THE BUILDING ESTIMATOR'S REFERENCE BOOK

A Reference Book Setting Forth Detailed Procedures And Cost Guidelines For Those Engaged in Estimating Building Trades

Editor
Scott Siddens

Technical Editors:
Jerrold Ratner, C.P.E., C.C.A.
William H. Spradlin, Jr., C.P.E.

Published By
FRANK R. WALKER COMPANY

Eugene R. Callahan, President

P.O. Box 3180
Lisle, Illinois 60532
(708) 971-8989

We will be grateful to readers of this volume who will kindly call our attention to any errors, typographical or otherwise, discovered therein. We also invite constructive criticism and suggestions that will assist in making future editions of this book more complete and useful.

FRANK R. WALKER COMPANY

BY
FRANK R. WALKER COMPANY

Publishers of
Walker's Practical Accounting and Cost Keeping for Contractors
Walker's Insulation Techniques and Estimating Handbook
Walker's Quantity Surveying & Basic Construction Estimating
Walker's Remodeling Estimator's Handbook
Walker's Manual for Construction Cost Estimating
Walker's Pocket Estimator
Walker's "Practical" standard forms for Contractors

ISBN: 0-911592-24-5
Library of Congress Catalog Card Number 15-23586

Printed in the U.S.A.

IMPORTANT

How to Obtain the Best Results From the Estimating and Cost Data in This Book

In a book of estimating cost data that is used in all parts of the U.S. and in many foreign countries, it is impossible to quote material prices and labor costs that apply in every locality. It takes some computation on the part of the user.

For this reason, all labor cost data give the quantity of work that a worker or crew should perform per hour or per day, together with the number of hours required to complete a certain unit of work. In each case, a certain basic price has been used.

The wage rates shown in the book are prevailing U.S. averages during the past year. Each rate is an average of the total wage and benefits that resulted from collective bargaining agreements and were reported to the Construction Labor Research Council, Washington, DC.

The user is urged to insert local wage scales and material prices to ensure accurate estimates. A standard method of tabulation has been used throughout this book, an example of which is given below.

When framing and placing wood floor joists up to 2"x8" in buildings of regular construction, a carpenter should frame and place 550-600 b.f. per 8-hr. day at the following labor cost per 1,000 b.f.:

	Hours	Rate	Total	Rate	Total
Carpenter	13.9	$....	$....	$24.04	$334.16
Labor	4.5	....	....	19.13	86.09
			$....		$420.25

The blank spaces are for you to insert your local wage rates and extend the prices as shown below:

	Hours	Rate	Total	Rate	Total
Carpenter	13.9	$15.22	$211.56	$24.04	$334.16
Labor	4.5	12.11	54.50	19.13	86.09
			$266.06		$420.25

Note that there is no factor included for overhead and profit on any of the costs in this book, because this is a variable item that each contractor must establish for each individual project.

Because of wide variance in state sales tax laws, this cost has not been included, and the user must apply the prevailing rate to obtain the total material cost.

To accommodate the maximum number of contractors, the insurance and taxes applicable to labor costs have not been included in any of the labor unit costs contained in this book, because some contractors prefer to incorporate these costs as part of overhead, while others maintain them as direct costs.

Neither the Frank R. Walker Company nor the editors guarantees or warrants the correctness or sufficiency of the information contained in this book; nor do either assume any responsibility or liability in connection with the use of this book or of any products, process or pricing described in this book.

TABLE OF CONTENTS

1. GENERAL REQUIREMENTS

2. SITE WORK

3. CONCRETE

4. MASONRY

5. METALS

6. CARPENTRY

7. THERMAL AND MOISTURE PROTECTION

8. WINDOWS AND DOORS

9. FINISHES

10. SPECIALTIES

11. EQUIPMENT

12. FURNISHINGS

13. SPECIAL CONSTRUCTION

14. CONVEYING SYSTEMS

15. MECHANICAL

16. ELECTRICAL

CHAPTER 1

GENERAL REQUIREMENTS

THE ROLE OF THE CONTRACTOR
CONSTRUCTION FINANCE
DRAWINGS AND SPECIFICATIONS
SETTING UP THE ESTIMATE
CONSTRUCTION SCHEDULING
TOOLS AND EQUIPMENT

Buildings are built by the efforts of several parties. Traditionally, these have been the owner, who either is or represents the user of the building, the architect who designs the building, and the contractor who constructs the building. The owner might have as consultants a realtor for land purchase and building management and financiers who provide short term cash for building construction and for the final long term mortgage when the project is completed.

The architect consults with structural and mechanical engineers, site planners, and interior space planners, any or all of whom might be part of his immediate office. He also consults with the governmental agencies that control zoning, building codes, and city planning, though he usually is not the one who takes out the building permit. He also relies on the cooperation of the producers who manufacture the products that he designs into a structure.

The contractor has on his team the various subcontractors and the producers and suppliers of the equipment, materials, services, appliances, tools, and machinery called for by the architect and required to complete the job. The contractor must also cooperate with the governing officials and must take out all building permits and conduct his operation without violating local laws that govern everything from blocking traffic to waking up the neighbors. And then there are the trade unions, the local utilities, and the ever-necessary bankers.

As land becomes scarce, when money rates soar, as ecological concerns are enacted into regulations, the exclusive club of owner-architect-contractor is forced to let in others, some welcome, some not, so that the triangle has now become more of a circle. On large projects today the owner-user may be the large landowner who wishes to develop his holdings, a government that is working to rebuild the inner city, or a large insurance company that needs to keep its investments active. The architect's concern may no longer be the individual building but an entire complex, a neighborhood, or a whole new town. And the general contractor may find himself in demand less for craftsmanship abilities than for managerial capabilities. But no matter how small or how complex the project, the owner has a need, the architect develops that need into satisfactory plans, and the contractor interprets the plans into an actual building.

THE ROLE OF THE CONTRACTOR

Erecting a building is a very complex undertaking, and seldom is one firm capable of doing all phases of the work. Yet the owner or developer usually prefers to let one contract and make one firm responsible for the completion of the project. That firm is then known as the *General Contractor* (or *general*). Usually, he assumes this role when the owner asks for firm price bids. He is the one who will compile all material, labor and service, and other costs called for on the architect's drawings and specifications. Where his own firm is not able to do work called for, he will turn to other contractors who have the expertise in those fields. These firms are then known as *Subcontractors* (or *subs*), and if the project is awarded to the general, they will enter into contracts with him and be responsible to him rather than directly to the owner. The general will ask for bids from many subcontractors in order to put together not only the best price but also the best team.

The amount of work performed by specialty type subcontractors varies to suit each project's needs, as determined by the general contractor. This major decision is not predicated solely upon economic reasons. Each time some portion of the work is assigned to a subcontractor, there is introduced into the construction team another member, whose strengths and weaknesses translate directly into the total performance plan. The general who attempts to "broker" a project by maximum use of subcontractors in order to limit financial responsibility will find that other hazards have been created. There will be a considerable loss in coordination and production, because the general has relinquished the right to and ability of direct control in the assignment of personnel, materials, and equipment. He might be subjected to a subcontractor who does not have his same dedication to the project.

Initially, it may appear that the large dollar value assigned for subcontractor performance insulates the general from financial loss on that portion of the project, but this is usually found to be false economy. Upon signing the subcontract document, the general contractor has inherited any faults and deficiencies that exist in the subcontractor's company. As the extent and seriousness of these faults become apparent during the life of the contract, they adversely impact the project schedule, subjecting the general to delay claims from other subcontractors, as well as from the owner. In addition, his own direct costs will increase in direct proportion to delayed completion.

In the most severe instances, the general may be forced to terminate the subcontract, and here again, excess costs will be incurred due to reprocurement costs and associated delays. It is an accepted practice to employ certain subcontractors who possess expertise in a particular discipline. However, they must be carefully selected on the basis of financial capacity, technical ability and the proven ability to perform consistent with the requirements of the contract documents. This policy can ensure a successful project, reflecting credit on the owner, architect/engineer, and the contracting team. The average percentage of work that is performed by subcontractors cannot be precisely determined, but surveys conducted by *The Associated General Contractors of America* indicate from 40% to 70%.

The general contractor must retain the financial ability to perform the work volume he has under contract, including allowances for unanticipated delays in receiving payments for work performed. Such delays seem to be increasing at an alarming rate, placing an additional burden on the financial resources of the general. There must be in-house personnel who are experienced in the particular type of projects in which the firm is engaged. Of equal importance, there must exist managerial skills to coordinate the field construction activities and the administrative functions into a total effort. The general is expected to provide many services to the owner and the architect prior to award of the contract as well as during construction. His knowledge and contribution may be the deciding factor in his receiving a contract award or in his expanding a project under construction.

How much of the contract will be sublet will vary with each general contractor. The *American Subcontractor's Association* claims that 90% of the work force in the building construction industry is employed by subcontractors. It is certainly quite possible for a subcontractor to erect a greater portion of the project than the general, and in some cases, the general is forced to sublet work. Some manufacturers insist that their products be installed by contractors licensed to them. Often, such as in curtain wall construction, the specifications will call for the supplier to install his product in order to put the responsibility for any corrective work on one source. Obviously, if anything like 90% of the contract is to be sublet to others, the general contractor must have the managerial skills to schedule and coordinate all the firms involved so that the work proceeds without delay.

In addition to building and managerial skills, he must be financially responsible. Not only must he meet his own payroll and overhead costs, but he also must pay his subcontractors and suppliers in advance of being reimbursed, hopefully with some profit, by the owner. Usually the general will put in a *request for payment* each month for that work completed the month before. These requests may have to be accompanied by *waivers of lien* for the amount requested from each sub and supplier. In submitting such a waiver, the sub gives up his right to file a claim against the property. To obtain a waiver, the general will have had to pay the sub or supplier at least the major portion of his request. Usually there is a *retainage* of at least 5% to 10% by the general contractor to his subs and in turn by the owner to the general contractor. Retainage is not completely released until the project is substantially complete, but it has become common practice to reduce the retainage by 50% when the project is 50% complete.

A contractor whose financial affairs are in good order and who has a contract with a responsible client, who in turn has secured adequate construction and mortgage loans, should have little trouble in securing a loan from his bank to finance the month-to-month payments. The contract will serve as the loan security. The amount needed for this interim financing is often stated as one dollar for every ten dollars in the contract. But before quoting a final price for a project, the contractor should determine the length of the period he must finance, estimate the amount involved, and add the cost of the money to the job cost. It should not have to come out of the profit.

The successful general contractor is one who has a broad knowledge of building construction. His initial contact with a prospective client may be about site selection, sources of construction finance, or the advantages of one type of construction over another. He must develop contacts with architects, bankers, realtors, mortgage brokers, soil engineers, and the local enforcers of building and zoning codes. His knowledge in these allied fields may be the deciding factor in his being considered for the job.

Most municipalities and many states require subcontractors and general contractors to be licensed. There is no general rule governing licensing in all jurisdictions. A state license may not fully qualify a contractor to practice in a municipality within that state. Many suburban areas will require additional fees from out-of-town contractors. One must always inquire about the laws governing the site of construction.

Often a license or permit bond is required guaranteeing a public body that the contractor will comply with all applicable statutes and ordinances. If land development is part of the contract, subdivision bonds may be required guaranteeing proper installation of roads, sewers and other utilities, today even extending to provision of schools and park systems.

And to practice, one must also enter into proper relationship with the unions who may also require a wage bond guaranteeing union scale and welfare fund payments.

Finally, one must also consider the type of business relationship that defines the firm; that is, whether to practice as an individual, within a partnership, or as a corporation. Consult your lawyer and banker.

Construction Management

In *Construction Management (CM)*, a general contractor or engineering company enters into a contract with the owner prior to the bidding period and acts in an advisory role. Bids are taken under the construction manager's supervision, but separate contracts are let for each phase directly between the contractor for that phase and the owner.

The construction manager is paid a flat fee by the owner, and this fee is completely separate from other job expenses. CM fees are determined in one of several ways:

a) A stated number of dollars plus expenses. The expense items would include the hourly wages paid the principals billed at an agreed upon fixed rate, plus employees' time billed at an agreed upon multiplier to cover overhead and profit, plus a multiple of expense for any outside consultants.

b) Same as above but without a fixed fee, all billing being on a multiple of direct expense.

c) An agreed upon fixed fee.

d) A percentage of the total construction cost.

Services rendered vary, but generally the CM is hired early in the planning and will work with the architect in setting up job budgets, bidding and construction schedules, site development, and selecting building systems. He might make detailed value engineering studies. Not only will he determine what material or procedure is least costly, but which ones will give optimum quality within the budget. He will cooperate with the architect in arranging specifications to include the proper extent of work for each phase to be bid. He will prepare prequalification criteria for the bidders and will conduct prebid conferences to acquaint the bidders with the extent of work covered under other contracts--what facilities are available to them at the job site and what they must furnish themselves, the time scheduling, and special procedures to be followed.

When the bids are received, the CM will evaluate them and recommend that one be accepted or that the phase should be modified and rebid.

When the construction begins, the CM will coordinate all work on the site to coincide with the agreed to schedules, budgets, and quality standards. He will recommend corrective procedures and job changes that he feels are necessary, but his decisions are subject to the approval of the architect and owner. He will maintain an office at the job site with sufficient personnel to inspect the work, record job progress, keep up-to-date records of drawing and specification changes, shop drawings, samples, operating manuals and other records which will be delivered to the owner upon completion of the project. He will set up and maintain the project accounting system, which on larger jobs might be computer based and which will give instant reports on the project cost status, payment status, and an analysis of each contract and the project cash flow. The construction management approach has gained wide acceptance and might be encountered in federally sponsored construction. It gives the owner access to complete data on all phases of the contract.

It frees the contractor from many of the risks of a lump sum contract. But if it limits his losses, it also limits his profits, because any job savings that the CM's expertise may bring about accrue to the owner.

The CM format is not attractive to all owners, especially not to the ones who desire to minimize their direct activity with the project during the construction period, in which case it is more desirable to employ the services of a general contractor on a lump sum contract basis.

Bidding for a Contract

Construction contracts are usually awarded in one of two ways--*competitive bidding* or *negotiation.* Competitive bidding is the method used for most contracts, but it is only truly effective when complete working drawings and specifications are available and when contractors are screened so all who bid are of the same caliber. Bids based on sketchy plans and vague specifications are a waste of time. You will almost always be underbid by those who sacrifice quality for price. One of the most valuable contacts an aspiring contractor can make is to get on the bidding list of an architect who produces detailed drawings and tight specifications and insists on their being carried out. The architect has already spent many hours working with the owner to make most of the critical decisions, and by the time bids are taken, the contractor's responsibilities are clearly set forth. He can proceed, if awarded the contract, knowing that the finished project will most probably satisfy the owner, and satisfied owners are probably the most valuable asset a contractor can have.

In private work the architect, often with the owner's help, will limit the list of bidders. You can judge your competition and know that if your price is low and reasonable you will get the contract. Public work is usually open to any "qualified" bidder, but qualifications are generally concerned more with financial status than reputation for quality. In order to propose a low bid against unknown competition, a good contractor must be certain his organization can bring to the project unique experience and efficiencies. Otherwise, he most certainly will be underbid by those desperate enough for work to bid without profit just to get a job.

Private work is sometimes announced in the newspaper or listed in trade publications. A real estate agent or a banker can often steer a contractor to jobs. Public work is almost always publicly advertised. Often, to get on a bidding list, a contractor must be prequalified. There are various ways this is done, by organizations, governmental bodies and banks, but in general, the information that must be submitted will follow that contained in *AIA Document A-305 Contractor's Qualifications Statement.* This requires that a firm--whether corporation, partnership, or individual--provide details on: length of time in business; names of principals; percent of work performed by own staff; record of and reasons for any past failures to complete a job; major projects now under construction; major projects completed in past five years; a record of experience of firm employees; bank references; trade references; name of bonding company; and a statement of financial conditions including current assets, liabilities, and the auditor's name. However, often just being a member in good standing of a local contractor's association and a bank reference will satisfy an inquiry.

Once on a bidding list, a contractor will receive an *Invitation to Bidders* for each prospective job. This form will include such information as job location, type of proposal, the number of sets of drawings and specifications to be furnished to bidders, what to do should discrepancies or errors be found in drawings and specifications, bulletins to be issued later, when and how bids will be opened, and when and where bids are due. A proposal sheet is usually attached so only the blanks need be filled out, but sometimes the contractor must furnish his own.

Often the invitation to bid is accompanied by an *Instruction to Bidders* further defining the job restrictions such as completion dates, visiting job site, special conditions, etc.

Bid deposits may be required, the amount being either a lump sum or a percentage of the proposal. The *bid bond* guarantees that the bidding contractor, if awarded the contract, will enter into the contract and furnish a *performance bond,* if required. If he does not honor his bid, he forfeits the amount of the bond, or the difference between his bid and the amount the owner can contract the work for with another contractor, whichever sum is less. Bid bonds, and the later performance, material, labor, maintenance, completion, supply, and subcontractor bonds, are often encountered in public work, but often are waived in private work, where the contractor's reputation is deemed sufficient and the cost of bonds unwarranted. The cost of the bid bond is a cost borne by the contractor. Performance bonds, unless made a part of the contract, are usually billed to the owner. Bonds are further discussed under "Definitions" in this chapter.

Negotiating a Contract

The other type of contract, one that is *negotiated,* allows the owner control over the selection of his general contractor. It is often used on complex work, for example, in remodeling jobs where the space is occupied and must be done piecemeal; interior work where strict control of subcontractors is desired; contracts in which mechanical equipment is a major factor and the owner may prefer one manufacturer over another; work which must proceed in great haste where the architect may not be able to produce complete plans in the time allotted and work will be let in sections (i.e., foundation work, then steel framing, then masonry). Often the contractor will proceed as with bid jobs in submitting a lump sum price, usually on the safe side and known as an *upset* or *guaranteed figure.* This will sometimes include the contractor's profit, but often he will be reimbursed for that which he has spent in completing the work. His profit and overhead will be a stated fixed dollar amount or a percentage of the job total.

A variation of the negotiated contract is *phased construction.* In this arrangement the project may be started before all the plans are fully developed. Each phase of the job, such as foundation, masonry, carpentry, etc., is bid separately, just before the phase is required to be installed. The advantages are:

a) The project can be started earlier, rather than wait until all the design details are settled.

b) The bidders need not add contingencies to cover unknowns such as wage and material cost rises or adverse job conditions. The work will proceed immediately, and these factors will already be known.

c) The bidding time is increased from three or four weeks to several months, and more "shopping around" is possible.

d) There is better control of the budget. If any phase comes in over the budget, it can be redesigned and rebid. If an earlier phase is contracted over the budget, later stages can be modified so that the total job can still conform to the budget.

As the phases are bid, the successful contractors may be assigned to a general contractor in the same manner as a lump-sum contract; or the general may act in the role of a project manager, in which case each contractor for each phase will have a direct contract with the owner. This variation is referred to as *multiple bidding.*

Contract Documents

Once it is determined which contracting firm is to do the job, a formal contract will be drawn up. The Contract Documents usually should include the Owner-Contractor Agreement; the General Conditions of the contract; Supplementary Conditions of the contract (if any); the Working Drawings, giving all sheet numbers with revisions; Specifications, giving page numbers; and Addenda or Bulletins issued prior to contract. After the contract is signed, any changes are added as Modifications to Contract and would take the form of written amendments, Change Orders or architectural interpretations.

The *Owner-Contractor Agreement* is a legal matter, but for many jobs standard forms are used. If the architect furnishes the Owner-Contractor Agreement, he will most probably use one printed by the American Institute of Architects. These are:

a) The Stipulated Sum Agreement A-101, four pages long or A-107, a shortened form agreement for small construction contracts. A special form A-10-111/CM is a stipulated sum agreement for construction management.

b) The Cost of the Work Plus a Fee Agreement, Form A-111, eight pages long.

c) Agreement Between Contractor and Subcontractor, Form A-401, seven pages long.

Agreements A-101 and A-111 assume that the *AIA Document A-201 General Conditions of the Contract for Construction*, 18 pages long, is made a part of the contract. These 14 articles are discussed later in this chapter; a shortened version of the General Conditions is included in Document A-107.

The Stipulated Sum Agreement will include the date of the agreement, the owner's name, contractor's name, the description of work, the architect's name, the contract sum, the time of commencement and completion, final payment, and a list of the contract documents.

The Cost Plus a Fee Agreement will include date of agreement, owner's name, contractor's name, description of work, architect's name, the contractor's duties and status, time of commencement and completion, cost of work and guaranteed maximum cost (this may be deleted and if included may be supplemented by a provision for distribution of savings), the contractor's fee, provisions for changes in the work, reimbursable costs, non reimbursable costs, discounts, rebates and refunds, subcontractors, accounting records, applications for payment, payments to contractor, termination of contract, and a list of the contract documents.

For less complicated *stipulated sum* contracts, where the contractor furnishes the form, one of the forms published by the Frank R. Walker Company may be useful. These include a single-page proposal Form #147, which includes an acceptance paragraph for the owner to sign to make a short contract, and a one-page contract agreement Form #158. If the contract is for a construction manager, AIA Form B-801 may be used.

PROPOSAL

________________ 19___

____

____

Dear Sir:

The undersigned proposes to furnish all materials and perform all labor necessary to complete the following:

____

All of the above work to be completed in a substantial and workmanlike manner for the sum of __________ (\$________) Dollars

Payments to be made each ________ as the work progresses to the value of ________ (____%) per cent of all work completed. The entire amount of contract to be paid within ________ days after completion.

Any alteration or deviation from the above specifications involving extra cost of material or labor will only be executed upon written orders for same, and will become an extra charge over the sum mentioned in this contract. All agreements must be made in writing.

The Contractor agrees to carry Workmen's Compensation and Public Liability Insurance, also to pay all Sales Taxes, Old Age Benefit and Unemployment Compensation Taxes upon the material and labor furnished under this contract, as required by the United States Government and the State in which this work is performed.

Respectfully submitted,

____
Contractor

By ____________________

ACCEPTANCE

You are hereby authorized to furnish all materials and labor required to complete the work mentioned in the above proposal, for which the undersigned agrees to pay the amount mentioned in said proposal, and according to the terms thereof.

____

Date ________________ 19___ ____________________

PRACTICAL Form 147

Short Form Proposal
Frank R. Walker Form 147

Contract Agreement

THIS AGREEMENT, made this __________ — __________ day of __________ A.D. 19____, by and between __________ hereinafter called the Owner, and __________ hereinafter called the Contractor.

For the consideration hereinafter named, the said Contractor covenants and agrees with said Owner, as follows:

FIRST. The Contractor agrees to furnish all material and perform all work necessary to complete the __________.

for the above named structure, according to the plans and specifications (details thereof to be furnished as needed) of __________ Architect, and to the full satisfaction of said Architect or Owner.

SECOND. The Contractor agrees to promptly begin said work as soon as notified by said Architect or Owner, and to complete the work as follows: __________

THIRD. The Contractor shall take out and pay for Workmen's Compensation and Public Liability Insurance, also Property Damage and all other necessary insurance, as required by the Owner, Architect or by the State in which this work is performed.

FOURTH. The Contractor shall pay all Sales Taxes, Old Age Benefit and Unemployment Compensation Taxes upon the material and labor furnished under this contract, as required by the United States Government and the State in which this work is performed.

FIFTH. No extra work or changes under this contract will be recognized or paid for, unless agreed to in writing before the work is done or the changes made.

SIXTH. This contract shall not be assigned by the Contractor without first obtaining permission in writing from the Architect or Owner. All Sub-contracts shall be subject to the approval of the Architect or Owner.

IN CONSIDERATION WHEREOF, the said Owner agrees that he will pay to the said Contractor, in __________ payments, the sum of __________ Dollars for said materials and work, said amount to be paid as follows: __________ per cent (______%) of all labor and material which has been placed in position by said Contractor, to be paid on or about the __________ of the following month, except the final payment, which the said Owner shall pay to the said Contractor within __________ days after the Contractor shall have completed his work to the full satisfaction of the said Architect or Owner.

The Contractor and the Owner for themselves, their successors, executors, administrators and assigns, hereby agree to the full performance of the covenants of this agreement.

IN WITNESS WHEREOF, they have executed this agreement the day and date written above.

Witness:

__________ Owner.

__________ Contractor.

PRACTICAL MFD. IN U.S.A. FORM 158

Short Form Contract Agreement
Frank R. Walker Form 158

Large corporations and governmental bodies will usually have contracts drawn up by their legal staffs. The contractor may want his own lawyer to review these contracts.

A third type of contract, but one only for both experienced owners and contractors, is the contract for a *cost plus a variable sum*. Here, a fixed cost for the building is agreed upon. If the contractor is able to complete a project at less than the agreed upon figure, the savings is split between owner and contractor; if the cost runs higher, the overage is also split. Often, not only cost is included with penalties and rewards, but also completion dates.

A bonus is given the contractor for each day of delivery earlier than the specified date, but a penalty is deducted for each day of failure to deliver on the specified date. Not only must a contractor be knowledgeable about his ability to perform under these conditions, he should also be well acquainted with the owner's reputation. What constitutes "completion" by a certain date can be a matter for the courts.

The General Conditions of the Contract usually refer to the American Institute of Architects Document A-201. These are often accompanied by Supplementary Conditions which modify and extend the General Conditions. Sometimes these two are combined and rewritten in a form more exactly fitting the special job conditions. But in any case, they set forth the rights, responsibilities, and relations of all parties to the contract and every contractor should be thoroughly acquainted with them.

The latest edition of the AIA form contains fourteen articles. Briefly they cover the following:

Article 1. Defines the contract documents, the work, the project, the execution, correlation, intent and interpretations of the contract documents, and sets forth the copies and ownership of these documents.

Article 2. Defines the architect and his administration of the contract. (A further discussion of administration is included later in this chapter.)

Article 3. Defines the owner, the information and services required of him, and his right to stop work and to carry out work if contractor defaults.

Article 4. Defines the contractor; his responsibility to review the contract documents; his supervision and construction procedures; his furnishing and control of labor and materials; his responsibility for a warranty, for taxes, for permits, fees, and notices; cash allowances, if any; the furnishing of a superintendent; his responsibility for those performing the work; the preparation of a progress schedule; the maintenance on the job site of drawings and specifications; the preparation of shop drawings and furnishing of samples; the use of the site; cutting and patching of work; cleaning up; proper channel for communications, and the indemnification of both owner and architect against all claims, damages, losses, and expenses within the limits set forth.

Article 5. Defines the subcontractors, the award of subcontracts, subcontractual relations, and payments to subcontractors.

Article 6. Sets forth the owner's right to award separate contracts, the mutual responsibilities of contractors, the problems of cutting and patching under separate contracts, and the owner's right to clean up.

Article 7. Includes miscellaneous provisions such as the law of the place, successors and assigns, what constitutes written notice, how claims for damages are to be made, right of owner to require performance and labor and material bonds and how they will be paid for, rights and remedies, payments of royalties and patents, tests that may be required, interest on money not paid when due, and arbitration of claims and disputes.

Article 8. Defines contract time, progress and completion, and delays and extensions.

Article 9. Discusses payments and completion sum, the filing of a schedule of values to be used as a basis for contractor's applications, progress payments, certificates for payment, payment withheld, failure of payment, and substantial completion and final payment.

Article 10. Covers protection of persons and property including safety precautions and programs, safety of persons and property, and emergency precautions.

Article 11. Lists the insurance requirements including contractor's liability insurance, owner's liability insurance, property damage insurance, and loss of use insurance.

Article 12. Covers changes in the work, including change orders, how to file claims for additional cost, and minor changes in the work and field orders.

Article 13. Discusses uncovering and correction of work and acceptance of defective or nonconforming work by the owner.

Article 14. Discusses termination of the contract by the contractor or by the owner.

The Supplementary Conditions may include two lists: those that amend the General Conditions and those that make additions to it. A special edition AIA Document A-201/CM is available for jobs using Construction Management. For federally assisted construction projects, the standard A-201 form is supplemented by form A-201/SC, which adds three articles that modify the 14 conditions and include conditions such as equal opportunity employment and special insurance requirements.

Setting Up the Job

Having been awarded the contract, the general contractor must then proceed to sign up his subcontractors. The standard AIA Owner-Contractor Agreement calls for the successful bidder to submit a list of subcontractors for approval prior to award of contract, and the owner and architect must object at that time to any firm on the list. If there is an objection, the contractor has two alternatives: submit an acceptable substitute with the change in cost, if any, in which case the owner may at his discretion accept or reject the bid; or the contractor may withdraw his bid without penalties.

Additional expenses occasioned by owner's or architect's rejection of subcontractors after award of contract will be added by change order to the contract sum and paid for by the owner. There is conflict on this issue, and extreme caution is to be observed.

The contractor must also submit a progress schedule and a schedule of values based on his subcontracts.

The AIA agreement further sets forth the contents of agreements between subcontractors and the general contractor and the method of paying the subcontractors. AIA Document A-401 may be used, or the shorter Frank R. Walker Forms #133 or #144.

The general contractor, with the cooperation of his subs, the architect, and owner, will file for and usually pay for all permits. Often it is necessary to have selected certain subcontractors before filing for permit, because the permit may require the license number of certain trades.

The Supplementary General Conditions will usually set forth the initial site requirements--office, telephone, toilets, water, light, heat, barricades, temporary partitions, etc., which the general contractor must provide. He must also make arrangements with local utilities.

Carrying Out the Work

The administration of the contract is clearly set forth in the AIA standard contractor agreements. The general contractor is responsible for executing the work in strict accordance with the contract documents and for turning the completed project over to the owner in full conformance with them. He must provide the continual supervision necessary to check all labor, materials, and procedures to accomplish this.

The architect serves in an advisory capacity to both contractor and owner, interpreting drawings, and specifications, approving shop drawings and samples, checking conformity of the job with contract documents, making periodic inspections to determine progress and approving payments to the contractors, and approving final acceptance of the contract. The architect will also have his consultants visit the site at certain intervals.

The owner often wants more supervision than the architect allows for in the standard contract and a "clerk of the works" is hired. He is usually paid by the owner but works under the architect. He keeps complete job records and reports daily progress, but he does not supplant the architect, whose decisions are the final ones, and does not relieve the contractor of responsibilities of supervision.

In addition, public building inspectors will probably visit the site periodically. Any objections they raise to apparent contradictions between codes and the contract documents should be reported in writing to the architect. At certain periods the contractor will be required to make tests, and he must notify the proper authorities and the architect when these are to occur.

Getting Paid

Applications for payment are usually made monthly. The Frank R. Walker Form #591 is an excellent one for this. It provides a column for the list of subcontractor names, what work the contract is for, the total amount of each contract, the amount previously requested, and finally the balance to complete. The bottom of the form provides a summary of total amounts, less percentages retained, and a space for notarizing. Waivers of lien showing that the contractor has paid for all materials, labor, and subcontractor billings are usually required to accompany the request for payment. The Frank R. Walker Company has printed forms for a Contractor's Affidavit (#592) and Waiver of Lien (#593).

The architect then checks the application, and if it agrees with what he feels to be proportionate to the amount of the contract, he will issue a certificate of payment. His decision in this may be reviewed by either the company writing the performance bond or the company supplying the interim financing. All of this takes time for the contractor to prepare, the architect to check and approve, and the owner to process, so the contractor must keep all his billings and payments current. With the high cost of money today, short term loans to tide a contractor over for a month because he didn't get around to filing a proper or complete request for payment can eat away at profits as surely as having to correct mistakes on the job.

When standard AIA contract forms are used, then the AIA *application for payment* will also be used to retain the same document format. Upon final completion and acceptance, the contractor should be prepared to file final waivers along with any written guarantees, certificates of inspections, operating instructions, etc., that may have been called for.

Sub-Contract Agreement

THIS AGREEMENT, made this ______ day of ______ A.D. 19____, by and between ______ hereinafter called the Contractor, and ______ hereinafter called the Sub-contractor.

For the consideration hereinafter named, the said Sub-contractor covenants and agrees with said Contractor, as follows:

FIRST. The Sub-contractor agrees to furnish all material and perform all work necessary to complete the ______

for the above named structure, according to the plans and specifications (details thereof to be furnished as needed) of ______ Architect, and to the full satisfaction of said Architect.

SECOND. The Sub-contractor agrees to promptly begin said work as soon as notified by said Contractor, and to complete the work as follows: ______

THIRD. The Sub-contractor shall take out and pay for Workmen's Compensation and Public Liability Insurance, also Property Damage and all other necessary insurance, as required by the Owner, Contractor or by the State in which this work is performed.

FOURTH. The Sub-contractor shall pay all Sales Taxes, Old Age Benefit and Unemployment Compensation Taxes upon the material and labor furnished under this contract, as required by the United States Government and the State in which this work is performed.

FIFTH. No extra work or changes under this contract will be recognized or paid for, unless agreed to in writing before the work is done or the changes made.

SIXTH. This contract shall not be assigned by the Sub-contractor without first obtaining permission in writing from the Contractor.

IN CONSIDERATION WHEREOF, the said Contractor agrees that he will pay to the said Sub-contractor, in ______ payments, the sum of ______ Dollars for said materials and work, said amount to be paid as follows: ______ per cent (____%) of all labor and material which has been placed in position by said Sub-contractor, to be paid on or about the ______ of the following month, except the final payment, which the said Contractor shall pay to the said Sub-contractor within ______ days after the Sub-contractor shall have completed his work to the full satisfaction of the said Architect or Owner.

The Contractor and the Sub-contractor for themselves, their successors, executors, administrators and assigns, hereby agree to the full performance of the covenants of this agreement.

IN WITNESS WHEREOF, they have executed this agreement the day and date written above.

Witness:

______ ______
Sub-Contractor

______
By

______ ______
Contractor

______
By

PRACTICAL MFD. IN U.S.A. FORM 133 FRANK R. WALKER CO. PUBLISHERS, CHICAGO

Short Form Sub-Contractor Agreement
Frank R. Walker Form 133

FORM 591

SWORN STATEMENT FOR CONTRACTOR AND SUBCONTRACTOR TO OWNER

State of ______________________ } ss.

County of ______________________

The affiant, ______________________ being first duly sworn, on oath deposes and says that he is (1) ______________________

contract with (2) ______________________ owner ___ for

(3) ______________________

on the following described premises in said County, to-wit: ______________________

That, for the purpose of said contract, the following persons have been contracted with, and have furnished, or are furnishing and preparing materials for, and have done or are doing labor on said improvement. That there is due and to become due them, respectively, the amounts set opposite their names for materials or labor as stated. That this statement is made to said owner for the purpose of procuring from said owner (4) Partial—Final Payment on said contract, and is a full, true and complete statement of all such persons, and of the amounts paid, due and to become due them.

(1) A member of the firm of, or officer of the corporation of, naming same. If a subcontractor so state and name the contractor. (2) Name of the owner or owners. (3) What the contract or subcontract is for. (4) Partial or Final Payment.

NAME AND ADDRESS	CONTRACT FOR	AMOUNT OF CONTRACT	TOTAL PREVIOUS REQUESTS	AMOUNT OF THIS REQUEST	BALANCE TO COMPLETE

MOUNT OF ORIGINAL CONTRACT	$	TOTAL AMOUNT REQUESTED	$
XTRAS TO CONTRACT	$	LESS % RETAINED	$
OTAL CONTRACT AND EXTRAS	$	NET AMOUNT EARNED	$
REDITS TO CONTRACT	$	AMOUNT OF PREVIOUS PAYMENTS	$
ET AMOUNT OF CONTRACT	$	AMOUNT DUE THIS PAYMENT	$
		BALANCE TO COMPLETE	

It is understood that the total amount paid to date plus the amount requested in this application shall not exceed % of the cost of work completed to date.

I agree to furnish Waivers of Lien for all materials under my contract when demanded.

Signed ______________________

____

Subscribed and sworn to before me this ______________ day of ______________ 19___

PRACTICAL

© FRANK R. WALKER CO., PUBLISHERS CHICAGO

The above sworn statement should be obtained by the owner before each and every payment.

Notary Public

Application For Payment
Sworn Statement For Contractor and Subcontractor To Owner
Frank R. Walker Form 591

Definitions

Standard contract forms use certain terms that have special meanings within the profession. Some of these are explained below:
Addenda: Modifications to the contract documents issued by the architect/engineer's office during the bidding period. Sometimes referred to as "bulletins". The addendum can be written, graphic, or both.

Alternates: Additions or subtractions to a contract sum for substitutions asked for by the architect, which the contractor must submit with his proposal.

Approved Equal: A contract term that allows the contractor to substitute another product for one specified by the architect, providing it is "equal" in all respects. Sometimes, the "or equal" must be accompanied by a proposal with a list of substitutions. The owner or the architect has the final say as to what is equal.

Arbitration: A method of settling a dispute whereby parties in disagreement submit their differences to a neutral third party. It is an alternative to formal litigation.

Bid: A proposal stating the sum for which the contractor will complete a project.

Bid Bond: A bond furnished by the contractor that guarantees he will sign a contract on the basis of his bid.

Cash Allowance: Sums set forth by the architect for specific items that the contractor must include in his proposal. Upon completion of a job, the contractor must make an accounting of these sums, returning any unspent amounts to the owner. Often used for buying hardware, face brick, light fixtures, and other special items.

Certificate of Occupancy: Issued by a government authority, authorizing occupation of a building.

Change Order: A written authorization by the architect to change the basic contract, plans, or specifications because of additions or deductions from it.

Contract Time: The period of time established in the contract to complete the contract or specific tasks. It can be expressed in calendar days, weeks, months, or final completion date.

Cost Breakdown (or Schedule of Values): A schedule prepared by the contractor showing how the contract sum is divided among the various divisions of the work, which is filed before work is started. The architect will use this as a basis for approving applications for payment and waivers. It should be accurate and kept up to date, reflecting any changes in the contract.

Extras: Construction costs outside the original contract amount. A change order usually is for additional work, or reduction of work scope.
Final Acceptance: When the owner accepts a project as complete.

Payment and Performance Bond: A bond assuring that the contractor will complete a project in accordance with contract documents including discharging all financial obligations.

Letter of Intent: Written prior to drawing up complete contract documents to tell the contractor a formal agreement is forthcoming. Often this authorizes certain preliminary work such as filing for building permits and site clearance.

Liens: Legal claims against an owner for labor and materials for construction of a project.

Liquidated Damages: A sum established by the owner, and stipulated in the contract as a charge to the contractor for damages suffered by the owner because of the contractor's failure to fulfill the contract obligations.

Maintenance Bond: Guarantees owner that contractor will rectify defects in workmanship or materials reported to him within a specified period following final acceptance of the work under contract. Often this is included as part of the Performance Bond and usually runs for one year. It may also be issued for longer periods covering specific phases of the work, such as roofing, curtain wall, paving, etc.

Punch List: A list, established by the owner or the owner's representative, of work yet to be done or to be corrected by the contractor.

Retainage: A percentage set forth in the Owner-Contractor Agreement to be withheld from each payment to the contractor. Usually 10%, reduced to a total of 5% upon substantial completion.

Separate Contract: A contract let by the owner directly to a contractor other than the general contractor. Usually for special equipment, landscaping, interior furnishings, etc.

Shop Drawings: Drawings prepared at the contractor's expense showing how special items shall be fabricated or installed. These usually are prepared by the subcontractors and are standard for such items as structural and ornamental steel, reinforcement steel, cut stone, millwork, partitions, door frames, fire protection, electrical fixtures, HVAC, etc. Shop drawings must be submitted to the architect or engineer for approval.

Subcontractor Bonds: A performance bond given by a subcontractor to the general contractor guaranteeing performance of contract and payment of labor and material bills.

Substantial Completion: When the project reaches a state where sufficiently complete to allow the owner to occupy it but not necessarily finally accept it.

Superintendent: Contractor's representative at the job site. This individual should not to be confused with the architect's superintendent (administration) or owner's clerk of the works.

Supplier: Sometimes referred to as a vendor. An organization which supplies materials, fabrication, or equipment for the project.

Supply Affidavit (Bond): Written notarized statement to the owner by the manufacturer that the materials delivered to the project comply with contract documents.

Unit Prices: Amounts asked for on the proposal for furnishing materials per unit of measurement. That is, cost of concrete per cu. yd., floor tile per sq. ft., or steel per ton.

Upset Price: An amount agreed to by the contractor as a maximum cost to perform a specific project. Used on Cost Plus a Fee jobs.

Warranty: A guarantee furnished by the contractor or a manufacturer covering quality, workmanship, and performance for a specified period.

Value Engineering

A primary objective of management is to achieve the maximum *value* for every expenditure. Many cost control programs have been devised to assist in this quest. None has been as easy to apply, and with as sure an expectation of substantial cost reduction, as *value engineering (VE).*

The value methodology is based on identification of function. In essence, it asks, What does the client really need in order to accomplish an objective? This is more difficult to quantify than it might appear. In order to be sure that the basic function is uncovered and clearly defined, a multidisciplinary team is assembled. The team should be composed of persons who can address all sides of an issue and usually includes the following:

a) certified value specialist
b) architect
c) structural engineer
d) mechanical engineer
e) electrical engineer
f) cost estimator

In the VE process, a specific time is allotted for study. A typical project might require one week for the *pre-workshop* phase, one week for the *workshop* phase, and another week for the *post-workshop* phase.

During the pre-workshop phase, the design documents are studied. When practical, a site visit is made. If a cost estimate exists, and if in the opinion of the team it does not represent the true cost of the design as it then exists, it is modified. If there is no cost estimate, then one is produced by the team cost estimator.

Cost/Worth Model. A *cost model* is produced by the cost estimator and the certified value specialist. This gives high cost visibility to the functions that compose the total project. Sometimes these models follow the CSI format--site, foundations, superstructure, etc. At other times, they might highlight design features, such as penthouse, annex, or parking. The purpose of the cost model is to graphically show how the costs are distributed by function.

The team next introduces a subjective factor called *worth.* This factor may modify the value of each *cost* by providing a comparative cost based on the experience of the team. For example, the team may perceive that the cost shown for HVAC in the cost model is much higher than for comparable projects. They determine what they consider the lowest cost possible to accomplish the basic function.

The cost model is then modified by giving each entry a worth, and it is then a *cost/worth model,* which will enable the team to see where the cost exceeds the worth by a significant amount. These become *targets* for the VE workshop.

VE Workshop. After a thorough review of the design documents, a site visit, and the preparation of a cost/worth model, the VE team convenes a workshop. Because this usually lasts a full week, it is called a *40-hour VE workshop.*

The workshop is the key intervention of value engineering. A multidiscipline team of experienced design and construction professionals work to seek cost effective alternatives to the present design. Each discipline sees the project from a different perspective, one of the important advantages of VE. However, it also means a potential problem in communication. This is solved by having a strong team leader, the certified value specialist as the *VE team coordinator (VETC).* To lead the team, the VETC uses the *VE job plan.*

VE Job Plan. The VE job plan is a logical series of phases that guides the team:

a) information phase
b) functional phase
c) creative phase
d) analytical phase
e) development phase
f) presentation phase

Information Phase. The information phase asks two questions of a component or system: What is it? What does it cost? During this phase the cost estimator works with each team member to produce a cost estimate for every target of investigation. Only the components involved in the system to be investigated are included. Simple forms are completed to clearly establish the present design for each target to be challenged.

Functional Phase. Now that the team has picked its targets and determined the costs, the functions are carefully analyzed. The questions asked in this phase are: What does the system do? What must the system do? The differences between answers to these two questions lead the team to fresh solutions that eliminate the functions that may be desirable but are not essential to the design.

Creative Phase. Each function is addressed to seek cost effective alternatives. Often this can be done by discussions among the members, and an effective technique is "brainstorming", where ideas are thrown out by any member as they occur, by association, contiguous thoughts, opposite concepts or the like.

What is sought in this phase is a great number of rough ideas, not engineered solutions, the reason being that an idea advanced by one member automatically triggers other ideas from other members of the team. Creativity seems to work best when negative expressions are not allowed to surface. The subconscious mind is at its best with positive direction, and it is this subconscious resource that we are trying to tap.

Analytical Phase. The long list of ideas generated during the creative phase is examined. Similar ideas are grouped so that a comparative analysis can be made to determine those ideas that must be discarded and those that show merit and should be developed. Several techniques are used to assist in this comparison. The experience of the team enables them to grade ideas with reasonable accuracy. The forms keep the team on track and expedite the comparative analysis.

Development Phase. One or two alternatives are selected by the team for development into the preferred solution. Closer cost estimates are made to compare with the present design.

The team must be aware of its responsibility. As professionals they are expected to propose concepts, systems and components that fulfill the basic functions required to properly satisfy the project. This is *not* a cheapening process! Anyone can use lower cost materials to reduce the cost. The goal of value engineering is to propose alternatives that completely fulfill the *needed* functions.

Often the savings are in *life cycle cost (LCC)*. The alternative may have the same or higher initial cost but may have significant cost savings over the life of the facility, product, etc. These LCC are highly desirable for most clients, though in some projects the "first costs" are the primary consideration. The VE team seeks this type of preliminary guidance during the pre-workshop phase to ascertain the real interest of the decision makers in LCC.

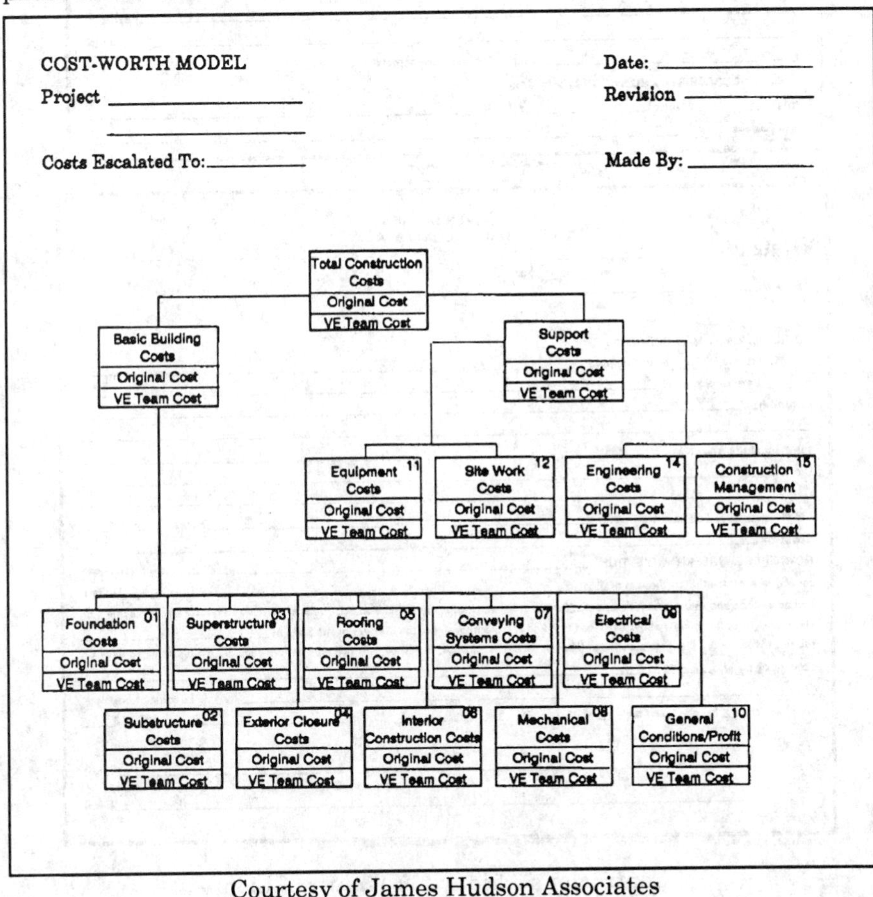

Courtesy of James Hudson Associates
Spotsylvania, VA

CONTRACTOR'S AFFIDAVIT

State of ______________________ ss. Office of ______________________

County of ______________________

being duly sworn on his oath deposes and says that he is ______________________

the contractor for the ______________________

for the building erected for ______________________, owner,

on the premises described as follows, to-wit: ______________________

All the bills for labor and material are fully paid and discharged ______________________

That ____ he ____ makes this affidavit for the purpose of procuring from ______________________

______________________ a ______________________ payment

of ______________________ Dollars and ______________________ Cents

upon ______________________ contract for said labor or material or both.

SUBSCRIBED AND SWORN to before me

this ____________ day of ______________ A.D., 19____

PRACTICAL ______________________ NOTARY PUBLIC

MFD. IN U.S.A. FORM 592 FRANK R. WALKER CO., PUBLISHERS, CHICAGO

WAIVER OF LIEN
MATERIAL OR LABOR

State of ______________________ } ss. ______________________ 19____

County of ______________________

TO ALL WHOM IT MAY CONCERN:

Whereas ________ the undersigned ______________________

ha ____ been employed by ______________________

to furnish ______________________

for the Building known as ______________________

City of ______________________

Lot No. __________ Section __________ Township __________ Range __________

County of ______________________ State of ______________________

NOW, THEREFORE, KNOW YE, That ______________________ the undersigned

for and in consideration of the sum of ______________________ Dollars

and other good and valuable considerations, the receipt whereof is hereby acknowledged, do hereby waive and release any and all lien, or claim or right to lien on said above described building and premises under the Statutes of the State of ______________________ relating to Mechanics' Liens, on account of labor or materials, or both, furnished or which may be furnished, by the undersigned to or on account of the said ______________________

______________________ for said building or premises.

Given under ____ hand ____ and seal ____ this ______________ Day of ______________ A.D., 19____

______________________ ______________________ (Seal)

______________________ ______________________ (Seal)

PRACTICAL

MFD. IN U.S.A. FORM 593 FRANK R. WALKER CO., PUBLISHERS, CHICAGO

Contractor's Affidavit & Waiver of Lien
Frank R. Walker Forms 592 & 593

It is the duty of the VE team to make very specific proposals. The designers do not need a "laundry list" of good ideas. Only a clear, well-defined alternative can be of use in modifying the design. The proposal must be clearly written, with sketches, to convey the concept.

Presentation Phase. *Alternatives are useless unless they are implemented.* It is imperative that the VE team prepare a written report, complete with all of the back-up worksheets prepared during the workshop, so that the decision makers fully understand the reasoning. In addition, whenever possible a verbal presentation should be given.

While the project is fresh in their minds, members of the VE team can give a complete oral recitation of their findings, proposals and emotional reaction to the present design and to the proposed design. Often, on the last day of a five-day workshop, the owner, designer, user, and other decision makers are asked to attend a presentation. During the presentation, each team member is given the opportunity to express without reservations the feelings that underlie the proposals. This might include minority opinions on some of the proposals. A properly led workshop seeks truth, not spectacular cost savings. This usually results in many cost effective alternatives, but it might also reveal serious deficiencies in the design. Great care is taken to divulge such findings with diplomacy and compassion.

Effectiveness of Value Engineering. VE has been used in construction for many years. When properly conducted, it has produced average savings of more than 10% of the estimated cost without sacrificing quality. VE has proven effective in controlling the cost of design and construction and is being used widely by several federal government agencies.

CONSTRUCTION FINANCE

Before getting into the details of estimating the materials and labor that go into a project, let us take a quick look at the "material" that makes it all possible--money.

The Cost of Money

Let's say a house that cost $45,000 ten years ago now costs $125,000. But add finance charges, based on a 20% down payment and a mortgage for twenty-five years, with interest charges of 9% ten years ago and 10% today. The house really cost $99,640 ten years ago and costs $272,700 today. The increase in the initial cost of the house in ten years was 74%, but the increase in the cost of financing was 120%. And in tight money markets of recent years, mortgages are not at 10% but at 16% and more! Obviously, the point is that money must be used as efficiently as any other commodity. And by efficiently we mean not only spent wisely but carefully planned for, shopped for, drawn upon only at the exact time it is needed, and paid for on time.

Our concern here is not with the money that the contractor needs to run his own business affairs, but the money to pay for the project itself. This breaks down into two types of financing--immediate cash to pay for the labor and materials as they are placed on the job, and once the project is completed, a loan against the project with repayment scheduled over many years. These are known as the *short term* or the construction loan and the *long term* or mortgage loan.

Project financing is not usually the direct concern of the contractor. It is the owner's or project developer's responsibility to secure both the short and long term loans. But it is essential that a contractor understand all phases of the project financing and his relationship to it.

First, few projects today are paid for from funds the owner already has accrued. He must find someone willing to loan the money to make the project possible, and that someone must have enough faith in the project to be willing to wait for repayment of his loan from money generated by the success of the project. In order to "sell" his project to a loan source, the owner must prepare a complete analysis of all aspects, including reliable cost projections, and will often turn to a contractor for this information. It is important that the contractor understands what such an estimate could commit him to, for what the lender "sees" is what the contractor later "gets" if the project loan is approved.

Second, the construction loan is the source from which the contractor will be paid, and he will have to work closely with the lender and his job inspectors to get his money on time. If the contractor is inefficient in compiling his requests for payment, his men and suppliers will go unpaid unless the contractor borrows the money until the next period the lender makes his payouts. The interest the contractor will have to pay will cut into his profit as surely as if he had underestimated the materials for the job.

Third, the owner may be very naive in construction finance. He may build only one project in his entire life. The local bank may be his only financial contact. The local banks are an excellent starting place to seek construction funds; they know the owner and the community and the owner's best interests are presumably the bank's best interests. But the bank may have other business loans with the owner, and they may not be willing, or even legally able, to make an additional loan of the size, at the rate, and for the length of time the owner requires. Thus the project may die in the bank's board room. The owner may lack the time and the sophistication to search further, and so the contractor loses the job as surely as if he had been underbid by a competitor. But there are other money sources besides banks, and an aggressive contractor will develop these other contacts so that he can steer the owner to them and salvage a good job that otherwise might never get off the drawing board.

Fourth, as we mentioned at the very beginning of this chapter, the triangle of owner-architect-contractor is nowadays often more of a circle, with each overlapping the other's territory. The contractor might also be the owner, either as the project developer or as the official landlord who then leases the project with perhaps an option to buy. Such arrangements can be very profitable and can be used in slack times to keep a contractor's force busy. But a good contractor does not always make a good landlord or even a good client for his company.

Sources of Money

The main sources of money are:

a) Commercial Banks--short and intermediate term financing
b) Savings and Loans--small residential loans
c) Insurance Companies--restricted long term loans
d) Pension and Trust Funds--long term loans. These funds are often administered by commercial banks.
e) Real Estate Investment Trusts (REITS)--long and short term loans, land loans, second mortgages, sell and leaseback deals.

f) Governmental Agencies--FHA and VA loans for residential work (guarantee loans)
g) Government Sponsored Bonds--industrial revenue bonds for projects that increase employment. Local governments issue tax-exempt bonds secured by the project income.
h) Mortgage Banking Firms--long and short term loans, land loans, second mortgages, sell and leaseback deals.

Each source has an area it is most interested in, but they compete to a degree. Many projects will draw on two or more sources. Short term loans are for three years or less. Long term loans are seldom for less than ten years and may run up to thirty years.

One thing all loan sources have in common is that they are "selling" their money, not "buying" your project. When the project is presented to them, they not only want to know the physical details but how the project will generate the money to pay back the loan. The essence of a good presentation is the proof that the loan can be repaid.

Mortgage Loans

While it is the construction, or short term, loan that is drawn on first, and land and "front money" loans may precede it, the mortgage, or long term, loan is shopped for first. Once the owner has the mortgage loan lined up, the construction loan is more easily secured, because it will eventually be paid by the mortgage loan when the project is completed. Further, the interest charged for the mortgage loan may decide whether the project can generate the income to make the necessary payments. If one fails to get a commitment for the mortgage interest and interest rates soar, the debt service may take all the projected profits.

Small projects can usually be financed both long and short term at local banks and savings and loan associations. These institutions make no direct charge for reviewing proposals and can usually give a yes or no answer in a few days, so one can shop several sources at the same time with no obligation. Once a loan is approved and accepted, there are certain charges called closing costs, and a part of these costs, known as *points, origination fees,* or *the discount,* cover the cost of setting up the loan. These points are a one-time charge made as a percentage of the loan and are deducted from the amount of the loan. If the points are 2%, then the amount of two dollars is deducted from every hundred dollars of the loan. The usual range of points is from 1% to 3%, but in states where there are legal limits set on the interest that may be charged, points have been quoted as high as 7% as a way to get around the usury laws. Such points are common to all loan sources and are one of the important items to shop for.

Mortgage Banker

For larger projects and projects that generate income, one should check out all the money sources. It isn't likely that even the most gregarious contractor knows all the institutional sources, or even if he did, could keep abreast of their current interests and capabilities. One must consult a specialist, and that specialist is the mortgage banker, as we shall call him here to distinguish him from a commercial banker.

Constant Annual Percent

Monthly Payment In Arrears

Description: This table shows the percent of the principal amount of a loan needed each year to pay off the loan when the actual payments are monthly and are paid in arrears. Divide the percent by 12 to get the level monthly payment per $100 that includes both interest and principal.

Example: The Constant Annual Percent needed to pay off a 15%, 30 year loan if payments are made monthly and in arrears is 15.18%. Divide by 12 to get the actual monthly payment. The constant annual payment for a $50,000 loan is $7,590. The monthly payment is $632.50.

Interest Rate	5 yr	6 yr	7 yr	8 yr	9 yr	10 yr	11 yr	12 yr	13 yr	14 yr	15 yr	16 yr	17 yr	18 yr	19 yr	20 yr
10.00	25.50	22.24	19.93	18.21	16.90	15.86	15.03	14.35	13.78	13.30	12.90	12.56	12.26	12.00	11.78	11.59
10.25	25.65	22.39	20.08	18.37	17.06	16.03	15.20	14.52	13.96	13.48	13.08	12.74	12.45	12.20	11.98	11.78
10.50	25.80	22.54	20.24	18.53	17.23	16.20	15.37	14.69	14.14	13.67	13.27	12.93	12.64	12.39	12.17	11.99
10.75	25.95	22.69	20.39	18.69	17.39	16.37	15.54	14.87	14.31	13.85	13.46	13.12	12.84	12.59	12.37	12.19
11.00	26.10	22.85	20.55	18.86	17.56	16.54	15.72	15.05	14.50	14.03	13.64	13.31	13.03	12.79	12.57	12.39
11.25	26.25	23.00	20.71	19.02	17.72	16.71	15.89	15.23	14.68	14.22	13.83	13.51	13.23	12.98	12.78	12.60
11.50	26.40	23.15	20.87	19.18	17.89	16.88	16.07	15.40	14.86	14.41	14.02	13.70	13.42	13.18	12.98	12.80
11.75	26.55	23.31	21.03	19.34	18.06	17.05	16.24	15.58	15.04	14.59	14.21	13.89	13.62	13.39	13.18	13.01
12.00	26.70	23.47	21.19	19.51	18.23	17.22	16.42	15.77	15.23	14.78	14.41	14.09	13.82	13.59	13.39	13.22
12.25	26.85	23.62	21.35	19.67	18.40	17.40	16.60	15.95	15.42	14.97	14.60	14.29	14.02	13.79	13.60	13.43
12.50	27.00	23.78	21.51	19.84	18.57	17.57	16.78	16.13	15.60	15.16	14.80	14.49	14.22	14.00	13.80	13.64
12.75	27.16	23.94	21.67	20.01	18.74	17.75	16.96	16.32	15.79	15.36	14.99	14.68	14.42	14.20	14.01	13.85
13.00	27.31	24.09	21.84	20.17	18.91	17.92	17.14	16.50	15.98	15.55	15.19	14.88	14.63	14.41	14.22	14.06
13.25	27.46	24.25	22.00	20.34	19.08	18.10	17.32	16.69	16.17	15.74	15.39	15.09	14.83	14.62	14.44	14.28
13.50	27.62	24.41	22.16	20.51	19.26	18.28	17.50	16.87	16.36	15.94	15.58	15.29	15.04	14.83	14.65	14.49
13.75	27.77	24.57	22.33	20.68	19.43	18.46	17.68	17.06	16.55	16.13	15.78	15.49	15.25	15.04	14.86	14.71
14.00	27.93	24.73	22.49	20.85	19.61	18.64	17.87	17.25	16.75	16.33	15.99	15.70	15.45	15.25	15.08	14.93
14.25	28.08	24.89	22.66	21.02	19.78	18.82	18.05	17.44	16.94	16.53	16.19	15.90	15.66	15.46	15.29	15.15
14.50	28.24	25.05	22.83	21.19	19.96	19.00	18.24	17.63	17.14	16.73	16.39	16.11	15.87	15.68	15.51	15.36
14.75	28.40	25.22	22.99	21.37	20.14	19.18	18.43	17.82	17.33	16.93	16.60	16.32	16.09	15.89	15.72	15.59
15.00	28.55	25.38	23.16	21.54	20.31	19.37	18.62	18.02	17.53	17.13	16.80	16.53	16.30	16.11	15.94	15.81
15.25	28.71	25.54	23.33	21.71	20.49	19.55	18.80	18.21	17.73	17.33	17.01	16.74	16.51	16.32	16.16	16.03
15.50	28.87	25.71	23.50	21.89	20.67	19.73	18.99	18.40	17.93	17.53	17.21	16.95	16.72	16.54	16.38	16.25
15.75	29.03	25.87	23.67	22.06	20.85	19.92	19.19	18.60	18.12	17.74	17.42	17.16	16.94	16.76	16.60	16.48
16.00	29.19	26.04	23.84	22.24	21.04	20.11	19.38	18.79	18.33	17.94	17.63	17.37	17.16	16.98	16.83	16.70
16.25	29.35	26.20	24.01	22.42	21.22	20.29	19.57	18.99	18.53	18.15	17.84	17.58	17.37	17.20	17.05	16.93
16.50	29.51	26.37	24.18	22.59	21.40	20.48	19.76	19.19	18.73	18.36	18.05	17.80	17.59	17.42	17.27	17.15
16.75	29.67	26.53	24.35	22.77	21.58	20.67	19.96	19.39	18.93	18.56	18.26	18.01	17.81	17.64	17.50	17.38
17.00	29.83	26.70	24.53	22.95	21.77	20.86	20.15	19.59	19.14	18.77	18.47	18.23	18.03	17.86	17.72	17.61
17.25	29.99	26.87	24.70	23.13	21.95	21.05	20.35	19.79	19.34	18.98	18.69	18.45	18.25	18.08	17.95	17.84
17.50	30.15	27.04	24.88	23.31	22.14	21.24	20.54	19.99	19.55	19.19	18.90	18.66	18.47	18.31	18.17	18.06
17.75	30.31	27.21	25.05	23.49	22.33	21.43	20.74	20.19	19.75	19.40	19.11	18.88	18.69	18.53	18.40	18.29
18.00	30.48	27.37	25.23	23.67	22.51	21.63	20.94	20.39	19.96	19.61	19.33	19.10	18.91	18.76	18.63	18.52
18.25	30.64	27.54	25.40	23.86	22.70	21.82	21.14	20.60	20.17	19.82	19.55	19.32	19.13	18.98	18.86	18.76
18.50	30.80	27.71	25.58	24.04	22.89	22.01	21.34	20.80	20.38	20.04	19.76	19.54	19.36	19.21	19.09	18.99
18.75	30.97	27.89	25.76	24.22	23.08	22.21	21.54	21.01	20.59	20.25	19.98	19.76	19.58	19.44	19.32	19.22
19.00	31.13	28.06	25.93	24.41	23.27	22.41	21.74	21.21	20.80	20.47	20.20	19.98	19.81	19.67	19.55	19.45
19.25	31.30	28.23	26.11	24.59	23.46	22.60	21.94	21.42	21.01	20.68	20.42	20.21	20.03	19.89	19.78	19.69
19.50	31.46	28.40	26.29	24.78	23.65	22.80	22.14	21.63	21.22	20.90	20.64	20.43	20.26	20.12	20.01	19.92
19.75	31.63	28.57	26.47	24.96	23.84	23.00	22.34	21.84	21.43	21.11	20.86	20.65	20.49	20.35	20.24	20.16
20.00	31.80	28.75	26.65	25.15	24.04	23.20	22.55	22.04	21.65	21.33	21.08	20.88	20.72	20.58	20.48	20.39
20.25	31.96	28.92	26.83	25.34	24.23	23.39	22.75	22.25	21.86	21.55	21.30	21.10	20.94	20.82	20.71	20.63
20.50	32.13	29.10	27.01	25.52	24.42	23.59	22.96	22.46	22.08	21.77	21.53	21.33	21.17	21.05	20.95	20.88
20.75	32.30	29.27	27.20	25.71	24.62	23.80	23.16	22.68	22.29	21.99	21.75	21.56	21.40	21.28	21.18	21.10

Interest Rate	21 yr	22 yr	23 yr	24 yr	25 yr	26 yr	27 yr	28 yr	29 yr	30 yr	31 yr	32 yr	33 yr	34 yr	35 yr	40 yr
10.00	11.41	11.26	11.13	11.01	10.91	10.82	10.73	10.66	10.59	10.54	10.48	10.44	10.39	10.36	10.32	10.19
10.25	11.62	11.47	11.34	11.22	11.12	11.03	10.95	10.88	10.82	10.76	10.71	10.66	10.62	10.58	10.55	10.43
10.50	11.82	11.68	11.55	11.43	11.34	11.25	11.17	11.10	11.04	10.98	10.93	10.89	10.85	10.81	10.78	10.67
10.75	12.03	11.88	11.76	11.65	11.55	11.46	11.39	11.32	11.26	11.21	11.16	11.12	11.08	11.05	11.02	10.91
11.00	12.23	12.09	11.97	11.86	11.77	11.68	11.61	11.54	11.48	11.43	11.39	11.35	11.31	11.28	11.25	11.14
11.25	12.44	12.30	12.18	12.08	11.98	11.90	11.83	11.77	11.71	11.66	11.62	11.58	11.54	11.51	11.48	11.38
11.50	12.65	12.51	12.40	12.29	12.20	12.12	12.05	11.99	11.94	11.89	11.85	11.81	11.77	11.74	11.72	11.62
11.75	12.86	12.73	12.61	12.51	12.42	12.35	12.28	12.22	12.16	12.12	12.08	12.04	12.01	11.98	11.95	11.87
12.00	13.07	12.94	12.83	12.73	12.64	12.57	12.50	12.44	12.39	12.35	12.31	12.27	12.24	12.22	12.19	12.11
12.25	13.28	13.16	13.05	12.95	12.87	12.79	12.73	12.67	12.62	12.58	12.54	12.51	12.48	12.45	12.43	12.35
12.50	13.50	13.37	13.26	13.17	13.09	13.02	12.96	12.90	12.85	12.81	12.78	12.74	12.71	12.69	12.67	12.59
12.75	13.71	13.59	13.48	13.39	13.31	13.24	13.18	13.13	13.09	13.05	13.01	12.98	12.95	12.93	12.91	12.84
13.00	13.93	13.81	13.71	13.62	13.54	13.47	13.41	13.36	13.32	13.28	13.25	13.22	13.19	13.17	13.15	13.08
13.25	14.14	14.03	13.93	13.84	13.77	13.70	13.64	13.59	13.55	13.51	13.48	13.45	13.43	13.41	13.39	13.32
13.50	14.36	14.25	14.15	14.07	13.99	13.93	13.87	13.83	13.79	13.75	13.72	13.69	13.67	13.65	13.63	13.57
13.75	14.58	14.47	14.37	14.29	14.22	14.16	14.11	14.06	14.02	13.99	13.96	13.93	13.91	13.89	13.87	13.81
14.00	14.80	14.69	14.60	14.52	14.45	14.39	14.34	14.30	14.26	14.22	14.19	14.17	14.15	14.13	14.11	14.06
14.25	15.02	14.92	14.82	14.75	14.68	14.62	14.57	14.53	14.49	14.46	14.43	14.41	14.39	14.37	14.36	14.30
14.50	15.24	15.14	15.05	14.98	14.91	14.86	14.81	14.77	14.73	14.70	14.67	14.65	14.63	14.61	14.60	14.55
14.75	15.47	15.37	15.28	15.21	15.14	15.09	15.04	15.00	14.97	14.94	14.91	14.89	14.87	14.86	14.84	14.80
15.00	15.69	15.59	15.51	15.44	15.37	15.32	15.28	15.24	15.21	15.18	15.15	15.13	15.12	15.10	15.09	15.04
15.25	15.92	15.82	15.74	15.67	15.61	15.56	15.51	15.48	15.45	15.42	15.40	15.38	15.36	15.34	15.33	15.29
15.50	16.14	16.05	15.97	15.90	15.84	15.79	15.75	15.72	15.69	15.66	15.64	15.62	15.60	15.59	15.58	15.54
15.75	16.37	16.28	16.20	16.13	16.08	16.03	15.99	15.95	15.93	15.90	15.88	15.86	15.85	15.83	15.82	15.79
16.00	16.59	16.50	16.43	16.37	16.31	16.27	16.23	16.19	16.17	16.14	16.12	16.10	16.09	16.08	16.07	16.03
16.25	16.82	16.74	16.66	16.60	16.55	16.50	16.47	16.43	16.41	16.38	16.36	16.35	16.33	16.32	16.31	16.28
16.50	17.05	16.97	16.89	16.83	16.78	16.74	16.71	16.67	16.65	16.63	16.61	16.59	16.58	16.57	16.56	16.53
16.75	17.28	17.20	17.13	17.07	17.02	16.98	16.94	16.92	16.89	16.87	16.85	16.84	16.82	16.81	16.80	16.78
17.00	17.51	17.43	17.36	17.31	17.26	17.22	17.19	17.16	17.13	17.11	17.10	17.08	17.07	17.06	17.05	17.02
17.25	17.74	17.66	17.60	17.54	17.50	17.46	17.43	17.40	17.38	17.36	17.34	17.33	17.32	17.31	17.30	17.27
17.50	17.97	17.90	17.83	17.78	17.74	17.70	17.67	17.64	17.62	17.60	17.59	17.57	17.56	17.55	17.55	17.52
17.75	18.20	18.13	18.07	18.02	17.97	17.94	17.91	17.88	17.86	17.85	17.83	17.82	17.81	17.80	17.79	17.77
18.00	18.44	18.37	18.31	18.26	18.21	18.18	18.15	18.13	18.11	18.09	18.08	18.06	18.05	18.05	18.04	18.02
18.25	18.67	18.60	18.54	18.49	18.45	18.42	18.39	18.37	18.35	18.34	18.32	18.31	18.30	18.29	18.29	18.27
18.50	18.91	18.84	18.78	18.73	18.69	18.66	18.64	18.61	18.60	18.58	18.57	18.56	18.55	18.54	18.54	18.52
18.75	19.14	19.07	19.02	18.97	18.94	18.90	18.88	18.86	18.84	18.83	18.81	18.80	18.80	18.79	18.78	18.77
19.00	19.37	19.31	19.26	19.21	19.18	19.15	19.12	19.10	19.09	19.07	19.06	19.05	19.04	19.04	19.03	19.02
19.25	19.61	19.55	19.50	19.45	19.42	19.39	19.37	19.35	19.33	19.32	19.31	19.30	19.29	19.28	19.28	19.26
19.50	19.85	19.79	19.74	19.69	19.66	19.63	19.61	19.59	19.58	19.56	19.55	19.55	19.54	19.53	19.53	19.51
19.75	20.08	20.02	19.98	19.94	19.90	19.88	19.86	19.84	19.82	19.81	19.80	19.79	19.79	19.78	19.78	19.76
20.00	20.32	20.26	20.22	20.18	20.15	20.12	20.10	20.08	20.07	20.06	20.05	20.04	20.03	20.03	20.02	20.01
20.25	20.56	20.50	20.46	20.42	20.39	20.36	20.34	20.33	20.32	20.30	20.30	20.29	20.28	20.28	20.27	20.26
20.50	20.80	20.74	20.70	20.66	20.63	20.61	20.59	20.57	20.56	20.55	20.54	20.54	20.53	20.53	20.52	20.51
20.75	21.03	20.98	20.94	20.90	20.88	20.85	20.84	20.82	20.81	20.80	20.79	20.78	20.78	20.77	20.77	20.76

The mortgage broker is an intermediator between borrowers and lenders, although his firm may have some funds of its own to invest. He is a "matchmaker" who brings parties together. To do this he must carefully evaluate both sides, shopping for the best terms for the owner and for projects most compatible with the portfolio of the lender. For consummating the transaction, he charges a flat fee of around 2% of all the loans placed by him. This fee is in addition to the usual closing costs that are charged by the lender. Both expenses are paid by the borrower. Many lenders require a 1% fee at the time the loan application is filled. This fee is refundable if the lender rejects the loan but becomes earned once the lender approves the application and is credited to the fee due the lender at the time of settlement.

There are other benefits in working with a mortgage broker. He can advise on the techniques of financing as well as the sources. There are numerous options available, and the owner will want to review them all with his accountant and lawyer and select what may suit his particular situation best. Decisions to be made would include the desired length of the mortgage; whether to select a loan at lower rates but "locked in" for many years or one at higher rates but with prepayment provisions; what personality--that is, items not part of the building itself such as equipment, carpet, drapes and the like--should be included in the mortgage; what owner assets should be pledged; what existing property should be made part of the new mortgage (wrap-around mortgage); what "kickers" might be offered, such as sharing income or equity with the lender if certain income or worth is exceeded; what advantages second and third mortgages might have; what method of repayment of the mortgage loan should be adopted (usually a constant covering interest and principal); payments of interest only, with the principal paid when the note is due; a constant payment of principal with interest figured on the declining balance; no payments on either interest or principal until the note is due but with the interest compounding; or a combination of the preceding with refinancing at certain periods to fit the owner's needs. There are no fixed answers that cover all situations. These decisions can contribute as much to the success of the project as decisions about room arrangements or types of building materials.

Having determined the best package for the owner, the mortgage banker will then shop his various money sources for the best deal. Each source will be looking for certain features, and the final deal probably will be a compromise.

The mortgage banker's opinions are useful in other ways. Having reviewed numerous similar projects at the preliminary stage, he can often suggest refinements in the physical arrangements, as they relate to potential income, that he has seen work out well in other situations. And finally, in accepting or refusing a project, he gives his opinion as to its probable success. If two or three mortgage bankers turn a project down when they are placing others, the owner should review his figures. The risk elements must be greater than those in the normal market. The mortgage broker need not be just a necessary evil. He can be a creative part of the building team.

Selling the Lender

Whether one works directly with a loan organization or through a broker, one must be prepared to make an effective presentation of his project. While each type of project will emphasize different elements, all should include the following:

a) physical analysis of the project including a surveyor's plot plan, an architect's sketch plans and specifications, an artist's rendering of the developed site, and a contractor's estimate of the cost;

b) a market feasibility study;

c) a financial analysis of the ownership including personal statements;

d) a financial analysis of the project.

A great deal of thought must go into the presentation of the physical plant. Once the project is approved, it is difficult to make major changes. A too glamourous presentation may backfire. It is utility not beauty that sells the lender. An impressive entrance framed by a tree lined drive circling a large fountain may be just an artist's window dressing and will be the first thing abandoned when the final budgets are reviewed. But that may be the feature a lender remembers and insists be built. Another lender, with his eye on the cash flow, will see such embellishments as so much maintenance expense and as a detriment rather than asset.

In addition to the contractor's estimate, the lender will make one of his own and may ask the borrower to have an independent source make up a third. This is usually done by a member of the American Institute of Real Estate Appraisers, whose members carry the title MAI (Member Appraisal Institute), or senior member of an appraisal society such as the American Society of Appraisers.

The owner's financial statement may be prepared by his certified public accountant, but often the lender will have his own standard form to fill out. The proposed contractor may also have to file a financial statement. Some loans will only be made if performance bonds are posted. Before presenting his proposal an owner will want to check with his lawyer and accountant as to the best form of ownership for the project--individual, corporate, or partnership. Some buildings are owned "at arms length". A third party actually owns the building, and the promoter of the project becomes the tenant.

The market feasibility study shows the compatibility of the project with the surroundings, means of public access, the effect on the environment, and compliance with zoning regulations. Similar successful projects in the area should be noted, and the continuing need for additional facilities proven. And finally, the adaptability to another occupancy and resale value should be evaluated.

The most carefully scrutinized section of the presentation will probably be the projected value of the project based on income, or whether the project will generate enough income to repay the loan on time with interest. What is included here will vary with the type of project. Where rents are involved, rent rolls should be projected. The lender in such cases may stipulate that a certain rent level be reached, usually around 80% of the projection, before the mortgage becomes final. If not, the mortgage will be for a lower amount. This is called *setting a floor* for the mortgage. The spread may be as much as 20%, and if the owner does not reach the agreed upon rent level, he must find another source to cover this amount. This will be discussed later.

The Commitment

Once a project is accepted by a lending organization, a commitment paper is issued stating that the lender will lend a certain amount of money for a certain length of time at a certain rate of interest, providing the borrower fulfills certain stated conditions and closes the loan by a certain date. This commitment, if the owner accepts, will become the basis for the first mortgage on the project. There may be other mortgages either on the project as a whole, or on some phase of it, such as furnishings, equipment, or the land itself. But the holder of the first mortgage takes precedence over all other claims except taxes, and he has the authority to take possession of the project upon proven default and to sell it and deduct the amount due him from the selling price, which is usually determined by a public auction. If any excess cash remains, it then goes to the holder of the second mortgage, then to the third, if any, and so on, with the original developer the last to receive any distribution. Holders of mortgages other than the first cannot instigate the sale of the project. They are subordinate to the first mortgage holder and therefore carry a higher risk and charge a higher rate of interest.

The commitment will be quite specific and should not be thought of as a preliminary proposal, but rather, as the basis of a contract that one must live with for many years. This is why the presentation of the project must be carefully thought out and based on realistic goals.

Short Term Loans

Having received a commitment, the owner must then work fast to finalize his plans. The commitment will carry a deadline date for acceptance that may be only a couple of weeks away. It will also carry an expiration date by which time all closing documents shall be submitted to the loan company so that the final mortgage loan may go into effect. If this date is not met, the owner forfeits an agreed upon deposit, which is usually around 1% of the mortgage amount and is deposited with the lender's agent upon accepting the commitment. One must then proceed immediately to secure a construction loan, gap financing (if it is needed), and "front money".

The construction loan is a short term loan to cover the building costs during the erection of the project. With the mortgage loan commitment in hand, the construction loan repayment is guaranteed providing that the project is completed in a satisfactory manner, so it is not necessary to sell the feasibility of the project all over again. But the construction loan lender does take the risk that the owner or the contractor will not get in financial difficulties during the construction period, or that the projected budget will not prove inadequate. The lender may then have to complete the project himself. Because the mortgage amount may be less than the stated amount if certain income levels are not reached, the construction loan will be based on the mortgage "floor" rather than the "ceiling" figure, which may deduct as much as 20% from the amount needed to complete the construction. Further, the construction loan is seldom for more than 80% of the floor figure, and like all loans, is subject to points and other closing costs, so perhaps another 2% will be deducted from the loan.

While shopping around might improve one's position, the construction loan will still fall short of the amount needed to complete the project. The missing funds must be supplied by the owner and represent his *equity* in the project. If the borrower can pledge sufficient liquid assets, that is cash, stocks or bonds, or other unencumbered property, he might be able to raise the additional funds from the same source as the construction loan. If not, he will have to take in a partner or borrow from those who take high risks. One such source is those who loan the difference between the floor and the ceiling of the mortgage. This is known as *gap financing*. They will issue a commitment to pay this difference if the income generated by the project is less than that specified to qualify for the full or ceiling amount. To obtain such a commitment one must pay in advance a flat fee, usually around 5% of the amount to be loaned. If the project reaches the income level to qualify for the full mortgage, and the gap loan is not needed, the fee is not refundable. If it is needed, the loan becomes a second mortgage against the project and is repaid with interest.

In setting up the project budget, there are other costs to be met besides construction costs. First, the land must be obtained. While this is sometimes leased or paid for under a mortgage arrangement, the first mortgage commitment may ask that the land title be *unencumbered*. In addition to land costs, there may be land development expenses such as surveys, soil investigation borings, utility extensions and connections, and access roads or sidings. There also may be purchasing expenses, including attorney and real estate agent fees, taxes and permits, and zoning fees. If the zoning for the land is to be changed, it can be both time-consuming and expensive.

And the architect must be paid. His fee will run from around 4% for a large industrial plant up to 15% for a fine residence or interior remodeling job. Some of this fee will not become due until the job is completed.

When the loan commitments are accepted, they may require a good faith deposit of around 1% of the loan. Consulting fees must be paid to accountants and lawyers. At loan closings legal and recording fees and transaction taxes may be charged, and if asked for, a performance bond obtained.

When the work is started, there is always a lapse of time between paying for permits, materials, and salaries and being reimbursed for these items from the construction loan. Usually, each request for payment from the construction loan must be accompanied by waivers of lien. In actual practice the contractor may finance this period rather than the owner, but the lender may still demand that the owner show he has the money to cover these start-up costs.

How much front money, or equity, an owner will need to launch his project will vary with the type of project, the money market, and the owner's reputation. It is often said that an owner with a proven need, a piece of property free of debt, and an architect's set of plans can obtain all the financing he will need. We have used the term "owner" in our preceding discussions as if the one who builds the project were the one who will occupy it. Today even large corporations may choose to be tenants rather than owners and will rely on developers to put the project together. The contractor may be asked to be the developer or at least to become a partner in it. Often the owner of the real estate and the architect will be included as *limited partners*, their limits of both liability and equity being proportional to the cost of the services or land contributed by them. When the construction loan is paid off by the mortgage loan, they are also often bought out and the ownership reverts to the general partner. However, some mortgage commitments guard against such arrangements by adding a penalty fee if there is any change of ownership within a stipulated time.

Rates of interest on construction loans vary from week to week and community to community but are usually tied to the prime rate, which is the interest charged by commercial banks to their preferred customers for short term loans. Currently construction loans are quoted at 2 ½% to 4% over the prime. High risk loans will have considerably higher rates.

It is not our intention to encourage the contractor to become the project entrepreneur. But with the cost of money tied so closely to the success of the project, from an income-producing standpoint a contractor must know all the procedures in order to follow the time and budget schedules, to process his monthly draws on time, and to turn the completed project over to the owner in such condition as to be completely acceptable to the mortgage lender.

Interim Financing

This particular type of financing has nothing to do with paying for the construction costs of a project. Nor does it have to do with long-term mortgage financing, which is repaid over an extended period from the proceeds of the revenue generated from the project.

The definition of the word interim is *provisional, intervening, interlude or relating to a time period.* This is the type of financial assistance required by virtually all contractors, regardless of size, for short-term working capital to compensate for unanticipated delays in receiving payments from owners for work performed and properly invoiced.

The source is usually a commercial bank that is responsive to the needs of the contractor. Not all banks are receptive to this type of business, and the contractor must establish banking relations with a bank that has personnel who possess intimate knowledge of the financial structure found in the construction industry. Once an acceptable financial institution has been selected, then the contractor should identify a particular person within the bank with whom a personal relationship can be established.

Next, it is the obligation of the contractor to prove, to the complete satisfaction of the bank, the credibility of the technical and financial assets of the company. This can be accomplished by furnishing *certified (CPA) financial statements* of the company along with a written detailed resume of the qualifications of each person directing the destiny of the company. Progress, at this point, should place the contractor in a position to obtain from the bank a general commitment as to the limit and terms under which they would participate in granting short term loans (usually 30 to 90 days). The ideal situation is to obtain from the bank a letter of credit outlining its position so that at any time during the next 12 months, a loan could be arranged without further review from a committee. This avoids days, or even weeks, of delay from the time the loan application is presented until the proceeds are available.

At the time a loan request is made, the position of the contractor can be significantly enhanced by presentation of a cash flow analysis of all current projects, projecting anticipated expenditures and income on a monthly basis so that the amount of money required and the availability of sufficient income to repay the loan can be clearly defined.

Expect the interest on this type of loan to be from 1% to 6% above the current prime rate. Banks consider this high risk business. If the current prime rate is 16% and the loan is assigned 3% over prime, this means the contractor is paying 19% for the use of the money. Obviously, a contractor is forced to impose restrictions on the use of such costly money. It cannot be used as continuous working capital, which at best has an earning capacity of about 10%. It has to be justified by large discounts. For example, prompt payment of equipment invoices would affect an early delivery date that would equate with a time and cost saving justifying the interest rate.

The availability of interim financing provides a measure of security for the contractor, who is assured that external financial assistance is available to augment working capital during periods of depressed cash flow.

DRAWINGS AND SPECIFICATIONS

The owner's dreams and the architect's visions must eventually be put down in a form the contractor can use as a basis for bidding the project and for constructing in enduring materials. This form has developed into the dual one of *working drawings* and *specifications*. Basically, the drawings present the design, the location, and the dimensions of a project. The specifications give the quality required.

The drawings are done in pencil or ink on tracing paper or cloth, which is easily reproducible, and prints are delivered to the contractor. The architect retains the original drawings in his office. The specifications, except on very small jobs for which they may be incorporated on the drawings, are printed and bound separately. Both are made available to each contractor bidding the job. As bid periods are usually quite limited, it is the contractor's ability to interpret them that enables him to present a realistic and complete bid. This chapter presents an outline of what information a contractor should expect to find in a bid set and where it can be found.

The Working Drawings

Arrangement. A typical complete project will find several drawing sheets, usually all of uniform size, bound into one job set. The first sheet should include a site plan, sometimes incorporating the roof plan, plus the complete job title, and sometimes, an index of the drawings. This will be followed by the sheets showing architectural details, usually in the following order: basement plan, first floor plan, upper floor plans, exterior elevations, sections, interior elevations and details.

The architectural sheets are followed by the structural set including footing and foundation plans, basement framing plans, floor framing plans, roof framing plans, and then the structural details. These are followed by the heating and ventilating plans; then the plumbing plans, and finally, the electrical plans, each starting with the lowest floor and working upward.

Each sheet should have a title block in the lower right-hand corner with the sheet number (usually "A-" for architectural, "S-" for structural, "M-" for heating and ventilating, "P-" for plumbing and "E-" for electrical); the number of sheets in each set (i.e., A-1 of 7); the date made plus each date it has been revised, and the initials of the person or persons who drew and approved the sheet.

Revisions are sometimes listed outside the title block. Some architects, rather than revising a drawing once it is printed, will issue supplementary drawings which must be attached to the appropriate set as they are issued.

Type of Drawings. Most working drawings for building construction are based on *orthographic projection,* which is a parallel projection to a plane by lines perpendicular to the plane. In this way, all dimensions will be true. If the plane is horizontal, the projection is a plan; if vertical, it is an elevation for outside the building, or a sectional elevation if through the building.

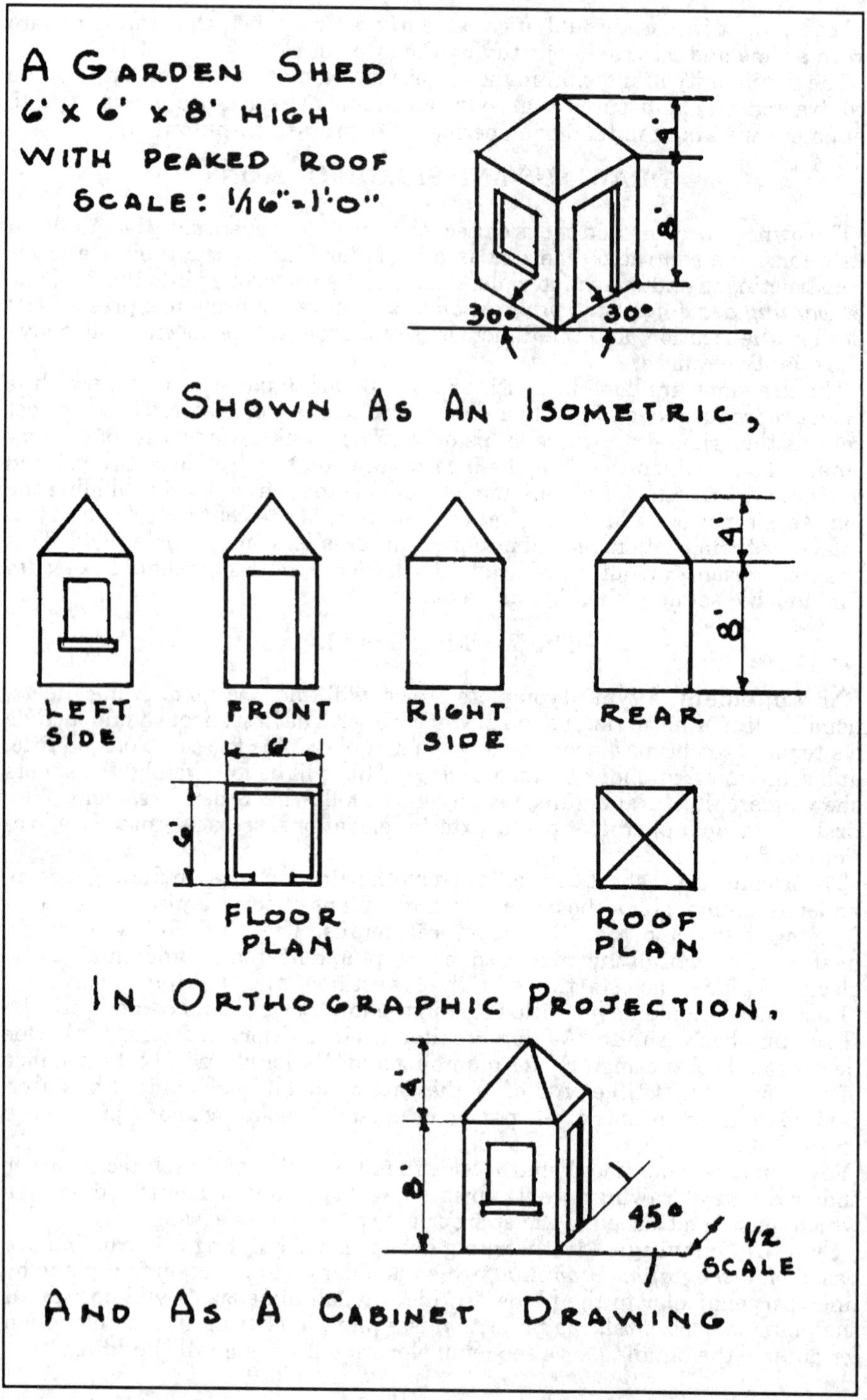
A GARDEN SHED
6' x 6' x 8' HIGH
WITH PEAKED ROOF
SCALE: 1/16" = 1'-0"
4'
8'
30°
30°
SHOWN AS AN ISOMETRIC,
4'
8'
LEFT SIDE
FRONT
RIGHT SIDE
REAR
6'
6'
FLOOR PLAN
ROOF PLAN
IN ORTHOGRAPHIC PROJECTION,
4'
8'
45°
1/2 SCALE
AND AS A CABINET DRAWING

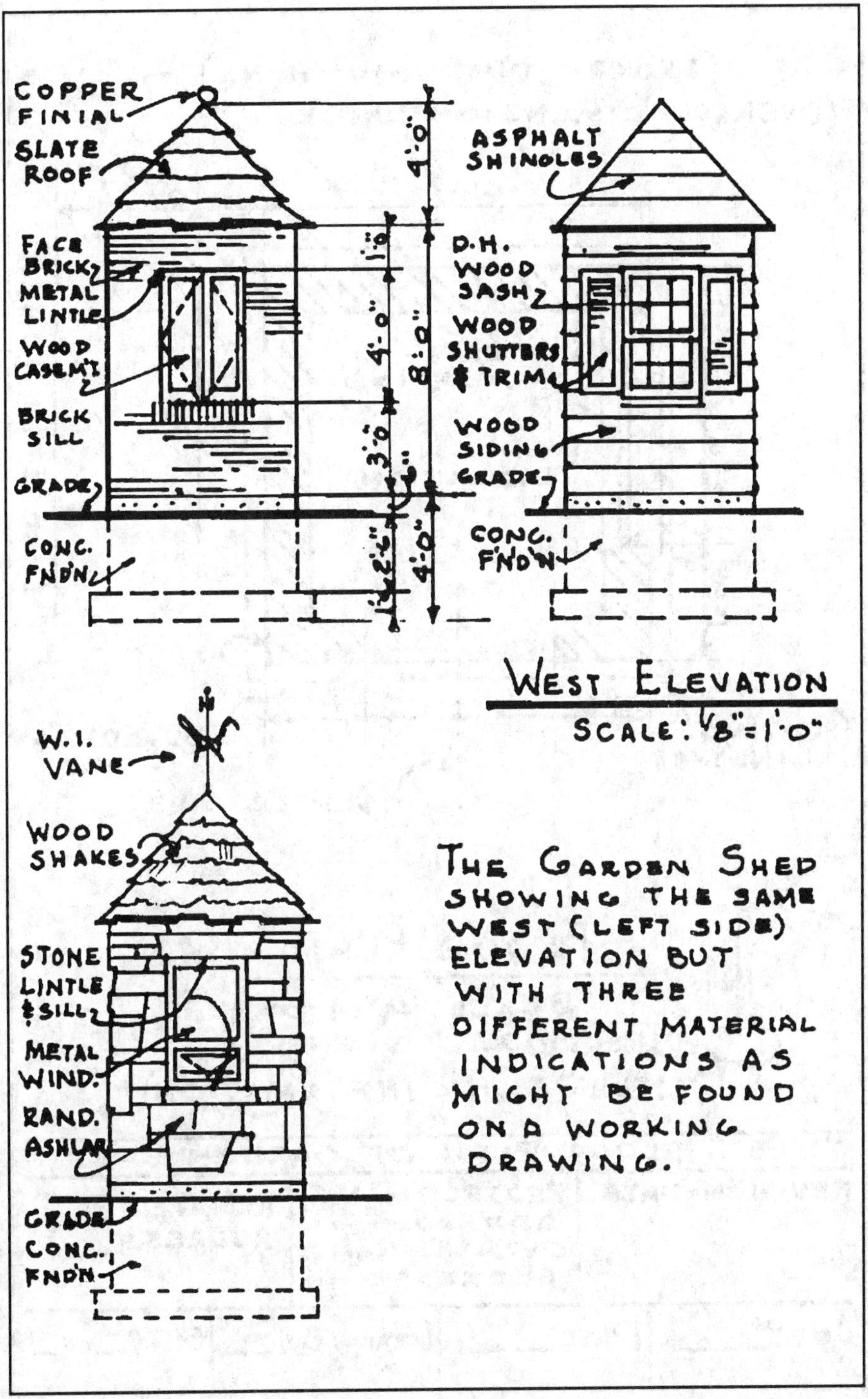
COPPER FINIAL
SLATE ROOF
FACE BRICK
METAL LINTLE
WOOD CASEM'T.
BRICK SILL
GRADE
CONC. FND'N.
4'-0"
1'-0"
4'-0"
8'-0"
3'-0"
1'-2"
2'-6"
4'-0"
ASPHALT SHINGLES
D.H. WOOD SASH
WOOD SHUTTERS & TRIM
WOOD SIDING
GRADE
CONC. FND'N
WEST ELEVATION
SCALE: 1/8"=1'-0"
W.I. VANE
WOOD SHAKES
STONE LINTLE & SILL
METAL WIND.
RAND. ASHLAR
GRADE
CONC. FND'N
THE GARDEN SHED SHOWING THE SAME WEST (LEFT SIDE) ELEVATION BUT WITH THREE DIFFERENT MATERIAL INDICATIONS AS MIGHT BE FOUND ON A WORKING DRAWING.

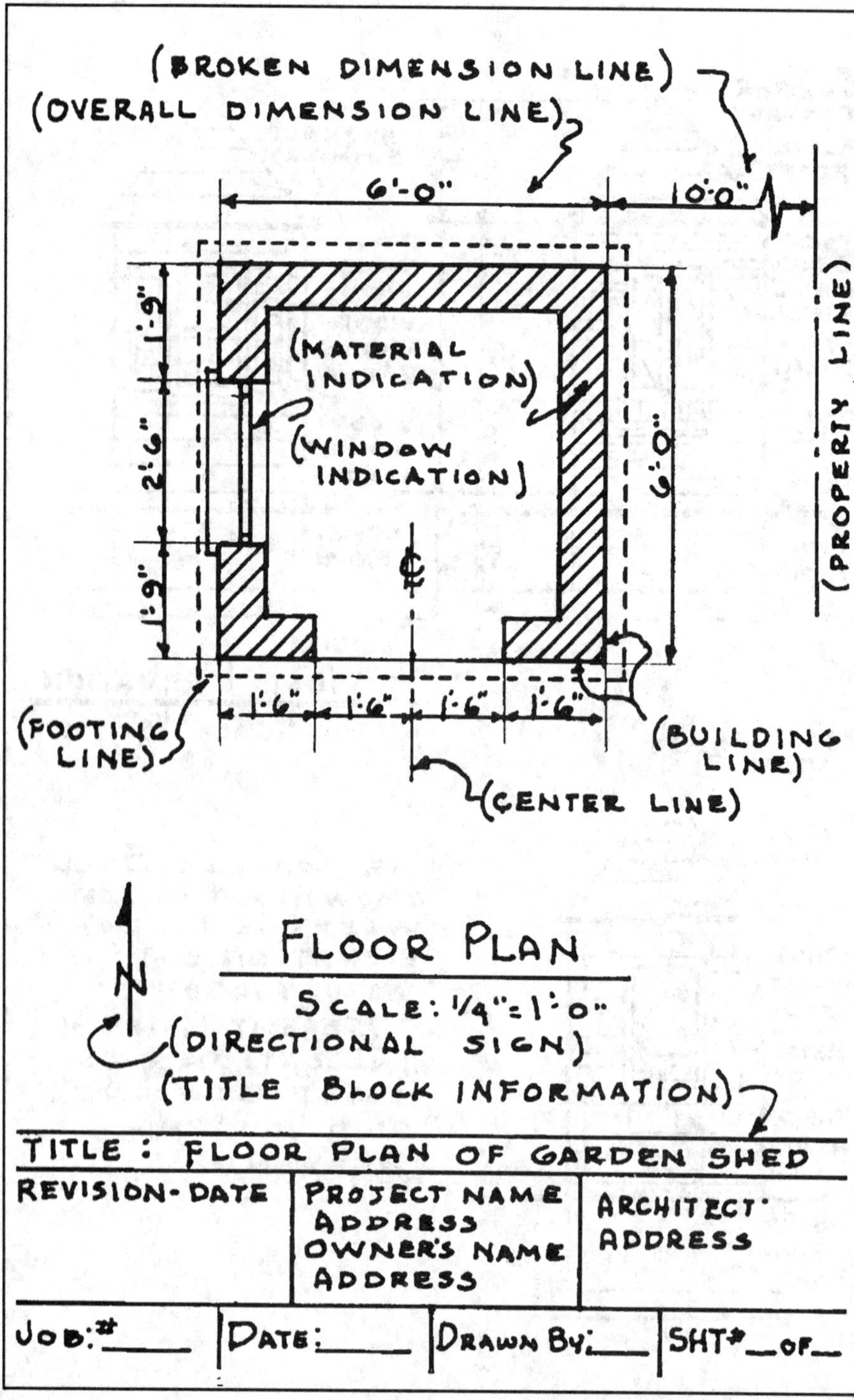
(BROKEN DIMENSION LINE)
(OVERALL DIMENSION LINE)
6'-0"
10'-0"
1'-9"
2'-6"
1'-9"
6'-0"
(MATERIAL INDICATION)
(WINDOW INDICATION)
(PROPERTY LINE)
1'-6"
1'-6"
1'-6"
1'-6"
(FOOTING LINE)
(BUILDING LINE)
(CENTER LINE)
N
FLOOR PLAN
SCALE: 1/4"=1'-0"
(DIRECTIONAL SIGN)
(TITLE BLOCK INFORMATION)
TITLE: FLOOR PLAN OF GARDEN SHED
REVISION-DATE
PROJECT NAME
ADDRESS
OWNER'S NAME
ADDRESS
ARCHITECT
ADDRESS
JOB:#
DATE:
DRAWN BY:
SHT#__OF__

ROOM FINISH SCHEDULE

JOB NO.:
SHEET NO.:
DATE:

ROOM	#	FLOORS													BASE AND/OR WAINSCOT									WALLS					CEILING				TRIM										DECORATING							NOTES
		CEMENT	ASPHALT TILE	RUBBER TILE	VINYL ASBESTOS T.	VINYL	LINOLEUM	CERAMIC TILE	WOOD	SLATE	MARBLE	TERRAZZO	SEAMLESS	CARPET	CEMENT HIGH	RUBBER HIGH	VINYL HIGH	WOOD HIGH	CER. TILE HIGH	SLATE HIGH	MARBLE HIGH	TERRAZZO HIGH	SEAMLESS HIGH	PLASTIC	WALLBOARD	CER. TILE	CONC. BLK	WOOD	PLASTER	WALLBOARD	ACOUSTICAL	WOOD	WOOD CASING	METAL CASING	WOOD SILLS	METAL SILLS	MARBLE SILLS	PLASTIC SILLS	CHAIR RAIL	CORNICE	BUILT-IN CABINET	PAINT FOR FLOORS	PAINT FOR WALLS	PAINT FOR CEILINGS	PAINT FOR TRIM	PAINT FOR DOORS	PAPER FOR WALLS	PAPER FOR CEILINGS		

PRACTICAL
MFG. IN U.S.A.
FORM 491

FRANK R. WALKER CO. PUBLISHERS CHICAGO

The only descriptive drawing that presents a building as the eye sees it is the *perspective*. But perspectives are used mainly to study the building and present it to the client in a form he can easily understand. A perspective is seldom useful for presenting information on working drawings.

However, other pictorial presentations are helpful to the builder. Two of these are the *isometric* and the *cabinet projection*. Isometrics are drawings in which all horizontal and vertical lines have a true length, and those lines parallel on the object are also parallel on the drawing. Vertical lines are vertical, but horizontal lines are set at 30 or 60 degrees. Isometrics are often used for piping diagrams and to show complicated intersections, such as on sloped roofs.

Cabinet drawings are those with the front face shown in true shape and size, as if it were an orthographic projection, but they simulate a perspective. The sides are shown receding at 45 degrees and at ½ scale. Variations of this are *oblique drawings*, where the angle and side scale may be anything that best shows the object, and *cavalier drawings*, where the side scale is at the same scale as the front. Cabinet drawings get their name from the fact that they are often used for cabinet work.

Scale. The architect's scale, with the inch divided into ⅛S, 1⁄16S, 1⁄32S, is standard for building construction in the United States. The engineer's scale, with the inch divided into tenths, is sometimes used in structural work or on site plans. However, it is advantageous to have architectural, structural, and mechanical drawings all at the same scale for a job.

The metric scale is divided into centimeters and millimeters, 2.54 centimeters equalling one inch. It is the common scale in Europe.

Graphic scales are those devised by the draftsman and shown on the drawings with subdivisions comparable to the other scale. This is convenient if the drawing is to be reduced in size in reproduction and is meant to be scaled rather than dimensioned. It is seldom used on working drawings.

Plans and elevations are usually drawn at no less than ⅛" = 1'-0"; plot plans at 1⁄16" = 1'-0". Complicated plan areas, such as toilet rooms, are usually repeated at ¼" to ½" scale. Wall sections and cabinet work must show more detail and are from ¾" to 1 ½" scale. Certain areas of sections will be further detailed at 3" scale, while moldings and decorative items might even be drawn at half to full scale.

Dimensioning. While scales should be clearly identified for each drawing, either in the title block or beneath each individual layout, contractors should never scale drawings but should insist that the architect furnish them with complete stated dimensions. This is the duty of the architect, and the contractor is foolish to take responsibility for it. Blueprints are subject to stretching and shrinking with humidity and some reproductive methods reduce the drawings in size.

Notes on Drawings. The drawings should not be cluttered with notes, which should be in the specifications, but architects will vary on this. The draftsman will often put notes on drawings to alert his own specification writer rather than the contractor. It is the specification that should govern, and all notes on drawings should be checked against the specifications and cross-referenced. If a contractor finds discrepancies during bidding, he should report this to the architect and ask for a written statement, known as an *Addendum*, to clarify it.

Finish Schedules. Schedules may appear either as part of the drawings or the specifications. Room finish schedules are often on the drawings next to the floor plan to which they refer. Schedules for lintels, columns, footings, doors, and frames are usually on the drawings, while hardware schedules are usually in the specifications. Fixture schedules for mechanical and electrical work may be found either place. A room finish schedule, Frank R. Walker Form 401 is shown.

Reproduction. *Blueprints,* with lines showing in white on a blue background, may be required by many building departments for permits, because they are difficult to alter. Notes can only be added in colored pencil, either red or yellow. The contractor should limit the notes he makes on his construction set and authorize only one person to add them so that somebody's doodles don't end up in brick and mortar.

Today, many architects have their own printing equipment. These are usually of the ammonia process which produce black, blue, or maroon lines on a white background. They are much more easily marked on, although the added note will not be as obvious as on a blueprint, unless made in color. Small prints may be run off on one of the many types of office duplicating machines.

If it is desirable to reduce or enlarge a drawing, a *photostat,* which is a photographic process that produces negatives as positives, can be used.

There are processes that transfer blueprints to cloth drawings or to *sepia prints,* which can be altered, added to, and printed just like an original tracing.

Interpreting Lines. The architect will use various lines in making his drawings. Thick solid lines are used to outline the edges of the plans and elevations. Lines consisting of long and short dashes in sequence denote center lines. Short dashes usually indicate hidden lines but are also used on windows to show where they are hinged if they are operable. Thin lines are used for dimensions. These terminate in arrows to show the limit of the given dimensions. Repeat marks or the notation "do" indicates the repetition of a given dimension. Dimension lines are often broken by a zigzag which shows the object in the drawing has been shortened.

Symbols

Material indications and items related to plumbing, heating and ventilation, and electrical work have been standardized. Some architects may have a symbol schedule on their drawings, but most no longer do. The following are many of the usual symbols a contractor may encounter, but always check with the specifications for building materials, because symbols may vary from draftsman to draftsman.

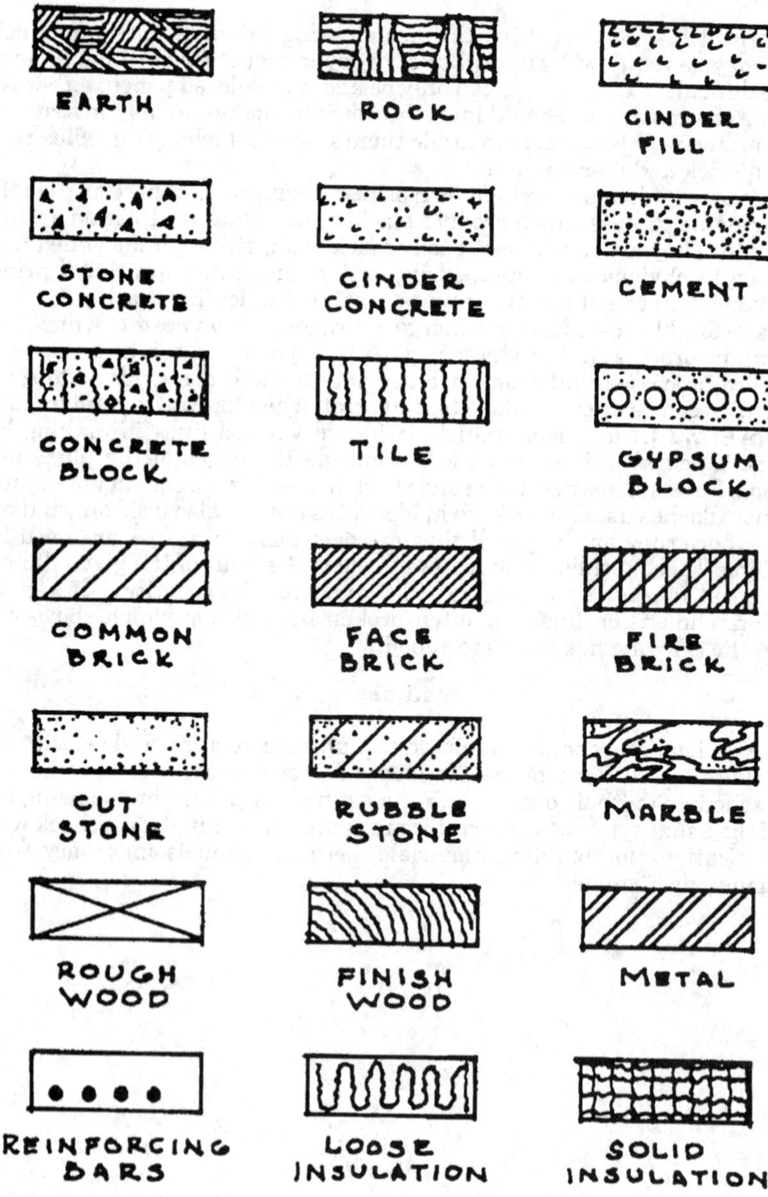
EARTH
ROCK
CINDER FILL
STONE CONCRETE
CINDER CONCRETE
CEMENT
CONCRETE BLOCK
TILE
GYPSUM BLOCK
COMMON BRICK
FACE BRICK
FIRE BRICK
CUT STONE
RUBBLE STONE
MARBLE
ROUGH WOOD
FINISH WOOD
METAL
REINFORCING BARS
LOOSE INSULATION
SOLID INSULATION

GENERAL OUTLETS:

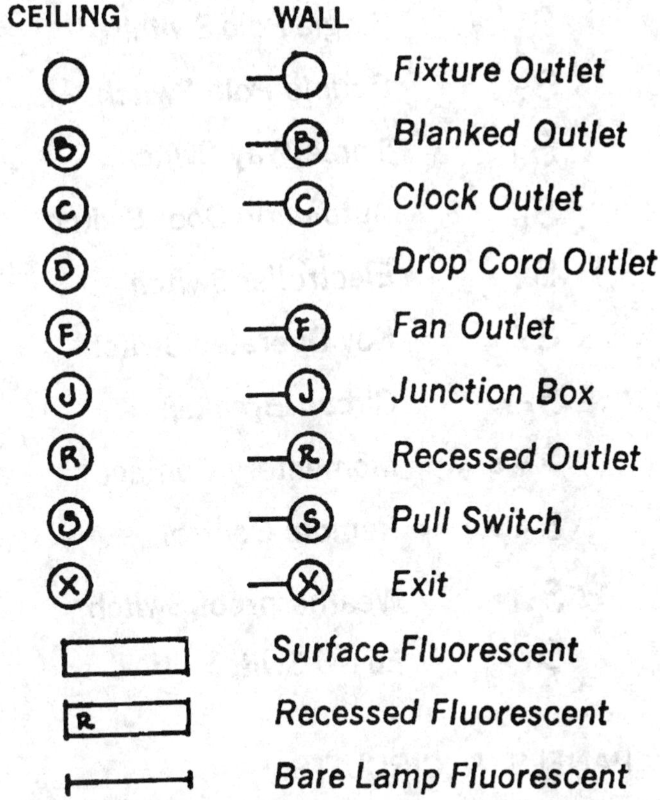

CONVENIENCE OUTLETS:

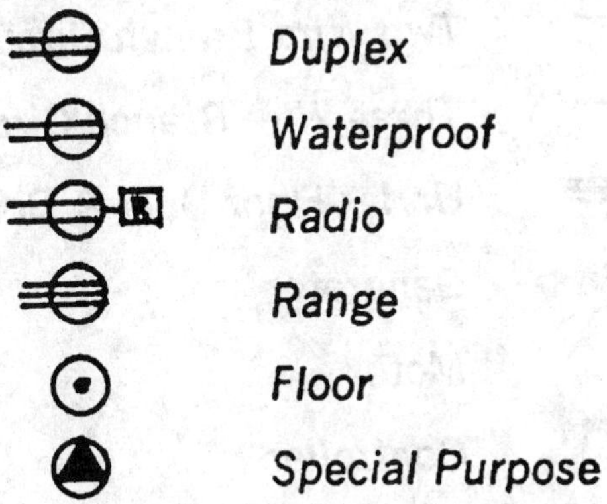

SWITCH OUTLETS:

S	*Single Pole Switch*
S_2	*Double Pole Switch*
S_3	*Three Way Switch*
S_D	*Automatic Door Switch*
S_E	*Electrolier Switch*
S_K	*Key Operated Switch*
S_{CB}	*Circuit Breaker*
S_{MC}	*Momentary Contact*
S_{RC}	*Remote Control*
S_{WP}	*Weatherproof Switch*
S_F	*Fused Switch*

PANELS & CIRCUITS:

 *Lighting Panel*

 *Power Panel*

 Two Wire Branch Circuit

 *Three Wire Branch Circuit*

 *Under Floor Duct & Box*

Generator

Motor

 *Controller*

SIGNALLING SYSTEMS:

	*Push Button*
	*Buzzer*
	*Bell*
	*Chime*
	*Annunciator*
	*Electric Door Opener*
	*Telephone Switchboard*
	*Interconnecting Telephone*
	*Outside Telephone*
	*Fire Alarm Bell*
	*Fire Alarm Station*
	*City Fire Alarm*
	*Automatic Fire Alarm*
	*Watchman's Station*
	*Horn*
	*Nurse's Signal Plug*
	*Maid's Signal Plug*
	*Radio Outlet*
	*Television Outlet*
	*Interconnection Box*

PIPE FITTING SYMBOLS:

————————	Low Pressure Steam
—/—/—/—	Medium Pressure Steam
—//—//—//—	High Pressure Steam
--- — — — —	Return Line
—o—o—o—	Condensate Line
·—·—·—·—	Make Up Water
—— — — — —	Air Relief Line
——FOF——	Fuel Oil Flow
——FOR ——	Fuel Oil Return
——FOV ——	Fuel Oil Vent
——A ——	Compressed Air
——RL——	Refrigerant Liquid
——RD ——	Refrigerant Discharge
——C ——	Condenser Water Flow
- - - -CR- - -	Condenser Water Return
——CH ——	Chilled Water
- - -CHR- - -	Chilled Water Return
—·—H —·—	Humidification Line
——D ——	Drain Line
——B ——	Brine Supply

HEATING SYMBOLS:

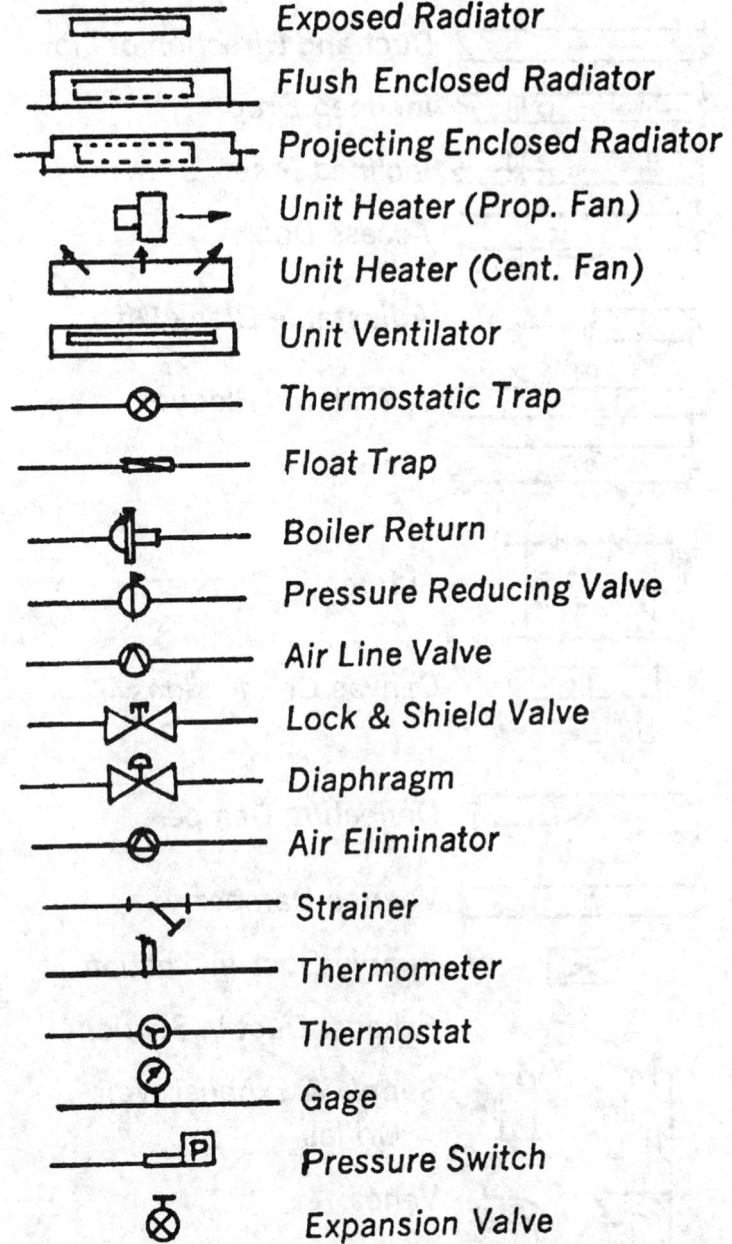

 Exposed Radiator

 Flush Enclosed Radiator

 Projecting Enclosed Radiator

 Unit Heater (Prop. Fan)

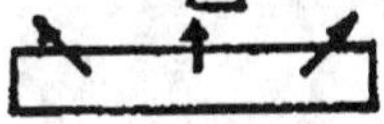 Unit Heater (Cent. Fan)

 Unit Ventilator

 Thermostatic Trap

 Float Trap

 Boiler Return

 Pressure Reducing Valve

 Air Line Valve

 Lock & Shield Valve

 Diaphragm

 Air Eliminator

 Strainer

 Thermometer

 Thermostat

 Gage

Pressure Switch

 Expansion Valve

DUCTWORK SYMBOLS:

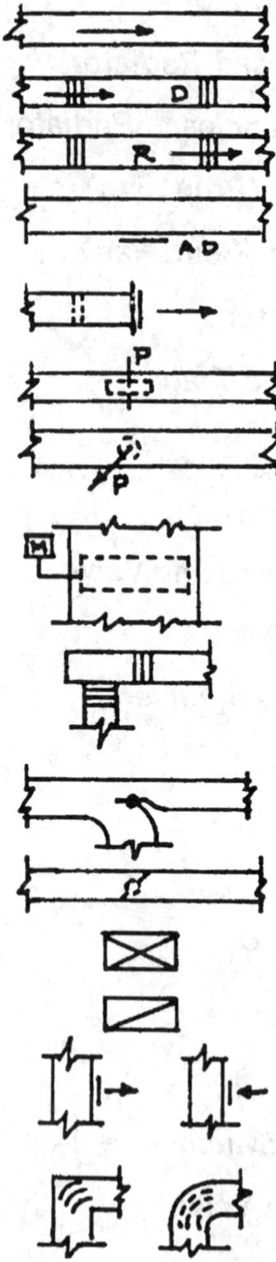

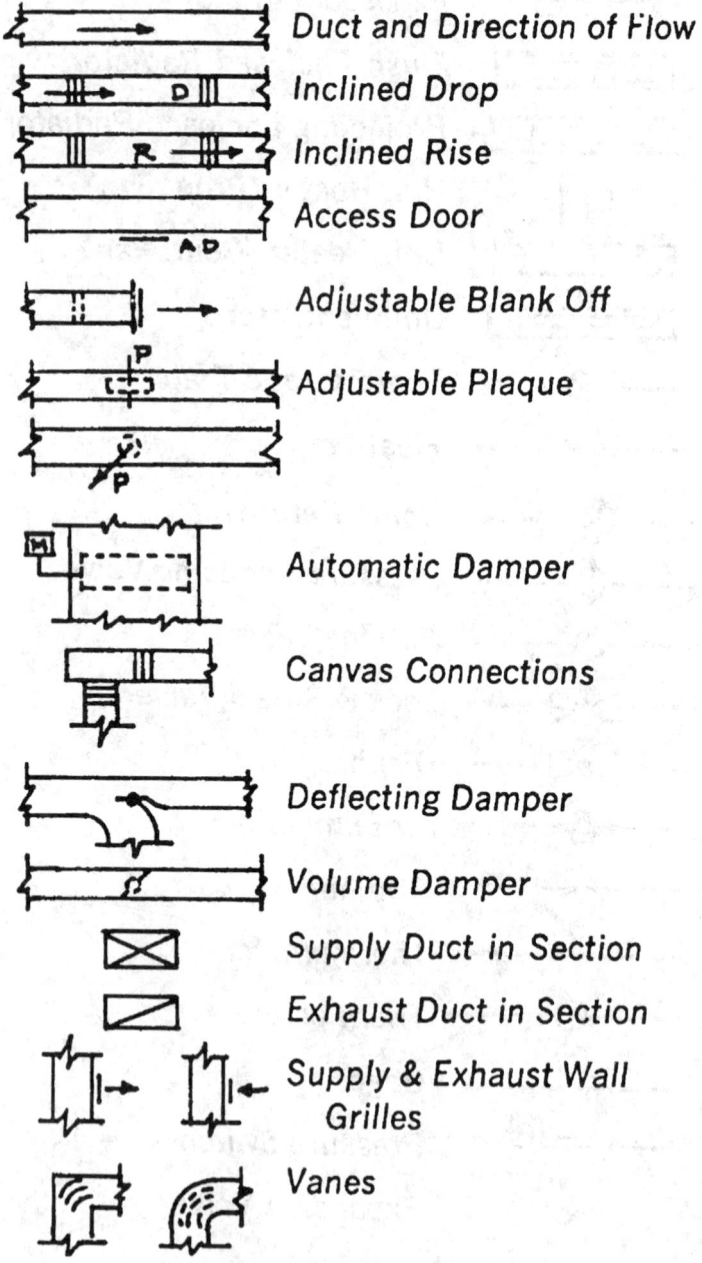

Duct and Direction of Flow

Inclined Drop

Inclined Rise

Access Door

Adjustable Blank Off

Adjustable Plaque

Automatic Damper

Canvas Connections

Deflecting Damper

Volume Damper

Supply Duct in Section

Exhaust Duct in Section

Supply & Exhaust Wall Grilles

Vanes

PIPING SYMBOLS:

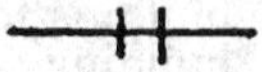	*Flanged Joint*
	*Screwed Joint*
	*Bell & Spigot Joint*
	*Welded Joint*
	*Soldered Joint*
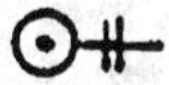	*Flanged Elbow-Up*
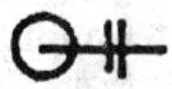	*Flanged Elbow-Down*
	*Flanged Tee-Up*
	*Flanged Tee-Down*
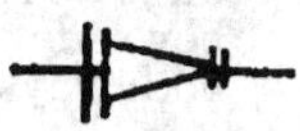	*Flanged Reducer-Concentric*
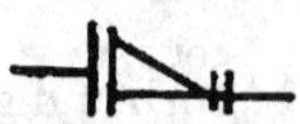	*Flanged Reducer-Eccentric*
	*Flanged Gate Valve*
	*Flanged Globe Valve*
	*Flanged Check Valve*
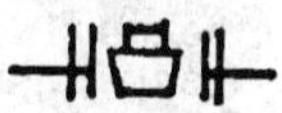	*Flanged Cock Valve*
	*Flanged Safety Valve*
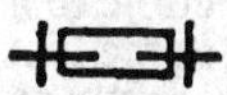	*Flanged Expansion Joint*
	*Reducing Flange*
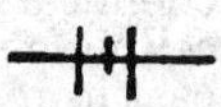	*Flanged Union*
	*Flanged Sleeve*

Abbreviations. The following is a list standard abbreviations used within the construction industry. The list is alphabetized by the abbreviations.

A	ampere; air
AB	anchor bolt
ABAND	abandoned
ABT	about
ABUT	abutment
ABV	above
AC	alternating current; acoustical
A/C	air conditioning
ACC	access
ACFL	access floor
ACPL	acoustical plaster
ACR	acrylic plastic
ACS	access
ACT	acoustical tile
AD	area drain; access door
ADD	addendum; additional
ADH	adhesive
ADJ	adjacent; adjustable
ADJT	adjustable
ADMIN	administration
AFF	above finished floor
AGG	aggregate
AL	aluminum
ALT	alternate
AMP	amperage; ampere
ANC	anchor
ANOD	anodized
AP	access panel
APPR	approved
APPROX	approximate
APX	approximate
ARCH	architect; architectural
ASB	asbestos
ASC	above suspended ceiling
ASPH	asphalt
ASSY	assembly
AT	asphalt tile
AUTO	automatic
AUX	auxiliary
AVG	average
AWG	American Wire Gauge
BAL	balance
BBD	bulletin board
BBL	barrel
BC	bottom of curb
BD	board
BDL	bundle
BEL	below
BET	between
BF	board foot, board feet
BH	baseboard heater
BIT	bituminous
BJT	bed joint
BL	building line
BLDG	building
BLK	block
BLKG	blocking
BLW	below
BLW FL	below floor
BM	bench mark; beam
BOT	bottom
BP	back plaster; base plate; bearing pile
BPL	bearing plate
BRG	bearing
BRK	brick
BRKR	breaker
BRZ	bronze
BS	both sides
BSMT	basement
BT	bent
BTM	bottom
BTU	British Thermal Unit
BTUH	British Thermal Unit per hour
BU	built up
BUR	built up roofing
BVL	beveled
BW	both ways
C	course
C & G	curb and gutter
CA	compressed air
CAB	cabinet
CAD	cadmium
CANTIL	cantilever
CAP	capacity
CB	catch basin
CBLSTN	cobblestone
CCF	hundred cubic feet
CEM	cement
CENT	centrifugal
CER	ceramic
CF	cubic foot; cellular floor
CFL	counterflashing
CFM	cubic foot per minute
CFS	cubic foot per second
CG	corner guard
CHAM	chamfer
CHBD	chalkboard
CHEM	chemical
CHT	ceiling height
CI	cast iron
CIP	cast in place; cast-iron pipe
CIPC	cast-in-place concrete
CIR	circle
CIRC	circ; circumference
CJT	control joint

CK	caulk
CLG	ceiling
CLJ	controlled joint
CLL	contract limit line
CLOS	closure
CLR	clear
CLS	closure
CM	centimeter
CMP	corrugated metal pipe
CMT	ceramic mosaic tile
CMU	concrete masonry unit
COL	column
COMB	combination
COMP	compression
COMPO	composition
COMPT	compartment
CONC	concrete
CONST	construction
CONT	continuous
CONTR	contract
COORD	coordinate
CORR	corrugated
CPM	critical path method
CPR	copper
CPT	carpet
CR	chromium
CRG	cross grain
CRS	course
CS	countersink; combined sewer
CSMT	casement
CST	cast stone
CT	ceramic tile
C to C	center to center
CTR	counter
CTSK	countersunk screw
CTWT	counter weight
CU	cubic
CUFT	cubic foot
CULV	culvert
CUYD	cubic yard
CWT	hundred pounds
CY	cubic yard
CX	connection
D	drain; depth; discharge
d	penny (nail size, i.e., 10d = 10 penny nail)
DA	doubleacting
DB	direct burial
DBL	double
DC	direct current
DEFL	deflect; deflection
DEG	degree
DEM	demolish
DEMOB	demobilization
DEP	depressed
DEPT	department
DET	detail
DF	drinking fountain
DH	double hung
DHW	domestic hot water
DI	drop inlet; ductile iron
DIA	diameter
DIAG	diagonal
DIAM	diameter
DIFF	diffuser
DIM	dimension
DIP	ductile iron pipe
DIR	direction
DISC	disconnect
DISCH	discharge
DIV	division
DK	deck
DL	dead load
DMH	drop manhole
DML	demolition; demolish
DMT	demountable
DN	down
DP	dampproofing
DPR	damper or dispenser
DR	door; drive
DRB	drainboard
DS	downspout
DT	drain tile
DTA	dovetail anchor
DTL	detail
DTS	dovetail anchor slot
DUP	duplicate
DW	dumbwaiter
DWG	drawing
DWL	dowel
DWR	drawer
E	east
EB	expansion bolt
EE	each end
EF	each face
EJ	expansion joint
EL	elevation
ELEC	electric
ELEV	elevator; elevation
EM	expanded metal
EMER	emergency
EMH	electrical manhole
EMT	electrical metallic conduit; thin wall conduit
ENC	enclose
ENG	engine
ENGR	engineer
ENT	entering
ENTR	entrance
EP	electrical panelboard; explosion proof; edge of pavement
EQ	equal
EQPT	equipment

EQUIP	equipment
ESC	escalator
EST	estimate
E to E	end to end
EW	each way
EWC	electric water cooler
EXC	excavation; excavate
EXCA	excavate
EXG	existing
EXH	exhaust
EXMP	expanded metal plate
EXP	exposed
EXS	extra strong
EXT	exterior
EXTN	extension
F	female; fill
f	fahrenheit
FA	fire alarm
FAB	fabricate
FAS	fasten
FB	face brick
FBD	fiberboard
FBO	furnished by others
FBRK	fire brick
FC	foot-candle
FD	floor drain; fire damper
FDN	foundation
FDR	feeder
FE	fire extinguisher
FEC	fire extinguisher cabinet
FF	factory finish; factory fresh; far face
FFE	finished floor elevation
FFL	finished floor line
FGL	fiberglass
FH	fire hydrant
FHC	fire hose cabinet
FHMS	flathead machine screw
FHS	fire hose station
FHWS	flathead wood screw
FIG	figure
FIN	finish
FJT	flush joint
FL	floor
FLCO	floor cleanout
FLG	flashing; flange
FLR	floor
FLSH	flashing
FLUR	fluorescent
FLX	flexible
FN	fence
FND	foundation
FO	fiber optics
FOC	face of concrete
FOF	face of finish
FOM	face of masonry
FOS	face of studs
FP	fireproof
FPL	fireplace or floor plate
FPM	foot (feet) per minute
FR	frame
FRA	fresh air
FRC	fire-resistant coating
FRG	forged
FRT	fire-retardant
FS	full size; far side
FT	foot; feet
FTG	footing
FT LB	foot-pound
F to F	face to face
FUR	furred
FURN	furnish; furniture
FUT	future
FW	fire water
G	gauge; gas; gas main or service
g	gram
GA	gauge
GAL	gallon
GALV	galvanized
GB	grab bar
GC	general contractor
GCMU	glazed concrete masonry unit
GD	grade
GEN	generator; general
GF	ground face
GI	galvanized iron
GKT	gasket
GL	glass or glazing
GLB	glass block
GLF	glass fiber
GOVT	government
GP	galvanized pipe
GPD	gallons per day
GPDW	gypsum dry wall
GPH	gallons per hour
GPL	gypsum lath
GPM	gallons per minute
GPPL	gypsum plaster
GPT	gypsum tile
GR	grade
GRN	granite
GSS	galvanized steel sheet
GST	glazed structural tile
GT	grout
GV	galvanized
GVL	gravel
GYP	gypsum
H	height
HB	hose bibb
HBD	hardboard
HC	handicapped; hollow core; high capacity
HD	heavy duty
HDR	header

HDW	hardware
HES	high early-strength cement
HH	handhole
HJT	head joint
HK	hook
HM	hollow metal
HOR	horizontal
HORIZ	horizontal
HP	horse power; high pressure
HPR	high pressure return
HPS	high pressure steam
HR	hour
HT	height
HTG	heating
HV	high voltage; heating/ ventilating unit
HVAC	heating/ventilating/ air conditioning
HW	hot water
HWD	hardwood
HWH	hot water heater
HWR	hot water return
HX	hexagonal
HYD	hydrant
HZ	hertz
ID	inside diameter; identification
ILK	interlock
IN	inch
INCIN	incinerator
INCL	inclusive
INS	insulation
INSC	insulating concrete
INSF	insulating fill
INSTL	install
INSUL	insulation; insulated; insulator; insulating
INT	interior
INTM	intermediate
INV	invert
IP	iron pipe
IPS	iron pipe size
IR	irrigation
J	joist
JB	junction box
JC	janitor's closet
JCT	junction
JF	joint filler
JT	joint
KCPL	Keene's cement plaster
KG	kilogram
KIT	kitchen
KLF	kips per lineal foot
KO	knockout
KPL	kickplate
KSF	kips per square foot
KSI	kips per square inch
KW	kilowatt
L	length; left; liter; long
LAB	laboratory
LAD	ladder
LAM	laminated
LAV	lavatory
LB	lag bolt
LBL	label
LBS	pounds
LBS/HR	pounds per hour
LBS/LF	pounds per lineal foot
LBS/ SQIN	pounds per square inch
LC	light control
LF	lineal foot; lineal feet
LG	long; length
LH	left hand
LIQ	liquid
LL	live load
LLH	long leg horizontal
LLV	long leg vertical
LMS	limestone
LOC	locate; local
LONG	longitudinal
LP	lightproof; light pole
LPT	low point
LS	lump sum
LT	light
LTL	lintel
LV	low voltage
LVR	louver
LW	lightweight
LWC	lightweight concrete
M	meter; thousand
MACH	machine
MAINT	maintenance
MAS	masonry
MATL	material
MAX	maximum
MB	machine bolt
MBF	thousand board feet
MBH	thousand BTU per hour
MBR	member
MC	medicine cabinet
MCC	motor control center
MECH	mechanical
MED	medium
MET	metal
MFD	metal floor deck
MFG	manufacture
MFR	manufacturer
MGD	million gallons per day
MH	manhole

MHW mean high water
MI mile; malleable iron
MID middle
MIN minimum
MIR mirror
MISC miscellaneous
MJ mechanical joint
MK mark
MLD moulding
MLW mean low water
MM millimeter
MMB membrane
MO month; masonry opening
MOB mobilization
MOD modular
MON monument
MOV movable
MPH miles per hour
MR mop receptor
MRB marble
MRD metal roof deck
MSL mean sea level
MTD mounted
MTFR metal furring
MTHR metal threshold
MTL material
MULL mullion
MWK millwork
MWP membrane waterproofing

N north
NA not applicable
NAT natural
NEC national electrical code
NF near face
NI nickel
NIC not in contract
NL nailable
NMT nonmetallic
NO. number
NOM nominal
NR noise reduction
NRC noise reduction coefficient
NTS not to scale

OA overall
O & P overhead and profit
OBS obscure
OC on center
OD outside diameter
OF outside face
OH overhead
OHMS ovalhead machine screw
OHWS ovalhead wood screw
OJ open-web joist
OP opaque
OPNG opening
OPH opposite hand
OPP opposite
OPS opposite surface
O to O out to out
OZ ounce

P pole; power; pump; page
PAR parallel
PB panic bar; push button
PBD particle board
PCC precast concrete
PCF pounds per cubic foot
PE porcelain enamel
PED pedestal
PEJ premolded expansion joint
PERF perforated
PERI perimeter
PFB prefabricated
PFL pounds per lineal foot
PFN prefinished
PG plate glass
PK parking
PL plate or property line
PLAM plastic laminate
PLAS plaster
PLY plywood
PLYWD plywood
PNL panel
PNT painted
PR pair
PRC precast reinforced concrete
PREFAB prefabricated
PRESS pressure
PRF preformed
PROV provide
PSC prestressed concrete
PSF pounds per square foot
PSI pounds per square inch
PSIG pounds per square inch gaug
PT point; point of tangent
PTC post-tensioned concrete
PTD paper towel dispenser
PTN partition
PTR paper towel receptor
PV paving
PVC polyvinyl chloride
PVMT pavement
PWR power

QT quarry tile

R riser; radius
RA return air
RAD radius
RB rubber base
RBL rubble stone
RBT rabbet; rebate; rubber tile
RBR rubber
RCP reinforced concrete pipe

RCPT receptacle
RD roof drain
RE reinforced
REF reference
REG register; regular
REINF reinforcement
REM remove
REQD required
RES resilient
RET return
REV revision or revised
RFG roofing
RFH roof hatch
RFL reflected
RH right hand
RL railing; roof ladder
RM room
RND round
RO rough opening
ROW right of way
RR railroad
RVS reverse
RVT rivet
R/W right of way
RWC rainwater conductor

S south; sanitary sewer; supply; sign
SAN sanitary
SBSTA substation
SC solid core
SCH schedule
SCHED schedule
SCN screen
SCT structural clay tile
SD storm drain
SEC section
SF square foot
SFGL safety glass
SG sheet glass
SH shelf
SHO shoring
SHT sheet
SHTH sheathing
SIM similar
SK sketch
SKL skylight
SL sleeve; slope
SMH sanitary manhole
SNT sealant
SP soundproof
SPC spacer
SPEC specification
SPK speaker
SPL special
SPR sprinkler
SQ square; 100 square feet
SQ FT square foot; square feet
SQ YD square yard
SS sanitary sewer
SSK service sink
SST stainless steel
ST steel
STA station
STD standard
STG seating
STL steel
STO storage
STR structural
STRL structural
STY story
SUBS subcontractors
SURF surface
SUS suspended
SVCE service
S/W sidewalk
SWGR switchgear
SY square yard
SYM symmetrical
SYMM symmetrical
SYN synthetic
SYS system

T tread; top; tangent; telephone
T & B top and bottom
T & G tongue and groove
TB towel bar
TC terra cotta; top of curb
TCP terra cotta pipe
TEL telephone
TEMP temperature
TERM terminal
THHN nylon jacked wire
THK thickness
THR threshold
THRU through
THW insulated strand wire
THWN nylon jacked wire
TKBD tackboard
TKS tackstrip
TOC top of concrete
TOL tolerance
TOS top of steel
TOT total
TPC top of pile cap
TPD toilet paper dispenser
TPTN toilet partition
TR transom
T/R top of rail
TRANS transformer
TRK track
TS top of steel; top of stone
TSL top of slab
TV television
TW top of wall

TYP	typical
TZ	terrazzo
UC	undercut
U/C	under construction
UCI	uniform construction index
UG	underground
UL	underwriters laboratory
UNF	unfinished
UNIF	uniform
UNO	unless otherwise noted
UR	urinal
USGS	United States Geological Survey
UTIL	utility
V	voltage; volt; vent; valve
VAC	vacuum
VAR	varnish
VAT	vinyl asbestos tile
VB	vapor barrier or vinyl base
VC	vertical curve
VEL	velocity
VENT	ventilate
VERT	vertical
VF	vinyl fabric
VG	vertical grain
VIB	vibration
VIN	vinyl
VJ	v-joint
VLF	vertical lineal feet
VNR	veneer
VOL	volume
VRM	vermiculite
VT	vinyl tile
W	west; wide; watt; water main; wire
W/	with
WB	wood base
WC	water closet
WD	wood
WDO	window
WE	water elevation
WF	wide flange
WG	wired glass; water gage
WH	wall hung
WHB	wheel bumper
WI	wrought iron
WIN	window
WM	wire mesh; water meter
WO	without
W/O	without
WP	waterproofing
WPT	working point
WR	water repellent
WS	waterstop
WSCT	wainscot
WST	waste
WT	watertight; weight
WTR	water
WTW	wall to wall
WV	water valve
WWF	welded wire fabric
XARM	cross arm
XFR	transfer
XFMR	transformer
XHD	extra heavy duty
Y	wye
YD	yard
YHYD	yard hydrant
YI	yard inlet
YR	year

Specifications

Arrangement. There has been much study of the proper way to set up specifications, and the industry has made progress toward standardizing. The *Uniform System* developed by the Construction Specifications Institute (CSI) has become the accepted standard. This system not only sets forth an arrangement for specifications, but it is equally applicable to data filing (most of the manufacturers' new catalogs today carry the CSI symbol, a rectangle enclosing three ellipses, and the Uniform System number) and to cost accounting systems. The CSI form divides the specifications into four sections: bidding requirements, contract forms, general conditions, and specifications. The specifications section is broken down into 16 divisions. A typical specification prior to the CSI format might have run to 60 sections. The present suggested divisions, under the fourth section *Specifications*, are as follows:

Division 1. General Requirements, including summary of work, schedules and reports, samples and shop drawings, temporary facilities, cleaning up, project closeout, allowances, and alternates.

Division 2. Site Work, including clearing of site, earthwork, piling, caissons, shoring and bracing, site drainage, site utilities, roads and walks, site improvements, lawns and planting, railroad work, and marine facilities.

Division 3. Concrete, including formwork, reinforcement, cast-in-place concrete, precast concrete, and cementitious decks.

Division 4. Masonry, including unit masonry, stone and masonry restoration.

Division 5. Metals, including structural, joists, decking, lightgauge framing, miscellaneous metals, ornamental metals, and special formed metals.

Division 6. Carpentry, including rough, finish, glue laminated, and custom woodwork.

Division 7. Moisture Control, including waterproofing, dampproofing, insulation, shingles and roof tiles, preformed roofing and siding, membrane roofing, sheet metal work, wall flashing, roof accessories, and caulking and sealants.

Division 8. Doors, Windows and Glass, including metal doors and frames, wood doors, special doors, metal windows, wood windows, finish hardware, operators, weatherstripping, glass and glazing, curtain walls, and store fronts.

Division 9. Finishes, including lath and plaster, gypsum, tile, terrazzo, veneer stone, acoustics, wood flooring, resilient flooring, special flooring, special coatings, painting and wall coverings.

Division 10. Specialties, including chalk and tack boards, chutes, compartments and cubicles, demountable partitions, disappearing stairs, firefighting devices, fireplace equipment, flagpoles, folding gates, identifying devices, lockers, mesh partitions, postal specialties, retractable partitions, scales, storage shelving, sun control devices, telephone booths, toilet and bath accessories, vending machines, wardrobe specialties, and waste disposal units.

Division 11. Equipment, including that for banks, commerce, darkrooms, churches, education, food service, gymnasiums, industry, laboratories, laundries, libraries, medicine, mortuaries, parking, prisons, residences, and stage.

Division 12. Furnishings, including artwork, light control, cabinets, carpets, drapes, furniture, and seating.

Division 13. Special Construction, including audio, bowling, broadcasting, clean rooms, conservatory, incinerator, integrated ceiling, observatory, pedestal floor, prefabricated, chimney, vault, swimming pool, and zoo.

Division 14. Conveying Systems, including elevators, dumbwaiters, hoists and cranes, lifts, material handling, escalators, and tubes.

Division 15. Mechanical, divided into general provisions, basic materials, and methods; and plumbing, heating, ventilating, and cooling, covering water supply, soil and waste, roof drainage, plumbing fixtures, gas piping, special piping, fire extinguishing, fuel handling, steam, hot water, chilled water, dual temperature air tempering, refrigeration, and controls and instruments.

Division 16. Electrical, including general provisions, basic materials and methods, electrical service, distribution, fixtures, communication, TV, power, heating, controls, and lighting protection.

This system is the generally accepted form for specifications today. This book is organized according to the system. The CSI Uniform System listing of 16 divisions and its subdivisions is as follows:

01	**General Requirements**
01010	Summary Of Work
01019	Contract Considerations
01020	Allowances
01100	Alternatives
01200	Project Meetings
01300	Submittals
01390	Coordination & Meetings
01400	Quality Control
01500	Construction Facilities & Temporary Controls
01600	Material & Equipment
01650	Starting Of Systems
01700	Contract Closeout
01999	Miscellaneous
02	**Site Work**
02000	Alternatives
02010	Subsurface Exploration
02011	Borings
02012	Core Drilling
02013	Standard Penetration Tests
02014	Seismic Exploration
02060	Building Demolition
02072	Minor Demolition For Remodeling
02100	Clearing
02102	Clearing & Grubbing
02103	Tree Pruning
02104	Shrub & Tree Relocation
02110	Site Clearing
02120	Structure Moving
02200	Earthwork
02210	Site Grading
02211	Rough Grading
02212	Embankment
02220	Excavating & Backfilling
02222	Excavation
02223	Backfilling
02224	Pipe Boring & Jacking
02225	Trenching
02226	Structure Backfill & Compaction
02227	Waste Material Disposal
02229	Rock Removal
02230	Soil Compaction Control
02231	Aggregate Base Course
02240	Soil Stabilization
02242	Soil Cement Stabilization
02252	Vegetation Control
02281	Termite Control
02300	Pile Foundations
02350	Caissons
02351	Drilled Caissons
02352	Excavated Caissons
02356	Pile Load Test
02361	Wood Friction Piles
02364	Concrete Filled Steel Piles
02365	Pressure Injected Footings
02367	Prestressed Concrete Piles
02368	Rolled Steel Section Piles
02371	Drilled Concrete Piers
02400	Shoring
02420	Underpinning
02500	Site Drainage
02510	Asphaltic Concrete Paving
02511	Crushed Stone Paving
02516	Asphaltic Block Pavers
02517	Stone Pavers
02518	Interlocking Pavers
02519	Brick Pavers
02520	Portland Cement Concrete Pavers
02550	Site Utilities
02600	Paving & Surfacing
02610	Paving
02620	Curbs & Gutters
02630	Walks
02640	Synthetic Surfacing

02675 Disinfection Of Water Distribution System
02700 Site Improvements
02710 Fences & Gates
02712 Foundation Drainage System
02720 Road & Parking Appurtenances
02722 Site Storm Sewerage Systems
02730 Playing Fields
02732 Site Sanitary Sewerage Systems
02740 Fountains
02750 Irrigation System
02760 Site Furnishings
02800 Landscaping
02810 Soil Preparation
02811 Landscape Irrigation
02820 Lawns
02850 Railroad Work
02851 Trackwork
02852 Ballasting
02900 Marine Work
02910 Docks
02920 Boat Facilities
02923 Landscape Grading
02930 Protective Marine Structures
02931 Fenders
02932 Seawalls
02933 Groins
02934 Jettys
02936 Seeding
02938 Sodding
02940 Dredging
02950 Trees, Plants, & Ground Cover
02960 Tunneling
02970 Tunnel Grouting
02980 Support Systems
02999 Miscellaneous

03 Concrete

03000 Alternatives
03100 Concrete Formwork
03150 Expansion & Contraction Joints
03200 Concrete Reinforcement
03210 Steel Bar & Welded Wire Reinf.
03230 Stressing Tendons
03300 Cast-in-place Concrete
03305 Concrete Curing
03310 Concrete
03320 Lightweight Concrete
03321 Insulating Concrete
03322 Lightweight Structural Concrete
03330 Heavyweight Concrete
03340 Prestressed Concrete
03346 Concrete Floor Finishing
03350 Specially Finished Concrete
03351 Exposed Aggregate Concrete
03352 Bushhammered Concrete
03353 Blasted Concrete
03354 Heavy-duty Concrete Floor Finish
03355 Exposed Aggregate Concrete Finish
03360 Specially Placed Concrete
03361 Shotcrete
03366 Post Tension Structural Concrete
03370 Concrete Curing
03400 Precast Concrete
03410 Precast Concrete Panels
03411 Structural Precast Panel
03415 Precast Concrete Hollow Core Planks
03420 Precast Structural Concrete
03430 Precast Prestressed Concrete
03451 Architectural Precast Concrete
03455 Glass Reinforced Concrete
03470 Tilt-up Precast Concrete
03505 Self-leveling Underlayment
03500 Cementitious Decks
03510 Gypsum Concrete
03520 Cementitious Wood Fiber Deck
03521 Insulating Concrete Fill
03721 Preparation For Resurfacing Concrete
03732 Concrete Repair
03999 Miscellaneous

04 Masonry

04000 Alternatives
04100 Mortar & Masonry Grout
04150 Masonry Accessories
04160 Joint Reinforcement
04170 Anchors & Tie Systems
04180 Control Joints
04200 Unit Masonry
04210 Brick Masonry
04220 Concrete Unit Masonry
04230 Reinforced Unit Masonry
04236 Shop Fabricated Masonry Panels
04240 Clay Backing Tile
04245 Clay Facing Tile
04250 Ceramic Veneer

04270	Glass Unit Masonry
04280	Gypsum Unit Masonry
04300	Unit Masonry System
04310	Single Wythe Masonry System
04320	Veneer Masonry System
04330	Cavity Wall Masonry System
04340	Reinforced Unit Masonry System
04400	Stone
04410	Rough Stone
04420	Cut Stone
04422	Marble
04430	Simulated Masonry
04435	Cast Stone
04440	Flagstone
04450	Cut Masonry Veneer
04500	Masonry Restoration & Cleaning
04510	Masonry Cleaning
04550	Refractories
04999	Miscellaneous
05	**Metals**
05000	Alternatives
05100	Structural Metal Framing
05120	Structural Steel
05130	Structural Aluminum
05200	Metal Joists
05210	Steel Joists
05300	Metal Decking
05311	Steel Roof Deck
05313	Steel Floor Deck
05400	Cold Formed Metal Framing
05500	Metal Fabrications
05510	Metal Stairs
05520	Handrails & Railings
05521	Pipe & Tube Railings
05530	Gratings
05531	Gratings & Floor Plates
05540	Castings
05700	Ornamental Metal
05710	Ornamental Stairs
05715	Prefabricated Spiral Stairs
05720	Ornamental Handrails & Railings
05730	Ornamental Sheet Metal
05800	Expansion Control
05805	Expansion Joint Assemblies
05999	Miscellaneous
06	**Wood & Plastic**
06000	Alternatives
06100	Rough Carpentry
06110	Framing & Sheathing
06111	Light Wooden Structures - Framing
06112	Framing & Sheathing
06113	Sheathing
06114	Wood Blocking & Curbing
06125	Wood Decking
06130	Heavy Timber Construction
06131	Heavy Timber Framing
06132	Mill-framed Structures
06133	Pole Construction
06136	Heavy Timber Trusses
06150	Trestles
06151	Wood Chord Metal Joists
06170	Prefabricated Structural Wood
06180	Glued-laminated Construction
06181	Glue-laminated Structural Units
06182	Glue-laminated Decking
06190	Wood Trusses
06193	Plate Connected Wood Trusses
06196	Plywood Web Joists
06200	Finish Carpentry
06220	Millwork
06240	Laminated Plastic
06300	Wood Treatment
06400	Architectural Woodwork
06410	Custom Casework
06411	Wood Cabinets: Unfinished
06420	Paneling
06421	Arch. Hardwood Plywood Paneling
06422	Softwood Plywood Paneling
06423	Wood Veneer Faced Paneling
06430	Stairwork
06431	Wood Stairs & Railings
06500	Prefab. Structural Plastics
06600	Plastic Fabrications
06610	Glass Fiber & Resin Fabrications
06620	Cast Plastic Fabrications
06999	Miscellaneous
07	**Thermal & Moisture Protection**
07000	Alternatives
07100	Waterproofing
07105	Bituminous Membrane Waterproofing
07110	Sheet Membrane Waterproofing
07115	Elastomeric Sheet Waterproofing
07120	Fluid Applied Waterproofing
07121	Liquid Waterproofing

07130 Bentonite Waterproofing
07140 Metal Oxide Waterproofing
07150 Dampproofing
07160 Bituminous Dampproofing
07170 Silicone Dampproofing
07175 Water Repellent Coating
07180 Cementitious Dampproofing
07181 Water Repellant Coating
07190 Vapor & Air Barriers
07200 Insulation
07210 Building Insulation
07211 Loose Fill Insulation
07212 Board Insulation
07213 Batt & Blanket Insulation
07214 Foamed-in-place Insulation
07215 Sprayed-on Insulation
07216 Loose Fill Insulation
07217 Blown Insulation
07218 Sprayed Insulation
07219 Sprayed Glass Fiber Insulation
07230 High & Low Temperature Insul.
07240 Coated Insulation System
07250 Perimeter & Under-slab Insul.
07253 Cellulose Fireproofing
07255 Cementitious Fireproofing
07270 Firestopping
07300 Shingles & Roofing Tiles
07310 Shingles
07311 Asphalt Shingles
07313 Wood Shingles & Shakes
07320 Roofing Tiles
07400 Preformed Roofing & Siding
07410 Preformed Wall & Roof Panels
07411 Preformed Metal Siding
07420 Composite Building Panels
07421 Composite Metal Building Tiles
07440 Preformed Plastic Panels
07460 Cladding/siding
07461 Wood Siding
07462 Composition Siding
07463 Asbestos-cement Siding
07464 Plastic Siding
07465 Preformed Metal Siding
07500 Membrane Roofing
07510 Built-up Bituminous Roofing
07512 Built-up Coal Tar Roof
07514 Built-up Asphalt Roofing
07520 Prepared Roll Roofing
07530 Elastic Sheet Roofing
07531 Elastomeric Sheet Roofing - Fully Adhered Conventional
07532 Elastomeric Sheet Roofing - Loose Laid Ballasted Conventional
07533 Elastomeric Sheet Roofing - Mechanically Attached Conventional
07536 Modified Bitumen Roofing - Conventional
07540 Fluid Applied Roofing
07545 Coated Foamed Roofing
07551 Protected Membrane Built-up Asphalt Roofing
07553 Elastomeric Sheet Roofing - Protected Membrane
07555 Modified Bitumen Roofing - Protected Membrane
07556 Rubberized Asphalt Roofing - Protected Membrane
07565 Preparation For Re-roofing
07570 Traffic Topping
07575 Traffic Membrane
07600 Flashing & Sheet Metal
07610 Sheet Metal Roofing
07611 Custom Sheet Metal Roofing
07620 Sheet Metal Flashing & Trim
07630 Roofing Specialties
07631 Gutters & Downspouts
07660 Gravel Stops
07700 Flashing
07710 Manufactured Roof Specialties
07724 Roof Hatches
07800 Roof Accessories
07810 Plastic Skylights
07820 Metal-framed Skylights
07830 Hatches
07840 Gravity Ventilators
07850 Prefabricated Curbs
07860 Prefabricated Expansion Joints
07900 Joint Sealers
07950 Gaskets
07999 Miscellaneous

08 Doors & Windows

08000 Alternatives
08100 Metal Doors & Frames
08110 Hollow Metal Work
08111 Standard Steel Doors
08112 Standard Steel Frames
08114 Custom Steel Doors
08115 Custom Steel Frames
08120 Aluminum Doors & Frames

08130	Stainless Steel Doors & Frames
08140	Bronze Doors & Frames
08200	Wood & Plastic Doors
08211	Flush Wood Doors
08212	Stile & Rail Wood Doors
08300	Special Doors
08305	Access Doors
08310	Sliding Metal Fire Doors
08311	Sliding Glass Doors
08320	Metal-clad Doors
08330	Coiling Doors
08331	Overhead Coiling Doors
08341	Overhead Coiling Grilles
08350	Folding Doors
08351	Folding Slat Doors
08355	Flexible Doors
08360	Section Overhead Doors
08370	Sliding Glass Doors
08375	Safety Glass Doors
08380	Sound Retardant Doors
08390	Screen & Storm Doors
08391	Aluminum Storm Doors & Frames
08400	Entrances & Storefronts
08410	Aluminum Entrances & Storefronts
08450	Revolving Doors
08470	Revolving Entrance Doors
08500	Metal Windows
08510	Steel Windows
08511	Rolled Steel Windows
08512	Sheet Steel Windows
08520	Aluminum Windows
08530	Stainless Steel Windows
08540	Bronze Windows
08600	Wood & Plastic Windows
08610	Wood Windows
08620	Plastic Windows
08631	Tubular Plastic Windows
08550	Special Windows
08700	Hardware & Specialties
08710	Door Hardware
08720	Operators
08721	Automatic Door Equipment
08725	Window Operators
08730	Weatherstripping & Seals
08740	Thresholds
08800	Glazing
08810	Glass
08811	Plate Glass
08812	Sheet Glass
08813	Tempered Glass
08814	Wired Glass
08815	Rough & Figured Glass
08820	Processed Glass
08821	Coated Glass
08822	Laminated Glass
08823	Insulating Glass
08830	Mirror Glass
08840	Glazing Plastics
08850	Glazing Accessories
08900	Window Walls-curtain Walls
08920	Glazed Aluminum Curtain Wall
08952	Translucent Panel System
08960	Sloped Glazing
08971	Suspended Glass
08975	Structural Sealant Glazing System
08999	Miscellaneous
09	**Finishes**
09000	Alternatives
09100	Lath & Plaster
09110	Furring & Lathing
09111	Metal Stud Framing System
09160	Plaster
09167	Gypsum Plaster
09180	Cement Plaster
09190	Acoustical Plaster
09206	Metal Furring & Lathing
09210	Gypsum Plaster
09215	Veneer Plaster
09220	Portland Cement Plaster
09250	Gypsum Wallboard
09260	Gypsum Board System
09280	Accessories
09300	Tile
09310	Ceramic Tile
09311	Ceramic Tile Floor Finish
09312	Ceramic Tile Wall Finish
09320	Ceramic Mosaics
09330	Quarry Tile
09331	Quarry Tile Floor Finish
09332	Quarry Tile Wall Finish
09340	Marble Tile
09350	Glass Mosaics
09360	Plastic Tile
09370	Metal Tile
09400	Terrazzo
09410	Portland Cement Terrazzo
09420	Precast Terrazzo
09430	Conductive Terrazzo
09440	Plastic Matrix Terrazzo
09500	Acoustical Treatment
09510	Acoustical Ceilings
09511	Suspended Acoustical Tile
09512	Adhesive Applied Acoustical Ceilings
09520	Acoustical Wall Treatment
09530	Acoustical Insulation & Barriers
09540	Suspension Systems
09550	Wood Flooring

09560	Wood Strip Flooring
09561	Hardwood Flooring - Adhesive Applied
09562	Hardwood Flooring - Nailed
09570	Wood Parquet Flooring
09580	Plywood Block Flooring
09590	Resilient Wood Floor System
09591	Resilient Hardwood Flooring
09600	Stone Flooring
09635	Brick Flooring
09650	Resilient Flooring
09651	Cementitious Underlayment
09660	Resilient Tile Flooring
09665	Resilient Sheet Flooring
09670	Fluid Applied Resilient Flooring
09680	Carpeting
09681	Carpet Cushion
09682	Carpet
09683	Bonded Cushion Carpet
09684	Custom Carpet
09686	Carpeting With Cushion
09688	Carpet - Glue Down
09690	Carpet Tile
09700	Special Flooring
09705	Resinous Flooring
09710	Magnesium Oxychloride Floors
09720	Epoxy-marble-chip Flooring
09730	Elastomeric Liquid Flooring
09731	Conductive Liquid Flooring
09735	Neoprene - Hapalon Flooring
09740	Heavy-duty Concrete Toppings
09741	Armored Floors
09750	Brick Flooring
09760	Fluid Applied Stone Chip Flooring
09765	Fluid Applied Vinyl Chip Flooring
09800	Special Coatings
09810	Abrasion Resistant Coatings
09815	High Build Glazed Coatings
09820	Cementitious Coatings
09825	Aggregate Wall Surfacing
09830	Elastomeric Coatings
09840	Fire-resistant Coatings
09841	Sprayed Fireproofing
09850	Aggregate Wall Coatings
09900	Painting
09950	Wall Covering
09951	Vinyl-coated Fabric Wall Cover.
09952	Vinyl Wall Covering
09953	Cork Wall Covering
09954	Wallpaper
09955	Vinyl Coated Fabric Wall Covering
09960	Flexible Wood Sheets
09970	Prefinished Panels
09990	Adhesives
09999	Miscellaneous

10	**Specialties**
10000	Alternatives
10100	Chalkboards & Tackboards
10105	Visual Display Boards
10150	Compartments & Cubicles
10151	Hospital Cubicles
10160	Metal Toilet Compartments
10161	Laminated Plastic Toilet Partitions
10162	Metal Toilet Partitions
10163	Stone Partitions
10165	Plastic Laminated Toilet Compartments
10170	Shower & Dressing Compartments
10196	Cubical Curtains
10200	Louvers & Vents
10211	Fixed Metal Wall Louvers
10240	Grilles & Screens
10260	Wall & Corner Guards
10270	Access Flooring
10280	Specialty Modules
10290	Pest Control
10300	Fireplaces
10301	Prefabricated Fireplaces
10302	Prefabricated Fireplace Forms
10305	Manufactured Fireplaces
10310	Fireplace Accessories
10350	Flagpoles
10400	Identifying Devices
10410	Directories & Bulletin Boards
10411	Directories
10420	Plaques
10440	Signs
10441	Plastic Signs
10450	Pedestrian Control Devices
10500	Lockers
10508	Metal Wardrobe Lockers
10522	Fire Extinguisher Cabinets & Accessories
10530	Protective Covers

10532 Car Shelters
10536 Awnings
10550 Postal Specialties
10551 Mail Chutes
10552 Mail Boxes
10600 Partitions
10601 Mesh Partitions
10605 Wire Mesh Partitions
10610 Demountable Partitions
10616 Demountable Gypsum Board Partitions
10620 Folding Partitions
10623 Accordion Folding Partitions
10650 Scales
10652 Folding Panel Partitions
10655 Accordion Folding Partitions
10660 Sliding Panel Partitions
10670 Storage Shelving
10700 Sun Control Devices (exterior)
10750 Telephone Enclosures
10800 Toilet & Bath Accessories
10900 Wardrobe Specialties
10999 Miscellaneous

11 Equipment
11000 Alternatives
11050 Built-in Maintenance Equipment
11051 Vacuum Cleaning System
11052 Powered Window Washing
11100 Bank & Vault Equipment
11150 Parking Control Equipment
11161 Dock Levelers
11164 Seal & Shelters
11165 Dock Bumpers
11170 Checkroom Equipment
11172 Waste Compactors
11180 Darkroom Equipment
11200 Ecclesiastical Equipment
11300 Educational Equipment
11400 Food Service Equipment
11401 Custom Food Service Equip
11410 Bar Units
11420 Cooking Equipment
11430 Dishwashing Equipment
11435 Garbage Disposers
11440 Food Preparation Machines
11450 Food Preparation Tables
11460 Food Serving Units
11470 Refrigerated Cases
11480 Vending Equipment
11500 Athletic Equipment
11550 Industrial Equipment
11600 Laboratory Equipment
11630 Laundry Equipment
11650 Library Equipment
11700 Medical Equipment
11800 Mortuary Equipment
11830 Musical Equipment
11850 Parking Equipment
11860 Waste Handling Equipment
11861 Packaged Incinerators
11863 Bins
11864 Pulping Machines & Systems
11870 Loading Dock Equipment
11872 Leveling Platforms
11873 Portable Ramps & Bridges
11880 Detention Equipment
11900 Residential Equipment
11970 Theater & Stage Equipment
11990 Registration Equipment
11999 Miscellaneous

12 Furnishings
12000 Alternatives
12100 Artwork
12110 Murals
12120 Photo Murals
12300 Cabinets & Storage
12301 Metal Casework
12370 Residential Casework
12500 Window Treatment
12512 Horizontal Louver Blinds
12531 Drapery Track
12550 Fabrics
12600 Furniture
12670 Rugs & Mats
12692 Floor Mats
12700 Seating
12710 Auditorium Seating
12730 Stadium Seating
12735 Telescoping Bleachers
12800 Furnishing Accessories
12999 Miscellaneous

13 Special Construction
13000 Alternatives
13010 Air-supported Structures
13025 Integrated Ceilings
13038 Cold Storage Rooms
13050 Integrated Assemblies
13052 Saunas
13054 Steam Rooms
13095 X-ray Radiation Protection
13100 Audiometric Room
13121 Pre-engineered Buildings
13152 Swimming Pools
13156 Hot Tubs
13250 Clean Room
13350 Hyperbaric Room

13400	Incinerators
13440	Instrumentation
13450	Insulated Room
13540	Nuclear Reactors
13550	Observatory
13600	Prefabricated Buildings
13700	Special Purpose Rooms & Bldgs
13750	Radiation Protection
13770	Sound & Vibration Control
13800	Vaults
13805	Building Management System
13999	Miscellaneous
14	**Conveying Systems**
14000	Alternatives
14100	Dumbwaiters
14120	Electric Dumbwaiters
14200	Elevators
14201	Elevator Hoisting Equipment
14202	Elevator Operation
14203	Elevator Cars & Entrances
14215	Electric Elevators - Passenger
14220	Electric Elevators - Freight
14245	Hydraulic Elevators - Passenger
14250	Hydraulic Elevators - Freight
14300	Hoists & Cranes
14310	Escalators
14400	Lifts
14430	Platform & Stage Lifts
14500	Material Handling Systems
14550	Conveyors & Chutes
14570	Turntables
14600	Moving Stairs & Walks
14700	Pneumatic Tube Systems
14800	Powered Scaffolding
14999	Miscellaneous
15	**Mechanical**
15000	Alternatives
15010	Basic Mechanical Requirements
15050	Basic Materials & Methods
15060	Pipe & Pipe Fittings
15075	Hose
15080	Piping Specialties
15100	Valves & Cocks (manual)
15120	Control Valves
15121	Expansion Compensation
15140	Supports And Anchors
15170	Motors
15175	Tanks
15180	Insulation
15190	Mechanical Identification
15200	Water Supply & Treatment
15220	Pumps & Piping
15230	Booster Pumping Equipment
15240	Water Reservoirs & Tanks
15242	Vibration Isolation
15250	Water Treatment
15260	Piping Insulation
15270	Distribution & Metering Systems
15280	Equipment Insulation
15290	Ductwork Insulation
15300	Waste Water Disposal & Treatment
15310	Fire Protection Piping
15320	Fire Pumps
15325	Sprinkler Systems
15330	Basins & Manholes
15340	Sewerage
15350	Lift Stations
15360	Septic Tank Systems
15365	Halon Agent Extinguishing System
15375	Standpipe & Fire Hose System
15380	Sewage Treatment
15400	Plumbing
15410	Plumbing Piping
15420	Equipment
15430	Plumbing Specialties
15440	Plumbing Fixtures
15450	Plumbing Equipment
15451	Water Coolers
15452	Washfountains
15476	Swimming Pool Equipment
15482	Medical Gas Systems
15484	Fuel Oil Piping Systems
15500	Fire Protection
15510	Hydronic Piping
15515	Hydronic Specialties
15520	Steam & Steam Condensate Piping
15525	Steam & Steam Condensate Specialties
15535	Refrigeration Piping & Specialties
15540	Hvac Pumps
15545	Chemical (water) Treatment
15550	Fire Extinguisher Cabinets
15556	Cast Iron Boilers
15558	Finned Water Tube Boilers
15559	Steel Water Tube Boilers

15560 Hood & Duct Fire Protection
15561 Fire Tube Boilers
15565 Electric Boilers
15575 Breeching, Chimneys, & Stacks
15600 Power Or Heat Generation
15610 Fuel Handling Equipment
15611 Oil Storage Tanks & Piping
15612 Bottled Gas Tanks & Piping
15613 Oil Piping
15614 Gas Piping
15615 Stokers & Conveyors
15616 Ash Removal System
15617 Breechings
15618 Exhaust Equipment
15619 Draft Control Equipment
15623 Forced Air Furnaces
15624 Direct Fired Air Units
15625 Fuel Fired Unit Heaters
15630 Solar Energy Devices
15640 Boiler Feedwater Equipment
15650 Refrigeration
15658 Reciprocating Water Compressors
15660 Compressors
15670 Condensing Units
15671 Air Cooled Condensing Chillers
15680 Chillers
15681 Reciprocating Water Chillers
15682 Air Cooled Water Chillers
15684 Centrifugal Water Chillers
15686 Absorption Water Chillers
15690 Evaporators
15698 Commercial Ice Making Equipment
15699 Refrigeration Accessories
15700 Liquid Heat Transfer
15710 Hot Water Specialties
15712 Induced Draft Cooling Tower
15714 Blow Thru Cooling Tower
15716 Ejector Cooling Tower
15720 Steam Specialties
15730 Heat Exchangers
15740 Terminal Units
15742 Air Cooled Refrigerant Condensers
15755 Heat Exchangers
15760 Packaged Heat Pump
15770 Packaged Heating & Cooling
15780 Humidity Control
15781 Packaged Roof Top Air Conditioning Units
15782 Packaged Terminal Air Conditioning Units
15784 Unit Air Conditioners
15785 Computer Room Air Conditioning Units
15786 Computer Room Air Conditioning Units - Unitary Ceiling
15790 Air Coils
15800 Air Distribution
15810 Furnaces
15811 Evaporative Humidifiers
15812 Evaporative Pan Humidifiers
15813 Steam Grid Humidifiers
15820 Fans
15821 Sprayed Coil Humidifiers
15835 Terminal Heat Transfer Units
15840 Induction Units
15850 Special Ductwork Equipment
15855 Air Handling Units
15860 Centrifugal Fans
15865 Axial Fans
15870 Power Ventilators
15880 Air Treatment Equipment
15885 Air Cleaning
15886 Dust Collection System
15887 Carbon Monoxide Exhaust System
15890 Ductwork
15900 Controls & Instrumentation
15910 Ductwork Accessories
15920 Sound Attenuators
15930 Air Terminal Units
15936 Air Inlets & Outlets
15950 Sequential Controls
15960 Recording Devices
15970 Alarm Devices
15973 Pneumatic Controls
15975 Direct Digital Control Systems
15980 Instrumentation
15985 Sequence Of Operation
15990 Testing, Adjusting, & Balancing
15999 Miscellaneous

16 Electrical

16000 Alternatives
16010 Basic Electrical Requirements
16060 Minor Electrical Demolition For Remodeling
16100 Basic Materials & Methods
16110 Raceways
16111 Conduit
16112 Surface Raceways
16113 Underfloor Ducts

16114 Cable Trays
16115 Indoor Service Poles
16120 Conductors
16121 Medium - Voltage Cable
16123 Building Wire & Cable
16125 Undercarpet Cable Systems
16130 Boxes
16133 Cabinets
16134 Panelboards
16140 Switches & Receptacles
16141 Wiring Devices
16149 Low Voltage Switching
16150 Manufactured Wiring Systems
16160 Cabinets & Enclosures
16170 Grounding & Bonding
16180 Equipment Wiring Systems
16190 Supporting Devices
16195 Electrical Identification
16199 Electronic Devices
16200 Power Generation
16210 Generator
16220 Engine
16230 Cooling Equipment
16240 Exhaust Equipment
16250 Starting Equipment
16260 Automatic Transfer Equipment
16300 Power Transmission
16310 Substation
16311 Secondary Unit Substation
16320 Switchgear
16321 Distribution Transfers
16330 Transformer
16340 Vaults
16350 Circuit Breaker Switchgear
16352 Medium Voltage Motor Controllers
16360 Rectifiers
16361 Air Interruption Switches
16362 Oil Interruption Switch
16370 Overhead Power Distribution
16380 Capacitors
16400 Service & Distribution
16410 Electric Service
16411 Power Factor Capacitors
16420 Service Entrance
16421 Utility Service Entrance
16426 Distribution Switchboards
16430 Service Disconnect
16435 Converters
16440 Metering
16441 Enclosed Switches
16445 Peak Load Controllers
16450 Grounding
16460 Transformers
16461 Dry Type Transformers
16466 Feeder & Plug - In Busway
16470 Panelboards
16476 Enclosed Circuit Breakers
16481 Enclosed Motor Controllers
16482 Motor Control Center
16483 Variable Frequency Controllers
16485 Contactors
16490 Converters
16491 Rectifiers
16496 Enclosed Transfer Switch
16500 Lighting
16510 Interior Luminaires
16512 Luminous Ceiling
16515 Signal Lighting
16530 Site Lighting
16531 Stadium Lighting
16532 Roadway Lighting
16535 Emergency Lighting Equipment
16550 Accessories
16551 Lamps
16552 Ballasts & Accessories
16570 Poles & Standards
16577 Obstruction Lights
16580 Theatrical Lighting
16600 Special Systems
16610 Lighting Protection
16611 Static Uninterruptible Power Supply
16612 Rotary Uninterruptible Power Supply
16615 Emergency Power Supply
16620 Emergency Light & Power
16622 Packaged Engine Generator Systems
16640 Cathodic Protection
16670 Lightning Protection Systems
16680 Unit Power Conditioners
16700 Communications
16710 Radio Transmission
16720 Alarm & Detection Equipment
16721 Fire Alarm & Smoke Detection Systems
16722 Intrusion Detection & Security Access Systems
16730 Clock & Program Systems
16740 Clock & Program Equipment
16741 Telephone Service Entrance
16750 Nurse Call System
16760 Intercom System
16770 Public Address Equipment
16780 Television Systems

16781	Television Distribution System
16782	Closed Circuit Television System
16850	Heating & Cooling
16855	Heating Cables & Mats
16858	Snow Melting Cable & Mat
16859	Heating Cable
16860	Electric Duct Heater
16864	Electric Space Heating Units
16865	Electric Baseboard
16870	Packaged Room Air Conditioners
16880	Radiant Heaters
16890	Electric Heaters (prop Fan Type)
16900	Controls & Instrumentation
16901	Electrical Sensing & Measurement
16902	Electrical Controls & Relays
16903	Programmable Logic Controllers
16910	Recording & Indicating Devices
16920	Motor Control Centers
16930	Lighting Control Equipment
16940	Electrical Interlock
16950	Control Of Electric Heating
16960	Limit Switches
16999	Miscellaneous

Subdivision. Each section of each division of the specifications is divided into the *Scope* or *Work Included, Materials* listing the quality of all materials to be furnished, and finally, *Fabrication and Erection*, defining the quality of workmanship desired.

Scope includes items to be furnished and installed, items to be furnished but installed by others, items furnished by others and installed under the section, and finally, related work in other divisions of the specifications. Scope will also list some general requirements such as those for shop drawings, samples, tests, methods of delivery, and storage that may be required. No matter how detailed the scope section, it will not include each and every item. The drawings must always be checked against scope, which often carries the phrase "including, but not limited to, the following".

Materials will list the materials to be used in one of several ways, often found in combination. The *closed specification* will list a single trade name, and the specified product must be furnished. The *contractor's option specification* (or bidder's choice) lists more than one trade name, and the contractor may choose from those listed. The *substitute bid specification* might list a choice of trade names and the bid must be based on one of the products included, but the contractor is allowed to suggest a substitute at the time of submitting a bid, naming the amount he would subtract for using the alternate product.

A variation is the *product approval specification* which asks the contractor to submit any substitutions prior to submitting a bid. If the architect approves the substitute, it will be put in an addenda sent to all contractors. This "or approved equal" type specification is the most common. It provides the widest competition and rewards the contractor who knows his products, prices, and sources.

However, it is the architect who decides whether the product is equal. If the contractor has any doubts about a substitute's being approved, then he should bring it to the architect's attention upon being selected as low bidder. In this way, any adjustments can be made before award of contract should the architect rule that the substitution is not equal.

The *product description specification* is used to describe bulk materials and some manufacturers' articles, the latter particularly in government work. The *performance specification* describes not the material but what work is required to produce strength, mechanical ability, or similar measurable results.

Specifications based on *reference standards* refer those which are published by industry associations and organizations. They are encountered often, are general in nature, and usually apply only in part to the specific job. A contractor should build a library of these published specifications.

Below is a list of construction associations, many of which offer commonly referred standards. In addition, many architects will refer to the published specifications of a named manufacturer. These are readily available and generally are included in installation procedures.

Fabrication and Erection will set performance standards for both shop and field work and will include, if required, sections on protection, cleaning, guarantees, warranties, maintenance, and operating instructions. The specifications of trade organizations and manufacturers' installation instructions are often made a part of this section.

Construction Organizations and Associations

ASA	Acoustical Society of America Woodbury, NY
ACCA	Air Conditioning Contractors of America Washington, DC
ARI	Air Conditioning & Refrigeration Institute Arlington, VA
ADC	Air Diffusion Council Chicago, IL
AMCA	Air Moving & Conditioning Association Arlington Heights, IL
AA	Aluminum Association Washington, DC
AAA	American Arbitration Association New York, NY
AACE	American Association of Cost Engineers Morgantown, WV
AASHTO	American Association of State Highway & Transportation Officials Washington, DC
ACAA	American Coal Ash Association Washington, DC
ACI	American Concrete Institute Detroit, MI
ACPA	American Concrete Paving Association Arlington Heights, IL
ACPA	American Concrete Pumping Association Vallejo, CA
ACEC	American Consulting Engineers Council Washington, DC

ACIL	American Council of Independent Laboratories Washington, DC
AFI	American Forest Institute Washington, DC
AHA	American Hardboard Association Palatine, IL
AIA	American Institute of Architects Washington, DC
AISC	American Institute of Steel Construction Chicago, IL
AITC	American Institute of Timber Construction Vancouver, WA
AIA	American Insurance Association Washington, DC
AISI	American Iron and Steel Institute Washington, DC
ALSC	American Lumber Standards Committee Germantown, MD
ANSI	American National Standards Institute New York, NY
APFA	American Pipe Fittings Association Springfield, VA
APA	American Plywood Association Tacoma, WA
ARBBA	American Railway & Bridge Building Association Homewood, IL
ARBA	American Road Builders Association Washington, DC
ASAHC	American Society of Architectural Hardware Consultants, San Rafael, CA
ASCET	American Society of Certified Engineering Technicians, El Paso, TX
ASCE	American Society of Civil Engineers New York, NY
ASHRAE	American Society of Heating, Refrigeration & Air-Cond Engineers Atlanta, GA
ASLA	American Society of Landscape Architects Washington, DC

ASME	American Society of Mechanical Engineers New York, NY
ASPE	American Society of Professional Estimators Wheaton, MD
ASSE	American Society of Sanitary Engineers Bay Village, OH
ASTM	American Society for Testing and Materials Philadelphia, PA
ASPA	American Sod Producers Association Rolling Meadows, IL
ASA	American Subcontractors Association Alexandria, VA
AWWA	American Water Works Association Denver, CO
AWS	American Welding Society Miami, FL
AWPA	American Wood Preservers Association Stevensville, MD
AWPI	American Wood Preservers Institute Vienna, VA
AAMA	Architectural Aluminum Manufacturers Association Chicago, IL
APA	Architectural Precast Association Indianapolis, IN
AWI	Architectural Woodwork Institute Arlington, VA
AI	Asphalt Institute College Park, MD
ARMA	Asphalt Roofing Manufacturers Association Rockville, MD
AABC	Associated Air Balance Council Washington, DC
AGC	Associated General Contractors of America Washington, DC
ALCA	Associated Landscape Contractors of America Falls Church, VA
ASC	Associated Schools of Construction Albuquerque, NM

AOSC	Association of Drilled Shaft Contractors Dallas, TX
AHAM	Association of Home Appliance Manufacturers Chicago, IL
BGA	Barre Granite Association Barre, VT
BIA	Brick Institute of America Reston, VA
BHMA	Builders Hardware Manufacturers Association New York, NY
BOCA	Building Official & Code Administrators Internationa Chicago, IL
BOMA	Building Owners and Managers Association Washington, DC
BRB	Building Research Board Washington, DC
BRI	Building Research Institute Washington, DC
BSI	Building Stone Institute New York, NY
CRA	California Redwood Association Novato, CA
CRI	Carpet and Rug Institute Dalton, GA
CISPI	Cast Iron Soil Pipe Institute Chattanooga, TN
CSSB	Cedar Shake and Shingle Bureau Bellevue, WA
CISCA	Ceilings & Interior Systems Construction Association Deerfield, IL
CTIOA	Ceramic Tile Institute of America Los Angeles, CA
CLFMI	Chain Link Fence Manufacturers Institute Washington, DC
CFFA	Chemical Fabrics and Film Association Cleveland, OH
CRSI	Concrete Reinforcing Steel Institute Schaumburg, IL

CSI	Construction Specifications Institute Alexandria, VA
CEMA	Conveyor Equipment Manufacturers Association Rockville, MD
DHI	Door and Hardware Institute McLean, VA
DORCMA	Door Operator & Remote Controls Manufacturers Association, Chicago, IL
ESCSI	Expanded Shale, Clay, & Slate Institute Rockville, MD
EJMA	Expansion Joint Manufacturers Institute Tarrytown, NY
FTI	Facing Tile Institute Canton, OH
FMERO	Factory Mutual Engineering & Research Organization Norwood, MA
FHVA/AWMA	Fine Hardwood Veneer Association American Walnut Manufacturers Association Indianapolis, IN
FGMA	Flat Glass Marketing Association Topeka, KS
GA	Gypsum Association Evanston, IL
HPMA	Hardwood Plywood Manufacturers Association Reston, VA
IESNA	Illuminating Engineering Society of North America New York, NY
ILIA	Indiana Limestone Institute of America Bedford, IN
ICA	Institute of Cost Analysis Alexandria, VA
IEEE	Institute of Electrical & Electronics Engineers New York, NY
IFRAA	Interfaith Forum on Religious Art and Architecture Philadelphia, PA
ICBO	International Conference of Building Officials Whittier, CA

IMI	International Masonry Institute Washington, DC
JCBC	Jute Carpet Backing Council New York, NY
LIA	Lead Industries Association New York, NY
LPI	Lightning Protection Institute Woodstock, IL
MSS	Manufacturers Standardization Society Vienna, VA
MFMA	Maple Flooring Manufacturers Association Northbrook, IL
MIA	Marble Institute of America Framington, MI
MBMA	Metal Building Manufacturers Association Cleveland, OH
ML/SFA	Metal Lath/Steel Framing Association Chicago, IL
NAA	National Aggregate Association Silver Spring, MD
NAPA	National Asphalt Pavement Association Riverdale, MD
NAAMM	National Association of Architectural Metal Manufacturers, Chicago, IL
NADAF	National Association of Decorative Architectural Finishes, Alexandria, VA
NAWIC	National Association of Women in Construction Fort Worth, TX
NBGQA	National Building Granite Quarries Association Barre, VT
NCMA	National Concrete Masonry Association Herndon, VA
NCAC	National Council of Acoustical Consultants Springfield, NJ
NCSA	National Crushed Stone Association Washington, DC
NECA	National Electrical Contractors Association Bethesda, MD

NEMA	National Electrical Manufacturers Association Washington, DC
NEII	National Elevator Industry, Inc. New York, NY
NEBB	National Environmental Balance Bureau Vienna, VA
NFPA	National Fire Protection Association Quincy, MA
NFPA	National Forest Products Association Washington, DC
NHLA	National Hardwood Lumber Association Memphis, TN
NKCA	National Kitchen Cabinet Association Falls Church, VA
NLA	National Lime Association Alexandria, VA
NMWIA	National Mineral Wool Insulation Association Alexandria, VA
NOFMA	National Oak Flooring Manufacturers Association Memphis, TN
NPCA	National Paint & Coatings Association Washington, DC
NPA	National Particleboard Association Gaithersburg, MD
NPCA	National Precast Concrete Association Indianapolis, IN
NRMCA	National Ready Mixed Concrete Association Silver Spring, MD
NRDCA	National Roof Deck Contractors Association Chicago, IL
NRCA	National Roofing Contractors Association Rosemont, IL
NSF	National Sanitation Foundation Testing Laboratory Ann Arbor, MI
NSSEA	National School Supply and Equipment Association Arlington, VA
NSPE	National Society of Professional Engineers Alexandria, VA

NSPI	National Swimming Pool Institute Alexandria, VA
NTMA	National Terrazzo and Mosaic Association Des Plaines, IL
MWWDA	National Wood Window and Door Association Des Plaines, IL
PDCA	Painting and Decorating Contractors of America Fairfax, VA
PI	Perlite Institute Chicago, IL
PPI	Plastic Pipe Institute New York, NY
PMDI	Plumbing Manufacturers and Drainage Institute Indianapolis, IN
PEI	Porcelain Enamel Institute Washington, DC
PCA	Portland Cement Association Skokie, IL
PCI	Prestressed Concrete Institute Chicago, IL
PG	Producers Group Washington, DC
RIS	Redwood Inspection Services Novato, CA
RFCI	Resilient Floor Covering Institute Rockville, MD
RDMI	Roof Drainage Manufacturers Institute Chicago, IL
SSA	Sauna Society of America Washington, DC
SAMA	Scientific Apparatus Makers Association Washington, DC
SIGMA	Sealed Insulating Glass Manufacturers Association Chicago, IL
SMACNA	Sheet Metal & Air Conditioning Contractors National Association, Marrifield, VA
SAME	Society of American Military Engineers Alexandria, VA

SPI	Society of the Plastics Industry Washington, DC
SWE	Society of Women Engineers New York, NY
SBCCI	Southern Building Codes Congress International Birmingham, AL
SFPA	Southern Forest Products Association New Orleans, LA
SPIB	Southern Pine Inspection Bureau Pensacola, FL
SGAA	Stained Glass Association of America Kansas City, MO
SDI	Steel Deck Institute Canton, OH
SDI	Steel Door Institute Cleveland, OH
SJI	Steel Joist Institute Myrtle Beach, SC
SSPC	Steel Structures Painting Council Pittsburgh, PA
SWI	Steel Window Institute Cleveland, OH
SCFPA	Structural Cement-Fiber Products Association Washington, DC
SMA	Stucco Manufacturers Association Sherman Oaks, CA
TCA	Tile Council of America Princeton, NJ
TPI	Truss Plate Institute Madison, WI
TRMI	Tubular Rivet and Machine Institute Tarrytown, NY
UL	Underwriters Laboratories Northbrook, IL
UBPVLS	Uniform Boiler & Pressure Vessel Laws Society Oceanside, NY
VA	Vermiculite Association Chicago, IL

WPCF	Water Pollution Control Federation Alexandria, VA
WSC	Water Systems Council Chicago, IL
WCLIB	West Coast Lumber Inspection Bureau Portland, OR
WRCLA	Western Red Cedar Lumber Association Portland, OR
WWMMP	Western Wood Moulding & Millwork Producers Portland, OR
WWPA	Western Wood Products Association Portland, OR
WRI	Wire Reinforcement Institute McLean, VA
WSFI	Wood and Synthetic Flooring Institute Hillside, IL

Alternates. At the time of bidding, the architect often wants to investigate the cost, either upward or downward, of a substitute product, material, or procedure, or the cost of adding or subtracting work from the contract. Such requests for alternate prices are usually noted at the end of the section of the specifications to which they refer, and are repeated at the end of Division 1 of the specifications and on the proposal form. On the typical lump sum proposal form, the Alternate follows the Statement of the Lump Sum price in a form such as the following:

If the following Alternates are accepted,	Add:	Deduct:
Alternate No. 1	$___	$___
Alternate No. 2	$___	$___

Obviously, a request for lots of alternates complicates bidding. Sometimes, the extent of the alternate is unclear. Since the architect will accept the price change for an alternate with the same responsibility as attached to the lump sum figure, the contractor must be certain of all the possible contingencies involved should the change be accepted.

Often an alternate is put only under the one section that makes the basic substitution, but other subcontractors' work is also affected by the change. It is the contractor's responsibility to foresee this. As an example, under Carpentry, an architect might include:

"Alternate No. 1--For substituting one 3'-0"x6'-8"x1 ¾" door with frame for the two 3'-0"x6'-8"x1 ¾" doors and frame shown between the living and dining rooms, deduct $____."

Obviously, the door cost will be less, but the wall construction will be more. Yet there might be no mention of this under the division for the latter.

Cash Allowances. Sometimes, the architect does not have a final decision from the owner on certain items, and rather than leave them out of the lump sum proposal, he gives a definite budget amount in the specification that is to be included in the bid. This is often encountered with the selection of brick, hardware, and other finish items. As an example:

"Facing brick shall be included under an allowance of $75 per M, fob, job site."

Every bidder will include the same material cost, rather than just guessing what type of face brick the owner may choose and using that cost. Cash allowances are almost always only for the cost of material and applicable taxes delivered to the site. Labor to set and overhead and profit on the material should be included in the lump sum price unless specifically noted otherwise.

Upon completion of the job, the actual cost of the items furnished under Allowance shall be tabulated with any savings credited to the owner, any overruns added to the contract cost.

Unit Prices. Where quantity of materials is in doubt, but quality is known, the specification may ask for unit prices. As opposed to allowances, unit prices are set by the bidder, not the architect; they are firm and binding on the bidder, the lump sum price being adjusted up or down, only according to variation in the quantity actually used on the job.

Unit Prices are included on the proposal form for each material per unit of measure. For example, unit prices are often asked for concrete per square yard, piling per lineal foot, partition block per square foot, etc. These prices should be complete with all costs, profit, and overhead included.

Unit Prices must be quoted with care, even if the amount of work covered by them is small, because they will be compared directly with those submitted by other bidders. The owner may expect the same unit prices to hold on additional work.

Addenda

Few plans and specifications are perfect when they leave the architect's office, and contractors bidding the job should call to the architect's attention all discrepancies they note. These, plus changes the architect and owner may wish to make after the plans and specifications have been issued but before bids are turned in, are incorporated in the *Addenda.* This usually takes the form of a printed sheet or sheets the same size as the specification to which it should be attached. Addenda eventually becomes a part of the Contract Documents.

Even changes to the drawings often can be described and included this way. But sometimes additional drawings or sketches must be issued, or a drawing is reissued with a revision date. Reissued drawings, however, usually should be avoided during bidding.

Each addendum is given a number, and the changes it covers are listed in the same order as the divisions in the specification, with drawing changes last.

Sometimes, addenda must be acknowledged by signing a receipt; often, they are sent by registered mail. Usually, they are acknowledged on the proposal form in a space provided for it. The single addendum or several addenda are also included in the Owner-Contractor Agreement.

Upon receipt of an addendum, the contractor should mark his specifications and drawings with appropriate notes, or with a stamp on each affected page saying "See Addendum #______." The Addendum will usually identify the change by division and page number and a reference to location on the page. As an example:

3:01--**Scope:** Add item "h) Remove existing wire fence along north edge of property."
6:02--**Materials:** Delete paragraph referring to concrete masonry lintels and add "...lintels in block work shall be units of sizes and shapes shown on the lintel schedule."
Dwg. Sheet A-2: Move rolling shutter door from center of Bay C-D to just north of Column B.
Dwg. Sheet A-4; delete ladder to roof.

Change Orders

Change orders are modifications issued after the contract is signed. Errors on drawings and specifications are usually uncovered during bidding. A contractor who sees where a drawing is not complete but does not include in his price what he knows must be furnished may very well end up having to pay for the item anyway to complete the job or face a lengthy lawsuit.

Some change orders are always necessary due to job conditions, changes required when plans are submitted for permits, substitutions made to satisfy the owner or his mortgage holder, products not being available, or because as the job progresses, certain changes for the benefit of the job become obvious.

The contractor must cooperate in making these changes. This is set forth in Article 12 of the General Conditions. Sometimes no cost is involved. Most often, an extra is called for. This may be handled by submitting a firm price change, or by order to proceed on a time and material basis. The change order is issued by the architect but signed by the owner and contractor. The format of the change order will follow closely that of the addendum, except that costs of changes and tabulation of the new contract amount are added. The Frank R. Walker Form #L-101 for contract change orders is shown.

Summary

As can be imagined, what appears on drawings will sometimes differ with what is stated in specifications. There is no legal precedent that one is more binding than the other, although some specifications will state which is to be followed in case of inconsistency.

In general, sizes, quantities, design, and location will be best taken from the drawings. Materials, quality, procedures, and general requirements are most reliably defined in the specifications. Never rely solely on one or the other. A draftsman's delineation of a stone wall in elevation may not show the coursing called for in the specifications at all, while the specifier may make no special mention of a stone sill condition carefully detailed on the drawings. One must always complement the other.

CONTRACT CHANGE ORDER

JOB
CONTRACT JOB NO.

FOR

CHANGE ORDER No.

DATE

TO

REVISED CONTRACT AMOUNT

PREVIOUS CONTRACT AMOUNT $

AMOUNT OF THIS ORDER $

TOTAL CONTRACT AND EXTRAS $

The work covered by this order shall be performed under the same Terms and Conditions as that included in the Original Contract.

CHANGES APPROVED

AMOUNT OF ORIGINAL CONTRACT $

CHANGE NO.

By____________

By____________

By____________

PRACTICAL MFD IN U.S.A. Form L-101 FRANK R. WALKER CO. PUBLISHERS CHICAGO

Contract Change Order
Frank R. Walker Form L-101

SETTING UP THE ESTIMATE

Estimating is one of the most important aspects of a building contractor's business. In nearly every instance it is necessary for a contractor either to estimate the cost of the work, or to bid in competition with others before a contract for the work is awarded. If a contractor's bid is too low and a contract is awarded on the basis of it, the contractor often must complete the work without profit and sometimes at a loss in time and money.

A contractor must know the size, complexity, and quality of work that his firm can efficiently handle. Exceeding these limits often leads to a poor contract, which can be costly not only monetarily but also in terms of dissatisfied customers and loss of future opportunities for bidding.

Most contractors fall into categories as to the type of work which their organizations can handle efficiently, expeditiously, and profitably. A heavy construction contractor specializing in structures such as airport runways, highways, bridge abutments, etc., usually should forego bidding on such work as churches, banks, hospitals, or modern office buildings. A contractor whose organization and personnel are geared to a particular category of construction can find himself in difficulty if he strays from his specialty.

Many contractors in business today are former employees of contracting firms who have gone into business for themselves. Many of these people doubtless were exceptional mechanics, foremen or superintendents, and had a thorough understanding of how the work should be carried out in the field, but sometimes they lack business ability and training to estimate costs accurately. Some of these contractors make a reasonable profit on their work and remain in business, but many more of them fail and eventually go back to work for someone else.

There are various reasons for these failures, but probably the most common one is the inability of the person estimating costs to come up with realistic and profitable estimates. The procedures involved in preparing an accurate, detailed cost estimate necessary for a realistic bid are relatively simple. Yet almost all instances of failure of contracting firms are the result of poor estimating practices, resulting in the submission of unsound and unprofitable bids. There is no substitute for complete and detailed material take-off and pricing and for cross-checking estimated costs with actual costs so that adjustments in pricing can be made for future bidding.

Another common reason for failure in the contracting business is overexpansion. Again, the contractor must know the capabilities of his office and supervisory staff and how much they can adequately handle at any one time. The successful contractor does not take on more work than his staff can take care of.

Bonding and insurance companies have certain minimum requirements that contractors applying for surety bonds or contractual liability insurance must meet. Some of the most important considerations that such companies make before bonding an applicant are:

a) the technical ability of the contractor applying for a bond, whether or not he understands his business thoroughly, and his capabilities in the preparation of detailed cost estimates;

b) the reputation of the contractor for honesty and his standing among those with whom he does business;

c) the financial ability of the contractor, and whether or not he has sufficient capital at his disposal to carry on his business, purchase materials and supplies, meet his payrolls and current expenses, etc.

Bonding companies seldom get the opportunity of bonding the best contracting firms. On public work projects the bidding is usually open to all bidders and in some instances as many as fifty or more bids may be received on one job. In most cases, the lowest bidder is awarded the contract, and he often may be the poorest risk because of substandard estimating procedures, lack of technical ability to perform the work as specified, and lack of financial responsibility. On privately owned projects, the risk to the bonding company may not be as great, because the bidding list can be controlled. Only contractors with the proper credentials are asked to submit proposals. Architectural and engineering firms that handle work for private owners are often asked to select contractors to bid on a project. They will attempt to choose bidders whom they know can perform the work equally well and will carry out the intent of the drawings and specifications in the best interest of the owner.

Others besides the contractor attach importance to a correct and detailed estimate. The success of a contractor depends as much on a well prepared estimate as it does on the manner in which he purchases materials, lets his subcontracts, or carries out his work on the job.

The Detailed Estimate

If twenty estimators were furnished the same set of plans and specifications and told to prepare a cost estimate, it is safe to say that not more than two of the estimates would be prepared on the same basis or using the same units. And an estimator doing the same estimate a year later would most likely arrive at different final costs the second time.

There are so many different method of estimating and a lack of detailed information regarding costs. Each contractor has developed a system of his own, making it difficult for one contractor to check another's figures. Some contractors use a *unit manhour* estimating procedure. A given unit of work takes a certain amount of time. For example, it might take 8 manhours to set 100 lineal feet of ½" rigid conduit. Other contractors use the *crew* method, i.e., a concrete pouring crew of 8 workers will place 100 cubic yards of concrete per day.

Each contractor develops his own unit prices from cost records that reflect his particular experience and expertise, which varies considerably among contractors. This is not to say that one estimate is correct and another is incorrect. In the final analysis of estimates and bids, they should be accepted as representing the projected final costs for performing certain prescribed work within a stipulated time schedule, taking into consideration all of the variables that a prudent contractor can anticipate during a construction contract based on information available at the time of bidding.

To prove the validity of the bid, one must consider the extreme risks that the contractor is willing to assume by furnishing payment and performance bonds and by entering into a legal contract dedicating the technical and financial resources of the firm to a successful completion of the contract in hopes of realizing a modest profit.

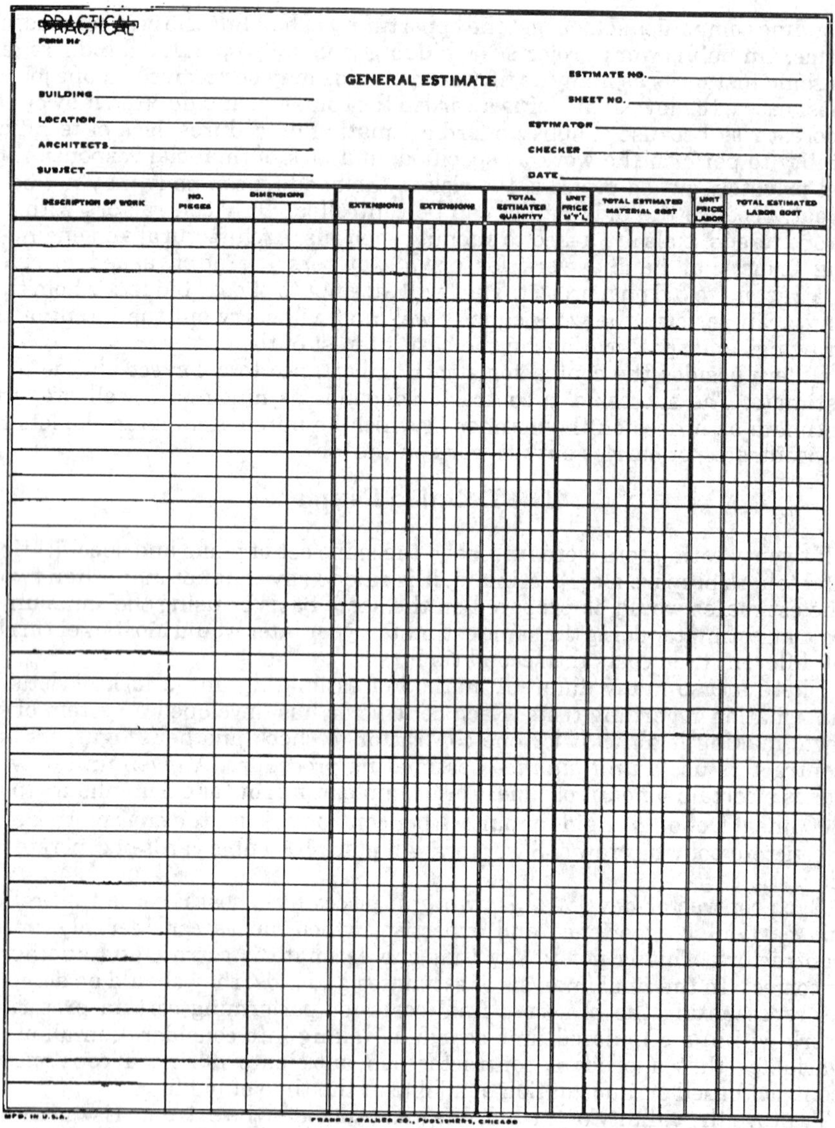
PRACTICAL
FORM 514

GENERAL ESTIMATE

BUILDING
LOCATION
ARCHITECTS
SUBJECT

ESTIMATE NO.
SHEET NO.
ESTIMATOR
CHECKER
DATE

DESCRIPTION OF WORK	NO. PIECES	DIMENSIONS	EXTENSIONS	EXTENSIONS	TOTAL ESTIMATED QUANTITY	UNIT PRICE M'T'L	TOTAL ESTIMATED MATERIAL COST	UNIT PRICE LABOR	TOTAL ESTIMATED LABOR COST

MFD. IN U.S.A.
FRANK R. WALKER CO., PUBLISHERS, CHICAGO

General Estimate
Frank R. Walker Form 514

Estimating is a laborious and costly operation and under no circumstances should an estimator or contractor produce a less than totally professional estimate that will stand the test of proving each element contained in the estimate.

There are many types of cost estimates that an owner may require before the bid estimate is needed. These estimates can be performed by or in conjunction with the owner's staff, contractors, professional estimators, or the estimating section of an engineering or architectural firm. Each type of cost estimate should serve the purpose for which it was intended.

Types of Cost Estimates

Budget Estimate. The budget estimate, once it has been developed, is effectively "cast in stone". No matter what happens throughout the life of the project, the costs stated in the budget estimate are what everyone will quote until the project is completed. If the budget estimate does not accurately represent the work to be performed, then it might mean one of two things.

On the one hand, if costs have been overstated, the project might be eliminated before it gets started. On the other hand, if the project starts with insufficient funding, the work might have to stop in midstream, or it will be necessary to borrow funds from other sources, delaying the start of other work.

While the budget estimate does not reduce the importance of the other types of cost estimates, it is usually the governing cost estimate of the project. At this stage the project either gets the green light or dies in committee. (Contingency percentage is 30% to 50%)

Pre-design Estimate. The purpose of this type of estimate is to provide approximate values for design work packages to be awarded A/E firms selected by the owner. At this stage of the process, there still has not been much increase in the design information available. (Contingency percentage is 30% to 50%)

Conceptual Estimate. For this level of estimating, the estimator has been supplied with sketches. These "drawings", in conjunction with discussions and meetings with the owner's representatives and the A/E, are used to form the estimate outline for costing.

The estimator formulates a list of pay items. These pay items have total unit costs applied to them. The conceptual estimate normally becomes the estimate that all other "working" estimates are compared with for a) design changes, b) scope changes, and c) cost changes. (Contingency percentage is 30%)

Preliminary Estimate 10%. At this stage the estimator has preliminary scope and drawing information. On the basis of information from the previous types of cost estimates, the estimator uses this new information, along with discussions with the owner's staff and the A/E, to expand to this level of estimating.

The item list generated in and expanded upon in the previous estimates is used. Individual items of work are broken down, where possible, into units of cost for labor, equipment, material, and subcontractor.

Prior to the presentation of the preliminary cost estimate, the estimator makes a site tour. The basic assumptions about how the work will be performed are reviewed and refined or revised as required. A written report on the site conditions, and on any effect that these conditions have on the estimate, is prepared for review by the owner.

At this stage of design development, cost items generally fall into the subcontract or "plug" cost column. As information becomes available during this or any later stage of the estimate, this "plug" cost can be expanded into a more detailed breakdown. All changes in quality, scope, etc. are fully identified and justified as the estimate progresses. (Contingency percentage is 25%)

Design Review Estimate 40%. This cost estimate is an expansion of the previous 10% cost estimate. Items of work are identified, to the extent possible for this level of estimating, in units of cost for labor, equipment, materials, and subcontract. In accordance with the details, as requested by the owner or his A/E, the 40% estimate provides backup to support labor rates, crew mixes, etc. Where possible, labor rates should include the contractor's indirect costs. Otherwise, indirect costs must be identified and enumerated separately. (Contingency percentage is 20%)

Pre-construction Estimate 70%. At the 70% estimate, the estimator completes the quantity take-offs on all defined items of work. Each identified item is quantified in terms of unit cost and total cost for material, labor hours and labor costs. The estimator adjusts the estimate to present either lump sum or unit price information, to conform to the type of contract that is being bid. This estimate uses the same format as the previous type of cost estimate. (Contingency percentage is 10%)

Construction Cost Estimate 100%. The estimator now starts an entirely new cost estimate based on the items of work identified in the contract specifications and drawings. All items of work are broken down into units of labor, equipment, material, and subcontract when possible. This is in the same format that has been used for all other cost estimates to date. Using a standard format allows for identification of cost and design changes when the 100% estimate is completed and compared with cost estimates from the previous stages.

The important difference between this estimate and those from all previous stages is that the estimator develops and completes the 100% estimate entirely separate from all the others. Only after the 100% estimate is complete is it then compared with estimates from the previous stages.

In addition, at this stage all vendor quotes, subcontractor costs and equipment prices are identified. All line items that the general contractor will perform are spread out into units of labor, equipment and material. (Contingency percentage is 5%)

The use of this detailed estimate extends far beyond being the basis of a bid. It serves as the guide for the purchase of materials, and the award of subcontracts. It also provides the information for the establishment of a cost control system so that project progress can be monitored to determine if the estimated material and the labor costs are consistent with the actual costs incurred during performance of the contract.

As an example, assume that the detailed estimate reflects the following labor cost allowances for each of the listed items of work:

	Labor Allowance	Quantity
Footing Excavation	$3,000.00	300 cu. yds.
Form Footings	$1,200.00	1,200 sq. ft.
Pour Footing Concrete	$1,500.00	300 cu. yds.

As each of these phases of work starts, the labor chargeable to each is allocated on a project daily time sheet (Walker Form 505) so that as work progresses, there is an accurate cost record that can be reviewed daily, weekly, or monthly, depending on the magnitude of the operation. In this way, periodic projections can be made. For example, the 50% completion review of each of the items reflects the following conditions:

Item	Labor Allowance	Expended To Date	Remaining Cost	+ - (loss)
Footing Excavation	$3,000.00	$1,800.00	$1,800.00	($600.00)
Form Footings	1,200.00	400.00	400.00	400.00
Pour Footings	1,500.00	600.00	600.00	300.00

The contractor should not view the combination of these three work items as reflecting a net $100.00 savings above the estimated costs but rather should be alerted to a serious overrun of 20% in the labor budget for footing excavation. This item should be reviewed in detail to determine if a remedy is available to improve labor production for the remaining 50% of work. Should it be found that there is no solution to the problem, then it is to be assumed that the project anticipated profits will be reduced by the $600.00 additional cost. However, valuable cost information has been developed, and these actual cost records become the basis of unit costs for estimating future projects. The above example indicates that excavation should be increased from $10.00 to $12.00 per cu. yd. and unit prices for each of the other items could be reduced, producing a more accurate and competitive bid.

An effective estimating format can be established using several pre-printed forms from the Frank R. Walker Company.

Post-estimating Requirements. An estimating system must be responsive to the construction industry. Every estimate should be compared with the ultimate cost of the work--the contract price--as determined by the successful bidder. It is normal practice for an estimator to review bids on a project and to compare them with the estimate of the work. He or she attempts, where possible, to identify variations from the estimate and variations among the bidders. Significant differences are investigated to check the estimating system and to identify possible alternative methods of performing the work. The usual output of such a review would be a recommendation to the owner to either award a contract or not award one. (Contingency percentage is 0%)

Review of Unit Price Estimates. In examining unit price contracts, the estimator examines the total price as well as the individual unit quantities and prices. Significant variations from the estimate, or from other bidders, should be further investigated by checking on estimated quantities, computations, and basic assumptions about construction methods. Where the bid of the apparent responsible low bidder is within an acceptable range of the estimate, the estimator should focus on the unit prices in order to identify imbalances in unit vs. time, "front loading", or in creating a "windfall" where there might be significant variations in the actual quantities of a particular item of work.

Review of Lump Sum Bids. In reviewing bids on lump sum contracts, the estimator should initially determine whether the apparent low bid varies significantly from his estimate, or from the other bidders. The estimator then prepares for the contractor's submission of his detailed breakdown, which is compared with the detailed estimate to verify the estimate and to identify possible imbalances. Where there is a significant variation between bids and the estimate, or where there are wide variations among the bidders, the estimator attempts to determine the cause.

DAILY TIME SHEET AND LABOR DISTRIBUTION

JOB. NO. REPORT NO.

NAME SHEET NO.

LOCATION DATE

REMARKS:

OCCUP-ATION	EMPLOYEE'S NAME	EMPLY. NUMBER	LABOR CLASSIFICATION														WEATHER / TEMPERATURE / 8 A.M. / 1 P.M.		
																	HOURS	RATE	AMOUNT
		TOTALS																	

PRACTICAL FORM 505

MFD IN U.S.A.

Daily Time Sheet & Labor Distribution
Frank R. Walker Form 505

PRACTICAL FORM 115

SUMMARY OF ESTIMATE

BUILDING	LOCATION	ESTIMATE NO.
ARCHITECT	OWNER	DATE
CUBICAL CONTENTS NO. OF STORIES	COST PER CUBIC FOOT	ESTIMATOR
FLOOR, AREA, SQUARE FEET	COST PER SQUARE FOOT	CHECKER

CLASSIFICATION	TOTAL ESTIMATED MATERIAL COST	TOTAL ESTIMATED LABOR COST	TOTAL SUB-BIDS	TOTAL	ADJUSTMENTS
1. GENERAL CONDITIONS AND OFFICE OVERHEAD					
2. JOB CONDITIONS AND JOB OVERHEAD					
3. CONSTRUCTION PLANT, TOOLS AND EQUIPMENT					
4. DEMOLITION AND SITE CLEARANCE					
5. EXCAVATION, GRADING AND DEWATERING					
6. SHEETING, SHORING AND BRACING					
7. PILING AND CAISSON WORK					
8. SITE DEVELOPMENT					
9. CONCRETE FORM WORK					
10. CAST IN PLACE CONCRETE					
11. PRECAST CONCRETE AND CEMENTITIOUS DECK					
12. BRICK, TILE, CONCRETE & GLASS UNIT MASONRY					
13. UNIT MASONRY PARTITIONS & FIREPROOFING					
14. CUT, ROUGH, NATURAL & SIMULATED STONE					
15. STRUCTURAL METALS					
16. OPEN WEB JOISTS AND METAL DECKING					
17. MISCELLANEOUS METALS					
18. ORNAMENTAL METALS					
19. ROUGH CARPENTRY AND ROUGH HARDWARE					
20. FINISH CARPENTRY					
21. CUSTOM MILLWORK					
22. WATERPROOFING AND DAMP-PROOFING					
23. BUILDING INSULATION					
24. SHINGLE AND ROOFING TILE					
25. MEMBRANE ROOFING					
26. PREFORMED ROOFING AND SIDING					
27. ROOFING ACCESSORIES					
28. SHEET METAL FLASHINGS					
29. SEALANTS AND CAULKING					
30. METAL DOORS AND FRAMES					
31. WOOD AND PLASTIC DOORS					
32. SPECIAL DOORS					
33. METAL WINDOWS					
34. WOOD AND PLASTIC WINDOWS					
35. FINISHED HARDWARE & WEATHER STRIPPING					
36. GLASS, GLAZING AND MIRRORS					
37. CURTAIN WALL AND STORE FRONT SYSTEMS					
38. LATHING AND FURRING					
39. PLASTERING AND STUCCO WORK					
40. GYPSUM DRYWALL					
41. CERAMIC, MOSAIC, QUARRY, MARBLE&SLATE TILE					
42. TERRAZZO AND SEAMLESS FLOORING					
43. WOOD FLOORING					
44. RESILIENT FLOORING					
45. ACOUSTICAL TILES AND PANELS					
46. SPRAY ON FIRE PROTECTION					
47. PAINTING AND SPECIAL FINISHES					
48. SPECIAL BUILDING PARTITIONING					
49. BUILDING SPECIALTIES, EQUIP. & FURNISHINGS					
50. ELEVATORS & MECHANICAL TRANSPORT					
51. PLUMBING					
52. FIRE PROTECTION & SPRINKLER SYSTEMS					
53. HEATING, VENTILATION & AIR CONDITIONING					
54. ELECTRICAL WORK					
55. COMMUNICATION SYSTEMS					
TOTALS					
56.		TOTAL COST			
57.		PROFIT			
58.		SURETY BOND			
59.		AMOUNT OF BID			
60.					

MFD. IN U.S.A.

FRANK R. WALKER CO., PUBLISHERS, CHICAGO

Summary of Estimate
Frank R. Walker Form 115

The Estimator

The estimator is the most important person in a contractor's organization, because it is up to him to prepare estimates that will either make or break the contractor. The most efficient job organization or the best purchasing department in existence cannot make money on contracts taken below cost.

On the other hand, it does not take a good salesman or contract man to obtain all the work you can handle if your estimates are way below your competitor's. Remember, the low man is wrong more often than he is right. The efficient organization is one that can get the work at the right price. It is not an easy task, but a business built upon the principles of service, dependability, good materials and workmanship, and fair price will be in business many years after the get-the-work-at-any-price contractor has made his last assignment in favor of his creditors.

The estimator should be practical and must possess a thorough understanding of job conditions, of how the work will be carried on in the field, and of the operations necessary to assemble the materials and put them in place in the structure. He must be able to visualize and must be able to take a set of blueprints and draw a mental picture of the building. He must know every branch of the work to be handled, the materials required, and the labor operations necessary to convert a pile of gravel, sand, brick, cement, lumber, steel, and glass into a completed building. The job superintendent or foreman may be able to take a set of blueprints and lay the entire job out in his mind and picture the building progressing from the foundation to the finished structure, but unless he possesses the knowledge necessary to compute the quantity of materials and the labor cost of putting them in place, he will never be able to prepare an accurate estimate of cost. The following requisites are essential for the making of a good estimator:

a) Able to read plans and measure them accurately.

b) Possess a good knowledge of arithmetic, including addition, subtraction, multiplication, and division, together with a knowledge of the decimal system. Nearly all measurements and computations in the average building estimate are made in lineal feet, square feet or yards, and cubic feet or yards. These material quantities are then usually multiplied at a certain unit price, so a knowledge of decimals is necessary to obtain the various costs. More and more drawings will carry dimensions based on the metric system, and one must be able to convert from one to the other.

c) Able to visualize. He should be able to take a set of blueprints and draw a mental picture of the building--how it will look during construction and at completion. He must be able to place his equipment so that the work can be handled economically. In short, he must first construct the building on paper.

d) Possess an intimate knowledge of job conditions, the most practical methods of handling materials and labor on the job and the most economical methods of construction. While the estimator is generally an office man, he must possess much the same knowledge of construction methods as the superintendent or foreman.

e) Know how and be able to assemble materials into workable units. A thousand brick in the wall means more than just the brick. It means a certain amount of sand, lime and cement, and possibly mortar color, to make the mortar. A cubic yard of concrete means that exact quantities of cement, sand, and gravel must be computed. The same applies to nearly every complete unit in the building. The competent estimator must have this knowledge at his finger tips.

f) Possess an intimate knowledge of labor performances and operations and be able to convert them into costs in dollars and cents. This requires a working knowledge of what a worker will do under given conditions and is obtained only by a careful study of different jobs completed under varying conditions.

g) Have at least an average amount of common sense.

The *American Society of Professional Estimators* (ASPE), with local chapters all over the country, has several categories of membership, including student and affiliate. Education programs are promoted, conventions are held on local and national levels, and newsletters are printed. ASPE has a certification program, and examinations for certified construction estimator are given periodically.

Office Overhead Expense

The item of overhead expense is neglected by many contractors, but it is of such importance that every contractor should give it careful consideration. Every contractor has a certain fixed expense that must be paid regardless of the amount of work done or contracts received, and these items should be charged to *office overhead.*

This includes such items as office rent, fuel, lights, telephone and telegraph, stationery, office supplies, advertising, trade journals and magazines, donations, legal and accounting expenses not directly chargeable to any one job, fire and liability insurance for the office, club and association dues, office employees, such as bookkeeper, stenographer, clerks, estimator, janitor services and salaries of executives, along with travel and entertainment expenses of a good will nature and depreciation on office furnishings and equipment. These items should be estimated for a year and then reduced to a certain percentage of the total annual business, as follows:

Non-reimbursable Salaries*	$.......	$70,000
Employee Benefits	....	21,000
Legal & Accounting Fees	....	5,000
Insurance	....	10,000
Office Rent	....	5,000
Office Utilities	....	3,000
Office Supplies	....	2,000
Job Procurement & Promotion	....	7,000
Depreciation	....	2,000
Maintenance	....	2,000
Job Management* (incl. transportation)	....	5,000
Education/Regulations	....	2,000
Miscellaneous	....	6,000
		$140,000

*Indirect expenses only, that is, those not reimbursed by fees from individual jobs.

The above list includes many more expenses than small contractor will have and considerably less than those carried by larger contractors, but it serves as an example of the items that should be included as overhead or general expense. For example, suppose you have an overhead expense of $140,000 per year, and do an annual business approximating $1,500,000. On this basis your overhead or general expense would average 9.3% of the total annual business.

Overhead may run 6% for some firms. For others, it might be 10% to 15% of the total annual business, but every contractor can approximate his overhead based on his average yearly business. Even an approximation is far better than completely ignoring office overhead.

Many smaller contractors do not maintain an office and probably attend to the greater part of their office duties themselves, so only such items of expense as they actually incur should be included.

Proposals for jobs taken on a time and material basis and provisions for extra work on jobs taken on a stipulated sum basis may require the contractor to state the percentage to be added for overhead, plus that for profit. In a recent job bid from five medium sized contractors in Chicago, these varied as follows: 8+4, 8+7, two at 10+5, and one at 10+10. While overhead percentage varied only from 8% to 10%, profit varied from 4 to 10%.

Whether these same percentages are used in figuring competitive bids is, of course, each contractor's secret. But certainly, some addition to cover office overhead should be made, unless the organization is so successful that the year's overhead has already been absorbed by other jobs, or the work is of such nature that it is primarily subcontracted and office expenses are minor.

All manufacturers, jobbers, and retailers add their overhead expense to the cost of their products when computing the selling price, and there is no reason why contractors should not do the same, always keeping their eye on the competition.

Office Furniture and Equipment. All expenditures for office furniture, typewriters, calculating machines, filing cabinets, etc., should be kept as a separate account and a certain percentage of this amount charged off each year as depreciation. The depreciation, however, should be charged as a part of the overhead expense of conducting the business, as the cash value of the equipment decreases each year.

In a completely efficient office, construction equipment would be constantly utilized on the job and would be charged as a job cost, not office overhead. However, there are always slack periods when equipment is idle, or it might have been bought for a specific job but still have a useful life. It can be depreciated and charged to overhead until it can be put to use again. The Internal Revenue Department publishes fixed schedules of allowed depreciation.

Some contractors set up a "rental" cost for equipment owned by the firm but chargeable to individual jobs. Today the trend is to rent tools and equipment from outside firms specializing in this, which simplifies bookkeeping--all costs are directly charged to the job and no depreciation need be carried--and frees the contractor's capital for more lucrative investment. A discussion of plant and equipment is included later in this chapter.

Insurance. Insurance costs directly chargeable to office overhead would include fire and theft for office property, liability for office premises, workmen's compensation for office employees, and automobile insurance for trucks and cars owned by the firm.

Liability insurance is a must in today's construction. This is often included as an office overhead, a necessary cost of doing business. However, if the policy has rather low limits, the contractor will find most specifications will require that he take out additional coverage, and this would be a job expense rather than overhead expense.

Surety bonds, performance bonds, property insurance for work under construction, and workmen's compensation for on-job employees can be charged directly to the job costs and are discussed under job expenses. The cost of bid bonds for work awarded to another contractor would obviously have to be absorbed by office overhead.

Job Overhead

Job overhead includes three divisions. First, those items the general contractor must furnish for the conduct of the job as a whole are usually known as *General Conditions*, often supplemented with modifications and additions under *Special Conditions*.

Secondly, there are items the general contractor may furnish to expedite the job which may benefit all or just a single trade. These are carried under *Plant and Equipment*.

Finally, the contractor may be asked to include in his work certain auxiliary functions which require outside advisors. This may range from a simple plot survey to complete design services. These expenses are carried under *Professional Services*.

Job Overhead: General Conditions

Each job will entail certain expenses that are chargeable to the job as a whole. These are set forth in the General Conditions, the 14 articles of the AIA usually made a part of the Contract, and the Supplementary Conditions and/or General Requirement Section of the Specifications. These will include such items as superintendence, temporary buildings, enclosures, barricades, offices, toilets, utilities, protection, clean-up, permits, surveys, photographs, tools and equipment, insurance and benefits, sales taxes, surety bonds, and warranties.

Sometimes, these are covered only in a general way, and it is up to the contractor to visit the site to determine what is needed. Often the unions and local laws may be more demanding than the architect's specifications, and the contractor must know what items he is liable for.

Job overhead and office overhead are best kept separately, although occasionally employees' insurance, tools and equipment, and other items may be divided between office expense and job expense.

Many contractors figure a certain percentage of the cost of the job to take care of all overhead expense and plant charges, but this is not the correct method to use and does not give the contractor or estimator a reliable basis for computing the cost of these expenses. Each item entering into the cost of job overhead expense should be estimated the same as any other branch of work.

Superintendence. This should include the salary and traveling expenses of the job superintendent, assistant superintendent, engineers, timekeepers, material clerks, watchmen, water boys, labor foremen, and all other foremen who are carried on the general contractor's payroll.

The proper method of computing these costs is to estimate the salaries of the various personnel required at a certain price per week or per month for the entire estimated time they will be required on the job.

Temporary Buildings and Enclosures. This should include temporary fences, bridges or platforms, temporary sheds and storerooms, temporary enclosures over hoisting engines, concrete mixers, temporary runways, ladders and stairs in the building before the permanent stairs are installed, temporary dustproof partitions to isolate work areas from building space still occupied, and temporary closures for doors and windows before the permanent doors and windows are in place.

Temporary Office. This should include the temporary job office for use by the contractor and architect during construction, together with all necessary expense in maintaining it, such as telephone, heat, light, stationery, and other items of expense. These are usually trailers, which may be rented. An 8' x 30' trailer, furnished, will cost around $5,000 but can be rented for around $200.00 per month.

Barricades and Signal Lights. This should cover the cost of building and maintaining barricades during construction, cost of maintaining signal lights, etc.

Temporary Toilet. Practically every job must have some provision for sanitary facilities. Local codes usually govern what must be supplied. Chemical, portable toilets will usually suffice. On large jobs, however, a more elaborate setup is usually specified requiring a complete water supply, drain, vent, and fixture installation. This work is sometimes included in the plumbing specifications. Chemical toilets can be rented for about $65.00 to $80.00 per month.

In the case of a remodeling job where permission is granted to use the existing facilities, an item should be included in the estimate for periodic and final cleaning of the toilet room and its appurtenances.

Temporary Water. On most jobs water for construction purposes must be provided by the general contractor for all trades. This item can become a high-cost operation, especially in barren, undeveloped localities where either a well must be drilled or water must be hauled into the job. When a water main is available, a plumber must be hired to tap it, install water meter, and run a water line to strategic locations on the job. This work is sometimes included in the plumbing specifications. Also included under this heading are the charges made by the local water department for the volume of water used. Often hydrants can be used at a set fee from the municipality.

Temporary Light and Power. This item can vary, from a small charge for tapping into existing adjacent power lines and running entrance cable into a meter setup from which extension cords may be run, to complex installations requiring long run pole lines, transformers, and switch gear, such as for electric mixers, hoists, and other heavy duty electric power equipment. This work is sometimes included in the electrical specifications. Also included in this item are the power company charges for energy consumed and rental of transformers, etc.

Temporary Heat. If a job is to run into or through the winter, it is very likely that some temporary heat will be required to maintain proper working conditions. On small jobs a few salamanders may be enough to fulfill the requirements. Large jobs, however, usually require temporary hookup of radiators to the new and as yet uncompleted heating system.

Repairing Streets and Pavements. This item should cover the repair and replacement of all sidewalks and pavements damaged or destroyed during construction.

Damage to Adjoining Buildings. There is always a chance of damage to adjacent structures, such as to windows, to foundations and walls improperly shored, or to skylights, and these items should be considered when preparing the estimate.

Protecting New Work from Damage During Construction. During construction there are certain classes of work that must be protected from damage, such as cut stone, marble, terra cotta, and granite. Cover newly-finished cement or wood floors to prevent damage.

Repairing New Work Damaged During Construction. This item should cover patching of damaged plaster, replacing broken lights of glass, etc. Quite often architects specify the cost of such repair work to be prorated among the various contractors working on the job unless specific responsibility for the damages can be charged to an individual contractor.

Cutting and Patching for Other Trades. Many specifications require the general or mason contractor to do all cutting and patching of walls and floors for the plumbing, heating, and electrical contractors, where it is necessary for the mechanical trades to place sleeves or pipes through the floors or walls after the permanent work is in place.

Removing and Replacing Public Utilities. In the larger cities it may be necessary to remove or relocate water hydrants, telephone and utility poles and wires, sewers and drains, or other underground pipes. In such instances it will be necessary for the contractor to pay for any damage incurred.

Removing Rubbish, Cleaning Floors, Windows, Etc. During the construction of every building there is always a certain amount of rubbish, such as old concrete, broken bricks, and tile, that must be removed at intervals. The most satisfactory method is to estimate so many loads of rubbish at a certain price per load.

If finish floors are to be swept or scrubbed before acceptance of the building by the owner, it is advisable to make a certain allowance to provide for this work. This should cost about $3.50 per 100 sq. ft.

Some architects require the contractor to clean all windows before the building will be accepted. To arrive at this cost, count the windows and allow a certain price for cleaning each window, generally around $.09 per sq. ft., 2 sides.

Building Permits. The cost of building permits varies from city to city. Some cities base the cost of the permit on the total cost of the work, or the cubic content of the building, while others charge for water consumed, street frontage occupied, and the number of plumbing fixtures, for which an inspection charge is made.

As a general rule a building permit will cost from 1/10 to 1/4 of 1 per cent of the total cost of the building. The following are examples of permit fees. Check with local governing agencies for costs in the area where you are building.

a) Single car garage, two car garage and shelter sheds: For each 1,000 cubic feet or fractional part, $4.00; minimum charge, $10.00.

b) New buildings or other structures, except those included in a): For each 1,000 cubic feet of volume or fractional part, $1.25; minimum charge, $20.00. Volume computations under a) and b) shall include every part of the building

from the basement to the highest point of the roof and include bay windows and other projections.
c) Alterations and repairs to any structures: For the first $1,000.00 of estimated cost, $7.75; for each additional $1,000.00 of estimated cost, $3.00; minimum charge, $20.00.

d) Shoring, raising, underpinning, or moving any building: For each 1,000 cubic feet of volume or fractional part, $0.40; minimum charge, $20.00.

e) Wrecking any building or other structure more than one story in height: For each 25 lineal feet of frontage or fractional part, $13.00; minimum charge, $20.00.

f) Fire Escape, erection or alteration: Four stories or less in height, $20.00; each story above four, $2.75.

g) Canopy or marquee, erection or alteration, $20.00.

h) Chimneys, isolated or over 50 feet above any roof, $20.00.

i) Tanks, above roof or tower: 400 gallon capacity or less, $13.00; over 400 gallon capacity, $25.00; structural supports for tank over 400 gallon capacity, $25.00.

j) Elevators and escalators, power operated, five floors or less in height, installed or altered, $20.00; each additional floor above five, $4.00; hand operated, $13.50; levelators, $13.50; theater curtains, $50.00; stage or orchestra platforms, $20.00. Dumbwaiters, power operated, five floors or less in height, installed or altered, $20.00; each additional floor above five, $4.00. Hand operated, five floors or less in height, $13.00; each additional floor above five, $2.75. Temporary construction towers over 50 feet in height $20.00.

k) Amusement devices, mechanical riding, sliding, sailing or swinging: portable, for each assembly or installation, $12.50. Permanent, installed or altered, $105.00. Temporary seating stands, $70.00.

l) Ventilating systems, mechanical supply or exhaust: Capacity, 3,000 cubic feet of air per minute, $10.50; for each additional 1,000 cubic feet per minute, $3.00; increase in capacity, each 1,000 cubic feet per minute, $3.00. Capacity shall be the sum of supply and exhaust. Warm air furnaces, $12.50 each.

m) Fences over 5 feet high, 100 lineal feet, $10.00; each additional 100 lineal feet or part thereof, $3.00.

n) Tanks for flammable liquids: Classes I and II, capacity 121 to 500 gallons, $21.00; each additional 1,000 gallons or fraction thereof, $0.65; Classes III and IV, capacity 551 to 1,000 gallons, $21.00; each additional 1,000 gallons or fraction thereof, $0.40.

o) Temporary platforms for public assembly units, $21.00.

p) Roof of any building, recoating or recovering, $6.50.

q) Billboards and signboards--erection, construction or alteration: up to 150 sq. ft. $5.00; 151 to 225 sq. ft. $7.75; 226 to 375 sq. ft. $13.00; for each additional 375 sq. ft. or part thereof $13.00.

A fee of $20.00 is paid to a city collector for the approval of plans and for inspection and test of any plumbing within any building containing not more than five plumbing fixtures. An additional fee of $3.00 shall be paid for every plumbing fixture in excess of five within such building.

Surveys. Where the contractor must furnish a survey of the property on which the building is to be erected, a charge of $500 to $1,000 is usually made for a lot up to 100' x 100'. In the downtown sections of large cities this charge will be substantially more. A topographical survey with 2' contours will cost around $300 for an acre; $1700 for 10 acres; and about $100 per acre from 10 to 100.

Photographs. Where it is necessary to have photographs taken at intervals to show job progress, the cost varies with the number of views made at one time. It will cost from $250 to $300 per visit, including shots of all building corners and two black and white 8" x 10" prints of each shot. Additional black and white 8" x 10" prints cost about $10.00 each. In smaller towns and cities, this cost may be reduced.

Social Security Tax. The Federal Insurance Contributions Act, F.I.C.A., requires contractors to pay a tax on all payrolls, both job and office. At the present time the employer's share of this tax is 6.20 percent of the first $53,400 of taxable wages paid each employee for the year.

The employer is also obliged to deduct 6.20 percent from each employee's salary or wages, but except for the cost of additional clerical help in maintaining the records, this portion of the tax does not increase the contractor's cost. In many organizations extra employees have been required to keep these records.

In addition to the social security tax, the employer must pay a medicare tax of 1.45% on wages or salary up to $125,000 and deduct 1.45% from the employee's wages or salary:

Regular FICA	6.20% of wages up to $53,400 annually
Medicare Tax	1.45% of wages up to $125,000 annually
Total	7.65%

Federal Unemployment Tax. This tax applies to the first $6,000 of wages paid each employee during a calendar year. The rate is 3.2 but you may take a credit not in excess of 2.5% of taxable wages for contributions you paid into State Unemployment Funds. Federal Unemployment Tax is imposed on employers and must not be deducted from wages of employees.

State Unemployment Tax. This will vary from state to state. Note above that a credit is allowed against the federal rate.

Sales Taxes. Many states have sales or use taxes applicable to building construction as well as to other commodities. This tax varies with the different states, and there are different methods of levying it. In some states, the tax is levied on all building materials and is paid by the contractor when he pays for his materials. In other states the tax is applicable to the entire cost of materials entering into the construction of the building and is paid by the contractor directly to the state. In either case, the contractor should add this cost to his estimate. Otherwise, it will make a big dent in estimated profits. Note that some jobs will be tax exempt, which could save as much as 6% or more on materials. However, it is up to the client to establish his tax exemption, and the contractor should qualify his bid accordingly.

Job Overhead: Plant and Equipment

Plant overhead includes equipment costs, the cost of transporting equipment to and from the job site, and the costs of operating it. These costs may be in the form of depreciation allowances if owned, or as actual monthly or daily charges if rented. Generally, equipment used by only one trade will be furnished as part of that subcontract and will not appear as general contractor's overhead.

At the other extreme, items such as outside hoists and elevators on a high-rise building will be used by all trades and would be charged similarly to a general conditions expense. But there are many items that will be used by several, though not all, trades.

As an example, a job may require a rolling scaffold that will be used by the mechanical trades and electricians installing above-ceiling work; by metal lathers and plasterers installing a fancy coved lobby ceiling; by acoustical tilers installing office ceilings; by drywall men installing storage room ceilings; and by painters decorating ceilings. If each trade brought its own scaffold, there would be 7 transportation charges, and individual trades would probably be renting on a daily rate. If the general contractor furnishes the scaffold, one transport charge is involved with a much cheaper monthly rate. This savings could lower the bid and win the job.

If the contractor owns the scaffold, he can rent to his subs at the going rate and make a profit for himself, as well as lowering the job cost. Another advantage in the general contractor's furnishing plant equipment is that he can control quality and safety on the job. With Occupational Safety and Health Administration (OSHA) rules in effect, it is crucial that the general contractor control the quality of equipment on his job or face stiff fines.

If the contractor owns the equipment, it may be depreciated for tax purposes. However, if the inflation rate is high, the depreciation allowed by the Internal Revenue Service might not be adequate to cover replacement costs, and the tax savings might not be so attractive. On the other hand, a high inflation rate might make it possible to buy equipment, use it, and resell it for what one paid for it.

The advantages in renting (a short term agreement) or leasing (a long term agreement) include:

a) eliminates tying up capital or borrowing against bank credit for equipment that most probably will not return as much money as if those assets were applied toward financing another job or taking advantage of cash discounts;

b) offers flexibility in choosing equipment; one can try one product on one job and another on a different job, until one finds just the right piece of equipment for each situation, or if the item does not perform to expectations, it can be returned;

c) eliminates personal property taxes on the equipment;

d) eliminates maintenance costs and the expense of training personnel to handle those tasks and minimizes possible losses due to obsolescence and vandalism;

e) eliminates costs of shipping, handling and storage, and tying up trucks and drivers to move this equipment, as this will be done by the lessor.

The question then becomes whether to rent or to buy and depreciate plant equipment. In a large community where several rental sources are available, renting is probably preferable, leaving the contractor's money for other forms of investment, such as building up capital in order to take on larger jobs or investing in land on which a contractor cannot only hope to make a profit on resale but also can insist his firm be used as contractor for any development.

If the contractor does elect to purchase equipment, he should set up separate records for each item. Be sure to consider the Internal Revenue Service's depreciation schedule allowed, the probable life, probable maintenance and storage cost, actual cost of transporting the item to and from the job, and the costs of operating the equipment on the job. Enter an hourly rate chargeable for the use of the item on each job estimate.

Charges should be sufficient to replace the item when it is worn out, as well as to cover interest charges if a loan is taken out to purchase it and a separate fund, often known as a *sinking fund,* set aside to cover future replacement. These charges should not be carried as profit. If the equipment is idle for long periods, the costs could be transferred to general overhead, but this is rather dodging the issue. If the equipment costs are accurately charged against each job, a contractor can easily tell whether his costs of ownership exceed current rental costs, always keeping in mind that the funds tied up in plant ownership could be making money elsewhere. Plant and equipment costs are discussed in greater detail later on in this chapter, under "Tool and Equipment".

Job Overhead: Prepurchase and Prepayment of Materials

With the cost of building materials being pushed ever upward, and with spot shortages of some raw materials a constant threat, a contractor may more and more be confronted with owners who want to protect themselves and order out materials months in advance of their being installed on the job. If this is a stated part of the contract, the estimator must make allowances for it.

In most cases the owner will reimburse the contractor upon his properly presenting his request for payment; but if the contractor is not to be paid until the material is installed, the estimator must add the costs of interim financing. Then the material must be suitably stored. Even if a material is weatherproof, few job sites today are vandalproof. Sometimes the material may arrive before the construction site is available. If the contractor must provide a shed, a watchman, or an off-site warehouse, or if the manufacturer is to store the material until it is needed at the site, these costs must also be figured and added to the bid or specifically excluded and made part of an extra work order. There are insurance and handling costs that must also be assigned either to the bid or to the owner. And it must be kept in mind that these costs will grow if the job schedule is not adhered to.

Job Overhead: Professional Services

Sometimes an owner may approach a contractor to act as his advisor on matters other than field construction. This can include finding a suitable site; lining up financing; handling rezoning or other legal and governmental problems such as sewer and road extensions, utilities and tax assessments; hiring of architects and engineers to design the project, landscapers to beautify the grounds, and decorators to furnish the interiors.

Some of the initial time spent on such consultation will be charged as "good will" under general office expense; then, if the job progresses, the contractor should be reimbursed not only for fees for outside services, but also for his and his staff's time spent in administering these services.

A contractor should move slowly in assuming these auxiliary functions. All too often they require much more time, responsibility and cash investment than anticipated, and his major responsibility, the contracting business, is left without adequate supervision and in a cash bind because reimbursement for preliminary phases has been held up until ground is broken and the interim loan is in effect.

Further, a contractor relies on the good will of the community for his future business. Choosing one real estate office, architect, lawyer, banker, or one decorator over another may limit his future prospects more than the immediate job can justify.

However, a good contractor does tend to be a good citizen, and as his experience grows he is a valuable source of information on building in the community. This knowledge is an asset, and it is entirely proper to charge for it on an hourly basis as for any other skill.

Bonds

Bonds are usually issued to cover a single job and are a direct job cost. Some cover only a single trade and may be included in a subcontract sum, such as warranties or roofing bonds. Bonds taken out by the general contractor are usually carried under *Overhead: General Conditions.* Some of the types of bonds encountered are discussed below.

Bonds are usually issued by insurance companies or surety companies specializing in such bonds. A division of the U.S. Department of Labor, Small Business Administration, has been set up to provide surety bonding for small and minority contractors. Surety bonds are credit guarantees of contractors and are issued after careful scrutiny by underwriters of a contractor's financial statements, skills, work in progress, past work, and general reputation. Bonds are not purchased by contractors, but rather, issued for the payment of a fee based on contract cost.

Bid Bonds. Bids are invited by advertisements, and the bidder may have to submit with his bid a certified check, usually for 5% of the bid, or a bid bond, usually for 10% of the bid. This requirement is made to insure that bids will be submitted only by qualified contractors who will not only enter into the contract but qualify for performance and payment bonds.

The bid bond guarantees that as the contract is awarded, the principal will, within a specified time, sign the contract and furnish the required performance and payment bonds. If the principal fails to furnish the required performance bond, there will be a default under the bid bond, and the damage will be the difference between the principal's bid and that of the next highest bidder, not exceeding the penalty of the bid bond.

Surety companies generally follow the practice of authorizing a bid bond only after the performance bond on the contract has been underwritten and approved, so there are few defaults under bid bonds.

Performance and Payment Bonds. Some owners, and most public work, require that each contractor to whom a contract is awarded shall furnish bonds which contain two obligations: a performance bond to indemnify the owner against loss resulting from the failure of the contractor to complete the work in accordance with the plans and specifications; and a payment bond to guarantee payment of all bills incurred by the contractor for labor or materials for the work.

The federal government, under the Miller act, requires that a contractor furnish two separate bonds, one for the performance and one for the payment of labor and materials. Some states follow the same procedure.

The risk of the surety under the construction contract bond is if the contractor fails in the performance of his obligation to complete the job, or to pay the bills for labor or materials, then it assumes that the responsibility for completing the project and paying the bills.

The underwriting skill is designed to weigh the hazards involved against the contractor's qualifications. If the contractor's ability is deemed sufficient to warrant the assumption that he will be able to complete the work and pay the bills, the bond is issued. If not, it is declined.

The general practice is to underwrite each bond on an individual basis. The contractor cannot assume that every request will be favorably received by underwriters. The contractor's volume of work that is still outstanding at the time of the bond request is a key factor.

Some of the most common causes of failure on the part of construction contractors are abnormal labor costs, fire or other loss not adequately covered by insurance, unforeseen restriction or withdrawal of credit, weather or subsurface conditions which increase the cost but not the contract price, over extension beyond the contractor's financial capacity, unavailability of materials, failure of subcontractors, death or disability of key men in the contractor's organization, a poor accident record, diversion of funds from one job to pay off overdue accounts on other jobs, inexperience in the type of work involved or dishonesty on the part of the contractor's employees.

Costs of performance bonds depend on amount of contract and length of coverage, as well as the contractor's reputation and whether the coverage is for a portion or complete cost of work.

Bonds are usually issued for a 24 month period and might run on an average job at $12 per $1,000 of contract cost for the first $500,000; $7.25 per $1,000 for the next $2,000,000; $5.75 per $1,000 for the next $2,500,000; $5.25 per $1,000 for the next $2,500,000 and $4.85 per $1,000 over $7,500,000. Should the bond need to be extended beyond two years, figure 1% of the regular premium per month.

Maintenance Bonds. These may be required to guarantee an owner that the general contractor agrees to correct all defects of workmanship and materials for a specified time following completion. Often the above performance bond can be arranged to cover this period which usually lasts one to three years.

License or Permit Bonds. These are bonds provided by a contractor to a public body, guaranteeing compliance with local codes and ordinances. If the contractor regularly operates within an area requiring such bonds, this cost should be carried under office overhead, because it is a normal cost of doing business. But if it is required for a specific job, it can be charged directly as a job expense.

Subdivision Bonds. With so many restrictions being applied to uses of sewers and water supplies, with shortages of natural gas, and with restrictions on fuel emissions, more and more public bodies require bonds be filed guaranteeing that all site improvements, utilities, and buildings themselves will be constructed in accordance with governing codes and ordinances. Unless the contractor is the developer, this is usually filed directly by the owner, who in turn is already protected by the contractor's performance bond to the extent that the drawings and specifications comply with the ordinances.

Specifications may include catchall phrases, such as "and in accordance with all local codes and ordinances...", which follow detailed instructions of procedures that may be contrary to such codes. In addition, there is a growing legal history on developments that have been refused licenses after completion because they do not comply. Consequently, the contractor should inform himself of all local requirements before entering into any contract.

Surety underwriters very carefully review requests for subdivision bonds, and these bonds are issued only when the underwriter is satisfied that the contractor is able to meet all obligations required by authorities.

Supply and Subcontractor Bonds. These are guarantees from manufacturers and installers for specific items and workmanship furnished, the cost of which is usually borne by them and included in the subcontract price for that trade rather than in overhead.

Insurance

A reliable insurance agent is one of the most valuable "subs" a contractor can add to his team. Almost everything can be insured and has been--at a price. Some types of insurance are required by law. Some are made part of the contract. Others may be indicated because of some job peculiarity. The following list discusses the more usual types encountered. Rates will obviously vary all over the country and from one part of town to the other, from one suburb to the next. There will be some overlap between insurance costs chargeable to office overhead and to job overhead, but it is best to keep them as separate as possible.

Insurance is a significant cost component in construction and is currently estimated to range between 3% and 5% of the "hard costs" of a contract. Construction contracts ordinarily contain a section titled *insurance* that specifies the insurance coverages that must be purchased by the contractor and those that are to be provided by the owner, usually coverage for damage to the project itself. This section of the contract usually sets insurance coverage conditions, terms, limits, and special wordings, such as for underground damage and contractual liability. Often, this section requires that the contractor name the owner as an additional insured and waives subrogation (the right to sue on one's behalf) in favor of the owner, its agents, and its employees. Sometimes, those who finance the project seek the same insurance protection. Insurance companies are reluctant to make such amendments, and if they do, the contractor can expect an additional premium charge.

Sometimes, construction contracts, particularly those for larger jobs, state that the owner will provide all insurance, except for specific types. Contractors will be required to bid *ex insurance* and the normal contractor overhead allowance for certain insurance will not be granted. This type of insurance is referred to as the *owner controlled program* and also as *wrap up insurance.*

Since its inception, owner controlled insurance has been the subject of much controversy and in many states is prohibited for public work. There are some notable exceptions. For instance, some states allow this type of insurance for certain *public authorities* but not for all. It is normally for private projects that exceed certain sums.

When faced with owner controlled insurance, a contractor should carefully review the insurance section and other pertinent parts of the contract in order to properly identify the risks and the extent to which the owner's program provides protection. It is not to recommend that the contractor exclude the project from an existing program, but rather to have a contractor's program, the excess over valid and collectible insurance provided by the owner. This is not automatic and must be specifically negotiated with insurance companies.

Workmen's Compensation. This insurance provides the contractor's employees with the benefits specified by the various state and federal compensation laws in the event of injury or death arising out of the employment. Most compensation acts contain at least some provisions specifying benefits for the most common occupational diseases.

Rates vary widely among the several states, the average being from less than 3% in Indiana to over 22% in the District Of Columbia. Sixteen states, primarily in the Southeast, average from 4% to 6%; eight states, primarily in the West, average over 10% and the remaining twenty-five average from 6 to 10%. In those states where medical benefits are limited, it is advisable to carry full or extra-legal medical coverage.

Certain compensation laws are restricted and provide benefits for only specified occupations. The workmen's compensation policy may be broadened to voluntarily pay compensation to employees injured in other occupations.

In addition to providing the statutory benefits, the standard workmen's compensation policy affords employer's liability insurance against claims for bodily injury or disease arising out of employment in the states named in the policy. This coverage against liability not within the scope of the compensation law is subject to $25,000 standard limit which can be increased for a moderate premium charge.

In the event of the death of an injured employee, his dependents generally have the right to choose between the benefits of the state where the accident occurred or the state where the employee was hired. In addition, a number of states impose penalties upon an employer who though subject to the compensation act has failed to provide insurance coverage. Accordingly, it is important that you discuss these points with your insurance representative before entering a state not covered by your policy and before hiring anyone in another state.

The general insurance rates in the preceding table are representative rates to be used as a guide only. Rates and experience modifications are subject to frequent and sometimes radical changes and should be verified with an insurance representative for each specific project. Major job classifications are listed in the table, but you must consult your insurance representative for specific classifications.

Reference has been made herein to base or manual rates and to experience modification of the manual rates or premium. A program of "retrospective rating" is available on an optional basis. This rating plan, superimposed on experience rated premiums, is usually attractive and appropriate to the larger contractors. Subject to certain state limitations as to the combinations permissible, it is available on an intrastate or interstate basis for worker's compensation and the liability lines, either separately or in combination.

Your insurance representative will be glad to explain the details of this rating plan, as well as any other features of coverage or rating on these various lines of insurance.

Workers Compensation Manual Rates for Certain States

Code No. & Type of Work	CA	DC	FL	IL	NJ	NY	TX
0042 Site Improvement	14.73	6.73	10.30	8.52	6.28	10.52	15.47
5022 Masonry	14.55	20.32	20.16	17.52	8.27	15.16	23.12
5040 Iron Steel Erection	24.99	54.46	36.02	70.09	20.44	23.34	50.47
5057 Structural Steel	31.33	53.40	43.58	102.24	9.81	25.67	31.01
5102 Misc. Ornamental Metal	12.92	23.67	16.80	12.95	....	12.80	14.29
5160 Elevators	6.27	10.75	10.16	15.75	2.45	8.31	4.67
5183 Plumbing	11.58	15.84	12.01	13.63	3.41	8.02	12.88
5188 Sprinklers	7.24	9.88	11.06	10.39	2.16	7.40	12.06
5190 Electrical Wiring	8.85	9.71	9.67	8.75	2.74	6.63	11.60
5213 Slab on Grade Foundation	12.23	31.58	33.24	25.80	6.81	15.36	25.59
5348 Ceramic Tile	8.49	26.72	9.53	12.67	3.14	9.63	9.94
5403 Carpentry	24.00	17.03	22.70	19.24	7.27	14.08	28.29
5445 Drywall	11.64	11.18	17.45	11.42	4.15	6.74	....
5474 Painting	17.17	10.62	22.64	15.10	7.68	10.75	18.29
5480 Lath & Plaster	18.12	12.14	23.70	14.16	6.88	12.02	20.99
5551 or 5595 (NY only) Roofing & Waterproofing	39.79	37.85	41.83	36.88	17.12	29.83	47.24
5550 (NY only) HVAC	....	....	....	....	....	7.48	....
5610 Gen. Con. & Cleaning	7.88	12.07	12.08	8.47	5.44	4.98	11.94
6217 Excavation	9.37	14.63	15.20	10.82	6.12	13.52	....
9521 Toilet Partitions	....	7.56	12.39	10.04	3.38	5.44	14.87
5606 Contractor Executive Supervisor	4.52	1.94	8.66	5.81	0.87	6.61	5.44
Effective Date:	1-1-91	5-15-90	9-1-90	1-1-91	1-1-91	7-1-90	1-1-90

Courtesy of William Cullen, Johnson & Higgins, New York, NY

Contractor's Public Liability Insurance. This policy protects against liability for bodily injury to a member of the public arising out of the contractor's operations and against property damage liability. It can be had under either a schedule general liability form policy or a comprehensive general liability form.

Under the schedule form, the several hazards to be insured are specifically named, and the coverage may be applied to specified locations and operations of the insured contractor. The comprehensive form is broader in nature and covers all the usual hazards of a general liability nature other than those specifically excluded.

Insurance companies are almost always willing to expand coverage conditions to cover excluded risks or to amend with broader terms the definition of an exclusion. The *broad form comprehensive general liability endorsement* is an example of a large package of additional coverage.

Under either form, consideration should be given to at least the lines of coverage briefly outlined below:

Owners, Landlords and Tenants, Premises Exposure: This hazard includes such exposures as offices, warehouses and fixed locations in contrast to specific contracting operations as such. The usual rating basis is a charge per 100 sq.

ft. of area, and normally, the premium for this hazard is relatively minor compared with that for the operations hazard.

Products Liability: Completed operations coverage applying to accidents which occur after the contractor's operations have been completed. Rates per $1,000 of receipts or contract price vary over a wide range, according to the operations being conducted. These rates are frequently the subject of individual account rate determination rather than the application of any base manual rates.

Sometimes, an owner specifically requires that the contractor carry product/completed operations coverage for a period of time after completion of a project. The completion of the work is the trigger that begins coverage, which applies to acts or errors, failures or negligence committed during the construction period. The need to provide such coverage after job completion is redundant. The key factor in the granting of coverage is when the negligent act or omission occurred, not when the accident happened. Of course, if the contractor has, or is required to provide, a new form of coverage called *claims made,* then the above can be ignored, because claims made coverage applies only for claims made against the insured during the term of the policy. With claims made insurance, the contractor is obligated to carry insurance protection for as many years subsequent to completion of the project or for as long as the contractor deems an exposure exists. It should be noted that claims made insurance has a like effect on all other coverage sections of the general liability policy, and contractors should carefully weigh all potential risks and exposures before accepting this type of coverage.

Owners Protective: Covering the liability of the owner for injuries to the public arising out of operations of independent contractors. Characteristic rates for general construction operations in most states are .018 bodily injury and .011 property damage per $100 of contract cost. These rates apply to the first $500,000 of cost on each specific project. Corresponding rates for the next $500,000 of contract price are .009 and .006; for contract cost over $1,000,000, it is .002 and .002. Rates in Illinois and New York are somewhat higher.

Contractor's Protective: Covering liability of the contractor for injuries to the public arising out of operations of subcontractors. Rating basis and rate levels stated above for owners protective are applicable under this line to sublet contract costs.

Additional features can be incorporated into the general liability policy, either for application on an individual exposure basis or to be applied on an automatic or blanket basis.

Contractual Coverage: Applying to the liability of others assumed by the contractor. Rates depend on the nature of exposure and degree of liability assumed.

Accident Defined Coverage: Broadening the coverage to apply to continuous or repeated exposure giving rise to injury or damage, in contrast to the "caused by accident" basis on which most General Liability Policies are written. Cost is usually expressed as a percentage of policy premium.

Personal Injury Coverage: Broadening the usual "bodily injury" nature to include such personal injury as libel, slander, false arrest, malicious prosecution, etc. Pricing varies with the degree of exposure and extent of coverage afforded.

Malpractice Coverage: Covering contractor for liability arising out of error or mistake in professional treatment in on-the-job hospital or medical facilities under his control. This line is individually rated.

Increased Limits of Liability: The general liability rates referred to herein and in subsequent pages are at standard limits of $5,000 per person, $10,000 per accident for bodily injury and at $5,000 per accident, $25,000 aggregate for property damage. The percentage increase factors for various higher limits can be secured from your insurance representative.

Automobile Liability Coverage: Bodily injury and property damage coverage for owned vehicles, non-owned vehicles, hired vehicles, or independent contractor's can be secured under a schedule form or a comprehensive form of automobile liability policy. Coverage can be selected either on the basis of a listing of owned automobiles or on an automatic basis whereby coverage applies to all owned vehicles. The rating basis for owned vehicles is based on size and weight, usage of the vehicle, and geographical territory involved.

Physical damage coverage (fire, theft, and collision) on owned vehicles may be either under the policy for automobile liability coverage or as a separate policy.

Property Damage Insurance: The contractor should make a careful study of what his responsibilities are in regard to damage to other property and to the work completed so that his policies will mesh with those carried by the owner.

If provision is not made in the specifications that fire insurance will be carried by the owner during construction, it is important that the contractor protect his interests by carrying some type of builders' risk insurance. This is normally written on a standard fire policy together with an extended coverage endorsement that extends the fire and lightning provisions of the standard fire policy to include windstorm, hail, explosion, riot, smoke damage, and damage from aircraft and vehicles. An additional endorsement, vandalism and malicious mischief, may be obtained. Some companies afford a broader coverage of an all risk nature although there are a number of exclusions.

There are two common builders' risk forms that may be written:

Builders' Risk Reporting Form: This form affords coverage for the contractor from the time material is placed on the job site until the building is completed. Under this form the contractor reports to the insurance company the actual value of the building at the end of each month as construction progresses and pays a premium each month on the additional values that go into the building.

Completed Value Builders' Risk Form: This form of coverage is normally preferred by a contractor, because it gives 100 percent coverage without the necessity of keeping records as to the value of the building as it progresses or reporting at the end of each month.

This form is written for the completed value of the building. The normal rate charged is about 55% of the rate on the basis that at the beginning of the job there is little value and at the end you have the full value. Normally the premium under the two types of policies is remarkably close.

The interests of a general contractor can be covered and he may exclude or include the interests of subcontractors who also may be employed in the construction of the building.

Temporary structures, tools, and equipment that belong to the contractor and are on the building site may also be included in the coverage for the hazards insured.

Today, many contracts that require the contractor to provide the coverage call for coverage on all risk of loss basis, naming the owner, lender, and numerous other parties, as well as the contractor, as insureds in the policy.

In addition, many contractors require the inclusion of specified exposures related to the project. Delay in opening and bank financing changes are two of the more frequently seen "soft cost" coverage requirements.

On occasion, contracts with clauses that are not contained in the insurance section of the contract but that are found in other sections, such as *force majeure*, risk of loss, indemnification, and work place safety, leave the contractor with liabilities that may or may not be covered by insurance. For example, the insurance section might specify limits of liability to be purchased for policies provided by the contractor, but another section of the contract or specification, contractual indemnification or risk loss, may specifically state that the contractor's liability is full and not limited to or by any insurance requirements otherwise stated in the contract.

Costs of such insurance vary by location, type of construction, and available public fire protection and may be obtained from your local insurance representative.

Contractor's Equipment Floater: Many contractors possess many thousands of dollars in contracting machinery and equipment. To properly protect these values, such equipment and machinery is usually insured under a contractor's equipment floater policy.

Insurance under this form is quite flexible so that many situations may be insured to fit the contractor's particular needs. The insurance is normally written on a named peril policy which includes loss or damage by fire or lightning, explosion, windstorm, flood, earthquake, collapse of bridges and culverts, theft, landslide, collision, and upset. Or if desired, a policy may be issued protecting from practically all risks of physical loss or damage. Such contracts usually write a deductible clause.

Installation Floater: The contractor who specializes in plumbing, heating, air conditioning, or electrical equipment may find an installation floater more suited to his needs. This type of policy may be written for an individual job or may be written to cover all the jobs he may secure during a year's time.

Coverage may be on a so-called "named peril" policy, usually including the hazards of transportation of his material to the job site and covering fire and lightning, windstorm and hail, explosion, riot and civil commotion, smoke damage, and aircraft and vehicle damage at the job site. He may also include coverage for vandalism and theft if desired.

Profit

The amount of profit to be added to the estimated cost of the work is a question that a contractor must answer individually for each bid. There is no set amount that can be added. It all depends on local conditions, competition, and how badly the job is wanted.

On small jobs, alterations, remodeling, and similar work, a contractor is justified in adding 20% to 25% profit to the actual cost, but he must ask himself whether he can actually obtain this amount.

On new work, where it is possible to estimate cost with a fair degree of accuracy, a contractor is entitled to 10% on the actual cost of the work (job overhead included in the actual cost of the job), but it is safe to say that competitive figures submitted for many jobs show a 5% instead of a 10% basis. A contractor is entitled to 10% profit, but getting it is another matter.

Another topic of considerable discussion is whether or not a contractor should add a certain percentage to his estimate to take care of contingencies. The answer is both yes and no. Yes, because it is not necessary to be quite so careful in the preparation of the estimate, and if any items have been overlooked, there is a fund to take care of omissions. But a contractor who is not careful in the preparation of his estimates never knows how much should be added to provide for forgotten items or contingencies. No, because if you add a percentage for contingencies, it is quite possible that one or more of your competitors has not, and another job is lost.

The only safe way is to be as careful as possible in listing all items from the plans and specifications. Include them all, and it will not be necessary to add for contingencies. This includes everything to be furnished and all labor costs, including job overhead expense and a percentage for profit.

An item often overlooked, but which can become a major expense, is the finishing up of a job and making corrections after the job is completed. Often, there is confusion on the job at the end and it is difficult to say who chipped the paint, scratched the door, broke the window, stole the light bulb, broke the door handle, or backed into the fencing. Most of the trades are on the job doing touch-up work, the owner is moving in, his decorator is hanging drapes, and his children are running wildly through the corridors. But the general contractor gets the complaint first, and in the interest of having a satisfied client, will probably assume the responsibility for correcting this. In a warehouse, such problems may be minimal; in a fine residence, you might be called back daily for weeks.

Other factors an estimator must evaluate in setting up his bid are the weather conditions that are apt to prevail at the time of construction; possible rises in labor rates and the possibility of strikes if labor contracts expire during the construction period; material shortages and strike actions at the manufacturers level; the money supply and its cost; and public relations with the community in which the project is located and possible job stoppage or slowdowns because of environmental or safety concerns which the drawings and specifications fail to cover.

Labor Inefficiencies on Construction Projects
Due to Overtime

Number of Days Per Week	Hours Per Day	Hours Per Week	Lost Efficiency Factors %
5.00	8.00	40.00	0
5.00	9.00	45.00	0-5
5.00	10.00	50.00	2-8
5.00	11.00	55.00	5-11
5.00	12.00	60.00	8-14

Number of Days Per Week	Hours Per Day	Hours Per Week	Lost Efficiency Factors %
6.00	8.00	48.00	2-3
6.00	9.00	54.00	10-13
6.00	10.00	60.00	16-17
6.00	11.00	66.00	20-21
6.00	12.00	72.00	22-25
7.00	8.00	56.00	8-15
7.00	9.00	63.00	16-16
7.00	10.00	70.00	21-22
7.00	11.00	77.00	24-25
7.00	12.00	84.00	28-29

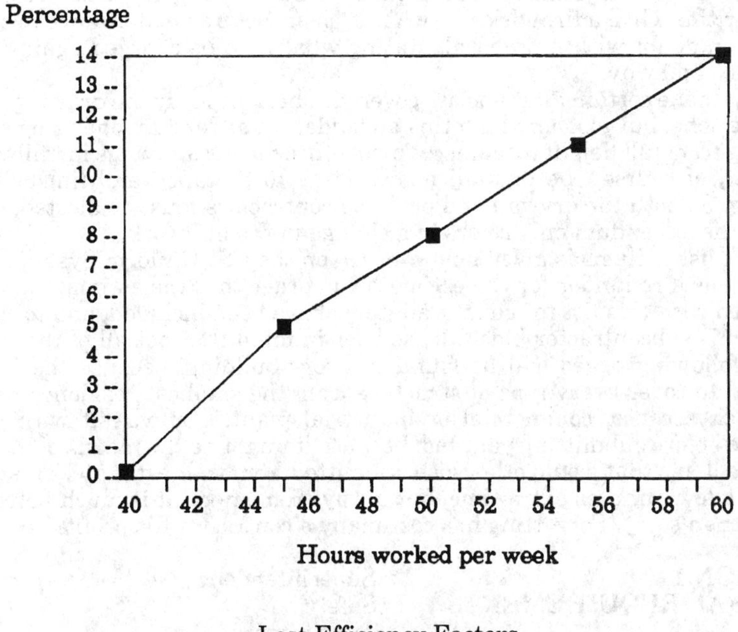

Lost Efficiency Factors

Estimate Check List

Too many estimates are done under the pressure of a deadline. If at all possible, a second party should check all arithmetic on the original estimate sheets and the original figures against those copied on the final bid sheet. It is not unusual to receive a bid from a subcontractor with an obvious error in the figures. This is an indication that the bid was not checked prior to tendering, and it could result in a severe impact on the company offering the bid and the general contractor utilizing the bid as part of the general estimate.

Storage, transportation, and handling costs are often overlooked. It is often a good idea to qualify a bid as to the time of year a job is to be started. A bid given in March anticipating an April start and October finish should say so. If the contract is delayed for any reason, so that the job starts in October and runs through the winter, allowances must be added for temporary enclosures and heat, idle days or weeks, and less efficient work conditions. Furthermore, inflation is a major factor today, and labor and material costs can escalate 4% to 6% in six months.

The estimate should be carefully checked for plant equipment costs, with each subcontractor being carefully informed as to what will be furnished to him. Architect specifications are usually very general on this matter, often stating in Work Included "...all labor, materials, and related items..." to cover plant equipment. It is left up to the general contractor to have a clear understanding with his subs on this.

Check your profit. How much of this is certain, and how much contingent on optimism? Set forth a definite percentage for contingencies and try to hold to it. Keep the profit apart. If there is a questionable item, alert the owner and/or architect, and then, qualify your bid accordingly. Most proposals provide a space for this. Give a firm price on what is clear. Set up an allowance to cover what is in doubt with a proposal, stating what is to be charged against the allowance and why.

Finally, make certain all items are covered. Check not only the drawings and specifications, but also any bulletins addenda. An addendum changing wood paneling from full height to wainscot in an office means a savings in millwork. But now, of course, the drywall above needs to be taped and finished for painting. So both the drywall and painting contractors must be alerted, even though the addendum only refers to a change in the millwork.

A check list of items is given below, based on the CSI Uniform System. It is intended as a reminder for the estimator, in order that the estimate may be compared with this list to see that all items have been included, and to check against his subcontractor bids. Please bear in mind that not all of the items on the following pages will be found in every building, because the list is intended to cover every type of structure from the smallest residence to the largest skyscraper, commercial or industrial plant. Compare it with your estimate before submitting your bid, because it might be the means of including some items that would otherwise be omitted from your estimate. This may require a few minutes extra time, but always remember, it is much better to be safe than sorry. Forgetting has cost many a contractor his profit.

DIVISION 1 GENERAL REQUIREMENTS

Summary of Work:
- Work under this Contract
- Work under Separate Contracts
- Work by Others
- Work by Owner
- Permits
- Taxes
- Warranties
- Fees and Notices
- Indemnification
- Insurance
  - Liability
  - Loss of Use
  - Property Damage
- Bonding
- Superintendence
- Safety
- Timekeepers
- Material Clerks
- Foremen
- Watchmen
- Coordination of Trades
- Cutting & Patching
- Allowances

Alternates
- Required
- Proposed

Project Meetings
- Preconstruction Conferences
- Progress Meetings
- Job Site Administration

Submittals
- Construction Schedules

Progress Reports
Survey Data
Shop Drawings
Samples
Photographs
Operation & Maintenance Data
Schedule of Values
As-built Drawings

Quality Control
Tests & Lab Services
Inspection Services
Quality Assurance/Control

Temporary Facilities
Temporary Utilities
Temporary Electricity
Temporary Lighting
Temporary Heat & Ventilation
Temporary Telephone Service
Temporary Water
Temporary Sanitary Facilities
Temporary First Aid Facilities
Construction Aids
Construction Elevators & Hoists
Temporary Enclosure
Swing Staging
Barriers
Fences
Tree & Plant Protection
Guardrails & Barricades
Security
Access Roads
Special Controls
Noise Control
Dust Control
Water Control
Pest Control
Rodent Control
Pollution Control
Maintenance & Protection of Traffic
Traffic Regulation
Traffic Signals
Flagmen
Flares & Lights
Parking
Project Identification
Temporary Field Offices

Products
Transportation
Handling
Storage at Job
Storage at Warehouse
Protection

Project Closeout
Cleaning Up
Record Documents
Start-up
Punch List
Touch-up & Repair

Transportation Allowance

DIVISION 2
SITE WORK

Subsurface Exploration
Borings
Core Drilling
Penetration Tests

Clearing
Structure Moving
Clearing & Grubbing
Tree Pruning
Shrub & Tree Relocation

Demolition
Buildings
Appurtenances
Interior

Earthwork
Containment
Site Grading
Rock Removal
Embankment
Excavating & Backfilling
Trenching, Backfilling, & Compacting
Structural Excavation, Backfill, and Compaction
Roadway Excavation, Backfill, and Compaction
Pipe Boring and Jacking
Waste Material Disposal
Soil Compaction Control
Soil Stabilization

Soil Treatment
Termite Control
Vegetation Control

Pile Foundations
Pile Driving
Pile Load Tests
Wood Piles
Precast Concrete Piles
Prestressed Concrete Piles
Compacted Concrete Piles
Steel H-Section Piles
Steel Pipe Piles
Concrete-Filled Steel Shell Piles

Caissons

Drilled
Excavated
Shoring
Steel Sheeting
Walers & Shores
Cribbing
Piling & Lagging
Underpinning
Site Drainage
Subdrainage Systems
Foundation Drainage
Underslab Drainage
Drainage Structures
Drainage Pipe
Dewatering
Wellpoints
Relief Wells
Erosion Control
Site Utilities
Gas Distribution System
Gas Transmission Lines
Oil Distribution System
Oil Transmission Lines
Water Distribution System
Water Transmission Lines
Telephone Distribution
Steam Distribution
Hot Water Distribution
Chilled Water Distribution
Water Wells
Sewage Lagoons
Paving & Surfacing
Mudjacking
Paving
Crushed Stone Paving
Asphalt Concrete Paving
Brick Paving
Portland Cement Concrete Paving
Bituminous Block Paving
Repair & Resurfacing
Pavement Sealing
Pavement Marking
Curbs & Gutters
Walks
Synthetic Surfacing
Site Improvements
Chain Link Fences
Wire Fences
Wood Fences
Guardrails
Signs
Traffic Signals
Culvert Pipe Underpasses
Playing Fields
Recreational Facilities
Fountains & Equipment
Irrigation System
Sprinkler Systems
Site Furnishings
Rubble Site Structures
Railroad Tie Structures
Lighting
Landscaping
Soil Preparation
Lawn Seeding & Sodding
Trees & Shrubs
Plants
Beds
Railroad Work
Trackwork
Ballasting
Service Facilities
Traffic Control
Marine Work
Docks
Boat Facilities
Protective Structures
Seawalls
Groins
Jetties
Dredging
Tunneling
Excavation
Grouting
Support Systems
Rockbolting

DIVISION 3
CONCRETE
Accessories
Inserts
Anchors
Ties
Hangers
Hardeners
Abrasives
Coloring
Formwork
Liners & Coatings
Wood Forms
Prefabricated Forms
Panel Forms
Pan Forms
Steel Forms
Fiberglass Forms
Stair Forms

Expansion & Contraction Joints
- Metal
- Plastic

Reinforcement
- Steel Bar
- Welded Fabric
- Stressing Tendons
- Epoxy Coating

Cast-In-Place Concrete
- Curing
- Concrete Types
  - Premixes
  - Lime
- Shrinkage Compensating
- Lightweight
- Insulating
- Lightweight Structural
- Heavyweight
- Prestressed
- Post-tensioned

Specially Finished Concrete
- Exposed Aggregate
- Tooled
- Blasted
- Heavy Duty Floor Finishes
- Grooved Surface

Specially Placed Concrete
- Shotcrete
- Grout
- Catalyzed Metallic Grout
- Non-shrink Grout
- Non-corrosive Grout
- Epoxy Grout

Precast Concrete
- Wall Panels
- Tilt-up Panels
- Structural
- Deck Panels
- Prestressed
  - Standard
  - Custom

Cementitious Decks
- Gypsum
- Wood Fiber

Installation Equipment
- Scaffolding
- Machinery
- Hoists
- Temporary Heat
- Tarpaulins
- Straw
- Special Handling & Storage

DIVISION 4
MASONRY

Mortar
- Cement & Lime
- Acid Resisting
- Premixed

Accessories
- Joint Reinforcement
- Anchors & Ties
- Control Joints
- Lintels

Unit Masonry
- Brick
  - Common
  - Face
  - Enamelled
  - Fire
  - Standard Sized
  - Special Sized
- Adobe
- Concrete Units
- Reinforced Units
- High-Lift Grouted
- Preassembled Panels
- Clay Backing Tile
- Clay Facing Tile
- Ceramic Veneer
- Terra Cotta Veneer
- Mechanically Supported Veneer
- Glass Unit
- Gypsum Unit
- Sound Absorbing

Stone
- Rough
- Cut or Carved
- Marble
- Limestone
- Granite
- Sandstone
- Simulated
- Cast
- Flagstone
- Natural Veneer

Restoration & Cleaning
- Cleaning
- Restoration
- Pointing
- Coating
- EPA Regulations

Refractories
- Flue Liners
- Corrosion Resisting

Combustion Chamber

Installation Equipment

Scaffolding
Machinery
Special Handling & Storage

DIVISION 5
METALS

Structural Metal Framing

Structural Steel
- Beams & Girders
- Roof Trusses
- Crane Tracks
- Grillage
- Bolts
- Welding

Structural Aluminum

Metal Joists

Standard Joists
Custom Joists
Aluminum
Framing Systems
Space Frames

Metal Decking

Roof
Floor

Lightgauge Framing

Metal Studs
Metal Joists

Metal Fabrications

Metal Stairs
Handrails & Railings
Pipe & Tube Railings
Gratings
Castings
Manhole Covers
Fireplace Accessories
Bumpers
Joist Hangers & Straps
Wall Plates
Package Receivers
Window Guards
Sidewalk Doors
Nosings
Rough Hardware For Above

Ornamental Metals

Stairs
Handrails & Railings
Sheet Metal Fabrications
Cast Iron Work
Bronze
Aluminum Stainless

Expansion Controls

Shop Painting

Field Painting

Epoxy Coating

Material Handling

Trucking
Unloading
Erecting Machinery
Storage/Protection

DIVISION 6
CARPENTRY

Rough Carpentry

Framing
- Trusses
- Columns
- Beams & Girders
- Rafters
- Floor & Ceiling Joists
- Porch Framing
- Rough & Exterior Stairs
- Stud Walls
- Bridging
- Lintels

Sheathing
- Roof
- Wall
- Insulating

Preassembled Components
Diaphragms
Accessories & Hardware

Heavy Timber Construction

Timber Trusses
Mill-Framed Structures
Pole Construction

Trestles

Prefabricated Structural Wood

Glue Laminated Structural Units
Glue Laminated Decking
Wood Trusses
Wood-Metal Joists

Finish Carpentry

Millwork
- Exterior
  - Facias, Cornices, & Trim
  - Columns
  - Shutters & Blinds
  - Railings
  - Stoops & Porch Work
  - Cupolas
- Interior
  - Base & Quarter Round

Chair & Picture Moldings
Cornices
Strip Paneling
Ceiling Beams
Closet Trim
Laminated Plastic

Wood Treatment
Pressure Treated Lumber
Preservative Treated Lumber
Fire Retardant Lumber

Architectural Woodwork
Custom Cabinet Work
Paneling
Solid Wood
Architectural Plywood
Ornamental Stair Work
Custom Trim
Decorative Wallboard

DIVISION 7
THERMAL & MOISTURE PROTECTION

Waterproofing
Membrane
Elastomeric
Fluid Applied
Liquid
Bentonite Clay
Metal Oxide
Epoxy

Dampproofing
Bituminous
Silicone
Water Repellant Coatings
Cementitious
Vapor Barriers
Bituminous
Laminated
Plastic

Insulation
Loose Fill
Rigid
Fibrous
Reflective
Foamed-in-place
Sprayed-On
High and Low Temperature
Roof and Deck
Perimeter
Under Slab

Shingles & Roofing Tiles
Asphalt
Wood
Slate
Porcelain Enamel
Metal
Clay Tiles
Cement Tile

Preformed Roofing & Siding
Metal Siding
Composite Panels
Preformed Plastic
Cladding/Siding
Wood
Composition
Plastic

Membrane Roofing
Built-Up Bituminous
Prepared Roll
Elastic Sheet
Fluid Applied

Traffic Topping

Flashing & Sheet Metal
Roofing
Flashing & Trim
Specialties
Gutters & Downspouts
Gravel Stops

Roof Accessories
Skylights
Plastic
Metal Framed
Hatches
Gravity Vents
Prefabricated Curbs
Expansion Joints

Sealants
Joint Fillers & Gaskets
Sealants & Caulking

Material Handling
Hoists
Scaffolds
Special Storage

DIVISION 8
DOORS & WINDOWS

Metal Doors & Frames
Hollow Metal
Aluminum
Stainless Steel
Bronze

Wood & Plastic
Flush
Panel

Plastic Faced
Plastic Formed

Special Doors
Sliding Metal Fire Doors
Metal Clad
Coiling Doors
Coiling Grilles
Folding
Accordion
Flexible
Overhead
Sliding Glass
Safety Glass
Sound Retardant
Screen & Storm

Entrances and Storefronts
Aluminum
Stainless
Bronze
Revolving Doors

Metal Windows
Steel
Aluminum
Stainless
Bronze
Operating Hardware

Wood & Plastic Windows
Milled Wood
Preglazed
Plastic Faced
Plastic Formed

Hardware & Specialties
Finish Hardware
Operating Hardware
Automatic Door Equipment
Weatherstripping & Seals
Thresholds

Glazing
Float Glass
Sheet Glass
Tempered
Wired
Rough & Figured
Processed
Coated
Laminated
Insulating
Mirror
Plastics
Accessories

Window Wall/Curtain Wall
Steel
Aluminum
Stainless
Bronze
Wood
Installation Machinery
Scaffolding

DIVISION 9
FINISHES

Lath & Plaster
Furring & Lathing
Gypsum Plaster
Cement Plaster
Acoustical Plaster

Gypsum Wallboard
System Types
Required Accessories
Metal Furring
Metal Stud Partitions

Tile
Ceramic
Mosaic
Quarry
Marble
Glass
Plastic
Metal

Terrazzo
Portland Cement
Precast
Conductive
Plastic Matrix

Acoustical Treatments
Ceilings
Panels
Tiles
Wall Treatments
Insulation and Barrier

Ceiling Suspension Systems

Wood Flooring
Strip
Parquet
Plywood Block
Resilient Systems
Industrial Block

Resilient Flooring
Cementitious Underlayment
Asphalt Tile
Vinyl
Rubber
Sheet
Fluid Applied

Carpeting

Cushioning
Wool
Synthetic
Blend
Custom
Tile Cut
Special Flooring
Magnesium Oxychloride
Epoxy-marble Chip
Elastomeric Liquid
Heavy-duty Concrete Toppings
Armored Floor
Brick
Laminated Plastic
Special Coatings
Abrasion Resistant
Cementitious
Elastomeric
Fire-resistant
Sprayed Fireproofing
Aggregate Wall Coating
Painting
Exterior
Flat
Enamel
Wall
Trim
Porches
Stairs
Windows
Doors
Railings
Sheet Metal Work
Steel & Cast Iron Work
Interior
Preliminary Preparation
Flat
Semi-gloss
Enamel
Varnish
Plastic
Graining
Sanding Between Coats
Washing
Touch-up
Wall Coverings
Vinyl-Coated Fabric
Cork
Paper
Fabric
Flexible Wood
Prefinished Panels
Adhesives

Temporary Equipment
Ladders
Scaffolds
Rolling Platforms

DIVISION 10
SPECIALTIES

Chalkboards & Tackboards
Compartments & Cubicles
Hospital
Toilet
Metal
Plastic
Stone
Shower & Dressing
Handicapped Accessories
Louvers & Vents
Grilles & Screens
Wall & Corner Guards
Access Flooring
Fireplaces, Prefabricated
Flagpoles
Identifying Devices
Lockers
Protective Prefab Covers
Walkway
Car
Postal Specialties
Prefab Partitions
Special Shelving
Scales
Sun Control Devices
Telephone Enclosures
Toilet & Bath Accessories
Wardrobe Specialties

DIVISION 11
EQUIPMENT
Built-In Maintenance Equipment

Vacuum
Window Washing
Bank & Vault
Commercial
Checkroom
Darkroom
Ecclesiastical

Educational
Food Service
Athletic
Industrial
Laboratory
Laundry
Library
Medical
Mortuary
Musical
Parking
Waste Handling
Loading Dock
Detention
Residential
Theater & Stage
Registration

DIVISION 12
FURNISHINGS

Artwork
Cabinet Furniture
Window Treatments
Fabrics
Rugs & Mats
Seating

DIVISION 13
SPECIAL CONSTRUCTION

Air Supported Structures
Integrated Assemblies
Audiometric Rooms
Clean Rooms
Hyperbaric Rooms
Incinerators
Instrumentation
Insulated Rooms
Integrated Ceilings
Nuclear Reactors
Observatory
Prefabricated Structures
Radiation Protection
Sound & Vibration Controls
Vaults
Swimming Pools
Saunas

DIVISION 14
CONVEYING SYSTEMS

Dumbwaiters
Elevators
Hoists & Cranes
Lifts
Material Handling Systems
Turntables
Moving Stairs & Walks
Pneumatic Tube Systems
Powered Scaffolding

DIVISION 15
MECHANICAL

General Provisions
Permits
Notices
Fees
Excavation & Backfill
Tests & Balances
Identification
Maintenance Contracts
Demonstration
Basic Materials & Methods
Steel Pipe
Cast Iron Pipe
Copper Pipe
Plastic Pipe
Glass Pipe
Stainless Steel Pipe
Aluminum Pipe
PVC Pipe
Hose
Piping Specialties
Gaskets
Swivel Joints
Strainers, Filters, & Driers
Vent Caps
Traps
Vacuum Breakers
Shock Absorbers
Supports, Anchors, & Seals
Anchors
Wall Seal
Flashing & Safing
Hangers & Supports
Gate Valves
Blowdown Valves
Butterfly Valves
Ball Valves
Globe Valves
Refrigerant Valves
Stop Cocks
Curb Stops
Hydrants

Check Valves
Swing Check Valves
Backwater Valves
Vertical Check Valves
Stop & Check Valves
Faucets
Washer Outlets
Self Contained Control Valves
Pressure Regulating Valves
Pressure Relief Valves
Automatic Temperature &
  Pressure Relief Valves
Solenoid Valves
Tempering Controllers
Photo Lab Tempering Controller
Mixing Station
Refrigerant Control Valves &
  Specialties
Feed Water Regulator
Pumps
  Centrifugal
  Rotary
  Turbine
  Reciprocating
  Sump Pump
  Submersible Pump
  Pneumatic Ejector
  Compressors
  Vacuum Pumps
  Air Compressors
Vibration Isolation & Expansion
  Compensation
  Vibration Isolation
  Expansion Joints
  Flexible Connections
Meters and Gauges
  Temperature Gauges
  Pressure Gauges
  Flow Measuring Devices
  Liquid Level Gauges
Tanks
  Steel Tanks
  Plastic Tanks
  Cast Iron Tanks

INSULATION
  General
  Cold Water Piping
  Chilled Water Piping
  Refrigerant Piping
  Hot Water Piping
  Steam & Condensate Return Piping
  Underground Piping
  Outside Piping
  Duct
  Breeching
  Equipment

Water Supply & Treatment
  Pump & Piping System
  Booster Pumping Equipment
  Water Reservoirs & Tanks
  Water Treatment
  Filtration Equipment
  Aeration Equipment
  Water Softening Equipment
  Chemical Feeding Equipment
  Chlorinating Equipment
  Metering & Related Piping

Waste Water Disposal & Treatment
  Sewage Ejectors
  Grease Interceptors
  Lift Stations
  Septic Tanks
  Drainage Fields
  Sewage Treatment Equipment
  Screens & Skimming Tanks
  Sedimentation Tanks
  Filtration Equipment
  Aeration Equipment
  Sludge Digestion Equipment

Plumbing
  Water Supply System
  Chilled Water Piping Systems
  Distilled Water Piping System
  Compressed Air Piping System
  Oxygen Piping System
  Helium Piping System
  Nitrous Oxide Piping System
  Vacuum Piping System
  Laboratory Gas Piping System
  Compressed Industrial Gas Piping
    Systems
  Central Soap Piping System
  Soil Piping System
  Waste Piping System
  Roof Drainage System
  Chemical Waste Drainage System
  Industrial Waste Drainage System
  Process Piping Systems
  Floor and Shower Drains
  Roof Drains
  Cleanouts & Cleanout Action Covers
  Domestic Water Heaters
  Aftercoolers & Separators

Stills
Anti-syphon Equipment
Sediment Interceptors
Laundry/Utility Units
Packaged Waste, Vent, or Water Piping Units
Domestic Water Conditioners
Special System Accessories
Gas Accessories
Plumbing Fixtures & Trim
Urinal Flush Valves
Special Fixtures & Trim
Fixture Carriers
Domestic Watercoolers
Wash Fountains
Showers
Receptors
Pool Equipment
Circulation & Filtration Equipment
Pool Drains, Inlets, & Outlets
Pool Cleaning Equipment
Chemical Treatment Equipment
Fountain Piping & Nozzles
Special Equipment

Fire Protection

Sprinkler Equipment
Foam Equipment
Halon Systems
CO^2 Equipment
Standpipe & Fire Hose Equipment
Fire Hose Connections
Fire Hose Cabinets & Accessories
Fire Hose Reels
Fire Hose
Portable Extinguishers
Fire Blankets
Fire Extinguisher Cabinets & Accessories
Hood & Duct Fire Protection
Non-electrical Alarm Equipment

Air Distribution

Furnaces
Direct Fired Furnaces
Cast Iron Furnaces
Steel Furnaces
Rooftop Furnaces
Direct Fired Unit Heaters
Direct Fired Duct Heaters/Reheaters
Fans
Centrifugal Fans
Propeller Fans
Attic Exhaust Fans
Fly Fans
Axial Flow Fans
Induced Draft Fans
Exhaust Fans
Power Roof Ventilators
Power Wall Ventilators
Roof Ventilators (connected to ductwork)
Air Handling Units (without coils)
Air Curtains
Ductwork
Low Pressure Steel Ductwork
High Pressure Steel Ductwork
Nonmetallic Ductwork
Special Ductwork
Prefabricated Insulated Ductwork
Flexible Ductwork
Duct Lining
Duct Hangers & Supports
Special Ductwork Systems
Tailpipe Exhaust Equipment
Dust Collection Equipment
Paint Spray Booth System Equipment
Fume Collection System Equipment
Breeching & Smokepipe
Duct Accessories
Manual Dampers
Gravity Backdraft Dampers
Barometric Dampers
Fire Dampers
Smoke Dampers
Turning Vanes
Distribution Devices
Duct Access Panels & Test Holes
Outlets
Wall & Floor Diffusers
Ceiling Diffusers
Ceiling Air Distribution System
Light Troffer-Diffusers
Warm Air Baseboard
Cabinet Diffusers
Air Floors
Roof Mounted Air Inlets & Outlets
Air Inlet & Outlet Louvers (connect to ductwork)
Air Treatment Equipment
Disposable Filters
Permanent Filters
High Efficiency Filters
Roll Filters
Oil Bath Air Filters

Electronic Air Filters
Air Washers
Dust Collectors
Fume Collectors or Dispensers
Sound Attenuators
Special Devices

Power or Heat Generation

Fuel Handling Equipment
Oil Storage Tanks, Controls, & Piping
L-P Gas Tanks, Controls, & Piping
Ash Removal System
Lined Breechings
Lined Prefabricated Chimneys & Stacks
Exhaust Equipment
Draft Control Equipment
Boilers
Cast Iron Boiler
Firebox Boiler
Scotch Marine Boiler
Water Tube Boilers
Absorption Boiler
Burners & Controls
Stokers
Fuel Preheaters
Boiler Accessories
Boiler Feedwater Equipment
Packaged Boiler Feed Pump System
Deaerators

Refrigeration

Refrigerant Compressors
Centrifugal Compressor
Rotary Compressor
Reciprocating Compressor
Condensing Units
Air Cooled Condensing Units
Water Cooled Condensing Units
Evaporative Condensing Units
Chillers
Reciprocating Chillers
Air Cooled Chillers
Ethylene Glycol Chillers
Centrifugal Chillers
Absorption Chillers
Rotary Chillers
Cooling Tower (Propeller type)
Cooling Tower (Centrifugal type)
Ice Bank
Special Ice Making Equipment
Commercial Ice Making Equipment
Evaporators
Unit Coolers
Condensers
Refrigeration Accessories

Liquid Heat Transfer

Hot Water Piping System
Chilled Water Piping System
Steam Supply & Return Piping System
Radiant Heat System
Snow Melting System
Hot Water Specialties
Steam Specialties
Condensate Pump & Receiver Set
Heat Exchangers
Storage Water Heater
Converter
Clean Steam Heat Exchanger
Water Heat Reclaim Equipment
Terminal Units
Induction Units
Radiant Panels
Coils
Baseboard Units
Finned Tube
Convectors
Radiators
Unit Heaters
Fan Coil Units
Unit Ventilators
Air Handling Units (with coils)
Packaged Heating and Cooling
Packaged Heat Pump
Humidity Control
Humidifiers
Centrifugal Type Humidifier
Dehumidifiers
Desiccant Dehumidifiers
Process Heating
Storage Cells
Special Devices

Controls and Instrumentation

Electrical and Interlocks
Identification
Inspection, Testing, & Balancing
Control Piping, Tubing, & Wiring
Control Air Compressor & Dryer
Control Panels
Instrument Panelboard
Primary Control Devices
Thermostats
Humidistats

Aquastats
Relays & Switches
Timers
Control Dampers
Control Valves
Control Motors
Sequence of Operation
Recording Devices
Alarm Devices
Special Process Controls

DIVISION 16
ELECTRICAL

General
Tests
Demonstration
Identification

Basic Materials & Methods
Raceways
Conduits
Bus Ducts
Underfloor Ducts
Cable Trays
Wires & Cables
Wire Connections & Devices
Pulling Cables
Outlet Boxes
Pull & Junction Boxes
Floor Boxes
Cabinets
Panelboards
Switches & Receptacles
Motors
Motor Starters
Disconnects (motor & circuit)
Overcurrent Protective Devices
Supporting Devices
Electronic Devices

Power Generation
Generator
Engine
Reciprocating Engine
Turbine
Cooling Equipment
Exhaust Equipment
Starting Equipment
Automatic Transfer Equipment

Power Transmission
Substation
Switchgear
Transformer
Vaults
Manholes
Rectifier
Converter
Capacitor

Service & Distribution
Electric Service
Underground Service
Service Entrance
Emergency Service
Service Disconnect
Primary Load Interrupter
Metering
Grounding
Transformers
Distribution Switchboards
Branch Circuit Panelboard
Feeder Circuit
Converters
Rectifiers

Lighting
Interior Lighting Fixtures
Luminous Ceiling
Signal Lighting
Exterior Lighting Fixtures
Stadium Lighting
Roadway Lighting
Accessories
Lamps
Ballasts & Accessories
Poles & Standards

Special Systems
Lightning Protection
Emergency Light & Power
Storage Batteries
Battery Charging Equipment
Cathodic Protection

Communications
Radio Transmission
Shortwave Transmission
Microwave Transmission
Alarm & Detection
Fire Alarm & Detection
Smoke Detector
Burglar Alarm
Clock & Program Equipment
Telephone
Telegraph
Intercommunication Equipment
Public Address Equipment
Television Systems
Master TV Antenna Equipment
Learning Laboratories

Heating & Cooling

Snow Melting Cable & Mat
Heating Cable
Electric Heating Coil
Electric Baseboard
Packaged Room Air Conditioners
Radiant Heaters
Duct Heaters
Electric Heaters (Prop Fan Type)

Controls & Instrumentation
Recording & Indicating Devices
Motor Control Centers
Lighting Control Equipment
Electrical Interlock
Control of Electric Heating
Limit Switches

Determining Costs

The detailed estimate is a projection into the future of the anticipated costs of a project. It reflects all of the known conditions at the time of preparation. It is important to maintain a cost system that uses as a base each of the components contained in the estimated cost. This is necessary in order to compare estimated costs with actual costs incurred during construction. If the estimated costs are realistic and accurate, they will produce the anticipated profit.

Equally important, the periodic review of actual cost versus estimated cost will serve as a data base that can be adjusted, either by increasing or decreasing the unit costs, so that future estimates will be produced more competitively.

Cost control systems can take many forms. They can be relatively simple and still produce accurate information, or they can be designed in an elaborate and detailed format to provide information relating to cost for every item of material and labor productivity for the entire project.

The selection of the system is one of personal choice, but it must be remembered that any cost system requires considerable expenditure of time in accumulating the material and labor records, in distributing to the proper categories of work, and then, in calculating the costs of each and recording this information on progressive cost record forms so that the actual costs can be monitored while the work is being performed. When this information is available, it is then possible, should an overrun appear, to review the situation to determine if a remedy is possible, such as adjusting crew balance, pursuing another method of construction, or allocating additional supervision to a specific operation. The system selected should be one that serves the needs of the company. Anything in excess of this is a burden and wasteful in both time and cost.

Walker Practical Forms 550, 505, and 506 are a suitable base to establish an effective cost system.

CONSTRUCTION SCHEDULING

In order to complete a construction project within the estimated cost, the contractor must insure that individual work activities are completed within the time allowed by the cost estimate. A schedule must be prepared and followed. The overall process can be separated into three distinct steps:

a) Planning
b) Scheduling
c) Monitoring

ESTIMATE AND COST SUMMARY

PROJECT

LOCATION

ARCHITECT ENGINEER

CLASSIFICATION

ESTIMATOR

EXTENSIONS

CHECKED

ESTIMATE NO.

SHEET NO.

DATE

			ESTIMATED COST								% COMPLETED	ACTUAL COST							
			MATERIAL			LABOR						MATERIAL			LABOR				
CODE	DESCRIPTION	UNIT	QUANTITY	UNIT COST	AMOUNT	HOURS	AMOUNT	SUB-CONTRACTS OTHER COSTS	TOTAL ESTIMATE			QUANTITY USED	UNIT COST	AMOUNT	HOURS	AMOUNT	SUB-CONTRACTS OTHER COSTS	TOTAL COST	PROFIT OR (LOSS)
	TOTALS																		

COMMENTS

FRANK R. WALKER CO., PUBLISHERS, CHICAGO

MFD. IN U.S.A. FORM 550

Frank R. Walker Form 550

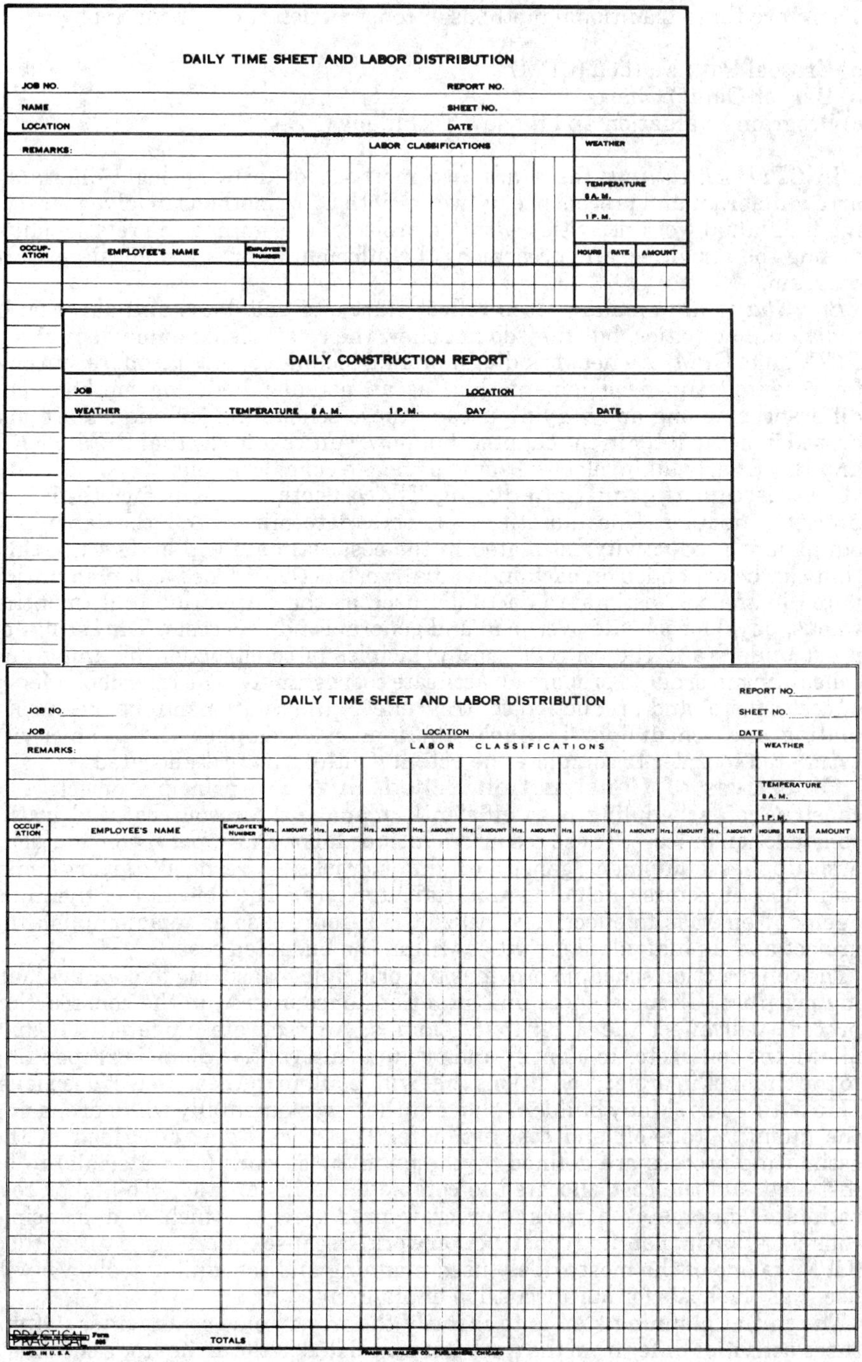

DAILY TIME SHEET AND LABOR DISTRIBUTION

JOB NO. REPORT NO.
NAME SHEET NO.
LOCATION DATE
REMARKS: LABOR CLASSIFICATIONS WEATHER
TEMPERATURE 8 A.M. 1 P.M.
OCCUPATION | EMPLOYEE'S NAME | EMPLOYEE'S NUMBER | HOURS | RATE | AMOUNT

DAILY CONSTRUCTION REPORT

JOB LOCATION
WEATHER TEMPERATURE 8 A.M. 1 P.M. DAY DATE

DAILY TIME SHEET AND LABOR DISTRIBUTION

JOB NO. REPORT NO.
SHEET NO.
JOB LOCATION DATE
REMARKS: LABOR CLASSIFICATIONS WEATHER
TEMPERATURE 8 A.M. 1 P.M.

OCCUPATION	EMPLOYEE'S NAME	EMPLOYEE'S NUMBER	Hrs.	AMOUNT	Hrs.	AMOUNT	Hrs.	AMOUNT	Hrs.	AMOUNT	Hrs.	AMOUNT	Hrs.	AMOUNT	Hrs.	AMOUNT	Hrs.	AMOUNT	Hrs.	AMOUNT	Hrs.	AMOUNT	HOURS	RATE	AMOUNT
	TOTALS																								

PRACTICAL Form 506 MFD. IN U.S.A.
FRANK R. WALKER CO., PUBLISHERS, CHICAGO

Frank R. Walker Forms 505, 506

There are three traditional methods of construction time scheduling:

a) Critical Path Method (CPM)
b) Bar (or Gantt) Charts
c) Program Evaluation and Review Technique (PERT)

In *CPM scheduling*, there are two methods of drawing logic diagrams: arrow-diagram and precedence network. Both CPM methods involve identifying individual work activities within a project, determining the relationships among the activities, and presenting the information in a graphical network diagram.

Bar Charts are similar. Bars reflect start and end dates and duration of individual activities, but they do not show the relationship among activities. *PERT*, like CPM, is a network diagramming technique, but it is more suitable for research and development than as an actual scheduling method. The differences among and varying uses of these scheduling techniques are discussed in detail later in the chapter. For now, suffice it to say that CPM, PERT, and Bar Charts all involve the same principle considerations.

Cost Estimating and Scheduling. The cost estimator is, in fact, the initial project scheduler. The quantities of work determined and the labor and equipment productivity estimated in the cost estimate will budget the time that can be expended on each individual work activity. The work plan anticipated in the cost estimate is actually used as the estimating unit (months, weeks, days) for job site overhead and general condition costs. The estimator must understand the concepts and principles of construction planning and scheduling in order to prepare an accurate cost estimate. The interdependency of cost control and production control reflects the relationship between estimating and scheduling. The field effort must accomplish the productivity estimated in order to complete the activity within the cost allocated.

Objectives of Construction Scheduling. The principle objective of construction scheduling is to efficiently manage the resources used in the construction process. These resources include labor and supervision, material and supplies, equipment, general condition or jobsite overhead support items, and subcontractors (including material suppliers). The objective of the management effort is to effectively utilize the resources so as to accomplish the project and its individual activities within the budgeted cost.

The construction schedule provides the principle measuring tool for evaluating progress. Circumstances and situations encountered in the construction process will affect the work that remains. An accurate, updated schedule allows the contractor to identify and evaluate alternative plans in responding to the changes in project conditions that will result in the best project decisions.

How to Plan. Many builders plan out their work mentally while preparing the quantity take-off and cost estimate. The work items described in the estimating process are defined by the quantity of work (or materials) to be installed and the cost allocated to complete the installation. The labor cost estimate anticipates a worker or crew productivity, which is a principle consideration in planning construction work. Because the critical path method (CPM) is one of the most widely used planning and scheduling tools, we will use it as the basis for our discussion of planning.

The initial planning step is to identify the work items or activities. Often, these activities differ from the work items as listed in the estimator's checklist. In the planning process the contractor determines the relationship between one activity and other activities, or series of activities. *Sequential* activities require that one activity be substantially complete before the next one begins. These are usually identified first. Then, *simultaneous* activities, or chains of activities, that are not directly dependent can be planned to take place at the same time. Planning also identifies the labor, equipment, and material resource requirements for each activity or chain of activities.

How to Schedule. An accurate construction schedule estimates a *duration* for each activity. When individual activity durations are totaled for a path of activities that occur in sequence, the duration for that chain of activities has been calculated. The highest sum of activity durations that form a continuous chain of sequential activities through the planned project, allowing contingency time for weather and other delaying factors, is the scheduled duration of the entire project. Then, when the start date is identified, the schedule can be converted from work days to calendar days, identifying the actual dates that each activity will begin and end, and indicating when each resource will be required.

The cost estimate for many equipment items, job site overhead items, and general condition items is often expressed as cost per unit of time (for example, dollars per month). The cost estimate assumes a time duration for the overall project and for certain chains of activity.

Monitoring and Reporting. Planning and scheduling are the only valid approach to understanding process and productivity of a project. Monitoring the progress of individual activities according to a valid construction schedule and promptly adjusting to changing conditions means maximum profit.

Most builders are capable of making good decisions when good, timely information is available. A system of reviewing and recording the cost and production of individual activities and chains of activities usually results in good control of construction time and costs. An active system of monitoring and reporting is even better.

Planning

The level of planning detail varies, depending on the size and type of project, on the amount of work that will be subcontracted, and on the method of scheduling used. The planning process described here is appropriate for Bar Chart, CPM, or PERT methods.

Planning is the most time consuming and critical element of the construction schedule process. The construction plan for a project requires an understanding of all the field requirements to complete the project. The sequence of the identified work activities must be planned in sufficient detail to account for all of the particular requirements of the contract documents, project site, anticipated availability of labor-equipment-material, anticipated weather conditions, and the anticipated performance of subcontractors and suppliers. Although some builders develop a checklist of activities for particular types of construction projects, it is generally recommended that the sequence of work for each portion of each project be considered unique.

When bar chart schedules are used, the description of work usually involves broad scope activities that have long durations. An individual activity, represented by one bar on the chart, usually includes a category of work, often categorized by craft or material, and does not always have a direct finish-start relationship with the other bars. It is not possible to develop accurate resource control based on typical bar charts. Their principle use is to plot progress for review.

CPM usually involves a more detailed breakdown of work items. There are no specific rules for the selection of activities, but the principle considerations are:

a) Activity Responsibility: Subcontract work is segregated into separate activities.

b) Activity Craft/Crew Requirements: The manpower requirements for individual activities must be identified.

c) Activity Material Requirements: The material requirements for individual activities must be identified.

d) Activity Equipment Requirements: The equipment requirements for an individual activity or series of activities must be identified.

e) Location of Work: The site conditions will often require similar work to be performed with different crews or at different times.

f) Subdivisions of Work: A particular project component often requires a breakdown of sequential activities to adequately plan and schedule the work. For example, structural slab is broken down into shoring, formwork, reinforcing, and other activities required to place structural slabs.

g) Cost Control Breakdown: The tracking of actual cost incurred is often linked to the planning and scheduling activities.

The level of detail of activities for a construction schedule depends on the anticipated use of the schedule. The *overall control* schedule defines the project by broad scope activities. *Short interval* schedules for production control are based on specific, short duration work activities. A successful schedule depends on an appropriate selection of activities.

Subcontract Activities. Numerous project activities will be completed under subcontract to trade and specialty contractors. The prime contractor spreads the financial risk and overhead cost by subcontracting work but often encounters difficulty managing the time aspect of the portions that have been subcontracted.

CPM has become the standard tool for resource control. One of the prime benefits of the initial CPM schedule is to identify which subcontractors are required on the project at specific times and to reveal the impact of subcontractor delays on the project schedule. The scheduling of subcontracted work requires both the subcontractors' input and their commitment to the work schedule. When using activity duration information furnished by a trade or specialty contractor, it is important for the general (prime) contractor to know the size and anticipated productivity of the trade crew in order to accurately monitor the progress of work. As an example, consider the CPM network in Figure 1.

The *Electrical Rough-in* Activity (33-34) on the sample project has an activity duration estimate of 3 days.

If the electrical subcontractor assumed using a crew of 4 electricians to complete the rough-in when the activity duration was established, then the prime contractor must anticipate a delay in the project if a smaller crew is available when it is time to perform the work.

The contractor cannot monitor the subcontract work and react promptly and effectively to prevent delay if the crew size and productivity assumed in the estimate of the activity duration is unknown.

The prime contractor should be certain that critical items and low-float activities that are subcontracted are identified in the performance terms of a subcontract. (*Float* measures the amount of time an activity can be delayed and still allow the project to finish on schedule; the concept of float is discussed in greater detail later in this chapter.) The construction schedule is a useful tool in providing for notification in advance of the start of subcontract activities.

Procurement Activities. The planning process must also recognize and identify procurement activities. The commencement of field work for a particular activity or series of activities is dependent on the delivery of the material. The delivery of material is often dependent upon other activities that involve the vendor, contractor, designer, or subcontractors.

Long-lead items and material procurement that require approval of submittals or shop drawings are often critical on the construction schedule. It is especially important that these activities be identified in the planning process. It must be noted that the monitoring and updating of procurement activities is often the most critical early project management task.

Network Sequence. Networking is an accurate, graphical method of recording the planned sequence of activities. A physical diagram of the work plan is a useful communication tool. Figure 1 is a typical network, with activity durations noted, for a construction project.

The network provides a visual reference for the flow of the work activities. CPM and PERT diagrams graphically display the finish to start relationship of activities that are sequential. The network is a communication tool that allows the construction planners to see the construction plan and to evaluate, critique, and revise it prior to preparing the schedule.

Construction Logic. One of the primary questions to be resolved in planning is to decide the logical progression of work. Although there may be numerous approaches to completing a given project, some are usually more efficient, and therefore more profitable, than others. The planning process involves evaluating alternate approaches for completing the work and determining reasonable alternatives that can be considered if the project schedule is disrupted during construction.

The relationship of one activity to another is often directly sequential. For example, concrete cannot be placed until the formwork and reinforcing activities have been completed. The logic of construction prohibits altering this sequence of activities. Likewise, the construction logic will often determine that there need not be a direct sequential relationship and that several activities can take place simultaneously. Separate chains of activities may have a direct, sequential relationship to the same starting and/or finishing activity.

Resource Constraints. Often, construction projects require multiple repetition of the same sequence of construction activities in different locations, as is usually the case in high-rise construction. Although sequences are independent according to construction logic, they are related if the same crews, equipment or supplies (e.g. formwork reuse) will be required on similar activities within repetitive work sequences.

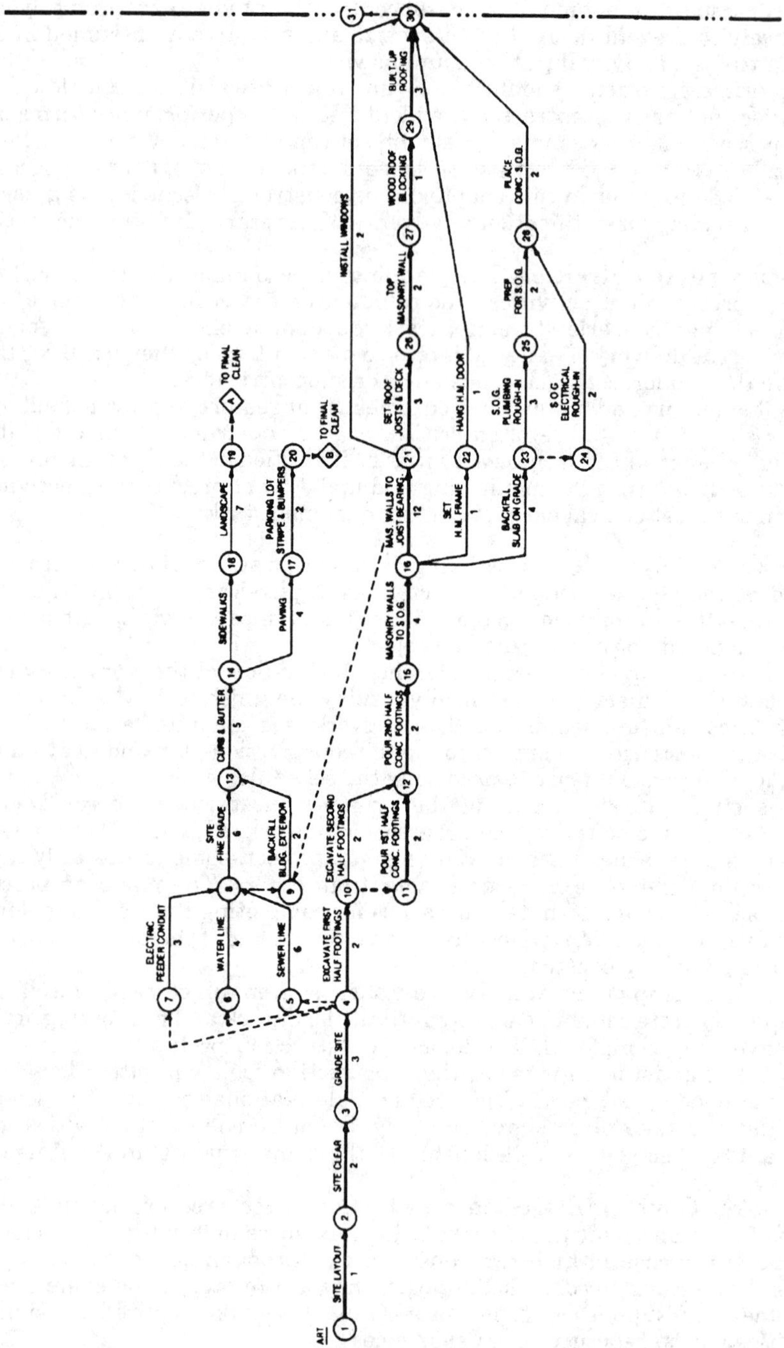

CPM Network
Figure 1

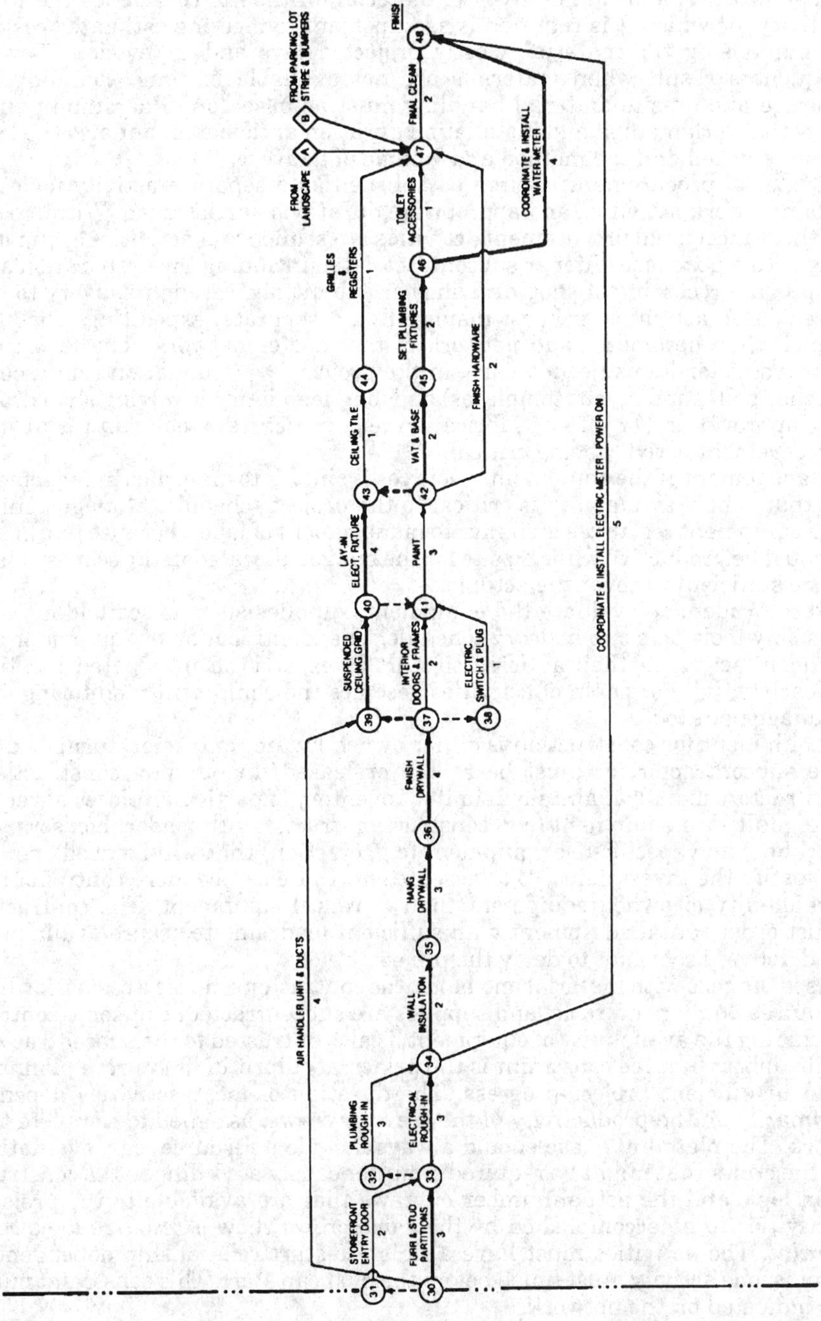

CPM Network
Figure 1 Cont'd

The procurement and delivery of material on time to commence the work activity for which it is required is as important to meeting estimated cost as is purchasing the material. Costly project delays and/or revision of work sequences result when materials are not available in time. Planning for storage and on-site material handling must be based on understanding not only the work method and installation duration anticipated, but also the lead time required and the method and volume of delivery.

Material procurement is often established as a separate activity for each chain of work activities and appropriate durations are assigned. Within each of these individual procurement activities is a sequence of activities commencing with a purchase order or subcontract, often including the preparation and approval processing of shop drawings or submittals, through delivery to the site. These activities are best managed on a separate, expediting schedule, which then becomes a sub network of the master network. The lead time constraint for items needed early in the project (e.g. foundation reinforcing, anchor bolts, and layout templates) and long-lead items involving shop drawing approval and/or off-site fabrication (e.g. structural steel) can result in a procurement activity being critical.

Management of the equipment resources is similar to materials management in that timely availability is critical to the project schedule. Management of the equipment resources is similar to management of labor, because like labor, it must be productive while on site for the budget allocated in the cost estimate to be sufficient to cover the actual cost.

It is prudent to evaluate the equipment, supplies such as scaffolding, and possibly tools on a *sub network* schedule. The identification of equipment-dependent activities, their anticipated start dates, and the anticipated duration of each activity or chain of activities becomes the equipment monitoring and management tool.

Equipment for construction is either owned by the contractor, furnished by the subcontractor, or must be rented or leased. In building construction, contractors usually maintain a limited inventory of particular pieces of versatile, multi-use equipment (e.g. construction tractor with loader, box scraper, backhoe) and specific-use equipment (e.g. erection) that will be used enough to justify the investment. To a large extent, there can be more control of the availability of owned equipment than of rental equipment. The contractor must order rental equipment with sufficient lead time to assure availability and delivery so as not to delay the project.

As is the case with material and labor, the contractor reduces a cost risk when required equipment, tools, and supplies are subcontracted. But some control, including the availability of equipment, is also entrusted to the subcontractor.

The labor resource constraint is the principal culprit in inaccurate planning and insufficient project progress. The duration of most activities depends primarily on the productivity of the crew, or crews, assigned to complete the work. The planning phase should always include a resource logic evaluation of the crews that might be required simultaneously according to the construction logic and the actual number of crews that are available to the project. Activities to be accomplished by the same crews show a *resource logic constraint*. The activities must have a finish-to-start relationship dependency; that is, one activity must finish before the next can start. This constraint must be indicated on the network.

Master Network/Sub Networks/Short Interval Plans. A bar chart is a *master network*. The direct logic and duration of individual activities are not shown. The diagram in Figure 2 identifies the anticipated start and finish dates of the principle work sequences required to complete the project. Milestone events identifying significant accomplishments in the progress--completion of the foundation, topping out of the structure, weatherproof enclosure--are easily identified and are the principle benefits of the bar chart.

Master networks are not developed to a sufficient level of detail to accomplish control of resources. The proper objective of a master network is to serve as a reference for those participants (owner, A/E, home office) not directly responsible for productivity and progress on the site.

Sub networks provide information for a particular task (e.g. expediting and material management, equipment management), a particular craft or subcontractor, or a specific sequence of work. Sub networks are developed to a level of detail sufficient to accomplish the start of work activities as scheduled.

Short interval plans are developed into short interval schedules during the actual construction phase. From one to several weeks in scope, the short interval plan and schedule provide an in-depth evaluation of the upcoming work and resource requirements. These networks have the greatest level of detail.

Network diagrams (CPM and PERT) are best at describing the interrelationship of individual project activities. However, these diagrams are not the most effective job-site communication tool. Supervisors, subcontractors, and craft personnel commonly use the Bar Chart as the schedule communication tool, because its format is easy and quick to use, but it is important to recognize that for a Bar Chart to be accurate, it should be based on CPM computations. Our discussion of scheduling begins with the complete development of CPM for a project.

Scheduling

Scheduling involves the development of duration estimates for individual activities and the mathematical application of those durations to the network. Often, the evolution of a valid and accurate construction schedule involves revisions in duration estimates and network logic.

The actual project scheduling is accomplished by mathematical computation and the application of the calendar in the CPM process. Computer programs that are based on CPM scheduling can do the computations, can facilitate development of sub networks and short interval schedules, can evaluate the impact of alternate methods or plans, and will ease schedule updating. Computer applications have made the scheduling process simpler and more accurate.

The scheduler must work to an appropriate level of detail to achieve the level of project control desired. The time (duration) required to complete an individual construction activity is based on the amount of work required and the productivity of the labor and equipment to be used. The following is taken from the sample project in Figure 1:

The *Masonry Walls to S.O.G. (Slab on Grade)* Activity (15-16), *Masonry Walls to Joist Bearing* Activity (16-21) and *Top Masonry Wall* Activity (26-27), in Figure 1 have durations of 4, 12, and 2 days respectively, or 18 days total duration. The activities are *critical;* that is, they are on the critical path of activities that must be completed in sequence and on time in order for the project as a whole to finish on schedule. The project schedule depends on accurate duration estimates for these activities.

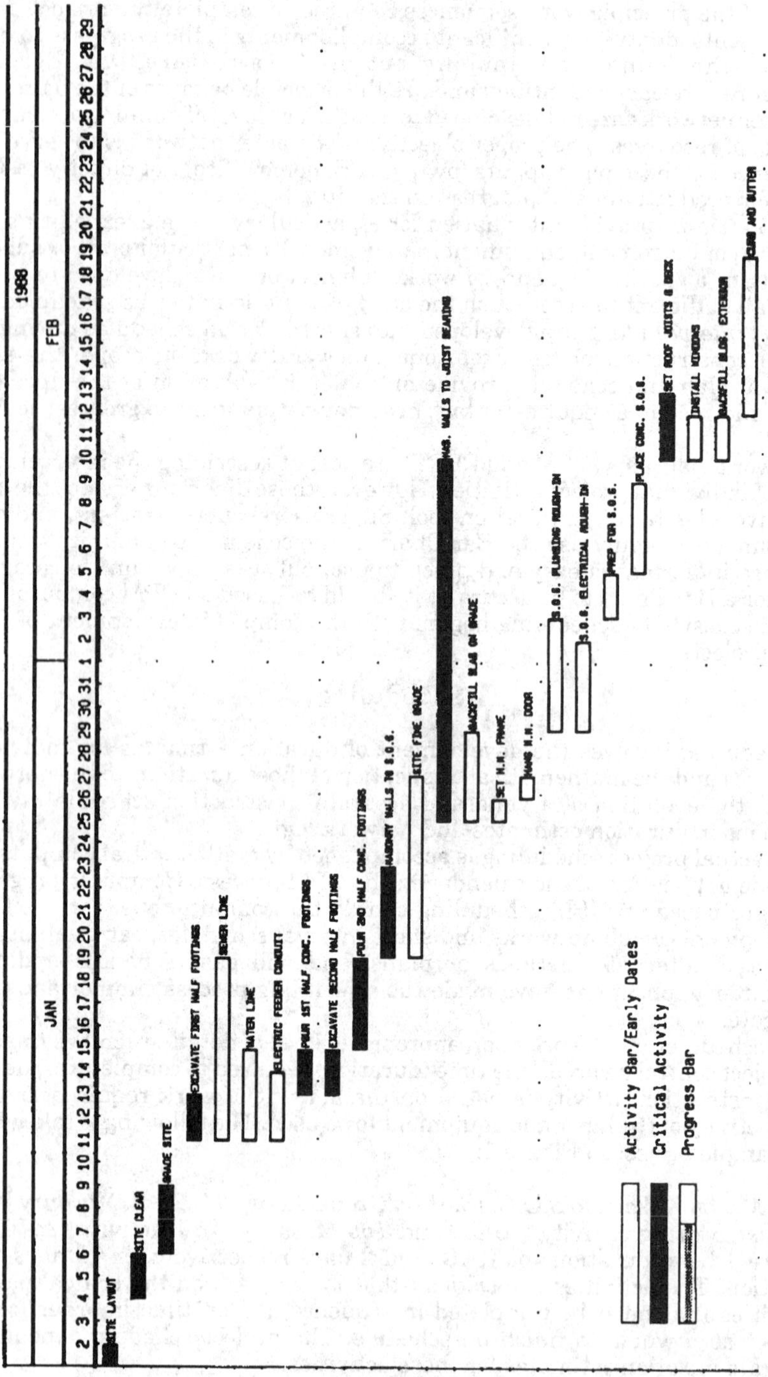
1988
JAN
FEB
2 3 4 5 6 7 8 9 10 11 12 13 14 15 16 17 18 19 20 21 22 23 24 25 26 27 28 29 30 31 1 2 3 4 5 6 7 8 9 10 11 12 13 14 15 16 17 18 19 20 21 22 23 24 25 26 27 28 29
SITE LAYOUT
SITE CLEAR
GRADE SITE
EXCAVATE FIRST HALF FOOTINGS
SEWER LINE
WATER LINE
ELECTRIC FEEDER CONDUIT
POUR 1ST HALF CONC. FOOTINGS
EXCAVATE SECOND HALF FOOTINGS
POUR 2ND HALF CONC. FOOTINGS
MASONRY WALLS TO S.O.G.
SITE FINE GRADE
MAS. WALLS TO JOIST BEARING
BACKFILL SLAB ON GRADE
SET H.M. FRAME
HANG H.M. DOOR
S.O.G. PLUMBING ROUGH-IN
S.O.G. ELECTRICAL ROUGH-IN
PREP FOR S.O.G.
PLACE CONC. S.O.G.
SET ROOF JOISTS & DECK
INSTALL WINDOWS
BACKFILL BLDG. EXTERIOR
CURB AND GUTTER
Activity Bar/Early Dates
Critical Activity
Progress Bar

Figure 2

The estimator can reference unit costs under "Section 04220 Concrete Unit Masonry" in this book. Using lightweight 12" concrete masonry units (CMU), the productivity of one mason with one laborer or hod carrier per 8 hour day is:

Below grade	120-160 units per day
Above grade	110-130 units per day

If the quantity take-off lists,

Below grade 12" CMU	4,480 units
CMU grade to joist bearing	11,520 units
CMU above joist	1,920 units

then the duration estimates require an 8 mason crew:

CMU		Crew x Blocks per day		Duration
4,480	divided by	(8 x 140)	=	4 days
11,520	divided by	(8 x 120)	=	12 days
1,920	divided by	(8 x 120)	=	2 days
Sequential Activity Duration =				18 days

If only 6 masons are available, the durations will be extended:

CMU		Crew x Blocks per day		Duration
4,480	divided by	(6 x 140)	=	5.3 days
11,520	divided by	(6 x 120)	=	16 days
1,920	divided by	(6 x 120)	=	2.7 days
Sequential Activity Duration =				24 days
Project Delay (24 minus 18) =				6 days

Alternately, if 12 masons are available, the durations will be shortened:

CMU		Crew x Blocks per day		Duration
4,480	divided by	(12 x 140)	=	2.7 days
11,520	divided by	(12 x 120)	=	8 days
1,920	divided by	(12 x 120)	=	1.3 days
Sequential Activity Duration			=	12 days
Project Acceleration (18 minus 12)			=	6 days

The construction schedule includes activities that will be completed by the contractor's own forces and activities that will be accomplished by subcontractors. The method of determining the duration of the individual activities is the same regardless of the craft, equipment, or material required. As in the cost estimate, an accurate and complete determination of work quantities is essential in order to prepare a valid construction schedule.

The productivity of the labor and equipment is the essential element in the activity duration estimate. Productivity is a contractor's prime project risk; when the field productivity is less than estimated, cost overruns occur. Many contractors use the schedule time duration as their primary check of the labor estimate. As an example, assume that the contractor's labor estimate for the sample project in Figure 1 is as follows (referencing "Section 09250 Gypsum Drywall"):

Item	Quantity	Unit Labor Cost	Total Labor Cost
⅝" Gypsum Board	19,360 SF	$0.11 per SF	$2,129
Fireproof Columns	1,430 SF	0.22 per SF	315
. . . .			$2,444

The contractor's schedule provides 3 days to complete the *Hang Drywall* Activity (35-36):

$2,444 divided by 3 days = $815 per Activity Day
$16.97 per manhour x 8 hours per day = $135.76 per Man Day
$815 per Activity Day divided by $135.76 per Man Day = 6 Man Crew

The total labor cost estimate budgets a crew of 6 carpenters given the 3-day duration. The contractor can evaluate the labor cost estimate by considering the amount of work normally accomplished by a 6-man crew in 3 days as compared to the amount of work required for the project.

The CPM process of transforming a plan into a schedule is based on computations summing the individual activity durations. An accurate estimate of each individual activity duration is critical to development of an accurate construction schedule. Since the activity duration is calculated based on the amount of work required to complete the activity and the anticipated productivity of the crew and/or equipment, the scheduler must have and use good information. The amount of work for each activity can be quantified by methods described elsewhere in this book. Likewise, anticipated productivity levels for a number of crews and tasks are also included.

CPM (and PERT) Computations. The computations used to develop construction schedules from the planning network are simply summations of the durations of interrelated activities. The early start time in work days or calendar dates is the earliest time that the activity can commence as soon as all preceding activities can be completed; that is, the early start time for each activity is dependent on the finish time of the last preceding activity to be completed.

As an example of how a CPM logic network is developed, a short interval job for reinforced concrete slab has been subdivided into nine activities with their respective durations and activity numbers:

Activity	Duration	Number
Start	0	10
Excavate	5	20
Build Forms	4	30
Procure Reinf. Steel	15	40
Fine Grade	2	50
Set Forms	3	60
Place Reinf. Steel	1	70
Place & Finish Concrete	2	80
Cure Concrete	7	90

The sequence of events is as follows. The procurement of reinforcing steel, the excavation, and the building of forms are all opening activities that can proceed at the same time and independently of one another. Fine grading follows excavation, but forms cannot be set until both the excavation and form building are completed. The placing of reinforcing steel cannot start until fine grading, form setting, and steel procurement have been completed. Concrete is placed and finished after the steel has been placed, and curing is the terminal operation. Figure 3 shows a precedence diagram for the job.

The scheduler must determine four limiting time for each network activity. The *early start* (ES) of an activity is the earliest time the activity can possibly start, allowing for the time required to complete preceding activities. The *early finish* (EF) of an activity is the earliest possible time by which it can be completed; it is determined by adding the activity's duration to its early start time. The *late finish* (LF) of an activity is the very latest it can finish and still allow the project to be completed by a designated time or date. The *late start* (LS) of an activity is the latest possible time that it can be started and still allow the targeted completion date of the project to be met; the LS is obtained by subtracting the activity's duration from its LF time.

The network computations to sum individual activity durations are known as *forward pass* and *backward pass*.

Forward Pass. The early start and early finish computations proceed from project start to project finish, left to right. Each activity starts just as soon as the last of its predecessor activities is finished. The ES of each activity is determined first. The project must start on Day Zero and that value is entered in the upper left hand corner of its activity box. (Notice in Figure 3 that each activity has its own box.) The EF is then obtained by adding the activity duration to its ES value; the EF is recorded in the upper right corner of the activity box. Because Activity 10 has a duration of zero, the EF value is zero. Activities 20, 30, and 40 can all start after Activity 10 is completed. Because Activity 10 has an EF equal to zero, each of the following three activities can start as early as Day Zero. THe EF of each of these three activities is then obtained by adding their respective durations to the ES value of zero. All remaining forward pass calculations for the job continue in the same manner, moving from left to right on the network diagram.

Activity 70 is an example of a *merge activity*. Its start depends on the completion of two or more antecedent activities; in this case, it is Activities 50 and 60. The computation rule for a merge activity on the forward pass is that its earliest possible start time is equal to the latest (or largest) EF value of the activities immediately preceding it. Figure 4 shows the completed forward pass.

Backward Pass. With the forward pass complete, project calculations are now "turned around" and a second set of calculations is performed to compute the late start (LS) and late finish (LF) times for each activity. These calculations start at the projected finish of the project and proceed backward through the network in a right to left manner. The supposition during backward pass is that each activity finishes as late as possible without delaying project completion. The LF value of each activity is obtained first and is entered into the lower right portion of the activity box. The LS is obtained by subtracting the activity duration from the LF value and is entered in the lower left corner of the activity box.

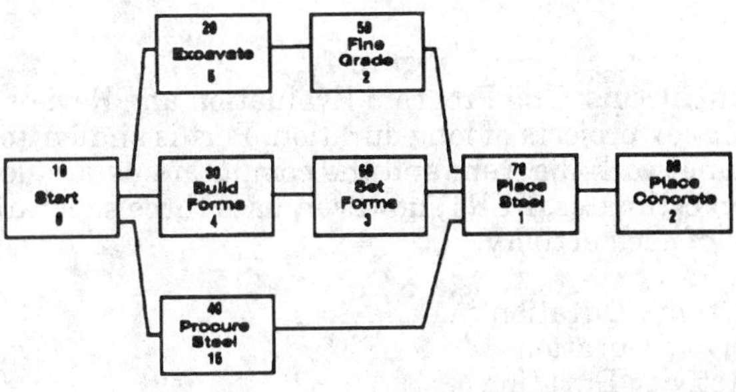

Figure 3

Activity 20 is known as a *burst activity,* which has more than one activity following it. The rule for burst activities is that the LF value for such an activity is equal to the earliest (smallest) LS for the activities that follow it. The backward pass computation proceeds until the start of the project is reached. A completed backward pass is shown in Figure 5.

Float. Time leeway exists in the scheduling of some activities but not in others. This leeway is a measure of the time available for a given activity above and beyond its estimated duration. This extra time is called *float.* There are two types of float.

The *total float* of an activity is obtained by subtracting its ES from its LS time. Subtracting the EF time from its LF time gives the same result. The total float of an activity is entered into the activity box in the space between ES and EF values. The importance of total float is with respect to a path of activities. The activities in a chain share total float. The use of total float for one activity will reduce it by a like amount in others.

The *free float* of an activity is found by subtracting its early finish time from the earliest start time of the activity or activities that directly follow it. Free float is the amount of time by which an activity can be delayed without affecting the early start of the following activity, and also affecting any other activities in the network.

An activity with zero float is called a *critical activity.* An activity with no float time cannot be delayed, or otherwise, the project is delayed. These activities are on the *critical path* of the schedule network. The significance of float is that it indicates the degree to which an activity is critical; that is, it measures the amount of time that an activity can be delayed without hindering scheduled completion of the project.

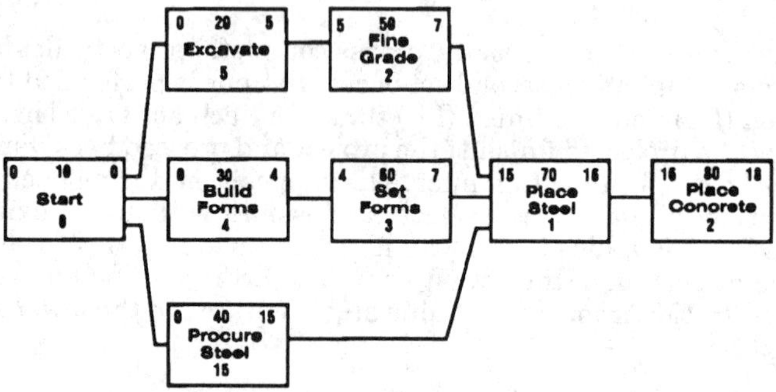

Figure 4

PERT Computations. The Program Evaluation and Review Technique is use on some complex projects of long duration. Pert is similar to CPM in that it is based on a network diagram and the computations are developed from activity duration estimates. PERT, however, uses three separate estimations for the duration of each activity:

a) Optimistic Activity Duration
b) Probable Activity Duration
c) Pessimistic Activity Duration

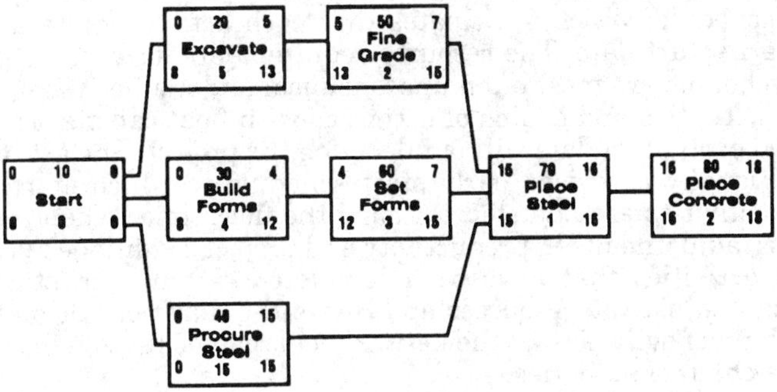

Figure 5

The project and activity chain duration forecasts can then be computed using a variety of anticipated conditions given a single network diagram, and a range of "what if" situations can be easily evaluated. However, the method is complex and is not necessary on less complex projects of shorter duration.

Project (Subproject) Durations. The principle objective of the planning and scheduling process is to forecast the time required to complete the project and portions of the project. As shown in the previous example, the duration of a particular chain of activities is the sum of the durations along the longest sequential path of activities.

Those activities are the critical activities in that if any delay occurs in any of the activities, the projected completion will be delayed. In the sample project, it was forecast that 18 days will be required from the beginning of excavation/form building/procurement through the placement and finishing of concrete.

Contingency Time. Planning and scheduling of a project must include consideration of the impact of weather and other potential delaying factors. The General and/or Supplemental Conditions of the Contract outline requirements for the performance and time extensions; these requirements are a principle consideration in preparing a realistic project schedule.

The development of individual activity durations is based on the concept that the actual productivity for each activity can be monitored and controlled to that duration. The actual time required for certain activities is affected by the time of year (and hence, weather conditions) when the work is scheduled for; this is not always known at the time of planning.

The best and most consistent method of planning for potential and anticipated extra days, for which contract time extensions will not be permitted, is to include time contingencies as a separate and specific consideration. The contractor should review a particular chain of activities that is likely to be delayed, review the overall duration of that chain of activities, anticipate the probable time impact, and include a time contingency as a separate activity at the end of the chain.

This process places float on the critical path to accommodate non-productive project time and assures that the scheduled start date for activities occurring later in the project are more accurate than if contingency time were not included.

Schedule Revisions. The initial CPM schedule is usually developed as the early start schedule, which anticipates that each activity will start on the computed early start date. The resource requirements for accomplishing an early start schedule are most often uneconomical and inefficient.

Individual activities and chains of activities with float can start later than the computed early start date without delaying the project. The actual project schedule is usually a revised early start schedule in which start dates of non-critical activities are adjusted by using the float time. When the project is in progress, adjustments in productivity and project conditions may require revisions in activities that have been completed or have not started. The contractor can adjust the resources and revise the duration estimate for an activity or chain of activities, or the network logic can be revised to shorten or lengthen the chain of activities.

The purpose of the initial schedule is to develop an effective resource utilization plan. The schedule updates provide the information to revise the schedule so that it accomplishes the specific needs and objectives of the contractor.

Milestone Schedules. A milestone schedule lists the dates anticipated for the start or completion of key and critical project activities and work sequences as a measurement of project progress. Although not valuable as a resource management tool, a milestone schedule is excellent for progress communication with concerned parties, such as architect or owner, who do not have frequent project interaction.

The milestones are events rather than activities, and therefore, the schedule date for the milestone most often indicates the beginning or end of some segment of the construction project.

Materials Management. Proper management of materials is essential to project progress. Whenever labor is employed on an hourly basis and short-term layoffs are undesirable, the lost productivity from waiting for or double handling of materials is expensive. Unless that loss is anticipated in the construction cost estimate, profits are reduced.

One category of materials that requires special attention is items that require the preparation, submission, and approval of shop drawings and submittals. Many contractors use a spreadsheet approach to management of procurement activities that involve an approval process. The process calls for all items requiring shop drawings or submittals to be identified as early in the construction phase as possible and for the procurement process to commence as soon as feasible. The expediting effort includes time management of the activities of both the vendor (preparation) and the project designer (approval).

The procurement activity for materials required in the initial segment of construction (e.g. reinforcing steel, anchor bolts and templates, approval of borrow/fill material) are often on the critical path. This portion of the work is also the most sensitive to adverse weather and the associated costs of the resulting delay. Attention to prompt ordering, fabrication and shipping time agreements in purchase orders, and expediting is essential to successful material management.

The site utilization plan, storage and security provisions, weather protection, and the cost and impact of material re-handling must be considered in sequencing delivery of material with the installation activities. Timely delivery is the arrival of materials on the project site so as to be installed undamaged at the appropriate time.

Monitoring

A common excuse to justify informal scheduling in lieu of a formal process is that construction schedules too often become invalid due to continuing changes in conditions. Until the recent availability of economical and efficient computer hardware and software packages, the updating process required for construction schedules prohibited prompt and frequent information for resource management needs. There continues to be some resistance in the field to formal scheduling.

The key to successful scheduling is prompt, accurate, and appropriate reporting. The simple form report shown in Figure 6 records the key schedule information needed. The essential data for scheduling falls under the following categories:

a) Delaying Factors
b) Activities Started
c) Activities Completed
d) Ongoing Activities
e) Crew Sizes

The data input and processing is most often accomplished weekly, after reviewing work progress over the past week. When combined with cost records, and more specifically with labor cost, the resulting information, reports, and forecasts can be integrated into a project management system for project control.

Material procurement and delivery is often the culprit in project delay. The contractor must anticipate and prepare for handling and storage in the event of early material arrival and additional labor costs in case of late arrival. Material management, including control of the shop drawing and submittal approval process, is a key element of project time management. It should be noted that off-the-shelf software for scheduling often requires the support of specific task electronic or manual spreadsheets for applications such as material management.

Collecting Information. The level of detail in the schedule indicates the level of detail of information to be collected for updating and monitoring the schedule. The activity durations are indicative of the frequency of information collection and reporting. Computers are beneficial in processing reported project progress and forecasting remaining durations. The principal time items to review at any point in the project are:

a) Anticipated Delays
b) Project Productivity
c) Time Remaining on Current Activities

The key to a successful schedule updating effort is to collect information on as simple a basis as possible and to collect only the information needed to process time and cost reports. Given the variety of information reporting efforts available from different project management teams, the collection system must be well-defined and consistent.

Reports. When CPM is a requirement of the general or supplemental conditions of a project, an updated report is required with each periodic, usually monthly, pay request. But a useful time management and resource control system as well as cost control reports must be more frequent, most often weekly. A good reporting system will accurately forecast the remaining duration of the ongoing activities.

A percent complete computed by time rather than cost for each activity is an essential element in the collection of information. A standard approach is to update the project schedule using the same scheduling system (Bar Chart, CPM, PERT) originated at the start of the project.

Schedule Updates. The concept of project time management is based on resource management to complete project activities in the sequence and within duration planned. A good system for resource control is one that provides for "management by exception" and for advance notice of delays or other time factors.

The principle complaint of many builders and owners is that the schedule is outdated before it is even seen. Infrequent updates are not useful to protect management. A bar chart allows for write-over updating on a current basis. CPM requires reprocessing. The update report provides vision of the impact of current progress on upcoming resource needs. Looking forward two weeks or even two months in advance will result in the opportunity to anticipate the delay impact of individual problems and to reduce the cost.

Changes and occurrences in the field that depart from the assumptions in project planning result in the need for continual updating of the project schedule. Ideally, the impact of variance from planned productivity would be dated on the short interval schedule and automatically integrated into a project milestone schedule.

Schedule Recovery Plan. The update anticipates the impact of recent project events on the scheduled completion of the project and on individual milestones. When delays are forecast, it is often desired, or even required, to accelerate the project so as to get back on schedule; the first step in the process is to identify, review, and make decisions based on alternate schedule recovery plans. The concept of schedule recovery is to complete work sequences within the time frame that was originally planned for certain chains of activities. Given that the duration is based on the amount of work required or remaining and the amount of available resources, the alternative duration reduction methods include:

a) Increase manpower or crews
b) Add crews
c) Add equipment
d) Work overtime--Extra hours or days
e) Multiple shifts

Obviously, the cost of completing the work at an accelerated rate often exceeds the cost of normal rate work. The schedule recovery plan is the method of comparing the increased cost of acceleration for individual activities and chains of activities. The ability to have control of scheduled completion at the most efficient cost is founded in the ability to promptly and accurately implement schedule recovery when actual project conditions differ from those anticipated.

Resource Planning. The objective of project management is the effective use of resources through the duration of the project. The resources include capital, labor, material, equipment, and subcontractors. The planning of resources for efficient use is required to successfully and profitably complete the project. The material management plan must control not only the quality and quantity of materials, but must identify when the materials are needed and how materials will be handled and stored on site. One of the principle causes of delay and reduced productivity is that materials are unavailable when needed. The project schedule is the control tool for material scheduling. Likewise, equipment must be available when required for the scheduled duration of the activity or chain of activities. The equipment plan is an essential cost control tool.

PRACTICAL
FORM 111
MFD IN U S A

DAILY MATERIAL REPORT

JOB NO. 103 DATE June 6, 19--
NAME OF WORK Smith Store Building WEATHER Clear
LOCATION 656 N Main St, Arlington, Ill TEMPERATURE 62 7 A.M. 83 2 P.M.

RAILROAD AND CAR NO	CLASS OF WORK	MATERIALS RECEIVED AND FROM WHOM	QUANTITY
	Cem Floors	Ready mixed concrete - Arlington Ready-Mix Co	9 cy
	Masonry	Common brick - Arlington Builders Supply Co	8000
	Rough Carpentry	62 Pcs · 2	
		31 " · 2	
		25 " · 1	

SUMM

KIND OF MATERIAL	Concrete Sand	Gravel	Crushed Stone
PREVIOUS			
TO-DAY			
TOTAL			

KIND OF MATERIAL	Fire Brick	Special Brick	Reinforcing Steel
PREVIOUS			
TO-DAY			
TOTAL			

KIND OF MATERIAL	Conc. Form Lumber	Framing Lumber	Ready-Mix Concrete
PREVIOUS	2500 FBM	—	56 CY
TO-DAY	—	3010 FBM	9
TOTAL	2500	3010	65

PRACTICAL
FORM 110
MFD IN U S A

DAILY CONSTRUCTION REPORT

BUILDING Smith Store Building DATE June 6, 19--
LOCATION 656 N Main St., Arlington, Ill. WEATHER Clear
ARCHITECTS Johnson and Anderson TEMPERATURE 62 7 A.M. 83 2 P.M.

CLASS OF WORK	Fore-men	Mech-anics	Labor-ers	Misc.
1. WRECKING—SHORING				
2. EXCAVATING—PUMPING		1	2	
3. PILING OR CAISSONS				
4. FOUNDATIONS—RETAINING WALLS				
5. WATER—DAMPPROOFING				
6. CEMENT FLOORS—WALKS—PAVEMENTS		2	2	
7. BRICK, TILE, CONCRETE MASONRY	1	4	6	
8. CAST STONE—CUT STONE—GRANITE				
9. TERRA COTTA				
10. FIRE-PROOFING, TILE—GYPSUM				
11. ARCHITECTURAL CONCRETE				
12. REINFORCED CONCRETE (a) Forms (b) Reinforcing Steel (c) Mix and Deposit Concrete				
13. ROUGH CARPENTRY	1	3	2	
14. FINISH CARPENTRY				
15. WOOD FLOORS				
16. INSULATION, SOUND DEADENING				
17. WEATHER STRIPS—CAULKING				
18. LATHING				
19. PLASTERING				
20. FIRE DOORS—WINDOWS				
21. STEEL DOORS—WINDOWS				
22. SHEET METAL WORK				
23. ROOFING				
24. TILE AND MOSAIC				
25. ASPHALT, CORK, LINOLEUM FLOORS				
26. ART MARBLE—SCAGLIOLA				
27. MARBLE—SLATE Exterior Interior				
28. GLASS AND GLAZING				
29. PAINTING AND DECORATING				
30. STRUCTURAL IRON—STEEL Erection Riveting		2		
31. MISC. IRON AND STEEL				
32. ORNAMENTAL IRON—STEEL				
33. PLUMBING—SEWERAGE—GAS FITTING		2	2	
34. HEATING AND VENTILATING				
35. AIR CONDITIONING				
36. ELECTRIC WIRING				
37. LIGHTING FIXTURES				
38. ELEVATORS—ESCALATORS				
39. SPRINKLER SYSTEM				
40. MAIL CHUTE				

REMARKS

2. Bulldozer operator and 2 laborers backfilling around foundation walls.

6. 2 cement masons and 2 laborers placing cement floor in basement.

7. Laying face brick and common brick back-up on east elevation. Laying solid common brick south wall.

13. Framing and placing joists for first floor.

30. 2 ironworkers finished erection of lally column and girders.

31. 2 plumbers working on connection to water main. 2 laborers backfilling sewer trenches.

Owner visited job today. Seems well satisfied with progress.

Must have 1x6" D&M for rough flooring immediately. Lumber yard says they are tracing a car of this material which was due in two days ago. Better check with them.

Also check with heating contractor. Should have started today.

FRANK R. WALKER CO., PUBLISHERS, CHICAGO

Daily Material and Construction Report
Frank R. Walker Form 110/111
Figure 6

Labor productivity is critical to the financial success of the project. The contractor, in establishing the estimate and activity duration, assumes productivity of the crew or crews. The resource control is based on accomplishing the work within the scheduled time with planned crew.

Subcontractors must often perform on a number of sites for several different contractors simultaneously. The contractor needs a commitment to the schedule from all subcontractors. The subcontractor needs advance notice of schedule requirements and any changes in the schedule. The management of the subcontract resources is critical to the successful accomplishment of schedule requirements.

TOOLS AND EQUIPMENT

There may be no other part of contracting that puts the estimator's ability to visualize future job conditions to test more than planning for equipment and plant. There was a time when the estimator would only include that equipment specified in the general and special conditions, such as barricades, scaffolds, dust chutes, and the like. It was assumed that each subcontractor would furnish the additional equipment he might need. This is still true in certain trades. The excavator bases his bid on furnishing and using specific equipment that allows him to complete the work expeditiously; the concrete firm furnishes its own mixers and vibrators; the mason provides his own saws and grinders.

But all trades will have certain needs in common. The estimator should foresee these needs and lump all shared equipment under the responsibility of the general. This eliminates duplication of machinery and operators, unnecessary transport to and from the site, and the cost of setup and dismantling. Further savings will be realized by taking advantage of reduced rental weekly and monthly rental rates that apply to extended periods. Further, the more equipment the general sets up, the more control he has over the safety conditions at the job site. Many a contractor has been fined, or worse, has had his job closed down, because a government inspector ruled job conditions unsafe.

Caution

Owning, renting, or using tools and equipment subjects the contractor to compliance with specific safety requirements, which may be imposed by cities, states, and the federal government. It is not unusual to detect considerable conflict or variance in the degree of protection required under the several sources of regulations. In such instances, it has been held that the contractor is required to comply with the maximum requirement.

A national safety guide is the Occupational Safety and Health Act (OSHA), established by the Department of Labor in 1970. All tools and equipment utilized on a construction project should meet OSHA standards. Purchase orders or rental agreements should contain a specific clause relative to this compliance. Failure to meet the provisions of the safety laws can subject the contractor to fines and costly litigation.

Determining Hourly Rate

The general should consult with his subs to determine what their needs will be and then tell them what he will be furnishing. He may then have the subs deduct these costs from their bids or make an agreed upon back charge for rental rates.

Whether the equipment is rented or owned, the estimator should charge for it at a fixed hourly rate. This rate would include the basic rental plus delivery, setup, and dismantling charges. It would then be divided by the working hours that the equipment stayed at the site. In some cases one might elect to charge only for the hours that the equipment was in use at the job site. The remainder of the time would then be charged to office overhead. Operating costs would be added as a separate item only for the actual in-use hours.

Equipment rental rates vary greatly throughout the country, depending on local practices and conditions. They might also vary according to how the equipment is used, such as for hard service in rocky soil or for use in sand.

It is standard practice to base rates upon one shift of 8 hours per day, 40 hours per week, or 176 hours per month for a 30 consecutive day period. In some urban areas, union building trades work 7 hours per day, 35 hours per week.

The lessor usually bears the cost of repairs due to normal wear and tear on non-tractor equipment. On rubber-tired hauling and tractor equipment, the lessee may be required to bear all costs for maintenance and repair.

The equipment rented should be delivered to the lessee in good operating condition and should be returned to the lessor in the same condition as delivered, less normal wear.

National average rates quoted herein are f.o.b. the lessor's warehouse or shipping point. The lessee pays all freight or delivery charges from shipping point to destination and return. Lessee also pays all charges for unloading, assembling, dismantling, and loading where required.

Rentals are usually paid in advance and subject to the terms and conditions of the lessor's rental contract. Rates do not include contractor's insurance, license fees, sales taxes, or use taxes.

Where the contractor owns the equipment, the estimator must arrive at an hourly cost to substitute for the basic rental. To do so, he must consider the following:

a) original capital investment
b) anticipated life
c) depreciation
d) estimate of maintenance
e) estimate of taxes, insurance, interest, and storage

Items c), d) and e) are expressed in percentages of the original capital investment, item a). The estimator will need the cooperation of the firm's accountant to determine the hourly cost figure, as well as some knowledge of the possible future use of the equipment. The capital investment cost should include all original delivery, test, and start-up charges. The expected life may be based on Internal Revenue tables. It is usually entered as 2000 hours per year (40 hour week, 50 week year).

Depreciation cost is the original capital investment less any resale or scrap value. This is often so small that depreciation on all but highway trucks may be figured at 100%. However, in times of high inflation, for equipment held only a short time, the inflation rate may approximate the depreciation rate. The equipment would appear as having zero depreciation.

Maintenance allowances will vary from 40% to 100% of the investment cost, but most equipment will fall in the 65% to 75% bracket. Those items falling below 50% would include air tools, gas engines, small motors, mortar mixers, vibrators, drills, screw jacks, small power tools, towers, small tractors, trailers, and trucks. Those items for which over 75% of investment cost should be used would include bituminous equipment, hoppers, crushers, derricks, electric generators, hydraulic jacks, pumps, and rollers.

Taxes, insurance, and interest percentages should be worked out with the accountant. Ten percent used to be an acceptable figure for these items, but with today's rates 15% would probably be more realistic, and 3% is generally added for storage. A formula to arrive at this entry would be:

$$\frac{p^{*}\text{(useful life)} + p^{**}}{2}\text{(useful life)} = \text{total percentage}$$

p^{*} is percentage for taxes, insurance and interest.
p^{**} is percentage for storage.

If the useful life is five years, p^{*} is 15% and p^{**} is 3%, then the total percentage to be entered under item e) would be:

$$\frac{15(5+1)}{2} + 3x5 = 45 + 15 = 60\%$$

Once items a) through e) are determined, the hourly rate can be figured as follows:

$$\frac{\text{capital investment}}{\text{anticipated life in hours}} \text{ x sum of \% for items c), d) and e)}$$

As an example, assume the contractor has purchased a $100,000 piece of equipment. The following is a reasonable cost projection:

a) capital investment	$100,000.00
b) anticipated life 5 yrs at 2000 hrs	10,000 hrs.
c) depreciation $2500 scrap value	.95%
d) maintenance	.70%
e) taxes, insur., int. & stor.	60%

Hourly Use Charge =

$100,000 divided by 10,000 hrs. = $10.00 per hr.
$10.00 x 95% + 70% + 60% = $22.50 per hr.

Material Handling

Manpower. Labor is too expensive today to use where machines can be used instead, but sometimes it is necessary. A laborer is generally rated as being able to handle a maximum of 100 lbs. For loads over that, or for oversized material, additional help must be scheduled.

One man should be able to unload and pile on a site 300 to 500 standard bricks per hour using his hands, or up to 1000 using special tongs. A forklift truck handling a palletted load can handle well over twenty times as many units in the same time and can provide a wider coverage than a man. One can allot considerable expense for temporary access for mechanical equipment, if much material is involved, and still show substantial savings over using manual labor.

Labor for unloading and piling other materials can be estimated as 80-120 per hour for sacks of cement; 100-200 per hour for block or tile; and 1000-2500 bf per hour for lumber.

Hand shovelling of bulk materials will vary as to whether the load is transferred at the same or lower elevation or raised to a higher one. The following table is based on the amount one man can handle in one hour under dry conditions.

	Even Transfer (tons/hr)	Upward Transfer (tons/hr)
Sand	3.00	2.60
Stone	2.12	1.80
Gravel	2.30	2.00
Top soil	2.60	2.35

Wheelbarrows. Distribution over the job site will still often require hand powered wheelbarrows, trucks, and buggies. Rental costs can be figured at $12.00 per day or $50.00 per month.

Power buggies are widely used on larger construction projects and they have proved to be a more economical method of transporting materials. The walk-behind models rent for $180.00 per day or $550.00 per month, not including gas.

Sizes, Weights and Capacities of Contractor's Wheelbarrows

Struck Capacity Cu. Ft.	Heaped Capacity Cu. Ft.	Approx. Weight Pounds	Type of Materials Used for
2 ½	3 ½	74	Dry Materials
3	4	78	Dry Materials
3 ½	4 ¼	75	Concrete-Mortar
4	5	79	Concrete-Mortar
4 ½	6	83	Concrete-Mortar

It requires about 150 shovels (No. 2 shovel) for one cu. yd.

Chutes and Conveyors. The cheapest energy the estimator can plan for is gravity. He should see that the job is laid out so that, as much as possible, material distribution will be by gravity. Straight line chutes can be rented in ten foot sections for $25.00 to $35.00 per week or may be built up out of plywood at the site. Prefabricated metal vertical chutes will cost $40.00 to $60.00 per foot erected. Gravity steel roller conveyors can be purchased in 10' lengths for around $550.00 for 18" widths. Powered conveyor belts will run two and a half times as much, plus one must add for operating costs.

Portable Belt Conveyors. Some applications require the use of portable belt conveyors of various sizes and capacities. The use of this equipment is flexible and does not limit the user to a fixed receiving or discharging point. When used in series, the flexibility to carry materials over long horizontal distances, over obstructions, and around corners is even greater.

Portable conveyors are available in boom lengths to 120 ft. and belt widths of 18", 24", 30", and 36". Swivel wheels permit building of high capacity radial stock piles or discharge into any one of several bins or hoppers. Both single- and multi-deck vibrating screens may be used at the discharge end to separate the material into two or three sizes. Feeder-trap and feeder-hopper units mounted at the tail end permit charging by front end loaders, dozers, cranes, draglines, shovels, or other conveyors.

Portable conveyors are used for conveying earth, clay, sand, gravel, stone, bulk cement, and concrete from stock piles or to convey and elevate materials to trucks. They can be used for conveying wet concrete from mixer to elevated structures where the incline is not too steep.

Conveyors are furnished with either plain or cleated belts, but the cleated belts should not be used with wet materials, such as wet concrete or clay.

The tables below list capacities of belt conveyors, wheelbarrows, mixers, and concrete carts.

Capacities of Standard Conveyors in Tons Per hour (Based on the Assumption of Uniform and Continuous Loading)

Belt Width	Belt Speed Ft. per Min.	Earth Dry	Clay Dry	Sand	Stone Gravel	Bulk Cement
18"	300	112	92	150	160	150
24"	300	215	185	290	315	290
30"	300	350	300	470	510	470
36"	300	520	450	690	750	690

Helicopters are playing an increasingly important role in modern construction, particularly in materials handling.

Hoisting Towers

Tubular steel hoisting towers are used extensively by contractors in all parts of the country for hoisting all types of building materials. They are also used for concreting because of the ease with which they are erected and dismantled, their great strength and stability, and their economy.

Tubular hoisting towers are designed for handling material and concrete, with or without chuting. They are built in any height and with one or more compartments. Their durability, low maintenance, and remarkable adaptability make them ideal as low cost equipment for contractors.

A tubular hoisting tower consists of a bottom section, the necessary number of intermediate tubular sections, each 6'-6" in height, and a top section containing the cathead with sheaves. All intermediate sections of a given type of tower are identical, and the top and bottom sections can be attached to any one of them.

The single and double (two compartment) towers are classified as follows:

	Heavy Type	Light Type
Size of legs	3-in.	2-in.
Maximum height	350'-0"*	201'-0"
Live load capacity	5,000 lbs.	3,200 lbs.
Maximum size concrete bucket	35 cu. ft.	....

*Higher towers require special material.

The light type towers are known as class 3200 and the heavy type towers are known as class 5000.

The size of the cage platform in the single tower is 5'-10 ½" x 7'-6 ½". For double towers the size of one cage platform will be 5'-10 ½" x 7'-6 ½" and the other platform will be 5'-10" x 7'-2".

Erecting and Dismantling Steel Tubular Towers. When estimating the cost of erecting and dismantling a steel tubular tower, the number of man-hours involved depends upon the height of the tower and the general conditions. For instance, it will take considerably longer to erect a tower 208'-0" high than one 65'-0" high, not only in time consumed but also in manhours per section, taking into consideration the amount of material involved as well as the height to which the equipment must be raised.

Time given includes erection, dismantling, and handling in manhours per lin. ft. of height.

Light Type Single Tubular Towers	1-1.5 hr.
Light Type Double Tubular Towers	1.5-2 hr.
Heavy Type Single Tubular Towers	1.5-2 hr.
Heavy Type Double Tubular Towers	2-2.5 hr.

Prices for light or heavy duty hoisting towers should be obtained from the manufacturer. Quotations for monthly rental charges can also be obtained for a specific project. A 50' tower will rent for about $1,500 per month, with each additional 10' section about $40 per month.

Scaffolding

Tubular Steel Scaffolding. The most versatile type scaffolding is galvanized steel tubing with a male fitting on one end and a female fitting on the other to form interlocking units. These units are then coupled together with a patented coupling device. It can be adapted to a variety of uses and its simplicity has found general acceptance among the building trades. Basically, it consists of these parts: interlocking steel tubes of various sizes, a base plate or caster, and an adjustable or right angle coupler. All parts are galvanized as a permanent protection against rust.

This galvanized carbon steel tubing or pipe is approximately 2" o.d. (1 ½" Standard Pipe) and weighs 2 ¾ lbs. per ft. Other sizes are 2 ½" o.d. (2" Standard Pipe Size) which weighs about 3 ¾ lbs. per ft.

Standard lengths for tube are 6'-0", 8'-0", 10'-0", 13'-0", 16'-0", and 20'-0", and for pipe are 6'-0", 8'-0", 10'-0", and 13'-0". Other sizes can be obtained, but those mentioned are the sizes generally used by the building trades. Each section is fitted with a bayonet locking device.

This locking device between members is permanently affixed to the tubes, which insures a perfectly rigid attachment that has been tested to a load of 20,000 lbs. in tension. The assembled tube and fittings make up a complete unit so designed as to have flush joints.

Standard couplers are made of drop-forged parts, except for the steel hinge pins on which the catch bolts hinge. The most widely used couplers are 2" x 2", both adjustable and standard, and 2" x 2 ½", both adjustable and standard, and 2 ½" x 2 ½", adjustable and standard. This type of scaffold is the most versatile of all built-up scaffolds.

Estimating costs of erecting scaffolding depend on many conditions: the type of job to be done, whether interior or exterior; ground conditions; height and width, as well as load to be carried; and length of time it will be in use. Costs of purchasing or renting "exterior tubular steel" scaffolding are quoted on the square footage of wall area actually covered and may be obtained from the equipment manufacturer. Low buildings will cost about $40.00 per hundred sq. ft. installed (rental). For buildings over 5 stories, it will cost about $60.00 per hundred sq. ft. of surface installed (rental).

Costs for purchasing or renting "interior tubular steel" scaffolding are quoted on the number of cubic feet actually occupied by the scaffolding and may be obtained from the manufacturer or supplier of the equipment.

Sectional Steel Scaffolding. Sectional steel scaffolding or prefabricated scaffolding is readily assembled and installed and is well adapted to various kinds of building and maintenance work. Various sizes of 5'-0" wide frame units, trusses, and ladder scaffold components make it simple to erect a rigid and safe scaffold under any conditions.

The scaffold components are manufactured to resist corrosion and rusting. Rugged construction prevents damage during assembly and dismantling. This is a lighter type scaffold than the tube-and-coupler type and has an approved working height limit of approximately 125'-0".

The tubing used in the frames and trusses is carbon steel, which reduces the weight without sacrificing strength.

Sectional steel scaffolding has found wide acceptance among bricklayers, plasterers, and other craft workers, because it adapts to rolling towers and to exterior and interior construction and maintenance.

Typical Materials and Manpower Lift Used in Present Day Construction

The spacing between frames may be either 3'-0", 5'-0", 6'-0", 7'-0", or 10'-0", and the frames vary from 3'-0" to 10'-0" in height, allowing for a wide variety of uses. Braces are attached to the frames by means of wing nuts or automatic locks. Costs of the component parts of sectional steel scaffolding may be obtained from the manufacturer or supplier.

Tubular Steel Scaffolding for Exterior Work

Suspended Scaffolding

The most economical method of scaffolding for masonry work on buildings with a structural steel or reinforced concrete skeleton framework is suspended scaffolding. The working platform is supported by regularly spaced scaffolding machines, each consisting of a pair of cable drum hoist mechanisms with connecting steel putlog, suspended by steel wire rope from steel or aluminum outrigger beams anchored to roof or floor framing above.

Standard width machines accommodate a 5'-0" width of solid planking. Machines are available to accommodate 8'-0" wide decking to permit wheeling of material on scaffold platform. Machines are also available for mechanized material handling with the inboard cable drums set back from the wall to provide a 5'-0" wide platform for material handling traffic and stockpiles on one side and a separate 20" wide working platform for masons on the other. All machines are available with or without steel frames to support overhead protective planking. For maximum safety, overhead planking should be used, together with toe boards and guard rails along outer edge of platform.

Scaffolding machines are usually spaced 7'-0" apart, permitting the use of 16'-0" scaffold plank. A ratchet device on the drum hoist mechanisms provides positive locking while still permitting easy raising or lowering of scaffold to maintain bench-high location of work for maximum efficiency and best workmanship.

Tubular Steel Scaffolding Used as an Outrigger in Curtain Wall Construction

Outrigger beams are usually anchored to the roof construction 7'-0" on centers, if building is not over 150'-0" high, the maximum lift capacity of scaffolding machines. For structures exceeding 150'-0" high, one or more additional installations will be required on lower floors. There are various methods of anchoring beams to building framework, such as securing with steel straps to anchor bolts set in concrete slabs or beams, bolting to inserts set in concrete, or anchoring to beams or joists with U-bolts and bars through holes in concrete slabs.

Conditions affecting the installation of suspended scaffolding vary so widely it is impossible to give unit costs that will be of any value in estimating. The manufacturer and erector will gladly quote a firm price on any specific job.

Ladder Type Scaffolding. For interior scaffolding, particularly for low to medium height work, ladder type scaffolds are considered very efficient. The type with horizontal bracing is most desirable for interior work, because it spans over furniture, machinery, or equipment. Scaffolding that can be assembled without bolts or nuts can be erected or dismantled more quickly and easily. Platforms that can be adjusted for height permit workers to stand at the most efficient distance from their work, a feature that is important for electricians, plumbers, heating and ventilating workers, acoustical erectors, painters, maintenance men, and others doing off-the-floor jobs.

Sectional Steel Scaffolding

Ladder type scaffolds are sold as complete scaffold units, which include two ladders, two platform support trusses, and one plywood platform (exterior type plywood, metal bound). The platforms are adjustable for height every three inches. Each unit is 6'-0" high by 6'-0" long (or longer if desired) by 2'-5" wide. They can be used separately, or in combinations to scaffold larger areas. Casters can be furnished either with or without locks.

Suspended Scaffolding for Masons

One of the outstanding advantages of scaffolds of this type is their maneuverability over permanent or semi-permanent objects, such as church pews, machinery, or furniture. It isn't necessary to scaffold an entire room or auditorium because the scaffolds can be readily moved from place to place.

Scaffolds of this type sell at approximately $160.00 to $215.00 per unit, including the platform, plus the casters.

Scaffolding Used to Erect Forms for Poured Concrete Decks

Sidewalk Bridges

Where construction is adjacent to heavy traffic areas and it is impracticable to reroute sidewalk traffic into the street, a sidewalk bridge must be erected to protect the public. Rigid standards must be met in the design and erection of these structures.

They are built of 4 ½" o.d. steel columns, set on heavy timber mudsills and supporting 8" steel I-beams bridging the sidewalk. Beams are securely attached to columns by means of special steel column head castings. Columns are tied together longitudinally with two rows of pipe girts and with an occasional diagonal pipe brace as required. Wood timbers, cut to a taper and set to slope towards the street, are fastened to the cross beams with U-bolts to support 6" steel beams, spaced approximately 16" on center, running the length of the sidewalk bridge. Two layers of 2" or 3" planking are then laid over the beams, the bottom layer running laterally and the top layer longitudinally. A parapet for the street side may be added if desired.

Most contractors lease sidewalk bridges on an erected and dismantled basis, and the supplier usually quotes a lump sum figure for a specific job. Before submitting a bid, a firm quote covering this work should be obtained.

For preliminary estimating purposes only, under average conditions sidewalk bridges can be figured as follows:

Erection and dismantling, per lin. ft.	$20.00-$36.00
Sidewalk bridge rental, per lin. ft., per month	4.75
Erect and dismantle parapet, per lin. ft.	9.00
Parapet rental, per lin. ft., per month	2.00

Air Compressors

One sales and rental source for air compressors advertises over 1800 models. Add to it all the various attachments that are made to work with the compressor, and one can see one must give some thought to matching the unit to the tools for the most efficient operation.

Two and more tools can be operated from the same compressor. The rating of the compressor should be the delivery, but leaking hoses, poor tools, and adverse job conditions could lower the cfm output. In some gang operations not all units will be in operation at the same time so that a unit sized for two may be adequate for three, providing the compressor is set up for more than two hose connections.

The following list will give some idea of air requirements for the more commonly used tools:

Breakers
35# - 30 cfm*
60# - 40
80# - 50

Diggers
20# - 20 cfm
25# - 25
35# - 35

Air Drills
1" - 35 cfm
2" - 50
4" - 60

Wagon Drills
3" - 160 cfm
3 ½" - 200
4" - 250

Jack Hammers
25# - 50 cfm
45# - 100
75# - 110

Chipping
light - 20 cfm
heavy - 30 cfm

Rivet
5/8" - 25 cfm
7/8" - 35

Circular Saws
12" - 50 cfm

Chain Saws
30" - 90
36" - 140
48" - 160

Tampers
35# - 30 cfm
60# - 40
80# - 50

Vibrators
3/4" - 35 cfm
1 3/4" - 50
2 1/2" - 25
3" - 45
5" - 80

Wrenches (Impact)
5/8" - 15 cfm
3/4" - 30
1 1/2" - 75

Drills
less than 1" - 50 cfm
more than 1" - 80

Air Nailers - 10 cfm

Grinders
small, light - 50 cfm
heavy duty - 80

Hoist
light duty - 150 cfm
heavy duty - 175

*Note: cfm = cubic foot per minute

Compressor rentals will vary as to the condition and rated size of the equipment, but should be available for following approximations:

fuel	cfm	per day	per week	per month
gas	85	$35	$130	$380
"	100	40	130	400
"	125	40	145	440
"	150	40	150	470
"	175	40	170	500
diesel	250	70	340	1,000
"	375	90	430	1,300
"	600	130	670	2,000
"	750	150	750	2,300
"	1,000	200	1,000	3,000
"	1,200	270	1,300	4,000
"	1,600	280	1,400	4,200

Tools, hoses, oilers, noise abatement controls, and other attachments must be added separately.

CHAPTER 2

SITE WORK

CSI DIVISION 2
02010 SUBSURFACE EXPLORATION
02100 CLEARING
02060 DEMOLITION
02200 EARTHWORK
02250 SOIL TREATMENT
02300 PILE FOUNDATIONS
02350 CAISSONS
02400 SHORING
02500 SITE DRAINAGE
02550 SITE UTILITIES
02600 PAVING AND SURFACING
02700 SITE IMPROVEMENTS
02800 LANDSCAPING

02010 SUBSURFACE EXPLORATION

When preparing estimates on any structures involving foundations or earthwork, there is a need to obtain all information available concerning the conditions and characteristics of the soils to be encountered in the subsurface. Frequently some of this information is made available on the working drawings as an aid to the contractors, but complete reliance on such limited information places the contractor at risk, especially when the drawings provide for a disclaimer of responsibility by the architect/engineer for the accuracy or completeness of the subsurface information. A site investigation is mandatory, because many conditions can be determined through adequate visual observations of the particular site involved and adjacent property. There can be discovery of previous stream beds, evidence of prior dumping of debris, existence of abandoned mining operations, existence of flood plains, appearance of rock outcroppings, and many other situations that would strongly influence the construction methods and the costing involved.

In urban areas records are usually available from utility companies and highway authorities concerning subsurface conditions affecting their installations.

It must be recognized that any penetration of the ground surface has attendant risk, and the accumulation of knowledge of the conditions to be encountered during excavation will tend to control the risk to a reasonable level thus avoiding exposure to a catastrophic situation.

Excavating test pits or caissons produces accurate information in cohesive soils, insofar as this method permits visual inspection of the soil in place, and samples for laboratory tests may be carved out of the pit walls in a relatively undisturbed condition. Costs of this work are high, however, and will be about the same as excavating costs for comparable pits and caissons.

However, most subsoil investigations require studies of the soil at greater depths than can be economically reached by test pits. For these projects, various methods of soil boring and sampling are used. A relatively small diameter hole is sunk into the ground from which soil samples are obtained.

Hand Auger Boring Method. The hand auger boring method, satisfactory for highway explorations at shallow depths, can be used only for preliminary investigations for foundations. Auger borings are used in cohesive soils and cohesionless soils above ground water. Depths up to 20 feet may be reached. Soil samples obtained through this method are disturbed to such an extent that little or no information is furnished about the character of the soil in its natural state. Where the auger hole is filled with water, sample disturbance is particularly acute. Auger holes are most useful for ground water determination.

Hand auger borings generally cost from $6.00 to $10.00 per lin. ft. depending on the soil, the amount of boring necessary, and the diameter of the hole, which will vary from 4" to 8".

Subsoil Investigations

Split Spoon Sampling. A split barrel sampler known as a split spoon is frequently used for obtaining representative samples. This method is described in detail in ASTM Method D-1586.

In securing split spoon samples, the drill hole is opened and cleaned to the desired sampling level by drilling bits and wash water, or alternatively, the hole may be drilled with a regular power auger. Drill casing or drilling mud is used when it is necessary to prevent the soil above the sampling level from closing the drill hole. A drive pipe is attached to the upper end of the drill rod and a slip jar weight placed on the drive pipe.

The sampling spoon is then driven into the soil by repeated blows of a drop hammer. The number of blows required to drive the spoon are recorded in the logs. The spoon is then brought to the surface, and the sample is removed, classified, and placed in a glass jar.

The ASTM method specifies a 2-inch OD x 1 3/8-inch ID split spoon driven 18" to 24" by a 140 lb. weight falling 30 inches. Various other sizes of split spoons and drop weights and lengths of drops are employed. The split spoon method is most useful in granular or dense soils.

Thin-Wall Tube Sampling. The method that yields the least disturbed samples is thin-wall tube sampling. This method is described in detail in ASTM Method D-1587.

In this method the hole is opened and cleaned as in the split spoon sampling method, but a thin-wall tube, usually with an outside diameter of 2", 3", or 5" is then attached to the end of the drill rod and lowered to the bottom of the boring. The tube is then pushed into the soil by a hydraulic piston arrangement with a rapid and continuous motion.

The thin-wall tube is then raised to the surface, cleaned, labeled, and sealed to prevent loss of moisture. It is then taken to the laboratory for testing. Here it is usually cut into short lengths, and the sample in each length is ejected and tested.

Samples are normally taken at 5-foot intervals or every time a soil change is expected, determined by watching the overflowing wash water and feeling the resistance to penetration of the drill bit.

Cost of drilling and sampling is usually between $6.00 and $16.00 per lin. ft. depending on the depth, diameter, and difficulty of drilling, plus on and off charges for the equipment and men. Laboratory testing is performed at an additional cost, depending on what tests are needed--for density, moisture content, compression, or mechanical analysis.

Core Borings. In addition to soil samples, information is frequently wanted on the nature or thickness of underlying rock. Core borings are made for this purpose with rotating coring tools such as diamond drills, shot drills, or carbide bits. Continuous samples are generally recovered for examination and further testing. ASTM Standard D-2113, Diamond Core Drilling for Site Investigation, should be called for in performing core borings.

The cost of making core borings will depend upon the location of the work, quantity, and the difficulty encountered in getting to the holes.

The drilling costs per lin. ft. should be estimated at from $18.00 to $22.00 per foot of drilling, with the lower costs on jobs of larger size. To the above prices, add the on and off charges for the equipment and men and the testing charges.

Other Methods. There are many methods used to investigate soil conditions for foundation purposes. Some are reliable and sufficient to use for design purposes while others produce inadequate or dangerously deceptive data. In problem soils the gathering of the basic information under controlled conditions is important, but to obtain definitive and factual information there is a need for professional engineering interpretation. In such instances the best approach is to obtain the services of a soils engineer with the laboratory facilities to perform a complete subsurface investigation and to provide a written report of the findings and recommendations for remedy of the problems to be encountered.

Engineering and laboratory services are cheap. However, when balanced against the magnitude of potential loss that could be encountered without this advance knowledge, the cost is usually justified.

Engineering costs vary widely by geographic location and services required. It is not practical to provide cost range figures here. There is enough competition to allow the contractor to solicit proposals from several engineering firms, who provide services that are tailored to the specific project.

Seismic resistivity, vane shear tests, cone penetration test, and nuclear methods are more frequently being used in subsurface investigations, performed by specially qualified engineers having the necessary equipment and regularly engaged in this type of work. Cost estimates will be tendered to the contractor predicated upon the services required.

02100 CLEARING

Clearing of a construction site by the removal of trees and vegetation has become more costly because of federal and state environmental protection regulations. Most areas prohibit the burning of trees. If the wood does not have a market value, the use of chipping machines is employed. This method requires additional personnel to cut the trees into manageable lengths, remove the branches, and transport them to the chipping machine. Some of the cost can be recovered by selling the by-product to nurseries and landscape companies to use as mulch. Stump disposal is very expensive, particularly where stumps cannot be buried at some location on the site and have to be transported to a distant disposal area.

Clearing, Cut & Chip	Unit	Avg. Price
Light -6" Diam.	Acre	$1,500.00-1,900.00
Medium - 12" Diam.	Acre	2,500.00-3,000.00
Heavy - 16" Diam	Acre	3,500.00-4,500.00
Grub Stump Plus Haul Charge	Ea	20.00-35.00

If chipping is not required, average prices can be reduced by 25%.

Selective Cutting Down and Removing Trees

The cost of cutting down and removing trees will vary with their size and the method used in cutting and disposing of them.

The following production times are based on a crew of 3 or 4 men working together, cutting down trees by hand, removing branches, and cutting the trunk into 4'-0" lengths. Labor costs of digging down around roots and removing a stump are given extra. It is assumed the limited quantity would not require chipping.

Approx. Dia. of Tree	Approx. Hgt of Tree	Labor Hours	Extra Labor Hrs Removing Stump
8" to 12"	20'-0" to 25'-0"	9 to 12	8
14" to 18"	30'-0" to 40'-0"	12 to 16	10
20" to 24"	45'-0" to 50'-0"	20 to 24	12

If necessary to cut trunks and stumps into small pieces and load into trucks, add extra for this work. Add also for removal from the site.

Where many trees are to be removed, a power or chain saw will speed the work and reduce costs considerably. A chain saw will cut within 2 to 4 inches of the ground, but if the stump must be removed, it is usually done by hand.

Chain saws are gas- , electric- , or air-powered; 24", 36", 48", and larger; and rent for about $40.00 per day.

Labor required using a chain saw, with 3 men in the crew cutting trees, removing branches, and cutting the trees into 2'-0" lengths, should average as follows:

Approx. Dia. of Tree	Approx. Hgt of Tree	Labor Hours	Extra Labor Hrs. Removing Stump
8" to 12"	20'-0" to 25'-0"	2 1/4 to 3	8
14" to 18"	30'-0" to 40'-0"	3 to 4	10
20" to 24"	45'-0" to 50'-0"	5 to 6	12

A faster and more economical method of removing trees is by using a bulldozer or loader, provided it does not cause damage to adjacent trees.

Trees up to 15 inches diameter or trees such as pine, which have small root systems, can be rapidly uprooted by raising the dozer blade or shovel bucket to maximum height and pushing against the tree. It may be necessary to strike the tree a few times to loosen it.

Trees with diameters of 24 inches or more, or trees like elms, maples, and oaks, which have heavy, wide-flaring root systems, usually require root cutting before they can be pushed over. First make a trench-cut with the dozer blade or shovel bucket across the roots on the side opposite the direction in which the tree must be felled. Backfill the trench and form a ramp up to the trunk so the machine can strike as high on the tree as possible. Unless the tree is exceptionally large or well-rooted, it can now be pushed over. If not, try cutting the roots on each side of the tree perpendicular to the first cut, and if necessary, the roots on the side toward which it will be felled.

Once the tree is felled, it can be pushed or dragged to a disposal area where it can be sawed into lengths for hauling or burning (where permitted).

A Word of Caution--Always watch for falling limbs. In the case of pines, if the whipping action is too great, the top may snap off and kick back on the machine or operator. Also, as the tree begins to fall, the machine must be backed up quickly so that the upturning root system will not come up under the machine, possibly damaging the unit or hanging it up.

An experienced operator, using a medium to large tractor (80-HP and up if crawler type, 160-HP and up if rubber-tired), with ground conditions affording good traction, and no problems regarding steep slopes, should fell trees at the following rates:

Approx. Diam. of Tree	Approx. Height of Tree	Minutes Req'd. for Felling
8" to 12"	20'-0" to 25'-0"	5 to 10
15" to 24"	30'-0" to 50'-0"	15 to 20
27" to 36"	55'-0" to 75'-0"	25 to 30

Add for pushing felled trees to disposal area and for disposing of them.

Stripping & Storing Topsoil

Following the site clearing operation, the stripping and storing of topsoil, if available, will be the next phase. In the interest of economy, the recovery of the maximum quantity of existing topsoil is important. However, any area cleared of trees and stumps will have a loss factor of approximately 15% to 50% depending on the number and size of stumps removed. The remaining topsoil should be stripped, cleaned of foreign materials, stored in a convenient location, and protected from erosion or contamination.

The depth of the topsoil influences the production rate. Less quantity would be available in a given area if it were 4" in depth as opposed to 6" and yet the same area would have to be stripped by the dozer. The selection of the storage area is important. It should not only provide protection for the topsoil material, but it also should minimize the haul time involved in the storing and redistribution of the material to the final location.

Strip & Store Topsoil

Description	Unit	Average Cost
With dozer, 4"-6" depth 100' haul-average conditions (erosion protection additional)	Cu. Yd.	$0.90-1.10
With dozer, 4"-6" depth 300' haul-average conditions (erosion protection additional)	Cu. Yd.	3.50-4.25

If adverse conditions are present increase unit price by 20% to 33%. Price range depends on terrain and capacity of equipment.

The redistribution of topsoil usually requires some preparation of the ground prior to placing the topsoil.

Construction activities cause rutting, depressions, and general unevenness of the site area, and to avoid excessive quantities of topsoil being used, the site should be graded to the designed contours less the depth of the topsoil. Any preparatory work such as scarifying, raking, or rolling should be accomplished before the introduction of the topsoil. Requirements for soil conditioning, fertilizer, etc., become cost items in addition to the basic spreading cost. When specifications have rigid conditions concerning the quality of topsoil, it may be necessary to submit samples to a laboratory for analysis to establish compliance.

Off-site purchase of topsoil is very expensive and can vary from $10.00 to $30.00 per cu. yd. depending on the quality and availability.

Spread and Prepare Topsoil

Description	Unit	Average Cost
Scarify subsoil-average conditions	Sq. Ft.	$0.02
Spread, 300' with dozer, average conditions	Cu. Yd.	1.50-2.50
Hand dress topsoil ($20.00 hourly rate)	Cu. Yd.	3.00-4.00
Rolling, push roller	Sq. Ft.	0.05-0.07

Caution is to be exercised in determining if the specified depth of topsoil is loose or compacted depth. Depending on the characteristics of the material involved, topsoil will compact from 33% to 50%. This could mean to obtain 4" compacted depth would require 6" of loose material. An estimator not taking this into consideration would discover a 50% shortage from the estimated quantities.

If topsoil is to be placed in a series of small restricted areas or at the perimeter of buildings, walks, drives, or other obstructions, additional hand labor will be required to place, grade, and hand compact the material.

It is absolutely essential that the estimator become familiar with local laws regarding disposal. Silt fence for containment during stripping and spreading operations costs about $0.75 to $3.00 per lin. ft. If salt hay is used, costs vary greatly depending on the availability of material, usually between $0.50 and $5.00 per lin. ft.

02060 DEMOLITION

Demolition can be divided into several main types--demolition of entire buildings or structures, controlled removals for historic preservation work, and general building demolition of items such as doors and frames, windows, floors, ceilings, and partitions.

In most areas of the country, there are specialty contractors whose sole business is the demolition of entire buildings and structures. These contractors will use a wide range of heavy equipment, even explosives, for the controlled demolition work. It is always best to solicit a quotation from one of these firms for this type of demolition. However, the following per cubic foot unit prices will provide some budget guidelines in the absence of quotes:

Building Type	Unit Price per Cu. Ft.
Low Rise-Steel Frame	$.25
Low Rise-Concrete Frame	.35
Low Rise-Masonry Wall Bearing	.30

Demolition associated with historical preservation work is highly specialized, and since each project has its own particular set of intricacies, it is extremely difficult to produce accurate historical cost data that can then be saved for use in future estimates.

Of less intricate nature than the controlled demolition of historical preservation work, but closely related, is general building demolition. It is this type of work that we will address for the various types and items of demolition and their appropriate labor production tables and costs.

The estimator must always take into consideration factors such as working space, is the building occupied or not, method and route in removing debris to the trash receptacle or truck, special equipment requirements, temporary bracing and shoring, dust and security control partitions, length of the haul to the dump site, and special use permits, EPA permits, and dump fees.

Demolition Using Compressed Air Tools

Compressed air is used extensively in all classes of construction work including such operations as concrete breaking, masonry chipping, excavating, compacting earth, and driving sheet pilings. To reduce the overall operating cost (ownership or rental rates), it is always wise to use an air compressor that is capable of handling at least two pneumatic tools. The use of two tools lowers the rental rate per hour for the machine down to a level half of what it would be if only one operator was using the machine to do an operation. For example, if you had 1,000 sq. ft. of concrete sidewalk to demolish, using one compressor with two hammers and two men would require the same total number of labor hours as one man would take. However, your equipment rental hours would be half of that required if you used only one man on the air compressor.

The following chart shows the classification of work to be demolished by an air hammer, the daily production of work per crew of two laborers at 8 hours each, and the total number of manhours required per 100 sq. ft. or lin. ft., whichever is applicable. Keep in mind that the production rates shown in the following table only include the material break-up, not transport/removal to the trash receptacle.

Item Description	Crew Daily Production	Man Hours Per 100 SF	Man Hours per 100 LF
Wall Footing-			
2' wide x 1' thick	125 LF	--	12.80
3' wide x 1'-6" thick	65 LF	--	24.62
Slab on Grade - 4" with mesh	450 SF	3.56	--
Slab on Grade - 8" with mesh	250 SF	6.40	--
Concrete Foundations Walls			
Lightly Reinforced:			
6" Thick	200 SF	8.00	--
8" Thick	160 SF	10.00	--
10" Thick	135 SF	11.85	--
12" Thick	110 SF	14.55	--
Header Curb	225 LF		7.11
Curb & Gutter	175 LF		9.14

Partition Removal

The following labor production table gives the daily crew (2 laborers, 8 hours each) output and the manhours required per 100 sq. ft. for the demolition of various types of partitions. The labor production hours do not include the transport/removal of demolished materials to the trash receptacle.

Partitions 10' High

Type Of Partitions	Daily Prod. SF	Man Hours Per 10(
Metal Stud & Drywall	850	1.88
Plaster & Lath on Wood Stud	650	2.46
4" Concrete Masonry Units	650	2.46
8" Concrete Masonry Units	500	3.20
12" Concrete Masonry Units	400	4.00
4" Brick	500	3.20

Partitions 15' High

Type Of Partitions	Daily Prod. SF	Man Hours Per 10(
Metal Stud & Drywall	800	2.00
Plaster & Lath on Wood Stud	600	2.67
4" Concrete Masonry Units	600	2.67
8" Concrete Masonry Units	450	3.56
12" Concrete Masonry Units	375	4.27
4" Brick	450	3.56

Door & Frame Removal

The following labor production rates are for the removal of doors and frames from walls to be demolished. Keep in mind that if a frame is to be removed from a wall that is to remain in place, more care must be used to prevent needless damage, and thus costly repairs, to the wall. As a guide the hours shown in the following table could be doubled for the removal of metal door frames from masonry walls that will remain in place. The production times shown in the table does not include any transport of the demolished materials to the trash receptacle.

Item Description	Man Hours Per Unit
Single wood door & frame in stud partition	1.00
Pair - wood doors & frame in stud partition	1.25
Single door and frame in masonry partition	1.25
Pair doors and frame in masonry partition	1.50
Single hollow metal door and frame in stud partition	1.50
Pair hollow metal doors & frame in masonry partition	2.50
Single hollow metal door & frame in masonry partition	2.25

Window Removal

The following table shows various types of window materials and the required man hours per unit for their removal. The production times shown in the table does not include transport of the demolished materials to the trash receptacle.

Type	Size	Man Hours Per Unit
Wood	up to 3' x 5'	.75
Wood	up to 5' x 6'	1.25
Aluminum	up to 3' x 5'	1.25
Aluminum	up to 5' x 6'	1.75
Steel	up to 3' x 5'	2.00
Steel	up to 5' x 6'	2.50

Floor Finishes Removal

The following table shows various types of finish flooring materials and the required man hours per 100 sq.ft. for their removal. The production times shown in the table does not include transport of the demolished materials to the trash receptacle.

Type of Flooring	Man Hours per 100 Sq. Ft.
Vinyl composition tile	2.00
Wood parquet	2.50
Wood strip	2.75
Carpet & pad - tack strip	0.40
Carpet - glued down	1.00
Ceramic tile	2.50

Ceiling Finishes Removal

The following table shows various types of ceiling finish materials and the required man hours per 100 sq. ft. for their removal. The production times shown in the table does not include transport of the demolished materials to the trash receptacle.

Type of Material	Man Hours per 100 Sq. Ft.
Drywall	1.00
Suspended plaster	3.00
Suspended acoustical tile and grid	1.25
Plywood	1.00

If the ceiling finishes are more than 10' above the finished floor, add a 10% labor factor multiplier for every 2' increment over 10'.

Wall Finish Removal

The following table shows various types of wall finish materials and the required man hours per 100 sq. ft. for their removal. The production times shown in the table does not include transport of the demolished materials to the trash receptacle.

Type Of Material	Man Hours per 100 Sq. Ft.
Drywall	.60
Plaster	1.00
Wood paneling	.85
Ceramic tile	2.25

Roofing Material Removal

The following table shows various types of roofing materials and the required man hours per 100 sq. ft. of their removal. The production times shown in the table do not include transport of the demolished materials to the trash receptacle.

Item Description	Man Hours per 100 Sq. Ft.
Built up roofing-4 ply w/gravel	3.50
Rigid insulation board	.75
Shingles - composition strip	.90
Shingles - wood	1.25
Shingles - slate	1.30
Sheathing - plywood	.95
Sheathing - 1" plank	1.10
Sheathing - 2" plank	1.50

Siding Removal

The following table shows various types of siding materials and the required man hours per 100 sq. ft. for their removal. The production times shown in the table do not include transport of the demolished materials to the trash receptacle.

Item Description	Man Hours per 100 Sq. Ft.
Metal - horizontal	2.75
Metal - vertical	3.00
Wood - horizontal	3.00
Wood - vertical	3.25
Shingle - wood or asbestos	3.50
Plywood	1.50

Core Drilling

One of the quickest and neatest ways of cutting a hole in a concrete slab or wall is by core drilling. This particular rig uses a circular-shaped diamond drill bit that is operated by a large drilling machine mounted on a stand. The drill requires continuous water circulated into the core drilling bit to provide lubrication and cooling. The water should not be a problem. The amount of water used is controllable, either by damming or suctioning, to prevent damage to surrounding finishes.

Core drill bits come in diameters from 1" to 18" in size and the depth of cut allowed by the average coring bit is about 12". However, special bits are made to allow boring through columns or walls up to 30" in thickness or more.

Core drilling is usually performed by specialty subcontractors whose only business is to provide this service to general contractors. It is best to obtain a firm price quote for any core drilling that is required for a project, but if a quick budget estimate is needed, core drilling is usually priced by the unit inch, which is the diameter in inches x the depth in inches. For example a hole 4" in diameter and 10" deep would be equal to 40 unit inches. Currently the going rate per unit inch is approximately $2.00 to $2.50 depending on the quantity of holes to be drilled, location, and set-up time required. Horizontal holes will cost about 10% to 15% more than a vertical hole. If reinforcement steel is to be encountered, add 50% to 100% more.

02200 EARTHWORK

It is not difficult to compute the number of cubic yards of earth to be excavated or removed from the premises on a specific project, but it is quite another matter to estimate the actual cost of performing this work, because of the many factors and possible unknown items entering into the cost. This includes such items as type of soil, whether or not water will be encountered, pumping, whether the banks of the excavated portion will be self-supporting or whether it will be necessary to brace and sheet them, disposal of excavated earth, length of haul to dump, type of material, etc.

The prudent estimator will examine the equipment that is available when preparing the cost estimate. The accurate choice of equipment at the bidding stage will determine the final costs. Type of equipment selected, as well as the size of equipment, should be based on several factors--subsurface information, site conditions, volume of material to be moved, and availability of specific equipment.

Estimating Quantities of Excavation. Excavating is measured by the cubic yard (cu.yd.). One cubic yard containing 27 cubic feet (cu. ft.).

General or Mass Excavation. When computing the volume or cubic contents of any basement or portion to be excavated, take all measurements from the outside face of wall footings, add 6" to 1'-0" on each side to allow for placing and removing forms, and then allow for sloping the sides of the excavation so they will stand reasonably stable without being supported by bracing or sheet piling. The amount of slope necessary to provide a "safe" hole will vary with the kind of soil to be excavated, depth of excavation, etc.

On construction projects the Federal Occupational Safety and Health Administration (OSHA), as well as state and local agencies, require for excavations of five feet deep and over soil support of sheeting or some other type of soil support. If none is used, then the side banks of the excavation shall be sloped back to the soils natural angle of repose so that the excavation sides will remain stable.

Digging in previously undisturbed material and assuming that no water or unstable conditions exist, most estimators, when taking off quantities for excavation, use a 1:1 slope; that is, one horizontal to one vertical, for sand and gravel; a 1:2 slope for ordinary clay and a 1:3 or 1:4 slope for stiff clay.

In some cases job conditions may not permit the sloping of banks and then it is usually necessary to sheet and brace the banks to prevent accidents that result in damage to adjoining property, or worse, worker injury or death.

On jobs with column footings that project beyond the wall footings, additional cuts must be made into the banks at the column locations to accommodate the additional width of the footings. The volume of this cut is computed by taking the column footing length plus 1'-0" or 2'-0" for work space plus "layback" of banks on both ends and multiplying by the amount of projection beyond the wall footing to obtain the area of the additional cut, which is then multiplied by the depth of the cut to obtain the volume of additional excavation required.

The depth of the general cut is usually taken from the underside of the topsoil, previously removed, to the underside of the floor slab. Where sand, gravel, or other fill material is to be placed under the concrete floor, the depth should be taken to the underside of the fill material.

Example: Assume you are figuring the general or mass excavation for a building 100'-0" x 100'-0" at grade. The wall footings extend 6" beyond the face of the foundation walls, and on each side there are 6 column footings that are 6'-0" long and project 2'-0" beyond the face of the wall footings.

Also assume the depth of the excavation to be 8'-0" and that the soil will stand at a slope of 1:2 (one horizontal to two vertical). The volume of the excavation should be figured as follows:

100'-0" building length plus 2 x 0'-6" for wall footings plus 2 x 0'-6" for work space plus an average of 4'-0" for slope equals 106' length. Since the building is square, the main volume would be 106'-0" x 106'-0" x 8'-0" or 89,888 cu. ft.

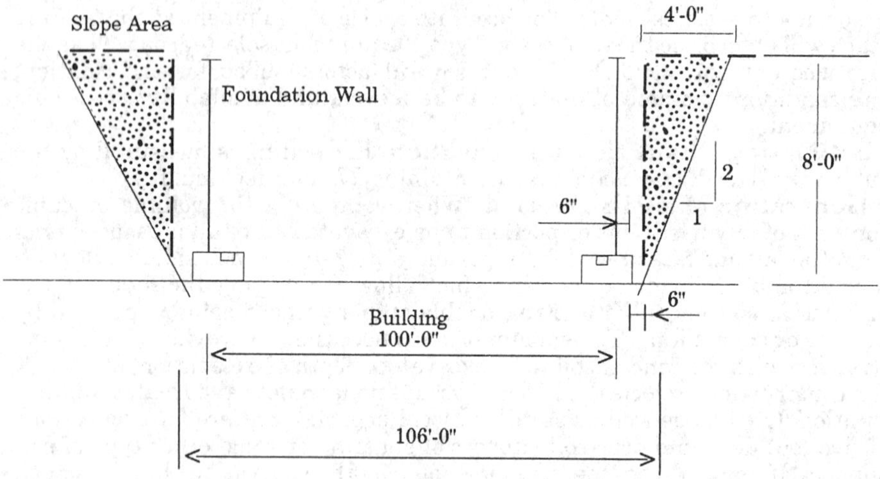

Elevation

Alternate example: 100'-0" x 100'-0" at grade. The wall footings extend 6" beyond the face of the foundation walls on each side. 100'0" building length + (2 x 6") for wall footings + (2 x 6") for work space equals 102'x102'. Next, add the slope area, remembering to add additional slope area at the corners.

Base Excavation: Length x Width x Height
Side Slope: Area of triangle = (base x height) divided by 2; multiply by 4 sides, plus 4 corners

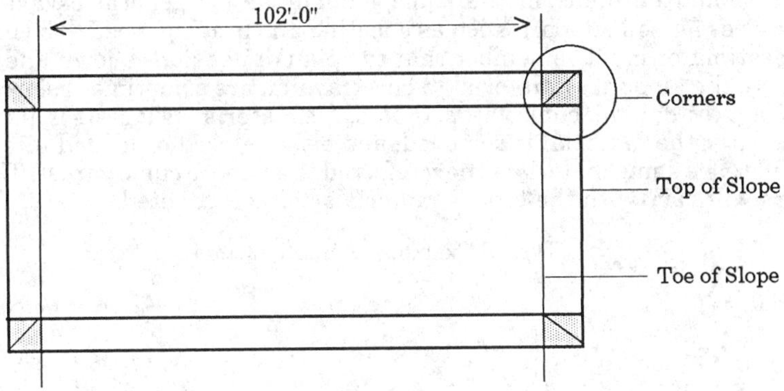

102' x 102' x 8' = 83,232
(4' x 8') divided by 2 x 102' x 4 = 6,528
(4' x 8') divided by 2 x 2 x 4 = 128
83,232 + 6,528 + 128 = 89,888

For the 24 column footings which project beyond the wall footings, the length of each additional cut would be the 6'-0" footing length plus 2 x 0'-6" for work space plus an average of 4'-0" for slope or 11'-0" and the width of each cut would be the 2'-0" projection. Therefore the additional excavation required to accommodate the column footing projections would be 24 x 11'-0" x 2'-0" x 8'-0" or 4,224 cu. ft.

The total volume required to be excavated would then be 89,888 cu. ft. plus 4,224 cu. ft. equals 94,112 cu. ft. or 3,486 cu. yds.

In some cases a portion of or the entire building site may be high and require a grading cut to produce the required contours. In these instances the grading cut should be figured first and the general or mass excavation computed afterwards taking the depth from the underside of the grading cut to the bottom of the floor slab or fill under the floor.

Trench and Pit Excavation. To complete the building excavation quantity take-off, there are usually additional items of excavation to figure for footings, foundations, pits, trenches, etc., which extend below the level of the general or mass excavation, or which occur in unexcavated areas. This kind of excavation is always estimated separately from the general work as it is usually a more expensive operation.

In many instances a good portion of this work may be done with power equipment, with some handwork required to clean out corners and grade the bottoms. On the other hand, there might be portions of the work that must be done entirely by hand, which would be much costlier than machine work. With such a big difference in unit cost between machine work and handwork, the estimator should decide, when taking off the quantities, into which category each portion of the work will fall and list the machine work separately from the handwork so that each may be properly priced.

Quantities for this work are measured and listed in the same manner as for general excavation, except the slope factor at times may be ignored, unless the additional depth exceeds 5'-0" or if the ground will not stand safely for this depth with vertical banks. Under these conditions, allowances for sloping the banks should be made in the same manner as for general excavation, or allowances for soil support, such as sheeting and bracing, need to be included.

The estimator must remember that the plan dimensions shown and used to compute the amount of material to be excavated are sometimes referred to as pit measured or bank cubic yards; that is, the material as it rests in its natural state. After the material is removed, and either stored or loaded into trucks, it swells. The same material is now referred to as "loose cubic yards". The swell factors will vary with the type of materials being excavated.

Typical Examples of Swell Factors

Material	Percentage Swe
Gravel	
Dry	15
Wet	15
Clay	
Dry	20
Wet	20
Rock, Blasted	
Blasted, well	20
Blasted, average	25
Blasted, poorly	50
Earth	
Dry	20
Wet	10
Gravel	
Dry	10
Dry, ¼" to 2"	15
Wet, ¼" to 2"	15
Topsoil	30

Backfill. After all excavating items have been figured, the amount of backfill should be computed. An easy method is to compute the displacement volume of the construction that is to be built within the limits of the excavation; that is, the volume of footings, piers, the basement volume figured from the underside of the fill under the floor to the elevation of the top of the general cut, etc., and deducting this volume from the sum of the general or mass excavation and the trench and pit excavation. The remainder is the volume of backfill required.

Another method of figuring backfill, usually more difficult, is to compute the volume from the actual dimensions of the spaces requiring backfill.

Some specifications require interior backfills to be made with selected materials, such as sand, gravel, bankrun gravel, etc., while the exterior backfill may be, if suitable and approved, the excavated material. In this case it is necessary to keep the two items separate as the interior backfill material will probably have to be purchased and brought in from outside sources.

Backfill very often is specified to be compacted by mechanical tampers and a degree of compaction is also specified. In these instances the backfill operations must be considered very carefully, because this is expensive work.

When calculations for the backfill are complete, the estimator should consider the compaction factor. This is the measured difference between the truck (loose cubic yardage) delivered quantity and the final in-place measured quantity. Backfill material purchased in loose cubic yardage after compaction will not yield the same in-place quantity. Therefore, additional yardage will be required. For example, assume a hole to be backfilled that is 9'0"x9'0"x3'0" deep; measured in place, 9 cu. yds. of material will be required. After compaction, assuming a compaction loss of 15%, the calculated quantity is 10.35 cu. yds. Order 10.50 cu. yds. to fill a hole requiring 9 cu. yds. in place measured.

Excavation and Fill for New Site Grades. Excavation and filling for new site grades, including cuts and fills for drives, walks, and parking areas, vary from simple to complex according to the site ground conditions. Estimating quantities for this work may be rather difficult and some surveying or engineering knowledge will greatly help at this point.

A number of methods are available for determining the volumes of cut and fill required, and the method which is most adaptable to the particular problem will have to be chosen by the estimator after considering the problem with the information given and the accuracy desired.

Cross Sectional Method

One of the easiest and most frequently used methods of computing grading cuts and fills when the plot plan shows both original and proposed contours is that of gridding or dividing up the area to be graded into squares, rectangles, triangles, or combinations of these figures, of regular and convenient dimensions, thus forming a series of truncated prisms. The depths of cut or fill at the corners of these prisms are computed and the volume of each prism may be determined by averaging these depths and multiplying the average by the cross sectional area of the prism. The results of these computations are totaled, keeping the cuts separate from the fills, and the total excavation and/or fill required is obtained.

Example: Assume the contour sketch shown on the next page to be a 400.0' x 200.0' portion of a site grading plan showing both present and proposed contour lines. Since both the present and proposed surfaces have fairly regular slopes, as indicated by the even spacing of contour lines, a grid spacing of 50 feet in both directions has been chosen and grid lines drawn as shown, labeled alphabetically for the vertical lines and numerically for the horizontal lines. The next operation is to figure the present ground elevation for each grid line intersection by interpolating between the present contour lines. The results are marked on the drawing in the upper right quadrant of each respective intersection.

Following this, the proposed ground elevations are similarly figured for each intersection, interpolating between the proposed contour lines and entering the results in the upper left quadrant of each respective intersection.

When this is completed, the depth of cut or fill at each intersection can be readily determined by subtracting one elevation from the other and noting whether a cut or fill is required. This determination should be made for each intersection and the results entered in the proper quadrant, using the lower left for cuts and the lower right for fills.

When all intersections have a depth value entered for either cut or fill, the plan should be examined closely to see where the grading operation will change from cut to fill or vice versa. In the sketch it will be seen that advancing from left to right on horizontal grid line "1" a change from fill to cut occurs between vertical grid lines "D" and "E" and by interpolating between the fill and cut depths the zero point "a" can be established approximately 36 feet to the right of intersection "D-1."

Similarly, zero points "b, d, e, and g" are established on the horizontal grid lines and zero points "c and f" on the vertical grid lines. These points are then connected by a line "a, b, c, d, e, f, g" and for all practical purposes it may be assumed that all points on this line are at the proposed elevation and require neither cut nor fill. In this case the line forms the boundary between cuts and fills. The change from cut to fill and vice versa could, of course, occur many times in actual problems depending on the irregularity of the present and proposed contour lines.

After the previously described operations have been completed, the actual volume of cut and/or fill required can be computed.

The volume of cut or fill in each prism in the grid system is obtained by averaging the corner depth values of each prism and multiplying by its cross sectional area. Thus, for the square prism "A-1, B-1, B-2, A-2" the volume of fill would be ¼ x (2.4' + 1.6' + 1.6' + 2.5') x 50.0' x 50.0' or 5,063 cu. ft. All grid prisms are similarly treated, keeping cut and fill results separated, except those prisms through which the zero line passes. So, for square prism "D-1, E-1, E-2, D-2" it will be seen that this volume has been divided into two trapezoidal prisms "D-1, a, b, D-2" and "a, E-1, E-2, b". To compute the respective fill and cut volumes of these prisms they must be divided into triangular prisms first and then the corner depths may be averaged and multiplied by their respective cross sectional areas. Thus for the triangular prism "D-1, a, b" the volume of fill would be ⅓ x (0.5' + 0.0' + 0.0') x ½ x 50.0' x 36.0' or 150 cu. ft. Similarly, the volume of fill for the triangular prism "D-1, b, D-2" would be ⅓ x (0.5' + 0.0' + 0.3') x ½ x 50.0' x 21.0' or 140 cu. ft. The volume of cut required for the trapezoidal prism "a, E-1, E-2, b" is figured in the same manner. In the case of square cross section "C-2, D-2, D-3, C-3" it will be seen that the zero line crosses one corner of the square and divides it into a triangle "c, D-3, d" and a five-sided figure "C-2, D-2, c, d, C-3". The five-sided figure should be further divided into three triangles "C-2, D-2, c", "C-2, c, C-3" and "C-3, c, d" and the volumes of fill for each triangular prism would be as follows:

⅓ x (0.7' + 0.3' + 0.0') x ½ x 50.0' x 30.0' or 250 cu. ft.
⅓ x (0.7' + 0.0' + 0.5') x ½ x 50.0' x 50.0' or 500 cu. ft.
⅓ x (0.5' + 0.0' + 0.0') x ½ x 36.0' x 20.0' or 60 cu. ft.

When there are several prisms with the same cross section to be either cut or filled the volume of such can be computed as one solid by assembling them as follows. Multiply each corner depth by the number of typical prisms in which it occurs and then add these results and divide by 4. This is then multiplied by the cross sectional area of one prism. For example the volume bounded by A-1, D-1, D-2, C-2, C-4, B-4, B-5, A-5, A-1 can be obtained by one computation because it is composed of a series of prisms having the same cross section. In the summation of depths, those at A-1, D-1, D-2, C-4, B-5 and A-5 are taken but once, those at B-1, C-1, C-3, A-4, A-3 and A-2 are multiplied by 2, at C-2 and B-4 the depths are multiplied by 3 and at B-2 and B-3 they are multiplied by 4. All of these values are then totaled and divided by 4 and then multiplied by the cross sectional area of one prism giving the volume of fill required.

When a volume has been computed for each grid division of the site, either individually or collectively, it is only a matter of summing up the cuts and fills to obtain the total excavation or fill required to bring the site to the proposed grades.

End Area Method

It often happens that a cut and fill operation is to be performed over a long and narrow area such as for roads, levees, ditches, etc. In this case the usual method of computing the earthwork quantities is by means of end areas.

In jobs of this sort a profile plan and a topographic plan is generally furnished to the contractor as part of the bidding documents. Using this information, sections are taken perpendicular to the longitudinal centerline of the work, usually at each 100 foot station, at each radical change in either present or proposed grades and where the cut changes to fill or vice versa. These sections are plotted to scale showing the original grades as obtained from the topographic plan and the proposed grades as obtained from the profile plan and design details of the surface required. After this has been done, it will be apparent that areas are enclosed by the plotted lines representing the present and proposed grades at each section and these areas should be computed.

To obtain the volume of cut and fill between each section, the usual and easiest method is to average the areas of two adjacent sections and multiply by the distance between them, keeping the cuts and fills separate. This method is not quite accurate, but the results are on the safe side and in practically every case are close enough for estimating purposes.

Example: Assume the following sketch to be a 300 foot portion of the topographic plan and profile plan of a proposed road location. The topographic plan shows the undulations of the present ground surface by the contours and the location of the road by the center and shoulder lines. The profile plan shows the relation between the present ground surface and the proposed road bed grade at the centerline of the road. Also assume the road bed is to be 20.0 feet wide and the shoulder slopes are to be 1 ½ : 1, that is 1 ½ horizontal to 1 vertical.

To compute the volumes of earthwork involved in this proposed grading operation, by means of the end area method, sections must be taken perpendicular to the center-line of the proposed road and the present and proposed grade lines plotted. In this case sections have been taken and plotted, as illustrated in the sketch, at even stations 15+00, 16+00, 17+00, and 18+00. Sections have also been taken at stations 15+07, 15+37, 16+50 and 16+65 because at these points the cut or fill at a shoulder of the proposed road is zero. A section is taken at station 17+65 because at this point the present grade makes a radical change in slope.

In plotting the sections any convenient scale may be used and if cross-section paper is used the job will be much easier.

If a greater degree of accuracy is required the prismoidal formula must be employed which necessitates figuring the middle area between two adjacent sections. This middle area is obtained by averaging the corresponding dimensions of both end areas and computing an area from the average dimensions; it should not be taken as the average of the two end areas. To obtain the volume between adjacent sections using the prismoidal formula, multiply the length between sections by one-sixth of the sum of the two end areas and 4 times the middle area combined.

Most specifications call for the fills to be placed in layers and then specify some degree of compaction to be attained. This should be considered by the estimator, because additional equipment such as sheepsfoot rollers, pneumatic rollers, or tandem rollers will be required plus labor for the operation.

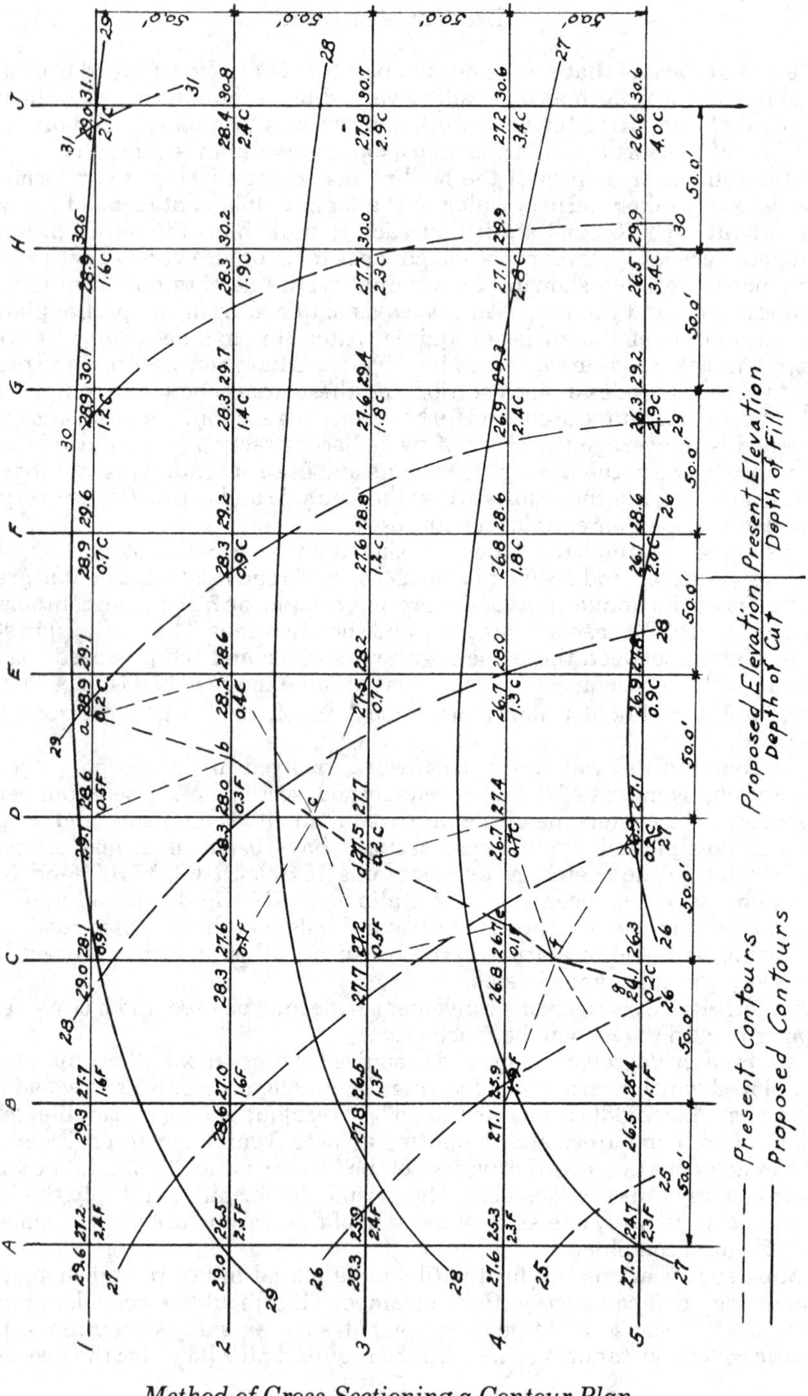

Method of Cross Sectioning a Contour Plan.

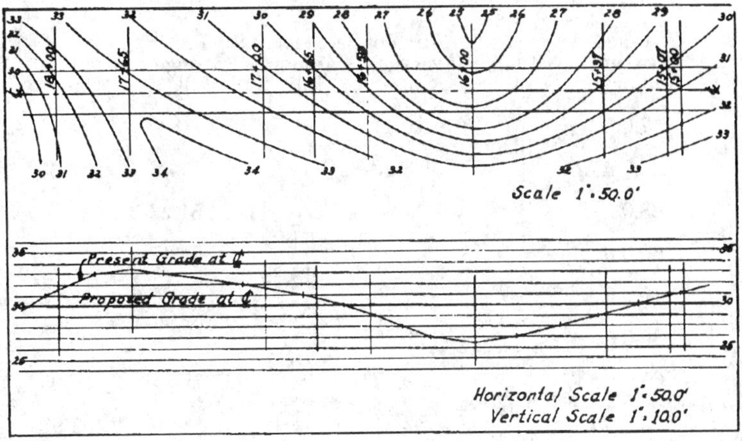

Topographic and Profile Plans

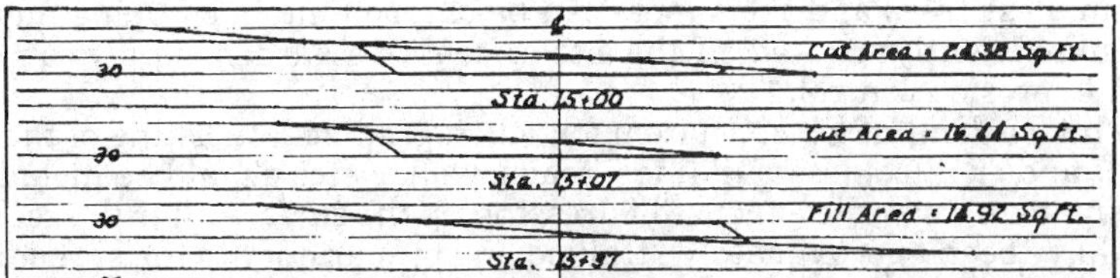

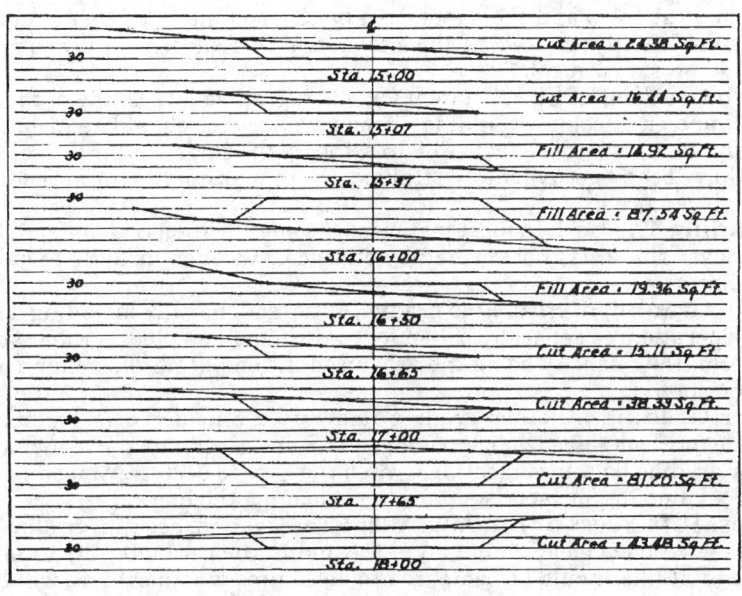

Method of Plotting End Areas

As each section is plotted, the area enclosed by the present and proposed grade lines is computed and the result tabulated. Final computations are made as follows:

		Area of Sections in Sq. Ft.					
Station	Distance Between Stations	Actual Cut	Average Cut	Actual Fill	Average Fill	Cu. Ft. Cut	Cu. Ft. Fill
15+00		24.38		--			
	7.0 ft.		20.41		--	142.87	--
15+07		16.44		--			
	30.0 ft.		8.22		7.46	246.60	223.80
15+37		--		14.92			
	63.0 ft.		--		51.23	--	3,227.49
16+00		--		87.54			
	50.0 ft.		--		53.45	--	2,672.50
16+50		--		19.36			
	15.0 ft.		7.56		9.68	113.40	145.20
16+65		15.11		--			
	35.0 ft.		26.72		--	935.20	--
17+00		38.33		--			
	65.0 ft.		59.77		--	3,885.05	--
17+65		81.20		--			
	35.0 ft.		62.34		--	2,181.90	--
18+00		43.48		--			
						7,505.02	6,268.99

Clear and Grub. Prior to stripping of top soil or to performance of rough grading, it may be necessary to clean and grub the area. This is the removal of all trees, bushes, and other growth, with roots to be grubbed out of the ground, in order for top soil to be removed in an efficient manner.

The smallest trees and their roots can be chipped into a chipping machine and then efficiently trucked off the site or sold to a garden supply company, or used in erosion control.

Rough Grading of Site. An item for rough grading, consisting of the total area cut and/or filled, less buildings, roads, walks, etc., should be included in the estimate to cover the expense of dressing up the surface after the required grades have been approximately attained and preparatory to the spreading of top soil, if required.

Spreading Top Soil. In most instances the specifications call for spreading top soil over site areas not covered by other construction, usually to a depth of 6 inches.

Finish Grading of Site. Fine grading of top soil, usually including a hand raking operation, is generally called for and should be listed in the estimate as an item, the quantity being the area used in computing the volume of top soil to be spread.

Other Grading. Another grading item to be considered is grading the subgrade for floor slabs, walks, paving, etc. Some contractors also include an item for grading bottoms of footings, when they are to be dug by machine, and include an amount for labor squaring and cleaning the holes.

Disposal of Excess or Borrow. After all excavation, backfill, grading cuts and fills, etc., have been computed and listed in the estimate, the total cuts and the total fills should be compared to determine whether there will be an excess of materials to dispose of or a deficiency of materials to be made up by borrowing or purchasing from outside sources. The results of this comparison should be listed in the estimate. The top soil comparison should be kept separate from the other materials, because the cost of any top soil that must be purchased usually is much more expensive than ordinary fill material. Also, any excess of top soil may find a ready market and be sold, reducing the overall cost of the earthwork. The difference between the cuts and fills will give the bank or compacted volume required. In case of a deficiency of materials, this compacted volume quantity must be adjusted for shrinkage due to compaction, because the material will probably be bought by loose measure. In the same manner all excess bank measurement volumes must be adjusted for swell due to bulking to obtain the true volume that must be handled in the disposal operation. The amount of increase for gravels and sand is from 5 to 12 percent and for clays and loams from 10 to 30 percent.

Pumping or De-watering. Finally, careful consideration should be given to the probability of pumping or de-watering operations. In some parts of the country this is no problem whatsoever, quite the reverse being the case. However, in most localities ground or climatic conditions make it necessary to include pumping or de-watering as an item in the estimate.

This item is quite variable, being affected by the season of the year as well as the locale. If only rainwater run-off is expected, an allowance for a few small pumps may be sufficient. If, however, the job is in close proximity to a body of water or springs are present, it might be necessary to de-water part or all of the operational area by means of a wellpoint or deep well system, in which case the contractor should consult a company that specializes in this work. (See the section on wellpoint systems in this chapter.)

Estimating the Labor Cost of Excavation. Before an excavating estimate is made up or a bid submitted, the estimator should visit the site and ascertain the exact conditions surrounding the work to be performed. The natural grade of the lot should be noted to compute the depth of the excavation from grade to underside of floor slab; the nature of the soil should be determined, whether sand, loam, clay, or rock, and whether or not water is likely to be encountered. Water usually means an additional cost--pumping out the hole day and night until the foundations are completed. Rock usually means blasting, with a greatly increased cost for drilling, dynamiting, excavating, and loading.

The safest method of determining the kind of soil is to dig a test hole the approximate depth of the basement, or take a set of borings. From this one can obtain a fairly accurate idea of the conditions to be encountered.

In most areas the soil strata, as well as the ground-water table, is fairly consistent, and its character may be ascertained by talking to contractors or engineers who have worked in the area for some time. In most larger jobs, the plans will show the results of test borings ordered by the owner or architect/engineer.

If the lot is low and the basement does not cover the entire lot, part of the excavated soil may be used for filling. On the other hand, if the natural grade of the lot is above the sidewalk or established grade, it may be necessary to remove a certain amount of soil from the entire lot.

In large structures having basements and sub-basements, the excavation may extend 40 to 50 feet below grade. This requires the construction of steep runways, haul roads, or ramps so that trucks may get in and out of the hole, and an extra truck or bulldozer may be required to help the loaded trucks out of the hole. These items must all be taken into consideration when pricing any excavating job.

If it is necessary to remove the excavated soil from the premises, locate the nearest available landfill, and ascertain just how the soil will be handled at the landfill (for example, whether it is necessary to spread it).

National and local environmental protection agency rules may require the contractor to test material to be excavated and disposed of, or even to be placed back into the same area, after excavation. Local EPA rules at times require local disposal site operators not to accept material without a certification from the contractor stating that no contaminants are in the material. Materials excavated that have contaminants must be disposed of at designated disposal sites. The types and amounts of contaminants will determine the amount of additional costs.

These are all items that affect the cost and should be given careful consideration by the estimator.

In most localities there are contractors who specialize in excavation and are familiar with the kind of soil encountered in different parts of the area. These contractors usually quote a price per cu. yd. for all of the general excavation, or they may quote a lump sum for the entire job.

Trench, pier, and all hand excavating is usually performed by the general contractor.

METHODS OF EXCAVATING

Excavating costs will vary with the kind of soil and the method used in removing and hauling to a final destination.

Except on very small jobs, practically all excavation today is performed by power operated equipment, i.e., power shovels, cranes with dragline or clamshell buckets, bulldozers, tractor excavators, front end loaders, scrapers, and trenching machines. Grading, spreading, and backfilling are performed by bulldozers, tractor excavators, and front end loaders.

In shallow excavations, where the excavated earth may be spread over the site not more than 50 feet from the excavation, bulldozers or loaders (a tractor equipped with a shovel that may be elevated to load directly into trucks or over the sides of the excavation) are frequently used. This type of excavator is also used for excavating basements, loading directly into trucks.

What was at one time referred to as a "back-digger" type of power shovel, is now on track or rubber tire and commonly called a "backhoe", which is extensively used for basement and trenching excavation. It is equipped with a bucket that "pulls" or "crowds" toward the machine as it excavates, which is always located on the bank above the excavation, working from the sides and ends of the excavation.

In large excavations, the use of a regular power shovel or hydraulic front shovel is the most satisfactory and economical method to use. The power shovel will excavate and load the trucks as fast as they can be handled in and out of the hole and will enable the job to proceed much faster.

If it is desirable to keep the backhoe on the bank above the excavation, but the depth is greater than the hoe can reach efficiently, a crane equipped with a dragline bucket or clam bucket can be used.

For individual pier footings, small pits, and hard-to-reach spots, a crane equipped with a clamshell bucket or a small combination backhoe/loader can be used.

On medium to large grading jobs, self-propelled or tractor-hauled scrapers are used quite often and perform cutting and filling operations very creditably.

Quantities and production times given in the tables below are average. Experienced excavators will perform more work than given, but an inexperienced worker will perform less, so average conditions have been used.

Hydraulic power excavating equipment has become very extensive. For the most part, hydraulic-operated excavating equipment has replaced the old cable-operated machines and is used almost 100 percent on the smaller excavation machines.

HAND EXCAVATION

Most construction projects require extensive use of hand labor to accomplish the selective types of excavation which cannot be performed by machine excavators.

In the excavation of footings, it is frequently found to be cost effective to combine the use of machines for the bulk excavation and then to "hand dress" the bottoms, sides, and corners to produce the precise horizontal and vertical planes required. Similarly, this is true in finalizing the subgrade elevation under slabs on grade, walks, and patio slabs.

The quantities and production rates given in the following table are for average conditions.

Material & Operation	Avg. No. Cu. Yd. per 8-Hr Day	Labor Hrs. per Cu. Yd.
Sandy loam		
Small footings, hand	6	1.33
Trenches 3' to 4' deep, hand	6	1.33
Piers 6' deep, hand 1 lift	5	1.60
Piers 12' deep, hand 2 lifts	3	2.66
Heavy soil and clay		
Small footings, hand	5	1.60
Trenches 3' to 4' deep, hand	5	1.60
Piers 6' deep, hand, 1 lift	4	2.00
Piers 12' deep, hand, 2 lifts	2	4.00
Excavation with ½ cu.yd. backhoe, casting to side		
Trench footing	300	--
Pier footing 6' deep	200	--
Pier footing 12' deep	150	--

Hand Trimming Trench, Piers And Slabs After Machine Excavation

Area	SF per 8-Hr Day	SF per Hour
Bottom and side areas	240	30
Fine grade under slabs	600	75

Backfilling By Hand With Compaction Material From Stockpile

Ordinary Light Soils	Cu. Yd. per 8-Hr Day	Labor Hrs per Cu. Yd.
Soil distribution laborer	12	.67
Compactor operator	12	.67

No allowance for compactor equipment cost.

MACHINE EXPLORATION

Excavating Using a 1 Cu. Yd. Tractor Shovel

Class of Work	Avg. No. Cu. Yds. per 8-Hr. Day	Tractor and Operator Hours Per 100 Cu. Yds.
Excavating ordinary soil and placing in piles on premises	325 to 375	2.30
Excavating heavy soil and clay and placing in piles on premises	250 to 300	2.90
Loading loose sand or gravel into trucks	400 to 450	1.90

Class of Work	Avg. No. Cu. Yds. per 8-Hr. Day	Tractor and Operator Hours Per 100 Cu. Yds.
Excavating ordinary soil and loading into trucks	400 to 450	1.90
Excavating heavy soil and clay and loading into trucks	300 to 350	2.50
Backfilling loose earth--Bulldozing	500 to 600	1.45

The above table is based upon a 50-foot haul from excavation to dump. For each additional 50 feet in length of haul, add 1 ½ to 2 hrs. tractor time per 100 cu. yds. or consider the use of trucks.

Excavating Using a 2 ¼ Cu. Yd. Tractor Shovel

Class of Work	Avg. No. Cu. Yds. per 8-Hr. Day	Tractor and Operator Hours Per 100 Cu. Yds.
Excavating ordinary soil and placing in piles on premises	600 to 650	1.30
Excavating heavy soil and clay and placing in piles on premises	450 to 500	1.70
Loading loose sand or gravel into trucks	775 to 825	1.00
Excavating ordinary soil and loading into trucks	775 to 825	1.00
Excavating heavy soil and clay and loading into trucks	575 to 625	1.35
Backfilling loose earth--bulldozing	925 to 975	0.85

The above table is based upon a 50-foot haul from excavation to dump. For each additional 50 feet in length of haul, add ¾ to 1 hrs. tractor time per 100 cu. yds. or consider the use of trucks.

Excavating Using Hydraulic Tractor (Crawler) Backhoe
1 Cu. Yd. Bucket

Type of Material	Avg. No. Cu. Yds. per 8-Hr. Day	Backhoe and Operator Hours per 100 Cu. Yds.
Sandy Clay/Moist Loam	720	1.10
Gravel/Sand	690	1.20
Clay, Hard	570	1.40
Rock		
Well Blasted	450	1.80
Poorly Blasted	300	2.70

Above production is based on the following: project conditions, average to above average; operator, average; obstructions, none; swing, to 60 degrees or ¼ turn.

Excavating Using Hydraulic Tractor (Crawler) Backhoe
¼-Cu. Yd. Bucket

Type of Material	Avg. No. Cu. Yds. per 8-Hr. Day	Backhoe and Operator Hours per 100 Cu. Yds.
Sandy Clay/Moist Loam	900	0.90
Gravel/Sand	840	0.95
Clay, Hard	570	1.40
Rock		
Well Blasted	450	1.80
Poorly Blasted	300	2.70

Above production is based on the following: project conditions, average to above average; operator, average; obstructions, none; swing, to 60 degrees or ¼ turn.

Excavating Using Hydraulic Tractor (Crawler) Backhoe
1 ½-Cu. Yd. Bucket

Type of Material	Avg. No. Cu. Yds. per 8-Hr. Day	Backhoe and Operator Hours per 100 Cu. Yds.
Sandy Clay/Moist Loam	1080	0.75
Gravel/Sand	1020	0.80
Clay, Hard	870	0.90
Rock		
Well Blasted	660	1.20
Poorly Blasted	420	1.90

Above production is based on the following: project conditions, average to above average; operator, average; obstructions, none; swing, to 60 degrees or ¼ turn.

Excavating Using Hydraulic Tractor (Crawler) Backhoe
1 ¾-Cu. Yd. Bucket

Type of Material	Avg. No. Cu. Yds. per 8-Hr. Day	Backhoe and Operator Hours per 100 Cu. Yds.
Sandy Clay/Moist Loam	1260	0.60
Gravel/Sand	1200	0.70
Clay, Hard	1020	0.80
Rock		
Well Blasted	750	1.05
Poorly Blasted	510	1.60

Above production is based on the following: project conditions, average to above average; operator, average; obstructions, none; swing, to 60 degrees or ¼ turn.

Excavating Using Hydraulic Tractor (Crawler) Backhoe
2 Cu. Yd. Bucket

Type of Material	Avg. No. Cu. Yds. per 8-Hr. Day	Backhoe and Operator Hours per 100 Cu. Yds.
Sandy Clay/Moist Loam	1440	0.55
Gravel/Sand	1380	0.60
Clay, Hard	1140	0.70
Rock		
Well Blasted	870	0.90
Poorly Blasted	720	1.10

Above production is based on the following: project conditions, average to above average; operator, average; obstructions, none; swing, to 60 degrees or ¼ turn.

Excavating Using Hydraulic Tractor (Crawler) Backhoe
3 Cu. Yd. Bucket

Type of Material	Avg. No. Cu. Yds. per 8-Hr. Day	Backhoe and Operator Hours per 100 Cu. Yds.
Sandy Clay/Moist Loam	1800	0.45
Gravel/Sand	1710	0.45
Clay, Hard	1440	0.55
Rock		
Well Blasted	1080	0.75
Poorly Blasted	720	1.10

Above production is based on the following: project conditions, average to above average; operator, average; obstructions, none; swing, to 60 degrees or ¼ turn.

ESTIMATING THE COST OF OWNING AND OPERATING CONSTRUCTION EQUIPMENT

The following information on the costs of owning and operating power shovels, hoes, cranes, truck cranes, tractors, and scrapers will give a general idea of the costs involved in making up an estimate for this type of work.

The equipment purchase prices given are only a relative guideline to be used in the tables. What is most importance is the format for determining these costs, not the price of the equipment listed, which will vary greatly depending on locality, method of purchase, number of pieces bought at one time, etc.

The price f.o.b. factory should cover the complete machine with all variable equipment and accessories, such as various attachments, light plant, magnet generators, and clamshell bucket.

Total Investment or Cost of Equipment. Depreciation, interest, taxes, and insurance are directly related to the initial investment in construction equipment. In addition, certain other costs may be estimated by their normal relationship to this figure. Therefore, the proper determination of the total cost is a basic requirement.

Approximate Prices of Power Shovels, Draglines, Clamshells, and Lift Cranes

The following are approximate prices, f.o.b. factory, and will give the estimator some idea of the investment required in shovels and cranes of various sizes:

Power Shovel
Crawler

Size of Shovel	
3/4 cu. yd	$91,000.00
1 cu. yd	96,000.00
1 1/2 cu. yd	146,000.00
2 cu. yd	310,000.00
3 1/4 cu. yd	547,000.00
4 1/4 cu. yd	672,000.00
5 1/2 cu. yd	800,000.00
6 cu. yd	950,000.00

Backhoe Crawler

Bucket Size of Hoe	
3/4 cu. yd.	$114,000.00
1 cu yd.	150,000.00
1 1/4 cu. yd.	155,000.00
1 1/2 cu. yd.	235,000.00
2 cu. yd.	271,000.00
2 1/2 cu. yd.	400,000.00
4 1/2 cu. yd.	925,000.00

Dragline Excavators Including a Medium Duty Bucket

Size of Dragline	
½ cu. yd	$81,000.00
¾ cu. yd	90,000.00
1 cu. yd	132,000.00
1 ¼ cu. yd	265,000.00
2 ½ cu. yd	470,000.00
3 cu. yd	582,000.00
3 ½ cu. yd	700,000.00

Clamshell Excavators Including a Medium Duty Bucket

Size of Clamshell	
½ cu. yd	$80,000.00
¾ cu. yd	90,000.00
1 cu. yd	132,000.00
1 ¼ cu. yd	262,000.00
2 ½ cu. yd	470,000.00
3 cu. yd	582,000.00
3 ½ cu. yd	700,000.00

Lift Cranes
Crawler Mounted

Capacity	
15 ton	$150,000.00
20 ton	185,000.00
30 ton	197,000.00
40 ton	210,000.00
45 ton	247,000.00
50 ton	275,000.00
70 ton	380,000.00
100 ton	543,000.00
150 ton	650,000.00
200 ton	800,000.00
400 ton	1,200,000.00

Approximate Rental Costs For Excavating Equipment With Operator

Type	Size	Day	Week	Month
Crawler Mounted Backhoe	½ cy	$453	$1,020	$6,770
	1 cy	600	2,240	8,000
	2 cy	1,020	3,700	12,500
	2 ½ cy	1,300	4,600	15,300
Compactor	1000# Ram	57	275	980
	1000# Vib	47	195	915
	5000# Vib	220	915	3,400
	2000#Drum	100	430	1,950
Crane-Cable Crawler				
	30 T	600	2,660	9,310
	40 T	660	2,920	10,100
	60 T	880	3,800	12,780
	100 T	1,060	4,570	15,100
	150 T	1,170	6,000	19,450
Crane-Cable Truck				
	40 T	790	3,260	10,950
	60 T	840	3,440	11,500
	80 T	1,080	4,350	14,150
	100 T	1,260	5,100	16,300

Type	Size	Day	Week	Month
Crane Hydro Truck				
	5 T	370	1,650	6,250
	15 T	500	2,170	7,440
	30 T	740	3,130	10,740
	60 T	1,140	4,790	15,600
Drill*	40# Rock	25	75	200
	65# Rock	30	85	240
Scraper	11 cy	500	2,250	9,000
	22 cy	850	3,900	15,800
Tractor - Dozer	105 hp	400	1,900	7,600
	180 hp	450	2,100	8,600
Loader				
	1 ½ cy	400	1,900	7,600
	2 ¼ cy	480	2,300	9,100
Trencher	16 cy	330	1,400	4,500
	24 cy	400	1,600	5,800

*No Operator

The above costs do not include move charges or permits, and they do not include fuel or other daily operating expenses.

Basic Power Shovel Unit and Front End Attachments. Naturally, the ideal situation would be to have the most suitable and economical power shovel, backhoe, crane, or dragline for each job, but this is seldom practical or profitable. More frequently, the contractor must take into consideration the handling of a wide range of work with varying conditions, requiring more or less frequent and easy conversion (for example, shovel to dragline or crane to clamshell) by simply changing attachments on the basic unit.

Power cranes and shovels are designed to accommodate a wide range of front-end attachments and working tools to meet the various job requirements. The attachments are classified into three basic groups: the shovel boom, the crane boom, and the hoe boom. These attachments generally are interchangeable in the field.

Since the basic purpose of the shovel boom is for shovel work and the hoe boom for hoe work, the crane-type boom serves for all other materials handling tools, such as dragline buckets, clamshell buckets, orange peel buckets, hooks, hook blocks, tongs, grabs, clamps, grapples, concrete buckets, skull crackers, and pile driving leads.

Crane type booms can be extended for extremely high lifts by inserting standard sections in the center of the boom and by boom tip extensions called jibs. These jibs are used primarily to extend horizontal reach with boom raised vertically close to maximum elevation. They also provide an increase in vertical lifting range. Since the jib adds weight at the outer lifting radius and extends the working range, it has a limiting effect on lifting capacity.

Average Useful Life of Power Shovels, Hoes, and Cranes. A reasonable allowance for the exhaustion, wear, and tear of property used in business, including a reasonable allowance for obsolescence, is permitted in figuring depreciation on your equipment.

The figures tabulated below are taken in part from Government Bulletin No. 456, "Depreciation Guide Lines and Rules", of the U. S. Treasury Dept. Internal Revenue Service.

Average Useful Life of Power Shovels and Hoes

3/8-3/4 cu. yds.	5 years or 10,000 hours
1-1 1/2 cu. yds.	6 years or 12,000 hours
2 cu. yds. and over	8 years or 16,000 hours

Average Useful Life of Draglines, Clamshells and Cranes

Cu. Yds.	Tons	Years	Hours
3/8-3/4	2 1/2-5	5	10,000
1-1 1/2	10-15	9	18,000
2 and over	20 and over	12	24,000

The grouping shown for lifting cranes is a separate grouping and does not necessarily relate to the machines by size in cu. yds. For example, many 3/4 cu. yd. excavators (clamshell or dragline) would have a greater rating than 5 tons and might be rated in tons to fall in a different group than they do as excavators.

Depreciation. The straight line method of figuring depreciation is used. This method is in general adequate, but if it is desired to take changing prices for equipment into account in estimating, an annual appraisal to determine a current value should be considered.

Therefore, for the years and hours used, the following percentage of depreciation for the various groups are used:

Years	Hours	% per Year	% per Hour
5	10,000	20%	.01%
6	12,000	16.67%	.00833%
8	16,000	12.5%	.00625%
9	18,000	11.11%	.00555%
10	20,000	10%	.005%
12	24,000	8.33%	.00416%

Interest, Taxes and Insurance. Interest, taxes, insurance, and storage are usually charged at 20% of the average investment, of which 15% is interest (to be adjusted for current rates) and 5% covers taxes, insurance, storage, and incidentals.

Average investment must be established, based on the number of years used for depreciation. Since the first year is considered as 100% and since the investment is considered at the beginning of each year, the method of calculating average investment for a 5-year depreciation period is as follows:

1st year	100% of total investment
2nd year	80% of total investment
3rd year	60% of total investment
4th year	40% of total investment
5th year	20% of total investment
	300% of total investment

Average investment equals 300% divided by 5 years equals 60% of total investment.

The percent of Total Investment to use for Average Investment for 5, 6, 8, 9, 10, and 12 year periods are as follows:

No. Years Depreciation	Percent Average Investment per Year
5	60%
6	58.33%
8	56.25%
9	55.56%
10	55%
12	54.17%

Therefore, the percentage of Total Investment per year and per hour to use for interest, insurance, taxes, and storage, based on 20 percent of average investment for the groups considered here, is as follows:

Interest, Insurance, Taxes, Storage
Based on 20 percent of Average Investment per Year

Percent Avg. Investment per Year	Years	Hours	Percent of Total Investment Per Year	Percent of Total Investment Per Hour
60%	5	10,000	12%	.006%
58.33%	6	12,000	11.666%	.005833%
56.25%	8	16,000	11.25%	.005625%
55.56%	9	18,000	11.112%	.005556%
55%	10	20,000	11%	.0055%
54.17%	12	24,000	10.834%	.005417%

Fixed Costs, Repairs, Maintenance, and Supplies. These are difficult to estimate. The figures used are not based on actual records but are believed to be representative. In the case of shovels and hoes, this cost is based on 100 percent of the total investment spread over the life of the machine and includes repairs, replacement parts and repair labor for normal operation, and such items as ropes, dipper and bucket teeth, oil and grease for lubricating machinery, as well as seasonal overhaul. It does not include engine fuel or lubricating oil.

In the case of draglines and clamshells, 80% of the total investment spread over the economic life of the machine is a fair average cost of these items.

In the case of lifting cranes, 60% of the total investment spread over the economic life of the machinery may be used.

The owner of an excavator should anticipate three things in regard to repairs and maintenance. First, the average figures used presume proper maintenance. Without this, repair bills will greatly exceed the estimates given here. Second, it is assumed the equipment will be used within its specified rating. Third, these expenses are not uniform each year.

Maintenance is then further defined in two major categories. The first is normal maintenance as a daily operating expense, which includes regular servicing on the job, such as fuel, changing oil, filters, greasing, brakes and hydraulic fluid levels, minor tuneups and the like.

The second category is defined as major maintenance, which requires that the equipment be idle for an extended period while repairs are made in the shop. An example of major maintenance would be the undercarriage of a dozer.

Provision should be made for a periodic general overhaul, and it should be expected that both the amount of repairs required and the losses caused by shutdowns will increase as the equipment becomes older.

Repairs, Maintenance, and Supplies, Including Labor
Associated With Them
Shovels and Hoes
Based on 100 percent of Total Investment Spread over Life of Machine.

			Percent Of Total Investment	
Cu. Yds.	Years	Hours	Per Year	Per Hour
3/8-3/4	5	10,000	20%	.01%
1-1 1/2	6	12,000	16.67%	.00833%
2-3 1/2	8	16,000	12.5%	.00625%
4 1/4 or more	10	20,000	10%	.005%

For example, assume a 3/4-cu. yd. shovel cost $125,000.00. Repairs, maintenance, and supplies equal $125,000.00 x .0001 = $12.50 per hr.

Draglines and Clamshells
Based on 80 percent of Total Investment Spread over Life of Machine.

			Percent Of Total Investment	
Cu. Yds.	Years	Hours	Per Year	Per Hour
3/8-3/4	5	10,000	16%	.008%
1-1 1/2	9	18,000	8.89%	.00445%
2-3 1/2	12	24,000	6.66%	.00333%

Lifting Cranes
Based on 60 percent of Total Investment Spread over Life of Machine.

			Percent Of Total Investment	
Tons	Years	Hours	Per Year	Per Hour
2 1/2-5	5	10,000	12%	.006%
10-15	9	18,000	6.67%	.00333%
20 and over	12	24,000	5%	.0025%

Operating Costs, Engine Fuel, and Lubricating Oil. A formula for estimating the approximate amount of diesel fuel consumption for equipment of this type is as follows:

$$\frac{\text{BHP x Factor x lbs. fuel per HP hr.}}{\text{Weight of fuel per gallon}}$$

= Gallons per Hour, where

BHP = Brake HP of engine--or rated HP
Factor = Factor for this use 50 to 60%
Diesel = 0.5 lbs. per brake horsepower hour.
Diesel = 7.3 lbs. per gallon (U.S.)

From this formula we obtain an approximate diesel fuel consumption of 0.034 to 0.041 gallons per horsepower hour (based on 50% to 60% factor respectively), and suggest using an average of about 0.040 gal. per horsepower hour.

Shovels normally will consume a greater amount of fuel than the other types of machines considered here. Therefore the larger consumption rate indicated should be used for estimating fuel for shovels and the smaller rate for draglines and clamshells. Machines used for lifting crane service only, usually operate intermittently and fuel consumption is difficult to estimate.

Example: A 100 horsepower engine in shovel service is estimated to consume per hour: 100 x .040 = 4.0 gals. of diesel fuel per hr.

Lubricating oil is considered with fuel, because it varies with the size and type of engine. It usually includes a complete change every 100 hours plus make-up oil between changes. Allow 15% of fuel costs.

Labor Operating Costs. The labor rates, as well as the number in the operating crew, vary in different parts of the country and on different jobs. The contractor must estimate this cost. Costs related to crew costs, besides rates of pay, are employer contributions for Federal Insurance Contribution Act, Workmen's Compensation Insurance, Unemployment Compensation, Overtime, Paid Holidays, and Contractor's Contributions to Union Welfare and Pension Funds.

Power Shovel Yardages. To estimate yardage production on a job is a real problem. If all conditions were the same on all jobs, it would be a simple matter, but jobs are as different as night and day. There are many factors which must be considered, many of which can be learned only through experience. The type of material, the depth of cut for maximum effect, delays in operation, a 90° swing for the shovel. All affect the output.

The following table gives an approximate idea of the difference in maximum output possible, subject to conditions such as those listed above.

Hourly Shovel Output in Cubic Yards

Class of Material	Shovel Capacity 3/8	1/2	3/4	1	1 1/4	1 1/2	1 3/4	2	2 1/2
Moist Loam, Light Sandy Soil	85	115	165	205	250	285	320	355	405
Sand and Gravel	80	110	155	200	230	270	300	330	390
Common Earth	70	95	135	175	210	240	270	300	350
Clay, Hard and Tough	50	75	110	145	180	210	235	265	310
Rock, Well Blasted	40	60	95	125	155	180	205	230	275
Common Earth, with Rocks and Roots	30	50	80	105	130	155	180	200	245
Clay, Wet and Sticky	25	40	70	95	120	145	165	185	230
Rock, Poorly Blasted	15	25	50	75	95	115	140	160	195

The quantities in the above table are based on bank measure, which means cubic yards removed from the bank rather than cubic yards in the hauling unit. There is a big difference between a yard of dirt in the bank and a yard of dirt in the truck. This is because of the swell of the material or its increase in volume due to voids when it is dug or loosened. For instance, common excavation will swell from 10 to 30 percent. Another condition is that the machine is working in a depth of cut suitable for maximum digging efficiency. The optimum depth of cut for various sizes of shovel may be defined as that depth which produces the greatest output and at which the dipper comes up with a full load.

The output figures are based on continuous operations, 60 minutes per working hours, without any delays for adjustments, lubrication, operator stopping for any reason, etc.

The figures are based on each dipper-full of material being swung through an arc of 90° before dumping. This is important, because a swing of either a lesser number or greater number of degrees than 90° will either save or consume time and affect output capacity.

Tables are based on all materials being loaded into hauling units, with the further understanding that these hauling units are properly sized to shovel capacity and are provided in sufficient quantities to take away all the material that the shovel can dig.

While these figures represent a very comprehensive job and as realistic a picture as can be presented on this involved subject, a word of caution is in order. These figures are general only and should not be considered as guaranteed outputs, or used by anyone as a basis for figuring and bidding jobs.

How Degrees of Swing and Depth of Cut Affect Shovel Yardage Output. The quantities in the above table are based on the optimum depth of cut and a swing of 90° was specified.

Because variations in either of these conditions will have considerable effect on the yardage output, the following chart shows what happens when swing and depth of cut vary from the conditions on which the previous basic hourly yardages were established.

Table Giving Effect of Depth of Cut and Angle of Swing on Power Shovel Output

Depth of Cut in % of Optimum	45°	60°	75°	90°	120°	150°	180°
40	.93	.89	.85	.80	.72	.65	.59
60	1.10	1.03	.96	.91	.81	.73	.66
80	1.22	1.12	1.04	.98	.86	.77	.69
100	1.26	1.16	1.07	1.00	.88	.79	.71
120	1.20	1.11	1.03	.97	.86	.77	.70
140	1.12	1.04	.97	.91	.81	.73	.66
160	1.03	.96	.90	.85	.75	.67	.62

Dragline Yardages. As in the case of power shovels and hoes, dragline yardage production is affected by the type of material to be excavated, the depth of cut, the swing before unloading, the type of unloading (unloaded into hauling units, cast onto spoil banks, etc.), the degree of continuity in the operation, etc.

The following table gives an approximate idea of the difference in maximum production possible subject to conditions listed:

Hourly Short Boom Dragline Output in Cubic Yards

Bucket Capacity	3/8	1/2	3/4	1	1 1/4	1 1/2	1 3/4	2	2 1/2
Class of Material									
Moist Loam, Light Sandy Clay	70	95	130	160	195	220	245	265	305
Sand and Gravel	65	90	125	155	185	210	235	255	295
Common Earth	55	75	105	135	165	190	210	230	265
Clay, Hard and Tough	35	55	90	110	135	160	180	195	230
Clay, Wet and Sticky	20	30	55	75	95	110	130	145	175

The dragline is working in the optimum depth of cut for maximum efficiency.

The dragline is working a full 60 minutes each hour--no delays
The dragline is making a 90° swing before unloading.
The bucket loads are being dumped into "properly sized" hauling units.

The proper type bucket is being used for the job.
The dragline is being used within the working radius recommended by the manufacturer for machine stability.

How Degrees of Swing and Depth of Cut Affect Dragline Yardage Output. The following table gives the effect of depth of cut and angle of swing on dragline yardage production. Variations in degrees of swing, in particular, have a marked effect on production and this is important to keep in mind when laying out a job as the shorter the swing the more the yardage.

Table Giving Effect of Depth of Cut and Angle of Swing on Dragline Output

Depth of Cut In % of Optimum	Angle of Swing in Degrees							
	30	45	60	75	90	120	150	180
20	1.06	.99	.94	.90	.87	.81	.75	.70
40	1.17	1.08	1.02	.97	.93	.85	.78	.72
60	1.24	1.13	1.06	1.01	.97	.88	.80	.74
80	1.29	1.17	1.09	1.04	.99	.90	.82	.76
100	1.32	1.19	1.11	1.05	1.00	.91	.83	.77
120	1.29	1.17	1.09	1.03	.985	.90	.82	.76
140	1.25	1.14	1.06	1.00	.96	.88	.81	.75
160	1.20	1.10	1.02	.97	.93	.85	.79	.73
180	1.15	1.05	.98	.94	.90	.82	.76	.71
200	1.10	1.00	.94	.90	.87	.79	.73	.69

Clamshell Production. The clamshell excavator is not to be considered a high production machine but rather a machine to be used where the work is beyond the scope of other types of equipment. For example, a condition which usually requires a clamshell is where digging is vertical or practically straight down as in digging pier holes or shafts. Digging in trenches that are sheathed and cross-braced generally calls for a clamshell, because the vertical action of the bucket enables it to be worked through the cross-bracing. Jobs requiring accurate dumping or disposal of materials are usually clamshell jobs. Also, for high dumping jobs, whether it be charging a bin, building a stockpile, or wherever the material must be dumped well above the machine level, the clamshell is well adapted. In general, the clamshell can operate vertically and dig or spot dump below, at, or above the level of the machine.

The clamshell is only effective where the materials to be handled are relatively soft or loose.

Conditions are so variable on clamshell operation that a table showing typical production has little value.

However, in an effort to establish some point to work from, the following table has been prepared showing maximum production to be expected from clamshell excavators operating under the following conditions:

a) The clamshell is engaged in open digging such as a basement or large footing, with the permissible cut at least a full bucket depth.
b) The depth of the excavation is not more than 10'-0".
c) The quantities are in terms of bank measure.
d) There is no wasted time, 60 minutes of digging each hour.
e) The clamshell is making a 90° swing before unloading.
f) The bucket loads are being dumped into "properly sized" hauling units.
g) The clamshell is being used within the working radius recommended by the manufacturer for machine stability.

Hourly Short Boom Clamshell Handling Capacity in Cubic Yards

Bucket Capacity	3/8	1/2	3/4	1	1 1/4	1 1/2	1 3/4	2	2 1/2
Class of Material									
Moist Loam or Sandy Clay	50	65	95	120	140	155	170	190	225
Sand and Gravel	45	60	85	110	130	140	160	175	205
Common Earth	40	55	70	95	115	125	145	160	185

It must be thoroughly understood that the above production is based on the most ideal job and management conditions.

How Degrees of Swing Affect Clamshell Yardage. The output of a clamshell operating at a steady pace is affected by the swing before unloading in the same manner as a dragline. The following table gives the effect of angle of swing on clamshell yardage production.

Angle of Swing in Degrees

30°	45°	60°	75°	90°	120°	150°	180°
1.32	1.19	1.11	1.05	1.00	.91	.83	.77

Job and Management Factors. Ideal conditions seldom, if ever, exist in the field, and therefore, it is necessary to rationalize these figures by some other factor to compensate for the fact that actual job conditions differ widely from the perfect conditions assumed so far.

There are two sets of factors on every job that have a great deal to do with the output of equipment on the job. They are the job factor and the management factor.

Job factors are the physical conditions pertaining to a specific job that affect the production rate, other than the class of material to be handled. These may be divided into three general headings.

1) topography and the dimensions of the work, which include the depth of cut and whether the work will require much moving within the cut or from one cut to another;
2) surface and weather conditions, which include in some cases the difference between summer and winter work, and also the question of drainage of either surface or underground water;
3) specification requirements that control the manner in which the work must be handled, or indicate the sequence of various operations, and which also show the amount of bank sloping or cutting close to finished line and grade that is required.

The above conditions are all inherent in the job itself and can be taken into account by the contractor in making up the bid.

Another group of factors which affect output are the management factors. These cover conditions which pertain to the efficiency of the operation and thus affect output. These are grouped under the head of the management factor because, in general, they are made up of items which management can determine and control. These include:

1) selection, training, and direction of personnel; the quantities given in the tables indicate what can be reasonably expected by an experienced operator who is willing to work; the availability of trained workers and the incentive to produce must be considered;

2) selection, care, and repair of equipment; the quality of the inspection and preventive maintenance program can have much to do with the lost time on the job;
3) planning, laying out the job, supervising, and coordinating the operations is a very important factor in increasing efficiency; providing the right number and size of hauling units is one of the first things to show its effect.

Dragline Excavator at Work

Foremanship and supervision are important. When everything is working smoothly, there is added incentive for high production rates.

Because both job and management factors must be considered jointly, that is, the output yardages must be modified by both a job and a management factor, the following table consolidates these factors.

To estimate the yardage, select the proper combined job and management factor from the following table. First, select a job factor classification of excellent, good, fair or poor, whichever applies to your job conditions and follow straight across the management factor classification of excellent, good, fair or poor, whichever applies to the management conditions. The resulting figure is the combination job management factor by which the yardage figures from the basic output tables (as modified by the conversion factors for depth of cut and swing) must be multiplied to give the yardage factor that can reasonably be expected under the job management conditions that will exist on the job.

Usually, pit and quarry and heavy construction jobs will fall into groups 1, 2, or 3, whereas highway grading will usually fall into groups 2, 3, or 4 because of the variations in cuts, required machine travel, close cuts, etc.

Combination Job and Management Factors

	Management Factors			
	1	2	3	4
Job Factors	Excellent	Good	Fair	Poor
1. Excellent	.84	.81	.76	.70
2. Good	.78	.75	.71	.65
3. Fair	.72	.69	.65	.60
4. Poor	.63	.61	.57	.52

Power Shovel Excavating

Modern shovels, powered by diesel engines, have tremendous capacity when conditions permit working without interruptions.

The cost of power shovel excavating will vary with the size of the job, size of the shovel, and the speed with which the trucks are handled in and out of the hole. Seldom can the shovel work to capacity on building work, due to the congestion of trucks waiting to get in or out of the hole or due to trucks being delayed at the dump.

The size of the job has considerable bearing on the costs, as it costs just as much to get the shovel to the job and remove it at completion for 1,000 cu. yds. as for a 10,000 cu. yd. job.

Where it is possible to use a ¾-cu. yd. shovel to capacity (figuring a 100% bucket load each cycle), it should excavate 135 cu. yds. an hr. but the bucket is not always filled to capacity and there are delays in loading, moving shovel from place to place, cleaning up, bad weather, etc., so the average on building work will probably run 50 to 60 per cent of the rated capacity of the shovel.

Basis for Computing Daily Output of Diesel Powered Shovel. The following method is used in figuring the output of a power shovel in basement excavation, based on a 1 cu. yd. shovel:

Bucket capacity	1 Cu. Yd.
Bucket efficiency, average material	90 percent
Load carried	0.9 Cu. Yds.
Average operating cycle	20 Seconds
Cycles per hour	180 Cycles
Cu. Yds. per hour, operating 100% time	162 Cu. Yds.

Operating efficiency in basement excavation due to haulage service, machine delays and cleanup work, approximately	50 to 60%
Average output per 8-hr. day	640 to 760 Cu. Yds.
Operating efficiency on heavy construction, roads, dams, open cut excavation, etc., including machine delays and cleanup work, approximately 66 ⅔ percent. Average output per 8-hr. day	864 Cu. Yds.

Dragline Excavating

The cost of dragline excavating will vary in the same manner as shovel excavating, according to the size of the job, size of the machine, and the capacity and speed of the hauling units. But the hauling hazards are reduced substantially by the trucks not having to be loaded in the hole, eliminating the necessity of trucking up a ramp.

As may be determined by comparing the basic output tables for shovels and draglines, the dragline has about 75 to 80 percent of the basic output capacity of shovels. However, the dragline definitely has a place in excavating for construction work by virtue of its generous digging reach, its ability to dig under extremely wet conditions (it can stand on the top on dry, firm footing), and the easier haul with no ramps to climb, all conditions that under certain circumstances would make shovel digging very expensive if not totally impossible.

In excavating for building construction, the dragline is affected by job and management factors the same as shovels and the same basis for computing daily output may be used as follows.

Basis for Computing Daily Output of Dragline Excavator. The following is based on a dragline having a 1 cu. yd. bucket.

Bucket capacity	1 Cu. Yd.
Bucket efficiency, average material	90%
Load carried	0.9 Cu. Yd.
Average operating cycle	24 Seconds
Cycles per hour	150 Cycles
Cu. Yds. Per hour, operating 100% time	135 Cu. Yds.
Operating efficiency in basement excavation due to machine delays, traffic delays to hauling units and clean up work, approximately	50 to 60%
Average output per 8-hr. day	540 to 650 Cu. Yds.
Operating efficiency on heavy construction, roads, dams, open cut excavation, etc., including machine delays and clean up work, approximately	66 ⅔%
Average output per 8-hr. day	720 Cu. Yds.

Power Hoe Excavating

The hoe may be said to be a cross between a dragline and a shovel, because it incorporates some of the characteristics of each and overcomes some of the limitations of each. The hoe is primarily a unit for digging below machine level, but it will dig harder material than the clamshell or dragline, because the weight of the boom itself may be used to force the dipper into the material. It is, however, limited in digging by the length of the boom and stick. The hoe dipper can be controlled more accurately than the dragline bucket and is better suited to close-limit work.

In building construction the hoe is used to dig trenches, footings, and basements. On small residence basements the hoe offers many advantages. It digs straight, vertical side walls (in soil which will stand); it cuts a level floor; it trims corners neatly and squarely; it can dig sewer and waterline trenches; it always works from the top on dry, safe ground; and it reduces hand trim to a minimum.

Clamshell Excavator

In general, the process of digging a basement with a hoe is to dig a trench around the four sides of the basement and scoop out the center as you go.

On small work it is necessary to have low-cost, simple and easy means of moving the shovel from job to job. This is best done by using a single-purpose trailer which can be loaded or unloaded in 15 to 20 minutes. The cost of moving the hoe from job to job will vary with the distance between jobs.

On time studies made on several small basement jobs, where the basements contained from 275 to 350 cu. yds. of excavation, a hoe power excavator equipped with a ¾-cu. yd. dipper averaged from 66 to 88 cu. yds. per hour.

An example of the cost of excavating a small basement containing approximately 350 cu. yds. of excavation, with the soil placed around the excavation, is as follows:

	Hours	Rate	Total	Rate	Total
Move-In Charge		$....	$....	$100.00	$100.00
Excavating					
Hoe operator	5	....	....	25.96	129.80
Labor, clean-up	5	....	....	19.13	95.65
Hoe charge	5	....	....	38.00	190.00
Total Direct Cost			$....		$515.45
Cost per cu. yd			....		1.47

The above costs do not include removal of excavated earth from site or backfilling.

Data on Diesel Power Shovel Costs and Operation

	Size of Shovel (in Cu. Yds.)						
Description	¾	1	1 ½	2	3 ¼	4 ¼	5 ½
Cu. Yds. per Hour Based on 100% Time	135	180	270	360	585	765	950
Cu. Yds. per Hour Based on 100% Time and 90% Dipper Efficiency	121	162	243	324	526	688	855
Output per Hour Based on 66 ⅔% Efficiency, Cu. Yds.	90	120	180	240	390	510	633
Output per Hour Based on 50% Efficiency, Cu. Yds.	68	90	135	180	293	383	475
Output per 8-Hour Day, Based on 66 ⅔% Efficiency, Cu. Yds.	720	960	1440	1920	3120	4080	5064
Output per 8-Hour Day, Based on 50% Efficiency, Cu. Yds.	544	720	1080	1440	2344	3064	3800
Average Cost--Diesel Shovel, Including Freight and Unloading	$91,000	$96,000	$146,000	$310,000	$547,000	$672,000	$955,000
Depreciation, Percent per Year	20	16.66	16.66	16.66	12.50	10	10
Depreciation per Hour Based on 2,000 Hours per Year	$9.10	$8.00	$12.16	$25.82	$34.19	$33.60	$47.75
Interest, Taxes, Insurance, 20% per Year of 2,000 Hours	$5.50	$5.80	$8.80	$18.60	$32.80	$33.60	$47.80
Hourly Consumption of Fuel, Approximate Gals. per Hour	4.5	4.5	5.2	6.4	12.0	15.5	18.0
Fuel Cost per Hour*	$6.75	$6.75	$7.80	$9.60	$18.00	$23.25	$27.00
Engine Lub. Oil Cost, 15% of Fuel	$1.01	$1.01	$1.01	$1.01	$1.01	$1.01	$1.01
Cost per Hour, Repairs, Maintenance and Supplies, Including Rope, Grease,etc	$8.00	$8.00	$8.00	$8.00	$8.00	$8.00	$8.00
Total Hourly Cost, Not Including Labor, Supervision, Compensation, Unemployment, Soc. Sec. etc.	$44.26	$44.56	$48.61	$60.21	$82.81	$88.86	$106.81

*Fuel costs based on diesel fuel at $1.50 per gal.

Below is a digging plan for excavating small basements using a hoe. The order of the letters indicates the order in which the sections are removed. The arrows indicate the direction the hoe travels in taking out each section.

Truck and trailer time to deliver the hoe and removing same at completion of job will vary with length of haul.

In medium to large work and at moderate depths, hoe production can approach that of shovels. Output falls off considerably, however, at greater depths.

Courtesy Caterpillar, Inc.

Power Hoe Being Used to Install Sewer Pipe

For the smaller machines, the size hoe is usually chosen for the width trench it will cut, but for the larger sizes, the capacity of the machine is chosen on a production basis.

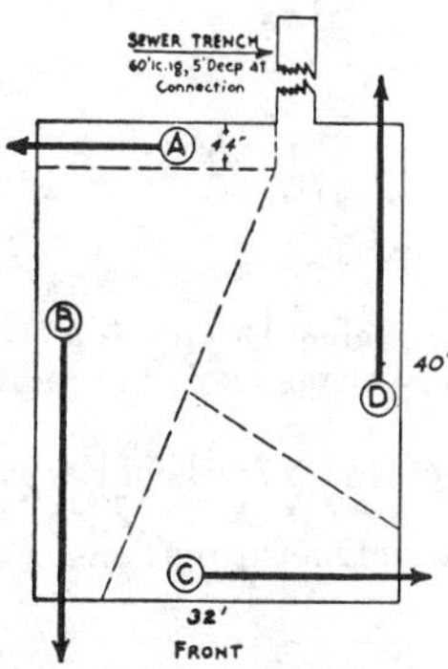

Courtesy Caterpillar, Inc.

Combination Backhoe/Loader

Data on typical hoe bucket dimensions and weight is given in the following table.

Dipper Capacity in Cu. Yds. Struck	Bucket Outside Width in Inches	Total Width Added for 2 Side Cutters in Inches	Weight In Lbs.
3/8	20 to 24	4 to 6	850
1/2	24 to 28	4 to 9	950
3/4	28 to 39	4 to 9	1,400
1	33 to 45	4 to 9	1,700
1 1/4	39 to 45	4 to 9	2,000
1 1/2	39 to 45	4 to 9	2,300
2	47 to 55	4 to 7	3,200

Production for Power Shovel Excavating Using a 3/8-Cu. Yd. Shovel Based on an Average Daily Output of 300 Cu. Yds.

	Number of Cubic Yards of Excavation in Job						
	1,000	1,500	2,000	2,500	3,000	4,000	5,000
	Actual Operating Time of Shovel						
Foreman*	27	40	54	67	80	107	134
Shovel Operator	27	40	54	67	80	107	134

*Foreman in charge of job, handling and expediting trucks, etc.

Production for Power Shovel Excavating Using a ½-Cu. Yd. Shovel

Based on an Average Daily Output of 400 Cu. Yds.

Number of Cubic Yards of Excavation in Job	1,000	1,500	2,000	2,500	3,000	4,000	5,000
Actual Operating Time of Shovel							
Foreman*	20	30	40	50	60	80	100
Shovel Operator	20	30	40	50	60	80	100

*Foreman in charge of job, handling and expediting trucks, etc.

Production for Power Shovel Excavating Using a ¾-Cu. Yd. Shovel

Based on an Average Daily Output of 600 Cu. Yds.

Number of Cubic Yards of Excavation in Job	1,000	1,500	2,000	2,500	3,000	4,000	5,000
Actual Operating Time of Shovel							
Foreman*	14	20	27	34	40	54	68
Shovel Operator	14	20	27	34	40	54	68

*Foreman in charge of job, handling and expediting trucks, etc.

Production for Power Shovel Excavating Using a 1-Cu. Yd. Shovel

Based on an Average Daily Output of 720 Cu. Yds.

Number of Cubic Yards of Excavation in Job	1,000	1,500	2,000	2,500	3,000	4,000	5,000
Actual Operating Time of Shovel							
Foreman*	11	17	22	28	34	45	56
Shovel Operator	11	17	22	28	34	45	56

*Foreman in charge of job, handling and expediting trucks, etc.

Production for Dragline Excavating Using a ½-Cu. Yd. Dragline

Based on an Average Daily Output of 320 Cu. Yds.

Number of Cubic Yards of Excavation in Job	1,000	1,500	2,000	2,500	3,000	4,000	5,000
Actual Operating Time							
Foreman*	25	38	50	63	75	100	125
Dragline Operator	25	38	50	63	75	100	125

*Foreman in charge of job, handling and expediting trucks, etc.

Production for Dragline Excavating Using a 1-Cu. Yd. Dragline

Based on an Average Daily Output of 560 Cu. Yds.

Number of Cubic Yards of Excavation in Job	1,000	1,500	2,000	2,500	3,000	4,000	5,000
Actual Operating Time							
Foreman*	18	27	36	45	54	72	90
Dragline Operator	18	27	36	45	54	72	90

*Foreman in charge of job, handling and expediting trucks, etc.

Sample Excavating Estimate

The following example is based on a basement containing 5,000 cu. yds. excavation, average conditions, using a 1 ¼-cu. yd. power shovel, loading into trucks, based on an average of 900 cu. yds. per 8-hr. day.

Bringing Shovel to Job	Hours	Rate	Total	Rate	Total
Move-In Charge		$....	$....	$250.00	$250.00
Excavating					
Foreman	48	....	....	26.74	1,283.52
Shovel operator	48	....	....	25.96	1,246.08
Shovel charge	48	....	....	46.00	2,208.00
Total Direct Cost			$....		$4,987.60
Cost per cu. yd			....		1.00

Add cost of trucks for removing excavated soil, overhead expense, and profit. The above method for estimating shovel excavation can also be used for hoe, dragline, and clamshell excavation by merely substituting the proper production values and equipment charges for those pertaining to shovels.

Caterpillar Power Shovel Excavating

Hauling Excavated Materials in Trucks

Where an excess of excavated materials over and above the amount required for backfill exists, it will have to be removed from the job by trucks.

The usual procedure followed in these cases is to determine beforehand the approximate quantity of material required for backfill, etc., and this amount subtracted from the estimated excavation quantity will give the volume of excavated material to be hauled away. The excess material should be loaded and hauled as the excavation progresses so that no rehandling is necessary.

Types of Trucks. Truck hauling units are mainly of two types: conventional highway hauling units and off-the-highway heavy duty hauling equipment.

The conventional highway hauling unit is the most frequently used. For the average contractor it is the most economical type of truck to use. It is adaptable to practically every job and may be used for other hauling purposes as well.

For the contractor specializing in large earth moving projects, the high production, heavy duty, off-the-highway hauling unit is the most economical equipment to use, because it is specifically designed for this purpose. Not limited by state highway regulations as to weight and size, it has a high ratio of net weight to payload, the additional weight being found in its more rugged construction, such as heavier frame, axles, body, hoist, and engine.

Hourly Cost of Ownership and Operation of Trucks. To estimate hauling costs, the contractor must know what the equipment costs to own and operate, and this can be determined by the same method previously described for excavating equipment.

Attention is called to the manner in which tire costs (original, replacement, and repair) are handled. Recognizing the fact that the economic life of tires will not be as long as the life of the vehicle, the original value of the tires is deducted from the total price before computing the hourly charge for depreciation and is then treated as an operating cost.

The hourly cost of tire replacement and tire repairs varies greatly according to operating conditions, road surfaces, and tire loadings. While tire costs in ordinary trucking operations normally are figured on a mileage basis, in construction work, with its rough roads, short hauls, and relatively low speeds (below 30 mph), they are figured on an hourly basis and are computed by dividing the replacement cost by the estimated tire life in hours. The average life expectancy of truck tires in construction work is from 3,000 to 3,500 hours. Tire repairs should run about 15 percent of the hourly tire cost.

The following tables give data on various sizes of rear dump trucks for both conventional highway hauling units and heavy duty off-highway equipment.

Table Showing How Hourly Cost of Ownership and Operation May Be Derived for Conventional Highway Rear Dump Trucks Used in Construction Work Based on an Economic Life of 3 Yrs. of 2,000 Hours Each

Capacity in Tons	5	12	15	20	25
Capacity in Cu. Yds., Struck	4	6	10	15	20
Approximate Delivered Price	$30,000	$60,000	$80,000	$90,000	$105,000
Original Tire Value	2,200	3,000	4,000	4,500	5,200
Depreciation, .0167% of Total Investment Less Tires	$5.01	$10.02	$13.36	$15.03	$17.54
Interest, Taxes and Insurance, .0067% of Total Investment	2.01	4.02	5.36	6.03	7.04
Hourly Tire Replacement Cost, Based on 3,000-Hr. Life	.70	.80	1.80	2.00	2.40

Capacity in Tons	5	12	15	20	25
Capacity in Cu. Yds., Struck	4	6	10	15	20
Hourly Cost of Tire Repairs, 15% of Hourly Tire Replacement Cost	.11	.12	.27	.30	.36
Maintenance and Repairs, .004% of Total Investment	1.20	2.40	3.20	3.60	4.20
Fuel Cost per Hour, .04 Gal. per HP	7.20	7.60	11.00	15.00	16.00
Oil and Grease, 15% of Fuel	1.08	1.14	1.65	2.25	2.40
Total Hourly Cost, Not Including Labor, Supervision, Unemployment, Soc. Sec., Overhead Expense & Profit	$17.31	$26.10	$36.64	$44.21	49.93

Table Showing How Hourly Cost of Ownership and Operation
May be Derived for Off-Highway Rear Dump
Trucks Based on an Economic Life of
3 Yrs. of 2,000 Hours Each

Capacity in Tons	15	20	35	45	50	62
Capacity in Cu. Yds., Struck	10	13	22.5	28	33	40
Horsepower	180	250	430	530	570	700
Approximate Delivered Price	$80,000	$100,000	$165,000	$230,000	$250,000	$310,000
Original Tire Value	4,000	8,200	10,500	17,500	21,000	32,000
Depreciation, .01% of Total Investment Less Tires	8.00	10.00	16.50	23.00	25.00	31.00
Interest, Taxes and Insurance, .0067% of Total Investment	5.36	6.70	11.06	15.41	16.78	20.77
Hourly Tire Replacement Cost, Based on 3,000-Hr. Life	1.33	2.73	3.50	5.83	7.00	10.67
Hourly Cost of Tire Repairs, 15% of Hourly Tire Replacement cost	.20	.41	.53	.88	1.05	1.60
Maintenance and Repairs, .004% of Total Investment	3.20	4.00	6.60	9.20	10.00	12.40
Fuel Cost per Hour, .025 Gal. per HP	4.50	6.25	10.75	13.25	14.25	17.50
Oil and Grease, 15% of Fuel	.68	.94	1.61	1.91	2.14	2.63
Total Hourly Cost	$23.27	$31.03	$50.54	$69.56	76.19	96.56

The foregoing tables are for the purpose of showing the items that must be taken into consideration in figuring hourly costs but do not necessarily reflect current prices of equipment, tires or fuel.

Cost of Hauling Excavated Material in Trucks. Hauling costs vary with the capacities of the trucks and time of the hauling cycle.

Truck capacity, for proper sizing, should be at least 4 times the dipper or bucket capacity of the excavator. This is important, because the efficiency of the excavator is seriously affected by undersized hauling units due to the increased hazards of truck delays.

The following table gives theoretical spotting time cycles for various sizes of power shovel excavators working at 100 percent efficiency and loading trucks of 4 times dipper size capacity.

Haul Units Needed to Spot Under Shovel per Hour in Medium Digging

Size Excavator Dipper	Minimum Haul Unit Capacity at 4 Times Dipper Size	Approximate Shovel Cycle in Seconds 90° Swing No Delays	Loading Time for 4-Dipper Truck in Seconds	Time of Spotting Cycle for Steady Operation in Minutes	Number of Spots Req'd per Hour for Steady Operation
3/8	1 1/2 Yd.	19	76	1.26	48
1/2	2 Yd.	19	76	1.26	48
3/4	3 Yd.	20	80	1.33	45
1	4 Yd.	21	84	1.40	43
1 1/4	5 Yd.	21	84	1.40	43
1 1/2	6 Yd.	23	92	1.53	39
2	8 Yd.	25	100	1.66	36
2 1/2	10 Yd.	26	104	1.73	35

For dragline operation the above time values may be increased from 20 to 30 percent.

The hauling time cycle consists of several operations--spotting of truck under excavator; loading truck by excavator; traveling to dumping area; dumping; returning to excavator.

Some of these operations, such as spotting, loading and dumping, can be minimized by good management and supervision. Traveling time, however, will vary with the haul distance, traffic conditions, road conditions, etc., and must be carefully analyzed for each job.

When the hauling time cycle has been estimated, the number of loads each truck is able to haul per day can be determined and multiplying this by the capacity of the truck will give the theoretical daily haul of each truck in cu. yds., assuming 100 percent job efficiency. This amount should be multiplied by the expected job efficiency factor to obtain the net yardage hauled. Divide the daily truck cost by the net yardage hauled and the result will be the hauling cost per cu. yd.

The number of trucks required to keep the excavator going can also be determined by dividing the hauling time cycle by the loading time cycle using 100 percent excavator efficiency as a basis. This is done to be sure that a hauling unit is under the excavator at all times. While an excavating job may actually work out to a 50 or 60 percent efficiency, there usually are times on the job when the excavator is working at full capacity and to take advantage of this, sufficient, properly sized hauling units must be on the job. In large operations many contractors maintain standby units to step in when one of the hauling fleet trucks breaks down or is otherwise unusually delayed.

In sizing trucks to conform to dipper or bucket capacities, the struck capacity of each is used, assuming that the heaping of loads on both trucks and dippers or buckets will offset each other and produce bank measure quantities.

However, all estimating should use actual job data rather than theoretical figures. Some materials do not follow the usual swell ratios and job efficiency ratings vary widely. Following book figures to the letter in excavating, or any other branch of construction work, can be disastrous. Book values, at best, can only give what can be done under a definite set of conditions. The contractor must exercise his judgment in estimating this work and base his figure on his own particular set of conditions, i.e., equipment, supervision, operators, local materials and local weather.

Courtesy Caterpillar, Inc.

Tandem Powered Wheel Tractor-Scraper

Tractors and Scrapers

The tractor is probably the most used piece of earth moving equipment on the construction job. It can perform a wide variety of operations, and equipped with the proper accessories, it is a high production and efficient tool for such jobs as site clearing, grading, shallow excavating, etc. In addition the tractor is often called upon to act as the "strong man" in freeing stuck trucks, hauling or dragging equipment, sheds, etc., into place, miscellaneous hoisting using a snatch block and cable, etc. In general, the tractor is almost indispensable on the job.

Equipped with bulldozer apparatus, the tractor, either rubber-tired or crawler type, has demonstrated its ability to do a wider variety of jobs than any other earth moving tool.

The bulldozer usually is employed to open up the job, which may consist of anything from clearing brush and vegetation to the removal of average size trees (up to 30 inches in diameter), stumps, rocks, paving, old floor slabs, old foundations, etc.

Stripping and stockpiling top soil on small quantity jobs with a haul distance from 200 to 400 feet or less is most efficiently done by bulldozer. On larger jobs this is done more economically by scrapers.

On small quantity, shallow excavation jobs with a short haul, tractor dozers are moderately efficient.

Crawler type bulldozers are reasonably economical on pushing distances up to 200 feet. Under the same conditions the rubber-tired dozer's economical push distance is about 400 feet.

Most backfilling jobs fall to the lot of the bulldozer where they out-perform any other piece of equipment.

Tractors also perform a valuable service when they are called upon to provide the additional power for push loading scrapers. On a production scraper job, dozers helping to load these units more than pay for themselves.

On residence and other small jobs, a loader usually can dig the basement and do all exterior grading cuts and fills with maximum efficiency.

Another application of the bulldozer is spreading fill hauled in and dumped by trucks or cast by other excavating equipment.

Tractors are also the ideal power source for hauling sheepsfoot rollers and pneumatic rollers used in compacting fills.

Scraper Application. Scrapers are high production earth moving units which are capable of digging their own load, hauling the load and then spreading the load in controlled layers. Furthermore, the scraper is a precision tool, as a skilled operator can cut a grade to within 0.1 ft. and can spread fill to the same degree of accuracy.

Scrapers may be powered by three types of prime movers: two wheeled or four wheeled rubber-tired units or crawler type tractors.

Scrapers, as a rule, are not used to any great extent in residence building or on any work in congested city areas but for jobs such as suburban and rural schools, hospitals, factories, housing projects, in fact any construction work with a medium to large site grading problem, scrapers are probably the most efficient pieces of excavating equipment for the job.

The scrapers can be used to strip and stockpile top soil, after the site has been cleared, and reclaim the top soil from the stockpile and spread it where required at the end of the job. Many times basement excavation may be carried on in conjunction with site grading cuts and fills by scrapers.

The basic efficiency of scrapers is relatively unaffected by depth of cut, length of haul or type of soil. Quantities being sufficient to warrant the use of scraper equipment, crawler tractor drawn scrapers are economical at haul distances up to 500 feet and with rubber-tired, high speed prime movers can compete with trucks on longer hauls.

Capacities of scrapers range from about 7 cu. yds. struck measure to approximately 40 cu. yds. heaped measure.

Scrapers powered by crawler tractors are capable of self-loading in most soils but generally it is more economical to use pusher assistance from a tractor to obtain heaped loads in the shortest time. With few exceptions, pusher assistance is essential for self-propelled, rubber-tired scrapers. Pusher loading is generally accomplished in an average of about 1 minute in 100 ft. of travel. To determine the number of scrapers one pusher can handle, divide the scraper cycle by the pusher cycle.

Estimating the Cost of Owning and Operating Tractors and Scrapers. In order to properly prepare an estimate on work to be done by tractors and/or scrapers, the contractor must know what to charge for the use of the equipment. He should work out an hourly charge for the equipment, to be used in pricing his estimates, so that the cost of ownership and operation is carried proportionately by each job and does not have to come out of profit.

Hourly Cost of Ownership. The cost of ownership has as its basis the total delivered cost of the piece of equipment and is composed of several items, namely, depreciation, interest, insurance and taxes.

In figuring the hourly depreciation cost, most contractors usually take the full delivered price of the machine, less cost of tires, and divide by the expected economic life of the machine in hours. The life expectancy figure commonly accepted is 5 years of 2,000 hours each or 10,000 hours.

Interest, insurance, taxes, and storage are computed on the basis of average investment. The average investment in a piece of equipment which is to be fully depreciated in 5 years is 60 percent of the full delivered cost, as is shown on previous pages under "Power Shovels and Cranes". The assumed rate for interest, insurance, taxes, and storage is a total of 20 percent, which when applied to the average investment and then divided by 2,000 hours, amounts to .006 percent of the full delivered cost to give the hourly cost of interest, insurance, taxes, and storage.

Total hourly ownership cost is then obtained by totaling the hourly cost of depreciation and the hourly cost of interest, insurance, taxes, and storage.

Hourly Cost of Operation. This cost includes tire replacement, tire repairs, fuel, lubrication, mechanical repairs, maintenance of blades and cable, etc.

The hourly costs of tire replacements and tire repairs are computed in the same manner as previously described, under "Hourly Cost of Ownership and Operation of Trucks", and are based on a tire life of 3,000 hours of operation with repair costs assumed to be 15 percent of the hourly tire cost.

Fuel costs may be figured on the following basis: Assuming an average machine in good repair, operating under average conditions, diesel fuel consumption of 0.04 gallons per horsepower per hour is used.

Experience has shown that lubricants and lubricating labor amount to approximately 50 percent of the fuel costs per hour.

Cost of repairs and maintenance, including major overhauls at the end of the working season, should be determined by the contractor from his own experience record.

Many contractors find that total repairs, maintenance and overhauls for the economic life of the machine runs from 70 percent of the full delivered price for the larger machines to 90 percent for the smaller machines. This percentage divided by the economic life of the machine in hours, 10,000, will give the hourly cost of mechanical repairs, maintenance, etc., and amounts from .007 percent to .009 percent of the total delivered cost.

The hourly cost of maintenance of blades and cables is figured separately from the cost of general maintenance and repair as this item will vary depending on the soil conditions encountered. This cost may be estimated by dividing the replacement cost of blades and cables by their expected life in hours. Under average conditions this life may be expected to be about 450 hours; under favorable conditions, such as digging soft loam, etc., this life may be increased to about 800 hours; under unfavorable conditions, where sharp rocks and shale are being loaded, the expected life should be reduced to about 250 hours.

Approximate Prices of Crawler Type Tractors with Bulldozer Equipment

Horsepower	Weight in Lbs.	Approx. Price
50	9,800	$36,000
65	14,600	54,000
75	19,500	72,000
105	30,000	100,000
200	45,900	200,000
400	90,000	291,000
460	112,800	400,000
520	122,000	490,000
770	200,000	1,800,000

Approximate Prices of Crawler Type Loaders

Capacity in Cu. Yds.	Horsepower	Approx. Price
½	75	$95,000
1	150	160,000
1 ½	180	168,000
2 ¼	260	250,000
3	300	500,000
4 ½	390	550,000

Approximate Prices of Self-Propelled Scrapers

Capacity in Cu. Yds. Struck	Heaped	Horsepower	Approx. Price	Orig. Tire Value
14	20	330	277,000.00	30,000.00
21	31	450	510,000.00	42,000.00
28	38	550	450,000.00	42,000.00
32	44	550	700,000.00	52,000.00

Approximate Prices of Tractor-Elevating Scrapers

Capacity, cu. yds.	Horsepower	Approx. Price F.O.B. Factory	Orig. Tire Value
11	150	$120,000.00	$12,000.00
16	265	250,000.00	25,000.00
22	330	290,000.00	30,000.00
34	450	530,000.00	35,000.00

Ownership and Operation of Bulldozers, Loaders, and Scrapers. The following data is the hourly cost of ownership and operation for various sizes of tractor and scraper equipment and is based on preceding price and horsepower data. It does not include operator's wages, supervision, compensation, and liability insurance or Social Security taxes.

Hourly Cost of Crawler Type Bulldozers

Horsepower	50	65	75	105	200	400	460
Approx. Delv'd. Cost	$36,000	$54,000	$72,000	$100,000	$200,000	$291,000	$400,000
Depreciation, .01%	3.60	5.40	7.20	10.00	20.00	29.10	40.00
Interest, Insurance and Taxes. .006%	2.16	3.24	4.32	6.00	12.00	17.46	24.00
Fuel, .04 Gal. per HP	3.00	3.90	4.50	6.30	12.00	16.00	18.40
Lubrication, Including Labor, 50% of Fuel	1.50	1.95	2.25	3.15	6.00	8.00	9.20
Mech. Repairs and Maintenance, .007%	2.52	3.78	5.04	7.00	14.00	20.37	28.00
Blade and Cable Replacement	.45	.47	.49	.51	.55	.60	1.00
Total Hourly Cost	$13.23	$18.74	$23.80	$32.96	$64.55	91.53	120.60

Hourly Cost of Crawler Type Loaders

Bucket Capacity, Cu. yds.	1	1 ½	2 ¼	3	4 1/2
Horsepower	150	180	260	300	390
Approx. Delv'd Cost	$160,000	$168,000	$250,000	$500,000	$550,000
Depreciation, .01%	16.00	16.80	25.00	50.00	55.00
Interest, Insurance and Taxes, .006%	9.60	10.08	15.00	30.00	33.00
Fuel, .04 Gal. per hp*	9.23	10.80	15.60	18.00	23.40
Lubrication, Including Labor, 50% of Fuel	4.61	5.40	7.80	9.00	11.70
Mech. Repairs and Maintenance, .007%	11.20	11.76	17.50	35.00	38.50
Blade Replacement	.80	.84	.88	.96	1.00
Total Hourly Cost**	$51.44	$55.68	$81.78	$142.96	$162.60

Hourly Cost of Self-propelled Scrapers

Capacity, Struck in Cu. Yds	14	21	28	32
Capacity, Heaped in Cu. Yds	20	31	38	44
Horsepower	330	450	500	550
Approx. Delv'd. Cost	$277,000	$510,000	$550,000	$700,000
Original Tire Value	27,000	40,000	40,000	45,000
Depreciation, .01% of Total Investment Less Tires	27.70	51.00	55.00	70.00
Interest, Insurance and Taxes, .006% of Total Investment	16.62	30.60	33.00	42.00
Tire Replacement, 3,000-Hr. Life	9.00	13.33	13.33	15.00
Tire Repairs, 15%	1.35	2.20	2.20	2.25
Fuel, .04 Gal. per hp*	13.20	18.00	22.00	22.00
Lubrication, Incl. Labor, 50% of fuel	6.60	9.00	11.00	11.00
Mech. Repairs and Maintenance, .007% of Total Investment	19.39	35.70	36.00	49.00
Blade and Cable Replacement	.70	.80	1.00	1.00
Total Hourly Cost**	$94.56	$160.63	$173.53	$215.25

*Based on $1.50 per gallon for diesel fuel.
**Not including Labor, Supervision, Unemployment, Soc. Sec., Overhead Expense

Bulldozer Production

In an average day's work, the bulldozer can be called upon to do several operations. Time will be lost in moving from one task to another. If this time loss is not considered in estimating bulldozer work, the job might not be profitable. Inasmuch as each job has its own peculiar problems, no cost table can reflect the time losses, and the estimator must analyze each job carefully and adjust the production rates to conform to the conditions present. The following table gives the approximate hourly production to be expected from various sized bulldozers and is based on good weather conditions, efficient and well-maintained machinery, and steady operation. It has been established that a well organized and supervised job will result in approximately 50 work-productive minutes each hour, a job efficiency of 83 percent.

Approximate Hourly Production of Bulldozers
Based on a Job Efficiency of 83 Percent

	Horsepower				
Description of Work	50	65	75	105	200
Simple clearing, vegetation, light brush, tree saplings, in sq. ft. per hour.	900 to 1,000	1,000 to 1,100	1,150 to 1,250	1,250 to 1,350	1,450 to 1,550
Moderately difficult clearing thick brush and tree saplings, in sq. ft. per hr Add extra for trees.	300 to 350	350 to 400	400 to 450	600 to 650	700 to 750
Strip top soil or shallow excavation of ordinary soil and placing in piles on premises, in cu. yds. per hr	15 to 20	17 to 22	20 to 25	22 to 27	27 to 32
Backfill ordinary soil placed slow enough to permit compacting, in cu. yds. per hr	33 to 38	37 to 42	43 to 48	50 to 55	60 to 65
Spread material dumped in piles by trucks or cast by other excavating equipment, in cu. yds. per hr	42 to 47	45 to 50	55 to 60	60 to 65	70 to 75

Caterpillar Dozer, D4 Series

Scraper Production

Relative scraper production is mainly governed by the cycle time of operation; that is, the length of time a scraper will take to load, haul, spread-dump, and return to start over again. In properly organized jobs with good, adequately-powered equipment, a portion of this cycle is usually termed the fixed time, because it is relatively unaffected by the length and grades of the haul and consists of the following: loading, approximately 1.0 to 1.5 minutes, spread-dumping, about 0.5 minutes, two turns of 0.25 minutes each, and shifting and acceleration of 0.5 to 1.0 minute or a total fixed time of 2.5 to 3.5 minutes. The other part of the cycle is the actual time of hauling the load to the fill area plus the time of returning empty to the beginning of a new cut. This traveling time is the big variable in the operation. It is affected by length of haul, road conditions, and grades to be negotiated. It must be estimated accurately. Scrapers powered by rubber-tired tractors can reach a speed of 30 m.p.h. when empty and on a good level road, but this will be reduced to 10 to 20 m.p.h. when loaded. It must also be remembered that adverse grades greatly reduce the speed, and favorable grades increase the speed. The traveling time plus the fixed time is the cycle time of operation.

Yardage production is figured in terms of bank measure. When estimating scraper production, the struck capacity of the machine is normally used as the pay yardage volume, anticipating the scraper will be heap loaded. The difference between struck and heaped capacity is generally assumed to be the amount the material will swell in being disturbed.

The following table gives the daily scraper production for various sized machines operating at several different time cycles and is based on ordinary soil conditions, good weather, good equipment, and an 83 percent job efficiency.

Daily Scraper Production for Various Operation Cycles
Based on a 50-Minute Hour and an 8-Hour Day

Operation Cycle in Minutes	Struck Capacity of Scraper in Cu. Yds. 7	14	24
3	930	1860	3200
4	700	1400	2400
5	560	1120	1920
6	465	930	1600
7	400	800	1370
8	350	700	1200
9	310	620	1070
10	280	560	960

Example Estimate for Scraper Excavation

Based on a site grading cut containing 8,000 cu. yds., excavated material to be spread-dumped in low areas requiring fill, good conditions permitting a 50-minute hour, an operation cycle of 6 minutes and using two 14 cu. yd., struck measure, self-propelled scrapers.

Bringing Scrapers to Job	Hours	Rate	Total	Rate	Total
Move-in charge 2 ea		$....	$....	$250.00	$500.00
Excavating					
Foreman	35	....	....	26.74	935.90
Scraper operators	70	....	....	25.96	1,817.20
Scraper Hours	70	....	....	94.56	6,619.20
Total Direct Cost			$....		$9,872.30
Cost per cu. yd			....		1.23

If a 105-Hp bulldozer is needed to give pusher assistance in loading, to maintain the operation cycle time of 6 minutes, the additional cost per cu. yd. would be as follows:

	Hours	Rate	Total	Rate	Total
Move-in charge		$....	$....	$250.00	$250.00
Excavating					
Bulldozer operator	35	....	....	25.96	908.60
Bulldozer charge	35	....	....	32.96	1,153.60
Additional Direct Cost			$....		$2,312.20
Additional cost per cu. yd.			....		.29

Trenching and Ditching Machine Excavation

Designed strictly for the purpose of cutting ditches, the trenching machine is the fastest earthmover for its weight and horsepower. Before a contractor can reap the benefits of this potential output, he must select a trenching machine of a type and size that fits the work to be done, and even more important, must put an operator on the machine who has both skill and experience.

There are two basic types of machine, the wheel type and the ladder type. The wheel type is generally considered to be the fastest and is by far the most common type in use today. Each type has advantages and disadvantages for performing certain classes of work.

Because most trenching or ditching operations in building construction will be foundations, service trenches, etc., the following data covers the cost and operation of a general purpose, crawler-mounted, wheel machine of the type and size generally used for this work.

This machine is equipped with shifting-tilting boom that permits it to dig close to obstructions, such as foundations or curbs, and to dig a vertical trench even when the machine is on a slope.

The cost of excavating with a trenching or ditching machine will vary with the size of the job, width and depth of trenches, kind of soil, accessibility, etc. Most machines are designed to dig certain standard width trenches, so that trenches a trifle narrower or wider will not make any appreciable difference in either output or costs.

While the following table gives the rated capacity of the machine, it will be necessary to allow for factors such as time lost moving machine and bad weather. It is not advisable to figure rated capacity as the average daily output of any machine.

Data on Crawler Mounted Ditching Machine
Operation and Cost

Wheel Size	Possible Cutting Widths	Maximum Depth Trench	Weight in Pounds	Approx. Cost FOB Factory
16"x11'-6"	16"-24"	11'-0"	27,000	$130,000
24"x12'-4"	24"-30"	15'-6"	43,000	$170,000

Maximum digging speed is 34 f.p.m. Forward and reverse travel speeds are 0.46, 0.95, 1.7, and 2.9 m.p.h. Digging speeds are controlled by a hydraulic transmission, permitting speed to be varied to match soil conditions. Wheel speeds vary from 0 to 9.2 r.p.m.

Crawler Mounted Ditching Machine Operation and Cost

When figuring operating cost on a ditching machine, include fuel and oil consumption, depreciation, interest on investment, repairs, replacements, and taxes.

Digging Width Description	12"	24"
Approximate Delivered Price	$130,000	$170,000
Depreciation per year, 18%	23,400	30,600
Interest 15% per year	19,500	25,500
Insurance and Taxes per year, 5%	6,500	8,500
Repairs and Replacement per year, 10%	13,000	17,000
Hourly Consumption of Fuel, Gals.	5	6
Daily Cost of Fuel, Oil, Grease, $1.50 gal. + 15%	69.00	82.80
Operating Cost per Mile (Based on 75 Mi. per Year, 200 Working Days per Year, Avg. Trench Depth, 5'-0")		
Depreciation	312.00	408.00
Interest	260.00	340.00
Taxes and Insurance	86.67	113.34
Repairs and Replacements	173.34	226.67
Fuel, Oil and Grease	184.00	220.80
Total machine cost per mile	$1,016.01	$1,308.81
Machine cost per lin. ft. trench	.193	.248

Cost of Digging 100 Lin. Ft. of Trench 5'-0" Deep,
Using a Crawler Mounted Ditching Machine,
Based on 350 Lin. Ft. per Hour

	Hours	Rate	Total	Rate	Total
Machine operator	0.29	$....	$....	$25.96	$7.53
Machine charge, 12" wide		....	....	.193	19.30
Machine charge, 24" wide		....	....	.248	24.80
Cost per 100 lin. ft., 12" wide					26.83
Cost per 100 lin. ft., 24" wide					32.33
Cost per lin. ft., 12" wide					.27
Cost per lin. ft., 24" wide					.32

To the above, add cost of trucking machine to job and removing same at completion, supervision, compensation and liability insurance, social security and unemployment taxes, overhead expense and profit. Add for backfilling trenches, if required.

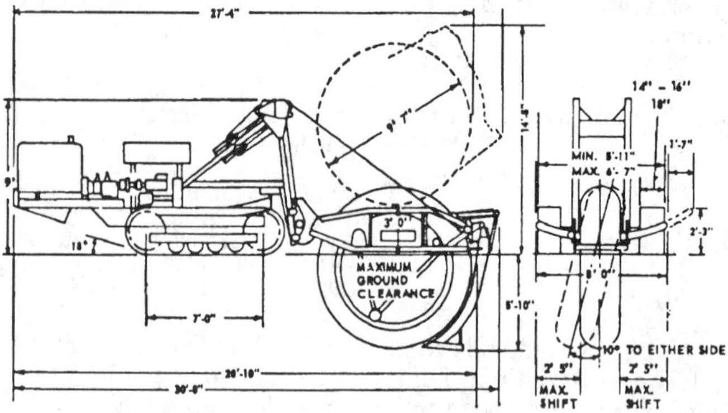

Dimensions of Crawler Mounted Wheel Ditcher

Volume of Trench Excavation
Cu. Yd. per 100 Lin. Ft.

Trench depth, in.	Trench width, in. 12	18	24	30	36	42	48
12	3.7	5.6	7.4	9.3	11.1	13.0	14.8
18	5.6	8.3	11.1	13.9	16.7	19.4	22.3
24	7.4	11.1	14.8	18.5	22.2	26.0	29.6
30	9.3	13.8	18.5	23.2	27.8	32.4	37.0
36	11.1	16.6	22.2	27.8	33.3	38.9	44.5
42	13.0	19.4	30.0	32.4	38.9	45.4	52.0
48	14.8	22.2	29.6	37.0	44.5	52.0	59.2
54	16.7	25.0	33.3	41.6	50.0	58.4	66.7
60	18.6	27.8	37.0	46.3	55.5	64.9	74.1
66	...	30.5	40.7	51.0	61.0	71.3	81.6
72	...	33.3	44.5	55.5	66.7	77.9	89.0
78	...	36.0	48.1	60.2	72.2	84.2	96.5
84	...	38.9	51.9	64.8	77.8	90.8	104.0
90	...	41.6	55.6	69.5	83.4	97.3	111.4
96	...	44.4	59.2	74.0	88.9	102.0	118.6

Drilling Holes Using an Earth Boring Machine. Where a large number of holes are required for line construction, foundation work, pre-boring, guard rails, piling, fencing, draining, or bridges, an earth boring machine will perform the work much faster and at lower cost than most other methods.

These machines will drill holes through loam, clay, hard pan, and shale and are furnished in various sizes by different manufacturers. Some machines will drill up to 8'-0" diameter holes to a depth of 120 ft. or more.

When drilling holes up to 18" in diameter in ordinary soil, a machine will drill at the rate of 1" depth in 1 ½ to 2 seconds, or drill a 3'-0" deep hole in 1 to 1 ¼ minutes. Moving the machine and spotting the holes takes some time, so this will govern the number of holes that may be drilled per day. A machine that will drill holes to 24" diameter and 9'-0" deep costs approximately $100,000.00 f.o.b. factory. This does not include cost of truck, which will have to be added. The larger machines which are used for drilling caisson foundations may cost as much as $300,000.00.

Based on 20 holes, 18" diam. and 3'-0" deep, an hour or 160 holes per 8-hr. day, the daily cost should average as follows:

	Hours	Rate	Total	Rate	Total
Operator	8	$....	$....	$25.96	$207.68
Machine charge	8	....	....	20.00	160.00
Truck charge	8	....	....	18.00	144.00
Total cost 160 holes			$....		$511.68
Cost per 3'-0" hole					3.20
Cost per lin. ft.					1.07

There are also small portable posthole diggers, mounted on two rubber tired wheels and equipped with a 5 to 6 Hp gasoline engine.

This machine can be operated by one man and in ordinary ground it will bore a 9" hole 3'-0" deep in 1 to 1 ½ minutes. Such machine costs $2000.00 to $3000.00.

Placing And Compacting Fills

Architects and engineers have become more rigid in their requirements as to the manner of placing and compacting fill materials. The usual specification for this work is to place the fill in from 6 to 12 inch layers and to compact each layer with some sort of roller equipment.

Placing and Compacting Site Grading Fills. Fill material in site grading work is placed either by scrapers spread-dumping their loads, trucks dumping their loads more or less in piles, or some sort of excavating equipment casting dipper or bucket loads as they are dug. In the case of the scraper spread-dumping fill material, no further spreading is necessary, because it can be controlled with precision by the scraper operator. In the other cases, however, a bulldozer will be required to spread out the fill material into layers of specified depth. Hourly production for this work can be found in the table under "Bulldozer Production".

When the fill has been placed and spread in the specified depth of layer, the compacting operation follows, and this is usually accomplished by rolling with a tractor-drawn sheepsfoot roller. The degree of compaction may be specified by the number of passes to be made by the roller or by a percentage, such as 95%, of maximum density of the material obtained at optimum moisture content in a soil mechanics laboratory. The latter specification is usually called for in government work and for fills under paving. Obtaining 95% of maximum density compaction may take as many as 12 passes of a sheepsfoot roller.

If rolling is not to interfere with production schedules, a rate must be established in cubic yards compacted per hour and applied to the rate of placing and spreading the fill material to determine the size and number of rollers required to keep the job moving. Where frequent turns are made or restricted areas require maneuvering these factors must be considered and the figures adjusted accordingly.

The following table gives rate of compaction, in cu. yds. per hr., for rolling fill with a 5'-0" wide sheepsfoot roller at 2.5 m.p.h. with the fill material placed in 12" layers for soil of various compaction factors and for 1 pass of the roller to 12 passes.

Rate of Sheepsfoot Roller Compaction in Cu. Yds. per Hr.
Based on 5'-0" Wide Roller, 2 ½ mph Speed,
12" Fill Layers and 100% Job Efficiency

Number of Passes of Roller	Percentage Factor of Pay Yd. to Loose Yd.		
	Hard Tough Clay 70%	Medium Clay 80%	Loam 90%
1	1,711	1,956	2,200
2	856	978	1,100
3	570	652	733
4	428	489	550
5	342	391	440
6	285	326	367
7	244	279	314
8	214	245	275
9	190	217	244
10	171	196	220
11	156	178	200
12	143	163	183

For rollers of different widths, adjust the above rates proportionately; for example, for a 10'-0" wide roller double the rates, and for a 15'-0" wide roller triple the rates.

For different depths of layers, adjust the above rates proportionately: i.e., for 6" layers reduce the rates to ½, and for 9" layers reduce the rates to ¾.

Courtesy Caterpillar, Inc.
After Fill has been Placed Self-powered Compactor is Used to Tamp Earth

The above rates must also be adjusted to normal job efficiency and further adjusted to take care of lost time in maneuvering and turning. In small jobs as much as 10 percent of the time could be lost.

Example: Assume a fill to be placed in 9" lifts and rolled 8 passes with a double drum sheepsfoot roller 10'-0" wide at 2.5 m.p.h. in soil with a compaction factor of 80%.

From the table, 8 passes in 80% material will compact 245 cu. yds. per hr. Adjusting this rate for a 10'-0" wide roller gives a rate of 490 cu. yds. per hr. and further adjusting for 9" fill layers results in a compaction rate of 368 cu. yds. per hr. Assuming that a 50-minute hour or a job efficiency of 83% is expected, the rate is reduced to 305 cu. yds. per hr. and with a time loss of 10% anticipated for turns and maneuvering the final result is a compaction rate of 275 cu. yds. per hr.

The approximate cost for compacting fill at rate of 275 cu. yds. per hr. is as follows :

	Hours	Rate	Total	Rate	Total
Hourly charge for 75-Hp. bulldozer with double drum sheepsfoot roller	1	$....	$....	$23.80	$23.80*
Bulldozer operator	1	....	....	25.96	25.96
Cost of Compacting 275 Cu. Yds			$....		$49.76
Cost per cu. yd			....		.18

*Does not include hauling equipment to and from job, compensation and liability insurance, social security taxes or overhead and profit.

The sheepsfoot roller method of compaction is applicable to all soils with cohesive qualities, such as loams and clays, and in very dry soils it may be necessary to add moisture to the material to obtain the correct degree of plasticity.

With non-cohesive materials, such as sands and gravels, the sheepsfoot roller usually is ineffective and rubber-tired pneumatic rollers are used instead.

Placing and Compacting Fills for Floors on Grade Inside Buildings. Fill material, necessary to raise the elevation of the subgrade for floors inside buildings, can be placed in a number of ways.

For small jobs this is usually done after the building is enclosed. Fill is brought in through door and window openings, and it is spread and tamped by hand. Three laborers working together should spread and tamp 35 to 40 cu. yds. of fill per 8-hr. day at the following cost per cu. yd.:

	Hours	Rate	Total	Rate	Total
Labor	0.6	$....	$....	$19.13	$11.48

On larger jobs it may be possible to truck the fill material directly into the space and spread-dump it, using a small bulldozer to further spread into even layers. Compaction of material in this case can usually be accomplished by using a single drum sheepsfoot roller or small pneumatic roller for the open areas and by hand tamping at walls and columns. Trucking traffic over the various layers of fill is a big help in compacting the material. The roller can be drawn by the same bulldozer that spreads the material and with 2 laborers spreading and tamping at "hard-to-get" places a daily production of 95 to 100 cu. yds. placed and compacted should be realized at the following cost per cu. yd.:

	Hours	Rate	Total	Rate	Total
50-Hp. bulldozer and single drum sheepsfoot roller	0.08	$....	$....	$13.23	$1.06
Bulldozer operator	0.08	....	....	25.96	2.08
Labor	0.16	....	....	19.13	3.06
Cost per Cu. Yd.			$....		$6.20

Does not include hauling equipment to and from job, compensation and liability insurance, social security taxes, or overhead and profit.

Grading

Grading consists of dressing up ground surfaces, either by hand or machine, to conform to specified contours or elevations and is usually preparatory to a subsequent operation such as placing a floor slab, walk or drive, or spreading top soil for planting areas.

Grading costs vary according to the degree of accuracy demanded and the method by which it is done. Following are approximate costs on some of the various types of grading usually encountered in construction work.

Rough Grading. After all backfill is in place and the site has been cut and/or filled to the approximate specified contours, rough grading of the site is done, by hand for small jobs and by machine for large areas, usually preparatory to the spreading of top soil. Tolerance for this type of grading is usually plus or minus 0.1 ft.

On small jobs a laborer should rough grade 800 sq. ft. of ground surface per 8-hr. day at the following cost per 100 sq. ft.:

	Hours	Rate	Total	Rate	Total
Labor	1.0	$....	$....	$19.13	$19.13
Cost Per Sq. ft.			....		.19

On large jobs a bulldozer is usually employed to do rough grading, and a laborer generally accompanies the machine to check the surface and direct the operation. A 75-Hp bulldozer should rough grade 6,500 to 7,500 sq. ft. per 8-hr. day at the following cost per 1,000 sq. ft.:

	Hours	Rate	Total	Rate	Total
75-Hp bulldozer charge	1.15	$....	$....	$23.80	$27.37
Bulldozer operator	1.15	....	....	25.96	29.85
Labor	1.15	....	....	19.13	22.00
Cost per 1,000 sq. ft.			$....		$79.22
Cost per sq. ft.			....		.08

Does not include hauling equipment to and from job, compensation and liability insurance, social security taxes, or overhead and profit.

Grading for Slabs on Ground. Grading for slabs on ground such as floors, walks, and driveways usually is done by hand unless the job is quite large and can be organized so that some of the work is done by machine. This work generally must be done accurately with tolerances of no more than ½". A laborer should grade 500 to 600 sq. ft. per 8-hr. day at the following cost per 100 sq. ft.:

	Hours	Rate	Total	Rate	Total
Labor	1.5	$....	$....	$19.13	$28.70
Cost per sq. ft.			....		.29

For sloping surfaces, add about 50% to the above cost, depending on steepness of pitch.

Finish Grading of Top Soil. After top soil has been spread over areas specified, a finish grading operation, usually including hand-raking, must be performed prior to seeding or sodding. The tolerance on this work usually is plus or minus 1 inch. A laborer should finish grade and hand-rake 600 to 700 sq. ft. per 8-hr. day at the following cost per 100 sq. ft.:

	Hours	Rate	Total	Rate	Total
Labor	1.25	$....	$....	$19.13	$23.91
Cost per sq. ft.			....		.24

For sloping surfaces, such as berms and ditch sides, add 50% to the above cost.

Grading for Footing Bottoms. When footing pits and trenches are dug by machine, there is always cleanup, squaring, and grading work to be done by hand. Many contractors price this work by the sq. ft. of footing bottom. A laborer should clean up, square, and grade 200 sq. ft. of footing bottom per 8-hr. day at the following cost per 100 sq. ft.:

	Hours	Rate	Total	Rate	Total
Labor	4.0	$....	$....	$19.13	$76.52
Cost per sq. ft.			....		.77

Costs Of Excavating

Digging Fence Post Holes. When digging 250 post holes about 3'-0" deep in black soil and sand, using an ordinary augur post-hole digger, a man will dig 4 holes an hour or 32 per 8-hr. day, at the following labor cost per hole:

	Hours	Rate	Total	Rate	Total
Labor	0.25	$....	$....	$19.13	$4.78

Loader Excavating under Favorable Conditions. The following costs are for a theoretical job performed where conditions are favorable, with no time lost because of bad weather or breakdowns.

This is a basement excavation 80'-0"x80'-0" and 13'-0" deep. The top 5'-0" consists of loose clay and loam, while the lower 8'-0" is blue grey dolomite limestone in beds varying from 6" to 24".

A 1 ½-cu. yd. loader is used, rented at $75.00 per hr. including operator and fuel.

Five 6-cu. yd. trucks transport the excavated earth and rock to a dump about ¼-mile from the job. Trucks are estimated to cost $50.00 per hr. including gasoline and driver.

A 5 ½"x5" portable air compressor, capacity 900 cu. ft. per minute, and a Type 12 standard I-R jackhammer is used for rock drilling. The rental of the compressor and jackhammer, including gasoline, operator, and repairs to drills, is estimated at $35.00 an hour.

Excavating Loose Clay and Loam. The area of loose clay and loam is 80'-0"x80'-0"x5'-0", a total of 1,185 cu. yds. of excavation.

The average on this job is 135 cu. yds. an hr. at the following cost per 100 cu. yds.:

	Hours	Rate	Total	Rate	Total
Loader and Operator	0.75	$....	$....	$75.00	$56.25
Cost per cu. yd.			....		.56

Five 10-cu. yd. trucks will haul 1,080 cu. yds. of excavated earth ¼-mile to a dump per day, or an average of 27 cu. yds. an hr. per truck, at the following cost per 100 cu. yds.:

	Hours	Rate	Total	Rate	Total
Truck and Driver	3.7	$....	$....	$50.00	$185.00
Cost per cu. yd.			....		1.85

Four men at the dump spread the excavated earth after it has been dumped by the trucks. Very little grading is necessary, because the earth dump is a high-terraced fill around a building. A worker will spread about 10 cu. yds. an hr. at the following labor cost per cu. yd.:

	Hours	Rate	Total	Rate	Total
Labor	0.10	$....	$....	$19.13	$1.91

The total cost of excavating, hauling and spreading one cu. yd. of loose earth is as follows:

	Rate	Total
Loader excavation	$....	$.56
Trucks hauling excavated earth	....	1.85
Labor spreading loose earth at dump	....	1.91
Cost per cu. yd.		$4.32

Note: Does not include equipment move-in charge.

Rock Excavation

After the top 5'-0" of soil has been removed, the excavation consists of blue grey dolomite limestone, hard and stratified in beds varying from 6" to 24". The depth of the cut was 8'-0", which was taken out in two lifts of 4'-0" each. The last lift was drilled 5'-0" deep or 1'-0" below the desired level. Total rock excavation is 80'-0"x80'-0"x8'-0", or 1,900 cu. yds.

Drilling Rock for Blasting. The area drilled is 80'-0"x80'-0", less a 5'-0" border on 3 sides and a 10'-0" border on one side. The rows were 5'-0" apart and the holes were 5'-0" on centers, staggered, and all of the holes will be 2" in diameter.

There will be 14 rows with 15 holes in a row or a total of 210 holes, as follows: 105 holes 4'-0" deep, containing 420 lin. ft. and 105 holes 5'-0" deep, containing 525 lin. ft., a total of 945 lin. ft. of 2" holes.

One man and a helper will drill 15 lin. ft. (possibly up to 60 lin. ft.) of 2" hole an hr. or 120 lin. ft. per 8-hr. day, at the following cost per 100 lin. ft.:

	Hours	Rate	Total	Rate	Total
Compressor, jackhammer and operator	7	$....	$....	$35.00	$245.00
Labor helping	7	....	....	19.13	133.91
Cost 100 lin. ft.			....		$378.91
Cost per lin. ft.			....		3.79
Cost per cu. yd.(1,900 cy)					.20

Explosives. The job will require 2.5 cases (50 lbs. each) of 40% dynamite, a total of 125 lbs., which is equivalent to 1/8 lb. per lin. ft. of hole or per cu. yd. of excavation. The cost of explosives is as follows:

	Hours	Rate	Total	Rate	Total
125 lbs. dynamite		$....	$....	$.85	$106.25
220 electric detonators with 6' wire lead		....	....	.45	99.00
75 lin. ft. No. 14, R.C.S.B. lead wire		....	....	.12	9.00
Total cost			$....		$214.25
Cost per hole			....		1.02
Cost per cu. yd. (1,900 cu. yds.)			....		.11

Rock Excavation. There will be 1,900 cu. yds. of rock excavation, taken out in two lifts of 4'-0" each, and the loader will handle 45 cu. yds. per hr. at the following cost per 100 cu. yds.:

	Hours	Rate	Total	Rate	Total
Loader and operator	2.3	$....	$....	$50.00	$115.00
Labor	2.3	....	....	19.13	44.00
Powderman	2.0	....	....	25.96	51.92
Labor assisting	5.0	....	....	19.13	95.65
Total cost 100 cu. yds.			$....		$306.57
Cost per cu. yd.			....		3.07

*One laborer on bull chain, placing chain on rocks too large for shovel to handle.

Two 6-cu. yd. trucks haul the excavated rock to a dump ¼ mile from the job, averaging about 23 cu. yds. per hr. per truck, at the following cost per 100 cu. yds.:

	Hours	Rate	Total	Rate	Total
Truck and driver	4.35	$....	$....	$45.00	$195.75
Cost per cu. yd.			....		1.96

Four men at the dump level 20 cu. yds. of excavated rock per hr. at the following labor cost per 100 cu. yds.:

	Hours	Rate	Total	Rate	Total
Labor	20	$....	$....	$19.13	$382.60
Cost per cu. yd.			....		3.83

The total cost of rock excavation, including drilling, blasting, excavating, hauling, and spreading, averages as follows per cu. yd.:

	Total	Total
Drilling	$....	$.20
Explosives	....	.11
Excavating	....	2.43
Hauling	....	2.11
Spreading and leveling	....	3.60
Cost per cu. yd.	$....	$8.45

Vehicle Mounted Hydraulic Hammer Used to Break Up Pavement

Rock excavation conditions and costs vary widely. Always obtain information from sources at or near the job site before bidding on this work. Check with one or more of the local powdermen--they will know what the conditions are and what production may be expected.

Drilling Rock Using Pneumatic Jackhammer Drills. The hand held jackhammer is a popular tool for rock drilling on all types of construction jobs. They are furnished in sizes from 30 to 80 lbs., and many rock drilling conditions can be met. Holes up to 20 ft. in depth are easily drilled with the heavier machines. Most sizes of jackhammer can be furnished in three styles, depending on the conditions of drilling and the depth of holes being drilled (dry style for shallow holes, blower style for deep holes, and wet for jobs where dust must be kept down to a minimum).

Hollow drill steel of various sizes and shapes are used, the most common being 7/8" and 1" hexagon. Bits are sometimes forged on the steel but the detachable bit is by far the most widely used.

When drilling holes in rock for work such as trenching and blasting, two men operating drills should drill 125 to 175 lin. ft. of holes per 8-hr. day, at the following cost per 100 lin. ft.:

	Hours	Rate	Total	Rate	Total
Jackhammerman	11.0	$....	$....	$19.13	$210.43
Compressor operator	5.5	....	....	25.96	142.78
Compressor expense	5.5	....	....	13.00	71.50
Cost per 100 l.f.			$....		$424.71
Cost per l.f.			....		4.25

Does not include bit charges.

Rock Drilling Using Wagon Drills. The wagon drill consists of a one-piece tubular frame on pneumatic or steel wheels and a tilting tower, on which is mounted a heavy pneumatic drill. It drills holes downward at any angle from horizontal to vertical, to depths of 25 to 30 feet. This type of drill is suitable for all kinds of excavation and trenching and usually does the job of at least 2 or 3 heavy handheld jackhammers. It requires only an operator, a helper, and a compressor operator.

Rock Drilling Using Crawler Mounted Drills. An innovation in rock drilling for blast holes, etc. is the completely mechanized, self-propelled drill unit. Many construction equipment manufacturers make these units, which can do the work of two or three wagon drills. The unit can move easily over rugged terrain and can tow its own air supply with it. The set-up time for each hole is greatly reduced because of the self-propelled features. It is not uncommon for a crawler mounted drill to drill more than 500 lin. ft. per 8-hour shift in hard sedimentary rock. It can be rented with operator for $800 to $1000 per day.

Excavating Using Pneumatic Diggers. These tools can be used to advantage on all classes of excavation that require picking of dirt or clay, such as trench work, tunneling, caisson sinking, and all kinds of building excavation, in fact, on all kinds of work in stiff clay or hard ground where power shovels, trenching machines, etc., cannot be used.

On work of this kind, two men using air picks and six men shoveling will loosen and shovel 35 to 40 cu. yds. per 8-hr. day, at the following labor cost per cu. yd.:

	Hours	Rate	Total	Rate	Total
Labor operating picks	0.4	$....	$....	$19.13	$7.65
Labor shoveling	1.2	....	....	19.13	22.96
Compressor operator	0.2	....	....	25.96	5.19
Compressor expense	0.2	....	....	13.00	2.60
Cost per cu. yd.			$....		$38.40

Breaking Up Old Curbing Using Pneumatic Jack Hammer

On large excavations providing plenty of room for workers, a worker with a pneumatic pick will loosen 25 to 30 cu. yds. per 8-hr. day. After the clay or earth is loosened, a worker will shovel 6 to 7 cu. yds. per 8-hr. day.

Comparative Cost By Hand

Excavating in stiff clay or tough ground, a worker using a hand pick and shovel will loosen 2 ½ to 3 cu. yds. of dirt per 8-hr. day. The same man, picking only, will loosen 3 ½ to 4 cu. yds. of dirt per 8-hr. day, and when shoveling only, will remove 6 to 7 cu. yds. per 8-hr. day.

Using a hand pick, a man will loosen and shovel one cu. yd. of dirt at the following cost per cu. yd.:

	Hours	Rate	Total	Rate	Total
Labor	3	$....	$....	$19.13	$57.39

Courtesy Caterpillar, Inc.

Excavating with Caterpillar Tractor Loader

Backfill Tamping Using a Pneumatic Tamper. A pneumatic tamper enables a workman to do nearly 10 times the amount of work he can do by hand. A backfill tamper strikes over 600 hard, snappy blows a minute, rams the fill hard, and works in and around pipe with far greater thoroughness than hand devices are able to do.

In piers, trenches, etc., where the excavated earth can be shoveled directly into the trench, a man should backfill 16 to 20 cu. yds. per 8-hr. day.

Where the trenches have to be tamped, one man with a pneumatic tamper will tamp as much as 2 men can backfill, or approximately 35 to 40 cu. yds. per 8-hr. day.

Courtesy Caterpillar, Inc.

Backfilling Using Caterpillar Tractors

The cu. yd. cost should average as follows:

	Hours	Rate	Total	Rate	Total
Labor backfilling	0.4	$....	$....	$19.13	$7.65
Labor tamping	0.2	....	....	19.13	3.83
Compressor operator	0.2	....	....	25.96	5.19
Compressor & tamper	0.2	....	....	15.00	3.00
Cost per cu. yd.			$....		$19.67

Comparative Cost By Hand

A man will tamp by hand 8 to 10 cu. yds. of backfill per 8-hr. day, at the following cost per cu. yd.:

	Hours	Rate	Total	Rate	Total
Labor backfilling	0.4	$....	$....	$19.13	$7.65
Labor tamping	0.8	....	....	19.13	15.30
Cost per cu. yd.			$....		$22.95

Breaking Pavement to Install Underground Electrical Cable

Oversize Hydraulic Tamper Finishing Off Sewer Trench

02250 SOIL TREATMENT

New regulations are being legislated almost on a daily basis. The estimator should become familiar with local EPA regulations and engage the services of a professional who specializes in this area.

02300 PILE FOUNDATIONS

The estimator needs to become familiar with the many types and models of pile hammers available for installing various pile types, such as wood, steel pipe, steel sheets, "H" piles, precast concrete (both round and square), combinations of concrete and steel, corrugated and step taper piles. Pile hammers available include air or steam, diesel, hydraulic, and drop hammers.

Drop Hammer. This is simply a heavy weight that is lifted and dropped on the pile. It operates within a guide that directs the impact. In the past the weight was lifted by human or horse power. Occasionally, this is still the case, but usually, a line on a crane or a winch raises the weight to a predetermined height, where a trip releases it to fall freely on the pile.

The weight of a drop hammer should be about one-third the weight of the pile to be driven. The fall must be regulated so that the pile is not damaged, and a suitable cap, with cushion material, is used on the pile to distribute the driving force. Drop hammers are used much more in other parts of the world than the U.S.

Air and Steam Hammers. *Single-acting pile hammers,* powered by steam or air, are essentially drop hammers with a short stroke and heavy ram. Steam or air is used to raise the weight much more rapidly than for drop hammers. The falling weight opens a port that admits steam or air under the ram to raise it for the next stroke. The rising elements cut off the pressure and open an exhaust port, permitting practically a free fall of the weight on the next stroke. It is essential that the weight be raised to its full height and fall free to develop the rated capacity. The hammer should be powered at the manufacturer's recommended pressures. It should operate at recommended blows per minute and height of fall. Energy ratings range from 7,000 ft. lbs. up to 60,000 ft. lbs., but they go up to even 1,800,000 ft. lbs. for big offshore hammers.

Double-acting pile hammers use steam or air both to raise the striking hammer and as a force to add energy on the downward stroke. They operate at about twice the speed of single-acting hammers of equal energy but require more steam or air at greater pressure.

A variation is the *differential-acting hammer,* where non-expansive use of the steam develops more effective pressure. The significant difference between double-acting and differential-acting is the manner and sequence of exhausting on the upward and downward strokes of the cycle. In the differential-acting hammer, there is no drop from the entering pressure to the main effective pressure moving the piston on the downward stroke.

Double-acting and differential-acting hammers usually give better results in granular non-cohesive soils or in soft clays. Used in proper soil conditions with the right pile, almost twice the production can be obtained as with a single-acting hammer.

Diesel Hammers. This type of pile hammer gets its energy from the compression blow of a falling weight and the reaction to controlled instantaneous burning and expansion of fuel, which raises the ram for the next stroke. The diesel hammer does not require steam or air under pressure for operation. The only input is a modest amount of diesel fuel, compressed and atomized to ignition readiness. The hammers have a total weight of one-third to one-half of that of comparable energy-rated single- or double-acting steam or air units. Diesel hammers are started by lifting the ram by outside means. In soft overburden there sometimes is a problem where piles move down without developing enough resistance to compress the fuel ignition heat and pressure for the next stroke. Diesels work better in hard driving.

Some diesel hammers have a closed top, designed to give added speed and energy to the down stroke. Contractors like diesel hammers because they are lighter and do not require exterior power.

Vibratory Driver/Extractor. These are like mini-stroke, high blow rate pile hammers, but the vibratory drivers/extractors in use today do not have linearly reciprocating weights or rams. Instead, they use rotating weights set eccentric from their centers of rotation. The result is a machine that is a mechanical sine wave oscillator. It is rigidly connected to the pile, usually by clamps, and it oscillates the pile through the soil. The vibratory driver/extractor incorporates the unique capability of converting from driver to extractor by pulling upon the vibratory hoist line.

Vibratory hammers ordinarily do not operate with leads, so other provisions must be made for holding the piles to plan position for driving. Equipment intended for vibratory or sonic installation or withdrawal of bearing or sheet piles should be capable of adjustment for frequency and amplitude to accommodate to different combinations of soils and piling. For efficient operation the machines must grip the pile firmly, with hydraulic or air actuated rams.

Verification of bearing capacity of piles installed by vibratory machines by load tests or final driving with an impact hammer usually is considered necessary.

Types and Specifications of Pile Hammers

Steam and Air Hammers

Rated Energy ft. lbs.	Rated Energy kg. m.	Manufacture	Model	Action	Average Blows per Minute	Striking Weight lbs.
1,800,000	248,868	Vulcan	6300	Single	40	300,000
1,200,000	165,912	Conmaco	2000	Single	40	200,000
750,000	103,695	Vulcan	5150	Single	46	150,000
350,000	48,391	Conmaco	700	Single	43	70,000
180,000	24,887	Vulcan	360	Single	60	60,000
150,000	20,739	Conmaco	5300	Single	46	30,000
120,000	16,591	Vulcan	340	Single	60	40,000
90,000	12,443	Vulcan	030	Single	55	30,000
90,000	12,443	Conmaco	300	Single	55	30,000
60,000	8,296	Vulcan	020	Single	60	20,000
60,000	8,296	Conmaco	200	Single	60	20,000
48,750	6,740	Vulcan	016	Single	60	16,250
48,750	6,740	Conmaco	160	Single	60	16,250
42,000	5,807	Vulcan	014	Single	60	14,000
42,000	5,807	Conmaco	140	Single	60	14,000
37,375	5,167	Conmaco	115	Single	50	11,500
32,500	4,493	Vulcan	010	Single	50	10,000
32,500	4,493	Conmaco	100	Single	50	10,000

Rated Energy ft. lbs.	Rated Energy kg. m.	Manufacture	Model	Action	Average Blows per Minute	Striking Weight lbs.
26,000	3,595	Vulcan	08	Single	50	8,000
26,000	3,595	Conmaco	80	Single	50	8,000
19,500	2,696	Vulcan	06	Single	60	6,500
19,500	2,696	Conmaco	65	Single	60	6,500
19,150	2,648	Mkt Mfg	11B3	Double	95	5,000
15,000	2,074	Vulcan	1	Single	60	5,000
15,000	2,074	Conmaco	50	Single	60	5,000
13,100	1,811	Mkt Mfg	10B3	Double	105	3,000
8,750	1,210	Mkt Mfg	9B3	Double	145	1,600

Diesel-Powered Hammers

Rated Energy ft. lbs.	Rated Energy kg. m.	Manufac	Model	Action	Average Blows per Minute	Ram Weight lbs.	Estimated Fuel Consumption gph
225,000	31,109	Delmag	D80-12	Single	36-45	19,500	6.5
165,000	22,813	Delmag	D62-23	Single	36-50	14,600	5.5
127,500	17,628	Mkt Mfg	DE150/110	Single	40-50	11,000	5.5
117,000	16,176	Delmag	D55	Single	36-50	12,100	5.5
105,600	14,600	Kobelco	K-60	Single	40-60	13,200	7.3
105,000	14,517	Delmag	D46-32	Single	37-53	10,140	4.5
100,000	13,826	Ice	200S	Single	53-70	20,000	5.0
91,000	12,595	Kobelco	K-45	Single	39-60	9,900	5.0
84,300	11,655	Mitsubishi	MH-45	Single	42-60	10,210	5.0
83,100	11,489	Delmag	D36-32	Single	37-53	7,900	3.0
70,800	9,789	Kobelco	K-35	Single	39-60	7,700	3.5
70,000	9,678	Ice	70S	Single	38-55	7,000	3.5
66,100	9,139	Delmag	D30-32	Single	38-52	6,600	2.0
65,600	9,070	Mitsubishi	MH-35	Single	42-60	7,720	4.5
59,500	8,226	Mkt Mfg	DE-70B	Single	40-50	7,000	3.3
50,700	7,010	Kobelco	K-25	Single	39-60	5,510	2.7
50,000	6,913	Ice	660	Double	84-88	7,564	3.3
48,500	6,706	Delmag	D25-32	Single	38-52	4,850	2.0
46,900	6,484	Mitsubishi	MH-25	Single	42-60	5,510	3.0
42,500	5,876	Mkt Mfg	DE-50B	Double	40-50	5,000	3.3
40,000	5,530	Ice	40S	Single	38-55	4,000	3.0
38,200	5,282	Mkt Mfg	DA-55B	Double	78-82	5,000	3.0
30,000	4,148	Ice	520	Single	80-84	5,070	2.0
28,100	3,885	Mitsubishi	MH-15	Single	42-60	3,310	2.0
27,100	3,747	Delmag	D16-32	Single	40-60	3,300	1.7
25,200	3,484	Kobelco	K-13	Single	40-60	2,860	1.5
23,800	3,291	Mkt Mfg	DE-30B	Double	40-50	2,800	2.0
22,500	3,111	Ice	30S	Single	44-67	3,000	1.5
21,000	2,903	Mkt Mfg	DA-35C	Double	78-82	2,800	2.7
18,100	2,503	Ice	440	Single	88-92	4,000	1.5
18,000	2,489	Delmag	D8-22	Single	38-52	1,760	1.0
17,000	2,350	Mkt Mfg	DE-20B	Double	40-50	2,000	2.0
9,350	1,293	Mkt Mfg	DA-15C	Double	80-92	1,100	1.5
9,100	1,258	Delmag	D5	Single	40-60	1,100	1.5
8,100	1,120	Ice	180	Double	90-95	1,725	0.7
3,630	502	Delmag	D4	Single	50-60	840	0.2
1,815	251	Delmag	D3	Single	60-70	485	0.1

Hydraulic Vibratory Drivers/Extractors

Manufac.	Model	Dynamic Force Tons	Line Pull Extraction Tons	Suspended Weight lbs.
Ice	1412	163	100	26,900
Ice	815	145	50	15,600
Mkt Mfg	V20	116	60	10,650
Ice	612	82	40	13,700
Mkt Mfg	V16	78	40	9,250
Ice	416L	65	40	9,900
Ice	216	35	30	4,825
Mkt Mfg	V5	30	15	5,100

When estimating the cost of piles, one must consider overhead expense before the actual cost of driving the piles can be computed. This includes cost of moving the equipment to the job, setting up the pile rig and being ready to operate, and dismantling and removing at completion.

This expense will always be pretty much the same, regardless of the size of the job. The equipment move charge is as much on a job with 100 piles as on one containing 1,000 piles.

The actual cost of driving the piles will vary with soil conditions, length of piles, and the amount of moving necessary to drive piles at the proper location on the site.

Setting up and Removing Equipment. The cost of moving the pile driving equipment to the job must be taken into consideration, and then after it arrives at the job, there is the task of setting up the pile driver and getting ready to operate.

After the pile driver has been delivered at the job it should take about one day for the crew to set it up ready to drive and this labor should cost as follows:

Setting up Equipment	Hours	Rate	Total	Rate	Total
Foreman	8	$....	$....	$26.74	$213.92
Pile driver engineer	8	....	....	25.96	207.68
Oiler	8	....	....	19.13	153.04
2 pile driver men	16	....	....	24.04	384.64
4 men, general labor	32	....	....	19.13	612.16
Total labor cost			$....		$1,571.44

The above item will remain practically the same whether there are 100 or 1,000 piles to be driven.

After the job has been completed, the pile driver must be dismantled and removed from the job and the dismantling should cost as follows:

Dismantling Equipment	Hours	Rate	Total	Rate	Total
Foreman	8	$....	$....	$26.74	$213.92
Engineer on pile driver	8	....	....	25.96	207.68
Oiler	8	....	....	19.13	153.04
2 pile driver men	16	....	....	24.04	384.64
4 men, general labor	32	....	....	19.13	612.16
Total Labor Cost			$....		$1,571.44

Various Types of Bearing and Friction Piles

Wood
- Treated
- Untreated (plain)

Concrete
- Cast-in-place
- Precast/prestressed
- Shell
- Shell-less

Steel
- HP shape
- Pipe
- Shell

Sheet Pile
- Wood
- Steel
- Precast Concrete

The above types can be used in various combinations, sometimes known as *composite piles*. A good example of composite pile would be heavy-walled pipe piles filled with concrete. Normally, the designer indicates on the drawings or in the written specifications, or on both, the type of pile and the size of hammer required to install.

Driving Wood Piles. After the pile driver is set up and ready to operate, the labor cost of driving the piles will vary with the length of the piles and job conditions.

The following table gives the approximate number of piles of various lengths that should be driven per hour and per day:

Length of Piles	1 Hour	8 Hours
24 feet	4.5	36
26 feet	4.2	34
28 feet	4.0	32
30 feet	3.7	29
32 feet	3.5	28
34 feet	3.3	26
36 feet	3.1	25
38 feet	3.0	24
40 feet	2.8	22
45 feet	2.5	20
50 feet	2.3	18
55 feet	2.1	17
60 feet	2.0	16

The labor operating the pile driver per 8 hr. day should cost as follows:

	Hours	Rate	Total	Rate	Total
Foreman	8	$....	$....	$26.74	$213.92
Pile driver engineer	8	....	....	25.96	207.68
Oiler	8	....	....	19.13	153.04
2 pile driver men	16	....	....	24.04	384.64
4 men, general labor	32	....	....	19.13	612.16
Labor cost per day			$....		$1,571.44

The above does not include time and costs for pile layout or resurvey, spotting piles, pumping, shoring, excavating, purchasing piles, pile points, or cutting off wood piles after they are driven. It only includes the actual time driving the piles. It also does not include the cost of equipment, fuel, oil, and other supplies for the pile driver. These costs should be added.

Driving Wood Piles on a Large Job

Cutting Off Wood Piles. After the wood piles have been driven, the tops projecting above the ground will have to be cut off to receive the foundation that is to be placed on them.

This cost will vary according to the size of the quarters in which workers must operate, but on the average job, it should require 1/6 to 1/4 hrs. to cut off each wood pile 12" or 14" in diameter with a chain saw, at the following labor cost per pile:

	Hours	Rate	Total	Rate	Total
Labor	0.22	$....	$....	$24.04	$5.29

To the above costs, add the hourly cost of chain saw, and the time for removing sawed-off ends, as well as proper removal and disposal.

Computing the Cost of Wood Piles. As described above, the cost of driving wood piles will vary with the number of piles, length, kind of soil, and other job conditions. Pile costs fluctuate so rapidly that current costs must be verified for each project.

The following is a sample estimate, showing a method for arriving at costs for driving 680 treated wood piles 40'-0" long:

	Local Rate	Total	Total
Trucking pile driver to job	$....	$....	$2,000.00
Setting up pile driver, 1 day for crew	....	....	1,571.44
680 wood piles, 40'-0" long @ $200.00	....	....	136,000.00
Driving piles (22 per 8-hr. day) 31 days @ $1,469.28	....	....	45,547.68
Fuel, oil, misc. supplies, etc. 31 days @ $175.00	....	....	5,425.00
Dismantling pile driver at completion, 1 day	....	....	1,571.44
Trucking pile driver, job to yard	....	....	2,000.00
Equipment rental or depreciation allowance (36 days)	....	....	18,000.00
Cost 680 piles		$....	212115.56
Cost per pile		....	311.93
Cost per lin. ft.		....	7.80

Add cost of cutting off piles after driving, and for overhead & profit.

Driving Steel HP Piles. Once the pile driver is set up and ready to operate, labor costs for actually driving the piles will vary with the length of piles and site conditions.

Rolled structural steel shapes are also used as bearing piles. The shape commonly used for this purpose is the HP. This type of pile has proved especially useful for trestle structures in which the pile extends above the ground and serves not only as a pile but also as a column.

Because of their small cross-sectional area, piles of this type can often be driven through dense soils to point bearing where it would be difficult to drive a pile of solid cross-section, such as a wood, cast-in-place, or precast concrete. But the feature that allows it to penetrate dense soils works to its disadvantage in other soil. Where piles are used to support loads by friction, or where they are used primarily for compaction, a considerably longer HP is required to carry the same load that can be supported by a pile of less length but greater cross-sectional area.

The upper ends of HP piles can be encased in concrete to prevent corrosion. Steel piles weighing more than 200 pounds per lin. ft. have been driven in lengths up to 130 feet. Lighter sections have been driven to even greater lengths.

Driving HP Section Piles. The following table gives the approximate number of HP piles of various lengths that can be driven per hour and per day, assuming average job site conditions:

Length of Piles	1 Hour	8 Hours
20 feet	4.00	32
30 feet	3.75	30
40 feet	3.00	24
50 feet	2.50	20
60 feet	2.25	18

The labor cost of operating the pile driver per 8-hr. day is as follows:

	Hours	Rate	Total	Rate	Total
Foreman	8	$....	$....	$26.74	$213.92
Pile driver engineer	8	....	....	25.96	207.68
Oiler	8	....	....	19.13	153.04
2 pile drivers	16	....	....	24.04	384.64
4 laborers	32	....	....	19.13	612.16
Cost per 8-hr. day					$1,571.41

The above does not include time and costs for pile layout or resurvey, spotting piles, pumping, shoring, excavation, purchase of piles, pile points and pile splices, or cutting off driven piles. It only includes the actual time driving piles.

It also does not include the cost of equipment, fuel, oil, and other supplies for the pile driver.

Prices of Steel HP Piles. Prices of steel HP piles vary considerably, mainly due to the material cost of the rolled steel sections. Steel sections commonly used for this purpose may range from an 8-inch HP section weighing 36 lbs. per lin. ft. to a 14-inch HP section weighing 117 lbs. per lin. ft.

Assuming a unit cost of $0.60 per lb., delivered to job site from warehouse, the HP beam material will cost from $21.60 to $35.10 per lin. ft. If the job is in an isolated location, additional transportation charges must be figured.

Cutting Off Steel HP Piles. After the steel HP section pile has been driven, the pile tops that project above the ground must be cut off to the required elevation to receive the foundation that will be placed on them.

The cost will vary according to the size of the pile and the quarters in which workers must work. A burning outfit, consisting of gauges, burning torch, hoses, acetylene, and oxygen is normally required to cut HP steel piles at the following cost per pile:

HP	Hours	Rate	Total	Rate	Total
8" section	.17	$....	$....	$24.04	$4.09
10"	.21	....	....	24.04	5.05
12"	.25	....	....	24.04	6.01
14"	.31	....	....	24.04	7.45

To the above, add the costs for a complete burning outfit, time for removing the cut off pile sections, and the cost for removal from the site and proper disposal.

Splicing Steel HP Piles. When splicing is necessary, steel HP piles can be spliced in the field. When butt welded, splices should be full penetration butt welds across both flanges and the web. Edges on the ends of the upper section, where welding is to be done, should be beveled for penetration welding. The top of the pile length, where driving will be done, should not be beveled.

Steel HP piles are spliced by butt welding at the following cost per pile:

HP	Hours	Rate	Total	Rate	Total
8"	1.33	$....	$....	$24.04	$31.97
10"	1.67	....	....	24.04	40.15
12"	2.00	....	....	24.04	48.08
14"	2.50	....	....	24.04	60.10

The costs vary according to the size of the piles and the quarters in which workers are obliged to operate. Add to the above all costs for additional material and equipment (such as plates, welding machine, and pile equipment), and standby time while welding is being performed, and all required testing.

Prefabricated splicers are available from Dougherty Foundation Products, Inc. of Franklin Lakes, New Jersey. These splicers temporarily hold a length of HP pile until the joint can be completed. The driver can set an extension and then move on to other work while the welding takes place, at the following labor cost per pile:

HP	Hours	Rate	Total	Rate	Total
8" to 14"	.50	$....	$....	$24.04	$12.02

Material Cost for DFP* Splicers

HP	
8"	$24.00 each
10"	29.00 each
12"	39.00 each
14"	48.00 each

*Dougherty Foundation Products, Inc., Franklin Lakes, NJ

The costs vary according to the size of the piles and the quarters in which workers must operate.

CONCRETE PILES

Due to the wide variety of conditions under which concrete piles are installed and used, it is difficult to give dependable cost figures without knowing the details of each installation. The different types of concrete piles vary considerably in cost. When estimating work containing concrete piles, it is always advisable to consult contractors specializing in this class of work. Most of them maintain offices in the principal cities and are usually willing to investigate and quote budget prices on anticipated work. More accurate figures can be obtained than can be given here.

There are two principal types of concrete piles, cast-in-place and precast. The cast-in-place pile is formed in the ground in the position where it is to be used in the foundation. The precast pile is cast above ground, and after it has been properly cured, it is driven or jetted just like a wood pile.

In ordinary building foundation work, the cast-in-place pile is more common, because the required length can be readily adjusted in the field as the job progresses. There is no need to predetermine pile lengths, and the required length is installed at each pile location.

In contrast, it is necessary to predetermine the length of precast piles, and to provide for contingencies, it is generally required that piles be ordered longer than the actual anticipated length. Also, it is not possible to determine the pile length at each location, because subsoil conditions at any construction site will show considerable variation. The cost of waste piling for precast piles can be quite high, as is the cost of cutting them off to the proper grade. Precast piles are prestressed and are manufactured at established plants. Job site casting is relatively rare. It is difficult and costly to handle and transport precast piles especially in long lengths, and a production plant must be relatively close to the job site.

In marine installations, either in salt water or fresh water, the precast pile is used almost exclusively, because of the difficulty of placing cast-in-place piles in open water. For docks and bulkheads, the cast-in-place pile is sometimes used in the anchorage system. On trestle type structures, such as highway viaducts, the precast pile is commonly used. A portion of the pile often extends above the ground and serves as a column for the superstructure. Precast concrete piles are always reinforced internally so that they resist stress from handling and driving. Cast-in-place piles are rarely reinforced, because lateral support of even the poorest of soils is sufficient to overcome any bending moments induced in the pile by column action.

Cast-in-place Concrete Pipe Piles. Pipe can be of any practical diameter. It is almost always filled with concrete, but with adequate wall thickness in non-corrosive ground, it can be used without concrete. Driving usually produces stresses greater than the working load on the pile. Pipe should conform to *ASTM A252, Specifications for Welded and Seamless Steel Pipe Piles.*

Pipe shell that does not contribute to the strength of the pile as a structural member may be of any metal that adequately resists installation stress and maintains an open shaft to receive the concrete, free from water or other foreign matter. Steel shell should meet the applicable requirements of *ASTM Specifications A-245, A-283, A-366,* or *A-415.*

The end closure for pipe or shell may be cast steel meeting requirements of *ASTM A-27 65/35* or structural steel meeting *ASTM A-36* or better.

Concrete for cast-in-place piles should conform to requirements for cement, aggregate, mixing, placing, and protection specified by ACI or PCA for quality concrete.

Cast-in-place concrete piles can be one of four types: open-end pipe, closed-end pipe, thin-cased and corrugated shell, large-base concrete piles. The closed-end pipe pile is simply a piece of steel pipe, closed at the bottom with a heavy boot, driven into the ground, and filled with concrete. The uses and allowable loads for these piles are about the same as for other types of cast-in-place piles with driven shells. On some jobs the pipe shell is driven all in one piece, but often it is driven in sections that are welded together or fitted together with internal sleeves as the driving progresses. Pipe wall thickness must be adequate to develop the required stiffness and driveability of the pipe.

Open-end steel pipe piles are usually driven to bearing on rock. Since the pipe is open at the bottom during driving, the interior fills with soil which must be removed before concreting. Cleaning is usually done with air and water jets, after which the pipe is re-driven to insure proper seating in the rock. Remaining water is pumped out and the pipe pile is filled with concrete. Open-end pipe installed in this manner usually carry relatively high working loads. They have been installed up to 36 inches in diameter and in lengths exceeding 200 feet.

Estimating the Cost of Concrete Pipe Piles. The cost of concrete pipe piles will vary greatly, depending upon the number of piles in the job and the length, because it is the total footage of piles in the job rather than the number of piles that controls the price. For example, the price per foot for 100 piles 35 feet long would be much greater than the price per foot for 1,000 piles 25 feet long.

For any type of pile, contractors who specialize in this work are usually willing to submit budget prices, which include all labor, materials, and equipment for the complete job. A detailed estimate would only be required when such budget prices are not available, or for some reason the estimator does not desire to seek out a budget price.

As with all pile driving, regardless of the size of the job, there is a fixed charge for moving equipment on and off the job that will cost from $5,000.00 to

$25,000.00, provided the equipment is located in the immediate vicinity of the work.

If it is necessary to ship plant and equipment from one city to another, it will be necessary to add additional freight and trucking amounting to $2,000.00 to $4,000.00, making a total fixed charge of $7,000.00 to $29,000.00, before figuring the cost of the piles.

Labor Driving Cast-in-place Concrete Pipe Piles. After the pile driver has been set up and ready to operate, the number of piles driven per day and the labor cost driving them will vary with the length of the piles and job conditions.

For instance, on a job where the piles were closely spaced and driven from a flat surface, a pile driver averaged 100 piles a day (which is exceptional), while on another job where the piles were more widely scattered and had to be driven from different elevations, it was with difficulty that 20 piles were driven per day.

Labor Placing Concrete For Piles. Practically 90 percent of all concrete used in concrete piles is ready-mix concrete, which is usually discharged directly into the piles. Ordinarily it requires 1/2 to 1 hour time for one man per cu. yd. of concrete.

Pipe Sizes Frequently Used for Piling and Cu. Yds. of Concrete Required per Lineal Foot of Pipe

Outside Dia. of Pipe, in.	Thickness of Pipe Wall, in.	Weight of Pipe per lin. ft.	Concrete Cu. Yds. per lin. ft. of Pipe
8 5/8	.125	11.35	.0142
	.188	16.90	.0137
	.219	19.64	.0135
	.250	22.36	.0133
	.312	27.74	.0129
	.375	33.00	.0125
10	.188	19.65	.0187
	.219	22.85	.0185
	.250	26.03	.0182
10 3/4	.188	21.15	.0217
	.219	24.60	.0215
	.250	28.04	.0212
	.307	34.24	.0207
	.365	40.50	.0203
	.500	54.70	.0192
12	.188	23.65	.0273
	.219	27.52	.0270
	.250	31.40	.0267
12 3/4	.188	25.16	.0309
	.219	29.28	.0306
	.250	33.40	.0303
	.312	41.50	.0297
	.375	49.60	.0291
	.500	65.40	.0279

Outside Dia. of Pipe, in.	Thickness of Pipe Wall, in.	Weight of Pipe per lin. ft.	Concrete Cu. Yds. per lin. ft. of Pipe
14	.188	27.70	.0375
	.219	32.20	.0372
	.250	36.70	.0368
	.312	45.70	.0362
	.375	54.60	.0355
	.500	72.10	.0341
16	.188	31.70	.0493
	.219	36.90	.0489
	.250	42.10	.0485
	.312	52.40	.0477
	.375	62.60	.0470
	.500	82.80	.0454
18	.219	41.50	.0623
	.250	47.40	.0619
	.312	59.00	.0610
	.375	70.60	.0601
	.500	93.50	.0584
20	.219	46.20	.0773
	.250	52.70	.0768
	.312	65.70	.0758
	.375	78.60	.0749
	.500	104.10	.0729
22	.250	58.10	.0934
	.375	86.60	.0912
	.500	114.80	.0891
24	.250	63.40	.1116
	.312	79.10	.1104
	.375	94.60	.1093
	.500	125.50	.1067
30	.375	118.70	.1728
	.500	157.50	.1700
36	.375	142.70	.2510
	.500	189.60	.2474

Quantity of Concrete Required in Various Sizes of Pipe Piles, in Cu. Ft. and Cu. Yds.

Length in Ft.	12" OD x 250 Wall		14" OD x .250 Wall		16" OD x .250 Wall		24" OD x .250 Wall		36" OD x .500 Wall	
	cu ft	cu yd	cu ft	cu yd	cu ft	cu yd	cu ft	cu yd	cu ft	cu yd
1	0.72	0.02	0.99	0.03	1.30	0.04	3.01	0.11	6.67	0.24
5	3.60	0.13	4.96	0.18	6.54	0.24	15.0	0.55	33.3	1.23
10	7.20	0.26	9.93	0.36	13.0	0.48	30.1	1.11	66.7	2.47
15	10.8	0.40	14.9	0.55	19.6	0.72	45.1	1.67	100	3.71
20	14.4	0.53	19.8	0.73	26.1	0.97	60.2	2.23	133	4.94
22	15.8	0.58	21.8	0.80	28.8	1.06	66.2	2.45	146	5.44
24	17.3	0.64	23.8	0.88	31.4	1.16	72.3	2.67	160	5.93
26	18.7	0.69	25.8	0.95	34.0	1.26	78.3	2.90	173	6.43
28	20.1	0.74	27.8	1.03	36.6	1.35	84.3	3.12	187	6.92
30	21.6	0.80	29.8	1.10	39.2	1.45	90.3	3.34	200	7.42
32	23.0	0.85	31.7	1.17	41.9	1.55	96.4	3.57	213	7.91
34	24.5	0.90	33.7	1.25	44.5	1.64	102	3.79	227	8.41

Length in Ft.	12" OD x 250 Wall cu ft	12" OD x 250 Wall cu yd	14" OD x .250 Wall cu ft	14" OD x .250 Wall cu yd	16" OD x .250 Wall cu ft	16" OD x .250 Wall cu yd	24" OD x .250 Wall cu ft	24" OD x .250 Wall cu yd	36" OD x .500 Wall cu ft	36" OD x .500 Wall cu yd
36	25.9	0.96	35.7	1.32	47.1	1.74	108	4.01	240	8.90
38	27.3	1.01	37.7	1.39	49.7	1.84	114	4.24	253	9.40
40	28.8	1.06	39.7	1.47	52.3	1.94	120	4.46	267	9.89
42	30.2	1.12	41.7	1.54	54.9	2.03	126	4.68	280	10.3
44	31.7	1.17	43.7	1.61	57.6	2.13	132	4.91	293	10.8
46	33.1	1.22	45.7	1.69	60.2	2.23	138	5.13	307	11.3
48	34.6	1.28	47.6	1.76	62.8	2.32	144	5.35	320	11.8
50	36.0	1.33	49.6	1.84	65.4	2.42	150	5.58	333	12.3
52	37.4	1.38	51.6	1.91	68.0	2.52	156	5.80	347	12.8
54	38.9	1.44	53.6	1.98	70.7	2.61	162	6.02	360	13.3
56	40.3	1.49	55.6	2.06	73.3	2.71	168	6.24	374	13.8
58	41.8	1.54	57.6	2.13	75.9	2.81	174	6.47	387	14.3
60	43.2	1.60	59.6	2.20	78.5	2.91	180	6.69	400	14.8

Thin Shell Pipe Piles. Monotube (Monotube Pile Company, Canton Ohio) piles are the only uniformly tapered steel shell pipe piles available. Monotubes are strictly a friction pile, where the shell is end driven with conventional pile driving equipment, visually inspected, and filled with concrete.

Monotubes are available in four diameters, four gauges, and three rates of taper, making them suitable for soils with considerable friction resistance, such as sand, silt, and clay. Because of the wedging action of the uniform taper and the factory attached conical nose, they are commonly driven to the same capacity as straight-sided friction piles with less depth. A combination of cold processing and vertical fluting permits the use of lighter gauges that are equal in strength to pipe piles with thicker walls. Tapered sections can be factory-welded to a consent diameter extension or spliced in the field. Field splices are made after the tapered portion has been driven into the soil. The extension is lifted into the leads and inserted 6 to 8 inches into the taper. Factory-crimped extension sections speed up alignment and field splicing.

Specifications for Monotube Piles
Tapered Sections

Type	Size Tip Point Dia. x Butt Dia. x Lngth.	Weight per Foot 9 Ga. .1495"	7 Ga. .1793"	5 Ga. .2092"	3 Ga. .2391"	Conc. Vol. Cu. Yds
	8 ½" x 12" x 25'	17	20	24	28	0.43
F	8" x 12" x 30'	16	20	23	27	0.55
Taper	8 ½" x 14" x 40'	19	22	26	31	0.95
.14 Inch	8" x 16" x 60'	20	24	28	33	1.68
per Foot	8" x 18" x 75'	--	26	31	35	2.59
J	8" x 12" x 17'	17	20	23	27	0.32
Taper	8" x 14" x 25'	18	22	26	30	0.58
.25 Inch	8" x 16" x 33'	20	24	28	32	0.95
per Foot	8" x 18" x 40'	--	26	30	35	1.37
Y	8" x 12" x 10'	17	20	24	28	0.18
Taper	8" x 14" x 15'	19	22	26	30	0.34
.40 Inch	8" x 16" x 20'	20	24	28	33	0.56
per Foot	8" x 18" x 25'	--	26	31	35	0.86

Extension Sections

Type	Dia. x Length	9 Ga. .1495"	7 Ga. .1793"	5 Ga. .2092"	3 Ga. .2391"	Cu. Yds. per Ft.
N12	12" x 12" x 20/40'	20	24	28	33	.026
N14	14" x 14" x 20/40	24	29	33	38	.035
N16	16" x 16" x 20/40'	27	32	38	44	.045
N18	18" x 18" x 20/40'	--	37	43	49	.058

Step-Taper Pile. Shells are made of sheet steel in gauges of 12 to 20 and in basic section lengths of 4, 8, 12, and 16 feet. Longer lengths are furnished for special conditions. Step-Taper shells are helically corrugated to provide greater strength against collapsing pressures. A driving ring is welded to the bottom of each shell section and joints between sections are screw-connected. The bottom section is closed with a flat steel plate welded to the driving ring. Step-Taper Pile shells are manufactured in standard nominal diameters ranging from 8 5/8 to 18 3/8 inches, but larger diameters can be made to meet special conditions.

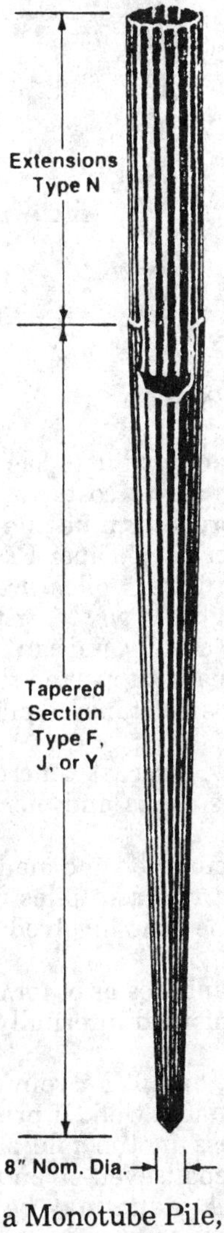

Cross-section of a Monotube Pile, Courtesy of Monotube Pile

When sufficient shell sections are joined together to make up the required pile length, the pile diameter increases from tip to butt at the rate of one inch per section length. Thus the rate of taper or pile shape will vary with the section lengths used. Within practical limits, different section lengths can be combined in a single pile, providing a wide range of available pile shapes. Nominal tip diameters usually range from 8 to 11 inches, but larger tip diameters can be used. With the variety of diameters and section lengths available, the Step-Taper pile provides maximum flexibility in the choice of pile size and configuration to best meet subsoil conditions and loading requirements.

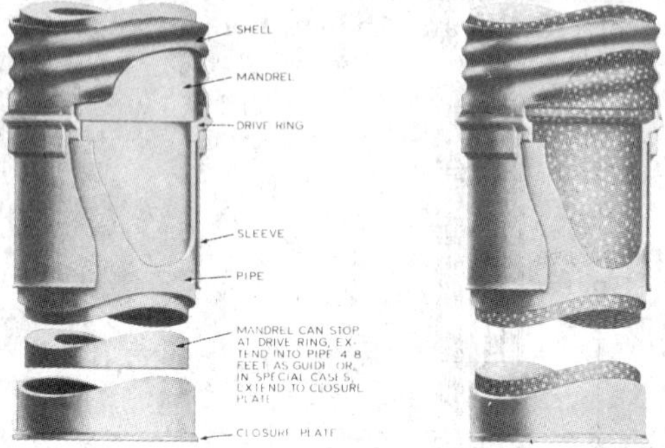

Step Taper Piles

The length of a Step-Taper Pile can be readily adjusted as the work proceeds to meet variable soil conditions, holding waste costs to a minimum. It is unnecessary to install drive-test piles to predetermine pile lengths.

Internal reinforcement is not required for Step-Taper Piles except to resist uplift loads or high lateral loads (where batter piles are not used) or for unsupported pile lengths extending through air, water, or very fluid soil.

Step-Taper Piles are suitable for all types of soils and can function as friction or point bearing piles. In most cases, piles are supported through a combination of friction and direct bearing. In many soils, tapered piles usually develop higher capacity than piles of no taper.

Precast/prestressed Concrete Piles. Precast concrete piles are commonly used in marine structures, such as docks and piers, where they are driven in open water.

Ordinarily they are designed for the particular job and made in a casting yard at or near the job site. The use of precast concrete piles usually delays the starting of actual pile driving because of the time involved in the casting and curing of this type of pile.

Precast concrete piles have the same advantages as other concrete piles over wood and steel piles. They are always reinforced internally to resist stresses produced by handling and driving.

Requirements for precast concrete piles, except for reinforcement requirements, generally apply to prestressed units as well. If prestressed piles are used, the minimum working net prestress in the pile should be 700 psi. Prestressing strands are of the ungalvanized, seven-strand type, conforming to the general requirements of ASTM A-416 and may be either regular or high-strength.

Precast piles may be moved when the compressive strength of concrete has reached 4,000 psi, but are not to be driven until 5,000 psi strength is attained.

Piles are to be lifted or supported only at designated points and handled and driven in such a manner as to avoid excessive bending stresses, cracking, or spalling.

Pile heads must be protected from direct impact of the hammer by a cushion head so arranged that strands or bars projecting from the pile head will not be displaced or deformed during driving. Minimum energy per blow of the hammer is established by the engineer. Jetting may be permitted, or required where necessary, to reach the desired depth.

Driving 20'-0" Concrete Piles. On a 12-story reinforced concrete building, requiring 280 concrete piles 20'-0" long, the following costs were obtained.

There was considerable broken time on this job, due to bad weather, and while the job required 17 ½ days, only 112 hrs. actual driving time was required. This included moving the pile driver from one place to another, because some of the piles were driven from the street level and others from the basement floor level.

The pile driving crew averaged two 20'-0" concrete piles per hr. or 16 piles (320 lin. ft.) per 8-hr. day, at the following labor costs:

	Hours	Rate	Total	Rate	Total
Setting up Pile Driver					
Foreman	32	$....	$....	$26.74	$855.68
Engineer	32	....	....	25.96	830.72
Oiler	32	....	....	19.13	612.16
Linesman	32	....	....	24.04	769.28
Four pile driver men	128	....	....	24.04	3,077.12
Driving Piles					
Foreman	140	....	....	26.74	3,743.60
Engineer	140	....	....	25.96	3,634.40
Oiler	140	....	....	19.13	2,678.20
Linesman	140	....	....	24.04	3,365.60
Four pile driver men	560	....	....	24.04	13,462.40
Dismantling Pile Driver					
Foreman	20	....	....	26.74	534.80
Engineer	20	....	....	25.96	519.20
Oiler	20	....	....	19.13	382.60
Linesman	20	....	....	24.04	480.80
Four pile driver men	80	....	....	24.04	1,923.20
Labor cost 280 piles	1,536		$....		36,869.76
Cost per 20'-0" pile			....		131.68
Cost per lin. ft			....		6.58

Does not include move charges for equipment, equipment costs, or pile material costs.

Various Types of Precast Concrete Piles

Squared Prestressed Piles

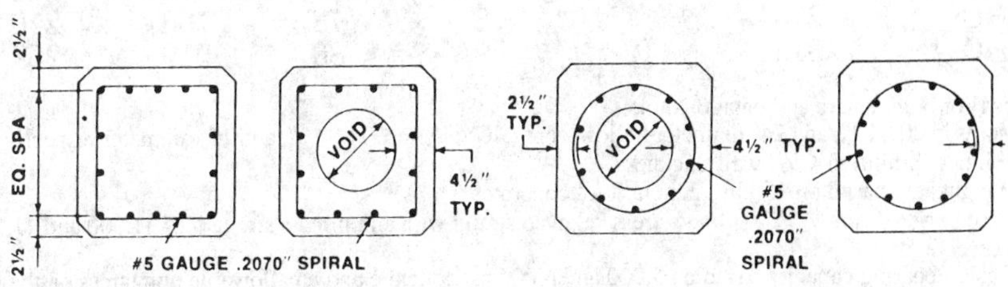

Cross-sections of Square Prestressed Piles

Pile Size(1) in.	Approx. Wgt.(2) lbs./l.f.	Strand per Pile Dia.(3) 7/16"	1/2"	Perimeter in.	Design Bearing Capacity Tons(4) Concrete Strength, psi 5000	6000
10	105	4	4	40	73	90
12	150	6	5	48	105	129
14	205	8	6	56	143	176
16	265	11	8	64	187	229
18	335	13	10	72	237	290
20	415	16	12	80	292	358
22	505	20	15	88	354	433
24	600	23	18	96	421	516
20 HC	320	13	10	80	223	273
22 HC	365	14	11	88	256	314
24 HC	415	16	12	96	291	357

Octagonal Prestressed Piles

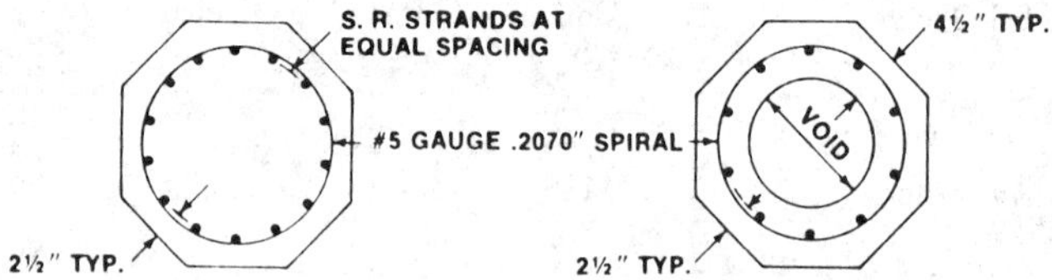

Cross-sections of Octagonal Prestressed Piles

Pile Size(1) in.	Approx. Wgt.(2) lbs./l.f.	Strand per Pile Dia.(3) 7/16"	1/2"	Perimeter in.	Design Bearing Capacity Tons(4) Concrete Strength, psi 5000	6000
10	85	4	4	33	61	74
12	125	5	4	40	87	107
14	170	7	5	46	118	145
16	220	9	7	53	155	190
18	280	11	8	60	196	240
20	345	14	10	66	242	296
22	420	16	12	73	293	359
24	495	19	15	80	348	427
20 HC	245	10	8	66	172	211
22 HC	280	11	8	73	196	240
24 HC	315	12	9	80	219	269

Notes (For both square and octagonal piles)

(1) Voids in 20", 22", and 24" diameter hollow-core (HC) piles are 11", 13", and 15" diameter respectively, providing a minimum 4 ½" wall thickness.

(2) Weights are based on 150 lbs. per cu. ft. of concrete.

(3) Based on 7/16" and ½" 270 Grade stress-relieved strand with an ultimate strength of 31,000 and 41,300 lbs. respectively.

(4) Design bearing capacity based on 5,000 and 6,000 psi concrete and an allowable unit stress on the full section of 0.33 ^{fce} - 0.27 ^{fce} where fce is the concrete stress in the pile due to prestressing, after all losses. These bearing capacity values may be increased if higher strength concrete is used.

Cutting Off Precast/Prestressed Square or Octagonal Piles. After concrete piles have been driven, it might be necessary to cut off the tops of piles to the proper elevation.

Costs for this work vary according to the size of piles and the space in which workers must operate. Before cutting a precast or prestressed pile, make a circumferential cut, with a diamond or carborundum saw, to score the pile and prevent spalling.

Manhours Required to Cut Off Precast/Prestressed Piles

Size	Square	Octagonal
10"	0.70	1.00
12"	1.00	1.30
14"	1.10	1.40
16"	1.30	1.70
18"	1.50	1.90
20"	1.70	2.10
22"	1.80	2.30
24"	2.00	2.50

To costs for the above, add the costs of cutting tools and equipment, as well as costs for removal and disposal of cut off sections of pile.

Prestressed Concrete Cylinder Piles. These are centrifugally cast in 16-foot sections, which are assembled end-to-end and post-tensioned, providing a wide range of available pile shapes. These piles are cast with zero-slump concrete, and strengths from 7,000 to 10,000 psi are attained. They are made with the Cen-Vi-Ro process, which is a combination of centrifugal casting, vibrating, and rolling. It produces an extremely dense concrete with an absorption factor about 40 percent less than bed-cast concrete piles. The use of zero-slump concrete prevents migration of the coarse aggregate during the spinning process so that the aggregates are uniformly distributed across the pile wall.

These piles are available in diameters from 36 to 90 inches, but the standard sizes are 36, 54, and 66 inch O.D. Wall thicknesses are 5 inches for the 36- and 54-inch piles and 6 inches for the 66-inch piles.

Cylinder Piles are driven either open or closed end. They are capable of carrying extremely high compression loads and resisting large lateral forces or bending moments. They are used for elevated highways, bridges, piers, offshore platforms, and breakwaters. For bridges and other trestle-type structures, the cylinder pile serves as both foundation pile and bridge pier or column.

Standard Sizes of Cylinder Piles

O.D. in.	I.D. in.	W in.	Weight/Foot lbs	Number of Cables
36	26	5	524	8 to 16
54	44	5	829	12 to 24
66	54	6	1217	16 to 32

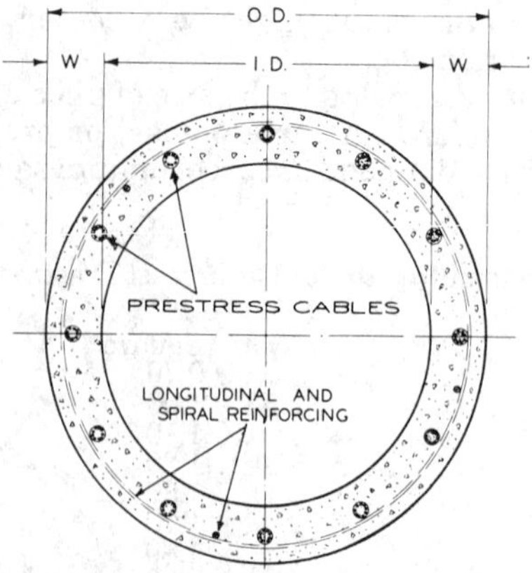

Prestressed Cylinder Piles

Pile Load Testing. When required, load testing might cost anywhere from $5,000 to more than $20,000. Because of the wide variation in piles, and in conditions under which they are tested, it is difficult to give dependable costs without knowing the details of an individual installation. Costs are affected by factors such as the information that is to be obtained and the quarters in which the test is to be undertaken.

There is a number of methods for determining the load-carrying capability of a pile: static load testing with intermittent addition of load; constant rate of penetration; dynamic testing while driving; and others. Before a pile load test is made, the proposed apparatus and structure for reactions and loading should be fully detailed and costs applied. Load tests are normally carried to twice the design load, and pile tests can be in excess of 500 tons.

For a load test, piles have to be driven. Often, reaction piles are also installed to counteract the test force. This requires the same materials and equipment as planned for production driving. Another pile or two, perhaps with a cast-steel point for comparison, can be added with little increase in the cost.

Typical Pile Load Test Setups. Supports for loads or reaction piles should be at least five pile diameters or 5 feet from the pile under test. In some soils, this should be as much as 10 diameters. The object is to keep the reaction far enough away that it will not affect the pile under test.

Pile load tests vary because of the different types of tests and the various types of piles. The test might vary from $5,000 to $50,000. In some rare cases, the cost has gone as high as $150,000. Not all piles require a load test. Usually, a pile of 20-ton capacity or less does not require tests.

Some local codes mandate a minimum length that must be driven before a pile is considered a pile (for example, 10 feet of pile driven tip to cut off). Pile driving contractors commonly quote projects giving several different unit costs:

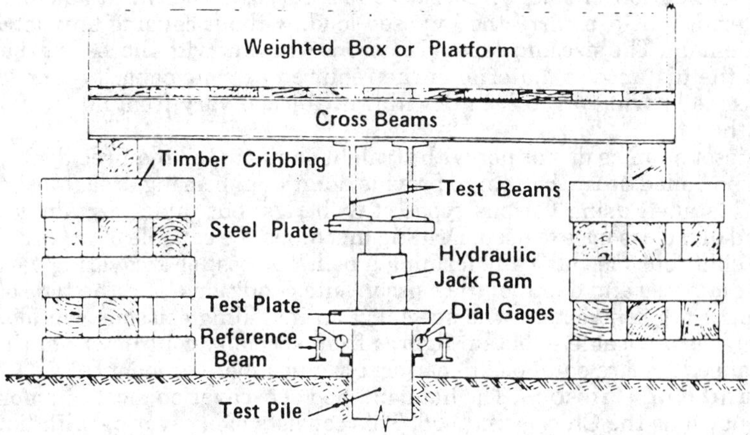

Typical setup for pile load testing in axial compression using a loading platform

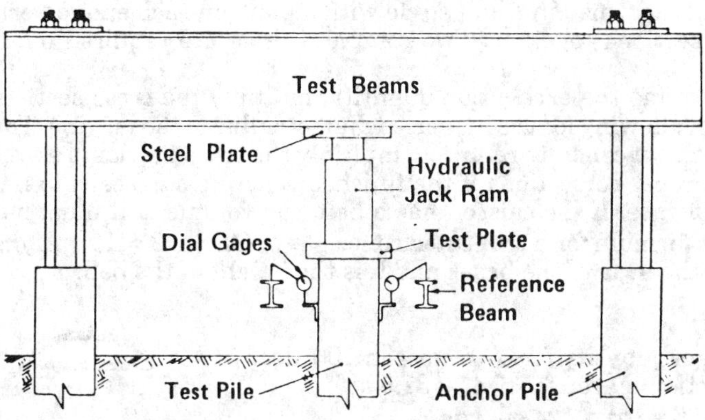

Typical setup for pile load testing in axial compression using anchor piles.

Item Description	Bid Unit
1) Mobilization	Lump Sum
2) Pile Load Test	Each
3) Pile Costs	Each or Lineal Foot
4) Demobilization	Lump Sum

Some pile driving contractors combine the mobilization and demobilization. On small jobs, where tests are not required, they will quote the project as a lump sum.

02350 CAISSONS

Caisson foundations are normally used to support heavy structures where soil conditions near grade cannot support them. Where such conditions exist, the foundation should be carried down through the unsatisfactory soil to material that can carry the imposed load, without causing any detrimental settlement. The size and depth of caissons will vary with the loads to be carried and the distance to material of the required bearing capacity. For example, caissons bearing on rock in downtown Chicago vary from 90 to 190 feet in depth.

Caissons can be dug either with straight shafts to the required depth, or they can be belled at the bottom to provide additional bearing area. Most caissons are installed using various types of drill rigs, but under certain conditions hand dug caissons are used. Drilling machines are capable of not only drilling the shaft but mechanically forming a bell. The cost of excavating varies with the diameter and depth of the caisson, site conditions, and the type of soil.

There are contractors who specialize in installing caissons. On jobs of any size, it is advisable to obtain figures from specialty contractors in this work, because they are equipped to perform it much more economically.

Hand Dug Caissons. The hand method of excavation most commonly used is known as the Chicago method. This consists of excavating with hand tools and lining the hole with wood lagging held in place by steel rings. The wood lagging has beveled edges and is tongue grooved. The lagging is milled from either 2"x6" or 3"x6" mixed hardwood in 16' lengths. The 16' lengths are commonly cut into thirds so that a set of lagging is 5'-4" long. Three inch steel channels are formed in a half circle with a plate on each end to permit bolting the half sections together. Two complete rings are required for each set of lagging.

To determine the excavation quantity, multiply the cross-sectional area by the length allowing for the increased diameter due to the lagging. To determine the quantity of concrete required, multiply the cross-sectional area by lengths making proper corrections if the finished concrete surface is to be below the ground surface. If the caisson has a bell, the volume of it must be computed using the formula for a truncated cone: $V = \frac{1}{3}(A1 + A2 + A1 + A2)h$, where A1 + A2 are the area of the bases and h is the height of the bell.

Excavation quantity for 6'-0" caisson 106' deep
Cross-sectional area for 6'-4" = 31.5 sq. ft.
31.5 x 106 x 1/27 = 124 cu. yds.
Concrete quantity:
Cross-sectional area for 6'-0" = 28.3 sq. ft.
28.3 x 106 x 1/27 = 111 cu. yds.
Allow 5% for waste or a total of 117 cu. yds.

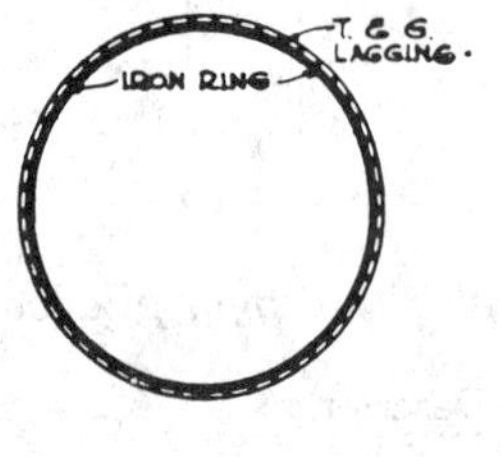

After determining the excavation quantities and the concrete quantities, it is necessary to determine the amount of lagging and rings. By cutting the bevel and tongue and grooves in the lagging, the width of the lagging is reduced, so that it requires 7 pieces of 2"x6" per foot of diameter of the caisson. For example, a 4' diameter caisson requires 28 pieces of lagging around the circumference. Therefore, using 2"x6" lagging, which is one board foot, it would require 28 board feet per foot of depth for a 4' caisson or 42 board feet using 3" thick lagging. To determine the amount of lagging required in a 6' diameter caisson, 106' deep using 2" lagging, multiply 6 x 7 or 42 board feet per foot of depth x the depth, which is 106' or 4,452 board feet. If the lagging had been 3", it would be 6,678. A caisson 106' deep would require 20 sections of 5'-4"

lagging. Using 2 rings for each section of lagging, the caisson would require 40 rings.

The rings are usually made from 3" channels, weighing 4.2 lbs. per foot. Each ring will weigh approximately 80 lbs. plus 10 lbs. for the plates at the ends of the half circles, for a total of 90 lbs.

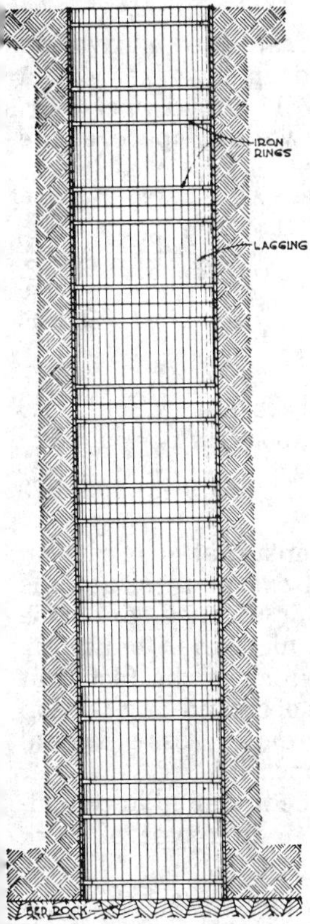

Section Through Caisson Showing Lagging in Place.

Labor Required for Caissons. Labor costs of caissons excavating varies with the diameter and depth of caissons, whether hand or pneumatic tools are used, the type of soil, and whether or not water or sand is encountered. It is generally assumed that one caisson digger can hand dig approximately 4 to 5 yards per 8 hour shift and place the lagging and rings.

Only one man can work in a 4' diameter shaft, but two men can work in a 5' diameter or larger shaft. In a 6' caisson, two men each taking out 4 yards should be able to excavate approximately 7' per shift. At the surface, a pneumatic tugger hoist and tripod would be required with two laborers to operate the hoist and dump the buckets. Generally, several caissons are constructed simultaneously, so the cost involved in furnishing compressed air for the hoist is spread over three or more caissons.

In addition to the above, it will be necessary to allow for time of foreman, timekeepers, material clerks, engineers, and cost of setting up and removing equipment, which will cost from $4.00 to $6.00 per cu. yd. It may be estimated separately under overhead expense and an amount allowed to cover the entire job.

To summarize the costs involved for constructing a caisson 6 ft. in diameter and 106 ft. deep (with two others being constructed simultaneously), time to excavate and lag:

124 cu. yds. @ 4 cu. yds./shift/digger = 31 shifts.
Using two diggers, this would be 15 ½ shifts.
Allow 16 shifts.

Labor	Hours	Rate	Total	Rate	Total
Caisson diggers	256	$....	$....	$19.13	$4,897.28
Hoist operator	128	....	....	25.96	3,322.88
Top laborers	256	....	....	19.13	4,897.28
Compressor operator	43	....	....	25.96	1,116.28
					14,233.72
Material					
Lagging	4.452 mbf	....	....	$760.00	$3,383.52
Rings	40	....	....	55.00	2,200.00
					$5,583.52
Equipment					
Hoist	16 shifts	....	....	$25.00	$400.00
Air tools	32 shifts	....	....	40.00	1,280.00
Compressor	5 ⅓	....	....	50.00	266.65
					$1,946.65
Removal of Excavated soil	124 cy	....	....	$10.00	1,240.00
Grand Total			$....		$23,003.89
Cost per cu. yd.			....		$185.52

To the above cost it will be necessary to add for setting up and removing necessary plant and equipment, electric current and wiring, concrete furnishing and placing, overhead expenses, contingencies, miscellaneous expenses, and profit.

Work of this type in the Chicago area generally costs from $200.00 to $275.00 per cu. yd. of completed caisson.

Excavating for Caissons with Pneumatic Diggers. When excavating for caissons 5'-0" to 8'-0" in diameter and up to 100'-0" deep, a worker with a pneumatic digger will loosen and load into buckets 9 to 11 cu. yds. per 8-hr. shift, depending upon the toughness of the clay and the working space, at the following cost per cu. yd.:

	Hours	Rate	Total	Rate	Total
Labor operating digger and loading buckets	0.90	$....	$....	$19.13	$17.22
Compressor operator	0.45	....	....	25.96	11.68
Compressor expense	0.45	....	....	11.25	5.06
Cost per cu. yd			$....		$33.96

Compressor and operator time at half time (assuming 2 men digging).

When excavating caissons by hand, two "diggers" will loosen and load into buckets, 6 to 7 cu. yds. per 8-hr. shift, at the following labor cost per cu. yd.:

	Hours	Rate	Total	Rate	Total
Labor excavating	2.5	$....	$....	$19.13	$47.83

Drilled Caissons. The use of drilling machines for the installation of caisson foundations has eliminated the hand digging method except in situations where adequate clearances to surrounding structures or utilities cannot be maintained. Drilled caissons have proved more economical than hand-dug caissons for several reasons, the most important of which is the fact that smaller shaft diameters may be used. The minimum size of a hand-dug caisson is 4' in diameter, which oftentimes provides a much greater cross-sectional area of concrete than is required for the load imposed upon it.

Another important reason is that drilled caissons can be installed in considerably less time. There are many firms specializing in this type of work throughout the country.

A typical job, consisting of 100 caissons with 2'-0" shaft diameters, 5'-0" bells, and 25' deep, might be estimated as follows, assuming normal clay digging. First, compute the total volume of excavation, which in this case would be 418 cu. yds. To this figure add about 10% overbreak, making a total of 460 cu. yds. Assuming each drill rig can complete 5 caissons per day, the job would take 20 days. Allow approximately 10% extra time for weather and total of 23 days to complete the job.

Labor	Hours	Rate	Total	Rate	Total
Drill Rig Operator	184	$....	$....	$25.96	$4,776.64
Oiler	184	....	....	19.13	3,519.92
Caisson Laborer	184	....	....	19.13	3,519.92
Laborer	184		....	19.13	3,519.92
					$15,336.40
Equipment					
Drill rig	23 days	$....	$....	$800.00	18,400.00
Material					
Concrete	460 cy	....	....	$55.00	25,300.00
Grand Total					59,036.40
Cost per cu. yd. (418 cu. yds.)					$141.24

These costs are for the complete caisson installation. Allowances must be added for reinforcing steel, additional labor based on union requirements, temporary or permanent casings if necessary, soil and concrete testing, surveying, disposal of excavated materials, and equipment move and set-up charges, if these items are included in the caisson contractor's contract, and overhead and profit.

Belling Tool for Hard-pan Caissons

Churn Drills. In areas where subsurface conditions are such that rotary drilling or augured caissons are difficult or impossible, the use of churn drills is usually effective. This caisson installation is sometimes known as the "drilled in" caisson method.

The cost for a 30" diameter drilled in caisson, complete in place, which includes the concrete, can range from $800.00 to $1,500.00 per lineal foot.

If bentonite is used in the installation of the caisson to produce the drillers mud, it is necessary to figure additional cost for the disposal of the bentonite. This can be an additional $200 to $500 per lin. ft. Labor cost for drilling 4 lin. ft. is as follows:

	Hours	Rate	Total	Rate	Total
Driller	8	....	....	$25.96	207.68
Helper	8	....	....	19.13	153.04
Cost for 4 lin. ft.					$360.72
Cost per lin. ft.					90.18
Equipment					
Churn Drill	8	....	....	100.00	800.00
Cost per lin. ft.					200.00

02400 SHORING

Wood sheet piling and bracing is usually estimated by the square foot, taking the number of sq. ft. of bank or trench walls to be braced, plus an additional amount for penetration at the bottom of the excavation and an allowance for extension above the top, and then estimating the cost of the work at a certain price per sq. ft. for labor and material required.

Sheet piling or bracing is required where the soil is not self supporting or where the excavation banksides cannot be sloped back. There is usually considerable vibration adjacent to railroad tracks and roadways that can cause the banks to cave in. Unless there is ample space on all sides of the excavation, the banks should be sheet piled to avoid damage to streets, alleys, or adjacent buildings.

Trenches or foundation piers 5'-0" to 8'-0" deep and 3'-0" to 4'-0" wide may require bracing or sheet piling, but this is much simpler than is ordinarily required for basement excavations.

The cost of this work will vary with the kind of soil, depth of excavation, amount of bracing or sheet piling required, and the method used to place. In ordinary excavations where no water is encountered, 2" or 3" square edge planks may be used, but if running sand or water is encountered, tongue and grooved planks should be used, as it is easier to keep them in line, and they are more water tight than square edge planks. It is customary to cut the bottom edge of each plank on a slight angle, so that in driving it is wedged against the preceding plank. The upper corners of the plank should also be cut off so that the effect of the driving will be concentrated along the vertical axis of the plank.

It requires 3 to 3 1/2 b.f. of lumber to sheet pile one sq. ft. of trench bank 5'-0" to 8'-0" deep, but on large deep excavations it will require 6 to 8 b.f. of lumber per sq. ft. of wall.

Sheet piling is usually driven with a pneumatic hammer designed especially for this purpose.

On very large jobs an ordinary pile driver is used for driving the sheet piling, but this is only on jobs of sufficient size to warrant the cost of a pile driver and crew.

Bracing and Sheet Piling Trenches and Piers. When excavating trenches and piers 5'-0" to 8'-0" deep, it is not always necessary to sheathe the banks solid but two or three lines of braces placed along the sides of the trench (as illustrated) will often be sufficient.

Where necessary to sheathe the banks solid, it will require about 3 to 3 ½ b.f. of lumber per sq. ft. of trench bank. A method of determining the quantity of lumber required for sheet piling a trench similar to the one illustrated is as follows:

Compute the quantity of lumber necessary to sheet pile a trench 50'-0" long and 7'-0" deep, sheathed on both sides. To the depth of 7'-0" must be added an amount for penetration. If 1'-0" will be sufficient, then figure 8'-0" lumber. If, however, 2'-0" penetration is required, then 10'-0" lumber must be figured leaving 1'-0" to extend above grade. Assuming 2'-0" penetration is required and 10'-0" lumber must be used, this is equivalent to 100 lin. ft. of sheathing 10'-0" high, or 1,000 sq. ft., and the following lumber will be required:

	B.f. of Lumber
150 pcs. 2"x8"-10'-0" sheathing	2,000
12 pcs. 4"x6"-16'-0" stringers or "wales"	384
26 pcs. 4"x6"-2'-0" braces spaced about 4'-0" apart	104
Total lumber required	2,488
Feet of lumber, b.f. per sq. ft. sheathing, (divide by 1,000)	2.49
Feet of lumber, b.f. per sq. ft. trench bank, (divide by 700)	3.55

This is based on using 2" plank for sheathing, 2 rows of 4"x6" stringers or "wales" placed near the top and bottom as illustrated, and 4"x6" or 6"x6" braces. Trenches over 10'-0" deep will require an additional line of stringers for each additional 4'-0" to 5'-0" depth.

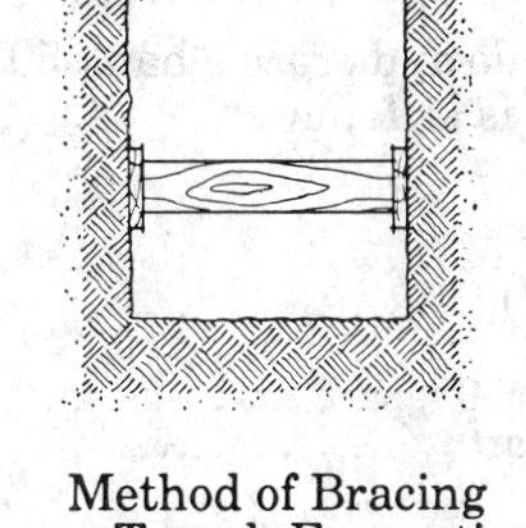

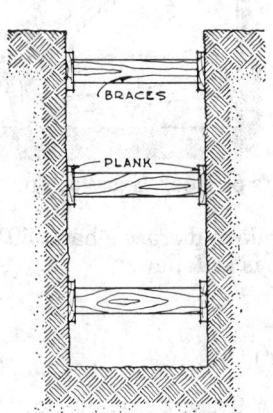

Method of Bracing Trench Excavation

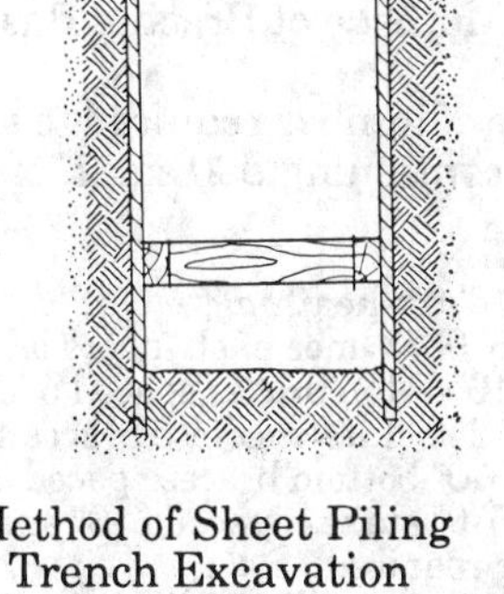

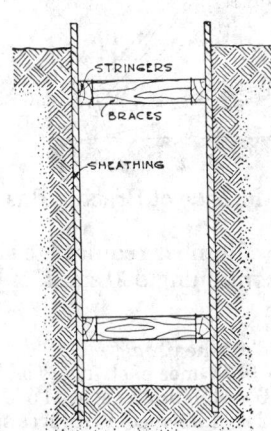

Method of Sheet Piling Trench Excavation

Most contractors who do a volume of this work use trench jacks instead of 4"x6" or 6"x6" braces. These jacks cost from $200.00 to $300.00 each and would not be economical for the contractor who does only an occasional job of this sort. The savings in material and labor is so slight that it is practically negligible.

Sheet Piling for Basements and Deep Foundations. On general basement excavation or large piers 8'-0" to 12'-0" deep, where it is necessary to brace any or all of the outside banks, the following lumber is required for a bank 50'-0" long, 8'-0" deep, and containing 400 sq. ft. of bank to be sheathed:

	B.f. of Lumber
80 pcs. 2"x8"-10'-0" sheathing	1,067
6 pcs. 6"x8"-16'-0" (2 lines of stringers or "wales")	384
10 pcs. 8"x8"-12'-0" top braces spaced about 5'-0" apart	640
10 pcs. 8"x8"-6'-0" bottom braces spaced about 5'-0" apart	320
10 pcs. 8"x8"-6'-0" bottom braces spaced about 5'-0" apart	320
Total lumber required	2,731
B.f. per sq. ft. of sheet piling, (divide by 500)	5.46
B.f. per sq. ft. of bank, (divide by 400)	6.83

On deep excavations and piers 14'-0" to 20'-0" below grade, 3" plank should be used for sheathing and either 6"x8" or 8"x8" lumber for stringers or "wales" and 8"x8" or 10"x10" timbers for bracing. Lines of stringers or "wales" should be spaced 4'-0" to 5'-0" apart, depending upon the earth pressure.

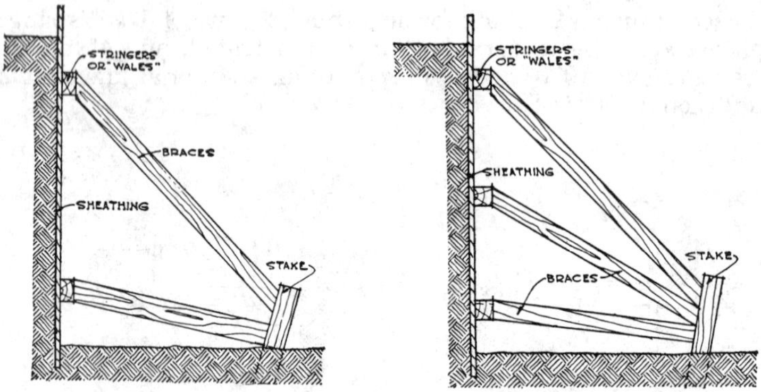

Methods of Bracing Basements or Deep Excavations

An example of lumber required to sheet pile and brace a bank 50'-0" long and 16'-0" deep, containing 800 sq. ft. of bank, is as follows:

	B.f. of Lumber
80 pcs. 3"x8"-18'-0" sheathing	2,880
9 pcs. 8"x8"-16'-0" (3 lines of stringers or "wales")	768
10 pcs. 8"x8"-16'-0" top braces spaced 5'-0" apart	853
10 pcs. 8"x8"-12'-0" Intermediate braces spaced 5'-0" apart	640
10 pcs. 8"x8"-8'-0" bottom braces spaced 5'-0" apart	427
10 pcs. 8"x8"-6'-0" stakes spaced 5'-0" apart	320
Total lumber required	5,888
B.f. per sq. ft. of sheet piling, (divide by 900)	6.54
B.f. per sq. ft. of bank (divide by 800)	7.36

Cost of lumber should be computed according to number of times it can be used on the job. Where labor hours are given, either carpenters or laborers may be used, as required.

Lumber Required for Sheet Piling

Size of Timbers	Width Sq. Edge	100 LF Covers SF	BF per 100 SF	Width T & G	100 LF Covers SF	BF per 100 SF
2"x8"	7 ½"	62 ½	215	7 ¼	60	220
3"x8"	7 ½"	62 ½	320	7 ¼	60	330
2"x10"	9 ½"	79	210	9 ¼	77	215
3"x10"	9 ½"	79	315	9 ¼	77	320

Wood Bracing and Sheet Piling Banks By Hand
Quantities and Production are given per 100 sq. ft. of bank braced

Class of Work	Depth in Feet	BF Lbr. Req'd 100 SF Bank	No. SF Placed 8-Hr. Day	Hours Placing 100 SF	Hours Removing 100 SF
Bracing Trenches	5-8	100-150	150-200	4-5	1-1 ½
Trench Sheet Piling	5-8	300-350	80-90	9-10	2 ½-3
Trench Sheet Piling	10-15	325-375	65-75	11-12	2 ½-3
Basement Sheet Piling	8-12	675-725	40-50	17-18	5-5 ½
Basement Sheet Piling	14-20	750-800	35-45	20-21	5 ½-6

Cost of Bracing 100 Sq. Ft. of Trench Bank With Plank Bracing

	Hours	Rate	Total	Rate	Total
125 b.f. lumber	..	$....	$....	$0.76	$95.00
Labor placing	4.50	....	....	19.13	86.09
Labor removing	1.25	....	....	19.13	23.91
Cost per 100 sq. ft.			$....		$205.00
Cost per sq. ft.			....		2.05

Cost of Sheet Piling 100 Sq. Ft. of Trench Walls
5'-0" to 8'-0" Deep

	Hours	Rate	Total	Rate	Total
325 b.f. lumber	..	$....	$....	$0.76	$247.00
Labor placing	9.50	....	....	19.13	181.74
Labor removing	2.75	....	....	19.13	52.61
Cost per 100 sq. ft.			$....		$481.35
Cost per sq. ft.			....		4.81

Cost of Sheet Piling 100 Sq. Ft. of Trench Walls
10'-0" to 15'-0" Deep

	Hours	Rate	Total	Rate	Total
350 b.f. lumber	..	$....	$....	$0.76	$266.00
Labor placing	11.50	....	....	19.13	220.00
Labor removing	2.75	....	....	19.13	52.61
Cost per 100 sq. ft.			$....		$538.61
Cost per sq. ft.			....		5.39

Cost of 100 Sq. Ft. of Sheet Piling for Basements or Foundations
8'-0" to 12'-0" Deep

	Hours	Rate	Total	Rate	Total
700 b.f. lumber		$....	$....	$0.76	$532.00*
Labor placing **	17.50	....	....	19.13	334.78
Labor removing	5.25	....	....	19.13	100.43
Cost per 100 sq. ft.			$....		$967.21
Cost per sq. ft.			....		9.67

Cost of 100 Sq. Ft. of Sheet Piling for Basements or Foundations
14'-0" to 20'-0" Deep

	Hours	Rate	Total	Rate	Total
775 b.f. lumber		$....	$....	$0.76	$589.00*
Labor placing **	20.50	....	....	19.13	392.17
Labor removing	5.75	....	....	19.13	110.00
Cost per 100 sq. ft.			$....		$1,091.17
Cost per sq. ft.			....		10.91

*Lumber costs may be reduced if it is removed & used again.
**Omit labor removing bracing or piling if left in place.

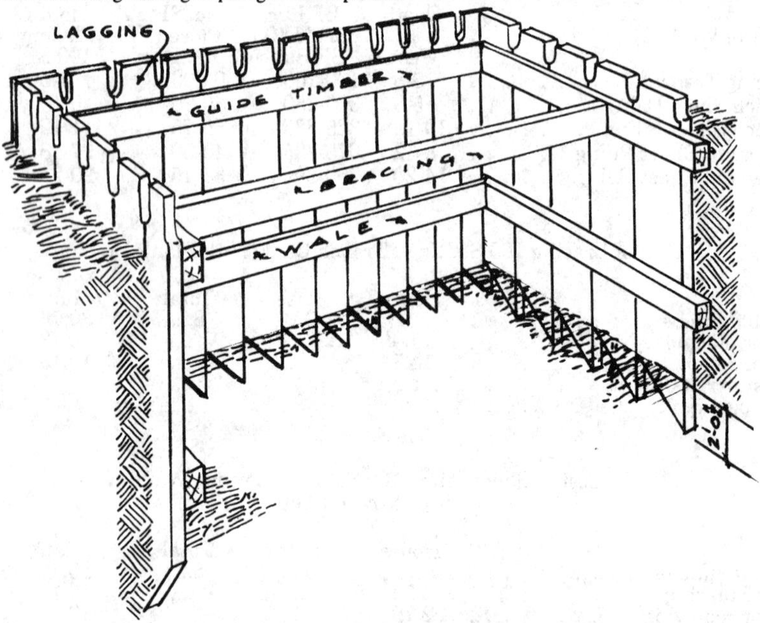

Method of Sheet Piling and Bracing Piers and Pits

Driving Sheet Piling by Compressed Air. The pneumatic pile driver is very effective for driving sheet piling. It is essentially a heavy paving breaker equipped with a special fronthead that adjusts for driving 2" to 3" piling. These machines weigh about 125 lbs. and will drive wood sheet piling into any ground that the piling can penetrate, such as sand, gravel, or shale, and any gradation between soils.

Driving speed varies from 2'-0" per minute in hard clay or shale to 9'-0" per minute in sand or gravel.

Under average conditions, two workers together on one machine should drive 100 lin. ft. (60 to 79 sq. ft.) per hr. at the following labor cost per 100 lin. ft.:

	Hours	Rate	Total	Rate	Total
Jackhammerman	1	$....	$....	$19.13	$19.13
Helper	1	....	....	19.13	19.13
Compressor & tools	1	....	....	13.50	13.50
Cost per 100 lin. ft			$....		$51.76
Cost per lin. ft			....		.52

Add for compressor engineer if required.

Cost of Sheet Piling 100 Sq. Ft. of Trench Banks 5'-0" to 8'-0" Deep, Using Pneumatic Hammers

	Hours	Rate	Total	Rate	Total
325 b.f. lumber		$....	$....	$0.76	$247.00*
Jackhammerman	2.00	....	....	19.13	38.26
Labor	4.00	....	....	19.13	76.52
Labor removing	2.75	....	....	19.13	52.61
Compressor & Tools	2.0	....	....	13.50	27.00
Compressor Engineer	2.0	....	....	25.96	51.92**
Cost per 100 sq. ft			$....		$493.31
Cost per sq. ft			....		4.93

Cost of Sheet Piling 100 Sq. Ft. of Trench Banks 10'-0" to 15'-0" Deep, Using Pneumatic Hammers

	Hours	Rate	Total	Rate	Total
350 b.f. lumber		$....	$....	$.0.76	$266.00*
Jackhammerman	2.00	....	....	19.13	38.26
Labor	4.50	....	....	19.13	86.09
Labor removing	2.75	....	....	19.13	52.61
Compressor & Tools	2.0	....	....	13.50	27.00
Compressor Engineer	2.0	....	....	25.96	51.92**
Cost per 100 sq. ft			$....		$521.88
Cost per sq. ft			....		5.22

*Lumber cost will vary with number of times it may be used.
**Add only if required.

STEEL SHEET PILING

Steel sheet piling is used for supporting soil in large excavations, cofferdams, caissons, and deep piers or trenches, where wood sheet piling is impractical. It is driven with a pile driver, the same as wood or concrete piles.

There are a number of different types of steel sheet piling on the market, but the following will give an idea of the common sizes and weights.

Sizes and Weights of United States Steel Sheet Piling

Section Number	Width Inches	Web Thick. Inches	Wt. Lbs. per Lin. Ft.	Wt. Lbs. per Sq. Ft. Wall
MZ38	18"	3/8"	57.0	38.0
MZ32	21"	3/8"	56.0	32.0
MZ27	18"	3/8"	40.5	27.0
MP102	15"	1/2"	40.0	32.0
MP101	15"	3/8"	35.0	28.0
MP113	16"	1/2"	37.3	28.0
MP112	16"	3/8"	30.7	23.0
MP110	16"	31/64"	42.7	32.0
MP116	16"	3/8"	36.0	27.0
MP115	19 5/8"	3/8"	36.0	22.0

Estimating the Quantity of Steel Sheet Piling. When estimating the quantity of steel sheet piling required for any job, take the entire girth of the basement, cofferdam, or piers to be sheet piled. This gives the number of lineal feet required.

Example: Suppose you have a basement 100' x 125' to be sheet piled. Adding the four sides of the excavation, 100 + 100 + 125 + 125 = 450 lin. ft. Using MZ-27 Carnegie Illinois sheet piling 18" (1.5') wide, 450'-0" divided by 1.50' = 300. It requires 300 pcs. to sheet pile the excavation. Using sheet piling 30'-0" long, 300 x 30 = 9,000 lin. ft. weighing 40.5 lbs. per lin. ft. = 364,500 lbs.

A shorter method is as follows: 450 lin. ft. around excavation multiplied by 30'-0" long, equals 13,500 sq. ft. of piling at 27 lbs. per sq. ft. equals 364,500 lbs.

Either method is satisfactory. But it is necessary to know the number of pieces to be driven in order to compute the labor costs accurately.

Labor Driving Steel Sheet Piling. The labor cost handling and driving steel sheet piling will vary with the size of the job, length of piling, and kind of soil, together with conditions encountered on the job, as described under "Wood Piles".

Example: Find the cost of driving 300 pcs. of MZ-27 sheet piling 30'-0" long.

	Total	Total
Trucking pile driver, yard to job	$....	$1,500.00
Setting up pile driver, 3 days for crew	....	3,173.04
300 pcs. MZ-27 piling, 364,500 lbs. at $0.40	....	145,800.00
Driving piling (30 per day) 10 days @ $1,057.68	....	10,576.80
Fuel, oil, misc. supplies, 10 days @ $150.00	....	1,500.00
Dismantling pile driver at completion, 2 days	....	2,115.36
Trucking pile driver, job to yard	....	1,500.00
Equipment Rental or depreciation allowance 16 days @ $500.00	....	8,000.00
Cost 300 pcs. sheet piling	$....	174,165.20
Cost per pc.	....	580.55
Cost per sq. ft. (13,500 sq. ft.)	....	12.90

Sq. ft. costs will vary with weight of piling used, size of job, etc.
Add for overhead and profit.

02500 SITE DRAINAGE

Site drainage includes such items as subdrainage, foundation and underslab drainage, drainage structures, sanitary and storm drainage piping, and dewatering and wellpoints. Some of these will be subcontracted directly with specialists. Most will probably be included in the plumbing subcontract, while others may involve local utilities. Some items the contractor may elect to handle directly.

Drainage trenches must have pitch. To find the cubic yards of earth excavation, first determine the average depth by multiplying the pitch by the required length of run, divide by two and add this to the depth at the start of the run. The total length x required width x the average depth = cubic feet. Divide by 27 to find cubic yards.

Subsoil drains are usually vitrified clay pipe, set with lowest point at the same elevation as the bottom of the footing, pitched a minimum of 6 in. in 100 lin. ft., laid 1/4" apart and joint covered with 15# asphalt felt or #12 copper or aluminum mesh. Four-inch pipe is minimum, but 6" pipe is preferred. One laborer can lay approximately 10 ft. of 4" pipe per hour, 8 ft. of 6" pipe. The 4" pipe costs about $1.00 per ft; the 6" pipe costs $1.40. Excavation is extra, and it is assumed the trench is properly sloped with a firm bottom before pipe is installed.

All sharp turns are formed with fittings, which cost $2.50 for 4" elbows, $4.75 for 6". Wide angles can be made by bevelling tile ends into easy radius bends.

The clay drain tile is laid on a thin bed of washed gravel. After the joints have been covered with felt or a filter cloth, more gravel is placed around and on top of the drain tile to allow water filtration into the drain tile itself. Approximately .75 to 1.0 cu.ft. of gravel per lineal foot of drain is required for proper filtration.

In recent years, vitrified clay drain tile is being displaced by the use of 4" plastic tubing that is already perforated by the manufacturer. This 4" tubing comes in 25', 50', and 100' rolls. It is very flexible, and since you are not handling small individual pieces as with the vitrified clay, this particular material is installed very rapidly. The joints are made by a coupling sleeve very much like that used with plumbing PVC piping. It is wise to lay a strip of filter cloth above the plastic tubing prior to installing the washed gravel to keep all sandy fines from entering the plastic tubing and creating a stoppage. All other grading preparation and gravels are installed around this particular material the same as for vitrified clay tile.

Subsoil drains, whether perimeter or underslab, can be connected to an existing storm sewer (in some localities, to a sanitary sewer), can be trenched to an outfall at grade level away from the structure (depending on the existing terrain), or connected to a sump pump area within the lower level of the building. If subsoil drainage piping terminates at a sump pump, a submergible pump with an automatic float control switch and discharge piping will be needed.

DRAINAGE STRUCTURES

There are a number of types of drainage structures required in storm drain piping systems. These structures can be of masonry, cast-in-place concrete, or precast concrete.

Curb inlets and yard drains are one type of structure that is used to collect surface water run-off and discharge it into the storm water piping system. The size of the inlet is determined by the area of run-off, as well as the size of the discharge pipe leading away from the inlet structure itself. An average price for a curb inlet drainage structure is $250 to $300 per vertical lineal foot, and the average price of a yard drainage structure is $175 to $210 per vertical lineal foot.

Storm drainage manhole structures are required at certain intervals along a storm pipe line, depending on the local codes, and whenever a storm drainage pipe system changes direction. The majority of manholes are precast concrete. It is much quicker to set an entire manhole in the course of the day instead of keeping the open pit shored while constructing a cast-in-place concrete manhole and bottom chamber. Once the manhole structure is in place and the piping connected, the bottom of the manhole structure is channeled using brick and mortar to aid the water flow from the inlet pipe to the outlet pipe.

Headwalls and endwalls are usually cast-in-place reinforced concrete walls at the ends of pipe, where either water enters the pipe (headwall) or where the water discharges from the pipe (endwall). Quite often at a headwall, there is a concrete flume entrance slab, and there might be a trash rack (a grillage screen) at the entrance to collect debris.

At the endwall an outfall slab or rip-rap area is installed that prevents the discharge water from eroding the soil away. The price of headwalls and endwalls varies tremendously, because each is sized individually depending upon the size of the inlet or discharge pipe and the amount of earth and height of the earth bank above the pipe that is being retained.

STORM PIPING

Concrete, asbestos cement, and corrugated metal are often used for piping other than sewer. The relative material costs per lineal foot are about as follows:

Size of Pipe	Concrete	Asbestos Cement	Corrugated Metal
6"	$3.00	$2.50	$....
8"	3.25	3.00	3.70
10"	3.75	4.25	4.30
12"	4.00	6.00	5.25
18"	7.00	11.50	7.35
24"*	11.00	19.00	12.50
30"*	15.00	27.00	15.00
36"*	22.00	36.00	23.00

*Reinforced

One cannot choose the type of pipe by price alone. While the specifications and job conditions usually dictate the selection, wage costs can also bear heavily. Lighter weight pipe will lay much faster and can be handled on the site much more easily.

For 12" pipe, a 4-worker crew will lay about 100 lin. ft. of concrete pipe, 150 ft. of asbestos cement, and 200 ft. of corrugated metal per day. For 36" pipe, a crew will lay about 20 ft. of concrete pipe, 30 ft. of asbestos, and 60 ft. of corrugated metal per day.

WELLPOINT SYSTEM OF DE-WATERING

The following is general information only. For specific applications consult a local de-watering contractor. Wellpoint systems are well known for their ability to de-water water bearing soils, such as sand and gravel, so that excavation and construction of foundations can proceed in the dry.

In addition, wellpoint systems are used extensively in such work as soil stabilization, pressure relief for dams and levees, and for water supply for municipalities and industrial plants.

A wellpoint system in construction work is usually a series of properly sized wellpoints, surrounding or paralleling the area to be de-watered and connected to a header pipe by means of risers and swing piping. Header piping, in turn, is connected to one or more centrifugal pumps depending on the volume of water that must be handled.

Wellpoints are usually self-jetted into place to the correct depth and at the proper spacing to meet requirements. Some soil conditions, however, require pre-drilling or "hole-punching" before the wellpoints can be installed.

The discharge from the pumps should be piped to an area where it will not interfere with construction. The entire installation should be located so that it will interfere as little as possible with other divisions of the work.

This system is usually installed in advance of the excavation, but where the water level is low, it is sometimes possible that a portion of the excavation can be at a cheaper rate when done prior to the installation of the system.

Any de-watering problem that is beyond the scope of ordinary pumping, or which the contractor or engineer thinks can be done more economically with a wellpoint system, should be referred to the engineering department of a wellpoint company. In this way the problem will receive expert analysis, at no cost to the contractor or engineer. It is essential for most contractors to obtain expert advice on the layout and installation of wellpoint systems, because frequently, a relatively small excavation with unusual soil presents greater de-watering difficulties than a larger, deeper job in a well-graded medium sand.

After carefully considering the factors involved--such as soil characteristics, site accessibility, hydrology of area, size and depth of area, total dynamic head involved, excavation and construction schedules, labor and working conditions, and power facilities, the wellpoint company can make a layout recommendation for the job.

Layout recommendations are usually accompanied by a quotation giving rental rates and charges for all equipment required, together with an estimate of the labor hours needed to install and remove the system, fuel and lubrication required per operating day, transportation charges for the equipment to and from the job, and daily rates for salary and expenses of company demonstrator. Companies specializing in this type of work will furnish rental quotations on a "sufficient equipment" basis, i.e., a fixed rental rate regardless of the amount of equipment actually required to do the job. Some of these companies will submit lump sum contract proposals for de-watering projects covering furnishing, installation operation, maintenance and removal of all de-watering equipment with a guaranteed result.

Method of De-watering Using Wellpoint System

It is sound practice to have a demonstrator from the wellpoint company on the job site to supervise the initial installation of the system. The time and expenses is more than compensated for by his working knowledge of wellpoint systems and their installation.

Operation of a wellpoint system is usually performed on a 24-hr. day, 7 day per week basis and must continue until all work, depending on dry conditions, is completed.

The labor cost of operating the system must be determined by the contractor and is based on the length of time operation is estimated to be required, figuring around-the-clock operation.

From the above information the contractor or estimator should be able to arrive at a lump sum figure for the required wellpoint system to be used in his estimate.

As an example, take the case of a job in which an underground tank was to be constructed. The test borings indicated the soil to be sand with a ground water level at 6'-0" below the surface. The excavation for the circular tank was an average of 84'-0" in diameter and 18'-6" deep. To maintain dry conditions, the water level had to be lowered at least 12'-6". It was also estimated that a dry condition would have to be maintained for a period of 3 months.

The following sample estimate shows how a detailed de-watering estimate might be put together.

	Hours	Rate	Total	Rate	Total
Equipment Rental, First Month					
52 wellpoints, 2"		$....	$....	$16.00	$832.00
260 l.f. header pipe, 6"		....	....	1.80	468.00
50 l.f. discharge pipe, 6"		....	....	1.00	50.00
2 valves, 6"		....	....	40.00	80.00
1 wellpoint pump		....	....	950.00	950.00
1 jetwell pump		....	....	1,025.00	1,025.00
Equipment Rental, Second Month					
52 wellpoints, 2"		....	....	10.00	520.00
260 l.f. header pipe, 6"		....	....	.90	234.00
50 l.f. discharge pipe, 6"		....	....	.50	25.00
2 valves, 6"		....	....	25.00	50.00
1 wellpoint pump		....	....	750.00	750.00
1 jetwell pump		....	....	850.00	850.00
Equipment Rental, Third Month					
52 wellpoints, 2"		....	....	10.00	520.00
260 l.f. header pipe, 6"		....	....	.90	234.00
50 l.f. discharge pipe, 6"		....	....	.50	25.00
2 valves, 6"		....	....	25.00	50.00
1 wellpoint pump		....	....	750.00	750.00
1 jetwell pump		....	....	850.00	850.00
2 weeks rental, 200 l.f. jetting hose		$....	$....	$200.00	$400.00
Labor	160	....	....	19.13	3,060.80
Supervisor, 5 days		....	....	200.00	1,000.00
Supervisor's expenses		....	....	500.00	500.00
Operating Expense					
Diesel fuel 91 days at 40 gals. per day		....	....	1.30	4,732.00
Lubricating oil, 91 days at 1 gal. per day		....	....	5.00	455.00
Pump operator, 3 shifts, regular time	1,560	....	....	25.96	40,497.60

	Hours	Rate	Total	Rate	Total
Pump operator, 3 shifts, overtime	624	....	....	38.94	24,298.56
Removal of Equipment Labor	60	....	....	19.13	1,147.80
Transportation to and from job, 16,000 lbs. at $4.00 per cwt		....	....		640.00
Total cost of wellpoint system			$....		$84,994.76

Add for insurance, social security, unemployment compensation, overhead and profit.

The jetwell pump is a double-duty unit and is connected into the system as a stand-by wellpoint pump after installation of wellpoints has been completed.

DEEP WELL SYSTEM OF DE-WATERING

A cheaper alternative to the wellpoint system of de-watering is the use of deep wells, which are 18" to 36" diameter shafts drilled into the ground and lined with a perforated, corrugated metal pipe in which an electric submergible pump is installed. This pump can be discharged directly onto the ground or into another collector pipe discharge system.

The size (diameter and depth) of the drilled shafts, the distance spaced apart, and the size of the electric submergible pump is determined by the depth of the planned excavation, the elevation of the water table, and the subsurface water flow.

When installing the submergible pump, the use of rigid PVC piping is preferred over flexible canvas discharge hoses. Depending on the depth of the shaft, a pump will be required to have sufficient head pressure to pump the water up to ground elevation. Therefore, any restrictions in collapsed discharge piping would create an additional burden for the pump and losing pump efficiency.

Pump costs vary widely depending upon the size of the pump, the head pressure rating, and the GPM capacity. Pumps can be purchased or rented, and the decision should be based on an evaluation of both methods for the amount of time that the pumps will be in operation.

EJECTOR SYSTEM OF DE-WATERING

The ejector method has a distinct advantage over the wellpoint system. Ejectors are not limited by depth, eliminating the need for multiple stages of wellpoints. When conditions are right, ejectors can cost less than deep wells and can be more effective, but to select the right system of de-watering for a specific project, one must have a good knowledge of the project and a thorough understanding of the various methods.

Installation of ejectors is the same as for wellpoints. The system works by supplying water down a pipe to create a vacuum and then pumping up the ejected water and the ground water. The *two-pipe ejector* system, which is the preferred method, works by pumping water into one pipe and drawing it out the other pipe.

One disadvantage with ejectors is the need for larger pumps with higher operating costs, because one is pumping out more water, both ejected water and ground water.

SUBMERSIBLE PUMP INSTALLATION

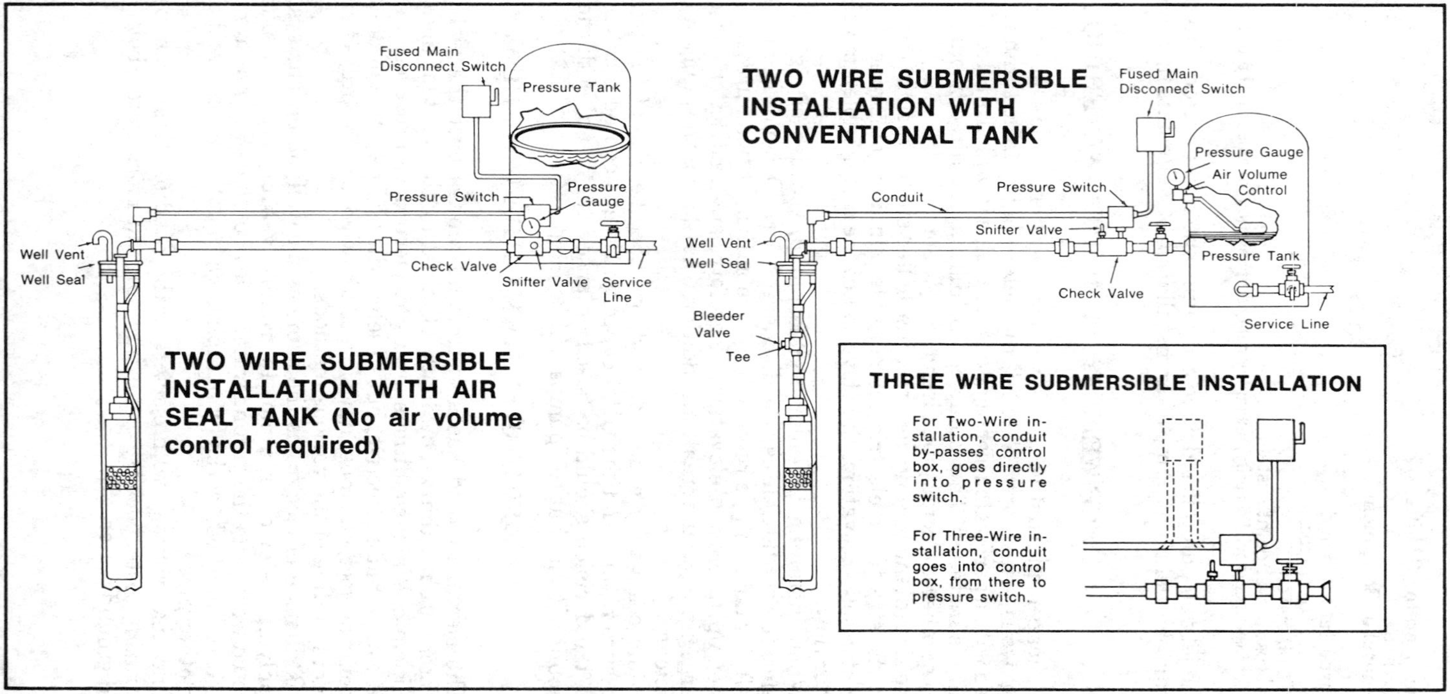
Fused Main
Disconnect Switch
Pressure Tank
Pressure Switch
Pressure
Gauge
Well Vent
Well Seal
Check Valve
Snifter Valve
Service
Line
TWO WIRE SUBMERSIBLE
INSTALLATION WITH AIR
SEAL TANK (No air volume
control required)
TWO WIRE SUBMERSIBLE
INSTALLATION WITH
CONVENTIONAL TANK
Fused Main
Disconnect Switch
Pressure Gauge
Air Volume
Control
Conduit
Pressure Switch
Snifter Valve
Well Vent
Well Seal
Pressure Tank
Check Valve
Service Line
Bleeder
Valve
Tee
THREE WIRE SUBMERSIBLE INSTALLATION
For Two-Wire installation, conduit by-passes control box, goes directly into pressure switch.
For Three-Wire installation, conduit goes into control box, from there to pressure switch.

Ground water quality can create discharge problems. It is always advisable to do a water analysis before deciding on and designing the de-watering system.

Upon completion of the project and termination of the need for the de-watering deep well system, the pump and piping are extracted from the hole and sand or gravel material is used to fill the shaft to prevent potential hazards of future cave-in.

The following two charts give the cost per vertical lineal foot for drilled shafts and perforated corrugated metal pipe liner.

Drilled Shafts		Installation Of Perforated Corrugated Pipe	
Dia. of Shaft	Price Per Lin. Ft.	Dia. of Pipe	Price per Lin. Ft.
18"	$12.00	18"	$14.50
24"	15.00	24"	22.00
30"	20.00	30"	27.75
36"	26.00	36"	42.00

02550 SITE UTILITIES

Gas and water are usually brought to the property line or building by the local utility company. It is then picked up by the piping contractor doing the building work. Costs on this work are discussed in Chapter 15, *Mechanical.* The distribution of utilities on new subdivision work is a specialized field subject to many code, health, and sanitation requirements.

Installing Service Pipes. Many public utility companies, instead of breaking up sidewalks and digging trenches, are installing service pipes by means of portable compressors and pneumatic tools. Where no rock is encountered, it is possible to drive pipe directly through the earth into cellars and basements.

This method consists of placing a pointed steel cap over the end of the pipe, which can be removed after the pipe holes through.

DRILLING WATER WELLS

Pollution of streams, rivers, and lakes has led many municipalities to look elsewhere for sources of water. Many have had to install wells as an alternate source of water. Industries sometimes also have to go to wells for their water supply where other sources are not available or are inadequate for their demands.

Large quantities of water are usually found in glacial deposits of sand and gravel. The quality of water from a glacial deposit is usually harder and of lower temperature than water derived from sand rock.

When it is necessary to locate large quantities of water for a town or city, it is customary to explore first by drilling a test hole to locate a water bearing strata, and to determine if there is an ample supply of water at that level. If a strata is found that is determined to be inadequate, blank pipe is then driven to the water bearing strata and a well screen is set in it.

Water travelling through a well screen has a tendency to precipitate the solids from the water on the screening, which eventually will restrict the water flow through it. Voids in the sand and gravel surrounding the pipe and screen may also become filled with the precipitates from the water. Water with a high mineral content may close the screen openings in a relatively short period of time, while water with low mineral content may not restrict the flow of water appreciably for many years. When screens do become clogged up with precipitates they sometimes can be opened by treating them with acid. If this process is not effective, it might be necessary to install a new well.

Rock wells are installed by driving a pipe through the overburden above the rock to the rock surface and continuing by drilling through the rock to the desired depth. It is sometimes possible to increase the quantity of water from a rock well in dense rock by shooting and fracturing the rock in the vicinity of the well. If the water bearing rock is already loose, porous, and creviced, shooting will probably not increase the water quantity.

Wells which extend into glacial sand and gravel deposits, will cost approximately $4.00 to $5.00 per inch in diameter per lineal foot of vertical depth. For example, a 12" diameter well would cost from $48.00 to $60.00 per lineal foot of depth.

Wells drilled into rock formations generally cost approximately $1.50 to $2.00 per inch of diameter per lineal foot of depth. A 6" diameter rock well would cost about $9.00 to $12.00 per lineal foot of depth. In the rock itself a pipe is not used, but if there is 100 ft. of earth or sand and gravel above the rock surface, it is necessary to include the cost of the pipe for this distance.

For wells with a capacity of 25 to 250 gallons per minute, submersible, multistage type pumps are generally used. Turbine type pumps are normally used for wells of more than 250 gallons per minute capacity.

Submersible pumps cost from $900.00 for a 10 gallon per minute pump to $10,000 for a 250 gallon per minute pump.

Turbine type pumps cost approximately $2,000 to $8,000 depending upon the capacity and the depth of the well.

It is recommended that quotations for drilling a well at a specific site be requested from local well drilling contractors who have knowledge of the area and know the approximate depth to the water bearing strata.

02600 PAVING AND SURFACING

Bituminous Roadways and Parking Lots. This section is confined to on-site roadways and parking lot construction for commercial and residential projects. Street and highway pavements are covered by the chapter on *Road and Highway Construction.*

Roadways and parking lots are constructed in any number of thicknesses, depending on the type of use, amount of traffic, gross vehicle weight anticipated, and the suitability of the subsoil conditions for the intended usage.

After careful attention is given to the designed grade contours and drainage patterns, and the rough grading operation is completed, the paving contractor performs the final grading with a motor grader. This machine is a rubber-tired tractor with a long cutting blade that is variable controlled for depths and angles of cut.

Once the motor grader has completed the final grading, the roadway or parking lot bed is rolled and compacted using a large rubber-tired compactor. While compacting the paving subgrade, the roller operator watches for any soft spots in the subgrade. If any are found, they are excavated and refilled with a select grade of backfill material. If these soft spots are missed and not repaired, the asphalt paving will flex and eventually break up over these soft areas causing costly repairs at a later date.

When the subgrade has been suitably prepared (graded and compacted), the base course of paving materials is laid to the proper thickness and compacted in preparation for the bituminous base course or bituminous wearing course.

This gravel base course can vary in thickness from 4" to over 20" depending on the paving composite design and is usually compacted to 98%. After the base course is installed, it may be sprayed with an asphalt tack coat just prior to the first bituminous paving course being laid.

The next step in the paving process is the laying of the bituminous paving materials to the proper thickness and rolling the surface to compact and smooth out the hot asphalt mix. The bituminous course can vary in thickness from 3" to 8" depending upon the requirements and can be laid in one course or can be split into two courses (binder course and wearing course). When the bituminous is laid in two courses, the top wearing course is usually a finer aggregate mix allowing for a much smoother compacted top surface.

Laying the bituminous paving in two courses will always cost a premium. However, there are some distinct advantages. It allows early paving of a work site without construction abuse of a finished surface. Should soft spots occur, they can be fixed prior to the finish wearing course installation. It allows sufficient time to completely check all areas for low spots (birdbaths) that collect water, and these can then be adjusted in the top wearing course thickness.

When estimating quantities of bituminous paving and base courses for a project, it is customary to figure all quantities in square yards.

Example: length in feet x width in feet = area in square feet divided by 9 = square yards.

Concrete Curb, Gutter & Paving

This section is not intended to cover road building but only small curb and gutter jobs, and small concrete paving projects. Large jobs of this kind require a vast amount of equipment, and the costs vary widely according to the amount of cut and fill required and the method of receiving and dispersing materials.

The item of plant and equipment should include the cost of transporting equipment to the job and removing it at completion and includes items such as bulldozers, loaders, power rollers, trucks, and wood or metal forms for curb and gutter. These items should be estimated on the basis of total transportation costs both ways and a rental charge for plant and equipment. The total should be divided by the number of sq. yds. of pavement in the job. It is customary to submit bids on paving at a price per sq. yd.

Rolling Streets and Roads. After the excavating and grading has been completed, the subgrade should be rolled to provide a solid foundation to receive the concrete base. A power roller should roll 130 to 150 sq. yds. of subgrade per hr. at the following cost per 100 sq. yds.:

	Hours	Rate	Total	Rate	Total
Operator	0.7	$....	$....	$25.96	$18.17
Power roller	0.7	....	....	22.00	15.40
Cost per 100 sq. yds.			$....		$33.57
Cost per sq. yd.			....		.34

Hauling Sand, Gravel, Crushed Stone, Etc. The cost of hauling paving materials will vary with the length of haul from pit or cars to point of delivery. Tables of hauling time are given under "Excavating".

Concrete Curb and Gutter

When estimating the cost of forms for concrete curb or curb and gutter, the size and shape must be considered and the amount of materials estimated.

Combined curb and gutter is usually 12" high at the outside edge (including pavement thickness), 24" to 30" wide and 6" to 7 ½" thick.

Separate concrete curb is usually 6" wide at the top, 8" to 9" wide at the bottom, and 24", 30", or 36" high.

Forms for Concrete Curb and Gutter. When placing wood forms for separate concrete curb, two mechanics and a helper should set forms for 100 to 110 lin. ft. of 24" curb per 8-hr. day, at the following labor cost per 100 lin. ft.:

	Hours	Rate	Total	Rate	Total
Mechanics	15.0	$....	$....	$24.04	$360.60
Helper	7.5	....	....	19.13	143.48
Cost per 100 lin. ft.			$....		$504.08
Cost per lin. ft.			....		5.04

Two mechanics and a helper should set forms for 90 to 100 lin. ft. of 30" concrete curb per 8-hr. day, at the following labor cost per 100 lin. ft.:

	Hours	Rate	Total	Rate	Total
Mechanics	17.0	$....	$....	$24.04	$408.68
Helper	8.5	....	....	19.13	162.61
Cost per 100 lin. ft.			$....		$571.29
Cost per lin. ft.			....		5.71

When setting wood forms for combined concrete curb and gutter, two mechanics and a helper should set forms for 80 to 90 lin. ft. of curb and gutter per 8-hr. day at the following labor cost per 100 lin. ft.:

	Hours	Rate	Total	Rate	Total
Mechanics	18.8	$....	$....	$24.04	$451.95
Helper	9.4	....	....	19.13	179.82
Cost per 100 lin. ft.			$....		$631.77
Cost per lin. ft.			....		6.32

Wood forms for concrete curb and gutter are ordinarily built of 2" lumber. Including all bracing, it requires 2 ¼ to 2 ½ b.f. of lumber per sq. ft. of forms. This cost will run from 15 to 20 cents per sq. ft. and should be added to the above setting costs, taking into consideration the number of uses expected from the forms. These forms can be used many times, if properly taken care of.

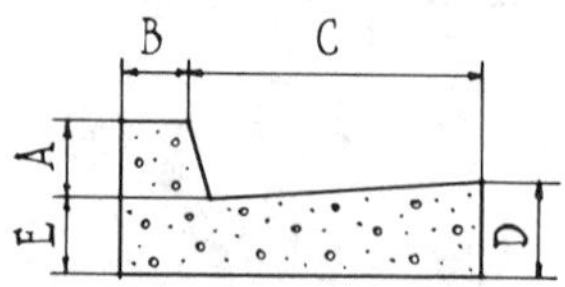

Combined Curb and Gutter

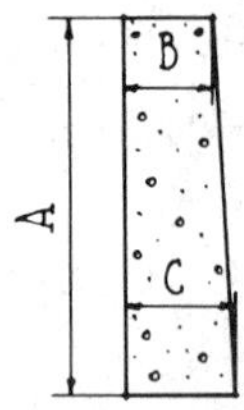

Separate Concrete Curb

Placing Ready-mix Concrete for Curb and Gutter. A two-worker crew should place about 1 cu. yd. of concrete an hr. at the following labor cost per cu. yd.:

	Hours	Rate	Total	Rate	Total
Labor	2	$....	$....	$19.13	$38.26

Concrete Roads and Pavements

Concrete roads and pavements could be thicker in the center than at the sides, so it will be necessary to obtain the average thickness of the pavement when estimating quantities of concrete materials.

Some pavements have uniform thickness. Others have uniform center section and thickened edges. In each case the area of the cross section must be determined and this is multiplied by the length to find the volume of concrete required.

Labor Placing Edge Forms for Concrete Paving. Wood edge forms or concrete paving usually consist of 2" plank of a width equal to the slab thickness, set to line and grade, secured to wood stakes, and braced.

Two mechanics and a helper should set to line and grade from 250 to 300 lin. ft. of edge forms per 8-hr. day at the following labor cost per 100 lin. ft.:

	Hours	Rate	Total	Rate	Total
Mechanics	3.0	$....	$....	$24.04	$72.12
Helper	1.5	....	....	19.13	28.70
Cost per 100 lin. ft.			$....		$100.82
Cost per lin. ft.			....		1.01

The above costs are based on the forms being straight or set to long radius curves. If sharp curves are required, the form must be either made from 1-inch boards, or the back of a 2" plank must be kerfed. In either case, the stakes must be set much closer together.

Concrete paving edge forms set to short radius curves will cost 50 to 100 percent more than straight forms.

Approximate Sq. Yds. of Concrete Pavement Obtainable from One Cu. Yd. of Concrete

Thickness Inches	Number Sq. Yds.	Thickness Inches	Number Sq. Yds.
4	9	8	4.5
4 ½	8	8 ½	4.25
5	7	9	4
5 ½	6.5	9 ½	3.75
6	6	10	3.5
6 ½	5.5	10 ½	3.5
7	5	11	3.25
7 ½	4.8	11 ½	3
		12	3

Placing Ready-mix Concrete Pavements. Using ready-mix concrete and wheeling the concrete into place with wheelbarrows, the labor per cu. yd. of concrete should cost as follows:

	Hours	Rate	Total	Rate	Total
Labor	1.2	$....	$....	$19.13	$22.96

If ready-mix concrete can be discharged from truck directly into place without wheeling, deduct 0.7-hr. labor time.

Larger jobs, with 100 cu. yds. or more of concrete, can realize better organization. Lower costs are the result. Assuming a pavement containing from 100 to 200 cu. yds. of ready-mixed concrete wheeled into place, the labor per cu. yd. of concrete should cost as follows:

	Hours	Rate	Total	Rate	Total
Labor	0.9	$....	$....	$19.13	$17.22

If ready-mix concrete can be discharged from truck directly into place without wheeling, deduct 0.4-hr. labor time.

Finishing Concrete Pavements. Finishing concrete pavements usually consists of a series of operations performed in the following sequence: strike-off and consolidation, straight-edging, floating, brooming, and edging.

Labor Finishing Concrete Pavements. Assuming an 8-inch concrete pavement being placed at the rate of 25 cu. yds. per hr. or 1,013 sq. ft. per hr., a crew of 10 cement masons should be able to keep up with the pour and perform all the above described finishing operations at the following labor cost per 100 sq. ft.:

	Hours	Rate	Total	Rate	Total
Cement Mason	1.0	$....	$....	$24.57	$24.57
Helper	0.4	....	....	19.13	7.65
Cement mason foreman	0.1	....	....	24.36	2.44
Cost per 100 sq. ft.			$....		$34.66
Cost per sq. ft.				....	.35

The above costs are based on a job that will run from 7,500 to 8,500 sq. ft. per 8-hr. day. On smaller jobs the costs will increase from 20 to 25 percent.

02700 SITE IMPROVEMENTS

02710 FENCES & GATES

Industrial chain link fences usually consist of 3" OD tubular end posts and 2" to 2 ½" OD intermediate posts depending upon the height of the fencing to be installed. Top and bottom rails are not always used, because having a top rail on a chain link fence gives much more stability for intruders to gain a handhold. Therefore, the spring wire is used for top and bottom stability in preventing chain link fabric sag. This spring wire is at least a 6 gauge material. The wire fabric is normally a 9 gauge wire and can be galvanized, aluminized steel, or vinyl coated in a number of colors.

For chain link fencing over 6' in vertical height, 6 gauge wire fabric will be used in lieu of the 9 gauge fabric.

Wire arms are available either in single or double sided for the installation of 2 to 4 strands of barbed wire per side.

The maximum line post or intermediate post spacing is 10' on center and these posts as well as in the corner posts are usually set in a 9" diameter by 3' deep concrete footing.

Industrial Chain Link Fencing - Cost Per Lineal Foot

Fabric	4' High 9 ga.	5' High 9 ga.	6' High 9 ga.	8' High 6 ga.
Galvanized	$7.00	$7.75	$8.25	$9.50
Aluminized Steel	7.25	8.50	9.00	10.50
Vinyl Coated	8.40	9.25	9.90	11.50

Add for barbed wire:

Single side - 3 strand	$.45/lin.ft.
Double side - 3 strand	.80/lin. ft.

Add for gates:

3' x 4'	$45.00 each
3' x 5'	60.00 each
3' x 6'	80.00 each
3' x 8'	105.00 each

RESIDENTIAL CHAIN LINK FENCING

The standard corner posts are 2" to 2 ½" diameter. The standard line posts are 1 ⅝" diameter set at 10' on center. Normally, with residential fencing, a top rail is added for stability.

Residential Chain Link Fencing - Cost Per Lineal Foot

Fabric	3' High 11 ga.	4' High 11 ga.
Galvanized	$4.15	$4.75
Aluminized	4.50	5.25
Vinyl Coated	4.70	5.50

Add for gates:

3' x 3'	$35.00 Each
3' x 4'	42.00 Each

Wood Residential Fences - Cost Per Lineal Foot

Type of Fence	Per LF	Per Each
Cedar Stockade Screen Fence 4' High	$8.50	--
Gate 3' Wide x 4' High	--	$55.00
Cedar Stockade Screen Fence 6' High	12.00	--
Gate 3' Wide x 6' High	--	80.00
Cedar Split Rail - 2 Rail 3' High	6.00	--
Cedar Split Rail - 3 Rail 3'6" High	6.75	--
Redwood or Cedar Board on Board 4' High	11.25	--
Redwood or Cedar Board on Board 6' High	14.00	--
Gates 3' Wide x 4' High	--	60.00
Gates 3' Wide x 6' High	--	85.00

02720 ROAD AND PARKING APPURTENANCES

Guard rails of formed steel are set on steel, wood, or concrete posts. Rail sections are set 18" to the centerline above grade. Posts are set every 12'-6" with rails lapped 11 ½" and attached with 7 ⅝" bolts. The 12 ga. sections weigh less than 100 lbs., so two workers can handle installation. Special terminal wings are furnished for exposed ends.

A budget price for this type of rail set on steel posts is $12.50 per lin. ft. and about $11.00 if set on wood posts.

The 4'x8' timber rails set on wood posts will cost about $6.00 per foot. They do not offer the protection of the metal rail but may better fit into the landscape.

Wheel stops for parking lots may be either wood or concrete. The concrete ones are shaped on the bottom to allow drainage; wood ones, unless set on standards, tend to collect water. Costs per lin. ft. installed are $3.25 for 6x6 wood dowelled into paving, $3.25 for concrete dowelled into paving and $2.40 for 4x4 wood set on metal brackets.

The 4" striping in parking lots will cost about $0.15 per foot or $3.00 per stall.

02760 SITE FURNISHINGS

Playground Equipment

Type of Equipment	Cost Each
Bike Rack, 10' long	$325.00
Horizontal Monkey Ladders, 12' long	575.00
Parallel Bars, 10' long	375.00
See-saw 2 Units	450.00
See-saw 4 Units	750.00
Slides - Stainless Steel, 12' long, 6' high	875.00
Swings - 4 seats, 8' high	800.00
Swings - 8 seats, 8' high	1,250.00

Benches

Type	Cost Each
Precast concrete with wood backs and seats, 8' long	$600.00
Fiberglass, 8' long	275.00
Fiberglass supports with wood backs and seats, 8' long	240.00
Wood, 8' long	200.00

Planters

Type of Planter	Cost Each
Precast concrete, 24" dia. x 18" high	$225.00
Precast concrete, 48" dia. x 30" high	375.00
Fiberglass, 36" dia. x 30" high	300.00
Fiberglass, 60" dia. x 36" high	525.00

02800 LANDSCAPING

Landscaping requirements are influenced to a considerable degree by the aesthetic appreciation of existing vegetation and the economics of retaining and protection costs during the construction period. Some owners and architects are dedicated to the preservation of the maximum number of trees that do not directly interfere with construction in an effort to blend the new construction into the existing environment. This theory produces exceptional immediate beauty to any new construction project but involves additional costs to construction by requiring extensive protective barriers for trees and vegetation and restricts work areas causing a lowering of productivity for automotive and heavy construction equipment.

In instances where existing trees and vegetation are of low quality, or the owner is unwilling to absorb the cost of retention, the site is cleared of all natural growth and provisions are made to provide new landscaping for the site at the conclusion of construction. The beauty of landscaping is subjective and may be limited to a combination of simple sodding, seeding, and minimal plantings. In other instances, it can progress to an elaborate extent, requiring the special talents of a landscape architect.

Most contractors lack the material sources, experienced personnel, and special equipment required for landscaping, and they rely upon landscape contractors to perform these services. Frequently, specifications require landscaping be performed by a certified contractor who is primarily engaged in this type of work and has the resources and capabilities to provide guarantees and certifications in compliance with specification requirements.

Estimators can develop a general knowledge of the costs involved in landscaping in order to evaluate the proposals received from the subcontractor.

Generally, nurseries providing service to homeowners and transit trade are not competitive in prices for commercial and large scale landscape projects. Soliciting proposals from qualified landscaping subcontractors presents a unique problem for the contractor. It has been found that these specialized subcontractors recognize that their performance will not be required until near the completion of a project, which may be many months in the future, and are reluctant to tender firm price proposals prior to the scheduled bid date.

Estimators must take a special effort to obtain these proposals in a timely manner to allow sufficient time to make an analysis of the bid before it is incorporated into the total project estimate.

The establishment of a work performance schedule to coincide with the accepted local planting season will tend to lower the subcontractor's price and will preserve the guarantee period. It is not unusual to provide for landscaping work slightly out of sequence in order to accommodate the planting season.

For the estimator to develop the proper unit of measure for pricing landscape work, or to determine the completeness of a proposal from a subcontractor, the following are required:

1) Area of seeding and fertilizing in square yards or acres on large projects.

2) Topsoil in cubic yards, with allowance for topsoil available from the site and for the amount that must be purchased from external sources.

3) Area of sodding in square yards.

4) Number, size, and species of shrubs.

5) Number, size and, species of trees.

6) Details of any special features, such as rock gardens.

7) Allowance for maintenance for specified periods.

8) Allowance for replacement guarantee, if specified.

After quantities, quality, and features are accurately determined, the estimator is prepared to develop a budget estimate that can be used as a cost comparison with the proposals received from landscaping subcontractors.

For general guidance the table below provides average cost ranges, but due to quantity, cost fluctuation, and geographical conditions, estimators should verify with local landscape subcontractors.

Item	Unit	In-Place Cost
Edging, 1x4 Redwood	Lin. Ft.	$1.15
Erosion Control - Plastic Netting	Sq. Yd.	1.00
Marble Stone Chips - Hand Placed	100#/Bag	9.20
Move Existing Plants - Replanting, 3'-0" high	Each	50.00
Seed, Fertilizer - Small Areas	Sq. Yd.	1.50
Mulch, Peat Moss 2" Deep	Sq. Yd.	1.40
Sodding	Sq. Yd.	2.50-4.00
Trees,		
Spruce 8'	Each	175.00
Birch 8'	Each	150.00
Pine-White 8'	Each	100.00
Dogwood 8'	Each	200.00
Shrubs,		
Honeysuckle 4'	Each	35.00
Forsythia 4'	Each	35.00
Plant Bed - By Hand	Sq. Ft.	3.75

Plant maintenance is necessary until rooting begins. The duration of this activity is determined by local weather conditions and the time of planting. It can extend for many months beyond the conclusion of the actual construction period, so allowances have to be made in the landscape estimate to accommodate this requirement. If the maintenance period extends beyond the conclusion of construction, it will necessitate the dispatch of personnel, including transportation of supplies and equipment, necessary to perform the maintenance work. The per trip cost will be considerably higher than if the work had been performed in the construction phase. These excess costs occur from the additional travel time involved for each trip, which would include the driver's wage cost plus a fair and reasonable allowance for truck rental. Also, the actual work performance would be in an unsupervised, which can mean low productivity.

PLANT GUARANTEE

Should the prime contractor assume the obligation of the guarantee, then the estimate should include an allowance for replanting shrubs and trees that fail to root and grow after transplanting. To arrive at a reasonable allowance for this cost, 10% to 20% of the original planting allowance should be sufficient.

CHAPTER 3

CONCRETE

CSI DIVISION 3
03100 CONCRETE FORMWORK
03200 CONCRETE REINFORCEMENT
03300 CAST-IN-PLACE CONCRETE
03400 PRECAST CONCRETE
03500 CEMENTITIOUS DECKS

When estimating the total cost of placing concrete for footings, foundations, walls, columns, slabs, and other items, there are five categories of labor operations to consider, in addition to the material and equipment costs:

1) *Form Preparation:* Concrete is placed in a fluid consistency. It must be contained or formed to the desired shape before its hardening process. The forming operation includes items such as: earth excavation to the desired shape for footings and foundations, together with supplemental wood forms when necessary for proper elevation, steps (changes in elevation), or unstable soil conditions; proper finish grading of earth surfaces for the placement of concrete slabs on grade to the correct thickness and elevation; and cutting, fabricating, placing, aligning, and bracing wood or metal forms for walls, columns, and slabs.

2) *Embedded Items Placement:* Labor will be required for the proper placement of all items to be cast into the concrete mix. Usually steel reinforcing (mesh or bars) will be necessary to insure the structural integrity of the concrete member. Non-structural items that frequently are installed into cast-in-place concrete include: anchor bolts, plates, or inserts for the future connection of other items of work, corner protection angles, and sleeves for the passage of pipes, conduit, and ductwork.

3) *Placement of Concrete:* The ready-mixed concrete that is delivered to the job site by truck must be physically moved from the truck to the location of placement by the use of a chute, wheelbarrow, buggy, or by some mechanical means such as a conveyor, pump, or crane. Any means of moving concrete mix involves a labor operation, whether it be a group of laborers, an equipment operator, or a combination. Once the concrete mix is in place, it is then necessary to tamp or vibrate the concrete for proper consolidation and "strike off" or screed the concrete to its designed level or elevation. In the event that only a small amount of concrete is required, or in remote areas where ready-mixed concrete is unavailable, it will be necessary to estimate the additional labor required to job mix the components (gravel, sand, cement, and water) for the quantity of concrete needed.

4) *Finish, Cure, & Protection:* Once the concrete mix is placed into the forms, vibrated, and screeded to the proper elevation, the next operation is the proper finishing of the exposed surfaces: a steel troweled finish to a slab surface to provide a smooth, hard and consolidated surface, free of voids and unevenness. Concrete requires a chemical action to change it from a fluid consistency to

the set or hardened condition, which generates a certain amount of heat. On hot days the heat generated by the setting process and the weather produces a concrete mix that sets too fast and hampers workability for the proper finish. It produces surface cracking, in some cases for the full depth of the slab. To help slow the setting process and to seal in moisture, slabs are covered with wet burlap, polyethylene, or a chemical curing compound. Slower curing time produces a stronger product.

Beyond proper curing and protection in hot weather, because of its water content prior to hardening, concrete must be kept from freezing by either covering with straw or some other insulating material to keep in its generated heat, or by a supplemental heating process such as covering the area and installing a temporary heat source. Concrete work, and in particular slabs, must be protected from any form of precipitation prior to the setting process, because the water droplets landing on the slab cause small craters in the finish surface, ruining the quality of finish, and dilute the cement ratio of the concrete mix, which weakens at least the top wearing surface of the slab.

5) *Form Removal:* After the concrete has hardened, the labor process of removing the forms begins. This entails stripping the nails from the forms, cleaning off the hardened concrete skim particles, preparing the forms for the next forming location by means of applying a form release agent (oiling), and physically moving the forms to the next area of the building for starting the forming process on another scheduled work item. The proper cleaning and oiling (coating the concrete contact surface) of form work is important, because it effects many reuses of form materials, reducing the costly expense of new materials and fabrication labor costs.

03100 CONCRETE FORMWORK

Concrete formwork is estimated by computing the actual surface area of the form that comes into contact with the concrete. This is referred to as the "contact surface". The formwork quantities are then categorized into specific groups (wall form, column form, etc.) and in the proper unit of measure (square foot or lineal foot) as follows:

Ribbon footing forms - Lineal Foot
Slab edge forms (by height) - Lineal Foot
Foundation edge forms - Square Foot
Wall forms - Square Foot
Pier forms - Square Foot
Column forms - Square Foot
Beam bottom forms - Square Foot
Beam side forms - Square Foot
Supported slab forms - Square Foot

Example: Compute the quantity of formwork required for a free standing wall 50'-0" long, 4'-0" high, and 1'-0" thick. Forms will be required for both sides of the wall. The formwork and quantity is obtained as follows:
Face Form: 50'-0" (length) x 4'-0" (height) x 2 (sides) = 400 sq. ft.
End Form: 1'-0" (width) x 4'-0" (height) x 2 (ends) = 8 sq. ft.
Total Contact Area = 400 + 8 = 408 sq. ft.

WOOD FORMS

Forms for Concrete Footings. There are two main types of footings in most construction work: column footings and wall footings. Column footings are usually square or rectangular masses of concrete designed to spread the column load over an area to accommodate the soil bearing value and are located at the column centers. Wall footings are usually ribbons of concrete, somewhat wider than the walls they carry, and are similarly designed to spread the loading over a larger area than the actual wall bottom thickness.

Forms for footings that are formed completely are usually simple and rough, using 2" planking for sides and 2"x4" stakes to hold the sides in place and 2"x4" struts for braces to the banks. This material may be reused numerous times in the course of the work. Plywood is sometimes substituted for the 2" planking for sides.

Ribbon forms for partly formed footings usually consist of 2" thick lumber in lengths as long as are practical, secured at the correct elevation to 2"x4" stakes, and tied across the tops at intervals of 4'-0" to 6'-0" with 1"x2" spreader-ties, except in the case of column footings, which are usually diagonally tied at the corners with 1-inch boards and tied transversely, if required, with strap iron.

The following are itemized costs of the column and wall footing forms found in the average simple job. Complicated or "cut-up" jobs require up to 25 percent more material and up to 50 percent more labor.

Material Cost of 100 Sq. Ft. of Column Footing Forms Based on Footings 5'-0"x5'-0"x1'-6" Completely Formed

Material	B.F.	Rate	Total	Rate	Total
Sides, 2" planking	200	$....	$....	$0.07*	$14.00*
Stakes and braces, 2"x4"	100	....	....	0.07*	7.00*
Nails,etc.		....	....	2.10	2.10
Cost per 100 sq. ft.			$....		$23.10
Material cost per sq. ft.			....		.23

Labor Cost of 100 Sq. Ft. of Column Footing Forms Based on Footings 5'-0x"5'-0"x1'-6" Completely Formed

Placing Forms--18-22 sq. ft. per hr.	Hours	Rate	Total	Rate	Total
Carpenter	5.0	$....	$....	$24.04	$120.20
Labor helping	2.5	....	....	19.13	47.83
Stripping Forms--90-110 sq. ft. per. hr.					
Labor	1.0	....	....	19.13	19.13
Cost per 100 sq. ft.			$....		$187.16
Labor cost per sq. ft.			....		1.87

Material Cost of 100 Lin. Ft. of Column Footing Ribbon Forms Based on Footings 5'-0"x5'-0"x1'-6" Partly Formed

Material	B.F.	Rate	Total	Rate	Total
Ribbons,2"x4",100 l.f.	67	$....	$....	$0.07*	$4.69*
Stakes, 2"x4", 3'-0" long	80	....	....	0.07*	5.60*
Corner ties, 1"x6", 2'-0" long	20	....	....	0.07*	1.40*
Nails, etc.		....	....	1.60	1.60
Cost per 100 Lin. Ft.			$....		$13.29
Material Cost per Lin. Ft.			....		.13

Labor Cost of 100 Lin. Ft. of Column Footing Ribbon Forms Based on Footings 5'-0"x5'-0"x1'-6" Partly Formed

Placing Forms--35 Lin. Ft. per hr.	Hours	Rate	Total	Rate	Total
Carpenter	3.0	$....	$....	$24.04	$72.12
Labor helping	1.5	....	....	19.13	28.70
Stripping Forms--450-500 Lin. Ft. per hr.					
Labor	0.21	....	....	19.13	4.02
Cost per 100 Lin. Ft.			$....		$104.84
Labor Cost per Lin. Ft.			....		1.05

Material Cost of 100 Sq. Ft. of Wall Footing Forms Based on Footings 1'-0" Deep Completely Formed

Material	B.F.	Rate	Total	Rate	Total
Sides, 2"x12" planks	200	$....	$....	$0.07*	$14.00*
Stakes and braces	75	....	....	0.07*	5.25*
Nails, etc.		....	....	1.30	1.30
Cost per 100 sq. ft.			$....		$20.55
Material cost per sq. ft.			....		.21

*Based on 6 uses of lumber.

Labor Cost of 100 Sq. Ft. of Wall Footing Forms Based on Footings 1'-0" Deep Completely Formed

Placing Forms--25-30 sq. ft. per hr.	Hours	Rate	Total	Rate	Total
Carpenter	3.50	$....	$....	$24.04	$84.14
Labor	1.75	....	....	19.13	33.48
Stripping Forms--70-75 ft. per hr.					
Labor	1.50	....	....	19.13	28.70
Cost per 100 sq. ft.			$....		$146.32
Labor cost per sq. ft.			....		1.46

Material Cost of 100 Lin. Ft. of Wall Footing Ribbon Forms Based on Footings 1'-0" Deep Partly Formed

Material	B.F.	Rate	Total	Rate	Total
Ribbons, 2"x4", 100 lin. ft.	67	$....	$....	$0.07*	$4.69*
Stakes, 2"x4", 2'-0" long	33	....	....	0.07*	2.31*
Ties, 1"x2"x2'-0" long	4	....	....	0.07*	.28*
Nails, etc.		....	....	1.30	1.30
Cost per 100 Lin. Ft.			$....		$8.58
Material Cost per Lin. Ft.			....		.09

Labor Cost of 100 Lin. Ft. of Wall Footing Ribbon Forms Based on Footings 1'-0" Deep Partly Formed

Placing Forms--40 Lin. Ft. per hr.	Hours	Rate	Total	Rate	Total
Carpenter	2.5	$....	$....	$24.04	$60.10
Labor helping	1.25	....	....	19.13	23.91
Stripping Forms--450-500 Lin. Ft. per Hr.					
Labor	0.21	....	....	19.13	4.02
Cost per 100 Lin. Ft.			$....		$88.03
Labor Cost per Lin. Ft.			....		.88

*Based on 6 uses of lumber.

On jobs where the wall footings are stepped down to a lower elevation, a riser form is required at each step. The cost of this riser form is about twice the cost of the regular footing side form.

Grooves are usually formed in the tops of footings and are called "footing keys". They are formed by forcing a beveled 2"x4" into the surface of the wet concrete. After the concrete has hardened, the "key" forms are pried loose, cleaned, and reused.

Keyways might be required by the contract drawings, specifications, or by local building codes. The two basic types of keyways are: a) female and b) male. Female keyways are usually either a 2x4 or 2x6, chamfered and placed in the middle of the new concrete pour. Male keyways are pre-formed with the lower pour and placed in the middle of the next pour location.

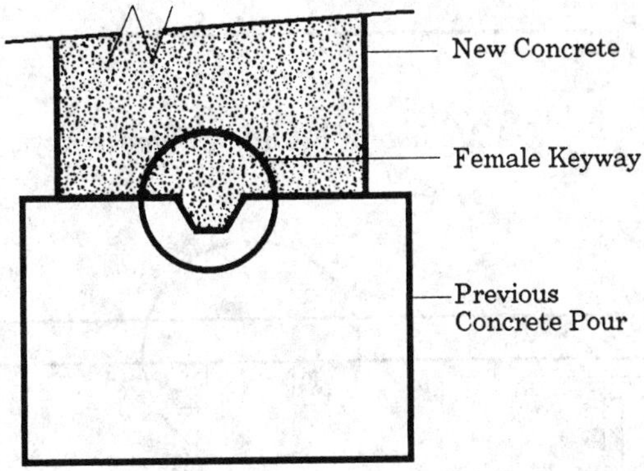

Female Keyway

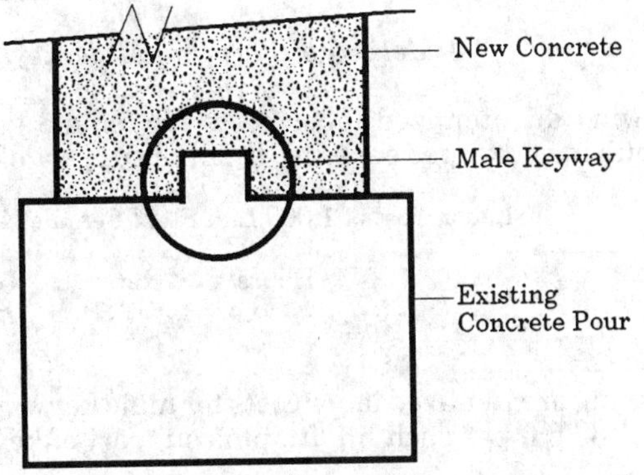

Male Keyway

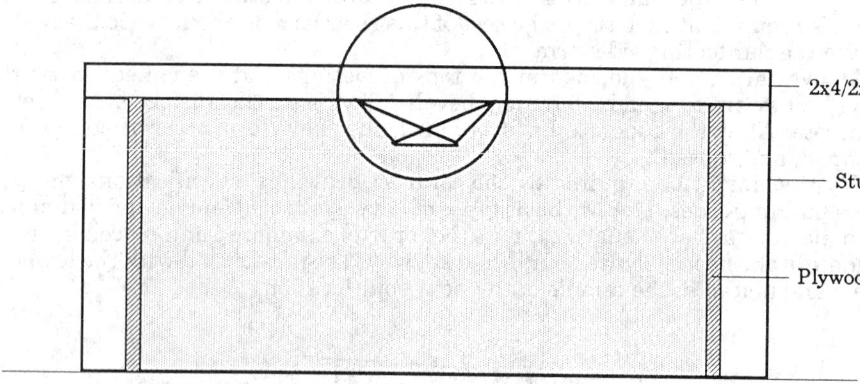

A Method of Forming Female Keyway 3.6

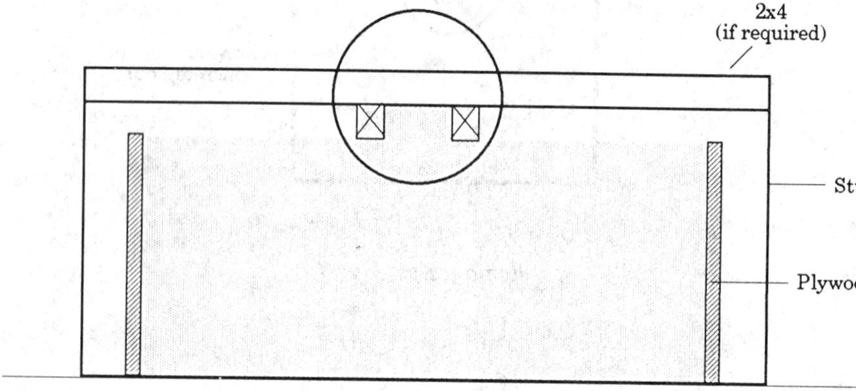

A Method of Forming Male Keyway

The following are itemized labor costs for female keyways, based on using 2"x6", requiring 1 b.f. per each lin. ft., plus support of keyway.

Labor Cost of 1,000 Lin. Ft. of Female Keyway

	Hours	Rate	Total	Rate	Total
Carpenter	8	$....	$....	$24.04	$192.32
Cost per lin. ft.					.19

The following are itemized labor costs for male keyways, based on using 2"x4", requiring 1.33 b.f. per each lin. ft., plus support of keyway.

Labor Cost of 400 Lin. Ft. of Male Keyway

	Hours	Rate	Total	Rate	Total
Carpenter	16	$....	$....	$24.04	$384.64
Cost per lin. ft.					.96

Labor Cost of 100 Lin. Ft. 2"x4" Footing Keys

	Hours	Rate	Total	Rate	Total
Carpenter placing	1.25	$....	$....	$24.04	$30.05
Labor stripping	.50	....	....	19.13	9.57
Cost per 100 lin. ft.			$....		$39.62
Labor cost per lin. ft.			....		.40

Forms for Pile Caps. After piles are driven and cut off at the required elevation, they are capped with masses of concrete that are designed to transmit the total loading of the structure to the piles. These pile caps vary in size and shape, from simple square caps covering one pile to complex many-sided caps covering clusters of twenty or more piles.

Forms for pile caps are usually prefabricated in panels at the bench and are then set up and put together in the pile location. Side forms are usually made of 3/4" plyform with 2"x4" stiffeners nailed to the back face. Five or 6 uses can usually be obtained from these forms.

After panels are set in place, they are secured with 2"x4" braces and tied with strap iron.

The following is an itemized cost of making average pile cap forms, based on 5'-6"x5'-6"x3'-0", containing 66 sq. ft.:

Material Costs

Making Pile Cap Side Forms	B.F.	Rate	Total	Rate	Total
3/4" Plyform s.f.	66	$....	$....	$0.60	$39.60
20 pcs. 2x4x3'-0"	41	....	....	0.40	16.40
8 pcs. 2x4x5'-0"	27	....	....	0.40	10.80
Bracing					
14 pcs. 2x4x8'-0"	75	....	....	0.40	30.00
Nails		....	....	1.30	1.30
Cost per cap, 66 sq. ft.			$....		$98.10
Material cost per sq. ft.			....		1.49

Divide above sq. ft. costs by number of times forms may be used on the job.

Labor Costs

Making Pile Cap Side Forms	Hours	Rate	Total	Rate	Total
Carpenter	4	$....	$....	$24.04	$96.16
Labor	2	....	....	19.13	38.26
Cost per cap, 66 sq. ft.			$....		$134.42
Labor cost per sq. ft.			....		2.04

Labor Cost Setting and Removing Pile Cap Forms

	Hours	Rate	Total	Rate	Total
Carpenter	3.0	$....	$....	$24.04	$72.12
Labor	1.5	....	....	19.13	28.70
Labor stripping, cleaning, moving	1.0	....	....	19.13	19.13
Cost per cap, 66 sq. ft.			$....		$119.95
Labor cost per sq. ft.			....		1.82

Forms for Concrete Piers. When column footing tops are lower than the bottom of the floor slab on the ground, piers are used to carry the column loading to the footings.

Forms for this work are usually made up in side panels at the job bench, hauled to the footings, erected, clamped, and braced.

Side panels are usually made of 3/4" plyform with 2"x4" stiffeners. Clamping may be accomplished by using either 2"x4" lumber or metal column clamps properly spaced to withstand the pressure developed when placing the concrete.

Bracing is required to hold the form in position and usually consists of 2"x4" or 2"x6" lumber, secured to the form at grade level, spanning the pit, and spiked to stakes driven into the ground. Bracing below this point is done by wedging 2"x4" struts between the form and the earth banks.

The following are itemized costs of building average pier forms, based on a pier 2'-0"x2'-0"x4'-0" containing 32 sq. ft.:

Material Costs

Making Pier Forms	B.F.	Rate	Total	Rate	Total
3/4" Plyform Sq. Ft.	32	$....	$....	$0.60	$19.20
12 pcs. 2x4x4'-0"	32	....	....	0.40	12.80
Clamps 8 pcs. 2x4x3'-0"	16	....	....	0.40	6.40
Nails, etc.		....	....	1.30	1.30
Cost 32 sq. ft.			$....		$39.70
Material cost per sq. ft.			....		1.24

Divide above sq. ft. costs by number of times forms may be used on the job.

Labor Costs

Making Pier Forms	Hours	Rate	Total	Rate	Total
Carpenter	1.0	$....	$....	$24.04	$24.04
Labor	0.5	....	....	19.13	9.57
Cost 32 sq. ft.			$....		$33.61
Labor cost per sq. ft.			....		1.05

Labor Cost Erecting Concrete Pier Forms

	Hours	Rate	Total	Rate	Total
Carpenter	1.0	$....	$....	$24.04	$24.04
Labor	0.5	....	....	19.13	9.57
Stripping, including cleaning and moving					
Labor	0.5	....	....	19.13	9.57
Cost per pier, 32 sq. ft.			$....		$43.18
Labor cost per sq. ft.			....		1.35

Forms for Concrete Walls

Forms for concrete walls must be designed and constructed to withstand the horizontal pressures exerted by the fluid concrete against them. This pressure against any given point of the form will vary and be influenced by any one or all of the following factors:

1) Rate of vertical rise in filling the forms (feet per hour).
2) Temperature of the concrete and the weather (ambient temperature).
3) Proportions of the concrete mix and its consistency.

4) Method of concrete placement and degree of vibration.

When wall forms are filled so rapidly that the concrete in the bottom of the form has not had sufficient time to start its initial setting process, then regardless of any other factor, the full height of the fluid concrete must be considered as a column of liquid with a weight of approximately 150 lbs. per cubic foot with a head pressure equal to its fluid height.

For example, at 10' below the top of the concrete, the pressure will amount to 10 feet x 150 lbs. = 1,500 lbs per sq. ft. exerted against the form sides at the bottom. If the wall form has not been adequately tied and braced to withstand this pressure, then the form will break open.

To avoid the necessity of overdesigning forms, which is a costly measure, or the possibility of form breakage, concrete is placed in layers or "lifts" at the rate of about 4 vertical feet per hour. This allows the concrete a chance to start the transformation from the fluid state to a hardened mass that supports itself and relieves pressure on the form surface.

The ambient temperature at which concrete is placed influences the time required for the setting process. Concrete placed at 40° F takes almost twice the time to set as concrete placed at 80° F. Therefore, concrete placed in wall forms during cold weather will require more control over the amount of concrete placed within a given time so that the forms will not be stressed beyond their design capability.

The proportions and consistency of the mix (water/cement ratio per cubic yard of concrete) has an effect on the rate of setting time as well as the strength of the concrete mix. The more the cement content and/or the less water needed to provide a workable consistency (slump) per cubic yard, the shorter the time for setting and the stronger the concrete.

When placing concrete into a form, care should be exercised to reduce the shock loading of the concrete hitting against the form material. This can be achieved by using a "tremie" (a flexible enclosed chute in the shape of an elephant's trunk) that confines the concrete until it reaches the bottom of the form or the level of the preceding lift. The tremie will also keep the concrete mix from segregating and should be used whenever placing concrete that could "free fall" a distance more than 4 feet. This segregation is caused by the large aggregate within the concrete mix hitting the reinforcing steel and shaking the cement/sand matrix coating off; the result being an improperly mixed concrete with inferior strength and possible voids (honeycomb).

Mechanical vibration of the concrete mix is accomplished after each lift is placed and is necessary for the proper consolidation (elimination of voids) and to ensure that the fine matrix of the mix reaches the form surface for a smooth wall finish. Although vibration of the concrete mix is desirable as well as necessary, it can have some adverse effects on formwork if done improperly. If the vibrator is allowed to pass through the just placed concrete lift into the preceding lift, it will disturb the preceding lifts "setting process" causing it to liquefy and increasing the head pressure against the formwork. Overvibrating the concrete will also imbalance the large aggregate to fine aggregate proportions within those areas of the mix.

Concrete Wall Forms Made in Panels or Sections. Many contractors, especially those performing smaller work, use foundation wall forms made up into panels 2'x6', 2'x8', 4'x6', 4'x8' and so on. The frame for these sections is usually 2"x4" lumber, with intermediate braces spaced 12" to 16" on centers. The corners are reinforced with galvanized straps or strap iron, and the entire frame is sheathed with 5/8" or 3/4" plyform plywood. Forms of this kind can be used a large number of times, especially where the walls are fairly straight.

The labor cost of making panel forms will vary with the method used and the number made up at one time. A carpenter should complete one panel (28 to 32 sq. ft.) in 3/4 to 1 hr. or 8 to 10 panels per 8-hr. day.

The cost of a panel 4'-0"x8'-0" in size, using 3/4" "Plyform" plywood should cost as follows:

Material Costs

	B.F.	Rate	Total	Rate	Total
2 pcs. 2"x4"-8'-0"	10.67	$....	$....	$0.40	$4.27
7 pcs. 2"x4"-4'-0"	18.67	....	....	0.40	7.47
Plyform plywood, 4'-0"x8'-0"-3/4" sq. ft.	32.00	....	....	0.60	19.20
Nails, etc.		....	....	.85	.85
Cost per panel 32 sq. ft.			$....		$31.79
Material cost per sq. ft			....		.99

Labor Costs

	Hours	Rate	Total	Rate	Total
Carpenter	0.75	$....	$....	$24.04	$18.03
Labor	0.25	....	....	19.13	4.78
Cost per panel			$....		$22.81
Labor cost per sq. ft.			....		.71

Many contractors specializing in concrete work, especially in foundations for residences, use panel forms exclusively and get up to 25 uses from the forms, with the usual amount of repairs. These conditions should be given consideration when pricing any wall work using panel forms.

Symons Steel-Ply Forms. Symons Steel-Ply Forms (from Symons Corporation, Des Plaines, IL) are lightweight panels with a rolled steel frame completely enclosing a plastic coated plywood face. The plywood has a high strength face, which is 12% stronger than standard plywood. Due to the plastic coating, effects of moisture are reduced 40%. The welded steel frames are painted to prevent rusting and with reasonable care should last for a number of years.

Panels are 2'-0" wide and come in lengths ranging from 2'-0" to 8'-0" high in one foot increments. Fillers are made in widths 4" to 22", from 2'-0" to 8'-0" high. All steel fillers are available in 1", 1 1/2", and 2" widths. The average weight of the Steel-Ply Forms is 6 lbs. per sq. ft. Outside corners are steel angles that lock adjoining forms together. Inside corners are 6"x6".

The side rails of the forms are rolled exclusively for Symons, and have a yield strength in excess of 55,000 psi. "L" shaped cross members on all forms are on one foot centers and have a yield strength in excess of 60,000 psi. The frame is 2 1/2" in width.

In the Symons system, only two basic pieces are necessary for erection--the tie and one-piece connecting hardware called the wedge-bolt. The same wedge-bolt is also used for connecting any Symons accessories to the form. In erecting forms, the panel tie is placed in the dado slot between the forms.

A wedge-bolt is inserted through the adjoining forms and tie loop. A second wedge-bolt is inserted through the slot of the first, automatically tying and joining the forms in one operation. The Symons System also uses an easy, efficient method for the erection of corners, pilasters, walers, bracing, and scaffolding. Standard panel ties have 1" breakbacks; 1 ½" and 2" are also available with wood cones. Safe load capacity of panel ties is 3,000 lbs. and 4,000 lbs.; flat ties 4,000 lbs. and 4,500 lbs. at 1.5 to 1.0 factor of safety. In stripping, panels may be quickly released from the finished wall. Wedge-bolts are easily loosened with a hammer tap, and panel ties then removed by giving tie loops a half twist, or in the case of flat ties, a vertical hammer blow.

Used panels may be rented for about 80 cents per sq. ft. for the first month and 70 cents per sq. ft. for each subsequent month. New panels rent for about 90 cents per sq. ft. for the first month and 80 cents per sq. ft. for each subsequent month. Rental costs should be checked with local suppliers. Minimum rental period is one month. Should the contractor wish to purchase the forms, a portion of the paid rentals are applicable to the purchase price. Rentals are charged from the date shipped from distribution center and stop on date of return. The contractor pays transportation charges both ways.

A typical foundation wall, 8" thick and 8' high containing 2500 sq. ft. of plywood faced steel wallforms and using a crew of three carpenters to four laborers would average as follows:

Labor Costs

Erecting, Aligning and Bracing Panels	Hours	Rate	Total	Rate	Total
Carpenter Foreman	4	$....	$....	24.77	$99.08
Carpenters	16	....	....	24.04	$384.64
Laborers helping	21	....	....	19.13	401.73
Stripping., Cleaning and Oiling for Reuse					
Carpenters	8	....	....	24.04	192.32
Laborers helping	16	....	....	19.13	306.08
Total Cost, 2,500 s.f.			$....		$1383.85
Cost Per Sq Ft					.55

To the above add form rental, form tie, and form oil costs.

The Symons Company has made on-the-job time studies of placing, aligning, and stripping its forms. A wall similar to the above example, with a crew of three lead and four supporting men, should work out to a productivity of 1.355 manhours per 100 sq. ft. In five hours this team should complete some 2500 sq. ft. of contact area.

A study of a curved wall forming a sewage treatment structure and using a crew of one carpenter to one laborer showed a productivity of 2.91 manhours per 100 sq. ft. of contact area. Setting of reinforcement and pouring of concrete are not included in the above figures.

Symons *Versiform*. This concrete forming system is a steel framed, plywood faced gang forming system for massive concrete structures and architectural concrete. It adapts to use in walls, columns, piers, culverts, etc., and can be had with four different standard facing materials.

Versiform panels are available for rental in 8'x4', 8'x2', 4'x4', 4'x2', 2'x4', and 2'x2' panels, with special sizes available for purchase only. A complete line of components and accessory hardware is available, for forming virtually any shape or size concrete structure. Versiform panels can be used in conjunction with Symons' other forming systems. A time study of Versiform panels and walers on continuous straight walls showed a productivity of 2.64 man hours per 100 sq. ft. of contact area.

Uni-Form Steel Panel Forms. Uni-Forms (from Universal Form Clamp Co., Chicago, IL) are prefabricated panels consisting of a steel frame 2'x1', 2'x2',

2'x3', 2'x4', 2'x5', 2'x6', 2'x7', and 2'x8', with a plywood facing. The panels are also furnished in fractional widths. The steel frame consists of a special rolled tee angle, with an equal legged tee strut to provide additional structural rigidity to the portion of the frame which acts as a plywood protector.

The plywood face is 5-ply plyform for use in concrete forms. The plywood facing provides a convenient nailing surface. It is advisable to oil the panels after each use. With proper care Uni-Forms have been used as many as 50 times before turning the plywood.

Uni-Form panels require alignment on one side only and employ a simplified method of forming inside corners and fillers.

Typical Method of Waling and Bracing Symons Steel-ply Forms

On a foundation wall job for 47 houses requiring 2,350 sq. ft. of forms per house, using the 7' Uni-Form panels, the following labor costs were obtained:

	Hours	Rate	Total	Rate	Total
Erection					
Carpenter foreman	6	$....	$....	$24.77	$148.62
Carpenters	30	....	....	24.04	721.20
Labor	18	....	....	19.13	344.34
Stripping, Cleaning and Moving					
Carpenters	10	....	....	24.04	240.40
Labor	8	....	....	19.13	153.04
Total cost 2,350 sq. ft.			$....		$1,607.60
Labor Cost per sq. ft.			....		.68

Ganged Symons Forms Showing Waling, Bracing, and Scaffold Attachment

Symons Forms Being Set to Form Room Walls in Circular High-rise Hotel

A carpenter erected and braced 78.3 sq. ft. of forms per hr. and required 0.6 hr. labor time per hr. of carpenter time. A carpenter stripped 235 sq. ft. of forms an hr. A laborer cleaned, oiled, and moved 294 sq. ft. of forms an hr.

On another foundation wall job for 50 houses, requiring 1,550 sq. ft. of forms per house (forms 2'-0"x7'-0" and 1'-0"x7'-0"), the following labor costs were obtained per 100 sq. ft. contact area of forms:

Uni-Form Concrete Wall Forms

	Hours	Rate	Total	Rate	Total
Erecting and Bracing Panels					
Carpenter Foreman	.65	$....	$....	$24.77	$16.10
Carpenter	2.6	$....	$....	$24.04	$62.50
Stripping Forms and Moving					
Carpenter	0.5	....	....	24.04	12.02
Labor	0.8	....	....	19.13	15.30
Oiling Panels					
Labor	0.5	....	....	19.13	9.57
Cost per 100 sq. ft			$....		$115.49
Labor Cost per sq. ft.			....		1.15

Uni-Form Concrete Wall Forms

A carpenter erected and braced 38.75 sq. ft. of forms per hr. A carpenter stripped 194 sq. ft. of forms per hr. A laborer oiled 194 sq. ft. of forms per hr. A laborer carried from one job to the other or loaded into trucks, 129 sq. ft. of forms per hr.

Based on the two preceding examples using Uni-Form prefabricated form panels, one finds that even though the panels, type of construction, and classification of labor are similar, the production time on the first example is considerably faster. There are many factors that affect the production time--weather, adaptability of materials used, accessibility to the work area, layout of the work, efficiency and experience of the workmen, and other factors. The estimator must evaluate all of these variables when preparing the estimate.

Built-In-Place Wood Forms For Concrete Walls

For those situations where the wall forms are to be built-in-place, the estimator should be familiar with form design parameters so that he can ascertain the correct quantity of form materials necessary for the work.

The following data and tables on form design have been furnished by the American Plywood Association (APA), Tacoma, Washington:

Grade-Use Guide for Concrete Forms*

Use these terms when you specify plywood	DESCRIPTION	VENEER GRADE	
		Faces	Inner Plies
APA B-B PLYFORM Class I & II**	Specifically manufactured for concrete forms. Many reuses. Smooth, solid surfaces. Mill-oiled unless otherwise specified.	B	C
APA High Density Overlaid PLYFORM Class I & II**	Hard, semiopaque resin-fiber overlay, heat-fused to panel faces. Smooth surface resists abrasion. Up to 200 reuses. Light oiling recommended between pours.	B	C Plugged
APA STRUCTURAL I PLYFORM**	Especially designed for engineered applications. All Group 1 species. Stronger and stiffer than PLYFORM Class I and II. Recommended for high pressures where face grain is parallel to supports. Also available with High Density Overlay faces.	B	C or C Plugged
Special Overlays, proprietary panels and Medium Density Overlaid plywood specifically designed for concrete forming**	Produce a smooth uniform concrete surface. Generally mill treated with form release agent. Check with manufacturer for design specifications, proper use, and surface treatment recommendations for greatest number of reuses.		

*Commonly available in 5/8" and 3/4" panel thicknesses (4'x8' size).
**Check dealer for availability in your area.

Concrete Pressures For Column And Wall Forms

Pressures of Vibrated Concrete (psf)*

Pour Rate, ft./hr.	50°F Columns	50° Walls	70°F Columns	70°F Walls
1	330	330	280	280
2	510	510	410	410
3	690	690	540	540
4	870	870	660	660
5	1050	1050	790	790
6	1230	1230	920	920
7	1410	1410	1050	1050
8	1590	1470	1180	1090
9	1770	1520	1310	1130
10	1950	1580	1440	1170

*Maximum pressure need not exceed 150h, where h is maximum height of pour. Based on concrete with density of 150 pcf and 4 in. slump.

Concrete Pressures For Slab Forms

Concrete Pressure (psf)

Depth of Slab (in.)	Non-Motorized Buggies*	Motorized Buggies**
4	100	125
5	113	138
6	125	150
7	138	163
8	150	175
9	163	188
10	175	200

*Includes 50 psf load for workmen, equipment, impact, etc.
**Includes 75 psf load for workmen, equipment, impact, etc.

Allowable Pressures On Plyform Class I For Architectural Applications

(deflection limited to 1/360th of the span)

Face Grain Across Supports Allowable Pressures (psf)*

Plywood Thickness (inches)

Support Spacing (in.)	1/2	5/8	3/4	7/8	1	1 1/8
4	2935	3675	4495	4690	5075	5645
8	970	1300	1650	1805	1950	2170
12	410	575	735	890	1185	1345
16	175	270	370	475	645	750
20	100	160	225	295	410	490
24			120	160	230	280
32					105	130
36						115

*Plywood continuous across two or more spans.

Face Grain Parallel To Supports Allowable Pressures (psf)*

Plywood Thickness (inches)

Support Spacing (in.)	1/2	5/8	3/4	7/8	1	1 1/8
4	1670	2110	2610	3100	4140	4900
8	605	810	1005	1190	1595	1885
12	215	360	620	740	985	1165
16		150	300	480	715	845
20		105	210	290	400	495
24			110	180	225	320

*Plywood continuous across two or more spans.

Allowable Pressures On Plyform Class I

(deflection limited to 1/270th of the span)

Face Grain Across Supports Allowable Pressures (psf)*

Plywood Thickness (inches)

Support Spacing (in.)	1/2	5/8	3/4	7/8	1	1 1/8
4	2935	3675	4495	4690	5075	5645
8	970	1300	1650	1805	1950	2170
12	430	575	735	890	1185	1345
16	235	325	415	500	670	770
20	135	210	285	350	465	535
24		110	160	215	295	340
32					140	170
36					105	120

*Plywood continuous across two or more spans.

Maximum Spans for Lumber Framing, Inches*
Douglas Fir-Larch No. 2 Or Southern Pine No. 2 (KD)

Equivalent Uniform Load (lb/ft)	Continuous Over 2 or 3 Supports (1 or 2 Spans)								Continuous Over 4 or More Supports (3 or More Spans)							
	Nominal Size								Nominal Size							
	2x4	2x6	2x8	2x10	2x12	4x4	4x6	4x8	2x4	2x6	2x8	2x10	2x12	4x4	4x6	4x8
400	36	53	70	90	109	50	79	101	41	60	78	100	122	62	91	118
600	30	43	57	73	89	44	66	88	31	49	64	82	99	51	74	98
800	24	38	50	63	77	39	58	76	25	39	52	66	81	44	64	85
1000	21	33	43	55	67	35	51	68	21	34	44	57	69	39	58	76
1200	19	29	38	49	60	32	47	62	19	30	39	50	61	35	53	69
1400	17	27	35	45	54	30	43	57	17	27	36	46	56	31	49	64
1600	16	25	32	41	50	27	41	54	16	25	33	42	52	28	44	58
1800	15	23	30	39	47	25	38	51	15	24	31	40	48	26	40	53
2000	14	22	29	37	45	23	36	48	14	22	29	38	46	24	37	49
2200	13	21	28	35	43	22	34	45	14	21	28	36	44	22	35	46
2400	13	20	26	34	41	20	32	42	13	20	27	34	42	21	33	44
2600	12	19	26	33	40	19	31	40	13	20	26	33	40	20	31	41
2800	12	19	25	32	38	19	29	38	12	19	25	32	39	19	30	39
3000	12	18	24	31	37	18	28	37	12	19	24	31	38	18	29	38
3200	11	18	23	30	36	17	27	35	12	18	24	30	37	18	28	36
3400	11	17	23	29	36	17	26	34	11	18	23	30	36	17	27	35
3600	11	17	22	29	35	16	25	33	11	17	23	29	35	16	26	34
3800	11	17	22	28	34	16	24	32	11	17	22	29	35	16	25	33
4000	10	16	22	28	34	15	24	31	11	17	22	28	34	15	24	32
4500	10	16	21	27	32	14	22	29	10	16	21	27	33	14	23	30
5000	10	15	20	25	31	13	21	28	10	16	20	26	32	14	22	28

*Spans are based on PS-20 lumber sizes, Single member stresses were multiplied by a 1.25 duration-of-load factor for 7-day loads. Deflection limited to 1/360th of the psan with ¼" maximum. Spans are center-to-center of the supports.

Maximum Spans for Lumber Framing, Inches*
Hem-Fir No. 2

Equivalent Uniform Load (lb/ft)	Continuous Over 2 or 3 Supports (1 or 2 Spans) — Nominal Size								Continuous Over 4 or More Supports (3 or More Spans) — Nominal Size							
	2x4	2x6	2x8	2x10	2x12	4x4	4x6	4x8	2x4	2x6	2x8	2x10	2x12	4x4	4x6	4x8
400	33	48	63	80	97	48	73	96	36	53	70	90	109	56	81	107
600	27	39	51	65	80	41	59	78	27	43	57	72	88	45	66	88
800	22	34	44	57	69	35	51	68	22	35	46	59	72	39	58	76
1000	19	29	39	50	60	31	46	61	19	30	40	51	62	35	51	68
1200	17	26	35	44	54	29	42	55	17	27	36	45	55	31	47	62
1400	15	24	32	41	49	27	39	51	16	25	33	42	51	27	43	57
1600	14	23	30	38	46	24	36	48	15	23	30	39	47	25	39	52
1800	14	21	28	36	43	22	34	45	14	22	29	36	44	23	36	47
2000	13	20	27	34	41	21	33	43	13	21	27	35	42	21	33	44
2200	12	19	26	33	40	19	31	40	13	20	26	33	40	20	31	41
2400	12	19	25	31	38	18	29	38	12	19	25	32	39	19	30	39
2600	12	18	24	30	37	18	28	36	12	18	24	31	38	18	28	37
2800	11	18	23	30	36	17	26	35	11	18	24	30	37	17	27	36
3000	11	17	23	29	35	16	25	33	11	17	23	29	36	17	26	34
3200	11	17	22	28	34	16	24	32	11	17	22	29	35	16	25	33
3400	10	16	22	27	33	15	24	31	11	17	22	28	34	15	24	32
3600	10	16	21	27	32	15	23	30	10	16	22	27	33	15	23	31
3800	10	15	20	26	32	14	22	29	10	16	21	27	33	15	23	30
4000	10	15	20	25	31	14	22	29	10	16	21	27	32	14	22	29
4500	10	14	19	24	29	13	21	27	10	15	20	26	31	13	21	28
5000	9	13	18	23	28	12	20	26	9	15	20	25	30	13	20	26

*Spans are based on PS-20 lumber sizes, Single member stresses were multiplied by a 1.25 duration-of-load factor for 7-day loads. Deflection limited to 1/360th of the psan with ¼" maximum. Spans are center-to-center of the supports.

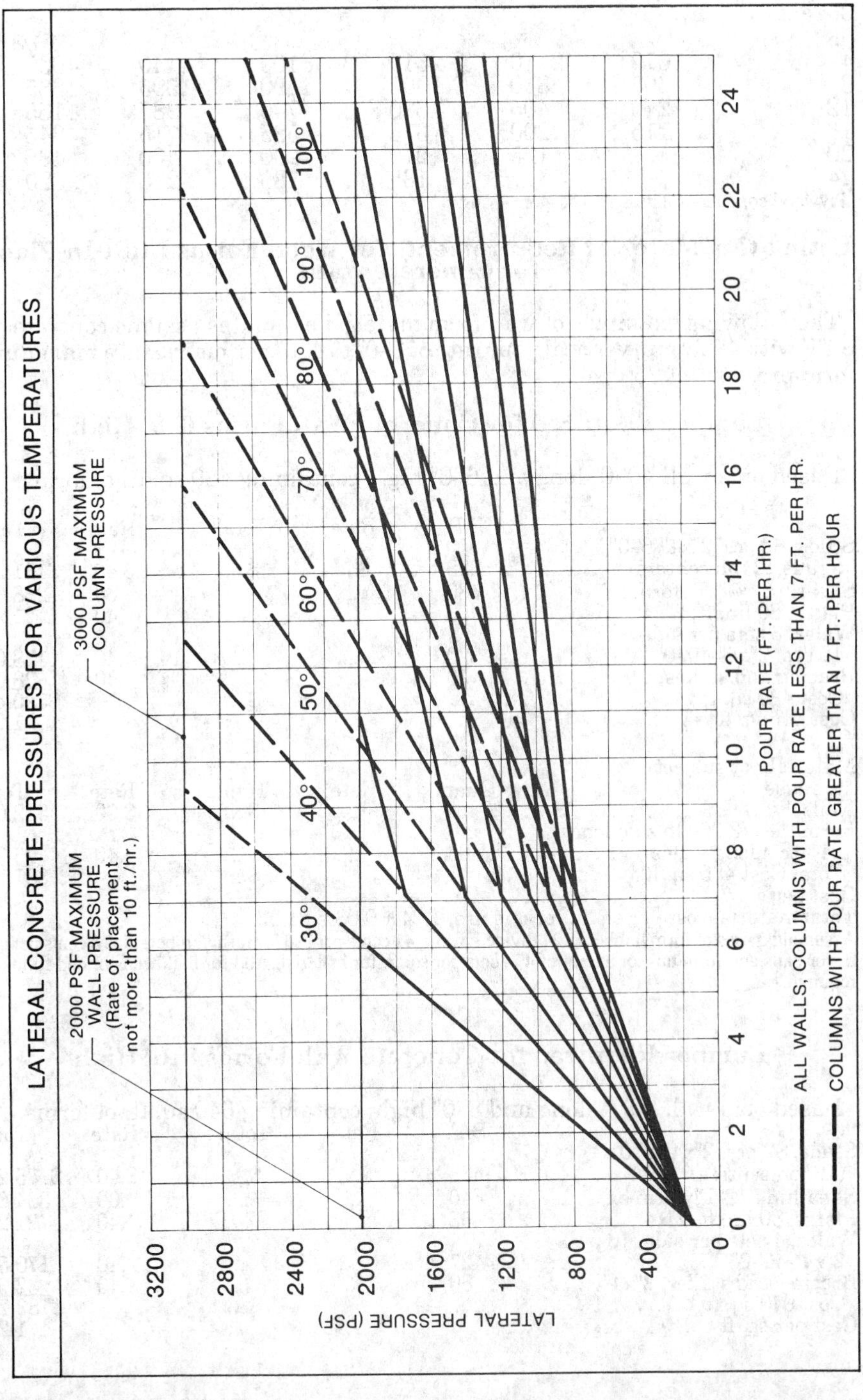
LATERAL CONCRETE PRESSURES FOR VARIOUS TEMPERATURES
2000 PSF MAXIMUM WALL PRESSURE (Rate of placement not more than 10 ft./hr.)
3000 PSF MAXIMUM COLUMN PRESSURE
30°
40°
50°
60°
70°
80°
90°
100°
3200
2800
2400
2000
1600
1200
800
400
0
LATERAL PRESSURE (PSF)
0
2
4
6
8
10
12
14
16
18
20
22
24
POUR RATE (FT. PER HR.)
ALL WALLS, COLUMNS WITH POUR RATE LESS THAN 7 FT. PER HR.
COLUMNS WITH POUR RATE GREATER THAN 7 FT. PER HOUR

Face Grain Parallel To Supports Allowable Pressures (psf)*

Support Spacing (in.)	Plywood Thickness (inches) ½	⅝	¾	⅞	1	1 1/8
4	1670	2110	2610	3100	4140	4900
8	605	810	1005	1190	1595	1885
12	270	405	620	740	985	1165
16	115	200	375	525	715	845
20		125	210	290	400	495
24			135	185	255	320

*Plywood continuous across two or more spans.

Estimating Material Requirements for Wood Forms Built-In-Place for Concrete Walls

The following examples of wall form material estimates assume concrete at 50°F with 4" slump, vertical pour rate of 3'-0" to 4'-0" per hour, and a maximum form pressure of 750 psf.

Lumber Required for Concrete Wall Forms 6'-0" High

Based on a wall 40'-0" long and 6'-0" high, containing 480 sq. ft. of forms.

	B.F.	Rate	Total	Rate	Total
Studs, 82 pcs. 2"x4"-6'-0" (studs 12" on centers)	328	$....	$....	$.40	$131.20
Sheathing ¾" Plyform s.f.	480	....	....	.60	288.00
Plates, 80 lin. ft. 2"x4"	53	....	....	.40	21.20
Wales, 3 sets per side 12 pcs. 2"x4"-40'-0"	320	....	....	.40	128.00
Bracing and stakes, 2"x4"	70	....	....	.40	28.00
Cost 480 sq. ft		....	$....		$596.40
Cost per sq. ft			....		1.24

Material not subject to reuse	Quan.	Rate	Total	Rate	Total
Nails	20 lbs	$....	$....	$.60	$12.00
Form ties, (3,000 lb safe load) 24" o.c. along wales	63	....	....	.40	25.20
Total Cost 480 sq. ft			$....		$37.20
Cost per sq. ft			....		.08

Total material cost per sq. ft. for one use, 1.24 + 0.08 = $1.32.

Assuming 3 uses of form lumber, 1.24 divided by 3 = 41 cents per use for form lumber, plus 8 cents per sq. ft. for nails and form ties, or a total of 49 cents per sq. ft. for material. Add for bulkheads or end forms if required.

Lumber Required for Concrete Wall Forms 8'-0" High

Based on a wall 40'-0" long and 8'-0" high, containing 640 sq. ft. of forms.

	B.F.	Rate	Total	Rate	Total
Studs, 82 pcs. 2"x4"-8'-0" (12" on centers)	438	$....	$....	$.40	$175.20
Sheathing ¾" Plyform s.f.	640	....	....	.60	384.00
Plates, 80 lin. ft. 2"x4"	53	....	....	.40	21.20
Wales, 4 sets per side, 16 pcs. 2"x4"-40'-0"	427	....	....	.40	170.80
Bracing and stakes, 2"x4"	80	....	....	.40	32.00
Cost 640 sq. ft			$....		$783.20
Cost per sq. ft			....		1.22

Material not subject to reuse:	Quan.	Rate	Total	Rate	Total
Nails	30 lbs	$....	$....	$.60	$18.00
Form ties (3,000 lb safe load) 24" o.c. wales	84	....	....	.40	33.60
Total cost 640 sq. ft			$....		$51.60
Cost per sq. ft			....		.08

Total material cost per sq. ft. for one use, 1.22 + 0.08 = $1.30.

Assuming 3 uses of form lumber, 1.22 divided by 3 = 41 cents per use of form lumber, plus 8 cents per sq. ft. for nails and form ties, or a total of 49 cents per sq. ft. for material. Add for bulkheads or end forms if required.

Lumber Required for Concrete Wall Forms 12'-0" High

Based on a wall 40'-0" long and 12'-0" high, containing 960 sq. ft. of forms.

	B.F.	Rate	Total	Rate	Total
Plates 4 pcs., 2"x4"-20'-0"	53	$....	$....	$.40	$21.20
Studs, 82 pcs. 2"x4"-12'-0" (12" on centers)	656	....	....	.40	262.40
Sheathing, ¾" plyform sf	960	....	....	.60	576.00
Wales, 6 sets per side wales @ 24" on centers 24 pcs. 2"x4"-40'-0"	640	....	....	.40	256.00
Bracing and stakes, 2"x4"	98	....	....	.40	39.20
Cost 960 sq. ft.			$....		$1,154.80
Cost per sq. ft.			....		1.20

Material not subject to reuse:	Quan.	Rate	Total	Rate	Total
Nails	40 lbs	$....	$....	$.60	$24.00
Form ties (3,000 lb. safe load) 24" o.c. along wales	126	....	....	.40	50.40
Total cost 960 sq. ft.			$....		$74.40
Cost per sq. ft			....		.08

Total material cost per sq. ft. for one use, 1.20 + 0.08 = $1.28.

Assuming 3 uses of form lumber, 1.20 divided by 3 = 40 cents per use for form lumber, plus 8 cents per sq. ft. for nail and form ties, or a total of 48 cents per sq. ft. for material.

In comparing the three preceding sample material estimates for wall forming on 6-, 8-, and 12-foot high walls, note that all three are in the same cost range. This holds true for even lower or higher walls, as long as the design parameters for temperature, slump, pour rate, and allowable form pressure remain the same.

If the temperature of the concrete is lower, the mix is wetter, or it is advantageous to place the concrete into the forms at a faster pour rate, then the forms will have to be designed to withstand the increased pressure per square foot on the form surface. The design changes would involve any of the following:

1) Stronger form tie (increased tensile strength).
2) Closer horizontal spacing for form ties.
3) Increased stud size to counteract bending under increased loads.
4) Closer on-center spacing of stud members to counteract deflection of the plyform, or use of thicker plyform material.
5) Increased size of wale members.
6) Additional bracing and stakes.

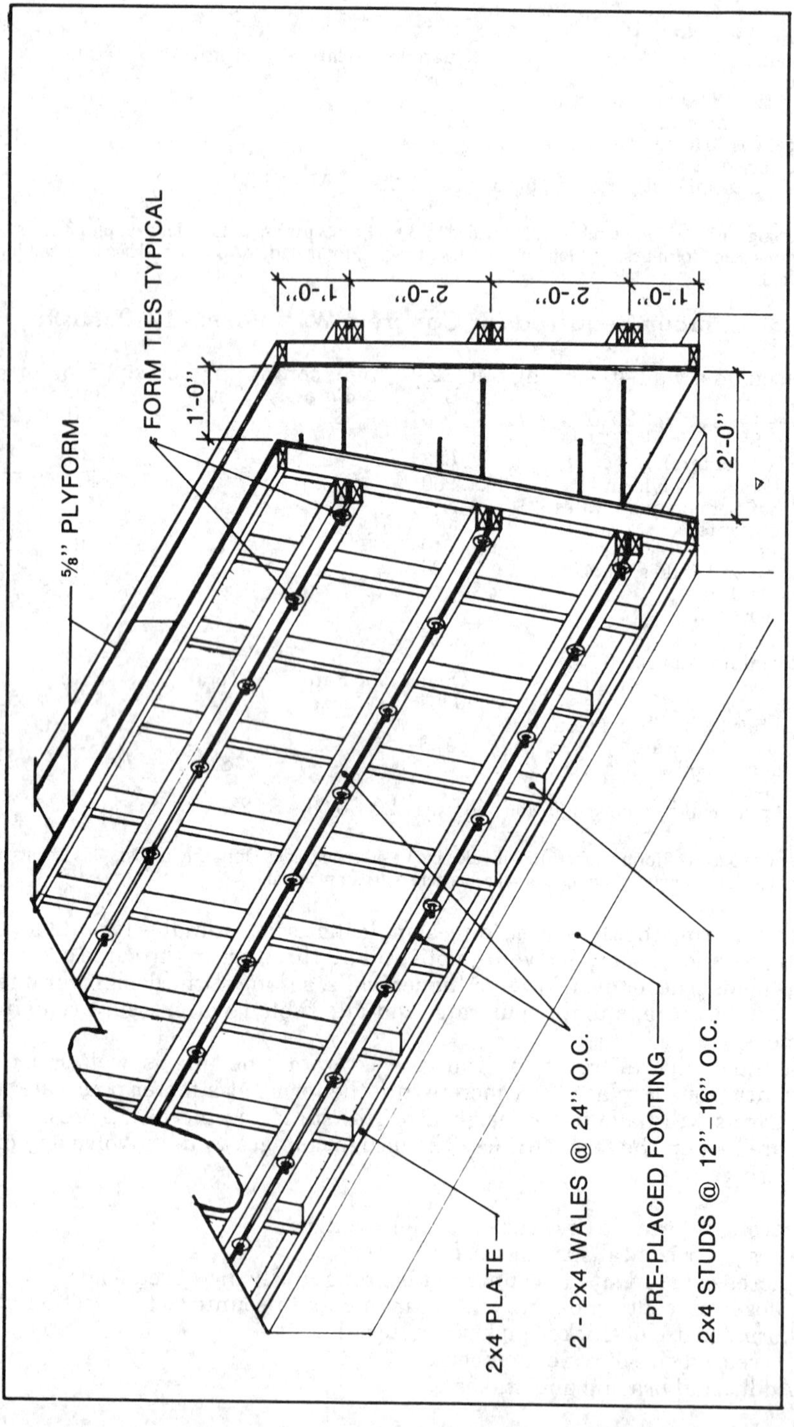
5/8" PLYFORM
FORM TIES TYPICAL
1'-0"
2'-0"
2'-0"
1'-0"
1'-0"
2'-0"
2x4 PLATE
2 - 2x4 WALES @ 24" O.C.
PRE-PLACED FOOTING
2x4 STUDS @ 12"–16" O.C.

Basis for Computing Labor Costs. After checking the labor costs on a large number of jobs over a period of years, it has been found that a carpenter can frame and erect about 325 to 400 b.f. of lumber per 8-hr. day and requires 3/8 to 1/2 hr. laborer time (carrying lumber, etc.) for each hour of carpenter time.

If there are no laborers employed on the job and all lumber is handled and carried by carpenters, a carpenter should handle, frame, and erect about 225 to 275 b.f. of lumber per 8-hr. day.

When removing or "stripping" concrete forms, a carpenter or laborer should remove 750 to 1,000 b.f. of lumber per 8-hr. day.

High walls necessitate working from a scaffold and require 10% to 50% more time per square foot of forms because of the additional handling, hoisting of lumber, and additional bracing. The sq. ft. costs on the following pages have been computed on this basis.

Production Times for 100 Sq. Ft. of Wood Forms for Concrete Walls

Class of Work	Height Of Wall in Feet	BF Lmbr per SF of Forms	SF Forms per 8-hr Day	Carpenter Hours	Labor* Hours	Removing Forms Sq. Ft. 8-Hr. Day	Removing Forms Labor Hours
Wall Forms Built-in-Place	4 to 6	2.6	190-210	4.0	2.0	360	2.25
Wall Forms Built-in-Place	7 to 8	2.6	165-190	4.5	2.3	350	2.20
Wall Forms Built-in-Place	9 to 10	2.6	150-160	5.3	2.5	325	2.50
Wall Forms Built-in-Place	11 to 12	2.6	125-135	6.0	3.0	300	2.70

*If Union rules do not permit laborer working with carpenter, figure labor time as additional carpenter time. Add foreman time as required. Normal practice on large concrete projects is to allow for one foreman when carpenter crew exceeds 3 carpenters.

When foundation walls rest directly on earth without a concrete footing, mud sills of 2" or 3" plank are usually employed to serve as layout lines and supports for the wall panel forms required. This will increase the cost of 4'-0" high panel wall forms about 10% for both material and labor. Similarly, 8'-0" high wall forms should be increased about 5 percent.

Labor Cost of 100 Sq. Ft. Built in Place Concrete Wall Forms 4'-0" to 6'-0" High, Requiring 2.6 BF of Lumber Per Sq. Ft. of Forms

	Hours	Rate	Total	Rate	Total
Carpenter	4.0	$....	$....	$24.04	$96.16
Labor	2.0	....	....	19.13	38.26
Labor stripping & cleaning	2.25	....	....	19.13	43.04
Cost 100 sq. ft			$....		$177.46
cost per sq. ft			....		1.77

Labor Cost of 100 Sq. Ft. Built-in-Place Concrete Wall Forms 8'-0" High, Requiring 2.6 BF of Lumber Per Sq. Ft. of Forms

	Hours	Rate	Total	Rate	Total
Carpenter	4.50	$....	$....	$24.04	$108.18
Labor helping	2.30	....	....	19.13	44.00
Labor stripping & cleaning	2.20	....	....	19.13	42.09
Cost 100 sq. ft			$....		$194.27
cost per sq. ft			....		1.94

Labor Cost of 100 Sq. Ft. Built-in-Place Concrete Wall Forms 10'-0" High, Requiring 2.6 BF of Lumber Per Sq. Ft. of Forms

	Hours	Rate	Total	Rate	Total
Carpenter	5.3	$....	$....	$24.04	$127.41
Labor helping	2.5	....	....	19.13	47.83
Labor stripping & cleaning	2.5	....	....	19.13	47.83
Cost 100 sq. ft			$....		$223.07
cost per sq. ft			....		2.23

Labor Cost of 100 Sq. Ft. Built-in-Place Concrete Wall Forms 12'-0" High, Requiring 2.6 BF of Lumber Per Sq. Ft. of Forms

	Hours	Rate	Total	Rate	Total
Carpenter	6.0	$....	$....	$24.04	$144.24
Labor helping	3.0	....	....	19.13	57.39
Labor stripping & cleaning	2.7	....	....	19.13	51.65
Cost 100 sq. ft			$....		$253.28
cost per sq. ft			....		2.53

Pilasters. Walls which are thickened for short intervals of 12" to 36" in length and then return to their former thickness are known as pilasters at that thickened area.

The cost of forming the face and sides of pilasters runs from 25% to 50% more than plain wall forms for both material and labor.

Radial Wall Forms. Wall forms that are built to radii or other curves require closer stud spacing, more bracing, and more labor laying out than straight wall forms. Formwork for radial walls will cost from 50% to 100% more than forms for straight walls.

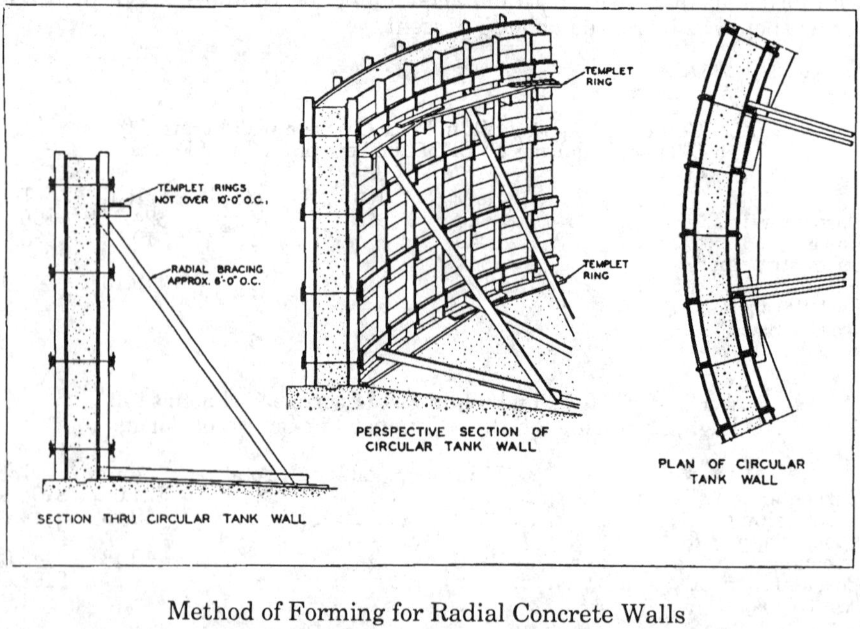

Method of Forming for Radial Concrete Walls

Forming Openings in Concrete Walls. Openings in concrete walls for doors, windows, and louvers are formed by placing bulkheads in the wall forms to form the sills, jambs, and heads of the openings required. These forms usually require from 2 ¼ to 3 b.f. of lumber per sq. ft. of contact area, depending upon the size of the opening.

The following is an itemized labor cost of boxed opening forms, based on 4'-0"x4'-0" openings in a 1'-0" wall requiring 2 ½ b.f. of lumber per sq. of contact area.

Labor Cost of 100 Sq. Ft. of Boxed Opening Forms

	Hours	Rate	Total	Rate	Total
Setting forms					
Carpenter	7.0	$....	$....	$24.04	$168.28
Labor Helping	3.5	....	....	19.13	66.96
Labor stripping forms	3.0	....	....	19.13	57.39
Cost 100 sq. ft			$....		$292.63
Labor cost per sq. ft			....		2.93

Forming Setbacks in Concrete Walls. When concrete foundation walls extend some height above ground level, it is often required that the portion above grade be faced with masonry. This necessitates the portion above grade being reduced in thickness sufficient to accommodate the facing material. This reduction in thickness is accomplished by placing a "setback" form in the regular wall forms.

Setback forms are usually made up of 2" framing lumber faced with ⅝" or ¾" plyform, and usually require 1 ¾ to 2 ¼ b.f. of lumber per sq. ft. of contact area, depending mainly upon the thickness of the setback.

The following is an itemized labor cost of 100 sq. ft. of setback forms requiring 2 b.f. of lumber per sq. ft. of contact area:

	Hours	Rate	Total	Rate	Total
Carpenter	6.5	$....	$....	$24.04	$156.26
Labor helping	3.3	....	....	19.13	63.13
Labor stripping	2.7	....	....	19.13	51.65
Cost 100 sq. ft			$....		$271.04
Labor cost per sq. ft			....		2.71

Forming a Continuous Haunch on Concrete Walls. When an intermediate support on a wall face is required to provide bearing for masonry, slabs, or future construction, and reducing the wall thickness is prohibited by the design, a continuous haunch is commonly used.

A continuous haunch is usually formed of 2" framing lumber faced with ⅝" or ¾" plyform and requires from 3 ½ to 4 b.f. of lumber per sq. ft. of contact area. Following is the itemized labor cost of 100 sq. ft. of continuous haunch forms:

	Hours	Rate	Total	Rate	Total
Carpenter	12.0	$....	$....	$24.04	$288.48
Labor helping	6.0	....	....	19.13	114.78
Labor stripping forms	2.5	....	....	19.13	47.83
Cost 100 sq. ft			$....		$451.09
Labor cost per sq. ft			....		4.51

Form Coating

For clean, easy stripping of forms and to prolong the life of form facings, the contact surface of all forms should receive a coating, which will prevent concrete from bonding to the form face.

The old method of brushing or spraying form faces with paraffin oil is still used by some contractors and gives adequate results for the average job, such as foundation walls, unexposed columns, beams, and slabs. Forms should always be coated before erection, but where this is impractical, such as for built-in-place wall forms, pan, and joist forms for combination slabs, oiling should be done in advance of setting reinforcing steel and care should be taken not to get any oil on concrete, masonry, or steel bearing surfaces.

On large areas, such as flat slab decks, a laborer can coat 400 to 500 sq. ft. of surface per hour, but production is considerably less when coating forms for columns, beams, combination slabs, etc. A fair overall average for coating forms is about 300 to 350 sq. ft. per hour.

For better class work, when using plywood, plastic-coated plywood, or lined forms, where smooth concrete surfaces are required, there are numerous patented coatings and lacquers available for this purpose. While the initial cost of these coatings is much higher than paraffin oil, better results are obtained and the forms may be used a couple of times without recoating, depending on the care used in applying the coating and in handling the forms when erecting or stripping.

Prices for these coating materials vary from $10.00 to $25.00 per gallon, but coverage and durability also vary. Some coatings require special thinners before application, while others can be thinned with relatively inexpensive mineral spirits.

Application may be either by brushing, spraying, or dipping. A worker can coat about 400 to 500 sq. ft. of form surface per hour.

Formfilm. Formfilm (from W.R. Grace & Co.) is a quick-drying plastic type plywood form coating with a high nitro-cellulose content. It is completely water-repellent, resistant to abrasion, and impervious to the alkaline action of concrete. It is recommended for use on untreated plywood forms to produce smooth concrete surfaces free from grain markings. It is not intended for use on metal or plastic forms.

Formfilm must be applied to concrete forms that are free from oil, grease, dirt, and foreign matter. It can be applied by brush, spray, roller, squeegee, or dipping. Formfilm is not ready mixed because job temperatures vary widely and affect the viscosity. It is usually thinned with *Formfilm Thinner* to the proper working consistency.

Concrete Column Forms

Reinforced concrete columns may be either square, rectangular, or round. The forms for square or rectangular columns are usually constructed of wood while metal or fiber molds are used for round columns.

Forms for reinforced concrete columns should be estimated by the sq. ft., obtained by multiplying the girth of the column by the height. Example: Obtain the number of sq. ft. of forms in a column 18" square and 12'-0" high. Proceed as follows: 18 + 18 + 18 + 18 = 72" or 6'-0", the girth. Multiply the girth by the height, 6'-0" x 12'-0" = 72 sq. ft. of forms.

The forms for all square or rectangular columns should be estimated in this manner.

To prevent the column forms from spreading while the concrete is placed and until it hardens, wood or metal clamps should be run around the column at intervals, depending upon the height of the column. In recent years the use of wood clamps has been almost entirely eliminated by the use of metal clamps.

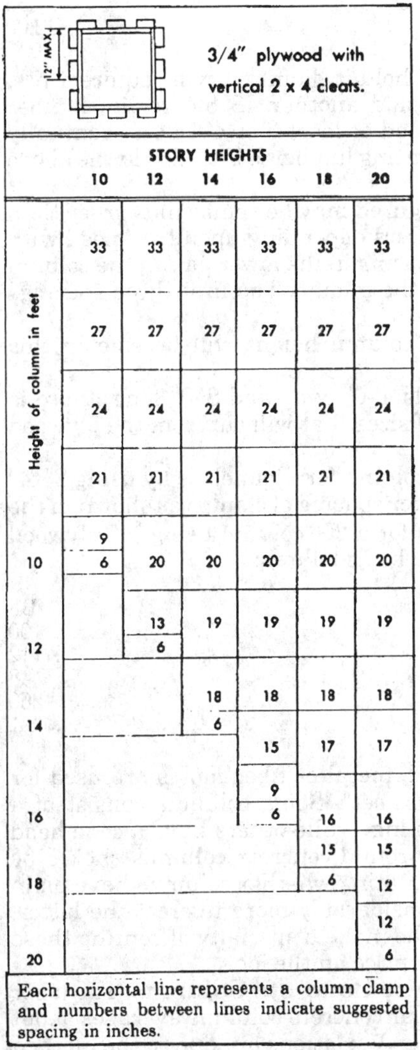

12" MAX
3/4" plywood with
vertical 2 x 4 cleats.
STORY HEIGHTS
10 12 14 16 18 20
Height of column in feet
10
12
14
16
18
20
33 33 33 33 33 33
27 27 27 27 27 27
24 24 24 24 24 24
21 21 21 21 21 21
9
6 20 20 20 20 20
13 19 19 19 19
6
18 18 18 18
6
15 17 17
9
6 16 16
15 15
6 12
12
6
Each horizontal line represents a column clamp and numbers between lines indicate suggested spacing in inches.

Column Clamp Spacing, in Inches, Based on Strength of Materials and Limited by 1/8" Deflection

The pressure of the fresh concrete has a greater effect on the design of column forms than on wall forms on account of the smaller amount of concrete to be placed. The contractor usually finds it more convenient and economical to fill a group of medium size columns in as short time as possible, perhaps 30 minutes, when the full liquid pressure of the fresh concrete will be acting, 150 lbs. per sq. ft. for every foot in depth of the concrete.

Column Form Design

There are two principal methods of erecting column forms; one is to unite all the four sides on two or three horses near the site where the form is to be erected and have 4 or 6 laborers hoist the complete form over the column dowels and stand it up temporarily; or in case of heavier columns, a simple 4"x6" pole derrick may be used for erection. Sometimes 3 sides of the column form are united and shoved into place without lifting the heavy box over the dowels, while the other and decidedly more expensive method is to erect each column form side separately.

As an example of the lumber required for a column form, take a concrete column 20" square and 11'-0" high underneath the lowest beam, and assuming a full liquid pressure of 150 lbs. per cu. ft. acting for the entire height of the column. Assume the clear story height as 12'-6". The area of the column form as usually taken off by the estimator will be 4 sides x 1.667 x 12'-6" high equals 83.33 sq. ft.

Using steel column clamps the following lumber would be required for a 20"x20" concrete column form 12'-6" high.

	BF
2 pcs. 3/4" plywood, 1'-9 1/2"x8'-0"	32
2 pcs. 3/4" plywood, 1'-9 1/2"x4'-6"	18
2 pcs. 3/4" plywood, 1'-8"x8'-0"	32
2 pcs. 3/4" plywood, 1'-8"x4'-6"	18
Total plywood required for 83.33 sq. ft. forms	100

20 pcs. 1x6x1'-9" cleats	18
Lumber required for 83.33 sq. ft. forms	118
Lumber required per sq. ft.	1.42
Bracing lumber per sq. ft. forms	.58
Total lumber required per sq. ft. forms	2.00

In order to brace the column form and hold it plumb, it will require 4 pcs. 2"x4", 16'-0" to 20'-0" long, which will add another 48 b.f. to the lumber required. This bracing lumber may be used at least 3 or 4 times, eventually for purposes other than column forms. Bracing lumber is included in the above table.

The number of steel column clamps required may be somewhat larger than usually recommended by the manufacturers depending upon the speed with which the column forms are filled. The spacing in the lower part of the column should be about 9" and about 1'-0" near the middle. The first three spacings from the top should be 1'-3" to 1'-8" apart.

An average of one clamp per foot of column height will be safe in this particular instance.

Because plywood is furnished in panels 4'-0" wide and 8'-0" long, there is considerable waste unless columns are of sizes that will cut from the plywood sheets without waste, which is unusual.

Another method of stiffening plywood column form panels is by using 2"x4" vertical cleats. This method permits greater spacing of clamps as shown in the table and the total lumber requirements for a 20" column using ¾" plywood sheathing and 2"x4" vertical cleats would be as follows:

	BF
Plywood sheathing	100
Cleats	112
Bracing	48
Lumber required for 83.33 sq. ft. forms	260
Lumber per sq. ft. of forms	3.1

When round concrete columns are used, metal or fiber molds are used for supporting the wet concrete until it has "set". Some columns consist of a straight shaft extending from floor to ceiling, while others have a cone head at the top of the column. The forms for round concrete columns should be estimated at a certain price per column, stating whether columns have plain shafts or cone heads, because it costs considerably more to erect the latter. There are several firms in the country who make a specialty of renting these column molds and will quote a lump sum price on the job.

Labor Cost of Wood Forms for Square Concrete Columns. The labor cost of wood forms for square or rectangular concrete columns averages about the same regardless of the type of building. Variations are due to the method used in assembling and building up the column section, their size, method of clamping, etc.

On jobs of any size it is customary to have a mill on the job, where the column forms are built to exact size, including cut-outs for beams or girders, and then sent to the floor on which they are to be used, ready for erection.

Here is an important item when figuring column form costs. How many times can they be used on the job without cutting down or extending their length or must they be reworked for each floor? This is a matter each estimator or contractor must decide from a study of the plans. For this reason separate labor costs are given for making up the column forms and for erection.

Roos Adjustable Metal Column Clamps
Hinged Bar Type—2 Hinged Units=1 Complete Clamp

Clamp Size	Bar Size	Weight Per Clamp	Net Concrete Sizes			
			With 1" Column Lumber		With 2" Column Lumber	
			Minimum	Maximum	Minimum	Maximum
36"	5/16" x 2 1/2" x 36"	40 lbs.	8 1/2" x 8 1/2"	26" x 26"	6 1/2" x 6 1/2"	24" x 24"
48"	3/8" x 2 1/2" x 48"	58 lbs.	12" x 12"	38" x 38"	10" x 10"	36" x 36"
60"	3/8" x 3" x 60"	85 lbs.	22 1/2" x 22 1/2"	48 1/2" x 48 1/2"	20 1/2" x 20 1/2"	46 1/2" x 46 1/2"
36" x 60"	1 Leg, 5/16" x 2 1/2" x 36" 1 Leg, 3/8" x 3" x 60"	65 lbs.	7 1/2" (36" Leg) 23 1/2" (Leg)	25" (36" Leg) 49 1/2" (60" Leg)	5 1/2" (36" Leg) 21 1/2" (60" Leg)	23" (36" Leg) 47 1/2" (60" Leg)

Labor Cost of 100 Sq. Ft. Concrete Column Forms Using Plywood Sheathing with Adjustable Steel Clamps and 2"x4" Vertical Cleats

Making Forms	Hours	Rate	Total	Rate	Total
Carpenter	4.0	$....	$....	$24.04	$96.16
Labor	2.0	....	....	19.13	38.26
Assembling, Erecting, Plumbing and Bracing					
Carpenter	3.0	....	....	24.04	72.12
Labor helping	3.0	....	....	19.13	57.39
Removing Clamps and Forms					
Labor	2.0	....	....	19.13	38.26
Cost per 100 sq. ft			$....		$302.19
cost per sq. ft			....		3.02

Metal Forms for Round Columns

The cost of metal forms for round concrete columns vary according to the diameter of the column, height, whether a plain shaft or one having a conical head, and whether it is necessary to cut out for beams and girders framing into the metal column molds.

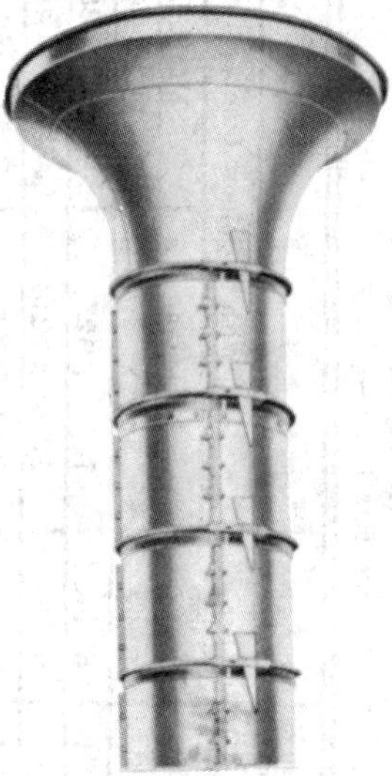

Round Metal Column Molds

Where it is necessary to cut the metal molds or "heads" to frame into wood forms for concrete beams and girders, an extra labor allowance should be made.

When plain round metal forms up to 24" in diameter are used without caps, which do not require cutting and fitting at head for concrete beams and girders, the labor erecting, removing, and oiling one column mold should cost as follows:

	Hours	Rate	Total	Rate	Total
Carpenter	2	$....	$....	$24.04	$48.08
Laborer	2	....	....	19.13	38.26
Total per column			$....		$86.34

Add for scaffolding and/or crane service if required.

If the column has a conical head, necessitating placing a metal cap in addition to the column shaft, the labor erecting, removing, and oiling one column should cost as follows:

	Hours	Rate	Total	Rate	Total
Carpenter	2.5	$....	$....	$24.04	$60.10
Laborer	2.5	....	....	19.13	47.83
Total per column			$....		$107.93

Add for scaffolding and/or crane service if required.

If necessary to cut and fit at column heads for concrete beams and girders, add 1 to 2 hrs. time per column.

The above labor costs are for columns up to 16'-0" in length. For columns 16'-0" to 20'-0" long, add 1 to 2 hrs. time.

Leasing or Renting Metal Column Molds. Where metal column molds are leased or rented by the manufacturer to the contractor to be used on a specified job and returned in good condition upon completion of the work, the following prices will prove a fair average for columns up to 24" in diameter.

For a plain metal mold for round concrete columns without cone head, the monthly rental should figure about $9.00 per lin. ft. of column form to be rented.

For plain round metal forms for concrete columns having cone head, add $55.00 per column per month rental.

To the above prices add delivery charges to the project site.

Cu. Ft. of Concrete and Sq. Ft. of Forms Required Per Lin. Ft. of Height for Round and Square Concrete Columns

	Round Columns			Square Columns			
Diam. in Inches	Area Sq. In.	Cu. Ft. Conc.	Sq. Ft. Forms	Area Sq. In.	Cu. Ft. Conc.	Sq. Ft. Forms	Square in Inches
12	113	0.78	3.16	144	1.00	4.00	12
13	133	0.92	3.40	169	1.18	4.33	13
14	154	1.07	3.67	196	1.36	4.67	14
15	177	1.22	3.93	225	1.57	5.00	15
16	201	1.40	4.20	256	1.78	5.33	16
17	227	1.57	4.45	289	2.05	5.67	17
18	255	1.77	4.71	324	2.25	6.00	18
19	284	1.97	4.97	361	2.52	6.33	19
20	314	2.18	5.24	400	2.78	6.67	20
21	346	2.40	5.50	441	3.06	7.00	21
22	380	2.64	5.76	484	3.36	7.33	22
23	416	2.88	6.10	529	3.67	7.67	23
24	452	3.14	6.30	576	4.00	8.00	24
25	491	3.41	6.54	625	4.34	8.33	25
26	531	3.69	6.81	676	4.70	8.67	26
27	573	3.97	7.08	729	5.05	9.00	27
28	616	4.27	7.33	784	5.43	9.33	28
29	661	4.59	7.60	841	5.85	9.67	29
30	707	4.91	7.86	900	6.25	10.00	30
31	755	5.24	8.08	961	6.68	10.33	31
32	804	5.58	8.38	1024	7.10	10.67	32
33	855	5.94	8.67	1089	7.56	11.00	33
34	908	6.30	8.90	1156	8.02	11.33	34
35	962	6.68	9.17	1225	8.50	11.67	35
36	1018	7.06	9.44	1296	9.00	12.00	36

Cu. Ft. of Concrete in Flared Caps For Round Concrete Columns

Diam. of Col. Cap at Top	Diam. of Col. Cap at Bottom	Cu. Ft. Conc. Including Column Diam.	Cu. Ft. Conc. Excluding Column Diam.
4'-0"	12"	9.75	8.5
4'-0"	14"	9.70	8.0
4'-0"	16"	9.60	7.6
4'-0"	18"	9.40	7.0
4'-0"	20"	9.30	6.5
4'-0"	22"	9.00	5.8
4'-0"	24"	8.80	5.3
4'-0"	26"	8.70	4.9
4'-6"	16"	13.6	11.2
4'-6"	18"	13.5	10.6
4'-6"	20"	13.3	10.0
4'-6"	22"	13.1	9.3
4'-6"	24"	12.9	8.6
4'-6"	26"	12.6	7.8
4'-6"	28"	12.3	7.1
4'-6"	30"	11.9	6.4
5'-0"	20"	18.2	14.3
5'-0"	22"	18.0	13.4
5'-0"	24"	17.8	12.6
5'-0"	26"	17.4	11.7
5'-0"	28"	17.1	10.9
5'-0"	30"	16.8	10.0
5'-0"	32"	16.4	9.1
5'-0"	34"	15.8	8.2
5'-0"	36"	15.3	7.3

Sonotube Fibre Forms. Sonotube fibre forms (from Sonoco Products Co., Hartsville, SC) are spirally wound, laminated fibre tubes that were developed to provide a fast and economical method of forming for round concrete columns, piers, and the like. They are particularly adaptable as forms for encasing wood or steel piling. They provide a fast, economical method of forming all types of round concrete columns. They also permit new design possibilities for architects, provide new structural features for engineers, and save time, labor, and money.

Sonotube fibre forms are available in standard sizes from 4" to 48" inside diameter, with wall thickness ranging from .200" to .500" They are available in lengths up to 48 ft. and can be easily cut to length on the job with an ordinary hand or power saw.

There are three types of Sonotube fibre forms to meet virtually any combination of job requirements. The seamless is the premium form with specially finished inner ply which produces a smoother, continuous concrete surface. Regular A-coated is the standard form for exposed columns. For exposed columns not exceeding 12 ft. in height, the light wall A-coated is recommended. Sonotube fibre forms are lightweight, easy to handle, and require few men for erection manually or with simple block and tackle. Minimum bracing is required, fixed in place at bottom and light lumber bracing to keep them erect and plumb. No clamping is necessary.

Under normal conditions, for average size columns, 12" to 18" diameter, 10-ft. to 14-ft. in length, a carpenter and a helper should handle, cut to length, erect, and brace about 12 Sonotube column forms per 8-hr. day at the following labor cost per column:

	Hours	Rate	Total	Rate	Total
Carpenter	0.7	$....	$....	$24.04	$16.83
Helper	0.7	....	....	19.13	13.39
Cost per column			$....		$30.22

Sonotube fibre forms strip easiest and most economically from two to ten days after concrete is placed. Stripping may be accomplished with the least effort by using an electric hand saw, with blade set to cut slightly less than the wall thickness of Sonotube, and make two or three evenly spaced vertical cuts from bottom to top, after which form may be pulled free from column. Sonotube forms are for one time use only.

Under normal conditions, a carpenter and a helper should strip 25 to 30 average size Sonotube column forms per 8-hr. day at the following labor cost per column:

	Hours	Rate	Total	Rate	Total
Carpenter	0.3	$....	$....	$24.04	$7.21
Helper	0.3	....	....	19.13	5.74
Cost per column			$....		$12.95

Approximate Lineal Foot Prices of Sonotube Fibre Forms with "A" Coating

Inside Dia. and Wall Thickness	Price Per Lin. Ft.	Approx. Wt. Per Foot
8" x .200"	$1.80	1.7
10" x .225"	2.45	2.3
12" x .225"	3.15	2.8
14" x .250"	3.95	3.6
16" x .300"	5.30	4.8
18" x .300"	6.10	5.4
20" x .375"	7.80	7.5
22" x .375"	8.75	8.2
24" x .375"	10.00	8.9
26" x .375"	11.60	9.6
28" x .375"	12.85	10.4
30" x .400"	14.40	11.8
36" x .400"	18.80	14.5
42" x .500"	35.00	18.6
48" x .500"	45.00	23.5

Sonotube forms are manufactured in plants across the U.S., and distributors maintain stocks in most principal cities. The above prices are f.o.b. the nearest shipping point.

Anchoring Brick, Stone, and Terra Cotta to Concrete Backing

There are a number of methods of anchoring brick, stone, and terra cotta to concrete, some of which consist of metal slots nailed to the column or beam forms, which form a slot in the concrete for placing anchors.

The dovetail anchor slot method of anchoring masonry to concrete consists of a metal slot dovetail in cross section and anchors with dovetail ends for cut stone, brickwork, and terra cotta.

The dovetail anchor slot is nailed to the form with the open side against the wood. The ends are then closed with a wood plug. Or they come with foam filled slots. After the concrete is poured and the forms removed, the slot remains in the concrete, available at any course height to receive the anchors.

The anchors are installed by merely inserting the dovetail end of the anchor edgewise in the slot, then turning the anchor crosswise of the slot with the hand, the other end of anchor being imbedded in the facing material or mortar joint. The dovetail slot holds the end of anchor securely, making any other means of fastening, drilling or boring unnecessary.

Courtesy Symons Corporation
Symons Adjustable Column Forms for Poured-concrete High-rise Construction

Forms for Concrete Beams, Girders, Spandrel Beams, and Lintels

Forms for reinforced concrete beams, girders, lintels, etc., should be estimated by the sq. ft., obtained by adding the dimensions of the three sides of the beam and multiplying by the length. For example, a beam or girder may be marked 12"x24" on the plans, but maybe 6" of this depth is included in the slab thickness, so the form area should be obtained by taking the dimensions of the beam or girder from the underside of the slab: 18" + 12" + 18" = 48" or 4'-0" the girth, which multiplied by the length gives the number of sq. ft. of forms in the beam, 4'-0" x 18'-0" = 72 sq. ft.

Spandrel beams, or those projecting above or below the slab, should be measured in the same manner. A good rule to remember when measuring beam forms is to take the area of all forms that come in contact with the concrete.

When preparing an estimate on formwork, beams and girders should always be estimated separately; that is, the columns should be taken off, then the beams, girders and lintels, followed by the floor slabs, stairs.

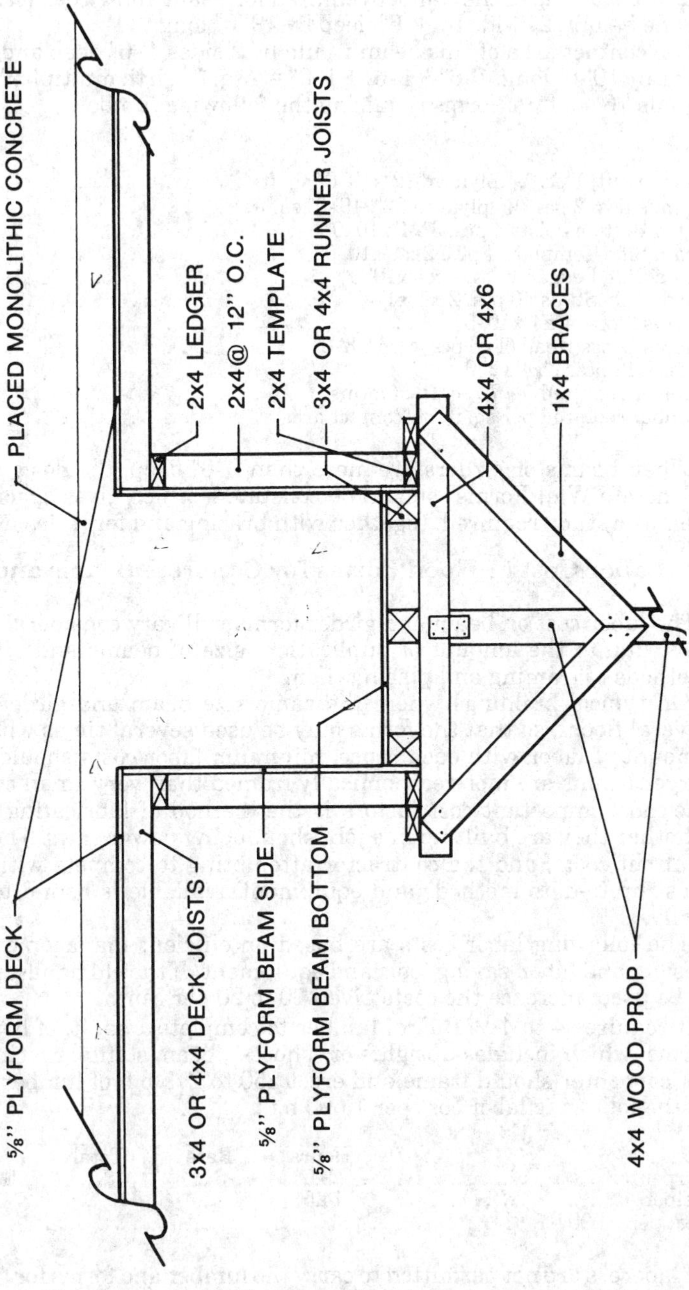

Method of Framing Wood Forms for Concrete Beams

As an example of the lumber required for beam and girder forms, assume an inside beam 12" wide by 1'-6" deep by 19'-0" long.

The contact area of this beam would be 2 sides 1'-6" deep and a bottom 1'-0" wide by 19'-0" long, 1'-6" + 1'-6" + 1'-0" = 4 sq. ft. girth multiplied by 19'-0" long, equals 76 sq. ft. of forms requiring the following lumber:

	B.F.
Beam soffit, 1 pc. 5/8" plyform 12"x19'-0" sq. ft.	19
Beam sides, 2 pcs. 5/8" plyform 22"x19'-0" sq. ft.	70
Beam Bottom Joists 4 pcs. 3"x4"x10'-0"	40
Beam Side Template 4 pcs. 2"x4"x10'-0"	27
Beam Side Ledger 4 pcs. 2"x4"x10'-0"	27
Beam Side Studs 40 pcs. 2"x4"x1'-4"	36
Shores 6 pcs. 4"x4"x10'-0"	80
Shores Cross Head (T) 6 pcs. 4"x4"x3'-0"	24
Braces 12 pcs. 1"x4"x2'-6"	10
Lumber required for 76 sq. ft. of forms	333
Lumber required per sq. ft. of "contact area"	4.4

When beams or girders are more than 2'-0" deep, the designs given under "Concrete Wall Forms" should be consulted, which gives spacing and size of cleats or studs required, together with bracing and form ties necessary.

Labor Cost of Wood Forms for Concrete Beams and Girders

The labor cost on beam and girder forms will vary considerably on different jobs, due to the amount of duplication, size of beams and girders, and the methods of framing and placing them.

On typical buildings where the same size beam and girders are used on several floors, so that the forms may be used several times with only a small amount of labor with each reuse, minimum labor costs should prevail. Also, larger beams are more economically framed than very small ones, but one of the most important cost factors is the method of fabricating the units and whether they are built up in a job "shop" using power saws. These are factors that cut costs, and the contractor attempting to compete without using the most up-to-date methods and equipment available is bound to suffer financially.

The following labor costs are based on efficient management, economical design, and labor-saving tools and equipment. If the old handsaw methods are to be used, increase the costs given 10 to 20 per cent.

It requires 4 to 4 1/2 B.F. of lumber to complete 1 sq. ft. of beam and girder forms, which includes uprights or "shores", beam soffits, bracing, etc.

A carpenter should frame and erect 250 to 275 b.f. of lumber per 8-hr. day, at the following labor cost per 1,000 b.f.:

	Hours	Rate	Total	Rate	Total
Carpenter	29.0	$....	$....	$24.04	$697.16
Helper	14.5	....	....	19.13	277.39
Cost per 1,000 b.f.					$974.55

If laborers are not permitted to carry the lumber and forms for the carpenters, the "helper" time given above should be figured as carpenter time.

A laborer should remove or "strip" 900 to 1,000 b.f. of lumber per 8-hr. day, at the following labor cost per 1,000 b.f.:

	Hours	Rate	Total	Rate	Total
Labor	9	$....	$....	$19.13	$172.17

Labor Cost of 100 Sq. Ft. of Inside Beam and Girder Forms Requiring 4 ½ BF of Lumber Per Sq. Ft. of Forms, Using 4"x4" Wood Shores

Making Beam Soffits and Sides	Hours	Rate	Total	Rate	Total
Carpenter	3.50	$....	$....	$24.04	$84.14
Labor helping	1.75	....	....	19.13	33.48
Assembling, Erecting, Shoring and Bracing					
Carpenter	5.50	....	....	24.04	$132.22
Labor helping	2.75	....	....	19.13	52.61
Removing Forms					
Labor	2.75	....	....	19.13	52.61
Cost per 100 sq. ft			$....		$355.06
cost per sq. ft			....		3.55

Labor Cost of 100 Sq. Ft. of Spandrel Beam or Lintel Forms Requiring 4 BF of Lumber Per Sq. Ft. of Forms, Using 4"x4" Wood Shores

Making Beam Soffits and Sides	Hours	Rate	Total	Rate	Total
Carpenter	3.50	$....	$....	$24.04	$84.14
Labor helping	1.75	....	....	19.13	33.48
Assembling, Erecting, Shoring and Bracing					
Carpenter	8.00	....	....	24.04	192.32
Labor helping	4.00	....	....	19.13	76.52
Removing Forms					
Labor	3.50	....	....	19.13	66.96
Cost per 100 sq. ft			$....		$453.42
cost per sq. ft			....		4.53

If the same forms can be reused without working over and cutting down, "Labor Making Forms" after first use may be reduced or omitted.

Forms for Upturned Concrete Beams. When the top of a beam is higher than the floor slab level, the beam side forms for that portion above the floor must be supported on temporary legs, which are removed after the concrete is poured and is still wet. It is also necessary to use spreaders to maintain the proper beam width.

While it is true that the additional form material required is negligible, the labor cost on this kind of beam side form will run 50% to 100% more than ordinary beam side forms.

Forms for Reinforced Concrete Floors

Forms for all types of reinforced concrete floors should be estimated by the sq. ft., taking the actual floor area or the area of the wood or metal forms that come in contact with the concrete.

Estimating the Cost of Form Lumber. When computing the lumber cost per sq. ft. of forms, the estimator should study the plans carefully to determine just how many times it is possible to use the same form lumber in the construction of the building.

On 8- to 12-story buildings, it may be possible to use the same lumber 4 to 5 times, while on a 3-story building it might be necessary to buy enough lumber to form almost the entire building. These are items that affect the material costs and can only be determined by the estimator or contractor, who should be in position to know just how soon the forms can be stripped and how fast they will be required to carry on the work.

On many jobs it will be more economical to use high early strength cement instead of ordinary portland cement, which will enable you to strip forms in 3 days instead of 7 to 10 days, saving in the number of sets of forms required.

Form lumber that has been used 3 or 4 times will have little salvage value, but each contractor should determine this for himself.

Types of Reinforced Concrete Floors. There are several types of reinforced concrete floor construction that require wood forms, centering, and shores. Because of the variations in costs, some of the more commonly used types will be described in detail.

Flat Slab Construction. Mill, factory, warehouse, and industrial buildings designed for heavy floor loads are usually of flat slab construction, which consists of floors without beams or girders, except spandrel beams between columns on outside walls, and around areas such as stair wells and elevator shafts. Buildings of this type usually have square or rectangular columns.

Beam and Girder Type Construction. This is the oldest type of concrete structure, having square or round columns with beams and girders running between each row of columns. The floor slabs is carried by the beams and girders and then by the columns.

Pan Construction with Concrete Joists. Apartment buildings, hotels, schools, hospitals, department stores, office buildings, and other structures requiring light floor loads ordinarily use this type of construction, which consists of metal pans in conjunction with concrete joists. This type of floor is used with either round or square columns and beams that run at right angles to the joists.

Forming Reinforced Concrete Floors Using Adjustable Shores

Designing Forms for Reinforced Concrete Floors and Estimating Quantity of Lumber Required

The process of designing forms for reinforced concrete floors may be divided into several parts, such as determining the span of the sheathing, the size and spacing of joists and girders or stringers, and the size and spacing of the upright supports or shores.

The spacing of joists for concrete floor forms is governed by the strength of the plywood used. When plywood panels are used for sheathing, the spacing of the joists must coincide with the sizes of the plywood panels, 4'-0"x8'-0", which means that the joists should be spaced 12", 16", or 19" on centers, depending on the load to be carried.

Weight of Concrete Floors of Various Thicknesses

In determining the weight per sq. ft. of floor, wet concrete is figured at 150 lbs. per cu. ft. Dead load of form lumber and live load on forms while concrete is being placed is figured at 38 lbs. per sq. ft., which is sufficient for temporary construction.

Floor Thickness	2"	3"	4"	5"	6"	7"	8"	9"	10"	11"	12"
Conc. Wt. per Sq. ft., Lbs.	25	37.5	50	62.5	75	87.5	100	112.5	125	137.5	150
Temp. Live Load per Sq. ft., lbs	40	37.5	38	37.5	38	37.5	38	37.5	38	37.5	38
Total Weight per Sq. Ft., Lbs.	65	75	88	100	113	125	138	150	163	175	188

Table Giving Spacing of Stringers and Shores When Supporting Floor Forms for Metal Pan and Concrete Joist Construction

Using 2" plank to support the concrete joists 2'-1" or 2'-11" on centers, with stringers spaced 4'-0" apart to support the 2" plank.

TABLE FOR DESIGNING SLAB FORMS FOR REINFORCED CONCRETE FLOORS BASED ON USING 3/4" SHEATHING, WITH JOISTS SPACED 2'-0" ON CENTERS

Thickness of Concrete Floor Slab and Total Dead and Live Load per Sq. Ft.

Thickness Concrete Slab Total Load per Sq. Ft., Lbs	2" 65	3" 75	4" 88	5" 100	6" 113	7" 125	8" 138	9" 150	12" 188
Distance Between Stringers When Joists are Spaced 2'-0" on Centers									
2"x4" Joists, S4S	6'-0"	5'-6"	5'-6"	5'-0"	5'-0"	4'-6"	4'-6"	4'-0"	4'-0"
2"x6" Joists, S4S	7'-6"	7'-0"	7'-0"	6'-6"	6'-6"	6'-0"	6'-0"	6'-0"	5'-6"
2"x8" Joists, S4S	9'-6"	9'-0"	8'-6"	8'-6"	8'-0"	8'-0"	7'-6"	7'-6"	7'-0"
2"x10" Joists, S4S	11'-0"	10'-6"	10'-0"	10'-0"	9'-6"	9'-6"	9'-0"	9'-0"	8'-6"
Distance Between Shores or Supports Using Stringers of Size and Spacing as Given Below									
4"x4" S4S Stringers, 5-ft. c.c.	5'-0"	4'-6"	4'-0"	4'-0"	4'-0"	3'-6"	3'-6"	....	....
4"x6" S4S Stringers, 6-ft. c.c.	6'-6"	6'-0"	6'-0"	6'-0"	6'-0"	5'-6"	5'-0"	5'-0"	4'-6"
3"x8" S4S Stringers, 6-ft. c.c.	8'-6"	7'-6"	7'-0"	7'-0"	7'-0"	6'-6"	6'-0"	6'-0"	5'-6"
2"x10" S4S Stringers, 7-ft. c.c.	7'-6"	6'-0"	5'-6"	4'-6"	4'-6"	4'-0"	....	....	....

The distance between shores or span of stringers is generally an even fraction of their length. Stringers 14'-0" long will permit a spacing of 4'-8" or 7'-0", while 18'-0" stringers will permit spans of 3'-7", 4'-6", 6'-0" and 9'-0".

Based on using yellow pine, Douglas fir or woods of equal strength.

The table gives the sizes of stringers and the distance between supports or shores when supporting various floor loads.

Size of Stringer	Thickness of Floor in Inches Weight of Floor per Sq. Ft.*	6+2½ 86 lbs.	8+2½ 92 lbs.	10+2½ 98 lbs.	12+2½ 105 lbs.	14+2½ 112 lbs.
		Distance Between Shores or Supports				
4"x 4"S4S		5'-0"	4'-9"	4'-8"	4'-8"	4'-6"
3"x 6"S4S		6'-6"	6'-5"	6'-3"	6'-1"	6'-0"
4"x 6"S4S		7'-0"	7'-0"	6'-8"	6'-7"	6'-6"
3"x 8"S4S		8'-0"	8'-0"	7'-6"	7'-6"	7'-6"
2"x10"S4S		8'-0"	8'-0"	7'-6"	7'-6"	7'-6"

The above tables are based on using yellow pine, Douglas fir, or woods of equal strength.

Data on Plywood Sheathing

When plywood is used for sheathing instead of 3/4" boards, the following loads may be safely carried.

Allowable Pressures on Plyform
Class I for Architectural Applications
(deflection limited to 1/360th of the span)

Face Grain Across Supports
Allowable Pressures (psf)*

Support Spacing (in.)	Plywood Thickness (inches) ½	5/8	¾	7/8	1	1 1/8
4	2935	3675	4495	4690	5075	5645
8	970	1300	1650	1805	1950	2170
12	410	575	735	890	1185	1345
16	175	270	370	475	645	750
20	100	160	225	295	410	490
24			120	160	230	280
32					105	130
36						115

*Plywood continuous across two or more spans.

Face Grain Parallel to Supports
Allowable Pressures (psf)*

Support Spacing (in.)	Plywood Thickness (inches) ½	5/8	¾	7/8	1	1 1/8
4	1670	2110	2610	3100	4140	4900
8	605	810	1005	1190	1595	1885
12	215	360	620	740	985	1165
16		150	300	480	715	845
20		105	210	290	400	495
24			110	180	255	320

*Plywood continuous across two or more spans.

Allowable Pressures on Plyform Class I
(deflection limited to 1/270th of the span)

Face Grain Across Supports Allowable Pressures (psf)*

Support Spacing (in.)	Plywood Thickness (inches) ½	5/8	¾	7/8	1	1 1/8
4	2935	3675	4495	4690	5075	5645
8	970	1300	1650	1805	1950	2170
12	430	575	735	890	1185	1345
16	235	325	415	500	670	770
20	135	210	285	350	465	535
24		110	160	215	295	340
32					140	170
36					105	120

*Plywood continuous across two or more spans.

Face Grain Parallel to Supports Allowable Pressures (psf)*

Support Spacing (in.)	Plywood Thickness (inches) ½	⅝	¾	⅞	1	1 1/8
4	1670	2110	2610	3100	4140	4900
8	605	810	1005	1190	1595	1885
12	270	405	620	740	985	1165
16	115	200	375	525	715	845
20		125	210	290	400	495
24			135	185	255	320

*Plywood continuous across two or more spans.

Carrying Capacity of 4"x4" Uprights or Shores. The carrying capacity of a given size upright or shore is generally limited by two principal factors. The first is compression that the shore exerts on the cross beam or stringer at right angles to the fibers . The stress per sq. in. on the bearing area of the stringer on the shore should not greatly exceed 500 lbs. per sq. in. for Yellow Pine, Western Fir, or similar lumber. Otherwise, there is a noticeable impression, often as much as ⅛", of the shore into the fibers of the stringer, especially when the lumber is green or water soaked after a long rain. A (4"x4", S4S), 3 ½"x3 ½" upright should never be loaded to more than 6,000 lbs., no matter how short it is, and a 4"x4" rough shore should not be loaded to more than 8,000 lbs.

The other principal limitation of strength is due to the length of the shore and the degree of crookedness in its length. Every experienced contractor knows that shores often show bows of 1" to 2", and if not cut off exactly square to their length, the carrying capacity of most shores are greatly limited by the eccentric loading impressed on them by the uneven bearing of the stringer and the bow in the shores.

The following table gives the permissible load on shores for various lengths and eccentricities:

Safe Load in Lbs. on Uprights or Shores of Various Lengths

Length of Upright or Shore	4"x4" S4S Yellow Pine or Fir 8'-0"	4"x4" S4S Yellow Pine or Fir 10'-0"	4"x4" Rough Yellow Pine or Fir 8'-0"	4"x4" Rough Yellow Pine or Fir 10'-0"	4"x4" Rough Wisconsin Hemlock 8'-0"	4"x4" Rough Wisconsin Hemlock 10'-0
Eccentricity						
(Bow+Top) 0"	6,000	6,000	8,000	8,000	8,000	8,000
(Bow+Top) 1"	5,700	4,800	8,000	7,100	7,300	5,600
(Bow+Top) 1 ½"	4,700	3,900	7,300	5,800	6,000	5,100
(Bow+Top) 2"	4,200	3,400	6,700	5,500	5,200	4,400
(Bow+Top) 2 ½"	3,600	3,000	5,300	4,500	4,800	4,100
(Bow+Top) 3"	3,400	2,800	5,100	4,300	4,500	3,800

Lintels as a rule produce quite large eccentricities on the shores due to the load from the adjoining slab, which comes through on the inside of the lintel forms onto the shore. Never allow a load larger than those given for 3" eccentricity for shores supporting lintels. One frequently sees lintels supported by shores spaced less than 2'-0" on centers, which is a clear waste of money.

Where the height of the uprights or shores are longer than given in the above table, it is advisable to cross brace the shores, in which case the vertical distance between braces plus 40% may be taken as the post height. Example:

You are using a 16'-0" shore, with 2 rows of cross bracing at 6'-0" and 12'-0" above the floor. The distance between braces 6'-0" plus 40%, or 2.40 ft., makes the capacity the same as a shore 8.40 or practically 8'-6" long. Thus an 8'-0" shore good for 8,000 lbs. would still be good for 8,000 lbs. when 16'-0" long, if cross braced by two rows of bracing 6'-0" and 12'-0" above the floor.

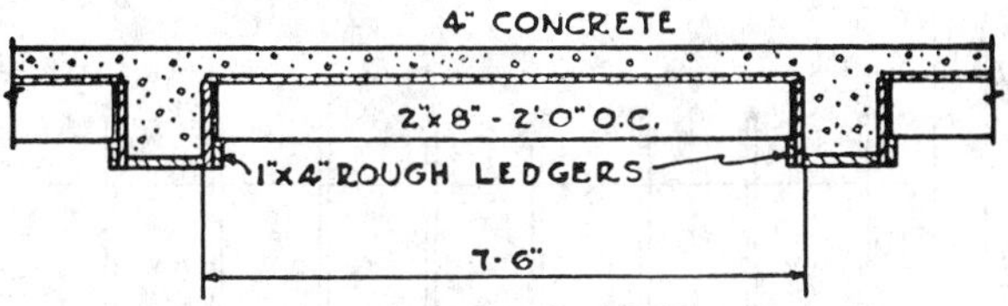

Method of Constructing Forms for Beam and Girder Type Solid Concrete Floor Slabs

Estimating the Quantity of Lumber Required for Forms for Beam and Girder Type Solid Concrete Floor Slabs. The accompanying illustrations give detailed designs, from which the lumber required per sq. ft. of forms may be computed.

The accompanying illustration shows a 4" concrete floor slab having a 7'-6" clear span between beams. Using a joist spacing of 2'-0" as given in the table for joist spacing, 2"x8" joists will carry the required load.

For a floor area 2'-0"x7'-6" containing 15 sq. ft. the following lumber will be required:

	BF
Sheathing, 2'-0" x 7'-6" = 15 sq. ft. plus 20% waste	18
Joist, 1 pc. 2"x8"-8'-0", plus 10% for joints	12
Ledger boards, 2 pcs. 1"x4"-2'-0"	1
Total lumber required for 15 sq. ft. forms	31
B.f. of lumber required per sq. ft. of forms	2.1

For a 6" solid concrete slab having a clear span of 11'-0" between girders, usually it is not economical to support joists with this span directly on the girders. It is a better practice to provide a stringer or girt halfway between the concrete beams so that the joists have a span of only 5'-6". Using a joist spacing of 2'-0", a 2"x6" joist is sufficiently strong, as shown by the table of joist spacings. The same table shows that a 4"x6" supported by shores 6'-0" on centers will suffice.

A floor area 2'-0"x11'-0" containing 22 sq. ft. will require the following lumber:

	BF
Sheathing, 2'-0" x 11'-0" = 22 sq. ft. plus 20% waste	27
Joists, 1 pc. 2"x6"-12'-0", plus 10% for joints	13
Stringer, 1 pc. 4"x6"-2'-0"	4
Shore, 1/3 of a 4"x4"-10'-0", S4S, plus sill, wedges, 18/3 =	6
Total lumber required for 22 sq. ft. forms	50
B.f. of lumber required per sq. ft. of forms	2.25

If shores require bracing, add 0.1 b.f per sq. ft. of slab.

Assume a solid concrete floor slab 16'-0" wide, 19'-0" long, and 12" thick, containing 304 sq. ft. of slab forms. Joists will be 2"x6" spaced 24" apart, supported by 3 lines of 3"x8"-16'-0" stringers, which in turn are supported by 4"x4" shores 5'-4" apart.

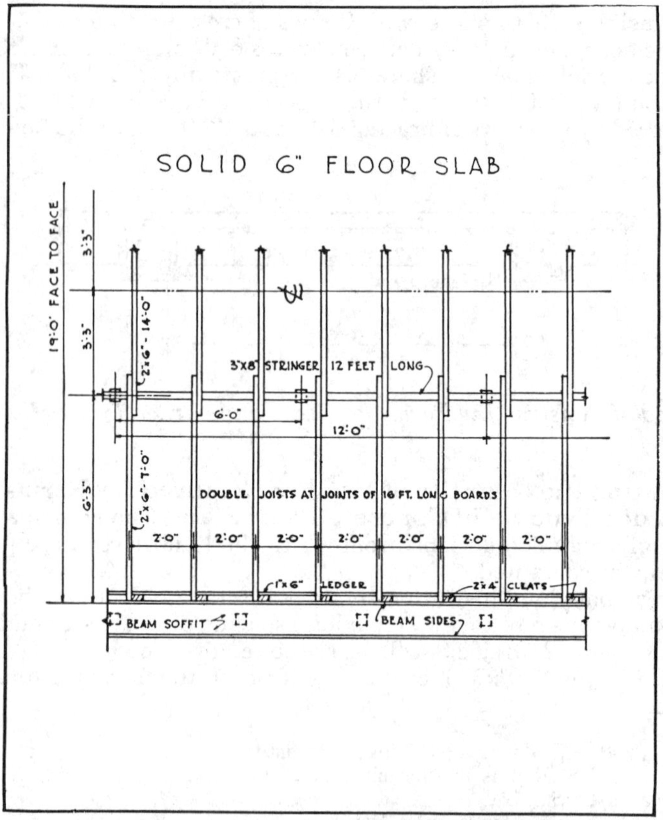

Method of Constructing Forms for Beam and Girder Type
Solid Concrete Floors

The following lumber will be required for 304 sq. ft. of forms:

	BF
Sheathing, 16'-0" x 19'-0" = 304 sq. ft. plus 20% waste	365
Joists, 9 pcs. 2"x6"-14'-0"	126
Joists, 9 pcs. 2"x6"-7'-0"	63
Stringers, 3 pcs. 3"x8"-16'-0"	96
Shores, 9 pcs. 4"x4"-10'-0", including sills and wedges	153
Braces, 6 pcs. 1"x6"-16'-0"	48
Joist support at girders, 2 pcs. 1"x6"-16'-0"	16
Total lumber required for 304 sq. ft. forms	867
B.f. lumber required per sq. ft. of forms	2.85

The above quantities are based on the assumption that the ends of joists and stringers are supported by the beam and girder forms. If a wall bearing job without beams and girders, the following additional lumber will be required:

	BF
Stringers, 2 pcs. 2"x8"-16'-0" at wall	43
Shores, 6 pcs. 4"x4"-10'-0", including sills and wedges	102
Braces, 2 pcs. 1"x6"-16'-0"	16
Additional lumber required for 304 sq. ft. forms	161
Additional lumber, b.f. required per sq. ft.	0.53

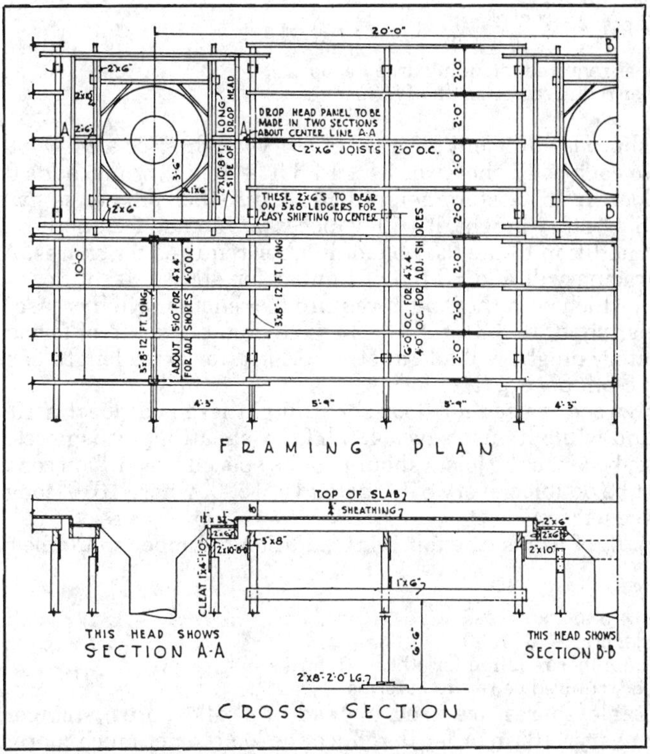

Method of Constructing Forms for Flat Slab Concrete Floors

Estimating the Quantity of Lumber Required for Flat Slab Concrete Floor Forms. Assume a flat slab having a 20'-0" span, with 7'-0" square drop heads, 8" thick as shown in illustration.

The maximum span of the joists is 5'-9". The table of joist spacing shows that 2"x6" joists and 3"x8" stringers are sufficient. The load on the shores of the center stringer will be 5'-9" x 6'-0" = 34.5 sq. ft. at 138 lbs. per sq. ft. = 4,761 lbs. Good practice requires a 4"x4" shore rough, though a 4"x4" S4S with bracing will do. The lumber required for a 20'-0"x20'-0" panel is as follows:

	BF
Sheathing, 400 sq. ft. plus 20% waste	480
Using 16'-0" long floor sheathing boards will require 9 joists (a double joist being placed at the joint) or for 20 ft. it will require 11 ¼ pcs 2"x6"-12'-0" joists (1 ¼ x 9)	135
Column Strips, 9 joists 2"x6"-10'-0"	90
Stringers, 3 pcs. 3"x8"-20'-0"	120
Stringers, 1 pc. 3"x8"-12'-0"	24
Shores, 13 pcs. 4"x4"-10'-0" plus sills, wedges, etc. @ 17 ft.	221
Horizontal bracing, 7 pcs. 1"x6"-20'-0"	70
X Bracing, not less than ⅓ horizontal bracing	23
Total lumber required for 400 sq. ft. forms	1,163

For Drop Heads	
Joists, 5 pcs. 2"x6"-14'-0"	70
Stringers, 2 pcs. 2"x8"-8'-0"	21
Shores, 4 pcs. 4"x4" at 17 b.f.	68
Total lumber required including drop heads	1,322
B.f. of lumber required per sq. ft. of forms	3.31

A 10'-0" shore is sufficient for a 12'-0" story height. For a 14'-0" story height, add 2'-0" to each of 17 shores or 34 x 1.33 ft. = 45 b.f., which adds 0.11 b.f. per sq. ft. of floor. A 16'-0" story height will require 0.23 b.f. per sq. ft. of floor.

For an 18'-0" story height, it will be necessary to add for another set of cross bracing in addition to the 0.33 b.f. of lumber required for shores. Another set of cross bracing will add 93 b.f. of lumber for 400 sq. ft. of floor or 0.23 b.f. additional, which with the 0.33 ft. required for shores, will increase the lumber required per sq. ft. to 3.31 + .33 + .23 = 3.87 b.f. per sq. ft. of forms.

A 20'-0" story height will add another 0.11 ft. for extra length of shores, or a total of 3.98 b.f. per sq. ft. of forms.

Where plywood is used for floor sheathing, there is at least a 10% waste in flooring, and while its use saves 480 b.f. for sheathing, it is necessary to use more joists, because the joists should not be spaced over 19" on center, and the joists must be doubled every 8'-0" instead of 16'-0" where 16'-0" long sheathing boards are used.

In 400 sq. ft. of forms, the following additional lumber would be required:

	BF
Extra joists, 4 pcs. 2"x6"-12'-0"	48
Extra joists, 2 pcs. 2"x6"-10'-0"	20
Total extra lumber required for 400 sq. ft. forms	68
Extra lumber required per sq. ft. of forms	0.17

If adjustable shores are used instead of 4"x4" shores, stringers must be supported every 4'-0" in order to reduce the load to not much more than 3,000 lbs. per shore although some adjustable shores will carry up to 6,000 lbs. This will require 17 to 23 adjustable shores instead of 17 pcs. of 4"x4" shores (247 b.f.) and 9 pcs. of horizontal bracing, 1"x6"-20'-0" (90 b.f.) instead of 7 pcs. of 1"x6"-20'-0" (70 b.f.) and the X bracing will also be increased by 7 b.f.

Using adjustable shores, 4"x4" stringers can be used instead of 3"x8" stringers, which will result in the following saving:

	BF
Shores, 17 pcs. 4"x4"-10'-0"	247
Stringers, 4"x4" instead of 3"x8", saving	44
Total saving using adjustable shores	291
Extra 1"x6" required for bracing	27
Extra sills, 6 pcs. 2"x10"-2'-0"	20
	47
Total saving on 400 sq. ft. forms	244
Total saving per sq. ft. of forms b.f.	0.61

On the other hand, it will be necessary to handle 17 to 23 shores weighing 1,000 to 1,400 lbs. while the 224 ft. saving weighs only 725 lbs.

For a 12" thick flat slab having a 20'-0" span, the lumber required can be approximated as follows:

	BF
Sheathing, per sq. ft forms	1.20
Joists, same spacing but using 2"x8" joists	0.75
Stringers, same as before	0.33
Shores, spacing will be 4'-6", requiring 21 shores	0.90
Bracing, 9 pcs. 1"x6"-20'-0"	0.23
X bracing, 1/3 of horizontal bracing	0.08
Drop heads, same as before	0.25
Total lumber, b.f. per sq. ft. for 12'-6" story height	3.74

For each 2 ft. story height, more or less, add or deduct per sq. ft..14

Weight of Combination Metal Pan and Concrete Joist Floors

Based on concrete joists 4", 5", and 6" wide, with a 2 ½" concrete slab on top, using metal pans of various depths and widths.

Concrete Thickness, in.	2 ½"	2 ½"	2 ½"
Width of Concrete Joists, in.	4"	5"	6"
		Weight per Sq. Ft. of Slab. Lbs.	
Using			
Metal Pans 6" deep, 20" wide	45	48	50
Metal Pans 6" deep, 30" wide	41	43	45
Metal Pans 8" deep, 20" wide	51	54	57
Metal Pans 8" deep, 30" wide	45	48	50
Metal Pans 10" deep, 20" wide	56	60	64
Metal Pans 10" deep, 30" wide	50	52	55
Metal Pans 12" deep, 20" wide	63	67	72
Metal Pans 12" deep, 30" wide	54	57	61
Metal Pans 14" deep, 20" wide	..	75	80
Metal Pans 14" deep, 30" wide	..	62	66

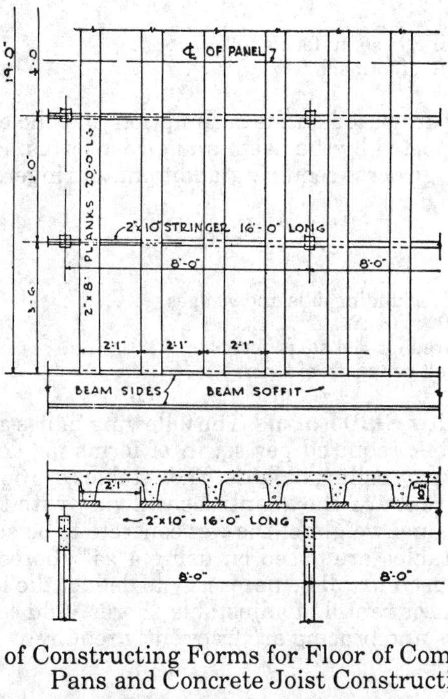

Method of Constructing Forms for Floor of Combination Metal Pans and Concrete Joist Construction

For each ½" variation in slab thickness, add or subtract 6.25 lbs. per sq. ft. of floor.

Add 38 lbs. per sq. ft. to the above weights to take care of temporary dead and live load on floors while concrete is being poured.

Lumber Required for Floor Forms Using Metal Pans and Concrete Joist Construction. Assume a slab of 19'-0" clear span and 16'-0" long, using 8" metal pans with 5" wide ribs spaced 2'-1" on centers and a 2 ½" concrete top over the entire slab.

According to the above table this construction weighs 52 lbs. per. sq. ft. and including a 38 lb. live load during construction, makes a total load of 90 lbs. per sq. ft.

It is customary to use 2"x8" S4S planks to support the concrete ribs or joists. If the planks are continuous without patching at the end spans, a span of 4'-0" will support a load of 265 lbs. per lin. ft. with a deflection of .235 in. In this instance the load is 2.09 x 90 lbs. per sq. ft. = 188 lbs. per lin. ft., which reduces the deflection to 0.166" or about ⅙", which is permissible.

It will require a stringer or girt every 4'-0" to support the planks and the load per lin. ft. on the stringers will be 360 lbs.

A 2"x10"-16'-0" S4S will easily carry this uniform load per lin. ft. when supported by 4"x4" shores spaced 8'-0" apart.

The following lumber will be required for 304 sq. ft. of forms:

	BF
Planks, 9 pcs. 2"x8"-20'-0"	240
Stringers, 4 pcs. 2"x10"-16'-0"	107
Shores, 8 pcs. 4"x4"-10'-0", including sills and wedges	136
Joist supports at girders, 2 pcs. 1"x6"-16'-0"	16
Bracing, 4 pcs. 1"x6"-16'-0"	32
Bracing, 2 pcs. 1"x6"-20'-0"	20
Total lumber required for 304 sq. ft. forms	551
Lumber required per sq. ft. of forms, b.f.	1.8

The above quantities are based on the assumption that the ends of the joists and stringers are supported by the beam and girder forms. If a wall bearing job without beams or girders, requiring additional stringers and shores at walls, add the following:

	BF
Stringers, 2 pcs. 2"x10"-16'-0"	54
Shores, 4 pcs. 4"x4"-10'-0", including sills and wedges	68
Bracing, 2 pcs. 1"x6"-16'-0"	16
Additional lumber required for 304 sq. ft. forms	138
Additional lumber required per sq. ft. of forms	0.45

Lumber Required for Slab Forms. The following tables give the approximate quantity of lumber required per sq. ft. of forms for floors of different thickness and with varying ceiling heights. These tables are only approximate. Every building is different and the quantities will vary with the span, ceiling height, slab thickness, and weight of the wet concrete to be supported.

Also, the following tables are based on using 4"x4" shores or uprights. If adjustable shores are used it will be necessary to deduct the lumber required for 4"x4"s and add cost or rental of adjustable shores. Add cost of additional bracing as more shores and bracing are frequently required.

The difference in the quantity of lumber required per sq. ft. for floors having 10'-0" ceiling heights and those having 20'-0" ceiling heights is principally in the shoring and bracing. If adjustable shores are used, deductions can be made on the following basis: floors 3" to 6" thick with 10'-0" ceiling heights require 1/3 to 3/8 ft. of lumber per sq. ft. for shores; 7" to 9" slabs require 1/2 to 5/8 ft.; and 10" to 12" slabs require 5/8 to 3/4 ft. of shoring lumber per sq. ft. of floor.

The above quantities can be used for all 10'-0" story heights, plus the additional lumber given in the tables for higher ceiling heights.

The forms are designed in yellow pine, Douglas fir, or woods of equal strength, with the loads computed as follows: wet concrete 150 lbs. per cu. ft., dead load of form lumber and live load on forms while concrete is being placed at 38 lbs. per sq. ft.

Number of Feet of Lumber Required for One Sq. Ft. of Flat Slab Concrete Floor Forms

Ceiling Height in Feet	Thickness of Slab in Inches 6" 113 lbs.	8" 138 lbs.	10" 163 lbs.	12" 188 lbs.
10	3.0	3.1	3.4	3.6
12	3.1	3.2	3.5	3.7
14	3.3	3.4	3.6	3.9
16	3.4	3.5	3.9	4.0
18	3.4	3.8	4.2	4.4
20	3.5	3.9	4.3	4.6

The above quantities include lumber required for forming drop heads. If adjustable shores are used, deduct lumber quantities given on previous pages.

Using Steel Truss Joists Instead of Shoring for Supporting Formwork for Reinforced Concrete Floors. In steel frame buildings where clear floor space is essential, steel truss joists may be used for supporting the wood formwork, as shown in the accompanying illustration.

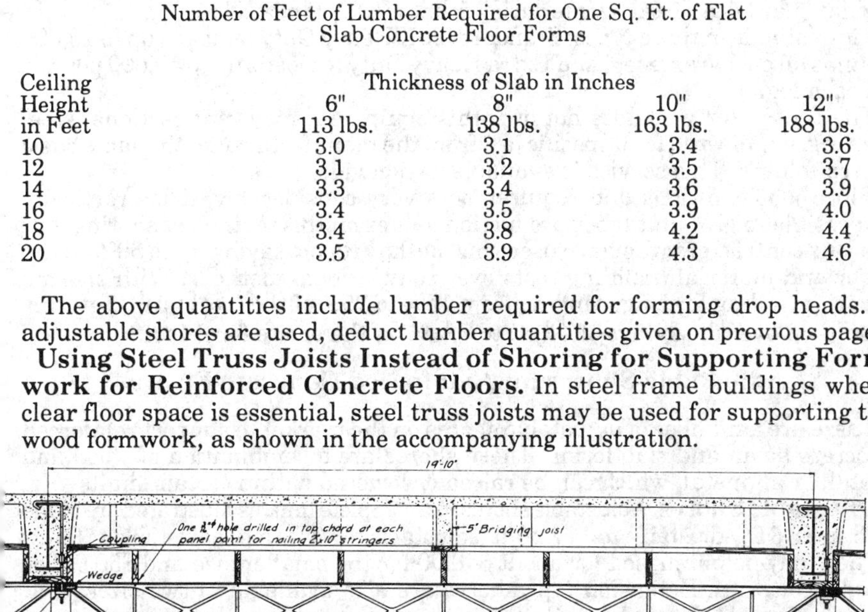

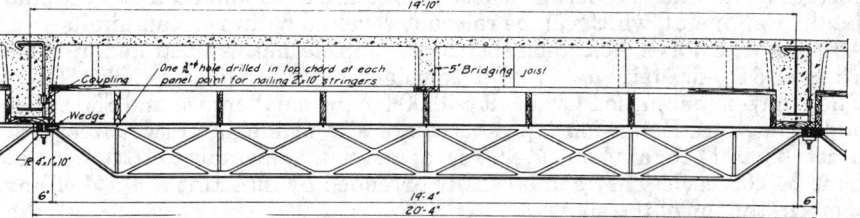

Steel Joists Used for Supporting Concrete Floor Forms

This is a simple method that is very effective as the salvage value of the truss joists is very high.

Sectional Steel Shoring. Many contractors now use their sectional steel scaffolding units for shoring formwork for reinforced concrete slabs and beams. Prefabricated sectional steel scaffolding sections are as simple to erect for shoring as they are for scaffolding. The same scaffolding components--base plates, adjustable extension legs, welded 5-ft. wide frames varying in height from 3'-0" to 6'-6" and diagonal braces to provide spacing between frames from 2'-6" to 7'-0"--are assembled quickly to provide free standing shoring sections suited to the easy placing of stringers, joists, and decking. In most instances, sectional steel shoring can be used wherever conventional shoring methods of adjustable shores or 4"x4" lumber are used, but is especially adaptable for shoring forms for large expanses of flat slab construction.

Height adjustments are quickly made by the use of 20" adjustable legs, which eliminates the need for cutting, fitting, and wedging. These legs also facilitate stripping operations by the ease with which the shoring units may be lowered. Sectional steel shoring also provides scaffolding within the shoring for forming and stripping, increasing job safety. This method also minimizes the use of wood, reducing the fire hazard and increasing the number of reuses obtainable from form lumber.

Safe working loads for sectional steel shoring depend on the spacing between frames in the same row, distance between rows of frames, and position of the stringers on the frames. Sectional steel shoring will support the largest loads when the stringers are placed directly over the frame legs by means of inverted base plates or U-heads as shown in the illustration. If this is not practical, stringers should be placed as close as possible to frame legs. If required, additional lateral rigidity may be attained by interlacing frames with standard bracing.

It is available in three types: Standard, 20K Heavy Duty for loads up to 10,000 pounds on each frame leg, and Extra Heavy Duty for loads up to 40,000 pounds on each leg.

The contractor who does not own this equipment may rent sectional steel shoring, engineered for a specific job, from the manufacturer on the same basis as scaffolding. This service is available nationwide.

Since job conditions and requirements vary considerably, it is practically impossible to give unit labor production values on this method of shoring, but various contractors who have used this method report saving up to 30% on the labor and material handling costs over conventional methods. With shoring labor, including bracing, ranging from 15% to 25% of the total form cost, this results in overall savings of 4 ½% to 7 ½% on the entire form cost.

Adjustable Shores for Reinforced Concrete Floors

There are a number of adjustable shores on the market to support reinforced concrete beam and slab forms. These shores are a combination of wood and metal, or all metal, which can be raised or lowered within certain limits.

Adjustable Shores. Adjustable shores in a complete unit, as illustrated are made in 3 sizes: 8 ft. adjustable to 14 ft.; 7 ft. adjustable to 13 ft.; 5 ft. adjustable to 9 ft.

They may be purchased for about $40.00 for the smaller size and $50.00 for the larger sizes. Extension type shores are also available in two sizes: 8 ft. adjustable to 14 ft., at about $45.00 each; and 6 ft. adjustable to 10 ft. 6 in., at about $43.00 each. They can be easily extended by inserting a 4"x4" of any length in the top of the shore.

Rental stocks are maintained in most principal cities and may be rented at about $2.50 per month, per section, f.o.b. warehouse and return.

Each shore will support about 3,000 lbs. with an adequate factor of safety. In some cases, however, shores are spaced to carry a load of not over 2,200 lbs., as some forming authorities consider it cheaper to use more shores than to use heavier lumber to prevent excessive deflection between supports.

For local prices and availability, the estimator should get a firm quote from the local supplier.

Adjustable Shores Versus 4"x4"s. It is much easier to shore up a floor using adjustable shores instead of 4"x4"s, which require wedging and cutting or adding to for each change in story height. This additional labor all costs money at present wage scales. However, to offset this, the original cost of the adjustable shores is much more than the 4"x4"s, but if they can be used often enough they will eventually show a saving. This will have to be considered when contemplating a purchase or rental basis.

Sectional Shoring Supporting Forms for Concrete Cap of Reservoir

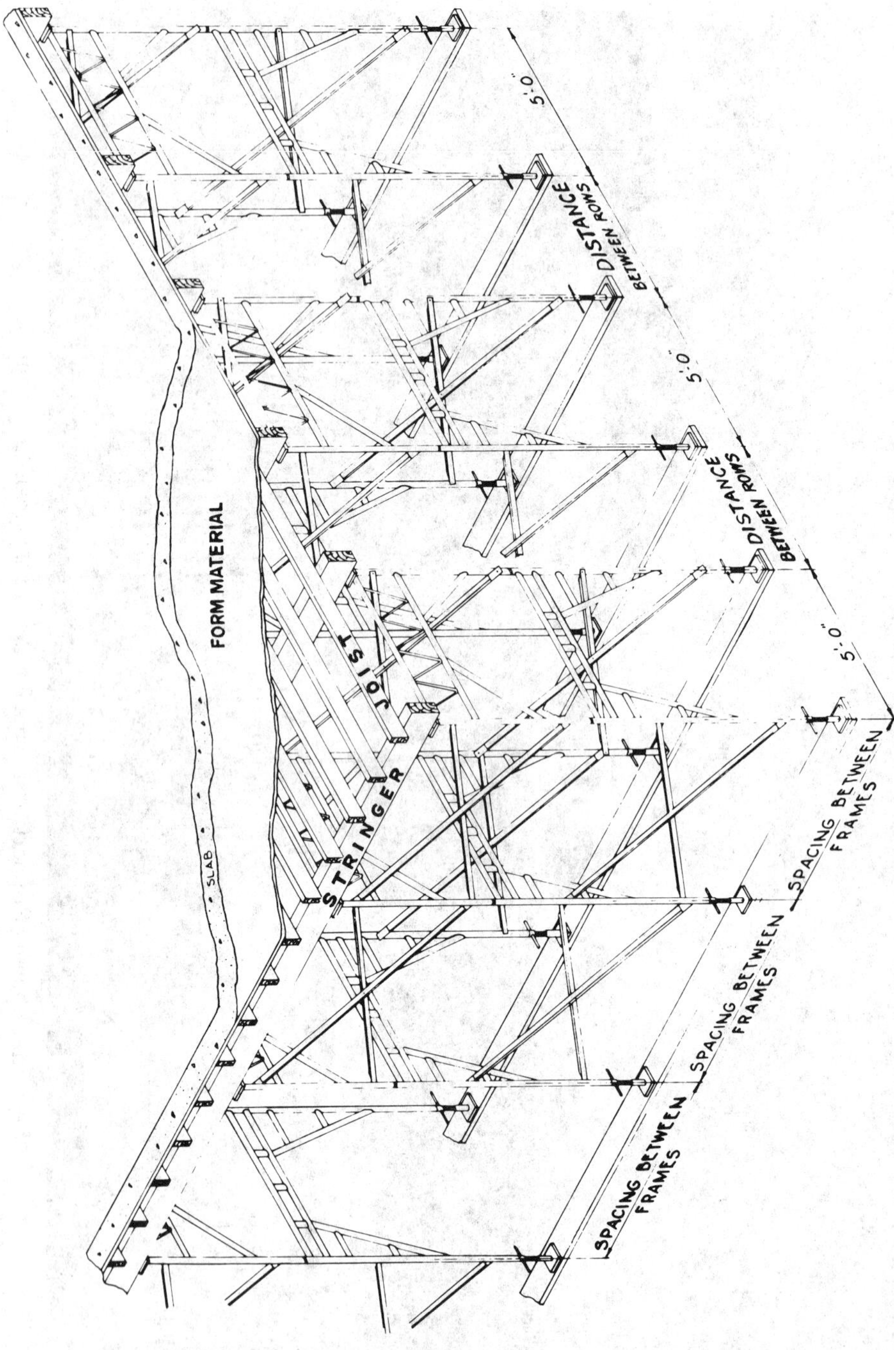

Method of Shoring for Reinforced Concrete Construction Using Sectional Steel Scaffolding Components

Courtesy Symons Corporation
Adjustable Shores Used to Support Reinforced Concrete Forms

Another thing that should be considered is the fact that a 4"x4" shore will usually carry a larger load than an adjustable shore, so in many cases it will require more adjustable shores than 4"x4"s to carry the same load. On the other hand, especially on light constructed floors, the adjustable shores may carry all the load required. This is an item that should be considered by the estimator before pricing the job.

The form costs on the following pages are computed on a rental basis, based on 10 days' time for each shore or 3 uses a month.

Labor Cost of Flat Slab Forms. Forms for flat slab floors are continuous--without breaks or offsets--except at column heads, stair wells, elevator shafts, etc., where it is necessary to frame for beams and girders.

The sq. ft. labor cost will vary with the amount of lumber required per sq. ft. of forms, as the heavier the slab and the higher the ceiling, the more lumber required per sq. ft.

Forms for exceedingly light floor slabs cost proportionately more per sq. ft. than heavy forms as the preliminary work is the same with both types. However, on work of this class a carpenter should frame and erect 375 to 425 BF of lumber per 8-hr. day, at the following labor cost per 1,000 BF:

	Hours	Rate	Total	Rate	Total
Carpenter	20	$....	$....	$24.04	$480.80
Labor	10	....	....	19.13	191.30
Cost per 1,000 BF		....	$....		$672.10

When removing or "stripping" forms, a laborer should remove 1,000 to 1,200 BF of lumber per 8-hr. day, at the following labor cost per 1,000 BF:

	Hours	Rate	Total	Rate	Total
Labor	7	$....	$....	$19.13	$133.91

Labor Framing for Slab Depressions at Column Heads. On flat slab jobs there may be a depression at each column head 5'-0" to 7'-0" square and 4" to 8" deep. The labor cost framing for these depressions will cost about 80% more than for straight slab forms.

A satisfactory method of estimating this additional labor is to figure the entire floor area at the regular price and then compute the area of the depressions and figure them at 80% additional labor.

Example: Floor contains 10,000 sq. ft. of slab forms, including 40 column head depressions 5'-0"x5'-0", containing 25 sq. ft. each.

10,000 sq. ft. slab forms $2.60 per sq. ft.	$26,000.00
Add for 40 depressions (5'-0"x5'-0") 1,000 sq. ft. at 80% more than plain slab forms or $2.08 per sq. ft.	2,080.00

This makes a total of $4.68 per sq. ft. for labor framing depressions at column heads.

Labor Cost of 100 Sq. Ft. of Flat Slab Concrete Floor Forms Requiring 3 BF of Lumber per Sq. Ft. of Forms

4"x4" Shores

	Hours	Rate	Total	Rate	Total
Carpenter	7.00	$....	$....	$24.04	$168.28
Labor	3.50	....	....	19.13	66.96
Labor removing forms	2.25	....	....	19.13	43.04
Cost per 100 sq. ft			$....		$278.28
cost per sq. ft			....		2.78

Adjustable Shores

	Hours	Rate	Total	Rate	Total
Carpenter	6.00	$....	$....	$24.04	$144.24
Labor	3.00	....	....	19.13	57.39
Labor removing forms	2.25	....	....	19.13	43.04
Cost per 100 sq. ft			....		$244.67
cost per sq. ft			....		2.45

For story heights over 16'-0", add 25% labor costs. Add labor for framing drop heads.

Labor Cost of 100 Sq. Ft. of Flat Slab Forms Requiring 4 Ft. of Lumber Per Sq. Ft.

4"x4" Shores

	Hours	Rate	Total	Rate	Total
Carpenter	9.0	$....	$....	$24.04	$216.36
Labor	4.5	....	....	19.13	86.09
Labor removing forms	3.0	....	....	19.13	57.39
Cost per 100 sq. ft			$....		$359.84
cost per sq. ft			....		3.60

Adjustable Shores

	Hours	Rate	Total	Rate	Total
Carpenter	8.0	$....	$....	$24.04	$192.32
Labor	4.0	....	....	19.13	76.52
Labor removing forms	3.0	....	....	19.13	57.39
Cost per 100 sq. ft			$....		$326.23
cost per sq. ft			....		3.26

For story heights over 16'0", add 25% to labor costs. Add labor for framing drop heads.

Forms for Beam and Girder Type Solid Concrete Floors

The labor cost of wood forms for beam and girder type floors will run somewhat higher than flat slab floors on account of the shorter spans, additional framing around beams, girders, etc.

On forms of this class, a carpenter should frame and erect 325 to 375 BF of lumber per 8-hr. day, at the following labor cost per 1,000 BF:

	Hours	Rate	Total	Rate	Total
Carpenter	23.0	$....	$....	$24.04	$552.92
Labor	11.5	....	....	19.13	220.00
Cost per 1,000 b.f.			$....		$772.92

An experienced laborer should remove or "strip" 1,000 to 1,200 BF. of lumber, per 8-hr. day, at the following labor cost per 1,000 BF.:

	Hours	Rate	Total	Rate	Total
Labor	7	$....	$....	$19.13	$133.91

Number of Feet of Lumber Required for One Sq. Ft. of Beam and Girder Type Solid Concrete Floor Forms

Ceiling Height in Feet	Thickness of Slab in Inches 3" 75 lbs	4" 88 lbs	5" 100 lbs	6" 113 lbs	7" 125 lbs	8" 138 lbs	10" 163 lbs	12" 188 lbs
10	2.0	2.1	2.4	2.5	2.6	2.6	2.7	2.7
12	2.1	2.2	2.4	2.5	2.6	2.6	2.7	2.8
14	2.1	2.2	2.4	2.5	2.6	2.6	2.7	2.8
16	2.2	2.3	2.5	2.6	2.7	2.7	2.8	2.9
18	2.4	2.5	2.6	2.7	2.8	2.8	2.9	3.0
20	2.5	2.6	2.7	2.8	2.9	2.9	3.0	3.1

Labor Cost of 100 Sq. Ft. of Slab Forms Between Beam and Girder Forms Requiring 2 Ft. of Lumber Per Sq. Ft. of Floor

4"x4" Shores

	Hours	Rate	Total	Rate	Total
Carpenter	5.0	$....	$....	$24.04	$120.20
Labor	2.5	....	....	19.13	47.83
Labor removing forms	1.5	....	....	19.13	28.70
Cost per 100 sq. ft			$....		$196.73
cost per sq. ft			....		1.96

Adjustable Shores

	Hours	Rate	Total	Rate	Total
Carpenter	4.6	$....	$....	$24.04	$110.58
Labor	2.3	....	....	19.13	44.00
Labor removing forms	1.5	....	....	19.13	28.70
Cost per 100 sq. ft			$....		$183.28
cost per sq. ft			....		1.83

Labor Cost of 100 Sq. Ft. of Slab Forms Between Beam and Girder Forms Requiring 2 ½ Ft. of Lumber Per Sq. Ft. of Floor

4"x4" Shores

	Hours	Rate	Total	Rate	Total
Carpenter	6.30	$....	$....	$24.04	$151.45
Labor	3.15	....	....	19.13	60.26
Labor removing forms	1.90	....	....	19.13	36.35
Cost per 100 sq. ft			$....		$248.06
cost per sq. ft			....		2.48

Adjustable Shores

	Hours	Rate	Total	Rate	Total
Carpenter	5.70	$....	$....	$24.04	$137.03
Labor	2.85	....	....	19.13	54.52
Labor removing forms	1.90	....	....	19.13	36.35
Cost per 100 sq. ft			$....		$227.90
cost per sq. ft			....		2.28

Labor Cost of 100 Sq. Ft. of Slab Forms Between Beam and Girder Forms Requiring 3 Ft. of Lumber Per Sq. Ft. of Floor

4"x4" Shores

	Hours	Rate	Total	Rate	Total
Carpenter	7.6	$....	$....	$24.04	$182.70
Labor	3.8	....	....	19.13	72.69
Labor removing forms	2.2	....	....	19.13	42.09
Cost per 100 sq. ft			$....		$297.48
cost per sq. ft			....		2.97

Adjustable Shores

	Hours	Rate	Total	Rate	Total
Carpenter	6.8	$....	$....	$24.04	$163.47
Labor	3.4	....	....	19.13	65.04
Labor removing forms	2.2	....	....	19.13	42.09
Cost per 100 sq. ft			$....		$270.60
cost per sq. ft			....		2.71

For story heights over 16'-0", add 25% to labor costs.

Forms for Floors of Metal Pan and Concrete Joist Construction

Wood forms for floors of metal pan and concrete joist construction may be of much lighter construction than solid concrete floors, as the dead load is less. Metal pans are usually furnished 20" to 30" wide, with 2"x6" or 2"x8" plank spaced 24" to 37" on centers, depending upon the width of the joists, which vary from 4" to 7". Fiberglass pans, sometimes used in place of steel pans, are usually furnished 41" to 52" wide. The open deck construction is ordinarily used owing to the wide spacing of the floor boards.

In buildings having one-way beams and long floor spans, a carpenter should frame and erect 350 to 400 BF of lumber per 8-hr. day, at the following labor cost per 1,000 BF:

	Hours	Rate	Total	Rate	Total
Carpenter	21.4	$....	$....	$24.04	$514.46
Labor	10.7	....	....	19.13	204.69
Cost per 1,000 BF			$....		$719.15

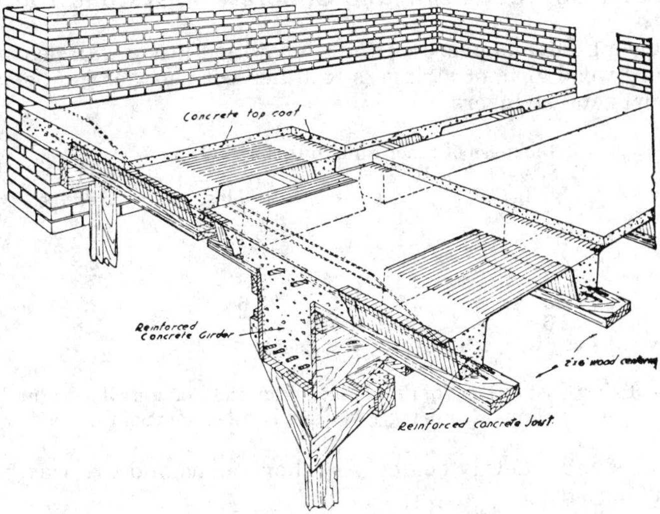

Method of Constructing Wood Slab Forms for Floors of Metal Pans and Concrete Joist Construction

Floors having beams and girders running in both directions with the floor panels averaging 16'-0"x16'-0" and smaller, a carpenter should frame and erect 325 to 375 BF of lumber per 8-hr. day, at the following labor cost per 1,000 BF:

	Hours	Rate	Total	Rate	Total
Carpenter	22.8	$....	$....	$24.04	$548.11
Labor	11.4	....	....	19.13	218.08
Cost per 1,000 BF			$....		$766.19

An experienced laborer should remove or strip 1,000 to 1,200 BF of lumber per 8-hr. day, at the following labor cost per 1,000 BF:

	Hours	Rate	Total	Rate	Total
Labor	7	$....	$....	$19.13	$133.91

Number of Feet of Lumber Required for One Sq. Ft. of Floor Forms for Metal Pan and Concrete Joist Construction

The following table is based on the assumption the ends of joists and stringers are supported by the beam and girder forms.

Ceiling Height in Feet	Thickness of Floor Slab in Inches 8 ½" 85 lbs.	10 ½" 90 lbs.	12 ½" 100 lbs.	14 ½" 105 lbs.	16 ½" 115 lbs.
10	1.7	1.7	1.7	1.8	1.8
12	1.7	1.7	1.8	1.9	1.9
14	1.7	1.7	1.8	1.9	2.0
16	1.8	1.8	1.9	2.0	2.1
18	2.1	2.1	2.2	2.3	2.4
20	2.2	2.2	2.3	2.4	2.5

Number of Feet of Lumber Required for One Sq. Ft. of Floor Forms For Metal Pan and Concrete Joist Construction

The following table is based on the assumption it is a wall bearing job and will require wood girts or stringers together with supporting shores at both ends of joists and stringers.

Ceiling Height in Feet	Thickness of Floor Slab in Inches 8 ½" 85 lbs.	10 ½" 90 lbs.	12 ½" 100 lbs.	14 ½" 105 lbs.	16 ½" 115 lbs.
10	2.1	2.1	2.1	2.2	2.2
12	2.1	2.1	2.1	2.2	2.2
14	2.2	2.2	2.2	2.3	2.4
16	2.3	2.3	2.3	2.4	2.4
18	2.5	2.5	2.6	2.7	2.8
20	2.6	2.6	2.7	2.8	2.9

Using yellow pine, Douglas fir, or woods of equal strength.

Labor Cost of 100 Sq. Ft. of Metal Pan and Concrete Joist Slab Forms Requiring 1 ¾ Ft. of Lumber Per Sq. Ft.

Long spans and one-way beams. For short spans and two-way beams, add 10% to labor costs.

4"x4" Shores

	Hours	Rate	Total	Rate	Total
Carpenter	4.2	$....	$....	$24.04	$100.97
Labor	2.1	....	....	19.13	40.17
Labor removing forms	1.3	....	....	19.13	24.87
Cost per 100 sq. ft			$....		$166.01
cost per sq. ft			....		1.66

Adjustable Shores

	Hours	Rate	Total	Rate	Total
Carpenter	3.8	$....	$....	$24.04	$91.35
Labor	1.9	....	....	19.13	36.35
Labor removing forms	1.3	....	....	19.13	24.87
Cost per 100 sq. ft			$....		$152.57
cost per sq. ft			....		1.53

For story heights over 16'-0", add 25% to labor costs.

Labor Cost of 100 Sq. Ft. of Metal Pan and Concrete Joist Slab Forms Requiring 2 Ft. of Lumber Per Sq. Ft.

Long spans and one-way beams. For short spans and two-way beams, add 10% to labor costs.

4"x4" Shores

	Hours	Rate	Total	Rate	Total
Carpenter	4.90	$....	$....	$24.04	$117.80
Labor	2.45	....	....	19.13	46.87
Labor removing forms	1.50	....	....	19.13	28.70
Cost per 100 sq. ft			$....		$193.37
cost per sq. ft			....		1.93

Adjustable Shores

	Hours	Rate	Total	Rate	Total
Carpenter	4.30	$....	$....	$24.04	$103.37
Labor	2.15	....	....	19.13	41.13
Labor removing forms	1.50	....	....	19.13	28.70
Cost per 100 sq. ft			$....		$173.20
cost per sq. ft			....		1.73

For story heights over 16'-0", add 25% to labor costs.

Labor Cost of 100 Sq. Ft. of Metal Pan and Concrete Joist Slab Forms Requiring 2 ½ Ft. of Lumber Per Sq. Ft.
4"x4" Shores

	Hours	Rate	Total	Rate	Total
Carpenter	6.0	$....	$....	$24.04	$144.24
Labor	3.0	....	....	19.13	57.39
Labor removing forms	1.8	....	....	19.13	34.43
Cost per 100 sq. ft			$....		$236.06
cost per sq. ft			....		2.36

Adjustable Shores

	Hours	Rate	Total	Rate	Total
Carpenter	5.4	$....	$....	$24.04	$129.82
Labor	2.7	....	....	19.13	51.65
Labor removing forms	1.8	....	....	19.13	34.43
Cost per 100 sq. ft			$....		$215.90
cost per sq. ft			....		2.16

For story heights over 16'-0", add 25% to labor costs.

Labor Cost of 100 Sq. Ft. of Metal Pan and Concrete Joist Slab Forms Requiring 3 Ft. of Lumber Per Sq. Ft.

4"x4" Shores

	Hours	Rate	Total	Rate	Total
Carpenter	7.0	$....	$....	$24.04	$168.28
Labor	3.5	....	....	19.13	66.96
Labor removing forms	2.2	....	....	19.13	42.09
Cost per 100 sq. ft			$....		$277.33
cost per sq. ft			....		2.77

Adjustable Shores

	Hours	Rate	Total	Rate	Total
Carpenter	6.4	$....	$....	$24.04	$153.86
Labor	3.2	....	....	19.13	61.22
Labor removing forms	2.2	....	....	19.13	42.09
Cost per 100 sq. ft			$....		$257.17
cost per sq. ft			....		2.57

For story heights over 16'-0", add 25% to labor costs.

Forms for Reinforced Concrete Stairs

The labor cost of framing and erecting concrete stair forms vary with the type of stair, open or box string, straight runs from floor to floor or having intermediate landing platforms, and whether the stair is straight or winding and has square or bull nose treads and risers. The story height also affects the labor costs on account of the shoring and bracing necessary to support the wet concrete.

It is rather difficult to estimate the cost of stair forms on a sq. ft. basis, because it costs as much to frame a 3'-0" stair as one 4'-0" wide, while the latter contains one-third more forms per lin. ft. The better method is to allow a certain price per flight of stairs of each type. This will result in more accurate estimates than the sq. ft. method. However, where it is desirable to estimate the stair forms by the sq. ft., take the area of the soffit of the stairs and platforms. For example, a stair 4'-0" wide and 18'-0" long, contains 72 sq. ft. of forms.

Some contractors and estimators estimate the cost of stairs at a certain price per lin. ft. of riser, which includes forms, reinforcing steel, concrete and finishing.

Wood Forms for Straight Concrete Stairs. On straight stairs extending from floor to floor without intermediate landing platforms and having an average story height of 10'-0" to 12'-0", two carpenters working together should lay out the stair, place rough stringers, mark off treads and risers on the rough string boards and set them in place in about 8 to 9 hours.

After the rough strings are in place, it will require another 8 to 10 hrs. for 2 carpenters to sheath the stairs, cut, bevel, and place risers, and place all necessary shoring and bracing ready for concrete.

The forms for an average flight of concrete stairs, 4'-0" wide and 18'-0" long, containing 72 sq. ft. of forms (soffit measurement) or 72 lin. ft. of risers, should cost as follows:

	Hours	Rate	Total	Rate	Total
Carpenter	36	$....	$....	$24.04	$865.44
Labor helping and removing forms	9	....	....	19.13	172.17
Cost per flight			$....		$1,037.61
Cost per sq. ft			....		14.41
per lin. ft. riser			....		14.41
Cost per riser (18)			....		57.65

The lumber cost is approximately as follows:

	BF	Rate	Total
Stringers, 2 pcs. 2"x12"-20'-0"	80	$0.40	$32.00
Risers, 18 pcs. 2"x8"-4'-0"	96	.40	38.40
Soffit sheathing, 4'-0"x18'-0"	72	.60	43.20
Joists supporting sheathing, 4 pcs. 2"x8"-18'-0"	96	.40	38.40
Shores or uprights, 6 pcs. 4"x4"-10'-0"	80	.40	32.00
Purlins or stringers, 3 pcs. 4"x6"-4'-0"	24	.40	9.60
Sills, wedges and bracing	30	.39	11.70
Cost per flight			$205.30
Cost per sq. ft. (72)			2.85

Wood Forms for Concrete Stairs Having Intermediate Landing Platforms. Where the concrete stairs consist of two short flights with an intermediate landing platform between floors, the labor cost per story will run somewhat higher than for straight run stairs, on account of laying out and framing for two short flights instead of one long one.

Stairs of this type up to 4'-0" wide and 8'-0" to 10'-0" long from floor to platform, with each short flight containing 36 to 40 sq. ft. of forms (soffit measurement), and 8 to 10 risers, should cost about as follows for labor:

	Hours	Rate	Total	Rate	Total
Carpenter	23	$....	$....	$24.04	$552.92
Labor helping and removing forms	7	....	....	19.13	133.91
Cost per flight			$....		$686.83
Cost per sq. ft. (36)			....		19.08
per lin. ft. riser (36)			....		19.08
per riser (9)			....		76.31

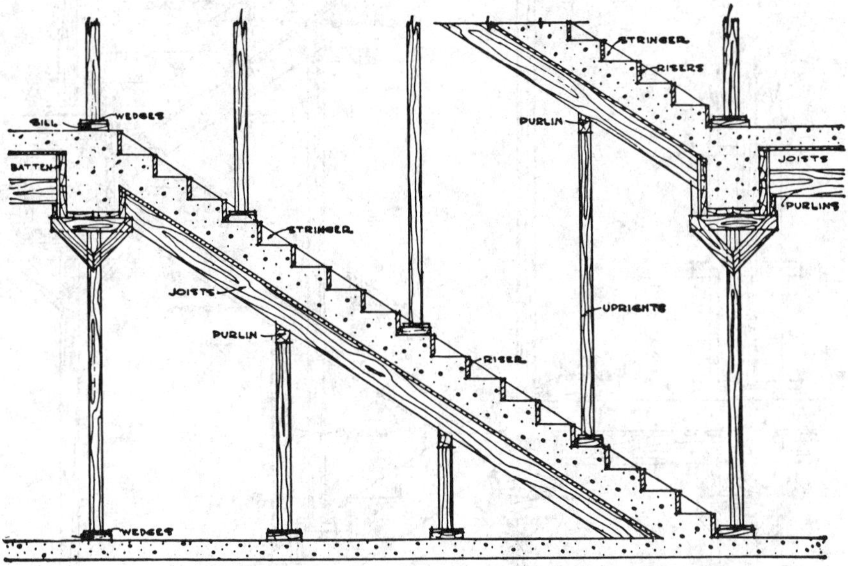

Method of Constructing Wood Forms for Concrete Stairs

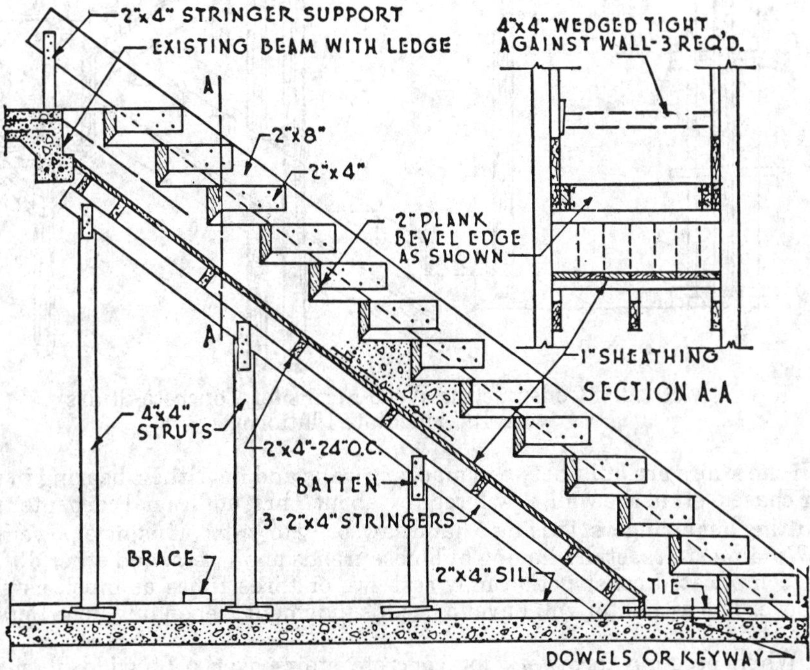

Another Method of Constructing Wood Forms for Concrete Stairs

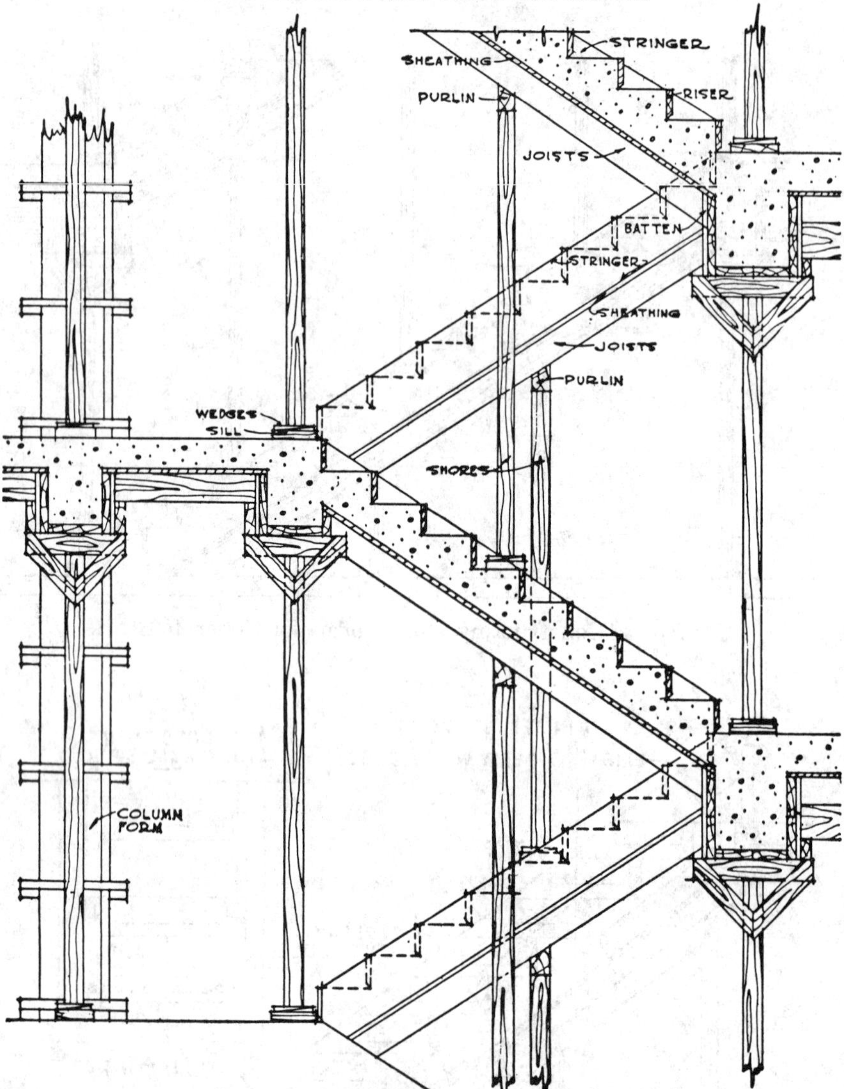

Method of Constructing Wood Forms for Concrete Stairs with Intermediate Platforms

If the stairs are built between masonry walls and have their bearing in slots or chases left in the wall, it will require about 2 hrs. additional carpenter time cutting out stringers, if using a hand saw, or 1/2 to 3/4-hr., if using a power saw.

Winding stairs, stairs having bull nose treads and risers, and other difficult or complicated construction may cost two or three times as much as given above. The estimator will have to use his judgment, depending upon detail of the stair.

When erecting formwork for concrete stairs over 4'-0" wide, it will be necessary to place additional "cut-out" stringers, to prevent the risers from bulging or giving way until the wet concrete has set. These "cut-out" stringers should be spaced 3'-0" to 4'-0" apart.

When figuring labor costs of landing platform, double the labor cost of plain slab forms.

STEELFORMS FOR JOIST CONSTRUCTED FLOORS AND ROOFS

Where the same size steelforms are used on a number of floors, they are removed after the concrete has attained the desired strength and reused again on the same job until the work is complete. The removable steelforms are available from and can be furnished by companies specializing in this type of work. These companies will erect and remove the steelforms with or without the supporting centering at the stipulated lump sum for the entire job.

Removable steelforms in flange and adjustable type are made from 14, 15, and 16 gauge metal that will stand hard use. The flange type is nailed through the flanges while the adjustables are nailed through the sides. The flange type steelforms are furnished in 20" and 30" standard widths in standard depths of 6", 8", 10", 12", 14", 16", and 20". The intermediate forms are in lengths of 1', 2', and 3' with end caps as closures. In addition to the straight ends, there are also single tapered end forms in 3' lengths for the various depths. Also available are adjustable straight side type forms that can be set to a depth of 8", 10", 12", 14", and 16". These are likewise furnished in standard 20" and 30" widths and have the same combination of 1", 2", and 3" intermediates, end forms, and the 3" single tapered ends. For filling out spaces, filler forms are furnished in 10" and 15" widths matching the steelform depths as described. Tapered lighter gauge metals do not withstand job abuse, particularly with men working over them before the concrete is placed, and as a result, often dent badly so that it requires considerable extra concrete and gives a poor surface finish. In additions, there is added labor expense in reconditioning the forms prior to the next use.

For two-way concrete joist construction, one-piece metal domes are also available. The depths furnished in the 30"x30" void are 8", 10", 12", 14", 16", and 20", and in the 19"x19" void are 4", 6", 8",10", 12", 16", and 20". Flanges for the 30/30 domes are 3" wide to make a 3 ft. module and for the 19/19 domes are 2 1/2" wide to make a 2 ft. module. The metal gauge ranges from 14 to 16 gauge.

The steelforms and steeldomes must be nailed to the supporting forms for centering, and to hold them rigid and in line while the concrete is placed.

It is common practice to oil the forms prior to the installation of the reinforcing steel so that they can be removed easier from the ceiling and without concrete adhering to their surfaces.

Estimating Quantities of Steelform. In estimating the area of floor and roof construction requiring removable or permanent forms, the gross floor or roof area is used. No deductions are to made for beams or for tees of beams or for wide joists. Major openings, 50 square feet or more, such as atriums, are deducted when applicable. Typical covering systems generally frame through smaller openings, so no area adjustments are made for these openings.

Concrete joist construction may be supported by any one of three types of structural systems. In addition to the aforementioned general estimating rules, the job area is subject to the following special rules. For a reinforced concrete frame, the areas are to be figured out-to-out of concrete frame. For a structural steel frame, the areas are to be figured center to center spandrel beams. For bearing wall construction, the areas are to be figured clear inside brick walls plus a bearing on all walls of 6".

Labor Handling and Placing Steelforms. The labor cost of placing steelforms will vary with the type of form, type of building, quantity of equipment furnished, amount of horizontal and vertical movement of equipment, depressions, openings in the slabs, headers, etc., and the locality in which the work is being executed.

Steelforms are placed usually by carpenters and laborers. In a few localities, iron workers handle the steelforms from the point of removal from the concrete ceiling to the next point of steelform installation on the job. In most localities, however, laborers assist the carpenters and handle the steelforms to the next point of erection on the job ready for the carpenter to install the steelform to the wood centering. Local practice of work assignment greatly affects the labor costs.

The following costs are based on steelforms being placed by experienced workers, such as ones employed by companies specializing in the erection and removal of steelforms.

On straight run work using flange type steelforms such as factory buildings, garages, warehouses, etc. having typical spans without much cutting for headers, openings, etc, an experienced crew consisting of two carpenters and four laborers should handle and place steelforms over approximately 3,500 sq. ft. of floor area in an 8-hr day at the following direct labor cost per 100 sq. ft.:

	Hours	Rate	Total	Rate	Total
Mechanic	0.46	$....	$....	$24.04	$11.06
Labor	0.91	....	....	19.13	17.41
Cost per 100 sq. ft.	1.37				$28.47
cost per sq. ft.					.28

For school buildings, hospitals, apartments, and office buildings, where the concrete joist require fitting and adjusting around pipes, pipe sleeves, chases and conduits, the handling and placing costs of the flange type equipment could increase as much as 40% to 50%, depending on the amount of special work required.

Labor Handling and Removing or Stripping Steelforms

In removing or stripping flange type steelforms after the concrete has set, an experienced crew, usually consisting of two carpenters and three laborers, can remove the steelforms from the ceiling over about 4,000 sq. ft. of floor per 8-hr day at the following direct labor cost per 100 sq. ft.:

	Hours	Rate	Total	Rate	Total
Mechanic	0.40	$....	$....	$24.04	$9.62
Labor	0.60	....	....	19.13	11.48
Cost per 100 sq. ft.					$21.10
cost per sq. ft.					.21

It should be noted that the above costs are based on flange type steelforms. The cost for handling and placing the one-piece dome is slightly higher. The cost for handling and placing adjustable type steelform is about 33 1/3 % higher.

Ceco Steelform Construction. Ceco steelform construction is a combination of concrete joist construction and thin top slabs. The steelforms are formed of 14, 15, or 16 gauge steel, depending on the width and depth of the form. These forms are placed on supporting centering and removed from the concrete after the concrete has reached sufficient strength. The type of forms made available to the industry by the Ceco Corporation are flange forms, the most commonly used type of form, followed by the one-piece metal dome, adjustable forms and the one-piece longforms.

The cost of the steelforms installed in a building will depend on a number of factors such as the type of steelform used, the size of the job, the number of reuses of the equipment, location, availability of the steelforms with respect to warehouse stocks, and so on. Local working conditions also play a very important part in job costs.

The Ceco steelform service includes supplying the necessary steelforms and the labor for their erection and removal. Also offered is a complete service of furnishing and erecting all centering for the support of the steelforms. In steeldome construction, centering is usually included for the concrete solid slab areas that are of the same depth as the dome construction and that are within the floor areas containing the steeldomes. The quotations and contracts are generally on a lump sum basis for specific projects. It is always recommended that the steelform prices be obtained from a steelform supplier before finalizing and submitting a bid on this portion of the job costs.

FORMS OTHER THAN WOOD

Lightweight Steel Forming Material for Concrete. High-strength corrugated steel forming material is often used for forming reinforced concrete floor and roof slabs. It comes either galvanized or uncoated and is manufactured from 100,000 psi tough-temper steel. It has a definite reliable structural strength nearly twice that of ordinary steel of equal weight. Used primarily in floor and roof systems having steel joists, junior beams, or purlins, it is also used over pipe tunnels or similar installations where economies of using permanent forms can be realized. Either structural grade concrete or lightweight insulating concrete may be used with corrugated steel forms, but in either case, reinforcing bars or wire mesh should be added to satisfy flexure and temperature steel requirements.

This material offers many advantages over other methods of flexible centering. The light, rigid sheets are quickly placed. No side pull is exerted on joists, and instead, joist top chords are given lateral support by the stiffness of the material. Sheets in place provide a safe working platform. Mesh is easily and effectively placed. Elimination of sag between joists reduces concrete quantity. Uniform thickness of slab over joists and mid span permits monolithic finish. Little cleanup is required underneath.

For exposed construction, it is available as galvanized. For unexposed joist construction, it can be purchased uncoated.

Sheets are placed with corrugations normal to the supporting joists or beams, with the end lap of sheets occurring over the joists or beams. The sheets are placed with edge lips up and are lapped a minimum of one corrugation with adjacent sheets. The ends of sheets should lap a minimum of 2". The sheets are fastened to the supporting members by clips, arc-welding, or nailing. For steel joists construction the most satisfactory method of attachment is arc-welding. The manufacturer can provide special curved washers for use in welding the sheets to the steel framework. These provide fast, high-strength welds.

Estimating. Corrugated steel forming material is sold by the square, with the area determination based on sheet width times actual sheet length.

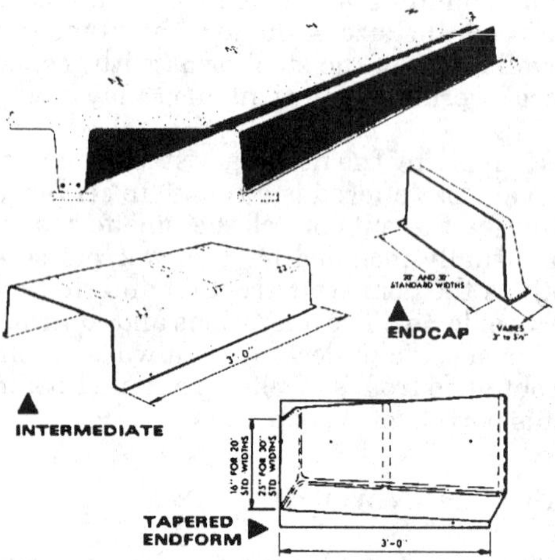

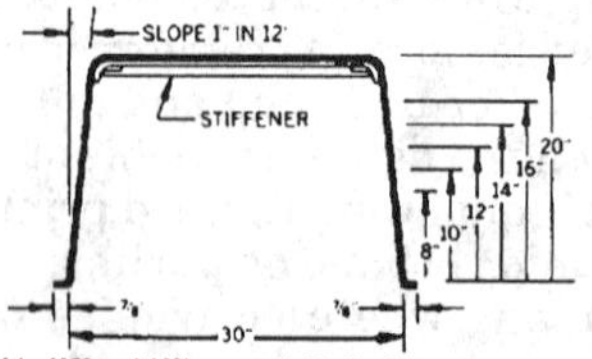

3E

Filler widths (10" and 15") are available for filling non-standard spaces only.

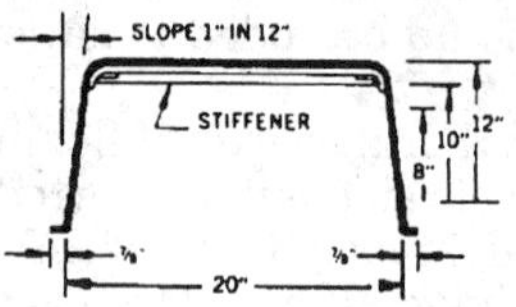

CONCRETE QUANTITIES/30" WIDTHS*

Depth of Steelform	Width of Joist	Cubic feet of concrete per square foot for various slab thicknesses 2½"	3"	4½"	Additional concrete for Tapered Endforms cu. ft. per lin. foot of bearing wall or beam (One side only)
8"	5"	.317	.359	.484	.143
	6"	.333	.374	.499	.139
	7"	.347	.389	.514	.135
10"	5"	.348	.390	.515	.179
	6"	.367	.409	.534	.174
	7"	.386	.427	.552	.169
12"	5"	.381	.422	.547	.214
	6"	.404	.445	.570	.208
	7"	.425	.467	.592	.203
14"	5"	.415	.456	.581	.250
	6"	.441	.483	.608	.243
	7"	.467	.508	.633	.236
16"	6"	.481	.522	.647	.278
	7"	.509	.551	.676	.270
	8"	.537	.578	.703	.263
20"	6"	.564	.606	.731	.347
	7"	.599	.641	.766	.338
	8"	.633	.675	.800	.329

**Apply only for areas over FLANGEforms and joists between them. Bridging joists, special headers, beam tees, etc., not included.*

CONCRETE QUANTITIES/20" WIDTHS*

**Depth of Steelform	Width of Joist	Cubic feet of concrete per square foot for various slab thicknesses 2½"	3"	4½"	Additional concrete for Tapered Endforms, cu. ft. per lin. foot of bearing wall or beam (One side only)
8"	4"	.339	.381	.506	.167
	5"	.361	.402	.527	.160
	6"	.380	.422	.547	.154
10"	4"	.377	.419	.544	.208
	5"	.404	.445	.570	.200
	6"	.428	.470	.595	.192
12"	4"	.418	.459	.584	.250
	5"	.449	.491	.616	.240
	6"	.479	.520	.645	.231

**Apply only for areas over FLANGEforms and joists between them. Bridging joists, special headers, beam tees, etc., not included.*

***14," 16" and 20" depths are also available. Contact your Ceco Construction Engineer.*

VOIDS CREATED BY VARIOUS SIZE FLANGEFORMS

Depth of Steelform	Cubic feet of void created per linear foot (for various widths of FLANGEforms) 30" Width	20" Width	15" Width	10" Width	**Cubic Feet per Tapered End 30" Width	20" Width
8"	1.628	1.072	.794	.516	4.465	2.882
10"	2.023	1.329	.982	.634	5.548	3.569
12"	2.414	1.581	1.165	.748	6.617	4.243
14"	2.801	1.829	1.343	.857	7.673	
16"	3.183	2.072	1.516	.961	8.715	
20"	3.933	2.544	1.850	1.155	10.756	

***Total void created by standard 3'-0" length Tapered Endform.*

Ceco Flange Type Forms

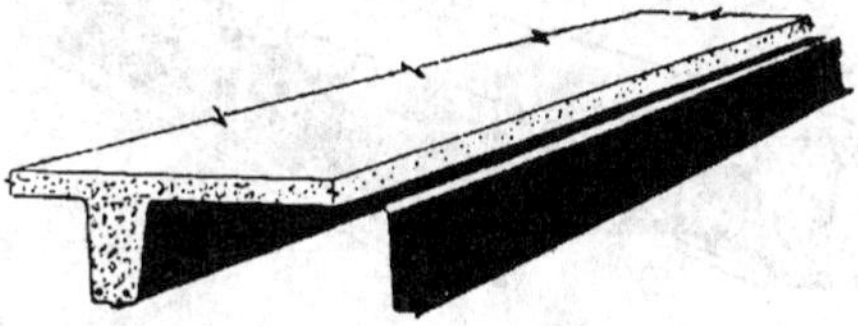

Concrete Quantities/30" Widths*

Depth of Steelform	Width of Joist	Cubic feet of concrete per square foot of floor for various slab thicknesses		
		2½"	3"	4½"
8"	6"	.333	.375	.500
10"	6"	.367	.409	.534
12"	6"	.403	.445	.570
14"	6"	.440	.482	.607
16"	6"	.479	.521	.646
20"	6"	.560	.602	.727

All sizes shown conform to the Concrete Reinforcing Steel Institute, CRSI, Code of Standard Practice, MSP-2-81.

**Apply only for areas over LONGforms and joists between them. Bridging joists, special headers, beam tees, etc., not included.*

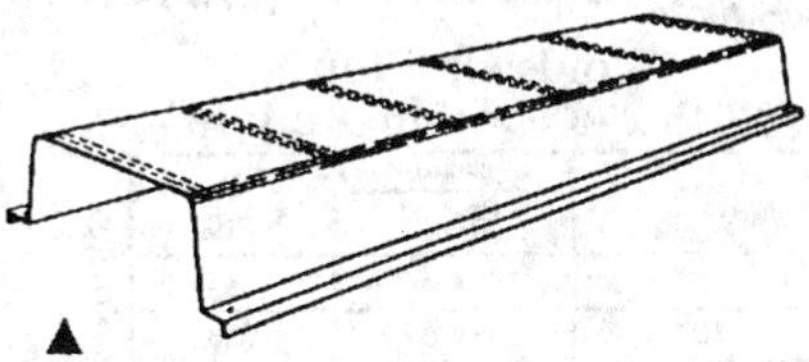

LONGFORM

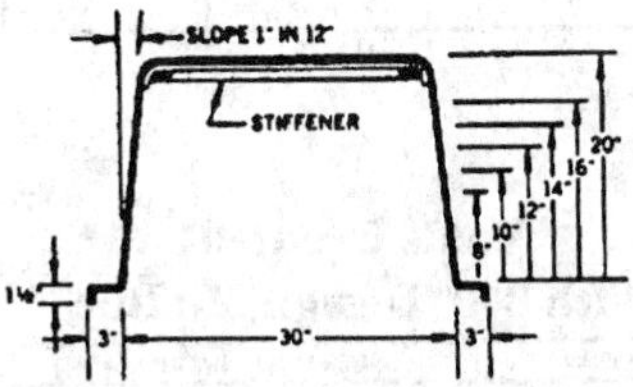

Filler widths (10" and 15") are available for filling non-standard spaces only.

Concrete Quantities/20" Widths*

Depth of Steelform	Width of Joist	Cubic feet of concrete per square foot of floor for various slab thicknesses		
		2½"	3"	4½"
8"	5"	.361	.403	.528
10"	5"	.404	.446	.571
12"	5"	.449	.490	.615

**Apply only for areas over LONGforms and joists between them. Bridging joists, special headers, beam tees, etc., not included.*

***14," 16" and 20" depths are also available. Contact your Ceco Construction Engineer.*

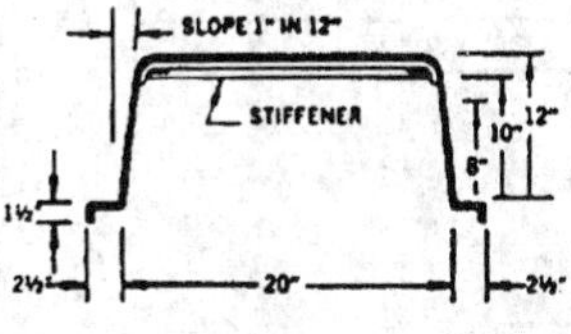

Voids Created by Various Size Longforms

Depth of Steelform	Cubic feet of void created per linear foot (for various widths of LONGforms)			
	30" Width	20" Width	15" Width	10" Width
8"	1.626	1.070	.793	.515
10"	2.023	1.329	.981	.634
12"	2.416	1.583	1.166	.749
14"	2.804	1.832	1.346	.860
16"	3.188	2.077	1.522	.966
20"	3.944	2.555	1.860	1.166

Shaded areas above indicate standard filler widths.

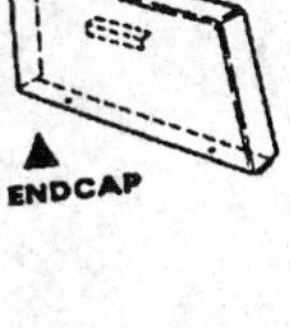

ENDCAP

Ceco Longform Type

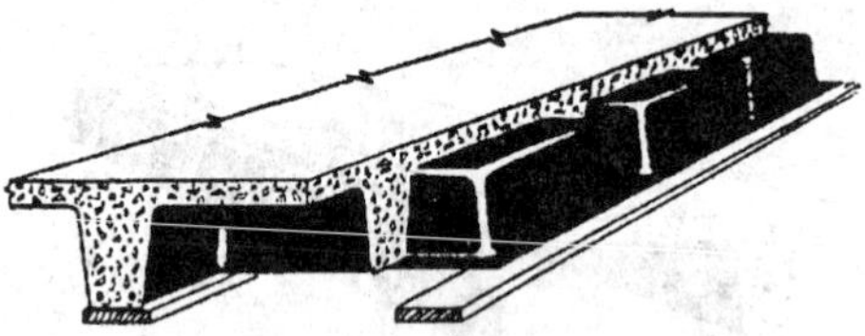

All 3'-0" module sizes (30" x 30" domes) and all 2'-0" module sizes (19" x 19" domes) except 14" depth conform to the Concrete Reinforcing Steel Institute, CRSI, Code of Standard Practice, MSP-2-81.

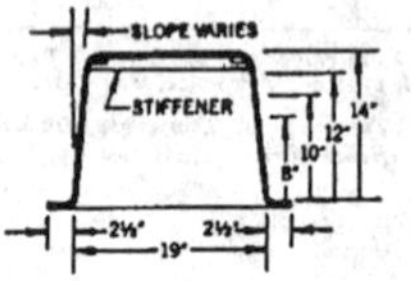

Voids Created With 2'-0" Design Module

Depth of Dome	OVERALL PLAN SIZE 24" x 24"	
	Plan Size of Void	Cu. Ft. of Void
8"	19" x 19"	1.410
10"	19" x 19"	1.862
12"	19" x 19"	2.200
14"	19" x 19"	2.447

3'-0" MODULE
(30"x30" Dome System)

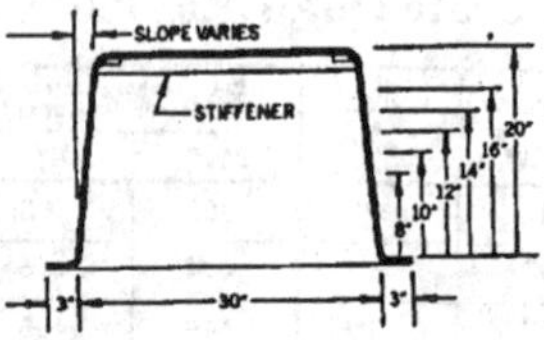

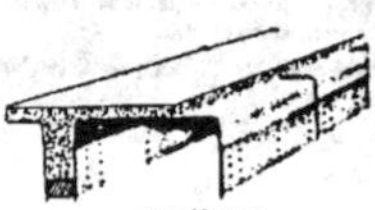

Voids Created With 3'-0" Design Module

STANDARD SIZES	OVERALL PLAN SIZE 36" x 36"	
Depth of Dome	Plan Size of Void	Cu. Ft. of Void
8"	30" x 30"	3.877
10"	30" x 30"	4.756
12"	30" x 30"	5.578
14"	30" x 30"	6.662
16"	30" x 30"	7.519
20"	30" x 30"	9.194

Filler Sizes for 3'-0" Module

Depth of Dome	OVERALL PLAN SIZE 26" x 36"		OVERALL PLAN SIZE 26" x 26"	
	Plan Size of Void	Cu. Ft. of Void	Plan Size of Void	Cu. Ft. of Void
8"	20" x 30"	2.535	20" x 20"	1.656
10"	20" x 30"	3.094	20" x 20"	2.011
12"	20" x 30"	3.607	20" x 20"	2.330
14"	20" x 30"	4.334	20" x 20"	2.817
16"	20" x 30"	4.873	20" x 20"	3.152
20"	20" x 30"	5.922	20" x 20"	3.808

U. S. Patent No. 2,850,785 applies to all sizes

Ceco Steel Dome Systems

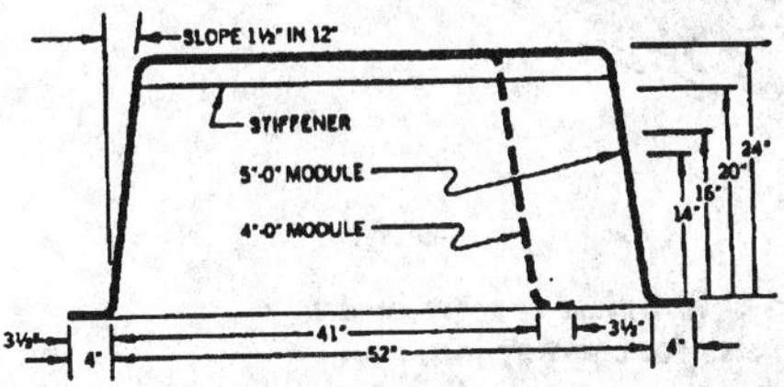

All sizes shown conform to the Concrete Reinforcing Steel Institute, CRSI, Code of Standard Practice, MSP-2-81.

VOIDS CREATED WITH 4'-0" DESIGN MODULE

Depth* of Dome	OVERALL PLAN SIZE 48" x 48"	
	Plan Size of Void	Cu. Ft. of Void
14"	41" x 41"	12.45
16"	41" x 41"	14.06
20"	41" x 41"	17.14
24"	41" x 41"	20.05

**18" and 22" depths are also available. Contact your Ceco Construction Engineer.*

VOIDS CREATED WITH 5'-0" DESIGN MODULE

Depth* of Dome	OVERALL PLAN SIZE 60" x 60"	
	Plan Size of Void	Cu. Ft. of Void
14"	52" x 52"	20.42
16"	52" x 52"	23.11
20"	52" x 52"	28.33
24"	52" x 52"	33.33

**18" and 22" depths are also available. Contact your Ceco Construction Engineer.*

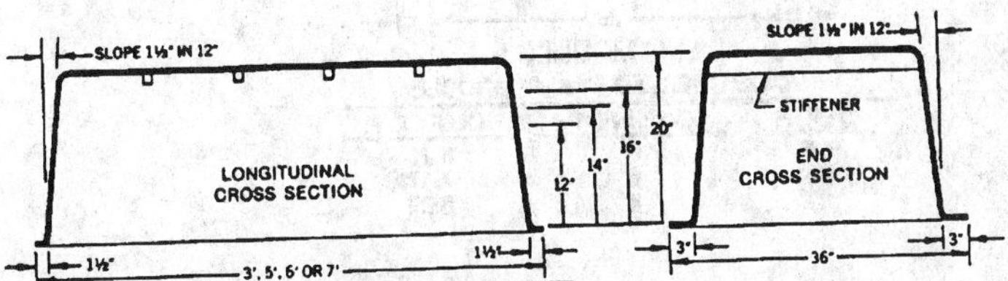

VOIDS CREATED

DEPTH OF LONGdome	CU. FT. OF VOID PER LONGdome UNIT			
	3'	5'	6'	7'
12"	6.208	10.950	13.321	15.692
14"	7.127	12.612	15.355	18.097
16"	8.014	14.228	17.335	20.442
20"	9.692	17.324	21.139	24.955

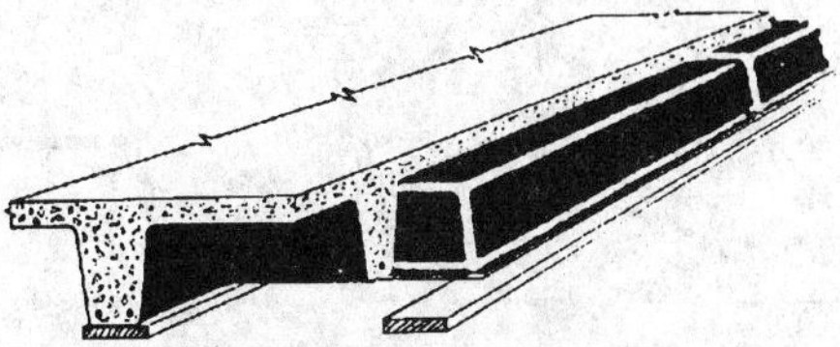

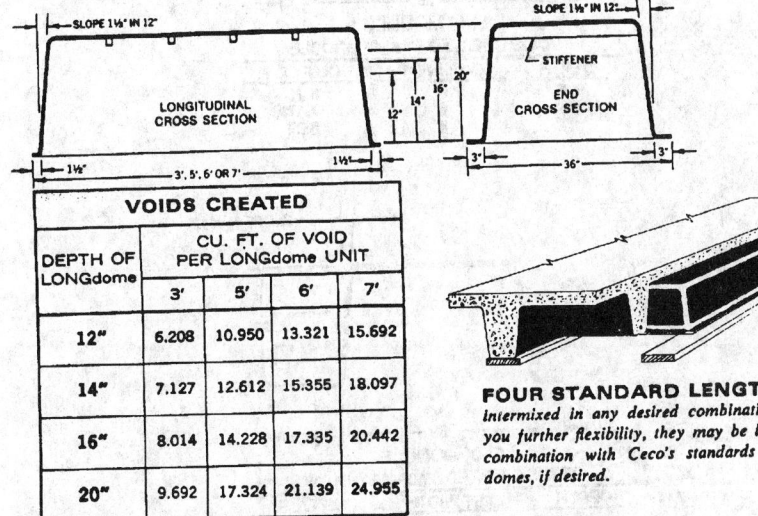

FOUR STANDARD LENGTHS *may be intermixed in any desired combination. To give you further flexibility, they may be installed in a combination with Ceco's standards 30" x 30" domes, if desired.*

Ceco Fiberglass Domes

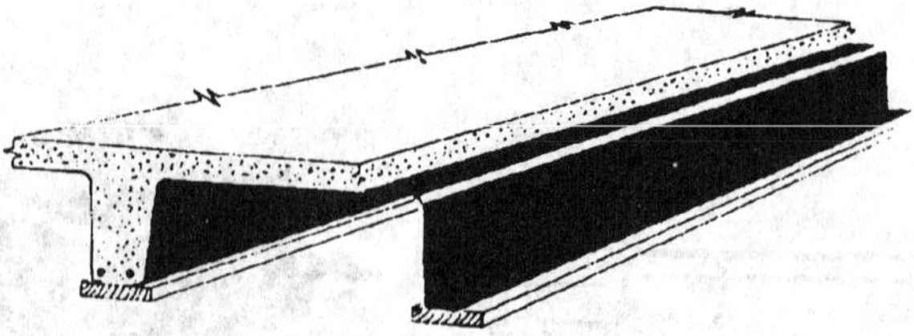

The reusable forms, constructed of fiberglass or segmented steel, make good use of the top slab for one-way joist construction and carry a 2-hour fire rating as required by fire codes.

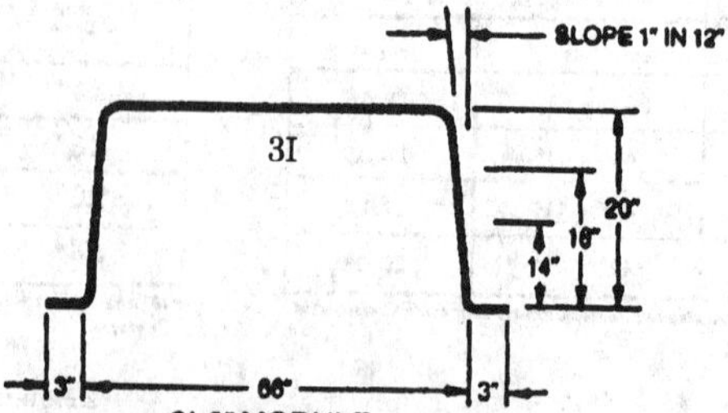

6'-0" MODULE

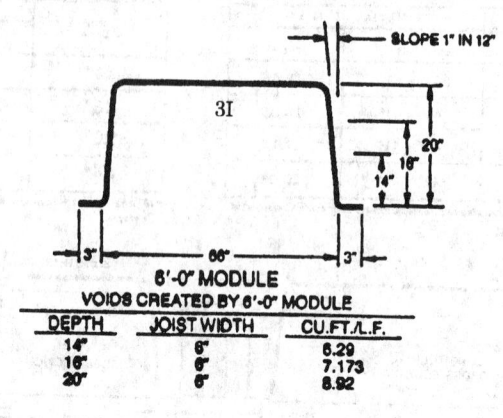

VOIDS CREATED BY 6'-0" MODULE

DEPTH	JOIST WIDTH	CU.FT./L.F.
14"	6"	6.29
16"	6"	7.173
20"	6"	8.92

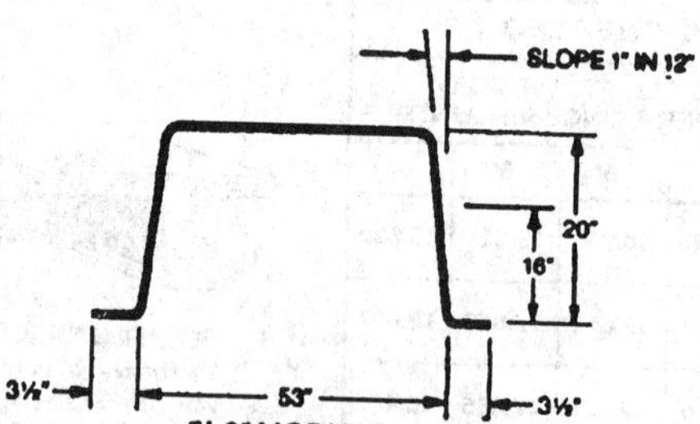

5'-0" MODULE

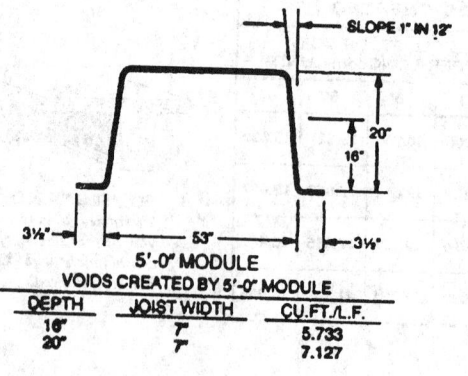

VOIDS CREATED BY 5'-0" MODULE

DEPTH	JOIST WIDTH	CU.FT./L.F.
16"	7"	5.733
20"	7"	7.127

Ceco Fiberglass Domes, Wide Module

Erection Costs. Corrugated steel forming material can be rapidly erected in all kinds of weather, thereby providing cover to floors below. The corrugated sheets are large in area but extremely light in weight and can be easily handled by one worker.

A well organized crew, working under favorable conditions, can place and fasten up to 10,000 sq. ft. of corrugated steel forming material per 8-hr day, but most jobs will average 3,000 to 5,000 sq. ft. per day.

Using an average figure of 4,000 sq. ft. per 8-hr day for a four-man crew, the labor costs for placing 1,000 sq. ft. of corrugated steel forming material would be as follows:

	Hours	Rate	Total	Rate	Total
Ironworker	8	$....	$....	$27.53	$220.24
Cost per sq. ft.					.22

The above figures are based upon one worker welding or fastening the deck in place and three workers handling and placing the material. Costs do not include equipment, overhead, and profit.

Computing Concrete Quantities. When computing volume of concrete placed over corrugated steel decking, subtract 1/4 in. from the slab thickness to allow for corrugations of standard and 3/8 in. for heavy duty corrugated steel forming. Size of corrugations: standard, 1/2-in. depth x 2 3/16-in. pitch; heavy duty, 3/4-in. depth x 3-in. pitch.

ARCHITECTURAL CONCRETE FORMWORK

Appearance is of paramount importance in architectural concrete, and every operation must be planned and executed with this in mind. The architect may design and specify, but regardless of how detailed the specifications, it is the contractor and the superintendent who must have the know-how to detail and construct the forms and be on constant watch during all operations to see that the correct procedures are followed.

Forms for Architectural Concrete

Form design and construction must be approached with a different viewpoint from that for ordinary structural concrete. In structural concrete some leakage and irregularities in formwork may be tolerated, so long as the forms are strong enough to carry the weight of the concrete and the imposed loads and are built fairly true to line and grade. In architectural concrete, there must be no leakage through the forms. Chamfer strips at corners and edges are helpful for this purpose. Wherever possible, the forms should be designed so that pressure of the fresh concrete will tighten joints rather than tend to loosen them.

A common source of trouble is bulging of forms at external corners, resulting in both poor alignment and leakage with subsequent ragged corners, sometimes to the extent of exposing the aggregate at the corners.

The accompanying illustration gives one method of constructing the forms to provide locking of the corner. The wales are extended beyond the intersection far enough to permit two vertical strips to be nailed in the corners. Wedges are driven between these strips and the wales to tighten the corner.

Method of Constructing Open Wood Centering and Placing Steelforms for Concrete Joist Floor Construction

Method of Constructing Open Wood Centering and Placing Steelforms for Concrete Joist Floor Construction

Corrugated Steel Floor Panels

Exposed Ceiling of Corrugated Steel Forms

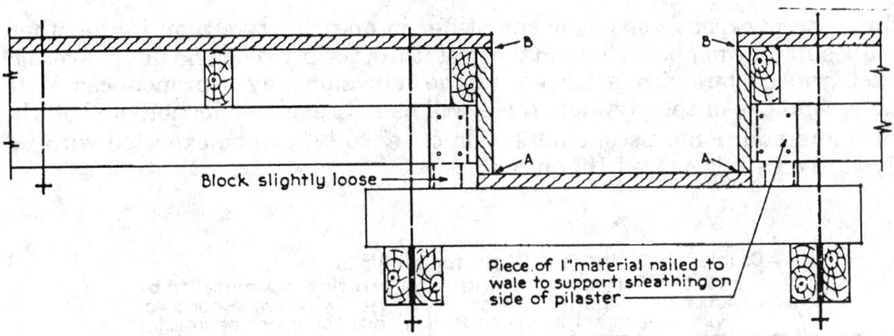

Form Designed so that Pressure of Fresh Concrete Tightens It to Prevent Leakage

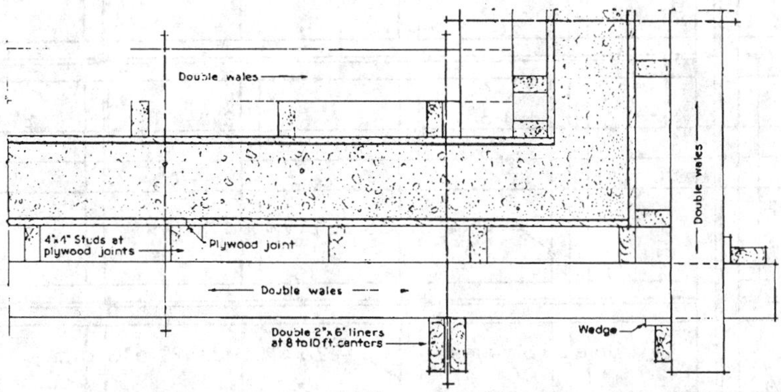

Form Construction to Provide Positive Corner Locking

Another method of locking the form corners is to use a tie instead of the wood strips and wedges. Since there are no wedges to work loose, a set of wales can be set and tightened on one level around the building without much chance of subsequent loosening when tightening the wales above or below.

Various materials are used for forming architectural concrete to eliminate imperfections and irregularities in the finished surface. Committee 622 of the American Concrete Institute has prepared a report entitled "Formwork for Concrete" that discusses these forming materials and their application. Plastic coated plywood, hardboard form lining, fiber or laminated pressed paper tubes, plaster waste molds, metal forms, and fiberglass forms are all used as forming material for architectural concrete. Fiberglass has been used on larger building projects with the forms being reused as many as 60 or 70 times.

Plywood Form Sheathing. Smooth surfaces required on much present day work are obtained with plywood form sheathing. This should be of structural grade, made especially for form use and preferably should be oiled at the factory. If oiled on the job, sufficient time should be allowed for the oil to penetrate, and any surplus oil should be wiped off before the plywood is used. Plywood 5/8" or 3/4" thick can be used directly against studs without backing when the outer plies are at right angles to the studding. Structural plywood usually has five plies with the fibers in the two outer and middle plies in one direction and the fibers of the other two plies in the opposite direction.

Thus, the plywood is stronger and stiffer in one direction than in the other. The result may not be serious on the first use of the plywood and under average placing speed, but with faster placing the deflection may be pronounced. With the second use of the plywood, there will be noticeable deflection, and on the third and additional uses, a deflection of 1/8" to 1/4" can be expected with 3/4" plywood on studs spaced 16" on centers.

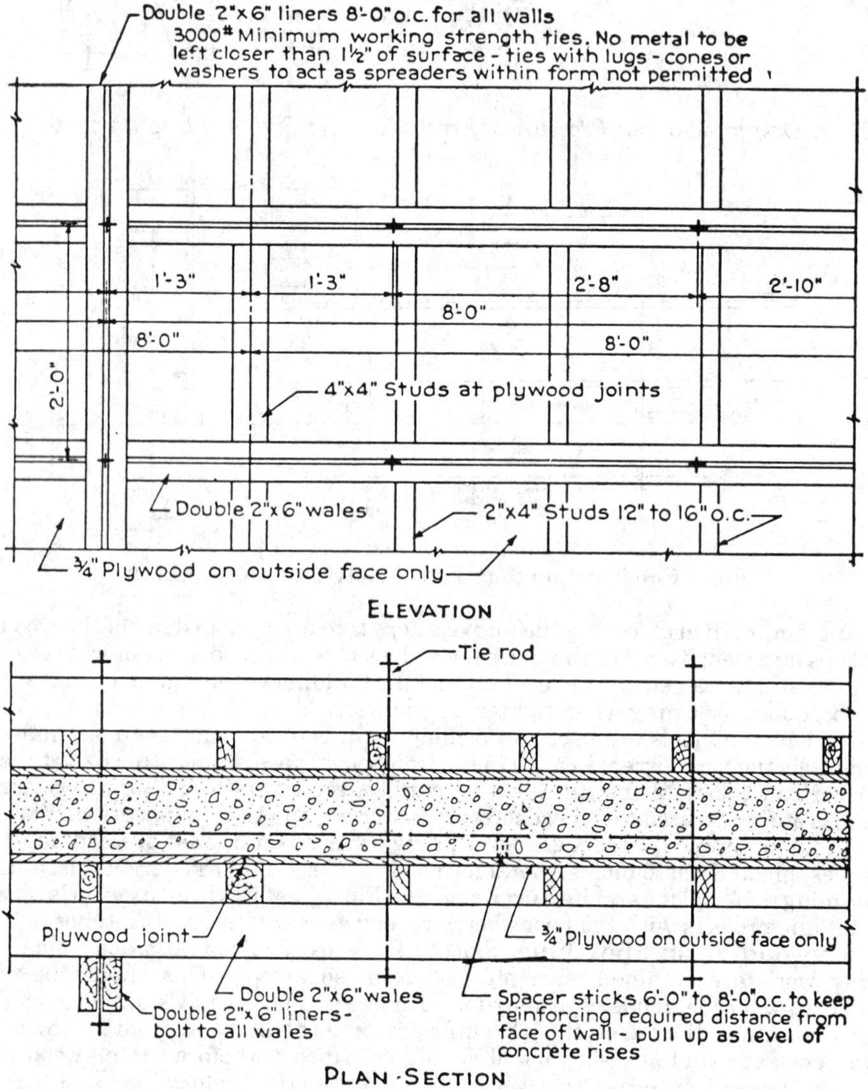

Form Layout for Use of Structural Plywood

Sometimes, concrete form *Presdwood* is used as a form liner, which is usually nailed to decking or sheathing lumber as a backing. Board 3/16" thick should be used over solid backing while 1/4" board may be used semi-structurally. Ordinary wood working tools are used in cutting and applying Concrete Form Presdwood.

Some manufacturers advertise that plywood may be used 10 times or more in the construction of formwork, but where appearance is important, it is seldom possible to use it more than 3 or 4 times. However, it may be used for other purposes where appearance is not so important.

When using plywood for the facade or facing forms, it must be thoroughly cleaned after each use. Where a special form coating is used to treat the plywood instead of oil, it should be applied with care and allowed to dry thoroughly.

Architectural concrete walls must be straight and plumb. This is essential. It means that the 2"x4" or preferably 2"x6" studs and wales must be of uniform depth and perfectly straight. Otherwise, the walls will be crooked, with the possibility of the work being rejected.

The top of the foundation walls on which the architectural concrete wall rests must be perfectly level and straight, using two 2"x6" wales or similar planks, braced at intervals of not less than 4'-0" in addition to all other bracing ordinarily required for foundation walls. It requires about 4 1/2 b.f. of lumber per lin. ft. of wall for these wales and bracing. In order to hold the studs of the form for the architectural concrete wall in line, it is necessary to insert 5/8" greased stud bolts, spaced about 3'-0" on centers, into the top of the foundation wall as illustrated. It requires about 1/8 hr. carpenter and 1/16 hr. labor time per lin. ft. of wall to brace the top of the foundation wall and to insert the bolts.

As a rule, it is necessary to obtain a perfect horizontal line of concrete at the top of the foundation wall, which will be assured by nailing a 1"x2" strip on the outside form of the foundation wall and placing the concrete about 1" higher than the bottom of this strip. About two hours after placing the concrete, the surplus concrete (which generally contains a good deal of laitance) is scraped off to the level of the underside of the strip.

After the forms for the foundation walls have been stripped, the wales should be installed and blocked about 3'-0" apart against the foundation wall. The 2"x4" or 2"x6" studs should now be installed between wales and the wall. The studs should have 8d nails (protruding about 1") driven into the outside edge exactly 8" from the lower end and 6d nails (protruding less than 3/4") exactly 8" from the lower edge. When inserting the studs between the wales and the foundation wall, the 8d nails form their support on the wales and the 6d nails furnish the support for the 3/4" plywood. The studs are usually spaced 12" or 16" on centers but at every vertical joint in the plywood (usually 8'-0") a double stud or a stud 3 1/2" wide should be used.

The studs are temporarily braced horizontally by nailing a 1"x6" board higher up and are braced vertically by wedging at the bottom between the studs and foundation wall. The first sheet of plywood is then inserted between the studs and the foundation wall and bolts, holding the wale tightened until the lowest part of the form is straight and plumb. The wales are spaced 24" vertically and the oiled tie bolts are spaced 2'-0", 2'-8" and 3'-4" apart, so that the holes do not interfere with the 16" stud spacing and are in proper relation to the ends of the 8'-0" long plywood sheets.

The arrangement of the holes for the tie bolts and the arrangement of the nails (6d box nails) by which the sheets of plywood are held to the studs are also shown in the accompanying illustration.

After 2 or 3 of the 4'-0" wide sheets of ¾" plywood have been fastened to the studs and the proper amount of horizontal walers put in place, the outside form is further strengthened by vertical wales consisting of 2 pcs. of 2"x6" spaced not more than 8'-0" on centers with about 2'-0" or more projecting over the top of the outside form. These wales are held in place and plumbed on top by two or more No. 9 or No. 10 wires, secured to column dowels, special bolts or other means provided in the foundations or in the floor. These vertical wales are bolted to every pair of horizontal wales they cross using ½" bolts. The form is now ready to support the vertical and horizontal reinforcing.

After the reinforcing steel has been placed, the inside form may be erected. The studs are nailed to a 1"x6" sill or plate, which has been nailed to the floor one day after concreting, and if the inside of the wall is to have a finished appearance, requiring the use of plywood, the plywood should be in sheets 24" wide and 8'-0" long in order to facilitate the nailing of the plywood to the studs and the insertion of the tie bolts and spreaders.

The joints of the plywood should be filled with a mastic consisting of a mixture of 50% tallow and 50% cement for summer use, while for winter work, the tallow should be replaced by pump grease. The surplus filler should then be scraped off using No. 0 sandpaper.

Estimating Quantities for Architectural Concrete Forms. When estimating the quantities for plain architectural concrete walls, do not make any deductions for ordinary door and window openings, because these must be formed with the wall and then boxes built in to allow for the door or window frame to be inserted at a later date.

The only exception to this rule is where there are large display windows in the first story of a commercial building or for skeleton type buildings where the piers and lintels are estimated at a higher rate than plain wall surfaces.

All ornamentation should be estimated separately, either by the lin. ft. or sq. ft. depending upon the class of work. Ornamental work will require the use of wood or plaster waste molds for forming, in addition to the structural backing, so that the entire wall area should first be figured as plain wall surfaces and then ornamentation figured extra.

Water tables, belt courses, cornices, door and window trim, copings, and the like should be estimated by the lin. ft. when less than 12" wide and by the sq. ft. when more than 12" wide. Always mention size and detail for each type of ornamentation, as this is necessary for pricing the work.

Columns, pilasters, etc. should be estimated separately, giving width, height, and area in sq. ft., together with a detail of the forming, such as whether fluted.

Sample Wall Form Calculation

This sample shows how to determine square footage of form contact surface area, size and number of studs and wales needed, and the number of steel ties required. The sample data, table, and charts are provided by Richmond Screw Anchor Company, Ft. Worth, Texas.

Given: Footing poured, 56'-0" long x 5'-0" wide x 3'-6" high

Required: Forms and material required for a concrete wall 50'-0" long x 1'0" wide x 12'-0" high

Assumptions:

1) Permanent or reusable formwork, Class "A".
2) ¾" thick Class I plyform with face perpendicular to supports (strong direction).
3) Studs vertical.
4) Wales horizontal.
5) Tie safe working load = 6,000 lbs.
6) Preferred tie spacing: vertical 2'-0"; horizontal 2'-0"; (2' x 2' = 4 sq. ft.).

We want to determine stud and wale sizes using ¾" plywood.

Determine Concrete Pressure:

Concrete Pressure = $\frac{\text{Tie safe working load}}{\text{Area}}$

Concrete Pressure = $\frac{6{,}000 \text{ lbs.}}{4 \text{ sq. ft.}}$

Concrete Pressure = 1,500 lbs./sq. ft.

Framing Material:

Stud Spacing: Use Table 2, "Spacing of Supports for Class I Plywood, (Strong Direction)--Face Grain Across Supports", or Chart A, "Spacing of Supports for Class I Plywood", which appear at the end of the sample calculation. For Class I plywood with a concrete pressure of 1,500 psf and plywood thickness of ¾", the support (tie) spacing is 7.6", so use 8" center to center (c.c.)

Stud Sizes:

Stud Loading = $\frac{\text{Stud Spacing}}{\text{12 in./ft.}}$ x Design Concrete Pressure

Stud Loading = $\frac{\text{8 inches}}{\text{12 in./ft.}}$ x 1,500 lbs/ft. sq.

Stud Loading = 1,000 lbs./sq. ft.
Assume studs are continuous or partially continuous.

Using Table 1 "Maximum Wale Spacing Along Stud/Joist Member", and Chart B, "Stud/Joist Loading", stud member sizes can be determined by beginning on Table 1 at 1,000 lbs./ft. and moving horizontal to 24" (2'-0") vertical tie spacing (wale spacing) and reading lumber sizes above, 2x6 or better.

Wale Size:

Wale Loading = $\frac{\text{Wale Spacing}}{\text{12 in./ft.}}$ x Design Concrete Pressure

Wale Loading = $\frac{\text{24 in. (2 ft.)}}{\text{12 in./ft.}}$ x 1,500 lbs./sq. ft.

Wale Loading = 3,000 lbs./ft.
Assume wales are continuous or partially continuous.

Using Table 4 and Chart C, begin on Table 4 at 3,000 lbs./ft. wale load and progress horizontally to a span of 24" (2'-0") horizontal tie spacing and read lumber sizes above, double 2x6's or better.

Summary of Calculations:
Ties - 6,000 lbs. safe working load
Tie Spacing - 2'-0" horizontal; 2'-0" vertical
Concrete Pressure - 1,500 lbs./ft. sq. (uniform)
Plywood - ¾" Class I perpendicular to supports
Stud Spacing - 8" center to center, perpendicular to plywood face grain
Stud Member Size - 2x6 or better, continuous or partially continuous
Wale Spacing - 24" (2'-0") center-to-center vertically
Wale Member Size - Double 2x6's or better, continuous or partially continuous

Material Calculation List:
¾" plywood (contact surface area) 50' x 12' = 60 sq. ft. one side x 2 sides = 1,200 sq. ft. of contact surface area (add end bulkheads if required)

Studs, 50 lin. ft. divided by 8 in. (.667 lin. ft.) c.c. = 75 each side x 2 sides = 150 pcs. 2x6, 12'0" long

Wales, 12' high divided by 2' c.c. = 6 each x 2 (doubles) = 12 each side; 12 each x 2 sides = 24 pcs. 2x6 @ 50' (5@10'=50') each

Ties, 50' x 12' divided by 4 = 150 each (plus hardware)

Notes:
1) Bottom row of ties, maximum 6" from bottom of form
2) Top row of ties, not to exceed 50% of vertical spacing, and 3" concrete cover required
3) First and last vertical row not to exceed 6" from edge of form
4) Concrete placement shall not exceed recommended placement rate of 7 ft./hr. in 4-foot layers.
5) Consult tie manufacturer for your special requirements

Table 1
Maximum Wale Spacing Along Stud/Joist Member
(Continuous or Partially Continuous Spans)

Load on Stud/Joist lb./ft.	Lumber Sizes 2x4	2x6	2x8	2x10	2x12	3x4	3x6	3x8	4x4	4x6	4x8
200	52	78 82	95 109	115 136	133 158	63 68	88 105	108 129	68 78	96 114	118 140
400	37	58	77	96 98	112 119	48	74 75	91 99	57	86 89	99 117
600	28	44	58	74	90	39	62	81	46	73	90 96
800	23	36	47	60	73	33	52	69	40	63	83
1,000	20	31	41	52	63	28	44	58	36	56	74
1,200	17	28	36	46	56	25	39	51	31	50	65
1,400	16	25	33	42	51	22	35	46	28	44	58
1,600	15	23	31	39	48	20	32	42	25	40	53
1,800	14	22	29	37	45	19	29	39	24	37	48
2,000	13	21	28	35	43	18	28	36	22	34	45
2,200	13	20	26	34	41	17	26	34	20	32	42
2,400	12	19	25	32	39	16	25	33	19	30	40
2,600	12	19	25	31	38	15	24	31	18	29	38
2,800	11	18	24	30	37	15	23	30	18	28	36
3,000	11	18	23	30	36	14	22	29	17	26	35

Table 2
Spacing of Supports for Class I Plywood (Strong Direction)--Face Grain Across Supports*

Pressure (psf)	1/4"	3/8"	1/2"	5/8"	3/4"	7/8"	1"	1 1/8"
100	10.0 12.2	14.5 16.2	18.3 19.3	21.7 21.9	24.9 24.3	28.5 26.9	32.1 29.4	35.2 31.5
200	7.9 9.3	11.5 13.1	14.5 16.1	17.2 18.4	19.8 20.4	22.6 22.6	25.5 24.7	27.9 26.4
300	6.9 7.6	10.1 10.7	12.7 13.1	15.0 15.1	17.1	19.3	21.7	23.4
400	6.3 6.6	9.1 9.3	11.4	13.1	14.8	16.8	18.8	20.3
500	5.8 5.9	8.3	10.2	11.8	13.2	15.0	16.8	18.1
600	5.4	7.6	9.3	10.7	12.1	13.7	15.3	16.5
700	4.6	7.0	8.6	9.9	11.2	12.7	14.2	15.3
800	4.0	6.6	8.1	9.3	10.5	11.8	13.3	14.3
900	3.6	6.2	7.6	8.8	9.9	11.2	12.5	13.5
1,000	3.2	5.6	7.2	8.3	9.4	10.6	11.7	12.8
1,100	2.9	5.1	6.9	7.9	8.9	10.1	11.3	12.2
1,200	2.7	4.6	6.5	7.6	8.5	9.7	10.8	11.7
1,300	2.5	4.3	6.0	7.3	8.2	9.3	10.4	11.2
1,400	2.3	4.0	5.6	7.0	7.9	9.0	9.8	10.8
1,500	2.2	3.7	5.2	6.7	7.6	8.5	9.2	10.2
1,600	2.0	3.5	4.9	6.3	7.4	7.9	8.6	9.6
1,700	1.9	3.3	4.6	5.9	7.2	7.5	8.1	9.0
1,800	1.8	3.1	4.4	5.6	6.9	7.0	7.6	8.5
1,900	1.7	2.9	4.1	5.3	6.5	6.7	7.2	8.0
2,000	1.6	2.8	3.9	5.0	6.2	6.3	6.9	7.6

*Plywood continuous across two or more spans.

Table 3
Spacing of Supports for Class I Plywood (Weak Direction)--Face Grain Parallel to Supports*,**

Pressure (psf)	1/2"	5/8"	3/4"	7/8"	1"	1 1/8"
100	10.7 12.9	14.5 16.2	19.0 19.7	22.4 22.4	26.2 25.2	29.5 27.6
200	8.5 10.5	11.5 13.6	15.0 16.6	17.7 18.8	20.8 21.2	23.4 23.2
300	7.4 8.5	10.1 11.7	13.1 15.0	15.5 17.0	18.2 19.1	20.5 21.0
400	6.7 7.4	9.1 10.1	11.9 13.2	14.1 15.4	16.5 17.8	18.6 19.5
500	6.2 6.6	8.5 9.0	11.0 11.8	13.0 13.8	15.3 16.1	17.2 18.4
600	5.9 6.0	8.0 8.2	10.4 10.8	12.3 12.6	14.4 14.7	16.2 16.4
700	5.5	7.6 7.6	9.9 10.0	11.7	13.7	15.2
800	4.8	7.1	8.8	10.9	12.8	14.2
900	4.3	6.3	7.9	10.3	12.1	13.4
1,000	3.9	5.7	7.1	9.6	11.4	12.7
1,100	3.5	5.2	6.4	8.7	10.3	12.1
1,200	3.2	4.7	5.9	8.0	9.5	11.2
1,300	3.0	4.4	5.4	7.4	8.8	10.4
1,400	2.8	4.1	5.1	6.8	8.1	9.6
1,500	2.6	3.8	4.7	6.4	7.6	9.0
1,600	2.4	3.5	4.4	6.0	7.1	8.4
1,700	2.3	3.3	4.2	5.6	6.7	7.9
1,800	2.2	3.2	3.9	5.3	6.3	7.5
1,900	2.0	3.0	3.7	5.0	6.0	7.1
2,000	1.9	2.8	3.5	4.8	5.7	6.7

*Plywood continuous across two or more spans.

**1/4" and 3/8" plywood not recommended for use in weak direction.

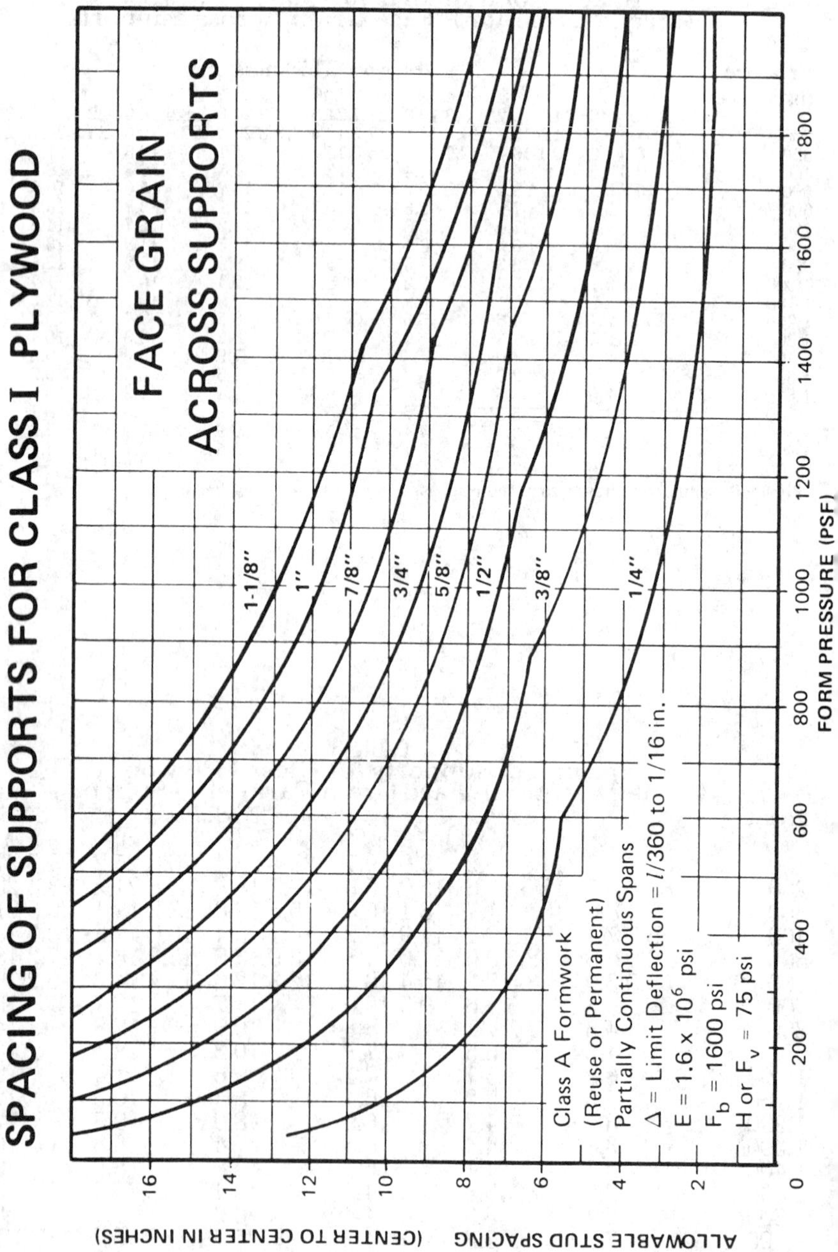

Spacing of Supports for Class I Plywood
Chart A

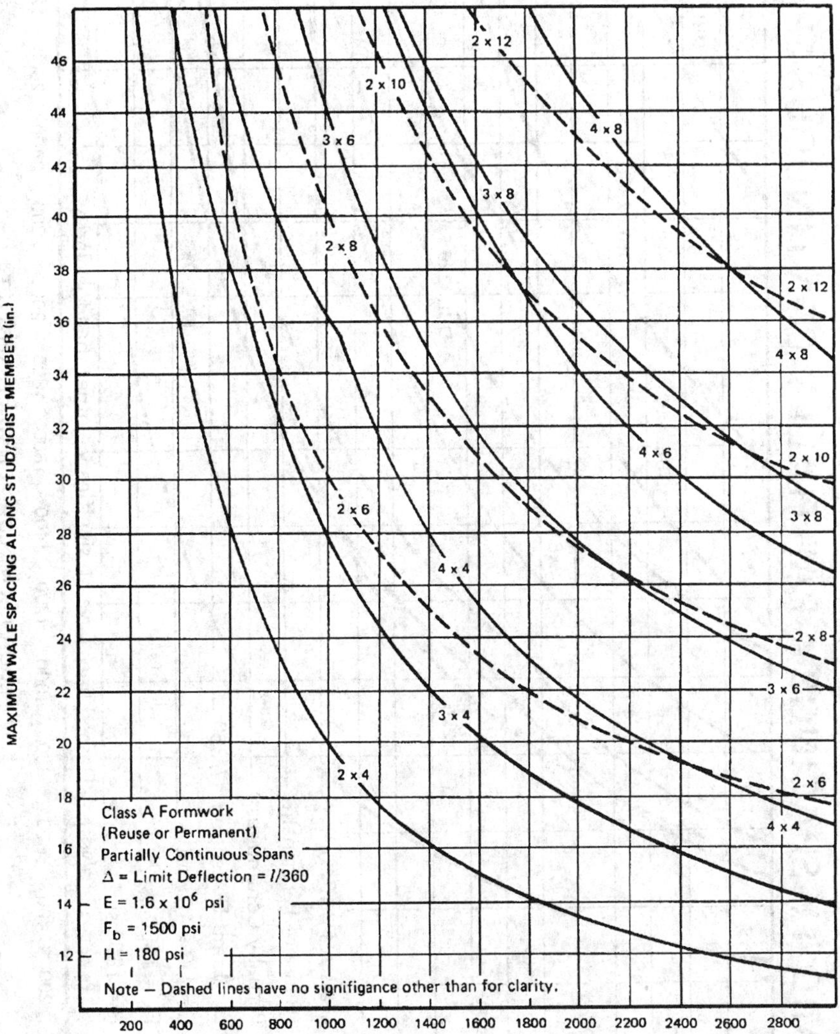
MAXIMUM WALE SPACING ALONG STUD/JOIST MEMBER (in.)
2 x 12
2 x 10
4 x 8
3 x 6
3 x 8
2 x 8
4 x 6
2 x 6
4 x 4
3 x 4
2 x 4
Class A Formwork
(Reuse or Permanent)
Partially Continuous Spans
Δ = Limit Deflection = l/360
E = 1.6 x 10^6 psi
F_b = 1500 psi
H = 180 psi
Note – Dashed lines have no signifigance other than for clarity.
200 400 600 800 1000 1200 1400 1600 1800 2000 2200 2400 2600 2800
12 14 16 18 20 22 24 26 28 30 32 34 36 38 40 42 44 46

Stud/Joist LoadingChart
Chart B

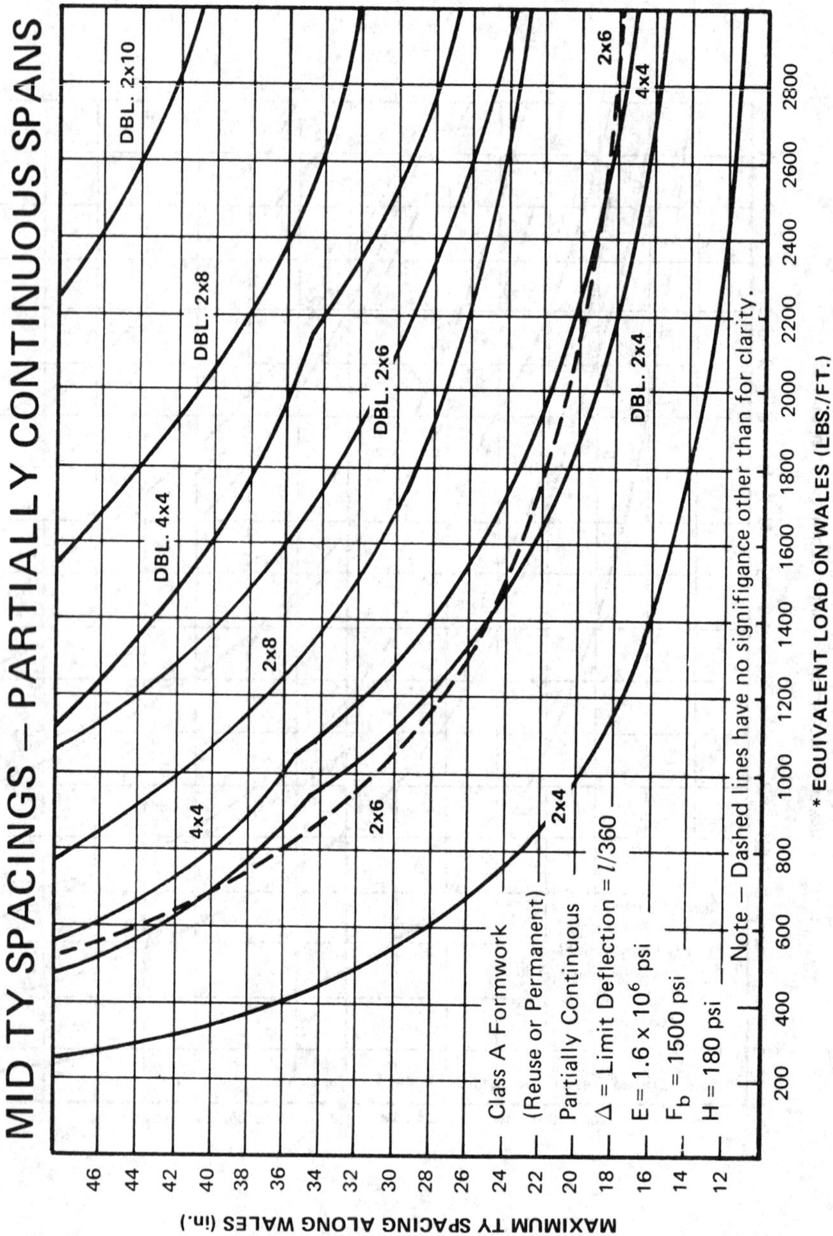

Mid Tie Spacings--Partially Continuous Spans
Chart C

Table 4
Maximum Tie Spacing for Class A (Permanent) Form Design

Load on Wales		Single Wales		Double Wales										
lbs./ft.	2x4	2x6	2x8	4x4	2x4	2x6	2x8	2x10	2x12	3x10	3x12	4x4	4x10	4x12
200	52	78 82	95 109	68 80	66 74	92 110	113 135	136 162	158 188	155 184	179 213	81 97	168 200	195 232
400	37	58	77	57	53	78 82	95 109	115 136	133 158	130 155	151 179	68 78	142 168	164 195
600	28	44	58	46	43	67	86 101	104 113	120 138	118 140	136 162	62 65	128 152	148 176
800	23	36	47	40	37	58	77	96 98	112 119	109 127	127 151	57	119 142	138 164
1,000	20	31	41	36	32	51	61	85	104	104 113	120 138	51	113 134	130 155
1,200	17	28	36	31	28	44	58	74	90	99 103	115 126	46	108 122	125 148
1,400	16	25	33	28	25	39	52	66	80	95 96	110 116	43	104 113	120 138
1,600	15	23	31	25	23	36	47	60	73	88	107 107	40	100 106	116 129
1,800	14	22	29	24	21	33	44	56	68	80	98	38	97 100	113 121
2,000	13	21	28	22	20	31	41	52	63	74	90	36	95 95	110 115
2,200	13	20	26	20	18	29	38	49	59	69	84	34	89	107 108
2,400	12	19	25	19	18	28	36	46	56	65	79	32	83	101
2,600	12	19	25	18	17	26	35	44	54	61	74	30	78	95
2,800	11	18	24	18	16	25	33	42	51	58	71	28	74	90
3,000	11	18	23	17	15	24	32	41	50	56	68	27	70	86

Lumber Size and Weight Table

Item	Nominal Size in Inches b x h	American Standard (In.) b x h	Board Feet per Lin. Ft.	Lbs. per Lin. Ft. Avg. Wgt.
1	1x4	¾ x 3 ½	0.33	0.8
2	1x6	¾ x 5 ½	0.50	1.2
3	1x8	¾ x 7 ¼	0.66	1.6
4	1x10	¾ x 9 ¼	0.83	2.0
5	1x12	¾ x 11 ¼	1.00	2.4
6	2x4	1 ½ x 3 ½	0.66	1.6
7	2x6	1 ½ x 5 ½	1.00	2.4
8	2x8	1 ½ x 7 ¼	1.33	3.2
9	2x10	1 ½ x 9 ¼	1.66	4.0
10	2x12	1 ½ x 11 ¼	2.00	4.8
11	3x4	2 ½ x 3 ½	1.00	2.4
12	3x6	2 ½ x 5 ½	1.50	3.6
13	3x8	2 ½ x 7 ¼	2.00	4.8
14	3x10	2 ½ x 9 ¼	2.50	6.0
15	3x12	2 ½ x 11 ¼	3.00	7.3
16	4x4	3 ½ x 3 ½	1.33	3.2
17	4x6	3 ½ x 5 ½	2.00	4.8
18	4x8	3 ½ x 7 ¼	2.66	6.4
19	4x10	3 ½ x 9 ¼	3.33	8.0
20	4x12	3 ½ x 11 ¼	4.00	9.7
21	6x6	5 ½ x 5 ½	3.00	8.0
22	6x8	5 ½ x 7 ½	4.00	10.7
23	6x10	5 ½ x 9 ½	5.00	13.3
24	6x12	5 ½ x 11 ½	6.00	16.0
25	8x8	7 ½ x 7 ½	5.33	14.2
26	8x10	7 ½ x 9 ½	6.66	17.8
27	8x12	7 ½ x 11 ½	8.00	21.3
28	10x10	9 ½ x 9 ½	8.33	22.2
29	10x12	9 ½ x 11 ½	10.00	26.7
30	12x12	11 ½ x 11 ½	12.00	32.0

Nominal Plywood Weights for Form Plywood

Item	Nominal Size in Inches	Weight in lbs. per Sq. Ft.	Weight in lbs. per 4'x8' Sheet (32 sq. ft.)
1	1/4	0.78	25
2	1/2	1.53	49
3	5/8	1.81	58
4	3/4	2.22	71
5	1 1/8	3.34	107

Nail Sizes and Weights

257 Common

Sizes	Gauge	Length in Inches	Approximate no. pcs/lb.
3d	14	1 1/4	543
4d	12 1/2	1 1/2	294
6d	11 1/2	2	167
7d	11 1/2	2 1/4	149
8d	10 1/4	2 1/2	101
10d	9	3	66
12d	9	3 1/4	61
16d	8	3 1/2	47
20d	6	4	30

Double Headers

Sizes	Gauge	Length in Inches	Approximate no. pcs/lb.
8d	10 1/4	2 1/4	8
16d	8	3	44

Concrete Quantities for Wall Forms per Square Foot Contact Area 50'-0" x 8'-0" (no bulkheads)

Wall Thickness in Inches	Sq. Ft. Contact Area, less Bulkheads	Cu. Ft. of Concrete	Cu. Yd. of Concrete	Sq. Ft. per Cu. Yd.	Cu. Yd. per Sq. Ft.
4	800	133.33	4.94	162	.006
6	"	200	7.41	108	.009
8	"	266.67	9.88	81	.012
10	"	333.33	12.35	64.8	.015
12	"	400	14.81	54	.019
14	"	466.67	17.28	46.29	.022
16	"	533.33	19.75	40.5	.025
18	"	600	22.22	36	.028
20	"	666.67	24.69	32.4	.031
22	"	733.33	27.16	29.46	.034
24	"	800	29.63	27	.037
26	"	866.67	32.10	24.92	.040
28	"	933.33	34.57	23.143	.043
30	"	1,000	37.04	21.6	.046
32	"	1,066.67	39.51	20.25	.049
34	"	1,133.33	41.98	19.06	.052
36	"	1,200	44.44	18	.056

Example:
50'-0" long x 8'-0" high wall, 12" thick, no bulkheads
50' x 8' = 400 sq. ft. x 2 sides = 800 sq. ft. of contact surface area.
50' x 8' x 1' = 400 cu. ft. divided by 27 = 14.81 cu. yds. concrete.
800 sq. ft. divided by 14.81 cu. yds. = 54 sq. ft. of forms per cu. yd.
14.81 cu. yd. of concrete divided by 800 = .019 cu. yd. concrete per sq. ft. of forms.

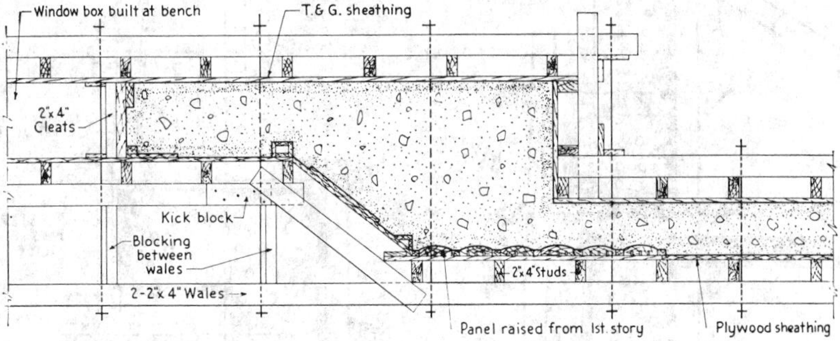

HORIZONTAL SECTION SHOWING WINDOW FORM

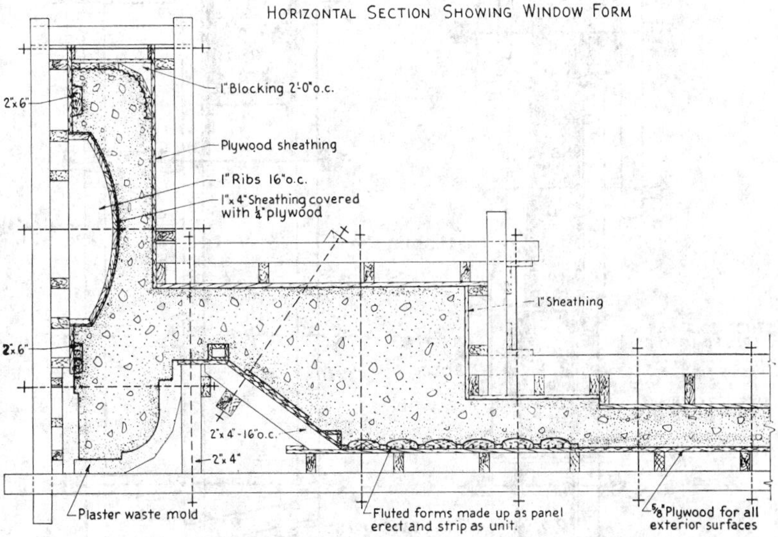

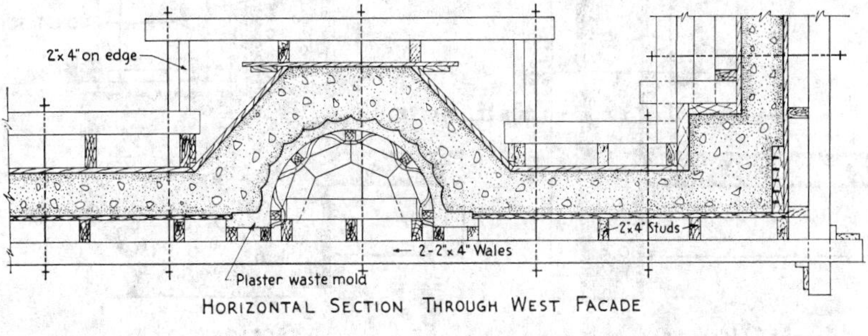

HORIZONTAL SECTION THROUGH WEST FACADE

Method of Forming Waste Molds and Other Ornamentation in Architectural Concrete Forms

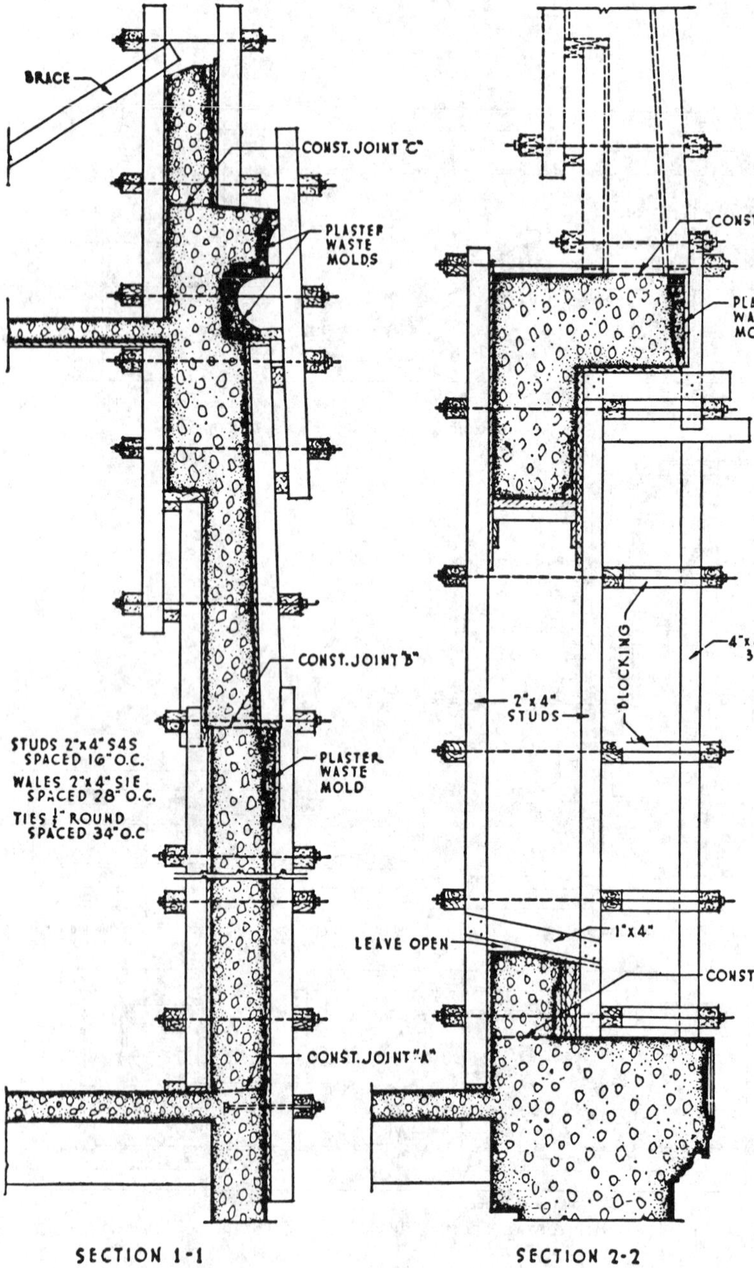

Method of Forming for Architectural Concrete Waste Molds for Belt Courses and Cornices

Window and Door Openings in Architectural Concrete Walls. It is necessary to form for all door and window openings in concrete walls. The outside wall form is built up solid. Then, boxes or "bucks" are constructed the exact size of the door and window openings. They should be made of 2" lumber and held in place by 1"x4" strips nailed to both the outside and inside wall sheathing and braced where necessary using 2"x4"s placed horizontally and diagonally. Where the opening is over 3'-0" wide, it will be necessary to leave an opening in the bottom of the frame so that concrete can be filled to the proper level. After filling to the correct level, the opening is closed by nailing a piece of plank to the bottom of the frame. A few hours after the concrete has set, the plank is removed and the surplus concrete scraped off. To do this, it is necessary to leave an opening in the inside form at every opening.

Reveals for door and window openings should be designed so that standard size 2" lumber may be used.

Rustication Strips. Rustication strips should be designed as narrow as possible and only about ¾" deep. When the strips are less than 1" wide, they should have one saw cut. When 2" wide, they should have 2 saw cuts in the back as illustrated. These strips should be nailed to a chalk line, using casing nails long enough to go through the strip and sheathing. It is recommended that the nails be withdrawn by pulling them through the sheathing before the form is removed and allow the wooden strip to remain in place a few days after the form is removed or long enough so that they can be withdrawn without injuring the edges of the concrete.

Ornamentation

Belt courses or other ornamentation where continuous members occur may be formed with wood moldings cut to the proper shape. While some contractors may prefer a waste mold for this particular detail, the wood mold has the advantage of being more easily held in alignment, because the various members are broken at different points, while in a plaster waste mold, the entire form would be cut in two at one point. Soft grained wood that does not warp or split easily should be used. Soft white pine is best, soft-grained Douglas fir is the second choice. In the cornice illustrated, the various members are narrow enough so that common stock sizes can be used throughout. Note the generous use of saw-kerfs in the back side of members to prevent swelling and warping of the lumber. The lumber should be thoroughly oiled on all sides before building the forms as further precaution against swelling and warping.

Large, heavy wood members should be avoided in detailing for ornamental work, because they will swell, and in stripping the corners of the projecting concrete will be fractured. Using several smaller pieces requires less lumber and the members may be arranged so that swelling is away from the concrete, making it easier to strip the form without damage to the concrete corners. Note the illustration where the left-hand detail shows a large member milled from a single wide board in contrast to the smaller pieces shown in the right-hand detail.

In assembling forms, one must constantly keep in mind the steps that can be taken to aid in removing them without injury to the concrete. Boxes, waste molds, wood molds, rustication strips, or anything applied to the face of the forms should be nailed lightly so that when the forms are removed, these members will pull loose from the wall form and remain in the concrete. After these materials have dried out thoroughly and have shrunk, they can be removed without much difficulty and without injury to the concrete corners or edges.

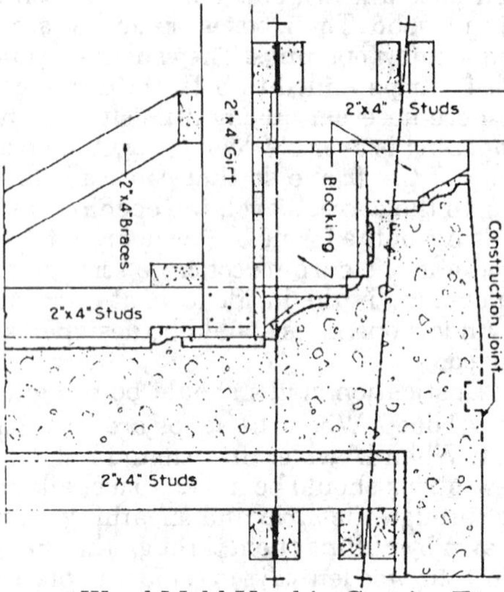

Wood Mold Used in Cornice Form

As previously indicated, solid strips of wood, even though well oiled, may swell and result in considerable breakage of concrete corners when the strips are removed. Saw-kerfs on the back of the member will prevent this trouble by relieving any pressure against the concrete. They should be approximately two-thirds the depth of the member, not more than 1 ½" apart, and in general, there should be one kerf within ¾" of each edge.

Waste molds are the only solution for some highly ornamental work. They should be made only by experienced ornamental plaster workers who have been given instructions as to how they are to be fitted into the forms. Ordinarily, the waste molds are about 2" thick, reinforced on the back with 2"x4"s that are attached to the mold with burlap dipped in plaster and then wound around the wood. The 2"x4"s permit easy handling of the mold and are also used to attach the mold to the form.

When waste molds are large and have deep undercuts, it is usually best to wire them back to the studs or wales with enough wire to be sure that all points will be pressed firmly against the form. Any openings between the form and waste mold should be pointed with plaster of Paris or a patching plaster and nailheads driven through the face and around the edge of the waste molds should be countersunk and similarly pointed.

Ornamental Concrete Cornices. Where it is necessary to use molds for belt courses, cornices, door and window reveals, columns, cornices, entrances, etc., it is more economical to build them of wood, if practicable. However, if plaster waste molds are required, the following will give some idea of their cost, although on this work, it is always advisable to obtain definite prices from a concern specializing in this work.

Plain molds consisting of a few straight lines, $10.00 per sq. ft.

Ordinary molds of stock design will cost from $10.00 to $14.00 per sq. ft.

Simple molds will cost from $10.00 to $14.00 per sq. ft. plus the cost of waste molds.

Cornice molds of stock design about 1'-6" deep and 1'-3" wide cost about $18.00 to $25.00 lin. ft.

Capitals for 1'-6" pilasters, stock designs, cost $170 to $280 each.

Joints in plaster waste molds should be patched with non-shrinking patching plaster and painted with shellac.

Individual pieces should not weigh more than 150 lbs. to be set without difficulty by 2 workers.

Plaster waste molds are usually given 2 coats of shellac in the shop to make them waterproof and non-absorbent. Before concrete is placed all molds should be greased with a light yellow cup-grease, which may be cut with kerosene if too thick. The grease should be wiped into all angles of the mold and every bit of surplus grease carefully wiped off. Care must be taken not to drop oil, grease or shellac onto hardened concrete or reinforcing.

Construction Joints. Much progress has been made on all types of concrete work in production of satisfactory construction joints, both from the standpoint of good appearance and good bond between successive lifts. On the other hand, some very unsightly joints, for which there is no excuse, occasionally are made on architectural concrete work. It is just as easy to produce a good joint as a poor one. The designer, of course, should indicate on his plans exactly where such joints should be located. Rustications or offsets will mask the joints and are advisable where they can be made to fit into the design.

One requirement that is sometimes neglected is to provide means of holding the form tightly against the hardened concrete. As a result, leakage occurs, discoloring the concrete below, and the upper lift will project slightly beyond the lower lift. The illustration shows the use of 5/8" bolts cast in the concrete for fastening the form for the next lift. Some types of form ties can be used instead of bolts. The bolt or form tie assembly can be used over and over, because only the nut is left in the concrete, if the bolt is used, or the insert, if a form tie is used. In re-erecting the forms for the next lift, the contact surface of the form sheathing should not overlap the hardened concrete more than 1". More overlapping than this only presents more opportunity for leakage due to irregularities in the wall surface against which the sheathing is to be held.

Since sufficient time must be allowed for the concrete between windows or other openings to settle and shrink before the concrete above is placed, the tops of such openings are "must" locations for construction joints. Similarly, it is desirable to locate a joint at the sill line. By so doing, the joints are broken up into short lengths by the opening, making them relatively inconspicuous. Moreover, cracking at corners of openings is thereby minimized.

The construction joint should be only a straight thin line at the surface. If a rustication strip is not used at the joint, a 3/4" wood strip to a point 1/2" above the bottom of this strip and just before the concrete becomes hard, the top should be lightly tamped to be sure that it is tight up against the strip after it has settled or shrunk. All surplus concrete or laitance should then be removed.

In resuming the placing of concrete in the new lift, steps should be taken to get good bond and to avoid honeycomb. The hardened concrete should be clean and thoroughly saturated. A 6" layer of concrete in which the coarse aggregate is reduced about 50%, or a 2" layer of cement-sand grout should then be placed on the hardened concrete followed by the regular concrete.

Control Joints. It is important, both for appearance and performance, that control joints be carefully installed so they will be perfectly straight and vertical. Care is required during placing of concrete to avoid knocking the joint out of alignment or otherwise injuring it. When control joints are to be caulked with a plastic compound, this should be done before the final clean down so that any compound smeared onto the surface is removed. Of course, excessive smearing should be avoided.

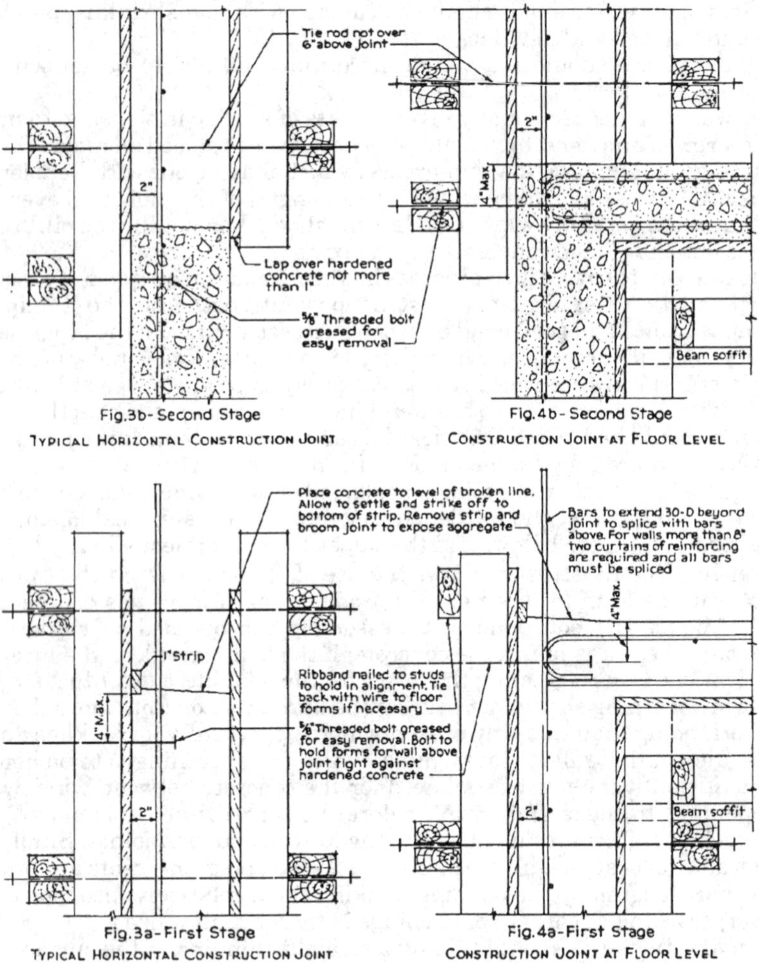

HORIZONTAL CONSTRUCTION JOINTS

Method of Embedding Bolts and Placing Forms for Successive Lifts

Form Ties. Various makes of form ties are available for architectural concrete construction as shown in the accompanying illustration. Important features to look for in selecting form ties are: removal of all metal to a depth of at least 1 1/2" from the face of the wall, minimum strength of 3,000 lbs. when fully assembled, adjustable length to permit tightening of forms around all openings and inserts and a design that will leave a hole at the surface not more than 7/8" in diameter and not less than 3/4" deep so that it will hold a patch. Ties should not be fitted with cones or washers acting as spreaders, because tapered and shallow holes cannot be patched properly.

Labor Cost of Architectural Forms. The labor fabricating, erecting, and stripping forms is usually estimated on the basis of sq. ft. of contact area. It is customary on a job having a normal amount of ornamentation to take off the total contact area as though the wall were plain. Window and other openings, unless very large, are figured solid. The area thus obtained is priced as though the wall were unornamented. Separate allowance is made for window frames or "bucks", ornamentation, etc.

The average amount of lumber required for plain wall forms is 2 1/4 to 2 1/2 b.f. of lumber per sq. ft. of contact area, plus the sheathing, plywood, or concrete form Presdwood, used for the facing materials. Figure one tie rod for approximately each 10 sq. ft. of form.

On plain walls up to 10 ft. story height, it requires about 11 to 13 hrs. carpenter time and 6 to 8 hrs. labor time per 100 sq. ft. of contact area. For walls 12 to 14 ft. high, add about 10% to these costs. This includes all time necessary for erecting and stripping the forms and hoisting it up for the next use. In addition, figure about 1 to 1 1/4 hrs. labor time for removing hardened concrete from the plywood and coating plywood between each use.

Door and window frame boxes or "bucks" require 5 to 8 hrs. carpenter time for each opening, depending upon size and detail of same.

Rustication strips should be placed at the rate of 25 to 30 lin. ft. an hr. for one carpenter.

Ornamental curtain walls, fluted pilasters and piers, reveals, etc., will require about 10 to 12 hrs. carpenter time and 6 hrs. labor time per 100 sq. ft. in addition to the cost of plain wall forms.

Ornamental wood cornices built to special detail will require 24 to 28 hrs. carpenter time and 12 to 14 hrs. labor time per 100 sq. ft. in addition to cost of plain wall forms.

Ornaments projecting from the face of plain walls, such as special sills, balconies, etc., will require about 36 to 40 hrs. carpenter time and 12 to 15 hrs. labor time per 100 sq. ft. in addition to the cost of plain wall forms.

Where plaster waste molds are used, they may be set either by carpenters or plasterers, depending upon local labor union regulations.

When setting simple belt courses of plaster waste mold, allow 12 to 14 hrs. mechanic time and 6 to 7 hrs. helper time per 100 sq. ft.

When erecting and blocking cornices and other elaborate moldings using plaster waste molds, figure 20 to 25 hrs. mechanic time and 10 to 12 hrs. helper time per 100 sq. ft.

When placing letters, a carpenter should set 2 to 3 letters an hr. depending upon size.

Labor Cost of 100 Sq. Ft. of Plain Architectural Concrete Walls Using Plywood for Facing Material

	Hours	Rate	Total	Rate	Total
Carpenter erecting forms	12.0	$....	$....	$24.04	$288.48
Labor helping	6.0	....	....	19.13	114.78
Labor stripping and cleaning forms	2.0	....	....	19.13	38.26
Labor coating plywood	0.33	....	....	19.13	6.31
Cost per 100 sq. ft.			$....		$447.83
cost per sq. ft			....		4.48

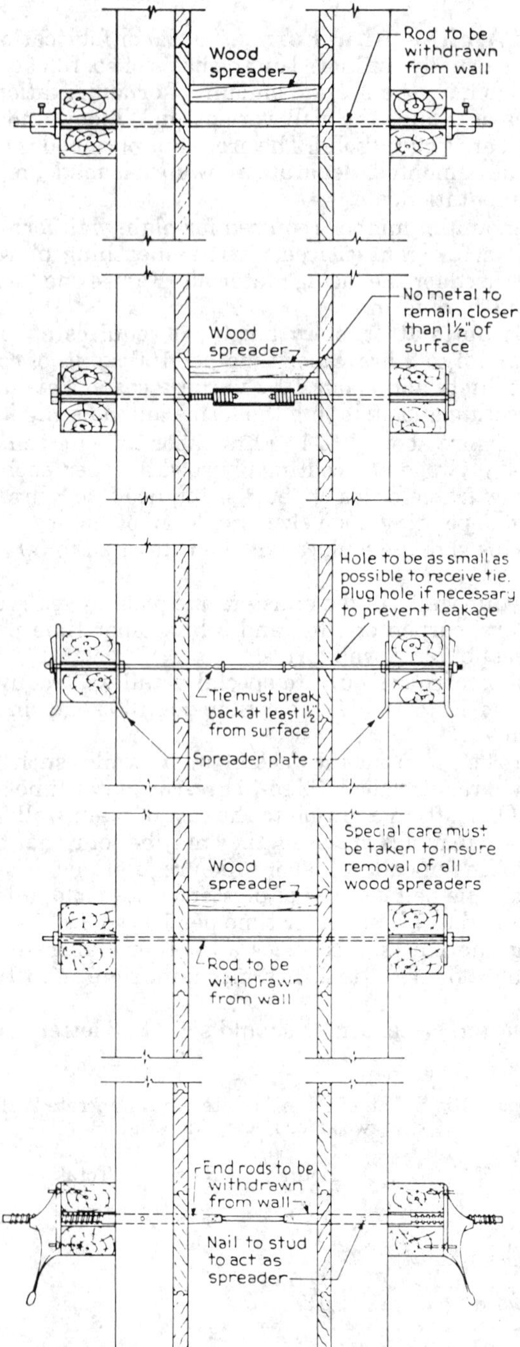

Types of Form Ties Available for Use in Architectural Concrete Construction

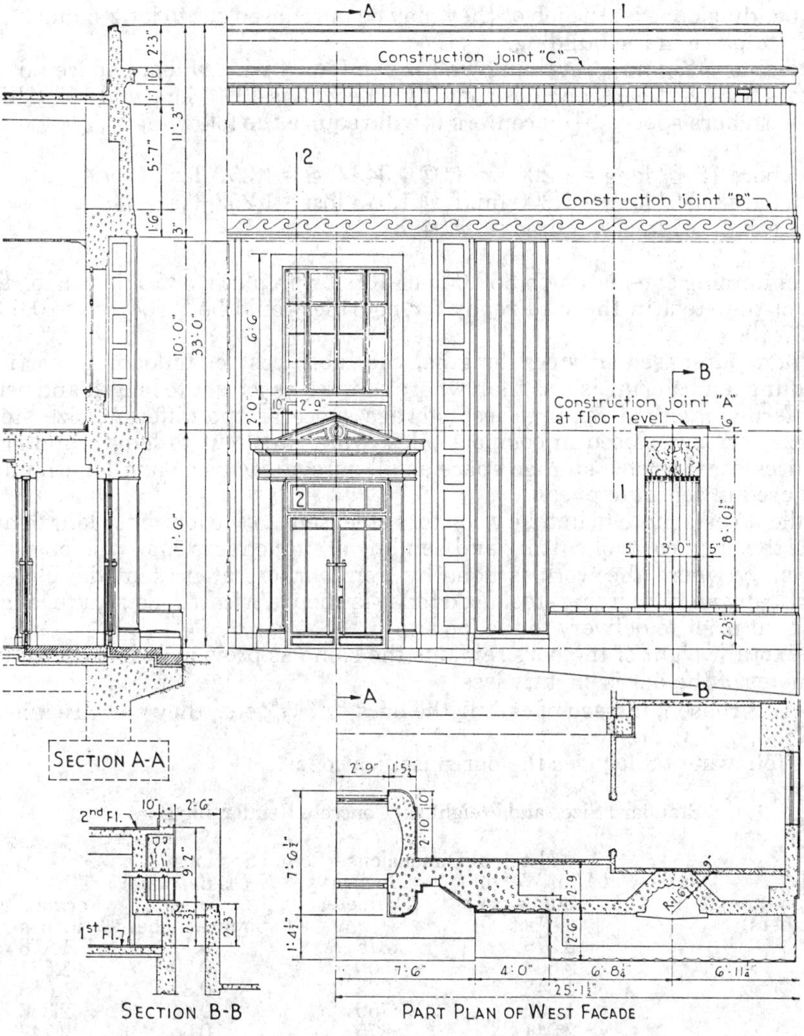

Details of Architectural Concrete Facade

Details for Architectural Concrete Forms. For the rapid and economical construction of architectural concrete forms, it is necessary to have complete working drawings that show the elevations in large scale, together with every piece of plywood, stud, wale, bolt, lumber for reveals, window boxes or "bucks", and so on, similar to the methods used for detailing cut stone, terra cotta, or structural steel.

03200 CONCRETE REINFORCEMENT

Reinforcing steel is estimated by the pound or the ton, obtained by listing all bars of different sizes and lengths and extending the total to pounds. Reinforcing bars may be purchased from warehouse in stock lengths and all cutting and bending done on the job or they may be purchased cut to length and bent, ready to place in the building.

Example: Assume a retaining wall is reinforced with #5 bars spaced 6" on centers both vertically and horizontally. The wall is 40'-0" long and 15'-0" high and with bars spaced 6" on centers it will require the following:

80 #5 bars 15'-0" long = 1,200 lin. ft.@ 1.043 lbs. = 1,252 lbs.
30 #5 bars 40'-0" long = 1,200 lin. ft.@ 1.043 lbs. = 1,252 lbs.
Total weight of steel required = 2,504 lbs.

If reinforcing steel is worth $60.00 per 100 lbs. in place, or $0.60 per lb., the cost of the steel in the wall ready for pouring would be 2,504 lbs. x 0.60 = $1502.40.

When purchased in stock lengths, the labor cost of unloading, sorting, handling, and storing is less than where sent to the job cut to length and bent. The former method requires less storage space as the different size stock length bars are placed in compact piles, while steel cut to length and bent requires considerable storage space as all bars of each size and length must be placed in separate piles.

While it costs more to handle and store steel sent to the job cut to length and bent, the shop costs of cutting and bending are much less than the job costs, especially where the work is done by iron workers at present day wages. Practically all reinforcing steel used today is shop fabricated, i.e., cut to length and bent prior to delivery to the job site.

The total weight of the bars remains the same as previously but the actual diameter of the bar is slightly less.

For this reason, instead of calling the bars ½", ¾", etc., they now are known as No. 3, 4, 5, 6, etc.

The following table gives the bar classifications:

Standard Sizes and Weights of Concrete Reinforcing Bars

	Nominal Dimensions,--Round Sections			
Bar Designation (Numbers)	Unit Weight Pounds per Foot	Inches Diameter	Cross Sectional Area Square Inches	Perimeter Inches
3	0.376	.375	0.11	1.178
4	0.668	.500	0.20	1.571
5	1.043	.625	0.31	1.963
6	1.502	.750	0.44	2.356
7	2.044	.875	0.60	2.749
8	2.670	1.000	0.79	3.142
9	3.400	1.128	1.00	3.544
10	4.303	1.270	1.27	3.990
11	5.313	1.410	1.56	4.430
14	7.650	1.693	2.25	5.320
18	13.600	2.257	4.00	7.090

Basis for Estimating Price of Reinforcing Steel. When estimating quantities and weights of reinforcing steel, all bars of each size should be listed separately to obtain the correct weight. No. 6 bars and over take the lowest price per lb. while bars smaller than No. 6 increase in price on a graduated scale depending upon the size of the bars. The smaller the bars the higher the price per lb. On large projects the normal practice is to quote the price per lb., or ton, based on all sites on the project as a whole.

Example: If base price is $60.00 per 100 lbs. what is price of No. 3 bars? The table shows that No. 3 bars take an increase of $5.75 per 100 lbs. over base or $65.75 per 100 lbs.

Prices are usually quoted f.o.b. warehouse, with freight allowed to destination.

Size Extras on Reinforcing Bars

Number	Extra per ton
14 and 18	$40.00*
7 to 11	20.00
6	25.00
5	30.00
4	40.00
3**	115.00

*Includes cutting extra.
**The #3 bars are not available in some areas.

Prices on rail steel are theoretically the same as for billet steel but will vary with the scrap market on rails.

There are four specifications covering the use of reinforcement steel:

1) ASTM A615: Deformed and plain billet-steel bars for concrete reinforcement.
2) ASTM A616: Rail steel deformed and plain bars for concrete reinforcement.
3) ASTM A617: Axle steel deformed and plain bars for concrete reinforcement.
4) ASTM A706: Low-alloy steel deformed bars for concrete reinforcement.

It is of utmost importance that the quantity take-off be correct. Normal practice is to allow between 2 to 5 percent of the calculated bar weight for waste and splices. Reinforcement steel up to No. 11 bars may be spliced by overlapping them and wiring them together.

Two standard grades of reinforcement steel available are, ASTM A615, Grade 40 and ASTM A615, Grade 60. Bars are of two minimum yield levels: 40,000 psi and 60,000 psi, designated as Grade 40 and Grade 60 respectively.

Both grades are standard in building codes and in the design recommendations of the American Concrete Institute.

Reinforcement Bar Supports. Chairs, sometimes referred to as reinforcement bar support, are required in order to accurately support and locate reinforcement steel in the concrete form. The bar support will hold the bar firmly in its proper location during placement of concrete.

Rust Prevention. Reinforcement steel supports are classified with respect to methods to diminish rust spots or similar imperfections from the concrete surface caused by reinforcement steel supports. There are three classes:

Class 1: Maximum protection; plastic protected reinforcement steel support which are intended for use in situations of moderate to severe exposure and/or situations requiring light grinding or sandblasting of concrete surface.

Class 2: Moderate protection; stainless steel protected reinforcement steel supports, which are intended for use in situations of moderate exposure and/or situations requiring light grinding or sandblasting of the concrete surface. Class 2 protection can be obtained by the use of either Type "A" or Type "B" stainless steel protection reinforcement steel supports. The difference between the two types is in the length of the stainless steel tip attached to the bottom of each leg support.

Type "A" - Stainless steel protected bar support; a tip of stainless steel is attached to the bottom of each leg so that no portion of the non-stainless steel wire lies closer than ¼" to the form face.

Type "B" - Stainless steel protected bar support, a tip of stainless steel is attached to the bottom of each leg so that no portion of the non-stainless steel wire lies closer than ¾" to the form face.

Class 3: No protection; bright basic reinforcement steel supports that have no protection against rusting and which are intended for use in situations where surface blemishes can be tolerated and/or where supports do not come in contact with the exposed concrete.

Costs for reinforcement steel supports can vary widely, from less than 1% to more than 10% of the reinforcement steel costs.

Bending Reinforcing Steel. The labor cost of bending reinforcing bars will vary with the size of the bars and the amount of bending necessary; the lighter the bars the higher the bending cost per ton.

Type L Bending. Bending all No. 3 bars; all No. 4 stirrups and column ties; truss bars and all sizes of bars bent at more than 6 points. It also includes all bars bent in more than one plane and all radius bending with more than one radius in any bar, approximately $90.00 per ton.

Type H Bending. Bending truss bars for beams and slabs, other than No. 3 and No. 4, radius bending and types not otherwise described, approximately $40.00 per ton.

When cut is made by means other than sawing or milling, $0.05 per end.

When cut is made by sawing or milling, $0.15 per end.

Spirals for Reinforced Concrete Columns. Spirals should always be coiled and fabricated in the shop as it is all machine work.

Prices of spirals vary throughout the country in the same manner and for the same reasons as reinforcing steel. Quantity extras are figured independently of reinforcing steel tonnages involved.

Cost of Engineering Service. The cost of making working drawings, bending details, setting plans, etc., will vary with the amount of work involved.

Where the engineers prepare the complete design for a reinforced concrete structure, including working drawings, placing plans, bar lists and bending details, add $45.00 per ton to the detailing charges given below.

For preparing order lists with or without bending details from plans on which all sizes and lengths are shown by others, $0.30 per 100 lbs. or $6.00 per ton.

Hauling Reinforcing Bars From Shops to Job. The cost of hauling reinforcing steel from warehouse to job will vary with the distance of the haul and the size of the city.

For example, in the Chicago area, the trucking charge for deliveries within city limits is about $1.50 per cwt. For jobs located in the Chicago suburbs, the trucking charge is about $1.75 per cwt. Minimum charge for both zones is $100.00.

Unloading Reinforcing Steel From Cars or Trucks. If steel can be unloaded direct from cars or trucks, without sorting or carrying to stock piles, a worker will unload 4 ½ to 5 ½ tons per 8-hr. day, at the following labor cost per ton:

		Hours	Rate	Total	Rate	Total
Iron Worker	. . . .	1.75	$....	$....	$27.53	$48.18

If it is necessary to carry the bars up to 60 ft., sort, and place in stock piles, figure 2 ½ to 3 tons per 8-hr. day per worker, at the following labor cost per ton:

		Hours	Rate	Total	Rate	Total
Iron Worker	. . . .	3	$....	$....	$27.53	$82.59

PERCENTAGES AND WEIGHTS OF STANDARD STEEL REINFORCING SPIRALS(1)

Weights are in pounds per foot of height exclusive of spacers and ends (2)(3)

(In each cell: % / Wt.)

¼" ∅ SPIRALS — PITCH IN INCHES

Core(3) Diam. Inches	1¾	2	2¼	2½	F
12	0.95 / 3.65	0.83 / 3.19			1.59
13	0.88 / 3.96	0.77 / 3.47			1.74
14	0.82 / 4.28	0.71 / 3.71	0.64 / 3.34		1.86
15	0.76 / 4.56	0.67 / 4.02	0.59 / 3.54	0.53 / 3.18	2.01
16	0.71 / 4.85	0.62 / 4.23	0.56 / 3.82	0.50 / 3.41	2.12
17	0.67 / 5.16	0.59 / 4.54	0.52 / 4.00		2.27
18	0.64 / 5.53	0.56 / 4.84			2.42
19	0.60 / 5.77	0.53 / 5.10			2.55
20	0.57 / 6.09	0.50 / 5.34			2.67
21	0.54 / 6.35				2.78
22	0.52 / 6.71				2.94
23	0.50 / 7.05				3.09
24					
25					
26					
27					
28					
29					
30					
31					
32					
33					

⅜" ∅ SPIRALS — PITCH IN INCHES

Core Diam. Inches	1¾	2	2¼	2½	2¾	3	3¼	F
12	2.09 / 8.02	1.83 / 7.03						3.55
13	1.93 / 8.71	1.69 / 7.63						3.84
14	1.79 / 9.35	1.57 / 8.21	1.40 / 7.32					4.14
15	1.67 / 10.00	1.47 / 8.82	1.30 / 7.80	1.17 / 7.02				4.44
16	1.57 / 10.71	1.37 / 9.35	1.22 / 8.33	1.10 / 7.51				4.73
17	1.48 / 11.40	1.29 / 9.93	1.15 / 8.85	1.03 / 7.93	0.94 / 7.24			5.03
18	1.40 / 12.09	1.22 / 10.52	1.09 / 9.42	0.98 / 8.47	0.89 / 7.68	0.81 / 7.00		5.32
19	1.32 / 12.70	1.16 / 11.16	1.03 / 9.91	0.93 / 8.95	0.84 / 8.09	0.77 / 7.41		5.61
20	1.25 / 13.34	1.10 / 11.73	0.98 / 10.47	0.88 / 9.40	0.80 / 8.85	0.73 / 7.80	0.68 / 7.26	5.91
21	1.20 / 14.11	1.05 / 12.33	0.93 / 10.93	0.84 / 9.88	0.76 / 8.94	0.70 / 8.22	0.65 / 7.64	6.21
22	1.14 / 14.70	1.00 / 12.90	0.89 / 11.48	0.80 / 10.32	0.73 / 9.42	0.67 / 8.65	0.62 / 8.00	6.50
23	1.09 / 15.38	0.96 / 13.53	0.85 / 11.98	0.77 / 10.86	0.70 / 9.88	0.64 / 9.02	0.59 / 8.32	6.80
24	1.05 / 16.12	0.92 / 14.13	0.81 / 12.43	0.73 / 11.20	0.67 / 10.29	0.61 / 9.37	0.56 / 8.60	7.10
25	1.00 / 16.65	0.88 / 14.65	0.78 / 13.00	0.71 / 11.82	0.64 / 10.67	0.59 / 9.83	0.54 / 9.00	7.39
26	0.97 / 17.46	0.85 / 15.30	0.75 / 13.50	0.68 / 12.23	0.62 / 11.17	0.56 / 10.08	0.52 / 9.36	7.68
27	0.93 / 18.08	0.82 / 15.93	0.72 / 14.00	0.65 / 12.63	0.59 / 11.47	0.54 / 10.50	0.50 / 9.72	7.97
28	0.90 / 18.81	0.79 / 16.52	0.70 / 14.63	0.63 / 13.18	0.57 / 11.92	0.52 / 10.88		8.27
29	0.87 / 19.51	0.76 / 17.05	0.67 / 15.03	0.61 / 13.68	0.55 / 12.33	0.51 / 11.43		8.57
30	0.84 / 20.15	0.73 / 17.52	0.65 / 15.60	0.59 / 14.16	0.53 / 12.72			8.86
31	0.81 / 20.78	0.71 / 18.20	0.63 / 16.15	0.57 / 14.62	0.52 / 13.33			9.16
32	0.79 / 21.60	0.69 / 18.86	0.61 / 16.67	0.55 / 15.03	0.50 / 13.67			9.46
33	0.76 / 22.02	0.67 / 19.42	0.59 / 17.11	0.53 / 15.38				9.76

½" ∅ SPIRALS — PITCH IN INCHES

Core Diam. Inches	2	2¼	2½	2¾	3	3¼	3½	F
12	3.33 / 12.79							6.30
13	3.08 / 13.89							6.82
14	2.86 / 14.92	2.54 / 13.26						7.35
15	2.67 / 16.00	2.37 / 14.20	2.13 / 12.78					7.87
16	2.50 / 17.05	2.22 / 15.14	2.00 / 13.64					8.40
17	2.35 / 18.10	2.09 / 16.10	1.88 / 14.48	1.71 / 13.18				8.93
18	2.22 / 19.18	1.97 / 17.02	1.78 / 15.38	1.62 / 14.00	1.48 / 12.79			9.45
19	2.10 / 20.21	1.87 / 18.00	1.68 / 16.16	1.53 / 14.72	1.40 / 13.47			9.97
20	2.00 / 21.36	1.78 / 19.00	1.60 / 17.09	1.45 / 15.48	1.33 / 14.20	1.23 / 13.12		10.50
21	1.90 / 22.32	1.69 / 19.85	1.52 / 17.85	1.39 / 16.34	1.27 / 14.92	1.17 / 13.75	1.09 / 12.81	11.02
22	1.82 / 23.48	1.61 / 20.79	1.45 / 18.71	1.32 / 17.02	1.21 / 15.61	1.12 / 14.45	1.04 / 13.42	11.54
23	1.74 / 24.55	1.54 / 21.70	1.39 / 19.60	1.26 / 17.76	1.16 / 16.35	1.07 / 15.09	1.00 / 14.10	12.08
24	1.67 / 25.64	1.48 / 22.72	1.33 / 20.42	1.21 / 18.59	1.11 / 17.04	1.02 / 15.67	0.95 / 14.59	12.60
25	1.60 / 26.62	1.42 / 23.62	1.28 / 21.30	1.16 / 19.32	1.07 / 17.82	0.98 / 16.31	0.91 / 15.15	13.12
26	1.54 / 27.72	1.37 / 24.65	1.23 / 22.14	1.12 / 20.18	1.03 / 18.55	0.95 / 17.10	0.88 / 15.84	13.64
27	1.48 / 28.79	1.31 / 25.42	1.18 / 22.90	1.08 / 21.00	0.99 / 19.24	0.91 / 17.69	0.85 / 16.52	14.17
28	1.43 / 29.90	1.27 / 26.58	1.14 / 23.82	1.04 / 21.77	0.95 / 19.86	0.88 / 18.40	0.82 / 17.14	14.70
29	1.38 / 30.99	1.22 / 27.39	1.10 / 24.65	1.00 / 22.41	0.92 / 20.63	0.85 / 19.08	0.79 / 17.72	15.22
30	1.33 / 32.00	1.18 / 28.32	1.07 / 25.70	0.97 / 23.27	0.89 / 21.38	0.82 / 19.69	0.76 / 18.23	15.73
31	1.29 / 33.08	1.15 / 29.48	1.03 / 26.39	0.94 / 24.10	0.86 / 22.04	0.79 / 20.25	0.74 / 18.98	16.28
32	1.25 / 34.19	1.11 / 30.30	1.00 / 27.33	0.91 / 24.89	0.83 / 22.69	0.77 / 21.05	0.71 / 19.40	16.80
33	1.21 / 35.10	1.08 / 31.34	0.97 / 28.15	0.88 / 25.51	0.81 / 23.49	0.75 / 21.75	0.69 / 20.00	17.32

⅝" ∅ SPIRALS(3) — PITCH IN INCHES

Core Diam. Inches	2	2¼	2½	2¾	3	3¼	3½	F
12	5.17 / 19.88							9.84
13	4.77 / 21.50							10.66
14	4.43 / 23.10	3.94 / 20.59						11.47
15	4.14 / 24.88	3.68 / 22.09	3.31 / 19.85					12.29
16	3.88 / 26.44	3.45 / 23.53	3.10 / 21.14					13.11
17	3.65 / 28.10	3.24 / 24.98	2.92 / 22.48	2.65 / 20.40				13.93
18	3.45 / 29.80	3.06 / 26.43	2.76 / 23.83	2.51 / 21.68	2.30 / 19.88			14.74
19	3.26 / 31.39	2.90 / 27.90	2.61 / 25.10	2.37 / 22.80	2.17 / 20.89			15.58
20	3.10 / 33.10	2.76 / 29.45	2.48 / 26.48	2.25 / 24.00	2.07 / 22.10	1.91 / 20.40		16.39
21	2.95 / 34.66	2.62 / 30.80	2.36 / 27.75	2.15 / 25.28	1.97 / 23.17	1.82 / 21.40	1.69 / 19.85	17.21
22	2.82 / 36.40	2.51 / 32.40	2.26 / 29.19	2.05 / 26.46	1.88 / 24.25	1.73 / 22.30	1.61 / 20.79	18.03
23	2.69 / 37.98	2.40 / 33.86	2.16 / 30.45	1.96 / 27.62	1.80 / 25.39	1.66 / 23.40	1.54 / 21.70	18.84
24	2.58 / 39.60	2.30 / 35.31	2.07 / 31.80	1.88 / 28.87	1.72 / 26.41	1.59 / 24.41	1.48 / 22.72	19.68
25	2.48 / 41.30	2.21 / 36.80	1.98 / 33.00	1.80 / 30.00	1.65 / 27.48	1.53 / 25.48	1.42 / 23.62	20.48
26	2.38 / 42.80	2.12 / 38.20	1.91 / 34.40	1.74 / 31.34	1.59 / 28.62	1.47 / 26.46	1.36 / 24.48	21.30
27	2.29 / 44.50	2.04 / 39.62	1.84 / 35.79	1.67 / 32.42	1.53 / 29.72	1.41 / 27.40	1.31 / 25.42	22.12
28	2.21 / 46.20	1.97 / 41.20	1.77 / 37.00	1.61 / 33.68	1.47 / 30.78	1.37 / 28.63	1.27 / 26.58	22.95
29	2.14 / 48.00	1.90 / 42.60	1.71 / 38.36	1.55 / 34.79	1.43 / 32.09	1.32 / 29.80	1.22 / 27.39	23.79
30	2.07 / 49.70	1.84 / 44.20	1.65 / 39.60	1.50 / 36.00	1.38 / 33.10	1.27 / 30.49	1.18 / 28.32	24.60
31	2.00 / 51.24	1.78 / 45.60	1.60 / 41.00	1.46 / 37.41	1.33 / 34.09	1.23 / 31.54	1.14 / 29.21	25.40
32	1.94 / 53.05	1.72 / 47.00	1.55 / 42.35	1.41 / 38.52	1.29 / 35.24	1.19 / 32.50	1.11 / 30.30	26.22
33	1.88 / 54.55	1.67 / 48.49	1.50 / 43.50	1.37 / 39.78	1.25 / 36.25	1.16 / 33.65	1.08 / 31.54	27.08

Spirals above the stepped line are not collapsible for shipment

1 Spirals are sold on the basis of theoretical weights.

2 F — Weight to Add for Finishing. (This includes 1½ turns at top and 1½ turns at bottom of spiral—ACI 318-56.) Total Wt. = (Wt. per Ft. x Height) + (F) + Wt. of Spacers.

3 Two spacers are required for Spirals 20" or less in diameter, 3 spacers for Spirals 20" to 30", and 4 spacers for Spirals over 30" in diameter and for ALL ⅝" or larger spiral rods, regardless of spiral diameter. Allow ¾ pound per lin. ft. per spacer.

4 The four spiral rod sizes shown in the table have been approved by the U. S. Department of Commerce Simplified Practice Recommendation R53 and are carried in stock ready for fabrication.

Labor Placing Reinforcing Bars. The labor cost of placing reinforcing steel will vary with the average weight of the bars, the manner in which it is placed; i.e., whether bars may be laid loose or whether it is necessary to tie them in place, and upon the class of labor employed.

In most of the larger cities there are concerns that make a specialty of setting reinforcing steel. These men become very adept at this work and under favorable conditions can place steel for 25% less than the average iron worker. For this reason costs are given both ways and the estimator should use his own judgment as to the class of workmen obtainable.

Setting Reinforcing Bars. On jobs using No. 5 bars and smaller, where it is not necessary to tie them in place, a worker should handle and place 700 to 900 lbs. per 8-hr. day, at the following labor cost per ton:

	Hours	Rate	Total	Rate	Total
Iron Worker	20.7	$....	$....	$27.53	$569.87

Experienced reinforcing steel setters should handle and place 900 to 1,100 lbs. of steel per 8-hr. day, at the following labor cost per ton:

	Hours	Rate	Total	Rate	Total
Iron worker	16	$....	$....	$27.53	$440.48

On jobs using No. 6 bars and heavier, where it is not necessary to tie the bars in place, a man should handle and place 900 to 1,100 lbs. of steel per 8-hr. day, at the following labor cost per ton:

	Hours	Rate	Total	Rate	Total
Iron worker	16	$....	$....	$27.53	$440.48

An experienced reinforcing steel setter should handle and place 1,400 to 1,600 lbs. of steel per 8-hr. day, at the following labor cost per ton:

	Hours	Rate	Total	Rate	Total
Iron worker	10.6	$....	$....	$27.53	$291.82

Setting Reinforcing Bars Tied in Place. On jobs having concrete floors where it is necessary to tie the bars in place with tie wire, and where lightweight bars, (No. 5 and under) are used, a worker should handle, place, and tie 600 to 800 lbs. per 8-hr. day, at the following labor cost per ton:

	Hours	Rate	Total	Rate	Total
Iron worker	23	$....	$....	$27.53	$633.19

An experienced reinforcing steel setter should handle, place and tie 800 to 1,000 lbs. of steel per 8-hr. day, at the following labor cost per ton:

	Hours	Rate	Total	Rate	Total
Iron worker	17.8	$....	$....	$27.53	$490.03

On jobs using heavy bars, (No. 6 and over) a worker should handle, set and tie 800 to 1,000 lbs. of steel per 8-hr. day, at the following labor cost per ton:

	Hours	Rate	Total	Rate	Total
Iron worker	17.8	$....	$....	$27.53	$490.03

An experienced reinforcing steel setter should handle, set and tie 1,300 to 1,500 lbs. per 8-hr. day, at the following labor cost per ton:

	Hours	Rate	Total	Rate	Total
Iron worker	11.5	$....	$....	$27.53	$316.60

Epoxy Coated Reinforcing Bars

Epoxy coated reinforcement steel is used in many concrete structures today. The bars are coated, cut, and bent to the approved configuration of shop drawings at the fabrication and coating shop and then are shipped to the project. Coated reinforcement steel shall meet ASTM A775 designation for the coating.

The estimator should allow as a minimum about 60% additional cost to the base price of reinforcement steel for purchasing epoxy coated reinforcement steel. Additional time in the field is required for the extra care needed to unload and handle the coated steel. Additional care refers to items such as nylon slings for unloading epoxy coated steel bars instead of steel wire slings, not walking on the bars after installation, and possibly covering the steel prior to placement into the form. It is suggested that approximately 5% to 25% additional time be added for the placement of epoxy coated reinforcement steel.

All coated reinforcement bars should be tied with coated tie wire. Coated bar supports should be non-corrosive chairs.

Welded Steel Fabric or Mesh Reinforcing

Welded steel fabric is a popular and economical reinforcing for concrete work of all kinds, especially driveways and floors. It may also be used for temperature reinforcing, beam and column wrapping, road and pavement reinforcing, etc. It is usually furnished in square or rectangular mesh.

It is usually sold at a certain price per sq. ft. or sq. yd. depending upon the weight.

Prices of mesh vary with quantity ordered and freight from mill to destination. Typical prices for some styles usually stocked, with truck job site delivery are as follows: Approximate prices are per 100 sq. ft.

Style	Wt. per 100 SF	Under 5 Tons
6x6-6x6 (W2.9)	42 lbs.	$17.00
6x6-8x8 (W2.1)	30 lbs.	12.00
4x4-8x8 (W2.1)	44 lbs.	17.00

Labor Placing Mesh Reinforcing In Walls. When mesh reinforcing is used, furnished in sheets or rolls, a worker should place 700 to 800 sq. ft. in walls per 8-hr. day, at the following labor cost per 100 sq. ft.:

	Hours	Rate	Total	Rate	Total
Steel setter	1.0	$....	$....	$27.53	$27.53
Cost per sq. ft			....		.28

Sizes and Weights of Welded Steel Fabric

Style	Wt. per 100 SF Based on	Spacing of Wires in Inches		A.S. & W. Co. Steel Wire Gauge Number		Sect. Areas Sq. In. per Ft.	
	60" Width	Longit.	Trans.	Longit.	Trans.	Longit.	Trans.
22-1616	13	2	2	16	16	.018	.018
22-1414	21	2	2	14	14	.030	.030
22-1313	28	2	2	13	13	.039	.039
22-1212	37	2	2	12	12	.052	.052
22-1111	48	2	2	11	11	.068	.068
22-1010	60	2	2	10	10	.086	.086
24-1414	16	2	4	14	14	.030	.015
24-1314	19	2	4	13	14	.039	.015
24-1212	28	2	4	12	12	.052	.026
212-38	105	2	12	3	8	.280	.021
212-06	166	2	12	0	6	.443	.029
216-812	46	2	16	8	12	.124	.007
216-711	55	2	16	7	11	.148	.008
216-610	65	2	16	6	10	.174	.011
216-510	75	2	16	5	10	.202	.011
216-49	89	2	16	4	9	.239	.013
216-38	104	2	16	3	8	.280	.015
216-28	119	2	16	2	8	.325	.015
216-17	139	2	16	1	7	.377	.018
33-1414	14	3	3	14	14	.020	.020
33-1212	25	3	3	12	12	.035	.035
33-1111	32	3	3	11	11	.046	.046
33-1010	41	3	3	10	10	.057	.057
33-99	49	3	3	9	9	.069	.069
33-88	58	3	3	8	8	.082	.082
316-812	32	3	16	8	12	.082	.007
316-711	38	3	16	7	11	.098	.009
316-610	45	3	16	6	10	.116	.011
316-510	52	3	16	5	10	.135	.011
316-49	61	3	16	4	9	.159	.013
316-38	72	3	16	3	8	.187	.015
316-28	83	3	16	2	8	.216	.015
316-17	96	3	16	1	7	.252	.018
316-06	113	3	16	0	6	.295	.022
44-1414	11	4	4	14	14	.015	.015
44-1313	14	4	4	13	13	.020	.020
44-1212	19	4	4	12	12	.026	.026
44-1010	31	4	4	10	10	.043	.043
44-88	44	4	4	8	8	.062	.062
44-77	53	4	4	7	7	.074	.074
44-66	62	4	4	6	6	.087	.087
44-44	85	4	4	4	4	.120	.120
48-1313	11	4	8	13	13	.020	.010
48-1214	12	4	8	12	14	.026	.008
48-1212	14	4	8	12	12	.026	.013
48-1112	17	4	8	11	12	.034	.013
48-1012	20	4	8	10	12	.043	.013
48-912	23	4	8	9	12	.052	.013
48-812	27	4	8	8	12	.062	.013
48-711	33	4	8	7	11	.074	.017

Style	Wt. per 100 SF Based on 60" Width	Spacing of Wires in Inches Longit.	Spacing of Wires in Inches Trans.	A.S. & W. Co. Steel Wire Gauge Number Longit.	A.S. & W. Co. Steel Wire Gauge Number Trans.	Sect. Areas Sq. In. per Ft. Longit.	Sect. Areas Sq. In. per Ft. Trans.
412-1212	13	4	12	12	12	.026	.009
412-1112	16	4	12	11	12	.034	.009
412-1012	19	4	12	10	12	.043	.009
412-912	22	4	12	9	12	.052	.009
412-812	25	4	12	8	12	.062	.009
412-711	31	4	12	7	11	.074	.011
412-610	36	4	12	6	10	.087	.014
412-510	42	4	12	5	10	.101	.014
412-57	45	4	12	5	7	.101	.025
412-49	49	4	12	4	9	.120	.017
416-1012	18	4	16	10	12	.043	.007
416-912	21	4	16	9	12	.052	.007
416-812	25	4	16	8	12	.062	.007
416-711	30	4	16	7	11	.074	.009
416-610	35	4	16	6	10	.087	.011
416-510	40	4	16	5	10	.101	.011
416-49	48	4	16	4	9	.120	.013
416-38	56	4	16	3	8	.140	.015
416-28	64	4	16	2	8	.162	.015
66-1212	13	6	6	12	12	.017	.017
66-1010	21	6	6	10	10	.029	.029
66-99	25	6	6	9	9	.035	.035
66-88	30	6	6	8	8	.041	.041
66-77	36	6	6	7	7	.049	.049
66-66	42	6	6	6	6	.058	.058
66-55	49	6	6	5	5	.067	.067
66-46	50	6	6	4	6	.080	.058
66-44	58	6	6	4	4	.080	.080
66-33	68	6	6	3	3	.093	.093
66-22	78	6	6	2	2	.108	.108
66-11	91	6	6	1	1	.126	.126
66-00	107	6	6	0	0	.148	.148
612-77	27	6	12	7	7	.049	.025
612-66	32	6	12	6	6	.058	.029
612-55	37	6	12	5	5	.067	.034
612-44	44	6	12	4	4	.080	.040
612-33	51	6	12	3	3	.093	.047
612-25	52	6	12	2	5	.108	.034
612-22	59	6	12	2	2	.108	.054
612-17	56	6	12	1	7	.126	.025
612-14	61	6	12	1	4	.126	.040
612-11	69	6	12	1	1	.126	.063
612-06	65	6	12	0	6	.148	.029
612-03	72	6	12	0	3	.148	.047
612-00	81	6	12	0	0	.148	.074

Sometimes, a style such as 66-44 is shown as 6x6-4/4.

Labor Placing Steel Fabric on Floors. When used for reinforcing floor slabs or for temperature reinforcing, a worker should place 1,400 to 1,600 sq. ft. of fabric per 8-hr. day, at the following labor cost per 100 sq. ft.:

	Hours	Rate	Total	Rate	Total
Iron worker	0.53	$....	$....	$27.53	$14.59
Cost per sq. ft			....		.15

Reinforcement For Architectural Concrete Walls. It is practically impossible to provide sufficient reinforcement to prevent cracking in a long wall, which is why control joints are used. But it is still necessary to provide adequate reinforcement between joints when properly spaced to resist volume change stresses. To be most effective, small bars with a type of deformation to give maximum bond should be placed relatively close together rather than using larger bars at wider spacings. Generally #3 and #4 bars are preferable to larger sizes and the horizontal reinforcement should be at least equal to 0.25 percent of wall area. In the vertical direction 0.15% reinforcement is sufficient.

The size and spacing of bars recommended for walls of various thickness are given below:

Wall Thickness (In.)	Horizontal reinforcement	Vertical reinforcement
(in outside face of wall)		
6	#3 @ 8"	#3 @ 8"
8	#3 @ 6"	#3 @ 8"
(in both faces of wall)		
10	#3 @ 10"	#3 @ 12"
12	#3 @ 8"	#3 @ 12"

Recommended Size and Spacing of Reinforcing Bars in Concrete Walls.

The bars should be tied at each intersection with a non-slip tie, and where two curtains of steel are used, it is advisable to include a detail drawing to show how the curtains are to be fastened together with tie spreaders at frequent intervals to make a rigid web of reinforcement.

Difficulty is sometimes experienced in placing concrete in walls because of a congestion of reinforcing bars and this condition should be avoided insofar as practicable.

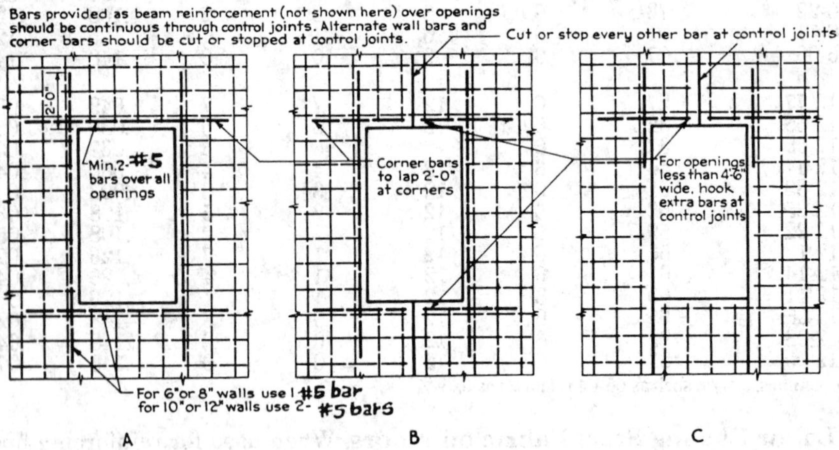

Alternate Methods of Placing Reinforcing around Openings

Labor Placing Reinforcing Bars. The labor cost of placing reinforcing steel in architectural concrete walls is expensive on account of the small size of the rods, the fact they run both horizontally and vertically, must be wired together, and the cramped quarters for the work.

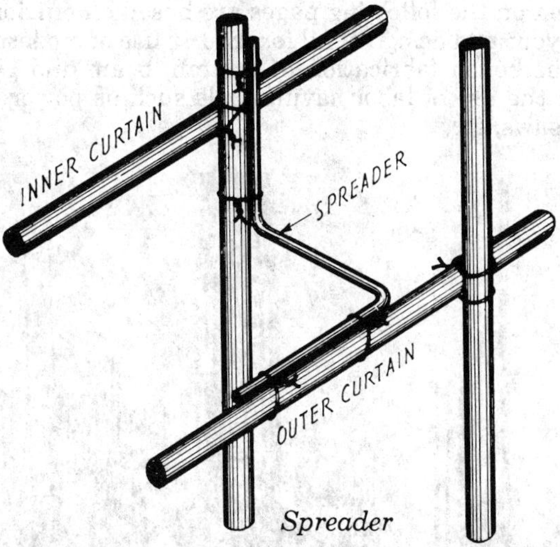

Spreader

On walls 8 to 12 inches thick, a reinforcing steel setter will handle, place, and wire 750 to 850 lbs. of reinforcing steel per 8-hr. day, at the following labor cost per ton:

	Hours	Rate	Total	Rate	Total
Iron worker	20	$....	$....	$27.53	$550.60

03300 CAST-IN-PLACE CONCRETE

Reinforced and architectural concrete are among our most common building materials and are used on virtually all types of construction work.

They are used in all types of structures, from residential, apartment, hotel and hospital buildings, requiring comparatively light floor loads, to the heaviest constructed warehouses, factory buildings, wharves, and industrial buildings.

In fact, entire buildings are constructed of concrete, including foundations, foundation walls, columns, beams and girders, floor slabs, ornamental exterior walls, cornices, etc.

Regardless of the fact that it is one of our most common building materials, costs on all types of reinforced concrete construction are subject to wide variation. This is due to a number of causes, such as the function and size of building, whether flat slab or beam and girder type, whether solid concrete construction or used in conjunction with steel, building codes, insurance requirements, etc.

The methods used in erecting and moving the formwork, handling, mixing and placing the concrete, and the ability of the superintendent or foreman, all affect the final cost of the work.

Regardless of how carefully or accurately the job has been estimated, if the superintendent or foreman in charge of the work are not proficient, the actual costs are almost certain to run higher than the estimate--not because the work was estimated too low but because it was poorly handled on the job.

All labor costs given on the following pages are based on efficient management; economical layout and design of all formwork; use of modern labor-saving methods, such as bench fabrication of column, beam and girder forms where possible, and the use of labor saving tools such as power saws, band saws, electric hand saws, etc.

Courtesy Symons Corporation

Setting Rebars Prior to Placing Forms for Cast-in-place Concrete Wall

In these days of high wage scales, the efficient contractor must use methods and tools that will reduce costs to a minimum, or else he may be out of the running when it comes to obtaining work in competition and of performing it economically after he has obtained the job. Remember, good management, economical design, labor-saving tools and methods are absolutely necessary to a profitable contract.

Items to be Included in the Estimate. Reinforced concrete should be estimated in the same manner and in the same units as the work is constructed on the job. This allows comparisons between estimated and actual costs and enables a contractor to tell where variations occur between estimated and actual costs.

The first operation on any reinforced concrete job is the erection of the temporary wood or metal forms to support the weight of the wet concrete until it has set sufficiently to be self-supporting. The members usually requiring forms are the columns, girders, beams, spandrel beams, lintels, floor slabs, stairs and platforms. Each class of formwork should be estimated separately and in detail.

All types of form work, including columns, beams, girders and the various type of floor slabs have been designed in the best engineering practice, giving lumber requirements, clamp spacing for columns, beam and girder lumber requirements, including shoring, together with a complete set of design tables for all types of concrete floor slabs, giving joist spacing, spans of stringers and shores required for any floor load.

After the forms are erected, it is necessary to place the reinforcing steel or wire mesh ready to receive the concrete. This may include unloading steel from trucks at the job, cutting steel to length and bending to special shapes for columns, beams, girders and floor slabs, and placing and tying it in position ready to receive the concrete.

After the reinforcing steel is in place, the job is then ready for concrete placement. This consists of preparing runways, unloading cement and other materials, mixing the cement, sand and gravel into the concrete mass, hoisting, wheeling and placing it in position, removing runways, removing old concrete drippings and spillage, patching, finishing, cleaning and washing wheelbarrows or buggies, and concrete mixer.

Each of these items should be estimated in detail to include all material and labor operations necessary to make up the completed unit.

Weights of Miscellaneous Building Materials
Weight of Cement, Sand, Gravel, Crushed Stone, Concrete, Etc.

The following table gives the approximate weights of cement, concrete, and concrete aggregates, and will assist the estimator in figuring freight or when purchasing materials by weight:

	Weight Lbs. per Cu. Yd.
Bank sand (dry)	2,500
Torpedo sand	2,700
Crushed stone	2,500
Crushed stone screenings	2,500
Gravel	2,700
Roofing gravel	2,700

Cement weighs 376 lbs. per bbl.
(1-sack weighs 94 lbs. and bulks 1-cu. ft.)

Cinder concrete, wt. per cu. ft. 112
Concrete of gravel, limestone, sandstone,
trap rock, etc., wt. per cu. ft. 150

Quantity of Pit Run Sand and Gravel Required for One Cubic Yard of Concrete. Pit run sand and gravel is not recommended for concrete work where strength is a factor, because it is impossible to grade the aggregate. There is either too much gravel and too little sand or too little gravel and too much sand, usually the latter, to make a dependable concrete.

An aggregate containing approximately 50% sand and 50% gravel (providing it does not contain any humic acid), may be considered acceptable aggregate, and the strength will probably be only 10% less than when using aggregates furnished by washing and grading plants.

It ordinarily requires about 1.25 cu. yds. of aggregate per cu. yd. of concrete, depending upon the coarseness of the sand.

CONCRETE

Concrete should be estimated by the cu. ft. or cu. yd. containing 27 cu. ft. The cu. yd. is the most convenient unit to use because practically all published tables giving quantities of cement, sand, gravel, or crushed stone are based on the cu. yd. of concrete. The cu. yd. is also a convenient unit to use when estimating labor costs.

It was formerly the practice to specify concrete mixtures as so many volumes of fine and coarse aggregate to each volume of cement, as for example 1:1:2, 1:1.5:3, 1:2:4, 1:3:5, etc. The condition of the aggregate when measured was not specified, nor were the consistency of the concrete or the water content.

As used on the job, sand nearly always contains some surface moisture, usually about 4 or 5 percent by weight of the sand, although it may be considerably more just after rain or if the sand is used soon after washing it. The moisture causes the sand to bulk up in volume, the amount of bulking depending on the amount of moisture and the grading. This bulking which may be 20 to 30 percent or more, reduces the true volume of surface dry sand in a given volume of damp sand. For example, if a 1:2:4 mix is specified and 2 volumes of damp sand are measured, the volume of sand on a surface dry basis would be less. Suppose the sand is bulked 25 percent due to the moisture. Then the volume of dry sand in each cubic foot of damp sand will be 100 divided by 125, or 0.8 cu. ft. of damp sand would give us 2 x 0.8 or 1.6 cu. ft. dry sand.

The coarse aggregate carries very little surface moisture and the moisture has little effect on the volume. The result was that undersanded mixes were often used. In the above example the 1:2:4 mix if based on damp materials would actually be a 1:1.6:4 mixture of surface dry materials. The amount of sand in this case would be less than 29 percent of the total volume of aggregate. For average mixtures 35 to 45 percent is usually desirable for good workability when concrete is placed by hand.

To correct for the bulking in the above example the amount of damp sand should be increased to 1.25 x 2 or 2.5 cu. ft. and the mix would then be 1:2.5:4 based on damp materials as measured in the field.

As a result of the increased knowledge that has been made available, other methods of specifying concrete have been adopted, such as a mix by weight, a minimum cement factor, a maximum water content or the strength at a given age, or a combination of these.

The Importance of Mixing Water. The hardening of concrete mixtures is brought about by chemical reactions between cement and water, the aggregates (sand, gravel, stone, etc.) being inactive ingredients used as fillers. So long as workable mixtures are used, the less water there is in the mix, the stronger, more watertight and more durable will be the concrete. Excess mixing water dilutes the paste made by the cement and water and makes weaker, more porous concrete. For any set of conditions of mixing, placing and curing and for given materials there is a definite relation between the strength of the concrete and the amount of water used in mixing. This relation can be determined by test and a water-cement ratio-strength curve can be developed representing the actual results on a specific job.

American Concrete Institute (ACI) Standard for Selecting Proportions for Concrete Mixes

Committee 211 of the American Concrete Institute recommends certain practices for selecting proportions for concrete mixes. Following are extracts from this Standard which should be of value to all users of concrete.

Basic Relationship

Concrete proportions must be selected to provide necessary placeability, strength, and durability for the particular application. Well established relationships governing these properties are discussed briefly below.

Placeability (including satisfactory finishing properties) encompasses traits loosely accumulated in the terms "workability" and "consistency". For the purpose of this discussion, *workability* is considered to be that property of concrete which determines its capacity to be placed and consolidated properly and to be finished without harmful segregation. It embodies such concepts as moldability, cohesiveness, and compactability. It is affected by the grading, particle shape and proportions of aggregate, the amount of cement, the presence of entrained air, admixtures, and the consistency of the mixture. Procedures in this recommended practice permit these factors to be taken into account to achieve satisfactory placeability economically.

Consistency, loosely defined, is the wetness of the concrete mixture. It is measured in terms of slump--the higher the slump the wetter the mixture--and it affects the ease with which the concrete will flow during placement. It is related to, but not synonymous with, workability. In properly proportioned concrete, the unit water content required to produce a given slump will depend on several factors. Water requirement increases as aggregates become more angular and rough textured (but this disadvantage may be offset by improvements in other characteristics such as bond to cement paste). Required mixing water decreases as the maximum size of well graded aggregate is increased. It also decreases with the entrainment of air. Mixing water requirement may often be significantly reduced by certain admixtures.

Strength. Strength is an important characteristic of concrete, but other characteristics such as durability, permeability, and wear resistance are often equally or more important. These may be related to strength in a general way but are also affected by factors not significantly associated with strength. For a given set of materials and conditions, concrete strength is determined by the net quantity of water used per unit quantity of cement. The net water content excludes water absorbed by the aggregates. Differences in strength, for a given water-cement ratio may result from changes in: maximum size of aggregate;

grading, surface texture, shape, strength, and stiffness of aggregate particles; differences in cement types and sources; air content; and the use of admixtures which affect the cement hydration process or develop cementitious properties themselves. To the extent that these effects are predictable in the general sense, they are taken into account in this recommended practice. However, in view of their number and complexity, it should be obvious that accurate predictions of strength must be based on trial batches or experience with the materials to be used.

Durability. Concrete must be able to endure those exposures which may deprive it of its serviceability--freezing and thawing, wetting and drying, heating and cooling, chemicals, de-icing agents, and the like. Resistance to some of these may be enhanced by use of special ingredients: low-alkali cement, pozzolans, or selected aggregate to prevent harmful expansion due to the alkali-aggregate reaction which occurs in some areas when concrete is exposed in a moist environment; sulfate resisting cement or pozzolans for concrete exposed to sea water or sulfate-bearing soils; or aggregate free of excessive soft particles where resistance to surface abrasion is required. Use of a low water-cement ratio will prolong the life of concrete by reducing the penetration of aggressive liquids. Resistance to severe weathering, particularly freezing and thawing, and to salts used for ice removal is greatly improved by incorporation of a proper distribution of entrained air. Entrained air should be used in all exposed concrete in climates where freezing occurs.

Background Data

To the extent possible, selection of concrete proportions should be based on test data or experience with the materials actually to be used. Where such background is limited or not available, estimates given in this recommended practice may be employed.

The following information for available materials will be useful:

a) Sieve analyses of fine and coarse aggregates.
b) Unit weight of coarse aggregate.
c) Bulk specific gravities and absorptions of aggregates.
d) Mixing water requirements of concrete developed from experience with available aggregates.
e) Relationships between strength and water-cement ratio for available combinations of cement and aggregate.

Estimates from Tables 3 and 4 may be used when the last two items of information are not available. As will be shown, proportions can be estimated without the knowledge of aggregate specific gravity and absorption (third item above).

Procedure

Estimating the required batch weights for the concrete involves a sequence of logical, straightforward steps, which in effect fit the characteristics of the available materials into a mixture suitable for the work. The question of suitability is frequently not left to the individual selecting the proportions. The job specifications may dictate some or all of the following:

Maximum water-cement ratio	Slump
Minimum cement content	Maximum size of aggregate
Air content	Strength

Other requirements relating to such things as strength over-design, admixtures, and special types of cement or aggregate

Regardless of whether the concrete characteristics are prescribed by the specifications or are left to the individual selecting the proportions, establishment of batch weights per cubic yard of concrete can best be accomplished in the following sequence:

Step 1. Choice of slump. If slump is not specified, a value appropriate for the work can be selected from Table 1. The slump ranges shown apply when vibration is used to consolidate the concrete. Mixes of the stiffest consistency that can be placed efficiently should be used.

Table 1--Recommended Slumps
For Various Types of Construction

Type of construction	Slump, in. Maximum	Minimum
Reinforced foundation walls and footings	3	1
Plain footings, caissons, and substructure walls	3	1
Beams and reinforced walls	4	1
Building columns	4	1
Pavements and slabs	3	1
Heavy mass concrete	3	1

*May be increased 1 in. for methods of consolidation other than vibration.

Step 2. Choice of maximum size of aggregate. Large maximum sizes of well graded aggregates have less voids than smaller sizes. Hence, concretes with the larger-sized aggregates require less mortar per unit volume of concrete. Generally, the maximum size of aggregate should be the largest that is economically available and consistent with dimensions of the structure. Ordinarily, the ratio of the nominal maximum aggregate size to the minimum dimension within which concrete must be placed should not exceed the value shown in Table 2.

Table 2--Recommended Maximum Ratio
of Nominal Aggregate Size to Minimum
Criterion Dimension

Criterion dimension		Recomm. max. ratio
Narrowest dimension between sides of forms		1/5
Depth of slab		1/3
Clear spacing between individual reinforcing bars, bundles of bars, or pretensioning strands	In cols.	2/3
	In all other members	3/4

These limitations are sometimes waived if workability and methods of consolidation are such that the concrete can be placed without honeycomb or void. When high strength concrete is desired, best results may be obtained with reduced maximum sizes of aggregate since these produce higher strengths at a given water-cement ratio.

Step 3. Estimation of mixing water and air content. The quantity of water per unit volume of concrete required to produce a given slump is dependent on the maximum size, particle shape and grading of the aggregates, and on the amount of entrained air. It is not greatly affected by the quantity of cement. Table 3 provides estimates of required mixing water for concretes made with various maximum sizes of aggregate, with and without air entrainment. Depending on aggregate texture and shape, mixing water requirements may be somewhat above or below the tabulated values, but they are sufficiently accurate for the first estimate. Such differences in water demand are not necessarily reflected in strength since other compensating factors may be involved. For example, a rounded and an angular coarse aggregate, both well and similarly graded and of good quality, can be expected to produce concrete of about the same compressive strength for the same cement factor in spite of differences in water-cement ratio resulting from the different mixing water requirements. Particle shape per se is not an indicator that an aggregate will be either above or below average in its strength-producing capacity.

Table 3 indicates the approximate amount of entrapped air to be expected in non-air-entrained concrete, and shows the recommended levels of average air content for concrete in which air is to be purposely entrained for durability. Air-entrained concrete should always be used for structures which will be exposed to freezing and thawing, and generally for structures exposed to sea water or sulfates. When severe exposure is not anticipated, beneficial effects of air entrainment on concrete workability and cohesiveness can be achieved at air content levels approximately half those shown for air-entrained concrete.

When trial batches are used to establish strength relationships or verify strength-producing capability of a mixture, the least favorable combination of mixing water and air content should be used. This is, the air content should be the maximum permitted or likely to occur, and the concrete should be gauged to the highest permissible slump. This will avoid developing an over-optimistic estimate of strength on the assumption that average rather than extreme conditions will prevail in the field. For information on air content recommendations, see ACI 201, 301, and 302.

Table 3--Approximate Mixing Water and Air Content Requirements for Different Slumps and Maximum Sizes of Aggregates*

Slump, in.	Water, lb. per cu. yd. of concrete for indicated maximum sizes of aggregate 3/8"	1/2"	3/4"	1"	1 1/2"	2"	3"
			Non-air-entrained concrete				
1 to 2	350	335	315	300	275	260	240
3 to 4	385	365	340	325	300	285	265
6 to 7	410	385	360	340	315	300	285
Air content, percent	3	2.5	2	1.5	1	0.5	0.3
			Air-entrained concrete				
1 to 2	305	295	280	270	250	240	225
3 to 4	340	325	305	295	275	265	250
6 to 7	365	345	325	310	290	280	270
Air content, percent	8	7	6	5	4.5	4	3.5

*These quantities of mixing water are for use in computing cement factors for trial batches. They are maxima for reasonably well-shaped angular coarse aggregates graded within limits of accepted specifications.

Step 4. Selection of water-cement ratio. The required water-cement ratio is determined not only by strength requirements but also by factors such as durability and finishing properties. Since different aggregates and cements generally produce different strengths at the same water-cement ratio, it is highly desirable to have or develop the relationship between strength and water-cement ratio for the materials actually to be used. In the absence of such data, approximate and relatively conservative values for concrete containing Type I portland cement can be taken from Table 4(a). With typical materials, the tabulated water-cement ratios should produce the strengths shown, based on 28-day tests of specimens cured under standard laboratory conditions. The average strength selected must, of course, exceed the specified strength by a sufficient margin to keep the number of low tests within specified limits.

For severe conditions of exposure, the water-cement ratio should be kept low even though strength requirements may be met with a higher value. Table 4(b) gives limiting values.

Table 4(a)--Relationships Between Water-Cement Ratio and Compressive Strength of Concrete

	Water-cement ratio, by weight	
Compressive strength* at 28 days, psi	Non-air-entrained concrete	Air-entrained concrete
6000	0.41	--
5000	0.48	0.40
4000	0.57	0.48
3000	0.68	0.59
2000	0.82	0.74

*Values are estimated average strengths for concrete containing not more than the percentage of air shown in Table 3. For a constant water-cement ratio, the strength of concrete is reduced as the air content is increased.

Strength is based on 6 x 12 in. cylinders moist-cured 28 days at 73.4 ± 3 F (23 ± 1.7 C) in accordance with Section 9(b) of ASTM C 31 for method of Making and Curing Concrete Compression and Flexure Test Specimens in the Field.

Relationship assumes maximum size of aggregate about 3/4 to 1 in.; for a given source, strength produced for a given water-cement ratio will increase as maximum size of aggregate decreases.

Table 4(b)--Maximum Permissible Water-Cement Ratios for Concrete in Severe Exposures*

Type of structure	Structure wet continuously or frequently and exposed to freezing† and thawing	Structure exposed to sea water or sulfates
Thin sections (railings, curbs, sills, ledges, ornamental work) and sections with less than 1 in. cover over steel	0.45	0.40**
All other structures	0.50	0.45***

*Based on report of ACI Committee 201, "Durability of Concrete in Service", previously cited.

**Concrete should also be air-entrained.

***If sulfate resisting cement (Type II or Type V of ASTM C 150) is used, permissible water-cement ratio may be increased by 0.05.

Step 5. Calculation of cement content. The amount of cement per unit volume of concrete is fixed by the determinations made in Steps 3 and 4 above. The required cement is equal to the estimated mixing water content (Step 3) divided by the water-cement ratio (Step 4). If, however, the specification includes a separate minimum limit on cement in addition to requirements for strength and durability, the mixture must be based on whichever criterion leads to the larger amount of cement.

Step 6. Estimation of coarse aggregate content. Aggregates of essentially the same maximum size and grading will produce concrete of satisfactory workability when a given volume of coarse aggregate, on a dry-rodded basis, is used per unit volume of concrete. Appropriate values for this aggregate volume are given in Table 6. It can be seen that, for equal workability, the volume of coarse aggregate in a unit volume of concrete is dependent only on its maximum size and the fineness modulus of the fine aggregate. Differences in the amount of mortar required for workability with different aggregates, due to differences in particle shape and grading, are compensated for automatically by differences in dry-rodded void content.

The volume of aggregate, in cubic feet, on a dry-rodded basis, for a cubic yard of concrete is equal to the value from Table 6 multiplied by 27. This volume is converted to dry weight of coarse aggregate required in a cubic yard of concrete by multiplying it by the dry-rodded weight per cubic foot of the coarse aggregate.

Table 6--Volume of Coarse Aggregate
Per Unit of Volume of Concrete

Maximum size of aggregate, in.	Volume of dry-rodded coarse aggregate per unit volume of concrete for different fineness moduli of sand			
	2.40	2.60	2.80	3.00
3/8	0.50	0.48	0.46	0.44
1/2	0.59	0.57	0.55	0.53
3/4	0.66	0.64	0.62	0.60
1	0.71	0.69	0.67	0.65
1 1/2	0.75	0.73	0.71	0.69
2	0.78	0.76	0.74	0.72
3	0.82	0.80	0.78	0.76

*Volumes are based on aggregates in dry-rodded condition as described in ASTM C 29 for Unit Weight and voids in Aggregate.

These volumes are selected from empirical relationships to produce concrete with a degree of workability suitable for usual reinforced construction. For less workable concrete such as required for concrete pavement construction they may be increased about 10 percent. When placement is to be by pump, they should be reduced about 10 percent.

Step 7. Estimation of fine aggregate content. At completion of Step 6, all ingredients of the concrete have been estimated except the fine aggregate. Its quantity is determined by difference. Either of two procedures may be employed: the "weight" method or the "absolute volume" method.

If the weight of the concrete per unit volume is assumed or can be estimated from experience, the required weight of fine aggregate is simply the difference between the weight of fresh concrete and the total weight of the other ingredients. Often the unit weight of concrete is known with reasonable accuracy from previous experience with the materials. In the absence of such information, Table 7 can be used to make a first estimate. Even if the estimate of concrete weight per cubic yard is rough, mixture proportions will be sufficiently accurate to permit easy adjustment on the basis of trial batches as will be shown in the examples.

If a theoretically exact calculation of fresh concrete weight per cubic yard is desired, the following formula can be used:

$$U = 16.85\, Ga\, (100 - A) + C(1 - Ga/Gc) - W(Gc - 1)$$

where

U = weight of fresh concrete per cubic yard, lb.
Ga = weighted average specific gravity of combined fine and coarse aggregate, bulk SSD*
Gc = specific gravity of cement (generally 3.15)
A = air content, percent
W = mixing water requirement, lb. per cu. yd.
C = cement requirement, lb. per cu. yd.

*SSD indicates saturated-surface-dry basis used in considering aggregate displacement. The aggregate specific gravity used in calculations must be consistent with the moisture condition assumed in the basic aggregate batch weights--i.e., bulk dry if aggregate weights are stated on a dry basis, and bulk SSD if weights are stated on a saturated-surface-dry basis.

A more exact procedure for calculating the required amount of fine aggregate involves the use of volumes displaced by the ingredients. In this case, the total volume displaced by the known ingredients--water, air, cement, and coarse aggregate--is subtracted from the unit volume of concrete to obtain the required volume of fine aggregate. The volume occupied in concrete by any ingredient is equal to its weight divided by the density of that material (the latter being the product of the unit weight of water and the specific gravity of the material).

Step 8. Adjustments for aggregate moisture. The aggregate quantities actually to be weighed out for the concrete must allow for moisture in the aggregates. Generally, the aggregates will be moist and their dry weights should be increased by the percentage of water they contain, both absorbed and surface. The mixing water added to the batch must be reduced by an amount equal to the free moisture contributed by the aggregate--i.e., total moisture minus absorption.

Table 7--First Estimate of Weight of Fresh Concrete

Maximum size, of aggregate, in.	First estimate of concrete weight, lb. per cu. yd. — Non-air-entrained concrete	Air-entrained concrete
3/8	3840	3690
1/2	3890	3760
3/4	3960	3840
1	4010	3900
1 1/2	4070	3960
2	4120	4000
3	4160	4040

*Values calculated by Eq. 7 for concrete of medium richness (550 lb. of cement per cu. yd.) and medium slump with aggregate specific gravity of 2.7. Water requirements based on values for 3 to 4 in. slump in Table 3. If desired, the estimated weight may be refined as follows if necessary information is available: for each 10 lb. difference in mixing water from the Table 3 values for 3 to 4 in. slump, correct the weight per cu. yd. 15 lb. in the opposite direction; for each 100 lb. difference in cement content from 550 lb., correct the weight per cu. yd. 15 lb. in the same direction; for each 0.1 by which aggregate specific gravity deviates from 2.7, correct the concrete weight 100 lb. in the same direction.

Step 9. Trial batch adjustments. The calculated mixture proportions should be checked by means of trial batches prepared and tested in accordance with ASTM C 192, "Making and Curing Concrete Compression and Flexure Test Specimens in the Laboratory", or full-sized field batches. Only sufficient water should be used to produce the required slump regardless of the amount assumed in selecting the trial proportions. The concrete should be checked for unit weight and yield (ASTM C 138) and for air content (ASTM C 138, C 172, or C 231). It should also be carefully observed for proper workability, freedom from segregation, and finishing properties. Appropriate adjustments should be made in the proportions for subsequent batches in accordance with the following procedure.

Re-estimate the required mixing water per cubic yard of concrete by multiplying the net mixing water content of the trial batch by 27 and dividing the product by the yield of the trial batch in cubic feet. If the slump of the trial batch was not correct, increase or decrease the re-estimated amount of water by 10 lb. for each required increase or decrease of 1 in. in slump.

If the desired air content (for air-entrained concrete) was not achieved, re-estimate the admixture content required for proper air content and reduce or increase the mixing water content of the above paragraph by 5 lb. for each 1 percent by which the air content is to be increased or decreased from that of the previous trial batch.

If estimated weight per cubic yard of fresh concrete is the basis for proportioning, re-estimate that weight by multiplying the unit weight in pounds per cubic foot of the trial batch by 27 and reducing or increasing the result by the anticipated percentage increase or decrease in air content of the adjusted batch from the first trial batch.

Calculate new batch weights starting with Step 4, modifying the volume of coarse aggregate from Table 6 if necessary to provide proper workability.

Sample Computations

Two sample problems illustrate application of the proportioning procedures. The following conditions are assumed:

1) Type I non-air-entraining cement will be used and its specific gravity is assumed to be 3.15. The specific gravity values are not used if proportions are selected to provide a weight of concrete assumed to occupy 1 cu. yd.
2) Coarse and fine aggregates in each case are of satisfactory quality and are graded within limits of generally accepted specifications, such as the "Specifications for Concrete Aggregates" (ASTM C 33).
3) The coarse aggregate has a bulk specific gravity of 2.68 and an absorption of 0.5 percent.
4) The fine aggregate has a bulk specific gravity of 2.64, an absorption of 0.7 percent, and fineness modulus of 2.8.

Example 1. Concrete is required for a portion of a structure which will be below ground level in a location where it will not be exposed to severe weathering or sulfate attack. Structural considerations require it to have an average 28-day compressive strength of 3500 psi.

This is not the specified strength used for structural design but a higher figure expected to be produced on the average. For the method of determining the amount by which average strength should exceed design strength, see "Recommended Practice for Evaluation of Compression Test Results of Field Concrete" (ACI 214-65).

On the basis of information in Table 1 as well as previous experience, it is determined that under the conditions of placement to be employed, a slump of 3 to 4 in. be used and that the available No. 4 to 1 ½-in. coarse aggregate will be suitable. The dry-rodded weight of coarse aggregate is found to be 100 lb. per cu. ft. Employing the sequence outlined in the Procedure section, the quantities of ingredients per cubic yard of concrete are calculated as follows:

Step 1. As indicated above, the desired slump is 3 to 4 in.

Step 2. The locally available aggregate graded from No. 4 to 1 ½ in., has been indicated as suitable.

Step 3. Since the structure will not be exposed to severe weathering, non-air-entrained concrete will be used. The approximate amount of mixing water to produce a 3-to 4-in. slump in non-air-entrained concrete with 1 ½-in. aggregate is found from Table 3 to be 300 lb. per cu. yd. Estimated entrapped air is shown as 1 percent.

Step 4. From Table 4(a), the water-cement ratio needed to produce a strength of 3500 psi in non-air-entrained concrete is found to be about 0.62.

Step 5. From the information derived in Steps 3 and 4, the required cement content is found to be 300/0.62 = 484 lb. per cu. yd.

Step 6. The quantity of coarse aggregate is estimated from Table 6. For a fine aggregate having a fineness modulus of 2.8 and a 1 ½ in. maximum size of coarse aggregate, the table indicates that 0.71 cu. ft. of coarse aggregate, on a dry-rodded basis, may be used in each cubic foot of concrete. For a cubic yard, therefore, the coarse aggregate will be 27 x 0.71 = 19.17 cu. ft. Since it weighs 100 lb. per cu. ft. the dry weight of coarse aggregate is 1,917 lb.

Step 7. With the quantities of water, cement, and coarse aggregate established, the remaining material comprising the cubic yard of concrete must consist of sand and whatever air will be entrapped. The required sand may be determined on the basis of either weight or absolute volume as shown below:

Weight basis. From Table 7, the weight of a cubic yard of non-air-entrained concrete made with aggregate having a maximum size of 1 ½ in. is estimated to be 4,070 lb. (For a first trial batch, exact adjustments of this value for usual differences in slump, cement factor, and aggregate specific gravity are not critical.) Weights already known are:

Water (net mixing)	300 lb.
Cement	484 lb.
Coarse aggregate	1917 lb. (dry)*
Total	2701 lb.

The weight of sand, therefore, is estimated to be 4070 - 2701 = 1369 lb. (dry)*

*Aggregate absorption is disregarded since its magnitude is inconsequential in relation to other approximations.

Absolute volume basis. With the quantities of cement, water, and coarse aggregate established, and the approximate entrapped air content (as opposed to purposely entrained air) taken from Table 3, the sand content can be calculated as follows:

Volume of water =	$\frac{300}{62.4}$ =	4.81 cu. ft.
Solid volume of cement =	$\frac{484}{3.15 \times 62.4}$ =	2.46 cu. ft.
Solid volume of coarse aggregate =	$\frac{1917}{2.68 \times 62.4}$ =	11.46 cu. ft.
Volume of entrapped air =	0.01 X 27 =	0.27 cu. ft.
Total solid volume of ingredients except sand =		19.00 cu. ft.
Solid volume of sand required =	27 - 19.00 =	8.00 cu. ft.
Required weight of dry sand =	8.00 x 2.64 x 62.4=	1318 lbs.

Batch weights per cubic yard of concrete calculated on the two bases are compared below:

	Based on estimated concrete weight, lb.	Based on absolute volume of ingredients, lb.
Water (net mixing)	300	300
Cement	484	484
Coarse aggregate (dry)	1917	1917
Sand (dry)	1369	1318

Note: These batch weights will require adjustment in the field to take into account moisture on aggregates. Also, some adjustment in proportions may be found desirable on the basis of actual field experience.

Example 2. Concrete is required for a heavy bridge pier which will be exposed to fresh water in a severe climate. An average 28-day compressive strength of 3000 psi will be required. Placement conditions permit a slump of 1 to 2 in. and the use of large aggregate, but the only economically available coarse aggregate of satisfactory quality is graded from No. 4 to 1 in. and this will be used. Its dry-rodded weight is found to be 95 lb. per cu. ft. Other characteristics are as indicated in the first part of this section.

The calculations will be shown in skeleton form only. Note that confusion is avoided if all steps of the Procedure section are followed even when they appear repetitive of specified requirements.

Step 1. The desired slump is 1 to 2 in.

Step 2. The locally available aggregate, graded from No. 4 to 1 in., will be used.

Step 3. Since the structure will be exposed to severe weathering, air-entrained concrete will be used. The approximate amount of mixing water to produce a 1 to 2-in. slump in air-entrained concrete with 1-in. aggregate is found from Table 3 to be 270 lb. per cu. yd. The recommended air content is 5 percent.

Step 4. From Table 4(a), the water-cement ratio needed to produce a strength of 3000 psi in air-entrained concrete is estimated to be about 0.59. However, reference to Table 4(b) reveals that, for the severe weathering exposure anticipated, the water-cement ratio should not exceed 0.50. This lower figure must govern and will be used in the calculations.

Step 5. From the information derived in Steps 3 and 4, the required cement content is found to be 270/0.50 = 540 lb. per cu. yd.

Step 6. The quantity of coarse aggregate is estimated from Table 6. With a fine aggregate having a fineness modulus of 2.8 and a 1 in. maximum size of coarse aggregate, the table indicates that 0.67 cu. ft. of coarse aggregate, on a dry-rodded basis, may be used in each cubic foot of concrete. For a cubic yard, therefore, the coarse aggregate will be 27 x 0.67 = 18.09 cu. ft. Since it weighs 95 lb. per cu. ft., the dry weight of coarse aggregate is 18.09 x 95 = 1719 lb.

Step 7. With the quantities of water, cement and coarse aggregate established, the remaining material comprising the cubic yard of concrete must consist of sand and air. The required sand may be determined on the basis of either weight or absolute volume as shown below.

Weight basis. From Table 7, the weight of a cubic yard of air-entrained concrete made with aggregate of 1 in. maximum size is estimated to be 3900 lb. (For a first trial batch, exact adjustments of this value for differences in slump, cement factor, and aggregate specific gravity are not critical.) Weights already known are:

Water (net mixing)	270 lb.
Cement	540 lb.
Coarse aggregate (dry)	1719 lb.
Total	2529 lb.

The weight of sand, therefore, is estimated to be 3900 - 2529 = 1371 lbs. (dry).

Absolute volume basis. With the quantities of cement, water, air, and coarse aggregate established, the sand content can be calculated as follows:

Volume of water =	$\frac{270}{62.4}$ =	4.33 cu. ft
Solid volume of cement =	$\frac{540}{(3.15 \times 62.4)}$ =	2.75 cu. ft.
Solid volume of coarse aggregate =	$\frac{1719}{(2.68 \times 62.4)}$ =	10.28 cu. ft.
Volume of air =	0.05 X 27 =	1.35 cu. ft.
Total volume of ingredients except sand =		18.71 cu. ft.
Solid volume of sand required =	27 - 18.71 =	8.29 cu. ft.
Required weight of dry sand =	8.29 X 2.64 X 62.4 =	1366 lbs.

Batch weight per cubic yard of concrete calculated on the two bases are compared below:

	Based on estimated concrete weight, lb.	Based on absolute volume of ingredients, lb.
Water (net mixing)	270	270
Cement	540	540
Coarse aggregate (dry)	1719	1719
Sand (dry)	1371	1366

See Notes for Example 1

READY-MIXED CONCRETE

In the United States more than 90 percent of all concrete being used in building projects is ready-mixed. Ready-mixed concrete may be defined as portland cement concrete manufactured for delivery to a purchaser in a plastic and unhardened state.

In some instances the concrete is mixed completely in a stationary mixer and then is transported to the job in trucks. This is known as central-mixed concrete. In other ready-mixed operations, the materials are dry batched and mixed enroute to the job in truck mixers. This is known as transit mixed concrete. In another procedure, the concrete is mixed in a stationary mixer at the central plant only enough to intermingle the ingredients, usually about ½ minute, and the mixing is completed enroute to the job. This is known as shrink-mixed concrete.

Truck mixers consist essentially of a mixer with a separate water tank and water measuring device mounted on a truck chassis. There are other truck conveyances which are similar but are without provisions for water.

ASTM C-94 (Ready-mix concrete) requires that when a truck mixer is used either for complete mixing or to finish the partial mixing, each batch of concrete is to be mixed not more than 100 revolutions of the drum or blades at the speed of rotation designated by the manufacturer as the mixing speed. Any additional mixing is to be done at the agitating speed. The specification also requires that the concrete must be delivered and discharged from the truck mixer or agitator truck within 1 ½ hours after introduction of the water to the cement and aggregate or the cement to the aggregate.

Quality of Ready-Mixed Concrete. ASTM C-94 states that in the absence of applicable general specifications, the purchaser shall select one of the alternate bases for specifying the quality of concrete.

Alternate No. 1: When the purchaser assumes responsibility for the proportioning of the concrete mixture, he shall also specify the following:

1) Cement content in bags or pounds per cubic yard of concrete, or equivalent units.

2) Maximum allowable water content in gallons per cubic yard of concrete, or equivalent units, including surface moisture on the aggregates, but excluding water of absorption.

3) If admixtures are required, the type, name, and dosage to be used. The cement content shall not be reduced when admixtures are used under Alternate No. 1 without the written approval of the purchaser.

Alternate No. 2: When the purchaser requires the manufacturer to assume full responsibility for the selection of the proportions for the concrete mixture, the purchaser shall also specify the following:

Requirements for compressive strength as determined on samples taken from the transportation unit at the point of discharge. The purchaser shall specify the requirements in terms of the compressive strength of standard specimens cured under standard laboratory conditions for moist curing. Unless otherwise specified the age at test shall be 28 days.

Alternate No. 3: When the purchaser requires the manufacturer to assume responsibility for the selection of the proportions for the concrete mixture with the minimum allowable cement content specified, the purchaser shall also specify the following:

1) Required compressive strength as determined on samples taken from the transportation unit at the point of discharge. The purchaser shall specify the requirements for strength in terms of tests of standard specimens cured under standard laboratory conditions for moist curing. Unless otherwise specified the age at test shall be 28 days.

2) Minimum cement content in bags or pounds per cubic yard of concrete.

3) If admixtures are required, the type, name, and dosage to be used. The cement content shall not be reduced when admixtures are used.

Note: Alternate No. 3 can be distinctive and useful only if the designated minimum cement content is at about the same level that would ordinarily be required for the strength, aggregate size, and slump specified. At the same time, it must be an amount that will be sufficient to assure durability under expected service conditions, as well as satisfactory surface texture and density, in the event specified strength is attained with it.

The proportions arrived at by Alternates 1, 2, or 3 for each class of concrete and approved for use in a project shall be assigned to a designation to facilitate identification of each concrete mixture delivered to the project. This is the designation required and supplies information on concrete proportions when they are not given separately on each delivery ticket. A certified copy of all proportions as established in Alternates 1, 2, and 3 shall be on file at the batch plant.

Curing. The protection of concrete during the early period to prevent loss of moisture at low temperatures is an important factor in the development of both strength and durability in concrete. Specifications require that the concrete be protected to prevent loss of moisture from the surface and to prevent temperatures at the surface from going below 50°F. for periods of 5 days where normal portland cement is used, and 3 days where high early strength portland cement is used.

In fixing these limits it is recognized that under average conditions, curing of concrete does not cease immediately upon removal of the protection against loss of moisture. Where the conditions are extremely severe, such as for thin sections in hot dry air or very low temperatures, it may be desirable to increase somewhat the protection periods specified.

High Early Strength Portland Cement. Most of the cement companies are now manufacturing what is designated as high early strength portland cement, which develops practically the same strength in 72 hours as is obtained with normal portland cement in 7 to 10 days.

High early strength cement is used in the same proportions and in exactly the same manner as ordinary portland cement and it is general knowledge with contractors that this cement is advantageous not only in special rush work but in every day reinforced concrete construction. High early strength cement at present costs $3.50 per barrel more than standard portland cement, or, if we assume that approximately 1 ½ barrels of cement per cu. yd. will be used, the cost per cu. yd. of concrete is increased by $5.25. If the average amount of concrete per sq. ft. of floor construction (including columns, beams, lintels, etc.) is 8", it takes 40 sq. ft. of floor for a yard of concrete, or the extra expense on account of the use of high early strength cement is 13.13 cents per sq. ft.

Under favorable conditions using high early strength portland cement, it is possible to strip part or an entire floor in three days so it may be possible to save half of the lumber required for a floor. Under "Concrete Formwork" it is shown that it takes not less than 3 to 4 BF of lumber for each sq. ft. of floor construction (including columns, girders, and lintels) in addition to the labor costs per sq. ft. of floor to make up the various sides, panels, shores, wedges, etc. required. If it is possible to save one-half of the lumber plus half the labor of the make up, the resulting saving per sq. ft. is greater than the additional 13.13 cents per sq. ft. for high-early cement.

During the winter there is the additional saving of two to three days for cold weather protection.

Air Entrained Concrete. Most of the highway departments are specifying air-entrained concrete for pavements and many require it for bridges and other structures. In this concrete 3 to 5 or 6 percent of air is incorporated in the concrete in the form of minute separated air bubbles. Such concrete is more resistant to freezing and thawing and to salt action than normal concrete. It is produced by using air-entraining portland cement or by the addition of an air-entraining agent at the mixer. Originally developed to prevent scaling of pavement where salts are used for ice removal, air-entrained concrete is being widely adopted for all types of work and in all locations because of its better workability as well as its better resistance to weathering even where salts are not used.

In air-entrained concrete the proportion of sand and the amount of water is generally reduced from the normal mix in sufficient quantities to make up for the increased volume of concrete produced by the air. Thus the same cement factor is maintained. In the tables given on the previous pages, the sand can be reduced about 5 lbs. for each 1 percent of air. Thus, if 4 percent air is incorporated, the quantities of sand shown in these tables would be reduced by 20 lbs. per cubic yard of concrete. For each 1 percent of air the amount of water would be reduced by about 6 to 7 lbs. per cubic yard.

Important Suggestions Regarding the Purchasing of Materials for Job Mixed Concrete. When purchasing sand, gravel or crushed stone for concrete, it is important to see that the aggregates and particularly the sand do not contain an abnormal amount of moisture, otherwise you will be short on materials, which will reduce profits or may even result in a loss, particularly if there are considerable quantities of aggregates to be purchased.

The following suggestions are important when placing orders:

1) Determine whether the sand has a high or low moisture content.
2) Determine whether the sand has a high or low specific gravity.
3) When sand is purchased at a certain price per cu. yd. make certain that you are not paying for water instead of sand, as it is not uncommon to see settlement of 4" in trucks or railroad cars from the time it is shipped to the time it arrives at the job. This is especially true where washed sand is loaded immediately after washing or after hard rains.
4) When sand is purchased by weight, also make certain you are not paying for water instead of sand. Ordinarily sand should not contain over 4 to 5 percent moisture, while wet sand may contain as high as 18 percent moisture. This makes a difference in the cost of sand.

5) Watch the grading of sand and percentages from fine to coarse.
6) Determine the grading of the gravel and its specific gravity.
7) When gravel is purchased by the cu. yd. stipulate measurement at job or destination, if possible.
8) When gravel is purchased by weight, weight at destination should govern instead of weight at loading point, when gravel may contain much moisture.
9) When purchasing crushed stone, obtain the grading and specific gravity.
10) When purchasing crushed stone, determine the weight per cu. yd. also the moisture content.

An easy method of determining the moisture content of sand, gravel or crushed stone, is to measure out one cu. ft. of each and weigh them. Then dry this same material and weigh it. The difference in weight is the moisture content. Example: If one cu. ft. of loose damp sand weighs 105 lbs. and after drying it weighs only 90 lbs. then the moisture content is 15 lbs. or 16 ⅔ percent.

Quantities of Materials Required for Job-Mixed Concrete. For estimating purposes, material quantities for job mixed concrete given in the following table may be used as a guide. The quantities shown in the table will produce concrete of medium consistency or with a slump of about 3" to 4".

Suggested Mixes for Non-Air-Entrained Concrete of Medium Consistency (3 to 4 in. slump)

					With fine sand—fineness modulus = 2.50**			With coarse sand—fineness modulus = 2.90**		
Water-cement ratio, lb. per lb.	Maximum size of aggregate in.	Air Content (entrapped air) per cent	Water, lb. per cu. yd. of concrete	Cement, lb. per cu. yd. of concrete	Fine aggregate, per cent of total aggregate	Fine aggregate, lb. per cu. yd. of concrete	Coarse aggregate lb. per cu. yd. of concrete	Fine aggregate, per cent of total aggregate	Fine aggregate, lb. per cu. yd. of concrete	Coarse aggregate, lb. per cu. yd. of concrete
0.40	⅜	3	385	965	50	1240	1260	54	1350	1150
	½	2.5	365	915	42	1100	1520	47	1220	1400
	¾	2	340	850	35	960	1800	39	1080	1680
	1	1.5	325	815	32	910	1940	36	1020	1830
	1½	1	300	750	29	880	2110	33	1000	1990
0.45	⅜	3	385	855	51	1330	1260	56	1440	1150
	½	2.5	365	810	44	1180	1520	48	1300	1400
	¾	2	340	755	37	1040	1800	41	1160	1680
	1	1.5	325	720	34	990	1940	38	1100	1830
	1½	1	300	665	31	960	2110	35	1080	1990

*Increase or decrease water per cubic yard by 3 per cent for each increase or decrease of 1 in. in slump, then calculate quantities by absolute volume method. For manufactured fine aggregate, increase percentage of fine aggregate, by 3 and water by 15 lb. per cubic yard of concrete. For less workable concrete, as in pavements, decrease percentage of fine aggregate by 3 and add water by 8 lb. per cubic yard of concrete.
**For definition of fineness modulus, see "Aggregates for Concrete", available from the Portland Cement Association.

Suggested Mixes for Air-Entrained Concrete Of Medium Consistency
(3 to 4 in. slump*)

0.50	3/8	3	385	770	53	1400	1260	57	1510	1150
	1/2	2.5	365	730	45	1250	1520	49	1370	1400
	3/4	2	340	680	38	1100	1800	42	1220	1680
	1	1.5	325	650	35	1050	1940	39	1160	1830
	1½	1	300	600	32	1010	2110	36	1130	1990
0.55	3/8	3	385	700	54	1460	1260	58	1570	1150
	1/2	2.5	365	665	46	1310	1520	51	1430	1400
	3/4	2	340	620	39	1150	1800	43	1270	1680
	1	1.5	325	590	36	1100	1940	40	1210	1830
	1½	1	300	545	33	1060	2110	37	1180	1990
0.60	3/8	3	385	640	55	1510	1260	58	1620	1150
	1/2	2.5	365	610	47	1350	1520	51	1470	1400
	3/4	2	340	565	40	1200	1800	44	1320	1680
	1	1.5	325	540	37	1140	1940	41	1250	1830
	1½	1	300	500	34	1090	2110	38	1210	1990
0.65	3/8	3	385	590	55	1550	1260	59	1660	1150
	1/2	2.5	365	560	48	1390	1520	52	1510	1400
	3/4	2	340	525	41	1230	1800	45	1350	1680
	1	1.5	325	500	38	1180	1940	41	1290	1830
	1½	1	300	460	35	1130	2110	39	1250	1990
0.70	3/8	3	385	550	56	1590	1260	60	1700	1150
	1/2	2.5	365	520	48	1430	1520	53	1550	1400
	3/4	2	340	485	41	1270	1800	45	1390	1680
	1	1.5	325	465	38	1210	1940	42	1320	1830
	1½	1	300	430	35	1150	2110	39	1270	1990

*Increase or decrease water per cubic yard by 3 per cent for each increase or decrease of 1 in. in slump, then calculate quantities by absolute volume method. For manufactured fine aggregate, increase percentage of fine aggregate, by 3 and water by 15 lb. per cubic yard of concrete. For less workable concrete, as in pavements, decrease percentage of fine aggregate by 3 and add water by 8 lb. per cubic yard of concrete.
**For definition of fineness modulus, see *Aggregates for Concrete*, available from the Portland Cement Association.

Water-cement ratio, lb. per lb.	Maximum size of aggregate in.	Air content, per cent	Water, lb. per cu. yd. of concrete	Cement, lb. per cu. yd. of concrete	With fine sand—fineness modulus = 2.50** Fine aggregate, per cent of total aggregate	Fine aggregate, lb. per cu. yd. of concrete	Coarse aggregate lb. per cu. yd. of concrete	With coarse sand—fineness modulus = 2.90** Fine aggregate, per cent of total aggregate	Fine aggregate, lb. per cu. yd. of concrete	Coarse aggregate, lb. per cu. yd. of concrete
0.55	3/8	7.5	340	620	54	1450	1260	58	1560	1150
	1/2	7.5	325	590	45	1250	1520	49	1370	1400
	3/4	6	300	545	39	1140	1800	43	1260	1680
	1	6	285	520	35	1060	1940	39	1170	1830
	1½	5	265	480	33	1030	2110	37	1150	1990
0.60	3/8	7.5	340	565	54	1490	1260	58	1600	1150
	1/2	7.5	325	540	46	1290	1520	50	1410	1400
	3/4	6	300	500	40	1180	1800	44	1300	1680
	1	6	285	475	36	1100	1940	40	1210	1830
	1½	5	265	440	33	1060	2110	37	1180	1990
0.65	3/8	7.5	340	525	55	1530	1260	59	1640	1150
	1/2	7.5	325	500	47	1330	1520	51	1450	1400
	3/4	6	300	460	40	1210	1800	44	1330	1680
	1	6	285	440	37	1130	1940	40	1240	1830
	1½	5	265	410	34	1090	2110	38	1210	1990
0.70	3/8	7.5	340	485	55	1560	1260	59	1670	1150
	1/2	7.5	325	465	47	1360	1520	51	1480	1400
	3/4	6	300	430	41	1240	1800	45	1360	1680
	1	6	285	405	37	1160	1940	41	1270	1830
	1½	5	265	380	34	1110	2110	38	1230	1990

* and **. See references bottom of previous table.

Labor Mixing and Placing Concrete

The labor cost of mixing and placing concrete will vary considerably according to the size of the job, the method used in mixing and placing the concrete, the proximity of the stock piles to the mixer, and the distance the concrete has to be transported after mixing.

On large jobs using conveyors and bins for feeding the aggregate to the mixer, the purchase or rental of this equipment, together with the labor cost of placing same, should be estimated separately and not included in the cu. yd. price of placing the concrete. This plant and equipment cost may be the same whether the job contains 1,000 or 10,000 cu. yds. of concrete, and unless it is kept as a separate item, the resulting labor costs will be misleading and valueless insofar as estimating the cost of future jobs are concerned.

Ready-mix concrete is used in all parts of the country and where the concrete is delivered to the job mixed and ready to place in the forms, it is often possible for the truck mixers to deposit the concrete directly into the trenches or wall forms without any handling on the job, other than spading, vibrating, spreading and leveling the concrete.

However, on many jobs this is not possible and it is necessary to have a receiving hopper where the truck mixers dump the concrete and it is then hauled from receiving hopper to place of deposit either by wheelbarrows, hand or power buggies. On large operations the concrete may be pumped from receiving hopper to place of deposit.

Another method of placing the concrete that is often used on large operations is to use a crane which picks up a batch of concrete in a bucket and places it directly over the footings, piers or forms.

Where any of the above methods are used, all mixer labor is eliminated but it usually requires one or two men at the receiving hopper or at the truck dump to assist the truck mixer, prevent spillage, cleanup, etc.

Most specifications prohibit the use of long spouts for chuting concrete from the mixer to the forms on account of the great danger of segregation. Buggies or buckets are generally desirable. Self-propelled buggies are available and are used on many of the larger jobs.

Weather conditions also influence the labor costs considerably because placing concrete in freezing weather usually necessitates heating the sand, gravel or stone and water, and protecting the concrete from freezing by the use of tarpaulins, straw, oil burning salamanders, or other artificial means, all of which add to the overall expense.

With so many small mixers on the market very little concrete is hand mixed today, but occasionally a job presents itself where it is more economical to mix by hand.

Mixes for Small Jobs

For small jobs where time and personnel are not available to determine proportions in accordance with the recommended procedure, mixes in Table 1 will usually provide concrete that is amply strong and durable if the amount of water added at the mixer is never large enough to make the concrete too wet. These mixes have been predetermined in conformity with the recommended procedure by assuming conditions applicable to the average small job, and for aggregate of medium specific gravity. Three mixes are given for each maximum size of coarse aggregate. For the selected size of coarse aggregate, Mix B is intended for initial use. If this mix proves to be oversanded, change to Mix C; if it is undersanded, change to Mix A. It should be noted that the mixes listed in the table are based on dry or surface-dry sand. If the sand is moist or wet, make the corrections in batch weight prescribed in the footnote.

The approximate cement content per cubic foot of concrete listed in the table will be helpful in estimating cement requirements for the job. These requirements are based on concrete that has just enough water in it to permit ready working into forms without objectionable segregation. Concrete should slide, not run, off a shovel.

Table 1--Concrete Mixes for Small Jobs

Approximate weights of solid ingredients per cu. ft. of concrete, lb.

Maximum size aggregate, in.	Mix designation	Cement	Sand.* Air-entrained concrete**	Sand.* Concrete without air	Coarse aggregate: Gravel or crushed stone	Coarse aggregate: Iron blast furnace slag
1/2	A	25	48	51	54	47
	B	25	46	49	56	49
	C	25	44	47	58	51

Maximum size aggregate, in.	Mix designation	Cement	Sand.* Air-entrained concrete**	Sand.* Concrete without air	Coarse aggregate: Gravel or crushed stone	Coarse aggregate: Iron blast furnace slag
3/4	A	23	45	49	62	54
	B	23	43	47	64	56
	C	23	41	45	66	58
1	A	22	41	45	70	61
	B	22	39	43	72	63
	C	22	37	41	74	65
1 1/2	A	20	41	45	74	65
	B	20	39	43	77	67
	C	20	37	41	79	69
2	A	19	40	45	79	69
	B	19	38	43	81	71
	C	19	36	41	83	72

*Weights are for dry sand. If damp sand is used, increase tabulated weight of sand 2 lb.; if very wet sand is used, 4 lb.

**Air-entrained concrete should be used in all structures that will be exposed to alternate cycles of freezing and thawing. Air-entrainment can be obtained by the use of an air-entraining cement or by adding an air-entraining admixture. If an admixture is used, the amount recommended by the manufacturer will, in most cases, produce the desired air content.

Power Buggy. A method of transporting construction materials, such as wet concrete, earth, sand, gravel, cement, etc., is through the use of a power operated buggy, which requires only steering on the part of the operator.

They are furnished in various sizes by different manufacturers and the cost varies according to size and capacity.

The power buggy is mounted on a power driven three-wheel chassis and is steered by means of a rear caster wheel. The maximum speed over the ground conforms to a fast walking pace either on the level or up to a 20 percent incline.

This machine has a capacity of 1,500 lbs.; a bucket volume of 10 cu. ft. and will handle 8 to 9 cu. ft. of wet concrete but a larger volume of light, dry materials. This buggy weighs 560 lbs., is powered by a 7 Hp. gasoline engine with a fuel capacity of 2 3/4 gals., which will operate the machine for eight hours. The machine is 5'-5 1/2" in length overall, 2'-7 1/2" wide and is furnished with either single or dual wheels and has a turning radius of 34 3/4".

Comparison of Hand and Power Buggy Operation. The smaller power buggy having a capacity of 8 to 9 cu. ft. of wet concrete will probably do the work of three workers pushing hand powered concrete buggies.

A large machine having a capacity of 10 to 13 cu. ft. of wet concrete will probably handle as much concrete as 4 to 5 men using hand operated buggies, but it may be that heavier runways will be required to handle the larger loads of concrete.

Sizes, Weights and Capacities of Concrete Carts or Buggies

Capacity Cu. ft.	Turning Radius	Width Overall	Length Overall	Maximum Height	Approx. Weight
12	72"	34"	87"	54"	950
15	82"	41"	94 ½"	54"	1020
21	91"	54"	106 ½"	56"	1400

The following is an approximate comparison of cost between the two methods, based on wheeling 100 cu. yds. of concrete from a 16-S mixer:

Hand Method	Quan.	Rate	Total	Rate	Total
5 Laborers (hours)	40	$....	$....	$19.13	$765.20
Cost per cu. yd			....		7.65
Power buggy method					
Power buggy, fixed charge (per) Day	1	$....	$....	$75.00	$75.00
Fuel, gals, and oil	5	....	....	1.25	6.25
Operator (hours)	8	....	....	19.13	153.04
Total cost per day			$....		$234.29
Cost per cu. yd			....		2.34

On some jobs, it may be that 2 power buggies would be required to keep the mixer operating at capacity, and in such instances would be double that given above--but in either instance the saving is considerable.

Capacities of Concrete Mixers Based on A. G. C. Standards
(in Sacks of Cement per Batch)

Size of mixer denotes number of cubic feet of mixed concrete, plus 10 percent excess, when mixer is level.

Concrete Proportions	Sizes of Standard Mixers 3 ½ - S	6 - S	11 - S	16 - S	28 - S
1:1 ½:3	½	1	3	4	8
1:1 ½:3 ½	½	1	2	3	7
1:2:3	½	1	2	3	7
1:2:3 ½	½	1	2	3	7
1:2:4	½	1	2	3	6
1:2 ½:4	½	1	2	3	6
1:3:5	½	1	1	2	5
1:3:6	½	1	1	2	4

Mixing Concrete in a One Sack (6-S) Mixer. Foundation concrete is usually mixed in the proportions 1:2:4, 1:3:5 or 1:3:6, which is a one sack mix using a 6-S mixer.

Often a 1:3:5 mix is used for foundation walls where watertight concrete is not necessary. This requires 1 sack of cement, 3 cu. ft. of dry sand and 5 cu. ft. of gravel or crushed stone, per batch. Wheelbarrows ordinarily used for dry materials hold 3 to 3 ½ cu. ft. loaded level full, so each batch will require 1 barrow of sand and 2 barrows of gravel or 9 cu. ft. of dry material.

Where possible to start mixing first thing in the morning and continue all day without delays, a mixer should average 30 batches (6 to 7 cu. ft. of concrete each) per hr. or 240 batches (54 to 62 cu. yds.), per 8-hr day. Runs like this on foundation work are exceptional, because in addition to the actual mixing time, it will be necessary to allow for delays, shutdowns, etc., as maximum output cannot be figured for the entire job, as the men must be paid for time getting ready to start work in the morning, cleaning and washing wheelbarrows or buggies and concrete mixer at noon and at night; cutting and removing old concrete drippings and spillage; building and removing runways; unloading cement; delays caused by non-delivery of materials, etc.

Based on 6 to 7 cu. yds. of concrete an hr. the following men will be required in the gang: 2 laborers loading and wheeling gravel or crushed stone; 2 laborers wheeling sand and cement; 1 laborer attending mixer; 3 or 4 laborers wheeling concrete from mixer to place of deposit; 2 or 3 laborers dumping barrows, spreading, leveling and tamping concrete; 1 mixer engineer (if required), and 1 mason foreman in charge of work (if required).

Allowing for average conditions, including runways, cleaning and repairing mixer, engine, etc., the labor per cu. yd. of concrete should cost as follows:

	Hours	Rate	Total	Rate	Total
Labor	2.5	$....	$....	$19.13	$47.83

Under extremely favorable conditions, this cost can be reduced 1/2 to 3/4-hr. per cu. yd. Add for mason foreman and mixer engineer, if required by union rules.

If a power buggy is used for wheeling concrete from mixer to place of deposit, one power buggy will take away all the concrete a 6-S mixer can turn out. This will eliminate 2 to 3 laborers wheeling concrete and result in a saving of about 1/4-hr. labor time per cu. yd. of concrete.

Mixing Concrete Using a 11-S Mixer. When using a 11-S mixer of 1 to 2 sack mix capacity, the costs will run about the same as given for 6-S or 16-S mixers, under the same conditions.

Mixing Concrete in a 16-S (3 Sack Batch) Mixer. A mixer of this size will turn out about 16 cu. ft. of mixed concrete at each batch, depending upon the proportions.

Opinions vary as to amount of mixing required. The time of mixing specified by architects and engineers usually varies between 1 1/2 and 2 minutes per batch. The time specified affects the labor costs to a considerable extent as the longer mixing time restricts the day's output.

If the mixer has a side-loader and is fed by wheelbarrows with the concrete wheeled direct from the mixer to place of deposit, in wheelbarrows or concrete buggies, the mixer should average one batch every 2 minutes or 30 batches (13 to 15 cu. yds.) an hr.

Using an average of 2 minutes per batch, the mixer should turn out 104 to 120 cu. yds. per 8-hr. day, depending upon the concrete proportions and the size of each batch. In addition to the actual mixing time, it will be necessary to allow for delays, shutdowns, etc., as maximum output cannot be figured for the entire job, as the men must be paid for time getting ready to start work in the morning, cleaning and washing wheelbarrows and concrete mixer at noon and at night; cutting and removing old concrete drippings and spillage; building and removing runways; delays caused by non-delivery of materials, etc.

Based on an average of 14 cu. yds. of concrete an hr. it will require 25 to 28 workers in the gang, distributed as follows: 6 laborers loading and wheeling gravel or crushed stone; 3 laborers wheeling sand; 1 or 2 laborers on cement; 1 laborer feeding mixer; 1 laborer dumping concrete from mixer into wheelbarrows or buggies; 5 to 7 laborers wheeling concrete from mixer to place of deposit, depending upon length of haul; 3 to 4 laborers dumping concrete carts or barrows, spreading, spading and tamping concrete in place; 3 to 4 laborers general labor, such as building and removing runways, cleaning concrete mixer, wheelbarrows or buggies, picking up or baling cement sacks and other miscellaneous labor; 1 engineer on mixer and 1 mason or concrete foreman in charge of gang.

On the above basis, the labor per cu. yd. of concrete should cost as follows:

	Hours	Rate	Total	Rate	Total
Labor	2.0	$....	$....	$19.13	$38.26
Foreman	0.13	....	....	19.79	2.57
Mixer engineer	0.13	....	....	25.96	3.37
Cost per cu. yd			$....		$44.20

In many instances concrete will be placed for 65% to 75% of the above costs, but this will merely represent a day's run and not the average for the entire job.

Using a power buggy to transport the concrete from the mixer or hopper to place of deposit, one or two power buggies should handle all the concrete a 16-S mixer can turn out. This will eliminate 3 to 4 laborers wheeling concrete and result in a saving of 1/4 to 1/3 hr. per cu. yd. of concrete.

Mixing Concrete in a 28-S (1 Cu. Yd.) Mixer. If the concrete is mixed in a 28-S mixer, fed by wheelbarrows and the concrete wheeled from the mixer to place of deposit the labor costs per cu. yd. will run approximately the same as given for 16-S mixers, under similar conditions.

While it is true the daily output will be greatly increased, the labor cost feeding the mixer and distributing the concrete will be increased in proportion.

Labor Mixing and Placing One Cu. Yd. of Concrete
for Foundations

	Hours per Cubic Yard			
Method of Mixing and Placing	Common Labor	Mixer Engr.	Hoist. Engr.	Foreman
Hand mixing using one mixing board	3.33*	..	..	0.33
Hand mixing using two mixing boards	3**	..	..	0.25
Using one sack (6-S) mixer and wheeling to place	2.5***	..	..	0.17
Using two sack (11-S) mixer and wheeling to place	2	0.125		0.125
Using 16-S mixer and wheeling to place	2	0.125	..	0.125
Using 28-S mixer and wheeling to place	2	0.083	..	0.08
Using conveyors, batcherplant, hoist and wheeling to place	1.25	0.06	0.06	0.06

*In extreme warm weather or southern states, add 1/2 to 2/3 hr.

**If necessary to load concrete into barrows and wheel from mixing board to place of deposit, add 1/2 to 3/4 hr. per. cu. yd.

***Under extremely favorable conditions, this may be reduced 1/4 to 1/2 hr.

Above time includes all lost time in connection with concrete foundations. Add for labor erecting and dismantling hoisting towers, spouts, hoisting engines, and mixers, as this may vary from $550.00 to $5,500.00, depending on the size of job.

If power buggies are used for wheeling concrete from mixer or receiving hopper to place of deposit, it usually results in a saving of 1/4 to 1/3 hrs. per cu. yd. of concrete.

Loading Mixer Using a Front-End Loader

Since labor costs are high, many new methods are being used to reduce costs.

One method employs a loader to load the sand and gravel from the stock piles and dump it into the loading skip at the mixer. This eliminates the workers loading wheelbarrows with sand and gravel by hand, but there is usually one worker necessary on the stock pile, plus the small loader and operator. When supplying a 6-S mixer, it would eliminate about 5 workers loading and wheeling sand and gravel but would add one worker for the stock piles, plus the loader and operator. When supplying a 16-S mixer with a loader, it would eliminate about 9 workers loading and wheeling sand and gravel but would add 1 worker on the sand stock pile; 1 worker on the gravel stock pile, plus the loader and operator.

When supplying a 11-S mixer in this manner, it would eliminate 6 to 7 workers loading and wheeling sand and gravel but would add 1 worker on the stock piles, plus the loader and operator.

Mixing and Placing Concrete Where Materials Are Delivered to Job Batch Proportioned. Another method of handling concrete work is to have the concrete aggregates delivered to the job in trucks, correctly proportioned for one batch of concrete, including cement, sand and gravel or crushed stone.

When this method is used, the concrete aggregates for each batch are proportioned in the material yard and delivered to the job ready to mix. The truck backs up to the mixer and discharges the cement, sand and gravel or crushed stone into the loading hopper. A paving mixer with a large, flat loading hopper is commonly used for this purpose as it facilitates dumping the aggregates from the trucks into the loading hopper.

This method eliminates the necessity of all men loading and wheeling cement, sand and coarse aggregates, and the time required per cu. yd. of concrete should average as given in the following table.

Labor Mixing and Placing One Cu. Yd. of Concrete for Foundations When Aggregates Are Yard Proportioned and Delivered to the Job a Batch to a Load

Hours per Cubic Yard

Method of Mixing and Placing	Common Labor	Mixer Engr.	Hoist. Engr.	Foreman
16-S mixer and wheeling to place	1.25	0.125	..	0.125
28-S mixer and wheeling to place	1.25	0.083	..	0.083

If power buggies are used for wheeling concrete from mixer to place of deposit, it will usually result in a saving of 1/4 to 1/3 hr. labor time per cu. yd. of concrete.

Above includes all lost time in connection with mixing and placing concrete.

Add for labor erecting and dismantling hoisting towers, hoisting engines, mixers, etc., as this may vary from $550.00 to $5,500.00, depending on the size of job.

Placing Concrete with a Crane and Bucket. The crane and bucket has become ever increasingly popular on jobs as a means for transporting mixed concrete from the mixer or ready-mix truck to the forms, thereby eliminating the necessity of building and removing runways, labor wheeling concrete, etc. This is especially true when the crane and bucket is used in conjunction with ready-mixed concrete as the amount of labor is reduced to a minimum.

Labor Placing One Cu. Yd. of Foundation Concrete
Using Crane and Bucket

Method	Qty. Placed per 8-Hr Day cu. yds.	Hrs. per Cu. Yd. Common Labor	Crane Operator	Mixer Engineer	Fore-man*
16-S mixer, batched material	70 to 90	.70	.10	.10	.10
28-S mixer, batched material	150 to 170	.50	.05	.05	.05
Ready-mix	150 to 170	.40	.05	...	.05

Add for equipment cost or rental of crane and bucket.

*Add only if required.

Cold Weather Concreting

During the late fall and early spring and during all of the winter months it is necessary to use special precautions to assure proper hardening of the concrete. Careful contractors will use protection whenever the temperature falls below 50°F. At such time the most important protection required is to heat the sand and mixing water.

During freezing weather an extra boiler (old hoisting engine boiler, for example) having a rating of 50 to 75 Hp according to the size of the job (10,000 to 25,000 sq. ft. floor area) with a steam gauge set for 2 or more lbs. pressure per sq. in. will be needed with at least 2 perforated pipes for heating the coarse aggregate in addition to that required for heating the sand. On larger jobs or for fast concreting where aggregate bins are used, a grillage of steam pipes 4'-0" apart at the bottom of the bins will be sufficient. These pipes should have 1/4" perforations about 4" on centers. A 1 1/2" steam line should be run from the boiler to the water tank and another line run up the side of the building with valves and connections for steam hose to be used for cleaning out ice and snow.

Sufficient steam should be available so the concrete has a temperature of 50 to 70 degrees when leaving the mixer. This will promote the setting of the cement and permit earlier removal of the forms.

On small jobs the flame of an oil heater is allowed to project into the mixer to heat the materials.

Footings are generally protected by covering the fresh concrete with 8" to 12" of hay or straw and with tarpaulins or polyethylene thrown over the hay.

Foundation walls should be protected in mild weather by covering the top of the wall with hay and the sides of the forms with tarpaulins. During more severe weather place oil burning salamanders every 30 to 40 ft. apart and cover with tarpaulins; care being taken to keep the tarpaulins sufficient distance from the salamanders.

In skeleton construction it will be necessary to curtain the walls with tarpaulins for the story being concreted and in very severe weather it may be necessary to curtain two stories with tarpaulins, and to cover the top floor with hay or straw, held down by planks or timbers, which in turn support a layer of boards and planks with tarpaulins thrown over them. Stair and elevator openings, and openings for steam and plumbing pipes will allow heat to penetrate beneath the protecting layers and to heat the top of the floor.

It will require one salamander for each 200 to 250 sq. ft. of floor surface according to the severity of the weather and sufficient fuel should be placed on the floor in the early afternoon to last until next morning. Heat should be supplied for 2 to 5 days or more depending upon whether high-early-strength or normal portland cement is used and on the temperature maintained. Barrels of water should be placed on each floor for fire protection.

Ready Mixed Concrete

When using ready-mixed concrete on all classes of reinforced construction, such as floors, beams, girders, columns, etc, the cost of handling and placing the concrete will run approximately the same as for job mixed concrete, except for the labor handling materials, supplying and operating mixer, etc., as the trucks dump the ready-mixed concrete into a receiving hopper or bucket, where it is distributed by wheelbarrows, concrete buggies or hoisted into place.

In addition to saving about 3/4-hr. labor time per cu. yd. of concrete for charging mixer, the use of ready-mixed concrete saves the cost of cement sheds and provides more room for storing other materials. It permits higher output per hour because ready-mixed concrete can usually be furnished in any quantity desired and it also saves the cost of installing and dismantling expensive equipment, which may easily cost $5,500.00 or more on a job of any magnitude. It is necessary to have a fairly large size hopper on the job to receive the concrete from the trucks, as this hopper must be of sufficient capacity to supply the hoisting bucket as fast as it can handle the concrete, otherwise the work will be slowed down and with the crew idle part of the time the labor costs go up.

The labor cost of handling and placing ready-mixed concrete will vary with the amount handled and placed per day, the method of handling, hoisting and placing the concrete, and the class of work.

The following tables gives the labor hours necessary to handle and place one cu. yd. of ready-mixed concrete, using various methods:

Labor Placing One Cu. Yd. of Ready-Mixed Concrete for Concrete Foundations

Hours per Cubic Yard			
Method and Quantity Placed per 8-Hr. Day	Common Labor	Hoisting Engineer	Foreman
Placing concrete directly in forms, 1 truck servicing job, 50 to 70 c.y.	0.35-0.50	....	0.13
Placing concrete directly in forms, 2 trucks servicing job, 100 to 150 c.y.	0.27-0.40	....	0.08
Using wheelbarrows, 50 to 70 c.y.	1.25-1.50	0.17	0.17
Using wheelbarrows, 100 c.y. & over	1.00-1.25	0.13	0.13
Using concrete buggies, 50 to 70 c.y.	1.00-1.25	0.17	0.17
Using concrete buggies, 100 c.y. & over	0.90-1.10	0.13	0.13

Using 13 c.f. power concrete buggies	0.65-0.75	0.13*	0.13**

*Where the concrete is hoisted and it is necessary to use a hoisting engineer, add as given in table.
**If union regulations require a mason or cement finisher foreman in charge of crew, add as given in table.

Placing Ready Mixed Concrete Using Bucket

Labor Placing One Cu. Yd. of Ready-Mixed Concrete for Reinforced Floors, Beams, Girders, Columns, Etc.

	Hours per Cubic Yard		
Method of Handling, Placing Concrete	Common Labor	Hoisting Engineer	Foreman
Hoisting concrete in wheelbarrows on a material hoist & wheeling into place. . . .	2.3	.11*	.11*
Hoisting concrete in concrete buggies on material hoist & wheeling into place. . . .	1.1	.08*	.08*
Thin floors 2 to 2 ½-in. thick. Hoisting concrete in wheelbarrows on material hoist & wheeling into place	2.8	.11*	.11*
Thin floors 2 to 2 ½-in. thick. Hoisting concrete in concrete buggies on material hoist & wheeling into place . . .	1.6	.08*	.08*
Hoisting concrete in concrete bucket, discharging into a floor hopper, & wheeling into place using wheelbarrows .	1.2	.05*	.05*
Hoisting concrete in concrete bucket, discharging into a floor hopper, & wheeling using concrete buggies	0.8	.04*	.04*
If power buggies are used for transporting concrete from hopper to place of deposit, deduct per cu. yd	0.25-0.33	...	...
Mobile crane with bucket on ground, hoisting concrete to floor or forms, & depositing directly into forms	0.5	...	.04*
Add for rental of mobile crane & operator	...	.04*	...
For thin concrete floors, 2 to 2 ½-in. thick, add to the above costs	0.5	...	...

*Time of hoisting engineer and foreman will vary according to the number of cubic yards of concrete placed per day.

The above costs do not include cost of erecting and dismantling material hoists, concrete hoisting towers, etc., which may run from $550.00 to $5,500.00, depending on size of job and type of equipment used.

PLACING CONCRETE BY PIPELINE

The development and perfection of pumps for transporting freshly mixed concrete through pipelines has revolutionized the placing of substantial yardages of concrete in dams, tunnels, retaining walls, bridge decks, mill, factory, warehouse and apartment buildings where concrete is required at rates varying from 15 to 80 cu. yds. an hr. These machines for pumping concrete force it through a pipeline in much the same manner that a piston pump pumps water. It will transport freshly mixed workable concrete to distances of 1,000 ft. and to heights in excess of 120 ft. Many instances are known where a concrete pump has delivered concrete to the forms 1,300 ft. away, or 200 ft. high where a well designed mix is used.

Concrete pumps can handle efficiently concrete of any slump from ½" to 7" but the most dependable slump for general conditions is around 3" to 5". However, plastic concrete of about 6" slump can generally be pumped the maximum distance and height. The rate of pumping depends considerably upon the consistency of the concrete.

Courtesy Portland Cement Association

Ready Mixed Concrete Placed Directly from Mixing Truck

The concrete is delivered to the forms in better condition than when discharged from the mixer due to the added mixing received in the remixing hopper which receives the concrete from the mixer and keeps it in a uniform condition as it is fed through the cylinder of the concrete pump to the pipeline. During the pumping operation, the pipeline is full at all times which keeps the concrete from segregating, thereby eliminating one of the major objections found in some methods used to transport concrete to the forms. The troweling action of the walls of the pipeline in contact with the concrete as it passes through improves the quality of the concrete.

The smooth, even flow of concrete into the forms reduces the shock loading on the bracing and ties. Danger of "honeycomb" is reduced as there is no "slug" of pebbles dumped in, as in the case of a buggy, at the end of a batch which might not be properly spaded. Walls 120 ft. long and 15 ft. high have been made at one pour in 3 ft. lifts without signs of longitudinal jointing or uneven surfaces due to the dumping of wheelbarrows or buggies in such a way as to splash the form above the point of placement.

Runways for wheelbarrows or concrete buggies are not required when placing concrete by pipeline, although it is necessary to provide supports for the pipelines, which usually consists of "horses" or "bridges".

Mechanical maintenance will vary with the efficiency of the operator, the type of work, abrasiveness of materials, etc., as with all construction machinery.

In addition to replacements for the machine there will be pipeline maintenance which will be dependent upon the abrasiveness of materials. The couplings, which represent approximately three-fourths of the cost of pipe sections, will last indefinitely. The straight pipe will last from 50,000 to

150,000 cu. yds. and the elbows will last from 30,000 to 70,000 cu. yds., depending upon the abrasiveness of the concrete. The piping can be replaced as it becomes worn out and the old couplings reused.

Placing Ready Mixed Concrete Using Truck Mounted Pumper

Modern Pumping Rigs Frequently Replace Bucket and Cranes

Cost of Placing Concrete by Pipeline. The cost of placing concrete by pipeline varies widely as do other methods of distribution, and is governed by the size of the job, area covered, the yardage of concrete that can be placed at one setup of the pipelines, distance from the pump, etc.

When placing concrete by ordinary methods, the cost of rental, erecting and dismantling the hoisting tower and spouts must be considered, also the cost of raising the concrete hopper and moving spouts for placing the concrete.

The same conditions must be considered with concrete pump distribution, which includes original cost of equipment, depreciation or salvage value, cost of setting up and removing pumping equipment, pipelines, etc.

Always bear in mind, the cost of mixing the concrete will be the same regardless of the method of distribution because where the concrete is mixed and dumped into a hoisting bucket using ordinary methods, it is deposited in the hopper. It is only the method of handling, hoisting and placing the mixed concrete that should be considered. This also applies where ready-mixed concrete is used.

The crew operating the pumping unit will vary with the size, area and yardage in the job. An operator is required on the concrete pump. A foreman and 6 to 12 laborers should place and remove pipelines, handle, spread and level the concrete, and the quantity will vary from 15 to 80 cu. yds. an hr. depending upon the size of the pumping equipment and the class of work.

Placing Concrete on a Commercial Building. Assume a 5-story steel frame commercial building, 140 by 250 ft. having floors of metal pan construction, with pans from 6" to 14" deep and 2 ½" of concrete over the pans, the entire job containing approximately 5,200 cu. yds. of concrete.

Assume the concrete is ready-mixed, using ¾" gravel aggregate and having a 5" slump.

The concrete is delivered to the job and deposited directly into the remixing hopper on the pumper and pumped to the various floors.

The labor per cu. yd. cost as follows:

	Hour	Rate	Total	Rate	Total
Pump operator	.07	$....	$....	$25.96	$1.82
Hopper man	.07	....	....	19.13	1.34
Labor vibrating, pipeline, etc	.49	....	....	19.13	9.37
Foreman	.07	....	....	19.79	1.39
Cost per cu. yd			$....		$13.92

Average rate of pour 15 cu. yds. per hr. Maximum rate of pour 30 cu. yds. per hr.

The above costs do not include the cost of the pumper, depreciation, parts and replacements, gas and oil, labor installing and removing equipment, etc.

Placing Concrete by Pipeline on a Sewage Disposal Plant. Assume a sewage disposal plant, size 370'-0"x380'-0", containing 15,000 cu. yds. of concrete consisting of heavily reinforced concrete in slabs and walls, all concrete to be placed by pipeline.

The mixer will discharge into the remixing hopper of a single pumper which is placed below grade in the excavation made for the plant addition. From this point the concrete is to be pumped through pipelines to the point of placement.

Triple Boom Pumper Placing Ready Mixed Concrete on Fifth Floor

Workmen Simultaneously Vibrate Concrete at Two Locations in Double Pumping Operation

The floors of the aerating basins, 30 ft. wide and 120 ft. long are poured by radial spouting from the end of the pipeline, which is supported on horses. The piping is taken up as the concreting progresses and is laid in the next basin.
Staging supported on the forms will permit erection of the pipeline, pouring 112 cu. yds. of concrete in the wall and removal of the pipe line in one day. The walls of the aerating and settling basins are 15 ft. high and are poured in 3 ft. lifts. Valves at 20 ft. intervals are inserted in the 120 ft. of distributing line which permits the withdrawal of concrete at these points. Elephant trunks (tremie) direct the concrete into specially constructed chutes leading the concrete down through the maze of steel to the level being poured where it is discharged without spattering the forms above that point. Pouring was started from the valve nearest the pumper. When enough concrete had been withdrawn at a valve, it was closed and the next valve opened for concreting at this point. This procedure is continued until the lift is completed at the end of the pipeline.
With the aid of internal vibration, these high long walls are poured relatively free of blemishes or jointing.
The average labor cost of placing one cu. yd. of concrete is as follows:

Placing Concrete by Pipeline	Hours	Rate	Total	Rate	Total
1 pumper operator	.071	$....	$....	$25.96	$1.84
1 foreman*	.071	....	....	19.79	1.41
8 laborers	.86	....	....	19.13	16.45
Cost per cu. yd			$....		$19.70

*Labor on pumper includes time placing and removing pipeline.

During the pouring of the concrete in the walls, 2 additional workers are required to operate vibrators. Vibrators would be required regardless of the distribution method used.

CAST IN PLACE CONCRETE

Concrete floors, sidewalks, and lightweight concrete floor fill are estimated by the sq. ft., taking the number of sq. ft. of any given thickness, as 2", 3",4", etc. to obtain the total cubic yards for placement.
The quantities are easily computed by obtaining the area of the various spaces to have concrete floors or walks and stating the number of sq. ft. of each thickness.
When preparing an estimate for concrete floors or sidewalks placed on the ground, there are distinct material and labor items to be considered as follows:

1. Labor grading and removing surplus earth under floors or walks or filling where present grade is too low.
2. Cost of sand, gravel or slag to be placed under concrete floor; labor spreading and tamping same in place.
3. Labor placing wood or metal screeds and forms.
4. Cost of vapor barrier and reinforcing material and placement.
5. Labor mixing and placing concrete.
6. Labor troweling finish surface and blocking off into squares, if required.
7. Labor and materials for curing and hardeners, if required.

Grading for Concrete Floors and Walks. On practically all cement floors and walks placed directly on the ground, it is necessary to do a certain amount of grading before the bed of sand, gravel or slag can be placed.

If it is not necessary to remove the excavated soil from the premises, a bulldozer or loader is probably the most economical method to use.

A medium size loader and operator should excavate and place in piles, 25 to 30 cu. yds. per hour at the following cost per cu. yd.:

	Hours	Rate	Total	Rate	Total
Loader w/Operator	0.045	$....	$....	$50.00	$2.25

This cost will vary with the kind of soil encountered, i.e., sand or loam, ordinary black soil, heavy soil or clay.

Number of Sq. Ft. of Soil of Various Thicknesses Obtainable From One Cu. Yd.

	Depth of Excavation In Inches									
One cu. yd. =	3"	4"	5"	6"	7"	8"	9"	10"	11"	12"
27 cu. ft.	108	81	65	54	46	40	36	32	29	27

To obtain the cost of grading one sq. ft. of any thickness, divide the cu. yd. cost by the number of sq. ft. obtainable from one cu. yd. and the result is the sq. ft. cost for grading. Example: With excavating at $4.50 per cu. yd., find the sq. ft. cost of removing 5" of top soil. One cu. yd. contains 65 sq. ft. of 5" fill. $4.50 divided by 65 = $0.069 per sq. ft. for grading.

Hand Grading. If the top soil is loaded by hand into wheelbarrows or power buggies, a worker should loosen, shovel and load 5 to 6 cu. yds. per 8-hr. day, at the following labor cost per cu. yd.:

	Hours	Rate	Total	Rate	Total
Labor	1.45	$....	$....	$19.13	$27.74

Grading and Tamping Sand, Gravel or Slag Fill Under Concrete Floors or Walks. Where sand, gravel or slag fill is placed under concrete floors or walks, the cost will vary with the thickness of the fill and job conditions.

It costs as much to grade a bed of slag or gravel 3" thick as one 9" thick, because it is only the surface that is graded and the only additional labor on the thicker fill is for handling, tamping, and spreading a larger quantity of material.

Where the fill varies from 3" to 4" thick, a man will spread, grade and tamp about 6 cu. yds. per 8-hr. day, at the following labor cost per cu. yd.:

	Hours	Rate	Total	Rate	Total
Labor	1.33	$....	$....	$19.13	$25.44

If the fill varies from 5" to 6" thick, a worker will spread, grade and tamp about 8 ½ cu. yds. per 8-hr. day, at the following labor cost per cu. yd.:

	Hours	Rate	Total	Rate	Total
Labor	0.94	$....	$....	$19.13	$17.98

If the fill varies from 7" to 9" thick, a worker will spread, grade, and tamp about 10 ½ cu. yds. per 8-hr. day, at the following labor cost per cu. yd.:

	Hours	Rate	Total	Rate	Total
Labor	0.76	$....	$....	$19.13	$14.54

A worker will spread, grade, and tamp about 12 ½ cu. yds. of 10" or 12" fill per 8-hr. day, at the following labor cost per cu. yd.:

	Hours	Rate	Total	Rate	Total
Labor	0.64	$....	$....	$19.13	$12.24

All fill materials will shrink under compaction--some more than others--and the additional material required to make up the deficiency due to shrinkage, together with the additional labor for handling same, must be provided for in the estimate.

The best and easiest way to do this is to increase the quantity before pricing by applying the proper shrinkage factor for the material to the net computed volume.

For example: If an area 100'-0" x 50'-0" is to receive a gravel fill 6-inches thick after tamping, the gross volume of gravel would be 100'-0" x 50'-0" x 0'-6" x 1.12 = 2,800 cu. ft. or approximately 104 cu. yds. In the computation, 1.12 is the shrinkage factor for gravel.

The following table gives approximate values for shrinkage percentages and shrinkage factors for various fill materials:

Material	Shrinkage Percentage	Shrinkage Factor
Cinder	30	1.30
Crushed Limestone	20	1.20
Granulated Slag	15	1.15
Gravel	12	1.12
Sand	8	1.08

To obtain the cost of one sq. ft. of fill of any thickness, divide the cost per cu. yd. by the number of sq. ft. obtainable from one cu. yd., multiply by the shrinkage factor and the result is the sq. ft. cost.

Forms or Screeds for Concrete Sidewalks. It is customary to use 2"x4" or 2"x6" lumber for forms at each side of a concrete sidewalk, the depth of the forms depending upon the thickness of the concrete. On ordinary sidewalk work, obtain the number of lin. ft. of walk and multiply by 2, which will be the number of lin. ft. of forms required. A sidewalk 4" thick will require 2"x4" lumber; 6" thick, 2"x6" lumber, etc.

Where concrete floors are placed on the ground it is customary to set wood screeds 6'-0" to 8'-0" on centers, so the floor will be level or have a uniform pitch. These screeds are removed as the work progresses and may be used many times. They are usually of 2"x4" lumber, with stakes placed 3 to 4 ft. apart to hold the screeds in place.

Labor Placing Forms and Screeds for Sidewalks and Floors. On all classes of sidewalk work where the forms are either 4", 6" or 8" wide, a carpenter should place and level 250 to 300 lin. ft. per 8-hr. day, at the following labor cost per 100 lin. ft.:

	Hours	Rate	Total	Rate	Total
Carpenter	3.0	$....	$....	$24.04	$72.12
Helper	1.5	....	....	19.13	28.70
Cost per 100 lin. ft			....		$100.82
per lin. ft			....		1.01
per lin. ft. walk			....		2.02

The above includes time removing forms.

When placing screeds for concrete floors on the ground, 2 carpenters working together should drive stakes, place and level 650 to 700 lin. ft. of screeds per 8-hr. day, at the following labor cost per 100 lin. ft.:

	Hours	Rate	Total	Rate	Total
Carpenter	2.4	$....	$....	$24.04	$57.70
Helper	1.2	....	....	19.13	22.96
Cost per 100 lin. ft			$....		$80.66
Cost per lin. ft			....		.81

Estimating the Quantity of Concrete Required for Floors and Walks. To compute the number of sq. ft. of concrete floor or sidewalk that may be obtained from one cu. yd. of concrete (27 cu. ft.), laid out into a cube 3'-0" wide, 3'-0" thick and 3'-0" high. The surface of this cube contains 9 sq. ft. and it is 36" high.

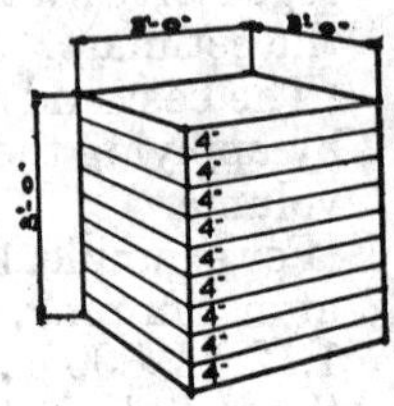

If the rough concrete is 4" thick, divide 36" (the height of the cube) by 4" (the thickness of the slab) and the result is 9; showing that it is possible to obtain 9 slabs, each containing 9 sq. ft. or 81 sq. ft. of 4" floor or walk per cu. yd. of concrete.

Number of Sq. Ft. of Concrete Floor of Any Thickness Obtainable From One Cu. Yd. of Concrete

Inches Thick	No. Sq. Ft.	Inches Thick	No. Sq. Ft.
1	324	7	46
1 ¼	259	7 ¼	44
1 ½	216	7 ½	43
1 ¾	185	7 ¾	42
2	162	8	40
2 ¼	144	8 ¼	39
2 ½	130	8 ½	38
2 ¾	118	8 ¾	37
3	108	9	36
3 ¼	100	9 ¼	35
3 ½	93	9 ½	34
3 ¾	86	9 ¾	33
4	81	10	32
4 ¼	76	10 ¼	31
4 ½	72	10 ½	31
4 ¾	68	10 ¾	30
5	65	11	29 ½
5 ¼	62	11 ¼	29
5 ½	59	11 ½	28
5 ¾	56	11 ¾	27 ½
6	54	12	27
6 ¼	52	12 ¼	26 ½
6 ½	50	12 ½	26
6 ¾	48	12 ¾	25 ½

Mixing and Placing Concrete for Floors and Walks

Mixing Concrete By Hand. In these days of high labor costs, concrete is seldom, if ever, mixed by hand but is usually mixed in a small mixer or ready-mix concrete is used and discharged direct into the forms or onto the floor.

However, for that one in a million job, here is some data on hand mixing, if you can find the men who will work that hard.

As with other classes of concrete work, the cost of mixing and placing the concrete will depend upon the number of times it is turned on the board and the method used in mixing and placing; but on the smaller jobs it will require 3 1/2 to 4 hrs. labor time to mix and place one cu. yd. of concrete, as this class of work does not proceed as fast as foundation work.

On work of this kind, the gang usually consists of 5 or 6 workers, so it is not possible for one individual to work at the same class of work all day but must alternate between wheeling sand, gravel, cement, and mixing concrete, as the job may require.

Five workers together should handle all materials, mix, place, and tamp 20 to 24 batches (1/2-cu. yd. each) or 10 to 12 cu. yds. per 8-hr. day, at the following labor cost per cu. yd.:

	Hours	Rate	Total	Rate	Total
Labor	3.64	$....	$....	$19.13	$69.63

Labor Mixing and Placing Concrete for Floors and Walks Using a Small Mixer. A small concrete mixer of 1/2 or 1 sack (3 1/2-S or 6-S) capacity is ordinarily used for floor and sidewalk work.

Where possible to use a full gang of men and keep the mixer in operation steadily, a mixer should discharge a batch every 2 minutes, but as in all other kinds of construction work there are always delays, such as engine adjustments, repairing mixer, delays caused by non-delivery of sand, gravel or cement, moving and placing runways, cleaning concrete barrows, etc., so the average for a job will be about 20 to 25 batches an hr. or 160 to 200 batches per 8-hr. day.

On the above basis, either a 3 1/2-S or 6-S concrete mixer, the labor cost per cu. yd. should run as follows:

	Hours	Rate	Total	Rate	Total
Labor	2.5	$....	$....	$19.13	$47.83

Using a larger mixer will increase the output but the labor cost will increase in direct proportion. Many contractors have found that a one sack mixer is the most economical unit for floor and sidewalk work.

Labor Placing Ready-Mixed Concrete for Floors. In order to place ready-mixed concrete for floors with the most economy it is necessary to have a crew large enough to handle the concrete quickly to eliminate waiting time charges by the material company and the supply of mixed concrete must be steady and of sufficient quantity to keep the crew busy.

A crew of 14 or 15 laborers, using buggies to wheel concrete, should place 22 to 27 cu. yds. of ready-mixed concrete per hr. at the following labor cost per cu. yd.:

	Hours	Rate	Total	Rate	Total
Labor	0.6	$....	$....	$19.13	$11.48

If the floor is small, such as for a residence, small addition, etc., the work must be organized in a different manner. It is obvious that a job of this sort is too small to keep a number of ready-mix trucks busy and therefore no more than 1 or 2 trucks will serve the placing operation.

Materials Required for 1 Cu. Yd. of Concrete

This table is based on use of average wet sand and on 3- to 4-in. slump. Quantities will vary according to grading of aggregate and workability desired. No allowance has been made for waste. 1 sack cement = 1 cu. ft. 4 sacks cement = 1 bbl.

Size of Aggregate	Mixes					Materials per Cu. Yd. of Concrete			
	Mixing Water* per sack of Cement	Portland Cement	Aggregates		Yield of concrete per Sack of Cement	Water	Portland Cement	Aggregates	
			Sand	Gravel or Crushed Stone				Sand	Gravel or Crushed Stone
	U.S. Gal.	Sacks	Cu. Ft.	Cu. Ft.	Cu. Ft.	U.S. Gal.	Sacks	Cu. Yd.	Cu. Yd.
Maximum size ¾ in.	5	1	2.50	2.75	4.3	31	6.25	0.58	0.64
Maximum size 1 in.	5	1	2.25	3.00	4.5	30	6.00	0.50	0.67
Maximum size 1½ in.	5	1	2.25	3.50	4.7	29	5.75	0.48	0.75

*Based on 6 gal. of water per sack of cement including water contained in average wet sand.

Materials Required for 100 Sq. Ft. of Surface Area for Various Thicknesses of Concrete*

Thickness	Quantity of Concrete	Mixes								
		1:2½:2¾			1:2¼:3			1:2¼:3½		
		Cement	Aggregate		Cement	Aggregate		Cement	Aggregate	
In.	Cu. Yd.	Sacks	Sand Cu. Yd.	Gravel or Crushed Stone Cu. Yd.	Sacks	Sand Cu. Yd.	Gravel or Crushed Stone Cu. Yd.	Sack	Sand Cu. Yd.	Gravel or Crushed Stone Cu. Yd.
3	0.93	5.8	0.54	0.60	5.6	0.47	0.62	5.4	0.45	0.70
4	1.23	7.7	0.71	0.79	7.4	0.61	0.82	7.1	0.59	0.92
5	1.54	9.6	0.89	0.99	9.2	0.77	1.03	8.9	0.74	1.16
6	1.85	11.6	1.07	1.18	11.1	0.93	1.24	10.6	0.89	1.39
8	2.46	15.4	1.43	1.57	14.8	1.23	1.65	14.1	1.18	1.85
10	3.08	19.3	1.79	1.97	18.5	1.54	2.06	17.7	1.48	2.31
12	3.70	23.1	2.14	2.37	22.2	1.85	2.48	21.3	1.78	2.78

No allowance has been made for waste.
*100 sq. ft. of portland cement plaster ½ in. thick requires 1.6. sacks of cement and 0.15 cu. yd. of sand.

Assuming a floor pour serviced by one 7 cu. yd. capacity ready-mix truck, delivering one load of concrete per hr., a crew of 6 or 7 laborers should be kept reasonably busy and still handle the ready-mixed load of concrete fast enough to prevent ready-mix concrete truck waiting time penalties.

Based on the above conditions the labor cost of placing one cu. yd. of floor concrete will be as follows:

	Hours	Rate	Total	Rate	Total
Labor	1.0	$....	$....	$19.13	$19.13

Estimating the Labor Cost Per Sq. Ft. of Floor or Sidewalk. Labor costs given for mixing and placing concrete are based on a cu. yd., but floors and sidewalks are usually estimated by the sq. ft.

Once the cost per cu. yd. has been obtained, divide it by the number of sq. ft. of floor obtainable from a cu. yd. of concrete. The result is the labor cost per sq. ft. of floor. Example: Find the labor cost per sq. ft. of 4" floor. The table on the previous page shows that 81 sq. ft. of 4" floor can be obtained from one cu. yd. of concrete. If the labor cost for concrete placement is $19.13 per cu. yd., $19.13 divided by 81 = $0.236 per sq. ft.

Labor Applying A Finish on Reinforced Concrete Floors. The cost of placing a finish on reinforced concrete floors is subject to wide fluctuations. One of the reasons is the custom of placing the concrete over as large an area as possible and then allowing the concrete to stand until the water disappears from the surface before finishing is started. However, the time required for cement to take its initial set depends upon the brand of cement used, as some cements are noted for being either "quick setting" or "slow setting".

The amount of water used in the concrete mix also has considerable bearing on the finishing cost because if "sloppy" concrete is used it requires much more time for the floor to set sufficiently to receive the finish than if concrete of a rather "stiff" consistency is used.

Architects and engineers specify the floor must be finished at the same time the concrete is placed, and in some instances it necessitates a large amount of overtime for finishers.

Where the gang places concrete right up until quitting time, it is a foregone conclusion that the finishers will have to work far into the night.

The item of overtime should be carefully considered by the estimator for as a usual thing the cement mason starts the finishing or floating work when all other trades have quit for the day. Therefore, provisions should be made and costs added, if required, for lights and for other trades necessary to support the finishers.

While under favorable conditions, some finishers may finish 175 to 200 sq. ft. of floor an hour on certain classes of work, the general average for a reinforced concrete job will vary from 70 to 80 sq. ft. an hr. or 560 to 640 sq. ft. per 8-hr. day.

The labor costs given are for straight time but the estimator should bear in mind that overtime costs time and a half or double time.

The labor per 100 sq. ft. of floor should cost as follows:

	Hours	Rate	Total	Rate	Total
Cement mason	1.3	$....	$....	$24.57	$31.94
Cost per sq. ft			....		.32

Finishing Concrete Floors by Machine. Finishing machines are used for screeding, floating, and troweling. When placing concrete floors, walks, and pavements on the ground, the wood or metal floor screeds are placed in the usual manner and the concrete is struck off by using a power operated screeding machine, which makes 2 ½" transverse strokes on the header boards, leveling and compacting the mix. A steady pull forward by the operator advances the machine at the rate of 10 ft. a minute or 600 lin. ft. per hour and will strike off a slab 6 to 20 feet in width.

A four-worker crew is usually required with the machine; two laborers spreading the concrete and two cement masons operating the screeding machine.

After the concrete slab has been screeded ready for troweling, allow sufficient time for the concrete to set hard enough to walk on before starting finishing.

Concrete Screeding Machine

For floating, the machine is equipped with 12 gauge steel trowels, which revolve on a 46" diameter. It usually requires 3 cement masons; 1 finisher to operate the machine and 2 finishers touching up corners, columns, edging, jointing, etc. On a well prepared slab, one finishing crew can float 4,000 sq. ft. of floor an hr. but this is not usually maintained due to job delays, etc.

A smaller finishing machine having a 34" diameter will float up to 3,000 sq. ft. of floor an hour under the same conditions as stated above.

After the floor has been floated, the heavy gauge steel floating trowels are removed and replaced with steel "finishing" trowels. The cement mason then guides the rotating, adjustable pitch trowels over the slab until a smooth, level surface is obtained.

Using the larger machines, and assuming a "four time over" was necessary to complete the job, it would require 4 hrs. of machine operation and assuming a ¾ to 1-hr. waiting period was necessary between each operation, it would require a total of 6 ¼ to 7 hrs. time to finish 4,000 sq. ft. of floor.

The labor screeding 4,000 sq. ft. of floor should cost as follows:

	Hours	Rate	Total	Rate	Total
Cement mason	3	$....	$....	$24.57	$73.71
Cost per sq. ft			....		.018

Add machine rental charges.

After the floor has been screeded, the labor floating and troweling (4 times over), 4,000 sq. ft. of cement finish floor should cost as follows:

	Hours	Rate	Total	Rate	Total
Cement mason	32	$....	$....	$24.57	$786.24
Cost per sq. ft			....		0.197

Add machine rental charges.

Color for Concrete Floors

Color for concrete is used for sidewalks, driveways, porches, tennis courts, floors, etc., the desired color being mixed into the cement topping which should be either ½, ¾, or 1-in. thick, depending upon service requirements. The amount required varies with the different manufacturers but the following proportions generally will produce good results.

Pigments for COLORED CONCRETE Floors

Color	Commercial names of colors for use with cement	Approximate quantities required--lb. per sack of cement: Light shade	Medium shade	
Greys,	Germantown lampblack*		½	1
blue-black	or carbon black* or	½	1	
and black	black oxide of manganese*		1	2
	or mineral black	1	2	
Blue	Ultramarine blue	5	9	
Brownish red to dull brick red	Red oxide of iron	5	9	
Bright red to vermillion	Mineral turkey red	5	9	
Red sandstone to purplish red	Indian red	5	9	
Brown to reddish-brown	Metallic brown (oxide)	5	9	
Buff, colonial tint and	Yellow ochre or	5	9	
yellow	yellow oxide	2	4	
Green	Chromium oxide or	5	9	
	greenish blue ultramarine		6	

*Only first-quality lampblack should be used. Carbon black is light in weight and requires very thorough mixing. Black oxide or mineral black is probably most advantageous for general use. For black use 11lb. of oxide for each sack of cement.

Approximate Quantities Required for 100 Sq. Ft. of Floor

	½" Top	¾" Top	1" Top
Gray portland cement	1.5 sacks	2.25 sacks	3.00 sacks
Clean sand	150 lbs.	225 lbs.	300 lbs.
Coarse aggregate, ⅜" max. size	263 lbs.	394 lbs.	525 lbs.
Conc. cement color	13.5 lbs.	20.3 lbs.	27 lbs.

For 1" topping, use no more than 5 gal. mixing water, including water in aggregates. Smaller mixes in proportion. Sand and coarse aggregate is based on 100 lbs. per cu. ft. Colors are usually packed in 9-lb. bags, 100-lb. drums and barrels. When mixed in the proportions of 1:2, use 9 lbs. of cement color to each bag of portland cement used in the topping.

It is advisable to establish the mixture from samples made under job conditions and allowed to dry. Careful measurement of every ingredient, including water, in each batch is essential to insure uniformity of color.

Final troweling should be done after the concrete has stiffened, as this assures a smooth, polished finish. Integral waterproofing or hardener can be added to mix materials, also polishing or a surface treatment can be applied, if desired.

Colorundum Surface Colorant and Hardener for Concrete. Colorundum (from W.R. Grace & Co.) is a ready-to-use powder containing non-crushable hardeners, dispersing agents and cementitious binders. It provides a durable, integral concrete color that does not wear off as paint and ordinary surface coatings do. When Colorundum is dusted on and trowelled into freshly poured concrete it produces bright, decorative surfaces with dense coloration and hard, abrasion resistant surfaces with long life.

Colorundum is for interior and exterior concrete floors and sidewalks. The hard wearing, highly decorative finish of Colorundum floors makes them ideal for use in factories, schools, hospitals, office buildings, showrooms, supermarkets, stores or any public building. Colorundum may be used inside or outside on traffic areas such as sidewalks, ramps, patios, terraces, sun decks and playgrounds. For best results, the slump of concrete monolithic slabs should not exceed 2 ½" and for toppings 3". Excess water causes laitance to rise to the surface, producing discoloration, and also causes separation, which delays the finishing operation.

Do not use air entraining agents or other admixtures in the mix. When excess water disappears and the topping will hold up knee boards, uniformly dust on ⅔ of specified amount of Colorundum per 100 sq. ft. of surface.

After spreading, allow the dry material to wet up. It should then be wood floated and worked into the slab. The first floating should be discontinued as soon as the surface becomes wet. Floating should be resumed when surface moisture has disappeared. Do not steel trowel. Immediately apply the remaining ⅓ of the specified amount per 100 sq. ft. and thoroughly wood float. (For heavy duty finish, the quantity of Colorundum may be increased for the second dust coat.) Apply Colorundum exactly as it comes from the container; do not mix with cement or concrete.

Steel trowel to an even plane free from surface marks and voids. When the floor has obtained its initial set, give it a final steel trowel burnishing. Leave exterior areas, such as sidewalks, under wood float finish after the trowelling.

To cure, use an approved curing and sealing compound according to manufacturer's instructions or cover the surface after finish trowelling with polyethylene. This seals against evaporation of the water so vitally necessary to hydrate the cement, thereby developing optimum strength and density throughout the mass. It also protects the colored surface against spillages.

Use approximately 40 lbs. per 100 sq. ft. for normal traffic areas. For areas subject to extra heavy traffic, the quantity of Colorundum may be increased proportionately.

Gilco Non-Shrink Grouting Compound. Gilco Grout (from Gifford-Hill & Co. Inc.) is a premixed compound containing iron particles, hardening and dispersing agents, binders, and oxidizing agents. When mixed with water, it produces a mass which has controlled expansion to overcome the inherent shrinkage in cementitious mixtures. The metallic aggregate also provides a ductile surface with the "give and take" properties concrete must have to absorb the vibration of machinery, equipment and columns.

Gilco grout can be used wherever non-shrink concrete is mandatory: setting machinery on concrete foundations; full bearing surface grouting for building columns; and grouting anchor bolts, floor grids, steel sash and jambs.

Grouting should be continuous once it has started. Do not mix more grout at one time than can be rodded and placed in a period of 20 minutes. Add only enough water to make mix placeable--avoid an excess of water.

Anchor forms securely to prevent movement during placing and curing. Intersection of form and base slab can be filled with a mix of equal parts of cement and sand before placing grout. Be sure to allow adequate clearance between forms and base plate.

Remove waste material and water from anchor bolt holes. All oil, grease and paint in contact with grout should be thoroughly removed from base plates before grouting.

It is important to place grout quickly and continuously to avoid all effects of overworking, resulting in segregation, bleeding, and break-down of initial set. Under no circumstances should the grout be retempered by introducing additional water after it has taken its initial set.

Lightweight Concrete Floor Fill

Concrete made of lightweight aggregate, such as cinders, slag, Haydite, pumice, vermiculite and other lightweight aggregates, is often used for floor fill under marble and tile floors, between wood floor sleepers, screeds, etc.

The labor cost of placing concrete floor fill runs considerably higher than for ordinary concrete because the fill is usually only 2" or 3" thick, which necessitates covering a large floor area to place a comparatively small amount of concrete. In many instances it is necessary to pump or hoist the concrete to the upper floors in wheelbarrows or concrete buggies and then wheel it to the different parts of the building where it is to be used.

When placing concrete fill between wood floor sleepers or screeds, it is important that every precaution be taken to prevent the sleepers from being knocked out of level when wheeling over them.

When placing concrete floor fill, it is not advisable to use a large mixer because of the necessity of covering a large floor area to place a small amount of concrete. It usually requires considerable leveling and tamping, making it impractical to handle the concrete in large quantities. A concrete mixer of 1 or 2 sacks capacity is best adapted for this class of work or partial loads of ready-mix can be ordered.

Mixing and Placing Concrete Floor Fill. Because of the additional labor required for spreading, grading and tamping concrete floor fill, the cu. yd. labor costs run higher than for foundation or sidewalk work.

There is also more labor required wheeling concrete from the mixer to the hoist and from the hoist to place of deposit, as floor fill is usually placed after the structural portions of the building have been completed, and it is not possible to carry on the work with much speed.

A crew of 13 workers with a 6-S mixer should handle materials, mix, and wheel into place 15 to 20 batches (containing 6 to 7 cu. ft. each) or 3 ½ to 5 cu. yds. of concrete an hour with the men distributed as follows : 2 workers wheeling coarse aggregate from stock pile to mixer; 2 workers wheeling sand and cement to mixer; 1 worker attending mixer; 3 workers wheeling concrete from mixer to hoist; 3 workers wheeling concrete from hoist to place of deposit; 2 workers dumping barrows and spreading concrete.

The labor per cu. yd. should cost as follows:

	Hours	Rate	Total	Rate	Total
Labor	3.00	$....	$....	$19.13	$57.39
Hoisting engineer	0.35	....	....	25.96	9.09
Cost per cu. yd			$....		$66.48

If union rules require a mixer engineer, deduct 0.35-hr. labor time and add 0.35-hr. engineer time per cu. yd.

If concrete is discharged from mixer directly into a hoist concrete bucket and wheeled from a floor hopper to place of deposit, deduct 0.50-hr. labor time.

If no hoisting is required, deduct 0.70-hr. labor time and omit hoisting engineer.

For concrete roof fill on flat roofs, deduct about 10% from the above costs. On pitch or gable roofs, add 25% to 50%, depending on steepness of slope, accessibility, etc.

After the cubic yard cost has been computed, to obtain the sq. ft. cost, divide the cu. yd. cost by the number of sq. ft. of fill obtainable from one cu. yd. and the result will be the labor cost per sq. ft. of fill. Example: Find the sq. ft. labor cost of 3" floor fill with the cu. yd. cost $43.90. The table on the previous pages gives 108 sq. ft. of 3" fill obtainable from one cu. yd. $43.90 divided by 108 = $0.406 per sq. ft.

To the above costs must be added the cost of setting screeds and all finishing operations.

Finishing Lightweight Concrete Floor and Roof Fill

All concrete floor and roof fills require some finishing labor, if only setting of screeds and striking off fill at screed height.

Where fill is placed between wood sleepers, screeds are not required but the fill must be struck off and darbied so that no part of it is higher than the top of the sleepers. In addition, the fill should be spaded along both sides of each sleeper so that good anchorage is obtained.

For 2" thick concrete fill between wood floor sleepers, 2 cement masons working with a helper should strike off and darby 5,000 to 6,000 sq. ft. per 8-hr. day or about all the fill a small-mixer crew can mix and place in an 8- hr. day. The labor per 100 sq. ft. should cost as follows:

	Hours	Rate	Total	Rate	Total
Cement mason	0.30	$....	$....	$24.57	$7.37
Helper	0.15	....	....	19.13	2.87
Cost per 100 sq. ft			$....		$10.24
Cost per sq. ft			....		.10

Finishing Concrete Floor Fill For Resilient Flooring. Where concrete fill is used as a base for cork, asphalt, vinyl, rubber or vinyl asbestos floor covering, it is necessary to obtain a smooth level finish on the concrete to receive the finish floor material or all irregularities will show through on the finished surface.

To obtain the required degree of smoothness, screeds must be placed 4'- 0" to 6'-0" on centers and concrete fill must be struck off, darbied, floated and troweled.

If the fill is not over 2" thick, 2 cement masons working together should strike off, darby, float and trowel about 1,600 sq. ft. of floor per 8-hr. day at the following labor cost per 100 sq. ft.:

	Hours	Rate	Total	Rate	Total
Cement mason	1.0	$....	$....	$24.57	$24.57
Cost per sq. ft			....		.25

Finishing Concrete Roof Fill. Concrete fill for flat roofs are usually struck off, darbied and floated. If fill is not over 2" to 3" thick, 2 cement masons with a helper should place screeds, strike off, darby and float 1,500 to 1,700 sq. ft. of roof fill per 8-hr. day at the following labor cost per 100 sq. ft.:

	Hours	Rate	Total	Rate	Total
Cement mason	1.0	$....	$....	$24.57	$24.57
Helper	0.5	....	....	19.13	9.57
Cost per 100 sq. ft			$....		$34.14
Cost per sq. ft			....		.34

Where concrete fill is placed on pitch or gable roofs, finishing is more difficult, screeds must be set closer together to help hold the fill from sliding down the slope and an additional helper is usually required. On work of this type, 2 cement masons and 2 helpers should place screeds, strike off, darby, and float 900 to 1,000 sq. ft. of fill per 8-hr. day at the following labor cost per 100 sq. ft.:

	Hours	Rate	Total	Rate	Total
Cement mason	1.60	$....	$....	$24.57	$39.31
Helper	1.60	....	....	19.13	30.61
Cost per 100 sq. ft			$....		$69.92
Cost per sq. ft			....		.70

CONCRETE SLAB ON GRADE CONSTRUCTION

Previously this type of floor was built without adequate test data or established construction standards relating to conditions involving comfort--dampness and floor temperature. Many such floors have proved to be cold, especially at the outer edge. Some have also been damp enough to damage floor coverings and wall construction.

To determine the most satisfactory floor from the standpoint of comfort, the University of Illinois studied the construction of concrete floor slabs which are laid on the ground and which are designed for use in climates where central heating is necessary (reprinted by permission of Small Homes Council from its circular F4.3, "Concrete Floors for Basementless Houses").

Nine different types of concrete floors were tested in an effort to determine the proper design for slab on grade houses with respect to:

1. Heat losses of the different types of floors.
2. Temperatures at various points throughout the floors. This information was sought for proper placement of floor insulation.
3. The amount of moisture passing from the ground to the top of the concrete slab.

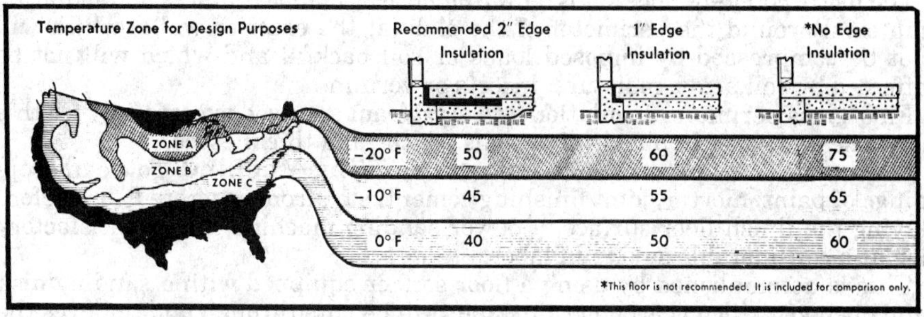

Insulation Requirements for Different Parts of the Country

The floors selected for tests represented standard construction practices and, at the same time, permitted the use of varying amounts and types of insulation along the edges. The floor slabs were tested simultaneously under similar conditions in a specially constructed laboratory.

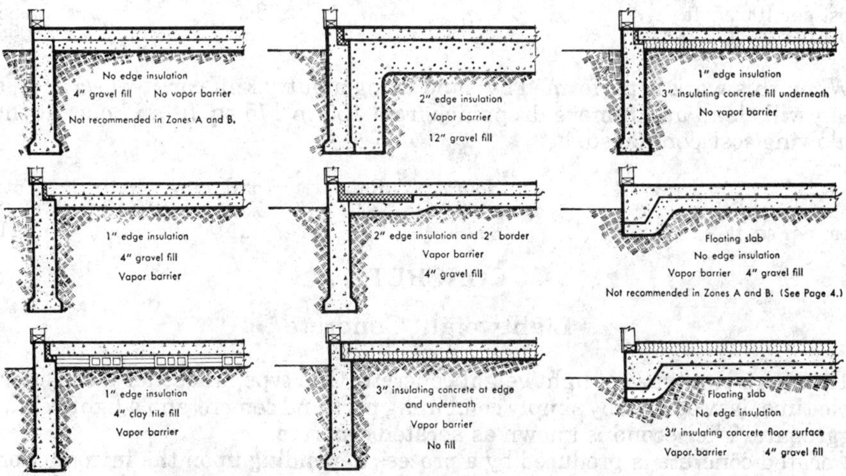

Methods of Constructing Concrete Floors for Basementless Houses

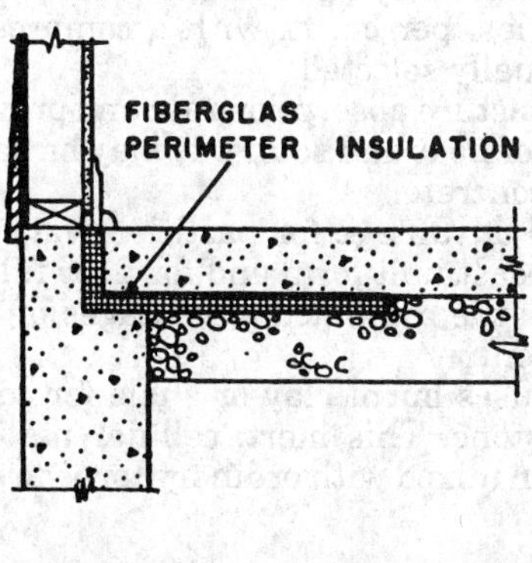

When panel heating systems are used with concrete floors, the heat loss to the ground and at the edge is increased due to the higher temperatures maintained in the floors. To prevent excessive heat losses in such floors, a minimum of 2 inches of rigid waterproof insulation should be provided at the edge. In addition, the use of insulation under the entire heated floor area is recommended. Gravel and rock fills are desirable for drainage controls, but they have no insulating value.

The use of concrete floor slabs for basementless homes requires an insulating material around the perimeter of the slab (at the exterior walls) which will not be compressed by imposed loads or soil backfill and which will not be affected by soil acids, moisture, insects or vermin.

Fiberglass perimeter insulation and styrofoam are used successfully for this purpose and come in panels of various widths and thicknesses.

Cleaning Concrete Floors Using a Sanding Machine. To clean droppings of paint, mortar, joint finishing cement, etc., from concrete floors before laying the finish floor surface, a power sanding machine does a satisfactory job at a considerable saving in labor.

This is accomplished by using a floor sander equipped with a sanding disk and open grit silicon carbon grain paper, which cuts through and removes the droppings satisfactorily.

A man operating a sanding machine will clean about 500 sq. ft. of floor an hour at the following cost per 100 sq. ft.:

	Hours	Rate	Total	Rate	Total
Machinery cost	0.2	$....	$....	$5.00	$1.00
Sanding disk		....	....	.75	.75
Labor	0.2	....	....	19.13	3.83
Cost per 100 sq. ft			$....		$5.58
Cost per sq. ft			....		0.056

When this work is performed by hand using a putty knife or steel scraper, a man will clean and remove droppings from 150 to 175 sq. ft. an hour, at the following cost per 100 sq. ft.:

	Hours	Rate	Total	Rate	Total
Labor	0.6	$....	$....	$19.13	$11.48
Cost per sq. ft			....		.11

CONCRETE

Lightweight Concrete

There are two types of lightweight concrete. One type, in use for a considerable time, is produced by simply combining portland cement and a lightweight aggregate. The second is known as aerated concrete.

Aerated concrete is produced by a process depending upon the introduction of certain chemicals to generate gases which cause the mass to expand. The weight of aerated concrete may be controlled with a variation of from 20 lbs. per cu. ft. to approximately the weight of ordinary lightweight concrete. For average conditions a weight of 40 to 50 lbs. per cu. ft. with a compressive strength of about 500 lbs. per sq. in. is usually selected.

Aerated concrete is fireproof, has a low moisture absorption rate and provides good insulation against the transmission of heat and sound. It may be sawed or nailed as readily as other lightweight concrete.

All lightweight concretes are of decided advantage for partition walls and fireproofing, whether in monolithic or precast unit form and for floor fill and roof slabs. The estimator should consider the combined advantages of each type of lightweight concrete before selection.

Haydite Concrete. Haydite concrete uses burnt clay or shale for coarse aggregate instead of gravel or crushed stone. This inert, cellular material weighs less than 50 lbs. per cu. ft. and when mixed with ordinary torpedo sand

and portland cement, produces concrete weighing about 98 lbs. per cu. ft. instead of 150 lbs. for ordinary concrete.

Full structural strengths can be obtained with Haydite concrete and in many instances a redesign based on lighter weight concrete and reduced dead load more than pays for the slight additional cost of the aggregates.

Perlite Lightweight Concrete. For lightweight concrete using Perlite as the aggregate, refer to "Concrete Floors and Walks".

Pottsco Lightweight Aggregate. Pottsco lightweight aggregate is an expanded blast furnace slag. By special processes, molten slag is converted into hard, cellular clinker, which is crushed and screened to commercial sizes, ready for use with portland cement and water, without the addition of sand or other materials. It is supposed to be chemically inert and has no corrosive effect on structural or reinforcing steel, plaster or paint.

Pumice Lightweight Concrete. For lightweight concrete using Pumice as the aggregate, refer to "Concrete Floors and Walks".

Vermiculite Lightweight Concrete. Refer to "Vermiculite Concrete" section.

Waylite Lightweight Concrete. For lightweight concrete using Waylite as the aggregate, refer to "Concrete Floors and Walks".

Reinforced Concrete

After the forms and reinforcing steel are in place, the job is ready for concrete. This should include the cost of gravel or crushed stone, sand and cement used in the concrete mixture as well as the labor cost of handling materials, mixing, hoisting, placing and curing the concrete.

For many years concrete used in reinforced concrete construction was specified to be mixed in the proportions of 1:1:2, 1:1 ½:3, 1:2:4, etc., depending upon the class of work for which it was used. More modern practice specifies that concrete used for specific purposes shall develop a compressive strength of 2,500, 3,000, 3,750 or 4,000 lbs. or more per sq. in. at 28 days.

On nearly all types of reinforced concrete building work, such as columns, beams, girders, floor slabs, etc., the coarse aggregate is graded from ¼" to 1" in size. Where conditions permit, however, coarse aggregate graded from ¼" to 1 ½" is used.

Estimating the Quantity of Concrete Required for Various Types of Concrete Floors

If all concrete floors were merely solid slabs of concrete, it would be an easy matter to compute the quantity of concrete required per sq. ft. of floor but with so many combination floors consisting of metal pans, etc, having joists 4" to 7" wide, 4" to 15" deep and from 16" to 35" on centers, it requires considerable figuring to obtain accurate concrete quantities.

It is practically impossible to estimate accurately the amount of concrete required per sq. ft. of floor for the various combination joist floors, as they invariably have T beam construction at all beams and girders, solid concrete slabs usually extend in 6" to 1'-6" from the outside walls, the joists vary in width and are often doubled under partitions, around stair wells, elevator shafts, etc. The only accurate method is to figure the floors as though they were solid concrete and then deduct for the displacement of the tile or pans.

Quantity of Concrete Required for Steelform and Concrete Joist Floor Construction. Floors of this type should be estimated by obtaining the entire slab volume as though of solid concrete and deduct for the displacement of the steelforms.

The deductions for steelform area in floors of this type is not simple since steelforms are furnished in a number of different sizes, with tapering sides and ends.

Steelforms furnished by different manufacturers vary slightly in shape but not sufficient to materially affect the concrete quantities. Ceco steelforms (The Ceco Corporation, Oak Brook, IL) have been used in the tables as being typical for general use.

Sizes and Cu. Ft. Displacement of Flange Type Steelforms

Ceco flange type steelforms slope 1" in each 12" in height. Example: a 12" pan is 20" wide at the bottom and 18" wide at the top, because of the 1" slope on each side. See illustrations on following pages and in the "Concrete Formwork" section.

The sides of single tapered endforms slope 1" for each 12" in height and one end slopes 1/4" for each 1" in height. For estimating purposes their displacement is the same as for standard steelforms in the following tables.

Sizes and Cu. Ft. Displacement of Tapered End Forms

Depth of Steelform Inches	Width at Wide End	Width at Narrow End	Average Width	Length in Feet	Cu. Ft. Displacement per 3-Ft. Section	Difference Between Tapered end forms and 3'-0" length of straight forms in Cu. Ft.
8"	20"	16"	18"	3'-0"	2.882	0.322
10"	20"	16"	18"	3'-0"	3.569	0.399
12"	20"	16"	18"	3'-0"	4.243	0.475
8"	30"	25"	27 ½"	3'-0"	4.465	0.406
10"	30"	25"	27 ½"	3'-0"	5.548	0.504
12"	30"	25"	27 ½"	3'-0"	6.617	0.602
14"	30"	25"	27 ½"	3'-0"	7.763	0.698
16"	30"	25"	27 ½"	3'-0"	8.715	1.188
20"	30"	25"	27 ½"	3'-0"	10.756	

Sizes and Cu. Ft. Displacement of Ceco Flange Type Steelforms

Depth of Steelform Inches	Width at Bottom	Width at Top	Average Width	Cu. Ft. Displacement per Lin. Ft.
8"	10"	8 ⅔"	9 ⅓"	0.516
10"	10"	8 ⅓"	9 ⅙"	0.634
12"	10"	8"	9"	0.748
14"	10"	7 ⅔"	8 ⅚"	0.857
16"	10"	7 ⅓"	8 ⅔"	0.961
20"	10"	--	--	1.155
8"	15"	13 ⅔"	14 ⅓"	0.794
10"	15"	13 ⅓"	14 ⅙"	0.982
12"	15"	13"	14"	1.165
14"	15"	12 ⅔"	13 ⅚"	1.343
16"	15"	12 ⅓"	13 ⅔"	1.516
20"	15"	--	--	1.850
8"	20"	18 ⅔"	19 ⅓"	1.072

Depth of Steelform Inches	Width at Bottom	Width at Top	Average Width	Cu. Ft. Displacement per Lin. Ft.
10"	20"	18 ⅓"	19 ⅙"	1.329
12"	20"	18"	19"	1.581
14"	20"	17 ⅔"	18 ⅚"	1.829
16"	20"	17 ⅓"	18 ⅔"	2.072
20"	20"	--	--	2.544
8"	30"	28 ⅔"	29 ⅓"	1.628
10"	30"	28 ⅓"	29 ⅙"	2.023
12"	30"	28"	29"	2.414
14"	30"	27 ⅔"	28 ⅚"	2.801
16"	30"	27 ⅓"	28 ⅔"	3.183

Sizes and Cu. Ft. Displacement of Adjustable Steelforms

Depth of Steelform Inches	Width at Bottom	Width at Top	Cu. Ft. Displacement per Lin. Ft.
8"	10"	8"	0.514
10"	10"	8"	0.653
12"	10"	8"	0.792
14"	10"	8"	0.931
15"	10"	8"	1.000
8"	15"	13"	0.792
10"	15"	13"	1.000
12"	15"	13"	1.208
14"	15"	13"	1.417
15"	15"	13"	1.521
8"	20"	18"	1.069
10"	20"	18"	1.347
12"	20"	18"	1.625
14"	20"	18"	1.903
15"	20"	18"	2.042
8"	30"	28"	1.625
10"	30"	28"	2.042
12"	20"	28"	2.458
14"	30"	28"	2.875
15"	30"	28"	3.083

Cu. Ft. of Concrete Displaced by Adjustable Steelforms and Tapered Endforms

Steelform Depth	Width	10'	11'	12'	13'	14'	15'	16'	17'	18'	19'
8"	30"	15.33	16.96	18.58	20.21	21.83	23.46	25.08	26.71	28.33	29.96
10"	30"	19.33	21.38	23.42	25.46	27.50	29.54	31.59	33.63	35.67	37.71
12"	30"	23.33	25.79	28.25	30.71	33.16	35.62	38.08	40.54	43.00	45.45
14"	30"	27.33	30.21	33.08	35.96	38.83	41.71	44.58	47.46	50.33	53.21
15"	30"	29.33	32.42	35.50	38.58	41.66	44.75	47.83	50.91	54.00	57.08

Steelform Depth	Width	20'	21'	22'	23'	24'	25'	26'	27'	28'	29'
8"	30"	31.58	33.21	34.83	36.46	38.08	39.71	41.33	42.96	44.58	46.21
10"	30"	39.75	41.80	43.84	45.88	47.92	49.96	52.01	54.05	56.09	58.13
12"	30"	47.91	50.37	52.83	55.29	57.74	60.20	62.66	65.12	67.58	70.03
14"	30"	56.08	58.96	61.83	64.71	67.58	70.46	73.33	76.21	79.08	81.96
15"	30"	60.16	63.25	66.33	69.41	72.49	75.58	78.66	81.74	84.83	87.91

Cu. Ft. of Concrete Displaced by Flange Type Steelforms and Straight Endforms

Steelform Depth	Width	1'	2'	3'	4'	5'	6'	7'	8'	9'	10'
		Length of Each Row of Steelforms in Feet									
8"	10"	0.52	1.04	1.55	2.07	2.59	3.11	3.63	4.14	4.66	5.18
10"	10"	0.64	1.27	1.91	2.54	3.18	3.82	4.45	5.09	5.72	6.36
12"	10"	0.75	1.50	2.25	3.00	3.75	4.50	5.25	6.00	6.75	7.50
14"	10"	0.86	1.72	2.58	3.44	4.30	5.15	6.01	6.87	7.73	8.59
16"	10"	0.96	1.92	2.88	3.84	4.80	5.75	6.71	7.67	8.63	9.59

		11'	12'	13'	14'	15'	16'	17'	18'	19'	20'
8"	10"	5.70	6.22	6.73	7.25	7.77	8.29	8.81	9.32	9.84	10.36
10"	10"	7.00	7.62	8.27	8.90	9.54	10.18	10.81	11.45	12.08	12.72
12"	10"	8.25	9.00	9.75	10.50	11.25	12.00	12.75	13.50	14.25	15.00
14"	10"	9.45	10.31	11.17	12.03	12.89	13.74	14.60	15.46	16.32	17.18
16"	10"	10.55	11.51	12.47	13.43	14.39	15.34	16.30	17.26	18.22	19.18

Steelform Depth	Width	1'	2'	3'	4'	5'	6'	7'	8'	9'	10'
		Length of Each Row of Steelforms in Feet									
8"	15"	0.80	1.59	2.39	3.18	3.98	4.78	5.57	6.37	7.16	7.96
10"	15"	0.98	1.97	2.95	3.94	4.92	5.90	6.89	7.87	8.86	9.84
12"	15"	1.17	2.33	3.50	4.67	5.84	7.00	8.17	9.34	10.50	11.67
14"	15"	1.35	2.69	4.04	5.38	6.73	8.07	9.42	10.76	12.11	13.45
16"	15"	1.48	2.96	4.44	5.92	7.40	8.88	10.36	11.84	13.32	14.80

		11'	12'	13'	14'	15'	16'	17'	18'	19'	20'
8"	15"	8.76	9.55	10.35	11.14	11.94	12.74	13.53	14.33	15.12	15.92
10"	15"	10.82	11.81	12.79	13.78	14.76	15.74	16.73	17.71	18.70	19.68
12"	15"	12.84	14.00	15.17	16.34	17.51	18.67	19.84	21.01	22.17	23.34
14"	15"	14.80	16.14	17.49	18.83	20.18	21.52	22.87	24.21	25.56	26.90
16"	15"	16.28	17.76	19.24	20.72	22.20	23.68	25.16	26.64	28.12	29.60

Steelform Depth	Width	1'	2'	3'	4'	5'	6'	7'	8'	9'	10'
		Length of Each Row of Steelforms in Feet									
8"	20"	1.07	2.15	3.22	4.30	5.37	6.44	7.52	8.59	9.67	10.74
10"	20"	1.33	2.66	3.99	5.32	6.66	7.99	9.32	10.65	11.98	13.31
12"	20"	1.58	3.17	4.75	6.33	7.92	9.50	11.08	12.66	14.25	15.83
14"	20"	1.83	3.66	5.49	7.32	9.16	10.99	12.82	14.65	16.48	18.31
16"	20"	2.07	4.14	6.21	8.28	10.35	12.42	14.49	16.56	18.64	20.70

		11'	12'	13'	14'	15'	16'	17'	18'	19'	20'
8"	20"	11.81	12.89	13.96	15.04	16.11	17.18	18.26	19.33	20.41	21.48
10"	20"	14.64	15.97	17.30	18.63	19.97	21.30	22.63	23.96	25.29	26.62
12"	20"	17.41	19.00	20.58	22.16	23.75	25.33	26.91	28.49	30.08	31.66
14"	20"	20.14	21.97	23.80	25.63	27.47	29.30	31.13	32.96	34.79	36.62
16"	20"	22.77	24.84	26.91	28.98	31.05	33.12	35.19	37.26	39.33	41.40

Steelform Depth	Width	1'	2'	3'	4'	5'	6'	7'	8'	9'	10'
		Length of Each Row of Steelforms in Feet									
8"	30"	1.63	3.26	4.89	6.52	8.15	9.77	11.40	13.03	14.66	16.29
10"	30"	2.03	4.05	6.08	8.10	10.13	12.15	14.18	16.20	18.23	20.25
12"	30"	2.42	4.83	7.25	9.67	12.09	14.50	16.92	19.34	21.75	24.17
14"	30"	2.80	5.61	8.41	11.21	14.02	16.82	19.62	22.42	25.23	28.03
16"	30"	3.30	6.60	9.90	13.20	16.50	19.80	23.10	26.40	29.70	33.00

		11'	12'	13'	14'	15'	16'	17'	18'	19'	20'
8"	30"	17.92	19.55	21.18	22.81	24.44	26.06	27.69	29.32	30.95	32.58
10"	30"	22.28	24.30	26.33	28.35	30.38	32.40	34.43	36.45	38.48	40.50
12"	30"	26.59	29.00	31.42	33.84	36.26	38.67	41.09	43.51	45.92	48.34
14"	30"	30.83	33.64	36.44	39.24	42.05	44.85	47.65	50.45	53.26	56.06
16"	30"	36.30	39.60	42.90	46.20	49.50	52.80	56.10	59.40	62.70	66.00

Cu. Ft. of Concrete Displaced by Adjustable Steelforms and Straight Endforms

Steelform		Length of Each Row of Steelforms in Feet									
Depth	Width	1'	2'	3'	4'	5'	6'	7'	8'	9'	10'
8"	10"	0.51	1.03	1.54	2.06	2.57	3.08	3.60	4.11	4.63	5.14
10"	10"	0.65	1.31	1.96	2.61	3.27	3.92	4.57	5.22	5.88	6.53
12"	10"	0.79	1.58	2.38	3.17	3.96	4.75	5.54	6.34	7.13	7.92
14"	10"	0.93	1.86	2.79	3.72	4.66	5.59	6.52	7.45	8.38	9.31
15"	10"	1.00	2.00	3.00	4.00	5.00	6.00	7.00	8.00	9.00	10.00
		11'	12'	13'	14'	15'	16'	17'	18'	19'	20'
8"	10"	5.65	6.17	6.68	7.20	7.71	8.22	8.74	9.25	9.77	10.28
10"	10"	7.18	7.84	8.49	9.14	9.80	10.45	11.10	11.75	12.41	13.06
12"	10"	8.71	9.50	10.30	11.09	11.88	12.67	13.46	14.26	15.05	15.84
14"	10"	10.24	11.17	12.10	13.03	13.97	14.90	15.83	16.76	17.69	18.62
15"	10"	11.00	12.00	13.00	14.00	15.00	16.00	17.00	18.00	19.00	20.00

Steelform		Length of Each Row of Steelforms in Feet									
Depth	Width	1'	2'	3'	4'	5'	6'	7'	8'	9'	10'
8"	15"	0.79	1.58	2.38	3.17	3.96	4.75	5.54	6.34	7.13	7.92
10"	15"	1.00	2.00	3.00	4.00	5.00	6.00	7.00	8.00	9.00	10.00
12"	15"	1.21	2.42	3.62	4.83	6.04	7.25	8.46	9.66	10.87	12.08
14"	15"	1.42	2.83	4.25	5.67	7.09	8.50	9.92	11.34	12.75	14.17
15"	15"	1.52	3.04	4.56	6.08	7.61	9.13	10.65	12.17	13.69	15.21
		11'	12'	13'	14'	15'	16'	17'	18'	19'	20'
8"	15"	8.71	9.50	10.30	11.09	11.88	12.67	13.46	14.26	15.05	15.84
10"	15"	11.00	12.00	13.00	14.00	15.00	16.00	17.00	18.00	19.00	20.00
12"	15"	13.29	14.50	15.70	16.91	18.12	19.33	20.54	21.74	22.95	24.16
14"	15"	15.59	17.00	18.42	19.84	21.26	22.67	24.09	25.51	26.92	28.34
15"	15"	16.73	18.25	19.77	21.29	22.82	24.34	25.86	27.38	28.90	30.42

Steelform		Length of Each Row of Steelforms in Feet									
Depth	Width	1'	2'	3'	4'	5'	6'	7'	8'	9'	10'
8"	20"	1.07	2.14	3.21	4.28	5.35	6.41	7.48	8.55	9.62	10.69
10"	20"	1.35	2.69	4.04	5.39	6.74	8.08	9.43	10.78	12.12	13.47
12"	20"	1.63	3.25	4.88	6.50	8.13	9.75	11.38	13.00	14.63	16.25
14"	20"	1.90	3.81	5.71	7.61	9.52	11.42	13.32	15.22	17.13	19.03
15"	20"	2.04	4.08	6.13	8.17	10.21	12.25	14.29	16.34	18.38	20.42
		11'	12'	13'	14'	15'	16'	17'	18'	19'	20'
8"	20"	11.76	12.83	13.90	14.97	16.04	17.10	18.17	19.24	20.31	21.38
10"	20"	14.82	16.16	17.51	18.86	20.21	21.55	22.90	24.25	25.59	26.94
12"	20"	17.88	19.50	21.13	22.75	24.38	26.00	27.63	29.25	30.88	32.50
14"	20"	20.93	22.84	24.74	26.64	28.55	30.45	32.35	34.25	36.16	38.06
15"	20"	22.46	24.50	26.55	28.59	30.63	32.67	34.71	36.76	38.80	40.84
		10'	11'	12'	13'	14'	15'	16'	17'	18'	19'
8"	30"	16.25	17.88	19.50	21.13	22.75	24.38	26.00	27.63	29.25	30.88
10"	30"	20.42	22.46	24.50	26.55	28.59	30.63	32.67	34.71	36.76	38.80
12"	30"	24.58	27.04	29.50	31.95	34.41	36.87	39.33	41.79	44.24	46.70
14"	30"	28.75	31.63	34.50	37.38	40.25	43.13	46.00	48.88	51.75	54.63
15"	30"	30.83	33.91	37.00	40.08	43.16	46.25	49.33	52.41	55.49	58.58
		20'	21'	22'	23'	24'	25'	26'	27'	28'	29'
8"	30"	32.50	34.13	35.75	37.38	39.00	40.63	42.25	43.88	45.50	47.13
10"	30"	40.84	42.88	44.92	46.97	49.01	51.05	53.09	55.13	57.18	59.22
12"	30"	49.16	51.62	54.08	56.53	58.99	61.45	63.91	66.37	68.82	71.28
14"	30"	57.50	60.38	63.25	66.13	69.00	71.88	74.75	77.63	80.50	83.38
15"	30"	61.66	64.74	67.83	70.91	73.99	77.08	80.16	83.24	86.32	89.41

Sizes and Cu. Ft. Displacement of Ceco Flange Type Single Tapered Endforms

Endforms are 20" wide at one end tapering to 16" at the other end. The 30" endforms are 30" wide at one end tapering to 25" wide at other end.

Sizes and Cu. Ft. Displacement of Ceco Adjustable Tapered Steel Endforms

Tapered endforms 3'-0" long are furnished for 30" wide steelforms only. Straight endforms 3'-0" long are furnished for the 30" standard widths and the 20", 15" and 10" special widths.

Depth of Steelform Inches	Width at Wide End	Width at Narrow End	Length in Feet	Cu. Ft. Displacement per 3-Ft. Section
8"	30"	25"	3'-0"	4.317
10"	30"	25"	3'-0"	5.483
12"	30"	25"	3'-0"	6.629
14"	30"	25"	3'-0"	7.775
15"	30"	25"	3'-0"	8.347

Another Method of Estimating the Concrete Required for Ceco Steelform Construction. The following method of estimating the concrete requirements for Ceco steelform construction (The Ceco Corporation, Oak Brook, IL) is that recommended by the manufacturers.

Figure the concrete as though it were a solid slab and then deduct for displacement of a certain number of lineal feet of straight metal steelforms and tapered endforms.

Ceco steelform construction is a combination of concrete joists and thin slabs. In computing the amount of concrete in the floor construction of a building designed for Ceco steelforms, it is more convenient, and there is less chance for error, to multiply the floor area by the equivalent thickness of the floor construction than to separately figure the concrete in the joists by multiplying the lineal feet of joists by their unit of volume and adding thereto the concrete in the thin slabs.

The cubic feet of concrete per square foot of floor (or roof) area of Ceco steelform construction, is a function of the thickness of the thin slab, the depth of the joists, and the width and spacing of the joists. The additional concrete in the joists at their ends as formed by tapered endforms cannot be pro-rated on a square foot basis. The concrete added to the joists by the tapered endforms must be figured per lineal foot of beam or wall into which the joists frame, and this unit of volume must be multiplied by the number of lineal feet of beams and walls along which the tapered endforms are set, to determine the total amount of concrete added by the use of the tapered endforms.

For convenience in estimating, an arrangement in tabular form of the quantities of concrete required for Ceco flange type steelform construction, is given on the following pages. In this connection it is well to bear in mind that the strength of the metal of which Ceco steelforms are made permits sharp angles in the steelforms and an absolutely flat top, thus requiring a minimum of concrete. These tables should, therefore, not be used in computing the concrete for other types of metal form construction.

Typical Cross-section of Flangeforms

All dimensions are outside to outside.

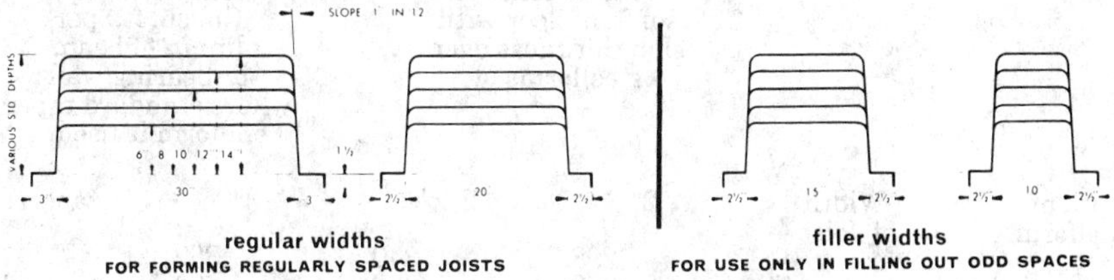

Sizes of Ceco Flange Type Steelforms

Typical Cross-sections of Longforms

All widths and depths are outside dimensions.

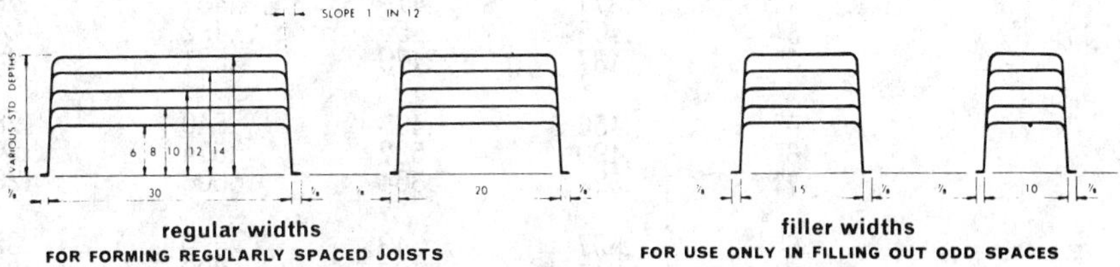

Sizes of Ceco Longform Steelforms

Typical Cross-sections of Adjustable Steelforms

All dimensions are outside to outside.

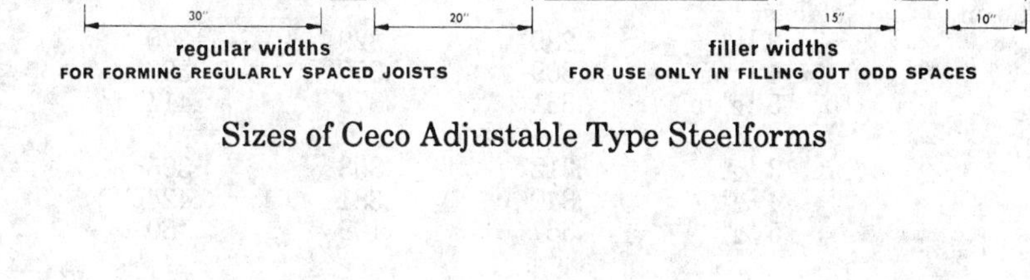

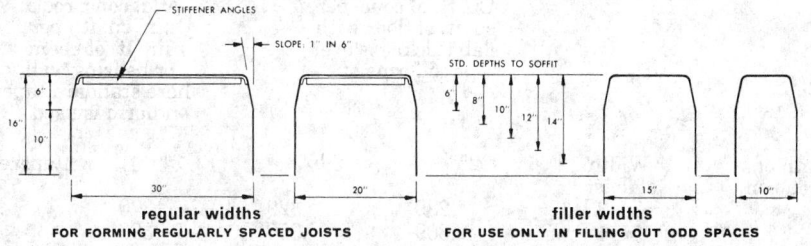

Sizes of Ceco Adjustable Type Steelforms

Concrete Required per Sq. Ft. of Floor Using Ceco Flange* Type Steelforms 20 Inches Wide

Depth of Steelform	Width of Joist	Cu. ft. of conc. per sq. ft. of floor with slab thickness over steelforms of			Add. conc. req'd. (in. cu. ft.) per lin. ft. of beam or bearing wall where standard taper endform is used**
		2"	2 ½"	3"	tapered
8"	4"	.298	.340	.382	.17
	5"	.320	.362	.404	.16
	6"	.339	.381	.423	.16
10"	4"	.336	.378	.420	.21
	5"	.363	.405	.447	.20
	6"	.387	.429	.471	.19
12"	4"	.377	.419	.461	.25
	5"	.408	.450	.492	.24
	6"	.437	.479	.521	.23
14"	5"	.456	.498	.540	.28
	6"	.490	.532	.574	.27
	7"	.522	.564	.606	.26
16"	5"	.507	.549	.590	.32
	6"	.545	.587	.629	.31
	7"	.581	.622	.664	.30

Concrete required per Sq. Ft. of Floor Using Ceco Adjustable Steelforms 20 Inches Wide

Depth of Steelform	Width of Joist	Cu. ft. of conc. per sq. ft. of floor with slab thickness over steelforms of			Add. conc. req'd. (in. cu. ft.) per lin. ft. of beam or bearing wall where standard taper endform is used**
		2"	2 ½"	3"	tapered
8"	3 ½"	.289	.329	.370	.17
	4 ½"	.309	.350	.393	.16
	5 ½"	.331	.372	.414	.16
10"	3 ½"	.312	.353	.395	.21
	4 ½"	.340	.381	.424	.20
	5 ½"	.367	.408	.450	.19
12"	3 ½"	.336	.378	.420	.25
	4 ½"	.371	.412	.455	.24
	5 ½"	.403	.444	.486	.23
14"	3 ½"	.361	.403	.444	.29
	4 ½"	.402	.443	.485	.28
	5 ½"	.438	.480	.522	.27

*The above concrete requirements are also applicable to one piece longforms.
**Amount of concrete given for standard tapered endforms is for one side of beam only.

Concrete Required per Sq. Ft. of Floor Using Ceco Flange* Type Steelforms 30 Inches Wide

		Cu. ft. of conc. per sq. ft. of floor with slab thickness over steelforms of			Add. conc. req'd. (in. cu. ft.) per lin. ft. of beam or bearing wall where standard taper endform is used**
Depth of Steelform	Width of Joist	2"	2 ½"	3"	tapered
	5"	.322	.364	.405	.15
8"	6"	.338	.380	.421	.15
	7"	.351	.392	.434	.14
	5"	.353	.395	.436	.19
10"	6"	.372	.414	.455	.18
	7"	.389	.430	.472	.18
	5"	.386	.427	.469	.22
12"	6"	.408	.450	.491	.22
	7"	.430	.470	.512	.21
	5"	.420	.460	.503	.26
14"	6"	.447	.488	.530	.25
	7"	.470	.511	.553	.24
	5"	.451	.493	.535	.29
16"	6"	.482	.523	.565	.28
	7"	.510	.552	.594	.27

*The above concrete requirements are also applicable to one piece longforms.
**Amount of concrete given for standard tapered endforms is for one side of beam only.

Concrete Required per Sq. Ft. of Floor Using Ceco Adjustable Steelforms 30 Inches Wide

		Cu. ft. of conc. per sq. ft. of floor with slab thickness over steelforms of			Add. conc. req'd. (in. cu. ft.) per lin. ft. of beam or bearing wall where standard taper endform is used*
Depth of Steelform	Width of Joist	2"	2 ½"	3"	tapered
	5 ½"	.326	.368	.408	.15
8"	6 ½"	.341	.383	.424	.15
	7 ½"	.355	.397	.438	.14
	5 ½"	.352	.394	.434	.19
10"	6 ½"	.371	.413	.454	.18
	7 ½"	.389	.430	.472	.18
	5 ½"	.378	.420	.461	.22
12"	6 ½"	.401	.443	.484	.22
	7 ½"	.422	.464	.505	.21
	5 ½"	.404	.446	.486	.26
14"	6 ½"	.430	.472	.514	.25
	7 ½"	.456	.497	.539	.24

*Amount of concrete given for standard tapered endforms is for one side of beam only.

Ceco One Piece Steeldomes for Two-Way Dome Slab Construction. To estimate the quantity of concrete required for steeldome construction, figure the concrete required for a solid slab of the depth of the dome plus the top slab, and subtract the voids created for each steeldome as given in the table below.

19" x 19" SYSTEM
REGULAR
SIZE

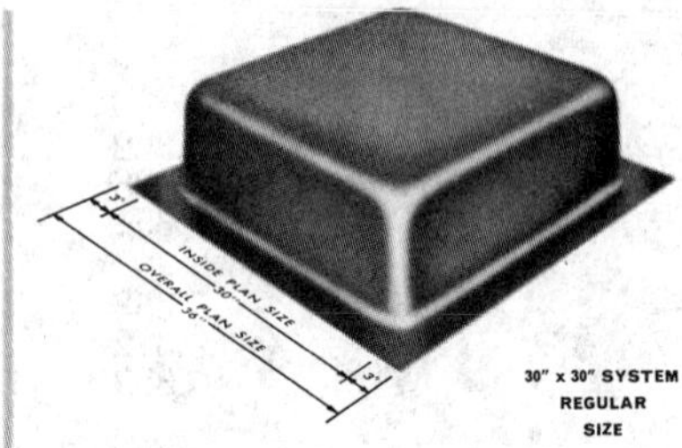

30" x 30" SYSTEM
REGULAR
SIZE

Ceco Steeldomes

Steeldome Sizes

	19" x 19" SYSTEM	30" x 30" SYSTEM	
Item	Standard Size	Standard Size	Filler Sizes
Plan size of void	19"x19"	30"x30"	20"x30" and 20"x20"
Overall plan size including flanges	24"x24"	36"x36"	26"x36" and 26"x26"
Width of flange	2 ½"	3"	3"
Depth of domes	4", 6", 8", 10", & 12"	8", 10", 12", 14", 16", & 20"	8", 10", 12", 14", 16", & 20"

Table of Concrete Voids
Voids Created By Steeldome, Cubic Feet

	19" x 19" SYSTEM	30"x30" SYSTEM		
	Regular	Regular	Fillers	
Depth of Steeldome	19"x19"	30"x30"	20"x30"	20"x20"
8"	1.41	3.85	2.54	1.65
10"	1.90	4.78	3.13	2.06
12"	2.14	5.53	3.63	2.41
14"		6.54	4.27	2.87
16"		7.44	4.85	3.14
20"		9.16	5.90	3.81

Lightweight Concrete, Insulating Concrete of Pumice, Perlite, Vermiculite and Other Lightweight Aggregates

Lightweight concrete of crushed slag, foamed slag, haydite, pumice, perlite, and vermiculite are used for floor and roof fill, structural roofs, floors in cold storage rooms, etc. A comparison of concrete aggregates is as follows:

Type of Aggregate	Aggregate Wt. per Cu. Ft. Lbs.	Lbs. per CF of Concrete Using Aggregate
Gravel	120	150
Sand	90-100	150
Crushed stone	100	145
Crushed Bank Slag	80	100-130
Haydite	40-60	100-120
Foamed slag	40-60	90-100
Cinders	40-50*	110-115
Pumice	30-60	60-90
Diatomite	28-40	55-70
Perlite	6-16	20-50
Vermiculite	6-10	20-40
Waylite	40-60	90-100
Expanded Shale	44-58	92-98

*Plus sand.

Vermiculite Insulating Concrete Aggregate. Vermiculite concrete aggregate is a type of mica which is mined, crushed and screened to size. When subjected to a temperature of about 2,000°F, it expands into a laminated granule containing millions of tiny dead air cells. This produces a lightweight aggregate weighing 6 to 10 lbs. per cu. ft., as compared to sand or crushed rock aggregate weighing about 100 lbs. per cu. ft. Vermiculite has high insulating value, is fireproof and will not rot or decay.

Vermiculite insulating concrete is a lightweight building material made like ordinary portland cement concrete, except vermiculite concrete aggregate is used instead of sand, gravel or crushed stone. Depending on the amount of vermiculite aggregate used, it is possible to make concrete weighing from 20 to 40 lbs. per cu. ft. Ordinary concrete weighs 145 to 150 lbs. per cu. ft.

Vermiculite concrete is used for poured concrete roof decks over a variety of forming materials; for roof insulation over such surfaces as concrete, steel, wood, etc.; and for poured concrete floor slabs.

Because of its high insulation value, vermiculite concrete is an excellent base for radiant heat floor installations and for grade level floors without radiant heat in homes, farm structures and all types of commercial buildings. At a density of approximately 25 lbs. per cu. ft., 1" of vermiculite concrete has an insulating value equal to approximately 16" of ordinary concrete. The density, insulating value and strength of vermiculite concrete can be varied to meet a wide range of design requirements.

Vermiculite concrete is incombustible and has earned the highest fire ratings attainable. It carries a 5-hour fire rating for spandrel wall construction and a 1 ½-hour rating on a steel deck covered with 1 ½" of vermiculite concrete and completely unprotected on the underside.

Mixing. Vermiculite concrete should be mixed in a mechanical mixer. The required amount of water and cement shall be placed in the mixer, and then the aggregate. Mixing shall be limited to the minimum time required to obtain a thorough mix and proper fluidity. (Maximum time recommended--5 minutes).

When transit-mixed vermiculite concrete is used, the operation shall be as follows:

a) Introduce water and cement into mixer. (Fill auxiliary water tank before leaving plant)
b) Rotate the mixer slowly until all aggregate has been added.

c) Continue to rotate mixer for approximately one minute after aggregate is in the mixer.
d) Do not rotate drum on way to job site. Mix concrete at job site at fastest speed until it is uniform and flows freely from the mixer.

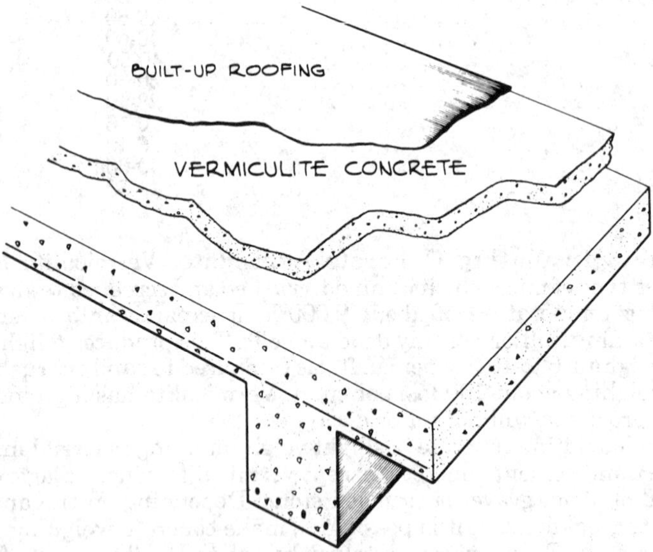

Vermiculite Concrete Roof Insulation Poured Over Structural Concrete Roof Decks

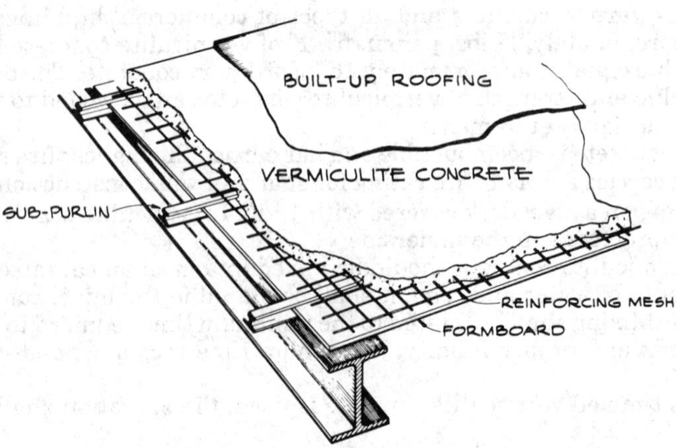

Vermiculite Concrete Roof Deck Poured in Place Over FormBoards

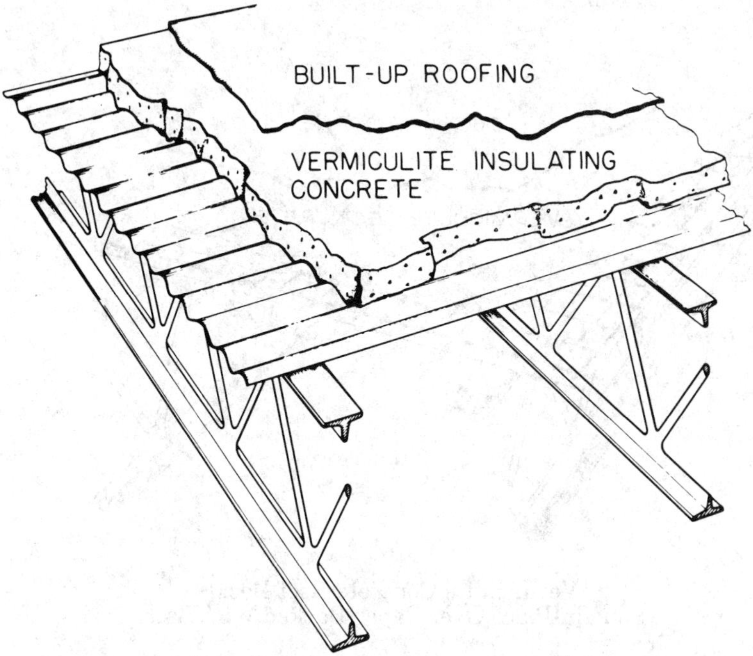

Vermiculite Insulating Concrete Over Vented Metal Roof Decks

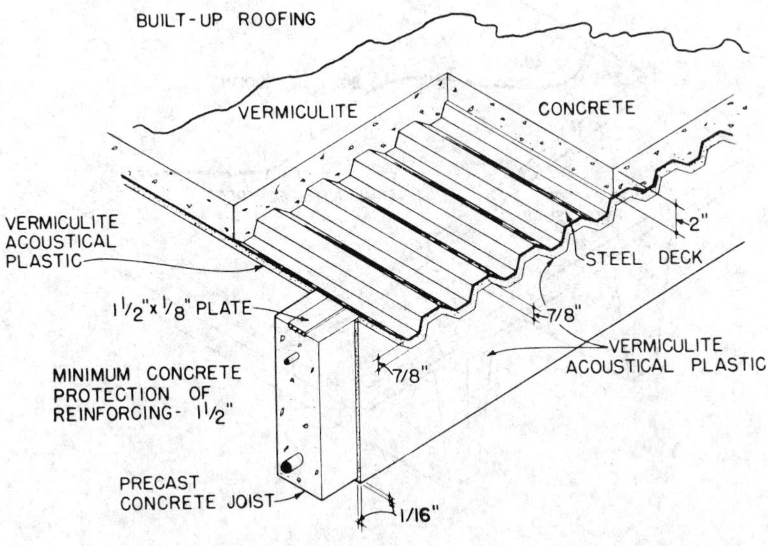

Vermiculite Concrete Insulation Over Vented Galvanized Steel Roof Decks, Precast Concrete Joists. (Two-hour Fire Rated)

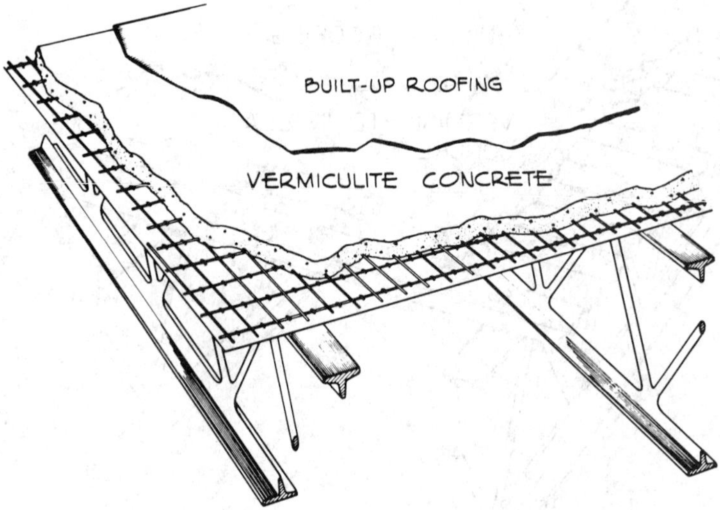

Vermiculite Concrete Roof Deck
Pour-in-Place Over Paper-Backed Wire Lath

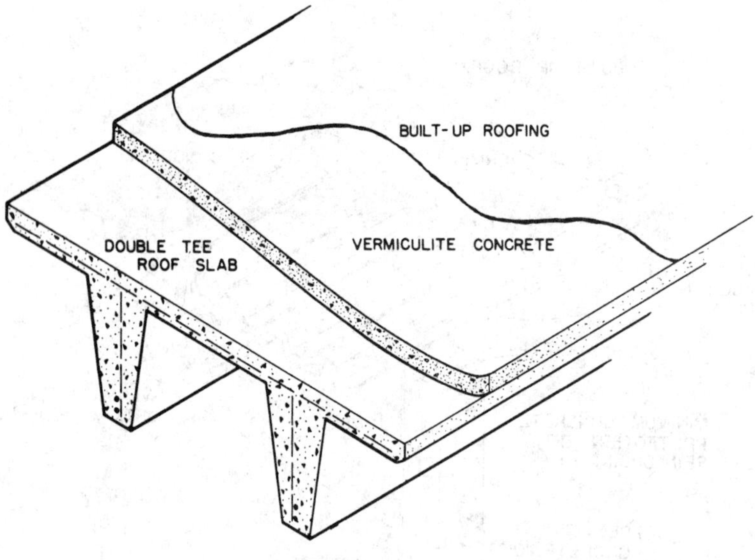

Vermiculite Concrete Insulation Over Precast Prestressed
Concrete Double Tee Roof Slabs

Placing. Vermiculite concrete shall be transported and placed immediately after mixing is completed, and the period between completion of the mixing and placing shall be of such short duration that the mixture does not appreciably change in consistency.

Vermiculite concrete shall not be placed when the temperature is less than 40°F. If the temperature expected after placing of the concrete is near or below 40°F., the mixing water shall be heated in the temperature range 75° to 100°F.

Vermiculite concrete may be placed by the use of standard concrete equipment such as bucket hoists, buggies or by specially designed pumping equipment.

Machine Placement. The use of high volume pumps to install cast-in-place vermiculite insulating concrete is an important development. Pumping or spraying the concrete is especially suited to roof decks with curved, sloped or irregular surfaces and is just as efficient on flat roof decks.

A pump is fast and simple. It can be set up or taken down in about 30 minutes. The long hose lengths that are possible enable the crew to get over obstructions on the job easily.

The Vermiculite Institute annually approves a national roster of roof deck applicators appointed by institute members to assure the highest quality of finished work.

Vermiculite concrete is covered by the United States of America Standards Institute Specification A122.1-1965.

Vermiculite Insulating Concrete Floors-on-Grade. A 1:4 mix of vermiculite concrete is recommended as an insulating base under a sand-gravel concrete topping for floors on grade. It is particularly effective in reducing heat loss where radiant heating coils or ducts are carried in the floor.

Minimum thickness of the slab should be 3". A reinforcing mesh of 6"x6"-10/10 ga. welded wire fabric, or equal, should be placed in the center of the vermiculite concrete, lapping the sides and ends one full mesh.

The sand-gravel topping should be designed to carry the maximum floor load. Minimum thickness should be 1 1/2" except when radiant heating coils are used. In that case minimum thickness over the top of the coils should be 1". The radiant heating coils should be supported on chairs to assure a minimum of 1/2" of sand-gravel concrete below the coils.

Proportion Table for Vermiculite Concrete

Description of Work	Vermiculite Aggregate Cu. Ft.	Portland Cement Cu. Ft.
Poured concrete roofs	4	1
Poured concrete roof insulation	6 or 8	1
Poured concrete slabs		
On grade insulating slab under 1 1/2 ordinary cement topping	4	1
Panel or spandrel walls	4	1

Mixing Table For Vermiculite Concrete

Mix by Vol. Vermiculite Aggregate Cu. Ft.	Portland Cement Sacks	Approx. Density in Lbs. per Cu. Ft.	Compressive Strength in Lbs. per Sq. In.	Thermal Conductivity (K in btu/SF/in./ hour/degree F
8	1	20 to 25	100 to 125	0.60
6	1	25 to 30	125 to 225	0.76
4	1	35 to 40	350 to 500	0.97

Approximate Quantities Required for One Cu. Yd. of Ready-Mixed Vermiculite Concrete

Mix	Bags (4 Cu. Ft.) Vermiculite Concrete Aggregate	Sacks (1 Cu. Ft.) Portland Cement	Water Gals.
8-1	7.5	3.75	85 to 95
6-1	7.5	5.0	85 to 95
4-1	7.5	7.5	85 to 95

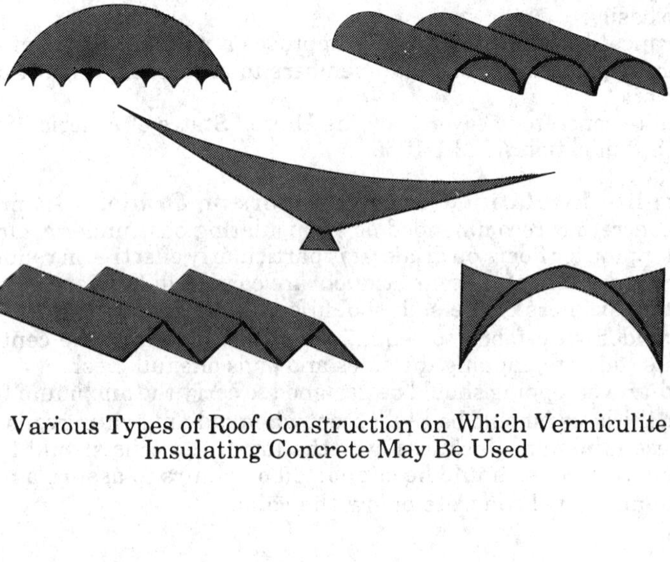

Various Types of Roof Construction on Which Vermiculite Insulating Concrete May Be Used

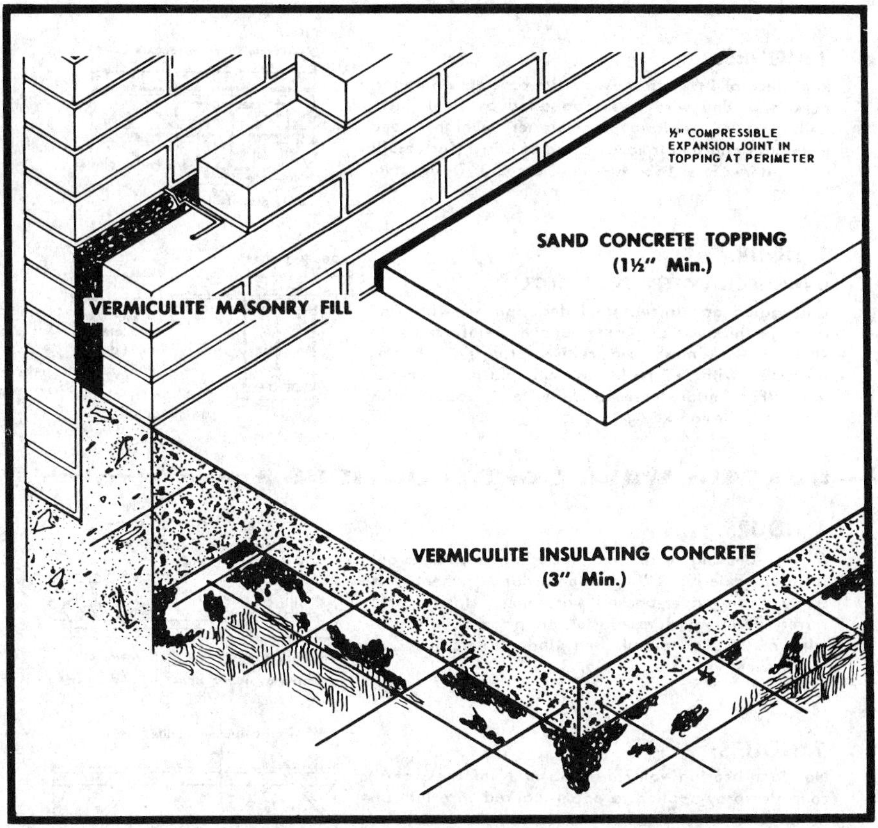

Vermiculite Insulating Concrete Floor

— ROOF CONSTRUCTIONS —

3 HOURS:

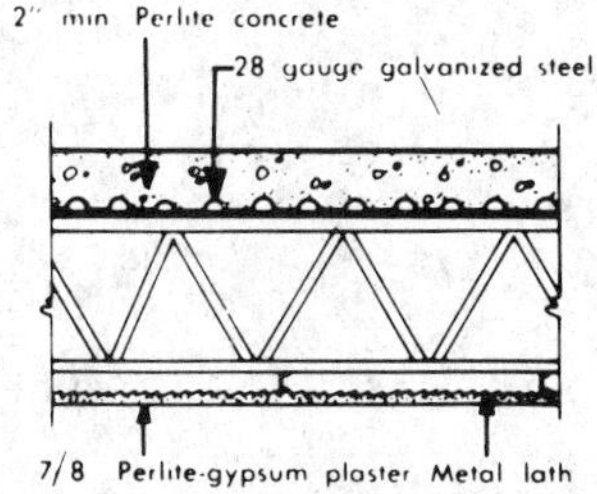

Roof deck of 2" minimum thickness perlite concrete on 28 gauge galvanized steel form units supported by steel joists 4' on center. Ceiling of ⅞" perlite-gypsum plaster on expanded metal lath attached to ¾" furring channels wire-tied to lower chord of joists. (For details see Underwriters' Laboratories Class C-3, Design No. RC1-3.)

2 HOURS:

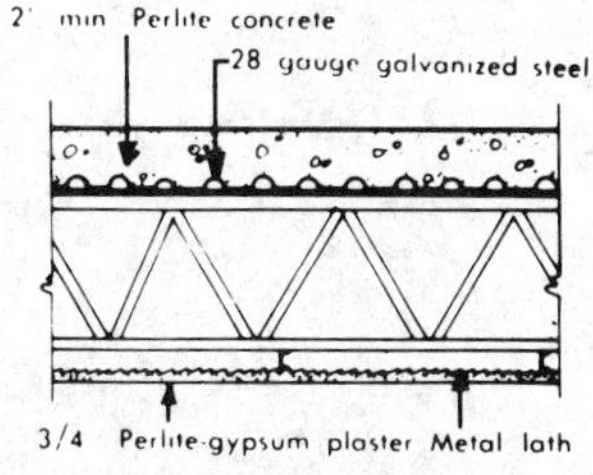

Same as 3 hour rated system above except that the perlite-gypsum plaster ceiling shall be ¾" thick. (For details see Underwriters' Laboratories Class D-2, Design No. RC11-2.)

3 HOURS:

Roof deck of 2½" thickness perlite concrete on paper-backed welded wire mesh supported by steel joists. Ceiling of ¾" perlite-gypsum plaster on high ribbed metal lath attached to lower chord of joists. (For details see Underwriters' Laboratories Class C-3, Design No. RC2-3.)

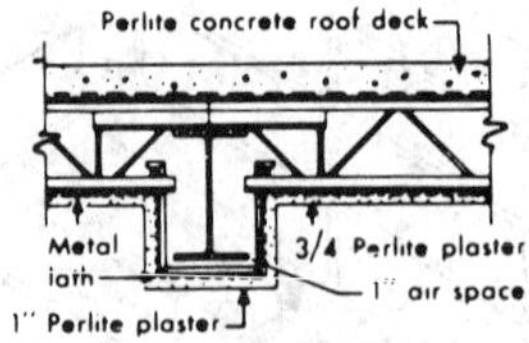

1 HOUR:

UNPROTECTED STEEL ROOF DECK

Corrugated galvanized steel deck topped with 2⅝" average thickness of perlite concrete reinforced with 48-1214 wire mesh. No ceiling protection. Beams protected with ⅞" perlite-gypsum plaster on metal lath. (For details see Underwriters' Laboratories Class E-I, Design RC2-1.)

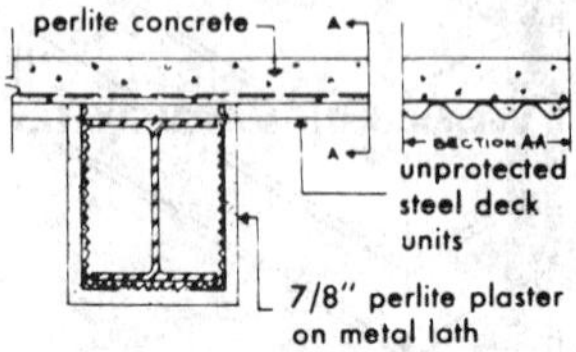

— CURTAIN WALL CONSTRUCTIONS —

4 HOURS:

6" non-bearing wall (not including weatherproof facade). Exterior: 4" perlite-portland cement mix sprayed on paper-backed wire mesh. Interior: 1" furred perlite-gypsum plaster on paper-foil backed wire mesh. (For details see Underwriters' Laboratories Class B-4, Design No. 3.)

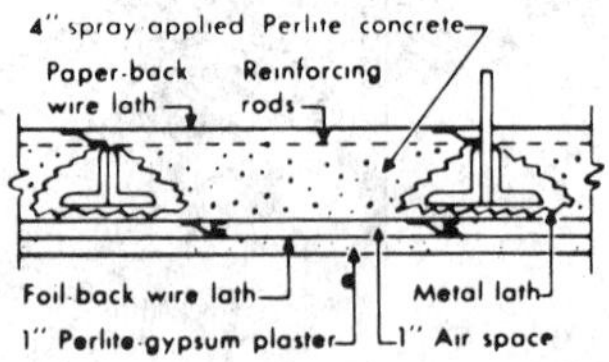

2 HOURS:

Non load-bearing wall consisting of 3" thickness perlite concrete spray-applied to paper- backed wire lath attached to 4" steel studs 16" on center. (For details see Underwriters' Laboratories Class D-2, Design No. 18.)

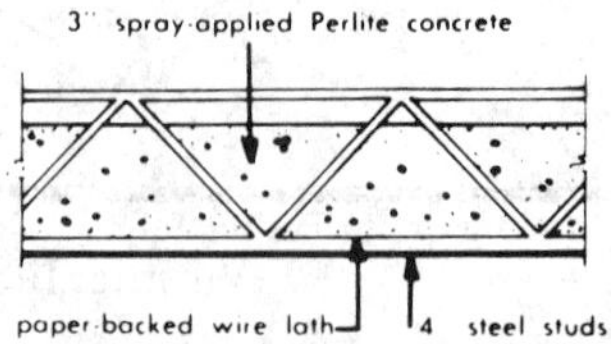

Perlite Concrete Aggregate

Perlite is a siliceous volcanic rock mined in western United States. When crushed and quickly heated to above 1500°F., it expands to form lightweight, non-combustible, glass-like particles of cellular structure. This material, white or light gray in color, weighs 7 ½ to 15 lbs. per cu. ft.

Perlite aggregate consists of expanded perlite sized for use in lightweight insulating concrete in place of sand or gravel. It is usually packed in 3 or 4 cu. ft. paper bags for easy handling.

The many, tiny glass sealed cells in each particle of expanded perlite make it highly insulating as well as comparatively nonabsorptive. Thus perlite mixes with about 30% less water than comparable lightweight aggregates.

Perlite concrete, in combination with portland cement forms a very lightweight concrete. Where ordinary concrete weighs 150 lbs. per cu. ft. the dry weight of perlite concrete can be designed from 20 lbs. to 40 lbs. per cu. ft. The extremely light weight and ease of handling make perlite concrete adaptable to practically any shape.

Perlite concrete has received up to 4-hour ratings in Underwriters' Laboratories fire tests. Its light weight makes it suitable for fireproof roof and floor fills, for thin concrete curtain walls (blocks, slabs or monolithic) and for many precast panel and block constructions.

For lightweight structural roof construction where perlite concrete is placed over galvanized steel forms, paper backed wire mesh, formboards, structural concrete, or other suitable form materials, use a 25 to 29 lb. per cu. ft. density perlite concrete. This offers an ideal balance of low dead weight, adequate compressive and indentation strengths, and good insulating value. For uses where higher strengths are more important than insulating value, i.e., floor fills and certain lightweight structural roof designs, use a 34 to 40 lb. per cu. ft. density perlite concrete.

Perlite concrete should be mixed in a paddle type plaster or a drum type concrete mixer. The required amount of water, air entraining admixture and portland cement should be placed in the mixer and mixed until a slurry is formed. The proper amount of perlite concrete aggregate should then be added to the slurry and all materials mixed until design wet density is reached. Perlite concrete may also be transit mixed.

Perlite concrete should be carefully deposited and screeded in a continuous operation until a panel or section is completed. Steel troweling should be avoided. Rodding, tamping, and vibrating should not be used unless so specified by the architect.

It is recommended that a 1" air space or expansion joint be provided through the thickness of the perlite concrete at the juncture of all roof projections, such as skylights, penthouses, and parapet walls. Expansion joints are recommended in accordance with good concrete and steel construction practice and are based upon the provision that suitable consideration has been given to the use of adequate through-building expansion joints. Other types of insulating concrete may not require expansion joints because of high initial shrinkage.

Perlite concrete, during the curing period, should be protected for at least the first three days to keep it from drying out too rapidly or freezing. Freshly poured concrete should also be given adequate protection from heavy rain. Allow no traffic until concrete can sustain a man's weight without indentation.

Perlite concrete should not be placed in temperatures under 40°F. nor on frosted surfaces. In near-freezing weather the mixing water should be heated to between 75° and 100°F.

Perlite concrete is also used as a base for radiant heating coils. For grade level floors a moisture resistant barrier should be used and a suitable wearing surface applied over the perlite insulating concrete.

Considerable savings in dead weight are possible by using perlite concrete as a floor fill over cellular steel or pan-type floors in multiple story buildings. The surface may be covered with concrete topping and other surfacing materials after the perlite concrete has cured several days. Another floor fill material, using perlite blended with other aggregates, is perlite/sand or perlite/expanded shale concrete. These concretes have higher compressive strengths and require only resilient flooring or carpeting as a wearing surface. Following are the results of tests conducted at nationally recognized laboratories. Due to the varying characteristics of naturally occurring aggregates in different geographical areas, trial mixes of aggregate blends are recommended for determination of mix proportions.

Density Wet	Density Dry	Compression 28 day, psi	Cement (Sacks)	Perlite (cf)	Heavy Aggregate (cf)	AEA† (oz)	Water U.S. Gal/Sack	Ct. Factor cu. yd.
82	72	1000	1	3	2*	6	8.5	5.82
66	54	1200	1	3	2**	4	8.5	6.12

*Sand. **Rotary Kiln Expanded Shale. †Air Entraining Agent.

Perlite concrete can be nailed, sawed, and worked with ordinary carpenter tools.

Mix Proportions By Volume for Lightweight Perlite Insulating Concrete

Cement (Sacks)	Perlite Cu. Ft.	Water-Gals. per Sack Cement	Air-Entraining Agent, Pints	Density Lbs. per Cu. Ft. Oven Dry	Comp. Str. psi-28 Days
1	4	9	1	36	400
1	5	11	1 ¼	30 ½	250
1	6	12	1 ½	27	170
1	7	14	1 ¾	24	120
1	8	16	2	22	95

Material Required for One Cu. Yd. of Placed Perlite Concrete

Proportions Cement Perlite	Cement Sacks	Perlite Cu. Ft.	Water Gallons	Air-Entraining Agent--Pints
1:4	6.75	27	61	6 ¾
1:5	5.40	27	59 ½	6 3/4
1:6	4.50	27	54	6 ¾
1:7	3.85	27	54	6 ¾
1:8	3.38	27	54	6 ¾

Pumice Concrete

Pumice is a natural white or gray glass foam. It is not a volcanic ash. It is mined extensively in the western part of the United States.

Uniform graded pumice is used as aggregate in monolithic concrete, concrete masonry units and precast products. It replaces sand and gravel in any concrete up to strengths of 3,000 lbs. per sq. in., where thermal insulation, acoustical and light weight properties are desirable.

All pumice aggregate shall be clean, free of all foreign matter and well graded so as to meet ASTM specifications C-330.

All water used in concrete work shall be free from oil, strong acids, alkali, or organic material.

Mixing. Concrete for columns, walls above ground, floors, window sills and all self-supporting floor and roof slabs, canopies, and other parts of structure not coming in contact with ground shall be of a mixture of water, portland cement, and pumice aggregate, and shall develop at 28 days a strength of not less than 2,500 psi. and with a slump of not more than 2" by standard test.

Water contained in the aggregate shall be deducted from the amount of water used in the mix except that the water needed to saturate the pumice aggregate shall not be deducted. Pumice aggregate shall be thoroughly saturated either in the stockpile or in the mixer prior to the addition of cement and water for mixing.

SUGGESTED MIX DESIGN DATA FOR PYRAMID PUMICE CONCRETE

Compression lbs/ sq. in	Design Weight lbs/ cu ft	Sand lbs (dry)	Pyramid Pumice per Cubic Yard Yield						Cement		Water				Recommended slump
			Fine		Coarse		Blend				to saturate		to mix		
			lbs (dry)	cu ft	lbs (dry)	cu ft	lbs (dry)	cu ft	sax/ cu yd yield	lbs/ cu yd yield	gal	lbs	gal	lbs	
150	40	0	0	0	792	36	0	0	4	376	36.1	300	Determine by sight		0
250	46	0	914	24	264	12	1180	36	3	282	34.8	290	Determine by sight		0
500	55	0	914	24	264	12	1180	36	4	376	34.8	290	Determine by sight		0
1000	60	0	914	24	264	12	1180	36	5	470	34.8	290	50	417	½"
1500	65	0	914	24	264	12	1180	36	6.5	611	34.8	290	52	433	½"
2000	70	0	914	24	264	12	1180	36	7.5	705	34.8	290	53	442	1"
2500	75	0	914	24	264	12	1180	36	8	752	34.8	290	48	400	2"

A Note on Pyramid Pumice Aggregate: It should be saturated prior to the addition of cement. This may be done in the stock pile or in the mixer. A 1" slump in pumice concrete will give a consistency equal to a 3" slump in hard rock concrete. Pumice concrete should be mixed with as little water as possible to get the required workability. It should be the consistency of thick mud and no free water should be in evidence. Air entraining agents are recommended to cut water-cement ratio and get greater workability. A combination of fine and coarse aggregate or the blended aggregate should be used. Volume measurements are based on loose volume non-compacted pumice.

Placing. All pumice concrete shall be thoroughly vibrated with a mechanical vibrator. Vibration shall be conducted in such a manner as to assure that the vibrator passes through the pumice mass with not more than 6 inches of space between the successive positions of the vibrator. In all cases where pumice is placed in layers the vibrator must penetrate into the underlying concrete layer after passing through the newly placed layer. Ample time shall be allowed for the settlement of the concrete before final topping or screeding to finished floor level or grade. The pumice aggregate shall be thoroughly saturated prior to mixing with the cement and water used for mixing and ample time shall be allowed for the saturation to take place.

Application. Pumice must be handled carefully if it is to be successful. Because it will float it must be pre-saturated before use. The stock pile should be sprayed for at least 48 hours or the pumice should be saturated in the mixer before use. Unless this is done, cement particles will be drawn into the cells of the pumice and excessive shrinkage may occur as a result of loss of water from the mix to the pumice particles. One half of the total water required should be added to the aggregate before the cement is introduced.

The concrete should be water cured for at least seven days after which time the outer fibers will have become strong and dense enough to protect against evaporation and resultant too rapid drying.

Gradations Available. Pumice aggregate is graded to meet ASTM specifications for properly graded lightweight aggregrates for concrete. Designation C-330 in sizes No. 4 to 0 for fine aggregate; 3/8 and 1/2 to 0 for blended aggregate; 1/2" to No. 4 for coarse aggregate.

Shipping Weight. Fine material or 1/4" to 0 aggregate will have an average weight of about 1,200 lbs. per cu. yd. Blended aggregate or 3/8" to 0 material will weigh about 1,100 lbs. per cu. yd. The coarse aggregate or 1/2" to No. 4 material will have an average shipping weight of about 800 lbs. per cu. yd. These weights may increase from 100 to 200 lbs. per cu. yd. during wet seasons. An open top, side dump, flat bottom (Caswell) car carries approximately 63 cu. yds. of material.

An open top, solid bottom, gondola car will carry approximately 63 cu. yds. of material.

An open top, hopper bottom car will vary in capacity from 60 to 100 cu. yds.

Waylite Lightweight Aggregate

A lightweight, cellular aggregate, refined and expanded by a process of agitating and rapidly cooling in an atmosphere of steam molten hot slag as it comes from the blast furnace at temperatures of 2500 to 3000°F. In this process the inherent gases expand and fill the material with minute air cells or bubbles completely sealed. These minute air cells are the active insulation and lightweight principle in all Waylite aggregates.

Waylite aggregate crushed and screened is available in three commercial sizes and average dry loose weights per cubic foot.

Aggregate	Grading	Average Weight
Coarse	7/16" to 3/16"	40 lbs.
Acoustical	3/16" to 3/32"	45 lbs.
Fine	1/8" to dust	60 lbs.

Waylite concrete requires a higher proportion of fine aggregate in relation to coarse than ordinary concrete. This is due to the lower top sizes of the coarse aggregate (7/16") and the additional angularity or harshness of the crushed fine and coarse aggregates.

The mixing of Waylite concrete requires a somewhat different manipulation procedure than for heavy aggregates. To obtain maximum workability with minimum weight, the aggregate and water should be mixed for one-half minute before the addition of the cement, and then mixed for at least an additional two minutes. A mix which appears harsh at first will improve in plasticity with continued mixing without additional water.

Waylite Concrete Mix Data

	Quantities Required per Cu. Yd., Loose Dry Volume				
Mix Loose Volume	Gals. Water per Sack	Cement Sacks	Fine Waylite Cu. Yds.	Coarse Waylite Cu. Yds.	Sand* Cu. Yds.
1-2 ¼-2 ½-¾*	9.4	6.4	0.53	0.60	0.17
1-2-2 ¼-¾*	8.2	7.2	.53	.60	.20
1-1 ¾-2½*	7.3	8.1	.54	.60	.15
1-1 ½-2½*	6.5	8.8	.49	.65	.17
1-1 ½-1 ¾-½*	5.9	9.6	.53	.62	.18

*Sand shall be a fine siliceous bank, lake, or mason sand.

The addition of approved concrete air entraining agents materially improve the workability of Waylite concrete mixes.

PLACING, PATCHING AND FINISHING ARCHITECTURAL CONCRETE

Architectural concrete demands the use of a plastic, workable mix that can be worked and molded readily in the forms. Surfaces of the hardened concrete should be free of defects such as sand streaking and honeycomb. It is important that sufficient fines be provided in the mixture, and for these reasons the fine aggregate should have at least 15% passing through a 50 mesh sieve and 3 percent passing the 100-mesh sieve, and the cement factor should not be less than 5 ½ sacks per cu. yd. of concrete. In other words, the concrete should not be richer than 1:2:3 by loose volumes and preferably 1:2 ⅓:3 ⅓, containing 6 ½ or 6 sacks of cement per cu. yd. of concrete respectively.

Not more than 6 ½ gals. of water per sack of cement should be allowed so that the concrete will have the durability required to withstand weathering.

Mixing and Placing Concrete. Concreting of architectural concrete walls should proceed at a rather slow rate allowing sufficient time for thorough vibrating at the finished surfaces so that the pressure exerted by the concrete on the forms does not exceed 600 lbs. per sq. ft.

A reasonable rate of placing the concrete should be considered when locating joints. Generally a rate of not more than 2 feet an hour is most conducive of good workmanship and lifts of 6 to 8 feet between joints are desirable and 12 feet should be the absolute maximum.

The concrete should be spaded or vibrated into all corners and along all form surfaces, but even if vibrated, some spading is necessary in the corners and angles as vibrators will not fill these places at all times. On many jobs small electric hammers moved along the forms at the point of placement have been found helpful. Vibration should be sufficient only for complete compaction, not to the point where segregation occurs and excess water is forced to the form faces. Such over-vibration may result in sand streaking and discoloration.

When concrete is splashed on the forms above the point of deposit and subsequently hardens, the wall in that area is likely to be rough and discolored. Splash boards are sometimes used but are not completely satisfactory, as they do not prevent segregation of the falling concrete. Tremies of various lengths equipped with a hopper at the top through which the concrete is placed will avoid both splash and segregation. The tremies should be spaced not more than about 9 feet apart. Sometimes splashing is the result of mortar being thrown from vibrators as they are lifted out of the concrete and raised to the placing platform. This can be prevented by shutting off the power as the vibrator is withdrawn from the concrete.

Labor Placing Concrete. The labor cost of placing architectural concrete will vary considerably, depending upon the size of job, size of sections, amount of ornamentation, quantity of reinforcing steel, etc., but on most work of this class, it will require 2 to 3 hours labor time per cu. yd. of concrete, plus the time for the hoisting engineer and foreman.

Based on a 64-cu. yd. pour per 8-hr. day, the labor cost should average as follows per cu. yd.:

	Hours	Rate	Total	Rate	Total
Foreman	0.13	$....	$....	$19.79	$2.57
Hoist engineer	0.13	....	....	25.96	3.37
Labor	2.50	....	....	19.13	47.83
Cost per cu. yd			$....		$53.77

Curing. On this type of work one of the surest and best ways of curing the concrete is by leaving the forms in place. Whenever it is practical to do so, they should be left in place 4 or 5 days. By that time the concrete is hard enough so that it is not scuffed nor are corners so likely to be broken off during form removal. If the forms are removed earlier, other means must be provided to keep the concrete wet or damp for at least 5 days if a durable concrete is to be obtained.

Patching and Finishing. When the forms are first stripped, the concrete does not always present a pleasing appearance. There may be pock marks, air holes, fins and some discoloration. Some of the latter will bleach out upon exposure but a simple clean down with cement grout will do wonders for the appearance. The projecting fins and other blemishes extending beyond the face of the wall are removed with a carborundum stone or abrasive hone and then a grout of 1 part portland cement to 1 ½ parts of very fine sand (sand passing through a fly screen will do) is applied. From ⅓ to ½ of the cement should be white cement.

The grout is mixed to the consistency of a heavy paste. After the wall is wet down, the grout is applied uniformly with a stiff fiber brush, completely filling all air voids and holes. Immediately after the grout has been applied, the wall is floated with a wood or cork float. A very fine abrasive hone can also be used. The grout is then allowed to harden partially, the exact time depending upon the weather. When it is hardened sufficiently so it will not be pulled out of the holes, all grout is scraped from the surface of the wall with the edge of a steel trowel. After the wall has dried thoroughly, it is rubbed with a dry piece of burlap to completely remove all of the dried grout or film. There should be no visible film left on the surface. Grout should not be left on the wall overnight. All cleaning of any one panel of concrete should be completed the same day.

Patching mortar used to fill tie holes and imperfections should be of the same mixture of gray and white cements, but should be a stiffer mixture and the cement and sand proportions should be the same as those used in the concrete. Patches should never be steel troweled, but may be finished with wood or cork floats. Allowing the mixed patching mortar to stand for an hour or two before using it reduces the amount of shrinkage but water must not be added in remixing it.

To correct blemishes appearing on the face of the walls, allow about 4 hrs. time for a cement finisher and a helper per 100 sq. ft. of wall. The labor cost per 100 sq. ft. of wall should average as follows:

	Hours	Rate	Total	Rate	Total
Cement mason	4	$....	$....	$24.57	$98.28
Helper	4	....	....	19.13	76.52
Cost per 100 sq.			$....		$174.80
cost per sq. ft.			....		1.75

Cutting Concrete with Portable Concrete Saw

Many states now specify that control joints in highway paving be cut with a concrete saw, rather than produced by forming. The advantages of this method are improved ridability of joint; less sealing material required to seal joint; better aggregate interlock at joint; and no spalling or cracking at joint, which reduces maintenance costs.

This technique is also widely used for cutting floors and paving for repair work, trenches for electric conduit, pipe lines, and sewers, and also for scoring before breaking out concrete for machinery bases and other foundation work.

Clipper concrete saws (from Norton Construction Products Division, Worcester, MA) are a good example of this type of saw. They are available in several electric and gasoline powered models. The most popular models are the 18 Hp model C-188, the 30 Hp model C-305, the 37 Hp model C-375, and the 65 Hp model C-655, all gasoline powered. Approximate prices range from $2,000.00 for the basic model C-188 manually propelled to $8,000.00 for the model C-655 with self propelling unit and electric starter. The water pump and spotlight are optional features. The self propelling unit and other accessories are optional on the C-188 and C-305 models. The larger C-655 with 65 Hp engine is generally used for sawing contraction joints in airports, highways, and turnpikes. The C-305 and C-188 are widely used in patching, trenching, and plant maintenance work. Electrically powered models are available from 3 Hp and up.

Two types of blades are used in cutting concrete with concrete saws as follows: 1) Green-Con wet-dry abrasive blades for cutting green concrete 2) diamond blades, of various specifications, for cutting cured or old concrete.

The cost of cutting concrete with a concrete saw varies considerably, depending on age of concrete, type of aggregate, blade specification used, accessibility of work, skill of operator, etc.

Assuming proper blade specification is used for the cutting job, green concrete 24 to 48 hours old may be cut 1-inch deep at the rate of 12 to 14 lin. ft. per minute and 3-inches deep at the rate of 6 to 8 lin. ft. per minute. These rates cover only actual cutting time. Overall production will be reduced considerably depending on length of each individual cut, number of machine moves, changes in direction of cuts, etc. Blade wear for a Green-Con blade of proper specification should cost from 1 to 2 cents per lin. ft. of cut 1-inch deep. For a 2-inch cut, blade cost should be 1 ½ to 3 cents per lin. ft.; 3-inch cut, 2 to 4 cents per lin. ft.

When cutting old or cured concrete, using a diamond blade of correct specification, cutting time may be double the amount given above for green concrete and cost of blade wear will probably run 5 to 6 times that given above.

CEMENT FLOOR HARDENERS

There are many preparations on the market for waterproofing, dustproofing and wearproofing cement finish floors. These preparations fall into two classifications, integral treatments in which the entire slab or topping is treated and surface treatments in which the surface is specially prepared to become waterproof, dustproof and wear resistant.

Integral Treatments. Integral treatments consist of powdered chemicals that are mixed with the cement before sand and water are added and become an integral part of the topping mix throughout its entire thickness. This type of treatment is said to densify and harden the topping and is generally recommended for areas subjected to light pedestrian traffic.

Surface Treatments. There are two types of surface treatments for cement finish floors, dust coat applications which must take place as floors are being finished and liquid chemical applications which may be done at any time after the floor finish is set hard enough for foot traffic.

METALLIC FLOOR TREATMENTS (HARDENERS)

By far the most wide spread surface treatment is the metallic dust coat. The term "hardeners" usually applied to metallic floor treatments is somewhat misleading since iron has a much lower rating on the scale of hardness than even the most common sand or stone aggregate. In this sense iron treatments could be described as "softeners", since the basic technique is to provide a floor surface where the hard, brittle sand and stone found in plain concrete is overlaid with a coating of soft, ductile iron. It is reasoned that where hard, brittle aggregates crush under impact and abrasion, iron particles are malleable under these forces and will not fracture or crush out of the floor surface.

Metallic floor treatments consist of a dust coat application of pulverized iron, size graded to pass through mesh screens in the following proportions:

	Percent Passing
4 mesh screen	100%
8 mesh screen	Not less than 90%
14 mesh screen	Not less than 70 nor more than 85%
28 mesh screen	Not less than 35 nor more than 50%
48 mesh screen	Not more than 10%
100 mesh screen	Not more than 5%

Iron particles of the above graduation are mixed with cement and dusted onto freshly floated concrete and then floated and troweled to a smooth even finish. This optimum grading produces a more workable material while plastic and provides a dense surface having great resistance to wear upon hardening.

The iron material should be free from oil and should contain no non-ferrous metals. Oil will prevent the cement paste from bonding properly to the iron particles and non-ferrous metals may react with the cement to form gases and cause serious surface blistering.

Manufacturers, architects, and engineers specify different methods and amounts of iron to be used, which is usually a certain number of pounds of iron mixed with one half that weight of cement. The material is dusted over the floor surface before the finish topping has taken its initial set. These quantities vary from 30 to 120 lbs. of iron per 100 sq. ft. of floor, depending upon the traffic the floor will have to withstand. When the heavier dust coats are used, water-reducing and plasticizing agents are required to produce a dust coat that can readily be worked onto the surface without resorting to an overly wet concrete mix. In wet mixes there is danger that the iron aggregate will sink into the concrete where it serves no benefit as surface armoring.

It is not necessary to add anything to the labor costs of applying the regular cement finish top. It merely means that the finisher dusts the floor with the metallic powder while it is still too wet to trowel. By adding this dust coat it is possible to trowel the floor sooner than it would be without it, as it absorbs a portion of the surface water.

Durocon Metallic Floor Hardener (Castle Chemical). A specially prepared metallic hardener for industrial and heavy duty concrete floors. Contains ocon, which provides concrete of increased strength, resisting acid, alkali, etc.

For average traffic, 30 to 40 lbs. hardener and 30 lbs. portland cement; for heavy traffic, 50 lbs. hardener and 30 lbs. portland cement; and for extra heavy traffic, 60 lbs. hardener and 30 lbs. portland cement per 100 sq. ft. floor.

Use 40 lbs. colored Durocon with 20 lbs. portland cement per 100 sq. ft.

Standard colors, natural, tile red, linoleum brown, French gray, Persian red, battleship gray, Nile green, black, russet and maroon.

Duroplate (Castle Chemical). Specially prepared, size-graded iron particles combined with dispersing agent. Applied as a dust coat while concrete is still in a plastic condition and machine floated into the surface. Produces a ductile, highly impact and abrasion-resistant, non-absorbent floor that will resist oils, grease, etc., and is easily cleaned and maintained. Floor life is extended up to ten times that of normal concrete floors when Duroplate aggregate is used. Use 60 to 120 lbs. per 100 sq. ft. of floor depending on degree of duty required.

Durundum Non-Slip Hardener (Castle Chemical). An aggregate floor hardener composed of powerful coloring mediums and hardening elements plus cementitious binders. Non-rusting and produces a non-slip surface. Entirely inert, very hard, tough and abrasive resistant. Apply as a dust coat from 30 to 60 lbs. per 100 sq. ft. as conditions require.

Ferro-Fax Metallic Floor Hardener (A. C. Horn, Inc.). Ferro-Fax metallic floor hardener meets government specifications. Free from oil and foreign metals. Produces wearproof, waterproof, dustproof concrete floors. Apply as a dust coat, using 30 lbs. per 100 sq. ft. for light traffic; 40 lbs. per 100 sq. ft. for moderate traffic; heavy-duty floors, 70 to 100 lbs. per 100 sq. ft.

Comes in two types: Ferro-Fax Standard, to be mixed with portland cement, and Ferro-Fax Ready-Mixed. Immediately following the leveling, deposit upon the surface ⅔ of a uniform dry mixture consisting of 2 sacks Ferro-Fax Standard and 1 sack portland cement.

Allow the Ferro-Fax to absorb the surface water. Float sufficiently to work the Ferro-Fax into the surface. Discontinue floating as soon as the surface becomes wet. Immediately after floating the first shake, apply the final ⅓ uniformly over the surface. Float just until moisture is brought to the surface. Trowel to the desired finish.

Mastercron Pre-Mixed (Master Builders). A specially graded hard silica aggregate, combined with light-fast and alkali-fast coloring pigments, selected cement and a water-reducing agent which will produce a tough, wear-resistant floor surface.

Ready to use as it comes from the bag, the material is applied as a dust coat to freshly floated concrete floors at the rate of 50 lbs. per 100 sq. ft. It is available plain and in colors.

Masterplate Pre-Mixed (Master Builders). Specially prepared metallic aggregate combined with a water reducing and cement dispersing agent. Makes possible the incorporation of 60 to 120 lbs. of iron on the surface to provide maximum resistance to abrasion under the heaviest of industrial traffic.

Average duty floors, 60 lbs. Masterplate plus 30 lbs. cement per 100 sq. ft.; heavy duty floors, 90 lbs. Masterplate plus 45 lbs. cement; extra heavy duty floors, 120 lbs. Masterplate plus 60 lbs. cement.

DPS Masterplate (Master Builders). A complete ready to use product containing specially processed iron aggregate, a conductive cement binder and a water- reducing and plasticizing agent. Applied as a dust coat at the rate of 180 lbs. per 100 sq. ft. of floor surface, DPS Masterplate meets federal and industrial requirements for conductive flooring. It retains its conductivity at temperatures from--20° to 150° F. and provides an extremely heavy duty floor that is spark-resistant and static-disseminating. Dark gray in color.

Sparkproof Duroplate Pre-Mixed (Castle Chemical). Static-disseminating and spark-resistant, meets all requirements of the Navy BuDocks Type Specifications, TS-F-15, Metallic Type (Superseding 48Y) for use in hazardous areas. Will not spark from friction with metal as do ordinary concrete aggregates. A special conductive cement binder is interground in this ready-to-use product. Cure with Conductive Kuraseal (200 sq. ft. per gallon). Available in the following colors: French grey, battleship grey, tile red, maroon, seal brown, black, Nile green, tan, and terra cotta.

Harcol R. M. (Sonneborn). A ready to use, rustproof, dry mix of abrasive resistant aggregates, specially selected for their flint-like hardness, portland cement and limeproof pigments of high and uniform tinctorial strength. Applied as a dust coat at the rate of 30 to 45 lbs. per 100 sq. ft., floated into the surface and troweled. Harcol produces an integrally colored surface highly resistant to abrasion.

Colors: bright red, dark red, gray, terra cotta, green and natural.

Ferrolith H. (Sonneborn). A finely ground metallic water absorbent hardener, free from oils and impurities.

Requires 60 to 180 lbs. per 100 sq. ft. for ordinary traffic, depending on planned use of floor.

Ferrocon Metallic (Castle Chemical). Finely powdered iron mixed with cement and water. Prevents absorption of moisture and seepage of water. For use on old and new structures. 3 to 5 coats recommended for pressure. Use 10 to 15 lbs. per 100 sq. ft. per coat.

Liquid Chemical Floor Treatments (Hardeners)

As concrete wears down, it is ground into dust. This condition has an especially detrimental effect in mills and factories where the dust from the floors has been carried into the bearings of machinery, motors, engines, etc., causing considerable damage and expense in the upkeep of the equipment.

To overcome this condition, there have been a number of chemicals in liquid form placed on the market that create a chemical action in the cement and lime in the floors and cause a complete transformation of the lime so that it solidifies the entire concrete aggregate into a hard, flint-like, homogenous mass that prevents dusting and wearing of the floors.

These liquid chemical compounds are sprayed over the floor until all of the pores in the surface are filled. It is usually necessary to treat the floors two or three times with the liquid to secure satisfactory results. When using these preparations, it is necessary to treat the floors at stated intervals, as the chemical action they create on the cement and lime is not permanent, although some manufacturers claim their product is permanent.

When using these liquid chemicals care must be exercised to see that the floor is absolutely clean and free from all grease, dust, dirt, oil, or other foreign matter.

Different manufacturers specify varying numbers of applications of their materials but all of them must be applied just as often as the floor shows signs of dusting.

Liquid Chemical Floor Hardeners

Kind of Hardener	No. Coats	Sq. Ft. per Gal.	Gals. Reqd. 100 SF	Approx. Price per Gal.	Labor Hours
Granitex, Devoe Paint	2	200	½	$7.80	0.8
Hornolith, A. C. Horn	3	150	⅔	5.00	0.8
Lapidolith, Sonneborn	3	150	⅔	4.25	1.0
Flintox, Toch	2	125	¾	3.75	0.8

Labor Cleaning Floors. The labor cost of cleaning floors is a matter that is entirely up to the judgment of the contractor or estimator. Floors that are full of oil and grease will require much more time to clean than a floor free of foreign matter.

The labor applying the liquid is also a variable item, but a man should spray 700 sq. ft. of floor an hr. on the first application and 800 sq. ft. and hr. on subsequent applications.

The labor applying the first coat should cost as follows per 100 sq. ft.:

	Hours	Rate	Total	Rate	Total
Labor	0.15	$....	$....	$19.13	$2.87
Cost per sq. ft			....		.029

On the second and third applications, the labor per 100 sq. ft. should cost as follows:

	Hours	Rate	Total	Rate	Total
Labor	0.13	$....	$....	$19.13	$2.49
Cost per sq. ft			....		.025

Concrete Accelerators and Densifiers

There are numerous preparations on the market for controlling the set of concrete, increasing its early strength, densifying and waterproofing the concrete mass, and for preventing freezing during winter weather by lowering the freezing point of the mixing water.

Anti-Hydro Accelerator and Anti-Freeze. (Anti-Hydro Co., Newark, N.J.) Can be mixed with the water or added to the wet mix. As a general rule, the standard proportion of 1.5 gals. of Anti-Hydro per cu. yd. of concrete, with reduced water content to compensate for increased slump, will give protection down to 23° F. on concrete deposited in forms except in the most exposed locations.

For cold weather work, be sure that sand and aggregate are free of ice.

To protect concrete or mortar against freezing, use the following table for anticipated outside air temperatures as follows:

32° to 25° F. use 1 part Anti-Hydro to 12 parts water.
25° to 15° F. use 1 part Anti-Hydro to 10 parts water.
15° or less, mechanical heat is necessary.

POZZOLITH-HE (Master Builders). An accelerator and water reducing agent recommended for use in concrete where accelerated set, increased early and ultimate strength and improved performance are required or desired: Concrete made with POZZOLITH-HE provides 3-day normal strength in 1 day, 7-day normal strength in 3 days, 28-day normal strength in 7 days and substantial increases in ultimate strength.

Also, non-chloride high early formulations are available for specialized concreting operations where calcium chloride is not permitted.

POZZOLITH-HE admixture is used at the dosage rate of anywhere from 16-64 fluid oz. per 100 lbs. (1040 ml to 4160 ml per 100 kg) of cement depending on formulation being used and the amount of set and strength acceleration needed or desired.

Plastiment (Sika). Plastiment is a powder added in the proportion of ½ to 1 lb. per sack of cement. It will increase workability, density, surface hardness and decrease shrinkage. It may be used for concrete or mortar.

Also available in a concentrated liquid form for use with automatic dispensing equipment. Use 2 to 4 fl. oz. per sack of cement.

Dehydratine No. 80 (W. R. Grace & Co.). Liquid integral cement floor hard- ener, accelerator and anti-freeze compound. Used in the proportions of 1 quart per sack of cement.

Sikacrete (Sika). Accelerating and plasticizing liquid for concrete or mortar, used as anti-freeze or rapid hardening compound. Used in the dilution of 1 part Sikacrete to 3 to 10 parts water. A dilution of 1 to 7 will reduce the hardening rate of mortar or concrete to one-half or protect against freezing to a temperature of 22° F.

Trimix (W. R. Grace & Co.). A chemical solution of uniform strength, which when used according to directions functions as an accelerator, integral hardener or anti-freeze compound, for use in mortar or concrete work. As an anti-freeze, use as follows:

25° to 32° F, 1 ½ qts. Trimix to each sack of cement.
20° to 24° F, 2 qts. Trimix to each sack of cement.
15° to 19° F, 2 ½ qts. Trimix to each sack of cement.
Below 15° F, heat is necessary.

BONDING OLD AND NEW CONCRETE

The problem of bonding new concrete to old, so the new concrete will remain permanently in place, is one that has received considerable attention among users of concrete. Unless the utmost care is exercised in preparing the old surface, the new concrete will invariably come loose where it is joined to the old concrete.

Different methods have been used to overcome this fault, but the basic condition of all of them is that the old surface must be thoroughly cleaned and washed, and the old aggregate must be exposed, which means that the thin film of cement that covers the surface of the concrete must be removed. One of the methods of preparing the old surface consists of hacking or picking it, and then washing or turning a steam hose under pressure to remove all dust and dirt from the old floor so as to present a perfectly clean surface.

Another method is to wash the concrete surface with a solution of muriatic acid and water to remove the old film of cement and then thoroughly wash the surface with clear water. After this has been done, the old concrete surface should be given a slush coat of neat portland cement and water, and the new concrete applied.

Hacking and Chipping Old Concrete. The cost of hacking and chipping old concrete surfaces in preparation for bonding new concrete to old, is a variable item, depending upon the hardness of the old concrete and the condition of the surface. Under average conditions, a laborer using a pick, should hack and roughen 175 to 200 sq. ft. of "green" concrete floor per 8-hr. day, at the following cost per 100 sq. ft.:

	Hours	Rate	Total	Rate	Total
Labor	4	$....	$....	$19.13	$76.52
Cost per sq. ft			....		.77

Where the concrete floors are of old concrete, a man using hand tools will do well to hack and roughen 8 to 10 sq. ft. an hour, and even when a compressor and jack-hammer or bush-hammering tool is used it is often possible to hack and roughen only 20 to 25 sq. ft. an hr. and the cost per 100 sq. ft. should average as follows:

	Hours	Rate	Total	Rate	Total
Compressor expense	2.25	$....	$....	$7.50	$16.88
Labor on jack-hammer	4.50	....	....	19.13	86.09
Cost per 100 sq. ft			$....		$102.97
Cost per sq. ft			....		1.03

Anti-Hydro Bonding Coat. Rough and clean the old concrete as previously described. Apply a coat of grout composed of ½ to ¾ sack of portland cement added gradually into a solution of 1 gal. of Anti-Hydro in 3 gals. of water until a thick, creamy consistency is obtained. After applying this grout to the prepared surface, the concrete or mortar should be placed while the grout is still wet.

Sufficient Anti-Hydro grout to cover 100 sq. ft. should cost as follows:

	Rate	Total	Rate	Total
0.25 gal. Anti-Hydro	$....	$....	$5.00	$1.25
0.25 sack portland cement	....	....	4.25	1.06
Cost per 100 sq. ft		$....		$2.31
Cost per sq. ft		....		.023

Daraweld-C (W. R. Grace & Co.) Daraweld-C is an emulsion of special internally plasticized high polymer resins uniformly dispersed in water. Mixed with cement mortars or grouts, it forms a durable, highly water-resistant bond. A Daraweld- C bond withstands water immersion without softening or disintegrating, resists the action of heat, cold, oil and gasoline, most acids and other corrosive materials. Its coupling action is so great that grouts containing Daraweld-C will bond not only to concrete and masonry, but to tile, wood, steel--even glass. Ready-to-use, non-settling, Daraweld-C will not re-emulsify when subjected to moisture. It is completely compatible with mixes containing calcium chloride.

Daraweld-C is used as a bonding agent admixture in grouts and mortars made from portland or Lumnite cement and assures their lasting adhesion to existing concrete. Formulated for exterior or interior use, it is especially useful for patching concrete floors, repairing cracks in walls, securing an integral bond between successive concrete pours, for waterproofing exteriors below and above grade, in Gunite applications, in bonding concrete to asphalt surfaces. Diluted with water, Daraweld-C is an excellent brush-on sealer for dustproofing concrete floors, and as a primer for painting concrete masonry.

Tremco Floor Bond. A chemically treated metallic powder to use in bonding new concrete toppings to set slabs, where a perfect bond must be secured to prevent cracking.

Old surface must be roughened and cleaned as described previously and the floor bond applied in 2 coats. Requires 25 lbs. of Metallic Floor Bond per 100 sq. ft. of surface.

When used for bonding vertical surfaces, 2 applications are required, using 15 lbs. of Floor Bond per 100 sq. ft.

Weld-Crete. (from Larsen Products Corp., Rockville, MD) A product that permanently bonds new concrete to old concrete. It also bonds new concrete to many other materials, such as brick, stucco, stone, etc.

For successful bonding of new concrete to old, the old concrete surface must be structurally sound and free from dust, dirt, loose material, grease, oil, wax, water soluble coatings, etc.

It is not necessary to chip, bush hammer or roughen the old surface in any way.

Weld-Crete may be used on either interior or exterior surfaces and may be applied over "green" concrete and damp or dry surfaces. May be applied with a brush, roller or spray. Spraying is most efficient and should be done with heavy industrial spray equipment.

Coverage, when sprayed, from 200 to 300 sq. ft. per gal. When brushed or rolled, coverage is somewhat less.

Repairing Leaks Against Hydrostatic Pressure

When repairing leaks in basement walls or floors, tunnels, pits, etc., where water pressure exists, the following methods and materials may be used.

Sika 2. Sika 2, red, fast-setting, sealing liquid is mixed with standard portland cement in small quantities. It is used to plug small infiltrations of water against high pressure. Sika 2 mortar will have an initial set of 15 seconds and a final set of 30 seconds. It is used after surrounding leaking concrete has been sealed with Sika 4A. Quantity required varies with degree of leakage.

Sika 4A. Sika 4A is a clear liquid mixed with neat portland cement in small quantities. The resulting mortar will have an initial set of less than 1 minute and a final set of approximately 5 minutes. Sika 4A may be diluted with water and the cement mixed with sand. This mortar is to be applied against damp surfaces or masonry over which water is running. If pressure is too heavy, bleeder pipes may be installed and after smaller leaks are sealed with Sika 4A, the bleeder holes may be plugged with Sika 2, liquid. Quantity required depends upon the degree of leakage.

CONCRETE ADMIXTURES

Since the introduction of the water-cement ratio law governing the strength and durability of concrete, the necessity for using comparatively low water ratio mixes has been generally recognized. Of greatest importance, however, has been the necessity of producing such mixes with a degree of workability which would insure their easy and economical placing and compacting.

In a cubic yard of concrete, approximately 2 ½ gallons of water per sack of cement are required to hydrate the cement. Any water in excess of that amount must be regarded as "placing" water, merely providing sufficient workability to make the mix placeable. This excess water occupies about 10% of the total space in concrete. As it evaporates, it causes the concrete to shrink. Excess water also reduces the strength of the concrete.

The purpose of admixtures is to improve the plasticity, workability, and finish of the concrete, to prevent segregation of the aggregates and to a certain extent provide integral waterproofing qualities by filling the small voids in the concrete mixture.

Pozzolith (Master Builders). A water reducing, set-controlling admixture available in several formulations to facilitate optimum performance benefits with the range of cements, sands and coarse aggregates used for making concrete.

In the plastic concrete, Pozzolith improves finishing characteristics for flat work and cast surfaces, reduces segregation of the mix, and enhances placement of low slump concrete. In the hardened concrete Pozzolith results in reduced cracking, increased compressive and flexural strengths, improved watertightness, and increased resistance of air entrained concrete to damage from freezing and thawing as well as scaling from deicing salts.

Pozzolith Normal and Retarding formulations are generally used at the rate of 5 ±2 fluid ounce per 100 pound (325 ±130 ml per 100 kg) of cement.

Plastiment (Sika). Plastiment, concrete densifier and retarding agent, is a liquid to be added to concrete or mortar at the rate of 2 to 4 oz. per sack of portland cement. For equal workability the water-cement ratio may be reduced approximately 10 percent. Results in better workability, density, adhesion to old concrete, surface hardness, and flexural and compressive strength. The delayed set results in elimination of cold joints and reduced shrinkage.

Plastocrete (Sika). A water-reducing admixture for ready-mix and performance concrete to produce higher strengths and workability. Improves structural quality of all concrete. Can be used throughout the year.

Trimix (Sonneborn). A multi-purpose concrete and mortar admixture which reduces water-cement ratio, produces higher early compression strengths, accelerates set and generally improves quality and workability of portland cement mixtures.

Used in the proportion of 1 qt. Trimix to each sack of cement in the mix.

Hydrated Lime as a Concrete Admixture. Hydrated lime is often used as an admixture to increase the plasticity and improve the workability of concrete. For this purpose 5 to 8 lbs. of hydrated lime are usually added to each bag of portland cement used in the concrete mixture.

Air Entraining Admixtures

In air entrained concrete, 3% to 6% of air is incorporated in the concrete in the form of minute separated air bubbles. Such concrete is more resistant to freezing and thawing and to salt action than normal concrete. Originally developed to prevent scaling of pavement where salts are used for ice removal, air entrained concrete is being widely used for all types of work and in all locations because of its better workability as well as its better resistance to weathering even where salts are not used.

MB-VR (Master Builders). MB-VR is a neutralized Vinsol resin type air entraining agent for concrete. It is furnished in water solution form and is ready to use as it comes from the drum. MB-VR is compatible for use with other admixtures commonly used in concrete; however, if more than one admixture is used, each should be dispensed into the mix separately. MB-VR is available via bulk tank delivery or 55-gallon steel drums.

Sika Aer. Air entraining resin solution of neutralized wood resin improves various properties of plastic and hardened concrete.

Add between 1/2 and 2 fl. oz. of Sika Aer solution to entrain between 3% and 6% air as air content is in direct proportion to quantity of Sika Aer added.

Darex AEA (W. R. Grace & Co.) Darex AEA is an aqueous solution of highly purified and modified salts of a sulfonated hydrocarbon. It contains a catalyst which promotes more rapid and complete hydration of portland cement. Darex AEA is specifically formulated for use as an air entraining admixture for concrete and is manufactured under rigid control which insures uniform, predictable performance. The addition of 3/4 fluid ounce of Darex AEA per sack of cement will generally entrain 4% to 7% air in a 5 1/2 sack concrete mix.

CONCRETE CURING PROCESS

Efficient methods of curing concrete are of vital importance to every contractor, because exposed surfaces such as concrete floors, roofs, road, pavements, airport runways, and other surfaces exposed to the sun require protection during the curing period. This was formerly accomplished by the use of damp burlap, sand, sawdust, dirt covering, paper, or ponding.

A method that is being used to a large extent involves curing compounds, which provides an air-tight seal over the surface of the concrete and prevents the evaporation of the water from the concrete until it has cured normally.

These compounds are usually applied with a brush or spray as soon as the concrete has set sufficiently to prevent injury.

Servicised/Horn Concrete Curing Compounds. (A. C. Horn, Inc.) Are scientifically formulated liquids for spray application to fresh concrete surfaces. The compound quickly forms a vapor-tight film which seals in 95% or more of the moisture in the concrete for 3 or more days, without changing color of the concrete. Can be applied with standard spray equipment. One application should control cure on any size surface area, from mass concrete to a few square feet.

Kure-N-Seal (Sonneborn). Kure-N-Seal cures, seals and dustproofs in one application, and it increases surface hardness when applied as a curing membrane after final troweling. By locking in essential curing moisture, the hydration process is refined and maximum hardness for the involved mix is assured.

Klearseal (Castle Chemical). A clear curing compound for use on surfaces where moderate traffic or corrosive conditions are anticipated, and where a completely clear surface is desired. One gallon covers 300 to 450 sq. ft. for smooth trowelled concrete; 250 to 350 sq. ft. for broom finished concrete, and 200 to 300 sq. ft. for rough finished concrete.

03366 POST-TENSIONING CONCRETE

Post-tensioning is the tensioning of concrete after it is placed. It has become increasingly popular in recent years to reduce shrinkage cracks and to enhance the strength, serviceability, and design flexibility of concrete structures. It can provide high quality, watertight floor systems for applications such as parking garage floors, and using corrosion protection systems that are available for post-tensioning tendons protect in corrosive environments.

Many concrete contractors think of post-tensioning as a type of specialty work, but actually, it is just an additional step in producing high-quality reinforced concrete. Of course, the concrete contractor will want the assistance of a qualified post-tensioning engineer and a reliable material supplier to produce a quality job at an economical price.

There are two basic types of post-tensioning: *multi-strand* and *mono-strand.* In the multi-strand type several strands of prestressed wire are placed in a duct, and after the strands are tensioned, the duct is grouted. This method is sometimes referred to as a *grouted* or *bonded* system.

In the mono-strand type, sometimes referred to as *unbonded,* individual strands are made up of seven prestressed wires. Strands are coated with a lubricant grease to prevent rusting and reduce friction during the stressing operation. The greased strand is covered with a plastic coat.

Mono-strand post-tensioning. In the estimate the contractor must allow for certain items that are unique to post-tensioning. Stressing access must be provided, usually a platform or floor foor extension at the perimeter of the structure.

Forming requirements must be coordinated with the stressing and anchorage of post-tensioning tendons. Horizontal supporting formwork cannot be removed until the stressing operation is completed in a given pour area.

Post-tensioning anchorages are typically designed to permit the stressing of tendons when the concrete reaches a strength of 3,000 psi. Check with the post-tensioning material supplier for actual concrete strength requirements for the systems.

The contractor must also make provisions for cutting off the stressing tails of the tendons and patching the stressing pockets.

The post-tensioning material supplier will provide the contractor with the placement drawings for the system and will coordinate the pour sequence with post-tensioning anchorage requirements.

Material for a mono-strand system is estimated by the pound. Contract drawings usually specify effective force requirements. The effective force translates into the number of single strands required by dividing the required force by the effective force of each tendon (about 26.5 kips per tendon for ½" strand).

The actual effective force provided by each tendon is calculated for each project and must consider seat loss, friction loss, elastic shortening of concrete, and time-dependent losses. The total length of a tendon is obtained by adding stressing tails to the length of each individual tendon (1'-6" at each stressing end). To obtain nominal strand weight, multiply the length by .525 lbs. per ft. of ½" diameter strand.

Material Costs. Post-tensioning material for mono strand averages about $0.80 per lb. The following are the standard basic steps for placement of post-tensioning tendons:

1) Form concrete slab as usual, but allow about 4'-0" for post-tensioning set up (jacking) adjacent to the concrete slab side forms.
2) Set side forms, drill holes for post tension wires (tendons), and set grommets.
3) Set reinforcement steel chairs about 3'-0" on center.

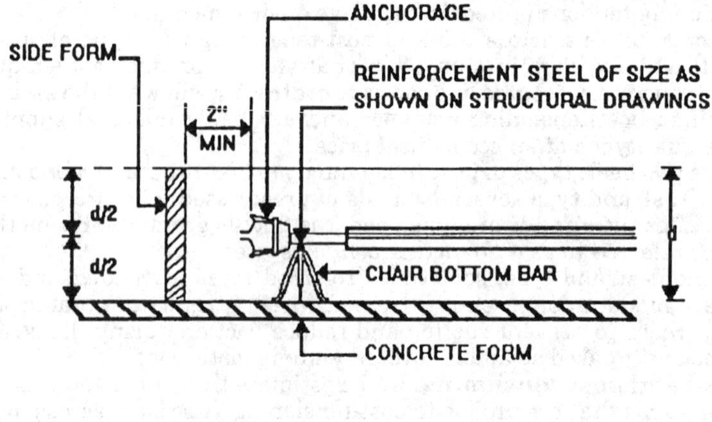

TYPICAL DEAD END POST TENSION ANCHORAGE

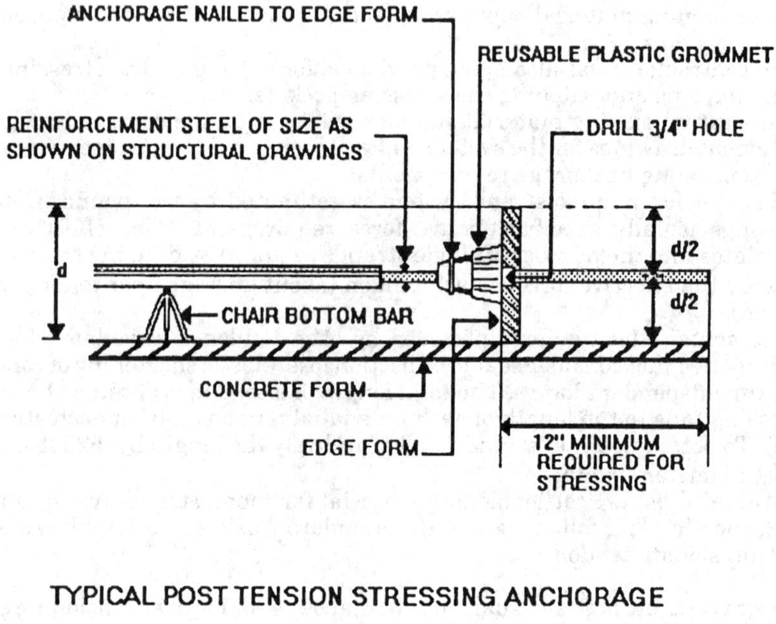

TYPICAL POST TENSION STRESSING ANCHORAGE

4) Set bottom reinforcement steel.
5) Set post-tension wire in one direction.
6) Set post-tension wire in the other direction.
7) Set top reinforcement steel.
8) Pour concrete.
9) Remove side forms, tension prestress wires (tendons) by jacking.
10) Remove grommet and finish concrete edge.

Labor Costs. The production average for post-tensioning is 40 to 50 sq. ft. per manhour at the following cost per 100 sq. ft.:

	Hours	Rate	Total	Rate	Total
Ironworker	2.22	$....	$....	$27.53	$61.12
Cost per sq. ft.			....		.61

03400 PRECAST CONCRETE

FLEXICORE* FLOOR AND ROOF SLABS

Prestressed Flexicore (*trademark of The Flexicore Co., Inc.) concrete floor and roof slabs, designed in accordance with the ACI 318 Building Code, are precast in central plants with patented equipment. The cross sections and lengths of these units are indicated below. Filler width slabs are also available.

	Span-Ft.	Lbs. per Sq. Ft. (Based on 150 Lb. Concrete)
6"x24"	15-25	43
8"x24"	20-33	57
10"x20"	28-40	61
10"x24"	28-40	72
12"x24"	35-50	79

The amount of deflection allowed shall be consistent with the finish to be applied to the underside of the flexicore units, and shall be within the limits specified by ACI 318 Building Code.

These units are manufactured in a rigid steel form made sufficiently strong to resist the pretensioning force applied by the seven wire strand reinforcing steel. Specially constructed rubber tubes are inflated to form circular voids so as to constitute approximately 50% of the cross section. The concrete is thoroughly vibrated to assure maximum denseness and strength.

The slabs may be cast of concrete made of gravel, crushed stone or lightweight aggregate. When lightweight aggregate is used, the unit weight is approximately 25% less. They are cured in a heated kiln.

Since all the Flexicore plants are not equipped to make all the sections listed in the above table, the availability for each section and available length should be checked locally.

Erection. Flexicore units are usually delivered to the job by truck. They are hoisted from the truck to the floor location and often placed directly into their final position by the crane. After the slabs are placed side by side, they are aligned and levelled. The keyways in the sides of the slab are then filled with a grout mixed in a ratio of one to three. The erection crew usually consists of six workers, including a crane operator, an oiler, a foreman, and three laborers. A crew will unload, erect and grout 2800 sq. ft. per day on smaller jobs to approximately 6000 sq. ft. on some larger jobs.

Cost of Flexicore Floor Slabs. Flexicore slabs are usually quoted on an erected basis by the manufacturer and the price will usually include the cost of grouting, and caulking of the joints on the ceiling side. The price will vary according to the type of building, size, live load, span, section specified, location, etc. but usually runs from $4.20 to $5.50 per sq. ft. erected, grouted, and caulked.

In localities distant from the manufacturing plant, the slabs are often sold on a "delivered only" basis and erection is performed by the contractor using his own personnel.

Caulking of the joints on the underneath is undertaken after the building has been enclosed. A non-staining type caulking material is used and will average about 20 cents per lin. ft. An underlayment which is not included in the bid is generally used over the Flexicore slabs and is provided by the general contractor.

CONCRETE PLANK

Cantilite (from Concrete Plank Co, Inc., North Arlington, NJ) is a lightweight high-strength concrete plank that is nailable and easily cut in the field. In laying, it is clipped to steel or wood beams. Floor or roof materials are nailed to it. Planks are factory-made in steel forms to give a smooth, even surface on all sides and to assure proper positioning of the reinforcing steel. Side edges are tongued and grooved.

Concrete plank are used for floors and roofs. The plank are 2", 2 ¾" and 3 ¾" thick, 16" wide and made in lengths up to 10'-0". The 2" plank are used on steel or concrete joists, 4'-0" on centers for floors. The 2 ¾" plank are placed on steel or concrete joists up to 5'-0" on centers for floors. The 3 ¾" plank are placed on steel or concrete joists up to 6'-0" on centers for floors. A 1" or 2" cement finish top is placed over the plank. Roofs having a span up to 7'-0" will require 2" plank, while 2 ¾" plank will be required for a roof span up to 8'-0". The 3 ¾" plank will be required for a roof span up to 10'-0".

Concrete plank costs $2.50 to $3.50 per sq. ft., f.o.b. factory, depending upon quantity in the job. The 2" plank weigh 13 lbs. per sq. ft.; the 2 ¾" plank, 18 lbs. per sq. ft.; the 3 ¾" plank, 28 lbs. per sq. ft.

Labor Erecting Concrete Plank. A crew consisting of 4 masons and 1 laborer can handle and lay approximately 2,000 sq. ft. of plank per 8-hr. day, on a flat surface.

Labor Cost of 100 Sq. Ft. of Concrete Plank Laid on a Flat Floor or Roof Surface

	Hours	Rate	Total	Rate	Total
Masons	1.6	$....	$....	$25.62	$40.99
Labor	0.4	....	....	19.13	7.65
Hoisting engineer	0.4	....	....	25.96	10.38
Cost per 100 sq. ft.			$....		$59.02
Cost per sq. ft.			....		.59

PRECAST CONCRETE ROOF SLABS

Precast concrete roof slabs are of three general types: rib, flat, and channel. They are adapted to all roof decks, flat or sloping, are fireproof, and with caulked joints, present a smooth surface for the application of built-up roofing.

Flexicore on Lightweight Steel Frame.

Where slate, ornamental tile or copper covering is specified, slabs with nailing surfaces can be furnished, permitting direct application.

Flexicore Slabs Erected Directly from Truck.

Precast concrete roof slabs are furnished made with regular concrete aggregates or where a lightweight slab is desired, they are made of lightweight aggregate.

The rib tile is self-weathering and its attractive red color provides an architectural feature that is very desirable and so often lacking in industrial buildings. Auxiliary pieces are furnished to suit the particular design required, as ridges, saw-tooth ridges, gable end finishing tile, monitor flashing tile and other specials such as would be required in connection with a hip or valley.

Tile of the same standard stock sizes, but having a glass insert, may also be furnished where additional light may be required. On this type of roof slab it is necessary to obtain prices from the manufacturer due to the fact that the amount of trim varies with each project.

Example of Architectural Concrete

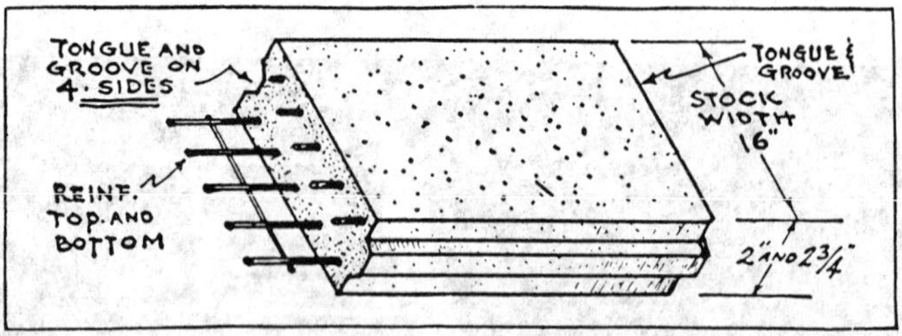

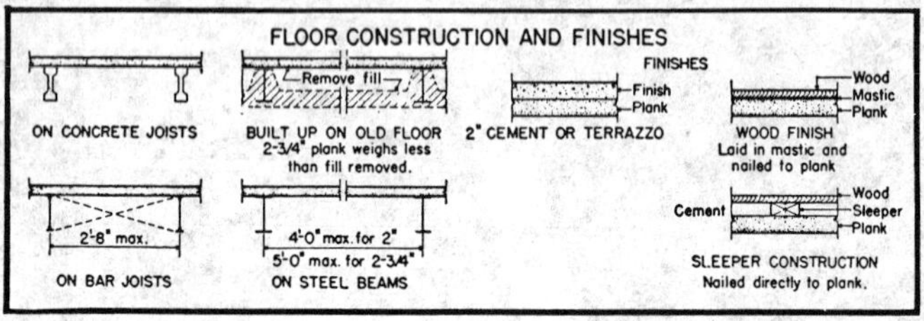

The other two types of precast concrete roof slabs, flat and channel, are used over all roof decks, whether flat or sloping, and present a smooth surface for the application of any type of built-up roofing. When used with a sloping roof and when slate or other types of ornamental covering are to be used, these slabs can have a concrete nailing surface superimposed and manufactured integrally with the structural slab.

The accompanying illustration gives the sizes and weights of the various types of tile.

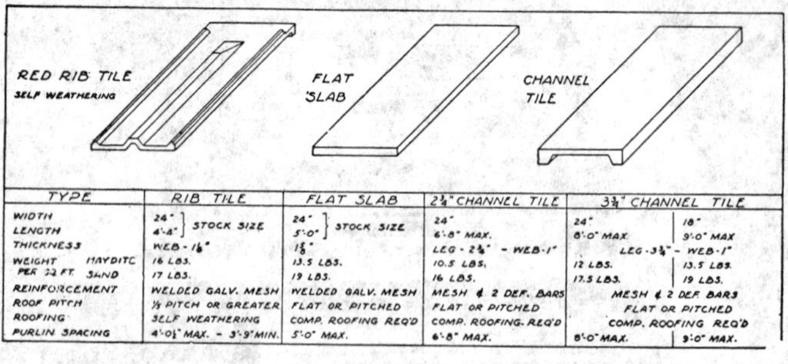

TYPE		RIB TILE	FLAT SLAB	2¼" CHANNEL TILE	3¾" CHANNEL TILE	
WIDTH		24" } STOCK SIZE	24" } STOCK SIZE	24"	24"	18"
LENGTH		4'-4"	5'-0"	6'-8" MAX.	8'-0" MAX	9'-0" MAX
THICKNESS		WEB - 1⅛"	1⅝"	LEG - 2¼" - WEB - 1"	LEG - 3¾" - WEB - 1"	
WEIGHT PER SQ. FT.	HAYDITE	16 LBS.	13.5 LBS.	10.5 LBS.	12 LBS.	13.5 LBS.
	SAND	17 LBS.	19 LBS.	16 LBS.	17.5 LBS.	19 LBS.
REINFORCEMENT		WELDED GALV. MESH	WELDED GALV. MESH	MESH & 2 DEF. BARS	MESH & 2 DEF. BARS	
ROOF PITCH		⅓ PITCH OR GREATER	FLAT OR PITCHED	FLAT OR PITCHED	FLAT OR PITCHED	
ROOFING		SELF WEATHERING	COMP. ROOFING REQ'D	COMP. ROOFING REQ'D	COMP. ROOFING REQ'D	
PURLIN SPACING		4'-0½" MAX. - 3'-9" MIN.	5'-0" MAX.	6'-8" MAX.	8'-0" MAX.	9'-0" MAX.

Sizes and Weights of Precast Concrete Roof Tile

Labor Placing Precast Concrete Roof or Floor Slabs. Because of their size and weight, it requires 2 workers to handle precast concrete roof or floor slabs. Each piece weighs from 125 to 200 lbs. Ordinarily it requires 2 workers on the ground handling the slabs and getting them to the hoist, 2 more on the floor or roof placing them for the mason, and a mason and helper to lay and caulk the joints.

A mason and helper should lay and caulk 960 to 1,050 sq. ft. of lightweight concrete tile per 8-hr. day on straight roofs.

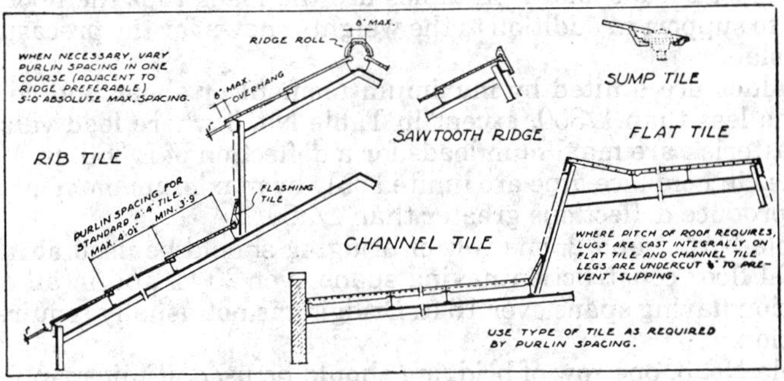

Details of Precast Concrete Roof Tile Construction

PRECAST CONCRETE JOIST FLOORS

Precast concrete joists are widely used to support concrete floors in homes, apartment buildings, office buildings, schools, hospitals and similar structures. Precast concrete joists are also extensively used for supporting concrete roof slabs for practically all types of buildings.

The design tables relate only to floors and roofs which carry relatively light loads. They are not intended for use in designing floors or roofs carrying heavy or concentrated loads, or where they are subject to heavy impact loads or to vibration of mounted machinery. These conditions impose special loading problems and to meet them floors must be designed accordingly.

The concrete joists are manufactured in a cement products plant and delivered to the job ready to set in place. The joists for each floor of a building should be laid out separately, showing size of joists, spacing, etc., which varies from 20 to 33 inches, depending upon depth of joists, span, floor load, etc.

To develop maximum strength, a reinforced concrete slab 2 to 2 ½" thick is placed over the tops of the joists with the joists embedded into the concrete slab to a depth of ½ to ¾ inch. Effective bond shall be obtained between the joists and slab. Concrete for both joists and slab shall have an average compressive strength of not less than 3,750 lbs. per sq. in. at age of 28 days. Floors made of precast joists mortised or embedded into a monolithic floor placed or poured on the job are called *precast joist cast-in-place concrete slab* floors.

Floors made of precast joist over which are laid precast slabs for flooring are called *precast joist and slab* concrete floors.

Concrete joists are manufactured 6, 8, 10, 12, and 14 inches deep and various strength requirements of each size may be obtained by increasing the amount of reinforcing steel. For instance, 8-inch concrete joists are manufactured containing from .22 to .88 sq. in. of reinforcing bars at the bottom of the joists and ⅜" round reinforcing bars at the top, with ¼" round steel stirrups spaced according to the load requirements of the particular job. The strength of 10, 12, and 14-inch joists is determined in the same manner.

Notes for Table Nos. 1, 2 and 3 on following pages:

Superimposed loads shown in tables are the loads that the floor or roof is designed to support in addition to the weights shown for the precast joists and concrete slab.

Load values are limited by maximum moment or shear and will produce deflections less than L/360, except in Table No. 3 where load values shown with an asterisk are maximum loads for a deflection of L/360. In Table No. 3 load values in bold face type are limited only by maximum moment and shear and will produce deflections greater than L/360.

For Table Nos. 1 and 2, one row of bridging should be used at midspan in residential floor construction having spans over 20 ft. and in all other floor construction having spans over 16 ft. Bridging is not usually required for roof construction.

For Table No. 3, one row of bridging should be used at midspan for all floor construction having spans over 16 ft. Bridging is usually not required for roof construction.

For floor or roof construction shown in Table Nos. 1 and 3, no shoring is required during construction.

For floor and roof construction shown in Table No. 2, shoring at midspan is required during construction and must remain in place until the concrete has reached its required strength.

Bar sizes are given in numbers based on the number of eighths of an inch included in the nominal diameter of the bars as specified in ASTM A305.

Concrete Joists Designed as Independent Beams. Where concrete joists are designed to be used with precast concrete slabs, where the slab does not form a part of the beam, it is necessary to use more reinforcing steel in the beams, which increases their cost considerably.

For example, 8" beams with 3/8" top bars and 5/8" bottom bars are used for spans 10 to 15 ft.; 1/2" top bars and 3/4" bottom bars for spans 11 to 16 ft.; and joists using 3/4" top bars and 7/8" bars for spans 12 to 16 ft. requiring heavier loads.

Ten inch joists with 3/8" top bars and 3/4" bottom bars are used for spans 12 to 19 ft.; 5/8" top bars and 7/8" bottom bars for spans 14 to 20 ft.; and 3/4" top bars and 1" bottom bars for spans 16 to 20 ft. required to carry heavier loads.

Twelve inch joists with 3/8" top bars and 3/4" bottom bars are used for spans 12 to 22 ft.; 3/8" top bars and 7/8" bottom bars for spans 14 to 24 ft.; and 1/2" top bars and 1" bottom bars for spans 16 to 24 ft. required to carry heavier loads.

These beams require from 1 to 2 lbs. of reinforcing steel per lin. ft. more than the T-beams and cost 30 to 40 cents per lin. ft. more than the T-beam joists.

Precast concrete joists are manufactured of Haydite, Waylite, Celocrete, and other lightweight aggregate, also of ordinary aggregate consisting of gravel or crushed stone and sand.

Weight of Precast Concrete Joists, Ordinary Aggregate

Size of Joist In.	Length of Joist in Feet 10	12	14	16	18	20
8	190	228	266	304	...	...
10	230	276	322	368	414	460
12	350	420	490	560	630	700

Table 1 • Joists and Slab Designed as T-Beams

No shoring used during construction

Joist size	Size of deformed reinforcing bars		Wt. of joist and slab	Slab thick-ness	Joist spac-ing c. to c.	Superimposed loads = psf/Clear span = ft.														
in.	Top	Bot	psf	in.	in.	10	11	12	13	14	15	16	17	18	19	20	21	22	23	24
8	#3	#5	34	2	20		113	88	69	53	40	30								
			32	2	24	118	90	69	53	40	29									
			32	2	27	101	76	57	42	31										
			37	2½	30	85	61	43	29											
			36	2½	33	74	52	36	23											
	#3	#6	34	2	20				113	91	73	59								
			32	2	24			113	90	72	57	45								
			32	2	27		122	96	72	59	46	36								
			37	2½	30	139	106	81	61	46	33	22								
			36	2½	33	123	93	70	53	38	26									
	#3	#7	34	2	20					137	113	94								
			32	2	24				134	110	90	74								
			32	2	27				115	93	76	61								
			37	2½	30			125	99	78	61	48								
			36	2½	33		128	111	87	68	53	40								
10	#3	#6	37	2	20					124	103	85	70	57	47	38				
			35	2	24				122	99	81	66	53	43	34	27				
			34	2	27			131	105	85	69	55	44	35	27					
			39	2½	30			113	89	70	54	41	31	22						
			38	2½	33		128	100	78	60	46	34	25							
	#3	#7	37	2	20							129	109	92	78	66				
			35	2	24						124	103	87	72	61	50				
			34	2	27					129	107	89	74	61	51	42				
			39	2½	30				137	111	90	73	59	47	37	28				
			38	2½	33				121	98	79	63	50	39	30	22				
	#3	#8	37	2	20									134	115	99				
			35	2	24								125	107	91	78				
			34	2	27							127	108	92	78	66				
			39	2½	30					138	125	110	91	76	63	52				
			38	2½	33					122	111	97	80	66	54	44				
12	#3	#6	44	2	20						126	104	86	71	59	48	39	31		
			41	2	24					123	100	83	68	55	44	35	27			
			39	2	27				131	106	87	70	57	46	36	28				
			43	2½	30				113	90	72	57	44	33	24					
			42	2½	33			126	99	79	62	48	36	27						
	#3	#7	44	2	20								135	115	97	83	70	59	50	42
			41	2	24							128	108	91	77	64	54	45	37	30
			39	2	27						133	111	93	78	65	54	45	37	30	23
			43	2½	30						115	95	78	63	51	41	32	25		
			42	2½	33					124	101	82	67	54	43	34	26			
	#3	#8	44	2	20											123	107	93	81	69
			41	2	24									133	114	98	85	73	62	53
			39	2	27								135	115	99	85	72	62	52	44
			43	2½	30								117	98	83	69	58	48	39	32
			42	2½	33							123	103	86	72	60	49	40	32	25
	#3	#9	44	2	20													131	115	102
			41	2	24											137	120	105	92	80
			39	2	27										137	119	103	90	78	68
			43	2½	30									138	118	101	87	74	63	54
			42	2½	33									122	104	89	75	64	54	45

Table 2 • Joists and Slab Designed as T-Beams

Shoring used at midspan during construction

Joist size (in.)	Size of deformed reinforcing bars: Top	Size of deformed reinforcing bars: Bot	Wt. of joist and slab (psf)	Slab thickness (in.)	Joist spacing c. to c. (in.)	Superimposed loads = psf/Clear span = ft. 10	11	12	13	14	15	16	17	18	19	20	21	22	23	24
8	#3	#5	34	2	20		124	99	79	64	51	41								
			32	2	24	128	100	80	63	50	39	30								
			32	2	27	111	86	67	52	41	31									
			37	2½	30	100	76	58	44	33	24									
			36	2½	33	89	67	50	38	28										
	#3	#6	34	2	20				123	102	84	70								
			32	2	24			123	100	82	67	55								
			32	2	27		133	106	86	70	57	46								
			37	2½	30		121	96	76	61	48	38								
			36	2½	33	138	108	85	67	53	42	32								
	#3	#7	34	2	20						124	105								
			32	2	24					120	101	85								
			32	2	27				126	104	87	72								
			37	2½	20				115	94	77	63								
			36	2½	33			126	102	83	68	55								
10	#3	#6	37	2	20					133	112	94	79	66	55	46				
			35	2	24				130	108	89	75	62	51	42	35				
			34	2	27				113	93	77	63	52	43	35	28				
			39	2½	30			125	101	82	66	53	43	34	26					
			38	2½	33			112	90	72	58	46	37	29						
	#3	#7	37	2	20								118	101	87	75				
			35	2	24						132	112	95	81	69	59				
			34	2	27					137	115	97	82	69	59	50				
			39	2½	30					123	102	85	71	59	49	40				
			38	2½	33				134	110	91	75	62	52	42	35				
	#3	#8	37	2	20										124	109				
			35	2	24								133	115	100	87				
			34	2	27							136	116	100	86	75				
			39	2½	30						137	122	104	88	75	64				
			38	2½	33					134	123	109	92	78	66	56				
12	#3	#6	44	2	20						134	112	95	80	67	56	47	39	32	26
			41	2	24					130	109	90	75	63	52	43	35	29		
			39	2	27					114	94	78	65	54	44	36	29			
			43	2½	30				124	101	83	67	55	44	35	28				
			42	2½	33			136	110	89	72	58	47	37	29					
	#3	#7	44	2	20									123	106	92	79	68	59	50
			41	2	24							136	116	99	84	72	62	53	45	38
			39	2	27							119	101	86	73	62	53	44	37	31
			43	2½	30						126	105	89	75	63	52	43	35	29	
			42	2½	33					135	112	93	78	65	54	44	36	29		
	#3	#8	44	2	20											133	116	102	90	79
			41	2	24										123	106	93	81	700	61
			39	2	27									123	106	92	80	69	61	52
			43	2½	30								128	109	94	81	69	59	50	43
			42	2½	33							134	114	97	83	71	60	51	43	36
	#3	#9	44	2	20														125	111
			41	2	24												128	113	100	88
			39	2	27											127	111	98	86	76
			43	2½	30										129	113	98	85	75	65
			42	2½	33									133	115	99	86	75	65	56

Table 3 • Joists Designed as Independent Beams

No shoring used during construction

Joist size in.	Size of deformed reinforcing bars Top	Size of deformed reinforcing bars Bot	Wt. of joist and slab psf.	Joist spac-ing c. to c. in.	Superimposed loads = psf/Clear span = ft. 10	11	12	13	14	15	16	17	18	19	20	21	22	23	24
8	#3	#5	34	20	112	86	67	52	40	31									
			32	24	89	68	52	40	30	21* 22									
			32	27	76	57	43	32	23										
	#4	#6	34	20		106	85	69	49* 56	23* 45									
			32	24	107	85	67	54	30* 43	8* 34									
			32	27	92	72	56	42* 44	16* 35										
	#6	#7	34	20					103	69* 85	39* 71								
			32	24				100	79* 82	46* 67	21* 55								
			32	27			106	86	60* 69	30* 56	8* 46								
10	#3	#6	37	20					100	82	68	56	46	26* 37					
			35	24				98	79	64	52	42	29* 34						
			34	27			104	84	68	54	44	35							
	#5	#7	37	20							104	88	70* 74	43* 63	21* 53				
			35	24						99	82	69	46* 58	23 48	5* 40				
			34	27					102	85	70	57* 59	32* 49	12* 40	— 33				
	#6	#8	37	20									85* 107	56* 93	32* 80				
			35	24								89* 100	59* 85	34* 73	14* 63				
			34	27							102	70* 86	43* 73	22* 62	4* 53				
12	#3	#6	44	20						105	87	72	60	49	40	32	26		
			41	24					102	83	68	56	45	36	29	23			
			39	27				109	88	72	58	47	38	30	23				
			43	30				90	71	57	45	35	26						
			42	33			100	79	62	49	38	29							
	#3	#7	44	20									96	81	69	58	49	39* 41	19* 34
			41	24							106	89	75	63	53	44	37	20* 30	
			39	27							92	77	65	54	45	37	28* 30	10* 24	
			43	30						91	75	61	50	40	32	21* 25			
			42	33					98	80	65	53	43	34	26				
	#4	#8	44	20											103	90	75* 78	50* 67	28* 58
			41	24										95	81	70	50* 608	28* 52	10* 44
			39	27									96	82	70	60	37* 51	18* 44	2* 37
			43	30								93	78	66	55	30* 46	9* 38	— 31	
			42	33							97	82	68	41* 57	19* 47	— 39	32		

Cost of Precast Concrete Joists. The cost of precast concrete joists will vary with the size of joists, amount of reinforcing steel required, size of job, locality, etc.

Inasmuch as precast concrete joists are always manufactured in a concrete products plant, it is important to obtain definite prices for each particular job.

Approximate Prices per Lin. Ft. for Precast Concrete Joists.

8-Inch	10-Inch	12-Inch	14-Inch
$4.75	$5.20	$7.00	$8.00

Where the concrete joists are designed as independent beams for use with precast concrete floor or roof slabs, add from 25 to 30 cents per lin. ft. to the above prices to allow for the extra reinforcing steel required.

Labor Setting Precast Concrete Joists. As long as the joist is not too heavy to handle by four workers or in some cases by light derricks, the precast joist is more economical than one which is cast in place. The economical limits of largeness of precast joists has not been established, although at the present time they are obtainable to a depth of 14" and 36'-0" long. Such a joist weighs approximately 1,150 to 1,330 lbs. if made of lightweight aggregate--they fall within the weight where erection by machinery is necessary.

On every job there should be a joist setting plan, prepared by an architect or engineer, as this will expedite setting the joists on the job.

Where smaller size joists are used, two workers can carry the joists from the stock pile to the building and set them in place, while with heavier joists it might require 4 to 6 workers to carry the joists to the building and place them on the wall.

Where joists are used on the second floor, two workers will probably be required to carry the joists into the building and 3 workers above to hoist the joists to the second floor. If the joists are not too heavy for two workers to handle, a rope can be slipped around each end of the joist and lifted to the second floor.

After the joists have been placed on the wall, it will require some additional time for exact spacing of the joists, placing headers, etc.

Where the joists can be handled and set by 2 laborers, it will require 6 to 8 min. to carry one joist from stock pile (not over 50 ft.) and set it on the wall, at the following labor cost per joist:

	Hours	Rate	Total	Rate	Total
Labor	0.23	$....	$....	$19.13	$4.40

If joists are hoisted to the second floor, double the above costs.

When handling and setting joists requiring 4 workers to carry them to the building and set them on the wall, it will require 6 to 8 min. for 4 workers, at the following labor cost per joist:

	Hours	Rate	Total	Rate	Total
Labor	0.47	$....	$....	$19.13	$8.99

If joists are hoisted to the second floor, double the above costs.

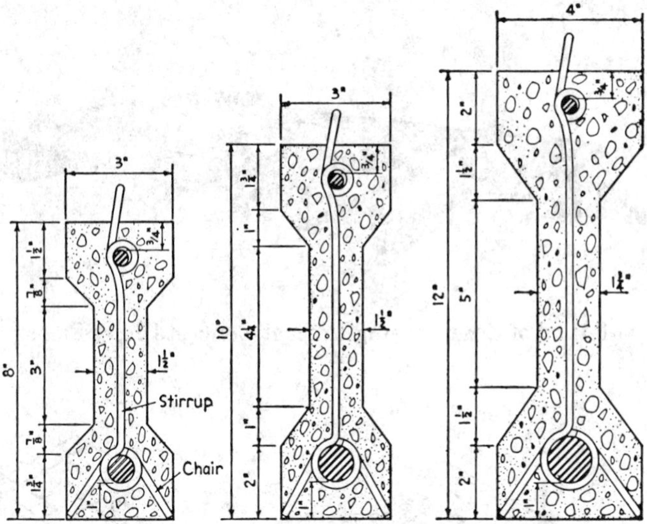

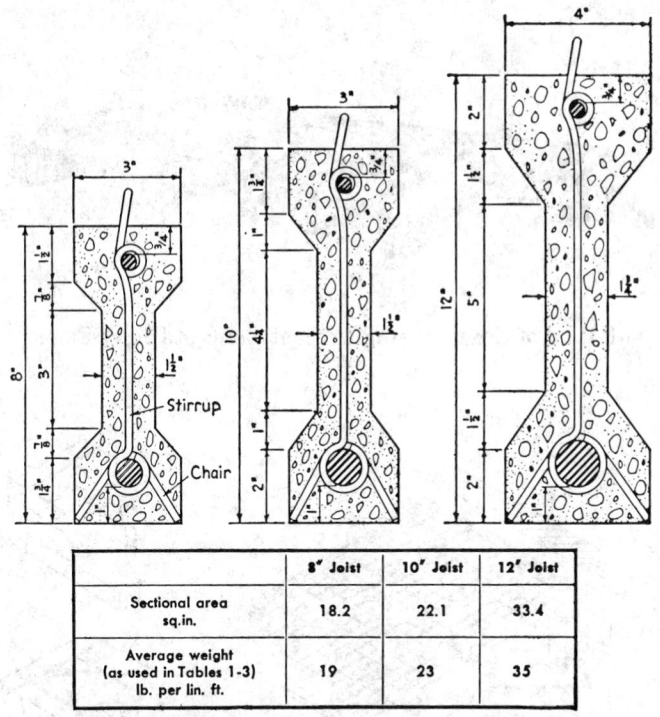

	8" Joist	10" Joist	12" Joist
Sectional area sq.in.	18.2	22.1	33.4
Average weight (as used in Tables 1-3) lb. per lin. ft.	19	23	35

Sizes and Weights of Precast Concrete Floor Joists

When handling and setting 12" joists up to 24-ft. long, requiring 6 workers to handle and set, it will require 6 to 8 min. at the following labor cost per joist:

	Hours	Rate	Total	Rate	Total
Labor	0.7	$....	$....	$19.13	$13.39

If joists are hoisted to the second floor, double the above costs.

On jobs having sufficient joists to warrant the use of a hydraulic crane, and using iron workers to set the joists instead of laborers, the cost of setting joists should average as follows:

Labor Setting One Concrete Joist

	Hours	Rate	Total	Rate	Total
Crane and operator	0.10	$....	$....	$52.00	$5.20
Iron workers	0.20	....	....	27.53	5.51
Cost per joist			$....		10.71

Wood Forms for Precast Concrete Joist Floors. The reinforced concrete floor slab used with precast concrete joists is usually 2 to 2 ½ inches thick, and the joists are spaced from 20 to 33 inches apart, so that very light formwork is sufficient.

The formwork usually consists of short pieces of 2x4-inch wood spreaders as shown in the illustration, cut to bear on the bottom flange of the concrete joists. These spreaders are placed between the concrete joists, and the ⅝" plywood are placed on top of the spreaders to receive the concrete.

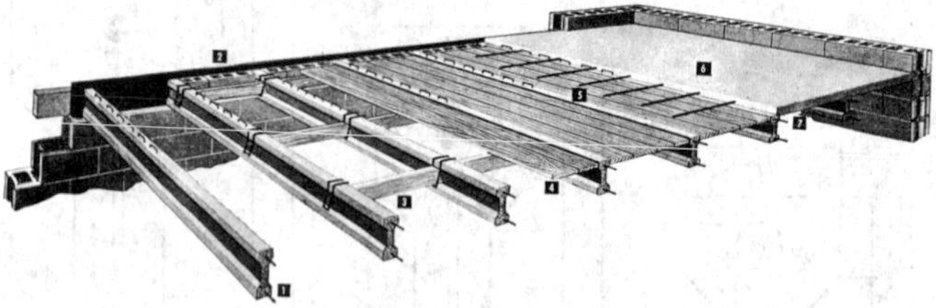

Method of Constructing Precast Concrete Joist Floors

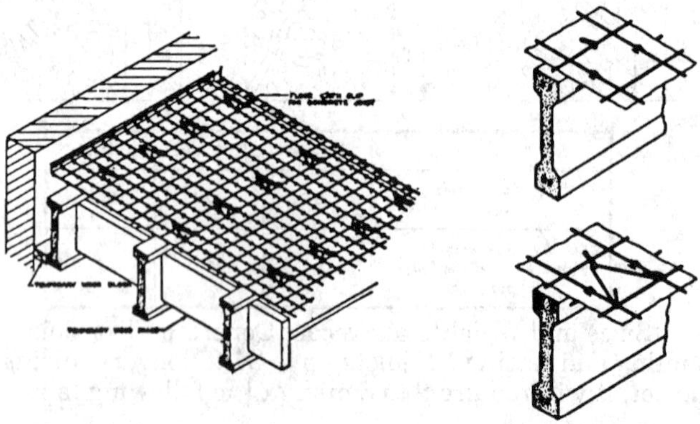

Method of Constructing Precast Concrete Joist Floors Using Paper-Backed Floor Lath for Permanent Form Material

On forms of this kind, it will require about 1 ¼ b.f. of lumber per sq. ft. of floor area.

When placing forms for precast concrete joist floors, a carpenter should frame and place forms for 175 to 225 sq. ft. of floor per 8-hr. day, at the following cost per 100 sq. ft.:

	Hours	Rate	Total	Rate	Total
Carpenter	4	$....	$....	$24.04	$96.16
Cost per sq. ft			....		.96

A laborer should remove forms from 600 to 750 sq. ft. of floor per 8-hr. day, at the following cost per 100 sq. ft.:

	Hours	Rate	Total	Rate	Total
Labor	1.2	$....	$....	$19.13	$22.96
Cost per sq. ft			....		.23

Permanent Forms for Precast Concrete Joist Floors. Permanent type forms, not requiring removal, are being widely used in precast concrete joist floor construction, especially for work such as floors over crawl spaces, tunnels, etc., where removal of wood forms would be difficult and expensive. Form materials usually used for this purpose are of two kinds--paper-backed floor lath, such as "Steeltex" (from National Wire Products Corp., Baltimore, MD) or corrugated steel sheets described in the first part of this chapter. With corrugated steel forms, reinforcing steel or wire mesh must be added to satisfy flexure and temperature steel requirements.

Steeltex Forms for Precast Concrete Joist Floors. Steeltex is a welded wire mesh with a waterproofed paper backing attached. The paper backing serves as the form and the mesh as the reinforcing for thin, short span concrete floor slabs. When concrete is placed, the paper backing sags away from the mesh to the limit of the stitch wires thereby placing the reinforcing mesh at the proper position for encasement in the slab.

Steeltex is furnished in rolls, 4'-0" wide and 125'-0" long, containing 500 sq. ft. Side laps should be at least 2" and end laps should be a minimum of 12" occurring only over a support with no adjacent splices across the same support. In estimating, compute area of slab and add about 5 percent for laps and waste. For floor slabs at or near grade level, two ironworkers working together should install about 4000 sq. ft. of Steeltex over steel joists per 8-hr. day. For slabs at higher levels, additional handling labor and hoisting must be added.

Labor Cost of 100 Sq. Ft. Steeltex Forms for
Precast Concrete Joist Floors

	Hours	Rate	Total	Rate	Total
Ironworker	0.4	$....	$....	$27.53	$11.01
Cost per sq. ft			....		.11

Corrugated Steel Forms for Precast Concrete Joist Floors. There are two methods of using corrugated steel forms for precast concrete joist floors. Corrugated steel forms may be laid continuous over joist tops with end laps occurring only over supports and no adjacent laps over same joist.

Corrugated steel forms also may be installed between joists, so that tops of joist and protruding stirrups are encased in slab concrete, providing T-beam action. For this method, joists should be precast with ledges on the top edges to furnish support for the corrugated sheets.

Side laps for corrugated steel forms should be one corrugation and end laps a minimum of 2". For estimating purposes, compute area of floor slab placed over corrugated steel sheets and add 5% to 10% for laps and waste, depending upon the method of installation.

Where corrugated steel forms are installed in continuous lengths over precast concrete joists for floors at or near grade level, with an area of 5,000 sq. ft. or more, a 4-worker crew should place and secure 2,500 to 2,700 sq. ft. per 8-hr. day. For areas less than 5,000 sq. ft., the crew should install 2,100 to 2,300 sq. ft. per 8-hr. day. For slabs at higher levels, additional handling labor and hoisting must be added.

Labor Cost of 100 Sq. Ft. of Corrugated Steel Slab Forms for
Precast Concrete Joist Floors in Residential Construction

	Hours	Rate	Total	Rate	Total
Ironworker	1.33	$....	$....	$27.53	$36.61
Cost per sq. ft.			....		.37

Where corrugated steel forms are installed between joists, the above costs should be increased about 10 percent.

Reinforcing Steel for Precast Concrete Joist Floors. A thin slab of concrete 2 to 2 ½-in. thick is used with concrete joists and the reinforcing steel consists of ¼-inch round rods spaced 10 in. on centers at right angles to the joists. It requires about .33 lbs. of steel per sq. ft. or 33 lbs. per 100 sq. ft. of floor. Steel reinforcing mesh of equal effective steel area may be used instead of reinforcing rods, if desired.

An ironworker should place about 400 lbs. of reinforcing steel per 8-hr. day and based on 33 lbs. per 100 sq. ft. the labor per 100 sq. ft. of floor should cost as follows:

	Hours	Rate	Total	Rate	Total
Ironworker	0.67	$....	$....	$27.53	$18.45
Cost per sq. ft.			....		.18

Where welded wire mesh is used for reinforcing concrete slabs for precast concrete joist floors, an ironworker should place 1,400 to 1,600 sq. ft. of mesh per 8-hr. day, at the following labor cost per 100 sq. ft.:

	Hours	Rate	Total	Rate	Total
Ironworker	0.53	$....	$....	$27.53	$14.59
Cost per sq. ft.			....		.15

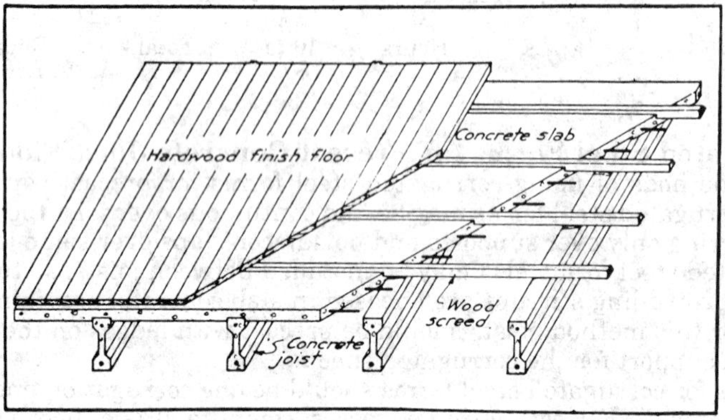

Precast Concrete Joist Floors with Wood Floor Screeds

Wood Sleepers For Use With Precast Concrete Joist Floors. Where wood floors are used with precast concrete joist floors, it will be necessary to place 2"x2" or 2"x3" wood strips or "sleepers" in the concrete for nailing the wood floors. These strips should be spaced about 16 in. on centers and should rest on top of the steel stirrups that project above the concrete joist. The sleepers are secured to the joists by wiring around the wood strip and through the top of the steel stirrups. Care must be taken when placing the concrete to see that the strips are level to receive the finish wood floors.

The floor strips should be beveled or nails driven into the sides at short intervals to provide anchorage into the concrete, otherwise after the concrete hardens and the wood strips dry out and shrink they become loose and cause squeaking floors.

Where wood strips are placed over the tops of the joists without leveling, a carpenter should place 350 to 450 lin. ft. per 8-hr. day, at the following cost per 100 lin. ft.:

	Hours	Rate	Total	Rate	Total
Carpenter	2.0	$....	$....	$24.04	$48.08
Cost per lin. ft.			....		.48
Cost per 100 sq. ft. (16-in. on centers):					
Carpenter	1.5	....	....	24.04	36.06
Cost per sq. ft.			....		.36

Labor Handling and Placing Precast Concrete Slabs. Precast concrete slabs are often used in conjunction with concrete joists instead of cast-in-place concrete slabs.

These slabs are ordinarily used for roofs or for floor slabs in unexcavated areas where it is either extremely difficult or impossible to remove the forms after the cast-in-place slab has been poured.

When precast slabs are used, it requires joists containing heavier reinforcing steel than where T-beam construction is used. This increases the cost of the concrete joists 10 to 15 cents per lin. ft.

Precast concrete slabs are furnished in several sizes, including, 2'-1"x 2'-6"x1 ¼"; 1'-2"x2'-1"x1 ¼"; 1'-0"x2'-6"x1 ¼" and 1'-6"x2'-1"x1 ¼". They are reinforced with 4"x4" No. 10 wire mesh.

When used as roof slabs, they are laid over the concrete joists with about a 1" space between the slabs. After the slabs are placed they are grouted in solid. Insulation board or roofing is applied directly over the slabs.

When used for floors, the precast slabs are placed over the joists and about 1 ½" of concrete is poured over the slab and either left rough or troweled to a smooth finish to receive the finish flooring.

This method eliminates form cost, but it requires heavier reinforced joists, and it is more difficult for the mechanical trades to place their pipes than where the cast-in-place floor is used.

On smaller jobs, the workers usually work in groups of two carrying and placing the precast slabs over the concrete joists.

On work on the first floor, where the precast slabs are on the floor or convenient to the place where they are to be used, two workers together should handle and set 50 to 60 pcs. an hour, depending upon the size of the slabs used, at the following labor cost per 100 pcs.:

	Hours	Rate	Total	Rate	Total
Labor	3.6	$....	$....	$19.13	$68.87
Cost per piece			....		.69

Add cost of carrying slabs from stock pile to floor. If slabs are to be hoisted to the second floor, double the above costs. On large jobs where it is possible to use a hydraulic crane to lift the slabs from the ground to the second floor or roof, a crane will hoist 7 slabs at one time and place them on the floor. To the above costs, it will be necessary to add cost of material and labor grouting joints between slabs.

PRESTRESSED CONCRETE

Within a relatively short period of time, prestressed concrete has taken a very significant place in the building industry. Firms producing prestressed concrete structural members number about 320 and are located in every part of the United States and Canada.

The use of prestressed concrete began in Europe some 50 years ago. A French engineer, Eugene Freyssinet, is generally regarded as the developer of the prestressing concept. Many spectacular prestressed concrete structures have been designed by Mr. Freyssinet and others in Europe and in Latin America and more recently in this country. The first major application of the use of prestressed concrete in the United States was the Walnut Lane Bridge in Philadelphia, Pa. This structure was completed in 1950, and since that time, thousands of bridges and buildings have been constructed with structural elements of prestressed concrete.

Prestressed concrete was developed partly out of necessity as scarcity and the consequent high price of conventional building materials was experienced in Europe during and after the war years. The concept of utilizing the available materials to their fullest capacities was experimented with until techniques were perfected which developed prestressing into a workable and practical method of construction.

The American Concrete Institute defines prestressed concrete as "concrete whose stresses resulting from external loadings are counterbalanced by prestressing reinforcement placed in the structure". This is accomplished by precompressing the concrete by means of internal high tensile strands stressed a predetermined amount. This prestressing introduces compressive stresses into the concrete which counteract the stresses induced when the external loading is applied. The high strength strands are restrained in the member by either bond between the concrete and the strands, or by end bearing devices. These two basic methods of restraining the strands are known as *pretensioning* and *post-tensioning*.

In pretensioning, the strands are tensioned prior to casting the concrete, and in post-tensioning the strands are tensioned after the concrete is cast. In general, the post-tensioning process is accomplished at the job site when casting large members. The member is cast with the strands properly positioned in the form, and after the concrete has reached sufficient strength, the tensioning members are stressed and secured by means of end bearing devices.

The pretensioning methods are particularly adaptable to the mass production of building elements such as double and single tees, channel sections, I joists and beams, T joists, hollow slabs and flat planks, among others. These members are usually cast in steel casting beds from 100 ft. to 600 ft. long. With the use of high quality concrete, low slumps and steam curing, a complete manufacturing cycle may be accomplished in a day.

The prestressed and precast industry is based upon the economics of prefabrication and maximum use of the casting facilities. The larger the project and the number of units of identical span and section, the lower the unit cost.

Building Elements of Prestressed Concrete. Precast/Schokbeton Inc. of Kalamazoo, Michigan has a modern prestressing plant for the manufacture of prestressed concrete building elements. The plant is equipped with a variety of casting beds capable of producing a full range of prestressed products, including both structural and architectural components.

Included in the structural items available in the United States are single and double tee slabs, "F" or Monowing slabs, channel slabs, and Flextee slabs for floor and roof construction. Columns, beams and girders are available in a wide range of sizes and shapes.

Architectural products made at the Precast/Schokbeton plant include precast concrete stadium seats, miscellaneous free form concrete made to order, and Schokbeton products produced under a license from N. V. Schokbeton of Zeist,

Walnut Lane Bridge, Philadelphia, Pa. The First Major Use of Prestressed Concrete
Structural Members in the Western Hemisphere

Holland. These include wall panels, window facades and frames, column shells, stair treads, etc.

Cost of Prestressed Concrete Structural Elements. As prestressed concrete building elements are now manufactured generally throughout the country, prices may usually be obtained from local sources. As stated previously, casting beds are in most cases long enough to cast any length member desired: length, width, height, and weight must be acceptable to local trucking regulations.

There are, of course, many variables that can influence the cost of the installed product, such as the size of the job; the number of identical units in a job; the distance from the plant to the job site; the length of the units, as extremely long units will probably require special handling at the plant, in transit and at the job site; accessibility at the site; the size and number of cranes or other equipment required for erection; the condition at the ground, etc.

Courtesy Portland Cement Association

Precast Concrete Tee Bars are Commonly Trucked to Site for Placement

Erection of Prestressed Concrete Building Units. A typical erection setup for an average job would include a crane, crane operator, oiler, five ironworkers, welding machines, and other miscellaneous equipment and tools. With move-in and move-out time for the cranes, dead time due to the weather or breakdowns, contingencies, etc., the daily cost of an erection setup such as this will average about $2,900 per day. On a large job and under good conditions an experienced crew can probably erect 6,000 to 10,000 sq. ft. or more per day. On smaller jobs or where working conditions are below average, a crew might only erect 3,000 to 4,000 sq. ft. per day.

The Prestressed Concrete Institute is a source of the latest reference material on prestressed concrete and can provide assistance to the design profession as well as to the purchaser of prestressed concrete products.

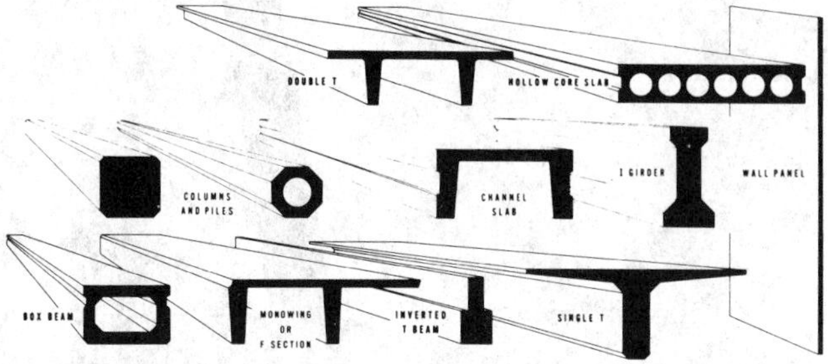

Typical Prestressed Concrete Products

03411 TILT-UP WALL PANELS

The tilt-up construction method can have some distinct advantages in reducing costs through utilization of semi-skilled work forces and producing a completed building project in a shorter number of construction days. These advantages provide a recognizable benefit to the contractor, and it would seem that there would result a major effort to adapt all possible structures to this method of construction. This has not been the case, because there are a sufficient number of incidents where contractors have failed to achieve the several advantages and have, in fact, experienced extreme difficulty in controlling costs, maintaining quality control and adhering to the construction schedule. The contractor/estimator should be aware of the disadvantages as well as the advantages and balance one against the other before reaching a final decision on the tilt-up construction method.

The most desirable project is one designed by the architect to adopt this construction method. Under this condition all of the engineering has been performed during the design stage thus requiring the contractor to cast and erect the panels in a sequence and quality established by the contract plans. All imbedded items, provisions for attachment to other structure members, erection details, allowances for flashing, water sealants and the interrelation with other structure members have been provided for in the design, leaving the contractor unobstructed concentration on scheduling, production, erection and finishes. These are the areas where the contractor has experience and expertise. Both office and field staff are accustomed to these situations and the attendant problems. Under these circumstances, the venture into a tilt-up construction project can be approached with a reasonable degree of assurance the project would succeed, but only after there is a complete understanding of the overall advantages and disadvantages.

Double Tee Being Hoisted into Position Where it will be Welded into Place

Small Crew Places LongSpan Prestressed Double Tee Roof Member for Column-free Interior Space

Erection of Roof Slabs

Converting Projects To Tilt-up Construction

Attempts to convert or adapt a project to incorporate the tilt-up technique are to be approached with extreme caution. Assuming there is sufficient background and experience to warrant the proposing of a design change, the contractor is exposed to a variety of issues, some of which may be obscure during the period of project planning when the compelling objective is developing methods and techniques to increase production, reduce costs, and thereby enjoy more profitability.

The realities of design change initially demand an indepth exploration of local building codes and restrictions or regulations that may be imposed by the lending institution for construction funding and permanent mortgage money. Some financial institutions have extreme objections to involvement with structures or projects utilizing other than conventional materials and construction methods. This is not because the changes are in any way inferior or detract from the value or stability of the project. More often the objection results from a lack of knowledge or prior experience with the particular construction technique. The same resistance can be encountered in city and county building inspection departments to the point where building permits will not be issued.

Owners usually are more receptive to change, especially in instances where they are the recipients of the ultimate lower cost for the same function level, as this fits with their objective for sponsoring the project. If a reduction in the square foot price is apparent to and understandable by owners, and so long as the project retains the same basic appearance and remains safe for occupancy without addition of excessive maintenance costs, most owners will be receptive to accepting a change in construction methods.

The most serious objection to the conversion process can be expected from the designing architect/engineer who will be reluctant to approve any attempt to modify the design concept which was adopted as the most acceptable solution available. Traditionally, aesthetics and design theory have been entrusted to the professional architect. For outsiders to tamper with these endowed rights is considered an intrusion by persons not meeting the standards, qualifications and professional status enjoyed by the architect. Often such proposed changes are considered a personal affront or even a challenge to ability, regardless of the motivation factor which inspired the contractor to initiate the action.

Should it be found that the designing architect/engineer is inclined to give some degree of consideration to the conversion, the contractor should be prepared to suffer some loss of time in obtaining approval and additional costs in generating engineering data upon which the architect would rely in making the final decision of acceptability. By attempting to modify the original architectural design, the contractor enters an unfamiliar role with many obligations and responsibilities which, if not recognized and accounted for, could transform an otherwise successful venture into a tragedy affecting the financial integrity of the company for many years.

It must be recognized when the contractor assumes even a partial responsibility for the structure design, the services of a registered architect/engineer are necessary to develop and produce the detailed drawings and engineering computations to demonstrate structural integrity and architectural compatibility to the satisfaction of the designing architect. Once design approval is obtained, production and erection drawings have to be provided for use by the field forces.

The loss of construction time resulting from delays in obtaining final approvals and additional engineering costs are strong influences on the feasibility of adapting a structure to the tilt-up method. Some manufacturers of inserts for slab lifting and connection devices offer information on design calculations for their products but this information is not to be relied upon as sufficient or complete in scope to satisfy the engineering responsibility assumed by the contractor. Exposure to design responsibility usually creates the need for a review of insurance coverage to provide adequate protection from the involved risks. Premiums may be exceptionally high or coverage may not exist in instances where this is a first time venture and an experience rating cannot be established.

Once the contractor is aware of the risk element involved, logical solutions can be developed and progress can proceed to accomplish the ultimate objectives of producing a project in a shorter period of time at a lower cost while retaining the functions and quality of the structure.

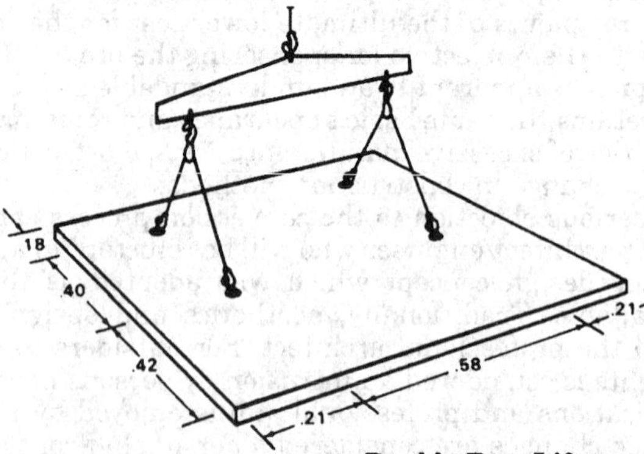

Double Row Lift

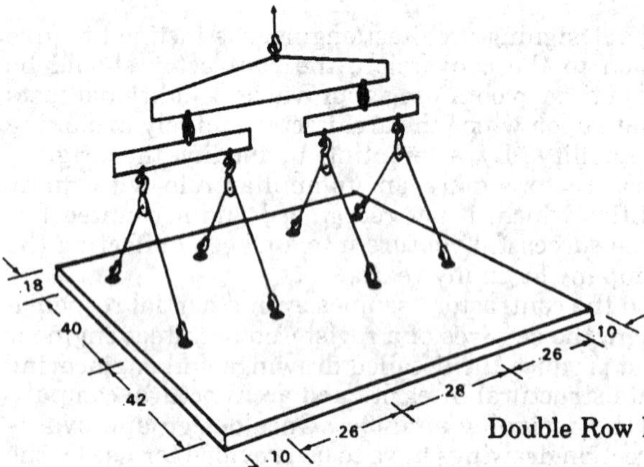

Double Row Lift

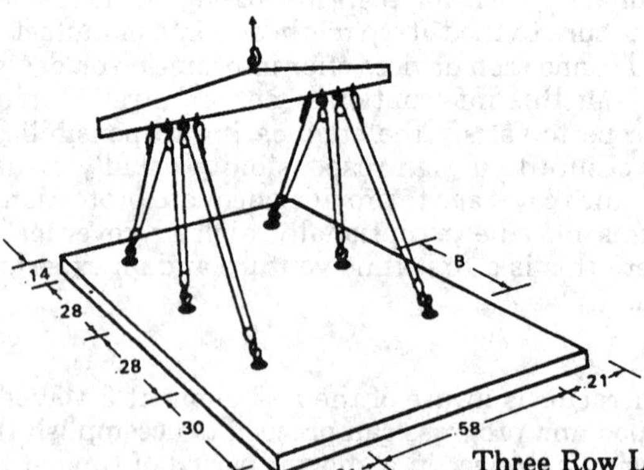

Three Row Lift

Courtesy of Dayton Superior Corp., Miamisburg, OH

Examples of Rigging Configurations for Tilt-up

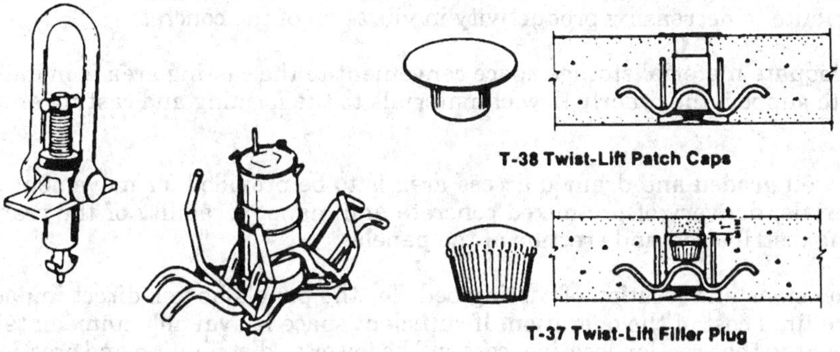

Courtesy of Dayton Superior Corp.

Examples of Hardware Accessories for Lifting Tilt-up Panels

Evaluation Of The Advantages & Disadvantages

Disadvantages. The disadvantages or special problems provide a base for the contractor/estimator to evaluate the unique situations involved in developing accurate unit costs to support the detailed estimate for tilt-up construction:

a) This type of concrete construction is special and unlike conventional poured-in-place method.

b) Because field personnel will be unfamiliar with the procedure, rejection and opposition undoubtedly will be encountered. To overcome this reaction and resulting low productivity, additional supervision should be assigned during the training period to avoid costly interruption in the work cycle due to indecision. Strong leadership by knowledgeable supervisors will produce coordinated work crews capable of lowering production and erection costs to an acceptable level.

c) Additional lay-out quality is required to insure precise location of all embedded items and openings within panel area.

d) Installation of reinforcing rods demand more time and precision to accommodate a panel type construction containing numerous embedded components and apertures for windows, doors, grilles, and duct, pipe and conduit penetrations.

e) Concrete pouring of horizontal thin (6") panels consisting of 400 square feet (20' x 20') requires the use of crane or chute facilities to economically distribute the concrete to all panel areas.

f) Vibration of the concrete to produce a homogeneous mass within the panel presents additional problems because the operators do not have free and equal access to all areas of the horizontal panel. Obstructions within the panel form such as opening forms, pipes, conduits, reinforcing and embedded items, all contribute to decreasing productivity in vibration of the concrete.

g) Adequate material storage space convenient to the casting area is mandatory to support the orderly flow of materials to the forming and casting area.

h) A well graded and drained access area is to be provided for movement of materials, delivery of pre-mixed concrete and maneuverability of the crane during distribution and erection of the panels.

i) Provisions and location of casting beds for the panels have a direct impact on the final cost of the operation. If sufficient space is available immediately adjacent to the erection location, cost will be lower as distribution and erection time will be less. However, if excessive time is consumed in distribution of the panels, the erection crew will be non-productive during this period thus escalating erection costs.

j) Development of a well defined casting schedule and erection drawings will eliminate confusion, delays and interruption to the overall pre-casting and erection operations.

Advantages.

a) Pre-casting of concrete panels provides an opportunity to develop work crews having limited prior experience into efficient groups at a reduced cost per hour.

b) This type of construction substantially reduces the investment in large quantities of forming materials and accessories.

c) Using the tilt-up method, most projects can be enclosed more rapidly providing access to interior finishes at an earlier date by avoiding prolonged weather delays.

d) Pre-planning during the casting operation encourages more efficient installation of window and door units in prepared openings.

e) Completion of panel erection supports availability of exterior site areas for storage or site development work.

f) Flexibility of the tilt-up system allows use of precast concrete columns, special wall surface finishes, incorporation of insulation in the wall panels and lightweight aggregate in the concrete.

g) Forming is basically limited to strip side forms at the perimeter of each panel. In instances where openings have to be formed within the panel are for doors and window areas, the cost for forming will be equal to the edge form cost.

Cost Development

Panel size and thickness can vary to accommodate specific job requirements, and engineering design will dictate special features to be found in all non-typical panels.

Cost development for a 20'x20'x6" thick solid panel (400 sq. ft), using 3000# pre-mixed concrete, #5 reinforcing rods 6" on center each way, would be as follows.

Preparing Pouring Beds. The pouring or casting bed for panels is usually provided by the surface of the concrete slab on grade of the structure. When there is insufficient surface area available to accommodate a reasonable form-pour-cure-erect cycle, a stack method can be employed where panels are cast on previously poured panels.

In both methods care is taken to prevent adherence between the two surfaces. Frequently, the cast bed surface is coated with a brush coat of a commercial release agent or a sheet of poly between pours. Cost for either method is minimal, and the allowances included in the table are adequate.

Edge Forms. Edge forms are constructed of 2x6, with a 1x4 kick strip secured at the bottom and with the top edge held in alignment by 2x4 braces 2'-0" long at 1'-0" intervals.

Material Cost

80 lin. ft. 2x6 =	80 b.f.
80 lin. ft. 1x4 =	26 b.f.
84 pcs. 2"x4"x2' =	112 b.f.
	218 b.f.

½" strip on edge of 2x6 not considered
Cost of Forming Materials: 218 b.f. x 0.40 = $87.20
Material Cost Per Sq. Ft. = $0.22

Labor Cost

Edge forms for one 20'x20' panel = 400 sq. ft.

	Hours	Rate	Total	Rate	Total
Carpenter	4	$....	$....	$24.04	$96.16
Helper	4	$....	....	19.13	76.52
			$....		$172.68
Cost per sq. ft. of panels					0.43
per b.f. of forms					0.79
per lin. ft. of edge forms					2.16

Labor cost for openings within the panel area to accommodate doors, windows, etc. would equal the labor cost of the edge forms.

Labor cost for stripping and cleaning forms for re-use would be minimal and are included in the above labor allowances.

Reinforcing. Reinforcing Rods - Plain 42 pcs. x 20'-0" = 840' x 1.043# = 876 lbs.

Material Costs

Reinforcing, 876 lbs. @ 0.60	$525.60
Accessories @ $60. per ton of Rebars - Allow	26.28
Tie Wire @ 5 lbs. per ton - Allow 2.5 lbs. @ 0.65	2.00
	$553.88
Material Cost Per Sq. Ft.	1.38

Labor Costs

	Hours	Rate	Total	Rate	Total
2 workers @ 4 hours	8	$....	$....	19.13	153.04
Cost per sq. ft.					$0.38

Panel Connectors & Inserts. These costs are predicated on the many available alternates. Design requirements and locality will determine the final cost for these items.

For the purpose of developing the average square foot panel cost, an allowance for material cost of $3.30 per lin. ft. of panel edge is considered adequate. Associated labor costs would be $1.15 per lin. ft. or $92.00.

Material & Labor Costs

80 lin. ft. @ $3.30	$264.00
Material Cost Per Sq. Ft.	0.60
Labor Cost Per Sq. Ft.	0.23

Concrete. The cost of concrete includes 5% waste factor. 20' x 20' x .5 x 1.05 = 210; 210 divided by 27 = 7.8 cu. yds. or 8 cu. yds.

Concrete, 8 cu. yds. @ $55.00 per cu. yd.	$440.00
Material Cost Per Sq. Ft.	1.10

Labor Costs

	Hours	Rate	Total	Rate	Total
5 laborers @ 1 hour	5	$....	$....	19.13	$95.65
Cost per sq. ft.					0.24

Finish. Finish is either float or broom finish.

Labor Costs

	Hours	Rate	Total	Rate	Total
2 laborers @ 4 hrs	8	$....	$....	19.13	$153.04
Cost per sq. ft.					0.38

Special surface finishes can increase finish cost an additional $1.10 to $4.50 sq. ft.

Material unit cost in table is sufficient to cover the minor material expenditures involved.

Cure. Curing of panel concrete can be accomplished in several ways:

a) Cover with burlap or poly, keeping surface moist and warm;
b) Use of commercial curing compound;
c) Providing a covering of sand or sawdust;
d) Construct canopy to protect from direct sun heat during summer months or to provide enclosure for temporary heat in winter months.

Erection Costs. The labor cost of panel erection is as follows:

	Hours	Rate	Total	Rate	Total
Crane with operator	8	$....	$....	250.00	2,000.00
Erection Supervisor	8	....	....	24.04	192.32
4 man crew	32	....	....	19.13	612.16
Crew Cost Per Day			$....		2804.48

Panel erection production includes lifting, distribution, positioning, alignment, and bracing:

Production: 1 completed panel per hour equals
400 sq. ft. x 8 = 3200 sq. ft. panels per day
Erection Cost Per Sq. Ft., $0.86

Frequently erection of panels is assigned to subcontractors having the capability to furnish a complete cost package for equipment and personnel.

Patch & Repair. During the stripping of forms, lifting, distribution, and erection of the panels, minor damage might require remedial work to maintain quality control. An allowance has been included in the panel cost development table.

Typical Cost Per Sq. Ft.
For Panels 20'-0"x20'-0"x6"

Item	Material	Labor	Total
1. Preparing Pouring Bed (Allowance)	.04	.04	.08
2. Edge Forms	.22	.43	.65
3. Reinforcing	1.38	.38	1.76
4. Panel Connectors and Inserts	.60	.23	.83
5. Concrete	1.10	.24	1.34
6. Finish - Float or Broom	.04	.38	.42
7. Cure (Allowance)	.04	.06	.10
8. Erection		.86	.86
9. Patch & Repair (Allowance)	.06	.12	.18
Total Panel Cost	3.48	2.74	6.22

Tilt panel method is compatible with structural steel, poured-in-place or precast concrete columns. Panels of irregular shape containing an unusual number of openings or requiring substantial modification to accommodate special features will inflict an increase in cost. Proper sequence planning to maximize use of equipment and personnel and work crew efficiency will produce lower cost per square foot.

03500 CEMENTITIOUS DECKS
USG METAL EDGE GYPSUM PLANK

Cementitious deck is lightweight, noncombustible construction for roofs and floors in connection with steel framing. USG (United States Gypsum Co.) Metal Edge Gypsum Plank are made in 2"x15"x10'-0" units and designed for roofs on spans up to 7'-0". The plank has a water resistant core.

All plank are molded and have steel tongue and groove edging on sides and ends, firmly anchored into the gypsum. They also have an electrically welded galvanized steel fabric as reinforcement for the gypsum. Weight 13.0 lbs. per sq. ft.

On jobs requiring approximately 3,000 sq. ft. or more, the plank cost about $1.35 per sq. ft. in the eastern zone of the United States.

s requiring approximately 3,000 sq. ft. or more, the plank cost about $1.35 per sq. ft. in the eastern zone of the United States.

Labor Laying Metal Edge Gypsum Plank.--One plank (12 ½ sq. ft. in area) will weigh 160 lbs. requiring 2 workers to handle a single plank. Usually, it will require 2 workers on the ground and 2 workers on the roof. This crew should lay approximately 1,200 sq. ft. per 8-hr. day at the following labor costs:

	Hours	Rate	Total	Rate	Total
Labor	2.7	$....	$....	$19.13	$51.65
Cost per sq. ft.			....		.52

The above costs contemplate ordinary flat slab work, on a job containing 3,000 to 4,000 sq. ft., with a minimum amount of cutting. Add for extra handling and cutting for pitched roof and double the above labor costs for pitched roofs having hips, valleys, dormers, etc. Add extra for hoisting engineer if required.

Monolithic or Poured-in-Place Gypsum Roof Construction

Monolithic or poured-in-place gypsum roof deck, with its continuous reinforcing and securely welded sub-purlins, becomes an integral part of the main steel construction of a building, adding rigidity to the entire structure and can be designed for any required loading condition.

Permanent formboards, which remain as the undersurface of the completed deck, may be ½" gypsum formboard; ¼" asbestos cement board; and 1" to 2" mineral fiber formboard.

Rolled steel sections designed to carry all loads between supports shall be laid on the top of walls, beams or roof supports at sufficient spacing to receive the formboards on the lower flanges; "reinforcing shall be 48-1214 galvanized welded wire fabric having 12 ga. longitudinal wire spaced 4" on centers and 14 ga. transverse wired 8" on centers or "Keydeck", a galvanized woven mesh as manufactured by Keystone Steel and Wire Co., having 16 ga. longitudinal wires and 19 ga. diagonal wires. The effective cross sectional area of reinforcing shall be not less than 0.026 sq. in. per ft. of slab width. Standard (mill mixed) gypsum composition containing at least 87 ½% calcined gypsum is mixed with water to a medium quaking consistency and poured or pumped in place, forming a solid reinforced slab 2" thick (or more, if desired), plus the board undersurface. The top surface of the slab shall be screeded smooth and left ready to receive the specified roofing.

The density of poured gypsum concrete is approximately 50 lbs. per cu. ft. When cast in the minimum thickness of 2", the total dead weight of the slab is about 10 lbs. per sq. ft.

The weights (excluding sub purlins) and insulating values of poured gypsum roof decks are as follows:

Description	Weight per SF	"U" Value BTU
2" gypsum concrete on ½" gypsum form boards or plastic faced board	11.0 lbs.	.34
2 ¼" gypsum concrete on ¼" asbestos cement board	11.0 lbs.	.37
2" gypsum concrete on 1" mineral fiber formboard	11.2 lbs.	.16
2" gypsum concrete on 2" mineral fiber formboard	11.2 lbs.	.10

Erection of Single Tee Roof Slabs

The accompanying illustration shows the type of roof construction that is widely used for schools, auditoriums, field houses, gymnasiums, factory, industrial and many commercial buildings, in fact, for most any steel skeleton structure where a permanent, noncombustible highly insulated roof is desired.

Practically all gypsum concrete is now mixed mechanically and pumped from ground to roof level as a means of further increasing the speed of operation and uniformity of quality, with corresponding reduction in cost.

The cost of poured-in-place gypsum roof decks will vary with the size and type of structure, availability of materials, freight charges, labor rates, etc., but the simplicity of design, lightweight and speed of erection, noncombustibility and finished undersurface make the overall cost of this construction very economical.

Poured-in-place gypsum roof decks are applied by contractors approved by the manufacturers of the materials.

The following are approximate prices only on the various types of poured-in-place gypsum roof decks. These prices may vary locally due to labor and material costs, freight rates or size of specific job. For definite prices on any project it is always advisable to consult a roof deck contractor specializing in this type of construction.

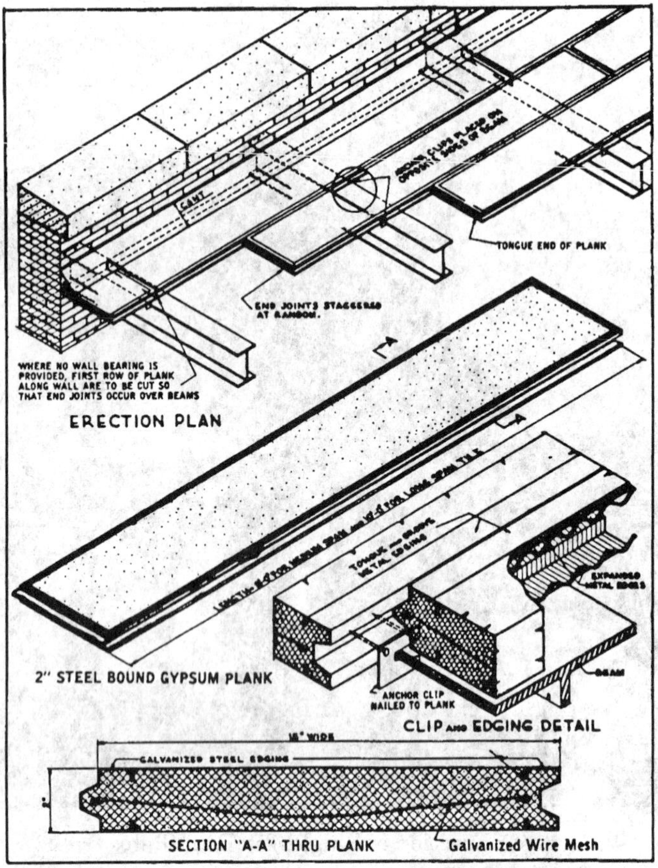

Details of USG Metal Edge Gypsum Plank

Type of Roof	Approx. Price Range per SF
2" gypsum concrete on ½" gypsum formboards	$1.25 - $1.40
2" gypsum on 1" mineral fiber formboard	1.45 - 1.60
For each additional ½" in thickness of gypsum concrete add	0.12 - 0.15

The accompanying illustration shows a poured gypsum roof on steel joists used in combination with acoustic ceiling boards of mineral fiber to comply with Underwriters Laboratories design No. P207 for a two hour fire rating. This assembly has a "U" value of 0.19, a noise reduction coefficient of .55 to .65, and a light reflectance of 85%. Other 2 hr. U.L rated systems are also available.

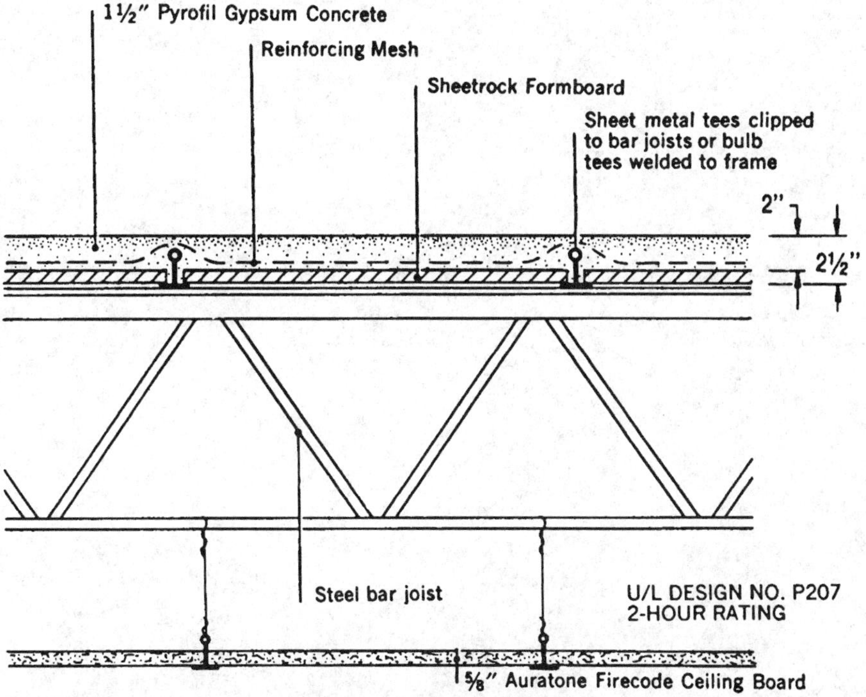

USG Floor Plank Systems *Courtesy of U. S. Gypsum Co.*

USG (United States Gypsum Co.) Floor Plank Systems are lightweight noncombustible assemblies for use with bar joists and floor loads of the magnitude generally found in residential buildings.

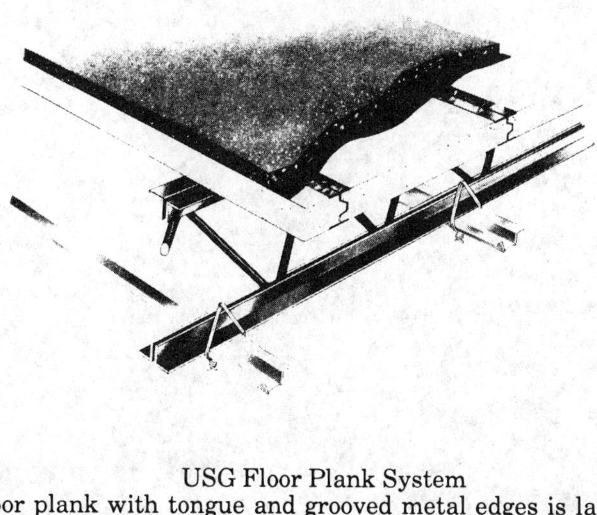

USG Floor Plank System

USG floor plank with tongue and grooved metal edges is laid dry without grout and welded to the joists. A topping of Mastical*, Flo-Fill, or ¼" plywood underlayment and a ceiling of ⅝" Sheetrock Firecode C* gypsum wallboard make up the complete floor system.

*United States Gypsum Co.2"

CHAPTER 4

MASONRY

CSI DIVISION 4
04100 MORTAR
04200 UNIT MASONRY
04210 BRICK
04220 CONCRETE
04240 CLAY TILE
04250 CERAMIC VENEER
04270 GLASS
04280 GYPSUM
04400 STONE
04500 MASONRY RESTORATION

To prepare an accurate estimate on the cost of masonry requires a knowledge of the various factors entering into the work. This includes the type of building, thickness of walls, kind of mortar, style of mortar joint, class of workmanship, ability of workmen, and last but not least, the ability of the foreman in charge.

Estimating masonry is more than a matter of referring to a table or chart to obtain the prevailing wage scale and then determining the labor cost of laying a thousand brick. Accuracy is important in the preparation of your estimates, and all of the above items should be considered when determining the unit costs.

The type of building plays an important part in the costs. A mason can lay far more brick on a factory or warehouse building having long straight walls than on a school or office building that is cut up with numerous windows, which require straight walls, plumb jambs, pilasters, etc.

A mason can lay far more brick on a 16" wall than on an 8" wall and can lay more brick in smooth working lime mortar than in coarse portland cement mortar. He can also lay more brick by merely cutting the mortar joints than by striking them with the point of his trowel.

The class of workmanship plays an important part. Many cheaply constructed, speculative buildings have crooked walls and open mortar joints, presenting an slovenly appearance, while the best grade of workmanship requires full mortar joints, straight walls, plumb jambs and corners--in other words, good workmanship.

Weather conditions also affect labor costs. A mason can lay more brick on a clear, dry day than when it is cold and wet, or when it is necessary to heat materials during freezing weather.

The demand for mechanics also affects their output. When there is an abundance of work, some (not all) mechanics take advantage of this condition and do just as little work as possible to hold their jobs. On the other hand, when work is scarce and there are two mechanics for every available job, they do a much better day's work when they know there is someone eagerly waiting to take their place.

The ability of the foreman in charge of the work affects the final costs; the ability to plan, schedule and lay out the work, along with obtaining a full day's work from the those under his supervision, are all important.

Masonry Specifications

In preparing his proposals from architectural plans and specifications, the estimator will usually find units referred to by specifications published by the American Society for Testing and Materials (ASTM). These specifications set limitations on requirements and properties of the various materials and are generally accepted by manufacturers.

ASTM specifications are available for all the following materials that the mason may encounter:

C 32* Sewer and Manhole Brick (made from clay or shale)
C 55 Concrete Building Brick
C 62 Building Brick (solid masonry units made from clay or shale)
C 73 Calcium Silicate Face Brick (sand-lime) brick)
C 91 Masonry Cement
C 105 Ground Fire Clay or Refractory Mortar for laying up fireclay brickwork
C 126 Ceramic Glazed Structural Clay Facing Tile, Facing Brick and Solid Masonry Units
C 144 Aggregate for masonry mortar
C 150 Portland Cement
C 207 Hydrated Lime for masonry purposes
C 270 Mortar for unit masonry
C 216 Facing Brick (solid masonry units made from clay or shale)
C 279 Chemical-Resistant Masonry Units
C 287 Chemical-Resistant Sulphur Mortar
C 331 Lightweight Aggregates for Concrete Masonry Units
C 476 Grout for Reinforced and Nonreinforced Masonry
C 652 Hollow Brick (hollow masonry units made from clay or shale)
C 902 Paving Brick

*Check for latest edition

While every estimator should have a general acquaintance with all the above specifications, he will deal primarily with face brick, building (formerly called common) brick and mortar for unit masonry. Building brick is subdivided into 3 grades: SW, MW, and NW, indicating severe, medium and negligible weathering. Face brick is subdivided into only SE and MW grades, but each of these grades has 3 types, FBX, FBS and FBA, which set forth requirements for appearance and size.

Mortar specifications cover both portland cement-lime and masonry cement mortars and are subdivided into types M, S, N, and O, based on 28 day compressive strengths of 2,500, 1,800, 750 and 350 lbs. per sq. in. ASTM specifications are also published for all the manufactured ingredients for the mortar.

An estimator should welcome a specification based on these ASTM designations, because it enables him to figure within well-defined limits, and he would do well to adopt the ASTM terminology in proposals he submits.

04100 MORTAR

Cements. Type I portland cement is the basic cement used in most mortars. Type II, for use where sulphate action is a problem, and Type III, a high early strength cement, are also included under ASTM specifications for mortar C 270.

Air entraining cements should usually not be substituted, because this affects bond strength. Full information on tests and specifications are available in ASTM publications and in information published by the Portland Cement Association.

Masonry cements are included in ASTM specifications, but their chemical makeup is not regulated and varies widely among the manufacturers. It may therefore take some experimenting to find the product most suited to the job.

Masonry cements do, however, contain additives that increase workability and water retentivity, and the single product saves handling both portland cement and lime on the job, and mortar can be used as soon as mixed. However, where a Type S mortar is selected for bonding strength, care should be taken not to select a masonry cement with air entrainment properties. Masonry cements are marketed under such names as Brixment, Atlas, Lehigh, Hy-Test, Medusa-Brisket, etc.

Lime. Lime for construction purposes is available in both the quick (unslaked) and hydrated (slaked) forms, though in recent years hydrated lime has been preferred to quicklime, because it is faster, easier and safer to use. Building lime is supplied in bulk or packaged in 50-lb. multiple-wall, moisture resistant bags. Barrels, once widely used to package lime, are no longer in use as containers for domestic lime shipments. Lime products sold in bulk or bags are generally quoted on a ton (2000 lbs. net) basis.

Quicklime may be purchased in the following forms, the principal difference being in the size of the particles: pebble lime, crushed lime, ground lime, and pulverized lime. All quicklime must be slaked prior to use and manufacturer's directions for slaking should be followed to secure best results. Due to the greater ease with which complete slaking is secured, the finer divided types of quicklime are preferred. Since quicklime is very reactive with water, care should be exercised at all times during the slaking operation to prevent splattering of the lime and potentially serious burns of the eyes and skin. Slaked quicklime putty should be screened and permitted to age until all the lime particles are completely slaked. The aging period may vary from a few hours to several days, depending on the chemical and physical properties of the quicklime used and on the skill of the operator in slaking the lime.

Hydrated lime for structural use is furnished in the dry powdered form usually packaged in 50 lbs. net paper bags and sold by the ton (2000 lbs. net). It may also be purchased in bulk, if proper facilities are available for handling and storage. In recent years a new special type of hydrated lime has been marketed known as Type "S" or pressure hydrated lime. This highly hydrated product contains a maximum of 8% combined unhydrated calcium and magnesium oxides. In addition, putties made from this lime develop high plasticities instantly upon mixing. Type "S" hydrated limes meeting all ASTM requirements are commercially available in both the dolomitic and high calcium varieties. Type "N" (normal) hydrated limes should be soaked in a paste or putty form for several hours or overnight prior to use in order to enhance their plasticity and workability. The preferred method for soaking is to sift the hydrated lime evenly into a watertight box or vat that has been previously half filled with clean water. Do not stir, mix, or agitate the mass, but allow the lime to settle naturally; continue to add the hydrate slowly and evenly over the entire surface of the water until the thick paste is formed, after which the mass is permitted to soak until required for use.

All building limes should comply with the requirements of the Standard Specifications for Hydrated Lime for Masonry Purposes (C-207) or Quicklime for Structural Purposes (C-5) as adopted by the American Society for Testing Materials (ASTM).

Storage of Lime and Lime Putty. Prior to slaking or use all quicklime and hydrated lime previously delivered to the job site should be stored in a clean, dry place.

Quicklime is shipped in bulk and in waterproofed, multi-wall paper bags. When furnished in bulk, it should be slaked immediately to assure maximum putty yield and to avoid loss. Quicklime furnished in bags may be stored for considerable time depending upon the humidity and storage conditions.

Hydrated lime in bulk or packed in bags may be stored for relatively long periods of time provided it is placed in a clean, dry, properly ventilated warehouse.

Lime putty should be aged and stored in clean, tight vats and should be well protected from direct contact with the atmosphere by maintaining a thin film of water over the surface.

Commercial plants for the production of scientifically prepared, well aged putty, are in operation in many markets. The use of commercially prepared lime putty eliminates all necessity for slaking and aging or soaking of lime on the job, and has the further advantage of providing the estimator with reliable figures for estimating purposes since purchase order is based on net quantity of lime putty required for the operation. Other advantages of commercially produced lime putty include flexibility during construction, eliminating lost time awaiting preparation of lime putty and admitting of expansion in operations on short notice, reduction in storage space required for materials at the site of the work and improved quality of putty due to careful supervision during slaking and aging of lime putty.

Lime Putty and Its Preparation. Job specifications generally require mortar to be mixed in definite predetermined proportions by volume and the lime proportion is usually determined in volumes of lime putty. Therefore, some knowledge of putty yield of lime is necessary in preparing estimates of quantity of lime required. The putty yield of different limes varies over a relatively wide range, according to chemical and physical properties of the lime, methods of processing, etc. However, the quantities given in the following table may be used as a sound, conservative average for estimating purposes.

Quantity of Lime Putty Obtainable from Various Types of Lime

Type of Lime	Average CF Putty per Ton of Lime	Lbs. Lime per CF Putty
Hydrated Lime	46	44
Pebble or Pulverized Quicklime	80	25

It takes 109 lbs. portland cement to produce 1 cu. ft. of cement paste of normal working consistency.

It takes 9 cu. ft. of putty or paste with 27 cu. ft. of ordinary building sand to produce 1 cu. yd. of 1:3 mortar.

Quantity of Materials Required for Masonry Using
Pebble or Pulverized Quicklime at 80 cu. ft. Lime Putty per Ton

	Proportions by Volume		Quantity of Materials Required For one cu. yd mortar			To lay 1000 brick		
Lime Putty	Portland Cement	Sand	Lime* lbs.	Cement Cu. Yd.	Sand lbs.	Lime* Sacks	Cement Cu. Yd.	Sand
1	0	3	225	0	1	150	0	.67
3	1	12	169	2.25	1	113	1.50	.67
2	1	9	150	3	1	100	2.00	.67
1 1/2	1	7.5	135	3.6	1	90	2.40	.67
1	1	6	112.5	4.5	1	75	3.00	.67
1/2	1	4.5	75	6	1	50	4.00	.67
0	1	3	0	9	1	0	6.00	.67
10%**	1	3	22.5	9	1	15	6.00	.67
15%**	1	3	34	9	1	23	6.00	.67

*Dry unslaked lime
**Based on volume of cement required. Add 5 to 10% for waste.

Where hydrated lime is used in dry powdered form, it is usually assumed that a given volume of dry hydrated lime will produce an equivalent volume of lime putty. A bag of dry hydrated lime (50 lbs. net) equals about 1.15 cu. ft. of lime putty.

Labor Slaking Lime and Making Mortar. When mixing mortar by hand, a good mortar maker should slake and sand about a ton of quicklime per 8-hr. day, and should make about 4 cu. yds. of mortar per 8-hr. day.

Quantity of Materials Required for Masonry Mortar Using
Hydrated Lime at 46 cu. ft. Lime Putty per Ton

	Proportions by Volume		Quantity of Materials Required For one cu. yd. mortar			For 1000 brick (18 CF)		
Lime Putty	Portland Cement	Sand	Lime lbs.	Cement Sacks	Sand Cu. Yd.	Lime lbs.	Cement Sacks	Sand Cu. Yd.
1	0	3	391	0	1	261	0	.67
3	1	12	293.5	2.25	1	196	1.50	.67
2	1	9	261	3	1	174	2.00	.67
1 1/2	1	7.5	235	3.6	1	157	2.40	.67
1	1	6	195.5	4.5	1	130	3.00	.67
1/2	1	4.5	130.5	6	1	87	4.00	.67
0	1	3	0	9	1	0	6.00	.67
10%*	1	3	39	9	1	26	6.00	.67
15%*	1	3	59	9	1	39	6.00	.67

*Based on volume of cement required. Add 5 to 10% for waste.

The actual labor cost of making 18 cu. ft. of mortar, sufficient for 1,000 common brick, should average as follows:

	Hours	Rate	Total	Rate	Total
Mortar maker	1.33	$....	$....	$19.13	$25.44

On jobs of any size, the mortar is mixed in a mixer, which produces a more uniform and more easily spread mortar. It not only enables a mason to lay more brick but also results in a saving of 10% to 20% in labor required to mix.

Quik-Slak Masons Lime. A lime furnished in pulverized form that slaks the instant it hits the water. Put up in 50 lb. paper sacks and the manufacturers say that 4 sacks are sufficient to lay, 1,000 brick under average conditions.

Aggregate. Aggregate may be either natural sand that is clean and sharp or sand manufactured by crushing stone, gravel or air-cooled iron blast furnace slag. Sand should be properly graded as set forth in ASTM Specification C 144, with all sand passing a No. 4 sieve and approximately 10% passing a No. 200 sieve for manufactured sand. Too coarse a sand decreases workability while too fine a sand decreases water retentivity.

Mortar Colors. Mortar coloring agents may be added, either natural, such as white sand, or pigments. While different color effects will require some experimentation, in general, 4 to 8 lbs. of color to one bag of cement, hydrated lime or cu. ft. of putty and 3 cu. ft. of sand can be figured, or about one 50-lb. bag per 1000 bricks with a 3/8" joint.

Waterproofing and Shrinkproofing Mortar

Leaky masonry walls have aroused considerable concern, and this has led to extensive investigation and tests by the National Bureau of Standards (Report BMS7), which investigated not only the materials such as brick, tile, concrete blocks, mortar, etc., but also the effects of different grades of workmanship.

After extensive tests, it was determined that workmanship affected the permeability of the walls more than any other factor. Walls with tooled joints were less permeable than similar walls with cut joints. But the quality of the workmanship inside the walls had a greater influence than the kind of surface finish on the joint. In other words, brick as laid on the average contract job, where mortar was spread and furrowed and the joints were not completely filled with mortar, leaked far more than bricks laid in a full bed of mortar and "shoved" into position with full bed and head joints and a complete filling of joints between all brick.

It was also determined that nearly all of the leakage was through the mortar joints regardless of the kind of brick used and that in practically every instance where the "commercial" grade of workmanship was used, leaky walls resulted.

Where brick of high absorptive properties were used, the brick absorbed the water from the mortar, making it difficult to work and resulting in small hair cracks between the mortar joints and the brick. This also caused leakage through the walls.

The use of lime with low plasticity, producing a mortar with low water retentivity, greatly increased the permeability of the walls. This effect was more pronounced when the mortar was used with high absorptive brick.

Water reduction in cement and lime mortars is a highly desirable objective because with less water there is less evaporation and mortar shrinkage is reduced below the critical point.

Since these investigations were made, numerous admixtures have been placed on the market. These are to be added to the mortar mixture and are said to give the mortar greater plasticity, to retain the water in the mortar, to prevent shrinkage and leakage in the mortar joints and to make a smoother mortar that will work better under the trowel. Some of these products are listed below:

Anti-Hydro. Cement mortar for use with brick, block or stone shall be of 1:3 mix, gauged and tempered with a solution of 1 gal. Anti-Hydro to 15 gals. of water, or use 1 qt. Anti-Hydro to each sack of cement.

Wet all brick. Push-lay and thoroughly grout all brick at every course. Parge with same mortar between the face brick and back-up masonry.

Castle Aquatite. Mortar to be composed of 1 part portland cement, 3 parts clean and mixed with a solution of 1 part Aquatite to 15 parts water.

Hydratite Plus. A powder in concentrated form, which when mixed with the mortar is dissolved and miscible in the water. It anchors the particle of water intimately to the particle of the cement, sand and lime with a bond greater than the suction of the brick. With less water it increases the flow or workability of the mortar. Use 2-lbs. per sack of cement in 1:1:6 mortar.

Hydrocide Powder. When used for waterproofing masonry mortar, use 1-lb. of powder to each sack of cement or per cu. ft. of lime putty used in the mixture.

Omicron Mortarproofing. A plasticizing and water reducing agent that increases the workability of mortar, increases water retentivity and minimizes shrinkage. Contains stearate for increased water repellency. Add 1 lb. of Omicron for each sack of cement, lime or putty in the mortar mix.

Requires 5 to 6 lbs. per 1,000 brick, depending upon mortar proportions and width of joints.

Plastiment (Sika). A powder used in the proportion of up to 1-lb. per sack of cement. Increases the workability of mortar, allows a reduction in the water-cement ratio, reduces bleeding and increases the bond between brick and mortar. Of particular advantage in hot weather construction, as it delays the setting time and preserves the workability of the mortar for a longer time.

Toxement IW. Toxement IW powder, paste or concentrated paste may be used for waterproofing and shrinkproofing masonry mortar.

Sikacrete. Accelerating and plasticizing liquid for use in brick mortar. Diluted with the gauging water 1 part Sikacrete to 3 to 7 parts water. Increases the bond and accelerates the hardening rate of mortar so that higher tiers of brick or glass block can be laid and the joints struck sooner.

Types of Mortar

The following mortars are based on proportion set forth in ASTM specifications. These mortars were designated Type A-1, A-2, B, and C prior to 1954.

Type M mortar is a high strength mortar used primarily in foundation masonry, retaining walls, walks, sewers and manholes. Its proportions and costs for a cement-lime mixture are:

1 cu. ft. portland cement	$5.75
¼ cu. ft. lime @ $4.25 per cu. ft.	1.06
3 cu. ft. sand @ $0.33 per cu. ft.	.99
	$7.80
Per cu. ft.	$2.60

Proportions and costs for a type M mortar using masonry cement are:

1 bag of portland cement	$5.75
1 bag Type II masonry cement	5.50
6 cu. ft. sand @ $0.33 per cu. ft.	1.98
	$13.23
Per cu. ft.	$2.21

Type S mortar also has a reasonably high compressive strength and develops maximum tensile bond strength between brick and cement-lime mortars. It is recommended for use in reinforced masonry and where flexural strengths are required, such as cavity walls exposed to strong winds, and for maximum bonding power, such as for ceramic veneers. Proportions and costs for cement-lime Type S mixtures are:

1 cu. ft. portland cement	$5.75
½ cu. ft. lime @ $4.25 per cu. ft.	2.13
4 ½ cu. ft. sand @ .33 cu. ft.	1.49
	$9.37
Per cu. ft.	2.08

Costs and proportions for cement masonry Type S Mortar are:

½ bag portland cement	$2.88
1 bag masonry cement	5.50
4 ½ cu. ft. sand @ $0.33 cu. ft.	1.49
	$9.87
Per cu. ft.	2.19

Type N mortar is a medium strength mortar most generally used in exposed masonry above grade. The proportions and cost of Type N cement-lime mortar is:

1 cu. ft. portland cement	$5.75
1 cu. ft. lime	4.25
6 cu. ft. sand @ $0.33 cu. ft.	1.98
	$11.98
Per cu. ft.	$2.00

Proportions and cost of Type N masonry cement mortar is:

1 cu. ft. masonry cement	$5.50
3 cu. ft. sand @ $0.33 cu. ft.	.99
	$6.49
Per cu. ft.	$2.16

Type O mortar is a low strength mixture for general interior use where compressive strengths do not exceed 100 psi. It may be used elsewhere where exposures are not severe and no freezing will be encountered. Proportions and costs of Type O cement-lime mortar is:

1 cu. ft. portland cement	$5.75
2 cu. ft. of lime @ per $4.25 cu. ft.	8.50
9 cu. ft. sand @ per $0.33 cu. ft.	2.97
	$17.22
Per cu. ft.	$1.91

Proportions and cost of Type O masonry cement mortar is:

1 cu. ft. masonry cement	$5.50
3 cu. ft. sand @ per $0.33 cu. ft.	.99
	$6.49
Per cu. ft.	$2.16

In the above costs, lime is figured in an exact proportion. As noted under "Quantity of Materials Required for Masonry Mortar Using Hydrated Lime...", a 50 lb. bag of hydrated lime will make 1.15 cu. ft. putty. The ASTM specifications for Types N and O mortar allow some leeway in lime proportions and a full bag of lime can be figured. Only with very high strength mortars might you have to figure splitting bags to achieve exact proportions.

Cu. Ft. of Mortar Required to Lay 1,000 Face Brick

Based on standard size brick having 1/4" to 3/8" end joints and bed joints as follows:

Width of Mortar Joints in Inches							
1/8"	1/4"	3/8"	1/2"	5/8"	3/4"	7/8"	1"
4	7	9	12	14	16	18	20

For Dutch, English, and Flemish Bond, add about 10% to the above quantities on account of additional head joints.

Cubic Feet of Mortar Required per Thousand Brick

For various thicknesses of walls and joints. No allowance for waste.

Joint	Wall Thickness					
Thickness	4"	8"	12"	16"	20"	24"
1/8"	2.9	5.6	6.5	7.1	7.3	7.5
1/4"	5.7	8.7	9.7	10.2	10.5	10.7
3/8"	8.7	11.8	12.9	13.4	13.7	14.0
1/2"	11.7	15.0	16.2	16.8	17.1	17.3
5/8"	14.8	18.3	19.5	20.1	20.5	20.7
3/4"	17.9	21.7	23.0	23.6	24.0	24.2
7/8"	21.1	25.1	26.5	27.1	27.5	27.8
1"	24.4	28.6	30.1	30.8	31.2	31.5

Labor cost to mix mortars is usually figured as part of labor cost setting brickwork. However, if figured separately, it will run about $0.92 per cu. ft., machine mixed, $1.38 per cu. ft. hand mixed.

Selecting Mortars. Often the estimator will have no choice in the makeup of his mortar, the architect's specifications setting forth exactly what is required. Because replacing masonry that is found not acceptable by an architect or owner is extremely expensive, it is false economy, not to mention the loss of reputation involved, to alter specifications without approval from the architect.

However, many jobs will leave considerable leeway in the selection of mortars, and an informed estimator may even perform a valuable service in suggesting changes to the mortar proportions that will either save money or make a better job. All too often a mortar may be selected on the basis of high compressive strength when the job may require only a 100 psi. This not only wastes cement, but such mortars do not have the adhesive and sealing powers of the weaker mortars, and these two characteristics are the primary functions required on most jobs.

Of further importance to the estimator is the workability of the mortar, which will enable the bricklayer to lay more units. A complete discussion of mortars and recommendations is contained in technical notes on brick construction from the Brick Institute of America (BIA) in Reston, VA.

Mortar for Concrete Masonry. For laying masonry walls subject to average conditions of exposure, use a mortar made in the proportions of 1 volume of masonry cement and between 2 and 3 volumes of damp, loose mortar sand; or 1 volume of portland cement and between 1 and 1 1/4 volumes of hydrated lime or lime putty and between 4 and 6 volumes of damp, loose mortar sand.

Walls which will be subjected to extremely heavy loads, severe winds, earthquakes, serious frost action or other conditions requiring extra wall strength, should be laid with a mortar made of 1 volume masonry cement plus 1 volume portland cement and between 4 and 6 volumes of damp, loose mortar sand; or 1 volume of portland cement, to which may be added up to 1/4 volume of hydrated lime or lime putty, and between 2 and 3 volumes of damp, loose mortar sand.

Cost of One Cu. Yd. Cement Mortar for Concrete Masonry
1 volume masonry cement and 3 volumes loose mortar sand

	Rate	Total	Rate	Total
9 sacks masonry cement	$....	$....	$5.50	$49.50
1.00 cu. yd. sand	....	....	8.90	8.90
Cost per cu. yd.		$....		$58.40
Cost per cu. ft.		....		2.16

1 volume portland cement:1 volume hydrated lime:6 volumes loose mortar sand

	Rate	Total	Rate	Total
4.5 sacks portland cement	$....	$....	$5.75	$25.88
196 lbs. hydrated lime	....	....	0.07	13.72
1.00 cu. yd. sand	....	....	8.90	8.90
Cost per cu. yd.		$....		$48.50
Cost per cu. ft.		....		1.80

1 volume portland cement:2 volumes hydrated lime or lime putty:9 volumes loose mortar sand

	Rate	Total	Rate	Total
3 sacks portland cement	$....	$....	$5.75	$17.25
261 lbs. hydrated lime	....	....	.07	18.27
1.00 cu. yd. sand	....	....	8.90	8.90
Cost per cu. yd.		$....		$44.42
Cost per cu. ft.		....		1.65

1 volume masonry cement:1 volume portland cement:6 volume loose mortar sand

	Rate	Total	Rate	Total
4.5 sacks masonry cement	$....	$....	$5.50	$24.75
4.5 sacks portland cement	....	....	5.75	25.88
1.00 cu. yd. sand	....	....	8.90	8.90
Cost per cu. yd.		$....		$59.53
Cost per cu. ft.		....		2.20

1 volume portland cement:1/4-volume hydrated lime:3 volumes loose mortar sand

	Rate	Total	Rate	Total
7.83 sacks portland cement	$....	$....	$5.75	$45.02
90 lbs. hydrated lime	....	....	.07	6.30
0.86 cu. yd. sand	....	....	8.90	7.65
Cost per cu. yd.		$....		$58.97
Cost per cu. ft.		....		2.18

Quantity of Mortar Required to Lay 1,000
Concrete Brick of Various Sizes

Width of Bed Joint	Thickness of Wall: 1 Brick or 4" Wall	2 Bricks or 8" Wall	2 Bricks or 8" Backup	3 Bricks or 12" Wall
Modular Size Brick, 2 ¼"x3 ⅝"x7 ⅝"				
5⁄12"	9.64 cu. ft.	12.19 cu. ft.	14.73 cu. ft.	13.01 cu. ft.
Jumbo Brick, 3 ⅝"x3 ⅝"x7 ⅝"				
⅜"	10.06 cu. ft.	13.88 cu. ft.	17.70 cu. ft.	15.15 cu. ft.
Double Brick, 4 ⅞"x3 ⅝"x7 ⅝"				
11⁄24"	12.68 cu. ft.	17.77 cu. ft.	22.86 cu. ft.	19.47 cu. ft.
Roman Brick, 1 ⅝"x3 ⅝"x11 ⅝"				
⅜"	11.79 cu. ft.	14.65 cu. ft.	....	.....
Roman Brick, 1 ⅝"x3 ⅝"x15 ⅝"				
⅜"	15.25 cu. ft.	19.07 cu. ft.	....	....

All end and back joints figured ⅜". Quantities include 10% for waste. Mortar for glass blocks shall be composed of 1 part waterproof portland cement, 1 part lime and 4 parts well graded sand, all well mixed to a consistency as stiff and dry as possible.

Cost of Mortar for Laying Glass Blocks

	Rate	Total	Rate	Total
25 lb. quicklime	$....	$....	$0.07	$1.75
1 bag w'p'f portland cement	....	....	7.50	7.50
4 cu. ft. sand	....	....	.33	1.32
Cost 4 cu. ft. mortar		$....		$10.57
Cost per cu. ft.		....		2.64

Mortar Required for Setting Stone. The quantity of mortar required for stone setting will vary with the size of the stone, width of bed, etc., but as a general thing it will require 4 to 5 cu. ft. of mortar per 100 cu. ft. of stone.

If the stone is to be back-plastered, the quantity of mortar will vary with the thickness of the plaster coat but as most ashlar is 4" to 8" thick, it will require the following cu. ft. of mortar per 100 cu. ft. of stone:

Thickness of Stone in Inches

Plaster Thickness	4"	6"	8"
¼"	6 ½	4 ¼	3 ¼
½"	13	8 ½	6 ½
¾"	19	13	9 1/2

It is customary to use white non-staining portland cement for setting and backplastering limestone or other porous stone to prevent stains from appearing on the face of the stone. Never use ordinary portland cement for setting limestone.

Cost of One Cu. Yd. of 1:3 Non-Staining Cement Mortar With 1/5 Part of Hydrated Lime Based on the Cement Volume

	Rate	Total	Rate	Total
2 bbls. Medusa Stone Set cement	$....	$....	$10.25	$20.50
100 lbs. hydrated lime	....	....	0.07	7.00
1 cu. yd. sand	....	....	8.90	8.90
Cost per cu. yd.		$....		$36.40
Cost per cu. ft.		....		1.35

If white sand is used for mortar, figure $18.25 per ton (3,000 lbs. per cu. yd.) If white cement is specified, figure $17.50 per bbl. If white cement (waterproofed) is specified, figure $18.75 per bbl.

Cost of One Cu. Yd. of Cement-Lime Mortar Consisting of 1/2 Part Lime, 1/2 Part Stone Set Cement, and 3 Parts Sand

	Rate	Total	Rate	Total
1.25 bbls. Stone Set cement	$....	$....	$10.25	$12.81
225 lbs. hydrated lime	....	....	2.00	9.00
1 cu. yd. sand	....	....	5.75	5.75
Cost per cu. yd		$....		$27.56
Cost per cu. ft		....		1.02

If white sand is used for mortar, figure $18.25 per ton (3,000 lbs. per cu. yd.)

If white cement is specified, figure $17.50 per bbl. If white cement (waterproofed) is specified, figure $18.75 per bbl.

04200 UNIT MASONRY

04210 BRICK

Information has been provided by the Brick Institute of America (BIA). The BIA separates brick walls into 3 general categories: conventional, bonded walls; drainage type walls; and barrier type walls.

The drainage type walls include brick veneer, the cavity type wall, and the masonry bonded hollow wall. Drainage type walls are recommended for walls subjected to the most severe exposures. The success of this type of wall, of course, depends on the care taken to provide continuous means for water to escape.

The barrier type walls include metal tied walls and reinforced brick masonry. The success of these walls will depend on a solidly filled collar joint. The BIA presently recommends using an 8" metal tied wall in preference to an 8" bonded wall for areas subjected to moderate exposure. Metal tied walls adapt well to brick and block backup.

When estimating the number of brick required for any job, obtain the length, height and thickness of each wall, stating the totals in cu. ft., or obtain the length and height, stating the totals in sq. ft.

In present day estimating, all openings should be deducted in full, regardless of their size, because the estimate should show as accurately as possible the exact number of brick required for the job. The old method of counting corners twice, doubling the number of brick required for pilaster, chimney breasts, etc., is no longer used by progressive builders, who know they would never get a job using obsolete methods--and if they did get it, they would have no idea of the actual number of brick required to build it. Actual quantities and costs is the only sure method of getting your share of work at a profit.

Brick walls are usually designated on the plans as 4" or 4 ½"; 8" or 9"; 12" or 13"; 16" or 17", etc., increasing by 4" or 4 ½" in width. This variation in thickness need not be considered by the estimator because a 4" or 4 ½" wall is 1 brick thick; and 8" or 9" wall is 2 bricks thick; a 12" or 13" wall is 3 bricks thick; a 16" or 17" wall is 4 bricks thick, etc. The easiest method for the estimator is to mark all walls, 4", 8", 12", or 16" and then reduce the totals to cu. ft. After the number of cu. ft. has been obtained, multiply by the number of brick required per cu. ft. of wall and the result will be the number of brick required for the job.

When brickwork is estimated on the above basis, it is advisable to add 1 ½ to 2% to the total to take care of salmon brick, broken brick, bats, etc. The waste should not exceed this amount unless very poor brick or poor job handling methods are used.

Years ago contractors figured their building brickwork on the basis of 7 or 7 ½ brick per sq. ft. of 4" or 4 ½" wall; 14 or 15 brick per sq. ft. of 8" or 9" wall; 21 or 22 ½ brick per sq. ft. of 12" or 13" wall, etc.

This method is no longer used, because modern business demands that actual quantities be figured. Work is being estimated too closely today to permit a 10% to 20% overrun on the brickwork, especially when many contractors are figuring on a 5% to 10% margin of profit.

After the number of cu. ft. of brickwork has been obtained, it will be necessary to determine the number of brick required per cu. ft. of wall. This will vary with the size of the brick and the width of the mortar joints. The standard non-modular size is 8"x2 ¼"x3 ¾". Modular brick is 8"x2 ⅔"x4". Sizes will vary somewhat depending upon the burning, the brick in the center of the kiln usually being more thoroughly burned.

When estimating the number of face brick required for any job, obtain the length and height of all walls to be faced with brick, and the total will be the number of sq. ft. of wall.

Always make deductions in full for all openings, regardless of size, because the estimate should show as accurately as possible the exact number of brick required to complete the job.

When making deductions for door and window openings, always note the depth or width of the brick jambs or "reveals". If they are only 4" deep, deduct the full size of the opening as the 4" end of the brick forms the "reveal" or jamb.

If the jamb is more than 4" wide, the full depth of the jamb or "reveal" should be deducted. Example: Deduct for a 5'-0"x7'-0" opening in a 12" brick wall where the face brick jamb or "reveal" returns the full thickness of the wall. It will be necessary to deduct for the 12" jamb or reveal on each side of the opening, i.e., 2 x 12"=24" or 2'-0". Deduct this from the width of the opening, i.e., 5'-0" - 2'-0" =3'-0". The opening deducted should be 3'-0"x7'-0" instead of 5'-0"x7'-0", as shown on the plan.

Standard Brick

A standard non-modular brick is 8" long, 2 ¼" high and contains 18 sq. in. on the face. A standard modular brick is 8" long, 2 ⅔" high and contains 21.28

sq. in. on the face. If using a standard non-modular brick, laid dry without a mortar joint, it would require 8 brick per sq. ft. of wall (144 divided by 18 = 8). However, the thickness of the mortar joint must be added, and this will vary from 3/8" to 5/8", with 1/2" the average width and 3/8" the normal thickness. A brick is 8" long, plus 3/8" for the vertical or end mortar joint makes a total length of 8 3/8". A brick is 2 1/4" high, plus 3/8" for horizontal or bed mortar joint, making the total height of each brick course 2 5/8", 8 3/8" x 2 5/8" = 21.984 or 22 sq. in. on the face.

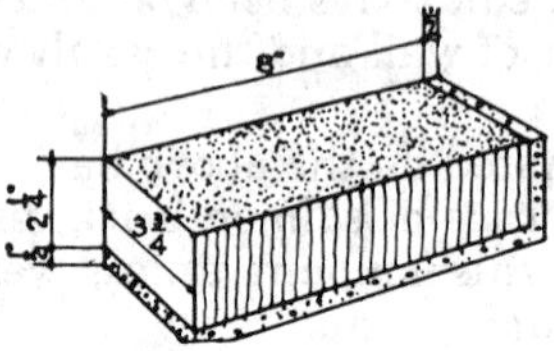

To obtain the number of brick required per sq. ft. of wall, divide 144 by 22 and the result is 6.54 or 6 1/2 brick per sq. ft. of 4" or 4 1/2" wall. If the wall is 8" or 9" (2 brick) thick, 2 x 6 1/2 = 13 brick per sq. ft. of 8" or 9" wall. If 12" or 13" (3 brick) thick, 6 1/2 x 3 = 19.5 brick per sq. ft. of wall, etc.

This same method should be used to obtain the number of brick of any size required for one sq. ft. of wall of any thickness.

Number of Standard Brick (8"x2 1/4"x3 3/4") Required for One Sq. Ft. of Brick Wall of any Thickness

Vertical or End Mortar Joints Figured as 1/4" Wide

Thickness of wall	Number of Brick Thick	Width of Horizontal or "Bed" Mortar Joints 1/8"	1/4"*	3/8"	1/2"	5/8"	3/4"
4" or 4 1/2"	1	7.33	7	6.67	6.33	6.08	5.8
8" or 9"	2	14.67	14	13.33	12.67	12.17	11.6
12" or 13"	3	22.00	21	20.00	19.00	18.25	17.4
16" or 17"	4	29.33	28	26.67	25.33	24.33	23.2
20" or 21"	5	36.67	35	33.33	31.67	30.42	29.0
24" or 25"	6	44.00	42	40.00	38.00	36.50	34.8

*Use this column for computing the number of brick required per cu. ft. of wall with any width mortar joint.

Variations in Face Brick Quantities. If the brick are laid in running bond without headers, the above quantities will prove sufficient, but if there is a full course of headers every 5th, 6th, or 7th course, it will be necessary to allow for the extra brick required. Also, if the brick are laid in English, Flemish, English Cross, or Dutch Bond, it will be necessary to make an allowance for extra brick where full header courses are required.

Number of Square Inches Occupied by One 8"x2 ¼" Face Brick
With Various Width Mortar Joints

All Vertical Mortar Joints Figured ¼" Wide
Width of Horizontal or "Bed" Mortar Joints in Inches

⅛"	¼"	⅜"	½"	⅝"	¾"	⅞"	1"
19.30	20.625	21.656	22.69	23.72	24.75	25.78	26.81

Number of 8"x2 ¼"x3 ¾" Face Brick Required Per Sq. Ft. of Wall
In Running Bond Without Headers

All Vertical Mortar Joints Figured ¼" Wide
Width of Horizontal or "Bed" Mortar Joints in Inches

⅛"	¼"	⅜"	½"	⅝"	¾"	⅞"	1"
7.46	7.0	6.65	6.35	6.08	5.82	5.60	5.38

The above table does not include any allowance for waste, breakage, nor for header courses extending into the building brick backing to form a bond, but is based on using all stretchers or blind headers with metal wall ties. On face brick work, 3% to 5% should be added to the net quantity for waste.

For face brick laid with a row of full "headers" every 5th, 6th or 7th course, add the following percentages to the above quantities:

Percentage to be Added for Various Brick Bonds

Common (full header course every 5th course)	20% or 1/8
Common (full header course every 6th course)	16 2/3% or 1/8
Common (full header course every 7th course)	14 1/3% or 1/7
English or English Cross* (full headers every other course)	50% or 1/2
English or English Cross* (full headers every 6th course)	16 2/3% or 1/6
Dutch or Dutch Cross* (full headers every other course)	50% or 1/2
Dutch or Dutch Cross* (full headers every 6th course)	16 2/3% or 1/6
Flemish (full headers every course)	33 1/3% or 1/3
Flemish (full headers every 6th course)	5.6% or 1/18
Double Header (two headers and a stretcher every 6th course)	8 1/3% or 1/12
Double Header (two headers and a stretcher every 5th course)	10% or 1/10
Double Flemish (full headers every other course)	10% or 1/10

*Add 10% to 15% extra brick for waste in cutting, unless a masonry saw is used.

Percentages To Be Added for Various Brick Bonds

Double Flemish (full headers every 3rd course)	6 2/3% or 1/15
3 Stretcher Flemish (full headers every other course)	7 1/7% or 1/14
3 Stretcher Flemish (full headers every 3rd course)	4.8% or 1/21
4 Stretcher Flemish (full headers every other course)	5.6% or 1/18
4 Stretcher Flemish (full headers every 3rd course)	3.7% or 1/27

For garden walls, porch walls, and other places where an 8" wall is used, with face brick on both sides of the wall, no additional brick are required for any type of bond.

For walks and floors with the brick laid on edge, in any pattern except diagonal, calculate as you would for face brick in common bond without headers. For herringbone or other diagonal work, additional brick will be required because of the waste in chipping or cutting the brick for the borders.

The additional number of brick will vary with the width of the walk or floor, as the wider the surface, the smaller the average waste per sq. ft. Walks and floors having the brick laid flat require one-third less brick than where they are laid on edge.

Coursing Tables

The following tables contain vertical coursing dimensions for both non-modular and modular brick. For the non-modular brick, vertical coursing dimensions are shown for both 3/8" and 1/2" mortar joints. In both tables the bricks are assumed to be positioned in the wall as stretchers.

Vertical Coursing Table for Non-modular Brick*

No. of	2 1/4" high		2 5/8" high		2 3/4" high	
Courses	3/8" joint	1/2" joint	3/8" joint	1/2" joint	3/8" joint	1/2" joint
1	0'-2 5/8"	0'-2 3/4"	0'-3"	0'-3 1/8"	0"-3 1/8"	0'-3 1/4"
2	0'-5 1/4"	0'-5 1/2"	0'-6"	0'-6 1/4"	0'-6 1/4"	0'-6 1/2"
3	0'-7 7/8"	0'-8 1/4"	0'-9"	0'-9 3/8"	0'-9 3/8"	0'-9 3/4"
4	0"-10 1/2"	0'-11"	1'-0"	1'-0 1/2"	1'-0 1/2"	1'-1"
5	1'-1 1/8"	1'-1 3/4"	1'-3"	1'-3 5/8"	1'-3 5/8"	1'-4 1/4"
6	1'-3 3/4"	1'-4 1/2"	1'-6"	1'-6 3/4"	1'-6 3/4"	1'-7 1/2"
7	1'-6 3/8"	1'-7 1/4"	1'-9"	1'-9 7/8"	1'-9 7/8"	1' 10 3/4"
8	1'-9"	1'-10"	2'-0"	2'-1"	2'-1"	2'-2"
9	1'-11 5/8"	2'-0 3/4"	2'-3"	2'-4 1/8"	2'- 4 1/8"	2'-5 1/4"
10	2'-2 1/4"	2'-3 1/2"	2'-6"	2'-7 1/4"	2'-7 1/4"	2'-8 1/2"
11	2'-4 7/8"	2'-6 1/4"	2'-9"	2'-10 3/8"	2'-10 3/8"	2'-11 3/4"
12	2'-7 1/2"	2'-9"	3'-0"	3'-1 1/2"	3'-1 1/2"	3'-3"
13	2'-10 1/8"	2'-11 3/4"	3'-3"	3'-4 5/8"	3'-4 5/8"	3'-6 1/4"
14	3'-0 3/4"	3'-2 1/2"	3'-6"	3'-7 3/4"	3'-7 3/4"	3'-9 1/2"
15	3'-3 3/8"	3'-5 1/4"	3'-9"	3'-10 7/8"	3'-10 7/8"	4'-0 3/4"
16	3'-6"	3'-8"	4'-0"	4'-2"	4'-2"	4'-4"
17	3'-8 5/8"	3'-10 3/4"	4'-3"	4'-5 1/8"	4'-5 1/8"	4'-7 1/4"
18	3'-11 1/4"	4'-1 1/2"	4'-6"	4'-8 1/4"	4'-8 1/4"	4'-10 1/2"
19	4'-1 7/8"	4'-4 1/4"	4'-9"	4'-11 3/8"	4'-11 3/8"	5'-1 3/4"
20	4'-4 1/2"	4'-7"	5'-0"	5'-2 1/2"	5'-2 1/2"	5'-5"
21	4'-7 1/8"	4'-9 3/4"	5'-3"	5'-5 5/8"	5'-5 5/8"	5'-8 1/4"
22	4'-9 3/4"	5'-0 1/2"	5'-6"	5'-8 3/4"	5'-8 3/4"	5'-11 1/2"
23	5'-0 3/8"	5'-3 1/4"	5'-9"	5'-11 7/8"	5'-11 7/8"	6'-2 3/4"
24	5'-3"	5'-6"	6'-0"	6'-3"	6'-3"	6'-6"
25	5'-5 5/8"	5'-8 3/4"	6'-3"	6'-6 1/8"	6'-6 1/8"	6'-9 1/4"
26	5'-8 1/4"	5'-11 1/2"	6'-6"	6'-9 1/4"	6'-9 1/4"	7'-0 1/2"
27	5'-10 7/8"	6'-2 1/4"	6'-9"	7'-0 3/8"	7'-0 3/8"	7'-3 3/4"
28	6'-1 1/2"	6'-5"	7'-0"	7'-3 1/2"	7'3 1/2"	7'-7"
29	6'-4 1/8"	6'-7 3/4"	7'-3"	7'-6 5/8"	7'-6 5/8"	7'-1- 1/4"
30	6'-6 3/4"	6'-10 1/2"	7'-6"	7'-9 3/4"	7'-9 3/4"	8'-1 1/2"
31	6'-9 3/8"	7'-1 1/4"	7'-9"	8'-0 7/8"	8'-0 7/8"	8'-4 3/4"
32	7'-0"	7'-4"	8'-0"	8'-4"	8'-4"	8'-8"
33	7'-2 5/8"	7'-6 3/4"	8'-3"	8'-7 1/8"	8"-7 1/8"	8'-11 1/4"
34	7'-5 1/4"	7'-9 1/2"	8'-6"	8'-10 1/4"	8'-10 1/4"	9'-2 1/2"
35	7'-7 7/8"	8'-0 1/4"	8'-9"	9'-1 3/8"	9'-1 3/8"	9'-5 3/4"

No. of Courses	2 ¼" high 3/8" joint	2 ¼" high ½" joint	2 5/8" high 3/8" joint	2 5/8" high ½" joint	2 ¾" high 3/8" joint	2 ¾" high ½" joint
36	7'-10 ½"	8'-3"	9'-0"	9'-4 ½"	9'-4 ½"	9'-9"
37	8'-1 1/8"	8'-5 ¾"	9'-3"	9'-7 5/8"	9'-7 5/8	10-0 ¼"
38	8'-3 ¾"	8'-8 ½"	9'-6"	9'-10 ¾"	9'-10 ¾"	10'-3 ½"
39	8'-6 3/8"	8'-11 ¼"	9'-9"	10'-1 7/8"	10'-1 7/8"	10'-6 ¾"
40	8'-9"	9'-2"	10'-0"	10'-5"	10'-5"	10'-10"
41	8'-11 5/8"	9'-4 ¾"	10'-3"	10'-8 1/8"	10'-8 1/8"	11'-1 ¼"
42	9'-2 ¼"	9'-7 ½"	10'-6"	10'11 ¼"	10'-11 ¼"	11'-4 ½"
43	9'-4 7/8"	9'-10 ¼"	10'-9"	11'-2 3/8"	11'-2 3/8"	11'-7 ¾"
44	9'-7 ½"	10'-1"	11'-0"	11'-5 ½"	11'-5 ½"	11'-11"
45	9'-10 1/8"	10'-3 ¾"	11'-3"	11'-8 5/8"	11'-8 5/8"	12'-2 ¼"
46	10'-0 ¾"	10"-6 ½"	11'-6"	11'-11 ¾"	11'-11 ¾"	12'5 ½"
47	10'-3 3/8"	10'-9 ¼"	11'-9"	12'-2 7/8"	12'-2 7/8"	12'-8 ¾"
48	10'-6"	11'-0"	12'-0"	12'-6"	12'-6"	13'-0"
49	10'-8 5/8"	11'-2 ¾"	12'-3"	12'-9 1/8"	12'-9 1/8"	13'-3 ¼"
50	10'-11 ¼"	11'-5 ½"	12'-6"	13'-0 ¼"	13'-0 ¼"	13'-6 ½"
100	21'-10 ½"	22'-11"	25'-0"	26'-0 ½"	26'-0 ½"	27'-1"

*Brick positioned in wall as stretchers. Vertical dimensions are from bottom of mortar joint to bottom of mortar joint.
Courtesy of Brick Institute of America

Vertical Coursing Table for Modular Brick*

No. of Courses	Nominal Height (h) of Unit 2"	2 ⅔"	3 ⅕"	4"	5 ⅓"
1	0'-2"	0'-2 ⅔"	0'-3 ⅕"	0'-4"	0'-5 ⅓"
2	0'-4"	0'-5 ⅓"	0'-6 ⅖"	0'-8"	0'-10 ⅔"
3	0'-6"	0'-8"	0'-9 ⅗"	1'-0"	1'-4"
4	0'-8"	0'-10 ⅔"	1'-0 ⅘"	1'-4"	1'-9 ⅓"
5	0'-10"	1'-1 ⅓"	1'-4"	1'-8"	2'-2 ⅔"
6	1'-0"	1'-4"	1'-7 ⅕"	2'-0"	2'-8"
7	1'-2"	1'-6 ⅔"	1'-10 ⅖"	2'-4"	3'-1 ⅓"
8	1'-4"	1'-9 ⅓"	2'-1 ⅗"	2'-8"	3'-6 ⅔"
9	1'-6"	2'-0"	2'-4 ⅘"	3'-0"	4'-0"
10	1'-8"	2'-2 ⅔"	2'-8"	3'-4"	4'-5 ⅓"
11	1'-10"	2'-5 ⅓"	2'-11 ⅕"	3'-8"	4'-10 ⅔"
12	2'-0"	2'-8"	3'-2 ⅖"	4'-0"	5'-4"
13	2'-2"	2'-10 ⅔"	3'-5 ⅗"	4'-4"	5'-9 ⅓"
14	2'-4"	3'-1 ⅓"	3'-8 ⅘"	4'-8"	6'-2 ⅔"
15	2'-6"	3'-4"	4'-0"	5'-0"	6'-8"
16	2'-8"	3'-6 ⅔"	4'-3 ⅕"	5'-4"	7'-1 ⅓"
17	2'10"	3'-9 ⅓"	4'-6 ⅖"	5'-8"	7'-6 ⅔"
18	3'-0"	4'-0"	4'-9 ⅗"	6'-0"	8'-0"
19	3'-2"	4'-2 ⅔"	5'-0 ⅘"	6'-4"	8'-5 ⅓"
20	3'-4"	4'-5 ⅓"	5'-4"	6'-8"	8'-10 ⅔"
21	3'-6"	4'-8"	5'-7 ⅕"	7'-0"	9'-4"
22	3'-8"	4'-10 ⅔"	5'-10 ⅖"	7'-4"	9'-9 ⅓"
23	3'-10"	5'-1 ⅓"	6'-1 ⅗"	7'-8"	10"-2 ⅔"
24	4'-0"	5'-4"	6'-4 ⅘"	8'-0"	10'-8"
25	4'-2"	5'-6 ⅔"	6'-8"	8'-4"	11'-1 ⅓"

No. of Courses	Nominal Height (h) of Unit 2"	2 ⅔"	3 ⅕"	4"	5 ⅓"
26	4'-4"	5'-9 ⅓"	6'-11 ⅕"	8'-8"	11'-6 ⅔"
27	4'-6"	6'-0"	7'-2 ⅖"	9'-0"	12'-0"
28	4'-8"	6'-2 ⅔"	7'-5 ⅗"	9'-4"	12'-5 ⅓"
29	4'-10"	6'-5 ⅓"	7'-8 ⅘"	9'-8"	12'-10 ⅔"
30	5'-0"	6'-8"	8'-0"	10'-0"	13'-4"
31	5'-2"	6'-10 ⅔"	8'-3 ⅕"	10'-4"	13'-9 ⅓"
32	5'-4"	7'-1 ⅓"	8'-6 ⅖"	10'-8"	14'-2 ⅔"
33	5'-6"	7'-4"	8'-9 ⅗"	11'-0"	14'-8"
34	5'-8"	7'-6 ⅔"	9'-0 ⅘"	11'-4"	15'-1 ⅓"
35	5'-10"	7'-9 ⅓"	9'-4"	11'-8"	15'-6 ⅔"
36	6'-0"	8'-0"	9'-7 ⅕"	12'-0"	16'-0"
37	6'-2"	8'-2 ⅔"	9'-10 ⅖"	12'-4"	16'-5 ⅓"
38	6'-4"	8'-5 ⅓"	10'-1 ⅗"	12'-8"	16'-10 ⅔"
39	6'-6"	8'-8"	10'-4 ⅘"	13'-0"	17'-4"
40	6'-8"	8'-10 ⅔"	10'-8"	13'-4"	17'-9 ⅓"
41	6'-10"	9'-1 ⅓"	10'-11 ⅕"	13'-8"	18'-2 ⅔"
42	7'-0"	9'-4"	11'-2 ⅖"	14'-0"	18'-8"
43	7'-2"	9'-6 ⅔"	11'-5 ⅗"	14'-4"	19'-1 ⅓"
44	7'-4"	9'-9 ⅓"	11'-8 ⅘"	14'-8"	19'-6 ⅔"
45	7'-6"	10'-0"	12'-0"	15'-0"	20'-0"
46	7'-8"	10'-2 ⅔"	12'-3 ⅕"	15'-4"	20'-5 ⅓"
47	7'-10"	10'-5 ⅓"	12'-6 ⅖"	15'-8"	20'-10 ⅔"
48	8'-0"	10'-8"	12'-9 ⅗"	16'-0"	21'-4"
49	8'-2"	10'10 ⅔"	13'-0 ⅘"	16'-4"	21'-9 ⅓"
50	8'-4"	11'-1 ⅓"	13'-4"	16'-8"	22'-2 ⅔"
100	16'-8"	22'-2 ⅔"	26'-8"	33'-4"	44'-5 ⅓"

*Brick positioned in wall as stretchers.
Courtesy of Brick Institute of America

Brick Dimensions and Nomenclature

The sizes of brick units shown in the following tables and figures are most typical of those currently being produced by the industry. However, few manufacturers produce all of the sizes shown; also, other sizes, produced by other manufacturers, will vary from those shown here. For these reasons, it is recommended that the designer consult with manufacturers or distributors in his area before proceeding with a design incorporating a specific size of brick that may not be readily available in that particular locality.

The nomenclature indicated, while typical, is not completely standard throughout the industry. Except for the *Standard, Roman,* and *Norman* sizes, individual manufacturers may have their own names for certain sizes listed here for a unit with different dimensions. It is suggested that in order to avoid confusion, the purchase or specifier first identify the brick by size.

Except for the non-modular Standard, Oversize, and 3-in. units, most bricks are produced in modular sizes. The nominal dimensions of modular brick are equal to the manufactured dimensions plus the thickness of the mortar joint for which the unit is designed. In general, the joint thicknesses used with brick are either ⅜ or ½ in.

The actual *manufactured* dimensions of the units may vary, of course, from the *specified* dimensions by not more than the permissible tolerances for variation in dimensions as prescribed in the applicable ASTM specifications (*Standard Specification for Facing Brick, ASTM Designation C 216; Standard Specifications for Building Brick, ASTM Designation C 62; Standard Specifications for Hollow Brick, ASTM Designation C 652*).

It should be noted that the designated manufacturers heights for the standard brick, the standard modular brick and all other modular brick designed to be laid 3 courses to 8", are the same (2 1/4"). There is a very practical reason for this. In 1946-47, with the adoption of modular coordination by the structural clay products industry, brick manufacturers who converted their production completely to modular sizes were faced with a problem in connection with supplying matching brick for additions to existing non-modular buildings. From the standpoint of appearance, most designers required that the vertical coursing in the addition built with modular brick match those in the existing building constructed with non-modular brick. It was agreed that the manufactured face height of the standard modular brick would remain at 2 1/4", even though the other dimensions (length and thickness) would become modular. Since that time, the custom has continued. The differences in mortar bed thickness required to maintain the modular coursing of 3 courses in 8" were considered minimal.

Size of Non-Modular Brick

Unit Designation	Manufactured Size, Inches t	h	l
Three-inch*	3	2 5/8	9 5/8
	3	2 3/4	9 3/4
	2 3/4	2 3/4	9 3/4
	2 5/8	2 3/4	8 3/4
Standard	3 3/4**	2 1/4	8
Oversize (Jumbo)	3 3/4	2 3/4	8

*In recent years, the so-called "three-inch" brick has gained popularity in certain areas. The term three-inch designates its thickness or bed depth. The sizes shown in the table are the ones most commonly produced. Originally developed primarily for use as a veneer unit, it is also used to construct 8-in. grouted walls.

**The manufactured thickness of standard or oversize non-modular brick will vary from 3 1/2 to 3 3/4 in. If other than a running bond is desired, the designer should check with the manufacturer of the brick selected.

Sizes of Modular Brick*

Unit Designation	Nominal Dimensions, inches t	h	l	Manufactured Dimensions, inches Joint, in.	t	h	l
Engineer	4	3 1/5	8	3/8	3 5/8	2 13/16	7 5/8
				1/2	3 1/2	2 11/16	7 1/2
Economy 8 or	4	4	8	3/8	3 5/8	3 5/8	7 5/8
Jumbo Closure				1/2	3 1/2	3 1/2	7 1/2
Double	4	5 1/3	8	3/8	3 5/8	4 15/16	7 5/8
				1/2	3 1/2	4 13/16	7 1/2
Roman	4	2	12	3/8	3 5/8	1 5/8	11 5/8
				1/2	3 1/2	1 1/2	11 1/2
Norman	4	2 2/3	12	3/8	3 5/8	2 1/4	11 5/8
				1/2	3 1/2	2 1/4	11 1/2

Sizes of Modular Brick*

Unit Designation	Nominal Dimensions, inches t	h	l	Manufactured Dimensions, inches Joint, in.	t	h	l
Norwegian	4	3 1/5	12	3/8	3 5/8	2 13/16	11 5/8
				1/2	3 1/2	2 11/16	11 1/2
Economy 12 or Jumbo Utility	4	4	12	3/8	3 5/8	3 5/8	11 5/8
				1/2	3 1/2	3 1/2	11 1/2
Triple	4	5 1/3	12	3/8	3 5/8	4 15/16	11 5/8
				1/2	3 1/2	4 13/16	11 1/2
6-in. Norwegian	6	3 1/5	12	3/8	5 5/8	2 13/16	11 5/8
				1/2	5 1/2	2 11/16	11 1/2
6-in. Jumbo	6	4	12	3/8	5 5/8	3 5/8	11 5/8
				1/2	5 1/2	3 1/2	11 1/2
8-in. Jumbo	8	4	12	3/8	7 5/8	3 5/8	11 5/8
				1/2	7 1/2	3 1/2	11 1/2
8-in.	8	4	8	3/8	7 5/8	3 5/8	7 5/8
15-in.	4	4	16	3/8	3 5/8	3 5/8	15 5/8

*Available as solid units conforming to ASTM C 216 or ASTM C 62, or in a number of cases, as hollow brick conforming to ASTM C652.

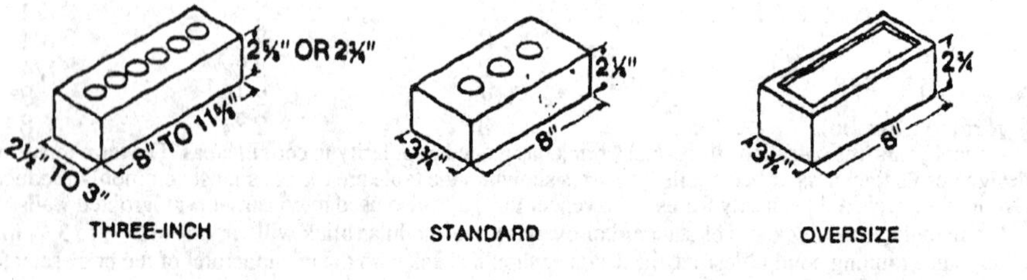

Non-Modular Brick (Nominal Dimensions)

Position of Brick in the Wall. Most brick are laid as *stretchers* so that the longer of the face dimensions is horizontal. The drawings below also illustrate terms applied to other brick positions as placed in the wall. The shaded areas indicate the surfaces of the brick that are exposed.

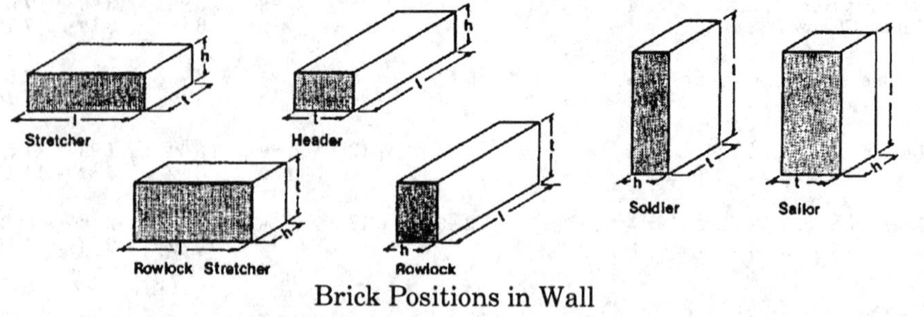

Brick Positions in Wall

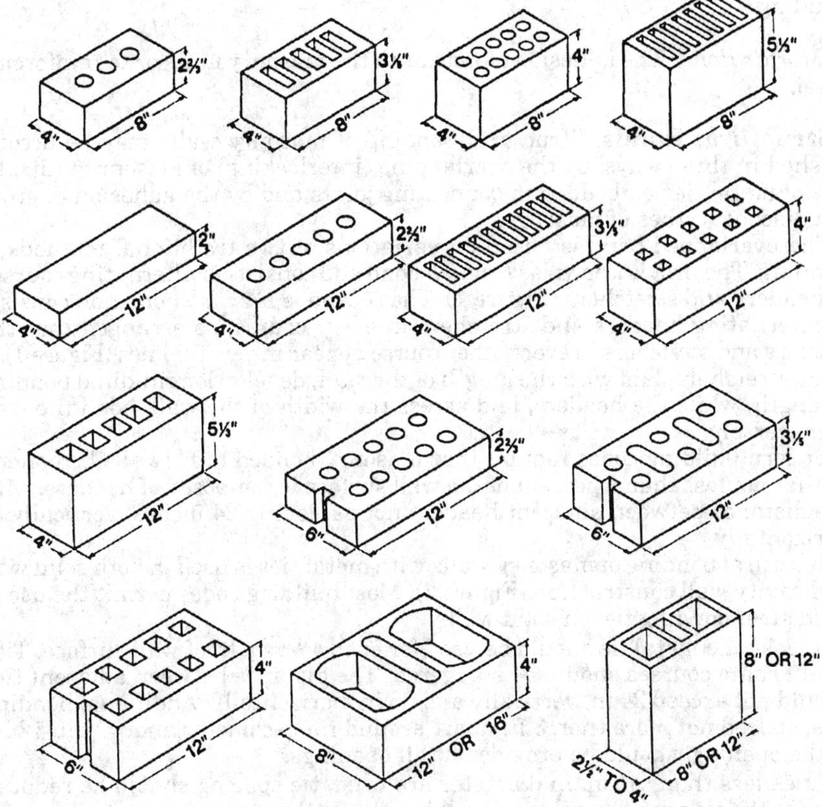

Modular Brick (Nominal Dimensions)

Brick Bonds and Patterns

Bond in brickwork is the overlapping of one brick upon the other, either along the length of the wall or through its thickness, in order to bind them together into a secure structural mass. Units are shifted back and forth so that the vertical joints in two successive layers or "courses" do not come into line; in other words, the brick are laid so as to break the joint, the whole forming a natural bond or a structural unit giving strength to the wall.

Bond. The word bond, when used in reference to masonry, has three meanings:

a) *Structural Bond:* The method by which individual masonry units are interlocked or tied together to cause the entire assembly to act as a single structural unit.

b) *Pattern Bond:* The pattern formed by the masonry units and the mortar joints on the face of a wall. The pattern may result from the type of structural bond used or may be purely a decorative one, unrelated to the structural bonding.

c) *Mortar Bond:* The adhesion of mortar to the masonry units or to reinforcing steel.

Structural Bonds. Structural bonding of masonry walls may be accomplished in three ways: by the overlapping (interlocking) of masonry units, by use of metal ties embedded in connecting joints and by the adhesion of grout to adjacent wythes of masonry.

The overlapped bond is based on variations of two traditional methods of bonding. The first is known as *English* bond and consists of alternating courses of headers and stretchers (Figure 1). The second is *Flemish* bond and consists of alternating headers and stretchers in every course, so arranged that the headers and stretchers in every other course appear in vertical lines (Figure 1).

The stretchers, laid with the length of the wall, develop longitudinal bonding strength, while the headers, laid across the width of the wall, bond the wall transversely.

Modern building codes require that masonry-bonded brick walls be bonded so that not less than 4 percent of the wall surface is composed of headers, with the distance between adjacent headers not exceeding 24 inches, vertically or horizontally.

Structural bonding of masonry walls with metal ties is used in both solid wall and cavity wall construction (Figure 2). Most building codes permit the use of rigid steel bonding ties in solid walls.

At least one metal tie should be used for each 4 ½ sq. ft. of wall surface. Ties in alternate courses should be staggered. The distance between adjacent ties should not exceed 24 in. vertically and 36 in. horizontally. Additional bonding ties, spaced not more than 3 ft. apart around the perimeter and within 12 in. of the opening, should be provided at all openings.

If ties less than 3/16 in. in diameter are used, tie spacing should be reduced so that the tie area per sq. ft. of wall is not less than specified above.

Structural bonding of solid and reinforced brick masonry walls is sometimes accomplished by grout, which is poured into the cavity or collar joint between wythes of masonry.

The method of bonding will depend on the use requirements, wall type and other factors. However, the metal tie method is generally recommended for exterior walls. Some of the advantages of this method are greater resistance to rain penetration and ease of construction. Metal ties also allow slight differential movements of the facing and backing which may relieve stresses and prevent cracking.

Pattern Bonds. Frequently, structural bonds, such as English or Flemish, or variations of these, may be used to create patterns in the face of a wall. However, in the strictest sense of the term, pattern refers to the change or varied arrangement of the brick texture or color used in the face. Therefore, it may be possible to secure many patterns using the same structural bond. Patterns also may be produced by the method of handling the mortar joint or by projecting or recessing certain brick from the plane of the wall, thus creating a distinctive wall texture that is not solely dependent upon the texture of the individual brick.

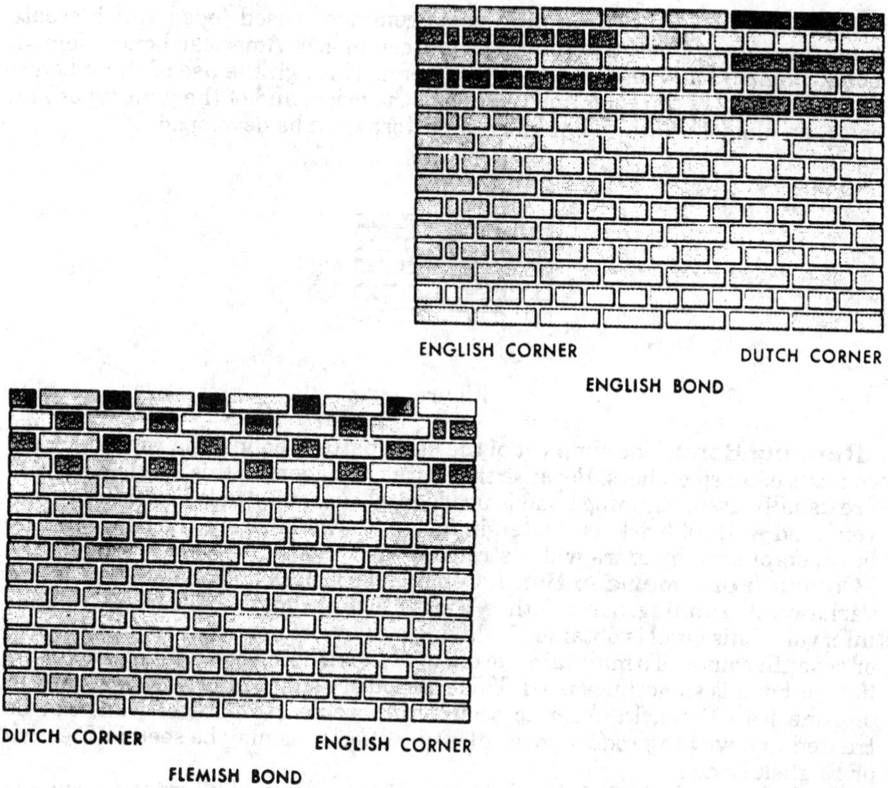
ENGLISH CORNER
DUTCH CORNER
ENGLISH BOND
DUTCH CORNER
ENGLISH CORNER
FLEMISH BOND

Figure 1

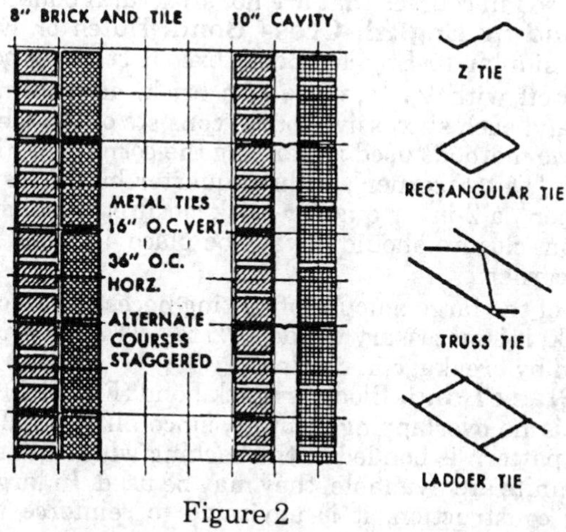
8" BRICK AND TILE
10" CAVITY
METAL TIES
16" O.C. VERT.
36" O.C.
HORZ.
ALTERNATE
COURSES
STAGGERED
Z TIE
RECTANGULAR TIE
TRUSS TIE
LADDER TIE

Figure 2

There are five basic structural bonds commonly used today which create typical patterns. These are running bond, common or American bond, Flemish bond, English bond and block or stack bond. Through the use of these bonds and variations of the color and texture of the brick, and of the joint types and color, an almost unlimited number of patterns can be developed.

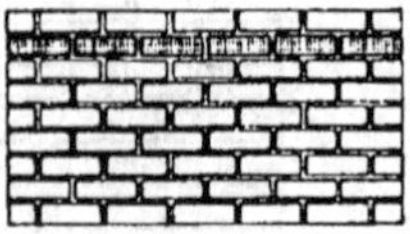

Figure 3

Running Bond. The simplest of the basic pattern bonds, the running bond, consists of all stretchers. Because there are no headers in this bond, metal ties are usually used. Running bond is used largely in cavity wall construction and veneered walls of brick, and often in facing tile walls where the bonding may be accomplished by extra width stretcher tile.

Common or American Bond. Common or American bond (Figure 4) is a variation of running bond with a course of full length headers at regular intervals. This bond is obtained by laying a course of headers every fifth, sixth or seventh course. To maintain the effect of the running bond, a special double header bond is sometimes used. This method of using headers, as in Common or American Bond, in order to secure transverse strength of wall, can be treated in a way to produce more pleasing effects, as may be seen in Flemish or English Bonds.

English Bond. English Bond (Figure 5) is made up of alternating courses of stretchers and headers. Ordinarily half brick are used for the header courses, except that every sixth course a full header course is used to tie the face and common brick walls together. The allowance for headers was given previously, under *Percentages to be Added for Various Brick Bonds*. Snap headers are used in courses that are not structural bonding courses.

Dutch Bond or English Cross Bond. Dutch or English Cross Bond (Figure 6) is similar to English Bond, except starting courses differ. Each course starts off with 1/4, 1/2, or 3/4 of a brick, alternating as shown in the illustration, and each successive course consists of stretchers and headers.

There are two methods used in starting the corners in Flemish and English bonds. In the "Dutch corner", a three-quarter brick closure is used. In the "English corner", a 2-in. or quarter brick closure, called a "queen closure", is used. The 2-in. closure should always be place 4 inches in from the corner, never at the corner.

On account of the large amount of cutting necessary to obtain the 1/4 and 3/4 pieces of brick, it is necessary to add 10% to 15% additional brick to allow for waste, caused by breakage in cutting.

Block or Stack Bond. Block or stack bond (Figure 7) is purely a pattern bond. There is no overlapping of units, since all vertical joints are aligned. Usually this pattern is bonded to the backing with rigid steel ties, but when 8-in. bonder units are available, they may be used. In large wall areas and in load bearing construction, it is advisable to reinforce the wall with steel reinforcement placed in the horizontal mortar joints. In stack bond it is imperative that prematched or dimensionally accurate masonry units be used if the vertical alignment of the head joints is to be maintained. The architect or contractor might want to use a "gauged" brick, allowing for the manufacturer's acceptable production tolerances for brick used in this type of bond. Check with the supplier for additional costs.

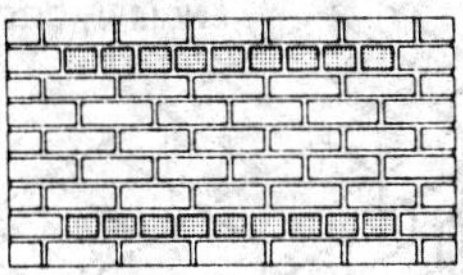

6th COURSE HEADERS

COMMON BOND

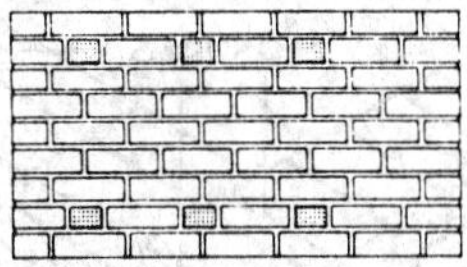

6th COURSE FLEMISH HEADERS

COMMON BOND

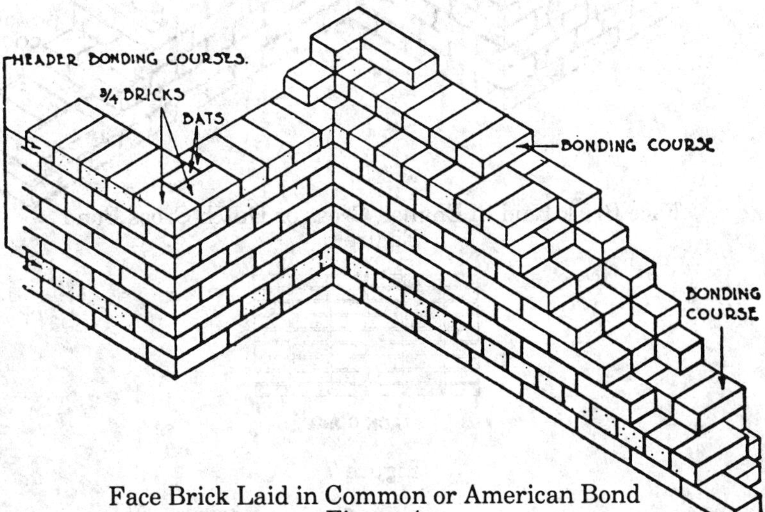

Face Brick Laid in Common or American Bond
Figure 4

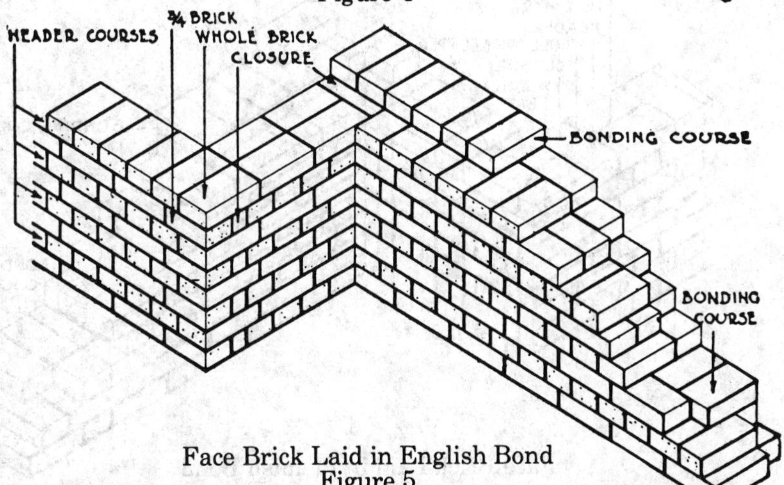

Face Brick Laid in English Bond
Figure 5

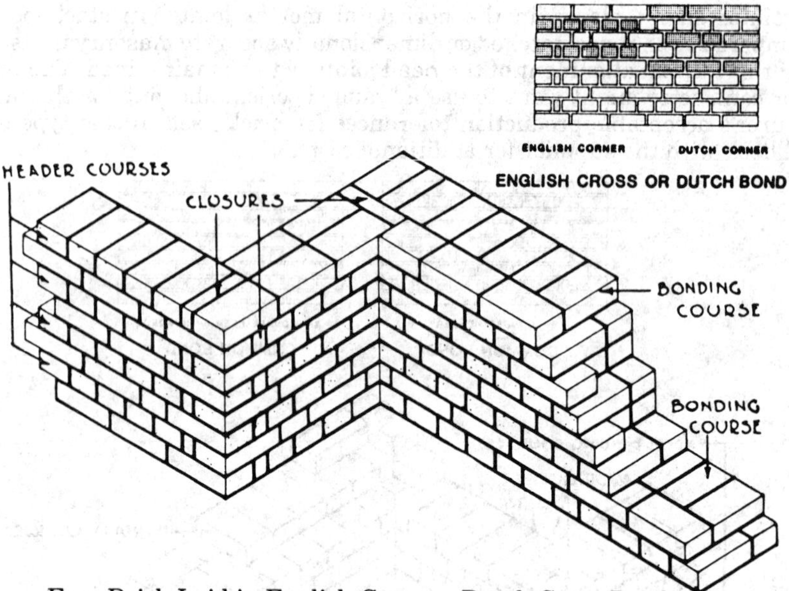

Face Brick Laid in English Cross or Dutch Cross Bond
Figure 6

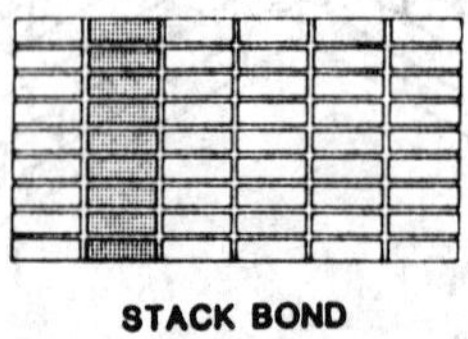

Figure 7

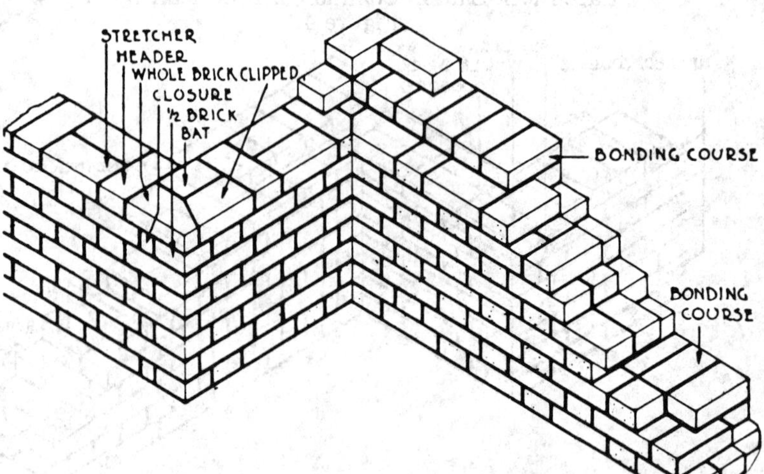

Face Brick Laid in Flemish Bond
Figure 8

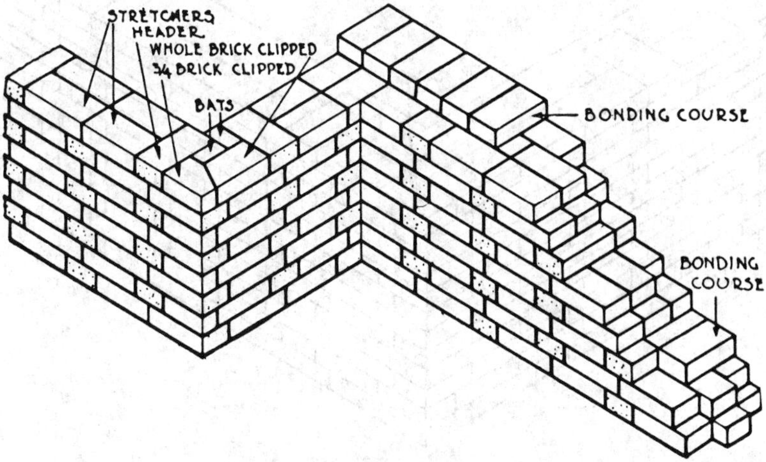

Face Brick Laid in Double Flemish or Garden Wall Bond
Figure 9

However, waste in cutting may be greatly reduced if a masonry saw is used for this purpose, because a brick can be cut in 10 to 15 seconds, and the resulting waste and spoilage of face bricks is greatly reduced.

If brick are laid with full header courses every second or sixth course, add the percentages given on the previous pages.

Flemish Bond. Flemish Bond (Figure 8) secures its effect by laying each course in alternate stretchers and headers, the header resting upon the middle of the stretcher in successive courses. When full brick are used for headers, it requires one-half more brick than where half brick or "blind" headers are used.

Double Flemish Bond. The arrangement of bricks along the course so that each header is preceded and followed by two stretchers for the entire course (Figure 9). On consecutive courses, the position of the header is directly over the joint between the two stretchers. Each stretcher has a lap of three-fourths of its length over the stretcher beneath.

Three Stretcher Flemish Bond. An arrangement (Figure 10) of bricks along the course so that a header is placed between each sequence of three stretchers. The header of one course is centered over the middle stretcher of the course below.

Four Stretcher Flemish Bond. In this bond (Figure 11), a header occurs between each sequence of four stretchers. The header of one course is centered between the headers of the course below. The allowances for header courses are given on the previous pages.

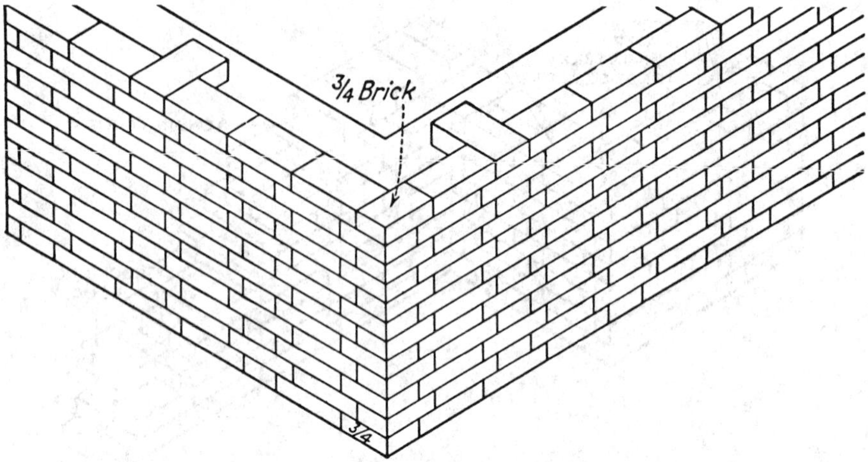

Face Brick Laid in Three Stretcher Flemish Bond
Figure 10

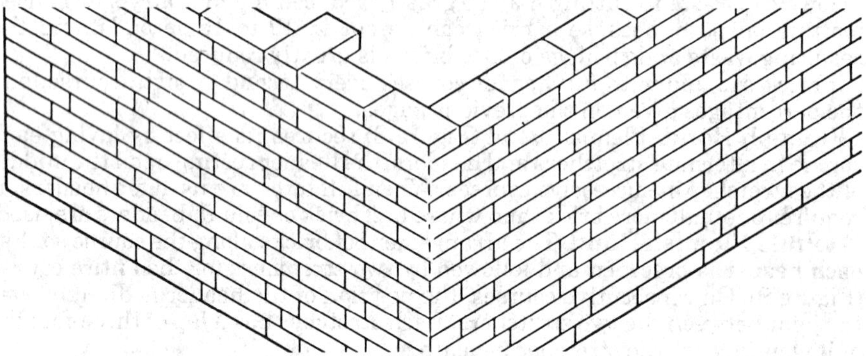

Face Brick Laid in Four Stretcher Flemish Bond
Figure 11

Garden Wall Bond. Garden Wall Bond (Figure 12) is merely a modification of Flemish Bond, secured by laying courses with two to four stretchers alternating with a header, and requires the same number of brick per square foot as Flemish Bond.

Diamond Bond Patterns. Diamond Bond Patterns (Figure 12) are secured by a modification of the Garden Wall Bond. It is, however, only in case of large wall surfaces that patterns of an elaborate character could be recommended, as any departure from simple bonds adds to the cost of bricklaying.

Face Brick Laid in Diamond Bond Patterns
Figure 12

Nominal Modular Sizes of Brick*

Unit Designation	Thickness, in.	Face Height, in.	Dimensions Length, in.	Number of Courses in 16 in.
Standard	4	2 2/3	8	6
Engineer	4	3 1/5	8	5
Economy 8 or Jumbo Closure	4	4	8	4
Double	4	5 1/3	8	3
Roman	4	2	12	8
Norman	4	2 2/3	12	6
Norwegian	4	3 1/5	12	5
Economy 12 or Jumbo Utility	4	4	12	4
Triple	4	5 1/3	12	3
6-in. Norwegian	6	3 1/5	12	5
6-in. Jumbo	6	4	12	4
8-in. Jumbo	8	4	12	4

*Available as solid units conforming to ASTM C 216 or ASTM C 62, or in some cases, as hollow brick conforming to ASTM C 652.

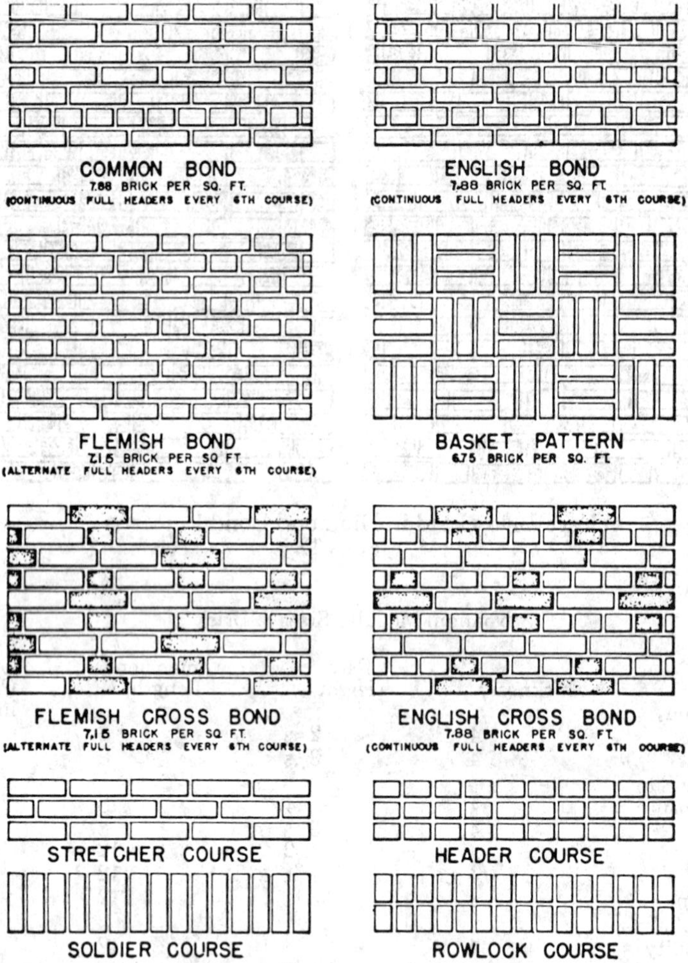

Elevations of Various Brick Bonds

Modular Size Brick

Economy is the prime objective of modular design. Its application has brought about economy of design in layout and in construction.

While modular design involves the coordination of building materials, this basic principle is employed only as a means to the objective of economy and is not carried beyond the point where its use would defeat the main purpose.

It is not to be considered a complex and revolutionary plan of coordinating building materials, but rather as a simplified and efficient system of design, layout and construction resulting in definite economies.

After an exhaustive study, a four-inch module or increment was agreed upon as a unit that would afford the maximum practical standardization and simplification, and at the same time, afford a sufficient flexibility in design.

On the basis of the four-inch module or increment, certain sizes of brick and tile have been recommended. In this connection, it must be remembered that the four-inch module, governing both the vertical and horizontal dimensions, is used to govern the layout of finished wall areas and to determine heights and widths of doors, windows, and other openings.

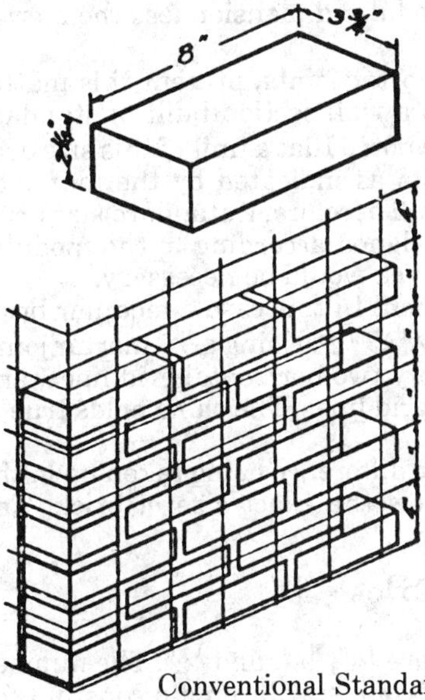

Conventional Standard Brick

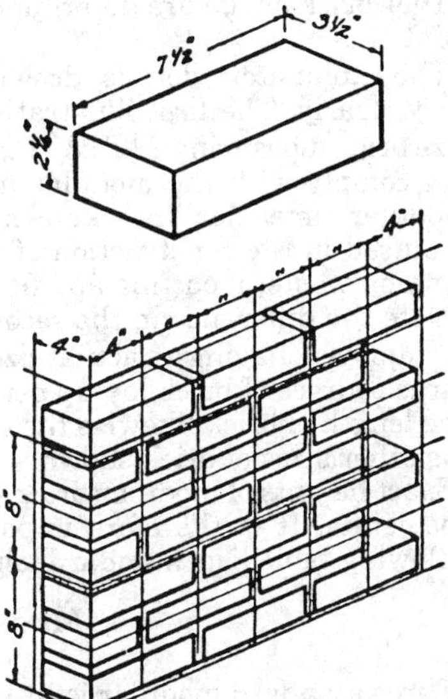

2-2/3" Modular Brick

3" Modular Brick

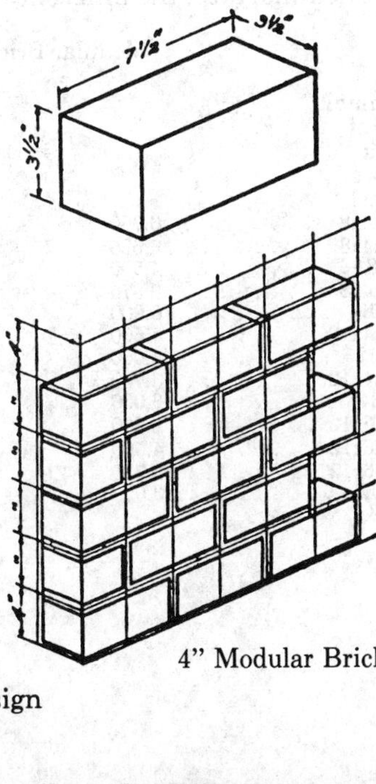

4" Modular Brick

Modular Design

Since the module is a unit of wall measurement, it cannot be applied to the individual brick or tile, but must be considered as governing the size of the brick or tile, plus the mortar joint.

In employing the modular system of measurement for masonry walls, the dimension is considered from center to center of the mortar joints. To fit properly into the modular system, the actual sizes of the individual units of either brick or tile are determined by the modular dimension less the mortar joint.

The explanatory details, drawn with ½" mortar joints, present this matter very clearly. The first illustration indicates a wall section built of standard size brick measuring 2 ¼"x3 ¾"x8". It will be noted that a unit of this size does not comply with the modular measurements as indicated by the four inch modular dimension lines, known as grid lines. Therefore, if standard size brick are used in the construction of a project designed according to the modular system, wasteful cutting and fitting of the brick would be necessary.

Note the difference in the second wall section. In this case a modular brick is employed, having an actual size of 2 ⅙"x3 ½"x7 ½". Using a ½" mortar joint, three courses of brick lay up perfectly to every two horizontal grid lines, and one length of brick likewise fits two vertical grid lines. The same holds true of the other sizes of brick shown.

Exact heights of modular brick will vary, the difference being taken up in the bed joints. It is therefore important to know the exact size of brick. The following table lists standard joint widths.

Modular Face Brick

Three standard modular joint thicknesses are ¼", ⅜", and ½". The number of modular masonry units per square foot, of course, will remain constant for each nominal size, the difference being taken up in the joint.

Modular Brick Walls Without Headers

Nominal Size of Brick, Inches h x t x l		Brick per Sq. Ft. of Wall	Cu. Ft. of Mortar per 100 Sq Ft of Wall Joint Thickness		Cu. Ft. of Mortar Per 1000 Brick Joint Thickness		
		¼"	⅜"	½"	¼"	⅜"	½"
2 ⅔x4x8	6.750	3.81	5.47	6.95	5.65	8.10	10.30
3 ⅕x4x8	5.625	3.34	4.79	6.10	5.94	8.52	10.84
4x4x8	4.500	--	4.12	5.24	--	9.15	11.65
5 ⅓x4x8	3.375	--	3.44	4.34	--	10.19	12.87
2x4x12	6.000	--	6.43	8.20	--	10.72	13.67
2 ⅔x4x12	4.500	3.52	5.06	6.46	7.82	11.24	14.35
3x4x12	4.000	--	4.60	5.87	--	11.51	14.68
3 ⅕x4x12	3.750	3.04	4.37	5.58	8.11	11.66	14.89
4x4x12	3.000	2.56	3.69	4.71	8.54	12.29	15.70
5 ⅓x6x12	2.250	--	3.00	3.84	--	13.34	17.05
2 ⅔x6x12	4.500	--	7.85	10.15	--	17.45	22.55
3 ⅕x6x12	3.750	--	6.79	8.77	--	18.10	23.39
4x6x12	3.000	--	5.72	7.40	--	19.07	24.67

Non-Modular Brick and Mortar Required for Single Wythe Walls in Running Bond (no allowances for breakage or waste)

Size of Brick, Inches t x h x l	3/8" Joints No. of Brick per 100 SF	Cu. Ft. Mortar per 100 SF	Cu. Ft. Mortar per 1000 Brick	1/2" Joints No. of Brick per 100 SF	CF Mortar per 100 SF	CF Mortar per 1000 Brick
2 3/4 x2 3/4x9 3/4	455	3.2	7.1	432	4.5	10.4
2 5/8x2 3/4x8 3/4	504	3.4	6.8	470	4.1	8.7
3 3/4x2 1/4x8	655	5.8	8.8	616	7.2	11.7
3 3/4x2 3/4x8	551	5.0	9.1	522	6.4	12.2

The above quantities include mortar for bed and vertical joints only. The following table gives allowances for backing mortar or collar joints.

Cubic Feet of Mortar Per 100 Sq Ft of Wall

1/4-in. Joint	3/8-in. Joint	1/2-in. Joint
2.08	3.13	4.17

Note: Cubic feet per 1000 units =

$$\frac{10 \times \text{cubic feet per 100 sq. ft. of wall}}{\text{number of units per sq. ft. of wall}}$$

Material Quantities per Cubic Foot of Mortar
Quantities by Volume

Material	Mortar Type and Proportions by Volume M 1:1/4:3	S 1:1/2:4 1/2	N 1:1:6	O 1:2:9
Cement	0.333	0.222	0.167	0.111
Lime	0.083	0.111	0.167	0.222
Sand	1.000	1.000	1.000	1.000

Material Quantities per Cubic Foot of Mortar
Quantities by Weight

Material	Mortar Type and Proportions by Volume M	S	N	O
Cement	31.33	20.89	15.67	10.44
Lime	3.33	4.44	6.67	8.89
Sand	80.00	80.00	80.00	80.00

Material Costs of Brick

Brick costs will vary widely around the country and within color ranges:

a) *Common Brick:* Standard modular (4x2 1/4x7 5/8) and standard non-modular (4x2 1/4x8) average between $240.00 and $350.00 per thousand.

b) *Jumbo Brick:* 4x2 3/4x8 averages between $290.00 and $390.00 per thousand.

c) *Utility Brick:* 4x3 5/8x11 5/8 averages between $600.00 and $1,100.00 per thousand.

d) *Economy Brick:* 4x7 5/8x7 5/8 averages between $420.00 and $700.00 per thousand.

e) *Paving Brick:* 4x2 1/4x8 or 4x3 5/8x7 5/8 averages between $315.00 and $460.00 per thousand.

f) Glazed Brick: 4x2 3/4x8 averages between $750.00 and $820.00 per thousand.

In general red tones are cheapest, earth tones fall in the middle range and whites and grays are the most expensive except for glazed units which run from $315.00 up in the more brilliant shades.

Brick Cavity Walls

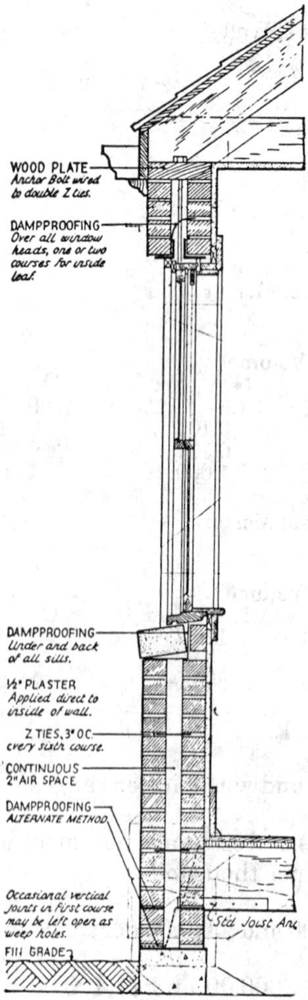

Brick cavity walls are distinguished from all other forms of hollow masonry by a continuous, vertical and horizontal air space, bridged by masonry ties. The primary purpose of this space is to provide an absolute barrier against the penetration of moisture to the inner side of the wall, including at heads and jambs of openings, joist bearing points, etc., assuring trouble-free masonry and permitting direct plastering without furring. It also increases resistance to heat-flow to about the same extent as the air space created by ordinary furring.

Since this construction affords no supporting material through which moisture penetrating the outer 4" wythe of the wall can soak to the inner wythe, it helps to insure that the inner wythe will remain dry, provided adequate precautions are taken to intercept and deflect moisture running down the inner side of the outer wythe at the heads of all openings, and provided the bottom of the wall is equipped with a suitable flashing to prevent absorption of soil water.

In place of the usual header courses, the inner and outer wythes of the cavity walls are connected by rust-proof metal ties--usually 3/16" diameter, square-end, Z bars--placed in every sixth course and 3'-0" on center horizontally.

Labor Laying Building Brick

There are so many factors entering into the labor cost of laying building brick that an explanation is necessary to use the following data intelligently. There are no set quantities of brick that a mason will lay per hour or day. With ten masons laying brick on the same wall and on exactly the same class of work, no two of them will lay the same number of brick. Averages must be taken.

Solid brick walls are used much less frequently now than in the past, and where brick is used as a facing, it is most often backed up with cement blocks. This is due to the lower labor costs in handling and setting the larger units of cement blocks in comparison with the smaller units of brick, as one 8"x8"x16" cement block usually displaces 12 building brick in the wall.

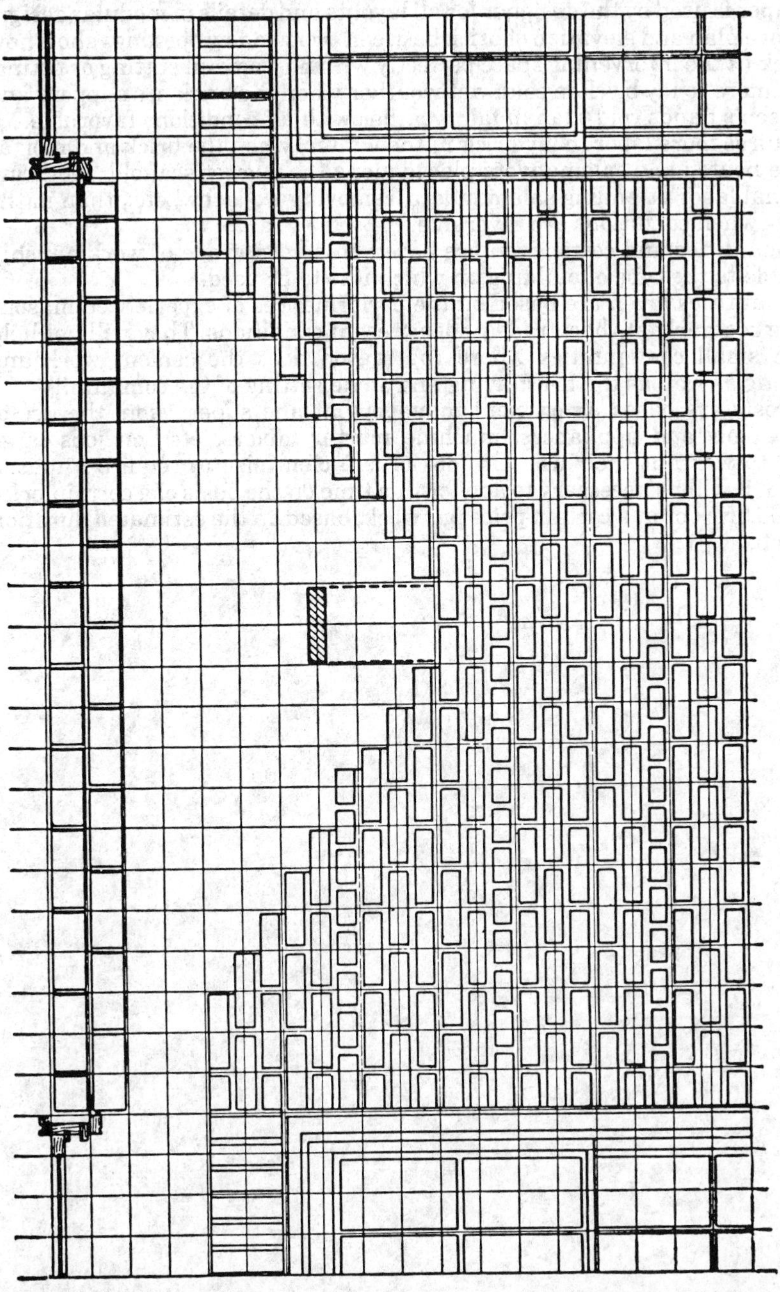

Grid Paper Is Used by the Designer for All Layouts and Details in Modular Design. The Above Plan and Elevation of Brick between Two Window Openings Show How the Brick Fit the Intervening Space Perfectly with No Wasteful Cutting or Fitting.

The class of work also has a great deal to do with the variation in labor costs. It costs much more to lay brick in an 8" wall with struck joints than in a 16" wall having cut joints on one side of the wall and struck joints on the other. Also, it costs more to lay brick in portland cement mortar than in lime mortar, because the lime mortar spreads far more easily than cement mortar.

Grid paper is used by the designer for all layouts and details in modular design. The above plan and elevation of brick between two window openings show how the brick fit the intervening space perfectly with no wasteful cutting or fitting. It costs more to lay brick in cold, wet weather when the brick are slippery and the mason's hands cold than in fair weather with all conditions favorable.

It requires more labor to lay brick in the winter, when the brick and mortar must be heated and salamanders placed along the floor or scaffold, requiring additional labor attending salamanders, removing ice, snow, etc., than in the summer when conditions are favorable.

The quantities and costs given are based on a good grade of workmanship applicable to the type of building on which it is to be used.

The quantities given are based on the performances of experienced masons working regularly at their trade under present conditions. They will not hold good in small communities where the mason does the cement work and plastering as well as the bricklaying and stone setting of the community.

The costs are based on efficient management and supervision, the use of modern tools and appliances, machine mortar mixers, etc., on jobs large enough to warrant their use. On jobs of sufficient size to require a mason foreman, it will be necessary to add for this time on the basis of a certain price per 1,000 brick or at a certain price per week, based on the estimated duration of the job.

Labor Production For Laying Building Brick

Class of Work	Mortar Joints Style	Kind of Mortar	Average Number Brick Laid per 8-hr. Day	Aver. Hrs. 1,000 Brick: Mason Hours	Aver. Hrs. 1,000 Brick: Labor Hours	Aver. Hrs. 1,000 Brick: Hoist Engr.*
8" walls, 1-story bungalows, garages,	Cut	Lime	750–800	10.5	9.0	....
two-flat buildings, residences, etc.	Struck	Lime	700–750	11.0	9.0	....
12" walls, ordinary construction, apart-	Cut	Lime	950–1,050	8.0	8.0	....
ment buildings, houses, garages, fac-	Struck	Lime	925–975	8.5	8.0	....
tories, stores, store and apartment	Cut	Cement	825–900	9.3	8.0	....
buildings, schools, etc.	Struck	Cement	750–850	10.0	8.0	....
Hodding brick to second story, add	....	....	....	....	2.0	....
16" walls, heavy warehouse, factory and	Cut	Lime	1,100–1,225	7.0	8.0	....
industrial work. Straight walls	Struck	Lime	1,000–1,125	7.5	8.0	....
	Cut	Cement	950–1,050	8.0	8.0	....
	Struck	Cement	900–1,000	8.4	8.0	....
Backing-up face brick, cut stone, terra cotta on ordinary wall bearing buildings, figure same as given above for 8" and 12" walls	....	....	....	....	....	

WINTER WORK. Heating brick, mortar, attending salamanders, removing snow and ice, add extra for this work.

*HOISTING. Add about ¼-hr. hoisting engineer time per 1,000 for reasonable size jobs using 35,000 or more brick per day; ½-hr. per 1,000 for jobs using 15,000 to 20,000 brick per day; and ⅓-hr. per 1,000 for jobs using 25,000 to 30,000 brick per day.

If automatic hoists are used, omit all time for hoisting engineer.

Labor Production For Laying Building Brick—Cont.

Class of Work	Mortar Joints Style	Kind of Mortar	Average Number Brick Laid per 8-hr. Day	Aver. Hrs. 1,000 Brick Mason Hours	Labor Hours	Hoist Engr.*
Backing-up face brick, cut stone, terra cotta on steel or concrete skeleton frame buildings. First Grade workmanship. Walls 8" to 12" thick.	Cut	Lime	825–900	9.3	9.0	
	Struck	Lime	750–850	10.0	9.0	
	Cut	Cement	725–800	10.5	9.0	
	Struck	Cement	700–775	10.8	9.0	
Public buildings, First Grade workmanship, Schools, college and university bldgs. Courthouses, state capitols, public libraries, etc. 12" to 20" walls.	Cut	Lime	850–950	9.0	8.0	
	Struck	Lime	800–900	9.4	8.0	
	Cut	Cement	750–850	10.0	8.0	
	Struck	Cement	700–800	10.6	8.0	
Powerhouses and other structures, having high walls 12" to 20" thick, without intermediate floors.	Cut	Lime	1,100–1,200	7.0	9.0	
	Struck	Lime	1,000–1,100	7.6	9.0	
	Cut	Cement	925–1,050	8.0	9.0	
	Struck	Cement	875–950	8.8	9.0	
Shoved joints. Brick laid in full shoved joints with all vertical joints slushed full of mortar, add to all of the above				1.5		
Basement foundation walls, paving brick.	Cut	Cement	900–1,000	8.5	8.5	
	Struck	Cement	850–950	9.0	9.0	

WINTER WORK. Heating brick, mortar, attending salamanders, removing snow and ice, add extra for this work.

*HOISTING. Add about ¼-hr. hoisting engineer time per 1,000 for reasonable size jobs using 35,000 or more brick per day; ½-hr. per 1,000 for jobs using 15,000 to 20,000 brick per day; and ⅓-hr. per 1,000 for jobs using 25,000 to 30,000 brick per day.

If automatic hoists are used, omit all time for hoisting engineer.

Labor Production For Laying Building Brick—Cont.

Class of Work	Mortar Joints Style	Kind of Mortar	Average Number Brick Laid per 8-hr. Day	Aver. Hrs. 1,000 Brick: Mason Hours	Aver. Hrs. 1,000 Brick: Labor Hours	Aver. Hrs. 1,000 Brick: Hoist Engr.*
Building brick foundation walls 8" to 12" thick, ordinary workmanship.	Cut	Lime	1,000–1,100	7.6	8.0	
	Struck	Lime	900–975	8.5	8.0	
	Cut	Cement	900–975	8.5	8.0	
	Struck	Cement	825–900	9.3	8.0	
Chimneys and stacks, building brick, 1'-4" to 2'-0" sq. 15'-0" above roof, 4" to 8" walls.	Struck	Lime	500–550	16.0	16.0	
Hodding brick, extra.					3.0	
Large chimney and stacks, 3'-0" to 4'-0" Sq. 15' to 30' high above roof, 8" to 12" walls.	Struck	Lime	550–600	14.0	16.0	
Large brick stacks 100' to 150' high. walls from 1'-8" at base to 12" top. Inside put-log scaffold. Outside scaffold extra.	Struck	Lime	650–750	11.5	14.0	
Bricking-in boilers, fire-boxes, etc.	Struck		600–650	12.5	10.0	

WINTER WORK. Heating brick, mortar, attending salamanders, removing snow and ice, add extra for this work.

*HOISTING. Add about ¼-hr. hoisting engineer time per 1,000 for reasonable size jobs using 35,000 or more brick per day; ½-hr. per 1,000 for jobs using 15,000 to 20,000 brick per day; and ⅓-hr. per 1,000 for jobs using 25,000 to 30,000 brick per day.

If automatic hoists are used, omit all time for hoisting engineer.

Labor Laying Face Brick

The labor cost of laying face brick will vary with the grade of workmanship, style of bond in which the brick are laid, width and kind of mortar joint, type of building, whether long, straight walls or walls cut up with pilasters, door and window openings, and last but not least, the ability of the foreman in charge.

Common bond is the most economical method of laying face brick. Other styles, such as English, Flemish Dutch and Garden Wall bond, require more labor on account of the various patterns and the additional cutting required. This is especially so on First Grade work where all vertical mortar joints must be plumb.

For instance, when laying Dutch, English and Flemish brick bonds, and the many variations of same, it is necessary to do a great deal of cutting, especially where half brick are used for headers, and it requires just as much time to cut the brick, spread mortar to lay one-half brick as for a full brick. The labor cost of laying 1,000 face brick in Dutch, English, or Flemish bond will be more if half-brick are used for headers than where full brick are used, because it will take just as much time to cut, spread mortar bed, apply mortar to brick, lay same and cut the joints whether a full or half brick is used.

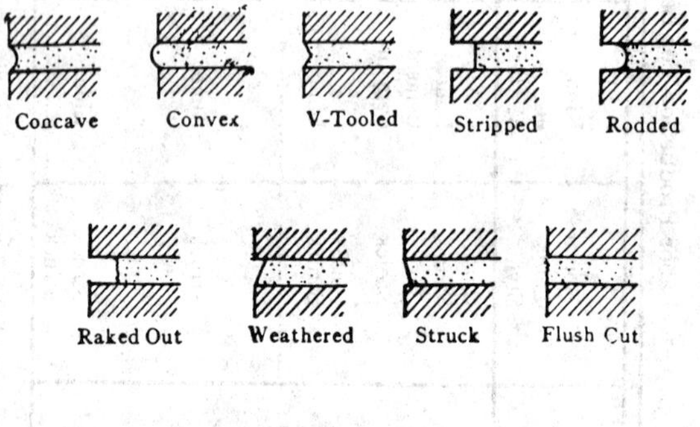

Styles of Mortar Joints

Flush Cut Joint. The style of mortar joints has considerable bearing on the labor costs. The *flush cut* joint, sometimes referred to as the *rough cut*, is the most economical, because the bricklayer cuts off the excess mortar with his trowel and the joint is finished. This produces an uncompacted joint with a small hairline crack where the mortar is pulled away from the brick by the cutting action. This joint is not always watertight.

Concave and V-Shaped Joints. These joints are normally kept quite small and are formed by the use of a steel jointing tool. These joints are very effective in resisting rain penetration and are recommended for use in areas subjected to heavy rains and high winds.

Weathered Joint. This joint requires care, as it must be worked from below. However, it is the best of the troweled joints, because it is compacted and sheds water readily.

Struck Joint. This is a common joint in ordinary brickwork. American mechanics often work from the inside of the wall, and this is an easy joint to strike with a trowel. Some compaction occurs, but the small ledge does not shed water readily, resulting in a less watertight joint than those discussed above.

Raked Joint. Made by removing the surface of the mortar while it is still soft. While the joint might be compacted, it is difficult to make weathertight and is not recommended where heavy rain, high wind or freezing is likely to occur. This joint produces marked shadows and tends to darken the overall appearance of the wall.

Grapevine Joint. This is a concave joint with a drip slot placed horizontal into the mortar, approximately midway between bricks. This joint is very effective in resisting rain penetration.

For concave, V-tooled, weathered and struck joints, it is necessary to cut off the excess mortar with the trowel and then another operation is required, using a trowel for the weathered or struck joints or a jointing tool for concave and V-tooled. A bricklayer must remove the mortar 1/4" to 3/8" back from the face of the brick, and for a rodded joint, the bricklayer must remove the excess mortar and then strike the joint with a jointer. These operations all require time and cost money.

The stripped joint is the most expensive of any illustrated, as it is necessary to place narrow wood strips in the mortar joints between each course of brick, and these strips must remain until the mortar has set, so it will not squeeze out toward the face of the brick.

If face brick are laid with 7/8" or 1" bed mortar joints, the work proceeds very slowly and requires a stiff mortar, because the weight of the top brick courses squeeze the mortar out of the lower mortar joints, producing an uneven appearance. For this reason the work cannot be carried up many courses at a time. Where the joints are struck, it is necessary for the mortar to take its initial set before the joints can be struck to produce a workmanlike job. Brick work of this kind is very expensive.

The quantities and costs given on the following pages are intended to cover all kinds of face brick work, whether laid in veneer in frame buildings, in solid brick or brick and tile walls.

Labor Laying Face Brick in English, Flemish, Dutch, Garden Wall, and Diamond Bonds

Face brick laid in any of the above bonds with the courses consisting of various arrangements of headers and stretchers requires considerable extra labor cutting brick on account of the large number of "bats" or headers required for each course. Ordinarily, a full header course extends through the wall every 5th or 6th course, and the other headers are only half brick. On First Grade workmanship, it is necessary to plumb all vertical mortar joints in order that the various patterns will work out accurately.

Labor Laying 1,000 Standard Size (8"x2¼"x3¾") or Modular Size (7⅝"x2¼"x3⅝") Face Brick in Common Bond

2:1:9 Lime-Cement Mortar				Ordinary Workmanship	Cement Mortar			
Mason Hours	Labor Hours	Hoist Engr.*	Number Laid per 8-Hr. Day	Style of Mortar Joint	Number Laid per 8-hr. Day	Mason Hours	Labor Hours	Hoist Engr.*
16.0	12	0.5	475-525	Flush Cut	425-475	17.8	13	0.5
17.8	12	0.5	425-475	V-Tooled or Concave	400-450	19.0	13	0.5
18.0	12	0.5	420-450	Struck or Weathered	390-435	19.5	13	0.5
17.8	12	0.5	425-475	Raked Out	400-450	19.0	13	0.5
19.0	12	0.5	400-450	Rodded	375-425	20.0	13	0.5
26.7	15	0.5	275-325	Stripped	275-300	27.8	16	0.5
				First Grade Workmanship				
19.0	13	0.5	400-450	Flush Cut	375-425	20.0	14	0.5
20.0	13	0.5	375-425	V-Tooled or Concave	360-400	21.0	14	0.5
21.0	13	0.5	360-400	Struck or Weathered	350-390	21.5	14	0.5
20.0	13	0.5	375-425	Raked Out	360-400	21.0	14	0.5
23.0	13	0.5	325-360	Rodded	300-340	25.0	14	0.5
32.0	16	0.5	225-275	Stripped	200-240	36.5	18	0.5
32.0	16	0.5	225-275	⅞" to 1" Flush Cut	200-240	36.5	18	0.5
36.0	16	0.5	200-250	⅞" to 1" Struck	190-230	38.0	18	0.5

*Add Hoisting engineer's time only if required.

Labor Laying 1,000 Standard Size (8"x2¼"x3¾") or Modular Size (7⅝"x2¼"x3⅝") Face Brick in English, Flemish**, Dutch and Garden Wall Bond, Laid with Full Headers Every 6th Course**

2:1:9 Lime-Cement Mortar				Ordinary Workmanship	Cement Mortar			
Mason Hours	Labor Hours	Hoist Engr.*	Number Laid per 8-Hr. Day	Style of Mortar Joint	Number Laid per 8-hr. Day	Mason Hours	Labor Hours	Hoist Engr.*
20.0	12	0.5	380-420	Flush Cut	360-400	21.0	13	0.5
21.3	12	0.5	350-400	V-Tooled or Concave	330-380	22.5	13	0.5
22.5	12	0.5	335-375	Struck or Weathered	320-360	23.5	13	0.5
21.3	12	0.5	350-400	Raked Out	330-380	22.5	13	0.5
22.5	12	0.5	335-375	Rodded	320-360	23.5	13	0.5
29.0	15	0.5	250-300	Stripped	240-285	31.0	15	0.5
				First Grade Workmanship				
22.5	13	0.5	335-375	Flush Cut	320-355	23.7	14	0.5
24.0	13	0.5	315-350	V-Tooled or Concave	300-335	25.0	14	0.5
25.0	13	0.5	300-340	Struck or Weathered	285-325	26.5	14	0.5
24.0	13	0.5	315-350	Raked Out	300-335	25.0	14	0.5
26.7	13	0.5	275-325	Rodded	260-310	29.0	14	0.5
35.5	16	0.5	200-250	Stripped	190-240	37.0	18	0.5
35.5	16	0.5	200-250	⅞" to 1" Flush Cut	190-240	37.0	18	0.5
40.0	16	0.5	175-225	⅞" to 1" Struck	165-215	42.0	18	0.5

*Add hoisting engineer's time only if required.
**If laid with full headers every course or every 2nd course, increase quantities laid approximately 8 percent.

Labor Laying 1,000 English Size 8⅞"x2⅞"x4" Face Brick

2:1:9 Lime-Cement Mortar				Common Bond	Cement Mortar			
Mason Hours	Labor Hours	Hoist Engr.*	Number Laid per 8-Hr. Day	Style of Mortar Joint	Number Laid per 8-hr. Day	Mason Hours	Labor Hours	Hoist Engr.*
22.5	17	0.6	335-375	Flush Cut	320-360	23.5	18	0.6
23.5	17	0.6	325-360	V-Tooled or Concave	310-340	24.6	18	0.6
26.3	17	0.6	285-325	Struck or Weathered	275-315	27.0	18	0.6
23.5	17	0.6	325-360	Raked Out	310-340	24.6	18	0.6
28.5	17	0.6	260-300	Rodded	250-290	29.5	18	0.6
				English, Flemish or Dutch Bond				
24.7	17	0.6	300-350	Flush Cut	285-325	26.3	18	0.6
26.3	17	0.6	285-325	V-Tooled or Concave	275-310	27.0	18	0.6
28.5	17	0.6	260-300	Struck or Weathered	250-290	29.5	18	0.6
26.3	17	0.6	285-325	Raked Out	275-310	27.0	18	0.6
31.0	17	0.6	240-275	Rodded	230-265	32.2	18	0.6

*Add hoisting engineer's time only if required.

Labor Laying 1,000 Standard Size (8"x2¼"x3¾") Vitrified Face Brick

2:1:9 Lime-Cement Mortar				Common Bond	Cement Mortar			
Mason Hours	Labor Hours	Hoist Engr.*	Number Laid per 8-Hr. Day	Style of Mortar Joint	Number Laid per 8-hr. Day	Mason Hours	Labor Hours	Hoist Engr.*
31.0	17	0.6	235-275	Buttered and Struck	225-265	32.7	18	0.6
22.5	16	0.6	335-375	Flush Cut	320-360	23.5	17	0.6
23.5	16	0.6	320-360	V-Tooled or Concave	300-340	25.0	17	0.6
26.3	16	0.6	285-325	Struck or Weathered	270-310	27.5	17	0.6
23.5	16	0.6	260-300	Rodded	250-285	30.0	17	0.6
28.5	16	0.6	320-360	Racked Out	300-340	25.0	17	0.6
				English, Flemish or Dutch Bond**				
34.8	17	0.6	210-250	Buttered and Struck	200-240	36.4	17	0.6
25.0	16	0.6	300-340	Flush Cut	285-325	26.3	17	0.6
26.3	16	0.6	285-325	V-Tooled or Concave	275-310	27.3	17	0.6
28.5	16	0.6	260-300	Struck or Weathered	250-290	29.7	17	0.6
26.3	16	0.6	285-325	Raked Out	275-310	27.3	17	0.6
31.0	17	0.6	240-275	Rodded	230-265	32.7	17	0.6

*Add hoisting engineer's time only if required.
**Full headers every sixth course. If full headers every other course, increase quantities laid approximately 8 percent.

Roman Face Brick

Roman size face brick, 12"x1½"x4" were extensively used some years ago for facing many types of buildings. Generally, they were laid with thin "buttered" joints and most of the original brick were made by the dry-pressed method and were relatively expensive.

Within the past few years, the modular Roman brick was introduced having a nominal size of 12"x2"x4", including mortar joint, while the actual brick size is 11⅝"x1⅝"x3⅝", requring a ⅜" mortar joint to lay 2" high and 12" long. These brick are furnished with either standard brick textured faces or a split (rock-like) texture. The split face is generally preferred as it greatly accentuates the horizontal bonding effect unobtainable in any other type of construction. These units are now being produced in practically all sections of the United States and are in great demand, particularly for single story ranch homes, although they are also used for commercial structures, store fronts and for finished masonry interiors.

When laid with a ⅜" mortar joint, it requires 6 brick per sq. ft. of wall. Add for headers, breakage and waste.

When laid with a ⅜" mortar joint, it requires 10¾ cu. ft. of mortar per 1,000 brick, not including any back-up mortar or any allowance for waste.

Labor Laying 1,000 Roman Face Brick 3⅝"x1⅝"x11⅝" Face Brick

Common Bond					Flemish Bond			
Mason Hours	Labor Hours	Hoist Engr.*	Number Laid per 8-Hr. Day	Style of Mortar Joint	Number Laid per 8-hr. Day	Mason Hours	Labor Hours	Hoist Engr.*
29.6	17	0.5	250-290	Buttered and Struck	210-240	35.5	17	0.5
20.5	17	0.5	370-410	Flush Cut	300-350	24.6	17	0.5
23.5	17	0.5	320-360	V-Tooled or Concave	265-300	28.3	17	0.5
25.0	17	0.5	300-340	Struck or Weathered	245-285	30.2	17	0.5
23.5	17	0.5	320-360	Raked Out	265-300	28.3	17	0.5
26.7	17	0.5	280-320	Rodded	230-270	32.0	17	0.5

*Add hoisting engineer's time only if required.

Using the labor laying brick in Common Bond as a base of 100, it will require approximately the following additional time to lay face brick in Dutch, English and the different variations of Flemish bond, where half brick are used as headers instead of full brick.

Kind of Bond	Percent
Common Bond using Full Headers	100
Dutch Bond using Full Headers every other Course	120
Dutch Bond using Full Headers every Sixth Course	129
English Bond using Full Headers every other Course	120
English Bond using Full Headers every Sixth Course	129
Flemish Bond using Full Headers every Course	120
Flemish Bond using Full Headers every Sixth Course	126
Double Flemish Bond using Full Headers every Course	100
Double Flemish Bond using Full Headers every Sixth Course	116
3-Stretcher Flemish Bond using Full Headers every Other Course	107
3-Stretcher Flemish Bond using Full Headers every Sixth Course	112
4-Stretcher Flemish Bond using Full Headers every Other Course	106
4-Stretcher Flemish Bond using Full Headers every Sixth Course	109

Double Size Building Brick

Double size building brick (3 3/4"x5"x8"), are used in different parts of the country. Their use effects a saving in both mortar and labor over standard size building brick (3 3/4"x2 1/4"x8").

Each brick contains 40 sq. in. on the face and when laid with a 1/2" bed or horizontal joint and a 1/4" vertical or end joint, covers 45.375 or 45 3/8 sq. in. and requires 3.2 brick per sq. ft. of 4" wall; 6.4 brick per sq. ft. of 8" wall and 9.6 brick per sq. ft. of 12" or 13" wall, not including any allowance for waste.

Mortar Required for Double Size Building Brick
Based on 1/2" bed or horizontal joints and 1/4" end or vertical joints.

Mortar Required per 1,000 Double Brick

4" Walls*	8" Walls	12" Walls
12 3/4 cu. ft.	19 1/4 cu. ft.	21 1/4 cu. ft.

Mortar Required per 100 Sq. Ft. of Wall

4" Walls*	8" Walls	12" Walls
4 cu. ft.	12 1/4 cu. ft.	20 1/2 cu. ft.

*Freestanding Walls.

No Allowance For Waste Included In Above Quantities.

Labor Laying Double Size Building Brick. A mason will lay about 60% as many double brick (3 3/4"x5"x8") as standard size building brick (3 3/4"x2 1/4"x8"), i.e., if a mason lays 800 building brick on a certain class of work, he would lay 480 double size building brick on the same class of work. In other words, if it required 10 hrs. mason time per 1,000 building brick, it would require 1 2/3 as many hours per 1,000 double brick, i.e., 10 x 1.667 = 16.67 or 16 2/3 hrs. mason time per 1,000 double brick.

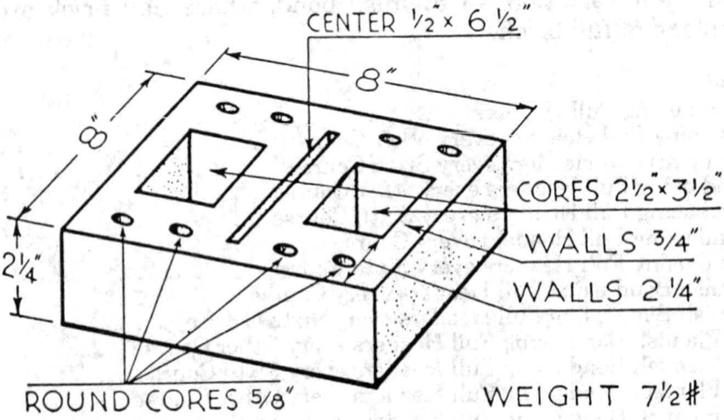

Twin Brick. Size 8"x8"x2 1/4"

Twin Brick

Twin brick is a double size brick that contains the face brick and a 4" backing brick in one unit, size 8"x8"x2 1/4". It is made with a hollow core so that the twin brick weighs only 7 1/2 lbs. and can be picked up and laid in a one-hand operation. It is designed so the wall has no continuous mortar joint from outer to inner wall faces, helping to avoid moisture penetration.

It is intended primarily for residential work having 8" walls that are to be furred and plastered. Extra face brick must be used at openings and corners to insure proper bond.

It requires about 25 cu. ft. of mortar per 1,000 units.

On four jobs on which these brick were used, requiring 50,000 units, a mason laid 60 twin brick an hour or 480 per 8 hr. day, at the following labor cost per 1,000 units:

	Hours	Rate	Total	Rate	Total
Mason	17	$....	$....	$25.62	$435.54
Labor	11.5		....	19.13	220.00
Cost per 1,000 units			$....		$655.54
Cost per brick					$.66

Skintled Brickwork

Skintled brickwork is used particularly in storefront work and interiors of the "Olde English Pub" variety.

Skintled brickwork consists of laying common brick irregularly, with random projecting brick at intervals and using clinkers or irregular and other off-colored brick to produce a variegated wall.

These brick are laid in a number of different ways to produce the desired effect. In some of them the mortar projects beyond the face of the brick just as it has been squeezed out of the joint; in others the joints are cut flush or raked out to produce shadows, etc.

The labor cost will vary with the design and care used in laying up the brick. When laying building brick without regard to color of wall and with uncut mortar joints or joints cut flush with the face of the wall, a mason should lay 350 to 400 brick per 8-hr. day, at the following labor cost per 1,000:

	Hours	Rate	Total	Rate	Total
Mason	21	$....	$....	$25.62	$538.02
Labor	15	....	....	19.13	286.95
Cost per 1,000 brick			$....		$824.97
Cost per brick					$.82

If brick joints are raked out, add 2.5 hrs. mason time per 1,000 brick. On skintled brickwork using clinkers and other dark or irregularly shaped brick to produce a textured wall, more care is required in laying the brick, working out the designs and in raking out the joints to produce shadows, etc., and a mason should lay 250 to 300 brick per 8-hr. day, at the following labor cost per 1,000:

	Hours	Rate	Total	Rate	Total
Mason	29	$....	$....	$25.62	$742.98
Labor	20	....	....	19.13	382.60
Cost per 1,000 brick			$....		$1,125.58
Cost per brick					$1.13

English Size Face Brick

English size face brick, 8 7/8"x2 7/8"x4" in size, are seldom used in this country any more and while the size is not standard, they are still furnished by a few manufacturers.

Inasmuch as they are practically half again as large on the face as the standard brick, a mason is unable to lay as many as where the standard size brick are used. They also require more labor time handling, hoisting and wheeling to the masons.

Number of English Size, 8 7/8"x2 7/8"x4", Face Brick Required
Per Sq. Ft. of Wall

All Vertical Mortar Joints Figured 1/4" Wide
Width of Horizontal or "Bed" Mortar Joints in Inches

Dry	1/8"	1/4"	3/8"	1/2"	5/8"	3/4"
5.65	5.34	5.06	4.88	4.67	4.52	4.37

Add for waste, headers, and the different brick bonds, as given under "Standard Brick".

Cu. Ft. of Mortar Required to Lay 1,000 English Size Face Brick
Based on 1/4" End Joints and "Bed" Joints as Given Below

1/8"	1/4"	3/8"	1/2"	5/8"	3/4"
5	9	11	14	17	20

Vitrified Face Brick

There are a number of vitrified face brick on the market that are so hard it is almost impossible to cut them. Because of their denseness and brittleness, they do not absorb water from the mortar, and for that reason, it requires more time for the mortar to take its initial set. It is impossible to lay up many courses of brick at one time, because the weight of the upper brick courses forces the mortar from the joints of the lower courses, causing an unsightly wall. It is also difficult to hold the brick in the wall and keep the wall plumb, because the brick slide out of place very easily.

Brick of this kind are ordinarily laid with "buttered" mortar joints or with 1/4" mortar joints struck flush. They are furnished in the standard size, 8"x2 1/4"x3 3/4" and require the same number of brick and the same amount of mortar as other standard size face brick.

Brick Floors and Steps

Face Brick Floors Laid in Basket Weave Pattern. Brick floors laid in square or basket weave pattern do not require any cutting and are much easier laid than herringbone designs.

On work of this kind, a mason should lay 180 to 225 brick per 8-hr. day, at the following labor cost per 1,000:

	Hours	Rate	Total	Rate	Total
Mason	40	$....	$....	$25.62	$1,024.80
Labor	20	....	....	19.13	382.60
Cost per 1,000 brick			$....		$1407.40
Cost per brick					$1.41

BRICK FLOOR PATTERNS

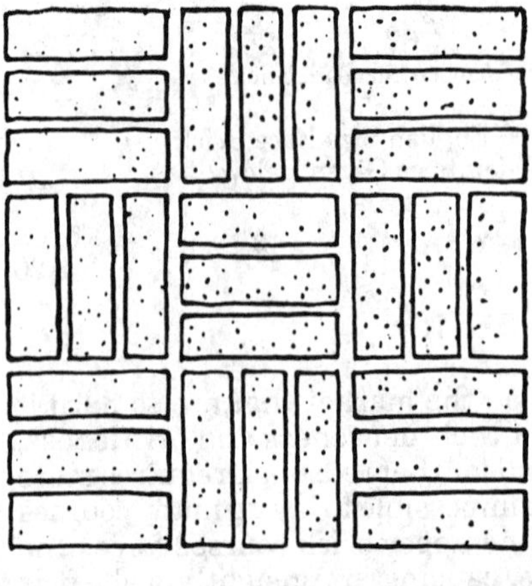

Brick Laid in Basket Weave Pattern.

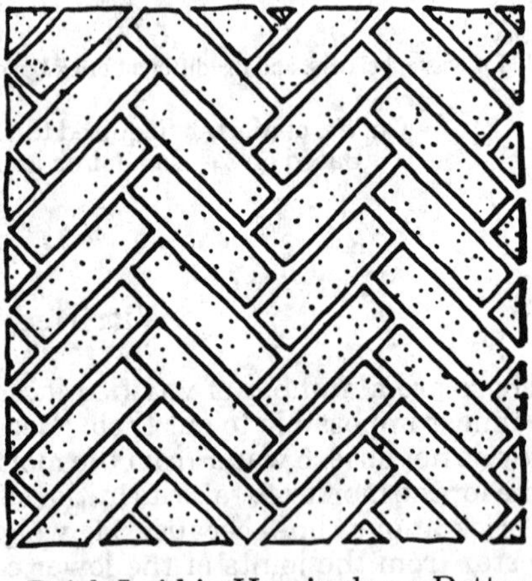

Brick Laid in Herringbone Pattern.

BRICK STAIR DETAILS

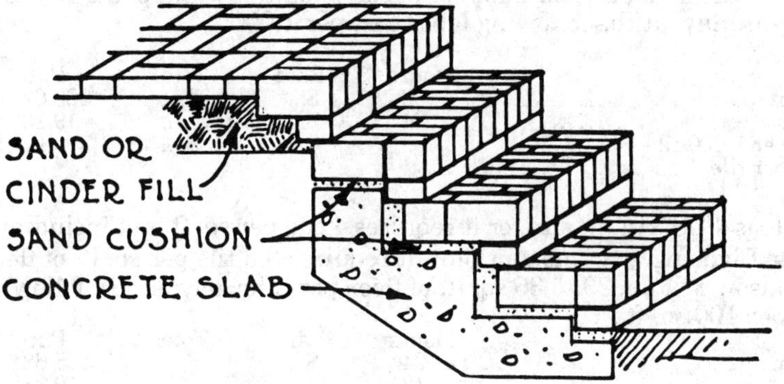

Brick Steps with End-Set Treads

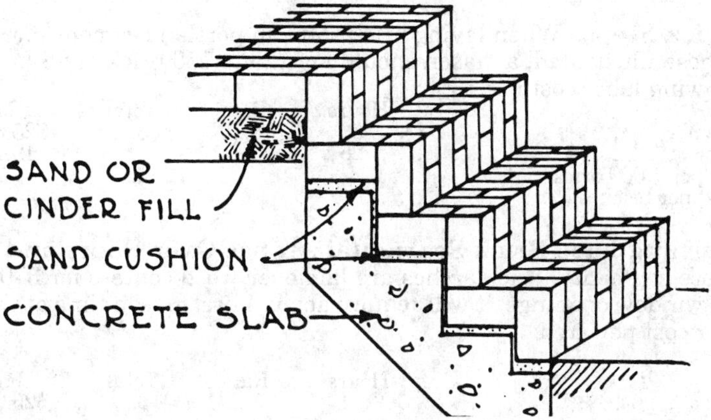

Brick Steps with Edge-Set Treads

When using brick for stairs, it is recommended that brick pavers be used instead of standard brick.

Face Brick Floors Laid in Herringbone Pattern. Brick floors are usually laid in portland cement mortar, and when laid in herringbone pattern require considerable time cutting and fitting the small pieces of brick at the edge or border of the pattern, as shown in the illustration.

A mason should lay 125 to 150 brick per 8-hr. day, at the following labor cost per 1,000:

	Hours	Rate	Total	Rate	Total
Mason	56	$....	$....	$25.62	$1,434.72
Labor	28	....	....	19.13	535.64
Cost per 1,000 brick			$....		$1,970.36
Cost per brick					1.97

If there is only one mason on the job, it will require a helper for each mason.

Promenade Tile Floor and Decks. When laying 4"x8"x½" promenade tile on porch floors and decks, with the tile laid in portland cement mortar and ½" mortar joints, struck flush, a mason should lay and point 100 to 125 tile per 8-hr. day, at the following labor cost per 1,000:

	Hours	Rate	Total	Rate	Total
Mason	68	$....	$....	$25.62	$1,742.16
Labor	34	....	....	19.13	650.42
Cost per 1,000 tile			$....		$2,392.58
Cost per tile					2.39

A tile is 4"x8"x½" in size, or it requires 4 tile per sq. ft. not including mortar joints. Figuring a ½" mortar joint, it requires 3.5 tile per sq. ft. of floor.

A mason will lay 30 to 36 sq. ft. of floor per 8-hr. day, at the following labor cost per 100 sq. ft.:

	Hours	Rate	Total	Rate	Total
Mason	24	$....	$....	$25.62	$614.88
Labor	12	....	....	19.13	229.56
Cost per 100 sq. ft			$....		$844.44
Cost per sq. ft			....		8.44

Brick Steps. When laying brick steps in portland cement mortar, similar to those illustrated, a mason should lay 125 to 150 brick per 8-hr. day, at the following labor cost per 1000:

	Hours	Rate	Total	Rate	Total
Mason	56	$....	$....	$25.62	$1,434.72
Labor	28	....	....	19.13	535.64
Cost per 1,000 brick			$....		$1,970.36
Cost per brick					1.97

Turning Face Brick Segmental Arches Over Doors and Windows. Where segmental brick arches are laid over wood centers for 3'-0" wide door and window openings, it will require about 4 hrs. mason time at the following labor cost per arch:

	Hours	Rate	Total	Rate	Total
Mason	4	$....	$....	$25.62	$102.48

For a 5'-0" arch, figure about 6.5 hrs. mason time, as follows:

	Hours	Rate	Total	Rate	Total
Mason	6.5	$....	$....	$25.62	$166.53

If the brick are cut or chipped to a radius, add cost of chipping as given below.

Laying Face Brick in Flat or "Jack" Arches. When laying face brick in flat or jack arches, a mason should complete one arch up to 3'-6" wide in about 5 hrs. at the following labor cost:

	Hours	Rate	Total	Rate	Total
Mason	5	$....	$....	$25.62	$128.10

For a 5'-0" jack arch, figure about 7 ½ hrs. mason time, as follows:

	Hours	Rate	Total	Rate	Total
Mason	7.5	$....	$....	$25.62	$192.15

The above costs do not include chipping or rubbing brick to a radius. If this is necessary, add cost of chipping as given below.

Chipping Brick. Where necessary to chip or cut brick for segmental or jack arches, a mason should chip 400 to 500 brick per 8-hr. day, depending upon the texture and density of the brick, at the following labor cost per 1,000:

	Hours	Rate	Total	Rate	Total
Mason	17.5	$....	$....	$25.62	$448.35

There is considerable lost motion in cutting brick, and unless there is a great deal of duplication in the cutting, where a mason can work continuously on a few special sizes, a mason will cut only 160 to 200 pieces per 8-hr. day at the following labor cost per 100 pieces:

	Hours	Rate	Total	Rate	Total
Mason	4.5	$....	$....	$25.62	$115.29
Labor	1.0	....	....	19.13	19.13
Cost per 100 cuts			$....		$134.42
Cost per cut			....		1.34

Fire Resistance

Fire Resistance Ratings. Tables and figures below list fire resistance ratings for various brick walls. Ratings listed are for load bearing walls tested with working loads of 160 psi of gross area, except the 4-in. wall shown was loaded to 92 psi.

The following table lists the fire resistance ratings for steel columns covered with brick:

Fireproofed Column Fire Ratings

Construction*	Ratings
Steel Columns, 6"x6" or larger	
3 ¾" brick with brick fill	4 hr.
2 ¼" brick with brick fill	1 hr.

*Column fire resistance varies with the cross-sectional area of solid material; the larger the area, the greater the fire resistance for a given thickness of protection around the structural steel. Column dimensions are outside dimensions. Smaller columns may require more cover to achieve equal ratings. For columns that are not square, protection should equal that of a square column having equal or lesser cross-sectional area. Thickness does not include plaster.

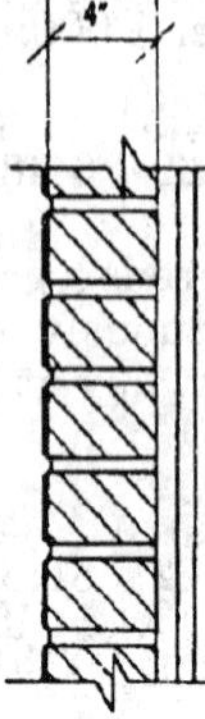

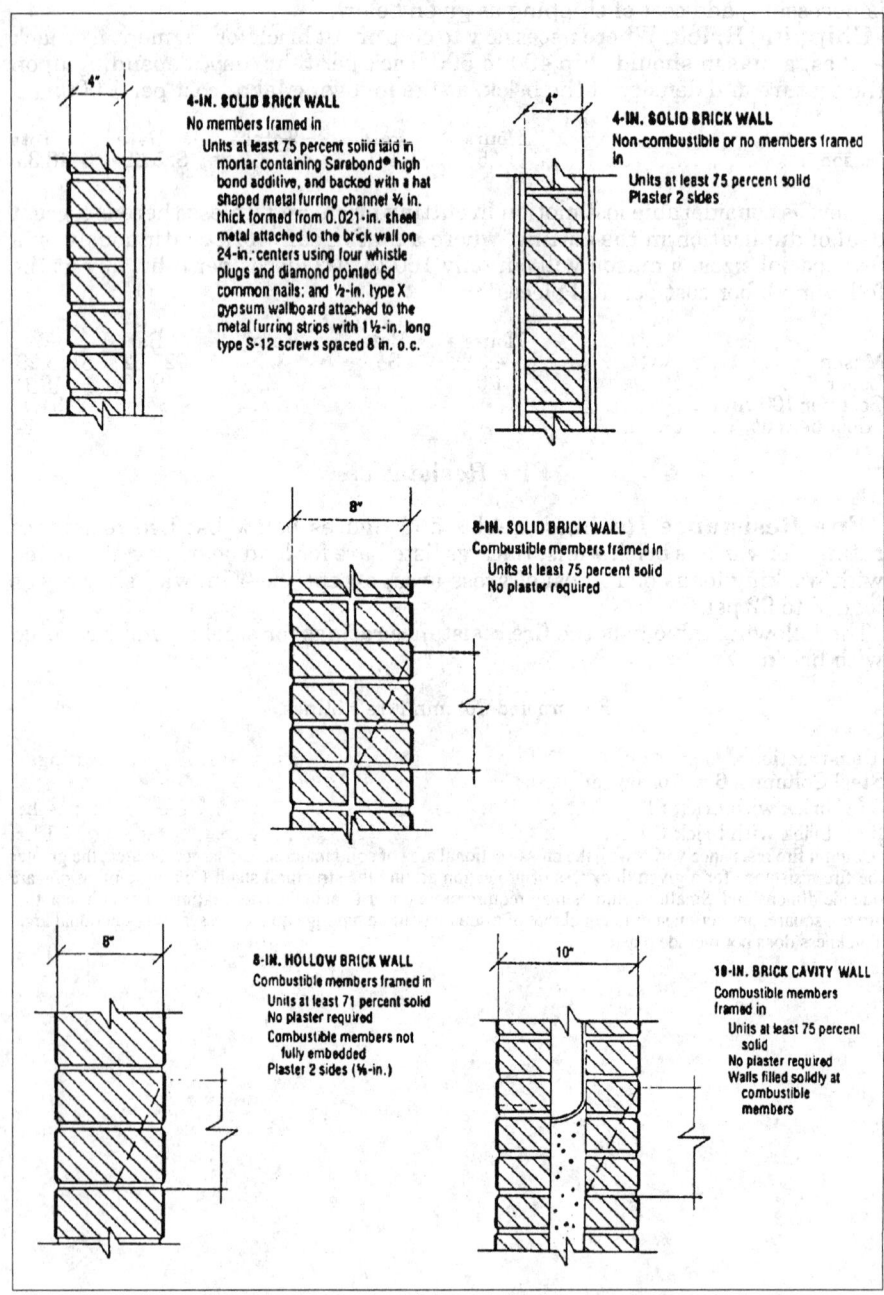

4-IN. SOLID BRICK WALL
No members framed in
Units at least 75 percent solid laid in mortar containing Sarabond® high bond additive, and backed with a hat shaped metal furring channel ¾ in. thick formed from 0.021-in. sheet metal attached to the brick wall on 24-in. centers using four whistle plugs and diamond pointed 6d common nails; and ½-in. type X gypsum wallboard attached to the metal furring strips with 1½-in. long type S-12 screws spaced 8 in. o.c.

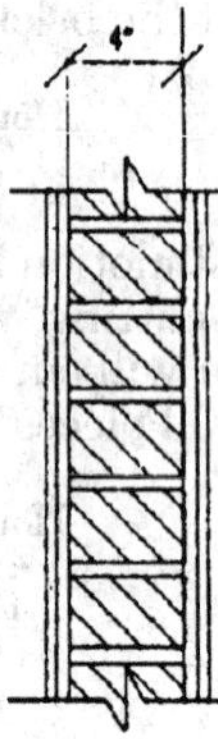

4-IN. SOLID BRICK WALL
Non-combustible or no members framed in
Units at least 75 percent solid
Plaster 2 sides

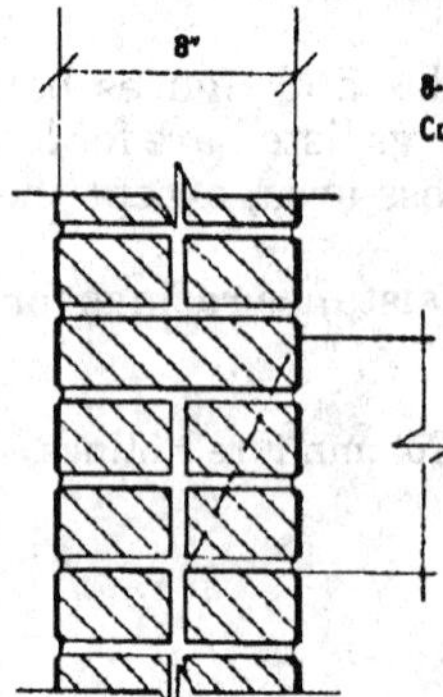

8-IN. SOLID BRICK WALL
Combustible members framed in
Units at least 75 percent solid
No plaster required

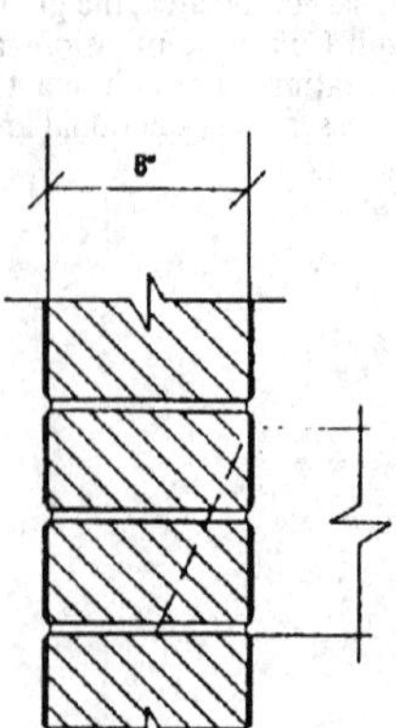

8-IN. HOLLOW BRICK WALL
Combustible members framed in
Units at least 71 percent solid
No plaster required
Combustible members not fully embedded
Plaster 2 sides (⅝-in.)

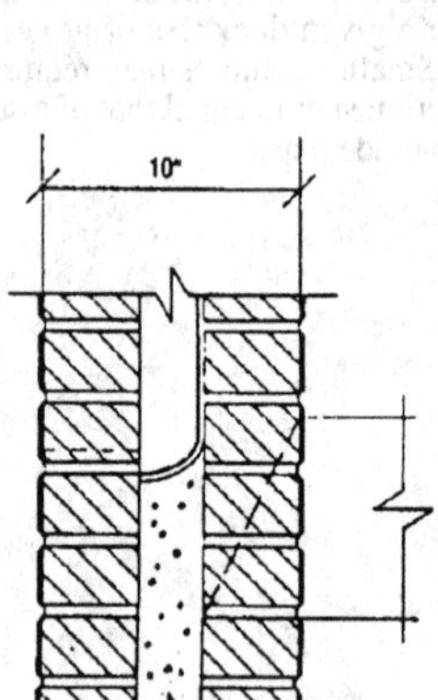

10-IN. BRICK CAVITY WALL
Combustible members framed in
Units at least 75 percent solid
No plaster required
Walls filled solidly at combustible members

Two-Hour Fire Ratings
Constructions Shown Are Loadbearing

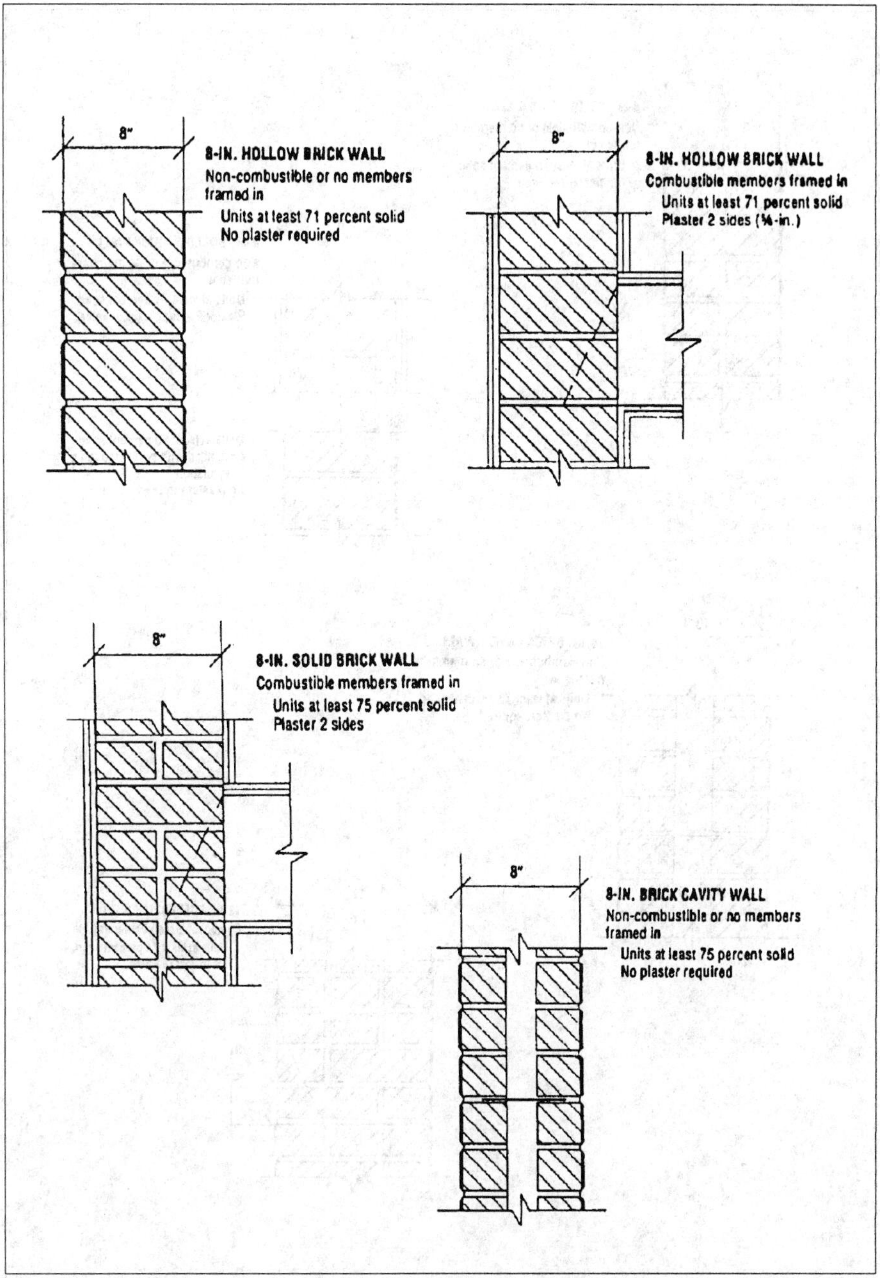

Three-Hour Fire Ratings
Constructions Shown Are Loadbearing

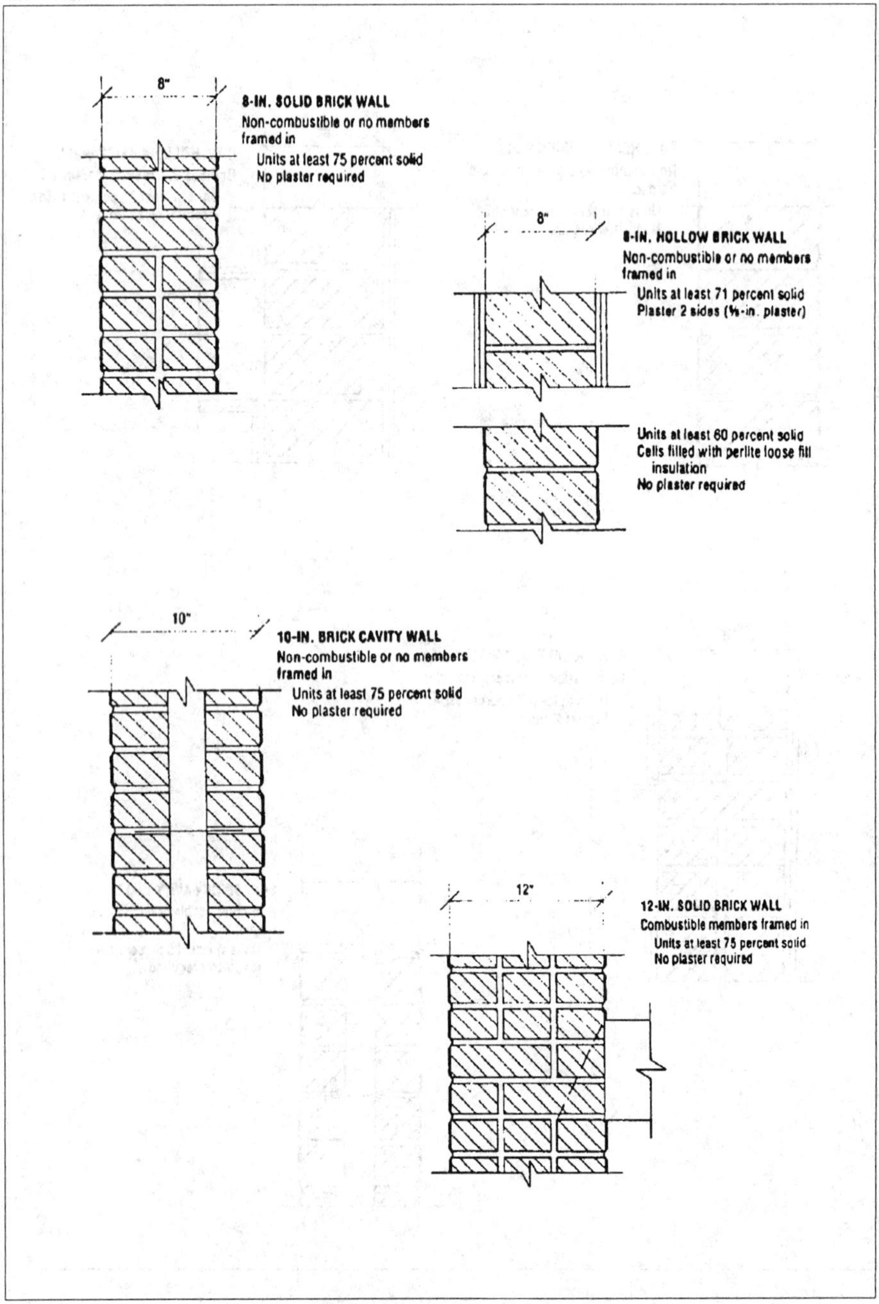
8"
8-IN. SOLID BRICK WALL
Non-combustible or no members framed in
Units at least 75 percent solid
No plaster required
8"
8-IN. HOLLOW BRICK WALL
Non-combustible or no members framed in
Units at least 71 percent solid
Plaster 2 sides (⅝-in. plaster)
Units at least 60 percent solid
Cells filled with perlite loose fill insulation
No plaster required
10"
10-IN. BRICK CAVITY WALL
Non-combustible or no members framed in
Units at least 75 percent solid
No plaster required
12"
12-IN. SOLID BRICK WALL
Combustible members framed in
Units at least 75 percent solid
No plaster required

Four-Hour Fire Ratings
Construction Shown Are Loadbearing

Cutting Masonry Products With Portable Masonry Saw

On practically every construction job there is a large amount of cutting necessary on masonry materials, and with the high cost of labor, this becomes an expensive item, especially when the cutting is done by hammer and chisel.

This applies particularly to the cross and diamond bonds in brick masonry, arch brick, radial brick, glazed tile, rotary kiln blocks, concrete blocks, natural stone, roofing tile, and conduits.

By using a masonry saw, this work can be done at a fraction of the cost of hand cutting plus the saving of materials.

Masonry saws are obtainable with 12", 14", 18" and 20" blade capacities, and are designed for both wet (dustless) and dry cutting. The advantage of wet cutting is in the confined areas where dust would be a problem. An exhaust assembly is available for use with dry cutting masonry saws. This assembly consists of a high velocity fan that draws the dust away from the machine and disposes of it through the nearest opening.

A smaller compact model masonry saw has been developed that is light enough for a man to carry and may be set up on the scaffolding, operated from a pickup truck or the back of a station wagon.

There is also a small, highly portable concrete saw that can do the work of many of the heavier models.

There are four types of blades that can be used on masonry saws: (1) wet-dry abrasive blades; (2) dry "break-resistant" abrasive blades; (3) wet "break-resistant" abrasive blades; (4) diamond blades. The wet abrasive and diamond blades should be used only with a masonry saw equipped for wet cutting. There are many specifications of each type of blade; it is very important that the correct specification be used for the particular material to be cut.

Compact Masonry Saw

Diamond blades for use on masonry saws vary in price from approximately $240.00 to $470.00 for 14" diameter blades and $425.00 to $905.00 for 18" diameter blades according to the specifications.

The cost of cutting masonry units varies greatly, depending upon the type of material, blade specifications and upon the operator.

Concrete products vary considerably in hardness, such as sand-gravel blocks compared with Haydite blocks. Therefore, blade wear and cutting time will vary accordingly. As an average, most 8"x8"x16" concrete blocks can be sliced in two in 14 to 16 seconds at a cost of less than one cent per unit.

Fire brick varies tremendously in hardness, as well as types. Ordinary dry press or silica can be cut in two very rapidly, while Super-duty refractories or Basic, Chrome, and Magnesite requires 15 to 18 seconds for each cut completely through, at a cost varying from 4 to 10 cents, depending upon actual density.

All the above costs and cutting speeds are based on cross cuts. Longitudinal cuts will approximate the same for equivalent area.

The following is an example of the approximate time required making cuts in various masonry products:

Material	Type of Cut	Length of Cut	Depth of Cut	Approximate Time Required
Acid Brick	Miter	4 ½"	2 ¼"	15 seconds
Concrete Block	Straight	8"	8"	19 seconds
Glazed Brick-Tile	Miter	6"	4"	15 seconds
Magnesite Brick	Miter	2 ¾"	4"	15 seconds
Rotary Kiln Block	Straight	8"	3"	11 seconds
Silica Brick	Straight	8"	2 ¼"	10 seconds
Roofing Tile	Miter	9"	2"	14 seconds
Clay Sewer Pipe	Cross	4" diam.	...	18 seconds
Concrete Cylinder	Cross	4" diam.	...	21 seconds

Cleaning Face Brick Work

The cost of cleaning face brick varies with the kind of brick. It is much easier to clean a smooth face vitrified brick than one having a rough texture. Where smooth face brick are used, and it is not necessary to do any pointing but merely wash the wall, the indiscriminate use of muriatic acid, or of the wrong proprietary compound, can cause unsightly, hard-to-remove stains. Also, chemical cleaning solutions are generally more effective when the outdoor temperature is 50° F or above.

An experienced mechanic should clean 5,000 to 5,500 brick (750 to 825 sq. ft. of wall) per 8-hr. day, at the following labor cost per 1,000 brick, or approximately 150 sq. ft. of wall:

	Hours	Rate	Total	Rate	Total
Mechanic	1.6	$....	$....	$25.62	$40.99
Helper	0.8	....	....	19.13	15.30
Cost per 1,000 brick			$....		$56.29
per sq. ft. wall			....		.38

The labor cost of cleaning and washing rough texture face brick will run higher than smooth brick on account of the cement and mortar getting into the surface of the brick.

When cleaning and washing rough textured brick, the same crew should clean and wash 3,500 to 4,000 brick (525 to 600 sq. ft. of wall) per 8-hr. day, at the following labor cost per 1,000 brick or per 150 sq. ft. of wall:

	Hour	Rate	Total	Rate	Total
Mechanic	2.2	$....	$....	$25.62	$56.36
Helper	1.1	....	....	19.13	21.04
Cost per 1,000 brick			$....		$77.40
per sq. ft. wall			....		.52

The labor cost of high pressure water cleaning face brick will vary considerably, depending on location and whether pointing and patching are required. When cleaning and washing textured brick, a crew can clean from 6,000 to 8,500 brick (900 to 1,275 sq. ft. of wall) per 8-hr. day at the following cost per 1,000 brick or per 150 sq. ft. of wall:

	Hour	Rate	Total	Rate	Total
Mechanic	0.55	$....	$....	$25.62	$14.09
Helper	0.55	....	....	19.13	10.52
Cost per 1,000 brick			$....		$24.61
per sq. ft. wall			....		.16

When estimating cleaning of masonry units, the following precautions apply, and costs must be added when applicable:

a) Saturation of the brick masonry wall surface with water before and after application of chemical or detergent cleaning solutions, add 0 to 10%.

b) Failure to properly use chemical cleaning solutions. This can result in recleaning or permanent staining.

c) Failure to protect windows, doors and trim. Cost will vary in accordance with the amount of material to be protected.

d) Costs of the chemical or detergents.

e) Classification and safe disposal of the chemicals.

f) Failure to protect adjacent property.

The following table is a guide for the estimator to determine the cleaning method(s) to be employed for various new masonry units.

Cleaning Guide for New Masonry

Brick Category	Cleaning Method
Red and red flashed	Bucket and brush hand cleaning High pressure water Sandblasting

Remarks: Proprietary compounds and emulsifying agents may be used. *Smooth texture:* Mortar stains and smears are generally easier to remove; less surface area exposed; easier to presoak and rinse; unbroken surface, thus more likely to display poor rinsing, acid stain, poor removal of mortar smears. *Rough texture:* Mortar and dirt tend to penetrate deep into textures; additional area for water and acid absorption; essential to use pressurized water during rinsing.

Brick Category	Cleaning Method
Red, heavy sand finish	Bucket and brush hand cleaning High pressure water

Remarks: Clean with plain water and scrub brush, or lightly applied high pressure and plain water. Excessive mortar stains may require use of cleaning solutions. Sandblasting is not recommended.

Brick Category	Cleaning Method
Light-colored units, white, tan, buff, gray, specks, pink, brown and black	Bucket and brush hand cleaning High pressure water Sandblasting

Remarks: *Do not use muriatic acid!* Clean with plain water, detergents, emulsifying agents or suitable proprietary compounds. Manganese colored brick units tend to react to muriatic acid solutions and stain. Light colored brick are more susceptible to "acid burn" and stains, compared to darker units.

Brick Category	Cleaning Method
Same as light-colored units, etc., plus sand finish	Bucket and brush hand cleaning High pressure water

Remarks: Lightly apply either method. (See notes for light-colored units, etc.) *Sandblasting is not recommended.*

Brick Category	Cleaning Method
Glazed Brick	Bucket and brush hand cleaning

Wiped glazed surface with soft cloth within a few minutes of laying units. Use soft sponge or brush plus ample water supply for final washing. Use detergents where necessary and acid solutions only for very difficult mortar stain. Do not use acid on salt glazed or metallic glazed brick. Do not use abrasive powders.

Brick Category	Cleaning Method
Colored mortars	Method generally controlled by the brick unit

Many manufacturers of colored mortars do not recommend chemical cleaning solutions. Most acids tend to bleach colored mortars. Mild detergent solutions are generally recommended.

Brick Fireplaces, Mantels, and Hearths

On account of the many sizes, shapes and styles of brick fireplaces, mantels and hearths, it is difficult to estimate the cost of the work at a certain price per 1,000 brick. The most satisfactory method is to figure a certain length of time to lay up each fireplace or mantel.

A mason should complete a plain brick fireplace 5'-0" to 6'-0" wide and 4'-0" to 5'-0" high, requiring 200 to 225 brick, not including hearth, in about 10 hrs. at the following labor cost:

	Hours	Rate	Total	Rate	Total
Mason	10	$....	$....	$25.62	$256.20
Labor	5	....	....	19.13	95.65
Cost per fireplace			$....		$351.85

On larger and more elaborate brick fireplaces and mantels having raked or rodded mortar joints, a mason should lay about 100 to 125 brick per 8-hr. day, at the following labor cost per 1,000:

	Hours	Rate	Total	Rate	Total
Mason	70	$....	$....	$25.62	$1,793.40
Labor	35	....	....	19.13	669.55
Cost per 1,000 brick			$....		$2,462.95

If there is only one mason working on the fireplace it will require one laborer with each mason.

Brick Fireplace Linings. A mason should complete the lining for a brick fireplace having an opening 3'-0" to 4'-0" wide, 2'-6" to 3'-0" high and 1'-4" to 1'-9" deep, in 6 to 8 hrs. all depending upon the class of work.

If the fireplace linings are laid in common bond and do not require an excessive amount of cutting and fitting, a mason should complete one fireplace lining in about 5 hrs. but where the back-hearths are of irregular shape and require considerable cutting and fitting at angles and at tops for dampers and smoke chambers, it will require about 8 hrs. mason time for each lining.

Fireplaces having an opening 3'-6" to 4'-0" wide, about 4'-0" high and 1'-9" deep, lined with firebrick laid in herringbone pattern, requiring cutting and fitting at all edges and corners, will require 20 to 24 hrs. mason time to line one fireplace.

Labor Laying Brick Hearths and Back-Hearths. If brick hearths and back-hearths are laid in square or basket weave pattern, a mason should complete one back-hearth in 3 to 4 hrs. and the front hearth will require the same length of time.

If laid in herringbone pattern with cut brick at all edges, figure 4 to 5 hrs. mason time for both hearth and back-hearth.

Figure about ½-hr. laborer time to each hour of mason time, except where only one mason is on the job when it will figure hour for hour.

Fire Brick Work

Fire brick and tile are furnished in innumerable sizes and shapes to meet the requirements of all classes of boilers, furnaces, stacks, etc. It is impossible to list them all, but the most common are listed below. Standard fire brick size is 8"x3 ½"x2 ¼" and will cost about $480.00 per 1,000 for low duty, $780.00 for high duty.

Fire Clay. When a close thin joint is desired, the clay should be soaked and mixed thin, and the brick should be dipped and rubbed. For work of this kind 300 lbs. of finely ground fire clay is required per 1,000 brick.

Sizes and Shapes of Standard Fire Brick
Number Required per Sq. Ft.

Name & Size	Laid Flat	Laid on Edge
8" Straight, 8"x3 ½"x2 ¼"	8.0	5.2
9" Straight, 9"x4 ½"x2 ½"	6.5	3.5
Small 9" Brick, 9"x3 ½"x2 ½"	6.5	4.5
Split Brick, 9"x4 ½"x1 ¼"	13.0	3.5
2" Brick, 9"x4 ½"x2"	8.0	3.5
Soap, 9"x2 ½"x2 ½"	6.5	7.0
Checker, 9"x2 ¾"x22 ¾"	6.0	6.0

Ordinary fire clay costs about $6.50 per 100 lbs. while high-grade refractory fire clay costs about $7.60 per 100 lbs.

Labor Lining Brick Chimneys and Stacks with Fire Brick. When lining small brick chimneys and stacks from 2'-0" to 3'-0" square, laid in fire clay, a mason should lay 450 to 525 brick per 8-hr. day, at the following labor cost per 1,000:

	Hours	Rate	Total	Rate	Total
Mason	16	$....	$....	$25.62	$409.92
Labor	12	....	....	19.13	229.56
Cost per 1,000 brick			$....		$639.48

On large brick stacks having an inside diameter of 4'-0" to 6'-0", a mason should lay 600 to 750 fire brick per 8-hr. day, at the following labor cost per 1,000:

	Hours	Rate	Total	Rate	Total
Mason	11.8	$....	$....	$25.62	$302.32
Labor	10.0	....	....	19.13	191.30
Cost per 1,000 brick			$....		$493.62

Number of Wedge and Arch Brick Required for Various Circles

No. 1 Wedge Brick 9"x4 ½"x (2 ½"-1 ⅞")

No. 2 Wedge Brick 9"x4 ½"x (2 ½"-1 ½")

No. 1 Arch Brick 9"x4 ½"x (2 ½"-2 ⅛")

No. 2 Arch Brick 9"x4 ½"x (2 ½"-1 ¾")

Labor Laying Fire Brick in Fireboxes, Breechings, Etc. When lining fireboxes, breechings, furnaces, etc., with fire brick, requiring arch and radial brick, special shapes, etc., a mason should lay 175 to 225 brick per 8-hr. day, at the following labor cost per 1,000:

	Hours	Rate	Total	Rate	Total
Mason	40	$....	$....	$25.62	$1,024.80
Labor	40	....	....	19.13	765.20
Cost per 1,000 brick			$....		$1,790.00

When estimating brick arches, it is advisable to allow additional labor. It is doubtful that a mason thoroughly experienced in boiler work will average more than 25 brick an hour for the actual arch and skew-backs, due to adverse conditions and close working space. One contractor specializing in this class of work checks the penetrations through the wall, and regardless of size, adds 4 hours mason time for each one.

Labor Bricking in Boilers. When bricking in steel boilers, fireboxes, breechings, etc., with common or fire brick, where the walls are 8" to 12" thick, a mason should lay 600 to 650 brick per 8-hr. day, at the following labor cost per 1,000:

	Hours	Rate	Total	Rate	Total
Mason	13	$....	$....	$25.62	$333.06
Labor	13	....	....	19.13	248.69
Cost per 1,000 brick			$....		$581.75

Table of Wedge Brick Required for Various Circles — Number of Arch Bricks Required for Various Circles

Diam. Ft.In.	No. 2 Wedge	No. 1 Wedge	Square	Total	No. 2 Arch	No. 1 Arch	9-inch	Total
2 0	...	...	...	...	42	...	...	42
2 6	63.	...	...	63.	10	40	...	50
3 0	48.	20.	...	68.	...	57	...	57
3 6	36.	39.	...	75.	...	57	7	64
4 0	24.	57.	...	81.	...	57	15	72
4 6	12.	78.	...	90.	...	57	22	79
5 0	...	94.	...	94.	...	57	29	86
5 6	...	94.	7.5	102.	...	57	37	94
6 0	...	94.	15.	109.	...	57	44	101
6 6	...	94.	23.	117.	...	57	52	109
7 0	...	94.	30.	124.	...	57	59	116
7 6	...	94.	39.	133.	...	57	67	124
8 0	...	94.	45.	139.	...	57	74	131
8 6	...	94.	53.	147.	...	57	82	139
9 0	...	94.	60.	154.	...	57	89	146
9 6	...	94.	68.	162.	...	57	97	154
10 0	...	94.	75.	169	...	57	104	161

Fire Clay Tile Flue Lining

Flue lining is estimated by the lin. ft. It is furnished in 2'-0" lengths in the following sizes:

Hours Required to Set 100 Lineal Feet of Flue Lining

Size of Flue Lining	No. Lin. Ft. per 8-Hr. Day	Mason	Labor
4"x8"	165-180	4.7	4.7
4"x12"	130-150	5.7	5.7
8"x8"	130-150	5.7	5.7
8"x12"	105-125	7.0	7.0
12"x12"	85-100	8.7	8.7
8"x18"	75-90	9.7	9.7
12"x18"	70-85	10.2	10.2
18"x18"	60-70	12.3	12.3
20"x20"	55-65	13.3	13.3
20"x24"	50-60	14.0	14.0
24"x24"	45-55	16.0	16.0
20" round	55-65	13.3	13.3

Approximate Prices of Tile Flue Lining

Size	Price per Lin. Ft.
8"x8"	$3.30
8"x12"	4.45
8"x18"	9.10
12"x12"	5.60
13"x13"	5.60
12"x16"	9.45
12"x18"	9.95
13"x18"	9.95
16"x16"	13.25
16"x20"	19.15
18"x18"	14.70
20"x20"	23.00
20"x24"	25.10
24"x24"	29.60
6" Round	$4.00
6" Round thimble	3.10
8" Round	4.45
8" Round thimble	6.15
10" Round	6.25
10" Round thimble	7.00
12" Round	6.60
12" Round thimble	8.50
15" Round	13.85
18" Round	16.00
20" Round	23.40
24" Round	37.30

Labor Setting Terra Cotta Wall Coping
Hours Required to Set 100 Lin. Ft.

Size of Coping	No. Lin. Ft. Set per 8-Hr. Day	Mason	Labor
9"	145-160	5.2	5.2
12"	115-130	6.5	6.5
18"	100-120	7.3	7.3

Approximate Lineal Foot Prices of Terra Cotta Wall Coping

Size	Double Slant	Single Slant
9"	2.50	2.75
13"	3.15	5.00
18"	6.30	12.00

Corners, ends and starters, four times price of straight coping; angles, six times the price of straight coping.

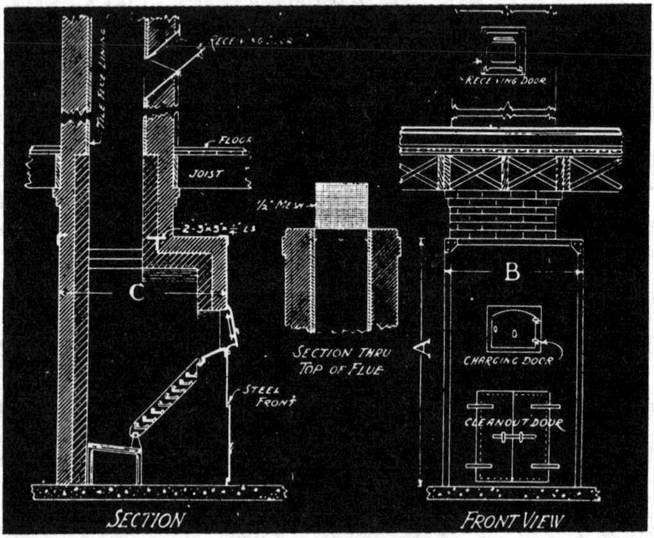

Illustrating Construction of One Type of Brick Incinerator

Brickwork for Incinerators

Incinerators have been used in residences, apartments, apartment hotels, hotels, hospitals, schools, and industrial plants for burning waste. The refuse might be fed into the chimney at each floor, dropped to the incinerator in the basement, or fed direct to the incinerator in the basement.

Practically all incinerators require a separate flue or stack, which should be estimated in the regular way, figuring fire brick, flue lining and common brick as shown on the plans.

The number of fire and common brick required for the various sizes and types of incinerators will vary with the different manufacturers.

However, the following data will assist the estimator in computing the number of brick and labor hours required to construct the brickwork for the different size incinerators, where they are not detailed on the plans.

If 4" outside walls are used, dimensions below will be reduced in length and height by 4" and in width by 8".

Model	Dimensions A	B	C	No. of Fire Brick	No. of Common Brick	Minimum Size of Flue	Mason Hours	Labor Hours
"A"	6'0"	3'5"	4'0"	300	900	12"x12"	20	14
"B"	5'7"	3'11"	4'7 1/2"	350	1100	14"x14"	24	18
"C"	6'0"	4'3"	4'7 1/2"	400	1300	16"x16"	28	20
"D"	6'10"	4'3"	5'11"	525	1500	20"x20"	33	24

Radical Brick Chimneys

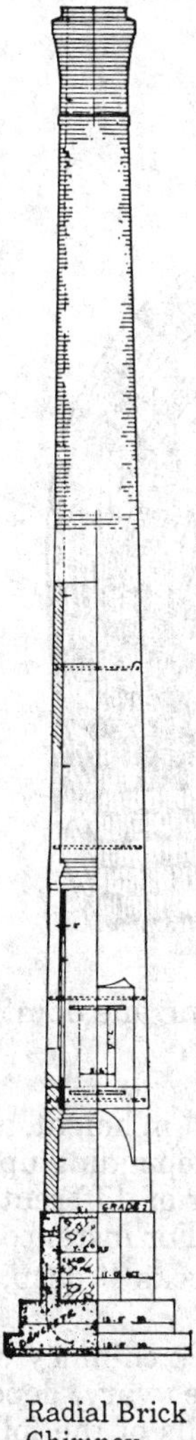

Radial Brick Chimney

Radial brick chimneys are usually constructed by contractors who specialize in chimney construction. These contractors employ workers with years of experience in this class of work and can build them much more economically than the general contractor. Also, most constructors of radial brick chimneys have their own designs and shapes of radial brick, manufactured especially for them, which are not available to the average contractor.

However, some information on this important subject is of value to all contractors. Radial brick chimneys are constructed of perforated radial brick formed to fit the circular and radial lines of the chimney. The brick are moulded with vertical perforations, and should be hard, well burned, acid proof, of necessary refractory powers and crushing strength and of maximum density.

The perforations serve to form a dead air space in the walls of the chimney that tends to prevent rapid heating and cooling of the walls by conserving the heat inside. Naturally, the higher the temperature of the gases the better the draft.

Four sizes of brick are ordinarily used in the construction of radial brick chimneys. All have the same face dimensions, approximately 6 1/2" wide by 4 1/2" high. The lengths of the blocks vary in order to make possible the breaking of the joints horizontally and vertically in the walls. The combination of bonds with this type of chimney provides for a lighter chimney than can be produced with ordinary building brick.

Radial chimney brick are generally either red or buff in color. Red radial brick are usually the most economical, because they are in greater demand and there are more sources of supply.

In laying the brick, the mortar is worked into the perforations, locking them together on the principle of a mortise and tenon joint. This produces the strongest bonded wall known to brick construction.

Sizes and Weights of Radial Brick for Radial Brick Chimneys

Type Brick	Outside Face	Inside Face	Depth Inches	Height Inches	Weight Lbs.
No. 4	6 ½"	5 ¾"	4"	4 ½"	7
No. 6	6 ½"	5 ½"	6"	4 ½"	9 ½
No. 7	6 ½"	5 ½"	7"	4 ½"	11
No. 8	6 ½"	5 ½"	8"	4 ½"	12
Pier	9"	9"	4"	4 ½"	9

Each radial brick with an outside face 6 ½" wide and 4 ½" high covers 29 ¼ sq. in. of surface, or it requires 5 (4.923) blocks per sq. ft. of wall, not including mortar joints. Allowing ½" for mortar joints, each brick covers 7"x5" or 35 sq. in. of surface, or it requires 4.12 blocks per sq. ft. of wall. Add about 5% for breakage and waste.

Perforated Radial Chimney Brick and Details of Bonding and Jointing of Perforated Radial Brick Construction

The walls of most radial chimneys taper about ¼" per 1'-0" in height, and the thickness of the walls vary from the bottom to the top, depending upon the height. The wall thickness must be made up of a number of different depth units to provide the proper bond and structural strength. For instance a wall 16 ¼" thick might be made up of 1 block 7" deep and 2 blocks 4" deep; a wall 8 ⅝" thick of 2 blocks 4" deep, etc.

Estimating Quantities of Radial Brick. Radial brick chimney work is frequently estimated by the ton, because wall thicknesses vary, depending upon the height. A chimney 125'-0" high might have walls of the following thicknesses:

First	25' height, 16 1/4" thick
Second	20' height, 13 1/4" thick
Third	20' height, 11 3/4" thick
Fourth	20' height, 10 1/4" thick
Fifth	20' height, 8 5/8" thick
Top	20' height, 7 1/8" thick

These wall thicknesses will be made up of radial brick of different depths to work out the correct wall thickness and proper bond, as follows:

Wall Thickness; Size of Units Used	Wgt. 29 1/4 Sq. In. Wall Lbs.*	Wgt per Sq. Ft., Lbs.*
20"-21"; 2 pcs. 8", 1 pc. 4"	31	128
18"-19"; 2 pcs. 7", 1 pc. 4"	29	120
16"-17"; 2 pcs. 6", 1 pc. 4" or 2 pcs. 8"	24-26	99-107
14"-15"; 1 pc. 6", 2 pcs. 4" or 2 pcs. 7"	22-23 1/2	91-97
12"-13"; 1 pc. 7", 1 pc. 4" or 1 pc. 8", 1 pc. 4"	18-19	75-79
10"-11"; 1 pc. 6", 1 pc. 4"	16 1/2	68
8"-9"; 2 pcs. 4" or 1 pc. 8"	12-14	50-58
7"-8"; 1 pc. 7" or 1 pc. 8"	11-12 lbs.	45 1/2-50

*Net weight of brick. Does not include mortar.

Figure radial brick chimney lining at about 30 lbs. per sq. ft. of wall surface.

Mortar Required Laying Radial Brick. It requires about two-thirds (1,334 lbs.) of a ton of mortar to lay one ton of radial brick. Mortar is usually mixed in the proportions of 1 volume portland cement, 1 volume lime putty, and 4 to 6 volumes clean mortar sand.

Unit	A	B	C	D	Inside Radius	Approx-imate Shipping Weight	Approx-imate Percent Voids
4 (N)	4 1/2"	6 1/2"	4"	5 20/32"	3'-4"	5.5	32
6 (A)	4 1/2"	6 1/2"	6"	5 21/32"	3'-4"	8.3	32
7 (B)	4 1/2"	6 1/2"	7"	5 1/2"	3'-8"	9.7	32
7-S.R.(B-S.R.)	4 1/2"	6 1/2"	7"	4 15/16"	2'-0"	9.7	32
8 (C)	4 1/2"	6 1/2"	8"	5 1/2"	3'-9"	10.8	32
8-S.R.(C-S.R.)	4 1/2"	6 1/2"	8"	5 1/16"	2'-6"	10.8	32
S (S)	4 1/2"	8"	4"	8"	Straight	7.0	32

Radial chimney brick is priced by the ton and will vary with type of brick and locality. Prices are f.o.b. factory, and the estimator must add freight and trucking from factory to job site and cost of unloading and storing brick at job.

Labor Laying Brick in Radial Brick Chimneys. The labor cost of laying radial brick chimneys will vary with the height of the chimney, thickness of walls, etc., but the following are actual production times on 6 radial brick chimneys constructed.

Labor Hours per Ton of Brick

Description	Tons of Radial Brick	Mason	Labor	Hoist Engr.
125'-0" high, Walls 16 1/4" to 7 1/8" thick. 12'-9" diam. at bottom, 7'-0" at top. Lined 50'-0"	157	2.7	4.7	0.9
80'-0" high, walls 11 3/4" to 7 1/8" thick. 8'-10" diam. at bottom, 5'-9" at top. Lined 35'-0"	66.5	3.1	5.3	1.0
60'-0" high, walls 10 1/4" to 7 1/8" thick. 7'-2" diam. at bottom, 4'-4" at top. Lined 21'-0"	31.1	4.7	8.2	1.9
46'-0" high, walls 10 1/4" to 7 1/8" thick. 6'-6" diam. at bottom, 4'-2" at top. Unlined	24.6	3.3	8.0	1.5
46'-0" high walls 10 1/4" to 7 1/8" thick. 6'-6" diam. at bottom, 4'-2" at top. Unlined (2)	49.2	3.0	5.0	1.5
55'-0" high, walls 10 1/4" to 7 1/8" thick. 6'-11" diam. at bottom, 4'-2" at top. Lined 20'-0"	31.6	3.0	5.0	1.5

The above does not include supervision time.

Labor time includes unloading materials from cars into trucks and unloading trucks at site. All mortar joints on the above stacks were struck flush with point of trowel.

Where stacks are lined, labor includes lining same with perforated radial fire clay brick, laid in high temperature cement. The lining is laid up from inside after stack is topped out.

Add for clean-out doors, band iron, ladders, lightning rods, cables and cable clips.

Miscellaneous Masonry Costs

Labor Costs on High Buildings. On buildings over 10 to 12 stories high, it will be necessary to add an extra allowance for labor handling and hoisting the bricks and mortar. The higher the building the longer it takes the hoist to make a trip from the street level to the floor on which the bricks are being laid and return, extra work raising and handling scaffolding, etc.

Taking a 12-story building, or 120 ft. in height, as normal or 100%, add about 1% to the labor cost for each story or each 10 feet in height above 120 ft.

For instance, a 12-story building or 120 feet in height is 100, the 13th story would be 101; the 14th story, 102; the 15th story, 103, etc. To find the average cost on a 15-story building, figure as follows:

The first 12 stories as 100 each or	1,200
The 13th story as 101	101
The 14th story as 102	102
The 15th story as 103	103
Total for 15 stories or 150 ft. in height	1,506

Dividing 1,506 by 15 gives 1.004 as the average cost of the masonry labor for a 15-story building. Other story heights would be computed in the same manner.

The following table applies to all classes of masonry work on high structures.

Height in Stories	Normal Base	Add Above Normal %	Average for Job
1-12	100.00	None	100.00
13	100.00	1%	100.08
14	100.00	2%	100.22
15	100.00	3%	100.40
16	100.00	4%	100.63
18	100.00	6%	101.17
20	100.00	8%	101.80
22	100.00	10%	102.50
24	100.00	12%	103.25
26	100.00	14%	104.04
28	100.00	16%	104.86
30	100.00	18%	105.70
40	100.00	28%	110.15
50	100.00	38%	114.82

Brick Catch Basins and Manholes

When laying building or sewer brick in catch basins, manholes, etc., from 3'-0" to 5'- 0" in diameter and having 8" to 12" walls, a mason should lay 600 to 750 brick per 8-hr. day, at the following labor cost per 1,000 :

	Hours	Rate	Total	Rate	Total
Mason	11.5	$....	$....	$25.62	$294.63
Labor	11.5	....	....	19.13	220.00
Cost per 1,000 brick			$....		$514.63

See *Highway & Road Construction, Standard Design Details* for block standards on masonry manholes and catch basins.

Pointing Around Steel Sash With Cement Mortar. All of the steel windows in a three story office building were pointed with portland cement mortar between the sash and the brick jambs after they had been set.

There were 39 windows 4'-0"x4'-10" and 96 windows 4'-0"x7'-0", making a total of 2,800 lin. ft. which were pointed at the following labor cost per 100 lin. ft.:

	Hours	Rate	Total	Rate	Total
Bricklayer	1.1	$....	$....	$25.62	$28.18
Labor	0.5	....	....	19.13	9.57
Cost per 100 lin. ft			$....		$37.75
Cost per lin. ft			....		.38

Reinforced Brick Masonry

Reinforced brick masonry has been widely used on the west coast because of its resistance to the lateral forces produced by earthquakes. It is now being used throughout the country (because most areas have been classified as seismic), where high winds or blast conditions may occur, and for retaining walls.

These walls are comparable to reinforced concrete construction with the masonry units serving as a form. The exterior wythes are built up in a conventional manner, the reinforcing is set in the interior joints and the whole is made homogeneous by filling the interior joints with grout.

No special skills or techniques are needed on the part of the mason, but certain care must be taken, including seeing that there is adequate lime in the mortar;

a sound base to start on with the aggregate in concrete exposed and wetted to saturation; that brick is wetted so grout will not dry out too quickly; that grout is sufficiently fluid so none will adhere to a trowel; that grout be poured in a layer of about three bricks high and be allowed to set for 15 minutes before another pour is made, and poured from the inside to avoid splashing the exposed surfaces; and that mortar cuts not be spilled into grout or grout allowed to accumulate on bed joints.

The amount of reinforcing will vary widely with the job. In lightly reinforced walls, the reinforcing will fit within the widths of normal brick walls. The more heavily reinforced walls may require wider inner joints or the use of soap courses to accommodate the steel. As headers are not used in reinforced walls, the contractor should check the drawings to see if, for aesthetic reasons, header rows are shown as this will require cutting all these brick. This is especially true when figuring reinforced lintels in conventional walls so that coursing can carry through.

Savings, as well as aesthetic advantages, often can be realized through the use of reinforced brick lintels. These are almost always poured in place with temporary shoring for the soffit. The economies result from savings in steel cost and elimination of painting. In better work it may call for the soffit brick joints to be filled temporarily with sand and then repointed later.

A variation on the above has been developed and is known as high lift grouted reinforced brick masonry. It varies in that it is essentially a cavity wall with a 2" or better inner joint. This allows the reinforcing to be preset and grouting delayed until a full 12' wall height is reached. Mechanical pumping can thus be used.

Precautions listed for standard reinforced walls also apply here with the following additions. Care must be taken to allow cleaning out of grout space by omitting every other unit on one side; brick must be allowed to set for at least 3 days before grouting; vibrating of grout is necessary; once grout pour is started, it must be continuous; and construction dams may be built every 20 ft. or 25 ft. to contain the grout laterally.

Handling of Masonry By Palletization

For years builders considered palletization for handling masonry materials, and many efforts were made in this direction with mixed results. Now, new systems have overcome most of the problems. Palletized masonry units, whether brick, block, or tile, can be transported from the manufacturing plant, through the job site, up the hoist, and around the scaffold to the mason's station, eliminating all hand handling of units until the mason picks them up for laying. Two companies that manufacture this equipment are the PCM Division of Koehring Corp., Port Washington, WI and Prime-Mover Co., Muscatine, IA.

Time studies on over 300 jobs using the PCM system indicate that with ten or more masons, the conventional ratio of 0.7 to 1.0 labor hour per mason hour can be reduced to 0.3 to 0.5 labor hours per mason hour--a savings of approximately 50% in labor time required for tending masons, or an overall savings on the total masonry labor cost of about 20 percent. These ratios include the labor time required to mix mortar, tend mortar, brick, tile, or block, and remove rubbish.

The success of these systems is mainly due to the equipment, which consists of properly sized wooden pallets, a tractor with specially design forklift mechanism, special brick buggies and special mortar buggies.

For efficient use of this system and equipment, certain job conditions must be established from the start and maintained throughout the life of the job.

All materials must be purchased on the basis of pallet delivery or palletized as soon as they arrive on the job. In the past, contractors were furnished the wood pallets, but dealers and manufacturers have shown an increasing preference for this type of delivery and in some cases furnish their own pallets. Contractor should specify the quantity of material to be placed on each pallet and the arrangement of same. Weight should not exceed 1,000 lbs. per pallet to stay within the capacity of scaffolding and buggies.

The effect on material prices for pallet delivery has been variable. Some suppliers have maintained their same prices, claiming the extra cost of labor and material for palletization is offset by savings in loading and unloading. Other suppliers have increased brick prices $2.00 to $3.50 per thousand. Actual cost of a 24"x32" pallet, if palletization is done on the job site, is around $4.00 per pallet. One laborer can handle 6 to 7 pallets per hour.

Access roads to job should be laid out and maintained to facilitate operation of the fork lift tractor. Area surrounding the building should be backfilled, consolidated and graded before masonry work starts and kept reasonably clear thereafter.

Exterior scaffolding must be wide enough to permit maneuvering brick and mortar buggies. Where tubular steel scaffolding is used, this may be accomplished with standard scaffolding components, using a 5-ft. width tubular scaffold with the addition of a 30" bracket on the inside for the masons and a 20" bracket on the outside for the extra width required for buggies and storage of pallets. No special planking is required, but plank laps should all run in the same direction and a wood easement strip placed at each lap to partially eliminate that irregularity.

Where suspended scaffolding is used, extra width can be obtained by using 8-ft wide Wheeling scaffold machines with the addition of 10-ft. pipe putlogs fastened directly under and through scaffold planking with U-bolts. This method moves the inside drum and cable 2'-0" out from the wall and creates a clear working space for the mason.

Scaffolding costs will be increased 5% to 15%, depending upon the type of scaffold used.

Standard Pallet 32"x24". A standard pallet size of 32" x 24" has been chosen, because it accommodates all common sizes of brick, block and tile, its loaded weight does not exceed safety regulations for scaffolding or light floor construction and it accommodates just enough material to build 10 lineal feet of wall, 4 feet high, 4 inches thick, allowing proper spacing for stacking on the scaffold according to standard masonry practice. In addition, its loaded weight can be handled by one man using hand powered equipment, and pallet load can be split into two 16" x 24" pallet loads, permitting passage through normal doorways, handling on occasional narrow scaffolds and maneuvering in confined spaces.

Pallets should be made to withstand hard use, preferably from a good grade of 1" lumber--1"x8" boards for the tops and 2"x4" for skids. Pallets made on the job, using power saws and a production line set-up, should cost about $2.25 per half pallet (16"x24") each, including material and labor.

One standard full size pallet is composed of two 16x24s strapped together with metal banding. When required, the metal banding can be removed, breaking the package into two halves of 16"x24" for use where operating space is limited.

Forklift Tractors. Manufacturers provide various capacities and sizes of rubber tired forklifts having lifting heights from 18'-6" up to 30'-6" with capacities from 2,500 lbs. to 4,000 lbs. The 3,000 lbs. unit can unload and stockpile as many as 4,000 brick in 10 minutes, and can be used to handle concrete planks and roofing materials as well as masonry.

The forklift tractor unloads pallets of material from delivery trucks and places them in stockpiles. From stockpiles, pallets are then transported to the masons' stations at ground level or on the scaffold up to maximum reach of tractor; to construction hoist for work on scaffolds or floors beyond tractor reach; to building floors within tractor reach for partition work or exterior masonry performed from inside the scaffold.

The forklift can handle as much palletized masonry material as can 20 laborers using conventional hand methods and wheelbarrows. It can also be equipped with other attachments to perform many other tasks on the job.

The fork lift also delivers mortar to the bricklayers from the mixing point by use of mortar buggies, which are mortar containers with a capacity of 7 cu. ft., equipped with wheels and casters so that they can be readily pushed on floors or scaffolds. The tractor handles these mortar containers to and from the mixer the same as packaged materials.

Prime Mover Mason Tender Transporting Brick From Stockpile to Hoist

Brick Buggies. Pallet loads are handled on the scaffold and building floors with brick buggies. These come in a variety of sizes, either hand propelled or power driven, with capacities of 1,000 lbs. for half (16"x24" pallets with 120 bricks) and 1,500 lbs. for full (24"x32" pallets with 240 bricks) loads. Lifting forks can be had for 10", 7'-6", and 10'-0" lifts, the latter being able to feed double deck scaffolding. Buggies can be operated under average outdoor conditions as well as on up and down ramps.

Cost of lift trucks for 24"x32" pallets is around $5,000. It can service 4 to 6 masons on a medium rise building instead of 1 laborer required to keep each mason supplied with brick. Two laborers, one on the ground and one delivering, should serve 5 masons.

The resulting savings in labor cost per day should amortize the lift and pallet costs in approximately 5 to 6 weeks under the best of conditions. Such handling also cuts down the time the hoist must be held for unloading by hand, which can be critical on larger jobs.

Courtesy J. Emil Anderson & Son, Inc. Photographer Ed Nastek

Large Mason Station Stocked with Brick and Mortar

04220 CONCRETE UNIT MASONRY

Concrete masonry units, commonly termed *concrete block or concrete brick,* are used extensively in all parts of the country for exterior and interior bearing walls, interior partitions, floor and roof fillers, and the like. Much of the information and data in this section comes from the National Concrete Masonry Association.

Concrete masonry units are widely used for backing up veneers. They are used both for bearing walls on wall bearing structures and for curtain walls in steel or reinforced concrete frame structures. The use of concrete masonry has increased greatly. This is partly due to the lower labor cost of handling and setting the larger concrete units in comparison with building brick, when used as a back-up material, but also because of increased use of architectural facing units.

The term concrete masonry is applied to block, brick or tile building units molded from concrete and laid by masons into a wall. The concrete is made by mixing portland cement with water and other suitable materials, such as sand, gravel, crushed stone, burned clay or shale, blast furnace slag and pumice.

Concrete masonry units should always be manufactured in a cement products plant where facilities for manufacturing and curing are uniform so as to obtain high quality products. The quality of all concrete masonry units should conform to the standards set by the American Society for Testing and Materials in their standard specifications for concrete masonry units and concrete brick.

Weight of Concrete Masonry. A 7 5/8"x7 5/8"x15 5/8" hollow loadbearing concrete block weighs from 35 to 45 lbs. when made from heavyweight aggregate and from 25 to 35 lbs. when made from lightweight aggregate. Heavyweight aggregates are sand, gravel, crushed stone, air-cooled slag, etc. Lightweight aggregates are coal cinders, expanded shale, clay or slag, pumice, etc.

Estimating Quantities of Concrete Brick. Concrete brick are manufactured, throughout the country generally, in the modular size of 2 1/4"x3 5/8" x 7 5/8". In some localities, additional sizes are available as follows: Jumbo brick, 3 5/8"x3 5/8"x7 5/8"; double brick, 4 7/8"x3 5/8"x7 5/8"; and Roman brick, 1 5/8"x3 5/8"x11 5/8" and 1 5/8"x3 5/8"x15 5/8".

They should be estimated by the sq. ft. of wall of any thickness or by the cu. ft., if all wall thicknesses are combined. For instance, if there are only 8" walls in the job, it is acceptable to take the number of sq. ft. of walls, but if the job contains 8", 12", and 16" walls, then it will be easier and more satisfactory to reduce the quantities to cu. ft. and multiply by a certain number of brick per cu. ft. of wall.

The number of concrete brick required per square foot or cubic foot of wall will vary with the size of brick and width of mortar joint used. Concrete brick sizes and widths of mortar joints have been standardized so that they will lay up to an even multiple of the 4-inch module.

All concrete brick units are 3 5/8" thick, which together with a 3/8" mortar back joint equal the modular thickness of 4 inches. Since there is always one less mortar joint than brick, in walls of multiple brick thickness, the actual wall thickness will always be 3/8" less than the modular 4 inches; i.e. 3 5/8", 7 5/8", 11 5/8", 15 5/8", etc.

In drawing plans, some architects give actual wall thickness dimensions, while others use nominal dimensions. For estimating purposes, nominal dimensions should be used, i.e. 4", 8", 12", 16", etc., to facilitate computations. When measuring the plans and listing the quantities of brick required, always take exact measurements for length and height and deduct all openings in full. Do not count corners twice but have your estimate show as accurately as possible the actual number of brick required to complete the job. For unexposed backup work, add 1% to 2% for waste. For exposed face work, add 3% to 5% for waste.

Modular Size Concrete Brick

Modular size concrete brick, 2 1/4"x3 5/8"x7 5/8" are laid so that 3 courses of brick, with 5/12" mortar joints, builds 8" in height, and the 7 5/8" length, plus 3/8" mortar joint, makes a total length of 8" laid in the wall. In other words, 3 brick lay 8"x8", or 64 sq. in., so it requires 6 3/4 or 6.75 brick per sq. ft. of wall of nominal 4" thickness.

Jumbo brick are 3 5/8"x3 5/8"x7 5/8" in size and are laid so that 3 courses of brick, with 3/8" mortar joints, builds 12" in height, and the 7 5/8" length, plus 3/8" mortar joint, makes a total length of 8" laid in the wall. One jumbo brick lays 32 sq. in. in the wall or it requires 4.5 brick per sq. ft. of wall of nominal 4" thickness.

Double brick are 4 7/8"x3 5/8"x7 5/8" in size, so that 3 courses of brick, with 11/24" mortar joint builds 16" high, and the 7 5/8" length, plus 3/8" mortar joint, makes a total length of 8" in the wall, or one brick lays 42.67 sq. in. in the wall, or it requires 3.4 brick per sq. ft. of wall of nominal 4" thickness.

Roman size brick are 1 5/8"x3 5/8"x11 5/8" and 1 5/8"x3 5/8"x15 5/8" and are laid so that 6 courses of brick, with 3/8" mortar joints, builds 12" in height, and the 11 5/8" or 15 5/8" lengths, plus 3/8" mortar joints makes a total length of 12" or 16" respectively laid in the wall. One nominal 12" brick lays 24 sq. in. in the wall or it requires 6 brick per sq. ft. of wall of nominal 4" thickness. One nominal 16" brick lays 32 sq. in. in the wall or it requires 4.5 brick per sq. ft. of wall of nominal 4" thickness.

The number of modular size brick per sq. ft. of wall of various thicknesses is as follows:

	Thickness of Wall		
	1-Brick (4")	2-Bricks (8")	3-Bricks (12")*
2 1/4"x3 5/8"x7 5/8"	6.75	13.50	20.25
3 5/8"x3 5/8"x7 5/8"	4.50	9.00	13.50
4 7/8"x3 5/8"x7 5/8"	3.40	6.80	10.20
1 5/8"x3 5/8"x11 5/8"	6.00	12.00	18.00
1 5/8"x3 5/8"x15 5/8"	4.50	9.00	13.50

*Use this column for number of brick required per cu. ft. of wall.
No allowance is included for waste in above quantities.

Labor Handling and Laying Concrete Brick

The labor cost of handling and laying concrete brick varies, depending on factors such as size of brick, kind of mortar, class of work, thickness of walls, and number of openings.

On buildings having long straight walls without many openings, such as basement walls, garages, and factories, a mason should lay 800 to 900 modular size concrete brick, 2 1/4"x3 5/8"x7 5/8", per 8-hr. day, while on dwellings and other structures having 8" walls, cut up with numerous openings and pilasters, a mason should lay 650 to 700 brick per 8-hr. day.

Labor Cost of 1,000 Modular Size Concrete Brick, 2 1/4"x3 5/8"x7 5/8",
Laid in 8" Walls

	Hours	Rate	Total	Rate	Total
Mason	11.8	$....	$....	$25.62	$302.32
Labor	8.8	....	....	19.13	168.34
Cost per 1,000 brick			$....		$470.66

Above costs based on flush cut mortar joints. For struck joints, add 1/2-hr. mason time per 1,000 brick.

Estimate the labor cost of laying concrete face brick the same as given for face brick under *Unit Masonry/Brick.*

On similar work, using jumbo brick, 3 5/8"x3 5/8"x7 5/8", a mason should lay 675 to 775 brick per 8-hr. day in 12" walls and 550 to 600 brick per day in 8" walls.

Using double brick, 4 7/8"x3 5/8"x7 5/8", a mason should lay 450 to 550 brick per 8-hr. day in 12" walls and 375 to 425 brick per day in 8" walls.

Roman size concrete brick, 1 5/8"x3 5/8"x11 5/8" or 1 5/8"x3 5/8"x15 5/8", are used only for facing or in 8" walls. The labor cost of handling and laying these brick will vary considerably, depending upon the type of joint treatment, flush cut, V-tooled or concave, struck or weathered, or raked out.

On average work with flush cut joints, a mason should lay 300 to 350 nominal 12" Roman brick per 8-hr. day or 225 to 265 nominal 16" Roman brick per 8-hr. day. For tooled, struck, weathered or raked out joints, reduce the above production quantities 8 to 12 percent.

Concrete Blocks and Partition Units

Concrete blocks and partition units are manufactured in weights of concrete. Specific information on density of concrete types and on weights of units should be obtained from local manufacturers. When specific information is not available, assume the following values for products:

Type	Lbs. per Cu. Ft.
Pumice	75
Expanded shale concrete	85
Expanded slag and cinder concrete	95
Air-cooled slag concrete	120
Crushed stone and gravel concrete	135-145

These types of units may be used interchangeably for all purposes, although lightweight units afford a savings of weight to secure economy in design, and they afford increased heat and sound insulation value.

Estimating the Quantity of Concrete Blocks and Partition Units. Concrete blocks and partition units should be estimated by the square foot of wall of any thickness and then multiplied by the number of blocks per 100 sq. ft. as given in the following tables.

When estimating quantities, always take exact measurements.

Make deductions in full for all openings, regardless of size. The result will be the actual number of sq. ft. or number of blocks required for the job.

Mortar. The same kind of mortar should be used in laying concrete blocks as given for *Modular Size Concrete Brick.*

When laying up exterior walls, face shell bedding shall be used with complete coverage of face shells. Furrowing of the mortar shall not be permitted.

Extruded mortar shall be cut off flush with face of wall and the joints firmly compacted, after the mortar has stiffened somewhat.

Tooling is essential in producing tight mortar joints. Mortar has a tendency to shrink slightly and may pull away from the masonry units causing fine, almost invisible cracks at the junction of mortar and masonry units.

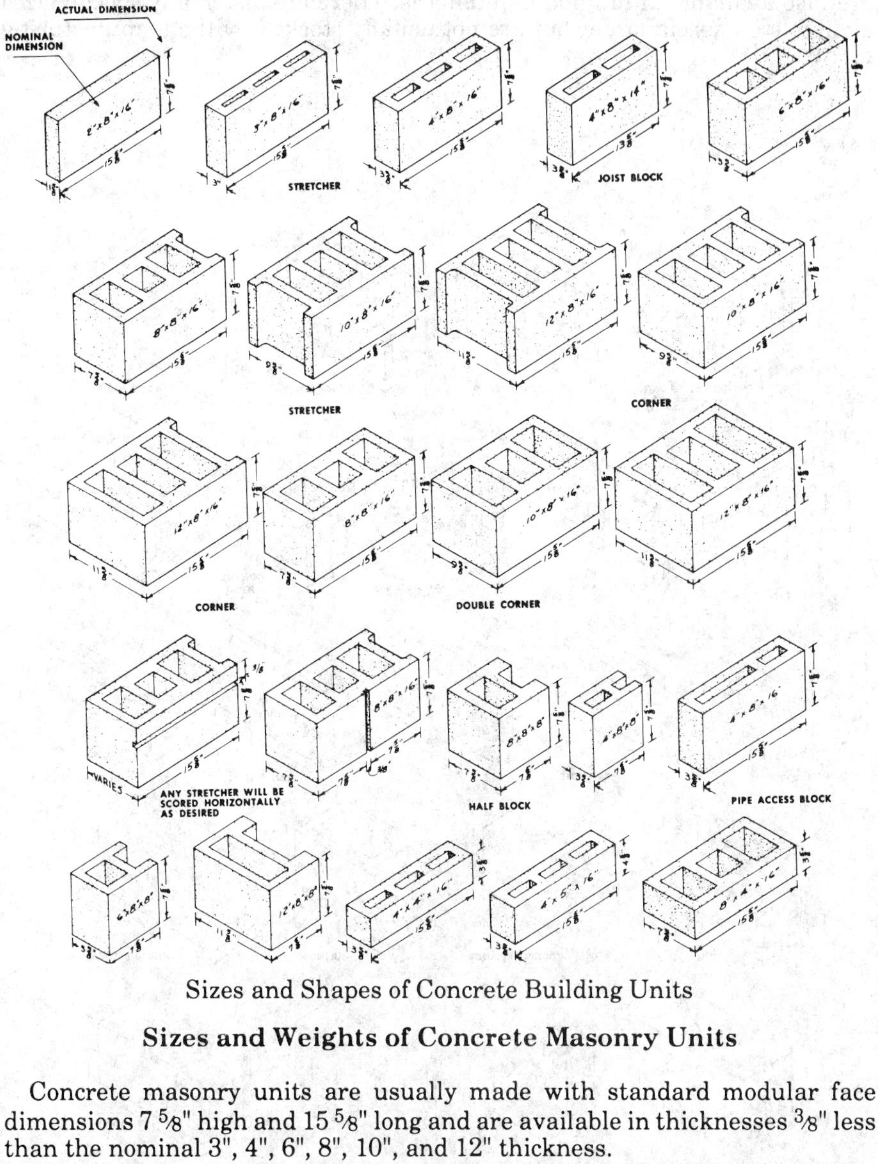

Sizes and Shapes of Concrete Building Units

Sizes and Weights of Concrete Masonry Units

Concrete masonry units are usually made with standard modular face dimensions 7 5⁄8" high and 15 5⁄8" long and are available in thicknesses 3⁄8" less than the nominal 3", 4", 6", 8", 10", and 12" thickness.

When laid up with 3⁄8" mortar joints, the units are 8" high and 16" long, requiring 112.5 units per 100 sq. ft. of wall, not including allowance for waste and breakage.

Hollow and solid partition and furring units have nominal thickness of 2", 3", 4", 6", and 8".

Hollow and solid loadbearing units have nominal thickness of 4", 6", 8", 10", and 12" and are also available in half lengths.

Standard specials such as steel and wood sash jambs, bullnose, and closures are also available in full and half lengths. There are also many special sizes available on special order but are not usually stocked by the manufacturing plants.

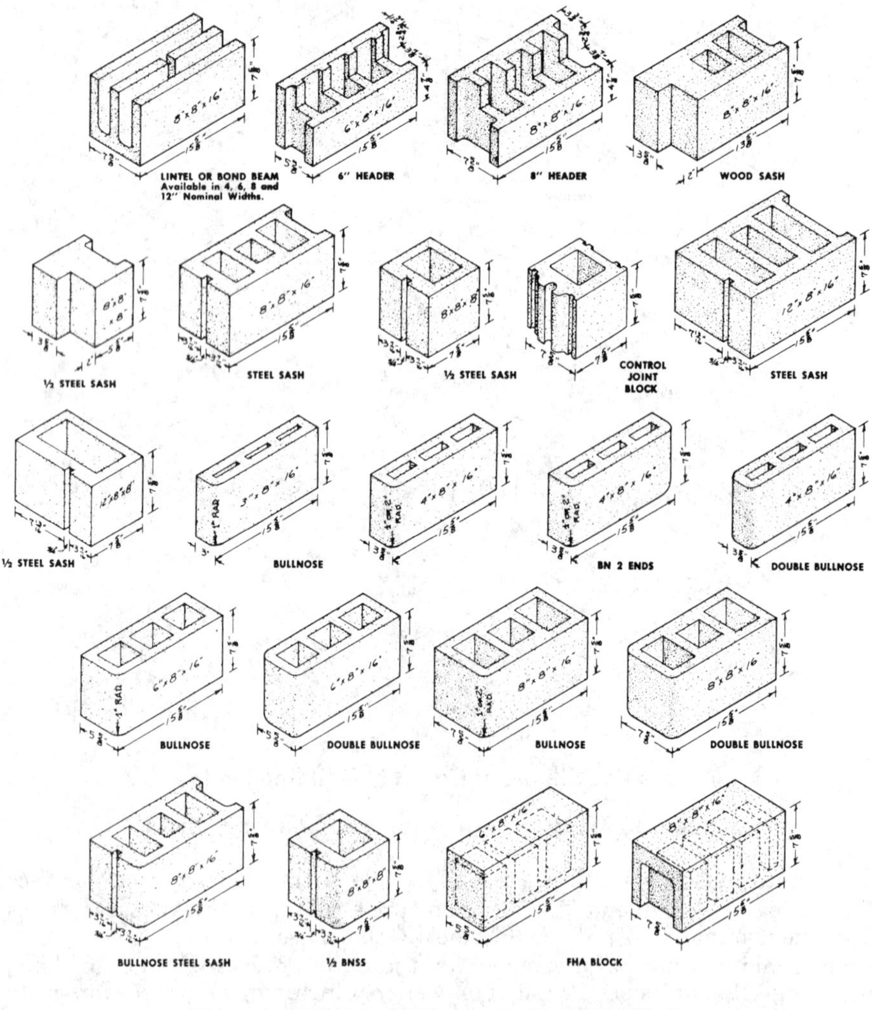

Sizes and Shapes of Concrete Building Units

Standard Block Nomenclature

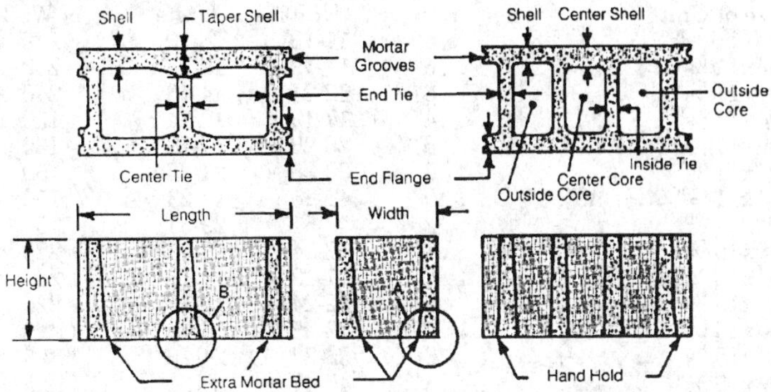

Note: The following nomenclature is the same for 2-core and 3-core block. The 2-core has been used here for illustration.

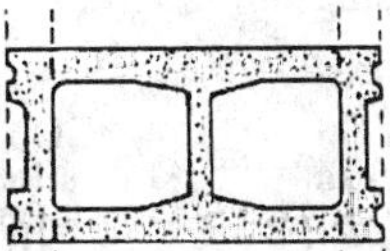

OBE with MG
Open Both Ends with Mortar Grooves

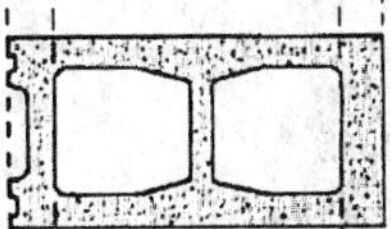

OE with MG-POE
Open End with Mortar Grooves – Plain Other End (Notice thickness of Plain End, unlike Permanent Plain End.)

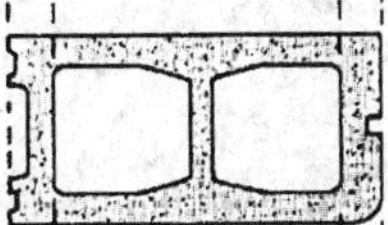

OE with MG-SSOE
Open End with Mortar Grooves – Steel Sash Other End (Also pictured is a Bullnose Corner: 1" & 2" radius available on any Plain End Corner.)

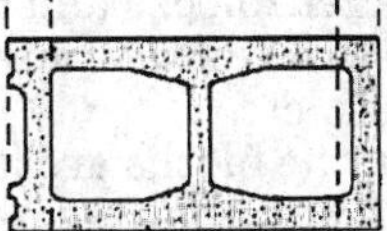

OE with MG-PPOE
Open End with Mortar Grooves – Permanent Plain Other End (Notice thickness of Permanent Plain End, unlike Plain End.)

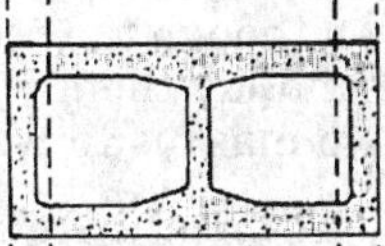

PPBE
Permanent Plain Both Ends.

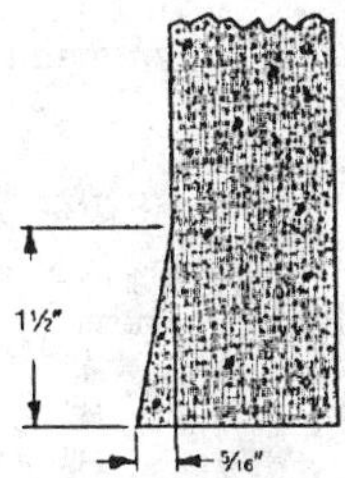

A Extra Mortar Bed

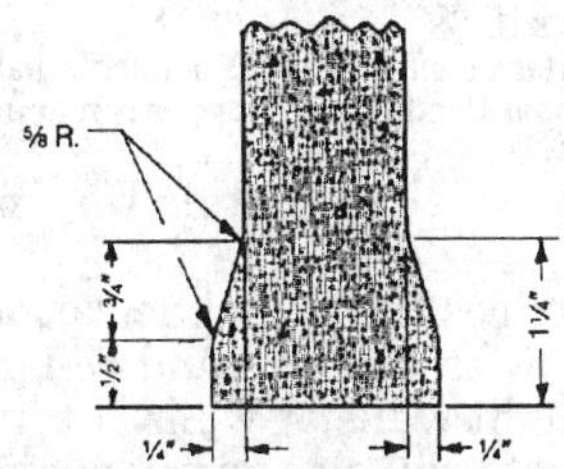

B Hand Hold

Hand Hold optional on 2 and 3 core block.

Dimensions of units: In practice, the first dimension of a concrete masonry unit represents the thickness; the second dimension, the height; the third, length.

Example: 8" x 8" x 16", or 200 mm. x 200 mm. x 400 mm.

Sizes, Weights and Quantities of Loadbearing Concrete Blocks and Tile

Actual Size of Units	Thick-ness	Approx. Weight in lbs. Hvy. Wt. Units	Approx. Weight in lbs. Light Wt. Units	No. of Units 100 SF of Wall	Cu.Ft. Mortar 100 SF of Wall*
3 5/8"x4 7/8"x11 5/8"	4"	11-13	8-10	240	8.0
5 5/8"x4 7/8"x11 5/8"	6"	17-19	12-14	240	8.5
7 5/8"x4 7/8"x11 5/8"	8"	22-24	14-16	240	9.0
3 5/8"x7 5/8"x11 5/8"	4"	17-19	12-14	150	6.0
5 5/8"x7 5/8"x11 5/8"	6"	26-28	17-19	150	6.5
7 5/8"x7 5/8"x11 5/8"	8"	33-35	21-23	150	7.0
9 5/8"x7 5/8"x11 5/8"	10"	42-45	27-29	150	7.5
11 5/8"x7 5/8"x11 5/8"	12"	48-51	29-31	150	8.0
3 5/8"x3 5/8"x15 5/8"	4"	12-14	9-10	225	9.0
5 5/8"x3 5/8"x15 5/8"	6"	17-19	11-13	225	9.5
7 5/8"x3 5/8"x15 5/8"	8"	22-24	14-16	225	10.0
3 5/8"x7 5/8"x15 5/8"	4"	23-25	16-18	112.5	5.0
5 5/8"x7 5/8"x15 5/8"	6"	35-37	24-26	112.5	5.0
7 5/8"x7 5/8"x15 5/8"	8"	45-47	29-31	112.5	6.0
9 5/8"x7 5/8"x15 1/2"	10"	57-60	36-38	112.5	6.5
11 5/8"x7 5/8"x15 5/8"	12"	64-67	40-42	112.5	7.0

Sizes, Weights and Quantities of Concrete Partition Tile

Actual Size of Units	Thick-ness	Approx. Weight in lbs. Light Wt. Units	No. of Units 100 SF of Wall	Cu.Ft. Mortar 100 SF of Wall*
3 5/8"x7 5/8"x11 5/8"	4"	12-14	150	6.0
5 5/8"x7 5/8"x11 5/8"	6"	17-19	150	6.5
7 5/8"x7 5/8"x11 5/8"	8"	21-23	150	7.0
1 5/8"x7 5/8"x15 5/8"	2"	11-12	112.5	3.0
3 5/8"x7 5/8"x15 5/8"	4"	16-18	112.5	5.0
5 5/8"x7 5/8"x15 5/8"	6"	24-26	112.5	5.0
7 5/8"x7 5/8"x15 5/8"	8"	29-31	112.5	6.0

*Actual mortar quantities figure out about half those given in the above table, but experience on the job shows that considerably more mortar is required, due to waste, droppings, etc.

Labor Laying Concrete Masonry

The labor cost of laying the various types and sizes of concrete blocks and tile will vary with the size and weight of the blocks, the class of work, whether long straight walls or walls cut up with numerous openings, etc.

Labor costs will also vary, depending upon whether or not the blocks are laid above or below grade, as basement walls usually proceed faster than exposed work above grade.

It is usually more economical to use lightweight units, even though they cost a few cents more per piece, because a mason can handle and lay them with less effort. In some localities union regulations require two masons to work together where the blocks weigh more than 35 lbs. each.

Where concrete blocks are used for exterior or interior facing walls, with the blocks carefully laid to a line and in various patterns and with neatly tooled mortar joints, the labor costs will run considerably higher than on straight structural walls.

The quantities given in the following tables are based on average conditions with blocks laid in 1:1:6 cement-lime mortar. If portland cement mortar is used above grade, reduce daily output about 5 percent.

All quantities are based on the output of one mason and one laborer or hod carrier per 8-hr. day. If hoisting engineer time is required, add about 1/4-hr. per 100 sq. ft.

Number of Concrete Partition Tile Laid Per 8-Hr. Day By One Mason

Actual Size of Units	Wall Thickness	No. of Light* Wt. Units per 8-Hr Day	Mason Hrs. per 100 Pieces	Labor Hrs. per 100 Pieces
3 5/8"x7 5/8"x11 5/8"	4"	195-215	3.9	3.9
5 5/8"x7 5/8"x11 5/8"	6"	175-195	4.3	4.3
7 5/8"x7 5/8"x11 5/8"	8"	155-175	4.8	4.8
1 5/8"x7 5/8"x15 5/8"	2"	180-200	4.2	4.2
3 5/8"x7 5/8"x15 5/8"	4"	190-210	4.0	4.0
5 5/8"x7 5/8"x15 5/8"	6"	170-190	4.5	4.5
7 5/8"x7 5/8"x15 5/8"	8"	150-170	5.0	5.0

Number of Concrete Building Units Laid Per 8-Hr. Day by One Mason

Actual Size of Units	Wall Thickness	No. of Light* Wt. Units per 8-Hr Day	Mason Hrs. per 100 Pieces	Labor Hrs. per 100 Pieces
3 5/8"x4 7/8"x11 5/8"	4"	215-235	3.5	3.5
5 5/8"x4 7/8"x11 5/8"	6"	195-215	3.9	3.9
7 5/8"x4 7/8"x11 5/8"	8"	175-195	4.3	4.3
3 5/8"x7 5/8"x11 5/8"	4"	195-215	3.9	3.9
5 5/8"x7 5/8"x11 5/8"	6"	175-195	4.3	4.3
7 5/8"x7 5/8"x11 5/8"	8"	155-175	4.8	4.8
9 5/8"x7 5/8"x11 5/8"	10"	135-155	5.5	5.5
11 5/8"x7 5/8"x11 5/8"	12"	115-135	6.4	6.4
3 5/8"x3 5/8"x15 5/8"	4"	215-235	3.5	3.5
5 5/8"x3 5/8"x15 5/8"	6"	195-215	3.9	3.9
7 5/8"x3 5/8"x15 5/8"	8"	175-195	4.3	4.3
3 5/8"x7 5/8"x15 5/8"	4"	190-210	4.0	4.0
5 5/8"x7 5/8"x15 5/8"	6"	170-190	4.5	4.5
7 5/8"x7 5/8"x15 5/8"	8"	150-170	5.0	5.0
9 5/8"x7 5/8"x15 5/8"	10"	130-150	5.7	5.7
11 5/8"x7 5/8"x15 5/8"	12"	110-130	6.7	6.7
7 5/8"x7 5/8"x15 5/8"	8"	225-250**	3.4	3.4
11 5/8"x7 5/8"x15 5/8"	12"	120-160**	5.7	5.7

*For heavyweight concrete units decrease above quantities and increase labor 10 percent.
**For walls below grade.

Mortar Joints. Hollow concrete blocks or tile should be laid with broken joints to secure the maximum resistance to moisture and heat penetration. Sufficient mortar is spread on the inner and outer shell bed and end joints to cover and to provide a joint of the required thickness when pressed into place. Solid units, brick, etc., are laid in solid beds.

Comparison of Concrete Masonry Sizes With Brick. Allowing for the mortar joints in brickwork, the various sizes of concrete masonry units are equivalent to the following number of modular size brick, i.e., 2 1/4"x3 5/8"x7 5/8".

Size of Blocks	Brick Equivalent
4"x8"x16"	6
8"x8"x16"	12
12"x8"x16"	18

Cost of Concrete Masonry Units. The prices quoted below are for index purposes only and are based on 8"x8"x16" block delivered to the job. As concrete masonry is widely manufactured, the estimator is always close to a source of supply. Costs on different size blocks are readily obtainable.

Prices for 8"x8"x16" Lightweight Block

City	Price Each
Baltimore	$0.85
Boston	1.05
Chicago	.80
Cincinnati	1.15
Cleveland	.70
Dallas	.80
Denver	.95
Detroit	1.05
Kansas City	1.10
Los Angeles	1.35
Minneapolis	1.05
New Orleans	1.05
New York	1.45
Philadelphia	.90
Pittsburgh	.85
San Francisco	1.85
St. Louis	.95
Seattle	1.10

Prices for 8"x8"x16" Heavyweight Block

City	Price Each
Baltimore	$0.85
Boston	.95
Chicago	.80
Cincinnati	.80
Cleveland	.70
Dallas	.80
Denver	.80
Detroit	1.05
Kansas City	1.00
Los Angeles	1.50
Minneapolis	.80
New Orleans	1.05
New York	2.05
Philadelphia	.80
Pittsburgh	.80
San Francisco	1.30
St. Louis	.85
Seattle	1.10

Precast reinforced concrete lintels can usually be obtained from concrete masonry manufacturers.

Control Joints in Concrete Masonry

To control movements in concrete masonry walls from various kinds of stresses, increasing use is being made of control joints. Control joints are continuous vertical joints built into the walls in such locations and in such a manner as to permit slight wall movement without cracking the masonry.

The spacing and location of control joints will depend upon a number of factors, such as length of walls, architectural details and especially on the experience records as to the need for control joints in the particular locality where the structure is to be built. Control joints should be placed at junctions of bearing as well as nonbearing walls, at junctions of walls and columns or pilasters, at construction joints in foundation, in roof and in floors, in walls weakened by chases and openings, columns, fixtures, etc. In long walls, joints are ordinarily spaced at approximately 20-ft. intervals, again depending upon local experience. At return angles in "L", "T", and "U" shaped structures, as a rule, control joint locations are determined by the architect or engineer and are usually indicated on the drawings.

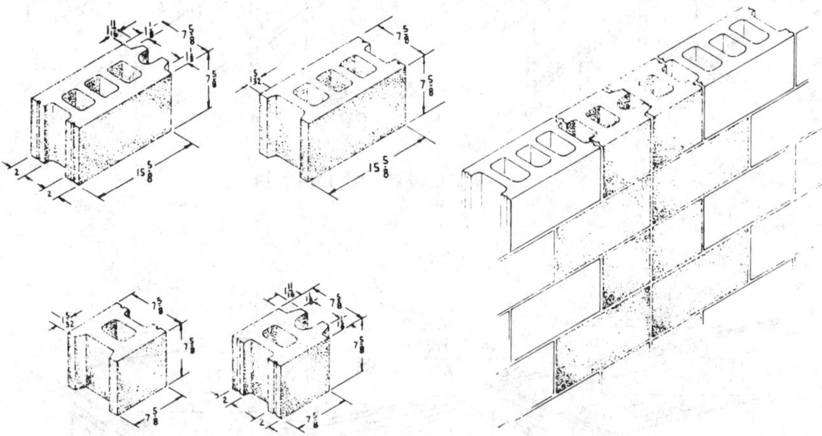

Method of Constructing Control Joints in Concrete Masonry Walls

Control joints can be built with regular full- and half-length stretcher block or full- and half-length offset jamb block. With this type of joint construction, a non-corroding metal Z-tiebar, placed in every other horizontal joint across the control joint will provide lateral support to wall sections on each side of the control joint.

In some localities, special control joint block are available. These block have tongue and groove ends, which provide the required lateral support. These block are also made in full- and half-length units.

Other common methods of constructing control joints are illustrated below. The joints permit free longitudinal movement, but they should have sufficient shear and flexural strength to resist lateral loads. They also must be weather tight when located in exterior walls.

Generally, a control joint is placed at one side of an opening less than six feet in width and at both jambs of openings over six feet wide. Control joints can be omitted if adequate tensile reinforcement is placed above and below wall openings.

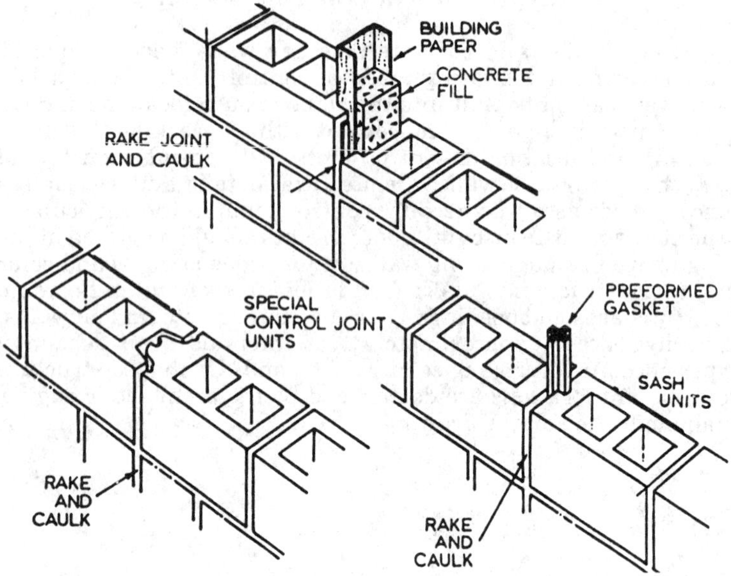

Typical Control Joint Details

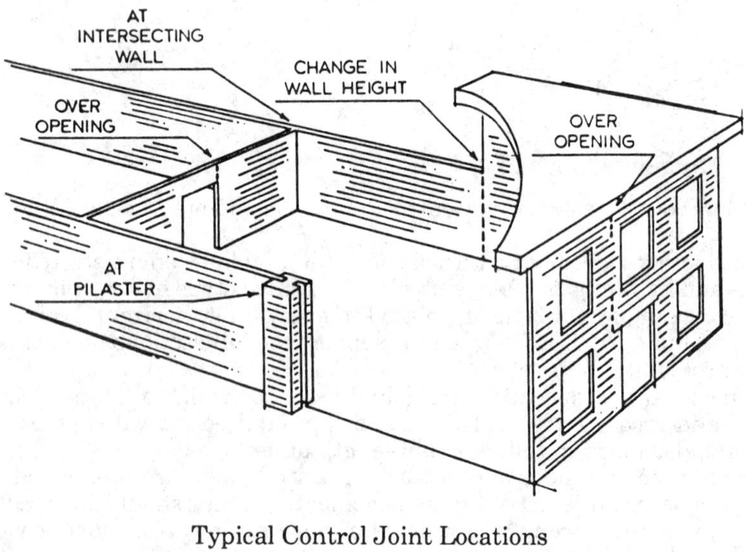

Typical Control Joint Locations

To keep control joints as unnoticeable as possible, care must be taken to build them plumb and of the same thickness as the other mortar joints. If joint is to be exposed to weather or view, it should be raked out to a depth of at least ¾" and sealed with knife grade caulking compound.

The additional cost of building control joints, over and above the regular wall cost, will run 25 to 30 cents per lin. ft. for material and 50 to 60 cents per lin. ft. for labor.

Reinforced Concrete Masonry

Local conditions frequently demand special construction methods. For example, in earthquake regions it is necessary to provide more than ordinary stability for all types of structures. This is also true in areas where severe wind storms occur or where foundation soils are unstable.

In almost any locality there are structures that are subject to excessive vibrations, very heavy loads, or other stresses. Any of these examples of unusual stress conditions require special attention in design and are often covered by local building codes.

In concrete masonry construction, additional strength is obtained by reinforcing the walls with steel reinforcing rods encased in grout. The walls may be reinforced horizontally, vertically or both.

Bond Beams. To reinforce concrete masonry walls horizontally, bond beams are frequently used at each story height. Under extreme stress conditions, it may be necessary to use them in every second or third course.

Bond beams may be constructed by forming and pouring a continuous ribbon of reinforced concrete at the course height required, but this method breaks up the continuity of the block pattern and may be objectionable.

In many localities, special bond beam blocks are available which are trough-shaped. When laid open side up, they form a continuous trough in which reinforcing steel and concrete encasement may be placed, eliminating the need for wood side forms, and the block pattern of the wall is preserved.

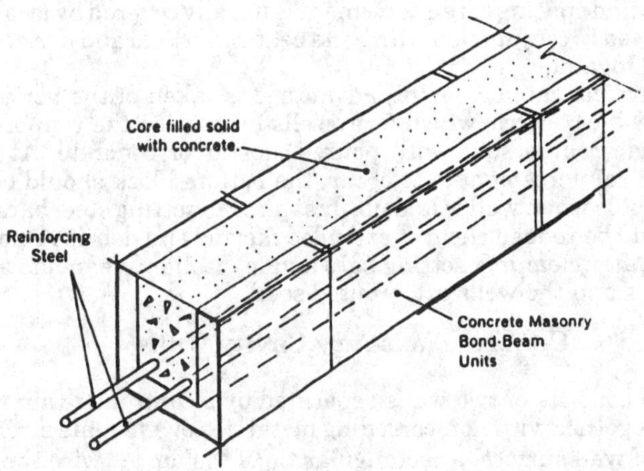

Typical Concrete Masonry Bond Beam

Bond beams serve both as structural elements and as a means of crack control. They are constructed with special shape masonry units that are filled with concrete or grout and reinforced with embedded steel. Their value in crack control is due to the increased strength and stiffness they give a masonry wall. Since they are capable of structural function as well as crack control, bond beams will be found serving the following functions:

a) As lintel beams over doors and windows. Lintels may be pre-constructed on the ground and set in place when they have attained sufficient strength or they may be built in place, using wood centers for support until mortar and concrete is strong enough to permit their removal.

b) Below the sill in walls with openings.

c) At the top of walls and at floor level to distribute vertical loads.

d) As horizontal stiffeners incorporated into masonry to transfer flexural stresses to columns and pilasters when unusually high lateral loading is encountered.

As a means of crack control, the bond beam's area of influence normally is presumed to extend 24 inches above and below its location in the wall. In walls without openings, they are spaced four feet apart and may be any length up to 60 feet maximum.

Reinforcement for bond beams must satisfy structural requirements but should not be less than two no. 4 steel bars. The beams are always discontinuous at expansion joints, and joints should be designed to transfer lateral force along the wall. Beams may be discontinuous at control joints; practice varies depending on structural requirements. Dummy joints are formed when a bond beam is continuous at a control joint.

Vertical Reinforcement in Concrete Masonry Walls. Where vertical reinforcement is required in concrete masonry walls, it is usually located at building corners, jambs of wall openings, and at regular intervals between wall openings. Size and spacing of reinforcement is usually covered by local building codes. When used in conjunction with bond beams, vertical and horizontal steel should be tied together.

In placing vertical reinforcement, advantage is taken of the vertical alignment of hollow block cores, which form wells into which the reinforcing bars are placed and grouted solid with poured mortar or concrete. At locations where vertical reinforcement is to occur, the bottom block should be left out for a cleanout hole when wall is laid up. Just prior to setting steel bars in place, the wells should be rodded clean of extruded mortar and debris removed from the cleanout. After cleaning, setting bars and inspection, cleanouts are closed with side-forms and the wells are grouted solid.

Concrete Masonry Cavity Walls

A cavity wall consists of two walls separated by a continuous air space and securely tied together with non-corroding metal ties of adequate strength. For each 3 sq. ft. of wall surface, a rectangular tie of 3/16" or 1/4" wire can be used. The ties are embedded in the horizontal joints of both walls. Additional ties are necessary at all openings, with ties spaced about 3 ft. apart around the perimeter and within 12 inches of the opening. Rectangular cavity wall ties are manufactured 2" and 4" wide and are either mill galvanized or hot dip galvanized. Costs for 2" hot dip galvanized range from $290.00 to 580.00 per

1,000 units. Costs for 4" hot dip galvanized range from $425.00 to 635.00 per 1,000 units.

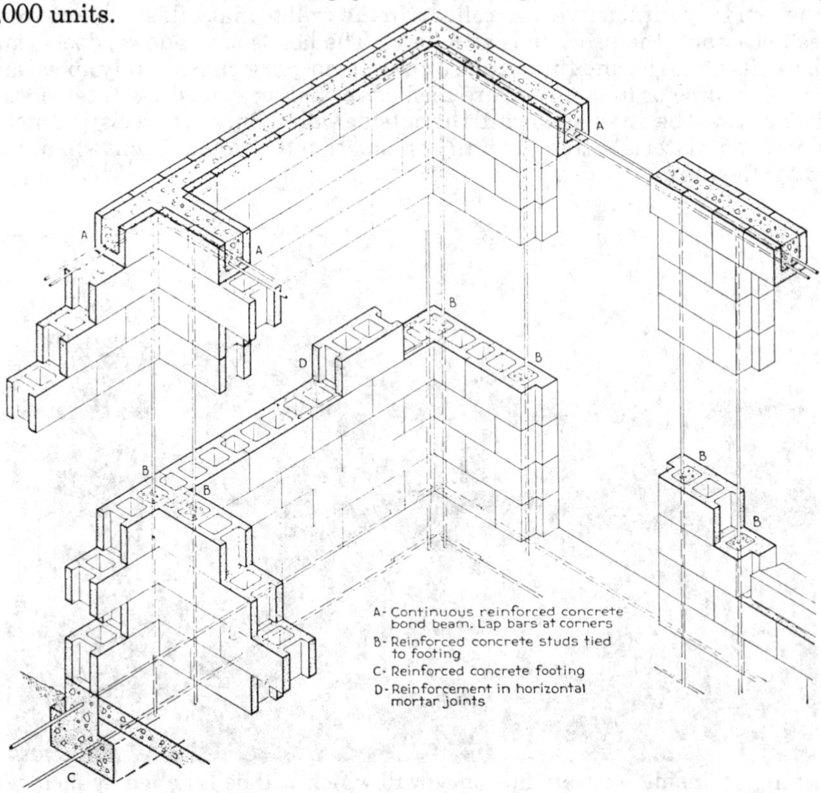

Method of Constructing and Reinforcing Concrete Masonry Walls

Typical codes require that 10-inch cavity walls should not exceed 25 ft. in height. In residential construction the overall thickness of concrete masonry cavity walls is nominally 10" or 12".

Neither the inner nor outer walls should be less than 4 inches thick (nominal dimension), and the space between them should not be less than 2 inches nor more than 3 inches wide. Usually the outer wall is nominally 4 inches thick and the remaining wall thickness made up by the air space and the inner wall. For example, a modular concrete masonry cavity wall nominally 12 inches thick is composed of a 4" outer wall, a 2" air space and a 6" inner wall.

A simple method of preventing the accumulation of mortar droppings between walls and maintaining a clear cavity is to lay a 1"x 2" wood strip across a level of ties to catch the droppings. As the masonry reaches the next level for placing ties, the strip is raised, cleaned and laid on the ties placed at this level.

The practice of providing special flashing and weep holes in masonry cavity walls primarily presumes that water will enter the wall from the outside. If concrete masonry walls are properly designed and built with well-compacted mortar joints and are painted or stuccoed, the walls should be weathertight. In the great majority of cases, there should be no need for special flashing and weep holes.

However, in limited areas subject to severe driving rains, or where experience has shown that sufficient water collects in the wall to make flashing and weep holes necessary, the practice is as follows. The heads of windows, doors, and other wall openings, and the bottom course of masonry immediately above any solid belt course or foundation, are flashed so that any moisture entering the wall cavity will be directed toward the outside walls. Only rust-resisting metal or approved materials treated with asphalt or pitch preparations should be used for flashing.

Laying up inside concrete masonry wall which will be finished by facing with brick

Weep holes are placed every 2 or 3 units apart in the vertical joints of the bottom course of the outside wall immediately above any solid belt course or foundation. In no case should the weep holes be located below grade. Weep holes can be formed by placing well-oiled rubber tubing in the mortar joint and then extracting it after mortar has become hard. The tubing should extend up into the cavity for several inches to provide a drainage channel through any mortar droppings that might have accumulated.

The labor cost of concrete masonry cavity walls varies the same as ordinary block walls according to the size of unit, pattern of laying, and nature of job.

Concrete Masonry Units Used for Exterior Facing

Concrete masonry walls are extensively used for exterior facing the same as face brick or stone. They may be laid up in any of the attractive designs shown in the accompanying illustrations, ranging from straight ashlar to the many variations of random or broken ashlar.

When used for exterior facing, the walls should be true and plumb and laid with full mortar coverage on vertical and horizontal face shells, (no furrowing permitted) with all vertical joints shoved tight. Mortar joints should be 3/8" thick.

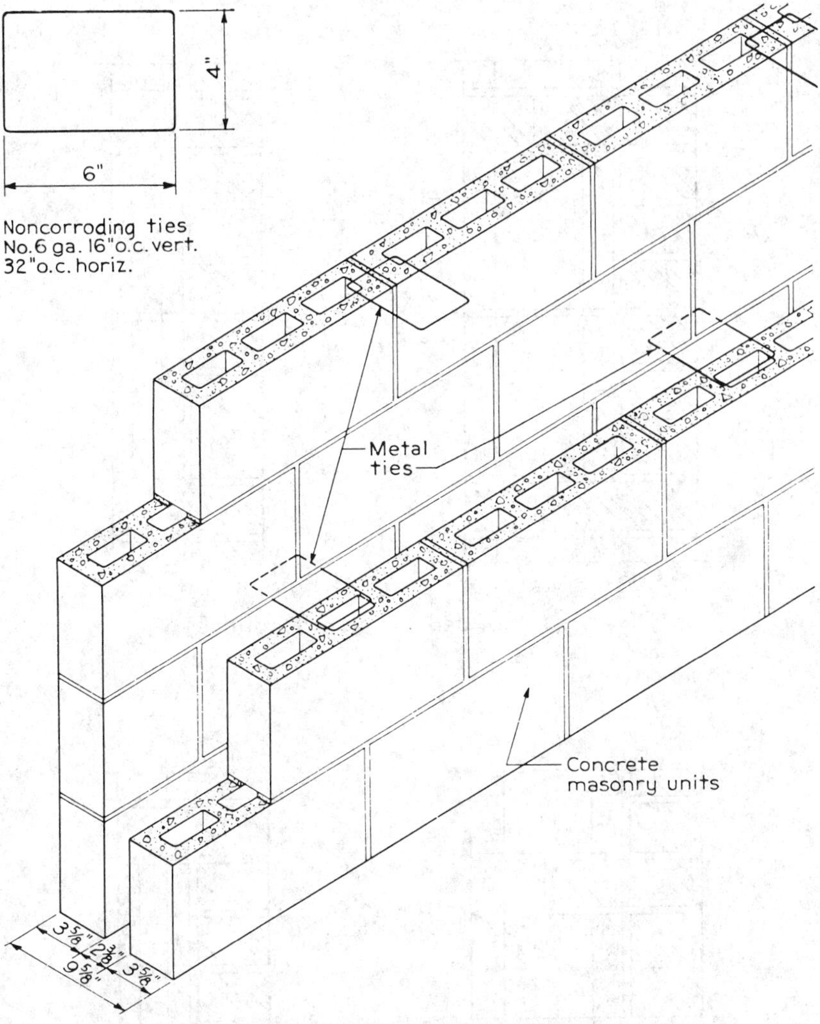

Method of Constructing Cavity Walls of Concrete Masonry

The mortar joints should be struck off flush with wall surface and when partially set shall be compressed and compacted with a rounded or V-shaped tool. This provides a more waterproof mortar joint. An attractive treatment is obtained by emphasizing the horizontal joints and obscuring the vertical joints in concrete walls. This is done by tooling the horizontal joints and striking the vertical joints flush with the wall surface and then rubbing with carpet or burlap to remove the sheen from the troweled mortar surface.

Varying the bond or joint pattern of a concrete masonry wall can create a wide variety of interesting and attractive appearances using standard units as well as sculptured-face and other architectural facing units. Due to the increased use of concrete masonry as the finished wall surface, the use of bond patterns other than the typical "running bond" has steadily increased for both loadbearing and nonloadbearing.

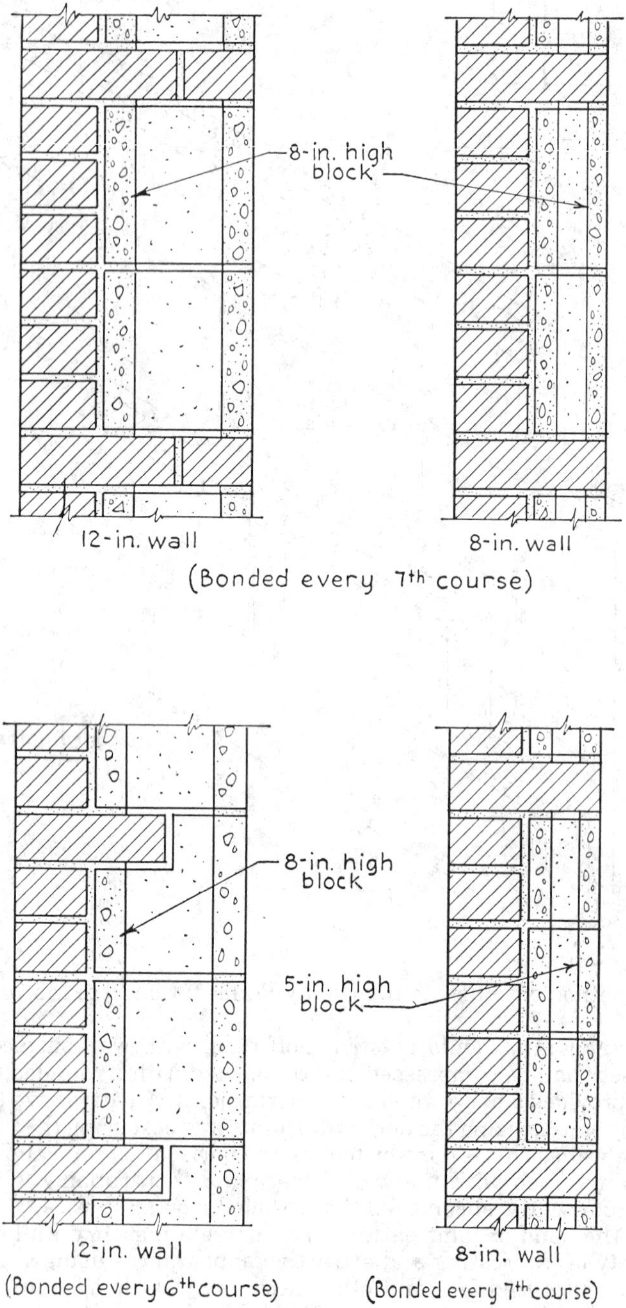

Examples of Concrete Masonry Used as Backup for Brick

After running bond construction, the next most widely used bond pattern with concrete masonry units is stacked bond. Lightweight concrete units should be used where obtainable. They provide better insulation than units made of ordinary concrete aggregates.

The labor costs given on the following pages are based on using lightweight units. If ordinary concrete units are used, add about 10% to labor costs given.

Diagonal Basket Weave
8"x16" units

Diagonal Bond
8"x16" units

Example of Horizontal and Vertical Face Shell Bedding

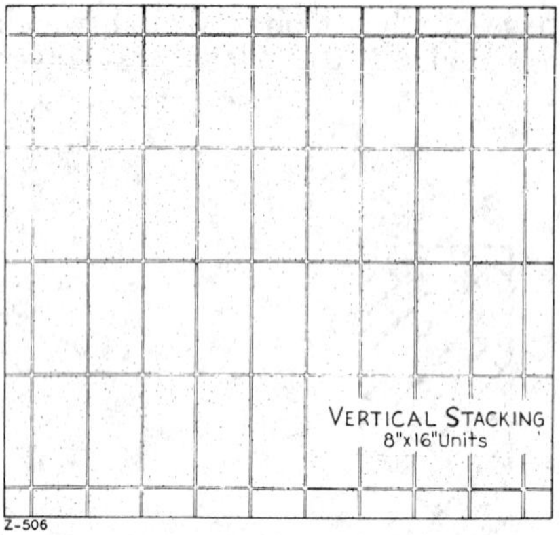

Z-506

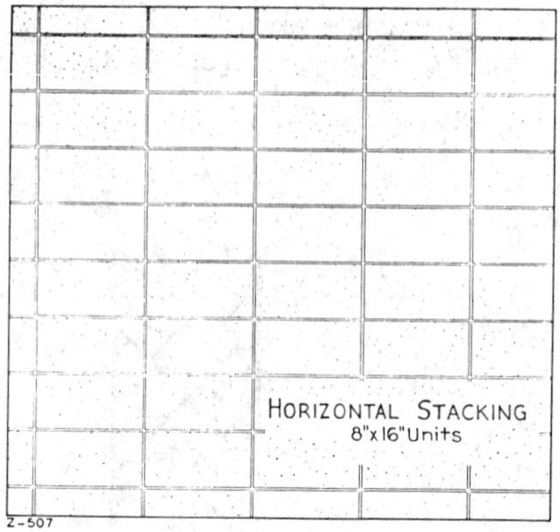

Z-507

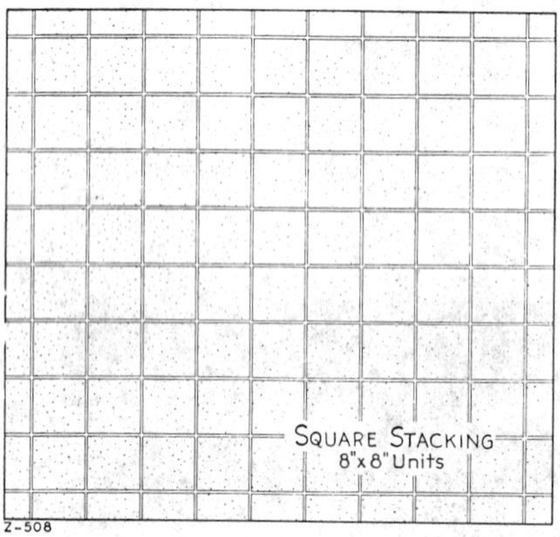

Z-508

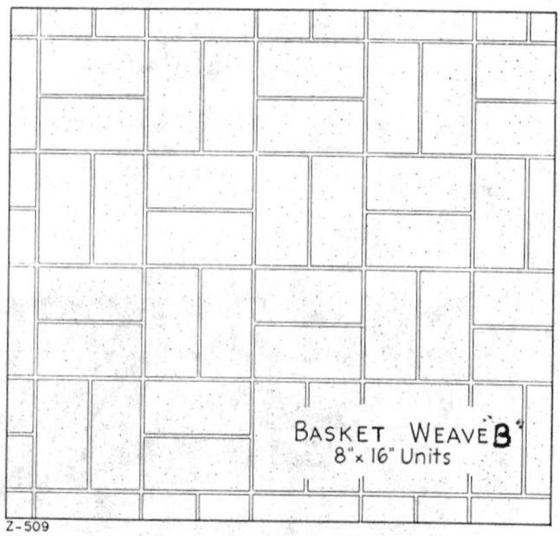

Z-509

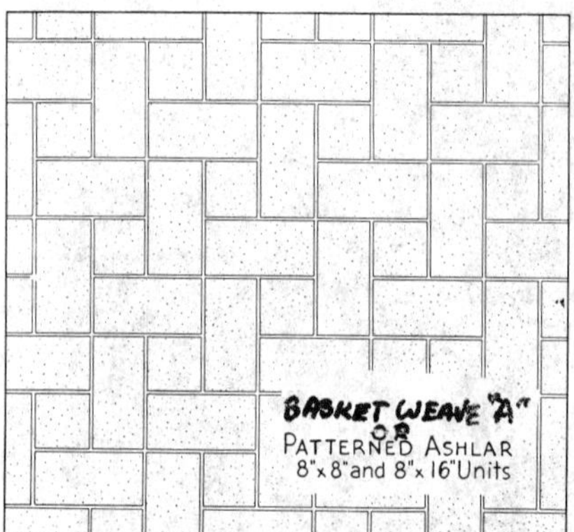

Z-510

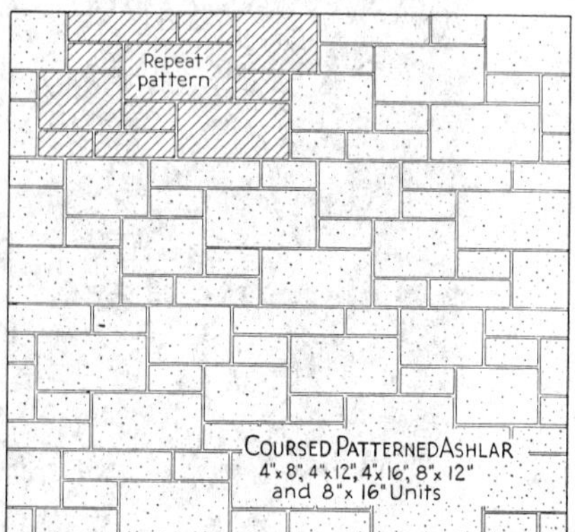

Z-511

Methods of Laying Concrete Masonry Units for Exterior Facing

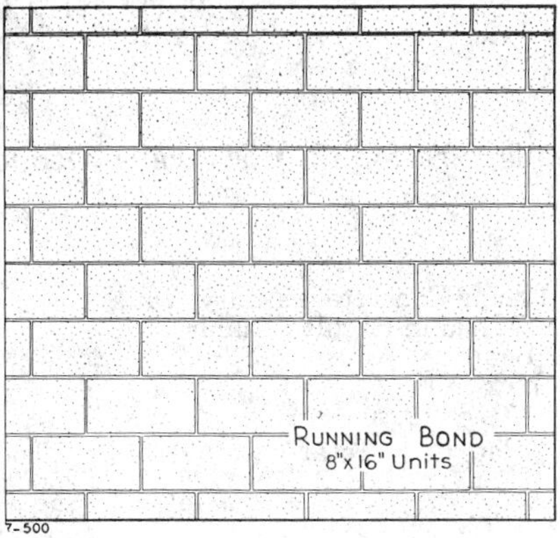

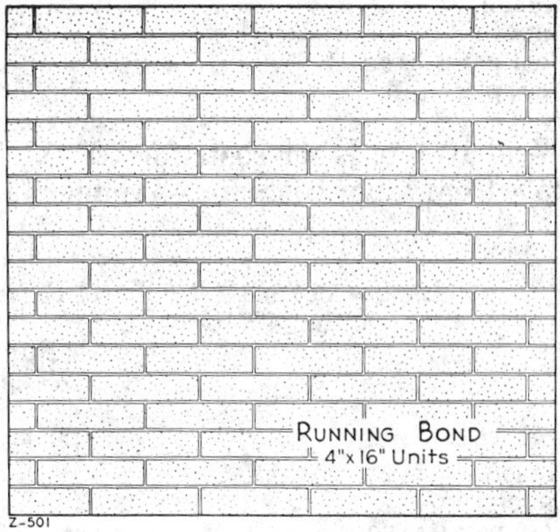

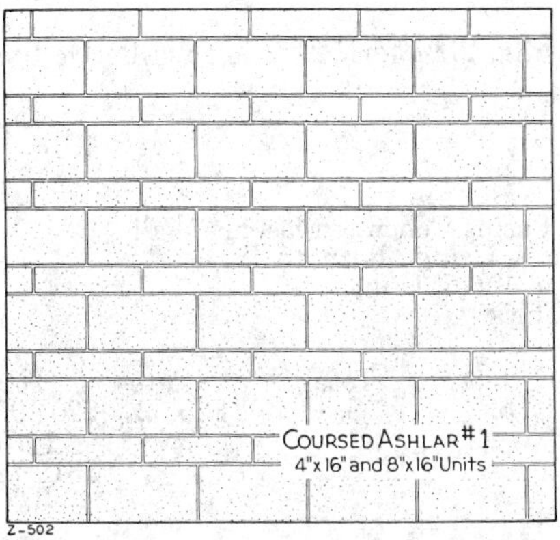

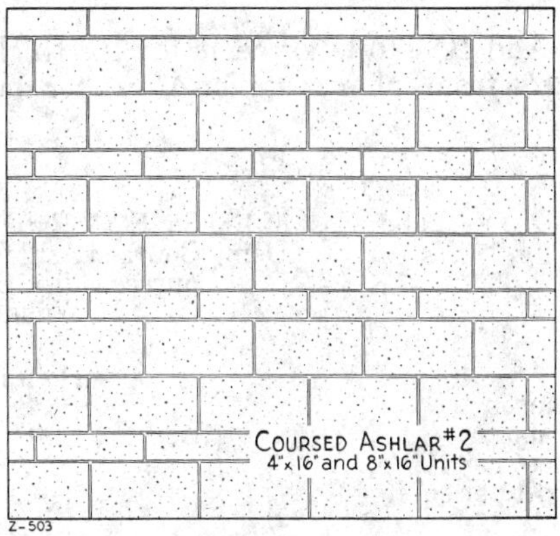

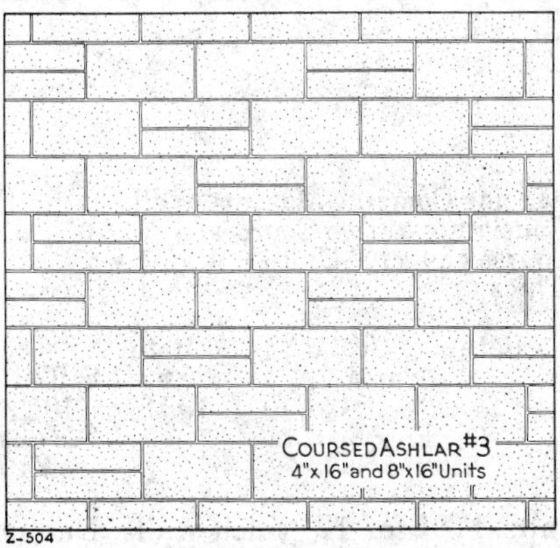

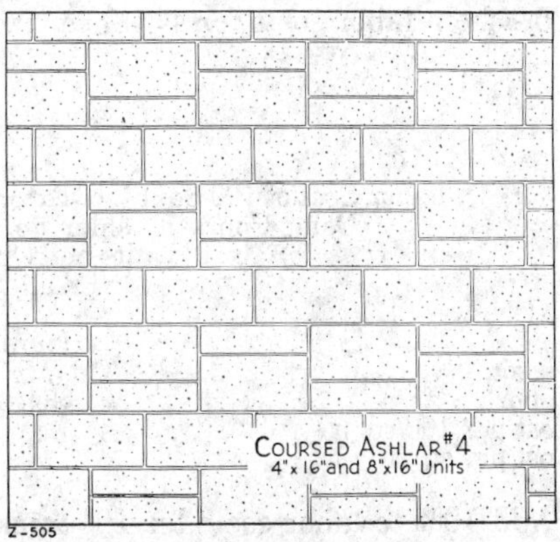

Methods of Laying Concrete Masonry Units for Exterior Facing

Labor Cost of 100 Sq. Ft. of Exterior Facing Concrete Masonry Walls
Using 8"x16" Units, Laid in Regular Ashlar or Running Bond
Walls 8" Thick

	Hours	Rate	Total	Rate	Total
Mason	7	$....	$....	$25.62	$179.34
Labor	7	....	....	19.13	133.91
Cost per 100 sq. ft.			$....		$313.25
Cost per sq. ft			....		3.13

Add 10% to labor costs for 10" thick walls, 20% for 12" thick walls. Deduct 25% from labor costs for 4" veneer facing.

Labor Cost of 100 Sq. Ft. of Exterior Facing Concrete Masonry Walls
Using 4"x16" or Half Height Units Laid in Regular Ashlar
or Running Bond
Walls 8" Thick

	Hours	Rate	Total	Rate	Total
Mason	10	$....	$....	$25.62	$256.20
Labor	7	....	....	19.13	133.91
Cost per 100 sq. ft			$....		$390.11
Cost per sq. ft			....		3.90

Add 10% to labor costs for 10" thick walls, 20% for 12" thick walls. Deduct 25% from labor costs for 4" veneer facing.

Labor Cost of 100 Sq. Ft. of Exterior Facing Concrete Masonry Walls
Using Coursed Ashlar No. 1, Using Alternate
Courses of 8"x16" and 4"x16" Units
Walls 8" Thick

	Hours	Rate	Total	Rate	Total
Mason	8	$....	$....	$25.62	$204.96
Labor	6	....	....	19.13	114.78
Cost per 100 sq. ft			$....		$319.74
Cost per sq. ft			....		3.20

Add 10% to labor costs for 10" thick walls, 20% for 12" thick walls. Deduct 25% from labor costs for 4" veneer facing.

Labor Cost of 100 Sq. Ft. of Exterior Facing Concrete Masonry Walls
Using Coursed Ashlar No. 2, Consisting of Two Courses
of 8"x16" Units and One Course of 4"x16" Units
Walls 8" Thick

	Hours	Rate	Total	Rate	Total
Mason	7.5	$....	$....	$25.62	$192.15
Labor	5.5	....	....	19.13	105.22
Cost per 100 sq. ft			$....		$297.37
Cost per sq. ft			....		2.97

Add 10% to labor costs for 10" thick walls, 20% for 12" thick walls. Deduct 25% from labor costs for 4" veneer facing.

Labor Cost of 100 Sq. Ft. of Exterior Facing Concrete Masonry Walls
Using Coursed Ashlar No. 3, Consisting of
8"x16" and 4"x16" Units
Walls 8" Thick

	Hours	Rate	Total	Rate	Total
Mason	8.25	$....	$....	$25.62	$211.37
Labor	6.50	....	....	19.13	124.35
Cost per 100 sq. ft			$....		$335.72
Cost per sq. ft			....		3.36

Add 10% to labor costs for 10" thick walls, 20% for 12" thick walls. Deduct 25% from labor costs for 4" veneer facing.

Labor Cost of 100 Sq. Ft. of Exterior Facing Concrete Masonry Walls
Using Coursed Ashlar No. 4, Consisting of
8"x16" and 4"x16" Units
Walls 8" Thick

	Hours	Rate	Total	Rate	Total
Mason	7.9	$....	$....	$25.62	$202.40
Labor	5.7	....	....	19.13	109.04
Cost per 100 sq. ft			$....		$311.44
Cost per sq. ft.			....		3.11

Add 10% to labor costs for 10" thick walls and 20% for 12" thick walls. Deduct 25% from labor costs for 4" veneer facing.

Labor Cost of 100 Sq. Ft. of Exterior Facing Concrete Masonry Walls
Using Vertical Stacking of 8"x16" Units
Walls 8" Thick

	Hours	Rate	Total	Rate	Total
Mason	9	$....	$....	$25.62	$230.58
Labor	7	....	....	19.13	133.91
Cost per 100 sq. ft			$....		$364.49
Cost per sq. ft			....		3.64

Add 10% to labor costs for 10" thick walls and 20% for 12" thick walls. Deduct 25% from labor costs for 4" veneer facing.

Labor Cost of 100 Sq. Ft. of Exterior Facing Concrete Masonry Walls
Using Horizontal Stacking of 8"x16" Units
Walls 8" Thick

	Hours	Rate	Total	Rate	Total
Mason	7.5	$....	$....	$25.62	$192.15
Labor	5.5	....	....	19.13	105.22
Cost per 100 sq. ft.			$....		$297.37
Cost per sq. ft			....		2.97

Add 10% to labor costs for 10" thick walls and 20% for 12" thick walls. Deduct 25% from labor costs for 4" veneer facing.

Labor Cost of 100 Sq. Ft. of Exterior Facing Concrete Masonry Walls
Using Square Stacking of 8"x8" Units
Walls 8" Thick

	Hours	Rate	Total	Rate	Total
Mason	12.50	$....	$....	$25.62	$320.25
Labor	8.75	....	....	19.13	167.39
Cost per 100 sq. ft.			$....		$487.64
Cost per sq. ft.			....		4.88

Add 10% to labor costs for 10" thick walls and 20% for 12" thick walls. Deduct 25% from labor costs for 4" veneer facing.

Labor Cost of 100 Sq. Ft. of Exterior Facing Concrete Masonry Walls
Using Basket Weave Design Consisting of 8"x16" Units
Walls 8" Thick

	Hours	Rate	Total	Rate	Total
Mason	7.9	$....	$....	$25.62	$202.40
Labor	5.7	....	....	19.13	109.04
Cost per 100 sq. ft.			$....		$311.44
Cost per sq. ft			....		3.11

Add 10% to labor costs for 10" thick walls and 20% for 12" thick walls. Deduct 25% from labor costs for 4" veneer facing.

Labor Cost of 100 Sq. Ft. of Exterior Facing Concrete Masonry Walls
Using Patterned Ashlar Consisting of 8"x16" and 8"x8" Units
Walls 8" Thick

	Hours	Rate	Total	Rate	Total
Mason	8.4	$....	$....	$25.62	$215.21
Labor	6.2	....	....	19.13	118.61
Cost per 100 sq. ft			$....		$333.82
Cost per sq. ft			....		3.34

Add 10% to labor costs for 10" thick walls and 20% for 12" thick walls. Deduct 25% from labor costs for 4" veneer facing.

Labor Cost of 100 Sq. Ft. of Exterior Facing Concrete Masonry Walls
Using Coursed Patterned Ashlar Consisting of 4"x8", 4"x12",
4"x16", 8"x12", and 8"x16" Units
Walls 8" Thick

	Hours	Rate	Total	Rate	Total
Mason	14	$....	$....	$25.62	$358.68
Labor	9	....	....	19.13	172.17
Cost per 100 sq. ft			$....		$530.85
Cost per sq. ft			....		5.31

Add 10% to labor costs for 10" thick walls and 20% for 12" thick walls. Deduct 25% from labor costs for 4" veneer facing.

Concrete block is often faced with brick. Three bricks equal the height of one block, and every seventh course of brick can be a header course, or in 12" walls every other block course can be a header block, that is, a block notched to receive a header brick. This allows every sixth course to be a bond.

Labor Cost of 100 Sq. Ft. of 8" Brick Faced Block Wall
Bonded Every Seventh Course

	Hours	Rate	Total	Rate	Total
Mason	18.5	$....	$....	$25.62	$473.97
Labor	13.5	....	....	19.13	258.26
Cost per 100 sq. ft			$....		$732.23
Cost per sq. ft			....		7.32

Labor Cost of 100 Sq. Ft. of 12" Brick Faced Concrete Block
Wall Bonded Every Sixth Course

	Hours	Rate	Total	Rate	Total
Mason	22	$....	$....	$25.62	$563.64
Labor	17	....	....	19.13	325.21
Cost per 100 sq. ft			$....		$888.85
Cost per sq. ft			....		8.89

Insulating Fill for Concrete Masonry Walls. For insulating masonry walls, a water-repellent vermiculite masonry fill is available. This is a free-flowing granular material, processed to provide a high degree of water-repellency. It is designed for the cores of masonry blocks and the voids of cavity walls. It does not bridge, and completely fills cores and cavities. Test installations show that the fill reduces heat loss up to 50% and air-conditioning costs about 25 percent. Masonry fill is also effective sound-deadening in masonry block partitions.

Vermiculite masonry fill is marketed in 4 cu. ft. bags and costs about $5.00 per bag. Approximate coverage is :

	Wall Area Sq. Ft.
2" cavity	24
2 ½" cavity	20
4 ½" cavity	11
6" block, 2 or 3 core	21
8" block, 2 or 3 core	14 ½
12" block, 2 or 3 core	8

Silicone treated perlite loose fill insulation insulates masonry walls to give greater comfort at lower cost. Silicone treated perlite pours readily into the cores of concrete block or cavity type masonry walls. Tests prove it reduces winter and summer heat transfer 50% or more. Silicone treatment of perlite provides lasting water repellency, prevents moisture penetration, and assures constant insulating efficiency in the most severe weather, including wind driven rain. Performance is rated excellent in accordance with test procedures developed by the National Bureau of Standards.

Insulating Cores of Masonry Units with Water-repellent Vermiculite Masonry Fill

Silicone treated perlite loose fill insulation is manufactured nationally and marketed in 4 cu. ft. bags. Coverage is approximately the same as shown for vermiculite. Installed costs will also be competitive with other loose fill materials.

Silicone Treated Perlite Fill in Masonry Walls

In addition to perlite and vermiculite, there are rigid insulating materials on the market. Some of these products are patten protected, such as *Korfil* and *Korfil II,* made of preformed expanded polystyrene and designed to fit two core masonry units from 6" to 12".

Other products available for insulation of masonry units are extruded and expanded polystyrene, as well as batt insulation. The estimator should be aware of the requirements of the owner and price the material that meets the engineers or owners needs and meets the local building code requirements both for insulation and fire. The following is a list of the common materials:

a) Vermiculite
b) 1" Extruded polystyrene
c) 2" Expanded polystyrene
d) 2" Extruded polystyrene

For *Korfil* it is suggested that the local masonry unit manufacturers be contacted for availability and costs.

Table A - R Values
Wall Constructed of 4" Hollow CMU

Details of Construction	Density, lb./cu. ft. 60	80	100	120	140
No insulation	2.8	2.4	2.2	1.9	1.7
Cores filled with vermiculite	4.9	4.1	3.4	2.7	2.2
Cores filled with perlite	5.4	4.4	3.5	2.8	2.3

Table B - R Values
Walls Construction of 6" Hollow CMU

Details of Construction	Density, lb./cu. ft. 60	80	100	120	140
No insulation	3.2	2.7	2.4	2.1	1.9
Cores filled with vermiculite	7.1	5.7	4.5	3.6	2.8
Cores filled with perlite	7.8	6.1	4.8	3.7	2.9
Rigid premolded insulation for two core CMU	7.7	6.7	5.9	5.0	4.2

Table C - R Values
Wall Constructed of 8" Hollow CMU

Details of Construction	Density lb./cu. ft. 60	80	100	120	140
No insulation	3.6	3.0	2.6	2.3	2.0
Cores filled with vermiculite	9.6	7.6	6.0	4.7	3.6
Cores filled with perlite	10.4	8.2	6.3	4.8	3.7
Rigid premolded insulation for two core CMU	8.3	7.7	6.3	5.3	4.2

Table D - R Values
Wall Constructed of 10" Hollow CMU

Details of Construction	Density, lb./cu. ft. 60	80	100	120	140
No insulation	3.8	3.2	2.7	2.4	2.1
Cores filled with vermiculite	11.8	9.3	7.2	5.5	4.2
Cores filled with perlite	12.9	10.0	7.6	5.7	4.3
Rigid premolded insulation for two core CMU	8.3	7.7	6.7	5.6	4.6

Table E - R Values
Wall Constructed of 12" Hollow CMU

Details of Construction	Density, lb./cu. ft. 60	80	100	120	140
No insulation	3.9	3.3	2.8	2.5	2.2
Cores filled with vermiculite	14.3	11.3	8.8	6.7	5.0
Cores filled with perlite	15.7	12.1	9.2	6.9	5.1
Rigid premolded insulation for two core CMU	9.1	8.3	7.1	6.3	5.3

Special Facing Blocks

In many areas, special block are available for the purpose of creating special facing effects. These block range from ordinary cement block, scored across the face to give the appearance of additional joints, to block made or faced with special aggregates which produce various textures and colors.

Scored Block. The simplest departure from ordinary block facing is obtained by scoring regular block one or more times across the face to give the illusion of shallower units and more joints. Most block manufacturers can furnish scored units for about 5 cents per score in addition to the regular block price.

Labor costs for laying up this type of facing is about the same as for a regular running bond facing except when scored grooves are pointed with mortar and tooled to match the actual joints.

Split-Block. Split-block afford another variation in wall finish. Split-block are made by splitting a hardened concrete unit lengthwise. The units are laid in the wall with the fractured face exposed. Many interesting variations can be obtained by introducing mineral colors and by using aggregates of different gradings and colors in the concrete mixture.

Split-block can be laid in simple running bond or in any of the other patterns used in concrete masonry construction. The fractured faces of the units produce a wall of rugged appearance.

Prices for 4" thick split-block veneer units run from $1.00 to $1.10 per sq. ft. of wall depending upon the size of units and pattern used.

Labor costs will be about the same as given on previous pages for the various patterns illustrated.

Slump Block. For a weathered stone effect, slump block are used. When manufacturing slump block, a special consistency mixture is used so that unit will sag or "slump" when released from the molds before complete "setting". This method produces many artistic irregularities with variations in height, texture, etc. Many are integrally colored in varying shades. Standard heights vary 1 5/8" to 3 5/8" and units are usually laid in a random ashlar pattern, although coursed patterns may be used if desired.

The cost of slump block varies with the locality and prices should be obtained locally when figuring this work. Prices range from $.95 to $1.25 per sq. ft. of wall facing.

Mortar requirements for slump block facing are high due to the irregularities of the material and approximate those of random ashlar stone work. As an average, it will require about 10 cu. ft. of mortar per 100 sq. ft. of 4" veneer facing.

When laying slump block in a random ashlar pattern, a mason and a helper should lay 45 to 55 sq. ft. of 4" veneer facing per 8-hr. day.

Glazed Concrete Block. Concrete block may be ordered with a factory applied glazed surface on one or more sides in 8"x16", 8"x8", and 4"x16" face sizes and in even thicknesses of 2" through 12". Standard cap, base and finished end units are also furnished. This glaze is a satin finish, factory applied to a 1/8" thickness and is available in a standard range of 48 colors and 11 scored patterns, with special colors and shapes a possibility. This glazing process is available to local sources through special licensing rather than being centrally manufactured and warehoused, and the local firm should always be consulted for availability and cost.

These units are especially useful where it is desired to have a block finish on one side and glazed on the other, as one unit will provide both, which results in considerable savings in both labor costs and floor space when compared with building a 4" block partition and adding a 2" glazed furring block. The glazed surface meets both USDA and OSHA requirements for sanitary surfaces. The high performance standards of concrete block for fire resistance and insulation value are retained. Tolerance on face dimensions and face distortions is limited to 1/16".

Units are set in standard mortar and struck with a groover but may be pointed or grouted for more finished work. Acid cleaning solutions should not be used, so units must be kept clean as the work progresses. A final cleaning with a masonry cleaning compound should be figured.

Adding the glazed surface to a standard block will add around $1.50 to $1.75 per square foot of glazed surface. Some colors will be more expensive than others, and the amount of scoring within the block surface will also affect the price. Two masons and one laborer will set 240-250 square feet of 4" partition in running bond per day.

Labor Cost of 100 Sq. Ft. of 4" Thick, 8"x16" Concrete
Block Glazed One Side Laid In Running Bond

	Hours	Rate	Total	Rate	Total
Mason	7	$....	$....	$25.62	$179.34
Labor	3.5	....	....	19.13	66.95
Cost per 100 sq. ft			$....		$246.29
Cost per sq. ft			....		2.46

Add to the above for pointing, grouting, and cleaning as required.

Tile Storage Bins

Storage bin and silo tile are clay, burned at a temperature of over 2,000 degrees and are salt glazed. They are made with a channel space at the top of each tile which permits the use of steel bands or reinforcing bars in each row of tile, amount of steel being proportioned to the amount of pressure, which is greatest at the bottom of the bin and decreases at the top.

These tile are furnished 12" high, 12" long and 6" and 8" thick, and it requires one tile per sq. ft. of wall, excluding the mortar joints which generally average ½" thick.

Mortar Required for Tile Storage Bins. Pure portland cement mortar should be used for laying all tile to make the strongest possible wall. Including mortar required to fill the channel space at the top of each tile, it will require approximately 9 cu. ft. of mortar for 6" thick tile and 12 cu. ft. for 8" thick tile per 100 sq. ft. of wall.

Labor Laying Silo Tile. When laying-up circular walls of 6"x12"x12" salt glazed tile, a mason should set 110 to 125 sq. ft. of tile per 8-hr. day, and will require one laborer to each mason.

When circular bin walls are laid up with 8"x12"x12" salt glazed tile, a mason should lay 90 to 100 sq. ft. of tile per 8-hr. day with a laborer helping full time.

Reinforcing Steel. Reinforcing steel will have to be added for each row of tile, according to size and weight of contents. It is impossible to give the amount of reinforcing steel required, as this will vary with the height of the storage bins and the pressure exerted by the contents and should be figured separately for each job.

Hoop Spacing for Concrete Stave Silos*

Based upon round rods of the sizes indicated of intermediate grade steel, with rolled threads. If cut threads are used the ends must be upset, or the spacing be closer. Unit tensile stress 20,000 psi (c)

Distance from top of silo in feet	Lateral Pressure of silo in psf (a) (c)	Hoop Spacing in Inches for 9/16 Inch Diameter Rods — Silo Diameter in feet						Hoop Spacing in Inches for 1/2 or 9/16 Inch Diameter Rds — Silo Diameter in feet					
		10	12	14	16	18	20	10	12	14	16	18	20
2½	12	30	30	30	30	30	30	30	30	30	30	30	30
5	33	30	30	30	30	30	30	30	30	30	30	30	30
7½	60	30	30	30	30	30	30	30	30	30	30	30	30
10	91	30	30	30	30	30	30	30	30	30	30	30	30
12½	125	30	30	30	30	30	30	30	30	30	30	30	30
15	162	30	30	30	30	30	30	30	30	30	30	30	30
17½	203	30	30	30	30	30	30	30	30	30	30	15	15
20	246	30	30	30	30	15	15	30	30	15	15	15	15
22½	294	30	30	30	15	15	15	30	15	15	15	15	15
25	340	30	30	15	15	15	15	15	15	15	15	15	15
27½	388	30	15	15	15	15	15	15	15	15	15	15	15
30	440	15	15	15	15	15	10	15	15	15	15	15	10
32½	495	15	15	15	15	10	10	15	15	15	15	10	10
35	550	15	15	15	10	10	10	15	15	15	10	10	10
37½	609	15	15	10	10	10	10	15	15	10	10	10	10
40	668	15	15	10	10	10	7½	15	15	10	10	10	7½
42½	725	15	10	10	10	7½	7½	15	10	10	10	7½	7½
45	793	15	10	10	7½	7½	7½	15	10	10	7½	7½	7½
47½	855	10	10	10	7½	7½	6	10	10	10	7½	7½	6
50	925	10	10	7½	7½	7½	6	10	10	7½	7½	7½	6
52½	990	10	10	7½	7½	6	6	10	10	7½	7½	6	6
55	1055	10	7½	7½	6	6	6	10	7½	7½	6	6	6
57½	1110	10	7½	7½	6	6	5	10	7½	7½	6	6	5
60	1200	7½	7½	6	6	5	5	7½	7½	6	6	5	5

For spacings above zig-zag line use 1/2" rods. For spacings below zig-zag line use 9/16" rods.
Psi denotes pounds per sq. in., psf denotes pounds per sq. ft. *A.C.I. 714

Concrete Stave Silo

Capacities of Storage Bins of Various Diameters—Per Foot of Height

DIAMETERS			10' Dia.	12' Dia.	14' Dia.	16' Dia.	18' Dia.	20' Dia.	22' Dia.	24' Dia.
Cubical Contents—		Cu. Ft.	78.54	113.10	153.94	201.06	254.47	314.16	380.13	452.39
Sand, Crushed Stone, etc		Cu. Yds.	2.91	4.18	5.70	7.44	9.42	11.63	14.08	16.75
Crushed Stone or Sand	(100 lbs. per cu. ft.)	Tons	3.93	5.65	7.70	10.05	12.72	15.71	19.01	22.62
Coal—Anthracite	(52 lbs. per cu. ft.)	Tons	2.04	2.94	4.00	5.23	6.62	8.17	9.88	11.76
Coal—Bituminous	(50 lbs. per cu. ft.)	Tons	1.96	2.82	3.85	5.02	6.36	7.85	9.50	11.31
Cinders	(45 lbs. per cu. ft.)	Tons	1.77	2.54	3.46	4.52	5.72	7.07	8.55	10.18
Salt	(48 lbs. per cu. ft.)	Tons	1.88	2.71	3.69	4.82	6.11	7.54	9.12	10.86
Lime—Unslaked	(53 lbs. per cu. ft.)	Tons	2.08	3.00	4.08	5.33	6.74	8.32	10.07	11.99
Lime—Hydrated	(40 lbs. per cu. ft.)	Tons	1.57	2.62	3.08	4.02	5.09	6.28	7.60	9.05
Cement in Bulk	(94 lbs. per cu. ft.)	Tons	3.69	5.32	7.23	9.45	11.96	14.76	17.87	21.26
Grain	(1 bu.=1.24445 cu. ft.)	Bushels	63.11	90.88	123.70	161.56	204.48	252.45	305.46	363.52
Liquid	(231 cu. in.)	Gallons	587.52	846.05	1151.55	1504.03	1903.57	2350.08	2843.57	3384.11
Interspace between 4 bins		Cu. Ft.	46.97	61.27	77.28	95.02	114.47	135.64	158.52	183.12
Interspace between 4 bins		Bushels	37.38	49.28	62.14	76.36	92.04	109.00	127.38	147.15

TO USE TABLE:—Example:

How many tons of hard coal will a bin 20' in diameter and 25' high hold—From table, 8.17 x 25'=204 tons capacity.

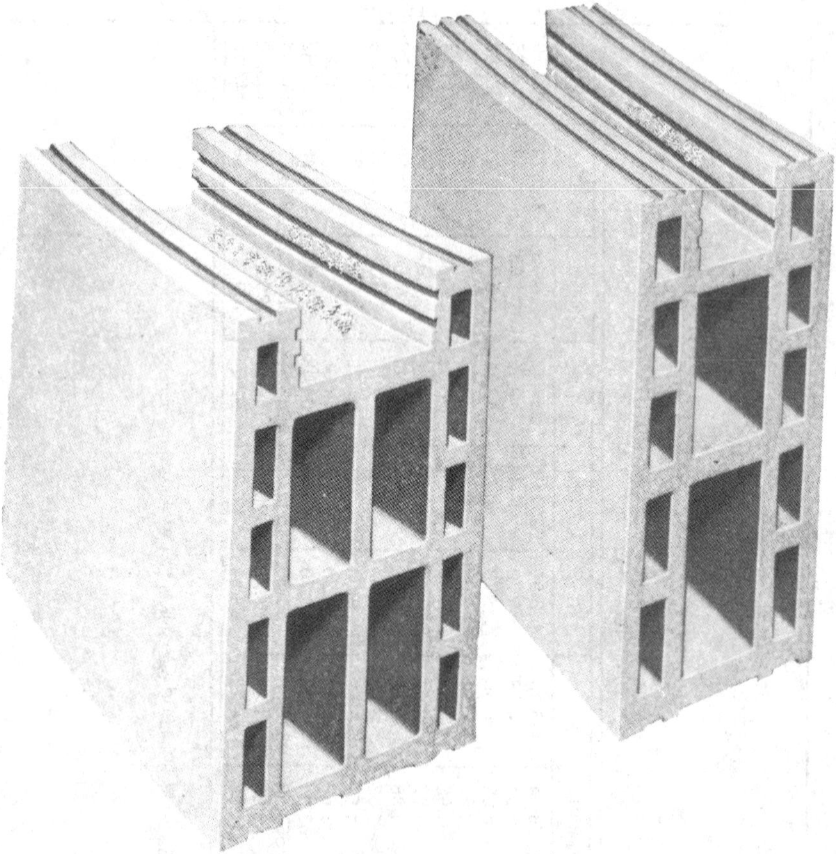

Number of 6"x7 3/4"x6 3/4" Double Grooved Concrete Manhole Blocks and Quantity of Mortar Required to Lay Up Manholes and Catch Basins of Various Sizes

Depth of Manhole		Internal Diameter of Manhole 30"	36"	42"	48"
	No. Block (Cone)	8	16	24	32
4 Ft.	No. Block (Bbl.)	35	32	27	20
	Total	43	48	51	52
	Cu. Ft. Mortar	1 1/2	2	2 1/2	3
	No. Block (Cone)	8	16	24	32
6 Ft.	No. Block (Bbl.)	56	56	54	50
	Total	64	72	78	82
	Cu. Ft. Mortar	2	3	4	5
	No. Block (Cone)	8	16	24	32
8 Ft.	No. Block (Bbl.)	77	80	81	80
	Total	85	96	105	112
	Cu. Ft. Mortar	2 1/2	3 1/2	4 1/2	5 1/2
	No. Block (Cone)	8	16	24	32

Depth of Manhole		Internal Diameter of Manhole 30"	36"	42"	48"
10 Ft.	No. Block (Bbl.)	99	104	108	110
	Total	107	120	132	142
	Cu. Ft. Mortar	3	4	5	6
	No. Block (Cone)	8	16	24	32
12 Ft.	No. Block (Bbl.)	119	128	135	140
	Total	127	144	159	172
	Cu. Ft. Mortar	3 ½	5	6 ½	8
	No. Block (Cone)	8	16	24	32
14 Ft.	No. Block (Bbl.)	140	152	162	179
	Total	148	168	186	202
	Cu. Ft. Mortar	4	5 ½	7 ½	9 ½

Concrete Manhole and Catch Basin Block

Concrete manhole and catch basin block are radial concrete units used for building circular manhole or catch basins in storm or sanitary sewers. The best quality units meet or exceed the specifications of the American Concrete Institute.

Size and Weight of Units. For manholes or catch basins having 48" inside diameter, 10 units are usually required for each course; for 42" 9 units; 36" 8 units, and 30" require 7 units. However, one popular type of block requires 12 units per course in 48" structures and 9 units in 36". A block 7 ¾" high and 6" thick weighs approximately 50 lbs. and a block 5 ¾" high and 6" thick weighs approximately 35 lbs.

Approximate Prices of Manhole Covers and Frames

Steel 18" Diameter	$200.00
Steel 24" Diameter	$260.00
Steel 36" Diameter	$575.00

Method of Construction. The concrete base is placed and then the first course of blocks is embedded at least 3" in the plastic concrete, leveling the units at once, as this construction insures a watertight joint at the base.

Labor Costs for Concrete Block Manholes and Catch Basins. The estimated cost of labor for a concrete block manhole, 48" inside diameter and 8'-0" deep, is as follows :

	Hours	Rate	Total	Rate	Total
Mason	7.5	$....	$....	$25.62	$192.15
Labor	7.5	....	....	19.13	143.48
Cost per manhole			$....		$335.63

Add for excavation, backfill, concrete base and cover.

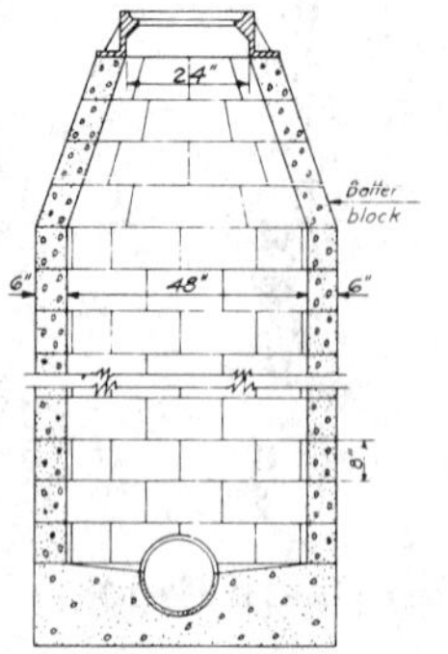

Note "Cone" construction consists of four courses of special batter block

CONCRETE BLOCK MANHOLE

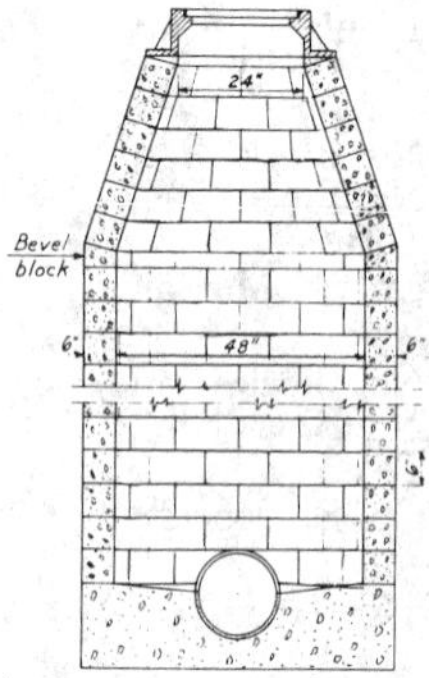

Note "Cone" construction consists of one course of bevel block and subsequent courses of standard units, using one less in each succeeding ring.

CONCRETE BLOCK MANHOLE

Seismic Concrete Masonry

In some regions of the U.S., there are seismic design requirements, and at times, additional material and labor are necessary to meet these local requirements. Additional design requirements should be included in the drawings and outlined in the specifications by the design engineer.

Reinforcement. In seismic zones 2, 3, and 4 on the accompanying map, the minimum area of reinforcement (in either direction) should not be less than 0.07% of the gross cross-sectional area of the wall. The sum of percents of horizontal and vertical reinforcement should be at least 0.2%. In other areas where reinforcement is used in selected portions of the wall as needed to resist tensile stresses, there is no requirement as to the minimum percentage of reinforcement. The size of bars used in reinforced masonry wall range from No. 4 through No. 11 bars.

The minimum and maximum vertical reinforcement in masonry columns is 0.25 and 4% respectively. Minimum size of bar is No. 4 and the minimum number of bars is four.

When a building experiences earthquake vibrations, its foundation will move back and forth with the ground. These vibrations can be quite intense, creating stresses and deformations throughout the structure. Flexible structures, such as frame buildings without shear wall stiffening elements, will endure large deflections often resulting in extensive non-structural damage. On the other hand, loadbearing concrete masonry buildings are much stiffer than frame buildings and resist deflection. Non-structural damage is minimized, as well as is hazard from falling debris. Loadbearing buildings are typically designed to resist twice the lateral force as buildings without stiffening elements. While limited research has been conducted on the damping properties of loadbearing buildings, its strength and structural field of performance have been demonstrated.

General. The latest edition of the *Uniform Building Code* requires every structure to be designed and constructed to resist minimum total lateral seismic forces assumed to act nonconcurrently in the direction of each of the main axes of the structure.

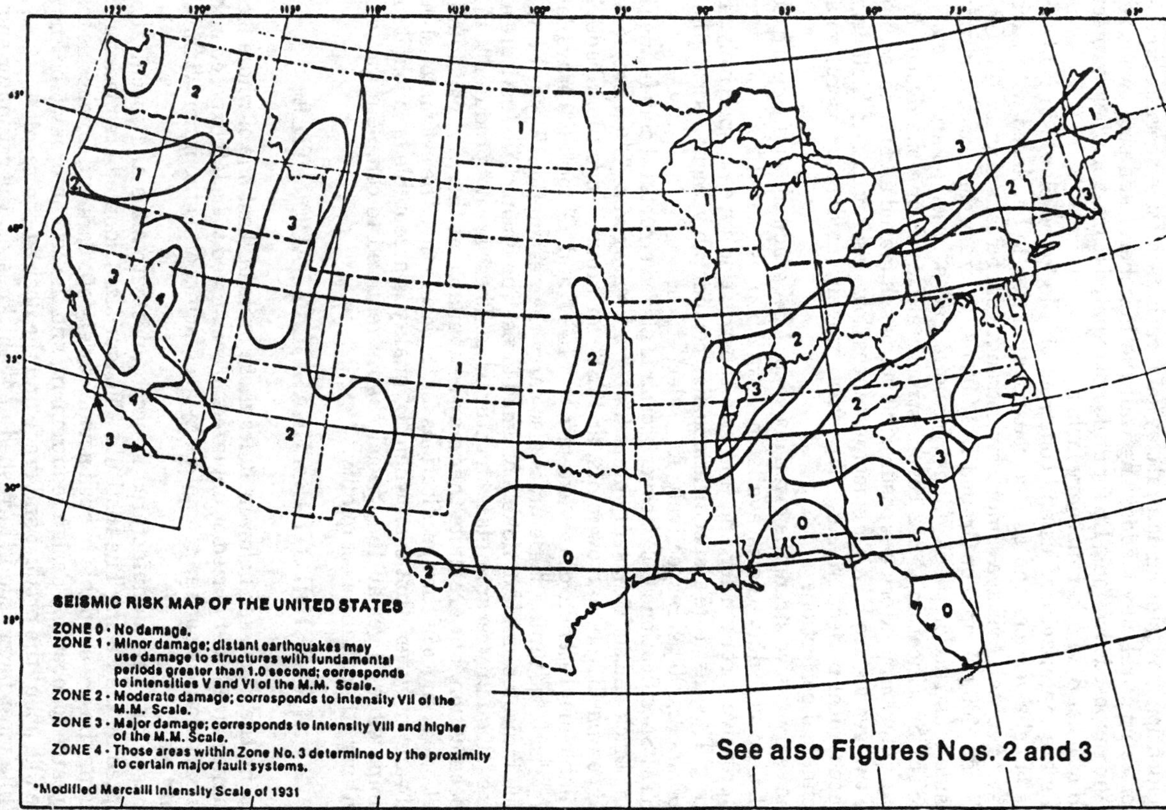
SEISMIC RISK MAP OF THE UNITED STATES
ZONE 0 - No damage.
ZONE 1 - Minor damage; distant earthquakes may use damage to structures with fundamental periods greater than 1.0 second; corresponds to intensities V and VI of the M.M. Scale.
ZONE 2 - Moderate damage; corresponds to intensity VII of the M.M. Scale.
ZONE 3 - Major damage; corresponds to intensity VIII and higher of the M.M. Scale.
ZONE 4 - Those areas within Zone No. 3 determined by the proximity to certain major fault systems.
*Modified Mercalli Intensity Scale of 1931
See also Figures Nos. 2 and 3

Figure 1—SEISMIC ZONE MAP OF THE UNITED STATES

LOADBEARING TILE

Loadbearing tile are used to carry part of the structural load of the building and are used for both interior and exterior walls. Some units have finished surfaces, some are scored for plaster, or other applied finishes.

The American Society for Testing and Materials (ASTM) has set up three specifications covering structural clay loadbearing wall tile. Specification C-34 covers structural clay loadbearing wall tile in general and sets two grades: LBX for exposed locations; LB for units not exposed to frost or where protected with at least 3" of other masonry facing.

Tile are manufactured as end construction tile, to be placed in walls with axes of cells vertical and as side construction tile, to be placed with axes of cells horizontal. Average compressive strength of end construction tile, Grade LBX, is 1,400 psi, and for Grade LB, 1,000 psi. Average compressive strength for side construction tile is 700 psi, for both grades, based on gross area.

ASTM Specification C-212 covers structural clay facing tile designed for either interior or exterior exposure, including salt glazed and unglazed units. The two grades are FTX for smoothest faces, low absorption and high degree of mechanical perfection, and FTS, which provides moderate absorption and mechanical perfection.

The compressive strengths given above for LBX grade tile apply here except that a special duty tile is available with 2,500 psi and 1,200 psi average compressive strength for end and side construction types respectively. The Facing Tile Institute's specifications list considerably higher strengths of 3,000 and 2,000 psi respectively, for "select" tiles.

ASTM Specifications C-126 covers ceramic glazed structural clay facing tile, facing brick and solid masonry units. Two grades are given: "S" (select) for use with comparatively narrow joints, and "SS" for ground edge tile. Grade S is available in all standard sizes except 7 ¾" x15 ¾" and "G" (or "SS") is available in 7 ¾"x15 ¾" units. These units also carry the designation, Types I and II, which indicates single or double finished faces. The Facing Tile Institute also lists a "B" quality ceramic glaze structural tile. These are not produced to meet a specification, but are units that may be available at the yard which fail to meet one or more of the specifications for "S" grade tile. The estimator should be aware, however, that even when "B" quality is acceptable on a job, it may be impossible to get all the special shapes required to complete the job in anything other than select quality.

INTERIOR STRUCTURAL CLAY FACING TILE

Facing tile discussed below includes those units at least one of whose faces, ceramic glazed, salt glazed or unglazed, are designed to be left exposed either as a partition, as backup to another type of facing, or as the inner wythe of a cavity wall.

These units are manufactured in accordance with sizes and specifications adopted by the Facing Tile Institute. The four sizes adopted are modular.

Facing tile are available in such a wide variety of colors and finishes as to be adaptable to most any design requirement, and the estimator must be knowledgeable about the product and grade and the desired end result.

While most facing tile is furnished in 4" thicknesses, the estimator should always check the cost of using 2" soaps in combination with a locally produced backing material against the cost of the full width tile, particularly if the wall carries no superimposed loads. Often, structural tile is used where its load-bearing capacity is never utilized.

Mortar design mixes for tile follow those listed in the previous chapter for brick, except interior non loadbearing partitions may utilize a low strength type "O" mortar that calls for .111 cu. ft. of cement and .222 cu. ft. of lime for each cubic foot of sand. Note, however, that in much tile work white portland cement and white sand may be called for. Most facing tile is figured for 1/4" joints, which limits sand to that passing a No. 16 sieve.

In the figures given below, mortar is figured as two separated 1" wide joints except for the bed joints of horizontal celled units which are figured as full. Where walls are built up of two units in thickness, metal ties should be figured 16" vertically and 36" horizontally, which will work out about 25 per 100 sq. ft. in all. Also if a solid filled collar joint is desired add 2.6 cu. ft. of mortar per 100 sq. ft. of wall.

Material Requirements for Facing Tile

8W Series.
Size: 8"x16"x2" or 4" thick
(7 3/4"x15 3/4" actual face size)
Tile per sq. ft. wall-1.125
Mortar in cu. ft.:
For vertical cell tile:
.77 per 100 sq. ft.
6.87 per 1000 units
For horizontal cell tile:
1.23 per 100 sq. ft.
10.92 per 1,000 units
Cost: 2" thick -$2,475 per M, or $2.79 per sq. ft., glazed 1 side
4" thick -$2,725 per M, or $3.07 per sq. ft., glazed 1 side
4" thick -$4,165 per M, or $4.69 per sq. ft., glazed 2 sides
6" thick -$3,865 per M, or $4.35 per sq. ft., glazed 1 side
8" thick -$4,625 per M, or $5.21 per sq. ft., glazed 1 side

6T Series.
Size: 5 1/3 x 12"x2", 4", 6" and 8" thick.
(5 1/16" x11 3/4" actual face size)
Tile per sq. ft. wall -2.250
Mortar in cu. ft.:
For vertical cell tile:
1.11 per 100 sq. ft.
4.94 per 1,000 units
For horizontal cell tile:
1.80 per 100 sq. ft. for 4" wall
7.98 per 1,000 units for 4" wall
2.58 per 100 sq. ft. for 6" wall
11.45 per 1,000 units for 6" wall
3.36 per 100 sq. ft. for 8" wall
14.93 per 1,000 units for 8" wall
Cost: 2" thick -$1,060 per M, or $2.39 per sq. ft., glazed 1 side
4" thick -$1,350 per M, or $3.04 per sq. ft., glazed 1 side
4" thick -$1,875 per M, or $4.22 per sq. ft., glazed 2 sides,
6" thick -$1,975 per M, or $4.44 per sq. ft., glazed 1 side,
8" thick -$2,365 per M, or $5.32 per sq. ft., glazed 1 side,

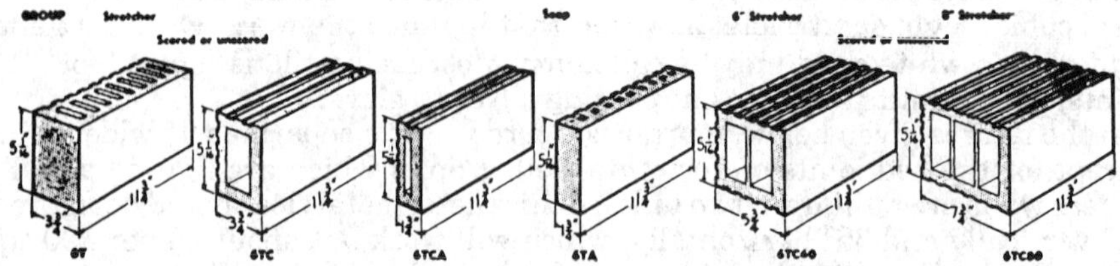

6T Series Structural Facing Tile, 5 ⅓"x12" Face

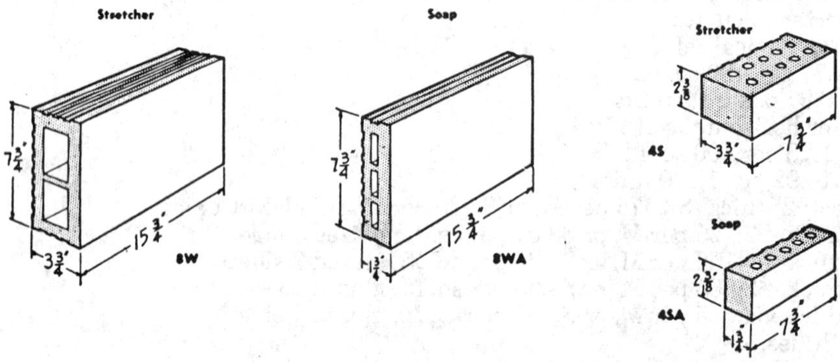

8W Series Structural Facing Tile, 8"x16" Face
4S Series 2 ⅔"x8" Face

Special Shapes. In addition to stretcher, or field units, these series also include many special shapes. These are given as belonging to groups; the higher the number, the more expensive the unit.

In general, the various groups include the following, although not all units will be available in all tile.

Group I:
4" bed return, bullnose or square
2" and 4" soap return, bullnose or square
coved internal corner
octagonal external corner
octagonal interior corner
cost: 8W--$2,690 per M
6T--$1,740 per M

Group II:

2" and 4" bed, bullnose cap
2" bed, cove base
cost: 8W--$4,525 per M
6T--$2,015 per M

Group III:

6" and 8" bed, bullnose cap
4" bed, cove base
2" bed, round top cove base
4" and 8" bed, full and half ends glazed both sides
4" bed, bullnose cap, glazed both sides
4" bed, sloped sill
Standard radial external or internal field units:
2" and 4" bed, 6 3/8" high, bullnose cap unit
2" and 4" bed, 7 3/4" high, bullnose cap unit
cost: 8W--$6,415 per M
6T--$3,175 per M

Group IV:

4" bed, miter unit
4" bed, miter unit with top return
4" bed, miter unit with finished end
2" and 4" bed, bullnose return
4" bed, bullnose return with starter
2" and 4" bed, cove base with return
4" bed, round top cove base
6" bed, bullnose glazed both sides
6" and 8" bed, sloped sill
4" bed, bullnose starter
2" bed, bullnose reveal
2" bed, bullnose coved internal corner
2" bed, bullnose octagonal internal and external corners
4" bed, 6 3/8" high slope sill
4" bed, 7 3/4" high slope sill
cost: 8W--$10,225 per M
6T--$6,675 per M

Group V:

4" bed, bullnose with 4" return
4" bed, miter 7 3/4" high with face return
4" bed, round top cove base with 4" return
2" bed, round top cove base with 4" return
2" bed, bullnose radial
2" bed, bullnose cove base radial
4" bed, bullnose internal corner with 4" reveal
6" bed, 6 3/8" high slope sill
2" bed, round top cove base with square corner
2" bed, round top cove base with coved interior corner
2" bed, round top cove base with octagonal exterior corner
2" bed, round top cove base with octagonal internal corner
4" bed, field radial starter
cost: 8W--$14,740 per M
6T--$7,165 per M

Group VI :
4" bed, bullnose full end, or half end glazed both sides
8" bed, bullnose half end, glazed both sides
8W--$18,226 per M
6T--$8,840 per M

Group VII :
4" bed, bullnose radial starter
4" bed, cove base radial starter

Estimating Facing Tile Quantities

The most widely used estimating procedure is the wall area method, which is as follows:

a) Compute gross wall area (height times length). Where coved base and/or bullnosed caps are specified, do not figure these in overall height.

b) Determine net area of wall; deduct wall openings. These are computed by using the actual width multiplied by the actual height plus the height of special sill and lintel units. Where openings overlap base or cap units, do not include this in the opening height as it was not included in the gross area. Also deduct for other special shapes by multiplying their height by the following factors:
For exterior bullnose or square corners--1.167 per ft. of height.
For Interior coved corners--.833 per ft. of height.
For bullnose or square jambs--.75 per ft. of height.

c) Determine number of stretchers required by multiplying net area by number of stretchers required per square foot:

30,58W:1.12
6T:2.25
4D:3.37

d) Compute number of standard fittings such as corners and jambs by the number of courses. Cove bases and caps, which should now be included as well as sills and lintels, are computed by the lineal foot. Miters, cove base corners, etc., should be taken off individually.

e) Summarize quantities and separate fittings in particular group.

f) Multiply by the unit cost figures.

g) Add 2% (or more, depending on job conditions) to cover breakage.

Example:

Step 1
Wall length = 29'-0"
Wall height, exclusive of the base course = 8'-0"
Gross area = 8.00 x 29 = <u>232 sq. ft.</u>

Step 2

Deductions:
Window area 4 x 4.89 = 19.56 sq. ft.
Door Area 2 x 6.67 = 13.33 sq. ft.
Exterior Corner 1.167 x 8.00 = 9.33 sq. ft.
Interior Coved Corner 0.833 x 8.00 = 6.66 sq. ft.
Jambs (4+4+6.22+6.22) x 0.75 = 15.33 sq. ft.
Total = 64.21 sq. ft.
Net wall area = 232 minus 64.21 = 167.79 sq. ft.

Step 3

Number of "6T" stretchers required 167.79 x 2.25 = <u>378</u>

Step 4

	Group I	II	III	IV	V
Exterior Bullnose corner					
18 courses of 5T4	18				
Interior coved corner					
18 courses of 4T8	18				
Jambs					
24 courses of 6T4	24				
24 courses of 3T4	24				
Lintels					
5 lin. ft. of 6T20		5			
1 miter 6T30R (door)				1	
1 miter 6T30L (door)				1	
1 miter 8T31R (window)					1
1 miter 8T31L (window)					1
Sill					
3 lin. ft. of 6T20		3			
1 miter 8T31R					1
1 miter 8T31L					1
Cove Base					
25 lin. ft. of 6T50			25		
1 outside corner 5T54L				1	
1 inside cove corner 4T58L				1	
1 bullnose jamb 6T504R				1	
1 bullnose jamb 6T504L				1	
Totals	84	8	25	6	4

Step 5 - Summary

Group	Series	Quantity
Stretcher	6T	378
I	6T	84
II	6T	8
III	6T	25
IV	6T	6
V	6T	4

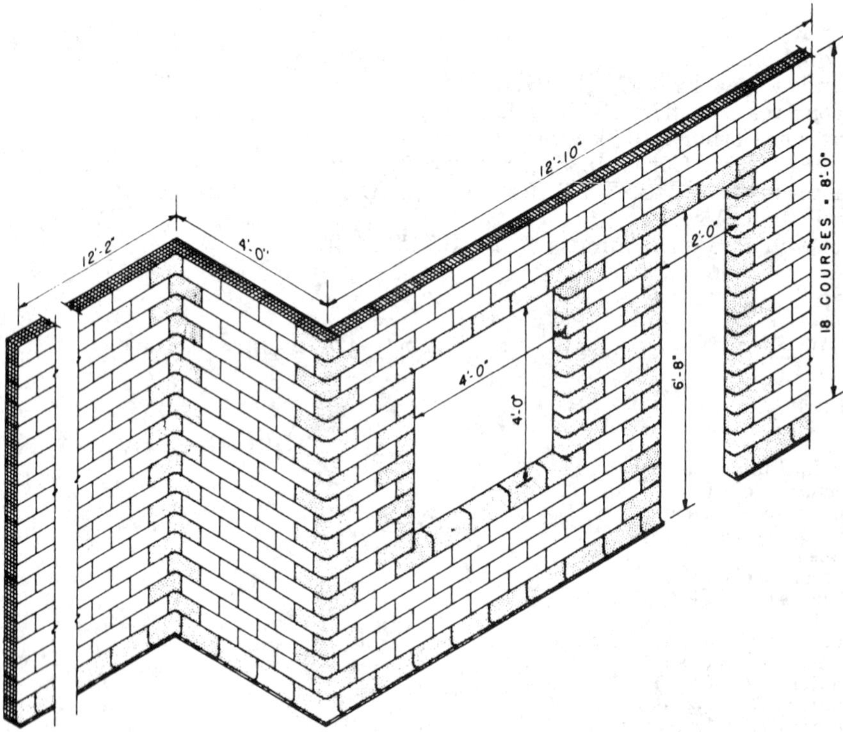

Example of Wall Area Method for Determining Facing Tile Quantities

Add a percentage for breakage between 2% and 5% depending on job conditions and distance of job from the plant. Chippage can frequently be used for special short lengths or other cuts.

A short cut method, devised by Stark Ceramics, is based on the following formula:

(Length x height minus actual openings) x (cost per sq. ft.) + (lineal feet x lineal foot cost factor) = total material cost.

In this method actual total heights are used to figure cross areas, and openings are figured using actual height and width with no allowances for sills, lintels, caps or bases. Lineal footage includes all special shapes: cove bases, internal coved corners, jambs, sills, lintels, caps, etc. The lineal foot cost factor is an average additional cost to cover these items. It is determined by taking an average of the group costs and deducting the stretcher unit cost already included in the net area figure.

Labor Laying Glazed Structural Facing Tile. To insure the most economical construction, as well as the most attractive appearance in the finished work, all cutting of tile on the job should be done with a power saw using a carborundum blade.

The following labor quantities are based on the average job requiring not more than 20% of special shapes, such as bullnose corners, cove base, cap, etc. If the percentage of special pieces vary from the above, increase or decrease labor time proportionately.

Labor Laying Glazed Structural Facing Tile
Labor Hours per 100 Pieces

Nominal Size of Tile*, Class of Work	No. of Pcs. per 8-Hr.Day	Mason	Labor
2"x5 1/3"x8", Glazed one side	155-175	4.8	4.8
4"x5 1/3"x8", Glazed one side	155-175	4.8	4.8
4"x5 1/3"x8", Glazed two sides	125-155	5.7	5.7
6"x5 1/3"x8", Glazed one side	155-175	4.8	4.8
6"x5 1/3"x8", Glazed two sides	125-155	5.7	5.7
8"x5 1/3"x8", Glazed one side	125-155	5.7	5.7
8"x5 1/3"x8", Glazed two sides	105-120	7.0	7.0
2"x5 1/3"12", Glazed one side	115-135	6.4	6.4
4"x5 1/3"x12", Glazed one side	115-135	6.4	6.4
4"x5 1/3"x12", Glazed two sides	90-110	8.0	8.0
6"x5 1/3"x12", Glazed one side	105-125	7.0	7.0
6"x5 1/3"x12", Glazed two sides	80-100	8.9	8.9
8"x5 1/3"x12", Glazed one side	85-100	8.7	8.7
8"x5 1/3"x12", Glazed two sides	65-85	10.6	10.6
2"x8"x16", Glazed one side	65-80	11.0	11.0
4"x8"x16", Glazed one side	55-65	13.3	13.3

Add for hoisting engineer if required.

*Nominal size includes thickness of standard mortar joint (1/4" for glazed tile) in length, height and thickness.

Tile 1 3/4" thick is usually used for wall furring or partition wainscoting. Tile glazed one side is usually used for partitions that are glazed on one side and plastered on the other side. Tile glazed two sides are usually used for partition walls glazed on both sides of wall.

Unglazed Structural Clay Tile

Backup tile are sized for backing up brickwork, but are also used with stone and other facing. These units are available with nominal face sizes of 5"x12", 8"x12", and 12"x12", in thickness of 4", 6", 8" and 12", and are usually furnished scored one side, smooth on the other. These units are often available, especially in larger sizes, in special designs with cast in handles for greater speed and ease in handling and non-continuous mortar joints to protect against moisture penetration. Generally, backup tile is sized to have 1/2" joints, which is the mortar amount given below, but certain units will be found to allow for less. Dimensional tolerances are wider in this type than in facing tile. The mortar figures below assume joints consist of two separated 1" wide joints and bed joints are two separated 2" wide joints.

Nominal 5"x12" tile is 4 7/8"x11 1/2"x3 1/2", 5 1/2", 7 1/2", and 11 1/2" actual, with 2.25 tile per sq. ft. wall, 3.75 cu. ft. of mortar per 100 sq. ft. of wall, and 16.69 cu. ft. of mortar per 1,000 units.

Nominal 8"x12" tile is 7 1/2"x11 1/2"x3 1/2", 5 1/2", 7 1/2", and 11 1/2" actual, with 1.5 tile per sq. ft. of wall, 2.73 cu. ft. of mortar per 100 sq. ft. of wall, and 18.23 cu. ft. of mortar per 1,000 units.

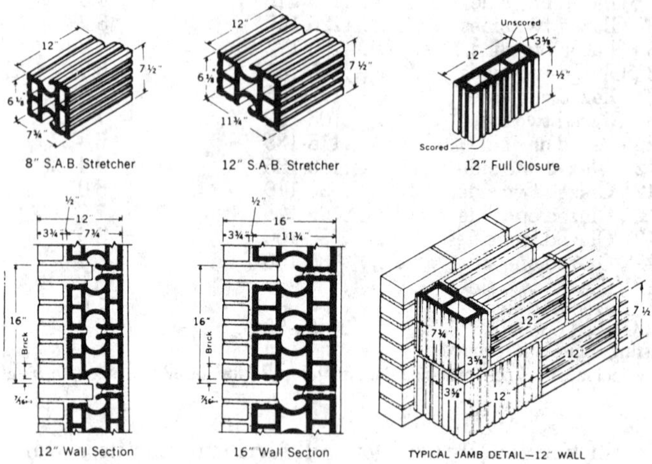

Speed-A-Backer Tile

Labor Laying Backup Tile. The following quantities are based on the average job, having the usual number of openings, pilasters, etc.

Size Tile	No. Pcs. Laid per 8-Hr. Day	Labor Hrs 100 Pcs. Mason	Labor	Engr.*	No. Sq. Ft. Laid per 8-Hr. Day	Labor Hrs. 100 Sq. Ft. Mason	Labor	Engr.*
4"x5"x12"	330-370	2.25	2.25	0.3	150-170	5.0	5.0	0.6
8"x5"x12"	250-290	3.00	3.00	0.3	115-130	6.5	6.5	0.7
4"x8"x12"	200-225	3.80	3.80	0.3	135-150	5.7	5.7	0.6
8"x8"x12"	155-185	4.70	4.70	0.5	110-132	6.6	6.6	0.7

*Add for hoisting engineer only if required.

Unit Wall Tile

Some tile units are designed to be used for a complete single unit, loadbearing, nominal 8" wall construction, with finished surfaces inside and out. These are usually manufactured under special patents. Some of these units may have one finished surface only and be used in combination with an inside backup.

Faces are available in buff unglazed, either smooth or Ruggtex finish; salt glazed, red textured (to give the appearance of high quality face brick); and ceramic glaze. Certain units are offered with supplemental full and half jamb corners, and lintel and pilaster units along with interior stretcher units to match the interior face of the unit tile.

DENISON BACK UP TILE

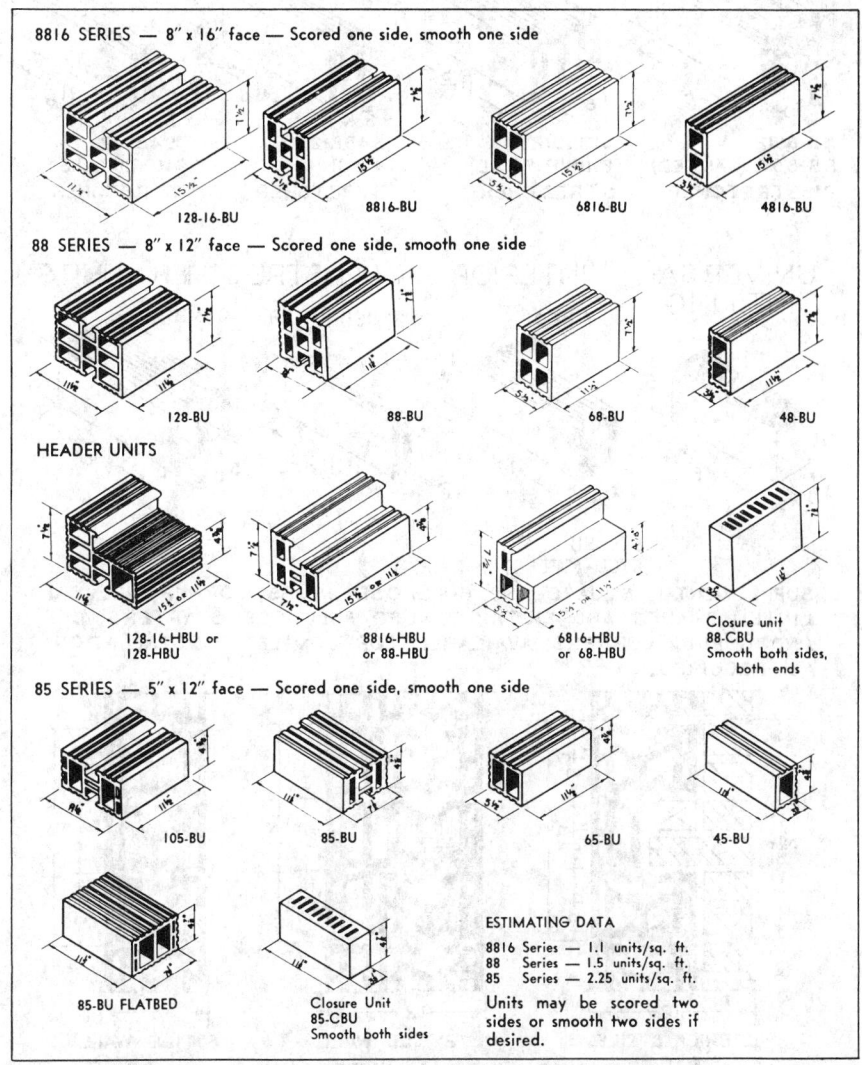

8816 SERIES — 8" x 16" face — Scored one side, smooth one side

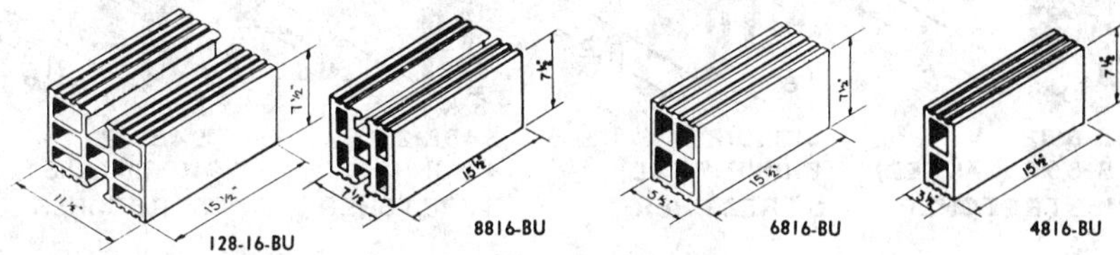

128-16-BU 8816-BU 6816-BU 4816-BU

88 SERIES — 8" x 12" face — Scored one side, smooth one side

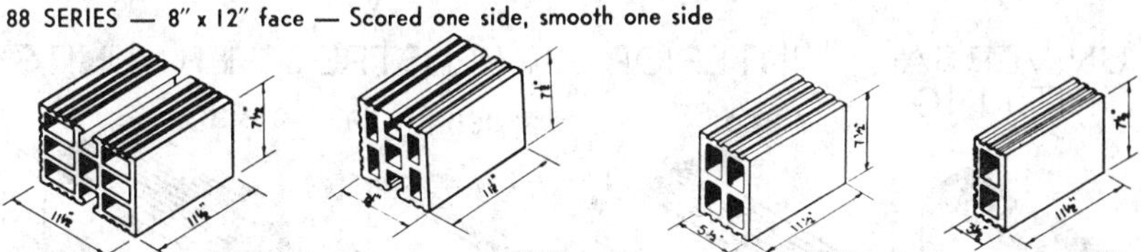

128-BU 88-BU 68-BU 48-BU

HEADER UNITS

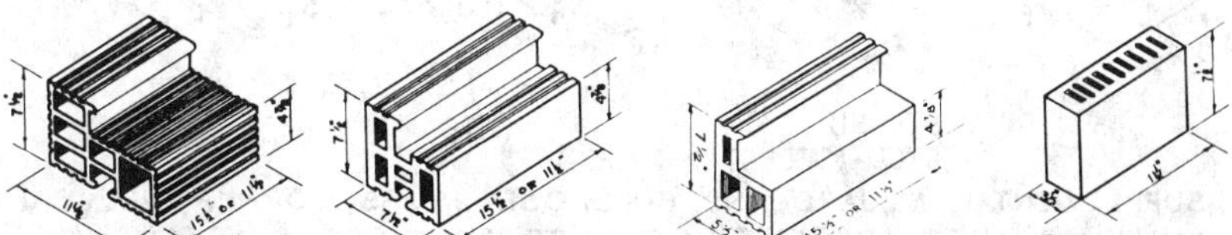

128-16-HBU or 128-HBU

8816-HBU or 88-HBU

6816-HBU or 68-HBU

Closure unit 88-CBU Smooth both sides, both ends

85 SERIES — 5" x 12" face — Scored one side, smooth one side

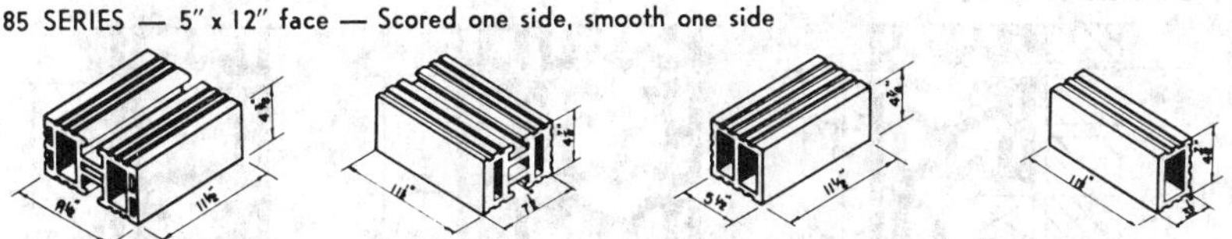

105-BU 85-BU 65-BU 45-BU

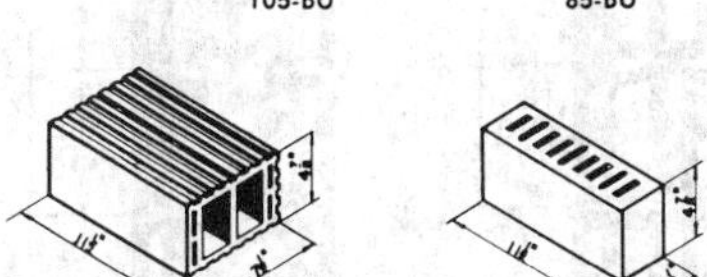

85-BU FLATBED

Closure Unit 85-CBU Smooth both sides

ESTIMATING DATA

8816 Series — 1.1 units/sq. ft.
88 Series — 1.5 units/sq. ft.
85 Series — 2.25 units/sq. ft.

Units may be scored two sides or smooth two sides if desired.

DENISON STRUCTURAL CLAY TILE
MASON CITY BRICK AND TILE COMPANY, MASON CITY, IOWA

Units are also available in 4" widths for use as exterior wythes in cavity wall construction. A wall of this type is highly impervious to water, thoroughly fire, termite and vermin proof, and cannot rot or decay. Textures offer a broad range including brick-like textures and colors suitable for homes and schools; glazed for situations where a sanitary finish is desired, and smooth faced units in natural range shades suitable for larger structures and areas.

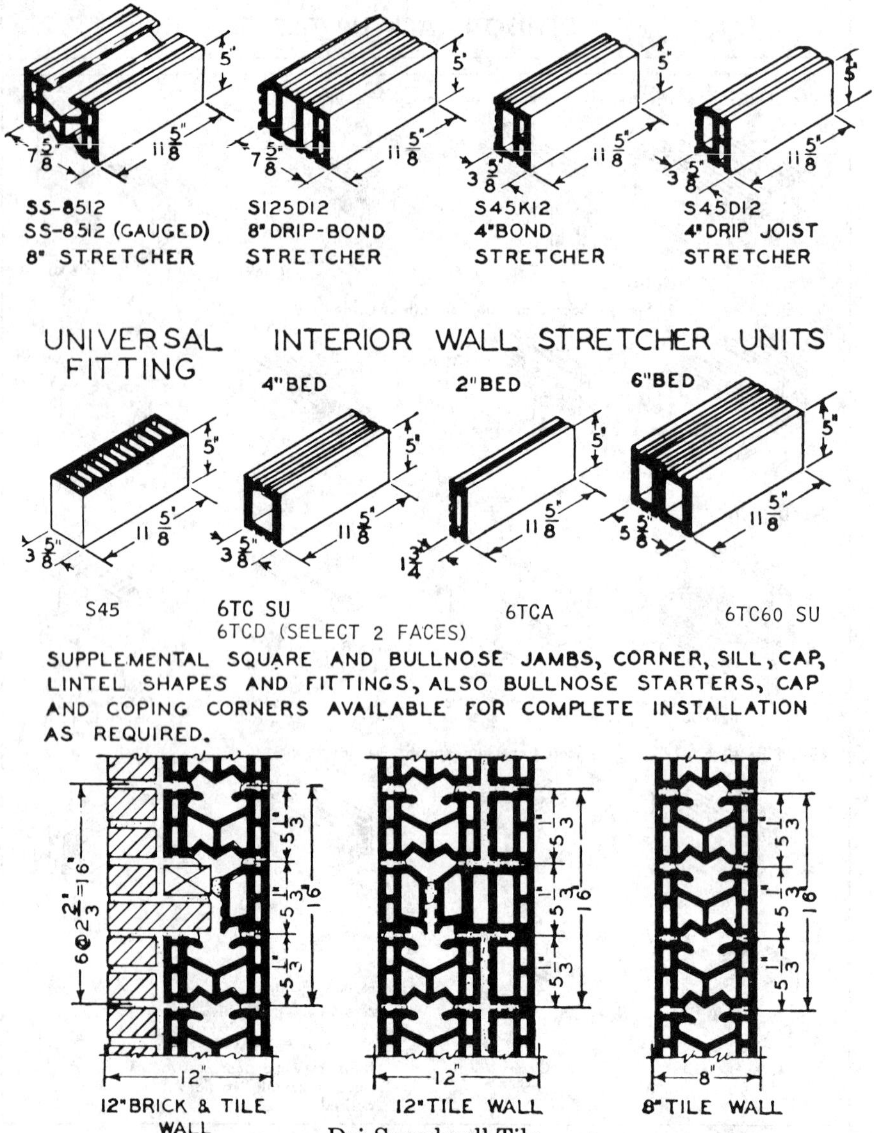

Dri-Speedwall Tile

The estimator must know exactly the quality of interior finish desired. While it is possible to obtain high grade finishes and workmanship both sides, such high quality is not always required, especially in larger warehouse and manufacturing buildings, and large economies can result where "select one face" units can be used. Often the lower wall may be constructed "select two faces" while "select one face" is used above. In general, though, the estimator must expect a variation of 1/8" in thickness, and if this is not acceptable, he should plan to use a built-up wall with a separate interior facing.

GLEN-GERY TEX DRI-WALL TILE (NOMINAL 4X12 FACE SIZE)

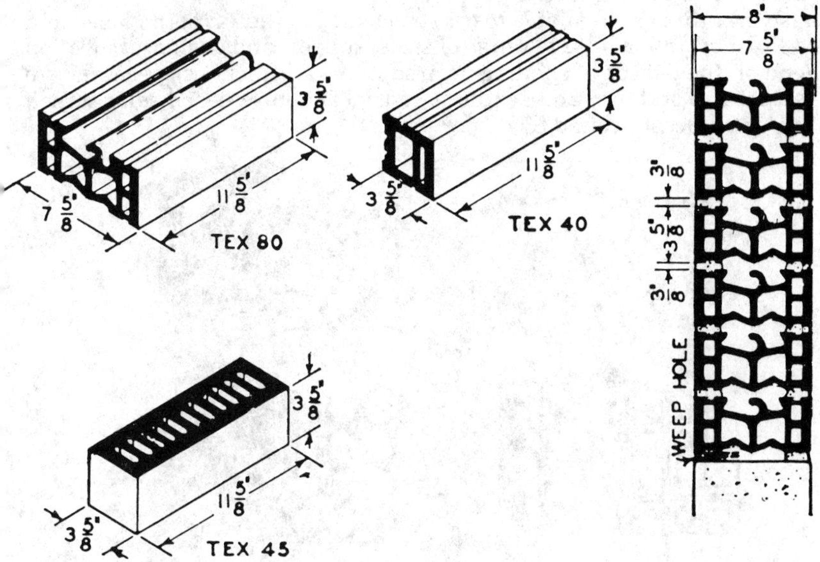

Unit wall tile comes in two nominal face sizes: 4"x12", and 5 ⅓"x12", in 4" and 8" widths.

4"x12" units, which include Uniwall and Tex Dri-wall, have an actual face dimension of 3 ⅝" x11 ⅝" with 3 ⅝" and 7 ⅝" thicknesses. These units will lay up 300 per 100 sq. ft. of wall and require 7.5 cu. ft. of mortar for 4" walls; 9.75 cu. ft. for 8" walls. Labor laying 4"x12" unit wall tile:

Size of Tile	No. Laid 8-Hr. Day	Labor Hours per 100 Pcs Mason	Labor Hours per 100 Pcs Labor	No. Sq. Ft. 8-Hr. Day	Labor Hours per 100 Sq. Ft. Mason	Labor Hours per 100 Sq. Ft. Labor
3 ⅝"x3 ⅝"x11 ⅝"	170-200	4.33	4.33	57-67	13.0	13.0
7 ¾"x3 ⅝"x11 ⅝"	150-170	5.0	5.0	50-57	15.0	15.0

The 5 ½"x12" units have actual face dimensions of 5"x11 ⅝" with 3 ⅝" and 7 ⅝" thicknesses, though they might vary somewhat in various parts of the country from 4 ⅞" to 5 1/16" in height and 11 ½" to 11 ¾" in length. Units can be figured to lay up 225 tile per 100 sq. ft. of wall and require 8.5 cu. ft. of mortar for 4" walls and 9.25 for 8" walls, the latter varying as to actual face size. Labor laying 5 ⅓"x12" unit tile:

Size of Tile	No. Laid 8-hr. Day	Labor Hours per 100 Pcs Mason	Labor Hours per 100 Pcs Labor	No. Sq. Ft. per Day	Labor Hours per 100 Sq. Ft. Mason	Labor Hours per 100 Sq. Ft. Labor
3 ⅝"x5"x11 ⅝"	175-200	4.25	1.90	80-90	9.50	4.12
7 ⅝"x5"x11 ⅝"	130-155	5.50	2.50	60-70	12.25	5.50

Clay Masonry Shading Devices

Architects have been experimenting with sun controls beyond the usual shades, blinds, and drapes. An effective way of sun control is the egg crate screen wall design which baffles both the low east and west rays of the sun as well as the high direct south rays.

Clay masonry units are available to construct such baffles in both glazed and unglazed finishes and in a large range of sizes, shapes, and designs. Selection will depend on the degree of shading desired, the wall thickness necessary to provide structural stability and aesthetic considerations. Glazed solar screen units 4" thick will cost around $7.00 per unit.

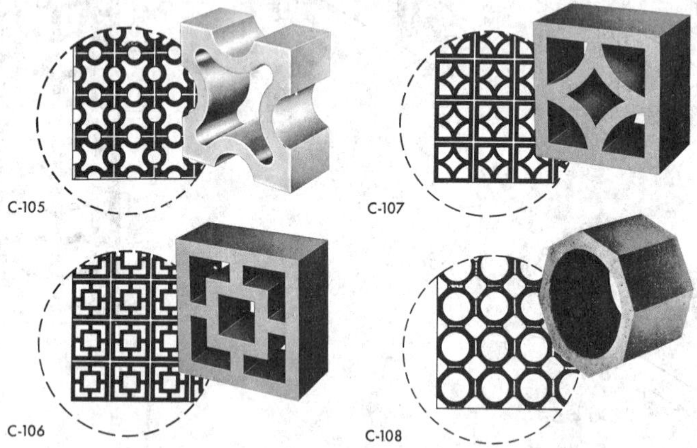

Four Patterns of Solar Screen Tile
Size: 7 5/8x7 5/8 for 3/8" mortar joints, thickness: 3 5/8", 5 1/16" and 7 3/4"

UNIT MASONRY/CERAMIC VENEER

Terra cotta is a building material made of a high grade clay (usually a mixture of several different kinds), to which is added a percentage of calcined clay to uniformly control the shrinkage and prevent undue warping.

Terra cotta is used for the same purposes as other building materials, like sandstone, limestone, granite, press brick, etc.; i.e., for interior or exterior trimmings or for entire fronts. It is also used extensively as glazed terra cotta for light courts, for the interior walls of swimming pools, operating rooms in hospitals, clinics and patients' rooms, corridors, vestibules and stair halls (full height of wainscot) and in large kitchens where a ceramic material is required for sanitary purposes as well as for fire resisting advantages. It is used for service stations, power houses and rooms housing machinery on account of the ease with which it can be kept free from dust, dirt and grease. Its unlimited possibilities in color effects make it an ideal material for store fronts. It is used as a lining for the interior of restaurants, palm courts, hotels, residences and railroad stations. In stations and public buildings it is also used in the construction of ceilings and domes. In fact, it is a universal building material and has the advantage over other materials in that it may be produced in almost any color and may be glazed or unglazed with any surface treatment desired. It furnishes unlimited possibilities to the architect not only in color effect, but in the beauty of the original modeling in the plastic clay, an effect that cannot be obtained in carved stone except at tremendous cost.

Method of Manufacturing Architectural Terra Cotta. After shop drawings have been approved, full size models are made of plaster to a shrinkage scale,that is, to an increased size to allow for the shrinkage in drying and burning. A model is made for each different shape on the shop drawings. If a piece is ornamental, such ornament as may be required is modeled in clay and attached to the plaster model. From the models, sectional molds of plaster are cast, from which the required number of pieces of terra cotta are produced. The mixture of clays and fusible materials used in forming terra cotta is carefully selected and proportioned to give the desired degree of plasticity when mixed with water. The composition must be such that when fired at high temperatures, it will produce a homogeneous body amply strong to support the required loads.

Forming terra cotta is a manual operation and consists of pressing the soft clay mixture into the molds. The walls of the pieces should be not less than one-inch thick following the contour of the mold, and the partitions should be properly sized and spaced to produce the strength required. The pressed piece remains in the mold until the clay stiffens, after which it is removed, retouched and placed in the dryer where most of the moisture in the clay is removed.

Following the drying process the exposed surfaces of each piece are sprayed with the ceramic mixture which, when fired, develops into the desired glaze color. The pieces are then placed in kilns and fired by gradually increasing the temperature to 2,000° F or more, depending on the temperature of maturity of the clay and glaze. After proper firing the kiln is allowed to cool slowly to normal temperature.

After firing, the terra cotta is fitted for alignment by cutting or grinding and marked to correspond with the piece numbers on the shop drawings.

Method of Manufacturing Ceramic Veneer. Ceramic veneer is the usual term applied to machine made architectural terra cotta and the process of manufacturing is the same as for handmade pieces except that the material is extruded instead of hand pressed into molds.

A machine mixes the proportioned clay with water and forces the plastic mixture through a steel die which forms it into the desired shape. In most cases, before entering the die, the plastic mixture passes through a de-airing chamber where it is subjected to a partial vacuum and most of the entrapped air is removed. The pieces are then dried and surface finished to a true plane. Subsequent operations are the same as for handmade pieces and include application of ceramic mixture, firing, fitting and marking.

Cost of Architectural Terra Cotta and Ceramic Veneer. In estimating the cost of architectural terra cotta and ceramic veneer, there is no unit that can be taken as a definite basis for estimating. All estimates are usually made by the manufacturer in a lump sum price for the entire job.

To be able to accurately arrive at the cost of these materials, an estimator must have an intimate knowledge of the manufacturing process. Each job is a custom job, there are no standard sizes, colors or shapes which are stocked. One manufacturer suggests that the cost of architectural terra cotta will vary from $440.00 to $630.00 and higher per ton and that the cost of ceramic veneer will run from $3.10 to $5.00 and higher per sq. ft. The extreme range of these costs serves to emphasize the necessity for obtaining firm quotations from the manufacturers.

Some of the conditions governing the costs of both architectural terra cotta and ceramic veneer are design of work, character and amount of ornamentation, amount of duplication, sizes of pieces, color of glaze, weight, supply and demand, shipping conditions, etc.

Estimating Quantities of Architectural Terra Cotta and Ceramic Veneer. When estimating quantities of terra cotta from the plans or details, all measurements should be squared; i.e., all molded courses, cornices, round columns, column caps, bases, coursed ashlar, sills, copings, etc., should be figured from the extreme dimensions and reduced to cubic feet.

The weight of terra cotta varies from about 80 lbs. per cu. ft. for plain 4" thick ashlar to 60 lbs. per cu ft. for large molded courses. For an average job about 72 lbs. per cu. ft., equivalent to 28 cu. ft. per ton of 2,000 lbs. may be used.

After the quantities have been taken off the plans and reduced to cu. ft., they may either be priced in that manner, or multiplying the number of cu. ft. by 72 lbs. will give the total weight of the terra cotta required for the job.

Ceramic veneer quantities are measured by the square foot area and should include all returns at doors, windows, soffits, etc.

Cost of Handling and Setting Architectural Terra Cotta. The labor cost of handling and setting architectural terra cotta will vary with the sizes of the different pieces, the amount of duplication and the method used in handling and setting same.

Terra cotta weighs about half as much as stone, so that it is seldom necessary to use derricks in handling and setting, as it is usually furnished in sizes that can be conveniently handled by 2 or 3 workers. Considerable space is usually required on the job for storing the terra cotta, as it is sent to the job each piece is marked to show its exact position in the building and must be carefully handled and placed in convenient piles for use as required. When handling the terra cotta on the job, it is customary to load it into wheelbarrows or stone barrows and wheel it to the location where it is to be used. However, if it is necessary to hoist it to the upper floors or roof of the building, the barrows can be placed on the material hoist, raised to the proper floor and then wheeled to the location where it is to be set.

Inasmuch as a great deal of terra cotta consists of trimmings or 4" ashlar, it requires considerable handling to set a cubic foot. For instance, most ashlar being 4" thick, it is necessary to set terra cotta in 3 sq. ft. of wall to set one cu. ft. This also applies to sills, lintels, etc.

Based on the average job consisting of ashlar, sills, lintels, ornamental trimmings, etc., a mason will set about 35 to 40 cu. ft. (105 to 120 sq. ft. wall) per 8-hr. day, at the following labor cost per 100 cu. ft.:

	Hours	Rate	Total	Rate	Total
Terra cotta setter	21	$....	$....	$25.62	$538.02
Helper	21	....	....	19.13	401.73
Labor sorting, handling wheeling, etc.	44	....	....	19.13	841.72
Cost per 100 cu. ft.			$....		$1,781.47
Cost per cu. ft			....		17.81
per sq. ft (300)			....		5.94
per ton (28 c.f.)			....		498.81

For labor cost of setting plain terra cotta wall ashlar, use same quantities and costs as given on the previous pages under "Labor Laying Glazed Structural Facing Tile".

On more complicated work consisting of gothic architecture, balusters and balustrade, glazed and enameled terra cotta, and other intricate work having a great deal of detail and requiring the utmost care in setting, a terra cotta setter will set only 20 to 25 cu. ft. (60 to 75 sq. ft. wall) per 8-hr. day, at the following labor cost per 100 cu. ft.:

	Hours	Rate	Total	Rate	Total
Terra cotta setter	36	$....	$....	$25.62	$922.32
Helper	36	....	....	19.13	688.68
Labor sorting, handling, wheeling, etc.	66	....	....	19.13	1,262.58
Cost per 100 cu. ft			$....		$2,873.58
Cost per cu. ft			....		28.74
per sq. ft. (300)			....		9.58
per ton (28 c.f.)			....		804.60

Small lintels and similar courses that have to be hung, may cost 25% to 50% higher than given above but these items are exceptional.

In general, the voids in architectural terra cotta should be completely filled with setting mortar or grout to prevent the possible accumulation of moisture in the cavities.

All supporting iron should be firmly embedded in masonry and completely covered with mortar to prevent rusting. Bronze or other non-corrosive metal or cadmium coated steel anchors should invariably be used for suspended features or for anchoring or bracing freestanding work.

Mortar Required for Setting Architectural Terra Cotta. Under average conditions it will require about 4 cu. ft. of mortar to set one ton of terra cotta, or about 1/2 cu. yd. of mortar per 100 cu. ft.

The following proportions are recommended as being suitable for making satisfactory mortar for setting terra cotta: One part standard portland cement, one part lime putty and six parts sand.

Do not under any circumstances use plaster of Paris or salt in the mortar.

All exposed joints in projecting and overhanging features, parapet work, and work standing free, should be raked out and pointed with a good asphaltic elastic cement.

Never use a cement that swells in setting for filling the voids in terra cotta. Use only the best quality of cement.

Handling and Setting Ceramic Veneer. Ceramic veneer is available in two types, namely, adhesion type, commonly called "thin" ceramic veneer, and anchored type.

Adhesion Type Ceramic Veneer. Adhesion type ceramic veneer is not over 1 1/8" in thickness, and the maximum face areas of individual slabs do not exceed 540 sq. in. Maximum overall face dimensions are 18"x30" or 20"x27". Actual face dimensions will vary with the manufacturer.

This type of ceramic veneer requires no metal anchorage. It is held in place by the adhesion of the mortar to the veneer body and the backing wall. The overall thickness from the face of the veneer to the face of the backing wall is 1 3/4" to 2".

Adhesion type ceramic veneer may be applied to a variety of backings, such as concrete, masonry, wood, and metal lath. Where the backing is wood, a sheet of waterproof building paper should be installed first, followed by a layer of wire mesh. In the case of metal lath backing or wire mesh on wood backing, 9 1/4" scratch coat of mortar should be applied in advance of the setting of veneer slabs.

Mortar Required for Adhesion Type Ceramic Veneer. Under average conditions, it will require 8 to 10 cu. ft. of mortar to set 100 sq. ft. of adhesion type ceramic veneer.

The following proportions are recommended as being suitable for making satisfactory mortar for setting adhesion type ceramic veneer: two parts portland cement, one part lime putty and eight parts sand. A stearate type admix may be added to the mix in the proportions recommended by the manufacturer.

Setting Adhesion Type Ceramic Veneer. Strict compliance with the manufacturer's recommendations for setting adhesion type ceramic veneer is necessary to assure the development of full shearing strength of veneer to backing. Step-by-step setting directions, recommended by manufacturers are as follows.

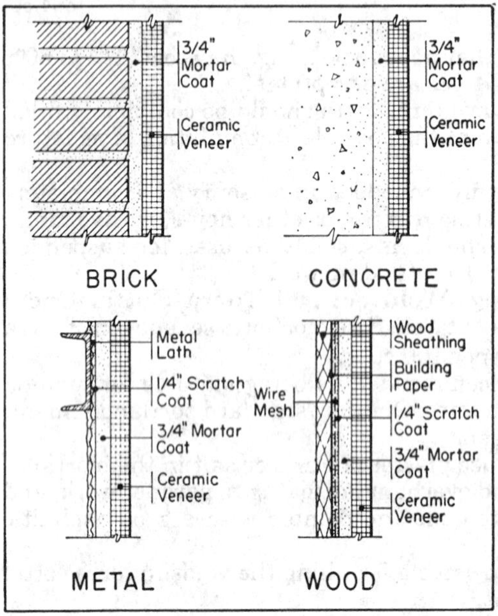

Method of Setting Adhesion Type Ceramic Veneer

All units must be soaked in clean water for at least one hour before setting. The backing wall surface should also be damp at the time of setting. Immediately before setting, the backing wall and the backs of veneer units should both receive a brush coat of portland cement and water. One-half of the setting mortar should be spread on the back of the unit to be set and the other half on the wall surface to be covered. The unit is then set in place, using wood wedges to maintain correct bed joint thickness. Sufficient mortar should be used to create a slight excess which will be forced out of the joints from the back when the unit is tapped into place, eliminating all air pockets and filling all voids. Face joints may be pointed as soon as mortar has set sufficiently to support the veneer slabs. As a final step, the surface of the veneer should be washed down with clean water.

Cost of Handling and Setting Adhesion Type Ceramic Veneer. On most jobs adhesion type ceramic veneer is delivered to the job in trucks and must be carefully unloaded, stored and protected in advance of the setting. Two laborers should do this work at the rate of 360 to 400 sq. ft. of veneer area per 8-hr. day.

Labor Cost Handling and Setting 100 Sq. Ft. Adhesion Type Ceramic Veneer on Concrete or Masonry Backing

Unloading and Storing	Hours	Rate	Total	Rate	Total
Labor	4.2	$....	$....	$19.13	$80.35
Setting					
Mason	11.5	....	....	25.62	294.63
Helper	11.5	....	....	19.13	220.00
Cost per 100 sq. ft			$....		$594.98
Cost per sq. ft			....		5.95

On the average job, where most of the veneer consists of typical panels, a mason and a helper should set 65 to 75 sq.ft. of veneer per 8-hr. day.

If veneer is to be applied to a backing of metal lath, add cost of 1/4" scratch coat of mortar.

If veneer is to be applied to a wood backing, add cost of one layer of waterproof building paper, one layer of wire mesh and 1/4" scratch coat of mortar.

Above costs do not include any allowance for scaffolding or hoisting. If hoisting is required, add 1 hr. hoisting engineer and 2 hrs. labor per 100 sq. ft. of veneer.

Anchored Type Ceramic Veneer. Anchored type ceramic veneer is recommended where a larger size slab is desired. Anchored type slabs are available in face sizes considerably larger than the adhesion type, again depending upon the manufacturer. Ribs or scoring are provided on the backs of such units and the overall thickness of the slabs range from 2" to 2 1/2". Depending on slab thickness, a total of 3" to 4 1/2" is required from rough wall to finished veneer surface to provide adequate grout space between the veneer and the backing. Anchor holes are provided in the bed edges of the slabs for the installation of loose wire anchors, which in turn, are fastened to pencil rods anchored to the backing. Once the wire anchoring is in place, the units are bonded to the wall by a poured grout core.

Grout Required for Anchored Type Ceramic Veneer. For ordinary work, it will require 20 to 25 cu. ft. of grout to fill the space back of 100 sq. ft. of anchored type ceramic veneer.

The following proportions are recommended for a suitable grout mix: One part portland cement, one part sand and five parts graded pea gravel passing a 1/4" sieve. Sufficient water should be added to cause the mixture to flow readily.

In addition to the above grout, it may require 1 to 2 cu. ft. of mortar per 100 sq. ft. of veneer for buttering the joints in setting to prevent leakage of grout. Any workable mortar mix may be used.

Setting Anchored Type Ceramic Veneer. The following directions for setting anchored type ceramic veneer are recommended by the manufacturers. As in the case of adhesion type veneer, both units and backing wall should be thoroughly soaked with clean water prior to setting.

Assuming the backing wall has had loop tie inserts or eye bolts properly located, the first step in veneer slab erection is the insertion of 1/4" diameter pencil rods into the loops projecting from the backing wall. The veneer slabs are then set in place and secured to the pencil rods with wire anchors inserted in the top of each unit. Wood wedges should be used to maintain the proper joint widths and the joints may be buttered with mortar or caulked with rope yarn or other material to prevent grout leakage. Temporary wood wedges should also be used to hold each piece in place at the required distance from the backing wall.

When the setting of each horizontal row of units is completed, grout is poured into the space between the veneer and the backing, puddled thoroughly and allowed to set. Wedges may be removed and joints pointed when grout has set sufficiently to hold slabs in place. To complete the operation the finished veneer surface should be washed down with clean water.

Cost of Handling and Setting Anchored Type Ceramic Veneer. Unloading, storing, and protecting anchored type veneer slabs is carried on in the same manner as for adhesion type slabs except that these units, being larger and heavier, are somewhat more difficult to handle. Two laborers should unload and store 320 to 360 sq. ft. anchored type slabs per 8-hr. day.

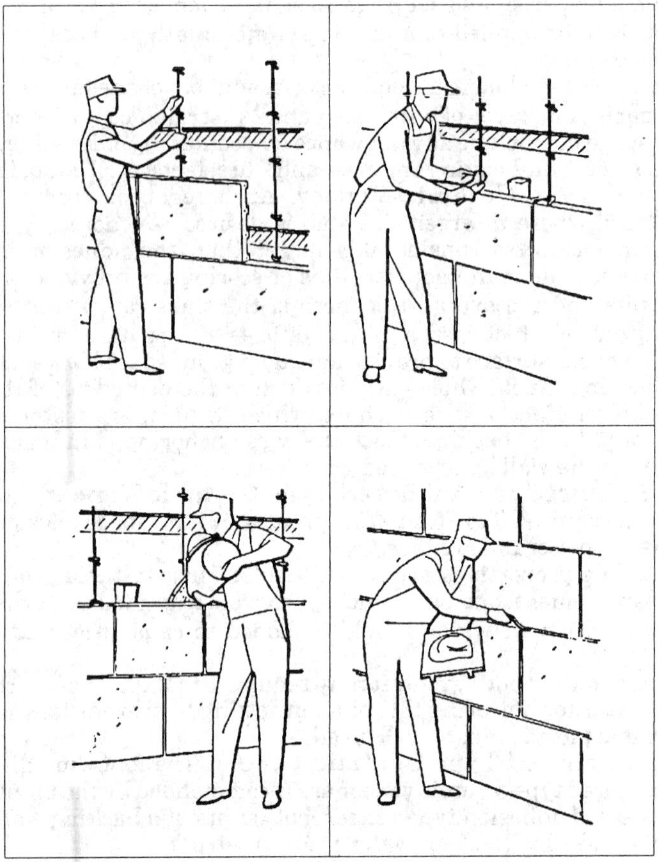

Steps in Setting Anchored Type Ceramic Veneer

Setting anchored type veneer slabs is dependent on the size and weight of slabs, nature of design, etc., but on an average facing job a mason and a helper should set, grout and point 60 to 70 sq. ft. of veneer per 8-hr. day.

Labor Cost Handling and Setting 100 Sq. Ft. Anchored Type Ceramic Veneer

Unloading and Storing	Hours	Rate	Total	Rate	Total
Labor	4.7	$....	$....	$19.13	$89.91
Setting					
Mason	12.3	....	....	25.62	315.13
Helper	12.3	....	....	19.13	235.30
Cost per 100 sq. ft			$....		$640.34
Cost per sq. ft			....		6.40

The above costs do not include any allowance for scaffolding or hoisting. If hoisting is required, add 1.5 hrs. hoisting engineer and 3 hrs. labor per 100 sq. ft. of veneer.

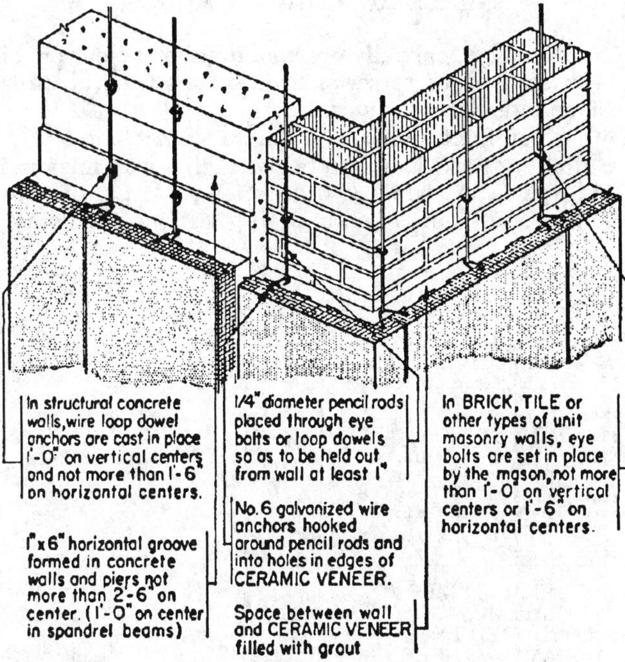
In structural concrete walls, wire loop dowel anchors are cast in place 1'-0" on vertical centers and not more than 1'-6" on horizontal centers.
1"x6" horizontal groove formed in concrete walls and piers not more than 2'-6" on center. (1'-0" on center in spandrel beams)
1/4" diameter pencil rods placed through eye bolts or loop dowels so as to be held out from wall at least 1"
No. 6 galvanized wire anchors hooked around pencil rods and into holes in edges of CERAMIC VENEER.
Space between wall and CERAMIC VENEER filled with grout
In BRICK, TILE or other types of unit masonry walls, eye bolts are set in place by the mason, not more than 1'-0" on vertical centers or 1'-6" on horizontal centers.

Method of Setting and Securing Anchored Type Ceramic Veneer Terra Cottato Brick and Concrete Backing

Cleaning Down or Washing Terra Cotta. Broadly speaking, there are three different surface finishes in terra cotta: lustrous glazed, matt glazed, and unglazed. The glazed surfaces are the easiest to clean.

For both lustrous and matt glazed finishes a good abrasive soap or washing powder is best. Where required, there can be added to the latter a slight proportion of sharp sand. A stiff lather should be made and the surface scrubbed hard, allowing the lather to remain long enough to soften the dirt and then rinse off with clean water. Acid is not necessary for cleaning glazed surfaces.

In the past, for cleaning unglazed terra cotta, the addition of a slight proportion of commercial muriatic acid was recommended. One quart of acid to 4 gallons of water was sufficient in ordinary cases. In no event did it exceed 1 1/2 pints of acid to one gallon of water and care was taken to rinse off the acid wash shortly after applying.

When acid was used, acid-resistant pails and fiber brushes were used. Metal pails caused a chemical reaction where the solution became a yellow stain instead of a cleaning fluid. Never use hydrofluoric acid. Many cleaners in order to save hand labor employ strong solutions of it, which is highly injurious.

Pointing Terra Cotta. Recommended practice today is to strike the joints as the work progresses. Pointing mortar applied afterward is usually of different composition and accordingly of questionable permanency.

Cold Weather Protection. Architectural terra cotta or ceramic veneer should not be set in freezing temperatures or when freezing can be expected within 30 days after setting, without providing enclosures and temporary heat.

GLASS BUILDING BLOCKS

Glass blocks are hollow, partially evacuated units of clear or colored glass. They have been engineered to serve several functions, all providing good thermal and sound insulating values.

Clear and other non-light directing units are available in 5 3/4", 8", and 12" square units, all 3 7/8" thick. These units give high light transmission and are available in clear glass for visibility, wide fluted patterns, and wavy, translucent surfaces. These units are often used in partitions and do not screen out sunlight.

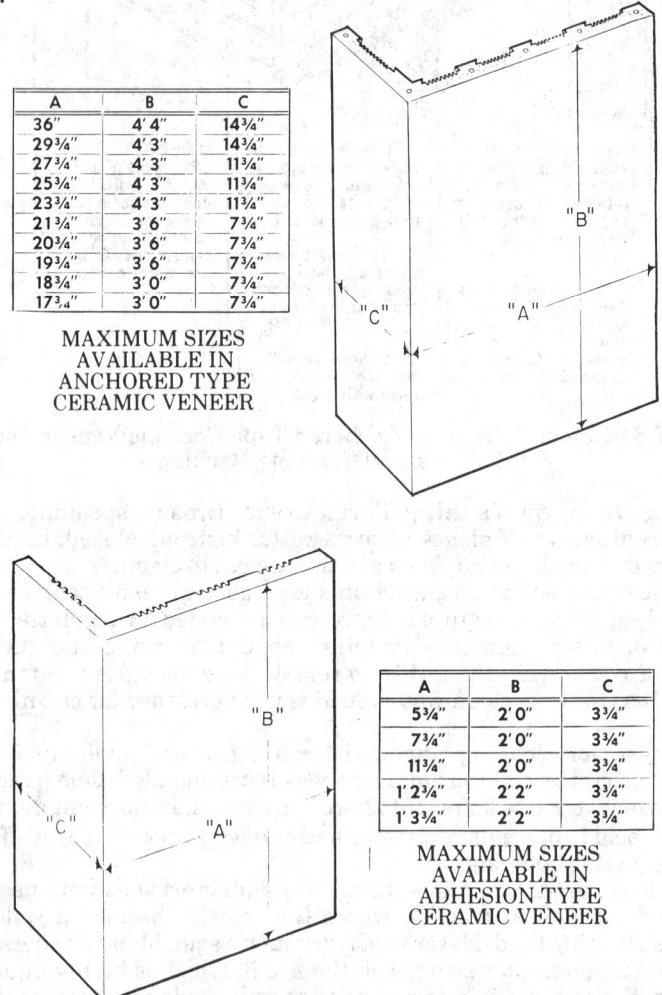

A	B	C
36"	4' 4"	14¾"
29¾"	4' 3"	14¾"
27¾"	4' 3"	11¾"
25¾"	4' 3"	11¾"
23¾"	4' 3"	11¾"
21¾"	3' 6"	7¾"
20¾"	3' 6"	7¾"
19¾"	3' 6"	7¾"
18¾"	3' 0"	7¾"
17¾"	3' 0"	7¾"

MAXIMUM SIZES AVAILABLE IN ANCHORED TYPE CERAMIC VENEER

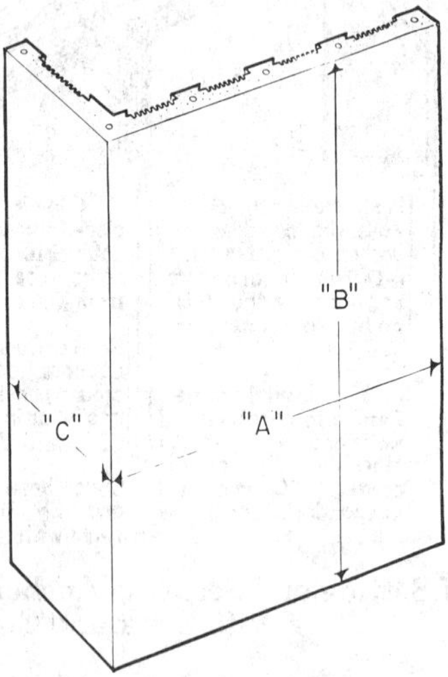

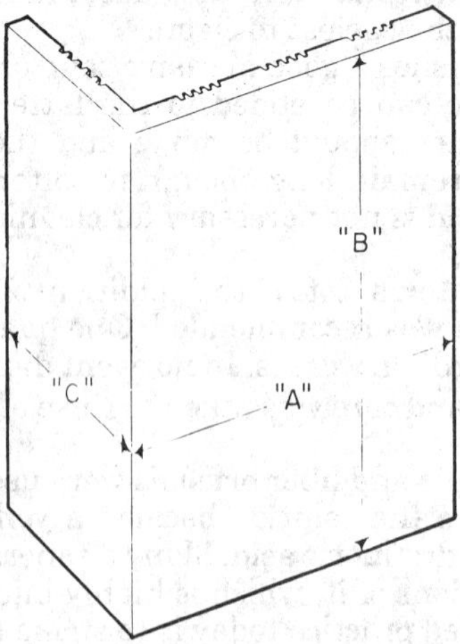

A	B	C
5¾"	2' 0"	3¾"
7¾"	2' 0"	3¾"
11¾"	2' 0"	3¾"
1' 2¾"	2' 2"	3¾"
1' 3¾"	2' 2"	3¾"

MAXIMUM SIZES AVAILABLE IN ADHESION TYPE CERAMIC VENEER

Light directing units are engineered for use on sun exposures of rooms having finished ceilings to provide efficient reflection where such indirect daylighting is desired, such as in schools and offices. The basic size is 7 3/4" x 7 3/4" x 3 7/8". These units can also be had with a white or green fibrous insert to reduce both glare and solar heat gain and are also available in 11 3/4" square size. The insert reduces the overall light transmission.

Light diffusing units diffuse light in all directions and give maximum light transmission. Size is 7 3/4" square. These units are also available with sun control inserts.

Color glass units have a ceramic coating fused to a single face of each block. They are used for decoration, often as accents with other blocks, and are available in 5 3/4" and 7 3/4" square units and 3 3/4" x 11 3/4" rectangles.

Sculptured glass modules have various geometric shapes pressed into the glass to a depth of approximately 1 1/2" on both sides of the unit. They are 11 3/4" square, and are available in clear glass and the lighter hues of the color glass units.

Grille wall units are patterned glass with a smooth exterior with an outline of grey colored ceramic frit fused at the edges leaving ovals, circles, dots, and hourglass designs exposed in the center. They are also available in solid grey frit. Sizes are 3 3/4"x7 3/4"x3 7/8" and 7 3/4" square. The finish is scratch, abrasive and chemical resistant and will not fade from exposure. They make handsome and practical screen walls.

Mortar for Glass Blocks. Materials used in making mortar for laying glass blocks shall be measured by volume. For this purpose 25 lbs. of quicklime or 40 lbs. of hydrated lime shall equal one cu. ft.

Lime shall be high-calcium or dolomitic hydrated lime. Dolomitic type lime must be pressure hydrated so that it does not contain more than 8% by weight of unhydrated oxides, in accordance with ASTM Specification C207 Type S or Type SA.

The mortar shall be composed of 1 part waterproof portland cement, 1 part lime, and 4 parts well graded sand. It shall be mixed to a consistency as stiff and dry as possible and still retain good working characteristics. Prepared masonry mortars of high strength and low volume change may be substituted, when approved. Setting accelerators or anti-freeze compounds are not to be used.

Expansion Joints. Expansion joints must be provided at the head and jambs of each glass block panel. This expansion area shall contain expansion strips as designated by the details or specifications. Caulk head and jamb expansion joints and at intermediate reinforcing members with sponge plastic strip or rope and a good grade of non-hardening mastic so that all expansion joints are waterproof.

Panels in residential construction which are not over 5'-0" in width or 7'- 0" in height, or 25 sq. ft. in area, may be mortared in solid without the use of expansion joints at the jambs. The expansion joint at the head must still be used to prevent loads from being carried by the panels. Reinforcing wall ties may be omitted in residential panels which do not exceed the above size limits.

Reinforcement. Reinforcing wall ties shall consist of two No. 9 wires spaced 2" apart to which are welded No. 14 gauge cross wires. Ties are to be 8'-0" long and not more than .20" thick at the weld. They shall be galvanized or treated with some other approved corrosion resisting coating. Ties are to run continuously with ends lapped 6" and are to be installed in horizontal mortar joints approximately 24" on centers.

Ties shall not be laid directly on the block, but shall be completely embedded in the center of the mortar joint.

Glass Building Units--Installation*

1. Sill area to be covered by mortar shall first have a heavy coat of PC Asphalt Emulsion and allowed to dry.

2. Adhere PC Expansion Strips to jambs and head with PC Asphalt Emulsion. Expansion strip must extend to sill.

3. When emulsion on sill is dry, place full mortar bed joint--do not furrow.

4. Set lower course of block. All mortar joints must be full and not furrowed. Steel tools must not be used to tap blocks into position. Mortar shall not bridge expansion joints. Visible width mortar joint shall be 1/4" or as specified.

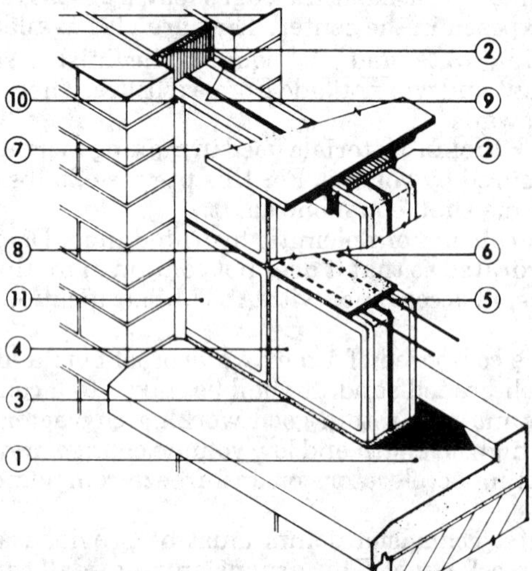

5. Install PC Panel Reinforcing in horizontal joints where required as follows:

a) Place lower half of mortar bed joint. Do not furrow.
b) Press panel reinforcing into place.
c) Cover panel reinforcing with upper half of mortar bed and trowel smooth. Do not furrow.
d) Panel reinforcing must run from end to end of panels and where used continuously must lap 6 inches. Reinforcing must not bridge expansion joints.

6. Place full mortar bed for joints not requiring panel reinforcing. Do not furrow.

7. Follow above instructions for succeeding courses. The number of blocks in successive lifts shall be limited to prevent squeezing out of mortar or movement of blocks.

8. Strike joints smoothly while mortar is still plastic and before final set. At this time rake out all spaces requiring caulking to a depth equal to the width of the spaces. Remove surplus mortar from faces of glass blocks and wipe dry. Tool joints smooth and concave, before mortar sets so that exposed edges of blocks have sharp clean lines.

9. After final mortar set pack PC Oakum tightly between glass block panel and jamb and head construction. Leave space for caulking.

10. Caulk panels as indicated on details.

11. Final cleaning of glass block faces shall not be done until after final mortar set.

Cleaning--Surplus mortar shall be removed and block faces wiped dry as joints are tooled. Final cleaning shall be done by others after mortar has attained final set.
Courtesy Pittsburgh Corning

Labor Laying Glass Blocks. In laying glass blocks, all mortar joints shall be completely filled with mortar. The joints shall not be less than 3/16" thick or more than 3/8" thick. Blocks shall not be hit with a metal tool but laid by the method known as "shoving", (working into place with the hands) thereby compressing the vertical joints.

When a bricklayer's trowel handle is capped with a common type of rubber ferule, such as used on the bottom of crutches, it prevents damage to the glass.

After the mortar has passed its initial set, the exposed edges of the joints shall be tooled and thoroughly compressed with a round jointer. The finished surface of the joint shall be slightly concave, smooth and nonporous. The final cleaning shall not be done until after the mortar has reached its final set.

If blocks are in any way disturbed after laying, during the plumbing, tooling, or cleaning process, such blocks shall be removed and relaid.

Inasmuch as glass blocks do not absorb water, the mortar must be stiff as possible and still be workable, otherwise the blocks will be inclined to slide and it will be difficult to lay up a straight wall and very few courses can be laid at a time as the weight of the glass blocks will squeeze the mortar out of the joints.

Basis for Estimating Labor Costs. The labor costs on the following pages are based on one laborer to two masons, which includes time mixing and supplying mortar, glass blocks, etc. It does not include time building and removing scaffolding, placing expansion joint strips, packing joints with sponge plastic rope and caulking, nor does it include time required cleaning glass blocks after erection. These are given as separate items.

Panels containing more than 144 sq. ft. and not more than 25'-0" in length or 20'-0" in height should be braced by and anchored to structural stiffeners, so a fair average for all classes of work would be panels 10'-0"x10'-0" in size, containing 100 sq. ft. Panels of this size have been used as a fair example for determining average costs.

Expansion joint strips 3/8" thick, 4 1/8" wide, and 25" long are used at the heads, jambs, and mullions of all openings. These strips are secured in place by an adhesive such as asphalt emulsion, and are butted together end to end to form a continuous cushion around the edges of the panels.

After the panels have been laid and the mortar has set, sponge plastic rope should be packed tightly between the side of the block and the sides of the "chase". Sponge plastic rope should be kept back 3/8" from the finished surface. The recess thus formed should then be caulked with a non-hardening waterproof caulking material to a depth of not less than 3/8".

In a 10'-0" x10'-0" opening, caulked on both sides of the blocks, there will be 40 lin. ft. of caulking on the inside of the wall and 30 lin. ft. of caulking on outside of the wall or 70 lin. ft. for each opening. A worker should pack rope into joints and caulk 30 to 40 lin. ft. (on one side) an hour. To clean both sides of 100 sq. ft. of smooth face glass block wall will require 2 to 2 1/2 hrs. labor time.

To clean both sides of 100 sq. ft. of ribbed face glass block (light directional type) will require 3 to 3 1/2 hrs. labor time.

Erecting and removing scaffolding on the average job will require 1 1/2 to 2 hrs. labor time per 100 sq. ft. of glass area.

Glass blocks must be carefully handled in unloading and handling on the job, as they require the same care as fine enameled brick.

Cost of Glass Blocks. The cost of glass blocks varies with the kind, size and quantity required. Always obtain delivered prices on blocks, expansion strips and wall ties for each job figured.

Approximate Prices of Standard Glass Blocks

Size of Blocks	Up to 1,000 Sq. Ft.	From 1,000 Sq. Ft.
5 3/4"x5 3/4"x3 7/8"	$2.20	$2.15
7 3/4"x7 3/4"x7 7/8"	3.00	2.90
11 3/4"x11 3/4"x3 7/8"	7.00	6.80

Special block runs considerably more and should be checked locally for both price and availability.

Estimating Data on Laying Glass Blocks Using 1/4-Inch Mortar Joints

	Size of Glass Blocks in Inches		
Cost per 100 Sq. Ft. of Wall	5 3/4x5 3/4x3 7/8	7 3/4x7 3/4x7 7/8	11 3/4x11 3/4x3 7/8
No. of Blocks per Sq. Ft. of Wall	4.00	2.25	1.00
No. of Blocks per 100 Sq. Ft. Wall	400	225	100
Cu. Ft. of Mortar Required	5	3.6	2.33
Mason Hours	21 to 23	14 to 16	10 to 12
Labor Hours	10 1/2 to 11 1/2	7 to 8	5 to 6
Labor on Scaffolding, Hours	2	2	2
Mason Ramming Oakum, Caulking, Hrs.	1 1/2	1 1/2	1 1/2
Mason Cleaning Blocks, Hrs.	2 1/2	2 1/2	2 1/2
Expansion Strips, 3/8"x4 1/8", l.f.	30	30	30
Wall Ties, l.f.	60	60	60
Sponge Plastic Rope Joints, l.f.	460	60	60
Caulking Joints, l.f.	70	70	70
Asphalt Emulsion, 40'-0"x4 1/2"	1/2 pt.	1/2 pt.	1/2 pt.

Cost per 1,000** Glass Blocks (Based on 10x10 Panels)

Cu Ft. of Mortar Required*	12.5	16	23.3
Mason Hours	53 to 57	62 to 71	100 to 120
Labor Hours	27 to 29	31 to 36	50 to 60
Labor on Scaffolding, Hours	5	9	20
Mason Ramming Oakum, Caulking, Hrs.	3	6	15
Mason Cleaning Blocks, Hrs.	6	11	25
Expansion Strips, 3/8"x4 1/8", l.f.	75	135	300
Wall Ties--Reinforcement, l.f.	110	200	440
Sponge Plastic Rope Joints, l.f.	150	266	600
Caulking Joints, l.f.	175	310	700
Asphalt Emulsion, 3 1/2" Wide	1 1/4 pt.	2 1/4 pt.	2 1/2 qts.
Number pcs. laid per 8-hr. day	140 to 150	115 to 125	70 to 80

*Does not include allowance for breakage. **Includes 10% for waste.

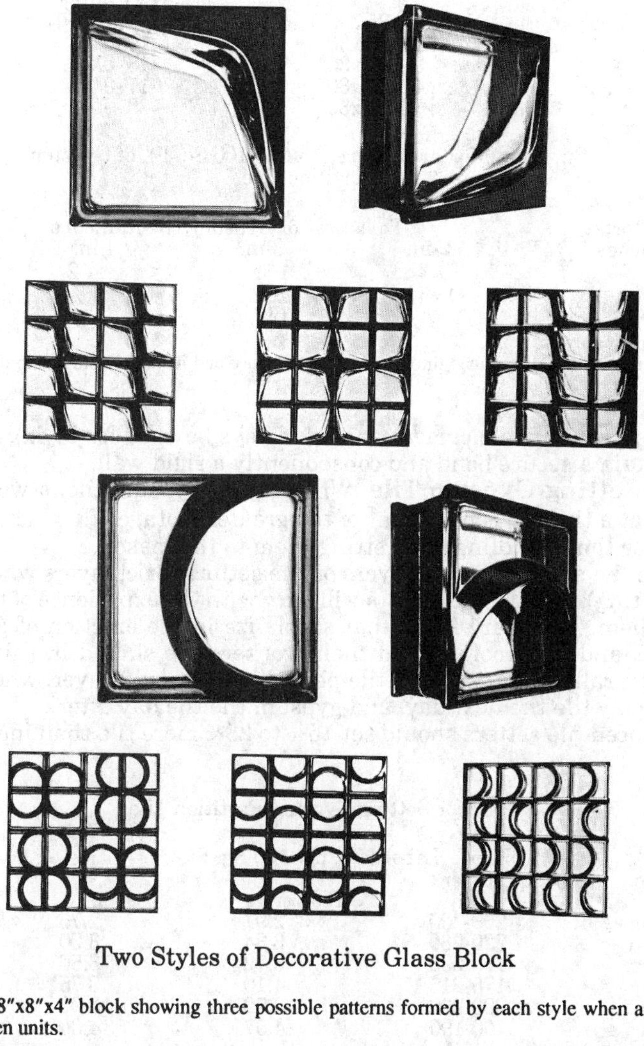

Two Styles of Decorative Glass Block

Two styles of 8"x8"x4" block showing three possible patterns formed by each style when arranged into panels of sixteen units.

GYPSUM TILE PARTITIONS AND FIREPROOFING

Gypsum tile are used for partitions and column fireproofing. They are 12"x30" in face dimensions. The large units are light in weight and require a minimum of labor for erection.

Estimating Quantities of Gypsum Tile. Gypsum tile should be estimated by the sq. ft. taking the area of all walls or partitions and making deductions in full for all openings. Gypsum tile are easily handled and may be cut with an ordinary hand saw, reducing waste to a minimum. The average mortar joint is 3/8" to 1/2" thick.

Sizes and Weights of Gypsum Partition and Furring Tile

	Size of Tile	For Ceiling Heights Up To	Weight per Sq. Ft. Lbs.
2" Solid	2"x12"x30"	10'-0"	10.0
3" Hollow	3"x12"x30"	13'-0"	10.0
4" Hollow	4"x12"x30"	17'-0"	13.0
6" Hollow	6"x12"x30"	30'-0"	19.0

Cu. Ft. of Mortar Required to Lay 100 Sq. Ft. of Gypsum Partition Tile

Width of Mortar Joint in Inches	Thickness of Gypsum Tile in Inches 2-in.	3-in.	4-in.	6-in.
1/4	1	1 1/2	2	3
3/8	1 1/2	2*	2 1/2	3 1/2
1/2	2*	2 1/2*	3*	4*

*Quantities ordinarily used.

Note: Mortar quantities determined through actual experience and include waste due to dropping, filling around pipe, etc.

Gypsum partition tile cement should always be used for laying up gypsum tile to obtain a secure bond and consequently a rigid wall.

Labor Setting Gypsum Tile. When laying gypsum tile, a worker sets 2 1/2 sq. ft. at a time but on account of the greater footage of tile set, it requires more labor time handling and getting them to the mason.

Gypsum tile are set by bricklayers or tile setters (bricklayers who specialize in tile setting), and the labor costs will vary with the experience of the workers setting them. There are firms that specialize in the erection of gypsum tile partitions and fireproofing, and their workers are skilled in this branch of work. Naturally, they set more tile per day than a bricklayer, who lays brick one day, clay tile the next day, and gypsum tile the day after.

Experienced tile setters should set 15% to 25% more tile than inexperienced workers.

Labor Setting Gypsum Partition Tile

Size of Tile in Inches	No. of Sq. Ft. per 8-hr. Day	Labor Hrs. per 100 Sq. Ft. B'layer	Labor	Engr.*
2" Partition	250-300	2.91	2.75	0.25
3" Partition	220-260	3.34	3.00	0.25
4" Partition	195-235	3.72	3.50	0.25
6" Partition	175-215	4.10	3.75	0.50
2" Column	200-230	3.72	3.00	0.25
3" Column	160-190	4.57	4.00	0.25

*Add hoisting engineer time only if required.

04400 STONE

Being a natural product, stone varies widely in type and in geographical distribution. Being a very heavy material, except for the lava type stones which have very limited use, the availability of stone locally is of utmost importance to the estimator as transportation costs can become a high %age of the total cost. Geologically, stones are classified as follows:

a) *Igneous:* those formed by cooling molten material inside the earth including granite, trap rock and lava.

b) *Sedimentary:* those consolidated from particles of decayed rocks deposited along streams and including limestone and sandstone.

c) *Metamorphic:* those either igneous or sedimentary which have been altered by deep-seated heat and pressure or intrusion of rock materials and including gneiss, marble, quartzite and slate.

An Indiana Limestone Quarry

The Building Stone Institute classifies stones as follows:

a) *Bluestone:* a hard sandstone, blue gray and buff in color, quarried in New York and Pennsylvania, used for paving, treads, copings, sills and stools, hearth and in random ashlar patterns for walls.

b) *Granite:* a fine to coarse grained igneous rock, grey, red, green, black or yellow in color, quarried in New England, New York, Virginia, Michigan, Minnesota and Colorado, used for paving, curbing and rubble, mosaic, ashlar and veneered wall construction for almost all building conditions.

c) *Greenstone:* a metamorphosed stone, a distinctive green in color, quarried in Virginia, and used inside and out for flooring, treads, spandrels and curtain walls, sills, copings, hearths.

d) *Limestone:* a sedimentary rock, oolitic, dolomitic and crystalline, grey, buff and pinkish in color, quarried in Indiana, Kentucky, Massachusetts, Minnesota, New York, Alabama, Kansas and Texas, used for wall facings in all patterns, from rubble through the various ashlars to smooth facings in large sheets and for all trim pieces including carved moldings, panels and sculpture.

e) *Marble:* a metamorphic rock, generally recrystallized limestone, available in a wide range of colors and variegated patterns, quarried in Georgia, Alabama and Vermont, used generally as interior facings and much of which is imported, but are increasingly used as exterior veneers on precast concrete panels and as brick sized and random ashlar units for exterior walls.

f) *Quartzite:* a compact granular rock composed of quartz crystals in a wide variety of colors, many of which are variegated, quarried in Colorado, Tennessee and Utah, and used for paving, ashlar and mosaic wall veneers, copings, sills, hearths and treads. Quartzite is also highly popular as aggregate for precast panels.

g) *Sandstone:* a sedimentary rock, in warm earth tones from near white to dark browns and reds, quarried in Colorado, Minnesota, Michigan, Indiana, Ohio, New York and Massachusetts, used for wall facings of all types rubble and mosaics, ashlars and smooth sheet facings.

h) *Slate:* a fine grained metamorphic rock derived from sedimentary rock shale, predominantly gray and blue-black in color but also available in red, green, mottled green, purple, mottled purple and weathering green, quarried in Pennsylvania, Virginia, New York and Vermont, and used in slabs for exterior facings, flooring, paving, treads, bases, sills and stools, copings, fireplace trim and, of course, roofing.

The stone pattern and the type of finish also are of utmost importance to the estimator. Patterns used in laying up stone work in walls include:

a) *Rubble:* including uncoursed field stone, polygonal or mosaic and random coursed.

b) *Ashlar:* including rubble field mixed with coursed ashlar, random range coursed ashlar, random range interrupted coursed, broken range ashlar and coursed ashlar.

c) *Panelling:* including spandrel and curtain wall units of both natural stone and stone aggregate precast with concrete.

Finishes for stone will vary as to the hardness and evenness of the material, and may be generally classified as natural, rough worked, and smooth.

Natural finishes may be weathered as in field stone, or seam or split faced, when the natural quarry texture is exposed.

Rough finishes are obtained mechanically by cutting with pitching chisels; gang sawing with steel shot or rough chat; plain sawn which still leaves saw marks; planed finish including those where small particles are plucked after planing to roughen the texture; hammered, hand and machine tooled, where

2 to 10 parallel concave grooved cuts per inch are cut into the exposed surface. A newly devised rough finish for granite is obtained by exposing it to flames.

Smooth finishes include those rubbed wet by hand with carborundum block and water; machine rubbed with carborundum disc; and honed, obtained with a polishing machine and which leaves the stone almost without texture. As this last finish is expensive, it should be limited to interior work or the finest exterior within 30' of eye range. Granite and marble may be polished smooth.

Rough faced stone is used in the construction of churches, college and university buildings, chapels, residences, and retaining walls. It is available in granite, quartzite and sandstone.

Rubble Stone Work

Rubble stone consists of irregularly shaped pieces, only partly trimmed or squared if at all. Finish is usually split faced or natural. Some specifications will restrict the size range. Strip rubble is ledge stone with beds of the natural cleft.

Rubble stone work is estimated by the cu. yd. containing 27 cu. ft. or by the perch containing 24 ¾ cu. ft. A perch of stone is nominally 16 ½ ft. long, 1 ft. high, and 1 ½ ft. wide, but is often computed differently in different locations. In many states, especially west of the Mississippi River, rubble work is figured by the perch containing 16 ½ cu. ft. Before submitting prices on rubble work by the perch, be sure to find out the common practice in your locality.

Cost of Rubble Stone. The cost of rubble stone will vary with the locality, kind of stone used, distance of quarry from the job, etc.

The method of buying and selling rubble stone also varies with the locality, as in some places it is sold by the ton containing 2,000 lbs. while in others it is sold by the perch containing 24 ¾ cu. ft. or by the cord containing 100 cu. ft. and weighing 13,000 lbs. which is about 130 lbs. per cu. ft. In some localities it is sold by the cu. yd. containing 27 cu. ft.

Mortar for Rubble Stone Work. The quantity of mortar required for rubble stone will vary with the size and shape of the stone and the thickness of the walls, but a fair average is 7 to 9 cu. ft. of mortar per cu. yd. of wall. Mortar costs are given under the mortar section early in this chapter.

Cost of Setting Rubble Stone. When handling and setting rubble stone in walls up to 1'-6" thick, a mason and laborer should handle and set 70 to 85 cu. ft. per 8-hr. day.

On heavy walls, 2'-0" to 3'-0" thick, a mason should set 100 cu. ft. per 8-hr. day, and will require about 1 ½ hrs. labor time to each hr. mason time.

Cost of One Cu. Yd. of Rubble Stone for Walls Over 2'-0" Thick

	Hours	Rate	Total	Rate	Total
Mason	3.0	$....	$....	$25.62	$76.86
Labor	4.5	....	....	19.13	86.09
Cost per cu. yd			$....		$162.95
Cost per cu. ft			....		6.04

Ashlar Stone Work

Ashlar Stone has a flat faced surface generally square or rectangular with sawed or dressed beds and joints. Wall patterns are referred to as "coursed"

when set to form continuous horizontal joints; "stacked" when set to form continuous vertical joints; and "random" when set with stones of various lengths and heights so that neither horizontal or vertical joints are continuous. The usual restrictions for random ashlar design are that no vertical joint should be higher than the highest course height being used; that no horizontal joint should be more than three stones long; and that no two stones the same height should be placed end to end. Random work will take more planning on the part of the mason than coursed work.

Finish for ashlar work is usually split face, which is the least expensive and brings out the character of the stone. It may also be ordered chat, sand, or shot sawn. The grade of stone for ashlar work is usually a mixture of all the grades available as a full range is part of the desired effect.

Many types of stone are furnished in ashlar stock including granite; limestone in the greys and buffs from Indiana, the creamy pinks from Minnesota and the popular lannon from Wisconsin; both Vermont and Georgia marble; Tennessee quartzite; and the local sandstones, those from Ohio and Pennsylvania being especially popular.

The common heights for ashlar are 2 ¼", 5", 7 ¾", and sometimes 10 ½", 13 ¼", and 16". Lengths are shipped from 1' to 6' for cutting on the job. Most ashlar work today is set as veneer and comes in thicknesses of 3 ½" to 4". Some ashlar comes in varying thicknesses to be set to a uniform back and to create deep shadow lines on the face. Some jobs will still specify that the ashlar be built as part of the back-up wall, not veneered to it. Ashlar must then be ordered with bond blocks. These units will be a nominal 8" deep and usually constitute around 10% of the stone furnished, or one stone in each 10 sq. ft. of wall.

Many producers offer stock patterns under their own trade names. These are less expensive than stone cut to specification, but are limited to veneer thickness and a fixed percentage of heights, usually around 15% of 2 ¼", 40% of 5" and 45% of 7 ¾".

Veneer ashlar must be tied to the back-up wall with noncorrosive ties not less than 18" apart horizontally and 24" vertically. The stock heights when used with ½" mortar joints level well with brick and block back-up, but adjustable anchors and stud anchors are also available.

Costs of ashlar stone will vary widely depending on the stone selected and the distance from the quarry as well as whether stock supplies or special cutting and finishes are required.

The cost of setting this stone also varies widely, depending upon the size of the stone and the amount of cutting required. On some of the more intricate jobs it requires as much mason's time trimming and cutting the stone as is required to set it. The entire success of the use of this random ashlar depends upon the artistic manner in which the work is performed, and the labor costs will be governed accordingly.

Labor time handling stone and helping masons varies widely, depending upon method of handling stone at job, access of material piles, size of stone, etc.

To show the variations in the cost of setting, six different classes of work are described.

Estimating Quantities of Random Ashlar. Split faced or other finishes of ashlar stone is usually sold by the ton and is furnished with rough beds, backs and joints. The stone weighs 160 to 165 lbs. per cu. ft. and there are 12 to 12 ½ cu. ft. of stone to the ton.

RUBBLE
Thickness: 2" and 2½"
Coverage: 70 sq. ft. per ton

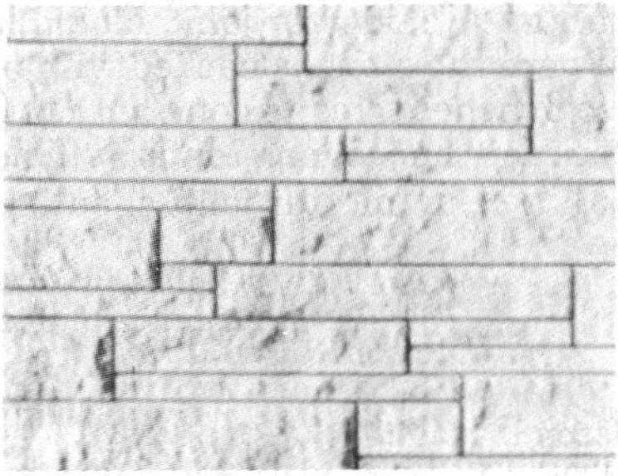

SPLITSTONE
Bed: 3⅝"
Height: 2¼", 5", 7¾"
Coverage: 48 sq. ft. per ton

WEBALL
Coverage: 30 sq. ft. per ton

CHAT SAWED ASHLAR
Bed: 3" and 4"
Height: 2¼", 5", 7¾", 10½"
Coverage: 3" bed, 60 sq. ft. per ton
4" bed, 40 sq. ft. per ton

HAND PITCHED ROCKFACE
Bed: 3"
Height: 3"
Coverage: 60 sq. ft. per ton

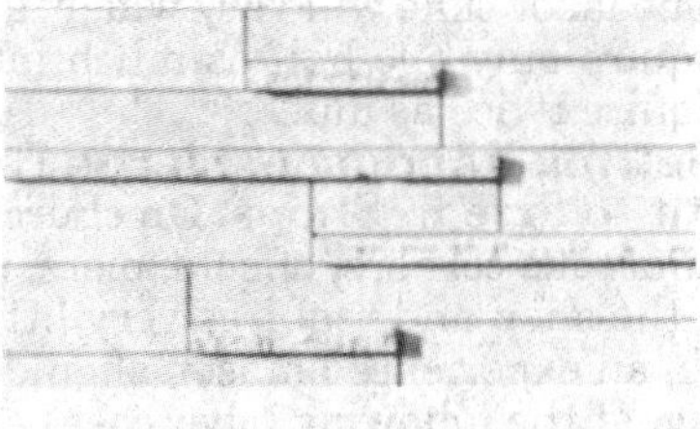

SPLITSTONE WITH
CHAT SAWED
Height: Splitstone–2¼"
Chat Sawed–5" and 7¾"
Coverage: 55 sq. ft. per ton

Limestone Veneer Patterns
Courtesy Victor Oolitic Stone Co., Bloomington, IN

Stone 4 inches thick produces 36 to 38 sq. ft. of wall to the ton and when laid with a ½-in. mortar joint should lay 40 to 42 sq. ft. of wall.

Stone 5 inches thick produces 29 to 30 sq. ft. of wall to the ton and when laid with a ½-in. mortar joint should lay 32 to 33 sq. ft. of wall.

Where the stone is backed with brick or block, the stone is usually furnished 4 to 8 inches thick to bond with the brick backing. Based on an average of 6-in. thick for the job, one ton is sufficient for 24 to 25 sq. ft. of wall and when laid with a ½-in. mortar joint should lay 26 to 28 sq. ft. of wall.

Cost of 100 Sq. Ft. of Random Ashlar Used as a Veneer Over Wood Framing and Sheathing, Stone Veneer 4-In. Thick

	Hours	Rate	Total	Rate	Total
Mason	24	$....	$....	$25.62	$614.88
Labor	12	....	....	19.13	229.56
Cost per 100 sq. ft			$....		$844.44
Cost per sq. ft			....		8.44

Cost Per Ton of Stone

	Hours	Rate	Total	Rate	Total
Mason	10	$....	$....	$25.62	$256.20
Labor	5	....	....	19.13	95.65
Cost per ton			$....		$351.85

On work above the second story or where large stones are used, it may require an extra laborer bringing stone to the mason.

Setting Random Ashlar in Straight Work Requiring Backing, Cutting and Fitting. When setting random ashlar in fairly straight walls as described above, but where the stone must be backed, and end joints cut on the job, an experienced mason should set stone in 17 to 25 sq. ft. of wall per 8-hr. day, at the following labor cost per 100 sq. ft.:

	Hours	Rate	Total	Rate	Total
Mason	38	$....	$....	$25.62	$973.56
Labor	38	....	....	19.13	726.94
Cost per 100 sq. ft			$....		$1,700.50
Cost per sq. ft			....		17.01

The labor time will vary widely according to job conditions, as some jobs may require only one-half as much labor time as mason time, while others may require twice as much.

Setting Random Ashlar in Churches, Chapels, Etc., Where Stone is Cut to Size in Shops. On churches, chapels, and similar structures having numerous corners, offsets, piers, pilasters, etc., where the stone is all backed in the shop and requires very little cutting and fitting by the masons on the job, an experienced mason should set stone in 25 to 30 sq. ft. of wall per 8-hr. day, at the following labor cost per 100 sq. ft.:

	Hours	Rate	Total	Rate	Total
Mason	29	$....	$....	$25.62	$742.98
Labor	29	....	....	19.13	554.77
Cost per 100 sq. ft			$....		$1,297.75
Cost per sq. ft			....		12.98

Setting Random Ashlar in Churches, Chapels, Etc., Where Cutting and Fitting is Done on the Job. On churches, chapels, and similar structures having numerous corners, offsets, piers, pilasters, etc., where the rough stone is sent to the job and all backing, jointing and fitting is done by the masons on the job, an experienced mason should set stone in 14 to 18 sq. ft of wall per 8-hr. day, at the following labor cost per 100 sq. ft.:

	Hours	Rate	Total	Rate	Total
Mason	50	$....	$....	$25.62	$1,281.00
Labor	50	....	....	19.13	956.50
Cost per 100 sq. ft			$....		$2,237.50
Cost per sq. ft			....		22.38

Detail of Random Ashlar Stonework

Random Ashlar in Residence Construction. Frank Walker made an actual study of a residence faced with 4500 square feet of 5" Lannon stone veneer. The stone varied from 1" to 10" high and consisted of one third weathered edge, one third bed face stone and one third rock face stone which was only one to two inches high. Care was taken to produce a natural looking pattern with very little cutting and fitting.

A mason set 34 sq. ft. of stone per 8-hr. day, at the following labor cost per 100 sq. ft.:

	Hours	Rate	Total	Rate	Total
Mason	24	$....	$....	$25.62	$614.88
Labor	14	....	....	19.13	267.82
Cost per 100 sq. ft			$....		$882.70
Cost per sq. ft			....		8.83

Random Ashlar in Memorial Chapel. The memorial chapel illustrated is also built of Lannon (Wisconsin) stone, but on account of the numerous corners, pilasters, piers, etc., and the fact that all backing, jointing and fitting of stone was done on the job, the labor costs ran rather high, as on work of this kind the time required cutting and fitting the stone is almost as much as the setting time.

On this job, a mason averaged 17 sq. ft. of 8" ashlar per 8-hr. day, at the following labor cost per 100 sq. ft.:

	Hours	Rate	Total	Rate	Total
Mason	47	$....	$....	$25.62	$1,204.14
Labor	40	....	....	19.13	765.20
Cost per 100 sq. ft			$....		$1,969.34
Cost per sq. ft			....		19.69

CUT STONE

The term "cut stone" refers here to stone that is cut to an exact dimension from selected quality stone, given a uniform texture and delivered to the site ready to set in place. In order to provide such quality control shop drawings must be made and approved in advance. Cut stone is used on the highest quality work and thus demands the highest quality of installation procedures.

Cut stone may be built up with and bonded to a masonry back-up, but today it is most often applied as a veneer or a facing panel sometimes preassembled with an insulating backing and frame.

In the past cut stone usually referred to granite or limestone or sometimes sandstone which was used to face the many "brownstones" that line the older streets of our northern cities. With the shift from slabs to thin veneers the denser marbles and slate are now often specified.

Shop Drawings. All cut stone work should be cut and set in accordance with approved shop drawings. These should show the bedding, bonding, jointing and anchoring details. Each stone is dimensioned and given a setting number. The producer will not cut the stone until all details and dimensions are agreed upon. As jointing in veneer work is generally ¼", extremely accurate measurements are required. Also, the contractor must plan ahead and have cutouts and holes required for lifting large slabs (called Lewis holes) as well as cutouts for posts, door checks, electrical fixtures, etc., as may be required for the work of other trades. Holes for anchorage are also cut by the contractor furnishing the stone.

Mortar for Cut Stone. The following three types of mortar are recommended for setting cut stone. The best type adapted to the particular work should be selected and included.

Non-Staining Cement Mortar. The cement mortar is 1 part non-staining cement to 3 parts sand, with the addition of 1/5 part of hydrated lime based on the cement volume.

Non-Staining Cement-Lime Mortar. Cement-lime mortar shall be composed of 1/2 part non-staining cement, 1/2 part of hydrated lime or quick lime paste, and not to exceed 3 parts of sand.

If quick lime paste is used, first thoroughly mix the quick lime paste and sand and stack to age; the cement shall be added and thoroughly worked into the lime and sand mixture in small batches just prior to use.

Quick lime shall be reduced to a paste by thorough and complete slaking with cold water and screening through mesh screen into a settling box. Lime putty shall stand not less than one week before use.

Sand shall be clean and sharp, free from loam, silt, vegetable matter, salts, and all other injurious substances with grains graded from fine to coarse. Screen through a 6-mesh screen.

Non-Staining Waterproof Cement Mortar. Recommended for use wherever parging is required or when any of the construction materials, such as water, sand, brick, tile, etc., are known to contain salts which may cause staining or efflorescence; and to walls exposed to excessive amounts of moisture.

Non-staining waterproof cement mortar should be composed of 1 part of non-staining waterproof cement to 3 parts of sand, with the addition of 1/5 part of hydrated lime based on the cement volume.

For quantity and cost of mortar required for setting limestone, refer to mortar section of this chapter.

Job Conditions. The cut stone contractor must check job conditions carefully. His product is expensive: defective or damaged work is almost always refused, and replacements can cause costly job delays.

Carefully unpack and inspect all material to assure it matches samples, is cut to dimensions of shop drawings and tolerances of specifications, and to check for shipping damages.

Carefully store all stone on non-staining platforms at least 4" off the ground. Plug all holes during freezing weather.

Check hoist and scaffolding conditions if they are involved. Dirty hoists and scaffolding can stain stone.

Check backup construction. Backup must be laid to a true line. Concrete and masonry backup should be dampproofed. If this is not in other contracts, this contractor should provide for it.

If work is to continue alongside or above stone after it is set, add an allowance for protection of exposed surfaces.

Estimating Cut Stone. The cost of cut stone will depend upon the kind of stone used (whether granite, limestone, or marble), the amount of cutting and hand work necessary, and freight from mill to destination.

Cut stone for exterior use is estimated by the cu. ft. (taking the cube of the largest dimensions), and any special work such as hand cutting, carving, etc., is estimated separately and added to the cost of the plain stone.

When estimating the cost of plain door and window sills, lintels, steps, platforms, copings, etc., the quantities are sometimes estimated by the lin. ft.

for sills and coping and by the sq. ft. for platforms, but the cu. ft. method is more generally used.

Setting Cut Stone. When estimating the labor cost of handling and setting cut stone, the size of the job should be considered, average size of stone, class of workmanship required and the method of setting, i.e., whether by hand, hand operated derricks or by using a mobile crane, stiff leg, guy or other types of derricks.

Setting Cut Stone by Hand. On small cut stone jobs consisting of door and window sills, lintels, wall coping, steps and other light stone that can be handled and set by two men without the use of derricks, a stone setter should set 50 to 60 cu. ft. of stone per 8-hr. day, at the following labor cost per 100 cu. ft.

	Hours	Rate	Total	Rate	Total
Stone setter	14	$....	$....	$25.62	$358.68
Labor	28	....	....	19.13	535.64
Cost per 100 cu. ft			$....		$894.32
Cost per cu. ft			....		8.94

Setting Cut Stone Using Hand Derricks. Where the stone consists of coursed ashlar, sills, lintels, lightweight cornices, copings, etc., which can be handled on stone barrows or rollers and set with a breast derrick, gin pole, or other hand operated derricks, a stone setter should set 80 to 100 cu. ft. of stone per 8-hr. day, at the following labor cost per 100 cu. ft.

	Hours	Rate	Total	Rate	Total
Stone setter	9	$....	$....	$25.62	$230.58
Helper	9	....	....	19.13	172.17
2 derrickmen	18	....	....	25.62	461.16
2 men handling stone	18	....	....	19.13	344.34
Cost per 100 cu. ft			$....		$1208.25
Cost per cu. ft			....		12.08

Setting Heavy Cut Stone Using Hand Derricks. On jobs requiring heavy stone, such as thick ashlar, heavy molded courses, cornices, copings, etc., where the stone is hoisted and set using hand operated derricks, a stone setter should set 115 to 140 cu. ft. of stone per 8-hr. day, at the following labor cost per 100 cu. ft.

	Hours	Rate	Total	Rate	Total
Stone setter	7	$....	$....	$25.62	$179.34
Helper	7	....	....	19.13	133.91
2 derrickmen	14	....	....	25.62	358.68
3 men handling stone	21	....	....	19.13	401.73
Cost per 100 cu. ft			$....		$1,073.66
Cost per cu. ft			....		10.74

Setting Cut Stone Using a Power Crane. On many large jobs it will be possible to set the stone using a mobile power crane to lift the stone from the ground to the wall.

This method is usually less expensive than setting up and dismantling stiff-leg or guy derricks, as the crane can cover the entire perimeter of the building without any setup or dismantling cost to speak of.

Where the stone can be set in this manner, it will require one operator for the crane, 2 laborers on the scaffold helping to land the stone and set it, 4 or

5 laborers on the ground sorting and handling stone from stock pile to derricks, plus the stone setter.

On a job of this kind, a crew should handle and set 200 to 240 cu. ft. of stone per 8-hr. day, depending upon size of stone being set, at the following labor cost per 100 cu. ft. (excluding crane and operator costs):

	Hours	Rate	Total	Rate	Total
Stone setter	4	$....	$....	$25.62	$102.48
2 helpers on scaffold	8	....	....	19.13	153.04
5 men handling stone	20	....	....	19.13	382.60
Cost per 100 cu. ft			$....		$638.12
Cost per cu. ft			....		6.38

The above does not include any allowance for scaffold, if a separate scaffold is required for stone setting, add for same.

Power Crane Used for Setting Stone

Setting Cut Stone Using Power Derricks. On large jobs where it is necessary to use stiff leg or guy derricks for hoisting and handling the stone, the costs will vary with the size and weight of the stone, quantity, and with the method of handling the stone on the ground.

On work of this kind, it is usually necessary to use stiff leg or guy derricks, or by having the boom of the derrick placed in a boom seat and bolted to the steel frame of the building. On jobs covering a large area, it will be necessary to use a number of derricks, so regardless of the method used in setting the stone, the cost of setting and removing derricks, hoisting engines, etc., should be estimated separately, as this charge will remain the same whether there are 5,000 or 50,000 cu. ft. of stone to be set.

After the derricks and equipment are in place, a stone setter should set 200 to 240 cu. ft. of stone per 8-hr. day. It will require about 2 laborers helping to land the stone and set it, 4 or 5 laborers on the ground sorting and handling the stone from stock piles to derricks, 3 derrickmen and one hoisting engineer.

In connection with work of this kind there is always a certain amount of lost time caused by delays in delivery of stone, repairs to derricks and hoisting engines, oiling and attending derricks, moving and raising derricks, inspecting cables, tightening guy lines and many other items that require attention and which must be charged against the cost of stone setting.

Pointing Cut Stone. Cut stone is often pointed at the time the stone is set. Recommended practice is to strike the joints as the work progresses. Pointing mortar applied afterward is usually of different composition and questionable permanency.

However, on some jobs it may be necessary to point all mortar joints with white or other non-staining cement after all stone has been set. The mortar joints may be struck flush or slightly rounded with a jointer as preferred.

The cost of this work will vary with the size of the individual pieces of stone. The fewer joints the more surface a tuckpointer will be able to point per day.

On an average job, a tuckpointer should point 500 to 600 sq. ft. of surface per 8-hr. day, at the following labor cost per 100 sq. ft.:

	Hours	Rate	Total	Rate	Total
Tuckpointer	1.50	$....	$....	$25.62	$38.43
Helper	0.75	....	....	19.13	14.35
Cost per 100 sq. ft			$....		$52.78
Cost per sq. ft			....		.53

Allow one helper on the ground to each 2 tuckpointers or to each scaffold. Add for mortar and scaffold.

Washing and Cleaning Cut Stone. After the stone work has been pointed it will be necessary to wash and clean it, using a stiff fiber brush and plenty of clean water or soapy water, if necessary, and then rinsed with clean water, which will remove the construction and mortar stains. A wire brush should never be used, as it tends to produce rust marks in the stone.

On new work or work that is fairly clean, an experienced tuckpointer should wash and clean 600 to 800 sq. ft. of surface per 8-hr. day, at the following labor cost per 100 sq. ft.

	Hours	Rate	Total	Rate	Total
Tuckpointer	1.1	$....	$....	$25.62	$28.18
Helper	0.6	....	....	19.13	11.48
Cost per 100 sq. ft			$....		$39.66
Cost per sq. ft			....		.40

Allow one helper on the ground to each 2 tuckpointers or to each scaffold. Add for scaffold.

Carving Cut Stone. The cost of carving cut stone is usually figured from full-size details or models furnished by the architect. This cost will vary with the amount of detail and the method used in carving, i.e., whether performed in the shop before setting or after the stone has been set on the job.

Where you are sure of a first class job of setting, it is more economical to have the carving done in the shop as all carved work consisting of a number of pieces is set up with wood strips forming the thickness of mortar joints and accurately fitted before shipping to the job.

Carving Stone by Hand in the Shop. When carving cut stone for medium priced buildings requiring just an average grade of workmanship and no fine detail work, a carver should complete 2 ½ to 3 ½ sq. ft. per 8-hr. day, at the following labor cost per sq. ft.

	Hours	Rate	Total	Rate	Total
Stone carver	2.7	$....	$....	$28.70	$77.49

On jobs having fine detail work and requiring first class workmanship, a carver should complete 1 to 1 ½ sq. ft. per 8-hr. day, at the following cost per sq. ft.

	Hours	Rate	Total	Rate	Total
Stone carver	7	$....	$....	$28.70	$200.90

Carving Stone in the Shop Using Pneumatic Tools. The use of air tools will greatly increase the daily output of a stone carver, regardless of the class of workmanship required.

On jobs requiring just an average grade of workmanship and no fine detail work, a carver should complete 5 to 8 sq. ft. per 8-hr. day, at the following labor cost per sq. ft.

	Hours	Rate	Total	Rate	Total
Stone carver	1.5	$....	$....	$28.70	$43.05

Where first class workmanship and fine detail is required, a carver should complete 2 to 2 ½ sq. ft. per 8-hr. day, at the following labor cost per sq. ft.

	Hours	Rate	Total	Rate	Total
Stone carver	4.0	$....	$....	$28.70	$114.80

Carving Stone by Hand After It Is Set in the Building. Where the stone is carved on the job after it has been set in place and where first class workmanship is required, a carver should complete 1 to 1 ¼ sq. ft. per 8-hr. day, at the following labor cost per sq. ft.:

	Hours	Rate	Total	Rate	Total
Stone carver	7	$....	$....	$28.70	$200.90

If the carving contains considerable fine detail, add ¾ to 1 hr. per sq. ft. If outside scaffold is required, add labor time to above.

Carving Stone on the Job Using Pneumatic Tools. If air tools may be used for carving the stone on the job, a carver should complete 1 ½ to 2 sq. ft. per 8-hr. day, at the following labor cost per sq. ft.:

	Hours	Rate	Total	Rate	Total
Stone carver	4.5	$....	$....	$28.70	$129.15

If the carving contains considerable fine detail, add 1/2 to 3/4 hr. per sq. ft. If outside scaffold is required, add labor time to above.

Cost of Models for Stone Carving. On the better class of work, clay or plaster models for all carving are usually furnished by the architect or a stipulated sum is allowed for models.

Clay models run about $40.00 to $100.00 per sq. ft. Plaster models cost about $50.00 to $125.00 per sq. ft. A small rosette or head cast in plaster will cost about $100.00. The models quite generally cost more than the carving, and often twice as much where models are made for an individual piece. It will often require a week's work for a skilled artisan to produce a carefully made model for a carved plaque or panel, 3'-6" to 4'-0" in diameter.

The sq. ft. price of models should be based on the sq. ft. of model required and not on the sq. ft. surface of carved stone, as one model may often be used for carving 10 or 15 pieces of stone, or one lin. ft. model of carved molding in plaster may often be used for carving hundreds of lin. ft. of stone.

Indiana Limestone

Indiana Limestone is probably the best known and most widely used building stone in the United States. This is doubtless due to its appearance, as well as the ease with which it is worked and its freedom from stratification and cleavage plane, and while it is a sedimentary stone, it is to all intents and purposes a free stone, having about equal strength in all directions.

It is used for the exterior facings of buildings, as well as all kinds of trimmings, such as sills, lintels, copings, steps, platforms, etc.

Indiana limestone can be produced in any reasonable quantity and size and is usually cut ready to place in the building by the mills located adjacent to the quarries. The creamy-pink limestone quarried in Minnesota is another high quality material. Limestone is furnished in a variety of different grades and color tones, a brief summary of which is given below:

Select stock is uniform in color and texture, more so than is required for ordinary building construction. It is recommended for entrance work and those portions of a building within ready range of vision, carving, interior, and other exceptional uses.

Standard stock is sound stone with a range of variation in color shades and texture not found in Select but necessarily confined within limits that make it impossible to determine at a distance of a few feet whether it is standard or select stock.

Rustic stock is of more or less coarse grain. Rustic stock is used to a considerable extent for sawed ashlar and various rough textured finishes.

Variegated stock in an irregular mixture of buff and gray produced from the blocks that are quarried where the buff and gray color tones adjoin in the quarry. It shows variation in texture as well as color. Variegated stone when cut into building units will exhibit, in these, units of each color tone and a small percentage of units with both color tones in one piece. Variegated stone is unusually effective in giving variety to plain surfaces and is a desirable class of material for trim as well as for facing.

There are certain grades of Indiana Oolitic limestone especially adapted for grade courses, steps, buttresses, floor tiling, etc., or any position in the building exposed to abrasion. Some are available in different color tones and may be identified by the following names: special hard gray and special hard buff Indiana limestone.

The following are the trade terms by which the different grades of Indiana limestone are classified: select buff, standard buff, select gray, standard gray, and variegated.

Some of the machine finishes suitable for limestone are sand sawed, chat or shot-sawed finish, smooth machine finish, wet rubbed finish, machine tooled finishes, carborundum finish, and plucked finish. There are also some rough textured finishes. The class of tooling in common practice is understood to cover what is properly termed fluting, that is, concave depressions, four and six parallel, concave grooves to the inch. Convex tooling is sometimes used effectively, but it is never applied unless specifically designated. Tooling and machinery has been developed to produce a variety of rough textured finishes.

It is generally recommended that all molded work be finished smooth. If it is tooled at all, the tooling should run in the direction of the mold and not across it, unless largely increased expense is of no consequence. As to tooled ashlar, some architects contend vertical tooling is preferable, others prefer the horizontal.

Indiana Oolitic limestone will take most hand finishes which under certain conditions are considered preferable to machine finish. Notable examples of this class of work are crandalling and tooth chiselling, but a number of the hand finishes can now be simulated by machine work.

A smooth finish is not commonly furnished other than in the form it comes from the planers. If sand rubbing is desired, it should be specified, and in such event the sand rubbing should be by the wet process or carborundum-and-water rubbed finish.

Indiana limestone weighs 200 lbs. per cu. ft. when furnished in rough quarry blocks and 180 lbs. per cu. ft. in scabbled quarry blocks. These figures include the weight of rough excess stone.

Sawed slabs or other squared material weighs 170 lbs. per cu. ft. The weight of the finished cut stone ready to set in the building is figured at 150 lbs. per cu. ft. If there are any large round columns in the job, they may be figured at 140 lbs. per cu. ft. of stock. The actual net weight of a cu. ft. of dry seasoned stone is 144 lbs.

Where the cut stone for any job consisting of ashlar, sills, steps, coping, plain machine moldings, etc., is furnished by any of the mills in the Bedford or Indiana limestone district, the price will vary from $12.00 to $20.00 per cu. ft. f.o.b. cars at the mills.

Considerable variation in price is due to the following reasons: Stone consisting of lightweight pieces, such as 4" or 6" ashlar, small molded courses, small turned columns, bases, sills, coping, etc., require a comparatively small amount of stone but involves considerable cutting and finishing expense, and even higher costs will prevail.

On other jobs consisting of fairly large stone, 8" or more thick, large molded pieces, bases, columns, etc., the cubical contents are a deciding factor, and the cu. ft. price of this stone, even for more elaborate work will usually average less in cost than the lighter stone described above.

Jobs containing many large pieces of stone will take the minimum price.

Always remember that the cube of the largest dimensions must always be used when computing the quantity, cost and weights of stone of any kind.

Indiana Limestone Sill Stock

Sill stock is available as standard items and sold by the lineal foot. The designation S4S means stock will be sent in random lengths of from 5' to 12'. The S6S designation means stock will be sawed to exact lengths. Sill stock is sold in rectangular shapes. If slopes, lugs, drips, or moldings are required, profile cuts will be made at additional cost. Sand and other special finishes are also extra.

STANDARD ITEMS

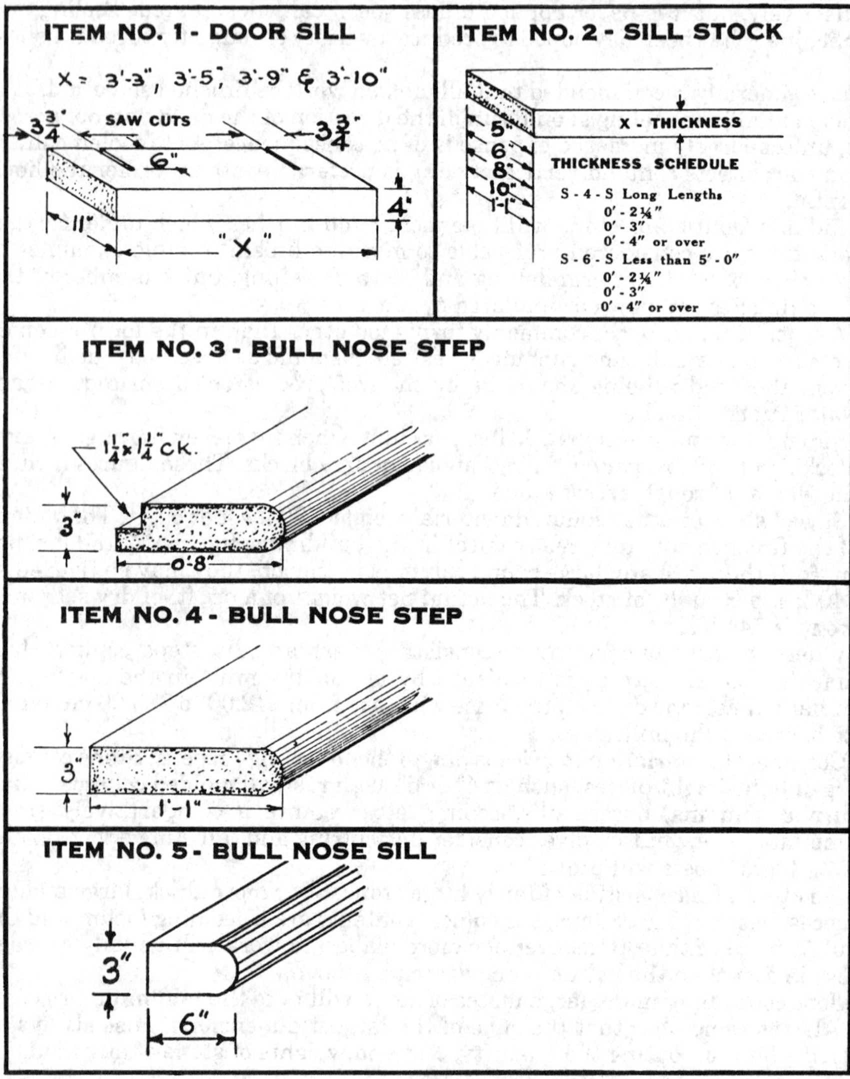

Sill stock will cost about $9.00 per cu. ft. If specific lengths are ordered add $2.00 per cu. ft. If custom cut to shop drawing details sills, copings, steps, and other simple trim will run around $25.00 per cu. ft. Belt courses, simple cornice molds and the like will run around $35.00 per cu. ft. Elaborately carved detail work will run $35.00 and up per cu. ft.

Cutting Beds and Joints. When cutting beds and joints the cost will vary with the size of the stone. The smaller the stone the higher the price per cu. ft. and the larger the stone the less the price per cu. ft. For instance, a piece of ashlar 8" thick contains twice as many cu. ft. of stone as a piece 4" thick, while the cost of cutting the joints will not be much more on the larger piece than on the smaller. Where the joints are cut by hand, the cost will vary from $1.00 to $1.50 per cu. ft., depending upon the conditions described above.

On the second sawing, where sawed slabs are used and it is necessary to run them through the saw the second time for top and bottom beds, it will cost $1.50 per cu. ft. for this work.

Where the jointing is done on the saw, figure $1.25 per cu. ft. for cutting joints.

Smooth or Planer Finish. When Indiana limestone is wanted with a smooth finish as it comes from the planer, the cost will vary from $0.75 to $1.00 per sq. ft. of surface planed.

The slabs are usually furnished sawed to the correct thickness and then the stone is run through the planer to remove the saw marks, which is not a very expensive process on straight work.

On the cheaper classes of work, and on some of the higher grade work where a sawed finish is desired, the sawed slabs are jointed and the stone is sent to the building without having been run through the planer. Work of this class is a little cheaper than stone having a first class finish.

Tooled Finish. If the stone has a tooled finish, having 4, 6, 8, or 10 bats to the inch, the cost of horizontal tooling on plain surfaces will cost $0.75 to $1.00 per sq. ft. (not cu. ft.) of surface tooled, or about the same as planer finish.

Hand vertical tooling in connection with machine molded work will add $0.75 to $1.00 per sq. ft. of surface finished.

Cross or vertical tooling on curved surfaces will add $1.50 per sq.ft. of surface finished.

Two bat tooling, which is always used vertically, can be done for almost the same cost as given above if it is not made to register. This tooling, however, usually has to be made to register and for that reason is generally more costly than 4, 6, 8, or 10 bat work.

Sand Rubbed Finish. If the stone is to have a sand rubbed finish, making it necessary to rub the surface with sand and water, add $0.75 to $1.00 per sq. ft. for all surface to be rubbed.

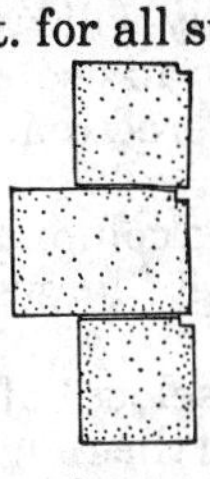

Rustication. If the stone is to be furnished with rustication or a "rabbet" along the edges of the stone, the cost will vary according to the depth of the rabbet. This is all planer work, so the width of the rabbet does not make as much difference as the depth to which it must be cut. The following will prove a fair average of the cost of rustication:

Depth in Inches	Price Per Lin. Ft.	Depth in Inches	Price Per Lin. Ft.
½	$0.75	2	$1.50
1	1.00	2 ½	1.80
1 ½	1.25	3	1.90

Rounded Edges on Stone. Add about $0.35 per lin. ft. for stone with edges slightly rounded.

Stone Moldings. The cost of stone moldings will vary greatly according to the size of the stone, the type of mold, breaks, heads, jointing, etc. The two kinds of moldings most commonly used are plain or Classic moldings and Gothic moldings.

When cutting stone moldings on a planer, the smaller the piece of stone the higher the price per cu. ft., on account of the additional labor involved and the small amount of stone required.

Tools for cutting moldings vary from 3" to 8" wide, so when the moldings are over 8" in width, it is necessary to run them through the planer a second time. The moldings that are the cheapest to run are similar to the illustration of plain moldings, where a large amount of stone is required on which no special work is necessary.

When estimating plain moldings, take the largest dimensions of the stone to obtain the cubical contents and figure at the average price as given on the previous pages, as this will cover most jobs where the molded courses do not form too large a proportion of the total stone in the job.

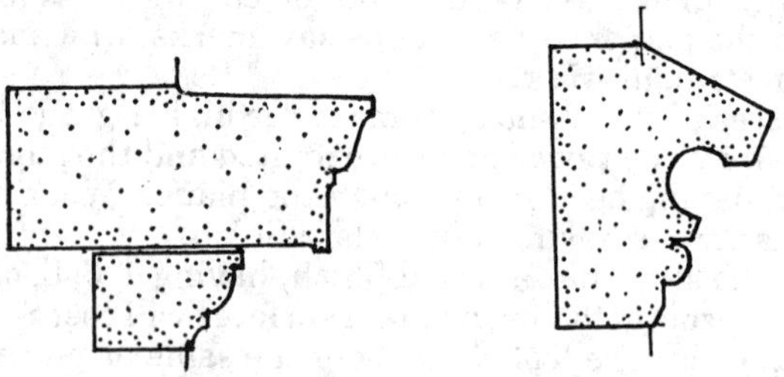

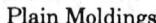

Plain Moldings Gothic Moldings

Gothic Moldings. Gothic moldings are much more expensive than plain or classic moldings, as in almost every instance it is necessary to run the stone through the planer twice or perform a certain amount of hand labor. Gothic moldings cost at least twice as much to run as plain or classic moldings. The cutting of heads for gothic moldings will run much higher than plain moldings, from $3.15 to $15.00 each, depending upon the size and character of the molding.

When an entire job consists of elaborate gothic stone, it frequently will run from $15.00 to $40.00 and more per cu. ft. On work of this kind it is advisable to obtain figures from a reputable concern covering the stone delivered ready to set.

Round Stone Columns. When estimating the cost of plain turned columns, the cubical contents should be obtained by taking the cube of the largest dimension (measured square) and price as follows:

Obtain the cu. ft. cost of the grade of stock to be used, whether select buff, standard buff, select gray, etc., and add seven times the cu. ft. cost of the rough quarry blocks. This will give the cu. ft. price at which the columns should be estimated. For instance, a turned column 3'-0" in diameter and 36'-0" long would contain 3'x3'x36' or 324 cu. ft. of stone. Select buff is worth $15.00 per cu. ft. in rough blocks. Multiply $15.00 x 7 gives $105.00 per cu. ft. 324 cu. ft. @ $105.00 equals $34,020.00, the cost of one column.

A stone column 38'-0" long is the usual limit of present day machine capacity, so when the column shaft is over 36'-0" long, it is best to get them out in three sections.

Fluted Stone Columns. Fluted stone columns should be figured on the same basis as plain turned columns described above. After the cost of the plain column has been obtained, the cost of the flutes should be estimated at $0.75 per lin. ft. of each flute in the column.

Cutting Raised Letters in Indiana Limestone. The cost of cutting raised letters in Indiana limestone will vary according to the size and style of the letters.

On average work, a stone carver should cut one 6" letter an hr. at the following labor cost per letter:

	Hours	Rate	Total	Rate	Total
Stone carver	1	$....	$....	$28.70	$28.70

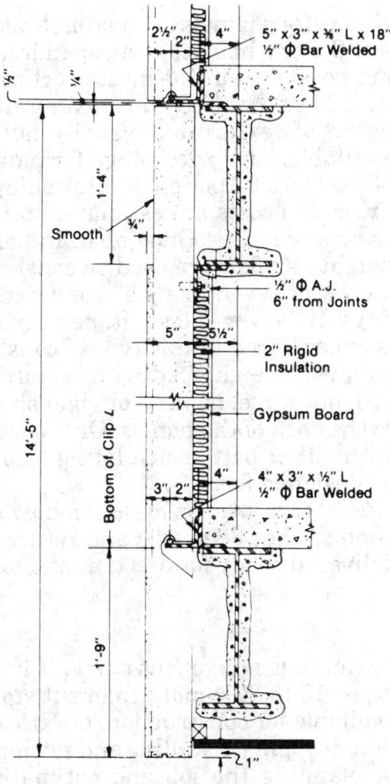

TYPICAL WALL SECTION

Courtesy Harding & Cogswell, Corp. Bedford, IN

Custom Fabricated Limestone Wall Panels

If the letters are 12" high, it will require about 1 ½ hrs. time per letter, at the following labor cost per letter:

	Hours	Rate	Total	Rate	Total
Stone carver	1.5	$....	$....	$28.70	$43.05

Cutting Sunk Letters in Indiana Limestone. The cost of cutting sunk letters is considerably less than raised ones.

When cutting letters up to 6" high, a cutter should cut one letter in about 20 minutes at the following labor cost per letter:

	Hours	Rate	Total	Rate	Total
Stone cutter	.33	$....	$....	$28.70	$9.47

If the letters are 12" high, a cutter will cut 1 letter an hr. at the following labor cost per letter:

	Hours	Rate	Total	Rate	Total
Stone cutter	1	$....	$....	$28.70	$28.70

Textured Indiana Limestone Facing Panels

Limestone panels are available as "curtain wall" units, as wall units spanning a full story height, or as spandrels spanning a full bay. Lengths of up to 15' are possible, but the most practical panels will fall in the range of 40 to 80 sq. ft., 120 sq. ft. being the maximum. Thicknesses will depend on panel size and depth of texture but usually fall in the 4" to 5" range. Nine textures are available. The price range for panels in the 40 to 60 sq. ft. size will run from $8.50 to 12.00 per sq. ft. plus shipping and erection costs.

Erection costs are estimated to run from $4.40 to $6.25 per sq. ft. including cranes or hoists. On a job in Indianapolis where 346 deep textured, full story height panels were used on construction four stories high as continuous piers a crew of seven using a crane and dual C-clamps erected 23,000 sq. ft. in 23 days. However 750 sq. ft. per day is probably a safer average unless the crew is experienced in this type of construction or the job very simple.

Panels are shipped with no back-up applied. Conventional masonry units can, of course, be used or rigid sheets of insulation board can be field applied to the back of the panels. Or the backing may be metal studs and plasterboard with either batt or insulating lath attached to the studs. Any "U" factor may be designed for.

Panels can also be preassembled at the plant from several stones with epoxy bond so that deep sills and returns or special pier or spandrel shapes can be delivered ready for erection in one piece at the job site.

Granite

Granite is an igneous stone. It is extremely hard and weighs 160 lbs. per cu. ft., or 15 to 20% more than cut stone. It is one of the most durable materials available for construction, but because of the high costs of quarrying, fabrication, shipping, handling and setting, considerable time and care must be taken in planning the job and estimating in order to make the best use of the material.

Granite is usually thought of as a facing material for the lower stories of buildings where durability is paramount. However, several tall buildings, such as the 38-story CBS Building in New York City and the Canadian National Bank in Montreal, have been faced with granite full height. It is also now popular as a sidewalk material for the large plazas around new skyscrapers, such as the John Hancock Center in Chicago.

Granite can be classified into three basic groups, each having its own design and specification requirements.

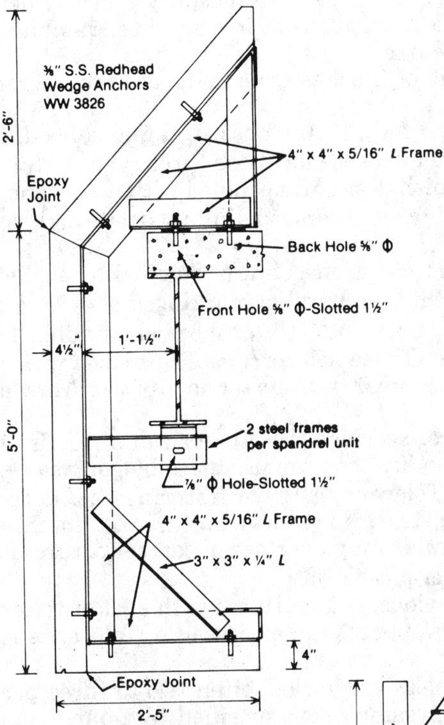

SECTION OF PREASSEMBLED SPANDREL
(3-pcs. stone 14'-0" length epoxied together and assembled on steel frames at our plant)

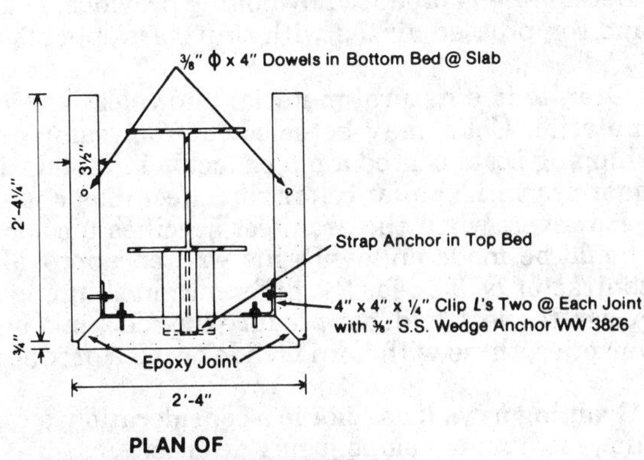

PLAN OF PREASSEMBLED COLUMN COVER
(3-pcs. plant assembled)

Courtesy Harding & Cogswell, Corp. Bedford, IN

Custom Fabricated Built-Up Limestone Wall Panels

Standard building granite is used structurally either self-supporting or as bonded masonry. Granite has a compressive strength of 20,000 psi and is well suited for bearing walls.

Granite veneer is a facing used for decoration or protection, rather than for its structural value and is applied to masonry backup with stainless steel anchors engineered for the job. Veneers will vary from 7⁄8" in thickness up to 8", the majority being in the 2" to 4" range. Also being developed today are precast "curtain wall" panels using granite facing. These are supported from the building framing rather than from the backup and present different construction problems altogether.

Masonry granite is granite used as an integral part of a masonry wall in which one of the more rugged finishes are used such as in churches, rambling residences, bridge abutments, and the like.

Granite is available in a wide range of finishes from split faced to mirror finished slabs.

Polished face is a mirror gloss; honed face, a dull gloss; and fine rubbed, a smooth finish with no gloss. Allowable tolerance for these three types is 3⁄64". Next are the rubbed finishes. Sand finish is slightly pebbled with a tolerance of 1⁄10"; spot finish has circular markings with an indefinite pattern and 3⁄32 surface tolerance.

Hammered surfaced granite comes in three grades: fine or 8-cut with parallel markings not over 3⁄32"; medium or 6-cut with parallel markings not over 1⁄8"; and coarse with parallel markings not over 3⁄32". Tolerances are limited to 7⁄32" for the first two, 1⁄8" for the third. These are corrugated finishes which become smoother near arris lines, that is, near the intersections of two dressed faces.

Sawn faces may be had with vertical scorings (VSM), horizontal scorings (HSM), or circular scorings (RSM). The first two are produced by gang saws, the second, by rotary or circular saws. Tolerance is 1⁄8" with scorings 3⁄32" deep. Thermal finishes are produced by flame texturing a sawn surface with torches to flake granite to a dull, uniform, regular rough surface of coarse texture. It is popular for large paving blocks. Tolerance is 1⁄8".

Jet honing is produced by honing previously cut surfaces with a jet of water and compressed air fed with tiny glass spheres under pressure. Tolerance is 1⁄8".

Granite is a natural material and colors are varied, often two to three per material. Color may be selected from samples or specified by quarry. As shipping costs can be a major factor in a material as heavy as granite, local quarries with similar colors can effect major savings.

However, when the architect specifies a color by quarry, any substitution should be made only with his written approval or as an alternate, lest the contractor be held for the difference in shipping costs.

Granites with feldspar produce the red, pink, brown, buff, grey and off-white varieties; those with horn blende or mica produce the dark greens and blacks.

Graining as well as color is a consideration; a uniformity of texture, whether fine, medium or coarse, being desirable.

The National Building Granite Quarries Assoc., Inc. publishes a list of granite suppliers with colors and graining available from each. Canada and Sweden are also major sources.

Material Cost of Granite. As discussed above, several factors determine the price of granite, including color, graining and transportation to the site.

The lower priced polished veneers will include the solid blacks and mottled greys. These will run around $10.00 to $13.00 per sq. ft. in 1" thickness, $13.00 to $15.00 in 2" thickness and $15.00 to $18.00 in 4" thickness. The variegated reds, pinks, and bluish and greenish blacks will cost $13.00 to $19.00 per sq. ft. Sawn finish will cost about $1.75 less per sq. ft. and thermal finish about $1.00 less.

Random ashlar granites are considerably cheaper, running from $5.00 to $10.00 in 4" thicknesses. This is also often priced as so much per cu. ft., the range being $15.00 to $31.00.

Dressed granite, machine cut in structural thicknesses, will run $15.00 to $50.00 a cu. ft. depending upon the number of openings, washes, and corners. Hand cut granite, as one might find on a mausoleum, can run to $125.00 or more a sq. ft.

Hand cutting, except on the most elaborate work, has been replaced by precision machines. Even surface carving and lettering is done today by sandblasting. Carved letters 6" to 10" high will cost around $20.00 per letter if cut into the granite, $31.00 if raised from the granite.

Cutting of granite into shapes other than square blocks adds quickly to the cost. One source suggests that if a simple square step, which might run $19.00 per cu. ft., were to have a wash or 1/8" slope on the top surface, the cost would increase 23%; a wash on two surfaces, 30%; a wash on two surfaces plus a drip, 42%; and an ogee molding plus a drip, 55%.

Labor Setting Granite Veneer. One mason and one helper should set from 30 to 40 sq. ft. of veneer per day, polished work with 1/4" joints taking more time than straight sawn work. Column enclosures can be figured at around 35 sq. ft. per day. Straight run work with no openings and no hoisting involved can be figured at 50 sq. ft. per day.

The following figures are based on one mason and one helper. On large jobs this ratio may be cut down with 3 laborers serving 4 masons providing scaffolding is arranged so workers can move from position to position. In addition, hoisting costs may have to be added.

On a low-rise building, the square foot labor cost for 100 sq. ft. of polished granite veneer would be figured as follows:

	Hours	Rate	Total	Rate	Total
1 mason	20	$....	$....	$25.62	$512.40
1 helper	20	....	....	19.13	382.60
Cost per 100 sq. ft.			$....		$895.00
Cost per sq. ft.					8.95

Work involving special trim at openings, copings and moldings would require more time:

	Hours	Rate	Total	Rate	Total
1 mason	26.5	$....	$....	$25.62	$678.93
1 helper	26.5	....	....	19.13	506.95
Cost per 100 sq. ft.			$....		$1,185.88
Cost per sq. ft.			....		11.86

Artificial Granite. Cast granite is manufactured in slabs 1 1/2" to 2 1/2" thick as a less expensive substitute for the natural material. It is not as durable and will not take or hold quite the high polish of the true stone. Still, it has been popular on remodeling and store front work. It is set and handled on the job like true granite. The material costs less than half, or in the $5.00 to $8.00 range. One mason and one helper should set 50 to 60 sq. ft. per day.

A Residence Constructed of Sawed Bed Rock Face Granite Ashlar

Labor Setting Granite Ashlar. The labor cost of setting granite ashlar will vary considerably, depending upon type of wall, height of rises and the skill of the mason. Being thin, sawed bed ashlar is light in weight and easy to handle and when finished or semi-finished, a mason should set 45 to 55 sq. ft. of wall consisting predominantly of low rises per 8-hr. day, and up to 70 sq. ft. of wall, where the ashlar consists principally of high rises.

Marble

Marble is classified according to soundness into A, B, C, and D grades. The A grade is in general the cheapest and also the best for exterior use.

Exterior marble veneer varies in thickness from 7/8" to 2". Curtain wall material is usually 1 1/4" thick, although if set in a metal frame or precast with a concrete backup 7/8" material is used.

The finish for exterior marble is either sawn, sand, or honed. A polished surface is not recommended.

Marble will vary widely in cost with the type selected and the distance it must be shipped. In Chicago 7/8" Class A marble will run from $9.00 to $15.00 per sq. ft.

Jointing for marble is 1/16" to 1/8" for thin veneer, 3/16" for the thicker stock. All bed and face joints must be solidly filled with either a non-staining mastic or mortar. Most work is either pointed or caulked.

Exterior veneer should be anchored to the back-up as follows: two anchors for up to 2 sq. ft.; four for up to 20 sq. ft.; and two for each additional 10 sq. ft.

Anchors may be dowels, straps, wire ties or dovetails, but all must be non-corrosive.

In addition to anchors marble veneer must be supported over all openings and at each story height or not more than 20' apart vertically.

Marble may be set against a dampproofed, prebuilt wall in which case it is set out 1" to 1 ½", anchored and spot mortared 18" on centers; or set along with the back-up in which case it is completely parged on the back with the anchors built into both facing and back-up as the work progresses; or it may be solidly grouted to the back-up with the grouting carefully paced so as not to dislodge the marble. In all cases the mortar must be non-staining. Soffit marble must have liners.

Marble is cleaned with clear water and fibre brushes. Acids will stain the surface.

Two masons and a laborer will set around 8 sq. ft. of 1" veneer per hour in straight runs. Pointing and caulking and hoisting costs must be added.

Labor Cost of 100 Sq. Ft. of 1 ¼" Marble Veneer Set Against a Prebuilt, Dampproofed Masonry Wall

	Hours	Rate	Total	Rate	Total
Mason	25	$....	$....	$25.62	$640.50
Labor	12.5	....	....	19.13	239.13
Cost per 100 sq. ft			$....		$879.63
Cost per sq. ft			....		8.80

Slate

Slate is sometimes used as an exterior veneer. When used in this way, it is usually 1" to 1 ¼" thick and furnished with a natural cleft finish. Panel sizes should not exceed 9'-6" in length, 5' in width, or about 25 sq. ft. in area.

Slate veneer is held in place with anchors and spot mortar. Relief angles should be provided at each story height. All angles and anchors must be non-corrosive. Anchors should be provided four to a slab or every 3 sq. ft. Round anchor holes cut in the edge of the slate to receive wire anchors is the recommended procedure. Mortar should be applied in 6" spots, 18" on center. All joints should be full and are usually ⅜" wide.

Slate will run around $10.00 per sq. ft. for material. Two masons and one laborer should set around 7 sq. ft. per hour.

Labor Cost of 100 Sq. Ft. Slate Veneer

	Hours	Rate	Total	Rate	Total
Mason	28	$....	$....	$25.62	$717.36
Labor	14	....	....	19.13	267.82
Cost per 100 sq. ft			$....		$985.18
Cost per sq. ft			....		9.85

04500 MASONRY RESTORATION

Cutting out Mortar Joints in Old Brickwork with an Electric Grinder. On old buildings, where it is necessary to cut out the mortar joints in the old brickwork to a depth of ½" and repoint them, the most efficient method is to an electric carborundum wheel, about 5" in diameter and ¼" thick, depending upon the width of the mortar joints. The motor is hung above the scaffolding with a flexible shaft connected to the carborundum wheel.

This method is nearly twice as fast as using an ordinary electric hand saw on account of the weight of the saw (approximately 17 lbs.), which is difficult and awkward to handle for workers on a swinging scaffold.

Either two or four workers work from each scaffold, depending upon the size of the job, and it will require one helper on the ground to each scaffold.

An experienced tuckpointer or operator should grind out lime or lime-cement mortar joints in 350 to 450 sq. ft. of brick wall per 8-hr. day, at the following labor cost per 100 sq. ft.:

	Hours	Rate	Total	Rate	Total
Tuckpointer	2.0	$....	$....	$25.62	$51.24
Helper	1.0	....	....	19.13	19.13
Cost per 100 sq. ft			$....		$70.37
Cost per sq. ft			....		.70

Using ordinary methods, with an electric saw or grinder, a mason will cut out about 25 sq. ft. of joints an hr. at the following labor cost per 100 sq. ft.:

	Hours	Rate	Total	Rate	Total
Mason	4	$....	$....	$25.62	$102.48
Cost per sq. ft			....		1.02

Repointing Mortar Joints in Old Brickwork. After the mortar joints have been cut out, two tuckpointers working on the scaffold, with one helper on the ground should point 400 to 500 sq. ft. of brick wall per 8-hr. day, at the following labor cost per 100 sq. ft.:

	Hours	Rate	Total	Rate	Total
Tuckpointers	3.5	$....	$....	$25.62	$89.67
Helper	1.5	....	....	19.13	28.70
Cost per 100 sq. ft			$....		$118.37
Cost per sq. ft			....		1.18

The ordinary hanging or swinging scaffold is 25 ft. long and the tuckpointers working from each scaffold can cover approximately 6 feet in height, so that each lift of the scaffold covers 150 sq. ft.

Cleaning Paving Brick Using Pneumatic Chipping Hammers. A satisfactory method of cleaning brick with pneumatic chipping hammers is to use a 4"x12" plank about 10'-0" long, placed on 2 carpenter horses and raised at one end 15 to 18 degrees. This is used to hold the brick which may be placed lengthwise across the plank, so that one side and one end can be cleaned without moving. A plank will hold about 56 brick. Using a pneumatic chipper with a hexagon nozzle and a 2" chisel bit, a worker should clean 150 to 180 brick an hr. or 1,200 to 1,440 per 8-hr. day.

Sandblast Cleaning of Buildings

The portable compressor and the sandblast are widely used for cleaning buildings, bridges and other structures, either preparatory to repainting or for the purpose of brightening up the surface.

The cost of this work, however, is contingent entirely upon the condition of the old buildings, as some of them have nothing more than many years' accumulation of dirt while others have been painted numerous times, making it necessary to remove several coats of paint before the surface of the brick or stone can be touched. Also, it is often necessary to repoint all of the old mortar joints. The estimator should consider all of these items when preparing the estimate.

The process of sandblasting is to force sand against the surface to be cleaned through a small hose connected with the air compressor, and the force of the sand striking the object to be cleaned removes the old surface. Where it requires a maximum pressure of sand and air to remove the old dirt, a 3/4" hose is used but where volume is desired more than high pressure a 1" hose is used.

It usually requires four workers to operate a sandblasting machine: one worker attending the machine, one at the nozzle, and two on the ground assisting with hose, scaffold, and sand.

When estimating quantities, take the entire area of the surface to be cleaned, make no deductions for door or window openings.

On court houses, state capitols, post offices, public libraries, and other ornamental structures, the quantities are computed as above and in addition, balustrades, balusters, etc., are measured on both sides, as it is necessary to go around each baluster separately to clean it.

With a portable compressor supplying air, the average operator can cover a strip 25'-0" wide and 60'-0" to 75'-0" high per 8-hr day. The exact area varies with the quality of the stone encountered but the following table shows the speed at which various stones can be cleaned.

	Sq. Ft. per Min.
Limestone	7-9
Terra Cotta	8-10
Marble	3-5
Brick or Brownstone	8-10
Granite	6-8
Sand stone	10-12

The quantity of sand required for sandblasting will vary with the method used but where a canvas screen or covering surrounding the scaffold catches most of the sand, it is then carried back to the tank on the ground. If this is done about 75% of the sand is saved for use again. This, however, is governed somewhat by the conditions of the work.

The following costs are for removing and cleaning only the natural accumulation of dirt, and do not include removing paint, oil stains, etc., unless otherwise mentioned.

Maximum quantities are not figured in the tables as average job conditions usually include delays, moving from one job to another, placing and removing scaffold, etc.

Labor Cost of Sandblasting 100 Sq. Ft.
of Old Brick or Limestone Buildings

	Hours	Rate	Total	Rate	Total
Compressor Operator	0.33	$....	$....	$25.96	$8.57
Nozzleman	0.33	....	....	19.79	6.53
Labor	0.67	....	....	19.13	12.82
Cost per 100 sq. ft			$....		$27.92
Cost per sq. ft			....		.28

Add for equipment rental, sand, and scaffold costs.

Sandblasting Old Brick Work That Has Been Painted. If the old brick work has been painted and the paint can be blasted off without the necessity of repointing the mortar joints, the costs will run about the same as given above for plain work.

Sandblasting Ornamental Buildings, Such as Post Offices, Courthouses, State Capitols, Public Libraries, Etc., Constructed of Granite or Limestone. The cost of sandblasting buildings as enumerated above, having exterior elaborations, such as balustrades, gables, pediments, plain or fluted columns, etc., will be higher than plain buildings on account of the extra labor required for scaffolding, sandblasting balusters, columns, caps, etc., that are not found on the average building.

Repointing Mortar Joints After Sandblasting. On many old buildings that are unusually dirty or that have been painted, it often happens that the mortar is blasted from the joints when removing the paint from the brick. After the job has been cleaned, the tuckpointers must go over the entire surface and repoint all of the brick joints.

CHAPTER 5

METALS

CSI DIVISION 5
05100 STRUCTURAL METAL FRAMING
05200 METAL JOISTS
05300 METAL DECKING
05400 LIGHTGAGE FRAMING
05500 METAL FABRICATIONS

05100 STRUCTURAL METAL FRAMING

The preparation of a structural steel estimate involves a great deal of work because of the large amount of detail required in listing the quantities from the plans, computing the weights of the various sections, making the shop drawings and detailing the steel, and estimating the cost of shop fabrication, freight, trucking, and erecting the steel on the job.

The structural steel framework of a building usually consists of anchor bolts, setting plates, base plates, columns, girders, beams, lintels, roof trusses, etc., which are fabricated from standard shapes, such as angles,"S" beams, channels,"W" beams and columns, plates, rods, etc., in combinations designed to give the required strength.

Estimating Quantities of Structural Steel

When estimating the quantity of structural steel required for any job, each class of work (column bases, columns, girders, beams, lintels, trusses) should be estimated separately, because each class involves different labor operations in fabrication and erection.

All standard connections can be taken from the *AISC Manual of Steel Construction.* Be sure to include rivets and bolts for connecting steel to steel. In addition to the structural framing of the building, it is often necessary for the structural steel contractor to furnish numerous miscellaneous items, such as bearing plates, loose lintels, and anchor bolts, even though they are set in place by other contractors.

Items to be Included in a Structural Steel Estimate. When preparing an estimate on structural steel, the following items should be included to obtain the erected price:

a) Cost of structural steel shapes at the mill or warehouse.
b) Freight or trucking of structural steel shapes from rolling mill or warehouse to fabricating shop.
c) Cost of making shop drawings and details for shop fabrication.
d) Cost of assembling and fabricating shapes into columns, girders, beams, trusses, etc. including templates in connection with these items.
e) Cost of applying one or more coats of shop paint.
f) Shop overhead expense and profit.
g) Freight on fabricated steel from shop to destination.
h) Cost of erecting steel, including equipment rental, setting up and removing cranes, hoisting equipment, unloading steel at site, erecting steel, and field riveting, and welded or bolted connections.

Estimating Cost of Structural Shapes at Mill or Warehouse. After the quantities have been listed from the plans, it will be necessary to compute the cost of the steel sections required.

All prices on structural steel are based on mill shipment from rolling mills in Pittsburgh, Birmingham, Chicago, etc., depending upon the location of the mill where the steel is purchased, plus freight or trucking from rolling mill to fabricating shop.

Where steel is wanted in a hurry, it is sometimes necessary to purchase it from warehouse stocks, which usually costs from 5 to 15 cents per lb. more than when purchased from mills, depending upon quantity.

Another item that must be considered when pricing steel work is the base price on standard sections and the additions to it. For example, assume the base price of structural shapes is $25.00 per 100 lbs. or $500.00 per ton. Find the cost of 4"x 4"x 1/2" steel angles. The standard classification of extras on the following pages shows that angles of this size cost $3.10 per 100 lbs. more than the base price or $28.10 per 100 lbs.

Warehouse prices vary from warehouse base depending upon size and quantity ordered. On orders of 30,000 lbs. or more, the decrease from base price is 70 cents per 100 lbs., while on orders from 100 to 399 lbs. the increase over base price is about $8.00 per 100 lbs.

Structural steel prices change so often that latest quotations from steel suppliers should always be obtained before figuring jobs of any size.

Cost of Making Working Drawings and Detailing Structural Steel. The cost of making shop or working drawings for steel structures is usually figured by the number of sheets estimated to be required for layout, details, erection diagrams, etc. Drawings, 24"x36" average about $300.00 to $500.00 per sheet, including drafting room overhead.

If drawing costs are reduced to a cost per ton, the unit costs will vary considerably, depending upon the type of building, weight of steel, amount of duplication, etc. The ton cost will run considerably less where fairly heavy shapes are used than on light structures using tube columns, large lightweight trusses, etc., as the latter type of construction requires considerable detailing without involving much tonnage.

Steel skeleton construction, such as office buildings, hotels, etc., without unusual features can usually be detailed for $100.00 to $125.00 per ton.

Ordinary mill and factory buildings consisting of columns, beams, light trusses, and crane runs, requiring considerable detailing without much duplication, will cost from $75.00 to $150.00 per ton.

Theaters, churches, and other structures of this class will cost from $100.00 to $200.00 per ton for details.

Shop Painting of Structural Steel. After the steel has been fabricated, it is customary to give it a coat of paint before it leaves the shop, except where steel is to be encased in concrete.

The cost of painting is usually estimated at a certain price per ton and the quantity of paint required and the surface area to be covered will vary with the class of steel. Ordinary structural beams contain 225 to 275 sq. ft. of surface per ton; plate girders 125 to 175 sq. ft. of surface per ton, while trusses contain 275 to 325 sq. ft. of surface per ton. The average for a job will vary from 175 to 225 sq. ft. per ton of steel.

A gallon of paint will cover 400 to 500 sq. ft. of surface or about 2 tons of steel, at a material cost of $15 per gallon, or $7.50 per ton.

The labor necessary to paint one ton of structural steel (200 sq. ft.) with one shop coat of paint should cost as follows:

	Hours	Rate	Total	Rate	Total
Labor (spray)	0.75	\$....	\$....	\$24.39	\$18.29

Basis for Estimating All Classes of Structural Steel Work

Every structural shop has constant or fixed costs that must be considered before estimating the detailed fabricating operations. The following items should be computed before adding the various shop operations required for fabricating the different members.

Approximate Constant or Fixed Costs

	Price per 100 lbs.
Base price of steel, f.o.b. Pittsburgh or Chicago	\$24-90.00
Freight or trucking on steel from rolling mill	2.00*
Extras	2.90*
Cost of structural steel, f.o.b. fabrication shop	\$29.80
Shop handling charge which includes unloading steel from cars or trucks upon arrival and loading cars or trucks after fabrication	1.25
Tonnage Charges--Painting structural steel in shop	.92
Handling structural steel in shop	.75
Cutting structural shapes to lengths	1.10
Shop Drawings (varies from \$30.00 to \$60.00 per ton) see explanation on previous pages, average cost	3.00**
Total Constant or Fixed Costs	\$36.82

*Varies. **Varies according to type of building. (In rare cases this could be as much as \$.10)

The above constant or fixed costs must be determined in every estimate before adding special fabrications necessary.

Cost of Fabricating Structural Steel

The cost of fabricating structural steel varies widely, depending upon the type of building, amount of duplication, weight of structural members, number of fabricating operations necessary, etc. The cost of fabricating light steel members usually runs considerably higher than heavier members on a per ton basis. The following gives approximate costs, including shop overhead, of the various shop operations and classifications.

Beam and Channel Punching Only. Where beams and channels are punched only, without any other operation, estimate as follows:

Beams and Channels	Price per 100 lbs.
6" to 10"	\$5.75
12" to 18"	4.50
Over 18"	4.25

Beams and Channels Framed with Connections. When beam and channel sections are punched and fabricated with the necessary connections, estimate as follows:

Beams and Channels	Price per 100 lbs.
Up to 10"	\$9.00
12" to 18"	7.50
Over 18"	6.00

Beam and Angle Framing with Shelf Angles. Where W-beams are fabricated with shelf angles as illustrated, estimate as follows:

Beams with Shelf Angles	Price per 100 lbs.
Up to 10"	$8.50
12" to 18"	6.50
Over 18"	6.00

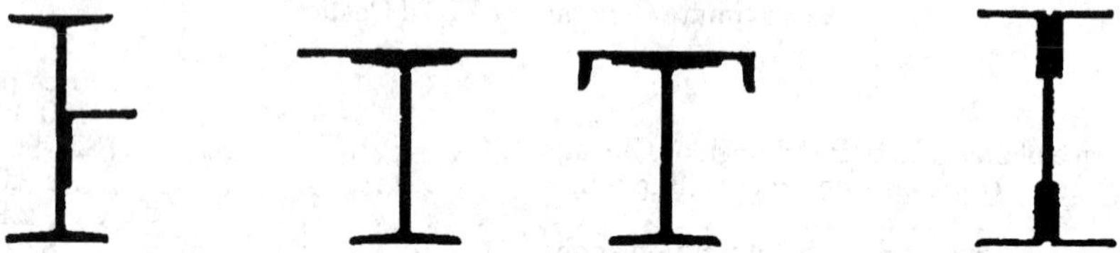

Beam Framing with Shelf Angles

Beam Framing with Plates or Channels

Built-up Plate Girder

Beam Framing with Plates or Channels. Where beams are fabricated with a plate or channel bolted or welded to same, as illustrated, estimate as follows:

Beams	Price per 100 lbs.
Up to 10"	$12.50
12" to 18"	9.50
Over 18"	8.00

Built-Up Plate Girders. When estimating the cost of built-up plate girders consisting of plates and angles as illustrated, estimate as follows:

. . . . $20.00

Built-Up Plate Girders with Cover Plates. Built-up plate girders consisting of plates, angles and cover plates, as illustrated, should be estimated, per 100 lbs:

. . . . $24.00

Plate and Angle Columns. Structural steel columns built-up of plates and angles as illustrated, should be estimated as follows:

Columns	Price per 100 lbs.
weighing up to 50 lbs. per lin. ft.	$24.00
weighing from 50 lbs. to 100 lbs. per lin. ft.	21.50
weighing over 100 lbs. per lin. ft.	20.00

Plate and Angle Columns with Cover Plates. Structural steel columns built-up of plates and angles with cover plates as illustrated, should be estimated as follows:

Not including Weight of Cover Plates Columns	Price per 100 lbs.
weighing up to 50 lbs. per lin. ft.	$28.00
weighing 50 lbs. to 100 lbs. per lin. ft.	25.00
weighing 100 lbs. per lin. ft.	23.50

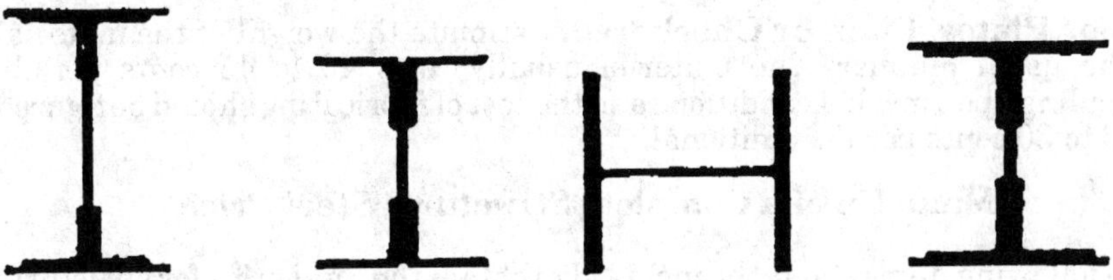

Built-up Plate Girder with Cover Plates — Built-up Plate and Angle Columns — Rolled W Sections — Built-up Plate and Angle Columns with Cover Plates

Rolled W Columns. Structural steel columns consisting of rolled wide flange and similar sections as illustrated, should be estimated as follows:

WF Columns	Price per 100 lbs.
Up to 6"	$20.00
8"	19.00
Over 8"	18.00

Crane Columns. Structural steel columns for supporting crane runs, etc.

. . . . $18.00

Steel Roof Trusses. When estimating the cost of structural steel roof trusses, the fabricating cost will vary with the weight of the trusses, as follows:

Steel Roof Trusses	Price per 100 lbs.
weighing up to 1,000 lbs. each	$36.00
weighing from 1,000 to 2,000 lbs. each	33.00
weighing from 2,000 to 3,000 lbs. each	31.00
weighing from 3,000 to 4,000 lbs. each	29.00
weighing over 4,000 lbs. each	28.00

Knee Bracing. Fabricating cost per 100 lbs.	$23.00
Angle Struts. Fabricating cost per 100 lbs.	23.00
"X" Framing and Bracing. Fabricating cost per 100 lbs.	14.00
Tie Rods. Tie rods 5/8" rounds, fabricating per 100 lbs.	45.00
Bracing Rods. For 3/4" to 1" rounds, fabricating per 100 lbs.	30.00

Fabricating Steel Bins, Hoppers and Similar Plate Work. When estimating the cost of fabricating steel plate work, such as steel bins, hoppers and work of this class, figure the fabricating cost, per 100 lbs.

. . . . 29.00

Floor Plates, Plain or Checkered. Estimate the weight of the material in the usual manner. The material usually costs 40 to 45 cents per lb. depending upon market conditions and the cost of fabricating should be figured at 23 to 30 cents per lb. additional.

Make Up of a Complete Structural Steel Price

The following items should be included to obtain the total price for structural steel delivered f.o.b. cars or job:

	Price per 100 lbs.
Total steel price, f.o.b. fabricating shop, including constant or fixed costs given on previous pages	$36.82*
Shop labor and fabrication cost (varies according to operations required, as given on previous pages)	12.00**
Total direct shop costs	$48.82
Business administration, overhead and sales expense, 20%	9.76
	$58.58
Profit on total cost, including overhead, 10%	5.86
	$64.44
Transportation from shop to job, varies	2.00
Price of structural steel, delivered at job	$66.44

*This item varies as described on previous pages.

**Fabricating costs vary according to amount of shop labor required as given under cost of the various fabricating operations.

Mill and Shop Inspection of Structural Steel. Where the structural steel is inspected by one of the testing laboratories, the usual charges for this work are as follows.

Mill inspection is usually figured at $3.00 to $3.50 per ton and shop inspection from $4.00 to $4.50 per ton. Where both mill and shop inspection are let to one firm, a common price is $6.00 to $7.00 per ton for both.

Field Inspection of Structural Steel. Where field inspection of the erection of the structural steel is required, figure $1,800.00 per week for salary and expenses of a field inspector, plus transportation to and from job.

Erection of Structural Steel

The field cost of handling and erecting structural steel is difficult to estimate for a number of reasons, such as weather conditions, strikes, delays in receipt of materials, storage facilities, size and type of building, equipment available, labor conditions, number and type of connections.

Any one of the above items may cause a considerable variation in the estimated labor costs and for that reason should be carefully considered when preparing the estimate.

Hauling Structural Steel From Shops to Job. The cost of hauling structural steel from the shop to the job will vary with the facilities for loading trucks, distance of haul, weight of steel members, etc.

The average cost of hauling structural steel is $30 per ton, but varies according to total tonnage.

Plant and Equipment for Steel Erection. The equipment required for steel erection will vary considerably with the size and type of building.

Skeleton frame buildings up to 100'-0"x100'-0" in size will require one derrick to unload the steel from the trucks and erect it in place, while buildings of irregular shape, such as "L" shape, "U" shape, etc., will probably require additional derricks. One derrick can handle a job 100'-0"x100'-0" but if the job is 75'-0"x125'-0", it will probably require two derricks to set the job, while a"U" shape building may require 3 derricks.

Where the buildings have setbacks at certain story heights, as required by many cities, it will often require 2 derricks on the job, one to unload the steel and hoist it from the street and another to set the steel.

It costs $5,000.00 to $10,000.00 to set up and remove a hoisting engine, derrick, compressor outfit, etc., for a one derrick job, with additional derricks at approximately the same cost.

On buildings covering considerable ground area, such as high schools, office buildings, auditoriums, mill and factory buildings up to 8 to 10 stories high, it is often more economical to use a crane, mounted on a truck or crawler, depending upon location of the job. Where a movable crane can be used to advantage, that is the most economical method to use, because the crane can be moved about at slight expense.

Steel Erection Contractors. Most structural steel is erected by specialty contractors. These contractors can erect structural steel much cheaper than the average general contractor, because they have well-organized crews specializing in this type of work. They also have on hand the most modern equipment, such as crawler cranes, truck cranes, lightweight and modern air tools, such as grinders, drills, rivet busters, impact wrenches, welding machines, and other special tools which are furnished to the job as needed and seldom figured in the job cost; to say nothing of the background and experience of its organization and personnel.

Erecting Structural Steel in Skeleton Frame Buildings. When figuring erection on a job of this type, consider carefully the amount of equipment, size of cranes, etc. required, as described previously.

On a one crane job, an erection gang should unload, handle, erect and connect 9 to 10 tons of steel per 8-hr. day, at the following labor cost per ton:

Erection gang	Hours	Rate	Total	Rate	Total
1 foreman	0.85	$....	$....	$28.36	$24.11
2 ir. wkrs."hooking on"	1.70	....	....	27.53	46.80
1 ir. wkr. giving signals	0.85	....	....	27.53	23.40
4 ir. wkrs."connecting"	3.40	....	....	27.53	93.60
Cost per ton			$....		$187.91

Add for workmen's compensation and liability insurance. If the structural connections are welded instead of bolted, add $26.00 to $31.00 per ton.

Erecting Structural Steel in Wall Bearing Buildings. On wall bearing buildings, the ends of all beams and channels usually rest on the exterior masonry walls while the interior framing is carried on masonry walls or structural steel columns. Field connections are usually bolted.

Eight workers together should handle and erect (including bolted connections) about 8 tons of steel per 8-hr. day, at the following labor cost per ton:

	Hours	Rate	Total	Rate	Total
1 Foreman	1	$....	$....	$28.36	$28.36
7 Iron Workers	7	....	....	27.53	192.71
Cost per ton			$....		$221.07

Add for worker's compensation and liability insurance.

Erecting Lightweight Steel for Mill Buildings, Foundries, Auditoriums, Etc. On light constructed steel buildings, such as machine shops, foundries, mill buildings, garages, auditoriums, etc., where the structural steel consists of steel columns and lightweight roof trusses, requiring considerable handling to set only a small tonnage, the labor costs run higher than on average skeleton frame construction.

Erecting Structural Steel Using Crawler Mounted Cranes

A crawler crane or a truck crane should be used for setting the steel. After the crane has been delivered at the job, the labor unloading, handling, erecting, and connecting one ton of steel should cost as follows:

Erection gang	Hours	Rate	Total	Rate	Total
1 Foreman	1.1	$....	$....	$28.36	$31.20
7 iron workers	7.7	....	....	27.53	211.98
Cost per ton			$....		$243.18

Add for worker's compensation and liability insurance.

Erecting Steel Bins, Hoppers and Similar Plate Work. When erecting steel bins, material hoppers and similar plate work, an erection crew consisting of 5 iron workers and an engineer should erect in place and bolt, 4 to 5 tons of steel a day at the following labor cost per ton:

	Hours	Rate	Total	Rate	Total
1 Foreman	2	$....	$....	$28.36	$56.72
5 iron workers	10	....	....	27.53	275.30
Cost per ton			$....		$332.02

Welding the joints or seams in hopper work will run from $60.00 to $100.00 per ton, additional. Add for workmen's compensation and liability insurance.

Riveting Steel After Erection. The labor cost of riveting steel after erection will vary with the type of construction and the weight of the structural members. Light constructed jobs usually require fewer rivets at each connection than heavy constructed jobs and require more moving of scaffold to drive the same number of rivets. The more rivets that can be driven from one scaffold, the greater the number driven per day.

On light constructed buildings, a riveting gang should drive 275 to 325 rivets per 8-hr. day, at the following labor cost per 100:

	Hours	Rate	Total	Rate	Total
Ir. wkr. heating rivets	2.6	$....	$....	$27.53	$71.58
Ir. wkr. catching rivets	2.6	....	....	27.53	71.58
Ir. wkr. driving rivets	5.2	....	....	27.53	143.16
Compressor engineer	1.3	....	....	25.96	33.75
Cost per 100 rivets			$....		$320.07
Cost per rivet			....		3.20

The cost of rivets will vary with their size and length, as steel of different thicknesses requires different size rivets. For example, when riveting steel 1/2" or less in thickness, it is customary to use 3/4" rivets, 3/4" steel requires 7/8" rivets, and so on.

Bolting Field Connections. Most structural steel projects have sections bolted together instead of riveted. On work of this kind, an iron worker should place 150 to 200 bolts per 8-hr. day, at the following labor cost per 100:

	Hours	Rate	Total	Rate	Total
Iron worker	5	$....	$....	$27.53	$137.65
Cost per bolt			....		1.38

Bolted connections normally are completed with a two-worker bolting crew. The cost of bolt units will vary with size, length, and grade of bolt. The most common bolts are ASTM A-325, high strength bolts for structural steel joints, 3/4" diameter x 1 3/4" long, at an approximate cost per bolt unit of $0.65 each. Each bolt unit consists of bolt, washers and nut. Other grades sometimes required for connections are ASTM 490, heat treated steel structural bolts (add up to 50% to the cost per bolt unit), and for secondary bolted connections, grade A307 carbon steel externally-threaded standard connection (decrease the cost about 15% per bolt unit).

If tension control bolts are used, add additional costs of about 65% per bolt unit.

Instead of counting individual bolts, some estimators use an average count per ton of steel. If one judges that the time involved in making an exact bolt count is not warranted by the cost, then use the following averages.

Average Amount of Field Bolts Required per Ton of Steel

Type of Work	Light	Average	Heavy
	30	25	20

Steel Infrastructure Repairs

Repair of steel infrastructure, which usually refers to bridge and elevated highway structures, present the estimator with many challenging problems that are not found in new construction. One major difficulty is that repairs are often undertaken in areas where the public has access to the work zone or to adjacent areas. There are many items that might not appear on contract drawings and documents, but which the estimator must consider:

a) Maintenance and protection of traffic - For costing and sample traffic patterns in accordance with federal regulations, see Chapter 22, *Road and Highway Construction.*

b) Maintenance and protection of railroad traffic - Normally, when working within 50 feet of the centerline of a rail track, special precautions must be taken. The estimator should contact local railroad authorities regarding the requirements for flagman, working hours, and the like.

c) Temporary support of existing structures - If temporary supports are not shown on the drawings, the contractor should enlist the services of a professional engineer to design them. The cost of these services will be in direct proportion to the complexity and risk involved. Allow from $65.00 to $100.00 per hour for design services.

d) Activity sequencing - A work sequence must be established. If one has already been set up by the designing engineer, then it is critical that it be followed. In pricing the infrastructure project, the estimator must carefully consider the schedule.

e) Safety - Provisions for netting, rigging, scaffolding, and platforms, as well as environmental conditions, must be considered and priced when required.

f) Work restrictions - Work hours might be restricted, such as during rush hours or holiday periods. In fact, it might be necessary to perform all work outside normal work hours. Area restrictions can result in productivity loss from 10 to 50 percent.

g) Hazardous material - Most older steel structures are covered with lead-based paints. Removal of these painted surfaces must comply with regulations set forth by federal, state, and local environmental protection agencies, or other agencies with jurisdiction in this area. For the cost of removing lead-based paints from structural steel, see the "Painting" section in Chapter 9, *Finishes.*

Rivet Removal and Replacement. Rivet heads are snapped off with a hand-held air scaling tool. The use of flame cutting is discouraged and usually not even permitted. The flat end of a chisel point is placed on top of the rivet against the steel member until the rivet head breaks off. The rivet is then hammered out, and a new bolt is placed in the existing hole. Two ironworkers can remove and replace bolts at the rate of about 20 bolts per hr.:

	Hours	Rate	Total	Rate	Total
Ironworkers	2	$....	$....	$27.53	$55.06
Cost per bolt			....		2.75

Drilling Holes into Existing Steel. The best tool for this purpose is a magnetic base drill with a hole bit. Gaining access to the work location and setting up usually take more time than the actual hole drilling.

Courtesy Karl Koch Erecting Co., Carteret, NJ

Traffic was maintained during this bridge repair work. Notice the scaffolding, netting, and siderails encapsulating the structure.

The set up will take about 15 minutes per hole. Drilling time is between 1 and 5 minutes for a 3/4" to 1 1/4" hole for steel up to 1 1/2" thick. Two iron workers can drill about 6 holes per hour:

	Hours	Rate	Total	Rate	Total
Ironworkers	2	$....	$....	$27.53	$55.06
Cost per hole			....		9.18

Cutting Existing Steel. When rusted or decayed areas need to be cut out and replaced with new steel, allow for two cuts. The first is cut rough, the second more carefully. After the second cut, additional time is needed for grinding. Production time for this work is difficult to estimate. There are so many types of steel and conditions of existing steel. Assume that 3 lin. ft. of burn can be accomplished in about 10 minutes.

Once repairs are complete, it will be necessary to paint the new steel and bolts, and to touch up areas adjacent to the repair work.

ARC WELDING IN BUILDING CONSTRUCTION

Arc welding structural steel in the construction of buildings and bridges has become an established practice. To aid the structural designer in its use, the American Welding Society, the American Institute of Steel Construction, and the American Association of State Highway Officials have established recommended codes and specifications governing design, fabrication, and erection of welded steel structures and applicable codes of most cities, counties and states are in accordance with them. In addition, most fabricating shops now have the equipment, personnel and experience required to fabricate steel members for welded construction, and in the field an increasing number of qualified operators are available for welding field joints during erection.

Advantages of Arc Welded Steel Design. Advantages of arc welded steel design in building construction are lower overall construction costs, less construction noise, particularly desirable in hospital zones, residential districts, etc., and smooth, clean appearance of erected members which permits leaving structural steel exposed where desirable or necessary.

Overall lower costs are realized through reduced steel material requirements and savings in foundation and exterior wall construction. When a structure with welded connections is properly designed, savings in steel material are gained from the following:

a) Gross section, rather than effective net section, is used for design, permitting lighter members.
b) Rigid connections provide continuous beam action, permitting a further reduction in member sections.
c) Connection steel, such as splice plates, gussets, etc., are greatly reduced, sometimes eliminated.
d) With beams of less depth, height of building may be reduced, saving column steel and exterior wall construction.
e) In addition, overall costs for detailing, fabrication, freight, trucking, handling and erection are lower due to reduced weight, fewer pieces and simplified connections.

f) On many structures, where designs and estimates were made for both riveted and welded construction, savings up to 15% on structural steel costs were obtained by using the welded method.

Conditions Affecting Arc Welding Costs

Best results and lowest costs are obtainable only when proper plate preparation and welding procedure is followed. The following conditions have considerable bearing on the cost and quality of welded joints.

Fit-Up. Care in cutting, forming, and handling shapes to be welded to avoid poor fit-up is a major factor in the cost and performance of welded joints. However, a gap of 1/64" to 1/32" is useful in preventing angular distortion and weld cracking. The accompanying graph shows how various sizes of gaps affect welding speeds for square or grooved butt welds. Fillet welds are affected by oversize gaps in a similar manner.

Position of Joints. The position of the joint has considerable effect on the speed and case of welding. Wherever practical, welds should be made in the downhand position with the joint level. Vertical or overhead welds require much more time and skill.

Foreign Matter in Joint. Excessive scale, paint, oil, or rust tend to interfere with welding and should be removed to obtain best speeds and results.

Build-Up or Overwelding. Any amount of weld metal in addition to that actually needed for the specified strength is useless, costly, wasteful and, in some instances, actually harmful. For butt welds, there should be just enough build-up (no more than 1/16") to make sure weld is flush with the plate. Excessive build-up not only wastes weld metal but increases welding time.

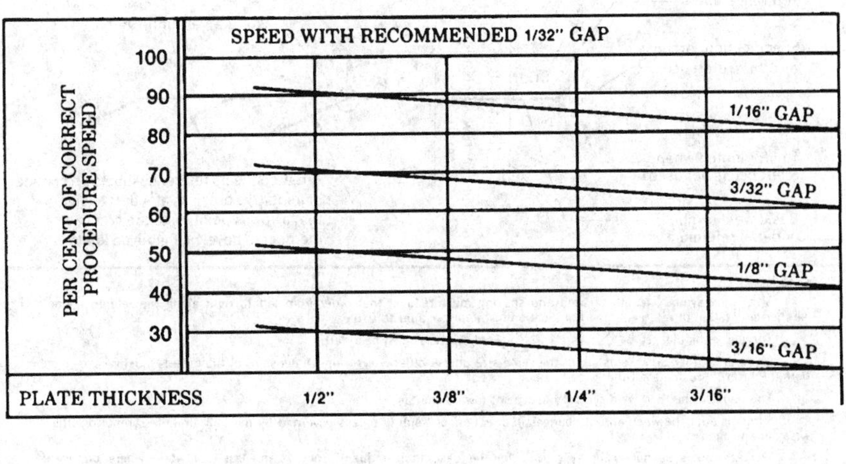

Effect of Gap Size on Welding Speeds for Flat Butt Joints Welded on Both Sides

WELDED JOINTS

Standard symbols

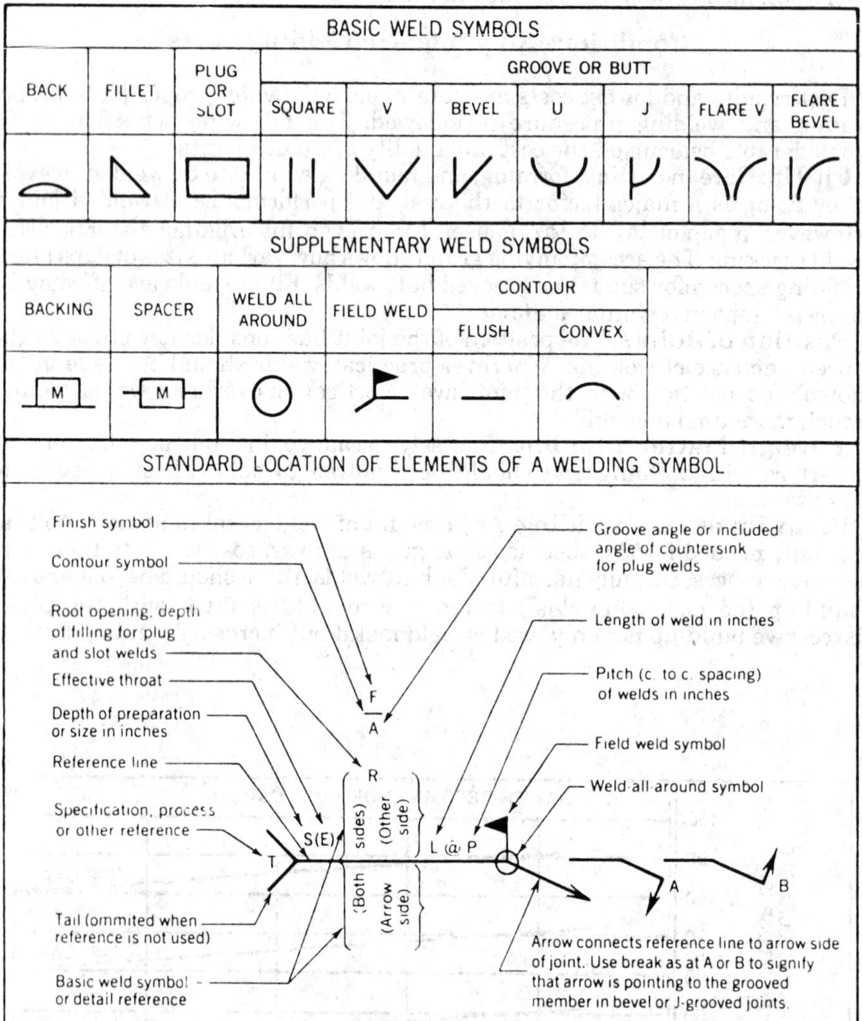

Note

Size, weld symbol, length of weld and spacing must read in that order from left to right along the reference line Neither orientation of reference line nor location of the arrow alter this rule.

The perpendicular leg of 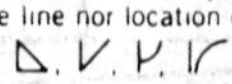 weld symbols must be at left.

Arrow and Other Side welds are of the same size unless otherwise shown. Dimensions of fillet welds must be shown on both the Arrow Side and the Other Side Symbol.

The point of the field weld symbol must point toward the tail.

Symbols apply between abrupt changes in direction of welding unless governed by the "all around" symbol or otherwise dimensioned.

These symbols do not explicitly provide for the case that frequently occurs in structural work, where duplicate material (such as stiffeners) occurs on the far side of a web or gusset plate. The fabricating industry has adopted this convention: that when the billing of the detail material discloses the existence of a member on the far side as well as on the near side, the welding shown for the near side shall be duplicated on the far side.

Courtesy of American Institute of Steel Construction, Inc.

Procedures for Making Various Types of Welded Joints

The following tables give procedures for making various types of welded joints usually encountered in field erection of structural steel and furnish information for calculating time and material required.

In the tables, electrodes specified are those which will make the joints at the lowest cost.

Arc speed in inches per minute is given for single pass welds and for the first pass in multiple pass welds, as the speed of the first pass is important in obtaining proper penetration.

Feet of joint welded per hour is based on actual welding time. No factor has been included for set-up, electrode changing, cleaning, moving scaffold, etc., which will vary with the job. For overall average quantities, these figures should be multiplied by an operating factor which may range from 25% to 50%, depending upon job conditions.

Pounds of electrode per foot of weld is the amount of electrode required to weld the joint with the recommended plate preparation, fit-up and a build-up of 1/16" or less--includes normal spatter loss and 2" stub ends. Any increase in gap or build-up will greatly affect amount of electrode required and time for depositing same.

Fillet and Lap Welds in Horizontal and Flat Positions

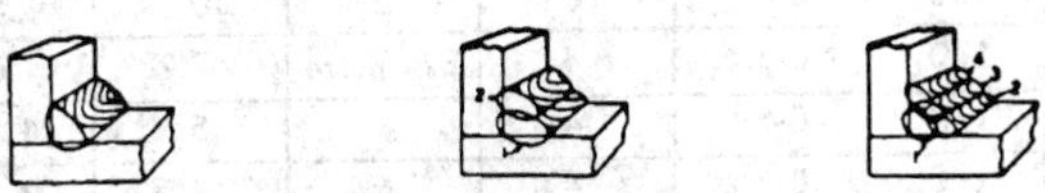

Horizontal Position

Flat Position

Plate Thick-ness (In.)	Gauge Size of Fillet (In.)	Electrode Size (In.)	Current (Amps.)	Electrode Melt-off Rate (In. per Min.)	Arc. Speed (In. per Min. for First Pass)	Passes or Beads	Ft. of Joint Welded Per Hr. (100% Operating Factor)	Lbs. of Electrode Per Ft. of Weld
3/16	5/32	1/8	170	15.0	15-16	1	75-80	.10
1/4	3/16	5/32	225	14.0	15-16	1	75-80	.15
5/16	1/4	3/16	275	12.0	14-15	1	70-75	.19
3/8	5/16	1/4	350-375	9.6-10.2	12-13	1	60-65	.30
1/2	7/16	1/4	350-375	9.6-10.2	10-11	2	28-31	.57
3/4	5/8	1/4	350-375	9.6-10.2	10-11	3-4	16-14	1.34
1	3/4	1/4	350-375	9.6-10.2	10-11	5	9-10	1.66

Preparation: square edge.
Fit-up: recommended gap, 1/32".
Electrode: iron powder E6024 AC. Use E6014 AC electrodes when weldment cannot be positioned close enough to flat for E6024 AC electrodes.

Fillet and Lap Welds in Vertical Position, Welded Up*

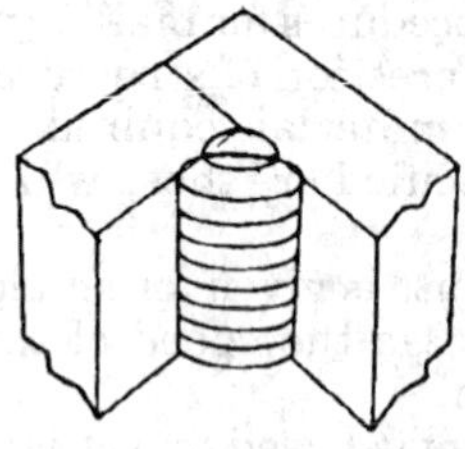

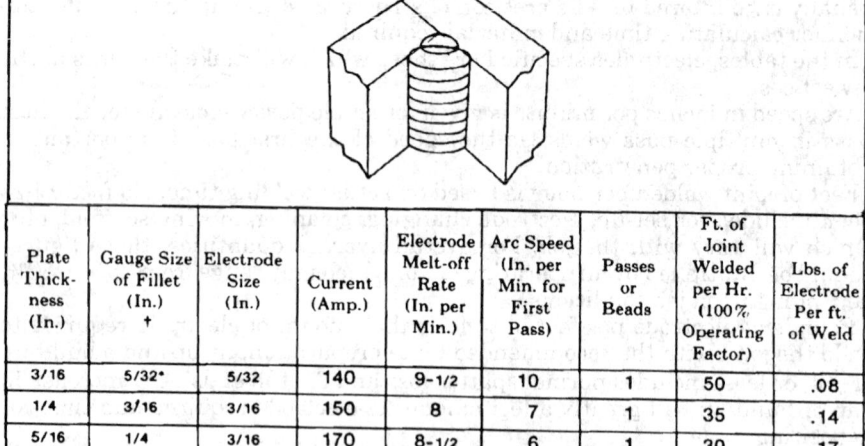

Plate Thick-ness (In.)	Gauge Size of Fillet (In.) †	Electrode Size (In.)	Current (Amp.)	Electrode Melt-off Rate (In. per Min.)	Arc Speed (In. per Min. for First Pass)	Passes or Beads	Ft. of Joint Welded per Hr. (100% Operating Factor)	Lbs. of Electrode Per ft. of Weld
3/16	5/32*	5/32	140	9-1/2	10	1	50	.08
1/4	3/16	3/16	150	8	7	1	35	.14
5/16	1/4	3/16	170	8-1/2	6	1	30	.17
3/8	5/16	3/16	170	8-1/2	3.9	1	19.5	.26
7/16	3/8	3/16	170	8-1/2	2.7	2	13.5	.38
1/2	7/16	3/16	170	8-1/2	6	2	10	.52
5/8	1/2	3/16	170	8-1/2	4	2 or more	7.7	.67
3/4	5/8	3/16	170	8-1/2	4	§	5	1.1
1	7/8	3/16	170	8-1/2	4	§	2.5	2.1

*Weld 5/32" fillet vertically down. T-joint welded both sides with this fillet will have plate strength. Total no. of passes varies with operator.

Preparation: square edge.
Fit-up: recommended gap, 1/32"; maximum gap 1/16". If greater gaps must be welded, use same procedure but add width of gap to fillet size.
Electrode: E6010
Polarity: electrode positive.

The following are actual production quantities of field welding on two different jobs located in different parts of the country and will give the estimator an idea of the amount of welding and labor required.

Description	9 Story Off Bldg.	18 Story Hospital
Size of building on the ground	100'x100'	140'x200'
Total tonnage of steel welded, tons	1,170	2,274
Fillet per ton of steel, lineal inches	63.5	50.68
Electrode used per ton of steel, lbs.	4.0	1.7
Electric current required per ton, kw. hrs.	8.1	6.05
Weld per lb. of electrode bought, lin.inches	16.00	21.34
Welding man hours per ton of steel, hours	1.51	1.80*
Steel welded per month, per machine, tons	128.00	....
Lin. ft. fillet per 8-hr. day, per welder	28.0	36.0
Lin ft. fillet per pound of electrode bought	1.33	1.78
Electric current required to melt one lb. of electrode, kw. hrs.	2.90	3.63

*Includes 0.26 hrs. foreman, 0.91 hrs. welder, 0.26 hrs. helpers, and 0.37 hrs. hoist. engr. time.

Data on Fillet and Lap Welds in Overhead Position

Estimating Data on Fillet and Lap Welds in Overhead Position

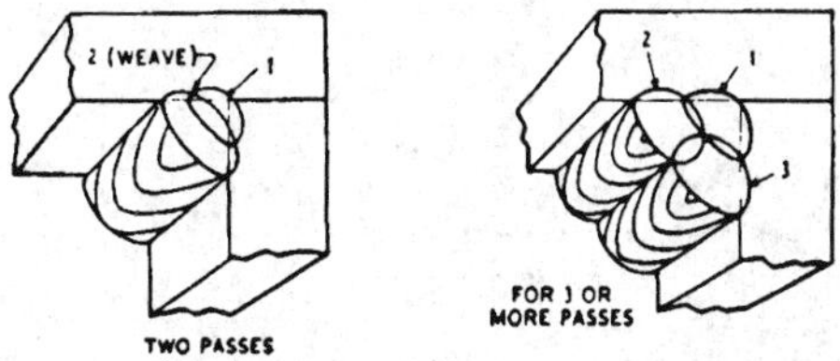

Plate Thick-ness (In.)	Gauge Size of Fillet (In.) †	Electrode Size (In.)	Current (Amp.)	Electrode Melt-off Rate (In. per Min.)	Arc Speed (In. per Min. for First Pass)	Passes or Beads	Ft. of Joint Welded per Hr. (100% Operating Factor)	Lbs. of Electrode per ft. of Weld
3/16	5/32	5/32	140	9-1/2	9	1	45	.09
1/4	3/16	3/16	160	8-1/4	7	1	35	.14
5/16	1/4	3/16	160	8-1/4	5.7	1	28.5	.18
3/8	5/16	3/16	160	8-1/4	3.7	2	18.5	.27
7/16	3/8	3/16	160	8-1/4	7-3/4	3	12.5	.40
1/2	7/16	3/16	160	8-1/4	7	§	9.2	.55
5/8	1/2	3/16	160	8-1/4	6	§	7.0	.72
3/4	5/8	3/16	160	8-1/4	6	§	4.5	1.2
1	7/8	3/16	160	8-1/4	6	§	2.2	2.3

T-joint welded both sides with this fillet will have plate strength. Total number of passes varies with operators.

Preparation: square edge.
Fit-up: recommended gap, 1/32", maximum gap, 1/16". If greater gaps must be welded, use same procedure but add width of gap to fillet size.

Approximate Prices of Iron and Steel Products

All prices for structural steel are based on carload shipments from mills, Pittsburgh, PA, Chicago, IL, or Birmingham, AL. Freight must be added from nearest mill point to destination.

Prices on steel should always be obtained from mills or nearest jobber, as steel prices are subject to frequent change.

Structural steel shapes are currently quoted at the basic price of $24.90 per 100 pounds for ASTM A36 steel.

The estimator should get a price quote from his or her steel supplier for each project.

Extras Over Base Price on Structural Shapes

S Shape Beams

Size Inches	Extra 100 lbs.
6"	$2.75
7"	2.75
8"	2.35
10"	2.25
12" 31.8 and 35 lbs.	2.25
12" 40.8 and 50 lbs.	2.25
15"	2.25
18"	2.25
20"	2.25
24"	2.25

W Shape Beams

Size Inches	Weight Pounds	Extra 100 lbs.
4"	13	$2.10
6"	9	2.10
6"	12 and 16	2.10
6"	20 and 25	2.10
8"	10	2.10
8"	13 and 15	2.10
8"	18 and 21	2.10
8"	24 to 28	2.10
8"	31 to 67	2.10
10"	12	2.10
10"	15 to 19	2.10
10"	22 to 30	2.10
10"	33 to 45	2.10
10"	49 to 112	3.60
12"	14	2.10
12"	16 to 22	2.10
12"	26 to 35	2.10
12"	40 to 50	2.10
12"	53 to 58	2.10
12"	65 to 190	4.50
14"	22 to 26	2.10
16"	36 to 57	2.10
16"	67 to 100	4.50
18"	35 to 46	2.10
18"	50 to 71	2.10
18"	76 to 119	4.50
18"	35 to 46	2.10
21"	44 to 57	4.50
21"	62 to 93	4.50
21"	101 to 147	4.50
24"	55 to 162	4.50
27"	84 to 178	4.50
30"	99 to 132	4.50
30"	173 to 211	4.50
33"	118 to 152	4.50
33"	201 to 241	4.50
36"	135 to 210	4.50
36"	230 to 300	4.50

M Shape Beams

Size Inches	Weight Pounds	Extra 100 lbs.
5"	18.9	$2.10
6"	20	3.10
8"	6.5	4.10

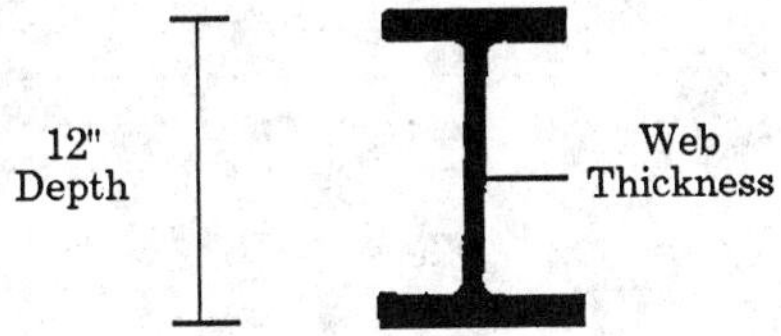

W12x45
W=shape
12=depth in inches
45=weight per lin. ft.

Angles

Size Inches	Extra 100 lbs.
3 x 2 x 3/16	$3.10
3 x 2 x 1/4 to 1/2	3.10
3 x 2 1/2 x 1/4 to 1/2	3.10
3 x 3 x 3/16	3.10
3 x 3 x 1/4 to 1/2	3.10
3 1/2 x 2 1/2 x 1/4 to 5/16	3.10
3 1/2 x 3 x 1/4 x 5/16	3.10
3 1/2 x 3 1/2 x 1/4 to 5/16	3.10
4 x 3 x 1/4	3.10
4 x 3 x 5/16	3.10
4 x 3 1/2 x 1/4 to 5/16	3.10
4 x 3 1/2 x 3/8 to 1/2	3.10
4 x 4 x 1/4 to 5/16	3.10
4 x 4 x 3/8 to 3/4	3.10
5 x 3 x 1/4	3.10
5 x 3 x 5/16	3.10
5 x 3 1/2 x 3/8 to 3/4	3.10
5 x 5 x 5/16	3.10
5 x 5 x 3/8 to 3/4	3.10
6 x 3 1/2 x 5/16 to 1/2	3.10
6 x 4 x 5/16	3.10
6 x 4 x 3/8 to 7/8	3.10
6 x 6 x 3/8 to 1	3.10
7 x 4 all thicknesses	3.10
8 x 4 all thicknesses	3.10
8 x 6 all thicknesses	3.10
8 x 8 all thicknesses	3.10
9 x 4 all thicknesses	3.10

Channels--Standard

Size Inches	Extra 100 lbs.
3"	$2.10
4"	2.10
5"	2.10
6"	2.10
7"	2.10
8"	2.10
9"	2.10
10"	2.10
12"	2.10
15"	3.10

Zees

Size	Extra 100 lbs.
Z3 x 6.7 to 12.6 lb.	$5.00
Z4 x 8.2 10.3 lb.	5.00

Estimating the Weight of Wrought Iron, Steel, or Cast Iron

When tables of weights are not handy, the following rules will prove of value to the estimator when computing the weights of wrought iron, steel, or cast iron.

Weight of Wrought Iron. One cubic foot of wrought iron weighs 480 pounds. One square foot of wrought iron 1 inch thick weighs 40 pounds. One square inch of wrought iron one foot long weighs 3 1/3 lbs.

To find the weight of one square foot of flat iron of any thickness, multiply the thickness in inches by 40, and the result will be the weight of the iron in pounds.

To find the weight of one lineal foot of wrought iron bar of any size, multiply the cross sectional area in square inches by 3 1/3, and the result will be the weight per lineal foot.

Weight of Steel. One cubic foot of steel weighs 489.6 pounds, or 2% more than wrought iron. One square foot of 1-inch thick steel weighs 40.8 lbs. A piece of 1-inch square by 1-foot long steel weighs 3.4 lbs.

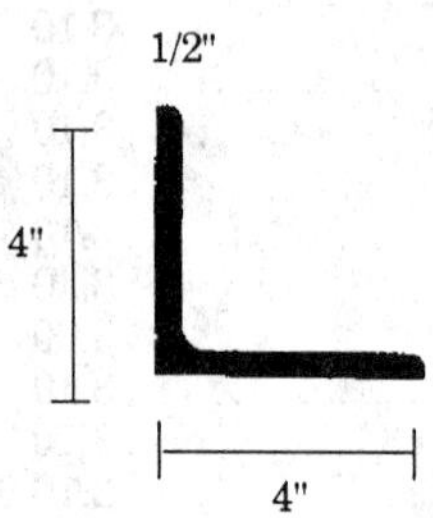

Equal Angles
4 x 4 x 1/2
Weight is listed
in AISC manual.

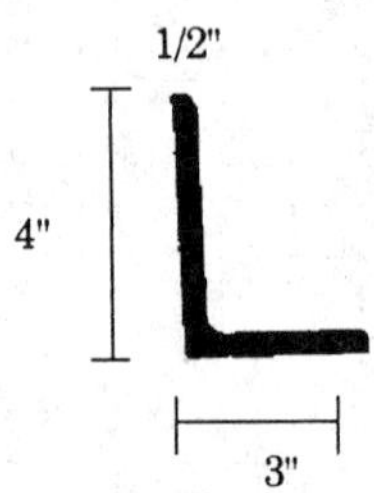

Unequal Angles
4 x 3 x 1/2
Weight is listed
in AISC manual

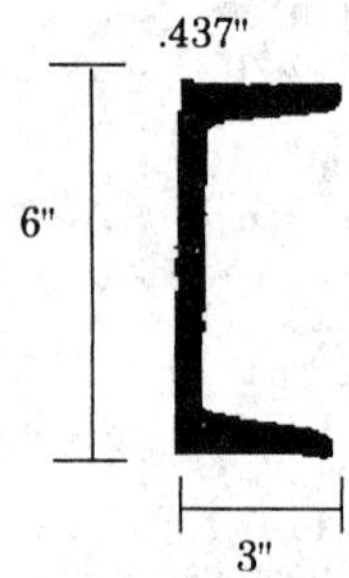

Standard Channel
6 x 13
6=channel size
13=weight per lin. ft.

To find the weight of one lineal foot of steel bar of any size, multiply the cross sectional area in square inches by 3.4, and the result will be the weight of the steel in pounds. If the weight per lineal foot is known, the exact sectional area in square inches may be obtained by dividing the weight by 3.4.

Weight of Cast Iron. One cubic foot of cast iron weighs 450 pounds. One square foot of cast iron, 1 inch thick, weighs 37 ½ lbs. A piece of cast iron 1 inch square and 1 foot long weighs 3 ⅛ lbs. One cubic inch of cast iron weighs .26 lb.

Weights of Square and Round Bars

Thickness or Diameter in Inches	Weight of Square Bar 1 foot long	Weight of Round Bar 1 foot long	Thickness or Diameter in Inches	Weight of Square Bar 1 foot long	Weight of Round Bar 1 foot long
¼	.213	.167	1 11/16	....	7.60
5/16	.332	.261	1 ¾	10.41	8.18
⅜	.478	.376	1 13/16	....	8.773
7/16	.651	.511	1 ⅞	11.95	9.39
½	.850	.668	2	13.60	10.68
9/16	1.076	.845	2 ⅛	15.35	12.06
⅝	1.382	1.04	2 ¼	17.21	13.52
11/16	1.607	1.26	2 ⅜	....	15.06
¾	1.913	1.50	2 ½	21.25	16.69
13/16	2.240	1.76	2 ⅝	23.43	18.40
⅞	2.603	2.04	2 ¾	25.71	20.20
15/16	....	2.35	2 ⅞	....	22.07
1	3.400	2.67	3	30.60	24.03
1 1/16	....	3.01	3 ⅛	....	26.08
1 ⅛	4.303	3.38	3 ¼	35.91	28.21
1 ¼	5.313	4.17	3 ⅜	....	30.42
1 5/16	....	4.6	3 ½	41.65	32.71
1 ⅜	6.428	5.05	3 ⅝	....	35.09
1 7/16	....	5.52	3 ¾	47.81	37.55
1 ½	7.650	6.01	4	54.40	42.73
1 ⅝	8.928	7.05			

Weights of Flat Steel Bars

Thickness in Inches	Width of Bars 1"	2"	3"	4"	5"	6"
3/16	.638	1.28	1.91	2.55	3.19	3.83
¼	.850	1.70	2.55	3.40	4.25	5.10
5/16	1.06	2.12	3.19	4.25	5.31	6.38
⅜	1.28	2.55	3.83	5.10	6.38	7.65
7/16	1.49	2.98	4.46	5.95	7.44	8.95
½	1.70	3.40	5.10	6.80	8.50	10.20
9/16	1.92	3.83	5.74	7.65	9.57	11.48
⅝	2.12	4.25	6.38	8.50	10.63	12.75
11/16	2.34	4.67	7.02	9.35	11.69	14.03
¾	2.55	5.10	7.65	10.20	12.75	15.30
13/16	2.76	5.53	8.29	11.05	13.81	16.58
⅞	2.98	5.95	8.93	11.90	14.87	17.85
15/16	3.19	6.38	9.57	12.75	15.94	19.13
1	3.40	6.80	10.20	13.60	17.00	20.40

Thickness in Inches	1"	2"	3"	4"	5"	6"
1 1⁄16	3.61	7.22	10.84	14.45	18.06	21.68
1 1⁄8	3.83	7.65	11.48	15.30	19.13	22.95
1 3⁄16	4.04	8.08	12.12	16.15	20.19	24.23
1 1⁄4	4.25	8.50	12.75	17.00	21.25	25.50
1 5⁄16	4.46	8.93	13.39	17.85	22.32	26.78
1 3⁄8	4.67	9.35	14.03	18.70	23.38	28.05
1 7⁄16	4.89	9.78	14.66	19.55	24.44	29.33
1 1⁄2	5.10	10.20	15.30	20.40	25.50	30.60
1 9⁄16	5.32	10.63	15.94	21.25	26.57	31.88
1 5⁄8	5.52	11.05	16.58	22.10	27.63	33.15
1 11⁄16	5.74	11.47	17.22	22.95	28.69	34.43
1 3⁄4	5.95	11.90	17.85	23.80	29.75	35.70
1 13⁄16	6.16	12.33	18.49	24.65	30.81	36.98
1 7⁄8	6.38	12.75	19.13	25.50	31.87	38.25
1 15⁄16	6.59	13.18	19.77	26.35	32.94	39.53
2	6.80	13.60	20.40	27.20	34.00	40.80

Weights and Dimensions of S Shape Beams

Size in Inches	Weight Per Ft. Lbs.	Thickness of Web Inches	Width of Flange Inches	Size in Inches	Weight Per Ft. Lbs.	Thickness of Web Inches	Width of Flange Inches
3	5.7	0.170	2.330	5	14.75	0.494	3.284
	7.5	0.349	2.509	6	12.5	0.230	3.330
4	7.7	0.190	2.660		17.25	0.465	3.565
	9.5	0.326	2.796	7	15.3	0.250	3.660
5	10	0.210	3.000		20	0.450	3.860
8	18.4	0.270	4.000	18	70	0.711	6.251
	23	0.441	4.171	20	66	0.505	6.255
10	25.4	0.310	4.660		75.0	0.641	6.391
	35	0.594	4.944		86	0.660	7.053
12	31.8	0.350	5.000		96	0.800	7.200
	35	0.428	5.078	24	80	0.500	7.000
	40.8	0.460	5.250		90.0	0.625	7.124
	50	0.687	5.477		100	0.745	7.247
15	42.9	0.410	5.500		106	0.620	7.875
	50	0.550	5.640		121	0.800	8.048
18	54.7	0.460	6.000				

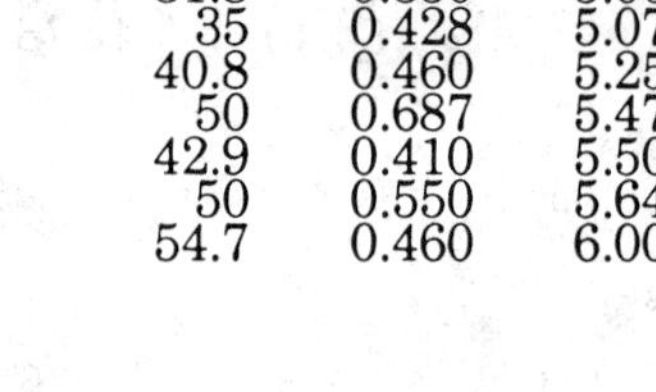

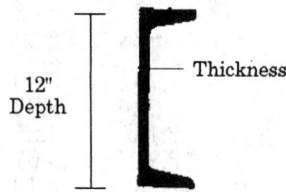

S24x100
S=shape of beam
24=depth in inches
100=weight per LF

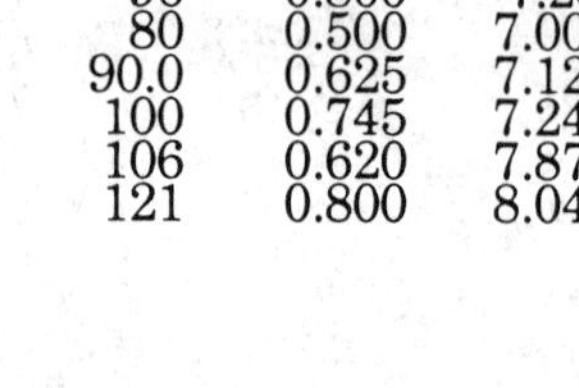

C12x20.7
C=shape (channel)
12=depth in inches
20.7=weight per LF

Weights of American Standard Channels

Depth of Channels Inches	Weight Per Ft., Lbs.	Thickness of Web, Inches	Width of Flange Inches	Depth of Channels Inches	Weight Per Ft., Lbs.	Thickness of Web, Inches	Width of Flange Inches
3	6.00	0.362	1.602		11.50	0.220	2.260
	5.00	0.264	1.504	9	20.00	0.452	2.652
	4.10	0.170	1.410		15.00	0.288	2.488
4	7.25	0.325	1.725		13.40	0.230	2.430
	5.40	0.180	1.580	10	30.00	0.676	3.036
5	9.00	0.330	1.890		25.00	0.529	2.889
	6.70	0.190	1.750		20.00	0.382	2.742
6	13.00	0.440	2.160		15.30	0.240	2.600
	10.50	0.318	2.038	12	30.00	0.513	3.173
	8.20	0.200	1.920		25.00	0.390	3.050
7	14.75	0.423	2.303		20.70	0.280	2.940
	12.25	0.318	2.198	15	50.00	0.720	3.720
	9.80	0.210	2.090		40.00	0.524	3.524
8	18.75	0.490	2.530		33.90	.400	3.400
	13.75	0.307	2.347				

Weights of Steel Angles

Size in Inches	Weight per Ft., Lbs.	Size in Inches	Weight per Ft., Lbs.
1/2 x 1/2 x 1/8	.38	2 x 1 1/2 x 5/16	3.39
5/8 x 5/8 x 1/8	.48	2 x 2 x 1/8	1.65
3/4 x 3/4 x 1/8	.59	2 x 2 x 3/16	2.44
3/4 x 3/4 x 3/16	.84	2 x 2 x 1/4	3.19
7/8 x 7/8 x 1/8	.70	2 x 2 x 5/16	3.92
1 x 5/8 x 1/8	.64	2 x 2 x 3/8	3.70
1 x 1 x 1/8	.80	2 x 2 x 1/2	6.00
1 x 1 x 3/16	1.16	2 1/4 x 1 1/2 x 3/16	2.28
1 x 1 x 1/4	1.49	2 1/4 x 2 1/4 x 1/8	1.86
1 1/8 x 1 1/8 x 1/8	.91	2 1/4 x 2 1/4 x 3/16	2.75
1 1/4 x 1 1/4 x 1/8	1.01	2 1/4 x 2 1/4 x 1/4	3.62
1 1/4 x 1 1/4 x 3/16	1.48	2 1/4 x 2 1/4 x 5/16	4.50
1 1/4 x 1 1/4 x 1/2	1.92	2 1/2 x 1 1/2 x 3/16	2.44
1 3/8 x 7/8 x 1/8	.91	2 1/2 x 1 1/2 x 1/4	3.19
1 1/2 x 1 1/2 x 1/8	1.23	2 1/2 x 2 x 3/16	2.75
1 1/2 x 1 1/2 x 3/16	1.80	2 1/2 x 2 x 1/4	3.62
1 1/2 x 1 1/2 x 1/4	2.34	2 1/2 x 2 x 5/16	4.50
1 1/2 x 1 1/2 x 5/16	2.86	2 1/2 x 2 x 3/8	5.30
1 1/2 x 1 1/2 x 3/8	3.35	2 1/2 x 2 x 1/2	6.74
1 3/4 x 1 3/4 x 1/8	1.44	2 1/2 x 2 1/2 x 1/8	2.08
1 3/4 x 1 3/4 x 3/16	2.12	2 1/2 x 2 1/2 x 3/16	3.07
1 3/4 x 1 3/4 x 1/4	2.77	2 1/2 x 2 1/2 x 1/4	4.10
1 3/4 x 1 3/4 x 5/16	3.39	2 1/2 x 2 1/2 x 5/16	5.00
2 x 1 1/4 x 3/16	1.96	2 1/2 x 2 1/2 x 3/8	5.90
2 x 1 1/4 x 1/4	2.55	2 1/2 x 2 1/2 x 1/2	7.70
2 x 1 1/2 x 1/8	1.44	2 x 2 x 3/16	3.07
2 x 1 1/2 x 3/16	2.12	3 x 3 x 1/8	2.50
2 x 1 1/2 x 1/4	2.77	3 x 3 x 3/16	3.71

Weights of Structural Steel Angles

Size in Inches	Weight per Ft., Lbs.
3 x 2 x 1/4	4.1
3 x 2 x 5/16	5.0
3 x 2 x 3/8	5.9
3 x 2 1/2 x 1/4	4.5
3 x 2 1/2 x 5/16	5.6
3 x 2 1/2 x 3/8	6.6
3 x 2 1/2 x 1/2	8.5
3 x 3 x 1/4	4.9
3 x 3 x 5/16	6.1
3 x 3 x 3/8	7.2
3 x 3 x 7/16	8.3
3 x 3 x 1/2	9.4
3 x 3 x 5/8	11.5
3 1/2 x 2 1/2 x 1/4	4.9
3 1/2 x 2 1/2 x 5/16	6.1
3 1/2 x 2 1/2 x 3/8	7.2
3 1/2 x 2 1/2 x 1/2	9.4
3 1/2 x 2 1/2 x 5/8	11.5
3 1/2 x 3 x 1/4	5.4
3 1/2 x 3 x 5/16	6.6
4 x 3 x 5/8	13.6
4 x 3 x 3/4	16.0
4 x 3 1/2 x 1/4	6.2
4 x 3 1/2 x 5/16	7.7
4 x 3 1/2 x 3/8	9.1
4 x 3 1/2 x 7/16	10.6
4 x 3 1/2 x 1/2	11.9
4 x 4 x 1/4	6.6
4 x 4 x 5/16	8.2
4 x 4 x 3/8	9.8
4 x 4 x 7/16	11.3
4 x 4 x 1/2	12.8
4 x 4 x 5/8	15.7
4 x 4 x 3/4	18.5
4 1/2 x 3 x 3/8	9.1
5 x 3 x 5/16	8.2
5 x 3 x 3/8	9.8
5 x 3 x 7/16	11.3
5 x 3 x 1/2	12.8
5 x 3 x 3/4	18.5
6 x 3 1/2 x 3/8	11.7
6 x 3 1/2 x 7/16	13.5
6 x 3 1/2 x 1/2	15.3
6 x 3 1/2 x 5/8	18.9
6 x 4 x 5/16	10.3
6 x 4 x 3/8	12.3
6 x 4 x 7/16	14.3
6 x 4 x 1/2	16.2
6 x 4 x 9/16	18.1
6 x 4 x 5/8	20.0
6 x 4 x 3/4	23.6
6 x 4 x 7/8	27.2
6 x 6 x 3/8	14.9
6 x 6 x 7/16	17.2
6 x 6 x 1/2	19.6
6 x 6 x 9/16	21.9
6 x 6 x 5/8	24.2
6 x 6 x 3/4	28.7
6 x 6 x 7/8	33.1
6 x 6 x 1	37.4
6 x 6 x 1 3/8	7.9
6 x 6 x 1 1/2	10.2
3 1/2 x 3 1/2 x 1/4	5.8
3 1/2 x 3 1/2 x 5/16	7.2
3 1/2 x 3 1/2 x 3/8	8.5
3 1/2 x 3 1/2 x 7/16	9.8
3 1/2 x 3 1/2 x 1/2	11.1
3 1/2 x 3 1/2 x 5/8	13.6
4 x 3 x 1/4	5.8
4 x 3 x 5/16	7.2
4 x 3 x 3/8	8.5
4 x 3 x 7/16	9.8
4 x 3 x 1/2	11.1
5 x 3 1/2 x 5/16	8.7
5 x 3 1/2 x 3/8	10.4
5 x 3 1/2 x 7/16	12.0
5 x 3 1/2 x 1/2	13.6
5 x 3 1/2 x 5/8	16.8
5 x 4 x 3/8	11.0
5 x 4 x 1/2	14.5
5 x 5 x 3/8	12.3
5 x 5 x 7/16	14.3
5 x 5 x 1/2	16.2
5 x 5 x 5/8	20.0
5 x 5 x 3/4	23.6
6 x 3 1/2 x 5/16	9.8
7 x 3 1/2 x 3/8	13.0
7 x 3 1/2 x 7/16	15.0
7 x 3 1/2 x 1/2	17.0
7 x 3 1/2 x 5/8	21.0
8 x 3 1/2 x 1/2	18.7
8 x 6 x 1/2	23.0
8 x 6 x 3/4	33.8
8 x 8 x 1/2	26.4
8 x 8 x 5/8	32.7
8 x 8 x 3/4	38.9
8 x 8 x 7/8	45.0
8 x 8 x 1	51.0
8 x 8 x 1 1/8	56.9

Weight of Steel Tees

Size Flange by Stem in Inches	Pounds Per Lineal Foot Thickness in Inches 1/8	3/16	1/4	5/16	3/8	1/2
3/4 x 3/4	0.60	....	....	....	....	....
7/8 x 7/8	.73	....	....	....	....	....
1 x 1	.90	1.20	....	....	....	....
1 1/4 x 1 1/4	....	1.55	1.98	....	....	....
1 1/2 x 1 1/2	....	1.88	2.43	....	....	....
1 3/4 x 1 3/4	....	2.30	2.90	....	....	....
2 x 1 1/2	....	....	3.10	....	....	....
2 x 2	....	....	3.60	4.40	....	....
2 1/4 x 2 1/4	....	....	4.10	4.90	....	....
2 1/2 x 2 1/2	....	....	4.60	5.50	6.4	....
2 1/2 x 2 3/4	....	....	....	5.9	6.8	....
2 1/2 x 3	....	....	....	6.1	7.2	....
3 x 2 1/2	....	....	....	6.1	7.2	....
3 x 3	....	....	....	6.7	7.8	....
3 1/2 x 3	....	....	....	....	8.5	....
3x4	....	....	....	....	9.3	....
3 x 3 1/2	....	....	....	....	8.5	....
3 1/2 x 3 1/2	....	....	....	....	9.2	11.7
3 1/2 x 4	....	....	....	....	9.8	....
4 x 2 1/2	....	....	....	....	8.5	....
4 x 3	....	....	....	....	9.2	....
4 x 4	....	....	....	....	10.5	13.5
4 x 5	....	....	....	....	11.9	15.3
4 1/2 x 2 1/2	....	....	....	....	9.2	....
4 1/2 x 3	....	....	....	8.6	10.0	....
5 x 3	....	....	....	....	11.0	13.6
6 1/2 x 6 1/2	....	....	....	....	....	19.8

Weights of Standard Diamond Steel Floor Plates

Thickness Inches	Weight per Sq. Ft., Lbs.	Thickness Inches	Weight per Sq. Ft., Lbs.
1/8	8.00	5/16	13.75
3/16	8.75	3/8	16.25
1/4	11.25	1/2	21.5

Weights of Steel M Shapes

Size	Weight Per Ft., Lbs.	Size	Weight Per Ft., Lbs.
M4x13	13.0	M8x6.5	6.5
M5x18.9	18.9	M10x9	9.0
M6x4.4	4.4	M12x11.8	11.8
M6x20	20.0	M14x18	18.0

Weights of Round Head Structural Rivets

Diam. and Length Inches	Weight per 100 Rivets Lbs.	Diam. and Length Inches	Weight per 100 Rivets Lbs.
1/2 x 1	9.49	3/4 x 6	87.61
1/2 x 1 1/2	12.24	3/4 x 6 1/2	93.80
1/2 x 2	14.99	3/4 x 7	99.97
1/2 x 2 1/2	17.74	3/4 x 7 1/2	106.15
1/2 x 3	20.49	3/4 x 8	112.33
1/2 x 3 1/2	23.24	7/8 x 1 1/2	46.67
1/2 x 4	25.99	7/8 x 2	55.10
5/8 x 1	16.38	7/8 x 2 1/2	63.53
5/8 x 1 1/2	20.68	7/8 x 3	71.96
5/8 x 2	24.98	7/8 x 3 1/2	80.39
5/8 x 2 1/2	29.28	7/8 x 4	88.82
5/8 x 3	33.58	7/8 x 4 1/2	97.25
5/8 x 3 1/2	37.88	7/8 x 5	105.68
5/8 x 4	42.18	7/8 x 5 1/2	114.11
5/8 x 4 1/2	46.76	7/8 x 6	122.54
5/8 x 5	51.10	7/8 x 6 1/2	130.97
5/8 x 5 1/2	55.45	7/8 x 7	139.40
5/8 x 6	59.80	7/8 x 7 1/2	147.83
3/4 x 1	25.81	7/8 x 8	156.26
3/4 x 1 1/2	31.99	1 x 2 1/2	86.82
3/4 x 2	38.17	1 x 3	97.80
3/4 x 2 1/2	44.35	1 x 3 1/2	108.79
3/4 x 3	50.53	1 x 4	119.77
3/4 x 3 1/2	56.71	1 x 4 1/2	130.76
3/4 x 4	62.89	1 x 5	141.74
3/4 x 4 1/2	69.07	1 x 5 1/2	152.73
3/4 x 5	75.25	1 x 6	163.71
3/4 x 5 1/2	82.05		

Weights of Steel Plates in Pounds per Square Foot

Thickness	lb./sq. ft.	Thickness	lb./sq. ft.
3/16	7.65	3/4	30.60
1/4	10.20	7/8	35.70
5/16	12.75	1	40.80
3/8	15.30	1 1/4	51.00
7/16	17.85	1 1/2	61.20
1/2	20.40	1 3/4	71.40
5/8	25.50	2	81.60

Weight of Corrugated Sheets in Pounds Per 100 Square Feet

	Black and Painted		Galvanized	
Gauge Nos.	2, 2 ½, & 3" Corrug.	1 ¼" Corrug.	2, 2 ½, & 8" Corrug.	1 ¼" Corrug.
16	271	...	286	...
18	217	...	232	...
20	163	170	178	185
22	136	142	151	157
24	110	114	125	129
26	83	86	98	101
27	76	79	91	94
28	68	72	85	87

JUNIOR STEEL BEAMS

Junior beams (M shapes) are lightweight, hot-rolled structural beams that are used as secondary floor and roof beams in schools, stores, apartments, hospitals, and other types of light occupancy buildings. They are well adapted for purlins in mill buildings.

They are rolled to ASTM specifications, and are made in 6", 8", 10", 12", and 14" sizes. Sizes and properties are given below:

Designation	Depth, Inches	Wt. per Ft., Lbs.	Web Thick. Inches	Flange Wdth Inches	Area Inches
M6 X 4.4	6	4.4	.114	1.844	1.29
M6 X 20	6	20.0	.250	5.938	5.89
M8 X 6.5	8	6.5	.135	2.281	1.92
M10 X 9	10	9.0	.157	2.690	2.65
M12 X 11.8	12	11.8	.177	3.065	3.47
M14 X 18	14	18.0	.215	4.000	5.10

The cost of cutting, punching holes, and coping junior beams will run approximately as given below, but when much fabricating is required, it is advisable to refer plans to a local fabricator or warehouse for a sub bid.

Top cope, per end	$4.00
Bottom cope and seat angles, per end	10.00
Cutting end on diagonal, per end	5.00
Anchor holes in web, per hole	1.00
Conduit holes in web, per hole	1.00

Junior steel beams can be used in conjunction with monolithically finished concrete floors or precast gypsum slabs laid over the top flange of the junior beams. By encasing the top flanges in concrete, the floor is made rigid and does not require bridging in light occupancy buildings.

Where the slab is poured on metal lath or where gypsum plank is used, adjustable steel rigid bridging is furnished.

When a poured-in-place concrete floor is used, the forms are easily constructed by cutting short pieces of lumber to rest on the lower flanges of the beams, which in turn support the floor sheathing.

The reinforcement of the concrete slab usually consists of welded wire fabric, the size and spacing of the wire depending upon the span of the slabs and the loads to be carried.

Where joists rest on walls or bearing partitions, anchors are provided. Where the tops of junior beams are flush with tops of supporting beams either clip angles or shelf angles are provided. Where the tops of beams are 2 ½" above the tops of supporting beams, part of the beam is coped out and an angle seat is welded to the junior beam which rests on the top flange of the supporting beam.

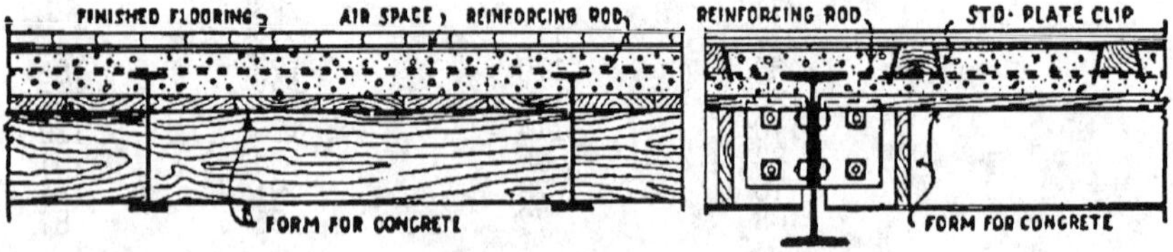

Method of Constructing Wood Forms for Concrete Floors
Supported by Junior Beams

050200 METAL JOISTS

Standard and longspan steel joists were developed primarily to take the place of wood construction. The fire-resistance and structural qualities of this type of floor and roof construction has meant that it has not only replaced wood but other types of fire-resistant floor construction as well. Steel joist construction is recognized and accepted by all building codes.

Standard open web and longspan steel joists are manufactured by welding chord members of hot rolled structural or cold formed sections to round bar or angle web members to form a truss. Information on the types of steel joists can be obtained from the Steel Joist Institute and the joist manufacturers.

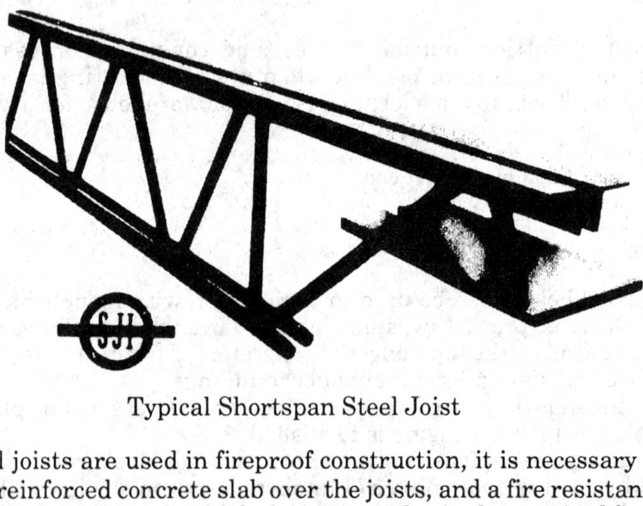

Typical Shortspan Steel Joist

When steel joists are used in fireproof construction, it is necessary to place a 2" to 2 ½" reinforced concrete slab over the joists, and a fire resistant ceiling of gypsum board or plaster beneath the joists to obtain the required fire rating.

Where a wood floor is required, wooden nailing screeds supported on screed clips are embedded in the concrete slab and the wood flooring is then nailed to the screeds.

Steel joists may also be used in semi-fireproof construction by nailing the wood sheathing directly to nailers attached to the joists without the use of a protective concrete slab. Some manufacturers of steel joists have met the demand for this type of construction by fabricating a nailer joist with the wood nailer attached in the shops by the use of lag screws or bolts.

Open web joists may be used to great advantage in all types of light occupancy buildings. The open web construction greatly facilitates the placing of plumbing, conduits, etc., making unnecessary the need of a furred ceiling or heavy fills on top of the structural slab.

Steel joists are also used in industrial construction to replace structural steel roof purlins with a resulting economy of materials. Metal roof deck is then usually welded to the joists and light weight insulating material either poured or rigid board is applied to the metal deck and the roofing is installed.

Estimating Steel Joists. Steel joists and related products are generally sold by the manufacturer directly to the contractor. When preparing general contract estimates the most satisfactory method is to send plans or sketches to the manufacturer or his agent thus assuring a sub-bid based on an economical layout and design and one which includes the necessary accessories.

In his quotation the manufacturer will note the tonnage of joists which will enable the contractor to figure the necessary handling and labor of erection. The cost of joist erection including placing of accessories varies from $150.00 to $250.00 per ton (excluding material costs) depending on labor conditions, location and type of building.

Labor Setting Steel Joists. The labor cost of handling and setting steel joists is usually estimated by the ton, based on the total weight of the joists to be placed.

This cost will vary considerably, depending upon the type of building, size of joists, length of spans, amount of handling and hoisting necessary, etc.

It is much cheaper to set joists on long straight spans and large floor areas, such as garage and factory buildings, office buildings, etc., than on small or other irregularly constructed buildings.

All steel joists must be bridged and connected according to the Steel Joist Institute standards.

On an ordinary job of steel floor joists of regular construction and fairly long spans, five workers together should handle and place about 4 ½ to 5 tons of joists (including bridging) per 8-hr. day at the following labor cost per ton:

	Hours	Rate	Total	Rate	Total
Ironworker	40	$....	$....	$27.53	$1,101.20
Cost per ton			....		231.83

On jobs of irregular construction, five men workers together should handle and place about 3 ½ to 4 tons of joists (including bridging), per 8-hr. day, at the following labor cost per ton:

	Hours	Rate	Total	Rate	Total
Ironworker	40	$....	$....	$27.53	$1,101.20
Cost per ton			....		293.65

Prices of Steel Joists. Costs of standard steel joists vary, depending on several factors, such as size of job, the number of joists of one particular size and length, and the job location. For an average job the cost of standard joists will be in the range of $600.00 to $700.00 per ton, which would include nominal bridging members and other accessories.

Costs of longspan steel joists will vary with the same factors as given above for standard joists. For an average job the cost of the longspan joists will range from $580.00 to $750.00 per ton including accessories.

05300 METAL DECKING

Steel roof deck is formed from steel sheets in 18, 20, or 22 gauge. Most commercial decks have a standardized cross section with longitudinal ribs spaced 6" on centers. Sections are 30" in width and are manufactured in various lengths to suit job conditions, usually in the 14-ft. to 31-ft. range, although some can be supplied in greater lengths.

Deck sections are usually 1 ½" deep, but some manufacturers supply a 1 ¾", 2", or 2 ½" deep deck. Sections have interlocking or nesting side laps and telescoping or nesting end laps.

Under ordinary roof live loads, 30 to 40 lbs. per sq. ft., 20-gauge deck is normally used for spans from 6'-0" to 7'-6" centers of purlins and 18 gauge deck for spans from 7'-0" to 8'-0". Specific information on allowable loads can be obtained from the various manufacturers. By extending a single sheet over two or more purlin spaces, structural continuity, increasing load carrying capacity, can be obtained. Dead load of deck, insulation and built-up roofing is approximately 7 lbs. per sq. ft.

Approximate cost of steel roof deck, delivered to job site in the eastern or northern states, on an average size job (10,000 sq. ft. or more) is as follows:

Gauge of Roof Deck	Price per Sq. Ft. 1 ½" Deck
22 Ga.	$.75
20 Ga.	.85
18 Ga.	1.10

Placing Metal Deck on Top of Steel Joists. When metal deck is placed over the top of steel joists to receive the concrete slab or lightweight insulating materials, the metal deck is usually spot welded to the joists.

On work of this type two workers together should place and weld about 200 sq. ft. of deck per hour at the following labor cost per square of 100 sq. ft.

	Hours	Rate	Total	Rate	Total
Iron Worker	1.0	$....	$....	$27.53	$27.53
Cost per sq. ft.			....		.28

Steel roof deck can be erected in almost any kind of weather, permitting other trades to work under cover much more rapidly. Deck must be welded to steel support members in accordance with the Steel Deck Institute recommendations. Two tack welds per 18" width of deck section at end laps and one weld on the outside rib on intermediate supports are recommended.

Longspan Deck

Longspan deck is formed from a single 14, 16, 18, or 20 gauge steel sheet into 4 ½", 6", or 7 ½" deep pans 12" wide. These are rolled sections with integral stiffening ribs formed into the top flange. Interlocking longitudinal side laps are provided on opposite edges for positive side lap attachment and are fastened by welding. Maximum spans allowable vary from 20'-0" to 30'-0".

Longspan deck is easily erected--on an average job, a 5-worker crew should erect 2,200 to 2,400 sq. ft. per 8-hr. day.

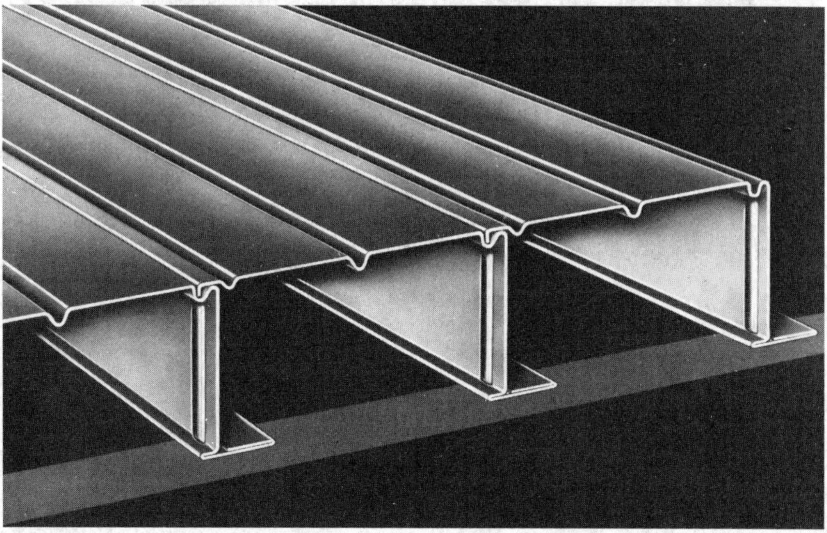

Longspan Roof Deck

Approximate cost of longspan steel roof deck, delivered to job sites in the eastern or northern states, is $2.00 to $3.50 per sq. ft., depending upon gauge and length of sections.

Reinforcing Floor Forms

Steel deck erected in an inverted position (with ribs up) can be used as a reinforcing form for concrete construction. It acts as a form to support the concrete and permits the ribs to act as reinforcing.

Steel deck reinforcing forms can be erected for an entire structure as soon as the steel framework is placed, providing an immediate working platform and protective staging for all trades. Concrete can be placed at any portion of the building without regard for removing, cleaning and re-setting temporary formwork. Steel deck forms usually provide all necessary reinforcing to satisfy flexure requirements--the only additional reinforcing ordinarily required is temperature mesh to minimize shrinkage cracking. Tight form joints prevent concrete from dripping to lower floors, saving clean-up time.

Steel deck reinforcing forms can also be used in conjunction with composite beam design. Ample area for concrete around stud shear connectors and between deck ribs permits full effectiveness of the connectors. Standard AISC Composite Design Procedure may be followed using the total slab depth in beam property calculations.

Reinforcing floor forms are installed by welding, the same as ordinary steel roof deck, and costs of materials and erection are about the same as given for "Metal Decking".

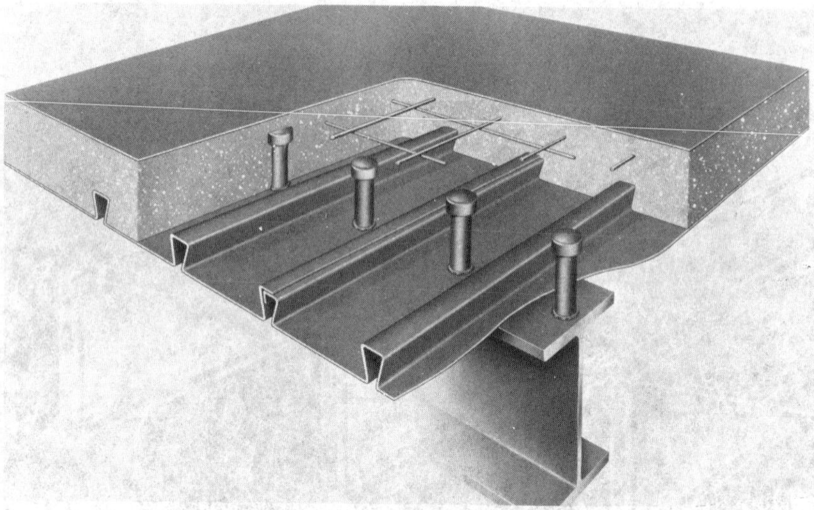

Concrete Slab With Steel Forms Used In Conjunction With Composite Beam Design

CELLULAR STEEL DECK AND FLOOR PANELS

Light gauge cellular steel structural panels are used for longspan structural floors and roofs. Some panels are composed of two identically formed beam sections and a flat plate--others are formed from two ribbed sections and a flat plate. All panels have interlocking side joints. Components are factory assembled by spot welding. Panels are manufactured in gauges from 20 to 13 and are normally available in 24" widths--some panels are fabricated in 12" widths. Lengths are available up to 40'-0", while depths vary from 1 ½" to 7 ½".

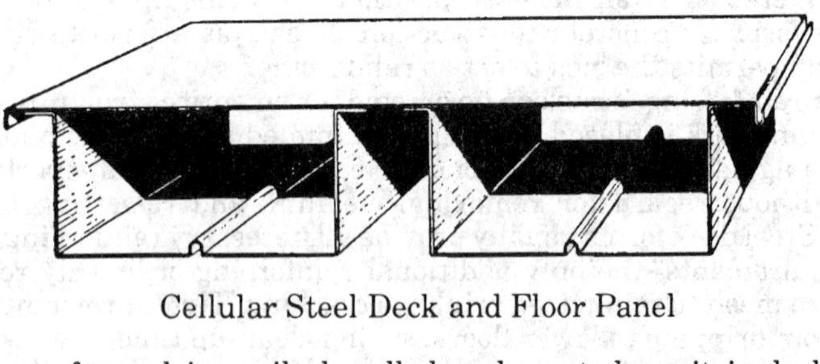

Cellular Steel Deck and Floor Panel

This type of panel is easily handled and erected, as it is deck and joist combined. Concrete floor forms are eliminated and panels provide a working platform for other trades.

A flat, plate down installation is often used to provide an attractive ceiling which needs only a finish coat of paint to complete the job. Voids between beam sections may be filled and covered with 2 ½" of concrete for floors or covered with rigid board insulation and built-up roofing for roofs. Installations with beam section forming the ceiling can be merely painted to give a finished fluted ceiling or a suspended ceiling can be erected to provide a flush type ceiling.

Where panels are covered with a concrete fill, the use of temperature mesh, to reduce shrinkage cracks, is recommended.

The cost of cellular panels, delivered to job site, ranges from $3.00 per sq. ft. for 20 ga. 2" to 3" thick panels, $3.25 to $4.00 for 3" to 4" 18 ga. units on up to $4.25 for 16 ga. deck.

Acoustically Treated Deck and Floor Panel

A crew of four structural iron workers plus a foreman should install around 1600 sq. ft. of light deck per 8 hour day, 1200 sq. ft. of 18 ga. deck, and around 1000 sq. ft. of the heavy gauge deck under normal job conditions.

For example, 100 sq. ft. of 18 ga. labor costs would be as follows:

	Hours	Rate	Total	Rate	Total
Foreman	0.65	$....	$....	$28.36	$18.43
Str'l Ironworkers	2.60	....	....	27.53	71.58
Cost per 100 sq. ft.					$90.01
Cost per sq. ft.					.90

05400 LIGHTGAGE FRAMING

The components involved are a complete range of studs, joists, and accessories for the steel framing of buildings. Sections are fabricated from structural-grade, high tensile strip steel by cold forming and are designed specifically for strength, light weight and low cost. Yet, structural framing carries all the benefits of conventional steel framing.

A nailing groove, developed for easy, economical attachment of other materials, is a feature of all double studs and joists. It is obtained by welding two cold-formed steel elements together, so designed that a nail driven into this space is not only held by friction, but is also deformed to provide maximum holding power.

The usual limitations imposed by prefabrication are avoided. Using nailable framing sections architects and engineers have unrestricted freedom in design. All sections are painted at the plant with a coat of oven-dried, rust-resisting red zinc chromate paint, or are galvanized.

Lightgage framing systems can supply complete wall, floor and roof construction for buildings up to four stories in height, or can be used in combination with other framing systems for interior, load bearing partitions, exterior curtain walls, fire separation walls, parapets, penthouses, trusses, suspended ceilings and mansard roofs.

Erecting Lightgage, Preassembled Wall

Advantages of using these systems include reduced dead loads, uniform fabrication, less on-the-job storage space, no warping, shrinkage or swelling, incombustibility, punched slots for passage of other trades and secure nailing to eliminate nail popping. When properly engineered and applied, this system may be cheaper than conventional wood or steel construction; but even when comparable in price, the advantages offered to other trades may yield savings on the overall job.

The components can be completely detailed, cut and assembled in the shop and delivered for erection to the job site. Where on-the-job cutting and assembly is preferred, the material can be cut with a radial saw fitted with a 1/8" high speed circular blade. Fastening can be by bolts, sheet metal screws or welding.

Joists come in 6", 8", 10", and 12" depths and in 12, 14, 16, and 18 gauge material. There are generally three styles: double nailable,"C" joists with 1 5/8" flanges and"C" joists with 2 1/2" flanges. Lengths from 6' to 40' are available. Joist webs may be solid, selectively punched or continuously punched for maximum raceway flexibility. Unnecessary punching should be avoided as the slotting lowers the structural value of the joist.

Joist bridging, which may be by stock "V" units or solid channels, must be supplied in the center of all spans up to 14'; at third points on spans from 14' to 20'; at quarter points on spans from 26' to 32'; and at eight foot centers on all spans over 32'.

Studs are available in 2 1/2", 3 5/8", 4", and 6" depths and in 14, 16, 18, and 20 gauges. Also available are 25 gauge studs, but they are suitable for non-load-bearing partitions only and are discussed under the drywall section of Chapter 14 "Finishes". Lengths from 7' to 40' are available with either slotted or solid webs. Studs may be of the double-nailable type, channels, or "C" type with knurled flanges.

Studs mount into standard tracks at floors, ceilings, sills, and facias. They should be stiffened with lateral bridging at the midpoint of all walls to 10' in height; at third points on walls from 10' to 14'; and every 4' on walls above 14'. Stock "V" bars, channels or special clipping systems may be ordered for this.

Costs will vary because of the various options available and each job should be priced with the supplier when all the design criteria are known. For general budget purposes joists can be figured at around 75 cents per pound galvanized. A punched, galvanized "C" joist will vary from 2.24 lbs. per ft. 6" deep to 5.85 lbs. 12" deep or from around $1.70 to $4.40 per foot. Carload lots will run less.

Studs 3 5/8" deep, punched and galvanized, will run 50 cents per foot in 20 gauge material in less than carload lots; 75 cents per foot in 18 gauge; and 85 cents per foot in 16 gauge. Tracks will approximate stud costs.

Attachment of Other Materials. Steel or aluminum roof decking, metal siding, or corrugated sheets can be quickly and economically attached to sections with self-tapping screws, or by other conventional methods.

When self-tapping screws are used, a hole slightly smaller than the screw is drilled through the material and the lightgage section. Then, the screw can be placed and tightened by one man. This is faster and more economical than most conventional methods of attachment which require two men. The 5/8" hex-head self-tapping, cadmium-plated screws are recommended.

Attachment of materials is generally with standard nails when the nailable sections are used. When channel studs or other non-nailable sections are used, the self-drilling screws are recommended for metal siding or plywood. When drywall is applied to non-nailable sections, self-drilling screws can be ordered from the drywall manufacturer.

Other types of materials can be readily attached to framing members by the use of bolts, clips, welds or other methods consistent with the finish desired.

DIMENSIONAL DATA – TABLE 1

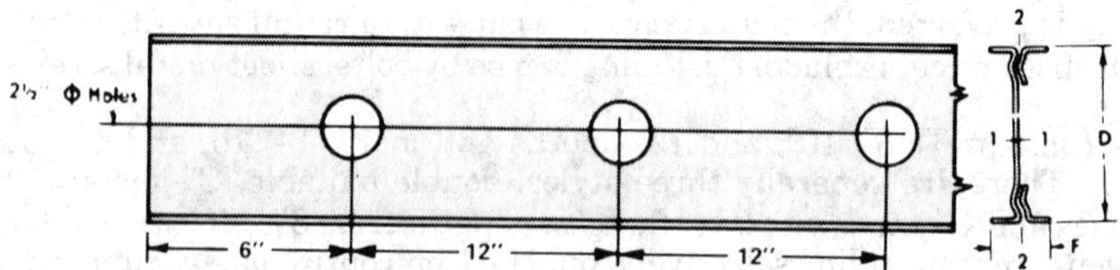

12", 10", 8" and 6" NAILABLE JOISTS

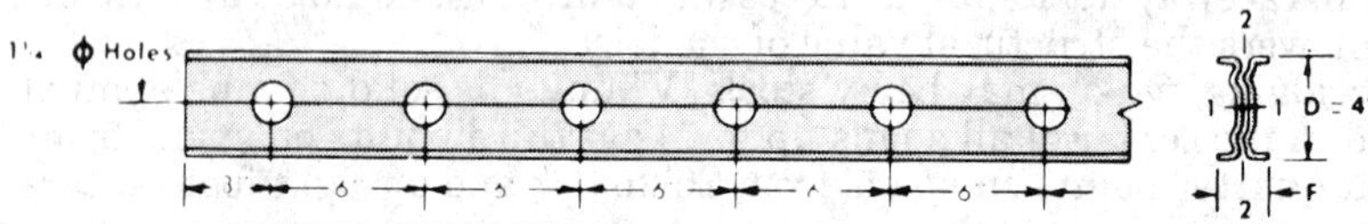

4" NAILABLE DOUBLE STUDS

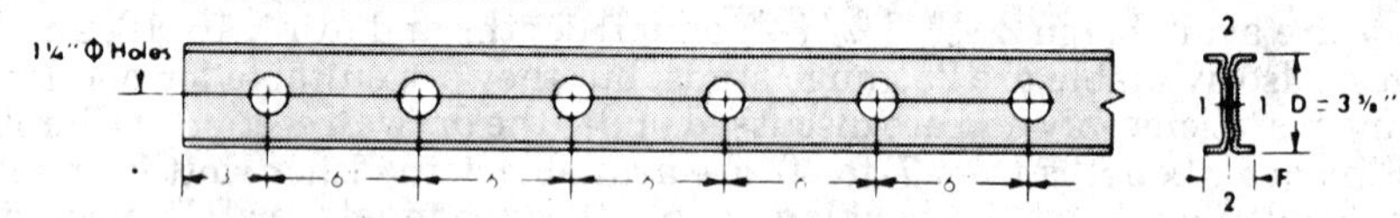

3⅝" NAILABLE DOUBLE STUDS

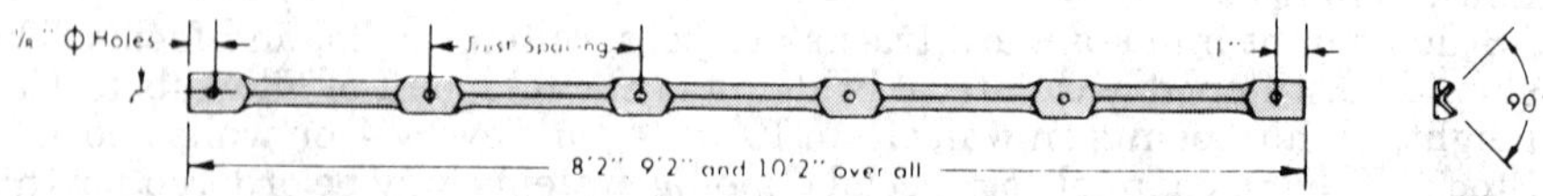

V-BAR BRIDGING*
(Weight = 0.26 lbs. ft., gage = 16, t = 0.0568, net area = 0.075 sq. in.)

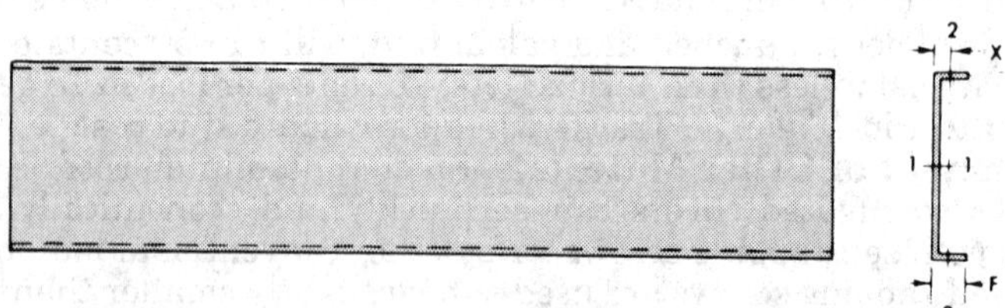

UNPUNCHED CHANNEL AND TRACK

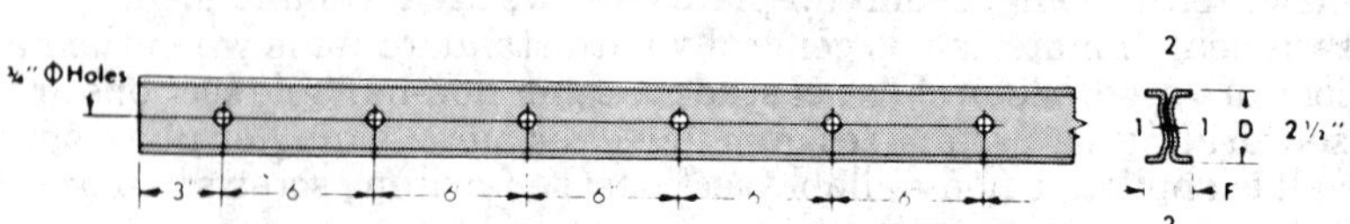

2½" NAILABLE DOUBLE STUDS

DIMENSIONAL DATA – TABLE 2

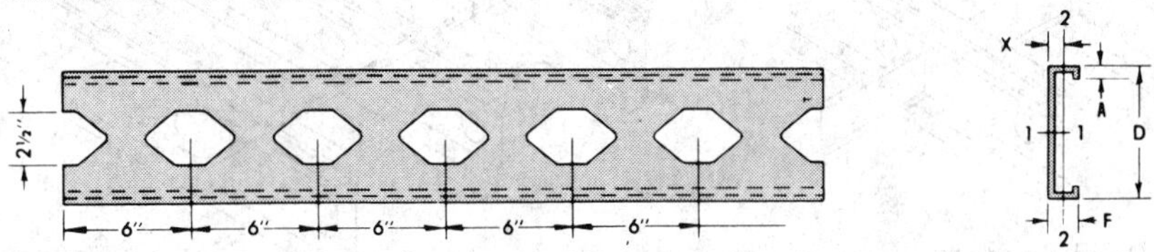

6" AND 8" CEE STUDS

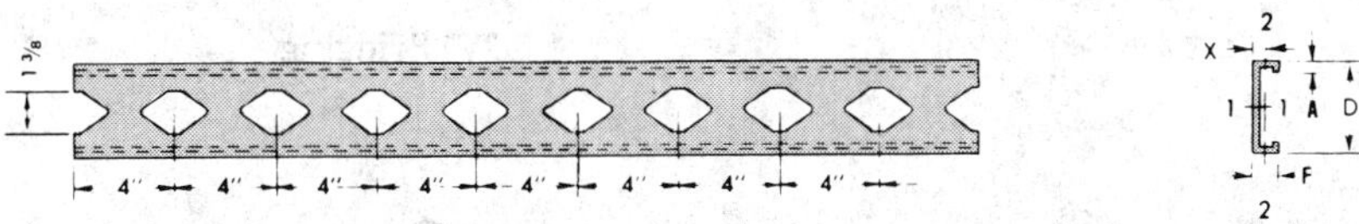

4", 3 5/8", 3 1/4" SCREW CEE STUDS

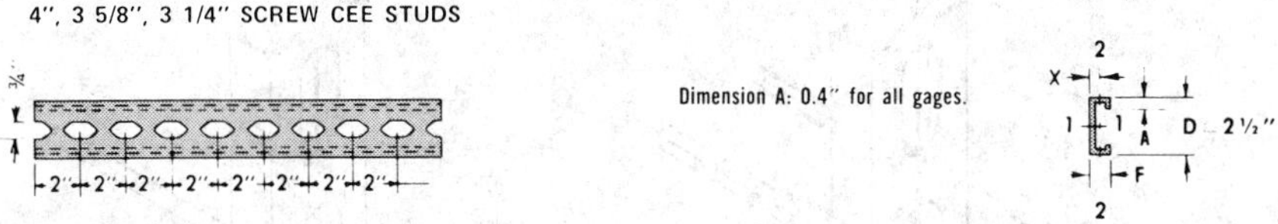

2 1/2" SCREW CEE STUDS

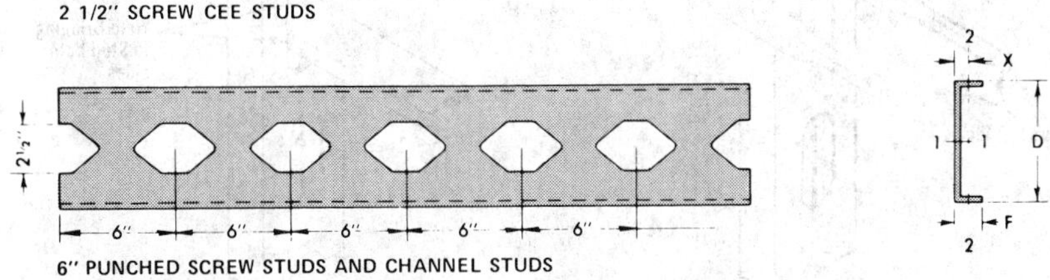

6" PUNCHED SCREW STUDS AND CHANNEL STUDS

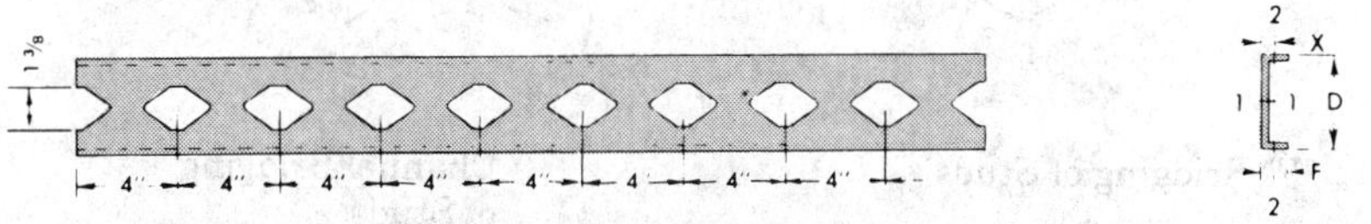

4", 3 5/8", 3 1/4" PUNCHED SCREW STUDS AND CHANNEL STUDS

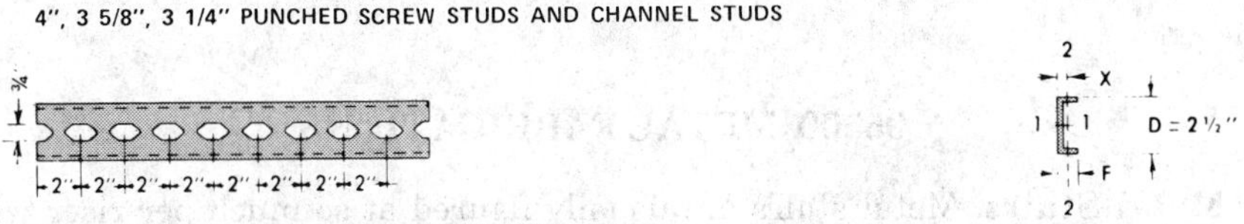

2 1/2" PUNCHED SCREW STUDS AND CHANNEL STUDS

NOTE: If holes are to line up, this specification must accompany order.

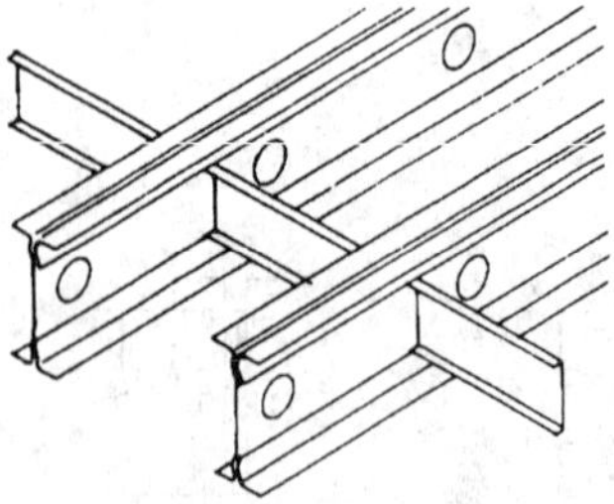

Solid Bridging

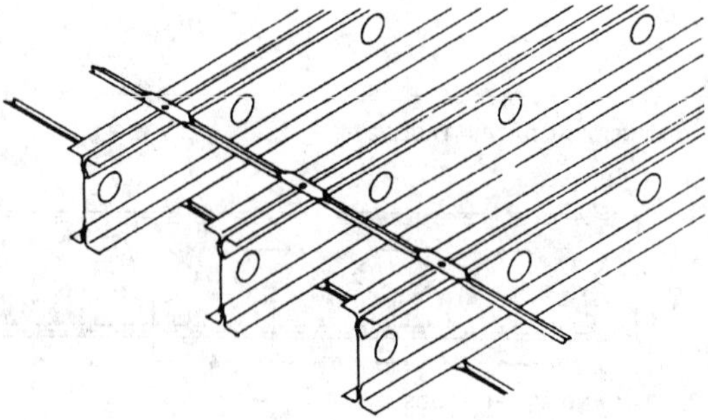

"V" Bridging

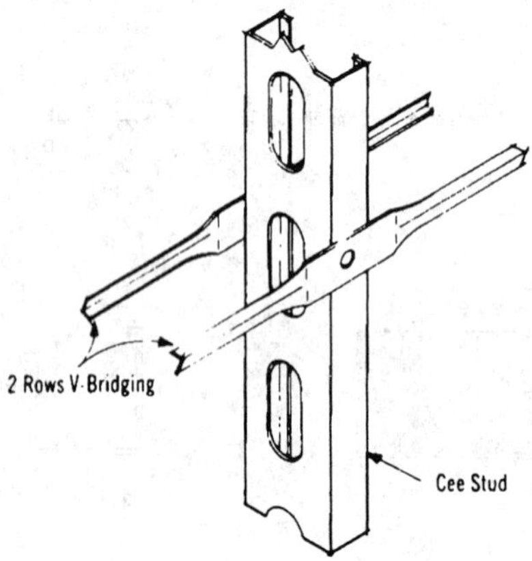

"V" Bridging of Studs

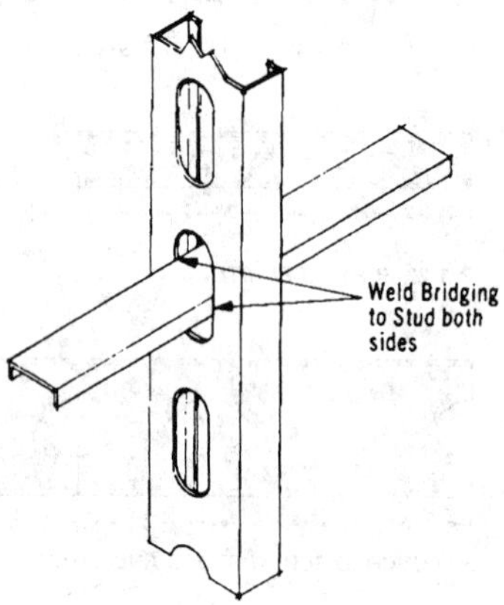

Channel Bridging of Studs

05500 METAL FABRICATIONS

Metal Stairs. Metal stairs are usually figured at so much per riser with stringers, treads, nosings, and railings included. A standard, 3-foot wide simple run metal pan stair can be budgeted at around $85.00 per riser, material cost. For each additional foot of width up to five feet add 10%. Custom stairs will run from $95.00 per riser to twice that amount where any of the stair components vary from the norm. Simple landings cost about $8.50 per square foot without an allowance for railings.

A four man crew should erect around 40 to 50 risers or 150 sq. ft. of landing per 8 hr. day.

The labor erection cost of a three foot wide steel stair consisting of three flights of sixteen risers and one 3' by 6' landing each run would figure as follows:

	Hours	Rate	Total	Rate	Total
Ironworkers					
Stairs - 48 risers	34.0	$....	$....	$27.53	$936.02
Landing 54 sq. ft.	11.1	....	....	27.53	305.58
Total Cost					$1,241.60
Cost per stair run (3)					413.87
Cost per riser incl.landings					25.87

Pipe Railings. When estimating plain pipe railing, 3'-6" high, consisting of two horizontal runs of pipe with uprights 6 ft. to 8 ft. on center, figure as follows for the various sizes, erected in place:

Description	Price per Lin. Ft.
1 ¼" pipe railing complete as described above per lin. ft.	$40.00
1 ½" pipe railing complete as described above per lin. ft.	$45.00
2" pipe railing complete as described above per lin. ft.	$60.00

For each curved section or termination point that is formed to radius, add $35.00 to $45.00.

For following the rake of a stair, add $8.50 per upright.

For each foot of wall railing, add $8.00.

Aluminum pipe railings can also be ordered from stock and will cost around $65.00 per foot, anodized. Aluminum wall rails will cost around $9.50 per foot. A two man crew should erect some 100 lineal feet of straight run railing per day, and 130 feet of wall railing.

Steel Ladders. Straight steel ladders cost from $40.00 to $60.00 per lin. ft. erected in place. Add $30.00 to $45.00 for curved handles and platform over coping walls. Add $40.00 per lin. ft. if cage is required.

Steel Gratings. Gratings may be welded or of expanded construction and made of steel, aluminum, or stainless steel.

Welded steel gratings weighing 4# will run about $4.50 per sq. ft.; weighing 5#, $4.50; 9#, $6.50; 12#, $8.50; and 16#, $9.25 to $10.25, all depending on the spacing and size of the bars. If the material is galvanized add 90 cents per sq. ft.

Aluminum grating will run from $11.00 per sq. ft. ¾" thick and weighing 1.5#, to $25.00 per sq. ft. 2 ¼" thick and weighing 5.9# in amounts of 75 sq. ft. or more.

Checkered plate is sometimes specified in place of or in combination with gratings and can be figured at about $9.50 per sq. ft. for ¼" thickness.

A typical crew laying grating will consist of four men who can install about 100 square feet per hour at the following cost per 100 sq. ft.:

	Hours	Rate	Total	Rate	Total
Ironworkers	4.0	$....	$....	$27.53	$110.12
Cost per sq. ft.					1.10

Steel Window Guards. Window guards made with ⅜" x 2" horizontal bars, top and bottom and round or square upright bars spaced 4" on centers. Delivered to job.

	Approx. Price per Sq. Ft.
Window guards with 1/2" round upright bars	$7.50
Window guards with 1/2" square upright bars	8.00
Window guards with 5/8" round upright bars	9.50
Window guards with 5/8" square upright bars	10.00
Window guards with 3/4" round upright bars	12.00
Window guards with 3/4" square upright bars	14.00

Do not figure any opening as containing less than 15 sq. ft.

Window Guards Made of Crimped Diamond Mesh. Window guards, made with 1-inch channel iron frames and crimped diamond mesh wire, per sq. ft. erected in place cost $12.00 to $20.00. Do not figure any opening as containing less than 20 sq. ft.

Curb Angles, Floor Frames, Trench Frames, and Covers. For miscellaneous items, such as curb angles, floor frames, trench frames, and covers, figure $1.50 per lb. delivered to job.

Ornamental Iron Entrance Rails. Fabricated from 1 1/4" x 1/4" or 1 1/4" x 3/8" bar stock, or 1 1/2" bar size channels for rails and 1/2" square bar pickets, stock designs for this item, delivered to job, cost about $50.00 per lin. ft. of rail, depending on height and amount of ornamentation. For rails following the rake of stairs, add 50 percent. Rails made to special design may cost 2 to 3 times the above prices.

Ornamental Porch Columns. Stock design are scrolled, wrought iron columns, with 3/4" x 3/4" bar stock, or 15/16" square tubing frames and 3/8" x 1/4" or 1/2" x 3/16" bar or strip stock scrolls. Flat columns, 8 1/2" to 12" wide x 8'-0" high cost $75.00 to $150.00 each. Corner columns to match flat columns cost $70.00 to $100.00 each.

Stock design columns, with 3/4" x 3/4" bar stock frames and cast iron panels, delivered to job, cost as follows: flat columns, 9 1/2" to 11" wide x 8'-0" high are $75.00 to $130.00 each; corner columns to match flat columns are $100.00 to $140.00 each. Brackets to match columns are $35.00 each.

Steel Stacks

Steel stack costs vary widely and are dependent on height, diameter, gauge of material, number of joints, transportation, etc. Steel stacks are usually fabricated by companies specializing in steel boiler and tank manufacturing.

When estimating a job that has a steel stack, always obtain a firm figure from a steel stack fabricator before submitting a bid on the work.

Labor Erecting Steel Stacks. When estimating the labor cost of erecting steel stacks for office buildings or other buildings of this type, where the stack extends from the ground to the top of the building and is either built into the building or placed on the outside of one of the rear walls, there are several items to be taken into consideration--the length and weight of each section of stack, the facilities for hoisting and placing, and the amount of rigging necessary.

Where each section of the stack weighs 500 to 1,000 lbs., it is customary to place a small gin pole on the roof of the building and lift each section in place.

Four iron workers should set up all necessary rigging in one day, ready to place the sections. After the derricks and rigging have been placed to start hoisting, the gang should handle, place and bolt one 15 to 20 foot section of stack in 4 hours, or 2 sections (30 to 40 lin. ft.), per 8-hour day.

The labor cost placing and removing derricks for erecting a steel stack should average as follows:

	Hours	Rate	Total	Rate	Total
Ir. wkrs. placing equip	32	$....	$....	$27.53	$880.96
Ir. wkrs. remove equip	32	....	....	27.53	880.96
Labor on equipment			$....		$1,761.92

The labor cost erecting a steel stack up to 30" in diam. and 60 feet long, should average as follows:

	Hours	Rate	Total	Rate	Total
Ir. wkrs.erecting stack	48	$....	$....	$27.53	$1,321.44
Cost per lin. ft.			....		22.02

To the above costs, add equipment rental and move charges to and from job for equipment.

CHAPTER 6

CARPENTRY

CSI DIVISION 6
06100 ROUGH CARPENTRY
06130 HEAVY TIMBER CONSTRUCTION
06170 PREFABRICATED STRUCTURAL WOOD
06200 FINISH CARPENTRY
06300 WOOD TREATMENT
06400 ARCHITECTURAL WOODWORK

06100 ROUGH CARPENTRY

To estimate carpentry quantities and costs accurately, an estimator should know how the various classes of work are constructed, where joists and studs should be doubled, where to place wood furring strips for walls and floors, openings requiring wood bucks, and where it is necessary to place wood grounds. These may appear to be small items but they are important, nevertheless.

For example, an inexperienced estimator taking off carpentry quantities might figure wood bucks for the exterior window frames because he is unfamiliar with how work is conducted in the field. He must know how a job goes together and what is required for each item of work. He is much like an insurance policy--an expense until it is needed--at which time it becomes the most valuable document in a contractor's possession.

When taking off lumber quantities from the plans, always estimate as closely as possible the exact quantity of each kind of lumber required. Some contractors estimate the cost of wood floors at a certain price per sq. ft. including joists, bridging, subfloor, deadening felt, furring strips, and the finish wood flooring. Entering into the construction of such a floor are materials and labor operations of widely varying costs, yet certain contractors are content to estimate their work in this haphazard manner, and at the completion of the job have no more idea of their labor costs than before they took the job.

When estimating labor costs of rough carpentry, bear in mind the costs will vary with the class of work performed, with the ability of the carpenters employed and with the experience and executive ability of the foreman or superintendent in charge of the work.

Modern tools, such as electric hand saws, electric drills, and sanders, play a big part in reducing labor costs. For instance an electric handsaw will cut off a 2"x12" plank in 4 or 5 seconds, while it takes about a minute to do it by hand; it takes 10 to 12 minutes to rip a 12'-0" plank 2" thick when sawing by hand while an electric handsaw will do the work in less than a minute. In other words, an electric saw will cut 10 to 20 times as fast as can be done by hand. The handling cost remains the same in either instance but when sawing for any length of time, a man can turn out three or four times as much work with an electric saw as by hand.

Another thing that should be considered is the cost of filing saws where there are a number of carpenters working. This will vary with the condition of the saw but on an average it will require ½ hr. to file a saw by hand, while an automatic saw filer will do the same work in ¼ hr. The same applies to circular saws used on saw rigs or electric hand saws, as these can be filed in ½ to ¾ hr. by machine, while hand filing requires nearly twice as much time.

The same thing applies to the use of electric drills, electric sanders, etc., so the contractor who is going to keep up with the procession must use modern, labor-saving tools and equipment wherever possible.

Here is another thing that should be remembered by every contractor. No matter how carefully the plans have been measured and how much consideration they have been given before pricing, if the work is not executed in an efficient manner, careful estimating will not avert losses.

Estimating Lumber Quantities

When estimating the quantity of lumber required for any job, the only safe method is to take off every piece of lumber required to complete that portion of the work. This is seldom done, however, and the following tables are given to simplify the work as much as possible and at the same time provide accurate material quantities.

Estimating Wood Joists. When estimating wood joists, always allow 4" to 6" on each end of the joist for bearing on the wall.

To obtain the number of joists required for any floor, take the length of the floor in feet, divide by the distance the joists are spaced and add 1 to allow for the extra joist required at end of span.

Example: If the floor is 28 ft. long and 15 ft. wide, it will require 16-ft. joists to allow for wall bearing at each end. Assuming the joists are spaced 16" on centers, one joist will be required every 16" or every 1 ⅓ ft. In other words it will require ¾ as many joists as the length of the span, plus one. Three-quarters of 28 equal 21, plus 1 extra joist at end, makes 22 joists 16 ft. long for this space.

The following table gives the number of joists required for any spacing:

Number of Wood Floor Joists Required for any Spacing

Distance Joists are Placed on Centers	Multiply Length of Floor Span by	Add Joists 1 inches
12"	1	
16"	¾ or .75	1
20"	⅗ or .60	1
24"	½ or .50	1
30"	⅖ or .40	1
36"	⅓ or .33	1
42"	2/7 or .29	1
48"	¼ or .25	1
54"	2/9 or .22	1
60"	⅕ or .20	1

Number of Feet of Lumber B.M. Required per 100 Sq. Ft. of Surface When Used for Studs, Joists, Rafters, Wall and Floor Furring Strips, etc.

The following table does not include any allowance for waste in cutting, doubling joists under partitions or around stair wells, extra joists at end of each span, top or bottom plates, etc. These items vary with each job. Add as required.

Lumber Size	12-Inch Centers	16-Inch Centers	20-Inch Centers	24-Inch Centers
1"x2"	16 ⅔	12 ½	10	8 ⅓
2"x2"	33 ⅓	25	20	16 ⅔
2"x4"	66 ⅔	50	40	33 ⅓
2"x5"	83 ⅓	62 ½	50	41 ⅔
2"x6"	100	75	60	50
2"x8"	133 ⅓	100	80	66 ⅔
2"x10"	166 ⅔	125	100	83 ⅓
2"x12"	200	150	120	100
2"x14"	233 ⅓	175	140	116 ⅔
3"x6"	150	112 ½	90	75
3"x8"	200	133 ⅓	120	100
3"x10"	250	187 ½	150	125
3"x12"	300	225	180	150
3"x14"	350	262 ½	210	175

Number of Wood Joists Required for any Floor and Spacing

Length of Floor	Spacing of Joists 12"	16"	20"	24"	30"	36"	42"	48"	54"	60"
6	7	6	5	4	3	3	3	3	2	2
7	8	6	5	5	4	4	3	3	3	2
8	9	7	6	5	4	4	3	3	3	3
9	10	8	6	6	5	4	4	3	3	3
10	11	9	7	6	5	4	4	4	3	3
11	12	9	8	7	5	5	4	4	3	3
12	13	10	8	7	6	5	4	4	4	3
13	14	11	9	8	6	5	5	4	4	4
14	15	12	9	8	7	6	5	5	4	4
15	16	12	10	9	7	6	5	5	4	4
16	17	13	11	9	7	6	6	5	5	4
17	18	14	11	10	8	7	6	5	5	4
18	19	15	12	10	8	7	6	6	5	4
19	20	15	12	11	9	7	6	6	5	5
20	21	16	13	11	9	8	7	6	5	5
21	22	17	14	12	9	8	7	6	6	5
22	23	18	14	12	10	8	7	7	6	5
23	24	18	15	13	10	9	8	7	6	6
24	25	19	15	13	11	9	8	7	6	6
25	26	20	16	14	11	9	8	7	7	6
26	27	21	17	14	11	10	8	8	7	6
27	28	21	17	15	12	10	9	8	7	6
28	29	22	18	15	12	10	9	8	7	7
29	30	23	18	16	13	11	9	8	7	7
30	31	24	19	16	13	11	10	9	8	7
31	32	24	20	17	13	11	10	9	8	7
32	33	25	20	17	14	12	10	9	8	7
33	34	26	21	18	14	12	10	9	8	8
34	35	27	21	18	15	12	11	10	9	8

Length of Floor	Spacing of Joists 12"	16"	20"	24"	30"	36"	42"	48"	54"	60"
35	36	27	22	19	15	13	11	10	9	8
36	37	28	23	19	15	13	11	10	9	8
37	38	29	23	20	16	13	12	10	9	8
38	39	30	24	20	16	14	12	11	9	9
39	40	30	24	21	17	14	12	11	10	9
40	41	31	25	21	17	14	12	11	10	9

One joist has been added to each of the above quantities to take care of extra joist required at end of span.

Add for doubling joists under all partitions.

Estimating Quantity of Bridging. It is customary to place a double row of bridging between joists about 6'-0" to 8'-0" on centers. Joists 10'-0" to 12'-0" long will require one double row of bridging or 2 pcs. to each joist.

Joists 14'-0" to 20'-0" long will require 2 double rows of bridging or 4 pcs. to each joist.

Bridging is usually cut from 1"x3", l"x4", 2"x2", or 2"x4" lumber.

The following table gives the approximate number of pcs. and the lin. ft. of bridging required per 100 sq. ft. of floor.

Joists Up to 12 Feet Long				Joists Up to 20 Feet Long			
12-In. Centers		16-In. Centers		12-In. Centers		16-In. Centers	
Pcs.	LF.	Pcs.	LF	Pcs.	LF	Pcs.	LF
20	30	16	24	40	60	32	48

Steel Bridging for Wood Joists. Several types of steel bridging are available, fabricated from de-formed strip steel varying from 16 to 20 gauge in thickness. Some are galvanized while others are painted with black asphaltum or have a baked enamel finish. Stock sizes fit all joist depths and spacings encountered in average construction work--special sizes are available made to order.

Prices vary according to type, size, finish and quantity, but a fair average for galvanized bridging is 30 to 40 cents per set.

Installation time is about 50 percent of that required for wood bridging and with many types can be done after subflooring is laid.

Estimating Number of Wood Studs. When estimating the number of wood partition studs, take the length of each partition and the total length of all partitions.

If a top and bottom plate is required, take the length of the wood partition and multiply by 2. The result will be the number of lin. ft. of plates required.

If a double plate consisting of 2 top members and a single bottom plate is used, multiply the length of the wood partitions by 3.

Example: Find the quantity of lumber required to build a stud partition 16'-0" long, 8'-0" high, with studs spaced l6" on centers and having single top and bottom plates. 16'-0" = 192". Next, 192" divided by 16" = 12 studs, plus 1 extra at the end equals 13 studs 8'-0" long. Two top and bottom plates 16'-0" long equal 32 lin. ft.

13 pcs. 2x4 @ 8'-0" = 104
2 pcs. 2x4 @ 16'-0" = 32
104 + 32 = 136 lin. ft.
136 lin. ft. x $^2/_3$ = 90.67 b.f.

After the total number of lin. ft. of lumber has been obtained, convert to board measure.

Board Feet of Lumber Required for Wood Stud Partitions
2"x4" Studs 16" on Centers, with Single Top and Bottom Plates

Length of Partition	Height of Partition 8'-0"	8'-6"	9'-0"	10'-0"	12'-0"
3'-0"	20	22	22	24	28
4'-0"	27	29	29	32	37
5'-0"	33	37	37	40	47
6'-0"	40	44	44	48	56
7'-0"	41	45	45	49	57
8'-0"	48	53	53	57	67
9'-0"	55	60	60	65	76
10'-0"	61	67	67	73	85
11'-0"	63	69	69	75	87
12'-0"	69	76	76	83	96
13'-0"	76	83	83	91	105
14'-0"	83	91	91	99	115
15'-0"	84	92	92	100	116
16'-0"	91	99	99	108	125
17'-0"	97	107	107	116	135
18'-0"	104	114	114	124	144
19'-0"	105	115	115	125	145
20'-0"	112	123	123	133	155
21'-0"	119	130	130	141	164
22'-0"	125	137	137	149	173
23'-0"	127	139	139	151	175
24'-0"	133	146	146	159	184
25'-0"	140	153	153	167	193
26'-0"	147	161	161	175	203
27'-0"	148	162	162	176	204
28'-0"	155	169	169	184	213
29'-0"	161	177	177	192	223
30'-0"	168	184	184	200	232
31'-0"	169	185	185	201	233
32'-0"	176	193	193	209	243
33'-0"	183	200	200	217	252
34'-0"	189	207	207	225	261
35'-0"	191	209	209	227	263
36'-0"	197	216	216	235	272
37'-0"	204	223	223	243	281
38'-0"	211	231	231	251	291
39'-0"	212	232	232	252	292
40'-0"	219	239	239	260	301

Add ⅔ b.f. of lumber for each lin. ft. of double top or bottom plate.

Number of Partition Studs Required for Any Spacing

Distance Apart Studs	Multiply Length of Partition by	Add Wood Studs
12 inches	1.0	1
16 inches	0.75	1
20 inches	0.60	1
24 inches	0.50	1

Add for top and bottom plates.

Number of Feet of Lumber Required Per Sq. Ft. of Wood Stud Partition Using 2"x4" Studs
Studs Spaced 16" on Centers, with Single Top and Bottom Plates.

Length Partition in Feet	No. Studs Req'd	Ceiling Height in Feet 8'-0"	9'-0"	10'-0"	12'-0"
2	3	1.25	1.167	1.13	1.13
3	3	0.833	.812	.80	.80
4	4	0.833	.812	.80	.80
5	5	0.833	.812	.80	.80
6	6	0.833	.812	.80	.80
7	6	0.833	.75	.75	.80
8	7	0.75	.75	.75	.70
9	8	0.75	.75	.75	.70
10	9	0.75	.75	.75	.70
11	9	0.75	.70	.70	.67
12	10	0.75	.70	.70	.67
13	11	0.75	.70	.70	.67
14	12	0.75	.70	.70	.67
15	12	0.70	.70	.70	.67
16	13	0.70	.70	.70	.67
17	14	0.70	.70	.70	.67
18	15	0.70	.70	.67	.67
19	15	0.70	.70	.67	.67
20	16	0.70	.70	.67	.67
For dbl. plate, add per sq. ft.		0.13	.11	.10	.083

For 2"x8" studs, double above quantities. For 2"x6" studs, increase above quantities 50%.

Example: Find the number of b.f. of lumber required for a stud partition 18'-0" long and 9'-0" high. This partition would contain 18x9 = 162 sq. ft.

The table gives 0.70 b.f.lumber of partition. Multiply 162 by 0.70 equals 113.4 b.f.

Quantity of Plain End Softwood Flooring Required Per 100 Sq. Ft. of Floor

Measured Size Inches	Actual Size Inches	Add for Width	BF Req. per 100 SF Surface	Weight per 1000 Ft.
1x3	¾x2 ⅜	27%	132	1800
1x4	¾x3 ¼	23%	128	1900

The above quantities include 5% for end cutting and waste.

Quantity of End Matched Softwood Flooring Required Per 100 Sq. Ft. of Floor

Measured Size Inches	Actual Size Inches	Add for Width	BF Req. per 100 SF Surface	Weight per 1000 Ft.
1x3	13/16x2 ⅜	27%	130	1800
1x4	13/16x3 ¼	23%	126	1900

The above quantities include 3% for end cutting and waste.

Quantity of Square Edged (S4S) Boards Required Per 100 Sq. Ft. of Surface

Measured Size Inches	Actual Size Inches	Add for Width	BF Req. per 100 SF Surface	Weight per 1000 Ft.
1x4	3/4x3 1/2	14%	119	2300
1x6	3/4x5 1/2	9%	114	2300
1x8	3/4x7 1/4	10%	115	2300
1x10	3/4x9 1/4	8%	113	2300
1x12	3/4x11 1/4	7%	112	2400

The above quantities include 5% for end cutting and waste.

Lineal Foot Table of Board Measure
Number of Board Feet per Lineal Foot of Any Size

2"x4"=0.667	4"x4"=1.333	8"x14"=9.333
2"x6"=1.	4"x6"=2.	8"x16"=10.667
2"x8"=1.333	4"x8"=2.667	10"x10"=8.333
2"x10"=1.667	4"x10"=3.333	10"x12"=10.
2"x12"=2.	4"x12"=4.	10"x14"=11.667
2"x14"=2.333	4"x14"=4.667	10"x16"=13.333
2"x16"=2.667	4"x16"=5.333	10"x18"=15.
2 1/2"x12"=2.5	6"x6"=3.	12"x12"=12.
2 1/2"x14"=2.917	6"x8"=4.	12"x14"=14.
2 1/2"x16"=3.333	6"x10"=5.	12"x16"=16.
3"x6"=1.5	6"x12"=6.	12"x18"=18.
3"x8"=2.	6"x14"=7.	14"x14"=16.333
3"x10"=2.5	6"x16"=8.	14"x16"=18.667
3"x12"=3.	8"x8"=5.333	14"x18"=21.
3"x14"=3.5	8"x10"=6.667	16"x16"=21.333
3"x16"=4.	8"x12"=8.	16"x18"=24.

Lengths of Common, Hip, and Valley Rafters Per 12 Inches of Run

1 Roof Pitch	2 Rise and Run or Cut	3 Length in In. Common Rafter per 12" of Run	4	5* % Increase Length Com. Rafter over Run	6** Length in In. Hip or Valley Rafters
1/12	2 and 12	12.165	.014	1.014	17.088
1/8	3 and 12	12.369	.031	1.031	17.233
1/6	4 and 12	12.649	.054	1.054	17.433
5/24	5 and 12	13.000	.083	1.083	17.692
1/4	6 and 12	13.417	.118	1.118	18.000
7/24	7 and 12	13.892	.158	1.158	18.358
1/3	8 and 12	14.422	.202	1.202	18.762
3/8	9 and 12	15.000	.250	1.250	19.209
5/12	10 and 12	15.620	.302	1.302	19.698
11/24	11 and 12	16.279	.357	1.357	20.224
1/2	12 and 12	16.971	.413	1.413	20.785
13/24	13 and 12	17.692	.474	1.474	21.378
7/12	14 and 12	18.439	.537	1.537	22.000
5/8	15 and 12	19.210	.601	1.601	22.649
2/3	16 and 12	20.000	.667	1.667	23.324
17/24	17 and 12	20.809	.734	1.734	24.021
3/4	18 and 12	21.633	.803	1.803	24.739

1 Roof Pitch	2 Rise and Run or Cut	3 Length in In. Common Rafter per 12" of Run	4	5* % Increase Length Com. Rafter over Run	6** Length in In. Hip or Valley Rafters
19/24	19 and 12	22.500	.875	1.875	25.475
5/6	20 and 12	23.375	.948	1.948	26.230
7/8	21 and 12	24.125	1.010	2.010	27.000
11/24	22 and 12	25.000	1.083	2.083	27.785
11/12	23 and 12	26.000	1.167	2.167	28.583
Full	24 and 12	26.875	1.240	2.240	29.394

*Use figures in this column to obtain area of roof for any pitch. See explanation below.
**Figures in last column are length of hip and valley rafters in inches for each 12 inches of common rafter run.

To Obtain Area of Roofs For Any Pitch. To obtain the number of square feet of roof area for roofs of any pitch, take the entire flat or horizontal area of the roof and multiply by the figure given in the fifth column (*), and the result will be the area of the roof. Always bear in mind that the width of any overhanging cornice must be added to the building area to obtain the total area to be covered. Example: Find the area of a roof 26'-0"x42'-0", having a 12" or 1'-0" overhanging cornice. Roof having a ¼ pitch. To obtain roof area, 26'-0" + 1'-0" + 1'-0" = 28'-0" width. 42'-0" + 1'-0" + 1'-0" = 44'-0" length. 28 x 44 = 1232 or 1,232 sq. ft. of flat or horizontal area.

To obtain area at ¼ pitch: multiply 1,232 by 1.12* = 1379.84 or 1,380 sq. ft. roof surface.

Add allowance for overhang on dormer roofs and sides.

Table of Board Measure
Giving Contents in Feet of Joists, Scantlings and Timbers

Size in Inches	Length in Feet 10	12	14	16	18	20	22	24	26	28	30
1x2	1 ⅔	2	2 ⅓	2 ⅔	3	3 ⅓	....	....	....	....	....
1x3	2 ½	3	3 ½	4	4 ½	5	....	....	....	....	....
1x4	3 ⅓	4	4 ⅔	5 ⅓	6	6 ⅔	....	....	....	....	....
1x6	5	6	7	8	9	10	....	....	....	....	....
1x8	6 ⅔	8	9 ⅓	10 ⅔	12	13 ⅓	....	....	....	....	....
1x10	8 ⅓	10	11 ⅔	13 ⅓	15	16 ⅔	....	....	....	....	....
1x12	10	12	14	16	18	20	....	....	....	....	....
1 ¼x4	4 ⅙	5	5 ⅚	6 ⅔	7 ½	8 ⅓	....	....	....	....	....
1 ¼x6	6 ¼	7 ½	8 ¾	10	11 ¼	12 ½	....	....	....	....	....
1 ¼x8	8 ⅓	10	11 ⅔	13 ⅓	15	16 ⅔	....	....	....	....	....
1 ¼x10	10 5/12	12 ½	14 7/12	16 ⅔	18 ¾	20 ⅚	....	....	....	....	....
1 ¼x12	12 ½	15	17 ½	20	22 ½	25	....	....	....	....	....
1 ½x4	5	6	7	8	9	10	....	....	....	....	....
1 ½x6	7 ½	9	10 ½	12	13 ½	15	....	....	....	....	....
1 ½x8	10	12	14	16	18	20	....	....	....	....	....
1 ½x10	12 ½	15	17 ½	20	22 ½	25	....	....	....	....	....
1 ½x12	15	18	21	24	27	30	....	....	....	....	....
2x2	3 ⅓	4	4 ⅔	5 ⅓	6	6 ⅔	....	....	....	....	....
2x3	5	6	7	8	9	10	11	12	13	14	15
2x4	6 ⅔	8	9 ⅓	10 ⅔	12	13 ⅓	14 ⅔	16	17 ⅓	18 ⅔	20
2x6	10	12	14	16	18	20	22	24	26	28	30
2x8	13 ⅓	16	18 ⅔	21 ⅓	24	26 ⅔	29 ⅓	32	34 ⅔	37 ⅓	40
2x10	16 ⅔	20	23 ⅓	26 ⅔	30	33 ⅓	36 ⅔	40	43 ⅓	46 ⅔	50
2x12	20	24	28	32	36	40	44	48	52	56	60
2x14	23 ⅓	28	32 ⅔	37 ⅓	42	46 ⅔	51 ⅓	56	60 ⅔	65 ⅓	70

Size in Inches	Length in Feet										
	10	12	14	16	18	20	22	24	26	28	30
3x4	10	12	14	16	18	20	22	24	26	28	30
3x6	15	18	21	24	27	30	33	36	39	42	65
3x8	20	24	28	32	36	40	44	48	52	56	60
3x10	25	30	35	40	45	50	55	60	65	70	75
3x12	30	36	42	48	54	60	66	72	78	84	90
3x14	35	42	49	56	63	70	77	84	91	98	105

Table of Board Measure
Giving Contents in Feet of Joists, Scantlings and Timbers

Size in Inches	Length in Feet										
	10	12	14	16	18	20	22	24	26	28	30
4x4	13	16	19	21	24	27	29	32	35	37	40
4x6	20	24	28	32	36	40	44	48	52	56	60
4x8	27	32	37	43	48	53	59	64	69	75	80
4x10	33	40	47	53	60	67	73	80	87	93	100
4x12	40	48	56	64	72	80	88	96	104	112	120
4x14	47	56	65	75	84	93	103	112	121	131	140
6x6	30	36	42	48	54	60	66	72	78	84	90
6x8	40	48	56	64	72	80	88	96	104	112	120
6x10	50	60	70	80	90	100	110	120	130	140	150
6x12	60	72	84	96	108	120	132	144	156	168	180
6x14	70	84	98	112	126	140	154	168	182	196	210
6x16	80	96	112	128	144	160	176	192	208	224	240
8x8	53	64	75	85	96	107	117	128	139	149	160
8x10	67	80	93	107	120	133	147	160	173	187	200
8x12	80	96	112	128	144	160	176	192	208	224	240
8x14	93	112	131	149	168	187	205	224	243	261	280
8x16	107	128	149	171	192	213	235	256	277	298	320
10x10	83	100	117	133	150	167	183	200	217	233	250
10x12	100	120	140	160	180	200	220	240	260	280	300
10x14	117	140	163	187	210	233	257	280	303	327	350
10x16	133	160	187	218	240	267	293	320	347	373	400
12x12	120	144	168	192	216	240	264	288	312	336	360
12x14	140	168	196	224	252	280	308	336	364	392	420
12x16	160	192	224	256	288	320	352	384	416	448	480
14x14	163	196	229	261	294	327	359	392	425	457	490
14x16	187	224	261	299	336	373	411	448	485	523	560
14x18	210	252	294	336	378	420	462	504	546	588	630
14x20	233	280	327	373	420	467	513	560	607	653	700
16x16	213	256	299	341	384	427	469	512	555	597	640
16x18	240	288	336	384	432	480	528	576	624	672	720
16x20	267	320	373	425	480	533	587	640	693	747	800
18x18	270	324	378	432	486	540	594	648	702	756	810
18x20	300	360	420	480	540	600	660	720	780	840	900
20x20	333	400	467	533	600	667	733	800	867	933	1000

Nails Required for Carpenter Work

The following table gives the number of wire nails in pounds for the various kinds of lumber per 1,000 b.f., or per 1,000 shingles and lath or per square (100 sq. ft.) of asphalt slate surfaced shingle, with the number of nails added for loss of material on account of lap or matching of shiplap, flooring, ceiling, and siding of the various widths. The table gives the sizes generally used for certain purposes with the nailing space 16" on centers, and 1 or 2 nails per board for each nailing space.

Description of Material	Unit of Measure	Size & Kind of Nail	No. of Nails Required	Lbs. of Nails Required
Wood Shingles	1,000	3d Common	2,560	4
Individual Asphalt Shingles	100 SF	7/8" Roofing	848	4
Three in One Asphalt Shingles	100 SF	7/8"Roofing	320	1
Wood Lath	1,000'	3d Fine	4,000	6
Wood Lath	1,000'	2d Fine	4,000	4
Bevel or Lap Siding, 1/2"x4"	1,000'	6d Coated	2,250	15*
Bevel or Lap Siding, 1/2"x6"	1,000'	6d Coated	1,500	10*
Byrkit Lath, 1"x6"	1,000'	6d Common	2,400	15
Drop Siding, 1"x6"	1,000'	8d Common	3,000	25 3/8"
Hardwood Flooring	1,000'	4d Common	9,300	16
25/32" Hardwood Flooring	1,000'	8d Casing	9,300	64
Subflooring, 1"x3"	1,000'	8d Casing	3,350	23
Subflooring, 1"x4"	1,000'	8d Casing	2,500	17
Subflooring, 1"x6"	1,000'	8d Casing	2,600	18
Ceiling, 5/8"x4"	1,000'	6d Casing	2,250	10
Sheathing Boards, 1"x4"	1,000'	8d Common	4,500	40
Sheathing Boards, 1"x6"	1,000'	8d Common	3,000	25
Sheathing Boards, 1"x8"	1,000'	8d Common	2,250	20
Sheathing Boards, 1"x10"	1,000'	8d Common	1,800	15
Sheathing Boards, 1"x12"	1,000'	8d Common	1,500	12 1/2
Studding, 2"x4"	1,000'	16d Common	500	10
Joist, 2"x6"	1,000'	16d Common	332	7
Joist, 2"x8"	1,000'	16d Common	252	5
Joist, 2"x10"	1,000'	16d Common	200	4
Joist, 2"x12"	1,000'	16d Common	168	3 1/2
Interior Trim, 5/8" thick	1,000'	6d Finish	2,250	7
Interior Trim, 3/4" thick	1,000'	8d Finish	3,000	14
5/8" Trim where nailed to jamb	1,000'	4d Finish	2,250	3
1"x2" Furring or Bridging	1,000'	6d Common	2,400	15
1"x1" Grounds	1,000'	6d Common	4,800	30

*Cement coated nails sold as 2/3 lb. = 1 lb. of common nails.

Nails Required for Subflooring. On wood floors up to 4" wide, quantities are based on 8d flooring nails. For flooring 6" and wider, 10d nails have been figured. The quantities given below are sufficient to lay 1,000 b.f. of flooring.

Width Flooring	Joist Spacing 12" on Centers	Joist Spacing 16" on Centers
2"	40 lbs. 8d flg.	30 lbs. 8d flg.
3"	30 lbs. 8d flg.	23 lbs. 8d flg.
4"	22 lbs. 8d flg.	17 lbs. 8d flg.
6"	24 lbs. 10d com.	18 lbs. 10d com.
8"	17 lbs. 10d com.	13 lbs. 10d com.

Data on Common Wire Nails

Size of Nails	Length of Nails Inches	Gauge Number	Approx. Number to Pound	Approx. Price Per 100 lbs.
4d	1 1/2	12 1/2	316	$36.00
5d	1 3/4	12 1/2	271	36.00
6d	2	11 1/2	181	36.00
8d	2 1/2	10 1/4	106	34.00
10d	3	9	69	34.00
12d	3 1/4	9	63	34.00
16d	3 1/2	8	49	34.00
20d	4	6	31	34.00
30d	4 1/2	5	24	34.00
40d	5	4	18	34.00
50d	5 1/2	3	14	34.00
60d	6	2	11	34.00

The following tables give the recommended nailing schedule for special (helically threaded) nails that are being used today throughout the building industry.

Recommended Nailing Schedule for Common Applications in Building Construction

Application	Nailed Into	Nail Size Inches	Nail Type	Head: Diameter Type	Point: Size	Point: Type	Nail-ing	Spacing o.c.	Nails Per Joint
Mudsill, partition plate, 2"	Concrete	2 1/2-2 3/4x0.148	Sc-1z	5/16" Checkered	Long	Dia.	Face	12"-24"	—
Ditto, in earthquake regions	Concrete	(3 1/4-) 3 1/2x0.250	Sc-1z	9/16" Checkered	Med.	Ndl.	Face	24"-48"	—
Ditto, 3"	Concrete	4 1/2x0.250	Sc-1z	9/16" Checkered	Med.	Ndl.	Face	24"-48"	—
Furring strips	Concrete	1 1/2-1 3/4x0.148	Sc-1z	5/16" Checkered	Long	Dia.	Face	12"-24"	—
Mudsill	Mudsill	2 1/2x0.120	Sc-2	9/32" Flat	Med.	Dia.	Toe	—	2
Sleepers	Mudsill	2 1/2x0.120	Sc-2	9/32" Flat	Med.	Dia.	Toe	—	2
Joists	Mudsill	3 1/4x0.135	Sc-2	5/16" Flat	Med.	Dia.	Toe	—	2-3
Subflrg., 1" lumber, plywood	Mudsill, sleeper, joist	2 1/8x0.105	St-14	1/4" Flat, Csk.	Med.	Dia.	Face	6" & 12"	2(3)
Subflrg., 2" lumber, plywood	Mudsill, sleeper, joist	2 7/8x0.120	St-14	9/32" Flat, Csk.	Med.	Dia.	Face	—	2(3)
Subflrg., 3/8"-1/2" plywood (dph.)	Mudsill, sleeper, joist	1 1/2x0.135	Hi-28	5/16" Flat, Csk.	Med.	Dia.	Face	6" & 12"	—
Subflrg., 5/8" plywood (dph.)	Mudsill, sleeper, joist	1 3/4x0.135	Hi-28	5/16" Flat, Csk.	Med.	Dia.	Face	6" & 12"	—
Subflrg., 3/4" plywd. (dph.), part bd.	Mudsill, sleeper, joist	2 x0.148	Hi-28	5/16" Flat, Csk.	Med.	Dia.	Face	6" & 12"	—
Subflrg., 1"-11/8" plywood (dph.) part. bd.	Mudsill, sleeper, joist	2 1/2x0.148	Hi-28	5/16" Flat, Csk.	Med.	Dia.	Face	6" & 12"	—
Underlayment, 1/4"-5/16" plywood	Subfloor	1 x0.083	St-16	3/16" Flat, Csk.	Med.	Dia.	Face	6" & 6"-12"	—
Underlayment, 3/8"-1/2" plywood	Subfloor	1 1/4x0.083	St-16	3/16" Flat, Csk.	Med.	Dia.	Face	6" & 6"-12"	—
Underlayment, 5/8" plywood	Subfloor	1 3/8x0.098	St-16	1/4" Flat, Csk.	Med	Dia.	Face	6" & 6"-12"	—
Underlayment, 3/4" plywood	Subfloor	1 1/2x0.098	St-16	1/4" Flat, Csk.	Med.	Dia.	Face	6" & 6"-12"	—
Underlayment, 7/8" plywood	Subfloor	1 5/8x0.098	St-16	1/4" Flat, Csk.	Med.	Dia.	Face	6" & 6"-12"	—
Underlayment, 3/16"-5/8" hardboard	Subfloor	1-1 3/8x0.083	St-15	3/16" Flat, Csk.	Med.	Dia.	Face	6" & 12"	—
Flooring, T & G hardwood	Subfloor, joist, sleeper	2-2 1/2x0.115	Sc-4	13/64" Casing	Blunt	Dia.	Toe	10"-18"	—
Flooring, T & G softwood	Subfloor, joist, sleeper	2-2 1/2x0.115	Sc-4	13/64" Casing	Blunt	Dia.	Toe	10"-18"	—
Flooring, T & G hardwood, 3/8" and 1/2"	Subfloor, joist, sleeper	1-1 1/4x0.072	Sc-4	9/64" Casing	Blunt	Dia.	Toe	10"-18"	—
Flooring, T & G parquet	Subfloor	1 1/2x0.105	Sc-4	9/64" Casing	Blunt	Dia.	Face	—	—
Framing plates	Stud	3 1/4x0.135	Sc-2	5/16" Flat	Med.	Dia.	Face	—	2
Framing studs	Stud, cripple, lintel, sill	2 1/2x0.120	Sc-2	9/32" Flat	Med.	Dia.	Face	16"-24"	—
Framing studs	Plate, cripple, lintel, sill	2 1/2x0.120	Sc-2	9/32" Flat	Med.	Dia.	Toe	—	3
Framing sole plate	Mudsill	3 1/4x0.135	Sc-2	5/16" Flat	Med.	Dia.	Face	16"	—
Framing top plate	Lower top plate	3 1/4x0.135	Sc-2	5/16" Flat	Med.	Dia.	Face	24"	—
Trussed rafter assembly		3 1/4x0.135	Sc-5	5/16" Flat	Med.	Dia.	Face	2 1/2"-3"	Given
Trussed rafter assembly		2 1/2x0.120	Sc-5	9/32" Flat	Med.	Dia.	Face	2 1/2"-3"	Given

See Notes, Key to Nail Types and Abbreviations on later page.

Application	Nailed Into	Nail Size Inches	Nail Type	Head Diameter Type	Point Size	Point Type	Nail-ing	Spacing o.c.	Nails Per Joint
Rafter, 4"	Top plate	3 1/4x0.135	Sc-2	5/16" Flat	Med.	Dia.	Toe	—	3
Rafter, 4"	Top plate	6 x0.177	St-34	7/16" Flat	Med.	Dia.	Face	—	2
Rafter, 4"	Top plate	7 x0.207	St-34	1/2" Flat	Med.	Dia.	Face	—	2
Rafter, 6", 8", 10"	Top plate	{4-6 x0.177 {7-9 x0.203	St-34	7/16" Flat	Med.	Dia.	Toe	—	2-3
Sheathing, 1" lumber	Framing, rafter	2 x0.120	St-3	9/32" Flat	Med.	Dia.	Face	—	2
Sheathing 3/8"-1/2" plywood	Framing, rafter	1 3/4x0.120	St-17	9/32" Flat	Med.	Dia.	Face	6" & 12"	—
Sheathing, 5/16"-1/2" plywood (dph.)	Framing, rafter	1 1/2x0.135	Hi-17	5/16" Flat, Csk.	Med.	Dia.	Face	6" & 12"	—
Sheathing, 5/8" plywood (dph.)	Framing, rafter	1 3/4x0.135	Hi-17	5/16" Flat, Csk.	Med.	Dia.	Face	6" & 12"	—
Sheathing, 3/4" plywood (dph.)	Framing, rafter	2 x0.148	Hi-17	5/16" Flat, Csk.	Med.	Dia.	Face	6" & 12"	—
Sheathing, 1"-11/8" plywood (dph.)	Framing, rafter	2 1/2x0.148	Hi-17	5/16" Flat, Csk.	Med.	Dia.	Face	6" & 12"	—
Sheathing, insulation board, gypsumboard	Framing, rafter	1 1/2-2x0.120	St-10g	3/8", 7/16" Flat	Blunt	Dia.	Face	3-4" & 6-8"	—
Sheathing, asbestosboard, 1/8"	Framing, rafter	1 1/4x0.083	St-6g	3/16" Flat, Csk.	Blunt	Dia.	Face	3-4" & 6-8"	—
Sheathing, asbestosboard, 1/4"	Framing, rafter	1 1/4x0.120	St-or Sc-6g	5/16" Flat, Csk.	Blunt	Dia.	Face	3-4" & 6-8"	—
Sheathing, hardboard, 3/8"-5/8"	Framing, rafter	2 x0.115	Sc-7g	13/64" Flat, Csk.	Med.	Ndl.	Face	3-4" & 6-8"	—
Building paper	Sheathing	1/2-3/4x0.105	Sq-30	15/16" Square	Med.	Dia.	Face	6"-12"	—
Stripping, 3/8"x35/8"	Framing, joist, rafter	2 x0.120	St-3	9/32" Flat	Med.	Dia.	Face	—	2
Stripping, 1"x4"	Framing, joist, rafter	2 1/2x0.135	St-3	5/16" Flat	Med.	Dia.	Face	—	2
Stripping, 2"x3"	Framing, joist, rafter	3 1/2x0.165	St-3	5/8" Flat	Med.	Dia.	Face	—	2
Siding, wood, 1"	Sheathing and framing	2 1/8x0.101-0.115	St-14g	1/4" Flat, Csk.	Med.	Dia.	Face	—	—
Siding, wood, 1"	Sheathing and framing	2 x0.120	Dr-14	5/32" Flat, Csk.	Med.	Ndl.	Face	—	1
Siding, wood, 2"	Sheathing and framing	3 x0.135	Dr-14	5/32" Flat, Csk.	Med.	Ndl.	Face	—	1
Siding, plywood	Sheathing and framing	1 7/8x0.109	Dr-8	5/32" Casing	Med.	Ndl.	Face	6" & 12"	1
Siding, T & G wood	Sheathing and framing	1 3/4x0.105	Sc-8	5/32" Casing	Med.	Dia.	Toe	6" & 12"	—
Siding, asbestos shingle	Sheathing and framing	1 1/2-1 3/4x0.105	Dr-33	3/16" Flat Button	Med.	Dia.	Face	—	Given
Siding, asbestos shingle	Sheathing and framing	1 1/2-1 3/4x0.083	St-19t	3/16" Flat	Med.	Dia.	Face	—	Given
Siding, asbestos shingle	Sheathing and framing	1 1/2-1 3/4x0.076	St-20	3/16" Flat	Med.	Dia.	Face	—	Given
Siding, insulated brick, wood shingle	Sheathing and framing	1 3/4x0.095	St-18ge	3/16" Flat	Med.	Dia.	Face	8"-12"	2
Siding, wood shingle	Insulating sheathing	1 3/4-2 x0.083	St-18ge	5/32" Finishing	Blunt	Dia.	Face	—	2
Siding, wood shingle	Insulating sheathing	1 3/4-2 x0.105	Dr-18	5/32" Finishing	Blunt	Dia.	Face	—	2
Siding, wood shingle	Plywood	1 1/8x0.102	Dr-18	3/16" Flat	Long	Dia.	Face	—	2
Siding, hardboard	Framing	2-2 1/2x0.115	Sc-7z	13/64" Casing	Long	Ndl.	Face	12"	—
Siding, hardboard battenboard	Framing	1 1/2x0.083	Sc-7z	9/64" Casing	Long	Ndl.	Face	12"	—

See Notes, Key to Nail Types and Abbreviations on later page.

Application	Nailed Into	Nail Size Inches	Nail Type	Head Diameter Type	Point Size	Point Type	Nailing	Spacing o.c.	Nails Per Joint
Fascia, 1"	Framing, rafter	2 1/2x0.120	Sc-2g	9/32" Flat	Med.	Dia.	Face	—	2
Fascia, 2" lumber	Framing, rafter	3 1/4x0.135	Sc-2g	5/16" Flat	Med.	Dia.	Face	—	2
Roofing, built-up	Sheathing	3/4-1 1/4x0.105	Sq-30	15/16" Square	Med.	Dia.	Face	10"	—
Roofing, built-up	Poured gypsum	1 1/2-1 3/4x0.120	Sq-31	15/16" Square	Med.	Dia.	Face	10"	—
Roofing, asphalt shingle	Sheathing	3/4-2x0.120	St-10g	3/8" Flat	Blunt	Dia.	Face	—	2-3
Roofing, asphalt shingle	Sheathing	3/4-2x.120-.135	Dr-10	3/8" Flat	Blunt	Dia.	Face	—	2-3
Roofing, wood shingle	Sheathing	3/4-2x.105-.120	Dr-18	3/16" Flat	Blunt	Dia.	Face	—	2
Roofing, wood shingle	Sheathing	1 3/4-2x0.083	St-18g	1/8" Flat, Csk.	Blunt	Dia.	Face	—	2
Roofing, asbestos shingle	Sheathing	As for siding							
Roofing, aluminum (corr. and flat)	Rafter, purlin	1 1/2-1 3/4x0.145	Dr-10	13/32" Flat ★	Long	Dia.	Face	12"	—
Roofing, sheet metal (corr. and flat)	Rafter, purlin	1-3x0.135	St-or	7/16" Flat ★	Long	Dia.	Face	12"	—
Roofing, glass fiber (corr. and flat)	Rafter, purlin	1 1/2-3x0.135	Sc-9g	7/16" Flat ★	Long	Dia.	Face	12"	—
Roofing, glass fiber (corr. and flat)	Rafter, purlin	1 1/2-3x0.148	Dr-9 or 10	7/16" Flat ★	Long	Dia.	Face	12"	—
				★ *With Neoprene washer attached*					
Lath, expanded metal, K-lath	Framing, joist	1 1/2x0.148	St-22g	L-Shaped	Med.	Dia.	Face	6" & 12"	—
Lath, gypsum plasterboard	Framing, joist	1 1/4x0.101	St-23b	19/64" Flat, Csk.	Long	Dia.	Face	5"	—
Gypsumboard, 3/8"	Framing, joist	1 1/4x0.098	St-24	1/4"-19/64" Flat, Csk.	Long	Dia.	Face	5"-8"	—
Gypsumboard, 1/2"-5/8"	Framing, joist	1 3/8x0.098	St-24	1/4"-19/64" Flat, Csk.	Long	Dia.	Face	5"-8"	—
Gypsumboard, prefinished	Framing, joist	1 3/8x0.083	K-32e	3/16" Flat, Csk.	Long	Dia.	Face	5"-8"	—
Paneling, trim		1-1 1/4x0.054	K-32e	3/32" Casing	Blunt	Dia.	Face	—	—
Paneling, trim, exterior		1-1 1/2x0.072	St-13	3/32" Casing	Blunt	Dia.	Face	—	—
Paneling, trim, exterior		1 x0.065	St-12	3/32" Casing	Blunt	Dia.	Face	—	—
Paneling, trim, exterior		1 1/2x0.076	Sc-12	3/32" Oval	Blunt	Dia.	Face	—	—
Paneling, trim		1 x0.072	Sc-11	3/32" Casing	Blunt	Dia.	Face	—	—
Paneling, trim		1 1/2 1 3/4x0.083	Sc-11	1/8" Casing	Blunt	Dia.	Face	—	—
Paneling, trim		2 1/2x0.105	Sc-11	9/64" Casing	Blunt	Dia.	Face	—	—
Acoustic tile		1-1 3/4x0.062	St-25z	—	Blunt	Dia.	Face	—	—
Electric conduit	Wood	1 1/2-2 x0.162	St-26z	1" Hook	Blunt	Ndl.	Face	—	—
Electric conduit	Masonry	1 1/2-2 x0.162	St-27z	1" Hook	Blunt	Ndl.	Face	—	—
Fencing wire	Softwood (treated)	1 1/2x0.148	St-22g	L-Shaped	Med.	Dia.	Face	—	—
Fencing wire	Hardwood	1 1/2x0.148	St-21g	L-Shaped	Med.	Dia.	Face	—	—

See Notes, Key to Nail Types and Abbreviations on next page.

Key to Nail Types

Sc-iz	Screw-Tite Masonry Nail, hardened HCS, zinc plated
Sc-2	Screw-Tite Framing Nail, hardened HCS
Sc-2g	Screw-Tite Framing Nail, hardened HCS, galvanized
Sc-3	Screw Tite Framing Nail, bright LCS
Sc-3g	Screw-Tite Framing Nail, bright LCS, galvanized
Sc-4	Screw-Tite Flooring Nail, hardened HCS
Sc-5	Screw-Tite Trussed Rafter Nail, hardened HCS
Sc-6g	Screw-Tite Asbestosboard Nail, hardened HCS, galvanized
Sc-7g	Screw-Tite Exterior Hardboard Nail, hardened HCS, galvanized
Sc-7z	Screw-Tite Exterior Hardboard Nail, hardened HCS, zinc plated
Sc-8	Screw-Tite Casing Nail, silver bronze
Sc-9g	Screw-Tite Roofing Nail, hardened HCS, galvanized
Sc-10g	Screw-Tite Roofing Nail, bright LCS, galvanized
Sc-11	Screw-Tite Finishing Nail, bright LCS,
Sc-12	Screw-Tite Finishing Nail, stainless steel
St-3	Stronghold Framing Nail, bright LCS
St-4	Stronghold Parquet Flooring Nail, hardened HCS
St-6g	Stronghold Asbestosboard Nail, hardened HCS, galvanized
St-9g	Stronghold Roofing Nail, hardened HCS, galvanized
St-10g	Stronghold Roofing Nail, bright LCS, galvanized
St-12	Stronghold Finishing Nail, stainless steel
St-13	Stronghold Finishing Nail, monel metal
St-14	Stronghold Sinker Nail, bright LCS
St-14g	Stronghold Sinker Nail, bright LCS, galvanized
St-15	Stronghold Underlay Nail, Hardened HCS
St-16	Stronghold Underlay Nail, bright LCS
St-17	Stronghold Sheathing Nail, bright LCS
St-18g	Stronghold Shingle Nail, bright LCS, galvanized
St-18ge	Stronghold Shingle Nail, bright LCS, galvanized and enameled
St-19t	Stronghold Shingle Nail, bronze, tin plated
St-20	Stronghold Shingle Nail, stainless steel
St-21g	Stronghold Fence Staple, hardened HCS, galvanized
St-22g	Stronghold Fence Staple, bright LCS, galvanized
St-22z	Stronghold Fence Staple, bright LCS, zinc plated
St-23b	Stronghold Lath Nail, bright LCS, blued
St-24	Stronghold Drywall Nail, bright LCS
St-25z	Stronghold Kollarnail, hardened HCS, zinc plated
St-26z	Stronghold Conduit Staple, bright LCS, zinc plated
St-27z	Stronghold Knurled Conduit Staple, hardened HCS, zinc plated
St-34	Stronghold Spike, hardened HCS
Hi-28	"Hi-Load" Shear-Resistant Nail, bright LCS
Hi-17	"Hi-Load" Sheathing Nail, bright LCS
Sq-30	Squarehed Annular Thread Cap Nail, bright LCS
Sq-31	Squarehed Spiral Thread Cap Nail, bright LCS
K-32e	Annular Thread Kolorpin, bright LCS, enameled
Dr-8	Drive-Rite Spiral Thread Casing Nail, aluminum
Dr-9	Drive-Rite Screw Thread Roofing Nail, aluminum
Dr-10	Drive-Rite Spiral Thread Roofing Nail, aluminum
Dr-14	Drive-Rite Spiral Thread Sinker Nail, aluminum
Dr-18	Drive-Rite Screw Thread Shingle Nail, aluminum
Dr-20	Drive-Rite Spiral Thread Shingle Nail, aluminum
Dr-33	Drive-Rite Knurled Asbestos-Cement Shingle Face, aluminum

NOTES: For fastening redwood, use only aluminum or stainless steel nails.

Local conditions, customs and popular usage may dictate minor variations in length and gauge of nails. Consult the Technical Service Department of Independent Nail & Packing Company, Bridgewater, Mass.

The above chart is based on a table appearing in Bulletin No. 38 (Revised Edition). "Better Utilization of Wood Through Assembly with Improved Fasteners," a study undertaken at Wood Research Laboratory, Virginia Polytechnic Institute, under the sponsorship of Independent Nail & Packing Company, Bridgewater, Mass., manufacturers of Stronghold® Annular Thread and Screw-Tite® Spiral Thread Nails and other improved fasteners.

ABBREVIATIONS USED IN THIS TABLE:

Corr.-Corrugated
Dph. Diaphragm
Part. bd—Particle board
LCS—Low Carbon Steel
HCS—High Carbon Steel
Plywd.—Plywood
Csk.—Countersunk
Flrg.—Flooring
Shgl.—Shingle
Med.—Medium
Dia.—Diamond
Ndl.—Needle

STRONGHOLD ® ANNULAR THREAD NAIL

STRONGHOLD ® SCREW THREAD NAIL

SCREW-TITE ® SPIRAL THREAD NAIL

SCREW-TITE ® KNURLED MASONRY NAIL

Aluminum Nails

Aluminum nails are excellent for use when the nailhead is exposed to the atmosphere or corrosive conditions. In other words, use aluminum nails wherever there is the possibility of rust or nail stain.

The manufacturers do not ordinarily recommend aluminum common nails for ordinary framing, because there is no economic advantage over steel wire nails for most common purposes, and the bending resistance of aluminum nails is less than that of steel nails despite the larger diameter of the aluminum nails.

Aluminum nails weigh about one-third as much as steel wire nails, but they are more expensive nail for nail; however, they do save labor and painting where rust or nail stain is a factor.

Common 3d to 20d nails will run $2.30 per lb., 30d to 60d around $2.05 per lb.

Hardware Accessories Used for Wood Framing

The following items may be used to advantage in all types of wood construction. Developed primarily to provide better joints between wood framing members, in many cases their use has resulted in lower overall costs--the additional material cost being more than offset by increased labor efficiency.

Steel Joist Hangers. Used for framing joists to beams and around openings for stair wells, chimneys, hearths, ducts, etc. Made of galv. steel, varying from 12 ga. to 3⁄16" in thickness, with square supporting arms, holes punched for nails and with bearing surfaces proportioned to size of lumber. Approximate prices for sizes most commonly used are as follows:

Joist Size	Gauge	Depth of seat	Opening in Hanger	Price Each
2"x6"	12	2"	1 5⁄8"x5"	$0.70
2"x8"	12	2"	1 5⁄8"x5"	.78
2"x10"	12	2"	1 5⁄8"x8 1⁄2"	.81
2"x12"	10	2 1⁄2"	1 5⁄8"x8 1⁄2"	1.00
4"x8"	11	2"	3 5⁄8"x5 1⁄4"	.94
4"x10"	9	2"	3 5⁄8"x5 1⁄4"	1.12
4"x12"	3⁄16	2 1⁄2"	3 5⁄8"x8 1⁄2"	1.73

Framing Accessories. These are used in light wood construction to provide face-nailed connections for framing members. Adaptable to most framing connections, they eliminate the uncertainties and weaknesses of toe-nailing. Manufactured of zinc-coated sheet steel in various gauges and styles. Framing accessories are designed to provide nailing on various surfaces. Special nails, approximately equal to 8d common nails, but only 1 1⁄4" long, to prevent complete penetration of standard nominal 2" lumber, are furnished with anchors. Approximate prices are as follows:

Type	Price per 100
Trip-L-Grips	$25.00
Du-Al-Clip	20.00
Nail-On Plates	30.00
Post Caps	66.00
H Clips	4.00

Angles 45.00

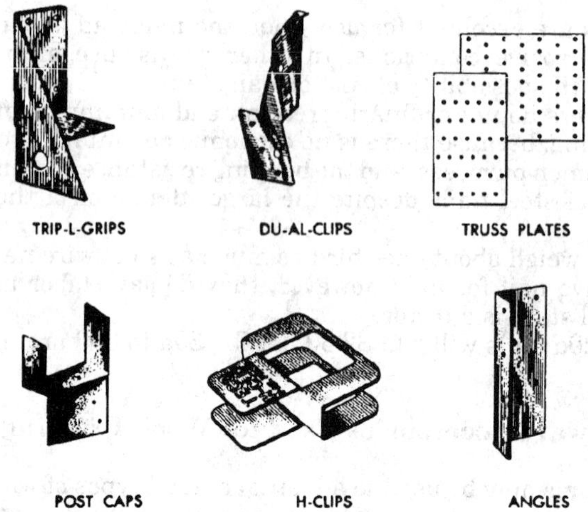

Timber Connectors. Timber connectors are devices for increasing the strength of bolted joints in timber construction. They are placed between adjacent faces of overlapping members, embedded to half-depth in each, to increase the bearing area at the connection. Several types of timber connectors are available, each for a particular purpose, as follows:

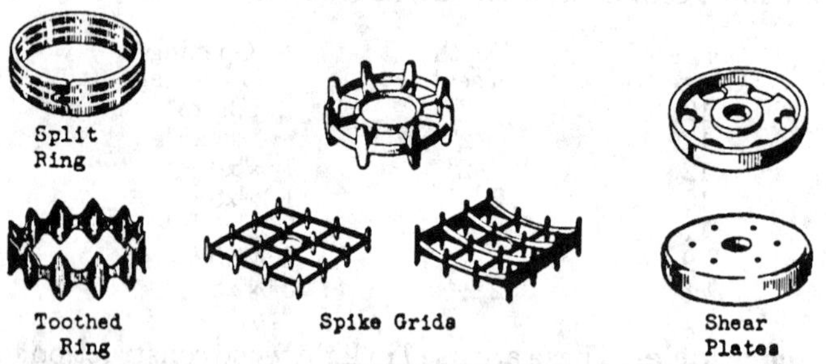

Timber Connectors

Metal Connector Plates (Truss Plates). These are made from structural quality galvanized sheet steel and have integral teeth protruding from one face perpendicular to the plate. They are made in various sizes, thicknesses, and gauges and are designed to laterally transmit load in wood. Metal connector plates are installed by applying uniform pressure over the plate surface, which requires special equipment such as hydraulic or roller presses.

Wedge-Fit Split Rings. Split ring connectors are hot-rolled carbon steel bands, bent to form a ring with a tongue and groove meeting joint. The cross section of the steel band is tapered both ways from center to edges, providing a wedge-fit when connectors are inserted into tapered grooves, conforming to ring section and precut to half-depth in each contact face of overlapping members.

Grooves in wood members must be cut with a special grooving tool, using a power drill press or portable drill. Grooves and bolt holes may be cut in a single operation, using a combination grooving tool and drill bit. If bolt holes are bored first, a smooth shank pilot is used to guide the grooving tool. Split rings are available in two sizes: 2 1/2" diameter rings are used for wood framing involving lumber of 2" nominal dimension and are used in trussed rafter construction for spans up to 50 ft., and 4" diameter rings are used for medium to heavy construction, generally with members 3" or more in nominal thickness. Approximate prices of split rings and grooving tools are as follows:

	Approximate Prices	
Description	2 1/2" Dia.	4" Dia.
Split rings, per 100 pcs.	$42.00	$68.00
Grooving tool cutter head, complete	120.00	175.00
Replacement cutters,		
Set of 4 for 2 1/2" dia.	69.00	....
Set of 6 for 4" dia.	....	93.00
Smooth shank pilot,		
9/16" dia.	10.00	....
11/16" dia.	10.00	....
13/16" dia.	....	12.00
15/16" dia.	....	15.00

Shear Plates. Shear plates are used for wood-to-steel connections and for wood structures that are frequently disassembled--used singly for wood-to-steel joints and in pairs for wood-to-wood assemblies. They are also often used in place of split rings for field connected joints. Shear plates are placed in daps precut in the contact faces of wood members with a special grooving tool. Load is transferred from wood member to shear plate, from shear plate to bolt and thence to opposing member, which may be another shear plate in wood, a steel gusset plate, or rolled steel shape. Drilling, grooving and dapping of wood members are accomplished in the same manner as described for split ring connectors.

Shear plates are available in two sizes--2 5/8" diameter, made of pressed steel, and 4" diameter made of malleable iron. Approximate prices of shear plates and grooving tools are as follows:

	Approximate Prices	
Description	2 5/8" Dia.	4" Dia.
Shear plates, per 100 pcs.	$62.00	$220.00
Grooving tool cutter head, complete	147.00	235.00
Replacement cutters,		
Set of 3 for 2 5/8" dia.	79.00	....
Set of 5 for 4" dia.	....	98.00
Smooth shank pilot,		
13/16" dia.	12.00	11.00
15/16" dia.	....	15.00

Toothed Rings. Toothed ring connectors are toothed metal bands with each tooth corrugated or curved along its cross section for greater rigidity. Used for wood-to-wood connections between lighter structural members, they are also used for strengthening structures in place and structural repairs.

The function of toothed rings is similar to that of split rings, but the method of installation is not as efficient and requires more labor. Toothed rings are embedded in wood members by applying pressure. Where a large number of joints are to be assembled, a hydraulic jack setup can be used to advantage. It requires a 5- to 7 ½-ton jack for single bolt connections and a 10-ton jack for joints with two bolts. Where fewer joints are involved, the high strength rod assembly is recommended. The assembly consists of a high strength rod with Acme threads on one end, double depth nuts, ball bearing thrust washer, heavy plate washers, lock washer, and nut. Metal sleeve adaptors permit the use of ½", ⅝", or ¾" diameter rods with a standard ball bearing washer. The length of rod required equals the total thickness of wood members plus one inch for each layer of toothed rings plus 5 inches for nuts and washers. Ratchet wrenches help speed assembly--impact wrenches have also been used with success. After joint is drawn closed, high strength rod assembly is replaced with ordinary bolt and washers.

Spike Grids. Spike grid connectors are malleable iron castings of a grid-like structure with teeth protruding from both faces. Available in three styles: square flat type, 4 ⅛"x4 ⅛"; circular type, 3 ¼" dia., used between sawn timbers; and single curve type, 4 ⅛"x4 ⅛", used between curved faces of poles or piling and sawn timbers. Spike grids are embedded under pressure, using the high strength rod assembly, as previously described for toothed rings. Approximate prices are as follows:

Type of spike grid	Square Flat	Single Curve	Circular
Price per 100 pcs.	$245.00	$245.00	$190.00

Installation tool for spiked grids, complete with adaptor - $185.00.

WOOD FLOOR AND ROOF TRUSSES

Some advantages of wood floor trusses are speed of construction and the ability to install utility lines, such as HVAC ducts, water lines, sprinkler lines, and electrical cables so that they do not interfere with one another. Easy of installation combines with economy.

Floor trusses are used for spans from 12'-0" to as large as 70'-0" and are found in all conventional types of construction, as well as in factories, storage facilities, and any other structure where a large, uninterrupted area is required.

Wood roof trusses, either shop or field fabricated, are used in buildings where clear floor space is a requirement and the width of the building exceeds the economical span of roof joists. Some of the advantages of trussed roof construction are fast erection, early availability of unobstructed weather-protected space, simplified installation of ceiling, floor, mechanical and electrical systems, and use of non-loadbearing movable partitions.

Wood roof trusses are used for spans as short as 25'-0" and can be used up to 200'-0". They are widely used in garages, factories, hangars, gymnasiums, auditoriums, bowling alleys, dance halls, supermarkets, and other structures of this type.

When using wood roof trusses, it is common practice to span the shorter dimension of the building. One might use any of several truss types, such as mechanically connected or bolted plywood gusset. The type of truss depends on the degree of economy desired, building use, and the general architectural effect desired. Prices of roof trusses are governed by the following conditions:

1) Cost of material and labor.
2) Loading conditions and spacing.
3) Difficulties of installation.
4) Requirements of local building ordinances.

The last condition causes the price of trusses to vary considerably in different parts of the country. Some cities may require only a 25-lb. live load while others require a 50-lb. live load, necessitating heavier construction.

The spacing of wood roof trusses that directly support roof sheathing is usually 2'-0". Where roof loads are light and the installation of a ceiling is not required, spacings of 4'-0" to 4'-10" are advantageous. Where snow loads are especially heavy, spacings of 16" and even 12" have been used.

Prices given on following pages are only approximate. The estimator must obtain definite prices, for trusses either delivered at the building site or installed in place, from the manufacturer. Truss suppliers are familiar with conditions in all parts of the country and can quote a definite price.

Wood Truss Configurations

Bowstring Bolted Roof Truss. For industrial and many commercial structures, the bowstring truss is popular. This type of truss has a curved top chord that starts from a point approximately 10 inches high at the end of the truss, rising to a maximum height at centerline of about 1/8 of the total span. This type of truss gives a curved roof shape. Because of its low height at the ends of the truss, a minimum of masonry is required for parapet or fire walls.

The truss may be left exposed, carrying roof load only, as in garages or industrial plant, or it can be designed to carry monorail, floor systems, or other concentrated loads. The bowstring truss is also designed to carry ceiling loads for use in stores, automobile agency salesrooms, and other commercial structures. Bowstring trusses provide wind bracing for walls through diaphragm action and can be used with "knee braces" to develop portal action in mill type buildings. They are efficiently fabricated with bolts, resulting in little reduction in cross-sectional areas of the pieces connected.

Approximate Prices of Wood Bowstring Truss. They are designed with glued laminated or nailed 2"x3" or 2"x4" top chords, built to conform to a parabolic curve, furnished in lots of 5 or more, spaced 16'-0" on centers, with 25 lb. live load and 40 lb. total load.

Span	Height	Price	Span	Height	Price
20'-0"	2'-6"	$315.00	90'-0"	11'-3"	1295.00
25'-0"	3'-1"	320.00	95'-0"	11'-10"	1445.00
30'-0"	3'-9"	325.00	100'-0"	12'-6"	1540.00
35'-0"	4'-4"	340.00	105'-0"	13'-1"	1630.00
40'-0"	5'-0"	375.00	110'-0"	13'-9"	1700.00
45'-0"	5'-7"	380.00	115'-0"	14'-4"	1705.00
50'-0"	6'-3"	420.00	120'-0"	15'-0"	1880.00
55'-0"	6'-10"	525.00	125'-0"	15'-7"	2020.00
60'-0"	7'-6"	595.00	130'-0"	16'-3"	2170.00
65'-0"	8'-1"	680.00	135'-0"	16'-10"	2300.00
70'-0"	8'-9"	810.00	140'-0"	17'-6"	2440.00
75'-0"	9'-4"	895.00	145'-0"	18'-2"	2500.00
80'-0"	10'-0"	1045.00	150'-0"	18'-9"	2630.00
85'-0"	10'-7"	1170.00			

Crescent Type Roof Truss. This is increasingly popular and is identified by its curved lower chord, which affords a higher ceiling at the mid-span of the truss. It produces a curved roof, similar to the bowstring truss. This truss presents a curved ceiling effect and is used in gymnasiums, auditoriums, low-cost churches, and more elaborate stores and restaurants. Recommended span is from 20'-0" to 85'-0".

Approximate Prices of Crescent Type Wood Roof Trusses. They are furnished in lots of 5 or more, spaced 16'-0" on centers, 25 lb. live load, and 40 lb. total load.

Span	Height	Camber	Price
20'-0"	5'-0"	1'-0"	$380.00
25'-0"	5'-6"	1'-3"	395.00
30'-0"	6'-0"	1'-6"	420.00
35'-0"	6'-6"	1'-9"	435.00
40'-0"	7'-0"	2'-0"	460.00
45'-0"	7'-6"	2'-3"	500.00
50'-0"	8'-0"	2'-6"	525.00
55'-0"	8'-6"	2'-9"	635.00
60'-0"	9'-0"	3'-0"	700.00
65'-0"	9'-6"	3'-3"	825.00
70'-0"	10'-0"	3'-6"	890.00
75'-0"	10'-6"	3'-9"	1050.00
80'-0"	11'-0"	4'-0"	1200.00
85'-0"	11'-6"	4'-3"	1325.00

Belgian Roof Truss. This type is used where it is desirable to give a conventional building a more pleasing appearance through the use of a peaked roof. It is recognized by its sloping top chord and horizontal lower chord. In most instances, a ceiling is applied to the lower chord of these trusses, while the top chord may have shingles or slate as a covering. Used on some higher class store buildings and low-cost churches. Recommended for spans from 20'-0" to 85'-0".

Belgian roof trusses are less efficient than the bowstring type, because the connections generally govern the member sizes and therefore cost approximately 50% more than bowstring type trusses.

The Double fink truss is also referred to as a belgian truss and is used for spans from 36'-0" to 60'-0".

Flattop Roof Truss. Industrial plants, warehouses, sheds, etc., when built of frame construction, often find use for a truss that absorbs wind loads, in addition to carrying roof and concentrated loads. The flattop truss is ideal for this purpose. It has parallel top and bottom chords and presents the overall appearance of a large rectangle. Spans should not exceed 65 feet where costs are an important factor.

Howe Truss. The Howe truss can be used for spans from 16' to 18'. It is efficient for loading conditions that balance the top and bottom chords and is often used when the truss is designed as a girder, with loads applied for the bottom chord.

Parallel Chord 4x2 Truss. Normally used in floor truss design, it has parallel chords of stress related 2"x4" lumber set flat so the wider faces are the bearing surface of the truss. The web members are also cut from 2"x4" and positioned so that they support the chord across their full widths. This type of truss can be manufactured with duct chase openings so wiring, piping, and ducts can run within the chords.

Parallel Chord 2x4 Truss. Generally used in roof truss design, it is gaining popularity because of the long, clear spans and shallow depth. The slope can be built into the the truss, or drainage can be provided by using different opposite side wall heights to slope the truss. Roof slopes should be at least ¼" per foot of span.

Modified Queen Post Truss. This type of truss is often used for spans the same as those requiring double W but that have load conditions that require fewer bottom chord panels.

Fink Truss. The fink truss is generally suitable for spans from 16' to as long as 46' and for all classes of construction. It is an efficient and cost effective truss configuration, 50% to 60% the cost of a comparable steel truss.

Three-Hinged Arch. This type of structure provides both good overhead clearance and appearance. However, high forces develop in some joints in the longer spans. It is recommended that the designer consult with the local fabricator before specifying this type.

Cantilever Truss. Trusses with single or double cantilever sections are possible. Cantilevers can approach one-fourth of the distance of the main interior truss span.

Clerestory Truss. This type of truss is used extensively in industrial and agricultural buildings, in spans up to 60 feet.

Inverted Truss. This truss is used for variations in interior and exterior roof profile.

Vaulted Ceiling Truss. A partial scissors is used to achieve a vaulted ceiling over a portion of a residence, such as the living room, while maintaining flat ceilings in the other rooms. Designers must consider horizontal movement at the bearing points.

Mono-Pitch Truss. Often used to form a shed roof structure on a residence or garden apartment building where a particular architectural effect is desired. It is also popular and economical for use in back-to-back construction with three or more bearing walls. This gives the same architectural effect as a gabled roof but uses two shorter span mono-pitch trusses. It might vary in configuration but offers a simple solution to some design problems.

Dual-Pitch Truss. Some design solutions require an asymmetric roof pitch. This type allows different pitch on each side of the roof.

Pitched Warren Truss. This is often specified when a long span and low pitch is required. It is well suited where natural lighting is sought, because vertical windows can be designed into the top section of the loadbearing walls between the truss supports. This truss form is most economical in spans from 30' to 70', on center spacings from 2' to 8'.

Types of Mechanically Connected Truss

Designers and builders of residential construction, multiple unit housing projects, and light commercial construction use trusses to reduce costs and to shorten construction time. Most small home constructors can benefit from this type of framing. Some of the advantages to using mechanically connected trusses are: roof framing and ceiling framing are accomplished at the same time; members can be made of lighter stock, saving material and reducing weight; trusses can be pre-assembled; trusses can be erected rapidly and the job put under cover quickly; and trusses permit the use of non-bearing partitions for all interior walls.

Trussed rafters used in house building and similar construction vary as to truss pattern and method of construction, depending on design requirements and individual preference of the builder. Some examples of truss patterns are:

Types of Trusses

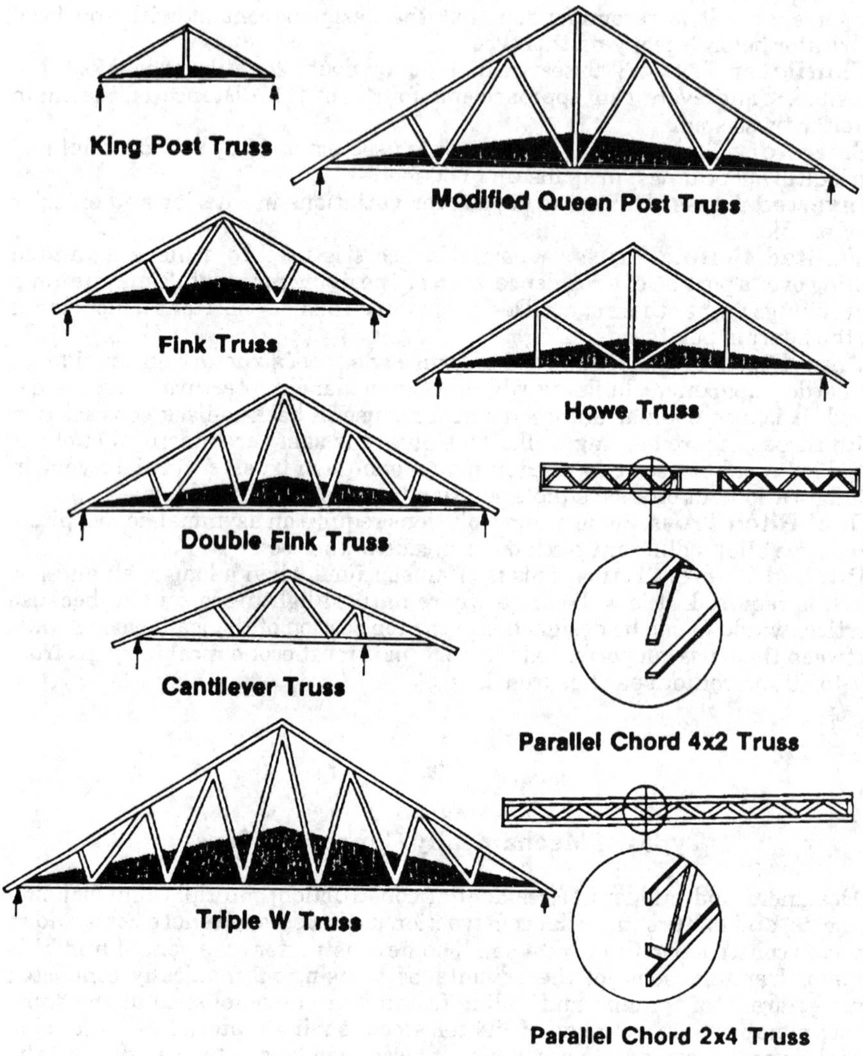
King Post Truss
Modified Queen Post Truss
Fink Truss
Howe Truss
Double Fink Truss
Cantilever Truss
Parallel Chord 4x2 Truss
Triple W Truss
Parallel Chord 2x4 Truss

Types of Trusses

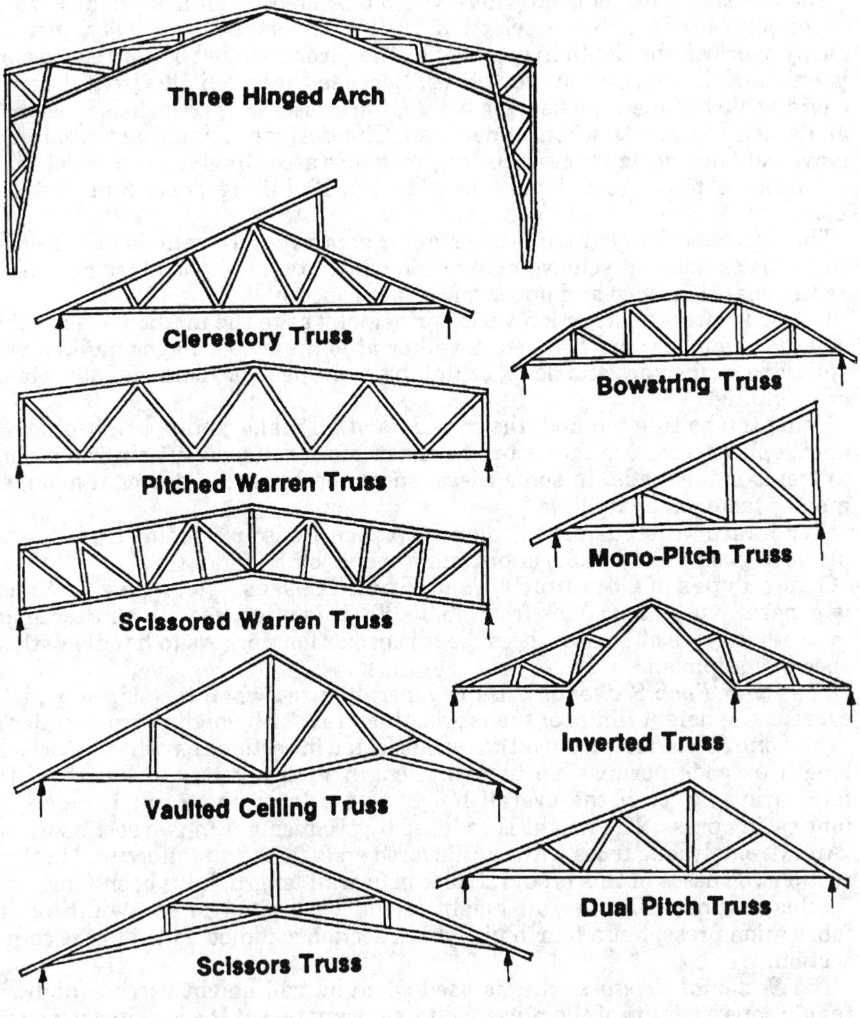
Three Hinged Arch
Clerestory Truss
Bowstring Truss
Pitched Warren Truss
Mono-Pitch Truss
Scissored Warren Truss
Inverted Truss
Vaulted Ceiling Truss
Dual Pitch Truss
Scissors Truss

The *W-Type* is the most popular type and is adaptable for spans from 18' up to 40'; roof slope from 2 in 12 to 6 in 12 and higher.

The *Triple-W* is used for spans up to 80' with slopes of 3 in 12 and higher. Centerline spacings can be from 2' to 20', depending on requirements.

The *Kingpost* truss is usually recommended for shorter spans. The economical range is up to 26' under most loading conditions. It has wide application, for residential and commercial garages, carports, and short span storage buildings.

The *Scissors* truss is used where vaulted or sloped ceilings are desired for architectural effect. Slope selection is very important, and a designer can easily overlook the depth in span ratio. The recommended bottom chord slope is one-half the slope of the top chord. Because there is no horizontal bottom chord or tie between the bearing points, the scissor truss develops movement at its bearing points when under load. The designer should anticipate this movement and design the supporting structure accordingly. Economical spans are from 18' to 36'; roof slopes 5 in 12 to 6 in 12; ceiling slopes 2 in 12 to 3 in 12.

The *Scissored Warren* truss has similar depth requirements as the regular chord truss and can achieve many of the advantages of portal frames, as far as internal clearance and appearances are concerned.

Trussed rafter construction varies principally as to the method of assembly, whether members are fastened together at joints by bolting or nailing, with split ring connectors and bolts, or flat, bent, pronged or toothed metal plates, or by gluing.

Using the bolting method, the members at all major joints, i.e., heel joints, apex, splices, etc., are connected by one or more bolts, or by using split rings to transfer the loads. In some cases, one or both ends of diagonal members may be fastened by nailing.

For nailed trussed rafters, all joint connections use metal truss plates, plywood gussets and nails, or ordinary nailed joints.

Other Types of Commonly Used Wood Trusses. The *Double Fink* truss is generally used for spans from 36' to 60'. It is often used in shorter spans over other types of trusses, because it is easier for workers to handle without special equipment.

The *Piggy Back Sectional* truss is generally used when the slope (or pitch) exceeds the height limits of the fabricating press or by highway restrictions.

The *Long Sectional* is a truss that is fabricated in sections for when the overall length exceeds permissible highway length restrictions, the length of the fabricating jig, or if the overall height is too large to be put through the fabricating press. The height is reduced by producing it in two sections.

An extremely long truss with cantilevered ends can be manufactured in three sections. Trusses of this type, 128 feet in overall length, have been fabricated in this manner. The overall height would be limited to the width of the fabrication press, but a fourth piggyback section could be added to the center section.

The *Sectional Scissors* truss is used when overall height exceeds highway regulations or limits of the press. Splices are applied at the job site by general contractor or a subcontractor. Due to its very high center of gravity and unstable configuration, this type requires extreme care in erection.

There are alternate methods of splicing sectional scissors trusses. One such method, when it is feasible, has the advantage of providing manufacturing symmetry.

Roof trusses offer numerous advantages over conventional roof framing: a truss can span greater roof distances; longer clear spans allow complete freedom in the layout of interior partitions; the bottom chord is available for ceiling finish material; and the cost is usually less, when on-site labor is considered.

Special Conditions Associated with Roof Truss. The *parallel chord* roof truss provides a flat roof that can be sloped to one side to drain by using wedge pieces added to the top chord. An alternate method is a slight variation in the bearing plate heights. This technique is acceptable where a slight variation in the interior room height is acceptable or where a hung ceiling will conceal the variation. When there is significant slope and no horizontal bottom chord, one must allow for horizontal movement at the wall.

Truss Openings. The two most widely used floor truss configurations are *Warren* and *Fan* trusses. Each type of truss allows openings for through-truss duct work. The table below shows the allowable openings for each type of truss. Note that the warren and fan trusses both allow a 7" opening when the depth of the truss is 12 inches. However, if the truss is 16" deep, the warren truss allows a 10¼" opening, the fan truss an 11" opening.

Center Duct Depth/Opening

Truss Depth in Inches	Maximum Opening Height in Inches	Width in Inches
12	9	18
14	11	22
16	13	24
18	15	24
20	17	24
22	19	24
24	21	24
26	23	24
28	25	24
30	27	24

Warren Truss Depth/Opening

Overall Truss Depth in Inches	Maximum Round Opening Diameter in Inches
12	7
14	8¾
16	10¼
18	11½
20	12½
24	14½
30	16½

The truss fabricator has dimensions of openings for intermediate depths of trusses.

Overall Truss Depth in Inches	Maximum Round Opening Diameter in Inches
12	7
14	9
16	11
18	13

The designer is cautioned to subtract insulation thickness and working tolerances from these dimensions to determine maximum net duct dimensions.

Installation of Wood Trusses

An estimator must consider the entire process of handling and installing wood trusses. The Truss Plate Institute recommends specific procedures, and the information provided below is excerpted from its publication, *Commentary and Recommendations for Handling Installing & Bracing Metal Plate Connected Wood Trusses (HIB-91).*

Unloading and Lifting. During the unloading process, and throughout all phases of construction, care should be taken to avoid lateral bending, which can cause joint and lumber damage. Use proper equipment to lift trusses. A crane with spreader bar and cables is strongly recommended for trusses with spans greater than 30 feet. For spans 30' or less, bridle and chokers may be used.

Do not lift bundles by the strapping, which is not strong enough to safely support the weight of the trusses. It is only meant for shipping purposes. Do not attach cables, chains, or hooks to web members. Thread cables under the top chord of a bundle of trusses near panel points closest to the quarter or third points of the trusses. Lift the bundle slowly in order to determine the load balance. If it is unbalanced, then lower the bundle, readjust the cable, and lift again.

Lifting with a bridle or spreader bar requires the same care in balancing the load. Never lift an unbalanced bundle of trusses. Do not lift single trusses with spans greater than 30' by the peak.

Storage. Whenever possible, trusses should be unloaded in bundles, on a relatively smooth ground, and picked up by the top chord in a vertical position only. It is common to unload trusses from a trailer one at a time and store them in a stable position to prevent toppling or shifting. When they are stored horizontally, blocking or sleepers should be maximum 10' centers to prevent lateral bending. Do not store unbraced bundles upright.

If trusses are stored vertically, extreme care should be taken to insure that they are braced and blocked in a stable manner that will prevent toppling. Pitched trusses should be stored with the peak up.

Installation. An open web metal plate connected truss is a manufactured product, comprised of various lumber components that are precisely cut and fitted together with metal connector plates. A metal plate connected wood truss should be installed with greater care than a monolithic item such as sawn wood joist, steel channel, or a beam. Without bracing, or some other type of restraining device, a truss can be unstable.

The installer is responsible for selecting the most suitable method and sequence of installation that is available and which is consistent with the plans and specifications and such information as may be furnished prior to installation. Trusses can be installed by hand or by mechanical means, depending on truss span, installed height above grade, and accessibility by equipment (such as crane or forklift).

The installer must understand truss design drawings, placement plans, and all notes and cautions on them. He should clarify any and all questions with the manufacturer before starting the installation. Use particular care when installing under adverse conditions, such as high winds, uneven or sloping terrain, proximity of high voltage powerlines, or constricted job site.

Never walk on trusses that are lying flat. Under no circumstances should construction loads of any description be placed on unbraced trusses.

Hand Installation. Installation by hand should be limited to trusses of a size and configuration that can be carried, raised up to bearing support height, and rotated into position without excessive lateral deflection (bow), which produces strain in the lumber or metal connector plates and weakens the joints. Lateral deflection greater than 3" in 10' of span is excessive. Trusses should be handled so as to ensure support at intervals of 25' or less.

When installed by hand, trusses are positioned over the side walls and rotated into position using a fork-like lifting pole. The longer the span, the more workers will be necessary to avoid lateral strain on the truss. Depending on length, the truss should be supported at the peak for spans less than or equal to 20' and at quarter points for spans less than or equal to 30 feet.

Mechanical Installation. Trusses that are installed by mechanical means should be handled so as to ensure support at intervals of 25' or less. The installer should provided adequate rigging (crane, fork lift, slings, tag lines, and spreader bars) for sufficient control during lifting and placement to assure safety to personnel and to prevent damage to trusses and property. Slings, tag lines, spreader bars, etc. should be used in a manner that will not cause any damage to metal connector plates.

Trusses that are lifted in place in banded bundles should be securely supported by temporary means, permitting the safe removal of banding and the sliding of individual trusses. Do not lift bundled trusses by their strapping.

Take care to position truss bundles so that the supporting structure is not overloaded. The bundle straps (banding) can be broken after the bundle has been placed on the supporting structure and prior to the release of the lifting cables.

Under certain conditions, several trusses may be assembled on the ground into structural subcomponents, complete with temporary bracing members and portions of the roof deck or subfloor, and then carefully lifted into place as a self supporting unit. These units are assembled as needed until the roof or floor structure is complete. Usually some decking is omitted so that the stagger-lap is provided between adjacent units for diaphragm continuity. This method requires substantial planning beforehand and special engineering for the trusses, such as location and design of pick-up or liftpoints. This method has been found to be a safe solution to the installation of especially long spans or complex shapes.

Trusses that are mechanically installed one at a time should be held safely in position with the erection equipment until such time as all temporary bracing has been installed. Lines from the end of the spreader bar should "toe-in". Do not permit lines to "toe-out", which will tend to cause trusses to buckle.

For truss spans greater than 60 feet, the suggested lift procedure requires the use a strong-back. The strong-back should be attached to the top chord and web members at intervals of approximately 10 feet and should be at or above mid height of the truss to prevent overturning. The strong-back should be monolithic with sufficient strength to safely carry the weight of the truss and rigid enough to resist truss bending.

Each truss should be set in position in accordance with the designer's plan and held with the hoisting equipment until the ends of the truss are securely fastened and temporary bracing is installed. Only then should the hoisting equipment be released from the braced truss and used for lifting, placing, and securing the next truss.

Installation tolerances are critical to achieving an acceptable roof or floor line and for effective bracing. A stringline, plumb bob, level, or transit is recommended to attain acceptable installation tolerances.

Trusses should not be installed with an overall bow, or bow in any chord or panel, that exceeds the lesser of L/200 or 2 inches, where L is the span of the truss, chord, or panel length in inches.

L (inches)	L/200 (inches)	L (feet)
25	1/8	2.1
50	1/4	4.2
75	3/8	6.3
100	1/2	8.3
125	5/8	10.4
150	3/4	12.5
175	7/8	14.6
200	1	16.7
225	1 1/8	18.8
250	1 1/4	20.8
275	1 3/8	22.9
300	1 1/2	25.0
350	1 3/4	29.2
400	2	33.3

Trusses should not be installed with a variation from plumb (vertical tolerance) at any point along the length of the truss from top to bottom chords that exceeds 1/50 of the depth of the truss at that point (D/50) or 2", whichever is less. This does not apply to trusses specifically designed to be installed out of plumb.

D (inches)	D/50 (inches)	D (feet)
12	1/4	1
24	1/2	2
36	3/4	3
48	1	4
60	1 1/4	5
72	1 1/2	6
84	1 3/4	7
96	2	8
108	2	9

Location of trusses along the bearing support should be within ±1/4" of plan dimensions. Special hangers or supports should be located to support trusses within ±1/4" of plan dimensions. Trusses are to be located at on-center spacing specified by the design engineer.

Top chord bearing parallel chord trusses should have a maximum gap 1/2" between the inside of the bearing and the first diagonal or vertical web.

Correction of Errors in Truss Installation. Errors in building lines or dimensions, or by subcontractors and suppliers, must be corrected before installation begins. Alterations, such as cutting overhangs to proper length, might be required and should be made by the builder of record. Any correction that involves the cutting, drilling, or relocation of a truss member or metal connector plate is considered major and should not be made without notifying the truss manufacturer.

A major correction calls for an engineering analysis by qualified registered engineer prior to the field work. Cutting or drilling trusses in the field without the approval of this engineer should be strictly prohibited. Trusses might be designed by the truss design engineer for specific reasons to be field cut by the installer at the designated locations. Such modifications should be indicated on the engineering drawing.

Field Assembly. Trusses that are too long for delivery to the jobsite in one piece can be designed to be delivered in two or more parts and then spliced together at the jobsite. The installer should carefully follow the splicing specifications shown on the truss drawings. Splicing may be performed on the ground before installation, or the truss sections may be supported by temporary shoring and splices installed by workers on a safe working surface.

Trusses that are too high for delivery to the jobsite in one piece may be manufactured in two or more sections and piggy-backed at the jobsite. The installer should observe the permanent bracing and connection details shown on the truss or piggyback design drawings. The supporting trusses should be completely installed, including the temporary and permanent bracing and sheathing when required, before installing top or supported truss sections.

Installation Sequence for Temporary and Permanent Bracing of Wood Truss. Install the first truss with a ground bracing system, which is constructed as follows:

a) Attach verticals to the end wall at the same spacing as the top chord lateral bracing.
b) Attach diagonals to the vertical and ground stake.
c) Attach horizontal ties or backup ground stakes if required.
d) Attach struts to the diagonals if required.
e) Attach lateral braces to the diagonals if required.
f) Attach end braces to the diagonals and lateral braces.

For the first group of three to six trusses, the sequence is as follows:

a) Install three to six trusses with top chord lateral braces affixed to the first (ground-braced) truss. The number of trusses to be initially installed should be specified in the bracing design.
b) Install top chord diagonal bracing between top chord lateral braces at intervals specified in the bracing design.
c) Install bottom chord lateral braces at no greater than 15 feet on center.
d) Install web lateral bracing as specified on the truss design.
e) Install bottom chord diagonal braces between bottom chord lateral braces at the same interval as top chord diagonal bracing.
f) Install web cross bracing on web(s) corresponding to the nearest adjacent panel point(s) to the bottom chord lateral brace(s).

Continue installation of trusses with top chord lateral braces as specified on the bracing design or truss layout plan. The number of trusses to be installed between sets of top chord diagonal braces should be specified on the bracing design or truss layout plan.

Repeat b) and d) above for each group of trusses and then install sets of web cross bracing at 20-ft. intervals. Install sets of bottom chord diagonal braces at the same intervals as top chord diagonal bracing and at each end bay. Top chord diagonal braces may remain as permanent bracing per the building design. Install permanent sheathing materials immediately as the temporary bracing is removed. Ground bracing may be removed only after the top chord plane is completely sheathed. Install any other permanent bracing as specified by the building design.

Note that all bracing lumber should be no less than 2'x4'x10'. A minimum of two 16d double head nails should be used at each connection.

Labor Erecting Wood Roof Trusses. Erection of wood roof trusses in confined areas, or where the supporting structure is not heavy enough to support new loads, has usually been accomplished using a gin pole. However, most truss erection today uses a special hydraulic crane mounted on a truck chassis, with lifting capacity up to 12 tons.

Before the crane is brought in, area around the building should be clear and all overhead electrical wires removed. Sizes of door openings must be predetermined to permit entrance of the crane.

Where a gin pole is used, the labor cost of erecting will vary with job conditions. Deep excavations, piles of brick, mortar boxes, lumber piles, etc. in the way of the derricks prevent erecting crews from making as good time as where a clear, unobstructed space is available.

Mechanical Erection of Trusses. Under average working conditions on jobs requiring 5 or more trusses, a crew consists of 5 workers with an erection machine. Sometimes, the truss fabricator sends trusses with a truck-mounted crane and the cost of unloading and erecting is included in the quoted price. If not, the estimator must add the cost of an erecting machine, such as a rubber-tired cherry picker or a crawler crane. The crew should erect one roof truss of given length at the following rate and costs:

Erecting Roof Trusses Using a Machine

Length of Truss	Hrs. to Erect	Labor Hrs. per Truss	Rate	Total	Rate	Total
30'-0"	0.52	2.62	$....	$....	$24.04	$62.98
40'-0"	0.75	3.75	....	....	24.04	90.15
50'-0"	1	5	....	....	24.04	120.20
60'-0"	1	5	....	....	24.04	120.20
75'-0"	2	10	....	....	24.04	240.40
90'-0"	3	15	....	....	24.04	360.60
100'0"	4	20	....	....	24.04	480.80
120'-0"	5	25	....	....	24.04	601.00

Erecting Floor Trusses Using a Machine

Length of Truss	Hrs. to Erect	Labor Hrs. per Truss	Rate	Total	Rate	Total
30'-0"	.15	.75	$....	$....	$24.04	$18.03
40'-0"	.20	1.00	....	....	24.04	24.04
50'-0"	.25	1.25	....	....	24.04	30.05
60'-0"	.35	1.75	....	....	24.04	42.07
75'-0"	.45	2.25	....	....	24.04	54.09

Hand Erection of Trusses. Wood roof trusses are erected by hand using a gin pole and normally requires a crew of four workers. The maximum length is 30'-0", and the contractor must careful to avoid "bowing" or deflection when erecting trusses by hand.

Erecting Roof Trusses by Hand

Length of Truss	Hrs. to Erect	Labor Hrs. per Truss	Rate	Total	Rate	Total
15'-0"	0.25	1.00	$....	$....	$24.04	$24.04
20'-0"	0.30	1.20	....	....	24.04	28.85
25'-0"	0.30	1.20	....	....	24.04	28.85
30'-0"	0.45	1.80	....	....	24.04	43.27

Erecting Floor Trusses by Hand

Length of Truss	Hrs. to Erect	Labor Hrs. per Truss	Rate	Total	Rate	Total
15'-0"	.15	.60	$....	$....	$24.04	14.42
20'-0"	.20	.80	....	....	24.04	19.23
25'-0"	.30	1.20	....	....	24.04	28.85
30'-0"	.45	1.80	....	....	24.04	43.27

Fabrication of Trussed Rafter Members. Fabrication of trussed rafter members is ideally suited for mass production methods, due to the large amount of repetition possible. In most cases, trussed rafters are symmetrical about their center lines so that corresponding members of each half are identical.

Since every member of a truss must do its share of the work, each one must be fabricated to exacting specifications. Design dimensions must be rigidly followed and all joints must have a good mechanical fit.

For efficient fabrication, a full-size layout should be made, either in the shop or on the subfloor of the building, after which one trussed rafter is carefully patterned and constructed with temporary connections. The unit may then be disassembled and the members used as templates for the balance of the trusses. On large projects, templates may be made of plywood or sheet metal.

To insure smooth roof and ceiling lines after erection, layout measurements should be made to upper edges of top chords and lower edges of bottom chords to eliminate variations in stock lumber widths.

A word of caution--attention is called to the fact that the joints and fabrication of various sized trusses are similar so it is important to exercise caution and see that the appropriate drawings are consulted and followed when trusses are fabricated.

Split-Ring Connected Trussed Rafters

As previously mentioned, members are connected by means of bolts and split rings at all major joints as the split rings transfer the loads and the bolts hold the members in contact.

In order that each member may function properly, the bolt holes must be located exactly in accordance with design dimensions and grooves for split rings must be of correct diameter, width and depth to provide a snug fit. Grooves must be cut with a grooving tool designed especially for this purpose.

Proper procedure for cutting grooves is as follows: after bolt locations have been established, drill bolt holes 1/16" larger in diameter than bolt; grooving tool is then fitted with a pilot mandrel of same diameter as drill and cutters are set to cut to a depth equal to half the width of the split ring; using bolt holes as pilot holes, concentric grooves are then cut in the faces of members which will contact each other. The grooving tool may also be fitted with the correct size drill bit, instead of a pilot mandrel, permitting both drilling and grooving operations to be done at one setup. For best results, drilling and grooving should be done with a bench drill or portable drill mounted in a drill stand.

Material Requirements for Split-Ring Connected Trussed Rafters. The material requirements for split-ring connected trussed rafters are based on the following design conditions:

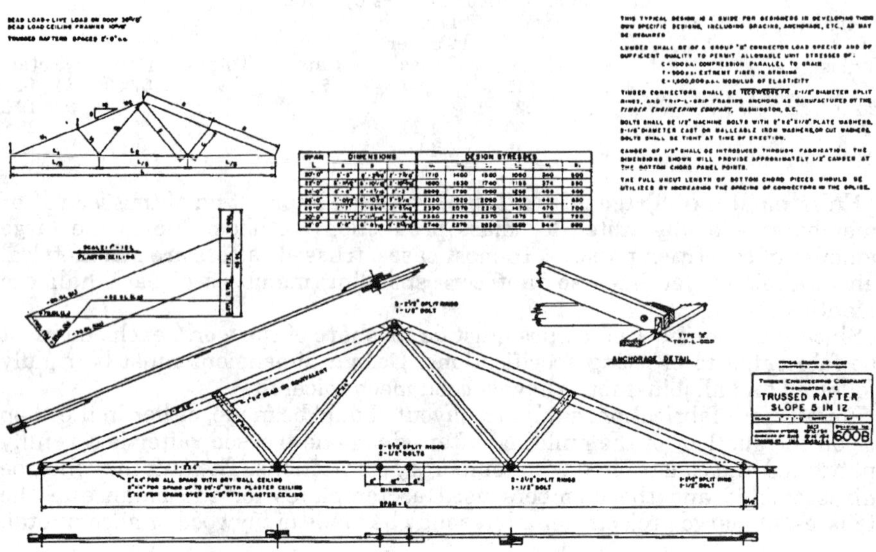

Design for Split-Ring Connected Trussed Rafter

Trussed rafters are designed to support a dead load plus live load on roof of 35 lbs. per sq. ft. and a ceiling load of 10 lbs. per sq. ft. with rafters spaced at 2'-0" centers.

Lumber shall be good grade of sufficient quality to permit the following unit stresses:

c = 900 lbs. per sq. in. Compression parallel to grain
f = 900 lbs. per sq. in. Extreme fiber in bending
E = 1,600,000 lbs. per sq. in. Modulus of elasticity

All split rings to be 2 ½" diameter. All bolts to be ½" diameter machine bolts. Washers may be 2"x2"x ⅛" plate washers, 2 ⅛" diameter cast or malleable iron washers or cut washers.

No allowances have been included for roof overhangs.

Quantities listed for diagonals are most economical lengths to produce a long and short member from a single length.

Material Requirements for One Split-Ring Connected Trussed Rafter of Various Spans for a 4 in 12 Roof Slope

Span	20'-0"	24'-0"	26'-0"	28'-0"	30'-0"	32'-0"
Top Chords	2-2x6	2-2x6	2-2x6	2-2x8	2-2x8	2-2x8
	12'-0"	14'-0"	16'-0"	16'-0"	18'-0"	18'-0"
Bottom Chords	2-2x4	2-2x4	2-2x4	2-2x4	2-2x4	2-2x4
	12'-0"	14'-0"	16'-0"	16'-0"	18'-0"	18'-0"
Diagonals	2-2x4	2-2x4	2-2x4	2-2x4	2-2x4	2-2x4
	8'-0"	10'-0"	10'-0"	12'-0"	12'-0"	12'-0"
Joint Scabs	1-1x4	1-1x4	1-1x4	1-1x4	1-1x4	1-1x4
	2'-0"	2'-0"	2'-0"	3'-0"	3'-0"	3'-0"
Total B.f.	51 1/3	60 2/3	67 1/3	81	89	89
Split Rings	11	11	11	13	13	13
Bolts,						
1/2"x4"	4	4	4	6	6	6
1/2"x6"	2	2	2	2	2	2
1/2"x7 1/2"	1	1	1	1	1	1
Washers	14	14	14	18	18	18
Nails, 8d	16	16	16	16	16	16

Material Requirements for One Split-Ring Connected Trussed Rafter of Various Spans for a 5 in 12 Roof Slope

Span	20'-0"	24'-0"	26'-0"	28'-0"	30'-0"	32'-0"
Top Chords	2-2x6	2-2x6	2-2x6	2-2x6	2-2x6	2-2x6
	12'-0"	14'-0"	16'-0"	18'-0"	18'-0"	20'-0"
Bottom Chords	2-2x4	2-2x4	2-2x4	2-2x4	2-2x4	2-2x4
	12'-0"	14'-0"	16'-0"	16'-0"	18'-0"	18'-0"
Diagonals	2-2x4	2-2x4	2-2x4	2-2x4	2-2x4	2-2x4
	10'-0"	12'-0"	12'-0"	12'-0"	14'-0"	14'-0"
Joint Scabs	1-1x4	1-1x4	1-1x4	1-1x4	1-1x4	1-1x4
	2'-0"	2'-0"	2'-0"	2-0"	2'-0"	2'-0"
Total B.f.	54	63 1/3	70	74	79 1/3	83 1/3
Split Rings	11	11	11	11	11	11
Bolts,						
1/2"x4"	4	4	4	4	4	4
1/2"x6"	2	2	2	2	2	2
1/2"x7 1/2"	1	1	1	1	1	1
Washers	14	14	14	14	14	14
Nails, 8d	16	16	16	16	16	16

Material Requirements for One Split-Ring Connected Trussed Rafter of Various Spans for a 6 in 12 Roof Slope

Span	20-0"	24'-0"	26'-0"	28'-0"	30'-0"	32'-0"
Top Chords	2-2x6	2-2x6	2-2x6	2-2x6	2-2x6	2-2x6
	14'-0"	16'-0"	16'-0"	18'-0"	18'-0"	20'-0"
Bottom Chords	2-2x4	2-2x4	2-2x4	2-2x4	2-2x4	2-2x4
	12'-0"	14'-0"	16'-0"	18'-0"	18'-0"	18'-0"
Diagonals	2-2x4	2-2x4	2-2x4	2-2x4	2-2x4	2-2x4
	10'-0"	12'-0"	14'-0"	14'-0"	16'-0"	16'-0"
Joint Scabs	1-1x4	1-1x4	1-1x4	1-1x4	1-1x4	1-1x4
	2'-0"	2'-0"	2'-0"	2'-0"	2'-0"	2'-0"
Total B.f.	58	67 1/3	72 2/3	76 2/3	82	86
Bolts,						
1/2"x4"	4	4	4	4	4	4
1/2"x6"	2	2	2	2	2	2
1/2"x7 1/2"	1	1	1	1	1	1
Washers	14	14	14	14	14	14
Nails, 8d	16	16	16	16	16	16

Material Requirements for One Split-Ring Connected Trussed Rafter of Various Spans for a 7 in 12 Roof Slope

Span	20'-0"	24'-0"	26'-0"	28'-0"	30'-0"	32'-0"
Top Chords	2-2x6	2-2x6	2-2x6	2-2x6	2-2x6	2-2x6
	14'-0"	16'-0"	16'-0"	18'-0"	20'-0"	20'-0"
Bottom Chords	2-2x4	2-2x4	2-2x4	2-2x4	2-2x4	2-2x4
	12'-0"	14'-0"	16'-0"	16'-0"	18'-0"	18'-0"
Diagonals	2-2x4	2-2x4	2-2x4	2-2x4	2-2x4	2-2x4
	12'-0"	14'-0"	14'-0"	16'-0"	16'-0"	18'-0"
Joint Scabs	1-1x4	1-1x4	1-1x4	1-1x4	1-1x4	1-1x4
	2'-0"	2'-0"	2'-0"	2'-0"	2'-0"	2'-0"
Total B.f.	60 2/3	70	72 2/3	79 1/3	86	88 2/3
Split Rings	11	11	11	11	11	11
Bolts,						
1/2"x4"	4	4	4	4	4	4
1/2"x6"	2	2	2	2	2	2
1/2"x7 1/2"	1	1	1	1	1	1
Washers	14	14	14	14	15	14
Nails, 8d	16	16	16	16	16	16

Plywood Gusset Trusses

Many contractors prefer plywood gusset trusses (PGT) for their simple construction. For small one-house jobs, containing only 20 to 30 rafter units, they can be easily fabricated and assembled on the job using simple hand tools--saw and hammer. They are also adaptable to mass production methods using portable power tools or shop equipment for projects involving thousands of trussed rafters.

Another advantage of the PGT is the fact that only standard dimension lumber and nails are used, in the case of "W" and the scissor-type, and standard plywood for gusset and splice plates.

To function properly, PGT must be fabricated and assembled with the same degree of accuracy as bolted and mechanically-connected trusses. Fabrication and assembly of truss units must adhere closely to design dimensions and joint details. It is important to have good bearing fits for top chord and compression diagonal member joints. Location, number, and size of nails must be in accordance with joint details.

For fast, accurate marking of nail locations, use nailing templates and spray colored paint or shellac through template openings with a paint gun or an aerosol-type container. To permit nailing from one side without nail clinching, use helically threaded nails. For dry, dense lumber, use hardened nails, which are more slender and less likely to split material.

While the use of seasoned lumber is not absolutely essential in the construction of PGT, better results are obtained with dry lumber and its use is recommended. This is particularly true for splice plates over 4" wide and is a design requirement for 8" wide material. The 1"x8" splice plates made from green lumber tend to split along nail rows, when undergoing shrinkage, causing the joint to fail.

Material Requirements for Plywood Gusset Trusses. The material requirements for PGT are furnished through the courtesy of Dr. E. George Stern, Research Professor of Wood Construction at the Virginia Polytechnic Institute, Blacksburg, Va. PGT designs for the various spans and roof slopes are based on experimental data obtained in the Wood Research Laboratory of the Virginia Engineering Experimental Station under the direction of Dr. Stern. PGT constructed in accordance with these designs have been subjected to all types of tests and have proved conclusively to be of ample strength for the use intended.

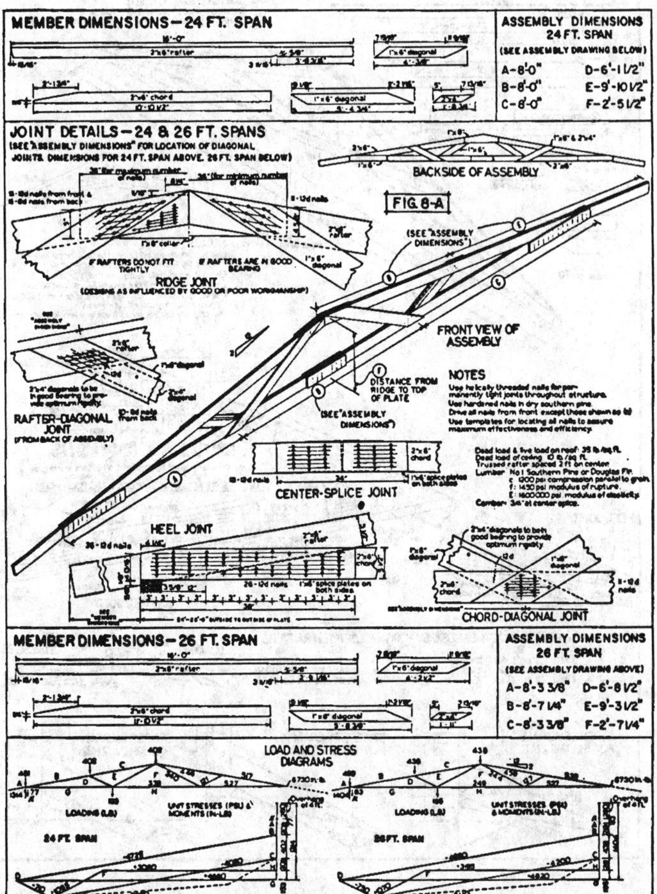

Design and Details for a Low Rise Plywood Gusset Truss

Material requirements are based on the following design conditions: PGT are designed to support a dead load plus live load on roof of 35 lbs. per sq. ft. and a ceiling load of 10 lbs. per sq. ft. with rafters spaced at 2'-0" centers. Lumber shall be stress-graded No. 1 southern pine or Douglas fir, preferably seasoned 12% to 15% moisture content, permitting the following unit stresses:

c = 1350 lbs. per sq. in. Compression parallel to grain
f = 1500 lbs. per sq. in. Extreme fiber stress
E = 1,760,000 lbs. per sq. in. Modulus of elasticity

Quantities listed for diagonals are most economical lengths to produce a long and short member from a single length.

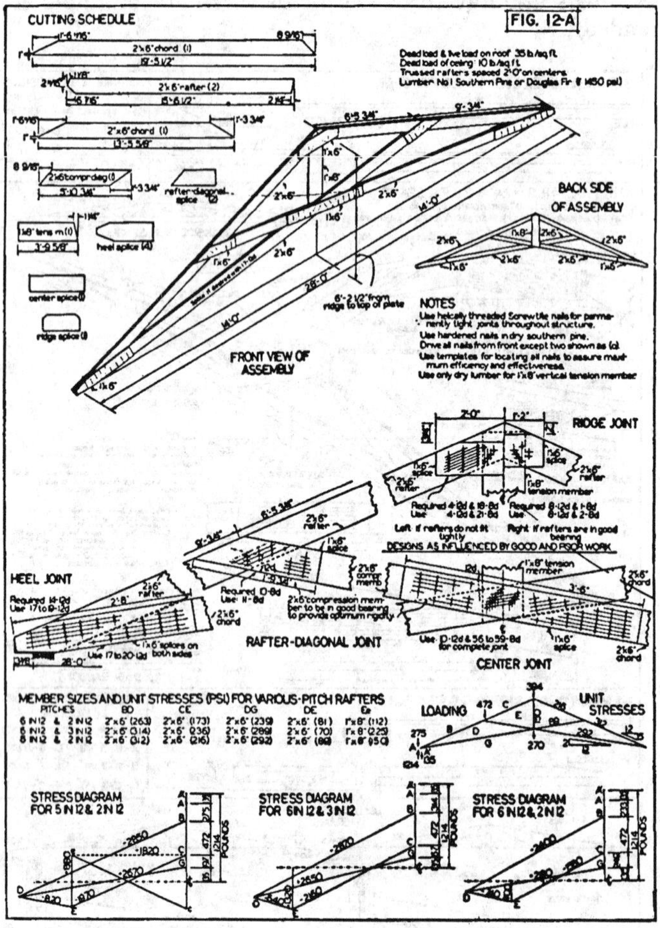

Design and Details for a Scissors Type Plywood Gusset Truss

Material Requirements for One Plywood Gusset Truss of Various Spans for a 2 in 12 Roof Slope

Span	18'-0"	20'-0"	24-0"	26'-0"	28'-0"	30'-0"
Approx. Overhang	3'-0"	2'-0"	4'-0"	3'-0"	4'-0"	3'-0"
Top Chords	2-2x6	2-2x6	2-2x6	2-2x6	2-2x6	2-2x6
	12'-0"	12'-0"	16'-0"	16'-0"	18'-0"	18'-0"
Bottom Chords	2-2x4	2-2x4	2-2x6	2-2x6	2-2x6	2-2x6
	8'-0"	10'-0"	12'-0"	12'-0"	14'-0"	14'-0"
Diagonals	2-1x6	2-1x6	2-1x6	2-1x6	2-1x6	2-1x6
	8'-0"	10'-0"	10'-0"	10'-0"	12-0"	12'-0"
	1-2x4	1-2x4	1-2x4	1-2x4	1-2x4	1-2x4
	4'-0"	4'-0"	4'-0"	4'-0"	6'-0"	6-0"
Splice Plates	4-1x4	4-1x4	4-1x6	4-1x6	4-1x6	4-1x6
	3'-0"	3'-0"	3'-6"	3'-6"	3'-6"	3'-6"
	2-1x6	2-1x6	2-1x6	2-1x6	2-1x6	2-1x6
	3'-0"	3'-0"	2'-0"	2'-0"	3'-0"	3'-0"
	1-1x8	1-1x8	1-1x8	1-1x8	1-1x8	1-1x8

	4'-0"	4'-0"	4'-0"	4'-0"	4'-0"	4'-0"
Total Board Feet	55	59 2/3	80 1/3	80 1/3	92 2/3	92 2/3
Nails,						
12d	144	144	198	198	242	242
8d	40	40	58	58	72	72

Material Requirements for One Plywood Gusset Truss of Various Spans for a 3 in 12 Roof Slope

Span	24'-0"	26'-0"	28'-0"	30'-0"	32'-0"	34'-0"
Approx. Overhang	3'-6"	4'-0"	3'-6"	4'-0"	3'-6"	2'-6"
Top Chords	2-2x6	2-2x6	2-2x6	2-2x6	2-2x6	2-2x6
	16'-0"	18'-0"	18'-0"	20'-0"	20'-0"	20'-0"
Bottom Chords	2-2x4	2-2x4	2-2x4	2-2x4	2-2x4	2-2x4
	12'-0"	14'-0"	14'-0"	16'-0"	16'-0"	18'-0"
Diagonals	2-1x6	2-1x6	2-1x6	2-1x6	2-1x6	2-1x6
	10'-0"	10'-0"	12'-0"	12'-0"	12'-0"	14'-0"
	1-2x4	1-2x4	1-2x4	1-2x4	1-2x4	1-2x4
	6'-0"	6'-0"	6'-0"	6'-0"	8'-0"	8'-0"
Splice Plates	4-1x4	4-1x4	4-1x4	4-1x4	4-1x4	4-1x4
	3'-0"	3'-0"	4'-0"	4'-0"	4'-0"	4'-0"
	3-1x6	2-1x6	2-1x6	2-1x6	2-1x6	2-1x6
	3'-0"	3'-0"	3'-0"	3'-0"	3'-0"	3'-0"
	.....	1-1x6	1-1x6	1-1x6	1-1x8	1-1x8
	.....	4'-0"	4'-0"	4'-0"	4'-0"	4'-0"
Total Board Feet	70 1/2	77 2/3	81	87 2/3	89 2/3	94 1/3
Nails,						
12d	138	138	146	146	168	168
8d	38	46	58	60	56	56

Material Requirements for One Plywood Gusset Truss of Various Spans for a 5 in 12 Roof Slope

Span	24'-0"	26'-0"	28'-0"	30'-0"	32'-0"	34'-0"
Approx. Overhang	3 1/4	3 1/4	3 1/4	3 1/4	3 1/4	3 1/4
Top Chords	2-2x4	2-2x4	2-2x6	2-2x6	2-2x6	2-2x6
	14'-0'	16'-0"	16'-0"	18'-0"	18'-0"	20'-0"
Bottom Chords	2-2x4	2-2x4	2-2x4	2-2x4	2-2x4	2-2x4
	12'-0"	14'-0"	14'-0"	16'-0"	16'-0"	18'-0"
Diagonals	2-1x6	2-1x6	2-1x6	2-1x6	2-1x6	2-1x6
	12'-0"	12'-0"	14'-0"	14'-0"	14'-0"	16'-0"
	1-2x4	1-2x4	1-2x4	1-2x4	1-2x4	1-2x4
	6'-0"	8'-0"	8'-0"	8'-0"	8-0"	10'-0"
Splice Plates	6-1x6	6-1x6	4-1x4	4-1x4	4-1x4	4-1x4
	2'-0"	2'-0"	3'-0"	3'-0"	3'-0"	3'-0"
	1-1x6	1-1x6	2-1x6	2-1x6	2-1x6	2-1x6
	3'-0"	3'-0"	2'-0"	2'-0"	2'-0"	2'-0"
	.....	....	1-1x6	1-1x6	1-1x8	1-1x8
	.....	....	3'-0"	3'-0"	3'-0"	3'-0"
Total Board Feet	57 5/6	64 5/6	77 1/2	84 1/6	84 2/3	94 2/3
Nails,						
12d	100	112	110	110	124	124
8d	26	18	42	44	34	34

Material Requirements for One Scissors Type Plywood Gusset Truss Rafter of Various Spans for a 5 in 12 Roof Slope and 2 in 12 Ceiling Slope

Span	18'-0"	20'-0"	24'-0"	26'-0"	28'-0"	30'-0"
Overhang	2"	2"	3 3/8"	3 3/8"	3 3/8"	3 3/8"
Top Chords	2-2x4	2-2x4	2-2x4	2-2x6	2-2x6	2-2x6
	10'-0"	12'-0"	14'-0"	16'-0"	16'-0"	18'-0"
Bottom Chords	2-2x4	2-2x4	2-2x4	2-2x4	2-2x6	2-2x6
& Compr. Diag	12'-0"	14'-0"	18'-0"	18'-0"	20'-0"	22'-0"

Vert. Member	1-1x8	1-1x8	1-1x8	1-1x8	1-1x8	1-1x8
	3'-0"	3'-0"	4'-0"	4'-0"	4'-0"	5'-0"
Splice Plates	4-1x4	4-1x4	4-1x4	4-1x4	4-1x6	4-1x6
	2'-6"	2'-6"	3'-0"	3'-0"	3'-0"	3'-0"
	2-1x4	2-1x4	2-1x4	2-1x4	3-1x6	3-1x6
	1'-6"	1'-6"	1'-6"	1'-6"	2'-0"	2'-0"
	1-1x6	1-1x6	1-1x6	1-1x6	1-1x6	1-1x6
	2'-0"	2'-0"	2'-0"	2'-0"	4'-0"	4'-0"
	1-1x6	1-1x6	1-1x6	1-1x6	....	....
	3'-0"	3'-0"	4'-0"	4'-0"	....	....
Total Board Feet	38 1/6	43 1/2	53 1/3	66 2/3	85 2/3	94 1/3
Nails,						
12d	63	63	78	80	104	109
8d	106	106	115	120	14	147

Plywood Gusset Kingpost Truss. PG kingpost trusses are the result of research conducted at Virginia Polytechnic Institute under the sponsorship of the Independent Nail & Packing Company of Bridgewater, Mass. These studies have brought about a new era in nail technology and wood construction.

Test data are available on request from the V.P.I. Wood Research Laboratory at Blacksburg, Va. Samples of the hardened-steel spiral-thread Hi-Load nails used to assemble these plywood gusset trusses can be obtained from the manufacturer.

Material Requirements for Plywood Gusset Kingpost Truss
with 1/2" Plywood Gusset and Splice Plates

Span	20'-8"	24'-8"	28'-8"	32'-8"	36'-5 1/2"	40'-0"
Roof Slope	2/12	2/12	3/12	2/12	5/12	4/12
Overhang	19"	18"	16"	18"	5 1/2"	12"
Top Chord, 2 ea.	2x4x12'	2x4x14'	2x4x16'	2x6x18'	2x6x20'	2x6x22'
Bottom Chord, 2 ea.	2x4x10'	2x4x12'	2x4x14'	2x6x16'	2x6x18'	2x6x20'
Post, 1 or 2 ea.	2x4x2'	2x4x2'3 5/8"	2x4x3'10 1/2"	2x4x3'1 7/8	2x6x8'	2-2x4x6'
Heel Gusset, 4 ea.	2'-9"x1'3/4"	4'x1'-3 1/4"	3'x1'-6 1/4"	4'x1'-7"	4'x2'-7"	4'x2'-3 1/2"
Ridge Gusset, 2 ea.	2'-9"x7"	4'x7 5/8"	3'x10 1/8"	5'x1'-0"	4'x1'-4"	4'x1'-5 1/4"
Center Splice, 2 ea.	3'x4"	4'x4"	4'x4"	4'x6"	4'x5 3/4"	4'x9"
Lumber, b.f.	31 1/3	36	55	71	100	109
Plywood, sq. ft.	16	24	22	36	43	44
(2 1/2"x0.135")						
Hardened Steel Nos.	140	168	218	304	308	342
Spiral Thread	1.39	1.66	2.16	3.01	3.05	3.40
Hi-Load Nails, lbs.						

Labor Required to Fabricate and Assemble Plywood Gusset Trusses. The labor required to fabricate and assemble PGT will vary with the scope of the job, number of truss units, production methods used, and the size of truss units.

On an average one-house job, two carpenters will require about 6 hours to make a layout on the subfloor, place jig stops to guide assembly work, and construct a pilot truss with temporary connections, at the following labor cost:

	Hours	Rate	Total	Rate	Total
Carpenter	12	$....	$....	$24.04	$288.48

For the balance of the truss units, a 2-worker crew, using portable power equipment for cutting, drilling, etc., should fabricate, assemble, and stockpile

10 to 12 24' to 28' span trussed rafters per 8-hr. day, at the following labor cost per truss:

	Hours	Rate	Total	Rate	Total
Carpenter	1.5	$....	$....	$24.04	$36.06

For larger spans up to 36', add about 5% to above costs. For smaller spans, reduce costs about 5%.

On large projects, containing several thousand PGT units, where the expense of setting up a shop for fabrication and assembly is warranted, a 2-worker crew should complete 20 to 24 units per 8-hr. day, at the following labor cost per truss:

	Hours	Rate	Total	Rate	Total
Carpenter	.75	$....	$....	$24.04	$18.03

To the above, add equipment charges and cost of setting up and dismantling fabrication shop. The above costs may be applied to either bolted trussed rafters or nailed trussed rafters.

Labor Erecting Trusses. Erection of trusses for one-story structures can usually be accomplished with a 4-worker crew: 2 laborers carrying units from stockpile to setting location, placing units in an inverted position on the bearing wall plates, and tilting up into an erect position ready for 2 carpenters to anchor and brace in the final position. For efficient operation, the layout of truss locations should be marked on the wall plates and metal framing anchors set in place before erection is started. The 4-worker crew should lay out, set anchors, and erect into place 22 to 26 trussed rafters, up to 28'-0" span, per 8-hr. day, at the following labor cost per truss:

	Hours	Rate	Total	Rate	Total
Carpenter	0.67	$....	$....	$24.04	$16.11
Labor	0.67	....	....	19.13	12.82
Cost per truss			$....		$28.93

For spans greater than 28'-0", it might be necessary to add an extra worker to the crew.

Labor Framing Lumber in Building Construction

The labor quantities and costs given in the following pages are intended to include all classes of frame construction, such as residences, ranch houses, barns, stables, apartment buildings, combined store and apartment buildings, country clubs, schools, and in fact, all buildings that are constructed entirely of wood or have brick, stone, tile or cement block walls and wood joists, rafters, stud walls, partitions, wood subfloors, wall and roof sheathing, finish wood floors, etc.

Costs are given on two classes of workmanship. The classification "ordinary workmanship" covers the grade of workmanship encountered in most buildings, where price is a factor. "First grade workmanship" is where quality is a factor as well as price, usually found in high-grade residences, apartments, hotel and office buildings, public buildings, high school, and college and university buildings.

The estimator must use his or her own judgment as to the grade of workmanship required, depending upon the type of building, the reputation of the architect, and the requirements of the specifications.

Another thing that is going to govern labor costs is the kind of equipment used. A contractor who uses electric handsaws and drills will easily outdistancing competitors who are still working on the old handsaw basis.

All costs given on the following pages are based on using an electric saw for cutting all joists, studs and rafters to length, cutting off ends of subflooring, roof sheathing, etc. If an ordinary handsaw is used for this purpose, add 1 to 1 ½ hrs. carpenter time per 1,000 b.f.

All of the labor quantities and costs on the following pages are based on laborers or helpers handling and carrying the lumber from the stock piles or benches where the lumber is being cut to length to the building. If the lumber is handled and carried by carpenters, figure labor time at carpenter wages.

Framing and Placing Foundation Wall Plates. Where 2"x4" or 2"x6" wood plates are placed on top of foundation walls or concrete slab to receive the floor joists and exterior wall studs, it is customary to place anchor bolts in the walls and floors and the plates are then bored to receive the bolts, the plates are placed on the wall (and usually wedged with shingles) ready to receive the joists and exterior wall studs. Special purpose nails may be used for anchoring wall plates to concrete.

Where just an ordinary grade of workmanship is required, two carpenters working together should handle, frame and place 225 to 275 lin. ft. of 2"x4" or 2"x6" plates per 8-hr. day, at the following cost per 100 lin. ft.:

	Hours	Rate	Total	Rate	Total
Carpenter	6.4	$....	$....	$24.04	$153.86
Cost per lin. ft.			....		1.54

First Grade Workmanship

On jobs where first grade workmanship is required, with the foundation wall plates drilled for bolts, plates set and bedded absolutely level in a bed of cement mortar, two carpenters working together should handle, frame, and place 175 to 225 lin. ft. of 2"x4" or 2"x6" wall plates per 8-hr. day at the following labor cost per 100 lin. ft.:

	Hours	Rate	Total	Rate	Total
Carpenter	8	$....	$....	$24.04	$192.32
Cost per lin. ft.			....		1.92

Framing and Placing Box Sills and Plates. On platform framing, where a wood box sill and plate is formed by using a 2"x4" or 2"x6" plate and a 2"x8" or 2"x10" on the end of the joists to form a box sill, the wall plate should be drilled, leveled, and set the same as described above, and the side or end piece is nailed after the joists are set.

This work should be estimated as given above for "Foundation Wall Plates" and the 2"x8" or 2"x10" end pieces should be figured in with the floor joists.

Framing and Erecting Exterior Stud Walls for Frame Buildings. The labor cost of framing and erecting stud walls is subject to wide variation, depending upon the type of building, height, regularity of the walls, etc.

The framing on square or rectangular buildings, such as Cape Cod and colonial type houses and buildings of similar shape and construction will cost much less to frame than English type houses, having walls of irregular shape and height.

When you consider the average house requires only 1,500 to 2,500 ft. of lumber for outside stud walls, it shows just how much labor is required to frame it.

On square or rectangular shaped buildings, such as colonial, Georgian houses, etc., a carpenter should frame and erect 350 to 400 b.f. of lumber per 8-hr. day, at the following labor cost per 1,000 b.f.:

	Hours	Rate	Total	Rate	Total
Carpenter	21.4	$....	$....	$24.04	$514.46
Labor	6	....	....	19.13	114.78
Cost 1,000 b.f.			$....		$629.24

On English type and other buildings having irregular wall construction, a carpenter should frame and erect 250 to 300 b.f. of lumber per 8-hr. day, at the following labor cost per 1,000 b.f.:

	Hours	Rate	Total	Rate	Total
Carpenter	29.1	$....	$....	$24.04	$699.56
Labor	6	....	....	19.13	114.78
Cost 1,000 b.f.			$....		$814.34

First Grade Workmanship

In high class wood constructed buildings, where every precaution is taken to prevent settling due to shrinkage, where the wood studs rest on masonry walls or steel I beams and are not set on top of the floor joists, and where it is necessary to bridge or truss between all studs and over door and window openings, a carpenter should frame and erect 225 to 275 b.f. of lumber per 8-hr. day, on square or rectangular type buildings, at the following labor cost per 1,000 b.f.:

	Hours	Rate	Total	Rate	Total
Carpenter	32	$....	$....	$24.04	$769.28
Labor	10	....	....	19.13	191.30
Cost 1,000 b.f.			$....		$960.58

On English type and other buildings having irregular wall construction, a carpenter should frame and erect 150 to 200 b.f. of lumber per 8-hr day, at the following labor cost per 1,000 ft.:

	Hours	Rate	Total	Rate	Total
Carpenter	45.7	$....	$....	$24.04	$1,098.63
Labor	10	....	....	19.13	191.30
Cost 1,000 b.f.			$....		$1,289.93

Framing Interior Stud Partitions. When framing and setting interior stud partitions set on top of rough wood floors that require just the ordinary amount of framing for door openings, a carpenter should frame and erect 375 to 425 b.f. per 8-hr. day, at the following labor cost per 1,000 b.f.:

	Hours	Rate	Total	Rate	Total
Carpenter	20	$....	$....	$24.04	$480.80
Labor	6	....	....	19.13	114.78
Cost 1,000 b.f.			$....		$595.58

First Grade Workmanship

Where the wood partition studs are set on masonry walls or steel I beams instead of the wood subfloor, and where it is necessary to brace between studs and truss over all door openings, a carpenter should frame and erect 250 to 300 b.f. of lumber per 8-hr. day, at the following labor cost per 1,000 b.f.:

	Hours	Rate	Total	Rate	Total
Carpenter	29.1	$....	$....	$24.04	$699.56
Labor	8.0	....	....	19.13	153.04
Costs 1,000 b.f.			$....		$852.60

Framing and Setting Floor Joists. When framing and placing wood floor joists up to 2"x8" in buildings of regular construction, a carpenter should frame and place 550 to 600 b.f. per 8-hr. day, at the following labor cost per 1,000 b.f.:

	Hours	Rate	Total	Rate	Total
Carpenter	13.9	$....	$....	$24.04	$334.16
Labor	6	....	....	19.13	114.78
Cost 1,000 b.f.			$....		$448.94

If 2"x10" or 2"x12" joists are used, a carpenter should frame and erect 600 to 650 b.f. per 8-hr. day, at the following labor cost per 1,000 b.f.:

	Hours	Rate	Total	Rate	Total
Carpenter	12.8	$....	$....	$24.04	$307.71
Labor	6	....	....	19.13	114.78
Cost 1,000 b.f.			$....		$422.49

First Grade Workmanship

On jobs where the wood joists must be set with the crowning edge up and it is not permissible to block up under the joists with shingles or wood wedges, a carpenter should frame and erect 500 to 550 b.f. (sizes up to 2"x8") per 8-hr. day, at the following labor cost per 1,000 b.f.:

	Hours	Rate	Total	Rate	Total
Carpenter	15.2	$....	$....	$24.04	$365.41
Labor	4.5	....	....	19.13	86.09
Cost 1,000 b.f.			$....		$451.50

Using 2"x10" or 2"x12" joists, a carpenter should frame and erect 550 to 650 b.f. of lumber per 8-hr. day, at the following labor cost per 1,000 b.f.:

	Hours	Rate	Total	Rate	Total
Carpenter	13.9	$....	$....	$24.04	$334.16
Labor	4.0	....	....	19.13	76.52
Cost 1,000 b.f.			$....		$410.68

Framing Panel and Girder Floor Systems. When framing and placing 4"x6" wood beams and 2"x4" spacers to form 4-ft. square grids for panel and girder floor systems, 2 carpenters and a helper should frame and place 1,500 to 1,700 b.f. of lumber per 8-hr. day, at the following labor cost per 1,000 b.f.:

	Hours	Rate	Total	Rate	Total
Carpenter	10	$....	$....	$24.04	$240.40
Labor	5	....	....	19.13	95.65
Cost per 1,000 b.f.			$....		$336.05

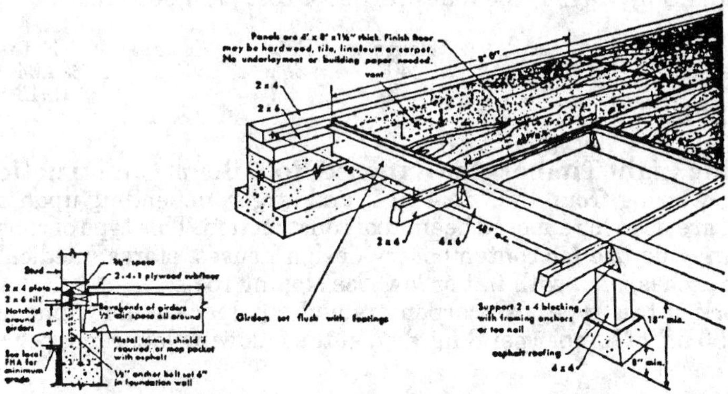

Method of Framing Panel and Girder Floor Systems

Installing Cross Bridging. Under average conditions a carpenter should cut and place 90 to 110 sets (2 pieces) of crossbridging per 8-hr. day at the following labor cost per 100 sets:

	Hours	Rate	Total	Rate	Total
Carpenter	8	$....	$....	$24.04	$192.32
Cost per set			....		1.92

Framing and Erecting Rafters for Gable Roofs. When framing and erecting rafters for plain double pitch or gable roofs, without dormers or gables, and where 2"x6" or 2"x8" lumber is used, a carpenter should frame and erect 285 to 335 b.f. per 8-hr. day, at the following labor cost per 1,000 b.f.:

	Hours	Rate	Total	Rate	Total
Carpenter	24.6	$....	$....	$24.04	$591.38
Labor	7.0	....	....	19.13	133.91
Cost 1,000 b.f.			$....		$725.29

On buildings having double pitch or gable roofs, cut up with dormers and gables framing into the main roof, a carpenter should frame and erect 250 to 300 b.f. of lumber per 8-hr. day, at the following labor cost per 1,000 b.f.:

	Hours	Rate	Total	Rate	Total
Carpenter	29.1	$....	$....	$24.04	$699.56
Labor	8.0	....	....	19.13	153.04
Cost 1,000 b.f.			$....		$852.60

Framing and Erecting Rafters for Hip Roofs. When framing and erecting plain hip roofs, without dormers or gables framing into the main roof, a carpenter should frame and erect 250 to 300 b.f. of lumber per 8-hr. day, at the following labor cost per 1,000 b.f.:

	Hours	Rate	Total	Rate	Total
Carpenter	29.1	$....	$....	$24.04	$699.56
Labor	8.0	....	....	19.13	153.04
Cost 1,000 b.f.			$....		$852.60

On difficult constructed hip roofs, where it is necessary to frame for dormers, gables, valleys, etc., a carpenter should frame and erect 180 to 220 b.f. of lumber per 8-hr. day, at the following labor cost per 1,000 b.f.:

	Hours	Rate	Total	Rate	Total
Carpenter	40	$....	$....	$24.04	$961.60
Labor	8	....	....	19.13	153.04
Cost 1,000 b.f.			$....		$1,114.64

Framing Light Timbers for Exposed Roof Beam Construction. Light timbers, varying from 4"x6" to 4"x14" and larger, depending upon span and spacing, are used for exposed beam roof construction. This type of construction is now in wide use for contemporary design houses, stores, medical groups, small churches, etc., with flat or low rise sloping roofs.

On work of this type, two carpenters and a helper should frame and place 750 to 850 b.f. of lumber per 8-hr. day, at the following labor cost per 1,000 b.f.:

	Hours	Rate	Total	Rate	Total
Carpenter	20	$....	$....	$24.04	$480.80
Labor	10	....	....	19.13	191.30
Cost per 1,000 b.f.			$....		$672.10

Framing for Roof Saddles on Flat Roofs. On flat roofs where it is necessary to frame saddles to pitch the water toward the drains, and where the saddles are built up of 2"x4" framing covered with 1" sheathing, a carpenter should frame and erect 400 to 450 b.f. of lumber per 8-hr. day, at the following labor cost per 1,000 b.f.:

	Hours	Rate	Total	Rate	Total
Carpenter	19	$....	$....	$24.04	$456.76
Labor	6	....	....	19.13	114.78
Cost per 1,000 b.f.			$....		$571.54

Placing Wood Cant Strips. When cant strips, diagonally cut from 4"x4" or 6"x6" lumber, occur in long runs with few breaks a carpenter should place 475 to 525 lin. ft. per 8-hr. day at the following labor cost per 100 lin. ft:

	Hours	Rate	Total	Rate	Total
Carpenter	1.6	$....	$....	$24.04	$38.46
Cost per lin. ft.			....		.38

If cant strips occur in short runs, such as around roof opening curbs, or if the runs are cut up with pilasters and other breaks, a carpenter should place 300 to 325 lin. ft. per 8-hr. day at the following labor cost per 100 lin. ft.:

	Hours	Rate	Total	Rate	Total
Carpenter	2.6	$....	$....	$24.04	$62.50
Cost per lin. ft.			....		.63

Laying Rough Wood Floors. When laying 1"x6" or 1"x8" rough wood subflooring, a carpenter should lay 825 to 925 b.f. per 8-hr. day, at the following labor cost per 1,000 b.f.:

	Hours	Rate	Total	Rate	Total
Carpenter	9.2	$....	$....	$24.04	$221.17
Labor	5.3	....	....	19.13	101.39
Cost 1,000 b.f.			$....		$322.56

Laying Rough Wood Floors Diagonally. When laying rough wood subflooring diagonally, making it necessary to cut both ends of the flooring on a bevel, a carpenter should lay 650 to 750 b.f. of flooring per 8-hr. day, at the following labor cost per 1,000 b.f.:

	Hours	Rate	Total	Rate	Total
Carpenter	11.5	$....	$....	$24.04	$276.46
Labor	5.5	....	....	19.13	105.22
Cost 1,000 b.f.			$....		$381.68

Laying Wood Sheathing on Flat Roofs. Figure same as given above for Rough Wood Floors.

Labor Laying Plywood Subflooring. The labor cost of handling and laying plywood subflooring can vary widely, depending upon the size of the floor area, the regularity of the joist spacing and upon the method of laying.

When plywood is used as subflooring, nailed directly to the joists and with the finish floor applied directly to the plywood, all edges (both longitudinal and crosswise) should be nailed about 6" on centers. Otherwise, the finish floor may develop squeaks caused by deflection of the long edges of the sheets when walked upon.

In other words, with the joists spaced 16" on centers, it is necessary to cut pieces of 2"x4" approximately 14 3/8" long and the ends of these 2"x4"s must be nailed into the sides of the joists to provide nailing for the long edges of the plywood sheets. As the plywood is usually 4'-0" wide, it means a row of 2"x4" blocking must be run at right angles to the floor joists every 4'-0" apart to provide nailing for the long edges of the plywood sheets. This requires a lot of extra work and costs a lot of money.

In buildings with regular spans, where an underlayment is to be provided, and where the long edges of the plywood sheets are not nailed, two carpenters working together should handle, fit, lay and nail 52 to 60 sheets (1,664 to 1,920 sq. ft.) of 4'-0"x8'-0" plywood 5/8" to 3/4" thick per 8-hr. day, at the following labor cost per 100 sq. ft.:

	Hours	Rate	Total	Rate	Total
Carpenter	0.9	$....	$....	$24.04	$21.64
Labor	0.3	....	....	19.13	5.74
Cost per 100 sq. ft.			$....		$27.38
Cost per sq. ft.			....		.27

If 2"x4" wood blocking is required between the joists, as described above, and the plywood sheets must be nailed both longitudinally and crosswise, two carpenters working together should cut and nail blocks, and handle, fit, lay and nail 26 to 30 sheets of 4'-0"6x8'-0" plywood (832 to 960 sq. ft.) 5/8" to 3/4" thick per 8-hr. day, at the following labor cost per 100 sq. ft.:

	Hours	Rate	Total	Rate	Total
Carpenter	1.8	$....	$....	$24.04	$43.27
Labor	0.5	....	....	19.13	9.57
Cost per 100 sq. ft.			$....		$52.84
Cost per sq. ft.			....		.53

Laying Plywood Decking for Panel and Girder Floor Systems. Plywood decking for panel and girder floor construction should be 1" or 1 ¼" thick and laid with staggered joints. To minimize waste, inside dimensions of building should be laid out on a 4-ft. module, which also reduces cutting of panels to starters in every other row. In some localities, 4'x4' plywood panels are available as a stock size.

On large projects, where the work may be highly organized, a carpenter and 2 helpers should lay 4,500 to 4,700 sq. ft. of plywood decking per 8-hr. day, or enough for 3 to 4 average one-story houses, at the following labor cost per 1,000 sq. ft.:

	Hours	Rate	Total	Rate	Total
Carpenter	1.75	$....	$....	$24.04	$42.07
Labor	3.50	....	....	19.13	66.96
Cost per 1,000 sq. ft.			$....		$109.03
Cost per sq. ft.			....		.11

For smaller jobs, where only 1 or 2 houses are involved, increase above costs about 25% to absorb lost motion in laying out work and getting started.

Where plywood decking is to be covered with linoleum or other resilient flooring, requiring a fairly smooth surface, a carpenter should spackle and sand joints, surface cracks and other imperfections at the rate of 200 sq. ft. per hr.

Laying Wood Sheathing on Pitch or Gable Roofs. When laying roof sheathing on plain hip or gable roofs, a carpenter should lay 575 to 615 b.f. of lumber per 8-hr. day, at the following labor cost per 1,000 b.f.:

	Hours	Rate	Total	Rate	Total
Carpenter	13.5	$....	$....	$24.04	$324.54
Labor	6.5	....	....	19.13	124.35
Cost 1,000 b.f.			$....		$448.89

On very steep roofs and roofs cut up with dormers, hips, valleys, etc., such as English type houses, etc., a carpenter will lay only 275 to 325 b.f. of sheathing per 8-hr. day at the following labor cost per 1,000 b.f.:

	Hours	Rate	Total	Rate	Total
Carpenter	26.5	$....	$....	$24.04	$637.06
Labor	7.5	....	....	19.13	143.48
Cost 1,000 b.f.			$....		$780.54

Labor Placing Sidewall Sheathing. When sheathing sidewalls of frame buildings with 1"x6" or 1"x8" boards laid horizontally, a carpenter should handle and place 625 to 700 b.f. per 8-hr. day, at the following labor cost per 1,000 b.f.:

	Hours	Rate	Total	Rate	Total
Carpenter	12	$....	$....	$24.04	$288.48
Labor	5	....	....	19.13	95.65
Cost 1,000 b.f.			$....		$384.13

Labor Placing Diagonal Sidewall Sheathing. When sidewall sheathing is placed diagonally, it usually requires two carpenters working together--one at each end of the board. It also usually requires scaffolding the outside walls before the sheathing is placed, as the man at the top must have a scaffold from which to work.

On work of this class, a carpenter should place 450 to 525 b.f. of 1"x6" or 1"x8" sheathing per 8 hr. day, at the following labor cost per 1,000 b.f.:

	Hours	Rate	Total	Rate	Total
Carpenter	16.5	$....	$....	$24.04	$396.66
Labor	5.5	....	....	19.13	105.22
Cost 1,000 b.f.			$....		$501.88

WOOD BLOCKING, FURRING AND GROUNDS

Placing Wood Furring Strips on Masonry Walls. Where it is necessary to place wood furring strips on brick or tile walls before lathing and plastering, allowing the strips to follow the line of the walls without wedging or blocking out to make them straight or plumb, with the nails driven into dry joints in the brickwork, a carpenter should place 500 to 550 lin. ft. of furring per 8-hr. day, at the following labor cost per 100 lin. ft.:

	Hours	Rate	Total	Rate	Total
Carpenter	1.5	$....	$....	$24.04	$36.06
Cost per lin. ft.			....		.36

Cost per Square (100 Sq. Ft.)

Strips 12" centers	1.7	$....	$....	$24.04	$40.87
Strips 16" centers	1.3	....	....	24.04	31.25

First Grade Workmanship

Where it is necessary to plug the masonry walls and place all furring strips absolutely straight and plumb to produce a level surface to receive lath and plaster, a carpenter should plug walls and place 200 to 250 lin. ft. of strips per 8-hr. day, at the following labor cost per 100 lin. ft.:

	Hours	Rate	Total	Rate	Total
Carpenter	3.5	$....	$....	$24.04	$84.14
Cost per lin. ft.			....		.84

Cost per Square (100 Sq. Ft.)

Strips 12" centers	3.9	$....	$....	$24.04	$93.76
Strips 16" centers	2.9	....	....	24.04	69.72

Placing Wood Grounds. Where wood grounds are nailed direct to wood furring strips, door and window openings, etc., allowing them to follow the rough furring, partitions and wood bucks, without wedging or blocking to make them absolutely straight, a carpenter should place 540 to 590 lin. ft. per 8-hr. day, at the following labor cost per 100 lin. ft.:

	Hours	Rate	Total	Rate	Total
Carpenter	1.4	$....	$....	$24.10	$33.66
Cost per lin. ft.			....		.34

First Grade Workmanship

Where it is necessary to keep all wood grounds absolutely straight, plumb and level for the finish plaster and to receive the interior wood finish, and where it is necessary to plug the masonry walls with wood plugs to hold the nails, a carpenter should place 200 to 250 lin. ft. of grounds per 8-hr. day, at the following labor cost per 100 lin. ft.:

	Hours	Rate	Total	Rate	Total
Carpenter	3.5	$....	$....	$24.04	$84.14
Cost per lin. ft.			....		.84

Wood Blocking and Grounds Around Window Openings. Where the masonry walls are 12" or more in thickness, requiring a wood jamb lining, a carpenter should place all necessary wood blocking and grounds for a window up to 4'-0"x6'-0" in size in 3/4 to 1 1/4 hr. at the following labor cost:

	Hours	Rate	Total	Rate	Total
Carpenter	1	$....	$....	$24.04	$24.04

Windows 4'-0"x7'-0" to 5'-0"x8'-0" in size will require 1 1/8 to 1 3/8 hrs. to place blocking and grounds, at the following labor cost per window:

Carpenter	1.25	$....	$....	$24.04	$30.05

First Grade Workmanship

In the better class of buildings where all grounds must be absolutely straight and plumb and it is necessary to plug the brick and block walls for securing the grounds and blocking, a carpenter should complete one window up to 4'-0"x6'-0" in size in 1 1/4 to 1 3/4 hrs. at the following labor cost:

	Hours	Rate	Total	Rate	Total
Carpenter	1.5	$....	$....	$24.04	$36.06

A carpenter should plug walls, place blocking and grounds for one window 4'-0"x7'-0" to 5'-0"x8'-0" in 1 1/2 to 2 hrs. at the following labor cost per window:

	Hours	Rate	Total	Rate	Total
Carpenter	1.75	$....	$....	$24.04	$42.07

Wood Blocking for Cabinet and Case Bases. A carpenter should locate, frame and set 115 to 135 b.f. of 2"x4" or 2"x6" blocking for cabinet or case bases per 8-hr. day at the following labor cost per 1,000 b.f.:

	Hours	Rate	Total	Rate	Total
Carpenter	64	$....	$....	$24.04	$1,538.56

Wood Furring Strips Over Wood Subfloors. When placing 1"x2" or 2"x2" wood furring strips over wood subfloors to receive finish flooring, a carpenter should place 550 to 600 lin. ft. (without wedging or leveling) per 8-hr. day, at the following labor cost per 100 lin. ft.:

	Hours	Rate	Total	Rate	Total
Carpenter	1.4	$....	$....	$24.04	$33.66
Labor	0.3	....	....	19.13	5.74
Cost per 100 lin. ft.			$....		$39.40
Cost per lin. ft.			....		.39

Cost per Square (100 Sq. Ft.)

Strips 12" centers	1.9	$....	$....	$24.04	$45.68
Strips 16" centers	1.4	....	....	24.04	33.66

First Grade Workmanship

In the better class of buildings where the wood furring strips must be wedged up and blocked to produce an absolutely level surface to receive finish flooring, a carpenter should place 250 to 300 lin. ft. of furring strips per 8-hr. day, at the following cost per 100 lin. ft.:

	Hours	Rate	Total	Rate	Total
Carpenter	3.0	$....	$....	$24.04	$72.12
Labor	0.3	....	....	19.13	5.74
Cost per 100 lin. ft.			$....		$77.86
Cost per lin. ft.			....		.79

Cost per Square (100 Sq. Ft.)

Strips 12" centers	3.6	$....	$....	$24.04	$86.54
Strips 16" centers	2.8	....	....	24.04	67.31

Placing Strip Deadening Felt and Wood Furring Strips Over Wood Subfloors. Frequently a narrow strip of deadening felt ½"x3" is placed under the wood furring strips and both are nailed to the wood subfloor. This provides better insulation and deadening than where the furring strips are nailed directly to the subfloor.

On work of this kind, a carpenter should place 425 to 475 lin. ft. of deadening felt and furring strips per 8-hr. day, at the following labor cost per 100 lin. ft.:

	Hours	Rate	Total	Rate	Total
Carpenter	1.8	$....	$....	$24.04	$43.27
Labor	0.4	....	....	19.13	7.65
Cost 100 lin. ft.			$....		$50.92
Cost per lin. ft.			....		.51

Cost per Square (100 Sq. Ft.)

Strips 12" centers	2.4	$....	$....	$24.04	$57.70
Strips 16" centers	1.9	....	....	24.04	45.68

First Grade Workmanship

Where it is necessary to wedge and block up the strips to produce an absolutely level surface to receive the finish flooring, a carpenter should place 220 to 250 lin. ft. of deadening and furring strips per 8-hr. day, at the following labor cost per 100 lin. ft.:

	Hours	Rate	Total	Rate	Total
Carpenter	3.5	$....	$....	$24.04	$84.14
Labor	0.4	....	....	19.13	7.65
Cost 100 lin. ft.			$....		$91.79
Cost per lin. ft.		....			.92

Cost per Square (100 Sq. Ft.)

Strips 12" centers	4.3	$....	$....	$24.04	$103.37
Strips 16" centers	3.3	....	....	24.04	79.33

Placing Deadening Quilt Over Rough Wood Floors. Where deadening felt or quilt is laid over wood subfloors, a man should handle and lay 200 to 250 sq. ft. of felt an hr. at the following labor cost per 100 sq. ft.:

	Hours	Rate	Total	Rate	Total
Labor	0.5	$....	$....	$19.13	$9.57
Cost per sq. ft.			....		.10

The above costs are based on deadening felts or quilts in rolls or sheets that can be laid on the rough floor and the furring strips placed directly over them.

Cost of Placing Deadening Felt and Wood Furring Strips Over 100 Sq. Ft. of Floor
Strips 12" on Centers--Ordinary Workmanship

	Hours	Rate	Total	Rate	Total
Carpenter	2.0	$....	$....	$24.04	$48.08
Labor	0.3	....	....	19.13	5.74
Cost 100 sq. ft.			$....		$53.82
Cost per sq. ft.			....		.54

Strips 16" on Centers--Ordinary Workmanship

	Hours	Rate	Total	Rate	Total
Carpenter	1.6	$....	$....	$24.04	$38.46
Labor	0.3	....	....	19.13	5.74
Cost 100 sq. ft.			$....		$44.20
Cost per sq. ft.			....		.44

Strips 12" on Centers--First Grade Workmanship

	Hours	Rate	Total	Rate	Total
Carpenter	3.8	$....	$....	$24.04	$91.35
Labor	0.3	....	....	19.13	5.74
Cost 100 sq. ft.			$....		$97.09
Cost per sq. ft.			....		.97

Strips 16" on Center--First Grade Workmanship

	Hours	Rate	Total	Rate	Total
Carpenter	3.0	$....	$....	$24.04	$72.12
Labor	0.3	....	....	19.13	5.74
Cost 100 sq. ft.			$....		$77.86
Cost per sq. ft.			....		.78

Placing Wood Floor Sleepers. When placing 2"x3" or 2"x4" wood floor screeds or sleepers over rough tile or concrete floors, to receive finish flooring, a carpenter should place 225 to 275 lin. ft. per 8-hr. day, at the following labor cost per 100 lin. ft.:

	Hours	Rate	Total	Rate	Total
Carpenter	3.2	$....	$....	$24.04	$76.93
Labor	0.8	....	....	19.13	15.30
Cost 100 lin. ft.			$....		$92.23
Cost per lin. ft.			....		.92

First Grade Workmanship

In the better class of buildings, 2"x3" or 2"x4" beveled floor sleepers are placed over the rough concrete floors and wedged or blocked up to provide a perfectly level surface to receive the finish flooring.

The screeds are usually held in place by metal clips placed in the rough concrete, or anchored with special purpose nails.

On work of this class, a carpenter should place, wedge up, and level 130 to 170 lin. ft. of sleepers per 8-hr. day, including setting sleeper clips, at the following labor cost per 100 lin. ft.:

	Hours	Rate	Total	Rate	Total
Carpenter	5.3	$....	$....	$24.04	$127.41
Labor	0.8	....	....	19.13	15.30
Cost per 100 lin. ft.			$....		$142.71
Cost per lin. ft.			....		1.43

Lumber Required for Wood Door Bucks of 2"x4" Lumber

Size of Opening	Lin. Ft. of Lumber Req'd.	Board Feet of Lumber Req'd.	Transom Add Lineal Ft.	BF of Lumber
2'-0"x4'-0"	12	8	6'-0"	4
2'-0"x4'-6"	12	8	6'-0"	4
2'-0"x5'-0"	13 ½	9	6'-0"	4
2'-6"x4'-0"	12	8	6'-6"	4 ⅓
2'-6"x4'-6"	13	8 ⅔	6'-6"	4 ⅓
2'-6"x5'-0"	14	9 ⅓	6'-6"	4 ⅓
2'-6"x5'-6"	15	10	6'-6"	4 ⅓
2'-6"x6'-0"	16	10 ⅔	6'-6"	4 ⅓
2'-6"x6'-6"	17	11 ⅓	6'-6"	4 ⅓
2'-6"x7'-0"	18	12	6'-6"	4 ⅓
2'-8"x6'-8"	17	11 ⅓	7'-0"	4 ⅔
2'-8"x7'-0"	18	12	7'-0"	4 ⅔
3'-0"x7'-0"	19	12 ⅔	7'-0"	4 ⅔
3'-6"x7'-0"	19	12 ⅔	7'-6"	5
4'-0"x7'-0"	20	13 ⅓	8'-0"	5 ⅓
5'-0"x7'-0"	21	14	9'-0"	6
6'-0"x7'-0"	22	14 ⅔	10'-0"	6 ⅔
7'-0"x7'-0"	24	16	11'-0"	7 ⅓
3'-6"x7'-6"	20	13 ⅓	7'-6"	5
4'-0"x7'-6"	20	13 ⅓	8'-0"	5 ⅓
4'-6"x7'-6"	21	14	8'-6"	5 ⅔
5'-0"x7'-6"	21	14	9'-0"	6
6'-0"x7'-6"	22	14 ⅔	10'-0"	6 ⅔
7'-0"x7'-6"	24	16	11'-0"	7 ⅓
3'-6"x8'-0"	21	14	7'-6"	5
4'-0"x8'-0"	22	14 ⅔	8'-0"	5 ⅓
5'-0"x8'-0"	22	14 ⅔	9'-0"	6
6'-0"x8'-0"	23	15 ⅔	10'-0"	6 ⅓
7'-0"x8'-0"	24	16	11'-0"	7 ⅓
8'-0"x8'-0"	25	16 ⅔	12'-0"	8
9'-0"x9'-0"	28	18 ⅔	9'-0"	6

The lumber quantities given in the table are estimated in lengths that cut with the least waste.

Lumber Required for Wood Door Bucks of 2"x6" Lumber

Size of Opening	Lin. Ft. of Lumber Req'd.	Board Ft. of Lumber Req'd.	Transom Add Lineal Ft.	BF of Lumber
2'-0"x4'-0"	12	12	6'-0"	6
2'-0"x4'-6"	12	12	6'-0"	6
2'-0"x5'-0"	13 ½	13 ½	6'-0"	6
2'-6"x4'-0"	12	12	6'-6"	6 ½
2'-6"x4'-6"	13	13	6'-6"	6 ½
2'-6"x5'-0"	14	14	6'-6"	6 ½
2'-6"x5'-6"	15	15	6'-6"	6 ½
2'-6"x6'-0"	16	16	6'-6"	6 ½
2'-6"x6'-6"	17	17	6'-6"	6 ½
2'-6"x7'-0"	18	18	6'-6"	6 ½
2'-8"x6'-8"	17	17	7'-0"	7
3'-0"x7'-0"	19	19	7'-0"	7
3'-6"x7'-0"	19	19	7'-6"	7 ½
4'-0"x7'-0"	20	20	8'-0"	8
5'-0"x7'-0"	21	21	9'-0"	9
6'-0"x7'-0"	22	22	10'-0"	10
7'-0"x7'-0"	24	24	11'-0"	11
3'-6"x7'-6"	20	20	7'-6"	7 ½
4'-0"x7'-6"	20	20	8'-0"	8
4'-6"x7'-6"	21	21	8'-6"	8 ½
5'-0"x7'-6"	21	21	9'-0"	9
6'-0"x7'-6"	22	22	10'-0"	10
7'-0"x7'-6"	24	24	11'-0"	11
3'-6"x8'-0"	21	21	7'-6"	7 ½
4'-0"x8'-0"	22	22	8'-0"	8
5'-0"x8'-0"	22	22	9'-0"	9
6'-0"x8'-0"	23	23	10'-0"	10
7'-0"x8'-0"	24	24	11'-0"	11
8'-0"x8'-0"	25	25	12'-0"	12
9'-0"x9'-0"	28	28	9'-0"	9

Wood Door Bucks

Wood door bucks are made from 2"x4" or 2"x6" lumber, depending upon the thickness of the partition in which they are to be used.

Considerable saving may be realized by using bucks made from 2"x6" lumber, having the back grooved out the thickness of the partition and ½" to ¾" deep. For a 5 ½" finished partition, this allows for a 4" tile or gypsum partition block and 1 ½" for plaster on both sides of the partition. The partition tile fit into the groove in the wood buck and the ¾" extension on each side provides a ground for nailing the trim.

The mills furnishing the lumber usually charge $15.00 to $20.00 per 1,000 b.f. for grooving the backs of 2"x6"s.

The lumber quantities given above are estimated in lengths that cut with the least waste.

Labor Making Rough Door Bucks. A carpenter should make and brace a rough door buck for an opening up to 3'-0"x7'-0" in about l-hr. If the door has a transom, add ½-hr. carpenter time.

When making rough door bucks for door openings 5'-0"x7'-0" to 6'-0"x9'-0", a carpenter should make and brace one door buck in 1 ¼ to 1 ½ hrs. If the door has a transom, add ½ to ¾ hr. carpenter time.

Rough Wood Bucks for Borrowed Lights. When making rough wood bucks for borrowed lights, a carpenter should make and brace one buck for an average size opening in 1 to 1 ½ hrs.

Setting Rough Door Bucks. After the rough door buck has been made, a carpenter should set, plumb, and brace on door buck for an opening up to 3'-0"x9'-0" in about 1 to 1 ¼ hr.

For double doors requiring an opening from 5'-0"x7'-0" to 6'-0"x9'-0", a carpenter should set, plumb and brace one large door buck in 1 ¼ to 1 ½ hrs.

Fireproofing Wood. Where framing lumber and timbers are chemically treated to fireproof the wood, the cost of this treatment practically doubles the cost of the untreated wood.

BUILDING AND INSULATING SHEATHING

There are any number of insulating sheathing boards on the market, consisting of felted wood fiber or vegetable fiber products treated to make them moisture resistant.

Some of these sheets are coated with high melting point asphalt or are asphalt impregnated to form a moisture resistant surface which retards moisture penetration.

Insulating sheathing is furnished in sheets 4 ft. wide and 6, 7, 8, 9, 10 and 12 ft. long, and ½" and 25/32" thick, the same thickness as wood sheathing.

Most manufacturers furnish this type of sheathing 25/32" thick, 2 ft. wide and 8 ft. long, with the long edges V-jointed or shiplapped.

Insulating sheathing is usually applied to wood studs under wood siding, shingles, stucco or brick veneer and because of the large size sheets used and the asphalt treatment, building paper is not ordinarily used with it, except under stucco.

Use 2 in. galvanized nails with ⅜" or ½" heads for insulating sheathing.

Place nails 3 inches apart on all outside edges of board and 6 inches apart for all intermediate nailing.

Insulating sheathing 25/32" thick costs $170 to $200 per 1,000 sq. ft.

Labor Placing Insulating Sheathing. When placing insulating sheathing on square or rectangular houses of regular construction, a carpenter should place 700 to 900 sq. ft. per 8-hr. day, at the following labor cost per 100 sq. ft.:

	Hour	Rate	Total	Rate	Total
Carpenter	1.0	$....	$....	$24.04	$24.04
Labor unloading and carrying sheets	0.3	....	....	19.13	5.74
Cost per 100 sq. ft			$....		$29.78
Cost per sq. ft			....		.30

On buildings of irregular construction, requiring a great deal of cutting and fitting, it is not possible to take advantage of the large size sheets. Too much cutting and fitting is required. A carpenter will place only 350 to 450 sq. ft. of sheathing per 8-hr. day, at the following labor cost per 100 sq. ft.:

	Hours	Rate	Total	Rate	Total
Carpenter	2.0	$....	$....	$24.04	$48.08
Labor unloading and carrying sheets	0.3	....	....	19.13	5.74
Cost per 100 sq. ft			$....		$53.82
Cost per sq. ft			....		.54

Insulating Roof Decking

Several manufacturers produce insulating roof decking made of multiple layers of ½" insulating board laminated together with vapor resistant cement and fabricated into tongue and groove planks, 2'-0" wide, 8'-0" long, and 1 ½", 2", or 3" thick, with a finish painted undersurface. It provides a roof deck, insulation, vapor barrier and interior ceiling finish using only one material, and is especially adaptable for use in contemporary design ranch houses with exposed beam ceilings.

Insulating roof decking can be used for flat, pitched or monosloped roofs. Flat roofs and surfaces with a slope of 3" in 12" or less are usually covered with built-up roofing. Steeper roofs may be covered with rigid shingles, slate or tile roofing, providing wood nailing strips are fastened through decking to supporting beams below.

Decking should be laid so that cross joints are staggered and occur only over supports. Decking should be face nailed to all framing members, spacing nails 4" to 6" apart and keeping back ¾" to 1" from edges of plank. Nails should be galvanized common of sufficient length to pass through decking and penetrate supports at least 1 ½" and should be driven flush but not countersunk. Where underside of decking will be exposed, planks must be handled and laid with care to prevent finished ceiling surface from being marred or damaged. Avoid excessive sliding of plank on roof beams.

Estimating Data on Insulating Roof Decking

Thickness	Max. Distance Between Supports	Size & Type of Nails	Nails Req'd. Per 100 SF	Approx. Price Per SF
1 ½"	24"	10d Com. Galv.	6.0 lbs.	$0.50
2"	32"	16d Com. Galv.	3.5 lbs.	.60
3"	48"	30d Com. Galv.	3.0 lbs.	.70

Labor Placing Insulating Roof Decking. The labor cost of handling and placing insulating roof decking will be about the same for the various thicknesses available, as the saving in handling the lighter material is offset by the increased nailing required due to closer allowable spacing of supports.

On simple, rectangular, flat or low-rise roofs, two carpenters and a helper should place 750 to 850 sq. ft. of decking per 8-hr. day, including stapling paper strips on roof beams for added protection against marring finished surface, at the following labor cost per 100 sq. ft.:

	Hours	Rate	Total	Rate	Total
Carpenter	2	$....	$....	$24.04	$48.08
Labor	1	....	....	19.13	19.13
Cost per 100 sq. ft.			$....		$67.21
Cost per sq. ft			....		.67

For steep roofs, with a slope greater than 6" in 12", add 15 to 25 percent.

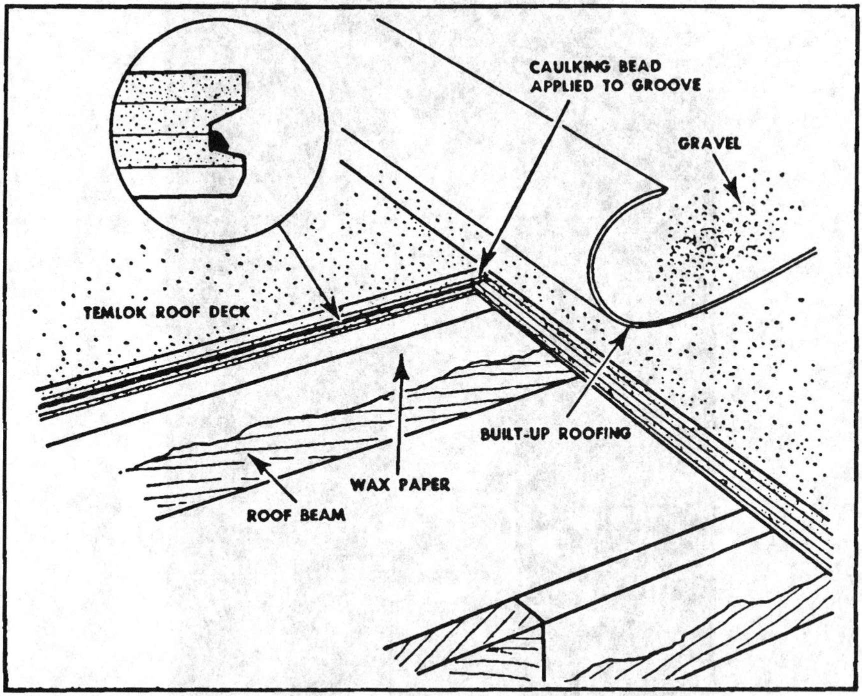

Method of Installing Insulating Roof Decking

PLYWOOD ROOF AND WALL SHEATHING, SUBFLOORING AND UNDERLAYMENT

Roof and Wall Sheathing. Plywood for use in roof and wall sheathing is readily available, easily worked with ordinary tools and skills, and is adaptable to almost any light construction application. Because of the large size sheets, installation time is less than for other types of sheathing, and waste is minimal. It may be used under any type of shingle or roofing material, or any type of siding.

Standard size sheets are 4"x8" but other sizes are available by special order. Most common sheathing thicknesses are 5⁄16", 3⁄8", 1⁄2", 5⁄8", and 3⁄4", which are unsanded, and are available with interior, intermediate or exterior glue.

For most roof and wall sheathing installations, where sheathing is to be covered, interior type plywood is used. Where installations are to be exposed to the weather, such as for roof overhangs or service building siding, Exterior type plywood should be used.

Nailing of plywood sheathing should be at 6" o.c. along panel edges and 12" o.c. at intermediate supports. 6d common nails should be used for panels 1⁄2" or less in thickness, and 8d for greater thickness.

Where plywood is used for wall sheathing, corner bracing is not required because of the rigidity of the plywood. Where other types of sheathing are used, plywood of the same thickness as the sheathing is sometimes used at corners to take the place of corner diagonal bracing as shown in the photograph.

Method of Applying Plywood Sheathing

The American Plywood Association has prepared maximum load and span tables for the various thicknesses of plywood for use in roof decking and wall sheathing installations as follows:

TABLE E Plywood Wall Sheathing [a] (Plywood continuous over 2 or more spans)

Panel Identification Index	Panel Thickness (inch)	Maximum Stud Spacing (inches) Exterior Covering Nailed to: Stud	Sheathing	Nail Size (b)	Nail Spacing (inches) Panel Edges (when over framing)	Intermediate (each stud)
12/0, 16/0, 20/0	5/16	16	16 (c)	6d	6	12
16/0, 20/0, 24/0	3/8	24	16 24 (c)	6d	6	12
24/0, 30/12, 32/16	1/2, 5/8	24	24	6d	6	12

NOTES:

(a) When plywood sheathing is used, building paper and diagonal wall bracing can be omitted.

(b) Common smooth, annular, spiral-thread, or galvanized box, or T-nails of the same diameter as common nails (0.113" dia. for 6d) may be used. Staples also permitted at reduced spacing.

(c) When sidings such as shingles are nailed only to the plywood sheathing, apply plywood with face grain across studs.

TABLE A Plywood roof decking (a) (b) (c)

(Plywood continuous over 2 or more spans; grain of face plys across supports)

Panel Ident. Index	Plywood Thickness (inch)	Max. Span (inches) (d)	Unsupported Edge—Max. Length (inches) (e)	Allowable Roof Loads (psf) (f) (g) (Spacing of Supports (Inches) Center to Center)										
				12	16	20	24	30	32	36	42	48	60	72
12/0	5/16	12	12	100 (130)										
16/0	5/16, 3/8	16	16	130 (170)	55 (75)									
20/0	5/16, 3/8	20	20		85 (110)	45 (55)								
24/0	3/8, 1/2	24	24		150 (160)	75 (100)	45 (60)							
30/12	5/8	30	26			145 (165)	85 (110)	40 (55)						
32/16	1/2, 5/8	32	28				90 (105)	45 (60)	40 (50)					
36/16	3/4	36	30				125 (145)	65 (85)	55 (70)	35 (50)				
42/20	5/8, 3/4, 7/8	42	32					80 (105)	65 (90)	45 (60)	35 (40)			
48/24	3/4, 7/8	48	36						105 (115)	75 (90)	55 (55)	40 (40)		
2·4·1	1-1/8	72	48							175 (175)	105 (105)	80 (80)	50 (50)	30 (35)
1-1/8" Grp 1&2	1-1/8	72	48							145 (145)	85 (85)	65 (65)	40 (40)	30 (30)
1-1/4" Grp 3&4	1-1/4	72	48							160 (165)	95 (95)	75 (75)	45 (45)	25 (35)

NOTES:

(a) These values apply for STANDARD C-D INT-DFPA, STRUCTURAL I AND II C-D INT-DFPA, C-C EXT-DFPA and STRUCTURAL I C-C EXT-DFPA grades only.

(b) For application where the roofing is to be guaranteed by a performance bond, recommendations may differ somewhat from these values. See Table D.

(c) Use 6d common, smooth, ring-shank or spiral-thread nails for 1/2" thick or less, and 8d common, smooth, ring-shank or spiral-thread for plywood 1" thick or less. Use 8d ring-shank or spiral-thread or 10d common smooth shank nails for 2·4·1, 1-1/8" and 1-1/4" panels. Space nails 6" at panel edges and 12" at intermediate supports, except that where spans are 48" or more, nails shall be 6" at all supports.

(d) These spans shall not be exceeded for any load conditions.

(e) Provide adequate blocking, tongue and grooved edges or other suitable edge support such as Plyclips when spans exceed indicated value. Use two Plyclips for 48" or greater spans and one for lesser spans.

(f) Uniform load deflection limitation: 1/180th of the span under live load plus dead load, 1/240th under live load only. Allowable live load shown in boldface type and allowable total load shown within parentheses. The allowable live load should in no case exceed the total load less the dead load supported by the plywood.

(g) Allowable roof loads were established by laboratory test and calculations assuming uniformly distributed loads.

Subflooring and Underlayment. Plywood subflooring, underlayment, or combined subfloor-underlayment, are products available for use in most areas of the country. Some particular panels may not be stocked in all areas, so it is advisable to check with local plywood sources as to their availability.

Subflooring panels are engineered grades of plywood marked by the American Plywood Association for maximum spans for the various thicknesses. (See table below) Deflection of the panels under concentrated loads at panel edges is the limiting factor in the design of plywood floors. Panel edges should be supported by solid blocking, or tongue and grooved panels should be used, unless a separate layer of underlayment is installed with its joints offset from the joints in the subfloor.

Nailing of subflooring should be at 6" o.c. along the edges and 10" o.c. at intermediate supports. Use 6d common nails for ½" plywood, and 8d for ⅝" to ⅞" thick plywood. For 1 ⅛" thick panels, use 8d ring shank or 10d common nails spaced 6" o.c. at both panel edges and intermediate supports.

Underlayment grades of plywood are touch sanded panels of ¼" and ⅜" thickness with a smooth, solid surface for application of nonstructural flooring finished directly to them. This grade or plywood is made with special inner-ply construction which resists punch-through by concentrated point loading.

Joints in the underlayment should be staggered with respect to the joints in the subfloor. Nailing should be done with 3d ring shanked underlayment nails at 6" o.c. along the edges and 8" each way in the interior of the panels. Nails should be countersunk if subfloor and joists are not completely dry, to prevent nail popping. See table below for American Plywood Association recommendations on Plywood Underlayment, and fastening.

TABLE G **Plywood subflooring** (a) (c) (d) For direct application of T&G wood strip and block flooring and lightweight concrete. (Plywood continuous over two or more spans, face grain across supports)

Panel Identification Index (b)	Plywood Thickness (inch)	Maximum Span (e) (inches)	Nail Size & Type	Nail Spacing (inches)	
				Panel Edges	Intermediate
30/12	5/8	12 (f)	8d common	6	10
32/16	1/2, 5/8	16 (g)	8d common (h)	6	10
36/16	3/4	16 (g)	8d common	6	10
42/20	5/8, 3/4, 7/8	20 (g)	8d common	6	10
48/24	3/4, 7/8	24	8d common	6	10
1-1/8" Groups 1 & 2	1-1/8 (i)	48	10d common	6	6
1-1/4" Groups 3 & 4	1-1/4	48	10d common	6	6

NOTES:

(a) These values apply for STANDARD C-D INT-DFPA, STRUCTURAL I AND II C-D INT-DFPA, C-C EXT-DFPA and STRUCTURAL I C-C EXT-DFPA grades only.

(b) Identification Index appears on all panels except 1-1/8" and 1-1/4" panels.

(c) In some nonresidential buildings, special conditions may impose heavy concentrated loads and heavy traffic requiring subfloor constructions in excess of these minimums.

(d) Edges shall be tongue and grooved or supported with blocking for square edge wood flooring, unless separate underlayment layer (1/4" minimum thickness) is installed.

(e) Spans limited to values shown because of possible effect of concentrated loads. At indicated maximum spans, floor panels carrying Identification Index numbers will support uniform loads of more than 100 psf.

(f) May be 16" if 25/32" wood strip flooring is installed at right angles to joists.

(g) May be 24" if 25/32" wood strip flooring is installed at right angles to joists.

(h) 6d common nail permitted if plywood is 1/2".

(i) For 2•4•1 application details see page 12.

TABLE H **Plywood underlayment** / For application of tile, carpeting, linoleum or other non-structural flooring.

Plywood Grades and Species Group	Application	Minimum Plywood Thickness	Fastener size (approx.) and Type (set nails 1/16")	Fastener Spacing (Inches)	
				Panel Edges	Inter-mediate
Groups 1, 2, 3, 4 UNDERLAYMENT INT-DFPA (with interior, intermediate or exterior glue) UNDERLAYMENT EXT-DFPA C-C Plugged EXT-DFPA	over plywood subfloor	1/4"	18 Ga. staples or 3d ring-shank nails (a) (b)	3	6 each way
	over lumber subfloor or other uneven surfaces	3/8"	16 Ga. staples (a)	3	6 each way
			3d ring-shank nails (b)	6	8 each way
Same Grades as above, but Group 1 only.	over lumber floor up to 4" wide. Face grain must be perpendicular to boards.	1/4"	18 Ga. staples or 3d ring-shank nails	3	6 each way

(a) Crown width 3/8" for 16 ga., 3/16" for 18 ga. staples; length sufficient to penetrate completely through, or at least 5/8" into, subflooring.

(b) Use 3d ring-shank nail also for 1/2" plywood and 4d ring-shank nail for 5/8" or 3/4" plywood.

Installation: Apply UNDERLAYMENT just prior to laying finish floor or protect against water or physical damage. Stagger panel end joints with respect to each other and offset all joints with respect to the joints in the subfloor. Space panel ends and edges about 1/32". For maximum stiffness, place face grain of panel across supports and end joints over framing. Unless subfloor and joists are thoroughly seasoned and dry, countersink nails 1/16" just prior to laying finish floor to avoid nail popping. Countersink staples 1/32". Fill any damaged, split or open areas exceeding 1/16". Do not fill nail holes. Lightly sand any rough areas, particularly around joints or nail holes.

Combined Subfloor-Underlayment Plywood. This is a single application installation combining both subfloor and underlayment, thus eliminating the expense of installing a separate underlayment. Panel edges perpendicular to the joists must be supported by solid blocking if tongue and grooved panels are not used.

Panels should be nailed at 6" o.c. at edges and 10" at intermediate supports with ringshank or spiral-thread nails. 6d deformed-shank nails may be used for panel thickness up to ¾" and 8d for thicker panels.

Approximate costs of plywood sheathing, subflooring and underlayment per 1000 sq. ft. are as follows:

Types of Plywood

Interior Type Standard (C-D INT)		Interior Type Structural 1 C-D		Exterior Type C-C EXT	
5/16"	$240.00	½"	$310.00	5/16"	$290.00
⅜	260.00	⅝	390.00	⅜	350.00
½	310.00	¾	470.00	½	390.00
⅝	360.00			⅝	430.00
¾	420.00			¾	500.00

Plywood Underlayment
Group 1, Interior

½"	$300.00
⅝	390.00
¾	450.00

Following is a table showing recommended support spacing for combined subfloor-underlayment.

TABLE I Combined subfloor-underlayment / For direct application of tile, carpeting, linoleum or other non-structural flooring. (Plywood continuous over two or more spans; grain of face plys across supports. Seasoned lumber is recommended.)

Plywood Grade (c)	Plywood Species Group	Maximum Support Spacing (a) (b): 16" o.c. Panel Thickness	16" o.c. Deformed Shank Nail Size (d)	20" o.c. Panel Thickness	20" o.c. Deformed Shank Nail Size (d)	24" o.c. Panel Thickness	24" o.c. Deformed Shank Nail Size (d)	32" (e) or 48" o.c. Panel Thickness	32" (e) or 48" o.c. Deformed Shank Nail Size (d)	Nail Spacing (inches): Panel Edges	Nail Spacing (inches): Intermediate
UNDERLAYMENT INT-DFPA (with interior, intermediate or exterior glue) UNDERLAYMENT EXT-DFPA C-C Plugged EXT-DFPA	1	1/2"	6d	5/8"[f]	6d	3/4"[g]	6d	–	–	6	10
	2 & 3	5/8"[f]	6d	3/4"[g]	6d	7/8"	8d	–	–	6	10
	4	3/4"[g]	6d	7/8"	8d	1"	8d	–	–	6	10
2·4·1	1, 2 & 3	(2·4·1 specifications are so written that panels from all groups have equal properties.)						1-1/8"	8d (or 10d common smooth shank if supports well-seasoned)	6	(h)

(a) Edges shall be tongue & grooved, or supported with framing.
(b) In some non-residential buildings, special conditions may impose heavy concentrated loads and heavy traffic requiring subfloor-underlayment constructions in excess of these minimums.
(c) For certain types of flooring such as wood block or terrazzo, sheathing grades of plywood may be used.
(d) Set nails 1/16" (1/8" for 2·4·1 panels) and lightly sand subfloor at joints if resilient flooring is to be applied. Don't fill nail holes.
(e) 2" wide supports for 32" o.c.; 4" wide supports for 48" centers.
(f) May be 19/32".
(g) May be 23/32".
(h) 10" for 32" o.c.; 6" for 48" o.c. supports.

Prices on all grades and types of plywood should be verified with local suppliers prior to submitting a proposal for work which includes these materials.

Approximate costs of various materials and labor required for sidewall sheathing are given below.

Labor Cost of 100 Sq. Ft. of Structural Insulating Sidewall Sheathing Applied to Square or Rectanglar Buildings of Regular Construction

	Hours	Rate	Total	Rate	Total
Carpenter sheathing	1.0	$....	$....	$24.04	$24.04
Helper unloading and carrying sheets	0.3	....	....	19.13	5.74
Cost per 100 sq. ft			$....		$29.78
Cost per sq. ft			....		.30

No paper required as insulating sheathing is treated to resist moisture.

If used on buildings of irregular construction, as described above, add 1 hr. carpenter time.

Labor Cost of 100 Sq. Ft. of 1"x6" Wood Sidewall Sheathing Applied Horizontally to Square or Rectangular Buildings of Regular Construction

	Hours	Rate	Total	Rate	Total
Carpenter sheathing	1.5	$....	$....	$24.04	$36.06
Labor unloading and carrying lumber	0.5	....	....	19.13	9.57
Carpenter applying building paper	0.4	....	....	24.04	9.62
Cost per 100 sq. ft			$....		$55.25
Cost per sq. ft			....		.55

If 1"x8" shiplap is used instead of 1"x6" D&M lumber, deduct 5 b.f. lumber and 0.1 hr. carpenter time per 100 sq. ft. of wall.

If used on buildings of irregular construction, as described above, add 1 hr. carpenter time.

Labor Cost of 100 Sq. Ft. of 1"x6" Wood Sidewall Sheathing Applied Diagonally to Square or Rectangular Buildings of Regular Construction

	Hours	Rate	Total	Rate	Total
Carpenter sheathing	2.0	$....	$....	$24.04	$48.08
Labor unloading and carrying lumber	0.5	....	....	19.13	9.57
Carpenter applying paper	0.4	....	....	24.04	9.62
Cost per 100 sq. ft			$....		$67.27
Cost per sq. ft			....		.67

If 1"x8" shiplap is used instead of 1"x6" D&M lumber, deduct 5 b.f. lumber and 0.1 hr. carpenter time per 100 sq. ft. of wall.

If used on buildings of irregular construction, as described above, add 1 hr. carpenter time.

Labor Cost of 100 Sq. Ft. of ½" Plywood Sidewall Sheathing Applied to Square or Rectangular Buildings of Regular Construction

	Hours	Rate	Total	Rate	Total
Carpenter sheathing	1.0	$....	$....	$24.04	$24.04
Helper unloading and carrying sheets	0.3	....	....	19.13	5.74
Carpenter applying paper	0.4	....	....	24.04	9.62
Cost per 100 sq. ft			$....		$39.40
Cost per sq. ft			....		.39

Paper optional; plywood comes in large sheets of 48"x96" treated with asphalt.

If used on buildings of irregular construction, as described above, add 1 hr. carpenter time.

Gypsum Sheathing. Gypsum sheathing is a fireproof solid sheet of gypsum encased in a tough, fibrous, water-resisting covering. The sides and ends are treated to resist moisture. Some manufacturers blend asphalt emulsion into the wet gypsum core mix to provide additional resistance to moisture. It is used for sheathing on frame structures under siding, shingles, stucco, and brick veneer.

Gypsum sheathing has V-joint edges on the long dimension to provide positive assurance of a tight fit at the unsupported joints. It is applied with its length at right angles to the studs.

Gypsum sheathing is ½" thick, 2 ft. wide, and 8 and 10 ft. long, and 4 ft. wide and 8 and 9 ft. long to fit supports 16" on centers. Nails should be 1 ¾ in. long, No. 10 ½ gauge, galvanized flat head roofing nails, spaced 4" on centers, except under wood siding and stucco, 8". Requires 14 to 21 lbs. per 1,000 sq. ft.

Price $85.00 to $125.00 per 1,000 sq. ft.

Labor cost same as given for insulating or plywood sheathing.

Labor Cost of 100 Sq. Ft. of Gypsum Sidewall Sheathing Applied to Square or Rectangular Buildings of Regular Construction

	Hours	Rate	Total	Rate	Total
Carpenter sheathing	1.0	$....	$....	$24.04	$24.04
Helper unloading and carrying sheets	0.3	....	....	19.13	5.74
Carpenter applying paper	0.4	....	....	24.04	9.62
Cost per 100 sq. ft			$....		$39.40
Cost per sq. ft			....		.39

Paper is optional. Gypsum sheathing is waterproofed sides and edges. Some manufacturers mix asphalt emulsion with wet gypsum to provide a waterproof board.

If used on buildings of irregular construction, as described above, add 1 hr. carpenter time.

Insulating Shingle Backer Strips

A number of manufacturers produce a shingle backer strip, of the same material as insulating sheathing board, to be used as the undercourse in double course shingle work. These strips are 4'-0" long and either 5⁄16" or 3⁄8" thick. They are available in two widths, 13 1⁄2" for 12" shingle exposure and 15 1⁄2" for 14" shingle exposure. They provide added insulation, improve the shadow line appearance of the finished wall, eliminate the necessity of building paper over wood sheathing and are more economical to use than second grade wood shingle undercoursing. Insulating shingle backer strips cost approximately 30 cents per sq. ft.

Labor Placing Insulating Shingle Backer Strips. On straight run jobs, with an average number of openings, a carpenter should place 500 to 600 sq. ft. of shingle backer strips per 8-hr. day at the following labor cost per 100 sq. ft.:

	Hours	Rate	Total	Rate	Total
Carpenter	1.5	$....	$....	$24.04	$36.06
Labor	0.5	....	....	19.13	9.57
Cost 100 sq. ft			$....		$45.63
Cost per sq. ft			....		.46

On complicated jobs, requiring a large amount of cutting and fitting, a carpenter should place 400 to 500 sq. ft. per 8-hr. day at the following labor cost per 100 sq.ft.:

	Hours	Rate	Total	Rate	Total
Carpenter	1.8	$....	$....	$24.04	$43.27
Labor	0.5	....	....	19.13	9.57
Cost 100 sq. ft			$....		$52.84
Cost per sq. ft			....		.53

Insulating Lath

Insulating lath is made by several manufacturers of insulating building sheathing. It is made of the same materials as the building boards previously described.

Insulating lath is usually furnished in the 18"x48" size, 1⁄2", 3⁄4", and 1" thick, although they are obtainable from some sources in 16"x48".

Some manufacturers furnish lath both plain and with the back coated with asphalt to form a vapor barrier.

Most insulating lath are manufactured with the long edges shiplapped and all edges beveled to reinforce the plaster against cracking and eliminate unsightly lath marks. Some of these insulating lath are metal reinforced along the edges to reinforce the unsupported horizontal joint.

Insulating gypsum lath consists of regular gypsum board lath with a thin sheet of aluminum attached to one side to provide insulating value. They are furnished in sheets 16"x32", 16"x48", and 3/8" and 1/2" thick.

Estimating Quantities of Insulating Lath

Size of Lath in Inches	No Sq. Ft. per Lath	No. Lath per 100 Sq. Ft.
16x48	5.33	18.76
18x48	6.00	16.67

Approximate prices of insulating lath are as follows per 1,000 sq. ft.: 1/2", $200.00; 5/8", $250.00.

Labor Applying Insulating Lath. The labor cost of applying insulating lath will vary with the locality and labor restrictions. Where there are no labor restrictions, an experienced man should place 125 sq. yds. per 8-hr. day, while in the larger cities a lather will place only 100 sq. yds. per 8-hr. day, at the following labor cost per 100 sq. yds.:

	Hours	Rate	Total	Rate	Total
Lather	8	$....	$....	$24.16	$193.28
Cost per sq. yd			....		1.93

WOOD AND WIRE FENCE

Setting Wood Fence Posts. When placing wire fence around the lots of 16 houses, it required 218 cedar posts, 5" in diameter and 7'-0" long, also 32 clothes posts, 7" in diameter and 10' -0" long.

After the holes had been dug, a man set 7 posts an hr. at the following labor cost per post:

	Hours	Rate	Total	Rate	Total
Carpenter	0.5	$....	$....	$24.04	$12.02

Treated pine posts and locust posts may also be used for fencing. Locust and cedar posts should be all heartwood.

Placing 2"x4" Top Bottom Fence Rail Ready to Receive Wire. The posts for these fences were spaced about 6'-0" apart and a 2"x4" top rail was nailed flatwise to the top of the post and a 2"x4" bottom rail was notched into the side of the post so that the wire could be stapled to it.

There was a total of 1,400 lin. ft. of fence or 2,800 lin. ft. of top and bottom rail. A carpenter placed rail for 110 lin. ft. of fence or 220 lin. ft. of single rail per 8-hr. day, at the following labor cost per 100 lin. ft.:

	Hours	Rate	Total	Rate	Total
Carpenter	7.3	$....	$....	$24.04	$175.49
Labor	0.9	....	....	19.13	17.22
Cost per 100 l.f.			$....		$192.71
per lin. ft. fence			....		1.93
per lin. ft. rail			....		.97

Stringing Fence Wire. Cyclone wire fencing in 42" rolls was used for these fences and was stapled to the top and bottom rails and posts about 4" on centers.

There were 1,400 lin. ft. of 42" wire fencing in the job and a carpenter would string and staple about 280 lin. ft. per 8-hr. day, at the following labor cost per 100 lin. ft.:

	Hours	Rate	Total	Rate	Total
Carpenter	2.9	$....	$....	$24.04	$69.72
Cost per lin. ft.			....		.70

Setting Lally Columns in Basements. In a group of brick houses, 3" iron columns were used to support the 6"x8" wood girders carrying the floor joists.

These columns were 3" in diameter and 7'-0" long, and a carpenter set, plumbed and braced 1 column an hr. at the following labor cost per column:

	Hours	Rate	Total	Rate	Total
Carpenter	1.0	$....	$....	$24.04	$24.04
Labor	1.0	....	....	19.13	19.13
Cost per column			$....		$43.17

MISCELLANEOUS HARDWARE AND ACCESSORIES

Steel Area Walls and Gratings

These are used around basement windows when first floor is at grade level and are made of 16-ga. and 20-ga. galvanized steel, with stiffening ribs and rounded tops. They are attached to masonry walls by screws or bolts.

Size Width Depth	For Use with Basement Windows	Price 16 ga.	Each 20 ga.
3'-2"x0'-11 ½"	15"x12"-2-Lt.	$10.25	$7.80
3'-2"x1'-5 ½"	15"x16"-2-Lt.	12.75	9.50
3'-2"x1'-11 ½"	15"x20"-2-Lt.	17.00	12.75

Steel gratings for use with above walls. Frames made of 1 ¼"x ¼" steel bars with cross bars 1"x 13/16" welded into one piece. Supports on the grating hold it flush with the top of the area wall. Price each painted, $23.50.

Metal Foundation Wall Ventilators

Brick Ventilators. Made of aluminum exactly the size of a brick (2 ½"x4"x8"). The bottom edge is flanged both front and back to insure a positive mortar lock. Louvered face allows 13 sq. in. of free area. Screened back. Price each, $9.25.

Frame Ventilators. Nailed to sheathing just above the foundation or to ends of joists spaced 16" on centers. Made of sheet aluminum, size 16"x8", providing 89 sq. in. of free area. Screened back. Price each, $23.50.

Concrete Block Ventilators. Made of cast aluminum 16" wide x 8" high, providing 89 sq. in. of free area. Screened back. Price each, $25.00.

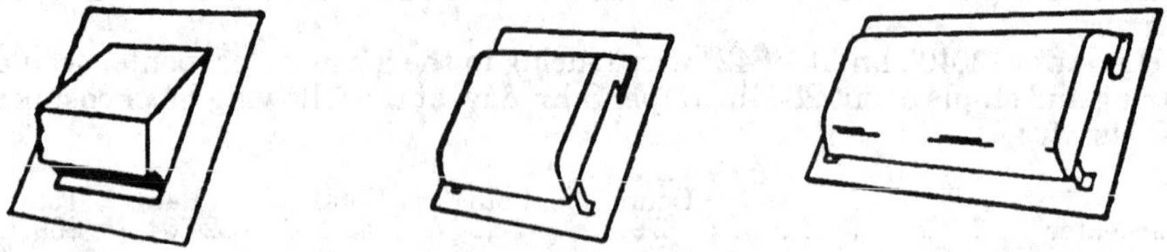

Metal Roof Ventilators

Metal roof ventilators provide air in the attic space under the roof. They are made of steel or aluminum and usually provide 18 to 75 sq. in. of free area.

Roof Opg. Inches	Free Area Sq. Inches	Price Each Aluminum
12x18	30	$7.50
15x20	50	10.00
16x16	65	16.00

Steel Clothes Chute Doors

	Price Each
Furnished in steel, white enameled, opening 9"x12"	$15.00

Steel Access Doors

For access to piping or wiring installations. Furnished with a flanged frame that may be nailed to studs 16" on centers. Size of door 16" x 24", overall size 17 1/2"x25 1/2".

. . . . $24.00

ROUGH HARDWARE

While it is possible to include the cost of rough hardware such as nails and screws in the material unit for each carpentry item figured, most contractors and estimators find it more convenient to make an allowance for this item at the end of the carpentry estimate based on the amount of lumber and other material involved. The usual method of arriving at a rough hardware allowance is to total all the material in the carpentry estimate taking, b.f. of dimension lumber and boards, sq. ft. area of plywood, paneling, etc., and lin. ft. of grounds, furring strips, trim, etc., and then figure from 25 to 35 lbs. of nails per 1,000 b.f. The weight of nails multiplied by the cost per lb. will give the rough hardware allowance required.

06130 HEAVY TIMBER CONSTRUCTION

The quantities and costs included under this classification cover types such as heavy constructed mill, warehouse, and factory buildings.

Framing and Erecting Heavy Wood Columns or Posts. The labor cost of erecting and framing heavy wood columns or posts, 10"x10" or larger, will vary according to the number of operations necessary, such as chamfering corners, cutting out and framing for post caps and bases, boring holes through the center, and upon the distance the posts must be hoisted before placing.

On the average mill constructed job requiring 10"x10" or 12"x12" posts 10'-0" to 16'-0" long, it will require 3 ½ to 4 hrs. time to chamfer corners, frame for bases and caps, and erect each column, at the following labor cost:

	Hours	Rate	Total	Rate	Total
Carpenter	3.7	$....	$....	$24.04	$88.95
Labor	0.5	....	....	19.13	9.57
Cost per column			$....		$98.52

If hoisted above the second floor, add ½ hr. of labor time per column. It is more difficult to estimate the cost of this work by the 1,000 b.f. of lumber, because it costs just as much to chamfer the 4 corners and frame for post caps and bases for an 8"x8" as a 12"x12" column, while the former contains less than half as much lumber per lin. ft. as the latter. The following costs are based on an average 10"x10"-12'-0" timber and should be increased or reduced accordingly for larger or smaller timbers.

Based on a 10"x10"-12'-0" column, a carpenter should chamfer corners, frame for post caps and bases, handle and erect 200 to 250 b.f. of lumber per 8-hr. day, at the following labor cost per 1,000 b.f.:

	Hours	Rate	Total	Rate	Total
Carpenter	35	$....	$....	$24.04	$841.40
Labor	8	....	....	19.13	153.04
Cost 1,000 b.f.			$....		$994.44

Rounding Corners on Large Timbers. Where the corners of wood columns, beams or girders must be rounded to a small radius, the yards or mills furnishing the lumber usually charge $16.50 to $21.50 per 1,000 b.f. for performing this work. **Boring Holes Through Center of Large Timbers.** Where the wood columns or posts have a hole bored through the center for their entire length, the yard or mill furnishing the lumber usually charges $32.00 to $37.00 per 1,000 b.f. for boring these holes.

Framing and Erecting Heavy Wood Beams and Girders for First Floor. The labor cost of framing and erecting heavy wood beams and girders will vary according to the amount of framing necessary and the method used in placing them; i.e., whether they set on top of the wood columns or whether it is necessary to frame for post caps, stirrups, etc.

Placing heavy beams and girders for the first floor of a building does not require as much labor as for the upper floors, because the timbers can be rolled to the place they are to be used without hoisting.

Where heavy wood girders are set on concrete or masonry piers without framing for post caps, joist hangers, etc., a carpenter should frame and erect 1,000 to 1,250 ft. per 8-hr. day, at the following labor cost per 1,000 ft.:

	Hours	Rate	Total	Rate	Total
Carpenter	7	$....	$....	$24.04	$168.28
Labor	7	....	....	19.13	133.91
Cost 1,000 b.f.			$....		$302.19

Where necessary to frame for post caps and bases, stirrups, etc., a carpenter should frame and erect 500 to 600 b.f. of lumber per 8-hr. day, at the following labor cost per 1,000 b.f.:

	Hours	Rate	Total	Rate	Total
Carpenter	14.5	$....	$....	$24.04	$348.58
Labor	7.0	....	....	19.13	133.91
Cost 1,000 b.f.			$....		$482.49

Framing and Erecting Heavy Wood Girders Above First Floor. Where heavy wood girders must be hoisted above the first floor, using a timber hoist or derrick, the carpenter time framing and placing the timbers will remain practically the same but the labor time handling, hoisting and skidding the timbers into place will be increased.

On work of this kind, a carpenter should frame and erect 450 to 550 b.f. of lumber per 8-hr. day, at the following labor cost per 1,000 b.f:

	Hours	Rate	Total	Rate	Total
Carpenter	16	$....	$....	$24.04	$384.64
Labor	10	....	....	19.13	191.30
Cost 1,000 b.f.			$....		$575.94

Framing and Erecting Heavy Wood Joists. When placing heavy wood joists 4"x12" or 4"x16" in size, where the joists set on top of wood girders or in iron stirrups, a carpenter should frame and erect 1,000 to 1,200 b.f. per 8-hr. day, at the following labor cost per 1,000 b.f.:

	Hours	Rate	Total	Rate	Total
Carpenter	7.3	$....	$....	$24.04	$175.49
Labor	3.5	....	....	19.13	66.96
Cost 1,000 b.f.			$....		$242.45

Where smaller joists are used, ranging from 3"x10" to 4"x10" in size, a carpenter should frame and erect 800 to 1,000 b.f. per 8-hr. day, at the following labor cost per 1,000 b.f.:

	Hours	Rate	Total	Rate	Total
Carpenter	9	$....	$....	$24.04	$216.36
Labor	4	....	....	19.13	76.52
Cost 1,000 b.f.			$....		$292.88

Framing and Placing Oak Bumpers at Edges of Loading Docks. Oak timbers, varying from 4"x6" to 12"x12", are usually called for at loading dock edges. The timbers generally are secured to the edge of the dock slab by bolts set in the concrete or with cinch anchors.

With bolts already set in place, a carpenter should lay out, drill and countersink holes and secure in place 150 to 175 b.f. of oak bumper per 8-hr. day at the following labor cost per 1,000 b.f.:

	Hours	Rate	Total	Rate	Total
Carpenter	48	$....	$....	$24.04	$1,153.92
Labor	24	....	....	19.13	459.12
Cost 1,000 b.f.			$....		$1,613.04

A carpenter should locate, set and secure in place in the slab edge or wall form about 25 anchor bolts per 8-hr. day at the following labor cost per bolt:

	Hours	Rate	Total	Rate	Total
Carpenter	0.3	$....	$....	$24.04	$7.21

Laying 2"x6" or 3"x6" Tongue and Groove Timber Subfloors or Roof Sheathing. When laying 2"x6" or 3"x6" tongue and groove flooring or roof decking in large areas without much cutting and fitting, a carpenter should lay 800 to 1,000 b.f. per 8-hr. day at the following labor cost per 1,000 b.f.

	Hours	Rate	Total	Rate	Total
Carpenter	8.9	$....	$....	$24.04	$213.96
Labor	6.0	....	....	19.13	114.78
Cost per 1,000 b.f.			$....		$328.74

If tongue and groove flooring or roof sheathing is laid in small spaces requiring considerable cutting and fitting around openings, stair wells, elevator shafts, skylights, etc., a carpenter should lay 700 to 800 b.f. per 8-hr. day at the following labor cost per 1,000 b.f.

	Hours	Rate	Total	Rate	Total
Carpenter	10.6	$....	$....	$24.04	$254.82
Labor	6.0	$....	....	19.13	114.78
Cost per 1,000 b.f.			$....		$369.60

LAMINATED WOOD FLOORS

Laminated wood floors consisting of 2" or 3" plank set on edge and spiked together, are used where a floor is desired capable of carrying heavy loads or supporting heavy machinery with a minimum amount of vibration.

Sometimes the floors are designed to use certain length plank reaching from bearing to bearing while on others they are spiked together with broken joints, regardless of bearing.

Quantity of Lumber Required for Laminated Wood Floors

Measured Size Inches	Finished Thickness Inches	Add for Waste Percent	To Obtain Qty of Lumber Reqd Mult. Area by	No. Ft. Lumber Per 100 Sq. Ft. of Surface
2x4	1 5/8	23	4.90	490 (74 pcs.)
2x6	1 5/8	23	7.40	740 (74 pcs.)
2x8	1 5/8	23	9.84	984 (74 pcs.)
2x10	1 5/8	23	12.30	1230 (74 pcs.)
2x12	1 5/8	23	14.76	1476 (74 pcs.)
2x14	1 5/8	23	17.22	1722 (74 pcs.)
3x6	2 3/4	10	6.60	660 (44 pcs.)
3x8	2 3/4	10	8.80	880 (44 pcs.)
3x10	2 3/4	10	11.10	1110 (44 pcs.)
3x12	2 3/4	10	13.20	1320 (44 pcs.)
3x14	2 3/4	10	15.40	1540 (44 pcs.)
4x6	3 3/4	7	6.42	642 (32 pcs.)
4x8	3 3/4	7	8.56	856 (32 pcs.)
4x10	3 3/4	7	10.70	1070 (32 pcs.)
4x12	3 3/4	7	12.84	1284 (32 pcs.)

The above quantities do not include any allowance for waste due to end matching, but are based on spans taking lumber of even lengths or laying flooring with broken joints.

Nails Required for 100 Sq. Ft. of 2" Laminated Flooring

Distance Spikes Are Spaced Apart

Size Spikes	12 In. No. Spikes	No. Lbs.	16 In. No. Spikes	No. Lbs.	20 In. No. Spikes	No. Lbs.	24 In. No. Spikes	No. Lbs.	30 In. No. Spikes	No. Lbs.
16d	740	15	600	12 ¼	518	10 ½	450	9 ¼	370	7 ½
20d	740	24	600	19 ¼	518	16 ¾	450	14 ½	370	12
30d	740	31	600	25	518	21 ½	450	18 ¾	370	15 ½
40d	740	41	600	33 ¼	518	28 ¾	450	25	370	20 ½
50d	740	53	600	43	518	37	450	32	370	26 ½
60d	740	67	600	55	518	47	450	41	370	33 ¾

Nails Required for 100 Sq. Ft. of 3" Laminated Flooring

Distance Spikes Are Spaced Apart

Size Spikes	12 In. No. Spikes	No. Lbs.	16 In. No. Spikes	No. Lbs.	20 In. No. Spikes	No. Lbs.	24 In. No. Spikes	No. Lbs.	30 In. No. Spikes	No. Lbs.
16d	440	9	352	7 ¼	308	0 ¼	264	5 ½	220	4 ½
20d	440	14 ¼	352	11 ½	308	10	264	8 ½	220	7
30d	440	18 ½	352	14 ¾	308	12 ¾	264	11	220	9 ¼
40d	440	24 ½	352	19 ½	308	17	264	14 ¾	220	12 ¼
50d	440	31 ½	352	25 ¼	308	22	264	18 ¾	220	15 ¾
60d	440	40	352	32	308	28	264	24	220	20

Nails Required for 100 Sq. Ft. of 4" Laminated Flooring

Distance Spikes are Spaced Apart

Size Spikes	12 In. No. Spikes	No. Lbs.	16 In. No. Spikes	No. Lbs.	20 In. No. Spikes	No. Lbs.	24 In. No. Spikes	No. Lbs.	30 In. No. Spikes	No. Lbs.
30d	320	13 ½	256	10 ¾	224	9 ½	192	8	160	6 ¾
40d	320	17 ¾	256	14 ¼	224	12 ½	192	10 ¾	160	9
50d	320	23	256	18 ¼	224	16	192	13 ¾	160	11 ½
60d	320	29	256	23 ¼	224	20 ½	192	17 ½	160	14 ½

06170 PREFABRICATED STRUCTURAL WOOD

Glued Laminated Beam Construction. Beams of glued laminated construction are popular where price is not the controlling factor.

They are made in uniform or tapered shapes. In the beam shape, similar to illustration, they can be used in spans from 20'-0" to 80'-0".

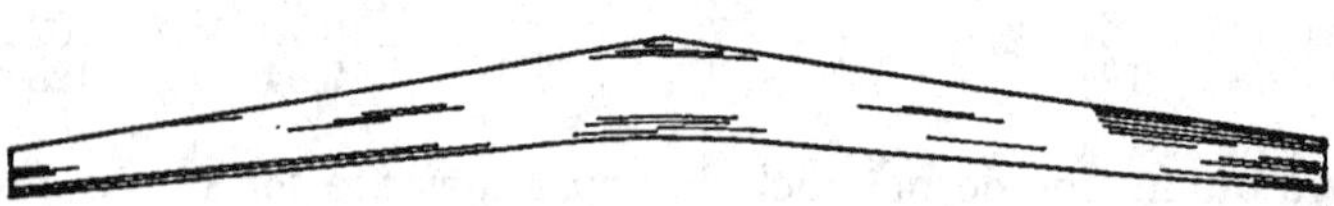

They are used in schools, auditoriums, churches, stores, and ranch-style homes and are made of kiln-dried structural woods bonded together by glue, applied under controlled conditions of temperature and pressure. They are planed and sanded to a very smooth surface and later treated with a transparent preservative, making their overall appearance very pleasing. With this type of construction, the ceilings are usually omitted, leaving the beams exposed to the room below. Roof insulation is accomplished through the use of standard insulating boards placed upon the roof deck and covered with roofing material. Because of the purlin construction generally used, two or four inch decking is recommended.

Span Feet	Center Height	End Height	Camber	Price
20'-0"	2'-4"	0'-9 ½"	0'-9"	$290.00
30'-0"	2'-10"	0'-11"	1'-0"	700.00
40'-0"	3'-11"	1'-2 ½"	1'-6"	1,020.00
50'-0"	5'-1"	1'-6 ½"	2'-0"	1,620.00
60'-0"	6'-1"	1'-9 ½"	2'-6"	2,330.00

Glued Laminated Three Hinged Arch

Another type of glued laminated construction is the three-hinged arch, which gains its support from floor level, incorporating column and beam in one compact design. Fabricated and finished in the manner of the glued beam, they are used where good appearance and function are the major factors and are used quite extensively in churches and schools.

Purlins are generally used to span the resulting bays and are covered with two inch decking and suitable insulating material.

This is the most expensive of the trusses discussed and costs from 5 to 6 times that of the bowstring type truss.

General Notes Regarding Trusses. Where rafters are used between trusses, 1" roof sheathing or ½" plywood is generally used. Where purlins are spaced 4'-0" to 6'-0" apart, 2" sheathing is common practice. Purlins may be eliminated and wood decking of 4 to 6 inches may be used to span up to 18'-0" between trusses.

The use of ceilings supported by the lower chord of the truss should be called to the attention of the truss designer, as it requires a heavier load.

Trusses may bear upon pilasters, steel columns, wood posts and plates, grillages or steel beams. A suitable steel bearing plate should be placed under each end of the truss to distribute the load throughout the masonry. Trusses may afford lateral support through angle irons fastened to truss and column.

Labor Framing Wood Roof Trusses. When framing wood roof trusses, a carpenter experienced in truss construction will handle and frame approximately the following amounts of lumber per hour on the various types of trusses:

Type of Truss	B.f. per Hr.
Bowstring Type Trusses	40
Scissors and Gothic Type Trusses	30
Glued Laminated Type Trusses	15

PREFABRICATED STRUCTURAL TIMBER

Structural timber decking in thickness of 2" to 4", generally 6" wide and double tongue and grooved is available in hemlock cedar, Douglas fir, and spruce. It can be used for roof decks, structural flooring and for planked walls. Approximate costs are as follows:

	Hemlock Per Sq. Ft.	Cedar or Fir Per Sq. Ft.	Spruce Per Sq. Ft.
4"x6" Select	$2.15	$2.45	$3.05
3"x6" Select	1.80	2.05	2.25

Glued Laminated Beams in a Church

All prices should be verified with local distributors prior to submitting proposals that include these materials.

06200 FINISH CARPENTRY

The estimator figuring finish carpentry work for a job whose specifications are arranged in accordance with the Construction Specification Institute (CSI) format must check at least five divisions to be certain that all of this type of work is included in the bid. Division 6, the subject of this chapter, covers rough and finish carpentry. But wood and related sidings are in Division 7, *Thermal and Moisture Protection.* Wood doors and windows are in a section of their own, Division 8. Division 10, *Specialties,* includes such items as tackboards, laminated plastic toilet partitions, signs, folding partitions, telephone booths, and toilet and wardrobe accessories, all of which may be furnished or set by the carpenter. Division 11, *Equipment,* covers kitchen and laundry cabinets and counters for residential work and such unrelated work as wood bank counters, church pews, bars, lab furniture, dock bumpers and library shelving. Finally Division 12, *Furnishings,* includes wood seating, class room furniture and similar semi-attached items. In addition, it is prudent to check all the allowances set forth in the specifications as often the furnishing of some of the items in Divisions 10, 11 and 12 will be covered under a cash allowance but the installation is to be included in the bid submitted by the carpenter.

EXTERIOR FINISH CARPENTRY

Wood siding and wood doors and windows are discussed under their own CSI divisions. The following items cover other adjacent exterior work.

Placing Corner Boards, Fascia Boards, Etc. When placing wood fascia boards, corner boards, etc., on houses, cottages, etc., a carpenter should place 175 to 225 lin. ft. per 8-hr. day, at the following labor cost per 100 lin. ft.:

	Hours	Rate	Total	Rate	Total
Carpenter	4	$....	$....	$24.04	$96.16
Cost per lin. ft.			....		.96

Placing Exterior Wood Cornices, Verge Boards, Etc. When placing exterior wood cornices, verge boards, fascia, soffits, etc., consisting of 2 members, two carpenters working together should place 150 to 175 lin. ft. per 8-hr. day, at the following labor cost per 100 lin ft.:

	Hours	Rate	Total	Rate	Total
Carpenter	10	$....	$....	$24.04	$240.40
Cost per lin. ft.			....		2.40

When the exterior cornices consist of 3 members (crown mold, bed mold, fascia, etc.), two carpenters working together should place 100 to 125 lin. ft. per 8-hr. day, at the following labor cost per 100 lin ft.:

	Hours	Rate	Total	Rate	Total
Carpenter	14.2	$....	$....	$24.04	$341.37
Cost per lin. ft.			....		3.41

If a 4-member wood cornice is used, two carpenters working together should place 60 to 75 lin. ft. per 8-hr. day, at the following labor cost per 100 lin. ft.:

	Hours	Rate	Total	Rate	Total
Carpenter	23.7	$....	$....	$24.04	$569.75
Cost per lin. ft.			....		5.70

The above quantities and costs do not include time blocking out for fascia boards, cornices, etc. An extra allowance should be made for all blocking required.

Placing Brick Mouldings. A carpenter should fit and set around 32 lin. ft. of brick moulding per hour at the following cost per 100 lin. ft.:

	Hours	Rate	Total	Rate	Total
Carpenter	3.2	$....	$....	$24.04	$76.93
Cost per lin. ft.			....		.77

Brick moulding costs about 50 cents per foot in random lengths and 75 cents in specified lengths.

Placing Wood Cupolas. One carpenter should set a prefabricated pine cupola in around two hours.

	Hours	Rate	Total	Rate	Total
Carpenter	2	$....	$....	$24.04	$48.08
Labor	2	....	....	19.13	38.26
Cost per each			$....		$86.34

Porch Work. The labor placing porch work is a variable item owing to the vast difference in the style and construction of porches and the amount of detail involved.

Placing Plain Porch Columns. When placing plain square or turned porch columns, such as commonly used for rear porches and other inexpensive work, a carpenter should place one post in about 1 hr., at the following labor cost:

	Hours	Rate	Total	Rate	Total
Carpenter	1.0	$....	$....	$24.04	$24.04

Placing Porch Top and Bottom Rail and Balusters. When placing wood top and bottom rail and wood balusters, such as used on front porches, a carpenter should complete 15 to 20 lin. ft. of rail per 8-hr. day, at the following labor cost per lin. ft.:

	Hours	Rate	Total	Rate	Total
Carpenter	0.46	$....	$....	$24.04	$11.06

When placing top and bottom rails with open balusters or using matched and beaded ceiling, such as is often used for the cheaper grades of work, a carpenter should complete 35 to 45 lin. ft. of rail per 8-hr. day, at the following labor cost per lin. ft.:

	Hours	Rate	Total	Rate	Total
Carpenter	0.2	$....	$....	$24.04	$4.81

Framing and Erecting Exterior Wood Stairs for Rear Porches. When framing and erecting outside wood stairs for rear porches on apartment buildings, etc., where the stringers are 2"x10" or 2"x12" lumber, with treads and risers nailed on the face of the stringers, it will require 18 to 22 hrs. carpenter time per flight of stairs.

This is for an ordinary stair having 14 to 18 risers extending from story to story, and the labor per flight should cost as follows:

	Hours	Rate	Total	Rate	Total
Carpenter	20	$....	$....	$24.04	$480.80

If the stair consists of 2 short flights with an intermediate landing platform between stories it will require 12 to 13 hrs. carpenter time per flight or 24 to 26 hrs. per story, including platform, at the following labor cost:

	Hours	Rate	Total	Rate	Total
Carpenter	25	$....	$....	$24.04	$601.00

Framing and Erecting Wood Stairs Having Winders. Where the wood stairs to rear porches have 4 to 6 winders in each story height, figure 24 to 28 hrs. carpenter time, at the following labor cost per story:

	Hours	Rate	Total	Rate	Total
Carpenter	26	$....	$....	$24.04	$625.04

up 1st set of illus. page 6.63

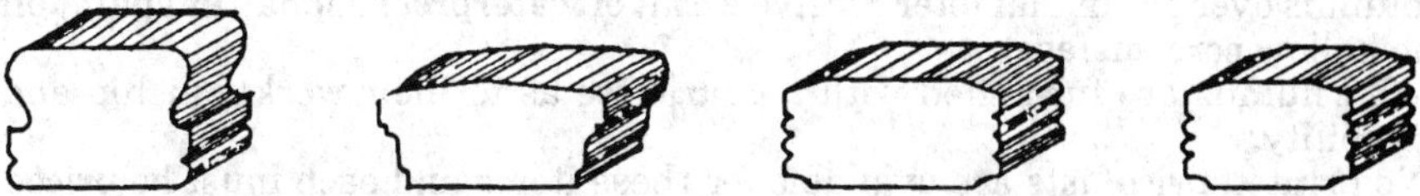

Porch Top Rail and Balustrade Cap

up 2nd set of illus. page 6.63

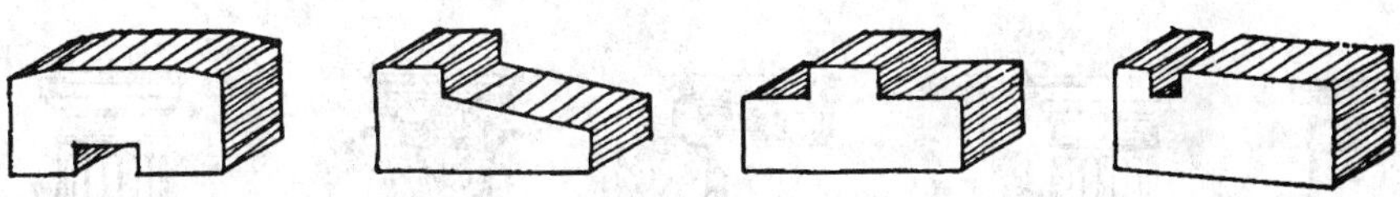

Rabbeted Porch Jamb, Plowed Shoe, and Bottom Rail

Clear Ponderosa Pine Porch Material

Kind of Molding	Size in Inches	Price per 100 Lin. Ft.
Square Baluster Stock	1 1/8 x 1 1/8	$50.00
Square Baluster Stock	1 3/8 x 1 3/8	64.00
Square Baluster Stock	1 5/8 x 1 5/8	75.00
Balustrade Cap	3 5/8 x 1 5/8	130.00
Balustrade Shoe	3 5/8 x 1 5/8	130.00
Rabbeted Porch Jamb	1 1/2 x 3 1/2	130.00
Plowed Porch Shoe	1 1/2 x 3 1/2	130.00
Top Rail	3 x 2 1/4	145.00
Top Rail	3 3/4 x 1 3/4	130.00
Bottom Rail	3 1/2 x 1 3/4	133.00

Cornice Boards*
Select Grade

Measured Size	to S4S	Price
1" x 3"	3/4" x 2 5/8"	$39.00
1" x 4"	3/4" x 3 1/2"	46.00
1" x 6"	3/4" x 5 1/2"	78.00
1" x 8"	3/4" x 7 1/2"	93.00
1" x 10"	3/4" x 9 1/2"	123.00
1" x 12"	3/4" x 11 1/2"	170.00

*Prices are per 100 lin. ft. in random lengths 6'-0" to 16'-0".

Wood Columns

The finest columns for exterior work are made from clear heart redwood staves glued together under clamp pressure with cold water waterproof glue. Shafts are turned in the lathe with the correct entasis, and to insure uniform stave thickness after turning, the staves are made straight for the lower third of the shaft and swelltapered for the upper two-thirds thus giving the correct entasis in the rough. Any style of fluting or reeding may be had.

The columns receive a prime coat of lead and oil on the exterior surface. Columns over 14" in diameter receive a coat of waterproof asphaltum paint on the hollow core surface.

All columns are furnished with a guarantee as to their workmanship and durability.

No printed price lists are available for these items and each must be priced at the factory. In order of price, the Roman is cheapest followed by the unfluted Ionic, the fluted Attic, the fluted Doric and finally the fluted Corinthian. A 12" diameter Doric style fluted column 10' high with a plinth and cap is currently quoted at around $318.00 at the factory.

Stock, tapered column shafts in pine can be figured at around $27.00 per vertical lineal foot in 12" diameter; $50.00 in 14"; $56.00 in 18"; and $70.00 in 24".

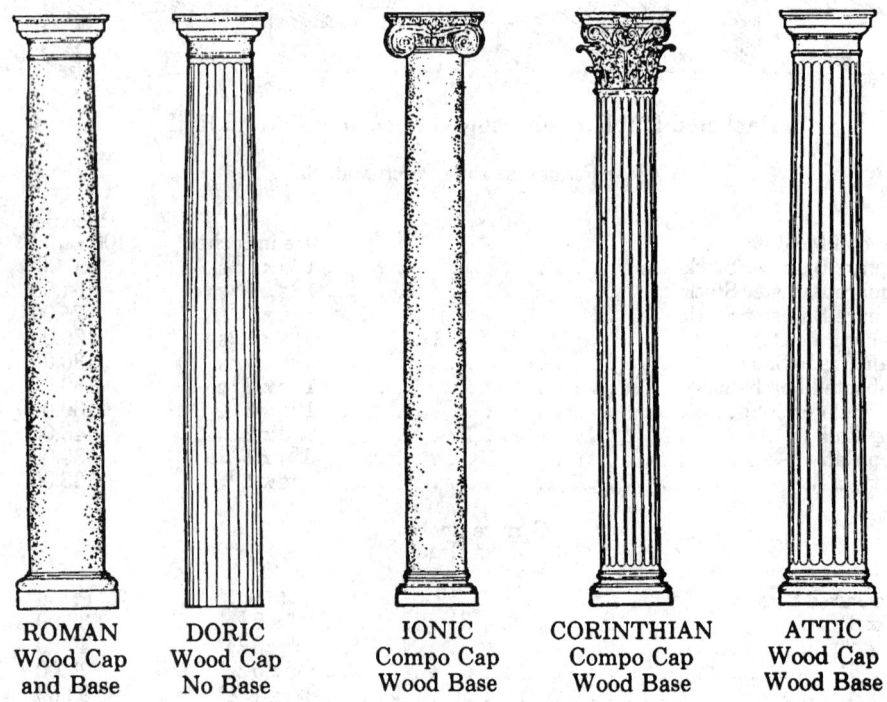

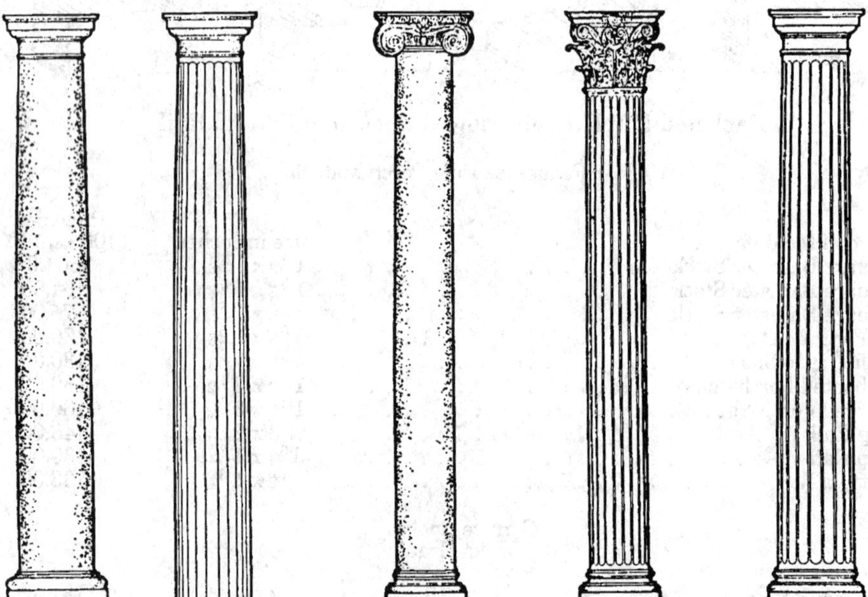

ROMAN Wood Cap and Base

DORIC Wood Cap No Base

IONIC Compo Cap Wood Base

CORINTHIAN Compo Cap Wood Base

ATTIC Wood Cap Wood Base

Fiberglass columns cast in a colonial design will run $6.00 per lin. ft. in 6" diameter; $7.50 in 8"; and $9.00 in 10".

Aluminum columns extruded to a fluted design are stocked in assorted sizes from 6" diameter in 8' lengths to 12" diameter in 18' lengths. They may be used as loadbearing members and come factory finished in a high-gloss, white, baked enamel. An 8" diameter shaft 12' long will run around $49.50. If no base or cap is used it may be attached to the slab and ceiling by cutting 2" lumber to fit the column diameter and bolting it to the adjacent construction. If a base and cap is ordered they will be furnished in aluminum castings and will run around $11.50 each.

Square porch columns costs about $2.00 per foot in 4x4 fir. Turned colonial posts in 8' lengths will about $15.00 in 4x4 stock, $22.00 in 5x5 stock. It takes two carpenters to set columns, and they should erect a medium sized column in an hour.

Cast Iron Column Bases. Cast iron column bases keep columns off the floor and permit air to circulate on the inside of the column.

Prices of Cast Iron Column Bases

Base Size	Height	For Col. Diam.	Wt., Lbs.	Approx. Price
11" x 11"	2"	8"	18	$32.00
13 ½" x 13 ½"	2 ½"	10"	24	47.00
16" x 16"	2 ½"	12"	32	62.00
18 ½" x 18 ½"	2 ½"	14"	35	90.00
21 ¼" x 21 ¼"	2 ½"	16"	54	120.00
24" x 24"	3 ¼"	18"	87	140.00
26 ½" x 26 ½"	3 ¼"	20"	96	160.00

*For Doric columns without base use one size smaller.

Cast iron bases for pilasters are ⅔ the price of bases for corresponding columns.

Placing Plywood Soffits. Two carpenters should set 75 sq. ft. of ¼" and 64 sq. ft. of ½" plywood on soffits per hour at the following labor cost per 100 sq. ft.:

	Hours	Rate	Total	Rate	Total
Carpenter	2.8	$....	$....	$24.04	$67.31
Cost per sq. ft			....		.67

Placing Matched and Beaded Ceiling. Matched and beaded ceiling is usually furnished 4" and 6" wide and ⅜", ½", and ¾" thick. Where the 4" material is used for porch ceilings, overhang soffits, etc., a carpenter should place 285 to 325 b.f. per 8-hr. day, at the following labor cost per 1,000 b.f.:

	Hours	Rate	Total	Rate	Total
Carpenter	26	$....	$....	$24.04	$625.04

The labor cost per square (100 sq. ft. surface) requiring 128 b.f. of ceiling should run as follows:

	Hours	Rate	Total	Rate	Total
Carpenter	3.33	$....	$....	$24.04	$80.05

When placing 6" wide matched and beaded ceiling, a carpenter should place 400 to 450 b.f. of ceiling per 8-hr. day, at the following labor cost per 1,000 b.f.:

	Hours	Rate	Total	Rate	Total
Carpenter	18.8	$....	$....	$24.04	$451.95

The labor cost per square (100 sq. ft. surface) requiring 120 b.f. of ceiling should run as follows:

	Hours	Rate	Total	Rate	Total
Carpenter	2.25	$....	$....	$24.04	$54.09

Placing Exterior Door Trim. Much of the architectural type door trim today is of molded material rather than milled wood. It is pre-engineered to fit together with little or no cutting or fitting, providing the door frame itself is set plumb and true. Stock trim can be ordered for widths of 3', 5', and 6' and for either brick or wood wall openings. Side trim is delivered for 7' door height but designed so that 4" may be cut off to fit a 6'-8" door without altering the design. Pilasters are nominally 8" in width. Cornices are 6" to 8" high and available in several styles. In addition pediments may be added. A basic cornice piece costs around $53.00 for a 3' opening, $80.00 for the larger ones. Pediments will run from $53.00 to $80.00 extra, for 3'-0" openings and up to $125.00 for the double door openings. Pilaster assemblies run $69.00 per pair. Two carpenters should set a cornice in 1/2 hour, the pediment in 1/2 hour, and the pilasters in one hour.

Outside Blinds or Shutters

Shutters are available in wood, aluminum, and molded plastic. Because of the popularity of the latter, it is not always easy to find wood stocked in all sizes. Custom made units will run considerably more than the prices given below.

All shutters have stiles, top and middle rails 2 1/4" wide; bottom rail 4 1/2" wide. All slats are stationary.

Shutters with raised panels can be fitted with decorative cutouts in the upper panel at an extra cost of $1.55 per pair. Designs include pine trees, cloverleafs, squirrels, crescents, and roosters.

A carpenter should hang 8 to 10 pair a day depending on size and height off the ground.

Price Per Pair, Hemlock, Preservative Treated

Window Size	w/Fixed Slats	w/Raised Panel
2'-0"x3'-3"	$14.00	$18.00
2'-0"x3'-11"	16.50	20.00
2'-0"x4'-7"	19.00	24.00
2'-0"x5'-3"	21.00	29.00
2'-4"x3'-3"	14.50	17.25
2'-4"x3'-7"	15.25	18.50
2'-4"x3'-11"	16.50	19.75
2'-4"x4'-3"	17.50	21.25
2'-4"x4'-7"	20.00	22.50
2'-4"x4'-11"	21.50	24.00

Window Size	w/Fixed Slats	w/Raised Panels
2'-8"x3'-3"	15.00	17.75
2'-8"x3'-7"	15.50	19.00
2'-8"x3'-11"	17.00	20.00
2'-8"x4'-3"	18.50	22.00
2'-8"x4'-7"	20.00	23.00
2'-8"x4'-11"	21.00	24.50
2'-8"x5'-3"	23.00	26.00
3'-0"x3'-3"	20.00	22.00
3'-0"x3'-7"	21.00	23.75
3'-0"x3'-11"	22.00	25.00
3'-0"x4'x3"	23.00	27.00
3'-0"x4'-7"	24.00	28.00
3'-0"x4'-11"	25.00	29.50
3'-0"x5'-3"	26.00	30.00
3'-4"x3'-3"	21.00	23.50
3'-4"x3'-11"	23.00	26.00
3'-4"x4'-7"	25.00	29.00
3'-4"x4'-11"	26.00	30.00
3'-4"x5'-3"	28.00	32.00
3'-8"x4'-7"	23.00	30.50
3'-8"x4'-11"	25.00	32.00
3'-8"x5'-3"	27.00	33.75

Door Blinds

Same as above, except 3 panels of slats in height.

Size	Price
2'-8"x6'-9", per pair	$29.00
3'-0"x6'-9", per pair	30.00

Style "A" Style "B" Style "C"
Outside Blinds or Shutters

Aluminum Shutters

Metal shutters of aluminum are designed to be screwed on the wall permanently. The louver section is stamped out of one piece of metal. They are furnished in white, black, green or brown and come packaged with mounting screws.

Size	Price per Pair
2'-4"x34"	$15.50
2'-4"x36"	16.00
2'-4"x39"	16.75
2'-4"x43"	17.50
2'-4"x48"	18.00
2'-4"x55"	19.00
2'-4"x64"	22.00
2'-4"x80"	26.00
2'-8"x34"	16.50
2'-8"x36"	17.00
2'-8"x39"	17.75
2'-8"x43"	18.50
2'-8"x48"	19.00
2'-8"x55"	20.50
2'-8"x64"	23.00
2'-8"x80"	28.00

One carpenter should hang one pair in .8 hour.

Molded Shutters

Shutters are now molded of thermo-formed polymer. They are prefinished in either black or white, but may be painted with an exterior latex paint. They are screwed to siding with 2" aluminum screws, 4 per unit to 55", 6 per unit over 55".

Size Per Panel	Price per Pair
14"x35"	$23.00
14x39	23.75
14x43	25.00
14x47	26.25
14x51	27.00
14x55	28.50
14x59	29.50
14x63	33.00
14x67	34.00
14x71	35.00
14x75	36.50
16"x80"	37.50

Labor Costs on Half Timber Work. The following costs on half timber work are taken from a college building. The timber was white oak 3" thick, 10" and 11" wide, and various lengths to meet the requirements of the job.

Nothing was done to the face of the timber and it was left just as received. The backs of all timbers were beveled, the thickness remaining 3 inches. A shop was set up on the job where the timbers were cut, beveled, and fit ready for erection.

There were 14,500 b.f. of lumber used in this half-timber work and the labor cost per 1,000 b.f. was as follows:

	Hours	Rate	Total	Rate	Total
Carpenter	66.3	$....	$....	$24.04	$1,593.85
Labor helping	12.0	....	....	19.13	229.56
Cost per 1,000 b.f.			$....		$1,823.41

Half Timber Work on Commercial Building

After the timbers were cut, beveled, and fit in the shop, the labor cost of placing and bracing the timbers in the building was as follows per 1,000 b.f.:

	Hours	Rate	Total	Rate	Total
Carpenter	41.5	$....	$....	$24.04	$997.66
Labor helping	2.2	....	....	19.13	42.09
Cost per 1,000 b.f.			$....		$1,039.75

INTERIOR FINISH CARPENTRY

When estimating the labor cost of interior finish, it is advisable to ascertain the grade of workmanship required by the architect or owner, because the quantity of work a carpenter will perform per hour or per day will vary considerably with the grade of workmanship required.

The costs given on the following pages embody two distinct grades of workmanship. The grade called "ordinary work" is by far the most common and is usually found in medium priced residences, cottages, apartment buildings, apartment buildings with stores underneath, factory and warehouse buildings, non-fireproof store, office and school buildings, and other buildings of this class.

The other grade, known as "first grade workmanship", is usually required in finishing fine residences, high-class fireproof apartments, hotel, bank and office buildings, high grade store buildings, fireproof school and university buildings, courthouses, city halls, state capitols, post offices, and other buildings of this type. The interior finish in buildings of this class is usually selected birch, selected gumwood, plain or quarter-sawed oak, mahogany, walnut or other first-class hardwoods, and is discussed later in this chapter under "Architectural Woodwork".

Placing Wood Base. This cost will vary with the size of the rooms and whether a single-, two- or three-member base is specified.

Where there are 55 to 60 lin. ft. of 2 member base in each room without an unusually large amount of cutting and fitting, a carpenter should place 125 to 150 lin. ft. per 8-hr day, at the following labor cost per 100 lin. ft.:

	Hours	Rate	Total	Rate	Total
Carpenter	5.8	$....	$....	$24.04	$139.43
Labor	1.0	....	....	19.13	19.13
Cost per 100 lin. ft			$....		$158.56
Cost per lin. ft			....		1.59

If there are an unusually large number of miters to make, such as required in closets and other small rooms, increase the above costs accordingly.

A carpenter should place almost as many lin. ft. of 3-member base (consisting of 2 base members and a carpet strip), as 2-member base (consisting of one member and carpet strip), as it is much easier to fit a small top member against the plastered wall than it is to nail a wide piece of base so that it will fit snug against the wall and follow the irregularities in the plaster.

Where there are 50 to 60 lin. ft. of base in a room, a carpenter should place 110 to 130 lin. ft. per 8-hr. day, at the following labor cost per 100 lin. ft.:

	Hours	Rate	Total	Rate	Total
Carpenter	6.7	$....	$....	$24.04	$161.07
Labor	1.0	....	....	19.13	19.13
Cost per 100 l.f.			....		$180.20
Cost per lin. ft.			....		1.80

First Grade Workmanship

In average size rooms, a carpenter should place 100 to 115 lin. ft. of 2-member hardwood base per 8-hr. day at the following labor cost per 100 lin. ft.:

	Hours	Rate	Total	Rate	Total
Carpenter	7.4	$....	$....	$24.04	$177.90
Labor	1.0	....	....	19.13	19.13
Cost per 100 l.f.			$....		$197.03
Cost per lin. ft.			....		1.97

Where 3-member hardwood base is used in average size rooms, a carpenter should place 85 to 100 lin. ft. (2 ordinary rooms) per 8-hr. day, at the following labor cost per 100 lin. ft.:

	Hours	Rate	Total	Rate	Total
Carpenter	8.7	$....	$....	$24.04	$209.15
Labor	1.0	....	....	19.13	19.13
Cost per 100 l.f.			$....		$228.28
Cost per lin. ft.			....		2.28

On work of this class, the wood grounds should be straight, so that it will not be necessary to "force" the wood base to make it fit tight against the finished wall.

Where a single 1"x4" pine base is to be fitted to straight runs, a single carpenter should set around 200 lin. ft. per day at the following cost per 100 lin. ft.:

	Hours	Rate	Total	Rate	Total
Carpenter	4	$....	$....	$24.04	$96.16
Cost per lin. ft					.96

Placing Wood Picture Molding. Where just an ordinary grade of workmanship is required, a carpenter should place picture molding in 5 or 6 ordinary sized rooms per 8-hr. day. This is equivalent to 250 to 275 lin. ft. at the following labor cost per 100 lin. ft.:

	Hours	Rate	Total	Rate	Total
Carpenter	3.0	$....	$....	$24.04	$72.12
Labor	0.5	....	....	19.13	9.57
Cost per 100 l.f.			$....		$81.69
Cost per lin. ft.			....		.82

First Grade Workmanship

Where the wood picture molding must fit close to the plastered walls with perfect fitting miters, a carpenter should place molding in 4 to 5 ordinary sized rooms per 8-hr. day. This is equivalent to 175 to 200 lin. ft., at the following labor cost per 100 lin.ft.:

	Hours	Rate	Total	Rate	Total
Carpenter	4.4	$....	$....	$24.04	$105.78
Labor	0.5	....	....	19.13	9.57
Cost per 100 l.f.			$....		$115.35
Cost per lin. ft			....		1.15

If the picture molding is placed in fireproof buildings having tile or brick partitions, it will be necessary to place wood grounds for nailing the picture molding but in non-fireproof buildings the nails may be driven into the plaster, as the nails will obtain a bearing in the wood studs or wall furring.

Placing Wood Chair or Dado Rail. In large rooms, long, straight corridors, etc., a carpenter should fit and place 275 to 300 lin. ft. of wood chair rail per 8-hr. day, at the following labor cost per 100 lin. ft.:

	Hours	Rate	Total	Rate	Total
Carpenter	2.8	$....	$....	$24.04	$67.31
Labor	0.5	....	....	19.13	9.57
Cost per 100 l.f.			$....		$76.88
Cost per lin. ft			....		.77

In small kitchens, pantries, closets, bathrooms, etc., a carpenter will place only 160 to 180 lin. ft. of chair rail per 8-hr. day, at the following labor cost per 100 lin. ft.:

	Hours	Rate	Total	Rate	Total
Carpenter	4.7	$....	$....	$24.04	$112.99
Labor	0.5	....	....	19.13	9.57
Cost per 100 l.f.			$....		$122.56
Cost per lin. ft.			....		1.23

First Grade Workmanship

Where first grade workmanship is required, a carpenter should place 200 to 225 lin. ft. of chair rail per 8-hr. day, at the following labor cost per 100 lin. ft.:

	Hours	Rate	Total	Rate	Total
Carpenter	3.8	$....	$....	$24.04	$91.35
Labor	0.5	....	....	19.13	9.57
Cost per 100 l.f.			$....		$100.92
Cost per l.f.			....		1.01

In small rooms, such as kitchens, pantries, bathrooms, etc., requiring considerable cutting and fitting around medicine cabinets, wardrobes, kitchen cases, etc., a carpenter should place 120 to 135 lin. ft. of chair rail per 8-hr. day, at the following labor cost per 100 lin. ft.:

	Hours	Rate	Total	Rate	Total
Carpenter	6.3	$....	$....	$24.04	$151.45
Labor	0.5	....	....	19.13	9.57
Cost per 100 l.f.			$....		$161.02
Cost per lin. ft			....		1.61

Placing Wood Cornices. Where 3- or 4-member wood cornices are placed in living rooms, reception rooms, dining rooms, etc., a carpenter should place cornice in one average sized room per 8-hr. day, which is equivalent to 50 to 60 lin. ft., and the labor cost per 100 lin. ft. would be as follows:

	Hours	Rate	Total	Rate	Total
Carpenter	14.5	$....	$....	$24.04	$348.58
Labor	2.0	....	....	19.13	38.26
Cost per 100 l.f.			$....		$386.84
Cost per lin. ft			....		3.87

First Grade Workmanship

Where it is necessary that the wood members fit the plastered walls and ceilings closely, with all miters true and even, two carpenters working together should complete one to one and a quarter rooms per day.

This is at the rate of 35 to 40 lin. ft. per 8-hr. day for one carpenter, at the following labor cost per 100 lin. ft.:

	Hours	Rate	Total	Rate	Total
Carpenter	21	$....	$....	$24.04	$504.84
Labor	2	....	....	19.13	38.26
Cost per 100 l.f.			$....		$543.10
Cost per lin. ft			....		5.43

Placing Vertical Wood Panel Strips. When vertical wood panel strips or "battens" are nailed to plastered walls to produce a paneled effect, a carpenter should place 22 to 28 pcs. (175 to 225 lin. ft.) per 8-hr. day, at the following labor cost per 100 lin. ft.:

	Hours	Rate	Total	Rate	Total
Carpenter	4.0	$....	$....	$24.04	$96.16
Labor	0.5	....	....	19.13	9.57
Cost per 100 l.f.			$....		$105.73
Cost per lin. ft			....		1.06

Placing Wood Strip Paneling. Where panels are formed of wood molding 1 ½" to 2 ½" wide, making it necessary to cut and miter both ends of each panel strip, the lin. ft. cost will vary with the size of the panels and the amount of cutting and fitting necessary, as there is almost as much labor required on a panel 2'-0"x3'-0" as one 3'-0"x6'-0", although there are only half as many lin. ft. in the former as in the latter.

On small sized panels up to 2'-0"x4'-0", requiring 12 lin. ft. of molding, a carpenter should complete 9 to 11 panels, containing 110 to 135 lin. ft. of molding per 8-hr. day, at the following labor cost per 100 lin. ft.:

	Hours	Rate	Total	Rate	Total
Carpenter	6.5	$....	$....	$24.04	$156.26
Labor	0.5	....	....	19.13	9.57
Cost per 100 l.f.			$....		$165.83
Cost per lin. ft			....		1.66

On larger panels 3'-0"x5'-0" to 4'-0"x6'-0" in size, where each panel contains 16 to 20 lin. ft. of molding, a carpenter should complete 7 to 9 panels, containing 140 to 180 lin. ft. of molding per 8-hr. day, at the following labor cost per 100 lin. ft.:

	Hours	Rate	Total	Rate	Total
Carpenter	5.0	$....	$....	$24.04	$120.20
Labor	0.5	....	....	19.13	9.57
Cost per 100 l.f.			$....		$129.77
Cost per lin. ft.			....		1.30

First Grade Workmanship

Where wood panel moldings are used over canvassed or burlap walls, with all strips plumb and level, fitting closely to the plastered walls with perfect fitting miters, a carpenter should complete 7 to 9 small panels, requiring 90 to 115 lin. ft. of molding per 8-hr. day, at the following labor cost per 100 lin. ft.:

	Hours	Rate	Total	Rate	Total
Carpenter	7.8	$....	$....	$24.04	$187.51
Labor	0.5	....	....	19.13	9.57
Cost per 100 l.f.			$....		$197.08
Cost per lin. ft			....		1.97

On larger panels 3'-0"x5'-0" to 4'-0"x6'-0" in size, where each panel contains 16 to 20 lin. ft. of panel molding, a carpenter should complete about 6 to 8 panels, containing 120 to 150 lin. ft. of molding per 8-hr. day, at the following labor cost per 100 lin. ft.:

	Hours	Rate	Total	Rate	Total
Carpenter	6.0	$....	$....	$24.04	$144.24
Labor	0.5	....	....	19.13	9.57
Cost per 100 l.f.			$....		$153.81
Cost per lin. ft			....		1.54

Placing Wood Ceiling Beams. In buildings where built-up ceiling beams are used, the labor costs will vary according to the number of intersections of beams in each room and the length of the beams, as it is just as easy to erect a 12'-0" built-up beam as an 8'-0" one.

On average work, a carpenter should place 35 to 45 lin. ft. of built-up wood beams per 8-hr. day, at the following labor cost per 100 lin. ft.:

	Hours	Rate	Total	Rate	Total
Carpenter	20	$....	$....	$24.04	$480.80
Labor	3	....	....	19.13	57.39
Cost per 100 l.f.			$....		$538.19
Cost per lin. ft			....		5.38

First Grade Workmanship

In the better class of buildings using wood ceiling beams, a carpenter should place 30 to 35 lin. ft. per 8-hr. day, at the following labor cost per 100 lin. ft.:

	Hours	Rate	Total	Rate	Total
Carpenter	25	$....	$....	$24.04	$601.00
Labor	3	....	....	19.13	57.39
Cost per 100 l.f.			$....		$658.39
Cost per lin. ft			....		6.58

Ponderosa Pine Moldings

Prices are per 100 lin. ft. unless noted otherwise.

Aprons
- 11/16"x2 1/4" . . . $46.00
- 11/16"x2 1/4" ogee . . . 46.00
- 9/16"x3 1/2" . . . 85.00

Astragals
- 1 3/8"x7'-0" . . . 3.15 ea.
- 1 1/4"x7'-0" . . . 3.50 ea.

Bases
- 9/16"x3" . . . 40.00
- 1/2"x3 1/2" . . . 42.00
- 1/2"x4 1/4" . . . 48.00

Base Cap
- 11/16"x1 3/8" . . . 20.00

Backband
- 11/16"x1 1/16" . . . 16.00

Balusters
- 3/4"x 3/4" . . . 25.00
- 1 1/16"x1 1/16" . . . 40.00
- 1 5/16"x1 5/16" . . . 65.00
- 1 5/8"x1 5/8" . . . 70.00

Base Shoe Combination
- 3/4"x2 5/8" . . . 30.00

Bed Molding
- 11/16"x1 3/4" . . . 35.00
- 11/16"x2 1/4" . . . 50.00

Base Shoe
- 1/2"x 3/4" . . . 20.00

Blind Stop
- 3/4"x1 3/8" . . . 40.00

Brick Molding
- 1 5/16"x2" . . . 60.00

Casings
- 11/16"x2 1/4" . . . 37.00
- 11/16"x3 1/2" . . . 68.00

Chair Rails
- 11/16"x2 1/8" . . . 50.00
- 7/16"x2 1/2" . . . 56.00

Corner Guards
- 3/4"x 3/4" . . . 21.00
- 1 1/8"x1 1/8" . . . 42.00
- 1 5/16"x1 5/16" . . . 46.00

Coves
- 11/16"x3 1/4" . . . 70.00
- 11/16"x2 1/4" . . . 48.00
- 11/16"x1 3/4" . . . 34.00
- 3/4"x 7/8" . . . 26.00
- 3/4"x 3/4" . . . 24.00
- 1/2"x 1/2" . . . 20.00
- 11/16"x1 1/8" . . . 27.00

Crowns
- 11/16"x4 1/4" . . . 102.00
- 11/16"x3 1/4" . . . 68.00
- 11/16"x2 1/4" . . . 48.00

Drip Caps
- 1 1/16"x1 5/8" . . . 48.00
- 3/4"x1 5/8" . . . 42.00

Full Rounds
- 1 5/8" . . . 75.00
- 1 5/16" . . . 50.00

Glazing Beads
- 1/2"x 9/16" . . . 18.00
- 3/8"x 3/8" . . . 15.00

Half Rounds
- 5/16"x 5/8" . . . 20.00

Hand Rails
- 1 1/2"x1 3/4" . . . 58.00

Hook Strips
- 9/16"x2 7/16" . . . 42.00

Jamb Extensions

$^3/_4$"x1 $^{15}/_{16}$" 37.00
$^3/_4$"x $^7/_8$" 28.00

Lattice
$^1/_4$"x1 $^1/_8$" 13.00
$^1/_4$"x1 $^3/_8$" 15.00
$^1/_4$"x1 $^3/_4$" 19.00
$^1/_4$"x2 $^1/_4$" 26.00

Mullion Casings
$^9/_{16}$"x5 $^1/_2$" 85.00
$^3/_{16}$x2 . . . . .42.00

Parting Strip
$^1/_2$"x $^3/_4$" 16.00

Picture Molding
$^3/_4$"x1 $^3/_4$" 39.00

Quarter Rounds
$^1/_4$"x $^1/_4$" 6.00
$^1/_2$"x $^1/_2$" 12.00
$^3/_4$"x $^3/_4$" 20.00
1 $^1/_{16}$"x1 $^1/_{16}$" 38.00

Screen Moldings
$^5/_{16}$"x $^5/_8$" 12.00
$^1/_4$"x $^3/_4$" 10.00

Screen Stock
$^3/_4$"x1 $^3/_4$" 40.00
$^3/_4$"x2 $^3/_4$" 61.00
1 $^1/_{16}$"x1 $^3/_4$" 63.00
1 $^1/_{16}$"x2 $^3/_4$" 75.00

Shelf Cleat
$^{11}/_{16}$"x1 $^1/_2$" 20.00

Stools
1 $^1/_{16}$"x3 $^1/_4$" 170.00
1 $^1/_{16}$"x3 $^5/_8$" 185.00
1 $^1/_{16}$"x3 $^1/_4$" 95.00
1 $^1/_{16}$"x2 $^3/_4$" 70.00
1 $^1/_{16}$"x2 $^1/_4$" 65.00
1 $^1/_{16}$"x2 $^1/_4$" 90.00

Stops
$^7/_{16}$"x2 $^1/_8$" 48.00
$^7/_{16}$"x1 $^5/_8$" 42.00
$^7/_{16}$"x1 $^3/_8$" 37.00
$^7/_{16}$"x1 $^1/_8$" 31.00
$^7/_{16}$"x $^7/_8$" 30.00
$^7/_{16}$"x $^{15}/_{16}$" 31.00
$^7/_{16}$"x $^5/_8$" 24.00

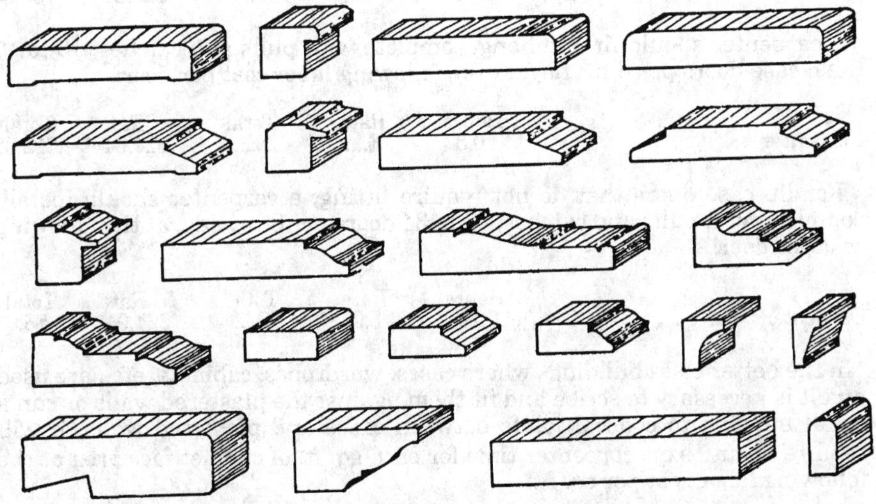

Wood Interior Trim, Casings, Back-band, Stops, Stool, Etc.

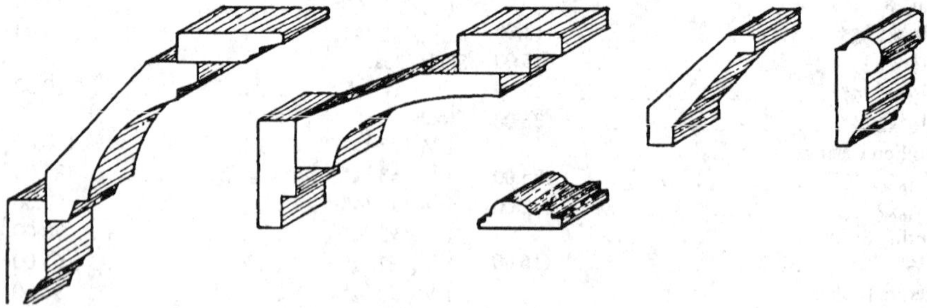

Wood Ceiling Cornice and Picture Mold

PLASTIC LAMINATE COUNTERTOPS

Plastic laminate for countertops will vary in price depending on the number of cutouts, corners, and whether the fabricator must include the cost of taking job measurements. A typical L-shaped kitchen countertop covering 18 ft. of inside wall, all edges plastic, laminated to 7/8" particle board, with 4" splashes back and ends, will cost around $35.00 per lineal foot in a standard pattern. Rounded edges will cost considerably more; metal banded ones somewhat less. For each cut out add $15.00. For each end splash add $10.00.

Setting Cabinets and Cases. When setting mill assembled cabinets and cases, such as used in the average residence or apartment building, it will require about 1/6 hr. carpenter time for each sq. ft. of cabinet face area at the following labor cost per sq. ft.:

	Hours	Rate	Total	Rate	Total
Carpenter	0.16	$....	$....	$24.04	$3.85

A carpenter should fit and hang, complete with pulls and catches, 15 to 17 flush case doors per 8-hr. day, at the following labor cost per door:

	Hours	Rate	Total	Rate	Total
Carpenter	0.5	$....	$....	$24.04	$12.02

For lip case doors that do not require fitting, a carpenter should install, complete with pulls and catches, 30 to 34 doors per 8-hr. day, at the following cost per door:

	Hours	Rate	Total	Rate	Total
Carpenter	0.25	$....	$....	$24.04	$6.01

In the better class buildings where cases, wardrobes, cabinets, etc., are used and it is necessary to scribe and fit them against the plastered walls or run a small molding to conceal joints between cases and plastered walls, it will require about 1/3 hr. carpenter time for each sq. ft. of cabinet face area at the following labor cost per sq. ft.:

	Hours	Rate	Total	Rate	Total
Carpenter	0.33	$....	$....	$24.04	$7.93

A carpenter should fit and hang, complete with pulls and catches, 11 to 13 flush case doors per 8-hr. day, at the following labor cost per door:

	Hours	Rate	Total	Rate	Total
Carpenter	0.67	$....	$....	$24.04	$16.11

Often, in restricted areas, such as adding cabinets in narrow lavatory areas, it is necessary to have cabinet work delivered in sections so that it may be turned and properly set within the space. Seldom is a job so organized that cabinet work is delivered before wall finishes are up, so the fit must be preplanned and exact.

Setting Factory Assembled and Finished Kitchen Cabinets. When setting factory assembled and finished kitchen cabinets, such as used in the average apartment or residence, it will require about 0.1 hr. carpenter time for each sq. ft. of cabinet face area at the following labor cost per sq. ft.:

	Hours	Rate	Total	Rate	Total
Carpenter	0.1	$....	$....	$24.04	$2.40

First Grade Workmanship

In better class apartments and residences where a high class job of installation is required it will take about 1/6 hr. carpenter time for each sq. ft. of cabinet face area at the following labor cost per sq. ft.:

	Hours	Rate	Total	Rate	Total
Carpenter	0.16	$....	$....	$24.04	$3.85

Setting Wood Fireplace Mantels. If wood fireplace mantels are factory assembled, merely requiring fitting and setting, a carpenter should install 2 to 3 mantels per 8-hr. day at the following labor cost per mantel:

	Hours	Rate	Total	Rate	Total
Carpenter	3.2	$....	$....	$24.04	$76.93
Labor	1.0	....	....	19.13	19.13
Cost per mantel		....	$....		$96.06

Fitting and Placing Closet Shelving. A carpenter should fit and place 115 to 135 sq. ft. of closet shelving per 8-hr. day including setting of shelf cleats at the following labor cost per 100 sq. ft.:

	Hours	Rate	Total	Rate	Total
Carpenter	6.4	$....	$....	$24.04	$153.86
Cost per sq. ft			....		1.54

Installing Hanging Rods in Closets. A carpenter should install about three hanging rods, including supports, per hr. at the following labor cost per rod:

	Hours	Rate	Total	Rate	Total
Carpenter	0.33	$....	$....	$24.04	$7.93

Setting Metal Medicine Cabinets. A carpenter should set 10 to 12 average size medicine cabinets per 8-hr. day at the following labor cost per cabinet:

	Hours	Rate	Total	Rate	Total
Carpenter	0.75	$....	$....	$24.04	$18.03

Setting Bathroom Accessories. In an 8-hr. day a carpenter should locate and set 30 to 34 bathroom accessories, such as towel bars, soap dishes, paper holders, etc. at the following labor cost per accessory:

	Hours	Rate	Total	Rate	Total
Carpenter	0.25	$....	$....	$24.04	$6.01

Installing Chalkboards, Corkboards, and Accessories. To install chalkboards, before starting, all grounds should be in place and final coat of finish should be thoroughly dry.

Chalkboard should be set so that all edges will be adequately covered by trim moldings. The writing surface should be plumb and in a true plane with 1/4" clearance on all sides. Chalkboard panels should be nailed at the top edges only and the balance secured to the wall with an even coat of spotting cement.

Two carpenters working together should install 125 to 150 sq. ft. of chalkboard per 8-hr. day at the following labor cost per 100 sq. ft.:

	Hours	Rate	Total	Rate	Total
Carpenter	11	$....	$....	$24.04	$264.44
Cost per sq. ft			....		2.64

Installation of corkboard is similar to that of chalkboard except nailing is eliminated. The labor cost of installing corkboard is approximately the same as for chalkboard.

Installation of aluminum trim moldings in conjunction with chalkboards and corkboards may be accomplished either with exposed wood screws or with concealed snap-on clips. A carpenter should install 90 to 110 lin. ft. of molding per 8-hr. day at the following labor cost per 100 lin. ft.:

	Hours	Rate	Total	Rate	Total
Carpenter	8	$....	$....	$24.04	$192.32
Cost per lin. ft			....		1.92

Labor Placing Finish Hardware

Description of Work	No. Set per 8-Hr. Day	Carp. Hrs. Each
Rim locks or night latches, cheap work	20-24	0.3-0.4
Mortised locks in soft or hardwood doors, Ordinary Workmanship	12-16	0.5-0.7
Mortised locks in hardwood doors, First Grade Workmanship	6-8	1.0-1.3*
Cylindrical locks	17-23	0.3-0.5
Front entrance cylinder locks in hardware doors, Ordinary Workmanship	6-8	1.0-1.3**
Front entrance cylinder locks in hardwood doors, First Grade Workmanship	4-5	1.6-2.0***
Panic bolts, First Grade Workmanship	2-3	2.7-4.0
Door closers, exposed	7-9	0.9-1.1
Door closers, concealed	2-3	2.7-4.0
Door holders	20-24	0.3-0.4
Sash lifts and locks, No. of windows	20-24	0.3-0.4
Cremone bolts for casement doors, windows	20-24	0.3-0.4
Kickplates, Ordinary Workmanship	8-10	0.8-1.0
Kickplates, First Grade Workmanship	6-8	1.0-1.3

*On this grade of work the lock must be flush with the face of the door and must operate perfectly.

**On this class of work the architect or owner must expect to have more or less trouble with hardware because the time given is not sufficient to adjust lock and see that it operates properly, etc.
***This includes front door locks having fancy escutcheon plates, knobs, knockers, etc.

Labor Placing Cabinet Hardware

Description of Work	No. Set per 8-Hr. Day	Carp Hrs. Each
Surface butts	180-200	.040-.045
Offset butts	90-100	.080-.090
Elbow catches	90-100	.080-.090
Friction catches	90-100	.080-.090
Drawer and door pulls	140-150	.053-.057
Knobs	180-200	.040-.045
Rim latches	90-100	.080-.090
Mortised locks	20-24	.3-.4

Mantels

There are many stock mantel designs from which to choose. A simple molding 6" wide extending out from the wall 2" on 3 sides will run around $55.00 for an outside dimension of 4'x7'. A full mantel with 7" shelf and of colonial design will cost from $100.00 to $150.00. More elaborate designs in Georgian or French provincial and with end returns to fit over the brick projection will run from $175.00 to $500.00. These costs are for wood members only. Brick or marble trim around the fireplace opening are, of course, extra. It will take a carpenter 3 to 4 hours to set a prefabricated unit in place.

China Cabinets

Corner cabinets can be purchased from stock to add architectural interest to dining, breakfast, or recreation rooms. Mostly of Ponderosa pine with either Colonial or Georgian design, they will vary from $100.00 for a unit 2'-8" wide with single doors top and bottom to $235.00 for a unit 3'-6" wide with double doors. Cabinets are from 7' to 7'-6" high. Bottom doors are usually paneled, tops factory glazed. Shipped with front assembled and body knocked down, the unit includes 3 shelves for the upper portion, 1 for the lower. Hardware is not included. Allow three hours of one carpenter's time to fit and set each unit, and apply the hardware.

Ironing Board Cabinet

A built-in feature that appeals to many people is an ironing board cabinet unit complete with adjustable board plus sleeve board, all operating hardware, and flush door which is sized to fit between wall studs 16" on centers. It is shipped in a carton and costs around $80.00. The door is available in either birch or Philippine mahogany. No casing is included, but if the unit is set in a matching paneled wall, it is almost inconspicuous.

Louvered Pine Panels

Shuttered panels are available in stock sizes, complete with hardware, or customized to fit exact openings. They are not only used inside window openings, but for cabinet doors and room separators.

Cost of Stock Single Panel Horizontal Slat Unit

	Widths				
Height	6"	7"	8"	10"	12"
20	$2.90	$3.20	$3.60	$4.45	$5.30
23	3.45	3.70	4.10	4.95	5.80
29	3.90	4.25	4.90	5.60	6.75
32	4.45	4.75	5.40	6.35	7.60
36	5.10	5.30	6.05	7.15	8.80

A four panel hardware kit costs $1.50. If no trimming is required one carpenter should set 2 panels per hour or 2 window openings consisting of 8 panel units each in a day.

Custom shutters in lengths of 6" to 18", and for openings from 12" to 24", 2 panels will cost $25.30 unfinished and $34.50 prefinished; for an opening 24" to 30", four panels same height, $42.60 and $60.00; for 30" to 42" opening, $39.00 and $64.00; in lengths of 18" to 24", openings 12" to 24", 2 panels will run $29.50 and $41.50; openings 24" to 30", 4 panels, $48.20 and $69.60; openings 30" to 42", 4 panels, $55.20 and $78.00. Custom door panels to 7' will run $98.00 unfinished, $137.00 finished for 2 panel for an opening to 24" wide. For openings to 30", 2 panels, $101.10 and $140.00; openings 30" to 42", 4 panels, $197.70 and $274.20.

06300 WOOD TREATMENT

Wood and plywood are often given special treatments at the mill to give additional protection against fire, rot, and termites.

Fire protected or fire retardant wood is pressure-impregnated with mineral salts that will react chemically at a temperature below the ignition point of the wood. An Underwriter's Laboratory Flame Spread Rating of 25 can be obtained, and no periodic maintenance is required to keep the rating. However, the lumber must also be kiln dried to 12% moisture content if it is to be painted. Also, wood is not suitable for direct exposure to weather and ground conditions unless further treatment is given. A fire retardant treatment will add 12 to 15 cents per board foot to lumber costs. If it is necessary to kiln dry the lumber add 6 cents for soft woods, 8 cents for hard woods.

Rot and termite resistance can be obtained by several methods. Creosote, when permitted by local environmental regulations, is impregnated at the rate of around 8# per cubic foot at a cost of 20 to 50 cents per board foot. This finish is not readily paintable and can stain adjacent finishes such as plaster and wallboard. Pentachlorophenol is a popular oil treatment and costs 20 to 30 cents per sq. ft. and leaves the wood with an oil residue. This residue can be eliminated and the wood left clean and paintable if the pantachlorophenal is impregnated into the wood by liquid petroleum. This will add 10 cents to the cost.

06400 ARCHITECTURAL WOODWORK

This section covers the assembly and installation of cabinets, wood and wallboard paneling, wood stairs, and railing from components made up in a shop or factory off the job premises.

If the items under this heading are not stock items, it will be necessary to include the cost of making on-the-job measurements and preparing shop drawings. Most cabinet makers are equipped to do this and will include such costs in their proposals. Often they will also install their work, but most usually the general contractor will handle this.

There is considerable difference between the grades of workmanship for architectural woodwork and the estimator must be able to determine the grade required.

These same differences apply to the erection and installation of all classes of millwork and interior finish.

The Architectural Woodwork Institute publishes a manual of "Illustrated Quality Standards" in which the work is divided into three grades: premium, custom, and economy. Premium is the highest grade available in both material and workmanship, intended for the finest work. Sometimes it is used throughout an entire building, but most often only in selected spaces or items within a building. Custom is the middle or normal grade in both material and workmanship, and is intended for high quality regular work. Economy is the lowest grade and intended for work where price outweighs quality considerations. An estimator may find references to these three grades in an architect's specification, and this manual sets forth very exactly what is to be expected with both drawings and notes.

Methods of Estimating the Labor Cost Millwork and Interior Finish. Different contractors use different methods of estimating the labor cost of erecting and setting millwork and interior finish, but the system generally used is to add a certain percentage of the cost of the millwork to cover the labor cost of erection. While this method is the easiest, it does not produce the best estimates. For example, suppose you are estimating the cost of a building on which you receive four millwork bids, ranging from $4,000 to $5,000. After receiving the bids you add the customary percentage to the lowest acceptable bid to take care of the labor cost of handling and erecting the millwork. This percentage usually runs from 40% to 50% of the cost of the millwork. Taking 40% of $4,000.00 gives $1,600.00 for labor to handle and erect all millwork and interior finish. Suppose a competitor received a bid for $3,500.00 from a mill which did not quote you. If your competitor added 40% to his bid, he would have only $1,400.00 to erect the same amount of millwork on which you figured $1,600.00.

Why this difference of 12 ½% on such a small item? While the method saves a great deal of time and is almost universally used, it does not produce the most accurate estimates. The only accurate method is to list each item separately; i.e., the number of lin. ft. of wood base, chair rail, picture mold, etc., the number of door jambs to set, the number of sash to fit and hang, together with the number of window openings to be cased or trimmed; the number of wardrobes, linen cases, vanity and kitchen units to set, etc. Estimating labor costs by this method produces more uniform and accurate estimates, reduces risk to a minimum, and enables the contractor to check actual and estimated costs during construction, but it does require a lot more work.

INTERIOR PANELING

Interior paneling used to mean solid wood with stiles and rails and elaborate mouldings for the mansion or bank job; knotty pine for the suburban retreat; and beaver board for the cottage. Today the selection is much wider.

Solid wood paneling is still with us, but it is more apt to be in the form of planks with face-applied mouldings rather than as elaborate paneling. Plywood veneer with hard wood is often favored over solid wood not only for cost savings but because the veneer exposes the beauty of the wood and offers a more easily cleaned surface.

Most paneling today though is sheet board faced to look like wood grain, marble, tile, stucco, leather, or a decorative pattern all its own. The designs come with a selection of moldings that blend into the overall wall.

Insulating wallboards and tiles are still sometimes used as exposed finishes but are selected more for their structural or insulating value rather than their decorative contribution.

Estimating Paneling Quantities. Estimating paneling is more of a task than merely computing the number of sq. ft. of surface to be covered. Each room should be laid out to obtain certain panel effects if the design of the panels will govern to any extent the amount of cutting and fitting necessary. Most paneling will come in stock lengths and widths. The job conditions must be carefully studied to minimize cutting, to make maximum use of studs for attachment, and to eliminate any awkward jointing.

The labor cost of applying wallboard depends upon three things: the cost of placing headers and nailing strips between wood studs to which wallboard and decorative strips are nailed; the cost of installing the wallboard; and the cost of placing decorative panel strips or other means of concealing joints. When installing a ceiling, furring must be applied to provide nailing pieces for all edges of the decorative strips.

Wood decorative or panel strips are usually 1/4" to 1/2" thick and of widths to meet architectural requirements.

There are several ways of covering joints. Today much paneling is grooved so joints are butted and appear as an overall pattern. Most flush panels have cover strips, caps, etc. available to match the basic material or to contrast with it and make a decorative feature of the joint.

A sketch of each room, should be made giving an elevation of each of the four walls and a plan showing the ceiling, similar to that illustrated. From this it is possible to obtain the number of lin. ft. of "headers" or nailing strips required, the size of each piece of paneling as well as the number of lin. ft. of panel strips and moldings necessary to complete the job.

Placing "Headers" or Nailing Strips. When placing "headers" or nailing strips to receive paneling it is necessary to cut them to length and place them flush with the face of the studs, as the paneling must be nailed top and bottom to the "headers" and at the sides to the wood studs.

If the studs are not wide enough to form nailing pieces for the panel strips, a piece of wood, usually 1"x1", must be nailed to either side of the stud along the entire length and flush with the face of the stud.

When estimating this class of work, care should be taken to obtain the exact number of lin. ft. of "headers" required as this is an important item in the cost of the work.

With studs spaced 16" on centers, a carpenter should cut and place 225 to 275 lin. ft. per 8-hr. day, at the following labor cost per 100 lin. ft.:

	Hours	Rate	Total	Rate	Total
Carpenter	3.2	$....	$....	$24.04	$76.93
Cost per lin. ft			....		.77

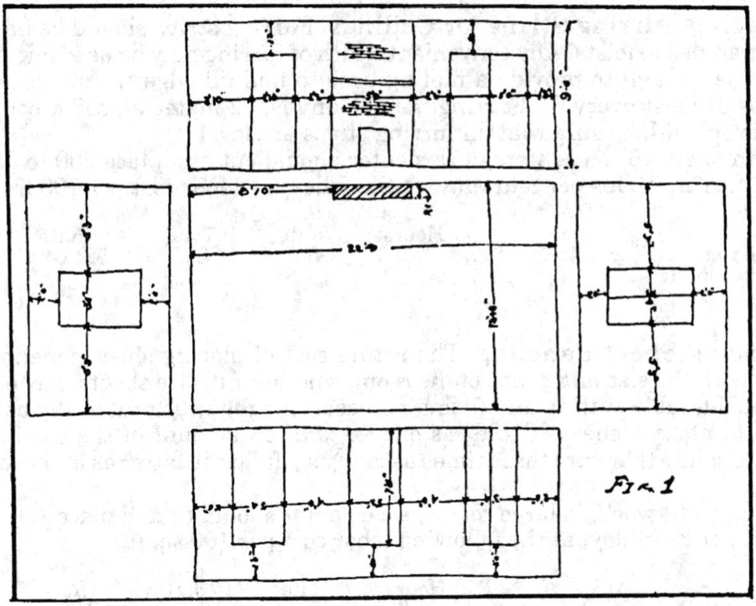

Rough Sketch of Room Showing Location of Doors and Windows

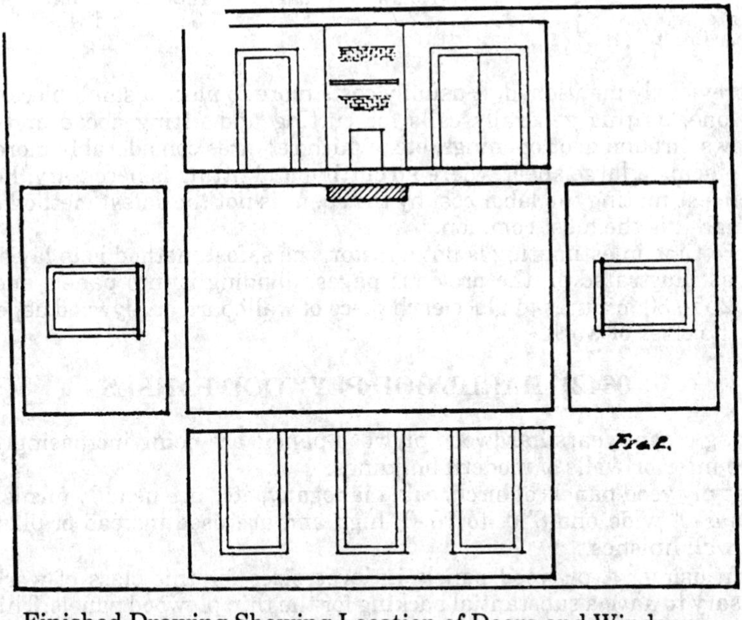

Finished Drawing Showing Location of Doors and Windows

Placing Furring Strips for Ceilings. Furring strips should be nailed at right angles to joists. Any convenient width of furring may be used but it must be wide enough to provide a nailing base to nail all edges of the decorative strips at least every 6". Furring may be any No. 2 material, soft wood being preferable. Make sure that all furring strips are level.

With joists 16" on centers, a carpenter should cut and place 300 to 350 lin. ft. of furring strips per 8-hr. day, at the following labor cost per 100 lin. ft.:

	Hours	Rate	Total	Rate	Total
Carpenter	2.5	$....	$....	$24.04	$60.10
Cost per lin. ft			....		.60

Placing Sheet Paneling. The actual cost of placing sheet paneling will vary with the size and shape of the room, whether full size sheets may be used or considerable cutting and fitting necessary. It requires practically as much time to place a sheet 16"x96" as one 48"x96" on account of the cutting and fitting, while there are three times as many sq. ft. in the latter as in the former.

On straight work, in large rooms, a carpenter should fit and place 450 to 550 sq. ft. per 8-hr. day, at the following labor cost per 100 sq. ft.:

	Hours	Rate	Total	Rate	Total
Carpenter	1.6	$....	$....	$24.04	$38.46
Cost per sq. ft			....		.38

When applying sheets in smaller rooms requiring considerable cutting and fitting around doors, windows, etc., a carpenter should fit and place 275 to 300 sq. ft. per 8-hr. day, at the following labor cost per 100 sq. ft.:

	Hours	Rate	Total	Rate	Total
Carpenter	2.7	$....	$....	$24.04	$64.91
Cost per lin. ft			....		.65

As previously mentioned, it usually costs more to place a small piece than a large one, as this generally calls for cutting and fitting above and below windows, around door openings, etc., which requires considerably more work than placing a large sheet where no cutting and fitting is necessary. For this reason, estimating the labor cost by the sq. ft. is not the safest method to use, although it is the most common.

Where time in estimating is not a factor, the safest method is to lay out the work as illustrated on the previous pages, dividing it into panels and then allow 20 to 30 minutes to place each piece of wallboard or plywood depending upon the class of work.

06421 HARDWOOD PLYWOOD PANELS

During recent years hardwood plywood panels are being increasingly used for the interior walls of modern buildings.

These plywood panels of birch, oak, mahogany, etc., are usually furnished in sheets 4'-0" wide and 8'-0" to 10'-0" high and are used instead of plaster or other wall finishes.

When using 1/4" plywood panels in large sizes for this class of work it is necessary to have a substantial backing for the thin plywood panels. This may be obtained by applying the plywood direct to a brown or finish plaster coat wall; a ½" gypsum board backing, or better still, a ½" thick plywood backing with the sheets running in the opposite direction to that of the finish plywood panels.

These plywood panels are usually applied with the panel running from floor to ceiling in one length, i.e., 4'-0"x8'-0" or 4'-0"x10'-0", with the vertical joints butted flush or having a very slight bevel at the edges.

To obtain a first class job, it is usually necessary to apply adhesive to the back of the plywood panels and also use small brads to secure the plywood in place until the adhesive sets.

This is particular work, because the adjoining sheets must be accurately fitted together if a first class job is to result.

If wood panel strips are to be used to conceal the joints, it is not necessary to use so much care in placing the panels.

Hardwood laminated veneer panels cover a wide variety of products, from the ¼" thick "grooved" veneers so popular now to ¾" thick flush panels with or without sequence matching of the flitches on up to special burl veneers suitable for the tops of the finest furniture.

Plywood may have a veneer core with a center and at least two intermediate cross banded veneers, a particleboard core, or a lumber core with strips of lumber edge glued into a solid slab with cross banding veneers either side. The back and face of each sheet then receive additional veneers at right angles to the cross banding.

The face veneers may be rotary cut with the log's annular rings exposing a bold variegated grain; flat sliced, which produces a variegated figure similar to that of plain sawn lumber; quarter sliced with growth rings of the log at right angles to the knife, producing a series of stripes in the veneer; and rift cut, which is used with oak logs to produce fairly uniform grains with a combed look caused by the medullary rays.

Then there are various grades of face veneers: premium, or architectural panels which have veneers sequence matched and no patching; custom, which allows no patching on face veneers but are not sequence matched; and economy, which allows "good" face veneers, but no matching of grain or colors.

And, finally, there is the type of tree itself which affects the price. The range from natural yellow birch to Brazilian rosewood is around 800%.

Costs should always be determined locally, since not all veneers are available at all times. The following will serve as a general outline of costs. All types are 4'x8'x ¼" custom.

Natural yellow birch	$30.00
Natural hard maple	30.00
Natural gum	30.00
Select white birch	38.00
Popular	34.00
Select birch heartwood	35.00
Select white maple	38.00
Plain sawn red oak	45.00
African mahogany	45.00
White ash	48.00
Limba	50.00
Rift sawn white oak	60.00
Pecan	53.00
Cherry	67.00

Butternut	58.00
Quarter sawn white oak	65.00
Walnut	5.00
Primavera	75.00
Avodire	80.00
Teak	120.00
Brazilian rosewood	240.00

If matched sequence panels are ordered, the cost will be about 50% more. Add around $12.00 per sheet for 3/4" thickness.

The stock V-grooved prefinished 4'x8'x 1/4" panels are considerably cheaper. Although the range of woods is not so extensive, the finishes vary widely and approximate the colors of more expensive woods still using the grains of the more common species. A random sampling of costs indicate the following range.

			Clear	Colonial
Birch	4'x8'x 1/4"	V-grooved	$18.00	$13.00
Oak	4'x8'x 1/4"	V-grooved	28.00	16.00
Walnut	4'x8'x 1/4"	V-grooved	32.00	19.00
Pecan	4'x8'x 1/4"	V-grooved	27.00	25.00

Inlaid paneling in 4'x8'x 1/4" sheets consisting of pecan panels striped with 1 1/4" wide walnut stripes runs $30.00 per sheet; with walnut striped in pecan the cost is $35.00 per sheet.

For lower budget jobs, simulated wood grain finishes on 1/4" plywood will run $7.00 per 4'x8' sheet.

For the lowest budget jobs, simulated wood finishes printed on hardboard are available in the $6.00 to $7.00 range per 4'x8'x 1/4" sheet.

Labor Placing Flush Plywood Wall Panels. When applying plywood panels 4'-0"x8'-0" to 4'-0"x10'-0", with the abutting joints fitted tightly together, using adhesive and small brads to secure the plywood to the backing, two carpenters working together should handle, fit, and place one plywood panel in 1 1/2 to 2 hours, or they should place 4 to 5 panels per 8-hr. day.

Many of the panels will have to be cut to half or three-quarter size, and in some instances cut out for doors and windows, so a certain length of time should be allowed for each piece of plywood, rather than figuring on a square foot basis.

06422 SOFTWOOD PLYWOOD PANELING

Softwood plywood consists of an odd number of thin sheets or veneers of Douglas fir of 3, 5, 7, or 9 standard thicknesses. These are laminated with alternating grain direction. Because the grain of each ply composing the material is at right angles to the grain of adjacent plies, the strength of Douglas fir plywood panels is approximately equal in both directions.

All Douglas fir plywood from American Plywood Association mills is made to conform to the moisture-resistance requirements set forth in the revised standards established by the National Bureau of Standards, as U. S. Commercial Standards CS45-55.

Plywall wallboard is the grade of plywood commonly used for interior walls and ceilings instead of lath and plaster or other types of wallboard. It is furnished in sheets 24" to 48" wide and 5 to 12 ft. long.

Plywall wallboard is furnished 1/4", 3/8", 1/2", 5/8", 3/4", 1", and 1 1/8" thick. The 1/4" thickness is most widely used for walls and ceilings, although for more substantial construction 3/8" or 1/2" thick is often preferred. Thicker material is used for shelving, wardrobes, cases, etc.

Practically any design may be obtained by the use of plywood as with other types of wallboard. Four penny casing or finishing nails are recommended for paneling up to 1/2" thick with the nails spaced 6 to 10 in. apart. One horizontal or fire stop should be cut in between top and bottom plates as a nailing piece.

Plywood panels may be finished by staining, painting, wall papering or mechanical surfacing.

Labor Applying Plywood Wallboard. The labor cost of applying plywood wallboard will run about the same as given for fiber wallboard covering the same class of work.

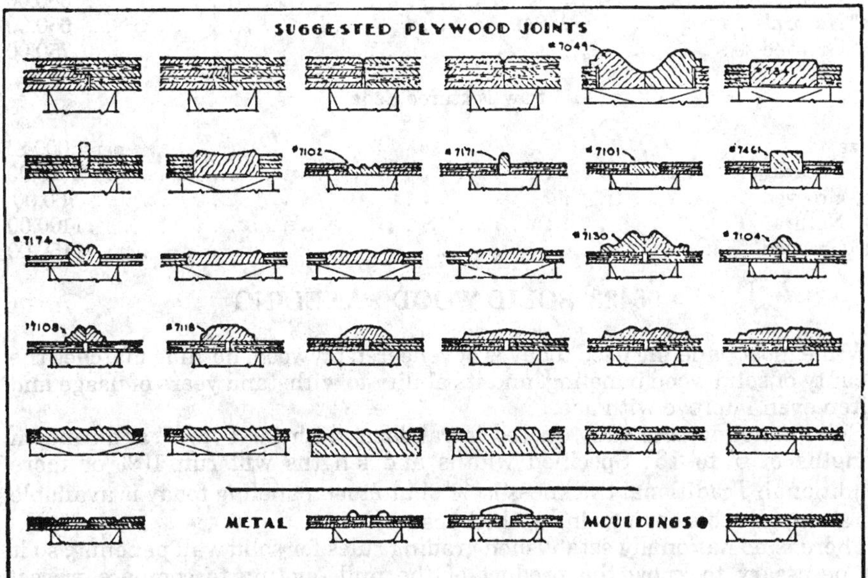

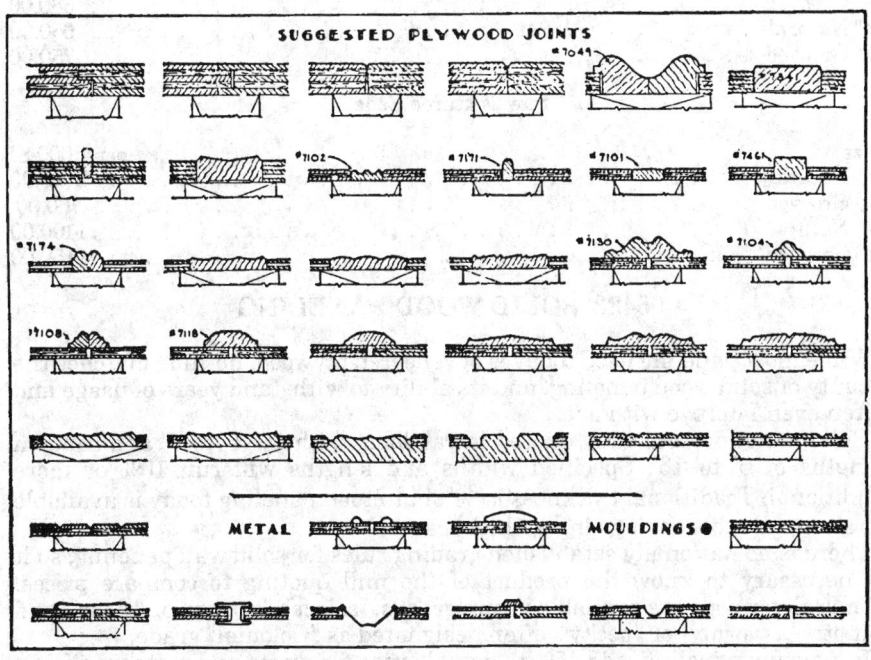

Types of Joints Used with Plywood Wallboard

Where special jointing of the panels is required to work out wall and ceiling panels, allow extra for grooving, beveling and any other designs worked into the wallboard. Also add extra for applying panel strips as given on previous pages.

Prices of Douglas Fir Plywood

Made with water resisting glue. All panels sanded 2 sides to net thickness shown. As plywood prices have been very volatile recently, check local sources.

Size of Panel	Thick, Inches	No. of Plies	Price per 1,000 SF Good(A-D) 1 Side	Good(A-A) 2 Sides
24"x60" to 48"x96"	1/4	3	$280.00	$380.00
24"x60" to 48"x96"	3/8	3	350.00	450.00
24"x60" to 48"x96"	1/2	5	390.00	490.00
24"x60" to 48"x96"	5/8	5	430.00	530.00
24"x60" to 48"x96"	3/4	5	525.00	580.00

Saw textured plywood siding is popular and relatively inexpensive. It is readily available in Douglas fir and cedar.

hcSaw Textured Douglas Fir Plywood

Size	per 1000 SF
3/8" Natural	$520.00
3/8" Grooved	580.00
5/8" Natural	650.00
5/8" Grooved	750.00

Saw Textured Cedar

Size	per 1,000 SF
3/8" Natural	$720.00
3/8" Grooved	760.00
5/8" Natural	1100.00
5/8" Grooved	1150.00

06423 SOLID WOOD PANELING

While most paneling used today is of veneered plywood, nothing matches the beauty of solid wood paneling, and its ability to withstand years of usage and often even improve with age.

Solid paneling is usually quoted in random widths of 4" to 8" and random lengths of 6' to 16'. Specified widths and lengths will run 10% or more additional. Traditional thickness is 3/4", but much paneling today is available at a considerable savings in 1/2" thickness.

There is no nationally established grading rules for solid wall paneling, so it is necessary to know the product of the mill quoting to compare prices. Generally, woods are available in two grades: "select", "luxury" or "clear", and "knotty", "wormy" or "pecky", often designated as "Colonial" grade.

In figuring paneling, add 15% to actual areas for waste and matching.

Paneling is available with V-joint tongue and groove, square joint, shiplap and Colonial beaded.

The most commonly encountered paneling is knotty pine. Cost is around $1.00 per sq. ft. Clear pine will run around $1.30.

Clear birch in 3/4" thickness will run around $1.85 per b.f.in lots over 1000 b.f., random width and length. Birch is an excellent cabinet wood with a firm texture, but has a rather strong grain pattern.

White ash is an open grained wood, light in color with contrasting light brown heartwood. It is easy to work and somewhat resembles flat grained white oak. It will cost $1.95 for 3/4" select grade.

Butternut, often called white walnut, is a soft-textured, easy-to-nail wood which takes a beautiful natural finish. It will run $2.20 per b.f. select, $1.15 character marked, 3/4" thick.

Cherry, cut from the wild black cherry, is a traditional Colonial paneling. It is a firm textured cabinet wood, but is available only in a rather narrow range of sizes. It runs around $2.10 for 3/4" select down to $1.50 for 3/4" Colonial.

Oak is available from either white or red varieties cut on the quarter grain. Oak is a very hard wood and subject to shrinkage and splitting in adverse conditions, but well worth the extra effort for the right place. Oak runs $1.70 for the red and $2.10 for the white in 3/4" select, $1.15 and $1.00 in 3/4" Colonial.

Walnut has become so popular and the sources for it so restricted that it is now a high priced wood and will run $3.25 or better per sq. ft. for 3/4" select. Because it is a hard wood, 1/2" thickness is practical and will run $2.75 for select and $2.30 for colonial.

Other domestic woods available are maple, around $1.40 for 3/4" select; wormy chestnut, at $1.50 for 3/4"; and cypress at 90 cents for 3/4" select. Imported woods are also available. The so-called "exotic" woods are more often found in the architectural grade, book-matched plywood where their veneers can be shown off to better advantage.

Some foreign woods stocked in the Chicago area include paldao, a firm textured striped wood for medium brown from the Philippines; African mahogany and the less costly Philippine mahogany which is more difficult to finish; avodire, a light creamy-colored wood which comes from Africa with a roll figure somewhat on the diagonal; limba, or korina, African wood which is cream or tan in color with a darker heartwood, and tigerwood or African lovea, a dullish brown to golden colored wood with a wide or ribbon stripe.

These are all special woods for special jobs. Prices and availability should always be checked, because they vary considerably, but all are expensive woods. **Placing Wood Wainscot or Paneling.** The labor cost of placing wood paneling or wainscoting will vary with the height of the wainscot and the manner in which it is delivered, i.e., whether assembled or knocked down.

It takes nearly as long to place wood wainscoting 4'-0" high as 7'-0" high, as the top and bottom members are identical and the only difference is the height of the panels.

If the wood wainscot is put together in the mill and the carpenters on the job do what little fitting is necessary, set it against the walls and nail it to the grounds, a carpenter should place 12 to 18 lin. ft. of wainscot per 8-hr. day, at the following labor costs for the different heights:

Height of Wainscot	No. Sq. Ft. 8-Hr. Day	Carp. Hrs. 100 Sq. Ft.
4'-0"	60-70	11-13
5'-0"	70-85	9-11
6'-0"	80-100	8-10
7'-0"	90-110	7-9
8'-0"	95-115	7-8

Placing Wood Wainscot or Paneling (Knocked Down). If wood wainscot or paneling is sent to the job knocked down, making it necessary to assemble the pieces and put them together on the job, a carpenter should place top and bottom rails, cap and intermediate panels for 10 to 15 lin. ft. of wainscoting per 8-hr. day, at the following costs for the different heights:

Height of Wainscot	No. Sq. Ft. 8-Hr. Day	Carp. Hrs. 100 Sq. Ft.
4'-0"	50-60	13-16
5'-0"	60-70	11-13
6'-0"	70-80	10-11
7'-0"	80-90	9-10
8'-0"	90-100	8-9

06424 SHEET BOARD PANELING

An inexpensive prefinished paneling is plastic coated hardboard. This comes in 1/8" and 1/4" thick, 4'x8' panels, and 16" wide, 1/4" thick grooved planks. The thinner boards should have solid backing for first class work. These panels have a wide range of printed designs from wood grains, solid colors, scenic designs, and marble textures. A full line of aluminum and presdwood moldings is available colored to match or compliment the selected material.

A recent innovation is vinyl surfaced gypsum wallboard. This comes in 1/2" thickness 8', 9', and 10' lengths. It can be applied with direct nailing using colored nails; glued to furring with a combination adhesive nail-on system, or laminated to a base layer of gypsum backed board. An advantage of this finish is that it can be patched and eventually painted.

Open Work Panels. A popular addition to stock partitioning available today are panels with open work. These vary from 1/8" thick prefinished masonite to solid walnut grilles ranging as high as $20.00 a square foot. The amount of openings can vary from 15% to 50%. These materials make good room dividers, help control sun and glare at windows and cut air conditioning and drapery costs.

The masonite product is known as Filigree and is available in a series of sizes up to 4'x8' sheets. Prefinished frame sections 3/4" thick and 1 1/2" wide, factory grooved to receive the panel are available.

Sculptured hardwood panels 3/4" thick are available in many designs and woods including walnut, maple, ash, oak and poplar, in stock sizes from 2'x4' to 4'x8'. These may be ordered sanded for finishing on the job or factory finished. Woods available listed in order of ascending cost include sycamore or gumwood; poplar, maple, red oak, ash, and walnut. Some designs, nine are available, are also made in 5/8" clear acrylic.

The least expensive design in the least expensive wood, sycamore, will run $57.00 for a 2'x6' unfinished, unframed panel. Framed and finished it will cost $180.00, plus shipping charges. The most expensive designs will run $108.00 unfinished and unframed, $235.00 framed and finished.

Wallboard–Kinds, Sizes and Costs. There are so many kinds and sizes of wall and insulating board, it is impossible to list all of them.

The thinnest boards are made of wood fiber and are usually furnished in sheets 4 ft. wide, 4 to 12 ft. long, and 1/8", 3/16", 1/4", and 3/8" thick. Prices vary from 20 cents per sq. ft. for the 1/8-in. up to 35 cents per sq. ft. for the 3/8" board.

Insulating and building board is usually furnished 3/8" and 1/2" thick, 4 ft. wide and 6, 7, 8, 9, 10, and 12 ft. long, all having square edges. Board of this type will cost 22 cents per sq. ft. for the 3/8" and 28 cents per sq. ft. for the 1/2" thickness.

Insulation Board Finish Plank. Used for interior finish to combine insulation and decoration in one material.

Insulating plank are ½" thick and furnished in random widths 8, 10, 12, and 16 in. wide and 8, 10, and 12 ft. long. The long edges may be shiplapped with either a plain bevel or bead and bevel design, or have a blind nailing flange.

Plank ½" thick cost about 50 cents per sq. ft. Wall plank with acoustical properties is also now available in 8' and 10' lengths.

Plank Effect Board. The plank effect board is ½" thick, 4'-0" wide, and 8'-0" high. A series of 5 score lines on each edge and through the center of the board gives the effect of 3 planks 16" wide. It may be applied vertically or horizontally either over old plaster or directly to studs. All edges are flush for butt jointing. The board is painted ivory color on one side and costs 60 cents per sq. ft.

Concealed Nailing Clips. Provide an invisible means for attaching tile or plank to nailing base. Clips fit into tongue and groove. Usually packed 1,000 to a carton. Four pieces are required for each 16"x16" tile. Prices per 1,000 pcs. $6.50.

Use 2d common coated nails with clips. One pound required per 1,000 clips.

Cement or Mastic. Waterproof cement or mastic for installing insulating plank. One gallon required for 80 to 100 sq. ft. Price per gal. is $3.75.

Nails Required for Applying Fiberboard, Tile, Plank, Wainscot and Moldings

For Material ½" Thick	Gauge of Nails	Length of Nails	Spacing Directions
Board	No. 17	1 ¼"	3"
Tile	No. 17	1 ¼"	6"-10"
Plank	No. 17	1 ¼"	6"
Wainscot	No. 17	1 ¼"	6"
Moldings	No. 15	1 ¾"	6"

As a general rule, board, plank and wainscot require about 6 brads per sq. ft. or 4 to 5 lbs. of 1 ¼" brads per 1,000 sq. ft. Galvanized wire brads are recommended. When they are not available, bright wire brads may be used.

Placing Insulating Plank. The labor cost of placing insulating plank will vary with the surface to which it is applied, length of walls, amount of cutting and fitting necessary, etc. Ceiling work costs more than wall work on account of working on a scaffold and placing the plank overhead.

The most economical method of application is to a solid backing of plywood, wood boards or to other satisfactory nailing surface, as it is necessary to have nailing facilities every 8, 10, 12, or 16 inches, depending upon width of the plank. In the event a solid backing is not available, it is necessary to place wood furring strips to provide for nailing the boards at edges and at ends, also for chair rails, wainscot cap and other ornamental strips.

For plank without blind nailing flange, nail through face or bead of the plank, never through bevel. Bury nail heads below surface with properly gauged hammer blow or with a nail set.

The cost of wood blocking and furring is given on previous pages of this chapter.

Number of Sq. Ft. of Insulating Plank of Various Widths
Placed Per 8-Hr. Day by One Carpenter, When Nailed In Place*

Plank Width	Height of Wainscoting in Feet 4'-0"	6'-0"	8'-0"	10'-0"
8"	210-260	275-315	315-365	325-390
10"	240-300	300-360	360-420	375-450
12"	275-340	350-410	410-475	425-500
16"	300-360	375-440	450-525	465-525

*Add for wood furring strips or other backing. When applied to ceilings, reduce above quantities approximately 25 percent.

Carpenter Hours Required to Place 100 Sq. Ft. of Insulating
Plank On Walls When Nailed In Place

Width of Plank, In.	Height of Wainscoting in Feet 4'-0"	6'-0"	8'-0"	10'0"
8	3.5	2.8	2.4	2.3
10	3.0	2.5	2.0	1.9
12	2.6	2.1	1.8	1.8
16	2.5	2.0	1.7	1.7

Add for wood furring strips or other backing.

Panel Strips and Moldings For Use With Sheet Paneling. Strips for working out decorative effects on walls or ceilings and for concealing the joints may be of the same material as the sheet or they may be wood or metal, depending upon the effect desired.

The labor cost of placing wood or composition decorative strips over the joints will vary with the design of panel strip used, whether it consists of one, two or three members, the size of the panels and the design of the walls and ceilings, as the more cutting and fitting required the fewer lin. ft. of strips a man will place per day.

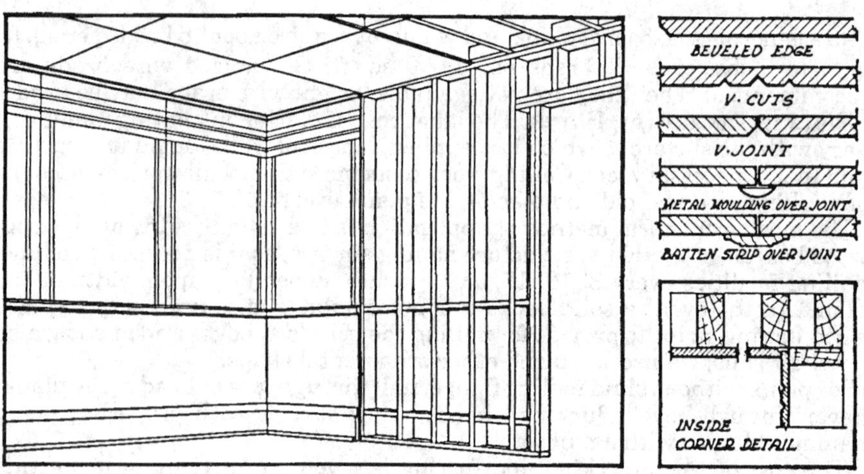

Method of Applying Wallboard and Treating Joints

On straight work, using an ordinary panel strip ¼"x2" to ½"x3", a carpenter should fit and place 350 to 450 lin. ft. per 8-hr. day, at the following labor cost per 100 lin. ft.:

	Hours	Rate	Total	Rate	Total
Carpenter	2	$....	$....	$24.04	$48.08
Cost per lin. ft			....		.48

On more complicated work, using a 1-member panel or decorative strip, a carpenter should place 275 to 325 lin. ft. per 8-hr. day, at the following labor cost per 100 lin. ft.:

	Hours	Rate	Total	Rate	Total
Carpenter	2.7	$....	$....	$24.04	$64.91
Cost per lin. ft			....		.65

Placing 3-Member Decorative or Panel Strips. When the panel strips form a plain wall and ceiling design but consist of 3 members (a strip 3" or 4" wide with a small molding on each side) a carpenter should fit and place 150 to 175 lin. ft. per 8-hr. day, at the following labor cost per 100 lin. ft.:

	Hours	Rate	Total	Rate	Total
Carpenter	5	$....	$....	$24.04	$120.20
Cost per lin. ft			....		1.20

Placing Wood Ceiling Beams, Cornices, Etc. In rooms having beam ceiling effects, wood ceiling cornices, etc., a carpenter should fit and place 175 to 200 lin. ft. of each member per 8-hr. day, at the following labor cost per 100 lin. ft.:

	Hours	Rate	Total	Rate	Total
Carpenter	4.3	$....	$....	$24.04	$103.37
Cost per lin. ft			....		1.03

Three member cornices would cost 3 times the above price, 4 member, 4 times, etc.

Applying Wallboard Over Existing Plastered Walls and Ceilings. In remodeling work where the wallboard is applied over existing walls and ceilings, it is advisable to remove the existing plaster wherever possible. To obtain a first class job it is recommended that all interior trim, such as door and window casings, base, etc., be removed before applying the wallboard. It is also recommended that furring strips be placed over all existing plastered walls, as this will straighten the walls and provide a much better base for securing the wallboard.

After the furring strips have been placed, the cost of applying the wallboard will run about the same as given on the previous pages.

The labor cost of placing wood furring strips is given on the previous pages under "Placing Furring Strips for Ceilings".

Prefinished Panels. New developments in wallboards have been most noticeable in prefinishing and the consequent care needed in concealing nailing.

One of the most popular items is V-grooved hardwood, prefinished with a baked-on finish. Grooves are generally random but with a groove every 16" on center to hit a stud. Most boards are 1/4" thick in sheets 4' wide by 8', 10' or 12' long. These sheets should be applied over a backing board for first class work. Colored nails are available as well as prefinished moldings for cornices, trim and base. Similar paneling is also available in 7/16" thickness for application directly to studs. A wider range of finishes is generally available in the heavier material.

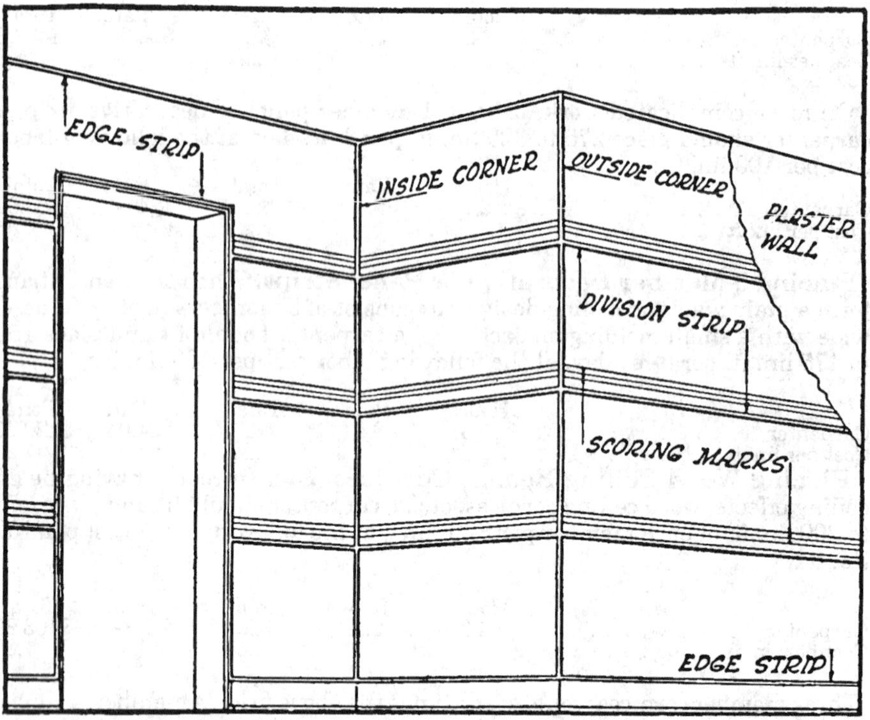

Method of Placing Wallboard and Horizontal Scoring Marks and Division Strips

Insulating Tile. Tile made of Celotex, Fir-Tex, Insulite, Nu-Wood, Temlok, and other fiber composition are still sometimes used for wall and ceiling finish. They are furnished with beveled edges either square or with a blind nailing or stapling flange. They come in sizes from 12"x12" to 16"x32" and may be used for obtaining ashlar effects on walls and tile effects on ceilings.

They may be applied to wood backing or to furring strips with adhesive, nails, staples, or special clips on the walls or ceilings. When furring strips are used, the strips must be spaced to work out with the tile sizes.

In applying square edged tiles, nails should be driven through the tile at an angle to the surface of approximately 60 degrees. Reverse the angle of half the nails and nail through the face of the tile, never through the bevel. Bury nail heads below the surface with properly gauged hammer blow or with nail set.

When applying tiles with blind nailing or stapling flanges, drive the nails or staples in the thick part of the flange perpendicular to the surface.

Insulating fiber tile ½" thick costs $10.50 per carton of 64.

Nailing Recommendations for Square Edged Fiber Tile. Where nailing is to be exposed, galvanized wire brads are recommended. When they are not available, bright wire brads may be used. 1 ¼" brads or finishing nails should be used for exposed nailing of ½" thick tile.

Nailing Recommendations for Square Edged Fiber Tile

Tile Size in Inches	Location of Brad	No. Brads per Tile	Approx. Lbs. of Brads per 1,000 SF Tile
12x12	1 in each corner	4	3 ½
12x24	4 in each long edge	8	3 ½
16x16	3 on 2 opposite edges 3 on center line	9	4
16x32	5 on each long edge 5 on center line	15	3 ½

Nailing Recommendations for Flanged Tile. Where attachment of flanged tile is to be by nails, 3d blued lath nails 1 ⅛" long should be used. Where application is by means of staples, rust resistant staples ½" or 9/16" in length should be used in conjunction with a gun type stapler.

Tile Size in Inches	No. Nails or Staples per Tile	Approx. No. Staples per 1,000 SF	Approx. Lbs. Nails per 1,000 SF
12x12	4	4,000	6.0
12x24	6	3,000	4.5
16x16	4	2,250	3.4
16x32	6	1,690	2.6

Labor Placing Fiber Tile When Nailed In Place. The labor cost of handling and placing fiber tile will vary considerably with the size of the tile and the skill of the mechanic placing same, the amount of nailing necessary and the surface to which the tile is applied. A man will place more tile on a good solid backing than on furring strips. To the quantities given in the following table, add for blocking and furring where necessary.

Quantities and Labor Costs of Applying Fiber Tile to Walls and Ceilings Where Tile are Nailed In Place

Tile Size, Inches	No. Sq. Ft. per Tile	No. Tile Req'd per 100 Sq. Ft.	No. Placed per 8-Hr. Day	No. Sq. Ft. per 8-Hr. Day	Carp. Hrs. 100 Sq. Ft.
12x12*	1.0	100	180-200	180-200	4.2
12x24*	2.0	50	95-105	190-210	4.0
16x16*	1.78	56	120-130	215-235	3.6
16x32*	3.56	28	83-90	295-315	2.7

*Stock sizes.
Add for wood furring strips or other necessary backing.

Applying Insulating Tile Using Adhesives. Apply the adhesive in spots or dabs about 2" in diameter, ¼" to ⅜" thick, one on each corner of the square tileboard and two or more additional spots on the rectangular sizes. Keep adhesive about 1" from edges of fiberboard units and use approximately 4 spots per sq. ft. Adhesive spots should not be spaced over 10" on centers in any direction.

After dabs of adhesive have been applied, place the tile in its final position. Slide back and forth for a distance of about ¾" inch under uniform pressure of the hands to obtain good contact. When working units larger than 16"x16", two men should exert pressure simultaneously for best results. If the surface of the unit is below the level of adjoining pieces, do not pull the unit away from the base to bring surface level but remove it and add more adhesive. Again work into position as described above to obtain correct surface level.

Brad or shore the first units on either walls or ceilings to prevent sliding out of position during application of surrounding units. Do not bend units over high spots.

Tile and plank may be applied to plasterboard, sound plaster, plywood, or lumber. In each case the use of adhesive or a combination of adhesive and brads is recommended.

Coverage Chart of Adhesive

Size of Tile in Inches	No. Spots per Tile	Sq. Ft. per Gal.	Gals. per 1,000 Sq. Ft.
12x12	4	40	13
12x24	6	40	13
16x16	4	60	10
16x32	6	60	10

Price, $3.50 per gal.

Labor Placing Fiber Tile Using Adhesive. The labor cost of placing fiber tile using adhesive or adhesive and nails will vary greatly with the skill and experience of the workman applying same, size of tile, etc. For instance, where an experienced man is placing the tile, and has a helper to build scaffold, unpack tile and apply the adhesive to the backs of the tile and the mechanic merely slides them in place, an experienced mechanic and helper should place 40 to 45 sq. ft. of 12"x12" tile an hour but if one man is working alone and he must stop to apply the adhesive to each tile, set it in place and then drive several brads through them, 15 to 20 sq. ft. an hour would probably be the average. For this reason, conditions under which the tile are to be installed must be carefully considered before arriving at the labor cost.

The design in which the tile are laid also affects the labor costs. For instance when tile are laid in herringbone pattern, it requires about 30% to 40% more time than regular square or common bond. Diagonal patterns require about 20% to 25% more time than regular squares or common bond while tile laid in mixed ashlar designs require about 50% more time than straight squares or common bond.

The quantities and costs given in the following table are based on using two workers--an experienced mechanic to place the tile and a worker to build scaffold, unpack the tile and place them on the scaffold, and place the adhesive on the back of the tile ready for the mechanic.

If the work is performed by inexperienced workers, the labor costs might be as much as double those given here.

Quantities and Labor Costs of Applying Fiber Wall Tile to Walls and Ceilings Using Adhesive and Nails Where Necessary

Add for cleaning plaster or other work necessary for applying tile. The quantities and costs given in the following table are based on using a 4- or 5-ft. scaffold, consisting of horses and planks, and laying the tile in plain squares or common bond.

Tile Size, Inches	No. of Sq. Ft. per Tile	No. Tile Req'd per 100 Sq. Ft.	No. Placed per 8-Hr. Day	Sq. Ft. per 8-Hr. Day	Hrs. per 100 Sq. Ft.
12x12*	1.0	100	165-185	165-185	4.5
12x24*	2.0	50	88-98	175-195	4.4
16x16*	1.78	56*	112-125	200-220	3.9
16x32*	3.56	28*	65-70	230-250	3.3

*Standard sizes

If tile are laid in herringbone pattern, add 30% to 40% to the above labor costs; if laid in diagonal patterns, increase the above labor costs 20% to 25%; and if laid in mixed ashlar designs, add about 50% to the above labor costs.

When placing tile in auditoriums and other spaces requiring high scaffolding, add cost of scaffolding. Large jobs are applied faster proportionately than small jobs.

Presdwood, Hardboard, Panelboard, Temwood. There are a number of makes of board made of wood fiber and pressed into board form under flat bed hydraulic presses that are much thinner and possess much greater strength than ordinary wallboard and are used for both interior and exterior work.

These hardboards are used for applications such as interior finish of walls and ceilings, cupboards, lining of closets, and clothes chutes.

They are usually furnished 1/8", 3/16", 1/4", and 5/16" thick, 4 ft. wide and 4 to 12 ft. long, although they vary somewhat among manufacturers.

Tempered hardboard, Temwood, or Presdwood are the same materials as described above, except they have been subjected to a tempering treatment which further increase the strength, resistance to abrasion, durability, and ease of painting. Tempered hardboards, 1/4" and 5/16" thick are used for the exterior of homes and other small buildings. It is also used quite extensively for concrete forms.

These hardboards or tempered boards are in no way comparable with ordinary wallboard. It is a far denser material, with greater tensile and transverse strength, greater resistance to abrasion, lower moisture absorption, and when properly applied, no warping or buckling.

The labor cost of placing hardboard will run approximately the same as placing other types of board under the same conditions.

Approximate Square Foot Prices of Hardboard

Type	Thickness	Price
Untempered, Plain	1/8"	$0.13
Untempered, Plain	1/4"	.18
Untempered, Pegboard	1/8"	.16
Untempered, Pegboard	1/4"	.21
Tempered Plain	1/8"	.15
Tempered Plain	1/4"	.23
Tempered Pegboard	1/8"	.17
Tempered Pegboard	1/4"	.24
Tempered Plastic Face	1/8"	.72
Tempered Plastic Face	1/4"	.80
Tempered Pl. F. Pegb'd	1/8"	.72
Tempered PL. F. Pegb'd	1/4"	.82

Tileboard

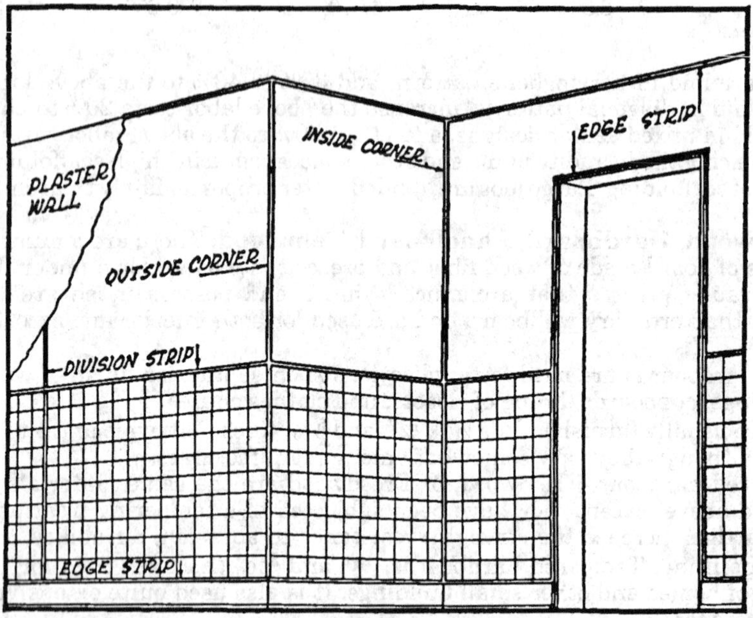

Method of Placing Tileboard in Bathrooms, Kitchens, Etc.

Tileboard has score lines impressed in one surface forming 4"x4" squares. The board is denser than Tempered Hardboard and has a lower rate of water absorption.

This material is used for walls in bathrooms, kitchens, lavatories, restaurants, barber shops, and similar places. It is furnished in sheets 4'-0" wide and 2'-0" to 12'-0" long.

Tempered board or tileboard should be cemented to the walls, using *Armstrong's Panelboard Cement* or a cement of equal quality. The adhesive should be applied to the back of the boards with a saw-tooth trowel, the boards having first been cut to fit. Use supplementary nailing or shoring to hold the boards firmly until the adhesive takes a partial set. Where it is necessary to nail the board to the walls, small brads should be used where scored lines intersect.

Figure about 1 gal. waterproof cement per 50 sq. ft. of wall.

Presdwood Temprtile 1/8" thick costs about 25 cents per sq. ft. unpainted.

Prefinished Tileboard. Tileboard 5/32" thick, made of masonite tempered Presdwood, is furnished scored 4"x4" to represent tile and finished with multiple coats of enamel or lacquer. Tileboard is also furnished without scoring, if desired.

These boards are furnished with dull or velvet finish, or a high gloss finish in a variety of colors with contrasting joints.

Tileboard is furnished in sheets 4 ft. wide and 4, 5, 6, 7, 8, 9, 10, and 12 ft. long.

Tileboard having a semi-gloss or high gloss finish costs about 60 to 70 cents per sq. ft. depending upon size of board.

Cap molding 1 3/4" wide for use at the top of the wainscoting is furnished in strips 4' and 8' long, scored to match tile or unscored, costs 35 cents per lin. ft.

Base molding 3 3/4" high, scored or unscored, in colors to match tileboard costs 40 cents per lin. ft.

Placing Tempered Tileboard. Where tempered tileboard is used in kitchens, for sink splashboards, bathrooms, shower stalls, and other small places, there is usually considerable cutting and fitting required to place a comparatively small square footage of board.

The safest method of estimating is to lay out each wall, showing number and sizes of pieces required, as it will cost almost as much to place a small panel as a large one.

Where the tileboard is secured by nails (without the use of adhesive), figure 3/4 to 1 hr. to fit and place each panel. Taking an average of 12 to 16 sq. ft. per panel, this will equal 125 to 150 sq. ft. per 8-hr. day.

However, if the tileboard is applied to walls, using an adhesive, a man should fit and place one panel in 1 to 1 1/4 hrs. or 100 to 125 sq. ft. per 8-hr. day. This applies to both plain and prefinished tileboard.

One gallon of adhesive will cover about 100 sq. ft. of surface if "spot" glued or about 65 sq. ft. of surface if "spread" glued.

Price per Gallon $10.00

Labor Cost of 100 Sq. Ft. Tempered Tileboard, Secured to Walls With Nails

	Hours	Rate	Total	Rate	Total
Carpenter	4	$....	$....	$24.04	$96.16
Cost per sq. ft			....		.96

Labor Cost of 100 Sq. Ft. Tempered Tileboard, Secured to Walls Using Adhesive

	Hours	Rate	Total	Rate	Total
Carpenter	3.5	$....	$....	$24.04	$84.14
Cost per sq. ft			....		.84

Wainscot Trim for Wallboard and Wall Tile

Moldings are made of stainless steel, polished finish in plain designs similar to the illustrations, punched for nails, and furnished in 4'-0", 6'-0", and 8'-0" lengths.

Description	for Wallboard	per Ft.
Cap Mold	1/8"	$0.22
Joint Mold	1/8"	.27
Outside Corner	1/8"	.32
Inside Corner	1/8"	.32
Tub edging or cove base	1/8"	.32

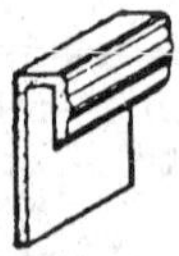
Cap Mold

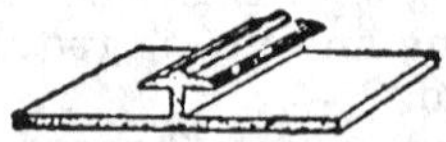
Joint Mold

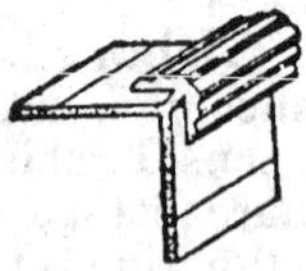
Outside Corner

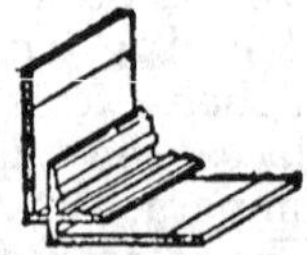
Inside Corner

Metal covered wood cap, base or trim, size 5/16" x 1 1/2", chrome zinc, polished finish. Price per lin. ft, $0.35.

Labor Placing Metal Moldings. The labor cost of placing metal moldings will vary with the design and size of the panels, cutting and fitting necessary, etc., but on the average job, a carpenter should place 100 to 150 lin. ft. per 8-hr. day, at the following labor cost per 100 lin. ft.:

	Hours	Rate	Total	Rate	Total
Carpenter	6	$....	$....	$24.04	$144.24
Cost per lin. ft			....		1.44

Items to Include When Making Up Wallboard Estimates. When making up estimates on wallboard construction, the following items should be included to insure a complete and dependable estimate:

1) Headers or nailing strips to receive wallboard.
2) Layout.
3) Furring strips on walls and ceilings where necessary.
4) Wallboard and labor placing same.
5) Wood or metal panel strips and labor placing same.
6) Labor filling nail holes with wood putty or crevice filler.
7) Beam ceilings, ceiling cornices, metal moldings, etc.
8) Painting and decorating wallboard and panel strips.
9) Nails and adhesive.
10) Overhead expense and profit.

06430 STAIR WORK

Framing and Erecting Wood Stairs of Unassembled Stock Material. When framing and erecting ordinary wood stairs having closed stringers where the treads and risers are not housed out, it will require 3 to 4 hrs. carpenter time to lay out the work, cut out for treads and risers and place wood stringers for a flight of ordinary wood stairs up to 4'-0" wide and 11'-0" story height, at the following labor cost:

	Hours	Rate	Total	Rate	Total
Carpenter	4	$....	$....	$24.04	$96.16
Labor	2	....	....	19.13	38.26
. . . .			$....		$134.42

After the wood stringers are in place, it will require 4 hrs. carpenter time to place treads and risers for each flight of stairs (providing it is a plain box stair without nosings to form and without newels, balusters or handrail), at the following labor cost per flight:

	Hours	Rate	Total	Rate	Total
Carpenter	4	$....	$....	$24.04	$96.16

Figure 3 hrs. carpenter time to place 4"x4" newels, 2"x4" rails and 1"x2" balusters for one side of this type stairs, at the following labor cost per flight:

	Hours	Rate	Total	Rate	Total
Carpenter	3	$....	$....	$24.04	$72.12

To lay out the work and erect the stairs complete, including plain wall hand rail secured by brackets (instead of newels, hand rails and balusters), figure 8 hrs. carpenter time, at the following labor cost per flight:

	Hours	Rate	Total	Rate	Total
Carpenter	8	$....	$....	$24.04	$192.32

For the above stair complete, including newels, handrail and balusters, one side, figure 10 hrs. carpenter time, at the following labor cost per flight:

	Hours	Rate	Total	Rate	Total
Carpenter	10	$....	$....	$24.04	$240.40

Erecting Wood Stairs Made in Shop. Where wood stairs are made in the shop with stringers housed out to receive treads and risers, it will require 8 hrs. stair builder's time to lay out the work, set stringers, treads and risers (with plain wall rail attached to brackets), at the following labor cost per flight:

	Hours	Rate	Total	Rate	Total
Stair builder	8	$....	$....	$24.04	$192.32
Labor	2	....	....	19.13	38.26
			$....		$230.58

If the stair consists of two short flights, having an intermediate landing platform between floors, it will require 10 to 12 hrs, stair builder's time to lay out work, place stringers, treads, risers and plain wall handrail, at the following labor cost:

	Hours	Rate	Total	Rate	Total
Stair builder	11	$....	$....	$24.04	$264.44
Labor	2	....	....	19.13	38.26
			$....		$302.70

Where a plain box stair (3'-0" to 4'-0" wide and 9'-0" to 10'-0" story height) with open stringers is used, it will require 10 to 12 hrs. stair builder's time to lay out work, place stringers, treads and risers (not including newels, balusters or handrail) for each flight of stairs, at the following labor cost:

	Hours	Rate	Total	Rate	Total
Stair builder	11	$....	$....	$24.04	$264.44
Labor	2	....	....	19.13	38.26
			$....		$302.70

If the stairs have newels, wood handrail and balusters, add 6 to 8 hrs. stair builder's time, at the following labor cost per flight:

	Hours	Rate	Total	Rate	Total
Star builder	7	$....	$....	$24.04	$168.28

Plain box stairs consisting of two short flights having an intermediate landing platform between stories will require 6 to 8 hrs. stair builder's time to lay out the work, place stringers, treads and risers for each short flight of stairs at the following labor cost per flight:

	Hours	Rate	Total	Rate	Total
Stair builder	7	$....	$....	$24.04	$168.28
Labor	2	....	....	19.13	38.26
			$....		$206.54

To place newels, balusters and handrail on each short flight of stairs will require about 4 hrs. stair builder's time, at the following labor cost per flight:

	Hours	Rate	Total	Rate	Total
Stair builder	4	$....	$....	$24.04	$96.16

The labor cost complete for two short flights of stairs, including newels, handrails and balusters should cost as follows per story height:

	Hours	Rate	Total	Rate	Total
Stair builder	22	$....	$....	$24.04	$528.88

Erecting Wood Stairs Having Open Stringers. Wood stairs of the open stringer type having treads with a return nosing projecting beyond the face of the stringer, will require about 12 to 14 hrs. stair builder's time to lay out the work and place stringers, treads and risers for one flight of stairs containing 16 to 18 risers, at the following labor cost per flight:

	Hours	Rate	Total	Rate	Total
Stair builder	13	$....	$....	$24.04	$312.52
Labor	2	....	....	19.13	38.26
			$....		$350.78

If necessary to place newels, handrails and baluster, add about 8 to 9 hrs. stair builder's time, at the following labor cost per flight:

	Hours	Rate	Total	Rate	Total
Stair builder	8.5	$....	$....	$24.04	$204.34

If the stair consists of two short flights having an intermediate landing platform between stories (each short flight contains 8 to 10 risers), it will require about 8 to 9 hrs. stair builder's time per flight or 16 to 18 hrs. per story, at the following labor cost:

	Hours	Rate	Total	Rate	Total
Stair builder	17.0	$....	$....	$24.04	$408.68
Cost per short flight	8.5		....		204.34

To set starting and landing newels, handrails and wood balusters will require 7 to 9 hrs. stair builder's time per flight or 14 to 18 hrs. per story, at the following labor cost:

	Hours	Rate	Total	Rate	Total
Stair builder	16	$....	$....	$24.04	$384.64
Cost per short flight	8		....		192.32

The above does not include carpenter time framing and erecting intermediate landing platform.

Placing Wood Handrail on Metal Balustrades. A carpenter should place 45 to 55 lin. ft. of wood handrail on metal balustrades per 8-hr. day at the following labor cost per 100 lin. ft.:

	Hours	Rate	Total	Rate	Total
Carpenter	16	$....	$....	$24.04	$384.64
Cost per lin. ft			....		3.85

Placing Wood Handrail on Wall Brackets. A carpenter should set brackets and install 65 to 70 lin. ft. of wall hung wood handrail per 8-hr. day at the following labor cost per 100 lin. ft.:

	Hours	Rate	Total	Rate	Total
Carpenter	12	$....	$....	$24.04	$288.48
Cost per lin. ft			....		2.88

Inside Stair Material

Starting Steps			
Quarter circle	Red Oak	4' long	$33.00
Half circle	Red Oak	4'-6" long	42.00
Bull nose	Red Oak	4'-6" long	38.00
Scroll end	Red Oak	4'-6" long	42.00

Stringers (not housed)		
3/4"x11 1/4"x8'-0" long	Red Oak	$28.00
3/4"x11 1/4"x10'-0" long	Red Oak	33.00
3/4"x11 1/4"x12'-0" long	Red Oak	50.00
3/4"x11 1/4"x14'-0" long	Red Oak	56.00
3/4"x11 1/4"x16'-0" long	Red Oak	64.00
Treads (not returned; for returns add $1.75)		
1 1/6"x10 1/2"x3'-0" long	Red Oak	$11.00
1 1/6"x10 1/2"x3'-6" long	Red Oak	12.50
1 1/6"x10 1/2"x4'-0" long	Red Oak	14.00
1 1/16"x11 1/2"x3'-0" long	Red Oak	14.50
1 1/16"x11 1/2"x3'-6" long	Red Oak	16.00
1 1/16"x11 1/2"x4'-0" long	Red Oak	19.00
1 1/16"x11 1/2"x5'-0" long	Red Oak	23.00
1 1/16"x11 1/2"x6'-0" long	Red Oak	29.00
Return Nosing		
1 1/8"x1 3/4"x1'-2" with 2" return	Red Oak	$5.00
Tread Nosing		
1 1/8"x1 3/4"x1'-2" long	Red Oak	1.90

1 1/16"x1 1/8"x1'-2" long	Red Oak	1.75

Landing Tread Nosing

1 1/16"x3 1/2"x3'-0" long	Red Oak	2.50
1 1/16"x3 1/2"x3'-6" long	Red Oak	2.75
1 1/16"x3 1/2"x4'-0" long	Red Oak	3.10

Tread Brackets

12"x7 3/4"x 1/4" 3 ply	Nat. Birch	2.00

Newel Posts	Birch	Red Oak
3 1/4"x3 1/4"x3'-2" tapered starting	$14.00	$13.50
3 1/4"x3 1/4"x3'-10" tapered, starting	16.00	16.50
3 1/4"x3 1/4"x5'-3" tapered, landing	22.00	22.00
3 1/4"x3 1/4"x7'-0" tapered, angle	26.00	26.00

Balusters		Birch	Red Oak
1 5/16" round	30"	$2.20	$2.00
or square	33"	2.30	2.10
	36"	2.40	2.20
	39 or 42"	2.75	2.50

Railings--Natural birch, random length

1 3/4"x1 5/8"	$1.75
1 3/4"6x1 11/16"	2.00
2 5/8"x1 11/16"	2.15
2 1/4"x2 3/8"	4.25

Rail Bolts and Plug	.40

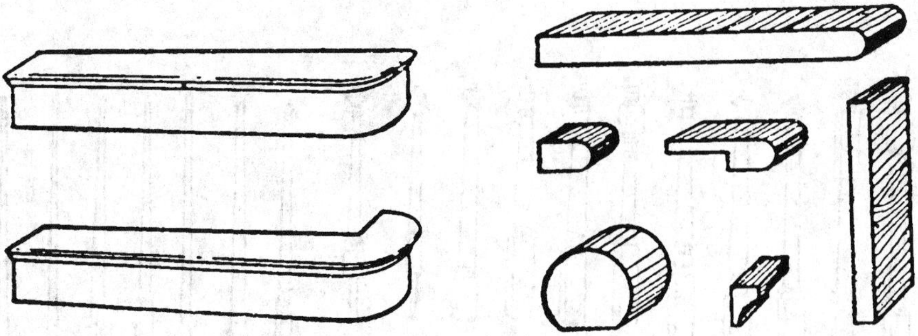

Inside Stair Materials

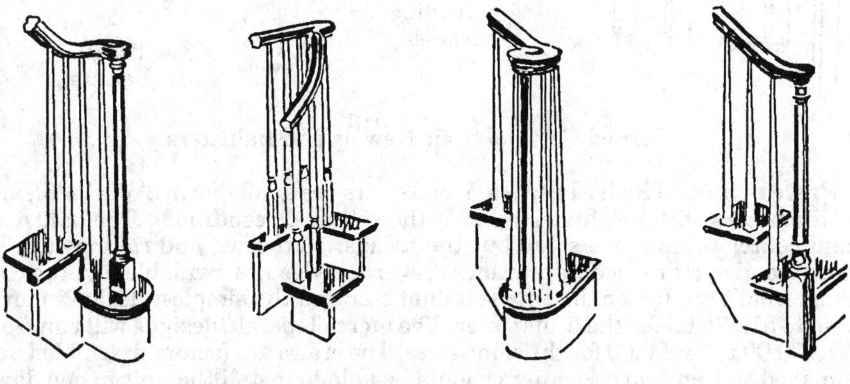

Starting Steps and Colonial Newels

Prefabricated Folding Stairways

For 9' finished floors to finished ceiling heights:

Premium grade, for 2'-6"x6'-0" opng	$255.00
Medium grade, for 2'-2"x4'-6" opng	80.00
Economy grade, for 2'-2"x4'-6" opng	55.00

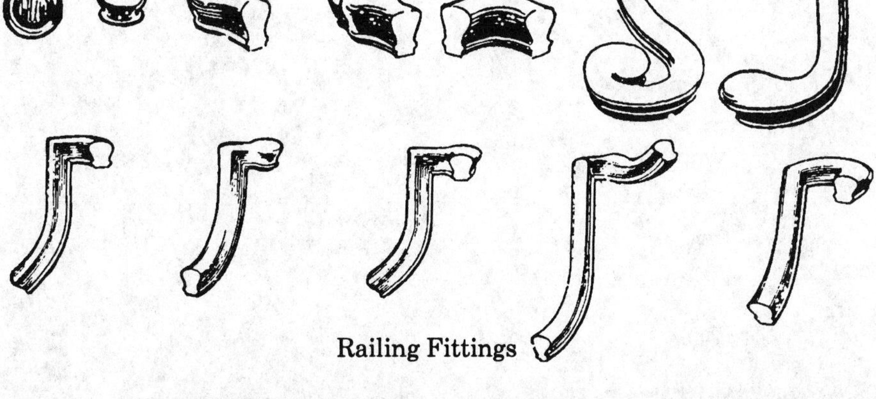

Railing Fittings

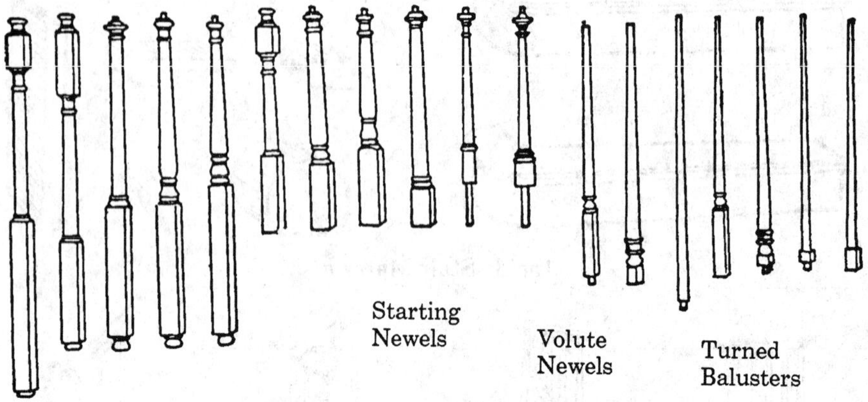

Turned Colonial Stair Newels and Balusters

Prefabricated Spiral Stairs. Wood stairs are available in 3'6", 4', 4'6", and 5' diameters with heights made to fit the actual job conditions. The balusters and center column are steel, but the treads, platforms, and railings can be wood to give a finished appearance. Several styles are available and the cost runs from $860.00 for the smallest diameter and the simplest design in an 8' rise to $1,275.00 for the 5' diameter. The more elaborate designs will run from $1,170.00 to $1500.00 for the same sizes. The stairs are factory assembled and finished so that two carpenters should be able to install the unit in one day.

CHAPTER 7

THERMAL AND MOISTURE PROTECTION

CSI DIVISION 7
07100 WATERPROOFING
07150 DAMPPROOFING
07200 INSULATION
07300 SHINGLES AND ROOFING TILES
07400 PREFORMED ROOFING AND SIDING
07500 MEMBRANE ROOFING
07600 FLASHING AND SHEET METAL
07800 ROOF ACCESSORIES
07900 SEALANTS

07100 WATERPROOFING

Every building requires water and damp resisting material, plaster bonds, or floor hardeners of one kind or another, and with so many of them on the market, it is impossible even for the best informed contractor to be familiar with all of them.

No attempt has been made here to pass on the relative merits of the various preparations. The data deals strictly with information that is of interest to the estimator--use of the material, covering capacity, and cost of applying. The prices quoted are approximate and may vary considerably in different locations. Always check with local distributors for current prices.

Water and Damp Resisting Methods. There are several methods used by the different manufacturers for mixing and applying their products to produce a water resisting wall or surface. In some instances, the water repelling compounds are incorporated with the cement, sand and gravel while the concrete is being mixed, and by filling the voids in the cement and sand produce a waterproof mixture. Where the various powders or pastes are mixed in the concrete mass, it is called the *integral method* of waterproofing.

Some water and dampness resisting preparations are furnished in mastic form and are applied with a trowel. Others are mixed with portland cement and sand and applied as a plaster coat to the surfaces to be treated.

There are also numerous water and dampness resisting compounds in the form of heavy paints, which are applied to the concrete or masonry walls, floors, and ceilings with a brush or mop. One or more applications of these materials are supposed to penetrate and fill the pores in the concrete or masonry surfaces to such an extent that the treated surfaces will be impervious to dampness or moisture.

Where there is considerable pressure or a large head of water, one of the most satisfactory methods is known as membrane waterproofing. This method consists of applying 2 to 7 hot moppings of pitch or asphalt on 1 to 7 plies of saturated felt or fabric. Waterproofing of this kind is applied in much the same manner as built-up roofing. After the last ply of saturated felt or fabric is applied, the entire surface is mopped with a heavy application of hot pitch or asphalt.

Estimating Quantities of Water and Damp Resisting Preparations. When estimating quantities of the various plaster coats, paints and membrane waterproofing compounds required for any job, the entire area of the walls or surfaces to be treated should be measured and the quantities stated in sq. ft., sq. yds. or by the square (100 sq. ft.).

Integral water resisting compounds are usually estimated by the cu. ft. or cu. yd. of concrete to which the liquid, paste, or powder is added.

Water Required for Mixing Concrete. When mixing concrete, sufficient water should be used to produce a plastic mix--no more and no less. A plastic mix is concrete that can be readily molded, and when the mold is removed, will flow sluggishly without segregation of the water or the fine materials from the coarse. Concrete should be mixed at least 2 minutes after all materials have been placed into the mixer. Remember there is nothing more injurious to good concrete than too much mixing water.

Refer to water required for mixing concrete as given in Step 4. "Selection of water-cement ratio" under *Cast-in Place Concrete.*

Waterproofing by the Integral Method

When using the integral method, the admixtures are usually mixed with the cement and aggregate and form a part of the concrete mass. The quantities are based on the specifications of the various manufacturers.

Material prices given are approximate, as it is impossible to quote prices that will apply to every locality. The contractor or estimator should obtain local prices on the materials specified. Prices are based on material being purchased in drum or barrel quantities. When purchased in one gallon or one pound quantities the cost will be increased considerably.

Integral Liquid Admixtures

Integral liquid admixtures consist of calcium chloride solutions, oil water repellent preparations, and other integral mixtures that flow freely from a drum.

The cost of hauling the barrel to the mixer, setting it up on blocks and drawing off the liquid by spigot into a measuring container will cost from 10 to 15 cents per gallon of waterproofing on small jobs. On large jobs this cost may be halved.

Concrete mixed in the proportions of 1:2:4 or 1:2 ½:3 ½ requires 5 ½ to 6 sacks of cement per cu. yd. of concrete.

Trade Name	Qts. per Sack Cement	Concrete Proportion	Gals. per Cu. Yd. Concrete	Price per Gal.
Anti-Hydro*	1	1:2:4	1 ½	$3.00
Aquatite	⅔	1:2:4	1	1.60
Aquatite**	1	1:2:4	1 ½	1.60
Hydratite	1	1:2:4	1 ½	2.35
Toxement IW	1	1:2:4	1 ½	2.30
Sikacrete	1	1:2:4	1 ½	3.30
Trimix	1	1:2:4	1 ½	1.50

*A 5-year maintenance guarantee furnished by manufacturer where supervision is supplied at additional cost.

**Use where there is water pressure.

Powdered Integral Admixtures

This material should be added to the portland cement before the addition of aggregates and then dry mixed thoroughly before adding gauging water. Add about 15 to 20 cents per lb. to the cost of material for labor handling and mixing.

Trade Name	Lb. per Sack Cement	Concrete Proportion	Lb. per Cu. Yd. Concrete	Price per Lb.
Hydratite Powder	1	1:2:4	6	$.30
Hydrocide Powder	1	1:2:4	6	.30
Plastiment	½-1	1:2:4	3-6	.55
Toxement IW	1 ½	1:2:4	7 ½	.45

Water and Damp-Resisting Plaster Coats

There are numerous preparations on the market to be mixed with portland cement and sand and applied as a plaster coat for waterproofing concrete and masonry walls. There are other preparations in mastic form that are applied with a trowel, the same as ordinary cement mortar.

The materials described on the following pages include many of the better known preparations used for this purpose.

Estimating Quantities of Water or Damp-Resisting Plaster Coats. Where water or damp-resisting plaster coats are applied to brick or concrete walls, the water resisting ingredient is usually in liquid, paste or powder form.

This compound is usually added to the mortar on the basis of a certain quantity to each sack of cement used and practically all manufacturers specify a mortar mixed in the proportions of 1 part portland cement to 2 parts sand, adding only sufficient water to make a workable mortar.

Quantity of Cement and Sand Required for One Cu. Yd. of Cement Mortar

Based on damp loose sand containing 5 percent of moisture and weighing 2,565 lbs. per cu. yd.

Proportions by Volume Cement Sand	Packed Cement Sacks	Loose Sand Cu. Yds.
1:1	18.0	0.70
1:1 ½	15.2	0.84
1:2	12.5	0.95
1:2 ½	11.0	1.03
1:3	9.5	1.10

Above quantities of sand include 5% for waste.

Number of Sq. Ft. of Cement Plaster Coat Obtainable from One Cu. Yd. of Cement Mortar

Thickness in Inches	Square Feet
¼	1,296
½	648
¾	432
1	324

Cleaning and Washing Old Masonry or Concrete Surfaces. Where old masonry or concrete surfaces are to be given a waterproof plaster coat, it will first be necessary to prepare the old surface to receive the cement plaster.

On some jobs it will be necessary to rake out the old brick joints to form a key for the plaster, and on others it will be necessary to thoroughly wire brush the old surface to remove all surface dirt and then wash the surface with clean water or a mixture of chemical cleaner and water.

Where the surfaces are to be brushed and washed with water, the labor cost of cleaning 100 sq. ft. of surface should average as follows:

	Hours	Rate	Total	Rate	Total
Labor	1.5	$....	$....	$19.13	$28.70
Cost per sq. ft.			....		.29

Roughing and Hacking Old Concrete Walls and Other Vertical Surfaces. Where it is necessary to roughen old, hard concrete surfaces to prepare a suitable bond for the new waterproof plaster coat, this can prove to be very expensive, especially where the old concrete surfaces are very hard and must be hacked and roughened using hand tools, a jackhammer, or bush-hammering tool. Workers often work in cramped quarters, which prevents the best daily output.

On work of this kind, a worker will roughen about 20 sq. ft. of surface per hour. The following table indicates the labor cost per 100 sq. ft. to roughen and clean the surface:

	Hours	Rate	Total	Rate	Total
Labor on Jackhammer	5	$....	$....	$19.13	$95.65
Labor for cleaning	2	....	....	19.13	38.26
Cost per 100 sq. ft.			$....		$133.91
Cost per sq. ft.			....		1.34

Cost of tools (purchase or rental), fuels, etc. has to be estimated, in addition to the labor requirements. These costs will vary, depending on the accessibility, depth to be roughened, and quantity of work. For example, light work or small areas can be accomplished with an electric bush-hammer, while heavier work will require an air compressor, hose and light jackhammer.

Equipment rental and fuel costs for an air compressor and tools to roughen the 100 sq. ft. of surface (as given in the preceding table) will be as follows:

	Hours	Rate	Total	Rate	Total
Air Compressor & Tools (Rental)	5	$....	$....	$7.00	$35.00
Fuel Costs 10 gal.		....	....	1.35	13.50
Cost per 100 sq. ft			$....		$48.50
Cost per sq. ft.			....		.49

Applying Slush or Grout Coat of Neat Cement. After the old surfaces have been cleaned, it will be necessary to give them a slush coat of neat portland cement thoroughly brushed into the pores of the wall, which acts as a bond for the subsequent plaster coats. The labor cost per 100 sq. ft. of surface should run as follows:

	Hours	Rate	Total	Rate	Total
Labor	2	$....	$....	$19.13	$38.26
Cost per sq. ft.			....		.38

The material needed for a slush coat of neat portland cement will require one bag of portland cement, mixed with water, per 100 sq. ft. of surface to be coated. A bag of cement costs about $6.00 each which amounts to the following material costs: $5.50 per 100 sq. ft. or $.06 per sq. ft.

Labor Applying Waterproof Plaster Coats to Vertical Surfaces. When applying waterproof plaster coats to vertical surfaces, the costs run high for a number of reasons:

1) Interference of other trades.
2) Difficulty of mixing and handling.
3) Delay in setting of coats.
4) Difficulty in finishing vertical surfaces of this type with plaster or cement finish.
5) Contingency cost of return to job for breaks and cracks.
6) Waste and spillage of materials.

This work is usually performed by cement masons, plasterers, or bricklayers.

When applying waterproof plaster coats, it is not necessary to give the surface a float or brush finish, but it must be reasonably smooth and without uneven projections that would prevent the water from running down the sides of the walls.

Most manufacturers specify the mortar to be mixed in the proportions of 1 :1 ½ or 1 :2, and the material is usually applied in one or two coats.

A mason should apply about 350 sq. ft. of first coat per 8-hr. day, at the following labor cost per 100 sq. ft.:

	Hours	Rate	Total	Rate	Total
Plasterer	2.3	$....	$....	$21.59	$49.66
Helper	2.3	....	....	19.13	44.00
Cost per 100 sq. ft			$....		$93.66
Cost per sq. ft.			....		.94

On the second or finish coat, a mason should apply about 400 sq. ft. per 8-hr. day, at the following labor cost per 100 sq. ft.:

	Hours	Rate	Total	Rate	Total
Plasterer	2.0	$....	$....	$21.59	$43.18
Helper	2.0	....	....	19.13	38.26
Cost per 100 sq. ft			$....		$81.44
Cost per sq. ft.			....		.81

If necessary to clean the old masonry or concrete surfaces, or to roughen the old concrete, add for this work as given on previous page.

Water Resisting Plaster Coats Using
Powdered Integral Water Repellents, Per 100 Sq. Ft.

Trade Name	Inches Thick	Mix Prop.	Number Coats	Sack Cement	Cu.Ft. Sand	Quan. W'pf'g	Approx Price/lb.	Mech. Hrs.	Lab. Hrs.
Hydratite	1	1 :2	2	4	8	8 lbs.	$0.30	8	8
Hydrocide Powder	1	1 :2	2	4	8	4 lbs.	.30	8	8
Toxement IW	¾	1 :2	2	3	6	4 ½	.35	7	7

Water Resisting Plaster Coats Using Liquid Integral Admixtures, Per 100 Sq. Ft.

Trade Name	Inches Thick	Mix Prop.	Number Coats	Sack Cement	Cu.Ft. Sand	Quan. W'pf'g	Approx Price/lb.	Mech. Hrs.	Lab. Hrs.
Anti-Hydro	3/4	1:2	2	3	6	3/4 gal	$3.00	7	7
Anti-Hydro	1	1:2	2	4	8	1 gal.	3.00	8	8
Aquatite	3/4	1:2	2	3	6	3/4 gal	1.60	7	7
Aquatite	1	1:2	2	4	8	1 gal.	1.60	8	8
Hydratite	3/4	1:2	2	3	6	3/4 gal	2.35	7	7
Hydratite	1	1:2	2	4	8	1 gal.	2.35	8	8
Toxement IW	3/4	1:2	2	3	6	3/4 gal	2.30	7	7
Toxement IW	1	1:2	2	4	8	1 gal.	2.30	8	8
Sikacrete	3/4	1:2	2	3	6	3/4 gal	3.30	7	7
Trimix	3/4	1:2	2	3	6	3/4 gal	1.50	7	7

Membrane Waterproofing

Membrane waterproofing is constructed in place by building up a strong, waterproof and impermeable blanket with overlapping plies of tar, asphalt saturated open mesh cotton fabric, or rag felt. The plies are coated and cemented together with hot coal tar pitch or waterproofing asphalt. There is always one or more applications of pitch or asphalt than plies of felt or fabric.

Estimating the Quantity of Felt or Fabric Required for Membrane Waterproofing. When estimating the quantity of felt or fabric required for waterproofing masonry or concrete walls, pits, floors, etc., it will be necessary to add a certain percentage for lap and waste. If the felt is applied to walls, it will be necessary to allow about 6" to lap over the footings at the bottom, also a small allowance at the top of the walls above grade. This will average from 7 1/2 to 10% of the total area to be waterproofed.

If the walls or floors are covered with a single thickness of felt or membrane, it will be necessary to lap each strip of felt 4" to 6", depending upon the specifications. As most felts are put up in 36" rolls it will be necessary to add about 12% waste for a 4" lap, and 17% for a 6" lap.

If the lap is 4" wide and only a single thickness of felt is used, it will be necessary to add about 20% to the actual area to allow for laps and the additional felt required at the top and bottom of the wall. On floors it is customary to run the felt or membrane up the sides of the walls 4" to 6".

If the lap is 6" and only a single thickness of felt is used, add about 25% to the actual wall or floor area for laps, etc.

When 2, 3, 4 or more plies of felt are used, it will not be necessary to allow for laps except at the ends of the wall or floor, as one thickness of felt overlaps the next without waste. However, extra material will be required at the tops and bottoms of the walls and where the paper runs up the sides of the walls 4" to 6". On work of this kind, add about 10% to the actual floor or wall area.

Fabric or Felt Required to Cover 100 Sq. Ft. of Surface

Class of Work	1 Ply 4" lap	1 Ply 6" lap	2-Ply	3-Ply	4-Ply	5-Ply	Ea. Add. Ply
Add percent	.20	25	10	10	10	10	10
Sq. Ft. Felt	120	125	220	330	440	550	110

Weight of Tar or Asphalt Felt for Membrane Waterproofing. Tar or asphalt felt for waterproofing is currently furnished in 4 sq. rolls of 432 sq. ft. weighing 60 lbs. per roll.

Double thickness asphalt felt is also furnished in 60 lb. per roll containing 216 sq. ft. and this felt is known as No. 30.

When specifying the grade or weight of felt to be used, it is customary to state that "felt shall weigh not less than 15 lbs. per 108 sq. ft." This is known as No. 15 felt. Felt is furnished in 4 sq. rolls of 432 sq. ft., so there are 32 sq. ft. per roll or 8 sq. ft. per 100 sq. ft. allowed for laps.

Tar or asphalt saturated fabric is usually sold by the roll containing 50 sq. yds. or by the sq. yd.

Estimating the Quantity of Pitch or Asphalt Required for Membrane Waterproofing. The quantity of pitch or asphalt required for mopping 100 sq. ft. of surface with one application is approximately 30 to 35 lbs.

Many special coatings are sold under trade names. Some are furnished solid, making it necessary to heat and melt them before using, while others are furnished in the form of a heavy black paint that can be either mopped or brushed on cold. The manufacturers usually specify how their materials shall be applied.

Engineers and architects usually specify the quantity of asphalt or compound to be used for each 100 sq. ft. of surface.

Weight of Pitch or Asphalt Per 100 Sq. Ft. of Surface

No. Plies of Saturated Fabric or Tarred Felt	Alternate Moppings of Pitch or Asphalt Required	Lbs. of Pitch or Asphalt Required 100 SF W'p'fing
2	3	90 to 105
3	4	120 to 140
4	5	150 to 175
5	6	180 to 210
6	7	210 to 245

Applying Membrane Waterproofing. Whether to use saturated fabric or saturated felt in constructing membrane waterproofing depends entirely upon local conditions. Saturated fabric weighing 12 oz. per sq. yd. has approximately three times the tensile strength of 15-lb. saturated felt, but it costs considerably more per sq. yd. than saturated felt. On flat surfaces where the waterproofing will be covered with a protection course immediately after it is applied and where it will not be subjected to any unusual strain, saturated felts can be used. On any surface where greater strength is required, fabric should be used. It is always desirable to provide a protection course over the waterproofing construction. While the labor operations involved in applying membrane waterproofing are similar to those followed in applying built-up roofing, the work is usually more expensive than roofing because of the limited working space and the necessity for using scaffolds on vertical surfaces over 8 feet high.

On vertical surfaces the work can be expedited if the plies of felt or fabric are installed up and down the surface like wallpaper rather than across. The pieces of felt or fabric can be cut to the required length and folded so that after the surface has been mopped the top edges can be pushed into the hot bitumen and the remainder of the piece will then fall into place where it can be rubbed into the hot pitch or asphalt. After the required number of plies have been embedded in alternate moppings of hot bitumen all angles at corners, walls, etc., should be reinforced with at least two additional plies of felt or fabric and alternate moppings of bitumen. A protection course of beadboard, fiberboard, or other material should immediately be placed over the waterproofing for best results.

Prior to the application of the first hot bitumen mopping coat, the wall surface is primed, uniformly and completely, with one gallon of primer (asphalt primer for asphalt specifications or tarbase primer for pitch specifications) per 100 sq. ft. of surface area.

For the application of the primer to the surface, one man should coat about 100 sq. ft. per hour, at the following labor cost:

	Hours	Rate	Total	Rate	Total
Labor	1.0	$....	$....	$22.72	$22.72
Cost per sq. ft.					.23

Labor Applying Membrane Waterproofing. When applying membrane waterproofing to walls and floors, it takes three or four workers together. It requires one worker to attend the fire and heat the bitumen, while two or three are mopping the walls and applying the felt or fabric.

If only one application of hot bitumen is applied, two workers together should heat materials and mop 2,000 to 2,200 sq. ft. per 8-hr. day, at the following labor cost per 100 sq. ft.:

	Hours	Rate	Total	Rate	Total
Roofer	0.8	$....	$....	$22.72	$18.18
Cost per sq. ft.			....		.18

Water Pressure Table and Waterproofing Required for Varying Heads of Water

Hydrostatic Head in Feet	Pressure Lbs. per Sq. In.	Lifting Pressure Lbs per SF	Wall Pressure Lbs per SF	Plies of Saturated Felt/Fabric	Asphalt Mastic Thickness
0.5	0.21	39.2	15.6	2-Ply	1/8"
1.0	0.43	62.5	31.2	2-Ply	1/8"
2.0	0.86	125.0	62.5	2-Ply	1/4"
3.0	1.30	187.5	93.7	2-Ply	1/4"
4.0	1.73	250.0	125.0	3-Ply	5/8"
5.0	2.17	312.5	156.2	3-Ply	5/8"
6.0	2.60	375.0	187.5	4-Ply	5/8"
8.0	3.47	500.0	250.0	4-Ply	5/8"
10.0	4.34	625.0	312.5	5-Ply	5/8"
12.0	5.21	750.0	375.0	6-Ply	5/8"
15.0	6.51	937.5	468.7	6-Ply	3/4"
20.0	8.68	1250.0	625.0	7-Ply	3/4"
25.0	10.85	1562.5	781.2	8-Ply	3/4"
30.0	13.02	1875.0	937.5	10-Ply	3/4"
40.0	17.36	2500.0	1250.0	11-Ply	3/4"

After the surface has been mopped with hot bitumen, it will be necessary for two men to cut the felt to lengths and place it on the wall. Usually it is cut long enough to lap over the top of the wall a few inches, so that it can be held in place by brick or stone until it is self-supporting.

Two men placing felt or fabric have to work along with the men mopping, so they will place felt or fabric on 2,000 to 2,200 sq. ft. of surface per 8-hr. day, at the following labor cost per 100 sq. ft.:

	Hours	Rate	Total	Rate	Total
Roofer	0.8	$....	$....	$22.72	$18.18
Cost per sq. ft.			....		.18

Labor Applying 100 Sq. Ft. of Membrane Waterproofing

Desc. of Work	Number of Plies	Mopping Hours	Felt Hours	Total Hours
1-ply fabric, 2 moppings	1	1.6	0.8	2.4
2-ply fabric, 3 moppings	2	2.4	1.6	4.0
3-ply fabric, 4 moppings	3	3.2	2.4	5.6
4-ply fabric, 5 moppings	4	4.0	3.2	7.2
Each additional ply of felt and mopping	1	0.8	0.8	1.6

Prices of Waterproofing Materials

Prices on waterproofing materials must always be checked, because prices vary considerably due to market conditions, geographical location and with the various manufacturers.

Asphalt primer costs about $2.25 per gal. in 55-gal. drums; $2.35 per gal. in 30-gal. drums and $2.50 per gal. in 5-gal cans.

Waterproofing pitch or asphalt costs about $270.00 per ton in carload lots; $280.00 per ton in less than carloads.

No. 15 tar or asphalt saturated felt costs $12.00 per roll of 432 sq. ft. in carloads and $12.50 per roll in less than carloads.

Tar or asphalt saturated cotton fabrics costs $1.00 to $1.25 per sq. yd. in carloads and $1.25 to $1.30 per sq. yd. in less than carloads.

Material Cost of 100 Sq. Ft. of 1-Ply Membrane Waterproofing
Consisting of 1-ply 15-lb. felt and 2 moppings of hot bitumen

	Rate	Total	Rate	Total
1 gal. asphalt primer*	$....	$....	$2.25	$2.25
70 lbs. asphalt or pitch	....	....	.13	9.10
.28 rolls No. 15 felt (120 s.f.)	....	....	12.00	3.36
Fuel, mops, etc.	....	....	1.00	1.00
Cost per 100 sq. ft.		$....		$15.71
Cost per sq. ft.		....		.16

Labor Cost of 100 Sq. Ft. of 1-Ply Membrane Waterproofing
Consisting of 1 ply of felt and 2 moppings of hot bitumen

	Hours	Rate	Total	Rate	Total
Roofer*	2.6	$....	$....	$22.72	$59.07
Roofer placing felt	0.8	....	....	22.72	18.18
Cost per 100 sq. ft.			$....		$77.25
Cost per sq. ft.			....		.77

*Includes labor cost for primer coat.

Material Cost of 100 Sq. Ft. of Each Additional Ply of Felt and Hot Bitumen

	Rate	Total	Rate	Total
35 lbs. asphalt or pitch	$....	$....	$.13	$4.55
.24 rolls No. 15 felt (100 sf)	....	....	12.00	2.88
Fuel, mops, etc.	....	....	.50	.50
Cost per 100 sq. ft.		$....		$7.93
Cost per sq. ft.		....		.08

Labor Cost per 100 Sq. Ft. for Each Additional Ply of Felt and Hot Bitumen

	Hours	Rate	Total	Rate	Total
Roofer	1.6	$....	$....	$22.72	$36.35
Cost per sq. ft.			....		.36

Material Cost of 100 Sq. Ft. of 1-Ply Membrane Waterproofing Using Tar or Asphalt Saturated Fabric and Hot Pitch or Asphalt
Consisting of 1-ply of Saturated Fabric and 2 Moppings of Hot Bitumen

	Rate	Total	Rate	Total
1 gal. asphalt primer	$....	$....	$2.25	$2.25
70 lbs. asphalt or pitch	....	....	.13	9.10
13.4 sq. yds. satur. fabric	....	....	1.25	16.75
Fuel, mops, etc.	....	....	.85	.85
Cost per 100 sq. ft.		$....		$28.95
Cost per sq. ft.		....		.29

Material Cost of 100 Sq. Ft. Additional Ply of Saturated Fabric and Hot Pitch or Asphalt

	Rate	Total	Rate	Total
35 lbs. asphalt or pitch	$....	$....	$.13	$4.55
12 sq. yds. satur. fabric	....	....	1.25	15.00
Fuel, mops, etc.	....	....	.50	.50
Cost per 100 sq. ft.		$....		$20.05
Cost per sq. ft.		....		.20

The labor costs for placing fabric and moppings of hot bitumen are the same as used in placing felts and moppings of hot bitumen.

The Iron Method of Waterproofing

The iron method of waterproofing uses an exceedingly fine metallic powder, containing no grease, asphalt, oil or other substances subject to disintegration.

The waterproofing is applied to the inside or outside surfaces of walls and to the tops of rough footings and floor slabs, in the form of brush coats, or in a combination of brush and plaster coats.

Applied either upon the inside or outside of walls, this type of waterproofing will resist hydrostatic pressure to a considerable degree.

When preparing old concrete vertical surfaces for iron waterproofing, the vertical surfaces should be given an entirely new bonding exposure by cutting not less than 1/16" with bush-hammers or other suitable tools. See cost of roughing old surfaces on previous pages. The new surface shall then be thoroughly cleaned by brushing with wire brushes. Surfaces of brick and stone need only thorough cleaning with the wire brush and washing.

Holes, cracks, and soft or porous spots in vertical or horizontal surfaces shall be cut out and pointed with one part iron floor bond, 2 parts portland cement and 3 parts sand, all by weight. These materials shall be mixed dry, screened and enough water added to make a stiff mix. Particular care must be taken at the intersections of all walls and floors, in corners, around pipes and other projections through the waterproofing and at construction joints in the concrete.

After the wall surfaces have been cleaned and pointed, they shall be thoroughly cleaned again by hosing. Excess water shall then be removed.

It usually requires 2 to 6 coats of iron waterproofing, depending upon the density of the surface and the conditions encountered.

Some of these materials are mixed with sand and water and applied as a brush coat while others are furnished already mixed ready to apply by the addition of water.

1-Coat Waterproofing Using the Iron Method, Brushed On, Per 100 Sq. Ft.

Trade Name	Number Coats	Quan. Req'd 100 Sq. Ft.	Approx. Cost Lb.	Labor Hrs. Applying
Ferrolith W	1	13-15 lbs.	.30	2
Kemox H (Sika)	1	60-100 lbs.	.16	2
Irontox	1	10-12 lbs.	.25	2

For each additional application, figure same as given above. Three to five applications are required where there is water pressure to be overcome.

Liquid Membrane Waterproofing

Liquid membrane (elastomeric) waterproofing is a cold, fluid applied, synthetic rubber, seamless coating system. It is available in single or two-component form and can be applied by trowel or spray. The material forms a continuous, flexible, water impervious film that bonds tightly to a wide range of surfaces, including concrete, stone, masonry, wood, and metal. On a troweled surface, an application rate of five gallons per 125 sq. ft. of surface will produce a cured coating thickness of approximately 55 mils.

Two workers should coat about 650 sq. ft. per day at the following labor cost:

	Hours	Rate	Total	Rate	Total
Labor	16.0	$....	$....	$22.72	$363.52
Cost per sq. ft.			....		.56

Cost of 100 Sq. Ft. Mastic Waterproof Coatings

The following materials are furnished in the form of a mastic or plastic cement and are applied with a trowel.

Kind of Waterproofing	Inches Thick	No. of Coats	Cover Cap. Sq. Ft. Gal.	Quan. Req'd 100 Sq. Ft.	Approx. Price per gal.	Labor Hours
Dehydratine No. 10	1/16	1	26	4 Gal.	$1.65	3
Hydrocide Mastic	1/16	1	20-25	5 Gal.	1.13	3
Hydrocide 700	1/16	1	20-25	5 Gal.	1.00	3
Tremco 103 Mastic						
Asphalt base	1/8	1	20	5 Gal.	1.00	3
Tar base	1/8	1	20	5 Gal.	1.70	3

07150 DAMPPROOFING

Dampproof Paints for Exterior Concrete or Masonry Surfaces

Where a heavy head of water is to be overcome, the concrete or masonry should preferably be waterproofed by the membrane method. Where only a dampproofing material is required, there are a number of paints on the market for this purpose. The application of one, two, or more coats of paint is intended to make the walls impervious to dampness.

Number of coats required and covering capacity will vary with porosity and surface of wall.

Heavy Dampproof Paints, Per 100 Sq. Ft.

Material	Use on	No. Cts.	Sq. Ft. per Gal.	Gals. Reqd. 100 Sq. Ft.	Price per Gal.	Lab. Hrs. Reqd.
Dehydratine 4	Fndtn.	2	33	3	$1.43	2.2
Dehydratine 10	Masonry	1	30	3 1/3	1.65	1.2
Hydrocide Semi-Mastic	Fndtn.	1	30-50	3	1.05	1.2
....	Fndtn.	2	15-18	6	1.05	2.2
Hydrocide 600	Fndtn.	1	75-100	1 1/3	1.13	1.2
Hydrocide 648	Fndtn.	2	50	2	1.13	2.2
Marine Liquid	Blw Grd	1	50-75	1 3/4	1.24	1.2
Sikaseal	Fndtn.	2	60-80	1 1/2	2.20	2.2
Tremco 110	Fndtn.	1	100	1	.95	1.2

Labor Applying Water and Dampproofing Paint to Exterior Surfaces Below Grade. Where the water or dampproofing consists of applying one or more coats of heavy paint to exterior concrete or masonry surfaces below grade, it is customary to use an ordinary roofer's brush or a mop. On foundation walls below grade, the men are usually obliged to work in cramped quarters which reduces output.

To make these paints effective, they must be thoroughly brushed into the surface and all voids and pin holes filled, otherwise water and moisture will seep through.

On work of this kind, a man should apply 600 to 700 sq. ft. per 8-hr. day, at the following labor cost per sq. ft.:

	Hours	Rate	Total	Rate	Total
Roofer	1.3	$....	$....	$22.72	$29.54
Cost per sq. ft.			....		.30

On the second and third coats, the labor cost will be slightly less than the first coat on account of the smoother surface to work on.

Dampproof and Plaster Bond Paints

Labor Applying Dampproofing and Plaster Bond Paints. Before applying the first coat of plaster to the interior of exterior brick or masonry walls, the walls are often painted with one or more coats of heavy black paint, which acts as a dampproofing and plaster bond.

The labor cost will vary with the consistency of the paint and the condition of the walls, as the paint must be well brushed in and all pin holes and porous places thoroughly covered with paint, otherwise moisture will seep through the walls.

Where the paint is applied by hand, a worker should cover 700 to 800 sq. ft. per 8-hr. day, at the following labor cost per 100 sq. ft. :

	Hours	Rate	Total	Rate	Total
Roofer	1.1	$....	$....	$22.72	$24.99
Cost per sq. ft.			....		.25

When the dampproof paints are applied with an air spray, a man should cover 300 to 350 sq. ft. per hr. at the following labor cost per 100 sq. ft.:

	Hours	Rate	Total	Rate	Total
Roofer	0.33	$....	$....	$22.72	$7.50
Cost per sq. ft.			....		.08

Dampproof and Plaster Bond Paints, Per 100 Sq. Ft.

Heavy black paints are used for dampproofing interior of exterior walls and to form a bond for plaster. Number of coats required will depend upon porosity of wall on which it is used.

Labor costs in the following table are based on hand application. If an air spray is used, deduct as given above.

Name of Material	No. Coats	Sq. Ft. per Gal.	Gals. Reqd. 100 SF	Price per Gal.	Labor Hrs. Applying
Dehydratine 4	1	70-80	1 ⅓	$1.43	1.1
Dehydratine 4	2	45-55	2 ⅓	1.43	2.0
Plasterbond 232	1	75-80	1 ⅓	1.47	1.1
Plasterbond 232	2	35-40	2 ½	1.47	2.0

Transparent Dampproofing of Exterior Masonry Walls

During the past few years, there has been a big change in the use of transparent dampproofings and the method of application due to an investigation by the National Bureau of Standards, Publications BMS7.

From the results of this investigation, it was determined that "leaky" walls were mainly caused by poor workmanship. In other words, unless the bricks were laid in a full bed of mortar (not furrowed), with all bed, end and vertical joints completely filled with mortar, the walls were likely to leak within a short period after the completion of the structure. A hairline crack between the brick and mortar joints was also given as a cause of leakage.

From these tests it was determined that painting the surfaces of masonry walls with transparent "paints" or "waxes" alone did not stop these leaks. One method that was really efficient was to cut out the old mortar joints to a depth of ½" to ¾" and repoint the joints with new mortar.

This class of work is quite expensive, because the cost depends upon the hardness and tenacity of the mortar joints that must be cut out.

The basis for the use of exterior waterproofing, labor and material, is calculated on the ordinary wall constructed with common brick of high absorption with ⅜" to ½" lime-cement mortar joints.

On lime-cement mortar joints, two masons can rake out and repoint about 12 sq. ft. of brick masonry surface an hour, removing joints to a depth of ¾ in. Allow about ¼ hr. helper's time for moving scaffolds, mixing, etc.

On the above basis, the labor cost of cutting out old mortar joints and repointing mortar joints to 100 sq. ft. of wall surface will cost as follows:

	Hours	Rate	Total	Rate	Total
Mason	16.0	$....	$....	$25.62	$409.92
Helper	4.0	....	....	19.13	76.52
Cost per 100 sq. ft.			$....		$486.44
Cost per sq. ft.			....		4.86

On hard portland cement mortar joints, two masons using a power saw can cut out and repoint joints in about 8 sq. ft. of wall per hour and requires 1/4 hr. helper time handling tools and mixing mortar.

On the above basis, the labor cost of cutting out old portland cement mortar joints, and repointing joints to 100 sq. ft. of wall will cost as follows:

	Hours	Rate	Total	Rate	Total
Mason	24	$....	$....	$25.62	$614.88
Helper	6	....	....	19.13	114.78
Cost per 100 sq. ft.			$....		$729.66
Cost per sq. ft.			....		7.30

Add cost of scaffold.

Colorless or Transparent Liquid Water Repellent Treatments

There are a number of colorless or transparent liquid preparations on the market used on brick, stone, stucco, cement, and concrete surfaces to render them water repellent and help to prevent rain absorption.

These liquids are colorless and do not affect the original color of the surface to which they are applied.

The covering capacities and quantities given in the following table were furnished by the various manufacturers.

It will be noted that some manufacturers specify one coat and others two coats. Regardless of the material used, sufficient liquid must be applied to completely seal the pores of the surface, or otherwise, the results will be unsatisfactory.

Transparent Liquid Water Repellents, Per 100 Sq. Ft.

Transparent liquids used for dampproofing brick, stone, stucco, concrete, and cement surfaces.

Name of Material	No. Coats	Sq. Ft. per Gal.	Gals. Reqd. 100 SF	Price per Gal.	Labor Hrs. Applying
Daracone	1	75-100	1 1/4	$4.05	1.6
Supertox	1	125-150	3/4	3.50	0.8
Tremco 141 Invisible	2	75-100	1 1/4	3.25	1.6

Colorless or Transparent Silicone Base Liquid Water Repellent Treatments

Recent developments in liquid water repellent treatments are the new silicone base products, which are truly colorless and permit the masonry to "breathe" while rendering the surface water repellent. They penetrate deeply into the cement or masonry surfaces coating pores, cracks, and fissures with an insoluble, nonoxidizing film of silicone that effectively stops capillary action by which water is absorbed.

Usually one flooding coat is sufficient and may be applied by either brushing or spraying. The covering capacities and quantities given in the following table are furnished by the various manufacturers.

Silicone Base, Transparent Liquid Water Repellents, Per 100 Sq. Ft.

Name of Material	No. Coats	Sq. Ft. Per Gal.	Gals. Req'd. 100 SF	Price Per Gal.	Labor Hours Applying
Dehydratine 22	1	75-100	1 ¼	$3.40	0.8
Hydrocide Colorless SX	1	75-100	1	4.50	0.8
Sika Transparent	1	80-200	⅔	2.70	0.8
Tremco 147-3%	1	100-200	⅔	4.70	0.8
Tremco 147-5%	1	100-200	⅔	6.40	0.8
Devoe Super-Por-Seal	1	150	⅔	5.29	0.8

Labor Applying Transparent Liquid Waterproofing. When applying transparent liquid dampproof paints by hand, a man should cover 900 to 1,100 sq. ft. of surface per 8-hr. day, at the following labor cost per 100 sq. ft.:

	Hours	Rate	Total	Rate	Total
Roofer	0.8	$....	$....	$22.72	$18.18
Cost per sq. ft.			....		.18

On the second and third coats, a man should apply 1,000 to 1,200 sq. ft. per 8-hr. day, at the following labor cost per 100 sq. ft.:

	Hours	Rate	Total	Rate	Total
Roofer	0.75	$....	$....	$22.72	$17.04
Cost per sq. ft.			....		.17

When applied by spray, a man should cover 300 to 350 sq. ft. per hr. at the following labor cost per 100 sq. ft.:

	Hours	Rate	Total	Rate	Total
Roofer	0.33	$....	$....	$22.72	$7.50
Cost per sq. ft.			....		.08

Add cost of compressor and spray equipment.

Back-Plastering and Back-Painting Cut Stone and Marble to Prevent Staining

When marble, limestone, brownstone, or sandstone are used in building work, considerable difficulty is encountered in preventing the moisture in the brick or concrete backing from penetrating the surface of the stone or marble and staining or discoloring the face. This is due to the open grain or porous nature of the stone or marble.

To overcome this condition, it is customary to paint the back of the stone, beds and end joints up to within 1" of the face to prevent staining and discoloration. White or non-staining cement is also used for this purpose. The stone is set in non-staining cement mortar, and the back of the stone is plastered or parged with the same mortar.

In many instances the backs, beds, and joints of the stone are painted and then back-plastered with non-staining cement, which acts as a double preventative.

Back-painting of Indiana limestone is not recommended by quarrymen and producers, who say that it tends to create conditions it is supposed to counteract. A full explanation is given under *Cut Stone.*

Where cut stone comes in contact with concrete, it is recommended that the concrete be painted instead of the stone and a one-inch space left between the concrete and cut stone.

Estimating the Cost of Back-Painting Cut Stone and Marble. It is customary to estimate the cost of cut stone by the cu. ft., so the cost of back-painting should be estimated in the same manner.

Plain stone ashlar is usually 4" to 8" thick or in alternating courses of 4" and 8" stone so as to provide a bond with the brick or masonry backing. Where just a veneer of cut stone is required, the stone is usually 4" thick and is bonded into the masonry backing by metal anchors.

If cut stone is furnished in alternating courses of 4" and 8", the stone for the entire job will average 6" thick. It will therefore be necessary to back-paint 2 sq. ft. of surface to each cu. ft. of stone. Including beds and joints, it will double the above quantities or it will be necessary to back-paint 4 1/3 sq. ft. of surface to one cu. ft. of stone.

If the stone averages 4" thick, it will be necessary to back-paint about 5 sq. ft. of surface to each cu. ft. of stone, including the backs, beds, and joints.

If the stone averages 8" thick, it will be necessary to back-paint about 4 sq. ft. of surface to each cu. ft. of stone.

Back-Painting Cut Stone and Marble. When back-painting limestone, sandstone, marble, etc., the stone is often painted in the yard before delivery, although on many jobs it is painted after delivery. It is customary to paint the stone a few days before setting, although the backs are frequently painted after the stone has been set.

Care must be exercised to thoroughly brush the paint into all the pores and not drop any paint on the face of the stone.

When painted at the building site, a man should paint 600 to 700 sq. ft. of surface per 8-hr. day, at the following labor cost per 100 sq. ft.:

	Hours	Rate	Total	Rate	Total
Mason	1.3	$....	$....	$25.62	$33.31
Cost per sq. ft.			....		.33

Back-Painting 100 Sq. Ft. Cut Stone or Marble to Prevent Staining

Heavy black paints applied to backs and sides of stone to prevent cement and mortar stains. Quantities based on 1 coat.

Name of Material	Sq. Ft. per Gal.	Gals. per 100 SF	Price per Gal.	Labor Hours
Dehydratine 4	100	1.00	$1.43	1.3
Hydrocide No. 648	125-175	0.75	1.13	1.3
Sikaseal	120	0.80	2.20	1.3

Number of Cu. Ft. of Stone to be Painted Per 100 Sq. Ft. of Surface
Includes back, beds and end joints
Average Thickness of Stone

4"	6"	8"	10"	12"
20	23	25	28.5	31

Number of Sq. Ft. of Painted Surface in One Cu. Ft. of Stone
Includes back, beds and end joints
Average Thickness of Stone

4"	6"	8"	10"	12"
5	4.3	4	3.5	3.2

Back-Plastering Cut Stone and Marble. After the stone or marble has been set, it is customary to plaster the back of the stone with non-staining cement, using an ordinary trowel and applying it 3/16" to 1/4" thick.

Inasmuch as the stone must be back-plastered as the setting progresses, so as not to delay the bricklayers in backing-up the stone, the time required is rather indefinite. Working steadily, however, a mason should back-plaster 65 to 75 sq. ft. an hr. at the following labor cost per 100 sq. ft.:

	Hours	Rate	Total	Rate	Total
Mason or stone setter	1.5	$....	$....	25.62	$38.43
Cost per sq. ft.			....		.38

07190 VAPOR BARRIERS/RETARDANTS

The combination of high indoor relative humidities and low outside temperatures causes condensation to form within the structure. Serious damage can result from this moisture in the form of rotting framing members, paint deterioration, and wet walls and ceilings.

Vapor barriers are recommended where the above conditions exist. Vapor seal paper should be installed on the warm side of wall, floor, ceiling, or roof. Paper should be installed with joints running parallel to and over framing members. All joints should be lapped about 2 inches. Where paper is to be exposed, nail wood lath strips over paper along framing members to provide a neat and permanent job.

A satisfactory vapor proof paper consists of a 50 lb. continuous asphalt film, faced on both sides with 30 lb. basis kraft paper. Other vapor proof papers consist of two 30 lb. basis sheets of kraft paper cemented together with asphalt and reinforced with strong jute cords, spaced 1/2 to 1 in. on centers and running in both directions. This same reinforced sheet is furnished with two sheets of asphalt coated paper cemented together and reinforced as described above. Another very satisfactory vapor proof paper consists of a heavy kraft paper coated on one side with a thin sheet of aluminum or copper, which prevents the penetration of moisture. Approximate prices on the various types of vapor seal paper are as follows:

Description	Price per Roll 500 Sq. Ft.
2 sheets 30-lb basis kraft paper, cemented together with asphalt	$5.00
2 sheets asphalt coated 30-lb kraft paper, cemented together with asphalt and reinforced with jute cords	$9.00
1 sheet heavy kraft paper coated two sides, with a reflective surface	$12.00

Sisalkraft. A strong, waterproof, windproof building paper consisting of 2 sheets of pretreated kraft paper, cemented together with 2 layers of special asphalt and reinforced with 2 layers of crossed sisal fibers. Costs about 2 cents per sq. ft.

Copper-Armored Sisalkraft is a combination of Sisalkraft and Anaconda electro-sheet copper bonded under heat and pressure. It is available in 3 weights, 1, 2, or 3 oz. copper per sq. ft.

It may be used for waterproofing, flashings, ridge roll flashing, etc. The approximate prices are as follows: 1-oz. costs 30 cents per sq. ft.; 2-oz. costs 50 cents per sq. ft. and 3-oz. costs 70 cents per sq. ft. Above prices subject to discount on large orders.

Polyethylene. Film comes in 100' rolls of various widths in thicknesses of .002", .004", .006", and .008". Costs vary from 1 1/2 to 4 cents per sq. ft. Tape for sealing joints costs $3.25 for a roll 2" wide by 100' long in the .004" thickness.

Labor Placing Vapor Seal Paper. Where vapor seal paper is applied to the interior of exterior stud walls, using a stapling machine, to prevent the penetration of moisture and condensation, a carpenter should handle and place 2,000 to 2,500 sq. ft. per 8-hr. day, at the following labor cost per 100 sq. ft.:

	Hours	Rate	Total	Rate	Total
Carpenter	0.4	$....	$....	$24.04	$9.62
Cost per sq. ft.			....		.10

07200 INSULATION

To insulate is defined as "to make an island of". This "island" not only excludes the outside elements but retains those created within. Today with energy costs soaring, and traditional energy sources drying up, the use of insulation becomes part of a larger, more inclusive concern--energy conservation. Building design must go beyond the casual approach of letting aesthetic and use-function considerations dictate the design and then turning to technology to provide a satisfactory interior climate regardless of the initial and operating cost. The building envelope itself must be modified to contribute in every possible way to interior conditions of comfort--excluding the extremes of weather but welcoming those that pleasantly modify; retaining our manufactured interior weather but also providing means to allow it naturally to renew and maintain itself without resorting exclusively to mechanical means.

There are three stages where crucial decisions are made concerning the building and its environment. First, in the preliminary planning; second, in material selection; and finally, in the actual construction.

Preliminary planning considers the basics. First the building form should approximate a cube and have a minimum exposure to the elements or sprawl and take maximum advantage of natural light, solar heat and breezes. Buildings should be sited to work with nature, not fight it. Fenestration should be concentrated to the south and tempered with overhangs engineered to exclude the sun when it is high in the summer sky, but to allow the rays to penetrate and add their warmth when the sun is low in the winter. Overhangs need not always be eaves but can be louvres, screens, balconies, porches, or a few well-spaced deciduous trees. Roof forms should not be chosen solely to exclude the elements but also to shade and vent the space above the top habitable floor. Wind conditions should be considered, doors protected, and windows placed to provide the optimum natural ventilation supplementing them with clear stories and skylights if need be. The earth itself is both a very cheap and a very good insulator and intelligent use of banking can pay large dividends.

Once the general arrangement has been determined, the second consideration is to select the best wall, ceiling, and floor materials for coping with the conditions they will face, and the best methods of installation to assure that the selected materials will perform to their maximum capabilities.

This discussion has concerned itself with new construction. One of the largest opportunities developing for the contractor is bringing existing buildings up to the energy efficient standards acceptable today. Most of the materials and procedures discussed in the following text apply to both new and remodeling work. However, the latter presents its own problems, and for that type of work, the reader is also referred to the chapter on *Remodeling Work* which deals with work on existing buildings.

The exterior and interior faces of a building will be chosen for appearance and wearability, and will by nature be hard and dense and therefore good conductors of heat. Also by nature insulating materials will be full of air pockets and be soft and easily damaged and need protection. Therefore most exterior envelopes will consist of three plies: an outside wearing surface, and intermediate space designed to interrupt the heat flow, and an interior wearing surface.

Even the exterior and interior plies will contribute some insulating quality. Materials are rated for their thermal resistance or their "R" value, the temperature difference between two exposed faces required to cause one BTU to flow through one square foot of the material per hour. "R" values can be added, one to the other, to arrive at a rating for the total wall, ceiling, or floor construction. The following list gives the "R" ratings for some of the more common building materials. Ratings for insulating materials will be given for each as it is discussed in the following text.

Material	R Value
4" Face brick	.44
4" Common brick	.80
8" Poured concrete	.64
8" Cinder block	1.11
8" Light weight block	2.00
4" Concrete slab	.32
½" Plywood	.47
½" Gypsum board	.45
½" Gypsum plaster	.32
1x8 Drop siding	.79

Even from the above small sampling, one can see that facing materials can be chosen one over the other to contribute to the insulating quality of the envelope. In addition, air spaces of over ¾" should also be added at an average of R = .91, depending on the position of the air space and whether the air movement is up or down. Also in choosing exterior, facings colors should be considered for their ability to reflect; the darker the color, the more heat will be absorbed. And finally the jointing of all materials must be carefully studied to eliminate infiltration. A study of one apartment tower found 50% of the heat loss was due to infiltration around the windows!

Heat transfer through the building enclosure is by three means: convection, conduction, and radiation.

Convection is the thermally produced upward and downward movement of air. Warm air rises, cold air falls. The concern here is to construct a total building envelope that blocks air currents. Hollow walls should be solidly blocked top and bottom at each story, and floor and roof construction should be continuously sealed from wall construction so air conditions in the exterior walls will not be able to spread out over the interior. This can be done either by extending the interior facing or by adding insulation or a combination of both.

Conduction is the transmission of heat through a material. The rate of conductance of a material or combination of materials, known as its "U" factor, is the BTU per hour per inch of thickness per square foot per degree temperature difference. The "R" ratings discussed previously are the reciprocals of conductance or 1 divided by the "U" factor. Most common insulating materials are manufactured to be as poor a conductor of heat as possible. However, their positioning within the wall can also make them block convection and when enclosed in reflective coverings and separated from adjacent construction by at least ¾" they also can cut down on the third means of heat transfer, radiation.

Radiation is the emission of energy from a surface. Bright surfaces, such as aluminum foil, are good reflectors and have low emissive coefficients and are therefore poor absorbers of heat. The interior faces of the two outer plies of building envelope tend to be dark, and the heat will radiate from one surface to the other constantly unless interrupted. By inserting one or more layers of reflective surfaces, the heat will be reflected back to the surface from which it escaped. Installation is crucial to the success of a radiant barrier, because the reflective surface will be highly conductive. If allowed to come into contact with either side of the enclosing construction, it will speed heat transfer rather than retard it.

Infiltration, as discussed above, will generally be solved by good detailing at the juncture of one material to another and by caulking. Strips of insulation, such as under sill plates, and the stuffing in of loose wools can also be used to plug air leaks.

There is no simple answer as to how much insulation should be added. Both comfort and the additional costs must be considered. Minimum accepted "R" values in the northern regions have been R19 for ceilings, R13 for floors and R11 for walls. These drop to R13, R9, and R8 in milder areas but should never be less than R9 for ceilings and R7 elsewhere. With the recent increase in fuel costs, many consider these figures far too low and recommend R24 for ceilings, R19 for floors and R13 for walls. If fuel costs keep advancing, some project an "R" value of 49 for ceilings in the Chicago area, over two and a half times today's standard. If it is a matter of fuel supplies actually running out, then perhaps such increases will come about. Based on whether it would be a good financial investment, it is doubtful, as each additional inch of insulation beyond a certain point reduces the heat loss less than the one before it. There is a point of sharply diminishing returns.

To determine the best "R" values for a particular project, one must consider the following:

a) Determine all wall, ceiling, and floor construction types and give each its proper "R" value without insulation.
b) Determine the heat loss (or heat gain for cooling) without insulation through each of the above.
c) Choose an "R" rating for walls, ceiling, and floors based on averages for existing climate and consistent with the type of construction contemplated.
d) Refigure heat loss with new "R" values.
e) Figure fuel savings of d) over b). Several alternate fuels might be considered and local utilities can be helpful in determining these costs.
f) Figure the difference in cost of the insulated construction over the uninsulated, remembering to deduct for savings in size of the mechanical equipment, if any.
g) Determine how long it will take savings in fuel to pay for insulation costs. This is the payback period and will have to take many things into consideration such as future fuel costs, and the interest on the additional mortgage to finance

the insulation. If the payback period looks attractive, the whole procedure can be repeated with a higher "R" value assigned, and this can be repeated until a point of diminishing returns is evident.

Once the desired "R" value is established, the selection of the best type of insulation can begin. The possibilities include:

a) Blankets and batts, with or without reflective and/or vapor barriers.
b) Rigid board type which may also serve as sheathing or lathing.
c) Sprayed on or foamed in place, which may also serve as sound or fire retardants.
d) Poured fill
e) Reflective barriers

Many of these may be used in combination with the other.

Method of Placing Insulating Wool Batts

Method of Placing and Stapling Paper-Backed Insulating Wool Batts

07210 BUILDING INSULATION BLANKET AND BATT INSULATION

Insulation formed into batts or rolls is made in a number of materials such as stone, slag, glass wools, vegetable and cotton fibers and suspended pulps. They are all light in weight, easily cut and handled and need support to stay in place. They are usually encased in either kraft paper or foil and can be used as a vapor barrier. The enclosing material is usually made with flapped edges for easy attachment to studs and joists. Units may be ordered unfaced and fitted between studs by friction. Unfaced units are most often used as a second layer of insulation in attic spaces as they do not form a second vapor barrier if the original layer has one. A narrow form of unfaced blanket 1" thick is made for inserting under sill plates. It will compress to 1/32".

Method of Placing Strip Insulating Wool

Most batt and blanket materials have "R" values of around 3.5 to 3.7 per inch of thickness. The 3" units will provide an "R" of 11, 3 ½" of 13 and 6" of 19. The 6" units are commonly used in attics and fitted between the joists; but more and more, they are used in walls with 2x6 studs placed 24" o.c. in place of the standard 2x4s 16" o.c. If foil faced units are used and an air space of at least ¾" maintained between the foil and the warm side of the room, the "R" rating can be increased some 15%.

Most glass fiber insulation is distributed on a nationwide basis, but other materials are often produced regionally and prices may vary widely. Widths are 15" and 23" to fit normal 16" and 24" stud and joist spacing. Batts come 48" and 96" long, blankets in rolls of from 24' to 40', often two rolls per package. The square foot cost for foil faced units 3 ½" thick is 16 cents; 6" thick is 26 cents. Units bound in kraft paper may run a couple of cents less and unfaced units even less. Sill sealer rolls come in 50' rolls 1" thick and in 3 ⅝" and 6" widths and cost 10 and 15 cents per lin. ft.

Because these materials are light and precut to fit standard construction, one carpenter can handle the installation. Friction fit type, with no attachments, will install at the rate of 2000 sq. ft. per day. Flanged type bound in kraft paper or foil will install at 1800 sq. ft. per day.

Labor Cost of 100 sq. ft. Faced Insulation

	Hours	Rate	Total	Rate	Total
Carpenter	0.45	$....	$....	$24.04	$10.82
Cost per sq. ft.			....		.11

07211 LOOSE FILL INSULATION

Loose fill insulation includes mineral wool, which is molten rock extruded by air and steam into fibers, known as blowing wool for machine applications, or nodules for pouring or spreading by hand; and expanded volcanic rocks such as vermiculite or perlite. The latter are most often specified for filling concrete block or cavity walls.

Loose Insulating Wool. Is suitable for any purpose where insulation can be packed by hand between ceiling joists or side wall studding. For maximum results, it is recommended that wool be applied full thickness of side wall studding and approximately 4" to 16" over ceiling areas.

The covering capacity varies considerably with the density to which it is packed. The National Bureau of Standards conductivity figure for glass or rock wool is 0.27 Btu at a density of 10 lbs. per cu. ft.

The covering capacity of bulk wool as given by most manufacturers is based on a density of 6 to 8 lbs. per cu. ft.

Number of Sq. Ft. of Surface Covered By One Bag of Loose Insulating Wool Weighing 40 Lbs. and Containing 4 Cu. Ft. Not Including Area Covered By Studs or Joists

Density, Lbs. per CF	Actual Cu. Ft.	Thickness of Loose Insulating Wool in Inches 1	2	3	3 1/2	3 5/8	4
6	6.67	80.0	40.0	26.6	22.8	22.0	20.0
7	5.72	68.5	34.4	22.9	19.7	18.8	17.2
8	5.00	60.0	30.0	20.0	17.2	16.6	15.0
9	4.45	53.4	26.7	17.8	15.2	14.7	13.4
10	4.00	48.0	24.0	16.0	13.7	13.3	12.0

Number of Sq. Ft. of Surface Covered By One Bag of Loose Insulating Wool Weighing 40 Lbs. and Containing 4 Cu. Ft. Including Area Covered By Studs or Joists

Density, Lbs. per CF	Actual Cu. Ft.	Thickness of Loose Insulating Wool in Inches 1	2	3	3 1/2	3 5/8	4
6	6.67	85.0	42.5	28.4	24.3	23.5	21.2
7	5.72	73.0	36.5	24.3	20.8	20.0	18.3
8	5.00	63.8	31.9	21.3	18.3	17.6	16.0
9	4.45	56.7	28.4	18.9	16.2	15.7	14.2
10	4.00	51.0	25.5	17.0	14.5	14.1	12.7

Loose or bulk rock wool is usually furnished in bags weighing 40 lbs. and containing 4 cu. ft. Average price $3.00 to $4.50 per bag.

Granule or Pellet Type Insulating Wool. Granule or pellet type insulating wool is furnished in particles sufficiently large that the material will not sift or dust through wall or ceiling cracks and is used for insulating ceiling areas between supporting joists, or void wall spaces where accessible and which permit pouring the pellets into place, particularly in old house construction.

Granule or pellet type insulating wool is usually furnished in bags weighing 40 lbs. and is placed at a density of 7 to 8 lbs. per cu. ft.

The tables for loose or granular insulating wool, based on 7 to 8 lbs. per cu. ft. may be used for estimating quantities obtainable per bag.

Granule or pellet type insulating wool in bags or cartons weighing 40 lbs. cost about $4.00 each.

Labor Placing Loose or Bulk Insulating Wool. When placing loose or bulk insulating wool between wood studs, a man should place about 350 to 450 sq. ft. per 8-hr. day, at the following labor cost per 100 sq. ft.:

	Hours	Rate	Total	Rate	Total
Labor	2	$....	$....	$24.04	$48.08
Cost per sq. ft.			....		.48

Labor Placing Granule or Pellet Type Insulating Wool Between Ceiling Joists. Where granule or pellet type insulating wool is poured between ceiling joists and other open spaces to a thickness of 3 1/2 or 3 5/8 in., a worker should place 700 to 900 sq. ft. per 8-hr. day, at the following labor cost per 100 sq. ft.:

	Hours	Rate	Total	Rate	Total
Carpenter	1	$....	$....	$24.04	$24.04
Cost per sq. ft.			....		.24

Vermiculite Or Perlite Loose-fill Building Insulation. Vermiculite loose- fill insulation is used to insulate attics, lofts, and side walls. It is fireproof, not merely "fire resistant". The fusion point is 2200° to 2400°F. It is completely mineral and does not decompose, decay, or rot. Because of its granular structure, vermiculite is free-flowing, assuring a complete insulation job without joints or seams. Rodents cannot tunnel into it, and it does not attract termites or other vermin. Vermiculite is a non-conductor and is excellent protection around electrical wiring.

Vermiculite loose-fill is marketed in 4 cu. ft. bags. Approximate coverage per bag, based on joists spaced 16" o.c., is 14 sq. ft. for a 3 5/8" thickness; 9 sq. ft. for 5 1/2" thickness. A bag will cost about $7.00.

Approximate Coverage of Vermiculite
Fill in Cavity and Block Walls

Sq. ft.	No. of Bags Required for Various Wall Types				
wall area	1" cavity	2" cav.	2 1/2" cav.	8" block	12" block
100	2	4	5	7	13
500	10	20	25	34	63
1,000	21	42	50	69	125
5,000	104	208	250	545	625
10,000	208	416	500	1,090	1,250

1 bag=4 cu. ft.

Applying Fiberglass Batts over Cellulose in Attics

One mason can pour around 50 bags or 200 cu. ft. per 8 hr. day, at the following labor cost:

	Hours	Rate	Total	Rate	Total
Mason	8	$....	$....	$25.62	$204.96
Cost per cu. ft.					1.02

When adequate ventilation is provided in attics or similar spaces, it is not necessary to use a vapor barrier in the ceiling construction below vermiculite fill insulation. A vapor barrier is recommended on the warm side of insulated exterior walls, except when the insulated space is ventilated to permit movement of water vapor and air from the space.

Reinsulation. Where an existing thickness of blanket or other insulation is inadequate, vermiculite can be poured over it to obtain the thickness desired.

07212 RIGID INSULATION

Rigid insulation boards are made of expanded polystyrene, often called "beadboard"; extruded polystyrene sold as *Styrofoam;* urethane; glass fibers; glass foams; and wood and vegetable fibers. Many rigid boards are made to serve dual purposes such as sheathing, lath and even interior finish. Many are also used for roof decks and perimeter insulation, which are discussed separately below.

Urethane boards have the highest "R" value per inch of thickness, 7.14, but are flammable and must be covered over, or treated. The cost is around 30 cents per sq. ft. plain, 40 cents treated. The sizes available are many, including the standard 4' x 8' sheet and in thicknesses from ½" to 24". Because of its high insulating value per thickness, it is ideal for use in cavity walls and for applying to the inside face of an exterior masonry wall. In cavity walls it is applied against the outside face of the inside wythe leaving a clear air space for any penetrating water to drain out.

Where urethane board is used against exterior masonry walls, the surface must be clean and even. Nailers must be applied wherever wood trim will be added, such as at the base and around doors and windows. The board is then applied to the wall with sufficient mastic to cover at least 50% of the back when shoved in place. Some manufacturers supply special channel systems for mechanical attachment as an alternative. The urethane board can then be covered with plasterboard or may be plastered direct. Because of the fire hazard, which ever method is chosen, the interior finish coat must be run up above the finished ceiling to cover all portions of the urethane as it is the fireproofing for it. Extruded polystyrene, or *Styrofoam,* under which name it is marketed by Dow Chemical Company, also has a high "R" value, 5.4 per inch of thickness. It costs about 35 cents per sq. ft. in 1" thickness. It is installed in the same locations and by the same methods as the urethane board above, and like urethane board, it should have a continuous interior finish that will protect it from fire.

Polystyrene, which is molded, is known as beadboard. It has a lower "R" rating, 3.85, and is absorbent, but it is much cheaper, about 15 cents per sq. ft. in 1" thickness. It too must be covered with a fireproofing finish.

Glass fiber boards come in various densities. A 1" board in a medium weight gives an "R" of 4.35 and costs around 35 cents per sq. ft.

The stronger rigid boards may also be used for sheathing on wood frame construction. But a 1" board in combination with standard 3 ½" batts will give a wall with the highly desirable rating of R19 or better and may be preferable to using 6" batts and 6" studs to gain such a rating.

Rigid board is light, easily cut and can be installed by one man who should erect about 1,500 sq. ft. per day depending on the amount of fitting.

Labor Cost of 100 sq. ft. of 1" Rigid Board
Applied to a Masonry Wall

	Hours	Rate	Total	Rate	Total
Mechanic	0.54	$....	$....	$24.04	$12.98
Cost per sq. ft.					.13

07213 REFLECTIVE INSULATION

Some reflective insulation can be incorporated as the enclosing sheet on batt and blanket insulation and as a backing on rigid insulation. Also available are sheets of insulating material depending entirely on the reflective principle and consisting of alternate layers of trapped air and foil, the outer layers of which are usually laminated to a backing sheet of kraft paper for extra strength and which are extended at the edges to form flaps for stapling. Another type depends on an inner layer or layers of accordion foils. Such insulation must be held away from the inner and outer wall finishes at least ¾" so that there will be an uninterrupted air space which serves the dual purpose of allowing the foil surface to reflect escaping heat (or entering cold) back to where it came from and keeping the highly conductive foil from transmitting the heat by contact.

A typical lamination would consist of three layers: the two outer sheets of foil laminated to a backing sheet and separated from a center sheet of pure foil, usually aluminum, with sealed air spaces both sides. The unit is held together with flanges extended to allow stapling to studs and joists. The "R" value will depend on the placement of the material, because reflective insulation is more effective with the heat flow down than with heat flow up. Where the heat flows down from a hot roof to a ceiling below, 93% of the heat transfer is by radiation and only 7% by conduction. As heated air always travels up, no convection is directed toward the ceiling. Considering the reverse travel, heated interior air travelling through the ceiling to the cold roof above, 50% will be transmitted by radiation, the same 7% by conduction and the remainder by convection. Heat transfer through the side walls will be 65% to 80% by radiation, 15% to 28% by convection, and 5% to 7% by conduction. The "R" rating then on a three-ply, two-inch thick reflective sheet will be 26.18 down, 9.52 up and 14.65 through the wall for a mean "R" of 17.85. When we speak of this as 2" thick, it is the distance from foil to foil. Another ¾" minimum must be added to each side to isolate the foil from other wall construction so that a full 3 ½" must be reserved for the reflective insulation.

One-inch thick units with just one inner and one outer layer of reinforced foil separated by a 1" air space will give an "R" value of 21.79 down, 7 up, and 10.72 through the wall for a mean "R" of 14.4.

Single sheets with aluminum foil either side of a reinforcing sheet with no integral air space if properly installed to preserve air spaces to adjacent construction will give "R" values of 14.28 down, 3.45 up and 7 through the wall, for a mean "R" of 9.

The cost of two-ply material is 12 cents per sq. ft.; three-ply is 14 cents; and five-ply is 25 cents.

The material is preformed to fit average stud and joist spacing and is very light; one man can install 2200 sq. ft. per day if the wall or ceiling space is clear, at the following labor cost per 100 sq. ft.:

	Hours	Rate	Total	Rate	Total
Carpenter	0.66	$....	$....	$24.04	$15.87
Cost per sq. ft.					.16

07214 FOAMED INSULATION

Foamed in place insulations are very efficient and are usually sublet to an experienced applicator. Foamed insulation will adhere well to most surfaces, will conform to irregular forms and will set up quickly. The plastic foams flow better into restricted areas than the sprayed on fibers, which have longer fibers and tend to clog in small spaces.

Application will require special equipment, and the space must be well ventilated. Some foams expand and will set up stresses in restricted areas if care is not taken. Despite these drawbacks, the resulting "R" can be as much as 7.7 per inch and will apply at the rate of 800 sq. ft. per day with a three-man crew. Material costs for urethane will run about 22 cents per inch per sq. ft. If subcontracted, a budget figure of $1.00 per sq. ft. for 1" and $1.65 for 2" should be adequate.

07215 SPRAYED INSULATION

Sprayed fibers of rock wool, glass, gypsum, perlite and cellulose will have "R" values in the 4 to 5 range. Special binders are required as well as special equipment, so this work too will be subcontracted. A two-worker crew will apply some 1200 sq. ft. of 1" thickness per day, possibly more on wide open ceilings. Material costs will be around 20 to 25 cents per sq. ft. per 1" of material. If subcontracted, a budget figure of $1.20 per sq. ft. should be adequate.

07230 HIGH AND LOW TEMPERATURE INSULATION

Rigid foam insulation, composed of expanded polystyrene, glass or polyurethane bound together into rigid board, has largely replaced corkboard for low temperature insulation. It is also used in general building for roofs, because it is quite strong, tough and resilient; for curtain wall back up because it is lightweight; for core walls as it is self-supporting; and for perimeter and crawl space insulation as it has a low water absorption and a zero rating for capillary action. In addition, it forms a vapor barrier, is dimensionally stable, incombustible, and easily cut.

Rigid foam insulation can be placed on concrete and masonry surfaces by adhering units, with hot asphalt; or with gobs of cement.

Units come in 1", 1 ½", 2", 3", 4", and 6" thicknesses and sizes from 12"x18" to 36"x36" in packaged forms and up to 9' lengths when shipped in carload lots. There are no standard sizes within the industry, and sizes, costs and insulating values must be carefully checked with the various manufacturers. **Back Plaster.** Where insulation is to be applied against masonry walls in hot asphalt, the walls must be made true and even with portland cement plaster.

The parge coat or back plaster may be omitted if the wall is of poured concrete.

Back plaster should be made by mixing 1 part portland cement and 3 parts clean sharp sand. Back plaster ½" thick requires 1.5 bags portland cement and 4.6 cu. ft. of sand per 100 sq. ft.

Material Cost of One Cu. Yd. of 1:3 Portland Cement Back Plaster

	Rate	Total	Rate	Total
9.5 sacks portland cement	$....	$....	$4.00	$38.00
1.10 cu. yds. sand	....	....	6.50	7.15
Cost per cu. yd.		$....		$45.15
Cost per cu. ft.		....		1.67

Asphaltic Priming Coat. All concrete or plaster surfaces to which insulation is to be applied in hot asphalt must be primed with asphaltic priming coat made especially for this purpose.

Asphalt. The proper grade of asphalt to be used is very important. It must be odorless, as any material with an odor will taint foodstuffs that may be stored in insulated rooms. It should be an oxidized asphalt of 180°F to 200°F melting point and should preferably be made from a Mexican base oil. The hot asphalt in which corkboard is dipped should be held between 350° and 375°F. Do not heat asphalt to higher temperatures than these. Most manufacturers recommend an asphalt for use with their product.

On dipping floor, wall and ceiling insulation, allow .4 lbs. of asphalt per sq. ft. of insulation per layer. Mop coating on floors or top of ceiling requires .8 lbs. per sq. ft.

Cold Erection Plastic. Where the use of hot asphalt for the erection of foamboard is not practical, use cold erection plastic on floors, walls and ceilings. A coat 1/16" thick requires 1 gallon of cold erection plastic per 25 sq. ft.

Nails or Impregnated Skewers. Use nails for the attachment of the first layer of insulation to wood and frame construction, for the attachment of the second layer to wood stripped concrete ceilings, or for toenailing the first layer of solid cork walls. All nails should be of galvanized wire and should be 1" longer than the thickness of the insulation. Where two courses of insulation are erected in hot asphalt or cold erection plastic, impregnated hardwood skewers should be used to additionally secure the second course in position. Do not use either nails or skewers when erecting the first course against brick, stone or concrete walls. Hot asphalt or cold erection plastic is all that is required.

Erection. The requirements (material and labor) for installing one layer of 2" thick insulation, using the hot asphalt dip method of setting, onto a masonry wall surface are given in the following tables. Bear in mind that the labor times given are for straight, flat surfaces; and that any irregularities such as columns, pilasters or beams will require additional labor time due to cutting and setting smaller pieces. If the insulation is thicker than 2", then adjust for the difference in material costs and add about ¼ hour to both mechanic and helper time per 1" thickness increase over 2".

2" Thick Single Insulation Layer, Hot Dip Method
Requirements Per 100 Sq. Ft.

Operation	Material Required	Mechanic Time	Helper Time
Masonry Wall			
Parge Coat- ½"	5.5 cu. ft. mortar	3.0 hrs.	3.0 hrs.
Asphalt Prime Coat	1 gallon primer	1.0	...
Hot Asphalt	40 lbs.	...	...
Applying Insulation	100 sq. ft.	1.5 hrs.	1.5 hrs.

For applications against concrete walls, the parge coat is not required.

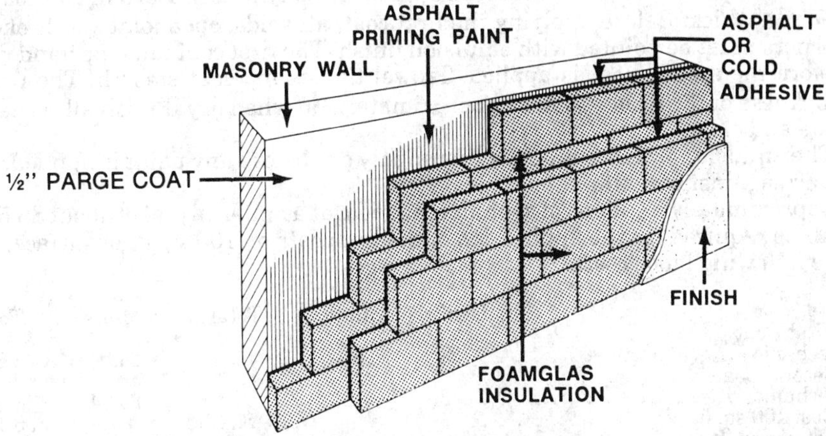

Low Temperature Insulation Using Two Layers Of Pittsburgh Corning Foamglas

Material Cost of 100 Sq. Ft. of Single Layer Insulation,
Set In Hot Asphalt on Masonry Walls

	Rate	Total	Rate	Total
5.5 cu. ft. cement plaster	$....	$....	$1.67	$9.19
1 gal. asphalt primer	....	....	2.00	2.00
40 lbs. asphalt	....	....	.13	5.20
100 sq. ft. 2" corkboard	....	....	.97	97.00
Cost 100 sq. ft.				$113.39
Cost per sq. ft.				1.13

Labor Cost of 100 Sq. Ft. of Single Layer Insulation,
Set in Hot Asphalt on Masonry Walls

	Hours	Rate	Total	Rate	Total
Plaster Coat					
Mechanic	3.0	$....	$....	$21.59	$64.77
Helper	3.0	....	....	19.13	57.39
Prime Coat					
Mechanic	1.0	....	....	24.04	24.04
Insulation					
Mechanic	1.5	....	....	24.04	36.06
Helper	1.5	....	....	19.13	28.70
Cost 100 sq. ft.			$....		$210.96
Cost per sq. ft.			....		2.11

Finishes. Corkboard insulation should receive a finish that will give long service under the temperature and moisture conditions common to cold storage rooms. The finish should bond securely to the corkboard and be reasonably resistant to bumps and abrasion. Cost, good appearance and availability should also be considered. These requirements are best met by asphalt emulsion finishes or portland cement plaster. Asphalt emulsion finishes are generally used on ceilings and portland cement plaster on walls. Rigid foam insulations may be sprayed or brushed with special white vinyl emulsion paints, aluminum paint, tile set with adhesive or cement plaster over mesh. Direct plastering should not be considered.

Asphalt emulsion finishes shall be applied directly to corkboard in two coats, each 1/8" thick. Before applying the first coat, all voids, open joints, or broken corners must be pointed with emulsion finish. The first coat must be hand dry before the second coat is applied. Trowel the second coat smooth. The final thickness of the finish shall be approximately 1/8" when dry. Finish all corners true and clean.

The emulsion finishes must be thoroughly dry before any paint is applied, to guard against cracking.

Approximately 5 gallons of asphalt emulsion, at a material cost of about $8.50, will be required for the application of the two coats on 100 sq. ft. of surface, at the following labor costs:

	Hours	Rate	Total	Rate	Total
First Coat					
Mechanic	2	$....	$....	24.04	$48.08
Second Coat					
Mechanic	2	....	....	24.04	48.08
Cost 100 sq. ft.			$....		$96.16
Cost per sq. ft.			....		.96

The labor cost of 100 sq. ft. of 1/2" of 1:3 portland cement plaster should average as follows:

	Hours	Rate	Total	Rate	Total
Mechanic	3	$....	$....	$21.59	$64.77
Helper	3	....	....	19.13	57.39
Cost per 100 sq. ft.	6		$....		$122.16
Cost per sq. ft.			....		1.22

Portland cement plaster should be mixed in the proportions of 1 part portland cement, 3 parts clean, screened sand, and 5% hydrated lime.

To reduce cracking to a minimum, finish coat of plaster may be scored each way approximately 4'-0" on centers.

Some manufacturers state that plaster can be applied directly to insulation; others specify wire mesh must be applied, particularly on ceiling work. As rigid foam insulation does not absorb water, plaster coats will dry very slowly. It is also necessary to add the cost of nailer strips for casings, trim, and bases, if any occur. Light trim such as beads may be stapled on.

07240 ROOF AND DECK INSULATION

Roof and deck insulation is often of the same material as has been discussed under rigid board insulation, but the units are generally furnished in smaller sizes more suitable for handling in exposed conditions. In addition to glass fiber, perlite, urethane and polystyrene boards, wood and mineral fiber boards costing as little as 25 cents per sq. ft. per 1" of thickness, and foam glass blocks are often specified. Poured decks are also common, but these have been covered at the end of the *Concrete* chapter. *Foamglas* (from Pittsburgh Corning) is a unique material that is inorganic, incombustible, dimensionally stable, and water and vapor proof and weighs only 8.5 lb. per cubic foot but is strong enough to be used under parking and promenade decks. It can also be ordered with an 1/8" per foot slope. Minimum thickness is 1 1/2", and it is available in 1/2" increments up to 4". The 1 1/2" material has an "R" value of 4.2 and costs about 60 cents per sq. ft. in 1 1/2" thickness. One roofer will install some 800 sq. ft. per day at the following labor cost:

Labor Cost of 100 sq. ft. of 1" Deck Insulation

	Hours	Rate	Total	Rate	Total
Roofer	1.0	$....	$....	$22.72	$22.72
Cost per sq. ft.			....		.23

07250 PERIMETER INSULATION

Board forms of polystyrene, urethane, fiberglass, and cellular glass are commonly employed as insulation on foundation walls and under slab edges to insulate an on-grade floor construction at the outer wall. Material costs are from 15 to 25 per sq. ft. in 1" thickness.

As perimeter insulation must fit the outline of the interior face of the outside wall and must be run continuously around offsets, perimeter ducts, column foundations and other irregularities, unit costs will vary widely. In general, insulation is run down 24" below outside grade. In addition, it may be run back 24" under the outer edge of the slab. One carpenter should apply around 800 sq. ft. per day on straight run work, at the following labor cost per 100 sq. ft.:

	Hours	Rate	Total	Rate	Total
Carpenter	1.14	$....	$....	$24.04	$27.41
Cost per sq. ft.			....		.27

07300 ROOF SHINGLES AND ROOFING TILES

Roofing is estimated by the square, containing 100 sq. ft. The method used in computing the quantities will vary with the kind of roofing and the shape of the roof.

The labor cost of applying any type of roofing will be governed by the pitch or slope of the roof, size, plan of same (whether cut up with openings, such as skylights, penthouses, gables, dormers, etc.), and upon the distance of the roof above the ground, etc.

Rules for Measuring Plain Double Pitch or Gable Roofs. To obtain the area of a plain double pitch or gable roof as shown in Figure 1, multiply the length of the ridge (A to B), by the length of the rafter (A to C). This will give the area of one-half the roof. Multiply this by 2 to obtain the total sq. ft. of roof surface.

Example: Assume the length of the ridge (A to B), is 30'-0" and the length of the rafter (A to C), 20'-0". By multiplying (A to B), 30'-0" by (A to C), 20'-0" equals 600. The area of one-half the roof is 600 sq. ft. 600 x 2 = 1,200 sq. ft. of roof.

Rules for Measuring Hip Roofs. To obtain the area of a hip roof as shown in Figure 3, multiply the length of the eaves (C to D) by 1/2 the length of the rafter (A to E). This will give the number of sq. ft. of one end of the roof, which multiplied by 2 gives the area of both ends. To obtain the area of the sides of the roof, add the length of the ridge (A to B) to the length of the eaves (D to H). Divide this sum by 2 and multiply by the length of the rafter (F to G). This gives the area of one side of the roof and when multiplied by 2 gives the number of sq. ft. on both sides of the roof.

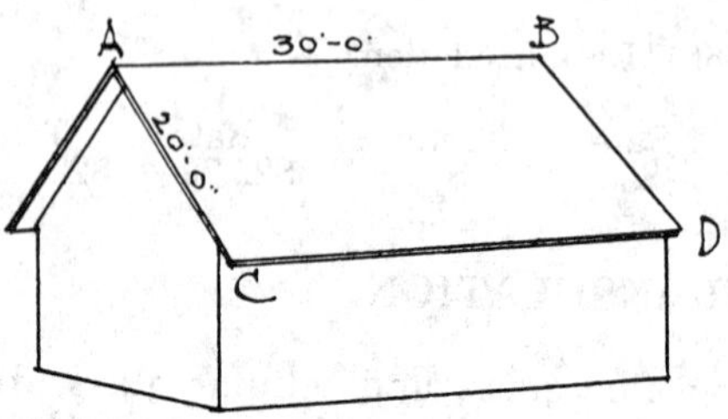

Fig 1. Plain Double Pitch or Gable Roof

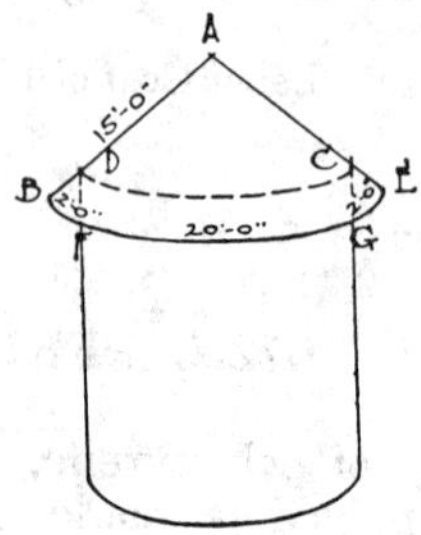

Fig 2. Conical Building and Roof

To obtain the total number of sq. ft. of roof surface, add the area of the two ends to the area of the two sides. This total divided by 100 equals the number of squares in the roof.

Example: Assume the length of the eaves (C to D) is 20'-0" and the length of the rafter (A to E) is 20'-0". Multiply (C to D) 20'-0" by ½ the length of the rafter (A to E), or 10'-0" equals 200 sq. ft., the area of one end of the roof. To obtain the area of both ends, 200 x 2 = 400 sq. ft.

To obtain the area of the sides of the roof, the length of the ridge (A to B), is 10'-0" and the length of the eaves (D to H), is 30'-0". Add (A to B) 10'-0" to (D to H) 30'-0", and the result is 40'-0". By taking ½ the combined length of the ridge and eaves, ½ of 40'-0" = 20'-0", the average length of the roof. Assuming the length of the rafter (F to G) as 20'-0", 20'-0"x20'-0" = 400 sq. ft., the area of one side of the roof, which multiplied by 2 equals 800 sq. ft., the area of both sides of the roof.

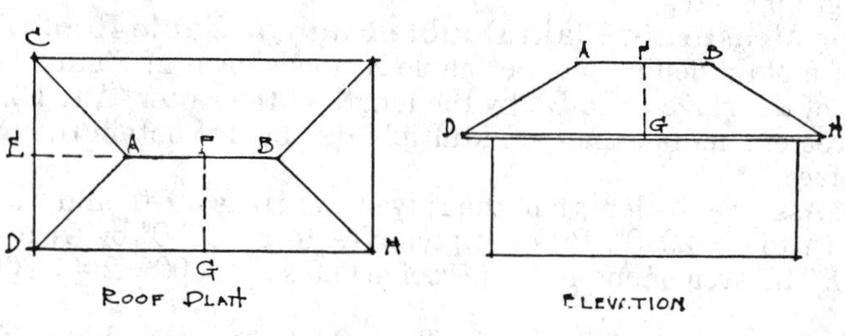

Fig 3. Hip Roof

Adding the area of the two ends to the area of the two sides gives 1,200 sq. ft. of roof area.

The area of a plain hip roof (Fig. 4) running to a point at the top is obtained by multiplying the length of the eaves (B to E) by ½ the length of the rafter (A to F). This gives the area of one end of the roof. To obtain the area of all four sides, multiply by 4. Example: Multiply the length of the eaves (B to E), which is 30'-0", by ½ the length of the rafter (A to F), 10'-0", and the result is 300 sq. ft., the area of one end of the roof. 300 x 4 = 1,200 sq. ft., the area of the 4 sides.

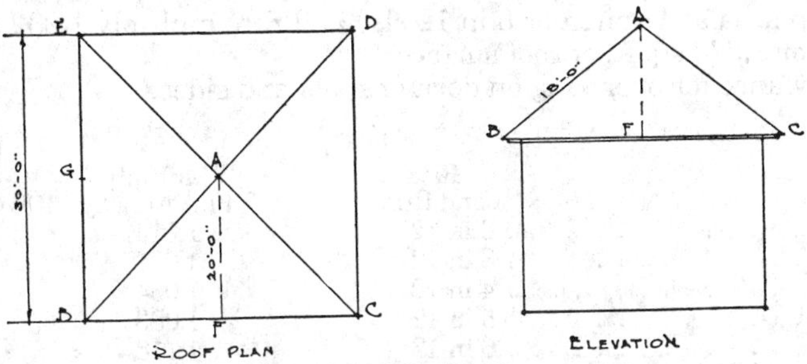

Fig 4. Hip Roof

Rules for Measuring Conical Tower Roofs and Circular Buildings. To obtain the area of a conical tower roof as shown in Figure 2, multiply ½ the length of the rafter (A to B) by the distance around the eaves at B. For example, the length of the rafter (distance from A to B) is 15'-0" and the diameter of the building at B is 20'-0".

To obtain the distance around the building, multiply the diameter by 3.1416. If the eaves project beyond the outside walls, and the diameter is given only to the outside walls of the building, add the length of the roof projection on both sides of the building to obtain the correct diameter. Example: If the diameter of the building (C to D) is 20'-0" and the eaves project 2'-0" on each side, the diameter of the building at the eaves would be 24'-0".

Multiplying 24'-0" (B to E), the diameter of the building at eaves, by 3.1416, gives 75.3984, or approximately 75'-5" around the eaves at projection (B to E).

To obtain the area of the roof, multiply ½ the length of the rafter (A to B), which is 15'-0", (½ of 15 is 7 ½) or 7'-6", by the distance around the eaves at B and E, which is 75'-5" or 75.4 feet, and the result is 565.5, or 565 ½ sq. ft., the area of the roof.

To obtain the wall area of a cylindrical or circular building, multiply the height by the circumference, or the distance around the building, and the result will be the number of sq. ft. to be covered. The circumference is obtained by multiplying the diameter (C to D) of the building by 3.1416. To obtain the area of the outside walls of a cylindrical building whose diameter is 20'-0" and the height 15'-0", 20 x 3.1416 = 62.832, or 62'-10". Multiply 62.832 ft. (the distance around the building) by 15 ft. (the height of the building) = 942.48, or 942 ½ sq. ft., the area of the outside walls.

A Short Method of Figuring Roof Areas

To obtain the number of square feet of roof area, where the pitch (rise and run) of the roof is known, take the entire flat or horizontal area of the roof and multiply by the factor given below for the roof slope applicable and the result will be the area of the roof.

Always bear in mind, the width of any overhanging cornice must be added to the building area to obtain the total area to be covered.

Example: Find the area of a roof 26'-0"x42'-0", having a 12" or 1'-0" overhanging cornice. Roof having a 1/4 pitch or a 6 in 12 rise and run.

To obtain the roof area, 26'-0" + 1'-0" + 1'-0" = 28'-0" width. 42'-0" + 1'-0" + 1'-0" = 44'-0" length. 28 x 44 = 1232 or 1,232 sq. ft. flat or horizontal area.

To obtain area at 1/4 pitch or 6 in 12 rise and run: multiply 1,232 by 1.118 = 1377.376 or 1,378 sq. ft. of roof surface.

Add allowance for overhang on dormer roofs and sides.

Pitch of Roof	Rise and Run	Multiply Flat Area by	LF Hips or Valleys per LF Common Run
1/12	2 in 12	1.014	1.424
1/8	3 in 12	1.031	1.436
1/6	4 in 12	1.054	1.453
5/24	5 in 12	1.083	1.474
1/4	6 in 12	1.118	1.500
7/24	7 in 12	1.158	1.530
1/3	8 in 12	1.202	1.564
3/8	9 in 12	1.250	1.600
5/12	10 in 12	1.302	1.641
11/24	11 in 12	1.357	1.685
1/2	12 in 12	1.413	1.732

Hips and Valleys. The length of hips and valleys, formed by intersecting roof surfaces, running perpendicular to each other, and having the same slope is also a function of the roof rise and run. For full hips or valleys, i.e. where both roofs intersect for their full width, the length may be determined by taking the square root of the sum of the rise squared plus twice the run squared.

Using the factors given in the last column of the above table, the length of full hips or valleys may be obtained by multiplying the total roof run from eave to ridge, (not the hip or valley run), by the factor listed for the roof slope involved.

07311 ASPHALT SHINGLES

Asphalt shingles are furnished in several styles, and the labor cost of laying them will vary with the style of shingle used and the type of roof to which they are applied.

Individual asphalt shingles are furnished and laid in exactly the same manner as wood shingles, the only difference being their size.

Strip asphalt shingles are furnished in strips, each containing 3 shingles, so when laying them a roofer handles and lays 3 shingles at a time.

Hexagon asphalt shingles are furnished in strips 11 1/3"x36".

Strip Asphalt Shingles

Estimating Quantities of Asphalt Shingles. Asphalt shingles are sold by the square, containing sufficient shingles to cover 100 sq. ft. of roof.

When measuring roofs of any shape, always allow one extra course of shingles for the "starters" at the eaves, as the first or "starting" course of shingles must always be doubled.

Obtain the number of lin. ft. of hips, valleys and ridges to be covered with asphalt shingles and compute same as 1'-0" wide.

Many roofing contractors do not measure hips, valleys and ridges, preferring to add a percentage of the roof area to cover these items and waste. The following percentages are commonly used: Gable roofs, 10%; hip roofs, 15%; hip roofs with dormers and valleys, 20%.

Asphalt shingles must be properly nailed--6 nails to a strip, and nailed low enough on the shingle (right at the cut-out), otherwise they will blow off the roof.

Most manufacturers produce a shingle, designed for high wind areas, that interlocks in such a manner that all of the shingles are integrated into a single unit. Interlocking shingles are available in single coverage for re-roofing and double coverage for new construction.

Self-sealing shingles with adhesive tabs are produced by most manufacturers.

When using asphalt shingles for roofs, roll asphalt roofing of the same materials as the shingles is often used for forming valleys, hips and roof ridges.

Nails Required for Asphalt Shingles. When laying individual asphalt shingles, use a 12 ga. aluminum nail, 1 1/2" long, with a 7/16" head. For laying over old roofs, use nails 1 3/4" long.

When laying square butt strip shingles, use 11 ga. aluminum nails, 1" long, with a 7/16" head. For laying over old roofs, use nails 1 3/4" long. It will require about 1 lb. of nails per sq. of shingles laid.

Sizes and Estimating Data on Asphalt Shingles

Kind of Shingle	Size	No. of Shingle per Sq.	Expos. Inches	Length Nails	Nails per Shingle	Lbs. Nails per Sq.
3-in.-1 strip	12"x36"	80	5	1	4	1

Approximate Prices on Mineral Surfaced Asphalt Shingles, Seal Tab

Kind of Shingle	Size	Weight per Square	Price per Square
3-in-1 strip	12"x36"	235 lbs.	$26.00
3-in-1 strip	12"x36"	340 lbs.	55.00
3-in-1 strip	12"x36"	350 lbs.	65.00

Most asphalt shingles carry an Underwriters Class C rating. Incorporating a glass fiber layer results in an "A" rating. Class A shingles weighing 225 lbs. will run $30.00 per square and weighing 300 lbs., $59.00 per square.

A roofer will lay one square (100 sq. ft.) of asphalt shingles in 1 hour on double pitched roofs with no hips, valleys or dormers. Figure about 35 lin. feet per hour for fitting hips, valleys and ridges. Hip and ridge roll material will run around 16 cents per lin. ft.; valley, 30 cents per lin. ft.

Labor Cost of 100 Sq. Ft. (1 Sq.) Strip Asphalt Shingles on Plain Double Pitch or Gable Roofs

	Hours	Rate	Total	Rate	Total
Roofer	1.0	$....	$....	$22.72	$22.72

On roofs with gables, dormers, etc. add 0.2 hr. per square. On difficult constructed hip or English type roofs, add 0.5 hr. per square.

07312 ASBESTOS CEMENT SHINGLES

This section is only for historical information. Asbestos cement shingles were widely used as a roofing and wall siding material in the 1950s. With the advent of improved construction materials technology, materials that were less brittle, lighter in weight, easier to install, and cheaper to manufacture replaced the asbestos cement materials. Eventually, regulation and banning of asbestos use has meant that this product is unavailable.

Even though the labor cost tables for installing this type of material have been updated to reflect current wage rates, the information that follows in this section is only included so that estimators who require information on this material may be familiar with the patterns and labor requirements.

Asbestos shingles were made of asbestos fiber and portland cement, united under hydraulic pressure, and cut to single shapes in a variety of sizes, shapes, and colors. The cost depended upon the style and color of shingles used and method of laying.

When estimating quantities of asbestos shingles, the measurements were taken in the same manner as for shingle or slate roofs. The length of the eaves were measured to obtain the number of lin. ft. of starters required, as well as the length of hips and ridges, because either shingles or a special ridge roll would be required. Asbestos shingles were sold by the square, made in hexagonal and Scotch or Dutch Lap styles, individual shingles, and slates and strip shingles, among others.

Preparation of Roof. Asbestos shingles were for use on pitched roofs only. For strip shingles the minimum allowable slope of roof was 4" per foot; for the hexagonal and Dutch Lap Method shingles, 5" per foot. A slope of 3" per foot was permissible with adequate underlayment procedure.

The roofing boards were well seasoned, of narrow width and laid in the usual manner breaking joints, and nailed securely in place with at least two nails at each rafter or purlin, leaving no loose ends. Plywood could be used for sheathing. A cant for the eaves shingles was to be provided by the projection of the cornice molding 1/4" above the roof boarding or by a lath (1/4"x1 1/2") nailed parallel and flush with lower edge of roof boarding. Over the roof boarding, one thickness of 15-lb. or heavier asbestos roofing felt was laid, horizontally with a 4" lap and with a 12" lap on hips, ridges, and valleys. Ridge shingles should be laid in the direction away from prevailing winds.

Strip Asbestos Shingles. These were applied over the roofing felt, one course of starter (9"x16") shingles at eaves lengthwise and parallel with same, overhanging the eaves about 1". The second course was applied using main body shingles entirely covering the first course, breaking joints, and then proceeding in the regular manner as with wooden shingles or slate exposing 7" to the weather, with a minimum headlap of 2". Each shingle was fastened in place with at least 2 galvanized iron or copper needle pointed shingle nails, never driving nails down tight, but driving them firmly. Over the ridges and hips were applied asbestos hip and ridge shingles 4 ¾" to 5 ⅜" wide by 14" long.

All chimneys, valleys, etc., were flashed with copper or other approved material. American Method shingles are 8"x16" in size, ¼" thick, and about 260 shingles per square, or 16x16 in size, ¼" thick, and 130 shingles per square.

For new work it was figured to take 3 lbs. of 1 ½" nails, and for re-roofing 4 lbs. 2" galvanized nails per square.

Hexagonal Method Asbestos Shingles. The Hexagonal or "Honeycomb" Method of applying asbestos shingles appears to show 6 sides of the shingle, thus overcoming the objections to severely straight lines and producing a pleasing effect.

One course of starter (4"x16") shingles were applied end to end, parallel with and overhanging the eaves 1", over which was applied one course of half shingles, entirely covering the starter and breaking all joints.

The balance of roof was covered with body shingles 16"x16", exposing 13"x13" to the weather, securely fastening all shingles in place with galvanized or copper needle pointed nails and fastening the points of the main body shingles with special copper storm nails. All the main body shingles were to be laid with the diagonal lines on a 45° angle with the eaves. Asbestos ridge and hip rolls, with not less than 3" lap, were fastened in place with special roll fasteners furnished for the purpose.

Scotch or Dutch Lap Method of Laying Asbestos Shingles. The Scotch or Dutch Lap Method of applying asbestos shingles closely resembles the American Method in appearance yet approximates the cost of the Hexagonal Method.

Over the felt at eaves, one course of eave starter shingles (4"x16") was applied parallel with it and overhanging eaves 1", with each shingle overlapping preceding shingle approximately ⅓ its length so that nail holes occurred one over the other. The first starter was cut to ⅔ its original length and with a new hole punched 2 13/16" from the cut end and applied at the end of the roof to overhang the gable 1". A nail is fastened through the new hole, a full length starter applied to overlap the first starter one-half its length, and then both fastened with a nail through coinciding holes. The roofers proceeded in a similar manner to end of roof.

Starters and balance of roof were covered with body shingles (16"x16"), exposing 10 ⅔"x13" to the weather, each shingle secured with 2 shingle nails and 1 storm anchor. The shingles were aligned on nail and storm anchor holes, one-third side lap and 3" head lap. If the shingles were laid with one-quarter side lap, 12"x13" is exposed. Ridges, hips and exterior corners were finished with asbestos ridge and hip shingles or asbestos ridge roll in the same manner as with other types of asbestos shingles.

Data on Scotch or Dutch Lap Method Asbestos Shingles

Laid with 4" or 1/4 Side Lap

Size	Thickness, Inches	Expos.	No. per Sq.	Lbs. per Sq.
16"x16"	5/32"	12"x13"	92	265

Laid with 5 1/2" or 1/3 Side Lap

16"x16"	5/32	10 2/3"x13"	104	300

When laid with 4" or 1/4 side-lap, it required (184) 0.8-lb. of 1 1/4" needlepoint nails and 92 storm buttons per square. When laid with 5 1/3" or 1/3 side-lap, it required 104 shingles (216) 1 lb. of 1 1/4" needlepoint nails and 104 storm buttons per square. Starter shingles for the above could be 4"x16" laid lengthwise.

Data on Hexagonal Method Asbestos Shingles

Size	Thickness Inches	Expos. Inches	No. Shingles per Sq.	Wt. Lbs. per Sq.	Nails Reqd. Lgth.	Lbs.
16"x16"	5/32"	13"x13"	86	240	1 1/4"	1*

*Also requires 86 storm anchors per square.

Requires 172 needlepoint nails and 86 storm buttons per square, starter shingles for above could be 4"x16" laid lengthwise, and eaves shingle were for use with hexagonal shingles 21" long requires 58 pcs. per 100 lin. ft.

Laying Asbestos Shingles by the Hexagonal Method.--When laying 16"x16" asbestos shingles by the hexagonal method on fairly plain roofs, a crew of two roofers and one helper handled and laid 600 to 700 sq. ft. (6 to 7 squares) per 8-hr. day.

On more complicated roofs, the labor cost was higher due to the necessity of cutting and fitting shingles. This additional labor varied, depending on how broken up to roof, but a crew consisting of two roofers and a helper handled and lay 400 to 500 sq. ft. (4 to 5 sqs.) per 8-hr. day, unless very unusual conditions exist.

Laying Asbestos Shingles by the Dutch Lap Method. Laying 16"x16" asbestos shingles by the Dutch lap method on either plain or complicated roofs required about 15% more labor than for the hexagonal method.

Labor Cost of 100 Sq. Ft. (1 Sq.) 8"x16" Roof Shingles Laid on Plain Roofs

	Hours	Rate	Total	Rate	Total
Roofer	4	$....	$....	$22.72	$90.88
Helper	2	....	....	19.13	38.26
Cost per 100 sq. ft.			$....		$129.14

If laid on a complicated or "cut-up" roof, it meant 2.5 hrs. roofer and 1.25 hrs. helper time per 100 sq. ft.

Labor Cost of 100 Sq. Ft. (1 Sq.) 16"x16" Asbestos Shingles Laid Hexagonal Method on Plain Roofs

	Hours	Rate	Total	Rate	Total
Roofer	2.4	$....	$....	$22.72	$54.53
Helper	1.2	....	....	19.13	22.96
Cost per 100 sq. ft			$....		$77.49

If laid on a complicated or "cut-up" roof, it meant 1.2 hrs. roofer and 0.6 hr. helper time per 100 sq. ft. The estimator added for starters, eaves starters, and hip and ridge shingles.

Labor Cost of 100 Sq. Ft. (1 Sq.) 16"x16" Asbestos Shingles Laid Dutch Lap Method with 4" or ¼ Side Lap on Plain Roofs

	Hours	Rate	Total	Rate	Total
Roofer	2.8	$....	$....	$22.72	$63.62
Helper	1.4	....	....	19.13	26.78
Cost per 100 sq. ft.			$....		$90.40

If laid on a complicated or "cut-up" roof, it took 1.4 hrs. roofer and 0.7 hr. helper time per 100 sq. ft. The estimator needed to add for starters and hip and ridge shingles.

Labor Cost of 100 Sq. Ft. (1 Sq.) 16"x16" Asbestos Shingles Laid Dutch Lap Method with 5 ⅓" or ⅓ Side Lap on Plain Roofs

	Hours	Rate	Total	Rate	Total
Roofer	3.2	$....	$....	$22.72	$72.70
Helper	1.6	....	....	19.13	30.61
Cost per 100 sq. ft.			$....		$103.31

If laid on a complicated or "cut-up" roof, it took 1.6 hrs. roofer and 0.8 hr. helper time per 100 sq. ft.

Applying Asbestos Shingles over Old Roofs. When asbestos shingles was applied over old roofs, it was necessary to place wood strips or laths over the old roof. Figure about 300 lin. ft. of strips per 100 sq. ft. of roof, and 0.5 to 1 hr. time to place same.

Asbestos Cement Sidewall Shingles and Siding. Are manufactured from asbestos fibers and portland cement, which are formed into various shingle shapes under hydraulic pressure.

They are furnished in various colors and white, are wood grained to represent wood shingles and siding and are used on both new and old work.

Asbestos cement siding shingles are furnished in strips 12"x24", 12"x27", or 32"x14 ⅝", and may be obtained with either staggered or thatched butts or with a wavy edge.

Estimating Data on Asbestos Cement Shingle Siding

Description Textured or Smooth	Dimensions	No. Reqd. per Sq.	Exposure Inches
Rock Shakes	24x12	54	11
Rock Shakes	32x14 ⅝	33	13 ⅝
Rock Shakes	32x9	57	8
Textured Only			
Wavy Butt	24x12	54	11
Smooth Only			
Sheet	32x96	4.7	*

*Use 3 ½" backer strips. Nails are furnished with shingles.

Special alloy nails for exposed nailing and asphalt joint strips were usually furnished by the manufacturers and are included in the price of the siding.

Labor Placing Asbestos Shingles on Sidewalls. The labor cost of placing asbestos shingle siding varied with the size of the shingle and the type of building, whether it had fairly long straight walls or was cut up, with numerous openings requiring considerable cutting and fitting.

Where the old walls were crooked and it was necessary to straighten the old walls and place furring strips on them ready to receive the shingles, an extra allowance was made for this work. It was figured that about 1.25 lin. ft. of furring was needed for each square foot of siding.

Before placing asbestos shingles on sidewalls, the walls were first covered with a heavy insulating and sheathing felt to keep out air and wind and prevent the penetration of water.

After the walls were covered with paper, the courses of shingles were marked with a chalk line, with the tops of the shingles following the chalk line. This resulted in straight courses and a satisfactory job.

An asbestos siding cutter was used for cutting siding for lower labor costs and to prevent breakage.

On straight sidewalls using 24"x12" shingles, a carpenter could place 250 to 300 sq. ft. per 8-hr. day; on complicated work requiring considerable cutting, 150 to 200 sq. ft. per 8-hr. day. Where 32"x14 5/8" shingles were used, the rate was 375 to 425 sq. ft. per 8-hr. day for straight walls and 225 to 275 sq. ft. per 8-hr. day for cut-up work. If wood stripping was required for nailing siding, extra was added for material and labor.

Labor Cost of 100 Sq. Ft. (1 Sq.) 24"x12" Asbestos Sidewall Siding on Straight Walls

	Hours	Rate	Total	Rate	Total
Carpenter	3	$....	$....	$24.04	$72.12
Cost per sq. ft.			....		.72

On jobs that required considerable cutting and fitting, 1.5 hrs. carpenter time was added.

Colorbestos. This product is no longer manufactured. It is an asbestos-cement material in sheets 32" wide, 96" long, and 3/16" thick. It was intended mainly as an exterior wall finish to be applied vertically from sill to plate and was also adaptable for soffits and exterior ceilings.

The face of this material is textured with striations running the long dimension of the sheet accomplished by embedding colored granules in the asbestos-cement sheet. Although the sheet appears to have an ingrained texture, the surface is smooth to the touch.

Labor Placing Colorbestos. Colorbestos was intended primarily for use on new construction. The large sheets were applied over a continuous backing of wood or non-wood sheathing and with the vertical joints butted, horizontal joints lapped or butted when flashed with a "Z" type, non-corroding metal flashing. Exterior corners could be finished with either wood or metal trim.

Each sheet has 33 nail holes, 1/8" in diameter, pre-punched at the factory, along the sides and down the center. Where non-wood sheathing was used, studs had to be accurately spaced 16" on center to permit driving nails through the sheathing into the studs, with studs centered behind all vertical joints. Both wood and non-wood sheathing was first covered with No. 15 asphalt saturated felt lapped 2" at horizontal joints and 6" at vertical joints. Nails, for fastening sheets, were of non-corroding metal such as aluminum.

Field cutting Colorbestos was usually by the "score and snap" method. Using a carbide tipped blade, a score mark was made in the surface of the sheet at the location of the cut, and with the sheet supported so that the edge of the support occurred under the cut-off line, the piece was snapped off cleanly. Cutting could also be done with a power saw equipped with a carbide tipped blade. To keep field cutting to a minimum, shorter factory-cut sizes were ordered along with full size sheets.

If additional nailing holes were required, they could be field punched using a sharpened nail set. However, holes close to the edge would be drilled.

Handling and placing full sheets of Colorbestos at ground level could be done by one worker. A 32"x96" sheet weighs only 36 lbs. Each full sheet covers 21 1/3 sq. ft. or 4.7 sheets provided 100 sq. ft. of finished surface. On ground story wall with a minimum of cutting, a carpenter placed 400 to 450 sq. ft. per 8-hr. day. Where the walls are cut up with door and window openings, a worker placed 250 to 300 sq. ft. per 8-hr. day.

For gable ends, where each piece had to be cut to fit the rake of the roof and placing was done from a scaffold, two carpenters together placed 375 to 425 sq. ft. per 8-hr. day. For 1 story soffit and exterior ceiling work, a carpenter placed 300 to 350 sq. ft. per 8-hr. day.

Labor Cost of 100 Sq. Ft. (1 Sq.) Colorbestos Siding on Ground Story Walls with a Minimum of Cutting

	Hours	Rate	Total	Rate	Total
Carpenter	1.9	$....	$....	$24.04	$45.68
Cost per sq. ft.			....		.46

For walls requiring cutting and fitting around door and window openings, 1 hr. carpenter time was added. Cost of metal or wood corner trim would also have been added to the above.

Labor Cost of 100 Sq. Ft. (1 Sq. Colorbestos Siding on Gable Ends)

	Hours	Rate	Total	Rate	Total
Carpenter	4	$....	$....	$24.04	$96.16
Cost per sq. ft.			....		.96

Add cost of scaffolding.

Labor Cost of 100 Sq. Ft. (1 Sq.) Colorbestos on 1-Story Soffits or Ceilings

	Hours	Rate	Total	Rate	Total
Carpenter	2.5	$....	$....	$24.04	$60.10
Cost per sq. ft.			....		.60

07313 WOOD SHINGLES AND SHAKES

The labor cost of laying wood shingles will vary with the type of roof, whether a plain gable roof, a steep roof, or one cut up with gables, or dormers, and with the manner in which they are laid, whether with regular butts, irregular or staggered butts, or thatched butts.

The costs given on the following pages are based on the actual number of shingles a roofer can lay per day and not upon the number of squares covered, which will vary with the spacing of the shingles. It does not make any difference to the roofer whether the shingles are laid 4", 4 ½", or 5" to the weather. He will lay practically the same number either way. It does make considerable difference in the number of sq. ft. of surface covered, which will vary from 10 to 40 percent.

The number of shingles laid will vary with the ability of workers and the class of work. Ordinary carpenters cannot lay as many shingles as carpenters who specialize in shingle laying. On the other hand, experienced shinglers usually demand a higher wage rate than ordinary carpenters.

Some carpenters claim to be able to lay 16 bundles (3,200 shingles) per 8-hr. day, but this is unusual and is generally found only on the cheapest grade of work, where only one nail is driven into each shingle instead of two that are necessary for a craftsman-like job.

Estimating the Quantity of Wood Shingles. Ordinary wood shingles are furnished in random widths, but 1,000 shingles are equivalent to 1,000 shingles 4" wide. Dimension shingles are sawed to a uniform width, being either 4", 5", or 6" wide.

Wood shingles are usually sold by the square based on sufficient shingles to lay 100 sq. ft. of surface, when laid 5" to the weather, 4 bundles to the square.

When estimating the quantity of ordinary wood shingles required to cover any roof, bear in mind that the distance the shingles are laid to the weather makes considerable difference in the actual quantity required.

There are 144 sq. in. in 1 sq. ft. and an ordinary shingle is 4" wide. When laid with 4" exposed to the weather, each shingle covers 16 sq. in. or it requires 9 shingles per sq. ft. of surface. There are 100 sq. ft. in a square. 100 x 9 = 900, and allowing 10% to cover the double row of shingles at the eaves, waste in cutting, narrow shingles, etc., it will require 990 shingles (5 bundles) per 100 sq. ft. of surface.

Number of Shingles and Quantity of Nails Required Per 100 Sq. Ft. of Surface

Distance Laid to Weather	Area Covered by One Shingle Sq. In.	Percent for Waste	Actual No. per Square without Waste	No. per Square with Waste	No. of 4-Square Bundles Required	Pounds 3d Nails Req'd
4"	16	10	900	990	5.0	3.2
4 ½"	17	10	850	935	4.7	2.8
4 ½"	18	10	800	880	4.4	2.5
5"	20	10	720	792	4.0	2.0
5 ½"	22	10	655	720	3.6	1.6
6"	24	10	600	660	3.3	1.5

How to Apply Shingles for Different Roof Slopes. Roof pitches are computed in fractions, such as ⅛, ⅓, ½ pitch. In the following cross section, the steepness of distances AB and BC constitutes *pitch*. Distance AC, extending from one eaveline to the other, is known as the *span*. One-half of this span, distance AD or DC, is called the *run*, and distance BD is called the *rise*. The relationship of the rise to run obviously affects the slope of AB or BD; in fact, roof pitches are computed from the ratio of rise to run. Therefore, the first step is to determine length of the run (AD or DC) and the rise (BD).

Wood shingles are manufactured in three lengths, 16", 18", and 24". The standard weather exposure (portion of shingle exposed to weather on roof) for 16" shingles is 5", for 18" shingles it is 5 ½", and for 24" shingles is 7 ½". These

Labor Laying One Square (100 Sq. Ft.) Wood Shingles

Class of Work	Mechanic	Number Laid per 8-Hr. Day	Distance Shingles are Laid to Weather					
			4"†	4¼"	4½"	5"	5½"	6"
Plain Gable or Hip Roofs	Carpenter	2,000–2,200	3.8 hrs.	3.6 hrs.	3.4 hrs.	3.0 hrs.	2.8 hrs.	2.5 hrs.
	Shingler	2,750–3,000	2.8	2.6	2.5	2.3	2.0	1.9
Difficult Gable Roofs, cut up with gables, dormers, hips, valleys, etc.	Carpenter	1,700–1, 900	4.5	4.2	4.0	3.6	3.3	3.0
	Shingler	2,200–2,500	3.4	3.2	3.0	2.7	2.5	2.3
Difficult Hip Roofs, Steep English Roofs, hips, valleys, etc.	Carpenter	1,300–1,500	5.7	5.4	5.1	4.6	4.2	3.8
	Shingler	2,000–2,200	3.8	3.6	3.4	3.0	2.8	2.5
Shingles Laid Irregularly or with Staggered Butts on Plain Roofs.	Carpenter	1,700–1,900	4.5	4.2	4.0	3.6	3.3	3.0
	Shingler	2,400–2,700	3.1	3.0	2.8	2.5	2.3	2.1
Shingles Laid Irregularly or with Staggered Butts on Difficult Constructed Roofs.	Carpenter	1,100–1,300	6.7	6.3	6.0	5.4	4.9	4.5
	Shingler	1,600–1,800	4.7	4.5	4.2	3.8	3.4	3.1
Shingles with Thatched Butts.*	Shingler	800–1,000	8.9	8.4	8.0	7.1	6.5	6.0
Plain Sidewalls.	Carpenter	1,300–1,500	5.7	5.4	5.1	4.6	4.2	3.8
	Shingler	1,700–1,900	4.5	4.3	4.0	3.6	3.3	3.0
Difficult Sidewalls, having bays, windows, breaks, etc.	Carpenter	1,100–1,250	6.8	6.4	6.0	5.4	5.0	4.5
	Shingler	1,400–1,650	5.2	5.0	4.6	4.2	3.8	3.5

To obtain number of bundles of shingles required, divide number of shingles as given above by 200, and the result will be the number of bundles required, i.e, 2,200 shingles ÷ 200 equals 11 bundles; 1,000 ÷ 200=5 bundles; 800 ÷ 200=4 bundles, etc.

*Shingles with thatched butts require 25% more shingles than when laid regularly.

†Use 4" column for carpenter or shingler time per 1,000 shingles. (5 bundles).

The above table is based on using 2 nails to each shingle and 10 per cent waste.

standard exposures are recommended on all roofs of 1/4 pitch and steeper (6" rise in 12" run). On flatter roof slopes, the weather exposure should be reduced to 3 3/4" for 16" shingles, 4 1/4" for 18" shingles, and 5 3/4" for 24".

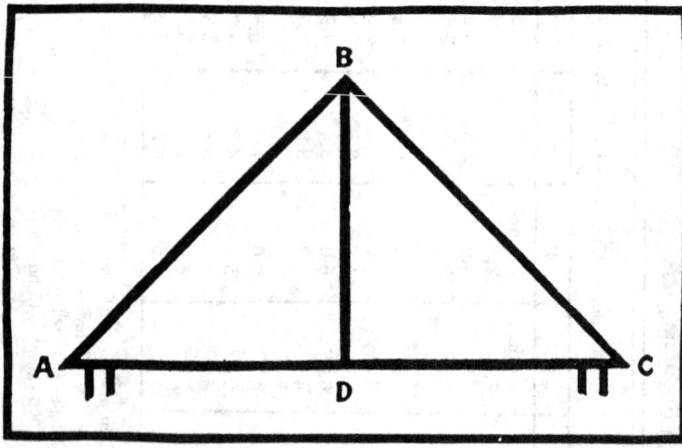

The following diagram shows the weather exposure to be used for various roof pitches. For example, if a roof has a rise of 8" in a run of 12", it is 1/3 pitch and that an exposure of either 5", 5 1/2", or 7 1/2" should be employed, depending on the length of the shingles used.

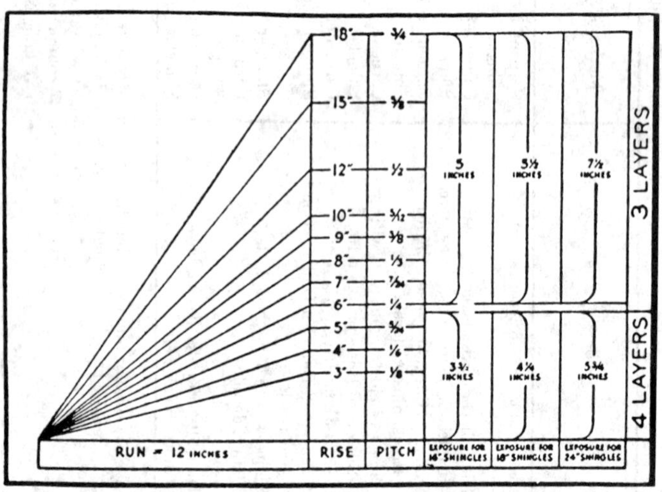

Wood shingles and shakes may be laid over open or solid sheathing, with solid preferred where wind driven snow conditions prevail. If open sheathing is selected, use 1"x4"s at centers equal to the weather exposure at which the shakes are to be laid, but not over 10". Building paper must be added for additional weather protection.

Western Red Cedar Shingles

144# 16" #1-5X Shingles 4 Bdl. per Sq.	$100.00
144# 16" #2-5X Shingles 4 Bdl. per Sq.	90.00
60# 16" Undercourse 14/14 2 Bdl. per Sq.	25.00
158# 18" #1 Perfection 4 Bdl. per Sq.	100.00
350# 24" #1 3/4" to 5/4 Handsplit Shakes, 5 Bdl. per Sq.	110.00

For fire retardant shingles add 100%

07314 SLATE ROOFING

Roofing slate is furnished in a number of sizes, thicknesses, and finishes to meet architectural requirements of the buildings on which they are to be used. The four principal classifications are standard, textural, graduated, and flat slate roofs.

Standard Roof. Standard slate roofs are composed of slate approximately 3/16" thick (commercial standard slate), of one uniform standard length and width; also one length and random widths or random lengths and widths. Any width less than one-half the length is not recommended. Standard roofs are suitable for any building where a permanent roofing material is desired at a minimum cost. If desired, the butts or corners may be trimmed to give a hexagonal, diamond, or "gothic" pattern for all or part of the roof.

Textural Roof. The term "textural" is used to designate slate of rougher texture than the standard, with uneven tails or butts, and with variations of thickness or size. In general, this term is not applied to slate over 3/8" in thickness.

Graduated Roof. The graduated roof combines the artistic features of the textural slate roof with additional variations in thickness, size, and exposure. The slate is so arranged on the roof that the thickest and longest occur at the eaves and gradually diminish in size and thickness until the ridges are reached. Slate for roofs of this type can be obtained in any combination of thicknesses from 3/16" to 3/4" and heavier when especially desired.

In addition to the usual standard sizes, slate above 1/2" thick are produced in lengths up to 24". The graduations in lengths generally range from 24" to 12". The variations in length will at once provide a graduation in exposure by using the standard 3" lap. To illustrate, a suitable range of lengths might be:

	Thickness	Length	Exposure
Course Under Eaves	3/8"	14"	No exposure
First Course	3/4"	24"	10 1/2"
1	3/4"	24"	10 1/2"
2	1/2"	22"	9 1/2"
2	1/2"	20"	8 1/2"
2	3/8"	20"	8 1/2"
4	3/8"	18"	7 1/2"
5	1/4"	16"	6 1/2"
3	1/4"	14"	5 1/2"
3	3/16"	14"	5 1/2"
8	3/16"	12"	4 1/2"

Random widths should be used and so laid that the vertical joints in each course are broken and covered by the slate of the course above.

Flat Roof. Flat roofs offer a wide field for roofing slate and are designated whether or not they are used for "promenade" purposes.

Slate of any thickness may be used instead of slag or gravel as a surfacing material for the usual built-up type of roof. For ordinary roofs the standard 3/16" slate are used, but for promenade or extraordinary service, the slate should be 1/4" or 3/8" thick.

Slate for flat roofs are furnished in the following standard sizes: 6"x6", 6"x8", 6"x9", 10"x6", 10"x7", 10"x8", 12"x6", 12"x7", and 12"x8".

Estimating Quantities of Roofing Slate. Obtain the net area of the roof in sq. ft. as described earlier in this chapter. Add 6" to rafter length to allow for waste of normal roof.

Deduct one-half the area of chimneys and dormers if over 20 sq. ft. and less than 80 sq. ft. Make no deductions if less than 20 sq. ft. and deduct 20 sq. ft. less than actual area if more than 80 sq. ft.

Include area of dormer sides, sides of dormers if slated, slate saddles, and other places where slate is used in addition to the main roof area. Include overhanging parts of dormers, etc.

Add 1 sq. ft. for each lin. ft. of hips and valleys, for loss in cutting and fitting.

Allow 2% to 15% additional slate, depending upon the extent to which the roof is intersected by other roofs, dormers, walls, etc.

Divide the total of the above by 100, which will give the number of "squares" of roofing required.

Slate is always sold at the quarry on the basis of the quantity required to cover 100 sq. ft. or a "square" of roof when slate is laid with a 3" head lap. If the roof is flat or other than 3" lap is used, the quantity must be corrected to the equivalent amount required as though the 3" lap were used.

Items to be Included in an Estimate for Slate Roofing. The following items should be included to make a complete and accurate estimate on slate roofing:

1) Cost of slate (punched) on cars or trucks at the quarry.
2) Freight or trucking from quarry to destination.
3) Unloading from cars and hauling to job, also unloading and piling at job, unless delivered to the job by truck. Trucking from quarry direct to job within a radius of 500 miles is now general practice in the Vermont-New York district.
4) Placing on roof and laying, a) roofing felt, b) elastic cement, c) nails, d) snow guards or snow rails, e) sheet metal, f) labor, g) waste in handling, cutting and fitting.

Size of Slate. The following table gives the different sizes in which roofing slate is furnished, number required per square when laid with 3" lap, and number of nails required per square based on 2 nails to each piece of slate.

Size of Slate Inches	Number per Square	Number of Nails Reqd. per Square	Size of Slate Inches	Number per Square	Number of Nails Reqd. per Square
10x6	686	1,372	16x12	185	370
10x7	588	1,176	16x14	159	318
10x8	515	1,030	18x9	213	426
10x10	414	828	18x10	192	384
12x6	533	1,066	18x11	175	350
12x7	457	914	18x12	160	320
12x8	400	800	18x14	137	274
12x9	355	710	20x10	169	338
12x10	320	640	20x11	154	308
12x12	267	534	20x12	141	282
14x7	374	748	20x14	121	242
14x8	327	654	20x16	106	212
14x9	290	580	22x11	138	276
14x10	261	522	22x12	126	252
14x12	218	436	22x14	109	218
14x14	188	376	22x16	95	190
16x8	277	554	24x12	115	230
16x9	246	492	24x14	98	196
16x10	221	442	24x16	86	172

Slate 3⁄4" and thicker and 24" or over in length should have 4 nail holes.

Weight of Slate Roofing. A square of slate roofing, i.e., sufficient slate to cover 100 sq. ft. of roof surface with a standard 3" lap, will vary from 650 to 8,000 lbs. for thicknesses from the commercial standard 3⁄16" to 2".

The weight of slate varies with the size of the slate, color and quarry, and even sometimes in the same quarry. The variation may be from 10% above to 15% below the weights given in the following table:

Average Weight of Slate per Square (100 sq. ft.)

	Slate Thickness, Inches	Sloping Roof, allowing 3" Lap	Flat Roof without Lap
Standard	3⁄16	700	240
Selected full	3⁄16	750	250
	1⁄4	900	335
	3⁄8	1,400	500
	1⁄2	1,800	675
	3⁄4	2,700	1,000
	1	4,000	1,330
	1 1⁄4	5,000	1,670
	1 1⁄2	6,000	2,000
	1 3⁄4	7,000	
	2	8,000	

Nails Required for Slate Roofing. The quantity of nails required for slate roofing will vary with size, kind, etc., but the following table gives quantities of those commonly used:

Approximate Number of Nails to the Pound

Length in Inches	"Copperweld" Slating Nails	Copper Wire Slating Nails	Cut Copper Slating Nails	Cut Brass Nails	Cut Yellow-Metal Slat. Nails
1	386	270	...	...	...
1 ¼	211	144	190	164	154
1 ½	176	134	135	140	140
1 ¾	133	112	100	108	...
2	87	104	...	88	...
2 ¼	...	46	...	80	...
2 ½	...	...	...	64	...
2 ¾	...	...	...	52	...
3	...	...	...	48	...

Labor Punching Slate. Because of high labor costs on the job, it is much cheaper to have slate punched at the quarry. The labor cost of punching slate on the job will vary with the size of the slate, as the larger the slate the fewer pieces required per square.

If one slater is working on the job with a helper, the helper should punch the slate and carry them to the slater as fast as required. In instances of this kind, there should be no additional charge for punching. If, however, there is only one worker on the job who must punch the slate and lay them, he should punch one square of slate in 30 to 40 min. or 12 to 16 sqs. per 8-hr. day, at the following labor cost per 100 sq. ft. (sq.):

	Hours	Rate	Total	Rate	Total
Slater	0.6	$....	$....	$22.72	$13.63

Labor Cost of Laying Roofing Slate. The labor cost of laying roofing slate will vary considerably with the size of the slate, thickness, type of roofs, etc., but when large thick slate are used for graduated roofs, the labor costs will run higher than when the same size slate are used on standard roofs.

The size of slate best adapted for plain roofs are large, wide slates, such as 16"x12", 18"x12", 20"x12", or 24"x12". Slate from 16"x8" to 20"x10" are also popular sizes. For roofs cut up into smaller sections, the 14"x7" and 16"x8" are very popular.

A roofing crew consisting of two slaters and one helper can work to advantage and perform the work more economically than one man working alone. In some localities the helper is considered an apprentice and is permitted to lay slate, in others he can only punch and carry the slate to the slaters.

Laying 16"x8" Roofing Slate. When working on plain roofs that do not require much cutting, 2 slaters and a helper should lay felt, handle and lay 350 to 450 sq. ft. (3 ½ to 4 ½ sqs.) of 16"x8" slate per 8-hr. day. On more complicated hip or gable roofs, requiring considerable cutting and fitting for hips, valleys, dormers, etc., 2 slaters and a helper should handle and lay 175 to 225 sq. ft. (1 ¾ to 2 ¼ sqs.) of 16"x8" slate per 8-hr. day.

Laying 18"x9" Roofing Slate. When laying 18"x9" slate on plain roofs, 2 slaters and a helper should lay felt, handle and lay 400 to 500 sq. ft. (4 to 5 sqs.) per 8-hr. day.

On more complicated hip or gable roofs, requiring considerable cutting and fitting for hips, valleys, dormers, etc., 2 slaters and a helper should lay 250 to 300 sq. ft. (2 ½ to 3 sqs.) per 8-hr. day.

Laying 20"x10" Roofing Slate. When laying 20"x10" slate on plain roofs, 2 slaters and a helper should lay felt, handle and lay 500 to 600 sq. ft. (5 to 6 sqs.) of roof per 8-hr. day, but on more complicated roofs, requiring considerable cutting and fitting for hips, valleys, dormers, etc., the same crew would lay only 300 to 350 sq. ft. (3 to 3 ½ sqs.) per 8-hr. day.

Laying 22"x12" Roofing Slate. On straight roofs, 2 slaters and a helper should lay felt, handle and lay 550 to 650 sq. ft. (5 ½ to 6 ½ sqs.) of 22"x12" slate per 8-hr. day, but on the more complicated roofs, requiring cutting and fitting for hips, valleys, dormers, etc., the same crew would lay only 350 to 400 sq. ft. (3 ½ to 4 sqs.) per 8-hr. day.

Laying Graduated Roofing Slate. When laying graduated roofing slate from 3/16" to 3/4" thick and 12" to 24" long, considerable care is required in selecting the right slate for each course and laying them to obtain the desired effect. On work of this kind 2 slaters and a helper should lay felt, handle and lay 150 to 200 sq. ft. (1 ½ to 2 sqs.) of roof per 8-hr. day.

Labor Cost of 100 Sq. Ft. (1 Sq.) 16"x8" Standard Slate Roofing

	Hours	Rate	Total	Rate	Total
Slater	4	$....	$....	$22.72	$90.88
Helper	2	....	....	19.13	38.26
Cost per 100 sq. ft.			$....		$129.14

For complicated hip or gable roofs, double the time given above.

Labor Cost of 100 Sq. Ft. (1 Sq.) 18"x9" Standard Roofing Slate

	Hours	Rate	Total	Rate	Total
Slater	3.50	$....	$....	$22.72	$79.52
Helper	1.75	....	....	19.13	33.48
Cost per 100 sq. ft.			$....		$113.00

For complicated hip or gable roofs, add 2.5-hrs. slater and 1.25-hrs. helper time.

Labor Cost of 100 Sq. Ft. (1 Sq.) 20"x10" Standard Roofing Slate

	Hours	Rate	Total	Rate	Total
Slater	3.0	$....	$....	$22.72	$68.16
Helper	1.5	....	....	19.13	28.70
Cost per 100 sq. ft.			$....		$96.86

For complicated hip or gable roofs, add 2 hrs. slater and 1 hr. helper time.

Labor Cost of 100 Sq. Ft. (1 Sq.) 22"x12" Standard Roofing Slate

	Hours	Rate	Total	Rate	Total
Slater	2.6	$....	$....	$22.72	$59.07
Helper	1.3	....	....	19.13	24.87
Cost per 100 sq. ft.			$....		$83.94

For complicated hip or gable roofs, add 1.6 hrs. slater and 0.8 hr. helper time.

Labor Cost of 100 Sq. Ft. (1 Sq.) Rough Texture Slate Roof Random Widths and 3⁄16" to 3⁄8" Thick

	Hours	Rate	Total	Rate	Total
Slater	7.50	$....	$....	$22.72	$170.40
Helper	3.75	....	....	19.13	71.74
Cost per 100 sq. ft.			....		$242.14

Labor Cost of 100 Sq. Ft. (1 Sq.) Graduated Slate Roof, Slate 12" to 24" Long and 3⁄16" to 3⁄4" Thick

	Hours	Rate	Total	Rate	Total
Slater	9.0	$....	$....	$22.72	$204.48
Helper	4.5	....	....	19.13	86.09
Cost per 100 sq. ft.			$....		$290.57

Prices on Slate Roofing Material

Slate shingles come from various parts of the country, and the estimator must obtain a quotation from the quarries in the geographical location, not only for price and shipping costs, but for availability of certain sizes and colors.

Vermont slate shingles in 3⁄16" thicknesses will cost as follows: black or gray, $290.00 per square; green and gray, $300.00; and purples, $375.00. The Pennsylvania black #1 clear and the Buckingham, Virginia black slates will cost about $375.00 per square. Slates thicker than 3⁄16" will cost appreciably more.

Copper nails are usually used for nailing the slates. Use 1 1⁄4" long nails for the 3⁄16" thick slates and 1 3⁄4" long nails for the rough texture and graduated slate roofs where thicker slates are encountered.

07315 PORCELAIN ENAMEL SHINGLES

Porcelain enamel shingles are manufactured to have an exposed surface of 10"x10" with 144 shingles per square. Weight is 225 lbs. per 100 sq. ft. Finish is fused on at 1500°F and provides a long lasting, self-cleaning finish that will not peel or blister.

In ordering these units, it is important to have complete shop drawings based on actual job conditions so that field cutting can be held to a very minimum. Shingles are nailed to the roof deck over 2 layers of 30 lb. felt. Special nails, ridge and hip sections are furnished by the manufacturer. Valley flashings and drips should be of non-corrosive metal and furnished by the sheet metal contractor.

Material costs will run in the vicinity of $280.00 per square, depending on the number of roof irregularities which run up the cost considerably, as all special fittings should be made up in the shop. One roofer can lay a square in about 6 1⁄2 hours.

07316 METAL SHINGLES

Metal shingles formed with a wood grain texture and available in a wide range of colors can be used for a lightweight, decorative roof, or false mansard. They can be applied in a conventional way on a solid nailable roof deck over 30 lb. felt underlayment or supported and clipped to sub-purlins on steel frames. Weights are 36 lbs. per square for .020 aluminum, 54 lbs. for .030 aluminum, and 88 lbs. for 30 ga. steel.

The shingles are not individual units, but made in strips either 10"x60", 12"x36" or 12"x48".

The shingle manufacturer can provide anchor clips, drip caps, starter strips, trim, ridge, and hip laps. Material Costs will run as follows:

	Per Sq.
Aluminum, .020", mill finish	$72.00
Aluminum, .030", mill finish	85.00
Aluminum, .020", anodized color	130.00
Aluminum, .030", anodized color	150.00
Aluminum, .020", bonderized	200.00
Aluminum, .030", bonderized	210.00
Steel, 26 ga., galvanized	33.00
Steel, 24 ga., galvanized	36.00
Steel, 26 ga., color galvanized	75.00
Steel, 24 ga., color galvanized	78.00

Ridge and cap sections cost $1.80 per lin ft. for .020" aluminum, $2.10 for 030". Valley sections cost $1.40 and $1.70. One carpenter will take about 4 to 5 hours to lay one square.

07321 CLAY ROOFING TILE

The following information on clay roofing tile has been furnished through the courtesy of the Ludowici-Celadon Company, Chicago, IL.

When estimating the quantities of clay roofing tile, the roof areas are obtained in the same manner as described at the beginning of this chapter.

After the total number of sq. ft. of surface has been obtained, add about 3 percent to the net measurement to take care of cutting and waste. Material is sold only in full pallets which vary from 3 to 8.7 squares.

Hip and valley fittings should be figured by the piece. Closure and finishing pieces, such as hip starters, closed ridge ends, terminals, etc., should also be figured by the piece. Fittings are sold only in full boxes.

Where interlocking shingle tiles are used, detached gable rake pieces are required for all gable roofs not intersecting with vertical surfaces. These pieces are measured by the total lineal feet of gable rake, keeping separate quantities for lefts and rights as determined by the roof eave to be covered.

For all shingle tile roofs, end bands or half-tile are required on each course, alternating from one end to the other to break joints. End bands are figured by the lineal feet of gable rake, taking half the length at each gable rake.

Laying Interlocking Shingle Tile. On a roof of average difficulty, a crew consisting of 2 roofers and an apprentice or helper should handle and lay 500 to 600 sq. ft. of interlocking shingle tile per 8-hr. day, at the following labor cost per 100 sq. ft. (1 sq.):

	Hours	Rate	Total	Rate	Total
Roofer	3.0	$....	$....	$22.72	$68.16
Helper	1.5	....	....	19.13	28.70
Cost per 100 sq. ft			$....		$96.86

On steep or cut-up roofs, add 15 to 25 percent. For roofs over 3 stories high to eaves, add 1.5 hrs. helper time.

Laying Tile Shingles (Architectural Patterns). On a roof of average difficulty, 2 roofers and a helper should handle and lay 325 to 375 sq. ft. of tile shingles per 8-hr. day, at the following labor cost per 100 sq. ft. (1 sq.):

	Hours	Rate	Total	Rate	Total
Roofer	4.6	$....	$....	$22.72	$104.51
Helper	2.3	....	....	19.13	44.00
Cost per 100 sq. ft			$....		$148.51

On steep or cut-up roofs, add 15 to 25 percent. For roofs over 3 stories high to eaves, add 2.3 hrs. helper time.

Laying Miscellaneous Tile Pattern Roofs. Where Spanish or French pattern tile is used on new work, figure same labor costs as for interlocking shingle tile.

Labor laying Mission tile will vary according to size and exposure of units. On new work of average difficulty, using 15"x8" tile with 12" exposure, 2 roofers and a helper should handle and lay 350 to 400 sq. ft. per 8-hr. day.

Items to Be Included When Estimating Tile Roofs. The following items should be included when estimating the cost of a clay shingle tile roof.

One roll coated felt or roll roofing weighing 40 lbs. per sq. and 3 lbs. roofing nails, 1 3/4" long.

Allow 1 lb. colored elastic cement for pointing joints for each 20 lin. ft. of hips and ridges.

Cost of Clay Tile Roofing Materials. The Luduwici-Celadon Company offers a number of different patterns and colors in each of their "Standard" and "Special" series product line. They will also produce custom shapes and colors, if given enough lead time.

Each of the patterns are produced with field tiles and special shapes for the ridge, hip, rake, etc.

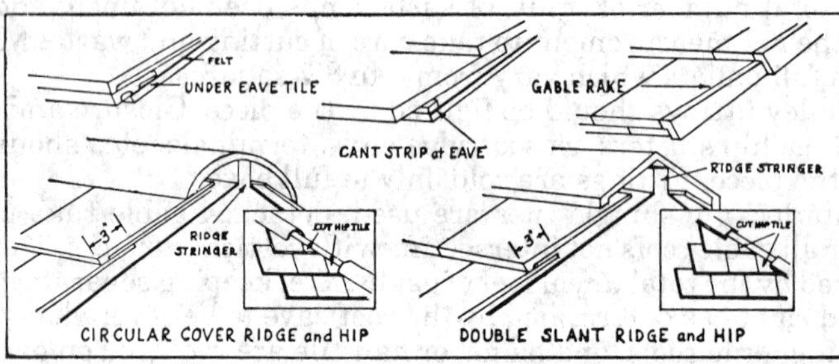

Details of Clay Interlocking Shingle Tile Roofing

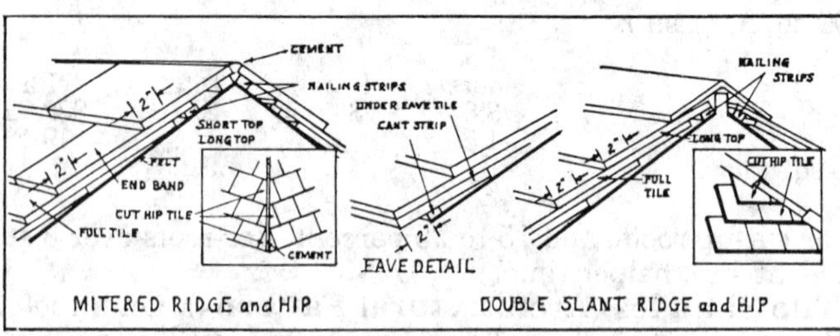

Details of Clay Tile Shingle Roofing

Approximate material prices for different patterns and colors are as follows:

	Price per 100 SF
Spanish Red	$180.00
Spanish Buff	350.00
Spanish Blue	550.00
Mission Granada Red	350.00
Americana Gray or Green	225.00
Lanai Black or Brown	225.00
Classic Red	225.00
French Red	300.00
French Blue	650.00

07400 PREFORMED ROOFING & SIDING

07411 PREFORMED METAL SIDING

Corrugated steel is used extensively for roofing and siding steel mills, manufacturing plants, sheds, grain elevators, and other industrial structures.

It is made with various corrugations, varying in width and depth, but the 2 ½" corrugation width is the most commonly used.

The sheets are usually furnished 26" wide and 6'-0" to 30'-0" long. Some manufacturers offer siding protected with vinyl coatings.

When estimating quantities of corrugated siding or roofing, always select lengths that work to best advantage and be sure and allow for both end and side laps.

Corrugated siding or roofing is nailed to the wood framework or siding when used on wood constructed buildings. When used as a wall and roof covering on structural steel framing, the corrugated sheets are fastened to the steel framework, using clip and bolts, which are passed around or under the purlins (which usually consist of channels, angles or Z bars). When angles are used for purlins, clinch nails are sometimes used for fastening the corrugated sheets.

When used for siding one corrugation lap is usually sufficient, but for roofing two corrugations should be used and if the roof has only a slight pitch, the lap should be three corrugations.

When used for siding, a 1" to 2" end lap is sufficient, but when laid on roofs it should have an end lap of 3" to 6" depending upon the pitch of the roof. For a ⅓ pitch, a 3" lap is sufficient; for a ¼ pitch, a 4" lap should be used; and for a ⅛ pitch, a 5" end lap is recommended.

When applying to wood sheathing or strips, the nails should be spaced about 8" apart at the sides. When applied to steel purlins, the side laps should extend over at least 1 ½ corrugations, and the sheets should be riveted together every 8" on the sides and at every alternate corrugation on the ends.

Number of Sq. Ft. of Corrugated Sheets Required to Cover 100 Sq. Ft. of Surface, Using 26"x96" Sheets

2 ½" Corrugation Width	Length of End Lap in Inches 1	2	3	4	5	6
	Square Feet of Corrugated Metal					
Side lap, 1 corrugation	110	111	112	113	114	115
Side lap, 1 ½ corrugations	116	117	118	119	120	121
Side lap, 2 corrugations	123	124	125	126	127	128
Side lap, 2 ½ corrugations	130	131	132	133	134	135
Side lap, 3 corrugations	138	139	140	141	142	143

If shorter sheets are used the allowance should be slightly increased.

Material Prices. Corrugated steel siding and roofing panels are manufactured in a number of metal gauges, widths and depths of corrugations, and in galvanized and prefinished painted surfaces. Average material prices, per 100 sq. ft., for sheeting with a 2 ½" corrugation width are as follows:

Metal Gauge	Galvanized	Painted
28	$45.00	$60.00
26	55.00	70.00
24	80.00	100.00
22	90.00	105.00
20	95.00	110.00

Placing Corrugated Steel Roofing or Siding on Wood Framing. If corrugated sheets used for roofing or siding are nailed to wood strips or framing, a carpenter and helper should place 100 sq. ft. (1 sq.) of 26"x96" or larger, in ⅞ to 1 ⅛ hr. or 7 to 9 sqs. per 8-hr. day, at the following labor cost per 100 sq. ft. (1 sq.):

Labor Cost of 100 Sq. Ft. (1 Sq.) Corrugated Metal Roofing on Wood Framing

	Hours	Rate	Total	Rate	Total
Carpenter	1	$....	$....	$24.04	$24.04
Labor	1	....	....	19.13	19.13
Cost per 100 sq. ft			$....		$43.17

Labor Cost of 100 Sq. Ft. (1 Sq.) Corrugated Metal Siding on Wood Framing

	Hours	Rate	Total	Rate	Total
Carpenter	1.2	$....	$....	$24.04	$28.85
Labor	1.2	....	....	19.13	22.96
Cost per 100 sq. ft			$....		$51.81

Placing Corrugated Steel Roofing on Steel Framing. The labor cost of applying corrugated steel roofing to structural steel framing, where the sheets are fastened with clips or bolts, will vary with the size of the roof area, regularity of roof, height above ground, etc., but on a straight job, 2 sheeters and 2 laborers should place 550 to 650 sq. ft. per 8-hr. day, at the following labor cost per 100 sq. ft.:

	Hours	Rate	Total	Rate	Total
Labor handling, hoisting	2.67	$....	$....	$19.13	$51.08
Sheeters	2.67	....	....	26.25	70.09
Cost per 100 sq. ft			$....		$121.17

On irregular roofs, above costs may be increased 25 to 50 percent.

Placing Corrugated Steel Siding on Steel Framing. The labor cost of applying corrugated steel siding to walls will vary greatly according to length and height of walls, quantity and regularity of openings, etc., but on a straight job, 2 sheeters and 2 laborers should place 450 to 550 sq. ft. per 8-hr. day, at the following labor cost per 100 sq. ft.:

	Hours	Rate	Total	Rate	Total
Labor handling, hoisting	3.2	$....	$....	$19.13	$61.22
Sheeters	3.2	....	....	26.25	84.00
Cost per 100 sq. ft			$....		$145.22

On coal tipples, grain elevators and other structures having high and irregular walls, the above costs may be doubled.

Protected Metal Roofing and Siding

Protected metal panels consist of a steel core sheets to which asbestos felt is bonded by hot molten zinc. When the zinc has cooled, the asbestos felt and the steel core are metallurgically bound together. The felt is then impregnated with asphalt saturant. Finally, a tough, thick, waterproof outer coating is applied. All material is made with protective coating both sides.

There are several colors to choose from. Protected metal panels are manufactured in 4 gauges: 18, 20, 22, and 24, and are available in lengths up to 30 feet.

Estimating and Cost Data. Panels are available in several profiles. The H.H. Robertson Company, Pittsburgh, Pa., has the following:

Huski-Rib has a deep fluted profile, designed specifically for high strength long span roofing and siding installations. The Huski-Rib sheet is 31 5/32" wide x 1 1/2" deep, and provides a coverage of 28". It is manufactured in 18, 20, 22, and 24 gauge metal, a maximum of 30' long, and is suitable for spans up to 15'-9" for siding, and up to 14'-14" for roofing depending upon local design load requirements.

Box-Rib has a strong, precise vertical texture, best suited for sidings only, either in standard or inverted positions. The standard sheet is 29 1/4" wide x 1 1/2" deep and provides a coverage of 28". It is manufactured in 18, 20, 22 and 24 gauge metal and is available in lengths up to 30 ft.

Sturdi-Rib sheets come 34" wide x 9/16" deep providing a coverage of 29 7/16", and are made in lengths up to 30 feet. They are made of 18, 20, 22, and 24 gauge metal and are capable of spanning up to 10'-1" for siding and 8'-9" for roofing installations, depending upon local design load requirements.

The *Magna-Rib* profile sheets are 26 5/8" wide x 4" deep providing a coverage of 24" installed. They come in 18, 20, and 22 gauge metal in lengths up to 38 feet. Magna-Rib sheets are capable of spanning up to 22'-8" for siding in one span.

Corrugated *Galbestos* is the original Robertson protected metal profile and comes in sheets 33" wide x 9/16" deep. Coverage is 29 3/4". Corrugated sheets are available in 18, 20, 22 and 24 gauge metal in lengths up to 30 feet. They are capable of spanning up to 9'-3" for siding and 7'-11" for roofing installations.

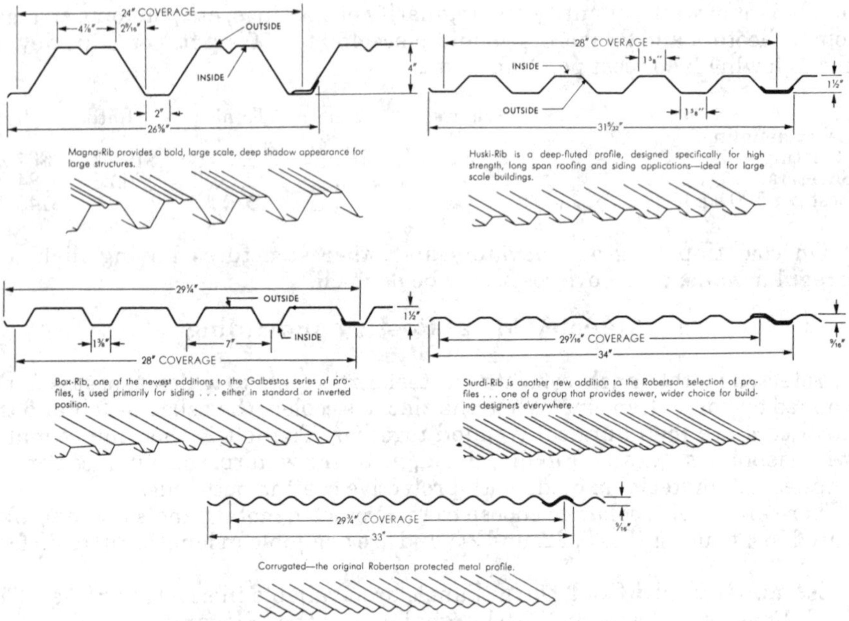

Profiles of Robertson Galbestos Roofing and Siding

Approximate Price per Sq. Ft. of Corrugated Protected Metal Panels

Description	Gauge of Sheets 18	20	22	24
Galbestos C2S (Asphalt) Black	$2.50	$1.90	$1.90	$1.70
Galbestos C2S (Resin) Color	3.00	2.50	2.20	2.00

After figuring the price of sheets from the above, approximately 25% should be added to cover cost of flashings, fastenings, closures, corners, etc.

True cost comparisons must include the cost of supporting steel girts or purlins. Longer spans may save up to $0.25 per sq. ft. or more.

Matching profiled translucent fiberglass reinforced plastic daylight panels are available.

07415 CORRUGATED ALUMINUM ROOFING AND SIDING

When using aluminum roofing and siding, it is very important when applying the sheets that they be insulated against electrogalvanic action. Ordinarily aluminum has a high resistance to corrosion, but when it is contiguous to steel or copper in the presence of moisture, an electrolytic cell is formed, and the resulting flow of current dissolves the aluminum, causing severe pitting. This condition can be avoided by preventing metal-to-metal contact.

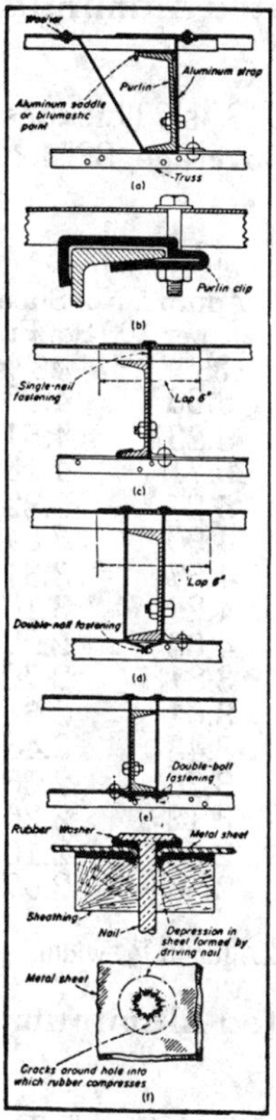

Types of Fasteners Avail. for Corrugated Alum. Roof & Siding

Another significant requirement in using aluminum sheet is to seal the openings made for the fastening devices. Methods for achieving this objective effectively include the use of special washers and waterproof roofing compounds. Washers should be used with fasteners to distribute stresses and prevent tearing of the sheet.

To provide adequate drainage, the roof surface should never have a slope less than 2 ½ in. per ft. and preferably not less than 3 in. per ft.

For roofing, sheets should have a side lap of 1 ½ corrugations. For siding, sheets should be lapped 1 corrugation.

A 6 in. overlap at the ends is recommended for roofing and 4 in. for siding. When corrugated aluminum sheets are used without sheathing on a steel frame, the method shown in (a) is suggested for fastening to purlins and girts. This method utilizes aluminum straps, 12 in. apart, with aluminum bolts and nuts, aluminum rivets or cadmium-plated steel bolts.

To prevent galvanic action, direct contact between the top of the purlin and roofing should be prevented preferably by insertion of an aluminum saddle or by painting the flange of the purlin with aluminum or bitumastic paint. Roofing paper may also be used to separate the metals.

Purlin clips (b) are also widely used as a fastening device. The U bend is slipped over the purlin flange and a leg hammered down against the web to lock the clip in place. The roofing is bolted to the clip. If standard steel purlin clips are used, they should be hot-dip galvanized.

Purlin nails, either single (c) or double (d) are also used for anchoring sheeting. A variation of this method employing aluminum washers, nuts and base plate under the purlin is shown in (e).

The best method of nailing to sheathing is to use galvanized roofing nails about 1 ¾-in. long with a head about ⅜-in. in diameter, spaced at 8-in. intervals. These nails should have a hot-dip zinc coating.

Corrugated sheets should always be riveted, bolted or nailed through the top of a corrugation. The reason is that rain running down the roof tends to collect in the bottom of corrugations and would penetrate imperfectly sealed holes there. At the top, water runs away from the holes.

A washer of non-metallic material, such as zinc chromate impregnated fabric, neoprene or other rubber should be employed under a nailhead. The purpose of this washer is to prevent metal-to-metal contact between the underside of the head and the roofing sheet, to seal the opening made in the sheet and to permit thermal expansion and contraction of the sheeting.

Hot-dip galvanized nails without washers may be used when washers are not available. Roofing nails with cast-on lead heads can be used in other than industrial or seacoast corrosive atmospheres.

Sizes, Weights, and Coverage Data on Corrugated Aluminum Sheets Used for Roofing

Lengths 5'-0" to 12'-0" in 6" increments. Widths, 35" and 48 3/8". Thicknesses, .024" and .032". Corrugation, 2.67" pitch and 7/8" deep. Coverage, 32" for 35" width and 45 3/8" for 48 3/8".

Data on .032" Corrugated Aluminum Roofing Sheets

Sheet Length (Ft.)	SF per Sheet 35"	SF per Sheet 48 3/8"	Wgt per Sheet, Lbs 35"	Wgt per Sheet, Lbs 48 3/8"	Approx. No. Sheets* per 100 Sq. Ft. 35"	Approx. No. Sheets* per 100 Sq. Ft. 48 3/8"
5	14.58	20.16	8.04	11.03	6.86	4.96
5 1/2	16.04	22.17	8.85	12.13	6.23	4.51
6	17.50	24.19	9.65	13.24	5.71	4.13
6 1/2	18.96	26.20	10.46	14.34	5.27	3.82
7	20.42	28.22	11.26	15.44	4.90	3.54
7 1/2	21.88	30.23	12.07	16.55	4.57	3.31
8	23.33	32.25	12.87	17.65	4.29	3.10
8 1/2	24.79	34.27	13.68	18.75	4.03	2.92
9	26.25	36.28	14.48	19.86	3.81	2.76
9 1/2	27.71	38.30	15.29	20.96	3.61	2.61
10	29.17	40.31	16.09	22.06	3.43	2.48
10 1/2	30.63	42.33	16.90	23.16	3.26	2.36
11	32.08	44.34	17.70	24.27	3.12	2.26
11 1/2	33.54	46.36	18.51	25.37	2.98	2.16
12	35.00	48.37	19.31	26.47	2.86	2.07

For .024" thickness use data given above except reduce weights 25 percent.
*For side and end lap allowance, add approximately 16% for 35" width and 12% for 48 3/8" width.

Sizes, Weight, and Coverage Data on Corrugated Aluminum Sheets Used for Siding

Lengths, 5'-0" to 12'-0" in 6" increments. Widths, 33 3/4" and 47 1/8". Thicknesses, .024" and .032". Corrugation, 2.67" pitch and 7/8" deep. Coverage, 32" for 33 3/4" width and 45 3/8 for 47 1/8" width.

Data on .032" Corrugated Aluminum Siding Sheets

Sheet Length in Feet	SF per Sheet 33 3/4"	SF per Sheet 47 1/8"	Weight per Sheet, Lbs 33 3/4"	Weight per Sheet, Lbs 47 1/8"	Approx. No. Sheets* per 100 Sq. Ft. 33 3/4"	Approx. No. Sheets* per 100 Sq. Ft. 47 1/8"
5	14.06	19.64	7.76	10.75	7.11	5.09
5 1/2	15.47	21.60	8.54	11.82	6.46	4.63
6	16.87	23.56	9.31	12.90	5.93	4.24
6 1/2	18.28	25.53	10.09	13.97	5.47	3.92
7	19.69	27.49	10.87	15.05	5.08	3.64
7 1/2	21.09	29.45	11.65	16.12	4.74	3.40
8	22.50	31.42	12.42	17.20	4.44	3.18
8 1/2	23.91	33.38	13.20	18.27	4.18	3.00
9	25.31	35.34	13.97	19.35	3.95	2.83
9 1/2	26.72	37.31	14.75	20.42	3.74	2.68
10	28.12	39.27	15.52	21.50	3.56	2.55
10 1/2	29.53	41.23	16.30	22.57	3.39	2.43
11	30.94	43.20	17.08	23.65	3.23	2.31
11 1/2	32.34	45.16	17.86	24.72	3.09	2.21
12	33.75	47.13	18.63	25.80	2.96	2.12

For .024" thickness use data given above, except reduce weights 25 percent.
*For side and end lap allowance, add approximately 9% for 33 3/4" width and 6% for 47 1/8" width.

Data on Nails Required for Roofing

Kind of Nail	Length Inches	Number per Lb.	Price per Lb.
Galvanized Needle Point	1 ¾"	83	$.85
Galvanized Needle Point	2"	77	.85
Aluminum Nails--Neoprene Washers	1 ¾"	318	2.90
Aluminum Nails--Neoprene Washers	2"	285	2.90

Approximate Prices Per Sq. Ft. of Corrugated Aluminum Roofing and Siding Material

Thickness	Finish	Price
.0175"	Mill	$0.40
.0175"	Painted	.45
.024"	Mill	.75
	Painted	1.00
.032"	Mill	1.00
	Painted	1.25

After figuring the price of sheets from the above, approximately 20% should be added to cover cost of closures, flashings, fillers, corners, etc.

Labor Placing Corrugated Aluminum Roofing and Siding. Estimate about the same labor cost for erecting aluminum corrugated roofing and siding as given on the previous pages under *Corrugated Steel Roofing and Siding.*

Aluminum is much lighter than steel, but the labor operations are practically the same.

Labor Cost of 100 Sq. Ft. (1 Sq.) Corrugated Aluminum Roofing on Wood Framing
Based on using sheets 35" wide, .032" thick, weighing .552 lbs. per sq. ft.

	Hours	Rate	Total	Rate	Total
Carpenter	1	$....	$....	$24.04	$24.04
Labor	1	....	....	19.13	19.13
Cost per 100 sq. ft.			$....		$43.17

Cost of 100 Sq. Ft. (1 Sq.) Corrugated Aluminum Siding on Wood Framing
Based on using sheets 33 ¾" wide, .032" thick, weighing .552 lbs. per sq. ft.

	Hours	Rate	Total	Rate	Total
Carpenter	1.2	$....	$....	$24.04	$28.85
Labor	1.2	....	....	19.13	22.96
Cost per 100 sq. ft.			$....		$51.81

Cost of 100 Sq. Ft. (1 Sq.) Corrugated Aluminum Siding on Steel Framing, Straight Work
Based on using sheets 33 ¾" wide, .032" thick, weighing .552 lbs. per sq. ft.

	Hours	Rate	Total	Rate	Total
Labor handling, hoisting	3.2	$....	$....	$19.13	$61.22
Sheeters	3.2	....	....	26.25	84.00
Cost per 100 sq. ft.			$....		$145.22

Add extra labor for high and irregular walls such as grain elevators, conveyor housing, etc. Add for closures, flashings, etc.

Cost of 100 Sq. Ft. (1 Sq.) Corrugated Aluminum Roofing on Steel Framing, Straight Work
Based on using sheets 35" wide, .032" thick, weighing .552 lbs. per sq. ft.

	Hours	Rate	Total	Rate	Total
Labor handling, hoisting	2.67	$....	$....	$19.13	$51.08
Sheeter	2.67	....	....	26.25	70.09
Cost per 100 sq. ft			$....		$121.17

Add extra labor for high or irregular roof areas. Add for closures, flashings, etc.

07460 CLADING/SIDING

Wood siding is available in plain bevel, plain shiplap, and tongue and groove patterns to be used horizontally in heights from 4" to 12". Bevel siding is either ½" or ¾" thick, T&G, and shiplap nominal 1". Cedar is the most common wood for siding but redwood, fir, hemlock, and spruce are also stocked. Wood may also be installed vertically with wood battens at the joints. Bevel siding ½" thick in clear "A" grade cedar will run about $1.25 a board foot, in redwood about $1.50. T&G red cedar 1x8s "D" and better will run $1.00 a board foot; T&G redwood, clear and better, runs $1.70.

Hardboard siding 7⁄16" thick and factory primed, costs 50 cents a square foot.

Plywood siding in 4x8 sheets, rough textured and grooved to look like individual boards ⅝" thick costs about 85 cents a square foot in select grade Douglas fir, $1.50 in redwood.

Aluminum siding, factory finished white in 8" widths, will run 75 cents uninsulated, $1.05 insulated per sq. ft. Flashing strips will cost 20 cents a foot; inside corners 55 cents; outside corners $1.10 per lin. ft. Colored nails cost $2.50 per pound.

Vinyl siding, factory finished in 8" wide strips, costs about 75 cents a square foot uninsulated, $1.05 with insulating backup. Corner boards run $1.10 for outside positions, 55 cents for inside. Trim moldings run 35 cents per lin. ft., while soffits run $1.15 per sq. ft.

Placing Bevel and Drop Siding. The labor cost of placing bevel or drop siding will vary with the class of work and the method of placing same.

On the less expensive type of construction, it is customary to square only one end of the siding, and the ends at the corners are left rough and are later covered with metal corner pieces. This is the cheapest method of placing siding. Metal corners are extensively used in the South, because the long summers and intense heat cause mitered wood corners to open up, making a very unsightly appearance.

On more expensive buildings, the door and window casings and corner boards are placed, and then it is necessary to cut and fit each piece of siding between the corner boards or between the casings and corner boards. On work of this class, two carpenters usually work together, because it is necessary to square one end of the siding and then measure and cut each board separately to insure a snug fit. This increases labor costs considerably but makes a nicer appearing job than the metal corners.

This also applies where the siding is mitered at the corners. Two carpenters working together must measure and miter each piece of siding so that it will fit the adjoining corner.

The following quantities and production time are based on using ordinary brackets with plank scaffolding. If a more elaborate scaffold is required for placing siding, add extra for same.

Labor Placing Drop Siding

Measured Size Inches	Actual Size Inches	Class of Work	Feet B.F. Placed per 8-Hr. Day	Carpenter Hours per 1000 BF
6	5 1/4	Rough Ends	525-575	14.5*
6	5 1/4	Fitted Ends	350-400	21.4
6	5 1/4	Mitered Corners	285-325	26.3
8	7 1/4	Rough Ends	600-650	12.8*
8	7 1/4	Fitted Ends	415-460	18.2
8	7 1/4	Mitered Corners	325-375	23.0

*Where an electric saw is used to square both ends of the siding before placing, leaving the exposed corners rough to be covered with metal corner pieces, deduct 1 to 1 1/2 hours time per 1,000 BF from the time given above.

Quantity of Bevel Siding Required Per 100 Sq. Ft. of Wall

Measured Size Inches	Actual Size Inches	Exposed to Weather	Pattern	Add for Lap	BF Required per SF Surface
1/2x4	1/2x3 1/4	2 3/4	Regular	46%	151
1/2x5	1/2x4 1/4	3 3/4	Regular	33%	138
1/2x6	1/2x5 1/4	4 3/4	Regular	26%	131
1/2x8	1/2x7 1/4	6 3/4	Regular	18%	123
5/8x8	5/8x7 1/4	6 3/4	Regular	18%	123
3/4x8	3/4x7 1/4	6 3/4	Rabbetted	18%	123
5/8x10	5/8x9 1/4	8 3/4	Rabbetted	14%	119
3/4x10	3/4x9 1/4	8 3/4	Rabbetted	14%	119
3/4x12	3/4x11 1/4	10 3/4	Rabbetted	12%	117

The above quantities include 5% for end cutting and waste.

Quantity of Drop Siding Required Per 100 Sq. Ft. of Wall

Measured Size Inches	Actual Size Inches	Exposed to Weather	Add for Lap	BF Required per 100 SF Surface
1x6	3/4x5 1/4	5 1/4	14%	119
1x8	3/4x7 1/4	7 1/4	10%	115

The above quantities include 5% for end cutting and waste.

Quantity of Shiplap Required Per 100 Sq. Ft. of Surface

Measured Size Inches	Actual Size Inches	Add for Lap	BF Required per 100 SF Surface
1x8	3/4x7 1/4	10%	115
1x10	3/4x9 1/4	8%	113

The above quantities include 5% for end cutting and waste.

Quantity of Bead Ceiling and Partition Required Per 100 Sq. Ft. of Surface

Measured Size Inches	Actual Size Inches	Add for Width	BF Required per 100 SF Surface
1x4	5/8x3 1/4	23%	128
1x4	3/4x3 1/4	23%	128

The above quantities include 5% for end cutting and waste.

Quantity of Dressed and Matched (D & M) or Tongued and Grooved (T & G) Boards Required Per 100 Sq. Ft. of Surface

Measured Size Inches	Actual Size Inches	Add for Width	BF Required per 100 SF Surface
1x6	¾x5 ¼	14%	119
2x6	1 ⅝x5 ¼	14%	238

The above quantities include 5% for end cutting and waste.

Labor Placing Bevel Siding

Measured Size Inches	Actual Size Inches	Exposed to Weather, In.	Class of Workmanship	BF placed per 8-Hr. Day	Carpenter Hrs. per 1000 BF
4	3 ¼	2 ¾	Rough Ends	350-400	21.3*
4	3 ¼	2 ¾	Fitted Ends	240-285	30.5
4	3 ¼	2 ¾	Mitred Corners	200-240	36.3
5	4 ¼	3 ¾	Rough Ends	415-460	18.2*
5	4 ¼	3 ¾	Fitted Ends	285-330	25
5	4 ¼	3 ¾	Mitred Corners	240-285	30.5
6	5 ¼	4 ¾	Rough Ends	475-525	16*
6	5 ¼	4 ¾	Fitted Ends	325-375	23
6	5 ¼	4 ¾	Mitred Corners	265-310	28
8	7 ¼	6 ¾	Rough Ends	570-620	13.4*
8	7 ¼	6 ¾	Fitted Ends	375-415	20
8	7 ¼	6 ¾	Mitred Corners	300-350	24
10	9 ¼	8 ¾	Rough Ends	650-700	12*
10	9 ¼	8 ¾	Fitted Ends	440-480	17.5
10	9 ¼	8 ¾	Mitred Corners	375-425	20
12	11 ¼	10 ¾	Fitted Ends	475-525	16
12	11 ¼	10 ¾	Mitred Corners	400-450	19

*Where an electric handsaw is used to square both ends of the siding before placing, leaving the exposed corners rough to be covered with metal corner pieces, deduct 1 to 1 ½ hours time per 1,000 b.f.

07463 ASBESTOS CEMENT ROOFING AND SIDING

Asbestos cement roofing and siding, like asbestos cement shingle, is no longer used. This section remains here for historical information only. The product is described so that estimators requiring information on this type of material can be familiar with its patterns of use and labor requirements.

Nowadays, the only handling of this material is when it has become friable and must be encapsulated or removed. It must be handled in strict accordance with OSHA requirements (see the *Hazard Waste* chapter).

Corrugated asbestos roofing and siding was used over skeleton frame construction. It is composed of asbestos fiber and portland cement combined under hydraulic pressure into homogeneous sheets of a light gray color, structurally strong, and relatively light in weight.

The same materials that were used in forming the corrugated sheets were molded at the factory into special standard shapes to be used as ridge roll, corner roll, louvers, and ventilators.

Standard sheets have corrugations with a 4.2" pitch and a depth of 1 ½". The thickness is approximately 7/16" at ridge and valley of corrugations and approximately 5/16" on tangent, an average thickness of ⅜". Sheets were furnished 42", or ten corrugations, wide. Standard lengths are 0'-6" to 12'-0" in 6" increments. Special curved sheets were furnished with the following limitations: sheets could be curved longitudinally, the arc being formed along the direction of the corrugation; minimum radius when so curved 60". Sheets could be curved crosswise, the arc being formed across the corrugations, minimum radius when so curved 24". Sheets could be curved either way but not in combination. The sheets were furnished in three types, as illustrated. The cut corner construction eliminates a thickness of material in the corner lap, thus allowing the sheets to nest perfectly at these points and permitting them to be laid with side laps forming vertical lines. The inside and outside radii of the corrugations are the same, which allowed the sheets to nest perfectly at all laps.

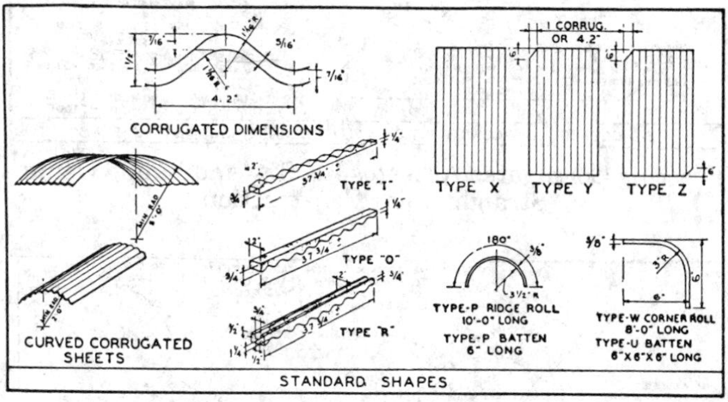

Types of Corrugated Asbestos Roofing and Siding

Corrugated asbestos roofing was not be laid on roofs having a pitch of less than 3" per foot; a pitch of 4" or more was preferred. Plastic cement or caulking was used in all laps.

Supporting purlins or girts were to be spaced not to exceed 54" center to center for roofing, and 66" for siding, depending upon the design loading.

Method of Applying Corrugated Roofing and Siding

Corrugated asbestos roofing and siding was applied directly over purlins and girts of skeleton frame construction or applied over solid wood sheathing.

When used as either roofing or siding, all sheets were to be lapped one corrugation at the sides and 6" at the ends. The size of the end lap was governed by the cut corner construction. The width of weather exposure of sheets for estimating purposes was 37.8".

Sheets were to be of proper length so that all end laps would occur over a purlin or girt and so that fasteners at ends of sheets would pass through both upper and underlying sheets.

Fasteners especially designed for use in connection with this material could be furnished by the manufacturer. The style of fastener to be used would be governed by the type, shape and position of purlins and girts. Sheets were to be secured at all purlins and girts, spacing fasteners as follows:

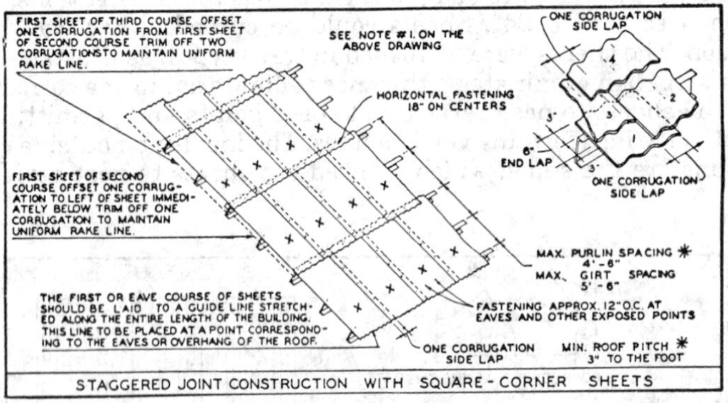

Erecting Corrugated Asbestos Roofing and Siding Using Straight Joint Construction

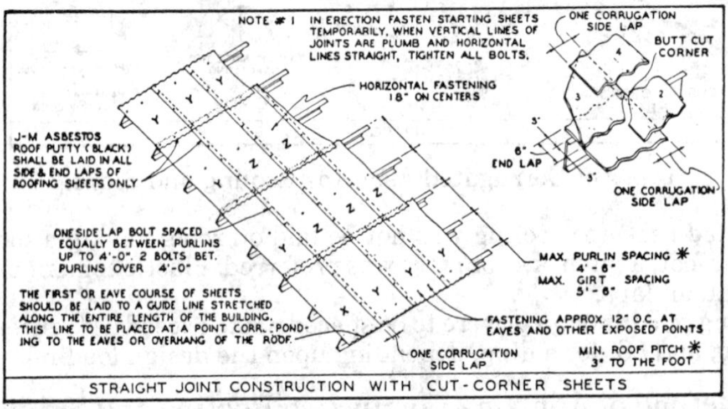

Erecting Corrugated Asbestos Roofing and Siding Using Staggered Joint Construction

The percentage of the area to be covered that is absorbed by the end and side laps of the corrugated asbestos sheets would vary with the length of that area and the number of courses of sheets required. A rough general layout of sheets would be made and area of laps determined before any estimating figures made.

Area of 10 side laps	(8'-6" + 4'-0").35' =	43.75 sq. ft.
Area of 1 end lap	(35'-0" x .5) =	17.5 sq. ft.
. . . .		61.25 sq. ft.
Area to be covered	12'-0" x 35'-0" =	420.00 sq. ft.
Sq. ft. of corrugated asbestos roofing or siding required		481.25 sq. ft.

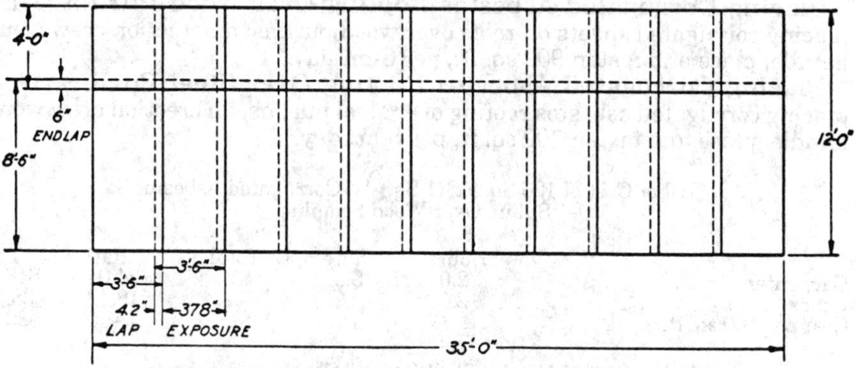

Representative Diagram of Roofing or Siding

Approximate Prices of Corrugated Asbestos Roofing and Siding

Weight per SF	Price per 100 SF Roof/Siding Area
4.1 lb.	$200.00

Prices of Ridge, Corner Roll and Louver Blades

	Approx. Wt. Lin. Ft.	Price per Lin. Ft.
Ridge and corner roll, including battens and fasteners	4 lbs.	$1.75
Louver blades	3.3 lbs.	.60

Fasteners for Asbestos Roofing and Siding

	Approx. Wt. 100 Sq. Ft.	Price per 100 Sq. Ft.
Lead headed fasteners (galv. iron) for use over wood framing	4 lbs.	$3.20
Lead headed fasteners (galv. iron) for use over steel framing	6 lbs.	4.80

Labor Handling and Placing Corrugated Asbestos Roofing and Siding. The labor cost of handling and placing asbestos siding and roofing varied according to the class of work, whether placed over wood or steel purlins or girts, size of sheets used, and conditions under which the work is performed.

The most economical organization for applying asbestos roofing and siding consisted of 2 mechanics and 2 laborers.

The labor costs would include all time necessary for cementing joints, and placing ridge roll, corner roll, louver blades, etc. Add for scaffolding, if necessary.

Placing Corrugated Asbestos Siding Over Wood Framing. When placing corrugated asbestos siding on wood framing, a crew would handle, place and fasten 800 sq. ft. per 8-hr day.

Placing Corrugated Asbestos Siding on Steel Framing. When placing corrugated asbestos siding on steel framing, a crew would handle, place and fasten 600 sq. ft., per 8-hr. day.

Placing Corrugated Asbestos Roofing Over Wood Purlins. When placing corrugated sheets on roofs over wood purlins, an erection crew would handle, place and fasten 900 sq. ft., per 8-hr. day.

Placing Corrugated Asbestos Roofing Over Steel Purlins. When placing corrugated asbestos roofing over steel purlins, an erection crew would handle, place and fasten 700 sq. ft. per 8-hr. day.

Labor Cost of 100 Sq. Ft. (1 Sq.) 3/8" Corrugated Asbestos Siding Over Wood Framing

	Hours	Rate	Total	Rate	Total
Carpenter	2.0	\$....	\$....	\$24.04	\$48.08
Labor	2.0	....	....	19.13	38.26
Cost per 100 sq. ft.			\$....		\$86.34

Labor Cost of 100 Sq. Ft. (1 Sq.) 3/8" Corrugated Asbestos Siding Over Steel Framing

	Hours	Rate	Total	Rate	Total
Sht. Met. wkr.	2.7	\$....	\$....	\$26.25	\$70.88
Labor	2.7	....	....	19.13	51.65
Cost per 100 sq. ft.			\$....		\$122.53

Labor Cost of 100 Sq. Ft. (1 Sq.) 3/8" Corrugated Asbestos Roofing Over Wood Framing

	Hours	Rate	Total	Rate	Total
Carpenter	1.8	\$....	\$....	\$24.04	\$43.27
Labor	1.8	....	....	19.13	34.43
Cost per 100 sq. ft.			\$....		\$77.70

Labor Cost of 100 Sq. Ft. (1 Sq.) 3/8" Corrugated Asbestos Roofing Over Steel Framing

	Hours	Rate	Total	Rate	Total
Sht. met. wkr.	2.3	\$....	\$....	\$26.25	\$60.38
Labor	2.3	....	....	19.13	44.00
Cost per 100 sq. ft.			\$....		\$104.38

Asbestos Cement Insulating Panels. This is an insulating unit consisting of a core of insulating board surfaced on both sides with asbestos cement board using a moisture-proof bituminous adhesive as a bond between the asbestos cement and core material. It is found in all types of buildings in exterior walls, roof decks, partitions, and as a lining material. It might be applied to steel or wood framing using bolts and clips, screws, or nails. Nailing would not fracture the surface if reasonable care were used in driving.

Panels are in sheets 4'-0" wide and from 4'-0" to 12'-0" long, in 2'-0" increments. Panel thickness varies depending on the insulating board core.

Wood fiber core	11/16", 1 1/8", 1 9/16", 2"
Polystyrene Core	1", 1 1/8", 1 9/16", 2"
Perlite Core	1", 1 1/8", 1 9/16", 2"

Material costs increase with the thickness and the type of insulating core used. Wood fiber core 11/16" thick costs about \$1.70 per sq. ft., wood fiber core 2" thick costs about \$2.75 per sq. ft., polystyrene core 2" thick costs about \$2.85 per sq. ft., and Perlite core 2" thick will cost about \$3.25 per sq. ft.

Labor Installing Insulating Board. When this material was used, it required more labor time than the corrugated sheets because of the increased weight per sq. ft. and the fastening time (this material was laid with butt joints, doubling the number of fasteners required around the edges). Using the same size crew as on the corrugated panels, about 40% more time was needed for the corrugated panel installation time, as previously given, to install the insulating board sandwich panels.

Labor Installing Insulating Board as Roof Decking. When 1 9/16" or 2" board was used for roof decking over wood framing, 2 carpenters and 2 laborers together could place 600 to 650 sq. ft. of decking per 8-hr. day at the following labor cost per 100 sq. ft.:

	Hours	Rate	Total	Rate	Total
Carpenter	2.5	$....	$....	$24.04	$60.10
Labor	2.5	....	....	19.13	47.83
Cost 100 sq. ft.			$....		$107.93
Cost per sq. ft.			....		1.08

Add for caulking joints with plastic roofing cement and covering joints with 6" wide strips of saturated fabric or 30-lb. roofing felt.

Labor Installing Asbestos Insulating Board as Siding. When 1 1/8", 1 9/16", or 2" board was applied as curtain walls on wood framing, 2 carpenters and 2 laborers together would place 550 sq. ft. of curtain wall per 8-hr. day at the following labor cost per 100 sq. ft.:

	Hours	Rate	Total	Rate	Total
Carpenter	2.8	$....	$....	$24.04	$67.31
Labor	2.8	....	....	19.13	53.56
Cost 100 sq. ft.			$....		$120.87
Cost per sq. ft.			....		1.21

Add for caulking joints and covering joints with batten strips or metal plates.

Labor Installing Asbestos Insulating Board as Partitions. When used for partitions, several methods of framing were employed. However, very little difference in labor placing the panels would be experienced.

In general, floor and ceiling plates, either rabbeted or built up to form a retaining slot, was installed and panels set in place. Joints between panels would be covered with battens, concealed with metal flush panel mouldings or splined together. In some instances 2"x4" studs and plates were used, the studs being spaced to allow one full panel to be installed between them and stops being used to hold the panels in place.

After framing is in place, 2 carpenters an 1 laborer working together would place 1,024 sq. ft. of partition panels per 8-hr. day at the following labor cost per 100 sq. ft.:

	Hours	Rate	Total	Rate	Total
Carpenter	1.6	$....	$....	$24.04	$38.46
Labor	0.8	....	....	19.13	15.30
Cost 100 sq. ft.			$....		$53.76
Cost per sq. ft.			....		.54

07464 PLASTIC SIDING

Fiberglass panels are made to most of the usual configurations and widths and in lengths up to 30', although stock lengths are in the 3' to 12' range.

Fiberglass panels may be used to enclose an entire building, walls and roof, or in combination with similarly configured panels of steel, aluminum, protected metal or asbestos. The standard panels are translucent and can be inserted as skylights or windows. They can be either clear or colored, smooth or embossed. Fiberglass has an ignition point of 850 to 900°F. Where greater fire protection is required, they may be ordered to be rated "fire retardant" and cost about 75 cents per square foot extra. The panels can be further treated to resist erosion or corrosion, or reinforced to be resistant to breakage. The weight of the sheet determines the cost. Standard sheets weighing 4 oz. per sq. ft. will cost around 80 cents per sq. ft.; 5 oz., 90 cents; 6 oz., $1.10; and 8 oz., $1.35. The material is light, and one carpenter should set about 100 sq. ft. per hour at the following cost per 100 sq. ft.:

	Hours	Rate	total	Rate	Total
Carpenter	1.0	$....	$....	$24.04	$24.04
Cost per sq. ft.			....		.24

07500 MEMBRANE ROOFING

Built-up roofing consists of alternate plies of saturated felt and moppings of pitch with tar-saturated felt, or asphalt with asphalt-saturated felt, covered with a top pouring of pitch or asphalt into which slag or gravel is embedded. On flat roofs with slopes of less than ¼" per foot, on which water may collect and stand, coal tar pitch and felt or a low melting point asphalt bitumen and asphalt felt are generally used. Coal tar pitch is not recommended for roofs having an incline in excess of 1" per foot.

On built-up roofs, where slope is over 2" per foot up to 4" per foot, steep asphalt (180°-200°F melting point) is more suitably used and slag is embedded in preference to gravel, because it remains embedded better than well rounded gravel. Where slag is not available and where a light gray or white surface is desired, hard limestone chips, angular pieces ¼" to ¾" in size, are embedded in the top pouring of bitumen. On slopes over 2" per foot, double coverage mineral surfaced roll roofing with a 19" selvage edge may be used as the top finish.

Smooth surface built-up roofing consists of alternate plies of asbestos felt cemented solid to the base sheet and to each other with asphalt, and a top coating of hot or cold asphalt, uniformly distributed.

The specifications on the various types of roofing vary widely, depending upon the surface to which the built-up roofing is applied and the service required.

The cost of built-up roofing is governed by the incline of the roof, size, plan (whether cut up with openings, skylights, penthouses, irregular roof levels, etc.), and the distance of the roof above the ground; the higher the roof, the higher materials must be hoisted, which increases the labor costs.

The term, "built-up roof" will be commonly heard in the trade, but it is more properly "built-up roofing". Technically, the "roof" is the supporting structure over which the "roofing" is applied. Although the term "built-up roof" may be loosely used in the trade, and seems to flow more easily in conversation, the specification writer, designer or their attorneys will be careful to distinguish between "roof" and "roofing".

Rules for Measuring Flat Roof. When measuring flat roof surfaces that are to be covered with composition, tar and gravel, tin, metal, or prepared roofing, the measurements should be taken from the outside of the walls on all four sides to allow for flashing up the side of each wall. The flashing is usually 8" to 1'-0" high. This applies particularly to brick, stone or tile buildings having parapet walls above the roof level.

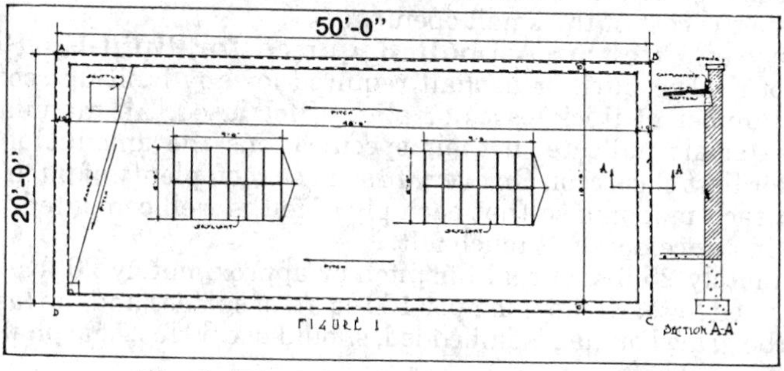

Fig. 1. Flat Roof with Parapet Walls

On flat roofs projecting or overhanging beyond the walls of the building, the measurements should cover the outside dimensions of the roof and not merely to the outside of the walls.

When estimating the area of flat roof surfaces, do not make deductions for openings containing less than 100 sq. ft. and then deductions should be made for just one-half the size of the opening.

Make deductions in full for all openings having an area of 500 sq. ft. or more.

Example: The length of the building (as shown in Figure 1), is 50'-0", which is the distance from A to B. The width of the building from B to C is 20'-0". Multiplying the length by the width, 50 x 20 = 1,000 sq. ft. Referring to the figure, note that the building is only 48'-0" long and 18'-8" wide between walls. The method of measuring the full length and width of the building is to allow for flashing up the side of the walls, which usually extends 8" to 12" high.

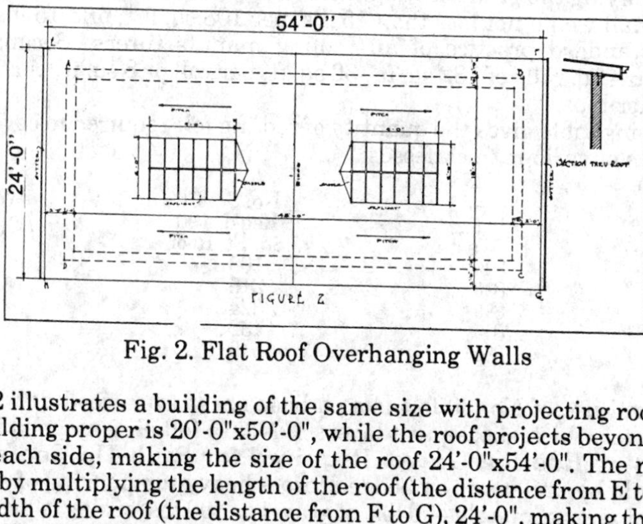

Fig. 2. Flat Roof Overhanging Walls

Figure 2 illustrates a building of the same size with projecting roof. The size of the building proper is 20'-0"x50'-0", while the roof projects beyond the walls 2'-0" on each side, making the size of the roof 24'-0"x54'-0". The roof area is obtained by multiplying the length of the roof (the distance from E to F), 54'-0", by the width of the roof (the distance from F to G), 24'-0", making the total roof area, 54'-0" x 24'-0" = 1,296 sq. ft.

The skylights on both roofs measure 6'-0"x9'-0" and contain only 54 sq. ft. each, so no deduction should be made for these openings, as they contain less than 100 sq. ft. each and the extra labor flashing up the sides of the walls more than offsets the cost of the small openings.

Quantity of Pitch or Asphalt Required for Built-Up Roofs. The quantity of roofing pitch or asphalt required for any built-up roof will vary with the number of thicknesses or plies of felt used. All manufacturers of roofing materials indicate in their specifications the amount of materials required for first class roof. On every first class roof, plenty of pitch or asphalt is used in each mopping so that each ply of felt is well cemented to the next and in no instance does felt touch felt.

Approximately 25 lbs. of coal tar pitch or approximately 20 lbs. of asphalt should be used for each mopping per 100 sq. ft. of surface and the last pouring, in which the gravel or slag is imbedded, should use 60 lbs. of asphalt or 75 lbs. of pitch per 100 sq. ft.

Quantity of Asphalt Required For 100 Sq. Ft of Roof

Surface to Which Roof is Applied	No. of Plies	Dry	Mopped	Lbs. per 100 SF
Wood, plywood, structural wood fiber	5	2	3	120
Poured gypsum, lightweight concrete	4	1	3	120
Concrete, precast concrete or gypsum	4		4	140

For roofing pitch add approximately 5 lbs. per mopping each ply per 100 sq. ft.--add 15 lbs. for top pouring per 100 sq. ft.

Estimating the Quantity of Roofing Felt. Asphalt or tarred felt for built-up roofing is furnished in 4 sq. rolls containing 432 sq. ft. and weighing from 56 to 62 lbs. per roll for No. 15 felt. No. 30 felt, which is generally used as the base sheet over wood construction when an asphalt specification is used, is furnished in 2 sq. rolls containing 216 sq. ft. and weighing 60 lbs. These weights are those used on a bonded job in the best type of building construction.

When specifying the grade or weight of felt to be used it is customary to state that "felt shall weigh not less than 15 lbs. per 108 sq. ft.", and 15 lbs. per 108 sq. ft. is standard practice of all roofing manufacturers. Because felt is furnished in 4 sq. rolls of 432 sq. ft., 32 sq. ft. per roll or 8 sq. ft. per 100 sq. ft. is allowed for laps.

The following table gives the quantity of roofing felt required to cover 100 sq. ft. of surface of various thicknesses:

Number of Plies	Add for Waste	No. Sq. Ft. Req'd. 100 Sq. Ft. Roof	Weight per Sq. of Roof Using No. 15 Felt
1	8%	108	15
2	8%	216	30
3	8%	324	45
4	8%	432	60
5	8%	540	75

The general use of No. 30 felt is as a base sheet over wood decks in 1-ply thickness over which subsequent layers of No. 15 felts are mopped in.

Quantity of Roofing Gravel Required for Built-Up Roofs. Roofing gravel should be uniformly embedded into a heavy top pouring of asphalt or pitch so that approximately 400 lbs. of gravel or 300 lbs. of slag is used per 100 sq. ft. of roof area.

Labor Applying Built-Up Roofing. The labor cost of handling materials and applying built-up roofing will vary with the specification used and the type of building to which the roofing is applied. On the low type of building, 1, 2 or 3 stories high, where there is no great distance from the ground to the roof deck, the labor for hoisting materials, and supplying hot asphalt or pitch from the ground to the roof deck is considerably less than on a high building, where there is a greater distance from the kettle to the roof.

On low type buildings, a 5-worker crew composed of 1 worker at the kettle, 1 worker carrying "hot stuff", 1 worker rolling in felts, 1 worker nailing in and 1 worker mopping in, can lay the following areas per 8-hr. day:

18 sqs. of 5-ply asphalt and gravel or pitch and gravel over wood roof deck, consisting of 1 sheathing paper, 2 dry felts and 3 plies of felt mopped in with a top pouring of pitch or asphalt and gravel surfacing.

18 sqs. of 4-ply asphalt and gravel or pitch and gravel over a concrete deck, all mopped including a top pouring of pitch or asphalt and gravel or slag surfacing.

22 sqs. of 4-ply asphalt and gravel or pitch and gravel over wood roof deck, consisting of 1 sheathing paper, 2 dry felts and 2 plies of felt mopped in with a top pouring of pitch or asphalt and gravel or slag surfacing.

22 sqs. of 3-ply asphalt and gravel or pitch and gravel over concrete roof deck, all mopped and including a top pouring of pitch or asphalt and gravel or slag surfacing.

24 sqs. of 3-ply asphalt and gravel, or pitch and gravel, over wood roof deck, consisting of 1 dry felt, and 2 plies of felt mopped in, with a top pouring of pitch or asphalt, and gravel or slag surfacing.

The above figures apply to buildings up to 3 stories in height.

If the roof surfaces are broken up with skylights, irregular roof level, etc., the above quantities should be reduced about 15 to 20 per cent.

A 5-worker crew should apply the following number of squares of roof per 8-hr. day, based on first class workmanship throughout:

Type of Roofing	Type of Structure "A"	"B"
3-ply roof over wood roof deck	24	18
3-ply roof over concrete roof deck or insulation	22	17
4-ply roof over wood roof deck	22	17
4-ply roof over concrete roof deck or insulation	18	14
5-ply roof over wood roof deck	18	14

Type "A" buildings are the most convenient type of structure for application of built-up roofing. A low type of building from 1 to 3 stories in height, straight, practically flat area not broken up by very many skylights or variations in the deck elevation.

Type "B" buildings are not convenient type of structures for the application of built-up roofing. These would include high buildings which require considerable handling of materials, and buildings with roofs broken up by skylights or penthouses, sawtooth construction, monitors, or considerable variation in deck levels.

An easy and accurate method of computing the labor cost on any type of roof is to take the cost of a 5-worker crew for an 8-hr. day, and divide this by the number of squares of roof applied per day, as follows:

	Hours	Rate	Total	Rate	Total
Foreman (working)	8	$....	$....	$24.05	$192.40
Roofers (4)	32	....	....	22.72	727.04
Crew cost per 8-hr. day			$....		$919.44
Cost per sq. 14 sqs. per day			....		65.67
Cost per sq. 17 sqs. per day			....		54.08
Cost per sq. 18 sqs. per day			....		51.08
Cost per sq. 22 sqs. per day			....		41.79
Cost per sq. 24 sqs. per day			....		38.31

On the average roof where the mopping between layers consists of mopping the width of the lap, a roofer should mop 100 sq. ft. of surface in 6 to 7 minutes.

On first grade work where the entire roof surface is mopped between each layer of felt, a man should mop 100 sq. ft. of roof in about 10 minutes.

After the roof surface has been mopped, a roofer should place roofing felt over 100 sq. ft. in about 6 minutes.

Where first grade workmanship is required, a roofer should handle pitch and gravel, pour hot pitch or asphalt and spread gravel over one square of roof in 35 to 40 minutes.

Prices on Roofing Materials. Prices on roofing materials such as coal tar pitch, asphalt, felt, etc., vary according to market fluctuations, location of job, etc. Current material prices should always be obtained, before submitting bid.

Material Cost of 100 Sq. Ft. (1 Sq.) 3-Ply Tar and Gravel Built-Up Roof Applied Over Wood Roof Deck

Maximum incline 2-in. per ft. One ply No. 30 felt, 2 plies No. 15 felt (1 dry, nailed and 2 mopped), coal tar pitch surfaced with gravel or slag.

	Rate	Total	Rate	Total
1.1 sqs. No. 30 felt	$....	$....	$6.00	$6.60
2.2 sqs. No. 15 felt	....	....	3.00	6.60
125 lbs. coal tar pitch	....	....	.13	16.25
400 lbs. roofing gravel	....	....	.50	2.00
Nails, fuel, mops, etc.	....	....	2.50	2.50
Cost per 100 sq. ft.		$....		$33.95

Material Cost of 100 Sq. Ft. (1 Sq.) 3-Ply Tar and Gravel Built-Up Roof Over Poured Concrete, Poured Gypsum Roof Decks, or Insulation Board

Maximum incline 2-in. per ft. Roof consisting of 3 plies of felt, coal tar pitch surfaced with gravel or slag.

	Rate	Total	Rate	Total
3.3 sqs. No. 15 felt	$....	$....	$3.00	$9.90
150 lbs. coal tar pitch	....	....	.13	19.50
400 lbs. roofing gravel	....	....	.50	2.00
Fuel, mops, etc.	....	....	3.00	3.00
Cost per 100 sq. ft.		$....		$34.40

Material Cost of 100 Sq. Ft. (1 Sq.) of Each Additional Ply of No. 15 Felt and Hot Mopping of Pitch

	Rate	Total	Rate	Total
1.1 Sqs. No. 15 Felt	$....	$....	$3.00	$3.30
25 lbs. Pitch	....	....	.13	3.25
Fuel, Mops, etc.	....	....	.50	.50
Cost per 100 sq. ft.		$....		$7.05

07520 PREPARED ROLL ROOFING

Ready roofing can be used on small buildings where an inexpensive, yet satisfactory, roofing is required. It is furnished in various grades and weights and is sold by the roll or in flat sheets packed with the nails and cement necessary for application.

Laying Ready-to-Lay Roll Roofing. When laying ready-to-lay roll roofing, a roofer should handle, lay and cement or nail in place about 800 sq. ft. per 8-hr. day, at the following cost per 100 sq. ft. (1 sq.):

	Hours	Rate	Total	Rate	Total
Carpenter	1.0	$....	$....	$24.04	$24.04

Asphalt Slate Surfaced Ready-to-Lay Roofing

A ready-to-lay roofing consisting of several plies of asphalt impregnated felt, surfaced with a thick layer of slate granules, either red or green. For roofs having a pitch of 1 1/2" or more per foot.

Furnished in rolls 36" wide and 36'-0" long, containing 108 sq. ft., sufficient for 100 sq. ft. of roof, at $13.00 per 90-lb. roll.

Smooth Surfaced Ready-to-Lay Roofing

Mica surface asphalt roofing consisting of good quality felt saturated with pure asphalt and both sides covered with flake mica. Rolls 36" wide and 36'-0" long.

Description	Weight per roll	No. Sq. Ft. in Roll	Price per Roll
Mica surfaced asphalt roofing	55 lbs.	108	$11.50
Mica surfaced asphalt roofing	65 lbs.	108	13.25

7530-40 ELASTIC SHEET AND FLUID APPLIED ROOFING

These types of elastomeric coatings include neoprene, Hypalon, urethane, butyl, and silicone. Some are applied by spraying, others in sheet form, and all are subcontracted to firms licensed by the manufacturer.

Since these coatings are thin, the roof surface to which they are to be applied must be firm, continuous, smooth, clean, and dry. New concrete should be sealed with a primer.

Because of their thinness and ability to conform to any shape and the fact that they can be had in most any colors including white they are often chosen as the roofing surface for decorative and fluid roof forms. Some materials are also used as traffic decks.

A 1/16" thick butyl sheet will cost around 65 cents a sq. ft. and one roofer can install some 225 sq. ft. per day for a total cost of around $1.25 a sq. ft.

A 1/16" neoprene sheet will have a material cost of some $1.25 a sq. ft. and will install at the same rate as the butyl for an installed price of around $1.80 a sq. ft.

Fluid applied Hypalon-neoprene .02" thick will be applied at the rate of some 100-110 sq. ft. per day and have a material cost of 85 cents for a total figure of around $2.00 a sq. ft.

07600 FLASHING AND SHEET METAL

07610 SHEET METAL ROOFING

Metal roofing includes galvanized steel, copper, lead, stainless steel and aluminum plus the many combinations and alloys of these metals such as lead-coated copper, terne (80% lead, 20% tin over copper-bearing carbon steel) microzinc and terne coated stainless. Terne and aluminum are the least expensive and copper and lead the costliest. Metal prices tend to fluctuate broadly as many of the ores are imported and at the mercy of the value of the dollar and the political climate of the country they are mined in.

Metal roofs may be applied in many ways. The simplest is the flat seam roof which may be used on slopes as low as ¼" to the foot. Standing seams and battened seams generally need a slope of at least 2 ½". They are more decorative, and in fact batten designs are often selected solely on their decorative value. Some materials may be ordered with prefabricated battens, others are formed in the traditional way over wood strips. Another decorative roof is the "Bermuda" type, where the metal is applied over wood "steps" provided by the carpenter. The step is based on the width of the metal roll to be used and is sloped at least 2 ½" to the foot. The roofer interlocks the rolls at each step edge giving a sharp shadowline.

For custom work terne and copper, with or without special coatings, are the usual choices. Many batten roofs today have the batten stamped into the metal and are usually factory finished aluminum or steel.

When estimating quantities of metal roofing, the measurements are taken in the same manner as for other types of roofs.

Where the metal roof is applied over flat roofs having parapet walls, an allowance of 1'-0" on each side and end of the wall should be made for flashing up the sides of the walls.

If metal roofing is applied to a flat surface or pitch roof with overhanging eaves, the measurement is taken to cover the entire roof surface.

Do not make deductions for openings containing less than 25 sq. ft. as the extra flashing costs as much as the roofing omitted.

Deduct in full for all openings containing over 50 sq. ft. and add for flashing.

Flat Seam Metal Roofing. The common sizes of tin plates are 10"x14" or multiples of that size. The sizes generally used for roofing are 14"x20" and 20"x28". The larger sizes are more economical to lay, and for flat roofs the 20"x28" size is preferable.

For a flat seam roof, the edges of the sheets are turned in about ½", locked together and well soaked with solder. The sheets should be fastened to the roof boards by cleats locked in the seams. These cleats are usually spaced about 8" apart and 2 nails are used in each cleat.

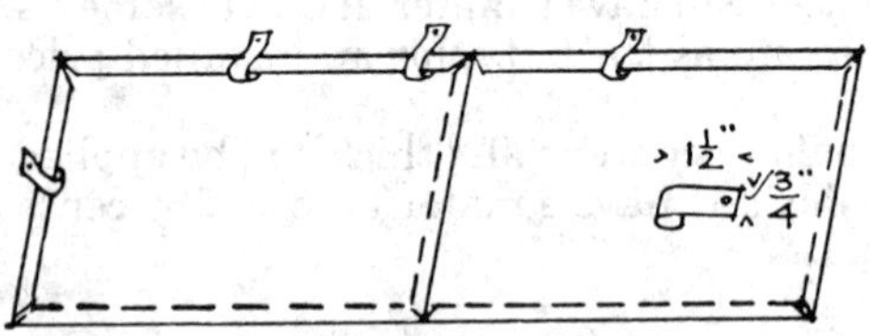

Method of Laying Flat Seam Metal Roofing

For flat seam roofing, a 14"x20" sheet of tin, with edges turned up ½" on each side and end, measures 13"x19" and contains 247 sq. in.; but the covering capacity when locked to other sheets is only 12 ½"x18 ½" or 231 ¼ sq. in. It requires 62 ½ sheets of 14"x20" tin to cover 100 sq. ft. of roof.

Sheets measuring 20"x28" with the edges turned for flat seam roofing measure 19"x27", and when locked to other sheets, have a covering capacity of 18 ½"x26 ½" or 490 ¼ sq. in. It requires 20 ½ sheets to cover 100 sq. ft. of roof.

Tin for roofs may also be obtained in rolls 14", 20", 24", and 28" wide. When used for flat seam roofing the loss due to turning edges amounts to 1 ½" on the width of any sheet.

Tin in 14" rolls loses 1 ½" or 11.2 percent of its width due to turning edges and laps. Another method is to take the length of the roof and divide by the net width of each sheet and the result will be the number of strips of tin required. Multiply the number of strips by the length of each strip to obtain the quantity of tin required for any roof.

The following table gives the covering capacity and allowances for edging and laps to be figured when using tin roofing in rolls for flat seam roofs:

Width of Sheets Inches	Allowance for Edge and Laps	Actual Width Inches	Add for Loss Due to Edges and Laps	No. SF Required Per Sq.
14	1 ½	12 ½	12.0%	112
20	1 ½	18 ½	8.0%	108
24	1 ½	22 ½	7.0%	107
28	1 ½	26 ½	6.0%	106

The quantities of galvanized iron or steel sheets required for flat seam roofing will depend upon the size of the sheets, as they may be obtained in sheets from 24"x96" to 48"x120" in size.

The actual measurements of the sheets will be 1 ½" less in width and length than the size of the sheet used. For instance, the actual covering capacity of a 24"x96" sheet will be only 22 ½" x 94 = 2,115 sq. in. or 14.76 sq. ft.

The following table gives the approximate covering capacity of one sheet of the different sizes, allowing 1 ½" in width and length for turning edge and lapping, together with the number of sheets of the different sizes required to cover one sq. (100 sq. ft.), of surface:

Size of Sheet, Inches	Allowance for Edge of Lap Length	Width	Actual Size Sheet, In.	Cover Cap. One Sheet	No. Sheets Per Sq.
24x96	1 ½	1 ½	22 ½x94 ½	14.76	6.8
24x120	1 ½	1 ½	22 ½x118 ½	18.51	5.4
26x96	1 ½	1 ½	24 ½x94 ½	16.08	6.2
26x120	1 ½	1 ½	24 ½x118 ½	20.16	5.0
28x72	1 ½	1 ½	26 ½x70 ½	13.00	7.7
28x84	1 ½	1 ½	26 ½x82 ½	15.18	6.6
28x96	1 ½	1 ½	26 ½x94 ½	17.40	5.8
28x108	1 ½	1 ½	26 ½x106 ½	19.60	5.1
28x120	1 ½	1 ½	26 ½x118 ½	21.80	4.6
30x96	1 ½	1 ½	28 ½x94 ½	18.70	5.4
30x120	1 ½	1 ½	28 ½x118 ½	23.45	4.3

Standing Seam Metal Roofing. Standing seam roofing requires a larger allowance for waste than flat seam roofing. The standing seam, edged 1 ¼" and 1 ½" takes 2 ¾" off the width and the flat cross seams edged ⅜", take 1 ⅛" off the length.

If 14"x20" tin is used, each sheet covers 11 ¼"x18 ⅞" or 212.34 sq. in. It requires 68 sheets of 14"x20" tin per 100 sq. ft. of roof.

If 20"x28" tin is used, each sheet covers 17 ¼x26 ⅞" or 463.59 sq. in. It requires 32 sheets of 20"x28" tin per 100 sq. ft.

If roll tin is used, 2 ¾" should be allowed for the two standing seams. The end waste will vary with the length of the sheets.

Tin in 14" rolls loses 2 ¾" off the width due to the standing seams, making it necessary to add 20 percent for waste. Another method is to take the width of the roof and divide by 11 ¼" (the net width of the sheet), and the result is the number of strips of tin required. Multiply the number of strips by the length of the roof to obtain the quantity of tin required.

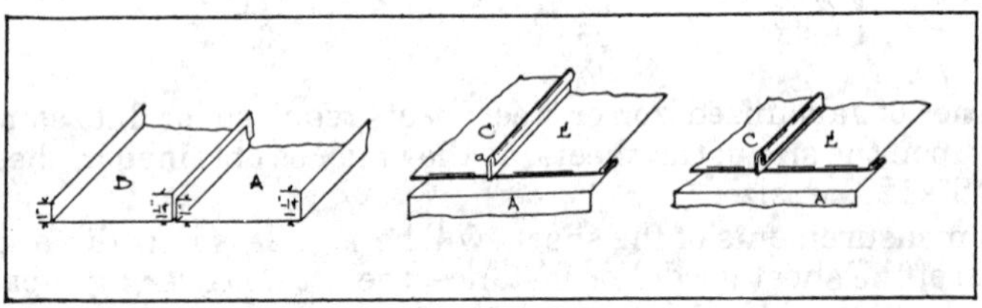

Method of Laying Standing Seam Metal Roofing

The following table gives the covering capacity and allowances for standing seams, when estimating roll roofing:

Width of Sheets Inches	Allowance for Standing Seams	Actual Width Inches	Add for Loss Due to Standing Seams	No. Sq. Ft. Required Per Square
14	2 ¾	11 ¼	25%	125
20	2 ¾	17 ¼	16%	116
24	2 ¾	21 ¼	13%	113
28	2 ¾	25 ¼	11%	111

If galvanized steel or any of the special process metals, such as Armco, Toncan, etc., are used for standing seam roofs, an allowance of 1 ¼" and 1 ½" or a total of 2 ¾" should be taken from the width of the sheets to allow for standing seams. The cross seams edged ⅜" take 1 ⅛" off the length of each sheet. For instance, where 24"x96" sheets are used, the actual size of the sheets is 21 ¼"x94 ⅞", or each sheet will cover 2,016 sq. in. or 14 sq. ft. The following table gives the covering capacity of painted or galvanized steel sheets of the different sizes:

Size of Sheet Inches	Standing Seam & End Lap Inches Width	Length	Actual Size Cover Cap. Sheet Inches	No. Sheets One Sheet Sq. Ft.	Per Square (100 SF)
24x96	2 3/4	1 1/8	21 1/4x94 7/8	14.00	7.2
24x120	2 3/4	1 1/8	21 1/4x118 7/8	17.54	5.7
26x96	2 3/4	1 1/8	23 1/4x94 7/8	15.33	6.6
26x120	2 3/4	1 1/8	23 1/4x118 7/8	19.20	5.3
30x96	2 3/4	1 1/8	27 1/4x94 7/8	17.92	5.6
30x120	2 3/4	1 1/8	27 1/4x118 7/8	22.50	4.5
36x96	2 3/4	1 1/8	33 1/4x94 7/8	22.00	4.6
36x120	2 3/4	1 1/8	33 1/4x118 7/8	27.45	3.7
42x96	2 3/4	1 1/8	39 1/4x94 7/8	25.85	3.9
42x120	2 3/4	1 1/8	39 1/4x118 7/8	32.40	3.1
48x96	2 3/4	1 1/8	45 1/4x94 7/8	29.80	3.4
48x120	2 3/4	1 1/8	45 1/4x118 7/8	37.40	2.7

V-Crimped Roofing. This type of roofing is used the same as standing seam roofing, the V-crimp answering the same purpose as the standing seam. It is usually furnished in sheets covering 24" in width and 6'-0" to 10'-0" long.

When estimating quantities of V-crimped roofing, allow for the end lap but there is no waste in the width as only the actual covering capacity is charged for by the manufacturers. For instance, a sheet 26" wide before crimping is 24" wide after crimping, but the sheet is sold as 24" wide.

The following table gives the quantity of V-crimp roofing required to cover 100 sq. ft. of roof with end laps 1" to 6":

End Laps in Inches

Length of Sheet Feet	1	2	3	4	5	6
	Square Feet of V-Crimp Roofing Required					
6	102	103	105	106	108	109
7	102	103	104	105	106	108
8	101	102	103	104	106	107
9	101	102	103	104	105	105
10	101	102	103	104	105	105

Labor Placing 100 Sq. Ft. Flat and Standing Seam Tin and Metal Roofing

Description of Roof	SF per 8-Hr. Day	Hours per 100 SF Tinner	Helper
Flat seam metal roofing, using 24"x96" sheets or larger	300-350	2.5	2.5
Flat seam tin roofing, using 14"x20" plates	175-225	4.0	4.0
Flat seam tin roofing, using 20"x28" plates	225-275	3.3	3.3
Flat seam tin roofing, using 20" tin in rolls	325-375	2.3	2.3
Flat seam tin roofing, using 14" tin in rolls	275-325	2.8	2.8
Flat seam tin roofing, using 28" tin in rolls	425-475	1.8	1.8
Standing seam metal roofing, using 24"x96" sheets	250-300	3.0	3.0
Standing seam tin roofing, using 14"x20" plates	125-175	5.3	5.3
Standing seam tin roofing, using 20"x28" plates	175-225	4.0	4.0
Standing seam tin roofing, using 14" tin in rolls	200-225	3.8	3.8

Standing seam tin roofing, using 20" tin in rolls	.250-300	3.0	3.0
Standing seam tin roofing, using 24" tin in rolls	.275-325	2.8	2.8
Standing seam tin roofing, using 28" tin in rolls	.300-350	2.4	2.4
V-Crimped metal roofing, using 24"x96" sheets or larger	.425-475	1.8	1.8

Material Cost of 100 Sq. Ft. (1 Sq.) Flat Seam Tin Roofing Using 14"x20" Plates

	Rate	Total	Rate	Total
1.1 sq. No. 15 felt	$....	$....	$3.00	$3.30
63 shts. 14"x20" 40 lb. coated	....	....	1.50	94.50
6 lbs. solder	....	....	1.00	6.00
252 cleats	....	....	.03	7.56
2 lbs. nails	....	....	.80	1.60
Flux and charcoal	....	....	1.30	1.30
*Cost per 100 sq. ft.		$....		$114.26

*Add for ridges, hips, valleys, flashing etc.

Material Cost of 100 Sq. Ft. (1 Sq.) Flat Seam Tin Roofing Using 20"x28" Plates

	Rate	Total	Rate	Total
1.1 sq. No. 15 felt	$....	$....	$3.00	$3.30
29 shts. 20"x28" 40 lb. coated	....	....	3.00	87.00
5 lbs. solder	....	....	1.00	5.00
145 cleats	....	....	.03	4.35
1 lb. nails	....	....	.80	.80
Flux and charcoal	....	....	1.30	1.30
*Cost per 100 sq. ft.		$....		$101.75

*Add for ridges, hips, valleys, flashing etc.

Material Cost of 100 Sq. Ft. (1 Sq.) Flat Seam Tin Roofing Using Roll Roofing

	Rate	Total	Rate	Total
1.1 sq. No. 15 felt	$....	$....	$3.00	$3.30
108 sq. ft. 20" roll rfg.	....	....	1.20	129.60
4 lbs. solder	....	....	1.00	4.00
110 cleats	....	....	.03	3.30
1 lb. nails	....	....	.80	.80
Flux and charcoal	....	....	1.00	1.00
*Cost per 100 sq. ft.		$....		$142.00

*Add for ridges, hips, valleys, flashing, etc.

Material Cost of 100 Sq. Ft. (1 Sq.) Flat Seam Roofing Using 24"x96" Galvanized Steel Sheets

	Rate	Total	Rate	Total
1.1 sq. No. 15 felt	$....	$....	$3.00	$3.30
7 shts. 26 ga. steel, 105 lbs.	....	....	.80	89.60
3 lbs. solder	....	....	1.00	3.00
100 cleats	....	....	.03	3.00
1 lb. nails	....	....	.80	.80
Flux and charcoal	....	....	.90	.90
*Cost per 100 sq. ft.		$....		$100.60

*Add for ridges, hips, valleys, flashing, etc.

Material Cost of 100 Sq. Ft. (1 Sq.) Standing Seam Roofing Using 14"x20" Plates

	Rate	Total	Rate	Total
1.1 sq. No. 15 felt	$....	$....	$3.00	$3.30
68 shts. 14"x20" 40-lb. coated	....	....	1.50	102.00
100 cleats	....	....	.03	3.00
2.5 lbs. solder	....	....	1.00	2.50
1 lb. nails	....	....	.80	.80
Flux and charcoal	....	....	.90	.90
*Cost per 100 sq. ft.		$....		$112.50

*Add for ridges, hips, valleys, flashing, etc.

Material Cost of 100 Sq. Ft. (1 Sq.) Standing Seam Roofing Using 20"x28" Plates

	Rate	Total	Rate	Total
1.1 sq. No. 15 felt	$....	$....	$3.00	$3.30
32 shts. 20"x28" 40-lb. coated	....	....	3.00	96.00
80 cleats	....	....	.03	2.40
2 lbs. solder	....	....	1.00	2.00
1 lb. nails	....	....	.80	.80
Flux and charcoal	....	....	.90	.90
*Cost per 100 sq. ft.		$....		$105.40

*Add for ridges, hips, valleys, flashings, etc.

Material Cost of 100 Sq. Ft. (1 Sq.) Standing Seam Roofing Using Roll Roofing

	Rate	Total	Rate	Total
1.1 sq. No. 15 felt	$....	$....	$3.00	$3.30
116 sq. ft. 20" roll rfg 40 lb.	....	....	1.20	139.20
80 cleats	....	....	.03	2.40
1 lb. solder	....	....	1.00	1.00
1 lb. nails	....	....	.80	.80
Flux and charcoal	....	....	.70	.70
*Cost per 100 sq. ft.		$....		$147.40

*Add for ridges, hips, valleys, flashing, etc.

Material Cost of 100 Sq. Ft. (1 Sq.) Standing Seam Roofing Using 24"x96" Galvanized Steel Sheets

	Rate	Total	Rate	Total
1.1 sq. No. 15 felt	$....	$....	$3.00	$3.30
8 shts. 26 ga. steel, 120 lbs.	....	....	.80	102.40
60 cleats	....	....	.03	1.80
1 lb. solder	....	....	1.00	1.00
1 lb. nails	....	....	.80	.80
Flux and charcoal	....	....	.70	.70
*Cost per 100 sq. ft.		$....		$110.00

*Add for ridges, hips, valleys, flashing, etc.

Material Cost of 100 Sq. Ft. (1 Sq.) V-Crimped Metal Roofing

	Rate	Total	Rate	Total
1.1 sq. No. 15 Felt	$....	$....	$3.00	$3.30
1 sq. 26 ga. V-crimped rfg	....	....	95.00	95.00
5 lbs. lead headed nails	....	....	1.50	7.50
60 lin. ft. wood strips	....	....	.20	12.00
*Cost per 100 sq. ft.		$....		$117.80

*Add for ridges, hips, valleys, flashing, etc.

COPPER ROOFING

Copper may be used as roofing for any type of roof. There are various methods of applying copper for roofing, but most frequently used are the flat seam and standing seam methods. The batten seam method is also used, where durability is of more importance than cost, e.g., cathedrals, government buildings, monumental structures, etc.

Flat Seam Roofing. Flat seam roofing is commonly used on dead level or flat surfaces, but is adaptable to any slope. This method is also adaptable for the construction of water cooling roof panels, permitting water to stand up to a depth of 3 inches.

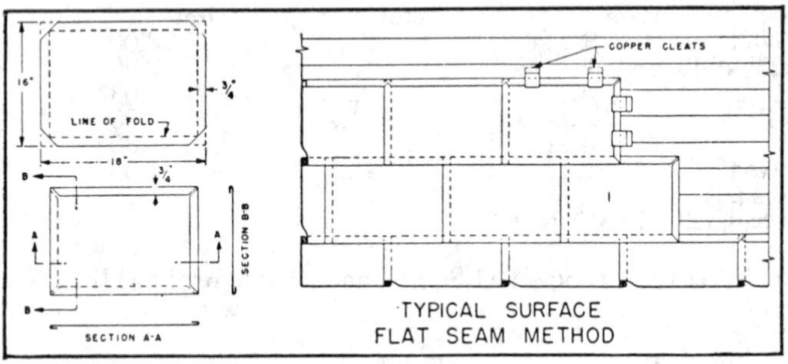

Method of Placing Flat Seam Copper Roofing

For flat seam roofing, the use of small sheets of 20-oz, cold-rolled copper, not larger than 16"x18", with ¾" seams flat-locked and soldered, is recommended. The sheets are tinned on all edges for soldering and are then formed to lock ¾" with adjacent sheet. Corners must be clipped to permit folding as shown. Opposite sides of the sheets are folded in opposite directions so they will hook into adjoining sheets.

The sheets are held down by 2" wide copper cleats, two on each of two adjacent sides. The other two sides are held by the edges of adjacent sheets already cleated. For small areas, it is general practice to use only three cleats, two on a long side and one on a short side.

The cross seams are folded in the direction of flow. All seams should be flattened with a mallet and solder sweated to completely fill the seam.

Each 16"x18" sheet when folded covers 14 ½"x16 ½" or 239 ¼ sq. in. It requires 61 sheets of 16"x18" copper per 100 sq. ft. of roof.

Standing Seam Roofing. Standing seam copper roofing may be used for surfaces with a minimum slope of 3" per foot. Standing seams are usually made to finish 1" high. The spacing of the seams is a matter of scale and architectural effect, but for economy, a choice should be made that will use sheets of stock sizes. The use of 20"x96" sheets is recommended as maximum for ordinary purposes.

For standing seams to finish 1" high, the vertical bends are made on the long edges of the sheets--1 ¾" on one side and 1 ½" on the other; the short bend on the one sheet always adjoins the long bend on the adjacent sheet. Recommended weights of material are: 16-oz. for widths between seams up to 20"; 20-oz. for widths over 20".

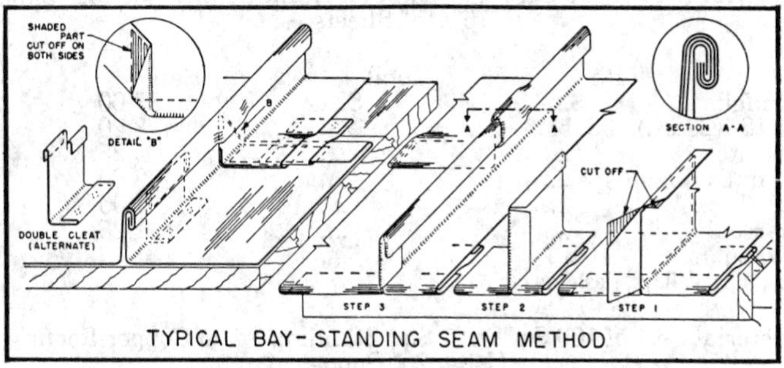

Method of Placing Standing Seam Copper Roofing

Formed pans are held to the roof deck with 2"x3" cleats spaced 12" apart and locked into the standing seam as shown in the accompanying illustration. Transverse joints should be staggered and each secured with one cleat locked into the flat seam.

Standing seams should not be riveted or soldered--cross seams should be left unsoldered whenever conditions permit.

A sheet loses 3 ¼" in width on account of the standing seam, i.e. a sheet 20" wide would cover only 16 ¾". Where 20" sheets are used, add 20% to roof area for standing seams.

The loss at the end of the sheet will depend upon the length of the roof and the length of copper sheet used, but allow 1 ½" for each flat seam.

Sizes and Weights of Copper Sheets Commonly Used in Building Work

Weight in Pounds per Sheet of Various Weight Copper

Size of Sheet Copper	16-Oz.	20-Oz.	24-Oz.	32-Oz.
20"x96"	13.33	16.67	20.00	26.67
24"x96"	16.00	20.00	24.00	32.00
30"x96"	20.00	25.00	30.00	40.00
36"x96"	24.00	30.00	36.00	48.00
20"x120"	16.67	20.83	25.00	33.33
24"x120"	20.00	25.00	30.00	40.00
30"x120"	25.00	31.25	37.50	50.00
36"x120"	30.00	37.50	45.00	60.00

Labor Placing Flat Seam Copper Roofing. When using 16"x18" copper sheets for flat seam roofing, a tinner and helper should place 175 to 225 sq. ft. of roof per 8-hr. day, on average size roofs.

Labor Placing Standing Seam Copper Roofing. When using copper sheets 20" wide and up to 8'-0" long for standing seam roofing, a tinner and helper should place 200 to 250 sq. ft. per 8-hr. day, on average size roofs.

Material Cost of 100 Sq. Ft. (1 Sq.) Flat Seam Copper Roofing Using 16"x18" Sheets

	Rate	Total	Rate	Total
1.1 sq. No. 15 felt	$....	$....	$3.00	$3.30
61 shts. 16"x18" (20-oz.) 153 lbs.	....	....	3.20	195.20
250 copper cleats	....	....	.25	62.50
2 lbs. copper nails	....	....	2.85	5.70
6 lbs. solder	....	....	1.00	6.00
Flux and charcoal	....	....	1.25	1.25
*Cost per 100 sq. ft.		$....		$273.95

*Add for ridges, hips, valleys, flashing, etc.

Material Cost of 100 Sq. Ft. (1 Sq.) Standing Seam Copper Roofing Using 20" Copper

	Rate	Total	Rate	Total
1.1 sq. No. 15 felt	$....	$....	$3.00	$3.30
120 sq. ft. (16-oz.) 120 lbs.	....	....	1.60	192.00
80 copper cleats	....	....	.25	20.00
1 lb. copper nails	....	....	2.85	2.85
*Cost per 100 sq. ft.		$....		$218.15

*Add for ridges, hips, valleys, flashing, etc.

Stainless steel, type 304, monel and lead are also used for flat, standing and batten seams.

07620 SHEET METAL FLASHING AND TRIM

Sheet metal work in its many branches is a highly specialized business, because the greater part of the work is performed in the shop, and the fabricated materials are sent to the job ready to erect. For this reason, an estimate on sheet metal work must take into consideration the cost of the finished materials at the shop or delivered to the building site, plus the labor cost of erection.

Estimating Quantities of Sheet Metal Work

The data given on the following pages will assist the estimator in measuring quantities from the plans and listing them on the estimate sheet.

When preparing the quantity survey for the different types, shapes, and sizes of metal items, be sure to identify the type and weight, or gauge, of metal that is to be used, e.g., galvanized steel, aluminum, stainless steel, copper, lead, terne, etc. Each one will have its own factors for waste, difficulty in shaping, method of jointing, etc., all of these affecting the production and installation costs, in addition to the sheet goods cost delivered to the shop.

Metal Gutters and Eave Trough. Metal gutters and eave trough are furnished in numerous designs, from 2" to 12" in size. When estimating quantities, obtain the number of pieces of each size gutter and the length of same, as the estimate should state the lin. ft. of each size.

Metal Conductor Pipe or Downspout. Metal conductor pipe is furnished both round and square and from 2" to 6" in size.

When estimating quantities, measure the distance from the roof to the ground, which will be the length of each downspout. After the quantities have been obtained in this manner, list the total number of lin. ft. of each size.

Metal Conductor Heads. Metal conductor heads are made in various styles, shapes and sizes.

Conductor heads are generally used on buildings having flat roofs where a cast iron conductor pipe is placed inside the building and installed by the plumbing contractor. When placed on the outside of a wall at the roof level, an opening is usually left in the masonry wall at the low point of the roof to allow the water to pass through the wall opening into the conductor head and down the conductor pipe.

When estimating conductor heads always note style and size of each head and the material specified.

Metal Flashing and Counterflashing. Flashing and counterflashing are furnished in tin, stainless steel, galvanized steel, aluminum or copper. They are measured by the lin. ft. when less than 12" wide and by the sq. ft. when over 12" wide.

Metal flashing used in connection with a flat roof deck covered with felt or built-up roofing usually extends 6" to 12" up the side of the brick fire wall and projects under the roofing the same distance. A counterflashing is then placed over and above this flashing. The upper edge of the counterflashing is inserted in a reglet (slot) in the brick or stone wall. After the counterflashing has been placed, it is caulked with caulking compound to prevent water from getting in back of the flashing and running under the roofing.

On many jobs using felt or built-up roofing, the felt is run up the wall about 12" and mopped. The metal counterflashing is then placed to extend down over the felt, which prevents water from getting in back of the flashing and running under the roofing.

Where metal counterflashing is used, it is necessary to cut an open joint or reglet in the mortar joint of the masonry wall.

The cost of cutting reglets should be estimated by the lin. ft.

Metal Valleys and Hips. Metal valleys and hips are frequently used with certain types of shingle and slate roofs. They are either continuous metal strips of a certain size or small shingles of tin, galvanized steel, aluminum, stainless steel or copper.

They should be estimated by the lin. ft. stating the width of the valley or hip and the kind of metal used.

Metal Ridge Roll, Hip Roll and Cap. Metal ridge roll, hip roll and cap should be estimated by the lin. ft. giving the width, gauge and kind of metal used.

Metal Ventilators. Metal ventilators for roofs and skylights should be estimated at a certain price for each ventilator, f.o.b. factory or erected. Always state size and type of ventilators required, as there are numerous kinds on the market at varying prices.

Metal Skylights. When estimating metal skylights, consider the following: size, type of skylight, whether single or double pitch or hip skylight; with or without ventilators or side sash; kind of glass, and whether a metal curb flashing is required.

Labor Placing Sheet Metal Work

The costs given on the following pages are for the erection of various kinds of sheet metal work on the job. Add cost of the fabricated material, contractor's overhead expense and profit.

Placing Hanging Gutter or Eave Trough. This cost will vary with the slope of the roof, distance above the ground and the method used in hanging the gutter.

On wood constructed buildings having hip or gable roofs, metal hangers or supports are placed over the gutters or eave trough and fastened to the eaves under the shingles or other roofing. The eave trough or gutter must always slope in the direction of the downspout, in order to drain satisfactorily.

On hip or gable roofs having eaves 20'-0" to 25'-0" above the ground, a tinner and helper should place 180 to 220 lin. ft. of gutter per 8-hr. day, at the following labor cost per 100 lin. ft.:

	Hours	Rate	Total	Rate	Total
Sht. met. wkr	4	$....	$....	$26.25	$105.00
Labor	4	....	....	19.13	76.52
Cost 100 lin. ft.			$....		$181.52
Cost per lin. ft.			....		1.82

Placing Metal Conductor Pipes or Downspouts. Metal conductor pipes or downspouts are furnished in both round and square and from 1 ½" to 6" diameter. The conductor pipe extends down the side of the building from the eave trough to grade, where it is connected to an elbow that throws the water away from the building or is connected with the drainage system. Metal conductor pipe is usually held in place by hooks fastened to the walls.

On one- or two-story buildings where the conductor pipe is 12'-0" to 25'-0" long, a tinner and helper should place 200 to 250 lin. ft. per 8-hr. day, at the following labor cost per 100 lin. ft.:

	Hours	Rate	Total	Rate	Total
Sht. met. wkr	3.5	$....	$....	$26.25	$91.88
Labor	3.5	....	....	19.13	66.96
Cost 100 lin. ft.			$....		$158.84
Cost per lin. ft.			....		1.59

Where outside conductor pipe is installed on buildings 3 to 4 stories high, a tinner and helper should place 160 to 200 lin. ft. per 8-hr. day, at the following labor cost per 100 lin. ft.:

	Hours	Rate	Total	Rate	Total
Sht. met. wkr.	4.5	$....	$....	$26.25	$118.13
Labor	4.5	....	....	19.13	86.09
Cost 100 lin. ft.			$....		$204.22
Cost per lin. ft.			....		2.04

Placing Metal Flashing. When placing metal roof flashing around parapet walls, etc., it is customary to have the flashing extend under the roofing and up the side of the wall 6" to 12", and then place the counterflashing above to lap over the flashing.

Where it is not necessary to cut reglets in the masonry wall, a tinner should place 140 to 160 lin. ft. of flashing per 8-hr. day, at the following cost per 100 lin. ft.:

	Hours	Rate	Total	Rate	Total
Sht. met. wkr.	5.4	$....	$....	$26.25	$141.75
Cost per lin. ft.			....		1.42

On frame buildings where the flashing extends under the wood wall siding to protect porch roofs, etc., it is only necessary to place the flashing or metal shingles on the roof and tack them against the sheathing. A tinner should place 165 to 185 lin. ft. per 8-hr. day, at the following labor cost per 100 lin. ft.:

	Hours	Rate	Total	Rate	Total
Sht. Met Wkr	4.6	$....	$....	$26.25	$120.75
Cost per lin. ft.			....		1.21

Placing Metal Counterflashing. When placing metal counterflashing over felt or metal flashing that extends up on the side of the wall, it will be necessary to place the counterflashing in a reglet or open joint in the masonry that can be caulked or sealed to prevent the water getting in back of the counterflashing and under the roof.

Where reglets have previously been cut, as is usually the case on stone copings, requiring only the sealing of the joint, a tinner should place 140 to 160 lin. ft. per 8-hr. day, at the following labor cost per 100 lin. ft.:

	Hours	Rate	Total	Rate	Total
Sht. Met. Wkr.	5.4	$....	$....	$26.25	$141.75
Cost per lin. ft.			....		1.42

Cutting Reglets in Masonry Walls. The labor cost of cutting reglets in brick walls will depend entirely upon the kind of mortar used and the condition of the mortar at the time of cutting the reglets.

If the brick are laid in portland cement mortar, it will cost more to cut the reglet than when laid in lime mortar. Also, if the reglets are cut within 24 hrs. after the brick are laid, the labor cost will be much less than cutting them two or three weeks later.

If reglets are cut before the mortar has had sufficient time to set, a worker should cut 175 to 225 lin. ft. per 8-hr. day, at the following labor cost per 100 lin. ft.:

	Hours	Rate	Total	Rate	Total
Labor	4	$....	$....	$19.13	$76.52
Cost per lin. ft.			....		.77

If the reglets are cut several days after the brick have been laid, giving the mortar sufficient time to harden, it will require a hammer and chisel to cut out the joints. Under these conditions a man should cut 65 to 80 lin. ft. per 8-hr. day, at the following labor cost per 100 lin. ft.:

	Hours	Rate	Total	Rate	Total
Labor	11	$....	$....	$19.13	$210.43
Cost per lin. ft.			....		2.10

If a portable electric saw, having an abrasive blade or cutting wheel is used for cutting the reglets, a man should cut 10 to 15 lin. ft. of reglet an hour.

Cutting reglets in stone coping or balustrade is usually performed by stone cutters at about the same cost as given above.

Placing Metal Valleys. When placing metal shingles or valleys in connection with gable or hip roofs, dormers, etc., a tinner should place 115 to 135 lin ft. per 8- hr. day, at the following labor cost per 100 lin. ft.:

	Hours	Rate	Total	Rate	Total
Sht. Met. Wkr.	6.4	$....	$....	$26.25	$168.00
Cost per lin. ft.			....		1.68

Placing Metal Ridge Roll of Cap. If metal ridge roll is placed on the ridge of a hip or gable roof, a tinner should place 90 to 110 lin. ft. per 8-hr. day, at the following labor cost per 100 lin. ft.:

	Hours	Rate	Total	Rate	Total
Sht. Met. Wkr.	8	$....	$....	$26.25	$210.00
Cost per lin. ft.			....		2.10

Placing Metal Conductor Heads. On buildings having flat roofs where the conductor pipe or downspout is placed inside of the building, requiring a conductor head at the low point of the roof which is connected to the downspout, a tinner should place one conductor head (made complete in the shop) in 1 ½ to 2 hrs. at the following labor cost:

	Hours	Rate	Total	Rate	Total
Sht. Met. Wkr.	1.8	$....	$....	$26.25	$47.25

07630 ROOFING SPECIALTIES

Flashings can be formed from a wide variety of materials of varying permanence, as follows:

Material		Cost per Sq. Ft.	Sq. Ft. Installed per Day
Aluminum	.013" thick	$0.24	150
(Mill)	.016"	.28	150
(Finish)	.019"	.38	150
	.032"	.72	145
	.040"	.86	145
	.050"	1.05	145
Copper	16 oz.	2.20	140
	20 oz.	2.70	140
	24 oz.	3.10	140
Stainless Steel	26 ga.	2.00	150
Asph. C't'd Cotton	17 oz.	.15	350
	40 oz.	.22	350
Lead 2.5 lb.		1.65	140
Zinc/Copper Alloy		1.35	140
Butyl	1/32"	.52	275
	1/16"	.75	275
Neoprene	1/16"	1.25	275
Polyvinyl	.02"	.14	275
	.03	.22	275

Gravel stops may be formed on the job or ordered out in extruded form. If extruded be sure to check if any special finish is specified as this will raise costs considerably. A few costs and rates for installation are given below:

Material	LF Cost 4" high	6" High	LF per Day
Aluminum .050" mill finish	$2.50	$2.90	135
Duranodic	3.00	3.40	135

Preformed gutters can be installed by one sheet metal worker at the rate of from 110' to 120' per day. Material costs are:

Material	Cost Per Lin. Ft. 4"	5"	6"
Aluminum .027" plain	$0.60	$0.65	$0.80
enam.	.65	.73	.90
Copper 16 oz.	3.40	3.80	4.70
Stainless 22 ga.	3.40	3.70	4.00
Vinyl	1.00	1.10	1.25
Galvanized 28 ga.	.55	.70	.90

Downspouts can be installed by one sheet metal worker at the rate of 175 lin. ft. for small sizes, 150 for medium and 125 for the larger sizes per day. The material costs will run as follows:

Material	Cost Per Lin. Ft. 3" dia.	4" dia.	5" dia.	2"x3"	3"x4"
Aluminum .025" plain	$0.50	$0.80	$1.10	$0.75	$0.95
enam.	.58	.90	1.15	0.85	1.10
Copper 16 oz.	3.00	3.50	4.30	3.30	4.05
Stainless	7.00	9.00	14.00	7.00	9.00
Vinyl	.60	--	--	.65	--
Galv. Steel 28 ga.	.45	.60	1.05	.50	.65

For elbows for the above work add the following:

3" diam. or 2"x3" in plain aluminum	$1.00
3" diam. or 2"x3" in enam. aluminum	1.25
3" diam. or 2"x3" in 16 oz. copper	5.00
4" diam. or 3"x4" in plain aluminum	1.25
4" diam. or 3"x4" in enam. aluminum	1.50
4" diam. or 3"x4" 16 oz. copper	8.00

07800 ROOF ACCESSORIES

Erecting and Glazing Metal Skylights. Metal skylights are manufactured in the shop and sent to the job "knocked down" or in sections ready to erect and glaze.

An average skylight (single pitch, double pitch or hip) up to 8'-0"x12'-0" in size, containing 100 sq. ft. of area, should be erected and glazed complete by a sheet metal worker and glazier in 8 to 10 hrs. time, at the following labor cost:

	Hours	Rate	Total	Rate	Total
Sheet metal worker	5	$....	$....	$26.25	$131.25
Glazier	4	....	....	23.58	94.32
Cost per skylight			$....		$225.57

When erecting skylights 10'-0"x10'-0" to 10'-0"x15'-0", containing 100 to 150 sq. ft., a sheet metal worker and glazier should erect and glaze one skylight in 14 to 16 hrs. at the following labor cost:

	Hours	Rate	Total	Rate	Total
Sheet metal worker	9	$....	$....	$26.25	$236.25
Glazier	6	....	....	23.58	141.48
Cost per skylight			$....		$377.73

When erecting skylights 10'-0"x16'-0" to 10'-0"x20'-0" in size, containing 160 to 200 sq. ft., it will require about 24 hrs. for a sheet metal worker and glazier to erect the skylight and glaze it complete, at the following labor cost:

	Hours	Rate	Total	Rate	Total
Sheet metal worker	16	$....	$....	$26.25	$420.00
Glazier	8	....	....	23.58	188.64
Cost per skylight			$....		$608.64

The above costs do not include overhead and profit or the cost of placing a metal curb flashing for skylight. For costs on this work, see below.

Placing Metal Skylight Curb Flashing. Metal skylights are often placed on top of a wood curb projecting 6" or more above the roof level. The space from the top of the curb to the roof level is covered with metal flashing.

When erecting metal skylight curb flashing, a tinner should place 190 to 210 sq. ft. of metal per 8-hr day, at the following labor cost per 100 sq. ft.:

	Hours	Rate	Total	Rate	Total
Sheet metal worker	4	$....	$....	$26.25	$105.00
Cost per sq. ft.			....		1.05

Erecting Skylights With Side Sash. If the skylights have side sash, the erection cost will vary with the number of sash in the skylight and whether stationary or pivoted. On an average it will require 1 to 1 ½ hrs. labor time for each sash in the skylight, at the following labor cost per sash:

	Hours	Rate	Total	Rate	Total
Sheet metal worker	1.25	$....	$....	$26.25	$32.81

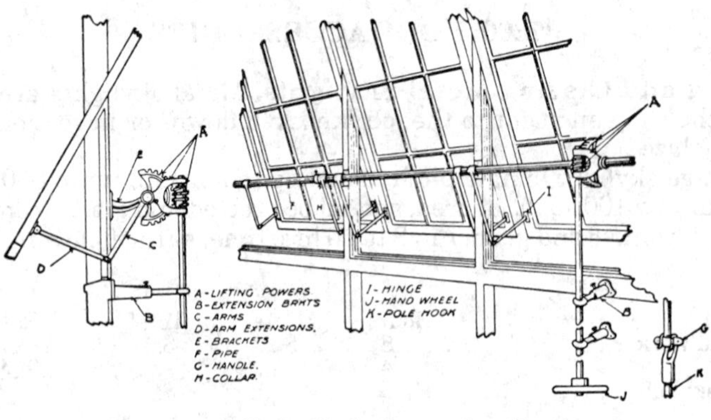

Worm and Gear Sash Operating Device

Erecting Sash Operating Devices for Movable Sash. Skylights furnished with hinged or pivoted sidewall sash require an operating device to open and close them.

An operating device usually consists of a mechanism similar to that illustrated, which is operated from the floor by a worm and gear or a chain attachment, from which all sash on each side of the skylight can be opened and closed.

The labor cost of erecting sash operating devices will vary with the length of each run and the number of runs in each skylight but on an average, a man should erect 50 to 75 lin. ft. per 8-hr. day, at the following labor cost per 100 lin. ft.:

	Hours	Rate	Total	Rate	Total
Sheet metal worker	13	$....	$....	$26.25	$341.25
Cost per lin. ft.			....		3.41

Ventilating Skylights. Ventilating skylights are used in factories, foundries, machine shops, garages, dairies, laundries and in any other style of building requiring economical light and ventilation. They are furnished in any type of metal desired, galvanized copper-bearing steel, aluminum or copper, depending upon the nature of the building on which they are to be used. Normally, for building of ordinary construction, the standard 18-gauge galvanized copper-bearing steel is furnished. Should the building be of reinforced concrete or steel and concrete construction, the non-corrosive aluminum or copper construction should be furnished.

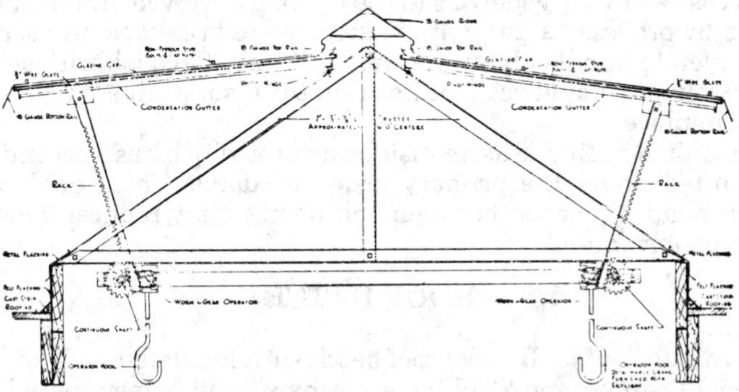

Section Through Ventilating Skylight

These skylights are of gable type as illustrated and are furnished in standard widths of 4'-0" to 20'-0" and any desired length in multiples of 2'-0". It is standard practice to interpret the size of a skylight as being outside of curb dimensions.

The length of the ventilating sections depend upon the width of the skylights; 4'-0", 6'-0" and 8'-0" widths having maximum ventilating sections of 40'-0" each controlled from one operating station; 10'-0" and 12'-0" wide skylights have maximum ventilation sections of 30'-0" in length; 14'-0" and 16'-0" widths have 40'-0" maximum lengths, and 18'-0" and 20'-0" widths have 30'-0" maximum ventilating sections.

A mechanical operating device consisting of either loose rod or chain control is furnished as required, so that the skylight can be readily controlled from the floor. All skylights under 30'-0" in length are arranged for full length ventilation. Longer skylights are arranged with ventilating sections as required.

The cost of ventilating skylights vary with their size and number of ventilating sections in each skylight. Prices should be secured from the manufacturer.

In estimating the cost of ventilating skylights, the principal thing to remember is: The smaller the area of the skylight, the higher the price per sq. ft.

Plastic Dome Skylights. This type skylight consists of a thermo-formed acrylic plastic sheet free blown into square, rectangular and circular domes.

This type skylight is furnished as a factory assembled unit; the plastic is mounted in an extruded aluminum frame, and prefabricated curbs are available. The aluminum frame is generally designed to form counterflashing.

Plastic Domes

Size	Cost Per Roof Dome
16"x16"	$40.00
24"x24"	50.00
36"x36"	65.00
48"x48"	115.00
24" Round	85.00
48" Round	150.00

Erection cost will vary widely as to the number involved and the preparatory work done by others. As units are vulnerable to breakage and scratching, it takes two men to handle all but the smallest unit. On a job with several units involved two men should set 2 units of up to 10 sq. ft. in a day if the curbing is already in place.

The dome unit is available as an insulating unit which has a secondary acrylic dome mounted under the primary (exterior) dome. This double dome unit provides a dead air space between the domes that reduces heat loss and condensation formation.

ROOF HATCH

A standard size 2'-6"x3'-0" steel roof hatch with insulated curb and cover and hardware will run around $275.00. A mechanic and helper should set this in 2 hrs.

GRAVITY VENTILATORS

The 12" stationary ventilators of the syphon type with a 24" base will cost around $40 and take two hours to set. Stationary ventilators of the mushroom type with a 24" base will cost $130.00 with the same labor cost.

07900 SEALANTS

No one sealant can solve the requirements across the full range of construction applications. There are sealants for interior use, exterior use, gun or pour grade, ability to expand and contract, service temperature range, paintability, and for compatability with the material to be sealed.

An oil base caulking compound has a life expectancy of about 4 or 5 years. Its restrictions list a joint maximum size of ½" wide x ¾" deep and is incapable of withstanding any joint movement without rupture. Therefore, using this material for exterior applications may not be prudent or cost effective.

Acrylic latex caulk is probably a good starting point for exterior use. It is considered to be the best value among sealants in the middle performance range due to; relatively inexpensive, life expectancy of 8 to 12 years, cures quickly, excellent paintability, non-staining, and has reasonable elongation and recovery for movement of joints. However, this material has a maximum joint size of ½" wide and ½" deep and should not be used for expansion joints.

Polysulfide base sealants, available in one or two part compounds, are among the premiere compounds on the market. It is reasonably expensive and requires a little more labor time to install, however its movement capabilities, bonding strength, wide use range (including expansion joints) and a life expectancy of 15 pm years makes it cost effective while providing excellent joint sealing characteristics.

It is important to have a properly designed joint, both in width and depth. Good sealants will expand and contract with the movement of the joint. As the joint moves, the sealant changes shape while the volume remains constant. It is critical that the width-to-depth ratio be designed to withstand the constant elongation and compression cycles over long periods of time.

Deep beads of sealant wastes material and are more prone to failure than shallow sealant beads. Another important factor to consider is that sealant works best when adhering to the two opposing faces and isolated from the third (back) side of the joint to be sealed. This is accomplished by filling the joint with a material such as oakum or polyethylene foam rod to within ⅜" or ½" of the face of the joint. In shallow depth joints, use a non-adhering tape at the back of the joint to prevent the sealant from bonding.

Sealants are manufactured in colors from white to black and some are clear (silicones). This is necessary to be architecturally compatible with the adjoining surfaces, in addition paint does not adhere at all to some sealants.

The following chart will provide a guide for the amount of sealant required to fill various size joints; expressed in linear feet of joint obtained per gallon of material.

DEPTH, INCHES	WIDTH, INCHES							
	1/16	1/8	1/4	3/8	1/2	5/8	3/4	1
1/16	4928	2464	1232	821	616	493	411	307
1/8	—	1232	616	411	307	246	205	154
3/16	—	—	411	275	205	164	137	103
1/4	—	—	307	205	154	123	103	77
3/8	—	—	—	137	103	82	68	51

In using the above chart, if the size of the joint to be sealed is not given on the drawings, use the principle that the depth of the sealant is ½ of the width of the joint to be sealed.

When estimating the labor required to caulk a joint, it will depend upon the width of the joint, the material used, and the location or accessibility of the joint. One worker should caulk about 600 lin. ft. per 8 hr. day of joints around doors or window frames when they are close to the ground or accessible without the use of scaffolding. This time may be reduced in half if working off ladders or scaffolding. The same person, working on sealing wide expansion joints, will only finish about 200 lin. ft. per 8 hr. day.

CHAPTER 8

DOORS AND WINDOWS

CSI DIVISION 8
08100 METAL DOORS AND FRAMES
08200 WOOD AND PLASTIC DOORS
08300 SPECIAL DOORS
08400 ENTRANCES AND STOREFRONTS
08500 METAL WINDOWS
08600 WOOD AND PLASTIC WINDOWS
08700 HARDWARE AND SPECIALTIES
08800 GLAZING
08900 WINDOW WALLS/CURTAIN WALLS

Most of the items covered in this chapter are furnished by material distributors, but set by the carpenter on the job; in the estimate, the cost of labor would be included under the proper trade.

Glazing is an exception, and both material and labor are included under a single item. However, it is more and more prevalent these days for sash and doors to come to the job factory glazed. The contractor must always check this out so as to eliminate any duplication in the estimate and to make certain the pre-glazing complies with both the specifications and local union rules.

Curtainwall material and labor are usually let to a single subcontractor. It is advantageous to have the metal erection, sealants and glazing all in one contract so that one source will be responsible for the weather tightness of the wall.

The estimator who is to perform the quantity survey for materials covered by this chapter should become very familiar with the various "Schedules" that the project architect includes on the plans.

These schedules are usually laid out in block column form consisting of a listing of the material involved and all of the particulars for each item, and may show architectural details for each type of listed item.

Door Schedules will normally list the door openings by "mark" numbers or letters, e.g., #1, #2, #3 or A, B, C. After the mark designation, there will be the size, material of the door and the frame, a code designation for the type and size of the frame, fire rating requirements (if any), a code designation for any louver or glass panels required, finish hardware set designations, and a "remarks" column for any special instructions. Bear in mind that the "mark" designation refers to that particular design type and may have nothing to do with the total quantity of doors and frames of that design type required for the project.

Window schedules are laid out in much the same fashion, showing: mark designation, size, material, glass, type, etc.

Finish hardware schedules are normally set up in the hardware section of the specifications, rather than on the drawings. The hardware schedule is a listing of the hardware items required to fit a particular door opening, e.g., hinges, latchset or lockset, closer, kickplates, etc.

Each type of opening, such as for toilets, closets, offices, etc., will have its own "HW Set #" designation; and each will vary from the other in some way. For example, a door that leads to a stairway in an office building could list under its HW Set #: "1½ pair hinges, latchset, and closer". A door that leads to an office in the same building could list under its HW Set #: "1½ pair hinges, lockset, and closer". As you can see from the two examples, they are very similar, but because of the lock requirements for the office door opening, a separate HW Set # will be given.

Here again, keep in mind that the HW Set # has nothing to do with the total quantity of finish hardware required, i.e., one set number may be applied to 10 door openings and the next set number applies to only one or two door openings.

08100 METAL DOORS AND FRAMES

Each hollow metal manufacturer produces their stock design doors and frames as well as a custom line to suit the project architect's requirements. Bear in mind that the "stock" line will vary between manufacturers, e.g., sizes, width of frames, hardware preparation, etc. For example, a hollow metal door from one manufacturer usually can not be hung in the hollow metal frame from another manufacturer, due to unmatched hinge or lockset locations.

"Stock" will also refer to the cut-out preparation for hinges, and in particular, for locksets. Most manufacturers prepare the frames for 3 hinge locations each sized 3½" or 4½" in height and for the standard cylindrical lockset. If the finish hardware specified for the project varies from the manufacturers' standards, the hollow metal manufacturer will consider the order "custom" and increase the price accordingly, as well as increase the delivery time to the project.

Material prices for hollow metal items are quoted inclusive of all mortising, reinforcing, drilling and tapping for mounting the door into the frame and preparation for items of finish hardware. Also included are frame anchors, and shop prime coat of paint.

Hollow Metal Frames. Frames are made of 18, 16, and 14 gauge metal and formed to receive a 1⅜" or 1¾" thick door, depending on which side of the stop the hinge and strike cut-outs are placed. See the following section through a standard hollow metal frame.

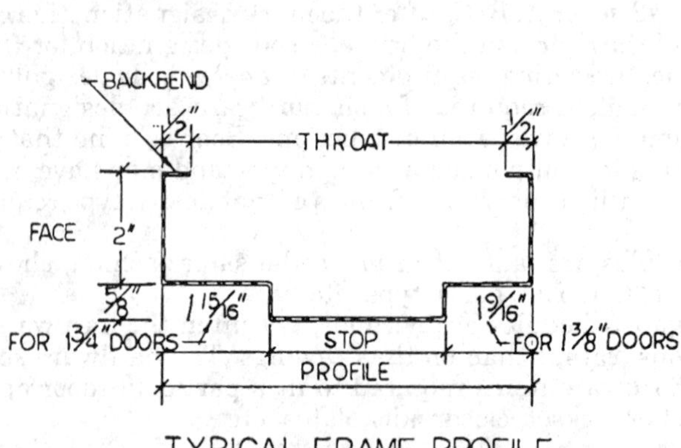

When the manufacturer varies the overall width of the frame to suit the different wall conditions that may be encountered, he simply increases the width of the stop. This allows them to set up any width frame for either thickness of door in lieu of manufacturing frames for 1⅜" doors and 1¾" doors.

As stated, frames are made in various widths (profile) to suit the thicknesses of the walls to which they are mounted. The standard widths with most manufacturers are 4¾", 5¾", 6¾" and 8¾".

In addition to the frame width dimensions, frames are ordered according to the door opening size, the standard being 2'-0", 2'-4", 2'-6", 2'-8", 3'-0", 3'-4", 3'-6", 3'-8", and 4'-0" wide for single swing and 4'-0", 4'-8", 5'-0", 5'-4", 6'-0", 6'-8", 7'-0", 7'-4", and 8'-0" wide for double swing by 6'-8", 7'-0", 7'-2", and 8'-0" in height.

Frames are available from the factory in welded joint construction (all 3 sides put together to form a unit) or in knocked down (K.D.) form for use on drywall partitions. The K.D. frame is considered to be a pressure-fit frame that once in place (capping the wall) it exerts a gripping action onto the wall. The welded frame is stronger and must be erected prior to the wall erection, whereas the K.D. frame is erected after the wall is built.

Special frames can be ordered which will incorporate a transom panel above the door and/or sidelites of various heights and widths beside the door.

Average material prices for hollow metal frames of various sizes are as follows:

18 Ga. Welded Hollow Metal Frames

Size	4¾"	5¾"	6¾"	8¾"
2'-6"x7'-0"	60.00	63.00	65.00	67.00
3'-0"x7'-0"	62.00	64.00	66.00	68.00
5'-0"x7'-0"	77.00	80.00	82.00	84.00
6'-0"x7'-0"	80.00	82.00	84.00	86.00

Deduct $10.00 each for knocked down (K.D.) frames.
Add $10.00 for "B" Label, $15.00 for "A" Label.

Labor Erecting Hollow Metal Frames. Two carpenters should erect about 16 welded frames per 8 hour day at the following labor cost per frame:

	Hours	Rate	Total	Rate	Total
Carpenter	1.0	$....	$....	$24.04	$24.04

One carpenter should erect about 12 K.D. frames per 8 hour day at the following labor cost per frame:

	Hours	Rate	Total	Rate	Total
Carpenter	0.67	$....	$....	$24.04	$16.11

Hollow Metal Doors. Doors are made with 20, 18, or 16 gauge face sheets and in 1⅜" and 1¾" thicknesses. Door sizes will correspond to the frame opening sizes stated previously in the Hollow Metal Frame data.

Doors may be ordered in a variety of styles including: small vision lites, half glass, full glass, dutch, louvered, and flush. Actually, when a door is ordered with a vision lite, what is received from the manufacturer is a flush door that has been cut out and prepared for future glass installation at the jobsite; the glass is not supplied by the door manufacturer.

The physical construction of the door itself is where the manufacturers really differ. They each have their own techniques for assembly, i.e. location of seams, inverted channel or flush top and bottom edges, gauge of hinge and closer reinforcements, leading edge reinforcement, core materials, etc. The contractor is obligated to determine the compliance of the manufacturers product with the requirements contained in the specification. Therefore, be sure of compliance with specification requirements when pricing hollow metal doors, because prices vary widely.

Average material prices for hollow metal doors are as follows:

20 Ga. Hollow Metal Doors

Size	1⅜"	1¾"
2'-6"x7'-0"	132.00	143.00
3'-0"x7'-0"	143.00	154.00
3'-6"x7'-0"	159.00	170.00

Add $20.00 for "B" Label, $45.00 for "A" Label
Add $60.00 for louver, $50.00 for lite cut-out.
Add $40.00 for mineral core (for temperature rise rating).

Labor Erecting Hollow Metal Doors. Two carpenters should hang, on hinges, about 12 doors per 8 hour day at the following labor cost per door:

	Hours	Rate	Total	Rate	Total
Carpenter	1.34	$....	$....	$24.04	$32.21
Laborer (distribution)	.5	....	....	19.13	9.57
			$....		$41.78

Fire Label Requirements. According to building codes, certain door locations within a building, such as at stairs, trash rooms, furnace rooms, garage entries, storage rooms, etc., will require a fire rating for the door and frame. This rating is expressed as "C Label" for ¾-hour rating, "B Label" for 1½- to 2-hour rating and "A Label" for 3-hour rating.

The rating is obtained by the manufacturer subjecting their product to actual fire testing by either Underwriters Laboratory or a Factory Mutual testing laboratory. The door and frame is then approved based on the physical construction of the unit. If the manufacturer changes the design after the approval, they must submit for a new test based on the re-designed unit. The approved designed and tested door and frame then qualifies to have a label affixed to it, which bears proof that it is a fire rated assembly.

When a door is to be fire rated, it also carries other restrictions, such as the latch or lockset must be installed with a UL throw bolt, glass is either not allowed (as in "A Label" rating) or restricted in size (as in "B or C Label" ratings), if a louver is needed in the door, it must have a fusible link that will melt and allow the louver blades to close when the fire reaches the door, and the door must have a closer device or spring hinges.

Metal Covered Jambs and Trim

Wood jambs and trim covered with No. 24 gauge zinc coated sheet steel, for use with kalamein doors.

Jambs are 1¾"x5½" rabbeted for 1¾" doors. Plain casing is ⅞"x3½". Round edge and brick mold is ⅞"x⅞".

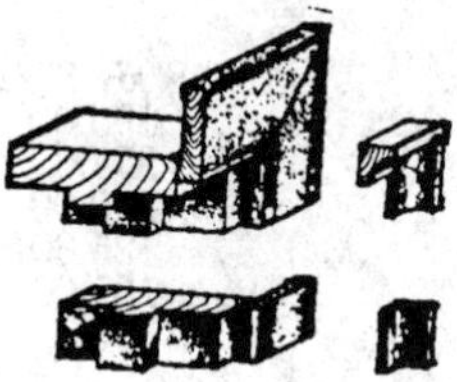

Size Up to	Prices Jamb Width To: 6 3/4	8 3/4	12
3'-0"x7'-0"	$74.00	$80.00	$85.00
4'-0"x7'-0"	80.00	85.00	90.00
5'-0"x7'-0"	85.00	90.00	95.00
6'-0"x7'-0"	90.00	95.00	100.00

For frames over 7'-0" add $5.00 for each 6" or less.

Exterior Metal Residential Door Assemblies

Prehung metal doors complete with moldings of the styles usually found on traditional wood doors and with a frame, threshold, weatherstripping, and a variety of locks are now available in over fifty stock designs. Prices for a 3' wide unit will vary from $190 to $730.00 depending on the type of lock and the type of glazing.

A plain, flush unit with a double cylinder lock and no glazing runs $210.00; a unit with raised panels runs $240.00; if the upper section is glazed with insulating glass the cost will raise to $270.00. Units similar to the above but in pairs for a 5' opening will run $400, $465 and $520 respectively. If moldings are ordered for the inside face of the door add $18.00.

08200 WOOD AND PLASTIC DOORS

Flush Veneer Slab Doors for Exterior Use

Furnished with water and weather resistant birch face veneers both sides of the door. All doors 1¾" thick.

Size	Hol. Core	Sol. Core
2'-6"x6'-8"	$40.00	$60.00
2'-8"x6'-8"	42.00	62.00
3'-0"x6'-8"	44.00	67.00
3'-0"x7'-0"	49.00	70.00
3'-0"x8'-0"	57.00	102.00

Extras for Altering Solid Core Flush Doors

For V-grooves, add per groove	$ 7.50
For cutting in one square light, add	18.00
For one circle light, approx. 12" diam., add	22.00
For one circle light, approx. 18" diam., add	26.50
For one peek light, add	18.00
For one diamond light, add	18.00

White Pine Entrance Doors

Type	2'-8"x6'-8"x1¾"	3'-0"x6'-8"x1¾"
6 Panel Colonial	$95.00	$100.00
4 Panel & Light	97.00	102.00
X Buck & 9 Light Rect.	110.00	115.00
X Buck & 12 Light Diag.	125.00	130.00
12 Panel Spanish	135.00	140.00
21 Panel Spanish	--	150.00

These doors can usually be purchased prehung in rabbeted frames of standard widths with 1½ pair of butts, without sill, lock or weatherstripping, for about $80.00 extra.

Exterior trim sets for front entrances are available; the price range is great, depending on depth of members, elaborateness and sharpness of carvings and flutings. One manufacturer offers trim resembling entrance "A" above at $160.00 in the deluxe model, $90.00 for a modified design; trim such as doorways "C", "D" and "E" around $90.00. These may also be purchased as complete entrance sets with frames, thresholds and sidelights if desired.

Western Pine Colonial Front Entrances

Colonial Front Entrances

Exterior wood doors are usually fitted with combination storm and screen units. These are usually of 1⅛" thick Ponderosa pine with aluminum wire. They run as follows for 6'-9" height:

	2'-6"	2'-8"	3'-0"
4 Light	$42.00	$44.00	$48.00
X Buck & 1 Light	--	54.00	57.00

Combination doors are also available in aluminum, 1" thick for either 6'-8" or 7'-0" openings in various qualities.

Standard quality, 2'-6", 2'-8" & 3'-0" wide	$42.00
Deluxe quality, 2'-6", 2'-8" & 3'-0" wide	52.00
Deluxe quality with white finish, 3'-0" wide	58.00
X-Buck type with white finish, 3'-0" wide	67.00

Soft Wood French Doors

Made of clear Ponderosa pine. Stiles and top rails are 4¾" and bottom rails are 9⅝" overall with bead and cove sticking. Muntins are ½" between glass. Wood stops or glazing beads are mitred and tacked in place.

	2'-6"x6'-8"		2'-8"x6'-8"		3'-0"x6'-8"	
Type	1⅜"	1¾"	1⅜"	1¾"	1⅜"	1¾"
1 Light French	$70.00	$74.00	$70.00	$74.00	$89.00	$95.00
5 Light French	71.00	75.00	71.00	75.00	90.00	97.00
15 Light French	72.00	76.00	72.00	76.00	91.00	100.00

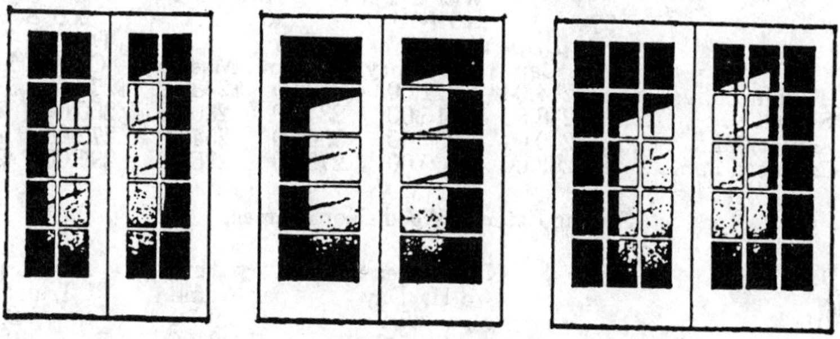

French Doors

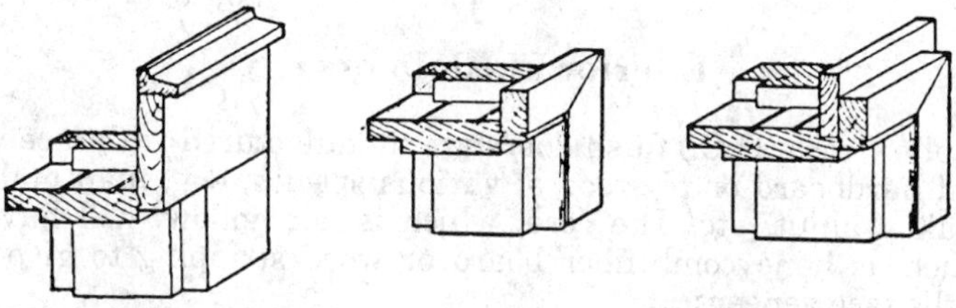

Stud Wall Drip Cap Frame.

Brick Wall Frame.

Brick Veneer Wall Frame.

Prices of Western Pine Exterior Door Frames

	Drip Cap 4⅝"	Drip Cap 5¼"	Masonry 4⅝"	Masonry 5¼"
2'-6"x6'-8"	$34.00	$35.00	$29.00	$30.50
2'-8"x6'-8"	35.00	36.00	30.00	31.00
2'-8"x7'-0"	36.00	37.00	30.50	31.50
3'-0"x6'-8"	37.00	38.00	30.50	32.50
3'-0"x7'-0"	37.50	38.00	31.50	32.50
4'-0"x6'-8"	40.00	42.00	32.50	33.50
5'-0"x6'-8"	43.00	45.00	33.50	34.50
6'-0"x6'-8"	48.00	49.00	35.50	36.50

Door Frame Prices with Preglazed Sidelight One Side for 3'-0"x6'-8" Doors

Glass	With 1-1'-0" x6'-8" S.L. Drip Cap	With 1-1'-0" x6'-8" S.L. Masonry	With 1-1'-4" x6'-8" S.L. Drip Cap	With 1-1'-4" x6'-8" S.L. Masonry	With 1-1'-8" x6'-8" S.L. Drip Cap	With 1-1'-8" x6'-8" S.L. Masonry
Saf. Gl.-1 Lite	$120.00	$114.00	$133.00	$128.00	$147.50	$142.00
3 Light Saf. Gl.	128.00	123.00	136.00	131.00	144.00	139.00
3 Light & Flat Pan.	143.00	138.50	150.50	145.00	158.50	153.00
3 Light & Raised Pan.	154.00	148.00	161.50	156.50	169.50	164.30

Also available in 4 and 5 light with or without flat or raised panel and with fluted, latticed rondel or loxenged, or insulating glass.

Door Frame Prices with Preglazed Sidelights Both Sides for 3'-0"x6'-8" Doors

Glass	With 2-1'-0" x6'-8" S.L. Drip Cap	With 2-1'-0" x6'-8" S.L. Masonry	With 2-1'-4" x6'-8" S.L. Drip Cap	With 2-1'-4" x6'-8" S.L. Masonry	With 2-1'-8" x 6'-8" S.L. Drip Cap	With 2-1'-8" x 6'-8" S.L. Masonry
Saf. Gl.-1 Light	$214.00	$211.00	$242.00	$238.50	$270.50	$267.00
3 Light Saf. Gl.	218.50	215.00	234.50	230.50	250.00	246.50
3 Light & Flat Pan.	240.00	237.00	250.00	254.50	272.00	268.00
3 Light & Raised Pan.	263.00	260.00	278.50	275.50	295.00	291.00

Setting Exterior Wood Door Frames

Size of Door Opening	No. Frames Set 8-Hr. Day	Carp. Hrs. per Frame	Add Hrs. for Transom
3'-0"x7'-0"	7-9	0.9-1.1	0.3
6'-0"x7'-0"	5-6	1.3-1.6	0.3
8'-0"x8'-0"	5-6	1.3-1.6	0.3
First Grade Workmanship			
3'-0"x7'-0"	5-7	1.1-1.6	0.3
6'-0"x7'-0"	4-5	1.6-2.0	0.3
8'-0"x8'-0"	4-5	1.6-2.0	0.3

Interior Flush Doors

Interior hollow core (H.C.) flush doors are manufactured with face veneers of tempered hardboard or plywood of various species, e.g., luan mahogany, birch, or oak, walnut, etc. The core, which is not hollow, will have some material such as honeycomb fiber board or wood stripping to give lateral support to the face veneers.

Interior Hollow Core Flush Doors

Doors $1\frac{3}{8}$" Thick	Plain Birch	Luan
2'-0"x6'-8"	$28.50	$16.00
2'-4"x6'-8"	29.50	17.00
2'-6"x6'-8"	31.50	22.00
2'-8"x6'-8"	35.00	24.00
3'-0"x6'-8"	36.00	25.00

For doors 7'-0" in height, add $6.00 per door.
For doors $1\frac{3}{4}$" thick, add $5.00 per door.

Interior solid core (S.C.) doors have face veneers of plywood in various species, the same as hollow core doors. The cores of the solid door are usually constructed of solid wood blocks (stave) or particleboard.

Interior Solid Core Flush Doors, Particleboard Core

	Rotary		Select
Size	Birch	Red Oak	Birch
2'-8"x6'-8"x$1\frac{3}{8}$"	$51.00	$54.00	$60.00
3'-0"x6'-8"x$1\frac{3}{8}$"	53.00	56.00	67.00
2'-8"x6'-8"x$1\frac{3}{4}$"	54.00	58.00	70.00
3'-0"x6'-8"x$1\frac{3}{4}$"	56.00	60.00	74.00

Pre-Hung Interior Door Units

Pre-hung door units, for interior use, consist of the door (usually $1\frac{3}{8}$" H.C.), frame, stops, face trim, and hinges all assembled together as a unit. The door is bored and mortised for the lockset and bolt and the frame is mortised for the strike plate.

The frame, which is referred to as "split jamb", parts into two pieces at the stop for installation in the wall opening.

Prices are for units with adjustable jambs, from $4\frac{3}{8}$" to $5\frac{1}{2}$" jambs.

Door Size	Luan Doors	Birch Doors
1'-6"x6'-8"	$63.50	$76.50
2'-0"x6'-8"	64.50	78.50
2'-4"x6'-8"	66.50	80.50
2'-6"x6'-8"	68.50	83.00
2'-8"x6'-8"	71.00	85.00

Interior Paneled Doors

Pine panel doors built up of stiles and rails have largely been supplanted by molded doors consisting of a single piece of resin-impregnated wood fibre. The doors are factory primed for finish painting on the job site and are joint-free and shrink and swell resistant. Colonial style 6 panel doors will cost as follows:

Size	
2'-0"x6'-8"x$1\frac{3}{8}$"	$32.00
2'-4"x6'-8"x$1\frac{3}{8}$"	34.00
2'-6"x6'-8"x$1\frac{3}{8}$"	35.00
2'-8"x6'-8"x$1\frac{3}{8}$"	37.00
3'-0"x6'-8"x$1\frac{3}{8}$"	40.00

Sliding and Bi-Folding Closet Doors

With the popularity of sliding doors for closets, many different combinations of doors, materials and sizes are on the market. With some of them, the doors are the same height as the other doors in the room or building, i.e. 6'-8" or 7'-0", while in others the doors extend to the ceiling, so that when opened, they provide access to the storage shelves at the top.

While there are many different types of doors, hardware, etc., the ones described below are of the type usually found in moderate class housing and apartments.

Frames for sliding doors may be obtained in kits containing two side jambs, head assembly with track attached and necessary hardware for hanging doors. Available with 5⅜" jambs for lath and plaster construction, or with 4⅝" jambs for dry wall construction. Openings may be arranged for two, three, or four by-passing doors. Any 1⅜" thick doors can be used.

The following table lists bypassing units, doors, and hardware only for a variety of 1⅜" thick doors.

2 Panel Opng.	H. C. Luan	H. C. Birch	W. P. Louver	W. P. Half Louver	6 Pan. W. P. Col.
4'-0"x6'-8"	$47.50	$58.00	$100.50	$127.00	$180.00
5'-0"x6'-8"	52.50	63.00	105.00	132.00	185.00
6'-0"x6'-8"	58.00	68.00	110.50	137.00	190.00

Bi-folding doors offer the advantage of full access to the closet behind. They are available in kits with doors, hinges, pivots, knobs, and guide assembly and track. Jambs and casing not included. Openings to 3'-2" panels hinged one side. 4'-0" and over 4 panels hinged two sides.

Sliding Closet Doors.

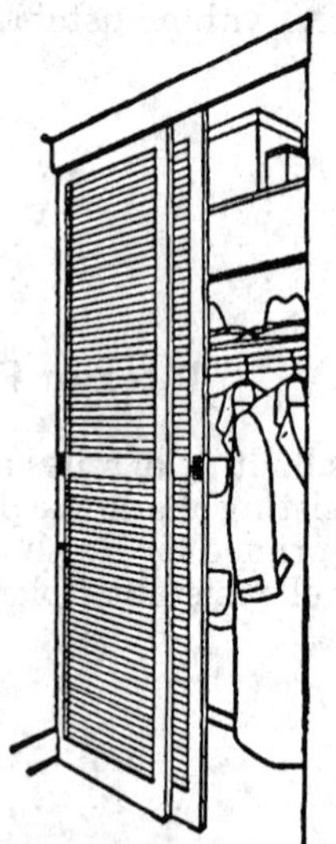

Louver Sliding Closet Doors.

Opng.	Panels per Opng.	$1\frac{3}{8}$" Hard-board	$1\frac{3}{8}$" H.C. Birch	$1\frac{1}{8}$" Louver	$1\frac{1}{8}$" Half Louver	$1\frac{3}{8}$" W. P. Col.
2'-0"x6'-8"	2	$26.50	$37.00	$47.50	$53.00	$58.00
2'-6"x6'-8"	2	29.50	42.00	52.00	58.00	63.00
3'-0"x6'-8"	2	33.00	44.00	54.50	62.00	68.00
4'-0"x6'-8"	4	39.00	69.00	78.00	88.00	113.00
5'-0"x6'-8"	4	46.00	75.00	83.50	93.00	118.00
6'-0"x6'-8"	4	52.00	78.00	85.50	98.00	125.00

PLASTIC FACED WOOD DOORS

For institutional use some doors are veneered on all surfaces with plastic laminate. The laminate can be used on both solid and hollow core doors as well as fire, acoustical, x-ray and other special units. Any standard plastic finish may be applied.

Unless otherwise specified, doors are edged as well as faced with plastic. The doors will be shipped in a carton prefinished and pre-mortised for all hardware and cannot be trimmed on the job. Therefore, special care must be taken in checking all details on the shop drawings to see that dimensions are accurately followed on the job.

Since all mortising and fitting is eliminated, hanging of the doors is simplified. One carpenter and one laborer should hang two doors an hour.

These doors are fabricated by manufacturers on special order. The veneers are not applied by contact adhesives, but are bonded under pressure. Costs should be obtained from the manufacturer or his distributor.

For budget purposes a door 2'-8"x6'-8" plastic faced both sides of a $1\frac{3}{4}$" solid core will run around $145.00. For one face plastic, one face birch around $110.00.

Interior Door Jambs

The following prices are for plain door jambs $\frac{3}{4}$"x$4\frac{1}{2}$". Side jambs are dadoed at top for head jambs. The prices are for jambs only, not including stops or trim which must be added. Furnished knocked down.

Price Per Set, Dadoed and Sanded, Not Including Stops

Size	Ponderosa Pine	Natural Birch
2'-8"x6'-8" or smaller	$16.00	$25.50
3'-0"x7'-0" or smaller	17.00	26.50
6'-0"x6'-8" or smaller	20.00	30.00

Add for Door Stops $\frac{7}{16}$"x$1\frac{3}{8}$", Cut to Length

	Pond. Pine	Nat. Birch
3'-0"x7'-0" or smaller	$1.20	$3.20
6'-0"x7'-0" or smaller	1.35	4.00

If door jambs are put together in the mill, add $3.00 per set to the above prices.

Setting Interior Wood Door Jambs

Door Size, Desc. of Jamb	No. Set 8-Hr. Day	Carp. Hrs. per Jamb	Add for Transom
3'-0"x7'-0", Plain door jambs	8-10	0.8-1.0	0.6
3'-0"x7'-0", Paneled door jambs	5-6	1.3-1.6	0.6
6'-0"x7'-0", Plain door jambs	5-6	1.3-1.6	0.6
First Grade Workmanship			
3'-0"x7'-0", Plain door jambs	7-9	0.9-1.1	0.6
3'-0"x7'-0", Paneled door jambs	4-5	1.6-2.0	0.6
6'-0"x7'-0", Plain door jambs	4-5	1.6-2.0	0.6

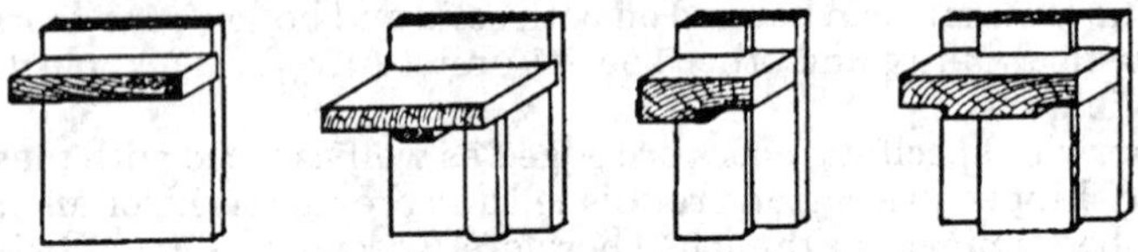

Door Jambs

Stock door casings per 100 lin. ft. cost $40.00 in Ponderosa pine and $117.00 in birch. Stops will run $13.00 and $26.00.

Oak thresholds, 3⅝"x⅝" will run $111.00 per 100 ft. and oak sills, 1$\frac{5}{16}$"x7¼" will run $318.00 per 100 ft. Pre-cut sills will run $3.50 for 2'-8" opening, $4.00 for 3' opening; in the narrow size, $10.00 and $12.50 for the wider.

Interior Door Trim

The labor cost of placing interior door trim will vary with the type of trim or casings used and the class of workmanship. Production time is given on 4 different types of interior door trim, as follows:

Style "A" consists of 1-member casing with either a mitred or square cut head or "cap". This is the simplest trim to use.

Style "B" consists of 1-member trim for sides and a built-up cap trim for heads. Caps may either be put together in the mill or assembled and put together on the job.

Style "C" consists of 2-member back-band trim, consisting of a one-piece casing with a back-band of thicker material.

Style "D" consists of any type of door trim that is assembled and glued-up in the mill, ready to set in place as a unit.

The class of workmanship will have considerable bearing on the labor costs. In ordinary workmanship the casings are nailed to the rough bucks or grounds and the trim may show some hammer marks, but where first grade workmanship is required, the utmost care must be used in cutting and fitting all casings, miters, etc. All nails must be carefully driven and set, the casings must show the same margin on all sides of the jamb and trim must fit close to the walls. In other words, it must be a first class job in every respect.

Fitting and Placing Stationary Transom Sash. Where just an ordinary grade of workmanship is required, a carpenter should fit and place about 12 to 14 stationary transom sash per 8-hr. day, at the following labor cost per sash:

Labor Erecting Interior Door Casings and Trim

Style Trim	Kind of Trim	Size of Opening	No. Sides Trim per 8-Hr. Day	Carp. Hrs. per Side	Carp. Hrs. per Opening	Add for Transom†
"A"	Single Casing	3'-0"x7'-0"	14–16	0.5–0.6	1.0–1.2	0.6
"B"	Cap Trim	3'-0"x7'-0"	14–16	0.5–0.6	1.0–1.2	0.6
"B"	Cap Trim*	3'-0"x7'-0"	7– 8	1.0–1.1	2.0–2.2	0.6
"C"	Back-band Trim	3'-0"x7'-0"	9–10	0.8–0.9	1.6–1.8	0.6
"D"	Mill Assembled**	3'-0"x7'-0"	14–16	0.5–0.6	1.0–1.2	0.6
"A"	Single Casing	6'-0"x7'-0"	9–10	0.8–0.9	1.6–1.8	0.8
"B"	Cap Trim	6'-0"x7'-0"	9–10	0.8–0.9	1.6–1.8	0.8
"B"	Cap Trim*	6'-0"x7'-0"	5– 7	1.1–1.6	2.2–3.2	0.8
"C"	Back-band Trim	6'-0"x7'-0"	7– 8	1.0–1.1	2.0–2.2	0.8
"D"	Mill Assembled**	6'-0"x7'-0"	9–10	0.8–0.9	1.6–1.8	0.8
		First Grade Workmanship				
"A"	Single Casing	3'-0"x7'-0"	9–10	0.8–0.9	1.6–1.8	0.8
"B"	Cap Trim	3'-0"x7'-0"	9–10	0.8–0.9	1.6–1.8	0.8
"B"	Cap Trim*	3'-0"x7'-0"	5– 7	1.1–1.6	2.2–3.2	0.8
"C"	Back-band Trim	3'-0"x7'-0"	8– 9	0.9–1.0	1.8–2.0	0.8
"D"	Mill Assembled**	3'-0"x7'-0"	11–12	0.7–0.8	1.4–1.6	0.8
"A"	Single Casing	6'-0"x7'-0"	7– 8	1.0–1.1	2.0–2.2	1.0
"B"	Cap Trim	6'-0"x7'-0"	7– 8	1.0–1.1	2.0–2.2	1.0
"B"	Cap Trim*	6'-0"x7'-0"	5– 6	1.3–1.6	2.6–3.2	1.0
"C"	Back-band Trim	6'-0"x7'-0"	6– 7	1.1–1.3	2.2–2.6	1.0
"D"	Mill Assembled**	6'-0"x7'-0"	7– 8	1.0–1.1	2.0–2.2	1.0

*Cap trim put together by carpenters on the job.
**Trim assembled and glued together in the mill.
†Includes transom jambs, stops and casings but no transom sash.

	Hours	Rate	Total	Rate	Total
Carpenter	0.6	$....	$....	$24.04	$14.42

First Grade Workmanship

Figure about 9 to 11 transom sash per 8-hr. day, at the following labor cost per sash:

	Hours	Rate	Total	Rate	Total
Carpenter	0.8	$....	$....	$24.04	$19.23

Fitting and Hanging Hinged or Pivoted Transom Sash. If the transom sash are hinged or pivoted, a carpenter should fit and hang 7 to 9 sash per 8-hr. day, at the following labor cost per sash:

	Hours	Rate	Total	Rate	Total
Carpenter	1.0	$....	$....	$24.04	$24.04

First Grade Workmanship

Figure about 5 to 7 hinged or pivoted transom sash per 8-hr. day, at the following labor cost per sash:

	Hours	Rate	Total	Rate	Total
Carpenter	1.3	$....	$....	$24.04	$31.25

Fitting and Hanging Wood Doors. The labor cost of fitting and hanging doors will vary with the class of workmanship, weight of door, whether soft or hard wood, and upon the tools and equipment used. There are power planes, hinge butt routers and lock mortisers that enable a carpenter to prepare about 3 times as much work per day as where the doors are fitted by hand. This must be considered when making up the estimate.

Where just an ordinary grade of workmanship is required, the doors do not always show the same margin at the top and sides, the screws are frequently driven instead of placed with a screw driver and in many instances the doors are "hinge-bound" so they will not open and close freely.

Ordinary Workmanship. Where doors are fit and hung by hand, a carpenter should fit and hang 8 doors per 8-hr. day, at the following labor cost per door:

	Hours	Rate	Total	Rate	Total
Carpenter	1.0	$....	$....	$24.04	$24.04

On large production line jobs where power tools can be used to advantage and where the same class of work is performed using a power plane, an electric hinge butt router for door and jambs and an electric lock mortiser, a carpenter should fit and hang 12 to 16 doors per 8-hr. day, at the following labor cost per door:

	Hours	Rate	Total	Rate	Total
Carpenter	0.6	$....	$....	$24.04	$14.42

The above costs will vary with the ability of the carpenter and the job setup, as a power planer should plane off a door in 5 to 10 minutes; a hinge butt router should do all the way from 60 to 75 openings per 8-hr. day, while a lock router should mortise out for a lock in about ½ minute, after it has been set on the door. The labor handling the doors is the same for both hand and machine work.

First Grade Workmanship

In the better class of buildings, the doors must show the same margin between the door and jamb at the top and sides, the hinges must be set even and flush with the jambs and the edge of the door and the doors must be free from "hinge binding" so they will fit well and work easily at the same time.

On this grade of work a carpenter should fit and hang about 4 doors per 8-hr. day, at the following labor cost per door:

	Hours	Rate	Total	Rate	Total
Carpenter	2.0	$....	$....	$24.04	$48.08

Where power planers, hinge butt routers and lock mortisers are used, a carpenter should fit and hang 7 doors per 8-hr. day, at the following labor cost per door:

	Hours	Rate	Total	Rate	Total
Carpenter	1.1	$....	$....	$24.04	$26.44

Labor Setting Jambs, Fitting and Hanging Doors, Placing Stops, Hardware and Casing One Door Opening Complete

Size of Door Opening	Style Trim	Kind of Trim	Carp. Hrs. per Opening	Add for Transom†
		Ordinary Workmanship		
3'-0"x7'-0"	"A"	Single Casing	3.1– 3.7	2.2
3'-0"x7'-0"	"B"	Cap Trim	3.1– 3.7	2.2
3'-0"x7'-0"	"B"	Cap Trim*	4.1– 4.7	2.2
3'-0"x7'-0"	"C"	Back-band Trim	3.7– 4.3	2.2
3'-0"x7'-0"	"D"	Mill Assembled**	3.1– 3.7	2.2
6'-0"x7'-0"	"A"	Single Casing	5.6– 6.5	2.8
6'-0"x7'-0"	"B"	Cap Trim	5.6– 6.5	2.8
6'-0"x7'-0"	"B"	Cap Trim*	6.2– 7.9	2.8
6'-0"x7'-0"	"C"	Back-band Trim	6.0– 6.9	2.8
6'-0"x7'-0"	"D"	Mill Assembled**	5.6– 6.5	2.8
		First Grade Workmanship		
3'-0"x7'-0"	"A"	Single Casing	5.5– 6.2	2.7
3'-0"x7'-0"	"B"	Cap Trim	5.5– 6.2	2.7
3'-0"x7'-0"	"B"	Cap Trim*	6.1– 7.6	2.7
3'-0"x7'-0"	"C"	Back-band Trim	5.7– 6.4	2.7
3'-0"x7'-0"	"D"	Mill Assembled**	5.3– 6.0	2.7
6'-0"x7'-0"	"A"	Single Casing	9.2–10.3	3.4
6'-0"x7'-0"	"B"	Cap Trim	9.2–10.3	3.4
6'-0"x7'-0"	"B"	Cap Trim*	9.8–11.3	3.4
6'-0"x7'-0"	"C"	Back-band Trim	9.4–10.7	3.4
6'-0"x7'-0"	"D"	Mill Assembled**	9.2–10.3	3.4

*Cap trim put together by carpenters on the job.
**Trim assembled and glued together in the mill.
†Above is for hinged or pivoted transom. For stationary transom deduct 0.4-hr.

For front door hardware, add 0.5-hr. for ordinary work and 1 to 2 hrs. for first grade work.

Where power planers, hinge butt routers, and lock mortisers are used, deduct 0.4 hr. for single door and 0.8-hr. for double door openings on ordinary workmanship; for first grade workmanship, deduct 0.9-hrs. for single door and 1.8 hrs. for double door openings.

Labor Installing Pre-Hung Door Units. Pre-hung door units are used extensively in the construction of houses and apartments ranging from "economy" grade to middle class.

In the completely assembled package, jambs are assembled, door is hung in place with hinges applied and stops are mitred and nailed into place.

Where assembled pre-hung door units are used, a carpenter should install 16 units per 8-hr. day, at the following labor cost per unit:

	Hours	Rate	Total	Rate	Total
Carpenter	0.5	$....	$....	$24.04	$12.02

Fitting and Hanging Hardwood Acoustical Doors. The labor cost of fitting and hanging hardwood acoustical doors is always figured on a first grade workmanship basis, because they must be perfectly installed to insure their efficient operation.

Assuming a carpenter, experienced in hanging acoustical doors, is employed to do the work, the following production should be obtained:

Assuming a carpenter, experienced in hanging acoustical doors, is employed to do the work, the following production should be obtained:

Type of Door	Approximate Weight per Sq. Ft.	Carpenter Hrs per Opng	Helper Hrs per Opng
Hardwood Class 36	4 lbs.	6	3
Hardwood Class 41	7 lbs.	8	4

Fitting and Hanging Toilet Stall Doors. A carpenter should fit and hang 9 to 11, ⅞" or 1⅛" toilet doors per 8-hr. day, at the following labor cost per door:

	Hours	Rate	Total	Rate	Total
Carpenter	0.8	$....	$....	$24.04	$19.23

First Grade Workmanship

Where hardwood stall doors are used and carefully hung, a carpenter should fit and hang 7 to 9 doors per 8-hr. day, at the following labor cost per door:

	Hours	Rate	Total	Rate	Total
Carpenter	1	$....	$....	$24.04	$24.04

Fitting and Hanging Heavy Wood Swinging Doors Up to 4'-0"x8'-0". When hanging heavy wood swinging doors, such as used in garages, stables, mill and factory buildings, etc., where the doors are 3'-0" to 4'-0" wide and 7'-0" to 8'-0" high, it will require about 4 to 4½ hrs. time to fit and hang one door or 8 to 9 hrs. per pair, at the following labor cost per door:

	Hours	Rate	Total	Rate	Total
Carpenter	4.3	$....	$....	$24.04	$103.37

Fitting and Hanging Heavy Wood Swinging Doors Up to 5'-0"x10'-0". To fit and hang extra heavy wood swinging doors up to 5'-0"x10'-0"x2½" or 3" thick, will require about 4 to 5 hrs. time for 2 carpenters, at the following labor cost per door:

	Hours	Rate	Total	Rate	Total
Carpenter	9	$....	$....	$24.04	$216.36

Fitting and Hanging Wood Sliding Doors Up to 4'-0"x8'-0". When fitting and hanging heavy wood sliding doors, such as used in garages, factories, etc., two carpenters working together should place tracks and hang one door complete in 3 to 3½ hrs. at the following labor cost per door:

	Hours	Rate	Total	Rate	Total
Carpenter	6.5	$....	$....	$24.04	$156.26

To place track and hang a pair of double sliding doors for and opening up to 8'-0"x8'-0", will require 10 to 12 hrs. carpenter time, at the following labor cost per opening:

	Hours	Rate	Total	Rate	Total
Carpenter	11	$....	$....	$24.04	$264.44

The above does not include cutting or drilling in masonry walls to place bolts or anchors but is based on this work being done when the walls are built.

Fitting and Hanging Heavy Wood Sliding Doors Up to 10'-0"x10'-0". To place track, fit and hang one sliding door up to 10'-0"x10'-0" in size and 2½" to 3" thick, will require about 12 hrs. carpenter time, at the following labor cost per opening:

	Hours	Rate	Total	Rate	Total
Carpenter	12	$....	$....	$24.04	$288.48

The above does not include cutting or drilling masonry walls for bolts or anchors.

Fitting and Hanging Heavy Wood Sliding Doors Up to 12'-0"x18'-0". To place track and hang one large door for an opening up to 12'-0"x18'-0", will require about 24 to 28 hrs. carpenter time, at the following labor cost per door:

	Hours	Rate	Total	Rate	Total
Carpenter	26	$....	$....	$24.04	$625.04

It will require 4 to 6 men to handle a door of this size.

Making and Setting Sliding Door Pockets. Where sliding door pockets are made on the job for door openings up to 3'-0"x7'-0", a carpenter should make the pocket, fit track and set the pocket in about 4 hrs. at the following labor cost:

	Hours	Rate	Total	Rate	Total
Carpenter	4	$....	$....	$24.04	$96.16

Where double doors are used and it is necessary to make and set 2 pockets, double the above costs.

Setting Sliding Door Pockets. If sliding door pockets are sent to the job put together with track in place or steel pocket door T-frames are used, and the carpenters on the job set the pocket or steel frame ready to receive the door, it will require about 1 hr. carpenter time, at the following labor cost per opening:

	Hours	Rate	Total	Rate	Total
Carpenter	1	$....	$....	$24.04	$24.04

If double sliding doors are used and it is necessary to set 2 pockets for each opening, double the above costs.

Fitting and Hanging Sliding Doors. To fit an ordinary sliding door up to 3'-0"x7'-0", place hangers and hang the door complete, not including finish hardware, will require about 1 hr. carpenter time, at the following labor cost per door:

	Hours	Rate	Total	Rate	Total
Carpenter	1	$....	$....	$24.04	$24.04

For double sliding doors, double the above costs.

Labor on Sliding Door Openings Complete with Pockets. Where necessary to make the sliding door pocket, hang the track, set the pocket and fit and hang the door complete, exclusive of finish hardware, figure about 5 hrs. carpenter time, at the following labor cost per opening:

	Hours	Rate	Total	Rate	Total
Carpenter	5	$....	$....	$24.04	$120.20

For double sliding door openings, double the above costs.

If pockets are sent to the job with track fitted in place or steel pocket door T-frames are used, and the job carpenters set the box or steel frame and fit and hang the sliding door, it will require about 2 hrs. carpenter time per opening, exclusive of finish hardware, at the following labor cost per opening:

	Hours	Rate	Total	Rate	Total
Carpenter	2	$....	$....	$24.04	$48.08

For double sliding door openings, double the above costs.

Labor Fitting and Hanging Sliding Wood Closet Doors. When hanging and fitting wood closet doors that slide past each other without the necessity for pockets, a carpenter should place track, attach hangers, fit and hang two doors complete for an opening 3'-0" to 5'-0" wide and 6'-8" to 8'-0" high in about 2 hours.

This does not include time setting door jambs and casing openings on two sides, as this should be added according to time given on previous pages.

The labor cost per opening should average as follows:

	Hours	Rate	Total	Rate	Total
Carpenter	2	$....	$....	$24.04	$48.08

08300 SPECIAL DOORS

TIN CLAD FIRE DOORS

Tin clad fire doors are used in factory, mill and warehouse buildings and may be either sliding or swinging, having fusible links and automatic closing devices.

Sliding doors and lap type swinging doors in brick walls do not ordinarily require a steel frame but they may be necessary with tile walls. Flush type swinging doors require a frame.

Standard tin clad fire doors are usually manufactured in two thicknesses depending on the fire rating required. For Class "A" rating, the thickness is 2⅝" consisting of 3-plies of tongue-and-groove lumber clad with tin sheeting on all sides. For Class "B" or "C" rating, the thickness is 1¾" consisting of 2-plies of tongue-and-groove lumber with cladding.

Method of Calculating the Area of Fire Doors. For flush doors, multiply the width of the opening in feet by the height of the opening in feet.

Square Top Lap Doors. Obtain size of door by adding 8" to width and 4" to height of opening, then multiply width of door in feet by height of door in feet.

Inclined Top Lap Doors. Obtain width of door by adding 8" to width of opening and obtain height by adding 8" to height of door for doors up to 4'-0" wide and 12" for doors up to 6'-0" wide. Multiply width of door in feet by height of door in feet.

Prices of Tin Clad Fire Doors and Shutters

Price includes door only, covered both sides with tin of a quality and in a manner specified by the National Board of Fire Underwriters.

	Price per Sq. Ft.
Standard 3-ply (2⅝" thick), Wt. 13 lbs./SF	$10.00
Standard 2-ply (1¾" thick), Wt. 10 lbs/SF	9.50
Standard 2-ply (1¾" thick), Shutters, Wt. 10 lbs./SF	9.00

Above prices include doors only, no hardware or fixtures of any kind or delivery costs.

Hardware for Swinging, Automatic Closing Tin Clad Doors

Fixture sets include hinges, pintles, weights, chains, catches, latches, complete automatic device with bolts and screws for attaching. Installation not included. Door frames not included.

Door Width & Height Up to and Including	Single Swinging Doors Flush Type	Single Swinging Doors Lap Type	Double Swinging Doors Flush Type	Double Swinging Doors Lap Type
3'-0"x6'-6"	$63.00	$86.00	$146.00	$169.00
3'-0"x8'-6"	79.00	108.00	165.00	197.00
3'-0"x10'-6"	94.00	128.00	185.00	221.00
4'-0"x6'-6"	67.00	89.00	148.00	171.00
4'-0"x8'-6"	82.00	111.00	168.00	199.00
4'-0"x12'-0"	98.00	133.00	188.00	225.00
5'-0"x6'-6"	70.00	91.00	150.00	174.00
5'-0"x8'-6"	85.00	115.00	171.00	203.00
5'-0"x12'-0"	102.00	137.00	192.00	232.00
6'-0"x6'-6"	74.00	94.00	153.00	176.00
6'-0"x8'-6"	89.00	118.00	174.00	206.00
6'-0"x12'-0"	107.00	141.00	196.00	233.00

If automatic controls are not wanted with swinging tin clad doors, deduct from the above prices, $8.00 for single doors and $12.00 per opening for double swinging doors.

Hardware for Sliding Tin Clad Doors

Prices include non-adjustable hangers, track, weights and bolts for attaching hardware to doors. Through wall bolts not included.

Hardware for single sliding door, 3'-0" wide opening	$191.00
For each 6" additional width of opening, add	16.00
Openings over 6'-0" wide require 3 hangers on each door, add for each additional hanger	11.00
Hardware for double sliding doors, 6'-0" wide opening	325.00
For each 6" additional width of opening, add	16.00

Labor Erecting Tin Clad Doors

The following schedules, furnished by the Richmond Fireproof Door Co., are guides for estimating labor for erection.

The hours of labor are based on brick walls 13" thick. For walls greater in thickness, add ¼ hour for one worker for each 4" in thickness per bolt involved. Concrete and stone walls require slightly more time drilling; terra cotta, slightly less. Where "doors both sides" are noted, this assumes doors back to back and tracks bolted together. When tracks cannot be bolted together use "one side" schedule multiplied by two.

Type	Width	Door One Side Mech.	Door One Side Help.	Door Both Sides Mech.	Door Both Sides Help.
Single Slide	2'-0" to 4'-0"	7	7	10	10
	4'-0" to 5'-2"	8	8	12	12
	5'-3" to 7'-8"	9	9	14	14
	7'-9" to 8'-8"	11	11	17	17
	8'-9" to 12'-0"	12	12	18	18
Slide in Pairs	4'-0" to 4'-5"	9	9	14	14
	4'-6" to 9'-8"	11	11	16	16
	9'-9" to 10'-6"	13	13	18	18
	10'-7" to 12'-0"	15	15	20	20
	Width				
Vertical	to 4'-10"	13	13	21	21
Sliding	4'-11" to 7'-6"	15	15	24	24
	7'-7" to 10'-2"	17	17	26	26
	10'-3" to 12'-0"	19	19	28	28
	Door Height				
Single Swing	to 5'-9"	5	5	7	7
(Flush or	5'-10" to 8'-9"	5½	5½	8	8
Lap)	8'-10" to 12'	6	6	9	9
	Door Height				
Swing in	to 5'-9"	8	8	12	12
Pairs (Flush	5'-10" to 8'-9"	9	9	14	14
or Lap)	8'-10" to 12'-0"	10	10	16	16

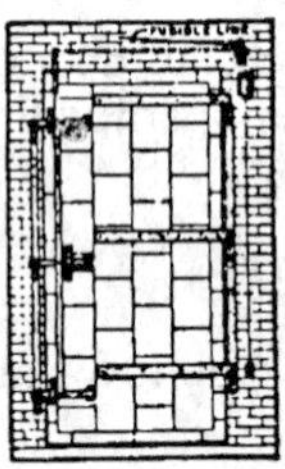

Swinging Type Tin Clad Door

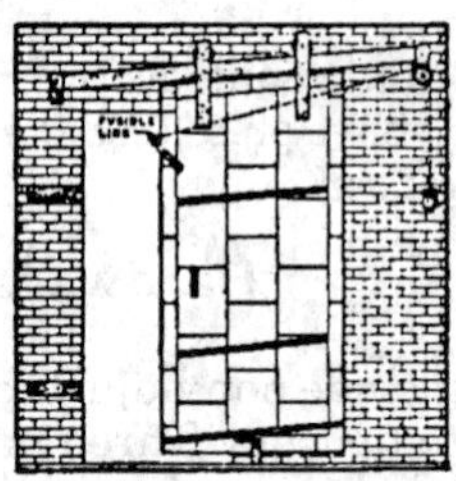

Gravity Sliding Type Tin Clad Door

Labor Cost of Installing One Single 3-Ply Sliding Tin Clad Fire Door For 4'-0"x7'-0" Wall Opening

The following estimate includes door, track, hangers, and necessary hardware installation.

	Hours	Rate	Total	Rate	Total
Mechanic	7	$....	$....	$24.04	$168.28
Helper	7	....	....	19.13	133.91
Cost per opening			$....		$302.19

Labor Cost of Installing One Pair Double Sliding 3-Ply Tin Clad Fire Doors for 6'-0"x7'-0" Wall Opening

The following estimate includes doors, track, hangers, and necessary hardware installation.

	Hours	Rate	Total	Rate	Total
Mechanic	11	$....	$....	$24.04	$264.44
Helper	11	....	....	19.13	210.43
Cost per opening			$....		$474.87

Labor Cost of Installing One Lap Type Single Swinging 2-Ply Tin Clad Fire Door for 3'-0"x7'-0" Wall Opening

The following estimate includes door, hardware and automatic controls installation.

	Hours	Rate	Total	Rate	Total
Mechanic	5	$....	$....	$24.04	$120.20
Helper	5	....	....	19.13	95.65
Cost per opening			$....		$215.85

Labor Cost of Installing One Pair Lap Type Double Swinging 2-Ply Tin Clad Fire Doors for 6'-0"x7'-0" Wall Opening

The following estimate includes doors, hardware, and automatic controls installation.

	Hours	Rate	Total	Rate	Total
Mechanic	9	$....	$....	$24.04	$216.36
Helper	9	....	....	19.13	172.17
Cost per opening			$....		$388.53

Labor Cost of Installing One Flush Type Single Swinging 2-Ply Tin Clad Fire Door for 3'-0"x7'-0" Wall Opening

The following estimate includes door, hardware, and automatic controls installation.

	Hours	Rate	Total	Rate	Total
Mechanic	5	$....	$....	$24.04	$120.20
Helper	5	....	....	19.13	95.65
Cost per opening			$....		$215.85

Labor Cost of Installing One Pair Flush Type Double 3-Ply Swinging Tin Clad Fire Doors for 6'-0"x7'-0" Wall Opening

The following estimate includes doors, hardware, and automatic controls installation.

	Hours	Rate	Total	Rate	Total
Mechanic	9	$....	$....	$24.04	$216.36
Helper	9	....	....	19.13	172.17
Cost per opening			$....		$388.53

Frames for tin clad doors are made of angles and channels usually furnished under metal fabrications and set by the mason. A 3x3 angle frame will run around $7.00 a foot installed. An 8" channel frame will run around $15.00 per foot installed.

08330 COILING DOORS

Coiling doors are usually called rolling shutters and are fully factory assembled units made of steel or aluminum. Often, they are labeled fire doors and installation is governed by Underwriter's specifications. Steel mounting jambs and heads are usually furnished by the metal fabrications contractor.

Costs for a manually operated, 20 gauge, door up to 12'x12' will cost around $7.00 to $8.00 a sq. ft., including hood, guides and operating hardware.

Where class A fire ratings are required, doors will cost $6.50 a square foot more. If 18 gauge construction, it will add another $1.00 a sq. ft. Motor operation will cost $800.00 for smaller sizes, $900 for medium. Each pass door will add another $800.00.

Two mechanics should hang and adjust a rolling shutter in size up to 12'x12' in one 8 hour day, at the following cost per shutter:

	Hours	Rate	Total	Rate	Total
2 Mechanics	16.0	$....	$....	$24.04	$384.64

Coiling Grilles

Rolling grilles are especially popular now for protection of store fronts, for open fronted stores in enclosed shopping areas and for closing off counter areas.

Stock commercial grilles will cost around $12.00 per sq. ft. with all hardware. As these are often part of a decorative front, they are often made of anodized aluminum, and cost around $14.00 a sq. ft. Exact prices can be obtained from the manufacturer, who will also furnish shop drawings upon request to show all the necessary anchoring involved. Steel grilles run around $8.00 a sq. ft. Motorizing and installation costs are the same as for rolling shutters.

08350 FOLDING DOORS

Steel Bi-fold Closet Doors. These may be installed in openings with drywall, wood, or steel jambs and head frame. Each bi-fold door package comes complete with door panels, track, guides, and pulls for a complete installation.

Doors are available in a variety of styles including flush, panel, half louver, full louver, etc. The doors are shipped "factory finished" with a baked-on paint finish.

A two panel unit is used for openings up to 3'-0" wide and will cost about $50.00 each. A four panel unit is used for openings from 4'-0" to 6'-0" wide, with the 6'-0" unit costing about $95.00 each. Units are sold in 6'-8" and 8'-0" heights, with the 8'-0" heights costing an additional $11 to $22 more per unit.

One carpenter should install a complete bi-fold unit and all hardware in about 45 minutes at the following labor cost per unit:

	Hours	Rate	Total	Rate	Total
Carpenter	0.75	$....	$....	$24.04	$18.03

Folding Doors. These are often folding partitions, can be ordered completely furnished and with all hardware in wood or vinyl. Folding doors for a 3'x6'-8" opening will run around $64.00 in vinyl, $95.00 in wood slats. Doors for 6'x6'-8" openings will run $150 for vinyl, $180 for wood. Deluxe and custom models will run somewhat more. All the above prices include hardware but the frame must be furnished by others. When the opening is over 150 sq. ft. the units are considered "partitions" rather than doors and will run around $5.00 a sq. ft. for light models to $7.00 a sq. ft. for heavy duty units.

08360 OVERHEAD DOORS

Overhead Garage Doors. Overhead garage doors are furnished in several different styles, including the one-piece overhead door; self-balancing overhead door and the spring balanced overhead doors.

For an average opening up to 8'-0"x7'-0", two carpenters should place tracks, fit and hang doors and apply the necessary hardware in 4 hours, at the following labor cost per door:

	Hours	Rate	Total	Rate	Total
Carpenter	8	$....	$....	24.04	$192.32

For larger openings up to 16'-0" wide and 7'-0" high, where one wide overhead folding door is used, two carpenters should place tracks, fit and hang doors and apply the necessary hardware complete in 5 hours, at the following labor cost per door:

	Hours	Rate	Total	Rate	Total
Carpenter	10	$....	$....	$24.04	$240.40

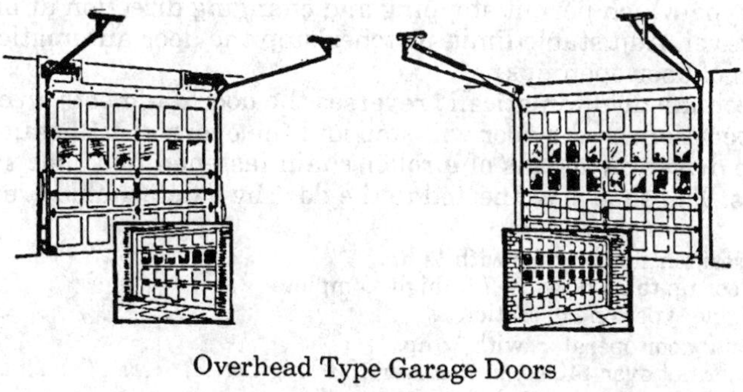

Overhead Type Garage Doors

GARAGE, FACTORY AND WAREHOUSE DOORS

Overhead Type Garage Doors

Overhead type garage doors are counter-balanced and controlled by oil tempered torsion springs. Cables are secured at the bottom of the door on each side and are directly connected with the pair of balancing springs located immediately over the door.

Overhead Type Garage Doors

Ball bearing rollers carry all sections of door over a continuous channel track. The track has a flange on it which keeps the rollers in the track.

Doors have kiln-dried fir or spruce sections, with hardboard panels. All joints are mortised, tenoned or doweled with steel pins and glued. All sections are fitted to size, squared for opening and drilled for hardware.

Doors over 12'-0" wide have steel struts in intermediate sections to prevent sagging. Doors over 15'-0" wide have steel struts in all sections.

Standard Size, Sectional Overhead Type, Garage Doors

Furnished complete with hardware, including cylinder lock, springs and track. Doors may be furnished with all wood panels or top section open for glass. Glass not included. Approximate prices:

Size Opening	1⅜" Thick	1¾" Thick
9'-0"x7'-0"	240.00	265.00
14'-0"x7'-0"	500.00	530.00
15'-0"x7'-0"	530.00	580.00
16'-0"x7'-0"	560.00	640.00
10'-0"x10'-0"	....	425.00
12'-0"x10'-0"	....	510.00
10'-0"x12'-0"	....	520.00
12'-0"x12'-0"	....	610.00

Electric Door Operators

Operators may be used with any type of overhead doors. Motors are suitable for single phase, 60 cycle, 110-120 Volt AC power supply, and are of the reversing type which permit stopping and changing direction at any point of the door travel. Adjustable limit switches stop the door automatically at top and bottom of door openings.

Safety mechanism automatically reverses the door if it meets an obstruction while in operation, or the door will stop and the operator will shut off.

Doors are driven by means of a roller chain fastened to a shoe sliding on 2 steel tracks. The shoe is connected to the door by a detachable arm.

Single door residential operator with ¼-hp. motor for door up to 22 ft. wide, 7 ft. high, complete with one toggle type control station	$190.00
For commercial door operator, with ¼-hp. motor for door not over 140 sq. ft. nor over 10 ft. high	265.00
Commercial door operator, with ⅓-hp. motor for door not over 168 sq. ft. or 12 ft. high	290.00
Commercial door operator, with ½-hp. motor for door not over 224 sq. ft. or 14 ft. high	350.00
Extra toggle type control station for commercial operator	11.00
Extra key switch (for residential operator only)	22.00

An experienced mechanic should install an electric door operator in 1 to 2 hours after the necessary wiring is in place.

Electronic Door Operators

For use where a transmitter is located in each automobile using the garage. A push of the button operates the door from the car.

For residential doors up to 16'-0"x8'-0" approx. price installed	$265.00

An experienced mechanic should install an electric door operator in 2 hours after the necessary wiring is in place.

Special Hardware--Add per Door

Low headroom, 6" minimum, for doors not over 132 sq. ft.	$16.00
Low headroom, 6" minimum, for doors not over 133-175 sq. ft.	$28.00
Low headroom, 6" minimum, for doors over 16'-0" wide or over 12'-0" high	$40.50
High lift (for lubritoriums) up to 2'-0" above head jamb	$23.50
High lift (for lubritoriums) up to 5'-0" above head jamb	$48.00

On commercial jobs steel doors are often specified. A 10'x10' door will cost about $425.00 with manual operator, $750.00 with electric operator.

Fiberglass and aluminum doors are light in weight and let in daylight. A stock door will cost about $240.00 for a single car size, $400.00 for a double car size.

Doors of Special Designs

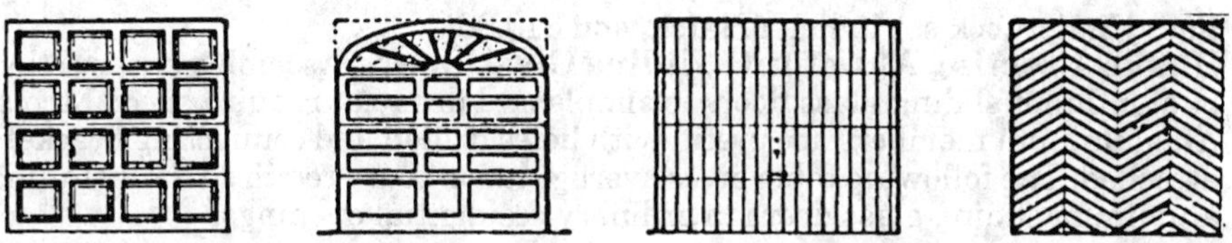

Colonial　　Fan Top　　Vertical Battens　　Diagonal Battens

Colonial doors, heavily molded outside with raised panels, add per sq. ft. $1.50.

For fan top on door, add $130.00 extra.

Vertical battens, 2-ply V-jointed or beaded, add per sq. ft. $1.50.

Diagonal battens, 2-ply V-jointed or beaded, add per sq. ft. $1.75.

08370 SLIDING GLASS DOORS

Aluminum sliding glass doors are available in several types, as illustrated, for glazing with crystal glass or 5/8" insulated glass. "X" indicates sliding panel; "O" indicates a fixed panel.

Frame, glazing bead and door members are heavy aluminum extrusions. Door sections move upon bottom mounted, ball bearing, grooved brass rollers. Frames are designed with provisions for installing bottom roller, horizontally sliding type screens. Weatherstripping is mohair pile, factory mounted in channels to form a continuous double seal. Hardware includes full-grip lucite pulls and touch latch. Cylinder lock sets are available at extra cost.

Screens are made from aluminum extrusions of tubular box type section with bottom rollers and fitted with fiberglass screen.

Prices are for door units set up. Factory assembly includes attachment of all accessory parts, such as weatherstripping, roller mechanisms, latch, corner connecting brackets, bumpers, etc. Screen units are assembled. Glass and glazing, is included.

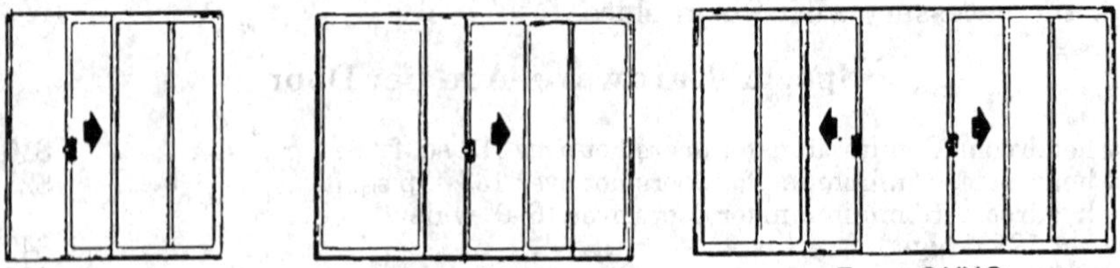

Sliding Glass Doors

Material Prices of Aluminum Sliding Doors

Type	Frame Size	Economy Crystal Gl.	Deluxe Insulating Gl.
XO	6'-0"x7'-0"	$265.00	$530.00
OXO	9'-0"x7'-0"	315.00	630.00
OXXO	12'-0"x7'-0"	415.00	888.00

For cylinder lock set in lieu of latch, add $10.00.

Labor Erecting Aluminum Sliding Glass Doors. Assembly and erection of aluminum sliding glass doors is simple and does not require special tools. All aluminum members are prefit, with holes drilled and connecting brackets attached. The following table gives average labor hours required for installing aluminum sliding glass doors in ordinary residential openings.

Type	Number of Doors	Number of Fixed Panels	Mechanic Hours Required for Assembly	Mechanic Hours Required for Erection
XO	1	1	2	6
OXO	1	2	3	8
OXXO	2	2	4	10

Performance Standards. There are many suppliers of aluminum windows and sliding doors, and it is often difficult to compare the quality of one manufacturer with that of another. To aid one's selection, the Architectural Aluminum Manufacturers Association publishes quality specifications, and manufacturers belonging to that association affix the AAMA "Quality-Certified" label to their products attesting that they complies with the designated specification. These specifications include the following for sliding glass doors:

SGD-B1 - For residential and limited commercial applications

SGD-A2 - For commercial and residential application requiring increased size or performance characteristics.

HP Series - Either of the above designations followed by "HP" indicates that the door complies with High Performance requirements.

The specifications for aluminum windows are as follows:

DH-B1	Double hung residential
DH-A2	Double hung commercial
DH-A3 & A4	Double hung monumental
C-B1	Casement residential
C-A2	Casement commercial
C-A3	Casement monumental
P-B1	Projected residential
P-A2	Projected commercial
P-A3	Projected monumental
A-B1	Awning residential
A-A2	Awning commercial
HS-B1 & B2	Horizontal sliding residential
HS-A2	Horizontal commercial
HS-A3	Horizontal monumental
J-B1	Jalousie residential
JA-B1	Jalousie-awning residential
VS-B1	Vertical sliding residential
TH-A2	Top hinged commercial
TH-A3	Top hinged monumental
VP-A2	Vertically pivoted commercial
VP-A3	Vertically pivoted monumental
HP Series	Any of the above followed by "HP" indicates the window complies with High Performance requirements.

The AAMA also publishes a certified products directory with a current listing of all certified windows and sliding glass doors by company, model and maximum size approved.

Wood Sliding Or Patio Doors

Sliding doors are now available from most of the nationally distributed millwork houses and follow the general designs established for aluminum sliding units. These doors can usually be purchased in two basic widths per panel and in various arrangements including two panels, one fixed, one operating; three panels, two sides fixed and center panel operable; and four panels, two end panels fixed and two center panels operable.

The following price schedule is based on units preglazed with 5/8" safety glass and complete with operating hardware, weatherstripping, sill, and screens for operable sash.

Glass Size	Nominal Door Size	Type	Material Cost
28" wide	5'-0"x6'-8"	Double XO	$500.00
34" wide	6'-0"x6'-8"	Double XO	538.00
46" wide	8'-0"x6'-8"	Double XO	698.00
34" wide	9'-0"x6'-8"	Triple OXO	830.00
46" wide	12'-0"x6'-8"	Triple OXXO	1,050.00

Add for keyed lock, $12.00. Grilles that can be removed for cleaning are available in either diamond or rectangular patterns. Costs are $43.50 each for the 28" width, $44.50 each for the 34" width, $53.00 for the 46" width in the rectangular pattern; and $45.00, $47.50 and $65.00 each for the diamond patterns.

08380 SOUND RETARDANT DOORS

Hardwood Sound Insulating Doors

Acoustical doors are for use in consultation rooms, conference rooms, hospitals, clinics, music rooms, television and radio studios, and other areas where acoustical levels must be closely maintained.

Dependent on the desired decibel reduction, or acoustical efficiency, doors are identified and furnished in several models.

Units 1¾" thick are rated in sound transmission class 36. Units 2½" thick are rated in sound transmission class 41. A door 3" thick is also made which has superior acoustical qualities.

Acoustical doors are made in a wide selection of face woods. Custom finishing of doors is available. These are veneered doors and, as such, can be furnished with veneers to match those of the balance of the doors in the building. Any standard hardware may be used for door erection.

The following prices include bottom closers, stops, gaskets, and stop adjusters and are for standard rotary cut veneers of natural birch and red oak.

DETAILS OF HPC ACCOUSTICAL DOORS

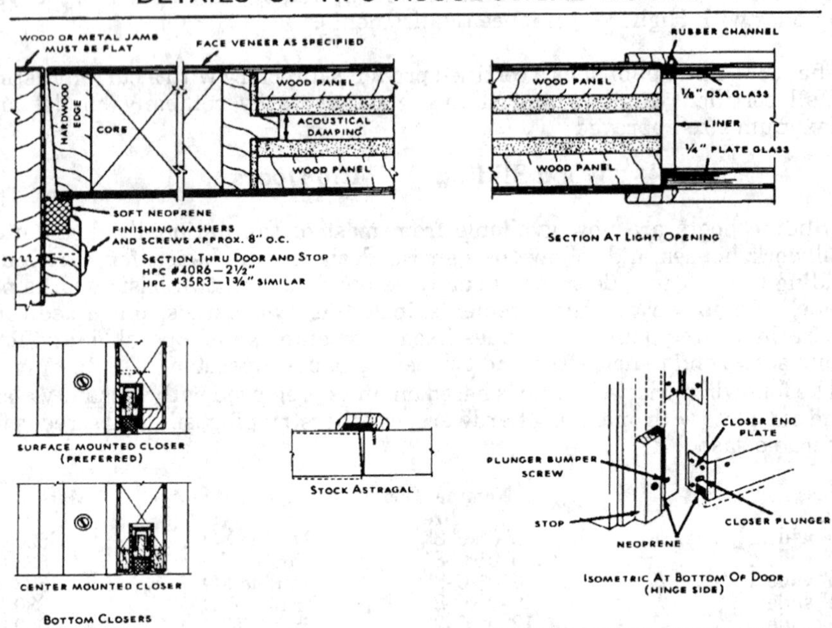

Material Prices of Hardwood Sound Insulating Doors*

Size	1¾" 35 Decibel	2½" 40 Decibel	3" 50 Decibel
2'-6"x6'-8"	$265.00	$345.00	$455.00
2'-8"x6'-8"	275.00	355.00	465.00
3'-0"x6'-8"	285.00	365.00	475.00
2'-6"x7'-0"	275.00	355.00	465.00
2'-8"x7'-0"	285.00	365.00	475.00
3'-0"x7'-0"	295.00	375.00	485.00

*f.o.b. factory, uncrated.

Light openings:	1¾"	2½"	3"
Rectangular vision light not over 18" square	$35.00	$37.00	$40.00

Astragals. Astragals for pairs of doors, 2½"x1" in any wood to match face veneer with 1"x½" rubber gasket, $30.00 per pair. For rounding meeting edges and grooving one edge with Eveleth bumper strip, add $30.00.

Other woods. For other woods, add to the total price the following percentages:

Sliced natural birch and sliced red oak	8%
Rotary cut red or white birch and rotary cut plain white oak	15%
Sliced red or white birch and sliced plain white oak	20%
Rift sliced white oak and sliced African mahogany	30%
Sliced American black walnut	50%

08400 ENTRANCES AND STORE FRONTS

METAL STORE FRONT CONSTRUCTION

Aluminum storefront and entrance doors are manufactured by a number of firms, in a wide variety of styles, to suit the architectural aesthetics desired or a particular design function for most any type of building.

The glass and glazing subcontractor is the firm to contact for this type of work, and they typically represent more than one manufacturer of these materials. This subcontractor will handle all the necessary functions of bidding, shop drawing preparation, ordering, and erection of the storefront materials, as well as the glass, for a complete installation.

Materials are made from an aluminum alloy and extruded into many different shapes and sizes of various thicknesses. The extrusions are offered in stock lengths of 24' long, however, longer lengths are available upon special order.

To prevent oxidation of the aluminum surfaces, the extrusions are anodized in a number of colors as well as the clear anodized finish. In recent years this material has become available with a paint coating that offers exceptional durability in addition to a controlled color gradation that is difficult to obtain with the color anodized finishes.

Entrance doors are available in numerous sizes and designs (narrow stile, medium stile, and wide or monumental stile) with various styles of applied panels and finish hardware.

The doors can be hung into the frame by using butt hinges, offset pivots, or center pivots. The controlled automatic closing of the doors can be achieved by installing an exposed overhead closer, a closer concealed in the horizontal head frame member, or by a floor closer that is set into a pocket in the floor under the hinge point of the door.

Installation. Once the subcontractor has a contract agreement to provide the storefront construction, he will prepare shop drawings based on the architectural drawings for the project. The shop drawings clearly define all dimensions and jointing details necessary for the installation, and the materials are ordered based on these drawings.

Prior to fabrication of the exact lengths necessary for erection, the subcontractor will verify the shop drawing dimensions with actual field dimensions taken at the jobsite.

After the extrusions are cut to the proper length for field erection, they are delivered to the jobsite, distributed to the proper location, and installation begins.

The extrusions are fastened to each other with concealed, extruded aluminum clips at all joints. The pre-fabricated frames are then set into the opening and fastened, through the glazing pocket, to the substrate material with screws or expansion anchors. Proper assembly allows the vertical extrusions to run top to bottom, with all horizontal extrusions fitted between the verticals.

Method Of Estimating Materials. Measure the length of each type and size extrusion and list into proper groupings. Since this material is sold in stock lengths, the estimator will have to combine the measured lengths together to arrive at the total number of stock lengths needed for the job with the least amount of waste. Count the joints to ascertain the required number of concealed clips (one per joint). It will also be necessary to measure the total length of vinyl gasket material needed for glazing, since this material is ordered separately from the aluminum extrusions. After all of the material quantities are measured, grouped and totaled, the appropriate material prices can be applied for the extension of the material costs.

Doors are listed by size and style, and depending upon the architectural requirements, they may be stock units or custom units. The pricing manual, supplied by the aluminum manufacturer, will list the material pricing strategy for the stock units as well as for the component parts necessary for the custom units.

Installed Costs Of Storefront Construction. Since there are so many different sizes and shapes of extrusions available for the installation of this type of construction, it would be inappropriate to try to list the material prices and labor time required for the installation, as the labor hours are directly related to the size and shapes of the extrusions to be installed. For instance, a simple storefront and entrance door installation, 40'-0" wide x 10'-0" high, in one system might take 72 man hours, while a more sophisticated system would require 100 man hours.

For budget purposes only, the following installed costs of aluminum storefront framing (glass not included) are given on a per square foot basis for the opening to be enclosed:

a) Windows without intermediate vertical mullions up to 5'-0"x5'-0" in size, using 1¾"x4½" tube - $10.00 per sq. ft.

b) Same as above, but with one vertical mullion - $11.00 per sq. ft.

c) Windows 40'-0"x10'-0" in size with vertical mullions spaced 5'-0" on center, using 1¾"x4½" tube - $9.00 per sq. ft.

d) Same as above, but with 3"x6" tube - $12.50 per sq. ft.

e) For a narrow stile single door and frame, 3'-0"x7'-0" - $580.00 each.

f) For a pair of narrow stile doors and frame, 6'-0"x7'-0" - $900.00 each.
Note:
Add 15% for bronze anodized finish.
Add 20% for black anodized finish.
Add for transom over single door, $85.00; over pair of doors, $150.00.

Revolving Doors. Revolving doors are made in a variety of styles and sizes in aluminum, bronze and stainless steel. The doors are fitted with a mechanism to control the travel speed and are pivoted so that they may collapse together to provide an unobstructed opening. The semi-circular wings at each side of the four leaf revolving door provides a surface for the door sweep to ride against, and therefore, at no point during normal operation is there a clear path to the exterior for the escape of the building heat or air conditioning to the outside atmosphere.

Average installed prices for revolving doors are as follows: Aluminum, $15,000; bronze, $26,500; stainless steel, $22,000.

Storefront Members. The following detail drawings will show some typical cut sections through aluminum storefront extrusions:

08500 METAL WINDOWS

Steel and aluminum windows are available in a great variety of types and sizes to meet almost all building requirements. These products have been standardized by the manufacturers to reduce costs and promote quicker deliveries.

The illustrations included in this chapter are for reference when selecting requirements. For full description and specifications of available products refer to manufacturers' catalogs.

All metal window and door products are either classified as "Standard", "Commodity" or "Special".

"Commodity Units" are those selected from "Standards" which are subject to the greatest demand and are available from manufacturers' warehouse stocks or through dealers engaged in the distribution of these products. "Standard Units" and "Special Units" are not available from stock and must be fabricated and shipped from factory. The trend is for distributors to send all orders to the factory for both pricing and fabrication.

The following information will deal with "Commodity Units" only. For "Standard Units" and "Special Units" obtain quotations from manufacturers or their dealer representatives.

Labor Installing Steel Windows and Doors. Installation costs will vary with the class of labor used, type and size of individual units and job working conditions. However, a reasonably accurate estimate of installation labor costs can be made by multiplying the number of man hours required to install each unit by the prevailing hourly rate of pay, plus rates for pension, welfare, etc., in effect in the area in which job is located.

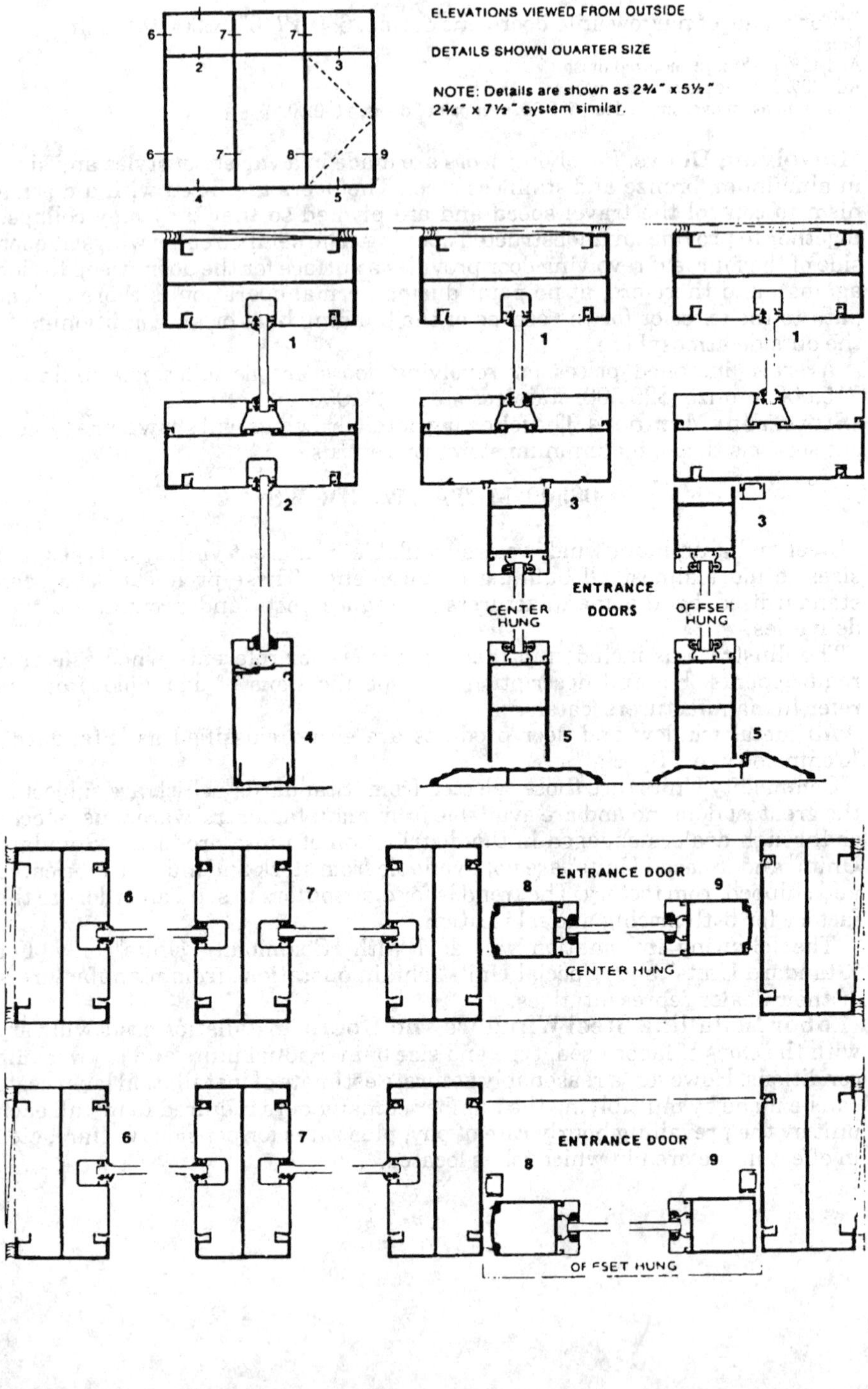
ELEVATIONS VIEWED FROM OUTSIDE
DETAILS SHOWN QUARTER SIZE
NOTE: Details are shown as 2¾" x 5½"
2¾" x 7½" system similar.
ENTRANCE DOORS
CENTER HUNG
OFFSET HUNG
ENTRANCE DOOR
CENTER HUNG
ENTRANCE DOOR
OFFSET HUNG

Product	Window Opening	Manhours Each Unit
Pivoted, Commercial	Single Unit*	2.25
Projected and Architectural	Two or more Units*	
Projected Windows	(a) Precast Sills	1.25
	(b) Poured Sills	1.50
Fixed Industrial	Single Unit	1.75
	Two or More Units	
	(a) Precast Sills	.75
	(b) Poured Sills	1.00
Fixed Architectural	Single Unit	2.00
	Two or More Units	
	(a) Precast Sills	1.00
	(b) Poured Sills	1.25
Intermediate Windows	In Wood	
(Casement, Combination	(a) Single Unit	2.75
and Projected)	(b) Two or More Units	2.25
	In Stone	
	(a) Single Unit	3.00
	(b) Two or More Units	2.50
	In Masonry	
	(a) Single Unit	2.25
	(b) Two or More Units	1.25
Mullions Only		.50
Mullions with Covers		.75
Residence Casements (Roto)	Single Unit	
	(a) In Wood	1.75
	(b) In Masonry or Brick Veneer	1.25
	Two or More Units	
	(a) In Wood	1.50
	(b) In Masonry or Brick Veneer	1.00
Residence Casements (Fixed)	Single Unit	
	(a) In Wood	1.50
	(b) In Masonry or Brick Veneer	1.00
	Two or More Units	
	(a) In Wood	1.25
	(b) In Masonry or Brick Veneer	.75
Casements in Casings	Single or Multiple Units	1.50
Continuous Top Hung, Fixed	Per Lineal Foot	.50
Continuous Top Hung, Swing	Per Lineal Foot	.75
Mechanical Operators, Rack and Pinion	Per Lineal Foot	.75
Mechanical Operator, Lever Arm	Per Lineal Foot	.50
Mechanical Operators, Tension	Per Lineal Foot	1.50
Industrial Doors and Frames	Up to 35 Sq. Ft.	
	Per Sq. Ft.	.125
	Over 35 Sq. Ft.	
	Per Sq. Ft.	.10
Double Hung Windows	Single Units	1.50
	Two or More Units	1.25
Screen Installation		
For Projected Windows	Net per Screen	1.10
For Basement and Utility	Net per Screen	.75
For Security Windows	Net per Screen	1.10
For Casement Windows	Net per Screen	.75
For Pivoted Windows	Net per Vent	1.50

*Single ventilator units. For each additional ventilator, add .25.

Prices of Commercial Projected Steel Windows

Window Type	Ventilated Window Price	Screen Price	Fixed Window Type	Window Price
*A12120	$58.00	$11.00	B-22	$31.00
*A13121	59.00	11.00	B-23	40.00
*B22140	70.00	15.00	B-24	45.00
*B23141	77.00	15.00	B-25	57.00
*B24141	85.00	15.00	B-26	69.00
*B25141	94.00	15.00	A-32	40.00
B2522402	99.00	**23.50	A-33	47.00
*B26141	104.00	15.00	A-34	59.00
B2622403	113.00	**23.50	A-35	70.00
*A32160	85.00	18.00	A-36	82.00
*A33161	93.00	18.00	A-42	86.00
*A34161	104.00	18.00	A-43	102.00
*A35161	115.00	18.00	A-44	114.00
*A36161	126.00	18.00	A-45	131.00
A3523602	115.00	**27.50	A-46	145.00
A3623603	126.00	27.50		
A43141	101.00	16.00		
A44141	115.00	16.00		
A45141	132.00	16.00		
A46141	138.00	16.00		

*These types available with either project-out or project-in vents.
**Price includes wicket screen for project-out vent and fixed screen for project-in vent.

Prices include cam handle and pole ring for project out ventilators and cam handle with strike or spring latch with strike for project in ventilators. Hardware is cadmium plated.

Prices also include standard installation fittings, provisions for standard screens, bonderizing and one factory coat of paint. Glass, putty, glazing and glazing clips not included.

Prices of Mullions and Covers

Lights High	2	3	4	5	6
Mullions	$7.00	$9.00	$12.00	$15.00	$17.00
Covers	8.00	10.00	13.00	16.00	19.00

Prices of Architectural Projected Steel Windows

Window Type	Window Price	Screen Price	Window Type	Window Price	Screen Price
252A	$69.00	$12.00	215	$137.00	$21.00
252B	69.00	9.00	216	140.00	21.00
253	85.00	12.00	218	145.00	21.00
261	$73.00	$9.00	221	$85.00	$10.00
262A	75.00	13.00	222A	94.00	15.00
262B	77.00	10.00	222B	98.00	13.00
263	88.00	13.00	223	100.00	15.00
264	104.00	13.00	224	112.00	15.00
265	123.00	19.00	225	132.00	24.00
266	174.00	19.00	226	142.00	24.00
268	188.00	19.00	228	147.00	24.00
211	81.00	9.00	273	115.00	18.00
212A	92.00	14.00	274	124.00	18.00
212B	93.00	12.00	275	164.00	26.00
213	96.00	14.00	276	181.00	26.00
214	100.00	14.00	278	191.00	26.00

Prices include bronze operating hardware for ventilated units. Cam handle, strike, and pole ring for project out vents and cam handle with strike or spring latch with strike for project in vents. Also included are standard installation fittings, bonderizing, and one factory coat of paint. Glass, putty, glazing and glazing clips not included.

Screen prices include screens for all vents in unit. Wicket screens are furnished for project out vents and outside fixed screens for project in vents. Prices also include clips for attachment.

Polished bronze hardware in place of standard bronze hardware, add per vent . $2.75

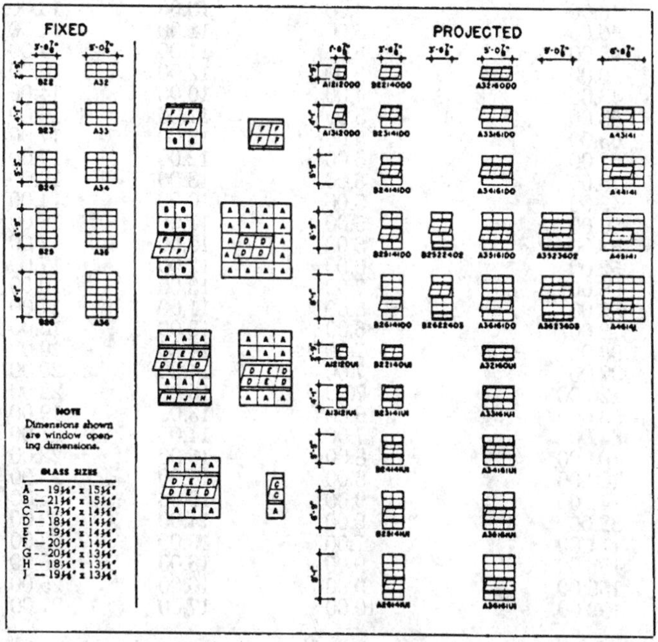

Types and Sizes of Commercial Projected Windows

Prices of Security Steel Windows

Window Type	Window Price	Screen Price	Mullion Price
32130	$80.00	$9.00	$4.00
33160	98.00	12.00	5.00
62 Fixed	77.00	....	4.00
62160	109.00	12.00	4.00
631120	112.00	16.00	5.00
641121	140.00	16.00	6.00
931180	177.00	20.00	5.00
941181	198.00	20.00	6.00

Prices include standard operating hardware, installation, fittings, bonderizing, and one factory coat of paint. Glass, putty, glazing, and glazing clips are not included.

Anchors

Anchors for securing frames to wood or masonry walls are included with window units. Four anchors per opening are furnished for heights up to and including 4'-5½". Six anchors per opening to and including 5'-9½". Eight anchors per opening to and including 8'-1½".

Prices of Residential Steel Casements Ventilated Casements

Window Type	Window Price	Head and Jamb Fins Per Set	Inside Casing Price	In/Out Trim Price	Screen Price
1212	$38.00	$4.00	$8.00	$12.00	$6.00
1313	43.00	4.00	10.00	13.00	7.00
1413	50.00	5.00	11.00	14.00	7.00
1414	48.00	5.00	11.00	14.00	8.00
1514	56.00	6.00	12.00	16.00	8.00
2212	43.00	4.00	10.00	14.00	6.00
2313	53.00	5.00	11.00	16.00	7.00
2414	65.00	5.00	12.00	17.00	8.00
2413	62.00	5.00	12.00	17.00	7.00
2514	78.00	6.00	13.00	19.00	7.00
2222	43.00	5.00	10.00	14.00	8.00
2323	50.00	5.00	11.00	16.00	10.00
2424	53.00	6.00	12.00	17.00	11.00
2423	52.00	6.00	12.00	17.00	10.00
2524	56.00	7.00	13.00	19.00	11.00
3222	52.00	5.00	11.00	17.00	8.00
3323	72.00	6.00	13.00	18.00	10.00
3424	96.00	7.00	14.00	20.00	11.00
3423	92.00	7.00	14.00	20.00	10.00
3524	120.00	7.00	15.00	22.00	11.00
4222	56.00	6.00	13.00	19.00	8.00
4323	84.00	7.00	14.00	21.00	10.00
4424	109.00	8.00	15.00	22.00	11.00
4423	108.00	8.00	15.00	22.00	10.00
4524	142.00	9.00	16.00	25.00	11.00
5222	82.00	6.00	14.00	22.00	8.00
5323	114.00	7.00	15.00	23.00	10.00
5424	158.00	9.00	16.00	25.00	11.00
5423	156.00	9.00	16.00	25.00	10.00
5524	192.00	10.00	17.00	27.00	11.00

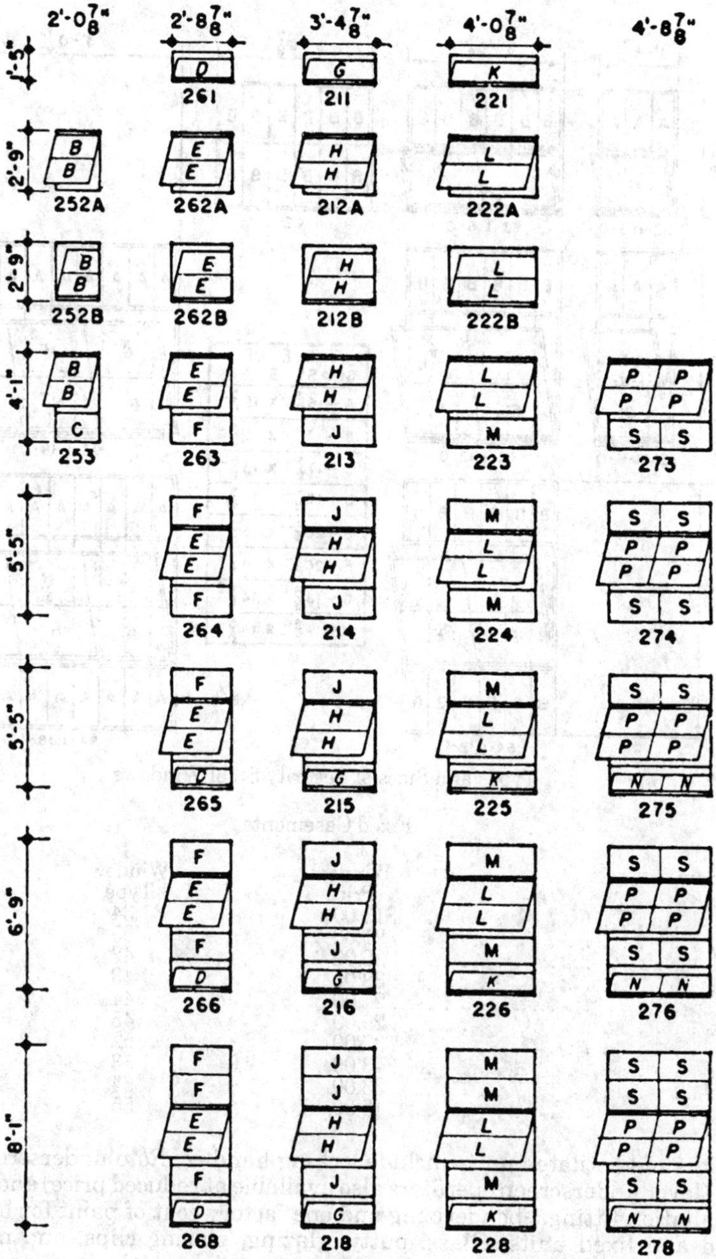

Types and Sizes of Architectural Projected Steel Windows

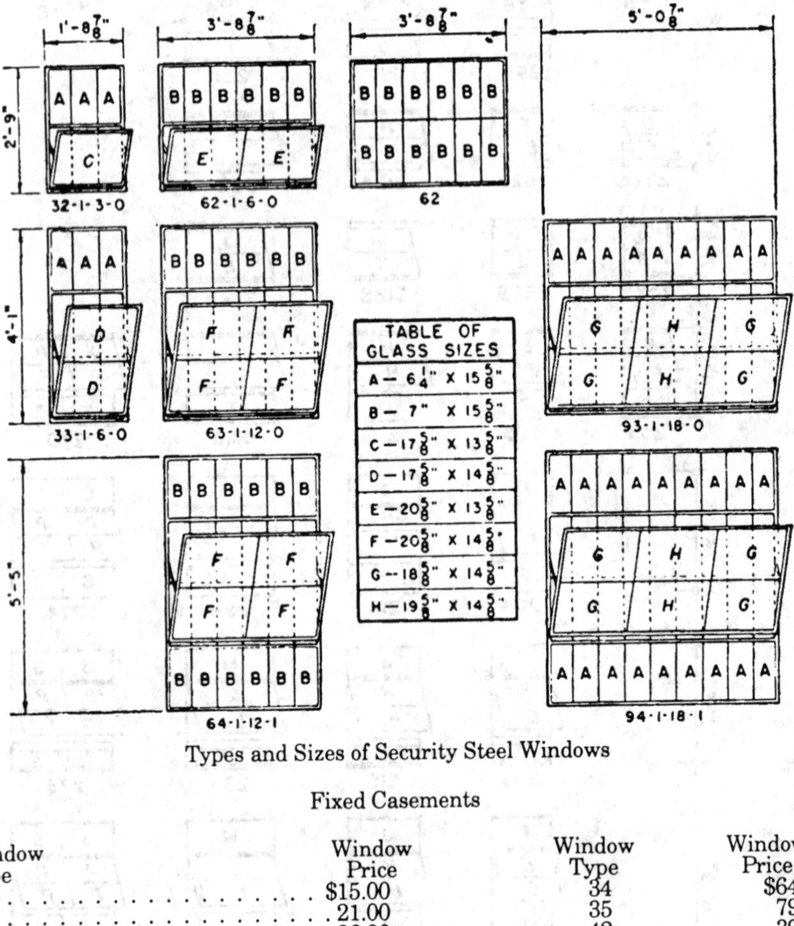

Types and Sizes of Security Steel Windows

Fixed Casements

Window Type	Window Price	Window Type	Window Price
12	$15.00	34	$64.00
13	21.00	35	79.00
14	26.00	42	39.00
15	32.00	43	58.00
22	21.00	44	78.00
23	29.00	45	94.00
24	37.00	52	52.00
25	43.00	53	78.00
32	33.00	54	100.00
33	45.00	55	122.00

Prices of ventilated units include locking handles, Roto underscreen operators (lever underscreen operators also available at reduced price) and standard installation fittings, bonderizing and one factory coat of paint for both ventilated and fixed units. Glass, putty, glazing, glazing clips, and mastic not included. Screen prices include screens for all vents in unit and clips for attachment.

Prices of Extras

Bronze hardware for roto casements instead of bronze lacquer finish hardware, add per side hinged vent	$6.50
Mastic, use 1 quart for each 40 lin. ft. of application	
Per quart (4 lbs.)	4.75
Per gallon ($16\frac{1}{2}$ lbs.)	14.00
5 gallons (85 lbs.)	63.50

Prices of Mullions

Vertical Mullions

2-Lt	$3.50
3-Lt	4.50
4-Lt	5.50
5-Lt	6.50

Horizontal Mullions

1-Lt	$3.50
2-Lt	4.50
3-Lt	5.50
4-Lt	6.50
5-Lt	7.50

Pipe Corners

2-Lt	$6.50
3-Lt	8.50
4-Lt	9.50
5-Lt	11.50

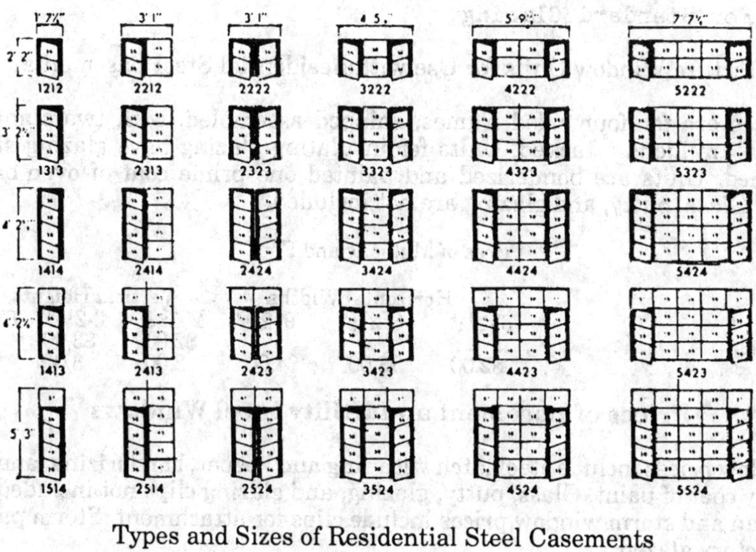

Types and Sizes of Residential Steel Casements

Prices of Residential Steel Casement Picture Windows

For Standard Glazing Window Type	Window Price
PW33	$20.00
PW34	22.00
PW35	24.00
PW43	23.00
PW44	26.00
PW45	28.00

For Insulating Glazing Window Type	Window Price
DG34	$24.00
DG35	27.00
DG44	28.00
DG45	30.00

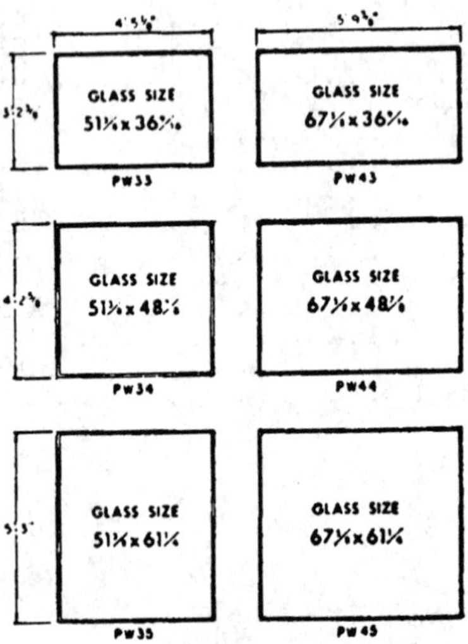

For Standard Glazing

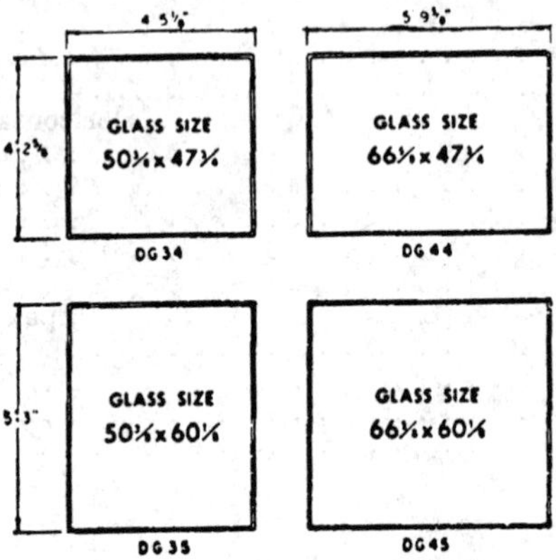

For Insulating Glazing

Picture Window Units for Use with Residential Steel Casements

Prices are for four sided frames, shipped assembled, with two standard vertical mullions attached. Units for insulating glazing have glazing strips attached. Units are bonderized and painted one prime coat of oven-baked paint. Glass, putty, and glazing are not included.

Prices of Mullions and Fins

	Horizontal Widths			Vertical Heights		
	3'-1¾"	6'-2½"	9'-2½"	2'-1¾"	4'-2⅛"	6'-2½"
Mullions				$2.00	$3.00	$4.00
Fins	$2.00	$3.00	$4.00	2.00	3.00	3.00

Prices of Basement and Utility Steel Windows

Window prices include steel latch with ring and keeper, bonderizing, and one factory coat of paint. Glass, putty, glazing, and glazing clips not included.

Screen and storm window prices include clips for attachment. Storm panels are factory glazed.

Basement Windows

Opening Dimensions	Window Type	Window Price	Mullion Price	Screen Price	Storm Sash Price
2'-8½"x1'-2⅝"	2-Lt.	$27.50	$2.50	$5.00	$10.00
2'-8½"x1'-6⅝"	2-Lt.	31.50	2.50	5.00	11.00
2'-8½"x1'-10⅝"	2-Lt.	39.50	2.50	6.00	13.00

Utility Windows

Window Type	Opening Dim. Width	Height	Window List Price	Mullion List Price	Screen List Price
22121	2'-8½"	3'-6⅝"	$28.00	$3.50	$6.50

08520 ALUMINUM CASEMENT WINDOWS

Prices include aluminum die cast roto operator with aluminum arm. Operator case and crank are brushed to a satin finish and given one coat of clear Duranite phenolic resin baked-on plastic coating. Casements are etched in a hot acid bath and receive a dip coat of methacrylate lacquer. Also included are wood screws and sill anchors when specified. Prices include ½" insulating glass and screens.

Prices of Aluminum Residential Casements
Ventilated

Window Type	Window Price	Window Type	Window Price
1212	$38.00	3222	90.00
1313	40.00	3323	77.00
1413	48.00	3423	101.00
1414	48.00	3424	101.00
1514	58.00	3524	124.00
2212	44.00	4222	63.00
2313	52.00	4323	110.00
2413	68.00	4423	117.00
2414	68.00	4424	120.00
2514	82.00	4524	149.00
2222	55.00	5222	82.00
2323	65.00	5323	120.00
2423	72.00	5424	159.00
2424	72.00	5524	197.00
2524	88.00		

Prices of Picture Windows for Use with Aluminum
Residential Casements
Glazed With ½" Insulating Glass

Window Type	Window Price	Window Type	Window Price
33MO	$69.00	43MO	79.00
34MO	80.00	44MO	94.00
35MO	90.00	45MO	106.00

Price of Aluminum Awning Windows

Window Type	Window Price	Window Type	Window Price
1722	$95.00	3142	222.00
1732	149.00	3153	258.00
1742	185.00	4522	176.00
1753	215.00	4532	218.00
3122	118.00	4542	248.00
3132	149.00	4553	308.00

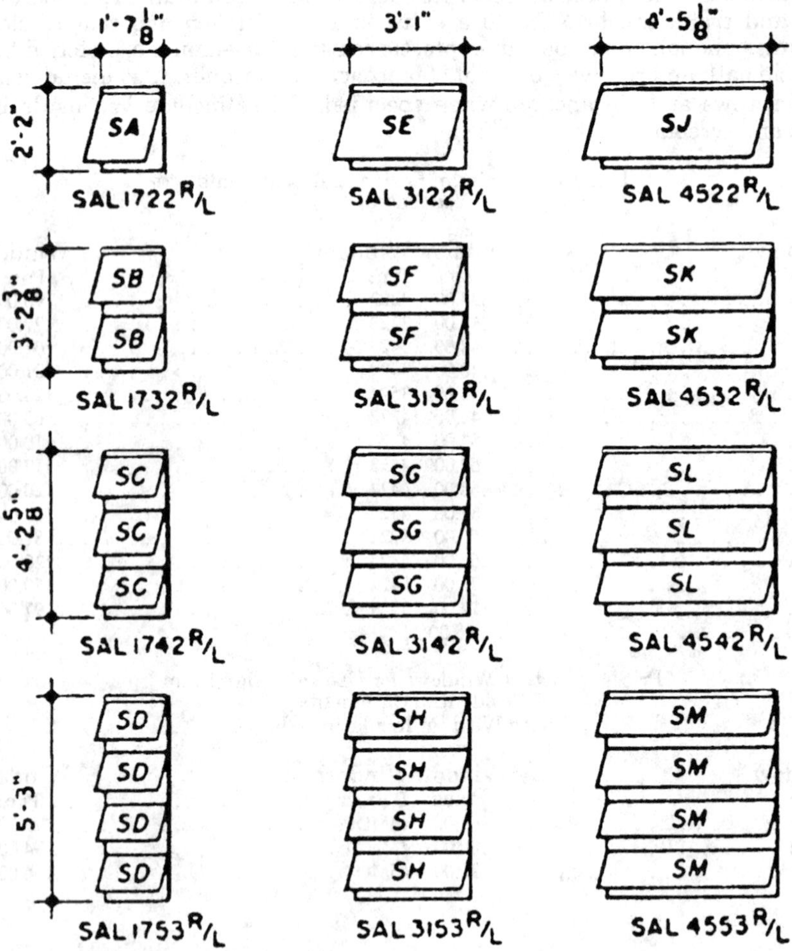

Aluminum Awning Windows

Prices of Mullions and Fins

Lts. Wide or High	Vertical Mullions	Head Fins Each Head	Jamb Fins Each Jamb
1	$...	$2.50	$...
2	6.00	2.50	2.00
3	9.00	3.50	2.50
4	11.00	4.50	3.50
5	14.00	5.50	3.50

Price includes a mechanical operator with crank handle. Windows are glazed by using snap-in aluminum beads. Price includes integral fin, and glazing with ½" insulating glass.

Windows are weatherstripped and receive a dip coat of methacrylate lacquer.

Screen prices include screens for all vents in unit. Screens have aluminum frames and 18x14 mesh aluminum cloth. Storm sash are factory glazed.

Prices of Aluminum Double Hung Windows

Prices of windows include silicone-treated wool-pile weatherstripping, spiral, sash balances, glazing clips, standard anchors, mullion screws, and window hardware consisting of sash lock, keeper, and pull down for upper sash. Prices include ½" insulating glass and screens.

All windows are finished with a coat of water clear lacquer.

Screens have extruded or rolled aluminum frame sections and are wired with 18x14 mesh aluminum cloth. Screens consist of full or half panels to allow for full or half-length screening as desired. Prices include all necessary installation hardware.

Storm sash are furnished glazed with single strength glass set in a plastic seal. All points of contact with window frame are insulated. Prices include all necessary hardware for installation.

Picture windows are shipped knocked down. Picture windows are designed to receive ¼" single glass or ½", ¾" or 1" insulating glass if specified. Prices include necessary assembly and installation hardware.

Prices of Aluminum Double Hung Windows

Window Type	2-Lt. Top 2-Lt. Bot.	6-Lt. Top 6-Lt. Bot.	Two Light
2030	$80.00	$85.00	$78.00
2830	85.00	90.00	83.00
3030	88.00	93.00	86.00
3430	91.00	96.00	89.00
3830	97.50	103.00	95.00
2038	83.00	88.00	81.00
2838	87.00	93.00	85.00
3038	95.00	100.00	93.00
3438	98.00	103.00	95.00
3838	117.00	127.00	115.00
2044	85.00	90.00	83.00
2844	88.00	93.00	86.00
3044	96.00	101.00	94.00
3444	98.00	103.00	96.00
3844	101.00	106.00	99.00
2050	90.00	95.00	88.00
2850	92.00	95.00	90.00
3050	95.00	100.00	93.00

Window Type	2-Lt. Top 2-Lt. Bot.	6-Lt. Top 6-Lt. Bot.	Two Light
3450	110.00	115.00	108.00
3850	115.00	120.00	113.00
2060	100.00	105.00	108.00
3060	105.00	110.00	103.00
3860	110.00	115.00	108.00

Prices of Aluminum Fins and Trim or Brick Molds

	Height of Window				
	3'-1⅝"	3'-9⅝"	4'-5⅝"	5'-1⅝"	5'-9⅝"
Jamb fin and trim or brick mold, per pair	$5.00	$6.00	$6.50	$7.00	$7.50

	Width of Window					
	2'-0½"	2'-4½"	2'-8½"	3'-0½"	3'-4½"	3'-8½"
Head Section	$0.60	$0.65	$0.75	$0.80	$0.85	$0.90
Sill Section	.50	.55	.60	.65	.75	.80

Prices of Aluminum Screens, Storm Sash, and Inside Casings for Aluminum Double Hung Windows

Window Type	Half Screen	Full Screen	Storm Sash 2-Piece	Inside Casing
2031	$6.50	$10.00	$39.50	$8.50
2431	7.00	11.00	41.00	8.70
2831	7.50	12.00	43.00	9.10
3031	8.75	14.50	45.00	9.50
3431	9.25	15.50	46.00	9.70
3831	10.00	16.00	50.00	10.10
2039	6.75	10.50	41.00	9.20
2439	8.00	13.00	43.00	9.50
2839	9.00	15.00	45.50	9.80
3039	10.00	17.00	46.50	10.20
3439	11.00	19.00	50.00	10.40
3839	12.00	21.00	54.00	10.80
2045	8.00	13.00	43.00	9.60
2445	9.00	15.00	45.50	9.80
2845	10.00	17.00	46.00	10.20
3045	11.00	19.00	50.00	10.50
3445	12.50	22.00	53.00	10.80
3845	13.50	24.00	55.00	11.20
2051	9.00	15.00	45.50	10.10
2451	10.00	17.00	46.50	10.30
2851	11.00	19.00	50.00	10.70
3051	13.00	23.00	53.00	11.00
3451	14.00	25.00	55.00	11.20
3851	15.00	27.00	64.00	11.60
2059	10.00	17.00	46.50	10.70
2459	11.00	19.00	50.00	10.90
2859	12.00	21.00	53.00	11.30
3059	14.00	25.00	55.00	11.60
3459	15.00	27.00	64.00	11.90
3859	16.00	29.00	65.00	12.30

Prices of Aluminum Picture Windows for Use with Aluminum Double Hung Windows, Glass Not Included

Window Type	Price	Window Type	Price
40P31	$39.00	44P45	$47.00
44P31	40.00	*48P45	49.50
40P39	40.50	*410P45	50.00
44P39	42.50	*50P45	50.50
48P39	43.00	*56P45	51.50
50P39	44.50	*62P45	52.50
40P45	45.50	*610P45	53.50
*42P45	47.00		

*These sizes are set up (standard) for insulated glazing.

Prices are for units for single glazing. For double glazing, add $3.00 per unit.

Prices of Aluminum Horizontally Sliding Windows

Prices of windows include silicone-treated wool-pile weatherstripping, and integral fin. Prices are for frames and sash assembled and glazed with ½" insulating glass.

All frame and sash members are cleaned, caustic etched, and coated with methacrylate-type lacquer at the factory.

Screens are aluminum, covering half of window opening, and are easily installed and removed from inside.

Storm sash are installed same as screens and are two piece SSB glazed.

Window Type	Window Price	Half Screen	Inside Casing
2020	$34.00	$6.00	$9.60
3020	36.00	7.00	10.50
4020	39.00	6.00	10.10
5020	41.00	7.00	11.20
6020	44.00	7.00	12.20
3030	48.00	7.00	9.80
4030	50.00	7.00	10.90
5030	54.00	8.00	12.00
6030	62.00	8.00	12.90
4040	60.00	8.00	12.00
5040	72.00	9.00	13.00
6040	83.00	9.00	14.50

Prices of Aluminum Picture Windows with Horizontally Sliding Flankers

Picture window prices are for frames and sash assembled and glazed.

Screen prices include two per unit. Storm sash prices are for three piece glazed.

Window Type	Window Price	2-Piece Screens	3-Piece Storm Sash	Inside Casing
7020	$70.00	$11.00	$55.00	$16.40
8020	73.00	11.00	57.00	17.00
9020	80.00	11.00	63.00	19.40
7030	86.00	12.00	67.00	17.90
8030	90.00	13.00	71.00	19.20
9030	97.00	13.00	76.00	20.60
7040	101.00	13.00	77.00	19.40
8040	104.00	14.00	81.00	20.50
9040	114.00	14.00	86.00	22.00

08600 WOOD AND PLASTIC WINDOWS

Wood windows generally are sold as complete units with frame, sash, operating hardware, weatherstripping and glazing assembled at the factory. Frames are usually treated. The exterior of wood windows is generally covered with rigid vinyl (PVC). The interior portion is unfinished, to be finished by painters or others. The following sizes are based on units manufactured by Andersen Window Company, Bayport, Minn. Sizes for other manufacturers will vary slightly. Prices, of course, should be checked locally, because they will vary considerably, depending on geographic location, on size of the order and on whether delivery is to a local dealer or to job site.

The contractor should add or deduct for items such as insect screens, divided light grilles, hardware, window blinds or window treatments when comparing individual window costs. The prices quoted will be helpful, however, for comparing the various window and glazing types.

Double Hung Window Units

Frame. Wood members are treated with a water repellent preservative and covered with a rigid vinyl (PVC) sheath. Sill ends are prefinished with polyurea or polyester urethane, depending on whether white or color.

Sash. The exterior of the wood sash is protected with a long-lasting patented polyurea finish or a polyester urethane finish coat, depending on color. Interior face of sash is clear for stain or paint finish.

Glazing. Select quality high-performance or high-performance sun insulating glass. Double-pane insulated glass is available for white only. (For high altitude and other special glazings, contact manufacturer).

Glazing System. Sash has a rigid vinyl snap-in-bead.

Weatherstripping. Foam type weatherstripping is applied to top and bottom rails. Rigid vinyl rib on head jamb liner and sills fits into vinyl covered foam weatherstripping on sash. Rigid vinyl leaf weatherstripping is applied to check rail. Polypropylene leaf weatherstrip with foam inserts contacts side jamb liner ribs.

Anchoring Flange and Windbreak. Factory applied rigid vinyl flanges at head and sides of outer frame members. A flexible vinyl sill windbreak is factory applied to the bottom of sill as flashing.

Sill. Wood core treated with a water repellent preservative, covered with PVC sheath.

Jamb Liner. White PVC for unit in white and complementary color PVC for unit in color.

Sash Lock and Lift. Stone color decorator finish, sash lock, keeper and lift are factory applied.

Courtesy Andersen Window Company

Frame and Sash Section, Sash Lock, and Sash Lift

Double Hung Units High Performance (H.P.) Insulating Glass

Window Type	Glass Size		Rough Opening Width x	Height	H.P. Insulating	Insect Screen Add	Grille Add	Combination Screen/Storm Add
1842W	16½"x	22"	1'10⅛" x	4'5¼"	$189.00	$16.40	$13.30	$72.00
1846W		24"		4'9¼"	193.00	17.30	13.50	74.00
2042W	20½"x	22"	2'2⅛" x	4'5¼"	197.00	18.00	15.10	77.00
2046W		24"		4'9¼"	205.00	18.80	15.60	80.00
2052W		28"		5'5¼"	223.00	20.50	16.70	85.00
2062W		68"		6'5¼"	251.00	22.00	22.60	104.00
20210W		14"		3'1¼"	168.00	14.40	13.30	69.00
2032W		16"		3'5¼"	172.00	15.30	13.60	70.00
20310W		20"		4'1¼"	191.00	16.60	14.20	75.00
2442W	24½"x	22"	2'6⅛"	4'5¼"	211.00	19.40	15.70	84.00
2446W		24"		4'9¼"	220.00	20.40	17.00	88.00
2452W		28"		5'5¼"	238.00	22.20	17.80	94.00
24210W		14"		3'1¼"	175.00	15.40	14.00	72.00
2432W		16"		3'5¼"	184.00	16.60	14.40	72.00
24310W		20"		4'1¼"	200.00	17.50	14.80	78.00
2842W	28½"x	22"	2'10⅛" x	4'5¼"	226.00	20.80	17.00	91.00
2846W		24"		4'9¼"	235.00	21.80	18.50	93.00
2852W		28		5'5¼"	255.00	24.20	19.40	100.00
2856W		60"		5'9¼"	270.00	26.00	22.30	104.00
2862W		68"		6'5¼"	283.00	30.00	25.20	110.00
28210W		14"		3'1¼"	188.00	16.70	15.20	75.00
2832W		16"		3'5¼"	194.00	17.50	15.70	76.00
28310W		20"		4'1¼"	215.00	19.80	16.00	83.00

Window Type	Glass Size		Rough Opening Width x	Height	H.P. Insulating	Insect Screen Add	Grille Add	Combination Screen/Storm Add
3042W	32½"x	22"	3'2⅛" x	4'5¼"	240.00	23.00	18.50	96.00
3046W		24"		4'9¼"	250.00	24.10	19.40	96.00
3052W		28"		5'5¼"	271.00	26.80	21.40	109.00
3056W		60"		5'9¼"	284.00	28.00	24.20	114.00
3062W		68"		6'5¼"	305.00	30.30	28.30	124.00
30210W		14"		3'1¼"	199.00	18.60	16.20	80.00
3032W		16"		3'5¼"	209.00	20.60	16.70	82.00
30310W		20"		4'1¼"	228.00	22.20	17.80	89.00
3442W	36½"x	22"	3'6⅛" x	4'5¼"	254.00	24.80	20.00	105.00
3446W		24"		4'9¼"	264.00	25.90	20.50	107.00
3452W		28"		5'5¼"	285.00	28.90	21.90	119.00
3456W		60"		5'9¼"	300.15	30.10	25.60	126.00
3432W		16"		3'5¼"	222.00	21.00	18.10	88.00
34310W		20"		4'1¼"	240.00	23.70	19.50	95.00
3846W	40½"x	24"	3'10⅛" x	4'9¼"	280.00	27.00	11.20	118.00

For double hung units in pairs, triple, or quadruple arrangements, multiply cost by two, three, or four respectively.

Double Hung Units High Performance (H.P.) Sun Insulating Glass

Window Type	Glass Size		Rough Opening Width x	Height	H.P. Sun Insulating	Insect Screen Add	Grille Add	Combination Screen/Storm Add
1842W	16½"x	22"	1'10⅛" x	4'5¼"	$204.00	$16.40	$13.30	$72.00
1846W		24"		4'9¼"	213.00	17.30	13.50	74.00
2042W	20½"x	22"	2'2⅛" x	4'5¼"	227.00	18.00	15.10	77.00
2046W		24"		4'9¼"	244.00	18.80	15.60	80.00
2052W		28"		5'5¼"	260.00	20.50	16.70	85.00
2062W		68"		6'5¼"	275.00	22.00	22.60	104.00
20210W		14"		3'1¼"	208.00	14.40	13.30	69.00
2032W		16"		3'5¼"	222.00	15.30	13.60	70.00
20310W		20"		4'1¼"	237.00	16.60	14.20	75.00
2442W	24½"x	22"	2'6⅛" x	4'5¼"	254.00	19.40	15.70	84.00
2446W		24"		4'9¼"	270.00	20.40	17.00	88.00
2452W		28"		5'5¼"	285.00	22.20	17.80	94.00
24210W		14"		3'1¼"	303.00	15.40	14.00	72.00
2432W		16"		3'5¼"	242.00	16.60	14.40	72.00
24310W		20"		4'1¼"	257.00	17.50	14.80	78.00
2842W	28½"x	22"	2'10⅛" x	4'5¼"	275.00	20.80	17.00	91.00
2846W		24"		4'9¼"	293.00	21.80	18.50	93.00
2852W		28"		5'5¼"	335.00	24.20	19.40	100.00
2856W		60"		5'9¼"	292.00	26.00	22.30	104.00
2862W		68"		6'5¼"	307.00	30.00	25.20	110.00
28210W		14"		3'1¼"	356.00	16.70	15.20	75.00
2832W		16"		3'5¼"	271.00	17.50	15.70	76.00
28310W		20"		4'1¼"	306.00	19.80	16.00	83.00
3042W	32½"x	22"	3'2⅛" x	4'5¼"	330.00	23.00	18.50	96.00
3046W		24"		4'9¼"	181.00	24.10	19.40	96.00
3052W		28"		5'5¼"	190.00	26.80	21.40	109.00
3056W		60"		5'9¼"	202.00	28.00	24.20	114.00
3062W		68"		6'5¼"	215.00	30.30	28.30	124.00
30210W		14"		3'1¼"	185.00	18.60	16.20	80.00
3032W		16"		3'5¼"	200.00	20.60	16.70	82.00
30310W		20"		4'1¼"	210.00	22.20	17.80	89.00
3442W	36½"x	22"	3'6⅛" x	4'5¼"	225.00	24.80	20.00	105.00
3446W		24"		4'9¼"	240.00	25.90	20.50	107.00
3452W		28"		5'5¼"	206.00	28.90	21.90	119.00
3456W		60"		5'9¼"	216.00	30.10	25.60	126.00
3432W		16"		3'5¼"	232.00	21.00	18.10	88.00
34310W		20"		4'1¼"	245.00	23.70	19.50	95.00
3846W	40½"x	24"	3'10⅛"	4'9¼"	260.00	27.00	11.20	118.00

For double hung units in pairs, triple, or quadruple arrangements, multiply costs by two, three, or four respectively.

Double Hung Units Double Pane Insulating Glass

Window Type	Glass Size		Rough Opening Width x	Height	Pane Insulating	Double Screen Add	Insect Grille Add	Combination Screen/Storm Add
1842W	16½"x	22"	1'10⅛" x	4'5¼"	$173.00	$16.40	$13.30	$72.00
1846W		24"		4'9¼"	180.00	17.30	13.50	74.00
2042W	20½"x	22"	2'2⅛" x	4'5¼"	192.00	18.00	15.10	77.00
2046W		24"		4'9¼"	207.00	18.80	15.60	80.00
2052W		28"		5'5¼"	220.00	20.50	16.70	85.00
2062W		68"		6'5¼"	233.00	22.00	22.60	104.00
20210W		14"		3'1¼"	176.00	14.40	13.30	69.00
2032W		16"		3'5¼"	188.00	15.30	13.60	70.00
20310W		20"		4'1¼"	201.00	16.60	14.20	75.00
2442W	24½"x	22"	2'6⅛" x	4'5¼"	215.00	19.40	15.70	84.00
2446W		24"		4'9¼"	228.00	20.40	17.00	88.00
2452W		28"		5'5¼"	242.00	22.20	17.80	94.00
24210W		14"		3'1¼"	257.00	15.40	14.00	72.00
2432W		16"		3'5¼"	205.00	16.60	14.40	72.00
24310W		20"		4'1¼"	218.00	17.50	14.80	78.00
2842W	28½"x	22"	2'10⅛"	4'5¼"	233.00	20.80	17.00	91.00
2846W		24"		4'9¼"	248.00	21.80	18.50	93.00
2852W		28"		5'5¼"	260.00	24.20	19.40	100.00
2856W		60"		5'9¼"	247.00	26.00	22.30	104.00
2862W		68"		6'5¼"	275.00	30.00	25.20	110.00
28210W		14"		3'1¼"	230.00	16.70	15.20	75.00
2832W		16"		3'5¼"	260.00	17.50	15.70	76.00
28310W		20"		4'1¼"	280.00	19.80	16.00	83.00
3042W	32½"x	22"	3'2⅛" x	4'5¼"	154.00	23.00	18.50	96.00
3046W		24"		4'9¼"	160.00	24.10	19.40	96.00
3052W		28"		5'5¼"	172.00	26.80	21.40	109.00
3056W		60"		5'9¼"	182.00	28.00	24.20	114.00
3062W		68"		6'5¼"	157.00	30.30	28.30	124.00
30210W		14"		3'1¼"	168.00	18.60	16.20	80.00
3032W		16"		3'5¼"	177.00	20.60	16.70	82.00
30310W		20"		4'1¼"	190.00	22.20	17.80	89.00
3442W	36½"x	22"	3'6⅛" x	4'5¼"	203.00	24.80	20.00	105.00
3446W		24"		4'9¼"	175.00	25.90	20.50	107.00
3452W		28"		5'5¼"	183.00	28.90	21.90	119.00
3456W		60"		5'9¼"	197.00	30.10	25.60	126.00
3432W		16"		3'5¼"	208.00	21.00	18.10	88.00
34310W		20"		4'1¼"	220.00	23.70	19.50	95.00
3846W	40½"x	24"	3'10⅛" x	4'9¼"		27.00	11.20	118.00

For double hung units in pairs, triple, or quadruple arrangements, multiply costs by two, three, or four respectively.

Double Hung Window Units - Typical Installation Details

2" x6" WALL WITH AUXILIARY CASING

Unit installed in frame using 2" x6" studs. Note special width extension jambs furnished by others.

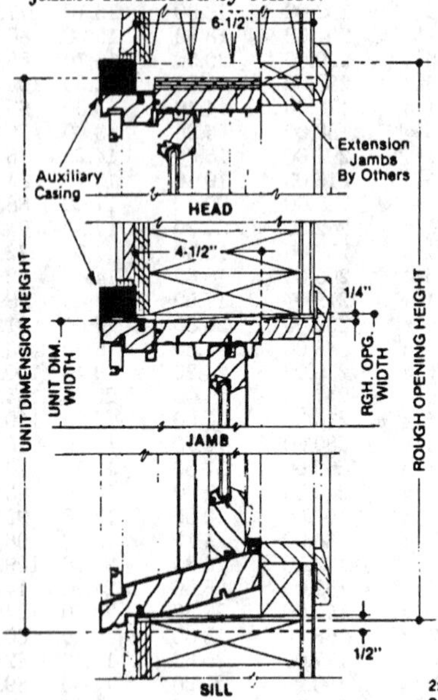

SOLID MASONRY WITH TREATED SUB FRAME

Unit installed in masonry wall with stucco on exterior. Interior finish applied on furring strips.

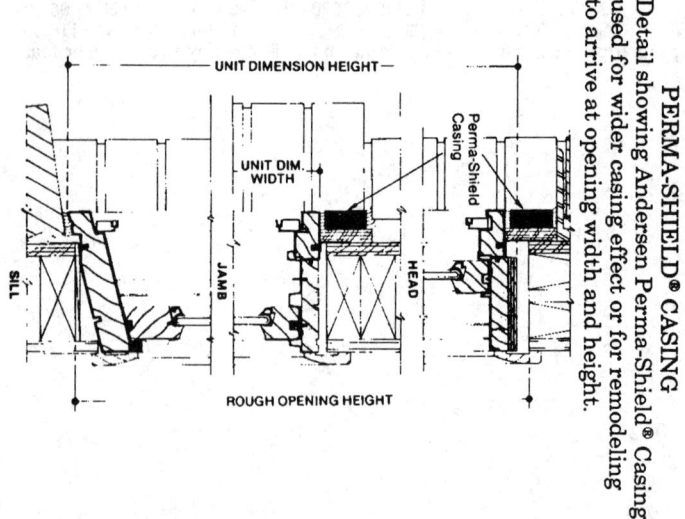

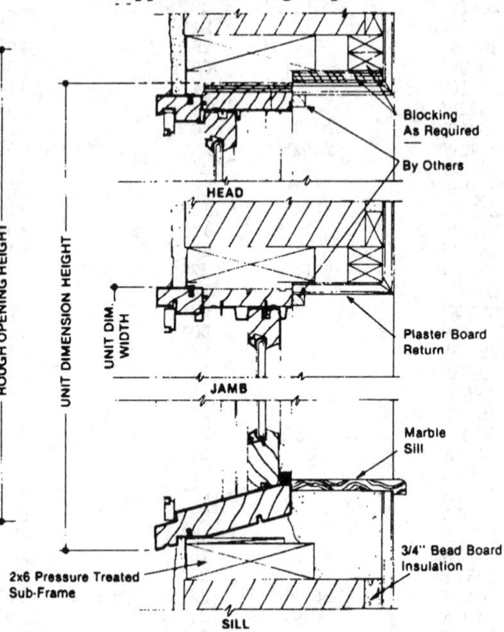

Courtesy Andersen Window Company

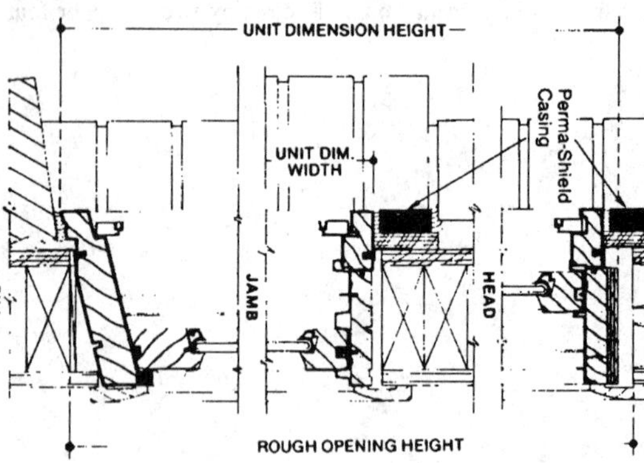

PERMA-SHIELD® CASING

Detail showing Andersen Perma-Shield® Casing used for wider casing effect or for remodeling to arrive at opening width and height.

Courtesy Andersen Window Company

Typical Pane Configuration

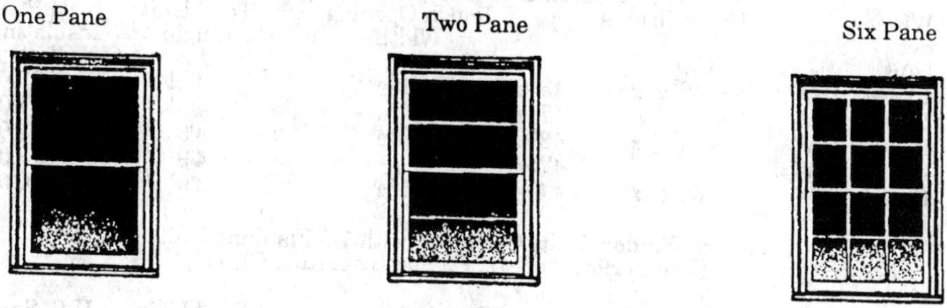

Typical Combination Units

STATIONARY PICTURE WINDOW UNITS

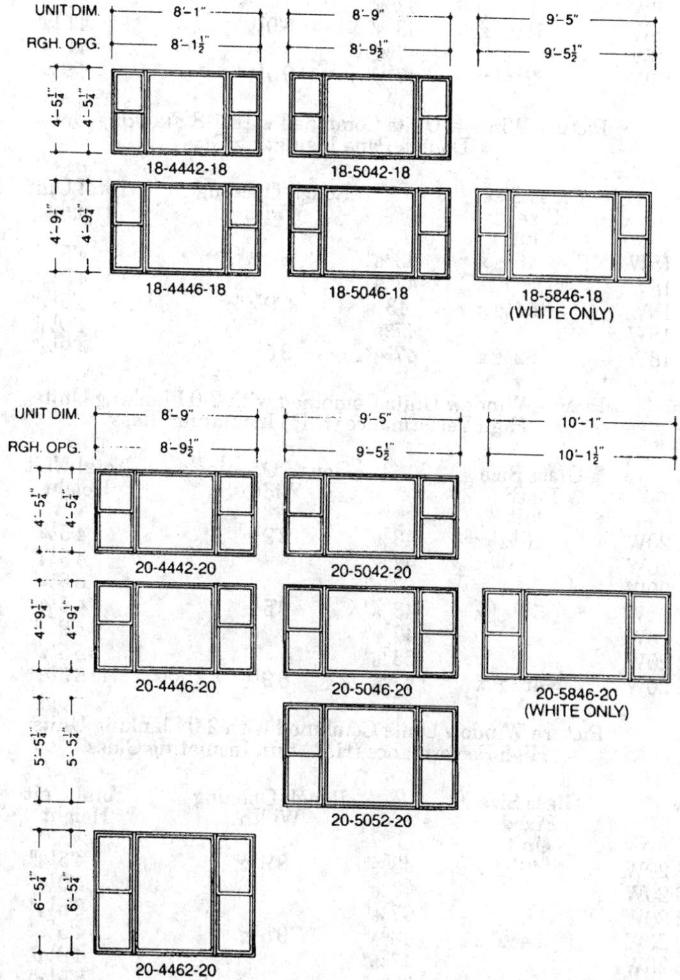

Picture Window Units Combined with 1-8 Flanking Units
High Performance (H.P.) Insulating Glass

Window Set Up Unit	Glass Size Fixed Unit		Rough Opening Width	Total Unit Height	H.P. Insulating Glass
18 4442 18W	46½" x	43⅝"	8'1½"	4'5¼"	$779.00
18 4446 18W		47⅝"		4'9¼"	801.00
18 5042 18W	54½" x	43⅝"	8'9½"	4'5¼"	843.00
18 5046 18W		47⅝"		4'9¼"	873.00
18 5846 18W	62½" x	47⅝"	9'5½"	4'9¼"	920.00

Picture Window Units Combined with 1-8 Flanking Units
High Performance (H.P.) Sun Insulating Glass

Window Set Up Unit	Glass Size Fixed Unit		Rough Opening Width	Total Unit Height	H.P. Sun Insulating Glass
18 4442 18W	46½" x	43⅝"	8'1½"	4'5¼"	$838.00
18 4446 18W		47⅝"		4'9¼"	862.00
18 5042 18W	54½" x	43⅝"	8'9½"	4'5¼"	907.00
18 5046 18W		47⅝"		4'9¼"	939.00
18 5846 18W	62½" x	47⅝"	9'5½"	4'9¼"	990.00

Picture Window Units Combined with 1-8 Flanking Units
Double Pane Insulating Glass

Window Set Up Unit	Glass Size Fixed Unit		Rough Opening Width	Total Unit Height	Double Pane Insulating Glass
18 4442 18W	46½ x	43⅝"	8'1½"	4'5¼"	$705.00
18 4446 18W		47⅝"		4'9¼"	732.00
18 5042 18W	54½ x	43⅝	8'9½"	4'5¼"	745.00
18 5046 18W		47⅝"		4'9¼"	765.00
18 5846 18W	62½ x	47⅝"	9'5½"	4'9¼"	803.00

Picture Window Units Combined with 2-0 Flanking Units
High Performance (H.P.) Insulating Glass

Window Set Up Unit	Glass Size Fixed Unit		Rough Opening Width	Total Unit Height	H.P. Insulating Glass
20 4442 20W	46½" x	43⅝"	8'9½"	4'5¼"	$795.00
20 4446 20W		47⅝"		4'9¼"	826.00
20 4462 20W		67⅝"		6'5¼"	1,068.00
20 5042 20W	54½" x	43⅝"	9'5½"	4'5¼"	858.00
20 5046 20W		47⅝"		4'9¼"	898.00
20 5052 20W		55⅝"		5'5¼"	1,000.00
20 5846 20W	46½" x	67⅝"	8'9½"	6'5¼"	945.00

Picture Window Units Combined with 2-0 Flanking Units
High Performance (H.P.) Sun Insulating Glass

Window Set Up Unit	Glass Size Fixed Unit		Rough Opening Width	Total Unit Height	H.P. Sun Insulating Glass
20 4442 20W	46½" x	43⅝"	8'9½"	4'5¼"	$855.00
20 4446 20W		47⅝"		4'9¼"	888.00
20 4462 20W		67⅝"		6'5¼"	1,149.00
20 5042 20W	54½" x	43⅝"	9'5½"	4'5¼"	924.00
20 5046 20W		47⅝"		4'9¼"	966.00
20 5052 20W		55⅝"		5'5¼"	1,077.00

Picture Window Units Combined with 2-0 Flanking Units
Double Pane Insulating Glass

Window Set Up Unit	Glass Size Fixed Unit		Rough Opening Width	Total Unit Height	H.P. Sun Insulating Glass
20 4442 20W	46½" x	43⅝"	8'9½"	4'5¼"	$719.00
20 4446 20W		47⅝"		4'9¼"	755.00
20 4462 20W		67⅝"		6'5¼"	939.00
20 5042 20W	54½" x	43⅝"	9'5½"	4'5¼"	759.00
20 5046 20W		47⅝"		4'9¼"	787.00
20 5052 20W		55⅝"		5'5¼"	874.00
20 5846 20W	46½" x	67⅝"	8'9½"	6'5¼"	825.00

Picture Window Units
High Performance (H.P.) Insulating Glass

Window Set Up Unit	Glass Size Fixed Unit		Rough Opening Width	Total Unit Height	H.P. Insulating Glass
4442W	46½" x	43⅝"	4'6⅛"	4'5¼"	$356.00
4446W		47⅝"		4'9¼"	420.00
4462W		67⅝"		6'5¼"	369.00
5042W	54½" x	43⅝"	5'2⅛"	4'5¼"	440.00
5046W		47⅝"		4'9¼"	488.00
5052W		55⅝"		5'5¼"	502.00
5846W	62½" x	47⅝"	5'10⅛"	4'9¼"	510.00

Stationary Picture Window Units

Picture Window Units (Stationary)

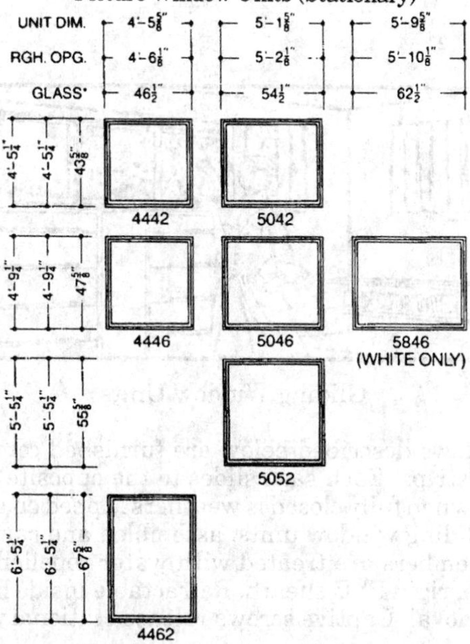

Unobstructed glass sizes shown in inches.

Stationary Picture Window Units
High Performance (H.P.) Sun Insulating Glass

Window Set Up Unit	Glass Size Fixed Unit		Rough Opening Width	Total Unit Height	H.P. Sun Insulating Glass
4442W	46½" x	43⅝"	4'6⅛"	4'5¼"	$384.00
4446W		47⅝"		4'9¼"	398.00
4462W		67⅝"		6'5¼"	550.00
5042W	54½" x	43⅝"	5'2⅛"	4'5¼"	453.00
5046W		47⅝"		4'9¼"	475.00
5052W		55⅝"		5'5¼"	543.00
5846W	62½" x	47⅝"	5'10⅛"	4'9¼"	475.00

Stationary Picture Window Units
Double Pane Insulating Glass

Window Set Up Unit	Glass Size Fixed Unit		Rough Opening Width	Total Unit Height	H.P. Sun Insulating Glass
4442W	46½" x	43⅝"	4'6⅛"	4'5¼"	$313.00
4446W		47⅝"		4'9¼"	332.00
4462W		67⅝"		6'5¼"	423.00
5042W	54½" x	43⅝"	5'2⅛"	4'5¼"	353.00
5046W		47⅝"		4'9¼"	365.00
5052W		55⅝"		5'5¼"	413.00
5846W	62½" x	47⅝"	5'10⅛"	4'9¼"	403.00

Gliding Window Units

When using the high narrow gliding windows, furniture may be placed under them making much better interior arrangements possible.

Gliding Window Units

The gliding windows described below are furnished complete with frames, sash, and weatherstrips. Each sash slides to the opposite side of the opening for ventilation but when fully closed is weatherstripped on all four sides. Prices are for complete gliding window units, assembled and carton packed.

Frame. Wood members are treated with water repellent preservative and covered with white, rigid PVC sheath. Retractable inside head stop allows for operating sash removal. Captive screws release stationary sash.

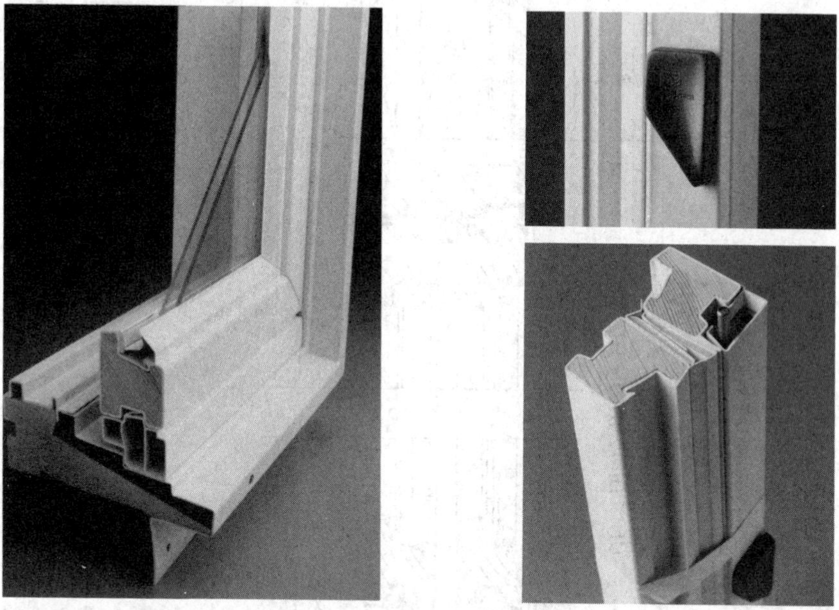

Gliding Windows Frame and Sash, Locking Latch, and Lock Rods

Sash. Wood core treated with a water repellent preservative and completely covered with white PVC sheath.

Glazing. Select quality double-pane insulating window glass. (Installations over 3500' altitude require high altitude glass.)

Glazing Bead. Rigid vinyl snap-in glazing bead with flexible vinyl tip against glass.

Weatherstripping. Rigid vinyl spring tension type, factory applied for a tight seal between sash and frame.

Flashing. Continuation of the rigid vinyl sheath on frame forms perimeter flashing and anchoring fin.

Corner Seal. Vinyl frame sheath eliminates corner joints. Sash corner joints are welded to form waterproof joints.

Sill Tank. For added tightness under severe conditions of exposure, a molded vinyl tank has been built in to drain any moisture to the exterior.

Glides. Adjustable chrome-plated steel glides, factory applied.

Locking Latch. Center interlock and two-way spring loaded locking mechanism secures sash at head and sill. Steel locking rods and zinc die-cast handle case with stone color decorator finish. Flexible baffles cover meeting stiles.

Andersen Combination Unit. A unit for triple glazing. Features aluminum frame and 1/8" tempered glass in storm panels and glass fiber screen cloth. Storm panels and screen with baked-on white decorator finish.

Vertical Gliding Window Detail

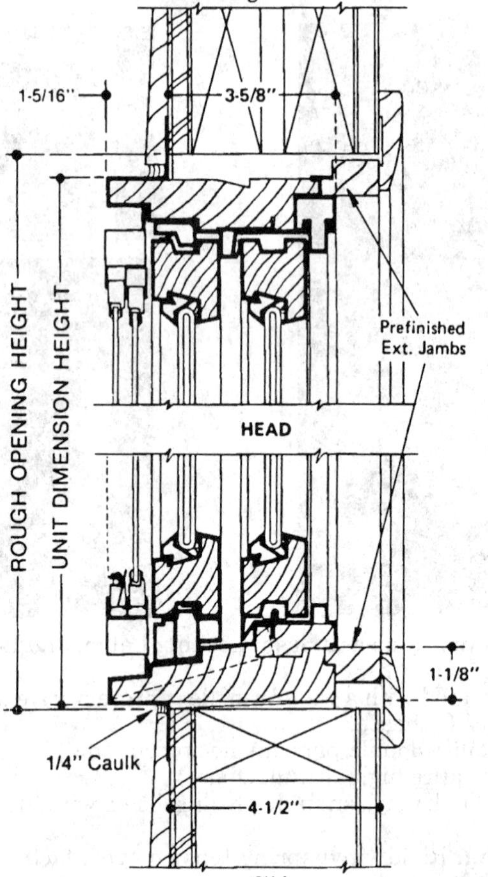

Courtesy Andersen Window Company

Vertical Gliding Window Detail

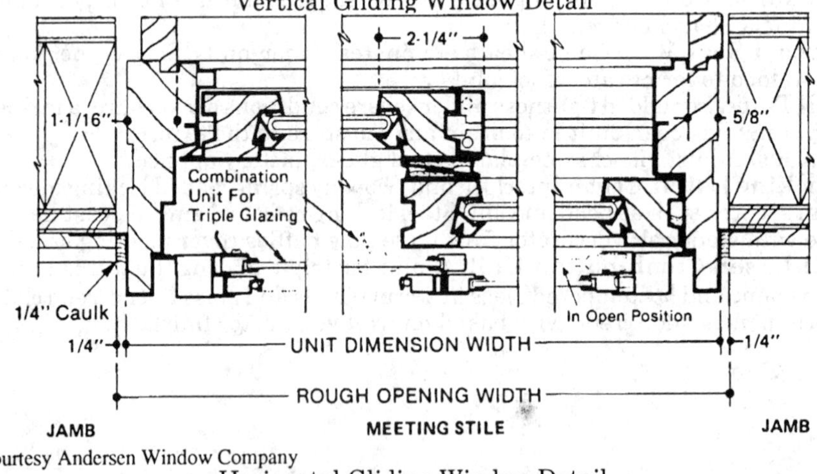

Courtesy Andersen Window Company

Horizontal Gliding Window Detail

Gliding Windows
Double Pane Insulation Glass

Window Unit Order No	Glass	Size Fixed Unit	Rough Opening Width	Total Unit Height	Double Pane Insulating Glass	Insect Screen
G33	14" x	29⅜"	3'0½"	3'0½"	$234.00	$14.00
G336		35⅜"		3'6½"	260.00	15.00
G436	20" x	35⅝"	4'0½"	3'6½"	291.00	16.00
G44		41⅜"		4'0½"	315.00	18.00
G536	26" x	35⅜"	5'0½"	3'6½"	330.00	19.00
G54		41⅜"		4'0½"	349.00	20.00
G55		53⅝"		5'0½"	402.00	23.00
G64	32" x	41⅜"	6'0½"	4'0½"	395.00	23.00
G65		53⅝"		5'0½"	506.00	25.00

Gliding Window Units - Typical Installation Details

2" X 6" WALL
Unit installed in frame wall using 2 x 6 studs. Note special width extension jambs furnished by others.

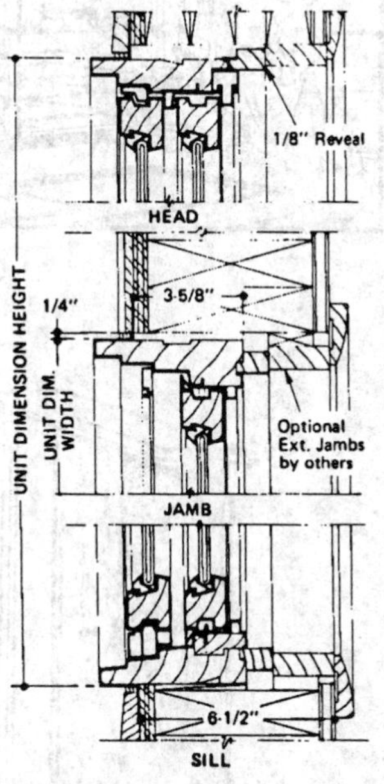

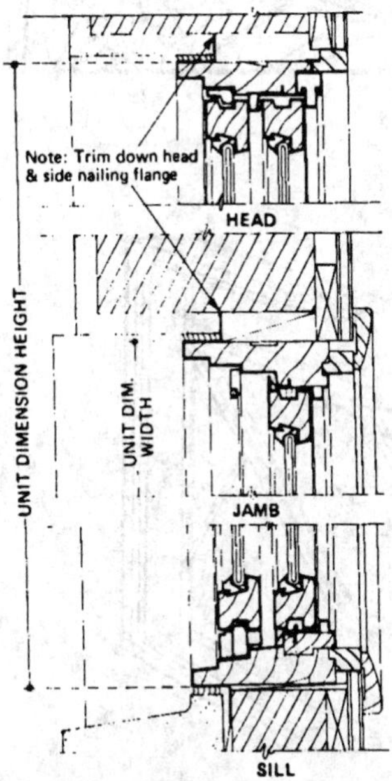

SOLID MASONRY
Unit installed in solid 8" masonry wall with stucco applied to exterior. Interior wall applied to furring strips.

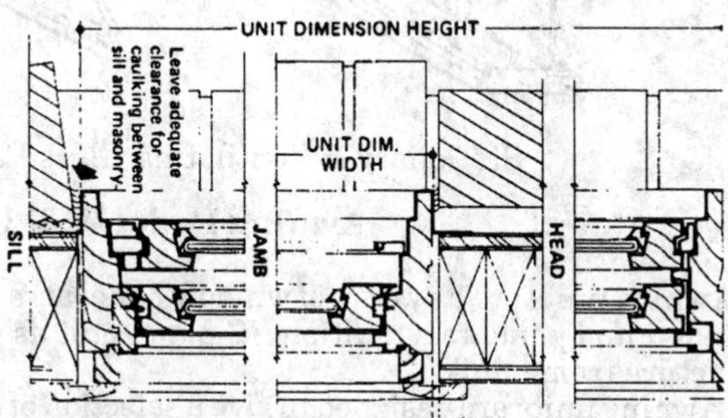

Courtesy Andersen Window Company

FRAME/BRICK VENEER
Unit installed in brick veneer wall with 1/2" sheathing and interior drywall.

Gliding type window units may be ordered with a fixed glass in the center and two gliding sash on either end, which when open, pass in front of the fixed panel.

Sash Extras. Glazing with obscure glass, add $3.00 to list, per sq. ft. Rectangular grilles same as double hung. Estimator should verify that it is available.

Glass Panel Wall Units

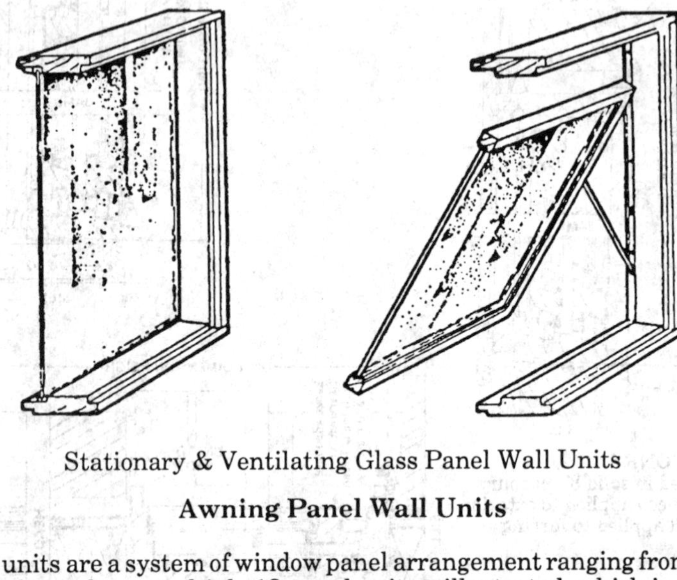

Stationary & Ventilating Glass Panel Wall Units

Awning Panel Wall Units

Awning units are a system of window panel arrangement ranging from a single individual panel to a multiple 12-panel unit as illustrated, which is used as a picture frame unit.

Awning units are designed to give a selection of either stationary glass panels, ventilating sash panels or a combination of both. Both come with high performance (H.P.) sun insulating glass and H.P. insulating glass.

Ventilating units have 2" sash, hinged at the top, which can be opened about 35 degrees, closed or held in intermediate position by rotary operators. Operators have worm gears, twin arms of cadmium plated steel and removable handles, and they are self-locking.

Insect screens, grilles and interior blinds are available for all units.

Glass panel wall units come completely set up and glazed. The ventilating units are weatherstripped, and operators are applied.

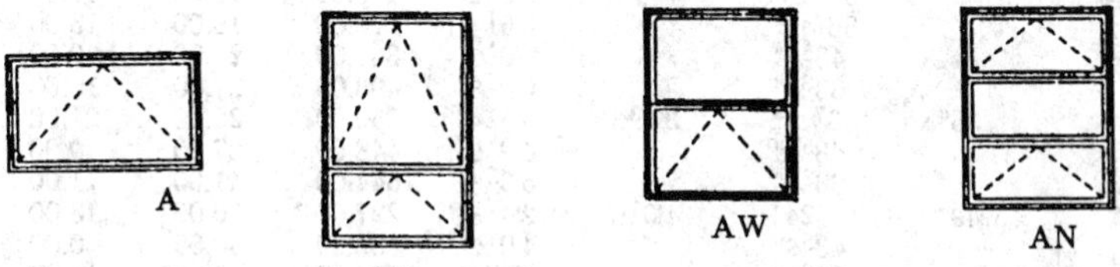

A AP AW AN

Awning Windows with High Performance (H.P.) Sun Insulating Glass

Window Type	Glass Size		Rough Opening Width	Height	H.P. Sun Insulating	Insect Screen Add	Grille Add	Interior Blinds Add
A21W	19 5/16" x	19 3/4"	2'0 5/8"	2'0 5/8"	$154.00	$10.00	$8.00	$58.00
AN31W	31 1/8" x	16 1/8"	3'0 1/2"	1'9"	160.00	11.00	9.00	56.00
A31W		19 3/4"		2'0 5/8"	180.00	12.00	10.00	65.00
AN351W	36" x	16 1/8"	3'5 3/8"	1'9"	177.00	13.00	11.00	59.00
A351W		19 3/4"		2'0 5/8"	194.00	16.00	13.00	63.00
AN41W	43 3/16" x	16 1/8"	4'0 1/2"	1'9"	194.00	22.00	18.00	72.00
A41W		19 3/4"		2'0 5/8"	210.00	11.00	9.00	59.00
AW31W	31 1/8" x	24"	3'0 1/2"	2'4 7/8"	189.00	14.00	10.00	67.00
A330W		31 9/16"		3'0 1/2"	196.00	18.00	17.00	72.00
A335W		36 7/16"		3'5 1/4"	278.00	19.00	18.00	75.00
AP32VW		43 5/8"		4'0 1/2"	306.00	25.00	0.00	80.00
AP321VW		63 3/8"		6'0 3/8"	493.00	37.00	10.00	145.00
A3535W	36" x	36 7/16"	3'5 3/8"	3'5 1/4"	256.00	23.00	21.00	85.00
AP352VW		43 5/8"		4'0 1/2"	250.00	27.00	0.00	91.00
AP3521VW		63 3/8"		6'0 3/8"	448.00	41.00	11.00	164.00
AW41W	43 3/16" x	24"	4'0 1/2"	2'4 7/8"	226.00	16.00	13.00	81.00
AP42VW		43 5/8"		4'0 1/2"	382.00	30.00	0.00	98.00
AP421VW		63 3/8"		6'0 3/8"	598.00	45.00	12.00	175.00
AP530VW	55 1/16" x	31 9/16"	5'0 3/8"	3'0 1/2"	362.00	29.00	0.00	109.00

Awning Windows with Double Pane Insulating Glass

Window Type	Glass Size		Rough Opening Width	Height	Double Pane Insulating	Insect Screen Add	Grille Add	Interior Blinds Add
A21W	19 5/16" x	19 3/4"	2'0 5/8"	2'0 5/8"	$143.00	$10.00	$8.00	$58.00
AN31W	31 1/8" x	16 1/8"	3'0 1/2"	1'9"	150.00	11.00	9.00	56.00
A31W		19 3/4"		2'0 5/8"	167.00	12.00	10.00	65.00
AN351W	36" x	16 1/8"	3'5 3/8"	1'9"	163.00	13.00	11.00	59.00
A351W		19 3/4"		2'0 5/8"	179.00	16.00	13.00	63.00
AN41W	43 3/16" x	16 1/8"	4'0 1/2"	1'9"	194.00	22.00	18.00	72.00
A41W		19 3/4"		2'0 5/8"	195.00	11.00	9.00	59.00
AW31W	31 1/8" x	24"	3'0 1/2"	2'4 7/8"	175.00	14.00	10.00	67.00
A330W		31 9/16"		3'0 1/2"	183.00	18.00	17.00	72.00
A335W		36 7/16"		3'5 1/4"	263.00	19.00	18.00	75.00
AP32VW		43 5/8"		4'0 1/2"	294.00	25.00	0.00	80.00
AP321VW		63 3/8"		6'0 3/8"	468.00	37.00	10.00	145.00
A3535W	36" x	36 7/16"	3'5 3/8"	3'5 1/4"	239.00	23.00	21.00	85.00
AP352VW		43 5/8"		4'0 1/2"	231.00	27.00	0.00	91.00
AP3521VW		63 3/8"		6'0 3/8"	415.00	41.00	11.00	164.00

Awning Windows with High Performance (H.P.) Insulating Glass

Window Type	Glass Size	Glass Size (Height)	Rough Opening Width	Rough Opening Height	H.P. Insulating	Insect Screen Add	Grille Add	Interior Blinds Add
A21W	19 5/16" x	19 3/4"	2'0 5/8"	2'0 5/8"	$154.00	$10.00	$8.00	$58.00
AN31W	31 1/8" x	16 1/8"	3'0 1/2"	1'9"	160.00	11.00	9.00	56.00
A31W		19 3/4"		2'0 5/8"	180.00	12.00	10.00	65.00
AN351W	36" x	16 1/8"	3'5 3/8"	1'9"	177.00	13.00	11.00	59.00
A351W		19 3/4"		2'0 5/8"	194.00	16.00	13.00	63.00
AN41W	43 3/16" x	16 1/8"	4'0 1/2"	1'9"	194.00	22.00	18.00	72.00
A41W		19 3/4"		2'0 5/8"	210.00	11.00	9.00	59.00
AW31W	31 1/8" x	24"	3'0 1/2"	2'4 7/8"	189.00	14.00	10.00	67.00
AW330W		31 9/16"		3'0 1/2"	196.00	18.00	17.00	72.00
A335W		36 7/16"		3'5 1/4"	278.00	19.00	18.00	75.00
AP32VW		43 5/8"		4'0 1/2"	306.00	25.00	0.00	80.00
AP321VW		63 3/8"		6'0 3/8"	493.00	37.00	10.00	145.00
A3535W	36" x	36 7/16"	3'5 3/8"	3'5 1/4"	250.00	23.00	21.00	85.00
AP352VW		43 5/8"		4'0 1/2"	448.00	27.00	0.00	91.00
AP3521VW		63 3/8"		6'0 3/8"	544.00	41.00	11.00	164.00
AW41W	43 3/16" x	24"	4'0 1/2"	2'4 7/8"	227.00	16.00	13.00	81.00
AP42VW		43 5/8"		4'0 1/2"	382.00	30.00	0.00	98.00
AP421VW		63 3/8"		6'0 3/8"	598.00	45.00	12.00	175.00
AP530VW	55 1/16" x	31 9/16"	5'0 3/8"	3'0 1/2"	362.00	29.00	0.00	109.00

Andersen Perma-Shield[r] awning windows combine slim line sash and frame for maximum exterior view. The vinyl will not chip, flake, peel, blister, rust, pit or corrode, and will not need painting every few years.

Frame. Exterior wood members are treated with a water repellent preservative and covered with a pre-formed rigid vinyl (PVC) sheath in white or color.

Sash. Wood core treated with a water repellent preservative and completely covered with a PVC sheath in white or color.

Glazing. Select quality high-performance, high-performance sun, or double-pane insulating window glass. (For high altitude and other special glazings, contact manufacturer.)

Glazing Bead. Rigid vinyl; snap-in feature with flexible vinyl tip against the glass.

Weatherstripping. Flexible PVC bulb type; factory applied for tight seal between sash and frame.

Flashing. Continuation of the pre-formed rigid vinyl sheath on frame forms perimeter flashing and anchoring fin.

Corner Seal. Pre-formed vinyl frame sheath eliminates corner joints. Sash corner joints are welded to form waterproof joint.

Inside Stops. Western cedar pine natural wood, can be finished to match interior decor.

Awning Roto-Lock Operator. Self-locking underscreen operator. Dual arms pull corners in tightly for maximum weathertightness. Stone color decorator finish. Sash disengages from operator without tools. Picture window units include same materials as ventilating units except no operating hardware. Recommended additional parts include extension jambs of Western clear pine. Sizes are available for 4 9/16", 5 1/4", 6 9/16", 7 9/16" wall thickness. They are pre-drilled for easy application (except for 7 9/16" size).

Optional Accessories. Two different Perma-Fit Divided Light Grilles are available. The white grille is rigid vinyl. Exterior is white and the interior surface has a pine color stainable/paintable polymer finish. The grille in color is polycarbonate. Exterior is color and interior surface has pine color stainable/paintable polymer finish. Perma-Fit grilles without the finish are available in white or color both sides. Grille fits tight against glass and is easily removed for cleaning. Grilles not available for picture window sizes.

Casement Window Units

Casement sash may be ordered in gangs of two, three, four, or five. Prices for each additional unit in a gang arrangement should be quoted at the time of estimate for possible gang arrangement discount. It may be cheaper to use casements in groups than as individual units. Additional savings will also be realized in gang arrangements because of fewer jambs to form and caulk, providing the lintel construction does not become more costly due to the longer span. If some of the sash are ordered as stationary units, a further savings of $30.00 on the basic price per opening can be made. Casements can also be used in combination with larger fixed units or "picture windows".

Casement units swing outward for top-to-bottom ventilation. They can be combined with other styles of windows and patio doors, or can be arranged in bow, angle bay and box bay windows.

Frame. Wood members are treated with a water repellent preservative and covered with a rigid PVC sheath in white or color.

Sash. Wood core treated with a water repellent preservative and completely covered with a rigid PVC sheath in white or color.

Glazing. Select quality high-performance, high-performance sun or double-pane insulating window glass.

Glazing Bead. Rigid vinyl; snap-in features with flexible vinyl tip against glass.

Weatherstripping. Flexible PVC bulb type; factory applied for tight seal between sash and frame.

Flashing. Continuation of the pre-formed rigid vinyl sheath on frame forms perimeter flashing and anchor fin.

Corner Seal. Pre-formed vinyl frame sheath eliminates corner joints. Sash corner joints are welded to form waterproof joint.

Inside Stops. Western clear pine natural wood, can be finished to match interior decor.

Sash Lock. Lock pulls sash firmly closed or ejects for opening. Reach-out action eliminates binding of locks and relieves stress on operator. Stone color decorator finish.

Casement Roto Operator. Under-screen operator with arm that attaches to bracket applied to sash. Either left- or right-hand operating. Stone color decor finish. Sash disengages from operator without tools. Picture window units include same materials as ventilating units except no operating hardware. Recommended additional parts include extension jambs of Western clear pine. Sizes available from Andersen for $4\frac{9}{16}$", $5\frac{1}{4}$", $6\frac{9}{16}$" or $7\frac{9}{16}$" wall thickness. Pre-drilled for easy application (except for $7\frac{9}{16}$" unit).

Optional Accessories. Two different Perma-Fit Divided Light Grilles are available. The white grille is rigid vinyl. Exterior is white and the interior surface has a pine color stainable/paintable polymer finish. The grille in color is polycarbonate. Exterior is color and interior surface has pine color stainable/paintable polymer finish. Perma-Fit grilles without the finish are available in white or color both sides. Grille fits tight against glass and is easily removed for cleaning. Grilles not available for picture window sizes.

Casement Windows with High Performance (H.P.) Insulating Glass

Window Type	Glass Size		Rough Opening Width x	Height	H.P. Insulating	Insect Screen Add	Grille Add	Combination Screen/Storm Add
CR12W	12 5/8" x	19 5/16"	1'5 1/2"	2'0 5/8"	$136.00	$10.00	$8.00	$56.00
CR13W		31 1/8"		3'0 1/2"	158.00	11.00	9.00	56.00
CR135W		36"		3'5 3/8"	172.00	12.00	10.00	58.00
CR14W		43 3/16"		4'0 1/2"	186.00	13.00	11.00	59.00
CR15W		55 1/6"		5'0 3/8"	226.00	16.00	13.00	63.00
CR16W		67 1/16"		6'0 3/8"	245.00	22.00	18.00	72.00
CN13W	16 1/8" x	31 1/8"	1'9"	3'0 1/2"	160.00	11.00	9.00	59.00
CN135W		36"		3'5 3/8"	174.00	13.00	10.00	61.00
CN14W		43 3/16"		4'0 1/2"	189.00	14.00	11.00	63.00
CN15W		55 1/16"		5'0 3/8"	226.00	17.00	14.00	67.00
CN16W		67 1/16"		6'0 3/8"	245.00	22.00	18.00	74.00
C13W	19 3/4" x	31 1/8"	2'0 5/8"	3'0 1/2"	158.00	12.00	10.00	62.00
C135W		36"		3'5 3/8"	186.00	13.00	11.00	64.00
C14W		43 3/16"		4'0 1/2"	202.00	15.00	12.00	66.00
C15W		55 1/16"		5'0 3/8"	226.00	17.00	14.00	71.00
C16W		67 1/16"		6'0 3/8"	245.00	22.00	18.00	76.00
CW12W	24" x	19 5/16"	2'4 7/8"	2'0 5/8"	158.00	12.00	9.00	65.00
CW13W		31 1/8"		3'0 1/2"	186.00	14.00	10.00	68.00
CW135W		36"		3'5 3/8"	200.00	15.00	11.00	68.00
CW14W		43 3/16"		4'0 1/2"	227.00	16.00	13.00	71.00
CW15W		55 1/16"		5'0 3/8"	270.00	19.00	15.00	77.00
CW16W		67 1/16"		6'0 3/8"	313.00	24.00	20.00	83.00
CR23W	12 5/8" x	31 1/8"	2'10 1/4"	3'0 1/2"	285.00	12.00	9.00	65.00
CR235W		36"		3'5 3/8"	321.00	25.00	20.00	116.00
CR24W		43 3/16"		4'0 1/2"	365.00	26.00	22.00	119.00
CN235W	16 1/8" x	36"	3'5 1/4"	3'5 3/8"	330.00	25.00	20.00	122.00
C23W	19 3/4" x	31 1/8"	4'0 1/2"	3'0 1/2"	477.00	24.00	19.00	124.00
C235W		36"		3'5 3/8"	365.00	27.00	21.00	128.00
C24W		43 3/16"		4'0 1/2"	398.00	29.00	23.00	133.00
C25W		55 1/16"		5'0 3/8"	460.00	35.00	28.00	142.00
C26W		67 1/16"		6'0 3/8"	527.00	44.00	36.00	152.00
CW24W	24" x	43 3/16"	4'9"	4'0 1/2"	445.00	32.00	25.00	142.00
CW25W		55 1/16"		5'0 3/8"	532.00	38.00	29.00	153.00
CW26W		67 1/16"		6'0 3/8"	527.00	48.00	39.00	165.00

Casement Windows with High Performance (H.P.) Sun Insulating Glass

Window Type	Glass Size		Rough Opening Width x	Height	H.P. Sun Insulating	Insect Screen Add	Grille Add	Interior Blinds Add
CR12W	12 5/8" x	19 5/16"	1'5 1/2"	2'0 5/8"	$136.00	$10.00	$8.00	$56.00
CR13W		31 1/8"		3'0 1/2"	158.00	11.00	9.00	56.00
CR135W		36"		3'5 3/8"	172.00	12.00	10.00	58.00
CR14W		43 3/16"		4'0 1/2"	186.00	13.00	11.00	59.00
CR15W		55 1/16"		5'0 3/8"	226.00	16.00	13.00	63.00
CR16W		67 1/16"		6'0 3/8"	245.00	22.00	18.00	72.00
CN13W	16 1/8" x	31 1/8"	1'9"	3'0 1/2"	160.00	11.00	9.00	59.00
CN135W		36"		3'5 3/8"	174.00	13.00	10.00	61.00
CN14W		43 3/16"		4'0 1/2"	189.00	14.00	11.00	63.00
CN15W		55 1/16"		5'0 3/8"	229.00	17.00	14.00	67.00
CN16W		67 1/16"		6'0 3/8"	249.00	22.00	18.00	74.00
C13W	19 3/4" x	31 1/8"	2'0 5/8"	3'0 1/2"	169.00	12.00	10.00	62.00
C135W		36"		3'5 3/8"	186.00	13.00	11.00	64.00
C14W		43 3/16"		4'0 1/2"	202.00	15.00	12.00	66.00
C15W		55 1/16"		5'0 3/8"	233.00	17.00	14.00	71.00
C16W		67 1/16"		6'0 3/8"	274.00	22.00	18.00	76.00
CW12W	24" x	19 5/16"	2'4 7/8"	2'0 5/8"	158.00	12.00	9.00	65.00
CW13W		31 1/8"		3'0 1/2"	186.00	14.00	10.00	68.00
CW135W		36"		3'5 3/8"	200.00	15.00	11.00	68.00
CW14W		43 3/16"		4'0 1/2"	227.00	16.00	13.00	71.00
CW15W		55 1/16"		5'0 3/8"	270.00	19.00	15.00	77.00
CW16W		67 1/16"		6'0 3/8"	313.00	24.00	20.00	83.00
CR23W	12 5/8" x	31 1/8"	2'10 1/4"	3'0 1/2"	158.00	12.00	9.00	65.00
CR235W		36"		3'5 3/8"	321.00	25.00	20.00	116.00

Window Type	Glass Size	Rough Opening Width x		Height	Double Pane Insulating	Insect Screen Add	Grille Add	Interior Blinds Add
CR24W		43 3/16"		4'0 1/2"	365.00	26.00	22.00	119.00
CN235W	16 1/8" x	36"	3'5 1/4"	3'5 3/8"	330.00	25.00	20.00	122.00
C23W	19 3/4" x	31 1/8"	4'0 1/2"	3'0 1/2"	335.00	24.00	19.00	124.00
C235W		36"		3'5 3/8"	365.00	27.00	21.00	128.00
C24W		43 3/16"		4'0 1/2"	398.00	29.00	23.00	133.00
C25W		55 1/16"		5'0 3/8"	460.00	35.00	28.00	142.00
C26W		67 1/16"		6'0 3/8"	527.00	44.00	36.00	152.00
CW24W	24" x	43 3/16"	4'9"	4'0 1/2"	445.00	32.00	25.00	142.00
CW25W		55 1/16"		5'0 3/8"	532.00	38.00	29.00	153.00
CW26W		67 1/16"		6'0 3/8"	604.00	48.00	39.00	165.00

Casement Windows with Double Pane Insulating Glass

Window Type	Glass Size	Rough Opening Width x		Height	Double Pane Insulating	Insect Screen Add	Grille Add	Interior Blinds Add
CR12W	12 5/8" x	19 5/16"	1'5 1/2"	2'0 5/8"	$130.00	$10.00	$8.00	$56.00
CR13W		31 1/8"		3'0 1/2"	146.00	11.00	9.00	56.00
CR135W		36"		3'5 3/8"	159.00	12.00	10.00	58.00
CR14W		43 3/16"		4'0 1/2"	171.00	13.00	11.00	59.00
CR15W		55 1/16"		5'0 3/8"	209.00	16.00	13.00	63.00
CR16W		67 1/16"		6'0 3/8"	227.00	22.00	18.00	72.00
CN13W	16 1/8" x	31 1/8"	1'9"	3'0 1/2"	147.00	11.00	9.00	59.00
CN135W		36"		3'5 3/8"	160.00	13.00	10.00	61.00
CN14W		43 3/16"		4'0 1/2"	173.00	14.00	11.00	63.00
CN15W		55 1/16"		5'0 3/8"	212.00	17.00	14.00	67.00
CN16W		67 1/16"		6'0 3/8"	230.00	22.00	18.00	74.00
C13W	19 3/4" x	31 1/8"	2'0 5/8"	3'0 1/2"	157.00	12.00	10.00	62.00
C135W		36"		3'5 3/8"	170.00	13.00	11.00	64.00
C14W		43 3/16"		4'0 1/2"	187.00	15.00	12.00	66.00
C15W		55 1/16"		5'0 3/8"	215.00	17.00	14.00	71.00
C16W		67 1/16"		6'0 3/8"	254.00	22.00	18.00	76.00
CW12W	24" x	19 5/16"	2'4 7/8"	2'0 5/8"	146.00	12.00	9.00	65.00
CW13W		31 1/8"		3'0 1/2"	172.00	14.00	10.00	68.00
CW135W		36"		3'5 3/8"	185.00	15.00	11.00	68.00
CW14W		43 3/16"		4'0 1/2"	210.00	16.00	13.00	71.00
CW15W		55 1/16"		5'0 3/8"	250.00	19.00	15.00	77.00
CW16W		67 1/16"		6'0 3/8"	300.00	24.00	20.00	83.00
CR23W	12 5/8" x	31 1/8"	2'10 1/4"	3'0 1/2"	257.00	12.00	9.00	65.00
CR235W		36"		3'5 3/8"	295.00	25.00	20.00	116.00
CR24W		43 3/16"		4'0 1/2"	336.00	26.00	22.00	119.00
CN235W	16 1/8" x	36"	3"5 1/4"	3'5 3/8"	304.00	25.00	20.00	122.00
C23W	19 3/4" x	31 1/8"	4'0 1/2"	3'0 1/2"	309.00	24.00	19.00	124.00
C235W		36"		3'5 3/8"	334.00	27.00	21.00	128.00
C24W		43 3/16"		4'0 1/2"	367.00	29.00	23.00	133.00
C25W		55 1/16"		5'0 3/8"	424.00	35.00	28.00	142.00
C26W		67 1/16"		6'0 3/8"	484.00	44.00	36.00	152.00
CW24W	24" x	43 3/16"	4'9"	4'0 1/2"	410.00	32.00	25.00	142.00
CW25W		55 1/16"		5'0 3/8"	491.00	38.00	29.00	153.00
CW26W		67 1/16"		6'0 3/8"	578.00	48.00	39.00	165.00

Circle Top Window Units

Circle top windows are specially designed with a frame/sash profile. The windows are compatible with casement, awning and double hung units and can be joined to these, or can be installed as single units. When combined, the glazing in the circle top window closely aligns with the window unit below it.

Frame. Wood members are treated with a water repellent preservative and covered with a pre-formed rigid PVC sheath in white or color. The vinyl sheath will not chip, flake, peel, blister, rust, pit or corrode, and it will not need painting every few years.

Flashing. Continuation of the pre-formed rigid vinyl sheath on frame forms a full perimeter flashing and anchoring fin.

Glazing. Select quality high-performance or high-performance sun insulating glass.

Recommended Optional Accessories. Interior arch casing is unfinished wood for painting or staining. Maple or oak is used for laminated portion, which forms the curved piece. Interior casing is available in ranch or colonial style. Divided light grilles are available in two different types. The white grille is rigid vinyl. Exterior is white and the interior surface has a pine color stainable/paintable polymer finish. The grille in color is glass fiber polycarbonate. Exterior is color and interior surface has pine color stainable/paintable polymer finish. Perma-fit grilles without the finish are available in white or color both sides. Grille fits tight against glass and can be removed for cleaning. Grilles not available for picture window sizes.

Plinth Blocks. Solid oak or maple plinth block with decorative radial sunburst pattern or reversible flush face. Plinth blocks come with arch casing or can be purchased separately.

up set of halftones on page 8.65

Courtesy Andersen Window Company

Cutaway View of Circle Top Window with Arch Casing and Plinth Blocks Shown at Right

Extension Jambs. Laminated maple curved extension jambs are available for wall thickness of 4$^{9}/_{16}$", 5$^{1}/_{4}$", 6$^{9}/_{16}$" and 7$^{9}/_{16}$" for casement and awning windows and 5$^{1}/_{4}$", 6$^{9}/_{16}$" and 7$^{9}/_{16}$" for double hung windows. They are pre-drilled, except for casement and awning 7$^{9}/_{16}$" size. One-piece pine straight extension jamb for sill is available in same sizes as maple parts for casement or awning. Both laminated maple and clear pine parts give uniform appearance when stained.

Circle Top Windows with High Performance (H.P.) Insulating Glass

Window Type	Glass Size		Rough Opening Width x	Height	H.P. Insulating	Grille Add
CTQC1W	19⅝" x	19⅝"	2'0⅝"	2'0⅝"	301.00	15.00
CTC1W	19½" x	9¾"	2'0⅝"	1'2⅞"	270.00	14.00
CTCW1W	23¾" x	11⅞"	2'4⅞"	2'2$^{3}/_{16}$	298.00	18.00
CTC2W	43⅜" x	21$^{11}/_{16}$"	4'0½"	2'2$^{13}/_{16}$"	414.00	27.00
CTCW4W	51⅞" x	25$^{15}/_{16}$"	4'9"	2'7$^{1}/_{16}$"	489.00	34.00
CTC3W	67⅛" x	35$^{9}/_{16}$"	6'0⅜"	3'2¾"	789.00	89.00
CTN20W	20⅜" x	10$^{3}/_{16}$"	2'2⅛"	1'3¾"	278.00	15.00
CTN24	24⅜" x	12$^{3}/_{16}$"	2'6⅛"	1'5¾"	322.00	18.00
CTN28	28⅜" x	14$^{3}/_{16}$"	2'10⅛"	1'7¾"	348.00	21.00
CTN30W	32⅜" x	16$^{3}/_{16}$"	3'2⅛"	1'9¾"	372.00	23.00
CTN34W	36⅜" x	18$^{3}/_{16}$"	6'6⅛"	1'11¾"	410.00	25.00

Circle Top Windows with High Performance (H.P.) Sun Insulating Glass

Window Type	Glass Size		Rough Opening Width x	Height	H.P. Sun Insulating	Grille Add
CTQC1W	19⅝" x	19⅝"	2'0⅝"	2'0⅝"	$301.00	$15.00
CTC1W	19½" x	9¾"	2'0⅝"	1'2⅞"	270.00	14.00
CTCW1W	23¾" x	11⅞"	2'4⅞"	1'5"	298.00	10.00
CTC2W	43⅜" x	21$^{11}/_{16}$"	4'0½"	2'2$^{13}/_{16}$"	414.00	27.00
CTCW4W	51⅞" x	25$^{15}/_{16}$"	4'9"	2'7$^{1}/_{16}$"	489.00	34.00
CTC3W	67⅛" x	35$^{9}/_{16}$"	6'0⅜"	3'2¾"	789.00	89.00
CTN20W	20⅜" x	10$^{3}/_{16}$"	2'2⅛"	1'3¾"	278.00	15.00
CTN24	24⅜" x	12$^{3}/_{16}$"	2'6⅛"	1'5¾"	322.00	18.00
CTN28	28⅜" x	14$^{3}/_{16}$"	2'10⅛"	1'7¾"	348.00	21.00
CTN30W	32⅜" x	16$^{3}/_{16}$"	2'2⅛"	1'9¾"	372.00	23.00
CTN34W	36⅜" x	18$^{3}/_{16}$"	6'6⅛"	1'11¾"	410.00	25.00

Bow, Box Bay and Angle Bay Window Units

30 Degree Angle Bay Unit Sizes with High Performance (H.P.)
Sun Insulating Glass

Note: Sizes of bay windows are the same with different glazing.

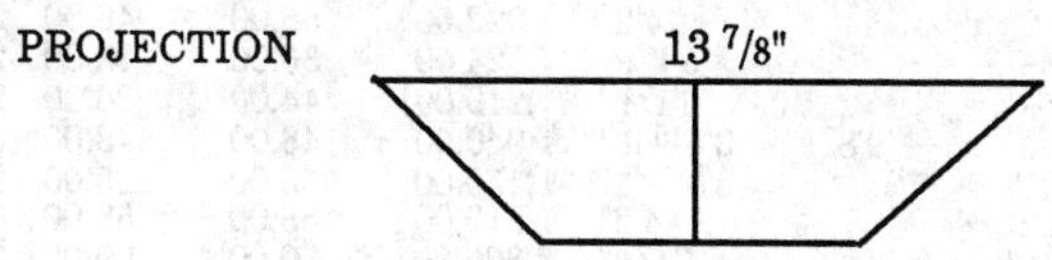

Window Type	Glass Size		Rough Opening Width x	Height	H.P. Sun Insulating	Insect Screens Add	Grille Add	Blinds with Hardware
30 C13 20W	5'10" x	3'1 15/16"	5'9 1/8"	3'1 15/16"	$683.00	$24.00	$29.00	$184.00
30 C135 20W		3'6 13/16"		3'6 13/16"	741.00	27.00	32.00	190.00
30 C14 20W		4'2"		4'2"	795.00	29.00	35.00	197.00
30 C15 20W		5'1 7/8"		5'1/78"	900.00	35.00	41.00	212.00
30 C16 20W		6'1 7/8"		6'1 7/8"	1,042.00	44.00	55.00	225.00
30 C23 20W	7'9 7/8" x	3'1 15/16"	7'9 7/8"	3'1 15/16"	876.00	48.00	39.00	247.00
30 C235 20W		3'6 13/16"		3'6 13/16"	949.00	53.00	42.00	256.00
30 C24 20W		4'2"		4'2"	1,022.00	58.00	46.00	265.00
30 C25 20W		5'1 7/8"		5'1 7/8"	1,158.00	70.00	55.00	284.00
30 C26 20W		6'1 7/8"		6'1 7/8"	1,326.00	88.00	73.00	303.00
30 CP23 20W	7'9 7/8" x	3'1 15/16"	7'9 7/8"	3'1 15/16"	754.00	24.00	19.00	207.00
30 CP235 20W		3'6 13/16"		3'6 13/16"	949.00	53.00	42.00	256.00
30 CP24 20W		4'2"		4'2"	1,022.00	58.00	46.00	265.00
30 CP25 20W		5'1 7/8"		5'1 7/8"	1,024.00	35.00	28.00	245.00
30 CP26 20W		6'1 7/8"		6'1 7/8"	1,219.00	44.00	36.00	265.00
30 C33 20W	9'9 3/4" x	3'1 15/16"	9'8 7/8"	3'1 15/16"	1,040.00	48.00	48.00	307.00
30 C335 20W		3'6 13/16"		3'6 13/16"	1,130.00	53.00	53.00	318.00
30 C34 20W		4'2"		4'2"	1,212.00	58.00	58.00	330.00
30 C35 20W		5'1 7/8"		5'1 7/8"	1,366.00	70.00	69.00	354.00
30 CP33 20W	9'9 3/4" x	3'1 15/16"	9'8 7/8"	3'1 15/16"	874.00	24.00	19.00	236.00
30 CP335 20W		3'6 13/16"		3'6 13/16"	958.00	26.00	21.00	247.00
30 CP34 20W		4'2"		4'2"	1,065.00	29.00	23.00	259.00
30 CP35 20W		5'1 7/8"		5'1 7/8"	1,288.00	34.00	28.00	284.00

30 Degree Angle Bay Unit Sizes with High Performance (H.P.) Insulating Glass

PROJECTION 13 7/8"

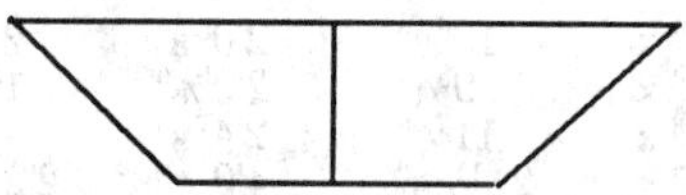

Window Type	Glass Size		Rough Opening Width x	Height	H.P. Insulating	Insect Screens Add	Grille Add	Blinds with Hardware
30 C13 20W	5'10" x	3'1 15/16"	5'9 1/8"	3'1 15/16"	$683.00	$24.00	$29.00	$184.00
30 C135 20W		3'6 13/16"		3'6 13/16"	741.00	27.00	32.00	190.00
30 C14 20W		4'2"		4'2"	795.00	29.00	35.00	197.00
30 C15 20W		5'1 7/8"		5'1 7/8"	900.00	35.00	41.00	212.00
30 C16 20W		6'1 7/8"		6'1 7/8"	1,042.00	44.00	55.00	225.00
30 C23 20W	7'9 7/8" x	3'1 15/16"	7'9 7/8"	3'1 15/16"	876.00	48.00	39.00	247.00
30 C235 20W		3'6 13/16"		3'6 13/16"	949.00	53.00	42.00	256.00
30 C24 20W		4'2"		4'2"	1,022.00	58.00	46.00	265.00
30 C25 20W		5'1 7/8"		5'1 7/8"	1,158.00	70.00	55.00	284.00
30 C26 20W		6'1 7/8"		6'1 7/8"	1,326.00	88.00	73.00	303.00
30 CP23 20W	7'9 7/8" x	3'1 15/16"	7'9 7/8"	3'1 15/16"	754.00	24.00	19.00	207.00
30 CP235 20W		3'6 13/16"		3'6 13/16"	949.00	53.00	42.00	256.00
30 CP24 20W		4'2"		4'2"	1,022.00	58.00	46.00	265.00
30 CP25 20W		5'1 7/8"		5'1 7/8"	1,024.00	35.00	28.00	245.00
30 CP26 20W		6'1 7/8"		6'1 7/8"	1,219.00	44.00	36.00	265.00
30 C33 20W	9'9 3/4" x	3'1 15/16"	9'8 7/8"	3'1 15/16"	1,040.00	48.00	48.00	307.00
30 C335 20W		3'6 13/16"		3'6 13/16"	1,130.00	53.00	53.00	318.00
30 C34 20W		4'2"		4'2"	1,212,00	58.00	58.00	330.00
30 C35 20W		5'1 7/8"		5'1 7/8"	1,366.00	70.00	69.00	354.00
30 CP33 20W	9'9 3/4" x	3'1 15/16"	9'8 7/8"	3'1 15/16"	874.00	24.00	19.00	236.00
30 CP335 20W		3'6 13/16"		3'6 13/16"	958.00	26.00	21.00	247.00
30 CP34 20W		4'2"		4'2"	1,065.00	29.00	23.00	259.00
30 CP35 20W		5'1 7/8"		5'1 7/8"	1,288.00	34.00	28.00	284.00

30 Degree Angle Bay Unit Sizes with High Double Pane Insulating Glass

PROJECTION 13 7/8"

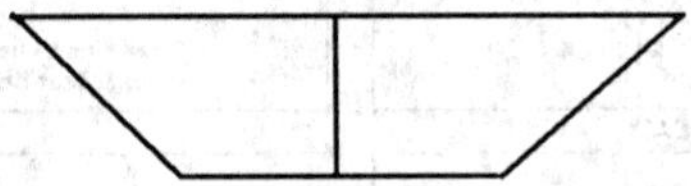

Window Type	Glass Size		Rough Opening Width x	Height	Double Pane Insulating	Insect Screens Add	Grille Add	Blinds with Hardware
30 C13 20W	5'10" x	3'1 15/16"	5'9 1/8"	3'1 15/16"	$644.00	24.00	29.00	184.00
30 C135 20W		3'6 13/16"		3'6 13/16"	694.00	27.00	32.00	190.00
30 C14 20W		4'2"		4'2"	750.00	29.00	35.00	197.00
30 C15 20W		5'1 7/8"		5'1 7/8"	847.00	35.00	41.00	212.00
30 C16 20W		6'1 7/8"		6'1 7/8"	979.00	44.00	55.00	225.00
30 C23 20W	7'9 7/8" x	3'1 15/16"	7'9 7/8"	3'1 15/16"	825.00	48.00	39.00	247.00
30 C235 20W		3'6 13/16"		3'6 13/16"	887.00	53.00	42.00	256.00
30 C24 20W		4'2"		4'2"	961.00	58.00	46.00	265.00
30 C25 20W		5'1 7/8"		5'1 7/8"	1,087.00	70.00	55.00	284.00
30 C26 20W		6'1 7/8"		6'1 7/8"	1,241.00	88.00	73.00	303.00
30 CP23 20W	7'9 7/8" x	3'1 15/16"	7'9 7/8"	3'1 15/16"	711.00	24.00	19.00	207.00
30 CP235 20W		3'6 13/16"		3'6 13/16"	784.00	53.00	42.00	256.00
30 CP24 20W		4'2"		4'2"	859.00	58.00	46.00	265.00
30 CP25 20W		5'1 7/8"		5'1 7/8"	964.00	35.00	28.00	245.00
30 CP26 20W		6'1 7/8"		6'1 7/8"	1,129.00	44.00	36.00	265.00
30 C33 20W	9'9 3/4" x	3'1 15/16"	9'8 7/8"	3'1 15/16"	975.00	48.00	48.00	307.00
30 C335 20W		3'6 13/16"		3'6 13/16"	1,052.00	53.00	53.00	318.00
30 C34 20W		4'2"		4'2"	1,135.00	58.00	58.00	330.00
30 C35 20W		5'1 7/8"		5'1 7/8"	1,276.00	70.00	69.00	354.00
30 CP33 20W	9'9 3/4" x	3'1 15/16"	9'8 7/8"	3'1 15/16"	825.00	24.00	19.00	236.00
30 CP335 20W		3'6 13/16"		3'6 13/16"	901.00	26.00	21.00	247.00
30 CP34 20W		4'2"		4'2"	986.00	29.00	23.00	259.00
30 CP35 20W		5'1 7/8"		5'1 7/8"	1,188.00	34.00	28.00	284.00

Add for casing 3 1/2" (not installed):

45-C1-20-W	45-C2-20-W	45-C3-20-W
$+18.83	$+30.43	$+30.43

Deduct for no top auxiliary casing:

45-C1-20-W	45-C2-20-W	45-C3-20-W
$-20.64	$-20.64	$-20.64

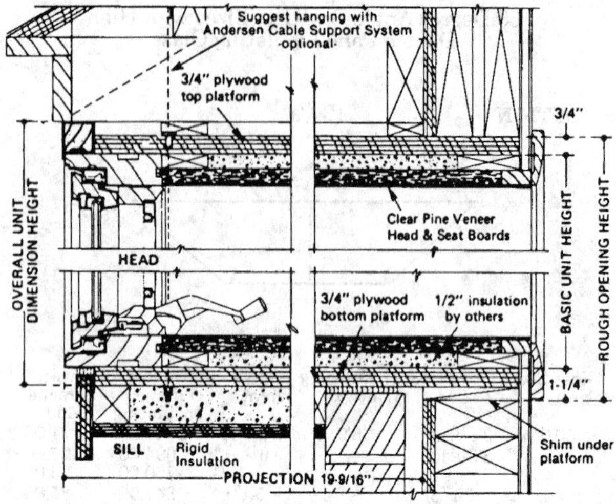

45° Angle Bay Section

45 Degree Angle Bay Unit Sizes with High Performance (H.P.) Insulating Glass

PROJECTION 19 9/16"

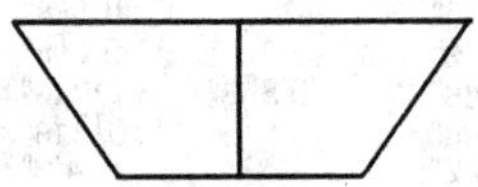

Window Type	Glass Size		Rough Opening Width x	Height	H.P. Insulating	Insect Screens Add	Grille Add	Blinds with Hardware
45 C13 20W	5'4 1/8" x	3'1 15/16"	5'2 5/8"	3'1 15/16"	$702.00	$24.00	$29.00	$184.00
45 C135 20W		3'6 13/16"		3'6 13/16"	760.00	27.00	32.00	190.00
45 C14 20W		4'2"		4'2"	823.00	29.00	35.00	197.00
45 C15 20W		5'1 7/8"		5'1 7/8"	936.00	35.00	41.00	212.00
45 C16 20W		6'1 7/8"		6'1 7/8"	1,079.00	44.00	55.00	225.00
45 C23 20W	7'3 7/8" x	3'1 15/16"	7'2 1/2"	3'1 15/16"	906.00	47.00	39.00	247.00
45 C235 20W		3'6 13/16"		3'6 13/16"	979.00	53.00	42.00	256.00
45 C24 20W		4'2"		4'2"	1,060.00	273.00	58.00	265.00
45 C25 20W		5'1 7/8"		5'1 7/8"	1,205.00	70.00	55.00	284.00
45 C26 20W		6'1 7/8"		6'1 7/8"	1,373.00	88.00	73.00	303.00
45 CP23 20W	7'3 7/8" x	3'1 15/16"	7'2 1/2"	3'1 15/16"	783.00	23.00	19.00	207.00
45 CP235 20W		3'6 13/16"		3'6 13/16"	864.00	27.00	21.00	215.00
45 CP24 20W		4'2"		4'2"	950.00	29.00	23.00	226.00
45 CP25 20W		5'1 7/8"		5'1 7/8"	1,071.00	35.00	28.00	245.00
45 CP26 20W		6'1 7/8"		6'1 7/8"	1,267.00	44.00	36.00	265.00
45 C33 20W	9'3 3/4" x	3'1 15/16"	9'2 3/8"	3'1 15/16"	1,074.00	47.00	48.00	307.00
45 C335 20W		3'6 13/16"		3'6 13/16"	1,165.00	53.00	53.00	318.00
45 C34 20W		4'2"		4'2"	1,255.00	58.00	58.00	330.00
45 C35 20W		5'1 7/8"		5'1 7/8"	1,417.00	70.00	69.00	354.00
45 CP33 20W	9'3 3/4" x	3'1 15/16"	9'2 3/8"	3'1 15/16"	909.00	24.00	19.00	236.00
45 CP335 20W		3'6 13/16"		3'6 13/16"	993.00	27.00	21.00	247.00
45 CP34 20W		4'2"		4'2"	1,108.00	29.00	23.00	259.00
45 CP35 20W		5'1 7/8"		5'1 7/8"	1,338.00	35.00	28.00	284.00

45 Degree Angle Bay Unit Sizes with High Performance (H.P.) Sun Insulating Glass

PROJECTION 19 9/16"

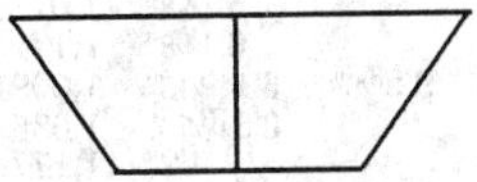

Window Type	Glass Size		Rough Opening Width x	Height	H.P. Sun Insulating	Insect Screens Add	Grille Add	Blinds with Hardware
45 C13 20W	5'4 1/8" x	3'1 15/16"	5'2 5/8"	3'1 15/16"	$702.00	$24.00	$29.00	$184.00
45 C135 20W		3'6 13/16"		3'6 13/16"	760.00	27.00	32.00	190.00
45 C14 20W		4'2"		4'2"	823.00	29.00	35.00	197.00
45 C15 20W		5'1 7/8"		5'1 7/8"	936.00	35.00	41.00	212.00
45 C16 20W		6'1 7/8"		6'1 7/8"	1,079.00	44.00	55.00	225.00
45 C23 20W	7'3 7/8" x	3'1 15/16"	7'2 1/2"	3'1 15/16"	906.00	47.00	39.00	247.00
45 C235 20W		3'6 13/16"		3'6 13/16"	979.00	53.00	42.00	256.00
45 C24 20W		4'2"		4'2"	1,060.00	273.00	58.00	265.00
45 C25 20W		5'1 7/8"		5'1 7/8"	1,205.00	70.00	55.00	284.00
45 C26 20W		6'1 7/8"		6'1 7/8"	1,373.00	88.00	73.00	303.00
45 CP23 20W	7'3 7/8" x	3'1 15/16"	7'2 1/2"	3'1 15/16"	783.00	23.00	19.00	207.00
45 CP235 20W		3'6 13/16"		3'6 13/16"	864.00	27.00	21.00	215.00
45 CP24 20W		4'2"		4'2"	950.00	29.00	23.00	226.00
45 CP25 20W		5'1 7/8"		5'1 7/8"	1,071.00	35.00	28.00	245.00
45 CP26 20W		6'1 7/8"		6'1 7/8"	1,267.00	44.00	36.00	265.00
45 C33 20W	9'3 3/4" x	3'1 15/16"	9'2 3/8"	3'1 15/16"	1,074.00	47.00	48.00	307.00
45 C335 20W		3'6 13/16"		3'6 13/16"	1,165.00	53.00	53.00	318.00
45 C34 20W		4'2"		4'2"	1,255.00	58.00	58.00	330.00
45 C35 20W		5'1 7/8"		5'1 7/8"	1,417.00	70.00	69.00	354.00
45 CP33 20W	9'3 3/4" x	3'1 15/16"	9'2 3/8"	3'1 15/16"	909.00	24.00	19.00	236.00
45 CP335 20W		3'6 13/16"		3'6 13/16"	993.00	27.00	21.00	247.00
45 CP34 20W		4'2"		4'2"	1,108.00	29.00	23.00	259.00
45 CP35 20W		5'1 7/8"		5'1 7/8"	1,338.00	35.00	28.00	284.00

45 Degree Angle Bay Unit Sizes with High Double Pane Insulating Glass

PROJECTION 19 9/16"

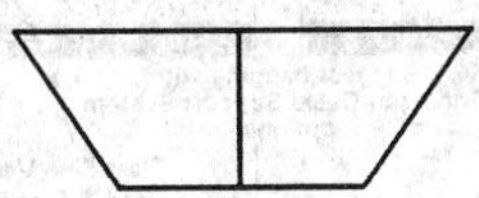

Window Type	Glass Size		Rough Opening Width x	Height	Double Pane Insulating	Insect Screens Add	Grille Add	Blinds with Hardware
45 C13 20W	5'4 1/8" x	3'1 15/16"	5'2 5/8"	3'1 15/16"	$663.00	$24.00	$29.00	$184.00
45 C135 20W		3'6 13/16"		3'6 13/16"	713.00	27.00	32.00	190.00
45 C14 20W		4'2"		4'2"	777.00	29.00	35.00	197.00
45 C15 20W		5'1 7/8"		5'1 7/8"	883.00	35.00	41.00	212.00
45 C16 20W		6'1 7/8"		6'1 7/8"	1,017.00	44.00	55.00	225.00
45 C23 20W	7'3 7/8" x	3'1 15/16"	7'2 1/2"	3'1 15/16"	855.00	47.00	39.00	247.00
45 C235 20W		3'6 13/16"		3'6 13/16"	917.00	53.00	42.00	256.00
45 C24 20W		4'2"		4'2"	999.00	273.00	58.00	265.00

Window Type	Glass Size		Rough Opening Width x	Height	Double Pane Insulating	Insect Screens Add	Grille Add	Blinds with Hardware
45 C25 20W		5'1 7/8"		5'1 7/8"	1,134.00	70.00	55.00	284.00
45 C26 20W		6'1 7/8"		6'1 7/8"	1,290.00	88.00	73.00	303.00
45 CP23 20W	7'3 7/8" x	3'1 15/16"	7'2 1/2"	3'1 15/16"	741.00	23.00	19.00	207.00
45 CP235 20W		3'6 13/16"		3'6 13/16"	815.00	27.00	21.00	215.00
45 CP24 20W		4'2"		4'2"	898.00	29.00	23.00	226.00
45 CP25 20W		5'1 7/8"		5'1 7/8"	1,011.00	35.00	28.00	245.00
45 CP26 20W		6'1 7/8"		6'1 7/8"	1,177.00	44.00	36.00	265.00
45 C33 20W	9'3 3/4" x	3'1 15/16"	9'2 3/8"	3'1 15/16"	1,009.00	47.00	48.00	307.00
45 C335 20W		3'6 13/16"		3'6 13/16"	1,085.00	53.00	53.00	318.00
45 C34 20W		4'2"		4'2"	1,177.00	58.00	58.00	330.00
45 C35 20W		5'1 7/8"		5'1 7/8"	1,327.00	70.00	69.00	354.00
45 CP33 20W	9'3 3/4" x	3'1 15/16	9'2 3/8"	3'1 15/16"	860.00	24.00	19.00	236.00
45 CP335 20W		3'6 13/16"		3'6 13/16"	935.00	27.00	21.00	247.00
45 CP34 20W		4'2"		4'2"	1,030.00	29.00	23.00	259.00
45 CP35 20W		5'1 7/8"		5'1 7/8"	1,239.00	35.00	28.00	284.00

Add for casing 3 1/2" (not installed):

45-C1-20-W	45-C2-20-W	45-C3-20-W
$+18.83	$+30.43	$+30.43

Deduct for no top auxiliary casing:

45-C1-20-W	45-C2-20-W	45-C3-20-W
$-20.64	$-20.64	$-20.64

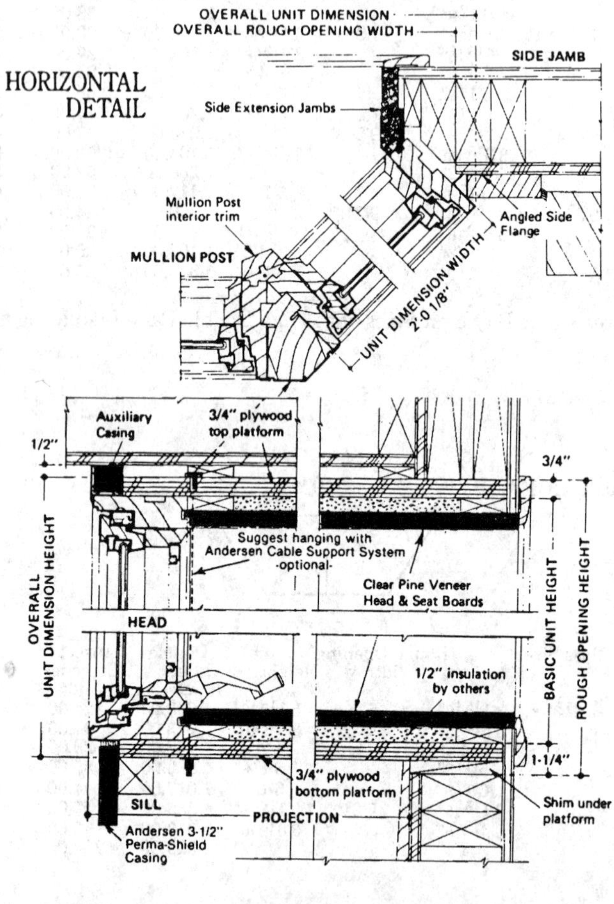

90 Degree Box Bay Window Sizes with High Performance (H.P.) Insulating Glass

PROJECTION 23"

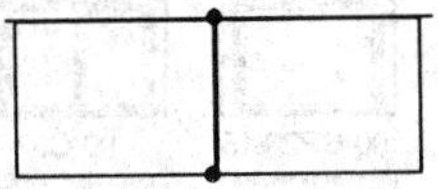

Window Type	Glass Size		Rough Opening Width x	Height	H.P. Insulating	Insect Screens Add	Grille Add	Blinds with Hardware
90 C23 15W	4'8 3/8" x	2'11 15/16	4'2"	3'1 15/16"	$883.00	$46.00	$38.00	$236.00
90 C235 15W		3'4 13/16"		3'6 7/8"	955.00	51.00	41.00	244.00
90 C24 15W		4'0"		4'2"	1,033.00	55.00	45.00	251.00
90 C25 15W		4'11 7/8"		5'1 7/8"	1,204.00	68.00	54.00	267.00
90 C26 15W		5'11 7/8"		6'1 7/8"	1,332.00	88.00	72.00	296.00
90 CP23 15W	4'8 3/8" x	2'11 15/16"	4'2"	3'1 15/16"	760.00	22.00	18.00	196.00
90 CP235 15W		3'4 13/16"		3'6 7/8"	840.00	25.00	20.00	203.00
90 CP24 15W		4'0"		4'2"	922.00	26.00	22.00	212.00
90 CP25 15W		4'11 7/8"		5'1 7/8"	1,070.00	33.00	27.00	228.00
90 CP26 15W		5'11 7/8"		6'1 7/8"	1,225.00	43.00	35.00	258.00
90 C33 15W	6'8 1/4" x	2'11 15/16"	4'2"	6'1 7/8"	1,068.00	46.00	47.00	296.00
90 C335 15W		3'4 13/16"		3'6 7/8"	1,159.00	51.00	51.00	306.00
90 C34 15W		4'0"		4'2"	1,246.00	55.00	57.00	315.00
90 C35 15W		4'11 7/8"		5'1 7/8"	1,436.00	68.00	68.00	337.00
90 CP33 15W	6'8 1/4" x	2'11 15/16"	4'2"	6'1 7/8"	903.00	22.00	18.00	225.00
90 CP3335 15W		3'4 13/16"		3'6 7/8"	987.00	25.00	20.00	235.00
90 CP34 15W		4'0"		4'2"	1,099.00	26.00	22.00	246.00
90 CP35 15W		4'11 7/8"		5'1 7/8"	1,357.00	33.00	26.00	267.00

90 Degree Angle Bay Unit Sizes with High Performance (H.P.) Sun Insulating Glass

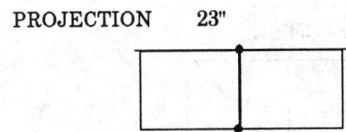

PROJECTION 23"

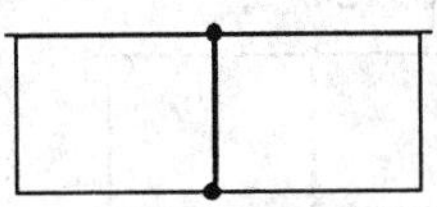

Window Type	Glass Size		Rough Opening Width x	Height	H.P. Sun Insulating	Insect Screens Add	Grille Add	Blinds with Hardware
90 C23 15W	4'8 3/8" x	2'11 15/16"	4'2"	3'1 15/16"	$883.00	$46.00	$38.00	$236.00
90 C235 15W		3'4 13/16"		3'6 7/8"	955.00	51.00	41.00	244.00
90 C24 15W		4'0"		4'2"	1,033.00	55.00	45.00	251.00
90 C25 15W		4'11 7/8"		5'1 7/8"	1,204.00	68.00	54.00	267.00
90 C26 15W		5'11 7/8"		6'1 7/8"	1,332.00	88.00	72.00	296.00
90 CP23 15W	4'8 3/8" x	2'11 15/16"	4'2"	3'1 15/16"	760.00	22.00	18.00	196.00
90 CP235 15W		3'4 13/16"		3'6 7/8"	840.00	25.00	20.00	203.00
90 CP24 15W		4'0"		4'2"	922.00	26.00	22.00	212.00
90 CP25 15W		4'11 7/8"		5'1 7/8"	1,070.00	33.00	27.00	228.00
90 CP26 15W		5'11 7/8"		6'1 7/8"	1,225.00	43.00	35.00	258.00
90 C33 15W	6'8 1/4" x	2'11 15/16"	4'2"	6'1 7/8"	1,068.00	46.00	47.00	296.00
90 C335 15W		3'4 13/16"		3'6 7/8"	1,159.00	51.00	51.00	306.00
90 C34 15W		4'0"		4'2"	1,246.00	55.00	57.00	315.00
90 C35 15W		4'11 7/8"		5'1 7/8"	1,436.00	68.00	68.00	337.00
90 CP33 15W	6'8 1/4" x	2'11 15/16"	4'2"	6'1 7/8"	903.00	22.00	18.00	225.00
90 CP335 15W		3'4 13/16"		3'6 7/8"	987.00	25.00	20.00	235.00
90 CP34 15W		4'0"		4'2"	1,099.00	26.00	22.00	246.00
90 CP35 15W		4'11 7/8"		5'1 7/8"	1,357.00	33.00	26.00	267.00

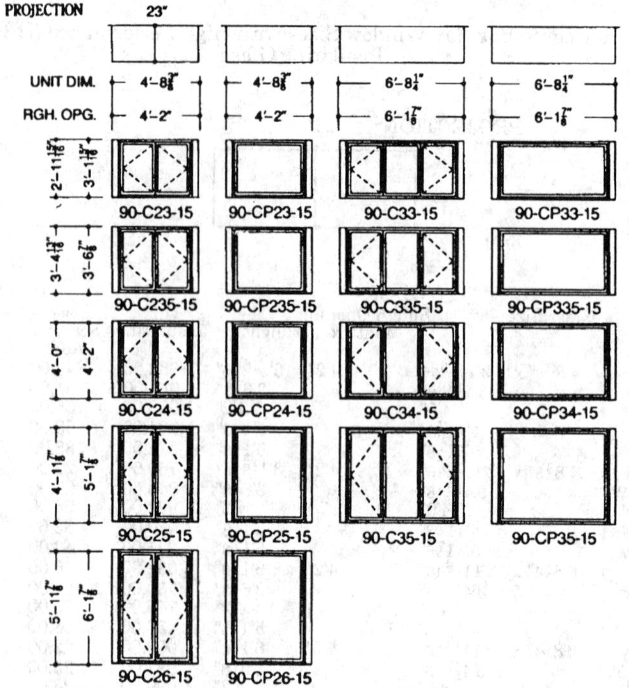

90 Degree Angle Bay Unit Sizes with High Double Pane Insulating Glass

PROJECTION 23"

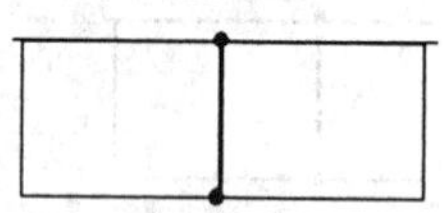

Window Type	Glass Size		Rough Opening Width x	Height	Double Pane Insulating	Insect Screens Add	Grille Add	Blinds with Hardware
90 C23 15W	4'8 3/8" x	2'11 15/16"	4'2"	3'1 15/16"	$883.00	$46.00	$38.00	$236.00
90 C235 15W		3'4 13/16"		3'6 7/8"	897.00	51.00	41.00	244.00
90 C24 15W		4'0"		4'2"	972.00	55.00	45.00	251.00
90 C25 15W		4'11 7/8"		5'1 7/8"	1,135.00	68.00	54.00	267.00
90 C26 15W		5'11 7/8"		6'1 7/8"	1,253.00	88.00	72.00	296.00
90 CP23 15W	4'8 3/8" x	2'11 15/16"	4'2"	3'1 15/16"	720.00	22.00	18.00	196.00
90 CP235 15W		3'4 13/16"		3'6 7/8"	795.00	25.00	20.00	203.00
90 CP24 15W		4'0"		4'2"	870.00	26.00	22.00	212.00
90 CP25 15W		4'11 7/8"		5'1 7/8"	1,011.00	33.00	27.00	228.00
90 CP26 15W		5'11 7/8"		6'1 7/8"	1,141.00	43.00	35.00	258.00
90 C33 15W	6'8 1/4" x	2'11 15/16"	4'2"	6'1 7/8"	1,005.00	46.00	47.00	296.00
90 C335 15W		3'4 13/16"		3'6 7/8"	1,085.00	51.00	51.00	306.00
90 C34 15W		4'0"		4'2"	1,169.00	55.00	57.00	315.00
90 C35 15W		4'11 7/8"		5'1 7/8"	1,347.00	68.00	68.00	337.00
90 CP33 15W	6'8 1/4" x	2'11 15/16"	4'2"	6'1 7/8"	856.00	22.00	18.00	225.00
90 CP335 15W		3'4 13/16"		3'6 7/8"	935.00	25.00	20.00	235.00
90 CP34 15W		4'0"		4'2"	1,021.00	26.00	22.00	246.00
90 CP35 15W		4'11 7/8"		5'1 7/8"	1,259.00	33.00	26.00	267.00

Add for casing 3½" (not installed):

90-C2-15W	90-C3-15W
$+60.86	$+79.69

Roof Window Units (Skylights)

Roof windows, also called "skylights", can enhance the appearance of a living space. However, the estimator must be cautious. Roof window installation is unique and usually requires additional costs. On the exterior, there might be flashing for ice build-up, curbs around the window on a flat roof, or on a sloped roof that is relatively flat, or deflector shield for maximum weather tightness. On the interior, additional finish work might be required.

Roof Windows, Basic and Stationary Unit with High Performance (H.P.) Insulating Laminated Tempered Glass

Window Type	Glass Size		Rough opening Width x	Height	H.P. Insulating Lam/Temp	Set Shingle Flashing	Incline Curb Flashing	Set Tile Flashing	Pleated Shade with Hardware
RW2133S	85 1/16" x	30 5/16"	1'10"	2'10¼"	$299.00	$68.00	$269.00	$79.00	$104.00
RW2144S		40 15/16"		3'8¾"	350.00	68.00	274.00	89.00	118.00
RW2172S		68⅞"		6'0¾"	511.00	0.00	123.00	139.00	119.00
RW2944S	26" x	40 15/16"	2'5⅞"	3'8¾"	419.00	82.00	290.00	97.00	131.00
RW2957S		53 15/16"		4'9¾"	513.00	82.00	320.00	107.00	155.00
RW2972S		68⅞"		6'0¾"	597.00	199.00	0.00	133.00	177.00
RW4144S	38 3/16" x	40 15/16"	3'6"	3'8¾"	526.00	91.00	324.00	111.00	166.00
RW4157S		53 15/16"		4'9¾"	640.00	91.00	343.00	119.00	189.00

Roof Windows, Venting Unit with High Performance (H.P.) Sun Insulating Tempered Glass

Window Type	Glass Size		Rough opening Width x	Height	H.P. Sun Insul/Tmp	Set Shingle Flashing	Incline Curb Flashing	Set Tile Flashing	Pleated Shade with Hardware
RW2133V	15 1/16" x	27⅛"	1'10"	2'10¼"	$444.00	$68.00	$270.00	$79.00	$94.00
RW2144V		37¾"		3'8¾"	506.00	68.00	274.00	89.00	107.00
RW2944V	23" x	37¾"	2'5⅞"	3'8¾"	565.00	82.00	290.00	97.00	119.00
RW2957V		50¾"		4'9¾"	656.00	82.00	320.00	107.00	139.00
RW4144V	35 3/16" x	37¾"	3'6"	3'8¾"	681.00	91.00	323.00	112.00	149.00
RW4157V		50¾"		4'9¾"	816.00	91.00	343.00	119.00	170.00

Courtesy Andersen Window Company

Skylight Frame, Optional Shade, and Extension Pole

Roof Windows, Basic and Stationary Unit with High Performance (H.P.) Sun Insulating, Laminated Tempered Glass

Window Type	Glass Size		Rough opening Width x	Height	H.P. Sun Insul Laminated	Set Shingle Flashing	Incline Curb Flashing	Set Tile Flashing	Pleated Shade with Hardware
RW2133S	85 1/16" x	30 5/16"	1'10"	2'10 1/4"	$299.00	$68.00	$269.00	$79.00	$104.00
RW2144S		40 15/16"		3'8 3/4"	350.00	68.00	274.00	89.00	118.00
RW2172S		68 7/8"		6'0 3/4"	511.00	0.00	123.00	139.00	119.00
RW2944S	26" x	40 15/16"	2'5 7/8"	3'8 3/4"	419.00	82.00	290.00	97.00	131.00
RW2957S		53 15/16"		4'9 3/4"	513.00	82.00	320.00	107.00	155.00
RW2972S		68 7/8"		6'0 3/4"	597.00	199.00	0.00	133.00	177.00
RW4144S	38 3/16" x	40 15/16"	3'6"	3'8 3/4"	526.00	91.00	324.00	111.00	166.00
RW4157S		53 15/16"		4'9 3/4"	640.00	91.00	343.00	119.00	189.00

Roof Windows, Venting Unit with High Performance (H.P.) Insulating Laminated Tempered Glass

Window Type	Glass Size		Rough opening Width x	Height	H.P. Insulating Lamin/Temp	Set Shingle Flashing	Incline Curb Flashing	Set Tile Flashing	Pleated Shade with Hardware
RW2133V	15 1/16" x	27 1/8"	1'10"	2'10 1/4"	$484.00	$68.00	$270.00	$79.00	$94.00
RW2144V		37 3/4"		3'8 3/4"	557.00	68.00	274.00	89.00	107.00
RW2944V	23" x	37 3/4"	2'5 7/8"	3'8 3/4"	622.00	82.00	290.00	97.00	119.00
RW2957V		50 3/4"		4'9 3/4"	756.00	82.00	320.00	107.00	139.00
RW4144V	35 3/16" x	37 3/4"	3'6"	3'8 3/4"	794.00	91.00	323.00	112.00	149.00
RW4157V		50 3/4"		4'9 3/4"	965.00	91.00	343.00	119.00	170.00

Window Screens

Wood window screens have been largely replaced by aluminum and vinyl units. Most prefabricated windows have insect screens available to fit them and can be ordered as part of the window unit or separately later. The following list of sample prices is based on prefabricated window glass sizes.

Glass Size		Price	Glass Size		Price
16 7/16" x	22"	$16.40		60"	26.00
	24"	17.30		68"	30.10
20 7/8" x	22"	18.00	32 7/16" x	22"	23.00
	24"	18.80		24"	24.20
	28"	20.60		28"	26.80
	68"	22.00		60"	28.00
24 7/16" x	22"	19.40		68"	30.30
	24"	20.50	36 7/16" x	22"	24.80
	28"	22.30		24"	26.00
28 7/16" x	22"	20.80		28"	28.90
	24"	21.80		60"	30.10
	28"	24.20	40 7/16" x	24"	27.00

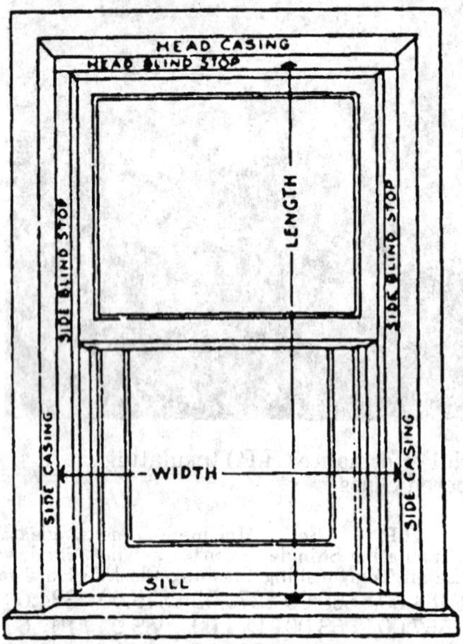

Full screen (order by screen sizes).
Length: Measure the distance from the bottom of the head casing to the sill of the window.
Width: Measure the distance between the side casing of each window.

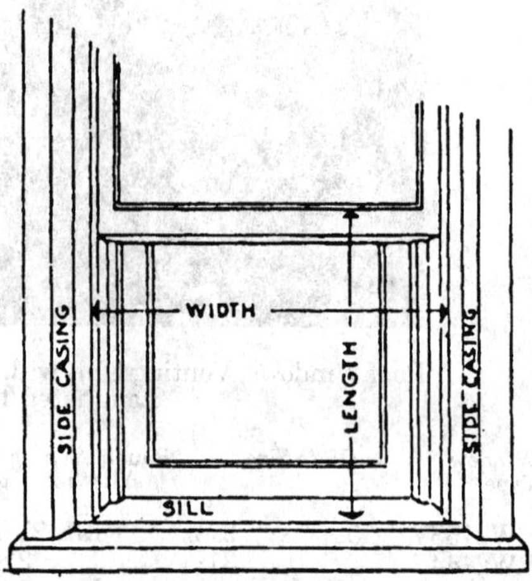

Half-Screen (order by screen sizes).
Length: Measure the distance from the top of the bottom rail of the top sash to the sill of the window.
Width: Measure the distance between the blind stop of each window.

Nailing Wood Window Frames Together. When wood window frames are sent to the job knocked down, it will require 1 to 1¼ hr. carpenter time to assemble and nail each frame together, at the following labor cost per frame:

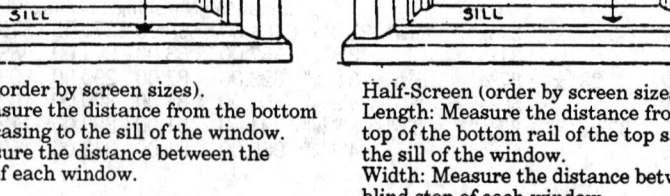

	Hours	Rate	Total	Rate	Total
Carpenter	1.2	$....	$....	$24.04	$28.85

It will require $1\frac{3}{4}$ to $2\frac{1}{4}$ hrs. carpenter time to assemble and nail together, a double window frame up to 6'-0" wide and 5'-6" or 6'-0" high, at the following labor cost per frame:

	Hours	Rate	Total	Rate	Total
Carpenter	2	$....	$....	$24.04	$48.08

Setting Single Wood Window Frames. The labor cost of setting wood window frames will vary with the size of the frame, amount of bracing necessary, and the distance they have to be carried or hoisted.

After a frame has been delivered at the job, it will require 10 to 15 minutes sorting frames and carrying them to the building, and it will then require 20 to 30 minutes for 2 carpenters to set, plumb, and brace each frame at the rate of 8 to 12 frames per 8-hr. day. The labor handling and setting each frame should cost as follows:

	Hours	Rate	Total	Rate	Total
Carpenter	0.8	$....	$....	$24.04	$19.23
Labor	0.2	....	....	19.13	3.83
Cost per frame			$....		$23.06

Setting Double or Triple Window Frames. It will ordinarily require 10 to 15 minutes for 2 workers to carry a double or triple window frame from the stock pile to place where it is to be used and it will then require 30 to 35 minutes for 2 carpenters to set, plumb and, brace the frame on the wall, at the following labor cost per frame:

	Hours	Rate	Total	Rate	Total
Carpenter	1.1	$....	$....	$24.04	$26.44
Labor	0.4	....	....	19.13	7.65
Cost per frame			$....		$34.09

Labor Handling and Setting Complete Window Units. When handling and setting complete window units consisting of frame, sash, balances, weatherstrips, etc., all installed complete, figure about $\frac{1}{4}$ hr. more than for handling and setting window frames of the same size.

Fitting Wood Sash. When fitting wood sash for double hung windows, a carpenter should fit 24 to 28 single sash per 8-hr. day, at the following labor cost per pair:

	Hours	Rate	Total	Rate	Total
Carpenter	0.6	$....	$....	$24.04	$14.42

First Grade Workmanship

Where the sash must fit perfectly, with neither too much nor too little play, a carpenter should fit 18 to 22 single sash per 8-hr. day, at the following labor cost per pair:

	Hours	Rate	Total	Rate	Total
Carpenter	0.8	$....	$....	$24.04	$19.23

Setting Roof Window (Skylight). When placing frame and window, two workers are usually required for ease of installation at the following cost per frame:

	Hours	Rate	Total	Rate	Total
Carpenter	3.0	$....	$....	$24.04	$72.12

It is important that the estimator include any additional roofing material and labor required.

Hanging Wood Sash. Including the time required placing sash cord, chain and weights or counterbalances, a carpenter should hang 24 to 28 single sash per 8-hr. day, at the following labor cost per pair:

	Hours	Rate	Total	Rate	Total
Carpenter	0.6	$....	$....	$24.04	$14.42

First Grade Workmanship

Where care must be exercised to see that all sash work fits perfectly and the cord or chain is the exact length with the correct weights or counterbalances, a carpenter should hang 22 to 26 single sash per 8-hr. day, at the following labor cost per pair:

	Hours	Rate	Total	Rate	Total
Carpenter	0.7	$....	$....	$24.04	$16.83

Fitting and Hanging Wood Sash. A carpenter should fit and hang, complete with sash cord, chain and weights or counterbalances, 12 to 14 single sash per 8-hr. day, at the following labor cost per pair:

	Hours	Rate	Total	Rate	Total
Carpenter	1.2	$....	$....	$24.04	$28.85

First Grade Workmanship

Where the sash must be fitted to have neither too much nor too little play and hung so they will work with ease and not be too loose, a carpenter should fit and hang (including sash cord, chain and weights or counterbalances), 10 to 12 single sash per 8-hr. day, at the following labor cost per pair:

	Hours	Rate	Total	Rate	Total
Carpenter	1.5	$....	$....	$24.04	$36.06

Fitting Casement Sash. A carpenter should fit 22 to 26 average size wood casement sash per 8-hr. day, at the following labor cost per pair:

	Hours	Rate	Total	Rate	Total
Carpenter	0.7	$....	$....	$24.04	$16.83

First Grade Workmanship

Where first class workmanship is required with all sash fitted to work easily without binding and hung so that each sash shows the same top, bottom and side margin, a carpenter should fit 14 to 18 single sash per 8-hr. day, at the following labor cost per pair:

	Hours	Rate	Total	Rate	Total
Carpenter	1.0	$....	$....	$24.04	$24.04

Hanging Casement Sash. The labor cost of hanging casement sash varies, depending on whether they are hung with 2 or 3 butts, style of butts used, etc. Where the sash are hung using 2 butts per sash, a carpenter should hang 20 to 24 single sash per 8-hr. day, at the following labor cost per pair:

	Hours	Rate	Total	Rate	Total
Carpenter	0.75	$....	$....	$24.04	$18.03

First Grade Workmanship

In fine residences, apartments and other buildings where the casement sash must be hung without binding and to show the same margin at the top, bottom and sides, a carpenter should hang 12 to 14 single sash per 8-hr. day, at the following labor cost per pair:

	Hours	Rate	Total	Rate	Total
Carpenter	1.2	$....	$....	$24.04	$28.85

Fitting and Hanging Casement Sash. Under average conditions, a carpenter should fit and hang 10 to 13 single casement sash per 8-hr. day, at the following labor cost per pair:

	Hours	Rate	Total	Rate	Total
Carpenter	1.4	$....	$....	$24.04	$33.66

First Grade Workmanship

Where casement sash must be fitted and hung to be neither too loose nor so tight as to cause binding when opening and closing and show the same margin between the sash and frame on all sides, a carpenter should fit and hang 7 to 9 single sash per 8-hr. day, at the following labor cost per pair:

	Hours	Rate	Total	Rate	Total
Carpenter	2.2	$....	$....	$24.04	$52.89

Fitting and Hanging Outside Window Shutters. When fitting and hanging outside window shutters on medium priced residences, cottages, etc., a carpenter should fit and hang 7 to 8 pairs per 8-hr. day, at the following labor cost per pair:

	Hours	Rate	Total	Rate	Total
Carpenter	1.1	$....	$....	$24.04	$26.44
Cost per single shutter	0.55		....		13.22

First Grade Workmanship

On the better class of buildings, where more care must be used in fitting and hanging outside window shutters, a carpenter should fit and hang 5 to 6 pairs per 8-hr. day, at the following labor cost per pair:

	Hours	Rate	Total	Rate	Total
Carpenter	1.4	$....	$....	$24.04	$33.66
Cost per single shutter	0.7		....		16.83

Fitting Roof Windows. The labor costs of setting wood, polyvinyl, or steel roof windows will vary with the size of the frame, amount of bracing necessary, and the distance that the window must be hoisted or carried.

After frames have been delivered to a project, it will require 10 to 15 minutes for sorting and carrying them to the building, and it will take 20 to 45 minutes for two carpenters to set, plumb and brace each frame, a rate of 6 to 12 units per day. The most important variable affecting the production rate is the slope of the roof that carpenters are working on.

	Hours	Rate	Total	Rate	Total
Carpenter	1.1	$....	$....	$24.04	$26.44
Labor	0.4			19.13	7.65
					$31.69

Add, if required, curbs around window unit at approximately 45 minutes per window unit.

Interior Window Trim

The labor cost of placing interior window trim will vary considerably with the type of building, thickness of walls, style of trim or casing, etc. Frame buildings, stucco buildings, brick veneer buildings, etc., having walls 6" to 8" thick will not require inside wood jamb linings, as the wood window frames usually extend to the inside face of the wall and the wood casings, window stools, aprons, stops, etc., are nailed directly to the wood frame.

Buildings having walls 10", 12", 16" or more in thickness will require a wood jamb lining to extend from the inside of the wood window frame to the face of the finished wall. It will also require wood blocking and grounds to nail the jamb linings, window stools, aprons, etc.

The style of trim will affect labor costs considerably and for that reason costs are given on 4 distinct types of interior casings, as follows:

Style "A" consists of 1-member trim with either a mitred or square cut head or "cap". This is the simplest trim it is possible to use.

Style "B" consists of 1-member trim for sides and a built-up cap trim for window heads. Caps may be put together in the mill or assembled and put together on the job.

Style "C" consists of 2-member back-band trim, consisting of a one-piece casing with a back-band of thicker material.

Style "D" consists of any type of window trim that are assembled and glued together in the mill and set in the building as a unit.

The class of workmanship will also have considerable bearing on the labor costs, as in ordinary structures the casings are nailed to the frames or jamb linings, and the trim may show some hammer marks but where first grade workmanship is required, the utmost care must be used in cutting and fitting all casings, miters, etc. All nails must be carefully driven and set, casings must show the same margin on all sides and must fit closely to the walls. In other words, it must be a first class job in every respect.

Labor Time Erecting Interior Window Trim

Style & Kind of Trim	No. Windows 8-Hr. Day	Carp. Hrs. per Window	Add for Wood Jamb Lining
"A" Single Casing	78	1.0-1.1	0.6
"B" Cap Trim	57	1.1-1.6	0.6
"C" Back-band Trim*	56	1.3-1.6	0.6
"D" Mill Assembled**	8-10	0.8-1.0	0.6
First Grade Workmanship			
"A" Single Casing	5-7	1.1-1.6	0.8
"B" Cap Trim	4-6	1.3-2.0	0.8
"C" Back-band Trim*	4-5	1.6-2.0	0.8
"D" Mill Assembled**	7-9	0.9-1.1	0.8

*For cap trim put together on job.
**Trim assembled and glued up in the mill.

On complicated window trim, having recessed jamb linings and window trim that has both a casing and a subcasing and a subcasing or panel under the stools, a carpenter will complete only about 1 window per 8-hr. day. This class of work is exceptional and is found only in high grade work.

Labor Setting Window Frames, Fitting and Hanging Wood Sash, and Trimming Interior Window Openings Complete

The following table includes all labor required to set window frame, fit and hang wood sash, place jamb linings, stops, window stool, apron, interior casings, hardware, etc., for one complete single window opening.

Window Frames, Sash, and Trim

Trim Style & Kind	Windows per 8-Hr Day	Hrs. per Window, Carpenter	Add for Wood Labor	Jamb Lining
"A" Single Casing	2½-3	3.0	0.3	0.6
"B" Cap Trim	2½-3	3.0	0.3	0.6
"B" Cap Trim*	2	3.5	0.3	0.6
"C" Back-band Trim	2	3.5	0.3	0.6
"D" Mill Assembled**	2½-3	3.0	0.3	0.6
First Grade Workmanship				
"A" Single Casing	2	3.5	0.3	0.8
"B" Cap Trim	2	3.5	0.3	0.8
"B" Cap Trim*	2	4.1	0.3	0.8
"C" Back-band Trim	2	4.4	0.3	0.8
"D" Mill Assembled**	2¼-2½	3.3	0.3	0.8

*For cap trim put together on job.
**Trim assembled and glued up in the mill.

08700 HARDWARE AND SPECIALTIES

08710 FINISH HARDWARE

Finish hardware refers to any item that is usually fitted to a door and frame to perform a specific function, e.g., hinges, latch or locksets, closers, stops, bolts, etc.

Most items of finish hardware can be obtained in various finishes and each finish has a U.S. Code Symbol Designation as follows:

US P - Prime Paint Coat
US 3 - Polished Brass
US 4 - Satin (dull) Brass
US 9 - Polished Bronze
US 10 - Satin (dull) Bronze
US 10B - Satin Bronze - Oil Rubbed
US 14 - Polished Nickel
US 15 - Satin (dull) Nickel
US 20 - Statuary Bronze
US 26 - Polished Chrome
US 26D - Satin (dull) Chrome
US 28 - Satin Aluminum - Anodized
US 32 - Polished Stainless Steel
US 32A - Satin (dull) Stainless Steel

Keep in mind that the finish will affect two areas of concern for the estimator: cost and delivery time. The US 28 finish on finish hardware is one of the cheapest to buy and readily available from most suppliers, compared with US 32, for example, which will be more expensive and a special order from the manufacturer.

Another item that will affect the cost and the delivery time will be the knob and rose trim-ring specified for latch and locksets. There are numerous styles. However, most suppliers only stock the most popular 3 or 4 patterns.

The specification writer will refer to a specific Federal Specification Series Designation when specifying the type of lockset required for the project, and each of these series relates to the construction and design function of the lockset. For example:

ANSI* Series 1000 - Mortise type latchset or lockset and keyed deadbolt housed in one casing.

ANSI Series 2000 - Mortise type latchset or lockset with integral deadbolt operated by the key in the knob.

ANSI Series 4000 - Cylindrical latch and locksets:
- Grade #1 - Heavy Duty
- Grade #2 - Medium Duty
- Grade #3 - Light Duty

*American National Standards Institute

The following listing of finish hardware material prices should be used as a guide only; always obtain a quotation from a material supplier for the specific type, function, finish and style of finish hardware specified for the project.

Material Prices For Finish Hardware

Item	Price Range
Cylindrical Locksets	
Std. Duty	$20.00-70.00
Heavy Duty	60.00-100.00
Mortise Locksets	
Heavy Duty	60.00-110.00
Push-Pull Bars	50.00-150.00
Anti-Panic Device, Single	175.00-400.00
Hinges (per pair)	
Steel Plain	10.00-20.00
Steel B.B.	25.00-70.00
Bronze B.B.	75.00-150.00
Door Closers	
III	45.00-125.00
IV	55.00-100.00
V	65.00-115.00
Surface Bolt	6.00-25.00
Kick Plate	
Alum	5.00-25.00
Bronze	15.00-50.00
Door Stops:	
Floor Bumper	5.00-10.00
Wall Bumper	10.00-15.00
Bumper Plus Holder	16.00-25.00
Door Plunger	15.00-25.00
Thresholds	
Alum.	10.00-25.00
Bronze	25.00-100.00

The following table will give average production time, in carpenter hours, necessary for the installation of finish hardware items:

Items of Finish Hardware	Hrs. to Install
Cylindrical Locksets (door prepared)	0.5
Mortise Locksets (door prepared)	0.7
Push-Pull Bars	0.4
Anti-Panic Device	1.5
Door Closers	1.0
Surface Bolts	0.3
Flush Bolts	1.0
Door Stops - Wall Type	0.3
Floor Type	0.5
Thresholds	0.5

08730 WEATHERSTRIPPING AND SEALS

The cost of metal weather strips will vary with the type of window, whether double hung or casement, size of opening, thickness of sash, etc., as they all affect the material and labor costs. It will make considerable difference in the labor costs whether the weather strips are installed in an occupied or unoccupied building. Ordinarily a workman must use more care when working in an occupied building, and for this reason, it will require 1/4 to 1/2 hour more time per opening than when working in new or unoccupied buildings.

The following quantities and costs are based on using men experienced in placing weather strips.

Metal Weather Strips for Double Hung Windows. When weatherstripping double hung windows, standard specifications require a track strip at least 3/4" wide at the top and bottom of the window, side strips for the lower sash not less than 1/16" wider than the width of the sash runway and side strips for the upper sash not less than 1/16" narrower than the width of the sash runway. Meeting rails to receive two interlocking hook strips or interlocking hook and flat strips.

When estimating the quantities of weather stripping required take the distance around the window, for the sides, head and sill strips plus the width, to take care of the meeting rail between the top and bottom sash. A 4'-0"x6'-0" double hung window will require 20 lin. ft. of weather strips for the sides, head and sill and 4 lin. ft. of interlocking material for the meeting rails.

The lineal foot cost of metal weather strip for double hung windows will average as follows in zinc:

	Sash Thickness		
Description	1-3/8"	1-3/4"	2-1/4"
Side, head and sill strips	$0.30	$0.35	$0.45
Meeting rail strips	.70	.75	1.15

The weather strip for a 4'-0"x6'-0"x1-3/8" double hung window would cost 20 x .30 plus 4 x .70 or $8.80. Bronze will run about twice as much.

Labor Installing Metal Weather Strips for Double Hung Windows. When installing metal weather strips on 1-3/8" sash for windows up to 3'-0"x5'-6", an experienced carpenter should weatherstrip 1 window in 3/4 to 7/8 hr. or 9 to 10 windows per 8-hr. day, at the following labor cost per window:

	Hours	Rate	Total	Rate	Total
Carpenter	0.9	$....	$....	$24.04	$21.64

When installing metal weather strips on 1¾" sash for windows up to 4'-0"x7'-0", an experienced carpenter should complete one window in ⅞ to 1 hr. or 8 to 9 windows per 8-hr. day, at the following labor cost per window:

	Hours	Rate	Total	Rate	Total
Carpenter	1	$....	$....	$24.04	$24.04

If installed in an occupied building, add ¼ hr. per window.

Installing Metal Weather Strips for Double Hung Windows Glazed with Plate Glass. When installing metal weather strips on 1¾" or 2¼" sash, glazed with plate glass, for openings up to 4'-0"x7'-0", a carpenter should weatherstrip one window in 1⅛ to 1¼ hrs. or 6 to 7 windows per 8-hr. day, at the following labor cost per window:

	Hours	Rate	Total	Rate	Total
Carpenter	1.25	$....	$....	$24.04	$30.05

If installed in occupied buildings, add ¼ to ½ hr. per window.

Weatherstripping In-Swinging Casement Windows. Single or double casement windows that swing in require a ¾" wide track strip on the hinged side, running in a groove in the sash. At the top, on the latch side, and where the sash meet (in the case of double casement windows) interlocking hook and flat strips are generally used. A trough strip and a hook strip is generally used at the bottom of the sash.

Material for jambs, head, and center stiles, in-swinging casement windows costs 40 cents per lin. ft. Brass trough section and interlocking sillstrip, $1.20 per lin. ft.

When estimating the number of lin. ft. of strip required for single casement windows, measure the entire distance around the window, but price the sill section separately.

For double casement windows, take the distance around the window, plus the height, which allows for the strip where the two sash meet. A 4'-0"x6'-0" double casement window will require 22 lin. ft. of weather strips for jambs, head and center stiles and 4'-0" of sill section.

Labor Weatherstripping In-Swinging Single Casement Windows. An experienced carpenter should weatherstrip one single casement window in 1 to 1⅓ hrs. or 6 to 8 windows per 8-hr. day, depending upon the size, at the following labor cost per window:

	Hours	Rate	Total	Rate	Total
Carpenter	1.1	$....	$....	$24.04	$26.44

If installed in an occupied building, add ¼ hr. per window.

Labor Weatherstripping In-Swinging Double Casement Windows. When weatherstripping in-swinging double casement windows, an experienced carpenter should weatherstrip one window in 1¾ to 2 hrs. or 4 to 5 windows per 8-hr. day, depending upon the size, at the following labor cost per window:

	Hours	Rate	Total	Rate	Total
Carpenter	2	$....	$....	$24.04	$48.08

If installed in an occupied building, add 1/4 hr. per window.

Weatherstripping Out-Swinging Casement Windows. Out-swinging single casement windows should have 3/4" wide track strip on the hinged side, running in a groove in the sash. At the top, on the latch side and where the sash meet (in the case of double casement sash), interlocking hook and flat strips are required. A heavy sill section with interlocking hook strip is generally used at the bottom of the sash. The material for jambs, heads, and center stiles will cost about 40 cents per lin. ft., while the sill section will cost about $1.00 per lin. ft.

Labor Weatherstripping Out-Swinging Single Casement Sash. An experienced carpenter should place weather strips on one single casement sash in 3/4 to 1 hr. or 8 to 10 windows per 8-hr. day, at the following labor cost per window:

	Hours	Rate	Total	Rate	Total
Carpenter	1	$....	$....	$24.04	$24.04

If installed in an occupied building, add 1/4 hr. per window.

Labor Weatherstripping Out-Swinging Double Casement Sash. An experienced carpenter should weatherstrip one double casement window in 1 1/4 to 1 1/2 hrs. or 5 to 6 windows per 8-hr. day, at the following labor cost per window:

	Hours	Rate	Total	Rate	Total
Carpenter	1.4	$....	$....	$24.04	$33.66

If installed in an occupied building, add 1/4 hr. per window.

Weather Strips for Single Wood Doors. Consisting of spring bronze strips for jambs and heads and 1 1/2" interlocking brass threshold for sill.

Door Size Up To	Thick	Price
2'-8"x6'-8"	1 3/8"	$11.20
3'-0"x7'-0"	1 3/8"	11.90
2'-8"x6'-8"	1 3/4"	12.00
3'-0"x7'-0"	1 3/4"	12.80

If aluminum interlocking thresholds are used instead of brass, deduct $1.75 for 2'-8" openings and $2.00 for 3'-0"x7'-0" openings.

If interlocking bronze is used instead of spring bronze, add $7.00 per door to the above prices.

Extruded aluminum and felt or neoprene sponge 1 1/8" wide and 1/4" thick to be set on stops at heads and jambs runs $1.10 per foot. Extruded aluminum with neoprene gasket 1 1/4" wide by 25/32" thick to be set on stops runs $2.20 per foot. A similar gasket for mounting on bottom rail of door runs $1.20 per foot.

A complete extruded aluminum channel with integral drip and vinyl insert at still runs $1.00 per foot. A 6" wide extruded aluminum thresholds for use with above rail strips runs $2.25 per foot. A complete extruded aluminum threshold 3 3/4" wide by 3/4" thick with integral latch track, interlock and hook stop for mounting on outside of door runs $3.00 per foot complete. A two-piece extruded aluminum with neoprene gasket astragal for surface mounting on double doors runs $2.60 per lineal foot.

A one-piece extruded aluminum and neoprene gasket designed for recessing in meeting rail will run $1.40 per lineal foot. Adjustable two-piece astragals with pile will run $3.50 per foot per each door leaf in extruded aluminum, $4.00 in dull bronze, and $4.80 in polished bronze or dull chrome.

Labor Weatherstripping Single Wood Doors. An experienced carpenter should weatherstrip one 3'-0"x7'-0"x1¾" door in 1 hour or about 8 doors per 8-hr. day, at the following labor cost per door:

	Hours	Rate	Total	Rate	Total
Carpenter	1.0	$....	$....	$24.04	$24.04

Where interlocking weather strip is used for head and jambs of doors instead of spring bronze, add ½ to ¾ hr. carpenter time to the time given above.

Metal Weatherstripping for French Doors. The material for weatherstripping double doors or French doors consists of spring bronze strips for jambs and head, spring bronze interlocking strip and threshold for sill. For doors up to 5'-0"x7'-0"x1⅜" thick, the material should cost about $13.00; for a 5'-0"x7'-0"-1¾ door, about $14.00.

Labor Weatherstripping French Doors. An experienced carpenter should weatherstrip one pair of 1¾" French doors in 2 hrs. or 3 to 4 pair per 8-hr. day, at the following labor cost per pair:

	Hours	Rate	Total	Rate	Total
Carpenter	2.0	$....	$....	$24.04	$48.08

Weatherstripping Double Doors. Where double exterior doors are used for openings up to 5'-0"x7'-0", using spring bronze weather strip for the head, jambs and between doors, with an interlocking threshold at the bottom of the door, figure 3 to 4 hrs. carpenter time per opening, as follows:

	Hours	Rate	Total	Rate	Total
Carpenter	2.0	$....	$....	$24.04	$48.08

If interlocking weather strip is used for the head, jambs and meeting rail, add 1 hour per opening.

Metal Thresholds

Price per lin. ft. or fraction thereof. Drilled and countersunk, with screws, ready for installation.

Metal	Size	Price per Lin. Ft.
Brass	1½"x¼"	$5.00
Brass	3½"x¾"	9.00
Brass	4½"x¾"	12.00
Brass	5"x⅞"	13.50
Aluminum	1½"x¼"	$1.50
Aluminum	3½"x¾"	2.80
Aluminum	4½"x¾"	3.90
Aluminum	5"x⅞"	5.50

All the above thresholds are furnished with interlocking strip.

08800 GLAZING

During the late 1970's, the glass manufacturers changed the process for making certain types of flat glass. Before that time, vertically drawn sheet glass was manufactured in various qualities (AA, A and B) and thicknesses (S.S., D.S., 3/16", and 7/32"). The 3/16" and 7/32" glass was further known as crystal sheet. Plate glass was made by grinding and polishing both surfaces to a level, parallel plane, thereby removing any distortion lines from the glass.

The new process of making glass consists of floating molten glass over large pools of molten tin. Through automated production stages, the floated glass is fire polished to remove all distortion, allowed to cool, and fed onto the tables in a continuous ribbon where it is cut to sheet size for packing into shipping crates. The glass is classified as "float glass" and is made in glazing quality and mirror quality of various thicknesses.

Clear Float Glass - Glazing Quality

Product	Thickness	Wt. in lb. per Sq. Ft.	Standard Max.Size	Material Cost per Sq. Ft.
Float	3/32"	1.22	40"x100"	$1.25
	1/8"	1.62	80"x120"	1.50
	3/16"	2.43	120"x212"	1.75*
Float/Plate	1/4"	3.24	130"x212"	2.00*
	3/8"	4.92	124"x204"	2.75*
	1/2"	6.56	124"x204"	4.00
	3/4"	9.85	124"x204"	6.25

*Note: Add $0.75 per sq. ft. for gray or bronze tint.

Float glass is also available in glare and heat reducing tints of gray and bronze, heat absorbing (blue-green), tempered, coated with thin metallic coverings for reflective qualities, laminated, and in insulating glass units consisting of two sheets of glass separated by an internal air space.

Tempered Glass. Tempered glass is a heat treated annealed glass that has high mechanical strength. It is 4 to 5 times as strong as annealed glass of the same thickness. When it is broken, it disrupts into innumerable small fragments of a cubical shape. In recent years, tempered glass has become much more reasonable in cost and should be given more consideration in general construction. However, the nature of its manufacture does not allow cutting or drilling on the job and all fabricating must be done prior to heat treatment. It is available in clear, tinted and heat absorbing. One-quarter inch thick units are available in sizes to 72"x120". Because this glass is all custom ordered, costs should be checked locally, but tempering will add around $2.00 a sq. ft. premium to base glass prices.

Laminated Glass. Laminated or "safety" glass is composed of two or more lights of glass with a layer or layers of tough, transparent vinyl plastic sandwiched between glass under heat and pressure to form a single unit. Laminated glass will crack but almost always will hold together. It is available in a wide range of thicknesses and sizes from 7/32" thick to 1" thick in sheets up to 80"x120". One-quarter inch laminated glass will cost about $4.25 a square foot, 1/2" will cost about $10.00 per sq. ft.

Insulating Glass. Glass units made from two sheets of glass with a sealed air space between are used to decrease heat loss through windows. The seal can be either metal or glass, and air space will vary from 3/16" to 1/2" in thickness.

Insulating glass units must be made to order in the exact size to be used, although standard sizes are manufactured.

They require a slightly larger allowance for clearance in setting them into the rabbet than single lights of glass.

Each piece weighs about 2½ times what a single lite of the same thickness would weigh and the glazing cost is about double that of single lites.

The price per square foot varies widely, depending upon kind of glass, size, etc., so it is always advisable to obtain definite prices on the sizes required.

For quick estimates two sheets of ⅛" for ½" thick unit will cost $3.50 a square foot.

Two sheets of ¼" for 1" thick unit will cost about $4.50 per sq. ft.

For tinted glass one side, add 15%.

Spandrel Glass. Heat-strengthened glass is used for spandrels and other opaque panels of curtain walls. The glass is coated on the back with a ceramic frit to give opacity and color.

It is manufactured in several standard colors, plus black and white, but can be ordered in almost any color desired.

Spandrel glass is factory cut to size (job cutting is not possible due to the heat-strengthening process in manufacturing). It is usually ¼" in thickness but may be ordered up to ¾" thick. Sealing of spandrel glass is usually accomplished by the use of liquid polysulfide sealants and neoprene gaskets. Material prices of ¼" spandrel glass are about $4.00 per sq. ft. for standard colors, and about $4.50 per sq. ft. for non-standard colors.

Wired Glass. Wire glass is a protective type glass for use in fire doors and windows, skylights and other places where breakage is a problem. It is available in ¼" thicknesses in sheets 60"x144" maximum size in hexagonal, square, diamond and pin stripe patterns, and comes in clear, hammered and many other patterned glass. Patterned wire glass will cost around $3.75 a sq. ft. Clear wire glass will cost $4.50 a sq. ft. in diamond pattern, $6.75 in pin stripe.

Patterned Glass. Glass can be rolled or figured for decoration and light diffusion. Patterns vary from manufacturer to manufacturer but usually include sandblasted, hammered, ribbed, fluted, pebbled, and rough hammered designs. It is usually available in two thicknesses: ⅛" in 48"x132" maximum size and 7/32" in 60"x132" maximum size.

Some patterns are illustrated on the accompanying pages. Cost of material are about $1.65 per sq. ft. for ⅛" thicknesses, $1.50 for 7/32".

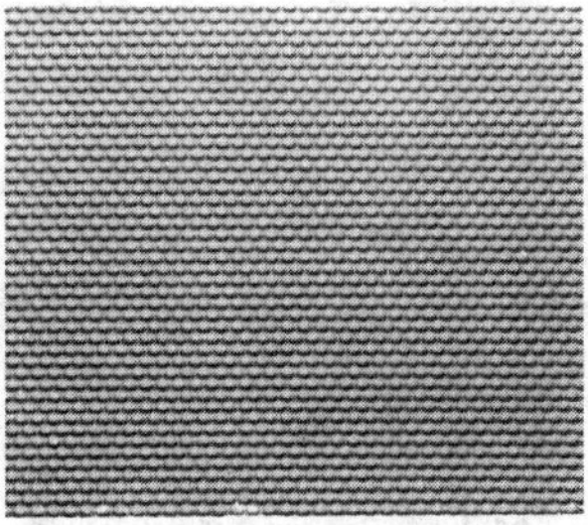

Crossnet
⅛" and 7/32" thick in sheets to 96"x132"
90% light transmission

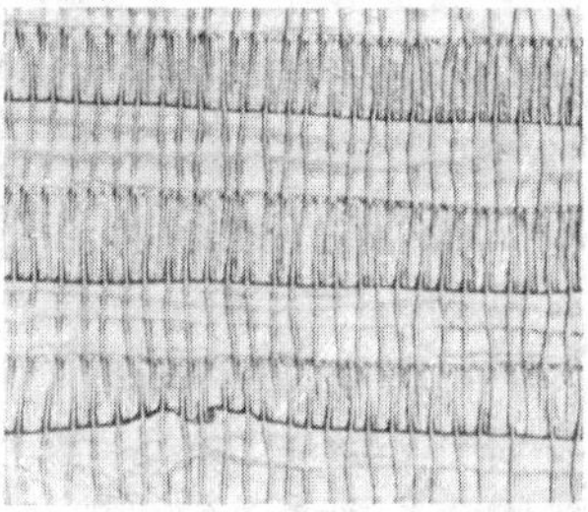

Bamboo
7/32" thick in sheets to 96"x132"
90% light transmission

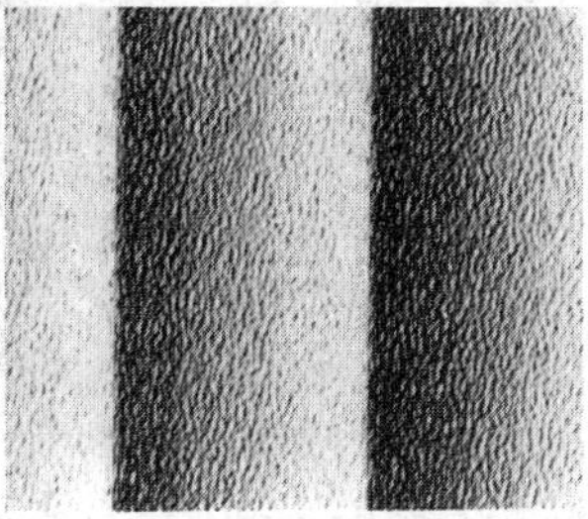

Textured Colonnade
7/32" thick in sheets to 96"x132"
90% light transmission

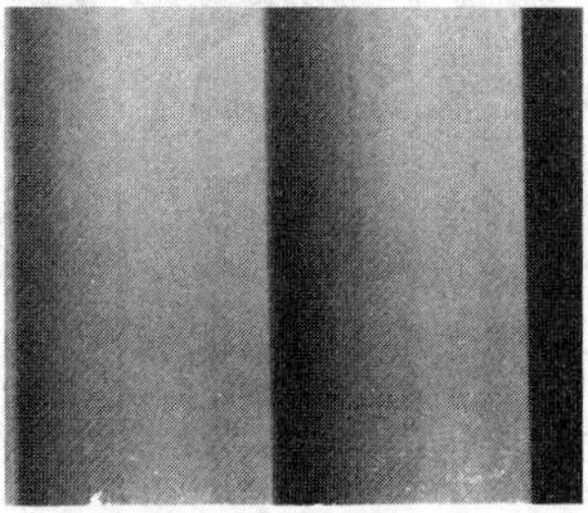

Colonnade
7/32" thick in sheets to 96"x132"
90% light transmission

Mirrors Mirrors are manufactured by taking a sheet of "mirror quality" float glass, in ⅛ or ¼ thickness, and hermetically sealing a silver coating with a film of electrolytic copper plating onto one side of the glass. The coating is then given a protective paint coat to seal out moisture and reduce the potential of damaging the silver coating.

Average material prices for mirrors will be about $3.50 per sq. ft.

Any edge work associated with mirrors will cost extra, as follows: plain polished edges at $0.25 per lineal inch, bevelled edges at $0.75 per lineal inch.

Estimating The Quantity Of Glass Required. When measuring and listing the quantities of glass for a project, list the glass by:

1) Type of glass and thickness
2) Size, measuring to the next even inch dimension, listing the width first and then the height
3) Type of frame the glass is placed in
4) Note if required to glaze from the inside or outside of the building. If from the outside, ladders or scaffolding may be necessary.
5) Height from the ground (as in multi-story buildings) for additional labor requirements of distributing the materials.

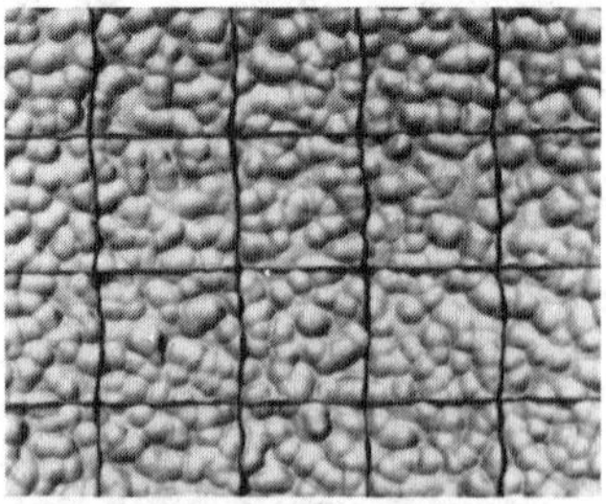

Crosswell Mottled Wired
¼" thick in sheets to 72"x130"
86% light transmission

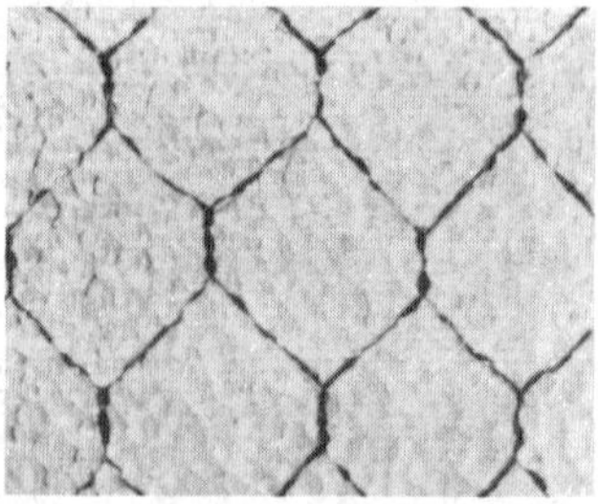

Hex Mottled Wired
¼" thick in sheets to 72"x130"
86% light transmission

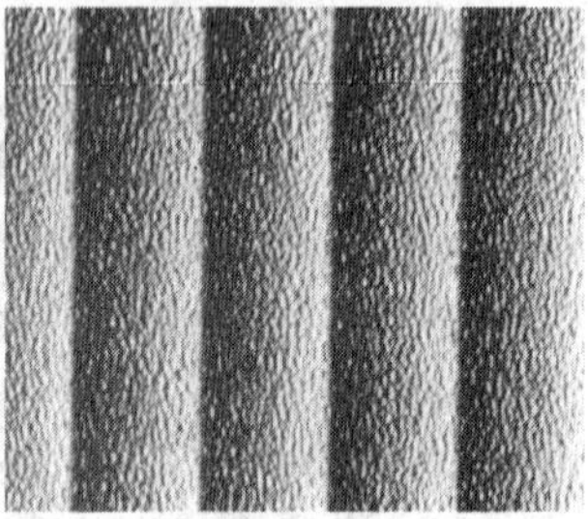

Textured Doric
⅛" and 7/32" thick in sheets to 96"x132"
90% light transmission

Doric
⅛" and 7/32" thick in sheets to 96"x132"
90% light transmission

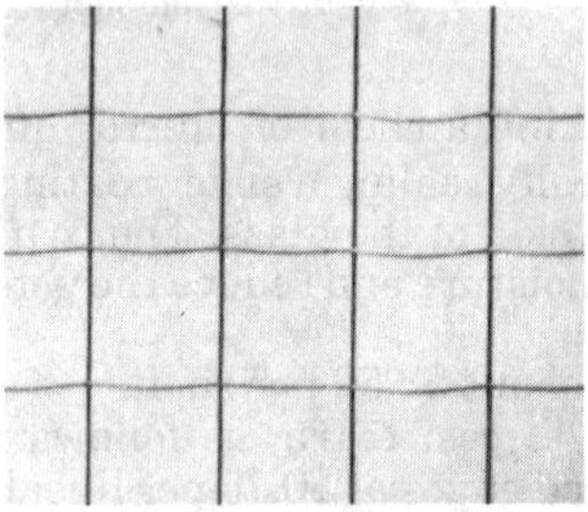

Crossweld Polished Wired
¼" thick in sheets to 72"x98"
86% light transmission

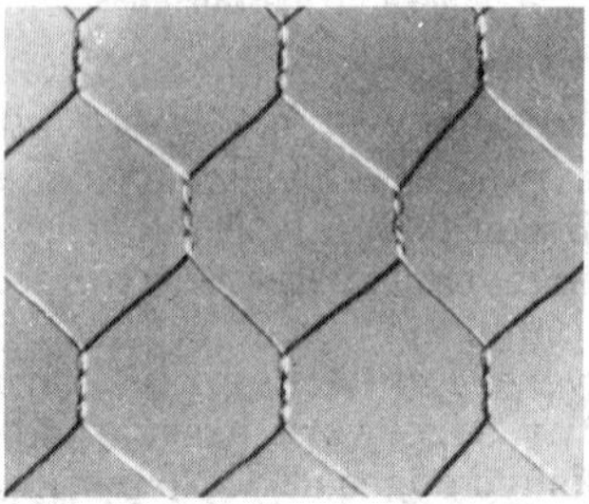

Hex Polished Wired
¼" thick in sheets to 72"x98"
86% light transmission

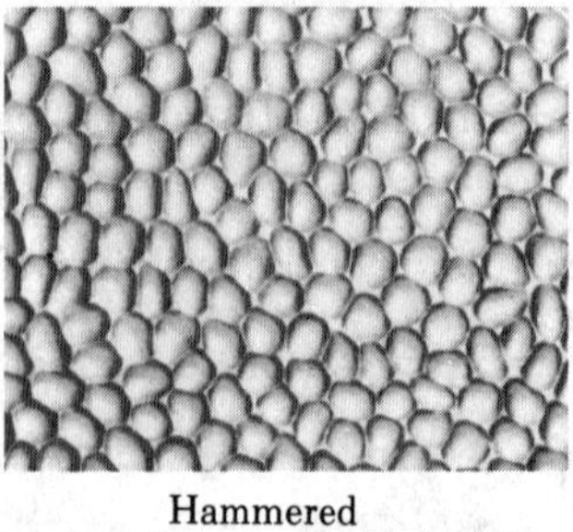

Hammered
⅛"and 7/32" thick in sheets to 96"x132"
90% light transmission

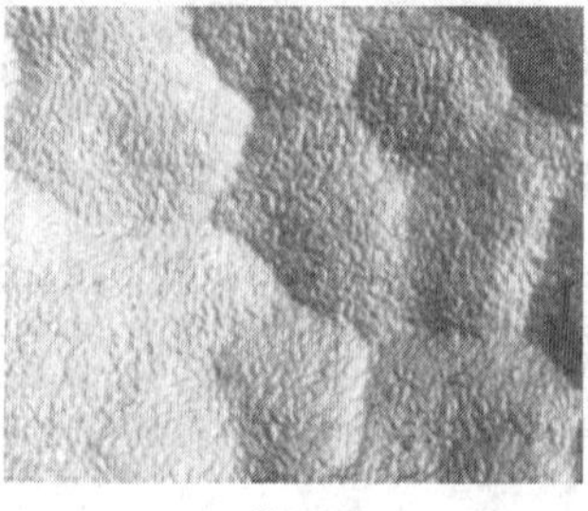

Surf
⅛" and 7/32" thick in sheets to 96"x132"
90% light transmission

Be sure not to list any glass for products that will be delivered to the job pre-glazed. It is common practice for sliding door units, aluminum sliding and single or double hung sash and some wood sash to be glazed at the factory.

Labor Glazing Window Glass. The labor cost of glazing will vary with the size and kind of glass, and whether set in putty or glazing compound, a Thiokol or Silicone sealant, or with wood or metal stops. Putty is now seldom recommended for glazing except on a very small light under 50 united inches (length plus width).

The estimator should have in his reference library the "Glazing Manual of the Flat Glass Jobbers Association", which reviews in detail recommended glazing procedures.

Labor Setting Window or Plate Glass in Wood Sash, Using Putty or Glazing Compound

Approximate Size of Glass	Number Lights Set 8-Hr. Day	Glazier Hours per 100 Lights
12"x14"	65	12.5
20"x28"	40	20.0
30"x40"	30	26.7
40"x48"	20	40.0

Labor Setting Window or Plate Glass Using Wood Stops

Approximate Size of Glass	Number Lights Set 8-Hr. Day	Glazier Hours per 100 Lights
12"x14"	40	20.0
20"x28"	20	40.0
30"x40"	15	53.3
40"x48"	12	66.7

Putty Required Setting Glass in Wood Sash. Window glass set in 1⅜" wood sash requires 1 lb. of putty for each 8 to 8½ lin. ft. of glass or sash rabbet; when set in 1¾" sash, it requires 1 lb. of putty to each 7¼ to 7½ lin. ft. of glass or sash rabbet.

The following table gives the quantity of putty (in lbs.) required for glazing various size glass:

Sash Thick	Size of Glass Inches 12"x14"	14"x20"	20"x28"	30"x36"	40"x48"
1⅜"	½	⅔	1	1⅜	1¾
1¾"	⅗	¾	1⅛	1½	2

A good grade of wood sash putty will cost about 50 cents per lb.

Labor Setting Glass in Metal Doors. When glazing metal doors and enclosures, the stops are usually steel and are fastened to the sash with small screws, and a glazier must remove all the screws from the stops, place the glass, and then replace the screws for each small light of glass. On work of this kind a glazier should set about 15 lights per 8-hr. day, at the following labor cost:

	Hours	Rate	Total	Rate	Total
Glazier	8.0	$....	$....	$23.58	$188.64
Cost per light			....		12.58

Labor Glazing Steel Sash. The labor cost of glazing steel sash will vary with the size of the glass, amount of putty required per light, and the class of work; i.e., whether sidewall sash, monitor sash, etc., also whether it is necessary to hoist the glass and erect special scaffolding to set it.

The time of year in which the work is performed will also influence the labor costs, as a glazier can set more glass in warm weather than during the winter months. The labor costs given below are based on summer glazing. For winter work add up to 25% to costs given.

Putty Required for Glazing Steel Sash. Putty used for glazing steel sash is a special mixture, part of which is red or white lead. One lb. should be sufficient to glaze 5 to 5½ lin. ft. of sash rabbet. For instance, a 16"x20" light would contain 72" or 6'-0" to be puttied.

Elastic glazing compound is displacing the regular steel sash putty to a great extent for use in steel sash and must be used in aluminum sash. This compound costs about 60 cents per lb. Sufficient putty to glaze 100 lin. ft. of steel sash should cost as follows:

	Pounds	Rate	Total	Rate	Total
Putty	20	$....	$....	$0.50	$10.00
Cost per lin. ft.			....		.10

	Pounds	Rate	Total	Rate	Total
Elastic compound	20	$....	$....	$0.60	$12.00
Cost per lin. ft.			....		.12

Labor Glazing Steel Sash. On jobs where there are no unusual window heights or other abnormal conditions, a glazier should set the following number of lights of glass in the various types of steel sash per 8-hr. day.

In clerestory, monitor or outside setting (from scaffold or swing stage), it will require one man on the ground for each two men on the scaffold to keep them supplied with glass, putty, etc. This adds approximately 50 percent to the costs as given below.

Add up to 25% to the following costs for winter work.

Glass Size Inches	Number Lights per 8-Hr. Day	Glazier Hrs. per 100 Lts.	Lbs. Putty per Light
Pivoted Steel or Commercial Projected Steel Windows*			
22"x16"	.50	26.7	1.25
32"x16"	.40	26.7	1.50
32"x22"	.30	32.0	1.67
Housing and Residence Steel Casements**			
9"x12"	.60	20.0	0.75
16"x12"	.60	23.0	1.0
18"x12"	.60	23.0	1.0
Architectural Projected Steel Windows**			
18"x16"	.30	23.0	1.25
20"x16"	.30	23.0	1.25
38"x16"	.25	40.0	1.75
40"x16"	.25	50.0	2.00
46"x16"	.25	50.0	2.00
48"x16"	.25	50.0	2.25

*Based on setting glass from inside and working from floor.
**Based on setting glass from outside, using glazing compound or steel sash putty.

Labor Setting Plate Glass in Store Fronts. The labor cost of handling and setting plate glass in store fronts will vary with the size of the glass, the length of the haul, etc.

The average size plate glass used in modern store fronts requires 4 to 6 workers to handle and set each lite of glass. No matter how many lites of glass there are in the job, it will be necessary to allow time for all workers going to the job and returning to the warehouse after completion, and time for the truck hauling glass from the warehouse to the job and return.

Unless the job is a considerable distance from the warehouse, an allowance of 1 hour time for each worker going to the job and returning to the warehouse should prove sufficient.

It is customary for the truck to remain on the job until the glass is set and return to the warehouse with the workers, so truck time should be figured on this basis.

Number of Workers Required for Various Sizes of Plate Glass in Store Fronts

From 100 united inches up to 140 united inches	2
From 140 united inches up to 160 united inches	3
From 160 united inches up to 180 united inches	4
Not over 210 united inches	5
Not over 240 united inches	6

On sizes larger than the above, some thought should be given to a mechanical means of hoisting the glass, in addition to the labor requirement.

It is further required that a) when glass is set 6 feet or more over street level, b) when weather conditions require it or c) when insulated glass (such as Thermopane or Twindow) is installed, additional labor shall be used on the following basis:

First size enumerated in the schedule, add. workers	1
Next two sizes enumerated in the schedule, add. workers	2
Next two sizes enumerated in the schedule, add. workers	3

LEADED AND ART GLASS

When estimating leaded or art glass, odd or fractional parts of inches are charged as even inches of the next larger size; for example, a 12¼"x13¼" lite will be charged as 14"x14".

Circles, ovals, or irregular patterns will be charged at the rate of a square or rectangle which will contain them.

The following prices are based on leading, using zinc cames not over ⅜" wide:

	Price per Sq. Ft.
Rectangles and squares, Clear sheet glass	$5.75
Rectangles and squares, Cathedral glass, white or colored	6.00-25.00
Rectangles or squares, Plate glass	8.00
Diamonds, Clear sheet glass	6.00
Diamonds, Cathedral glass, white or colored	7.00-35.00
Diamonds, Plate glass	10.00

Colored Art Glass Leaded. Leaded art glass furnished in colors may be had in many designs at $6.00 to $25.00 a square foot, which includes only straight work.

Landscape and water scenes in colored art glass will run from $20.00 to $40.00 a square foot.

Very elaborate designs of art glass in colors with flowers, fruits, etc., can be purchased from $50.00 up per sq. ft.

The architect usually specifies an allowance of so much per square foot to cover the cost of all art glass in the building.

08900 WINDOW WALLS/CURTAIN WALLS

INSULATED METAL WALL PANELS

Insulated steel, aluminum, and aluminum-steel wall panels were developed to provide the answer to dry wall construction, enabling the builder to enclose large areas in a minimum of time. They are designed to serve as exterior walls and partitions.

One typical wall panel consists of two members pressed together at the side laps to form a structural unit. This type has a strip of felt inserted the full length of the joint to prevent metal-to-metal contact between the two members. These panels are factory filled with borosilicate glass fiber insulation, 2½-lb. density. An end closure is attached to both ends of the panel to insure the proper retention of insulation and to seal the panels. Panels are manufactured in standard widths of 12", 16", 24", or 32" and thicknesses vary from 1⁹⁄₁₆" to 3⅜", depending upon whether surfaces are flat or fluted and thickness of insulation used.

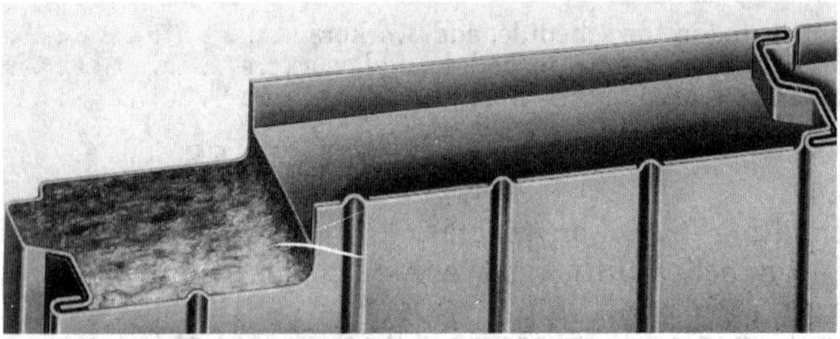

Insulated Metal Wall Panel

Lengths vary from 6'-0" to 30'-0". Maximum allowable clear span for an all galvanized steel panel, based on a wind load of 20 lbs. per sq. ft., is 14'-0"; for an all aluminum panel, 10'-6"; and for an aluminum-galvanized steel panel, 12'-0".

The interlocking tongue and groove between adjacent panels offers a joint that unites the series of units into a continuous wall surface. When using the interlocking tongue and groove connection the panels can be erected either vertically or horizontally in walls and partitions. This double-acting tongue and groove joint has three positive bearing surfaces that permit a series of panels to act as a structural unit in resisting loads from either side of a wall.

With panels in the horizontal position, the double tongue and groove joint forms a shiplap construction, adding to the weather-resisting properties of the complete wall unit. Where panels are used as semi-permanent constructions enclosing the face of a building to which a future extension is contemplated, details can allow for 100% salvaging of the panels.

Steel Wall Panels. Steel type panels are manufactured of 18 to 26 gauge sheet steel, with either a baked-on coat of shop primer or galvanized finish. The all-steel panel has the advantage of high structural strength and low initial cost. The panels are caulked during erection by buttering a mastic in the female joints prior to putting the panels in place. Welding of panels to supporting steel is the most common method of attachment. The material cost of shop assembled steel wall panels is approximately $2.50 per sq. ft. For two sheets of 26 ga. metal and 1" insulation to $2.75 per sq. ft. for two sheets of 22 ga. If the outer sheet is to have a porcelain enamel finish, add $1.25 to the sq. ft. cost.

Aluminum Wall Panels. The extreme light weight of the long length aluminum panels allows easier and faster handling and erecting in the field. A typical aluminum panel, 12'-0" long, 24" wide, weighs only 60 lbs. The erection procedure is the same as for steel panels except that direct contact between the aluminum and steel must be prevented. A coating of bituminous paint is recommended for this purpose. Maintenance costs become negligible with an all-aluminum construction. The original mill finish will eventually darken with weathering. Bolting of panels to supporting steel and use of pre-formed sections are the most common methods of attachments. Aluminum wall panels cost about $3.50 per sq. ft.

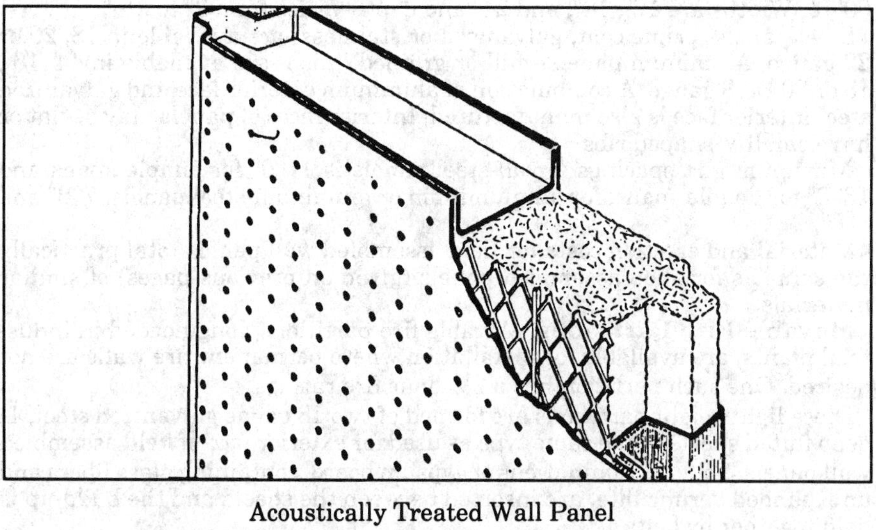

Acoustically Treated Wall Panel

Aluminum-Steel Wall Panels. To supply the need for an economical panel which would have an aluminum exterior surface for high resistance to corrosion and weather, but did not require this type surface on the interior, the aluminum-steel panel was developed.

For particular use in conjunction with aluminum-steel panels and all aluminum panels, extrusions have been developed which enable the panel to be placed in position and held through the clamping action of the extrusion at the base of the panel. These extrusions eliminate bolting and accelerate erection. They also provide 100% salvageability of the panel walls when placed in buildings which are to be expanded later. At the head of the panels there is usually no extrusion, but rather a toggle bolt is used to fasten at this position. The material cost of aluminum-steel wall panels is about $3.50 per sq. ft., depending on the gauge of metals used.

Labor Erecting Metal Wall Panels. A normal wall panel job with panel lengths ranging between 6 and 8 ft. should be constructed at the rate of 12 manhours per 100 sq. ft., which includes handling all work from unloading to actual fastening of the panels in place. A difficult job, involving the use of toggle bolts, aluminum panels, short runs, either long or short panels, will take somewhat longer--up to 20 manhours per 100 sq. ft. A typical crew would consist of one worker to caulk and hoist, a worker to receive and distribute the panel, a worker to place and weld the bottom, one to plumb and weld the top, and a worker to do the cutting, attach closures, place flashings, bitumastic, etc. The erection costs for a normal job are about $1.20 per sq. ft. and $2.25 per sq. ft. for a difficult job. On large jobs with no openings, erection costs of $0.75 per sq. ft. can be achieved, with 180-190 sq. ft. erected per man per 8-hour day.

Field Assembled Wall Panels. Steel, aluminum and combination aluminum-steel wall panels are also available for field assembly--to be assembled and erected simultaneously. Manufactured with a deeply fluted or ribbed exterior surface, they are particularly suited to modern architectural designs.

Panel lengths generally range from 6'-0" to 40'-0"--some are available up to 60'-0". Widths are 12", 18", and 24" and depth varies from 3" to 3½".

Steel panels--prime coat, galvanized or stainless--are available in 18, 20 or 22 gauge. Aluminum panels--mill or grained finish--are available in 14, 16 , 18 or 20 B&S gauge. A combination of aluminum exterior face and galvanized steel interior face is also manufactured. Interior faces of panels may be flat or have small V-shaped ribs.

Maximum girt spacings for all-steel panels is 11'-0" for simple spans and 13'-0" for double spans; for all-aluminum or aluminum-steel panels, 7'-6" and 8'-6".

Material and erection costs for field assembled wall panels total practically the same as for shop assembled panels (given on previous pages) of similar materials.

Movable Fire Partitions. Movable fire partitions, long needed in industrial plants, are available for installation where permanent fire walls are not desired. One such partition has a 2½ hour fire rating.

These lightweight partitions are formed of two 18 gauge galvanized steel, 1½" deep fluted sheets of the same type as used for exterior face of field assembled wall panels. Four ½" laminations of gypsum board, containing glass fibers and unexpanded vermiculite, are inserted between the sheets and the build-up is held together by bolts.

Depth of the fire partition is 5" and coverage width per section is 24". Length of section varies with ceiling to floor dimension.

Fire partitions are lightweight and easily erected. Portable attachment to structural steel is by bolting--permanent attachment is by welding.

Cost in place of movable fire partitions is approximately $4.00 per sq. ft.

Curtain Wall Construction. Buildings of all types are now using the curtain wall method of exterior wall construction, in preference to the conventional masonry, for the following reasons.

Curtain walls are lightweight, permitting savings in structural framing and foundations.

Units are thin--varying from 1¼" to 4½"--increasing ratio of usable floor area to total building area.

They provide good insulation with the use of efficient insulating filler materials and smooth reflective exterior and interior surfaces.

The large prefabricated and shop assembled panels, containing window units, spandrels, mullions, etc., are erected rapidly, reducing labor costs and permitting earlier completion and occupancy of building.

At the present time, curtain walls must be designed and fabricated to order to fit a particular job. Each job is custom built and costs vary considerably depending upon design, materials used, size of units, amount of duplication, type of windows, secondary framing requirements, etc.

Materials commonly used for curtain wall panels are painted, galvanized or stainless steel, aluminum and porcelain enameled steel, or aluminum. Various combinations of these materials are also used.

Various types of windows are used, including intermediate projected, casement or combination types, awning type, reversible windows, and others.

Panel sizes vary considerably, but most economical designs for multiple story structures have a coverage width of about 4'-0" and a story to story height of 10'-0" to 12'-0". For 2 and 3 story structures, full height panels have been used successfully.

The following average curtain wall costs are given for preliminary estimating purposes only. Always obtain firm figures for curtain walls before submitting bid, unless an allowance has been set up for this work.

Description	Approx. Cost Per Sq. Ft.
1½" Steel frames and sash, insul. panels	$10.50-20.00
Same in aluminum	13.50-25.00
Same in stainless steel	23.00-35.00
Same in bronze	30.00-50.00

The curtain wall can be priced out after four elements are agreed on: the framing system, the window system, the spandrel material, and the type of glazing.

CHAPTER 9

FINISHES

CSI DIVISION 9
09100 LATH AND PLASTER
09250 GYPSUM DRYWALL
09300 TILE
09400 TERRAZZO
09500 ACOUSTICAL TREATMENT
09550 WOOD FLOORING
09650 RESILIENT FLOORING
09680 CARPETING
09900 PAINTING
09950 WALL COVERING

09100 LATH AND PLASTER

Plastering costs vary widely depending on the kind of materials used and the class of workmanship performed, and the estimator should consider these items carefully before pricing. There are two distinct grades of plastering, and the estimator should possess a fair knowledge of the class of work required and estimate accordingly.

The first grade of plastering, the one used on most types of buildings, has the walls rodded and angles and corners fairly straight and true. Work of this class permits variations and "waves" in the finish up to 1/8" to 3/16". This is the regular commercial grade of plastering and on the following pages is headed "Ordinary Work".

The highest grade of work is found only in high grade residences, hotels, schools, courthouses, federal government buildings, state capitol buildings, public libraries, and the like, where all walls and ceilings together with all interior and exterior angles must be absolutely straight, and plumb and not permit variations or "waves" exceeding 1/32" to 1/16". This is termed "First Grade Workmanship".

Estimating Quantities of Lathing and Plastering. Lathing and plastering are estimated by the square yard, but the method of deducting openings, etc., varies with the class of work and the individual contractor.

Many plastering contractors make no deductions for door or window openings when estimating new work, while others deduct for one-half of all openings over a certain size.

Both lathing and plastering are estimated by the sq. yd., which is obtained by multiplying the girth of the room by the height, plus the ceiling area. For example: Obtain the number of sq. yds. of plastering in a room 12'-0"x18'-0", having 9'-0" ceilings. 18 + 18 + 12 + 12 = 60 lin. ft. of wall.

60 x 9 =540 Sq. Ft. Wall Area
12 x 18 = 216 Sq. Ft. Ceiling Area
540 + 216 = 756 Sq. Ft. Area of Walls and Ceiling
756 divided by 9 Sq. Ft. = 84 Sq. Yds.

Openings should be measured and no deductions should be made for openings less than 2'-0" wide, and then only one-half the area of each opening should be deducted.

The complete contents of large openings should be deducted, and the contractor should figure 1'-6" wide for each jamb by the height, when the return or "reveal" is less than 1'-0".

For all beams or girders projecting below ceiling line, allow 1'-0" in width by the total length as an extra for each internal or external angle, in addition to the actual area.

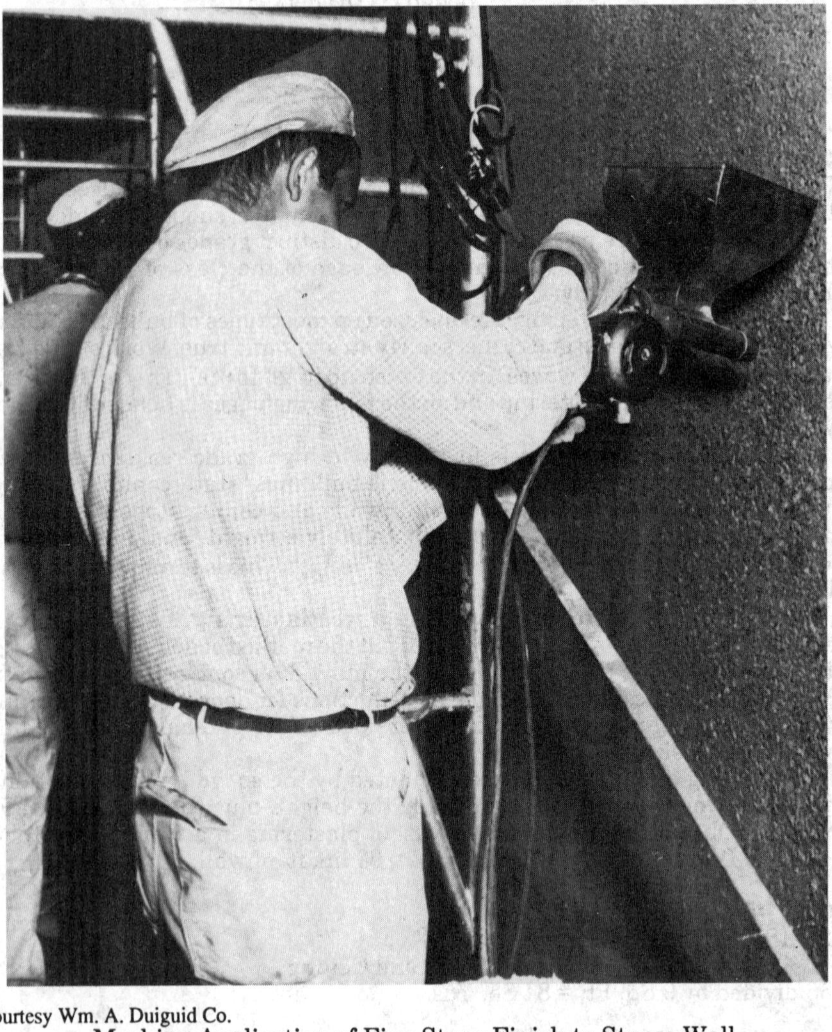

Courtesy Wm. A. Duiguid Co.

Machine Application of Fine Stone Finish to Stucco Wall

Corner Beads, Moldings, Etc. All corner beads, quirks, rule joints, moldings, casings, cornerites, screeds and other lathing trim accessories necessary to complete a plastering job are measured by the lin. ft., from their longest extension with 1'-0" extra added for each stop or miter.

Plaster Cornices. Plaster cornices should be measured by the total length of the walls. If the cornices are 1'-0" or less in girth, measurement should be made by the lin. ft., stating the girth of each cornice. If the cornices are over 1'-0" in girth, measurement may be made either by the sq. ft. of exposed face or by the lin. ft., stating the girth of each cornice.

Allow one lin. ft. for each internal miter and 3 lin. ft. for each external miter. Enriched cornices (cast work) should be measured by the lin. ft. for each enrichment.

Arches, Corbels, Brackets, Ceiling Frieze Plates, Etc. Arches, corbels, brackets, rings, center pieces, pilasters, columns, capitals, bases, rosettes, pendants, and niches are estimated by the piece. Ceiling frieze plates should be estimated by the sq. ft.

Columns. Plain plastered columns should be measured by the lin. ft. or by the sq. yd., multiplying the girth by the height.

Cement Wainscot. Cement wainscot should be measured by the sq. ft. or sq. yd.

Circular or Elliptical Work. Circular or elliptical work on a small radius should be estimated at double the rate of straight work.

Gypsum Lath

Gypsum lath is manufactured in 3 types--plain, perforated, and insulating. Each type is made of a core of gypsum with a tough, fibrous surfacing material. For the usual surfaces, either plain or perforated gypsum lath is used. Where maximum fire protection is required or desired, perforated gypsum lath should be employed. For insulation against heat and cold, and as a vapor barrier, insulating gypsum lath having an aluminum foil back should be used on the inside face of exterior walls and for top floor ceilings.

Gypsum lath is furnished 3/8" thick and face dimension is approximately 16"x48", 6 laths per bundle. Each bundle of laths contains 32 sq. ft. and weighs about 50 lbs. These sizes fit the usual stud or joist spacings without cutting.

Sizes and Weights of Gypsum Lath

Size and Thickness	Wt. Per Sq. Yd.	Approx. Price 1,000 Sq. Ft.
16"x48"x3/8"	14 Lbs.	$110.00
16"x48"x1/2"	19 Lbs.	$115.00

Insulating or aluminum foil backed gypsum lath costs about 8 cents per sq. ft. more than plain lath.

Gypsum lath is also made 1/2"x16"x48" for framing over 16" on center but not exceeding 24" on center 3/8"x24"x job length for exterior wall furring with or without foil backing; and 3/8"x16"x96" plain or perforated.

For use in 2" solid studless partitions, gypsum lath is made 1/2" thick, 24" wide, and in lengths up to 12 ft. These laths are erected with the long dimension vertical.

Nails Required for Gypsum Lath. Nails shall be 1-1/8" long for 3/8" lath and 1-1/4" long for 1/2" lath, 13 gauge with 19/64" flathead and blued. For 3/8" lath on 16" spacing, nails shall be spaced approximately 5" apart using 4 nails per stud and for 1/2" lath on 24" spacing, nails shall be spaced approximately 4" apart, using 5 nails per framing member. Lath should be butted lightly together. Gaps over 1/2" wide should be reinforced with metal lath.

With studs spaced 16" on centers it will require 6 to 7 lbs. of nails per 100 sq. yds. of gypsum lath and with studs spaced 24" on centers it will also require 6 to 7 lbs. per 100 sq. yds. of 1/2" lath. Lath may also be attached by staples, screws or clips.

Labor Placing Gypsum Lath. A lather should place 90 to 105 sq. yds. of gypsum lath per 8-hr. day on average work, at the following labor cost:

Courtesy Wm. A. Duguid Co.

Mechanically Fastening Metal Lath

Labor Cost of 100 Sq. Yds. of 3/8" Gypsum Lath Applied to Wood Studs or Joists with Nails

	Hours	Rate	Total	Rate	Total
Lather	8	$....	$....	$24.16	$193.28
Cost per sq. yd.			....		1.93

Mechanical Stapling of Gypsum Lath. Increased production in placing gypsum lath can be gained where labor conditions permit the use of air-driven staplers for attaching lath to wood studs or joists.

Air-driven stapler heads, using galvanized frozen staples 7/8" long with a 3/8" crown, cost about $250.00 each. Staples cost $.90 to $1.00 per thousand depending upon quantity purchased and are packed 12,000 to the carton. An air compressor able to handle one stapler head will cost about $370.00 and a 2-stapler head capacity machine will cost about $740.00.

For this type of attachment, it is general practice to tack lath into place, using 2 nails per piece for wall lath and 4 nails per piece for ceiling lath, until enough area is covered to permit continuous stapling. Attachment is completed by stapling 4" on center to all wood studs or joists. Approximately 3,500 staples and 1-lb. of nails are required per 100 sq. yds. of gypsum lath.

Labor Cost of 100 Sq. Yds. of 3/8" Gypsum Lath Applied to Wood Studs or Joists Using Air-Driven Staples

	Hours	Rate	Total	Rate	Total
Lather	5.3	$....	$....	$24.16	$128.05
Cost per sq. yd.			....		1.28

Two (2") Inch Solid Gypsum Long Length Lath and Plaster Partition

A studless, non-loadbearing partition consists of 1/2" thick, 24" wide, long length (up to 12'-0") gypsum lath with V'd tongue and groove edges, held vertically in floor and ceiling runners and plastered to 3/4" grounds each side.

Temporary bracing consists of 3/4" cold rolled channels horizontally (one channel for heights up to 9'-0" and two channels at the third points for heights up to 12'-0") is required with adequate vertical braces on partitions over 6'-0" long, and every 6'-0" or fraction thereof during initial plastering stages.

A metal ceiling runner is used for attachment at the ceiling and either a 2 1/2" flush metal base or a routed out wood floor runner may be used at the floor to secure the long sheets of gypsum board lath.

On work of this class, it will require 20 to 24 hours lather time to set and place floor and ceiling runners and gypsum lath for 100 sq. yds. of partition, after the material has been hoisted and stocked by laborers, which will average about 4 hrs. per 100 sq. yds.

Material Cost of 100 Sq. Yds. 2" Solid Gypsum Long Length Lath and Plaster Partition
(Partition only, not including plastering)

	Rate	Total	Rate	Total
115 lin. ft. ceiling runner	$....	$....	$.25	$28.75
115 lin. ft. 2 1/2" flush metal base*	....	....	.48	55.20
945 sq. ft. 1/2" long length gypsum lath	....	....	.12	113.40
50 lin. ft. 3" strip lath and paper backing	....	....	.13	8.00
155 lin. ft.** temp. bracing channels	....	....	.09	13.95
100 temp. bracing clips	....	....	.05	5.00
Cost per 100 sq. yds.		$....		$224.30
Cost per sq. yd.		....		2.24
Cost per sq. ft.		....		.25

*Floor runner may be a special wood or steel runner channel at correspondingly lower material and installation cost in carpentry labor, but will require wood base at additional cost.

**115 lin. ft. of horizontal temporary bracing channels, plus five 8-ft. verticals figured on a minimum of 3 re-uses.

Labor Cost of 100 Sq. Yds. 2" Solid Gypsum Long Length Lath and Plaster Partition
(Partition only, not including plastering)

	Hours	Rate	Total	Rate	Total
Lather	22	$....	$....	$24.16	$531.52
Labor	4	....	....	19.13	76.52
Cost per 100 sq. yds			$....		$608.04
Cost per sq. yd.			....		6.08
Cost per sq. ft.			....		.68

Add for corner beads, cornerites and other accessories required to complete lathing work. Also available are 1" thick gypsum lath either solid or laminated which reduce the requirements for bracing and grounds.

Floating Systems of Attachment of Gypsum Lath. In order to eliminate the sources of sound transmission in walls and ceilings, the principle of resiliently isolating the interior walls and ceilings from their structural elements (studs and joists) has proven extremely effective as a means of confining the sound at its source. It interposes a shock absorber that reduces the energy transmitted to the structural elements of the building. Interior wall and ceiling finish should be completely and thoroughly isolated from all structural members surrounding the source of sound.

Laying Out and Placing Floor Runners for Long Length Gypsum Lath Partitions

Attachment devices for gypsum lath are designed to isolate the lath from framing members. They are made from sheet metal or wire, and generally include field and corner clips for the attachment to wood, metal furring and masonry.

In addition to reducing sound transmission, the source of plaster cracking is also often eliminated when occasioned by movement of the structural elements which are stronger than the plastered interior. Floating systems of attachment permit a limited movement of the structural members without imposing undue stresses in the lath and plaster interior.

The spacing of framing members should not exceed 16" o.c. for clip attachments. Plain gypsum lath should be used on ceilings, and either plain or perforated lath may be used on walls.

Suspended Gypsum Lath Ceilings. Suspended gypsum lath ceilings consist of metal hangers, runners and furring channels, which in turn support gypsum lath as a base for plaster, or the adhesive application of acoustical tile.

The hangers, runners and furring channels for ceilings are approximately the same as required for metal lath. Furring channel spacing must not exceed 16". Some fire resistive ratings require 12" spacing. Starter channels along walls should be within 2" of wall. A ½" space should be provided between ends of channels and masonry walls. Each lath is supported transversely under the channels by a long wire clip, with not less than 3 such clips per full length of lath.

Stagger the end joints by starting alternative rows with half lengths of lath, allowing end joints to fall between channels.

Material Cost of 100 Sq. Yds. Suspended Gypsum Lath Ceilings Using 16"x48"x⅜" Gypsum Lath with Main Runners 3'-0" on Centers and Cross Furring Channels 16" on Centers

	Rate	Total	Rate	Total
77 pcs. 3/16"x3'-0" hangers, 231'-0"	$....	$....	$.04	$9.24
364 lin. ft. 1½" channels	....	....	.25	91.00
700 lin. ft. ¾" channels	....	....	.17	119.00
945 sq. ft. 16"x48"x⅜" gypsum lath	....	....	.11	103.95
500 Brace-tite field clips	....	....	.09	45.00
340 Bridjoint clips, B-1	....	....	.06	20.40
Cost 100 sq. yds.		$....		$388.59
Cost per sq. yd.		....		3.89

Labor Cost of 100 Sq. Yds. Suspended Gypsum Lath Ceilings Using 16"x48"x⅜" Gypsum Lath with Main Runners 3'-0" on Centers and Cross Furring Channels 16" on Centers

	Hours	Rate	Total	Rate	Total
Lather	20	$....	$....	$24.16	$483.20
Labor	4	....	....	19.13	76.52
Cost 100 sq. yds.			$....		$559.72
Cost per sq. yd.			....		5.60

The above ceiling has a fire rating of 1 hour. If the job specification calls for a fire resistance rating of two hours, 14 ga. diagonal wire reinforcement must be added below the lath, to obtain a 4 hour rating, 1" 20 ga. galvanized 1" hex wire mesh must be added. All the above ratings are dependent on the type of floor construction above.

Hollow Steel Stud Partitions Using Gypsum Lath as a Plaster Base. A non-loadbearing partition of trussed steel studs, track and shoes to which gypsum lath is applied to both sides by means of metal clips, nailing or stapling.

The construction is similar to hollow steel stud non-bearing partitions except that gypsum lath is applied to the studs by means of special metal clips and the studs are spaced 16" on centers to receive the 16"x48" gypsum lath.

Material Cost of 100 Sq. Yds. of Hollow Steel Stud Non-Bearing Plastered Partitions Using Gypsum Lath on Both Sides of Partitions, 2½" Studs 16" on Centers, Ready to Receive Plaster

	Rate	Total	Rate	Total
105 lin. ft. top track	$....	$....	$.25	$26.25
105 lin. ft. bottom track	....	....	.25	26.25
76 9'-0" studs, use 750 lin. ft.*	....	....	.34	255.00
1,900 sq. ft. gypsum lath. ⅜"	....	....	.11	209.00
1,000 Trus-Lok field clips	....	....	.08	80.00
680 Bridjoint clips	....	....	.06	40.80
Cost 100 sq. yds.		$....		$637.30
Cost per sq. yd.		....		6.37

*Allowance made for opening framing.

Perforated Gypsum Lath on Steel Studs

Labor Cost of 100 Sq. Yds. of Hollow Steel Stud Non-Bearing Plastered Partitions Using Gypsum Lath on Both Sides of Partitions, 2½" Studs 16" on Centers, Ready to Receive Plaster

	Hours	Rate	Total	Rate	Total
Lather	28	$....	$....	$24.16	$676.48
Labor	4	....	....	19.13	76.52
Cost 100 sq. yds.			$....		$753.00
Cost per sq. yd.			....		7.53

Add for corner beads, cornerites and other accessories required to complete lathing work.

Gypsum lath is also used for exterior wall furring and column fireproofing. Special framing support systems are manufactured for this. Foil backed lath may be used for exterior walls.

Wood Lath

Wood lath for plastering is a thing of the past. Gypsum lath and metal lath are commonly used. The following data is included for that one in a thousand job where wood lath may be used for patching in remodeling work.

Wood lath should be estimated by the sq. yd. Because it is customary to deduct only one-half the area of openings, it is not necessary to add any allowance for waste. When placing wood lath it is customary to allow ⅜" between each lath for a plaster key. A standard wood lath is 1½" wide and 48" long, as follows:

Size of Lath	No. Req'd. 100 Sq. Yds.
1½"x48"	1,425-1,450

Labor Placing Wood Lath. A lather should place 1,400 to 1,600 1½"x48" wood lath, (95 to 110 sq. yds.) per 8-hr. day.

Metal Lath and Furring

Where metal lath is used as a plaster base in partition work, it may be nailed or stapled to wood or steel studs, nailed to masonry, or wire-tied to steel channels and prefabricated steel studs. It may also be used without studs or back-up when attached to floor and ceiling runners and temporarily braced during application and curing of first coat of plaster. Where used as a plaster base in ceiling work, it may be nailed to wood joists, wire-tied to various metal support systems either in contact with or suspended from the building structure, or wire-tied directly to the bottom chords of steel joists. It may also be attached directly to the underside of reinforced concrete floors or joists by means of attachments previously embedded in the concrete.

In addition to its utility as a plaster base, metal lath is also used to reinforce plaster applied to other surfaces. Prefabricated, right-angled strips of metal lath are attached to masonry or other rigid plaster bases, prior to plastering, at the interior angels formed by intersecting walls and along the junction of walls and ceilings. Flat strips of metal lath are attached to solid plaster bases to reinforce the joints in such bases, or to bridge potential crack lines at the corners of door and window openings.

Metal lath is used in the individual fireproofing of beams and columns. It may be adapted to decorative shapes for cornices, coves, curved walls, convoluted ceilings and various acoustic shapes such as inverted pyramids. In specialty classifications, it is used to form the membrane of hyperbolic paraboloids, or the web section of concrete beams and girders.

Metal Lath Construction. The costs of metal lathing will vary according to the complexities of the work in which it is used, the size of the rooms, the height above floors, and the method of attachment. In the sections which follow, attempt is made to cover only the type of lathing which is usual and normal in building construction.

Materials. A number of metal and wire laths, and welded wire fabrics are available to suit the various conditions which might be encountered in building construction. In general, there are sheets of slitted and expanded metal, or perforated metal, commonly called metal laths; rolls of netting formed of 19 ga. wire woven into a pattern which provides 2½ strands per inch, commonly called wire laths; and sheets of open mesh laths formed of strands of 16 gauge wire welded at right angles to one another and spaced 2" apart, commonly called welded wire fabric. Laths may be of galvanized metal or painted with a rust-inhibitive paint. Some metal laths and all welded wire fabrics are available with a paper backing. Also, some laths may be obtained with a backing of aluminum foil laminated with paper. The backings sometimes perform useful functions insofar as utility and plastering are concerned, but many such laths are often more difficult to install.

Expanded metal lath can be either flat dimpled (self-furring), or ribbed. The flat expanded type is classed as small mesh, having approximately 1000 openings per square foot. The perforated metal lath, commonly called sheet lath, is fabricated either with furring indentations or with channel-shaped ribs. The openings in sheet lath and in rib lath are classified as small. Although they are somewhat larger than those of the expanded metal laths, they are separated by larger areas of unperforated metal. Stucco mesh has diamond-shaped openings which are approximately 1½"x3". Wire laths are flat when installed and the openings are classified as of medium size. The welded wire fabrics are either flat or, in the crimped, self-furring variety, all the wires laying in one direction are bent approximately ¼" out of a flat plane, at their intersection with every third wire lying in the other direction.

Installation Procedures. Rather rigid industry standards govern the installation of metal laths, particularly with respect to ceiling work. Complete specifications are prepared by associations of the manufacturers, by individual manufacturers, or by sectional committees of the American National Standards Institute. The building construction estimator will require a brief outline of the critical factors of installation in order to derive a realistic estimate of costs.

Spans of Lath. In all work except where laths are supported by solid surfaces, and except in the case of studless solid partitions, the weight of metal lath must be suitable to the distance between studs, joists or other supporting members. The laths are attached to these members by nails, wire-ties, staples or clips, at intervals not exceeding 6". The following chart gives maximum spans, or distance between supports for various weights of metal and wire laths and welded wire fabric.

Type of Lath	Maximum Allowable Spacing of Supports in Inches					
	Min. Wt. Lath Lbs. per Sq. Yd.	Vertical Supports			Horizontal Supports	
		Wood	Metal Solid Partit.	Others	Wood or Concrete	Metal
Flat Expanded	2.5	16	16	12	0	0
Metal Lath	3.4	16	16	16	16	13½
Flat Rib	2.75	16	16	16	16	12
Metal Lath	3.4	19	24	19	19	19
⅜" Rib	3.4	24	..	24	24	24
Metal Lath	4.0	24	..	24	24	24
Sheet Lath*	4.5	24	..	24	24	24
Wire Lath	2.48	16	16	16	13½	13½
V-Stiffened						
Wire Lath	3.3	24	24	24	19	19
Wire Fabric	..	16	0	16	16	16

*Used in studless solid partitions.

Laps of Lath. Industry standards require overlapping at ends and sides of sheets of lath. Most sheets are manufactured slightly larger than nominal size and the estimator need not allow extra quantities on this account. A few extra ties are required in all laps: one between each supporting member, and where laps of the ends of sheets fall between supports, such laps should be laced with tie wire. These extra ties are considered part of normal lathing work and no extra allowance need be made for them.

Attachment of Lath. Nail-on attachment of metal lath to horizontal or sloping surfaces (ceilings and soffits) requires the use of No. 11 ga. barbed, galvanized roofing nails with a head diameter of 7/16". The length of such nails must permit a minimum penetration into the wood of 1⅜". In nail-on attachment of metal lath to vertical surfaces (walls and partitions), 4d common nails are permitted. They should penetrate the wood at least ¾". However, and the remainder should be bent over the strands of lath. Also, in vertical work, 1" long roofing nails with a head diameter of 7/16" are permitted. They are driven home without crushing the strands of metal. In direct attachment to masonry, concrete stub nails with ⅜" flat heads are generally used.

Tie-on attachment of metal lath is generally done with 18 ga. galvanized annealed wire, a strand of wire passing through the mesh, around the support member, and back through the mesh with the two ends twisted together approximately three turns. This operation is more difficult for the lather when paper or foilbacked lath is being used, unless adequate openings in the backing have been provided. Therefore, when estimating hand tying of lath having heavy backing, a reduction in labor productivity should be allowed. In certain cases 18 ga. tie wire is not permitted. Where metal lath is attached directly to structural elements of the building, such as concrete slabs or joists or the bottom chord of open web steel joists, wire of heavier gauge, or more than one loop of 18 ga. wire, is required (16 ga. or 2-18 ga. for steel joist work; 14 ga.

twisted or 10 ga. bent over for concrete work). The extra difficulty of twisting and cutting the heavier gauge wires has been considered in the productivity rates suggested below.

Special devices for attachment of laths are permitted, where they develop a fastening of strength equal to the accepted standards. Such devices often result in noticeable labor savings but the estimator should not neglect to adjust his material costs as necessary.

Horizontal Support Systems. Horizontal supports for metal lath fall under the general classifications of contact systems, where the lath is fastened in direct contact with the structural elements of the building, furred systems, or suspended systems. In both of the latter systems, the laths are attached in the usual manner to a framework of pencil rods or channel shapes which have been fastened directly to, or suspended from, structural elements of the building. In a furred system, the channels or pencil rod furring are attached to the structural elements of the building, by means of 16 ga. wire saddle ties--in open web steel joist construction, or by 18 ga. wire saddle ties supported by nails driven in or through wood joists. The forming of saddle ties is considered normal lathing work and the productivity rates suggested below contemplate this type of tying.

In a suspended system, the channels or pencil rods to which the lath is attached, are called cross-furring. These members are saddle-tied with 16 ga. wire to the underside of an intersecting framework of heavier channels, which have been hung from the structural elements of the building by means of wire, rod or flat steel hangers. Wire hangers and rods are saddle-tied around these support channels, which are called main runners, and flats may be welded or bolted to them.

A combination of the furred and suspended systems, is usually found where furring members are installed in contact with concrete joists. The furring members which support the lath are in contact with the bottoms of the joists, but they are supported by transverse channels, or runners, which are suspended between and run parallel to the joists. The furring members are saddle-tied with 16 ga. wire to the runners, and the runners are suspended by means of 10 ga. wire hangers embedded in the slab above. The wire hangers are saddle-tied to the runners.

In contact, furred or suspended systems, the lath must be supported at spans not exceeding those given in the table of maximum spans, above. In furred and suspended systems, all pencil rods and channels are limited as to distance between supports and center to center spacing, and--in suspended systems--the size of each hanger is dependent upon the area of ceiling which it supports. The following tables list the usual elements and arrangements to be found in furred and suspended ceilings.

Hanger Sizes for Suspended Ceilings

Ceiling Area, SF Max.	Hanger Size Min.	Hangers for Supporting up to 25 SF Ceiling	
8	12 ga. wire	6 ga. wire	1"x3/16" flat
12	10 ga. wire	5 ga. wire	1 1/4"x1/8" flat
12 1/2	9 ga. wire	3/16" rod	1 1/4"x3/16" flat
16	8 ga. wire	7/32" rod	1 1/2"x1/8" flat
17 1/2	7 ga. wire	1/4" rod	1 1/2"x3/16" flat

All wire hangers should be galvanized steel, all flat and rod hangers should be coated with rust-inhibitive paint.

Spans and Spacing for Main Runners in Suspended Ceilings

Min. Size and Type	Max. Span Between Hangers or Supports	Max. CtoC Spacing of Runners
¾" channel, .3 lb.	3'-0"	2'-4"*
¾" channel, .3 lb.	2'-6"	2'-6"
¾" channel, .3 lb.	2'-0"	3'-0"
1½" channel, .475 lb.	5'-0"	2'-0"
1½" channel, .475 lb.	4'-0"	3'-0"
1½" channel, .475 lb.	3'-6"	3'-6"
1½" channel, .475 lb.	3'-0"	4'-0"
2" channel, .59 lb.	7'-0"	2'-0"
2" channel, .59 lb.	5'-0"	3'-6"
2" channel, .59 lb.	4'-6"	4'-0"

All weights of channel are for cold-rolled members.
*This spacing for runner channels supporting furring members against concrete joists, only.

Spans and Spacing for Cross Furring

Min. Size and Type	Max. Span Between Runners or Supports	Max. CtoC Spacing of Cross Furring
⅜" pencil rod	2'-6"	12"
⅜" pencil rod	2'-0"	19"
¾" channel, .3 lb.	4'-0"	16"
¾" channel, .3 lb.	3'-6"	19"
¾" channel, .3 lb.	3'-0"	24"

All weights of channel are for cold-rolled members.

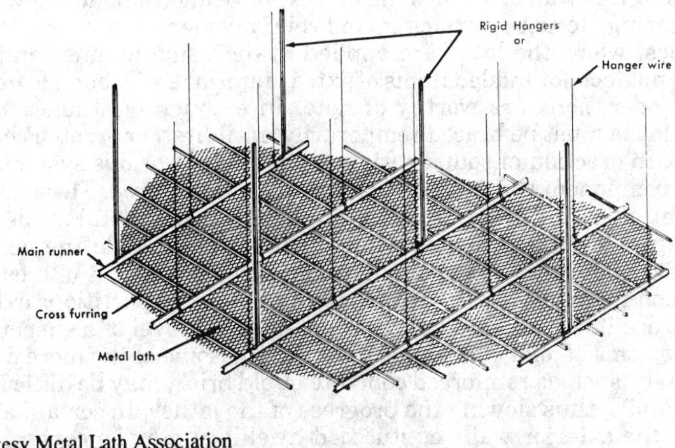

Courtesy Metal Lath Association

Details of Suspended Ceiling

In most horizontal work, the laths, framing members, hangers, and inserts are installed in the usual manner. The estimator should look for the presence of large ducts, which might require additional framing or hangers not detailed on the drawings. He should allow for hung or tied supports at the perimeter of rooms and not plan to set channels into wall pockets. The estimator's checklist of materials should usually contain allowances for lath, furring channels, carrying channels, tie wire, hanger wire, nails, special inserts if any, corner reinforcing, and accessory items, if required, such as expansion bead or casing bead.

Vertical Support Systems. Metal lath may be used in a number of wall and partition framing systems. A generalized classification would include hollow walls and partitions, solid plaster partitions, vertical furring and wall cladding. Hollow assemblies may have either metal or wood studs. The metal studs are generally used where non-combustible construction is required, where concealed horizontal pipe must be accommodated, or where sound resistive qualities are needed. Although the majority of such studs are rated as non-loadbearing, because they are fabricated from light-gauge steel or rod, tests of completed walls have indicated surprising load carrying capacity and resistance to horizontal impact.

Thin solid plaster partitions may be formed over a core of metal lath alone, or an assembly of metal lath and metal studs. The studless solid partitions are generally 2" thick and may be as much as 10' in height. The solid partitions with 3/4" coldrolled channels may range between 1½" thickness and 8½" height to 2½" thickness and 16' height. Heavier channels may be used for thicker and higher solid partitions.

Vertical, or wall, furring systems may be braced from exterior walls, pilasters, or interior columns, or may be free-standing. Their utility is sometimes decorative, at other times functional. Functional aspects include concealment, fireproofing, acoustic treatment, condensation and temperature control. The lathing installation is essentially that of a partition with studs, lathed on one side only. However, horizontally-placed 3/4" channels are required for stiffening steel studs, and above certain heights--dependent upon size of these studs--the framing must be braced from the structure behind.

Exterior and interior wall claddings include overcoating (remodeling work), decorative surfacing, acoustic surfacing, and reinforcement.

In most vertical work, the laths are applied in the usual manner and the estimator need not look for hidden items of extra material and labor. However, a few unusual operations are worthy of note. In erection of studless solid partitions, the laths must be braced temporarily until first, or scratch coat of plaster has set. In erection of sound resistive partitions, various systems for separating the components of the partition may be encountered. These might include resilient clips and pencil rod furring between the metal studs and metal lath, staggered stud arrangements which require double the normal number of studs, or double rows of small studs with short pieces of cross-tie channels between them. Note also that most soundproofing partitions extend up to the structure itself, and are not terminated at the level of a suspended ceiling or the bottoms of joists. In wall cladding work, some of the more dense back-up materials, such as reinforced concrete or old brick, may be difficult to penetrate with nails, thus slowing the progress of the lather. In certain areas of the country, the exterior walls of stuccoed dwellings may be constructed without sheathing. In this style of erection, known as line wire construction, the lathing work sometimes includes installation of 18 ga. wires, stretched taut completely around the building, and fastened securely to the studs at 6" intervals vertically. This operation is generally performed only when woven wire or stucco mesh is the plaster base.

The estimator's checklist of materials for vertical work should usually contain allowances for lath, studs if any, four shoes for each prefabricated metal stud (2 top and 2 bottom), ceiling and floor tracks for holding studs and shoes, nails for attachment of either the tracks or bent-over ends (of small studs) to ceiling and floor, nails for attachment of lath to wood or solid back-up, wire for tying lath and shoes to studs, and shoes to tracks, if required, metal base or screeds, if any, and accessory items such as corner, casing or expansion beads, and picture molds, if required. In estimating requirements for studs, allow for double-studding at door frames where necessary, and where splicing is required, allow for an 8" lap and double wire-tying.

Installation--Labor

Nail-On Attachment of Metal Lath. For either horizontal or vertical work, the lather should apply 90 to 100 sq. yds. of metal lath per 8-hr. day, depending on size of rooms.

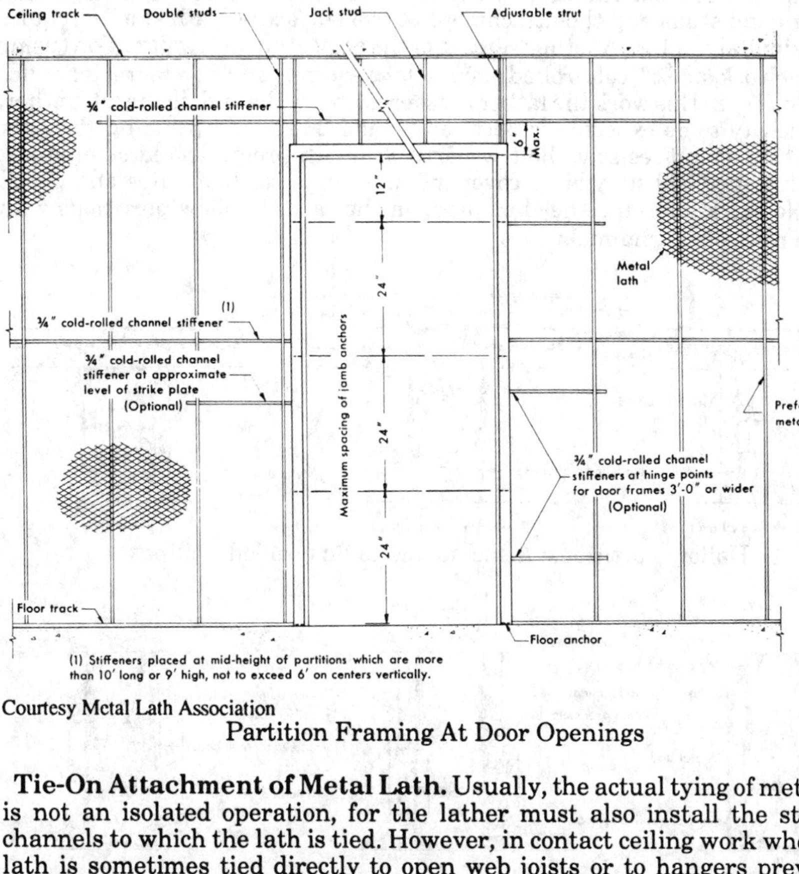

Courtesy Metal Lath Association

Partition Framing At Door Openings

Tie-On Attachment of Metal Lath. Usually, the actual tying of metal lath is not an isolated operation, for the lather must also install the studs or channels to which the lath is tied. However, in contact ceiling work where the lath is sometimes tied directly to open web joists or to hangers previously embedded in concrete, the lather should apply 100 sq. yds. of metal lath per 8-hr. day.

Furred and Suspended Horizontal Systems. In furred ceiling work, including installation of furring members and lath but not including drilling for and placement of inserts in concrete, the lather should apply 50 sq. yds. per 8-hr. day. In suspended ceiling work, including installation of wire and rod hangers (but not concrete inserts for hangers), main runner channels, cross furring channels, and lath, the lather should apply 40 sq. yds. per 8-hr. day. Where strap hangers are used (bolted construction), the lather should complete 30 sq. yds. per 8-hr day. In drilling concrete and placing hanger inserts, the lather should cover 25 sq. yds. of area per 8-hr. day. In closets, toilet rooms and other small spaces, a lather may complete only 30 sq. yds. of suspended ceiling per 8-hr. day.

Beam Fireproofing. In furring and lathing around structural beams, the lather should complete 9 lin. ft. of beam per hour.

Vertical Systems Other Than Nail-On Work: Hollow Partitions. In hollow partition work, not including sound-resistive construction, a lather should install top and bottom tracks, tie shoes on studs, tie studs and shoes to tracks, apply lath, and complete 30 sq. yds. of partition per 8-hr. day. In 100 sq. yds. of such partitions, approximately 16 hrs. are required to install runners and studs and the attachment of two surfaces of lath should require approximately 11 hrs. Another version of the steel stud hollow partition (using a framework of 3/4" cold-rolled channels) requires slightly more labor for installation. In this work the lather must erect two walls and tie them together. Productivity to be expected is between 24 and 28 sq. yds. per 8-hr. day. For estimating sound-resistive hollow partitions with prefabricated steel studs, add 5 hrs. per 100 sq. yds. to cover installation of resilient clips and pencil rods. No extra allowance need be made in the case of hollow partitions with double rows of 3/4" channels.

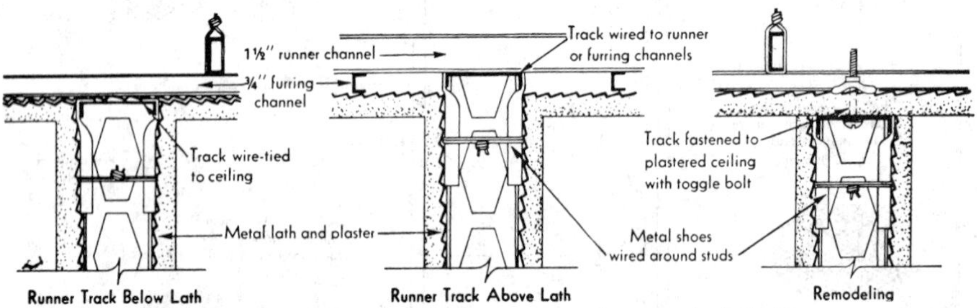

Hollow Partitions: Attachments to Suspended Ceilings

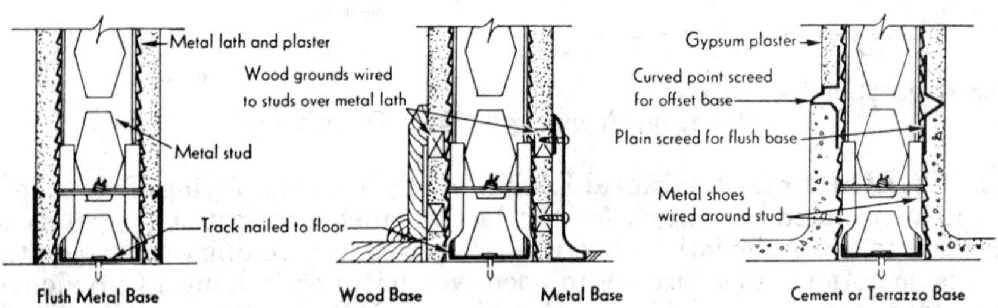

Courtesy Metal Lath Association

Hollow Partitions: Attachments to Floor

Vertical Systems, Solid Partitions, and Vertical Furring. In erecting solid partitions with 3/4" cold rolled channels, a lather should complete 22 sq. yds. per 8-hr. day, depending on spacing of channels and method of attaching channels at floor and ceiling. Studless solid partitions should be erected at the rate 27 sq. yds. per 8-hr. day. Labor savings are realized in the studless system due to elimination of wire tying the lath to channel studs. Vertical furring systems should generally be erected at the same rate as solid partitions with channel studs. However, where wall braces are required, add 4 hrs. per 100 sq. yds.

Vertical Systems, Wall Cladding. In overcoating work, and in other direct-applied lath attachments, a lather should nail metal lath to surfaces of wood and concrete block at the same rate as for nail-on work to wood studs and joists. Lathing should proceed at the rate of 160 sq. yds. per 8-hr. day. In attachment to old brick or reinforced concrete, however, a lather should be expected to complete only 125 sq. yds. per 8-hr. day.

Accessories. Many metal lathing accessories are available for a variety of functional or trimming purposes. Only a few of the more common items are covered herein. The labor rates given below for erection of some of the functional accessories are inclusive of items of work covered as part of the total assembly. The estimator should, therefore, use the highest productivity rate for erection of an entire assembly, when he lists these functional accessories as separate items. For example, installation of metal base sometimes includes layout of partition and attachment of clips to floor--items of work which are also contemplated in erection of some partition types.

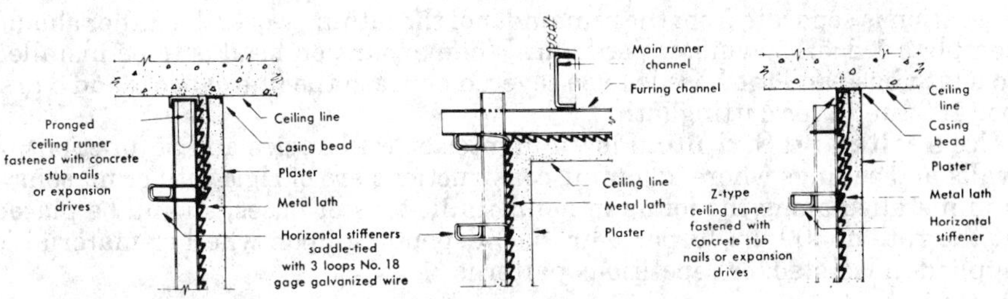

Details At Ceiling

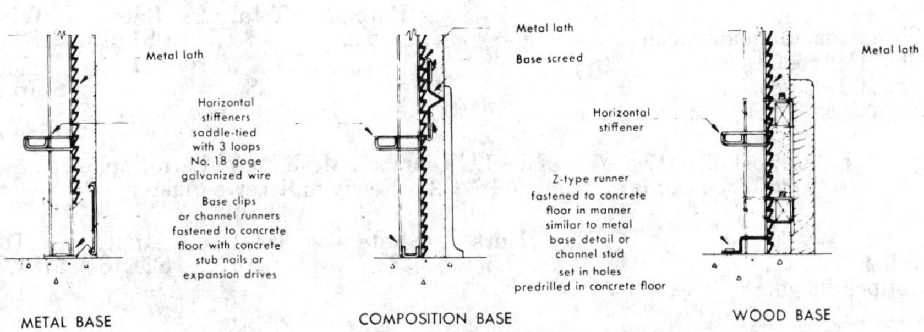

Floor Attachments

Courtesy Metal Lath Association

Flush Metal Base. Double for solid partitions--including layout of partition, attachment of clips to floor, cutting and fitting, attachment of side plates: 5 hrs. per 100 lin. ft. of partition.

Single for Masonry Trim. Including drilling for and setting masonry clips, cutting and fitting, attachment of side plates: 3 hrs. per 100 lin. ft.

Single for Wall Furring. Including layout of furring, attachment of clips to floor, cutting and fitting, attachment of side plates: 4 hrs. per 100 lin. ft.

Flush or Applied Metal Base. Single for snap-on to metal studs or for screwing to wood grounds--including cutting and fitting, attaching side plates, and attaching clips to studs, but not including wood grounds: 3 hrs. per 100 lin. ft.

Beads and Molding. Including corner and bullnose beads, picture molds, base screeds, casing bead (except when used for door and window trim), and expansion bead (except when lath and/or framing is severed beneath line of bead):

	Hrs. per 100 Lin. Ft.
wired or stapled to lath bases	2½
nailed or stapled to wood construction	2
nailed to brick, tile or block construction	3
stub-nailed to concrete construction	3½

If on-the-job mitering is required, add up to ½ hr. per 100 lin. ft. to above rates.

When windows and doors are being trimmed with casing bead, and the operation is separate from the remainder of the lathing work, the lather should complete 1 door or window per hour. When expansion bead is to be installed to a lath base and the base is to be severed beneath the line of bead, add 3 hrs. per 100 lin. ft. for cutting lath.

Cornerite and Stripite. Placed in interior wall angles and at junctions of walls and ceilings where adjoining constructions are of rigid lath or masonry, and installed along the joints in nonmetallic plaster bases, should be placed at the rate of 500 lin. ft. per 8-hr. day, depending upon whether material is applied in isolated, or repetitious patterns.

Material Cost of 100 Sq. Yds. of 3.4 Lb. Diamond Mesh Metal Lath Applied to Wood Studs or Joists @ 16" O.C., Ready to Receive Plaster

	Rate	Total	Rate	Total
105 sq. yds. diamond mesh	$....	$....	$1.59	$199.95
9 lbs. 1½" roofing nails	....	....	1.06	9.54
Cost 100 sq. yds.		$....		$176.49
Cost per sq. yd.		....		1.76

Labor Cost of 100 Sq. Yds. of 3.4 Lb. Diamond Mesh Metal Lath Applied to Wood Studs or Joists @ 16" O.C., Ready to Receive Plaster

	Hours	Rate	Total	Rate	Total
Lather	7.5	$....	$....	$24.16	$181.20
Cost per sq. yd.			....		1.81

Material Cost of 100 Sq. Yds. of 3.4 Lb. ⅜" Rib Metal Lath Applied to Wood Studs or Joists @ 24" O.C., Ready to Receive Plaster

	Rate	Total	Rate	Total
105 sq. yds. ⅜" rib mtl lath	$....	$....	$2.33	$244.65
6 lbs. 1½" roofing nails	....	....	1.06	6.36
Cost 100 sq. yds.		$....		$251.01
Cost per sq. yd.		....		2.51

Labor Cost of 100 Sq. Yds. of 3.4 Lb. 3/8" Rib Metal Lath Applied to Wood Studs or Joists @ 24" O.C., Ready to Receive Plaster

	Hours	Rate	Total	Rate	Total
Lather	7.0	$....	$....	$24.16	$169.12
Cost per sq. yd.			....		1.69

Material Cost of 100 Sq. Yds. of 3.4 Lb. 3/8" Rib Metal Lath Applied Directly to Ties Embedded in Concrete or to Steel Joists at 24" on Centers

	Rate	Total	Rate	Total
105 sq. yds. 3/8" rib metal lath	$....	$....	$2.73	$244.65
15 lbs. 18 ga. tie wire	....	....	.95	14.25
Cost 100 sq. yds.		$....		$258.90
Cost per sq. yd.		....		2.59

Labor Cost of 100 Sq. Yds. of 3.4 Lb. 3/8" Rib Metal Lath Applied Directly to Ties Embedded in Concrete or to Steel Joists at 24" on Centers

	Hours	Rate	Total	Rate	Total
Lather*	9	$....	$....	$24.16	$217.44
Cost per sq. yd.			....		2.17

*Lather hours listed are for attachment of lath only. For concrete joist construction, add for placing wire in forms before pour, or if necessary to drill holes in concrete.

Material Cost of 100 Sq. Yds. of Furred Ceiling Using 3.4 Lb. Flat Rib Metal Lath Applied to 3/8" Pencil Rods 19" on Centers Tied to Structure at 2 Ft. Intervals

	Rate	Total	Rate	Total
105 sq. yds. flat rib metal lath	$....	$....	$2.33	$244.86
660 lin. ft. 3/8" pencil rods	....	....	.08	52.80
6 lbs. 18 ga. tie wire (lath to rods)	....	....	.95	5.70
2 lbs. 16 ga. wire for splicing rods	....	....	.85	1.70
4 lbs. 16 ga. wire for attaching rods	....	....	.85	3.40
Cost 100 sq. yds.		$....		$308.46
Cost per sq. yd.		....		3.08

Labor Cost of 100 Sq. Yds. of Furred Ceiling Using 3.4 Lb. Flat Rib Metal Lath Applied to 3/8" Pencil Rods 19" on Centers Tied to Structure at 2 Ft. Intervals

	Hours	Rate	Total	Rate	Total
Lather*	16	$....	$....	$24.16	$386.56
Cost per sq. yd.			....		3.87

*Lather hours listed are for attachment of lath only. For concrete joist construction, add for placing wire in forms before pour, or if necessary to drill holes in concrete.

Material Cost of 100 Sq. Yds. Suspended Metal Lath Ceiling Using 3.4 Lb. 3/8" Rib Metal Lath With Furring Channels 24" on Centers Runner Channels 36" on Centers, Hangers 48" on Centers

	Rate	Total	Rate	Total
100-9 ga. x 3'-0" hangers 300'-0" or 19 lbs.	$....	$....	$.74	$14.06
345 lin. ft. 1½" c.r. channels	....	....	.25	86.25
490 lin. ft. ¾" c.r. channels	....	....	.17	83.80
3 lbs. 16 ga. wire	....	....	.85	2.55
105 sq. yds. 3/8" rib metal lath	....	....	2.33	244.65
5 lbs. 18 ga. wire	....	....	.95	4.28
Cost per 100 sq. yds.		$....		$435.09
Cost per sq. yd.		....		4.35

Labor Cost of 100 Sq. Yds. Suspended Metal Lath Ceiling Using 3.4 Lb. 3/8" Rib Metal Lath With Furring Channels 24" on Centers Runner Channels 36" on Centers, Hangers 48" on Centers

	Hours	Rate	Total	Rate	Total
Lather*	50	$....	$....	$24.16	$1,208.00
Cost per sq. yd.			....		12.08

*For erection under concrete, add for placing hangers in forms before pour; or, if necessary to drill holes in concrete or tile, and place inserts.

Material Cost of 100 Sq. Yds. Suspended Metal Lath Ceiling Using 3.4 Lb. Diamond Mesh Metal Lath With Furring Channels 13½" on Centers, Runner Channels 48" on Centers, Hangers 48" on Centers

	Rate	Total	Rate	Total
85-8 ga. x 3'-0" hangers 255'-0" or 19 lbs.*	$....	$....	$0.74	$14.06
280 lin. ft. 1½" c.r. channels	....	....	.25	70.00
850 lin. ft. ¾" c.r. channels	....	....	.17	144.50
4 lbs. 16 ga. wire	....	....	.85	3.40
105 sq. yds. diamond mesh	....	....	1.59	166.95
8 lbs. 18 ga. wire	....	....	.95	7.60
Cost per 100 sq. yds.		$....		$406.51
Cost per sq. yd.		....		4.07

*See above for erection under concrete.

Labor Cost of 100 Sq. Yds. Suspended Metal Lath Ceiling Using 3.4 Lb. Diamond Mesh Metal Lath With Furring Channels 13½" on Centers, Runner Channels 48" on Centers, Hangers 48" on Centers

	Hours	Rate	Total	Rate	Total
Lather*	56	$....	$....	24.16	$1,352.96
Cost per sq. yd.			....		13.53

*See above for erection under concrete.

Material Cost of 100 Lin. Ft. Beam or Girder Fireproofing Using 3.4 Lb. Self-Furring Diamond Mesh Metal Lath Tied to Adjacent Ceiling Lath and Wrapped Downwards Around 12" WF Beam

	Rate	Total	Rate	Total
33 sq. yds. 3.4 lb. self-furring diamond mesh	$....	$....	$1.70	$56.10
2.5 lbs. 18 ga. wire	....	....	.95	2.38
Cost per 100 lin. ft.		$....		$58.47
Cost per lin. ft.		....		.58

Labor Cost of 100 Lin. Ft. Beam or Girder Fireproofing Using 3.4 Lb. Self-Furring Diamond Mesh Metal Lath Tied to Adjacent Ceiling Lath and Wrapped Downwards Around 12" WF Beam

	Hours	Rate	Total	Rate	Total
Lather	12	$....	$....	$24.16	$289.92
Cost per lin. ft.			....		2.90

Material Cost of 100 Sq. Yds. of Hollow Steel Stud Partitions; Lathed Both Sides, Studs 12" on Centers, Using 2.5 Lb. Diamond Mesh Metal Lath

	Rate	Total	Rate	Total
200 lin. ft. snap-in-runner track	$....	$....	$.25	$50.00
Nails or expansion drives-allowance	....	....	26.50	26.50
101--9'-0" 3⅝" galv. steel studs (909 lin. ft.)	....	....	.37	336.33

	Rate	Total	Rate	Total
210 sq. yds. diamond mesh	....	....	1.27	266.70
17 lbs. 18 ga. wire	....	....	.95	16.15
Cost per 100 sq. yds.		$....		$695.68
Cost per sq. yd.		....		6.96

*Add necessary extras for framing around openings, returns,etc.

Labor Cost of 100 Sq. Yds. of Hollow Steel Stud Partitions; Lathed Both Sides, Studs 12" on Centers, Using 2.5 Lb. Diamond Mesh Metal Lath

	Hours	Rate	Total	Rate	Total
Lather	27	....	....	$24.16	$652.32
Cost per sq. yd.			....		6.52

Material Cost of 100 Sq. Yds. of Sound Insulation Partition Using Prefabricated Steel Studs 24" on Centers, Resilient Clips and Pencil Rods, Lathed Both Sides With 3.4 Lb. 3/8" Rib Metal Lath

	Rate	Total	Rate	Total
200 lin. ft. plain runner track	$....	$....	$.25	$50.00
Nails or expansion drives-allowance	....	....	26.50	26.50
51--9'-0" 3⅝" galv. steel studs (459 lin. ft.)	....	....	.37	169.83
204 shoes	....	....	.08	16.32
210 sq. yds. 3/8" rib metal lath	....	....	2.33	489.30
11 lbs. 18 ga. wire	....	....	.95	10.45
Add for sound treatment:				
100 lin. ft. cork, 3½" wide x 1" thick	....	....	.53	53.00
1020 resilient clips	....	....	.08	81.60
918 lin. ft. ¼" rod	....	....	.07	64.26
Cost per 100 sq. yds.		$....		$961.26
Cost per sq. yd.		....		9.61

*Add necessary extras for framing around openings, etc.

Labor Cost of 100 Sq. Yds. of Sound Insulation Partition Using Prefabricated Steel Studs 24" on Centers, Resilient Clips and Pencil Rods, Lathed Both Sides With 3.4 Lb. 3/8" Rib Metal Lath

	Hours	Rate	Total	Rate	Total
Lather	27	$....	$....	$24.16	$652.32
Add for sound treatment:					
Lather	5	....	....	24.16	120.80
Cost per 100 sq. yds					$773.12
Cost per sq. yd.			....		7.73

Material Cost of 100 Sq. Yds. Metal Lath Solid Partitions, Lathed One Side Only, Using Channels 16" on Centers, Metal Floor and Ceiling Runners and 2.5 Lb. Diamond Mesh Metal Lath

	Rate	Total	Rate	Total
690 lin. ft. ¾" cold rolled channel	$....	$....	$.17	$117.30
200 lin. ft. Z-type runner, slotted	....	....	.19	38.00
Nails or screws--allowance	....	....	26.50	26.50
105 sq. yds. diamond mesh	....	....	1.27	133.35
8 lbs. 18 ga. wire	....	....	.95	7.60
Cost per 100 sq. yds.		$....		$322.75
Cost per sq. yd.		....		3.23

Labor Cost of 100 Sq. Yds. Metal Lath Solid Partitions, Lathed One Side Only, Using Channels 16" on Centers, Metal Floor and Ceiling Runners and 2.5 Lb. Diamond Mesh Metal Lath

	Hours	Rate	Total	Rate	Total
Lather	37	$....	$....	$24.16	$893.92
Cost per sq. yd.			....		8.94

Material Cost of 100 Sq. Yds. of Hollow Partitions (Concealment) Using Double Rows of ¾" Channel Studs at 16" on Centers and 3.4 Lb. Diamond Mesh Metal Lath

	Rate	Total	Rate	Total
152 pcs. ¾" studs x 9'-2" 1390 lin. ft.: (tops of studs tied to ceiling furring channels)*	$....	$....	$.17	$236.30
208 lin. ft. ¾" stiffener	....	....	.17	35.36
152 pcs. ¾" spacer 512 lin. ft. : (based on 2' between rows of studs	....	....	.17	87.04
200 lin. ft. snap-on metal base	....	....	.48	96.00
152 single base clips (furring clip)	....	....	.08	12.16
Nails or expansion drive-allowance	....	....	26.50	26.50
210 sq. yds. diamond mesh	....	....	1.59	333.90
17 lbs. 18 ga. wire	....	....	.95	16.15
Cost per 100 sq. yds.		$....		$843.41
Cost per sq. yd.		....		8.43

*Add necessary extras for framing around opening, returns, etc.

Labor Cost of 100 Sq. Yds. of Hollow Partitions (Concealment) Using Double Rows of ¾" Channel Studs at 16" on Centers and 3.4 Lb. Diamond Mesh Metal Lath

	Hours	Rate	Total	Rate	Total
Lather	32	$....	$....	$24.16	$773.12
Cost per sq. yd.			....		7.73

Material Cost of 100 Sq. Yds. of Vertical Furring Using 2.75 Lb. Flat Rib Metal Lath and ¾" Channels 16" on Centers

	Rate	Total	Rate	Total
76 pcs. ¾" studs x 9'-0" 684 lin. ft.*	$....	$....	$.17	$116.28
104 lin. ft: ¾" stiffener	....	....	.17	17.68
51 pcs. ¾" braces x 2'-0" 102 lin. ft.	....	....	.17	17.34
200 lin. ft. double-pronged ceiling-floor runner	....	....	.25	50.00
Nails or expansion drives-allowance	....	....	26.50	26.50
105 sq. yds. flat rib metal lath	....	....	1.59	166.95
8 lbs. 18 ga. wire	....	....	.95	7.60
Cost per 100 sq. yds.		$....		$402.35
Cost per sq. yd.		....		4.02

*Add necessary extras for framing around opening, for soffits, returns, etc.

Labor Cost of 100 Sq. Yds. of Vertical Furring Using 2.75 Lb. Flat Rib Metal Lath and ¾" Channels 16" on Centers

	Hours	Rate	Total	Rate	Total
Lather*	22	$....	$....	$24.16	$531.52
Cost per sq. yd.			....		5.32

Material Cost of 100 Lin. Ft. of Metal Corner Bead Attached to Lath Bases

	Rate	Total	Rate	Total
100 lin. ft. corner bead	$....	$....	$.16	$16.00
Staples or tie wire	....	....		1.06
Cost per 100 lin. ft.		$....		$17.06
Cost per lin. ft.		....		.17

Labor Cost of 100 Lin. Ft. of Metal Corner Bead Attached to Lath Bases

	Hours	Rate	Total	Rate	Total
Lather	2.5	$....	$....	$24.16	$60.40
Cost per lin. ft.			....		.60

In estimating for other beads, molding or screeds, substitute proper material price, and for other bases consult discussion of labor on beads and molding for necessary additions to hours required. Also, if beads or molding are to serve as plaster grounds and all plumbing or straightening work is assigned to the installation of the accessory, add ½ hrs. per 100 lin. ft.

ESTIMATING PLASTER QUANTITIES

The following pages contain covering capacities and quantities of the different kinds of plastering materials required and are based on manufacturer specifications.

Gypsum plasters are sold by the ton and are usually packed in 100 lb. sacks. When sand is mixed with plaster in the proportion of 1 to 1, 1 to 2, or 1 to 3, it means 100, 200, or 300 lbs. of sand respectively are to be added to each 100 lb. sack of plaster.

Certain fire-resistant ratings may require an increase in the ratio of plaster to aggregate (such as 1:1; 1:2 in lieu of 1:2, 1:3) or an increase in the thickness of the plaster over normal grounds or both, and the estimator must take this into consideration in the use of the following data.

Covering Capacity. The covering capacity of plaster necessarily depends upon conditions under which the plaster is applied--upon the mechanic, thickness of grounds, kind and trueness of the surface to be plastered, type of lath, quality of sand, etc.

The covering capacity given in the following paragraphs for each type of material and for the various surfaces is based on average conditions and may be accepted as a safe basis for calculating the quantity of each class of material required when applied according to the manufacturer's specifications.

Thickness of grounds will have considerable bearing on the covering capacity of plaster and the quantities given are based on the following: ¾" grounds for metal lath, or about ⅝" of plaster over the face of the lath; ⅞" grounds for ⅜" gypsum lath or ½" of plaster; 1" for ½" insulating board lath or ½" of plaster and ⅝" for unit masonry.

Number of 100 Lb. Sacks of Gypsum Cement Plaster Required Per 100 Sq. Yds.

Kind of Plastering Surface		
Metal Lath	Gypsum Lath	Unit Masonry
18 to 20 (Sanded 1:2;1:3)	9 to 11 (Sanded 1:2;1:3)	10 to 12 (Sanded 1:3)
For each ⅛" in Thickness, Proportions 1:3; Add or Deduct Sacks		
2.5	2.5	2.5

Hollow Steel Stud Non-Bearing Partition

For insulation board lath, use same plaster quantities as given for gypsum lath.

Standard specifications specify that the first (scratch) coat on all types of lath shall be mixed in the proportions of 1 part gypsum neat plaster to not more than 2 parts of sand by weight (or 100 lbs. of gypsum neat plaster to 2 cu. ft. of perlite or vermiculite aggregate).

First coat on masonry surfaces (except monolithic concrete) and second (brown) coat on all lath bases shall be mixed in the proportions of one (1) part gypsum neat plaster to not more than three (3) parts of sand by weight (or 100 lbs. of gypsum neat plaster to 2 cu. ft. of perlite or vermiculite aggregate).

The sacks of gypsum cement plaster shown in the table above per 100 sq. yds. for various bases are based on these proportions when used with sand aggregate.

Perlite and Vermiculite Plaster

Vermiculite is a mineral that expands when heated at about 2000°F. Each ore particle expands to about 12 times its original size. It will not rot, decay, burn or be attacked by vermin or termites.

Perlite is a siliceous volcanic rock mined in western United States. When crushed and quickly heated to above 1500°F., it expands to form lightweight, noncombustible, inert glass-like particles of cellular structure. This material, white in color is about 1/10 the weight of sand.

Perlite or vermiculite plaster aggregate is a lightweight material. Compared with 100 lbs. per cu. ft. for sand aggregate, perlite ranges from 7½ to 15 pounds per cu. ft., and vermiculite 6 to 10 lbs. per cu. ft. Plaster using this aggregate will weigh approximately 30 to 45 lbs. per cu. ft. (depending upon proportions used), against 100 lbs. per cu. ft. where sand aggregate is used.

When perlite or vermiculite is expanded each granule has countless numbers of small dead air spaces, which in addition to making the material very light, gives it a high insulating value. It has approximately three and one half times as much insulating value as ordinary plaster.

It is this insulating quality that provides good fire resistance when combined with gypsum plaster. These aggregates are frequently called for when specific fire ratings are required for the protection of steel framing. As such fire ratings usually require greater plaster thicknesses than normal grounds provide, the estimator is cautioned to determine the proper quantities required per 100 sq. yds.

Perlite or vermiculite plaster aggregate is usually packed in multi-wall paper bags containing 3 or 4 cu. ft. Since there may be some compression or packing in shipping or storage, to insure correct proportioning do not use the original package but a cubic foot measuring box.

Machine-applied vermiculite portland cement plaster is used for panel or spandrel walls and exterior columns. The contractor can apply it to exterior columns simultaneously with the wall, since the paper-backed wire fabric used in wall construction is also wrapped around the columns. The protection is also used for interior columns where more than ordinary wear and tear is expected due to the type of occupancy.

The following table gives the approximate quantities of gypsum plaster and perlite or vermiculite for 100 sq. yds. of various mixtures.

Recommended Proportions and Quantities

In the following table giving recommended proportions for gypsum plaster and perlite or vermiculite, the proportions 1-2, means 100 lbs. of gypsum plaster to 2 cu. ft. of perlite or vermiculite; 1-3 means 100 lbs. of gypsum plaster to 3 cu. ft. of perlite or vermiculite.

Number of 100 Lb. Sacks or Wood Fiber Plaster Required
Per 100 Sq. Yds.

Kind of Plastering Surface		
Metal Lath	Gypsum Lath	Unit Masonry
32 to 34 unsanded	17 to 18 unsanded	12 to 14 Equal parts of sand by weight to be added
For each 1/8" in Thickness, Add or Deduct Sacks		
5	4 1/2	3

For insulation board lath, use same plaster quantities as given for gypsum lath.

For two and three-coat work on all types of lath, gypsum wood fiber plaster may be used without the addition of aggregate.

For two and three-coat work on masonry surfaces and as scratch coat on lath (except monolithic concrete), gypsum wood fiber plaster shall be mixed in the proportions of 1 part of plaster to 1 part of sand, or perlite by weight.

Type of Construction	Grounds (See Note)	Actual Plaster Thickness	Coat	Recommended Volume Proportions	Required per 100 Sq. Yds. Bags Plaster	Required per 100 Sq. Yds. Cu. Ft. Aggregate
Metal Lath	3/4"	1/4"	Scratch	1—2	10	20
		7/16"	Brown	1—2	12½	45
Gypsum Lath	7/8"	3/16"	Scratch	1—2	5	10
		1/4"	Brown	1—2	10	20
2" Solid Partition	2"	3/8"	Scratch (One side only)	1—2	15	30
		3/8"	Brown (One side only)	1—2	15	30
		1 1/8"	Brown 2nd Side	1—2	45	90
Unit Masonry	5/8"	3/16"	Scratch	1—3	4	12
		3/8"	Brown	1—3	8	24

Grounds attached to studs, joists or channels--not the lath. 1/16" allowed for finish coat thickness.

Gypsum Ready-Mix Plasters Made with Lightweight Aggregate

Gypsum ready-mix plasters are generally made with perlite. The regular formula for lath bases is packed in 80-lb. sacks and the masonry formula for application over all masonry bases is packed in 67-lb. sacks. The following table gives the number of sacks required per 100 sq. yds. applied over various surfaces:

Kind of Plastering Surface		
Metal Lath	Gypsum Lath	Unit Masonry
32 to 34*	17 to 19*	24 to 26**
For Each 1/8" in Thickness, Add or Deduct Sacks		
5	5	5

*80-lb. sacks. **67-lb. sacks.

Quantity of Putty Obtained from Various Plastering Materials

The quantity of lime putty obtainable from one ton of lime will vary with the kind of lime used, whether pebble quicklime, pulverized quicklime or hydrated lime, also upon the quality of the lime, as lime produced in different parts of the country will produce varying amounts of lime putty. Under average conditions the following quantities should be obtained.

Plastering Material	Cu. Ft. of Putty per Ton of Material	Lbs. Req'd for One Cu. Ft. Putty
Pebble quicklime	80	25
Pulverized quicklime	80	25
Hydrated lime	46	43½
Gypsum plaster	37	54
Keene's cement	30	66⅔
Portland cement	18½	109

Base Coat

Covering Capacity of Lime Plaster. There has been some confusion in the past regarding the covering capacity of lime plaster due to the lack of definite specifications covering the method of mixing and proportioning the materials.

Lime plaster is usually proportioned by volume, using 1 volume of stiff, well aged lime putty to 2 volumes of dry sand, plus necessary fiber for scratch coat.

For the brown coat, use 1 volume of lime putty to 3 volumes of dry sand, plus fiber as needed.

One cu. ft. of sand is considered 100 lbs. or 2,700 lbs. per cu. yd.

Covering Capacity of One Cu. Yd. of Lime Plaster

Based on same thickness of grounds as given on previous pages.

	Plastering Surface Metal Lath	Unit Masonry ⅝-in. Grounds
Scratch Coat	50-60 sq. yds.	....
Brown Coat	70-80 sq. yds.	60-75 sq. yds.

Bondcrete For Interior Concrete Surfaces. Bondcrete is used as a bond coat on rough concrete surfaces only. If the concrete surfaces are smooth, they must be roughened by bush-hammering or hacking or a dash coat of portland cement grout, composed of one part portland cement and one and one-half parts of fine sand mixed to a mushy consistency, shall be applied. Use a stiff fiber brush forcibly dashing the mix onto the surface and damp cure for two days and then allow to dry.

Bondcrete adheres to dry concrete surfaces because of its cementitious qualities and because it maintains a practically unchanging volume during the process of setting and hardening.

On properly prepared rough concrete ceilings, apply a coat of bond plaster scratched in thoroughly, doubled backed, and filled out to a true, even surface and left rough, ready to receive the finish coat. Total thickness of the plaster not to exceed ⅜".

On properly prepared rough concrete columns or walls, scratch in a minimum ¼" coat of bond plaster and then follow with a brown coat of gypsum plaster (one part gypsum neat plaster to not more than three parts sand, by weight) troweled into the scratch coat before it sets. Brown coat shall be brought out to ground, straightened to a true surface and left rough ready to receive the finish coat. Total thickness of plaster not to exceed ⅝".

It usually requires 1,700 to 2,200 lbs. of Bondcrete per 100 sq. yds. depending upon the regularity of the concrete surface and thickness applied.

Plaster-Weld. Plaster-Weld permanently bonds new gypsum plaster (finish or base coat), or acoustical plaster to most surfaces, but cannot be used over calcimine or other water soluble coatings.

No chipping or roughening of surfaces is required. The surface must be structurally sound and free from dust, dirt, grease, oil or wax.

It may be applied by brush, roller or spray equipment. After allowing Plaster-Weld to become dry to the touch (one hour on porous surfaces and two hours on non-porous surfaces), plaster can be applied either immediately or later. Too much time lag in excess of a week or ten days may require recoating. Plaster-Weld covers up to 500 sq. ft. per gal. when sprayed on, which gives the most satisfactory results. When applied by brush or roller, coverage is reduced. When sprayed on the labor cost per 100 sq. yds. should average as follows:

	Hours	Rate	Total	Rate	Total
Labor	2.25	$....	$....	$21.59	$48.58
Cost per sq. yd.			....		.49
Cost per sq. ft.			....		.05

Plaster-Weld costs approximately $20.00 per gal. in 1-gal. cans.

Finishing Lime. Finishing lime is essential to two common types of job mixed finish coats--the white putty coat trowel finish and the sand float finish.

For the trowel finish coat, the proportion is generally 1 part of gypsum gauging plaster to 2 parts of hydrated lime on a dry weight basis, or 100 lbs. of gypsum gauging plaster to 4½ cu. ft. of lime putty. One ton of gypsum gauging plaster and 2 tons of hydrated finishing lime, or 90 cu. ft. of lime putty will finish approximately 1,200 sq. yds.

While a sand float finish may be mixed with 1 part of lime putty and 2½ to 3 parts of sand, it is generally necessary to add Keene's cement or gypsum gauging plaster to give the finish early hardness and strength. A standard mix for a Keene's cement-lime putty sand float finish is 2 parts of lime putty to 1½ parts of Keene's cement and 4½ parts of sand by volume. This is equivalent to 1 ton of hydrated lime, or 45 cu. ft. of putty, to 1¼ tons of Keene's cement and 4¾ tons of sand, and this quantity will cover approximately 1,600 sq. yds.

If type "S" (fully) hydrated limes are used, they must be added to the mix in dry form, as they form immediate plasticity on wetting. If quicklimes are used, they must be slaked and properly aged and added to the mix in the volume indicated above for putty.

Prepared Gypsum Trowel Finishes. There are a number of prepared gypsum trowel finishes on the market which require only the addition of water to make them ready for use. A few of these are Universal White Trowel Finish, Red Top Trowel, and Red Top Imperial (from Larson Products Corp., Bethesda, MD).

Under average conditions, if mixed and applied in accordance with standard specifications, one ton will cover 350 to 400 sq. yds. of surface or it requires 500 to 575 lbs. per 100 sq. yds.

Prepared Gypsum Sand Float Finishes. There are a number of gypsum sand float finishes, such as Red Top Sand Float Finish (Larson Products).

These finishes are ready for use when water is added and under average conditions one ton is sufficient for 250 to 275 sq. yds. or it requires 725 to 800 lbs., per 100 sq. yds.

Estimating Labor Costs of Plastering

Before it is possible to prepare an accurate estimate on the cost of any plastering job, the quality of work must be considered, as labor costs may vary up to 25 or 30 percent.

The costs given on the following pages cover two grades of workmanship, which are fully described at the beginning of this chapter.

Applying Scratch Coat of Plaster. A plasterer should apply 140 to 165 sq. yds. of scratch coat per 8-hr. day, where ordinary grade of workmanship is required, and 135 to 155 sq. yds. per 8-hr. day where first grade workmanship is required.

Applying Brown Coat. Where an ordinary grade of workmanship is required, a plasterer should apply 90 to 110 sq. yds. of brown coat per 8-hr. day.

Where first grade workmanship is required, a plasterer should brown out 70 to 85 sq. yds. per 8-hr. day.

Applying "Scratch–Double-Back" Basecoat. Over masonry bases and in many markets over gypsum lath base it is common practice to apply a single plaster mix in a "scratch-in" and "double-back" operation to fill out to grounds.

For ordinary work, a plasterer should apply 90 to 110 sq. yds. per 8-hr. day, and for first grade workmanship, a plasterer should apply 70 to 85 sq. yds. per 8-hr. day in this manner.

Applying White Finish or "Putty" Coat. On an ordinary job of white coating, a plasterer should apply 90 to 110 sq. yds. per 8-hr. day.

Where first grade workmanship is specified, a plasterer will apply 60 to 80 sq. yds. per 8-hr. day.

General Notes Applying to Finish Plaster Coats

All of the tables on the following pages give costs on a finish white coat. If wanted with sand float finish, deduct quantities given in tables for Finish White Coat and add cost of Sand Float Finish as given on the following pages under "Sand Float Finish".

Applying Sand Float Finish. A plasterer should apply 100 to 120 sq. yds. of simple sand float finish per 8-hr. day, where an ordinary grade of workmanship is required or 70 to 90 sq. yds. when first grade workmanship is required.

Labor Applying 100 Sq. Yds. of Plaster to Various Plaster Bases
Ordinary Workmanship

Description of Work	Sq. Yds. 8-Hr Day	Plasterer Hours	Tender Hours
Gypsum Mortar on Gypsum lath--½" Plaster--2-Coat Work 1:2½ Brown Coat	90-110	8	5
Gypsum Mortar on Unit Masonry--⅝" Plaster--2-Coat Work--1:3 Brown Coat	80-100	9	6
Lime Mortar on Unit Masonry--⅝" Plaster--2-Coat Work Brown Coat 1:3	90-100	8.5	5.5
Lime Mortar on Metal Lath--¾" Grounds--⅝" Plaster--3-Coat Work Scratch Coat 1:2	150-160	5	5
Brown Coat 1:3	90-110	8	5
Lime Mortar on Unit Masonry--⅝" Plaster--3-Coat Work Scratch Coat 1:2	150-160	5	5
Brown Coat 1:3	90-110	8	5
Gypsum Mortar on Unit Masonry Tile--⅝" Plaster--3-Coat Work Scratch Coat 1:3	150-160	5	5
Brown Coat 1:3	90-110	8	5
Gypsum Mortar on Gypsum Lath--½" Plaster--3-Coat Work Scratch Coat 1:2	150-160	5	5
Brown Coat 1:3	90-110	8	5

Description of Work	Sq. Yds. 8-Hr Day	Plasterer Hours	Tender Hours
Gypsum Mortar on Metal Lath--¾" Grounds--⅝" Plaster--3-Coat Work			
Scratch Coat 1:2	150-160	5	5
Brown Coat 1:3	85-100	9	6
Two (2") Inch Solid Plaster Partitions--Metal Lath One Side--Back Up--Finished 2 Sides--Gypsum Mortar--¾" Grounds--2" Plaster			
Scratch Coat 1:2	150-160	5	5
Back Up Coat 1:2	150-160	5	5
2 Brown Coats 1:2	40-50	18	12

Labor Applying 100 Sq. Yds. of Plaster to Various Plaster Bases
Ordinary Workmanship

Description of Work	Sq. Yds. 8-Hour Day	Plasterer Hours	Tender Hours
Hollow Stud Partitions--Lathed 2 Sides--Plastered to ¾" Grounds			
Scratch Coat 1:2--2 Sides	75-80	10	10
Brown Coat--2 Sides	40-50	18	12
Finish White--All Classes of Undercoat	90-110	8	4
Sand Finish--All Classes of Undercoat	100-120	7.25	5

First Grade Workmanship--Where First Grade Workmanship is Required as described on theprevious pages, add to Hours Given above

Brown Coat		2.25	1
Finish White Coat		3.50	1
Sand Finish		2.75	1

Applying Finish Coat of Cement Wainscoting. Where cement plaster finishes are used in kitchens, bathrooms, laundries, corridors, etc., using Keene's cement, or similar finishes applied over a brown coat of gypsum cement plaster or portland cement, a plasterer should apply 50 to 60 sq. yds. per 8-hr. day.

If the wainscoting is blocked off into squares, 3"x6", 4"x4" or 6"x6" to represent tile, a plasterer should block off 45 to 55 sq. yds. per 8-hr. day.

To apply the finish coat of plaster and block same off into squares to represent tile, a plasterer should complete 25 to 30 sq. yds. per 8-hr. day.

Where first grade workmanship is required, a plasterer should apply 35 to 45 sq. yds. of cement wainscoting per 8-hr. day, and block off 40 to 45 sq. yds. or he will apply cement wainscoting and block off 18 to 22 sq. yds. per 8-hr. day.

Itemized Plastering Costs

The itemized cost given on the following pages are intended to furnish the estimator and contractor detailed information regarding material quantities and labor costs of all kinds of plastering.

All plaster quantities are based on thickness of grounds as given. If the work is "skimped" and the plaster does not fill out the grounds or is applied thinner than specified, then the quantities in the tables will not be correct. If sand is not added in accordance with manufacturer's specifications, the quantities may be more or less than given. Rates given for quantities in lbs. are usually for a 100-lb. bag.

Labor hours include time required for building and removing ordinary scaffolding, but will not cover scaffolding for high work in churches, auditoriums, etc.

2-Coat Plastering

Material Cost of 100 Sq. Yds. 2-Coat Gypsum Cement Plaster, White Finish on Gypsum Lath
Ordinary Workmanship--3/8" Grounds--1/2" Plaster

Add for gypsum lath.

Brown Coat--1:2.5	Rate	Total	Rate	Total
1,000 lbs. gypsum plaster	$....	$....	$6.40*	$64.00
25 cu. ft. sand	....	....	.48	12.00
Finish White Coat				
340 lbs. fin. hydrated lime	....	....	8.50*	28.90
170 lbs. gauging plaster	....	....	9.55*	16.24
Cost 100 sq. yds.		$....		$121.14
Cost per sq. yd.		....		1.21

*100 lb. bag.

Labor Cost of 100 Sq. Yds. 2-Coat Gypsum Cement Plaster, White Finish on Gypsum Lath
Ordinary Workmanship--7/8" Grounds--1/2" Plaster

Add for gypsum lath.

Brown Coat--1:2.5	Hours	Rate	Total	Rate	Total
Plasterer	8	$....	$....	$21.59	$172.72
Labor	5	....	....	19.13	95.65
Finish White Coat					
Plasterer	8	....	....	21.59	172.72
Labor	4	....	....	19.13	76.52
Cost 100 sq. yds.			$....		$517.61
Cost per sq. yd.			....		5.18

Material Cost of 100 Sq. Yds. 2-Coat Gypsum Cement Plaster, White Finish on Unit Masonry
Ordinary Workmanship--5/8" Grounds

Brown Coat--1:3	Rate	Total	Rate	Total
1,100 lbs. gypsum plaster	$....	$....	$6.40	$70.40
33 cu. ft. sand	....	....	.48	15.84
Finish White Coat				
340 lbs. fin. hydrated lime	....	....	8.50	28.90
170 lbs. gauging plaster	....	....	9.55	16.24
Cost 100 sq. yds.		$....		$131.38
Cost per sq. yd.		....		1.31

Labor Cost of 100 Sq. Yds. 2-Coat Gypsum Cement Plaster, White Finish on Unit Masonry
Ordinary Workmanship--5/8" Grounds

Brown Coat--1 :3	Hours	Rate	Total	Rate	Total
Plasterer	9	$....	$....	$21.59	$194.31
Labor	6	....	....	19.13	114.78
Finish White Coat					
Plasterer	8	....	....	21.59	172.72
Labor	4	....	....	19.13	76.52
Cost 100 sq. yds.			$....		$558.33
Cost per sq. yd.			....		5.58

Material Cost of 100 Sq. Yds. 2-Coat Wood Fiber Plaster, White Finish on Gypsum Lath
Ordinary Workmanship--7/8" Grounds--1/2" Plaster

Add for gypsum lath.

	Rate	Total	Rate	Total
Brown Coat				
1,700 lbs. wood fiber	$....	$....	$6.40	$108.80
Finish White Coat				
340 lbs. fin. hydrated lime	....	....	8.50	28.90
170 lbs. gauging plaster	....	....	9.55	16.24
Cost 100 sq. yds.		$....		$153.94
Cost per sq. yd.		....		1.54

Labor Cost of 100 Sq. Yds. 2-Coat Wood Fiber Plaster, White Finish on Gypsum Lath
Ordinary Workmanship--7/8" Grounds--1/2" Plaster

Add for gypsum lath.

	Hours	Rate	Total	Rate	Total
Brown Coat					
Plasterer	9	$....	$....	$21.59	$194.31
Labor	5	....	....	19.13	95.65
Finish White Coat					
Plasterer	8	....	....	21.59	172.72
Labor	4	....	....	19.13	76.52
Cost 100 sq. yds.			$....		$539.20
Cost per sq. yd.			....		5.39

Bondcrete plaster can only be applied to rough concrete surfaces. If the concrete surfaces are smooth, they must be roughened by bush-hammering or hacking or a dash coat of portland cement grout, composed of one part portland cement and one and one-half parts fine sand, mixed to a mushy consistency. Use a stiff fiber brush forcibly dashing the mix onto the surface and damp cure for two days and then allow to dry. For costs on this work, refer to section on Waterproofing.

Material Cost of 100 Sq. Yds. 2-Coat Bondcrete Plaster, White Finish on Concrete Ceilings
Ordinary Workmanship--Plaster 3/8" Thick

	Rate	Total	Rate	Total
Brown Coat				
2,000 lbs. Bondcrete	$....	$....	$7.42	$148.40
Finish White Coat				
340 lbs. fin. hydrated lime	....	....	8.50	28.90
170 lbs. gauging plaster	....	....	9.55	16.24
Cost 100 sq. yds.		$....		$193.54
Cost per sq. yd.		....		1.94

Labor Cost of 100 Sq. Yds. 2-Coat Bondcrete Plaster, White Finish on Concrete Ceilings
Ordinary Workmanship--Plaster 3/8" Thick

	Hours	Rate	Total	Rate	Total
Brown Coat					
Plasterer	8	$....	$....	$21.59	$172.72
Labor	5	....	....	19.13	95.65
Finish White Coat					
Plasterer	8	....	....	21.59	172.72
Labor	4	....	....	19.13	76.52
Cost 100 sq. yds.			$....		$517.61
Cost per sq. yd.			....		5.18

3-Coat Plastering

Material Cost of 100 Sq. Yds. 3-Coat Gypsum Cement Plaster, White Finish on Gypsum Lath
Ordinary Workmanship--7/8" Grounds--1/2" Plaster

Add for gypsum lath.

	Rate	Total	Rate	Total
Scratch Coat--1:2				
400 lbs. gypsum plaster	$....	$....	$6.40	$25.60
8 cu. ft. sand	....	....	.48	3.84
Brown Coat--1:3				
700 lbs. gypsum plaster	....	....	6.40	44.80
21 cu. ft. sand	....	....	.48	10.08
Finish White Coat				
340 lbs. fin. hydrated lime	....	....	8.50	28.90
170 lbs. gauging plaster	....	....	9.55	16.24
Cost 100 sq. yds.		$....		$129.46
Cost per sq. yd.		....		1.29

Labor Cost of 100 Sq. Yds. 3-Coat Gypsum Cement Plaster, White Finish on Gypsum Lath
Ordinary Workmanship--7/8" Grounds--1/2" Plaster

Add for gypsum lath.

	Hours	Rate	Total	Rate	Total
Scratch Coat--1:2					
Plasterer	5	$....	$....	$21.59	$107.95
Labor	5	....	....	19.13	95.65
Brown Coat--1:3					
Plasterer	8	....	....	21.59	172.72
Labor	5	....	....	19.13	95.65
Finish White Coat					
Plasterer	8	....	....	21.59	172.72
Labor	4	....	....	19.13	76.52
Cost 100 sq. yds.			$....		$721.21
Cost per sq. yd.			....		7.21

If first grade workmanship is required, add 5.75 hrs. plasterer time and 2 hrs. labor time per 100 sq. yds.

Material Cost of 100 Sq. Yds. 3-Coat Gypsum Cement Plaster, White Finish on Unit Masonry
Ordinary Workmanship--5/8" Grounds

	Rate	Total	Rate	Total
Scratch Coat--1:3				
500 lbs. gypsum plaster	$....	$....	$6.40	$32.00
15 cu. ft. sand	....	....	.48	7.20
Brown Coat--1 :3				
700 lbs. gypsum plaster	....	....	6.40	44.80
21 cu. ft. sand	....	....	.48	10.08
Finish White Coat				
340 lbs. fin. hydrated lime	....	....	8.50	28.90
170 lbs. gauging plaster	....	....	9.55	16.24
Cost 100 sq. yds.		$....		$139.22
Cost per sq. yd.		....		1.39

Labor Cost of 100 Sq. Yds. 3-Coat Gypsum Cement Plaster, White Finish on Unit Masonry
Ordinary Workmanship--⅝" Grounds

	Hours	Rate	Total	Rate	Total
Scratch Coat--1:3					
Plasterer	5	$....	$....	$21.59	$107.95
Labor	5	....	....	19.13	95.65
Brown Coat--1:3					
Plasterer	8	....	....	21.59	172.72
Labor	5	....	....	19.13	95.65
Finish White Coat					
Plasterer	8	....	....	21.59	172.72
Labor	4	....	....	19.13	76.52
Cost 100 sq. yds.			$....		$721.21
Cost per sq. yd.			....		7.21

If first grade workmanship is required, add 5.75 hrs. plasterer time and 2 hrs. labor time per 100 sq. yds.

Applying Plaster Scratch Coat to Metal Lath Walls and Ceilings

Material Cost of 100 Sq. Yds. 3-Coat Gypsum Cement Plaster, White Finish on Metal Lath Walls and Ceilings
Ordinary Workmanship--¾" Grounds--⅝" Plaster

Add for metal lath and furring.

	Rate	Total	Rate	Total
Scratch Coat--1:2				
800 lbs. gypsum plaster	$....	$....	$6.40	$51.20
16 cu. ft. sand	....	....	.48	7.68
Brown Coat--1 :3				
1,200 lbs. gypsum plaster	....	....	6.40	76.80
36 cu. ft. sand	....	....	.48	17.28
Finish White Coat				
340 lbs. fin. hydrated lime	....	....	8.50	28.90

	Rate	Total	Rate	Total
Scratch Coat--1:2				
170 lbs. gauging plaster	....	....	9.55	16.24
Cost 100 sq. yds.		$....		$198.10
Cost per sq. yd.		....		1.98

Labor Cost of 100 Sq. Yds. 3-Coat Gypsum Cement Plaster, White Finish on Metal Lath Walls and Ceilings
Ordinary Workmanship--3/4" Grounds--5/8" Plaster

Add for metal lath and furring.

	Hours	Rate	Total	Rate	Total
Scratch Coat--1:2					
Plasterer	5	$....	$....	$21.59	$107.95
Labor	5	....	....	19.13	95.65
Brown Coat--1:3					
Plasterer	9	....	....	21.59	194.31
Labor	6	....	....	19.13	114.78
Finish White Coat					
Plasterer	8	....	....	21.59	172.72
Labor	4	....	....	19.13	76.52
Cost 100 sq. yds.			$....		$761.93
Cost per sq. yd.			....		7.62

If first grade workmanship is required, add 5.75 hrs. plasterer time and 2 hrs. labor time per 100 sq. yds.

Material Cost of 100 Sq. Yds. 2" Solid Plaster Partition on Metal Lath
Ordinary Workmanship

Measurement taken on one side only. Add for metal lath and furring.

	Rate	Total	Rate	Total
Scratch Coat--1:2				
2,000 lbs. gypsum plaster	$....	$....	$6.40	$128.00
40 cu. ft. sand	....	....	.48	19.20
Brown Coat--1:3				
2,800 lbs. gypsum plaster	....	....	6.40	179.20
84 cu. ft. sand	....	....	.48	40.32
Finish White Coat				
680 lbs. fin. hydrated lime	....	....	8.50	57.80
340 lbs. gauging plaster	....	....	9.55	32.47
Cost 100 sq. yds.		$....		$456.99
Cost per sq. yd.		....		4.57

Labor Cost of 100 Sq. Yds. 2" Solid Plaster Partition on Metal Lath
Ordinary Workmanship

Measurement taken on one side only. Add for metal lath and furring.

	Hours	Rate	Total	Rate	Total
Scratch Coat--1:2					
Plasterer	10	$....	$....	$21.59	$215.90
Labor	10	....	....	19.13	191.30
Brown Coat--1:3					
Plasterer	18	....	....	21.59	388.62
Labor	12	....	....	19.13	229.56
Finish White Coat					
Plasterer	16	....	....	21.59	345.44
Labor	8	....	....	19.13	153.04
Cost 100 sq. yds.			$....		$1,523.86
Cost per sq. yd.			....		15.24

If first grade workmanship is required, add 11.5 hrs. plasterer time and 4 hrs. labor time per 100 sq. yds.

Material Cost of 100 Sq. Yds. 2" Solid Gypsum Long Length Lath and Plaster Partition

Add for gypsum lath.

Ordinary Workmanship
Partition 2" Thick; ½" Gypsum Lath; ¾" of Plaster Each Side of Lath

Scratch Coat--1:2	Rate	Total	Rate	Total
800 lbs. gypsum plaster	$....	$....	$6.40	$51.20
16 cu. ft. sand	....	....	.48	7.68
Brown Coat--1:3				
2,400 lbs. gypsum plaster	....	....	6.40	153.60
72 cu. ft. sand	....	....	.48	34.56
Finish White Coat				
680 lbs. fin. hydrated lime	....	....	8.50	57.80
340 lbs. gauging plaster	....	....	9.55	32.47
Cost 100 sq. yds.		$....		$337.31
Cost per sq. yd.		....		3.37

Labor Cost of 100 Sq. Yds. 2" Solid Gypsum Long Length Lath and Plaster Partition

Add for gypsum lath.

Ordinary Workmanship
Partition 2" Thick; ½" Gypsum Lath; ¾" of Plaster Each Side of Lath

Scratch Coat--1:2	Hours	Rate	Total	Rate	Total
Plasterer	10	$....	$....	$21.59	$215.90
Labor	10	....	....	19.13	191.30
Brown Coat--1:3					
Plasterer	16	....	....	21.59	345.44
Labor	10	....	....	19.13	191.30
Finish White Coat					
Plasterer	16	....	....	21.59	345.44
Labor	8	....	....	19.13	153.04
Cost 100 sq. yds.			$....		$1,442.42
Cost per sq. yd.			....		14.42

Material Cost of 100 Sq. Yds. 2" Solid Studless Metal Lath and Plaster Partition

Add for metal lath and furring.

Ordinary Workmanship
Partition 2" Thick; 3.4-lb. Metal Lath; Plaster 2" Thick

Scratch Coat--1:2	Rate	Total	Rate	Total
2,000 lbs. gypsum plaster	$....	$....	$6.40	$128.00
40 cu. ft. sand	....	....	.48	19.20
Brown Coat--1 :3				
2,800 lbs. gypsum plaster	....	....	6.40	179.20
84 cu. ft. sand	....	....	.48	40.32
Finish White Coat				
680 lbs. fin. hydrated lime	....	....	8.50	57.80
340 lbs. gauging plaster	....	....	9.55	32.47
Cost 100 sq. yds.		$....		$456.99
Cost per sq. yd.		....		4.57

Labor Cost of 100 Sq. Yds. 2" Solid Studless Metal Lath and Plaster Partition

Add for metal lath and furring.

Ordinary Workmanship
Partition 2" Thick; 3.4-lb. Metal Lath; Plaster 2" Thick

	Hours	Rate	Total	Rate	Total
Scratch Coat--1:2					
Plasterer	10	$....	$....	$21.59	$215.90
Labor	10	....	....	19.13	191.30
Brown Coat--1:3					
Plasterer	18	....	....	21.59	388.62
Labor	12	....	....	19.13	229.56
Finish White Coat					
Plasterer	16	....	....	21.59	345.44
Labor	8	....	....	19.13	153.04
Cost 100 sq. yds.			$....		$1,523.86
Cost per sq. yd.			....		15.24

Material Cost of 100 Sq. Yds. 3-Coat Wood Fiber Plaster, White Finish on Metal Lath Walls and Ceilings

Add for metal lath and furring.

Ordinary Workmanship--¾" Grounds--⅝" Plaster

	Rate	Total	Rate	Total
Scratch Coat				
1,100 lbs. wood fiber	$....	$....	$6.40	$70.40
Brown Coat				
2,300 lbs. wood fiber	....	....	6.40	147.20
Finish White Coat				
340 lbs. fin. hydrated lime	....	....	8.50	28.90
170 lbs. gauging plaster	....	....	9.55	16.24
Cost 100 sq. yds.		$....		$262.74
Cost per sq. yd.		....		2.63

Labor Cost of 100 Sq. Yds. 3-Coat Wood Fiber Plaster, White Finish on Metal Lath Walls and Ceilings

Add for metal lath and furring.

Ordinary Workmanship--¾" Grounds--⅝" Plaster

	Hours	Rate	Total	Rate	Total
Scratch Coat					
Plasterer	5.5	$....	$....	$21.59	$118.75
Labor	5.0	....	....	19.13	95.65
Brown Coat					
Plasterer	10.0	....	....	21.59	215.90
Labor	6.0	....	....	19.13	114.78
Finish White Coat					
Plasterer	8.0	....	....	21.59	172.72
Labor	4.0	....	....	19.13	76.52
Cost 100 sq. yds.			$....		$794.32
Cost per sq. yd.			....		7.94

If first grade workmanship is required, add 5.75 hrs. plasterer time and 2 hrs. labor time per 100 sq. yds.

On properly prepared rough concrete columns or walls, scratch in a minimum ¼" coat of bond plaster and then follow with a brown coat of gypsum plaster (one part gypsum neat plaster to not more than three parts sand, by weight) troweled into the scratch coat before it sets. Brown coat shall be brought out to grounds, straightened to a true surface and left rough ready to receive the finish coat. Total thickness of plaster not to exceed 5/8". Add for preparing concrete surfaces, if required.

Material Cost of 100 Sq. Yds. 3-Coat Bondcrete Plaster, White Finish on Concrete Walls and Columns

Ordinary Workmanship--Plaster 5/8" Thick

	Rate	Total	Rate	Total
Scratch Coat--1/4"				
1,500 lbs. Bondcrete	$....	$....	$7.42	$111.30
Brown Coat--1 :3				
800 lbs. gypsum plaster	....	....	6.40	51.20
24 cu. ft. sand	....	....	.48	11.52
Finish White Coat				
340 lbs. fin. hydrated lime	....	....	8.50	28.90
170 lbs. gauging plaster	....	....	9.55	16.24
Cost 100 sq. yds.		$....		$219.16
Cost per sq. yd.		....		2.19

Labor Cost of 100 Sq. Yds. 3-Coat Bondcrete Plaster, White Finish on Concrete Walls and Columns

Ordinary Workmanship--Plaster 5/8" Thick

	Hours	Rate	Total	Rate	Total
Scratch Coat--1/4"					
Plasterer	5	$....	$....	$21.59	$107.95
Labor	5	....	....	19.13	95.65
Brown Coat--1:3					
Plasterer	8	....	....	21.59	172.72
Labor	5	....	....	19.13	95.65
Finish White Coat					
Plasterer	8	....	....	21.59	172.72
Labor	4	....	....	19.13	76.52
Cost 100 sq. yds.			$....		$721.21
Cost per sq. yd.			....		7.21

If first grade workmanship is required, add 5.75 hrs. plasterer time and 2 hrs. labor time per 100 sq. yds.

Plastering Radiant Panel Heat Ceilings. Where radiant panel heating coils are embedded in ceiling plaster, base coat plastering costs will be higher than normal for both material and labor.

For this type of heating, 3/8" I.D. (1/2" O.D.) copper tubing or electric cable is secured to ceiling framing after lath is installed, in contact with the underside. For proper operation of the heating system, all ceiling coils must be completely embedded in plaster, in perfect contact with plaster (no air pockets permitted) and have a plaster cover of 3/8" or more. This means at least 7/8" of plaster below the face of the lath for copper tubing and 1/2" for electric cable.

Scratch coat application is more difficult, due to the care which must be taken to obtain proper contact with tubing. Brown coat application will require more labor, due to the increased thickness of material required and the difficulty in making it stick to the ceiling. In some cases, a second scratch coat may be necessary to obtain the required thickness.

There should be no change in the method or cost of applying the various plaster finish coats.

Use sand aggregate only for radiant heat ceilings and walls. Lightweight aggregate is an insulator.

Material Cost of 100 Sq. Yds. 3-Coat Gypsum Plaster ⅞" Thick, Applied to Gypsum Lath Ceilings with Radiant Heating Coils Embedded in the Plaster

Add for gypsum lath.

Ordinary Workmanship--⅞" Plaster

	Rate	Total	Rate	Total
Scratch Coat 1:2				
400 lbs. gypsum plaster	$....	$....	$6.40	$25.60
8 cu. ft. sand	....	....	.48	3.84
Brown Coat 1 :3				
1,400 lbs. gypsum plaster	....	....	6.40	89.60
42 cu. ft. sand	....	....	.48	20.16
Finish White Coat				
340 lbs. fin. hydrated lime	....	....	8.50	28.90
170 lbs. gauging plaster	....	....	9.55	16.24
Cost 100 sq. yds.		$....		$184.34
Cost per sq. yd.		....		1.84

Labor Cost of 100 Sq. Yds. 3-Coat Gypsum Plaster ⅞" Thick, Applied to Gypsum Lath Ceilings with Radiant Heating Coils Embedded in the Plaster

Add for gypsum lath.

Ordinary Workmanship--⅞" Plaster

	Hours	Rate	Total	Rate	Total
Scratch Coat 1 :2					
Plasterer	6	$....	$....	$21.59	$129.54
Labor	6	....	....	19.13	114.78
Brown Coat 1 :3					
Plasterer	10	....	....	21.59	215.90
Labor	6	....	....	19.13	114.78
Finish White Coat					
Plasterer	8	....	....	21.59	172.72
Labor	4	....	....	19.13	76.52
Cost 100 sq. yds.			$....		$824.24
Cost per sq. yd.			....		8.24

Material Cost of 100 Sq. Yds. of 3-Coat Gypsum Plaster ⅝" Thick, Applied to Metal Lath Ceilings With Radiant Electric Heating Cables Embedded in the Plaster

Add for metal lath.

Ordinary Workmanship--⅝" Plaster

	Rate	Total	Rate	Total
Scratch Coat--1:2				
800 lbs. gypsum plaster	$....	$....	$6.40	$51.20
16 cu. ft. sand	....	....	.48	7.68
Brown Coat--1 :3				
1,200 lbs. gypsum plaster	....	....	6.40	76.80
36 cu. ft. sand	....	....	.48	17.28
Finish White Coat				
340 lbs. fin. hydrated lime	....	....	8.50	28.90
170 lbs. gauging plaster	....	....	9.55	16.24
Cost 100 sq. yds.		$....		$198.10
Cost per sq. yd.		....		1.98

Labor Cost of 100 Sq. Yds. of 3-Coat Gypsum Plaster ⅝" Thick, Applied to Metal Lath Ceilings With Radiant Electric Heating Cables Embedded in the Plaster

Add for metal lath.

Ordinary Workmanship--⅝" Plaster

	Hours	Rate	Total	Rate	Total
Scratch Coat--1:2					
Plasterer	6	$....	$....	$21.59	$129.54
Labor	6	....	....	19.13	114.78
Brown Coat--1:3					
Plasterer	9	....	....	21.59	194.31
Labor	6	....	....	19.13	114.78
Finish White Coat					
Plasterer	8	....	....	21.59	172.72
Labor	4	....	....	19.13	76.52
Cost 100 sq. yds.			$....		$802.65
Cost per sq. yd.			....		8.03

If it is necessary to apply the plaster in four coats instead of three, add labor for a second scratch coat, as follows:

	Hours	Rate	Total	Rate	Total
Plasterer	6	$....	$....	$21.59	$129.54
Labor	6	....	....	19.13	114.78
Cost 100 sq. yds			$....		$244.32
Cost per sq. yd.			....		2.44

Labor Applying Perlite or Vermiculite Plaster

When using perlite or vermiculite plaster aggregate, a plasterer is applying material weighing 30 to 45 lbs. per cu. ft. against 100 lbs. per cu. ft. for gypsum-sand plaster, so that a plasterer should apply a greater yardage of light aggregate plaster.

Material Cost of 100 Sq. Yds. 3-Coat Plaster Applied to Gypsum Lath, Using Perlite or Vermiculite Aggregate

Add for gypsum lath.

Ordinary Workmanship--⅞" Grounds--1½" of Plaster

	Rate	Total	Rate	Total
Scratch Coat--1:2				
500 lbs. gypsum plaster	$....	$....	$6.40	$32.00
10 cu. ft. aggregate	....	....	3.18	31.80
Brown Coat--1:2				
1,000 lbs. gypsum plaster	....	....	6.40	64.00
20 cu. ft. aggregate	....	....	3.18	63.60
Finish White Coat				
340 lbs. fin. hydrated lime	....	....	8.50	28.90
170 lbs. gauging plaster	....	....	9.55	16.24
Cost 100 sq. yds.		$....		$236.54
Cost per sq. yd.		....		2.37

Labor Cost of 100 Sq. Yds. 3-Coat Plaster Applied to Gypsum Lath, Using Perlite or Vermiculite Aggregate

Add for gypsum lath.

Ordinary Workmanship--7/8" Grounds--1½" of Plaster

	Hours	Rate	Total	Rate	Total
Scratch Coat--1:2					
Plasterer	4.5	$....	$....	$21.59	$97.16
Labor	4.5	....	....	19.13	86.09
Brown Coat--1:2					
Plasterer	7.5	....	....	21.59	161.93
Labor	5.0	....	....	19.13	95.65
Finish White Coat					
Plaster	8.0	....	....	21.59	172.72
Labor	4.0	....	....	19.13	76.52
Cost 100 sq. yds.			$....		$690.07
Cost per sq. yd.		....	....		6.90

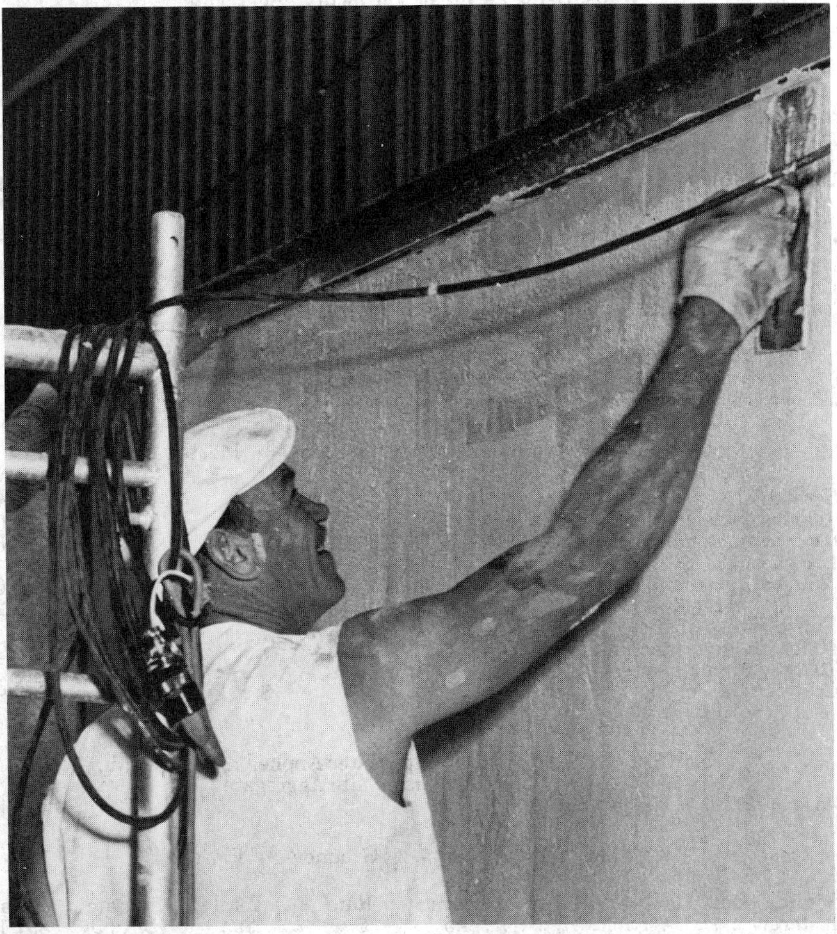

Courtesy Wm. A. Duguid Co.

Applying Finish Plaster Coat

Material Cost of 100 Sq. Yds. 3-Coat Plaster Applied to Unit Masonry Walls, Using Perlite or Vermiculite Aggregate

Ordinary Workmanship--⅝" Grounds--⅝" of Plaster

Scratch Coat--1:3	Rate	Total	Rate	Total
400 lbs. gypsum plaster	$....	$....	$6.40	$25.60
12 cu. ft. aggregate	....	....	3.18	38.16
Brown Coat--1:3				
800 lbs. gypsum plaster	....	....	6.40	51.20
24 cu. ft. aggregate	....	....	3.18	76.32
Finish White Coat				
340 lbs. fin. gypsum plaster	....	....	8.50	28.90
170 lbs. gauging plaster	....	....	9.55	16.24
Cost 100 Sq. yds.		$....		$236.42
Cost per sq. yd.		....		2.36

Labor Cost of 100 Sq. Yds. 3-Coat Plaster Applied to Unit Masonry Walls, Using Perlite or Vermiculite Aggregate

Ordinary Workmanship--⅝" Grounds--⅝" of Plaster

Scratch Coat--1:3	Hours	Rate	Total	Rate	Total
Plasterer	4.5	$....	$....	$21.59	$97.16
Labor	4.5	....	....	19.13	86.09
Brown Coat--1:3					
Plasterer	7.5	....	....	21.59	161.93
Labor	5.0	....	....	19.13	95.65
Finish White Coat					
Plasterer	8.0	....	....	21.59	172.72
Labor	4.0	....	....	19.13	76.52
Cost 100 Sq. yds.			$....		$690.07
Cost per sq. yd.			....		6.90

Material Cost of 100 Sq. Yds. 3-Coat Plaster Applied to Metal Lath, Using Perlite or Vermiculite Aggregate

Add for metal lath and furring.

Ordinary Workmanship--¾" Grounds--¾" Plaster

Scratch--1:2	Rate	Total	Rate	Total
1,000 lbs. gypsum plaster	$....	$....	$6.40	$64.00
20 cu. ft. aggregate	....	....	3.18	63.60
Brown Coat--1:2				
1,250 lbs. gypsum plaster	....	....	6.40	80.00
40 cu. ft. aggregate	....	....	3.18	127.20
Finish White Coat				
340 lbs. fin. hydrated lime	....	....	8.50	28.90
170 lbs. gauging plaster	....	....	9.55	16.24
Cost per 100 sq. yds.		$....		$379.94
Cost per sq. yd.		....		3.80

Labor Cost of 100 Sq. Yds. 3-Coat Plaster Applied to Metal Lath, Using Perlite or Vermiculite Aggregate

Add for metal lath and furring.

Ordinary Workmanship--¾" Grounds--¾" Plaster

Scratch Coat--1:2	Hours	Rate	Total	Rate	Total
Plasterer	4.5	$....	$....	$21.59	$97.16
Labor	4.5	....	....	19.13	86.09
Brown Coat--1:2					

Scratch Coat--1:2	Hours	Rate	Total	Rate	Total
Plasterer	8.0	....	....	21.59	172.72
Labor	6.0	....	....	19.13	114.78
Finish White Coat					
Plasterer	8.0	....	....	21.59	172.72
Labor	4.0	....	....	19.13	76.52
Cost per 100 sq. yds			$....		$719.99
Cost per sq. yd.			....		7.20

Material Cost of 100 Sq. Yds. 2" Solid Plaster Partition Applied to Metal Lath, Using Perlite or Vermiculite Aggregate (Measurement One Side Only)

Add for metal lath and furring.

Ordinary Workmanship--2" Partition--2" of Plaster

Scratch Coat--1:2	Rate	Total	Rate	Total
1,500 lbs. gypsum plaster	$....	$....	$6.40	$96.00
30 cu. ft. aggregate	....	....	3.18	95.40
Brown Coat--1:2				
6,000 lbs. gypsum plaster	....	....	6.40	384.00
120 cu. ft. aggregate	....	....	3.18	381.60
Finish White Coat				
680 lbs. fin. hydrated lime	....	....	8.50	57.80
340 lbs. gauging plaster	....	....	9.55	32.47
Cost 100 sq. yds.		$....		$1,047.27
Cost per sq. yd.		....		10.47

Labor Cost of 100 Sq. Yds. 2" Solid Plaster Partition Applied to Metal Lath, Using Perlite or Vermiculite Aggregate (Measurement One Side Only)

Add for metal lath and furring.

Ordinary Workmanship--2" Partition--2" of Plaster

Scratch Coat--1:2	Hours	Rate	Total	Rate	Total
Plasterer	5	$....	$....	$21.59	$107.95
Labor	5	....	....	19.13	95.65
Brown Coat--1:2					
Plasterer	23	....	....	21.59	496.57
Labor	17	....	....	19.13	325.21
Finish White Coat					
Plasterer	16	....	....	21.59	345.44
Labor	8	....	....	19.13	153.04
Cost 100 sq. yds.			$....		$1,523.86
Cost per sq. yd.			....		15.24

Material Cost of 100 Sq. Yds. 3-Coat Plaster (Scratch-Double-Back Basecoat) Applied to Gypsum Lath Using Pre-Mixed Perlite Plaster

Add for gypsum lath.

Ordinary Workmanship--7/8" Grounds--1/2" of Plaster

Scratch Coat	Rate	Total	Rate	Total
880-lb. sacks P. plaster	$....	$....	$6.90	$60.72
Double-Back Brown Coat				
1180-lb. sacks P. plaster	....	....	6.90	81.42
Finish White Coat				
340 lbs. fin. hydrated lime	....	....	8.50	28.90
170 lbs. gauging plaster	....	....	9.55	16.24
Cost 100 sq. yds.		$....		$187.28
Cost per sq. yd.		....		1.87

Labor Cost of 100 Sq. Yds. 3-Coat Plaster (Scratch-Double-Back Basecoat) Applied to Gypsum Lath Using Pre-Mixed Perlite Plaster

Add for gypsum lath.

Ordinary Workmanship--7/8" Grounds--1/2" of Plaster

	Hours	Rate	Total	Rate	Total
Scratch Coat					
Plasterer	4.5	$....	$....	$21.59	$97.16
Labor	4.5	....	....	19.13	86.09
Double-Back Brown Coat					
Plasterer	7.5	....	....	21.59	161.93
Labor	5.0	....	....	19.13	95.65
Finish White Coat					
Plasterer	8.0	....	....	21.59	172.72
Labor	4.0	....	....	19.13	76.52
Cost 100 sq. yds.			$....		$690.07
Cost per sq. yd.			....		6.90

Material Cost of 100 Sq. Yds. 3-Coat Plaster (Scratch-Double-Back Basecoat) Applied to Unit Masonry Walls Using Pre-Mixed Perlite Plaster

Ordinary Workmanship--5/8" Grounds--5/8" of Plaster

	Rate	Total	Rate	Total
Scratch Coat				
1080-lb. sacks P. Plaster	$....	$....	$6.90	$74.52
Double-Back Brown Coat				
1680-lb. sacks P. Plaster	....	....	6.90	115.92
Finish White Coat				
340 lbs. fin. hydrated lime	....	....	8.50	28.90
170 lbs. gauging plaster	....	....	9.55	16.24
Cost 100 sq. yds.		$....		$235.58
Cost per sq. yd.		....		2.36

Labor Cost of 100 Sq. Yds. 3-Coat Plaster (Scratch-Double-Back Basecoat) Applied to Unit Masonry Walls Using Pre-Mixed Perlite Plaster

Ordinary Workmanship--5/8" Grounds--5/8" of Plaster

	Hours	Rate	Total	Rate	Total
Scratch Coat					
Plasterer	4.5	$....	$....	$21.59	$97.16
Labor	4.5	....	....	19.13	86.09
Double-Back Brown Coat					
Plasterer	7.5	....	....	21.59	161.93
Labor	5.0	....	....	19.13	95.65
Finish White Coat					
Plasterer	8.0	....	....	21.59	172.72
Labor	4.0	....	....	19.13	76.52
Cost 100 sq. yds.			$....		$690.07
Cost per sq. yd.			....		6.90

Material Cost of 100 Sq. Yds. 3-Coat Plaster Applied to Metal Lath Using Pre-Mixed Perlite Plaster

Add for metal lath and furring.

Ordinary Workmanship--¾" Grounds--¾" Plaster

	Rate	Total	Rate	Total
Scratch Coat				
1580-lb. sacks P. Plaster	$....	$....	$6.90	$109.02
Brown Coat				
1880-lb. sacks P. Plaster	....	....	6.90	129.72
Finish White Coat				
340 lbs. fin. hydrated lime	....	....	8.50	28.90
170 lbs. gauging plaster	....	....	9.55	16.24
Cost 100 sq. yds.		$....		$283.88
Cost per sq. yd.		....		2.84

Labor Cost of 100 Sq. Yds. 3-Coat Plaster Applied to Metal Lath Using Pre-Mixed Perlite Plaster

Add for metal lath and furring.

Ordinary Workmanship--¾" Grounds--¾" Plaster

	Hours	Rate	Total	Rate	Total
Scratch Coat					
Plasterer	4.5	$....	$....	$21.59	$97.16
Labor	4.5	....	....	19.13	86.09
Brown Coat					
Plasterer	8.0	....	....	21.59	172.72
Labor	6.0	....	....	19.13	114.78
Finish White Coat					
Plasterer	8.0	....	....	21.59	172.72
Labor	4.0	....	....	19.13	76.52
Cost 100 sq. yds.			$....		$719.99
Cost per sq. yd.			....		7.20

Material Cost of 100 Sq. Yds. 2" Solid Plaster Partition Applied to Metal Lath Using Pre-Mixed Perlite Plaster (Measurement One Side Only)

Add for metal lath and furring.

Ordinary Workmanship--2" Partition--2" of Plaster

	Rate	Total	Rate	Total
1580-lb. sacks P. Plaster	$....	$....	$6.90	$109.02
Brown Coat				
7080-lb. sacks P. Plaster	....	....	6.90	488.52
Finish White Coat				
680 lbs. fin. hydrated lime	....	....	8.50	57.80
340 lbs. gauging plaster	....	....	9.55	32.47
Cost 100 sq. yds.		$....		$687.81
Cost per sq. yd.		....		6.88

Labor Cost of 100 Sq. Yds. 2" Solid Plaster Partition Applied to Metal Lath Using Pre-Mixed Perlite Plaster (Measurement One Side Only)

Add for metal lath and furring.

Ordinary Workmanship--2" Partition--2" of Plaster

	Hours	Rate	Total	Rate	Total
Plasterer	5	$....	$....	$21.59	$107.95
Labor	5	....	....	19.13	95.65
Brown Coat					
Plasterer	23	....	....	21.59	496.57

	Hours	Rate	Total	Rate	Total
Labor	17	....	....	19.13	325.21
Finish White Coat					
Plasterer	16	....	....	21.59	345.44
Labor	8	....	....	19.13	153.04
Cost 100 sq. yds.			$....		$1,523.86
Cost per sq. yd.			....		15.24

Machine Application of Plaster. Various types of machines are available for the spray application of base coat plaster mortar to walls and ceilings where it is then worked in the normal manner. They vary in size from large machines capable of spraying any plaster mortars, including base coat materials for ordinary plastering and membrane fireproofing, to small machines that handle only the finish coat of acoustical plaster or a sand float finish. Most models have self-contained pump units with gasoline or electric power air compressors. Several small hand gun-hopper units are available for attachment to separate air compressors.

When applying plaster mortars by machine, approximately 20% to 25% additional material will be required to provide equal coverage to hawk and trowel application. This is caused by material densification due to machine action and increased material loss from spattering.

Using the machine method, the labor cost of applying the various plaster coats will be considerably less than for the hawk and trowel method, but an allowance must be made for machine cost and maintenance and additional protection and clean-up labor will be required.

Where scratch coat is machine applied over metal lath, an experienced operator can apply 750 to 850 sq. yds. per 8-hr. day. For this area, an additional 4 hrs. of plasterer time will be required for touching up and scratching surface.

Material Cost of 100 Sq. Yds. of Machine Applied Scratch Coat on Metal Lath Using Perlite or Vermiculite Aggregate

Add for metal lath and furring.

	Rate	Total	Rate	Total
1,200 lbs. gypsum plaster	$....	$....	$6.40	$76.80
24 cu. ft. aggregate	....	....	3.18	76.32
Machine cost	....	....	5.30	5.30
Cost 100 sq. yds.		$....		$158.42
Cost per sq. yd.		....		1.58

Labor Cost of 100 Sq. Yds. of Machine Applied Scratch Coat on Metal Lath Using Perlite or Vermiculite Aggregate

Add for metal lath and furring.

	Hours	Rate	Total	Rate	Total
Plasterer	1.5	$....	$....	$21.59	$32.39
Labor	2.0	....	....	19.13	38.26
Cost 100 sq. yds.			$....		$70.65
Cost per sq. yd.			....		.71

Where brown coat is machine applied over scratch coat or gypsum lath, an experienced operator, with another plasterer rodding and leveling, can apply 375 to 425 sq. yds. per 8-hr. day.

Material Cost of 100 Sq. Yds. of Machine Applied Brown Coat,
7/16" Thick over Scratch Coat or Gypsum Lath,
Using Perlite or Vermiculite Aggregate

	Rate	Total	Rate	Total
1,550 lbs. gypsum plaster	$....	$....	$6.40	$99.20
50 cu. ft. aggregate	....	....	3.18	159.00
Machine cost	....	....	10.60	10.60
Cost 100 sq. yds.		$....		$268.80
Cost per sq. yd.		....		2.69

Labor Cost of 100 Sq. Yds. of Machine Applied Brown Coat,
7/16" Thick over Scratch Coat or Gypsum Lath,
Using Perlite or Vermiculite Aggregate

	Hours	Rate	Total	Rate	Total
Plasterer	4	$....	$....	$21.59	$86.36
Labor	4	....	....	19.13	76.52
Cost 100 sq. yds.			$....		$162.88
Cost per sq. yd.			....		1.63

Special Prepared Finishes

There are a number of prepared plaster finishes on the market that are ready for use with the addition of water. These may be applied over a brown coat of gypsum cement plaster.

Smooth trowel finishes such as gypsum gauging-lime putty require a much higher degree of rigidity and strength in the lath and basecoat plaster for good visual performance. When the use of high strength basecoat plasters is impractical, consideration should be given to the elimination of restraint at the perimeter angles or the selection of a more rigid lath, or both. They help to compensate for the lower strength of some basecoat plasters.

Where a special finish is specified, deduct the cost of the finish white coat as given under the ordinary plastering classification and add the cost of the special finish specified.

Material Cost of 100 Sq. Yds. Gypsum Trowel Finish
Ordinary Workmanship

	Rate	Total	Rate	Total
550 lbs. gyp. trowel finish	$....	$....	$15.90	$87.45
Cost per sq. yd.		....		.87

Labor Cost of 100 Sq. Yds. Gypsum Trowel Finish
Ordinary Workmanship

	Hours	Rate	Total	Rate	Total
Plasterer	8	$....	$....	$21.59	$172.72
Labor	4	....	....	19.13	76.52
Cost 100 sq. yds.			$....		$249.24
Cost per sq. yd.			....		2.49

For first grade workmanship, add 3.5 hrs. plasterer time and 1-hr. labor time per 100 sq. yds.

If plaster is wanted with a sand float finish, deduct quantities of plaster and labor given under finish white coat and add the following:

Material Cost of 100 Sq. Yds. Sand Float Finish
Ordinary Workmanship

	Rate	Total	Rate	Total
200 lbs. unfibered plaster	$....	$....	$6.90	$13.80
4 cu. ft. sand	....	....	.48	1.92
Cost per 100 sq. yds.		$....		$15.72
Cost per sq. yd.		....		.16

Labor Cost of 100 Sq. Yds. Sand Float Finish
Ordinary Workmanship

	Hours	Rate	Total	Rate	Total
Plasterer	7.25	$....	$....	$21.59	$156.53
Labor	5.00	....	....	19.13	95.65
Cost per 100 sq. yds			$....		$252.18
Cost per sq. yd.			....		2.52

If first grade workmanship is required, add 3 hours plasterer time and 1 hour labor time per 100 sq. yds.

Material Cost of 100 Sq. Yds. Keene's Cement Smooth Finish, (Medium Hard) Applied Over a Brown Coat of Gypsum Cement Plaster
Ordinary Workmanship

	Rate	Total	Rate	Total
530 lbs. Keene's cement	$....	$....	$16.96	$89.89
270 lbs. fin. hydrated lime	....	....	8.50	22.95
Cost 100 sq. yds.		$....		$112.84
Cost per sq. yd.		....		1.13

Labor Cost of 100 Sq. Yds. Keene's Cement Smooth Finish, (Medium Hard) Applied Over a Brown Coat of Gypsum Cement Plaster
Ordinary Workmanship

	Hours	Rate	Total	Rate	Total
Plasterer	14.5	$....	$....	$21.59	$313.06
Labor	5.0	....	....	19.13	95.65
Cost 100 sq. yds.			$....		$408.71
Cost per sq. yd.			....		4.09

If blocked off into squares to represent tile, add about 16 hrs. plasterer time per 100 sq. yds.

If first grade workmanship is required, add 5.5 hrs. plasterer time and 1 hr. labor time per 100 sq. yds.

Keene's cement sand float finish coat may be applied over a brown coat of gypsum cement plaster, either plain or blocked off into ashlar effects. If blocked, add extra for this work.

Material Cost of 100 Sq. Yds. Keene's Cement Sand Float Finish
Ordinary Workmanship

	Rate	Total	Rate	Total
160 lbs. Keene's cement	$....	$....	$16.96	$27.14
125 lbs. fin. hydrated lime	....	....	8.50	10.63
6 cu. ft. clean sand	....	....	.48	2.88
Cost 100 sq. yds.		$....		$40.65
Cost per sq. yd.		....		.41

Labor Cost of 100 Sq. Yds. Keene's Cement Sand Float Finish
Ordinary Workmanship

	Hours	Rate	Total	Rate	Total
Plasterer	13	$....	$....	$21.59	$280.67
Labor	6	....	....	19.13	114.78
Cost 100 sq. yds			$....		$395.45
Cost per sq. yd			....		3.95

If first grade workmanship is required, add 4 hrs. plasterer time and 1 hr. labor time per 100 sq. yds.

The resistance of gypsum gauging-lime putty trowel finishes to cracking, particularly to check cracking, can be increased by the addition of fine aggregate. The use of not less than 1/2 cubic foot of fine silica sand or perlite to each 100 pounds of gauging plaster or Keene's cement increases the factor of safety in trowel finishes.

Acoustical Plaster. Acoustical plaster is occasionally specified instead of trowel putty coat or sand float finishes for noise quieting in classrooms and corridors and for acoustical correction of churches and auditoriums. A number of prepared acoustical plasters are available having sound absorption values of 40% to 45% at 512 cycles and a noise reduction coefficient of 55 to 60 percent. There are two general types of acoustical plaster. Those with a gypsum or lime binder, such as Sabinite, Hi-Lite, Kilnoise, Gold Bond, Perfolite, etc., are for trowel application only with either a troweled, floated or stippled and perforated finish. Other acoustical plasters, such as Audicote, Sprayolite, Zonolite Acoustical Plaster, etc., are made with an adhesive type binder that shrinks on drying, providing the porosity for sound absorption, and may be applied by machine or trowel.

In general, acoustical plasters are for application over gypsum base coats (scratch and brown) and should be applied in two coats to a total thickness of 1/2"--usually a 3/8" first coat and a 1/8" finish coat.

Because of restrictions due to the asbestos fiber content, acoustical plaster is temporarily out of supply but will probably be back on the market with a new mixture. The prices below are for the old formula material.

Where applied by trowel, a plasterer should complete 65 to 75 sq. yds. of acoustical plaster, two coats to 1/2" total thickness, per 8-hr. day.

Material Cost of 100 Sq. Yds. of Acoustical Plaster
Applied by Trowel

Add for lath and base coats required.

	Rate	Total	Rate	Total
20 bags (5 sq. yds., 1/2" thick), acoustical plaster	$....	$....	$7.95	$159.00
Cost per sq. yd		....		1.59

Labor Cost of 100 Sq. Yds. of Acoustical Plaster
Applied by Trowel

Add for lath and base coats required.

	Hours	Rate	Total	Rate	Total
Plasterer	11.5	$....	$....	$21.59	$248.29
Labor	6.0	....	....	19.13	114.78
Cost 100 sq. yds			$....		$363.07
Cost per sq. yd			....		3.63

Where application of the two coat acoustical finish is by machine, a plasterer should complete 100 to 110 sq. yds. per 8-hr. day, but one extra hour of labor time must be figured for the additional cleanup and protection required, and an allowance included for cost and maintenance of equipment.

Material Cost of 100 Sq. Yds. of Acoustical Plaster Applied by Machine
Add for lath and base coats required.

	Rate	Total	Rate	Total
14 bags (7.2 sq. yds. ½" thick) acoustical plaster	$....	$....	$7.95	$111.30
Machine cost	....	....	39.75	39.75
Cost per 100 sq. yds		$....		$151.05
Cost per sq. yd		....		1.51

Labor Cost of 100 Sq. Yds. of Acoustical Plaster Applied by Machine
Add for lath and base coats required.

	Hours	Rate	Total	Rate	Total
Plasterer	8.0	$....	$....	$21.59	$172.72
Labor	7.0	....	....	19.13	133.91
Labor (masking)	7.0	....	....	19.13	133.91
Cost per 100 sq. yds			$....		$440.54
Cost per sq. yd			....		4.41

Vermiculite Direct-to-Steel Fireproofing. Vermiculite Type-MK direct-to-steel fireproofing is a new and economical method of protecting steel columns, beams, girders, and trusses, and steel floor and roof sections. It is applied by machine directly to the metal, without a primer, lath, or reinforcing. It presents a tough, fire-protective and insulating coating free of fissures, and can be applied in the early stages of construction when the building is free of other trades.

This is a mill-mixed material that requires only the addition of water. It is packaged in 25- and 50-lb. bags. Coverage varies with specified thickness.

Colored Interior Finish Plaster

There are a number of colored finish plasters on the market used for textured or float plaster finish.

These special finishes are applied over a brown coat of gypsum cement plaster and may be finished in a number of different ways, such as Spanish Palm Finish, Brush Swirl, Float Finish, English Trowel finish, etc., in addition to a simple float finish.

Labor Applying Special Finish Coats. For colored finish plaster with a simple float finish, a plasterer should apply 100 to 120 sq. yds. per 8-hr. day, where an ordinary grade of workmanship is permitted.

Oriental Interior Color Finish Plaster. From the U.S. Gypsum Co., this is used for textured or floated plaster finish. It is made in 13 standard shades and white and is ready for use with the addition of water only.

It requires approximately 8 lbs. of Oriental finish per sq. yd. depending upon the texture used. Price about $12.00 per 100 lbs.

Material Cost of 100 Sq. Yds. of Oriental Interior Colored Finish Plaster, with Simple Float Finish, Applied Over a Brown Coat of Gypsum Cement Plaster
Ordinary Workmanship

Finish Coat	Rate	Total	Rate	Total
800 lbs. Oriental finish	$....	$....	$12.72	$101.76
Cost per sq. yd		....		1.01

Labor Cost of 100 Sq. Yds. of Oriental Interior Colored Finish Plaster, with Simple Float Finish, Applied Over a Brown Coat of Gypsum Cement Plaster
Ordinary Workmanship

Finish Coat	Hours	Rate	Total	Rate	Total
Plasterer	7.25	$....	$....	$21.59	$156.53
Labor	4.50	....	....	19.13	86.09
Cost 100 sq. yds			$....		$242.62
Cost per sq. yd			....		2.43

If first grade workmanship is required, add 2.75 hrs. plasterer time and 1 hr. labor time per 100 sq. yds.

Material Costs
Colored Plaster with Smooth or Textured Finish

	Rate	Total	Rate	Total
350 lbs. Keene's cement	$....	$....	$16.96	$59.36
450 lbs. fin. hydrated lime	....	....	8.50	38.25
16 lbs. color	....	....	1.06	16.96
Cost 100 sq. yds		$....		$114.57
Cost per sq. yd.		....		1.15

Material Costs
Colored Plaster with Sand Float Finish

	Rate	Total	Rate	Total
100 lbs. Keene's cement	$....	$....	$16.96	$16.96
500 lbs. fin. hydrated lime	....	....	8.50	42.50
1,200 lbs. white silica sand	....	....	1.06	12.72
36 lbs. color	....	....	1.06	40.45
Cost 100 sq. yds		$....		$112.63
Cost per sq. yd.		....		1.13

The above formulas are standard mixture for permanent colors. Best results are obtained by soaking lime overnight or for 24 hours.

Ornamental Plastering

Ornamental plastering is not used as extensively as it once was, and mechanics who can do this type of work are not always available. It is best to contact a plastering subcontractor for firm prices on intricate ornamental plaster work.

Material Cost of 100 Lin. Ft. of Plaster Cornice Under 12" in Girth

	Rate	Total	Rate	Total
200 lbs. gypsum plaster	$....	$....	$6.36	$12.72
4 cu. ft. sand	....	....	.48	1.92
50 lbs. molding plaster	....	....	10.60	5.30
Cost 100 lin. ft.		$....		$19.94
Cost per lin. ft.		....		.20

Add for metal lath and furring as given on previous pages.

Labor Cost of 100 Lin. Ft. of Plaster Cornice Under 12" in Girth

	Hours	Rate	Total	Rate	Total
Plasterer	18	$....	$....	$21.59	$388.62
Labor	6	....	....	19.13	114.78
Cost 100 lin. ft.			$....		$503.40
Cost per lin. ft.			....		5.03

Add for metal lath and furring as given on previous pages.

Material Cost of 100 Sq. Ft. of Plaster Cornice Over 1'-0" in Girth

	Rate	Total	Rate	Total
200 lbs. gypsum plaster	$....	$....	$6.36	$12.72
4 cu. ft. sand	....	....	.48	1.92
50 lbs. molding plaster	....	....	10.60	5.30
Cost 100 sq. ft.		$....		$19.94
Cost per sq. ft.		....		.20

Add for metal lath and furring as given on previous pages.

Labor Cost of 100 Sq. Ft. of Plaster Cornice Over 1'-0" Girth

	Hours	Rate	Total	Rate	Total
Plasterer	17	$....	$....	$21.59	$367.03
Labor	6	....	....	19.13	114.78
Cost 100 sq. ft.			$....		$481.81
Cost per sq. ft.			....		4.82

Add for metal lath and furring as given on previous pages.

Material Cost of 100 Sq. Ft. of Portland Cement Cornice

	Rate	Total	Rate	Total
4 sacks portland cement	$....	$....	$4.24	$16.96
40 lbs. hydrated lime	....	....	8.50	3.40
10 cu. ft. sand	....	....	.48	4.80
Cost 100 sq. ft.		$....		$25.16
Cost per sq. ft.		....		.25

Add for metal lath and furring as given on previous pages.

Labor Cost of 100 Sq. Ft. of Portland Cement Cornice

	Hours	Rate	Total	Rate	Total
Plasterer	46	$....	$....	$21.59	$993.14
Labor	10	....	....	19.13	191.30
Cost 100 sq. ft.			$....		$1,184.44
Cost per sq. ft.			....		11.84

Add for metal lath and furring as given on previous pages.

Running Bull Nose Corners, Ceiling Coves, Etc. When running bull nose corners on plaster columns, pilasters, window jambs, etc., and plain ceiling coves and wall angles, a plasterer should run 80 to 100 lin. ft. per 8-hr. day, at the following labor cost per 100 lin. ft.:

	Hours	Rate	Total	Rate	Total
Plasterer	9	$....	$....	$21.59	$194.31
Labor	2	....	....	19.13	38.26
Cost 100 lin. ft.			$....		$232.57
Cost per lin. ft.			....		2.33

Portland Cement Plaster

Portland cement plaster is used for both interior and exterior work. It should not be mixed too rich. A too rich mixture has a tendency to check and shrink.

Proper attention to curing is important. After each coat has set, and before the following coat is applied, the mortar should be kept moist for at least three days and then permitted to dry out gradually.

When using portland cement mortar it is customary to add a small percentage of lime (usually about 10%) to make the mortar work easier, because a straight portland cement mortar is very difficult to apply.

For the cost of portland cement plaster applied over metal lath and the cost of metal lath and furring, see previous pages of this chapter.

Labor Applying 100 Sq. Yds. Portland Cement Plaster to Various Plaster Bases

Description of Work Ordinary Work*	Sq. Yds. per 8-Hr. Day	Hrs. 100 Sq. Yds. Plasterer	Tender
Interior Work on Brick, Clay Tile or Cement Block--3/4" Thick			
Scratch Coat	120-140	6	4
Brown Coat	75-85	10	5
Float Finish Coat	60-70	12	4
Trowel Finish Coat	45-55	16	4
Interior Portland Cement Plaster on Metal Lath			
Scratch Coat	120-140	6	5
Brown Coat	60-70	12	6
Float Finish Coat	60-70	12	4
Trowel Finish Coat	45-55	16	4
Exterior Portland Cement Stucco on Brick, Clay Tile or Cement Block			
Scratch Coat	120-140	6	5
Brown Coat	75-85	10	6
Scaffold	....	..	12
Exterior Portland Cement Stucco on Metal Lath on Frame Construction			
Scaffold	....	..	12
Scratch Coat	120-140	6	6
Brown Coat	60-70	12	8
Floated Finish Coat	60-70	12	6
Troweled Finish Coat	45-55	16	6
Textured Finish Coat	50-60	14	6
Blocking Off Portland Cement Plaster into Squares to Represent Tile	45-55	16	4
Portland Cement Straight Base	45-55**	16	4
Portland Cement Coved Base	30-35**	24	6

*For first grade workmanship as described at the beginning of this chapter, add 30% to 40% to plasterer time for brown and finish coats.

**Lineal feet.

Material Cost of 100 Sq. Yds. 2-Coat Portland Cement Plaster on Brick, Clay Tile or Cement Block
Ordinary Workmanship--5/8" Grounds

Brown Coat	Rate	Total	Rate	Total
14 sacks portland cement	$....	$....	$4.24	$59.36
140 lbs. hydrated lime	....	....	8.50	11.90
42 cu. ft. sand	....	....	.48	20.16
Finish Coat				
4 sacks portland cement	....	....	4.24	16.96
40 lbs. hydrated lime	....	....	8.50	3.40

Brown Coat	Rate	Total	Rate	Total
12 cu. ft. sand	....	....	.48	5.76
Cost 100 sq. yds.		$....		$117.54
Cost per sq. yd.		....		1.18

Labor Cost of 100 Sq. Yds. 2-Coat Portland Cement Plaster on Brick, Clay Tile or Cement Block
Ordinary Workmanship--5/8" Grounds

Brown Coat	Hours	Rate	Total	Rate	Total
Plasterer	13	$....	$....	$21.59	$280.67
Labor	6	....	....	19.13	114.78
Finish Coat					
Plasterer	16	....	....	21.59	345.44
Labor	4	....	....	19.13	76.52
Cost 100 sq. yds			$....		$817.41
Cost per sq. yd			....		8.17

Material Cost of 100 Sq. Yds. 3-Coat Portland Cement Plaster on Brick, Clay Tile or Cement Block
Ordinary Workmanship--5/8" Grounds

Scratch Coat	Rate	Total	Rate	Total
6 sacks portland cement	$....	$....	$4.24	$25.44
60 lbs. hydrated lime	....	....	8.50	5.10
18 cu. ft. sand	....	....	.48	8.64
Brown Coat				
8 sacks portland cement	....	....	4.24	33.92
80 lbs. hydrated lime	....	....	8.50	6.80
24 cu. ft. sand	....	....	.48	11.52
Finish Coat				
4 sacks portland cement	....	....	4.24	16.96
40 lbs. hydrated lime	....	....	8.50	3.40
12 cu. ft. sand	....	....	.48	5.76
Cost 100 sq. yds		$....		$117.54
Cost per sq. yd		....		1.18

Labor Cost of 100 Sq. Yds. 3-Coat Portland Cement Plaster on Brick, Clay Tile or Cement Block
Ordinary Workmanship--5/8" Grounds

Scratch Coat	Hours	Rate	Total	Rate	Total
Plasterer	6	$....	$....	$21.59	$129.54
Labor	4	....	....	19.13	76.52
Brown Coat					
Plasterer	10	....	....	21.59	215.90
Labor	5	....	....	19.13	95.65
Finish Coat					
Plasterer	16	....	....	21.59	345.44
Labor	4	....	....	19.13	76.52
Cost 100 sq. yds			$....		$939.57
Cost per sq. yd			....		9.40

If first grade workmanship is required, add 8 hrs. plasterer time and 2 hrs. labor time per 100 sq. yds.

If blocked off into squares to represent tile, add 16 hrs. plasterer time for ordinary work and 18 hrs. plasterer time for first grade workmanship per 100 sq. yds.

Material Cost of 100 Sq. Yds. 3-Coat Portland Cement Plaster on Metal Lath
Add for metal lath and furring.

Ordinary Workmanship--¾" Grounds

	Rate	Total	Rate	Total
Scratch Coat				
7 sacks portland cement	$....	$....	$4.24	$29.68
70 lbs. hydrated lime	....	....	8.50	5.95
21 cu. ft. sand	....	....	.48	10.08
Brown Coat				
11 sacks portland cement	....	....	4.24	46.64
110 lbs. hydrated lime	....	....	8.50	9.35
33 cu. ft. sand	....	....	.48	15.84
Finish Coat				
4 sacks portland cement	....	....	4.24	16.96
40 lbs. hydrated lime	....	....	8.50	3.40
12 cu. ft. sand	....	....	.48	5.76
Cost 100 sq. yds		$....		$143.66
Cost per sq. yd.		....		1.44

Labor Cost of 100 Sq. Yds. 3-Coat Portland Cement Plaster on Metal Lath
Add for metal lath and furring.

Ordinary Workmanship--¾" Grounds

	Hours	Rate	Total	Rate	Total
Scratch Coat					
Plasterer	6	$....	$....	$21.59	$129.54
Labor	5	....	....	19.13	95.65
Brown Coat					
Plasterer	12	....	....	21.59	259.08
Labor	6	....	....	19.13	114.78
Finish Coat					
Plasterer	16	....	....	21.59	345.44
Labor	4	....	....	19.13	76.52
Cost 100 sq. yds			$....		$1,021.01
Cost per sq. yd.			....		10.21

If blocked off into squares to represent tile, add 16 hrs. plasterer time per 100 sq. yds.

If ⅝" grounds are used instead of ¾", deduct 2 sacks portland cement, 20 lbs. lime and 6 cu. ft. sand.

If first grade workmanship is required, add 8 hrs. plasterer time and 2 hrs. labor time per 100 sq. yds.

Material Cost of 100 Lin. Ft. of 6" Portland Cement Straight Base

	Rate	Total	Rate	Total
2 sacks portland cement	$....	$....	$4.24	$8.48
20 lbs. hydrated lime	....	....	8.50	1.70
6 cu. ft. sand	....	....	.48	2.88
Cost 100 lin. ft.		$....		$13.06
Cost per lin. ft.		....		.13

Labor Cost of 100 Lin. Ft. of 6" Portland Cement Straight Base

	Hours	Rate	Total	Rate	Total
Plasterer	16	$....	$....	$21.59	$345.44
Labor	4	....	....	19.13	76.52
Cost 100 lin. ft.			$....		$421.96
Cost per lin. ft.			....		4.22

For coved base, add 50% to material and labor.

Exterior Stucco

Exterior stucco is generally composed of a portland cement base. The scratch and brown coats should be mixed in the proportions of 1 part of portland cement by weight to 3 parts of clean sand by weight. Do not add more than 8 lbs. of lime to each 100 lbs. of portland cement used in the mixture. Portland cement plaster should not be applied when the outside temperature is below 32°F.

The cost of portland cement stucco will vary according to the materials used and the grade of workmanship required.

Applying Scratch Coat of Portland Cement Stucco. Where just an ordinary grade of workmanship is required, a plasterer should apply 120 to 140 sq. yds. of scratch coat per 8-hr. day on metal lath.

Applying Brown Coat of Portland Cement Stucco. A plasterer should apply 75 to 85 sq. yds. of brown coat per 8-hr. day, on brick, tile or concrete block walls and 60 to 70 sq. yds. per 8-hr. day on metal lath.

Applying Trowel Finish Coat of Portland Cement. A plasterer should apply 45 to 55 sq. yds. of portland cement trowel finish per 8-hr. day, where just an ordinary grade of workmanship is required and 35 to 40 sq. yds. per 8-hr. day, where first grade workmanship is required.

Applying Wet Rough Cast Finish to Portland Cement Stucco. If a wet rough cast finish is used and the mortar and aggregate are thrown against the wall with a paddle or similar tool, a plasterer should complete 30 to 40 sq. yds. per 8-hr. day.

Applying Pebble Dash or Dry Rough Cast. When applying a pebble dash or dry rough cast finish where the aggregate is thrown against the wet cement or "butter" coat, a plasterer should apply 35 to 40 sq. yds. per 8-hr. day.

Washing Exterior Stucco With Acid to Expose Crystals. Where the finish coat of stucco contains granite, marble or crystal screenings, it is washed off with a solution of muriatic acid to expose the crystals. A man should wash 475 to 525 sq. ft. (53 to 58 sq. yds.) per 8-hr. day.

On some jobs it will be necessary to wash the walls two or three times to bring out the crystals satisfactorily, and in such instances the labor cost should be increased accordingly.

Material Cost of 100 Sq. Yds. 3-Coat 1:3 Portland Cement Stucco, Float Finish, Applied to Metal Lath Over Wood Framing

Add for metal lath.

Ordinary Workmanship

	Rate	Total	Rate	Total
Scratch Coat--1:3				
8 sacks portland cement	$....	$....	$4.24	$33.92
60 lbs. hydrated lime	....	....	8.50	5.10
24 cu. ft. sand	....	....	.48	11.52
Brown Coat--1:3				
16 sacks portland cement	....	....	4.24	67.84
120 lbs. hydrated lime	....	....	8.50	10.20
48 cu. ft. sand	....	....	.48	23.04
Float Finish--1:3				
5 sacks w'p'f. portland cement	....	....	5.30	26.50
15 cu. ft. sand	....	....	.48	7.20
Cost 100 sq. yds.		$....		$185.32
Cost per sq. yd.		....		1.85

Labor Cost of 100 Sq. Yds 3-Coat 1:3 Portland Cement Stucco, Float Finish, Applied to Metal Lath Over Wood Framing

Add for metal lath.

Ordinary Workmanship

	Hours	Rate	Total	Rate	Total
Scratch Coat--1:3					
Plasterer	6	$....	$....	$21.59	$129.54
Labor	6	....	....	19.13	114.78
Brown Coat--1:3					
Plasterer	12	....	....	21.59	259.08
Labor	8	....	....	19.13	153.04
Float Finish--1:3					
Plasterer	12	....	....	21.59	259.08
Labor	6	....	....	19.13	114.78
Cost 100 sq. yds.			$....		$1,030.30
Cost per sq. yd.			....		10.30

If a smooth troweled finish is desired, add 4 hrs. plasterer time.
If a textured finish coat is desired, add 2 hrs. plasterer time.
If first grade workmanship is required, add 8 hrs. plasterer time to the above.

Add for scaffold.

Material Cost of 100 Sq. Yds. 3-Coat 1:3 Portland Cement Stucco, Float Finish, Applied Over Brick, Clay Tile or Concrete Block Surfaces
Ordinary Workmanship

	Rate	Total	Rate	Total
Scratch Coat--1:3				
6 sacks portland cement	$....	$....	$4.24	$25.44
45 lbs. hydrated lime	....	....	8.50	3.83
18 cu. ft. sand	....	....	.48	8.64
Brown Coat 1:3				
10 sacks portland cement	....	....	4.24	42.40
75 lbs. hydrated lime	....	....	8.50	6.38
30 cu. ft. sand	....	....	.48	14.40
Float Finish 1:3				
5 sacks w'p'f. portland cement	....	....	5.30	26.50
15 cu. ft. sand	....	....	.48	7.20
Cost 100 sq. yds.		$....		$134.79
Cost per sq. yd.		....		1.35

Labor Cost of 100 Sq. Yds. 3-Coat 1:3 Portland Cement Stucco, Float Finish, Applied Over Brick, Clay Tile or Concrete Block Surfaces
Ordinary Workmanship

	Hours	Rate	Total	Rate	Total
Scratch Coat--1:3					
Plasterer	6	$....	$....	$21.59	$129.54
Labor	5	....	....	19.13	95.65
Brown Coat 1:3					
Plasterer	10	....	....	21.59	215.90
Labor	6	....	....	19.13	114.78
Float Finish 1:3					
Plasterer	12	....	....	21.59	259.08
Labor	6	....	....	19.13	114.78
Cost 100 sq. yds.			$....		$929.73
Cost per sq. yd.			....		9.30

For smooth troweled finish, add 4 hrs. plasterer time.
If a textured finish coat is desired, add 2 hrs. plasterer time.
If first grade workmanship is required, add 8 hrs. plasterer time to the above.

Add for scaffold.

Special Finishes for Portland Cement Stucco

If any of the following special finishes are wanted with portland cement stucco, deduct the finish coat given above and add cost of the finish coat desired.

Material Cost of 100 Sq. Yds. of White Cement Float Finish

Ordinary Workmanship	Rate	Total	Rate	Total
5 sacks w'p'f. white cement	$....	$....	$8.50	$42.50
1,500 lbs. white silica sand	....	....	1.06	15.90
Cost 100 sq. yds.		$....		$58.40
Cost per sq. yd.		....		.58

Labor Cost of 100 Sq. Yds. of White Cement Float Finish

Ordinary Workmanship	Hours	Rate	Total	Rate	Total
Plasterer	12	$....	$....	$21.59	$259.08
Labor	6	....	....	19.13	114.78
Cost 100 sq. yds.			$....		$373.86
Cost per sq. yd.			....		3.74

Material Cost of 100 Sq. Yds. Finish Coat Colored Stucco, Float Finish

Ordinary Workmanship	Rate	Total	Rate	Total
2 sacks w'p'f white cement	$....	$....	$8.50	$17.00
100 lbs. fin. hydrated lime	....	....	8.50	8.50
600 lbs. white silica sand	....	....	1.06	6.36
18 lbs. color	....	....	1.06	19.08
Cost 100 sq. yds.		$....		$50.94
Cost per sq. yd.		....		.51

Labor Cost of 100 Sq. Yds. Finish Coat Colored Stucco, Float Finish

Ordinary Workmanship	Hours	Rate	Total	Rate	Total
Plasterer	12	$....	$....	$21.59	$259.08
Labor	6	....	....	19.13	114.78
Cost 100 sq. yds.			$....		$373.86
Cost per sq. yd.			....		3.74

If a smooth troweled finish is desired, add 4 hrs. plasterer time.
If a textured finish is desired, add 2 hrs. plasterer time.
If first grade workmanship is required, add 8 hrs. plasterer time to the above.

Material Cost of 100 Sq. Yds. Wet Rough Cast Finish Coat

	Rate	Total	Rate	Total
4 sacks w'p'f. portland cement	$....	$....	$5.30	$21.20
12 cu. ft. sand	....	....	.48	5.76
500 lbs. aggregate	....	....	6.36	31.80
Cost 100 sq. yds.		$....		$58.76
Cost per sq. yd.		....		.59

Labor Cost of 100 Sq. Yds. Wet Rough Cast Finish Coat

	Hours	Rate	Total	Rate	Total
Plasterer	20	$....	$....	$21.59	$431.80
Labor	10	....	....	19.13	191.30
Cost 100 sq. yds.			$....		$623.10
Cost per sq. yd.			....		6.23

Material Cost of 100 Sq. Yds. of Dry Rough Cast Finish

	Rate	Total	Rate	Total
4 sacks w'p'f. portland cement	$....	$....	$5.30	$21.20
12 cu. ft. sand	....	....	.48	5.76
750 lbs. aggregate	....	....	6.36	47.70
Cost 100 sq. yds.		$....		$74.66
Cost per sq. yd.		....		.75

Labor Cost of 100 Sq. Yds. of Dry Rough Cast Finish

	Hours	Rate	Total	Rate	Total
Plasterer	20	$....	$....	$21.59	$431.80
Labor	10	....	....	19.13	191.30
Cost 100 sq. yds.			$....		$623.10
Cost per sq. yd.			....		6.23

A dry rough cast finish requires a finish or "butter" coat 1/4" thick applied directly over the brown coat. This is brought to a straight, smooth finish and then the aggregate is thrown onto the "butter" coat dry.

If coarse aggregate is used, it requires about 1,000 lbs. per 100 sq. yds.

If medium size aggregate is used, it requires 750 lbs. per 100 sq. yds.

If fine aggregate is used, it requires 500 lbs. per 100 sq. yds.

Approximate Prices on Lathing and Plastering Materials

Metal Lath

Description	Finish	Wt. Lbs. Sq. Yd.	Price Sq. Yd.
Expanded diamond mesh (flat)	Painted	2.5 lbs.	$1.30
Expanded diamond mesh (flat)	Painted	3.4 lbs.	1.60
Expanded diamond mesh (flat)	Galvanized	3.4 lbs.	1.85

Self-furring lath, add 10 cents per sq. yd. to the above prices.

Rib Metal Lath

Description	Finish	Height of Rib	Wt. Lbs. Sq. Yd.	Price Sq. Yd.
Flat rib lath	Painted	1/8"	2.75	$1.60
Flat rib lath	Painted	1/8"	3.4	1.70
Rib or rod ribbed lath	Painted	3/8"	3.4	2.30
Rib or rod ribbed lath	Painted	3/8"	4.0	2.55

Hot Rolled Channels

Width	Leg	Weight per 1,000 Ft.	Price per 1,000 LF
3/4" Standard	15/32"	300	$175.00
1" Standard	3/8"	410	210.00
1 1/2" Standard	19/32"	650	265.00
2" Standard	7/16"	1,260	330.00

Peforating, add $10.00 per 1,000 ft.

Cold Rolled Channels

Width	Leg	Weight per 1,000 Ft.	Price per 1,000 LF
3/4"	1/2"	300 lbs.	$170.00
1 1/2"	19/32"	475 lbs.	255.00
2"	19/32"	590 lbs.	320.00

Wire

	Per Cwt.
16 Ga	$85.00
18 Ga	$95.00

Studs & Tracks
20 Ga.

Size	Per 1,000 Lin. Ft.
1⅝"	$320.00
2½"	340.00
3¼"	370.00
4"	390.00
6"	500.00
Shoes (per 1,000 pcs.)	80.00

Corner Bead

Type	Per 1,000 Lin. Ft.
Regular	$160.00
Expanded	185.00
Bull Nose	370.00
Cornerite	105.00

Rods

Type	Per 1,000 Lin Ft.
3/16" plain	$48.00
3/16" galvanized	58.00
¼" plain	64.00
¼" galvanized	74.00
⅜" plain	80.00

Gypsum Lath

Description	Per 1,000 Sq. Ft.
Gypsum lath, plain or perforated, 16"x48"x⅜"	$110.00
Gypsum lath, foilback, 16"x48"x⅜"	195.00
Gypsum lath, plain, 16"x48"x½"	115.00
Gypsum lath, foilback, 16"x48"x½"	200.00

Nails and Staples

Description	Price per 100 Lbs.
Gypsum lath, 1⅛" blued	$95.00
Gypsum lath, 1¼" blued	95.00
Roofing, barbed, 1½" galv	105.00
Concrete stub	120.00
Annular ring nail (blued)	95.00
Brick--plain 2" and 2½"	105.00

Lime and Finishing Plaster

Description	Price per 100 Lbs.
Pulverized Quicklime	$7.40
Plasterer's Hydrated Lime	8.50
Gypsum Gauging Plaster	9.50
No. 1 Moulding Plaster	10.60
Keene's Cement, Regular or Fast	16.96
Portland Cement	4.24

Plastering Items

Liquid Bonding Agent	$26.50 per Gal.

Prepared Finishes

Hair (4 Lb. to Bu.)	$8.50 per Bu.
Fiber (2 Lb. to Bu.)	8.50 per Bu.

Gypsum Plaster

	Per 100 Lbs.
Gypsum Cement Plaster	$6.40
Wood Fiber (prepared)	6.40
Bondcrete (for concrete)	7.40
Perlite Plaster	12.70
Gypsum Trowel Finish	7.40
Silica Sand Float	9.50

09250 GYPSUM WALLBOARD

Gypsum wallboard, or drywall, is manufactured from a gray-white colored rock called gypsum, which is a nonmetallic mineral composed of calcium sulphate chemically combined with crystallized water.

After the gypsum ore is mined, it is crushed, dried, and ground to a fine powder. It is then heated to remove most of the chemically combined water. The calcined gypsum is then mixed with other ingredients and water and sandwiched between two sheets of treated paper to form a smooth gypsum wallboard panel. After the gypsum core has achieved a set, it is cut to length, dried, finished, and packaged for shipment.

Advantages. Gypsum wallboard is fire resistant and will not support combustion. When one surface is exposed to a flame, the opposite panel surface remains cool until the gypsum core is calcinized.

The panel is easily cut and quickly hung in large sheets. This tends to promote an earlier completion and occupancy of the building.

Since the wallboard is a dry panel, less moisture is introduced into the building, and cold weather is less of a factor for a completed finished surface. Only the joint finishing must be kept from freezing.

The strong, smooth face paper of the gypsum wallboard panel is suitable for any decorative treatment such as paint, wallpaper, and textured coatings.

Size and Thickness. Gypsum wallboard panels are manufactured in a number of thicknesses and sizes with various longitudinal edge designs.

Size	Thickness	Edge Design
4'-0"x8'-12'-0"	1/4"	S.E.*, T.E.**
4'-0"x8'-12'-0"	3/8"	S.E., T.E.
4'-0"x8'-12'-0"	1/2"	S.E., T.E.
4'-0"x8'-12'-0"	5/8"	S.E., T.E.

*Square Edge
**Tapered Edge

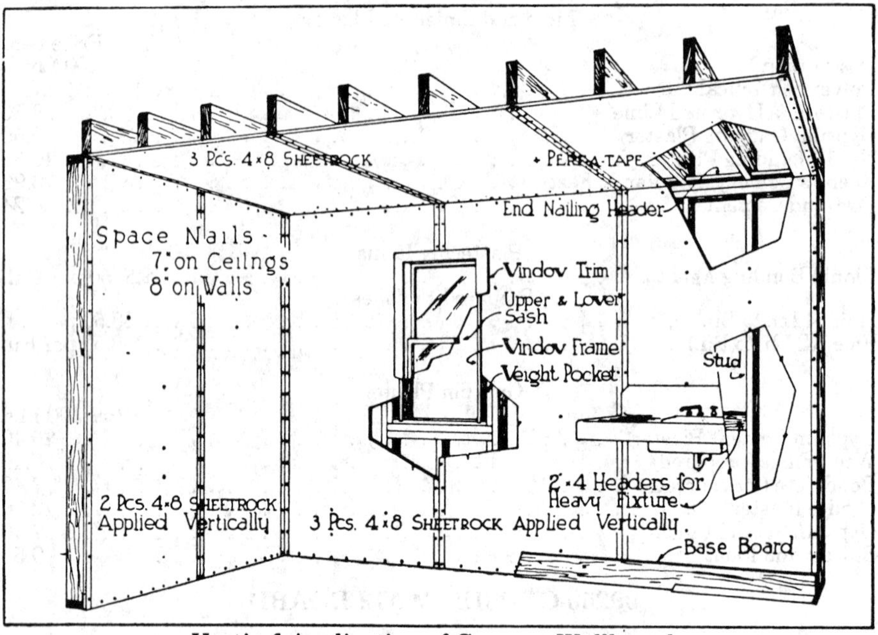

Vertical Application of Gypsum Wallboard

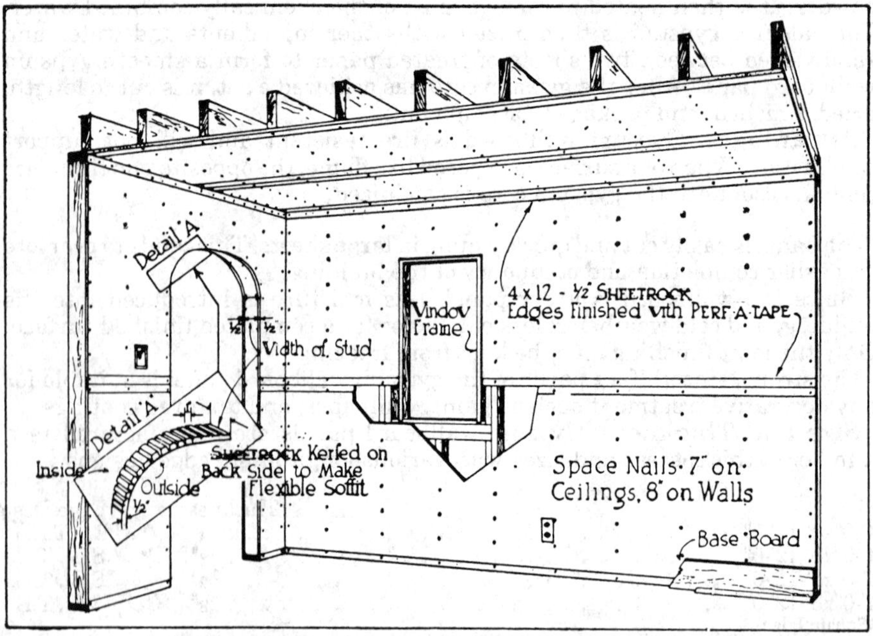

Horizontal Application of Gypsum Wallboard

Square edge panels may be used as the first layer of a two layer system or where the joints will be covered by a batten strip. The tapered edge panel should always be used on the finished surface layer, where the joints are to be finished with tape and compound. The tapered edge is a depression along the longitudinal edge that receives the tape and compound build-up to smooth and hide the butting edges of the gypsum panels.

The 1/4" gypsum wallboard is used as a lightweight, low cost, utility wallboard. It is also used for curved surfaces or to cover existing wall surfaces. The 3/8" gypsum wallboard is used principally for repair or remodeling work or in double wall construction. One-half inch gypsum wallboard is used in single layer new construction, and the 5/8" gypsum wallboard is used where a one-hour fire rating is required or where the framing is spaced in excess of 1/2" wallboard limitations.

Perforated Tape Joint System for Gypsum Wallboard. This method of concealing joints in gypsum wallboard is backed by over 20 years of joint reinforcing experience. It conceals the joints between boards and bonds the gypsum wallboard units together into a single, smooth, even wall and ceiling surface unit.

First, the hollow or channel at the edges of the wallboard is filled with compound, using a 4" or 5" joint finishing broadknife.

Apply the perforated tape immediately, directly over the compound and press it into place with the broadknife, squeezing excess compound out from under the tape but leaving enough compound under tape for proper bond. Apply thin covering coat of compound immediately and let dry.

Next, apply another thin coating of compound so that the tape will be completely hidden. Feather out edges beyond previous coat as smoothly as possible and let dry thoroughly. Intermediate nail heads should also be carefully filled with compound and brought flush with the surface of the board. Apply third and finish coat of compound, of thinner consistency to even up surface and edges of joint.

Sand, as necessary, between and after coats to insure smooth, inconspicuous joint.

Foil-Back Gypsum Panels. These panels are made by laminating a sheet of aluminum foil to the back surface of the gypsum wallboard. The foil reduces outward heat flow in winter and inward heat flow in summer, has a significant thermal insulating value if facing an air space of 3/4" or more, and is effective as a vapor barrier.

Moisture Resistant Gypsum Panels. MR board has a special asphalt composition gypsum core and is covered with chemically treated face papers to prevent moisture penetration. This board is a little harder to cut, has a brownish color core, and is usually covered with a light green finish face paper. These panels were developed for application in bathrooms, kitchens, utility rooms, and other high moisture areas.

Fire Rated Gypsum Panels. A specially formulated mineral gypsum core is used to make panels in 1/2" and 5/8" thicknesses for application to walls, ceilings, and columns where a fire rated assembly is needed. Based on tests by Underwriters' Laboratories, Inc., certain wall, floor/ceiling, and column assemblies give 45 minute to 4 hour fire resistance ratings.

Drywall Accessories. The following items will be necessary to provide a complete finish for most any drywall job:

Fastener Description	Nail Spacing c. to c.	Approx. Lbs. Nails Req'd per MSF Gypsum Panels
1¼″ GWB-54 Annular Ring Nail 12½ ga.; ¼″ dia. head with a slight taper to a small fillet at shank; bright finish; medium diamond point; meets ASTM C380 	7″ ceiling 8″ walls	5¼
1⅜″ Annular Ring Nail (Same as GWB-54 except for length) 	7″ ceiling 8″ walls	5¼

Fastening Application	Fastener Used
GYPSUM PANELS TO STANDARD METAL FRAMING	
½″ single-layer panels to standard studs, runners, channels	 ⅞″ Type S Bugle Head
⅝″ single-layer panels to standard studs, runners, channels	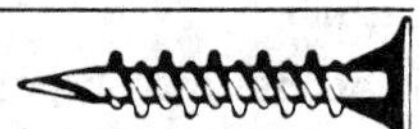 1″ Type S Bugle Head
½″ double-layer panels to standard studs, runners, channels	 1⁵⁄₁₆″ Type S Bugle Head
⅝″ double-layer panels to standard studs, runners, channels	 1⅝″ Type S Bugle Head
1″ coreboard to metal angle runners in solid partitions	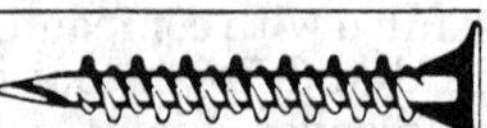 1¼″ Type S Bugle Head
½″ panels through coreboard to metal angle runners in solid partitions	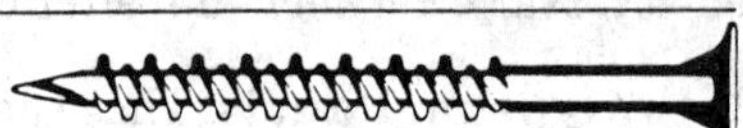 1⅞″ Type S Bugle Head
⅝″ panels through coreboard to metal angle runners in solid partitions	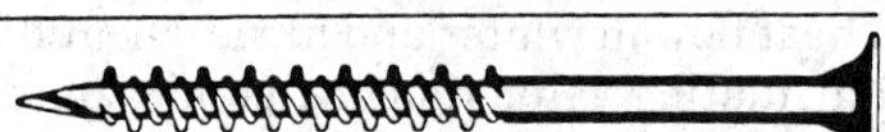 2¼″ Type S Bugle Head
GYPSUM PANELS TO 12-GA. (MAX.) METAL FRAMING	
½″ and ⅝″ panels and gypsum sheathing to 20-ga. studs and runners	 1″ Type S-12 Bugle Head
USG Self-Furring Metal Lath through gypsum sheathing to 20-ga. studs and runners	 1¼″ Type S-12 Bugle Head
½″ and ⅝″ double-layer gypsum panels to 20-ga. studs and runners	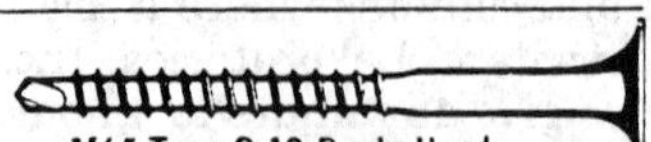 1⅝″ Type S-12 Bugle Head
Multi-layer gypsum panels to 20-ga. studs and runners	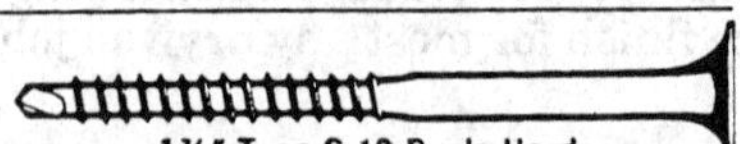 1⅞″ Type S-12 Bugle Head

1) Fasteners: Nails or screws may be used for attachment of gypsum panels to wood framing, however screws must be used for attachment to metal framing members.
2) Corner Bead: Metal corner bead, furnished in lengths of 8' and 10', are used at all external corners in drywall construction to protect and prevent damage to the gypsum wallboard. This metal bead also provides a true and straight finish line to the corner.

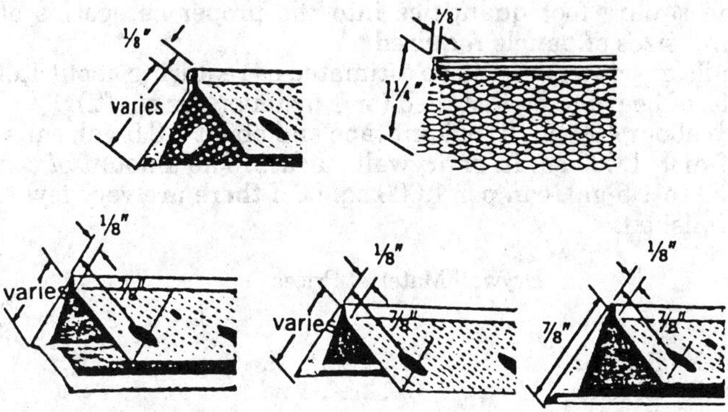

3) Stop Bead: Metal stop beads are used to cap the edge of a gypsum panel at its termination point, whenever that edge is exposed and not to be covered by either a tape/compound finish or other trim members.

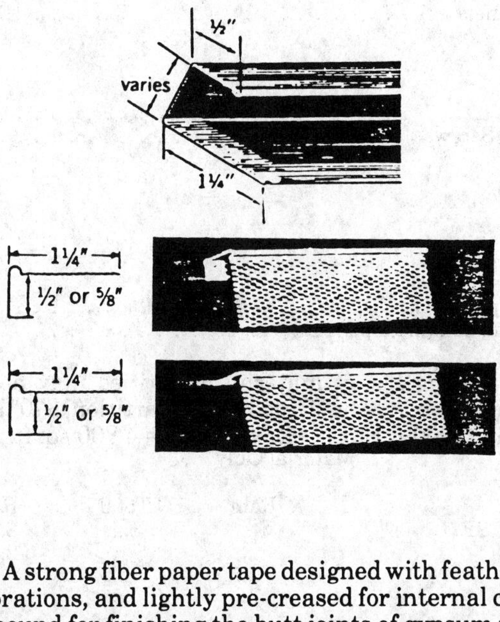

4) Joint Tape: A strong fiber paper tape designed with feathered edges, random pin-hole perforations, and lightly pre-creased for internal corners is used with the joint compound for finishing the butt joints of gypsum panels. This tape is packaged and sold in rolls of 250 and 500 lin. ft. per roll.
5) Joint Compound: This material is available in many different forms, e.g., powdered, ready-mixed, all-purpose, fast setting, exterior application, etc. The

ready-mixed all-purpose joint compound is used extensively for all interior applications of taping and joint finishing, and is readily available in 5 gallon cans weighing about 68 lbs. each.

Estimating Material Quantities. The estimator would first measure the wall and ceiling surfaces to ascertain the required square footage of surface to be covered. Do not deduct for openings smaller than 50 sq. ft.

Separate the square foot quantities into the proper categories of types, thicknesses, and sizes of panels required.

Fasteners (nails or screws) should be estimated as requiring about 1,000 each per 1,000 sq. ft. of board to be installed (or 1 fastener per sq. ft.).

It will require about 400 lin. ft. of joint tape and about 1.3 5-gal. cans of joint compound to finish 1,000 sq. ft. of drywall surface. The amount of compound may be reduced to 1 5-gal. can per 1,000 sq. ft. if there are very few external corners to be finished.

Drywall Material Prices

	Price per 1,000 SF
1/4" Gypsum Panels - Regular	$108.00
3/8" Gypsum Panels - Regular	113.00
1/2" Gypsum Panels - Regular	117.00
5/8" Gypsum Panels - Regular	133.00
1/2" Gypsum Panels - Fire Rated	131.00
5/8" Gypsum Panels - Fire Rated	143.00
1/2" Gypsum Panels - Moisture Resistant	170.00
5/8" Gypsum Panels - Moisture Resistant	182.00
Add: For Foil Back Panels	80.00

	Price per 1,000 Pcs.
1¼" GWB Ring Nails	$2.15
1" Type S Bugle Head Screws	5.80

	Price per Lin. Ft.
Corner Bead	$.10
Stop Bead	.12
Joint Tape - 250' roll (price per roll)	.95
Joint Compound - 5 gal. can (price per can)	8.50

Example: The following list of material quantities may be required for installation and finishing of 300 lin. ft. of drywall partition, 8'-0" high, covering both sides of partition (300 L.F. x 8' Ht. x 2 sides = 4,800 sq. ft.).

"Material Only"

Item	Rate	Total	Rate	Total
1/2" Drywall Reg. 4,800 SF	$....	$....	$.12	$576.00
1¼" Nails 4.8 M Pcs.	....	....	2.12	10.18
Corner Bead 96 L.F.	....	....	.08	7.68
Stop Bead 40 L.F.	....	....	.11	4.40
Joint Tape-8 rolls	....	....	.95	7.60
Joint Compound - 6 5-gal cans	....	....	8.50	51.00
Cost for 4,800 sq. ft.		$....		$656.86
Cost per sq. ft		....		.14

Labor Placing Gypsum Wallboard. When placing 1/2" gypsum wallboard in average size rooms, 2 carpenters experienced in wallboard erection should place about 3,000 sq. ft. of board per 8-hr. day, at the following labor cost per sq. ft.:

	Hours	Rate	Total	Rate	Total
Carpenter	16	$....	$....	$24.04	$384.64
Cost per sq. ft.			....		.13

When placing 1/2" gypsum wallboard in small rooms and spaces requiring considerable cutting and fitting in proportion to the number of square feet of board placed, 2 carpenters should place about 2,000 sq. ft. per 8-hr. day, at the following labor cost per sq. ft.:

	Hours	Rate	Total	Rate	Total
Carpenter	16	$....	$....	$24.04	$384.64
Cost per sq. ft.			....		.19

Where fire resistant 5/8" material is specified, 2 carpenters should place about 2,500 sq. ft. per 8 hr. day on straight wall and ceiling construction at the following labor cost per sq. ft.

	Hours	Rate	Total	Rate	Total
Carpenter	16	$....	$....	$24.04	$384.64
Cost per sq. ft.			....		.15

Where gypsum wallboard is used to fireproof beams and columns, 2 carpenters should place about 1,200 sq. ft. per 8 hr. day at the following labor cost:

	Hours	Rate	Total	Rate	Total
Carpenter	16	$....	$....	$24.04	$384.64
Cost per sq. ft.			....		.32

Labor Finishing Gypsum Wallboard. The finishing of gypsum wallboard is a multi-step process of tape coat, block coat, skim coat, and point-up coat. Sanding between the block and skim coats is not required if the finisher is careful and a first class mechanic. The complete finishing labor for 1,000 sq. ft. of wallboard is as follows:

	Hours	Rate	Total	Rate	Total
Tape Coat	2.0	$....	$....	$24.04	$48.08
Block Coat	1.6	....	....	24.04	38.46
Skim Coat	1.3	....	....	24.04	31.25
Sand	0.4	....	....	24.04	9.62
Point-Up	0.8	....	....	24.04	19.23
Cost per 1,000 sq. ft			$....		$146.64
Cost per sq. ft			....		.15

Metal Studs and Furring

Stud framing and wall or ceiling furring on commercial projects is normally accomplished with light gauge metal studs, runner channels, and furring members instead of wood framing.

This material is manufactured from cold rolled galvanized metal and is available in a number of sizes, shapes, and gauges.

Metal Stud Partitions. Runner channels are channel shaped members that are positioned at the top and bottom of the studs, as in top and bottom plates for a wood stud partition. The runner channels are attached to the floor and overhead structure with nails, screws, or powder actuated drive pins.

The metal studs, which are channel shaped with a backbend to give stiffness, are placed within the web of the runner channels, located for "on-center" spacing, plumbed, and screwed into place with a ⅜" pan head screw, top and bottom. The web of the metal studs have cutouts for the passage of conduit and piping.

Runner channels and studs are made in 1⅝", 2½", 3⅝", 4", 6" widths in 25 ga., 2½", 3⅝", 4", 6" widths in 20 ga., and 4", 6" widths in 18 ga. and 16 ga. metals.

The 25 ga. runners and studs are usually specified for non-loadbearing partitions up to 14'-16' in height. Over this height and up to 22', the 20 ga. material is used. The 18 ga. and 16 ga. materials are used for greater heights, loadbearing walls, and for exterior wall construction where lateral pressure (wind loads) will be imposed, therefore, requiring more stiffness and less deflection.

Metal Furring. This material is made from 25 gauge metal in the following shapes:

1) Z furring channel is used in conjunction with rigid insulation board when furring exterior wall surfaces.

2) The hat shaped furring channel is used for furring walls and ceilings when insulation board is not required.

3) The resilient channel is used primarily on ceilings to provide a separation between the gypsum panels and the framing members for the floor above. It is used extensively in wood framed garden apartments for noise dampening between apartment units.

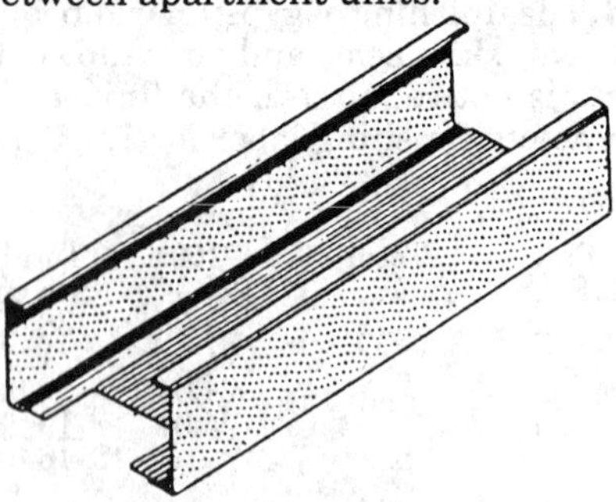

Metal Stud

Metal Runner

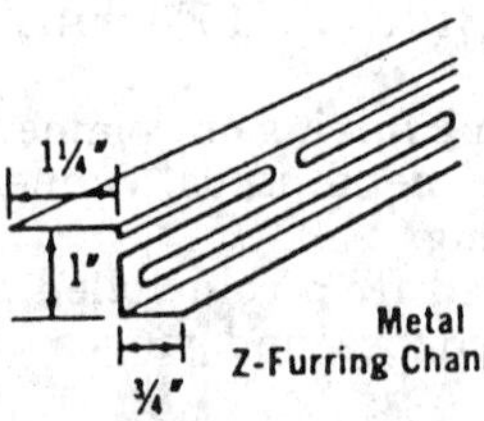

Metal Z-Furring Channel

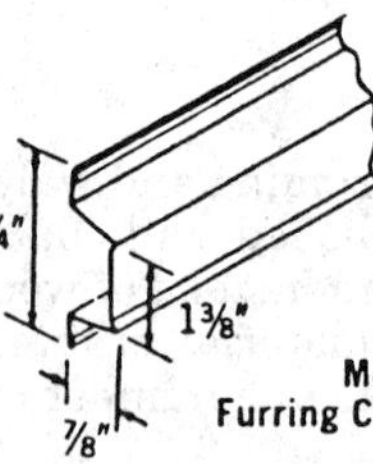

Metal Furring Channel

Estimating Material Quantities. To determine the quantity of runner channel, studs, and fasteners for a partition, the estimator would measure the total length of partition to be erected. Note the thickness and gauge of materials specified, ascertain the height of the partition for length of studs, ascertain the required spacing of studs and add extras for door or window openings and at corners, and determine what type and spacing of fasteners is required for attachment of the runner channels to the floor and overhead structure. Horizontal stud bridging or fire stops are usually not required as in wood frame partition, but be sure to check, because this item in metal is expensive to install.

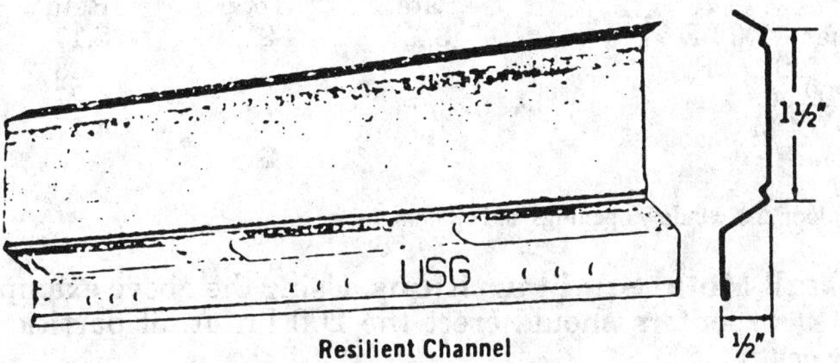

Resilient Channel

To determine the quantity of furring channel required, ascertain the on-center spacing of the members for quantity of pieces by the length of the members (height of wall, as furring members are usually vertically installed) to give total linear footage to be installed. Determine the fastener type and spacing for attachment to the furring channel.

Material Prices

	Per 1,000 Lin. Ft.
25 ga. - 1⅝" Runner Channel	138.00
25 ga. - 1⅝" Stud	148.00
25 ga. - 2½" Runner Channel	164.00
25 ga. - 2½" Stud	180.00
25 ga. - 3⅝" Runner Channel	201.00
25 ga. - 3⅝" Stud	212.00
25 ga. - 4" Runner Channel	217.00
25 ga. - 4" Stud	233.00
25 ga. - 6" Runner Channel	281.00
25 ga. - 6" Stud	302.00
25 ga. - ⅞" Hat Furring Channel	148.00
25 ga. - 1" Z Furring Channel	175.00
25 ga. - Resilient Channel	154.00
20 ga. - 2½" Runner Channel	323.00
20 ga. - 2½" Stud	339.00
20 ga. - 3⅝" Runner Channel	360.00
20 ga. - 3⅝" Stud	387.00
20 ga. - 4" Runner Channel	381.00
20 ga. - 4" Stud	403.00

20 ga. - 6" Runner Channel 488.00
20 ga. - 6" Stud 514.00

3/8" Pan Head Screws 4.75
Powder Actuated Drive Pins (avg. cost inc. shot) . . . . .40 ea.

Example. The following list of material quantities would be required for a metal stud partition 200 feet long, 10 feet high, stud at 24" on-center, 25 ga. metal, and installed with drive pins at 24" on-center, top and bottom to concrete slabs:

"Material Only"

Item	Rate	Total	Rate	Total
2½" 25 ga. Runner - 400 LF	$....	$....	$.17	$68.00
2½" 25 ga. Stud - 1,010 LF	....	....	.18	181.80
1" Drive Pins - 202 ea.	....	....	.27	54.54
3/8" Pan Screws - 0.4 M	....	....	4.77	1.90
Cost for 200 Lin. Ft		$....		$306.24
Cost per Lin. Ft.		....		1.53

Add extra studs for door and window openings and corners.

Labor to Install Metal Stud Partitions. Using the above example given for materials, 2 carpenters should erect the 200 lin. ft. of partition at the following labor cost:

	Hours	Rate	Total	Rate	Total
Layout Wall Line	1.0	$....	$....	$24.04	$24.04
Fasten Bottom Runner	4.0	....	....	24.04	96.16
Fasten Top Runner	4.0	....	....	24.04	96.16
Install Studs	3.0	....	....	24.04	72.12
Plumb & Screw Studs	4.0	....	....	24.04	96.16
Cost for 200 lin. ft			$....		$384.64
Cost Per Lin. Ft			....		1.92

09300 TILE
CERAMIC WALL AND FLOOR TILE

Ceramic tile provides a durable, colorful surface that is virtually maintenance-free. Its applications include interior and exterior finishes for functional and decorative purposes in all types of structures. Special uses include acid resistant and electrically conductive installations.

Ceramic tile is available in many sizes, shapes, and finishes. Terms of reference for the most commonly used ceramic tile are as follows.

Glazed Ceramic Wall Tile. Ceramic tile that has an impervious facial finish fused onto the body of the tile. The glazed surface comes in a wide variety of colors.

Ceramic Mosaic Tile. Ceramic tile, either glazed or unglazed, having a facial area of less than six square inches and which is usually mounted on sheets or mesh approximately 2'x 1' to facilitate setting.

Quarry Tile. A rugged ceramic tile used primarily as a finish flooring (interior and exterior) where a long wearing, easily cleanable surface is desired.

Recent developments in the ceramic tile industry make it necessary to stress the importance of relating tile costs to each individual job specification. Glazed ceramic wall tile, for example, can be backmounted or unmounted, and can be installed using conventional portland cement mortar, various types of adhesives or the more recently developed "dry-set" portland cement mortar. All of these variations can affect costs, both material and labor, and, consequently, the estimator should review job requirements carefully prior to making the estimate.

The above notwithstanding, the estimator's job has been simplified in recent years. Virtually the entire ceramic tile industry has adopted a "simplified practice", prepared under the auspices of the Tile Council of America, which has reduced the number of sizes and shapes and which has established generally recognized standards for the industry.

Estimating Quantities. Ceramic tile is estimated by the square foot, with trim pieces such as base, cap, etc. being estimated by the lineal foot. The estimator should deduct door and window openings, but the trim pieces necessary to finish the openings must be added. The quantities should be related to the type and size of ceramic tile, since these items will affect the cost of the ceramic tile when priced.

Estimate Composition. The finished ceramic tile estimate will include the following items:

a) Cost of ceramic tile delivered to the job site.
b) Cost of accessory materials such as wire mesh, sand and cement for floor fill under ceramic tile.
c) Cost of mixing and placing floor fill.
d) Direct labor cost of laying and cleaning the ceramic tile.
e) The ceramic tile contractor's overhead and profit.

Variable Factors Influencing Costs. As in the other construction trades, estimating the labor costs of setting ceramic tile requires an intimate knowledge of the labor market in which the work is to be performed. Wage rates vary throughout the country, and it is important the estimator determine the rate in his locality.

Ceramic tile is unique in that it is often necessary to install it in very small quantities (for example, in 1 or 2 bathrooms in a house), and therefore, the costs can vary greatly. However, that such small installations are the exception rather than the norm, and the costs given on the following pages are based on the assumption that the areas involved are large enough to permit the tile contractor to operate efficiently.

The costs developed on the following pages are based on $4\frac{1}{4}$"x$4\frac{1}{4}$" glazed wall tile, 1"x1" ceramic mosaic tile and 6"x6"x$\frac{1}{2}$" quarry tile, since these are the sizes most commonly employed. If other sizes are specified, substitute the material price only. For all practical purposes, the labor cost remains constant regardless of size. The developed costs do not include areas where an unusually large amount of trim is required. Therefore, the estimator should make an allowance for extra trim pieces in such special cases.

Setting Methods. Recent years have seen the development of many methods of adhering ceramic tile to a subsurface. There are, however, three methods that are most commonly used and accepted:

Conventional Portland Cement Mortar Method. This method is to bond each ceramic tile with a layer of pure portland cement paste to a portland cement setting bed. This is done while the setting bed is still plastic. Wall tile must be soaked in water so that the water needed for curing is not absorbed from the paste. While being the traditionally accepted method, it is also more costly than the other methods.

Dry-Set Portland Cement Mortar Method. This method utilizes a dry curing portland cement mortar (accomplished through the use of water retaining additives) and has made ceramic tile installation cheaper and simpler. "Dry-set" is ideally suitable for use with concrete masonry, brick, poured concrete and portland cement plaster. It should not be used over wood or gypsum plaster. Labor costs are appreciably reduced when this method is used.

Water-Resistant Organic Adhesive. Organic adhesives can be used over smooth base materials, such as wallboards, plywood, and metal. Labor productivity is comparable to that of "dry-set" mortar.

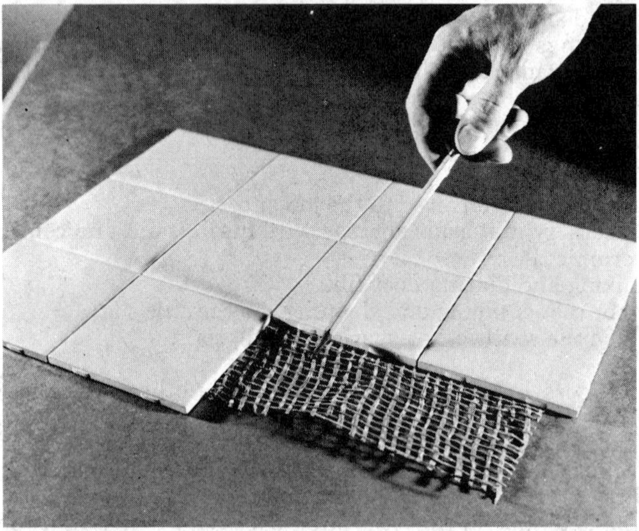

Glazed Wall Tile, $4^1/_4$ by $4^1/_4$ Inch, Mounted on Mesh Backing

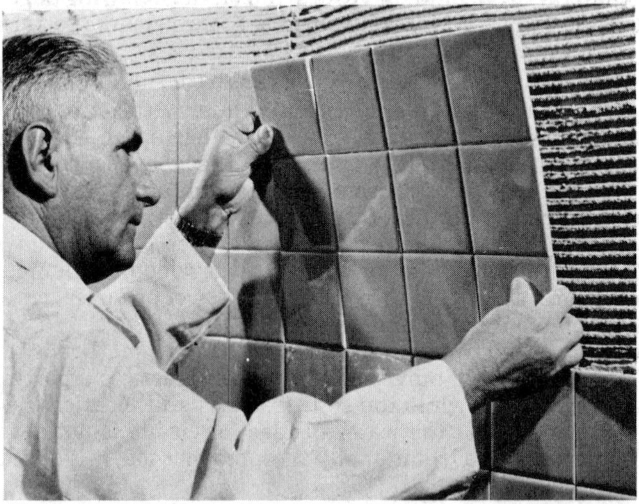

Dry-set Mortar and Back Mounted Tile Sheets

Labor Productivity
(sq. ft. per team day; team is 1 tile setter and 1 helper)

Conventional Mortar

Description	Face-mounted	Back-mounted	Unmounted
Glazed wall tile	....	75-90 SF	55-65 SF
Ceramic mosaic			
tile--walls	45-50 SF	55-65 SF	....
floors	100-125 SF	100-125 SF	....
Quarry tile floors	....	....	100-125 SF

"Dry-Set" Mortar

Description	Face-mounted	Back-mounted	Unmounted
Glazed wall tile	....	125-150 SF	100-125 SF
Ceramic mosaic			
tile--walls	100-125 SF	120-140 SF	....
floors	125-150 SF	125-150 SF	....
Quarry tile floors	....	....	125-150 SF

Organic Adhesive

Description	Face-mounted	Back-mounted	Unmounted
Glazed wall tile	....	150-175 SF	120-140 SF
Ceramic mosaic			
tile--walls	120-140 SF	125-150 SF	....
floors	150-175 SF	175-200 SF	....
Quarry tile floors	....	....	....
Cove Or Base	65-75 LF		
Cap	80-90 LF		

Approximate Prices of Ceramic Tile Materials
Ceramic Wall Tile

Description	Size	Price per sq. ft.
Flat tile, all colors, incl. white	4¼"x4½"	$1.35
	6"x6"	1.45
	6"x4¼"	1.50
	6"x9"	1.65
	8½"x4¼"	1.55

For use with 4¼"x4¼" tile	Size	Price per Piece*
Cove	6"x3¼"	$.85
Cove	4¼"x4¼"	.85
Bullnose	2"x6"	.85
Bullnose	4¼"x4¼"	.90
Double bullnose	4¼"x4¼"	1.30
For use with 6"x6" tile		
Cove	6"x6"	$1.15
Bullnose	6"x6"	1.20
Double bullnose	6"x6"	1.50
Miscellaneous trim shapes		
Bead	6"x¾"	$.85
Bullnose	6"x3¾"	1.15
Base	6"x4"	1.30
Base	6"x6"	1.35
Sink Trim	6"x2"	1.30
Window Sill	9"x5⅛"	1.50

*Prices are for stretchers; for angles, multiply prices by two.

Ceramic Mosaic Tile Description
Unglazed, Modular, Solid Color

Description White & Grays	Price per sq. ft.	Other Colors
1"x1"	$1.30	$1.50
2"x1"	1.40	1.60
2"x2"	1.50	1.70
¾"x¾"	1.30	1.50
1 9/16"x¾"	1.30	1.50
1 9/16"x1 9/16"	1.30	1.50
Trim Shapes (price per lineal foot)		
1"x1" Bullnose		$1.40
2"x1" Bullnose		1.60

¾" Series trim pieces same price as 1" Series

Various manufacturers furnish, as standard items, patterns made from combinations of modular tile sizes and colors. Prices vary depending on number of different sizes, number of colors, and depth of colors used. Patterns such as these are too numerous to itemize, but the general range is from $1.35 to $1.70 per sq. ft. There are exceptions to this range however, and when a specific pattern price is needed, the estimator should check with the specified supplier.

	Price per sq. ft.
Abrasive Ceramic Mosaic Tile	$1.45 to 1.80
Conductive Floor Tile	2.55 to 3.20
Granitized Ceramic Mosaic Tile	1.35 to 1.55
Abrasive Granitized Ceramic Mosaic Tile	1.45 to 1.70
Glazed Satin Finish Ceramic Mosaic Tile	1.40
Glazed Textured Finish Ceramic Mosaic Tile	1.60
Special Decorative Glazed Ceramic Mosaic Tile for Accent Effects	3.20 to 13.80

Venetian glass mosaics mounted on sheets 12¼"x12¼" square run $2.65 a sq. ft. for ¾"x¾", $2.90 for 1"x1" and $3.10 for 1½"x1½". Metallic finishes run considerably more. Beads and coves for glass mosaics will run $2.15 a lin. ft.

Sculptured tiles which can be used for wall surfacing, murals, screens, and sculpture are available in a wide range of ceramic designs and treatments. Design-Technics of New York have tiles in sizes 4¼"x8½"x¼" at approximately $4.25 to $5.30 a sq. ft., 12"x12"x⅜" at $5.30 to $6.65 a sq. ft. and 12"x12"x1⅜" at $6.90 to $8.20 a sq. ft.

Quarry Tile

Description, Size	Price per sq. ft.
Deep red, plain surface,	
6"x6"x½"	$1.30
6"x2¾"x½"	1.35
6"x6"x¾"	2.00
9"x9"x¾"	3.25
Deep red, abrasive,	
6"x6"x¾"	2.25
6"x6"x¾"	2.75

Note: Other colors, with the exception of green, are virtually the same price as the red shown above. Green in the 6"x6"x½" size costs $2.15 per sq. ft.

Extras, when specified, are:

	Price per sq. ft.
Tile ground square after burning	$0.10
Wax coating	0.22

	Price per M pieces
Floor brick, deep red, plain	
tray packed, 8"x4"x1⅜	$175.00
in sealed cartons	206.00
Floor brick, deep red, abrasive	
tray packed, 8"x4"x1⅜"	190.00
in sealed cartons	222.00

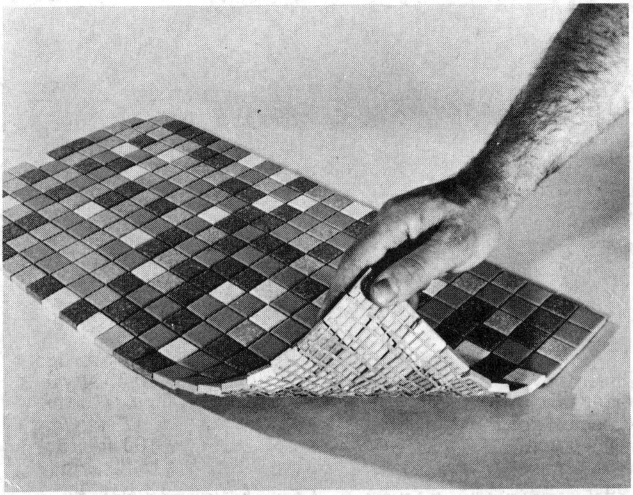

Ceramic Mosaic Tile, 1"x1" Tile, Back Mesh-Mounted

There are many specially cast heavy floor and wall tiles that compliment the popular provincial and country modes. Many of these are quite expensive and are not regularly stocked. Always check locally for availability and prices, and for similar tiles that might be allowed for substitution should this be necessary.

Ceramic Tile Bathroom Accessories

Numerous ceramic tile bathroom accessories are available in a variety of sizes and qualities. They may be recessed or surface mounted, and they come in the full range of tile colors. Some of the more commonly used items are as follows:

Description	Price per piece
Recessed soap holder	$12.75
Recessed glass holder	12.75
Roll paper holder	15.90
Robe hook	4.25
Double robe hook	4.75
Toothbrush holder	6.36
Tile bar brackets (pair)	8.50

Decorative Ceramic Wall Tiles

Glazed ceramic wall tile containing designs as an integral part of each piece of tile is also available. Normally supplied in 4¼"x4¼" and 6"x6" sizes, the price of this material will vary with the design selected and number of colors in the design. If the design is known, the estimator should obtain a firm price from a supplier. The general range is as follows:

	Price per sq. ft.
One color,	
4¼"x4¼"	$2.65
6"x6"	2.50
Two colors,	
4¼"x4¼"	3.20
6"x6"	3.00
Three colors,	
4¼"x4¼"	3.70
6"x6"	3.45

Ceramic Tile Adhesives and Accessory Materials

	Coverage	Price
Organic adhesive	40-50 sf/gal	$8.50/gal.
Dry-set mortar mix	35 lbs./100 sf	.50/lb.
Primer (for use in damp areas)	125 sf/gal.	6.50/gal.
Plastic underlayment	100 sf/unit*	8.50/unit
Wet tile grout	15-20 sf/lb.**	.30/lb.
Dry tile grout	12-15/sf/lb.**	.35/lb.

*Unit consists of 1 gallon liquid & 34 lbs. aggregate.

**When used with 4¼"x4¼" tile.

Material Cost of 100 sq. ft. of 4¼"x4¼" Glazed Ceramic Wall Tile Using Conventional Mortar Method of Installation (Unmounted Tile)

	Rate	Total	Rate	Total
2 sacks portland cement	$....	$....	$4.24	$8.48
4 cu. ft. sand	....	....	.48	1.92
100 sq. ft. glazed wall tile	....	....	1.33	133.00
7 lbs. wet tile grout mix	....	....	.27	1.89
Cost per 100 sq. ft		$....		$145.29
Cost per sq. ft		....		1.45

Labor Cost of 100 sq. ft. of 4¼"x4¼" Glazed Ceramic Wall Tile
Using Conventional Mortar Method of Installation (Unmounted Tile)

	Hours	Rate	Total	Rate	Total
Tile Setter	13.5	$....	$....	$19.13	$258.26
Cost per sq. ft			....		2.58

Material Cost of 100 sq. ft. of 4¼"x4¼" Glazed Ceramic Wall Tile
Using Water Resistant Organic Adhesive (Unmounted Tile)

	Rate	Total	Rate	Total
2.5 gallons adhesive	$....	$....	$8.50	$21.25
100 sq. ft. glazed wall tile	....	....	1.33	133.00
9 lbs. dry tile grout mix	....	....	.37	3.33
Cost per 100 sq. ft		$....		$157.58
Cost per sq. ft		....		1.58

Labor Cost of 100 sq. ft. of 4¼"x4¼" Glazed Ceramic Wall Tile
Using Water Resistant Organic Adhesive (Unmounted Tile)

	Hours	Rate	Total	Rate	Total
Tile Setter	8.0	$....	$....	$19.13	$153.04
Cost per sq. ft			....		1.53

Material Cost of 100 sq. ft. of 4{1/4"x4¼" Glazed Ceramic Wall Tile
Using "Dry-Set" Portland Cement Mortar (Unmounted Tile)

	Rate	Total	Rate	Total
35 lbs. "dry-set" mortar mix	$....	$....	$.48	$16.80
100 sq. ft. glazed wall tile	....	....	1.33	133.00
9 lbs. dry tile grout mix	....	....	.37	3.33
Cost per 100 sq. ft		$....		$153.13
Cost per sq. ft		....		1.53

Labor Cost of 100 sq. ft. of 4¼"x4¼" Glazed Ceramic Wall Tile
Using "Dry-Set" Portland Cement Mortar (Unmounted Tile)

	Hours	Rate	Total	Rate	Total
Tile Setter	10.0	$....	$....	$19.13	$191.30
Cost per sq. ft			....		1.91

Material Cost of 100 sq. ft. 1"x1" Ceramic Mosaic Tile Floors Using
Conventional Mortar Method of Installation (Face-Mounted Tile)

	Rate	Total	Rate	Total
2 sacks portland cement	$....	$....	$4.24	$8.48
4 cu. ft. sand	....	....	.48	1.92
100 sq. ft. ceramic mosaic tile	....	....	1.27	127.00
17 lbs. wet tile grout mix	....	....	.27	4.59
Cost per 100 sq. ft		$....		$141.99
Cost per sq. ft		....		1.42

Labor Cost of 100 sq. ft. 1"x1" Ceramic Mosaic Tile Floors Using
Conventional Mortar Method of Installation (Face-Mounted Tile)

	Hours	Rate	Total	Rate	Total
Tile setter	10.5	$....	$....	$19.13	$200.87
Cost per sq. ft			....		2.01

Material Cost of 100 sq. ft. of 1"x1" Ceramic Mosaic Tile Floors Using Water Resistant Organic Adhesive (Face-Mounted Tile)

	Rate	Total	Rate	Total
2.5 gallons adhesive	$....	$....	$8.50	$21.25
100 sq. ft. ceramic mosaic tile	....	....	1.27	127.00
23 lbs. dry tile grout mix	....	....	.37	8.51
Cost per 100 sq. ft		$....		$156.76
Cost per sq. ft		....		1.57

Labor Cost of 100 sq. ft. of 1"x1" Ceramic Mosaic Tile Floors Using Water Resistant Organic Adhesive (Face-Mounted Tile)

	Hours	Rate	Total	Rate	Total
Tile setter	8.0	$....	$....	$19.13	$153.04
Cost per sq. ft			....		1.53

Material Cost of 100 sq. ft. of 1"x1" Ceramic Mosaic Tile Floor Using "Dry-Set" Portland Cement Mortar (Face-Mounted Tile)

	Rate	Total	Rate	Total
35 lbs. "dry-set" mortar mix	$....	$....	$.48	$16.80
1 cu. ft. of sand	....	....	.48	.48
100 sq. ft. ceramic mosaic tile	....	....	1.27	127.00
23 lbs. dry tile grout mix	....	....	.37	8.51
Cost per 100 sq. ft.		$....		$152.79
Cost per sq. ft.		....		1.53

Labor Cost of 100 sq. ft. of 1"x1" Ceramic Mosaic Tile Floors Using "Dry-Set" Portland Cement Mortar (Face-Mounted Tile)

	Hours	Rate	Total	Rate	Total
Tile setter	9.0	$....	$....	$19.13	$172.17
Cost per sq. ft.			....		1.72

Material Cost of 100 sq. ft. of 1"x1" Ceramic Mosaic Tile on Walls Using Conventional Mortar Method of Installation (Face-Mounted Tile)

	Rate	Total	Rate	Total
2 sacks portland cement	$....	$....	$4.24	$8.48
4 cu. ft. sand	....	....	.48	1.92
100 sq. ft. ceramic mosaic tile	....	....	1.27	127.00
17 lbs. wet tile grout mix	....	....	.27	4.59
Cost per 100 sq. ft.		$....		$141.99
Cost per sq. ft.		....		1.42

Labor Cost of 100 sq. ft. of 1"x1" Ceramic Mosaic Tile on Walls Using Conventional Mortar Method of Installation (Face-Mounted Tile)

	Hours	Rate	Total	Rate	Total
Tile setter	11.0	$....	$....	$19.13	$210.43
Cost per sq. ft.			....		2.10

Material Cost of 100 sq. ft. of 1"x1" Ceramic Mosaic Tile on Walls Using Water Resistant Organic Adhesive (Face-Mounted Tile)

	Rate	Total	Rate	Total
2.5 gallons adhesive	$....	$....	$8.50	$21.25
100 sq. ft. ceramic mosaic tile	....	....	$1.27	127.00
23 lbs. dry tile grout mix	....	....	.37	8.51
Cost per 100 sq. ft.		$....		$156.76
Cost per sq. ft.		....		1.57

Labor Cost of 100 sq. ft. of 1"x1" Ceramic Mosaic Tile on Walls
Using Water Resistant Organic Adhesive (Face-Mounted Tile)

	Hours	Rate	Total	Rate	Total
Tile setter	9.0	$....	$....	$19.13	$172.17
Cost per sq. ft.			....		1.72

Material Cost of 100 sq. ft. of 1"x1" Ceramic Mosaic Tile on Walls
Using "Dry-Set" Portland Cement Mortar (Face-Mounted Tile)

	Rate	Total	Rate	Total
35 lbs. "dry-set" mortar mix	$....	$....	$.48	$16.80
1 cu. ft. sand	....	....	.48	.48
100 sq. ft. ceramic mosaic tile	....	....	1.27	127.00
23 lbs. dry tile grout mix	....	....	.37	8.51
Cost per 100 sq. ft.		$....		$152.79
Cost per sq. ft.		....		1.53

Labor Cost of 100 sq. ft. of 1"x1" Ceramic Mosaic Tile on Walls
Using "Dry-Set" Portland Cement Mortar (Face-Mounted Tile)

	Hours	Rate	Total	Rate	Total
Tile setter	9.0	$....	$....	$19.13	$172.17
Cost per sq. ft.			....		1.72

Material Cost of 100 sq. ft. of 6"x6"x½" Quarry Tile Floors
Using Conventional Mortar Method of Installation

	Rate	Total	Rate	Total
2 sacks portland cement	$....	$....	$4.24	$8.48
6 cu. ft. sand	....	....	.48	2.88
100 sq. ft. quarry tile plain surface	....	....	1.27	127.00
35 lbs. portland cement grout	....	....	.53	18.55
Cost per 100 sq. ft.		$....		$156.91
Cost per sq. ft.		....		1.57

Labor Cost of 100 sq. ft. of 6"x6"x½" Quarry Tile Floors
Using Conventional Mortar Method of Installation

	Hours	Rate	Total	Rate	Total
Tile setter	16.0	$....	$....	$19.13	$306.08
Cost per sq. ft.			....		3.06

Material Cost of 100 sq. ft. of 6"x6"x½" Quarry Tile Floors
Using "Dry-Set" Portland Cement Mortar

	Rate	Total	Rate	Total
35 lbs. "dry-set" mortar mix	$....	$....	$.48	$16.80
2 cu. ft. of sand	....	....	.48	.96
100 sq. ft. quarry tile plain surface	....	....	1.27	127.00
35 lbs. portland cement grout	....	....	.53	18.55
Cost per 100 sq. ft.		$....		$163.31
Cost per sq. ft.		....		1.63

Labor Cost of 100 sq. ft. 6"x6"x½" Quarry Tile Floors
Using "Dry-Set" Portland Cement Mortar

	Hours	Rate	Total	Rate	Total
Tile setter	12.0	$....	$....	$19.13	$229.56
Cost per sq. ft.			....		2.30

Material Cost of 100 lin. ft. of Cove or Base (Ceramic or Quarry Tile)

	Rate	Total	Rate	Total
100 lin. ft. cove	$....	$....	$1.70	$170.00
Adhesive or mortar--allow	....	....	6.36	6.36
Cost per 100 lin. ft.		$....		$176.36
Cost per lin. ft.		....		1.76

Labor Cost of 100 lin. ft. of Cove or Base (Ceramic or Quarry Tile)

	Hours	Rate	Total	Rate	Total
Tile setter	20.0	$....	$....	$19.13	$382.60
Cost per lin. ft.			....		3.83

Material Cost of 100 lin. ft. of Wainscot Cap (Glazed Ceramic Wall Tile)

	Rate	Total	Rate	Total
100 lin. ft. bullnose	$....	$....	$1.70	$170.00
Adhesive or mortar--allow	....	....	6.36	6.36
Cost per 100 lin. ft.		$....		$176.36
Cost per lin. ft.		....		1.76

Labor Cost of 100 lin. ft. of wainscot cap (Glazed Ceramic Wall Tile)

	Hours	Rate	Total	Rate	Total
Tile setter	20.0	$....	$....	$19.13	$382.60
Cost per lin. ft.			....		3.83

Placing Cement Floor Fill. Cement floor fill under ceramic tile floors is usually placed by tile setters and helpers (one tile setter and one or two helpers working together). The fill is placed one or two days in advance of the tile if the overall thickness from rough floor to finished tile surface is over 3". For 3" thickness and under, fill and setting bed may be placed in one operation. When such tile floors are to be placed over wood subfloors, it is necessary to first place a layer of waterproof building paper and a layer of wire mesh reinforcing before placing the fill.

A tile setter and two helpers should place 450 to 500 sq. ft. of fill per 8-hour day in areas large enough to permit efficient operations, at the following cost per 100 sq. ft.:

Material Costs

	Rate	Total	Rate	Total
6 bags portland cement	$....	$....	$4.24	$25.44
1 cu. yd. sand	....	....	12.75	12.75
Cost per 100 sq. ft.		$....		$38.19
Cost per sq. ft.		....		.38

Labor Costs

	Hours	Rate	Total	Rate	Total
Tile setter	1.5	$....	$....	$19.13	$28.70
Helper	3.0	....	....	19.13	57.39
Cost per 100 sq. ft.			$....		$86.09
Cost per sq. ft.			....		.86

Metal Wall Tile

Wall tile of aluminum, having a baked enamel finish in various colors, or tile of copper or stainless steel are used for the same purposes.

They require the same type of base and are applied with adhesive. A worker will apply about 100 sq. ft. per day.

Metal tile mastic weighs about 16 lbs. per gal. and it requires ½ lb. per sq. ft. or one gal. is sufficient for 32 sq. ft. of tile. Price about $8.50 per gal. Aluminum metal tile cost about $1.60 per sq. ft. Copper or stainless steel tile cost about $3.95 per sq. ft.

MARBLE TILE

To prepare an accurate estimate of cost on marble work, the estimator should understand the correct method of measuring and listing quantities from the plans, the manner in which the marble should be handled and set on the job, as well as the most economical and practical methods of handling the work.

When estimating marble wainscoting, it is incorrect to take the height of the wainscot from the floor to the cap and multiply by the number of lin. ft., because in all probability the wainscot is made up of three members, (base, die and cap) and each member should be estimated separately; the base and cap by the lin. ft. and the die by the square foot.

The length and size of the pieces of marble has considerable bearing on the labor cost, because it usually costs as much to set a 6" piece as one 2'-0" long, so that a job made up of short pieces will cost more per foot than a job having a larger percentage of long pieces of marble.

Trade Practice. Trade practice in the marble industry recognizes all marbles and stones commonly used for interior building purposes, as falling within one of four groups or classifications, according to their respective characteristics and working qualities. Therefore, for purposes of standardization and clearer definition, the Marble Institute of America has officially adopted these four classifications:

Group "A": Sound marbles and stones that require no "sticking", "waxing", or "filling"; characteristically uniform and favorable working qualities.

Group "B": Marbles and stones similar in character to the preceding group but somewhat less favorable working qualities; occasional natural faults and requiring a limited amount of "waxing" and "sticking".

Group "C": Marbles of uncertain variation in working qualities, geological flaws, voids, veins, and lines of separation common; common shop practice to repair nature's shortcomings by sticking, waxing and filling; "liners" and other forms of reinforcement freely employed when necessary.

Group "D": Marbles and stones similar to the preceding group and subject to the same methods of finishing and manufacture but embracing those materials that contain a larger proportion of natural faults and a maximum variation in working qualities, etc. This group comprises many of the highly colored marbles prized for their decorative qualities.

Below is a list of some of the more commonly used varieties of marbles and stones and the groups to which they belong:

Group "A"

Alabama--usual grades
Blanco P
Brocadillo, Vermont
Carthage
Georgia
Italian White
Italian English Vein
Napoleon Gray (Vt.)
Tennessee, Gray and Pink
Vermont, Black
Vermont, White Grades
Vermont Pavonazzo
Westland Green Vein Cream

Group "B"

Alabama Cream Veined A
Belgian Black
Champville
Cremo Italian
Imperial Black, Tenn.
Mankato Buff
Travertine (Italian)

Group "C"

Belgian Grand Antique
Bois Jourdan
Botticino
Breche Opal
Escalette
Hauteville
Mankato Pink
Ozark Rouge
Red Levanto
Red Verona
Rosato
Tavernelle, all types
Verdona
Verona, Yellow
Vermont Verde Antique
Westfield Green

Group "D"

Alps Green
Black and Gold
Bleu Belge
Breche Oriental
Breche Rosora
Forest Green
Grand Antique, French
Grand Antique, Italian
Onyx, Pedrara
Rouge Antique
Rouge Jascolin
Sienna
Verde Antico (Italian)
Vert St. Denis

Estimating Quantities of Marble Work. When measuring and listing quantities of marble work from the plans, there are several general rules that should be followed to insure accuracy.

To estimate accurately the quantity of marble base, obtain the number of lineal feet of base, listing the various heights separately. All base under 1'-0" high is extended and priced on a lineal foot basis. All base 1'-0" or over is extended and priced on a square foot basis. Always bear in mind that no piece of marble should be figured as being less than 1'-0" long, as it requires just as much time to set a 6" piece as one 2'-0" long. Short pieces are usually found around pilasters, door and window returns, and the like.

Marble die is estimated by the square foot, and the labor costs computed in the same manner. When listing quantities, obtain the length of each run and multiply by the height, and the result will be the number of sq. ft. of die. If marble pilasters project beyond the face of the wall, list each projection according to its actual dimensions, and if less than a 6" projection, it should be priced as 6" wide, but if the pilasters are over 6" wide, then use actual dimensions.

Marble wainscot cap under 1'-0" wide or any marble under 1'-0" wide, should be measured by the lineal foot.

Marble stair treads should be measured by the square foot, as they are ordinarily 12" or more in width. Be sure to mention length, width and thickness of all treads, type of nosing, etc.

Marble stair risers should be estimated by the lin. ft. giving length, height and thickness of each riser, as it costs practically as much to set a short riser as a long one.

When estimating toilet stall work consisting of toilet backs, partitions, stall fronts, stiles, etc., the fronts should be estimated by the square foot if over 12" wide and by the lin. ft. if less than 12" wide. Partitions and backs should be estimated by the square foot. Marble cap is usually estimated by the lin. ft. Some marble contractors figure toilet work at a certain price per stall, including all labor in connection with same, although most contractors recommend pricing on a square or lineal foot basis.

Marble column bases and caps formed of solid stock over 2" thick, should be estimated at a certain price for each base or cap, mentioning the size and number of cu. ft. in each.

Marble floor tile and border are usually estimated by the square foot, giving size of tile, such as 6"x6", 9"x9", 12"x12", etc. Small or irregular sized tile, such as dots, etc., are estimated in the same manner, except that a notation should be made of the class of work, as it costs more to lay small and irregular shaped tile.

Marble handrail, moldings, etc., should be estimated by the lin. ft., giving the size of each, in order that the quantities may also be stated in cu. ft. if over 2" thick.

All classes of circular work, such as base, wainscot or die, should be estimated in the same units as straight work, except that circular work should be estimated separately, as it usually requires considerable more cutting and fitting than straight work.

All marble work cut and set "on the rake", such as stair wainscoting, wainscoting under stairs, etc., should be estimated by the square foot, using the largest dimensions. Work of this kind should be estimated separately as it is more expensive to set than straight work.

Handling and Setting Interior Marble. Estimating the labor cost of handling and setting interior marble will depend as much upon the conditions under which the work is performed as any other class of construction work. The amount of labor handling, sorting, and distributing the marble on the job before it is set should be considered, and also, whether it is necessary to hoist the marble above the first floor, because this will require considerably more handling. The average size of the pieces should be considered, such as the average length of base, size and thickness of wainscot, length of thresholds, treads, risers, caps, etc., as it costs almost as much to handle and set a short piece of marble as a long one. This is especially true on stock that can be handled by one person.

The labor costs given on the following pages are based on the performance of the regular crew of any well organized marble company, but if it is necessary to hire "extra" workers or "floaters", the daily production may be decreased 25% and the labor costs increased accordingly.

On practically all jobs it is necessary to have laborers handle the marble and distribute it ready for the setters, although in the larger cities all marble must be handled by marble setters and helpers.

Hoisting Marble. The proposal sheet of practically all marble contractors states that the marble contractor shall have free use of the general contractor's hoisting facilities, runways, etc., so for this reason no allowance has been made for hoisting in the following estimates of cost.

However, if it is necessary for the marble contractor to pay for the use of hoist for hoisting marble to the upper floors of a building, figure about ½ hr. hoist time per 100 sq. ft. of marble.

Trucking on Marble From Mill or Cars to Job. If the marble is furnished by an out of town mill and it is necessary to haul it in trucks from cars to the job, the cost will vary with the length of the haul. If the marble is furnished by a mill located in the same town as the job, the price of the marble will probably include delivery to the job.

Marble Contractor's Foreman Expense. It is necessary to have a foreman on all jobs of any size to supervise the arrival, unloading, and distribution of the stock, as well as to lay out the work and supervise the setting. On an average allow 1/16 to ⅛ hr. foreman time to each hour setter's time, depending upon the size of the job and the number of setters employed.

Most small jobs will not require a foreman and no allowance should be made for same.

Average Labor Cost Setting Interior Marble. While the labor cost of marble setting will vary with the kind of material handled, some of the large producing marble companies estimate the setting cost of the entire job at a certain price per foot. The quantities for the entire job are taken, which include base, cap, wainscot or die, treads, risers, etc., and the labor priced per sq. ft., or 100 sq. ft. should cost as follows:

	Hours	Rate	Total	Rate	Total
Foreman	1.5	$....	$....	$26.59	$39.89
Marble setter	12.0	....	....	25.62	307.44
Helper	12.0	....	....	19.13	229.56
Labor or helper	8.0	....	....	19.13	153.04
Cost 100 sq. ft			$....		$729.93
Cost per sq. ft			....		7.30

The above price includes setting on jobs consisting principally of ⅞" and 1¼" stock.

If 2" marble is used, the handling and setting cost will be increased considerably, and the labor per 100 sq. ft. should cost as follows:

	Hours	Rate	Total	Rate	Total
Foreman	1.5	$....	$....	$26.59	$39.89
Marble setter	16.0	....	....	25.62	409.92
Helper	16.0	....	....	19.13	306.08
Labor or helper	12.0	....	....	19.13	229.56
Cost 100 sq. ft			$....		$985.45
Cost per sq. ft			....		9.85

On jobs having an average amount of base, wainscot or die, cap, toilet stalls, treads and risers, etc., the above prices will prove close enough, but where there are large quantities of any one class of work, it is advisable to refer to that particular class of work to obtain accurate quantities and costs.

Setting Marble Base. The cost of setting marble base will vary with the size of the rooms, whether straight walls or broken up with pilasters, piers, etc., and whether the job consists of long or short pieces of base.

In small corridors and other spaces having numerous pilasters, piers, etc., requiring short pieces of base, a setter and helper should set 60 to 75 lin. ft. per 8-hr. day, at the following labor cost per 100 lin. ft.:

	Hours	Rate	Total	Rate	Total
Foreman	1.5	$....	$....	$26.59	$39.89
Marble setter	12.0	....	....	25.62	307.44
Helper	12.0	....	....	19.13	229.56
Labor or helper	8.0	....	....	19.13	153.04
Cost 100 lin. ft			$....		$729.93
Cost per lin. ft			....		7.30

If the base can be set in reasonably long pieces, a marble setter and helper should set 80 to 100 lin. ft. per 8-hr. day, at the following labor cost per 100 lin. ft.:

	Hours	Rate	Total	Rate	Total
Foreman	1	$....	$....	$26.59	$26.59
Marble setter	9	....	....	25.62	230.58
Helper	9	....	....	19.13	172.17
Labor or helper	8	....	....	19.13	153.04
Cost 100 lin. ft			$....		$582.38
Cost per lin. ft			....		5.82

Setting Circular Base. Setting circular base will run about double the cost of straight work of the same kind and set under similar conditions. There is usually considerable cutting and fitting on all circular work.

Setting Marble Wainscot or Die. The labor cost of setting marble wainscot or die will vary with the size of the pieces, the height of the wainscot or die, etc., as a marble setter will set almost as many lin. ft. of 5'-0" wainscot as 3'-0" wainscot. The labor cost handling stock will be increased somewhat, as it will require 2 to 4 laborers or helpers to handle each piece of marble and place it on the floor ready for the setter, while lighter stock, such as base, risers, treads, etc., can be handled by one man.

Setting Marble Wainscot Up to 3'-0" High. When setting marble wainscot up to 3'-0" high, a marble setter and helper should set 22 to 26 lin. ft. containing 66 to 78 sq. ft. per 8-hr. day, at the following labor cost per 100 sq. ft.:

	Hours	Rate	Total	Rate	Total
Foreman	1.5	$....	$....	$26.59	$39.89
Marble setter	11.0	....	....	25.62	281.82
Helper	11.0	....	....	19.13	210.43
Labor or helper	8.0	....	....	19.13	153.04
Cost 100 sq. ft			$....		$685.18
Cost per sq. ft			....		6.85

The cost of setting circular wainscot or die will run about double the cost of straight work on account of the extra cutting and fitting.

Setting Marble Wainscot 3'-0" to 4'-0" High. A marble setter and helper should set 20 to 25 lin. ft. containing 75 to 90 sq. ft. of wainscot per 8-hr. day, at the following labor cost per 100 sq. ft.:

	Hours	Rate	Total	Rate	Total
Foreman	1.2	$....	$....	$26.59	$31.91
Marble setter	10.0	....	....	25.62	256.20
Helper	10.0	....	....	19.13	191.30
Labor or helper	8.0	....	....	19.13	153.04
Cost 100 sq. ft			$....		$632.45
Cost per sq. ft			....		6.32

The cost of setting circular wainscot or die will run about double the cost of straight work on account of the extra cutting and fitting.

Setting Marble Wainscot 4'-0" to 5'-0" High. Where the marble wainscot varies from 4'-0" to 5'-0" high, a marble setter and helper should set 20 to 23 lin. ft. containing 85 to 100 sq. ft. per 8-hr. day, at the following labor cost per 100 sq. ft.:

	Hours	Rate	Total	Rate	Total
Foreman	1	$....	$....	$26.59	$26.59
Marble setter	9	....	....	25.62	230.58
Helper	9	....	....	19.13	172.17
Labor or helper	8	....	....	19.13	153.04
Cost 100 sq. ft			$....		$582.38
Cost per sq. ft			....		5.82

The cost of setting circular wainscot or die will run about double the cost of straight work on account of the extra cutting and fitting.

Setting Marble Wainscot Over 5'-0" High. When setting marble wainscot 5'-0" to 7'-0" high, a marble setter and helper should set 90 to 110 sq. ft. per 8-hr. day, at the following labor cost per 100 sq. ft.:

	Hours	Rate	Total	Rate	Total
Foreman	1	$....	$....	$26.59	$26.59
Marble setter	8	....	....	25.62	204.96
Helper	8	....	....	19.13	153.04
Labor or helper	8	....	....	19.13	153.04
Cost 100 sq. ft			$....		$537.63
Cost per sq. ft			....		5.38

The cost of setting circular wainscot or die will run about double the cost of straight work on account of the extra cutting and fitting.

Wainscot over 7'-0" high will require working from a scaffold, which decreases the production considerably. In such cases the labor cost of that portion of wainscot over 7'-0" high should be increased 50% to 100%, and the cost of erecting and removing steel tubular scaffold should be added.

Setting Marble Wainscot Cap. When setting marble wainscot cap, a marble setter and helper should set 60 to 75 lin. ft. per 8-hr. day, at the following labor cost per 100 lin. ft.:

	Hours	Rate	Total	Rate	Total
Foreman	1.5	$....	$....	$26.59	$39.89
Marble setter	12.0	....	....	25.62	307.44
Helper	12.0	....	....	19.13	229.56
Labor or helper	8.0	....	....	19.13	153.04
Cost 100 lin. ft			$....		$729.93
Cost per lin. ft			....		7.30

Add for overhead and profit.

Setting Marble Stair Treads. The cost of setting marble stair treads will vary with the thickness and length of the treads. The shorter each piece of marble, the higher the cost per lin. ft.

If the treads are 3'-0" to 3'-6" long and 1¼" thick, a marble setter and helper should set 20 to 22 treads containing 60 to 77 lin. ft. per 8-hr. day, at the following labor cost per 100 lin. ft.:

	Hours	Rate	Total	Rate	Total
Foreman	1.5	$....	$....	$26.59	$39.89
Marble setter	11.0	....	....	25.62	281.82
Helper	11.0	....	....	19.13	210.43
Labor or helper	8.0	....	....	19.13	153.04
Cost 100 lin. ft			$....		$685.18
Cost per lin. ft			....		6.85

If the treads vary from 4'-0" to 6'-0" long, a marble setter and helper should set 80 to 90 lin. ft. per 8-hr. day, at the following labor cost per 100 lin. ft.:

	Hours	Rate	Total	Rate	Total
Foreman	1.1	$....	$....	$26.59	$29.25
Marble setter	9.5	....	....	25.62	243.39
Helper	9.5	....	....	19.13	181.74
Labor or helper	8.0	....	....	19.13	153.04
Cost 100 lin. ft		....	$....		$607.42
Cost per lin. ft			....		6.07

Setting Marble Stair Risers. The cost of setting marble stair risers will vary with the length of the pieces of marble, width of stairs, etc.

Where the stairs are 3'-0" to 3'-6" wide, a marble setter and helper should set 60 to 75 lin. ft. of risers per 8-hr. day, at the following labor cost per 100 lin. ft.:

	Hours	Rate	Total	Rate	Total
Foreman	1.5	$....	$....	$26.59	$39.89
Marble setter	11.0	....	....	25.62	281.82
Helper	11.0	....	....	19.13	210.43
Labor or helper	8.0	....	....	19.13	153.04
Cost 100 lin. ft			$....		$685.18
Cost per lin. ft			....		6.85

If the marble stairs are over 4'-0" wide, a marble setter and helper should set 80 to 90 lin. ft. of risers per 8-hr. day, at the following labor cost per 100 lin. ft.:

	Hours	Rate	Total	Rate	Total
Foreman	1.1	$....	$....	$26.59	$29.25
Marble setter	9.5	....	....	25.62	243.39
Helper	9.5	....	....	19.13	181.74
Labor or helper	8.0	....	....	19.13	153.04
Cost 100 lin. ft			$....		$607.42
Cost per lin. ft			....		6.07

Setting Marble Stair Wainscot on the Rake. Where marble wainscoting is set on the rake of the stairs, the labor cost will run considerably higher than straight work on account of the additional cutting and fitting necessary.

If the wainscoting is 3'-0" to 3'-6" high, a marble setter and helper should set 10 to 15 lin. ft. containing 35 to 45 sq. ft. per 8-hr. day, at the following labor cost per 100 sq. ft.:

	Hours	Rate	Total	Rate	Total
Foreman	2	$....	$....	$26.59	$53.18
Marble setter	20	....	....	25.62	512.40
Helper	20	....	....	19.13	382.60
Labor or helper	8	....	....	19.13	153.04
Cost 100 sq. ft			$....		$1,101.22
Cost per sq. ft			....		11.01

Where the wainscot is 5'-0" to 7'-0" high and follows the rake of the stairs, a marble setter and helper should set 50 to 65 sq. ft. per 8-hr. day, at the following labor cost per 100 sq. ft.:

	Hours	Rate	Total	Rate	Total
Foreman	1.5	$....	$....	26.59	$39.89
Marble setter	14.0	....	....	25.62	358.68
Helper	14.0	....	....	19.13	267.82
Labor or Helper	8.0	....	....	19.13	153.04
Cost 100 sq. ft			$....		$819.43
Cost per sq. ft			....		8.19

Setting Marble Toilet Stalls and Partitions. The cost of handling and setting marble toilet stalls, backs, partitions, stiles, etc., will vary greatly with the individual job. On work of this kind the marble partition slabs are seldom less than 5'-0" in height and width and each slab contains 25 to 30 sq. ft. of marble. The backs are usually the same height as the partitions but not so wide and each back contains 16 to 20 sq. ft. of marble. The stall fronts or stiles are usually 6" to 8" wide and 6'-0" to 7'-0" high and extend to the floor to support the dividing partitions where metal standards are not used.

As an average on toilet stall work, a good marble setter should complete a stall in 3 to 3½ hours, after the marble backs are in place. Since the average partition contains about 25 sq. ft. and each stile and cap about 7 sq. ft., a setter and helper should set 75 to 80 sq. ft. of marble per 8-hr. day, at the following labor cost for 1 complete stall exclusive of back:

	Hours	Rate	Total	Rate	Total
Foreman	0.4	$....	$....	$26.59	$10.64
Marble setter	3.3	....	....	25.62	84.55
Helper	3.3	....	....	19.13	63.13
Labor or helper	2.6	....	....	19.13	49.74
Setting cost 1 stall			$....		$208.06

Toilet partition work is estimated by the sq. ft. by most marble contractors.

Setting Marble Toilet Backs. The marble backs for toilet stalls are usually 6'-0" to 7'-0" high. A marble setter and helper should set 75 to 85 sq. ft. per 8-hr. day, at the following labor cost per 100 sq. ft.:

	Hours	Rate	Total	Rate	Total
Foreman	1.2	$....	$....	$26.59	$31.91
Marble setter	10.0	....	....	25.62	256.20
Helper	10.0	....	....	19.13	191.30
Labor or helper	8.0	....	....	19.13	153.04
Cost 100 sq. ft			$....		$632.45
Cost per sq. ft			....		6.32

Setting Marble Door and Window Trim. A marble setter and helper should set 45 to 55 lin. ft. of marble door and window trim or casings per 8-hr. day, at the following labor cost per 100 lin. ft.:

	Hours	Rate	Total	Rate	Total
Foreman	2	$....	$....	$26.59	$53.18
Marble setter	16	....	....	25.62	409.92
Helper	16	....	....	19.13	306.08
Labor or helper	8	....	....	19.13	153.04
Cost 100 lin. ft			$....		$922.22
Cost per lin. ft			....		9.22

Add for overhead and profit.

Setting Marble Countertops. When setting marble countertops, a marble setter and helper should set 75 to 85 sq. ft. per 8-hr. day, at the following labor cost per 100 sq. ft.:

	Hours	Rate	Total	Rate	Total
Foreman	1.2	$....	$....	$26.59	$31.91
Marble setter	10.0	....	....	25.62	256.20
Helper	10.0	....	....	19.13	191.30
Labor or helper	8.0	....	....	19.13	153.04
Cost 100 sq. ft			$....		$632.45
Cost per sq. ft			....		6.32

Setting Marble Thresholds. When setting marble thresholds 3'-0" to 3'-6" long, a marble setter and helper should set 12 to 14 thresholds per 8-hr. day, at the following labor cost each:

	Hours	Rate	Total	Rate	Total
Marble setter	.60	$....	$....	$25.62	$15.37
Helper	.60	....	....	19.13	11.48
Cost per threshold			$....		$26.85

Setting Marble Plinths. When setting marble plinths at door openings, a marble setter and helper should set 20 to 25 plinths per 8-hr. day, at the following labor cost per 100 plinths:

	Hours	Rate	Total	Rate	Total
Foreman	4.5	$....	$....	$26.59	$119.66
Marble setter	36.0	....	....	25.62	922.32
Helper	36.0	....	....	19.13	688.68
Labor or helper	8.0	....	....	19.13	153.04
Cost 100 plinths			$....		$1,883.70
Cost per plinth			....		18.84

Setting Marble Stair Railings and Balusters. On jobs having marble balusters up to 6" in diameter and 2'-6" long, a marble setter and helper should set about 20 balusters per 8-hr. day, at the following labor cost per 100 balusters:

	Hours	Rate	Total	Rate	Total
Foreman	5	$....	$....	$26.59	$132.95
Marble setter	40	....	....	25.62	1,024.80
Helper	40	....	....	19.13	765.20
Labor or helper	16	....	....	19.13	306.08
Cost 100 balusters			$....		$2,229.03
Cost per baluster			....		22.29

Setting Marble Ashlar. Where marble ashlar is used for the interior walls of buildings, it is customary to use 7/8" or 1¼" marble and the method of handling and setting is similar to that used for setting exterior work, as it is often necessary to use a light breast derrick to set the marble.

On work of this kind, a marble setter and helper should handle and set 60 to 80 sq. ft. of ashlar per 8-hr. day, at the following labor cost per 100 sq. ft.:

	Hours	Rate	Total	Rate	Total
Foreman	1.5	$....	$....	$26.59	$39.89
Marble setter	12.0	....	....	25.62	307.44
Helper	12.0	....	....	19.13	229.56
Labor or helper	12.0	....	....	19.13	229.56
Cost 100 sq. ft			$....		$806.45
Cost per sq. ft			....		8.06

Setting Marble Moldings, Handrail, Etc. On jobs having light marble moldings, stair handrail, well hole rail, etc., where the marble is about 6" square and furnished in reasonably long pieces, a marble setter and helper should set 75 to 85 lin. ft. per 8-hr. day, at the following labor cost per 100 lin. ft.:

	Hours	Rate	Total	Rate	Total
Foreman	1.2	$....	$....	$26.59	$31.91
Marble setter	10.0	....	....	25.62	256.20
Helper	10.0	....	....	19.13	191.30
Labor or helper	8.0	....	....	19.13	153.04
Cost 100 lin. ft			$....		$632.45
Cost per lin. ft			....		6.32

Setting Marble Column Bases. When setting marble column bases 2'-6" to 3'-0" square and 0'-9" to 1'-3" high, a marble setter and helper should handle and set about one base an hr. at the following labor cost per base:

	Hours	Rate	Total	Rate	Total
Foreman	0.1	$....	$....	$26.59	$2.66
Marble setter	1.0	....	....	25.62	25.62
Helper	1.0	....	....	19.13	19.13
Labor or helper	0.5	....	....	19.13	9.57
Cost per base			$....		$56.98

Laying Marble Floor Tile. There are three labor operations to be considered when estimating the labor cost of setting marble floor tile, viz., the cement bed under the marble floor, labor handling and setting floor tile and the cost of rubbing or smoothing the floors after they have been laid.

As a general rule, the cement bed under the marble floor is 2½" to 3" thick, composed of cement and sand mixed rather dry. It requires about 1½ bbls. of portland cement and one cu. yd. of sand to each 100 sq. ft. of floor.

Material and labor per 100 sq. ft. of floor fill should cost as follows:

Material Costs

	Rate	Total	Rate	Total
6 sacks portland cement	$....	$....	$4.24	$25.44
1 cu. yd. sand	....	....	12.72	12.72
Cost 100 sq. ft		$....		$38.16
Cost per sq. ft		....		.38

Labor Costs

	Hours	Rate	Total	Rate	Total
Foreman	0.2	$....	$....	$26.59	$5.32
Marble setter	1.5	....	....	25.62	38.43
Helper	3.0	....	....	19.13	57.39
Cost 100 sq. ft			$....		$101.14
Cost per sq. ft			....		1.01

If the marble floor tile vary from 6"x6" to 12"x12" in size, a marble setter and helper should handle and lay 90 to 110 sq. ft. of floor per 8-hr. day, at the following labor cost per 100 sq. ft.:

	Hours	Rate	Total	Rate	Total
Foreman	1	$....	$....	$26.59	$26.59
Marble setter	8	....	....	25.62	204.96
Helper	8	....	....	19.13	153.04
Labor or helper	8	....	....	19.13	153.04
Cost 100 sq. ft			$....		$537.63
Cost per sq. ft			....		5.38

When laying marble floors it is customary to place the cement bed just ahead of the tile, and the entire work is performed by the setter and helper laying the floor. If the subfloor is sound and even, marble in residential and light commercial work is only ⅜" thick and set in mastic. Setting costs would average about 25 cents per sq. ft. less, unless the area was small and irregular, or set with an intricate pattern.

Smoothing Marble Floors. After the marble floors have been laid, it is necessary to rub them with carborundum stone to remove the uneven spots at the joints. It is customary to use a rubbing machine powered by gasoline or electricity, which is operated by one man and runs back and forth over the floors and grinds them to an even surface. A machine rents for about $7.50 per hour.

The quantity of floor that can be surfaced per day will vary with the kind of marble used, as it is possible to surface considerably more soft marble than hard marble.

When rubbing and surfacing soft marble, a machine should surface 125 to 150 sq. ft. per 8-hr. day, at the following labor cost per 100 sq. ft.:

	Hours	Rate	Total	Rate	Total
Machine operator	6.0	$....	$....	$25.62	$153.72
Foreman	0.7	....	....	26.59	18.61
Cost 100 sq. ft			$....		$172.33
Cost per sq. ft			....		1.72

When rubbing and surfacing hard marble, a rubbing machine should surface 100 to 125 sq. ft. of floor per 8-hr. day, at the following labor cost per 100 sq. ft.:

	Hours	Rate	Total	Rate	Total
Machine operator	7.0	$....	$....	$25.62	$179.34
Foreman	0.8	....	....	26.59	21.27
Cost 100 sq. ft			$....		$200.61
Cost per sq. ft			....		2.01

Setting Marble Floor Border. Where the marble floor has a different colored marble border, the labor setting the border will cost about the same as the balance of the floor.

Setting Structural Slate. The labor cost of setting structural slate work, such as base, treads and risers, toilet stalls and partitions, etc., will run about the same as for the same class of marble work and should be estimated accordingly.

If, however, field cutting should be required, as is often the case with window stools and door thresholds, the labor costs should be increased 20% to 25% as slate is more difficult to work than marble.

Estimating Quantities of Interior Marble

When estimating quantities of interior marble, any fraction of an inch will be treated as a whole inch. No piece of marble will be considered as being less than 6" wide and 1'-0" long.

Thickness. Thicknesses given are approximate but not exact. For slabs gauged to a given thickness, an extra charge of 10% to the list price is made.

Extra Lengths. On slabs 10'-0" to 12'-0" long, add 10% to list price; on slabs over 12'-0" long, add 20 percent.

Extra Widths. On slabs over 5'-6" wide, add 10% to list price.

Polished Both Sides. Prices given are based on marble having one side polished unless otherwise noted. If both sides are to be polished, add $1.50 per sq. ft. for marbles in Group "A" and "B" and $1.25 per sq. ft. for marbles in Group "C" and "D".

Hone Finish. Hone finish takes the same price as polished work.

Sand Finish. If marble is finished with sand finish on one side only, deduct 15 cents per sq. ft. If sand finish on both sides, add 25 cents per sq. ft.

Polished Edges. For each polished edge of 7/8" and 1 1/4" thickness, add 50 cents per lin. ft. of edge; on 1 1/2" and 2" thickness, add 75 cents per lin. ft.

Beveled or Rounded Edges. For each beveled or rounded edge on 7/8" and 1 1/4" thickness, add 50 cents per lin. ft.; on 1 1/2" and 2" thickness, add 75 cents per lin. ft.

Ogee Molding. For single member ogee molding $3.25 per lin. ft. of edge on 7/8" thick; $3.50 per lin. ft. on 1 1/4", and $4.25 per lin. ft. on 1 1/2". For each additional member, add $2.25 per lin. ft. of edge on 7/8" and 1 1/4", and $2.50 on 1 1/2".

Countersinking. For countersinking 1/4" deep or less, $2.00 per sq. ft. of surface. For each additional 1/4" deeper, or fraction thereof, add extra $2.50 per sq. ft. of surface.

Grooving. For grooving 1/4" deep or less $1.50 per lin. ft. for each groove. For each additional 1/4" deeper, or fraction thereof, add extra $1.50 per lin. ft. of each groove.

Rabbeting. For rabbeting 1/4" deep or less, $1.50 per lin. ft. for each rabbet. For each additional 1/4" deeper, or fraction thereof, add extra $1.25 per lin. ft. of each rabbet.

Special Cutting, Polishing, Drilling, Etc. All special work other than listed above will be charged for extra.

Shipping Weights of Marble. The approximate shipping weights of marble, boxed for shipment, are as follows per square foot: 7/8" thick 15 lbs.; 1 1/4" thick 20 lbs.; 1 1/2" thick 24 lbs.; 2" thick 32 lbs.; per cu. ft. 192 lbs.

Approximate Material Prices per Square Foot of 7/8" Marble, Polished One Side

Always check local sources for price and availability. Foreign marbles are always affected by exchange rates. The trend has been to marbles of even color. Many of the quarries for the more colorful marbles have been shut down, and when a block of such marble is found, the price may be out of line because of its scarcity.

Type of Marble	
Alabama Clouded "A" or Cream "A"	$8.50
Brocadillo	10.60
Carthage Gray	7.95
French Gray	8.50
Georgia Creole, Cherokee or Mezzotint	8.50
Gravina	7.95
Kasota Pink	7.95
Kasota Yellow	9.55
Light Cloud Vermont	8.50
Napoleon Gray	8.50
Radio Black	9.55

Tennessee Marbles	8.50
Verde Antique	10.60
Westfield Green	11.65
York Fossil	9.55
Foreign Marbles	
Alps Green	14.30
Belgian Black	12.75
Black and Gold	16.95
Botticino	12.75
Bois Jordan	12.75
Breche Oriental	18.05
Grand Antique, French	21.20
Grand Antique, Italian	15.90

The above prices are for marble in less than carload lots, uncrated, f.o.b. mill. Crating 40 to 50 cents per sq. ft. for marble up to 2" thick.

Approximate Material Prices on Sand Finish Marble Floor Tile and Border 7/8" Thick

Tile sizes 6"x6" to 1'-0"x2'-0": Prices per Square Foot

Alabama Clouded "A"	$5.80
Carthage Gray	4.00
Georgia Creole	5.80
Georgia Cherokee	3.20
Italian White	3.70
Italian English Vein	3.70
Tennessee Pink	4.80
Tennessee Gray	4.80
Tennessee Dark	4.25
Verde Antique	7.45
Travernelle	5.85
Travertine	5.30

Approximate Material Prices on 3/8" Thinset Marble Tiles Per Square Foot

	Polished	Honed
Cremo	$2.70	$2.65
Negro-Marquina	3.20	3.15
Rose Aurora	2.55	2.50
Statuary Vein	2.25	2.20
Travertine Filled	2.15	2.10
Travertine Unfilled	2.05	2.00
White Italian	2.15	2.10

12"x12" tiles, 3/8" thick, 10 tiles per box.

SLATE TILE

Slate is a natural mica granular crystalline stone. Its main sources are Vermont, Pennsylvania and Virginia. Vermont produces colored slates, including greens; purples; mottled green, purple and red; as well as the standard dark grays.

Slate is usually graded into clear stock and ribbon stock. Ribbon stock has bands of a darker color running through it and is cheaper than clear stock, which has no ribbons but will have some spots and veining in it. Finishes available are as follows:

Natural Cleft: Natural split or cleaved face, moderately rough with some textural variations.

Sand Rubbed Finish: Wet sand on a rubbing bed is used to eliminate any natural cleft, and leave slate in an even plane.

Honed Finish: This is a semi-polished surface.

Slate is used for spandrels, sills, stools, treads and risers, toilet stalls, floors and walks, fireplace facings and hearths.

Spandrels are generally clear slate with either cleft, rubbed or honed finish. Thickness is generally 1", 1¼" or 1½". Sizes are somewhat limited, with a length of no more than 6 ft. and width of 2'-6" to 4' being standard recommendations, although larger sizes may be available for certain conditions. Many spandrels today are set in metal grid frames. Otherwise, they must be anchored with bronze or stainless steel anchors in each corner.

Sills and stools may be either ribbon or clear stock. Exterior sills are generally furnished with a sand rubbed finish in lengths up to 4' and 1" to 2" in thickness. Sills may have drip grooves cut in at additional cost.

Interior stools and fireplace surrounds are usually 1" thick with honed finish, a maximum length of 4' being most economical.

Treads and risers on the exterior are generally natural cleft finish. Interior work is usually sand rubbed rather than honed, because it is both cheaper and gives a good non-slip surface, although the front edges may be honed. Thickness is generally 1" or 1¼" for stairs, 1½" to 2" for landings.

The cost of slate is determined by the grade, the size involved and the finish.

Ribbon grade, often used on stairs, will run around $5.00 per sq. ft. in 1" thickness and $8.00 for 2" thickness, sand finished. Honed finishes will add $0.50 per sq. ft. to the cost.

Clear grade, 1" thick and sand finished, will run $5.50 per sq. ft. in sizes to 3 sq. ft., $7.75 to $8.75 in sizes 3 sq. ft. to 6 sq. ft., and $11.00 in sizes over 6 sq. ft. Lineal foot costs for sills and stools will vary with both length and width. Up to 6" wide stock will cost about $4.75 per lineal foot for 1" thickness; $7.50 for 2" thickness. The 10" wide stock will cost $7.50 per lineal foot for 1" thickness to $9.25 per lineal foot for 2" thickness.

MORTAR BED METHOD
SLATE FLOORING OVER METAL LATH

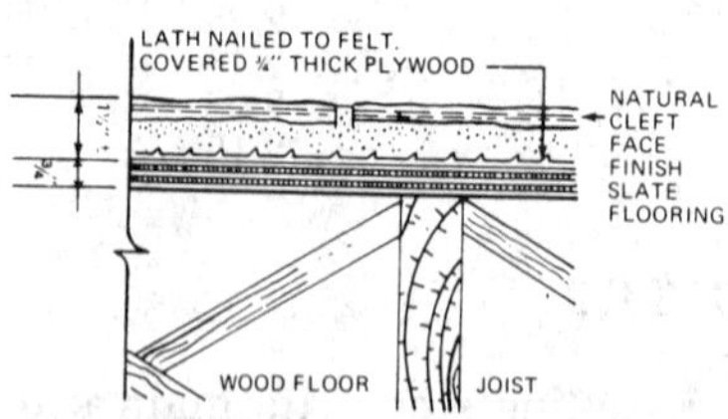

THINSET MASTIC METHOD
OVER PLYWOOD SUB-FLOOR

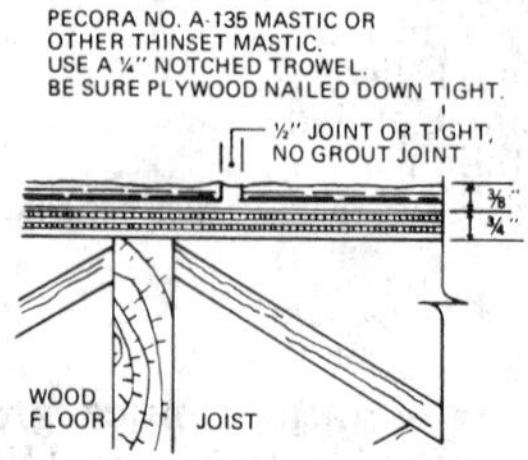

Slate for flooring is available for either mortar bed or thin set application. This slate is ¼" thick and available in 3" multiples 6"x6" to 12"x12". It weighs 3¾ lbs. per sq. ft. It is set in ¼" mastic applied with a notched trowel over concrete or plywood. It can have grouted joints or be ordered for butt joints. Cost will run $2.00 per sq. ft. for material and $2.50 for labor.

Slate for heavier duty floors and for outside applications is generally ½", ¾" or 1" thick, weighing 7½ lbs., 11¼ lbs. and 15 lbs. respectively. It is set in 1" mortar beds on concrete slabs or plywood. Plywood should be covered with felt and metal lath, nailed on. Minimum thickness is 1½". If plywood is set over joists without a subfloor, it should be at least ¾" thick. Sizes are 6"x6" to 24"x18", in multiples of 3". Cost runs from $3.75 per sq. ft. for ½" to $5.00 for 1".

09400 TERRAZZO

The cost of terrazzo work varies with the size of the job, size of rooms or spaces where the floors are to be installed, floor designs, strips of brass or other non-rusting material, and method of laying the floors.

Two methods are used in laying terrazzo floors over concrete construction. One is to bond it to the concrete and the other to separate it from the structural slab.

When the first method is used, the concrete fill is provided for by another contractor and should be left 2" below the finished floor. Before the terrazzo contractor installs his underbed, he must see that this concrete fill is thoroughly cleaned of plaster droppings, wood chips and other debris, and wetted to insure cohesion.

The second method is used in buildings where cracking is anticipated either from settlement, expansion and contraction or vibration. In this case the terrazzo contractor begins his work from the structural floor slab up. This method requires a thickness of at least 3". The concrete slab is covered with a thin bed of dry sand over which a sheet of tar paper is laid. Over the paper the underbed is installed as in the first method except that coarser aggregate can be used in the underbed, such as cinders or fine gravel where its thickness exceeds 2½". When this method is used, the cracks originating in the structural slab are not likely to appear on the surface but terminate at the sand bed.

When terrazzo is laid over wood floors a thickness of not less than 2" is required. The floor should first be covered with tarred paper. Over this paper, nail galvanized wire netting no. 14 gauge, 2" mesh. Then lay the concrete underbed as specified under the second method.

The underbed for terrazzo consisting of 1 part portland cement and 4 parts sharp, screened sand shall be spread and brought to a level not less than ½" nor more than ¾" below the finished floor.

Into this underbed, while still in a semi-plastic state, install metal strips or preassembled decorative units, having proper bonding features.

The terrazzo topping shall be not less than ½" nor more than ¾" in thickness and shall be of granulated marble of colors selected. The topping shall be uniform in composition and the marble granule that appears on the surface shall be used for its entire thickness. The granule to be composed of such proportions of Nos. 1-2-3 sizes as required.

Composition for terrazzo mixture shall be in the proportion of 200 lbs. of marble granules to 100 lbs. of gray or white portland cement, mixed dry, and water added afterward to make the mix plastic, but, not too wet. The mix shall then be placed in the spaces formed by the metal dividing strips and rolled into a compact mass by means of heavy stone or metal rollers until all superfluous cement and water is extracted, after which it must be hand troweled to an even surface, disclosing the lines of the metal strips on a level with the terrazzo filling.

When the floors have set sufficiently hard, they shall be machine rubbed, using coarse carborundum grit stones for the initial rubbing after which a light grouting of pure gray white portland cement shall be applied to the surface, filling all voids, and allowed to remain until the time of final cleaning.

Floors shall have the grouting coat removed by machine, using a fine carborundum grit, after which it must be thoroughly washed. Do not use acids in cleaning terrazzo floors.

Ramped or other surfaces in terrazzo floors so specified, shall be made anti-slip by the addition of an abrasive aggregate. For heavy duty floors, the proportion shall be 2 parts of abrasive to 3 parts of marble granule and the abrasive shall be mixed in the terrazzo topping for its entire thickness. For light traffic floors, the abrasive shall be sprinkled on the surface only, and show a proportion of 5 parts marble granule and 1 part abrasive aggregate.

Terrazzo Base. Terrazzo base is usually coved at the floor and may be any height desired. If the base is to finish flush with the finish plaster, a metal base bead shall be furnished and set by the plasterer. The base shall be divided approximately every 4 lin. ft. using metal base dividers. The walls back of the base shall have a scratch coat of cement and sand mortar brought to a line 3/8" back of the finished face of the base, into which the base dividers shall be set. Base shall be finished with a very fine stone, so as to leave the surface at a hone finish.

Estimating the Cost of Terrazzo Work. In addition to the direct cost of labor and materials entering into terrazzo work, an allowance should be made to provide for wood grounds, carborundum stone for rubbing, wiring for electric power, depreciation on machines, use of hoist, cleaning up rubbish, freight and trucking on machines and tools. This will vary with size and location of job.

The costs on the following pages are based on an average job containing about 3,000 sq. ft. The labor on a small job containing 100 to 200 sq. ft. will run twice as much per sq. ft. as on a job containing 5,000 sq. ft.

Metal Strips in Terrazzo Floors and Base. Brass strips or other non-rusting metals are used in practically all terrazzo floors and add to the cost in proportion to the quantity used.

Standard brass strips B & S gauge No 20 cost 75 cents per lin. ft. including waste.

Metal strips B & S gauge No. 14, cost 1.45 per lin. ft. including waste in brass and 50 cents per lin. ft. in zinc.

The cost of metal strips per sq. ft. of floor will vary with the size of the squares or design and should be estimated accordingly.

Terrazzo base is jointed every 4'-0" with vertical metal joint dividers.

Estimating Quantities of Terrazzo Work. Terrazzo floors are estimated by the sq. ft. and will vary from 2" to 3" thick, depending upon the subfloors, as described in the previous paragraphs.

Terrazzo floor borders are estimated by the sq. ft. when over 12" wide and by the lin. ft. when less than 12" wide.

Terrazzo base is estimated by the lin. ft. stating height. An allowance of 1'-0" should be made for each corner or miter.

Double cove terrazzo base having 2 finished faces, such as ordinarily used under narrow partitions, shower stalls, etc. are estimated by the lin. ft. and the cost is double that of single cove base.

Labor Placing Terrazzo, Floor and Base. The labor cost of placing terrazzo work will vary with the size of the rooms, design of the floor, size of the squares or pattern requiring brass strips, etc., as the smaller the squares, the more metal strip required and the higher the cost.

The following are approximate quantities of various classes of terrazzo work a crew of 2 mechanics and 3 helpers should install per 8-hr. day :

Description of Work	SF per 8-Hr. Day	Hrs. per 100 SF Mechanic	Helpers
Floor blocked off into 5'-0" squares	400-450	3.75	5.65
Floor blocked off into 4'-0" squares	375-425	4.00	6.00
Floor blocked off into 3'-0" squares	350-400	4.25	6.40
Floor blocked off into 2'-0" squares	325-375	4.60	6.90
Floor blocked off into 1'-0" squares	275-325	5.33	8.00
Floor border 12" to 24" wide	275-325	5.33	8.00
		Hrs. per 100 LF	
Terrazzo Cove Base			
Terrazzo cove base, 3" high	60-65	13	13
Terrazzo cove base, 6" high	5055	16	16

Double cove base, double time given for single base.

Rubbing and Finishing Terrazzo Work. Terrazzo floors are rubbed by machine and by hand only in corners where the machine cannot reach. On jobs consisting of large and small rooms, a worker rubs and completes about 100 sq. ft. of terrazzo floor per 8-hr. day. A worker should rub and finish about 80 lin. ft. of terrazzo base per 8-hr. day.

Cost of Concrete Subbase Under Terrazzo Floors. There are three methods of applying the concrete underbed for terrazzo floors, as described previously: bonded to the structural concrete slab, separate from the structural slab, or applied over wood floors.

Inasmuch as terrazzo topping is ½" thick for all types of floors, cost of finished floor varies with type of underbed.

Material Cost of 100 Sq. Ft. of Concrete Underbed 1¼" Thick Bonded To Structural Concrete Slab

	Rate	Total	Rate	Total
3 sacks portland cement	$....	$....	$4.24	$12.72
11 cu. ft. sand	....	....	.48	5.28
Cost 100 sq. ft.		$....		$18.00
Cost per sq. ft		....		.18

Material Cost of 100 Sq. Ft. of Concrete Underbed, Consisting of ¼" Dry Sand, 1 Thickness Tarred Felt and 2" of Concrete, Underbed Separated From Structural Slab

	Rate	Total	Rate	Total
2.5 cu. ft. dry sand	$....	$....	$.48	$1.20
1 sq. No. 15 felt	....	....	3.18	3.18
4.5 sacks portland cement	....	....	4.24	19.08
18 cu. ft. sand	....	....	.48	8.64
Cost 100 sq. ft		$....		$32.10
Cost per sq. ft		....		.32

Material Cost of 100 Sq. Ft. of Concrete Underbed 2" Thick, Applied Over Wood Floor

	Rate	Total	Rate	Total
1 sq. No. 15 felt	$....	$....	$3.18	$3.18
105 sq. ft. 14 ga. galv. wire netting	....	....	.13	13.65
4.5 sacks portland cement	....	....	4.24	19.08
18 cu. ft. sand	....	....	.48	8.64
Cost 100 sq. ft		$....		$44.55
Cost per sq. ft		....		.45

Material Cost of 100 Sq. Ft. of Terrazzo Floor ½" Thick, Applied Over Concrete Underbed

	Rate	Total	Rate	Total
3 sacks portable cement	$....	$....	$4.24	$12.72
600 lbs. marble granules	....	....	.13*	78.00
0.5 sack port. cement (grout)	....	....	4.24	2.12
Cost 100 sq. ft		$....		$92.84
Cost per sq. ft		....		.93

*Varies according to kind of granules used. Add for metal strips.

Material Cost of 100 Sq. Ft. Terrazzo Floors Blocked Into 5'-0" Squares, Including 1¼" Concrete Underbed

	Rate	Total	Rate	Total
Underbed 1¼" Thick				
3 sacks portland cement	$....	$....	$4.24	$12.72
11 cu. ft. sand	....	....	.48	5.28
Terrazzo Floor				
3 sacks portland cement	....	....	4.24	12.72
600 lbs. marble granules	....	....	.13*	78.00
0.5 sack cement (grout)	....	....	4.24	2.12
50 lin. ft. metal strips	....	....	.53**	26.50
Cost 100 sq. ft		$....		$137.34
Cost per sq. ft		....		1.37

Labor Cost of 100 Sq. Ft. Terrazzo Floors Blocked Into 5'-0" Squares, Including 1¼" Concrete Underbed

	Hours	Rate	Total	Rate	Total
Mechanic	3.75	$....	$....	$25.62	$96.08
Helper	5.65	....	....	19.13	108.08
Labor rub and finish	8.00	....	....	25.62	204.96
Cost 100 sq. ft			$....		$409.12
Cost per sq. ft			....		4.09

If terrazzo is placed over wood subfloor or separated from concrete floor, add difference in cost of underbed.

Material Cost of 100 Sq. Ft. Terrazzo Floors Blocked Into 4'-0" Squares, Including 1¼" Concrete Underbed

	Rate	Total	Rate	Total
Underbed 1¼" Thick				
3 sacks portland cement	$....	$....	$4.24	$12.72
11 cu. ft. sand	....	....	.48	5.28
Terrazzo Floor				
3 sacks portland cement	....	....	4.24	12.72
600 lbs. marble granules	....	....	.13*	78.00
0.5 sack cement (grout)	....	....	4.24	2.12
60 lin. ft. metal strips	....	....	.53*	31.80
Cost 100 sq. ft		$....		$142.64
Cost per sq. ft		....		1.43

Labor Cost of 100 Sq. Ft. Terrazzo Floors Blocked Into 4'-0" Squares, Including 1¼" Concrete Underbed

	Hours	Rate	Total	Rate	Total
Mechanic	4	$....	$....	$25.62	$102.48
Helper	6	....	....	19.13	114.78
Labor rub and finish	8	....	....	25.62	204.96
Cost 100 sq. ft			$....		$422.22
Cost per sq. ft			....		4.22

If terrazzo is placed over wood subfloor or separated from concrete floor, add difference in cost of underbed.

Material Cost of 100 Sq. Ft. Terrazzo Floors Blocked Into 3'-0" Squares, Including 1¼" Concrete Underbed

	Rate	Total	Rate	Total
Underbed 1¼" Thick				
3 sacks portland cement	$....	$....	$4.24	$12.72
11 cu. ft. sand	....	....	.48	5.28
Terrazzo Floor				
3 sacks portland cement	....	....	4.24	12.72
600 lbs. marble granules	....	....	.13*	78.00
0.5 sacks cement (grout)	....	....	4.24	2.12
80 lin. ft. metal strips	....	....	.53**	42.40
Cost 100 sq. ft		$....		$153.24
Cost per sq. ft		....		1.53

Labor Cost of 100 Sq. Ft. Terrazzo Floors Blocked Into 3'-0" Squares, Including 1¼" Concrete Underbed

	Hours	Rate	Total	Rate	Total
Mechanic	4.25	$....	$....	$25.62	$108.89
Helper	6.40	....	....	19.13	122.43
Labor rub and finish	8.00	....	....	25.62	204.96
Cost 100 sq. ft			$....		$436.28
Cost per sq. ft			....		4.36

*Varies according to kind of granules used. **Varies according to kind and gauge of metal.

If terrazzo is placed over wood subfloor or separated from concrete floor, add difference in cost of underbed.

Material Cost of 100 Sq. Ft. Terrazzo Floors Blocked Into 2'-0" Squares, Including 1¼" Concrete Underbed

	Rate	Total	Rate	Total
Underbed 1¼" Thick				
3 sacks portland cement	$....	$....	$4.24	$12.72
11 cu. ft. sand	....	....	.48	5.28
Terrazzo Floor				
3 sacks portland cement	....	....	4.24	12.72
600 lbs. marble granules	....	....	.13*	78.00
0.5 sack cement (grout)	....	....	4.24	2.12
120 lin. ft. metal strips	....	....	.53**	63.60
Cost 100 sq. ft		$....		$174.44
Cost per sq. ft		....		1.74

*Varies according to kind of granules used. **Varies according to kind and gauge of metal.

Labor Cost of 100 Sq. Ft. Terrazzo Floors Blocked Into 2'-0" Squares, Including 1¼" Concrete Underbed

	Hours	Rate	Total	Rate	Total
Mechanic	4.6	$....	$....	$25.62	$117.85
Helper	6.9	....	....	19.13	132.00
Labor rub and finish	8.0	....	....25.62		204.96
Cost 100 sq. ft			$....		$454.81
Cost per sq. ft			....		4.55

If terrazzo is placed over wood subfloor or separated from concrete floor, add difference in cost of underbed.

Material Cost of 100 Sq. Ft. Terrazzo Floors Blocked Into 1'-0" Squares, Including 1¼" Concrete Underbed

	Rate	Total	Rate	Total
Underbed 1¼" Thick				
3 sacks portland cement	$....	$....	$4.24	$12.72
11 cu. ft. sand	....	....	.48	5.28
Terrazzo Floor				
3 sacks portland cement	....	....	4.24	12.72
600 lbs. marble granules	....	....	.13*	78.00
0.5 sacks cement (grout)	....	....	4.24	2.12
210 lin. ft. metal strips	....	....	.53	111.30
Cost 100 sq. ft		$....		$222.09
Cost per sq. ft		....		2.22

Labor Cost of 100 Sq. Ft. Terrazzo Floors Blocked Into 1'-0" Squares, Including 1¼" Concrete Underbed

	Hours	Rate	Total	Rate	Total
Mechanic	5.35	$....	$....	$25.62	$137.07
Helper	8.00	....	....	19.13	153.04
Labor rub and finish	8.00	....	....	25.62	204.96
Cost 100 sq. ft			$....		$495.07
Cost per sq. ft			....		4.95

If terrazzo is placed over wood subfloor or separated from concrete floor, add difference in cost of underbed.

Material Cost of 100 Lin. Ft. of 6" Terrazzo Cove Base

	Rate	Total	Rate	Total
Scratch Coat				
1.5 sacks portland cement	$....	$....	$4.24	$6.36
3 cu. ft. sand	....	....	.48	1.44
Terrazzo Base				
1 sack portland cement	....	....	4.24	4.24
200 lbs. marble granules	....	....	.13*	26.00
25 metal dividers	....	....	.53**	13.50
Cost 100 lin. ft		$....		$51.54
Cost per lin. ft		....		.52

*Varies according to kind of granules used. **Varies according to kind and gauge of metal.

Labor Cost of 100 Lin. Ft. of 6" Terrazzo Cove Base

	Hours	Rate	Total	Rate	Total
Mechanic	16	$....	$....	$25.62	$409.92
Helper	16	....	....	19.13	306.08
Labor rub and finish	10	....	....	25.62	256.20
Cost 100 lin. ft			$....		$972.20
Cost per lin. ft			....		9.72

Double cove base, double cost of single cove base.

Abrasive Materials for Terrazzo Floors and Stair Treads. Where an anti-slip surface is required on terrazzo floors, platforms, landings and treads, mix 2 parts abrasive aggregates with 3 parts marble granule. For heavy duty floors, the abrasive aggregate shall be mixed in the terrazzo topping for its entire thickness. For light traffic floors, the abrasive shall be sprinkled on the surface only, to show in the proportion of 3 parts marble granule and 1 part abrasive aggregate.

Abrasive aggregates for 100 sq. ft. of heavy duty floors should cost as follows:

	Rate	Total	Rate	Total
240 lbs. abrasive aggregates	$....	$....	$0.53	$127.20
Cost per sq. ft		....		1.27

Deduct cost of 240 lbs. of marble granules included in itemized estimates. Abrasive aggregates for 100 sq. ft. of light duty floors should cost as follows:

	Rate	Total	Rate	Total
10 lbs. abrasive aggregates	$....	$....	$0.53	$5.30
Cost per sq. ft		....		.05

For prices on special terrazzo floors, such as venetian, mosaic and conductive, it is advisable to contact a local installation contractor.

There are several products on the market that resemble terrazzo floors and are installed by terrazzo mechanics. These are magnisite, latex, epoxy, and polyester resin type terrazzo floors. These floors can be installed over any sound existing floor in ½" thickness, with the exception of the epoxy and polyester resin types which can be installed as thin as the marble granules will allow. The costs of these floors are similar to that of regular terrazzo.

Precast terrazzo is available for base, floor, wainscots and stair treads. Straight 6" base runs about $4.00 per lin. ft., and one mechanic sets 125 lin. ft. per 8 hr. day. Floor tiles, 1" thick runs $4.30 per sq. ft. and one mechanic sets 60-70 sq. ft. per 8 hr. day. Stair treads cost around $10.50 per lin. ft., because these are heavy. On a stair of any width it takes a mechanic and helper to set 100 lineal ft. per 8 hr. day.

09500 ACOUSTICAL TREATMENT

There is a large variety of prefabricated acoustical products available on the market. In general, they fall into three categories: acoustical tiles that are mounted on ceilings, on suspended boards, or clipped or splined to suspension systems; acoustical panels for lay-in systems on exposed runners; and metal pan systems for mechanical suspension. Many systems today are also engineered to distribute air and light as well as to control sound; also, some systems are rated as fire resistive assemblies.

Acoustical tiles are applied to ceilings and walls for the purpose of absorbing and deadening sound in offices, banks, and other spaces where the utmost quiet is desired and in auditoriums to obtain optimum reverberation time.

Acoustical tiles are made of many different materials, such as wood fiber, cane fiber, mineral wool, cork, specially processed mineral filaments, perforated metal units, and other insulating materials.

They are made in a variety of sizes, such as 12"x12", 12"x24", 24"x24", 24"x48", etc., and vary from ½" to 2" thick, depending upon the materials used.

Acoustical tile may be installed over sound, dry plaster or concrete that is thoroughly seasoned, over gypsum board, or nailed or screwed to 1"x3" wood furring strips or by means of mechanical suspension systems.

Acoustical tile work is simple to estimate, but because it is almost always factory cut and prefinished, it is extremely important to see that job conditions are the very best. Once installed, there is little opportunity to "touch up" the work. Call backs can quickly eat up profits as well as ruin reputations.

One of the most important conditions to check is temperature and humidity. These should always approximate the interior conditions that will exist when the building is occupied.

All plastering, poured roof decks, concrete and floor work should be complete and dry; all windows and doors in place and glazed. If weather demands it, the building heating, rather than space heating, should be turned on and kept on as though the building were occupied.

The trend today is to square edge tile, which minimizes joint visibility. However, tile work will always have slight unevennesses in it, and if the job conditions include cove lights or high windows where light strikes the tile surface at a small angle, tile with beveled edges will help conceal any unevenness.

Budget Prices on Acoustical Tile. Most acoustical tile jobs are sublet to concerns specializing in this work. They have mechanics experienced in applying their particular brands of tile and can complete the job at less cost than a general contractor who purchases the tile and attempts to install them.

The following are approximate prices on the various classes of tile erected in place on fair size jobs as they might be quoted by a subcontractor including his overhead and profit.

Type of Installation	Price per Sq. Ft.
½" wood fibre tile surface applied	$1.15
⅝" mineral tile suspended*	1.50
¾" mineral tile suspended*	1.65
1" pad on steel pan suspended*	2.55
1" pad on alum. pan suspended*	3.70
⅝" mineral fibreboard suspended*	1.00
Air distributing ceilings-board* 2'x2'	1.45
Air distributing ceilings tile* 12"x12"	1.75

*Including hangers but not additional channel supports.

Acoustical Tile Directly Applied. When cemented, four spots of cement or adhesive should be applied to each tile on each corner of a 12"x12" tile. The spots should be placed as uniformly as possible with relation to the corners of the tile. A sufficient area on each corner of the tile should be primed prior to the application of the spot to receive the entire spot when it has been pressed to its final diameter. The spots of cement should be applied to these primed areas using a putty knife or trowel. The spot should be approximately the size and shape of a walnut, and should be nearly uniform in size and shape except where thicker spots are required to level surfaces. They should stand out from the tile about ½" to ¾" so as to contact the surface at about the same time. The diameter of the spot when pressed in place should be approximately 2½" to 2¾" and about ⅛" thick.

In placing, the tile should be held with both hands so that when the pressure is applied it will be between the spots rather than outside them. The tile should be held as nearly parallel to its final position as possible and pressed to position with a firm pressure. A slight lateral sliding motion of the hands while setting is necessary to settle the cement into the ceiling surface. It is important that the tile be pressed to level when setting.

A crew of at least two men is required to handle the work of installation efficiently, one man on the scaffold applying the adhesive to the tile and the other man placing the tiles on the ceiling.

One gallon of acoustical cement is sufficient for approximately 60 sq. ft. of tile under average conditions, although this varies somewhat with the surface to which it is applied. Adhesive costs about $10.00 per gallon.

The labor cost of placing acoustical tile on ceilings will vary with the size of the rooms or spaces where the tile are to be applied, pattern in which the tile are laid, height of scaffolding, etc.

The following quantities are based on 2 mechanics working together from a 4- to 6-ft. scaffold and laying tile in plain square or ashlar designs:

Size of Acoustical Tile	No. Pcs. Placed per 8-Hr. Day	No. SF per 8-Hr. Day	Mechanic Hrs. per 100 SF
12"x12"	600-700	600-700	2.5
12"x24"	400-450	800-900	2.0

When acoustical tile are laid in herringbone pattern, it requires 30% to 40% more time than regular squares or common bond. Diagonal patterns require 20% to 25% more time than regular squares or common bond, while mixed ashlar designs require about 50 percent more time than straight squares or common bond.

When placing tile in auditoriums and other spaces requiring high scaffolding, the additional cost of the scaffolding will have to be added to the cost of installing the tile. Large jobs are applied faster proportionately than small jobs.

Material Cost of 100 Sq. Ft. 12"x12" Acoustical Tile on Ceilings

	Rate	Total	Rate	Total
100 sq. ft. mineral tile 3/4"	$....	$....	$.70*	$70.00
1.75-gal. adhesive	....	....	7.95	13.91
Cost 100 sq. ft		$....		$83.91
Cost per sq. ft		....		.84

*Cost of tile varies according to kind and thickness.

Labor Cost of 100 Sq. Ft. 12"x12" Acoustical Tile on Ceilings

	Hours	Rate	Total	Rate	Total
Mechanics	2.5	$....	$....	$24.04	$60.10
Cost per sq. ft			....		.60

Acoustical Tile On Suspended Systems

Lay-in type systems for 2'x2' and 2'x4' grids will have a material cost of around 55 cents and 50 cents per sq. ft. respectively, not including any channel supports if they are needed. One man should install 700 sq. ft. per eight hour day of 2'x4' grid, 600 sq. ft. for 2'x2' grid.

Concealed spline systems are more time consuming. Material costs will run around 60 cents a sq. ft. and one man will install 500 sq. ft. per day.

If channel carriers are required, the material cost will be about 15 cents per sq. ft. and one man should install about 500 sq. ft. per 8 hr. day.

Labor Cost To Install 100 Sq. Ft. 2'x2' Lay-in Grid

	Hours	Rate	Total	Rate	Total
Mechanic	1.3	$....	$....	$24.04	$31.25
Cost per sq. ft					.31

Labor Cost To Install 100 Sq. Ft. 2'x4' Lay-in Grid

	Hours	Rate	Total	Rate	Total
Mechanic	1.2	$....	$....	$24.04	$28.85
Cost per sq. ft					.29

Labor Cost To Install 100 Sq. Ft. 12 x 12 Z Bar System Grid

	Hours	Rate	Total	Rate	Total
Mechanic	1.6	$....	$....	$24.04	$38.46
Cost per sq. ft					.39

Labor Cost To Install 100 Sq. Ft. Channel Ceiling Supports

	Hours	Rate	Total	Rate	Total
Mechanic	1.6	$....	$....	$24.04	$38.46
Cost per sq. ft					.38

Acoustical Tile For Suspended Grid Systems. This material is made "cut to size" ready to install into the grid system. The 24"x24" and 48" boards cost about 50 to 60 cents per sq. ft. depending on the type of tile, and one worker should install about 100 sq. ft. per hr. The 12"x12" tiles for the Z bar system costs about 75 to 85 cents per sq. ft., and one worker should install about 100 sq. ft. in 2 hrs.

Metal Pan Type Acoustical Tile. This type of acoustical treatment consists of metal units 12"x24" in size, having center grooves which result in the appearance of 12"x12" units when applied. The edges, which are beveled and returned vertically, are firmly held in place on the 12" sides by special tee runners. Within the metal units and supported on crimped, galvanized wire mesh, or on miniature channels, are mineral wool absorbent pads.

These systems will install at the rate of around 250 sq. ft. per day, including sound absorbing pads, but without channel supports if they are required. Material costs will run around $2.00 for steel, $3.00 for aluminum and $4.00 for stainless.

Labor Cost To Install 100 Sq. Ft. Suspended Steel Pan System

	Hours	Rate	Total	Rate	Total
Mechanic	3.2	$....	$....	$24.04	$76.93
Cost per sq. ft					.77

09550 WOOD FLOORING

Estimating Quantities of Wood Flooring. When estimating the quantity of board feet (b.f.) of wood flooring required for any job, take the actual number of sq. ft. in any room or space to be floored, and add allowances as given in the following tables:

Measured Size Inches	Finished Size Inches	No. Pcs. Pcs. in Bundle	Add for Waste Percent	To Obtain Quantity Flooring Repaired Multiply Area by	No. BF Flg. Required for 100 Sq. Ft. Floor
1x1	3/8x7/8	24	16 2/3	1 1/6 or 1.17	117
1x2	3/8x1 1/2	24	33 1/3	1 1/3 or 1.33	133
1x2 1/2	3/8x2	24	25	1 1/4 or 1.25	125
1x2 1/4	25/32x1 1/2	12	50 1/3	1 1/2 or 1.50	150
1x2 3/4	25/32x2	12	37 1/2	1 3/8 or 1.375	137 1/2
1x3	25/32x2 1/4	12	33 1/3	1 1/3 or 1.33	133
1x4	25/32x3 1/4	8	25	1 1/4 or 1.25	125

Amount of Surface 1,000 Board Feet of Flooring Will Cover and Quantity of Nails Required to Lay it

How Measured	Size Flooring	Covers SF Fl.	Nailed Every	Lbs. Nails Req'd. Cut Nails	Helically threaded nails
1x2	$^3/_8$x1$^1/_2$	750	8 in.	20 lbs. 4 "d" cut	
1x2$^1/_2$	$^3/_8$x2	800	8 in.	17 lbs. 4 "d" cut	
1x2$^1/_4$	$^{25}/_{32}$x1$^1/_2$	667	12 in.	70 lbs. 8 "d" cut	60 lbs 7 "d"
1x2$^3/_4$	$^{25}/_{32}$x2	727	12 in.	56 lbs. 8 "d" cut	47 lbs 7 "d"
1x3	$^{25}/_{32}$x2$^1/_4$	750	10 in.	64 lbs. 8 "d" cut	54 lbs 7 "d"
1x4	$^{25}/_{32}$x3$^1/_4$	800	10 in.	29 lbs. 8 "d" cut	24 lbs 7 "d"

Drill flooring for nails, if possible, for best results. No predrilling is required for helically threaded nails.

Note: The above figures are based on laying the flooring straight across a rectangular room without producing any design whatever.

Rules for Grading Hardwood Flooring. In many instances the specifications of the architects and owners are confusing and not in accordance with the standard rules for grading as adopted by the manufacturers. The rules adopted by the flooring manufacturers are given below so that they may be compared with the architect's specifications and the proper grade of flooring figured.

First Grade. $^{25}/_{32}$ inch and thicker, shall have the face practically free of all defects, but the varying natural color of the wood shall not be considered a defect. Standard lengths in all widths in this grade shall be in 2 foot bundles and longer as the stock will produce. Not over 30% of the total footage shall be in bundles under 4 feet, and not more than 17% of the total footage shall be in 2-ft. bundles.

Second Grade. $^{25}/_{32}$ inch and thicker, will admit tight, sound knots and slight imperfections in dressing, but must lay without waste. Standard lengths in all widths in this grade shall be in 2-ft. bundles and longer as the stock will produce. Not over 45% of the total footage shall be in bundles under 4 feet, and not more than 25% of the total footage shall be in 2-ft. bundles.

Third Grade. $^{25}/_{32}$ inch and thicker must be of such character as will lay and give a good serviceable floor. Standard lengths in all widths of this grade shall be in 1$^1/_4$-ft. bundles and longer as the stock will produce. Not over 65% of the total footage shall be in bundles under 4 feet, and not more than 27% of the total footage shall be in 1$^1/_4$-foot bundles.

Combination Grades

Second and Better Grade. In all thicknesses and widths, it is a combination of first and second grades developing in the strip without crosscutting for each grade. The lowest grade pieces admissible shall not be less than standard second grade. Standard lengths in all widths in this grade shall be in 2-ft. bundles and longer as the stock will produce. Not over 40% of the total footage shall be in bundles under 4 feet, and not more than 20% shall be in 2-ft. bundles.

Third and Better Grade. In all thicknesses and widths, a combination of first, second and third grades developing in a strip without crosscutting. The lowest grade pieces admissible shall not be less than standard third grade. Standard lengths in all widths of this grade shall be in 1$^1/_4$-foot bundles and longer as the stock will produce. Not over 50% of the total footage shall be in bundles under 4 feet.

Special Color Grades

(The Finest Grades Produced)

Selected First Grade Light Northern Hard Maple. This grade is selected for light color. The color tones in individual strips will vary somewhat, but after laying, this grade provides a luxurious "light" appearing floor. It costs around $4.00 per sq. ft. for material.

Selected First Grade Amber Northern Hard Maple. This grade is selected for amber color. The color tones in individual strips will vary somewhat, but after laying, this grade provides a luxurious "amber" appearing floor. The cost is $3.75 per sq. ft. for material.

Oak Flooring - Quarter Sawed

Clear. The face shall be practically clear, admitting an average 3/8" of bright sap. The question of color shall not be considered. Bundles to be 2 ft. and up. Average length is 3 3/4 feet.

Select. The face may contain sap, small streaks, pin worm holes, burls, slight imperfections in working, and small tight knots which do not average more than one to every 3 feet. Bundles to be 2 ft. and up. Average length is 3 1/4 feet.

Plain Sawed

Clear. The face shall be practically clear, admitting an average of 3/8" bright sap. The question of color shall not be considered. Bundles to be 2 foot and up. Average length 3 3/4 feet.

Select. The face may contain sap, small streaks, pin worm holes, burls, slight imperfections in working, and small tight knots which do not average more than one to every 3 feet. Bundles to be 2 ft. and up. Average length 3 1/4 feet.

No. 1 Common. Shall be of such nature that will lay a good residential floor and may contain varying wood characteristics, such as flags, heavy streaks and checks, worm holes, knots and minor imperfections in working. Bundles to be 2 foot and up. Average length 2 3/4 feet.

No. 2 Common. May contain sound natural variations of the forest product and manufacturing imperfections. The purpose of this grade is to furnish an economical floor suitable for homes, general utility use, or where character marks and contrasting appearance is desired. Bundles to be 1 1/4 foot and up. Average length 2 1/4 feet.

1 1/4 Shorts. Pieces 9 to 18 inches long are to be bundled together and designated as 1 1/4-foot shorts. Pieces grading No. 1 Common, Select and Clear to be bundled together and designated No. 1 Common and Better with pieces grading No. 2 Common bundled separately and designated as such. Although pieces 6" under and only 3" over the nominal length of the bundle may be included, the pieces must average 1 1/4" which is achieved through the natural preponderance of longer lengths.

Standard Thicknesses and Widths

25/32" thickness; widths 1 1/2" face, 2" face, 2 1/4" face, and 3 1/4" face.

$^3/_8$" thickness; width $1^1/_2$" face and 2" face.
$^1/_2$" thickness; width $1^1/_2$" face and 2" face.
Above tongue and groove and end matched.

Square Edge Strip Flooring

Grades as shown above but bundling and lineals as follows: $^5/_{16}$x2", $^5/_{16}$x$1^1/_2$", $^5/_{16}$x$1^1/_3$", $^5/_{16}$x$1^1/_4$", $^5/_{16}$x$1^1/_8$", $^5/_{16}$x1", $^5/_{16}$x$^7/_8$". Also made rough back in $^{11}/_{32}$x2" and $^{11}/_{32}$x$1^1/_2$".

Clear. Bundled 2 feet and up. Average length $5^1/_2$ feet.

Select. Bundled 2 feet and up. Average length $4^1/_2$ feet.

No. 1 Common. Bundled 2 feet and up. Average length $3^1/_2$ feet.

No. 2 Common. May contain defects of all characters, but will lay a serviceable floor.

Bundles to be $1^1/_4$ feet and up. Average length $2^1/_2$ feet. All faces shown above in $^5/_{16}$" square edge are finished $^1/_{64}$" over face.

Approximate Prices of Oak Flooring

Oak flooring prices will vary widely throughout the country. A recent quote gives $^{25}/_{32}$"x$2^1/_4$ oak at $3.25 per sq. ft., select, $3.00 common. Clear will run 30% more than select, if it is at all available.

Grading Rules on Prefinished Hardwood Flooring

Oak. White and Red Oak to be separated in each grade. Grades are established after the flooring has been sanded and finished.

Prime Grade. Face shall be selected for appearance after finishing, but sapwood and the natural variations of color are permitted. Minimum average length 4'. Bundles 2' and longer.

Standard Grade. Will contain sound wood characteristics which are even and smooth after filling and finishing and will lay a sound floor without cutting. Minimum average length 3'. Bundles $1^1/_4$ and longer.

Standard and Better Grade. A combination of Prime and Standard to contain the full product of the board except that no pieces are to be lower than Standard Grade. Minimum average length $3^1/_2$'. Bundles $1^1/_4$' and longer.

Tavern Grade. Shall be of such nature as will make and lay a serviceable floor without cutting, but purposely containing typical wood characteristics, which are to be properly filled and finished. Minimum average length $2^1/_2$'. Bundles $1^1/_4$' and longer.

Beech and Pecan

(Will be furnished only in a combination grade of Tavern and Better)

Tavern and Better. A combination of Prime, Standard, and Tavern to contain the full product of the board, except that no pieces are to be lower than Tavern Grade. Minimum average length 3'. Bundles $1^1/_4$' and longer.

General Rules for All Species

Hardwood flooring is bundled by averaging the lengths. A bundle may include pieces from 6" under to 6" over the nominal length of the bundle. No piece shorter than 9" admitted.

The percentages under 4' referred to apply on total footage in any one shipment of the item.

A 3/4" allowance shall be added to the face length when measuring the length of each piece of flooring.

Flooring shall not be considered of standard grade unless the lumber from which the flooring is manufactured has been properly kiln dried.

Laying and Finishing Wood Floors

The following pages contain detailed itemized costs of laying and finishing all kinds of soft and hardwood floors.

Labor Laying 100 Sq. Ft. of Hardwood Floors

Description of Work	SF Laid 8-Hr. Day	Carp. Hours	Labor Hours
	Ordinary Workmanship		
25/32"x3 1/4" face softwood floors for porches, kitchens, factories, stores, etc.	400-500	1.8	0.6
25/32"x2 1/4" face Third Grade Maple, for warehouse, factory and loft building floors	375-425	2.0	0.6
25/32"x2 1/4" face Oak or Birch in residences, apartments, stores, offices, etc	250-300	2.9	0.6
Same as above laid by experienced floor-layer	300-350	2.5	0.6
25/32"x1 1/2" face Third Grade Maple for warehouse, factory and loft building floors	300-325	2.3	0.7
25/32"x1 1/2" face Oak or Birch in residences, apartments, stores, offices, etc	150-200	4.5	0.8
Same as above laid by experienced floor-layer	225-275	3.2	0.8
	First Grade Workmanship		
25/32"x2 1/4" face Oak or Birch in fine residences, apartments, hotels, stores and offices	200-225	3.8	0.6
Same as above laid by experienced floor-layer	225-250	3.3	0.6
25/32"x1 1/2" face Oak or Birch, same class of work as described above	120-150	6.0	0.8
Same as above laid by experienced floor-layer	175-200	4.3	0.8

In larger cities there are carpenters who make a specialty of laying and finishing floors, and they seldom do any other kind of work. Due to their long experience, they not only do a better job but accomplish more work per day than the average carpenter, and they expect to be paid accordingly. For this reason, production times are given for floors laid and finished by both carpenters and floorlayers.

Another reason some floorlayers will lay more floors than carpenters is because they do not nail the flooring 8" or 16" on centers but will nail at every other bearing unless watched continually.

The first or Ordinary Workmanship includes the grade of work found in moderate priced homes, apartments, stores, etc., where the floor laying permits some hammer marks. Considerable savings can be obtained by laying the strip flooring all the way through prior to the erection of partitions.

First Grade Workmanship includes all classes of high class buildings where the best in workmanship is required, with all flooring closely driven up, laid free from hammer marks and with all nails set.

Labor Laying 3/8" or 1/2" Hardwood Flooring. Figure same costs as given for 25/32" flooring.

Setting Sleepers in Mastic on Concrete Floors. A sound, fast, economical way to provide nailing for strip wood flooring installed over concrete subfloors is the sleeper-in-mastic method, used extensively in the construction of slab on grade houses.

Assuming concrete slab is smooth, level and dampproofed to prevent moisture seepage, 2"x4" wood sleepers are laid flat in hot or cold mastic and are immediately ready to receive strip flooring. Installation is as follows: After priming concrete slab, hot or cold mastic is spread to a 3/32" depth over the entire slab or in 1/4" deep rivers along parallel lines in positions where sleepers are to be placed.

Wood sleepers should be straight, flat, dry 2"x4" lumber in random lengths from 18" to 48". For best results, sleepers should be treated with an approved non-creosote wood preservative. Lay sleepers in mastic, flat side down, in staggered rows on 12" centers at right angles to direction of flooring. Ends should overlap 4" and end joints should be staggered by alternating short and long lengths. Leave 1" clearance between sleepers and all vertical surfaces to allow for normal expansion.

Laying strip wood flooring over sleepers in mastic should follow the same general procedure as for over wood subfloors with the following additional precautions:

a) Use end-matched flooring only.
b) Individual strips of flooring should span at least two sleeper spacings, with no adjacent joints occurring in the same spacing.
c) Flooring strips passing over sleeper laps should be nailed to both sleepers.

Material coverages and approximate prices are as follows:

Asphalt primer, 200 to 400 sq. ft. per gal., depending upon porosity of concrete surface. Price $4.50 per gal.

Hot mastic, 25 sq. ft. per gal. Price $3.50 per gal.

Cold mastic, 40 to 50 sq. ft. per gal. Price $3.25 per gal.

For average size rooms, 100 sq. ft. of floor area will require 120 to 135 lin. ft. or 80 to 90 ft. b.m. of 2"x4" wood sleepers.

Under ordinary conditions, a floorlayer should prime concrete floor, spread mastic and place wood sleepers at the rate of 100 sq. ft. per hour.

Labor Laying 1,000 BF Soft and Hardwood Flooring

Ordinary Workmanship

Description of Work	B.f. Lumber Per 8-Hr. Day	Carp. Hours	Labor Hours
25/32"x3¼" face softwood floors for porches, kitchens, factories, stores, etc	500-600	14.5	4
25/32"x2¼" face Third Grade Maple, used in warehouses, factory and loft buildings, etc	500-575	15.0	4
25/32"x2¼" face Birch or Oak flooring in residences, apartments, stores, etc	350-400	21.0	4
Same as above laid by experienced floorlayer	400-450	19.0	4
25/32"x1½" face Third Grade Maple, used in warehouses, factory and loft buildings, etc	450-500	17.0	4
25/32"x1½" face Birch or Oak flooring in residences, apartments, store and office buildings, etc	250-280	30.0	5
Same as above laid by experienced floorlayer	350-400	21.0	5

First Grade Workmanship

Description of Work	B.f. Lumber Per 8-Hr. Day	Carp. Hours	Labor Hours
25/32"x1¼" face Oak or Birch flooring in fine residences, apartments, hotels, stores and offices	275-300	28	4
Same as above laid by experienced floorlayer	325-375	23	4
25/32"x1½" face Oak or Birch, same class of work as described above for 2¼" face flooring	175-225	39	5
Same as above laid by experienced floorlayer	275-300	28	5

Sanding Wood Floors by Machine

Practically all wood floors today are sanded and finished by machine. The cost varies, depending on the kind of flooring, whether old or new work, class of workmanship, size of rooms, etc.

The floors are first sanded with a floor sanding machine and then the edges are finished, using a disc type edging machine, which is capable of sanding up to the base shoe or quarter round.

The class of workmanship of number of operations governs production and cost, but where just an ordinary grade of workmanship is required in average size rooms in houses, apartments, offices, etc., a good machine and operator should finish and edge 800 to 900 sq. ft. per 8-hr. day, at the following labor cost per 100 sq. ft.:

	Hours	Rate	Total	Rate	Total
Operator-finisher	1	$....	$....	$24.04	$24.04
Cost per sq. ft			....		.24

When used in store rooms, auditoriums and other large floor areas, a machine and operator should finish and edge 1,000 to 1,250 sq. ft. of new floor per 8-hr. day, at the following labor cost per 100 sq. ft.:

	Hours	Rate	Total	Rate	Total
Operator-finisher	0.75	$....	$....	$24.04	$18.03
Cost per sq. ft			....		.18

Surfacing Floors by Machine, First Grade Workmanship. In residences, apartments, hotel, store and office buildings, etc., where first grade workmanship is required, eliminating all irregularities and waves in the floor, about four cuts are necessary; first with No. 1½ or 2 grit sandpaper, then with No. ½ or 1 grit paper and the last two cuts with No. 0 paper.

On work of this kind an experienced operator should finish and edge 400 to 500 sq. ft. of floor per 8-hr. day, at the following labor cost per 100 sq. ft.:

	Hours	Rate	Total	Rate	Total
Operator-finisher	1.75	$....	$....	$24.04	$42.07
Cost per sq. ft			....		.42

Resurfacing Old Floors by Machine. When resurfacing old wood floors, including the removal of old varnish, dirt, grease spots, etc., the cost will vary with the condition of the old floors. This variation will run from 300 to 1,000 sq. ft. per 8-hr. day and should be determined by the estimator after examining the old floors.

There is little or no difference in resurfacing an old floor or surfacing a new one except that under normal conditions one extra cut with a special open-faced abrasive is usually necessary to remove old varnish or paint.

Sanding Wood Floors

The following specifications list the operations necessary to produce a first class job. Cheaper grades of workmanship will require less sanding and consequently less labor--and lower costs.

Sanding Old Wood Floors Laid in Strips. The first operation is to remove the surface finish by sanding in the direction of the grain of the wood until all surface finish has been removed, using a drum type sanding machine, with extra heavy open coat sandpaper grit No. 3½ or 4.

The second operation is to remove the effects of the first operation and level the floor. Sand as in first operation, using same type machine with sharp sandpaper, grit No. 2 or 2½, to remove the effects of the coarse open coat sandpaper and level the floor surface.

The third operation is to use a spinner along the edges of the floor and hand scrape all corners where the machines could not reach. Use a No. 3½ or 4 grit paper and repeat, using No. 2 or 2½ grit paper.

The fourth operation is to sweep the entire floor to remove the coarse grits and abrasives left by the previous operations.

The fifth operation is to sand as in the first operation using the same type machine with No. 0 or No. ½ fine grit.

The sixth operation is the final smoothing of the edges. Sand as in third operation using same type of machine with fine sandpaper, grit 0 or ½.

After sweeping the entire floor clean it is ready for any kind of floor finish.

Sanding New Wood Floors Laid in Strips. The first operation is to sand to a level surface. Sand in the direction of the grain of the wood until all the high edges or over-wood on the floor has been sanded off to a level surface, using a drum type sanding machine with No. 1½ or 2 grit sandpaper.

The following chart gives the proper abrasives or grits for surfacing all classes of wood floors:

Chart for Floor Surfacing Operations

Kind of Floors	Operation	Floor Conditions	Proper Grit
		New Floors	
Oak, Maple and Close-Grained Hardwood Floors	Roughing	Ordinary Floors	2-2½
		Well Laid Floors	1½
		Very Uneven Floors	2½
		Very Hard Floors	2½
	Finishing	Ordinary Finish	½
		Fine Finish	0
		Extra Fine Finish	00
		Rough Finish	1
Parquet Floors	Roughing		1½
	Finishing	Semi-Finish	½
		Finishing	0
Cork Tile Floors	Roughing		1½
	Finshing	Semi-Finish	½
		Finishing	0
		Old Floors	
Resurfacing Old Floors— Soft and Hard Woods	Removing Varnish, Paints, etc.	Ordinary Conditions	3½
		Extra Heavy Coat Paint or Varnish	4
	Roughing	Ordinary Floors	2-2½
	Finishing	Ordinary Finish	½
		Fine Finish	0
		For Penetrating Sealer	00

The second operation is to remove the scratch pattern of the first sanding. Sand in the direction of the grain of the wood until the scratch pattern of the first operation is eliminated using the same type sanding machine with No. 1 or ½ sandpaper. Frequent changes of sandpaper must be made to insure cutting rather than sliding over the surface.

The third operation is to use a spinner around the edges and scrape by hand where it is impossible to use the machine, without crossing the grain of the wood. Use same grits as on drum sander.

The fourth operation is to sweep the entire floor free from any grit and abrasives and then sand in direction of the grain of the wood, using the drum type machine as in the first operation, with No. 0 sandpaper. Change the sandpaper frequently to insure cutting rather than sliding over the surface.

After sweeping the floor clean it is ready for any kind of floor finish.

Sanding Old Parquet and Herringbone Wood Floors. The first operation is to sand lengthwise of the room or at 45 degrees, using a drum type sanding machine with No. 3½ or 4 open coat grit, using a very light pressure on the drum. The second operation is to remove the coarse marks of the first operation by sanding crosswise of the room using same type machine with No.

1 or 1½ grit. Change paper frequently to assure cutting rather than sliding over the surface. Then sand lengthwise of the room using No. ½ or No. 1 grit.

The third operation is to use a portable edging machine or hand scrape all places where the drum machine and edger cannot reach. Sweep the entire floor to remove coarse grits and abrasives left by the previous operations.

The fourth operation is to use a large diameter rotary sanding machine across and then lengthwise of the room with No. ½ sandpaper to remove the cross marks of previous operations.

The fifth and final operation is to sweep the entire floor free from all grits, then sand using the large diameter rotary sander with No. 00 or No. 0 sandpaper, going over the floor once across and once lengthwise of the room.

After sweeping the floor clean it is ready for any kind of floor finish.

Finishing Hardwood Floors

The modern and approved method of finishing hardwood floors is to use wood sealers which are a special compound of ingredients blended so that they penetrate into the fiber and cells of the wood, sealing the pores with tough elastic materials, thereby building up a base resistance to water, dirt and stains. It wears only as the wood wears down.

Finishing Hardwood Floors in Natural Color. Before finishing hardwood floors, they should be sanded as stated in previous paragraphs.

After the floor has been completely sanded and swept clean of all dust, a good grade of permanent penetrating seal should be applied. One such seal on the market is Pentra-Seal (from the American-Lincoln Corp., Toledo, OH). May be applied quickly with a brush, rag, mop, lamb's wool applicator, or squeegee and does not leave brush or lap marks.

For an extra fast job, Fast-Drying Seal is recommended. Apply a thin even coat and after it has become tack free remove the excess with a No. 2 steel wool pad. Sweep floor clean and apply a second coat in the same manner. Where a highly glossy finish is desired, apply a third coat and do not use steel wool. This will remain on the surface, since the preceding coats have thoroughly filled every cell and pore, and result in a glossy surface finish.

The floor should now be given a final protective coat of a good grade of paste wax or Lustre Finish. Paste wax should be applied manually and Lustre Finish can be put on with a mop. Two coats are recommended, followed by buffing with a mixed fiber brush.

Fast-Drying Seal is available in natural as well as seven decorator type colors. Coverage ranges from 400 to 450 sq. ft. per gallon for the first coat and 650 to 750 sq. ft. per gallon for the second coat. Price, Fast-Drying Seal Natural, about $15.00 per gal.; colors, slightly higher.

Coverage of Lustre-Finish is about 1,250 sq. ft. per gallon. Price, about $10.50 per gal.

Labor Finishing Wood Floors. Assuming the floor has been sanded and is ready for the seal, an experienced worker can finish about 1,000 sq. ft. of natural finish floor per 8-hr. day. Colored seal takes more time as the excess pigment must be removed by hand wiping.

Labor Cost of 100 Sq. Ft. of Fast-Drying Seal and Lustre Finish
Floor Finish on Wood Floors, Natural Color

	Hours	Rate	Total	Rate	Total
Mechanic	0.8	$....	$....	$24.04	$19.23
Cost per sq. ft			....		.19

Prefinished Hardwood Plank Flooring

Plank or strip flooring in ¾" thickness is available in several widths and finishes, most furnished in random lengths and widths. It is generally of red oak and predrilled for nailing. Widths are in the 3" to 8" range. Special distressed finishes and pegging are also available.

Plain flooring in narrower widths will run around $3.00 per sq. ft. For pegging, add 10 cents per sq. ft. Wider planks will run $3.75 per sq. ft. Premium finishes can raise costs to as much as $4.00 a sq. ft.

One carpenter should lay at least 150 sq. ft. per day.

Wood Parquet Flooring

Wood parquet is available in stock designs of oak, walnut, teak, cherry, and maple. Standard thickness is 5/16", although heavier parquet is available on custom orders in 11/16" and 13/16" thicknesses. Sizes will vary with patterns running from 9"x9", 12"x12", 16"x16", on up to 36"x36".

Parquet is set in mastic on any level floor above grade. Most parquet is ordered factory finished, but may be unfinished if desired. Many patterns are available in "prime grade" and "character marked", square or bevelled edges.

Costs of all wood flooring are changing rapidly, but square foot material costs will approximate $1.70 for a standard 5/16" oak, $3.75 for walnut, and $3.75 for teak. The prime grades in more elaborate patterns will run much higher. If prefinished, add 50 cents per sq. ft.

Laying Wood Parquet Floors in Mastic Over Concrete. Concrete subfloor shall be primed with one coat of asphalt priming paint and allowed to dry before laying parquet. Average covering capacity of primer 200 to 400 sq. ft. per gal. depending upon porosity of cement floor finish. Price $3.25 per gal. in 5-gal. cans.

Where concrete floors are in direct contact with the ground, below grade, or where there is not sufficient ventilation to prevent condensation, a membrane waterproofing shall be applied to the slabs, extending same up the sides of all walls, 4" to 6", according to manufacturer's specifications. Where wood parquet is to be laid in hot mastic, a hot application, two-ply membrane shall be applied. Where parquet is to be laid in cold mastic, a cold application one-ply polyethylene film shall be applied.

Installation of polyethylene film is as follows: after concrete surface is primed, cold mastic is applied at the rate of 80 to 100 sq. ft. per gal., into which polyethylene film is rolled, lapping joints 4 inches. Laps do not have to be sealed.

Parquet shall be laid to a level surface bedded in a suitable mastic without the use of nails or screws. One gallon of mastic will cover 45-50 sq. ft. and will run around $15.00 per gallon. Expansion joints must be left around all edges of rooms according to manufacturer's specifications.

Labor Laying Wood Parquet Flooring. When laying wood parquet floors in mastic in ordinary size rooms such as residences, apartments, etc., an experienced floorlayer with a helper should lay 300 to 350 sq. ft. of floor per 8 hr. day, while in large spaces such as schools, auditoriums, gymnasiums, etc., an experienced floorlayer should lay 350 to 400 sq. ft. of floor per 8-hr. day. This applies to either finished or unfinished parquet, as the same care is required in laying either type of flooring. This does not include any time preparing or leveling old concrete floors, as this must be added extra if required.

Where unfinished parquet is used, add for sanding and finishing.

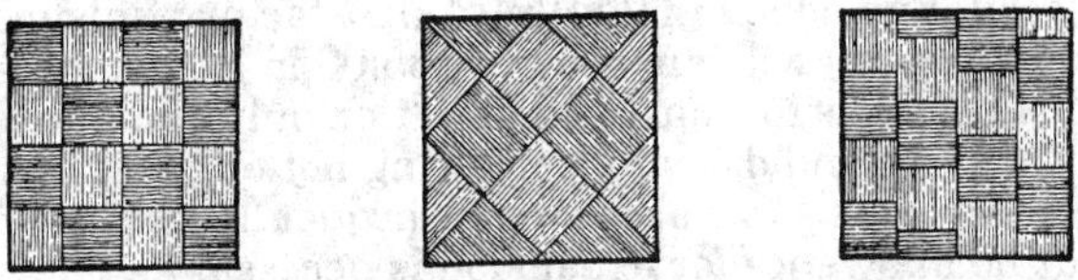

Various Designs in Wood Parquet Flooring

Laying Wood Parquet Floors in Mastic over Other Surfaces. Wood parquet flooring may also be laid in mastic over plywood or board subflooring, present wood floors, asphalt tile, etc., using the same general procedures as given above for over concrete, except subsurface will not require priming before applying mastic.

Subsurfaces must be sound, smooth, clean and dry--asphalt tile must be firmly bonded and in good condition--wood subflooring or present wood flooring may require re-nailing. Where installation is over square edged board subflooring, one layer of 15-lb. asphalt saturated felt should be laid to keep mastic from seeping through joints.

Prices of Wood Parquet Floors. Owing to the variation in freight rates, dealer's markup, etc., it is practically impossible to give prices that will apply to all parts of the country. Dealer's markup will probably vary from 25% on large jobs to 40% on smaller installations, plus freight from factory to destination.

Material Cost of 100 Sq. Ft. of 9"x9" 5/16" Oak Parquet Floors Laid in Mastic

	Rate	Total	Rate	Total
0.33 gal. asphalt primer	$....	$....	$2.65	$0.87
100 sq. ft. oak parquet	....	....	1.80	180.00
2.5 gal. hot mastic	....	....	12.75	31.88
Cost per 100 sq. ft		$....		$212.75
Cost per sq. ft		....		2.13

Cost of 100 Sq. Ft. of 9"x9" 5/16" Oak Parquet Floors Laid in Mastic

	Hours	Rate	Total	Rate	Total
Floorlayer	3	$....	$....	$24.04	$72.12
Helper	3	....	....	19.13	57.39
Cost per 100 sq. ft			$....		$129.51
Cost per sq. ft			....		1.30

If wood parquet is laid over wood or asphalt tile surfaces, omit asphalt primer.

If parquet floors are laid in large spaces, such as auditoriums, etc., deduct 1/4 to 1/2 hr. floorlayer time per 100 sq. ft.

Add for membrane waterproofing, if required.

If unfinished parquet is used, change price of blocks and add for sanding and finish.

Above costs do not include any allowance for preparing subfloor, if necessary.

Laminated Block Flooring. Laminated oak blocks are fabricated from 3-ply all oak plywood. Furnished in 9"x9"x1/2" size, factory finished.

A 27 sq. ft. carton (48 pcs.) will run around $35.00 in a light or dark finish.

Laminated block flooring is for mastic installation only and may be laid over any sound, smooth, clean and dry subfloor, using hot or cold mastic. General installation procedure is the same as for wood parquet floors given on previous pages, except that no allowance for expansion is necessary.

Labor costs for laying laminated block flooring should be about the same as for 9"x9"x5/16" wood parquet given above.

Wood Block Industrial Flooring. Creosoted wood blocks 2" to 3" thick will run from $1.70 to $3.80 per sq. ft. One carpenter will install around 200 sq. ft. per day.

09650 RESILIENT FLOORING

Resilient flooring includes asphalt tile, rubber, cork, vinyl asbestos, pure vinyl tile, and vinyl sheet flooring. Each type has its own qualities and should be selected on these considerations rather than cost alone. If material is to be installed on or below grade, cork tile and certain sheet vinyls are not suitable.

If grease is a problem, rubber, cork, and asphalt tile perform poorly, while if resilience and quietness are important, cork and rubber tile are excellent, but sheet vinyl is poor.

A quick guide for approximate installed costs per sq. ft. is as follows:

Type	Thickness	Cost
Asphalt tile	3/32"-1/8"	$0.70-0.95
Vinyl asbestos	1/16"	0.85-0.95
	3/32" & 1/8"	0.90-1.00
Cork tile-Prefinished	1/8", 3/16", 5/16"	1.30-2.00
Vinyl tile	.050"	1.40-2.00
Vinyl sheet	.070-.075"	1.70-2.65
Rubber tile	1/8", 3/16"	1.60-1.95

Asphalt Tile Floors

Asphalt tile is one of several types of resilient flooring that can be used satisfactorily on concrete floors that are in direct contact with the ground, at or below grade. It is suitable for basement rooms, because it is not affected by the moisture and alkali always present in concrete in contact with the ground. Asphalt tile is also used in residences, stores, offices and institutions where a low cost floor is desired.

They are furnished in a variety of colors, and tile sizes are 9"x9"x1/8" thick.

When applied over approved wood subfloors, the wood floor shall be finished smooth and level and covered with a layer of lining felt bonded to the wood with mastic and then rolled thoroughly with a linoleum roller. The asphalt tile is then laid over the lining felt, using an approved asphalt emulsion for bonding the tile to the felt.

When applied over concrete floors above, at, or below grade, the concrete floor must be smooth, dry, and even. The tile is laid directly on the concrete, using approved asphalt emulsion. At grade or subgrade, if dampness due to humidity is extreme, the concrete shall be given one coat of asphalt cut-back primer after which the tile is laid in asphalt (solvent type or cutback) cement.

How to Estimate Tile Floors

To Determine and Price the Quantity of Material Required. Compute the actual square footage of surface to be covered, and add a percentage for waste. The percentage will depend on the size of the area and regularity of outline. A fair percentage of waste to add to the actual area is shown in the accompanying table.

After the gross area has been determined, multiply it by the square foot price of tile. The result will be the price of the material.

The square foot price of the material will depend on the colors and quantity.

Table of Approximate Percentages of Allowable Waste for Various Areas

Up to 50 sq. ft.	14%
50 to 100 sq. ft.	10%
100 to 200 sq. ft.	7% to 9%
200 to 300 sq. ft	6% to 7%
500 to 1,000 sq. ft.	4% to 5%
1,000 to 5,000 sq. ft.	4%
5,000 to 10,000 sq. ft.	3%
10,000 and up	1½% to 3%

Quantity of Primer and Asphalt Cement Required. When priming concrete floors, one gallon of primer should cover 200 to 300 sq. ft. of surface, depending upon the porosity of the surface. Price $5.00 per gal.

One gallon of asphalt cement should lay 175 to 225 sq. ft. of asphalt tile over concrete floors. Costs about $12.00 per gal.

One gallon of asphalt emulsion for installing asphalt tile over lining felt on wood subfloors and direct to above grade concrete floors will cover approximately 140 to 175 sq. ft. Price $5.00 per gal.

Labor Laying Asphalt Tile. The cost of laying asphalt tile varies with the size and shape of the room or space where the tile are to be laid, size of tile, kind of subfloor, etc., as a tile setter will lay more sq. ft. of tile in a large room than one cut up with pilasters, offsets, etc., also a man will lay more sq. ft. of large size tile than of the smaller sizes.

When placing lining felt over wood subfloors, an experienced man should spread paste, lay and roll 250 sq. ft. of felt an hr. depending upon the room sizes.

After the lining felt has been laid (over wood floors), an experienced tile setter should lay the following quantities of tile per 8-hr. day:

Size of Tile	SF per 8-Hr. Day	Hours per 100 Sq. Ft.
9"x9"	400-425	2.0

The above quantities are based on fairly large rooms. If laid in small spaces, increase the labor costs accordingly.

Sq. Ft. Prices on Asphalt Floor Tile

The following are average prices on asphalt tile.

Group	1/8"
A colors	$0.50
B colors	.60
C colors	.75
D colors	.80

Diagonal half tile of 9"x9", add 3 cents per sq. ft. Feature strips in the standard colors are priced at their group price plus 10 cents per sq. ft. additional. Standard, die-cut design asphalt tile inserts, size 18"x18", cost about $3.00 each.

"A" colors--Dark red and black solid patterns.
"B" colors--Dark marbleized patterns.
"C" colors--Medium light marbleized patterns.
"D" colors--Light marbleized patterns.

Greaseproof Asphalt Tile

Greaseproof colors are also acid resistant. Standard size 9"x9".

	Price per Sq. Ft.
1/8"-Thick Greaseproof Asphalt Tile	$0.80

Vinyl Plastic Tile

Vinyl plastic tile is a more expensive resilient flooring material. It is very stable, will not shrink or expand and is highly resistant to oils, fats, greases, acids, alkalis, detergents, soaps and most solvents.

Furnished in 9"x9", 12"x12" and 18"x36" sizes and in 3/32" and 1/8" thicknesses.

Subfloor requirements are the same as for other resilient flooring. May also be laid over concrete floors, or on grade or below grade, using special adhesives.

Vinyl plastic tile is laid in the same manner as other resilient tile and labor costs are about the same.

Material costs vary according to pattern, ranging from 75 cents per sq. ft. for marbleized patterns to $3.00 per sq. ft. for translucent patterns, for 1/8" thick tiles. The 3/32" thick tiles cost about 15% to 20% less.

Rubber Tile Floors and Base

Rubber tile is furnished in 9"x9", 12"x12", and 18"x36" tile, 1/8" thick and in a variety of colors.

Rubber floor tile may be laid over wood or above grade concrete floors. It may also be laid over clean, dry, concrete floors on ground or below grade providing a special adhesive, recommended by the manufacturer of the tile, is used.

When laying rubber tile over an old wood floor, all loose, defective, or badly worn boards shall be replaced and face nailed. Any unevenness in the boards shall be planed or sanded smooth and even.

If the wood floors cannot be placed in proper condition by these methods, install Masonite or 1/4" plywood over the original wood floor.

Over the wood base there shall be installed semi-saturated lining felt which shall be securely affixed to the base with mastic. Depending upon the condition of the subfloor, the spreading capacity should be about 140 sq. ft. per gallon. After the felt is laid, it should be thoroughly rolled with a three-section 100-lb. linoleum roller.

The tile should be installed over the felt, using mastic.

The finished floor should be thoroughly rolled with a three-section 100-lb. linoleum roller and any high edges smoothed in place with a flat headed hammer.

Laying Rubber Tile. The cost of laying rubber floor tile varies with the size and shape of the room or space where the tile are to be laid, size of tile, kind of subfloor, etc., as it is easier to lay tile in a large square room than one cut up with offsets, pilasters, bays, etc., and a man will lay more sq. ft. of large size tile than of the smaller sizes.

When laying lining felt over wood subfloors, an experienced man should spread paste, lay and roll 250 sq. ft. of felt an hr. depending upon the room sizes.

After the felt has been laid (over wood floors), an experienced tile setter should lay and finish the following sq. ft. of tile of various sizes per 8-hr. day:

Size of Tile	SF per 8-Hr. Day	Hours per 100 Sq. Ft.
9"x9"	400-425	2.0
12"x12", 18"x36"	600-675	1.3

The above quantities are based on fair sized rooms. If laid in small spaces, increase labor costs accordingly.

Material Prices of Rubber Floor Tile, All Standard Colors

	Price per Sq. Ft.
1/8" Thick	$1.05-1.80

Rubber tile has lost some of its market to the vinyls but is still used quite often as a stair tread covering.

Stair Treads

Size	Marbleized	Plain Black
12½"x36"x¼"	$10.50	$10.00
12½"x42"x¼"	12.25	10.50
12½"x48"x¼"	14.25	13.00
12½"x54"x¼"	15.25	14.50
12½"x60"x¼"	16.50	16.00
12½"x72"x¼"	21.00	20.00

Landing tile will run around $1.10 per sq. ft. and 3/32" riser material about .65 cents per lin. ft.

Prices on Rubber Composition Top-Set Cove Base

Description	Price per Lin. Ft.
4" base	$0.50
6" base	.55

Cork Tile

Cork tile are made from high-grade pure cork shavings compressed and baked in molds. It is used for floors and wainscoting in offices, libraries, hospitals, schools, and other places where footsteps must be muffled.

They are furnished in 6"x12", 9"x9" and 12"x12" size 1/8", 3/16", and 5/16" thick. Cork tile are supplied beveled or unbeveled. With unbeveled cork tile, sanding the finished floor is usually necessary to offset unevenness of the subfloor. With beveled cork tile, the beveling conceals such irregularities, with the result that sanding is unnecessary and the smooth surface given at the tile factory is retained.

Cork tile should never be placed over a single wood subfloor but should be first covered with Masonite or 3/8" plywood. Never install on grade or subgrade floors.

The wood or above grade concrete subfloor must be smooth and even before cork tile can be placed. Lining felt is used over wood subfloors but this is not necessary when applied over smooth, level concrete floors.

After the floor has been laid and sanded (if unbeveled tile are used), the floor should be given a coat of cork tile filler, followed by two coats of cork tile finish (glossy or flat) and subsequent waxings with an approved water emulsion wax, or the floor may be finished by merely applying 3 applications of water emulsion wax and vigorously buffed by machine.

Approximate Square Foot Prices of Cork Tile

Kind of Tile	Tile Thickness 1/8"	5/16"
Color according to manufacturer's specs., regular sizes	$1.00	$1.50

Border feature strips: In any width under 3" and any length between 18" and 36" at 10 cents per sq. ft. additional.

Factory finished tile: Add 15 cents per sq. ft. for beveled cork tile finished at factory with sealer, undercoat, and one application of wax. After installation, apply one application of wax and buff.

Material Cost of 100 Sq. Ft. 9"x9"x1/8" Cork Tile Floors Over Wood Flooring

	Rate	Total	Rate	Total
0.75-gal. adhesive	$....	$....	$8.50	$6.38
12 sq. yds. lining felt	....	....	.32	3.84
105 sq. ft. cork tile	....	....	.95	99.75
0.75-gal. adhesive	....	....	8.50	6.38
Finishing Floor				
0.33-gal. cork tile sealer	....	....	3.18	1.05
0.25-gal. cork tile undercoat	....	....	3.18	.80
0.13-gal. water emul. wax	....	....	3.18	.41
Cost 100 sq. ft		$....		$118.61
Cost per sq. ft		....		1.19

Labor Cost of 100 Sq. Ft. 9"x9"x1/8" Cork Tile Floors Over Wood Flooring

	Hours	Rate	Total	Rate	Total
Mechanic	3.5	$....	$....	$24.04	$84.14
Sanding 100 sq. ft	1.0	....	....	24.04	24.04
Finishing Floor					
Clean and wax floor	2.0	....	....	24.04	48.08
Cost 100 sq. ft			$....		$156.26
Cost per sq. ft			....		1.56

Sheet Vinyl Flooring

Sheet vinyl can be installed on concrete floors, below and above grade, and on wooden floors that have a smooth surface. If the wood floor is rough, then install a layer of hardboard or 1/4" plywood to provide a good surface for the sheet vinyl.

This material is available in 6 and 12 foot wide rolls up to 120 feet in length, depending on the manufacturer. Thicknesses will vary depending on the product design, type of backing material, and the manufacturer; however, the most recognized gauges will fall in the .065", .080" and .090" range for non-cushion backed sheet vinyl.

Most sheet vinyl flooring materials are designed to require a minimum of maintenance, which means that the flooring cleans easily, is resistant to normal abuse, and does not require waxing to maintain the luster appearance.

Installation is by full adhesive coating on the 6 foot wide material and adhesive coating on seams only for the 12 foot wide rolls.

Material prices will vary from $1.25 to $2.50 per sq. ft. depending on the gauge and backing, but the estimator should be aware that this material is sold by the square yard and that proper floor layout should be determined to reduce the waste factor.

Labor for installation will vary depending on the size and shape of the area to be covered and the number of cuts to be made. One worker lays about 600 sq. ft. in large areas per 8 hour day, and about 350 sq. ft. in smaller areas per 8 hour day.

09680 CARPETING

With the development of man-made fibers special padding and backings, carpeting has become competitive with resilient tile in commercial construction. While almost always subcontracted to a firm specializing in the field, the general contractor should be acquainted with the basic choices available today.

Pure wool has set the high standards most people expect from carpet and no man-made fiber has yet equaled it in overall performance and appearance, although in some respects, such as abrasion resistance, texture retention and mildew resistance, man-made fibers excel. But wool carpeting is in the high cost range, running $16.00 to $27.00 per square yard in the 36 to 42 oz. commercial grades, without padding.

Nylon is the most often used of the man-made fibers. It has outstanding wear ability and resiliency, is mildew resistant and mothproof. But there are problems with static electricity unless it is specially treated or has grounding mesh woven into the backing. In the lighter weights, it is extremely economical, running around $7.00 per square yard for 15 oz.; $8.50 for 22 oz., both without padding.

The olefin (polypropylene) fibers are comparable in price to nylon, but they are found mostly in the indoor-outdoor type of carpet as they offer the lowest moisture absorbency and are exceptionally stain-resistant. Their appearance is dull, resiliency low; they have the lowest melting point, so are subject to cigarette burns. Their range is $5.50 to $9.75.

Acrylics surpass the other man-made fibers in looks and feel, most closely resembling wool, but they do not wear as well as nylon, and will pill some. They run $8.00 to $13.00 per square yard for 36 to 42 oz. weights, without padding.

Polyesters have poor resiliency and therefore are seldom used alone, except for shag-type carpets which are seldom encountered in contract work. They fall in a medium cost range. Padding may be laid separately or come with the carpet. It will add $1.10 to $2.75 to the material cost.

Carpets can be taped down or laid in mastic. Installation cost will run around $3.00 per square yard on open work. If not integral with carpet, add another $1.50 per square yard for laying padding.

Estimating Quantities. Carpet is estimated by the square yard measure (9 sq. ft. per sq. yd.). Since most carpeting is manufactured in 12 foot wide material, the estimator must be careful to determine how the carpet will lay on the floor to minimize the waste factor. Bear in mind that carpet has a definite "lay" to the nap and the "lay" must be in the same direction when joining pieces together, or else the carpet pieces will appear to have a different color.

09900 PAINTING

Estimates and bids on painting and interior decorating vary more than for almost any other construction trade. There are several kinds of painters--good, fair, poor, and just plain "daubers". There are painters who figure on "skinning the job" by applying one or two fewer coats of paint or varnish than called for in the specifications, others who figure on substituting cheap and inferior materials for those called for, and others who do not know how to do good work. Expect wide variation in figures. Of course, a cheap or inferior job costs less than a first class job, and the buyer usually gets what he pays for. When preparing painting estimates, consider the quality of materials and workmanship required.

Methods of Measuring and Listing Painting Quantities

When estimating quantities of painting or interior finishing, the actual surface to be painted should be measured as accurately as possible from the plans or from measurements taken at the building. From these measurements it should be possible to estimate the quantity of materials required, but the labor quantities present a much more difficult problem, owing to the different classes of work and the difficulties encountered in applying the paint. For instance, the covering capacities of paint will be the same for a plain surface or a cornice, but the labor cost painting the cornice will be considerably more owing to the height of the cornice above the ground and the amount of "cutting" or "trimming" necessary. Care must be used when pricing any piece of work, as conditions on each job are different.

Clapboard or Drop Siding Walls. Obtain actual area of all walls and gables. Add 10% to actual surface measurement to allow for painting under edge of boards. Do not deduct for openings less than 10'-0"x10'-0".

Shingle Siding. Obtain actual area of all walls and gables and multiply by 1½. Do not deduct for openings less than 10'-0"x10'-0".

Brick, Wood, Stucco, Cement and Stone Walls. Obtain actual area of all walls and gables. Do not deduct for openings less than 10'-0"x10'-0".

Eaves. Plain eaves painted same colors as side walls, obtain area and multiply by 1½. If eaves are painted a different color than side walls, obtain area and multiply by 2.

Eaves with rafters running through, obtain area and multiply by 3.

Eaves over brick, stucco or stone walls, obtain area and multiply by 3.

For eaves over 30'-0" above ground, add ½ for each additional 10'-0" in height.

Cornices, Exterior. Plain cornices, obtain area and multiply by 2. Fancy cornices or cornices containing dentils, etc., obtain area and multiply by 3.

Downspouts and Gutters. Plain downspouts and gutters measure the area and multiply by 2. For fancy downspouts and gutters, obtain area and multiply by 3.

Blinds and Shutters. Plain blinds or shutters, measure one side and multiply by 2. Slatted blinds or shutters, measure one side and multiply by 4.

Columns and Pilasters. Plain columns or pilasters, obtain area in sq. ft.; if fluted, obtain area and multiply by 1½; if paneled, obtain area and multiply by 2.

Lattice Work. Measure one side and multiply by 2 if painted one side only; if painted two sides, obtain area and multiply by 4.

Porch Rail, Balustrade, Balusters. If solid balustrade, add one foot to height to cover top and bottom rail. Multiply the length by the height. If painted two sides, multiply by 2.

If balustrade consists of individual balusters and handrail, multiply length by the height, then multiply the result by 4.

Handrail only. If under 1'-0" girth, figure as 1'-0" by length and multiply by 2.

Moldings. Cut in on both sides, figure 1 sq. ft. per lin. ft. if under 12" girth. If over 12" girth, take actual measurement.

Doors and Frames, Exterior. Inasmuch as it costs almost as much to paint a small door as a large one, do not figure any door less than 3'-0"x7'-0". Allowing for frame, add 2'-0" to the width and 1'-0" to the height. For instance, a 3'-0"x7'-0" door would be figured as 5'-0"x8'-0", or 40 sq. ft.

If a sash door, containing small lights of glass, add 2 sq. ft. for each additional light. A 4-lt. door would contain 8 sq. ft. additional; a 12-lt. door, 24 sq. ft. additional, etc.

If painted on both sides, obtain sq. ft. area of one side and multiply by 2.

Door frames only, where no door is hung, allow area of opening to take care of both sides.

Windows, Exterior. Inasmuch as it costs almost as much to paint a small window as a large one, do not figure any window less than 3'-0"x6'0". Add 2'-0" to both the width and height of the opening, to take care of the sides and head of the frame and the outside casing or brick mold, and multiply to obtain the area. For instance, a window opening 3'-0"x6'-0", add 2'-0" to both the width and height, making 5'-0"x8'-0", containing 40 sq. ft. of surface.

If sash contain more than one light each, such as casement sash, etc., add 2 sq. ft. for each additional light. A 6-lt. window would contain 12 sq. ft. additional; a 12-ft. window, 24 sq. ft. additional, etc.

Roofs. For flat roofs or nearly flat, measure actual area. For roofs having a quarter pitch, measure actual area and add 25%; roofs having a one-third pitch, measure actual area and add 33⅓%; roofs having a half pitch, measure actual area and add 50 percent.

Fences. Plain fences, measure one side and multiply by 2; picket fences, measure one side and multiply by 4.

Picture Mold and Chair Rail. On picture mold, chair rail, etc., less than 6" wide, obtain the number of lin. ft. to be painted or varnished and figure ¾ sq. ft. per lin. ft.

Wood and Metal Base. Wood or metal base 6" to 1'-0" high should be figured as 1'-0" high; under 6", figure at .5 sq. ft. per lin. ft.

Wood Panel Strips, Cornices, Etc. When estimating quantities of wood panel strips, wainscot rail and ceiling cornices, measure the girth of the member and if less than 1'-0" wide, figure as 1'-0".

If over 1'-0" wide, multiply the actual girth by the length, viz., 1'-3" x 200'-0" = 250 sq. ft.

Refer to rule for exterior cornices.

Interior Doors, Jambs and Casings. When estimating quantities for interior doors, jambs and casings, add 2'-0" to the width and 1'-0" to the height of the opening. This allows for painting or varnishing the edges of the door, the door jambs which are usually 6" wide, and the casings on each side of the door which average from 4" to 6" wide. Example: on a 3'-0"x7'-0" door opening, add 2'-0" to the width and 1'-0" to the height, which gives an opening 5'-0"x8'-0", containing 40 sq. ft. on each side. Some painters figure all single doors at 40 sq. ft. per side, or 80 sq. ft. for both sides, while others figure them at 50 sq. ft. per side or 100 sq. ft. for both sides. Do not deduct for glass in doors.

If a sash door, containing small lights of glass, add 2 sq. ft. for each additional light. A 4-lt. door would contain 8 sq. ft. additional; a 12-lt. door 24 sq. ft. additional, etc.

Interior Windows, Jamb Linings, Sash and Casings. When estimating painting quantities for windows and window trim, add 2'-0" to the sides and length to allow for jamb linings, casing at the top and sides, and window stool and apron at the bottom. Example: If the window opening is 3'-0"x6'-0", adding 2'-0" to both the width and length gives a window 5'-0"x8'-0", containing 40 sq. ft. of surface.

If sash contain more than one light each, such as casement sash, etc., add 2 sq. ft. for each additional light. A 6-lt. window would contain 12 sq. ft. additional; a 12 lt. window, 24 sq. ft. additional, etc.

Stairs. When estimating quantities of paint or varnish for wood stairs, add 2'-0" to the length of treads and risers to allow for stair strings on each side of the stairs. A stair tread is 10" to 12" wide, riser is about 7" high, and the average cove and underside of a stair tread 2" in girth, so each tread should be figured 2'-0" wide. Multiply the width by the length to obtain the number of sq. ft. of paint or varnish required for each step. Multiply the area of each step by the number of treads in the stair, and the result will be the number of sq. ft. of finish. Example: A flight of stairs containing 20 treads, 4'-0" long. Adding 2'-0" to the length of the treads to allow for the stringers, gives 6 lin. ft. in each tread, which, multiplied by the girth of 2'-0" gives 12 sq. ft. of surface in each step. Since there are 20 treads in the stairs, 12 x 20 = 240 sq. ft.

For painting the soffits of stairs, use same rules of measurement as given above.

Balustrades and Handrails with Balusters. When estimating balustrades around well holes or stair handrail and balusters, measure the distance from the top of the treads to the top of the handrail and add 6" for painting or varnishing the handrail. Multiply the height of the balustrade by the length and the result will be the number of sq. ft. of balustrade. An easy method for computing the length of the stairs is to allow 1 lin. ft. in length for each tread in the stairs. Example: If the handrail is 2'-6" above the treads and the stairs contain 20 treads, add 6" to the height of 2'-6" to take care of the extra work on the handrail proper, making a total height of 3'-0". 3 x 20 = 60 sq. ft. of surface for one side of the balustrade or 120 sq. ft. for two sides. The balustrade around the stair well hole is easily estimated by multiplying the height by the length.

On fancy balustrades having square or turned spindles, which require considerable additional labor, use actual measurements as given above and multiply by 4 to allow for extra labor.

Wood Ceilings. When estimating quantities for wood ceilings, multiply the length by the width. Do not deduct for openings less than 10'-0"x10'-0".

Wainscoting. Plain wainscoting, obtain actual area. Paneled wainscoting, obtain actual area and multiply by 2.

Floors. To compute the quantities of floors to be finished, multiply the length of each room by the width, and the result will be the number of sq. ft. of floor to be finished.

Plastered Walls and Ceilings. To obtain the area of any ceiling, multiply the length by the breadth and the result will be the number of sq. ft. of ceiling. When estimating walls, measure the entire distance around the room and multiply by the room height. The result will be the number of sq. ft. of wall to be decorated. For instance, a 12'-0"x15'-0" room has two sides 12'-0" long and two sides 15'-0" long, giving a total of 54 lin. ft. If the ceilings are 9'-0" high, 54 x 9 = 486 sq. ft. the area of the walls. Do not deduct for door and window openings.

Cases, Cupboards, Bookcases, Etc. When estimating surfaces of cupboards, wardrobes, bookcases, closets, etc., to be painted or varnished, obtain the area of the front and multiply by 3 if the cases do not have doors.

If the cases have doors, obtain the area of the front and multiply by 5.

This takes care of painting or varnishing doors inside and out, shelves 2 sides, cabinet ends and backs.

Radiators. For each front foot, multiply the face area by 7.

Sanding and Puttying Interior Trim. Sanding and puttying on high grade interior trim should be figured as one coat of paint or varnish; on medium grade trim, figure at 50% of one coat of paint or varnish. For industrial work, figure about 25% of the cost of one coat of paint or varnish.

Emulsified Asphalt

Emulsified asphalt may be used as a floor surfacer, a bonding coat, or a quicksetting, hard-drying and durable patching material. It may be used with satisfactory results over either old or new floors. The following procedure is generally used. First, clean the old surface free from dirt, grease and foreign matter; second, cut out the worn spots or cracks with perpendicular sides to avoid the formation of feather edges; third, a bonding coat of emulsified asphalt is applied, consisting of 1 part clean water to 4 parts of emulsified asphalt, dry overnight; fourth, the mixture of portland cement, sand, gravel and emulsified asphalt is applied and rolled.

When applied direct to a concrete surface, 1 gallon will cover about 75 sq. ft. one coat or 50 sq. ft. two coats.

When used instead of water as a material for binding cement, sand and gravel the quantity will vary with the thickness applied and will require 4 to 5 gallons per sack of cement.

Painting Concrete and Masonry Surfaces

Hydrocide Colorcoat (Sonneborn). An alkyd-epoxy-silicone-oil base formula, decorative coating especially designed for exterior, highly porous surfaces. Properly applied, it effectively seals and protects such surfaces with a color-fast, decorative coating, which will not run or sag, in one application. It will fill pores and bridge hairline cracks on concrete or masonry surfaces which are completely cured.

One gallon of material, used as it comes in the container, will cover 60 to 70 sq. ft. depending upon porosity of the surface.Price, $9.50 per gal.

Hydrocide Super Colorcoat (Sonneborn). Super Colorcoat is for use on cured, green, or slightly damp surfaces. It is water dispersed acrylic emulsion formulation. It is also mildew and fungus resistant and non-flammable.

Super Colorcoat will cover approximately 80 to 100 sq. ft. per gal. Cost per gal. is $10.50.

SikaGard Hi-bild (Sika). Tough 100% solids epoxy resin coating for concrete, block, steel, wood and other structural materials. Can be applied to dry, damp, or wet surfaces. Coverage with short nap roller will be approximately 175 to 200 sq. ft. per gal.; when spraying, 250 to 400 sq. ft. per gal. Coverage varies with temperature, substrate and environment. Available in colors. Price, $22.50 per gal.

Covering Capacity of Paints

It is difficult to say how many sq. ft. of surface one gallon of paint will cover, because there are several items that influence the covering capacity. By brushing the paint out very thin, it will cover more surface than when applied thick. Dark paint will hide the surface better than light paint, and for that reason it can be brushed out thinner than the lighter colors. Also, a rough surface will require considerable more paint than a smooth surface.

Soft and porous wood will absorb more oil and require more paint than close grained lumber.

Another thing that must be considered are the ingredients entering into the paint. Different brands of paint vary in covering capacity or hiding power.

Covering Capacity of Oil Base Paint on New Exterior Wood. In figuring the number of square feet a gallon of oil base paint will cover, a great deal depends upon the surface to be painted; that is, the kind of wood, the degree of roughness, etc. Some woods are more porous than others and absorb more paint. Much depends on the way the paint is brushed out as some painters brush the paint out more and thus cover more surface.

The priming coat, properly applied, should cover 450 to 500 sq. ft. per gallon. The second coat should cover 500 to 550 sq. ft. per gallon and the third coat should cover 575 to 625 sq. ft. per gallon.

When estimating exterior trim painting, with measurements taken in accordance with standard methods as given on previous pages, the priming coat should cover 750 to 850 sq. ft. per gallon; the second coat should cover 800 to 900 sq. ft. per gallon and the third coat should cover 900 to 1,000 sq. ft. per gallon.

Bear in mind the above coverages are based on surfaced lumber, as the covering capacity will be greatly reduced when applied to rough boards.

Painting Old Exterior Wood Surfaces. When estimating repaint work on old exterior wood surfaces, if the existing surface is sound and in good condition, two coats should be sufficient to give a good job and paint coverages should be about the same as for the second and third coats on similar new work.

If the old surface shows cracking, blistering, scaling, and peeling, the old paint should be removed by using either a paste or liquid paint remover of the slow drying type, or a blow torch and scraper.

If the old paint is removed completely, three coats of paint should be applied the same as recommended for new work.

Painting Wood Shingle Siding. Wood shingle siding should receive three coats of paint on new work. Paint coverages per gallon should be as follows: first coat, 300 to 325 sq. ft.; second coat, 400 to 450 sq. ft.; third coat, 500 to 550 sq. ft.

Wood shingles which have been previously painted with an oil paint and are in suitable condition for repainting, should receive two coats of paint. The first coat should cover 400 to 450 sq. ft. per gal. and the second coat 500 to 550 sq. ft. per gal.

Painting Exterior Wood Floors and Steps. When painting exterior wood floors and steps with oil base paint, material coverages per gallon should be as follows: first coat, 325 to 375 sq. ft.; second coat, 475 to 525 sq. ft.; third coat, 525 to 575 sq. ft.

Painting Interior Wood. New interior wood should receive three coats of paint, a priming coat, a second coat and a final or finish coat.

Paint coverage for interior wood surfaces will vary with the type of work, such as doors and windows, running trim, paneling, etc., in accordance with the standard method of measuring the quantities.

For doors and windows, coverage per gallon should be: first coat, 475 to 500 sq. ft.; second coat, 500 to 550 sq. ft.; third coat, 500 to 550 sq. ft.

For running trim up to 6" wide, coverage per gallon should be; first coat, 800 to 900 lin. ft.; second and third coats, 1,100 to 1,200 lin. ft.

For ordinary flat wood paneling, coverage per gallon should be: first coat, 450 to 500 sq. ft.; second and third coats, 500 to 550 sq. ft.

When painting interior woodwork that has been previously painted and is in good condition, one gallon of paint should cover approximately the same as given for the second and third coats on new wood.

Painting Brick, Stone, Stucco and Concrete. When painting masonry surfaces that are reasonably smooth, one gallon of oil base paint should cover 170 to 200 sq. ft. for the priming coat; 350 to 400 sq. ft. for the second coat and 375 to 425 sq. ft. for the third coat.

These quantities are subject to considerable variation due to the porosity of the surfaces to which the paint is applied, smoothness of walls, etc., as a rough porous surface will require considerably more paint than a smooth hard surface.

Painting Metal Work. The area that any paint may be expected to cover on metal work will vary with the surface to be painted. Badly pitted or rough metal will require more paint than a perfectly smooth surface. The covering capacity will also vary with the temperature, consistency of the paint and the effort behind the brush.

However, for ordinary smooth surfaces, one gallon of paint should cover 550 to 650 sq. ft., one coat.

Aluminum Paint. Aluminum paint is used for both priming and finishing coats on metal surfaces. The covering capacity varies with the condition of the surface, but on smooth surfaces, one gallon should cover 600 to 700 sq. ft. with one coat.

Painting Interior Walls. When painting interior walls, the covering capacity of the paint will depend upon the surface to be painted, whether smooth or sand finished plaster, porous or hard finished wallboard, etc.

Three coats are recommended for interior plaster that has never been painted: a priming coat, a second or body coat, and a third or finishing coat. However, if a two-coat job on unpainted plaster is desired, use a good wall primer or sealer for the first coat, followed by the second or finishing coat. To make two coats hide better, tint the first to nearly the same color as the second coat. If the surface has been painted before and the old paint is still in good condition, two coats are sufficient, no priming coat being required.

On smooth plaster or hard finished wallboard, one gallon of good wall primer or sealer should cover 575 to 625 sq. ft. for the priming coat. For the second coat, figure 500 to 550 sq. ft. per gallon and for the third coat, 575 to 625 sq. ft. gallon.

On rough, porous sand finished plaster, one gallon of wall primer or sealer may cover only 275 to 300 sq. ft. while on very porous wallboard it may be only 150 to 200 sq. ft. per gallon.

For the second and third coats on rough surfaced plaster or wallboard, figure 400 to 475 sq. ft. per gallon.

One Coat Wall Finishes. There are any number of "one-coat" wall finishes on the market, but when applied to new plastered surfaces or over surfaces that have been previously unpainted, it is advisable to use a primer or wall seal coat, which seals the pores in the plaster or other surface and provides a base suitable to receive the one coat finish. Fully described above.

Latex Base Paints. There are any number of latex base paints on the market that are compatible with most surfaces.

When used over smooth plastered surfaces, one gallon should cover 450 to 500 sq. ft. first coat and 550 to 650 sq. ft. second coat.

When applied over rough sand finish plaster one gallon should cover 325 to 350 sq. ft. per coat.

Covering Capacity of Wall Size. The covering capacity of wall size will vary greatly with the material used.

Water size consisting of ground or flake glue and water will cover 600 to 700 sq. ft. of surface per gallon.

Varnish size consisting of varnish, benzine or turpentine and a little paint, will cover 450 to 550 sq. ft. per gallon.

Hard oil or gloss oil size consisting of rosin and benzine will cover 450 to 500 sq. ft. per gallon.

The covering capacity of prepared wall primers and sealers is given on previous pages under Painting Interior Walls.

Shingle Stain. When staining wood shingles after they have been laid, one gallon of good stain should cover 120 to 150 sq. ft. of surface, for the first coat.

The second and third coats will go farther, as the wood does not absorb as much stain as on the first coat and should cover 200 to 225 sq. ft. per gallon. One gallon of good stain should cover about 70 sq. ft. of roof with 2 coats or 50 sq. ft. with 3 coats.

If the shingles are dipped in stain before laying, it will require 3 to 3½ gals, of stain per 1,000 shingles. When dipping shingles in stain, only two-thirds of the shingle need be dipped, as it is unnecessary to dip the end of the shingle that is not exposed.

Covering Capacity of Oil and Spirit Stains. When staining finishing lumber, such as birch, mahogany, oak, gum, etc., one gallon of oil stain should cover 700 to 725 sq. ft.

One gallon of spirit stain should cover 500 to 600 sq. ft.

Covering Capacity of Wood Fillers. Liquid filler is usually used with close grained woods, such as pine, birch, etc., and one gallon should cover 500 to 550 sq. ft.

It is necessary to use a paste filler to fill the pores of all open grained wood, such as oak, ash, walnut, mahogany, etc.

If paste filler is used on oak, one gallon of filler should cover about 450 sq. ft. of surface.

Covering Capacity of Shellac. The covering capacity of shellac will vary with its purity but one gallon of good shellac should cover 550 to 750 sq. ft. of surface, depending upon whether first, second, or third coat.

Covering Capacity of Varnish. One gallon of good varnish should cover 400 to 450 sq. ft. of softwood floors with one coat; 200 to 225 sq. ft. with 2 coats and 135 to 150 sq. ft. with 3 coats.

When applied to hardwood floors, one gallon of good varnish should cover 500 to 550 sq. ft. with one coat; 250 to 275 sq. ft. with 2 coats and 170 to 185 sq. ft. with 3 coats.

When applied to softwood interior finish, one gallon of good spar finishing varnish should cover 400 to 425 sq. ft. with one coat; 200 to 215 sq. ft. with 2 coats and 135 to 145 sq. ft. with 3 coats.

When applied to hardwood interior finish, one gallon of good spar varnish should cover 425 to 450 sq. ft. with one coat; 210 to 225 sq. ft. with 2 coats and 140 to 150 sq. ft. with 3 coats.

Covering Capacity of Wax. One gallon of good liquid wax should cover 1,050 to 1,075 sq. ft. of surface.

Covering Capacity of Varnish Remover. The amount of varnish remover required to remove old varnish from floors and interior finish will depend upon the condition of the old work, but one gallon of good varnish remover should be sufficient for 150 to 180 sq. ft. of surface.

Enamel Finish. Where interior woodwork is to receive an enamel finish, at least three coats are required, as follows: first coat, oil base paint primer and sealer; second coat, prepared enamel undercoat; third coat, enamel finish coat. If a four-coat job is specified, the third coat may be a mixture of ½ undercoat and ½ enamel, followed by the enamel finish coat.

For flat work, figure material coverages per gallon as follows: paint primer and sealer, 575 to 600 sq. ft.; undercoat, 375 to 400 sq. ft.; enamel finish, 475 to 500 sq. ft.

For running trim, one gallon of material should cover as follows: paint primer and sealer, 1,100 to 1,200 lin. ft.; undercoat, 775 to 800 lin. ft.; enamel finish, 775 to 800 lin. ft.

Glazed Finish. One gallon of good glazing liquid, colored with oil colors, should cover 500 to 550 sq. ft. of surface.

For glazed finish on interior running trim, not over 6" wide, one gallon should cover 1,050 to 1,100 lin. ft.

Estimating Labor Costs for Painting

Perhaps in no other trade will labor costs vary to a greater extent than in painting and decorating. On some classes of work the material costs are almost negligible, while the labor costs may make it one of the most expensive kinds of work.

Naturally, the cost of the work will depend upon the grade of workmanship specified or upon the class of work the painting contractor figures on doing.

In the large cities particularly, there are two classes of painters--those who make a specialty of ordinary or commercial work such as factory buildings, flat or apartment buildings, speculative work and work of this class, and those who specialize in high class work. Naturally, a painter who figures the lower grade of workmanship is bound to be lower in cost. The estimator must consider the class of work required.

Another thing to keep in mind is to be sure that all measurements and quantities are listed and computed in strict accordance with the methods of measurement as given at the beginning of this section. This applies particularly to the allowances for labor when estimating such items as sash doors, windows having divided lights, drop siding and shingle siding, fences, trellis and lattice work where the face and edges of all surfaces must be painted, stairs where it is necessary to paint all four sides or the entire circumference of newels, handrails, balusters, etc.

These are all exceedingly important, because if the quantities are not computed according to the methods given, then the labor costs given below will be far too low in many cases.

Itemized Costs of Painting

The labor and material tables on the following pages give detailed information on material coverages and production times for various types of paint on different kinds of surfaces.

In preparing an estimate for painting work, after obtaining quantities required for the project, the estimator need only to refer to the proper table to determine the required quantity of material and labor hours necessary to perform the work. Once the gallons of paint and man hours are computed, apply the appropriate unit costs applicable for your area.

Do not overlook the items of protection, scaffolding or other equipment costs in preparing the estimate. These are not given in the following tables.

For example, let's assume that the estimator is performing an estimate for staining shingle siding on a residence, and the quantity of work to be performed amounts to 3,300 sq. ft. of surface to be covered with two 2 coats of stain. Refer to the chart for exterior residential work and find the proper work classification, "Stain-Shingle Siding". The material coverage is stated as 120 to 150 sq. ft. per gallon for the first coat and 200 to 225 sq. ft. per gallon for the second coat. The labor production is 1.20 hours per 100 sq. ft. for first coat and 0.82 hours per 100 sq. ft. for the second coat. The sample estimate for this 3,300 sq. ft. of surface would be as follows:

Material Costs

	Rate	Total	Rate	Total
1st Coat Stain 27 gal.	$....	$....	$15.00	$405.00
2nd Coat Stain 16 gal.	....	....	15.00	240.00
Cost of 3,000 sq. ft.		$....		$645.00
Cost per sq ft.		....		.22

Labor Costs

	Hours	Rate	Total	Rate	Total
1st Coat Painter	39.6	$....	$....	$24.39	$965.84
2nd Coat Painter	27.1	....	....	24.39	660.97
Cost of 3,300 sq. ft.			$....		$1,626.81
Cost per sq. ft.			....		.49

Labor and Materials Required for Painting Operations

Exterior Work - Residential - Brush

Description of Work	Coat	No. of Sq. Ft. per Hr.	Hours per 100 Sq. Ft.	Material Coverage Sq. Ft. per Gal.
Sanding & Puttying, Plain Siding & Trim		200-210	0.50	....
Sanding & Puttying, Outside Trim Only		115-120	0.86	....
Burning Paint Off Plain Surfaces		40-50	2.20	....
Burning Paint Off Wood		30-40	2.85	....
Exterior Painting, Plain Siding & Trim	Priming	100-110	0.95	400-500
	Second	115-125	0.85	500-550
	Third	125-135	0.75	575-625
Exterior House Painting, Rubberized Wood Bond	One Coat	125-135	0.75	450-500
Exterior Trim Only	Priming	65-75	1.40	750-850
	Second	85-95	1.12	800-900
	Third	95-105	1.00	900-1,000
Shingle/Shake Siding* - Oil Paint	First	115-125	0.82	250-300
	Second	150-160	0.65	375-425
Shingle/Shake Roofs* - Oil Paint	First	100-120	0.90	120-150
	Second	155-175	0.60	220-250
Shingle/Shake Siding* - Stain	First	75-85	1.20	120-150
	Second	115-130	0.82	200-225
Shingle/Shake Roofs* - Stain	First	145-155	0.67	100-120
	Second	190-210	0.50	170-200
Shingle/Shake Roofs & Siding* - Sealer	First	200-210	0.49	125-150
	Second	220-250	0.43	150-200
*If surface is extremely dry, increase labor hours and decrease covering capacity.				
Brick Walls - Oil Paint	First	100-120	0.93	170-200
	Second	140-160	0.67	350-400
	Third	155-175	0.60	375-425
Brick Walls - Latex Paint	First	95-110	0.98	180-210
	Second	130-150	0.71	350-410
	Third	150-170	0.63	380-430
Porch Floors & Steps - Oil Paint	First	235-245	0.42	325-375
	Second	270-280	0.36	475-525
	Third	285-295	0.35	525-575
Smooth Face Brick - Waterproof Cement Paint	First	175-185	0.55	90-110
	Second	260-270	0.38	140-160
Asphalt Shingle - Clear Waterproof Paint	First	45-55	2.00	120-140

	Second	50-70	1.67	165-185
	Third	60-80	1.43	175-205
Smooth Face Brick - Clear Waterproof Paint	First	200-210	0.50	500-510
	Second	210-230	0.45	590-610
Stucco, Medium Texture - Oil Paint	First	90-100	1.10	140-160
	Second	150-160	0.65	340-360
	Third	150-160	0.65	340-360
Stucco, Medium Texture - Latex Paint	First	110-125	0.85	160-180
	Second	175-200	0.53	180-210
	Third	175-210	0.53	180-220
Stucco, Medium Texture - Waterproof Cement Paint	First	130-140	0.74	90-110
	Second	190-210	0.50	125-145
Exterior Masonry, Stucco, Shingle Siding - Latex Flat Finish	First	100-120	0.91	325-375
	Second	165-175	0.59	375-425
Concrete Walls, Smooth - Waterproof Cement Paint	First	175-180	0.57	110-130
	Second	225-285	0.36	150-170
Concrete Floors & Steps - Floor Enamel	First	250-280	0.38	440-460
	Second	190-210	0.50	575-625
	Third	200-220	0.48	575-625
Cement Floors - Color Stain & Finish	First	325-370	0.28	475-525
	Second	280-300	0.35	450-500
Cement Floors - Epoxy	Primer	100-125	0.89	175-200
	First	115-135	0.80	190-220
	Second	150-180	0.60	200-250
Fences, Plain, Average	First	125-135	0.77	530-550
	Second	190-200	0.51	630-650
Fences, Picket, Average	First	140-150	0.70	630-650
	Second	160-170	0.60	650-675
Fences, Wire Metal, Average	First	100-110	0.95	900-1,000
	Second	140-150	0.70	975-1,125
Shutters, Average, Each Coat	No. Shutter	2-3	0.40	11-13
Downspout & Gutters - Paint	First	170-180	0.57	540-560
	Second	185-195	0.53	575-600
Screens, Wood Only, Average, Each Coat	No. Screens	6-8	0.14	45-55
Storm Sash, 2 Light, Average, Each Coat	No. Sash	3-4	0.29	23-27
Sheet Metal Work	First	230-240	0.43	500-550
	Second	260-270	0.38	550-600
Steel Factory Sash	First	80-90	1.18	950-1,000
For old work, add 10 to 20% to prepare	Second	100-110	0.95	950-1,000
Surface	Third	125-135	0.77	1,250-1,350
Doors & Windows - Paint	First	175-185	0.55	475-500
	Second	130-140	0.75	500-550
	Third	165-175	0.60	500-550
Base, Chair Rail, Picture Mold, & Other Trim less than 6" wide (priced per lin. ft.)	First	275-300	0.35	800-900
	Second	175-200	0.55	1,100-1,200
	Third	150-165	0.65	1,100-1,200
Base, Chair Rail, Picture Mold, & Other Trim less than 6" wide - polyurethane (per lin. ft.)	First	260-280	0.37	525-575
	Second	280-300	0.34	550-600
	Third	280-300	0.34	550-600
Masonry Walls - Bonding & Penetrating Oil Paint	First	350-360	0.28	225-250
	Second	350-375	0.28	275-300

Common Brick, Smooth Brick, Concrete or Cinder Blocks, Cement Plaster - Waterproofing	First	100-110	1.00	90-100
	Second	180-195	0.53	175-200
Wire Cut Tapestry Brick or Very Rough Concrete Block or Stucco - Waterproofing	First	90-100	1.05	50-60
	Second	150-165	0.65	150-175
Smooth Plaster Walls - Flat Finish Paint	First	450-475	0.22	500-550
	Second	475-500	0.21	500-550
Plaster Walls - Sand Finish	First	400-425	0.25	450-500
	Second	425-450	0.24	400-450
Smooth Finish Plaster - Industrial Plaster	First	450-475	0.22	500-550
	Second	500-550	0.20	400-450
	Third	500-550	0.20	450-500

Exterior Work - Residential - Roller/Pad

Description of Work	Coat	No. of Sq. Ft. per Hr.	Hours per 100 Sq. Ft.	Material Coverage Sq. Ft. per Gal.
Shingle/Shake Siding* - Oil Paint	First	130-140	0.74	220-260
	Second	170-185	0.56	330-370
Shingle/Shake Roofs* - Oil Paint	First	115-135	0.80	105-130
	Second	180-200	0.53	190-215
Shingle/Shake Siding* - Stain	First	85-100	0.74	105-130
	Second	130-150	0.49	170-195
Shingle/Shake Roofs* - Stain	First	165-180	0.58	85-105
	Second	220-240	0.43	150-170
Shingle/Shake Roofs & Siding* - Sealer	First	230-240	0.43	110-130
	Second	250-285	0.37	130-170
*If surface is extremely dry, increase labor hours and decrease covering capacity.				
Brick Walls - Oil Paint	First	115-135	0.80	150-170
	Second	160-180	0.59	305-345
	Third	180-200	0.53	325-370
Brick Walls - Latex Paint	First	110-125	0.85	155-182
	Second	150-170	0.63	305-350
	Third	170-190	0.55	330-370
Porch Floors & Steps - Oil Paint	First	270-280	0.36	285-325
	Second	310-320	0.32	415-450
	Third	330-340	0.30	460-500
Smooth Face Brick - Waterproof Cement Paint	First	200-210	0.50	80-95
	Second	300-310	0.33	120-140
Asphalt Shingle - Clear Waterproof Paint	First	50-65	1.48	105-120
	Second	60-80	1.42	140-160
	Third	70-90	1.25	150-180
Smooth Face Brick - Clear Waterproof Paint	First	230-240	0.43	430-440
	Second	240-265	0.50	515-530
Stucco, Medium Texture - Oil Paint	First	100-115	0.93	120-140
	Second	170-185	0.55	295-310
	Third	170-185	0.55	295-310
Stucco, Medium Texture - Latex Paint	First	130-145	0.72	140-155
	Second	200-230	0.47	155-180
	Third	200-240	0.45	155-190
Stucco, Medium Texture - Waterproof Cement Paint	First	150-160	1.00	75-95
	Second	215-240	0.44	110-125
Exterior Masonry, Stucco, Shingle Siding - Latex Flat Finish	First	115-135	0.80	280-325
	Second	190-200	0.51	325-370

Concrete Walls, Smooth - Waterproof Cement Paint	First	200-210	0.49	95-115
	Second	260-320	0.48	130-150
Concrete Floors & Steps - Floor Enamel	First	290-320	0.37	380-400
	Second	220-240	0.43	500-540
	Third	230-250	0.42	500-540
Cement Floors - Color Stain & Finish	First	370-420	0.25	410-450
	Second	320-345	0.30	390-430
Cement Floors - Epoxy	Primer	115-140	0.78	150-170
	First	130-155	0.70	165-190
	Second	170-210	0.53	170-210
Fences, Plain, Average	First	145-155	0.67	460-480
	Second	220-230	0.45	550-565
Fences, Picket, Average	First	160-170	0.61	550-565
	Second	185-195	0.53	565-585
Fences, Wire Metal, Average	First	115-125	0.83	780-870
	Second	160-170	0.60	850-980
Downspout & Gutters - Paint	First	195-205	0.50	470-485
	Second	210-225	0.46	500-520
Masonry Walls - Bonding & Penetrating Oil Paint	First	400-410	0.25	200-220
	Second	400-430	0.24	240-260
Common Brick, Smooth Brick, Concrete or Cinder Blocks, Cement Plaster - Waterproofing	First	115-125	0.83	80-85
	Second	205-225	0.53	150-170
Smooth Plaster Walls - Flat Finish Paint	First	520-550	0.19	435-475
	Second	550-575	0.18	435-475
Plaster Walls - Sand Finish	First	460-490	0.21	390-430
	Second	490-515	0.20	350-390
Smooth Finish Plaster - Industrial Plaster	First	515-545	0.19	435-475
	Second	575-630	0.17	350-390
	Third	575-630	0.17	390-430

Spray Painting - Residential & Commercial Work

Description of Work	Coat	No. of Sq. Ft. per Hr.	Hours per 100 Sq. Ft.	Material Coverage Sq. Ft. per Gal.
Brick, Tile, & Cement - Latex Paint	First	500-525	0.20	90-110
	Second	525-550	0.19	140-160
Brick, Tile, & Cement - Oil Paint	First	475-500	0.21	250-275
	Second	500-525	0.20	400-450
Brick - Waterproofing	First	550-600	0.17	200-300
	Second	650-700	0.15	225-350
Note: For weathered brick, add 15% labor and 35% to 50% for material.				
Rough Brick, Tile, Cement & Stucco - Silicone	One Coat	800-900	0.13	80-90
Stucco - Exterior Latex Paint	First	400-425	0.25	90-110
	Second	450-500	0.22	125-150
Stucco - Oil Paint	First	300-325	0.34	200-225
	Second	325-350	0.30	350-400
Stucco - Waterproofing	First	550-600	0.17	150-180
	Second	650-700	0.15	160-190
Note: For weathered stucco, add 15% for labor and 35% to 50% for material.				
Shingle Roofs - Oil Paint	First	400-450	0.25	125-150
	Second	450-500	0.22	200-225
Shingle Roofs - Stain	First	450-475	0.22	125-150
	Second	475-500	0.21	200-225
Shingle Siding - Oil Paint	First	350-375	0.29	125-150
	Second	400-425	0.25	225-250
Shingle Siding - Stain	First	400-450	0.25	125-150
	Second	450-500	0.22	200-225

Smooth Plaster Walls - Flat Finish Paint	First	450-475	0.22	500-550
	Second	475-500	0.21	500-550
Plaster Walls - Sand Finish	First	400-425	0.25	450-500
	Second	425-450	0.24	400-450
Smooth Finish Plaster - Industrial Enamel	First	450-475	0.22	500-550
	Second	500-550	0.20	400-450
	Third	500-550	0.20	450-500
Fence, Picket - Oil or Latex Paint	First	475-500	0.20	500-525
	Second	475-500	0.20	550-600
Fence, Plain - Oil or Latex Paint	First	450-500	0.21	375-400
	Second	570-600	0.17	400-450
Doors, Garage - Oil or Latex Paint	First	540-550	0.22	550-575
	Second	540-550	0.22	560-600
Blinds or Shutters - Paint	First	120-140	0.77	120-140
	Second	130-170	0.67	140-180
	Third	150-190	0.59	150-190

Trim must be figured separately as hand work.

Finishing Interior Trim - Residential

Description of Work	Coat	No. of Sq. Ft. per Hr.	Hours per 100 Sq. Ft.	Material Coverage Sq. Ft. per Gal.
Preparatory Work for Painting	Sanding	290-300	0.35	....
	Sand & Putty	190-200	0.50	....
	Light Sanding	340-350	0.30	....
Back Priming Interior Trim up to 6" Wide	Lin. Ft.	475-500	0.21	1,100-1,200
Doors & Windows, Interior - Paint	First	140-150	0.70	475-500
	Second	115-125	0.85	500-550
	Third	115-125	0.85	500-550
Wood Sash - Oiling & Priming	No. of Sash	9-10	....	600-700
Preparatory Work for Enamel Finish	Sanding	290-300	0.35	....
	Sand & Putty	120-130	0.80	....
	Light Sanding	135-145	0.72	....
Doors & Windows - Enamel Finish	Paint First	140-150	0.70	575-600
	Undercoat Second	90-100	1.10	372-400
	Enamel Third	100-110	1.00	475-500
Four coat work, add 1/2 undercoat, 1/2 enamel	Add'l Coat	90-100	1.10	425-450
Base, Chair Rail, Picture Mold, & Other Trim up to 6" Wide (lin. ft.)	First	250-260	0.40	800-900
	Second	145-155	0.67	1,100-1,200
	Third	125-135	0.77	1,100-1,200
Base, Chair Rail, Picture Mold, & Other Trim up to 6" Wide - Polyurethane (lin. ft.)	First	260-280	0.37	525-575
	Second	280-300	0.34	550-560
	Third	280-300	0.34	550-560
Four coat work, add 1/2 undercoat, 1/2 enamel	Add'l Coat	135-145	0.72	1,100-1,200
Stain Interior Work	One Coat	215-225	0.45	700-725
Stain & Fill Interior Woodwork, Wipe Off	One Coat	65-70	1.50	400-450
Shellac Interior Woodwork	One Coat	215-225	0.45	700-725
Varnish, Gloss, Interior Woodwork	One Coat	165-175	0.60	425-450
Varnish, Flat, Interior Woodwork	One Coat	170-180	0.57	600-625
Standing Trim - Wax & Polish	One Coat	100-110	0.95	550-600

Standing Trim - Penetrating Stainwax	One Coat	170-180	0.57	525-550
	Second	190-200	0.51	600-625
	Polishing Coat	190-200	0.51	....
Sanding for Extra Fine Varnish Finish	Sanding	40-45	2.25	....
Synthetic Resin Finish, Requires Wiping	First	190-200	0.50	550-600
	Second	210-220	0.47	625-675
Spackling or Swedish Putty over Flat Trim	One Coat	60-65	1.60	140-150
Glazing & Wiping over Enamel Trim	One Coat	60-65	1.60	1,050-1,100
Brush Stippling Interior Trim - Painted	....	85-90	1.15	....
Flat Varnishing & Brush Stippling over Glazed Trim	Varnish	170-180	0.57	600-625
	Stipple	240-250	0.42	....
	Old Work			
Washing Average Enamel Finish	Washing	85-90	1.15	....
Washing Better Grade Enamel Finish	Washing	60-65	1.60	....
Polishing Better Grade Enamel Finish	Polish	170-180	0.57	1,700-1,800
Removing Varnish with Liquid Remover	Flat Surfaces	30-35	3.00	150-180
Wash, Touch Up, One Coat Varnish	Wash-Touch Up	165-175	0.60	....
	Varnish	100-110	0.95	575-600
Wash, Touch Up, One Coat Enamel	Wash-Touch Up	140-150	0.70	....
	Enamel	70-75	1.40	375-400
Wash, Touch Up, One Coat Undercoat and One Coat Enamel	Wash-Touch Up	140-150	0.70	....
	Undercoat	75-85	1.25	475-500
	Enamel	75-85	1.25	475-500
Burning Off Interior Trim	....	20-25	4.00	....
Burning Off Plain Surfaces	....	30-35	3.00	....

Finishing Interior Floors

Description of Work	Coat	No. of Sq. Ft. per Hr.	Hours per 100 Sq. Ft.	Material Coverage Sq. Ft. per Gal.
Wood Floors - Paint	First	290-300	0.35	525-550
	Second	260-270	0.38	425-450
Wood Floors - Filling, Wiping	Fill-Wipe	170-180	0.57	425-450
Hardwood Floors - Penetrating Stainwax	First	390-400	0.25	525-550
	Second	590-600	0.17	1,200-1,250
Floor Seal	First	490-500	0.20	600-625
	Second	590-600	0.17	1,200-1,250
Shellac	First	390-400	0.25	500-550
	Second	475-500	0.21	650-675
Stainfill, 1 Shellac, 1 Varnish	Stainfill	260-270	0.38	475-500
	Shellac	390-400	0.25	650-675
	Varnish	300-310	0.33	500-550
Stainfill, 1 Shellac, Wax & Polish	Stainfill	260-270	0.38	475-500
	Shellac	390-400	0.25	650-675
	Wax/Polish	200-210	0.50	1,050-1,075
Varnish, Each Coat over Shellac	Varnish	300-310	0.33	500-550
Buffing Floors by Machine	Buffing	390-400	0.25	....

Waxing over 2 Coats of Seal & Polish	Wax-Polish	200-210	0.50	1,050-1,075
	Old Work			
Removing Varnish with Liquid Remover	Flat Surfaces	40-45	2.35	170-180
Clean, Touch Up, & Varnish	Clean-Touch Up	125-135	0.77	....
	Varnish	300-310	0.33	252-550
Clean, Touch Up, Wax, & Polish	Clean-Touch Up	125-135	0.77	....
	Wax-Polish	200-210	0.50	1,050-1,075

Painting Interior Walls

Description of Work	Coat	No. of Sq. Ft. per Hr.	Hours per 100 Sq. Ft.	Material Coverage Sq. Ft. per Gal.
Taping, Beading, Spotting Nail Heads, & Sanding Gypsum Wallboard		90-110	1.00	....
Texture over Gypsum Wallboard per lb.	One Coat	250-270	0.39	8-10
Casein or Resin Emulsion over Textured Gypsum Wallboard	One Coat	300-320	0.32	325-350
Sizing New Smooth Finish Walls	Sizing	350-400	0.27	600-700
Wall Sealer or Primer on Smooth Walls	Sealer	200-225	0.45	575-625
Wall Sealer or Primer on Sand Finish Walls	Sealer	125-135	0.80	275-300
Smooth Finish Plaster - Flat Finish	Sealer	200-225	0.45	575-625
	Second	165-175	0.60	500-550
	Third	175-185	0.65	575-625
	Stippling	190-200	0.52	....
Sand Finish Plaster - Flat Finish	Sealer	125-135	0.80	275-300
	Second	135-145	0.70	400-450
	Third	125-135	0.80	425-475
Smooth Finish Plaster - Gloss or Semi-Gloss	Sealer	200-225	0.45	575-625
	Second	170-180	0.57	500-550
Third		140-150	0.70	500-550
	Stippling	85-95	1.10	....
Sand Finish Plaster - Gloss or Semi-Gloss	Sealer	125-135	0.80	275-300
	Second	115-125	0.83	275-400
	Third	125-135	0.80	400-450
Texture Plaster, Average - Semi-Gloss	Sealer	100-110	0.95	250-275
	Second	115-125	0.83	320-350
	Third	125-135	0.80	375-400
Smooth Finish Plaster - Latex Rubber Paint	First	140-160	0.67	300-350
	Second	140-160	0.67	300-350
Sand Finish or Average Texture Plaster - Latex Rubber Paint	First	110-130	0.83	250-300
	Second	110-130	0.83	250-300
Glazing & Motting over Smooth Finish Plaster		80-90	1.15	1,050-1,075
Glazing & Motting over Sand Finish Plaster		60-65	1.60	875-900
Glazing & Highlighting Textured Plaster		80-90	1.15	825-850
Starch & Brush Stipple over Painted Glazed Surface		115-125	0.83	....

Flat Varnish & Brush Stipple over Painted Glazed Surface		105-115	0.90	500-550
Textured Oil Paint over Smooth Finish Plaster	Size	215-225	0.45	675-725
	Texture	40-45	2.30	125-150
Water Textured over Smooth Finish Plaster	Size	215-225	0.45	700-725
	Texture	50-55	1.90	85-95
New Smooth Plaster - Latex Base Paint	First	190-200	0.50	500-550
	Second	210-220	0.47	650-700
on Rough Sand Finish	One Coat	150-160	0.65	325-350
on Cement Blocks	One Coat	135-145	0.70	300-325
on Acoustical Surfaces	One Coat	135-145	0.70	200-225
Casein Paint over Smooth Finish Plaster	First	270-280	0.37	500-550
	Second	300-310	0.33	500-550
over Rough Sand Finish Plaster	One Coat	250-260	0.40	325-350
over Cement Blocks	One Coat	165-175	0.60	300-325
over Acoustical Surfaces	One Coat	165-175	0.60	200-225
over Cinder Concrete Blocks	One Coat	125-135	0.70	125-150

When applying paint to walls and ceilings using a roller applicator, increase quantities 10% to 15% and reduce painter time about the same amount.

Old Work

Description of Work		No. of Sq. Ft. per Hr.	Hours per 100 Sq. Ft.	Material Coverage Sq. Ft. per Gal.
Washing Off Calcimine, Average Surfaces		115-125	0.83	....
Washing Smooth Finish Plaster Walls, Average		145-155	0.67	....
Washing Sand Finish Plaster Walls, Average		100-110	1.00	....
Washing Starched Surfaces & Restarching, Smooth Surfaces	Washing	145-155	0.67	....
	Restarching	130-140	0.75	....
Removing Old Wallpaper, 3 layers or less		65-75	1.45	....
Washing Off Glue after Removing Paper (including fixing average cracks & sizing)		125-135	0.77	....
Cutting Hard Oil or Varnish Size Walls (including fixing average cracks)		125-135	0.77	....
Cutting Gloss Painted Walls (incl. fixing cracks)		125-135	0.77	....
Wash, Touch Up, One Coat Gloss Paint to Smooth Plaster Surfaces	Wash-Touch Up	125-135	0.77	....
	One Coat	125-135	0.77	450-500
Synthetic-Resin Emulsion Paint over Old Painted Walls	One Coat	190-200	0.50	400-425

Wallpaper, Canvas, Coated Fabrics, Paper Hanging, Wood Veneer

Canvas Sheeting		50-60	1.82	....
Coated Fabrics	Single Roll	2-2 1/2	....	....
Wallpaper				
One Edge Work	Single Roll	3 1/2-4	....	....
Wire Edge Work	Single Roll	2 1/2-3	....	....
Butt Work	Single Roll	2 1/4-2 3/4	....	....

Scenic Paper 40"x60"	Single Roll	1-1 1/4	....	....
Wood Veneer Flexwood	Sq. Ft.	18	....	....
Lacquer Finish over Wood Veneer	First Coat	100	1.00	275-325
	Second	100	1.00	450-500
Penetrating Wax or Synthetic Resin Application over Wood Veneer	First	100	1.00	550-600
	Second	100	1.00	650-675

Colors in Oil

Colors Ground Fine in Pure Linseed Oil, One-half Pint Cans

Blacks	Price
Drop Black	$0.80
Lamp Black	1.40
Blue	
Palco Blue	3.20
Prussian Blue	2.60
Cobalt Blue	2.60
Ultramarine Blue	2.60
Brown	
Raw Turkey Umber	1.20
Burnt Turkey Umber	1.20
Raw Italian Sienna	1.30
Burnt Italian Sienna	1.30
Vandyke Brown	1.20
Green	
Palco Green	3.50
Chrome Green, L.	2.10
Chrome Green, M.	2.10
Chrome Green, D.	2.10
Red	
Vermilion	2.70
Rose Lake	2.20
Venetian Red	1.80
Yellow	
Chrome Yellow, L.	3.20
Chrome Yellow, M.	3.20
Chrome Yellow, O.	3.20
Ochre	3.20

Shellac and Varnish

Kind of Material	Per Gal.
Shellac, White, 4-lb. Cut	$18.00-22.00
Shellac, Orange, 4-lb. Cut	16.00-20.00
Cabinet Finish Varnish	20.00-26.00
Floor Varnish	19.00-25.00
Flat Varnish	19.00-25.00
Spar Varnish	19.00-29.00
Varnish Remover, Liquid	15.00-25.00
Varnish Remover, Paste	15.00-29.00
Wallpaper Lacquer	12.00-14.00
Wallpaper Lacquer Thinner	8.00-10.00
Lacquer	20.00-25.00
Lacquer Thinner	10.00-12.00

Enamels and Enamel Undercoating

Kind of Material	Per Gal.
Enamel, Non-Yellowing White	$13.00-15.00
Architectural Enamel Paint	13.00-18.00
Enamel Undercoater	11.00-15.00
Enamel, White	11.00-15.00
Enamel, Quick Drying	11.00-16.00
Enamel, Industrial	13.00-18.00

Wall Size, Sealer, Paste, Etc.

Kind of Material	Price
Sizing, prepared per gal.	$5.00-9.00
Glue, Dry for Sizing per lb.	.75-1.25
Wall Primer and Sealer per gal.	8.00-12.00
Wallpaper Cleaner per lb.	.75
Starch, gloss per lb.	1.00
Patching Plaster per lb.	1.00
Spackle per lb.	.75
Glazing Liquid per gal.	8.00-12.00
Pre-mixed Vinyl Adhesive per lb.	1.00

Wood Fillers and Stains

	Per Gal.
Oil Stains	$10.00-15.00
Acid Stains	10.00-15.00
Penetrating Stainwax	10.00-15.00
Stain, Varnish, all colors	12.00-30.00
Stain, Fill, and Seal	10.00-12.00
Paste Wood Fillers	8.00-11.00
Floor Seal	12.00-15.00

Floor Wax

Liquid Floor Wax per gal.	$12.00-20.00
Self-Polishing Wax per gal.	7.00-9.00
Paste Wax per lb.	6.00-8.00

Miscellaneous Paints

	Per Gal.
Ready Mixed House Paint - Latex	$10.00-15.00
House Paint--One Coat	9.00-12.00
Porch and Floor Paint	10.00-15.00
Flat Wall Paint - Latex	8.00-18.00
One Coat Wall Paint - Oil Base	9.00-15.00
Metal Paint for Structural Steel, Bridges, Stacks, etc.	15.00-25.00
Cement Floor Paint	8.00-16.00
Cement Floor Paint - Rubber Base	8.00-16.00
Aluminum Paint	15.00-18.00
Vinyl-Chloride Co-Polymer	25.00
Sand Finish	7.00-10.00
Polyurethane	10.00-18.00
Drywall Primer	6.00-10.00
Waterproof Sealer, Clear	9.00-20.00

Lettering On Glass or Wood

The cost of lettering will vary with the style and size of letters and whether painted or gold leaf. Costs given on the following page include only Roman or Gothic style letters, such as are ordinarily used for signs on doors, transoms, and show windows, and are based on all work being performed by experienced sign writers.

Plain Black Letters 2" to 4" High. For plain black Gothic letters and numerals, such as appear on windows, transoms, and office doors, an experienced sign writer should average about 25 letters an hr. at the following labor cost per 100 letters:

	Hours	Rate	Total	Rate	Total
Sign painter	4	$....	$....	$32.92	$131.68
Cost per letter			....		1.32

Plain Black Letters 6" to 12" High. An experienced sign writer should average 12 to 14 plain Gothic letters 6" to 12" high, an hr. at the following labor cost per 100 letters:

	Hours	Rate	Total	Rate	Total
Sign painter	7.7	$....	$....	$32.92	$253.48
Cost per letter			....		2.53

Gold Leaf Letters 2" to 4" High. An experienced sign writer should complete about 5 plain Roman or Gothic letters in gold leaf an hr. (this class of work proceeds much slower than painted letters) at the following labor cost per 100 letters:

	Hours	Rate	Total	Rate	Total
Sign painter	20	$....	$....	$32.92	$658.40
Cost per letter			....		6.58

Gold Leaf Letters 6" to 12" High. An experienced sign writer should complete 2 to 3 plain Roman or Gothic letters 6" to 12" high, in gold leaf, an hr. at the following labor cost per 100 letters:

	Hours	Rate	Total	Rate	Total
Sign painter	40	$....	$....	$32.92	$1,316.80
Cost per letter			....		13.17

Painting Pipe

The cost of painting pipe varies considerably, depending on accessibility of work, layout of piping, scaffolding required, type of paint, whether one color or more, and in the case of old work, the amount of surface preparation required and whether or not pipes are in service.

Quantities are expressed in square feet of surface. When pipe quantities are measured from drawings, determine whether or not pipes are covered and the covering thickness, which greatly affects the diameter and consequently the surface area to be painted, especially in the smaller sizes. Most estimators convert lin. ft. of pipe to sq. ft. of surface according to the following table:

O.D. Pipe or Covering	Sq. Ft. Area per Lin. Ft.
3" and under	1
4" to 7"	2
8" to 11"	3

12" to 15"	4
16" to 19"	5
20" to 22"	6
23" to 26"	7
27" to 30"	8
31" to 34"	9
35" to 38"	10
39" to 42"	11
42" to 45"	12
46" to 48"	13

Various types of paint materials are specified for pipes, depending on the function, exposure, whether bare or covered, etc. Average material coverages and approximate prices for a few of these materials are as follows:

Description	SF per Gal.	Price per Gal.
Bare Pipe		
Black asphalt paint, one coat	130-150	$9.00-15.00
Aluminum paint, each coat	550-600	15.00-18.00
Radiator enamel, each coat	550-600	18.00-25.00
Covered Pipe		
Sizing, one coat	350-400	5.00-9.00
Paint (oil base), each coat	450-500	9.00-19.00
Industrial enamel, 1st Coat	450-500	15.00-25.00
Industrial enamel, 2nd Coat	550-600	15.00-25.00

Labor Painting Pipe. Labor costs for painting pipes are extremely variable due to the many conditions that might be encountered.

For simple work, consisting of horizontal runs and risers within stepladder reach, all painted one color, a painter should paint 800 sq. ft. of pipe, one coat, per 8-hr. day.

For boiler room or steam power plant piping of average difficulty, where some high ladder work is required, a painter should paint 600 sq. ft. of pipe, one coat, per 8-hr. day.

For difficult industrial process piping or in very congested steam power plants, where most of the work requires high scaffolding, a painter will paint only 350 to 450 sq. ft. of pipe, one coat, per 8-hr. day.

If pipes are painted different colors for identification, reduce the above production 10 to 15 percent.

If pipes are coded by color banding at 10- to 15-foot intervals, consider this work as another coat for labor, but practically no material will be required.

If painted flow direction indicators are required after the pipes have been painted, then add $7.50 each.

Painting Structural Steel

The cost of painting structural steel after erection will depend upon a number of factors, i.e., the weight of the steel, whether light or heavy sections; access to work and method of painting, whether brush painting by hand or using a spray.

Lightweight members cost more to paint than heavy members, and the less climbing required, the more paint a man will apply, either by brush or when using a spray.

Structural steel may be estimated, either by the ton of steel or by the number of square feet of surface to be painted.

The ton method is much easier to compute but the square feet of surface method is much more accurate.

The following tables giving the approximate number of square feet of area on various size beams, angles and channels will assist the estimator in computing accurate quantities where the square foot method is used.

Number of Sq. Ft. Area Per Lin. Ft. of Structural "S" Beams of Various Sizes

Depth Beam Inches	Weight per Lin. Ft. Lbs.	Sq. Ft. Area per Lin. Ft.
3	5-8	1.25
4	7-10	1.40
5	10-15	1.90
6	12-17	2.00
7	15-20	2.60
8	18-25	2.70
9	21-35	2.90
10	25-40	3.40
12	31-55	3.70
15	42-75	4.50
18	54-70	5.00
20	65-100	5.40
24	80-120	6.40
27	90-135	7.50

Number of Sq. Ft. Area Per Lin. Ft. of Structural Channels of Various Sizes

Depth of Channel Inches	Weight per Lin. Ft. Lbs.	Sq. Ft. Area per Lin. Ft.
3	4-6	1.00
4	5-7	1.20
5	7-11	1.50
6	8-16	1.70
7	10-20	2.00
8	11-21	2.20
9	14-25	2.35
10	15-35	2.70
12	21-40	3.00
15	32-50	3.70
18	34-55	4.40
20	42-58	5.00

Number of Sq. Ft. Area Per Lin. Ft. of Structural Angles of Various Sizes

Size of Angle Inches	Weight per Lin. Ft. Lbs.	Sq. Ft. Area per Lin. Ft.
2x2	2-5	0.70
3x3	2-11	1.00
4x4	5-20	1.35
5x5	8-31	1.70

6x6	13-40	2.00
8x8	26-57	2.70
4x3	4-17	1.20
5x3	6-19	1.40
6x4	10-30	1.70
7x4	13-34	1.90
8x4	17-37	2.00
8x6	20-50	2.40

Painting Lightweight Structural Steel After Erection. When painting, by hand and using a brush, lightweight structural steel after erection, a painter should paint 600 to 650 sq. ft. per 8-hr. day on the first coat; 800 to 900 sq. ft. per 8-hr. day on the second and third coats.

On a tonnage basis, this would be at the rate of 2 to 3 tons per 8-hr. day on the first coat and 3 to 4 tons per 8-hr. day on the second and third coats.

When using a spray gun, a painter should paint 3,600 sq. ft. per 8-hr. day on the first coat and 4,800 sq. ft. on the second and third coats.

Painting Medium to Heavyweight Structural Steel After Erection. When painting, by hand using a brush, medium to heavyweight structural steel after erection, a painter should paint 750 to 850 sq. ft. on the first coat and 900 to 1,000 sq. ft. per 8-hr. day on the second and third coats.

On a tonnage basis, this would be at the rate of 4 to 5 tons per 8-hr. day on the first coat and 5 to 6 tons per 8-hr. day on the second and third coats.

Using a spray gun, a painter should paint 4,400 sq. ft. per 8-hr. day on the first coat and 5,600 sq. ft. per 8-hr. day on the second and third coats.

Labor Cost of 100 Sq. Ft. (1 Sq.) 2-Coat Oil Paint Applied to Lightweight Structural Steel

	Hours	Rate	Total	Rate	Total
First Coat					
Painter	1.3	$....	$....	$24.39	$31.71
Second Coat					
Painter	1.0	....	....	24.39	24.39
Cost 100 sq. ft.			$....		$56.10

For third coat, add same as second coat.

Spray Painting

	Hours	Rate	Total	Rate	Total
First Coat					
Painter	.23	$....	$....	$24.39	$5.61
Second Coat					
Painter	.17	....	....	24.39	4.15
Cost 100 sq. ft.			$....		$9.76

For third coat, add same as second coat.

Add for air compressor and engineer, depending upon number of spray guns operated.

Labor Cost of 100 Sq. Ft. (1 Sq.) 2-Coat Oil Paint Applied to Medium to Heavyweight Structural Steel

	Hours	Rate	Total	Rate	Total
First Coat					
Painter	1.0	$....	$....	$24.39	$24.39
Second Coat					
Painter	0.9	....	....	24.39	21.95
Cost 100 sq. ft.			$....		$46.34

For third coat, add same as second coat.

Spray Painting

	Hours	Rate	Total	Rate	Total
First Coat					
Painter	.18	$....	$....	$24.39	$4.39
Second Coat					
Painter	.15	....	....	24.39	3.66
Cost 100 sq. ft.			$....		$8.05

For third coat, add same as second coat.

Add for air compressor and engineer, depending on number of spray guns operated. Not included in above costs are equipment such as scaffolding, and costs of complying with hazardous waste regulations.

Preparing Steel Surfaces

Proper protection of steel depends upon proper adhesion of the protective paint. Thus, when rust, oils, grease, scale and other dirt are not completely removed, the paint coat or coats will fail no matter how skillfully they are applied.

The Steel Structures Painting Council (SSPC), Pittsburgh, PA publishes detailed specifications for cleaning methods.

The simplest method is hand cleaning using scrapers, sanders, wire brushes and chipping tools. These remove only loose rust and scale and effectiveness depends largely on the zeal of the worker.

Power tool cleaning, using power operated sanders and brushes is slightly more expensive and slightly more effective.

For a superior job, one of the blasting methods is suggested: commercial, near white or white. The last specifies the metal be blasted down to a uniform "white" surface and will provide a base good for 7 to 8 years. Near white specifies that at least 95% of the metal will meet the "white" surface test, and its durability almost equals white blast.

Commercial blast is difficult to specify, but the results should leave at least two-thirds of the metal in the "white" stage, with durability of from 6 to 7 years.

With all the above methods, it is essential to use a primer with a good "wetting" power.

In figuring all these methods, cleaning up and protection for adjacent surfaces can add materially to the cost.

Surface Preparation by Hand and Abrasive Method. A properly prepared surface can make a good painting job look even better. A poorly prepared one can make an otherwise good job look bad. Remember that paint is not only to make a surface attractive. It protects a material from exposure and prolongs the useful life of the material. Paint protects steel surfaces from oxidation that results in rust.

Estimators need to understand the types of paints and coatings that an architect or engineer might specify for steel. Knowing the proper procedures for applying a specific painting material will give the estimator a competitive edge. The following checklist indicates major items to consider:

a) Type of surface to be painted.
b) Type of surface preparation - scrape or sandblast.
c) Scaffolding and other equipment requirements (OSHA).
d) Federal, state, and local environmental protection agency requirements.
e) Proper applicators - brushes, rollers, or sprayers.

f) Priming requirements - compatible with final material.
g) Number of paint coats required.
h) Disposal and clean up requirements.

The method of cleaning the surface in preparation for painting is usually dictated by the specifications, which might refer one to the paint manufacturer's recommendations. Wood surfaces are usually limited to a few choices of methods, but steel and concrete offer a wide range of possible methods, depending on the condition of the existing surface. Scaled or rusted surfaces can be cleaned by hand, with power tools, or using abrasive methods. A required degree of cleanliness is often stated in the contract documents and is defined by SSPC specifications.

Cleaning Structural Steel with Hand Wire Brush (SSPC-SP 2). When cleaning structural steel surfaces by hand with a wire brush, a worker can clean about 300 sq. ft. of surface per 8-hr. day at the following labor cost per 100 sq. ft.:

	Hours	Rate	Total	Rate	Total
Painter	2.67	$....	$....	$24.39	$65.12
Cost per sq. ft.			....		.65

Cleaning Piping 14" Diameter or Less with Hand Wire Brush. When cleaning steel pipe surfaces by hand with a wire brush, a worker can clean about 225 sq. ft. of surface per 8-hr. day at the following labor cost per 100 sq. ft.:

	Hours	Rate	Total	Rate	Total
Painter	3.20	$....	$....	$24.39	$78.05
Cost per sq. ft.			....		.78

Cleaning Structural Steel with Power Tools (SSPC-SP 3). When cleaning rust and scale from structural steel members, bridges, gas holders, oil tanks, and other metal surfaces before repainting, the number of sq. ft. cleaned per 8-hr. day varies with the type of surface and accessibility. A worker using air-powered wire brushes should clean 1,000 to 1,200 sq. ft. per day at the following cost per 100 sq. ft.:

	Hours	Rate	Total	Rate	Total
Painter	.72	$....	$....	$24.39	17.56
Compressor Operator	.36	....	....	25.96	9.35
Compressor Expense & Tools	.36	....	....	35.00	12.60
Cost per 100 sq. ft.			$....		$39.51
Cost per sq. ft.			....		.40

Cleaning Piping 14" Diameter or Less with Power Tools. When cleaning rust and scale off structural steel pipe, the number of sq. ft. cleaned per day will vary with surface and accessibility. A worker should clean 750 to 900 sq. ft. per day using air-powered wire brushes at the following cost per 100 sq. ft.:

	Hours	Rate	Total	Rate	Total
Painter	.97	$....	$....	$24.39	23.66
Compressor Operator	.49	....	....	25.96	12.72
Compressor Expense & Tools	.49	....	....	35.00	17.15
Cost per 100 sq. ft.			$....		$53.53
Cost per sq. ft.			....		.54

Add scaffolding or rigging, clean up, and disposal.

Commercial Blast Cleaning of Structural Steel (SSPC-SP 6). A blast cleaned surface should be free of all visible (viewed without magnification) oil, grease, dirt, mill scale, rust, paint, oxides, corrosion products, and other foreign matter, except staining. This should provide a base good for 7 to 8 years when the surface is properly painted.

A worker should clean 500 to 600 sq. ft. of structural steel members, bridges, etc., per day using abrasive blasting methods.

	Hours	Rate	Total	Rate	Total
Painter	1.45	$....	$....	$24.39	$35.37
Compressor Operator	1.45	....	....	25.96	37.64
Compressor Expense & Tools	1.45	....	....	25.00	36.25
Sandblasting Equipment	1.45	....	....	30.00	43.50
Cost per 100 sq. ft.			$....		$152.76
Cost per sq. ft.			....		1.53

Add for scaffolding or rigging, abrasive material, clean up, and disposal.

Commercial Blast Cleaning of Piping, 14" Diameter and Less. A worker should clean 300 to 500 sq. ft. of steel pipe per day using abrasive blasting methods at the following cost:

	Hours	Rate	Total	Rate	Total
Painter	2	$....	$....	$24.39	$48.78
Compressor Operator	2	....	....	25.96	51.92
Compressor Expense & Tools	2	....	....	25.00	50.00
Sandblasting Equipment	2	....	....	30.00	60.00
Cost per 100 sq. ft.			$....		$210.70
Cost per sq. ft.			....		2.11

Add for scaffolding or rigging, abrasive material, clean up, and disposal.

Brush Off Blast Cleaning of Structural Steel (SSPC-SP 7). A brush off blast cleaned surface should be free of all visible (viewed without magnification) oil, grease, dirt, dust, loose mill scale, loose rust, and loose paint. Tightly adherent mill scale, rust, and paint that cannot be removed by lifting with a dull putty knife may remain on the surface.

On structural steel members, bridges, etc., a worker should clean 900 to 1,100 sq. ft. per day using this method:

	Hours	Rate	Total	Rate	Total
Painter	.80	$....	$....	$24.39	$19.51
Compressor Operator	.80	....	....	25.96	20.77
Compressor Expense & Tools	.80	....	....	25.00	20.00
Sandblasting Equipment	.80	....	....	30.00	24.00
Cost per 100 sq. ft.			$....		$84.28
Cost per sq. ft.			....		.84

Add for scaffolding or rigging, abrasive material, clean up, and disposal.

Brush Off Blast Cleaning Piping 14" Diameter or Less. On steel pipes, a worker should clean 680 to 820 sq. ft. per day using this method at the following cost:

	Hours	Rate	Total	Rate	Total
Painter	1.07	$....	$....	$24.39	$26.10
Compressor Operator	1.07	....	....	25.96	27.78
Compressor Expense & Tools	1.07	....	....	25.00	26.75
Sandblasting Equipment	1.07	....	....	30.00	32.10
Cost per 100 sq. ft.			$....		$112.73
Cost per sq. ft.			....		1.13

Add for scaffolding or rigging, abrasive material, clean up, and disposal.

Near White Blast Cleaning of Structural Steel (SSPC-SP 10). A near-white blast cleaned surface, when viewed without magnification, should be free of all visible oil, grease, dirt, dust, mill scale, rust, paint, oxides, corrosion products, and other foreign matter, except for staining.

On structural steel members, bridges, etc., a worker using this method should clean 300 to 400 sq. ft. per day at the following cost:

	Hours	Rate	Total	Rate	Total
Painter	2.29	$....	$....	$24.39	$55.85
Compressor Operator	2.29	....	....	25.96	59.45
Compressor Expense & Tools	2.29	....	....	25.00	57.25
Sandblasting Equipment	2.29	....	....	30.00	68.70
Cost per 100 sq. ft.			$....		$241.25
Cost per sq. ft.			....		2.41

Add for scaffolding or rigging, abrasive material, clean up, and disposal.

Near White Blast Cleaning Piping 14" Diameter and Under. On steel pipes, a worker should clean 225 to 300 sq. ft. per day:

	Hours	Rate	Total	Rate	Total
Painter	3.05	$....	$....	$24.39	$74.39
Compressor Operator	3.05	....	....	25.96	79.18
Compressor Expense & Tools	3.05	....	....	25.00	76.25
Sandblasting Equipment	3.05	....	....	30.00	91.50
Cost per 100 sq. ft.			$....		$321.32
Cost per sq. ft.			....		3.21

Add for scaffolding and rigging, abrasive material, clean up, and disposal.

White Metal Blast Cleaning Structural Steel (SSPC-SP 5). Viewed without magnification, a white metal blast cleaned surface should be free of all visible oil, grease, dirt, dust, mill scale, rust, paint, oxides, corrosion products, and other foreign matter.

On structural steel members, bridges, etc., a worker should clean 230 to 300 sq. ft. per day using this method:

	Hours	Rate	Total	Rate	Total
Painter	3.01	$....	$....	$24.39	$73.41
Compressor Operator	3.01	....	....	25.96	78.14
Compressor Expense & Tools	3.01	....	....	25.00	75.25
Sandblasting Equipment	3.01	....	....	30.00	90.30
Cost per 100 sq. ft.			$....		$317.10
Cost per sq. ft.			....		3.17

Add for scaffolding and rigging, abrasive material, clean up, and disposal.

White Metal Blast Cleaning Piping, 14" Diameter or Less. On steel pipes, a worker should clean 170 to 230 sq. ft. per day using this method:

	Hours	Rate	Total	Rate	Total
Painter	4.00	$....	$....	$24.39	$97.56
Compressor Operator	4.00	....	....	25.96	103.84
Compressor Expense & Tools	4.00	....	....	25.00	100.00
Sandblasting Equipment	4.00	....	....	30.00	120.00
Cost per 100 sq. ft.			$....		$421.40
Cost per sq. ft.			....		4.21

Add for scaffolding and rigging, abrasive material, clean up, and disposal.

Spray Painting. Almost any painting job can be done quicker with spray equipment. This is particularly true where large surfaces are to be covered or where there are angles and corners that cannot be reached with brushes or rollers.

Practically all surfaces that are now being painted with a bristle brush can be coated by spray painting. On rough surfaces such as cement, stucco, rough plaster, brick, tile, shingles and rough boards, spray painting affects a considerable saving in time and labor.

It is now possible to obtain mechanical painting equipment that will apply any paint, varnish, graphite, mineral oxides, epoxy, etc.

Painting Oil Tanks and other Large Metal Surfaces by Compressed Air. In painting oil tanks and other large metal surfaces, a man should cover 1,000 sq. ft. an hr. at the following cost per 100 sq. ft.:

	Hours	Rate	Total	Rate	Total
Painter	.10	$....	$....	$24.39	$2.44
Compressor operator	.05	....	....	25.96	1.30
Compressor expense	.05	....	....	15.00	.75
Cost per 100 sq. ft.			$....		$4.49
Cost per sq. ft.			....		.04

Does not include paint, scaffold, or preparation.

Comparative Cost by Hand

On the same class of work, a painter will brush about 200 sq. ft. of surface per hr. at the following cost per 100 sq. ft.:

	Hours	Rate	Total	Rate	Total
Painter	.5	$....	$....	$24.39	$12.20
Cost per sq. ft.			....		.11

Does not include paint, scaffold, or preparation.

09950 WALLCOVERING

Wallpaper is estimated by the roll containing 36 sq. ft. Most rolls are 18" or 20½" wide and contain 36 sq. ft. or 4 sq. yds. of paper. The double roll contains twice as much as the single roll: 72 sq. ft. or 8 sq. yds. of paper.

Estimating Quantity of Wallpaper for Any Room. When estimating the quantity of wallpaper required for any room, measure the entire distance around the room in lin. ft. and multiply by the height of the walls or the distance from the floor to the ceiling. The result will be the number of sq. ft. of surface in the four walls.

Make deductions in full for the area of all openings such as windows, doors, mantels, and built-in bookcases. The difference between the area of the four walls and the area of the openings will be the actual number of sq. ft. of surface to be papered.

When adding for waste in matching, cutting, fitting, etc., paper hangers use different methods and allow different percentages, contingent upon the height of the ceiling, the design of the paper, and the size of the pattern or figure, but an allowance of 15% to 20% is sufficient in nearly all cases.

When estimating the quantity of paper required for walls, bear in mind it is not necessary to run the side wallpaper more than one or two inches above the border, when a border is used. The height of the wood base should also be deducted from the total height of the wall.

In rooms having drop ceilings, the depth of the drop should be deducted from the height of the wall, which will decrease the area of side wallpaper in the same amount as it increases the quantity of ceiling paper.

If a border is required, measure the distance around the room and the result will be the number of lin. ft. of border. Dividing by 3, gives the number of yards of border required.

When estimating the quantity of wallpaper required for ceilings, multiply the width of the room by the length, and the result will be the number of sq. ft. of ceiling to be papered.

In rooms having a drop ceiling, add the depth of the drop or the distance the ceiling paper extends down the side walls to the length and breadth of the room and multiply as described above.

After the areas of both side walls and ceiling have been computed in sq. ft. divide by 36, and the result will be the number of single rolls of paper required, or by dividing by 72 gives the number of double rolls of paper required.

Paste Required Hanging Wallpaper. The quantity of paste required for hanging wallpaper will vary with the weight of the paper and the surface to which it is applied.

Where light or medium weight wallpaper is used, one gallon of paste should hang 12 single rolls of paper.

If heavy or rough texture paper is used, it will often be necessary to give it two or three applications of paste to obtain satisfactory results. On work of this kind one gallon of paste should hang 4 to 6 single rolls of paper.

There are prepared pastes on the market which require only the addition of cold water to make them ready for use.

One pound of prepared cold water dry paste should make 1½ to 2 gallons of ready to use paste.

Labor Hanging Wallpaper. Different methods of hanging wallpaper are used in different sections of the country. In some localities a paper hanger and helper work together, one worker trimming the paper and pasting while the other hangs the paper. However, in many cities the work is performed by one person, who cuts, trims, fits, pastes, and hangs the paper.

The following quantities and costs are based on the performance of one worker but will prove a fair average for use in any locality.

Hanging Wallpaper on Walls and Ceilings, One Edge Work. When hanging light or medium weight paper on ceilings and drops, a paper hanger should hang 28 to 32 single rolls of paper per 8-hr. day, at the following labor cost per roll:

	Hours	Rate	Total	Rate	Total
Paper hanger	0.27	$....	$....	$24.39	$6.59

Hanging Wallpaper on Walls, Butt Work. Where light or medium weight paper is used for bedrooms, halls, etc., and where a good grade of workmanship is required with all paper trimmed on both edges and hung with butt joints, a paper hanger should trim, fit and hang 20 to 24 single rolls of paper per 8-hr. day, at the following labor cost per roll:

	Hours	Rate	Total	Rate	Total
Paper hanger	0.36	$....	$....	$24.39	$8.78

Another thing that will slow up the work is where new paper is applied over old rough textured wallpaper, such as oatmeal designs, etc.

Hanging Wallpaper, First Grade Workmanship. Where a good grade of medium or heavy weight wallpaper is used, with all paper hung with butt joints, a paper hanger should trim, fit and hang 16 to 20 single rolls of paper per 8-hr. day, at the following labor cost per roll:

	Hours	Rate	Total	Rate	Total
Paper hanger	0.44	$....	$....	$24.39	$10.73

Labor Hanging Scenic Paper. Where scenic paper or paper having mural designs are used over the wall or up stairways, a paper hanger will hang only 8 to 10 single rolls per 8-hr. day, at the following labor cost per roll:

	Hours	Rate	Total	Rate	Total
Paper hanger	0.9	$....	$....	$24.39	$21.95

Placing Wallpaper Borders. Wallpaper borders are usually estimated by the lin. ft. or yd. and as the width varies from 3" to 18" there is never over one width required.

A paper hanger should place border at the rate of 100 to 125 lin. ft. an hr. The labor cost per 100 lin. ft. (33⅓ yds.) should run as follows:

	Hours	Rate	Total	Rate	Total
Paper hanger	0.9	$....	$....	$24.39	$21.95
per lin. ft.			....		.22
Cost per yd.			....		.66

Hanging Coated Fabrics. Where coated fabrics, such as Walltex, Sanitas, etc., are used, a paper hanger should hang 16 to 20 single rolls per 8-hr. day, at the following labor cost per roll:

	Hours	Rate	Total	Rate	Total
Paper hanger	0.44	$....	$....	$24.39	$10.73

Canvassing Plastered Walls. When applying canvas to plastered walls, an average mechanic should apply 50 to 60 sq. ft. an hour.

Labor Cost of 100 Sq. Ft. Canvassed Walls

	Hours	Rate	Total	Rate	Total
Paper hanger	1.82	$....	$....	$24.39	$44.39
Cost per sq. ft.			....		.44

Flexwood. Flexwood from U.S. Plywood, Louisville, KY, is a genuine wood veneer cut to $1/85$" thick and glued under heat and hydraulic pressure to cotton sheeting with a waterproof adhesive. A patented flexing operation breaks the cellular unity of the wood to produce a limp, pliable sheet which may be applied to any smooth surface, flat or curved. Waterproof Flexwood 710 Adhesive, which makes a permanent bond, is used to apply Flexwood. Standard sizes of stock material are 18" and 24" wide and 8' to 12' long.

The mechanics of installing Flexwood are as follows: the background is sized with Flexwood cement. Another coating of cement is brushed on the Flexwood which is then hung in the manner of any sheet wall covering. A stiff, broad knife, used with considerable pressure, smoothes out the Flexwood, removes air spaces and furnishes the necessary contact.

Some of the wood available in Flexwood are mahogany, walnut, oak, prima vera, knotty pine, orientalwood, satinwood, zebrawood, rosewood, English oak, maple, lacewood, etc. The cost of the material varies from $1.20 to $3.75 per sq. ft. depending upon the kind of wood. Most woods run from $1.50 to $1.75. Adhesive runs $9.50 per gallon.

It requires a skilled and thoroughly competent paper hanger to apply this material and the labor costs vary considerably with the design to be obtained.

Considerable time is usually required to lay out any room to obtain the desired effect, spacing of strips, etc. This is especially true where certain designs must be obtained on columns, walls and pilasters and where narrow strips of contrasting woods are used to obtain inlay effects.

On plain walls without inlays, a paper hanger experienced in this class of work should apply 150 sq. ft. of Flexwood per 8-hr. day, at the following labor cost per 100 sq. ft.:

	Hours	Rate	Total	Rate	Total
Paper hanger	5.4	$....	$....	$24.39	$131.71
Cost per sq. ft.			....		1.32

On pilasters, columns and walls inlaid with narrow strips of contrasting woods, the work is considerably slower, due to the additional cutting and fitting required. On work of this kind, a paper hanger, experienced in this class of work should apply 75 sq. ft. per 8-hr. day, at the following labor cost per 100 sq. ft.:

	Hours	Rate	Total	Rate	Total
Paper hanger	10.7	$....	$....	$24.39	$260.97
Cost per sq. ft.			....		2.61

Additional time for paper hanger or foreman will be required laying out the work into patterns, marking off strips, etc. This will vary with the size of the job.

After the Flexwood is applied, the joints should be sanded lightly to remove any imperfections in the joints.

Finishing Flexwood. Flexwood will take any wood finish but where the natural color of the wood is desired, the Flexwood is given one coat of lacquer sealer, sanded lightly between coats and then given one coat of lacquer for a finish coat.

One gallon of lacquer sealer will cover about 300 sq. ft. of surface, one coat.

After the sealer has been applied, one gallon of lacquer should cover 450 to 550 sq. ft.

An experienced painter should apply lacquer sealer, sand lightly between coats and apply one coat of lacquer to 90 to 110 sq. ft. of surface an hr. at the following labor cost per 100 sq. ft.:

	Hours	Rate	Total	Rate	Total
Painter	1	$....	$....	$24.39	$24.39
Cost per sq. ft.			....		.24

Vinyl Wall Covering. Vinyl wall covering is composed of a woven cotton fabric, to which a compound of vinyl resin, pigment and plasticizer is electronically fused to one side. It comes 24", 27", and 54" wide in three weights: heavy (36 oz. per lin. yd.); medium (24-33 oz. per lin. yd.); and light (22-24 oz. per lin. yd.).

Some of the patterns available simulate linens, silks, moires, grasses, tweeds, Honduras mahogany, travertine, damasks, and burlaps. Costs range from 35 to 90 cents per square foot.

Vinyl fabric is hung using regular wallpaper hanging procedures for hanging fabric baked wallcoverings. A broad knife is used to smooth out any wrinkles, air pockets and insures a good contact. Wash off any excess paste that remains on the surface of the material. No further finishing is necessary.

A competent paper hanger can apply approximately 600 sq. ft. per 8 hr. day at the following costs per 100 sq. ft.

	Hours	Rate	Total	Rate	Total
Paper hanger	1.34	$....	$....	$24.39	$32.68
Cost per sq. ft.					.33

If excessive cutting and fitting is required, the above figures should be increased to allow for the above conditions.

CHAPTER 10

SPECIALTIES

The Construction Specifications Institute (CSI), whose format we use in the general presentation of this text, follow sections covering the usual building trades with four sections covering allied fields, which may or may not be part of the typical general contract.

CSI Division 10 includes specialties that are prefabricated and added to the building, such as chalkboards, toilet partitions, decorative grilles, freestanding fireplaces, flagpoles, directories and signs, lockers, mailboxes and chutes, movable partitions, sun control devices, telephone booths, and the like. Many of these items will be subcontracted and installed under the general contractor's supervision. Some may be covered by allowances.

10100 CHALKBOARDS AND TACKBOARDS

Chalkboards are made of ¼" or ½" hardboard with special finishes applied. These include the following which are available in either black or green:

¼" Thick - Material Prices

	Price per Sq. Ft.
Standard, weighing 1.75 lbs. per sq. ft.	$1.60
Moisture sealed at 1.8 lbs. per sq. ft.	1.85
Porcelain steel at 2.1 lbs. per sq. ft.	4.80

These are available in 4' widths in sheets to 16' long and are applied to the wall with mastic at the rate of one gallon per 70 sq. ft. They can be set in custom trim, or the following stock aluminum trim, which comes complete with all necessary attaching clips:

	Price per Lin. Ft.
Perimeter trim 1½" wide	$2.00
Maprail headtrim	2.25
Joint strip	1.18
Chalktrough 2¼" wide	4.70

Bulletin board materials are ¼" cork and available as follows:

	Price per Sq. Ft.
Natural cork	$2.10
Vinyl cork	3.10
Cork on ⅜" fiberboard	3.15
⅛" on ⅜" fiberboard vinyl f'c'd	3.10

The cork comes in rolls 4' wide by up to 90' long. The panels are 4' wide up to 16' long. Stock trim costs $2.00 for ¼" stock, $2.15 for ½" stock.

10150 COMPARTMENTS AND CUBICLES

Toilet Partitions and Screens

Toilet partitions are most often metal units of the floor mounted type, with baked enamel finish and chrome hardware. One door and side partition for use in a corner or abutting other stall side partitions cost about $185.00. A porcelain enamelled finish adds $300.00 to the cost. Ceiling hung units cost around 10% more, plus of course the cost of ceiling reinforcement if needed. One carpenter and laborer will set a unit in about two hours. Urinal screens hung from the wall cost $115.00 in baked enamel and will take a two-person team half an hour to set. Wall hung toilet stalls will cost $355.00 per unit.

10301 PREFABRICATED FIREPLACES

Metal fireplaces are available in a wide range of styles for burning wood or gas. They are made for placing against walls, in corners and freestanding. However, there are requirements for clearances from walls and for floor protection. They can be vented to an existing flue or have a prefabricated smoke pipe terminating above the roof. A simulated masonry chimney top is available. Finishes are either matte black or porcelainized. A 24" wide unit costs $390.00 in matte, $500.00 in porcelain. A 45" wide unit costs $530.00 in matte and $645.00 in porcelain.

Freestanding units open on all sides will cost $475.00 for matte and $590.00 for porcelain in 36" diameter size; $550.00 and $675.00 in 42" diameter. If they are to rest on the floor, add $55.00 for the metal base. If hung from the ceiling, add $30.00 for chains.

The 8" metal flues cost about $22.00 per foot. Chimney top housings for single flues cost $85.00 to $115.00.

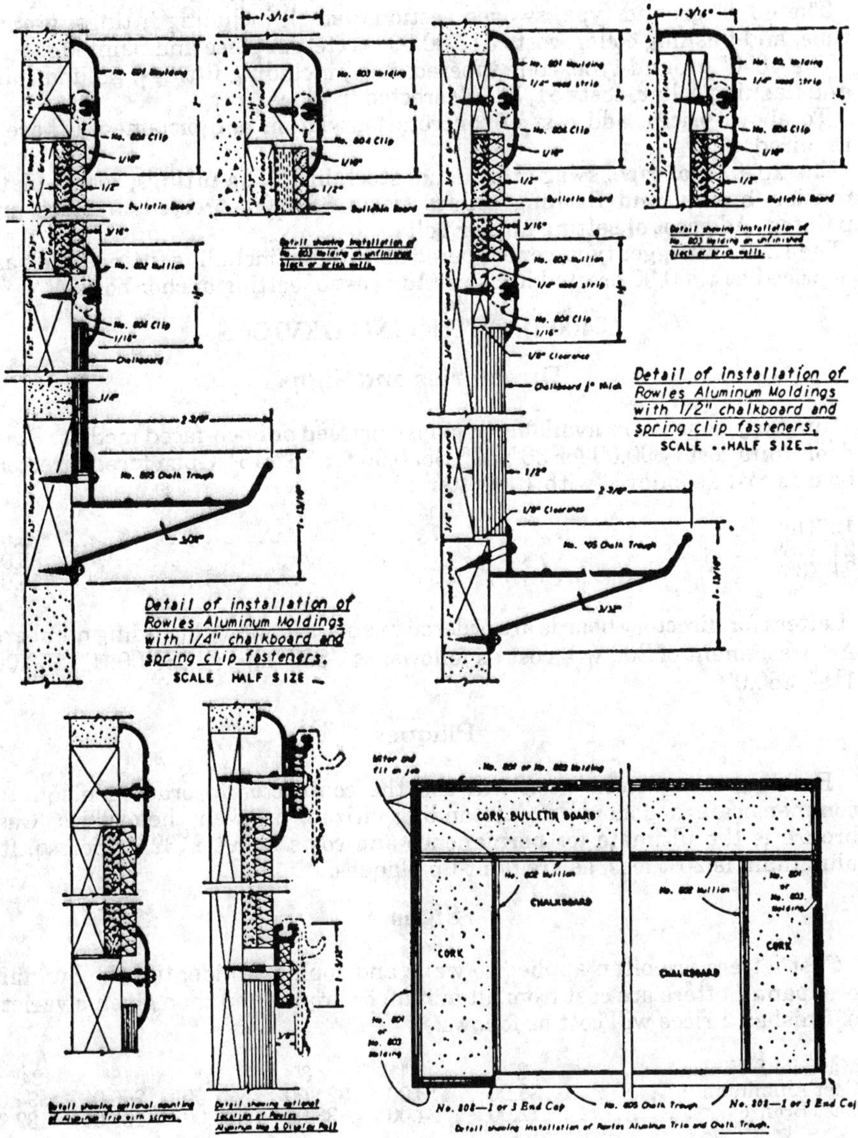

Details of Aluminum Moldings Used with Chalkboards and Corkboards

10350 FLAGPOLES

Typical prices for several types of flagpoles are as follows:

The 40'-0" ground type, swaged sectional steel, including fittings, ground tube, and flashing collar, costs $1,500.00 erected in place and painted.

The 40'-0" ground type, cone tapered steel, including fittings, ground tube, and flashing collar, costs $1,700.00 erected in place.

To above prices, add cost of concrete foundation and ornamental base if required.

The 20'-0" roof type, swaged sectional steel, including fittings, socket plate, tension braces, and flashing collars, costs $850.00, erected in place and painted. Add cost of setting anchor bolts.

The 12'-0" outrigger type, cone tapered aluminum, including fittings and base is priced at $900.00 erected in place. Add cost of setting anchor bolts.

10400 IDENTIFYING DEVICES

Directories and Signs

Directory boards are available in glass-enclosed or open-faced models. Glass door units cost $390.00 for 36"x30", $550.00 for 36"x48". Open faced directory boards cost as follows, with 100 letters:

12"X18"	$75.00
24"X18"	85.00
24"X36"	115.00

Letters for directory boards are ordered in an assortment, including numbers. An assortment of 200 will cost as follows: ½", $33.00; ¾", $39.00; 1", $50.00; 1½", $60.00.

Plaques

Building committees often will ask the contractor to order a plaque to commemorate a donor, an outstanding citizen, or even themselves. Cast bronze is the ultimate for permanence and costs about $132.00 per sq. ft., aluminum is 20% less, as are built-up plaques.

Signs

Cast letters are often applied to walls and copings to identify the building occupant. Letters are cast from aluminum or bronze and then given a variety of finishes. Prices will cost as follows:

Material & Finish	2"	4"	6"	12"	18"	24"
Cast Aluminum	$8.00	$13.00	$24.00	$37.00	$66.00	$94.00
Cast Bronze	12.00	24.00	33.00	66.00	100.00	132.00

A worker should install 3 to 4 letters per hour. Plastic letters will cost $12.00 each in 8" height.

10500 LOCKERS

Lockers are made in single, double and triple high units totaling 72" high (single tier also comes 60" high), in widths 9", 12", 15", and 18" and depths of 12", 15", and 18". There are also box lockers 6 units high.

A 12"x15"x72" unit in baked enamel will cost around $40.00 single tier; $55.00 double tier; $75.00 triple tier. For a sloping top add $2.50 per tier. For closed base, add $3.00 per ft. Open mesh or "athletic" types will cost $90 to $110 per tier 72" high. Basket racks 8 high, 5 wide will cost $125.00 plus $6.50 per basket. For a combination lock, add $20.00 per door, $10.00 for a key lock.

Benches for lockers are sold by the lineal foot. Benches made of 1¼" thick laminated maple cost $20.00 per foot plus $25.00 for each pedestal.

10550 POSTAL SPECIALTIES

Mail chutes with glass and aluminum enclosures 9"x3½" will cost $60.00 per lin. ft., and one worker can install two floor heights per day. A lobby collection box will cost from $1,200.00 on up depending on the size and design.

Keyed mail delivery boxes in gangs will cost around $55.00 per unit rear loading and $60.00 front loading. One worker can set around 30 units per day.

10600 PARTITIONS

Movable partitioning is available in metal in both flush and panel construction, in laminated gypsum board and made of open wire mesh. Solid units can be specially constructed for sound and fire isolation, and can be prefinished with baked enamel, vinyl, or wood veneers or painted in the shop or on the job.

Metal office type partitioning is usually a nominal 2¼" or 3½" thick with a height limit of 12 feet. Free standing units have a standard cornice height of 7'-1½". "Banker" types are usually 1¾" thick and 30" high. A standard glass panel can be mounted on the top rail to bring the height to 42" or 54".

The base for movable partitioning is designed to form a raceway for both electrical and telephone lines. Switches can usually be built into the door frame panels, which with doors, glazing panels and louver panels can all be ordered as part of the partitioning.

Mesh partitioning is usually made of 10 ga. wire worked into a 1½" diamond mesh set in a channel frame. It will cost around $2.00 per square foot on straight runs using standard sections. A 3'x7' door will cost $170.00. A 5'x7' sliding panel, $300.00. A shop coat of paint is standard. A baked enamel coat will add 20 cents per sq. ft. to costs. Two erectors can install some 100 lin. ft. per day. Add 1.5 hrs. for each door. Ceiling panels will install at the rate of 75 sq. ft. per hour two persons working together.

Office partitions come in so many grades and job conditions are so diverse that it is difficult to give any basic costs, but because the field is highly competitive, it is always a simple matter to get prices together. A standard 2⅜" 20 ga. flush steel floor to ceiling partition 8 ft. to 9 ft. high will cost around $5.00 per sq. ft., without installation. Two men should erect around 40 lineal feet per day. Each door with lock but no closer will cost $230.00. If partitions are glazed add for both labor and material.

Gypsum partitions, 2¼" thick, will run about $3.25 per sq. ft. installed but unfinished. If glazed panels are desired, add $13.00 per sq ft.

Freestanding "bank" type partitioning 3' high cost $30.00 to $40.00 per lin. ft. Two workers will set 70 lin. ft. per day.

10650 SCALES

Built-in scales must be very carefully set, adjusted, and maintained and should be set by a contractor licensed by the scale manufacturer. For general budget allowances the following is a guide:

Platform Size	Capacity	Cost of Mechanical	Cost of Electronic
7'x9'	5 ton	$9,400	....
10'x24'	10 ton	10,600	....
10'x50'	30 ton	12,500	....
10'x70'	50 ton	30,000	35,000
10'x70'	60 ton	31,250	37,500
10'x70'	80 ton	35,600	40,000
10'x70'	100 ton	....	45,000

10670 STORAGE SHELVING

Free standing, baked enamel industrial shelving comes knocked down in 3' sections, complete with all hardware for erection at the job site. Cost will approximate the following:

Depth	Price per Lineal Foot 4 Shelves	7 Shelves
12"	$29.00	$48.00
18"	34.00	56.00
24"	40.00	65.00

One worker can erect from 9 to 12 lin. ft. per day.

10700 SUN CONTROL DEVICES

Awnings and canopies of canvas, vinyl, and aluminum are prefabricated in the shop and delivered to the job site ready for erection.

Continuous, wall hung, aluminum canopies will cost from $4.75 to $7.00 per sq. ft., and two workers can erect about 250 sq. ft. per day.

Awnings over individual openings 3'-6" wide will cost $300.00 in aluminum, $200.00 in cloth. Awnings are often used as decorative features and take forms unrelated to sun control and can easily cost up to $35.00 a sq. ft. Two workers will take two hours to hang an individual awning.

10750 TELEPHONE ENCLOSURES

The trend in telephone noise control has been to rely on sound absorbtion rather than physical separation. Where booths are still used a simple steel and wood unit costs $1,500.00 while an extruded metal unit will cost twice that. Two workers will set three units in two days.

Outdoor post types will cost around $700.00 and take one worker 6 hours to set.

Indoor shelf types will average around $400.00 a station and take one worker 3 hours to install. Directory shelves will cost around $300.00 for a size that holds up to four books, and will take one worker one hour to set.

10800 TOILET AND BATH ACCESSORIES

The following list may serve as a guide in purchasing chrome plated bathroom accessories. Sometimes accessories are of ceramic tile, costs of which are discussed under *Finishes*. Also, some items may be furnished with toilet compartments or will be furnished by service companies who supply the owner. Check to avoid duplication.

Prices of Toilet and Bath Accessories

Ash tray bowl	
wall-mounted	$35.00
recessed	75.00
Clothes hook	
single	3.00
double	3.50
Grab bar	
12"	10.00
24"	16.00
Hand dryer, electric	225.00
Medicine cabinet, 18"x24", lighted	65.00
Paper towel dispenser	
wall-mounted	25.00
recessed with waste receptacle	240.00
Sanitary napkin dispenser	
coin operated & surface-mounted	200.00
Shelf, stainless steel, 24" long	18.00
Shower rod	
3'	9.00
5'	12.00
Soap dish, recessed	7.00
Soap dish with grab bar	10.00
Soap dispenser	
powder	15.00
liquid	22.00
Toilet tissue dispenser	8.00
Towel bar	
12"	12.00
18"	12.50
24"	14.00
Towel ring	9.00
Towel ladder, 16"x30"	42.00
Tumbler/toothbrush holder	8.00
Waste receptacle, 10 gal	85.00

10900 WARDROBE SPECIALTIES

Hat and coat checking storage units will cost approximately the following:

Material	Price per Lineal Feet Single Tier	Double Tier
Steel	$20.00	$22.00
Aluminum	29.00	35.00
Stainless	35.00	40.00

One worker can set 45 lin. ft. of single tier and 32 lin. ft. of double tier rack per day.

CHAPTER 11

EQUIPMENT

CSI DIVISION 11
11900 RESIDENTIAL EQUIPMENT

CSI Division 11 includes equipment such as vacuum systems, powered window washers, bank, checkroom, church, school, food service, vending, athletic, lab, laundry, library, medical, parking, waste handling, loading dock, detention and stage equipment.

These are often purchased by the owner separately and installed by the general contractor. Even if the general contractor is not involved with the installation of such equipment, he should have a list of all equipment going into the building, with cuts and shop drawings so that he can point out any conflicts between the structure and such equipment. A floor that has to be broken up to conceal a conduit or duct becomes a permanent scar that reflects on the general contractor, even though he might be an innocent partner.

11900 RESIDENTIAL EQUIPMENT

Kitchen Cabinets

The following cabinets are furnished completely assembled and finished. Frames are ¾" kiln dried hardwood; doors and drawer fronts are ⅞" thick solid lumber; shelves are ½" plywood or chipboard; backs of base cabinets are Masonite; backs of wall cabinets and drawer bottoms are plywood; ends of cabinets are flush with hardwood faces.

Cabinets are factory finished with a penetrating sealer coat and clear lacquer. Interiors are finished and drawer bottoms are clear lacquered. Door and drawer hardware come in a variety of styles.

These cabinets are the type found in most residential construction. Economy grade cases for cheaper construction will range from 25% to 50% less for finished units ready for installation. For top quality kitchen cabinets, the prices will run up to 50% higher for stock sizes and 100% higher or more for custom made units, depending upon the design, materials, hardware and finish specified. All prices are f.o.b. factory.

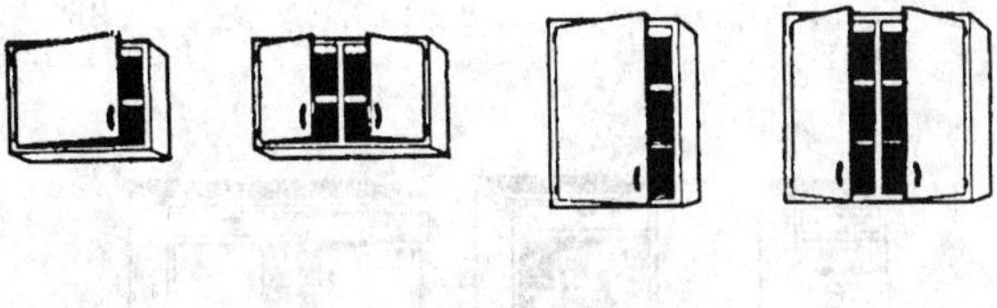

Wall Units

Wood Wall Units, 12" Deep

Width	No. Doors	12" High	15" High	24" High	30" High
12"	1	$....	$....	$....	$60.00
15"	1	....	....	....	66.00
18"	1	....	....	....	70.00
21"	1	....	....	....	74.00
24"	2	....	....	67.00	85.00
27"	2	....	....	70.00	95.00
30"	2	53.00	58.00	75.00	104.00
33"	2	58.00	60.00	81.00	109.00
36"	2	60.00	67.00	83.00	114.00
42"	2	....	....	89.00	120.00
48"	2	....	....	100.00	143.00

Corner Shelf and Wall Units, 30" High

Corner wall unit, 23" deep and wide, with 2 shelves	$100.00
Corner wall unit, 23" deep and wide, with revolving 3 shelf "Lazy Susan"	143.00

Utility Cabinet and Built-In Oven Cabinets

Utility cabinet, 84" high, 24" deep, 18" wide	$222.00
Oven cabinet, 84" high, 24" deep, 24" wide	265.00
Oven cabinet, 84" high, 24" deep, 27" wide	280.00

Wood Cabinet Base Units

Base units are 34½" high and 24" deep without tops.

Description	Width, Inches	Without Tops
4 drawers	15"	$127.00
4 drawers	18"	132.00
4 drawers	21"	140.00
4 drawers	24"	148.00
1 top drawer, 1 door below	12"	91.00
1 top drawer, 1 door below	15"	97.00
1 top drawer, 1 door below	18"	103.00
1 top drawer, 1 door below	21"	113.00
1 top drawer, 1 door below	24"	127.00
1 top drawer, 2 doors below	27"	138.00
2 top drawers, 2 doors below	36"	170.00
2 top drawers, 2 doors below	42"	180.00
2 top drawers, 2 doors below	48"	195.00
Corner base cabinet	42"	150.00
Corner base cabinet with revolving "Lazy Susan	35"	160.00
3" corner filler		6.00

Wood Cabinet Base Units

Wood Sink Fronts

Description	Width	Price
Sink front only, 1 pair doors	27"	$72.00
	30"	74.00
	33"	77.00
	36"	83.00
	42"	89.00
	48"	96.00
Sink end	24"	12.00

Wood Sink Base Units

Width, Inches	Price Without Tops or Sinks
27"	$117.00
30"	122.00
33"	127.00
36"	133.00
42"	143.00
48"	154.00

Steel Kitchen Units

Prices on steel case units vary considerably with different manufacturers, due to weight of metal used, method of manufacturing, kind of finish, sprayed or baked, and volume of production as many kinds of steel cases are on a mass production basis, while others are custom made to fit the job. Naturally, the custom cabinets cost more.

The cabinets priced below are completely finished and ready to be installed. They are fabricated from 18, 20, and 22 gauge cold rolled steel and are spot welded, bonderized and then finished with 2 coats of enamel baked on at high temperature. They are medium quality.

Base units are mounted on an integral black enameled subbase 4" high and recessed 3" in front for toe space.

Doors are 3/4" thick, double-walled, die formed of one piece of steel, and spot welded. Drawers have 3/4" fronts which make them flush with the doors. Drawers operate on roller bearings.

Hardware consists of semi-concealed hinges. Die-cast chromium plated handles are furnished on doors and drawers.

Steel Wall Cabinets

Wall units are 13″ deep, 18″ units have 1 removable shelf, and 30″ units have 2 shelves.

Width	No. Doors	18" High	30" High
12"	1	$....	$39.00
15"	1	....	41.00
18"	2	39.00	46.00
21"	2	42.00	51.00
24"	2	45.00	53.00
27"	2	....	57.00
30"	2	48.00	60.00
36"	2	55.00	68.00

Corner Shelf and Wall Units, 30" High

Description	Price Each
Corner shelf unit, 12¼ deep, 6" wide, with 2 fixed shelves	$20.00
Corner shelf unit, 12¼" deep and wide, with 2 fixed shelves	23.00
Corner wall unit, 21" deep and wide, with revolving 3-shelf "Lazy Susan"	110.00

Courtesy Bill Hedrich, Hedrich-Blessing

Steel Base Cabinets

Base units--34½" high and 24" deep

Description	Width, Inches	Without Top
1 drawer above, 1 door below	12"	$72.00
1 drawer above, 1 door below	15"	80.00
1 drawer above, 1 door below	18"	86.00
1 drawer above, 2 doors below	21"	90.00
1 drawer above, 2 doors below	24"	100.00
1 drawer above, 2 doors below	27"	108.00
1 drawer above, 2 doors below	30"	115.00
1 drawer above, 2 doors below	36"	129.00
3 drawers	18"	123.00
Corner base cabinet with revolving "Lazy Susan"	33"	133.00

Steel Sink Fronts

Description	Width	Price
Sink front only--one pair doors	24"	$44.00
Sink front only--one pair doors	30"	45.00
Sink front only--one pair doors	36"	51.00

Fillers and End Panels for Steel Cabinets

Description	Price Each
Corner wall filler, 30" high, 15" deep, 15" wide	$15.00
Straight wall filler, 18" high, 13" deep, 3" wide	7.00
Straight wall filler, 30" high, 13" deep, 3" wide	9.00
Corner base filler, 34½" high, 27" deep, 27" wide	22.00
Straight base filler, 34½" high, 3" wide	5.00
End panel, 34½" high, 25" deep	12.00

Steel Kitchen Sink Units

Sinks are 36" high, 25" deep. Constructed of 18, 20, or 22 ga. furniture steel with a baked on finish. Sink tops are 14 ga. deep drawn enameling steel to which porcelain enamel is bonded.

Tops have grooved drainboards and 4" back-splash, all insulated to deaden noise. Sink tops are one piece construction, roll rimmed.

Length Unit	Size Bowl	Single or Dbl. Bowl	Price Each
42"	20"x16½"x7"	Single	$220.00
54"	20"x16½"x7"	Single	250.00
66"	Two 14½"x 17½"x7"	Double	300.00

KITCHEN COUNTERTOP

Most all countertops for kitchen cabinets are plastic laminate over a plywood, particleboard or masonite subsurface.

The plastic decorative laminate is offered in a wide variety of colors and patterns including simulated wood grain and butcher-block styles.

The tops are usually made in self-edge design (square edges and 90 angle for the backsplash), however they are available in post-form design (rolled edges and angles) at a slight decrease in cost.

To estimate the total lineal feet of countertop, measure the length of the area to be covered along the longest side (usually the wall surface). Plastic laminate tops, made to size, ready to install will cost from $20.00 to $30.00 per lin. ft.

Dupont *Corian* is a synthetic countertop material that can be manufactured in one piece. Costs installed range from $110.00 to $200.00 per lin. ft.

CHAPTER 12

FURNISHINGS

CSI DIVISION 12

CSI Division 12 covers furnishings. Separate contracts are usually let for this work, but again, some coordination is necessary, and the general contractor should have a clear understanding regarding damages to the building by other parties working on the premises, especially if painting and decorating are part of the general work. Also, furnishings and draperies can play havoc with a carefully balanced air handling system. If the owner is uncomfortable, the general contractor is the one he will call, even though it is the 10-foot sofa that blocks the baseboard heater or the $5,500.00 chandelier that interrupts the ceiling outlet airflow and causes a draft on the boss's neck. The architect should be the coordinator of all this, but an alert contractor will try to forestall as many problems as he can.

Typical furnishings supplied by the contractor and coordinated with the architect and interior decorator are:

Modular furniture
Planters for plants
Benches
Bulletin boards
Cabinets
Booths
Trash receptacles
Fire extinguishers
Blinds, shades, or other window decorations
Refrigeration
Shredders
Shop equipment
Gym equipment
Recreation equipment
Mailroom equipment
Special lighting
Special sound system

Local labor practices often determine which trades unload and position special furnishings. Laborers might unload benches to be placed loose in an area, but skilled workers might be needed for benches to be positioned and secured.

Time and money should be allotted for protective floor runners when required to protect new construction from damage during installation of furnishings or other special equipment.

CHAPTER 13

SPECIAL CONSTRUCTION

CSI DIVISION 13
13700 SPECIAL PURPOSE ROOMS & BUILDINGS
CONCRETE STORAGE BINS AND SILOS

CSI Division 13 covers such items as air supported structures, audiometric rooms, and hyperbaric rooms. We have chosen to include in this chapter a section on concrete storage bins and silos.

13700 SPECIAL PURPOSE ROOMS

Sauna Rooms

The luxury residence or office is often fitted with a prefabricated sauna cabinet. These come in a wide variety of sizes and there are several manufacturers so that more than one source should be contacted.

The minimum sauna, for sitting position only will be around 4'x4', weigh 500 lbs., and will need a separate 120 volt 20 AMP service. The cost complete with door, floor, ceiling, bench, and equipment is about $2,000.00. A 4'x7' unit for one lying down or 2 sitting will weigh 750 lbs., will require 220 volt 30 AMP service and cost about $3,300.00. Units for several people measuring up to 7'x10' will weigh 1200 to 1500 lbs., require 220 to 250 AMP service and cost $4,000.00 to $7,500.00.

These units are prefabricated and simple to install but require a carpenter, a helper, and an electrician. Except for the largest units, or when preparatory work is required, the installation should be completed in a days' time.

CONCRETE STORAGE BINS AND SILOS

Concrete storage bins and silos are used for storing grain, coal, sand, gravel, etc., in addition to those used for the processing and preserving of forages for feed on the farm.

Silos are constructed of numerous types of materials such as concrete stave, cast-in-place concrete, concrete block, glazed tile, wood, and steel. The cost of storage silos varies with the storage capacity, the type of structure and the kind of materials to be stored.

Practically all silos are round or circular, which gives more capacity for the least amount of construction cost, and the stored material packs better.

Farm silos are built for storing corn, sorghum, hay crops, and miscellaneous forages. Research studies reveal that lateral pressures increase with the moisture content of the silage and the size of the silo. All silages exert approximately equal lateral pressures when of the same moisture content. Therefore, the reinforcing requirements are identical. However, hay crop silages are usually stored at excessive moisture content and more reinforcing is recommended.

Excavating and Concrete Foundations for Silos

Footings should be ample to carry the weight of the silo without settlement or tipping. The width and depth, therefore, are varied according to the height of the silo and the load-carrying capacity of the soil. Recommended footing sizes are given in the accompanying table. These take into consideration the bearing capacities of the different soils as well as the weights of silos of different heights, including the weights of the concrete chutes, roofs, and the weight of the silage carried by the walls.

The depth of the foundation wall below ground level will vary according to the location of the silo. The footings should extend below frost penetration and to firm soil. The bottom of the trench for the footing should be flat and even in width to assure uniform distribution of the load to the soil. Forms will not be required for construction of the footings if care is taken to excavate the trench so that the walls stand vertically. Wall forms are used from the footing up.

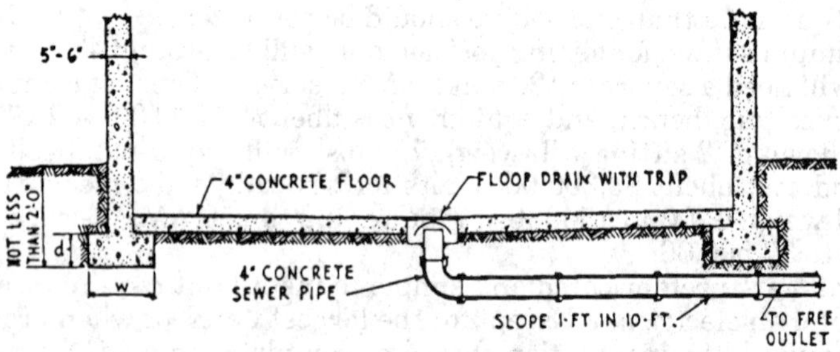

Concrete Foundation and Floor for Cast-in-Place Concrete Silos

Dimensions of Annular Footings for Silos with Walls 6" Thick*

Height of Silo ft.	Type of Soil** — Sand and Gravel (Type 1)		Firm clay, wet sand or clay and sand mixture (Type 2)		Soft Clay (Type 3)	
	Width Inches	Depth Inches	Width Inches	Depth Inches	Width Inches	Depth Inches
20	12	8	12	8	17	8
25	12	8	12	8	24	8
30	12	8	16	8	32	11
35	12	8	21	10		
40	13	9	26	13		
45	16	11	32	15		
50	19	13				
55	23	16				

*ACI-714.
Soil pressures in psf (Type 1) 8,000 (Type 2) 4,000 (Type 3) 2,000.
Silos higher than 45 ft. not recommended on Type 2 soil.
Silos higher than 30 ft. not recommended on Type 3 soil.

Cu. Ft. of Excavation or Concrete Footing Required for Silo Foundations of Various Sizes per Inch of Depth or Thickness

Footing Width	Diameter in Feet 10"	12"	14"	16"	18"	20"	22"	24"
12	2.67	3.17	3.67	4.17	4.75	5.25	5.75	6.33
13	2.89	3.43	3.97	4.51	5.14	5.69	6.23	6.86
16	3.55	4.22	4.88	5.55	6.33	7.00	7.67	8.44
17	3.77	4.48	5.19	5.90	6.72	7.44	8.14	8.97
19	4.21	5.01	5.80	6.59	7.51	8.31	9.10	10.03
21	4.66	5.53	6.40	7.28	8.30	9.19	10.06	11.08
23	5.10	6.06	7.01	7.97	9.10	10.06	11.02	12.15
24	5.33	6.33	7.33	8.33	9.50	10.50	11.50	12.67
26	5.55	6.86	7.94	9.03	10.29	11.37	12.45	13.72
27	5.77	7.12	8.25	9.37	10.68	11.81	12.95	14.25
32	7.10	8.44	9.77	11.10	12.65	14.00	15.33	16.90

Diameter and Circumference of Standard Size Silos

Diam. Silo, Ft.	Circum. Silo in Ft.	Nominal Circum. in Ft.	Diam. Silo in Ft.	Circum. Silo in Ft.	Nominal Circum. in Ft.
10	31.42	32	18	56.55	57
12	37.70	38	20	62.83	63
14	44.00	44	22	69.10	69
16	50.27	50	24	75.41	76

To obtain number of cubic feet or cubic yards of excavation or footings required for any job, proceed as follows. Find number of cubic feet of concrete required for a silo 16 feet in diameter, requiring a footing 19" wide and 13" thick. Refer to above table: a 16-ft. diameter silo requiring a footing 19" wide, requires 6.59 cu. ft. of concrete per inch in thickness. Multiply 6.59 by 13" (thickness of footing) and the result is 85.67 cu. ft. of concrete for the footing.

Silo Wall Forms. The use of commercial steel forms is recommended for the construction of cast-in-place silos. Such forms are quickly set and removed and result in true, smooth walls. Several types of commercial steel forms are available. They differ mainly in the size of the sections or panels and in the manner they are fitted together. In all types, the sections are curved and joined together snugly with clamps to form a complete circle or course. One course of forms is referred to as a lift. Each course or circle is 2 or more feet high, according to the forms used. From 6 to 8 feet of wall can be built in a day, depending upon the size of the crew and the outdoor temperature. In summer weather, forms can usually be removed in 24 hours. In building 6 or 8-ft. of wall per day, three or four complete courses of forms, using 2-ft. high sections, will be required. The usual practice is to set and fill one lift at a time. Where three or more courses of forms are in use it is the usual practice to remove the lower course and set it on top. This procedure is continued until the required height of wall is built. Form faces are cleaned and re-oiled for each resetting.

A plumb wall is assured when using commercial forms provided the first course is carefully leveled. Each manufacturer usually furnishes instructions covering the use of his forms. These companies also make forms for casting chutes and roofs and can furnish erection derricks, hoists and other construction accessories.

Constructing the Walls. Walls of well-built cast-in-place silos are smooth, plumb, and air and watertight. The walls are 4" to 6" thick. A 5" to 6" thickness is recommended for large silos.

Concrete for durable silo walls should be of high quality, using clean sharp sand and gravel. Sand should be from very fine up to particles which pass a No. 4 screen (4 openings per lin. in.). Gravel should be clean, hard, and range in size from 1/4" up to 1 1/2".

A workable concrete mix is one that is "mushy" but not "soupy". It should be somewhat sticky when worked with a shovel or trowel and should be stiff enough to require some spading in the forms. For machine mixing allow about 2 minutes mixing after all materials are in the mixer. Concrete for footing should be mixed in approximately the proportions of 1 bag air entrained portland cement; 2 3/4 cu. ft. sand and 4 cu. ft. gravel. Use about 5 gals. water per bag of cement. Concrete for walls, floor, chute, and roof should be mixed in the proportions of 1 bag of air entrained portland cement; 2 1/4 cu. ft. sand and 3 cu. ft. gravel. Use about 5 gals. water per bag of cement.

Courtesy A. O. Smith Harvestore Products, Inc.

Concrete Foundation Being Poured for Silo

Total Weight of Horizontal and Vertical Bars Required for Cast-in-Place Concrete Silos

Height of silo, ft.	Diameter of silo 10 ft. Lb.	12 ft. Lb.	14 ft. Lb.	16 ft. Lb.	18 ft. Lb.	20 ft. Lb.
20	547	652	....	....	....	....
24	675	796	968	1,263	....	....
28	787	935	1,174	1,492	....	....
32	910	1,047	1,381	1,721	2,014	2,609
36	1,034	1,213	1,587	2,017	2,351	3,112
40	....	1,400	1,816	2,339	2,836	3,649
44	....	....	2,044	2,702	3,291	4,317
48	....	....	2,351	3,066	3,780	4,986
52	....	....	....	3,430	4,386	5,654

No allowance made for lapping of bars. Add about 10% when three lengths of bars are used per circle.

Number and Length of Horizontal Reinforcing Bars Required for Cast-in-Place Concrete Silos with Continuous Doorway

Height of silo, ft.	Diameter of Silo 10 ft.	12 ft.	14 ft.	16 ft.	18 ft.	20 ft.
Length of bars*	31 ft. 10 in.	38 ft. 0 in.	44 ft. 4 in.	50 ft. 6 in.	56 ft. 10 in.	63 ft. 4 in.
	Number of bars required					
20	14	17	....	16	....	....
24	17	21	14	20	....	....
28	20	25	18	24	....	....
32	24	29	22	28	20	26
36	28	33	26	34	24	32
40	....	39	30	40	30	38
44	....	....	34	48	36	46
48	....	....	38	56	42	54
52	....	....	....	64	50	62
	No. 3 Bars		No. 4 Bars		No. 5 Bars	

*Bar lengths do not allow for laps. Add approximately 10% when three lengths of bar are used per circle.

Number and Length of Vertical Reinforcing Bars Required for Cast-in-Place Concrete Silos

Height of silo, Length of bars	Diameter of Silo 10'	12'	14'	16'	18'	20'	22'	24'
	Number of bars required							
20 ft., 20 ft. 0in.	22	26	....	....	....	....	....	....
24 ft., 13 ft. 0 in.	44	52	60	68	....	....	....	....
28 ft., 15 ft. 0 in.	44	52	60	68	....	....	....	....
32 ft., 17 ft. 0 in.	44	52	60	68	76	86	92	102
36 ft., 19 ft. 0 in.	44	52	60	68	76	86	92	102
40 ft., 14 ft. 8 in.	....	78	90	102	114	129	138	153
44 ft., 16 ft. 0 in.	....	78	90	102	114	129	139	153
48 ft., 17 ft. 4 in.	....	....	90	102	114	129	138	153
52 ft., 18 ft. 8 in.	....	....	90	102	114	129	138	153
60 ft., 16 ft. 6 in.	....	....	120	136	152	168	184	204

All bars are No. 3, spaced 18" apart. Bars are lapped 24 in. at splices.

Basis for Estimating Labor for Cast-in-Place Concrete Silos. The following estimating data is applicable to and based on circular type cast-in-place concrete silos.

Labor excavating and placing concrete footings for silos should be estimated separately, because it varies with the depth of excavation and size of silo.

After the footings are in place, an experienced crew of three should build 8 feet of 10'-0", 12'-0", or 14'-0" diameter silo per 8-hr. day, including the labor on feed room. It will also require about a day to place the roof, and after the concrete is placed, a day will be required to tear down and remove the forms and equipment.

Labor Required to Construct Cast-in-Place Concrete Silos
Labor Required to Construct Each 8'-0" of Wall Height

Diam. of Silo	Thickness of Wall	Size of Crew	No. of Hours for Crew	Labor Hours for Each 8'-0" Height
10'-0"	5"-6"	3	8	24
12'-0"	5"-6"	3	8	24
14'-0"	5"-6"	3	8	24
Roof	....	3	8	24
16'-0"	5"-6"	4	8	32
18'-0"	5"-6"	4	8	32
20'-0"	5"-6"	4	8	32
22'-0"*	5"-6"	4	8	32
24'-0"*	5"-6"	4	8	32
Roof	....	4	8	32

*Labor required to construct each 6 ft. of wall height.

A set of metal forms usually includes 4 sets of forms 2'-0" high. Reinforcing steel may be either rods or mesh but many silo builders prefer mesh reinforcing as it may be placed at lower cost than rods.

To the above, add labor cost excavating for footings, also labor placing concrete footings.

To the above, add time for 1 day for crew bringing tools and equipment to job and setting up; also 1 day tearing down and removing equipment at completion.

Silo Walls Using Slip Forms. For casting monolithic concrete structures of great height the method of slip forms is often used and this type of form is particularly adaptable to the construction of silos and storage bins. Despite the fact that the sliding forms must be built with more precision and of heavier material than fixed forms, slip forms have proven to be more economical in the long run for high structures. Naturally the higher the construction the less the cost per square foot of wall.

The forms are prefabricated of wood or metal and are usually raised by using screw jacks or a hydraulic jacking system. The yokes, which are usually made of wood and tied together with bolts, are suspended with a screw or jack arrangement on a steel jackrod. The raising is done by a crew, each worker operating from 1 to 5 jacks, depending on the size of the job. This operation must be done on a shift basis, continuing day and night until the structure is completed to prevent the forms from adhering to the concrete.

Concrete Stave Silos

Footings for concrete stave silos can be made smaller than those for cast-in-place silos, because staves are only 2½" thick. American Concrete Institute Bulletin 714 gives the following information for footings for concrete stave silos.

Concrete Staves. Concrete stave silos are built of cast concrete staves usually 10" wide, 30" long, and 2½" thick. Every other stave in the bottom circle is 24" high with a 6" stave at the top of the corresponding column.

The number of staves required may be computed by figuring the number of staves for each circle, with the height of the structure governing the number of circles required.

Labor Setting Staves. The setting of 1,000 10"x30" staves may be considered a fair day's work under average conditions for a 4-worker crew, consisting of 2 workers on the ground, 1 worker setting staves, and a foreman to follow up, leveling staves and inspecting the tension of hoops. This includes the raising of the necessary scaffolding, placing and tightening hoops, etc.

Dimensions of Annular Footings for Silos with Walls 2½" Thick

	Type of Soil*					
	Sand and Gravel (Type 1)		Firm clay, wet sand or clay and sand mixture (Type 2)		Soft Clay (Type 3)	
Height of Silo Silo ft.	Width Inches	Depth Inches	Width Inches	Depth Inches	Width Inches	Depth Inches
20	12	8	12	8	12	8
25	12	8	12	8	18	8
30	12	8	12	8	24	8
35	12	8	16	8	32	11
40	12	8	21	10		
45	13	9	26	13		
50	16	11	32	15		
55	19	13				
60	23	16				

*Silos higher than 50 ft. not recommended on Type 2 soil.
Silos higher than 35 ft. not recommended on Type 3 soil.

Number of Staves Required Per Circle for Concrete Stave Silos. The number of staves required for one complete circle of various size silos. Use last (*) column where door frames are included.

Diameter of Silo	Outside Diameter of Silo	Number of Staves per Circle	Number of Staves per Circle*
10'-0"	10'-4"	38	35
12'-0"	12'-6"	46	43
14'-0"	14'-4"	53	50
16'-0"	16'-5"	61	58
18'-0"	18'-6*"	69	66
20'-0"	20'-8"	77	74

Hoops for Concrete Stave Silos. It will be necessary to place steel rods or "hoops" around the concrete stave silo to reinforce the staves. The hoops spacing for either grass or corn silage of moisture content not over 75% should not exceed that given in the table of "Hoop Spacing for Concrete Stave Silos".

Lugs (connections for hoops of round cross section) should be of malleable cast iron or steel. The strength of the lug should be such that the ultimate strength of the hooping to be used with it can be developed by the lug in place. For silos up to and including 10'-0" in diameter, at least two lugs per hoop should be used; from 10'-0" to 16'-0", at least 3 lugs per hoop should be used; and for diameters from 16'- 0" to 22'-0", at least 4 lugs per hoop should be used.

On a new silo a final retightening of the reinforcing rods should be made 5 to 7 days after the silo is erected.

CHAPTER 14

CONVEYING SYSTEMS

CSI DIVISION 14
14100 DUMBWAITERS
14200 ELEVATORS
14400 LIFTS
14600 MOVING STAIRS & WALKS

14100 DUMBWAITERS

Dumbwaiters must be carefully selected to provide the maximum in convenience and efficiency for the minimum cost. They range from simple hand-operated to computerized models.

Standard speeds are 50 fpm, but speeds to 150 fpm are possible on some models. Where speed is not a consideration, lower speeds save some initial cost. Car sizes are directly related to capacities.

Size and Capacity of Dumbwaiters

Capacity, lbs.	Speed, fpm	Hp	Stops	Cab Size, Sq. Ft.	Price
50	50	½	2	3.33	$10,000
100	50	¾	2	4.00	12,000
200	50	1	2	4.70	14,000
300	50	1½	2	6.25	16,000
500	50	2	2	9.00	19,000

Dumbwaiters can be mounted on channel guides that are attached to the building construction, but a packaged, self-supporting tower is also available.

14200 ELEVATORS

Elevator construction is such a highly specialized business that it is out of the question to expect anyone except an experienced elevator contractor to prepare an estimate for the installation of such equipment.

There are many items that affect the cost of elevator construction. All elevator manufacturers submit quotations on an installed basis, and the location of the job must be taken into consideration, as well as the intended use, capacity, car speed, size of car, travel, number of stops and openings, design of the car, and kind of current available.

In addition there are many types of controls, such as car switch, continuous pressure push button, single call push button, up down collective (for operation with an operator or automatic operation) signal control, duplex, and others.

All modern passenger elevator installations are partly or completely computerized. In new office buildings computerized systems are available that recognize traffic demands and automatically program elevator service to these demands. With these programming systems, elevator service is made as efficient as possible during peak or down periods.

Elevator Safety Codes

Many states and cities have their own elevator codes, and all elevator installations must comply with the rules and regulations of that state or city, as well as with American National Standards Institute (ANSI) codes. In general, all municipal or state codes follow closely the Code for Elevators, Dumbwaiters, and Escalators (ANSI Standard A17.1 with supplement ANSI A17-1a) and the Code Rules for Moving Walks (ANSI A17.1.13).

It would be a good idea for anyone endeavoring to estimate the cost of an elevator to become familiar with these codes. For example, the capacity of a passenger elevator must be based upon the effective area of the car platform in accordance with the contact load graphs shown in the latest edition of the ANSI codes.

Essential Information for Estimating the Cost of Elevators. The following information is essential to enable the elevator manufacturer to prepare an intelligent estimate on the elevators required for any building.

1. New or renovation project
2. Location of building
3. Number of elevators
4. Use of elevator (passenger, freight, or general purpose)
5. Capacity of elevators
6. Car speed of elevators
7. Controls and operation of elevators
8. Type of motor (single speed, two speed, or variable voltage)
9. Total travel of elevator in feet
10. Number of stops and openings
11. Single or double entrance car
12. Location and size of elevator machinery room or penthouse
13. Kind of current available
14. Size of hatch or platform
15. Design of cab, doors, and frames
16. Car and hatch door power operators
17. Signals
18. Access door location
19. Federal, state, local, and ANSI codes

Costs of Electric and Hydraulic Elevators. The cost of an elevator is a function of its speed, capacity, and type of control, which in turn are determined by type of building to be served, the use of the elevator, and the height of the building or feet of travel and number of stops.

There are so many factors that enter into the cost that an estimator should get definite prices from an elevator contractor whenever possible. The following budget figures can be used for rough estimating only.

Electric Elevators

General Purpose Elevators. There are two basic types of general purpose elevator: gear and gearless. To choose the correct one for a specific need, the contractor should contact a specialist in vertical people moving.

Gearless elevators satisfy most high-rise passenger elevator requirements and are most cost effective for structures 10 floors and greater. The prices given below represent standard platform sizes, with center opening of doors, for a 12-story building.

Capacity, lbs.	Speed, fpm	No. of Passengers	Cab Size, Sq. Ft.	Price
3,000	500	20	60.40	$191,000
3,500	700	23	66.00	192,000
3,500	1,000	23	68.00	195,000
4,000	1,200	27	76.00	200,000

For each additional floor above 12 floors, add $4,200.

Many tall building elevators now operate at speeds up to 1,800 fpm, but these require special pricing, depending on express and local runs.

Geared elevators satisfy most low rise passenger elevator requirements. The prices given below represent standard platform sizes, with center opening of doors, for an 8-story building.

Capacity, lbs.	Speed, fpm	No. of Passengers	Cab Size, Sq. Ft.	Price
2,000	200	13	48.20	$112,000
2,500	200	16	54.80	113,000
3,000	200	20	59.00	115,000
3,500	200	23	66.00	123,000
4,000	300	27	74.00	130,000

For each additional floor above 8 floors, add $4,000.

For hospitals, special size elevators are available for increased length requirements and multiple openings necessary for hospitals and modern care facilities.

Capacity, lbs.	Speed, fpm	Cab Size, Sq. Ft.	Price
4,000	125	104	$200,000
4,500	200	115	210,000

The above prices are for 5 stops. For each additional stop, add $4,500.

Oil Hydraulic Elevators. There are two types of hydraulic elevators. One type requires the installation of a steel casing into the ground for the piston. Rise is limited to about 60 feet, so they can be used for 6- or 7-story buildings. They are economical to install and require no overhead penthouse. They are available in speeds up to 300 fpm and capacity to 5,000 lbs.

Capacity, lbs.	Speed, fpm	No. of Passengers	Cab Size, Sq. Ft.	Price
2,000	120	13	42.10	$85,000
2,100	120	13	42.10	90,000
2,500	120	16	48.00	95,000
3,000	120	20	52.00	102,000
3,500	120	23	57.60	115,000
4,000	200	13	42.00	120,000
4,500	250	15	45.00	130,000
5,000	300	16	48.70	140,000

The second type of hydraulic elevator, sometimes referred to as a holeless hydraulic, has limited capacity with a maximum travel of 15'-0" and maximum two stops. It is available in speeds up to 100 fpm and capacity to 2,500 lbs.

Capacity, lbs.	Speed, fpm	No. of Passengers	Cab Size, Sq. Ft.	Price
2,000	100	13	42.10	$50,000
2,100	100	13	42.10	55,000
2,500	100	16	47.90	60,000

The cost of an oil hydraulic installation must include the drilling of the hole for the piston. The cost of this hole depends on local conditions such as soil, water, and rock. Elevator prices depend on size, speed, control, and purpose.

Garage Elevator. Freight elevator capacities range from 3,500 to 10,000 lbs. Speeds vary from 50 to 100 fpm., and platforms are from 35 to 112 sq. ft. The following is a typical configuration for this type of equipment:

Freight elevator, single-call push-button control, two-speed motor, automatic leveling, capacity 6,000 lbs., speed 100 fpm; total rise 36'-0" with 4 stops and 4 openings; car platform 10'-0"x20'-0"; car built of steel wainscoting on 2 sides 6'-0" high, a steel car top and two wood or expanded metal lift-up gates with electric contacts. Price installed $75,000. No doors or hatchway gates included.

Residence Type Elevators. During recent years residence type elevators have become quite popular, especially where there are older people or invalids in the house. These elevators are easily installed both in new homes and homes already built. Residential elevator capacities range from 450 to 700 lbs., with speeds from 450 to 700 fpm.

A typical residential elevator with 450 lbs. capacity, 30 fpm speed, total travel 11'-0", 2 stops and 2 openings, 110 or 220 volts, single phase 60 cycles, and platform 36"x42", is priced installed approximately $35,000.00. For each additional stop, add $2,500.00.

The same elevator as above with 36"x48" platform is priced installed at about $40,000.00. For each additional stop, add $2,500.00.

14400 LIFTS

A wheelchair lift is designed to permit people with physical handicaps to go up and down stairs. The system is usually fastened to a wall but sometimes to the stair treads. Before installing a lift system, the details of walls or stairs must be reviewed to determine whether support members can support a wheelchair lift. A typical wheelchair lift has a travel distance of 27 ft., a speed of 25 fpm, and a 1 Hp motor. The chair is supported on 2" OD steel rails, with platform sizes from 2'-6"x3'-8" to 3'-2"x4'-2". For a single story lift adjacent to stairway, the cost is about $30,000.00.

14600 MOVING STAIRS

Escalators. Escalators move large volumes of people from one level to another efficiently. They are available in several widths and speeds.

Width, inches	Speed, fpm	People per Hr.	Cost
32	90	5,000	$100,000
32	120	6,500	110,000
40	90	7,000	104,000
40	120	9,000	114,000
48	90	8,000	116,000
48	120	10,000	120,000

To the above price, add about $1,200 per ft. of rise.

Horizontal People Movers. These are also called power walks, power ramps, or moving walks. Horizontal people movers provide high volume transportation of people within buildings where walking is not advantageous. People movers come in various sizes from 24" to 48" tread width. Larger sizes of 40" and 48" accommodate an adult with luggage. For the special requirements for this type of installation, a contractor should consult the manufacturer.

For a 40" wide, flat horizontal people mover with no curves or offsets, prices range from $120,000 to $150,000 per 50 ft.

CHAPTER 15

MECHANICAL

CSI DIVISION 15
15A PLUMBING AND SEWERAGE
15B HEATING AND AIR CONDITIONING
15C SPRINKLER SYSTEMS

15A PLUMBING AND SEWERAGE

To prepare an accurate detailed estimate on plumbing and sewerage requires a working knowledge of the trade. All work must be laid out on paper, and the number of lineal feet of each kind of pipe estimated separately, such as tile or cast iron sewers, soil and vent pipe, water pipe, gas pipe, drains, valves, and pipe fittings. Very few, if any, general building contractors or estimators possess this knowledge, so it makes it doubly difficult for them to prepare more than an approximate estimate on the various kinds of plumbing work.

Budget figures for average plumbing installations can be estimated in one of two ways: as a percentage of the total cost of a project, or as the grand total of a predetermined cost allowance for the installation of each fixture.

Plumbing estimates as a percentage of the total job will vary from a minimum of 3% on up to 12%, depending on the number and type of fixtures required and on their disposition within the building. Percentage estimates would be valid for general plumbing only. Process and other special condition piping would of course have to be entered as a separate item.

Those buildings which would require a high percentage allowance would include motels and nursing homes, which are generally spread out over a large area and require plumbing for each room, and apartment high rises with many studio units, which require full baths and kitchens for relatively small living areas. On the low side would be assembly, mercantile, warehouse, and industrial buildings where plumbing can be concentrated in a central core. Recent OSHA requirements limiting the travel distance to a rest room have raised the costs on some of these types.

The private residence is difficult to peg. Where kitchens and baths can be back to back, the percentage is on the low side. But in a large, rambling house with a central powder room area but with kitchens, family, and master bedrooms all in separate wings and with hose bibs and lawn sprinkling systems on all four sides and sunken bathtubs, shell-shaped lavatories and gold plated fixtures, even 12% may be too low.

However the average building types will fall in the following ranges:

apartments	9% to 12%	motels	9% to 12%
assembly	4% to 7%	office bldgs.	4% to 7%
banks	3% to 6%	shops	4% to 7%
dormitories	7% to 10%	schools	5% to 8%
factories	4% to 8%	stores	3% to 6%
hospitals	8% to 12%	warehouses	3% to 7%

The allowance per fixture estimate is generally more accurate. An experienced plumber can acquire very accurate prices but will have to use some judgment on how much to add for bringing in the main waste and supply lines, depending on how spread out the project may be.

The per fixture price will be the cost of the fixture itself plus the cost of the immediate piping connections and installation within the room. Some per fixture costs have added to them a prorated cost of the central piping system. It would be more accurate to add it, together with the cost of bringing waste and supply lines from the street to the building and permit costs and overhead and profit, each as separate items, because some buildings require a much more extensive central system to supply fewer fixtures than others. Per fixture allowances are discussed later in this chapter.

15400 PLUMBING SYSTEMS

Estimating Quantities and Costs of Sewer Work. Sewerage is estimated by the lineal foot, obtained by measuring the number of lineal feet of each size sewer pipe required, such as 4", 6", 8", and 12" pipe, and all fittings should also be listed in detail, giving the number of ells, Ys, Ts, etc.

Cast-iron and ductile iron sewer pipe is estimated in the same manner as tile sewer pipe. Most plumbers estimate their sewer work at a certain price per lineal foot, including excavating, sewer pipe, and labor. This is all right on ordinary jobs that do not require more than the usual amount of deep excavation, but if there is an exceptionally large amount of deep excavation, (from 5 to 15 feet deep), then the excavating, shoring, and bracing should be estimated separately.

The best method is to estimate the number of cubic yards of excavating required for sewers, based on a trench 1'-6" wider than the pipe diameter by the required depth. Refer to chapter on *Site Work* for labor costs excavating for trenches.

Where 4" to 6" sewer pipe is used, an experienced sewer layer should lay 12 to 15 lin. ft. of pipe per hour or 100 to 125 lin. ft. per 8 hour day, at the following labor cost per 100 lin. ft.:

	Hours	Rate	Total	Rate	Total
Sewer layer	7	$....	$....	$26.62	$186.34
Average per lin. ft.			....		1.86

Add for excavating and additional time for setting catch basins, gravel basins, triple basins, etc.

Estimating Quantities of Cast Iron or Ductile Iron Soil Pipe, Downspouts, Stacks, and Vents. When estimating soil pipe, stacks, vents, etc., note the number of stacks and length of each, listing the number of lineal feet of each size pipe, the number of pieces, size, and kind of fittings required, and price them at current market prices. The labor cost of placing cast iron pipe is given on the following pages.

Estimating Quantity of Pipe and Fittings. When estimating the quantity of black or galvanized pipe, list the number of lineal feet of each size pipe required, such as 1/2", 3/4", 1", 1 1/4", 1 1/2", 2", 2 1/2", 3", etc., and estimate at the current market price.

After the cost of the pipe has been computed, take 60% to 75% of the cost of the pipe to cover the cost of all fittings required.

Brass or copper pipe and fittings should be estimated in the same manner as galvanized pipe.

Estimating the Quantity and Prices of Valves. All valves of the different sizes should be listed on the estimate separately, stating the number of each size required, and pricing them at the current market price.

Estimating the Quantity of Fixtures. Each type of fixture such as sinks, lavatories, laundry trays, water closets, bathtubs, drinking fountains, shower baths, hot water tanks and heaters, house pumps, bilge pumps, etc., should be listed separately and priced at the current market prices.

Labor Roughing in for Plumbing. The usual practice among plumbers is to allow a certain percentage of the cost of the roughing materials to cover the labor cost of installing them. This will vary with the type of building and grade of work.

The total cost of the roughing in materials, such as cast-iron soil pipe, downspouts, stacks, vents, and fittings; also all black and galvanized pipe and fittings, all valves, increasers, tees, Ys, 1/8 bends; nipples; in fact, all pipe required to rough in the job is computed. Then the labor cost is estimated at a certain percentage of the cost of the pipe.

In medium-priced one and two story residences, the labor cost of roughing in the job will vary from 75% to 85% of the cost of the roughing materials, with 80% a fair average.

In apartment buildings of non-fireproof construction, the labor cost of roughing in will average about 90% to 100% of the cost of the roughing in materials, while on high grade fireproof construction, the labor cost roughing in will run from 100% to 120% of the cost of the roughing in materials.

In public garages, the labor cost roughing in the job will average about 90% to 110% of the cost of the roughing in materials.

Sewer work is estimated separately from roughing in.

Labor Handling and Placing Plumbing Fixtures. The labor cost handling and placing all kinds of fixtures, such as laundry tubs, kitchen sinks, lavatories, bathtubs, shower baths, water closets, drinking fountains, etc., is usually estimated at 25% to 30% of the cost of the fixtures.

Preparing Detailed Plumbing Estimates

When preparing detailed estimates on plumbing, it is necessary to list the quantity of each class of work separately, such as the number of lineal feet of 4", 6", and 8" sewers, and fittings; number of lineal feet of each size soil pipe, vents, and fittings; number of lineal feet of water pipe and fittings of different materials and sizes; number of valves of the various kinds and sizes; number of each type of fixture required, such as hot water heaters, laundry trays, kitchen sinks, slop sinks, bathtubs, lavatories, water closets, shower baths, etc., together with a list of fittings and supplies required for each.

The following is a list of items the plumbing contractor should include in his estimate:

1. Sewerage, including double sewer system
2. House tanks, foundations, etc
3. Compression tanks, foundations, etc
4. Sewer ejector and bilge pump and basins
5. Iron, gravel, catch and other basins
6. Water filters, foundations
7. Water meter
8. Soil and vent pipe and fittings
9. Shower and urinal traps

10.Closet bends
11.Drum traps
12.Lead, solder, sundries
13.Inside downspouts
14.Iron sewer and fittings
15.Floor drains
16.Back water gates
17.Service pipe to building
18.Mason hydrant
19.Stop cock and box
20.Galvanized pipe and fittings
21.Brass or copper pipe and fittings
22.Valves, check valves
23.Pet cocks, sill cocks, hose bibbs
24.Hot water tank and heater
25.Pipe covering, tank covering, heater covering, filter covering, painting
26.Gas fitting, mains, ranges, heaters, etc
27.Manholes and covers
28.Catch basins and covers
29.Surface drain basins and covers
30.Fire system, incl. pumps, hose, Siamese connection, meter, etc.
31.Roughing in material
32.Fixtures, such as water closets, lavatories, bathtubs, shower baths, sinks, laundry tubs, drinking fountains, N. P. fittings
33.Permits, insurance, trucking, freight, telephone, watchman, etc.
34.Overhead expense and profit

15060 PIPE AND PIPE FITTINGS

Approximate Lineal Foot Prices of Cast-Iron Soil Pipe

Prices on all types of pipe and pipe fittings and supplies should be checked with suppliers at the time of the job, because they are subject to continual change.

	Single Hub		Double Hub	
Size	Service	Extra Heavy	Service	Extra Heavy
2"	$3.10	$4.30	$4.25	$5.85
3"	4.00	5.05	4.94	6.85
4"	4.75	6.75	6.60	8.98
5"	6.20	8.65	8.40	11.80
6"	7.30	10.10	9.80	13.80
8"	11.25	15.70	15.40	21.35
10"	18.00	25.30	24.70	33.70
12"	25.85	36.30	35.95	48.90

Prices of Cast-Iron Soil Pipe Fittings

Description	Size in Inches					
of Fitting	2"	3"	4"	5"	6"	8"
Quarter bends, service	$3.40	$6.20	$9.00	$11.80	$15.70	$51.70
Extra heavy	3.90	7.30	10.30	14.00	17.60	58.40
Eighth bends, service	2.50	5.05	7.30	10.30	12.35	35.95
Eighth bends Extra heavy	2.80	5.70	8.40	11.80	14.00	41.60

Description of Fitting	2"	3"	4"	5"	6"	8"
Sanitary T & Y branches Extra heavy	7.00	13.50	16.85	24.70	38.20	78.65
Plain Traps S P or 3/4 Service	9.60	12.90	18.00	34.80	44.90	102.00
T cleanout Extra heavy	30.30	40.45	57.30	103.40	148.30	185.40
Comb. Y & eighth bends Extra heavy	10.10	13.50	19.10	37.10	45.00	104.50

Closet bends slip collar type: 4"x4"x12", $23.00; 4"x4"x18", $27.00.

Lineal Foot Prices of Buttweld Steel Pipe

Size	Standard Weight Black	Standard Weight Galvanized	Extra Heavy Black	Extra Heavy Galvanized
1/4"	$0.33	$0.50	$0.50	$0.67
3/8"	.45	.60	.65	.70
1/2"	.47	.65	.65	.80
3/4"	.67	.80	.85	1.00
1"	1.00	1.25	1.30	1.45
1 1/4"	1.20	1.60	1.65	1.90
1 1/2"	1.50	1.85	1.95	2.30
2"	2.00	2.50	2.65	3.10
2 1/2"	2.90	3.65	4.10	4.75
3"	3.80	4.75	5.40	6.30
3 1/2"	4.70	5.85	6.70	7.85
4"	5.60	6.85	8.00	9.30

Prices on Standard Malleable Iron Fittings

Type of Fitting	1/2"	3/4"	1"	1 1/4"	1 1/2"	2"	3"	4"	6"
Tee Black	$.62	$.90	$1.60	$2.60	$3.15	$4.65	$11.80	$24.15	$67.40
Galv.	.90	1.30	2.10	3.70	4.50	6.00	16.30	33.70	95.50
90° El Black	.50	.55	1.05	1.70	2.25	3.20	10.70	18.00	53.90
Galv.	.75	.80	1.50	2.35	3.15	4.50	14.60	25.30	73.00
45° El Black	.75	.95	1.25	2.00	2.50	3.40	11.80	19.10	61.80
Galv.	1.05	1.25	1.75	2.50	3.40	3.60	16.30	26.60	84.30

Lineal Foot Prices of Copper Water Tubing

Type	1/4"	3/8"	1/2"	5/8"	3/4"	1"	1 1/4"	1 1/2"	2"	3"
Light Type M	$.40	$.45	$.67	$.80	$.95	$1.30	$1.80	$2.35	$2.60	$6.75
Medium Type L	.40	.60	.85	1.00	1.25	1.70	2.20	2.80	4.20	8.15
Heavy Type K	.45	.70	.95	1.10	1.60	2.00	2.50	3.25	4.90	9.55

Prices of Wrought Copper Soldered Joint Fittings

Type	1/2"	3/4"	1"	1 1/4"	1 1/2"	2"	3"	4"
Tee	$.22	$.60	$1.70	$2.65	$3.37	$5.40	$17.65	$35.95
90° Elbow	.17	.38	.80	1.25	1.70	3.00	9.00	20.20
45° Elbow	.28	.50	.85	1.45	1.90	2.80	10.95	22.50

15080 PIPING SPECIALTIES

Cast-Iron Drum Traps

Drum trap with cover and gasket, 4"x8", 2 inlets	$10.5(
Drum trap with cover and gasket, 4"x8", 3 inlets	12.1(

Roof Drains

	3"	4"	5"	6"	8"
Cast Iron	$130.00	$135.00	$195.00	$215.00	$340.00
Galv.	220.00	225.00	280.00	300.00	425.00

Floor Drains

Type	Size 3" outlet	4" outlet	5" outlet
Flat round top cast iron	$40.00	$50.00	$85.00
Flat square top cast iron	55.00	55.00	95.00
Funnel type brass	80.00	80.00	
Drain with bucket cast iron	110.00	110.00	115.00
Trench drain 10"x24"	130.00	150.00	

Sizes and Weights of Steel Storage Tanks
Black, Heavy Weight, Welded, 100 Lbs. Working Pressure

Capacity in Gals.	Dimensions Diam. Length	Overall Length	Heavy Weight Welded
Tanks with One Convex and One Concave Head			
66	20" x 4'	52"	225
82	20" x 5'	64"	265
120	24" x 5'	64"	340
140	24" x 6'	76"	395
180	30" x 5'	66"	480
220	30" x 6'	78"	550
295	30" x 8'	102"	680
Tanks with Two Convex Heads			
69	20" x 4'	55"	220
85	20" x 5'	67"	260
125	24" x 5'	68"	320
147	24" x 6'	80"	375
200	30" x 5'	71"	450
235	30" x 6'	83"	520
305	30" x 8'	107"	650
350	36" x 6'	85"	850
450	36" x 8'	109"	1050
560	36" x 10'	133"	1250
550	42" x 7'	99"	1200
770	42" x 10'	135"	1550
900	42" x 12'	159"	1800
1050	42" x 14'	183"	2100
1000	48" x 10'	137"	2150
1200	48" x 12'	161"	2500
1500	48" x 15'	197"	3050
1800	48" x 18'	233"	3550

Septic Sewage Disposal Tanks

Septic tanks are used in rural or suburban districts where running water is available but not sewers and are used for buildings such as homes, hotels, summer resorts.

The capacity of a septic tank should be at least equal to the maximum daily flow of sewage. This is normally estimated at 100 gallons per person per day. For part time service in factories, churches, schools, etc., 50 gallons per person per day is generally satisfactory for estimating tank capacities. In some localities, health authorities require at least a 900 gallon tank for residential installation.

Tanks are furnished in 12 and 14 gauge copper bearing steel, electrically welded and covered with a thick coating of asphalt to protect the tanks against corrosive action.

Size of Tank	Capacity Gallons	Size Tile Connection	Weight Pounds	Home Capacity No. People	Price Each
46" dia.x48"	300	4"-6"	275	56	$67.50
52" dia.x 60"	500	4"-6"	380	67	102.00
46" dia.x 120"	750	6"	575	712	140.00
48" dia.x 144"	1000	6"	980	1215	195.00

Precast concrete tanks, not steel tanks, are now the standard in the U.S. The inner surface of the concrete tank is coated with a tar-base material to prevent gases from deteriorating the concrete. Some precast tank suppliers not only manufacture tanks but also set the tank on delivery into an excavated hole using a delivery truck with boom or other lifting device. Size and weight of the tank limits this service. Consult individual suppliers. Sizes range from 900 to 6,000 gallon capacity.

Capacity in gals.	Size	Price	Riser per ft.	Casting	Roof Slab
900	4'-6"x8'x6'	$290.00	$20.00	$150.00	
1,200	8' diameter	650.00	20.00	150.00	140.00

For larger sizes, consult local suppliers.

Prices of Brass Valves

Kind of Valve	Size in Inches ½"	¾"	1"	1¼"	1½"	2"	2½"	3"
Gate valve, 100 lbs. pressure	$8.00	. .$10.00	$13.00	$16.50	$19.25	$30.00	....	....
Gate valve, 125 lbs. pressure	11.00	. . 13.75	17.00	22.00	28.00	40.00	80.00	120.00
Globe valve, 100 lbs. pressure	9.50	. . 12.50	17.00	22.00	29.00	43.00	....	....
Globe valve, 125 lbs. pressure	10.00	. . 13.50	18.00	25.00	32.00	48.00	80.00	125.00
Check valve, horizontal, 125 lbs.	15.00	. . 20.00	28.00	36.00	45.00	74.00	100.00	140.00
Check valve, 125 lb. swing	12.00	. . 14.00	18.00	23.00	27.00	39.00	70.00	100.00

Angle valves, same price as Globe valves.

Gas Heaters

Capacity Gals.	Overall Dia.	Overall Ht.	Approx. Wt.,lbs.	Prices 5 Year	Prices 10 Year
20	14½"	61"	144	$100.00	$120.00
30	16¼"	64"	174	120.00	170.00
40	18½"	65"	195	150.00	185.00
50	20¼"	65"	274	190.00	240.00

Electric Heaters

Capacity Gals.	Overall Dia.	Overall Ht.	Approx. Wt.	Prices 10 Yr.
40	20	50	160	$170.00
52	23	50	181	200.00
66	23	62	227	220.00
82	25	63	270	250.00
110	28	69	430	290.00

15450 PLUMBING FIXTURES AND TRIM

Residential kitchens and bathrooms have tended to go full circle. When plumbing first moved indoors, it was generally installed in a full-sized room and was arranged about the walls as so much "furniture". Later sanitary concerns became paramount and kitchens and baths were designed as machines-- all chrome and tile and porcelain enamel and reduced to minimum dimensions to facilitate cleaning.

Eventually, the kitchen emerged as an entertainment center, where family and friends could gather. Today it is often difficult to tell where the family room ends and the kitchen begins. All this has had its effect on the materials used with appliances and fixtures made as inconspicuous as possible.

The bathroom now seems to be following suit, especially the bathing area which is often in a separate room together with built in equipment for exercise and relaxation and sometimes even a sauna, greenhouse and fireplace. Often the contractor will be called in to convert a spare bedroom into a bath and dressing room suite.

The following discussion of fixtures and their prices give figures which include the fittings. Also are included averages for the piping hook-ups to the main risers and stacks that might be added to arrive at per fixture installed budget figures.

Bathtubs

The least expensive, standard, recessed tub is made of enamelled steel. This will cost around $115 for the 4'-6" length and $145 for the 5' length. Units made of cast iron will run $225 and $250 in those sizes in white and $300 and $330 in color. If these units are to be recessed into the floor add another $50. Square recessed tubs 42"x48" will run $490 in cast iron.

Units complete with three enclosing walls can be had in fiberglass for around $400.

A plumber with a helper should set a tub with shower in about two hours. To arrive at a per-fixture installed budget figure add around $250.

Whirlpool bathtubs are not only found in luxury homes these days. They are being installed in place of regular units in average homes. Costs of these units vary depending on the number of jets and other options. Prices range from $1,500.00 to more than $5,000.00. It takes a plumber and helper about one day to install. Whirlpools usually require greater widths than regular tubs for the water jets. An additional cost is the 1-Hp electric motor needed to drive the jets.

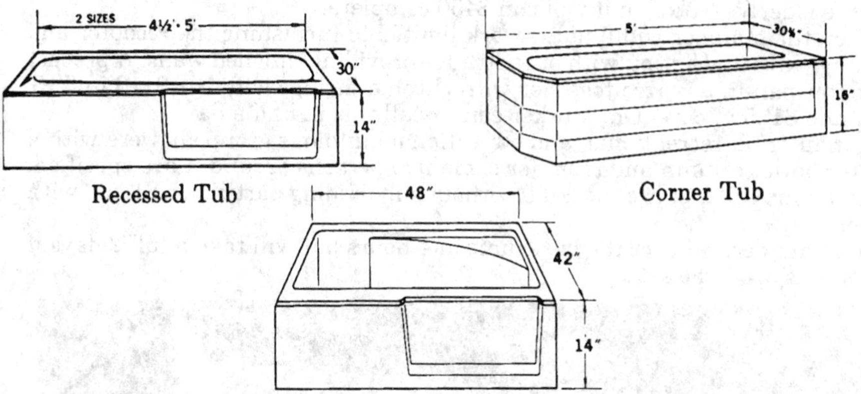

Recessed Tub

Corner Tub

Square Recessed Tub

Illustrations courtesy Eljer Plumbingware Division Wallace-Murray Corp.

Courtesy Kraft Hardware, New York, NY

Whirlpool Bathtub

Shower Stalls

These can range from simple baked enamel on steel units costing around $150 to custom marble enclosures costing six times that. Prefabricated fiberglass units are very popular and can be had with built-in seats and grab bars for the elderly. A 3'x3' unit will run $450 complete.

Often the plumber will find his work limited to furnishing the receptor and mixing valve and outlet, with other trades providing finished walls. A precast receptor pan in terrazzo together with chrome fittings will cost $110 in 3'x3' size and $125 in 3'x4'. One worker can install this in half a day.

In industrial, recreational, and educational buildings group showers with a center outlet column and room for some five persons around it are specified. Such a unit will cost around $750 without subdividing partitions, $1,300 with them.

Two plumbers will set the basic unit in 4 hours but will take a full 2 days if partitions are to be set.

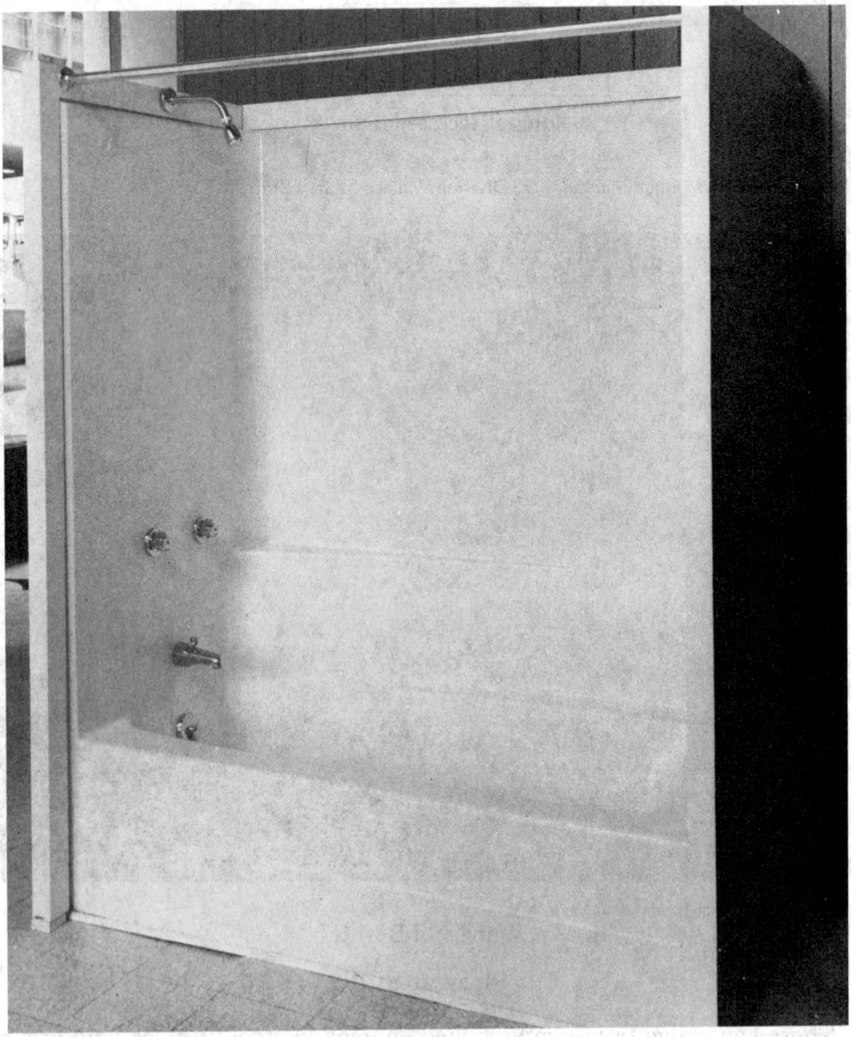

Prefabricated Fiberglass Tub and Shower

Lavatories

Wall hung lavatories are available in porcelain enamelled cast iron or vitreous china. A 19"x17" unit will run $95 in white, cast iron, $110 colored; $100 china white, $120 china colored. A 22" long unit will run $100 cast iron white, $115 cast iron colored and $130 china white and $150 china colored.

Allow two workers an hour to install. Add $230 to arrive at a per fixture installed price.

Drop-in, counter top units are mainly self-rimming today. They come in cast iron, china, steel, stainless and molded plastic. An 18" round unit in cast iron or vitreous will run around $65 in white, $75 in color; in stainless, $120 and in enameled steel, $50. A 20"x18" unit will cost some $10 more. Two workers will take 1 ½ hours to set per unit.

Washfountains

A 3-ft. round, freestanding unit will serve five to six persons at a cost of $600 per unit in precast terrazzo and $725 in fiberglass or stainless. A 4'-6" unit will serve up to 10 and costs $700 in terrazzo, $920 in fiberglass and $1,100 in stainless.

Half round units are also available. A 4'-6" unit in terrazzo will cost $600, in stainless $890. Since installation costs will remain almost the same as for circular units, the cost per person served is much more with these units.

While terrazzo units are cheapest, they are also the heaviest, and one should allow another hour to set these units. Otherwise, allow two workers 2½ hours to set a 36" unit, 3 hours to set a 54" unit.

Also available for vanity units are precast, one piece, bowl and countertops in a variety of marbleized colors and patterns. These are generally locally manufactured and prices should be checked but a 2' long top and bowl should run around $100 and take two workers an hour and a half to set and a 6' top with double bowl should cost around $225 and take 3 hours to set.

Water Closets

Water closets can be either wall hung or floor mounted, have tank or flush valve water supply and the tank can be a separate piece or the unit can be cast in one piece with the tank dropped behind the seat.

A floor mounted one piece will cost $275 in white, $325 in color. A floor mounted two piece unit will cost $125 in white, $165 in color. For areas where water supply is a problem, units designed to flush with as little as two quarts, aided by compressed air, cost about $600. (Some locales now mandate a flush of not greater than 1.6 gallons.

Floor mounted units will take two workers 1 ½ hours to set. For a per fixture installed budget figure add $200.

In commercial work wall hung, flush valve units are the norm. These will cost $175 for the fixture, $150 for a standard carrier and take two workers 2 hours to set.

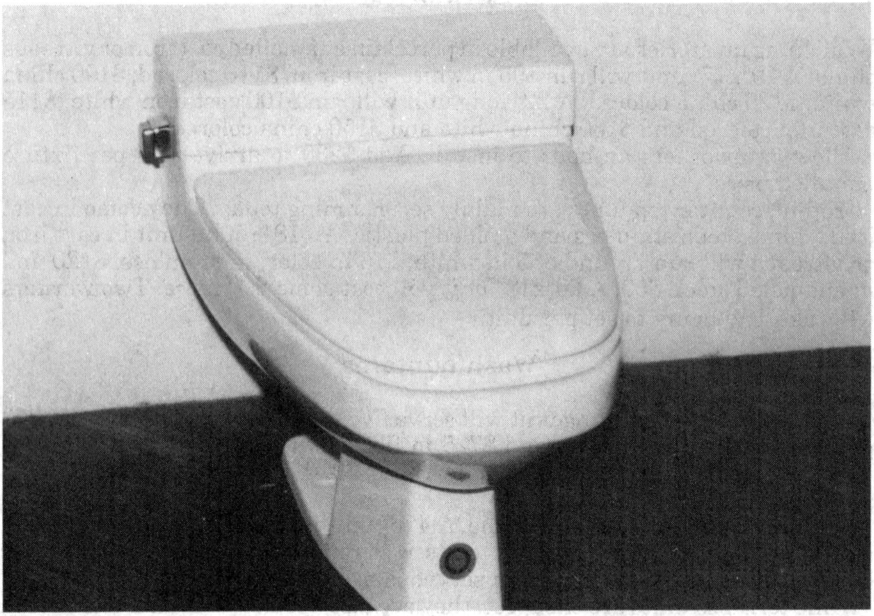

Courtesy Kraft Hardware

Typical Water Closet

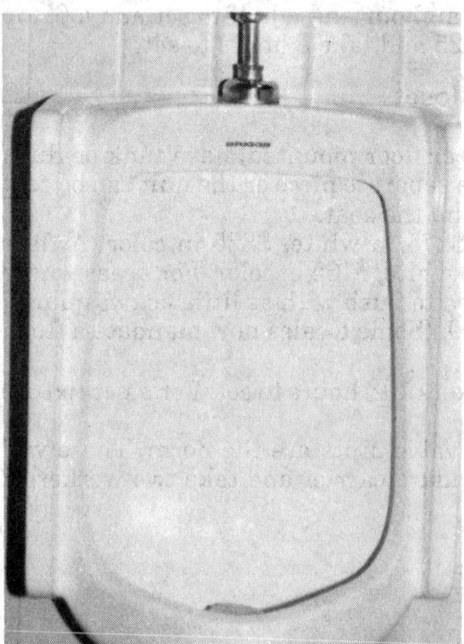

Urinal

Urinals

Urinals are available in cast iron or vitreous china, floor, pedestal or wall mounted, single or in gangs or as wall mounted troughs.

Vitreous china, floor mounted units 18" wide and 38" high with 4" recessed into the floor will cost around $250. If set in gangs add $60 for a spacer-cover 6" wide.

Wall hung china units 18" wide and projecting 12½" will cost $210.

A women's pedestal type will run $265, men's, $220.

Two workers should install a urinal in 2 ½ hours.

For a per fixture installed budget figure add $250 per unit.

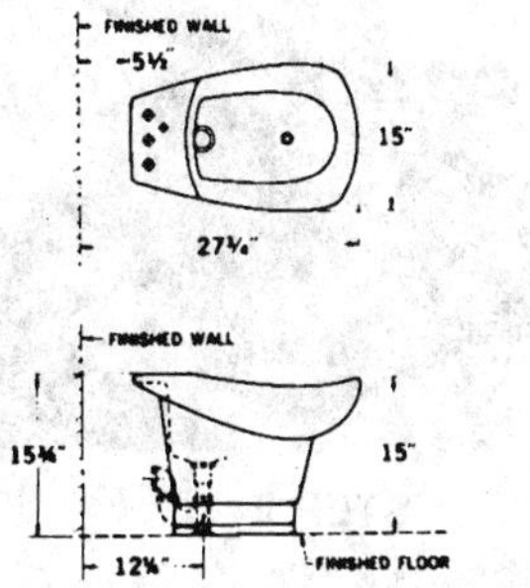

Floor Mounted Bidet

Courtesy Eljer Division, Wallace Murray Corp.

Bidets

For the ultimate touch in the luxury bath, install a bidet. It costs about $325 in white, $450 in color, and takes two workers 2 hours to set.

Sinks

Kitchen sinks are usually self-rimming and set into a cabinet top of plastic or tile. With electric dishwashers, many are only single bowl. Even in luxury kitchens, a single bowl may be set where the dishes are washed, with a second bowl near the refrigerator for washing vegetables, and a third near the ice making machine for making drinks.

A single bowl 24"x21" will cost $130 in enamelled steel, $160 in white cast iron, $250 in colored cast iron, and $300 in stainless. Figure two workers will take 3 hours to set. Add $600 for a per fixture installed price.

A double unit 42"x21" will cost $200 in steel, $230 in cast iron white, $320 in cast iron color, and $400 in stainless. Two men will take 3 ½ hours to set such a unit.

Cast iron sink tops with drain boards will run $375 for single bowl 42" unit, $500 for double bowl 60" unit.

Enamelled steel sink and drain board tops will run 150 for a 42" length, $380 for a double bowl 60" length.

Laundry sinks in enamelled cast iron will cost $150 for a single compartment, $225 for a double; in cast stone, $100 and $150; in fiberglass, $75 and $110; in stainless steel, $300 and $500. Two workers will set a single unit in 1 ½ hours; a double in 2 hours.

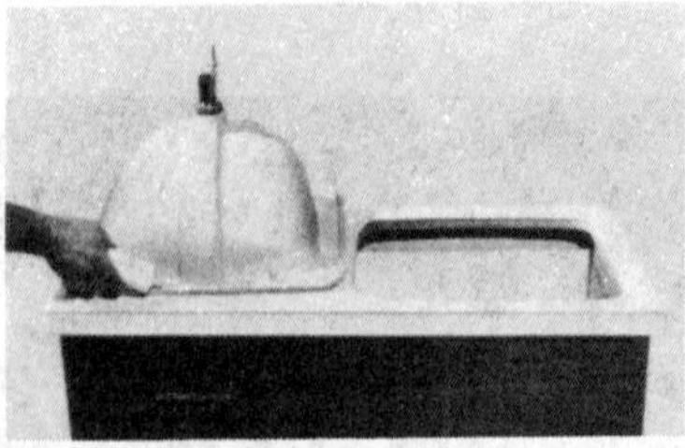

STEP 1

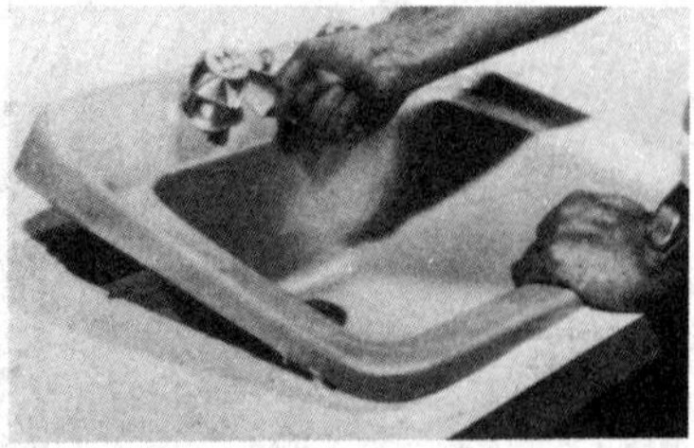

STEP 2

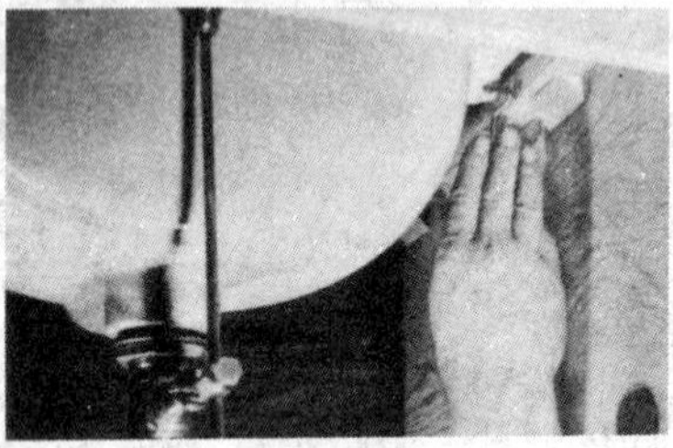

STEP 3

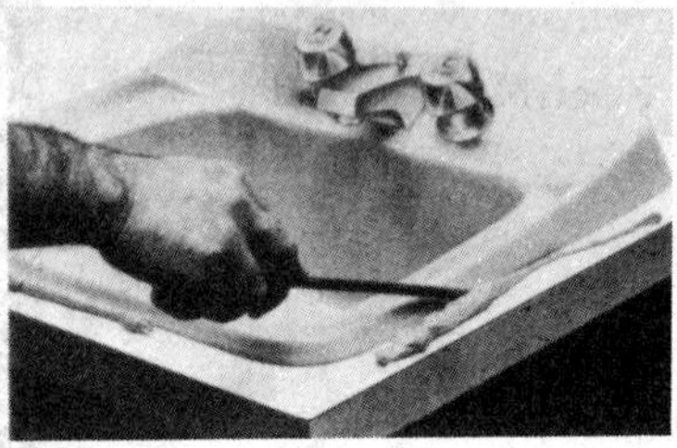

STEP 4

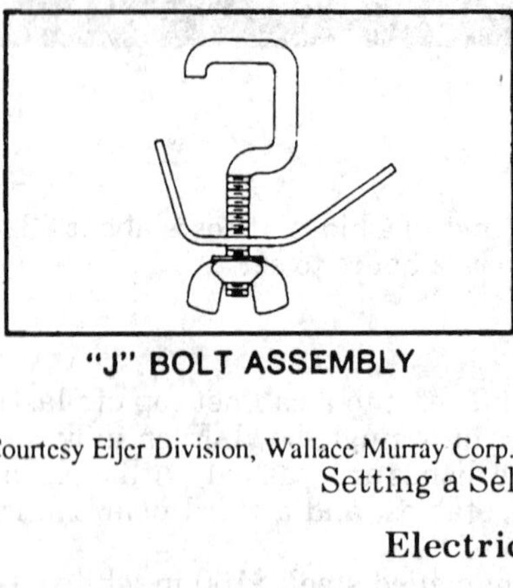

"J" BOLT ASSEMBLY

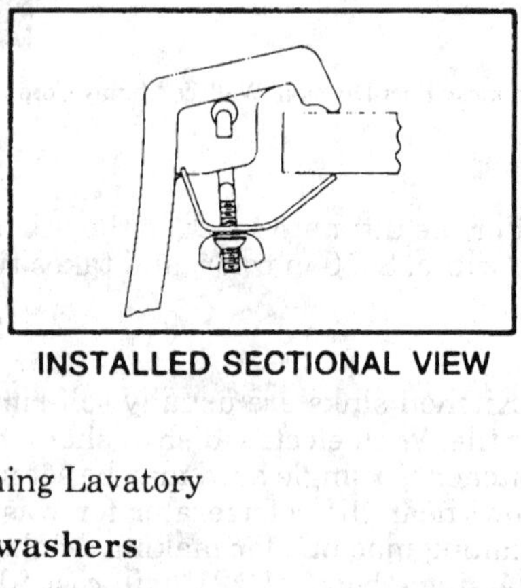

INSTALLED SECTIONAL VIEW

Courtesy Eljer Division, Wallace Murray Corp.

Setting a Self Rimming Lavatory

Electric Dishwashers

Dishwashers are installed as separate units, or they may form a part of a complete sink. Typical prices for various units are as follows:

Dishwasher unit only with floor cabinet but without top and backsplasher, where used under a continuous countertop, size 24"x25" and 34½" high will cost about $500.00.

Dishwasher unit with porcelain enamel top and backsplasher, cabinets made of electrically welded rust-resistant steel with baked on enamel finish, size 23"x27" and 39" high will cost about $475.00.

Add installation charges to all of the above prices.

Garbage Disposal Units. The disposal reduces solid waste but does not do away with the kitchen garbage can. It is a self-contained unit attached to the kitchen sink to form an enlarged drain into which all kitchen food wastes can be placed.

The unit consists of a housing in which is contained a propeller and shredding mechanism. A 1/2-hp motor is directly connected to the propeller and shredding mechanism and supplies the power for operating the waste unit.

Waste material can be accumulated in the upper receptacle of the unit until a normal charge is collected or it may be operated to dispose of the material immediately. The capacity of the waste receptacle is 1 gallon.

Waste material passes through a series of shredders where they are reduced to a fine pulp. This pulp then passes into a revolving strainer disc through which it is forced centrifugally into a chamber below and around the flywheel. Fins on the flywheel centrifugally force the pulp into the outlet passage connected to the drain line which carries the waste to the sewer. Cold water from the faucet flowing through the unit during the grinding operation thoroughly flushes the waste down the drain. The price of this unit averages about $250.00.

Add installation charge to above price.

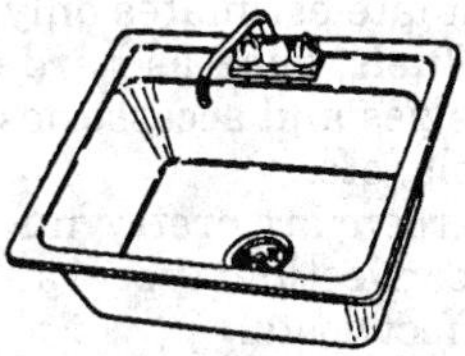

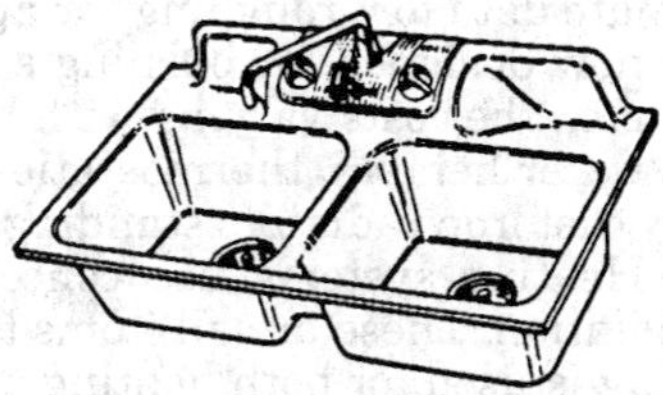

15B HEATING AND AIR CONDITIONING

Over the years methods of heating and cooling buildings has continually improved, especially in residential work due to the research programs conducted by heating and air conditioning associations, engineering societies, equipment manufacturers, and research departments of various universities.

Types of Conventional Heating Systems. There are four basic types of conventional heating systems--warm air, hot water, steam, and electric, including both resistance types and heat pumps, each of which may be used separately or in combination. Solar heating is still in the experimental stage, although in certain areas of the country, there are enough installations and research that reliable information is available.

For small to medium size residences, stores, churches, etc., of similar size and construction, warm air systems are most widely used. For medium size to large residences and non-residential structures of comparable size, hot water systems are usually selected. For large scale installations, steam systems are generally required. Because there are no definite rules stating just when to select one system in preference to another, there is much overlapping in their application. An owner's preference for a particular system may be governed by initial cost of installation; operation cost; convenience of operation; adaptability of system to other uses, such as summer cooling using ducts of a warm air system, snow melting using hot water piping connected to a hot water system; fuel availability; and aesthetics.

Estimating the Cost of Heating Systems. There is no fast and easy "rule of thumb" method for estimating the cost of a heating system. This requires expert knowledge, because most jobs must be designed and laid out before they can be estimated. In residential and small commercial work, architects seldom give more information than merely specifying the type of system desired, the performance expected, and the kinds of material to use. It is then up to the heating contractor to figure the heating loads and design a system to fit the conditions before estimating the job. On larger work, the heating system is usually designed and laid out diagrammatically, but the heating contractor must work out details for piping, equipment connections, controls, etc., which all influence the cost.

To acquire the necessary technical knowledge for this purpose requires everyday participation in this line of work plus constant study to keep up to date. Heating and air conditioning is too complex a subject to treat in detail in a book of this kind. However, much useful data and information can be gained from manuals published by the HVAC associations (see the List of Associations in Chapter 1). Here, we have provided only the most elementary and approximate data for "roughing" or approximate estimates only.

Boilers are now designed for burning special fuels, such as hard coal, soft coal, oil, gas, etc., the costs varying with boiler sizes and accessories, such as domestic hot water heaters, thermostatic controls, etc.

The ordinary cast iron radiators standing in the room are pretty much a thing of the past. Heating systems use concealed convection type radiators or baseboard radiation. These are all items that affect costs.

Copper piping is used for both heating and plumbing. The cost of this work depends to a large extent upon the experience of the contractor installing same. Mechanics familiar with sweating joints and installing copper piping can install this work rapidly, while those who are not accustomed to its use, will spend considerable time on it.

For the above reasons, the estimator should be familiar with the requirements of the trade in order to prepare intelligent estimates.

Determining Heating Loads. The first step in designing and estimating heating work is to determine the maximum heating load. For most residential work, the maximum heating load is the total heat loss of the structure figured at design temperatures. For commercial and industrial structures, special purpose heating loads and pick-up loads may be required which must be added to the total building heat loss to obtain maximum heating load.

Heating loads and heat losses are invariably expressed in British thermal units (Btu) per hour. A Btu is the quantity of heat required to raise the temperature of 1 lb. of water 1°F.

Building heat losses vary considerably depending upon climatic location of job, exposure, building size, architectural design, purpose of structure, construction materials used, quality of construction workmanship, and other factors.

Heat losses should be figured separately for each room so that the proper amount of radiation may be provided for comfort in all areas. Space limits do not permit giving detailed information and data on figuring building heat losses here. Anything less than a thorough treatment of this subject would be misleading. Complete information and data on this subject are contained in the manuals and guides from the HVAC associations.

Forced Warm Air Heating Systems

Forced warm air heating is the most widely used residential heating method in use at the present time. Generally speaking, the forced warm air heating method differs from the old gravity type by using blower equipment to circulate the air in the system.

It is practically impossible for the average general estimator to compute the cost of a modern forced warm air heating system, unless he has a working knowledge of this subject and is able to calculate building heat losses and design systems which will satisfy heating requirements under design conditions. In addition, the subject has become so complex it cannot be covered in a volume of this kind so as to provide adequate information for the preparation of a cost estimate, other than a rough approximation to be used for preliminary purposes only. Always obtain firm quotations from reputable warm air heating contractors before submitting bids on jobs containing this work.

Winter Air Conditioning or Forced Air Systems. Forced warm air heating systems that are equipped with automatic humidifying devices and air filters are popularly known as winter air conditioning systems. Most forced air installations are of this type.

Each system consists of a direct fired heating unit, a blower, a system of warm air ducts and return air duct system. Air filters are located in the return air duct system just ahead of the blower--humidifying equipment is placed in the warm air plenum chamber over the furnace unit. Air cooling equipment may be added to make this a year-round system.

The unit may be gas, oil, or coal fired. Gas or oil fired units designed for that particular fuel operate at high efficiency, although satisfactory, if less efficient results may be obtained through the use of conversion burners installed in units primarily designed for solid fuel.

Thermostatic control of heating apparatus is highly desirable and invariably used. Automatic controls vary the supply of heat in accordance with demand set up by a thermostat located in the living portion of the home. Actuated by limit controls, the blower commences operation automatically when the heated air in the generating unit has reached a predetermined temperature and cuts out after the source of heat has been shut down and the temperature in the bonnet dropped to a predetermined point.

Design of air conveying systems of this kind presents an engineering problem that can only be solved with a reasonable degree of accuracy by a competent designer. The heat loss of the building, room by room, is established and the volume of heated and conditioned air required for indoor comfort establishes data from which a duct system can be designed with due regard for velocities, static pressures and delivery temperatures.

Items to be Included in the Estimate. Costs vary considerably and methods of estimating also differ, but every estimate for a winter air conditioning system should include the following items:

1. Cartage of material and equipment from shop to job.
2. Winter air conditioning unit, complete with humidifier, air filters, and controls.
3. Labor assembling and setting up unit.
4. Oil storage tank installation, including fill and vent piping, oil booster pump if required, oil gauge, piping to unit, oil filter, etc., for oil fired units, or--piping for gas fired units.
5. Installation of smoke pipe, fittings and accessories.

6. Warm air and return air plenum chambers.
7. Warm air and return air duct systems.
8. Diffusers, registers, intakes, grilles, etc.
9. Special insulation.
10. Electrical work.
11. Labor starting plant in operation and balancing system.
12. Service allowance.
13. Miscellaneous costs.
14. Overhead and profit.

Cost of Winter Air or Forced Air Conditioning Units. Sizes, capacities and prices of units vary with the different manufacturers, but the following listings are representative of models available for residential use.

Approximate Net Prices of Winter Air Conditioning Units
Complete with Automatic Controls, Humidifier, and Air Filters

Bonnet Rating Btu per Hr.	Register Rating Btu per Hr.	Heating Only		Net Price With Cooling
	Gas Fired Units			
60,000	48,000	$500.00	24 MBH	$1870.00
76,000	60,800	530.00	29 MBH	2,295.00
90,000	72,000	560.00	36 MBH	2,465.00
108,000	86,400	640.00	47 MBH	2,975.00
120,000	96,000	820.00	47 MBH	3,145.00
150,000	120,000	1,060.00	58 MBH	3,910.00
180,000	144,000	1,190.00	....	....
		Oil Fired Units		
84,000	71,400	890.00	36 MBH	2,465.00
112,000	95,200	1,020.00	42 MBH	2,975.00
140,000	119,000	1,200.00	47 MBH	3,400.00
175,000	148,750	1,360.00	....	....
210,000	178,500	1,530.00	....	....

Assembling and Setting up Unit. Most gas fired units can be assembled and set up by a sheet metal worker and a helper in 3 to 4 hours. For an oil fired unit, figure 4 to 5 hours.

Oil Storage Tanks. Common practice for oil storage tanks is the use of one or two inside storage tanks with a capacity of 275 gallons each. Occasionally, outside buried tanks of 550 or 1,000 gallon capacity are used. Buried tanks must be double wall with fuel leak detectors and must be in accordance with local environmental regulations. On a contract basis, these tanks installed, including all piping, cost as follows:

1-275 gallon inside tank	$400.00
2-275 gallon inside tanks	780.00
1-550 gallon outside tank	600.00
1-1,000 gallon outside tank	1,000.00

In the use of an outside tank, a basement wall type pump costing from $120.00 to $160.00, must be used with some types of burner equipment.

Gas Piping. An allowance of $200.00 to $300.00 should cover cost of material and labor for connecting gas to unit within the heating room.

Smoke Pipe. Material and labor for smoke pipe connection from unit to flue should cost $80.00 to $120.00 for gas fired and oil fired units.

Plenum Chambers. Plenum chambers must be fabricated to order in the shop and are usually made from light gauge galvanized steel. They are mounted on top of the furnace and serve to connect the warm air and return air duct systems to the unit. Two plenum chambers are required for each installation.

Plenums will run $8.00 per sq. ft in 18 ga., $10.00 in 16 ga., $12.00 in 14 ga.

Warm Air and Return Air Duct Systems. The cost of duct work will vary with the size of job and type of distribution.

Under ordinary conditions, the warm air heating contractor will measure and list each type and size of duct, fitting and accessory. These quantities are priced for material and then an estimate is made of labor required for installation. This is a lengthy operation and requires an experienced heating contractor.

When pressed for time, the duct system cost may be approximated on an outlet basis, using a unit price which covers both material and labor. For ordinary installations, the following costs per outlet are about average where sheet metal workers' wages are from $20.00 to $25.00 per hr.:

Type of System	Cost per Outlet
Conventional system--warm air outlets located on inside walls.	
1st floor outlets	$100.00-125.00
2nd floor outlets	140.00-160.00
Capped stacks	40.00 to 50.00
Radial perimeter system in small 1-sty. buildings with basement or crawl space.	
Low velocity--6" to 8" dia. ducts	$100.00-125.00
High velocity--4" and 5" dia. ducts	95.00-120.00
Large central return air duct	130.00-140.00
Radial perimeter system in small 1-sty. basementless buildings.	
Galv. steel sheet metal pipe and fittings	$132.00-140.00
Fiber duct with galv. steel fittings	132.00-140.00
Trunk and branch perimeter system in medium-size, 1 sty. buildings w/basement or crawl space.	
Std., galv. steel, rectangular ducts	$150.00-165.00

Add for diffusers, registers, intakes, special insulation, etc. Excavation, backfill and concrete encasement where required, not included.

Diffusers, Registers, and Intakes. Diffusers and registers generally used are of pressed steel construction, with prime coat finish, are fully adjustable and are available in floor, baseboard or wall types. Return air intakes are of same construction and finish, but are non-adjustable.

Floor type diffusers cost $16.00 to $21.00 for sizes commonly used. Baseboard type diffusers cost $15.00 for 2'-0" lengths and $21.00 for 4'-0" lengths. Wall type registers cost $7.00 to $15.00. Wall type intake grilles cost $9.00 to $18.00.

Special Insulation. Ducts in unexcavated spaces or in unheated attic spaces, both supply and return, should be insulated with not less than 1" of adequate insulating material. This material is furnished in flexible form and averages $2.00 to $2.50 per sq. ft. applied.

Electrical Work. Custom varies with respect to the supply and installation of electrical circuit from meter to fused safety switch adjacent to unit. Likewise for the connection of blower motor to 110 volt controls and also for the 22 volt wiring to thermostat when used. Most often, the job electrician runs the 110 volt service to the safety switch adjacent to unit. From this point it is handled as a subcontract under the heating contractor and represents a cost from $90.00 to $225.00 for complete wiring installation of all controls and motors.

Labor Starting and Balancing System. It is customary for the heating contractor to start the plant in operation, test all controls and balance the air distribution system. For jobs completed during summer months, this means a come-back call at the beginning of the heating system.

Labor costs for this work will run $100.00 to $150.00 depending upon size of job. For commercial work, an independent certified testing and balancing contractor might be required. The cost can be up to $5,000.00.

Service Allowance. Most jobs carry a 1-year free service warranty and an allowance should be included in the estimate to cover this contingency. Average costs for service during the first year of operation are $200.00 for gas fired and $220.00 for oil fired systems.

Miscellaneous Costs. Under miscellaneous costs are classified such items as federal tax, sales tax, permits for oil burner or for installation, if required.

Approximate Cost of a Complete System Installed. Based on installations in residences of ordinary construction, located in the larger cities, the cost of average winter air conditioning systems ranges as follows for gas fired units. For oil fired units, add 10 to 15 percent.

Floor Area Sq. Ft.	Heat Loss Btu per Hr.	Price for Complete Gas Fired Installation
Conventional Systems, 1-Sty. Buildings with Basements or Crawl Space		
1,000	65,000	$2,550.00-3,400.00
1,500	80,000	3,400.00-3,910.00
2,000	100,000	3,910.00-4,250.00
Conventional Systems, 2-Sty. Buildings with Basements		
1,500	75,000	$3,400.00-3,910.00
2,000	90,000	4,250.00-5,100.00
Perimeter Systems, 1-Sty. Buildings with Basements or Crawl Space		
1,000	65,000	$2,550.00-3,400.00
1,500	80,000	3,400.00-3,910.00
2,000	100,000	4,250.00-4,760.00
Perimeter Systems, 1-Sty. Basementless Buildings		
1,000	65,000	$2,550.00-3,400.00
1,500	80,000	3,400.00-3,910.00

The above prices are for individual jobs. For multiple housing projects, where large numbers of similar dwelling units are involved, much better prices may be obtained.

As previously mentioned, equipment, design, and practices vary greatly even in localized areas. No safe guide may be set down for general use. So much depends upon correct design, proven equipment and experienced installation that it behooves the buyer to carefully investigate all of these items before awarding a contract for this important part of the building.

Summer Air Conditioning

Residential summer air conditioning is available at a price the average home owner can afford. This applies to both existing homes and new homes under construction and is especially true where forced warm air heating is used, because the same air distribution duct system can be used for both heating and cooling. Various warm air heating equipment manufacturers make cooling units for this purpose, and installation work is usually done by heating contractors.

The function of summer air conditioning equipment is the reverse of winter air conditioning--circulating air is cooled and de-humidified--but design problems are similar, though more complex. A basic requirement for designing a cooling system is calculating the total cooling load or total heat that must be removed from the structure to achieve a predetermined inside temperature and humidity under design conditions. Heat gain calculations are much more involved than those for heat losses, as many additional factors must be considered, such as sun effect on roof, wall and glass areas; internal heat gains from human occupancy, cooking, lighting, appliances, etc. and latent heat energy involved in de-humidification. This is a job for an expert.

Heat gain calculations are always expressed in Btu per hour. Cooling equipment capacities are usually stated per ton of refrigeration. A ton of refrigeration will remove heat at the rate of 12,000 Btu per hr. In a 24-hr. period, this is equivalent to the heat required to melt a ton of ice. With water-cooled compression refrigeration equipment, one horsepower usually equals about one ton of cooling capacity.

Types of Cooling Equipment for Residential Use. There are several types of cooling units available and in popular use for residential summer air conditioning. In general, equipment may be classified as water-cooled or air-cooled, with each type available for several methods of application.

One method employs a remotely located condensing unit, with refrigerant lines connected to a cooling coil installed in the warm air plenum chamber of a forced warm air heating system. This method uses the heating system blower and ducts for air distribution.

Another method uses a self-contained unit, with the condenser, cooling coil and blower combined in one cabinet. This type may be cut in to the heating ducts, or may have its own duct system where hot water or steam heat is used.

Net Prices of Remote System Cooling Units. Approximate prices are for units factory assembled with heating and cooling thermostat included. No pipe, duct materials or blower assembly included.

Capacity Tons	Rating Btu per Hr.	Net Price Air-cooled	Labor Hrs. To Install
2	24,000	$1,000.00	8
3	36,000	1,450.00	14
5	60,000	2,700.00	30

Where a complete air circulation duct system is required, figure same as previously given for winter air conditioning systems.

Approximate Cost of Complete Cooling Unit Installation. Based on conditions of average difficulty, complete residential cooling unit installations should cost approximately as follows:

Capacity Tons	Rating Btu per Hr.	Price for Complete Installation
	Remote, Water-cooled System	
2	24,000	$3,740.00-4,250.00
3	36,000	4,250.00-4,760.00
	Remote, Air-cooled System	
2	24,000	3,400.00-3,740.00
3	36,000	4,250.00-4,590.00
5	60,000	7,650.00-8,500.00
	Self-contained, Water-cooled System	
2	24,000	3,910.00-4,250.00
3	36,000	5,950.00-6,290.00
	Self-contained, Air-cooled System	
2	24,000	2,600.00-4,760.00
3	36,000	7,650.00-8,160.00
5	60,000	11,900.00-12,750.00

Room Size Air Conditioning Units. The cost of portable room size air conditioning units vary according to size, ranging from ½-Hp., which cools room areas up to 400 sq. ft., to 2-Hp. for spaces up to 1,200 sq. ft.

All models are for window installation and include automatic thermostat controls.

Net prices run from $400.00 for a 4,000 Btu unit to $1,500.00 for a 20,000 Btu unit.

If electric wiring is necessary, add extra for installation.

Commercial Installations. A number of mechanical equipment manufacturers are offering self-contained, one-piece rooftop electric heat pump units for the smaller commercial projects.

These units which are made in numerous heating and cooling capacities are pre-wired and precharged at the factory so that when they are delivered to the project site, the installation is relatively simple. The unit is set on a roof curb, power and control wiring connected and duct connections hooked up--and the unit is in operation.

Standard construction of these units includes a manual damper that can be preset to admit up to 25% outdoor ventilation air for year-round fresh air. Some units come equipped with an "economizer cycle" that has a thermostatically-controlled, fully-modulating damper which admits up to 100% outdoor ventilation air to provide low-cost cooling on mild days.

For quick approximations of air cooling costs cooling units can be estimated from $1,500.00 to $3,000.00 per ton depending on the ductwork, job conditions, types of grilles, etc. 12,000 Btus requires one ton of air cooling. The following table will give an approximation of final requirements:

Type of Space	Btu per Sq. Ft. Floor Area
Large Stores	40
Offices	45
Specialty Shops	50-70
Bars	80
Restaurants	100-120

Steam and Hot Water Heating

Because of different conditions surrounding the installation of heating apparatus, it is impossible to give any set rule that can be accepted, without modification, for all kinds of buildings to be heated. It is necessary to take into consideration all of the conditions in and around any building, and additions or deductions made to suit the requirements, no matter what rule may be used for figuring.

Methods of Computing Heating. The most advanced method of figuring heating, the one that is generally used by heating engineers, is the Btu Method. This method is based on replacing the heat loss through exposed walls, doors and windows, floors, ceilings, etc., and uses so many factors based on the type of construction that it is impossible to go into detail in a book of this kind. Intricate and elaborate formulas are used for both warm air, steam and hot water heating. The subject of heating requires a book as large as this one. For this reason only approximates are given here.

Flow Control Hot Water Heating Systems. This method of hot water heating has largely superseded the older gravity or "open" type of hot water heating system. It is a complete combined automatic water heating and domestic water supply system for year-round use. Its major advantages are instantaneous heating response, high efficiency, fuel economy, and completely automatic operation.

The flow control valve makes the boiler a reservoir of high temperature water, in that it prevents circulation through the system when heat is not required. A circulator or pump attached to the main near the boiler forces this hot water through the mains the instant the room thermostat calls for heat. When the thermostat is satisfied, the circulator ceases to operate, and the flow control valve stops all circulation of hot water, thus preventing overheating and needless heat losses.

The flow control hot water heating system requires approximately 20% more radiation than is figured for steam heat. The mains are smaller, reducing the cost of pipe, fittings and labor to install.

All in all, a flow control hot water heating system costs a trifle less than the older gravity systems requiring large radiators and piping. A gravity hot water heating system can be converted into a flow control system at a nominal expense.

Cost of Special Fittings for Flow Control Hot Water Heating Systems. The cost of fittings will vary with the amount of radiation in the job, and whether a complete flow control system is used, or merely a flow control valve and a circulating pump.

Where only the flow control valve and water circulating pump are used, the costs for the special equipment will run as follows based on the amount of radiation in the job.

Capacities and Prices of Flow-Control Valves

Valve Size	Radiation Capacity in Square Feet 150 Btu	200 Btu	240 Btu	Weight, lbs.	Price Each
1"	354	265	221	5	$20.00
1¼"	786	590	492	5	21.00
1½"	1170	875	730	10	24.00
2"	2140	1610	1340	21	29.00
2½"	3680	2750	2300	33	57.00
3"	6660	5000	4160	42	72.00
4"	14000	10000	8750	57	165.00

Sizes and Prices of Water Circulators or Booster Pumps

Recommended for use on modern forced circulation systems or for improving gravity hot water systems.

Size Inches	Motor Hp	Dely Gals. per Min.	Direct Radiation Capacity in Sq. Ft. 150 Btu	200 Btu	240 Btu	Price Each Iron Body	All Bronze
1	1/12	31	750	562	469	$160.00	$155.00
1¼	1/12	33	1000	750	675	160.00	180.00
1½	1/12	33	1000	750	675	160.00	180.00
2	1/6	75	3500	2625	2187	280.00	460.00
2½	¼	120	4500	3375	2820	405.00	735.00
3	¼	120	4500	3375	2820	405.00	840.00

Sizes and Prices of Relief Valves

ASME Capacity at 30 lbs. in Btu per Hr.	Size Inches	Net Price
175,000	3/4	$12.00
250,000	3/4	17.00
350,000	3/4	20.00
480,000	3/4	21.00
750,000	1	35.00
1,050,000	1 1/4	56.00

Sizes and Prices of Air Control Boiler and Tank Fittings

Air control fittings are used for the purpose of removing air from a hot water heating system and preventing its return.

Air control boiler fittings consist of a tube inserted into a T-shaped fitting. The tube, in effect, extends the supply main down into the boiler water, thus preventing air (which has accumulated at the top of the boiler) from rising into the piping and heating units. This air flows around the tube, up into the T-fitting and then to the pipe leading to the air control tank fitting.

The air control tank fittings provide two separate passages--one through which air can flow to the top of the tank, and the other for the water displaced by the air to flow back into the pipe connections to the system. These will cost from $22.00 to $45.00 in sizes 1" to 3" for boiler fittings, $22.00 for tank controls up to 24".

Coils for Heating Water. Where coils are inserted in the boiler for heating water for domestic purposes, the size of the boiler should be increased, figuring each gallon of water tank capacity as equivalent to 1 sq. ft. of steam radiation or 2 sq. ft. of hot water radiation. For example, a 30-gal. tank is equivalent to 30 sq. ft. of steam or 60 sq. ft. of hot water radiation.

Rule for Computing Boiler Sizes. After the total number of square feet of radiation of total heat loss in Btu per hour has been obtained, the size of the boiler should be computed.

Practically all boiler manufacturers rate their boilers, giving maximum heat output under full drive and also the recommended load representing the attached load, including piping, which may be placed on the boiler in accordance with accepted installation standards for economical operation.

Regarding Steam Boiler Installation. It is recommended on all installations of steam boilers that drain valves be placed on the returns and that the condensation from such returns be discharged into the sewer for a period of 3 days to one week after starting fire, thereby clearing system of grease and dirt.

At the end of this period the boiler should be thoroughly washed and blown out.

Boiler Capacity. Boiler output is defined as the number of sq. ft. of standard direct column radiation or its equivalent in Btu, which a boiler will supply at its outlet in one hour, under standard conditions, with 240 Btu representing one sq. ft. of standard direct column steam radiation and 150 Btu representing one sq. ft. of standard direct column water radiation.

Boiler Horsepower. One boiler horsepower is defined as the evaporation of 34½ lbs. of water per hour from a feed water temperature of 212° into steam at 212°. This is equal to 34.5 x 970.4 = 33478.8 Btu per hour. One boiler horsepower is equal therefore to 33478.8 divided 240 = 139 sq. ft of steam radiation. Note that boiler horsepower is not derived from engine horsepower. 1 horsepower hour = 2545 Btu.

To find the Btu output per hour of a steam boiler, multiply the output in sq. ft. by 240.

To find the Btu output per hour of a water boiler, multiply the output in sq. ft. by 150.

To find the output in sq. ft. of steam radiation, divide the output in Btu per hour by 240.

To find the output in sq. ft. of water radiation, divide the output in Btu per hour by 150.

Estimating Pipe and Fittings. To prepare an intelligent estimate on the cost of any heating plant, the entire job should be laid out using the quantity of pipe of each size required; connections, such as elbows, tees, nipples, valves, etc., should all be figured separately. This requires considerable labor and a man must thoroughly understand his business to figure with any degree of accuracy.

When estimating the quantities of piping required for any heating plant, the risers, mains and returns must be figured, together with all connections to and from the boilers. These will vary according to the system used and the type of building.

A one-pipe steam heating system should have the highest point in the main directly over the boiler, and from this point it pitches down around the circuit until it drops to the floor or enters the boiler. The more pitch the better, but there should be at least one inch in 20 feet. The lowest point in the steam main, where it drops to enter the boiler or to the floor, should not be less than 14 inches above the water line of boiler and the more the better.

Branches for radiators should be taken from top of main, using a nipple and elbow or a nipple and a 45-degree ell, and should pitch up to the radiator.

Connections for steam should never be taken from side of main, as this will cause water hammer and syphoning of the water of condensation in the steam main into the radiator.

Each radiator should be supplied with a good disc radiator valve and a good automatic air valve, and each return main should also have an automatic air valve where it drops to enter boiler.

If branches are longer than 8 to 10 feet, they should be one size larger than the upright connecting pipes. Where mains drop to floor, a pipe two sizes smaller can be used to return to boiler. Risers for either steam or water should be plumb and straight. If exposed in a room, and it is necessary to use couplings on same, the couplings should be at a uniform height from floor.

The valve on a steam radiator should be either wide open or tightly closed to prevent radiator filling with water.

In a two-pipe gravity hot water heating system the highest return rises from boiler to radiator, the lowest point being at the boiler and the highest point at the end of the main. These pipes may be reduced as the branches are taken off, care being taken not to reduce too rapidly. Each radiator should be fitted with a hot water radiator valve, union elbow and a compression air valve. All pipes should be well supported, so that no pockets or depressions will occur to retard or trap the circulation. All mains, both for steam and hot water, should run about 4 feet from wall to allow for expansion.

Table Giving Number of Square Feet of Radiation in Exposed Heating Pipes of Various Sizes and Lengths

Length of Pipe	Size of Pipe											
	¾	1	1¼	1½	2	2½	3	4	5	6	7	8
1	.275	.346	.434	.494	.622	.753	.916	1.175	1.455	1.739	1.996	2.257
2	.5	.7	.9	1.	1.2	1.5	1.8	2.4	2.9	3.5	4.	4.5
3	.8	1.	1.3	1.5	1.9	2.3	2.7	3.5	4.4	5.2	6.	6.8
4	1.1	1.4	1.7	2.	2.5	3.	3.6	4.7	5.8	7.	8.	9.
5	1.4	1.7	2.2	2.4	3.1	3.8	4.6	5.8	7.3	7.7	10.	11.3
6	1.6	2.1	2.6	2.9	3.7	4.5	5.5	7.	8.7	10.5	12.	13.5
7	1.9	2.4	3.	3.4	4.4	5.3	6.4	8.2	10.2	12.1	14.	15.8
8	2.2	2.8	3.5	3.9	5.	6.	7.3	9.4	11.6	13.9	16.	18.
9	2.5	3.1	3.9	4.4	5.6	6.8	8.2	10.6	13.1	15.7	18.	20.3
10	2.7	3.5	4.3	4.9	6.2	7.5	9.1	11.8	14.6	17.4	20.	22.6
11	3.	3.8	4.8	5.4	6.8	8.3	10.	12.9	16.	19.1	22.	24.9
12	3.3	4.1	5.2	5.9	7.5	9.	11.	14.1	17.4	20.9	24.	27.1
13	3.6	4.5	5.6	6.4	8.1	9.8	11.9	15.3	18.9	22.6	26.	29.4
14	3.8	4.8	6.1	6.9	8.7	10.5	12.8	16.5	20.3	24.3	28.	31.6
15	4.1	5.2	6.5	7.4	9.3	11.3	13.7	17.6	21.8	26.1	30.	33.9
16	4.4	5.5	6.9	7.9	10.	12.	14.6	18.8	23.2	27.8	32.	36.1
17	4.7	5.9	7.4	8.4	10.6	12.8	15.5	20.	24.7	29.5	34.	38.4
18	5.	6.2	7.8	8.9	11.2	13.5	16.5	21.2	26.2	31.3	36.	40.6
19	5.2	6.6	8.3	9.4	11.8	14.3	17.4	22.3	27.6	33.1	38.	42.9
20	5.5	6.9	8.7	9.9	12.5	15.	18.3	23.5	29.1	34.8	40.	45.2
21	5.8	7.3	9.1	10.4	13.	15.8	19.2	24.7	30.5	36.5	42.	47.4
22	6.	7.6	9.6	10.9	13.7	16.5	20.2	25.9	32.	38.3	44.	49.7
23	6.3	8.	10.	11.3	14.3	17.3	21.1	27.	33.5	40.	46.	52.
24	6.6	8.3	10.4	11.9	14.9	18.	22.	28.2	34.9	41.7	48.	54.2
25	6.9	8.6	10.9	12.3	15.6	18.8	22.9	29.3	36.3	43.5	50.	56.4
26	7.1	9.	11.3	12.8	16.2	19.5	23.8	30.5	37.8	45.2	52.	58.6
27	7.4	9.4	11.7	13.3	16.8	20.3	24.7	31.7	39.3	47.	54.	61.
28	7.7	9.7	12.2	13.8	17.4	21.	25.6	32.9	40.7	48.7	56.	63.2
29	8.	10.	12.6	14.3	18.	21.8	26.6	34.1	42.2	50.4	58.	65.5
30	8.3	10.4	13.	14.8	18.7	22.5	27.5	35.3	43.6	52.1	60.	67.7
31	8.5	10.7	13.5	15.3	19.3	23.3	28.4	36.4	45.1	53.9	62.	70.
32	8.8	11.1	13.9	15.8	19.9	24.1	29.3	37.6	46.5	55.6	64.	72.2
33	9.1	11.4	14.3	16.3	20.5	24.8	30.2	38.8	48.	57.4	66.	74.4
34	9.4	11.7	14.7	16.8	21.2	25.6	31.1	40.	49.5	59.1	68.	76.7
35	9.6	12.1	15.2	17.3	21.8	26.3	32.	41.1	50.9	60.8	70.	79.
36	9.9	12.5	15.6	17.8	22.4	27.	33.	42.3	52.4	62.6	72.	81.3
37	10.2	12.8	16.1	18.3	23.	27.8	33.9	43.5	53.8	64.3	74.	83.5
38	10.5	13.2	16.5	18.8	23.7	28.5	34.8	44.6	55.2	66.	76.	85.8
39	10.7	13.5	16.9	19.3	24.3	29.3	35.7	45.8	56.7	67.8	78.	88.
40	11.	13.8	17.4	19.8	24.9	30.1	36.6	47.	58.2	69.5	80.	90.2
41	11.3	14.2	17.8	20.3	25.5	30.8	37.6	48.2	59.6	71.3	82.	92.5
42	11.5	14.5	18.2	20.8	26.1	31.6	38.5	49.4	61.1	73.	84.	94.8
43	11.8	14.9	18.7	21.3	26.8	32.3	39.4	50.6	62.5	74.8	86.	97.
44	12.1	15.2	19.1	21.8	27.4	33.1	40.3	51.7	64.	76.5	88.	99.3
45	12.4	15.6	19.5	22.2	28.	33.8	41.2	52.9	65.5	78.2	90.	101.6
46	12.7	15.9	20.	22.7	28.6	34.6	42.2	54.	67.	80.	92.	103.8
47	12.9	16.3	20.4	23.2	29.2	35.3	43.	55.2	68.4	81.7	94.	106.
48	13.2	16.6	20.8	23.7	29.9	36.1	43.9	56.4	69.8	83.5	96.	108.4
49	13.5	17.	21.3	24.2	30.5	36.8	44.8	57.6	71.2	85.1	98.	110.5
50	13.8	17.3	21.7	24.7	31.1	37.6	45.8	58.7	72.7	87.	100.	112.8

Practically all hot water heating systems installed today are flow control systems where the hot water is circulated from the boiler through the mains and pipes, using a circulating or booster pump. When this method is used, the size of the mains can be greatly reduced, as will be noted from the following table. With the water being circulated through the system at 30 to 120 gallons a minute, it is very much faster than the old gravity systems. Also the radiators may be placed above or below the boiler level, as desired.

Sizes of Pipes. When estimating steam heating, using a rule of thumb method, the size of the main may be determined by taking the total amount of direct radiation to which add 25% for piping and from this total extract the square root, dividing same by 10, which gives the size of the main to use. This is for one pipe work.

For two pipe work one size less is sufficient and the return can be one or two sizes less than the supply. A steam main should not decrease in size but very little according to the area of its branches.

An example of the above method is as follows: A job contains 400 feet of direct steam radiation. By adding 25% for piping, the result is 500 square feet. Extracting the square root of 500, gives 22.4. Divide 22.4 by 10 equals 2.24 or it will require a 2½-inch steam main.

List of Sizes of Steam Mains. The following table gives the correct sizes of mains for both one and two pipe steam heating systems, as determined by the above method:

Radiation Square Feet	One-Pipe Work Inches	Two-Pipe Work Inches
125	1½	1¼x1
250	2	1½x1¼
400	2½	2x1½
650	3	2½x2
900	3½	3x2½
1,250	4	3½x3
1,600	4½	4x3½
2,050	5	4½x4
2,500	6	5x4½
3,600	7	6x5
5,000	8	7x6
6,500	9	8x6
8,100	10	9x6

List of Sizes for Hot Water Mains. The following table gives the correct sizes of mains for hot water heating systems:

Radiation Square Feet	Gravity Size Pipe Inches	Forced H.W. 2-Pipe System*
75 to 125	1¼	1
125 to 175	1½	1
175 to 300	2	1
300 to 475	2½	1
475 to 700	3	1¼
700 to 950	3½	1½
950 to 1,200	4	1½
1,200 to 1,575	4½	2
1,575 to 1,975	5	2
1,975 to 2,375	5½	2½
2,375 to 2,850	6	2½

*Based on 200 Btu emission and circulators having 1,725 rpm motors.

Hot water flow mains may be reduced in size in proportion to the branches taken off. They should, however, have areas as large as the sum of all branches beyond that point. Returns should be same as flows.

Tables of Mains and Branches. The following table gives the size of mains and the number of branches of different sizes that each main will supply:

Mains	Branches
1 in. will supply	2 3/4-in.
1 1/4 in. will supply	2 1-in.
1 1/2 in. will supply	2 1 1/4-in.
2 in. will supply	2 1 1/2-in.
2 in. will supply	2 1 1/2-in. and 1 1 1/4-in. or 1 2-in. and 1 1 1/4-in.
3 in. will supply	1 2 1/2-in. and 1 2-in. or 2 2-in. and 1 1 1/2-in.
3 1/2 in. will supply	2 2 1/2-in. or 1 3-in. or 1 2-in. or 3 2-in.
4 in. will supply	1 3 1/2-in. and 1 2 1/2-in. or 2 3-in. and 4 2-in.
4 1/2 in. will supply	1 3 1/2-in. and 1 3-in. or 1 4-in. and 1 2 1/2-in.
5 in. will supply	1 4-in. and 1 3-in. or 1 4 1/2-in. and 1 2 1/2-in.
6 in. will supply	2 4-in. and 1 3-in. or 4 3-in. or 10 2-in.
7 in. will supply	1 6-in. and 1 4-in. or 3 4-in. and 1 2-in.

Radiator Tappings. The following is a schedule of tappings or branches required for radiators of different sizes.

Steam	One Pipe
Radiators containing up to 24 sq. ft. of radiation	1"
Radiators containing 24 to 60 ft. of radiation	1 1/4"
Radiators containing 60 to 100 ft of radiation	1 1/2"
Radiators containing above 100 ft. of radiation	2"

Gravity Hot Water	Two Pipe
Radiators containing up to 40 ft. of radiation	1x1"
Radiators containing 40 to 72 ft. of radiation	1 1/4x1 1/4"
Radiators containing more than 72 ft. of radiation	1 1/2x1 1/2"

Forced Flow Hot Water

A 3/4" branch will take care of most any ordinary size radiator where forced flow is used.

Sizes and Net Prices of Standard Wrought Steel Pipe
All Weights and Dimensions are Nominal

Size	Price per 100 LF Random T. & C. Black	Galv.	Diameter External	Diameter Internal	Weight per foot Plain Ends	Weight per foot Threads and Couplings	No. of Threads Per In. of Screw
1/8	$50.00	$80.00	.405	.269	.24	.24	27
1/4	70.00	90.00	.540	.364	.42	.42	18
3/8	80.00	100.00	.675	.493	.57	.57	18
1/2	80.00	90.00	.840	.622	.85	.85	14
3/4	85.00	100.00	1.050	.824	1.13	1.13	14
1	115.00	135.00	1.315	1.049	1.68	1.68	11 1/2
1 1/4	150.00	175.00	1.660	1.380	2.27	2.28	11 1/2
1 1/2	170.00	200.00	1.900	1.610	2.72	2.73	11 1/2
2	235.00	280.00	2.375	2.067	3.65	3.68	11 1/2
2 1/2	325.00	375.00	2.875	2.469	5.79	5.82	8
3	425.00	500.00	3.500	3.068	7.58	7.62	8

Size	Price per 100 LF Random T. & C. Black	Galv.	Diameter External	Internal	Weight per foot Plain Ends	Threads and Couplings	No. of Threads Per In. of Screw
3½	540.00	625.00	4.000	3.548	9.11	9.20	8
4	625.00	750.00	4.500	4.026	10.79	10.89	8
5	1150.00	1250.00	5.563	5.047	14.62	14.81	8
6	1350.00	1550.00	6.625	6.065	18.97	19.18	8
8	2000.00	2200.00	8.625	8.071	24.70	25.55	8
8	2200.00	2400.00	8.625	7.981	28.55	29.35	8
10	2800.00	3000.00	10.750	10.136	34.24	35.75	8
10	3000.00	3200.00	10.750	10.020	40.48	41.85	8
12	3600.00	3800.00	12.750	12.090	43.77	45.45	8
12	3800.00	4000.00	12.750	12.000	49.56	51.15	8

For lengths cut to order, add 15% to random length prices.

Sizes and Net Prices of Extra Strong Wrought Steel Pipe

Price per 100 Lin. Ft.

Size	Random PE Black	Galv.	Diameter External	Internal	Thickness	Price per Ft. Plain Ends
1	$90.00	$154.00	1.315	.957	.179	$3.05
1¼	120.00	195.00	1.660	1.278	.191	4.20
1½	140.00	224.00	1.900	1.500	.200	5.10
2	185.00	310.00	2.375	1.939	.218	7.00
2½	270.00	455.00	2.875	2.323	.276	10.70
3	350.00	595.00	3.500	2.900	.300	14.35
3½	425.00	735.00	4.000	3.364	.318	17.50
4	500.00	875.00	4.500	3.826	.337	21.00
5	875.00	1540.00	5.563	4.813	.375	29.10
6	1050.00	1750.00	6.625	5.761	.432	40.00
8	1350.00	2310.00	8.625	7.625	.500	60.75
10	1850.00	3080.00	10.750	9.750	.500	76.60
12	2400.00	4200.00	12.750	11.750	.500	91.60

For lengths cut to order, add 33⅓% to random length prices.

Net Prices Per 100 Lin. Ft. of Youngstown "Yoloy" Alloy Steel

Size	Pipe Standard Random T. & C. Black	Galv.	Extra Heavy Random P.E. Black
½	$135.00	$160.00	$182.00
¾	160.00	189.00	196.00
1	224.00	266.00	301.00
1¼	294.00	350.00	400.00
1½	343.00	400.00	462.00
2	441.00	525.00	637.00
2½	665.00	770.00	997.00
3	861.00	1015.00	1197.00
3½	1064.00	1274.00	....
4	1295.00	1540.00	2100.00
5	1890.00	2240.00	2910.00
6	2450.00	2940.00	3780.00
8-28.55 lb.	3920.00	4400.00	4340.00
10-35 lb.	5600.00	6580.00	7560.00
12-45 lb.	7000.00	8400.00	11,200.00

For black pipe lengths cut to order, add to random length prices 25% for standard pipe and 35% for extra heavy pipe.

Size of Expansion Tanks Required for Hot Water Heating Systems

The following table gives the sizes of expansion tanks required for hot water heating systems, complete with trimmings:

Capacity Gallons	Size Inches	Sq. Ft. of Radiation	Price Each with Trimmings
15	12x30	350	$63.00
18	12x36	500	70.00
24	12x48	1000	77.00
30	12x60	2000	88.00
40	14x60	3000	112.00
80	20x60	4500	180.00

For two gauge glass tappings, add $15.00.

Radiation capacities are based on forced circulation and small pipes. On old systems use next larger size tank.

Heights, Sizes and Ratings of Thin Tube Cast Iron Radiators

Cast iron radiations are seldom used these days, but standard measures are 1¾" in length per section. Add ½" to the length for each bushing.

Sq. Ft. of Radiation per Section of Various Height Radiators

Number Tubes	Width Inches	19"	22"	25"	32"
3	3½	...	...	1.6	...
4	4¾	1.6	1.8	2.0	...
5	6	...	2.1	2.4	...
6	7¼	2.3	...	3.0	3.7

Approximate Prices of Thin Tube Radiators

Sold only in even number of sections. Price per 100 sq. ft.

No. Tubes	Height 19"	22"	25"	32"
3	$380.00			
4	380.00	371.00	371.00	
5		364.00	336.00	
6	385.00		364.00	560.00

Radiant Baseboard Panels

Radiant baseboard panels are produced and marketed by a number of concerns. Some are made of cast iron, others with a finned pipe covered with a steel facing or cover.

They are made in various heights from 7" to 9" and usually extend out from the wall 1¾" to 2½", depending on the manufacturer.

Rated Heat Emission of Radiant Baseboard Panels

Rating in sq. ft. is based on standard emission of 240 Btu per hour per sq. ft. at average temperature of 215°.

9⅞-Inch Radiant Panel Baseboard

Rating of Panel Sq. Ft.	Water Flow Lbs./Hr.	Heat Output Btu per Hour per Lin. Ft. 220°	210°	200°	190°	180°	170°	160°	150°
1.95	500	460	420	380	350	310	270	240	200
1.95	2,500	490	450	410	370	330	290	250	210

9⅞ Inch Radiant Convector Panel

Rating based on standards given for baseboard panels.

Rating of Panel Sq. Ft.	Water Flow Lbs./Hr.	Heat Output Btu per Hour per Lin. Ft. 220°	210°	200°	190°	180°	170°	160°	150°
3.25	500	780	720	660	600	530	470	420	350
3.25	2,500	830	760	700	640	570	510	440	380

Price of Baseboard Radiation and Accessories

Description	Weight Lbs.	Approx. Price
Radiant Baseboard Panels 9⅞", Price per lin. ft.	12.25	$9.00
Radiant Convector Panels 9⅞", Price per lin. ft.	14.00	18.00
Plain ends, 9⅞", Price each	2.50	6.00
Corner Plate, 9⅞", Price each	.50	6.00
Angle Radiator Valve, ¾", Price each	1.00	18.00

Convector-Radiators

Radiant convectors combine convection heating with quick-acting and comfortable radiant heating. They are formed of cast iron with finned sections, or of copper with or without finned sections. There are so many types of convectors it is impossible to describe all of them in detail.

Convectors described below are 6¼" deep overall and may be installed as freestanding or semi-recessed units. Maximum depth of recess should not exceed 4¾".

Size, Capacity, and Approximate Price of Convectors

Height Inches	Length Inches	Steam SF EDR	Forced Hot Water, 1,000 Btu per Hr. 180°	200°	220°	Price Ea.
20	16	13.0	1.74	2.17	2.60	$73.00
20	20	17.0	2.28	2.84	3.40	91.00
20	24	21.0	2.81	3.51	4.20	109.00
20	28	26.0	3.48	4.34	5.20	136.00
20	32	29.5	2.95	4.93	5.90	154.00
20	36	33.5	4.48	5.59	6.70	175.00
20	40	38.5	5.15	6.43	7.70	196.00
20	44	43.5	5.82	7.26	8.70	218.00
20	48	46.5	6.22	7.76	9.30	238.00
20	56	55.5	7.43	9.26	11.10	280.00
20	64	63.0	8.44	10.50	12.60	315.00

Height Inches	Length Inches	Steam SF EDR	Forced Hot Water, 1,000 Btu per Hr. 180°	200°	220°	Price Ea.
24	20	18.5	2.48	3.09	3.70	95.00
24	24	23.0	3.08	3.84	4.60	117.00
24	28	28.0	3.75	4.67	5.60	132.00
24	32	32.5	4.35	5.42	6.50	168.00
24	36	36.5	4.88	6.09	7.30	189.00
24	40	42.0	5.62	7.01	8.40	218.00
24	44	47.5	6.36	7.93	9.50	224.00
24	48	51.0	6.83	8.51	10.20	266.00
24	56	60.5	8.18	10.10	12.10	308.00
24	64	69.0	9.24	11.52	13.80	357.00

For dampers, add $10.00. For air chambers and accessories less air valve, add $12.00.

Types of Standard Radiant Convectors and Cabinets

Standard Convectors and Cabinets--Non-Ferrous

Convectors listed below are 4", 6", 8", and 10" deep plus ¼" overall, and may be installed as freestanding or semi-recessed units.

		Height 20"			24"			32"		
Dep.	Len.	EDR	Approx. Price	Shpg. wt. lbs.	EDR	Approx. Price	Shpg. wt. lbs.	EDR	Approx. Price	Shpg. wt. lbs.
4	24	13.1	$67.00	25	15.0	$70.00	25	--	$84.00	--
6	24	19.0	94.00	30	22.4	98.00	30	25.5	115.00	35
	32	27.0	98.00	40	31.6	85.00	35	36.0	126.00	50
6	36	30.8	105.00	40	36.2	112.00	45	41.2	133.00	55
	40	34.9	115.00	45	40.8	98.00	45	46.5	147.00	60
	44	38.8	126.00	50	45.4	140.00	55	51.7	168.00	65
	48	42.7	134.00	55	50.0	147.00	60	56.9	173.00	75
	56	50.6	143.00	60	59.2	154.00	65	67.4	193.00	75
	64	58.3	154.00	65	68.4	168.00	70	77.9	196.00	80
8	32	34.0	115.00	45	37.5	98.00	45	41.6	147.00	55
	36	39.1	123.00	45	43.1	133.00	45	47.8	157.00	60
	40	44.2	143.00	50	48.7	154.00	50	54.0	193.00	65
	48	54.3	168.00	60	59.8	173.00	65	66.4	207.00	75
	56	64.4	182.00	65	70.9	196.00	70	78.8	235.00	115
	64	74.4	203.00	70	82.0	231.00	70	91.0	266.00	120

		Height								
		20"			24"			32"		
Dep.	Len.	EDR	Approx. Price	Shpg. wt. lbs.	EDR	Approx. Price	Shpg. wt. lbs.	EDR	Approx. Price	Shpg. wt. lbs.
10	36	45.0	140.00	50	49.0	161.00	50	54.5	185.00	65
	40	50.8	137.00	55	55.3	175.00	55	61.5	207.00	70
	48	62.1	182.00	65	67.7	203.00	75	75.4	238.00	85
	56	73.6	203.00	70	80.2	231.00	80	89.3	280.00	120
	64	85.1	231.00	75	92.7	259.00	85	103.1	301.00	130

Stock Model Convector Accessories

Description	Price	Shpg. Wt. Lbs.
Damper Package (up to and incl. 64")	$17.00	5
Snap-in Inlet Grille for Stock Models Only	.20.00	5
CM Molding Trim--24" to 36" Long	.25.00	6
CM Molding Trim--40" to 64" Long	.32.00	8
Air Vent--Manual	.14.00	1
Air Vent--Auto	.14.00	1

Hot Water Rating in Btu per Sq. Ft. EDR

Temp. Drop	Average Water Temp. °F 180°	190°	200°
10°F	160	181	201
20°F	141	159	177
30°F	131	149	166

Sizes and Approximate Prices of Cast-Iron Sectional Boilers and Gas or Light Oil Burners With Jacket

Steam: Installed Radiation Sq. Ft.*	Steam: Price Each	Hot Water: Installed Radiation Sq. Ft.* For Oil Firing	Hot Water: Btu Rating	Hot Water: Price Each
845	$1220.00	1500	225,000	$1220.00
1090	1460.00	1910	287,000	1460.00
1340	1675.00	2320	348,000	1675.00
1600	1870.00	2735	410,000	1870.00
1860	1950.00	3145	472,000	1950.00
2120	2050.00	3560	534,000	2050.00
2380	2350.00	3960	594,000	2350.00
2650	2850.00	4375	656,000	2850.00
2920	2900.00	4785	718,000	2900.00
3200	3000.00	5205	781,000	3000.00
3480	3100.00	5620	843,000	3100.00

*Capacities are based on net installed radiation and on the presence of a quantity of net installed radiation sufficient for the requirements of the building. Nothing need be added for normal piping.
All ratings are based on radiation at 240 Btu per sq. ft. for steam and 150 Btu per sq. ft. for water.

Above prices are net to the heating contractor, f.o.b. factory. Prices include boiler trimmings.

Sizes and Approximate Prices of Oil Burning Boiler Units

Price includes boiler complete with oil burner, trombone tank type water heater or tankless type water heater, steel jacket with baked enamel finish, air-cell type insulation furnished with jacket, plain room thermostat, limit control, and combustion control.

In addition, steam boiler prices include: low water cut-off, pop safety valve, try cocks, water gauge, pressure and vacuum gauge and flue brush.

Water boiler prices include: combination altitude gauge and thermometer and flue brush.

Installed Radiation Sq. Ft.*	Steam Boiler Jacketed price with Trombone Heater	Steam Boiler Jacketed price with Tankless Heater	Installed Radiation Sq. Ft.*	Net Btu Rating	Water Boiler Jacketed price with Trombone Heater	Water Boiler Jacketed price with Tankless Heater
460	$1300.00	$1380.00	840	126,000	$1300.00	$1380.00
570	1581.00	1650.00	1030	155,000	1581.00	1650.00
680	1700.00	1785.00	1220	183,000	1700.00	1785.00
790	1870.00	1955.00	1405	211,000	1870.00	1955.00
900	1955.00	2040.00	1590	239,000	1955.00	2040.00

*Capacities are based on net installed radiation and on the presence of a quantity of net installed radiation sufficient for the requirements of the building. Nothing need be added for normal piping.
All ratings are based on radiation at 240 Btu per sq. ft. for steam and 150 Btu per sq. ft. for water.

Above prices are net to the heating contractor, f.o.b. factory. Prices include boiler trimmings. If boilers are furnished without tankless heater, deduct $45.00; less trombone tank type heater, deduct $25.00.

Water Heater Capacities

Tankless type	4 gals. per minute
Trombone tank type	66 gals. in 3 hrs.

Above capacities are based on 40° to 140° rise with boiler water at 180°. Above prices are f.o.b. factory. Prices include boiler trimmings as listed above.

Approximate Prices of Gas Fired Boilers for Hot Water Heating Systems

Furnished with finned cast iron sections to provide staggered heat travel, draft diverter, heavily insulated steel jacket with baked enamel finish.

Boiler price includes gas valve and hydraulic limit control, relay gas valve and recycling manual control switch, transformer, electric safety pilot, gas pressure regulator, manual gas shut-off valve, combination thermometer and altitude gauge, pilot pressure regulator, drain cock, pilot cock, draft diverter, gas manifold, tubing and fittings. Room thermostat not included. Prices are f.o.b. factory.

Radiation Sq. Ft.	Net I-B-R Rating Output Btu per hr.	A.G.A. Gross Output Btu per hr.	Price Boiler Complete
515	77,500	103,000	$924.00
860	129,000	172,000	1220.00
1205	181,000	241,000	1595.00
1550	233,000	310,000	1820.00
1720	258,000	344,000	2030.00
1960	294,000	392,000	2240.00
2240	336,000	448,000	2660.00
2520	378,000	504,000	2940.00
2800	420,000	560,000	3150.00
3080	462,000	616,000	3640.00
3350	502,500	672,000	4200.00

Radiation Sq. Ft.	Net I-B-R Rating Output Btu per hr.	A.G.A. Gross Output Btu per hr.	Price Boiler Complete
3920	588,000	784,000	4760.00
4480	672,000	896,000	5040.00
5040	756,000	1,008,000	5670.00
5600	840,000	1,120,000	6020.00
6160	924,000	1,232,000	6440.00
6666	1,000,000	1,344,000	7000.00
7222	1,083,350	1,456,000	7560.00
7777	1,166,650	1,568,000	7840.00
8333	1,250,000	1,680,000	8120.00
8888	1,333,300	1,792,000	8680.00

Approximate Prices of Gas Fired Boilers for Steam and Vapor Heating Systems

This is furnished with finned cast iron sections to provide staggered heat travel, draft diverter, and heavily insulated steel jacket with baked enamel finish.

Boiler price includes gas valve, relay gas valve and recycling manual control switch, transformer, electric safety pilot, gas pressure regulator, pilot pressure regulator, manual gas shut-off valve, electric hot water cut-off, steam pressure regulator, retard steam gauge, pop safety valve, water gauge, try cocks, drain cock, pilot and burner cocks, draft diverter, gas manifold, tubing and fittings. Room thermostat not included. Prices are f.o.b. factory.

Radiation Sq. Ft.	Net I-B-R Rating Output Btu per hr.	A.G.A. Gross Output Btu per hr.	Price Boiler Complete
280	66,750	103,000	$918.00
475	113,500	172,000	1,320.00
670	161,000	241,000	1,955.00
875	210,000	310,000	2,465.00
980	235,000	344,000	2,975.00
1121	269,100	392,000	3,400.00
1293	310,250	448,000	3,570.00
1466	351,700	504,000	3,740.00
1641	393,800	560,000	4,080.00
1996	478,950	672,000	4,420.00
2353	565,250	784,000	5,100.00
2717	652,100	896,000	5,780.00
3081	739,550	1,008,000	6,120.00
3449	827,800	1,120,000	6,460.00
3814	915,300	1,232,000	7,140.00
4167	1,000,000	1,344,000	7,820.00
4514	1,083,350	1,456,000	8,330.00
4861	1,166,650	1,568,000	8,840.00
5208	1,250,000	1,680,000	9,180.00
5555	1,333,300	1,792,000	11,220.00
5903	1,416,650	1,904,000	13,600.00

Net boiler is the amount of actual installed radiation which may be attached to the boiler, based on a heat emission rate from the radiators of 240 Btu per sq. ft. for steam and 150 Btu for gravity hot water. For forced circulation for hot water systems, select the boiler from the net Btu column.

Standard Black and Galvanized Steel Pipe

Inside Diameter Inches	Black Pipe Price Per Lin. Ft.	Galv. Pipe Price Per Lin. Ft.	Threads Extra Each End
1/4	$0.98	$1.26	$0.98
3/8	1.12	1.40	.98
1/2	1.12	1.26	.98
3/4	1.19	1.40	.98
1	1.69	1.89	.98
1 1/4	1.61	2.45	1.12
1 1/2	2.38	2.80	1.00
2	3.29	3.85	1.68
2 1/2	4.55	5.25	2.80
3	5.95	7.00	3.00

Above prices are for pipe cut to order. Add for threads as given.

Prices of Steam Radiator Valves

	Sizes in Inches					
	1/2	3/4	1	1 1/4	1 1/2	2
Angle steam radiator valve	$29.00	$36.00	$50.00	$69.00	$88.00	$140.00
Gate steam radiator valve	15.40	20.00	24.00	30.00	39.00	56.00

Prices of Water Radiator Valves

	1/2	3/4	1	1 1/4	1 1/2
Hot water radiator valve	$8.00	$11.00	$15.00	$25.00	$35.00
Hot water radiator union ell	3.50	4.00	6.50	8.50	11.00

Prices on Floor and Ceiling Plates

Pipe size, inches	1/2	3/4	1	1 1/4	1 1/2	2	2 1/2	3
Chrome plated, ea.	$0.64	$0.64	$0.70	$0.76	$0.81	$0.89	$1.37	$1.71

Floor and ceiling plates made of cold rolled steel, halves securely riveted by a concealed hinge. Can be opened or closed on pipe without effort.

Copper Pipe

The cost of copper pipe on heating work depends to a large extent upon the experience of the steam fitter installing it. The technique of installing copper pipe is entirely different from that of steel pipe. Inexperienced steam fitters figure more for labor than on a steel pipe job, while others figure the same for labor, whether steel or copper. Steam fitters who are thoroughly experienced in installing copper pipe and fittings and who are equipped to handle it efficiently say that they can install copper pipe for 15% to 20% less than steel pipe.

The price of the pipe itself is subject to wide fluctuations, due to the price of copper so that prices on copper pipe and fittings should *always* be obtained separately for each job. Copper pipe is furnished in 3 weights:

a) Heavy Copper Pipe (Government Type "K") for underground service, such as water, gas, steam, oil lines, industrial uses, also interior plumbing.

b) Standard Copper Pipe (Government Type "L") for interior plumbing and heating.

c) Light Copper Pipe (Government Type "M") for low pressure interior plumbing.

Sizes and Weights of Copper Pipe

Nominal Size Inches	Type "K" Heavy Wall Thickness Inches	Type "K" Heavy Weight per Lin. Ft. Lbs.	Type "L" Standard Wall Thickness Inches	Type "L" Standard Weight per Lin. Ft. Lbs.	Type "M" Light Wall Thickness Inches	Type "M" Light Weight per Lin. Ft. Lbs.
3/8	.049	.269	.035	.198	.025	.145
1/2	.049	.344	.040	.285	.028	.204
5/8	.049	.418	.042	.362	....	....
3/4	.065	.641	.045	.455	.032	.328
1	.065	.839	.050	.655	.035	.465
1 1/4	.065	1.04	.055	.884	.042	.681
1 1/2	.072	1.36	.060	1.14	.049	.940
2	.083	2.06	.070	1.75	.058	1.46
2 1/2	.095	2.93	.080	2.48	.065	2.03
3	.109	4.00	.090	3.33	.072	2.68
3 1/2	.120	5.12	.100	4.29	.083	3.58
4	.134	6.51	.110	5.38	.095	4.66
5	.160	9.67	.125	7.61	.109	6.66
6	.192	13.87	.140	10.20	.122	8.91
8	.271	25.90	.200	19.30	.170	16.46
10	.338	40.30	.250	30.10	.212	25.60
12	.405	57.80	.280	40.40	.254	36.70

Because of the smooth inside surface of copper pipe, usually one size smaller pipe can be used than where steel pipe is used. Outside diameter of pipe is 1/8" more than nominal size of tubing.

Approximate Prices on Copper Tubing

Prices on copper tube and fittings are subject to wide price fluctuations. Always check prices before submitting bids.

Copper Tubing

Nominal Size

Price per lin. ft.	1/4"	3/8"	1/2"	5/8"	3/4"	1"	1 1/2"	1 1/2"
Type "K"	$0.56	$0.91	$1.19	$1.40	$2.09	$2.52	$3.15	$4.06
Type "L"	.52	.70	1.05	1.26	1.54	2.10	2.03	3.50

Price per lin. ft.	2"	2 1/2	3"	3 1/2	4"	5"	6"
Type "K"	$6.09	$8.68	$11.90	$16.80	$21.00	$49.00	$70.00
Type "L"	5.25	7.56	10.15	13.72	16.80	39.20	53.20

Prices of Brass Solder-Joint Valves

These valves are made for use with copper tubing and can be used for steam or water service.

	Size in Inches					
Fitting	½"	¾"	1"	1¼	1½"	2"
Gate valve	$15.61	$19.95	$24.15	$32.20	$41.65	$58.50

Approximate Quantity of Solder and Flux Required to Make 100 Joints

Size, Inches	⅜"	½"	¾"	1"	1¼"	1½	2"	2½	3"	3½	4"	5"	6"	8"	10"
Solder, Lbs	½	¾	1	1½	1¾	2	2½	3½	4½	5	6½	9	17	35	45
Flux, Oz.	1	1½	2	3	3½	4	4	7	9	10	13	18	34	70	90

Estimating the Cost of Pipe and Fittings. To estimate the cost of pipe and fittings required for any heating plant, obtain the total number of square feet of radiation in the job (based on cast iron radiation) and figure at $2.00 to $2.15 per sq. ft. This is based on prices of pipe as given on the previous pages.

For one story residences having short pipe runs, figure the cost of the pipe and fittings at $2.00 per sq. ft. of radiation (based on standard cast iron radiation).

An example of the method used in estimating the cost of pipe and fittings is as follows: For a job containing 1,400 square feet of standard radiation, take 120% of 1,400 which equals 1,680, or the pipe and fittings would cost approximately $3,400.00.

The above allowances for pipe and fittings do not include radiator valves, floor and ceiling plates, boiler fittings, expansion tanks, etc., but only the pipe and fittings required for roughing-in the job.

Pipe Covering

Pipe covering is furnished in several kinds of materials. Fiberglass sectional pipe insulation covering is the most common and cheapest for covering steam and hot water pipes; 85 percent magnesia pipe covering is for high pressure steam work; and wool felt pipe covering is for protecting pipes against freezing and pipe covering in general.

Pipe covering is usually furnished in 3-ft. lengths, ¾" to 1" thick and is covered with a canvas or fire safety jacket.

Labor Covering Pipes Up to 3" in Diameter. When placing covering on pipes up to 3" in diameter, a pipe coverer should apply 10 to 15 lin. ft. of covering per hour, or 80 to 120 lin. ft. per 8-hr. day, and the labor cost per 100 lin. ft. should average as follows:

	Hours	Rate	Total	Rate	Total
Pipe coverer	8	$....	$....	$26.62	$212.96
Cost per lin. ft			....		2.13

Labor Covering Pipes 4" to 6" in Diameter. When placing covering on pipes 4" to 6" in diameter, a pipe coverer should apply 9 to 11 lin. ft. of covering per hour or 72 to 88 lin. ft. per 8-hr. day, and the labor cost per 100 lin. ft. should average as follows:

	Hours	Rate	Total	Rate	Total
Pipe coverer	10	$....	$....	$26.62	$266.20
Cost per lin. ft			....		2.66

When estimating quantities of pipe covering, add 1 lin. ft. of covering for each fitting for pipe up to 2" and 2 lin. ft. of covering for each fitting on pipe over 2" thick.

Temperature Regulators for Heating Plants

Temperature regulators and thermostats are furnished in a variety of different models for controlling the temperature.

There are so many different methods of controlling the inside temperature with straight thermostats, day and night thermostats, inside-outside controls, etc., that it is practically impossible to give even a fraction of them in a book of this kind.

Temperature Regulators With Clock. The thermostat with a clock attachment automatically lowers the temperature at night and raises it in the morning. Electric clocks are now usually furnished with all thermostats of this type.

Models of this kind are furnished with thermometer, clock, electric motor, chain, pulleys, transformer and fittings, at the approximate price of $150.00.

Temperature Regulators Without Clock. Many thermostats or regulators are used without the clock attachment and are either manually operated or the same temperature maintained day and night.

There are a number of different types of regulators on the market and they vary in price from $50.00 to $150.00, depending on the method of wiring and number of controls necessary.

Labor Installing Thermostats. An experienced mechanic should install any of the above thermostat outfits in 1 to 2 hours. If they are installed by the manufacturers in cities where they maintain offices, a charge of $150.00 to $200.00 is usually made.

Making Up a Complete Estimate on a Steam Heating Plant. When making up an estimate on a steam heating system, the following items should always be included in order that the estimate may be complete in every detail.

1. Boiler
2. Oil or gas burner
3. Radiators or convectors
4. Radiator Valves
5. Automatic air valves
6. Pipe and fittings
7. Smoke pipe
8. Special foundation, if required
9. Asbestos boiler covering
10. Pipe covering
11. Floor and ceiling plates
12. Painting radiators
13. Thermostat regulator
14. Freight and trucking
15. Water Connection
16. Labor installing system
17. Bond and insurance
18. Overhead and misc. expense
19. Profit

Making Up a Complete Estimate on a Hot Water Heating Plant. When making up an estimate on a hot water heating plant, the following items should always be included in order that the estimate may be complete in every detail.

1. Special boiler foundation, if required
2. Boiler
3. Oil or gas burner
4. Pipe and fittings
5. Radiators or convectors
6. Radiator valves-union ells
7. Circulating pumps
8. Compression air valves
9. Expansion tank and gauge
10. Altitude gauge
11. Thermometer
12. Smoke pipe
13. Painting radiators
14. Boiler covering
15. Pipe covering
16. Floor and ceiling plates
17. Thermostat regulator
18. Labor installing system
19. Freight and trucking
20. Water connection
21. Bond and insurance
22. Overhead and miscellaneous expense
23. Profit

Unit Heaters. Unit heaters are often used to heat commercial and industrial areas, garages, and to supplement other heating units at entrances. They may be direct-fired by gas, oil, or electricity, or have hot water or steam piped to them. Direct-fired are cheaper when only a few units are required and no central plant is available. Hot water and steam are the most common units in use where a central plant is available or its cost justified by the number of units it serves.

Units may have propeller or centrifugal fans blowing either horizontally, up or down. Air may be drawn through or blown through, the latter being used for direct-fired units.

The following costs do not include valves, fittings or piping which will vary widely with job conditions and are based on units served with 2 lbs. steam, 60° entering air, single phase 115 volts.

Cost of Unit Heaters

Type	H.P.	Btu	1 Speed	2 Speed	Hrs. to Install
Horizontal	1/40	25,000	$155.00	$--	1
Delivery	1/15	50,000	200.00	245.00	1½
	1/10	100,000	300.00	340.00	2½
	1/5	200,000	480.00	540.00	4
Vertical	1/40	50,000	190.00	--	1½
Delivery	1/12	100,000	260.00	300.00	2
	1/3	200,000	520.00	575.00	5
Cabinet	1/20	25,000	160.00	--	1½
	1/12	60,000	230.00	--	2
	1/8	120,000	475.00	--	3

For 230 volts, add $16.00. For ceiling suspension, add $25.00.

15C SPRINKLER SYSTEMS

The construction industry and the general public for many years have been aware of the need for and availability of fire protection systems. However, for a long time this protection was restricted to areas where local building codes or extreme business risks dictated a demand for their use. In general, there was a great insensitivity toward providing fire protection for the public, even in areas of high concentration such as restaurants, theaters, hotels, sporting events, etc.

This disregard for public safety changed abruptly because of an increase in the number of fire disasters that had received national publicity. They became the catalyst to inspire total remedial action by federal and state authorities, as well as by insurance companies, building owners, and professional architects.

The fire protection industry responded promptly to provide the specialized technical design data and to produce trained personnel with the capability of educating the designers and specification writers in selecting the particular system, or combination of systems, most appropriate to provide the degree of protection desired.

As the need has developed to provide special protection for costly equipment such as computer rooms and tape storage libraries, for irreplaceable documents, and art objects, for sophisticated research projects, and for the aged and the handicapped, new alarm, control and protection systems have been developed. Some of the systems are mechanically operated, while others are completely electronically controlled. In some cases, alarm and control systems are designed as a combination of mechanical and electrical devices. In these instances, the estimator must determine with great precision the cost and installation responsibility that each trade will assume in the overall function of the system.

Wet Pipe Sprinkler System

This type of system meets most fire protection requirements for normal hazards, where piping is not subject to freezing. Water is constantly supplied under pressure to each sprinkler head.

Dry Pipe Sprinkler System

This system can be used in areas subject to freezing, because the piping is filled with air under pressure, which contains the water at the supply source until an activated sprinkler head allows the air pressure to disburse and let the water supply move to the head.

Sprinkler systems are a specialized type of work, requiring the services of a subcontractor who has access to the necessary equipment, which in many instances is a sole source item and not available on the open market. One usually finds that the contract documents require the services of a specialist who has a certain number of years' experience in this type of work and who has the capabilities to design the sprinkler system.

For the wet and dry pipe systems, unit costs have been developed on a per head basis. It is assumed that an adequate water supply is available and that special alarm and control work is not included. Different heads are used depending on the degree of hazard, and local fire regulations are consulted for a specific project.

For high-rise buildings, booster pumps may be required, which will add from $20,000 to $50,000 to the cost of installation.

Maximum area coverage per head is 130 sq. ft. The average area covered per head is from 80 sq. ft. to 130 sq. ft. under normal hazard. Distance between sprinkler heads cannot exceed 15 ft.

Wet System*	Total Cost per Head
Warehouse	90.00
Hospital	150.00
Shopping Center	100.00
Municipal Buildings	125.00

*For dry system, add 15.00 20.00 per head.

From a look at the above costs, it might seem that a fire sprinkler system is costly to install. But an owner must look at all the benefits. The owner should consider it as an insulation against costly litigation resulting from fire loss. In a more identifiable form, it produces an immediate and continuing savings in fire insurance premiums. Most sprinkler installations return their initial cost in reduced insurance premiums within 3 to 10 years.

CHAPTER 16

ELECTRICAL

CSI DIVISION 16

Electricity plays an increasingly more important role in construction each day. Buildings towering 70 stories or more might depend completely on electricity, not only for light, power, and building services, but for electrified tenant equipment and for their heat. Heating and lighting depend on each other so that a maximum use of electric energy may be obtained at a minimum of expense.

Obviously such sophisticated designs are well beyond the scope of this book. But even a modest residence will involve electrical planning and power demands well beyond anything thought of 25 years ago.

An electrical estimate, whether commercial or residential, can be divided into four general categories: service, distribution, branch, and devices/loads. The service comprises the outdoor utility-supplied feeders and connection to the building, as well as the revenue metering and main disconnect switches. Distribution consists of the panelboards and switchgear that immediately follow the main switch and the feeders that serve them and emanate from them. The branch work, also referred to as lighting circuitry, is the raceway and wire between the panelboard breakers and the devices/loads. Finally there are the devices/loads, such as receptacles and switches, lighting outlets, fixtures, and heaters.

The electrical service is installed in two phases: the contractor's portion, consisting of the setting and wiring of the main disconnect switches and the revenue meter pans, followed by the utility companies portion, consisting of the installation of the feeder, either on poles or underground, its connection to the meter pan, and installation of the meter itself.

For residence construction, the typical service size is 200 amperes, 208 volts, single phase; that is, two phase wires and a neutral, making both 120 and 208 volts available. The main disconnect switch is usually incorporated in the main circuit breaker. The typical cost of such a panelboard, 30 circuits and the usual complement of circuit breakers, is $400.00 plus about 24 hours of labor. The matching meter pan would add $50.00 and an additional 5 hours for installation. Where a 100 ampere service is provided, the panelboard costs would be $200.00 plus about 15 hours of labor for installation, with the meter pan not substantially affected.

In commercial work, a separate disconnect switch connected to a distribution panelboard is common. A 200 ampere fused disconnect switch cost is $400.00 plus about 10 hours labor for installation. A matching distribution panel with typical breakers, 42 circuits, no main breaker, costs $800.00 plus about 36 hours of labor for installation. The wiring between the devices (3/0,2" emt) costs $6.00 plus about 0.30 hours of labor per lin. ft. for installation. Where necessary, a 400 ampere switch at $650.00 plus about 16 hours of labor for installation and a 400 ampere, 42 circuits no main breaker, panelboard at $500.00 plus about 40 hours of labor for installation may be used. In this case, the wiring between the service switch and the panel would be 500 mom, 3 ½" emt, $17.30 plus about 0.43 hours of labor per foot of installation.

With the service and distribution completed, the branch work can be analyzed. In residential work, most of the branch work will run as non-metallic sheath cable (romex) or metal sheath cable (bx). Costs are: Romex 2#12 plus ground wire, $0.30 plus about 0.035 hours of labor per lin. ft. of installation; 2#10 plus ground wire, $0.40 plus about 0.038 hours of labor per lin. ft. of installation; BX 3#12, $0.30 plus about 0.05 hours of labor per lin. ft. of installation; BX 3#10, $0.65 plus about 0.06 hours of labor per lin. ft. of installation. In commercial work, emt and building wire is often used. The equivalent costs are: ¾" emt 3#12, $0.45 plus about 0.138 labor hours per lin. ft. of installation; ¾" emt 3#10, $0.61 plus about 0.146 labor hours per lin. ft. of installation.

For very preliminary estimates, electrical work can be assumed to run from 6% to 12% of the total job cost for residences, 8% to 11% for office buildings, and 8% to 12% for industrial buildings.

An estimator should make a detailed electrical take off and must have a thorough knowledge of the working drawings and specifications from which the job will be built, as well as of the building code under which it will be inspected and regulations set by the local utility company.

A rotometer which will scale distances directly off the drawings is handy if drawings include sufficient interior room elevations so that not only horizontal but vertical distances can be measured.

The electrical take off sheet for residential work should have a column for each of the following:

a) Ceiling outlets
b) Bracket outlets
c) Duplex convenience outlets
d) Triplex convenience outlets
e) Waterproof convenience outlets
f) Range outlets
g) Lighting outlet with lampholder
h) Special purpose outlets (dishwashers, clothes dryers, etc.)
i) Single pole switches
j) Three-way switches
k) Length of wiring

The estimator then should check off each item and enter the total on the sheet. This is best done floor by floor, or in large one-story residences, wing by wing, so that checking and changes ordered on the job will be simplified.

Enter next the circuit and feeder control centers, listing them by type and size, and then the branch circuits. As mentioned above, a rotometer will be helpful in measuring raceway runs, but be sure all drawings used are at the same scale. List separately two-wire and three-wire runs and empty conduit for telephones and such other special items as intercoms, buzzer systems, television outlets, and the like.

A final list of such items as fixture hangers, studs, locknuts, bushings, cable clamps, and outlet covers complete the take off, and then, all these items can be transferred to the pricing sheet item by item.

A pricing sheet should include a line for each type of item; the quantity involved; the material cost by unit and in total; the hours estimated for installation per unit and hourly rate; the total labor cost; and finally, the total labor and material cost.

Service Entrance Equipment

The service entrance equipment consists of the service drop, service entrance conductors, the meter, main control center, and the service ground. The local utility will govern the details of this installation and may or may not furnish the meter. Underground service is highly desirable and may be required by local ordinances. Underground service for residential use will run about $8.00 per lin. ft. Overhead lines will run around $600.00 per pole. The main control center for residential work will consist of a circuit breaker-fuse panel of at least 100 amp. capacity.

Circuit Breakers

Capacity	Cost	Hours to set
100 amp. 3 pole, gen. purpose	$285.00	4
200 amp. 3 pole, gen. purpose	625.00	7
400 amp. 3 pole, gen. purpose	1000.00	11
600 amp. 3 pole, gen. purpose	1700.00	14

16110 RACEWAYS

Rigid galvanized steel conduit is the safest system and in some large cities, required by code. Note that in some ares, when reference is made to rigid pipe, it is understood to mean rigid galvanized pipe. Conduit systems may be galvanized steel rigid, aluminum, or electric metallic tubing, generally referred to as thinwall or EMT.

Material and Labor to Set 100 Lin. Ft. of Conduit (including supports and fittings, drag wire, raceway, terminations, elbows, and junction boxes)

	Galvanized		Aluminum		EMT	
Size	Cost	Hours to Set	Cost	Hours to Set	Cost	Hours to Set
½"	$60.00	15	$75.00	11	$25.00	6
¾	70.00	18	100.00	13	28.00	11
1"	101.00	19	150.00	14	35.00	12
1½"	160.00	26	240.00	19	58.00	16
2"	208.00	30	350.00	22	73.00	17
3"	435.00	38	700.00	28	208.00	19
4"	632.00	45	1100.00	33	338.00	23

Flexible conduit, or Greenfield, is easier to install.

Size	Conduit Cost	Connector Cost	Hours to Set
½"	$30.00	$1.35	5
1"	75.00	3.50	8
1½"	120.00	12.30	14
2"	160.00	10.40	18
3"	250.00	32.00	30

Where codes allow, the least expensive method of wiring is pre-wired armored cable (BX) or non-metallic sheathed cable such as Romex.

Material Cost to Install 100 Lin. Ft. of Armored Cable

Gauge of Wire	No. of Wires	Cost	Hours to Set
#10	2	$63.00	5
#10	3	82.00	6
#12	2	36.00	4
#12	3	53.00	5
#14	2	33.00	4
#14	3	42.00	5

Material and Labor to Install 100 Lin. Ft. of Non-Metallic Sheathed Cable with Ground

Gauge of Wire	No. of Wires	Cost	Hours to Set
#12	2	$26.00	3.5
#12	3	43.00	4.0
#14	2	16.00	3.2
#14	3	33.00	3.5

Duct System

Underfloor metal duct systems vary widely in their complexity. Some floor systems have raceways built in, others are added in the floor fill and may be a one or two level system. Standard duct is 1⅜"x3⅛". Super duct is 1⅜"x7¼". Lengths are 5', 6', 10', and 12'. Detailed take offs on these systems are beyond the scope of this book. The Square D Company catalog gives a suggested take off method with all the required fittings. Unit do not include receptacle, telephone, or other devices outlets or fittings.

	Material Cost	Standard Hours Per 100 LF	Material Cost	Super Hours per 100 LF
Plain Duct Per LF	$3.50	10	$6.00	14.0
Duct with Inserts per LF	3.75	11.5	6.25	16.0
Single Box	85.00	2 ea.	110.00	3.0 ea.
Double Box	100.00	2.5 ea.	200.00	3.5 ea.
Support	5.00	.25 ea.	6.00	0.5 ea.
Elbow	23.00	.75 ea.	44.00	1.0 ea.
Connector	4.00	.30 ea.	8.00	0.3 ea.

Material Cost and Labor to Set 100' of Fiber Duct

Size	Duct Cost	Labor to Set	Coupling Cost	Hours to Set
2"	$75.00	4	$12.00	.1
3"	110.00	5	13.00	.1
4"	115.00	7	15.00	.2
6"	330.00	10	20.00	.3

Material and Labor to Pull 1000 Lin. Ft. of Wire in Conduit Already in Place (600 V Type THHN/THWN)

Gauge	Material Cost	Hours to Pull
14	$50.00	6.00
12	60.00	7.25
10	100.00	9.00
8	180.00	12.00
6	220.00	13.00
4	330.00	16.00
2	560.00	18.00
1/0	880.00	26.00
2/0	1,020.00	29.00
4/0	1,580.00	38.00

Note: THW has been replaced by THHN/THWN

16130 METALLIC OUTLET BOXES

Type	Cost per Unit	Hrs to Install
Switch Box	$1.50	.30
3 Gang Box	10.00	.60
4" Octagon	1.20	.50
4" Square	1.50	.50
4 Concrete	2.20	.40
1 Gang Cast WP	9.00	.66
3 Gang Cast	20.00	1.00

16140 SWITCHES AND RECEPTACLES

Once the box has been set and connected to the conduit and the wiring pulled, the proper switch or receptacle can be installed which will run as follows:

Type	Cost per Unit	Hours to Install
15 amp single pole toggle	$3.00	.2
20 " " "	4.00	.3
15 amp 3-way toggle	5.00	.35
20 " "	6.00	.5
15 amp 4-way toggle	14.00	.6
20 " "	16.00	.75
600 watt single pole dimmer	7.00	.5
" " three " "	12.00	.66
1000 watt single pole dimmer	50.00	.5
" " three " "	75.00	.66
15 amp duplex receptacle	2.00	.2
20 " " "	4.00	.3
30 amp dryer	8.00	.5
50 amp range	11.00	.75

Cost of Toggle Switch and Duplex Receptacle Plates per 100

Description	Cost in Phenolic	Cost in Stainless
1 Gang Switch	$60.00	$142.00
2 Gang Switch	120.00	375.00
3 Gang Switch	230.00	600.00
4 Gang Switch	500.00	950.00
1 Gang Outlet	50.00	142.00
2 Gang Outlet	120.00	375.00
3 Gang Outlet	330.00	600.00
Blank-Single	60.00	142.00
Telephone	60.00	142.00

Weatherproof switches with grey Hypalon presswitch plates will run $600.00 a hundred for the plate not including the switch or box.

16500 LIGHTING

The following is a sampling of some fixtures an electrician might encounter on average work.

Type	Wattage	Size	Shielding	Cost
Industrial Fluorescent	2-40	4'	Open	$35.00
Pendant Mtg.	4-40	8'	Open	65.00
	2-40	4'	Louvers	40.00
	4-40	8'	Louvers	80.00
Commercial Fluorescent	2-40	4'	35°x35°	50.00
Surface Mounted	4-40	4'	35°x35°	80.00
Average Quality	4-40	8'	35°x35°	150.00
Commercial Fluorescent	8-40	8'	35°x35°	100.00
Surface Mounted	2-40	4'	Acrylic	65.00
Premium Quality	4-40	8'	Acrylic	100.00
Commercial Fluorescent	2-40	4'	Holophane	80.00
Recessed Troffers	4-40	4'	Holophane	190.00
Commercial Fluorescent	4-20	2x2	Acrylic	45.00
Square Surface	6-30	3x3	Acrylic	75.00
	6-40	4x4	Acrylic	175.00
Commercial Fluorescent				
Corridor Unit	1-40	4'	Acrylic	130.00
Incandescent Industrial	1-100	12" diam.	Open	50.00
Dome Bowl	1-150	14" diam.	Open	60.00
	1-200	16" diam.	Open	65.00
	300-500	18" diam.	Open	75.00
	750-1000	20" diam.	Open	125.00
High Bay Mercury	400	24" diam.	Open	170.00
	1000	24" diam.	Open	225.00
Commercial Incandescent				
Recessed Fixed Spot	150	10" diam.	Open	55.00
Adj. Spot	150	10" diam.	Open	65.00
Punch Spot	150	8" diam.	Open	65.00
Recessed Square	100	8x8	Opal	80.00
Alum. Wall Bracket	75	5" diam.	Glass	100.00
Pendant Metal Shade	100	11" diam.	Glass	50.00
Pendant Sphere	40	16" diam.	Glass	75.00
Drum	100	18" diam.	Opal	50.00
For Stems add				
6" Drop				$5.00
18" Drop				10.00
30" Drop				15.00
For plaster rings for spots add				$7.00
Track Lighting				
1 Circuit	...	8'	....	$55.00
3 Circuit	...	12'	....	120.00
150 W Spot	...	...	....	35.00 ea.

16720 ALARM AND DETECTION EQUIPMENT

The largest single threat to property and life associated with buildings is fire. While the primary concern is always potential loss of life, damage to or loss of buildings, equipment, materials, files, and records due to fire can put a company out of business. In the case of a museum or other places where historical data is stored, part of history could be lost forever. Consequently, life safety and fire codes are being enacted across the country for all types of buildings, from institutional to office buildings, from hotels to single family residences.

The threat of property damage and loss of life due to fire can be substantially reduced through the proper use of smoke, fire, and gas detectors. These detectors are either wired into the building's electrical system, battery operated, or both. The detector that is wired to the electrical system should be designed to receive backup power from a battery source in the event of a power failure.

Battery operated detectors normally have a small light that flashes when the battery is operative. Some battery systems have a flashing light, short beeps, or both, periodically indicating the battery is not functioning properly. Battery powered systems do not connect to the main electrical system, which makes this device ideal for use in existing buildings (primarily residences) where exposed wiring is objectionable and concealed wiring means costly "cut and patch" work. The one drawback to the battery operated detector is that while it will warn building occupants, it cannot be connected to an exterior alarm system.

Fire detectors and smoke detectors are different. The former can be adjusted to alarm at a particular temperature and the latter is set to alarm at a certain percentage of smoke in the air surrounding the detector. Because most fire fatalities are due to smoke inhalation, in most cases the smoke detector gives the best protection and the quickest alarm.

Placement of detectors will vary depending upon the type of building and the interior partitioning. In institutional and commercial buildings, a detector should be located in every room and storage closet, one for every 900 sq. ft. in large open areas, and at approximately 60 lin.ft. on center in corridors. In residential homes a detector should be placed near the kitchen, in hallways leading to bedrooms, and at the top of all stairways at each floor level.

The cost of fire and or smoke detectors varies widely and depends on the amount of sensitivity and sophistication of the alarm system. Residential detectors, either battery or hardwired type, range in price from $10.00 to $55.00 each. Detectors for institutional and commercial buildings will range from $65.00 to $130.00 each.

In addition to the cost of the detectors for non-residential use, there are other associated items of equipment that can be added to the alarm system, such as alarm station pull boxes, test panels, and control console panels, costing several thousand dollars each. Hardwired systems, both commercial and residential, can be provided with an automatic dialers. When the alarm is activated, the system will automatically notify the proper emergency agencies. The automatic dialer installation costs range from $75.00 to $500.00. The wide range of costs is due to the many expenses associated with an automatic dialer. Some emergency units prefer an alarm system wired into a private central station to weed out false alarms. The central control station costs are private and are normally billed directly to the user on an annual basis after installation is completed.

16850 ELECTRICAL HEATING

Heating electrically includes a broad range of methods--radiant units, convector units, baseboard units, electric furnaces with forced air, and heat pumps with forced air. Some of these methods are often included as simple supplemental heaters; others are full plants and involve quite sophisticated design in order to function most efficiently.

Wall heaters are the simplest, most commonly encountered means of heating electrically. Both natural and fan driven units are available and they are usually recessed and have built-in thermostats. A 1500 watt ceiling fan unit will cost around $90.00; a 1500 watt ceiling radiant unit around $150.00; a 1000 watt ceiling radiant unit, $130.00. A 1200 watt fan forced unit will run $100.00 with manual control, $125 with thermostat.

Baseboard heaters are available in standard and high wattage models in surface mounted and semi-recessed designs. The wiring serving these units may be part of the common wiring system or fed from separate branch control centers.

In some areas electricity for heating is separately metered and must be on a separate system. For preliminary figures, baseboard units will run about $30 a lin. ft. installed, not including separate wiring system. Add $32.00 for each thermostat. Baseboards, if used as the sole heating source, are often supplemented by fan units at entrances.

Electric furnaces are similar to conventional warm air furnaces but substitute electric resistance-heating units for flame. They are very compact, require no flue or vent, and can be placed anywhere so that zone heating is simplified. Also, air cooling systems can be added to provide a year-round system. A variation on this are through-the-wall units including both heating and cooling capacities. These are especially suitable for motels, dormitories, and other similar buildings where individual control is important. Electric furnaces will cost about $375 for a 35 MBH/hr unit up to $850.00 for a 140 MBH model. Assuming wiring is already in the furnace room, two workers can install the unit in half a day.

Heat pumps are more efficient than resistance type heaters, because they are designed to utilize existing outside conditions to supplement inside heating and cooling. In the winter, the heat pump extracts heat from outside air, ground, or water and conducts it to the inside. In summer, it removes heat from base to the outside. Heat pumps are generally placed outside to eliminate noise. Units are usually designed to handle air cooling loads. In areas of extreme cold where heating capacity of the pump is inadequate, it can be supplemented by electric resistance heaters. The unit is connected to a duct system within the house.

CHAPTER 17

REMODELING WORK

Remodeling, rehabilitation, and renovation work poses many problems not found in new work, but if estimated and executed properly, it can be quite profitable and lead to new clients and future work. It can also provide the carry-over jobs needed at times to keep crews busy during bad weather or slack periods. Many new structures abut existing structures, requiring a connection between the two and making renovation work necessary.

The following estimating suggestions are items encountered in remodeling work. The estimator must investigate and make allowances for all job conditions:

a) Need for protecting adjacent work.
b) Removal of rubbish and salvage.
c) Restrictions during normal working hours, or limited work hours.
d) Extent of daily cleaning up involved. (Will owner occupy work area at the end of the day?)
e) Is there enough work to keep the various trades busy for a full day's work, or will there be callbacks for only fractional periods?
f) Will owner be constantly supervising work?
g) Are plans of original structure, as well as previous renovations, that show mechanical and structural details available?
h) Is work governed by a strict building code that might require work beyond what is immediately ordered, such as strengthening structural members, adding additional exits to a remodeled third floor, or adding a fire suppression system?
i) Are workers who can form a cooperative team available? There are almost always changes made to fit actual job conditions as they are uncovered?
j) Are there materials present, such as lead paint, that are classified as industrial waste?
k) Are there materials present, such as asbestos floor tiles, exterior siding, or boiler covering, that are classified as hazardous waste?

A contractor is often asked whether he thinks it best for the owner to remodel, move, or build anew. On residential construction, a common rule is that if the cost of remodeling plus the original building cost is equal to or less than the cost of a comparable new house, it is best to remodel. Not only are moving costs eliminated, but so are the percentages charged to sell the old house and to buy the new one. Also, investments in landscaping, drapes, carpets, and the like are salvaged.

On commercial work there are other considerations that might be favorable to remodeling rather than rebuilding. First, remodeling may eliminate zoning and building code restrictions that apply to new construction. Second, the expense and time of wrecking is eliminated. Third, if the basic structure is sound, this may be considered as equity for a construction loan. If this is so, financing may be simplified by eliminating preliminary loans before final project is under way.

The savings in time will mean less chance of cost escalation and earlier occupancy. These savings can be reflected in the rental scale, making a remodeled structure more competitive with new. And finally, one might find that preservation of older structures, especially those that are registered with national or local historical societies, is a definite plus for one's public image.

A visit to the job site is mandatory for remodeling estimating. Existing conditions must be compared with drawings and any differences brought to the owner's attention prior to bid. A site visit is also important to determine the extent of alterations in existing mechanical and electrical work, which is costly, and to match existing materials, patterns, and colors.

Masonry Repairs

Removal of Brick by Hand Chipping. A mason should hand chip around 200 pieces per 8-hr day at the following labor cost per 100 pieces:

	Hours	Rate	Total	Rate	Total
Mason	4.5	$....	$....	$25.62	$115.29
Helper	1.0	....	....	19.13	19.13
Cost per 100 pieces			$....		$134.42
Cost per brick					1.34

Removal of Concrete Block. Block walls can usually be knocked down manually by laborers. For 12" or 8" block walls, figure per 100 sq. ft. as follows:

	Hours	Rate	Total	Rate	Total
Labor	8	$....	$....	$19.13	$153.04
Cost per sq. ft					1.53

To remove 4" partition block, figure per 100 sq. ft. as follows:

	Hours	Rate	Total	Rate	Total
Labor	5	$....	$....	$19.13	$95.65
Cost per sq. ft					.96

Where bearing walls are cut through, it will be necessary to provide needling and shoring, and to set in a new lintel. A 3' wide opening in an 8" wall will cost about $100.00 to shore and another $45.00 to build in lintels. A 6' wide opening will cost about $110.00 to shore and $55.00 to set a beam for a lintel.

To Patch Face Brick. To patch 4" face brick, figure one mason and one helper can lay about 65 sq. ft. per day at the following cost per 100 sq. ft.:

	Labor	Rate	Total	Rate	Total
Mason	12	$....	$....	$25.62	$307.44
Helper	12	....	....	19.13	229.56
			$....		$537.00
Cost per sq. ft					5.37

Material cost per sq. ft. will add another $1.45 for brick in the $130.00 per 1000 range.

To Patch Building Brick. Where openings are closed up or otherwise patched out in small areas, one mason and one helper should lay about 30 cu. ft. of building brick per day.

Labor Cost to Patch 100 Sq. Ft. of 12" Building Brick Wall

	Hours	Rate	Total	Rate	Total
Mason	27	$....	$....	$25.62	$691.74
Helper	27	....	....	19.13	516.51
			$....		$1,208.25
Cost per sq. ft					12.08

Material cost per cu. ft. will add another $1.50.

To Patch Concrete Block Partitions. One mason and a helper should patch out 150 sq. ft. of 4" block partitioning per day at the following cost per 100 sq. ft.:

	Hours	Rate	Total	Rate	Total
Mason	5.5	$....	$....	$25.62	$140.91
Helper	5.5	....	....	19.13	105.22
Cost per 100 sq. ft.			$....		$246.13
Cost per sq. ft					2.46

Material costs will add about $0.60 per sq. ft.

Removing and Replacing (Repointing) Brick Mortar Joint. The mortar joint is usually removed to a depth of ½". One mason and a helper should remove and replace about 500 lin. ft. of joint per day at the following cost per 100 lin. ft.:

	Hours	Rate	Total	Rate	Total
Mason	1.6	$....	$....	$25.62	$40.99
Helper	1.6	....	....	19.13	30.61
Cost per 100 lin. ft			$....		$71.60
Cost per lin. ft.			....		.72

To Rebuild Chimneys. On small chimneys one mason with a helper will lay about 500 bricks per day.

To Add Ceramic Brick to Existing Walls. One-inch thick ceramic units formed to the size of standard and Norman bricks may be applied over any nailable surface such as plywood, wood sheathing, furring strips, and various types of concrete and masonry. One sq. ft. of wall will require 6 of the standard units which cost $95.00 per 1000 in red, $105.00 in golden blend and $118.50 in gray. On small jobs one mason (check local jurisdiction on this, as it varies) working alone should install around 700 standard units per 8 hour day. Mortar requirements are 1½ cu. ft. per 100 sq. ft. of wall, and a tuckpointer working alone should point about 450 standard units a day. Special units are available for corners.

Siding

To determine siding costs, it is best to sketch an outline of each elevation. For straight walls, multiply the width by the height and deduct from the total the combined square foot area of the doors and windows. Gable area are found by multiplying the width by the height and dividing by two. Deduct window openings. Wall areas for dormers can be figured from the following table, the window areas being already deducted.

Side Wall Area in Sq. Ft. for Dormers

	Type of Dormer		
Dormer Width	Lift Type	Gable Type	Hip Type
5'-0" (One Window)	88	60	52
6'-0"	94	70	58
7'-0"	100	80	64
8'-0" (Two Window)	88	70	52
9'-0"	94	75	58
10'-0"	100	85	64
11'-0"	106	90	70
12'-0" (Three Window)	93	80	57

Add the net square feet of each elevation, gable and dormer together, add 8% for waste and figure to the nearest larger half or full square or nearest bundle of shingles for material costs.

Old siding may be stripped off before new is applied, or may be covered over either by applying an underlayment board or by applying wood stripping and felt over present siding. Underlayment board would figure the same quantity as siding. Wood stripping can be figured per square, depending on the exposure of existing siding, as follows:

Exposure, inches	4"	4½"	5"	5½"	6"	6½"	8"
LF of Stripping	300	280	240	224	200	118	152

For corners, use full height for metal corners. For corner boards figure one 3" and one 4" board for each other. Add molding, flashing and caulking costs at doors and windows.

Labor Cost To Add ½" Plywood Sheathing Over Existing Boards, per 100 Sq. Ft.

	Hours	Rate	Total	Rate	Total
Carpenter	1.0	$....	$....	$24.04	$24.04
Helper	0.3	....	....	19.13	5.74
Cost Per 100 Sq. Ft			$....		$29.78
Cost Per Sq. Ft					.30

Corner Boards and Beads

Allow ½ hour carpenter time to set one 1x3 and one 1x4 corner per story. Beads will take approximately the same time.

Often exterior moldings, corner boards and base molds must be replaced because of decay. Cedar, cypress or redwood, preferably the heartwood, will resist future decay.

Costs per lineal foot for #2 pine trim and labor would be $0.85 for 1x4; $1.30 for 1x6; $1.50 for 1x8; $2.35 for 1x12.

Remodeling For Energy Conservation

With the great rise in fuel costs, most buildings, if reappraised for their insulating value, would be found wanting. This has opened a whole new field, which is generally referred to as "retrofitting". In the *Thermal & Moisture Protection* chapter, we discussed how deciding whether to invest in better methods of conserving energy is a simple matter of comparing dollars spent on new materials and equipment against the dollars saved in fuel consumption. Obviously, if a gallon of oil increases from $.90 to $1.50 or more in a six year period, while a 3½" batt of insulation only goes from 10 cents a square foot to 14 cents, the balance swings in favor of adding more insulation. It will not only pay for itself in a few years but will make a solid return for years to come, will help to preserve our natural resources and will create a more comfortable environment.

As most insulation is hidden within walls and under roofing, one must become a "house detective" before one can become a "house doctor". One of the most reliable aids in determining just where the worst leaks are occurring is the infrared scanner, on which escaping heat shows up as red. The worse the leak the brighter the color. Such studies can be made into photographs and submitted in reports, and can be retaken after the corrective measures have been taken to show their effectiveness. Even when the original drawings and specifications are available to show what was originally installed such scans can be useful as installation may have been poorly done, condensation within the walls may have rendered the insulation useless, or the insulation may have packed down in the case of loose types or shrunk in the case of sprayed or foamed types, leaving substantial gaps for free air passage.

While it is air passage through the walls and roofs that is usually attacked first, infrared scans will often show the greatest heat loss from infiltration around openings, which is probably the first thing to correct.

Caulking

Most caulking materials are available in either knife grade for wide cracks and gun grade for narrow ones. They can be further classified into basic, intermediate and high performance groups lasting approximately up to five years, from five to twenty years and twenty years and over. On the low end are cord and rope types and those made of oil and resin. These are low in cost and fine for seasonal protection but will need constant replacement. The intermediate caulks are the rubbers, natural and synthetic, starting at the low end with latexes, with butyls in the middle and neoprene at the top. The high performance types are all synthetic including polyurethane, polysulfide and silicones. A building erected as late as 1980 and caulked with an intermediate grade could have major problems today with infiltration. All openings around windows, doors, trim, sills, cornice, and verge boards and all through the wall piping, air conditioners, vents, and intakes should be carefully checked.

When the space between the door and window frames and a masonry wall are caulked with oakum, an experienced caulker should crank 340 to 400 lin. ft. of opening per 8 hour day at the following labor cost per 100 lin. ft.:

	Hours	Rate	Total	Rate	Total
Caulker	2.1	$....	$....	$24.04	$50.48
Cost per lin. ft					.50

Where it is not necessary to pack in oakum, one worker using a pressure gun should do about 600 lin. ft. per 8 hour day at the following labor cost per 100 lin. ft.

	Hours	Rate	Total	Rate	Total
Caulker	1.33	$....	$....	$24.04	$31.97
Cost per lin. ft					.32

The cost of caulk sealants, intermediate grade, averages about $1.25 for a 10-oz. tube. High performance silicone caulking averages about $4.00 per 10-oz. tube.

Refer to the *Thermal & Moisture Protection* division for the quantity of material that will be required for various sizes of joints.

Porch Repairs

To replace an existing porch floor with new 25/32"x3¼" face soft wood, allow 2 carpenter hours and 1 laborer hour per 100 sq. ft. (painting not included). Cypress, Douglas fir, larch pine, and redwood resist decay and warping and offer medium resistance to wear.

To replace a simple porch column, allow ½ hour to remove and shore existing column, 1 hour to replace column.

To replace a simple wood porch railing, figure a carpenter can remove and replace 15 lin. ft. per 8 hour day.

To erect an exterior utility stair of 2x10 stringers, allow 20 carpenter hours per average story in one run, 25 hours if stair has intermediate landing.

Roof Repairs

To Remove Built-Up Roofing. To remove a 3- to 5-ply roof, figure ½ hour each of a roofer's and laborer's time per each square.

To Replace Built-Up Roofing. Figure a crew of 5 per 8-hour day can install 18 squares of 5 ply roofing, 22 squares of 4 ply, and 24 squares of 3 ply roofing over wood decks on low buildings.

To Replace Shingle Roofing. To remove asphalt roof shingles and underlayment, figure 1 roofer and 1 laborer working together can strip around 20 squares per 8 hour day. A carpenter can install a square of 12"x36" strip asphalt shingles on plain double pitch or gable roofs without hips, valleys or dormers in 2 hours. New shingles are often applied over existing.

To Rehabilitate Existing Roofing. The use of urethane asphalt roofing material is an approved method, but special care must be exercised to insure penetration into all voids, blisters, and cracks. One manufacturer recommends a coating on existing surfaces to be a minimum 50 mil. Coverage is approximately 1 gal. per 30 sq. ft. for 50 mil. thickness, or 3½ gals. per square. A roofer should apply 10 to 15 gals. per hour.

To replace sheet metal gutters, a tinner and a helper should replace 20 lin. ft. of gutter or 25 lin. ft. of downspout per hour.

To replace metal counterflashing, figure 1 tinner can replace 15 lin. ft. per hour providing a satisfactory reglet exists. If a reglet has to be chiseled into masonry, figure 1 laborer can cut 8 to 10 lin. ft. per hour. If an electric cutting saw can be used, figure 12 to 15 lin. ft. per hour.

Chimney Flashing Table

No. of Bricks		Inches		Pitch of Roof: Lin. Ft. of Flashing		
Wide	Long	Wide	Long	To ⅓	To ½	To ¾
2	2	18"	18"	11'	11'	12'
2	2½	18"	22"	12'	12'	13'
2	3	18"	26"	13'	13'	15'
2	3½	18"	30"	13'	14'	16'
2	4	18"	34"	14'	15'	17'
2½	4	22"	34"	15'	16'	18'
2½	4½	22"	38"	16'	17'	20'
2½	5	22"	43"	16'	18'	22'
2½	5½	22"	47"	17'	19'	23'
2½	6	22"	51"	18'	20'	24'

Material Requirements to Add a Gabled Dormer
(Roofing, Siding, Painting, and Interior Finishes Not Included)

Materials & Size	Width of Dormer 5'	6'	8'	10'	12'
Rafter Headers 2"x6"	2-10'	2-12'	2-16'	4-10'	4'-12'
Rafters Plates 2"x6"	2-16'	2-16'	2-16'	2-16'	2-16'
Studs and Plates					
Side 2"x4"	2-10'	2-10'	2-10'	2-10'	2-10'
Side 2"x4"	2-12'	2-12'	2-12'	2-12'	2-12'
Front 2"x4"	7-10'	8-10'	11-10'	13-10'	11-12'
Gable 2"x4"	1-12'	1-12'	3-10'	2-10'	2-12'
Gable 2"x4"	1-8'	1-8'	1-10'	1-14'	1-14'
Rafters 2"x4"	6-8'	6-10'	6-12'	6-12'	6-14'
Ridge Boards 1"x6"	1-10'	1-10'	1-10'	1-10'	1-10'
Framing Total Bf	179 Bf	200 Bf	237 Bf	264 Bf	283 Bf
Sheathing 1"x6"	60 Bf	70 Bf	70 Bf	85 Bf	80 Bf
Roof Boards 1"x6"	60 Bf	72 Bf	86 Bf	100 Bf	120 Bf
Cornice Fascia 1"x6"	22 Lf	24 Lf	26 Lf	28 Lf	30 Lf
Crown Mold 1"x3"	22 Lf	24 Lf	26 Lf	28 Lf	30 Lf
Nails	10 lb.	11 lb.	12 lb.	13 lb.	14 lb.
Flashing 12" Wide	33 Lf	35 Lf	39 Lf	42 Lf	46 Lf
Felt 30 lb.	1 RL	1 RL	1 RL	1 RL	2 RL
Window Frame $^{2}/_{6}$x$^{4}/_{10}$	1	1	2	2	3
Window Sash $^{2}/_{6}$x$^{4}/_{10}$	1	1	2	2	3
Labor To Install	26 Hrs.	27 Hrs.	32 Hrs.	36 Hrs.	40 Hrs.

Gabled Dormer Roof and Side Areas for 1/2 Pitch Roofs

	Width 5'	6'	8'	10'	12'
Roof	60 SF	72 SF	86 SF	100 SF	120 SF
Wall	60 SF	70 SF	70 SF	85 SF	80 SF
Starter	10 LF	10 LF	10 LF	10 LF	10 LF
Ridge	12 LF	12 LF	12 LF	12 LF	12 LF

Weatherstripping

All doors and windows should be tightly sealed including those leading to unheated attics, garages or basements. Check windows for sealing gaskets and tight closing latches. Interlocking devices at jambs and heads of doors are preferable to spring or pressure types and select a proper adjustable sill seal and threshold that will fit to floor finish.

Lineal foot cost of material weatherstrip for double hung windows will average as follows in zinc:

Description	Thickness 1 3/8"	1 3/4"	2 1/4"
Side, head & sill strips	$0.40	$0.45	$0.50
Meeting rail strips	.60	.65	.75

Bronze will run about 100% more.

Single or double casement windows that swing in require a 3/4" wide track strip on the hinged side, running in a groove in the sash. At the top, on the latch side, and where the sash meet, if in pairs, interlocking hook and flat strips are generally used. A trough strip and a hook strip is generally used at the bottom of the sash.

Weather strips for single wood doors will range from $4.75 for zinc to $7.40 for bronze per opening. If interlocking bronze rather than spring type is used, add $3.20 to the above figures. Extruded aluminum and neoprene spring stop mounted strips run $1.10 per foot for 1⅛"x¼" and $2.55 per foot for 1¼"x25/32". Bottom rail gasket runs $1.90 per foot.

Labor Installing Weather Strips

	Hours
1⅜" DH sash to 3x5'6"	0.9
1¾" DH sash to 4x7	1.0
2¼" DH sash to 4x7	1.25
Single inswinging casement	1.3
Double " "	2.0
Single outswinging casement	1.0
Double outswinging casement	1.4
3x7x1¾" wood door, spring bronze	1.6
3x7x2¼" wood door " "	2.0
3x7x1¾" wood door interlock	2.5
1 pr. 1¾" French doors	2.3

It may pay to weatherstrip a door at the head of a stair leading to the attic. Chimneys should also be checked for tight fitting dampers.

Doors and windows in exposed locations should be fitted with storm sash if operable and double glazing if fixed. Openings on the north side and on the side toward prevailing winter winds especially need this extra protection.

A carpenter with helper should set an average preglazed storm window in 1½ hours. Storm windows, up to 12 sq. ft. in area, will average about $55.00 per unit installed. Larger units cost around $75.00 installed. Smaller units can be handled by a carpenter working alone.

Doors in exposed locations are best protected by adding a vestibule rather than just a storm door. In commercial work with heavy traffic a revolving door will offer the best protection although the initial cost will be $15,900 or more.

Windows that are not used for ventilation may be permanently sealed, but it should be remembered that if sealing it means having to rely completely on running air cooling in the summer months this could be self-defeating.

Fixed windows can often be fitted with an interior storm window. This is especially adaptable to curtain wall construction where tubular frames can be fitted with a spacer, new metal sash bars, and an additional layer of glass, which can be of the solar type, further cutting down on heat penetration and glare. Such additions can be added entirely from the inside, but because the area between the old and new glass will be hermetically sealed, only experienced installers should be considered. Pittsburgh Plate Glass has a patented system which they will install on a custom basis.

Several manufacturers offer special replacement windows, which can be pivoted, projected, double hung, or casement types. To install these, one first removes the existing window, but not the frame. A factory assembled subframe is then fitted over the existing frame and secured to it. The new window sash is then installed within the new subframe. These units are usually aluminum, can be had preglazed in either single or double glazing or with part of the window blocked off with insulated panels, and often can be set, window by window, from the inside. A two-worker crew should remove the existing sash, install the new subframe, and add the new sash in approximately three hours per opening.

Unwanted windows had best be eliminated altogether by either bricking them up or covering them with insulated panels. A variation of this, which can retain partial light and ventilation, is to add an insulated panel over part of the window, usually the upper half. These are usually porcelainized metal laminated over a polystyrene core with an aluminum frame factory assembled to fit to the existing window frame. One such manufacturer is Mapes Industries, Lincoln, Nebraska, which rates its add-on panels 1¼" thick at an R value of 6.56.

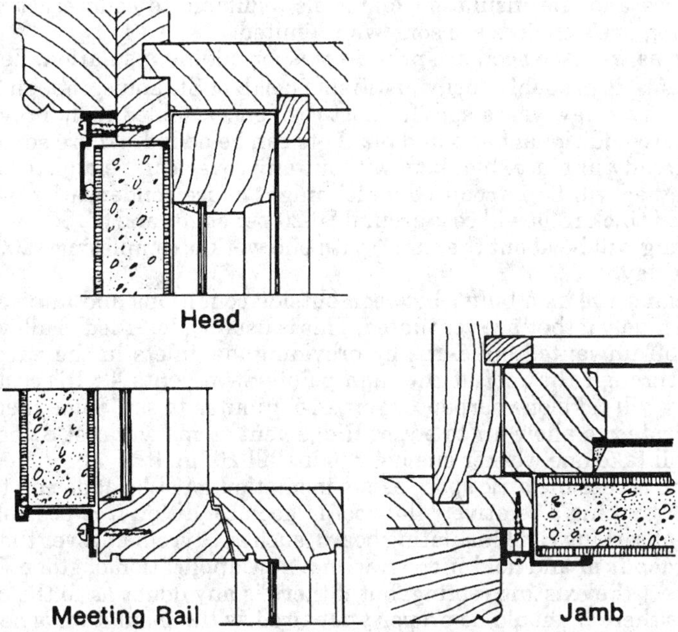

Courtesy Mapes Industries

Insulated Panel Over Double Hung Window

Solar Heat Controls

Heat gain by the rays of the sun penetrating into a building is a blessing in the winter but not in the summer. There are ways to take advantage of this by cooperating with the season. The sun is high in the summer, low in the winter and exterior screens and shades, such as canopies, awnings, operable shutters and even a few well placed deciduous trees, can be so placed that the summer rays are excluded while the low winter rays are welcomed in. Excluding the sun from the outside rather than blocking it once it has entered the inside should be the first consideration.

Continuous canopies will run around $5.30 per sq. ft. while individual window ones will cost $160 in canvas, $265 in aluminum per unit. With a three foot overhang two workers should erect 80 lin. ft. of running canopy or four individual ones per day.

If windows are to be reglazed the sun effect can be counteracted by using tinted or reflective glass. Glass with a fused reflective coating will cost about $3.00 per sq. ft. more than plain plate.

Existing glass can be coated with a solar rejecting film for around $1.60 per sq. ft.

Insulation

The chapter on *Thermal & Moisture Protection* has a discussion on the types of heat loss and the insulation materials available to counteract them. In retrofitting, one's choices are somewhat limited.

Ceilings as we have seen are prime areas for adding insulation. Few would argue that 6" is probably easily justifiable north of St. Louis, though 3½" was standard just a few years ago. In most residences gabled roofs provide attic space where additional batts and blankets can be added. Where some insulation is already in place, blankets without any cover can be added so that no vapor barrier will be introduced which might cause condensation problems. Three-inch thick rolls will cost around $0.20 per sq. ft. and 6", $0.26. The cost of installing will be about the same, with one worker completing 1200 to 1400 sq. ft. per day.

Attics can serve as a buffer between outside conditions and interior spaces most effectively if they are ventilated. This is usually designed to allow gravity to carry off unwanted build-ups by providing low inlets in the eaves which exhaust through the roof at the high point. Eave vents 4"x16" cost around $7.40 and will take one carpenter some 15 minutes to set. Some metal eaves are perforated to allow air to enter. Ridge vent strips will cost $1.55 per lin. ft. and will take one worker one hour to install 20 lin. ft.

Where no accessible attic space exists insulation must be added on top of the roofing. Assuming a relatively flat roof in good condition it is possible to add new non-absorbing rigid insulating board, such as Styrofoam, over the existing roofing, mop it in and hold it down with a new application of stone. This will help protect the existing roofing, but if there is any doubt as to the condition of what is there it should be properly repaired as the insulation is not to take the place of roofing membranes. The amount of insulation added can vary from 1½" to 3" in single sheets but can also be built up by adding additional plies. Should the old roofing eventually need replacement, the stone can be removed from the insulation and new roofing applied over it. Roof drains and flashings of course would have to be reset to the new roof level.

Another possibility for covering existing roofs, and this is not limited to level ones, is to spray on a polystyrene or urethane insulating coat and then cover this with a butyl or elastomeric roofing coat. Costs of applying such coats and roofing will be similar to new work as discussed in *Thermal & Moisture Protection* with the addition of costs to adapt adjacent drains and flashings to the new levels.

Walls present special problems. If the exterior and interior finishes are to remain, there is little alternative to adding insulation to whatever cavity space exists in the center of the wall. In wood stud construction this is relatively simple. To get into the cavity between the studs it is necessary to cut holes between each stud at least every story height. These are usually cut from the outside by removing a strip of siding. Bevel and shiplap can be pried up easily, but tongue and grooved will have to be broken out. Most walls will have a sheathing course behind the siding. This will have to be drilled with holes large enough to accommodate the nozzle for blowing in the insulation, usually 2" to

3". Once the holes are cut the cavity must be plumbed to determine if any obstructions are present, and if any are found additional holes will have to be drilled. Material costs for fiberglass or mineral wool will run from $0.42 to $0.55 per cu. ft. One carpenter can install some 400 cu. ft. per day. To this must be added the cost of removing and replacing siding.

Where an owner contemplates new interior finish in a building or space the existing walls can be covered with rigid insulation board and then faced with whatever new finish is contemplated. In older buildings where the plaster is in poor condition this may be the better alternative. It will mean replacing all trim, but in removing the baseboards it will also allow batts to be packed between the floor joists to keep air currents from flowing down the outside walls and out over the floor area. Rigid board of fiberglass or extruded polystyrene will cost around $0.25 per sq. ft. and will have R values of 4.3 and 5.4 respectively. One carpenter should install 700 to 800 sq. ft. per day.

Wood framed structures can also be resided with vinyl or aluminum siding with insulation fitted to the back. Adding the insulation will cost around $0.10 per sq. ft. so that vinyl will run $0.80 and aluminum $0.90 per sq. ft. and one carpenter can install 250 sq. ft. per day.

Masonry veneer walls can have the cavities filled much the same as wood sided walls, removing brick to allow access. Patching a masonry wall will be more obvious than a wood one. Where the brick has concrete block as a back-up it will be easier to add rigid insulation to the inside face. If the wall is at all irregular it will have to be furred, but the air space will contribute to the insulation value.

Block walls can also be foamed by drilling block every other unit on four foot vertical runs. Foaming is tricky and should be subcontracted to someone experienced in its application. Cost will be around $0.50 per inch of thickness installed per sq. ft.

Basement and crawl space walls, especially those with portions above grade should be insulated. Rigid board or sprayed on types are applicable here although the sprayed on type will need protection if the space is subject to much traffic. The same treatments along with batts or blankets can be continued under the floor construction above these spaces. Ducts and piping should also be insulated.

Interior Work

In remodeling work it is especially important that new wood used with old be dry so that shrinkage will be minimized. Wood should be ordered that has been kiln dried. Edge grained wood has a shrinkage just half that of flat grained wood, and should be used.

Shrinkage is considerable in birch, gum, maple and oak; is average in cherry, hemlock, pine and walnut; and minimal in cedar and redwood.

In extending or closing up doorways and openings in existing walls, often stud dimensions will not match the existing. Special lumber or furring will be required. New studs are 1/8" less in size than those in use a few years ago. In much older buildings studs will be found that are not even sanded. In demolishing old work, plan ahead to see if some of the old lumber should not be salvaged. It may be harder to work with, but the size may be irreplaceable.

Often contractors are asked to judge what caused cracks in walls and ceilings and whether to repair, replace or reface them. Many cracks are not attributable to the workmanship of the cracked surface, but to faults in the materials to which they are applied. Such cracks can be caused by lumber shrinkage, foundation settlement, inadequate bracing and expansion and contraction.

Plaster can crack, spall and disintegrate because of poor workmanship. A common condition occurs when the white coat is applied over a base coat that has been allowed to dry out too much. When the white coat is applied, the base coat soaks the moisture from the finish coat and a spidery pattern of cracking occurs. If the surface of the plaster seems soft, the plaster has probably been oversanded. If cracks occur at random on the white coat, it is probably too thin a coat.

Structural cracks usually show up in the first few years of occupancy. Unless major defects are involved, they reach an equilibrium and can be successfully patched if the plaster on either side is not loose. If the crack is less than ½" wide, dig out an inverted "V" joint, moisten the adjacent areas to reduce suction and fill with prepared patching compound in two stages. For larger cracks, the crack should be made wider to accommodate a piece of metal lath. The resulting channel is then filled with a three coat application.

When the plaster is soft or cracking is prevalent over a large area, it is best to cover the plaster. This can be done with plasterboard, canvas, acoustical tile, or any of the decorative hardboards and plywoods. If the plaster is over sound wood lath, the applied board can be nailed through the plaster into the lath or stud. If the backing is unsound, a new furring system should be applied. Furring strips are nailed at right angles to studs or joists at intervals of 12", 18", or 24", depending on module of new finish.

Walls. To strip a wall of plaster and leave ready to replaster if masonry, or relath and plaster, figure one laborer will strip from 60 to 70 sq. ft. per hour. Add for cost of clean up and debris removal if this is to be included.

To add ½" gypsum board over stripped wood studs, figure 1 carpenter will place around 50 sq. ft. per hour in the average room. If large unbroken expanses are involved, he may double this. If gypsum board is not a pre-finished type but is to be painted, it will require taping. Figure a carpenter will tape 100 lin. ft. of joint, which, in usual applications, would cover the needs of 300 sq. ft. of wall in about 4 hours.

When pre-finished plywood is specified for first class work, gypsum back-up board should be used as an underlayment.

To add wallboard over an existing wood stud and plaster wall, the wall should first be furred. Furring strips should be nailed at right angles to studs. Any convenient width may be used, but it must be wide enough to provide a nailing base to nail all edges of the decorative strips at least every 6". Furring may be of No. 2 soft wood. The success of the wall will depend largely on the success of leveling the furring strips.

With studs 16" on center, a carpenter should cut and place 450 to 500 lin. ft. of furring strips per 8 hour day at the following cost per 100 lin. ft.:

	Hours	Rate	Total	Rate	Total
Carpenter	1.8	$....	$....	$24.04	$43.27
Cost per lin. ft.					.43

When applying wallboard in smaller rooms requiring considerable cutting and fitting around doors, windows, etc., a carpenter should fit and place 400 sq. ft. per 8 hour day, at the following labor cost per 100 sq. ft.:

	Hours	Rate	Total	Rate	Total
Carpenter	2	$....	$....	$24.04	$48.08
Cost per sq. ft.					.48

In large rooms with unbroken walls, this cost would be reduced to about $0.25 per sq. ft.

To add thin board, such as 5⁄32" prefinished over an existing plaster base with adhesive, figure labor costs as follows per 100 sq. ft.:

	Hours	Rate	Total	Rate	Total
Carpenter	3.2	$....	$....	$24.04	$76.93
Cost per sq. ft.					.77

To remove ceramic tile from an existing wall, figure one laborer will remove around 30 to 40 sq. ft. per hour, not including removal of debris. As old existing tile is undoubtedly set in a mortar base, it will be necessary to set the new in the same way or to replaster the wall to provide a smooth base for applying tile with adhesive. 4"x4" tile applied with mortar will run about $5.00 per sq. ft. Adhesive set tile will run in the neighborhood of $4.25 per sq. ft. to which must be added to the cost of a new base.

Ceilings

To remove plaster from ceiling joists, figure one laborer will clear about 50 sq. ft. of area per hour. If the ceiling is suspended and the hangers are also to be removed, figure about 30 sq. ft. per hour.

To strip a plastered ceiling of acoustical tile, figure one laborer can clear 75 sq. ft. per hour. These figures do not include removal of debris.

Often a contractor is asked to patch plaster over concrete or old plastered ceilings. This can be accomplished with the use of a product such as *Plaster-Weld* from Larsen Products Corp. It is brushed, rolled or sprayed on, and in an hour or so, recoated with the desired plaster finish.

To add 12"x12" wood fiber tile to existing plaster ceiling, figure 2 mechanics working together on a scaffolding can set 800 tiles per day at the following labor cost per 100 sq. ft.:

	Hours	Rate	Total	Rate	Total
Mechanics	2	$....	$....	$24.04	$48.08
Cost per sq. ft					.48

Flooring

To Remove Wood Subflooring. Where existing subfloors are badly worn and must be removed, figure one carpenter and one laborer can remove 100 sq. ft. per hour at the following cost:

	Hours	Rate	Total	Rate	Total
Carpenter	1	$....	$....	$24.04	$24.04
Laborer	1	....	....	19.13	19.13
Cost per 100 sq. ft.			$....		$43.17
Cost per sq. ft.			....		.43

To Replace Wood Subfloors. In small cut-up areas, a carpenter should lay between 700 and 800 b.f per 8-hr day of 1"x6" or 1"x8" at the following labor cost per 1,000 b.f.:

	Hours	Rate	Total	Rate	Total
Carpenter	10.6	$....	$....	$24.04	$254.82
Laborer	6.0	....	....	19.13	114.78
			$....		$369.60

To Add Plywood Over Existing Subfloors. Labor for 100 sq. ft. of plywood added over existing subfloor would be as follows:

	Hours	Rate	Total	Rate	Total
Carpenter	1.0	$....	$....	$24.04	$24.04
Laborer	0.5	....	....	19.13	9.57
Cost per 100 sq. ft			$....		$33.61
Cost per sq. ft			....		.34

To Add New Finished Flooring. Once an existing subfloor is put in good condition, replaced, or given a good underlayment, the following installed costs per sq. ft. may be used for various floor finishes:

1x1 Ceramic Tile, Adhesive Set	$4.25-5.50
Asphalt Tile	.70-.95
Vinyl Asbestos 1/16"	.85-1.05
Cork Tile	1.30-2.00
Vinyl Tile .050"	1.05-1.15
Vinyl Sheet .070"	1.40-1.85
Rubber Tile	1.60-1.95
9x9x5/16" Oak Parquet	2.85-3.00

Trim

To Remove Existing Wood Base. A carpenter and helper working together should remove a one-piece wood base at the rate of 100 lin. ft. per hour.

	Hours	Rate	Total	Rate	Total
Carpenter	1	$....	$....	$24.04	$24.04
Laborer	1	....	....	19.13	19.13
Cost per 100 lin. ft			$....		$43.17
Cost per lin. ft			....		.43

To Replace Wood Base. To place a single member base plus carpet strip, allow $1.30 per lin. ft. for 3½, $1.60 for 4½. To add new interior trim, the following installed prices may be used.

	Per LF
Wood picture molding	$1.10
Wood chair rail	.95
Wood cornices 1"x8"	1.50
Panel strips	.60

To Replace Window Sash and Trim. A carpenter should fit and hang, complete with sash cord, chain and weights or counterbalances, 10 to 12 single sash per 8 hour day, at the following labor cost per pair:

	Hours	Rate	Total	Rate	Total
Carpenter	1.5	$....	$....	$24.04	$36.06

A carpenter should fit and hang 10 to 13 casement sash per 8-hr. day at the following labor cost per pair:

	Hours	Rate	Total	Rate	Total
Carpenter	1.4	$....	$....	$24.04	$33.66

A carpenter should set a one member window trim and cap at 1.5 hours per window at the following labor cost:

	Hours	Rate	Total	Rate	Total
Carpenter	1.5	$....	$....	$24.04	$36.06

To Replace Wood Doors and Trim. A carpenter should set a plain wood jamb for a 3'x7' wood door in one hour and trim both sides in 1.5 hours. To hang the door will take another hour for a total cost as follows:

	Hours	Rate	Total	Rate	Total
Carpenter	3.5	$....	$....	$24.04	$84.14

A frame set for 3'x6'-8" opening will run about $16.00 in pine, $21.00 in birch.

If door unit is prefit and knocked down, a carpenter should assemble and install it in one hour.

To Set Factory Assembled and Finished Kitchen Cabinets. When setting factory assembled and finished kitchen cabinets, about 0.15 hours of carpenter time for each sq. ft. of cabinet face area should be allowed.

To Fit and Place Closet Shelving. A carpenter should fit and place approximately 125 sq. ft. of closet shelving per 8 hour day at the following labor cost per 100 sq. ft:

	Hours	Rate	Total	Rate	Total
Carpenter	6.4	$....	$....	$24.04	$153.86
Cost per sq. ft					1.54

To Install Rod in Closet. A carpenter should install about 3 hanging rods per hour at the following labor cost per rod:

	Hours	Rate	Total	Rate	Total
Carpenter	0.33	$....	$....	$24.04	$7.93

To Set Medicine Cabinets. A carpenter should set 12 face mounted cabinets per 8-hr. day at the following labor cost per cabinet:

	Hours	Rate	Total	Rate	Total
Carpenter	0.66	$....	$....	$24.04	$15.87

To Set Bathroom Accessories. A carpenter should set 30 to 34 bathroom accessories per 8-hr. day at the following cost per accessory:

	Hours	Rate	Total	Rate	Total
Carpenter	0.25	$....	$....	$24.04	$6.01

To Set Finish Hardware

Type	Carpenter Hrs. per Each
Rim Lock	.5
Mortised Lock	1.0
Cylinder Lock	.5
Front Entrance Cylinder Lock	2.0
Surface Door Closer	1.0
Concealed Closer	3.0
Sash Lifts and Lock	.5
Kickplates	1.0

To Set Cabinet Hardware

Type	Carpenter Hrs. per Each
Surface Bolts	.05
Offset Bolts	.1
Catches	.1
Drawer Pulls	.06
Knobs	.05
Rim Catch	.1
Mortise Lock	.4

Painting

To repaint exterior wood siding and trim:

	SF/Hr.	Hrs./100 SF	SF/Gal.
First Coat	115-125	.85	500-550
Second Coat	125-135	.75	575-625

If existing surface is in poor condition, add:

	SF/Hr.	Hrs./100 SF	SF/Gal.
Burning Paint Off	30-40	2.75	
Prime Coat	100-110	.95	400-450

To add 2 coats to wood shingles:

	SF/Hr.	Hr./100 SF	SF/Gal.
First Coat, Oil	115-125	.82	250-300
Second Coat, Oil	150-160	.65	375-425
First Coat, Stain	75-85	1.2	120-150
Second Coat, Stain	115-130	.82	200-225

To prepare interior work:

Type of Work	SF/Hr.	Hrs./100 SF	SF/Gal.
Sanding & Putty	190-200	0.5	
Washing Enamel	85-90	1.15	
Removing Varnish with Liquid Remover	30-35	3.00	150-180
Wash & Touch Up Varnish	165-175	.60	
Add 1 Coat Varnish	100-110	.95	575-600
Burn Off Interior Trim	20-25	4.00	
Burn Off Flat Surfaces	30-35	3.00	
Spackle Over Flat Surface	60-65	1.60	140-150
Filling Wood Floors, Wiping	170-180	.57	425-450
Taping, Spotting & Sanding Wallboard	90-110	1.00	
Wash Off Calcimine	115-125	.83	
Wash Smooth Wall	145-155	.67	
Remove Old Wallpaper	65-75	1.45	
Wash Off Glue	125-135	.77	

To paint interior work:

Type of work	SF/Hr.	Hrs./100 SF	SF/Gal.
Wood Floors			
1st Coat Paint	290-300	.35	525-550
2nd Coat Paint	260-270	.38	425-450
1st Coat Stainwax	390-400	.25	525-550
2nd Coat Stainwax	480-490	.21	600-625
Varnish, Each Coat	300-310	.33	500-550
Buffing	390-400	.25	....
Waxing	200-210	.50	1,050-1,075
Walls			
Sizing	350-400	.27	600-700
Sealer	200-225	.45	575-625
Flat Finish			
1st Coat	165-175	.60	500-550
2nd Coat	175-185	.55	575-625
Latex			
1st Coat	140-160	.67	300-350
2nd Coat	140-160	.67	300-350
Canvas	50-60	1.82	
Wallpaper			
Per Single Roll	2½-3		
Trim			
Enamel			
Undercoat	90-100	1.10	375-400
Finish Coat	100-110	1.00	475-500
Stain Woodwork	215-225	.45	700-725
Shellac Woodwork	215-225	.45	700-725
Varnish Woodwork	165-175	.60	425-450
Wax and Polish	100-110	.95	550-600

Plumbing

The following may be used as a guide for replacing existing fixtures. No roughing in or pipe replacement is figured. Fixture cost and roughing in is discussed under *Division 15, Mechanical.*

	Hours
Replace Bathtub	8
Add Shower-Surface Mtd.	2
Replace Single Kitchen Sink	5
Replace Double Kitchen Sink	7
Replace Laundry Tub	5
Replace Lavatory	5
Replace Shower Stall	12
Replace Water Closet	7
Replace Water Heater	4
Replace Central Air Compressor	4
Replace Gas Dryer	2
Replace Automatic, Top-loading Washer	2

Furnish and connect:	Cost
40-gal. Natural Gas Water Heater	$330.00
50-gal. Natural Gas Water Heater	360.00
40-gal. LP Gas Water Heater	290.00

Electrical

To rewire, pulling in old conduit: per outlet, 2.5 hours.
To add new, using BX cable in open wall: per outlet, 1 hour.

Budget figures for remodeling residential work using armored cable and including labor and materials but not fixtures or patching, unless so noted, would run somewhat as follows:

To add a ceiling outlet	$75.00
To add a wall switch	60.00
To add a wall outlet	75.00
Wire and connect a dishwasher	90.00
Wire and connect a clothes washer	100.00
Wire and connect a disposal	110.00
Wire and connect wall or window air conditioner	110.00

To Furnish, wire, and connect the following :

40-gal. electric water heater	$290.00
52-gal. electric water heater	300.00
82-gal. electric water heater	350.00

CHAPTER 18

EPOXY SYSTEMS

The current emphasis on renovation of structures has reached enormous proportions, the result of many factors, one of which is the high cost of replacement. Efforts are underway to restore structures of all types to their original, or better, condition to forestall replacement and reduce overall costs of operation and maintenance.

This situation has created a need for new products and methods for the preservation of buildings, bridges, floors, tunnels, airports, runways--in short, all conceivable types of structures. These new products include, among others, polymers in general and epoxies in particular. The epoxy industry has grown substantially over the past 20 years and has matured to the extent that it is providing reliable products, supported by technically proficient manufacturers who provide architects, engineers, owners and contractors with a variety of services. These services include guide specifications, on-site technical assistance and even product design to suit the needs of specialized applications.

Epoxy Compounds

Epoxy compounds are used in a variety of applications, in construction, electronics, space and aircraft adhesives, and for structural components. Some aircraft are now being constructed almost entirely of fiber reinforced epoxies, because pound for pound, they are stronger than steel and offer greater design latitude. Before we can discuss the specific applications and uses of epoxies, one first needs a basic understanding of the nature and limitations of these materials.

Epoxies are plastics. More specifically, they are thermosetting resins. When resin and a *hardener* (curing agent) are mixed together, a chemical reaction that liberates energy in the form of heat takes place. This reaction is classified as *exothermic* (exo means "out"; therm means "heat").

Temperature Limitations. Epoxies require some warmth to cure or harden, which explains why low temperature limitations are imposed for application of these items. Conversely, as the temperature rises, they will harden more rapidly, and in some cases (for example, in the sun at 120°F), it might be impossible to install certain types of epoxy compounds, because they will harden too fast.

Pot Life Limitations. The pot life of epoxies, after the resin base and hardener have been mixed, depends on several factors:

a) The type of curing agent used. Some cause reactions to proceed faster than other.
b) The quantity of resin and hardener mixed at one time (total mass).
c) The temperature of epoxy at the time of mixing.
d) Atmospheric temperature.

If the quantity mixed is large (e.g., a 5-gallon container full), the exothermic chemical reaction proceeds more rapidly, because the heat cannot be readily dissipated to the atmosphere. The large mass "insulates" the compound and causes the reaction to build on itself and proceed more rapidly.

If the mix quantity is small (e.g., a half-pint container), the heat of the reaction is readily dissipated through the walls of the container to the atmosphere. The mass is small, and the mixture cannot effectively retain the heat of reaction. Also, if the air temperature is low, the reaction will proceed slowly. If the air temperature is high, it will proceed more rapidly.

As an example, let us assume that we mix 4 gallons of *Durabond* epoxy bonding compound and intend to brush it onto preexisting prepared concrete before pouring a new concrete overlay. Assume that only half, about 2 gallons, of bonding compound is applied to the concrete, while the rest stays in the mixing bucket. What happens?

The epoxy in the mixing bucket (large mass) will react fairly rapidly and harden in about one hour, while that applied to the concrete (a very small mass, about 10 mils thick) will take three to five hours, depending on the temperature of the deck and of the atmosphere.

Mixing Ratio Limitations. Unlike other types of polymer, the mixing ratio of epoxy items cannot and must not be altered under any circumstances. Neither the setting nor the curing time can be changed by adding more or less curing agent to the mix. When cold weather approaches (40 to 50°F), and an epoxy compound appears to be setting or curing too slowly, Do *not* attempt to speed the setting time by adding excess hardener. Also, in the summer when temperatures are high (90 to 100°F), Do *not* try to slow the setting time by using less hardener. Either of these actions will only create a soft and gummy compound.

The chemical ratio of resin to hardener is exact, and the resin requires a fairly precise quantity of curing agent with which to react. If the hardener quantity is insufficient, it is possible that some of the resin will not react and will remain in liquid or semi-liquid state. Also, if an excess of hardener is added, because the hardener can only react with a specific quantity of resin, the excess hardener might not react and possibly remain free in the compound as liquid or semi-liquid. In any event, the physical and chemical properties of the cured epoxy can be severely impaired by these actions.

In summary, it might be well to make an analogy with a familiar product--concrete. There are many similarities, such as in the mass vs. set time and in the exotherm relationship. Consider that cement and water provide the matrix or binder (adhesive) for the aggregate. This is similar to the resin and hardener in epoxies and other thermosetting resin systems.

Both concrete and epoxy are subject to limitations of temperature. Concrete normally cannot be poured when the weather is below freezing, or when it is very hot. Both cannot be placed without consideration of the mass and thickness of the pour. If the pour is excessively large, the heat generated could destroy or weaken the structural value if the exotherm is not controlled.

With concrete, the water/cement ratio is very important, because if it is excessively high, then the concrete will not develop adequate physical properties for a long time, if at all. If it is excessively low, hydration might not be complete, and again, a weak mix will result. The same can be said of the epoxy resin/hardener ratio described above.

Everyone is familiar with the fact that concrete once mixed (sand, stone, cement, and water) begins to cure or harden via chemical reaction and will set normally within a given period of time. Epoxy also has a limited working life after mixing and requires that it be applied within a given period of time.

Cure Time. One final but very important point in our comparison of epoxy and concrete is that concrete must be allowed to cure for a period of time before forms can be stripped and it can be opened to traffic or otherwise placed in service. Curing of concrete is critical.

Curing of epoxy coatings, adhesives, membranes, or patching compounds is also critical. Because temperature affects the cure rate, it is essential during cold weather to provide additional cure time before using the structure, acting on the epoxy or even just expecting it to perform properly. It is advisable to consult the technical data of the manufacturer for cure times before placing the compound in service.

If one applies the basic theory and ideas inherent in concrete technology, then one will not go far astray, providing that he or she does not attempt to add water to epoxy. Adding water to epoxy is a taboo and is like adding water to oil. Epoxy and water do not mix. There are, however, exceptions to every rule. Some epoxy systems are designed to be water dispersible, but these are very specialized products.

Shelf Life. Regardless of the limits imposed by some specifications, epoxies will have almost an unlimited shelf life in their separate containers. To illustrate this point, Dural has retained samples from production batches that are over five years old. Periodically, these samples are re-examined for compliance with specifications and been found to still comply fully with the requirements of the appropriate specification. Pre-dispersion (mixing) of any settled pigments and/or fillers is the only prerequisite for satisfactory performance of out-of-date or expired products.

Surface Preparation

Surfacing cleaning and/or preparation is frequently required to remove laitance or foreign matter so that concrete or masonry surfaces can be restored or prepared to receive some type of finish requiring a positive bond with the existing surface.

Sandblasting is also used extensively in producing an architectural finish for exposed concrete. Sandblasting and/or automatic shotblasting is considered the most effective and economical method of surface cleaning when using epoxy, because the quality of the finished work is predicated upon preparation of the existing surface. While sandblasting is usually performed by a specialty subcontractor, the general contractor and owner should be aware of the need for providing protection for persons, adjacent properties, glass and metal surfaces, ornamental structures, trees, power lines or other items subject to damage by the sandblasting operation.

Clean up at the conclusion of each work day is another issue that should be included. By endorsement on an existing property damage insurance policy, or the issuance of a special policy covering this specific risk, protection can be obtained from the hazards of this work.

Sandblasting

	Price per Sq. Ft.
Light penetration, including protection and cleaning	$0.55
Heavy penetration, including protection and cleaning	1.65 to 1.75
On metal surfaces, light penetration, including protection and cleaning	1.20
On metal surfaces, heavy penetration, including protection and cleaning	1.90

Sealers

The following are only some of the many products available to the industry today.

Epoxy Penetrating Sealers. This system is a two-component resin, available in 15%, 30%, and 50% solids. Its use is to waterproof concrete, brick mortar and cement block and to produce a protective zone that increases the substrate's resistance to damage from fuels, oils, alkali conditions, mild acids, de-icing chemicals, salts, and aggressive waters. When properly applied on concrete bridge decks, skid resistance is not altered and savings can be realized by reducing the amount of de-icing chemicals required per application.

This method is similar to the application of a coat of silicone, but with certain advantages. Epoxy offers a chemical as well as a mechanical adhesion. In addition, chemical resistance and film integrity are superior in epoxy. In general, epoxy provides greater resistance to intrusion of salts and other aggressive chemicals, as well as to water, gasoline, and oils.

Application to bridge decks, where salt applications are generally heavier, should be 2 coats rather than one. It might be desirable to broadcast a silica sand into the final coat to eliminate any skid potential when the surface is wet. This might be necessary where concrete is extremely dense and porosity is minimal.

Penetrating sealer can be applied by rollers, brush or spraying at the following cost:

Cost of Applying Penetrating Sealer

	Cost per Sq. Ft.
Clean surface by sandblasting methods	$0.55
Epoxy sealer material	0.15
Labor applying epoxy sealer	0.45
Total cost per sq. ft.	$1.15

Epoxy Injection Systems. This system uses a two-component epoxy resin that is injected under pressure into cracks in concrete decks, columns, piers, and walls. It restores the structural integrity of the concrete, and its principal uses are: bonding cold joints; repairing delaminations; stopping water seepage through foundation walls, floors, beams, and decks. Some injection epoxies are used underwater to repair pilings, water tanks, retaining walls and tunnel linings.

Cartridge Method: This method is generally used where cracking is extensive and it becomes desirable to use automatic mixing and application equipment.

	Cost per Lin. Ft.
Material - 8 oz. tube (5.5 lin. ft. per tube)	$4.00
Labor	22.00
Total cost per lin. ft.	$26.00

Injection Machine Method: This method is used when cracking is extensive, and it becomes desirable to use automatic mixing and application. It is strongly suggested that several sample areas first be repaired, and with the cost information developed from these tests, a unit price per gallon of epoxy or a lineal foot price can be established. General costs of "welding" cracks in concrete are as follows:

Material cost is $45.00 per gallon, which under average conditions will "weld" cracks 1/16" wide at the rate of 40 lin. ft. per gallon.

	Cost per Lin. Ft.
Material	$1.13
Injection machine rental, $500.00 per week, 100 l.f.	5.00
Injection machine operating costs, 40 hrs. @ 4.00 per hr.	1.60
Injection machine operator, 40 hrs. @ 17.05 per hr. = $682.00 per week (or per 100 lin. ft.)	6.82
Labor cost, 40 hrs. @ 20.50 per hr.	8.00
Total cost per lin. ft.	$22.55

Epoxy Mortar Systems. This system is a two-component resin with an aggregate added to the resin. The aggregate is used for a number of reasons, some of which are:

a) reduce the cost of application;
b) provide a textured surface;
c) increase abrasion resistance;
d) increase mortar compatibility with the physical properties of the concrete.

The epoxy resin, if low viscosity, can be extended with aggregate, up to 4 parts aggregate to 1 part mixed epoxy resin by volume. If the epoxy is a gel or paste for vertical or overhead applications, then up to 1 part aggregate to 1 part mixed epoxy may be added by volume.

Some products can be used as a thin overlayment, as thin as 1/8". Some can be placed 1/4" and thicker and are able to cure to 0°F (-18°C) on dry surfaces and 33°F (1°C) on moist surfaces. Some products can even cure below freezing without heaters or lamps to initiate the curing. All aggregates, whether silica sand, stone or blends of the two, must be bone dry. Sandblasting sand is often used for this purpose. Larger aggregates must be specially processed to remove latent moisture.

This system is effective as an overlay over existing concrete slabs and requires the following work procedure with the application costs as follows:

Costs of Applying 3/8" Mortar System

	Cost per Sq. Ft.
Sandblast or pressure clean surface	$0.55
Repair spall areas, material 2.00 and labor 3.00 (if req'd)	5.00
Crack welding*	25.00-45.00
Apply 3/8" overlay	4.50

*Price per lin. ft.

The actual extent of work required can only be determined by a detailed site investigation, and because of the variance that may be encountered in the condition of an existing concrete slab, the tendering of a fixed price proposal for work of this nature is very risky. It should not be attempted by an estimator who is not experienced in this type of work.

Epoxy Coatings. Coatings are generally two-component systems and are used in many applications. Some of the more popular uses are:

a) floor coatings
b) ramp coatings
c) pool coatings
d) tank coatings
e) acid coatings
f) food processing coatings
g) underwater coatings
h) chemical liner coatings
i) abrasion resistant coatings
j) bridge roadway coatings

Many coatings are available in a variety of colors, including green, red, tile red, blue, white, light blue, dark blue, dark gray, and concrete gray. Some epoxy manufacturers will even formulate colors to specification.

Some products can cure on moist surfaces down to 32°F (0°C) and on dry surfaces to 20°F (-7°C).

Epoxy Coatings

Epoxy coatings require basically the same surface preparation as epoxy mortars. If one knows what preparations are required for the latter (sandblasting, steel shotblasting, etc.), then one can use the same costs per sq. ft. for preparation and cleaning.

Because 100% solids epoxy coatings are free from any solvents, the thickness of coating applied as a wet film will be the same as the dry film thickness. For estimating purposes, 1 gallon of mixed epoxy, when applied 1 mil (.001") thick, will cover 1600 sq. ft., or at 10 mils thickness, it will cover 160 sq. ft.

To figure total gallons required to cover 1,000 sq. ft., simply determine the mil thickness specified or required, divide 1,600 by this thickness and then divide 1,000 sq. ft. by that number:

Area to be covered = 1,000 sq. ft.
Thickness specified = 10 mils
1,600 divided by 10 = 160 sq. ft. per gallon
1,000 sq. ft. divided by 160 = 6.25 gallons

Be aware that the coverage rates are theoretical and do not take into account waste or absorption into concrete.

Cost of Applying 1,000 sq. ft. of 5-mil Coating

Epoxy material, $45.00 per gal. x 3.12 gal*	$140.40
Labor, painting (with roller) 4 hrs. @ 20.00	80.00
Total cost per 1,000 sq. ft.	$220.40

*Coverage for 5 mils figured using coverage formula described above.

Epoxy Bonding Compounds. These products are two-compound systems whose principal use is to bond dissimilar materials, such as concrete, steel, and wood. Many other uses are possible, but one should always consult the manufacturer prior to using the adhesive for an unusual application.

Another primary construction application for epoxy bonding compounds is adhering new concrete to existing concrete. The aerospace industry uses these to bond aluminum aircraft skins to their structural members.

Product Selection

In order to assist and simplify the product selection process, ASTM C-881 should be consulted by the specifier or user. This specification sets forth a variety of applications and the parameters for the products suited to these applications. Once this is done, the manufacturer should be consulted to determine the product that best meets the specifications for the application intended. Some important factors to consider in product selection are:

a) Most epoxies do not cure properly below 50°F. Although the most current ASTM C-881 specifications provide for low temperature curing systems, these are generally limited to 40°F and rising. Applications at temperatures below 32°F are not recommended, because frost and ice crystals may be present on the surface, which would prevent adhesion of the epoxy.

On the other hand, if temperatures are above 85°F, a slower setting product should be selected, because most systems will set too rapidly and will not permit sufficient working time. Most reputable epoxy manufacturers provide products that cover the broadest range of temperatures, from -15°F to 140°F and above.

b) Moisture-insensitive epoxies (or for that matter, any polymer) should be considered. Moisture-sensitive epoxies will not bond well to damp or wet surfaces. Some manufacturers provide epoxies that will cure underwater as well as on damp and dry surfaces.

c) Modulus and flexibility. On large-scale applications on concrete and steel, particularly when using coatings, mortars or overlays, flexible products should be used, because the coefficient of expansion for these products is significantly greater than for concrete or steel. This means that they tend to expand and contract at a greater rate than the substrate, and in the case of concrete, could shear off at the concrete interface. Flexible, low modulus products will be able to relieve any stress set up between the epoxy and underlying surface.

Structural applications where the epoxy adhesive or epoxy mortar is to provide structural support calls for high modulus, high strength and low creep. Again, to assist in simplifying matters, ASTM C-881 should be consulted.

d) Viscosity. In performing coating or epoxy mortar work on horizontal surfaces, generally a low viscosity epoxy should be used. When applying epoxy mortars to overhead or vertical surfaces, a paste or gel version should be used to prevent sagging or slumping from the surface. It should be noted that low viscosity epoxies, when used to prepare a mortar, may accommodate 3 to 4 volumes of sand, while gel or paste systems may accommodate 1 volume of sand per volume of epoxy.

Equipment

The following is a general list of tools required for epoxy work. It is a guide for the most commonly used tools and should not be considered necessarily a complete list.

Basic Tool and Equipment Guide

Substrate Cleaning Equipment
Sandblaster
High pressure water blaster
Grinders
Scabblers
Broom, mop & pails
Automatic scrubbers

Measuring Tools
Tape measure
Empty 1-, 2-, & 5-gallon pails

Mixing Equipment
Muller type plastic mixer (stationary drum, blades turn)
Drill and mixing tool (200-600 rpm)
Wooden mixing paddles

Application Equipment and Tools
Airless sprayer
Air sprayer
Power screed
Manual screed
Squeegees
Trowels (flat, various sizes)
Coving trowels (½", ¾" and 1" radius)
Roller pans, rollers
Fiber bristle brushes with wood handles
Paint brushes

Miscellaneous Items
Hand tools, pliers, hammers, screwdrivers, etc.
Kneeling boards or pads
Plaster spiked shoes
Rags and paper towels
Plastic or masking paper and masking tape
Wet film gauge

Safety Equipment
Respirator for organic vapors
Goggles or face shield
Rubber or plastic gloves
Skin cream

Techniques

Injection. There are generally three accepted methods of injecting epoxies into concrete cracks, honeycombing, delaminations, cold joints, and the like. Selection of a method depends in great part on the quantity and quality of injection required.

The basic difference between quantity and quality referred to is that quantity is a primary consideration when filling voids, such as honeycombing, adequately enough to accomplish mass, while quality is important in filling fine hairline cracks (.002") with a 95% to 100% integrity requirement.

Using Cartridges. Some manufacturers provide an injection resin in plastic cartridges not unlike caulking tubes. They do differ from caulking tubes in that the "A" and "B" components must be mixed immediately within the tube itself prior to use.

Using Pressure Pot. A pressure pot is a paint type pressure vessel used to injection epoxies. The major drawback to this type of procedure is that the epoxies must be mixed in bulk prior to injection and that the injection must be completed prior to the epoxies gelling within the container and lines.

The pressure pot method is practical where it is known that no slowing of the injection will occur, and it is particularly adaptable to the filling of void areas, such as honeycombing, large cracks, and delaminations.

Using Injection Machine. An injection machine is suited to large-scale projects and to when the injection procedure is steady and slow. There are many different injection machines on the market. The use of any type machine with a any given epoxy product should depend on whether it is endorsed by the epoxy manufacturer.

The injection machine must be able to properly proportion the "A" and "B" components, to completely mix them and to inject the epoxies at the epoxy manufacturer's recommended sustained operating pressure.

Any injection machine considered should be as self contained as possible to avoid having to rent air compressors, power generators, etc.

Epoxy Mortars. The selection of an epoxy for a mortar application should be with the job site conditions in mind. Some job conditions that one should consider are:

a) temperature curing range of the epoxy;
b) moisture insensitivity or sensitivity of the epoxies;
c) whether local aggregates can be used, or will it be necessary to ship aggregates as a part of the epoxy system;
d) whether the aggregate/epoxy system actually performs as an overlay in the environment specified.

When selecting an epoxy for a mortar application, one must be aware that epoxies are available that are capable of curing at a temperature as low as -15°F and as high as 140°F. Some epoxies can be extended with an aggregate using local sands, and for thicker applications, using ¾" to 1" stone.

By knowing these things, an estimator can introduce economy into an estimate for materials. Some epoxy mortars have very poor workability and require power troweling, while others can be placed with a screed to line and grade. Some epoxy mortars can be self leveling, while others might require power trowels. Knowing this, the estimator can economize by reducing labor and equipment costs.

Estimating Quantities. The common denominator is to reduce all quantities to the cu. ft. (the epoxy manufacturer will provide the contractor with a "mix design", i.e., aggregate size and weights and the weights of epoxy). For example:

Aggregate	120.00 lb.
Epoxy Resin	24.00 lb.
Yield	144.00 lb. per cu.ft.

The epoxy weight is 9.2 lb. per gallon, and 1 cu. ft. = 2.6 gallons. So, coverage for a ¼-inch thick overlay is 12" divided by .25 = 48 sq. ft. per cu. ft.

If a prime coat is required, add the quantity of epoxy required for it. Usually the coating of epoxy, as a primer, is the mortar resin itself. If a separate resin is recommended as a primer, it should be 100% solids. This means that it should not contain any solvents.

A typical crew size for a mortar overlayment job might be as follows:

1 laborer for mixing
2 laborers for screed
1 finisher for finishing
2 laborers for priming and spreading
6 man crew = 48 manhours per day

Assuming all the substrate cleaning sandblasting, etc. has been completed, the crew can apply a ⅛" or ¼" overlayment as follows:

6'0" ribbons, 400 lin. ft. = 2,400 sq. ft.
2,400 sq. ft. divided by 48 manhours = 50 sq. ft. per manhour

Important Considerations in Applying Epoxy Mortars

Viscosity. This is a critical factor, normally not considered during applications. However, it partly controls the success of the project. Viscosity must be considered, especially when applications are made in the winter or during any weather that produces cold nights. The following figures indicate typical viscosity variations with temperatures.

Typical Initial Viscosities
"A" & "B" Components Mixed

°F	°C	Centipoises
90	32	300
77	25	500
68	20	1,000
60	15.5	3,000
50	10	4,000
33	1	7,500

One can see that the warmer the temperature, the lower the viscosity. The chief requirement for adding the filler or silica sand in the proper quantities is that the viscosity of the epoxy be such that all surfaces of the filler are evenly wetted and bonded together. When laboratory mortar formulations are designed, the epoxy is maintained at an even temperature of 72°F (22°C). This is the balancing point for filler addition. When the viscosity rises or becomes thicker, it is not possible to maintain the same filler balance, thus the mortar may exhibit characteristics that are not desirable. During cold weather applications, it is recommended that the resin and curing agent be stored in areas above 40°F (4.4°C) and be heated to a minimum temperature of 90°F (32°C) before use.

Epoxy heating jackets are an easy and fast method of warming the materials. Experience has shown that upon pouring the warm components into the mortar mixer, the components will drop immediately to a temperature of 60 to 70°F (17 to 21°C). This range of temperature at the beginning of the mixing procedure will continue to drop with the addition of the cold sand and gravel. Heating both epoxy components will result in mortar mixtures that provide proper aggregate proportions, giving the desired physical properties when cured. When temperatures are above 50 to 60°F (10 to 15.5°C), heating is generally not required. Of course, heated aggregate will also provide better workability and cure rate.

Reactivity. The reactivity of the mixed system determines such characteristics as pot life, tack free time, cure speed, working schedule, and method of application. The cure reaction starts as soon as components "A" and "B" are mixed. The colder the temperature of the resin system, the slower the reactivity. It shows a longer pot life time and longer time to fully cure.

In cold temperatures applications, slower reactivity can possibly cause application defects or failure, because the substrate can go through a thermal expansion or contraction movement prior to the mortar developing its tack-free stage. If temperatures become sufficiently cold, the reaction could stop entirely and the epoxy not cure at all. Effects could be loss of adhesion or cracks in the mortar, or in the case of very low temperatures, a gooey mess and uncured epoxy. By heating both components to a preselected temperature, desirable lower working viscosities are achieved and the product will reach a "tack-free" state more rapidly. The shorter the time required to reach this desirable stage, the less chance of mortar damage prior to initial curing. Once the tack-free stage develops, common physical properties are obtained in the epoxy mortar that are similar to those of concrete.

Sand. It may be necessary to blend different grades of sand at the job site to insure the proper aggregate blend. The sand should be bagged, clean and, dry. It should be of high quality silica or quartz, with a minimum hardness of 7.0 on the Mohs scale. Sandblasting sand is often used in this application.

Particle shape, as well as gradation, is important in order to have a workable mix at the required sand to binder ratio. All these factors must be checked carefully before specifying a particular sand to assure a mix that will give the greatest ease of application and performance. All sand should be tested by the manufacturer's laboratory prior to use. Remember, different applications require different graded blends to achieve the best results. The manufacturer's representative should assist you in selecting the best and most economical blends.

Preparation of Concrete and Other Substrates for Epoxy Floor Systems

Epoxy flooring systems require a certain amount of preparation of the substrate prior to actual application. The following are general instructions on floor and job preparation. However, each job should be considered individually, and preparation should be estimated for the specific conditions. Very few jobs are exactly alike. You might use only a few of the clean-up steps on many of your jobs.

Preparation

General. Careful attention should be paid to surface preparation of areas to be treated, to mixing and storing of materials area and to adjacent areas not being treated. Epoxy products have excellent adhesion, so it is extremely important to prevent waste and spills. Materials allowed to cure where they are not wanted will be extremely difficult and time-consuming to clean up later. It is a good practice to protect the flooring area where the mixing takes place, if the area is not going to receive a new overlay. The typical method of protecting the floor area would be to spread heavy cardboard or plastic over the floor.

Good working conditions and preparations are essential to a smooth operation and high quality job.

Work Areas. Plan your job well. Provide proper material storage and mixing area. Allow sufficient space and ease of access to the area where the floor is to be treated. Locate the storage and mixing area as close as possible to the area to be treated. On large jobs, mix at many different locations so that the hauling labor is reduced to a minimum. When mortar mixers are used, make sure the mixer is elevated high enough for the mixed to dumped into wheelbarrows. Set out your tools so that they can be cleaned as necessary in the solvent containers. Clean your mixer once every 30 to 40 minutes by adding a gallon of solvent mixture and rotating it until the drum cleans itself. The solvent may be used many times over, but remember to keep a cover on the solvent pail. The mixing area is always a "no smoking" area.

Moisture. Moisture creates no problem with most epoxy flooring systems. However, some require less labor to apply than others when the floor surfaces are dry.

Epoxy flooring systems may be applied to moist or damp surfaces down to 33°F in temperature. However, cure time might be severely retarded at these low temperatures. The practical lower temperature limit should be 40°F. Remember, do not apply epoxies on ice surfaces or frozen concrete.

Temperatures. As a general rule, air temperatures do not govern the concrete temperatures, which are normally related to soil and moisture temperatures. A room with an average air temperature of 70°F (20°C) could still have substrate concrete temperatures of 40 to 50°F (4.5 to 10°C). Cold temperatures will affect the application of some epoxy flooring systems. However, most epoxies require surface temperatures higher than 40°F (4.5°C) to apply their materials.

General Surface Preparation. Epoxy flooring systems may be applied over any sound substrate with the exception of asphalt and petroleum base surfaces. Surfaces must be cleaned and properly prepared, as outlined in this section. The ultimate success or failure of the job depends a great deal on proper bonding to the substrate.

Service life of mortars or coating systems primarily depends upon good surface preparation. The life will be extended considerably by following the suggested recommendations.

Steel Surfaces. These must be dry and clean in accordance with the following requirements:

a) Remove all debris from the surface to be treated.

b) Removal of grease, oil and contaminants can be accomplished by wiping the surface with a recommended cleaner. The cleaner will dissolve and dilute most contaminants sufficiently to permit them to be wiped or washed off. Before wiping, scrape or wire brush all heavier areas of contaminants to remove thick deposits. Rags or brushes may be required to adequately clean the surface. Wear protective clothing during the cleaning operation.

c) Sandblast the surface to white metal finish, removing all mill scale, rust, rust scale, paints or other foreign matter. Care must be taken to see that extreme peaks and valleys are not left on the surface, because they require

additional coating build-up. These peaks, if too high, also cause thin spots in the coating application and could lead to early coating failure. Vacuum all debris and sand from area to be treated. Immediately apply the epoxy system as recommended.
d) Remove or grind all sharp edges and welds to a rounded surface.
e) Follow manufacturer's recommendations for applying the epoxy system.

New or Non-contaminated Concrete Surfaces. These must be prepared in accordance with the following requirements:

a) Remove all debris from working surface.
b) Surface must be cleaned free of all curing compounds, release oils and dust. New concrete should be fully cured. Dry surfaces permit easier application, but most epoxy products will adhere to clean, damp surfaces.
c) Determine and record the pH factor of concrete surface.
d) Record the substrate and the ambient temperatures.
e) Etch concrete surface with a 10% solution of commercial hydrochloric acid. Apply 1 gallon of 10% solution per 40 sq. ft. The acid solution should be mechanically worked into the surface for 2 or 3 minutes or until bubbling stops. Wash surface with clean water to remove all acid. Rinse second time if any bubbling is noticed. Check pH after rinsing.
f) Sandblast or waterblast surface to remove contaminants, heavy laitance, sharp edges or protrusions that will interfere with the uniform application.

Old Concrete Surfaces. Old surfaces are usually contaminated or previously coated or sealed. The most economical means of preparing these surfaces is with automatic steel shotblasting equipment. These devices are neither self-propelled, nor are they pushed along the surface. They have self-contained vacuum units that automatically recycle the steel shot, while removing the dust and debris. The degree of shotblasting should be such that aggregate is exposed or at least 1/16" to 1/8" of concrete is removed. In cases where heavy oil or grease contamination has occurred, it may be necessary to follow shotblasting with degreasing, thorough rinsing and drying prior to application of the epoxy.

Tile Preparation. Epoxy flooring systems may be applied over old ceramic tile floors. The floor should be prepared by removing all grease, oil, paint and other contaminants. Scarify the surface of the floor with a sander or grinder. Remove old loose material in the joints. Vacuum all dust and then acid etch as described in the concrete preparation section. Conditions should now be ready for application of the epoxy floor system.

Wood Floor Preparation. Wood floors are the least desirable substrate of all surfaces, but if carefully prepared, an excellent floor system can be produced. Make sure the floor is sound and subject to a minimum of movement. Epoxy floor systems will provide an excellent surface and will hold the wood together. However, if there is a significant amount of floor movement, then cracking could develop.

It is very important to remove all greases, oils and any other contaminants, including oil paints. A floor sander is often the best method of cleaning the surface. Vacuum the entire area so that no dust is present. The floor must be free of all dust and dirt. Wood that has been subjected to weather elements must be well sanded before epoxy installation. Wood floors that are subject to heavy use must have a 1/2"x1/2" wire mesh hardware cloth or wire lath nailed to the floor after priming has been completed. The primer need not be dry to install the wire mesh or to start the application of epoxy flooring systems.

Other Coatings or Epoxy Surfaces. It is a must that all existing epoxy coatings and surfaces be removed by abrasive blasting or scarifying. Vacuum all dust and debris. If the previous system failed, then it is obviously unacceptable to have it between the substrate and your epoxy floor system. The adhesion will be no better than the surface that previously failed.

Epoxy Floor and Deck Coatings

Epoxy floor coatings are a two-component epoxy system designed to coat concrete, steel floor and deck areas for extra durability and wear resistance. They provide an attractive, durable finish for industrial space, warehousing areas and parking deck surfaces, where service is too severe for most paints.

The epoxy floor coating must be resistant to a variety of chemical spillage and exposures normally found in industrial plants and parking structures, such as alkalies, mild non-oxidizing acids, gasoline, de-icing salts and many solvents.

The epoxy floor coating must have an excellent resistance to abrasion and be available in smooth or non-skid formulations. Standard colors are clear amber, gray, tile red and green. One should be able to apply the epoxy floor coating on damp surfaces (no standing water) down to 35°F (2°C), or on dry surfaces down to 20°F (6.6°C).

Outstanding Features. The following is a list of some of the advantageous features of epoxy floor and deck coatings:

a) High film-build.
b) May be applied at low ambient temperatures. Applicable on moist substrates to 35°F (2°C). Applicable on dry substrates down to 20°F (-6.6°C).
c) Rapid drying time.
d) Can be applied in neat form or with a sand filler.
e) Both chemical and mechanical adhesion to concrete.
f) Excellent resistance to chemical and petroleum spillages.
g) Tough, flexible, abrasion resistance to hard rubber wheels and moving vehicular traffic.
h) Excellent colors provide attractive finish, hiding patches and stains.

Typical Uses. Some of the typical uses of these products are:

a) Bridge decks
b) Chemical plants
c) Equipment
d) Paper mills
e) Parking garage decks
f) Pharmaceutical plants
g) Machinery areas
h) Marine uses
i) Water pollution control plants
j) Water treatment plants

Resistance Properties. Epoxy coating products for floors and decks are generally resistant to:

a) Abrasion
b) Fuel spillages
c) Industrial fumes
d) Ultraviolet rays
e) Machine tool cutting oil
f) Natural and synthetic oils
g) Alkalies
h) Gases (hydrogen sulfide, carbon and sulphur dioxide)
i) Acid fumes
j) Fresh water
k) Salt water
l) Sour crude oil

Limitations. The epoxy floor coatings are usually not recommended where exposure will include direct contact with sulphuric, nitric, chromic, and acetic acids.

Preparation for Special Requirements

In addition to the typical preparation steps that were described earlier, the following are offered for special requirements of the floor coating system.

Concrete. Broom finished surfaces are the easiest to prepare. Lightly abrasive blast or mechanically abrade the surface to remove curing compounds and other contaminants. Remove dust from surface by vacuum or wash with water and air blow or sqeegee off puddles. Puddles indicate low spots and can create unsafe conditions during wet weather.

Aged Concrete. All older concrete must be thoroughly cleaned by an abrasion method, such as sandblasting or some other factory-approved method. All contaminants, such as old sealers, grease, oil, paints, asphalt, tars, and other coatings, must be completely removed before applying the epoxy floor coating. Wash or vacuum all loose debris and dust after abrasive cleaning of the surface (note puddles to locate low areas).

Acid Etching. All concrete should be checked after the cleaning process to determine the correct pH of the substrate. Acid etching is not required unless the concrete indicates a pH of 10 or higher.

Cleaning. Cleaning solutions that have not been approved in writing by the epoxy manufacturer should not be used.

Repair of Structural Cracks. The repair of structurally cracked and delaminated concrete is of paramount importance, because reflective cracking may occur. The crack may transmit contaminants and moisture from one level to another.

Hairline Cracks. Small fissures, often called *hairline cracks,* usually result from concrete shrinkage during curing, or from the lack of proper strength. Routine surface preparation will remove hairline cracks and expose the cracks that are truly a structural problem.

Non-moving Joint Sealing. Saw cuts and construction joints should be filled with a flexible resin (do not cover the exposed surface area of the flexible resin with the epoxy floor coating).

Spalled Areas. Repair all chuckholes and spalled areas with a low modulus, flexible epoxy mortar system.

Typical Application Tools. a) A 9" to 18" wide roller with extension handle. Use solvent-resistant short cover. b) Squeegee with flat edges. c) Airless sprayer with extension spray gun.

Typical Rates of Application for Neat Material. The *1st coat* is applied evenly over the surface at an average rate of 1 gallon per 160 to 250 sq. ft. with squeegees, rollers or airless sprayer.

The *2nd coat* is applied within 24 hours of applying the first coat and is applied at an average rate of 1 gallon per 160 to 250 sq. ft. with sqeegees, rollers or airless sprayer.

Typical Rates of Application for a Sand Seeded System for Skid Resistance. The *1st coat* is applied at the rate of 100 to 125 sq. ft. per gallon with sqeegees, rollers and or airless sprayer. Immediately seed dry, clean silica sand, sandblasting grade, evenly over the tacky surface. The sand should be seeded until a thin layer of dry sand is on the surface. Use approximately 30 lbs. of sand per 100 sq. ft. Allow to become tack-free.

After tack-free stage has developed, remove excess sand and vacuum free of all dust particles. Any areas where the sand is uneven or too high, grind the surface level to adjacent surfaces. Re-vacuum the area.

Immediately apply the *2nd coat* with squeegees, rollers and airless sprayer. The final anti-skid surface will be determined by the amount of epoxy applied to the surface. Typically, the 2nd coat will cover 80 to 100 sq. ft. per gallon. For actual quantities of epoxy to be used, a sample should be made, because sand gradations and surface areas vary widely.

CHAPTER 19

HANDICAPPED FACILITIES

The issue of providing facilities to enable the handicapped individual access to the built environment was presented to the construction industry and governmental agencies during the early 1950s in an unstructured and fragmented form by a number of civic groups. By 1961 a specification had been developed as a guide for making buildings and other facilities accessible to and usable by the physically handicapped. This standard was adopted by the American National Standards Institute (ANSI Standard A117.1 - 1986) and became the basis for future laws enacted to protect the interest of the handicapped. The first major federal law was the Architectural Barrier Act of 1968 (amended 36 CFR 1190), which provided that any building construction that was financed in any part with federal funds must make provisions for accessibility to the handicapped. Unfortunately, this law did not provide for enforcement, and it had little impact on changing the design of buildings to incorporate new features.

In 1973 The Rehabilitation Act was passed, and this law provided for the establishment of an Architectural and Transportation Compliance Board to oversee the application of the law. As a result, most cities and states have enacted similar laws requiring the inclusion of design features in buildings to insure barrier-free access and use by the physically handicapped. This also includes access to buildings, such as parking spaces assigned to the handicapped and sloped walkways for wheelchair access.

The magnitude of the handicapped population is brought into focus by studies showing that more than 12% of the entire population, or 30 million people, suffer some type of handicap. This figure continues to increase with the added longevity of 24 million aged persons who suffer some degree of disability associated with the normal process of aging, causing a reduction in perception, mobility, and coordination.

Contractors are subjected to providing the specialized facilities and equipment necessary to allow the handicapped to function and perform in a manner compatible with their peers.

Studies conducted on new projects indicate that less than 1% will be added to the construction costs to make facilities acceptable to accommodate the needs of the handicapped. When modifying existing structures to provide the support services for the handicapped, considerable expense can be encountered, especially if structural, electrical, and vertical lift equipment must be modified.

A classic example is a project to provide free access and all support facilities for the use of a toilet area by the handicapped. In this case, coordination between trades is most difficult. Considerable nonproductive time will be encountered simply because of a large number of trades in a restricted area attempting to perform relatively small but necessary items of work. Under these conditions, the contractor should not attempt to rely on the usual unit cost method, but would be more accurate in identifying each item of work by trade and develop a daily work schedule for each crew. An allowance should be included for interference from other trades, and the added cost of demobilization and remobilization of equipment and work crews to accommodate other trades. Work production will be very poor under these conditions. The development of and adherence to a work schedule is the most positive means of cost control.

Though clouded by the uncertainty of job conditions that could substantially alter them, the following unit prices have been used as base cost figures in a large number of cases and will serve as a guide.

Site Facilities

Concrete ramps on grade, 1:12 pitch, 4" thick with mesh, broom finish, 4" base	$3.00 SF
Steel pipe handrail, 1½" galv., 2 rails 36" high	31.00 LF
Non slip abrasives for walks and ramps	.40 SF
Parking signs for handicapped areas	31.00 EA
Pavement painted indicator	5.00 EA
Embedded abrasive warning strips	7.50 LF
Exterior hazardous area warning grooves 1" deep	1.90 LF
Depressed curb ramps	5.00 SF

Interior Stairs and Corridors

Hazardous area warning abrasive strips	$7.50 LF
6" curbs with pipe rail at dangerous areas	32.00 LF
Integrated stair nosings, non slip	30.00 EA
Corridor wall rails, aluminum pipe	18.00 LF
Balcony railings, aluminum 1½" post with pickets	37.00 LF
Stair hand rails, steel, with newel and brackets	31.00 LF
Floor grating, aluminum, ¼" maximum spacing, safety surface	10.20 SF
Vinyl tile for sound reflection	.80-1.10 SF
Ramps, wood not to exceed 12'-0" length and 4'-0" width	6.50 SF

For curved railings add 33%

Information Only

1. Corridors, minimum 36" width single with turn around area every 200 LF.

2. 360-degree wheelchair turnaround is 60" dia. or 19.62 SF.
3. Right angle grid pattern for corridors and public spaces are perceived more easily than curves or serpentines.
4. Diagonal reach from wheelchairs is 48" above floor level for telephones, etc., with forward reach of 18.5".
5. Overhead obstructions not less than 7'-0" above finish floor.
6. Parking spaces as specified in 36 CFR 1190.31.

Total Parking in Lot	Required Minimum No. of Accessible Spaces
1 to 25	1
26 to 50	2
51 to 75	3
76 to 100	4
101 to 150	5
151 to 200	6
201 to 300	7
301 to 400	8
401 to 500	9
501 to 1,000	2% of total
over 1,000	20 plus 1 for each add'l 100

7. Passenger loading zones shall provide an access aisle at least 60" wide by 20 ft. long, adjacent, parallel, and level with the vehicle standing space. Vehicle standing spaces and access aisles should have surface slopes not exceeding 1:48 (1/4" per ft.) in all directions.
8. Ramps and curb Ramps, Maximum running slope shall not exceed 1:12 or 8%; maximum rise for any run shall not exceed 2'-6".
9. Grab bars, like handrails, shall have an outside bar diameter of 1 1/4" to 1 1/2", and the space between the wall and the centerline of the pipe shall be 1 1/2 inches.

Case Work

Counter height should be adjustable, between 28" and 36", with minimum depth of 19" (per lin. ft.)	$42.00
General work surfaces 25" to 30" above finished floor with maximum depth of 25" (per lin. ft.)	50.00-65.00
Built-in case work, with toe clearance of 7" recess and 10" above finished floor (per lin. ft.)	42.00
Closets to be partially accessible for wheelchairs should have adjustable clothes rod (per set)	25.00

All case work to have recessed hardware.

Doors and Openings

Electric operated automatic doors with delay mechanism, complete including hardware (each)	$5,500.00
Access door adjacent to revolving door	1,200.00
Vision light in solid doors (each)	50.00
Lever type hardware (per set)	30.00
Knurled back handles for dangerous areas (per set)	30.00
Corner protection, stainless steel, 16 ga. (per lin. ft.)	17.00
Corner protection, vinyl, high impact, with stainless frame (per lin. ft.)	21.40

Identifying Devices

International handicapped sign (each)	$31.00
Plaques for rooms with raised letters (per letter)	2.75
interior sign character 2 1/2" to 4"	5.15
Exterior sign character 4" to 8"	10.50

Elevators

Hall lanterns with visual and audible signals indicating elevator direction and floor location (each)	$250.00
Elevator door jambs equipped with floor designations on each side having 2" characters (per pair)	15.00
Doors equipped with reopening devices not requiring contact to be activated (per car)	500.00
Interior car position indicator with visual and audible signals for each floor level (per car)	250.00
Emergency telephone located 48" above floor	N/C
Call buttons with raised characters (each)	40.00

Kitchen

Front loading dishwashers, with rack space accessible from the front	N/C
Front loading washers and dryers	N/C
Cooking surfaces with front controls	N/C
Wall ovens mounted with top of door 31" above floor	N/C
Storage and utility areas to have peg boards, shelving, etc. at convenient level	N/C
Refrigerator/freezer with 50% of freezer space below 54" and 100% of refrigerator controls below 54 inches	50.00-60.00

Mechanical and Toilet Facilities

Water closets, mounting height 17" to 19" above floor with 18" clearance each side (additional unit cost 75.00 each)

For complete installation, hook-up to existing waste & supply	$925.00
Stainless steel drinking fountain, dual level for handicapped installed complete, waterspouts to be no higher than 36" from floor to spout outlet	865.00
Drinking fountain, wheelchair handicapped unit installed complete, waterspout no higher than 36" from floor to spout outlet	950.00
Shower, fiberglass with cast base 32" x 32", complete hook-up to existing waste and supply	610.00
Recessed drinking fountains (add for alcove)	250.00
Water cooler, non-recessed stainless steel cabinet, wheelchair type complete with hook-up to existing waste and supply	780.00
Grab bars, 24" stainless steel	28.00
Handicap shower head controls, shower spray with a minimum 60" of hose (each)	75.00
Shower transfer seat (each)	70.00
Toilet partitions, handicap units, porcelain enamel (ea.)	1,775.00
Toilet partitions, handicap units, stainless steel (each)	1,715.00
Toilet screens, porcelain enamel (each)	120.00
Toilet screens, stainless steel (each)	110.00

Type of anchoring system for toilet partitions will influence cost. Unit prices quoted above are for ceiling hung units.

Tilting mirror, bracket only (each)	70.00
Medicine cabinet, useable bottom shelf no higher than 44"	N/C

Electrical

Electric automatic door opening control located near door jamb (per station)	$400.00
Audible signals at visual electric signals (each)	350.00
Electric door holder	125.00
Public Address System (per speaker)	225.00
Sound System, AM-FM, 12 outlets	8,000.00
Fire detection systems, including smoke and heat detectors, for complete installation of a 10-detector system	2,500.00
Master clock system, including clocks and bells, complete for a 20-unit system	14,250.00

Master TV antenna including VHF-UHF antenna and distribution
to a 10-outlet system 4,800.00
Locating receptacles and switches at lower height
for handicapped . . . . N/C

Elevator controls, all car buttons to be a minimum 3/4" in the least dimension, and all buttons shall be between 48" and 54" above the floor, depending on the type of door opening.

The following standard design details are taken directly from the *Federal Register,* Vol. 47, No. 150, August 4, 1982. We have included them here as examples of minimum guidelines and requirements for accessible design.

However, there is no guarantee that these are the latest requirements, and the contractor is strongly cautioned to check and make sure that these are the most current.

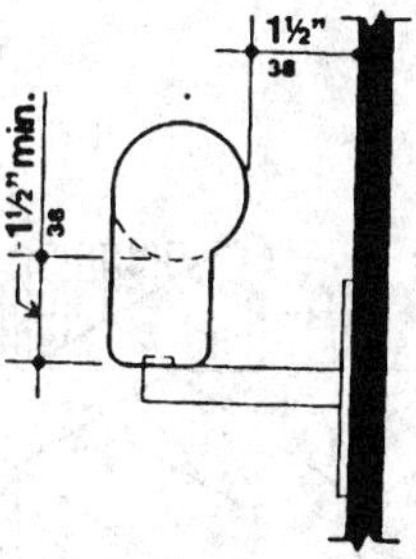

Handrail

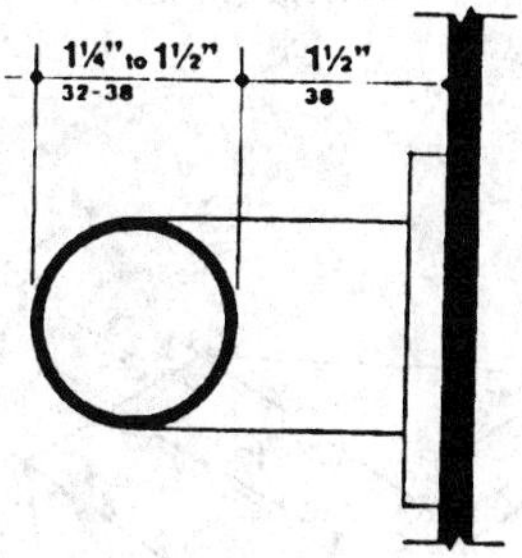

Grab Bar

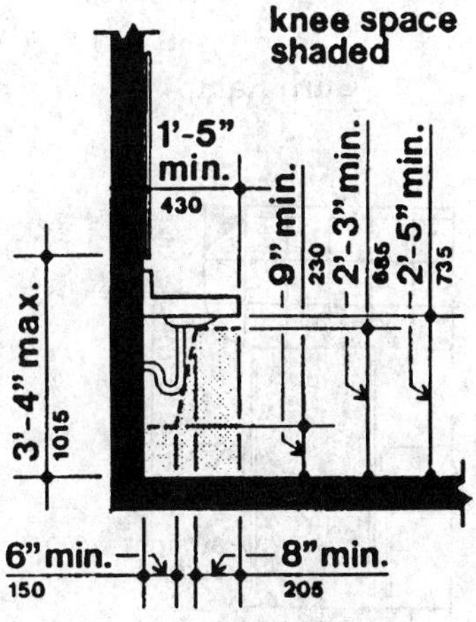

Lavatory

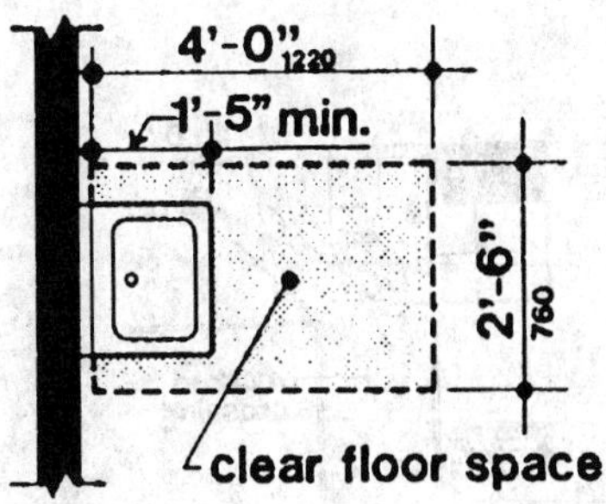

Lavatory

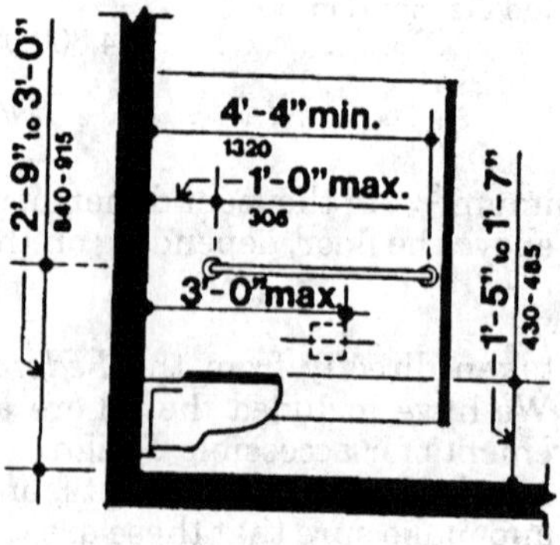

Water Closet

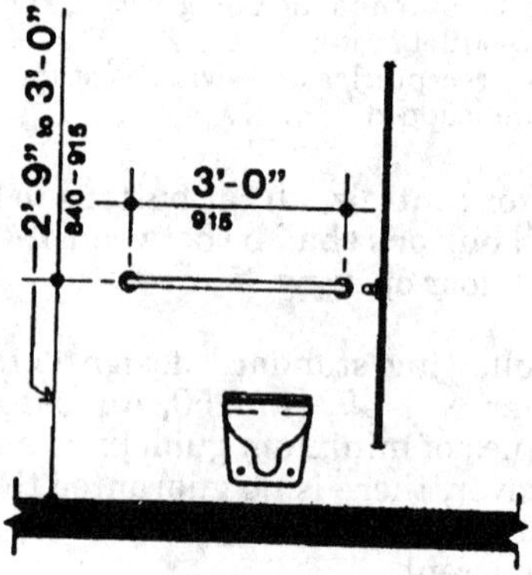

Water Closet

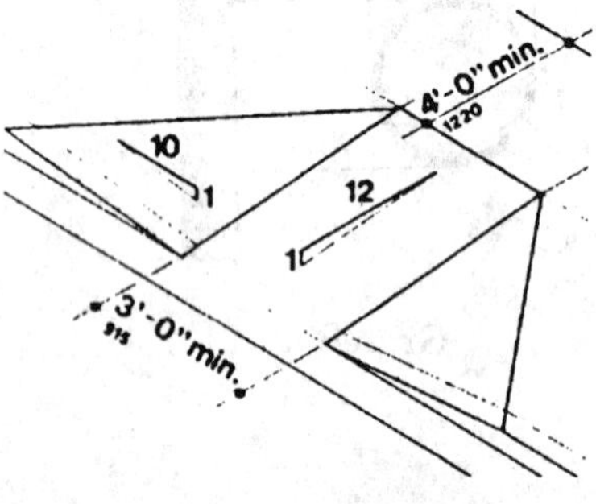

Curb Ramp

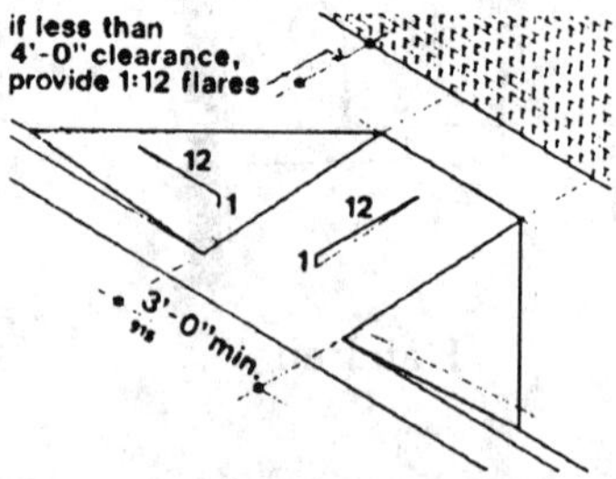

Curb Ramp

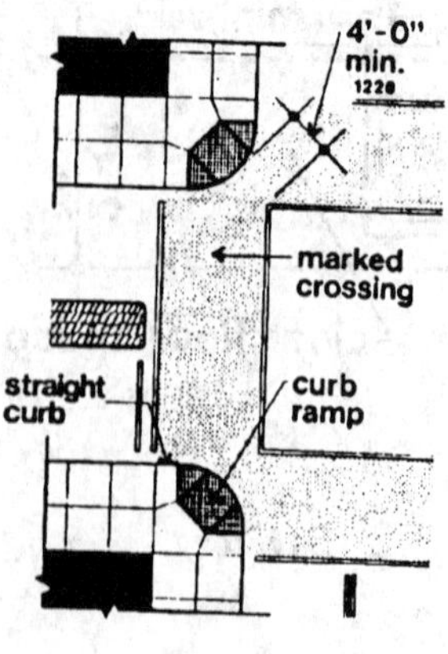

Crossing

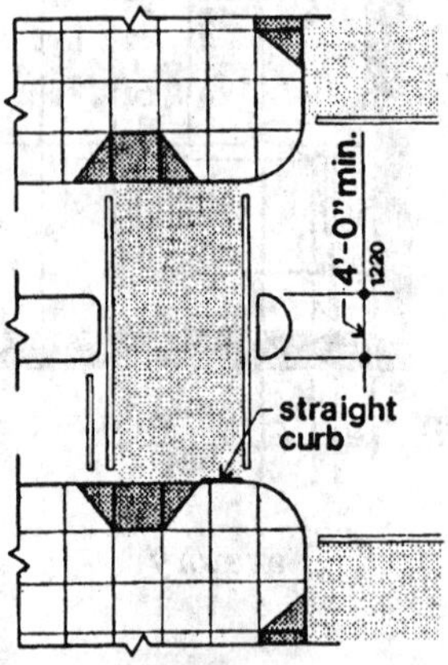

Crossing

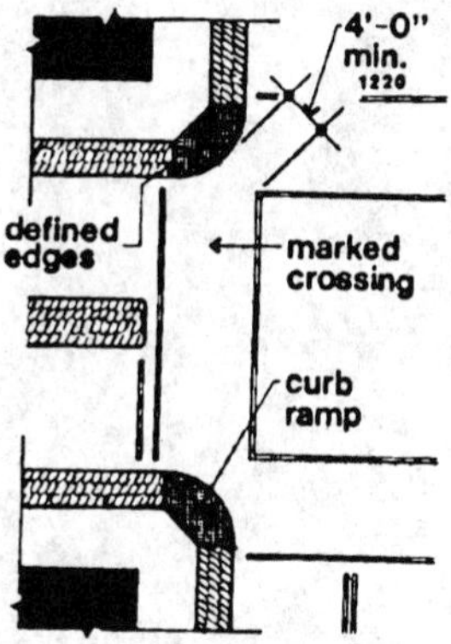

Crossing

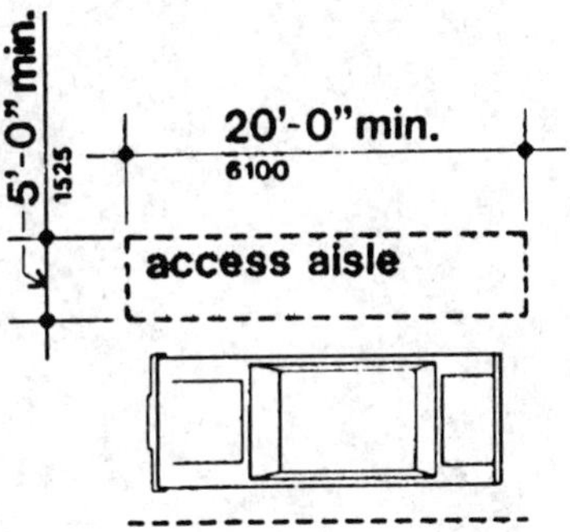

Unloading Zone

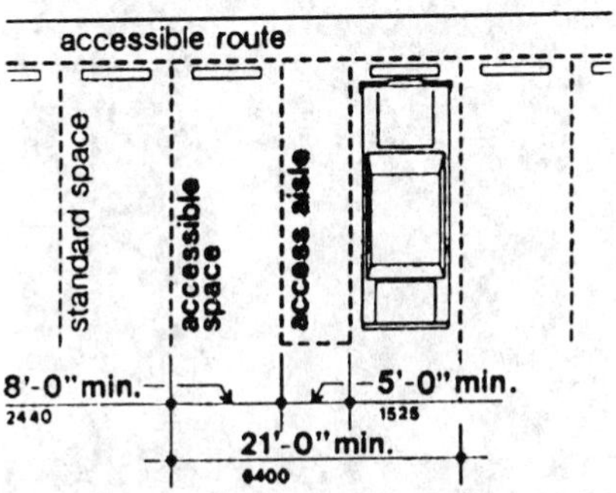

Accessible Parking

CHAPTER 20

ENERGY MANAGEMENT

Energy management systems can no longer be considered a luxury. The impact of energy costs has forced building owners to seek relief through management systems that control energy consumption, optimize comfort, interface with human performance and provide automatic data analysis.

This new market, which has only been lightly penetrated so far, has attracted a great variety of manufacturers, consultants, vendors and suppliers, each attempting to capture a share of this lucrative business, and there has resulted many cases where owners have experienced excessive costs due to over design, or where systems did not provide results because of a lack of knowledge or expertise on the part of the installer.

Although energy management cannot be considered an organized industry as yet, competition and advanced technology are stabilizing factors. It is now possible to make an analysis of the needs of a conservation package for a building and procure responsible bids from several specialists with a feeling of confidence that the system will be cost effective and will produce the desired results.

The first major thrust into energy management was made by the national control and computer manufacturers. Through their established marketing and advertising capabilities, they were able to rapidly develop the large client market of hospitals, commercial and industrial buildings, educational institutions and major office buildings. With the advance of microprocessor technology, which can sell for less than $1,000.00, it is possible to install a system that does not require a $50,000 to $250,000 computer, and the advantages of energy management are available for use in smaller buildings.

In spite of the wide variety of electronic energy management systems available today, there also exists for relatively simple applications the ability to design non-electronic control panels that can provide automatic setback of thermostats and restrictions on electrical or fuel consumption, both of which translate into operating economies of major significance. Other than the economy of initial installation costs, this simple system would be found attractive for use in the more remote areas, where service could be supplied by the ordinary electrical or air conditioning service company. It is estimated that such panels could be installed for approximately $3,000 to $4,000 for an open space store area of about 25,000 square feet with limited glass area and insulated roof area.

The costs involved in the more complex computer controlled energy management systems cannot be predetermined, because each system is specifically designed for the needs of the owner. It is not uncommon to include additional functions in the system, such as fire and smoke alarms, security controls, or other specialized features.

One of the hidden or ignored costs in building energy management systems is that they share a fundamental need with other complex contemporary technology, the continuous availability of maintenance service. Low cost microprocessors are reliable because of their relatively simple design, but these systems usually provide for service contracts that range in annual cost from 6% to 10% of the system purchase price. In the more complex computer controlled systems, service contracts are offered at approximately 10% of the purchase price with an annual escalation of 10% to 12%.

Consideration must be given to the costs associated with the training of personnel to operate the system, because the maintenance staff will have to take an active part in the installation and commissioning procedure and must actively participate with the operators in the initial training by the manufacturer of the system. While these costs are related to the operational costs that the owner must provide for, they are many times not given consideration by the owner as part of the installation cost of the energy management system. The estimator who is aware of these oftentimes overlooked expenses and their relation to the efficient operation of the system, including minimizing the need for maintenance service by the manufacturer, will be in a position to extend an additional service to the owner by offering certain guidelines for the type and duration of training for the operating personnel. In this way the owner will become conscious of the need for operational training and alerted to the associated costs.

In the absence of a highly developed and organized energy management industry, there is no single source to obtain historical data relating to installation costs, performance reports, reliability studies or cost effectiveness of the various systems available. Much of the information available to the public is obtainable from the large control and computer manufacturers and is essentially sales oriented to their particular equipment and the accomplishments they have achieved. This situation places a burden on the contractor involved in an energy management system for a new building, or in modifying the energy systems in a existing building, because he will be forced to evaluate technical proposals with extreme care before reaching a decision that the offer provides the maximum efficiency for the most reasonable price.

Even though there were some early financial disasters resulting from insufficient knowledge of the capabilities and limitations of energy management systems and from the lack of expertise on the part of some companies, the industry is providing a significant contribution.

It is not unusual to develop systems serving the needs of both small and large clients that will reduce energy consumption in a building by 25% and yield a payback of the initial installation cost in about two years. With such attractive figures as these, it must be contemplated that this industry will continue to mushroom and progressive contractors will recognize the potential of this market by initiating an in-house program to educate their personnel in the arts of this new industry.

From the position of establishing some basic method of determining a cost criteria to serve as an estimating guide, surveys indicate that it is inappropriate to attempt to develop such information at this time due to the absence of sufficient historical data cost bank. It has been found installations costs can be as low as $0.15 per square foot of floor area for industrial plants containing up to 2 million square feet, while in a 36,000 square feet store using a non-electronic system, the cost was found to be $0.10 per square foot. In the instance of the institutional and office complex, installations that are frequently loaded with a series of other functions, in addition to energy management, it is virtually impossible to develop meaningful conceptual costs because of the individuality of the design. Estimating techniques and technical information relating to energy management will eventually be provided to the estimator as it becomes available.

Energy management has rapidly become a major consideration of every business and industry as energy costs continue to rise. To lower energy usage, users are reducing the operation of lighting and HVAC equipment. However, additional help is often needed in the form of *load control*. Load control is defined as automatically controlling electrical equipment loads to reduce the electrical consumption and usage rate.

Most electric utilities base the monthly charges to commercial users on two separate items: a) consumption, the total kilowatt-hours of electricity used during any billing period; and b) demand charges, based on the rate at which electricity is used. Most utility companies measure the electricity being used during a short period of time called a *demand interval*. The monthly demand charge is then based on the maximum peak usage that occurred within the demand interval of each billing period.

Load control systems design requires the input of several important factors before the proper implementation can occur.

Every building has its own set of characteristics. Even for similar buildings, such as chain stores, each has unique situations that change the design and application of a load control system. Variables affecting the control system design include climate, building design and construction, type and capacity of HVAC and electrical equipment, business hours and traffic flow in and out of the building.

The building's electrical usage data (load profile) must be analyzed and all equipment items examined for their operating cycles in order to determine which items are in operation at peak demand times, lengths of continuous operation, etc. This information needs to be obtained for weekday, weekend and seasonal changes.

Local health codes are an important factor should the load control system be installed in food service facilities. It is possible for the load control system to reduce the operation of certain equipment to the point where a violation of the minimum health code requirements may occur. For example, if the load control system turns off the hot water heater too often or for too long a period of time, the temperature of the dishwasher water may fall below the minimum allowable by health codes.

Methods Of Load Control

Duty Cycling. Duty cycling accomplishes load control by turning on or off all loads on a programmed basis. The major function of this method is to reduce the kilowatt-hour consumption. Duty cycling equipment based on temperature (temperature selective duty cycling) increases the equipment's operational efficiency and longevity and lowers energy consumption. An example would be the shut-down of air conditioning equipment once the A/C unit has lowered the temperature to the prescribed level. Needless energy is wasted in allowing the unit to run nearly unloaded.

Each load (equipment item) that is connected to the duty cycler is programmed to turn off for a specific portion of time in a repeating time interval cycle. The amount of time a load is off is directly dependent upon the importance of that item to the overall operation of the facility.

Duty cycling can be accomplished by using two separate cycles, Long-Off-Time-Cycle (LOTC) and Short-Off-Time-Cycle (SOTC). When in the LOTC mode, the equipment load is turned off for longer periods of time in the repeating cycle. The duration of the LOTC, SOTC and the repeating time interval cycle is defined by the system user. Temperature set points are also used in conjunction with the LOTC and SOTC to give virtually any combination of temperature regulated on/off mode for the equipment load.

In applications where product temperatures are critical (food service), temperature override devices can be utilized to prevent the degree limit being reached.

The duty cycle function can be overridden by a demand limit control program should the actual demand level exceed the set points of the duty cycle load control center.

Demand Limiting. Demand limiting is a term that describes the process of turning off (shedding) loads temporarily to keep the kilowatt (KW) usage demand below a predetermined level. To accomplish this, the loads connected to the demand limit control are turned off, or shed, when the KW demand level reaches the set level, which is the lowest practical KW usage rate allowable to meet the facilities operating requirements.

The demand limit control is programmed to shed loads in a predetermined order when the demand set level is exceeded. This control will continue to shed loads until the KW demand level is less than the set level minus a *deadband* range. The deadband is intended to prevent unnecessary cycling of loads controlled by the demand limit control when the KW demand level is in an area around the set level. If the demand level is in the deadband range, additional loads are not shed or restored by the demand limit control.

Demand limit controls can monitor KW demand in several ways. One is the instantaneous rate method which compares the instantaneous rate of power usage to the predetermined set level. The other way is the average rate method which continuously samples and averages the KW demand level over a short period of time. The averaged usage range is then compared with the predetermined set level.

Time Of Day Programming. Another method of load control is time of day programming or load scheduling, which is the process of automatically turning on and off specific equipment loads based on the time of day.

Some loads are often left on because of failure to manually disengage the switch. These types of loads are the ones controlled by the load scheduling device.

Most all energy management load control systems incorporate all three methods of load control (duty cycling, demand limiting, and time of day programming) into one integrated package that gives the user a complete system for controlling all equipment loads in his facility.

CHAPTER 21

HAZARDOUS WASTE

This chapter mainly addresses issues of asbestos abatement and control in buildings, but we have also attempted to cover other areas of concern, such as disposing of contaminated soil and removing underground storage tanks. All of these items can directly affect commercial building projects.

ASBESTOS

For many years asbestos materials were used for acoustical insulation, fireproofing spray on structural steel framing members, fireproofing spray on ducts, pipe and duct insulation and boiler and breeching padding.

Early in the 1970s, scientists determined conclusively the extreme health hazard posed by asbestos fibers. The government, at that point, immediately imposed strict regulations on the manufacture and use of asbestos products. The Occupational Safety and Health Administration (OSHA) has sets of regulations that control such things as permissible asbestos exposure limits and methods of complying with them. Whenever work is performed involving the installation or removal of asbestos materials, it is required that instruments be used to measure and monitor the area's airborne fiber content. Special protective clothing shall be worn by all personnel exposed to the contaminated areas.

In 1978 asbestos was banned from all spray-on materials including those used for decorative purposes. The Environmental Protection Agency (EPA) contracted with Battelle Memorial Institute to conduct a test program to determine whether encapsulation or sealing of asbestos would be effective. As a result, several laws were proposed, and in June, 1980, the Asbestos School Hazard Detection and Control Act was passed. On October 22, 1986, Congress passed the Asbestos Hazard Emergency Response Act (AHERA) for removal of asbestos in private and public schools. Asbestos abatement projects now in progress are being funded at the federal, state, local and private levels.

Asbestos is considered to be hazardous in a friable state, which means that the material can be easily crushed or turned into powder by using hand pressure. The greatest hazard posed to humans by friable asbestos is airborne fibers. At present, OSHA has set levels of the allowable asbestos fibers in a certain quantity of air. The current Permissible Exposure Limit (PEL) is two fibers, 5 micrometers in length, per cubic centimeters of air. A person inhales about one cubic meter of air per hour which is 1 million cubic centimeters. This amounts to an intake of 2 million fibers per hour and that quantity does not include the fibers that are too small for detection by the current measuring equipment.

OSHA has proposed new standards on exposure to asbestos in the work place. Among other things, the new standard would set the *permissible exposure limit* at either 0.5 fiber or 0.2 fibers per cubic centimeter of air. However, at this writing, the new standard has not yet been accepted as the prescribed standard. Certain states, however, have taken it upon themselves to issue stricter requirements for airborne asbestos fibers. For example, Virginia has recently passed a new guideline that reduces the acceptable asbestos exposure limits from 2 fibers per cubic centimeter of air to 0.5 fibers.

Asbestos is facing ultimate market extinction under a sweeping ban being contemplated by the EPA, and alternative product development is already underway. In late 1983, EPA advised asbestos producers, product manufacturers and environmental groups that it planned to issue a ruling in July, 1984 banning six categories of asbestos containing products and to propose another rule in October 1984 setting production ceilings for remaining products. The limits would decline annually until all products containing asbestos, a known carcinogen, are eventually phased out. The products covered by the proposed ban were saturated asbestos roofing felt, unsaturated roofing felt, asbestos flooring felt, asbestos sheet flooring, vinyl asbestos floor tile, and cement asbestos pipe. A major factor in the schedule of planned phase-out of the remaining asbestos containing products were to be the availability of a suitable substitute material. These proposed rulings were a continuation of the rule making that was begun in 1979 when the EPA announced its intent to ban such products under the Toxic Substance Control Act. Presently, thirty-two states have passed regulations related to asbestos removal.

The estimator must be cautioned that abatement rules, regulations and restrictions are continually being altered and revised. The estimator must ascertain the latest rules and regulations as of the time of bid for each new project.

Abatement Methods. Once the problem of asbestos content had been discovered and analyzed, then the abatement program was initiated and continues to be expanded. There are presently three methods of controlling asbestos.

The first is *removal.* This is by far the most expensive method, because it takes more time for removal and controlled disposal for final deposit in limited designated disposal sites around the country. Also consider that the material was originally installed for a purpose, whether it be thermal or sound, and therefore, usually needs to be replaced with a new non-asbestos material that performs the function of the removed asbestos.

The second method is the *enclosure system* which is simply installing another material, such as drywall, over the existing asbestos materials.

The third method is known as *encapsulation* which requires the spraying of a sealant material onto the asbestos either to form a bridge over the asbestos insulation, or to penetrate into the asbestos to bind the mass together. The effectiveness of encapsulation depends largely on the condition of the existing asbestos. For example, if the asbestos is very loose and becoming delaminated from the surface, encapsulation will not work because the material will simply fall off the structure once it is agitated by the application process of encapsulation sealants. These sealants are applied with an airless spray. The EPA has determined that brushes and rollers should not be used, because they create excessive disturbance on the surface of the asbestos materials. Using the airless spray also disturbs the surface of the asbestos, but it is the lesser of evils.

To be effective, the encapsulation sealant should virtually eliminate any asbestos fallout into the atmosphere (airborne materials), provide some resistance to impact and abrasion and be reasonably flexible to handle any normal movement of the building. It needs to have a flamespread resistance for building code compliance, and the material, in the dry state, has to be free of odor.

As stated, the encapsulation sealants basically fall into two categories, *bridging* and *penetrating,* but one important fact to note is that the AHERA

Act of 1986 mandates the *removal* of all asbestos material for all school districts, public and private.

A *bridging sealant* is a coating that forms a barrier between the asbestos materials and its surrounding atmosphere. These sealants are applied at the rate of approximately 3 gallons per 100 sq. ft. for a dry film thickness of about 25 mils. There is very little surface penetration when using a bridging sealant. The bridging sealant is an application of two coats, with half of the material applied in the first coat, and the remaining half on the second coat. Normally, the coats will be color coded for coat identification. Proper coverage is very important; if cracks develop in the finished surface, there is an escape point for the asbestos fibers and, therefore, the encapsulation will be rendered ineffective and the surface area must be retreated.

The *penetrating sealant* is a very watery solution which penetrates into and hardens the asbestos materials. The penetrating sealant method is a little more difficult to control in the initial installation, since the purpose of the penetrating sealant is to absorb into the materials, cure and bind them together. If insufficient material is used on the first coat, and that coat is allowed to dry, a film is formed that interferes with the penetration of any subsequent coats. Therefore, the penetrating sealant is normally applied in a three coat application with each coat being applied prior to the drying of the previous coat. It is more difficult to estimate the quantity of materials to be used in the penetrating method because of the porosity and thickness of the asbestos material. Normal coverage can be obtained with 2 to 3 gallons per 100 sq. ft., but in extreme conditions, 10 to 15 gallons per 100 sq. ft. may be required. This greatly increases the material cost and the amount of labor time required to spray on the additional material. The easiest method to determine the actual quantity of penetrating sealant to be used is to spray a test area of the asbestos, allowing it to cure, and then coring the portion of asbestos and examine it for proper penetration and bonding characteristics.

In any asbestos abatement program, whether removal, enclosure, or encapsulation, the most expensive and time consuming job is the preparation of the area where the asbestos is to be removed or encapsulated. The walls and floors have to be completely covered with polyethylene. The entrance to the area has to be converted into an air lock (and preferably 2 air locks at least 6' apart). Light fixtures have to be removed, other types of equipment and items have to be covered or removed, the return and supply air grills and diffusers have to be removed and the openings sealed in the duct. The work area itself should maintain a negative air pressure which helps prevent asbestos fibers from escaping into other areas.

Special change rooms are required where the workers can change into their work clothes and exit through an air lock into the work area. Upon return from the work area, the workers will go into a room, strip, dispose of their work suits, enter the shower and wash, exit into the area containing their street clothes and exit the area. All workers are prohibited from smoking, eating, or drinking in the contaminated work area, or the contaminated portions of the change rooms. Thus the average worker will normally exit the work area, dispose of the work suit, shower, and dress 3 to 4 times each work day.

One of the manufacturers of asbestos encapsulating sealant materials is the Foster Products Division of the H.B. Fuller Co. in Springhouse, Pa. Its bridging encapsulant material is a latex elastomer, white in color, and has the consistency of heavy paint with a weight of 11.4 pounds per gallon. The material is applied by the airless spray method, and has a mild odor while wet but odorless when dry. The recommended coverages for this particular material is 3 gallons per 100 sq. ft. (or if applying in 2 coats, 1½ gallons per 100 sq. ft., each coat) and this will depend on the texture and porosity of the asbestos.

The *Protektor* sealant material is an acrylic copolymer, comes in colors of yellow, blue, and untinted, and has a consistency of a thin fluid for proper penetration into the asbestos. Its weight per gallon is approximately 8.4 lbs. It is applied with the airless spray method the same as the bridging encapsulant. Its odor is mild while wet, odorless when dry. The recommended coverage of this material is 2 to 10 gallons per 100 sq. ft. depending on the absorbency and thickness of the asbestos and is best applied in 2 to 3 coats.

During the process of asbestos removal or encapsulation, there are two main pieces of equipment that the contractor will need to have at his disposal during the entire process. The first is an exhaust filtration unit, available from a number of manufacturers, such as the Torit Division of the Donaldson Company, which has local representatives throughout the country. This unit is an exhaust fan made in varying sizes depending upon the amount of air, cubic feet per minute (CFM), that needs to be moved. The air that flows through this exhaust fan is filtered by means of a high efficiency particle air (HEPA) filter and deposits the asbestos fibers in a special drum below the exhaust fan. The typical drum size is a 55-gallon unit, which contains a disposable plastic bag with a special marking, warning the handler of the contaminated hazardous waste material.

The size of the exhaust fan required for the project is determined by the number of air changes per hour that is specified (normally four) and the size of the area that is to be exhausted. An example of this would be if you had a room 50' long x 50' wide x 15' high; the cubic content of this room would be 37,500 cu. ft. Because the exhaust unit is rated in cu. ft. of air moved per minute (CFM), you would divide 37,500 cu. ft. by 60 minutes, then multiply by 4 air changes, and the results would be 2,500 CFM capacity for the exhaust fan.

In quotations from the Torit Division, a 1,000 CFM exhaust filtration unit, complete with HEPA filter, costs approximately $5,000.00. A 2,000 CFM exhaust filtration unit, complete with HEPA filter, costs approximately $7,000.00. Units can be equipped with automatic alarms and other sensitive detection devices that can double the above prices.

The other item of equipment that will be required is the HEPA vacuum cleaner. One such unit is manufactured by the American Cleaning Equipment Corporation, in Addison, Ill., which is handled by numerous sales representatives throughout the country. This particular vacuum cleaner contains a special five-stage filter unit similar to the exhaust filtration unit. The filter unit alone will cost in excess of $500. The drum that collects the asbestos fibers as they are vacuumed from a surface comes in 15, 30, and 55 gallon sizes. A specially marked plastic bag that the asbestos fibers are deposited into fits in the drum. When the bag is full, it is removed, sealed and placed ready for transport to an approved disposal site.

To alert the contractor to the specific detailed procedures that all site personnel will be subjected to when involved with asbestos work the following standard "worker protection procedure" program has been included.

It is obvious that extremely low productivity will result from the work restrictions placed on employees. It is estimated that labor production might be reduced by 15% to 60% of normal production rates.

Worker Protection Procedure. *Arrival:* Each worker or authorized visitor arriving at the job site shall remove street clothes in the Clean Change

Room and put on a respirator with new filters and protective clothing and shoes before entering the Equipment Room or the Work Area.

Worker Decontamination: Each worker or authorized visitor shall, each time he leaves the Work Area, remove gross contamination from clothing, proceed to the Equipment Room, and remove all clothing except respirators. Still wearing the respirator, proceed naked to the Shower Room, clean the outside of the respirator with soap and water, remove the respirator, shampoo and wash themselves, remove the respirator filter, wet them and dispose the filters in the container provided for that purpose, and wash and rinse the inside of the respirator.

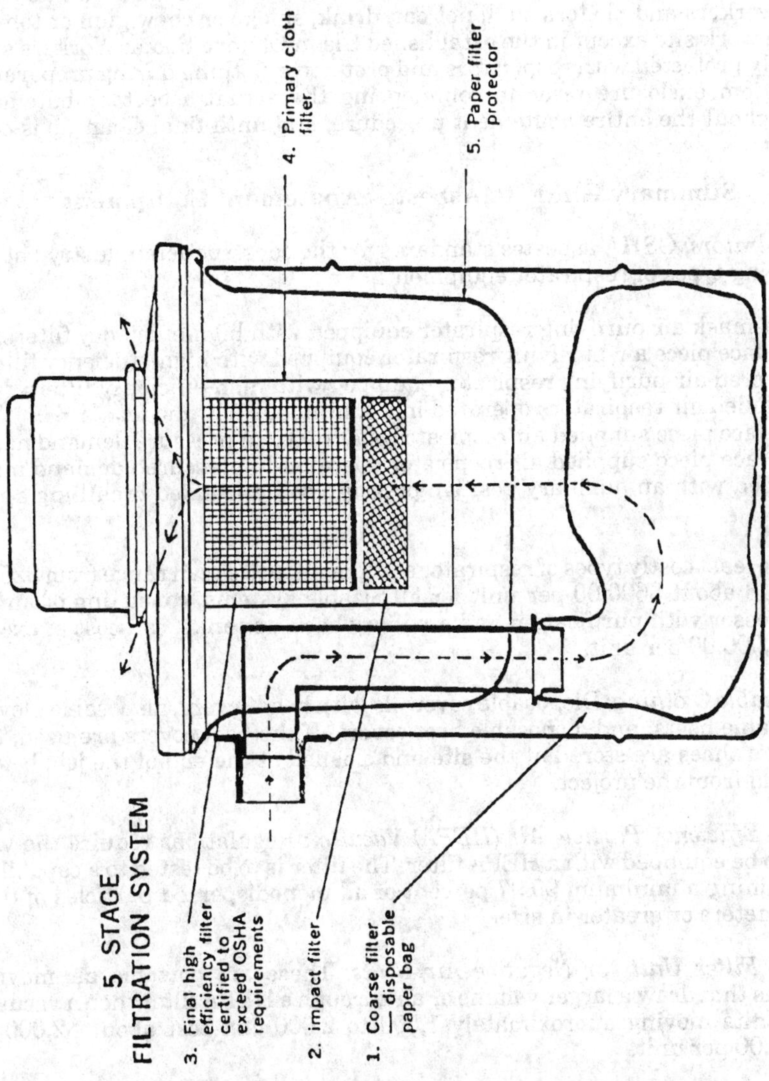

Components of a Typical Asbestos Vacuum

Following showering and drying off, each worker and authorized visitor shall proceed directly to the Clean Change Room and dress in street clothes at the end of each day's work, or in clean coveralls for eating, smoking or drinking. Contaminated work footwear should be stored in the Equipment Room when not in use. Store contaminated work suits in the Equipment Room for re-use or place in container for disposal as asbestos contaminated material.

Upon completion of the asbestos abatement work, dispose of footwear as contaminated waste or clean thoroughly inside and out with soap and water before removing from the Equipment Room. Workers removing waste containers from the Equipment Decontamination Enclosure shall enter the Shower Room from outside wearing a respirator and dressed in clean coveralls.

All workers and visitors shall not eat, drink, smoke or chew gum or tobacco at the work site except in the established Clean Change Room. Workers shall be fully protected with respirators and protective clothing during preparation of system enclosure prior to commencing the actual asbestos abatement, throughout the entire abatement procedure, and until final clean-up is completed.

Summary Guide to Asbestos Abatement Equipment

Respirators: OSHA asbestos standards for the construction industry list the following types of respirator equipment:

a) half-mask air purifying respirator equipped with high-efficiency filters;
b) full face piece air-purifying respirator equipped with high-efficiency filters;
c) powered air-purifying respirator equipped with high-efficiency filters;
d) supplied air respirators operated in a continuous flow mode;
e) full face piece supplied air respirator operated in a pressure demand mode;
f) full face piece supplied air respirator operated in pressured demand mode equipped with an auxiliary positive pressure self-contained breathing apparatus.

For the least costly types of respirator equipment, prices can range from $25.00 for a) to about $600.00 per unit for c). Sizable systems, consisting of an air compressor with purification and auxiliary back-up tanks, can cost in excess of $20,000.00 per unit.

Disposable Clothing: Disposable coveralls with head covers, disposable gloves, disposable boots, and disposable boot covers. If the boot covers are used, the workers shoes are stored at the site and cleaned at the end of the job, before removal from the project.

High Efficiency Particle Air (HEPA) Vacuum: Regulations require the vacuum to be equipped with a HEPA filter. The filter is to be tested to a capability of retaining a minimum 99.97 percent of all monodispersed particles of 0.30 micrometers or greater in size.

HEPA Filter Units or Negative Air Units: These are portable air moving systems that draw a larger volume or air through a HEPA filter then a vacuum can. Units moving approximately 1,500 to 2,000 cfm cost about $2,000 to $3,000.00 per unit.

Pressure Monitoring Equipment: HEPA cabinets or negative air units are used to provide a controlled flow of air and to prevent leakage from a work area to an occupied area.

Disposable Clothing for Hazardous Waste Removal

Wetting Agent: These are chemicals added to water, with concentrations stated by individual manufacturers.

Encapsulant: Encapsulation is the application of a liquid sealant onto sprayed or troweled asbestos. Sealants are usually water based material.

Air Monitoring: There are four basic analytical procedures:

a) phase contrast microscopy (PCM) is a method for counting fibrous particles of defined size, which are deposited on a filter by drawing air through the filter;
b) scanning electron microscopy (SEM) is an alternative to the PCM;
c) fibrous aerosol monitor (FAM) is a laser based method, developed and calibrated to give the same result as the PCM method.
d) transmission electron microscopy (TEM) analyzes and counts every airborne asbestos fiber.

Decontamination Units: Asbestos removal requires the use of decontamination units, both for workers and for waste. These units may be constructed on the site or prefabricated and moved to the site. All units are required to be made airtight with tape and plastic.

Glove Bags: These are plastic bags with at least two long-sleeve gloves fastened to them. The glove bag can be fastened onto an asbestos insulated pipe, and the asbestos insulation can be removed from the pipe within the sealed back. There are two types:

a) polyethylene (PE) with either *Tyvek*, rubber or other material for the sleeves and gloves. Normally used on horizontal pipes.

b) polyvinyl chloride (PVC) type of glove bag normally comes equipped with a zipper. The bags are designed for use on horizontal and vertical pipes, as well as on pipe tees and valves.

Air Monitoring. The contractor shall be responsible for monitoring and testing to determine the fiber count content of the air and on surfaces in and adjoining the work area, during the abatement process and the clean-up phase. All air and surface sampling and testing is normally the expense of the contractor, not the owner. Air and surface samples shall be taken in accordance with OSHA and EPA requirements during each work day. Sampling shall begin by taking samples to establish a "background level" before the start of any construction, including the temporary facilities. Sampling may stop only when final acceptance of the area has been made in writing by the owner.

The laboratory analyzing the samples must be accredited for asbestos analysis by the American Industrial Hygiene Association. Certified copies of the results must be delivered to the owner normally within 24 hours of taking the samples. Continuous air monitoring equipment is used by the contractor to supplement the testing program. However, this air monitoring is not allowed to substitute for the air and surface sampling that is required.

Area Protection. Most contractors use polyethylene to completely cover the walls and floors (assuming the asbestos is on the ceiling) of all areas for decontamination. Normally a 6 mil poly is used for floors, the air locks, and sealing doors and windows. A 4 mil poly is used against wall surfaces. The poly should be as wide as possible to minimize joints, thereby minimizing leakage of airborne asbestos fibers to the outside area.

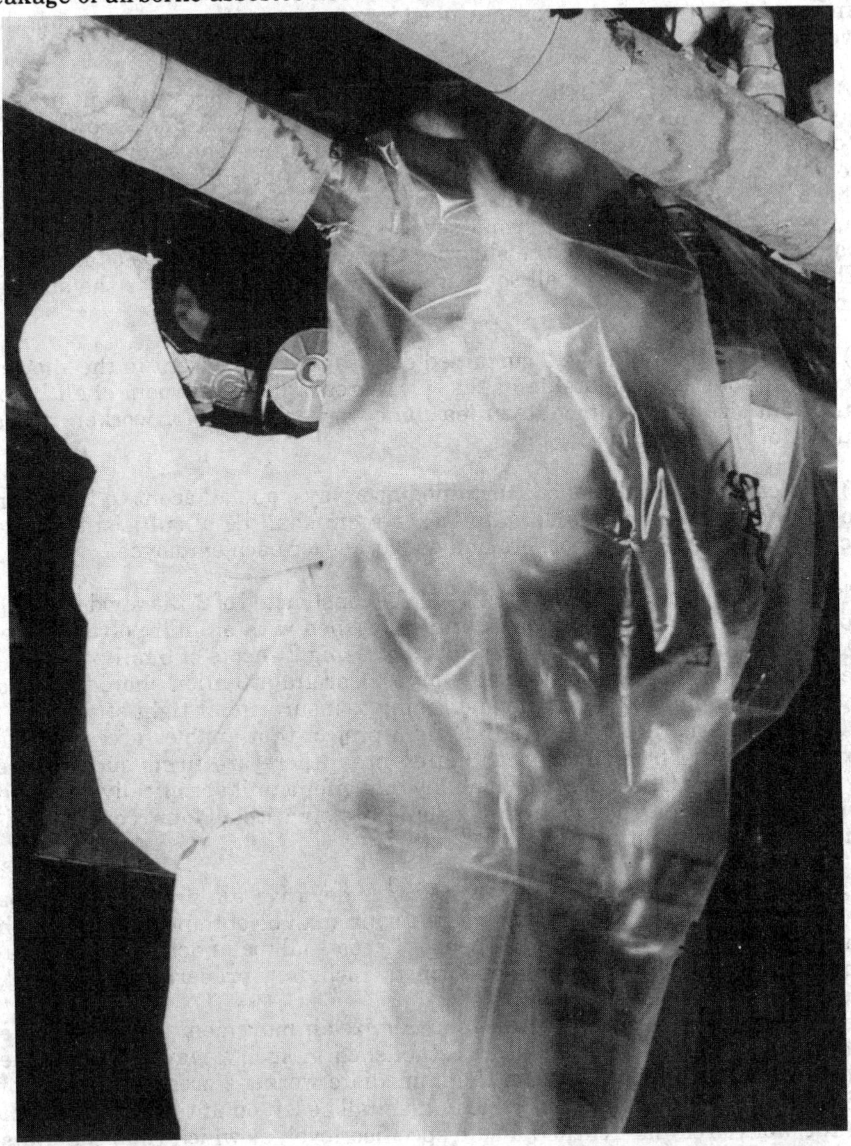

Worker Removing Asbestos Insulation from Pipes Using Glove Bag

Worker Decontamination Facilities. The contractor shall construct a worker decontamination facility consisting of 4 chambers as follows:

a) *Equipment Room:* Shall have two curtained doorways, one adjacent to the work area and one adjacent to the shower room. The equipment room shall be of sufficient size to accommodate at least one worker, allowing him enough room to remove his disposable uniform, as well as being large enough to house any equipment the contractor may need in the contaminated area during the work process.

b) *Shower Room:* Shall have 2 curtained doorways, one adjacent to the equipment room and one adjacent to the clean room. At least one shower with hot and cold water shall be installed in this room. The shower shall be constructed in such a manner that all water will be collected and pumped through a 5-micron filter. From the filter, water may be drained off in any conventional manner to a normal sewer system. Careful attention shall be paid to the construction of the shower to insure against leakage of any kind. The contractor maintains all soap and other drying supplies for the shower area.

c) *Clean Room:* Shall have 2 curtained doorways, one adjacent to the shower room and one to the outside of the work area. The clean room shall be of sufficient size to accommodate at least one worker, towels for workers and a supply of clean overalls.

d) *Locker Room:* Shall have 2 curtained doorways, one adjacent to the clean room and one to the outside. The locker room shall be of sufficient size to accommodate at least 2 workers and one locker for each employee.

This worker decontamination facility can be constructed of 2"x4" wood framing with plywood for sheathing and completely lined with a 6 mil polyethylene. Doorways and air locks can be constructed using 2 sheets of 6 mil poly hung between the head and the floor and jambs. Careful attention should be paid to the overlapping of poly joints and taping to insure an air tight structure.

On the occasion that you are bidding a project that enables access to the exterior directly from the contaminated area, there are firms such as the Williams Mobile Office Trailer Company that offer a unit specifically designed for asbestos abatement projects. It comes complete with all four rooms including the shower facilities.

Placement of Exhaust Air Filters. A negative air pressure shall be established in the work area and make-up air shall be obtained only through the decontamination chambers. The work area shall be under negative pressure whenever work is being performed, including preparation, removal, clean-up, or re-insulation.

Equipment shall be located so as to maximize air movement throughout the work area. Air movement shall be established in such a way that airborne fibers will be carried away from the immediate worker's area. At the end of each work day, the exhaust filtration unit shall be left on until the fiber count in the work area has reached a safe specified level, at which time the units can be turned off and the employee's decontamination disposal routes sealed. Note that an employee will need to be left at the project to accomplish this particular function.

Removal of Asbestos Materials. The asbestos material shall be sprayed with water containing a wetting agent (amended water) to reduce airborne fibers preceding the removal of the asbestos material. The material shall be sufficiently saturated to prevent emission of airborne fibers in excess of the exposure limits prescribed in the OSHA standards.

The asbestos material shall be removed in small sections by 2-worker teams. Asbestos contaminated materials shall be handled carefully and deliberately. Unnecessary agitation of the material is strictly prohibited. The use of any high rpm power equipment, pressure washers or hydroblasters is not acceptable. The contractor shall not allow asbestos material to dry out or collect on the floors.

Removed materials shall be immediately placed in 6 mil plastic bags, sealed and placed into metal drums for transport to the disposal site. Bags and drums shall be marked with the OSHA label prescribed by the OSHA regulation referenced by their specification. The outside of all containers shall be cleaned by wet sponging prior to leaving the work area.

After the wet removal process has been completed, the existing substrate and any surface from which asbestos has been removed shall be thoroughly vacuumed with a HEPA filtered vacuum cleaner. A note of caution: A HEPA filter will fail if used on wet materials.

Clean-up After Decontamination. The following work is performed after decontamination has been accomplished. All plastic coverings shall be vacuumed using a HEPA vacuum cleaner to remove the loose asbestos fibers. The poly is then removed, folded carefully and disposed of in a proper container. The work area is then again vacuumed with a HEPA vacuum cleaner, wet wiped, and in some instances, a second wet wiping is specified. During this particular process, the air monitoring equipment and the exhaust filtration units are in full operation. Once all of the poly is picked up and the room has been vacuumed and wet wiped, the process of re-installing the registers and grilles, the light fixtures, smoke detectors, and any other equipment that was removed prior to the decontamination process can begin.

Estimating Checklist. Because each asbestos removal project has unique job performance parameters, cost data is almost impossible to supply. We offer the following checklist items that the estimator needs to consider in preparing an estimate for asbestos removal work:

1. Work procedures submittal to required agencies.
2. Procedure approvals and permits.
3. Set-up of decontamination rooms.
4. Water, drain and electrical connections to the decontamination rooms.
5. Maintenance of decontamination rooms during the progress of work.
6. Dismantle and removal of decontamination rooms at completion of work.
7. Air monitoring equipment rental, set-up and dismantle.
8. Polyethylene covering for floors, walls and other large objects to remain in the work space; and the subsequent removal of this covering at completion of asbestos removal.
9. Removal or covering for lighting fixtures, smoke detectors, duct grills and registers, etc; and the subsequent reinstallation or uncovering at completion of asbestos removal.
10. Disposable clothing suits.
11. Respirators and filters.
12. Scaffolding, ladders, and work platforms.
13. Surfactant for amended water application.
14. Removal of asbestos materials from substrate.

15. HEPA vacuum equipment rental and use.
16. Exhaust filtration units rental, set-up, and dismantle.
17. Plastic disposal bags and transport drums for protection during haul to disposal site.
18. Transport costs to disposal site and dump fees.
19. Wet cleaning of all surfaces and fixtures.
20. Spray sealants to prevent release of residual asbestos fibers.
21. Replacement thermal or fireproofing materials where asbestos was removed.
22. Record keeping and final submittals.
23. Insurance requirements.
24. Legal liabilities.
25. Training.
26. Development and costs of emergency procedures.
27. Cost for industrial hygiene consultant.
28. Cost to implement an approved safety program.

Sample Costs for Asbestos Removal. Encapsulation of the area where contaminated material is to be removed by negative atmospheric controls; portable showers, portable clean rooms, and portable change rooms. Allow for installation and removal.

	Hours	Rate	Total	Rate	Total
Laborers	32	$....	$....	$19.13	$612.16

To the above, add costs for equipment.

Use of glove bags is an ideal method of removing sound asbestos material from around pipes and structural steel. Under normal conditions, a laborer can remove about 40 sq. ft. per hour:

	Hours	Rate	Total	Rate	Total
Laborer	1	$....	$....	$19.13	$19.13
Cost per sq. ft.			....		$0.48

Monitoring devices are portable but require constant calibration and an experienced operator. Cost per project can vary widely, from $1,000.00 to $5,000.00 per day.

Clean up costs will vary with the type of material being removed and its location. It is a standard procedure to check the atmosphere, as well as the surrounding materials (such as the walls), just after completion of the work, and then again 24 hours after the work is completed. The cost can run from $1,500.00 up to $15,000.00 per clean-up operation.

CONTAMINATED SOIL

When removing contaminated soil from any site, precautions are necessary to prevent hazards to human health and to the environment. Specific procedures are required to comply with EPA, DOT, and in some case OSHA regulations regarding hazardous waste site clean up.

Hazardous wastes refer to those listed in the EPA's 40 CFR Part 261 *Identification and Listing of Hazardous Wastes.* Requirements for the transportation of hazardous waste is outlined in DOT 49 CFR Parts 172, 173 and 178 and in some cases in OSHA 29 CFR Part 1910.120 for health and safety.

Estimating Amount to be Removed. When removing contaminated soil from a site, it is important to understand how to determine the magnitude of the work. At most sites, the area of contamination is already defined, and the information is usually available in a site work plan.

Soil samples (borings) are obtained from the site at different locations using split barrel (split spoon) sampling apparatus or a hand auger. The split barrel method is used to obtain deep samples and is usually operated from a drilling rig. Hand augers may be used for shallow soil samples, up to about five feet deep. Samples are then sent to a laboratory for analysis, and the results are used to determine the extent of contaminated soil to be excavated. A plot plan or other diagram may be used to define the length, width and depth. Calculations obtained in feet should be converted to cubic yards, and prices for disposal should be quoted in cubic yards.

Methods of Excavation, Storage, and Transportation. Common earthmoving equipment is used for excavation, including backhoes, bulldozers, front-end loaders, and cranes. Solid material vacuum trucks and front-end loaders are sometimes necessary to complete the work in a timely manner.

The backhoe is used to excavate and consolidate soil from the ground into a pile. Piles of soil should be consolidated on 6-mil polyethylene sheeting used as ground liner. The front-end loader is then used to load soil from the pile into a roll off container or a dump truck. A typical 30-cubic yard dump truck can only be filled with 15 cubic yards of soil (50% of normal capacity) to remain within weight limits mandated by the Interstate Commerce Commission and the Department of Transportation. But also, overboarding a roll off or dump truck can damage the vehicle.

Using this method, it takes approximately 1 hour to load each roll off or dump truck with about 15 cubic yards of soil.

Once the vehicle is loaded, the material is either stored on site, if roll offs are used, or if dump trucks are used, then it is moved off site as soon as loading is completed. Dump trailers should be lined with 6-mil polyethylene sheeting and covered with standard roll off or dump trailer tarpaulin. The tarpaulin must be securely fastened onto the vehicle and capable of preventing soil from blowing onto the roadway when moving or by the wind when standing or parked. Polyethylene sheeting can be purchased in rolls measuring 100'x20', which is enough to line two standard 30-cubic yard roll offs or trailers.

If the soil is a hazardous waste, an appropriate *hazardous waste manifest* must be used as the shipping paper, and as a minimum, all vehicles must have hazardous waste hauling permits.

When removing small amounts of contaminated soil, it might be appropriate to consolidate excavated soil into 55-gallon drums. It will take about 15 to 30 minutes to fill each drum, depending on soil compaction and on the level of personal protective equipment. Each drum will weigh between 600 and 800 lbs. In order to move and load these drums onto a trailer, a forklift equipped with a drum-moving attachment or a front-end loader is commonly used.

Drums are usually loaded onto a flatbed or 40-foot closed box trailer. Drums containing hazardous waste must be labeled according to EPA, DOT, and possible state regulations. One, two or more EPA hazardous waste labels and DOT hazardous class labels may be required. It will take between 5 and 10 minutes to label and load each drum, depending on the level of protective equipment.

Personal Protective Equipment. Anyone who enters a hazardous waste site must be protected against potential hazards. The purpose of personal protective clothing and other equipment (PPE) is to shield the individual from chemical, physical or biological hazards. A proper selection of PPE protects the respiratory tract, skin, eyes, face, hands, feet, head, body and hearing. It is important to note that when using PPE, workers are at considerable risk of developing heat stress.

There are four levels of PPE. Each worker at a hazardous waste site must have the appropriate level of PPE available for use when working in the contamination zone. The use of PPE can increase the work duration by two or three times. The following gives recommended protective equipment at each level.

Level A. Totally encapsulated chemical protective suit (TECP). Pressure demand full face self contained breathing apparatus (SCBA) or supplied air respirator with escape SCBA. Inner outer chemical resistant gloves. Chemical resistant safety boots/shoes. Two-way radio communications. Hard hat. This level will mean a loss of 60 to 80% in work efficiencies.

Level B. SCBA or supplied air respirator with escape SCBA. Inner outer chemical resistant gloves. Chemical resistant safety boots/shoes. Two-way radio communications. Hard hat. Disposable coveralls may be used. *Tyvek* coveralls, hooded one-piece chemical splash suit. Duct tape should be used to seal ankles, wrists and other openings. Loss of efficiency is 50 to 60%.

Level C. Full face air purifying cartridge filter respirator. Use organic vapor acid gas and high efficiency particulate air filter (HEPA) as necessary for appropriate personal protection. Disposable coveralls, hooded one-piece chemical splash suit. Duct tape should be used to seal ankles, wrists and other openings. Disposable chemical resistant boots to be worn over safety shoes. Hard hat. Loss of efficiency is 15 to 50%.

Level D. Work uniform and possibly dust respirator might be required. Hard hat.

Decontamination. All equipment that comes into contact with hazardous waste must be decontaminated before leaving the contamination zone. There are various methods, including solvent cleaning and high pressure steam cleaning with detergent.

When decontaminating large or heavy equipment, a sump lined with polyethylene sheeting should be constructed. The sump, used to collect cleaning solutions, prevents contaminated runoff from spreading. When all equipment has been decontaminated, the residue must be pumped into drums using a drum pump. In the case where large amounts of cleaning residue are generated, a vacuum trailer might be more appropriate.

Labor Removing Contaminated Soil. When removing large amounts of soil with heavy equipment, it will require one operator for each backhoe, etc. When removing small amounts of soil, a crew of at least three workers might be necessary because of the weight of the drums.

Qualified supervisors must be available. All personnel supervising hazardous waste operations or handling hazardous waste initially must receive training, which must be updated regularly in accordance with federal and state regulations. Federal regulations may also require other personnel to be on site, such as an emergency coordinator or health and safety supervisors, chemists, professional engineers and certified hazardous waste specialists. Specialized direct-reading air monitoring equipment may be needed.

Cost Estimating for Removing Soil. When estimating the cost of contaminated soil removal, the following equipment, labor, personal protective equipment, transportation and disposal costs should be considered.

Excavating, Loading Equipment
Backhoe
Front-end Loader
Solids Vacuum Trailers
Roll off Container, Tractor and Flatbed Trailer
Dump Trailer with Tractor Equipped with Dump Trailer Hydraulic Lines
Forklift

Miscellaneous Equipment
55-Gallon Steel Drums
6-mil Polyethylene Sheeting
Duct Tape to Seal Seams on Trailer Liners
EPA Hazardous Waste Labels
DOT Hazardous Class Labels
Shovels and Picks
Ratchet Wrench to Bolt Drums Closed
Shipping Paper (Hazardous Waste Manifest)
Felt Tip Markers for Drum Labeling
Decontamination Equipment (Pressure Steam Cleaner)

Personal Protective Equipment
(One Ensemble for Each Worker)
Level A
Level B
Level C
Level D
Air Monitoring Equipment

Transportation Costs
Cost per Load for Roll offs and Dump Trailers
Cost per 55-gallon Drums
Cost Vary Depending on Distance to Disposal Site

Disposal Costs
Cost for Disposal of Contaminated Polyethylene
Cost per Cubic Yard of Soil Disposal
Cost per 55-gallon Drum Soil Disposal
Cost per Gallon for Bulk Cleaning Residue from Pressure Washing and Decontamination Solids Cost per Drum for Cleaning Residues from Pressure Washing and Decontamination Solids.

Labor Costs
Heavy Equipment Operators
Laborers for Liner Placement and Drum Handling
Supervisors

Average Costs to Remove 100 Cu. Yds. of Contaminated Material
Loaded into Dumpsters or Dump Trucks

	Hours	Rate	Total	Rate	Total
Backhoe	6.66	$....	$....	$18.74	$124.81
Loader	6.66	....	....	18.74	124.81
Equipment operators	13.32	....	....	25.96	345.79
Laborers	13.32	....	....	19.13	254.81
Cost per 100 cu. yds.			$....		$850.22

Cost per cu. yd. 8.50

The estimator must review the above figures and might possibly want to add the following items: level of protective equipment and clothing required for the specific project and the "lost efficiency" percentage for that level. Trucking and disposal and special liability costs. Cost of permits is not included in the above.

Average Costs to Remove Contaminated Material in Approved 55-Gallon Steel Drums*

	Hours	Rate	Total	Rate	Total
Backhoe	2	$....	$....	$18.74	$37.48
Equipment operators	2	....	....	25.96	51.92
Laborers	3	....	....	19.13	57.39
Cost per cu. yd.					$146.79
Cost per drum					38.17

*Allow 7 cu. ft. per drum.

The estimator should add the following items to the estimate: lost efficiency factor for protective equipment and clothing level required. Trucking and disposal and special liability costs. Cost of permits is not included in the above costs.

Removal of contaminated materials in buildings requires special liability insurance coverage. The high liability risks involved, combined with high minimum limits of liability insurance required by owners and agencies, prohibits most contractors from performing this work. The prudent estimator will become knowledgeable about all requirements and ramifications before submitting a bid for contaminated material removal. It is suggested that the estimator leave this operation to a qualified and approved asbestos removal contractor.

Some regulatory agencies are requiring the removal of all asbestos material in public buildings to be completed by the year 1995. Different materials are being continually added, removed or revised from listings of contaminated material by federal, state, and local OSHA and environmental protection agencies. The estimator is cautioned to check with all agencies that have jurisdiction for a project. Listed below are some of the agencies, both federal and local, that may have jurisdiction over contaminated material removal:

a) Occupational Safety and Health Agency (OSHA)
b) Environmental Protection Agency (EPA)
c) conservation agencies
d) Department of Transportation (DOT) federal, state and local
e) health departments
f) United States Coast Guard (USCG)
g) U.S. Army Corps of Engineers (COE)
h) fish and wildlife agencies
i) Consumer Products Safety Commission (CPSC)
j) Mine Safety and Health Administration (MSHA)

UNDERGROUND STORAGE TANKS

Most states have adopted regulations for immediate testing of existing underground storage tanks for leaks and for necessary repairs to be made without delay. In fact, all single-wall steel storage tanks (tanks for fuel, gasoline, oil, heating oil, and industrial storage), even those that do not leak must be replaced by the end of the century.

Tank monitoring regulations for underground storage tanks mandate a secondary containment system, which requires a leak detection monitoring system. The program might employ a continuous system, connected to an audible or visual alarm system, or a manual inspection, performed on a daily basis.

Leak detection systems that have been approved by regulatory agencies include the following: liquid level indicators that detect a change in the height of a fixed volume of liquid in an annular space; liquid sensors that detect the presence of liquid in an annular space; vapor sensors for volatile hazardous substances; and pressure or vacuum loss sensors.

It is strongly recommended that the general contractor contact a specialist in industrial waste to perform the required testing. Testing for leaks varies according to the type, size, and material of the tank. Prices for tank testing range from $1,500 to $10,000.

If a leak is detected, the owner must remove the contents of the tank. It will be necessary to determine the source of the leak; properly remove all substances from the underground storage tank system; repair, replace, or close the system in accordance with the local governing regulatory agency mandates.

The possessor will then be required to determine the extend of the leak. An accepted method is to drive 2½" diameter pipe observation wells about 20' into the ground and take samples of the soil and water if present. The number of observation wells is normally 4 per tank or farm tank. Costs vary due to the many types of soil and subsurface conditions. Installations of wells continues until the extent of soil contamination is determined. The contaminated soil is removed, the tank cleaned and filled, with clean material, and then either left in place or removed.

Removal of the tank is the preferred method, but method of disposal can depend on many extenuating circumstances, such as location of tank with respect to other facilities, or it might be that removal is an insurmountable task. The final determination whether to leave it in place or remove it depends on local regulations.

In-place Disposal of Underground Storage Tanks. The following is a step-by-step procedure for the safe in-place disposal of underground tanks:

1. Drain product from piping into tank, disconnect all piping, and cap and remove all piping.
2. Remove liquids and residues from the tank using explosion-proof or air driven pumps. The last few inches might require a hand pump.
3. Excavate to the top of the storage tank.
4. Remove and cap drop tube, fill pipe, gauge pipe, vapor recovery connections, submersible pumps, and all other tank fixtures and connections, except for the vent line. The vent line remains connected until the tank is purged.
5. Purge the tank of flammable vapors. Vent all vapors a minimum of 12 feet above grade or 3 feet above adjacent roof lines.
6. Cut tank, if necessary, for the proper introduction of inert (clean fill) material to fill the tank.
7. Place suitable solid material, such as sand or other clean fill, through the openings in the top of the tank. It is important to fill the tank as full as possible.
8. Sand or other clean fill is suitable if it is free of rocks or stones, which might limit leveling out of the fill in the tank. Sand may be introduced dry, as long

as it flows freely. When the sand cones near the top of the tank, it can be washed into the tank with a nominal amount of water so that it flows to the ends. The use of large amounts of water must be avoided.
9. Fill the tank to about 80% of its calculated capacity. Mix soil and water to make a free-flowing mud and pour the mixture into the tank opening until the tank is full and the mixture overflows the opening.
10. Plug or cap all openings.
11. Disconnect the vent line, and remove and plug the vent line opening.
12. Restore ground surface.

Cost will vary for this type of work. Prices range from $1,500 to $3,000 per tank for tanks up to 2,500 gallon capacity, $3,000 to $5,000 for tanks to 6,000 gallon capacity.

Removal of Underground Storage Tanks. The following procedures are followed for the safe removal of underground tanks:

1. Drain product from piping into the tank, disconnect piping from the tank, and cap and remove all piping.
2. Remove liquids and residues from the tank using explosion-proof or air driven pumps. The last few inches might require a hand pump.
3. Excavate to the top of storage tank.
4. Remove and cap the drop tube, fill pipe, gauge pipe, vapor recovery connections, submersible pumps, and all other tank fixtures and connections. One plug shall have a 1/8" vent hole to prevent the tank from being subjected to excessive differential pressure caused by temperature changes.
5. Purge the tank of flammable vapors. Vent all vapors a minimum 12 feet above grade or 3 feet above adjacent roof lines.
6. Excavate around tank to uncover it for removal. Remove the tank from the excavation and place it on a level surface. Use wood wedges or blocks to prevent the tank from rolling.
7. After tanks are removed from the ground but before they are removed from the site, they must be labelled. For example, the label might read:

TANK HAS CONTAINED LEADED GASOLINE
NOT VAPOR FREE
NOT SUITABLE FOR STORAGE OF FOOD OR
LIQUIDS INTENDED FOR HUMAN OR ANIMAL
CONSUMPTION

or

TANK HAS CONTAINED LEADED GASOLINE
LEAD VAPORS MIGHT BE RELEASED IF HEAT
IS APPLIED TO THE TANK SHELL

8. Tanks shall be removed from the site as quickly as possible after the tank is vapor free.
9. Tank atmosphere shall be checked with a combustible gas meter to insure that the tank does not exceed 20% of the lowest flammable rate.
10. Tank should be loaded and secured on a truck for transportation to the storage or disposal site with the 1/8" vent hole located at the uppermost point

on the tank. The tank shall be transported in accordance with all applicable local, state, and federal regulations.

Costs will vary. Prices range from $2,500 to $6,000 per tank for tanks up to 2,500 gallon capacity, $3,500 to $15,000 for tanks up to 6,000 gallon capacity. Price also varies depending on whether there is one tank to fill or more than one tank at a location.

Removal of a single wall steel tank and replacement with an approved double wall tank, with all the necessary leak and fume detectors, averages about $50,000 for a 1,000 to 1,500 gallon tank; a 2,000 to 6,000 gallon tank costs from $75,000 to $100,000.

The contractor must follow all local, state, and federal regulations. Regulatory agency requirements can be supplemented with information published by the American Petroleum Institute (API).

CHAPTER 22

ROAD AND HIGHWAY CONSTRUCTION

This chapter serves the needs of the general contractor and the specialty subcontractor who prepare estimates for and engage in paving work associated with construction projects. It is a guide and a reference in solving many of the problems encountered by providing practical information to expand the basic knowledge and understanding of the methods and techniques.

Much of the information contained in this chapter will be applicable to construction of major highway works and interstate transportation systems. However, this sophisticated and specialized form of highway construction is not specifically addressed in this chapter, as it would produce information useful to relatively few construction organizations dedicated to this highly competitive and enormously complex segment of the construction industry.

Supplemental information is available through extensive publications devoted to research data, current practices, and standard specifications that have been produced through the efforts of the American Association of State Highway Officials, the Highway Research Board of the National Research Council and the Bureau of Public Roads. In addition, several material associations have provided factual publications for the benefit of the highway construction industry. The most notable and active associations are The American Concrete Institute, Association of Asphalt Paving Technologists, and The Asphalt Institute.

Road, highway and paving construction is a high risk business under the most favorable conditions, demanding a thorough knowledge of the technical skills involved, accompanied by managerial and administrative talent to support a totally efficient and coordinated operation. Like other types of contracting, the paving contractor is faced with the obligation of accurately evaluating the capabilities and resources of his organization. This specifically includes the financial capacity of the company and the availability of external sources for interim financing; availability of construction equipment through ownership, lease or rental with emphasis on condition, suitability and sufficiency, facilities for maintenance and repair; existence of systems and personnel to efficiently provide cost control, fiscal control, scheduling, purchasing, expediting; and qualified estimators capable of producing accurate estimates. Once this evaluation is complete, it can then be determined the type, size and complexity of construction projects that would best match the resources and capabilities of the company. This matching process is the basis of selection of projects for bidding.

It should be noted that firms having sufficient financial capacity to qualify for bonding on projects of a certain magnitude may have a deficiency in providing sufficient experienced personnel to efficiently accomplish the work. Thus, contractors would be well advised to follow the formula: financial assets + people assets = total capability.

Because of the impact of road and highway construction on the economic, social and political decisions of a community, there has to be a presumption that many laws and regulations affecting the conduct of a construction program involving roads and highways will exist. Each geographic location will have its own peculiarities requiring the contractor's investigation and familiarity with these requirements prior to becoming involved in the bidding process. Prequalification of contractors is often a prerequisite, as is compliance with local contractor license laws or registration as a foreign corporation if the company is not incorporated under that state's law.

The intent of this chapter is not to present extensive engineering and design theory, but rather, to provide sufficient examples of incidents where contractors, estimators, project managers, and other supervisory personnel have been confronted with problems relating to paving operations where without a source of some basic knowledge of design, they were placed at a disadvantage in the decision making process. This lack of knowledge could become an extremely important factor when attempting to establish a fair and reasonable cost factor in the initial estimating process and possibly have more impact when attempting to evaluate a remedial solution for defective paving which may be assigned as a contractor's responsibility.

Contractors will generally be working within the requirements of the project specifications. Strict compliance normally will protect the contractor from claims. However, not all specifications are complete and totally descriptive of the precise methods necessary in order to achieve the desired end results. In some instances the designing architect also lacks sufficient understanding of the capabilities and characteristics of the materials involved in the construction. Frequently, the problem is further clouded by including a general reference to the AASHO System (American Association of State Highway Officials) or an attempt to include elements extracted from the Unified Classification System for Soils and fortified with vague references to density, moisture content, compactness and gradation of materials.

While the road and paving may be a somewhat minor cost factor as it relates to an entire construction project, it would be unacceptable and unprofessional not to establish the most accurate cost possible for this work. Every estimator should be aware that failures are costly and properly developed estimates do not make provisions for re-performance of work. It is obvious that unacceptable work requiring remedial action escalates the cost of the project and lowers the profit margin.

Generally, it is found advantageous to utilize the talent, expertise and services of a specialty contractor in the paving operation. However, this subcontractor will demand certain supporting performances by the prime or general contractor. This may include the preparation of the subgrade, all line and grade engineering and any testing which may be required.

More extensive knowledge on the part of the contractor creates a position where methods, schedules, unit costs, and other specific details can be discussed with authority and confidence prior to entering a contract for paving work, with reasonable assurance the ultimate objective will be achieved with minimal problems. This same theory should be applied with the same force and effect on each of the other divisions of work included in the total project.

Construction Documents

Plans. Properly prepared plans provide sufficient information for an estimator to make a quantitative analysis of each discipline of work included in the project. This function should be performed even if the owner has provided the estimated quantities in the *bid form,* because accuracy cannot be determined except by comparison with the contractor's quantity survey.

Specifications. The technical specification for highway work is usually a "guide specification" making many references to the Standard Specification requirements where the details appear for the various types of work to be performed. As an example, the guide specifications may identify curb and gutters as Types A, C, D, and F at certain locations. To determine the specific design of each type, reference is made to the Standard Specification. Copies of Standard Specifications can be purchased from the agency issuing the plans and specifications.

Bid forms. In the case of unit price bidding, which is the usual format in road and highway construction, the form will provide for each classification of work with estimated quantities included. The contractor is required to insert a unit price, including allowance for overhead and profit on each item, and calculate the estimated quantity by the unit price to determine total cost on each line item. A typical example would be as follows:

Bid Form

Code, Item	Quantity	Unit	Unit price	Total
200 Clearing	6	Acres	$2,000.00	$12,000.00
303 Excavation, Class 2	100,000	Cu. Yds.	1.00	100,000.00
310 Seed	500	Lbs.	3.00	1,500.00
490 Asphalt Conc. Agg.	2,000	Ton	26.00	52,000.00
520 Cement Treated Base	2,000	Cu. Yds.	31.00	62,000.00
710 Modify Signs	11	Each	300.00	3,300.00
730 Pipe Culvert 24"	200	Lin. Ft.	40.00	8,000.00
732 RC Pipe 48"	200	Lin. Ft.	65.00	13,000.00
Total Price				$251,800.00

Should an error appear in the extension to the total line amount, it has been ruled many times that the owner has the right to use the unit price offered by the contractor and correct the mathematical error. Estimators should use extreme caution in developing the unit price bid item.

Instruction to Bidders. These documents provide the precise information relating to bid guarantee, number of copies of the bid to be furnished, establishment of date and time bids are to be received, documentation to be included in the bid package, and other instructions appropriate to the project. These requirements are to be fulfilled without deviation or the bid may be considered nonresponsive, making the contractor ineligible for award.

Special Instructions. Physically inspect every project site prior to bidding to determine working conditions. Avoid accepting oral interpretation or instructions as they cannot be enforced. If uncertainty or lack of confidence appears during the bidding process, the best decision would be to avoid bidding on the project rather than risk a serious mistake and possible loss.

Soil Stabilization

The need for soil stabilization is widespread in the construction industry to support the designed contour configuration that provides beautification in the landscaping of buildings, parks, and other green areas. In addition, there is an important functional aspect to soil stabilization. This applies to the prevention of soil erosion and providing a sound and stable base to support roads, walks, terraces, parking areas, and other surfaces subjected to concentrated loading and heavy traffic wear.

Without sufficient stability of the subgrade, all finishes placed upon it are, to some degree, influenced by the credibility of the subgrade. It is not uncommon to encounter complete failure of paved areas which can be traced to poor or ineffective preparation of the subgrade prior to the installation of properly designed and installed finish surface.

Soils exist in a broad range of types with wide variation in the value of their properties and they react differently to the several methods available to improve their degree of stabilization. It is generally recognized that the ultimate resistance of a soil to the penetration of concentrated load depends on the following two properties of the soil: 1) *friction* between soil particles and 2) the *cohesion* between the soil particles. Soil resistance due to cohesion is dependent on the external pressure on the soil. However, frictional resistance increases in direct proportion to the external pressure.

Example: Sand has a high frictional resistance without appreciable cohesion and the bearing capacity is increased considerably when a concentrated load is surrounded by a distributed load. Clay, having high cohesive resistance, but without appreciable frictional resistance, reacts quite differently, because the bearing capacity is not increased to any great extent by applying a concentrated load surrounded by a distributed load.

The stability of a soil for subgrade purposes also depends on its *elasticity* or the ability to rebound after a load has been applied and then removed. Excessive elasticity makes a soil less stable because it cannot be permanently compacted.

Water content in the soil can be a significant factor in stabilization, particularly in the expansive soils which are subject to change in volume, because they possess cohesion and capillarity in appreciable amounts.

Often the quality of subgrades can be improved by installing some type of artificial drainage to prevent the accumulation of excessive moisture from the subgrade materials. Such a provision would also provide protection from "pumping" and "heaving" action during freezing and thaw periods.

Excavation

Excavation, moving, and depositing of existing materials to a new location are major cost items for the contractor involved in road and highway construction. All encountered materials are basically rock. However, because of the forces of nature, there are great variances in the size of particles, from materials classified as solid rock continuing down the scale to those identified as sand or clay.

This variation in material is so extensive it would be presumptuous and extremely dangerous for the excavator to attempt calculating the cost involved without engaging in preliminary studies. Each classification retains its own characteristics. In some instances, blasting or ripping is required to separate the materials from the natural position, while other materials can be readily removed by loader or scraper. Dissimilar conditions will be encountered when attempting to compact the various types of materials.

On a given construction project, it would not be unusual to encounter several types or classifications of materials, requiring the contractor to modify methods and equipment to accommodate each condition. An understanding of geology and soils mechanics is necessary to effectively deal with the complexities of excavation. Experience in soils of a geographical location also contributes to the overall knowledge.

In excavating hard materials such as shales, sandstone, and limestone, the strata of rock may be hard and strong, but if lamination thickness and vertical joint spacing are limited in size, production will be improved over instances where these planes of weakness produce greater cubic content of the material.

The most common practices used to determine the blasting and ripping zones for rock excavation is the track drill because of it is widely available, is portable, and can produce data relative to drilling costs and blasting techniques. The seismic timer is an instrumental means to determining the degree of consolidation of the rock which can be related to the cost of excavation.

When excavating the usual non-cementitious materials, a variety of machines and methods are available to perform specific tasks at the most economical cost. Loaders, shovels, cranes, Gradalls, backhoes, draglines, and pans are the most common machines in use, with each one having certain advantages and limitations. It is the duty of the estimator to know the capabilities of these machines and to quantitatively prove their efficiency before making a judgment. The following formula will provide hourly production for a front-end loader with a 3 cubic yard capacity bucket working average soil. Loading time cycle through an angle of 120 degrees is a 0.5 minutes, comprised of the following four elements:

a) Loading - 7 seconds (0.12 minutes)
b) Swing-Position - 12 seconds (0.20 minutes)
c) Dumping - 3 seconds (0.05 minutes)
d) Return-Position - 8 seconds (0.13 minutes) or a total of 30 seconds (0.50 minutes)

Hourly production in cubic yards, bank measurement, per 50-minute working hour is determined by the following formula:

$$P = \frac{50}{CT} \times BF \times BC \quad \text{where}$$

P = Hourly Production, CYBM (Cu. Yds. Bank Measurement)
CT = Cycle Time in Minutes
BF = Bucket Factor
BC = Bucket Capacity

Bucket Factor Table

Material	Fill Factor LM* / BC	Conversion Factor BM** / LM	Bucket Factor BM / BC
Sand-Gravel	1.05	0.88	0.92
Weathered Rock-Earth	0.90	0.78	0.72
Blasted Rock	0.80	0.67	0.54

*Loose Measurement **Bank Measurement

The equation becomes:

$$P = \frac{50}{.50} \times BF \times BC$$

$$P = 100 \times BF \times BC$$
$$P = 100 \times 0.92 \times 3.0$$
P = 276 cybm per 50 min. workhour

Assume this 3 cu. yd. capacity front-end loader could be rented including operator, fuel, maintenance, and transportation charges for $80.00 per hour, then the cost for excavation would be:

$$\text{Cost/CYBM} = \frac{\text{Rental Cost Per Hour}}{\text{CYBM Per Hour}}$$

$$\text{Cost/CYBM} = \frac{\$80.00}{275}$$

$$\text{Cost/CYBM} = \$0.29$$

This cost is pure excavation and loading cost. To this must be added hauling, dumping charges, job supervision, establishment and maintenance of haul roads, compaction of fills, and other miscellaneous costs. In making cost production analysis on each type of machine planned for use on a project, other valuable information will be produced. The hourly production rate on the excavating machine will determine the number of trucks required to haul the excavation to other locations within or external to the construction site. Truck costs can be established in similar analysis predicated upon loading time, capacity of the truck and the haul time to the disposal area.

To determine the number of trucks required to service and support the production rate of the previous front-end loader, it can be established as follows:

P = 275 CYBM

PM = CYBM per minute of 50 min. work hour

WH = 50 minute/hour

$$PM = \frac{P}{WH}$$

$$PM = \frac{275}{50}$$

$$PM = 5.5 \text{ CYBM}$$

A 10-cu. yd. truck could be loaded every 2 minutes of work time, thus 25 trucks would be required to service the loader. Haul time for the trucks would have an influence on the number of trucks required.

Hauling

The hauling of excavated materials to a disposal area is the most important phase of excavation. It is usually the most costly and subject to many variables. In instances of off-site hauling by trucks, the traffic patterns will have a substantial influence on the haul time. More haul time will be consumed during morning and evening peak traffic periods than during the midday normal traffic flow.

On-site hauling by self propelled scrapers can be determined with more precision as the conditions can be controlled by the contractor.

After selection of scrapers of a suitable capacity, there is a need to determine the travel time involved from the point of loading to the disposal destination. This can be accomplished by using the following equation:

$$\text{Travel time (minutes)} = \frac{\text{Length of haul (feet)}}{88 \times \text{Average speed (mph)}}$$

Eighty-eight represents a factor to convert miles per hour to feet per minute. Speed factor tables are available from equipment manufacturers for the conversion of maximum speed to average speed, but these are somewhat complicated to use. On smaller projects, past experience would serve the needs of the contractor. In establishing an average speed factor, utilize the rated maximum speed of the machine and multiply by realistic percentages reflecting job conditions to determine average speed.

Short haul distances, 200 to 600 feet would be at a 50% level because of stop frequency and turnaround time loss. For haul distances of 2,000 to 3,000 feet, the percentage will increase to 75% to 80%. These percentages would be for average condition with level grades, no sharp turns or hazards, and without interference of cross traffic.

In using the equation to establish travel time after the loading operation, assume the length of haul to be 3,000 feet with an average speed of 12 mph.

$$\text{Travel time (mins.)} = \frac{3{,}000}{88 \times 12}$$

$$= \frac{3{,}000}{1{,}056}$$

$$= 2.84 \text{ Minutes}$$

To establish the number of haul cycles, divide the 50 minute workhour by the travel time:

$$\frac{50}{2.84}$$

$$\text{Number of Haul Cycles/Hour} = 17.6$$

Production could be calculated by allowing a 2 minute loading time plus the 2.84 minute haul time for a total 4.84 minute production cycle for a 24 cu. yd. scraper:

$$P = \frac{50}{4.84} \times 24$$

$$P = 10.33 \times 24$$

$$P = 247.92 \text{ cubic yards/hour}$$

Should materials or job conditions be encountered that would require a pusher machine, then it would be an additional cost to the basic scraper.

Fill Compaction

The distribution and compaction of fills to rigid specifications calls for the use of several machines designed for the densifying operation, which is accomplished by spreading-mixing, wetting or drying, and the use of compactors. Production and costs for spreading materials are complicated by the overlapping and sharing of work with other machines. The following general equations have been found acceptable for hourly production when using the horse power (HP) rating of the machine as a base:

Dozer	CYBM =	3 x HP
Motor Grader	CYBM =	3 x HP

The use of water wagons is a function of the haul cycle time and can be estimated by the following equation:

$$\text{Production (gal/h)} = \frac{50}{\text{Cycle time (min.)}} \times \text{Capacity (Gal.)}$$

Water wagons range in capacity from 1,500 to 14,000 gallons. The contractor must decide on the most economical based on the anticipated needs at the project. Laboratory test data provides the existing natural moisture content. The moisture-density curve indicates that the desired percent of relative density will be attained with a percentage of moisture by dry weight. With this information the estimator can produce an estimated need for water and project this toward a water wagon cost.

In instances where excessive moisture is encountered, this can be reversed by aeration of existing materials or the importation and mixing of dry materials. In each instance, it becomes a costly operation and delays job progress.

Calculation of Excavation Quantities

In establishing the quantities of excavation, which is subdivided into cuts and fills, the estimator is actually determining the volume or cubic yards involved in each operation through calculations involving areas, distance, and geometric figures.

Volume:	1 cu. yd.	= 27 cu. ft.
Distance:	1 lin. ft.	= 12 in.
	1 station	= 100 lin. ft.
	1 mile	= 5,280 ft.
Area:	1 sq. yd.	= 9 sq. ft.
	1 acre	= 43,560 sq. ft.
	1 sq. mile	= 640 acres

See Chapter 24, *Mensuration,* for tables that provide the formulas for calculating the quantities of the most frequently found geometric figures.

The usual methods of estimating are the *end area* method and the *cross section* Method. These two methods are described in Chapter 2, *Site Work.*

Bituminous Road Paving

This flexible paving method consists of several layers or courses of mineral materials placed upon a properly prepared subgrade of earthwork which performs as a foundation for the paving. Quality, quantity, and methods of performance are governed by the requirements of the contract specifications.

Subbase Course. This is a layer of selected granular material of pre-determined depth applied to the subgrade for the support of the paving materials. It also prevents or reduces frost damage and provides under pavement drainage.

Base Course. A high quality material such as crushed stone which is placed on the subgrade or the subbase. Gradation and depth controlled by specification. Also may be treated with portland cement or bituminous to reduce thickness and improve durability.

Leveling Course. Used to eliminate irregularities in existing paved surfaces prior to resurfacing.

Binder Course. In asphalt paving a binder course, which is often referred to as a leveling course, is installed at a prescribed thickness over the subbase. The strength of the binder course will be substantially equal to the designed surface or wearing course.

Surface Course. The uppermost or exposed surface of the paving designed to withstand wear and polishing by traffic. Smooth surfaces reduce noise level but increase skid action.

Pumping. Commonly referred to as mud pumping. Occurs when the subgrade contains excessive moisture and is a type containing a high percentage of fine particles. The surface becomes flexible and unstable to any load imposed upon it. This condition is unsuitable to support paving and must be removed and replaced with structural support for the paved surface.

Subgrade. Foundation on which the subbase and paving materials are placed. The utilization of locally available materials for use in the subgrade has important economic benefit, unless the material is of a poor quality requiring additives in the form of chemicals or modification in the soil content and gradation to achieve the desired quality of compaction. In such instances, consideration should be given to the economics of removal of a sufficient quantity of the existing soil and replacement by fill material requiring limited compaction effort to produce a suitable surface to receive the subbase material.

For various terms that are used in road and highway construction, including those that apply specifically to asphalt paving, see the Appendix A, *Glossary*.

Standard Design Details

For general guidance and reference, the following standard design details are included, all of which have been approved by the Federal Highway Administration.

Certain states, counties, or municipalities may require some slight modifications, but it is assumed that deviations will not significantly exceed the requirements of these standard details.

Index of Standards

1. Soils & Soil - Aggregate Mixtures, Characteristics, & Performance
2. Soils & Soil - Aggregate Mixtures, Treatment, & Use
3. Soils & Soil - Aggregate Mixtures, Guide to Classification
4. Standard Timber Barricade
5. Standard Berm Ditches
6. Standard Surface Drain Ditches, Concrete, & Sod

7. Standard Side Ditches Cut Areas
8. Standard Side Ditches, Standard Median Ditches - Flat
9. Standard Median Ditches, V Slope
10. Standard Pipe Outlets, Surface Drainage on Fills
11. Method of Depressing Pavement at Inlets
12. Standard End Support Wall, Round Pipe
13. Standard End Support Wall, Arch Pipe
14. Standard Headwall
15. Standard Headwall
16. Standard Type C Endwall, Round Pipe
17. Standard Type C Endwall, Arch Pipe
18. Standard Type E Endwall, Round Pipe
19. Standard Type E Endwall, Arch Pipe
20. Standard Type F Endwall, Round Pipe
21. Special Type F Endwall, Round Pipe
22. Standard Type G Endwall, Round Pipe
23. Standard Type H Endwall, Round Pipe
24. Standard Type K Inlet, Open-End Grate
25. Concrete Manholes, Catch Basins, and Inlets
26. Standard Underdrains
27. Sediment and Erosion Control, Fiberglass Erosion Stop
28. Standard Concrete Valley Gutter, Flues, Concrete Shoulder, & Rebut
29. Sediment and Erosion Control Placed Rip-Rap Ditch
30. Piled Rip-Rap Detail
31. Concrete Bridge Parapet and Bridge, Median Barrier Details
32. Reinforced Concrete Pavement, Types of Joints
33. Standard Bituminous Concrete Curb
34. Standard Types of Concrete Curb, Combination Concrete Curb & Gutter

GENERAL CLASSIFICATION	GRANULAR MATERIALS						SILT - CLAY MATERIALS							
GROUP CLASSIFICATION	A-2		A-3	A-2-4	A-4-2	A-2-7	A-7-2	A-4	A-4-7	A-7-4	A-7	A-6	A-5	A-8
GENERAL DESCRIPTION	SAND			SILTY-SAND	SANDY-SILT	CLAYEY-SAND	SANDY-CLAY	SILT	CLAYEY SILT	SILTY-CLAY	CLAY	COLLOIDAL CLAY	MICA, DIATOMS & SILT	SWAMP MUCK
STABILITY	WHEN N.P., HIGH	WHEN PLASTIC, GOOD WHEN DRY	IDEAL WHEN CONFINED	GOOD WHEN DRY	GOOD WHEN DRY	GOOD WHEN DRY & PROPERLY COMPACTED	GOOD WHEN PROPERLY COMPACTED OR UNDISTURBED				POOR		GOOD TO POOR	NONE
USE AS A BASE	FAIR	FAIR	EXCELL.	FAIR	FAIR	POOR	UNSATISFACTORY							
USE AS A SUB-BASE	EXCELL.	GOOD	EXCELL.	FAIR	FAIR	FAIR	UNSATISFACTORY							
USE AS A SUB-GRADE	EXCELL.	EXCELL.	EXCELL.	FAIR	FAIR	FAIR	POOR				VERY POOR	POOR		UNSATISFACTORY
FILLS UNDER 50'	EXCELL.	GOOD	GOOD	GOOD	POOR	FAIR	GOOD TO POOR				POOR	VERY POOR	POOR	UNSATISFACTORY
FILLS OVER 50'	GOOD	GOOD TO FAIR	GOOD TO FAIR	FAIR	POOR	FAIR	POOR TO VERY GOOD				VERY POOR	VERY POOR	POOR	UNSATISFACTORY
FROST ACTION	NONE TO LOW			MEDIUM	HIGH	MEDIUM	MEDIUM	HIGH	MEDIUM TO HIGH		MEDIUM		HIGH	MEDIUM
RANGE OF MAX. DRY DENSITY (AASHO T-180) (PCF)	115-135	115-135	105-130	110-130	110-135	115-135	115-130	110-135	110-135	105-130	100-120	90-115	100-135	LESS THAN 100
RANGE OF OPTIMUM MOISTURE CONTENTS (AASHO T-180X%)	9-12	8-12	8-15	8-15	9-15	6-12	9-15	8-15	10-15	10-17	12-25	14-30	11-18	——
REQUIRED COMPACTION (AASHO T-180) (%)	92-95	92-95	92-95	92-95	92-95	92-95	92-95			≤ 95 %				WASTE
COMPACTION METHODS	ROLLING WITH SMOOTH FACE TAMPING RUBBER TIRED ROLLER OR VIBRATORY COMPACTOR		TRACTOR DISKING VIBRATION	TAMPING OF RUBBER-TIRED ROLLER	TAMPING OR RUBBER-TIRED ROLLER	TAMPING OR RUBBER-TIRED ROLLER	TAMPING SHEEPFOOT ROLLER OR RUBBER-TIRED ROLLER						TAMPING OF RUBBER-TIRED ROLLER	WASTE
COMPACTION ABILITIES	GOOD WITH CLOSE CONTROL		GOOD	GOOD TO POOR	GOOD TO POOR	GOOD TO POOR	GOOD TO POOR				POOR	POOR	VERY POOR	WASTE
PUMPING ACTION	SLIGHT TO NONE						FAIR TO POOR				POOR			WASTE
BEARING VALUE	EXCELLENT TO FAIR			GOOD TO FAIR			FAIR TO POOR				POOR			WASTE
DRAINAGE	GOOD		DRAINS FREELY	FAIR TO PRACTICALLY IMPERVIOUS			FAIR TO IMPERVIOUS				POOR	IMPERVIOUS	FAIR TO IMPERVIOUS	POOR

NOTES:

A-2 TO A-3 SOILS: WHEN USED AS A BASE, PLASTICITY INDEX AND LIQUID LIMIT SHOULD NOT EXCEED 6 AND 25 RESPECTIVELY. BEST FOR SOIL-CEMENT STABILIZATION, GENERALLY 8 TO 12 % CEMENT BY WEIGHT WILL BE SUFFICIENT.

NON-PLASTIC A-2 TO A-3 SOILS: MAY REQUIRE VIBRATION FOR COMPACTION.

A-4 TO A-7 SOILS: FILLS SHOULD BE PLACED IN DRY SEASON.

A-4 SILTS: SUSCEPTIBLE TO SETTLEMENT AND EROSION.

A-5 SOILS: WHEN MICA IS PRESENT, VERY DIFFICULT TO COMPACT BECAUSE OF EXPANSION AND REBOUND.

A-6 SOILS (CLAY): WILL PUMP IN POROUS BASES FORMING CRACKS FILLS WILL SETTLE OVER LONG PERIODS OF TIME. HIGH BANKS IN CUTS AND FILLS VERY LIABLE TO SLIDE.

Soils & Soil - Aggregate Mixtures
Characteristics & Performance

TYPE	TREATMENT & USE
A-2 SOILS:	WELL GRADED TO POORLY GRADED SAND AND GRAVELS GOOD BASE FOR MODERATE FLEXIBLE OR THIN RIGID PAVEMENT. GOOD FILL. FROST HEAVE, BREAK-UP IF PLASTIC. SOFTENS WHEN WET IF PLASTIC. USE BASE COURSE WHEN SUB-GRADE P.I. IS GREATER THAN 6. SUB-DRAINAGE EFFECTIVE. STABILIZE; WITH BITUMEN, CHLORIDES, CEMENT OR ADMIXTURE SOIL.
A-3 SOILS:	CLEAN SANDS AND GRAVELS. IDEAL BASE FOR MODERATE FLEXIBLE OR THIN RIGID PAVEMENT. GOOD FILL. NO FROST HEAVE OR BREAK-UP. SUB-DRAINAGE ONLY THROUGH IMPERVIOUS SHOULDERS. STABILIZE; WITH SOIL BINDER, BITUMINOUS, OR CHEMICAL ADMIXTURES.
A-4 SOILS:	SILTY SOILS. NOT GOOD FOR SURFACE. POOR BASE. ABSORBS WATER. UNSTABLE WHEN WET. BAD FROST HEAVE AND BREAK-UP. USE SUB-DRAINAGE AND/OR BASE AND SUB-BASE WITH FLEXIBLE PAVEMENT. USE BITUMINOUS SUB-GRADE PRIME. USE THICK CONCRETE PAVEMENT (7" TO 10") WITH STEEL REINFORCEMENT AND CRACK CONTROL.
A-5 SOILS:	ELASTIC SILTS. USE SUB-DRAINAGE AND/OR GRANULAR BASE AND SUB-BASE WITH BITUMINOUS SUB-GRADE PRIME. USE THICK CONCRETE PAVEMENT, REINFORCED WITH CRACK CONTROL.
A-6 SOILS:	CLAYS. IMPERMEABLE AND STABLE WHEN DRY AND UNDISTURBED (HARD CLAY). PLASTIC AND ABSORBENT IF DISTURBED. BAD PUMPING INTO POROUS BASE. MACADAM OR PAVEMENT JOINTS. SHRINKS OR CRACKS WHEN DRY. USE GRANULAR BASE AND SUBBASE. USE SUB-DRAINAGE ONLY WHEN MADE PERVIOUS BY CRACKS, ROOT HOLES AND LAMINATIONS. FROST HEAVE SLIGHT WHEN IMPERMEABLE; BAD WHEN PERVIOUS. USE SUB-GRADE PRIME. USE THICK, STRONG, DENSE FLEXIBLE PAVEMENT OR REINFORCED CRACK CONTROLLED CONCRETE.
A-7 SOILS:	EXPANSIVE, PLASTIC CLAYS. EXCESSIVE VOLUME CHANGE. BAD FROST HEAVE AND BREAK-UP. SUB-DRAINAGE NOT EFFECTIVE. USE THICK, DENSE, FLEXIBLE PAVEMENT WITH BASE AND SUB-BASE OVER SUB-GRADE PRIME OR REINFORCED CRACK CONTROLLED CONCRETE PLACED ON IMPERVIOUS PAPER.
A-8 SOILS:	MUCK AND PEAT. UNFIT FOR CONSTRUCTION PURPOSES. EXCAVATE TO SOLID STRATUM AND REPLACE WITH SELECTED FILL. DISPLACEMENT BY SUPERIMPOSED FILL IS DOUBTFUL. DISPLACEMENT BY EXPLOSIVE UNDER SUPERIMPOSED FILL IS SOMETIMES EFFECTIVE.

Soils & Soil - Aggregate Mixtures
Treatment & Use

	SYMBOLS	TYPICAL GRADING	TYPICAL PHYSICALS	REMARKS
A-3 SAND	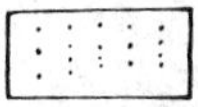	C.S. =22% F.S. =48% SILT =20% CLAY = 8% COLL. = 2%	L.L. = N.P P.I. = N.P.	SAND-53% MIN. %-# 200-20% MAX. P.I.- N.P. L.L.-MUST BE N.P
A-2 SAND & FINES	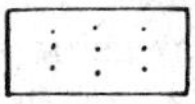	C.S. =20% F.S. =43% SILT =19% CLAY =10% COLL. = 8%	L.L. =22 P.I. = 2 S.L. =18	SAND-53% MIN. %-# 200 -20% MAX. P.I.-7 MAX. L.L.-34 MAX. (MUST HAVE L.L.)
A-2-4 SILTY SAND	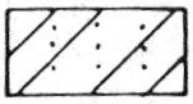	C.S. =25% F.S. =30% SILT =32% CLAY = 7% COLL. = 6%	L.L. =24 P.I. = 2 S.L. =21	SAND-53% MIN. % #200-21% MIN.-30% MAX. P.I.-7 MAX. L.L.-34 MAX. (MAY BE N.P.)
A-4-2 SANDY SILT		C.S. =23% F.S. =28% SILT =33% CLAY =10% COLL. = 6%	L.L. =25 P.I. = 3 S.L. =21	SAND-48% MIN %-#200-31% MIN. P.I.-7 MAX. L.L.-40 MAX. (MAY BE N.P)
A-2-7 CLAYEY SAND	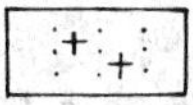	C.S =38% F.S. =31% SILT =15% CLAY = 8% COLL. = 8%	L.L. =31 P.I. =10 S.L. =18	SAND-48% MIN. CLAY-29% MAX. P.I.-8-14 L.L.-40 MAX.
A-7-2 SANDY CLAY	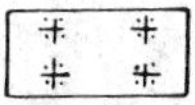	C.S. =20% F.S. =29% SILT =17% CLAY =21% COLL. =13%	L.L. =39 P.I. =17 S.L. =16	SAND-48% MIN. CLAY-17%-35% P.I.-15 MIN. L.L.-30 MIN.
A-4 SILT	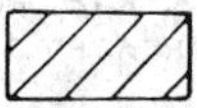	C.S. =20% F.S. =22% SILT =40% CLAY =10% COLL = 8%	L.L. =30 P.I. = 6 S.L. = 19	SAND-47% MAX. CLAY-29% MAX. P.I.-9 MAX. L.L.-40 MAX.
A-4-7 CLAYEY SILT	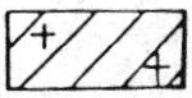	C.S. = 8% F.S. = 17% SILT = 40% CLAY = 23% COLL. = 12 %	L.L. =33 P.I. = 11 S.L. = 18	SAND-47% MAX. CLAY-25% MIN. P.I.-14 MAX. L.L.-40 MAX.
A-7-4 SILTY CLAY	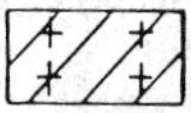	C.S. =18% F.S. =20% SILT =35% CLAY =12% COLL. =15%	L.L. =39 P.I. =15 S.L. =16	SAND-47% MAX. CLAY-29% MAX. P.I.-15 MIN. L.L.-30 MIN.
A-7 CLAY	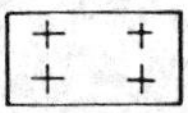	C.S. =18% F.S. =22% SILT =23% CLAY =22% COLL. =15%	L.L. =40 P.I. =17 S.L. =15	SAND-47% MAX. CLAY-30%-59% P.I.-15 MIN. L.L.-35 MIN.
A-6 COLLOIDAL CLAY	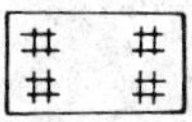	C.S. = 6% F.S. = 7% SILT = 18% CLAY =33% COLL. =36%	L.L. =50 P.I. =33 S.L. =14	CLAY-60% MIN. P.I.-25 MIN. L.L.-45 MIN.
A-5 MICA, DIATOMS, DECOMPOSED ROCK	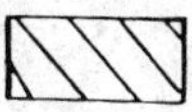	C.S. =15% F.S. =35% SILT =30% CLAY =15% COLL. = 5%	L.L. =35 P.I. = 4 S.L. =26	GRAD NOT SIGNIFICANT P.I.- LOW L.L.-HIGH S.L.-26 MIN. VISUAL INSPECTION NECESSARY TO DETERMINE TYPE
A-8 SWAMP MUCK	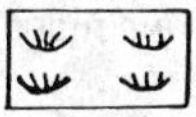	C.S. =18% F.S. =26% SILT =45% CLAY = 7% COLL. = 4%	L.L. =52 P.I. = 7 S.L. =38	ORGANIC CONTENT-4% MIN. P.I.-LOW L.L.-HIGH, WHEN OBTAINABLE S.L.-26 MIN.
ROCK REFUSAL	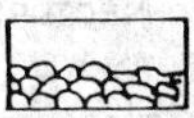			

Soils & Soil - Aggregate Mixtures
Guide to Classification

RAILS TO BE UNTREATED TIMBER.

POSTS TO BE TREATED TIMBER. WHERE PAINT IS CALLED FOR, TWO COATS TO BE APPLIED

ALL LUMBER TO BE SOUTHERN YELLOW PINE OF 1400# STRESS GRADE OR BETTER.

TO BE PAINTED IN A COLOR SCHEME AND STRIPING DETAILS IN ACCORDANCE WITH THE LATEST "MANUAL OF UNIFORM TRAFFIC CONTROL DEVICES."

TO BE USED WHERE NOTED ON PLANS

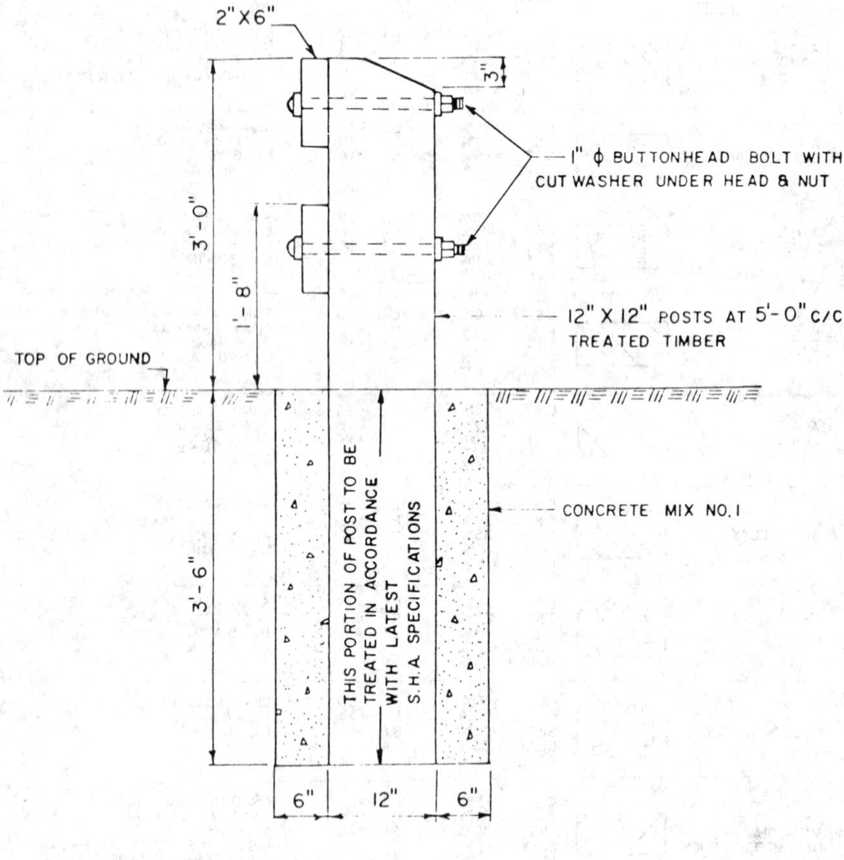

THE COST OF ALL EXCAVATION AND BACKFILL TO BE INCLUDED IN LUMP SUM PRICE BID FOR TIMBER BARRICADE.

THE LUMP SUM PRICE BID SHALL INCLUDE THE FURNISHING OF ALL MATERIALS, PRESERVATIVE TREATMENTS, TIMBER CONNECTORS AND HARDWARE, PAINTING, GALVANIZING AND CONCRETE AS WELL AS ALL LABOR, TOOLS AND EQUIPMENT AND ALL WORK INCIDENTAL THERETO.

Standard Timber Barricade

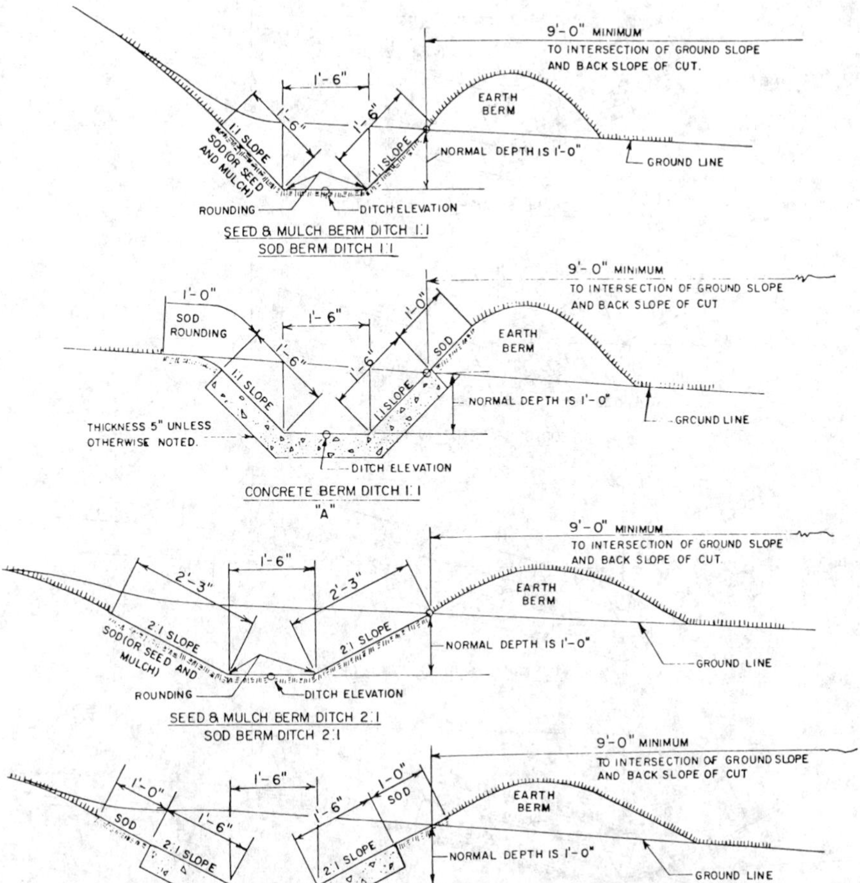

GENERAL NOTES

1. DITCH ELEVATIONS AS NOTED ON PLANS.
2. 2" TOP SOIL AND SEED & MULCH ALL CONSTRUCTION SLOPES NOT OTHERWISE TREATED.
3. WHERE DITCH LOCATION IS OTHER THAN SHOWN, DISTANCE FROM DITCH ℄ TO A REFERENCE POINT SHALL BE INDICATED ON THE PLANS.
4. EARTH BERM TO BE CONSTRUCTED FROM DITCH EXCAVATION AND COMPACTED AS DIRECTED BY THE ENGINEER. THIS WILL NOT BE A PAY ITEM.

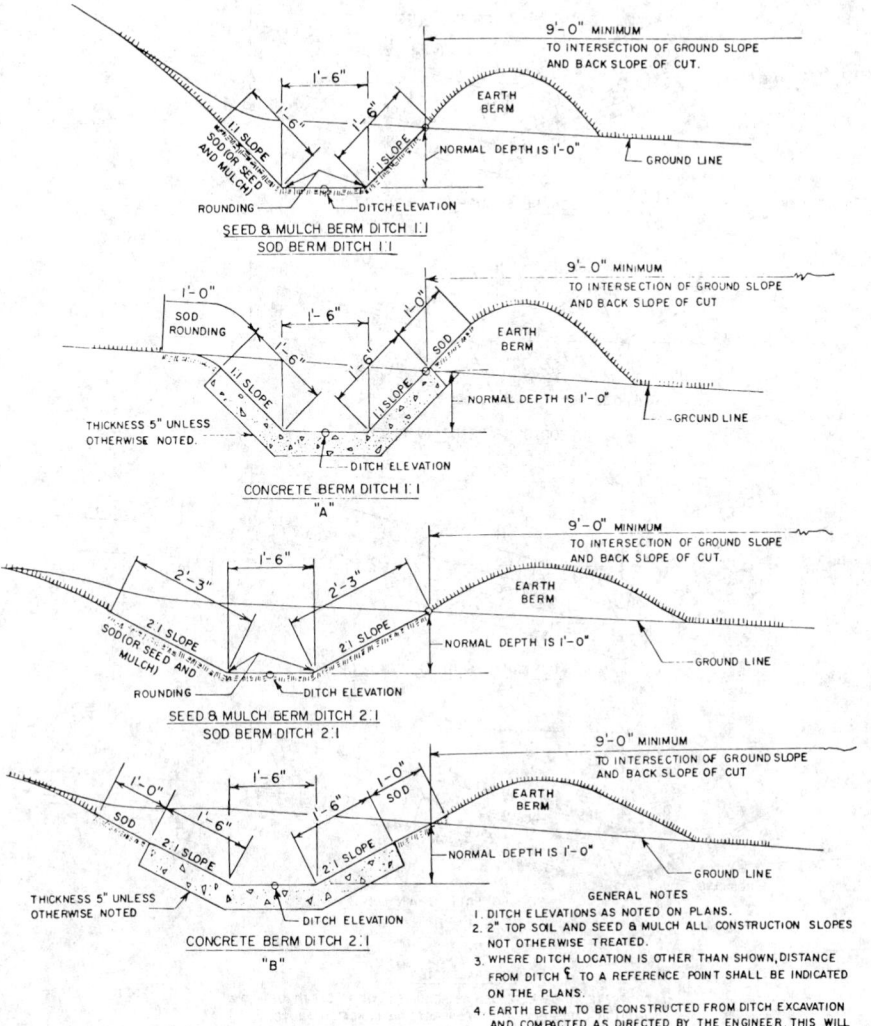

Standard Berm Ditches

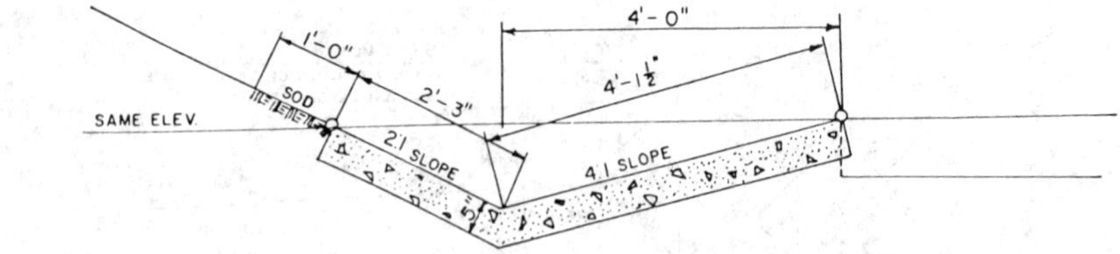

CONCRETE SURFACE DRAIN DITCH

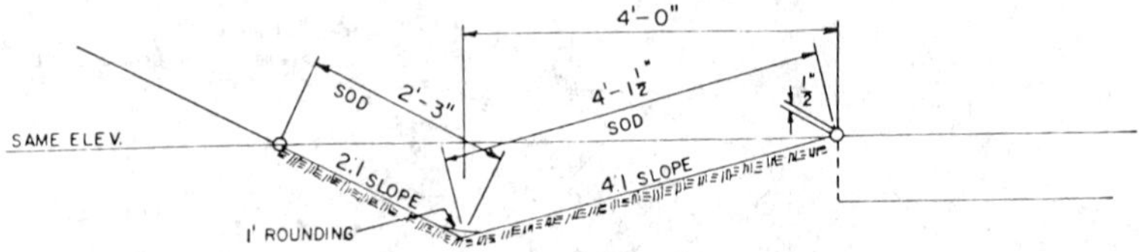

SOD SURFACE DRAIN DITCH
"A"

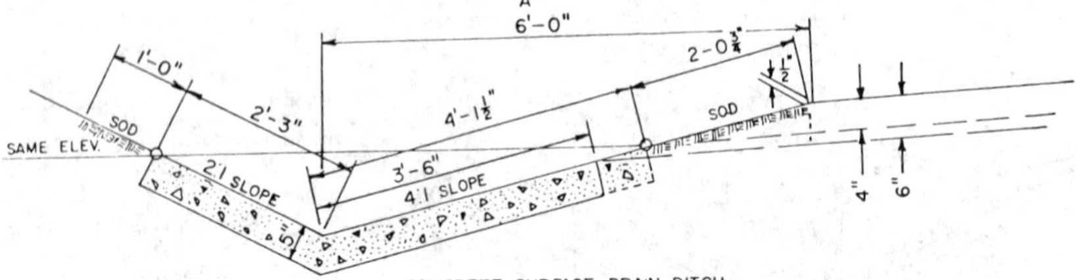

CONCRETE SURFACE DRAIN DITCH

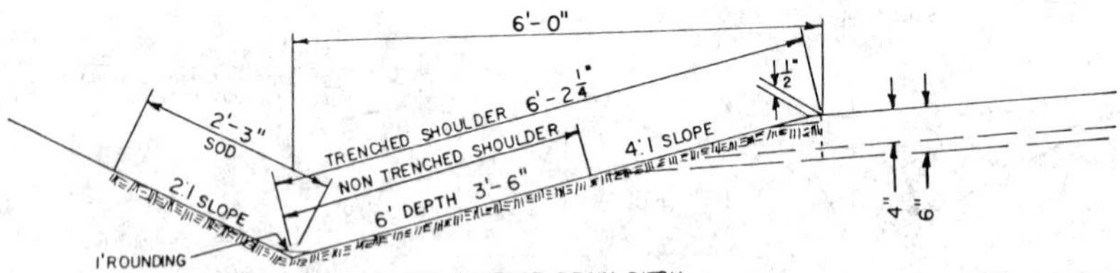

SOD SURFACE DRAIN DITCH
"B"

GENERAL NOTES

1. CONCRETE TO BE EXTENDED AS REQUIRED WHERE DEPTH OF FLOW IN DITCH EXCEEDS LIMIT OF CONCRETE AS INDICATED ABOVE.
2. OMIT SOD STRIP ON SHOULDER SIDE OF DITCH OF NON-TRENCHED SHOULDERS.

Standard Surface Drain Ditches
Concrete & Sod

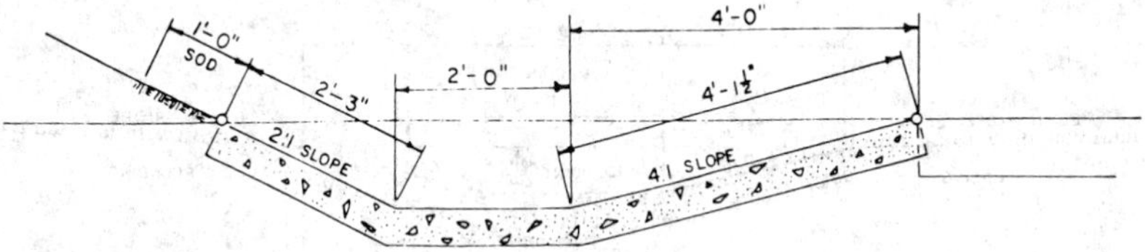

CONCRETE SIDE DITCH

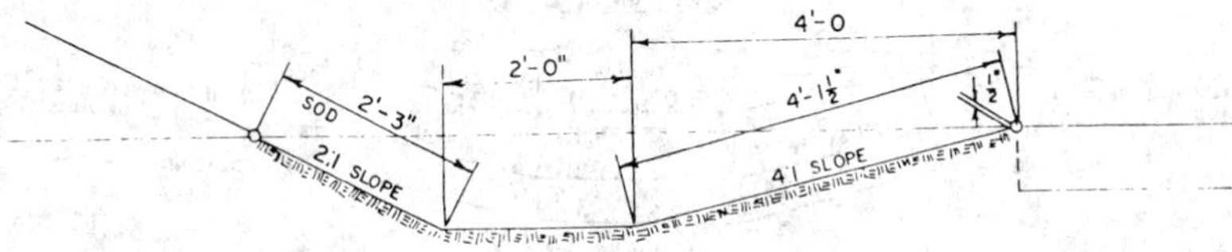

SOD SIDE DITCH
"A"

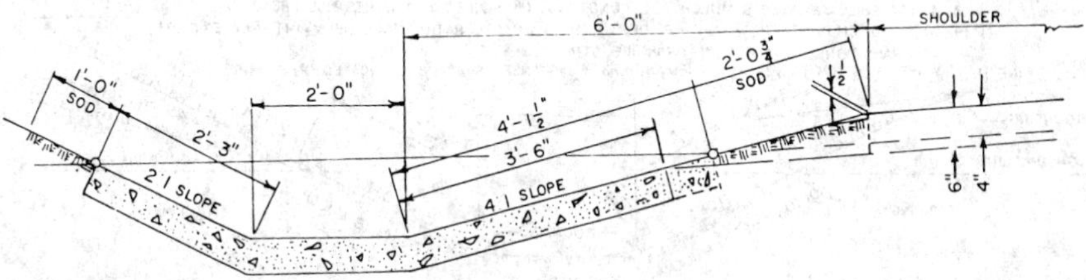

CONCRETE SIDE DITCH

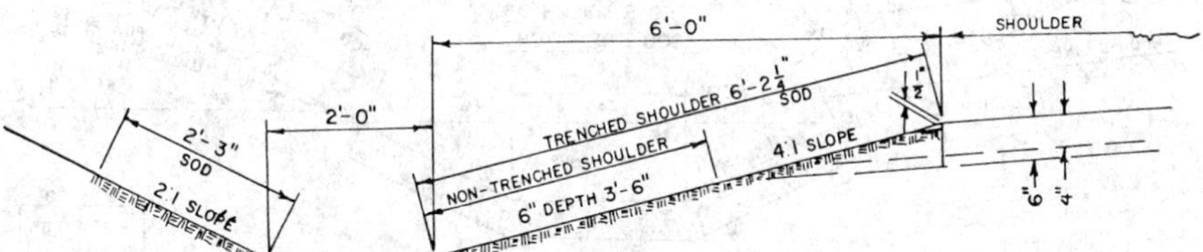

SOD SIDE DITCH
"B"

GENERAL NOTES

WHEN DITCHES ARE CONSTRUCTED OTHER THAN WHAT IS SHOWN, THE ELEVATIONS WILL BE NOTED ON THE PLANS.

Standard Side Ditches Cut Areas

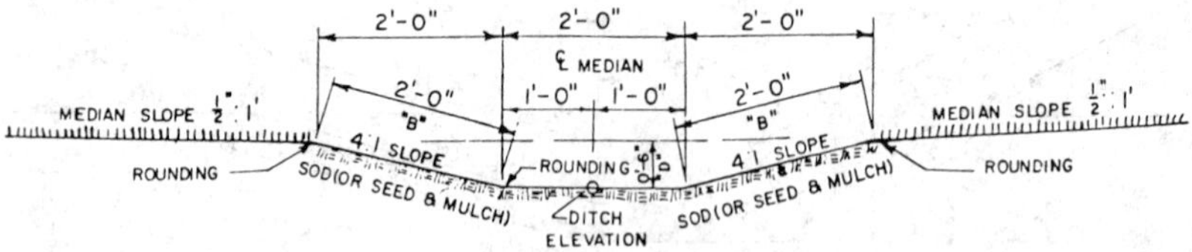

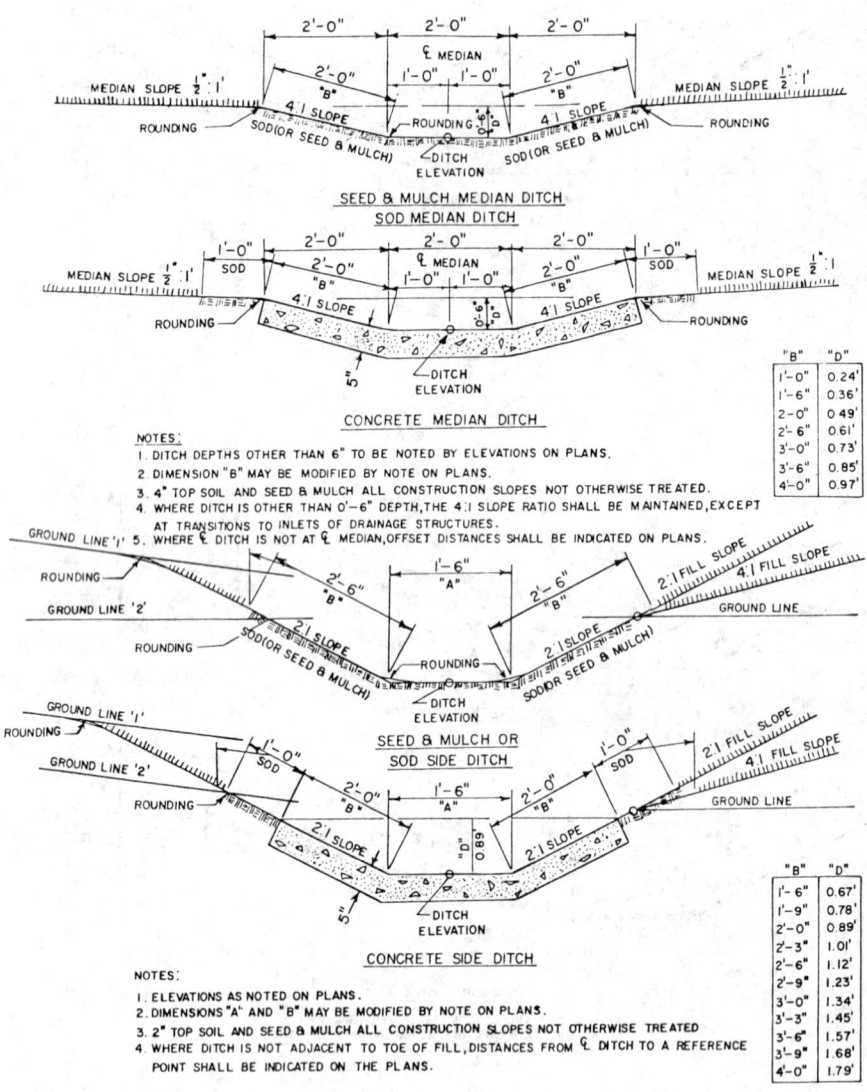

SEED & MULCH MEDIAN DITCH

SOD MEDIAN DITCH

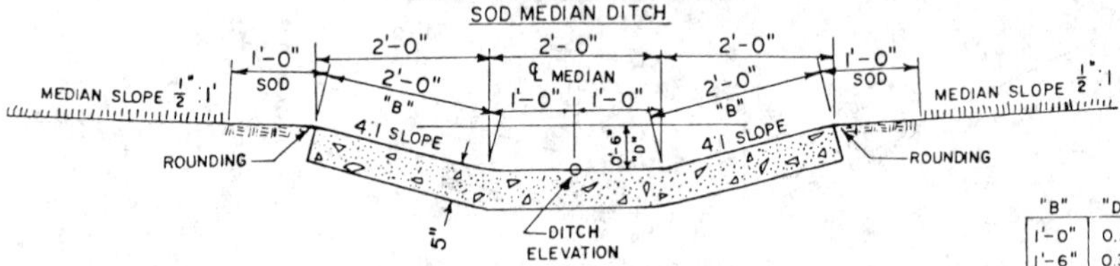

"B"	"D"
1'-0"	0.24'
1'-6"	0.36'
2-0"	0.49'
2'-6"	0.61'
3'-0"	0.73'
3'-6"	0.85'
4'-0"	0.97'

CONCRETE MEDIAN DITCH

NOTES:

1. DITCH DEPTHS OTHER THAN 6" TO BE NOTED BY ELEVATIONS ON PLANS.
2. DIMENSION "B" MAY BE MODIFIED BY NOTE ON PLANS.
3. 4" TOP SOIL AND SEED & MULCH ALL CONSTRUCTION SLOPES NOT OTHERWISE TREATED.
4. WHERE DITCH IS OTHER THAN 0'-6" DEPTH, THE 4:1 SLOPE RATIO SHALL BE MAINTAINED, EXCEPT AT TRANSITIONS TO INLETS OF DRAINAGE STRUCTURES.
5. WHERE ℄ DITCH IS NOT AT ℄ MEDIAN, OFFSET DISTANCES SHALL BE INDICATED ON PLANS.

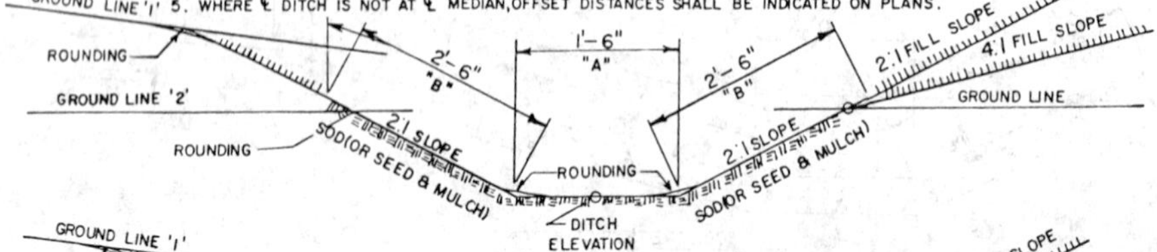

SEED & MULCH OR SOD SIDE DITCH

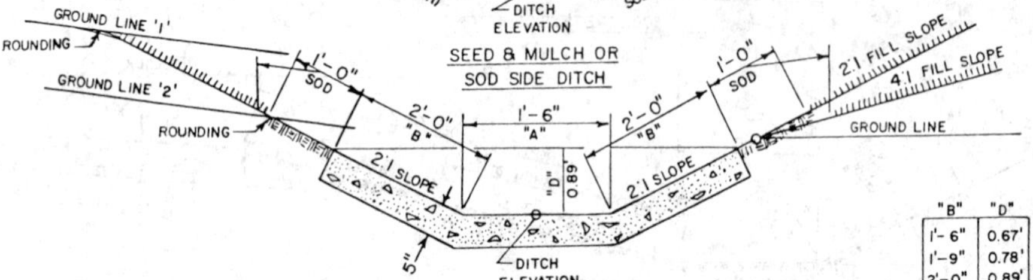

"B"	"D"
1'-6"	0.67'
1'-9"	0.78'
2'-0"	0.89'
2'-3"	1.01'
2'-6"	1.12'
2'-9"	1.23'
3'-0"	1.34'
3'-3"	1.45'
3'-6"	1.57'
3'-9"	1.68'
4'-0"	1.79'

CONCRETE SIDE DITCH

NOTES:

1. ELEVATIONS AS NOTED ON PLANS.
2. DIMENSIONS "A" AND "B" MAY BE MODIFIED BY NOTE ON PLANS.
3. 2" TOP SOIL AND SEED & MULCH ALL CONSTRUCTION SLOPES NOT OTHERWISE TREATED
4. WHERE DITCH IS NOT ADJACENT TO TOE OF FILL, DISTANCES FROM ℄ DITCH TO A REFERENCE POINT SHALL BE INDICATED ON THE PLANS.

Standard Side Ditches
Standard Median Ditches - Flat

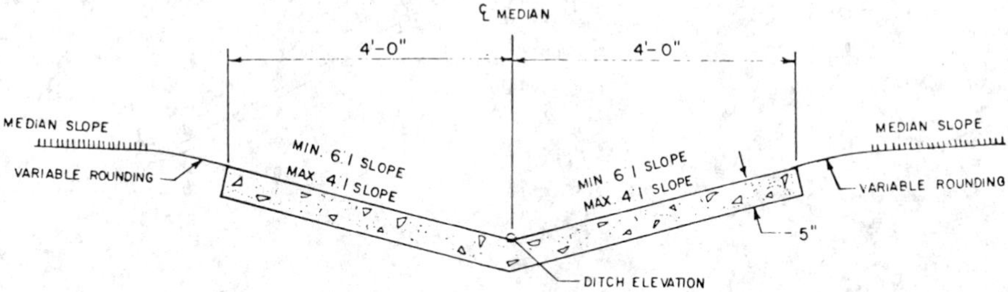

CONCRETE MEDIAN 'V' DITCH

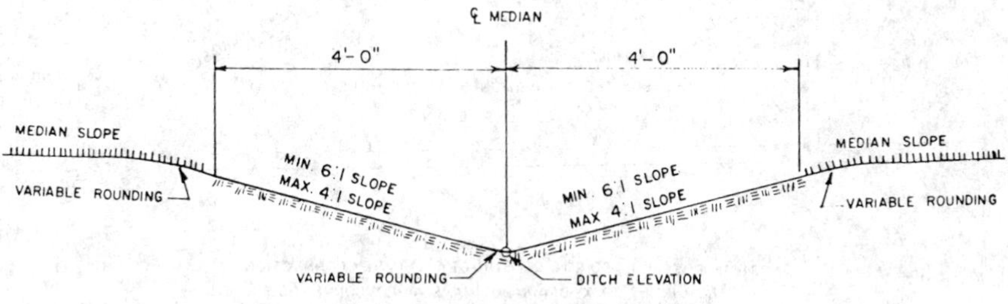

SOD MEDIAN 'V' DITCH

Standard Median Ditches - V Slope

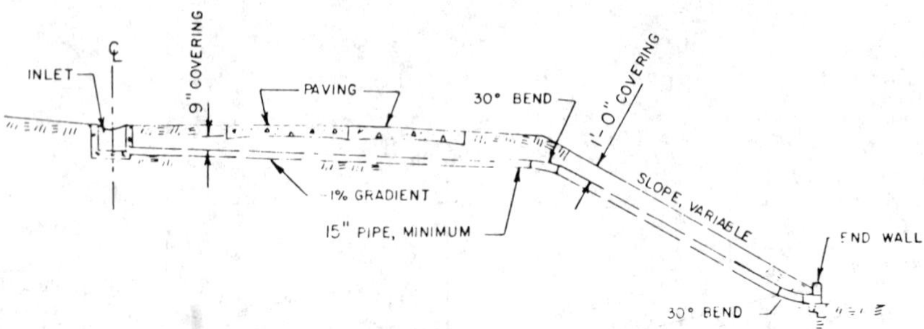

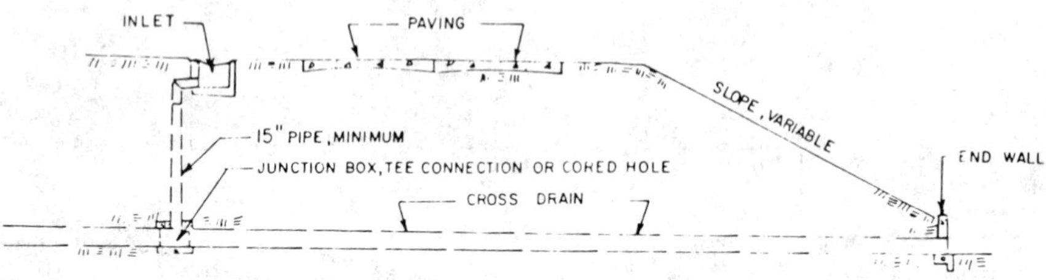

NOTE IF POSSIBLE, DO NOT CONSTRUCT A CHIMNEY CONNECTION FROM INLET TO X-DRAIN BUT MAKE A 45° CONNECTION TO X-DRAIN WHEN NECESSARY TO USE A CHIMNEY CONNECTION POUR A 6" CONCRETE COLLAR AROUND THE VERTICAL PIPE FOR ITS ENTIRE HEIGHT.

Standard Pipe Outlets
Surface Drainage on Fills

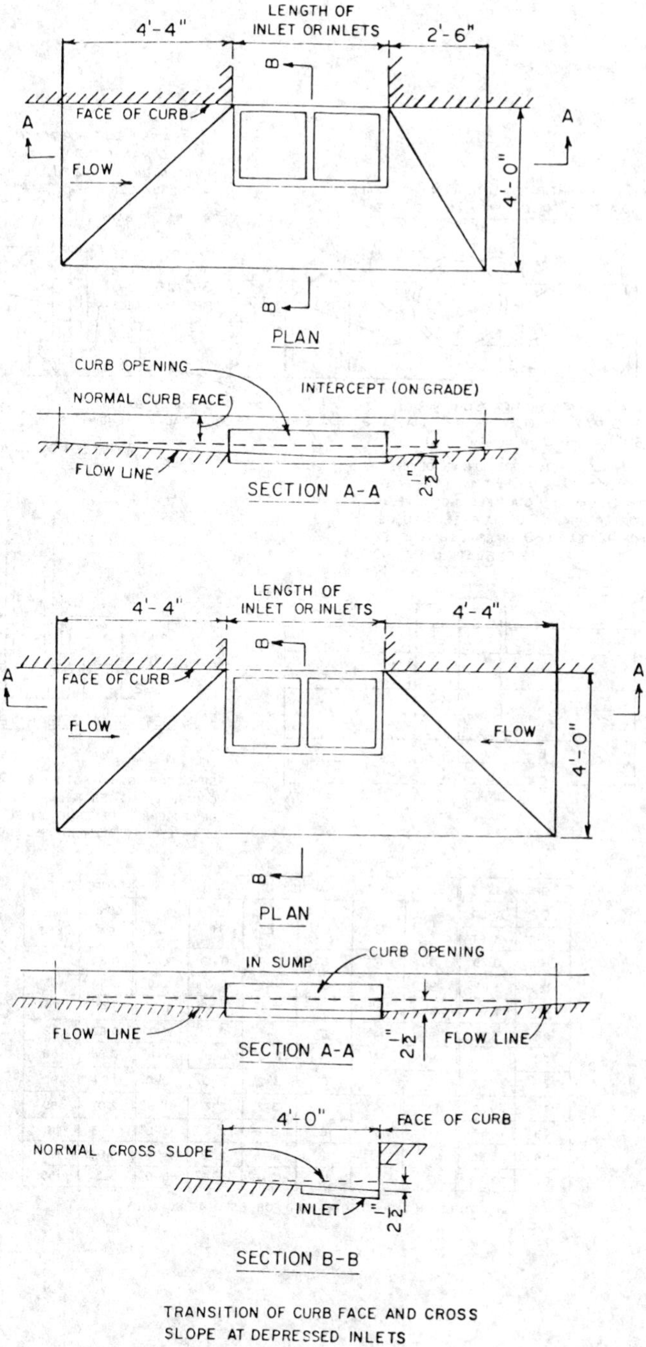

Method of Depressing Pavement at Inlets

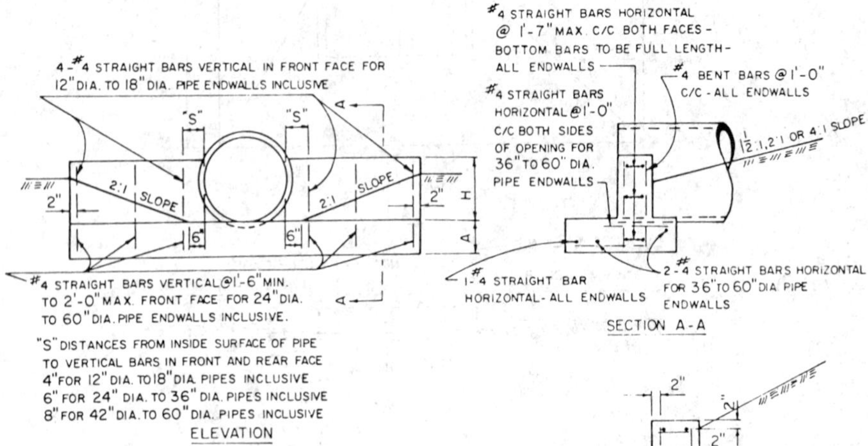

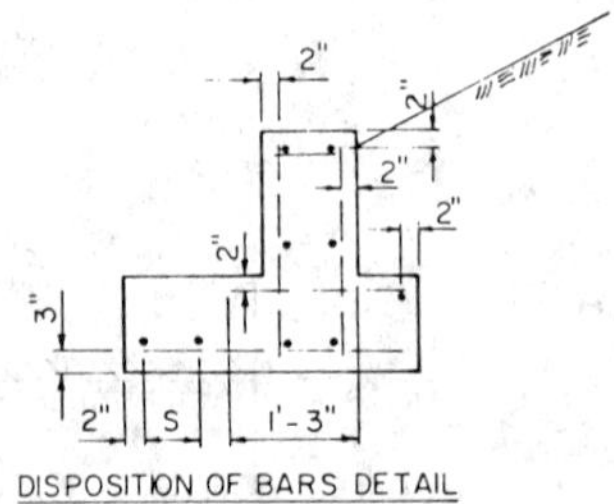

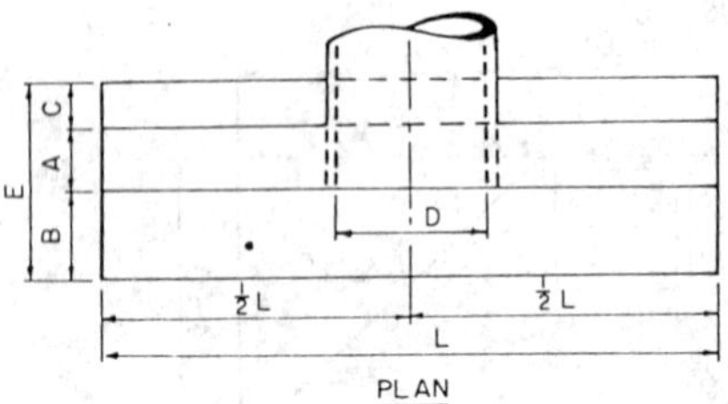

GENERAL NOTES:
SPECIFICATIONS : LATEST S.H.A.
CONCRETE : SEE S.H.A. SPECIFICATIONS
REINFORCING : DEFORMED STEEL BARS #4 BARS (1/2" ϕ)
CHAMFER : ALL EXPOSED EDGES 1"X1" OR AS DIRECTED

OPENINGS		DIMENSIONS						VOLUME	STEEL
D INCHES	AREA SQ. FT.	A	B	C	E	H	L	CONC. C.Y.	LBS.
12	0.79	9"	6"	6"	1'-9"	0'-10"	4'-0"	0.27	19
15	1.23	9"	6"	6"	1'-9"	1'-0½"	4'-9"	0.34	23
18	1.78	9"	6"	6"	1'-9"	1'-3"	5'-6"	0.41	29
24	3.14	9"	14"	6"	2'-5"	1'-6"	7'-0"	0.69	38
30	4.91	9"	14"	6"	2'-5"	1'-9"	8'-6"	0.88	58
36	7.07	12"	16"	10"	3'-2"	2'-0"	10'-0"	1.74	90
42	9.62	12"	16"	10"	3'-2"	2'-3"	11'-6"	2.07	99
48	12.57	12"	16"	10"	3'-2"	2'-6"	13'-0"	2.43	119
54	15.90	12"	20"	12"	3'-8"	2'-9"	14'-6"	3.08	132
60	19.64	12"	20"	12"	3'-8"	3'-0"	16'-0"	3.50	158

QUANTITIES IN TABLE TO BE USED FOR ESTIMATING ONLY

Standard End Support Wall
Round Pipe

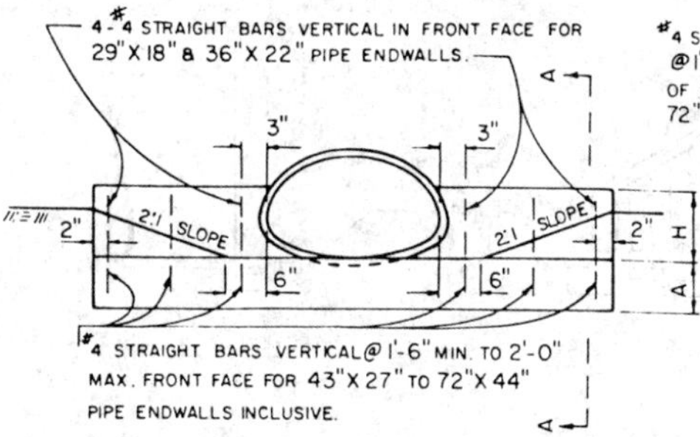

ELEVATION

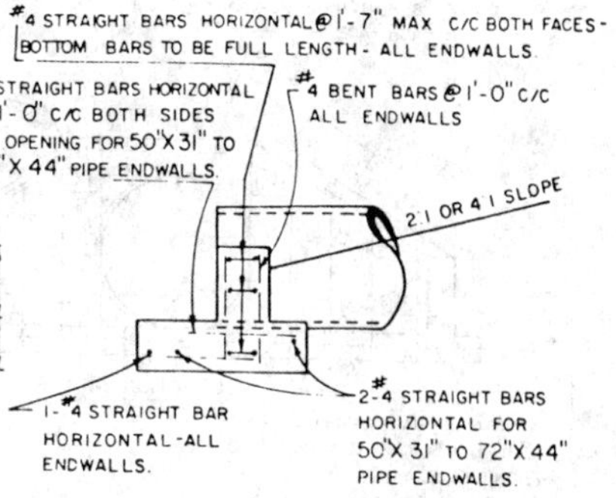

SECTION A-A

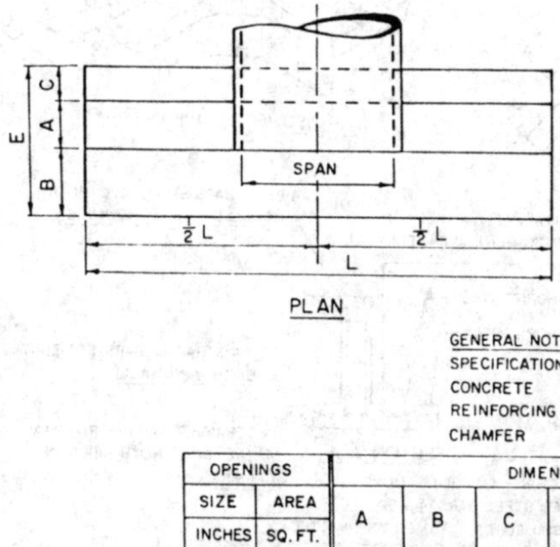

PLAN

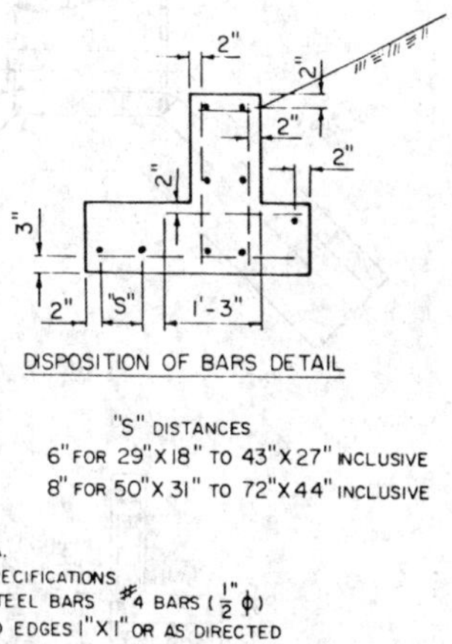

DISPOSITION OF BARS DETAIL

"S" DISTANCES
6" FOR 29"X18" TO 43"X27" INCLUSIVE
8" FOR 50"X31" TO 72"X44" INCLUSIVE

GENERAL NOTES:

SPECIFICATIONS : LATEST S.H.A.
CONCRETE : SEE S.H.A. SPECIFICATIONS
REINFORCING : DEFORMED STEEL BARS #4 BARS ($\frac{1}{2}$" ϕ)
CHAMFER : ALL EXPOSED EDGES 1"X1" OR AS DIRECTED

OPENINGS		DIMENSIONS						VOLUME	STEEL
SIZE INCHES	AREA SQ. FT.	A	B	C	E	H	L	CONC. C.Y.	LBS.
29"X18"	2.2	9"	14"	6"	2'-5"	1'-0"	6'-0"	.51	34
36"X22"	4.4	9"	14"	6"	2'-5"	1'-3"	7'-6"	.67	41
43"X27"	4.8	12"	16"	10"	3'-2"	1'-6"	8'-11"	1.37	59
50"X31"	6.5	12"	16"	10"	3'-2"	1'-9"	10'-5"	1.66	103
58"X36"	8.7	12"	16"	10"	3'-2"	2'-0"	12'-2"	2.01	133
65"X40"	11.0	12"	20"	12"	3'-8"	2'-3"	13'-8"	2.59	156
72"X44"	13.5	12"	20"	12"	3'-8"	2'-6"	15'-2"	2.97	176

QUANTITIES IN TABLE TO BE USED FOR ESTIMATING ONLY

**Standard End Support Wall
Arch Pipe**

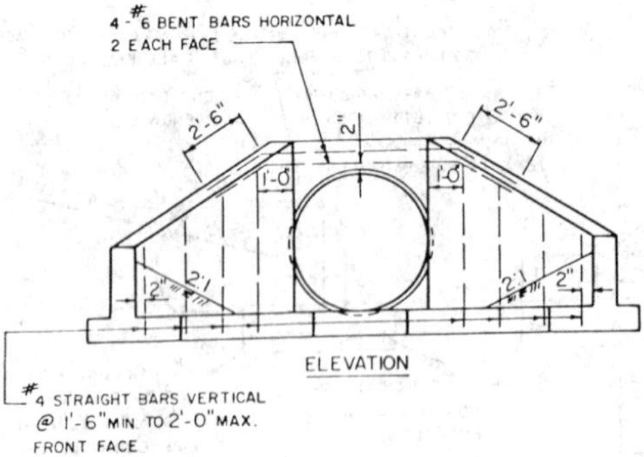

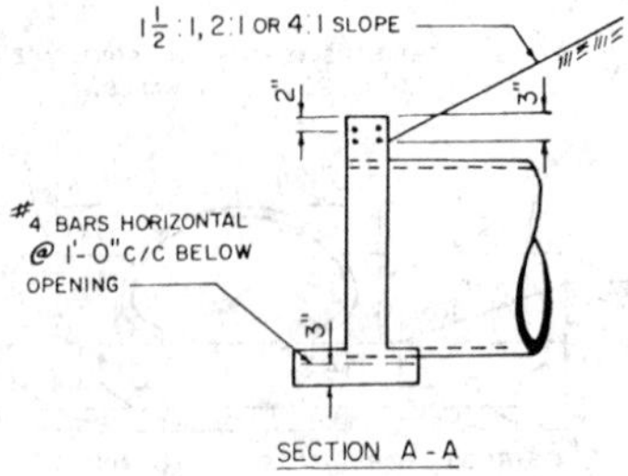

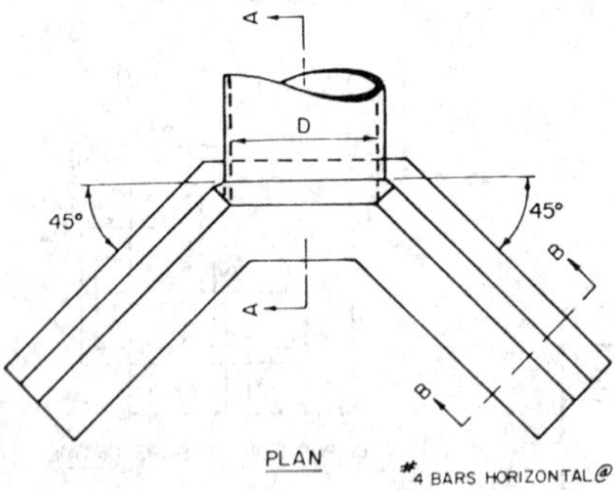

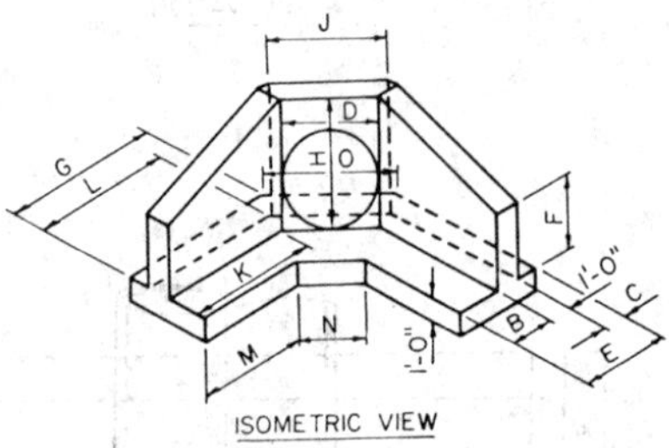

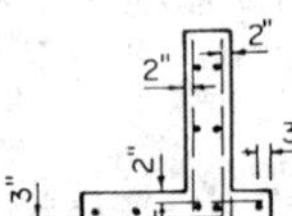

DISPOSITION OF BARS DETAIL

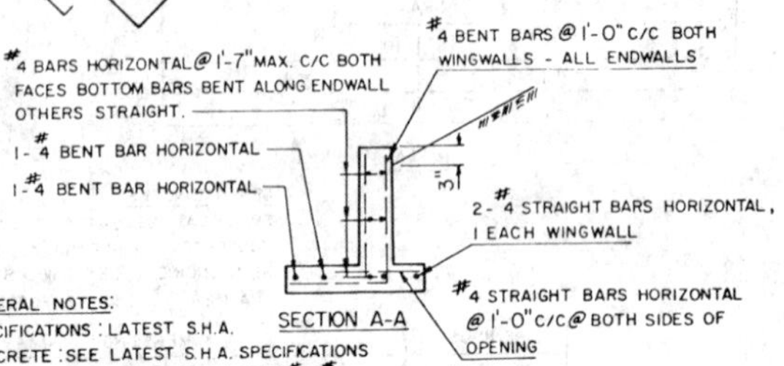

SECTION A-A

GENERAL NOTES:
SPECIFICATIONS : LATEST S.H.A.
CONCRETE : SEE LATEST S.H.A. SPECIFICATIONS
REINFORCING: DEFORMED STEEL BARS #4 & #6
CHAMFER: ALL EXPOSED EDGES 1" X 1" OR AS DIRECTED

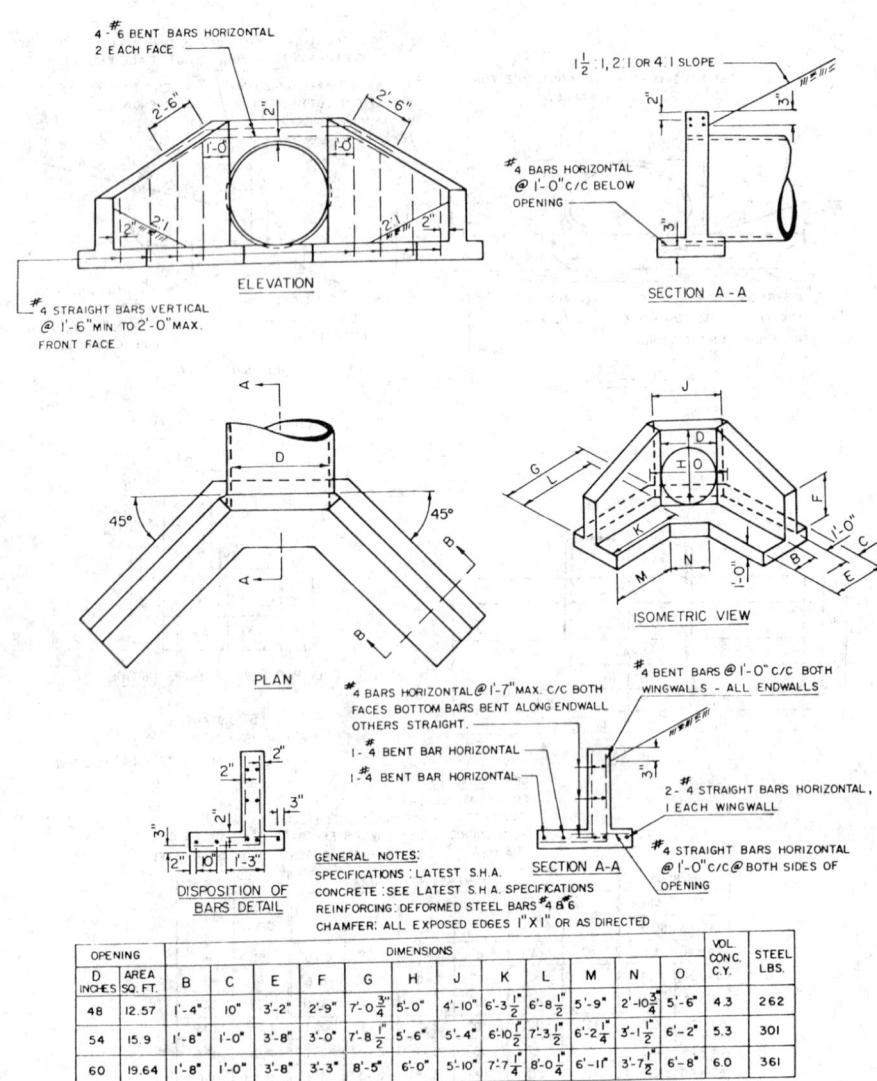

OPENING		DIMENSIONS												VOL. CONC. C.Y.	STEEL LBS.
D INCHES	AREA SQ. FT.	B	C	E	F	G	H	J	K	L	M	N	O		
48	12.57	1'-4"	10"	3'-2"	2'-9"	7'-0 3/4"	5'-0"	4'-10"	6'-3 1/2"	6'-8 1/2"	5'-9"	2'-10 3/4"	5'-6"	4.3	262
54	15.9	1'-8"	1'-0"	3'-8"	3'-0"	7'-8 1/2"	5'-6"	5'-4"	6'-10 1/2"	7'-3 1/2"	6'-2 1/4"	3'-1 1/2"	6'-2"	5.3	301
60	19.64	1'-8"	1'-0"	3'-8"	3'-3"	8'-5"	6'-0"	5'-10"	7'-7 1/4"	8'-0 1/4"	6'-11"	3'-7 1/2"	6'-8"	6.0	361

QUANTITIES IN TABLE TO BE USED FOR ESTIMATING ONLY

Standard Headwalls

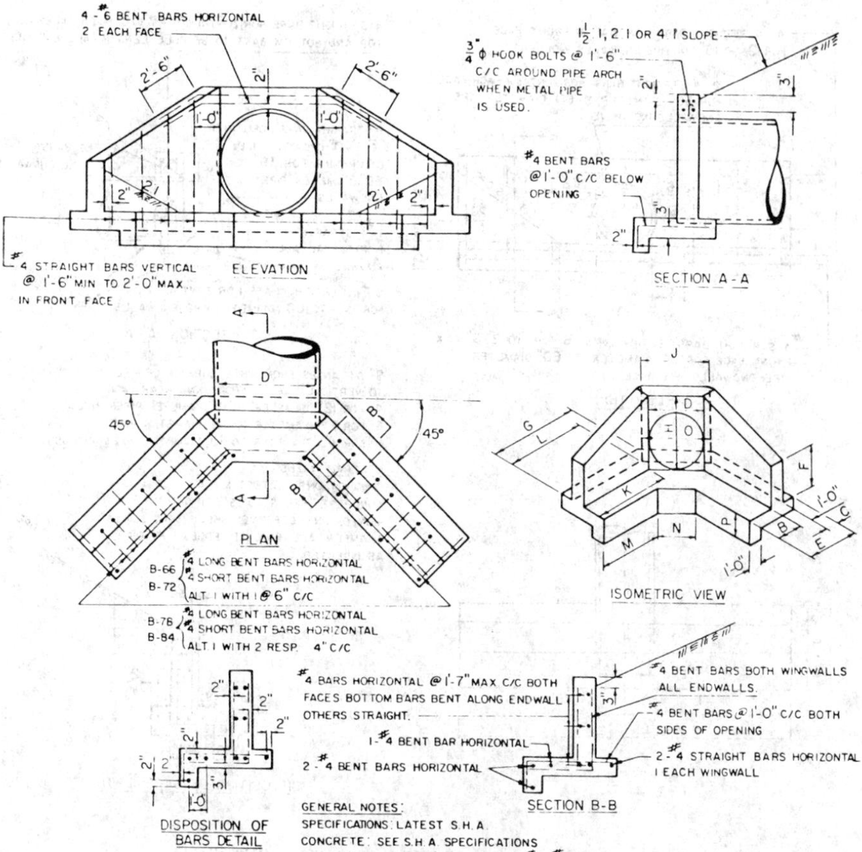

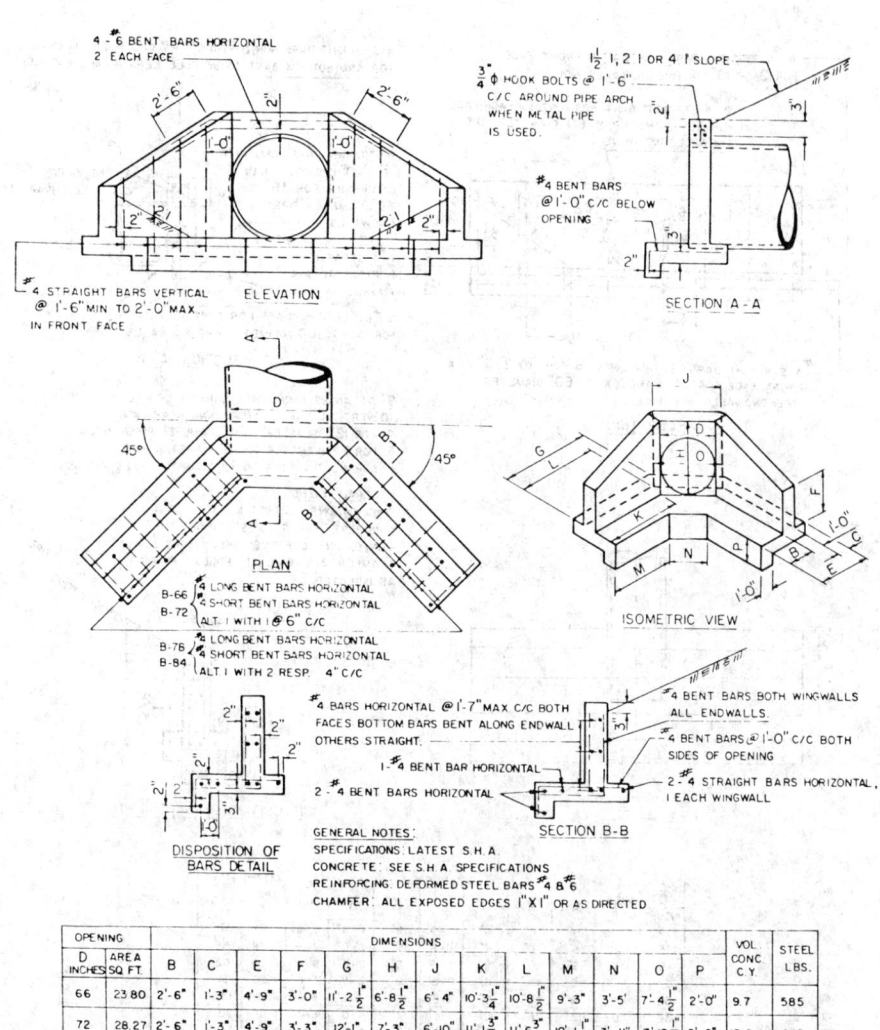

GENERAL NOTES:
SPECIFICATIONS: LATEST S.H.A.
CONCRETE: SEE S.H.A. SPECIFICATIONS
REINFORCING: DEFORMED STEEL BARS #4 & #6
CHAMFER: ALL EXPOSED EDGES 1"X1" OR AS DIRECTED

OPENING		DIMENSIONS													VOL. CONC. C.Y.	STEEL LBS.
D INCHES	AREA SQ. FT.	B	C	E	F	G	H	J	K	L	M	N	O	P		
66	23.80	2'-6"	1'-3"	4'-9"	3'-0"	11'-2 1/2"	6'-8 1/2"	6'-4"	10'-3 1/4"	10'-8 1/2"	9'-3"	3'-5'	7'-4 1/2"	2'-0"	9.7	585
72	28.27	2'-6"	1'-3"	4'-9"	3'-3"	12'-1"	7'-3"	6'-10"	11'-1 3/4"	11'-6 3/4"	10'-1 1/4"	3'-11"	7'-10 1/2"	2'-0"	10.9	645
78	33.20	3'-0"	1'-6"	5'-6"	3'-6	13'-0 1/2"	7'-9 1/2"	7'-4"	12'-0"	12'-5"	10'-9"	4'-0"	8'-6 3/4"	2'-6"	13.3	865
84	38.48	3'-0"	1'-6"	5'-6"	3'-9"	13'-10"	8'-4"	7'-10"	12'-9 1/2"	13'-2 1/2"	11'-6 1/2"	4'-6"	9'-0 3/4"	2'-6"	14.7	984

QUANTITIES IN TABLE TO BE USED FOR ESTIMATING ONLY

Standard Headwalls

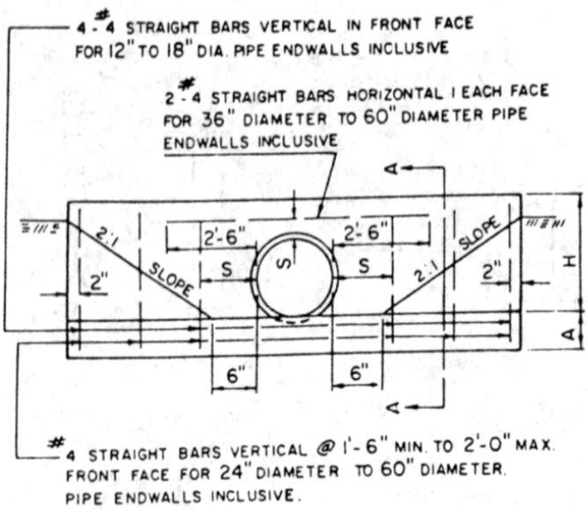

ELEVATION

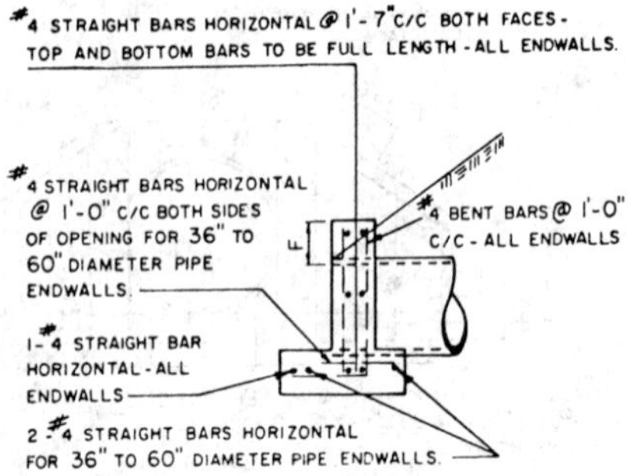

SECTION A - A

"S" DISTANCES FROM INSIDE SURFACE OF PIPE TO VERTICAL BARS IN FRONT AND REAR FACE.
4" FOR 12" DIAMETER TO 18" DIAMETER PIPES INCLUSIVE
6" FOR 24" DIAMETER TO 36" DIAMETER PIPES INCLUSIVE
8" FOR 42" DIAMETER TO 60" DIAMETER PIPES INCLUSIVE

GENERAL NOTES:
SPECIFICATIONS: LATEST S.H.A.
CONCRETE: SEE S.H.A. SPECIFICATIONS
REINFORCING: DEFORMED STEEL BARS #4 φ BARS
CHAMFER: ALL EXPOSED EDGES 1"X1" OR AS DIRECTED

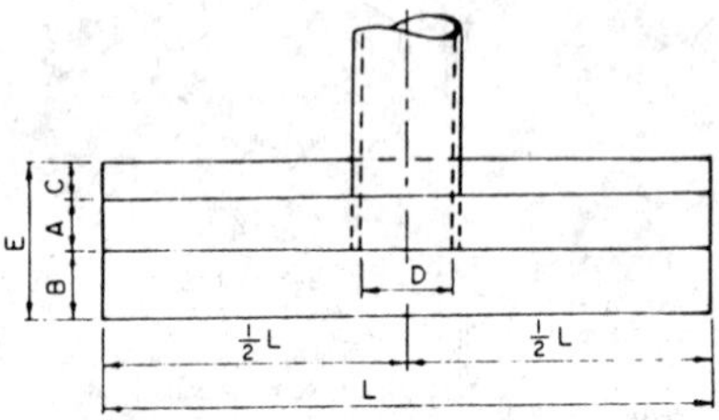

PLAN

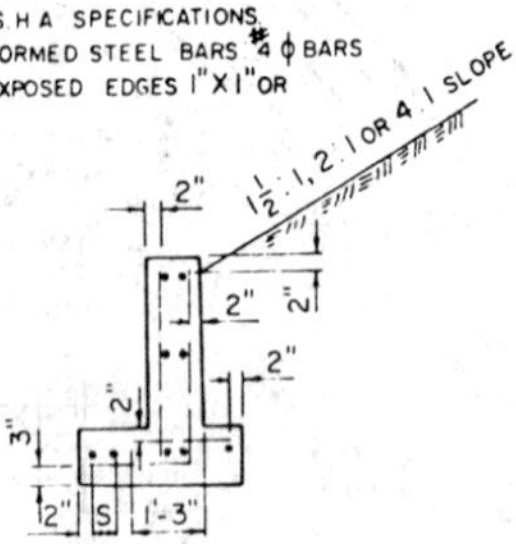

DISPOSITION OF BARS - DETAIL

OPENINGS D INCHES	AREA SQ. FT.	A	B	C	E	F	H	L	VOLUME CONC. C.Y.	STEEL LBS.
12	0.79	9"	6"	6"	1'-9"	9"	1'-9"	6'-6"	0.61	38
15	1.23	9"	6"	6"	1'-9"	9"	2'-0"	7'-9"	0.78	55
18	1.78	9"	6"	6"	1'-9"	9"	2'-3"	9'-0"	0.95	62
24	3.14	9"	14"	6"	2'-5"	9"	2'-9"	11'-6"	1.56	91
30	4.91	9"	14"	6"	2'-5"	12"	3'-6"	14'-2"	2.19	136
36	7.07	12"	16"	10"	3'-2"	12"	4'-0"	16'-8"	4.18	222
42	9.62	12"	16"	10"	3'-2"	12"	4'-6"	19'-2"	5.13	265
48	12.57	12"	16"	10"	3'-2"	12"	5'-0"	21'-8"	6.12	326
54	15.90	12"	20"	12"	3'-8"	12"	5'-6"	24'-2"	7.68	384
60	19.64	12"	20"	12"	3'-8"	12"	6'-0"	26'-8"	8.86	428

QUANTITIES IN TABLE TO BE USED FOR ESTIMATING ONLY

Standard Type C Endwall
Round Pipe

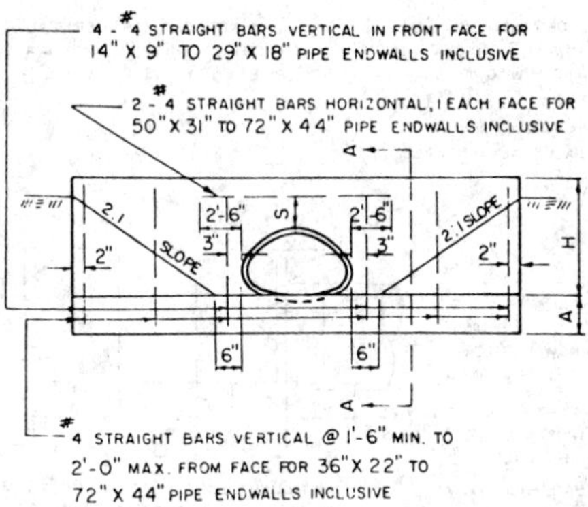

ELEVATION

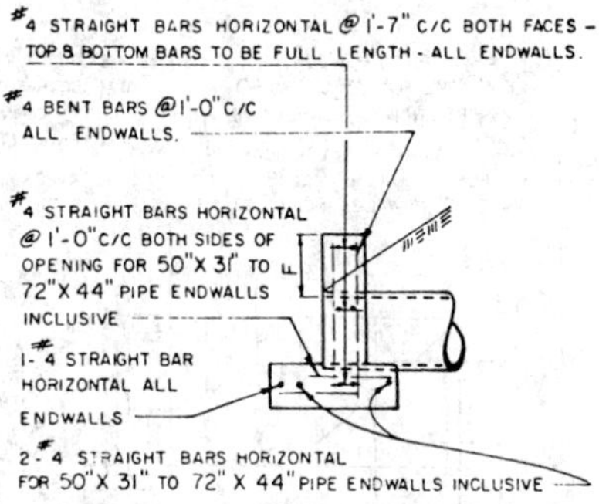

SECTION A - A

"S" DISTANCES
4" FOR 14" X 9" TO 22" X 13" INCLUSIVE
6" FOR 29" X 18" TO 43" X 27" INCLUSIVE
8" FOR 50" X 31" TO 72" X 44" INCLUSIVE

GENERAL NOTES:
SPECIFICATION : LATEST S. H. A.
CONCRETE : SEE S. H. A. SPECIFICATIONS
REINFORCING : DEFORMED STEEL BARS: #4 φ BARS
CHAMFER : ALL EXPOSED EDGES 1" X 1" OR AS DIRECTED.

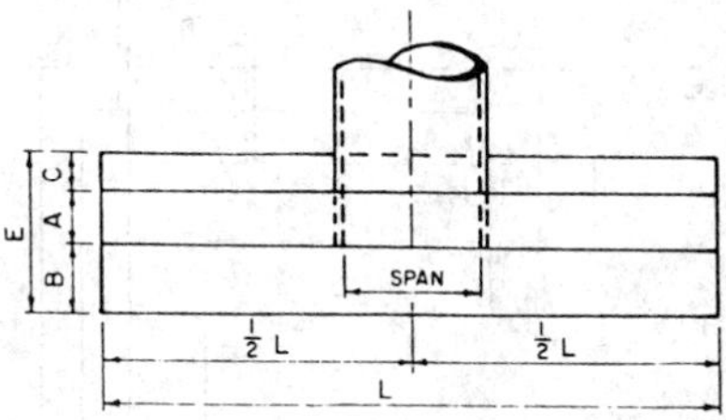

PLAN

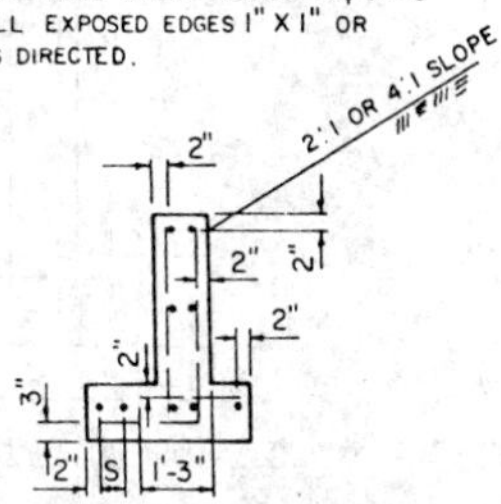

DISPOSITION OF BARS - DETAILS

OPENINGS		DIMENSIONS							VOLUME	STEEL
SIZE INCHES	AREA SQ. FT.	A	B	C	E	F	H	L	CONC. C.Y.	LBS.
14"X 9"	0.8	9"	6"	6"	1'-9"	8"	1'-5"	6'-4"	.54	41
18"X 11"	1.1	9"	6"	6"	1'-9"	6"	1'-5"	6'-4"	.53	41
22"X13"	1.6	9"	6"	6"	1'-9"	11"	2'-0"	9'-7"	.95	68
29"X18"	2.8	9"	6"	6"	1'-9"	6"	2'-0"	9'-7"	.92	67
36"X22"	4.4	9"	14"	6"	2'-5"	14"	3'-0"	13'-9"	1.95	120
43"X27"	6.4	9"	14"	6"	2'-5"	9"	3'-0"	13'-9"	1.89	126
50"X31"	8.7	12"	16"	10"	3'-2"	14"	3'-9"	18'-0"	4.29	226
58"X36"	11.4	12"	16"	10"	3'-2"	9"	3'-9"	18'-0"	4.19	226
65"X40"	14.3	12"	20"	12"	3'-8"	13"	4'-5"	21'-10"	6.01	310
72"X44"	17.6	12"	20"	12"	3'-8"	9"	4'-5"	21'-10"	5.89	310

QUANTITIES IN TABLE TO BE USED FOR ESTIMATING ONLY.

Standard Type C Endwall
Arch Pipe

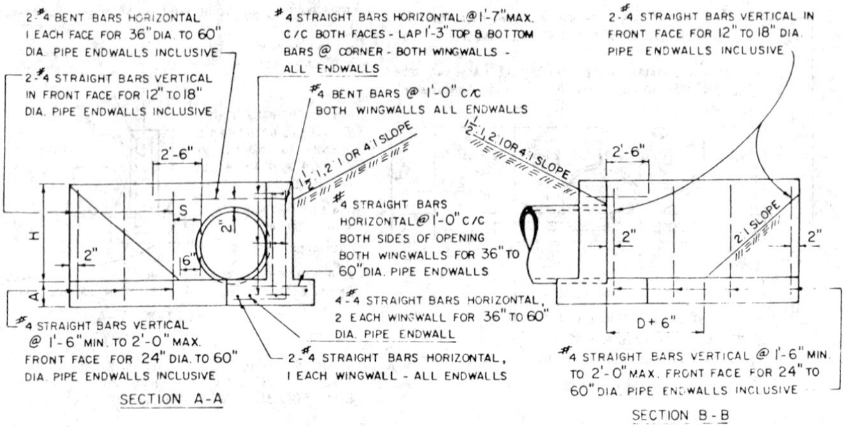

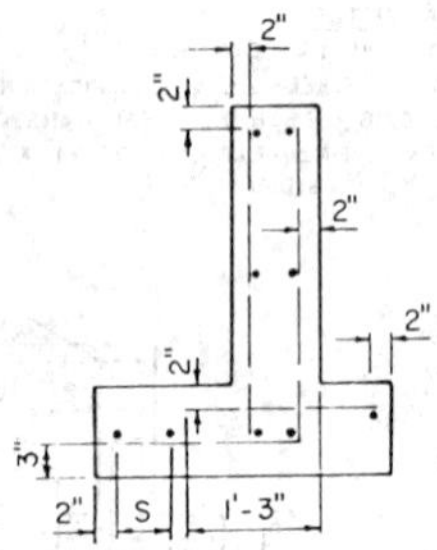

DISPOSITION OF BARS DETAIL

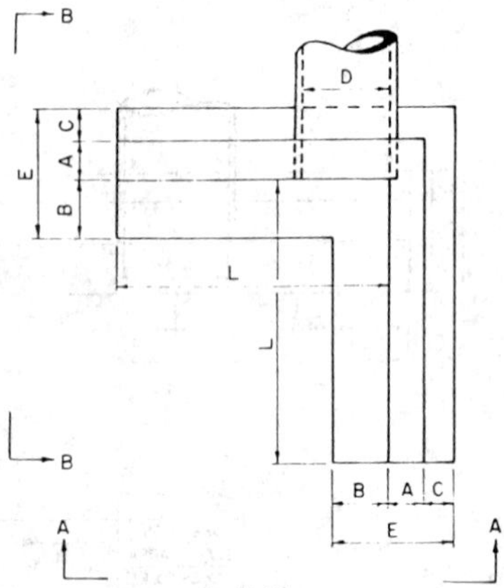

PLAN

OPENINGS		DIMENSIONS						VOL.	STEEL
D INCHES	AREA SQ. FT.	A	B	C	E	H	L	CONC. C.Y.	LBS.
12	0.79	9"	6"	6"	1'-9"	1'-9"	3'-6"	0.73	56
15	1.23	9"	6"	6"	1'-9"	2'-0"	4'-3"	0.93	77
18	1.78	9"	6"	6"	1'-9"	2'-3"	5'-0"	1.14	89
24	3.14	9"	14"	6"	2'-5"	2'-9"	6'-6"	1.8	124
30	4.91	9"	14"	6"	2'-5"	3'-6"	8'-0"	2.6	160
36	7.07	12"	16"	10"	3'-2"	4'-0"	9'-6"	5.0	287
42	9.62	12"	16"	10"	3'-2"	4'-6"	11'-0"	6.2	333
48	12.57	12"	16"	10"	3'-2"	5'-0"	12'-6"	7.4	387
54	15.90	12"	20"	12"	3'-8"	5'-6"	14'-0"	9.2	458
60	19.64	12"	20"	12"	3'-8"	6'-0"	15'-6"	10.7	519

QUANTITIES IN TABLE TO BE USED FOR ESTIMATING ONLY

"S" DISTANCES FROM INSIDE SURFACE OF PIPE TO VERTICAL BARS IN FRONT AND REAR FACE
4" FOR 12" DIA. TO 18" DIA. PIPES INCLUSIVE
6" FOR 24" DIA. TO 36" DIA. PIPES INCLUSIVE
8" FOR 42" DIA. TO 60" DIA. PIPES INCLUSIVE

GENERAL NOTES

SPECIFICATIONS : LATEST S.H.A
CONCRETE : SEE S.H.A SPECIFICATIONS.
REINFORCING : DEFORMED STEEL BARS - 4 φ BARS.
CHAMFER : ALL EXPOSED EDGES 1" X 1"

Standard Type E Endwall
Round Pipe

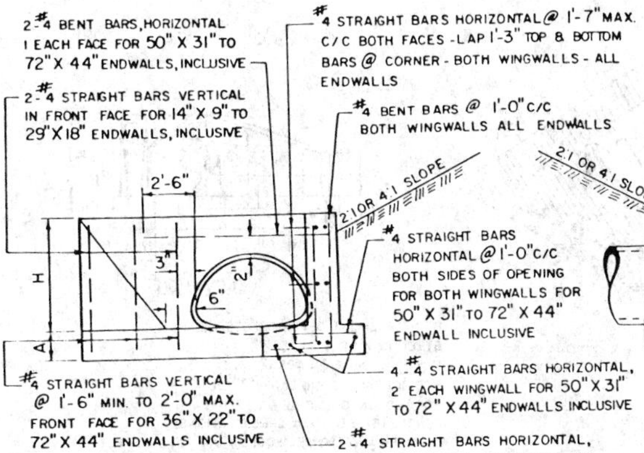

SECTION A-A

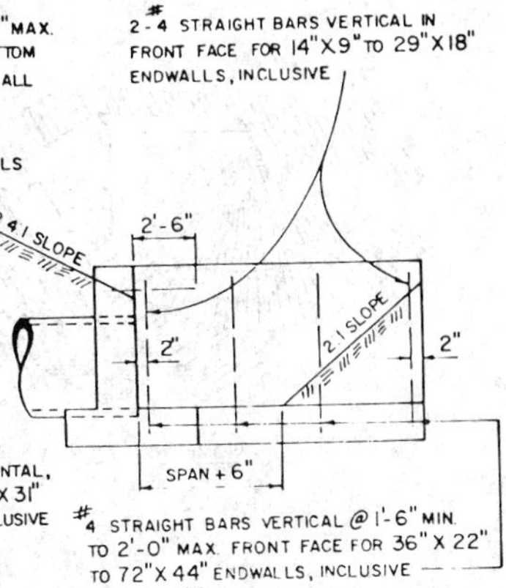

SECTION B-B

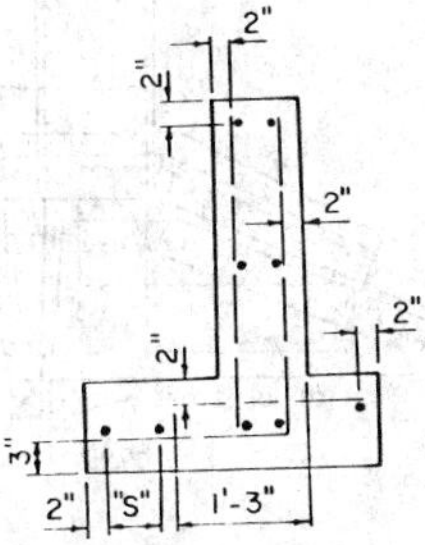

DISPOSITION OF BARS DETAIL

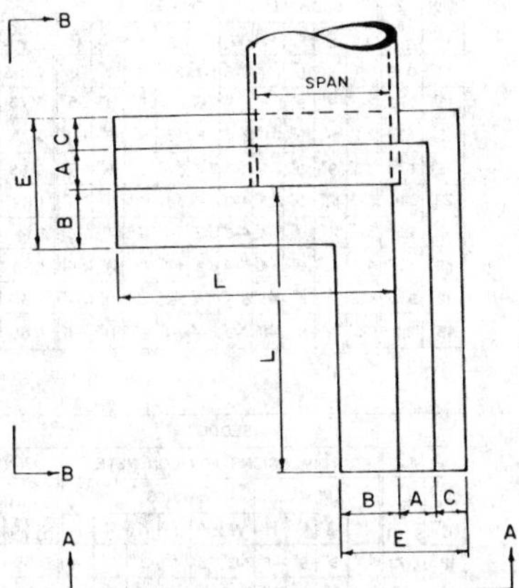

PLAN

OPENINGS		DIMENSIONS						VOL.	STEEL
SIZE	AREA	A	B	C	E	H	L	CONC. C.Y.	LBS.
INCHES	SQ. FT.								
14"X9"	0.8	9"	6"	6"	1'-9"	1'-5"	3'-10"	.72	56
18"X11"	1.1	9"	6"	6"	1'-9"	1'-5"	3'-10"	.71	56
22"X13"	1.6	9"	6"	6"	1'-9"	2'-0"	5'-11"	1.27	90
29"X18"	2.8	9"	6"	6"	1'-9"	2'-0"	5'-11"	1.23	87
36"X22"	4.4	9"	14"	6"	2'-5"	3'-0"	8'-7"	2.53	173
43"X27"	6.4	9"	14"	6"	2'-5"	3'-0"	8'-7"	2.47	167
50"X31"	8.7	12"	16"	10"	3'-2"	3'-9"	11'-4"	5.68	334
58"X36"	11.4	12"	16"	10"	3'-2"	3'-9"	11'-4"	5.58	331
65"X40"	14.3	12"	20"	12"	3'-8"	4'-5"	13'-10"	7.97	451
72"X44"	17.6	12"	20"	12"	3'-8"	4'-5"	13'-10"	7.85	442

QUANTITIES IN TABLE TO BE USED FOR ESTIMATING ONLY

"S" DISTANCES
4" FOR 14" X 9" TO 22" X 13" INCLUSIVE
6" FOR 29" X 18" TO 43" X 27" INCLUSIVE
8" FOR 50" X 31" TO 72" X 44" INCLUSIVE

GENERAL NOTES

SPECIFICATIONS : LATEST S.H.A.
CONCRETE : SEE S.H.A. SPECIFICATIONS
REINFORCING : DEFORMED STEEL BARS - #4 ϕ BARS
CHAMFER : ALL EXPOSED EDGES 1" X 1" OR AS DIRECTED

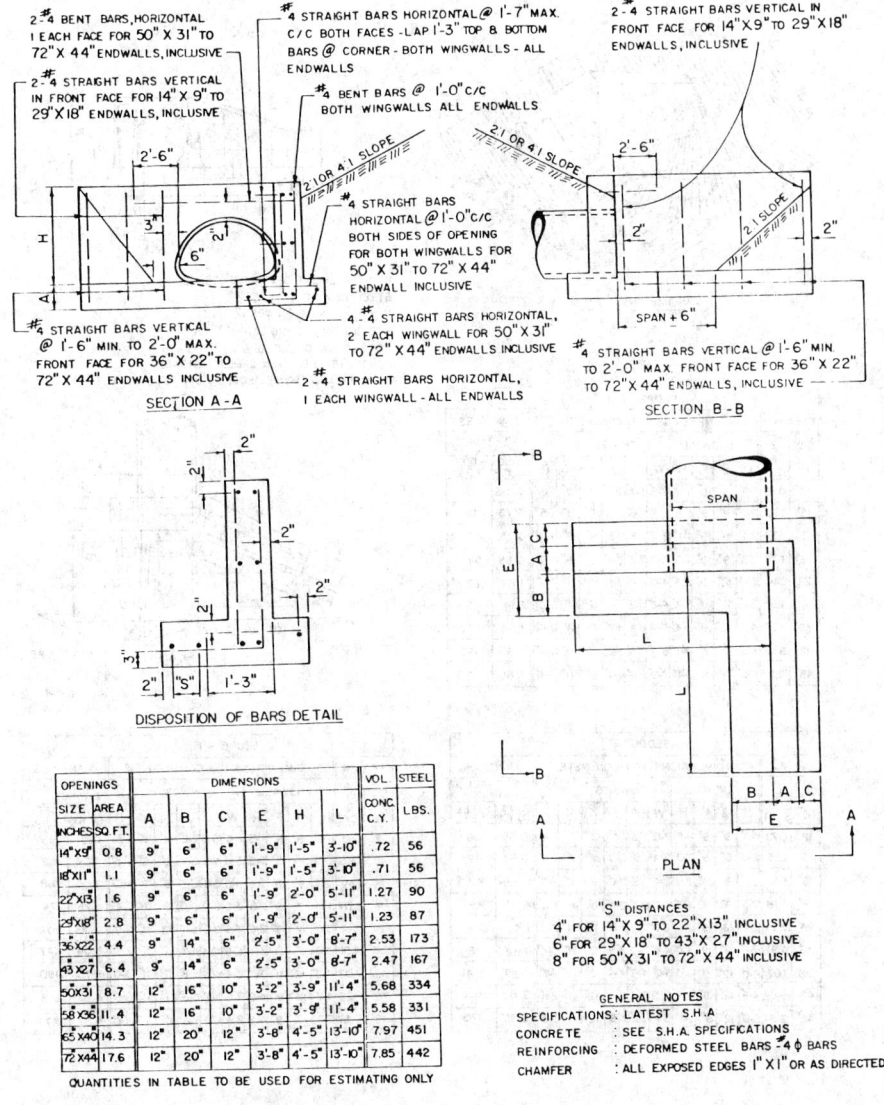

Standard Type E Endwall
Arch Pipe

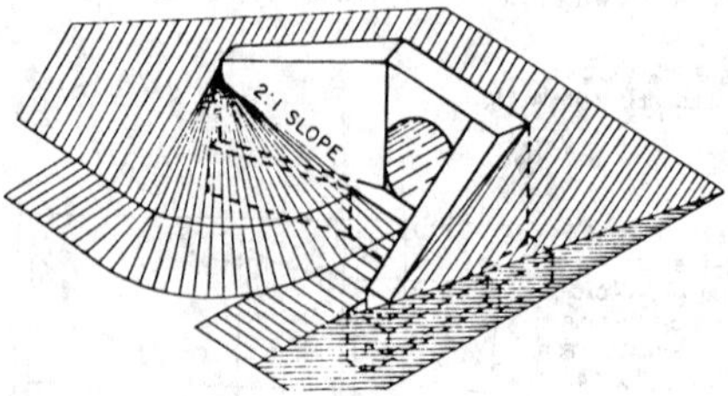

ISOMETRIC VIEW

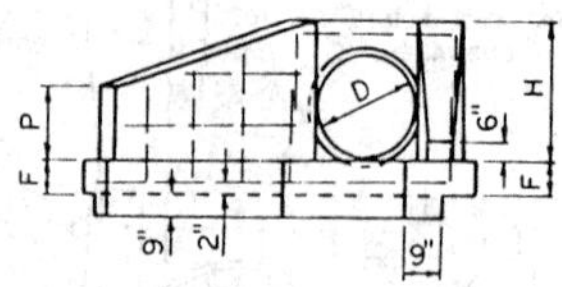

ELEVATION

REINFORCEMENT, DEFORMED STEEL BARS
STEEL IN WINGS:
VERTICAL NO. 6 ϕ BARS 12" C/C
HORIZONTAL NO. 4 ϕ BARS 12" C/C
HOOKED ON ONE END
CONCRETE (SEE S.H.A. SPECIFICATIONS)
CHAMFER ALL EXPOSED EDGES 1" X 1" OR AS DIRECTED

SLOPE 2:1														
OPENING		DIMENSIONS FOR CONCRETE											QUANTITIES	
DIAM. PIPE	AREA	ENDWALL				WINGS							1 ENDWALL 2 WINGS	
D	SQ.FT.	L	B	F	H	W	X	Z	K₁	K	P		CONC. CU.YDS.	STEEL LBS.
12"	0.79	1'-9"	9"	9"	1'-7"	2'-3"	1'-10"	2'-0"	9"	7"	1'-3"		0.55	38
15"	1.23	2'-0"	9"	9"	1'-11"	2'-6"	2'-6"	2'-2"	9"	8"	1'-5"		0.73	46
16"	1.39	2'-1"	9"	9"	1'-11"	2'-7"	2'-6"	2'-3"	9"	8"	1'-5"		0.75	46
18"	1.78	2'-3"	9"	9"	2'-2"	3'-0"	3'-0"	2'-7"	9"	8"	1'-7"		0.89	57
24"	3.14	3'-0"	12"	9"	2'-9"	4'-0"	4'-0"	3'-5"	12"	11"	2'-1"		1.64	82
30"	4.91	3'-6"	12"	9"	3'-4"	4'-8"	5'-2"	4'-1"	12"	11"	2'-6"		2.14	113
36"	7.07	4'-0"	12"	9"	3'-10"	5'-4"	6'-2"	4'-7"	12"	11"	2'-10"		2.48	132
42"	9.62	4'-6"	12"	9"	4'-5"	6'-4"	7'-4"	5'-6"	12"	11"	3'-2"		3.30	163
48"	12.57	5'-0"	12	9	4'-11"	7'-0"	8'-4"	6'-1'	12"	11"	3'-6"		3.85	193

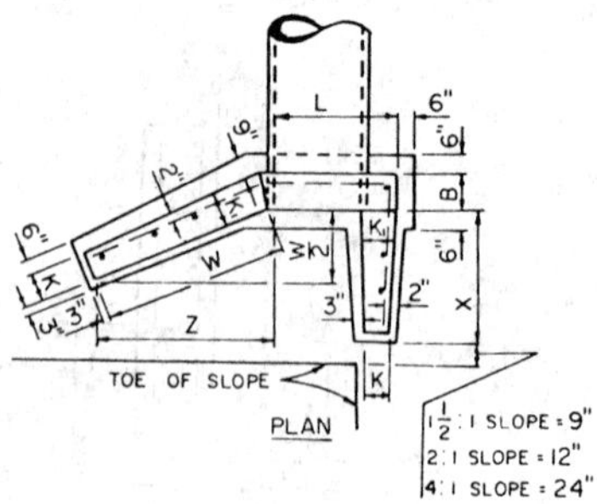

PLAN

SLOPE 1½:1													
OPENING		DIMENSIONS FOR CONCRETE										QUANTITIES	
DIAM. PIPE	AREA	ENDWALL				WINGS						1-ENDWALL 2-WINGS	
D	SQ.FT.	L	B	F	H	W	X	Z	K₁	K	P	CONC. CU.YDS.	STEEL LBS.
12"	0.79	1'-9"	9"	9"	1'-8"	2'-0"	1'-3"	1'-9"	9"	7"	1'-4"	0.51	38
15"	1.23	2'-0"	9"	9"	2'-0"	2'-4"	1'-9"	2'-0"	9"	7"	1'-5"	0.63	42
16"	1.39	2'-1"	9"	9"	2'-0"	2'-4"	1'-9"	2'-0"	9"	7"	1'-5"	0.65	42
18"	1.78	2'-3"	9"	9"	2'-3"	2'-8"	2'-1"	2'-4"	9"	8"	1'-6"	0.77	48
24"	3.14	3'-0"	12"	9"	2'-11"	3'-6"	2'-11"	3'-0"	12"	11"	2'-1"	1.43	73
30"	4.91	3'-6"	12"	9"	3'-6"	4'-4"	3'-9"	3'-9"	12"	11"	2'-4"	1.88	96
36"	7.07	4'-0"	12"	9"	4'-0"	5'-0"	4'-6"	4'-4"	12"	11"	2'-7"	2.32	118
42"	9.62	4'-6"	12"	9"	4'-7"	5'-9"	5'-4"	5'-0"	12"	11"	2'-11"	2.85	144
48"	12.57	5'-0"	12"	9"	5'-1"	6'-4"	6'-2"	5'-6"	12"	11"	3'-3"	3.37	170

SLOPE 4:1													
OPENING		DIMENSIONS FOR CONCRETE										QUANTITIES	
DIAM. PIPE	AREA	ENDWALL				WINGS						1 ENDWALL 2 WINGS	
D	SQ.FT.	L	B	F	H	W	X	Z	K₁	K	P	CONC. CU YDS.	STEEL LBS.
12"	0.79	1'-9"	9"	9"	1'-5"	2'-6"	3'-8"	2'-2"	9"	8"	1'-3"	0.81	45
15"	1.23	2'-0"	9"	9"	1'-9"	3'-0"	5'-0"	2'-7"	9"	8"	1'-5"	1.04	68
16"	1.39	2'-1"	9"	9"	1'-9"	3'-0"	5'-0"	2'-7"	9"	8"	1'-5"	1.06	68
18"	1.78	2'-3"	9"	9"	2'-0"	3'-6"	6'-0"	3'-0"	9"	8"	1'-7"	1.26	74
24"	3.14	3'-0"	12"	9"	2'-6"	4'-6"	8'-0"	3'-11"	12"	11"	2'-2"	2.23	104
30"	4.91	3'-6"	12"	9"	3'-1"	5'-6"	10'-4"	4'-9	12"	11"	2'-7"	3.04	147
36"	7.07	4'-0"	12"	9"	3'-7"	6'-6"	12'-4"	5'-8"	12"	11"	3'-0"	3.75	180
42"	9.62	4'-6"	12"	9"	4'-2"	7'-6"	14'-8"	6'-6"	12"	11"	3'-5"	4.67	218
48"	12.57	5'-0"	12"	9"	4'-8"	8'-3"	16'-8"	7'-2"	12"	11"	3'-10"	5.57	277

Standard Type F Endwall
Round Pipe

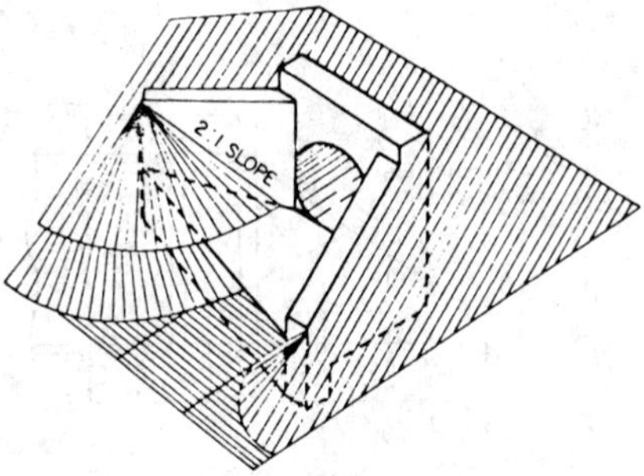

ISOMETRIC VIEW

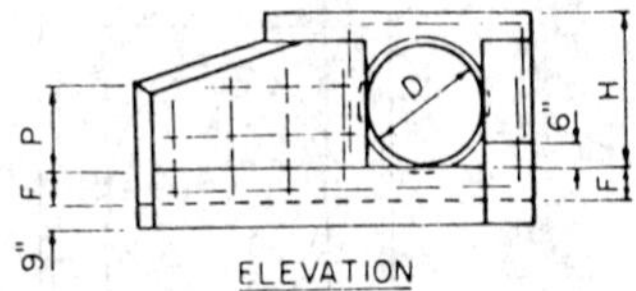

ELEVATION

CONCRETE: (SEE S.H.A. SPECIFICATIONS)
STEEL: DEFORMED BARS
ENDWALLS - NO 6 ϕ U-BEND
WINGS - VERTICAL NO.6 ϕ 12" C/C - 18" BEND ONE END
HORIZONTAL NO.4 ϕ 12" C/C - 6" HOOK ONE END
FLOOR - NO. 4ϕ 12" C/C, 2 WAYS
CHAMFER ALL EXPOSED EDGES 1"X1" OR AS DIRECTED

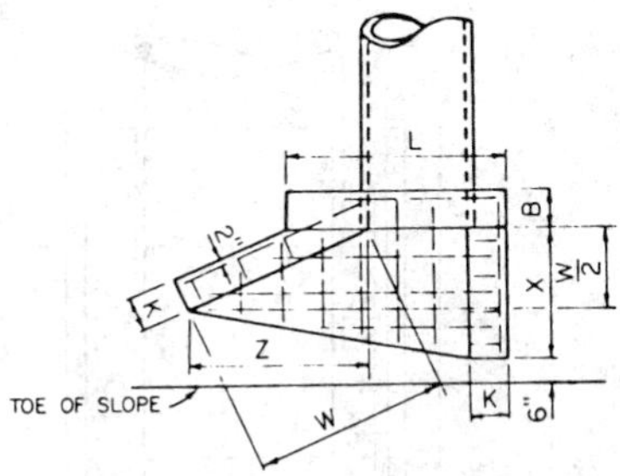

PLAN

SLOPE 1½:1

OPENING		DIMENSIONS FOR CONCRETE									QUANTITIES	
		ENDWALL				WINGS					1-ENDWALL 2-WINGS	
DIAM. PIPE D	AREA SQ.FT.	L	B	F	H	W	X	Z	K	P	CONC. CU.YDS	STEEL LBS.
12"	0.79	3'-3"	9"	9"	1'-8"	1'-11"	1'-3"	1'-8"	9"	1'-0"	0.47	33
15"	1.23	3'-6"	9"	9"	2'-0"	2'-5"	1'-8"	2'-1"	9"	1'-2"	0.61	50
16"	1.39	3'-7"	9"	9"	2'-0"	2'-5"	1'-9"	2'-1"	9"	1'-2"	0.63	50
18"	1.78	3'-9"	9"	9"	2'-3"	2'-8"	2'-1"	2'-4"	9"	1'-4"	0.75	53
24"	3.14	5'-0"	12"	9"	2'-11"	3'-5"	2'-11"	2'-11"	12"	1'-9"	1.44	89
30"	4.91	5'-6"	12"	9"	3'-6"	4'-2"	3'-8"	3'-7"	12"	2'-1"	1.93	112
36"	7.07	6'-0"	12"	9"	4'-0"	4'-10"	4'-6"	4'-2"	12"	2'-4"	2.51	154
42"	9.62	6'-6"	12"	9"	4'-[illegible]	5'-7"	5'-4"	4'-0"	12"	2'-8"	3.20	189
48"	12.57	7'-0"	12"	9"	5'-1"	6'-3"	6'-2"	5'-5"	12"	2'-11"	3.92	216

SLOPE 2:1

OPENING		DIMENSIONS FOR CONCRETE									QUANTITIES	
		ENDWALL				WINGS					1-ENDWALL 2-WINGS	
DIAM. PIPE D	AREA SQ.FT.	L	B	F	H	W	X	Z	K	P	CONC. CU.YDS	STEEL LBS.
12"	0.79	3'-3"	9"	9"	1'-7"	2'-1"	1'-10"	1'-10"	9"	1'-0"	0.53	37
15"	1.23	3'-6"	9"	9"	1'-11"	2'-7"	2'-5"	2'-3"	9"	1'-3"	0.70	55
16"	1.39	3'-7"	9"	9"	1'-11"	2'-7"	2'-6"	2'-3"	9"	1'-3"	0.72	55
18"	1.78	3'-9"	9"	9"	2'-2"	2'-11"	2'-11"	2'-6"	9"	1'-5"	0.82	68
24"	3.14	5'-0"	12"	9"	2'-9"	3'-7"	4'-0"	3'-1"	12"	1'-10"	1.63	92
30"	4.91	5'-6"	12"	9"	3'-4"	4'-5"	5'-1"	3'-10"	12"	2'-2"	2.24	125
36"	7.07	6'-0"	12"	9"	3'-10"	5'-2"	6'-2"	4'-6"	12"	2'-6"	2.88	166
42"	9.62	6'-6"	12"	9"	4'-5"	6'-0"	7'-3"	5'-2"	12"	2'-10"	3.74	221
48"	12.57	7'-0"	12"	9"	4'-11"	6'-9"	8'-4"	5'-10"	12"	3'-2"	4.61	262

SLOPE 4:1

OPENING		DIMENSIONS FOR CONCRETE									QUANTITIES	
		END WALL				WINGS					1-ENDWALL 2-WINGS	
DIAM. PIPE D	AREA SQ.FT.	L	B	F	H	W	X	Z	K	P	CONC. CU.YDS.	STEEL LBS.
12"	0.79	3'-3"	9"	9"	1'-5"	2'-4"	4'-2"	2'-0"	9"	1'-2"	0.76	61
15"	1.23	3'-6"	9"	9"	1'-9"	2'-11"	5'-3"	2'-6"	9"	1'-5"	1.03	81
16"	1.39	3'-7"	9"	9"	1'-9"	2'-11"	5'-6"	2'-6"	9"	1'-5"	1.10	81
18"	1.78	3'-9"	9"	9"	2'-0"	3'-1"	6'-4"	2'-11"	9"	1'-7"	1.33	98
24"	3.14	5'-0"	12"	9"	2'-6"	4'-0"	8'-6"	3'-6"	12"	2'-0"	2.45	161
30"	4.91	5'-6"	12"	9"	3'-1"	5'-0"	10'-8"	4'-4"	12"	2'-5"	3.44	225
36"	7.07	6'-0"	12"	9"	3'-7"	5'-11"	12'-10"	5'-2"	12"	2'-10"	4.59	246
42"	9.62	6'-6"	12"	9"	4'-2"	7'-0"	15'-0"	6'-1"	12"	3'-4"	5.98	367
48"	12.57	7'-0"	12"	9"	4'-8"	7'-11"	17'-2"	6'-10"	12"	3'-9"	7.46	499

Special Type F Endwall
Round Pipe

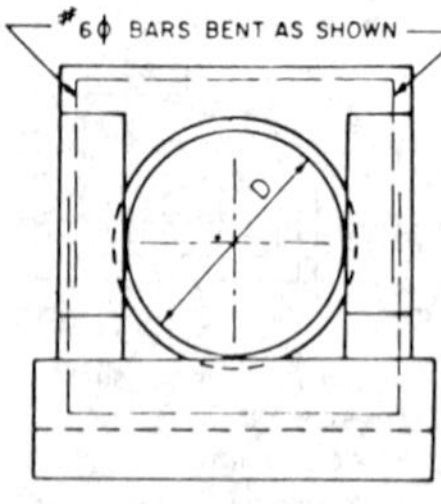

ELEVATION

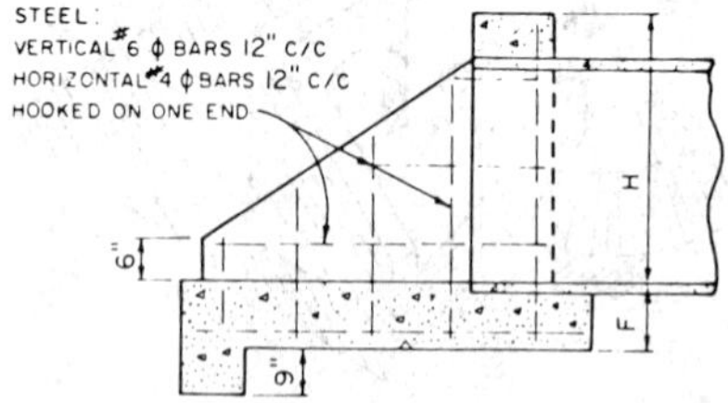

SECTION A-A

REINFORCEMENT, DEFORMED STEEL BARS
CONCRETE (SEE S.H.A. SPECIFICATIONS)
CHAMFER ALL EXPOSED EDGES 1"X1"
OR AS DIRECTED.

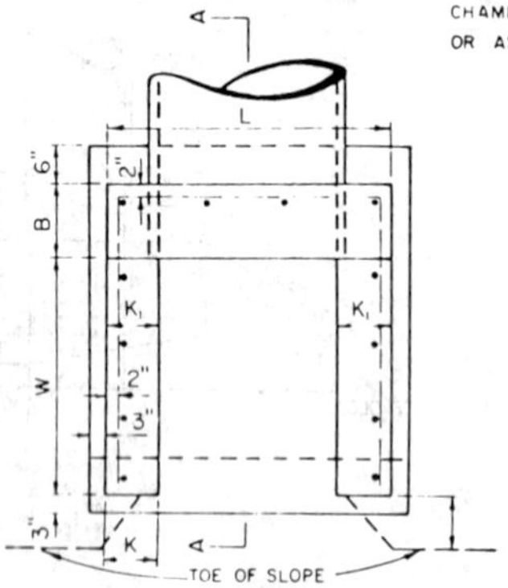

PLAN

$1\frac{1}{2}$: 1 SLOPE = 9"
2 : 1 SLOPE = 12"
4 : 1 SLOPE = 24"

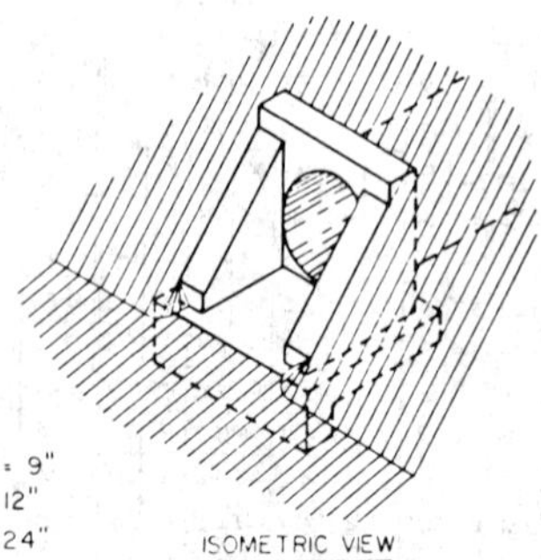

ISOMETRIC VIEW

QUANTITIES APPROXIMATE

SLOPE $1\frac{1}{2}$: 1

OPENING: DIAM OF PIPE D	OPENING: AREA SQ FT	DIMENSIONS FOR CONCRETE: L	B	F	H	W	K	K_1	QUANTITIES 1-ENDWALL 2-WINGS: CONC CU YDS	STEEL LBS
12"	0.79	2'-0"	9"	9"	1'-8"	1'-0"	6"	6"	0.33	28
15"	1.23	2'-3"	9"	9"	2'-0"	1'-5"	6"	6"	0.42	31
16"	1.39	2'-4"	9"	9"	2'-0"	1'-6"	6"	6"	0.44	35
18"	1.78	2'-6"	9"	9"	2'-3"	1'-10"	6"	6"	0.52	41
24"	3.14	3'-6"	12"	9"	2'-11"	2'-8"	9"	9"	1.02	53
30"	4.91	4'-0"	12"	9"	3'-6"	3'-6"	9"	9"	1.39	73
36"	7.07	4'-6"	12"	9"	4'-0"	4'-3"	9"	9"	1.77	104
42"	9.62	5'-6"	12"	9"	4'-7"	5'-1"	9"	12"	2.61	126
48"	12.57	6'-0"	12"	9"	5'-1"	5'-11"	9"	12"	3.21	155

SLOPE 2 : 1

OPENING: DIAM OF PIPE D	OPENING: AREA SQ FT	ENDWALL: L	ENDWALL: B	ENDWALL: F	ENDWALL: H	WINGS: W	WINGS: K	WINGS: K_1	QUANTITIES 1-ENDWALL 2-WINGS: CONC CU YDS	STEEL LBS
12"	0.79	2'-0"	9"	9"	1'-7	1'-4"	6"	6'	0.36	34
15"	1.23	2'-3"	9"	9"	1'-11"	1'-11"	6"	6"	0.47	37
16"	1.39	2'-4"	9"	9"	1'-11"	2'-0"	6"	6"	0.50	38
18"	1.78	2'-6"	9"	9"	2'-2"	2'-6"	6"	6"	0.61	53
24"	3.14	3'-6"	12"	9"	2'-9"	3'-6"	9"	9"	1.17	71
30"	4.91	4'-0"	12"	9"	3'-4"	4'-7"	9"	9"	1.62	91
36"	7.07	4'-6"	12"	9"	3'-10"	5'-8"	9"	9"	2.12	116
42"	9.62	5'-6"	12"	9"	4'-5"	6'-9"	9"	12"	3.15	141
48"	12.57	6'-0"	12"	9"	4'-11"	7'-10"	9"	12"	3.88	174

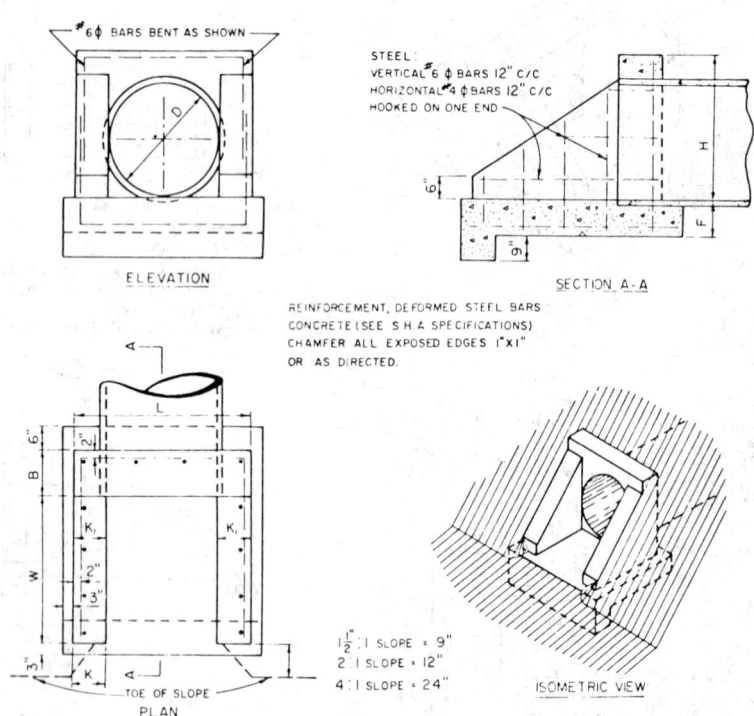

Standard Type G Endwall
Round Pipe

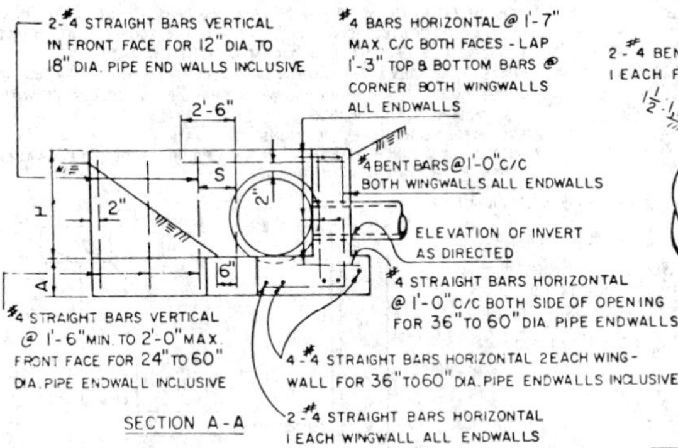

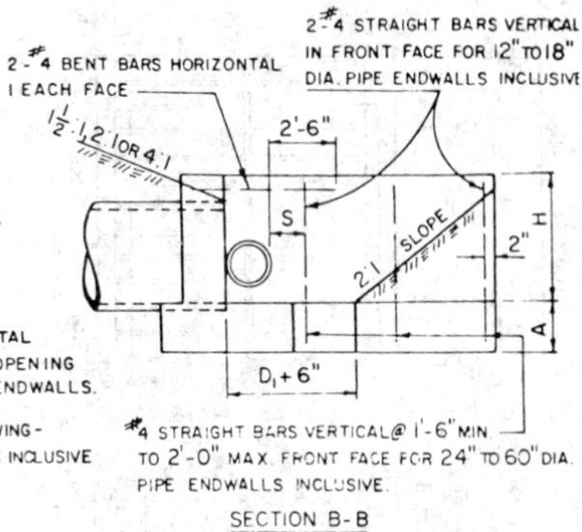

'S' DISTANCES FROM INSIDE SURFACE OF PIPE TO VERTICAL BARS IN FRONT & REAR FACE.
4" FOR 12" DIA. TO 18" DIA. PIPES INCLUSIVE
6" FOR 24" DIA. TO 36" DIA. PIPES INCLUSIVE
8" FOR 42" DIA. TO 60" DIA. PIPES INCLUSIVE

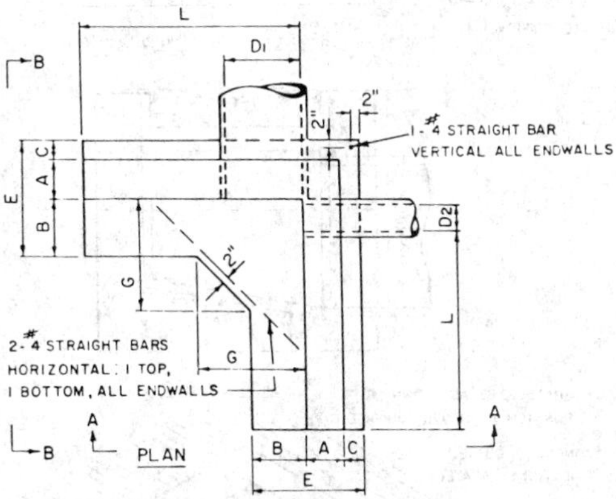

QUANTITIES IN TABLE TO BE USED FOR ESTIMATING ONLY

D	D	A	B	C	E	H	L	G	VOL. CONC. C.Y.	STEEL LBS.
12"	12"	9"	6"	6"	1'-9"	1'-9"	3'-6"	1'-6"	0.7	62
15"	12"	9"	6"	6"	1'-9"	2'-0"	4'-3"	2'-0"	0.9	76
15"	15"	9"	6"	6"	1'-9"	2'-0"	4'-3"	2'-0"	0.9	76
18"	12"	9"	6"	6"	1'-9"	2'-3"	5'-0"	2'-3"	1.1	88
18"	15"	9"	6"	6"	1'-9"	2'-3"	5'-0"	2'-3"	1.1	85
18"	18"	9"	6"	6"	1'-9"	2'-3"	5'-0"	2'-3"	1.1	85
24"	12"	9"	14"	6"	2'-5"	2'-9"	6'-6"	3'-0"	1.8	123
24"	15"	9"	14"	6"	2'-5"	2'-9"	6'-6"	3'-0"	1.8	122
24"	18"	9"	14"	6"	2'-5"	2'-9"	6'-6"	3'-0"	1.8	117
24"	24"	9"	14"	6"	2'-5"	2'-9"	6'-6"	3'-0"	1.8	116
30"	12"	9"	14"	6"	2'-5"	3'-6"	8'-0"	3'-9"	2.6	172
30"	15"	9"	14"	6"	2'-5"	3'-6"	8'-0"	3'-9"	2.6	171
30"	18"	9"	14"	6"	2'-5"	3'-6"	8'-0"	3'-9"	2.5	171
30"	24"	9"	14"	6"	2'-5"	3'-6"	8'-0"	3'-9"	2.5	166
30"	30"	9"	14"	6"	2'-5"	3'-6"	8'-0"	3'-9"	2.5	164
36"	12"	12"	16"	10"	3'-2"	4'-0"	9'-6"	4'-6"	5.0	279
36"	15"	12"	16"	10"	3'-2"	4'-0"	9'-6"	4'-6"	5.0	279
36"	18"	12"	16"	10"	3'-2"	4'-0"	9'-6"	4'-6"	5.0	274
36"	24"	12"	16"	10"	3'-2"	4'-0"	9'-6"	4'-6"	4.9	272
36"	30"	12"	16"	10"	3'-2"	4'-0"	9'-6"	4'-6"	4.9	262
36"	36"	12"	16"	10"	3'-2"	4'-0"	9'-6"	4'-6"	4.8	260
42"	12"	12"	16"	10"	3'-2"	4'-6"	11'-0"	5'-3"	6.1	328
42"	15"	12"	16"	10"	3'-2"	4'-6"	11'-0"	5'-3"	6.1	328
42"	18"	12"	16"	10"	3'-2"	4'-6"	11'-0"	5'-3"	6.1	327
42"	24"	12"	16"	10"	3'-2"	4'-6"	11'-0"	5'-3"	6.1	317
42"	30"	12"	16"	10"	3'-2"	4'-6"	11'-0"	5'-3"	6.0	315
42"	36"	12"	16"	10"	3'-2	4'-6"	11'-0"	5'-3"	5.9	308
42"	42"	12"	16"	10"	3'-2"	4'-6"	11'-0"	5'-3"	5.8	307
48"	12"	12"	16"	10"	3'-2"	5'-0"	12'-6"	6'-0"	7.4	404
48"	15"	12"	16"	10"	3'-2"	5'-0"	12'-6"	6'-0"	7.4	404
48"	18"	12"	16"	10"	3'-2"	5'-0"	12'-6"	6'-0"	7.4	393
48"	24"	12"	16"	10"	3'-2"	5'-0"	12'-6"	6'-0"	7.3	391
48"	30"	12"	16"	10"	3'-2"	5'-0"	12'-6"	6'-0"	7.2	383
48"	36"	12"	16"	10"	3'-2"	5'-0"	12'-6"	6'-0"	7.2	381
48"	42"	12"	16"	10"	3'-2"	5'-0"	12'-6"	6'-0"	7.1	373
48"	48"	12"	16"	10"	3'-2"	5'-0"	12'-6"	6'-0"	7.0	371

GENERAL NOTES:

SPECIFICATIONS:	LATEST S.H.A.
CONCRETE	: SEE S.H.A. SPECIFICATIONS
REINFORCING	: DEFORMED STEEL BARS #4 ϕ BARS
CHAMFER	: ALL EXPOSED EDGES 1"X1" OR AS DIRECTED
SOD	: 3' OF SOD WILL BE CALLED FOR ON INDIVIDUAL JOBS.

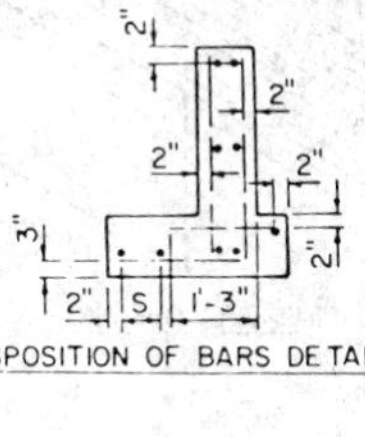

DISPOSITION OF BARS DETAIL

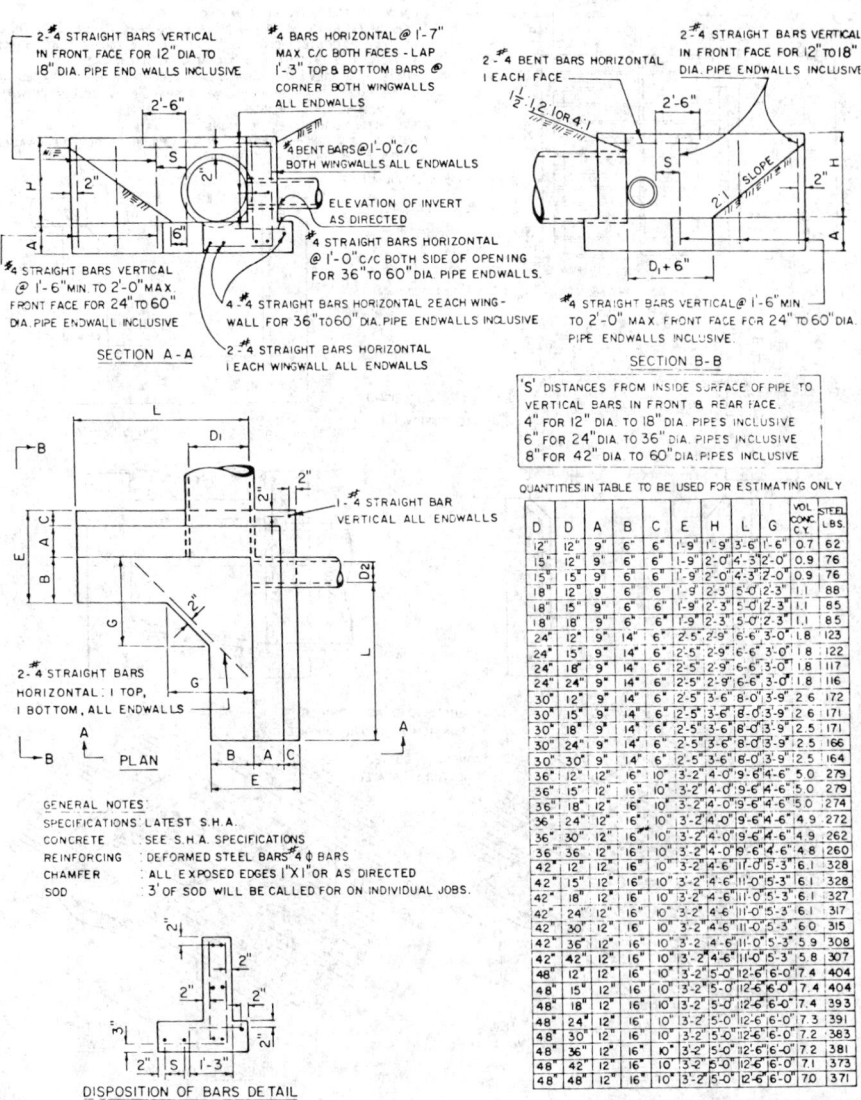

Standard Type H Endwall
Round Pipe

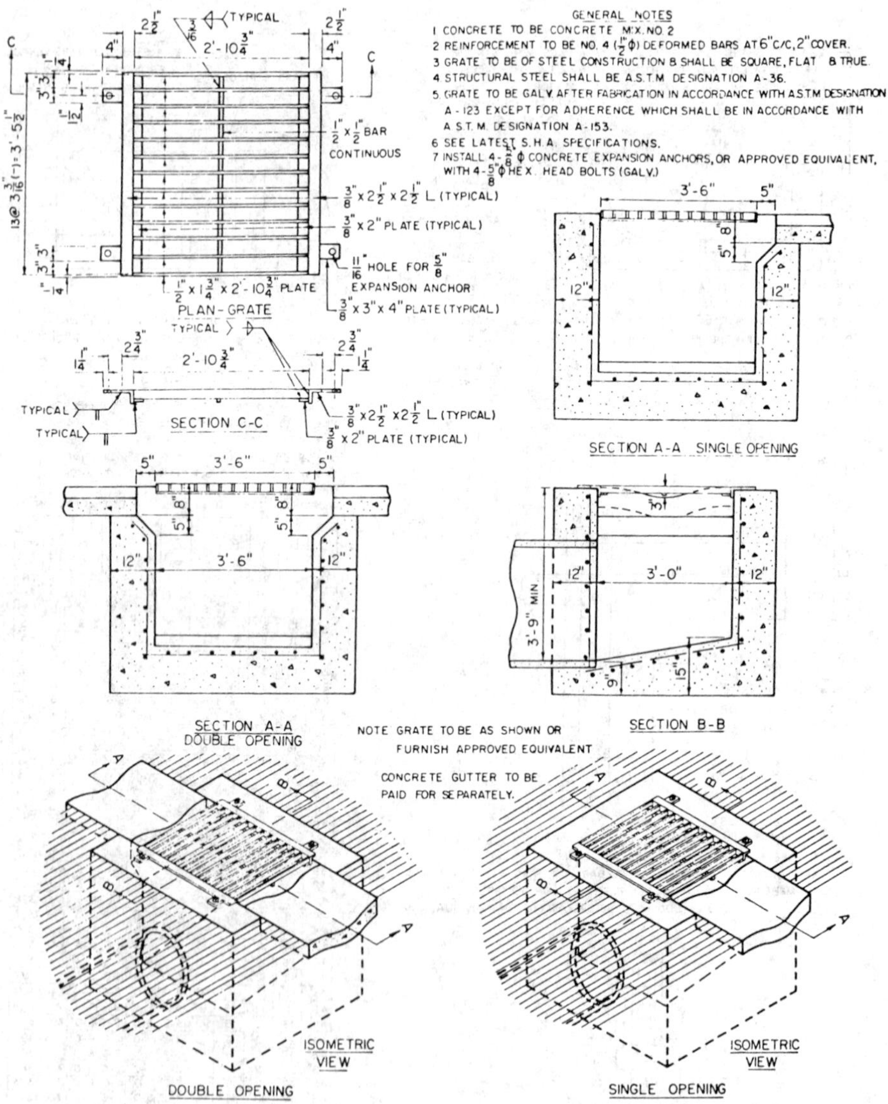
GENERAL NOTES
1 CONCRETE TO BE CONCRETE MIX NO. 2
2 REINFORCEMENT TO BE NO. 4 (1/2" ϕ) DEFORMED BARS AT 6" C/C, 2" COVER.
3 GRATE TO BE OF STEEL CONSTRUCTION & SHALL BE SQUARE, FLAT & TRUE.
4 STRUCTURAL STEEL SHALL BE A.S.T.M. DESIGNATION A-36.
5 GRATE TO BE GALV. AFTER FABRICATION IN ACCORDANCE WITH A.S.T.M. DESIGNATION A-123 EXCEPT FOR ADHERENCE WHICH SHALL BE IN ACCORDANCE WITH A.S.T.M. DESIGNATION A-153.
6 SEE LATEST S.H.A. SPECIFICATIONS.
7 INSTALL 4-3/4" ϕ CONCRETE EXPANSION ANCHORS, OR APPROVED EQUIVALENT, WITH 4-5/8" ϕ HEX. HEAD BOLTS (GALV.)
TYPICAL
2'-10 3/4"
1/2" x 1/2" BAR CONTINUOUS
3/8" x 2 1/2" x 2 1/2" L (TYPICAL)
3/8" x 2" PLATE (TYPICAL)
11/16" HOLE FOR 5/8" EXPANSION ANCHOR
3/8" x 3" x 4" PLATE (TYPICAL)
1/2" x 1 3/4" x 2'-10 3/4" PLATE
13@ 3 3/16"(-) = 3'-5 1/2"
PLAN-GRATE
SECTION C-C
SECTION A-A SINGLE OPENING
SECTION A-A DOUBLE OPENING
SECTION B-B
3'-6"
3'-0"
3'-9" MIN.
NOTE GRATE TO BE AS SHOWN OR FURNISH APPROVED EQUIVALENT
CONCRETE GUTTER TO BE PAID FOR SEPARATELY.
ISOMETRIC VIEW
DOUBLE OPENING
ISOMETRIC VIEW
SINGLE OPENING

Standard Type K Inlet
Open-End Grate

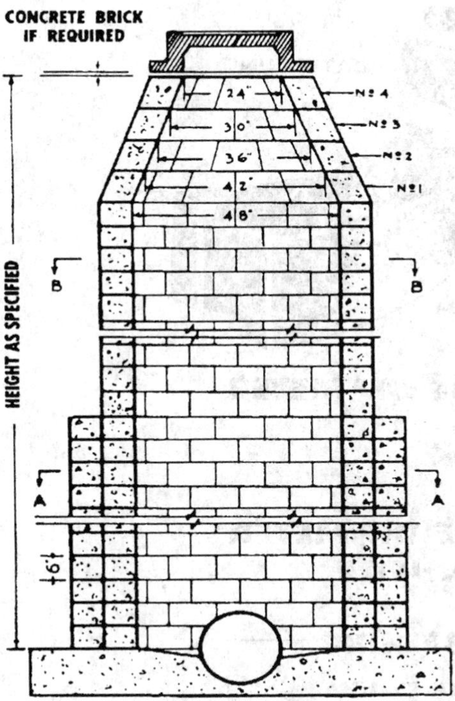

SHOWING 12" WALL

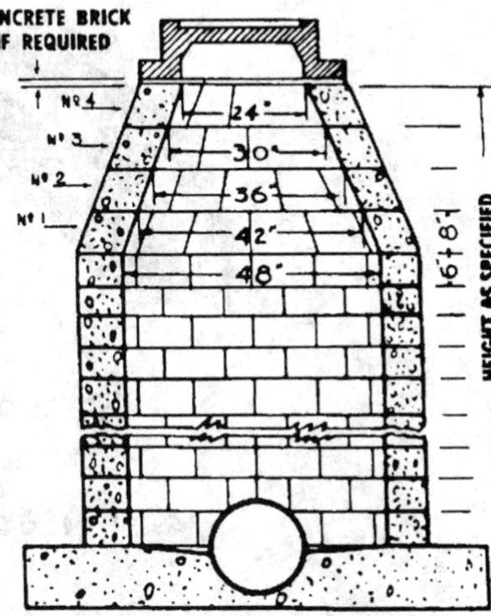

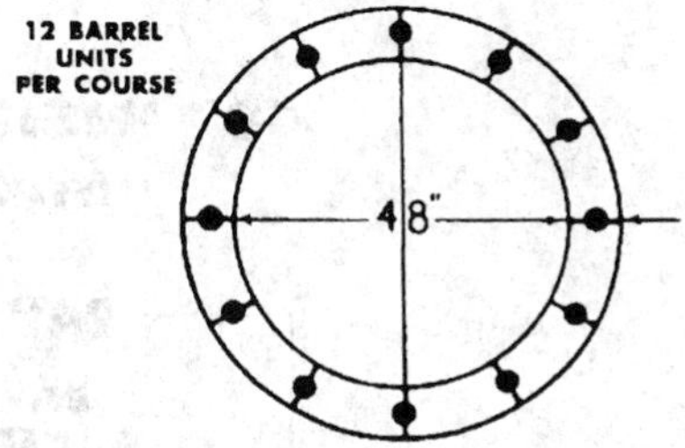

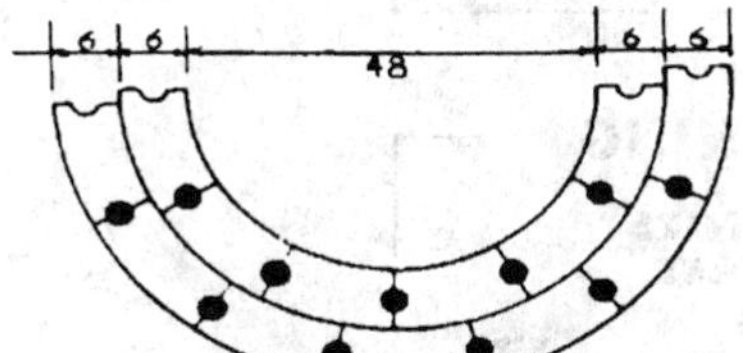

BARREL UNITS — SECTION AA

UNITS PER COURSE

Outside Barrels . . . 14
Inside Barrels 12

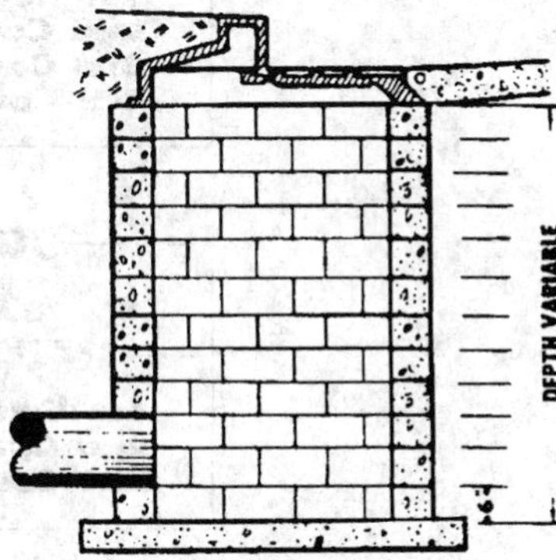

CROSS-SECTION

All dimensions may be varied in multiples of six inches.

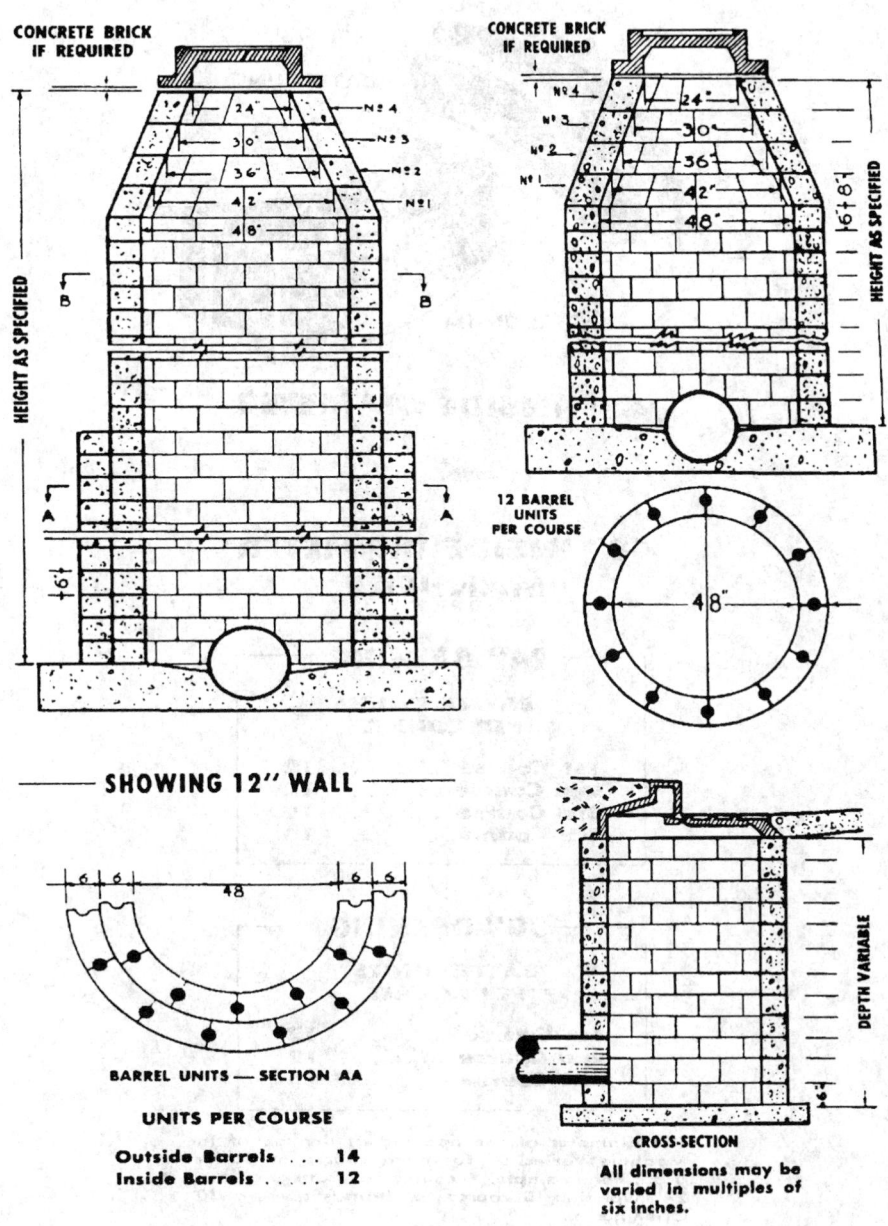

Concrete Manholes (Double Coursed)
Concrete Catch Basin (lower right)

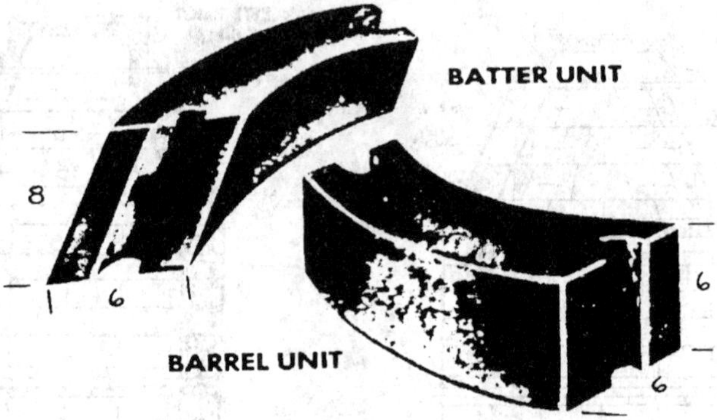

48" INSIDE DIAMETER

48" INSIDE DIAMETER MANHOLE

24" OPENING

BATTER UNITS PER COURSE

1st Course	12
2nd Course	12
3rd Course	10
4th Course	10

30" OPENING

BATTER UNITS PER COURSE

1st Course	12
2nd Course	12
3rd Course	10

The diameter of the opening at the top of the Manhole varies 6" for each course of Batters used. For example, 4 courses of Batters gives a 24" Opening: 3 courses of Batters gives a 30" Opening.

You will note the Barrel Units are 6" high and Batter Units are 8" high.

For Manhole depths of over 12 feet, double courses of Barrel Units are recommended --

Specifications for Concrete Manholes

UNITS PER COURSE FOR 42"x48" INSIDE DIMENSIONS

Catch Basin Stretchers 9
Catch Basin Corners 4

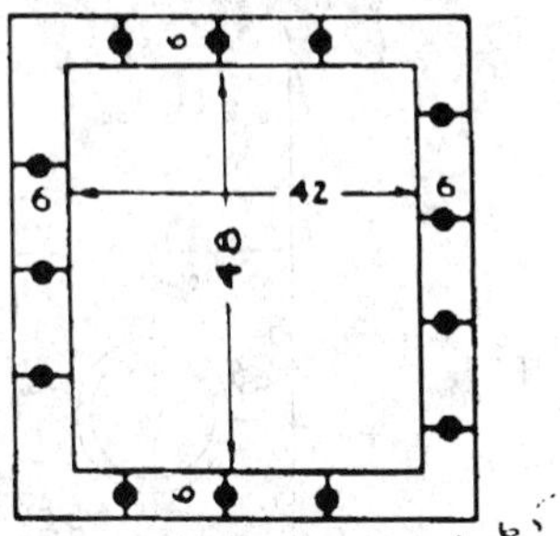

1st COURSE

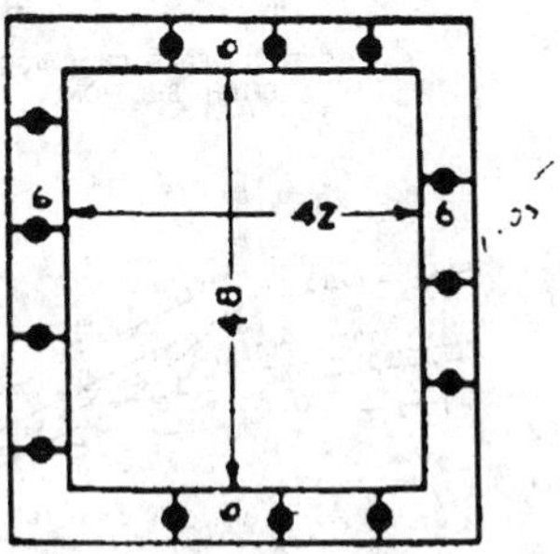

2nd COURSE

UNITS PER COURSE FOR 48"x48" INSIDE DIMENSIONS

Catch Basin Stretchers 12
Catch Basin Corners 4

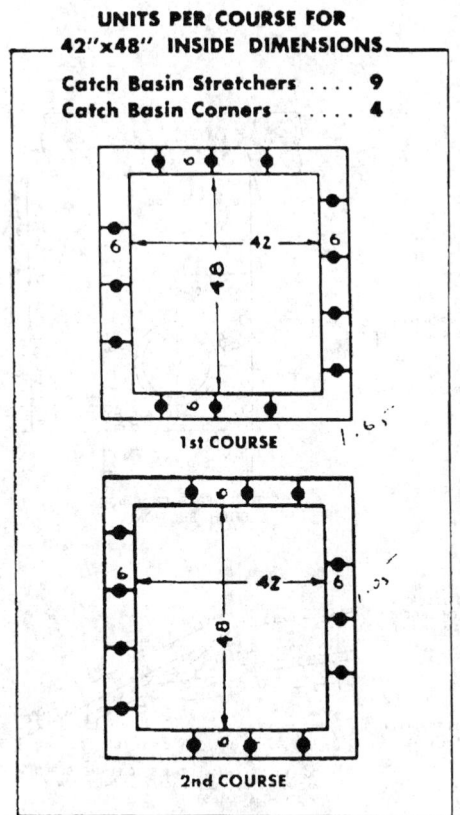

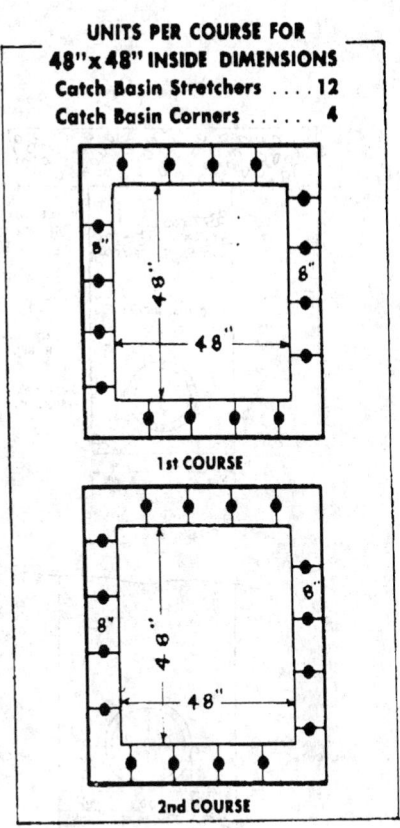

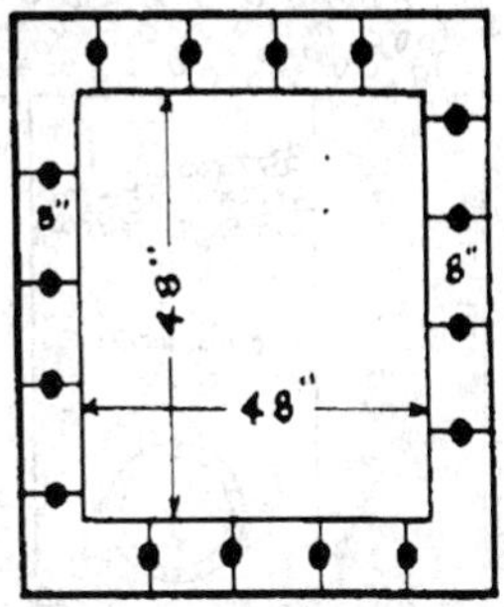

1st COURSE

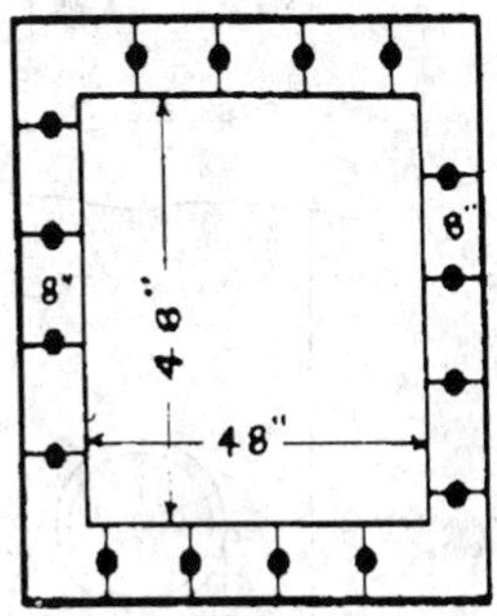

2nd COURSE

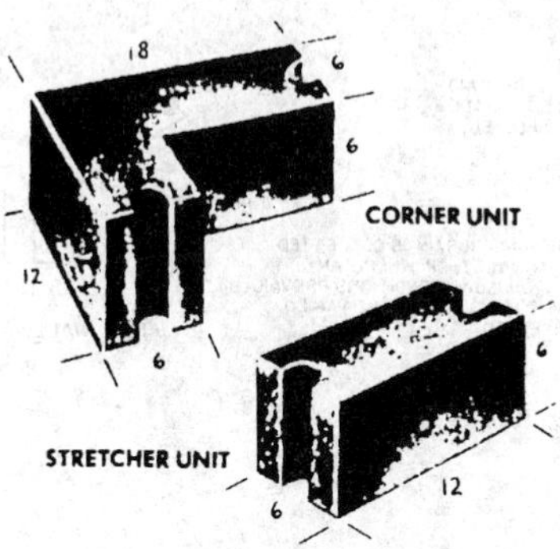

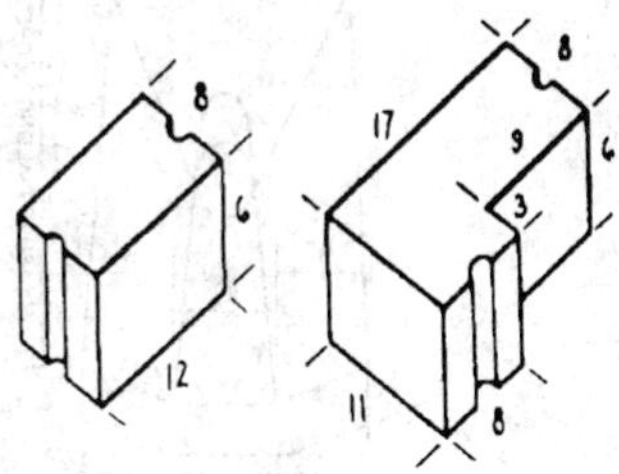

Catch Basins and Inlet Units

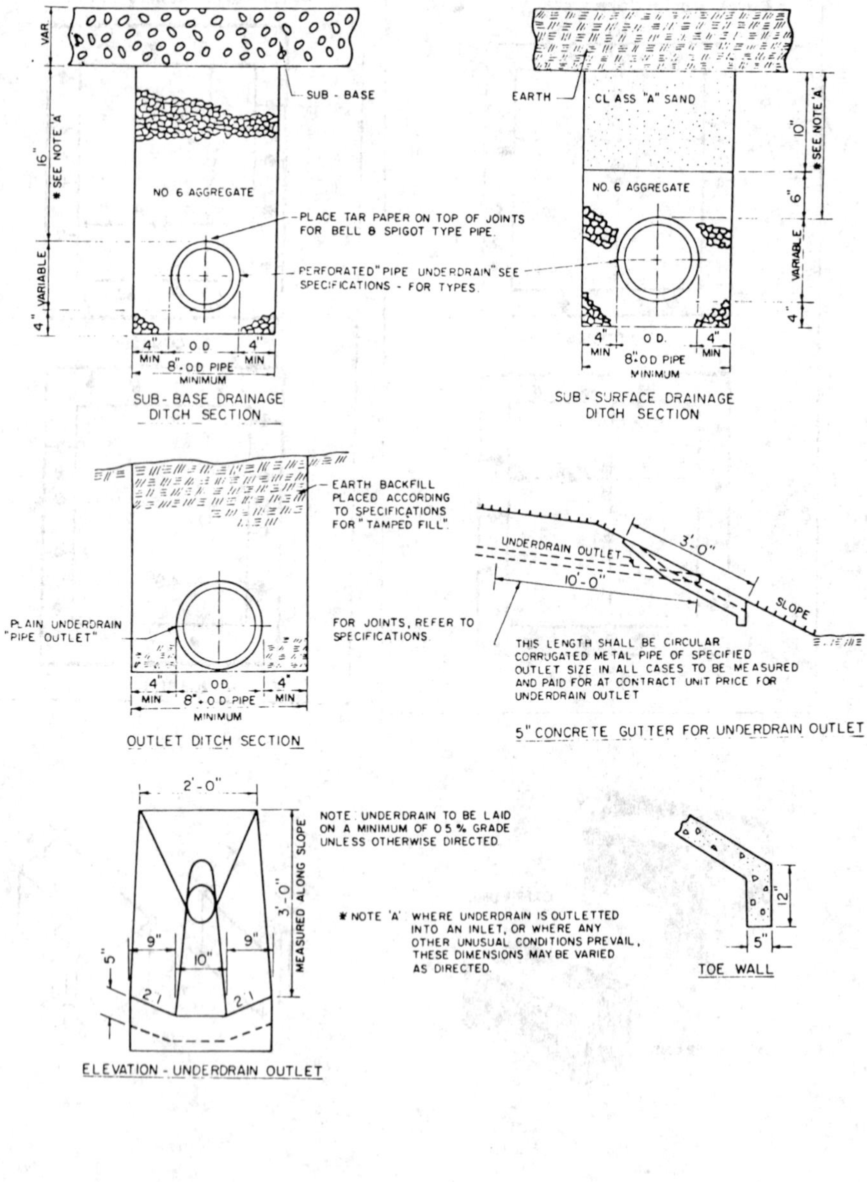

Standard Underdrains

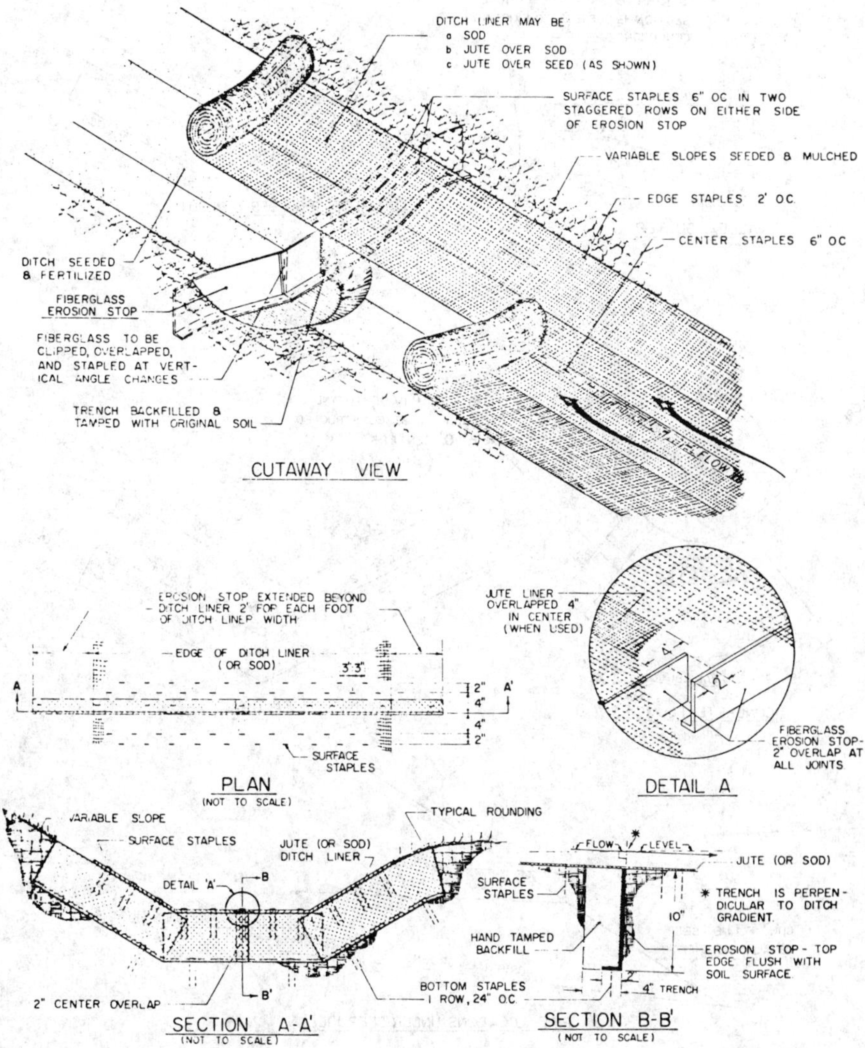

Sediment and Erosion Control
Fiberglass Erosion Stop

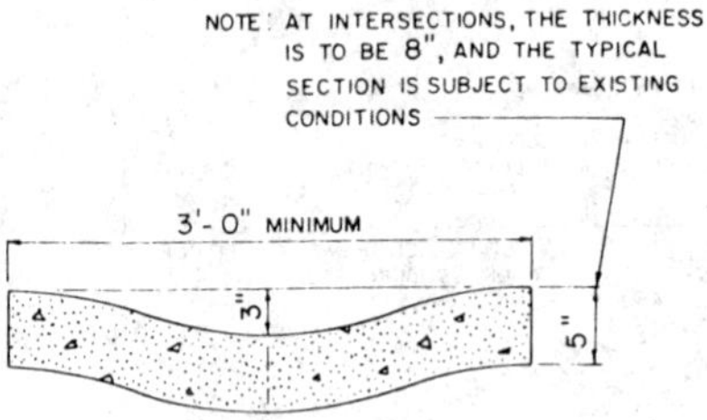

VALLEY GUTTER

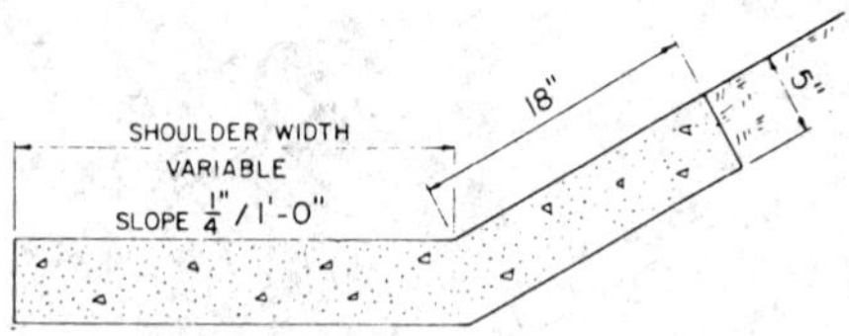

CONCRETE SHOULDER & REBUT

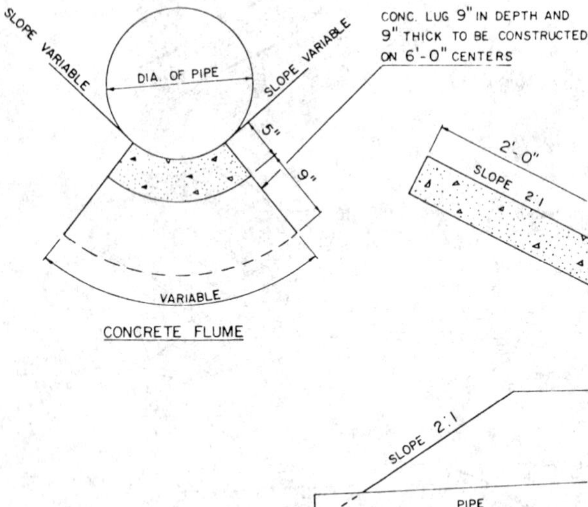

CONCRETE FLUME

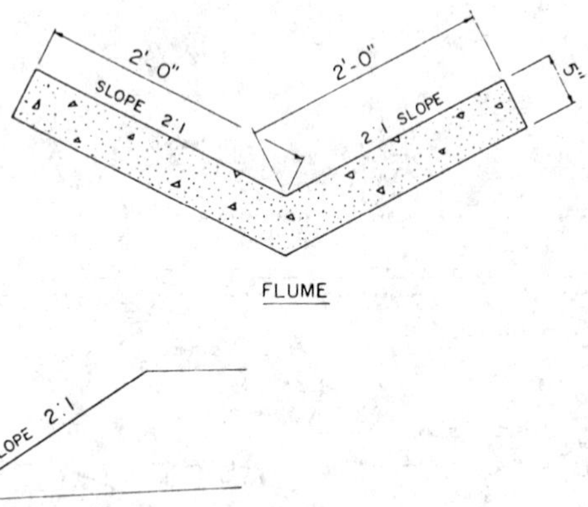

FLUME

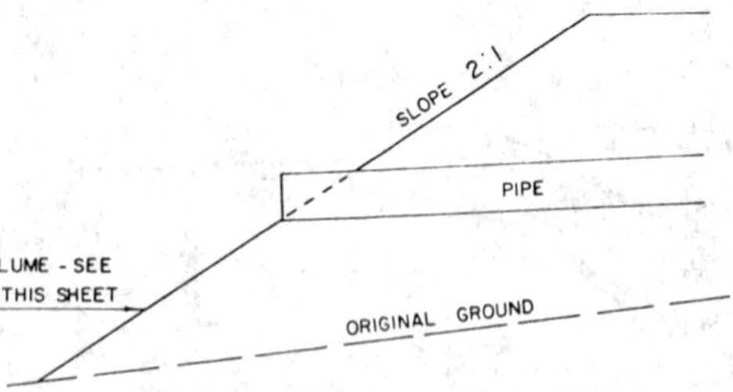

PIPE LOCATIONS UNDER DEEP FILL

Standard Concrete Valley Gutter, Flumes
Concrete Shoulder & Rebut

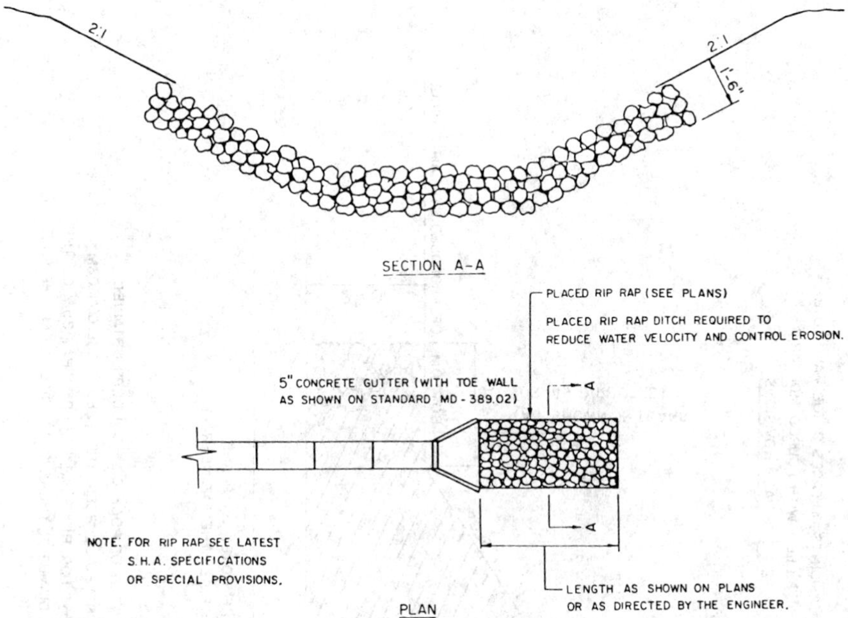

GENERAL NOTE:
ALL DIMENSIONS AND LOCATIONS
NOT INDICATED, FOR ITEMS APPEARING
ON THIS SHEET OR ON THE PLANS,
SHALL BE DIRECTED BY THE ENGINEER.

Sediment and Erosion Control
Placed Rip-Rap Ditch

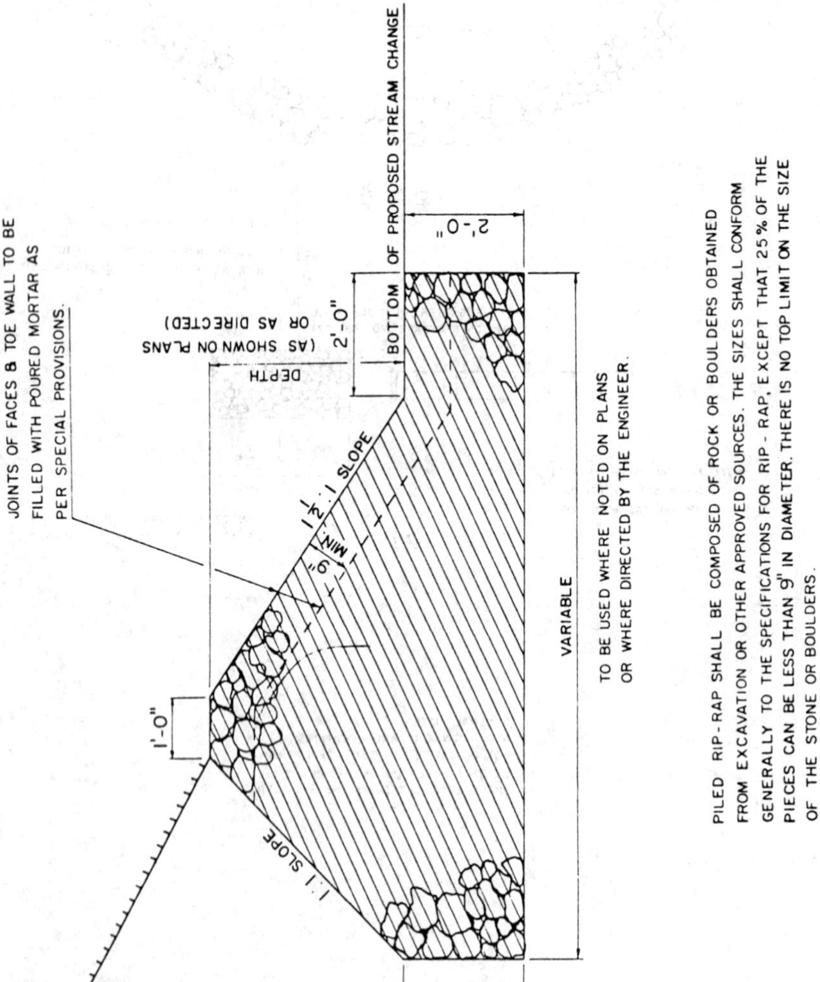
JOINTS OF FACES & TOE WALL TO BE FILLED WITH POURED MORTAR AS PER SPECIAL PROVISIONS.
DEPTH (AS SHOWN ON PLANS OR AS DIRECTED)
2'-0"
BOTTOM OF PROPOSED STREAM CHANGE
2'-0"
1'-0"
1½ : 1 SLOPE
9" MIN.
1:1 SLOPE
VARIABLE
2'-0"
TO BE USED WHERE NOTED ON PLANS OR WHERE DIRECTED BY THE ENGINEER.
PILED RIP-RAP SHALL BE COMPOSED OF ROCK OR BOULDERS OBTAINED FROM EXCAVATION OR OTHER APPROVED SOURCES. THE SIZES SHALL CONFORM GENERALLY TO THE SPECIFICATIONS FOR RIP-RAP, EXCEPT THAT 25% OF THE PIECES CAN BE LESS THAN 9" IN DIAMETER. THERE IS NO TOP LIMIT ON THE SIZE OF THE STONE OR BOULDERS.

Piled Rip-Rap Detail

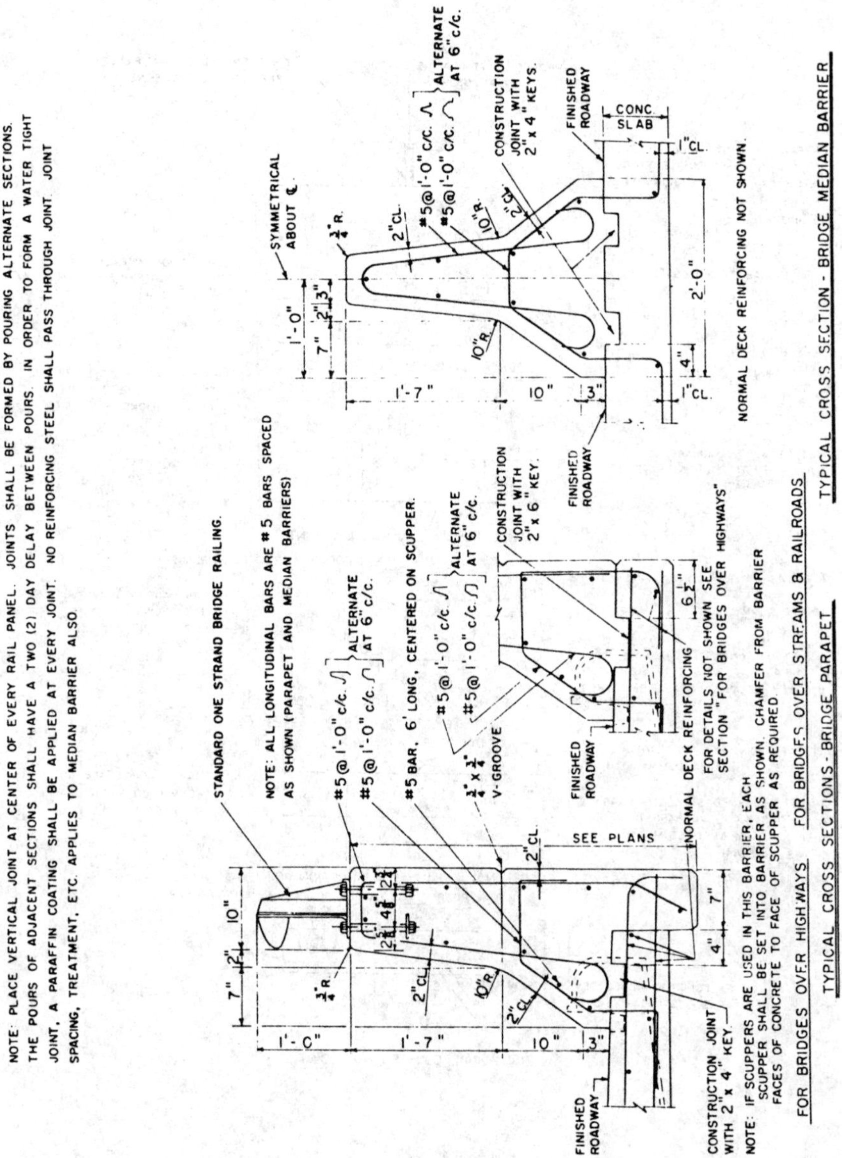
NOTE: PLACE VERTICAL JOINT AT CENTER OF EVERY RAIL PANEL. JOINTS SHALL BE FORMED BY POURING ALTERNATE SECTIONS. THE POURS OF ADJACENT SECTIONS SHALL HAVE A TWO (2) DAY DELAY BETWEEN POURS. IN ORDER TO FORM A WATER TIGHT JOINT, A PARAFFIN COATING SHALL BE APPLIED AT EVERY JOINT. NO REINFORCING STEEL SHALL PASS THROUGH JOINT. JOINT SPACING, TREATMENT, ETC. APPLIES TO MEDIAN BARRIER ALSO.
STANDARD ONE STRAND BRIDGE RAILING.
NOTE: ALL LONGITUDINAL BARS ARE #5 BARS SPACED AS SHOWN (PARAPET AND MEDIAN BARRIERS)
#5@1'-0" c/c.
#5@1'-0" c/c.
ALTERNATE AT 6" c/c.
#5 BAR, 6' LONG, CENTERED ON SCUPPER.
1/4" x 1/4" V-GROOVE
CONSTRUCTION JOINT WITH 2" x 6" KEY.
FINISHED ROADWAY
SEE PLANS
NORMAL DECK REINFORCING
FOR DETAILS NOT SHOWN SEE SECTION "FOR BRIDGES OVER HIGHWAYS"
CONSTRUCTION JOINT WITH 2" x 4" KEY.
NOTE: IF SCUPPERS ARE USED IN THIS BARRIER, EACH SCUPPER SHALL BE SET INTO BARRIER AS SHOWN. CHAMFER FROM BARRIER FACES OF CONCRETE TO FACE OF SCUPPER AS REQUIRED.
FOR BRIDGES OVER HIGHWAYS
FOR BRIDGES OVER STREAMS & RAILROADS
TYPICAL CROSS SECTIONS - BRIDGE PARAPET
SYMMETRICAL ABOUT ℄
CONSTRUCTION JOINT WITH 2" x 4" KEYS.
CONC. SLAB
NORMAL DECK REINFORCING NOT SHOWN.
TYPICAL CROSS SECTION - BRIDGE MEDIAN BARRIER

Concrete Bridge Parapet and Bridge Median Barrier Details

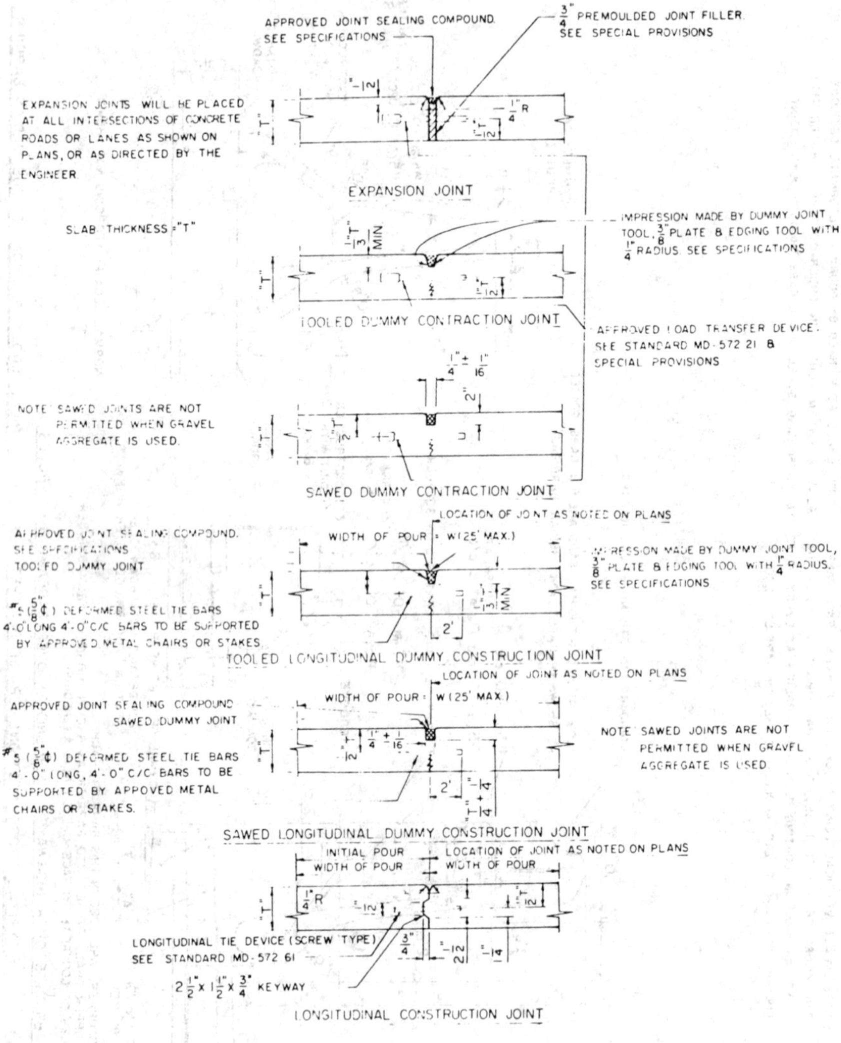
APPROVED JOINT SEALING COMPOUND. SEE SPECIFICATIONS
3/4" PREMOULDED JOINT FILLER. SEE SPECIAL PROVISIONS
EXPANSION JOINTS WILL BE PLACED AT ALL INTERSECTIONS OF CONCRETE ROADS OR LANES AS SHOWN ON PLANS, OR AS DIRECTED BY THE ENGINEER
EXPANSION JOINT
SLAB THICKNESS = "T"
IMPRESSION MADE BY DUMMY JOINT TOOL, 3/8" PLATE & EDGING TOOL WITH 1/4" RADIUS. SEE SPECIFICATIONS
TOOLED DUMMY CONTRACTION JOINT
APPROVED LOAD TRANSFER DEVICE. SEE STANDARD MD-572 21 & SPECIAL PROVISIONS
NOTE: SAWED JOINTS ARE NOT PERMITTED WHEN GRAVEL AGGREGATE IS USED.
SAWED DUMMY CONTRACTION JOINT
LOCATION OF JOINT AS NOTED ON PLANS
WIDTH OF POUR = W (25' MAX.)
APPROVED JOINT SEALING COMPOUND. SEE SPECIFICATIONS TOOLED DUMMY JOINT
#5 (5/8" ¢) DEFORMED STEEL TIE BARS 4'-0" LONG 4'-0" C/C BARS TO BE SUPPORTED BY APPROVED METAL CHAIRS OR STAKES
IMPRESSION MADE BY DUMMY JOINT TOOL, 3/8" PLATE & EDGING TOOL WITH 1/4" RADIUS. SEE SPECIFICATIONS
TOOLED LONGITUDINAL DUMMY CONSTRUCTION JOINT
APPROVED JOINT SEALING COMPOUND SAWED DUMMY JOINT
#5 (5/8" ¢) DEFORMED STEEL TIE BARS 4'-0" LONG, 4'-0" C/C BARS TO BE SUPPORTED BY APPOVED METAL CHAIRS OR STAKES.
NOTE: SAWED JOINTS ARE NOT PERMITTED WHEN GRAVEL AGGREGATE IS USED.
SAWED LONGITUDINAL DUMMY CONSTRUCTION JOINT
INITIAL POUR WIDTH OF POUR
WIDTH OF POUR
LONGITUDINAL TIE DEVICE (SCREW TYPE) SEE STANDARD MD-572 61
2 1/2" x 1 1/2" x 3/4" KEYWAY
LONGITUDINAL CONSTRUCTION JOINT

Reinforced Concrete Pavement
Types of Joints

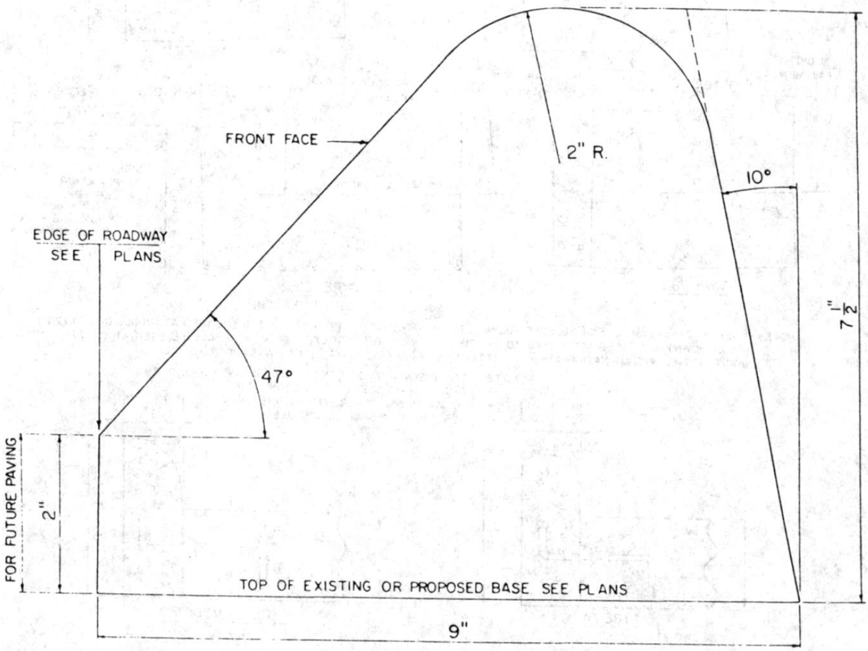

Standard Bituminous Concrete Curb

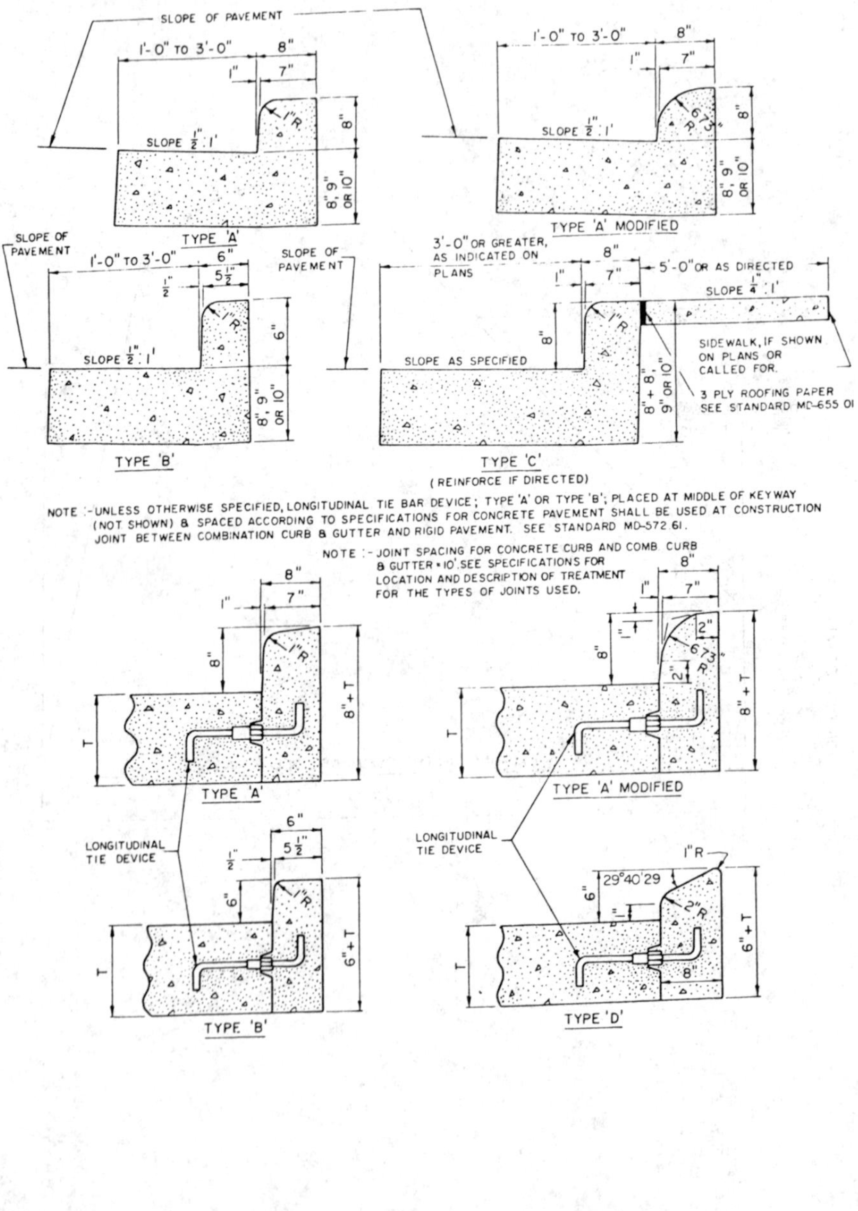
SLOPE OF PAVEMENT
1'-0" TO 3'-0"
8"
1"
7"
1"R
SLOPE 1/2" : 1'
8"
8", 9" OR 10"
TYPE 'A'
TYPE 'A' MODIFIED
6.73" R
SLOPE OF PAVEMENT
6"
5 1/2"
1/2"
TYPE 'B'
3'-0" OR GREATER, AS INDICATED ON PLANS
5'-0" OR AS DIRECTED
SLOPE 1/4" : 1'
SLOPE AS SPECIFIED
SIDEWALK, IF SHOWN ON PLANS OR CALLED FOR.
3 PLY ROOFING PAPER SEE STANDARD MD-655 01
8" + 8", 9" OR 10"
TYPE 'C'
(REINFORCE IF DIRECTED)
NOTE :- UNLESS OTHERWISE SPECIFIED, LONGITUDINAL TIE BAR DEVICE; TYPE 'A' OR TYPE 'B', PLACED AT MIDDLE OF KEYWAY (NOT SHOWN) & SPACED ACCORDING TO SPECIFICATIONS FOR CONCRETE PAVEMENT SHALL BE USED AT CONSTRUCTION JOINT BETWEEN COMBINATION CURB & GUTTER AND RIGID PAVEMENT. SEE STANDARD MD-572 61.
NOTE :- JOINT SPACING FOR CONCRETE CURB AND COMB. CURB & GUTTER = 10'. SEE SPECIFICATIONS FOR LOCATION AND DESCRIPTION OF TREATMENT FOR THE TYPES OF JOINTS USED.
8" + T
T
2"
LONGITUDINAL TIE DEVICE
6" + T
29°40'29
2"R
TYPE 'D'

Standard Types of Concrete Curb
Combination Concrete Curb and Gutter

Traffic Control

Sometimes, traffic must be moved through and around road and street construction, maintenance operations, utility work, curb and sidewalk installation, and any other type of work on or adjacent to roadways. Because there are so many different situations where one encounters the need for traffic control, there is no standard sequence of signs or controlling devices for this purpose, but the following provides a guide.

The general principles set forth here are applicable to both urban and rural areas. These guidelines are presented as a minimum of what is needed, and one cannot emphasize too strongly that the needs of each project must be reviewed carefully in order to provide workers and the public maximum protection.

In general, traffic conditions on city streets are characterized by low speeds, wide ranges in volume, limited maneuvering space, frequent turns and cross movements, and significant pedestrian movement. Rural highways tend to see lower volume of traffic but much higher speeds, less pedestrian traffic, and fewer turns and encroachments.

Motorists must be guided in a clear and positive manner when approaching and traversing construction work areas. Warning signs in construction areas should have a black legend on an orange background. All signs intended for use during hours of darkness should be reflectorized with a material that has a smooth, sealed outer surface, or they must be illuminated to show approximately the same shape and color day or night. Note that street or highway lighting is not considered sufficient to meet the illumination requirements.

As a general rule, signs should be located on the right side of the road or street. In rural areas the edge of the sign shall be a minimum 6'-0" from the roadway edge and a minimum 5'-0" from the ground to the bottom of the sign. In an urban area, the edge of the sign shall be a minimum 2'-0" from the outside face of the curb, and the bottom of the sign shall be at least 7'-0" from the top of the sidewalk.

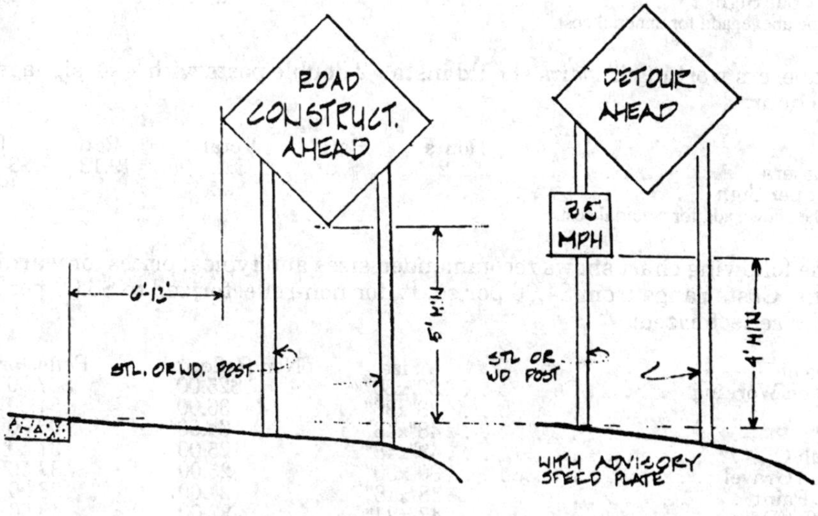

Roadside Sign Rural District

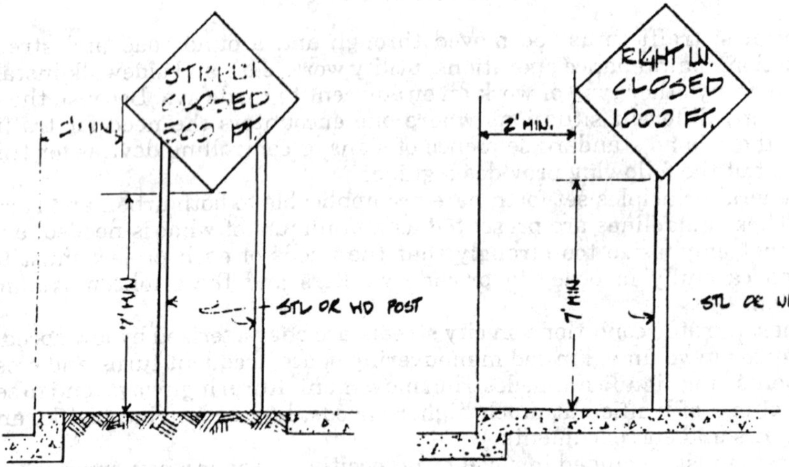

Roadside Sign Urban District

Height and Lateral Locations of Signs
Typical Installations

Temporary Sign Erection. Laborers working in pairs can install 3 single posts with 4'x4' signage in an hour:

	Hours	Rate	Total	Rate	Total
Laborers	2	$....	$....	$19.13	$38.26
Cost per Sign			....		12.75

To the above, add for material cost.

Laborers working in pairs should install 2 double posts with 4'x4' signage in one hour:

	Hours	Rate	Total	Rate	Total
Laborers	2	$....	$....	19.13	$38.26
Cost per Sign			....		19.13

To the above, add for material cost.

The following chart shows recommended sizes and typical prices for warning signs. Costs range from $4.00 per sq. ft. for non-reflectorized to $6.00 per sq. ft. for reflectorized:

Type of Use	Size	Non-Reflector	Reflector
Person Working	30"x30"	$25.00	$37.50
"	36"x36"	36.00	54.00
"	48"x48"	64.00	96.00
Fresh Oil	48"x48"	25.00	37.50
Loose Gravel	30"x30"	25.00	37.50
Wet Paint	48"x18"	24.00	36.00
Do Not Pass	48"x24"	32.00	48.00
Road Work Ahead	48"x36"	48.00	72.00

Road Work xx Feet	48"x48"	64.00	96.00
Detour xx Feet	48"x48"	64.00	96.00
Road Closed xx Feet	42"x42"	49.00	73.50
"	48"x48"	64.00	96.00
Right Lane Closed xx Feet	48"x48"	64.00	96.00
Single Lane	48"x48"	64.00	96.00
Arrow Boards	48"x24"	32.00	48.00
Flagman Ahead	36"x36"	36.00	54.00
"	48"x48"	64.00	96.00

Wood sign posts may be used in place of steel posts, but the portion of wood to be placed in the ground may require wood preservative.

Type of Post	Length	Price Each
Steel	12'-0"	$12.00
4x4 Wood	12'-0"	6.00

When working adjacent to a roadway, a flagman may be required to stop traffic intermittently as work progresses to protect the work crew and the travelling public. When flagging is required, the estimator needs to add flagman to the work crew.

Advanced Warning Arrow Panel. These are sign panels with a matrix of lights capable of either flashing or sequential display. Necessary signs, barricades, and other traffic control devices shall be used in conjunction with advance warning arrow boards. Arrow panels are generally used for day and night lane closures, roadway diversions, and slow construction on travelled roadways. Placement of arrows is as needed to achieve the desired recognition distance. The color of light emitted should be yellow. Prices range from $4,000 to $12,000 depending on engine, fuel capacity, and sign illumination.

Other miscellaneous items are:

Cones, 28" High	$8.00 each
Plastic Barrels	35.00 each
Blinking Lights	20.00 each

CHAPTER 23

CITY COST INDEXES

Obviously, wage rates vary from place to place, and from job to job. The hourly labor rates shown throughout the book are U.S. averages for prevailing wages as of mid 1991.

The following table shows labor cost adjustment factors for 97 major U.S. cities and is intended to be used in conjunction with the labor rates used on cost tables throughout this book.

The U.S. average is the base, so its cost factor is theoretically 1.000, and all of the cost adjustment factors below are figured in relation to it. The labor rates for any of these 97 cities are computed by multiplying the labor rates shown in cost tables throughout the book by the appropriate city cost factor from the table below.

For example, the wage rate for general construction labor is shown as $19.13. This is a 1991 U.S. average. Suppose that you want the wage rate for general construction labor in Akron, Ohio. Locate the cost factor below under for Akron and multiply: 19.13 x .995 = 19.03435 or $19.03

City	Index
Akron, OH	.995
Albany, NY	.869
Albuquerque, NM	.756
Anchorage, AK	1.404
Atlanta, GA	.751
Baltimore, MD	.863
Billings, MT	.987
Birmingham, AL	.633
Boise, ID	.989
Boston, MA	1.299
Buffalo, NY	1.055
Burlington, VT	.765
Charleston, SC	.894
Charleston, WV	.918
Charlotte, NC	.854
Cheyenne, WY	.826
Chicago, IL	1.037
Cincinnati, OH	.904
Cleveland, OH	1.053
Columbia, SC	.854
Columbus, OH	.880
Dallas, TX	.568
Dayton, OH	.895
Denver, CO	.683
Des Moines, IA	.922
Detroit, MI	1.105
Duluth, MN	1.021
El Paso, TX	.707
Erie, PA	.866
Fairbanks, AK	1.396
Fargo, ND	.819
Grand Rapids, MI	.754
Hartford, CT	1.236
Honolulu, HI	1.181
Houston, TX	.822
Indianapolis, IN	.827

City	Index
Jackson, MS	.517
Jacksonville, FL	.841
Kansas City, MO	.901
Knoxville, TN	.637
Lansing, MI	.862
Las Vegas, NV	1.042
Little Rock, AR	.911
Los Angeles, CA	1.113
Louisville, KY	.839
Madison, WI	.831
Manchester, NH	.765
Memphis, TN	.729
Miami, FL	.616
Milwaukee, WI	.944
Minneapolis, MN	.893
Montgomery, AL	.596
Nashville, TN	.663
Newark, NJ	1.287
New Haven, CT	1.294
New Orleans, LA	.863
New York, NY	1.491
Norfolk, VA	.560
Oakland, CA	1.478
Oklahoma City, OK	.608
Omaha, NE	.752
Philadelphia, PA	1.169
Phoenix, AZ	.858
Pittsburgh, PA	.899
Portland, ME	.840
Portland, OR	.887
Providence, RI	1.092
Raleigh, NC	.862
Reading, PA	.818
Richmond, VA	.793
Rochester, NY	.983
Sacramento, CA	1.042

City	Index
St. Louis, MO	.993
St. Paul, MN	1.049
St. Petersburg, FL	.817
Salt Lake City, UT	.887
San Diego, CA	1.175
San Francisco, CA	1.232
Santa Fe, NM	.925
Savannah, GA	.773
Scranton, PA	.857
Seattle, WA	.943
Sioux Falls, SD	.756
South Bend, IN	.818
Spokane, WA	.852
Stamford, CT	1.272
Syracuse, NY	.947
Tampa, FL	.847
Toledo, OH	.950
Topeka, KS	.594
Trenton, NJ	1.155
Tulsa, OK	.594
Washington, DC	.816
Wichita, KS	.820
Wilmington, DE	.977
York, PA	.764
Youngstown, OH	.904

CHAPTER 24

MENSURATION

The information in this chapter will enable contractors and estimators to estimate quantities and costs accurately and efficiently. A number of tables use decimals in the quantities, because by using decimals, it is possible to state the fractional parts of inches, feet, and yards in a smaller space than when regular fractions are used.

For the estimator who is not thoroughly familiar with the decimal system, complete explanations are given covering the use of all classes of decimal fractions so that they can be used rapidly and accurately.

Estimating is nearly all "figures" of one kind or another, so it is essential that the estimator possess a fair working knowledge of arithmetic for estimates to be accurate and dependable.

Most of the estimator's computations involve measurements of surface and volume and are stated in lineal feet (lin. ft.), square feet (sq. ft.), square yards (sq. yds.), squares (sqs.) containing 100 sq. ft., cubic feet (cu. ft.), and cubic yards (cu. yds.). These quantities are often further reduced to thousands of brick, board feet of lumber, etc.

The following abbreviations are used throughout the book to make all tables brief and concise and to insert them in the smallest possible space:

	Abbreviation	Symbols
Inches	= in.	= "
Lineal feet	= lin. ft. or l.f.	= '
Feet-inches	= ft. in.	= 2'-5"
Square feet	= sq. ft. or s.f.	
Square yards	= sq. yds. or s.y.	
Squares	= sqs.	
Cubic feet	= cu. ft. or c.f.	
Cubic yards	= cu. yds. or c.y.	
Board feet	= b.f.	

Linear Measures

12 inches	= 12 in.	= 12"	= 1 foot	= 1 ft.	= 1'-0"
3 feet	= 3 ft.	= 3'-0"	= 1 yard	= 1 yd.	
16½ feet	= 16½ ft.	= 16'-6"	= 1 rod	= 1 rd.	
40 rods	= 40 rds.	= 1 furlong	= 1 fur.		
8 furlongs	= 8 fur.	= 1 mile	= 1 mi.		
5,280 feet	= 1,760 yds.	= 1 mile	= 1 mi.		

Square Measure or Measure of Surfaces

144 square inches (sq. in.)	= 1 square foot (sq. ft.)	
9 square feet (sq. ft.)	= 1 square yard (sq. yd.)	
100 square feet (sq. ft.)	= 1 square (sq.)*	
30¼ square yards (sq. yds.)	= 1 square rod (sq. rd.)	
160 square rods	= 1 Acre = 1 A.	
43,560 sq. ft.	= 4,840 sq. yds.	= 1 Acre = 1 A.
640 acres	= 640 A.	= 1 square mile (sq. mi.)

*Architects' and builders' measure.

Water Quantities

1 cu. ft.	= 7.4805 gal.	= 62.4 lbs.
1 gal.		= 8.34 lbs.
1 acre-foot	= 325,850 gal.	
1 gal.	= 3.7854 liters	
1 cu. meter	= 264.17 gal.	
1 cu. ft. salt water	= 64.4 lbs.	

Power Units

1 hp	= 550 ft-lbs. per sec.
1 hp	= 0.746 kw

Cubic Measure or Cubical Contents

1,728 cubic inches (cu. in.)= 1 cubic foot (cu. ft.)
27 cubic feet (cu. ft.)= 1 cubic yard (cu. yd.)
128 cubic feet (cu. ft.) = 1 cord (cd.)
24¾ cubic feet (cu. ft.) = 1 perch (P.)*

*A perch of stone is nominally 16½ ft. long, 1 ft. high, and 1½ ft. thick, and contains 24¾ cu. ft. However, in some states, especially west of the Mississippi, rubble work is figured by the perch containing 16½ cu. ft. Before submitting prices on masonry by the perch, find out the common practice in your locality.

Computing Areas and Volumes

To Compute the Area of a Square, Rectangle, or Parallelogram. Multiply the length by the breadth or height. Example : Obtain the area of a wall 22 ft. (22'-0") long and 9 ft. (9'-0") high. 22 x 9 = 198 sq. ft.

To Compute the Area of a Triangle. Multiply the base by ½ the altitude or perpendicular height. Example: Find the area of the end gable of a house 24 ft. (24'-0") wide and 12 ft. (12'-0") high from the base to high point of roof. 24 ft. x 6 ft. (½ the height) = 144 sq. ft.

To Compute the Circumference of a Circle. Multiply the diameter by 3.1416. The diameter multiplied by $3\frac{1}{7}$ is close enough for all practical purposes. Example: Find the circumference or distance around a circle, the diameter of which is 12 ft. (12'-0"). 12 x 3.1416 = 37.6992 ft. the distance around the circle or 12 x $3\frac{1}{7}$ = 37.714 ft.

To Compute the Area of a Circle. Multiply the square of the diameter by 0.7854 or multiply the square of the radius by 3.1416. Example: Find the area of a round concrete column 24 in. (24" or 2'-0") in diameter. The square of the diameter is 2 x 2 = 4. 4 x 0.7854 = 3.1416 sq. ft. the area of the circle.

The radius is ½ the diameter. If the diameter is 2 ft. (2'-0"), then the radius would be 1 ft. (1'-0"). To obtain the square of the radius, 1 x 1 = 1. Multiply the square of the radius, 1 x 3.1416 = 3.1416, the area of the circle.

To Compute the Cubical Contents of a Circular Column. Multiply the area of the circle by the height. Example: Find the cubical contents of a round concrete column 2 ft. (2'-0") in diameter and 14 ft. (14'-0") long.

From the previous example, the area of a circle 2 ft. in diameter is 3.1416 sq. ft. 3.1416 x 14 ft. (the height) = 43.9824 cu. ft. or for all practical purposes 44 cu. ft. of concrete in each column.

To Compute the Cubical Contents of any Solid. Multiply the length by the breadth or height by the thickness. Computations of this kind are used extensively in estimating all classes of building work, such as excavating, concrete foundations, reinforced concrete, brick masonry, cut stone, granite, etc.

The following charts give illustrations and formulas for all the different geometric shapes of area and volume that the estimator is likely to encounter.

AREA (A : Area)

NAME FORMULA	SHAPE
Parallelogram $A = B \times h$	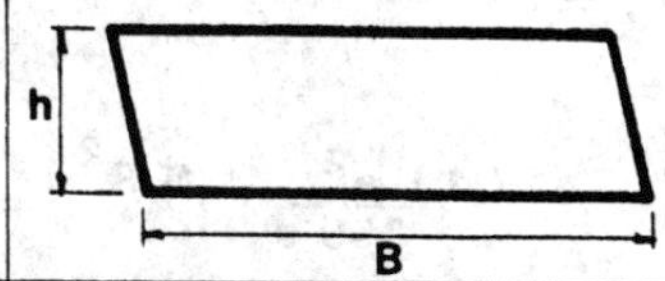
Trapezoid $A = \frac{B + C}{2} \times h$	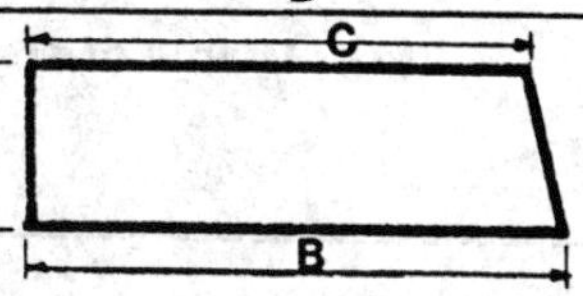
Triangle $A = \frac{B \times h}{2}$	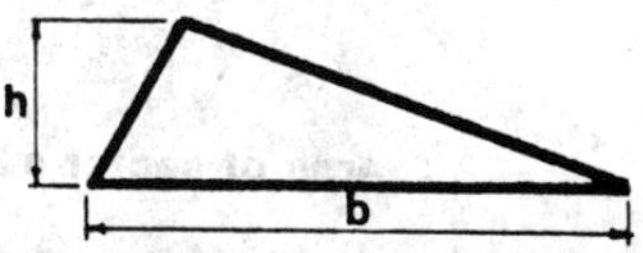
Trapezium (Divide into 2 triangles) A = Sum of the 2 triangles (See above)	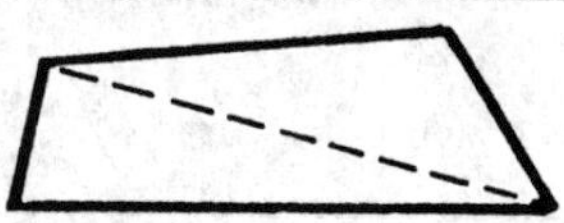
Regular Polygon A= $\frac{\text{Sum of sides (s)}}{2}$ x Inside Radius (R)	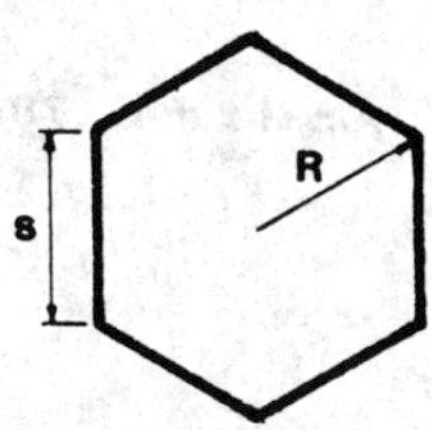
Circle (1) πR^2 A : (2) $.7854 \times D^2$ (3) $.0796 \times C^2$	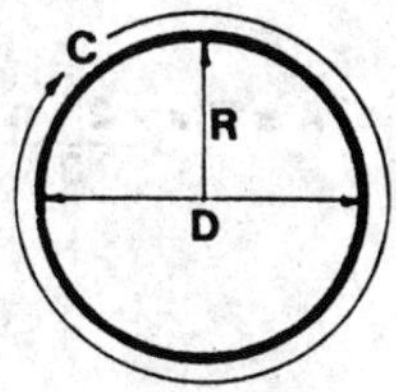

AREA (A : Area)

NAME FORMULA	SHAPE
Sector A : (1) $\frac{a^2}{360\ o} \times \pi R^2$ (2) Length of area x $\frac{R}{2}$	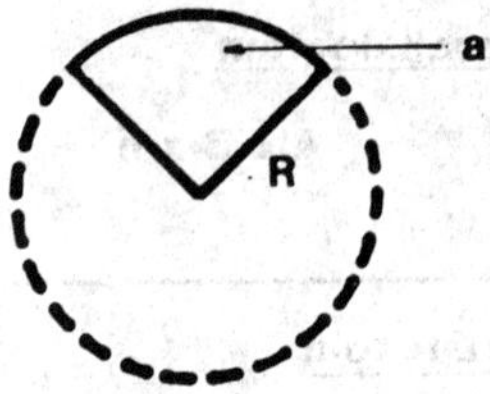
Segment A : Area of sector minus triangle (See above)	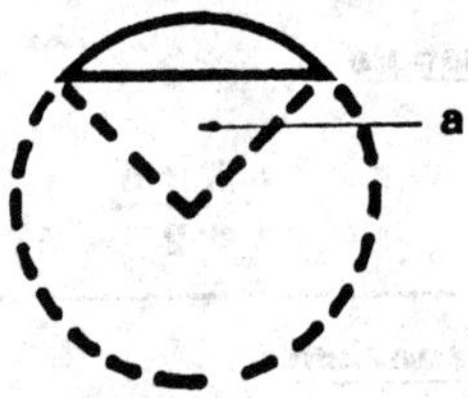
Ellipse A = M x m x .7854	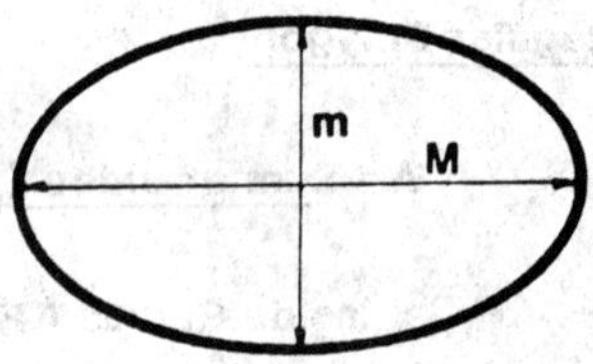
Parabola A = B x $\frac{2\ h}{3}$	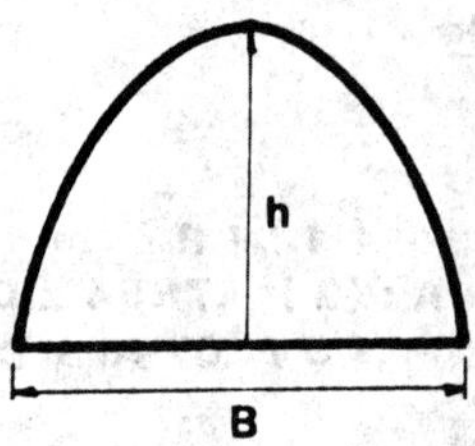

VOLUME (V : volume)

NAME FORMULA	SHAPE
Cube $V = a^3$ (in cubic units)	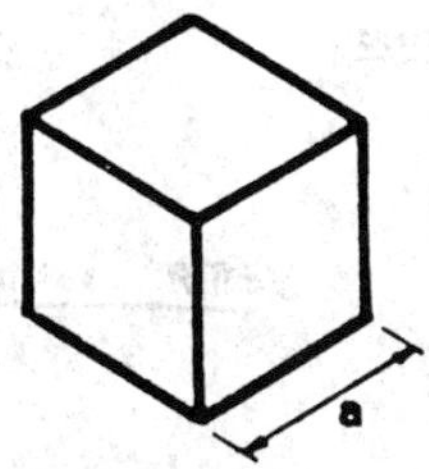
Rectangular Solids $V = L \times W \times h$	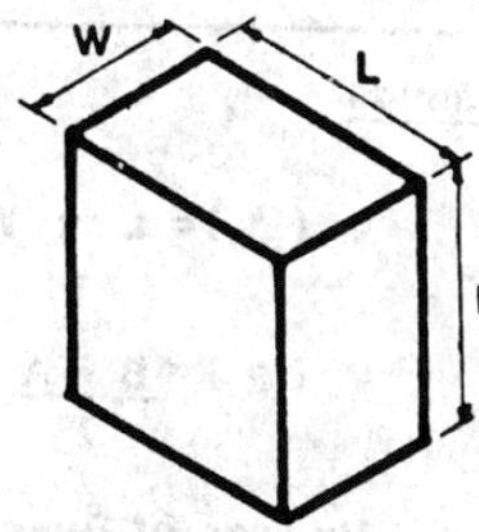
Prisms $V\ (1) = \frac{B \times A}{2} \times h$ $V\ (2) = \frac{s \times R}{2} \times 6 \times h$ V = Area of end × h	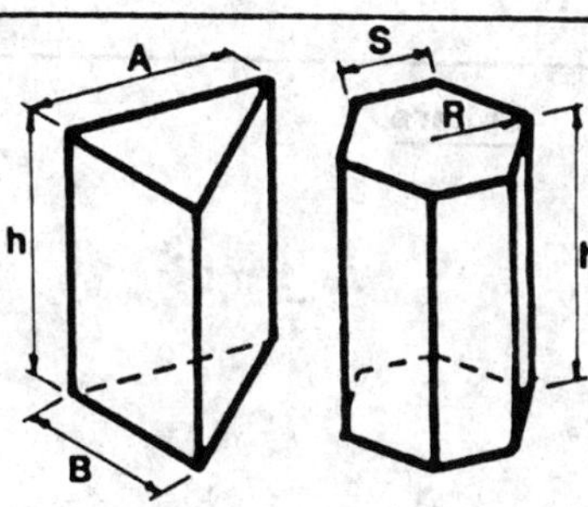
Cylinder $V = \times \pi R^2 \times h$	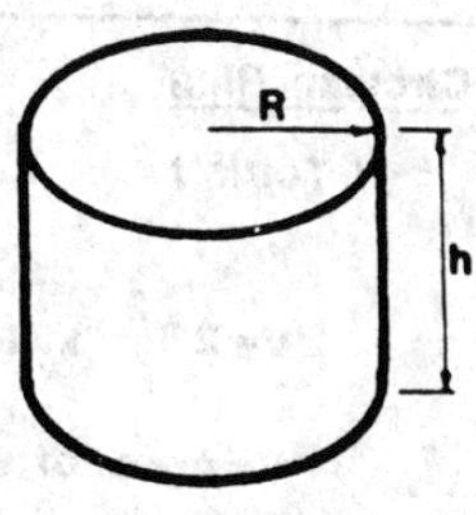

VOLUME (V : volume)

NAME FORMULA | **SHAPE**

Cone

$$V = \frac{\pi R^2 \times h}{3}$$

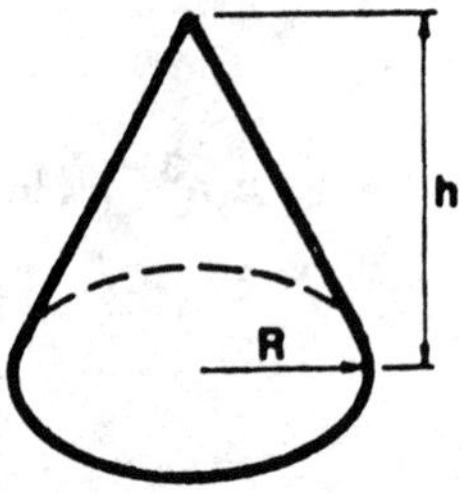

Pyramids

$$V(1) = L \times W \times \frac{h}{3}$$

$$V(2) = \frac{B \times A}{2} \times \frac{h}{3}$$

$$V = \text{Area of Base} \times \frac{h}{3}$$

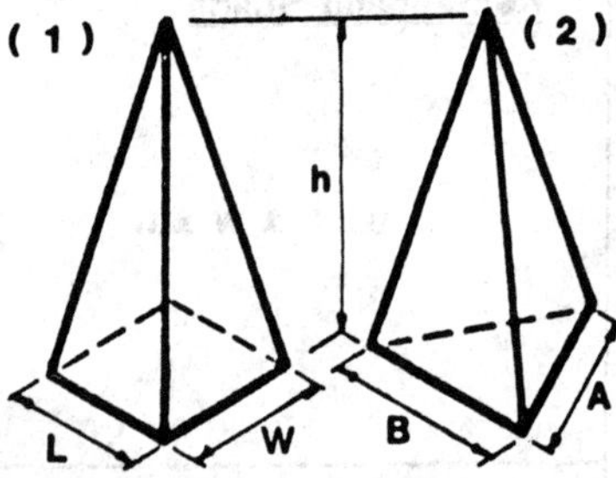

Sphere

$$V = \frac{1}{6} \pi D^3$$

Circular Ring

(Torus)

$$V = 2\pi^2 \times Rr^2$$

$$V = \text{Area of section} \times 2\pi R$$

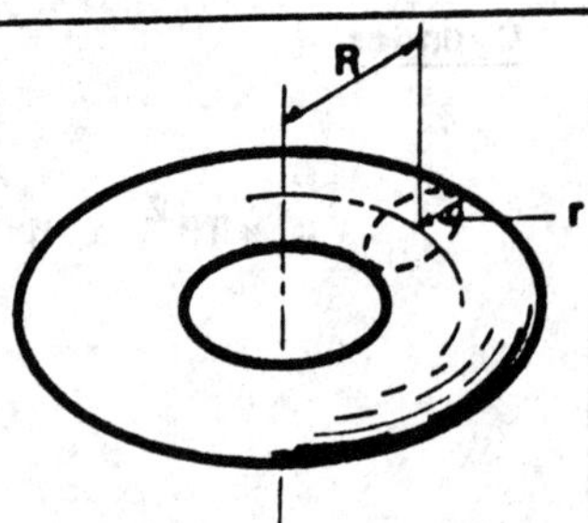

Example: Find the cubical contents of a wall 42 ft. (42'-0") long, 5 ft. 6 in. (5'-6") high, and 1 ft. 4 in. (1'-4") thick, 42'-0" x 5'-6" x 1'-4" = 308 cu. ft. To reduce cu. ft. to cu. yds., divide 308 by 27. The result is $11^{11}/_{27}$ or $11^{1}/_{2}$ or 11.41 cu. yds.

How to Use the Decimal System in Estimating.

A decimal is a fraction whose denominator is not written and is some power of 10. It is often called a decimal fraction but more often simply a decimal. Example: We know 50 cents is $^{50}/_{100}$ or $^{1}/_{2}$ of a dollar. Writing the same thing in decimals would be $0.50.

Numerator	50
Denominator	100

When written as .50 or .5, the denominator of the fraction is not written, but it is understood to be 10 from the fact that 5 occupies the first place to the right of the decimal point. Therefore we can have the following:

0.5 means $^{5}/_{10}$, for the 5 extends to the 10th's place;
0.25 means $^{25}/_{100}$, for the 25 extends to the 100th's place;
0.125 means 100, for the 125 extends to the 1000th's place.

The name of the places are, in part, as follows:

Thousands	Hundreds	Tens	Units	(Decimal Point)	Tenths	Hundredths	Thousandths	Ten-Thousandths
1	3	4	5		2	7	6	5

This number is read "one thousand three hundred forty-five and two thousand seven hundred and sixty-five ten thousandths". The orders beyond the ten thousandths are hundred thousandths, millionths, ten millionths, etc., but in figuring construction work it is seldom necessary to carry the figures beyond three or four decimal points.

A whole number and a decimal together, form a *mixed decimal.* Example: 2.25 is the same as $2^{25}/_{100}$ or $2^{1}/_{4}$.

The period written at the left of tenths is called the decimal point. Example: 0.5 = $^{5}/_{10}$; 0.25 = $^{25}/_{100}$ = $^{1}/_{4}$; 0.375 = 300 = $^{3}/_{8}$, etc.

It is not necessary to write a zero at the left of the decimal point in the above examples, because 0.5 means the same as .5. The zero is often written to call attention more quickly to the decimal point.

In the construction business, decimals are used chiefly to denote feet and inches, hours and minutes, the fractional working units of the various kinds of materials, and in money, which is dollars and cents or fractional parts of 100.

Table of Feet and Inches Reduced to Decimals

The following table illustrates how feet and inches may be expressed in four different ways, all meaning the same thing.

1 inch = 1" = 1/12th foot	=0.083
1½ inches = 1½" = 1/8th foot	=0.125
2 inches = 2" = 1/6th foot	=0.1667
2½ inches = 2½"= 5/24ths ft	=0.2087
3 inches = 3" = 1/4th foot	=0.25
3½ inches = 3½" = 7/24ths ft	=0.2917
4 inches = 4" = 1/3rd foot	=0.333
4½ inches = 4½"= 3/8ths ft	=0.375
5 inches = 5" = 5/12ths foot	=0.417
5½ inches = 5½" =11/24ths ft	=0.458
6 inches = 6" = ½ foot	=0.5
6½ inches = 6½" =13/24ths ft	=0.5417
7 inches =7" = 7/12ths foot	=0.583
7½ inches = 7½" = 5/8ths ft	=0.625
8 inches = 8" = 2/3rds foot	=0.667
8½ inches =8½" = 17/24ths ft	=0.708
9 inches = 9" = 3/4ths foot	=0.75
9½ inches = 9½" =19/24ths ft	=0.792
10 inches =10" = 5/6ths foot	=0.833
10½ inches = 10½" =7/8ths ft	=0.875
11 inches = 11" = 11/12ths foot	=0.917
11½ inches = 11½" =23/24ths	=0.958
12 inches =12"= 1 foot	=1.0

Example : Write 5 feet, 7½ inches, in decimals. It would be written 5.625, which is equivalent to 5⅝ feet.

Table of Common Fractions Stated in Decimals

The following table gives the decimal equivalents of common fractions frequently used in estimating:

1/16	= 0.0625	9/16	= 0.5625
1/8	= 0.125	5/8	= 0.625
3/16	= 0.1875	11/16	= 0.6875
1/4	= 0.25	3/4	= 0.75
5/16	= 0.3125	15/16	= 0.9375
3/8	= 0.375	13/16	= 0.8125
7/16	= 0.4375	7/8	= 0.875
1/2	= 0.5	8/8	= 1.0

Annexing zeros to a number does not change its value; 0.5 is the same as 0.500.

Table of Hours and Minutes Reduced to Decimals

The following table illustrates how minutes may be reduced to fractional parts of hours and to decimal parts of hours.

No. of Minutes	Fractional Part of an Hour	Decimal Part of an Hour	No. of Minutes	Fractional Part of an Hour	Decimal Part of an Hour
1	=1/60th	= 0.0167	31	=31/60ths	= 0.5167
2	=1/30th	= 0.0333	32	=8/15ths	= 0.5333
3	=1/20th	= 0.05	33	=11/20ths	= 0.55
4	=1/15th	= 0.0667	34	=17/30ths	= 0.5667
5	=1/12th	= 0.0833	35	=7/12ths	= 0.5833
6	=1/10th	= 0.10	36	=3/5ths	= 0.60
7	=7/60ths	= 0.1167	37	=37/60ths	= 0.6167
8	=2/15ths	= 0.1333	38	=19/30ths	= 0.6333
9	=3/20ths	= 0.15	39	=13/20ths	= 0.65
10	=1/6th	= 0.1667	40	=2/3rds	= 0.6667
11	=11/60ths	= 0.1833	41	=41/60ths	= 0.6833
12	=1/5th	= 0.20	42	=7/10ths	= 0.70
13	=13/60ths	= 0.2167	43	=43/60ths	=0.7167
14	=7/30ths	= 0.2333	44	=11/15ths	= 0.7333
15	=1/4th	= 0.25	45	=3/4ths	= 0.75
16	=4/15ths	= 0.2667	46	=23/30ths	= 0.7667
17	=17/60ths	= 0.2833	47	=47/60ths	= 0.7833
18	=3/10ths	= 0.30	48	=4/5ths	= 0.80
19	=19/60ths	= 0.3167	49	=49/60ths	= 0.8167
20	=1/3rd	= 0.333	50	=5/6ths	= 0.8333
21	=7/20ths	= 0.35	51	=51/60ths	= 0.85
22	=11/30ths	= 0.3667	52	=13/15ths	= 0.8667
23	=23/60ths	= 0.3833	53	=53/60ths	= 0.8833
24	=2/5ths	= 0.40	54	=9/10ths	= 0.90
25	=5/12ths	= 0.4167	55	=11/12ths	= 0.9167
26	=13/30ths	= 0.4333	56	=14/15ths	= 0.9333
27	=9/20ths	= 0.45	57	=19/20ths	= 0.95
28	=7/15ths	= 0.4667	58	=29/30ths	= 0.9667
29	=29/60ths	= 0.4833	59	=59/60ths	= 0.9833
30	=1/2	= 0.5	60	=1	=1.0

Example: Write 4 hours and 37 minutes in decimals. It would be written 4.6167. Now find the labor cost for 4.6167 hrs. at $7.50 an hr.:

Hours	4.6167
X Hourly Rate	$7.50
	2308350
	323169
	$34.625250
	or $34.63

Similar Decimals. Decimals that have the same number of decimal places are called similar decimals. Thus, 0.75 and 0.25 are similar decimals and so are 0.150 and 0.275; but 0.15 and 0.275 are dissimilar decimals.

To reduce dissimilar decimals to similar decimals, give them the same number of decimal places by annexing or cutting off zeros. Example, 0.125,

0.25, 0.375, and 0.5 may all be reduced to thousandths as follows: 0.125, 0.250, 0.375, and 0.500.

To Reduce a Decimal to a Common Fraction. Omit the decimal point, write the denominator of the decimal, and then reduce the common fraction to its lowest terms.

Example: 0.375 equals 300, which reduced to its lowest terms, equals 3⁄8:

25300 = 5 15⁄40 = 3⁄8

How to Add Decimals. To add numbers containing decimals, write like orders under one another, and then add as with whole numbers. Example, add 0.125, 0.25, 0.375, and 1.0. The total of the addition is 1.750, which equals 1 700, and which may be further reduced to 1 3⁄4

```
 0.125
 0.25
 0.375
 1.0
 1.750
```

How to Subtract Decimals. To subtract one number from another, when either or both contain decimals, reduce to similar decimals, write like orders under one another, and subtract as with whole numbers. Example: Subtract 20 hours and 37 minutes from 27 hours and 13 minutes, both being written in decimals. The difference is 6.6000 or 6.6 hours, which reduced to a common fraction is 6 3⁄5 hours or 6 hours and 36 minutes.

```
 27.2167
-20.6167
  6.6000
```

How to Multiply Decimals. The same method is used as in multiplying other numbers, and the result should contain as many decimal points as there are in both of the numbers multiplied. Example: Multiply 3 feet 6 inches by 4 feet 9 inches.

4 ft. 9 in. = 4'-9" = 4 3⁄4 ft. = 4.75 Multiplicand	4.75
3 ft. 6 in. = 3'-6" = 3 1⁄2 ft. = 3.5 Multiplier	3.5
To multiply, proceed as illustrated	2375
	1425
Product	16,625
	or 16.625

In the above example, there are 2 decimals in the multiplicand and 1 decimal in the multiplier. The result or product should contain the same number of decimals as the multiplicand and multiplier combined, which is three. Starting at the right and counting to the left three places, place the decimal point between the two sixes. The result would be 16.625, which equals 16600 = 16 5⁄8 sq. ft.

Practical Examples Using Decimals

The quantities of cement, sand and gravel required for one cu. yd. of concrete are ordinarily stated in decimals. For instance, 1 cu. yd. of concrete mixed in the proportions of 1 part cement:2 parts sand:4 parts gravel is usually stated 1:2:4 and requires the following materials:

1.50 bbls. cement = 1 50/100 bbls. = 1 1/2 bbls. = 6 sacks.
0.42 cu. yds. sand = 42/100 cu. yds. = 21/50 cu. yds. = 11 1/3 cu. ft.
0.84 cu. yds. gravel = 84/100 cu. yds. = 22 2/3 cu. ft.

There are 4 sacks of cement to the bbl., and each sack weighs 94 lbs. and contains approximately 1 cu. ft. of cement.

There are 27 cu. ft. in a cu. yd.; to obtain the number of cu. ft. of sand required for a yard of concrete, 0.42 cu. yds = 42/100 cu. yds. = 21/50 cu. yds. and 21/50 of 27 cu. ft. =

$$\frac{21 \times 27}{50} =$$

$$\frac{567}{50} =$$

= 11.34 or 11 1/3 cu. ft. sand.

To find the cost of a cu. yd. of concrete based on the above quantities, and assuming it requires 2 1/4 hrs. labor time to mix and place one cu. yd. of concrete, proceed as follows:

1 1/2 bbls. cement=1.50 bbls. @ $9.00 per bbl.	$13.5000
11 1/3 cu. ft. sand=0.42 cu. yd. @ 5.75 per cu. yd.	2.4150
22 2/3 cu. ft. gravel=0.84 cu. yd. @ 5.75 per cu. yd.	4.8300
2 1/4 hrs. labor=2.25 hrs. @ 9.80 per hr.	22.0500
Cost per cu. yd.	$42.7950

You will note the total is carried in four decimal places. This is to illustrate the actual figures obtained by multiplying. The total would be $42.80 per cu. yd. of concrete.

In making the above multiplications you will note there are 2 decimals in both the multiplicand and the multiplier of all the amounts multiplied, so the result should contain as many decimals as the sum of the multiplicand and multiplier, which is 4. You will note all the totals contain 4 decimal places. Always bear this in mind when making your multiplications, because if the decimal place is wrong it makes a difference of 90 percent in your total. You know the correct result is $42.7950, but suppose by mistake you counted off 5 decimal places. The result would be $4.27950 or $4.28 per cu. yd. or just about 1/10 enough; or if you made a mistake the other way and counted off just 3 decimal places, the result would be $427.950 or $427.95 per cu. yd., or just about 10 times too much. Watch your decimal places.

The same method is used in estimating lumber. All kinds of framing lumber are ordinarily sold by the 1,000 b.f., so 1,000 is the unit or decimal used when estimating lumber.

Suppose you buy 150 pcs. of 2"x8"-16'-0", which contains 3,200 b.f. This is equivalent to 3200 = 3 2/10 = 3 1/15 thousandths or stated in decimals it may be either 3.2 or 3.200. The cost of this lumber at $230.00 per 1,000 b.f., would be obtained by multiplying 3.2 ft. at $230.00 per 1,000 b.f. as follows: 3.2 mbf of lumber @ $230.00 = $736.00.

Things to Remember When Using Decimals

.6 indicates tenths, thus .6 = 6/10 = .60 = 60/100 = .600 = 600 .06 indicates hundredths, thus .06 = 6/100 = .060 = 60/1000
.006 indicates thousandths, thus .006 = 6/1000
.0006 indicates ten thousandths, thus .0006 = 6/10000

When multiplying decimals, always remember that the result or product must contain as many decimal places as the sum of the decimal places in both the multiplicand and the multiplier. Find the cost of 3 hrs. 20 min. laborer's time at $9.80 an hr.

3 hrs. 20 min. equal 3 1/3 hrs. = 3.3333
9 dollars 80 cents an hr. = 9.80

```
  2666640
 299997
 32666340
= 32.666340
= $32.67
```

There are 3 decimals in the multiplicand and 2 decimals in the multiplier, so the result or total should contain as many decimals as the sum of the multiplicand and multiplier, which is 5. Be sure your decimal point is in the *right* place and the rest is easy.

Conversion of Feet, Inches to Decimal and Metric Equivalents

Inches	Inches in Decimals	Feet in Decimals	Millimeters	Centimeters	Meters
1/16	0.0625	0.005	1.5875	0.15875	0.001588
1/8	0.1250	0.010	3.1750	0.31750	0.003175
3/16	0.1875	0.016	4.7625	0.47625	0.004763
1/4	0.2500	0.021	6.3500	0.63500	0.006350
5/16	0.3125	0.026	7.9375	0.79375	0.007938
3/8	0.3750	0.031	9.5250	0.95250	0.009525
7/16	0.4375	0.036	11.1125	1.11125	0.011113
1/2	0.5000	0.042	12.7001	1.27001	0.012700
9/16	0.5625	0.047	14.2876	1.42876	0.014288
5/8	0.6250	0.052	15.8751	1.58751	0.015875
11/16	0.6875	0.057	17.4626	1.74626	0.017463
3/4	0.7500	0.063	19.0501	1.90501	0.019050
13/16	0.8125	0.068	20.6376	2.06376	0.020638
7/8	0.8750	0.073	22.2251	2.22251	0.022225
15/16	0.9375	0.078	23.8126	2.38126	0.023813
1	1.0000	0.083	25.4001	2.54001	0.025400
1 1/8	1.1250	0.094	28.5751	2.85751	0.028575
1 1/4	1.2500	0.104	31.7501	3.17501	0.031750
1 3/8	1.3750	0.115	34.9251	3.49251	0.034925
1 1/2	1.5000	0.125	38.1002	3.81002	0.038100
1 5/8	1.6250	0.135	41.2752	4.12752	0.041275
1 3/4	1.7500	0.146	44.4502	4.44502	0.044450
1 7/8	1.8750	0.156	47.6252	4.76252	0.047625
2	2.0000	0.167	50.8002	5.08002	0.050800

Inches	Inches in Decimals	Feet in Decimals	Millimeters	Centimeters	Meters
2 1/8	2.1250	0.177	53.9752	5.39752	0.053975
2 1/4	2.2500	0.188	57.1502	5.71502	0.057150
2 3/8	2.3750	0.198	60.3252	6.03252	0.060325
2 1/2	2.5000	0.208	63.5003	6.35003	0.063500
2 5/8	2.6250	0.219	66.6753	6.66753	0.066675
2 3/4	2.7500	0.229	69.8503	6.98503	0.069850
2 7/8	2.8750	0.240	73.0253	7.30253	0.073025
3	3.0000	0.250	76.2003	7.62003	0.076200
3 1/8	3.1250	0.260	79.3753	7.93753	0.079375
3 1/4	3.2500	0.271	82.5503	8.25503	0.082550
3 3/8	3.3750	0.281	85.7253	8.57253	0.085725
3 1/2	3.5000	0.292	88.9004	8.89004	0.088900
3 5/8	3.6250	0.302	92.0754	9.20754	0.092075
3 3/4	3.7500	0.313	95.2504	9.52504	0.095250
3 7/8	3.8750	0.323	98.4254	9.84254	0.098425
4	4.0000	0.333	101.6004	10.16004	0.101600
4 1/4	4.2500	0.354	107.9504	10.79504	0.107950
4 1/2	4.5000	0.375	114.3005	11.43005	0.114300
4 3/4	4.7500	0.396	120.6505	12.06505	0.120650
5	5.0000	0.417	127.0005	12.70005	0.127001
5 1/4	4.2500	0.354	107.9504	10.79504	0.107950
5 1/2	5.2500	0.438	133.3505	13.33505	0.133351
5 3/4	5.7500	0.479	146.0506	14.60506	0.146051
6	6.0000	0.500	152.4006	15.24006	0.152401
6 1/4	6.2500	0.521	158.7506	15.87506	0.158751
6 1/2	6.5000	0.542	165.1007	16.51007	0.165101
6 3/4	6.7500	0.563	171.4507	17.14507	0.171451
7	7.0000	0.583	177.8007	17.78007	0.177801
7 1/4	7.2500	0.604	184.1507	18.41507	0.184151
7 1/2	7.5000	0.625	190.5008	19.05008	0.190501
7 3/4	7.7500	0.646	196.8508	19.68508	0.196851
8	8.0000	0.667	203.2008	20.32008	0.203201
8 1/4	8.2500	0.688	209.5508	20.95508	0.209551
8 1/2	8.5000	0.708	215.9009	21.59009	0.215901
8 3/4	8.7500	0.729	222.2509	22.22509	0.222251
9	9.0000	0.750	228.6009	22.86009	0.228601
9 1/2	9.5000	0.792	241.3010	24.13010	0.241301
10	10.0000	0.833	254.0010	25.40010	0.254001
10 1/2	10.5000	0.875	266.7011	26.67011	0.266701
11	11.0000	0.917	279.4011	27.94011	0.279401
11 1/2	11.5000	0.958	292.1012	29.21012	0.292101
12	12.0000	1.000	304.8012	30.48012	0.304801
18	18.0000	1.500	457.2018	45.72018	0.457202
24	24.0000	2.000	609.6024	60.96024	0.609602
30	30.0000	2.50	762.0030	76.20030	0.762003
36	36.0000	3.00	914.4036	91.44036	0.914404
42	42.0000	3.50	1066.8042	106.68042	1.066804
48	48.0000	4.00	1219.2048	121.92048	1.219205
54	54.0000	4.50	1371.6054	137.16054	1.371605

Inches	Inches in Decimals	Feet in Decimals	Millimeters	Centimeters	Meters
60	60.0000	5.00	1524.0060	152.40060	1.524006
66	66.0000	5.50	1676.4066	167.64066	1.676407
72	72.0000	6.00	1828.8072	182.88072	1.828807
78	78.0000	6.50	1981.2078	198.12078	1.981208
84	84.0000	7.00	2133.6084	213.36084	2.133608
90	90.0000	7.50	2286.0090	228.60090	2.286009
96	96.0000	8.00	2438.4096	243.84096	2.438410
102	102.0000	8.50	2590.8102	259.08102	2.590810
108	108.0000	9.00	2743.2108	274.32108	2.743211
114	114.0000	9.50	2895.6114	289.56114	2.895611
120	120.0000	10.00	3048.0120	304.80120	3.048012
126	126.0000	10.50	3200.4126	320.04126	3.200413
132	132.0000	11.00	3352.8132	335.28132	3.352813
138	138.0000	11.50	3505.2138	350.52138	3.505214
144	144.0000	12.00	3657.6144	365.76144	3.657614

Conversion Factors
S. I. Metric - English Systems

Multiply	by	to obtain
acres	0.404687	hectares
"	4.04687×10^{-3}	square kilometers
ares	1076.39	square feet
board feet	144 sq in. X 1 in.	cubic inches
" "	0.0833	cubic feet
bushels	0.3521	hectoliters
centimeters	3.28083×10^{-2}	feet
"	0.3937	inches
cubic centimeters	3.53145×10^{-5}	cubic feet
" "	6.102×10^{-2}	cubic inches
cubic feet	2.8317×10^{4}	cubic centimeters
" "	2.8317×10^{-2}	cubic meters
" "	6.22905	gallons, Imperial
" "	0.2832	hectoliters
" "	28.3170	liters
" "	2.38095×10^{-2}	tons, British shipping
" "	0.025	tons, U.S. shipping
cubic inches	16.38716	cubic centimeters

Conversion Factors (Cont'd)

Multiply	by	to obtain
cubic meters	35.3145	cubic feet
" "	1.30794	cubic yards
" "	264.2	gallons, U. S.
cubic yards	0.764559	cubic meters
" "	7.6336	hectoliters
degrees, angular	0.0174533	radians
degrees, F (less 32 F)	0.5556	degrees, C
" C	1.8	degrees, F (less 32 F)
foot pounds	0.13826	kilogram meters
feet	30.4801	centimeters
"	0.304801	meters
"	304.801	millimeters
"	1.64468×10^{-4}	miles, nautical
gallons, Imperial	0.160538	cubic feet
" "	1.20091	gallons, U. S.
" "	4.54596	liters
gallons, U.S	0.832702	gallons, Imperial
" "	0.13368	cubic feet
" "	231.	cubic inches
" "	0.0378	hectoliters
" "	3.78543	liters
grams, metric	2.20462×10^{-3}	pounds, avoirdupois
hectares	2.47104	acres
"	1.076387×10^{5}	square feet
hectares	3.86101×10^{-3}	square miles
hectoliters	3.531	cubic feet
"	2.84	bushels
"	0.131	cubic yards
hectoliters	26.42	gallons
horsepower, metric	0.98632	horsepower, U. S.
horsepower, U.S.	1.01387	horsepower, metric
inches	2.54001	centimeters
"	2.54001×10^{-2}	meters
"	25.4001	millimeters
kilograms	2.20462	pounds
"	9.84206×10^{-4}	long tons
"	1.10231×10^{-3}	short tons
kilogram meters	7.233	foot pounds
kilograms per m	0.671972	pounds per ft
kilograms per sq cm	14.2234	pounds per sq in.
kilograms per sq m	0.204817	pounds per sq ft
" " " "	9.14362×10^{-5}	long tons per sq ft
kilograms per sq mm	1422.34	pounds per sq in.
" " " "	0.634973	long tons per sq. in.
kilograms per cu m	6.24283×10^{-2}	pounds per cu ft
kilometers	0.62137	miles, statute
"	0.53959	miles, nautical
"	3280.7	feet
liters	0.219975	gallons, Imperial
"	0.26417	gallons, U.S.
"	3.53145×10^{-2}	cubic feet
"	61.022	cubic inches
meters	3.28083	feet
"	39.37	inches
"	1.09361	yards

Conversion Factors (Cont'd)

Multiply	by	to obtain
miles, statute	1.60935	kilometers
" "	0.8684	miles, nautical
miles, nautical	6080.204	feet
" "	1.85325	kilometers
" "	1.1516	miles, statute
millimeters	3.28083×10^{-3}	feet
"	3.937×10^{-2}	inches
pounds, avoirdupois	453.592	grams, metric
" "	0.453592	kilograms
" "	4.464×10^{-4}	tons, long
" "	4.53592×10^{-4}	tons, metric
pounds per ft	1.48816	kilograms per m
pounds per sq ft	4.88241	kilograms per sq m
pounds per sq in	7.031×10^{-2}	kilograms per sq cm
" " " "	7.031×10^{-4}	kilograms per sq mm
pounds per cu ft	16.0184	kilograms per cu m
radians	57.29578	degrees, angular
square centimeters	0.1550	square inches
square feet	9.29034×10^{-4}	ares
square feet	9.29034×10^{-6}	hectares
" "	0.0929034	square meters
square inches	6.45163	square centimeters
" "	645.163	square millimeters
square kilometers	247.104	acres
" "	0.3861	square miles
square meters	10.7639	square feet
" "	1.19599	square yards
square miles	259.0	hectares
" "	2.590	square kilometers
square millimeters	1.550×10^{-3}	square inches
square yards	0.83613	square meters
tons, long	1016.05	kilograms
" "	2240.	pounds
" "	1.01605	tons, metric
" "	1.120	tons, short
tons, long, per sq ft	1.09366×10^{-4}	kilograms per sq m
tons, long, per sq in	1.57494	kilograms per sq mm
tons, metric	2204.62	pounds
" "	0.98421	tons, long
" "	1.10231	tons, short
tons, short	907.185	kilograms
" "	0.892857	tons, long
" "	0.907185	tons, metric
tons, British shipping	42.00	cubic feet
" " "	0.952381	tons, U. S. shipping
tons, U. S. shipping	40.00	cubic feet
" " "	1.050	tons, British shipping
yards	0.914402	meters

Functions of Numbers, 1 to 49

No.	Square	Cube	Square Root	Cubic Root	Logarithm
1	1	1	1.0000	1.0000	0.00000
2	4	8	1.4142	1.2599	0.30103
3	9	27	1.7321	1.4422	0.47712
4	16	64	2.0000	1.5874	0.60206
5	25	125	2.2361	1.7100	0.69897
6	36	216	2.4495	1.8171	0.77815
7	49	343	2.6458	1.9129	0.84510
8	64	512	2.8284	2.0000	0.90309
9	81	729	3.0000	2.0801	0.95424
10	100	1000	3.1623	2.1544	1.00000
11	121	1331	3.3166	2.2240	1.04139
12	144	1728	3.4641	2.2894	1.07918
13	169	2197	3.6056	2.3513	1.11394
14	196	2744	3.7417	2.4101	1.14613
15	225	3375	3.8730	2.4662	1.17609
16	256	4096	4.0000	2.5198	1.20412
17	289	4913	4.1231	2.5713	1.23045
18	324	5832	4.2426	2.6207	1.25527
19	361	6859	4.3589	2.6684	1.27875
20	400	8000	4.4721	2.7144	1.30103
21	441	9261	4.5826	2.7589	1.32222
22	484	10648	4.6904	2.8020	1.34242
23	529	12167	4.7958	2.8439	1.36173
24	576	13824	4.8990	2.8845	1.38021
25	625	15625	5.0000	2.9240	1.39794
26	676	17576	5.0990	2.9625	1.41497
27	729	19683	5.1962	3.0000	1.43136
28	784	21952	5.2915	3.0366	1.44716
29	841	24389	5.3852	3.0723	1.46240
30	900	27000	5.4772	3.1072	1.47712
31	961	29791	5.5678	3.1414	1.49136
32	1024	32768	5.6569	3.1748	1.50515
33	1089	35937	5.7446	3.2075	1.51851
34	1156	39304	5.8310	3.2396	1.53148
35	1225	42875	5.9161	3.2711	1.54407
36	1296	46656	6.0000	3.3019	1.55630
37	1369	50653	6.0828	3.3322	1.56820
38	1444	54872	6.1644	3.3620	1.57978
39	1521	59319	6.2450	3.3912	1.59106
40	1600	64000	6.3246	3.4200	1.60206
41	1681	68921	6.4031	3.4482	1.61278
42	1764	74088	6.4807	3.4760	1.62325
43	1849	79507	6.5574	3.5034	1.63347
44	1936	85184	6.6332	3.5303	1.64345
45	2025	91125	6.7082	3.5569	1.65321
46	2116	97336	6.7823	3.5830	1.66276
47	2209	103823	6.8557	3.6088	1.67210
48	2304	110592	6.9282	3.6342	1.68124
49	2401	117649	7.0000	3.6593	1.69020

Functions of Numbers, 50 to 99

No.	Square	Cube	Square Root	Cubic Root	Logarithm
50	2500	125000	7.0711	3.6840	1.69897
51	2601	132651	7.1414	3.7084	1.70757
52	2704	140608	7.2111	3.7325	1.71600
53	2809	148877	7.2801	3.7563	1.72428
54	2916	157464	7.3485	3.7798	1.73239
55	3025	166375	7.4162	3.8030	1.74036
56	3136	175616	7.4833	3.8259	1.74819
57	3249	185193	7.5498	3.8485	1.75587
58	3364	195112	7.6158	3.8709	1.76343
59	3481	205379	7.6811	3.8930	1.77085
60	3600	216000	7.7460	3.9149	1.77815
61	3721	226981	7.8102	3.9365	1.78533
62	3844	238328	7.8740	3.9579	1.79239
63	3969	250047	7.9373	3.9791	1.79934
64	4096	262144	8.0000	4.0000	1.80618
65	4225	274625	8.0623	4.0207	1.81291
66	4356	287496	8.1240	4.0412	1.81954
67	4489	300763	8.1854	4.0615	1.82607
68	4624	314432	8.2462	4.0817	1.83251
69	4761	328509	8.3066	4.1016	1.83885
70	4900	343000	8.3666	4.1213	1.84510
71	5041	357911	8.4261	4.1408	1.85126
72	5184	373248	8.4853	4.1602	1.85733
73	5329	389017	8.5440	4.1793	1.86332
74	5476	405224	8.6023	4.1983	1.86923
75	5625	421875	8.6603	4.2172	1.87506
76	5776	438976	8.7178	4.2358	1.88081
77	5929	456533	8.7750	4.2543	1.88649
78	6084	474552	8.8318	4.2727	1.89209
79	6241	493039	8.8882	4.2908	1.89763
80	6400	512000	8.9443	4.3089	1.90309
81	6561	531441	9.0000	4.3267	1.90849
82	6724	551368	9.0554	4.3445	1.91381
83	6889	571787	9.1104	4.3621	1.91908
84	7056	592704	9.1652	4.3795	1.92428
85	7225	614125	9.2195	4.3968	1.92942
86	7396	636056	9.2736	4.4140	1.93450
87	7569	658503	9.3274	4.4310	1.93952
88	7744	681472	9.3808	4.4480	1.94448
89	7921	704969	9.4340	4.4647	1.94939
90	8100	729000	9.4868	4.4814	1.95424
91	8281	753571	9.5394	4.4979	1.95904
92	8464	778688	9.5917	4.5144	1.96379
93	8649	804357	9.6437	4.5307	1.96848
94	8836	830584	9.6954	4.5468	1.97313
95	9025	857375	9.7468	4.5629	1.97772
96	9216	884736	9.7980	4.5789	1.98227
97	9409	912673	9.8489	4.5947	1.98677
98	9604	941192	9.8995	4.6104	1.99123
99	9801	970299	9.9499	4.6261	1.99564

Functions of Numbers, 100 to 149

No.	Square	Cube	Square Root	Cubic Root	Logarithm
100	10000	1000000	10.0000	4.6416	2.00000
101	10201	1030301	10.0499	4.6570	2.00432
102	10404	1061208	10.0995	4.6723	2.00860
103	10609	1092727	10.1489	4.6875	2.01284
104	10816	1124864	10.1980	4.7027	2.01703
105	11025	1157625	10.2470	4.7177	2.02119
106	11236	1191016	10.2956	4.7326	2.02531
107	11449	1225043	10.3441	4.7475	2.02938
108	11664	1259712	10.3923	4.7622	2.03342
109	11881	1295029	10.4403	4.7769	2.03743
110	12100	1331000	10.4881	4.7914	2.04139
111	12321	1367631	10.5357	4.8059	2.04532
112	12544	1404928	10.5830	4.8203	2.04922
113	12769	1442897	10.6301	4.8346	2.05308
114	12996	1481544	10.6771	4.8488	2.05690
115	13225	1520875	10.7238	4.8629	2.06070
116	13456	1560896	10.7703	4.8770	2.06446
117	13689	1601613	10.8167	4.8910	2.06819
118	13924	1643032	10.8628	4.9049	2.07188
119	14161	1685159	10.9087	4.9187	2.07555
120	14400	1728000	10.9545	4.9324	2.07918
121	14641	1771561	11.0000	4.9461	2.08279
122	14884	1815848	11.0454	4.9597	2.08636
123	15129	1860867	11.0905	4.9732	2.08991
124	15376	1906624	11.1355	4.9866	2.09342
125	15625	1953125	11.1803	5.0000	2.09691
126	15876	2000376	11.2250	5.0133	2.10037
127	16129	2048383	11.2694	5.0265	2.10380
128	16384	2097152	11.3137	5.0397	2.10721
129	16641	2146689	11.3578	5.0528	2.11059
130	16900	2197000	11.4018	5.0658	2.11394
131	17161	2248091	11.4455	5.0788	2.11727
132	17424	2299968	11.4891	5.0916	2.12057
133	17689	2352637	11.5326	5.1045	2.12385
134	17956	2406104	11.5758	5.1172	2.12710
135	18225	2460375	11.6190	5.1299	2.13033
136	18496	2515456	11.6619	5.1426	2.13354
137	18769	2571353	11.7047	5.1551	2.13672
138	19044	2628072	11.7473	5.1676	2.13988
139	19321	2685619	11.7898	5.1801	2.14301
140	19600	2744000	11.8322	5.1925	2.14613
141	19881	2803221	11.8743	5.2048	2.14922
142	20164	2863288	11.9164	5.2171	2.15229
143	20449	2924207	11.9583	5.2293	2.15534
144	20736	2985984	12.0000	5.2415	2.15836
145	21025	3048625	12.0416	5.2536	2.16137
146	21316	3112136	12.0830	5.2656	2.16435
147	21609	3176523	12.1244	5.2776	2.16732
148	21904	3241792	12.1655	5.2896	2.17026
149	22201	3307949	12.2066	5.3015	2.17319

Functions of Numbers, 150 to 199

No.	Square	Cube	Square Root	Cubic Root	Logarithm
150	22500	3375000	12.2474	5.3133	2.17609
151	22801	3442951	12.2882	5.3251	2.17898
152	23104	3511808	12.3288	5.3368	2.18184
153	23409	3581577	12.3693	5.3485	2.18469
154	23716	3652264	12.4097	5.3601	2.18752
155	24025	3723875	12.4499	5.3717	2.19033
156	24336	3796416	12.4900	5.3832	2.19312
157	24649	3869893	12.5300	5.3947	2.19590
158	24964	3944312	12.5698	5.4061	2.19866
159	25281	4019679	12.6095	5.4175	2.20140
160	25600	4096000	12.6491	5.4288	2.20412
161	25921	4173281	12.6886	5.4401	2.20683
162	26244	4251528	12.7279	5.4514	2.20952
163	26569	4330747	12.7671	5.4626	2.21219
164	26896	4410944	12.8062	5.4737	2.21484
165	27225	4492125	12.8452	5.4848	2.21748
166	27556	4574296	12.8841	5.4959	2.22011
167	27889	4657463	12.9228	5.5069	2.22272
168	28224	4741632	12.9615	5.5178	2.22531
169	28561	4826809	13.0000	5.5288	2.22789
170	28900	4913000	13.0384	5.5397	2.23045
171	29241	5000211	13.0767	5.5505	2.23300
172	29584	5088448	13.1149	5.5613	2.23553
173	29929	5177717	13.1529	5.5721	2.23805
174	30276	5268024	13.1909	5.5828	2.24055
175	30625	5359375	13.2288	5.5934	2.24304
176	30976	5451776	13.2665	5.6041	2.24551
177	31329	5545233	13.3041	5.6147	2.24797
178	31684	5639752	13.3417	5.6252	2.25042
179	32041	5735339	13.3791	5.6357	2.25285
180	32400	5832000	13.4164	5.6462	2.25527
181	32761	5929741	13.4536	5.6567	2.25768
182	33124	6028568	13.4907	5.6671	2.26007
183	33489	6128487	13.5277	5.6774	2.26245
184	33856	6229504	13.5647	5.6877	2.26482
185	34225	6331625	13.6015	5.6980	2.26717
186	34596	6434856	13.6382	5.7083	2.26951
187	34969	6539203	13.6748	5.7185	2.27184
188	35344	6644672	13.7113	5.7287	2.27416
189	35721	6751269	13.7477	5.7388	2.27646
190	36100	6859000	13.7840	5.7489	2.27875
191	36481	6967871	13.8203	5.7590	2.28103
192	36864	7077888	13.8564	5.7690	2.28330
193	37249	7189057	13.8924	5.7790	2.28556
194	37636	7301384	13.9284	5.7890	2.28780
195	38025	7414875	13.9642	5.7989	2.29003
196	38416	7529536	14.0000	5.8088	2.29226
197	38809	7645373	14.0357	5.8186	2.29447
198	39204	7762392	14.0712	5.8285	2.29667
199	39601	7880599	14.1067	5.8383	2.29885

Functions of Numbers, 200 to 249

No.	Square	Cube	Square Root	Cubic Root	Logarithm
200	40000	8000000	14.1421	5.8480	2.30103
201	40401	8120601	14.1774	5.8578	2.30320
202	40804	8242408	14.2127	5.8675	2.30535
203	41209	8365427	14.2478	5.8771	2.30750
204	41616	8489664	14.2829	5.8868	2.30963
205	42025	8615125	14.3178	5.8964	2.31175
206	42436	8741816	14.3527	5.9059	2.31387
207	42849	8869743	14.3875	5.9155	2.31597
208	43264	8998912	14.4222	5.9250	2.31806
209	43681	9129329	14.4568	5.9345	2.32015
210	44100	9261000	14.4914	5.9439	2.32222
211	44521	9393931	14.5258	5.9533	2.32428
212	44944	9528128	14.5602	5.9627	2.32634
213	45369	9663597	14.5945	5.9721	2.32838
214	45796	9800344	14.6287	5.9814	2.33041
215	46225	9938375	14.6629	5.9907	2.33244
216	46656	10077696	14.6969	6.0000	2.33445
217	47089	10218313	14.7309	6.0092	2.33646
218	47524	10360232	14.7648	6.0185	2.33846
219	47961	10503459	14.7986	6.0277	2.34044
220	48400	10648000	14.8324	6.0368	2.34242
221	48841	10793861	14.8661	6.0459	2.34439
222	49284	10941048	14.8997	6.0550	2.34635
223	49729	11089567	14.9332	6.0641	2.34830
224	50176	11239424	14.9666	6.0732	2.35025
225	50625	11390625	15.0000	6.0822	2.35218
226	51076	11543176	15.0333	6.0912	2.35411
227	51529	11697083	15.0665	6.1002	2.35603
228	51984	11852352	15.0997	6.1091	2.35793
229	52441	12008989	15.1327	6.1180	2.35984
230	52900	12167000	15.1658	6.1269	2.36173
231	53361	12326391	15.1987	6.1358	2.36361
232	53824	12487168	15.2315	6.1446	2.36549
233	54289	12649337	15.2643	6.1534	2.36736
234	54756	12812904	15.2971	6.1622	2.36922
235	55225	12977875	15.3297	6.1710	2.37107
236	55696	13144256	15.3623	6.1797	2.37291
237	56169	13312053	15.3948	6.1885	2.37475
238	56644	13481272	15.4272	6.1972	2.37658
239	57121	13651919	15.4596	6.2058	2.37840
240	57600	13824000	15.4919	6.2145	2.38021
241	58081	13997521	15.5242	6.2231	2.38202
242	58564	14172488	15.5563	6.2317	2.38382
243	59049	14348907	15.5885	6.2403	2.38561
244	59536	14526784	15.6205	6.2488	2.38739
245	60025	14706125	15.6525	6.2573	2.38917
246	60516	14886936	15.6844	6.2658	2.39094
247	61009	15069223	15.7162	6.2743	2.39270
248	61504	15252992	15.7480	6.2828	2.39445
249	62001	15438249	15.7797	6.2912	2.39620

Functions of Numbers, 250 to 299

No.	Square	Cube	Square Root	Cubic Root	Logarithm
250	62500	15625000	15.8114	6.2996	2.39794
251	63001	15813251	15.8430	6.3080	2.39967
252	63504	16003008	15.8745	6.3164	2.40140
253	64009	16194277	15.9060	6.3247	2.40312
254	64516	16387064	15.9374	6.3330	2.40483
255	65025	16581375	15.9687	6.3413	2.40654
256	65536	16777216	16.0000	6.3496	2.40824
257	66049	16974593	16.0312	6.3579	2.40993
258	66564	17173512	16.0624	6.3661	2.41162
259	67081	17373979	16.0935	6.3743	2.41330
260	67600	17576000	16.1245	6.3825	2.41497
261	68121	17779581	16.1555	6.3907	2.41664
262	68644	17984728	16.1864	6.3988	2.41830
263	69169	18191447	16.2173	6.4070	2.41996
264	69696	18399744	16.2481	6.4151	2.42160
265	70225	18609625	16.2788	6.4232	2.42325
266	70756	18821096	16.3095	6.4312	2.42488
267	71289	19034163	16.3401	6.4393	2.42651
268	71824	19248832	16.3707	6.4473	2.42813
269	72361	19465109	16.4012	6.4553	2.42975
270	72900	19683000	16.4317	6.4633	2.43136
271	73441	19902511	16.4621	6.4713	2.43297
272	73984	20123648	16.4924	6.4792	2.43457
273	74529	20346417	16.5227	6.4872	2.43616
274	75076	20570824	16.5529	6.4951	2.43775
275	75625	20796875	16.5831	6.5030	2.43933
276	76176	21024576	16.6132	6.5108	2.44091
277	76729	21253933	16.6433	6.5187	2.44248
278	77284	21484952	16.6733	6.5265	2.44404
279	77841	21717639	16.7033	6.5343	2.44560
280	78400	21952000	16.7332	6.5421	2.44716
281	78961	22188041	16.7631	6.5499	2.44871
282	79524	22425768	16.7929	6.5577	2.45025
283	80089	22665187	16.8226	6.5654	2.45179
284	80656	22906304	16.8523	6.5731	2.45332
285	81225	23149125	16.8819	6.5808	2.45484
286	81796	23393656	16.9115	6.5885	2.45637
287	82369	23639903	16.9411	6.5962	2.45788
288	82944	23887872	16.9706	6.6039	2.45939
289	83521	24137569	17.0000	6.6115	2.46090
290	84100	24389000	17.0294	6.6191	2.46240
291	84681	24642171	17.0587	6.6267	2.46389
292	85264	24897088	17.0880	6.6343	2.46538
293	85849	25153757	17.1172	6.6419	2.46687
294	86436	25412184	17.1464	6.6494	2.46835
295	87025	25672375	17.1756	6.6569	2.46982
296	87616	25934336	17.2047	6.6644	2.47129
297	88209	26198073	17.2337	6.6719	2.47276
298	88804	26463592	17.2627	6.6794	2.47422
299	89401	26730899	17.2916	6.6869	2.47567

Appendix A

GLOSSARY OF CONSTRUCTION TERMS

This glossary is not intended as a complete and comprehensive dictionary of all the specialized terminology that one will encounter in construction estimating. It is to assist the reader in using this text.

We have drawn on many sources and gratefully acknowledge the Brick Institute of America, the National Concrete Masonry Association, and the Structural Steel Painting Council for permission to reprint entries from their glossaries.

abatement. Procedures to eliminate fiber release from asbestos containing building materials including encapsulation, enclosure or removal methods.

abrasive. A fine graded (sized) granular or spherical material that is used for cleaning, etc.

absorption. The weight of water a masonry unit absorbs when immersed in either cold or boiling water for a stated length of time, expressed as a percentage of the weight of the dry unit.

accelerator. Material such as calcium chloride and compositions predominately of calcium chloride which accelerate hardening and promote early strength development of concrete or mortar.

acceptance testing. Testing of received products to determine that the quality meets required specifications.

additive. Any substance added in small quantities to another substance, usually to improve properties.

admixtures. Materials other than water, aggregates, and hydraulic cement, used as an ingredient of concrete, mortar, or grout, and added to the batch immediately before or during its mixing.

adsorption. Concentration of a substance at a surface or interface of another substance.

aggregate. Granular material such as natural sand, manufactured sand, expanded clay, shale or slate, pumice, volcanic scoria, bituminous or anthracite cinders, gravel, crushed gravel, crushed stone, heavyweight aggregate such as magnetite or ilmenite, and air-cooled or expanded blast-furnace slag, which when bound together into a conglomerate mass by a matrix forms concrete, mortar, or grout.

air lock. A system for hazardous substance removal and abatement that permits ingress and egress from one room to another while permitting minimal air movement between rooms, typically constructed by placing two overlapping sheets of plastic over an existing or temporarily framed doorway. Two curtained doorways spaced a minimum of 6 ft. apart from an air lock.

air monitoring. The process of measuring asbestos fiber content of a specific volume of air in a stated period of time.

alkyd resins. Synthetic resins formed by the condensation of polyhydric alcohols with polybasic acids.

ambient air quality. Average atmospheric purity, as distinguished from discharge measurements taken at the source of pollution.

amended water. Water to which a surfactant is added.

anchor. A piece or assemblage, usually metal, used to attach building parts (i.e., plates, joists, trusses, etc.) to masonry or masonry materials.

anchor pile. A pile connected to a structure by one or more ties to furnish lateral support or resist uplift.

ANSI. American National Standards Institute

arch. A curved compressive structural member, spanning openings or recesses; also built flat.

back arch. A concealed arch carrying the backing of a wall where the exterior facing is carried by a lintel.

jack arch. One having horizontal or nearly horizontal upper and lower surfaces. Also called flat or straight arch.

major arch. Arch with spans greater than 6 ft. and equivalent to uniform loads greater than 1000 lb. per ft. Typically known as Tudor, semicircular, Gothic, or parabolic arch. Has rise to span ratio greater than 0.15.

minor arch. Arch with maximum span of 6 ft. and loads not exceeding 1,000 lb. per ft. Typically known as jack, segmental, or multicentered arch. Has rise to span ratio less than or equal to 0.15.

relieving arch. One built over a lintel, flat arch, or smaller arch to divert loads, thus relieving the lower member from excessive loading. Also known as discharge or safety arch.

trimmer arch. An arch, usually a low rise arch or brick, used for supporting a fireplace hearth.

ashlar masonry. Masonry composed of rectangular units of clay or shale, or stone, generally larger in size than brick and properly bonded, having sawed, dressed or squared beds, and joints laid in mortar. Often the unit size varies to provide a random pattern, random ashlar.

asphalt. Solid or semi-solid mixture of bitumens that are found in natural deposits or created as a petroleum by-product and used mainly for paving and roofing.

natural asphalt (or lake asphalt). Derived from petroleum by the natural process of evaporation leaving the asphalt material in open beds or lakes ready for extraction and processing.

petroleum asphalt: Asphalt refined from crude petroleum.

asphalt cement (AC). Refined asphalt for paving purposes.

asphalt joint filler. Liquid asphalt product used to seal cracks and joints in pavement to prevent moisture penetration.

asphalt joint filler, preformed. Asphalt mixed with various fibrous materials and manufactured in solid form to a specific thickness. Widely used as an expansion joint material.

asphalt primer. Liquid asphalt applied to a non-bituminous surface to promote adhesion between existing surface and new surface application.

ASTM. American Society for Testing and Materials

back filling. 1. Replacing of fill materials in an excavated area 2. Rough masonry built behind a facing or between two faces. 3. Filling over the extrados of an arch. 4. Brickwork in spaces between structural timbers, sometimes called brick nogging.

backup. That part of a masonry wall behind the exterior facing.

bat. A piece of brick.

batter. Recessing or sloping masonry back in successive courses; the opposite of corbel.

bed joint. The horizontal layer of mortar on or in which a masonry unit is laid; may cover entire masonry unit or face shell only.

belt course. A narrow horizontal course of masonry, sometimes slightly projected such as window sills, that is made continuous. Sometimes called string or sill course.

bevel. One side of a solid body which is inclined with respect to the other, with the angle between the two sides either greater or less than a right angle.

blast cleaning. Cleaning or roughing of a surface using metallic or non-metallic grit or shot projected with compressed air, centrifugal force, or water.

block, concrete. A hollow or solid unit consisting of portland cement and suitable aggregates combined with water. Other materials such as lime, fly ash, air-entraining agents, or other admixtures may be permitted.

blocking. A method of bonding two adjoining or intersecting walls, not built at the same time, by means of offsets whose vertical dimensions are not less than 8 in.

blown asphalt. Asphalt specially treated at high temperature levels with blown air to obtain specific characteristics necessary in products for roofing, pipe coating, membrane envelopes, and hydraulic application.

bond. 1. Tying various parts of a masonry wall by lapping units one over another or by connecting with metal ties. 2. Patterns formed by exposed faces of units. 3. Adhesion between mortar or grout and masonry units or reinforcement.

bond, grout. The adhesion to and/or interlocking of grout with the masonry units and reinforcement.

bond, mechanical. Tying masonry units together with metal ties or reinforcing steel or keys.

bond, mortar. The adhesion of mortar to masonry units and reinforcement.

bond beam. Course or courses of a masonry wall grouted and usually reinforced in the horizontal direction. Serves as horizontal tie of wall, bearing course for structural members or as a flexural member itself.

bond breaker. A material used to prevent adhesion between two surfaces.

bond course. The course consisting of units which overlap more than one wythe of masonry.

bond strength. Resistance to separation of mortar from concrete masonry units and of mortar and grout from reinforcing steel and other materials with which it is in contact.

bonder. A bonding unit. See Header.

bracing. Any type of horizontal or inclined structural member used to increase stability and to resist lateral forces.

breaking joints. Any arrangement of masonry units that prevents continuous vertical joints from occurring in adjacent courses.

brick. A solid masonry unit of clay or shale, formed into a rectangular prism while plastic, and burned or fired in a kiln.

acid-resistant brick. Brick suitable for use in contact with chemicals, usually in conjunction with acid-resistant mortars.

adobe brick. Large roughly-molded, sun-dried clay brick of varying size.

angle brick. Any brick shaped to an oblique angle to fit a salient corner.

arch brick. 1. Wedge-shaped brick for special use in an arch. 2. Extremely hard-burned brick from an arch of a scove kiln.

building brick. Brick for building purposes not especially treated for texture or color. Formerly called common brick.

clinker brick. A very hard-burned brick whose shape is distorted or bloated due to nearly complete vitrification.

common brick. See Building brick.

dry-press brick. Brick formed in molds under high pressure from relatively dry clay (5% to 7% moisture content).

economy brick. Brick whose nominal dimensions are 4"x4"x8".

engineered brick. Brick whose nominal dimensions are 4"x3.2"x8".

facing brick. Brick made especially for facing purposes, often treated to produce surface texture. They are made of selected clay, or treated, to produce desired color.

fire brick. Brick made of refractory ceramic material that resists high temperatures.

floor brick. Smooth dense brick, highly resistant to abrasion, used as finished floor surface.

gauged brick. 1. Brick that have been ground or otherwise produced to accurate dimensions. 2. A tapered arch brick.

hollow brick. A masonry unit of clay or shale whose net cross-sectional area in any plane parallel to the bearing surface is not less than 60% of its gross cross-sectional area measured in the same plane.

jumbo brick. A generic term indicating a brick larger in size than the standard. Some producers use this term to describe oversize brick of specific dimensions manufactured by them.

Norman brick. A brick whose nominal dimensions are 4"x2⅔"x12".

paving brick. Vitrified brick especially suitable for use in pavements where resistance to abrasion is important.

Roman brick. Brick whose nominal dimensions are 4"x2"x12".

salmon brick. Generic term for underburned brick that are more porous, slightly larger, and lighter colored than hard-burned brick. Usually pinkish-orange color.

SCR brick. See SCR.

sewer brick. Low absorption, abrasive-resistant brick intended for use in drainage structures.

soft-mud brick. Brick produced by molding relatively wet clay (20% to 30% moisture). Often a hand process. When insides of molds are sanded to prevent sticking of clay, the product is sand-struck brick. When molds are wetted to prevent sticking, the product is water-struck brick.

stiff-mud brick. Brick produced by extruding a stiff but plastic clay (12% to 15% moisture) through a die.

brick and brick. A method of laying brick so that units touch each other with only enough mortar to fill surface irregularities.

brick, concrete. A solid unit having a rectangular prismatic shape, usually not larger than 4x4x12 inches, made from portland cement and suitable aggregates, with or without the inclusion of other materials.

brick grade. Designation for durability of the unit expressed as SW for severe weathering, MW for moderate weathering, or NW for negligible weathering. See ASTM Specifications C 216, C 62, and C 652.

brick type. Designation for facing brick that controls tolerance, chippage, and distortion. Expressed as FBS, FBX, and FBA for solid brick, and as HBS, HBX, HBA, and HBB for hollow brick. See ASTM Specifications C 216 and C 652.

bullnose block: A unit having one or more rounded exterior corners.

buttress. A masonry pilaster decreasing in area from the base to the top, generally used to give greater lateral strength and stability to a wall.

buttering. Placing mortar on a masonry unit with a trowel.

caisson. Also caisson pile, a large-diameter shaft, hand or machine excavated to bearing stratum inside a protective casing.

capacity insulation. The ability of masonry to store heat as a result of its mass, density and specific heat.

cathodic protection. A technique to prevent corrosion of a metal surface by making that surface the cathode of an electrochemical cell.

cavity wall. A wall built of two or more wythes of masonry units separated by a continuous air space (with or without insulating materials) and in which the wythes are securely tied together with rigid corrosion resistant metal ties.

c/b ratio. The ratio of the weight of water absorbed by a masonry unit during immersion in cold water to weight absorbed during immersion in boiling water. An indication of the probable resistance of brick to freezing and thawing. Also called saturation coefficient.

cement paint. Supplied in dry powder form, it is based on portland cement to which pigments are sometimes added for decorative purposes.

centering. Temporary formwork for the support of masonry arches or lintels during construction. Also called center(s).

ceramic color glaze. An opaque colored glaze of satin or gloss finish obtained by spraying the clay body with a compound of metallic oxide, chemicals and clays. It is burned at high temperatures, fusing glaze to body and making them inseparable.

chase: A continuous recess built into a wall to receive pipes, ducts, etc.

class of unit. Distinguishes between masonry units of different grade or type in ASTM specifications, being manufactured from different raw materials, or having different specified compressive strengths.

clay. A natural, mineral aggregate consisting essentially of hydrous aluminum silicate; it is plastic when sufficiently wetted, rigid when dried and vitrified when fired to a sufficiently high temperature.

clay mortar-mix. Finely ground clay used as a plasticizer for masonry mortars.

cleanout. An opening in the first course of masonry for cleaning mortar droppings prior to grout placement in grouted masonry. Required in high lift grouting.

clean room. An uncontaminated area or room that is part of the worker decontamination enclosure system in asbestos abatement and removal, with provisions for storage of street clothes and protective equipment.

clear ceramic glaze. Same as Ceramic Color Glaze except that it is translucent or slightly tinted with a gloss finish.

clip. A portion of a brick cut to length.

closer. The last masonry unit laid in a course. It may be whole or a portion of a unit.

closure. Supplementary or short length units used at corners or jambs to maintain bond patterns.

collar joint. The vertical, longitudinal joint between wythes of masonry.

column. 1. A vertical member whose horizontal dimensions measured at right angles to the thickness does not exceed three times its thickness. 2. (in concrete masonry) a compression member, vertical or nearly vertical, the width of which does not exceed four times its thickness and the height of which exceeds four times its least lateral dimension.

composite wall. A multiple wythe masonry wall in which at least one of the wythes is dissimilar to the other wythe with respect to type or grade of unit or mortar.

compressive strength. The maximum compressive load in lbs. that a unit will support divided by the gross cross-sectional area of the unit in square inches.

concrete masonry unit, hollow. A unit whose net cross-sectional area in any plane parallel to the bearing surface is less than 75% of its gross cross-sectional area measured in the same plane.

concrete masonry unit, solid: A unit whose net cross-sectional area in every plane parallel to the bearing surface is 75% or more of its gross cross-sectional area measured in the same plane.

control joint. A continuous unbonded masonry joint to regulate the location and amount of separation resulting from the dimensional change of different parts of a structure so as to avoid the development of excessively high stresses.

coping: The material or masonry units forming a cap or finish on top of a wall, pier, pilaster, chimney, etc. It protects masonry below from penetration of water from above.

corbel: A shelf of ledge formed by projecting successive courses of masonry out from the face of the wall, pier, or column.

core: The molded open space in a masonry concrete unit.

corrosion. The deterioration of a material by direct or electrochemical reaction with its environment.

course: One of the continuous horizontal layers of units bonded with mortar in masonry.

cross bracing. Any type of bracing that is constructed at right angles to the main or longitudinal axis.

cross-sectional area, gross. In masonry, the total area of a section perpendicular to the direction of the load, including areas within the cells and within reentrant spaces unless these spaces are to be occupied in the masonry by portions of adjacent masonry. (The gross cross-sectional area of scored units is determined to the outside of the scoring.)

cross-sectional area, net. In masonry, the gross cross-sectional area of a section minus the average area of ungrouted cores or cellular spaces. (The cross-sectional area of grooves in scored units is not deducted from the gross cross-sectional area to obtain the net cross-sectional area.)

culls. Masonry units that do not meet the standards or specifications and have been rejected.

curing. To preserve or finish by chemical or physical process.

curing agent. An additive that promotes curing.

curtain wall. A non-loadbearing, usually prefabricated, wall between columns and piers.

curtained doorway. A device used in asbestos removal to allow ingress or egress from one room to another while permitting minimal air movement between the rooms, typically constructed by placing two overlapping sheets of plastic over an existing or temporarily framed doorway. Two curtained doorways spaced a minimum of 6 ft. apart from an air lock.

customized masonry. Architectural masonry units having textured or sculptured surfaces. Methods used to obtain different surface textures including splitting, grinding, forming vertical striations and causing the units to "slump". Sculptured faces are obtained by forming projecting ribs or flutes, either rounded or angular, as well as vertical and horizontal scoring, recesses and curved faces.

cutback asphalt. Asphalt cement blended with petroleum solvents to produce a more liquid or fluid product thus improving handling and workability qualities. Exposure to atmosphere evaporates the solvents and the remaining asphalt performs the intended function. Variation in the types of solvent used produces the following products: a) Rapid Curing Asphalt (RC), b) Medium Curing Asphalt (MC), and c) Slow Curing Asphalt (SC).

dampcheck. An impervious horizontal layer to prevent vertical penetration of water in a wall consisting of either a course of solid masonry, metal or a thin layer of asphaltic or bituminous material. Generally near grade to prevent upward migration of moisture by capillary action. See damp course below.

damp course. A masonry course or layer of impervious material that prevents capillary entrance of moisture from the ground or a lower course. Same as damp check.

dampproofing. Prevention of moisture penetration by capillary action, usually by addition of one or more coats of a compound that is impervious to water.

decontamination enclosure system. In asbestos abatement and removal, a series of areas or connected rooms, with curtained doorways between them, for decontamination of workers or materials and equipment. The system always contains a minimum of one airlock between two curtained doorways.

direct costs. Costs that are charged directly to a project, such as for labor, taxes, insurance, materials, scaffolding, equipment, and inspection.

disposal. Procedures necessary to transport and deposit hazardous waste containing material removed from a building or site in compliance with regulations.

dog's tooth. Brick laid with their corners projecting from the wall face.

double-walled tank. An underground storage tank in which a rigid secondary container is attached to the primary container and which has an annular space.

dovetail anchor. A splayed tenow shaped like a dove's tail, broader at its end than at its base, which fits into the recess of a corresponding mortise.

dowel. Straight metal bars used to connect two sections of masonry.

drilled pile. Also called an augured pile, it is a cast-in-place concrete pile i an augured hole, possibly belled at the bottom.

drip. A projecting piece of material, shaped to throw off water and prevent it running down the face of the wall or other surface.

dry stack. Masonry work laid without mortar.

EBM. See Engineered brick masonry.

eccentricity. The normal distance between the centroidal axis of a membe and the parallel resultant load.

e1/e2. Ratio of virtual eccentricities occurring at the ends of a column or wa under design. The absolute value is always less than or equal to 1.0.

effective area of reinforcement. The area obtained by multiplying th right cross-sectional area of metal reinforcement by the cosine of the angl between its direction and the direction for which the effectiveness of th reinforcement is to be determined.

effective height. The height of a member to be assumed for calculating th slenderness ratio.

effective thickness. The thickness of a member to be assumed for calculat ing the slenderness ratio.

efflorescence. A powder or stain sometimes found on the surface of masonry resulting from deposition of water-soluble salts.

emission standard. The maximum amount of a pollutant that is permitte to be discharged from a single source.

encapsulation. Procedures necessary to coat all spray- or trowel-applie asbestos containing materials with an encapsulant to control the release o asbestos fibers into the ambient air.

empirical design. Designed based on applying physical limitations based o experience or observations gained through experience without a structura analysis.

engineered brick masonry. Masonry in which design is based on a rationa structural analysis.

engineered design. Design based on a rational analysis considering th interrelationships of the various construction materials, their properties an actual design loads in lieu of empirical design procedures.

EPA. Environmental Protection Agency.

epoxy. Thermosetting resin that is mixed with a curing agent to produce adhesives, coatings, etc.

equipment decontamination system. A decontamination enclosure system for materials and equipment, typically consisting of designated area of the work area, a wash room and an uncontaminated area.

equipment decontamination enclosure system. A decontamination enclosure system in asbestos work for materials and equipment, typically consisting of designated area of the work area, a wash room and an uncontaminated area.

expansion joint. A separation between adjoining parts of a structure that is provided to allow small relative movements, such as those caused by thermal changes, to occur without one part affecting an adjacent part.

face. 1. The exposed surface of a wall or masonry unit. 2. The surface of a unit designed to be exposed in the finished structure.

face shell. The side wall of a hollow concrete masonry unit, generally between 3/4" and 1 1/2" thick.

face shell bedding. Mortars is applied only to the horizontal face of the face shells of hollow masonry units and in the head joints to a depth equal to the thickness of the face shell.

facing. Any material, forming a part of a wall, used as a finished surface.

field. The expanse of wall between openings, corners, etc., principally composed of stretchers.

filter block. A hollow, vitrified clay masonry unit, sometimes salt-glazed, designed for trickling filter floors in sewage disposal plants.

fire clay. A clay that is highly resistant to heat without deforming and used for making brick.

fire-resistive material. See non-combustive material.

fireproofing. Any material or combination protecting structural members to increase their fire resistance.

fire wall. Any wall that subdivides a building so as to resist the spread of fire, by starting at the foundation and extending continuously through all stories to, or above, the roof.

flashing. 1. A thin impervious material placed in mortar joints and through air spaces in masonry to prevent water penetration and/or provide water drainage. 2. Manufacturing methods to produce specific color tones.

flux oil. A thick petroleum product used to soften asphalt to a desired consistency.

frog. A depression in the bed surface of a brick. Sometimes called a panel.

foundation wall. A wall below the floor nearest grade serving as a suppo for a wall, pier, column, or other structural part of a building.

furring. Strips of wood or metal used to provide a level surface for finishi or to create an air space, or the installation of such material.

furrowing. The practice of striking a v-shaped trough in a bed of mortar (n recommended).

galvanized. Coating of zinc applied to steel in order to protect it fro corrosion.

glove bag system. A portable asbestos abatement system designed f isolated and small areas of pipe, fittings, etc. requiring asbestos removal.

grid pavers. Open type masonry units that allow the growing of grass whe employed for soil stabilization in parking areas, along the shoulders of hig ways and airport runways, embankment erosion control, fire engine lin while providing a base to support vehicular traffic.

grounds. Nailing strips placed in masonry walls as a means of attaching tri or furring.

grout. Mixture of cementitious material and aggregate to which sufficie water is added to produce pouring consistency without segregation of th constituents.

high-lift grouting. The technique of grouting masonry in lifts up to 12 ft.

low-lift grouting. The technique of grouting as the wall is constructed.

grout lift. The height to which grout is placed in a cell, collar joint, or cavit without intermission.

grout pour. The total grouted height between masonry lifts. A grout pou may consist of one or more grout lifts.

grouted masonry. Concrete masonry construction composed of hollow unit where hollow cells are filled with grout, or multi-wythe construction in whic space between wythes is solidly filled with grout.

hacking. 1. The procedure of stacking brick in a kiln or on a kiln car. 2. Layin brick with the bottom edge set in from the plane surface of the wall.

hand cleaning. Any surface preparation using hand tools such as wir brushes, scrapers, and chipping tools.

hard-burned. Nearly vitrified clay products that have been fired at hig temperatures. They have relatively low absorptions and high compressiv strengths.

head joint. The vertical mortar joint between ends of masonry units. Often called cross joint.

header. 1. Masonry unit that overlaps two or more adjacent wythes of masonry to tie them together. Often called bonder. 2. A floor or roof beam placed between joists or beams to support the ends.

blind header. A concealed brick header in the interior of the wall, not showing on the face.

clipped header. A bat placed to look like a header for purposes of establishing a pattern. Also called false header.

flare header. A header of darker color than the field of the wall.

heading course. A continuous bonding course of header brick. Also called header course.

HEPA filter. A high efficiency particulate absolute (HEPA) filter capable of trapping and retaining 99.9% of asbestos fibers greater than 0.3 microns in length. These filters are used in vacuum cleaners or in air transfer units (exhaust fans).

HEPA vacuum equipment. High efficiency particulate absolute filtered vacuuming equipment with a filter system capable of collecting and retaining asbestos fibers. Filters should be of a 99.9% efficiency for retaining fibers 0.3 microns (3⁄1000 of a millimeter) or larger.

hot-rolled steel. Steel that is hot reduced; that is, it is formed and shaped when it is hot.

hot specs. New and modified federal specifications that are issued every 15 days on microfilm.

indirect costs. Costs that are not directly attributable to just one job, such as office overhead, cost of capital, equipment depreciation.

initial rate of absorption. The weight of water absorbed expressed in grams per 30 sq. in. of contact surface when a brick is partially immersed for one minute. Also called suction. See ASTM Specification C 67.

interlocking block paver. Solid masonry units capable of transferring loads and stresses laterally by arching or bridging action between units when subjected to vehicular traffic.

IRA. See initial rate of absorption.

joint reinforcement. Steel wires placed in the mortar joint (over the face shell in hollow masonry) and having cross wires welded between them at regular intervals.

kiln. A furnace oven or heated enclosure used for burning or firing brick or other clay material.

kiln run. Brick from one kiln that have not been sorted or graded for size or color variation.

king closer. A brick cut diagonally to have one 2-in. end and one full-width end.

laitance. A milky white deposit on new concrete.

lateral support. Means whereby walls are braced either vertically or horizontally by columns, pilasters, cross walls, beams, floors, roofs, etc.

latex paint. A paint containing a stable aqueous dispersion of synthetic resin, produced by emulsion polymerization, as the principal constituent of the binder.

lead[1]. The section of a masonry wall built up and racked back on successive courses. A line is attached to leads as a guide for constructing a wall between them.

lead[2]. A heavy metal that can be hazardous to health if breathed or swallowed.

lime, hydrated. Quicklime to which sufficient water has been added to convert the oxides to hydroxides.

lime putty. Hydrated lime in plastic form ready for addition to mortar.

lintel. A beam placed over an opening in a wall to carry the superimposed weight of the construction and loads above the opening.

loadbearing. A structural system or element designed to carry loads in addition to its own dead weight.

masonry. Brick, stone, concrete, etc. or masonry combinations thereof, bonded with mortar.

masonry cement. 1. A mill-mixed cementitious material to which sand and water must be added. See ASTM C 91. 2. Hydraulic cement produced for use in mortars for masonry construction where greater plasticity and water retention are desired than is obtainable by the use of portland cement alone. Such cements always contain one or more of the following materials: portland cement, portland-pozzolan cement, natural cement, slag cement, hydraulic lime. They usual contain one or more of the following: hydrated lime, pulverized limestone, chalk, talc, pozzolan, clay or gypsum; many masonry cements also include air-entraining and water-repellent additions.

masonry unit. Natural or manufactured building units of burned clay, concrete, stone, glass, gypsum, etc.

hollow masonry unit. One whose net cross-sectional area in any plane parallel to the bearing surface is less than 75% of the gross.

modular masonry unit. One whose nominal dimensions are based on the 4-in. module.

solid masonry unit. One whose net cross-sectional area in every plane parallel to the bearing surface is 75% or more of the gross.

moisture content. The amount of water contained at the time of sampling expressed as a percentage of the total absorption.

mortar. A plastic mixture of cementitious materials, fine aggregate, and water. See ASTM Specifications C 270, C 476 or BIA M1-72.

fat mortar. Mortar containing a high percentage of cementitious components; it is a sticky mortar which adheres to a trowel.

high-bond mortar. Mortar that develops higher bond strengths with masonry units than normally developed with conventional mortar.

lean mortar. Mortar that is deficient in cementitious components; it is usually harsh and difficult to spread.

net section. The minimum cross-section of the member under consideration; in masonry, usually the mortar bedded area plus the grouted area.

nominal dimension. A dimension greater than a specified masonry dimension by the thickness of a mortar joint, but not more than ½ in.

non-combustible material. Any material that will neither ignite nor actively support combustion in air at a temperature of 1200°F when exposed to fire.

oil paint. A paint that contains drying oil, oil varnish, or oil-modified resin as the basic vehicle ingredient. Not that the common but technically incorrect definition is any paint soluble in organic solvents.

overhand work. Laying brick from inside a wall by workers standing on a floor or on a scaffold.

pargeting. The process of applying a coat of cement mortar to masonry. Also parging.

particulates. Fine liquid or solid particles, such as dust, smoke, mist, fumes, or smog, found in the air or in emissions.

panel wall. A non-loadbearing wall constructed between columns or piers and wholly supported at each story.

parapet wall. That part of a wall that extends above the roof level.

partition. An interior wall, one story or less in height.

party wall. A wall on an interior lot line, or any wall used to adapt for joint service between two buildings or adjacent living or work spaces.

phenolic resin. Resin made by the condensation of phenols and aldehydes.

pick and dip. A method of laying brick whereby the bricklayer simultaneously picks up a brick with one hand, and with the other hand, enough mortar on a trowel to lay the brick. Sometimes called the eastern or New England method.

pier. Any of various vertical supporting structures; an isolated column of masonry.

pilaster. A wall portion projecting from either or both wall faces and serving as a vertical column or beam.

plasticizer. A substance added to paint, varnish, or plaster to give it flexibility.

plumb rule. This is a combination plumb rule and level. It is used in a horizontal position as a level and in a vertical position as a plumb rule. They are made in lengths of 42" and 48" and short lengths from 12" to 24".

pointing. Troweling mortar into a joint after masonry units are laid.

prefabricated brick masonry. Masonry construction fabricated in a location other than its final inservice location in the structure. Also known as preassembled, panelized, and sectional brick masonry.

prism. 1. Any of a number of polyhedron shapes 2. A small masonry assemblage made with masonry units and mortar. Primarily used to predict the strength of full scale masonry members.

queen closer. A cut brick having a nominal 2" horizontal face dimension.

quoin. A projecting right angle masonry corner.

racking. A method entailing stepping back successive courses of masonry.

raggle. A groove in a joint or special unit to receive roofing or flashing.

RBM. Reinforced brick masonry.

reinforced masonry. Masonry units, reinforcing steel, grout, or mortar combined to act together in resisting forces.

return. Any surface turned back from the face of a principal surface.

reveal. That portion of a jamb or recess that is visible from the face of a wall.

rowlock. A brick laid on its face edge so that the normal bedding area is visible in the wall face. Frequently spelled rolok.

salt glaze. A gloss finish obtained by thermochemical reaction between silicates of clay and vapors of salt or chemicals.

saturation coefficient: See c/b ratio.

SCR. Structural Clay Research (trademark of the Structural Clay Products Institute, BIA).

SCR acoustile. A side-construction, two-celled facing tile, having a perforated face backed with glass wool for acoustical purposes.

SCR brick. Brick whose nominal dimensions are 6"x2⅔"x12".

SCR building panel. Prefabricated, structural ceramic panels, approximately 2½" thick.

SCR insulated cavity wall. Any cavity wall containing insulation that meets rigid criteria established by the Structural Clay Products Institute (BIA).

SCR masonry process. A construction aid providing greater efficiency, better workmanship, and increased production in masonry construction. It utilizes story poles, marked lines and adjustable scaffolding.

screen block. Open-faced masonry units used for decorative purposes or to partially screen areas from the sun or outside viewers.

sealant. 1. Any type of sealing agent. 2. A material spray applied to the substrate after the wet removal of existing asbestos material. The purpose of this is to prevent the release of residual asbestos fibers left after removal and cleaning operations.

shale. Clay that has been subjected to high pressures until it has hardened.

shear wall. A wall which, in its own plane, carries shear resulting from wind, blast, or seismic forces.

shower room. A room constituting an air lock between the clean room and the equipment room in the worker decontamination enclosure system, with hot and cold or warm running water suitably arranged for complete showering of workers during decontamination. Shower room always includes an air lock.

shoved joints. Vertical joints filled by shoving a brick against the next brick when it is being laid in a bed of mortar.

shrinkage. Volume change due to loss of moisture or decrease in temperature.

single wythe wall. A wall of only one masonry unit in thickness.

slenderness ratio. Ratio of the effective height of a member to its effective thickness.

slushed joints. Vertical joints filled, after units are laid, by "throwing" mortar in with the edge of a trowel (generally not recommended).

soap. A masonry unit of normal face dimensions, having a nominal 2" thickness.

soffit. The underside of a beam, lintel, or arch.

soft-burned. Clay products that have been fired at low temperature ranges, producing relatively high absorptions and low compressive strengths.

solar screen. A perforated wall used as a sunshade.

soldier. A stretcher set on end with face showing on the wall surface.

spall. A small fragment removed from the face of a masonry unit by a blow or by action of the elements.

stack. Any structure or part thereof that contains a flue or flues for the discharge of gases.

stacked bond. A bonding pattern where no unit overlaps either the one above or below, all head joints form a continuous vertical line.

story pole. A marked pole for measuring masonry coursing during construction.

stretcher. A masonry unit laid with its greatest dimension horizontal and its face parallel to the wall face.

stringing mortar. The procedure of spreading enough mortar on a bed to lay several masonry units.

struck joint. Any mortar joint that has been finished with a trowel.

suction. See Initial rate of absorption.

surfactant. In asbestos work, a chemical wetting agent added to water to improve penetration into the asbestos thus reducing the amount of water required for a given operation or area. This surfactant is spray applied to the asbestos to decrease the amount of airborne fibers while removing the asbestos material from the substrate. (Wetting agent-50% polyglycol ether and 50% polyoxyethlene ether or approved equal shall be mixed with water to provide a concentration of 1 oz. surfactant to 5 gallons of water or the manufacturer's recommended concentration.

temper. To moisten and mix clay, plaster, or mortar to a proper consistency.

tie. Any unit of material that connects masonry to masonry or other materials. See wall tie.

tooling. Compressing and shaping the face of a mortar joint with a special tool other than a trowel.

toothing. Constructing the temporary end of a masonry wall with the end stretcher of every alternate course projecting. Projecting units are toothers.

traditional masonry. Masonry in which design is based on empirical rules that control minimum thickness, lateral support requirements, and height without a structural analysis.

transformed section. An assumed section of one material having the same elastic properties as the section of two materials.

tuck pointing. The filling in with fresh mortar of cut-out or defective mortar joints in masonry.

veneer: 1. A thin layer of material that is bonded to and serves as the surface for another material. 2. A single wythe of masonry for facing purposes, not structurally bonded.

virtual eccentricity: The eccentricity of a resultant axial load required to produce axial and bending stresses equivalent to those produced by applied axial loads and moments. It is normally found by dividing the moment at a section by the summation of axial loads occurring at that section.

vitrification. The condition resulting when kiln temperatures are sufficient to fuse grains and close pores of a clay product, making the mass impervious.

wall. A vertical member of a structure whose horizontal dimension measured at right angles to the thickness exceeds three times its thickness.

apron wall. That part of a panel wall between window sill and wall support.

area wall. 1. The masonry surrounding or partly surrounding an area. 2. The retaining wall around basement windows below grade.

bearing wall. One that supports a vertical load in addition to its own weight.

cavity wall. A wall built of masonry units so arranged as to provide a continuous air space within the wall (with or without insulating material), and in which the inner and outer wythes of the wall are tied together with metal ties.

composite wall. A multiple-wythe wall in which at least one of the wythes is dissimilar to the other wythe or wythes with respect to type or grade of masonry unit or mortar.

curtain wall. An exterior non-loadbearing wall not wholly supported at each story. Such walls may be anchored to columns, spandrel beams, floors, or bearing walls, but not necessarily built between structural elements.

dwarf wall. A wall or partition that does not extend to the ceiling.

enclosure wall. An exterior non-loadbearing wall in skeleton frame construction. It is anchored to columns, piers, or floors but not necessarily built between columns or piers nor wholly supported at each story.

exterior wall. Any outside wall or vertical enclosure of a building other than a party wall.

faced wall. A composite wall in which the masonry facing and backing are so bonded as to exert a common reaction under load.

fire division wall. Any wall that subdivides a building so as to resist the spread of fire. It is not necessarily continuous through all stories to and above the roof.

fire wall. Any wall that subdivides a building to resist the spread of fire and that extends continuously from the foundation through the roof.

foundation wall. That portion of a loadbearing wall below the level of the adjacent grade, or below first floor beams or joists.

hollow wall. A wall built of masonry units arranged to provide an air space within the wall. The separated facing and backing are bonded together with masonry units.

insulated cavity wall. See SCR insulated cavity wall

loadbearing wall. A wall that supports any vertical load in addition to its own weight.

non-loadbearing wall. A wall that supports no vertical load other than its own weight.

panel wall. An exterior, non-loadbearing wall wholly supported at each story.

parapet wall. That part of any wall entirely above the roof line.

party wall. A wall used for joint service by adjoining buildings.

perforated wall. One that contains a considerable number of relatively small openings. Often called pierced wall or screen wall.

shear wall. A wall that resists horizontal forces applied in the plane of the wall.

single wythe wall. A wall containing only one masonry unit in wall thickness.

solid masonry wall. A wall built of solid masonry units, laid contiguously, with joints between units completely filled with mortar or grout.

spandrel wall. That part of a curtain wall above the top of a window in one story and below the sill of the window in the story above.

veneered wall. A wall having a facing of masonry units or other weather-resisting non-combustible materials securely attached to the backing, but not so bonded as to intentionally exert common action under load.

wall plate. A horizontal member anchored to a masonry wall to which other structural elements may be attached. Also called head plate.

wall tie. A bonder or metal piece that connects wythes of masonry to each other or to other materials.

wall tie, cavity. A rigid, corrosion-resistant metal tie that bonds two wythes of a cavity wall. It is usually steel, $3/16$" in a diameter and formed in a "Z" shape or a rectangle.

wall tie, veneer. A strip or piece of metal used to tie a facing veneer to the backing.

water retentivity. That property of a material that prevents the rapid loss of water.

water table. A projection of lower masonry on the outside of the wall slightly above the ground. Often a damp course is placed at the level of the water table to prevent upward penetration of ground water.

waterproofing. Prevention of moisture flow through masonry due to water pressure.

web. The cross wall connecting the face shells of a hollow concrete masonry unit.

weep holes. Openings placed in mortar joints of facing material at the level of flashing to permit the escape of moisture.

wet cleaning. The process of eliminating asbestos contamination from building surfaces and objects by using cloths, mops, or other cleaning tools that have been dampened with water and by afterwards disposing of these cleaning tools as asbestos contaminated waste.

with inspection. Masonry designed with the higher stresses allowed under EBM. Requires the establishment of procedures on the job to control mortar mix, workmanship, and protection of masonry materials.

without inspection. Masonry designed with the reduced stresses allowed under EBM.

worker decontamination enclosure system. A decontamination enclosure system for workers typically consisting of a clean room, a locker room, a shower room, and an equipment room.

wythe. 1. Each continuous vertical section of masonry one unit in thickness. 2. The thickness of masonry separating flues in a chimney. Also called withe or tier.

Appendix **B**

CONSTRUCTION SAFETY

Construction is the largest industry in the U.S. and one of the most dangerous. The issues of accident prevention and safety in construction are in the limelight, partly due to the number of avoidable disasters in recent years.

The most important reason for developing and carefully monitoring a safety program is that employees' lives depend on it, but there are economic advantages as well. Healthy workers are productive ones, and a good safety record can mean lower workers compensation rates, which for some contractors can be 5% to 10% of total contract amount.

There are many agencies and organizations that provide guidelines and regulations for construction safety. The most visible and probably most important is the federal Occupational Safety and Health Administration (OSHA). OSHA regulations are ever changing, and the contractor must keep abreast of the latest revisions in OSHA standards. This point cannot be stressed too much. Listed below are some of the key OSHA publications that concern construction safety, but it is the contractor's responsibility to determine the latest versions of these publications. The easiest way is by contacting the nearest OSHA field office.

At the same time, the contractor must fulfill state and local regulatory requirements regarding the health and safety of workers and the public as well. It might seem like a daunting task, but it is well worth the time and effort.

The list below outlines the subparts of the Code of Federal Regulations (CRF) Volume 29, Chapter XVII - Occupational Safety and Health Administration, Department of Labor. These are the OSHA Construction Industry Standards. The subparts are self-descriptive and cover all major areas of safety concern for the industry. The revisions date indicated below was the latest when the current edition of our book was published. Once again, the contractor must check for more current revisions of these standards:

Occupational Safety and Health Administration Construction Industry Standards (29 CFR 1926/1910) OSHA 2207 Revised 1991

Subpart	Topic
A	General
B	General Interpretations
C	General Safety and Health Provisions
D	Occupational Health and Environmental Controls
E	Personal Protective and Life Saving Equipment
F	Fire Protection and Prevention
G	Signs, Signals, and Barricades
H	Materials Handling, Storage, Use, and Disposal
I	Tools--Hand and Power
J	Welding and Cutting
K	Electrical

L	Scaffolding
M	Floors and Wall Openings
N	Cranes, Derricks, Hoists, Elevators, and Conveyors
O	Motor Vehicles, Mechanized Equipment, and Marine Operations
P	Excavations
Q	Concrete and Masonry Construction
R	Steel Erection
S	Underground Construction, Caissons, Cofferdams, and Compressed Air
T	Demolition
U	Blasting and Use of Explosives
V	Power Transmission and Distribution
W	Rollover Protective Structures; Overhead Protection
X	Effective Dates

In addition to OSHA 2207 - Construction Industry Standards described above, there are a number of other OSHA publications that pertain to the construction industry:

OSHA No.	Title
2098	OSHA Inspections
2201	General Industry Standards
2202	Construction Industry Digest
2209	Handbook for Small Business
2226	Excavation and Trenching Operations
2254	Training Requirements in OSHA Standards and Training Guides
3007	Ground Fault Protection on Construction Sites
3040	Health Hazards of Asbestos
3071	Job Hazard Analysis
3075	Controlling Electrical Hazards
3077	Personal Protective Equipment
3079	Respiratory Protection
3080	Hand and Power Tools
3088	How to Prepare for Workplace Emergencies
3096	Asbestos Standards for Construction Industry
3097	Electrical Standards for Construction
3100	Crane and Derrick Suspended Personnel Platforms
3106	Concrete and Masonry Construction

OSHA 2207 (11-14-90 Edition)

Construction Stds.

Subpt. P, App. B

Appendix B to Subpart P

Sloping and Benching

(a) *Scope and application.* This appendix contains specifications for sloping and benching when used as methods of protecting employees working in excavations from cave-ins. The requirements of this appendix apply when the design of sloping and benching protective systems is to be performed in accordance with the requirements set forth in § 1926.652(b)(2).

(b) *Definitions.*

Actual slope means the slope to which an excavation face is excavated.

Distress means that the soil is in a condition where a cave-in is imminent or is likely to occur. Distress is evidenced by such phenomena as the development of fissures in the face of or adjacent to an open excavation; the subsidence of the edge of an excavation; the slumping of material from the face or the bulging or heaving of material from the bottom of an excavation; the spalling of material from the face of an excavation; and ravelling, i.e., small amounts of material such as pebbles or little clumps of material suddenly separating from the face of an excavation and trickling or rolling down into the excavation.

Maximum allowable slope means the steepest incline of an excavation face that is acceptable for the most favorable site conditions as protection against cave-ins, and is expressed as the ratio of horizontal distance to vertical rise (H:V).

Short term exposure means a period of time less than or equal to 24 hours that an excavation is open.

(c) *Requirements*—(1) *Soil classification.* Soil and rock deposits shall be classified in accordance with appendix A to subpart P of part 1926.

(2) *Maximum allowable slope.* The maximum allowable slope for a soil or rock deposit shall be determined from Table B-1 of this appendix.

(3) *Actual slope.* (i) The actual slope shall not be steeper than the maximum allowable slope.

(ii) The actual slope shall be less steep than the maximum allowable slope, when there are signs of distress. If that situation occurs, the slope shall be cut back to an actual slope which is at least ½ horizontal to one vertical (½H:1V) less steep than the maximum allowable slope.

(iii) When surcharge loads from stored material or equipment, operating equipment, or traffic are present, a competent person shall determine the degree to which the actual slope must be reduced below the maximum allowable slope, and shall assure that such reduction is achieved. Surcharge loads from adjacent structures shall be evaluated in accordance with § 1926.651(i).

(4) *Configurations.* Configurations of sloping and benching systems shall be in accordance with Figure B-1.

TABLE B-1

MAXIMUM ALLOWABLE SLOPES

SOIL OR ROCK TYPE	MAXIMUM ALLOWABLE SLOPES (H:V)[1] FOR EXCAVATIONS LESS THAN 20 FEET DEEP [3]
STABLE ROCK	VERTICAL (90°)
TYPE A [2]	3/4 : 1 (53°)
TYPE B	1:1 (45°)
TYPE C	1½: 1 (34°)

NOTES:

1. Numbers shown in parentheses next to maximum allowable slopes are angles expressed in degrees from the horizontal. Angles have been rounded off.
2. A short-term maximum allowable slope of 1/2H:1V (63°) is allowed in excavations in Type A soil that are 12 feet (3.67 m) or less in depth. Short-term maximum allowable slopes for excavations greater than 12 feet (3.67 m) in depth shall be 3/4H:1V (53°).
3. Sloping or benching for excavations greater than 20 feet deep shall be designed by a registered professional engineer.

Figure B-1

Slope Configurations

(All slopes stated below are in the horizontal to vertical ratio)

B-1.1 Excavations made in Type A soil.

1. All simple slope excavations 20 feet or less in depth shall have a maximum allowable slope of ¾:1.

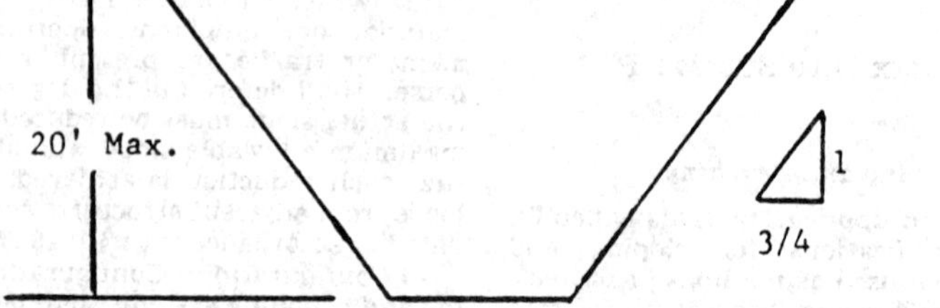

SIMPLE SLOPE—GENERAL

Exception: Simple slope excavations which are open 24 hours or less (short term) and which are 12 feet or less in depth shall have a maximum allowable slope of ½:1.

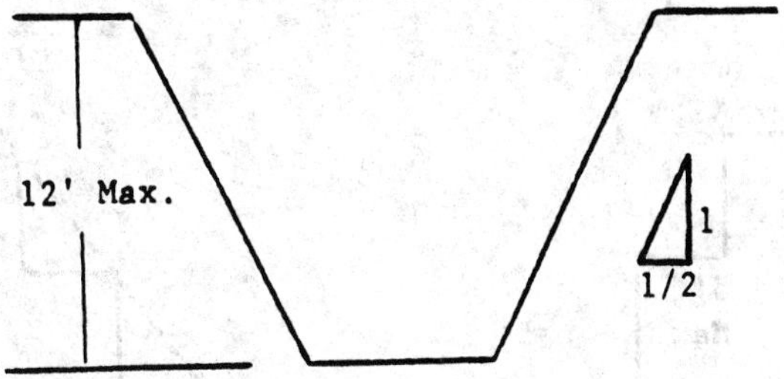

SIMPLE SLOPE—SHORT TERM

2. All benched excavations 20 feet or less in depth shall have a maximum allowable slope of ¾ to 1 and maximum bench dimensions as follows:

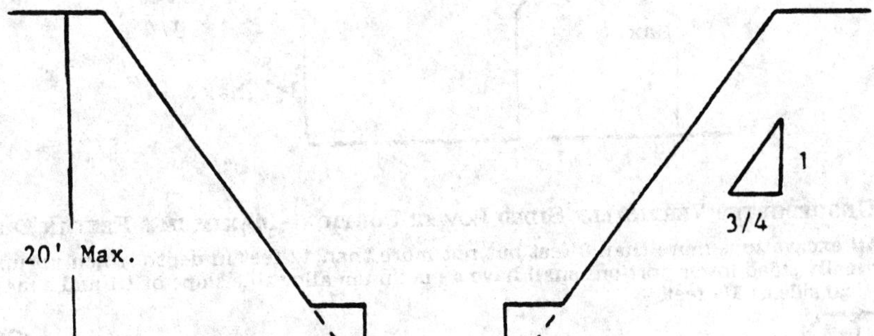

SIMPLE BENCH

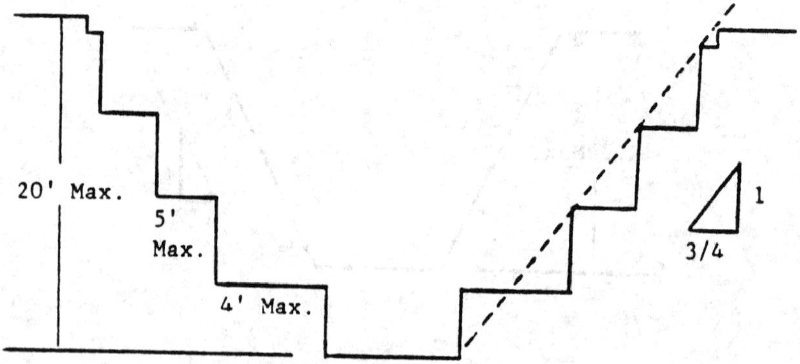

Multiple Bench

3. All excavations 8 feet or less in depth which have unsupported vertically sided lower portions shall have a maximum vertical side of 3½ feet.

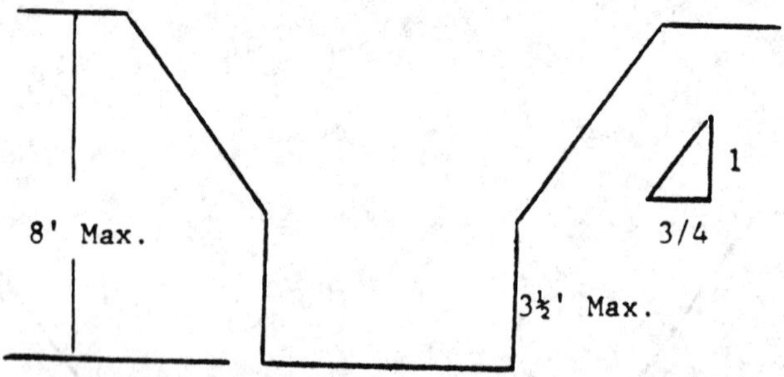

Unsupported Vertically Sided Lower Portion—Maximum 8 Feet in Depth

All excavations more than 8 feet but not more than 12 feet in depth which unsupported vertically sided lower portions shall have a maximum allowable slope of 1:1 and a maximum vertical side of 3½ feet.

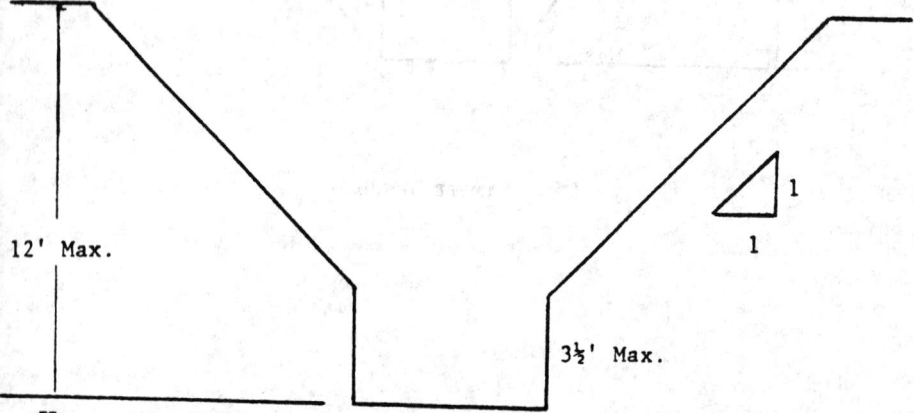

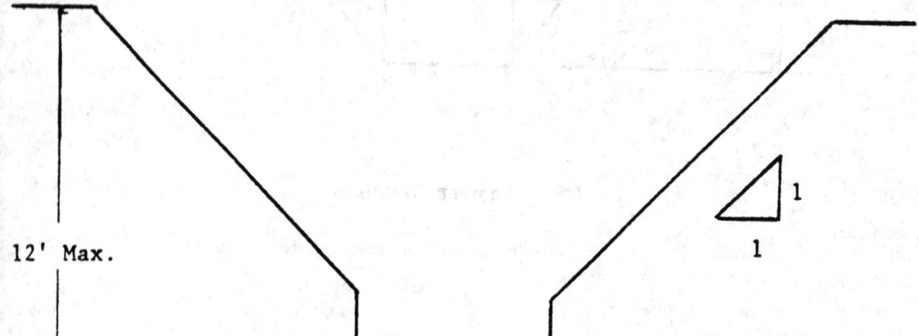

Unsupported Vertically Sided Lower Portion—Maximum 12 Feet in Depth

All excavations 20 feet or less in depth which have vertically sided lower portions that are supported or shielded shall have a maximum allowable slope of ¾:1. The support or shield system must extend at least 18 inches above the top of the vertical side.

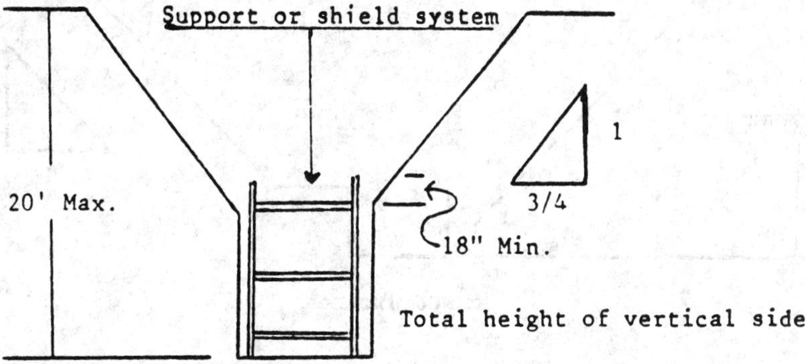

Supported or Shielded Vertically Sided Lower Portion

4. All other simple slope, compound slope, and vertically sided lower portion excavations shall be in accordance with the other options permitted under § 1926.652(b).

B-1.2 Excavations Made in Type B Soil

1. All simple slope excavations 20 feet or less in depth shall have a maximum allowable slope of 1:1.

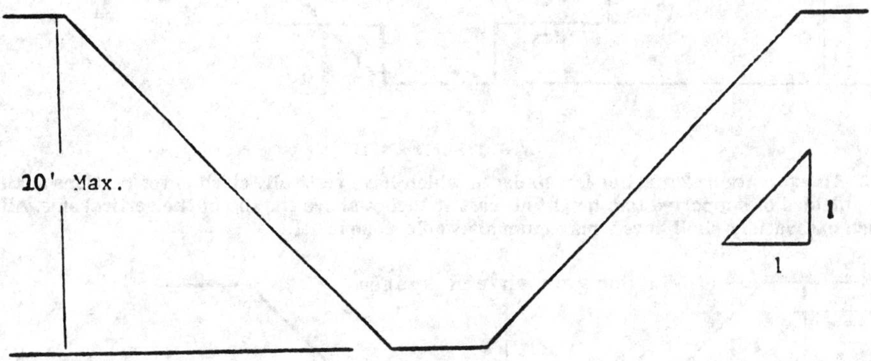

Simple Slope

2. All benched excavations 20 feet or less in depth shall have a maximum allowable slope of 1:1 and maximum bench dimensions as follows:

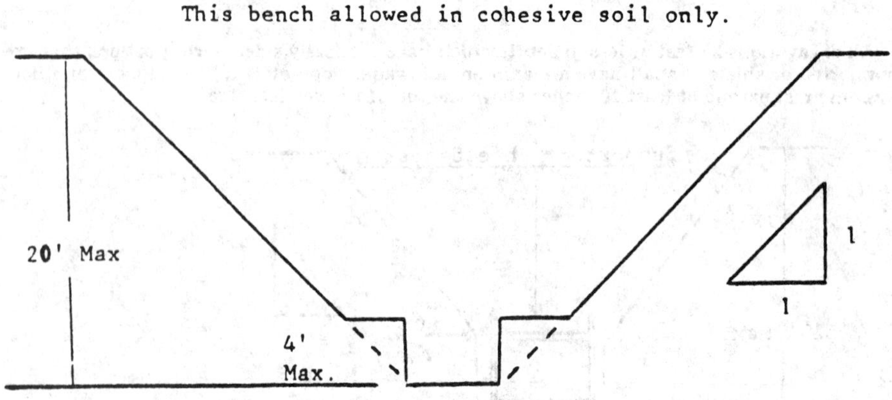

Single Bench

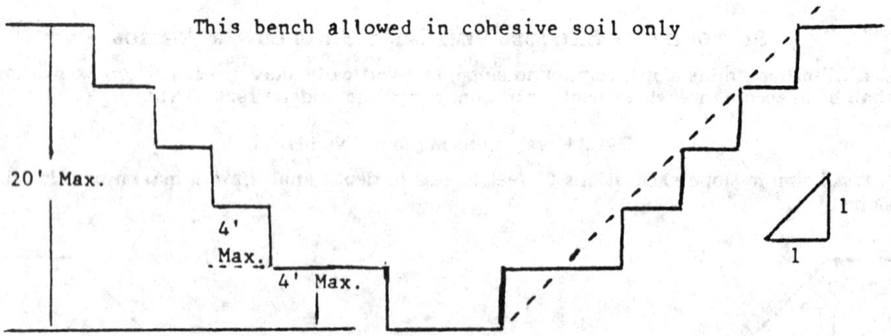

Multiple Bench

3. All excavations 20 feet or less in depth which have vertically sided lower portions shall be shielded or supported to a height at least 18 inches above the top of the vertical side. All such excavations shall have a maximum allowable slope of 1:1.

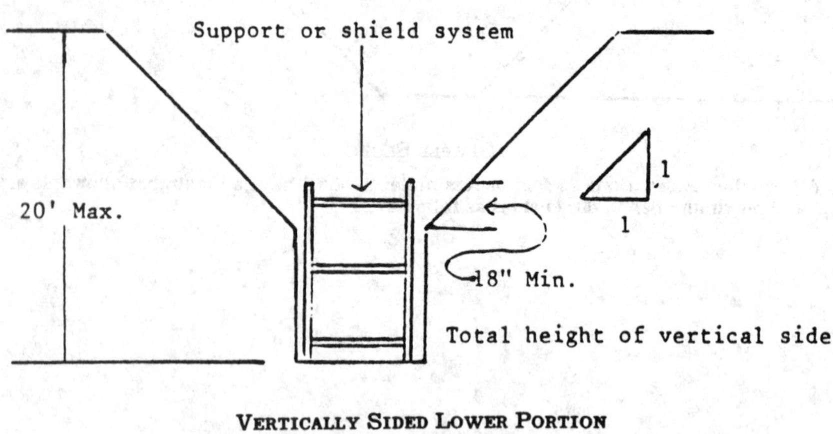

Vertically Sided Lower Portion

4. All other sloped excavations shall be in accordance with the other options permitted in § 1926.652(b).

B-1.3 Excavations Made in Type C Soil

1. All simple slope excavations 20 feet or less in depth shall have a maximum allowable slope of 1½:1.

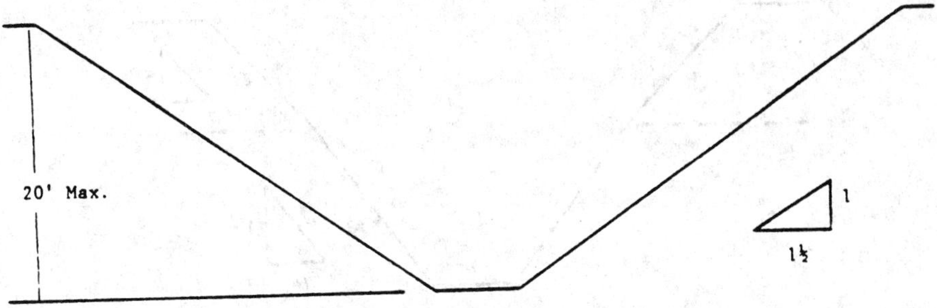

SIMPLE SLOPE

2. All excavations 20 feet or less in depth which have vertically sided lower portions shall be shielded or supported to a height at least 18 inches above the top of the vertical side. All such excavations shall have a maximum allowable slope of 1½:1.

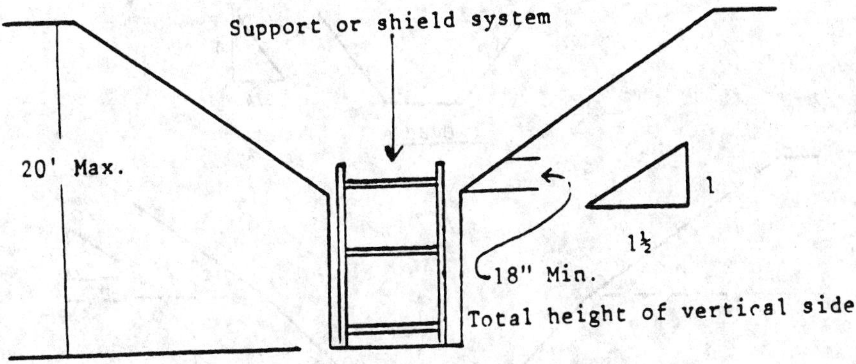

VERTICAL SIDED LOWER PORTION

3. All other sloped excavations shall be in accordance with the other options permitted in § 1926.652(b).

B-1.4 Excavations Made in Layered Soils

1. All excavations 20 feet or less in depth made in layered soils shall have a maximum allowable slope for each layer as set forth below.

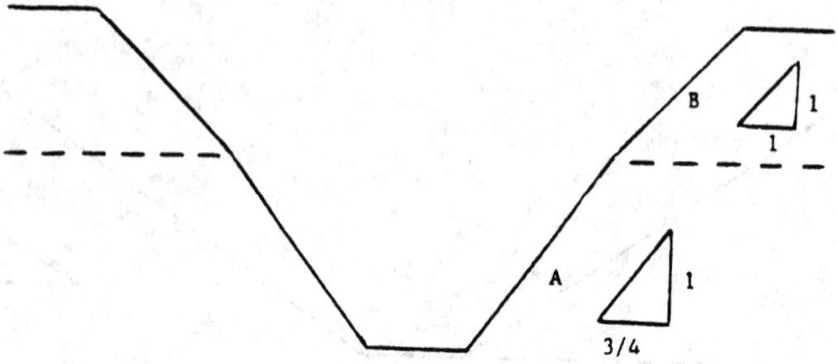

B OVER A

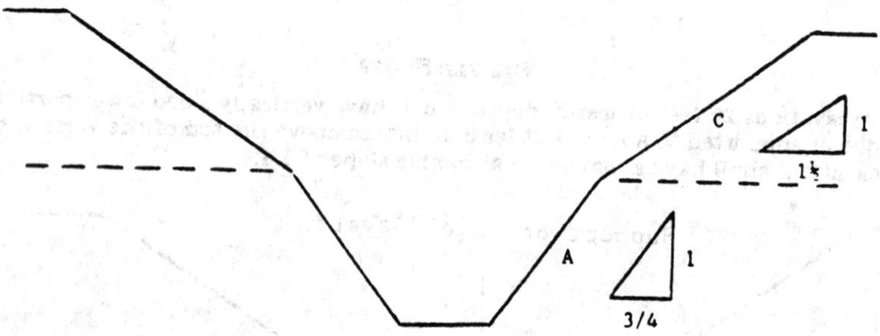

C OVER A

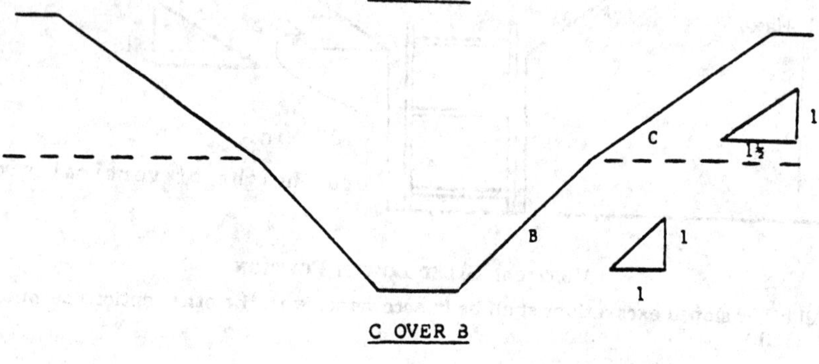

C OVER B

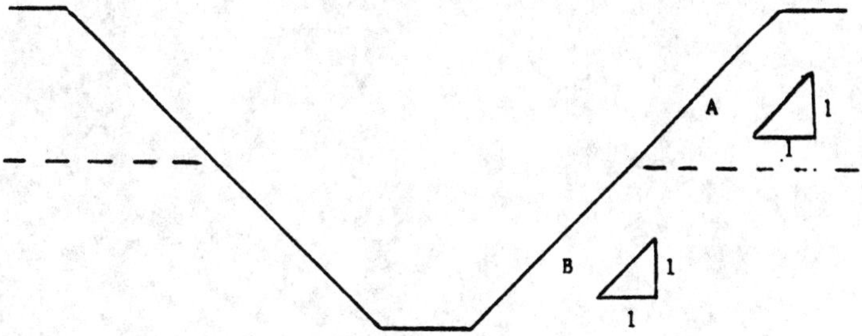

A OVER B

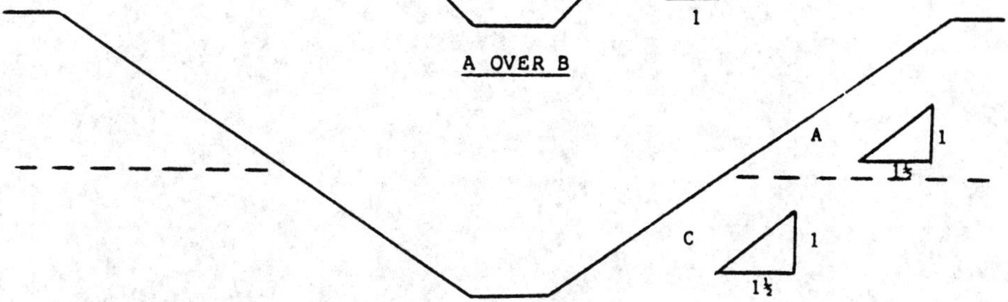

A OVER C

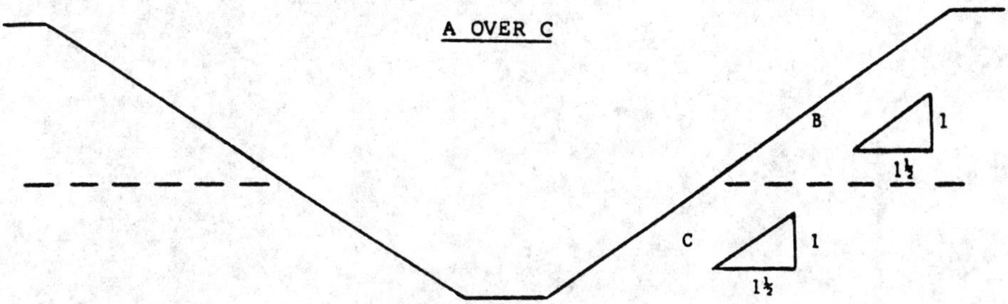

B OVER C

2. All other sloped excavations shall be in accordance with the other options permitted in § 1926.652(b).

C

D

E

F

H

I

N

U

V

W

XYZ